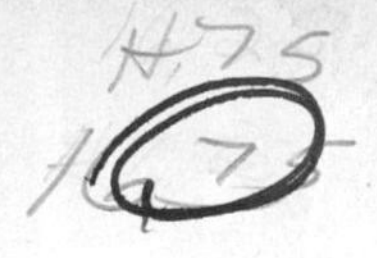



# ENGINEERING DRAWING AND GRAPHIC TECHNOLOGY

# ENGINEERING DRAWING AND GRAPHIC TECHNOLOGY

**Thomas E. French**
Late Professor of Engineering Drawing
The Ohio State University

**Charles J. Vierck**
Visiting Professor to the Graphics Division of the
Department of Mechanical Engineering
University of Florida

**Eleventh Edition**

**McGRAW-HILL BOOK COMPANY**

New York St. Louis San Francisco Düsseldorf Johannesburg Kuala Lumpur
London Mexico Montreal New Delhi Panama Rio de Janeiro Singapore
Sydney Toronto

Library of Congress Cataloging in Publication Data

French, Thomas Ewing, 1871–1944.
Engineering drawing and graphic technology.

First–10th editions published under title:
A manual of engineering drawing for students & draftsmen.
Bibliography: p.
1. Mechanical drawing. I. Vierck, Charles J.
II. Title.

T353.F85 1972 604'.2 70-38135
ISBN 0-07-022157-X

**ENGINEERING DRAWING AND GRAPHIC TECHNOLOGY**



3 4 5 6 7 8 9 0 DODO 7 9 8 7 6 5 4

This book was set in News Gothic by York Graphic Services, Inc., and printed and
bound by R. R. Donnelley & Sons Company. The designer was John Horton.
The editors were B. J. Clark, Peter Karsten, and J. W. Maisel.
John F. Harte supervised production.

## CONTENTS

**7 AUXILIARY VIEWS** *251*

*Auxiliary views are special projections used to clarify and complete orthographic descriptions of shape*

**8 SECTIONAL VIEWS AND CONVENTIONS** *281*

*These special views and practices are specific aids to complete and accurate orthographic representation*

**9 SURFACE INTERSECTIONS** *309*

*Geometric components of an object or assembly meet in intersections which must be shown in order to complete the graphic description*

**10 DEVELOPED VIEWS** *329*

*These show a representation of the shape and size of thin material used to make an object by folding, rolling, or forming*

**11 METHODS USED IN MANUFACTURE** *347*

*Concise and complete specification of size requires a knowledge of manufacturing procedures*

**12 DIMENSIONS, NOTES, LIMITS, AND PRECISION** *363*

*These are the elements used to describe size, which along with a description of shape comprise a complete graphic representation*

## ENGINEERING DRAWING FILMS

THE FOLLOWING 16-mm sound motion pictures are especially recommended for use with various chapters of this book:

**According to Plan.** An introduction to engineering drawing (9 min). *For Chap. 1.*

**Orthographic Projection.** Shape description and the principles of orthographic projection (18 min). *For Chap. 5.*

**Pictorial Sketching.** Basic principles of axonometric, oblique, and perspective pictorial sketching (11 min). *For Chap. 6.*

**Auxiliary Views: Single Auxiliaries.** Reviews orthographic projection and explains auxiliary projection (23 min). *For Chap. 7.*

**Auxiliary Views: Double Auxiliaries.** The theory and practice of double-auxiliary or oblique view (13 min). *For Chap. 7.*

**Sections and Conventions.** Theory and practice of sectioning and conventional principles (15 min). *For Chap. 8.*

**Simple Developments.** What simple developments are and how they are used (11min). *For Chap. 9.*

**Oblique Cones and Transition Developments.** Animated drawings illustrate oblique cones and transitions (11 min). *For Chap. 9.*

**Drawings and the Shop.** Relationships between the drawing and production operations in the shop; basic machines (15 min). *For Chap. 10.*

**Selection of Dimensions.** Principles governing choice of dimensions and their applications (18 min). *For Chap. 11.*

Prints of these films may be purchased from Text-Film Department, McGraw-Hill Book Company.

# PREFACE

The graphics of engineering design and construction may very well be the most important course of all studies for an engineering or technical career. The indisputable reason why graphics is so extremely important is that it is the *language of the designer, technician, and engineer,* used to communicate designs and construction details to others. No matter how knowledgeable an engineer may be concerning the highly complex technical and scientific aspects of his profession, without a command of graphics he would be completely ineffectual simply because he would fail miserably in transmitting his designs to others. All the technical people working under the direction of an engineer also must have the same command of the language. The language of graphics is written in the form of drawings which represent the shape, size, and specifications of physical objects. The language is read by interpreting drawings so that physical objects can be constructed exactly as originally conceived by the designer.

Sometimes the designer's drawings will be made freehand for technicians to follow in making final finished drawings. At other times, depending largely on complexity, the designer may make accurate instrument drawings to be followed in making final drawings. This, incidentally, does not mean that the engineer never makes a final drawing because, depending somewhat upon the size of the organization, he may do so. In any case, the engineer must be able to check and approve of the final drawings made by others. Thus, it is absolutely necessary for the engineer to have a comprehensive command of the language.

Technicians must also have a complete command of the language, because as a design is worked on and brought to completion, the sketches and drawings record ideas and really become an integral part of design thinking.

The study of design continues to increase in importance. Leading industries have practically demanded that the schools of engineering and technology include more design courses in the curriculum. There has also been much discussion of the importance, the trends, and the procedures for design in the meetings and publications of the ASME, SAE, ASEE, and other organizations. Chapter 14 is intended to fill the gap in available design information that exists between machine drawing and machine design. All texts on machine drawing with which the author is familiar are restricted to drawing techniques. In addition, all texts on machine design are limited to discussions of mechanisms and the calculations of the strength of parts. In neither area is anything said about definitions of design, explanation of design categories, how a designer thinks, procedures for design, materials, good practices and proportioning, construction and manufacturing methods, or aesthetics, all of which are included in Chapter 14. This material will be expanded in future revisions as the need for more information increases.

This textbook represents many years of study not only in teaching, engineering experience, and writing, but also painstaking attention to book design principles and usage. A carefully considered plan to make this book more readable and readily usable has been employed.

The two-column format promotes readability and also provides flexibility in the location and size of accompanying illustrations. The illustrations have been outlined to accent their area, allow the use of captions within the enclosed area, and positively relate caption and illustration. Running heads on all pages give the chapter number and title as well as the page number for easy reference and to facilitate the location of assigned material. Each chapter has an opening photograph typifying covered material. The accompanying opening page not only lists chapter title and number but also gives a stylized outline of the topics covered, matching the description given in the table of contents. The table of contents not only gives chapter titles and numbers, but briefly gives details of the chapter subjects. The new type faces for reading material and captions have been chosen for maximum clarity and readability. Figure numbers are individual numbers within the chapter, a simplification made possible because of chapter number and title identified on each page spread. This eliminates the necessity of identifying the chapter number with each figure number. Problem numbers are also continuous in each chapter.

The use of color is a strong part of the overall book design. The choice of color and the details of its use have been carefully worked out to give maximum readability and to promote reader interest. Research on teaching methods has proved that page makeup, effective use of color, illustration placement, and caption content aid greatly in the processes of learning and retention.

A book of such scope and completeness requires cooperation between the author and all other persons involved in bringing the work to fruition. The frank, honest, and sometimes extensive discussion with associates is highly valued, along with the assistance of Esther E. Vierck in the total effort necessary to bring this work to completion.

**Charles J. Vierck**

# ENGINEERING DRAWING AND GRAPHIC TECHNOLOGY

This Chapter Outlines the Graphic Language

- Basic Concepts—Writing and Reading
- Essentials—Lines and Lettering
- Methods of Expression
- Methods of Shape Description—Orthographic and Pictorial
- Auxiliaries, Sections, Intersections, Developments
- Methods of Size Description
- Charts and Graphs
- The Fundamentals of Design
- Basic Machine Elements—Screw Threads, Fasteners, Keys, Springs
- Welding and Riveting
- Jigs, Fixtures, Gears, Cams
- Piping; Drawings of Electrical Systems, Structures, Maps and Topography
- Drawings for Engineering Design and Construction

# 1

# INTRODUCTION

## This Chapter Outlines the Graphic Language in Theory and Practice

## 1. The Graphic Language—Theory and Practice.

In beginning the study of graphics, you are embarking upon a rewarding educational experience and one that will be of real value in your future career. When you have become proficient in it, you will have at your command a method of communication used in all branches of technical industry, a language unequaled for accurate description of physical objects.

The importance of this graphic language can be seen by comparing it with word languages. All who attend elementary and high school study the language of their country and learn to read, write, and speak it with some degree of skill. In high school and college most students study a foreign language. These word languages are highly developed systems of communication. Nevertheless, any word language is inadequate for describing the size, shape, and relationship of physical objects. Study the photograph at the opening of this chapter and then try to describe it verbally so that someone who has not seen it can form an accurate and complete mental picture. It is almost impossible to do this. Even a picture such as Fig. 1, although possibly easier to describe, presents an almost insurmountable problem. Furthermore, in trying to describe either picture, you may want to use pencil and paper to sketch all or a part in an attempt to make the word description more complete, meaningful, and accurate, or tend to use your hands, gesturing to aid in explaining shape and relationship. From this we can see that a word language is often without resources for accurate and rapid communication of shape and size and the relationships of components.

**FIG. 1.** Try to describe in words the shape, the relative size, and the position of the objects in this picture.

Engineering is applied science, and communication of physical facts *must* be complete and accurate. Quantitative relationships are expressed mathematically. The written word completes many descriptions. But whenever machines and structures are designed, described, and built, graphic representation is necessary. Although the works of artists (or photography and other methods of reproduction) would provide pictorial representation, they cannot serve as engineering descriptions. Shaded pictorial drawings and photographs are used for special purposes, but the great bulk of engineering drawings are made in line only, with separate views arranged in a logical system of projection. To these views, dimensions and special notes giving operations and other directions for manufacture are added. This is the language of graphics which can be defined as *the graphic representation of physical objects and relationships.*

As the foundation upon which all designing and subsequent manufacture are based, engineering graphics is one of the most important single branches of study in a technical school. Every engineering student must know how to make and how to read drawings. The subject is essential in all types of engineering practice, and should be understood by all connected with, or interested in, technical industry. All designs and directions for manufacture are prepared by draftsmen, professional writers of the language, but even one who may never make drawings must be able to read and understand them or be professionally illiterate. Thorough training in engineering graphics is particularly important for the engineer because he is responsible for and specifies the drawings required in his work and must therefore be able to interpret every detail for correctness and completeness.

Our object is to study the language of graphics so that we can write it, expressing ourselves clearly to one familiar with it, and read it readily when written by another. To do this, we must know its basic theory and composition, and be familiar with its accepted conventions and abbreviations. Since its principles are essentially the same throughout the world, a person who has been trained in the practices of one nation can readily adapt himself to the practices of another.

This language is entirely graphic and written, and is interpreted by acquiring a visual knowledge of the object represented. A student's success with it will be indicated not alone by his skill in execution, but also by his ability to interpret lines and symbols and to visualize clearly in space.

In the remainder of this chapter we shall introduce briefly the various aspects of graphics that will be discussed at length later. It is hoped that this preview will serve as a broad perspective against which the student will see each topic, as it is studied, in relation to the whole. Since our subject is a graphic language, illustrations are helpful in presenting even this introductory material; figures are used both to clarify the text and to carry the presentation forward.

## 2. Essentials of Graphic Writing: Lines and Lettering.

Drawings are made up of lines that represent the surfaces, edges, and contours of objects. Symbols, dimensional sizes, and word notes are added to these lines, collectively making a complete description. Proficiency in the methods of drawing straight lines, circles, and curves, either freehand or with instruments, and the ability to letter word statements are fundamental to writing the graphic language. Furthermore, lines are connected according to the geometry of the object represented, making it necessary to know the geometry of plane and solid figures and to understand how to combine circles, straight lines, and curves to represent separate views of many geometric combinations.

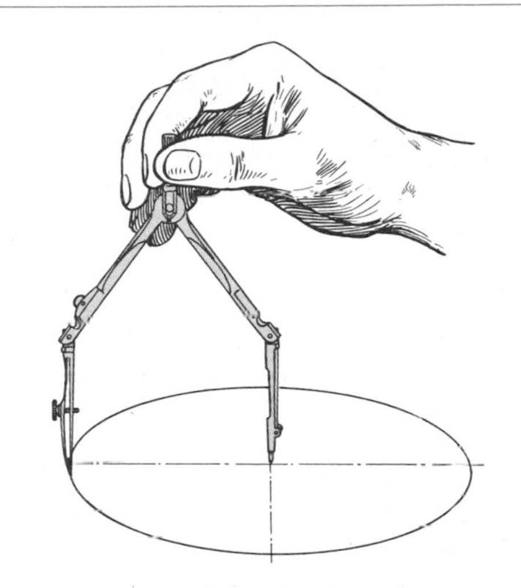

**THE USE OF GRAPHIC INSTRUMENTS.** Facility in the use of instruments makes for speed and accuracy.

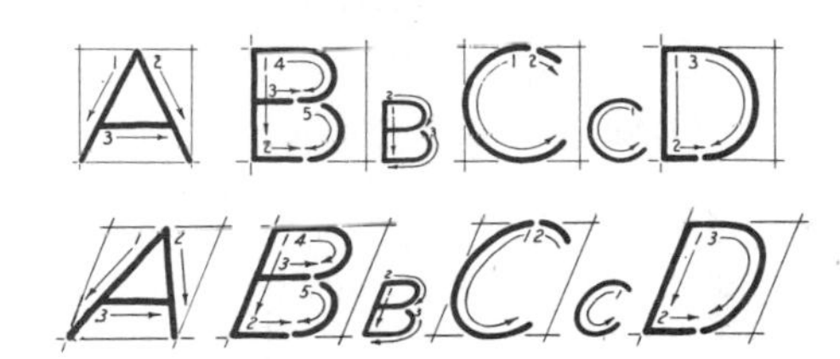


**LETTERING.** The standard lettering for engineering drawings is known as "commercial Gothic." Both vertical and inclined styles are used.

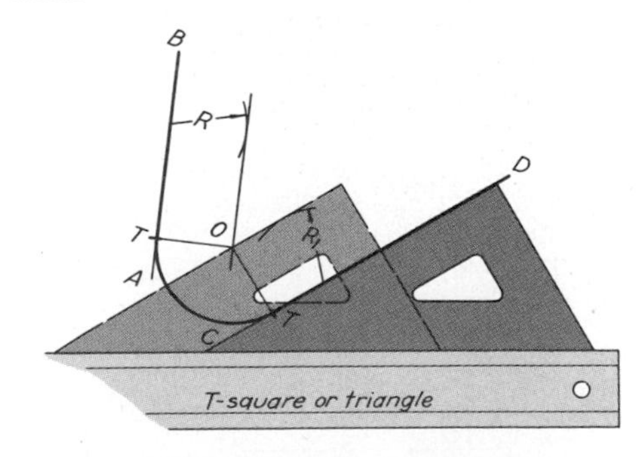


**GRAPHIC GEOMETRY.** A knowledge of, and facility in, the construction of lines and geometric figures promotes efficiency.

### 3. Methods of Expression.

There are two fundamental methods of writing the graphic language: freehand and with instruments.

Freehand drawing is done by sketching the lines with no instruments other than pencils and erasers. It is an excellent method during the learning process because of its speed and because at this stage the study of projection is more important than exactness of delineation. Freehand drawings are much used commercially for preliminary designing and for some finished work. Instrument drawing is the standard method of expression. Most drawings are made "to scale," with instruments used to draw straight lines, circles, and curves concisely and accurately. Training in both freehand and instrument work is necessary for the engineer so that he will develop competence in writing the graphic language and the ability to judge work done under his direction.

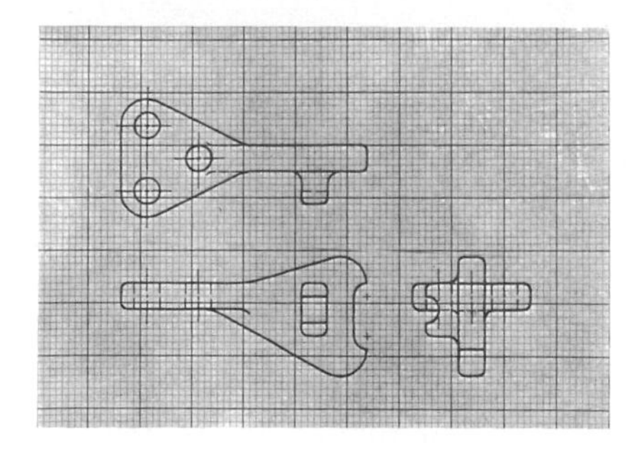

**FREEHAND DRAWINGS.** The freehand method is fine for early study because it provides training in technique, form, and proportion. It is used commercially for economy.

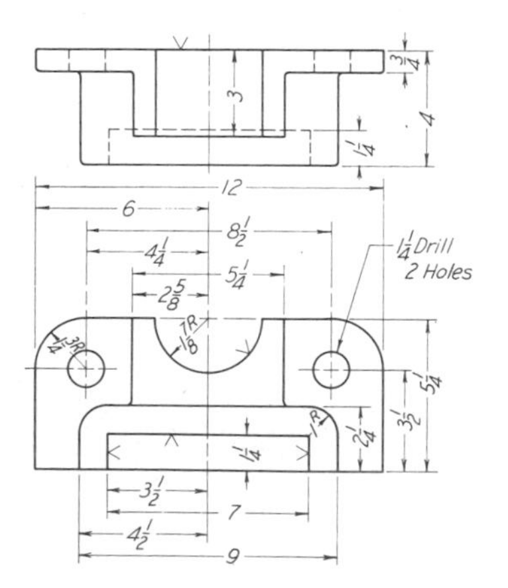


**INSTRUMENT DRAWINGS.** Because of the necessity of drawing "to scale," most drawings are made with instruments.

### 4. Methods of Shape Description.

Delineation of the *shape* of a part, assembly, or structure is the primary element of graphic communication. Since there are many purposes for which drawings are made, the engineer must select, from the different methods of describing shape, the one best suited to the situation at hand. Shape is described by projection, that is, by the process of causing an image to be formed by rays of sight taken in a particular direction from an object to a picture plane.

Following projective theory, two methods of *representation* are used: orthographic views and pictorial views.

For the great bulk of engineering work, the orthographic system is used, and this method, with its variations and the necessary symbols and abbreviations, constitutes an important part of this book. In the orthographic system, separate views arranged according to the projective theory are made to show clearly all details of the object represented. The figures that follow illustrate the fundamental types of orthographic drawings and orthographic views.

"Pictorial representation" designates the methods of projection resulting in a view that shows the object approximately as it would be seen by the eye. Pictorial representation is often used for presentation drawings; text, operation, and maintenance book illustrations; and some working drawings.

There are three main divisions of pictorial projection: axonometric, oblique, and perspective. Theoretically, axonometric projection is projection in which only one plane is used, the object being turned so that three faces show. The main axonometric positions are isometric, dimetric, and trimetric.

Oblique projection is a pictorial method used principally for objects with circular or curved features only on one face or on parallel faces; and for such objects the oblique is easy to draw and dimension.

Perspective projection gives a result identical with what the eye or a single-lens camera would record.

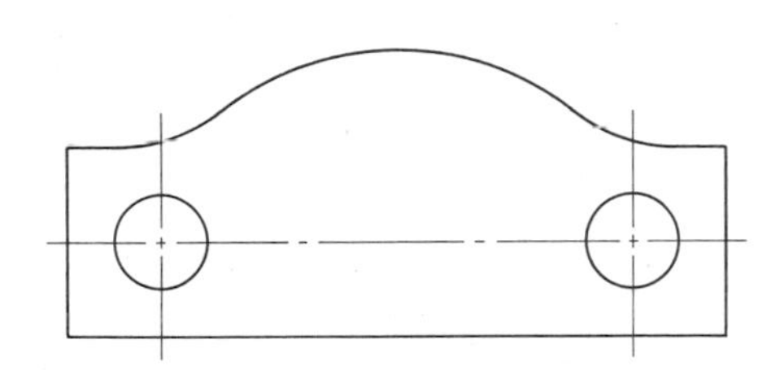

**ONE-VIEW DRAWINGS.** These are used whenever views in more than one direction are unnecessary, for example, for parts made of thin material.

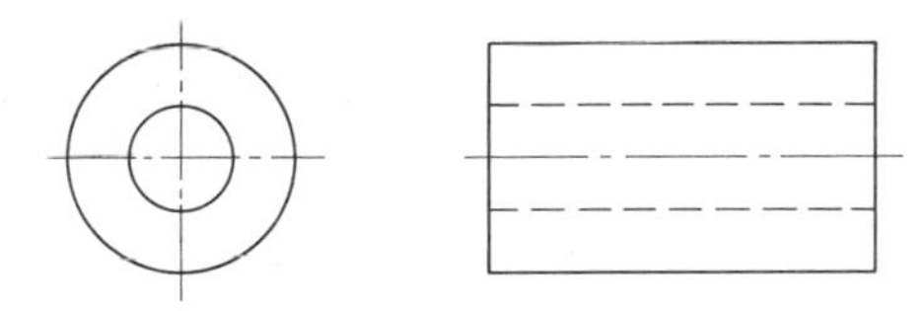

**TWO-VIEW DRAWINGS.** Parts such as cylinders require only two views. More would duplicate the two already drawn.

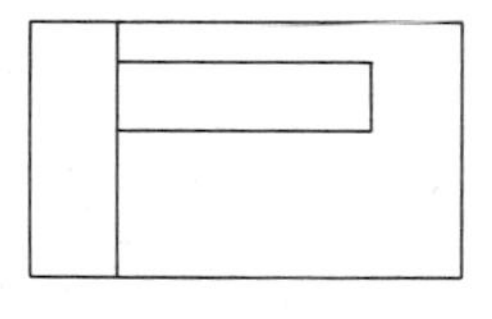

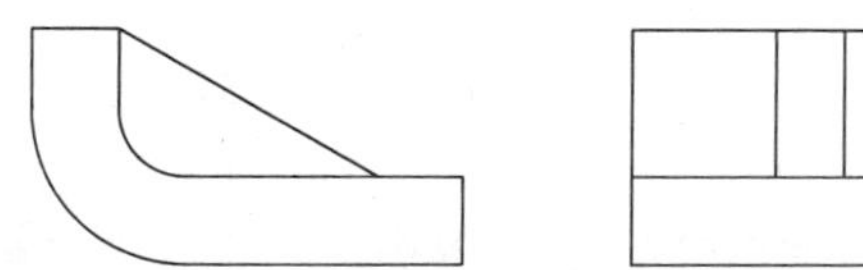

**THREE-VIEW DRAWINGS.** Most objects are made up of combined geometric solids. Three views are required to represent their shape.

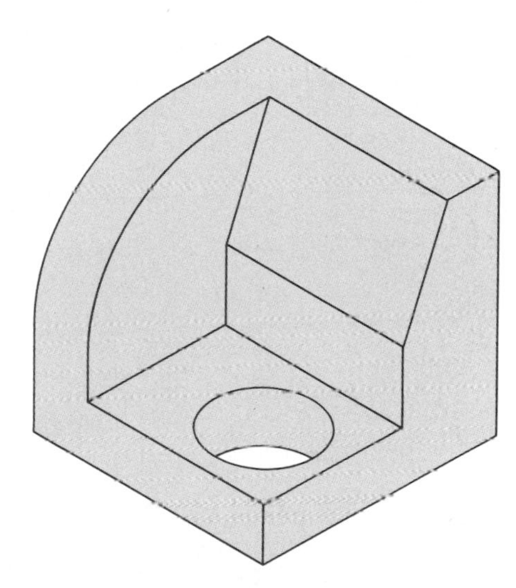

**ISOMETRIC DRAWING.** This method is based on turning the object so that three mutually perpendicular edges are equally foreshortened.

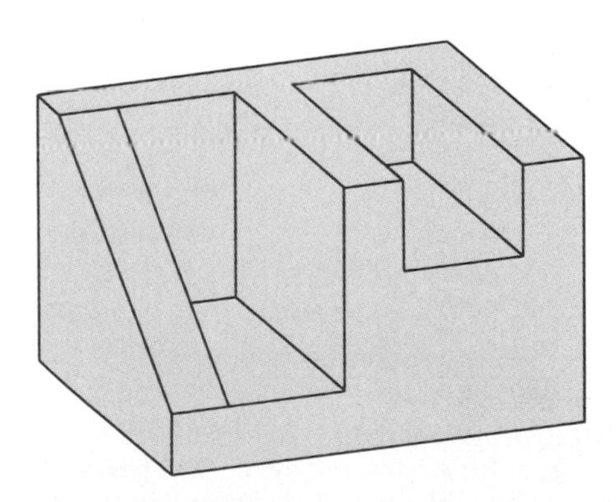

**DIMETRIC DRAWING.** This method is based on turning the object so that two mutually perpendicular edges are equally foreshortened.

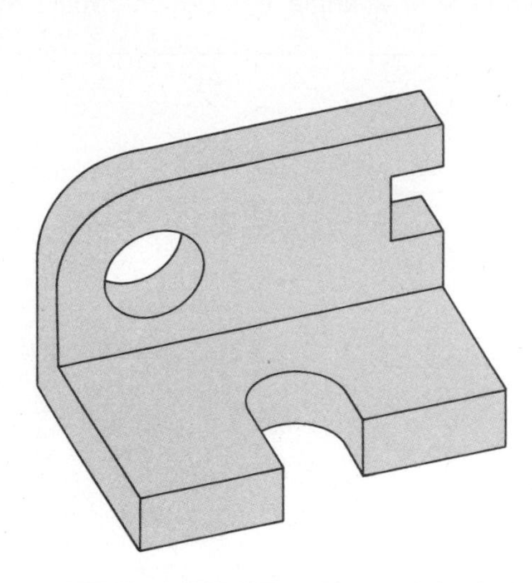

**TRIMETRIC DRAWING.** This method is based on turning the object so that three mutually perpendicular edges are all unequally foreshortened.

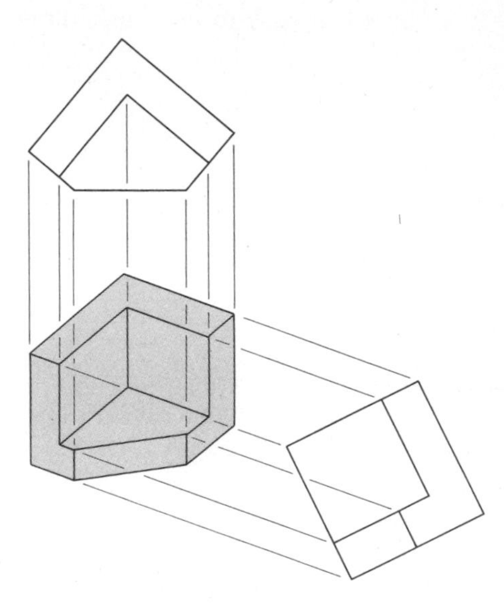

**AXONOMETRIC PROJECTION FROM ORTHOGRAPHIC VIEWS.** Pictorials—sometric, dimetric, or trimetric—may be obtained by projection from orthographic views. The views are located by a geometric method.

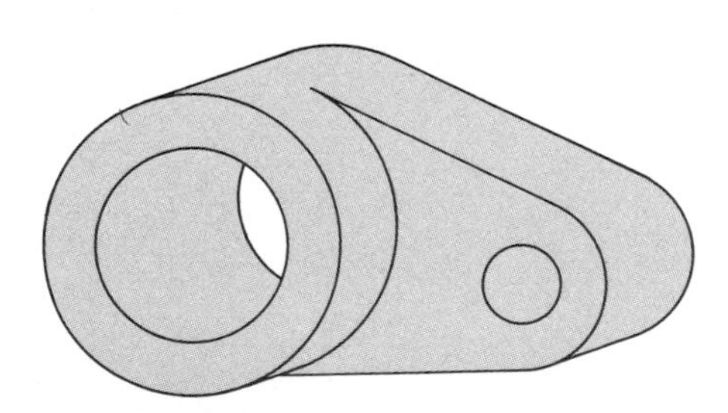

**OBLIQUE DRAWING.** This pictorial method is useful for portraying cylindrical parts. Projectors are oblique to the picture plane. Cavalier and cabinet drawings are specific forms.

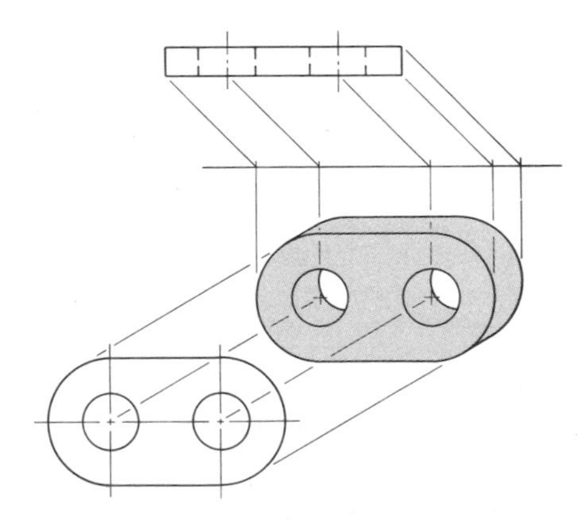

**OBLIQUE PROJECTION FROM ORTHOGRAPHIC VIEWS.** In this pictorial method views are arranged so that, by projection, an oblique drawing results.

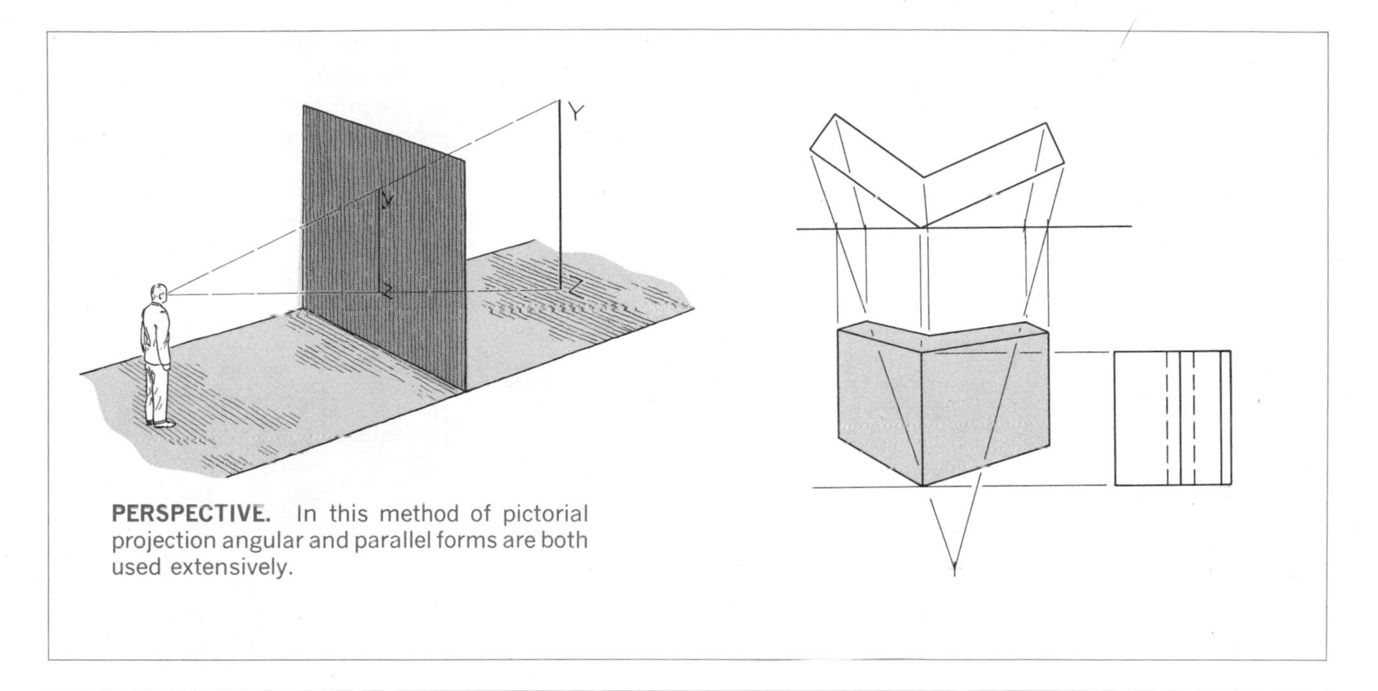


**PERSPECTIVE.** In this method of pictorial projection angular and parallel forms are both used extensively.

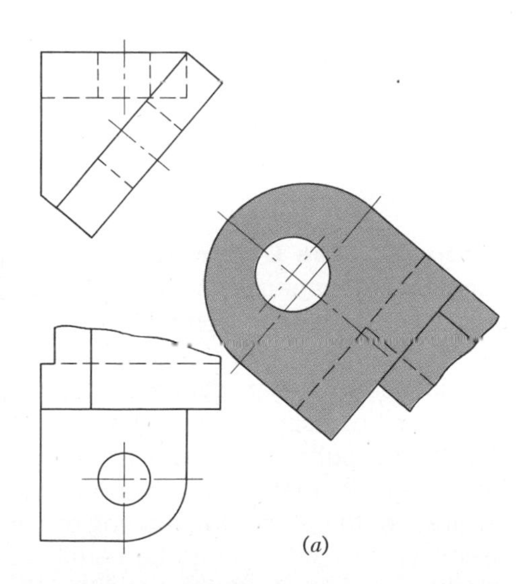

(*a*)

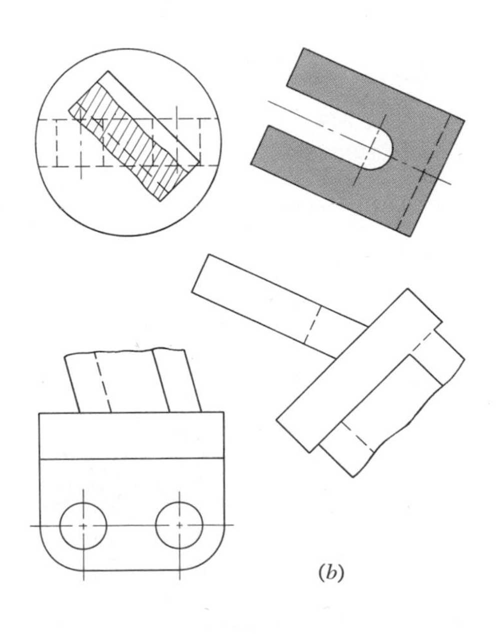

(*b*)

**AUXILIARIES: EDGE AND NORMAL VIEWS.** Auxiliaries are used to show the normal view (true size and shape) of (*a*) an inclined surface (at an angle to two of the planes of projection) or (*b*) of a skew surface (inclined to all three planes of projection).

**5. Special Practices.**
These are used to simplify and clarify.

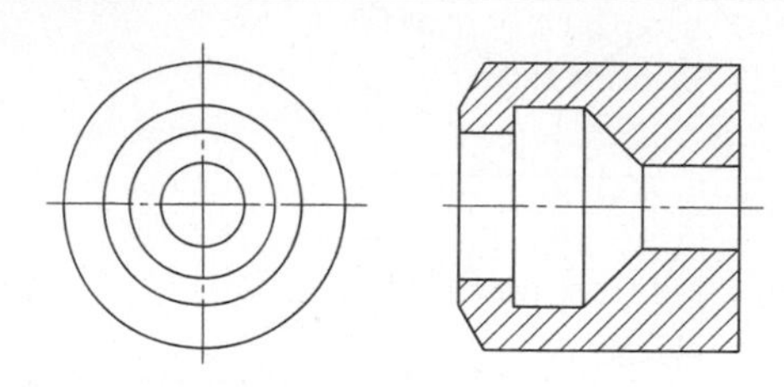

**SECTIONAL VIEWS.** These are used to clarify the representation of objects with complicated internal detail.

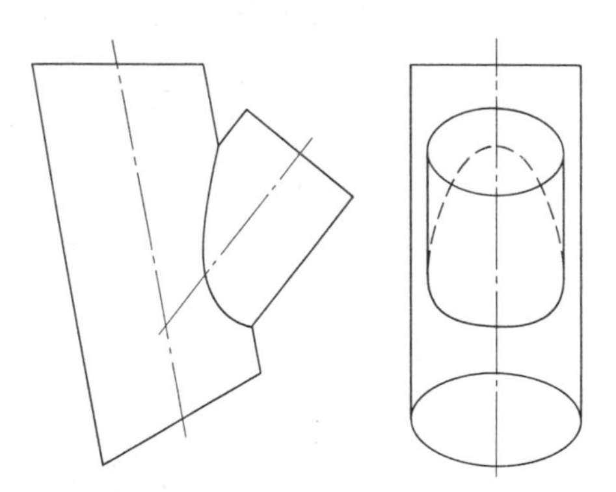

**INTERSECTIONS.** Geometric surfaces or solids often are combined so that additional projection is needed to determine the line of intersection between the parts.

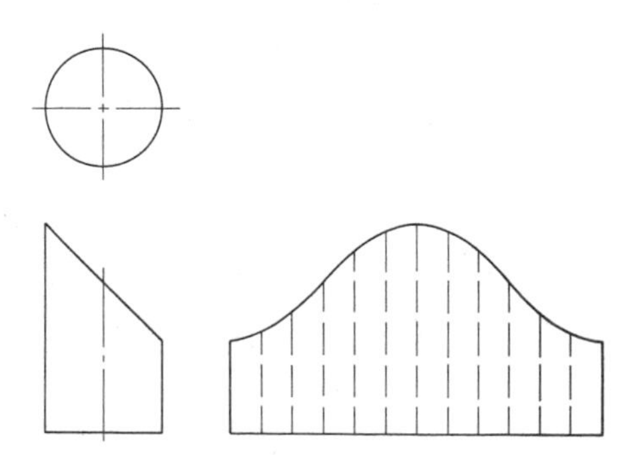

**DEVELOPMENTS.** These are the projections of geometric surfaces into a flat pattern.

**6. Methods of Manufacture.**
To delineate a drawing properly and give accurate size description and specifications for manufacture, a knowledge of processing methods is essential.

**7. Methods of Size Description.**
After delineation of shape, *size* is the second element of graphic communication, completing the representation of the object. Size is shown by "dimensions," which state linear distances, diameters, radii, and other necessary magnitudes.

The dimensions put on the drawing are *not necessarily* those used in making the drawing but are those required for the proper functioning of the part after assembly, selected so as to be readily usable by the workers who are to make the piece. Before dimensioning the drawing, study the machine and understand its functional requirements; then put yourself in the place of the patternmaker, diemaker, machinist, etc., and mentally construct the object to discover which dimensions would best give the information.

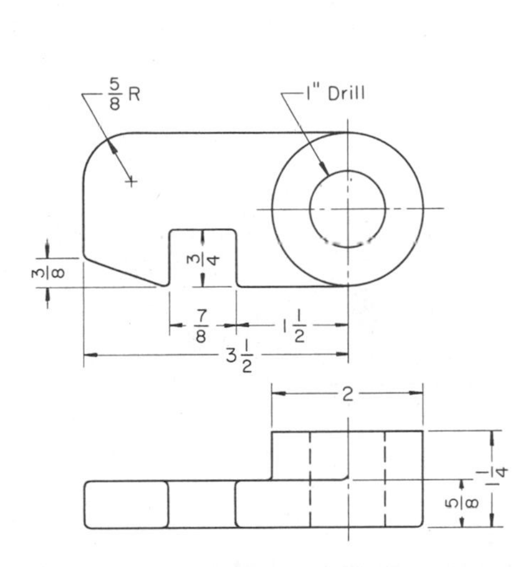


**DIMENSIONING ORTHOGRAPHIC DRAWINGS.** Dimensions showing the magnitude and relative position of each portion of the object are placed on the view where each dimension is most meaningful.

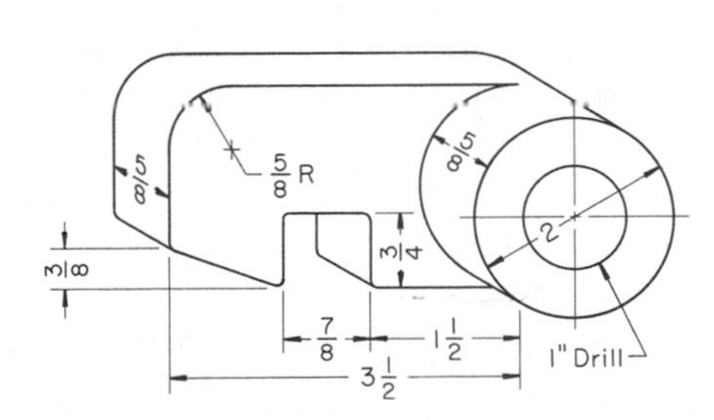


**DIMENSIONING PICTORIAL DRAWINGS.** The descriptions of magnitudes and positions are shown on the pictorial by dimensions placed so as to be easily readable.

## 8. Charts, Graphs, and Diagrams.

These are graphic plots for the analysis of engineering data or the presentation of statistics. Special standards designate preferred construction, lines, symbols, and lettering.

Charts, graphs, and diagrams fall roughly into two classes: (1) those used for purely technical purposes and (2) those used in advertising or in presenting information in a way that will have popular appeal. The engineer is concerned mainly with those of the first class, but he should be acquainted also with the preparation of those of the second and understand their potential influence. The aim here is to give a short study of the types of charts, graphs, and diagrams with which engineers and those in allied professions should be familiar.

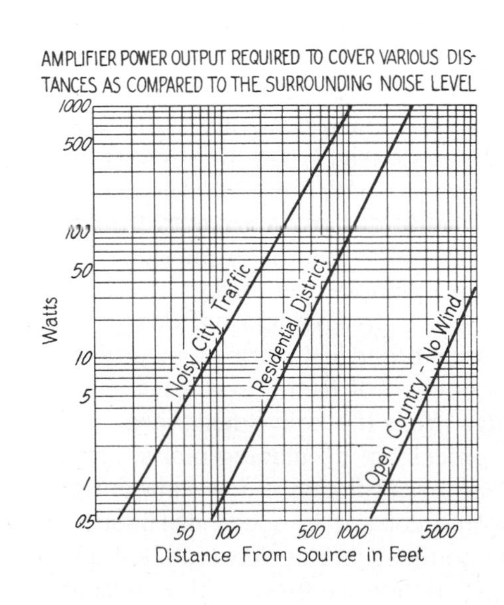

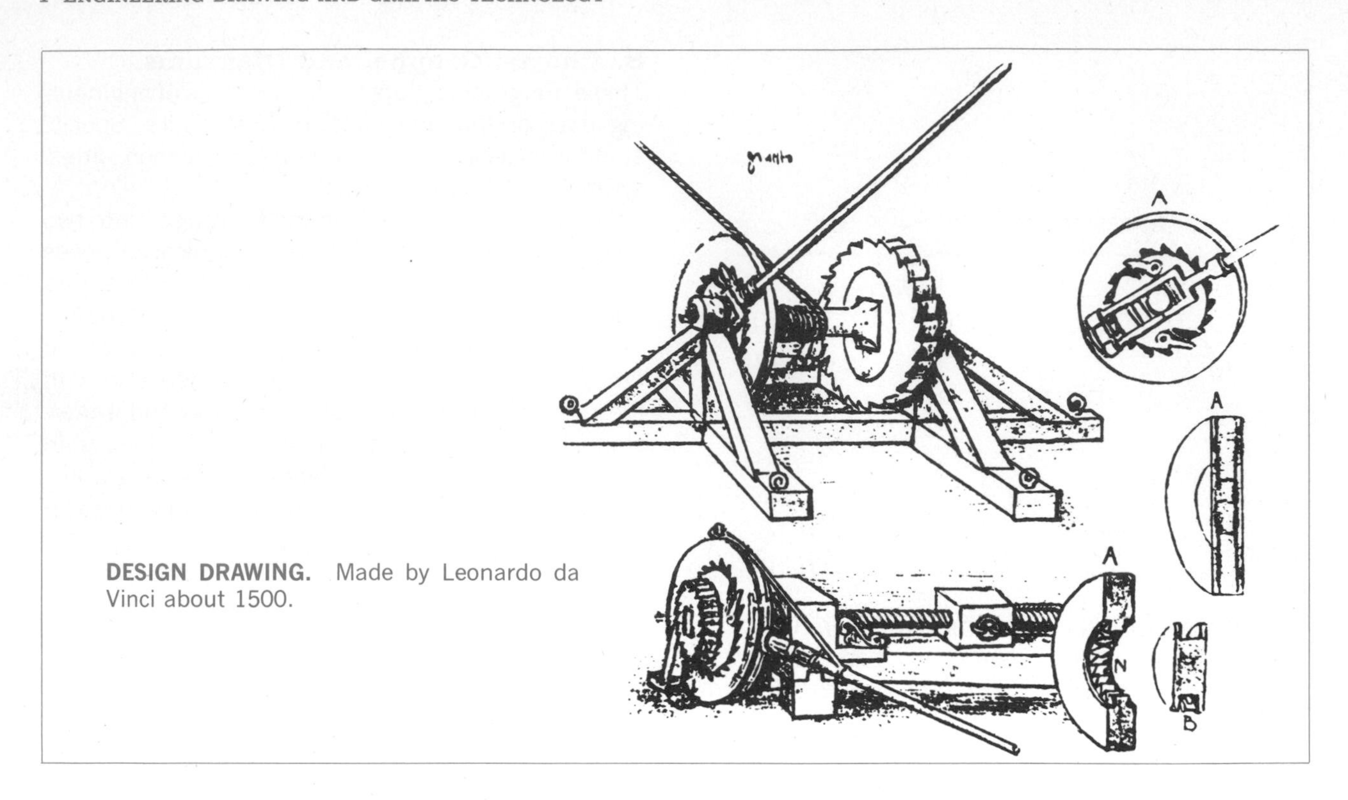


**DESIGN DRAWING.** Made by Leonardo da Vinci about 1500.

### 9. The Fundamentals of Design.

The chapter on design discusses materials, procedures, and the aspects of good mechanical design and explains how a designer thinks and works.

The word "design" has many meanings. A digest of various dictionary definitions is: to plan, conceive, invent, and to designate so as to transmit the plan to others. *Design* has many purely artistic connotations. For example, the design of fabrics, clothing, furniture, etc. In engineering, design has come to mean that broad category of invention leading to the production of useful devices.

*Design*, from the latin "designare" (to mark out), is the process of developing plans, schemes, directions, and specifications for something new. Thus it is within context to speak of Hitler's designs for world conquest; the design (conception) of a book, play, or motion picture; or the design of fabric, clothing, furniture, appliances, or other completely physical objects. Design is distinguished from production and craftsmanship: design is the creative original plan, and the production and craftsmanship are a part of the execution of the plan.

Design means creation in the purest sense. Specifically, design does not go beyond creation, and it logically follows that the execution of a design, that is, the carrying out of the plan by presentation, action, production, manufacture, craftsmanship, and use are not design at all but are simply and positively the products of the design. Also, when examining a finished product, it is proper to speak of the *design* of it, and by this term of reference we mean the original plan or scheme and not the product itself.

### 10. Basic Machine Elements.

Many machine elements occur repeatedly in all kinds of engineering work. Familiarity with these elements is necessary so that dimensioning and specifications on the drawings will be correct. The material that follows introduces basic machine elements and shop processes, illustrating the principles of graphics laid out in the previous sections.

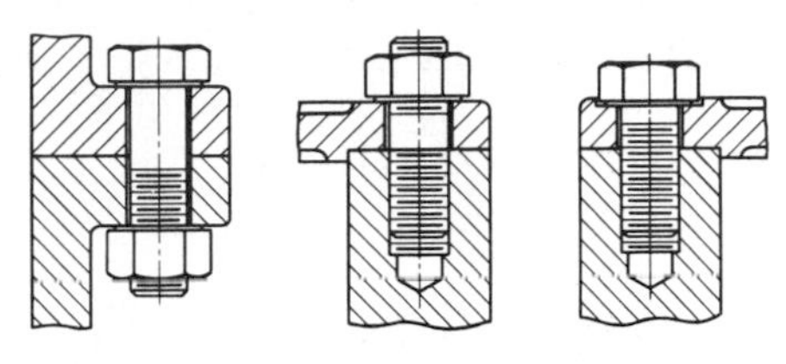

**FASTENERS.** Bolts and screws in a wide variety of forms are used for fastenings. Familiarity with the standard sizes available and their drawing and specification is fundamental.

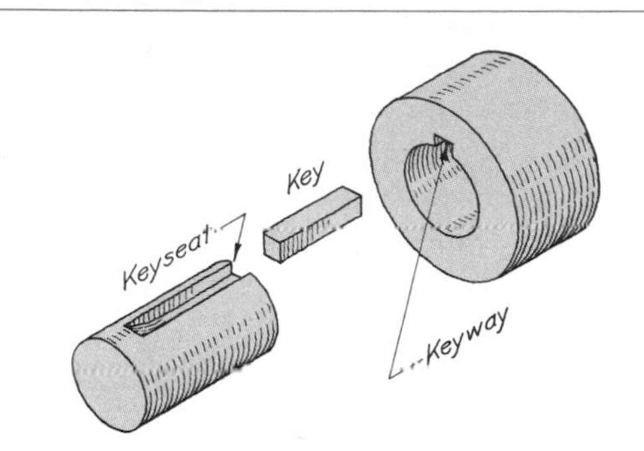


**KEYS.** Keys are used to prevent cylindrical elements such as gears and pulleys from rotation on their shafts.

The earliest records of the screw are found in the writings of Archimedes (278 to 212 B.C.). Although specimens of ancient Greek and Roman screws are so rare as to indicate that they were seldom used, there are many from the later Middle Ages; and it is known that crude lathes and dies were used to cut threads in the latter period. Most early screws were made by hand, by forging the head, cutting the slot with a saw, and fashioning the screw with a file. In America in colonial times wood screws were blunt on the ends; the gimlet point did not appear until 1846. Iron screws were made for each threaded hole. There was no interchanging of parts, and nuts had to be tied to their own bolts. In England, Sir Joseph Whitworth made the first attempt to set up a uniform standard in 1841.

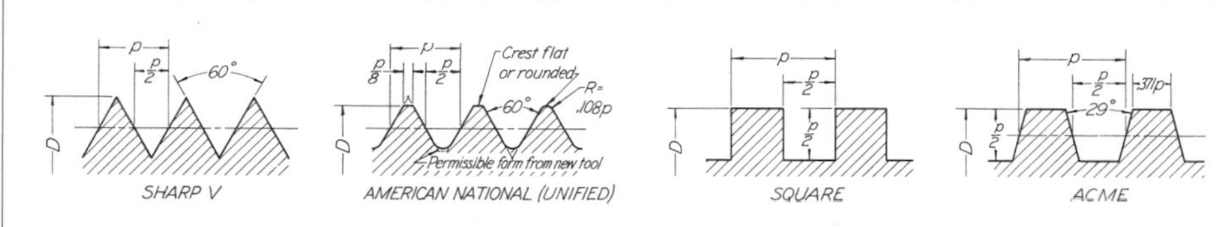


**SCREW THREADS.** Screw threads are used on fasteners, on devices for making adjustments, and for the transmission of power and motion. The screw thread is the most extensively used machine element.

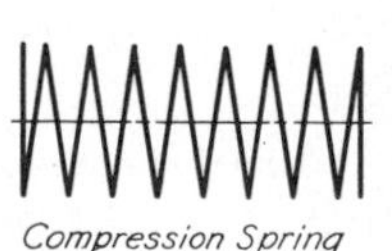


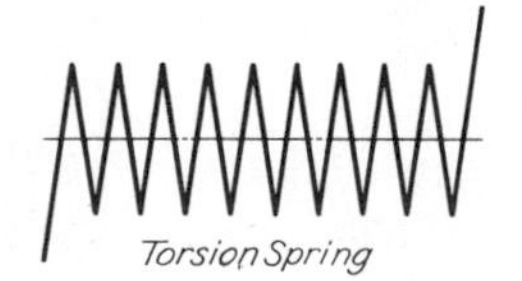


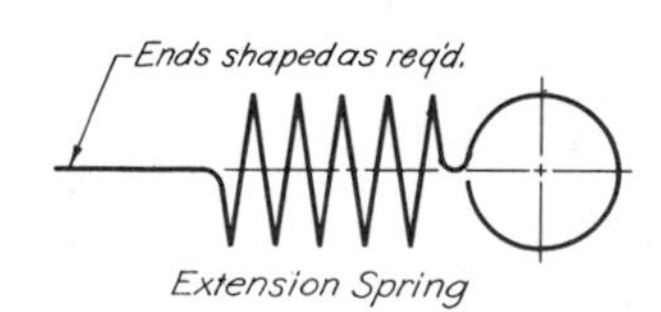


**SPRINGS.** A spring is an elastic body that stores energy when deflected. Basic forms are illustrated.

## 11. Special Fields and Practices.

The methods of projection, the choice of representational type (shape description), and the accompanying size description are more or less uniform in all fields because these are the elements of which all drawings are composed. However, for

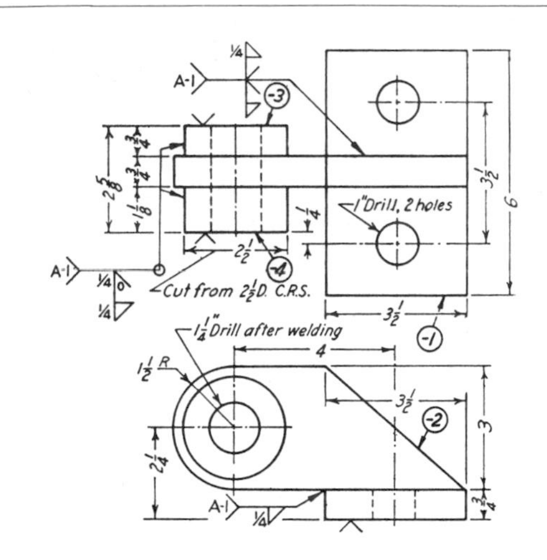


**REPRESENTATION OF WELDS.** Welds are represented by symbols standardized by the American Welding Society (AWS) and the American Standards Association (ASA).

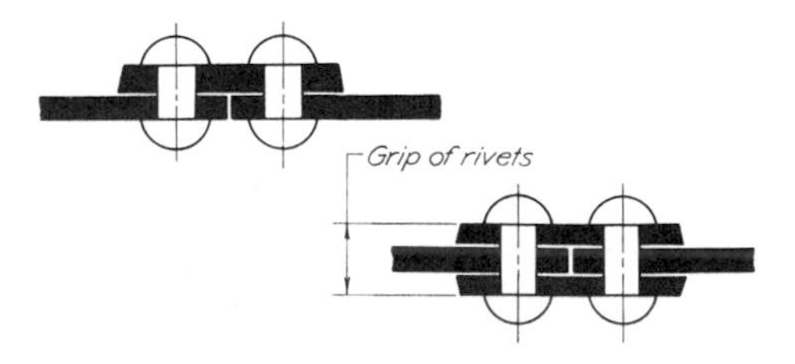


**RIVETS.** Rivets are permanent fasteners. They are cylinders of metal with a head on one end. When placed in position, the opposite head is formed by impact.

**JIGS AND FIXTURES.** In modern high-quantity production these devices are used for holding parts during machining. Drawings are made to conform to tool-engineering standards.

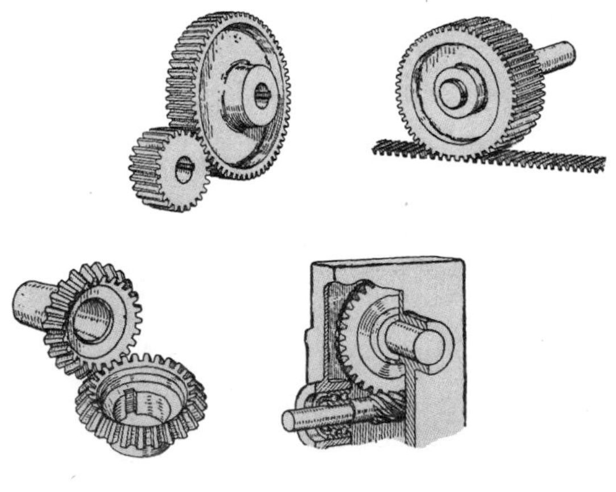

**GEARS.** Special practices include calculation of detailed sizes and their tabulation on drawings.

most special fields certain symbols and notation, drawing, and dimensioning practices have been standardized, and a knowledge of these is necessary so that drawings will be readily and efficiently understood.

The special practices or fields given in this book include: various portions of mechanisms and systems; welding; electrical drawing; structural and topographic practices; and commercial practices and economies.

A "production," or "working," drawing is any drawing used to give information for the manufacture or construction of a machine or the erection of a structure. Complete knowledge for the production of a machine or structure is given by a *set* of working drawings conveying all the facts fully and explicitly so that further instructions are not required.

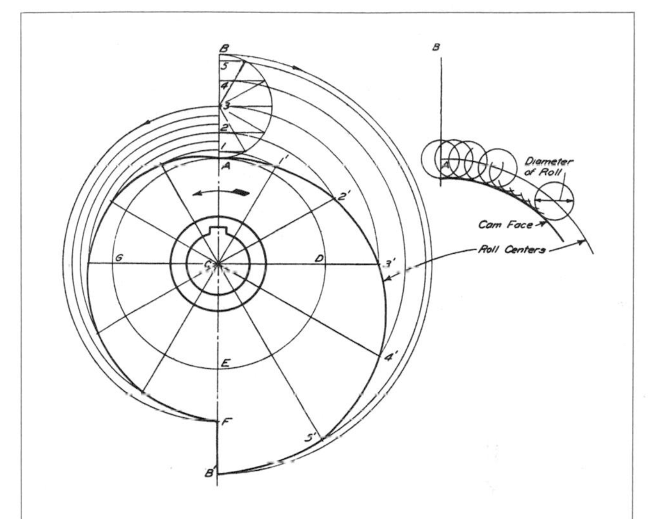


**CAMS.** Special practices include full-size layouts and also calculation and dimensioning of points on cam surfaces.

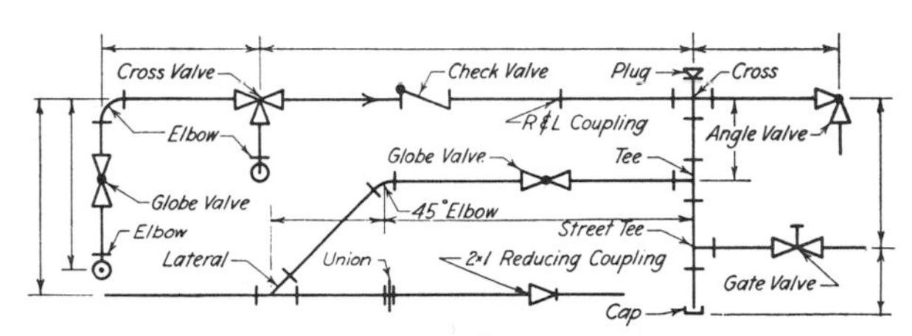


**PIPING.** Piping drawings are made to scale or symbolically. Symbols for pipe, valves, fittings, and other details are standardized by the ASA.

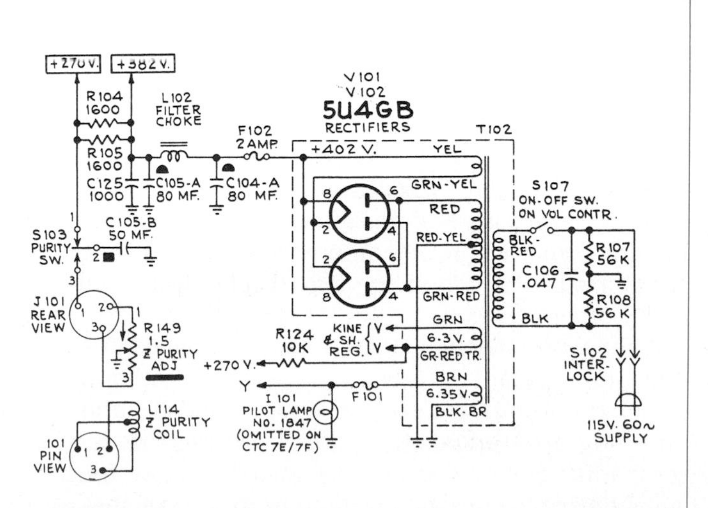


**DRAWINGS OF ELECTRICAL SYSTEMS.** Most electrical drawings are made symbolically, using ASA standards.

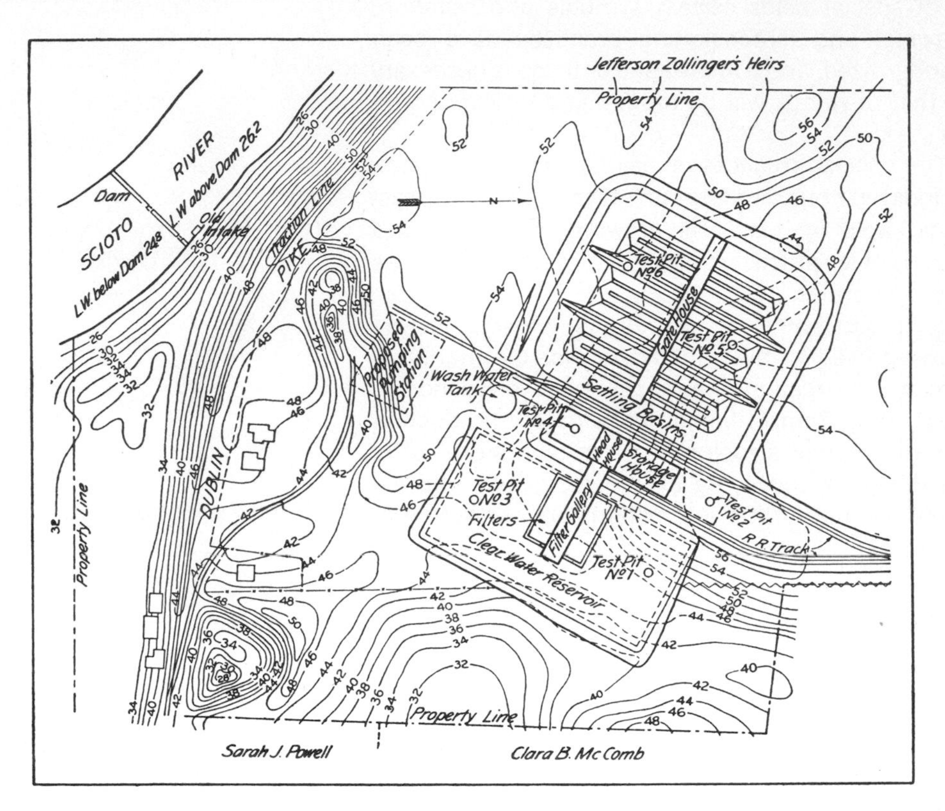


**MAPS AND TOPOGRAPHY.** Maps and topographic drawings are made up chiefly of symbols.

## 12. Drawings for Engineering Design and Construction.

There are many types of drawings used in a wide diversity of fields. Typical are drawings for presentation, proposal, and design; drawings for manufacture, construction, and assembly; drawings giving specialized information for a particular field; and many others.

When a new machine or structure is designed, the first drawings are usually in the form of freehand sketches on which the original ideas, scheming, and inventing are worked out. These drawings are accompanied or followed by calculations to prove the suitability of the design. Working from the sketches and calculations, the design department produces a *design* drawing. This is a preliminary pencil drawing on which more details of the design are worked out. It is accurately made with instruments, full size if possible, and shows the shape and position of the various parts. Little attempt is made to show all the intricate detail; only the essential dimensions, such as basic calculated sizes, are given. On the drawing, or separately as a set of written notes, will be the designer's general specifications for materials, heat-treatments, finishes, clearances, or interferences, etc., and any other information needed by the draftsman in making up the individual drawings of the separate parts.

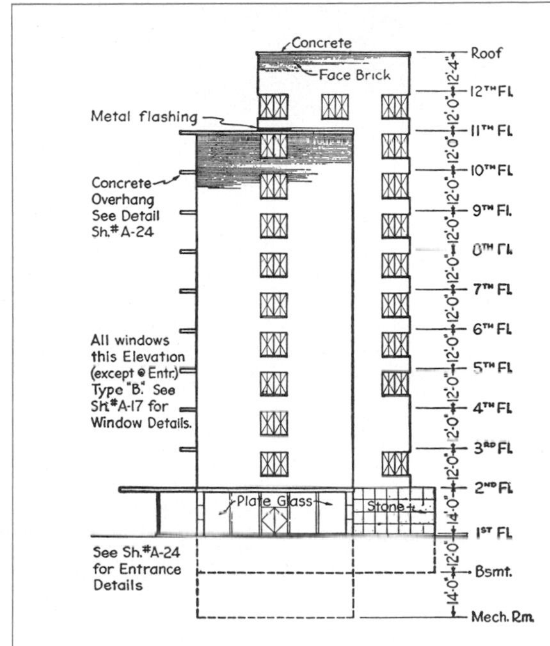


**STRUCTURES.** Architectural standards of the American Institute of Architects (AIA) and structural standards of the American Institute of Steel Construction (AISC) prevail in the building industry.

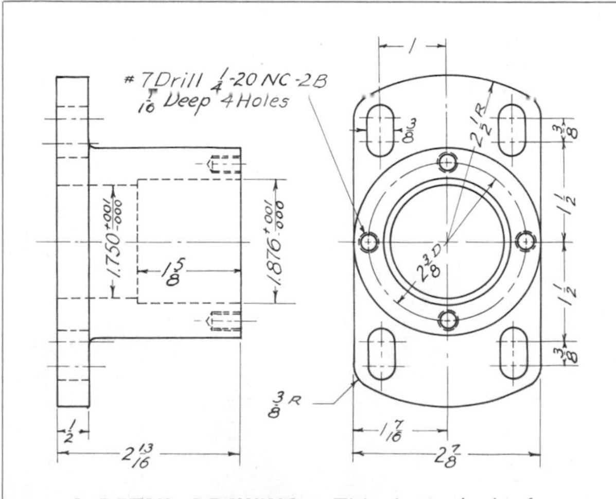


**A DETAIL DRAWING.** This is typical of a drawing used to manufacture a single part.

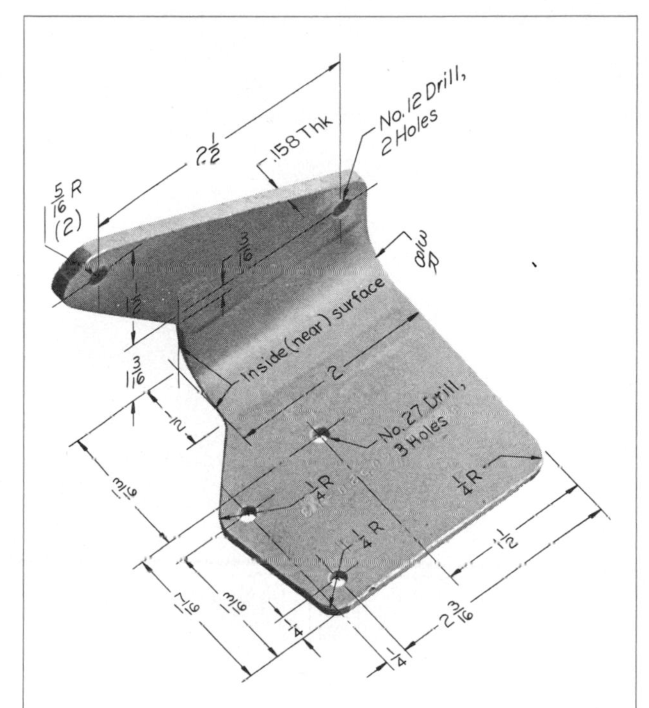


**SPECIALIZED PRACTICES.** These include simplified drawing practices, special uses of reproduction processes, photography, and modelmaking. The figure above is a photo drawing.

## 13. Terminology.

During any course of study in graphics the student must become thoroughly familiar with the terminology of design and construction. The glossary of technical, structural, architectural, and welding terms in the Appendix should be consulted as needed. Form the habit of never passing over a word without understanding its meaning completely.

## 14. Appendix Tables.

Many details of graphics depend upon specifications for drills, bolts and screws, fits, limits, pipe sizes, symbols, and other standards. It is important to become well acquainted with this material, which is given in the Appendix.

This Chapter Discusses:

Selection of Instruments

Boards, Paper, Tape, Thumbtacks

Pencils, Erasers, Pens, Ink

T square, Triangles, Scales, Curves

Case Instruments

The Use of T Square, Triangles, Scales, Dividers, Compasses, Ruling Pens, Curves, Erasers

The Alphabet of Lines

Tangents

Problems

# 2

# GRAPHIC INSTRUMENTS AND THEIR USE

**Accurate Representation of Shape, Relationship, and Size is Accomplished through the Use of Instruments**

## 1. Graphic Instruments—Why They Are Needed.

To record information on paper (or another surface), instruments and equipment are required. Even for drawings made freehand, pencils, erasers, and sometimes coordinate paper or other special items are used.

The lines made on drawings are *straight* or *curved* (including circles and arcs). They are made with drawing instruments, which are the necessary tools for laying down lines on a drawing in an accurate and efficient manner. In order to position the lines, a measuring device, a scale, is needed.

The various instruments will be described in detail later, but the opening of this chapter will serve as an introduction. To draw straight lines, the T square, with its straight blade and perpendicular head, or a triangle is used to support the stroke of the pencil. To draw circles, a compass is needed. In addition to the compass, the draftsman needs dividers for spacing distances and a small bow compass for drawing small circles. To draw curved lines other than circles, a French curve is required. A scale is used for making measurements. To complete drawings in ink, the draftsman will require pens for inking and an ink bottle in holder.

## 2. Selection of Instruments.

In selecting instruments and materials for drawing, secure the *best* you can afford. For one who expects to do work of professional grade, it is a mistake to buy inferior instruments. Sometimes a beginner is tempted to get cheap instruments for learning, expecting to buy better ones later. With reasonable care a set of good instruments will last a lifetime, whereas poor ones will be an annoyance from the start and worthless after short use. Since poor instruments can be difficult to distinguish from good ones, it is well to seek trustworthy advice before buying. Instruments have been greatly improved in recent years.

## 3. Drawing Boards.

The drawing surface may be the table top itself or a separate board. In either case the working surface should be made of well-seasoned clear white pine or basswood, cleated to prevent warping. The working edge must be straight and should be tested with a steel straightedge. Some boards and table tops are supplied with a hardwood edge or a steel insert on the working edge, thus ensuring a better wearing surface. Figure 1 illustrates a drawing board which, because of its design and the materials of which it is made, is rigid, sturdy, and light.

## 4. Drawing Paper.

Drawing paper is made in a variety of qualities and may be had in sheets or rolls. White drawing papers that will not turn yellow with age or exposure are used for finished drawings, maps, charts, and drawings for photographic reproduction. For pencil layouts and working drawings, cream or buff detail papers are preferred as they are easier on the eyes and do not show soil so quickly as white papers. In general, paper should have sufficient grain, or "tooth," to take the pencil, be agreeable to the eye, and have a hard surface not easily grooved by the pencil and good erasing qualities. Formerly imported papers were considered the best, but American mills are now making practically all the paper used in this country. Cheap manila papers should be avoided.

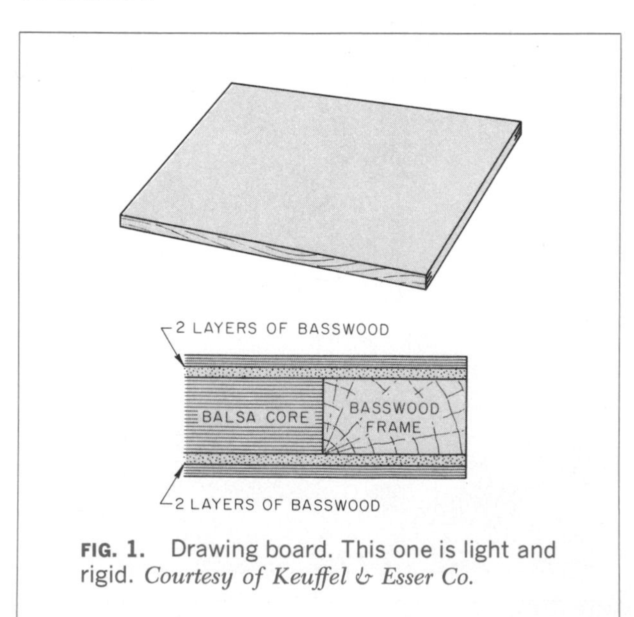


**FIG. 1.** Drawing board. This one is light and rigid. *Courtesy of Keuffel & Esser Co.*

## 5. Tracing Paper.

Tracing papers are thin papers, *natural* or *transparentized,* on which drawings are traced, in pencil or ink, and from which blueprints or similar contact prints can be made. In most drafting rooms original drawings are penciled on tracing papers, and blueprints are made directly from these drawings, a practice increasingly successful because of improvements both in papers and in printing. Tracing papers vary widely in color, thickness, surface, etc., and the grade of pencil and the technique must be adjusted to suit the paper; but with the proper combination, good prints can be obtained from such drawings.

## 6. Tracing Cloth.

Finely woven cloth coated with a special starch or plastic is used for making drawings in pencil or ink. The standard tracing cloth is used for inked tracings and specially made pencil cloth for pencil drawings or tracings. Cloth is more permanent than paper as it will stand more handling. Tracing and duplicating processes are described later.

## 7. Drafting Tape.

The paper to be used is usually attached to the drawing board by means of Scotch drafting tape, with a short piece stuck across each corner or with tape along the entire edge of the paper. Drafting tape is not the same as masking tape (made by the same company); the latter has a heavier coating of adhesive and does not come off the drawing paper so cleanly as the former.

## 8. Thumbtacks.

The best thumbtacks are made with thin heads and steel points screwed into them. Cheaper ones are made by stamping. Use tacks with tapering pins of small diameter and avoid flat-headed (often colored) map pins, as the heads are too thick and the pins rather large.

## 9. Pencils.

The basic instrument is the graphite lead pencil, made in various hardnesses. Each manufacturer has special methods of processing designed to make the lead strong and yet give a smooth clear line. Figure 2 shows five varieties of pencils. At the left are two

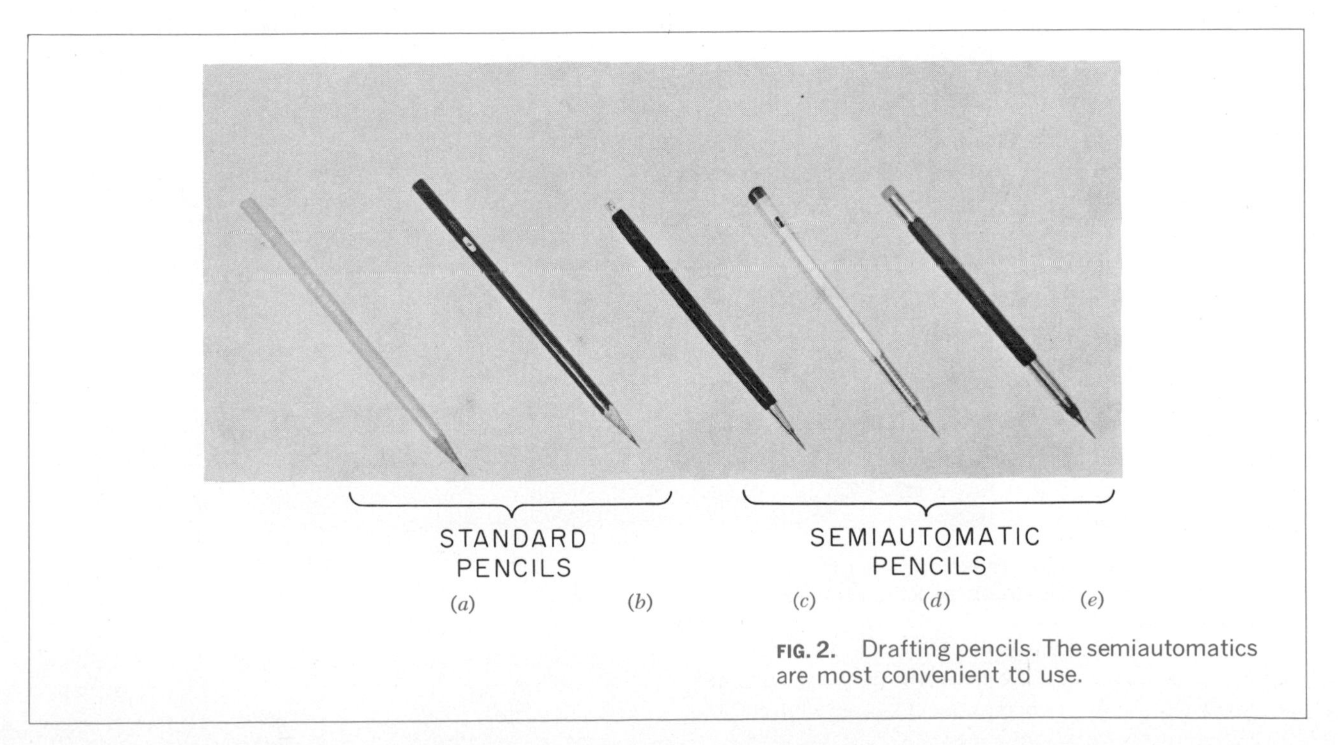


**FIG. 2.** Drafting pencils. The semiautomatics are most convenient to use.

ordinary pencils, with the lead set in wood. (*a*) is of American and (*b*) of foreign manufacture. Both are fine but have the disadvantage that in use, the wood must be cut away to expose the lead (a time-consuming job), and the pencil becomes shorter until the last portion must be discarded. Semiautomatic pencils, (*c*) to (*e*), with a chuck to clamp and hold the lead, are more convenient. (*c*) has a plastic handle and changeable tip (for indicating the grade of lead). (*d*) has an aluminum handle and indexing tip for indication of grade. (*e*) is the rather elegant Alteneder pencil with rosewood handle, hardened-steel chuck ring, and eraser. Drawing pencils are graded by numbers and letters from 6B, very soft and black, through 5B, 4B, 3B, 2B, B, and HB to F, the medium grade; then H, 2H, 3H, 4H, 5H, 6H, 7H, and 8H to 9H, the hardest. The soft (B) grades are used primarily for sketching and rendered drawings and the hard (H) grades for instrument drawings.

**10. Pencil Pointer.**

After the wood of the ordinary pencil is cut away with a pocketknife or mechanical sharpener, the lead must be formed to a long, conic point. A lance-tooth steel file, Fig. 3A, about 6 in. long is splendid for the purpose. Some prefer the standard sandpaper pencil-pointer pad, Fig. 3B.

**11. Erasers.**

The Ruby pencil eraser, Fig. 3C, large size with beveled ends, is the standard. This eraser not only removes pencil lines effectively but is better for ink than the so-called ink eraser, as it removes ink without seriously damaging the surface of paper or cloth. A good metal erasing shield, Fig. 3D, aids in getting clean erasures.

Artgum or a *soft*-rubber eraser, Fig. 3E, is useful for cleaning paper and cloth of finger marks and smears that spoil the appearance of the completed drawing.

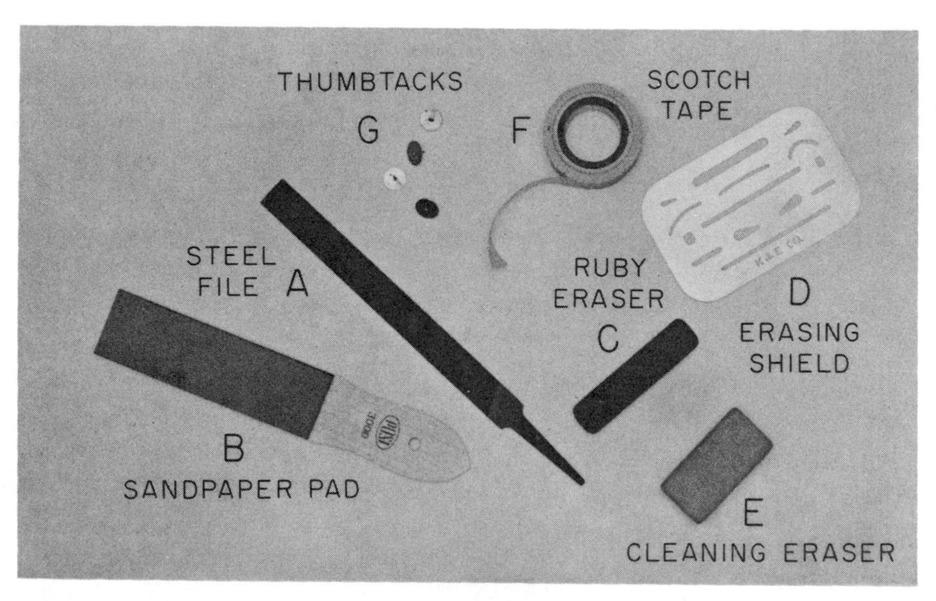


**FIG. 3.** Accessories. These are used for fastening paper, sharpening pencils, and erasing.

## 12. Penholders and Pens.

The penholder should have a grip of medium size, small enough to enter the mouth of a drawing-ink bottle easily yet not so small as to cramp the fingers while in use. A size slightly larger than the diameter of a pencil is good.

An assortment of pens for lettering, grading from coarse to fine, may be chosen from those listed in Chap. 4.

A pen wiper of lintless cloth or thin chamois skin should always be at hand for both lettering and ruling pens.

## 13. Drawing Ink.

Drawing ink is finely ground carbon in suspension, with natural or synthetic gum added to make the mixture waterproof. Nonwaterproof ink flows more freely but smudges easily. Drawing ink diluted with distilled water or Chinese ink in stick form rubbed up with water on a slate slab is used in making wash drawings and for very fine line work.

Bottleholders prevent the bottle from upsetting and ruining the drawing table or floor. They are made in various patterns; one is illustrated in Fig. 4. As a temporary substitute, the lower half of the paper container in which the ink is sold may be fastened to the table with a thumbtack, or a strip of paper or cloth with a hole for the neck of the bottle may be tacked down over the bottle. Several companies are now supplying ink in small-necked plastic bottles or tubes, Fig. 4, which are squeezed to supply the ink, a drop at a time. These containers prevent spillage and protect the ink from deterioration through evaporation or contamination.

## 14. The T Square.

The fixed-head T square, Fig. 5, is used for all ordinary work. It should be of hardwood, and the blade should be perfectly straight. The transparent-edged blade is much the best. A draftsman will have several fixed-head squares of different lengths and will find an adjustable-head square of occasional use.

## 15. Triangles.

Triangles, Fig. 5, are made of transparent celluloid (fiberloid) or other plastic material. Through internal strains they sometimes lose their accuracy. Triangles should be kept flat to prevent warping. For ordinary work, a 6- or 8-in. 45° and a 10-in. 30–60° are good sizes.

**FIG. 4.** Standard bottle in holder; Leroy pen-filler cartridge. The cartridge is particularly convenient.

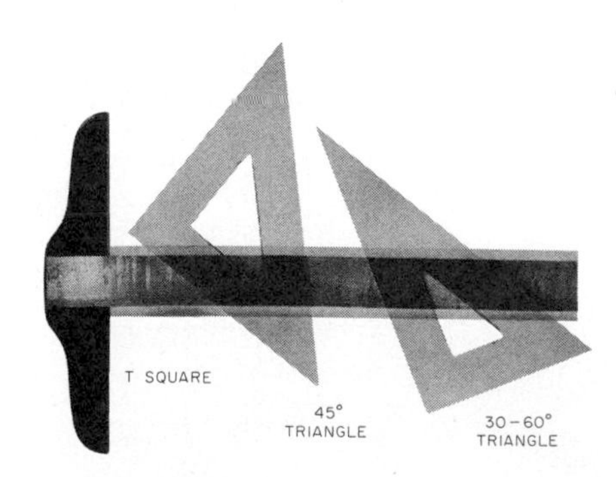


**FIG. 5.** T square and triangles. These are standard equipment for drawing straight lines.

## 16. Scales.

Scales, Figs. 6 to 9, are made in a variety of graduations to meet the requirements of many different kinds of work. For convenience, scales are classified according to their most-common uses.

*Mechanical Engineer's Scales.* These are divided and numbered so that fractions of inches represent inches. The most common ranges are ⅛, ¼, ½, and 1 in. to the *inch*. These scales are known as the *size* scales because the designated reduction also represents the ratio of size, as, for example, *one-eighth* size. A full- and half-size scale is illustrated in Fig. 6. Mechanical engineer's scales are almost always "full divided"; that is, the smallest divisions run throughout the entire length. They are often graduated with the marked divisions numbered from right to left, as well as from left to right, as shown in Fig. 6. Mechanical engineer's scales are used mostly for drawings of machine parts and small structures where the drawing size is never less than one-eighth the size of the actual object.

*Civil Engineer's Scales.* These are divided into decimals with 10, 20, 30, 40, 50, 60, and 80 divisions to the inch (Fig. 7). Such a scale is usually full divided and is sometimes numbered both from left to right and right to left. Civil engineer's scales are most used for plotting and drawing maps, although they are convenient for any work where divisions of the inch in tenths is required.

*Architect's Scales.* Divided into proportional feet and inches, these scales have divisions indicating ⅛, ¼, ⅜, ½, ¾, 1½, and 3 in. to the *foot* (Fig. 8).

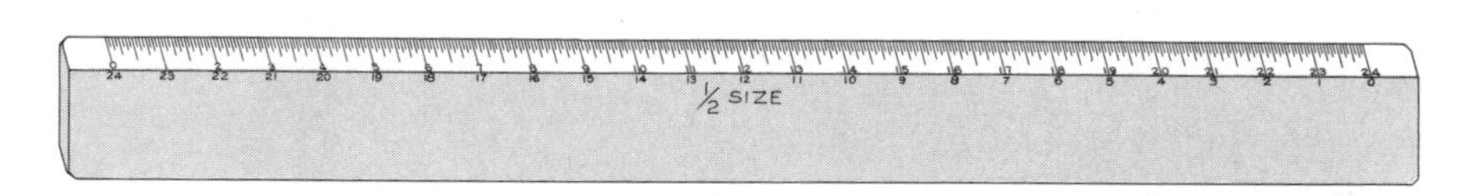


**FIG. 6.** A mechanical engineer's full- and half-size scale. Divisions are in inches to sixteenths of an inch. (The half-size scale is on the back of the full-size scale.)

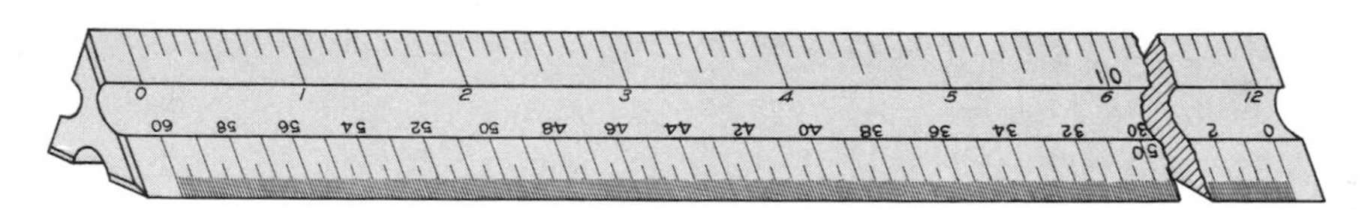


**FIG. 7.** A civil engineer's scale. Divisions are 10, 20, 30, 40, 50, and 60 parts to the inch.

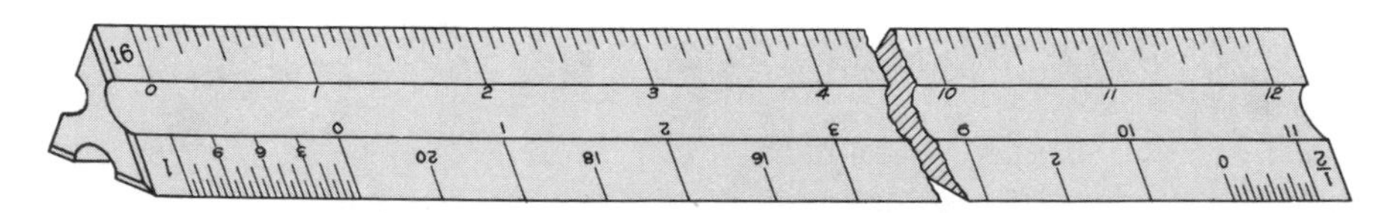


**FIG. 8.** An architect's (or mechanical engineer's) scale. Reduction in size is based on proportion of a foot.

They are usually "open divided"; that is, the units are shown along the entire length, but only the end units are subdivided into inches and fractions. These scales are much used by all engineers—mechanical, industrial, chemical, etc.—for both machine and structural drawings and are sometimes called *mechanical engineer's* scales.

A variety of special scales, made with divisions specified by the customer, are available from most instrument companies. Compared with standard scales, they are expensive.

Scales are made with various cross-sectional shapes, as shown in Fig. 9. The triangular form, (*a*) and (*b*), has long been favored because it carries six scales as a unit and is very stiff. However, many prefer the flat types as being easier to hold flat to a board and having a particular working scale more readily available. The "opposite-bevel" scale, (*c*) and (*d*), is easier to pick up than the "flat bevel" scale, (*f*); moreover, it shows only one graduation at a time. The "double-bevel" scale, (*e*), in the shorter lengths is convenient as a pocket scale, but it can be had in lengths up to 24 in.

Practically all drafting scales of good quality were formerly made of boxwood, either plain or with white edges of celluloid. Although metal scales have been available for more than 30 years, they were seldom used until about 1935, when drafting machines equipped with metal scales made rather important gains in popularity. Extruded medium-hard aluminum alloys are the preferred metals. The Second World War brought a period of considerable experimentation with various types of plastic for all kinds of drafting scales. Up to the present time the white-edge scale has retained a large measure of its popularity. Another type of scale is now getting considerable attention, a metal scale with a white plastic coating that carries the graduations. Both magnesium and aluminum have been used successfully. These scales have all the reading advantages of a white-edge boxwood scale together with the stability afforded by metal.

### 17. Curves.

Curved rulers, called "irregular curves" or "French curves," are used for curved lines other than circle

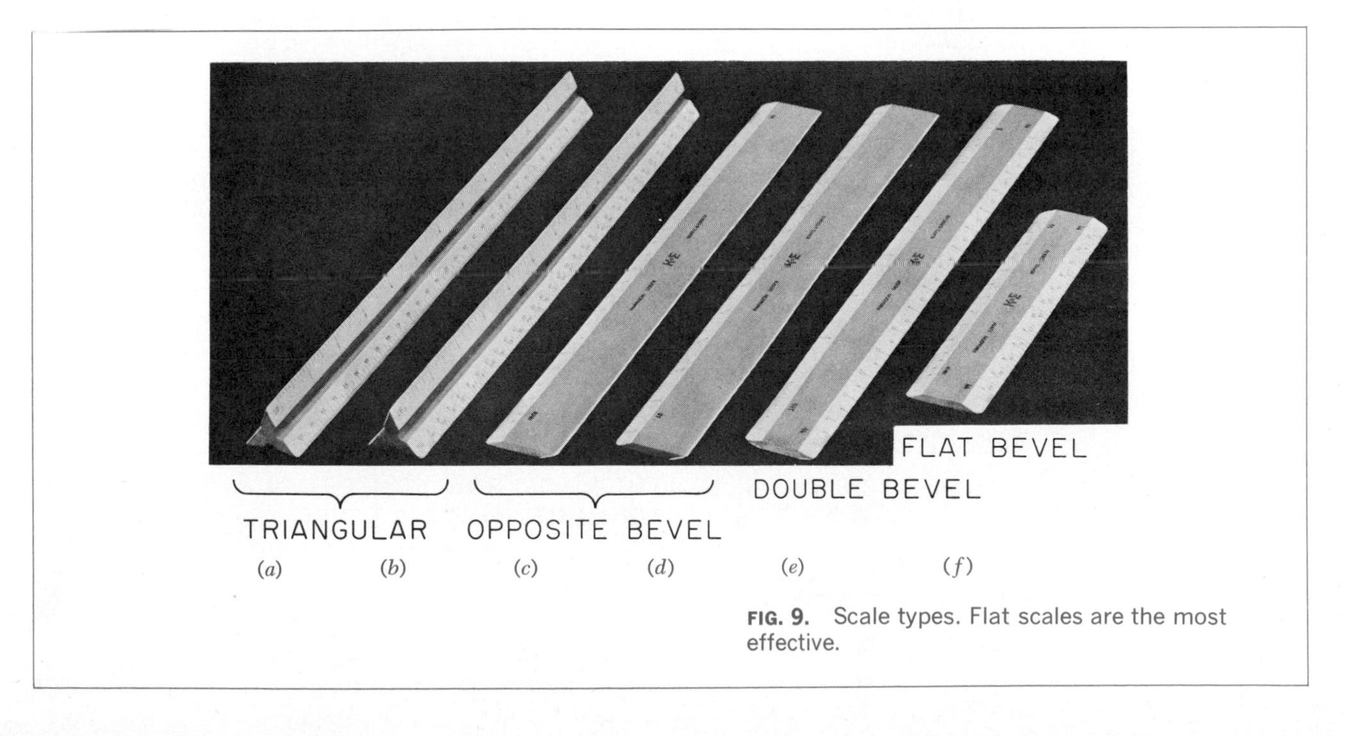


**FIG. 9.** Scale types. Flat scales are the most effective.

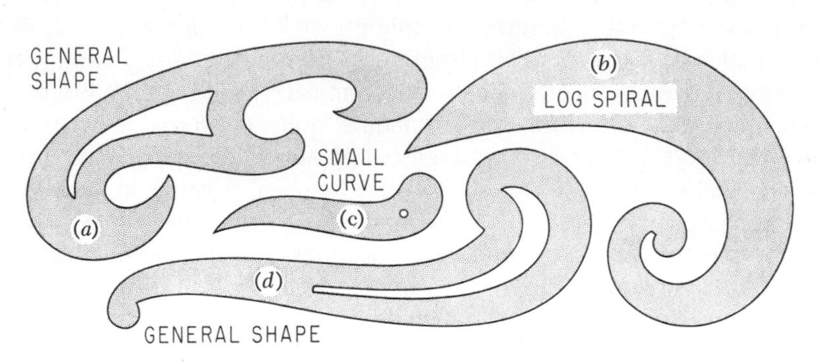


**FIG. 10.** Irregular curves. These are used for drawing curves where the radius of curvature is not constant.

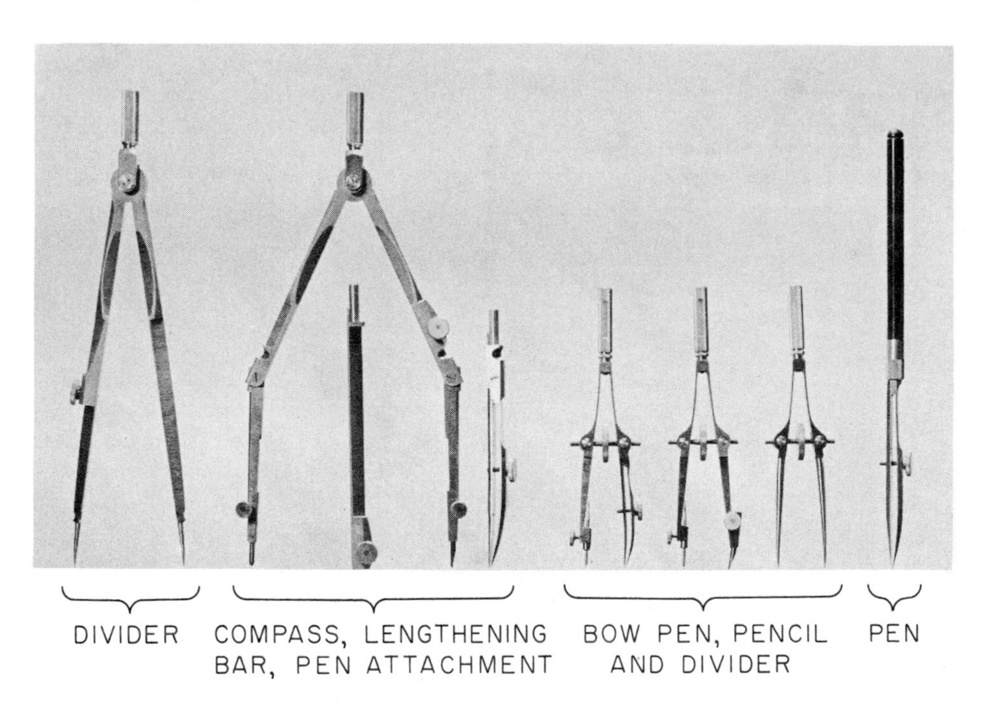


**FIG. 11.** The basic three-bow set. Large compass and dividers; bow pen, pencil, and dividers; ruling pen.

arcs. The patterns for these curves are laid out in parts of ellipses and spirals or other mathematical curves in various combinations. For the student, one ellipse curve of the general shape of Fig. 10*a* or *d* and one spiral, either a logarithmic spiral, (*b*), or one similar to the one used in Fig. 60, is sufficient. (*c*) is a useful small curve.

### 18. The "Case" Instruments.

We have so far, with the exception of curves, considered only the instruments (and materials) needed for drawing straight lines. A major portion of any drawing is likely to be circles and circle arcs, and the so-called "case" instruments are used for these. The basic instruments are shown in Fig. 11. At the left is a divider of the "hairspring" type (with a screw for fine adjustment), used for laying off or transferring measurements. Next is the large compass with lengthening bar and pen attachment. The three "bow" instruments are for smaller work. They are almost always made without the conversion feature (pencil to pen). The ruling pen, at the extreme right, is used for inking straight lines. A set of instruments of the type shown in Fig. 11 is known as a three-bow set.

Until recently the three-bow set was considered the standard, in design and number of pieces, for all ordinary drafting work. However, the trend now is for more rigid construction and fewer pieces. The 6-in. compass in Fig. 12 embraces practically a whole set in one instrument. Used as shown, the instrument is a pencil compass; with pencil replaced by pen, it is used to ink circles; with steel point installed in place of pencil point, it becomes a divider; the pen point, placed in the handle provided, makes a ruling pen; and there is a small metal container for steel points and lead. Similar instruments of other manufacture are shown in Figs. 13 and 14. The instrument in the upper right of Fig. 14 is a "quick-change" design (with vernier adjustment), a great convenience when changing from a very small setting to a large one.

Figure 13*b* shows a quick-action bow employing special nylon nuts in which the center screw operates. The design allows the legs to be moved in or out to change the setting, after which the center screw is used for fine adjustment.

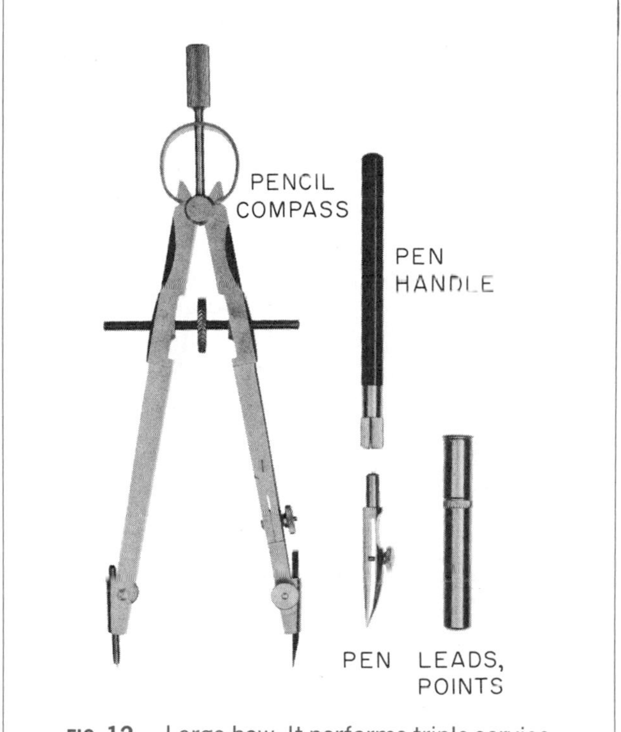


**FIG. 12.** Large bow. It performs triple service as dividers, pen, and pencil compass. Pen leg in handle makes ruling pen.

Even though it is possible to find an instrument that serves practically all purposes, it is convenient to have several instruments, thus saving the time required to convert from pencil leg to ink leg or divider points. For this reason the newer more rigid instruments are also made as separate pieces in different sizes. Figure 15 shows the 6-in. dividers and large 6-in. bow compass; three 3-in. bow compasses, two with pencil and one with ink leg, which can also be used as dividers by changing attachments; and pen points, which are used on the 6-in. compass or in a handle (not shown) to make a ruling pen.

The standard 6-in. compass will open to only approximately 5 in. The lengthening bar of the

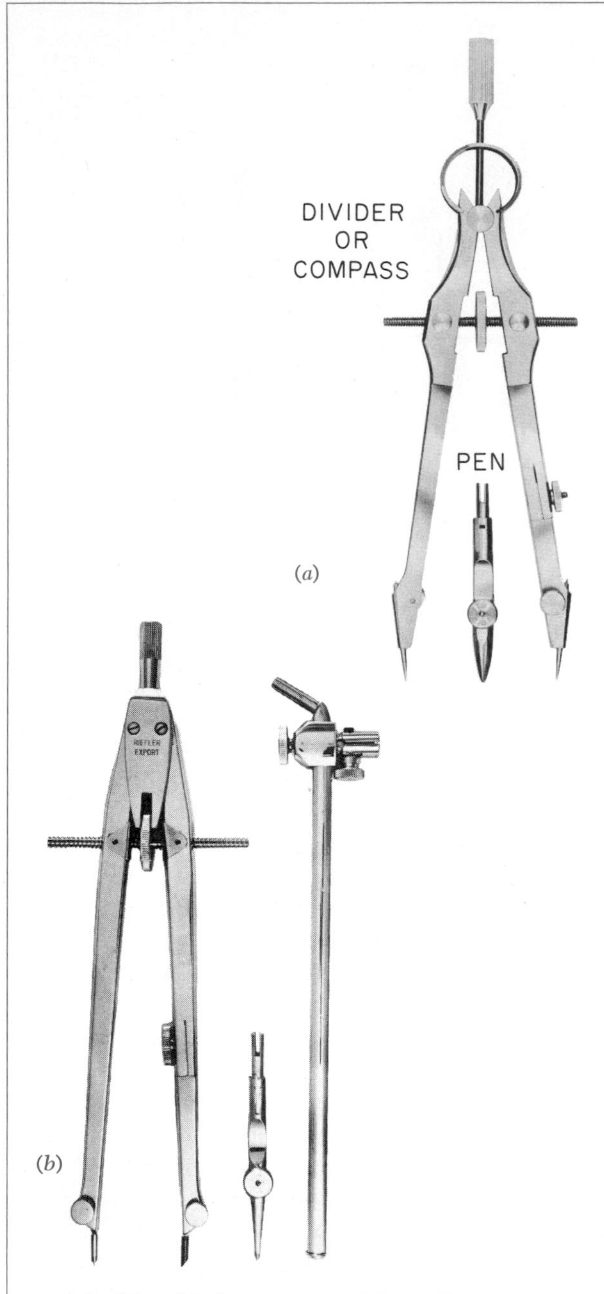


**FIG. 13.** (*a*) A very rugged bow. Serves as dividers, or pen or pencil compass. (*b*) A quick-action bow instrument, with pencil and pen legs and extension bar.

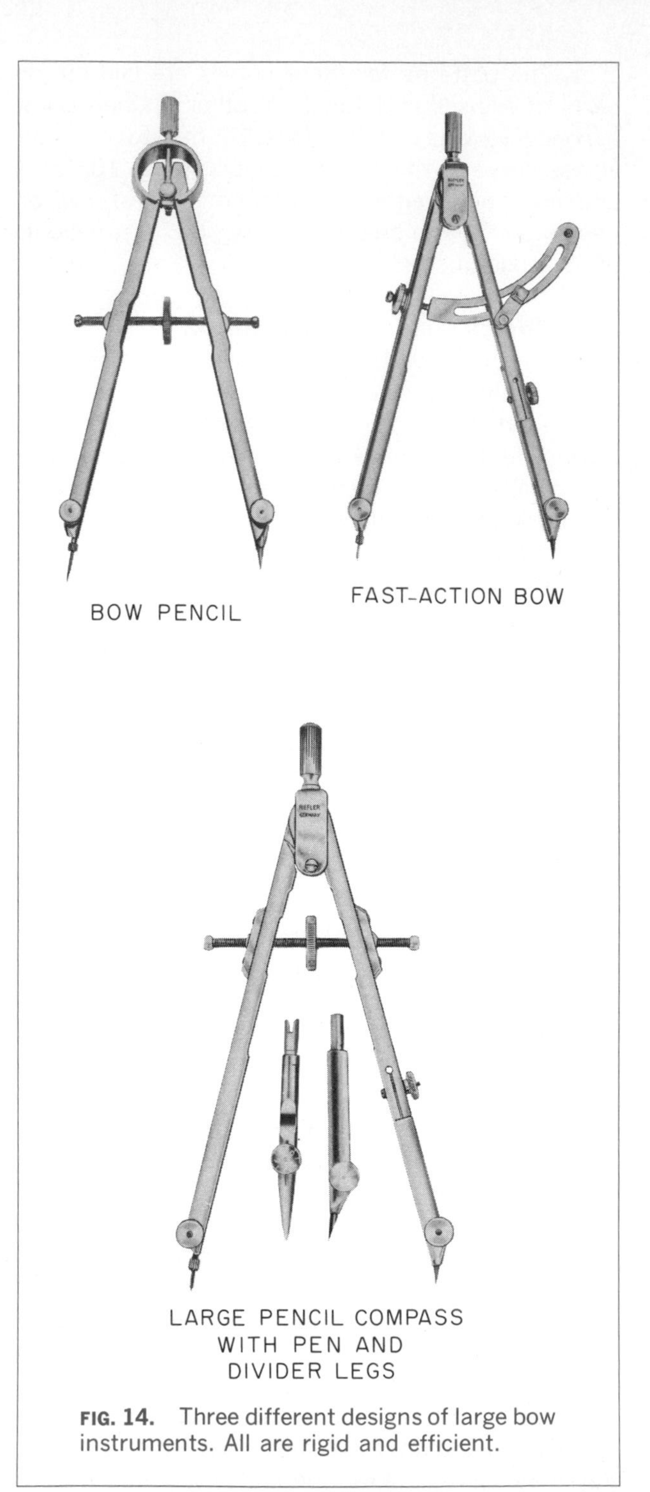


**FIG. 14.** Three different designs of large bow instruments. All are rigid and efficient.

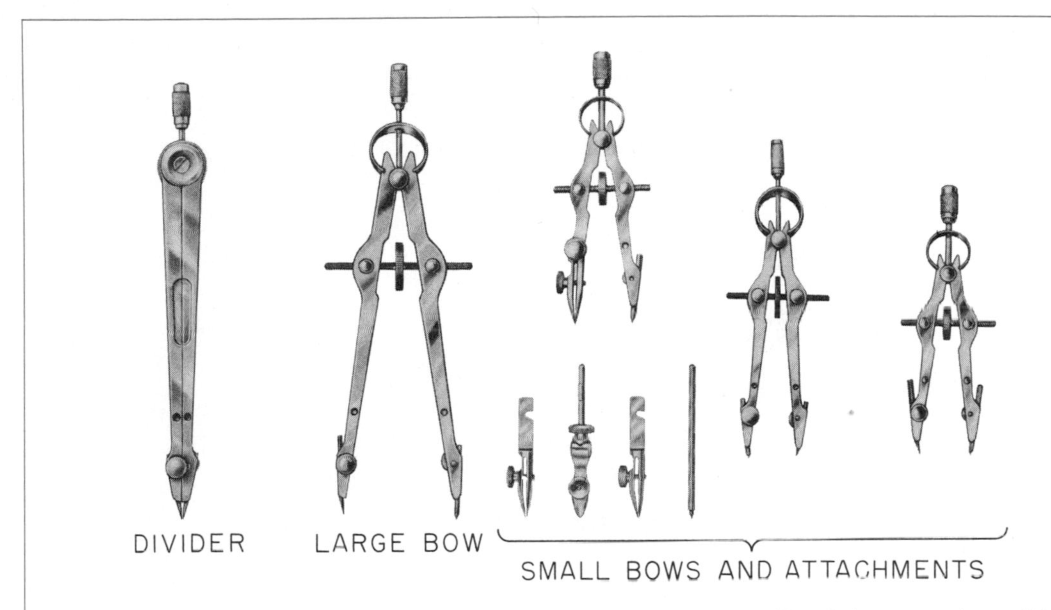


**FIG. 15.** A large- and small-bow set. Combines the advantages of both standard three-bow and large-bow designs.

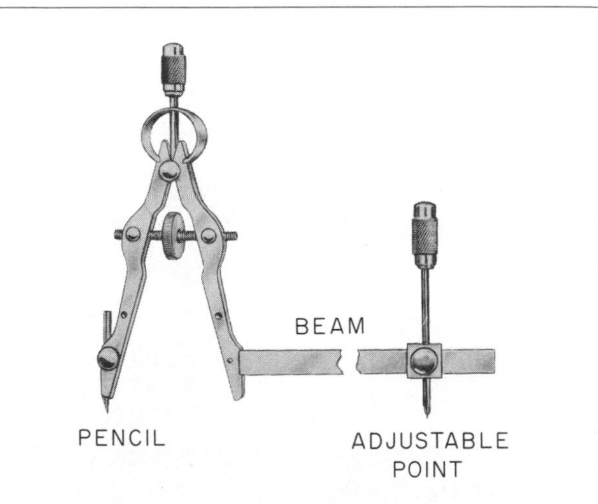


**FIG. 16.** Extender for large bow. This type extends the range of its center point, with the pen or pencil on the bow.

older-type compass (Fig. 11) will extend the radius to about 8 in. Lengthening bars for the newer spring-bow instruments are of the beam type. The instrument in Fig. 16 has a center point on the beam, thus employing the pencil and pen of the instrument itself, while the compass in Fig. 17 uses the center point of the instrument and separate pen and pencil attachments on the beam. A standard metal beam compass for drawing circles up to 16-in. radius is shown in Fig. 18.

Various ruling pens are shown in Fig. 19. (*a*) to (*e*) are standard types. Note that (*d*) uses the compass-pen leg in a handle. (*f*) is a contour pen, (*g*) is a border pen for wide lines, (*h*) and (*i*) are "railroad" pens for double lines, and (*j*) and (*k*) are border pens with large ink capacity.

All manufacturers supply instruments made up in sets, with a leather, metal, or plastic case. Figure 20 shows a standard three-bow combination and Fig. 21 a large-bow set with beam compass. The case in Fig. 22*a* is unique in that fillers may be removed

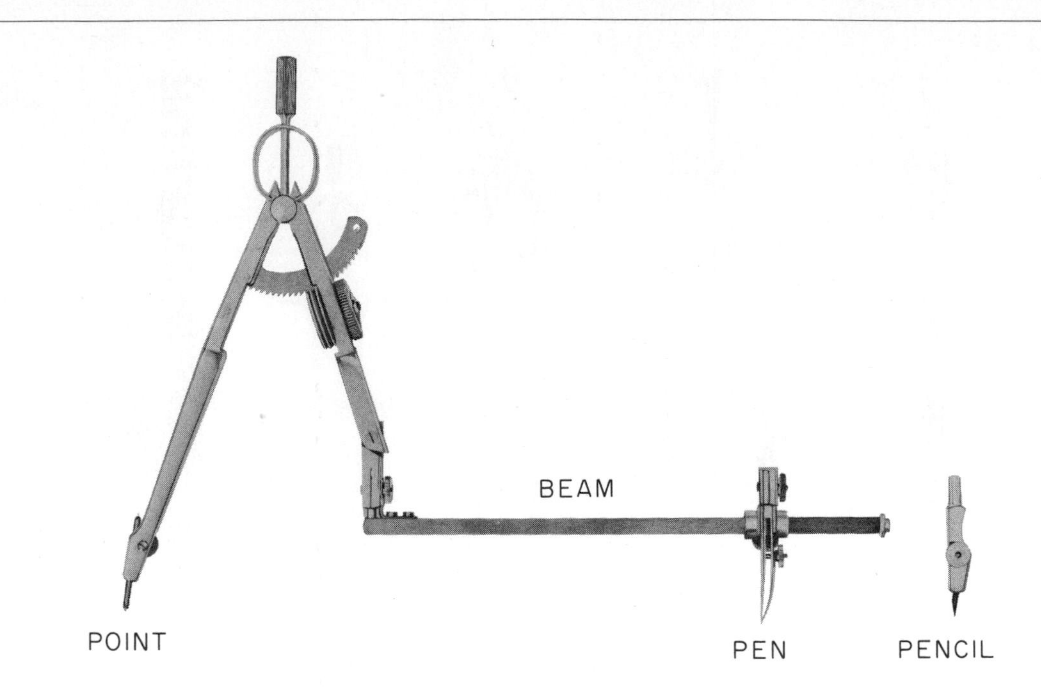


**FIG. 17.** Extender for large bow. This type uses the center point of the bow and has its own pen and pencil attachments.

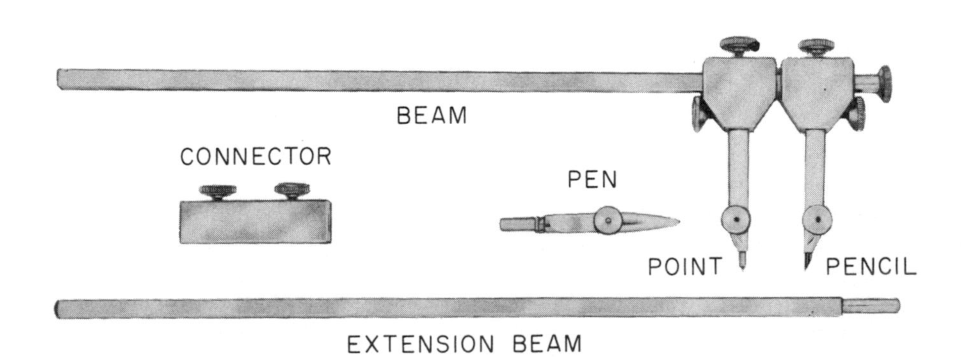


**FIG. 18.** A beam compass. This compass is adjustable for radii of 1 to 16 inches.

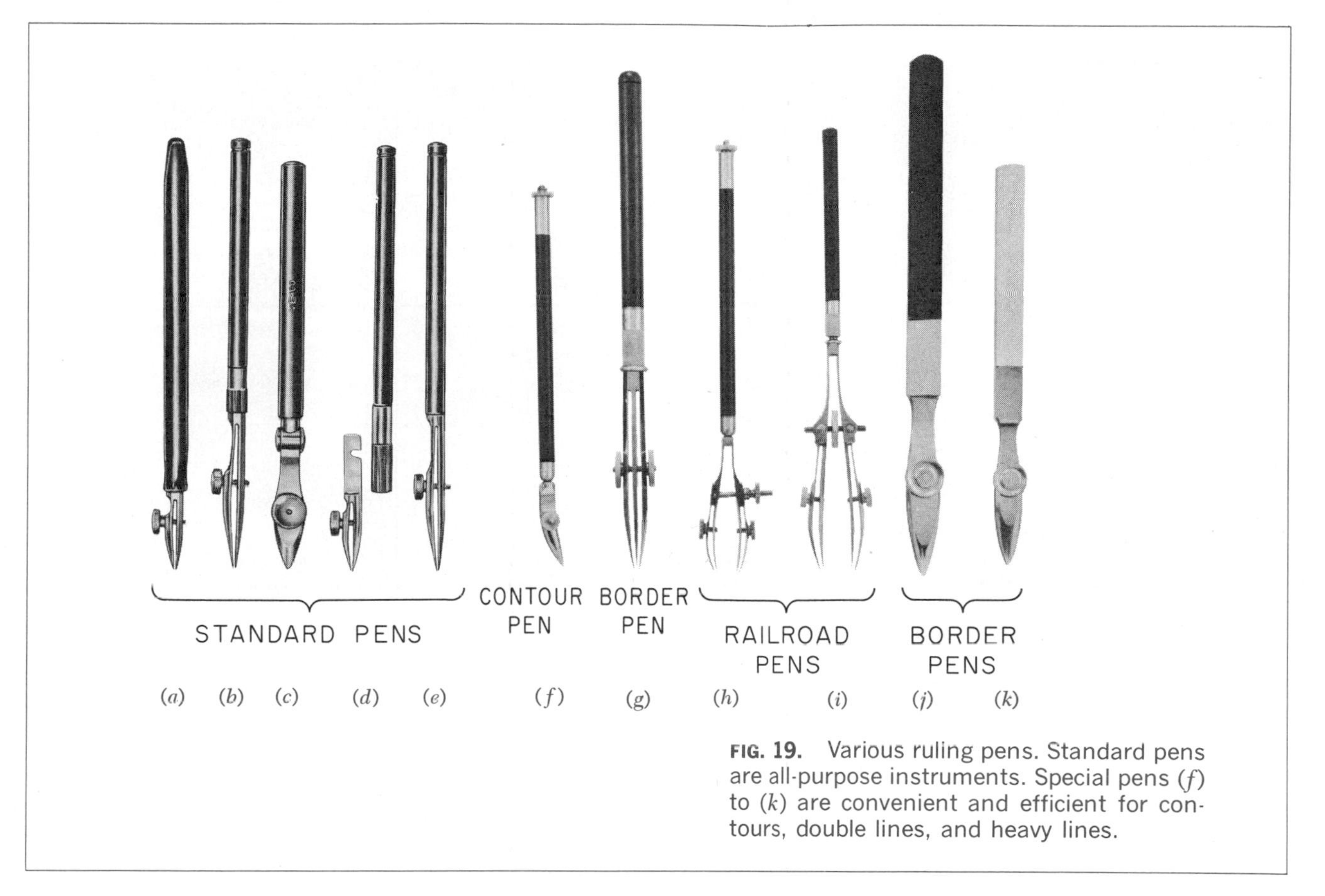


**FIG. 19.** Various ruling pens. Standard pens are all-purpose instruments. Special pens (*f*) to (*k*) are convenient and efficient for contours, double lines, and heavy lines.

to provide space for any combination of instruments desired.

Figure 22*b* shows a modern case made of plastic with a magnetic fastener. This set of instruments includes the newer technical fountain pen with several sizes of points.

## 19. Lettering Devices.

The Braddock-Rowe triangle and the Ames lettering instrument (Figs. 2*a* and *b*) are convenient devices used in drawing guide lines for lettering.

## 20. Checklist of Instruments and Materials.

Set of drawing instruments, including: 6-in. compass with fixed needle-point leg, removable pencil and pen legs, and lengthening bar; 6-in. hairspring dividers; 3½-in. bow pencil, bow pen, and bow dividers; two ruling pens; box of leads. Or large-bow set containing 6½-in. bow compass; 4½-in. bow compass; 6½-in. friction dividers and pen attachment for compass; 5½-in. ruling pen; beam compass with extension beam; box for extra leads and points

Drawing board
T square
45° and 30–60° triangles
Three mechanical engineer's scales, flat pattern, or the equivalent triangular scale
Lettering instrument or triangle
French curves
Drawing pencils, 6H, 4H, 2H, H, and F
Pocketknife or pencil sharpener
Pencil pointer (file or sandpaper)

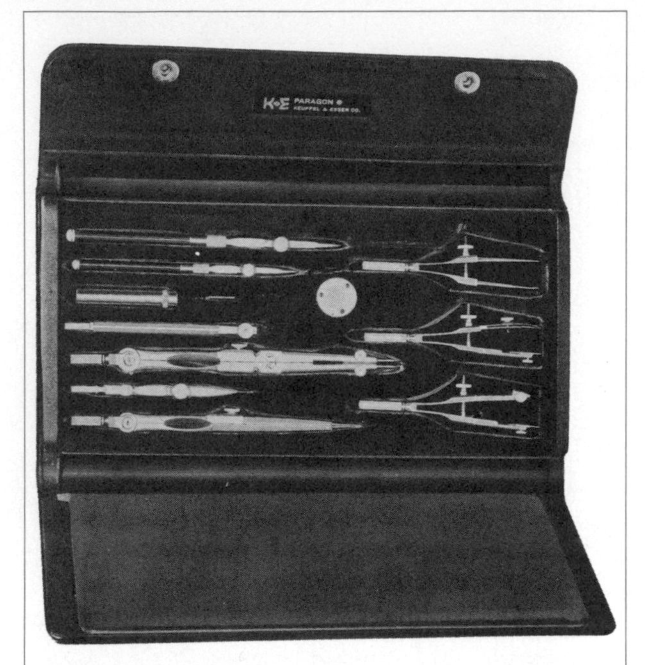

FIG. 20. A three-bow set, in case. It contains large dividers, large compass with extension bars and pen attachment, two ruling pens, bow pen, bow pencil, bow dividers, and bone center.

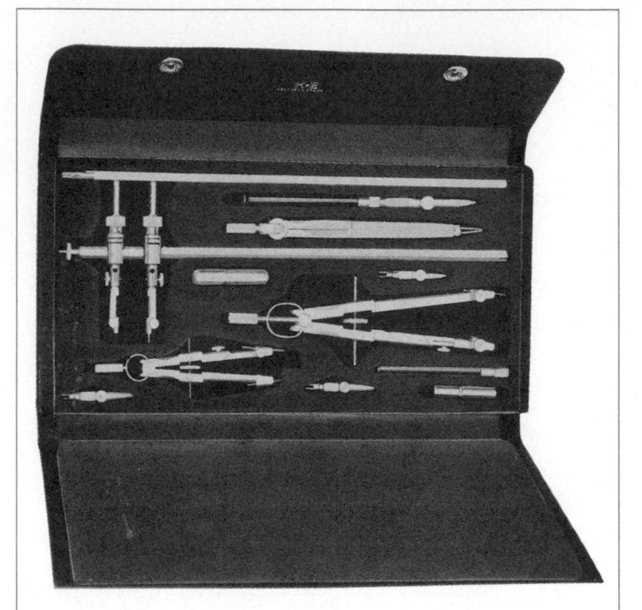

FIG. 21. A large-bow set, in case. It contains large- and small-bow compasses with attachments, dividers, ruling pen, and beam compass.

Pencil eraser (Ruby)
Artgum or cleaning rubber
Penholder, pens for lettering, and penwiper
Bottle of drawing ink and bottleholder
Scotch drafting tape or thumbtacks
Drawing paper to suit
Tracing paper and cloth
Dustcloth or brush

To these may be added:

Civil engineer's scale
Protractor
Erasing shield
Slide rule
Six-foot steel tape
Clipboard or sketchbook
Hard Arkansas oilstone
Piece of soapstone
Cleaning powder or pad

The student should mark all his instruments and materials plainly with his initials or name as soon as they have been purchased and approved.

**21. Additional Instruments.** The instruments and materials described in this chapter are all that are needed for ordinary practice and are, with the exception of such supplies as paper, pencils, ink, and erasers, what a draftsman is as a rule expected to take with him into a drafting room.

There are many other special instruments and devices that are not necessary in ordinary work but with which the draftsman should be familiar, as they may be convenient in special cases and are often found as a part of drafting-room equipment. Consult manufacturers' catalogues.

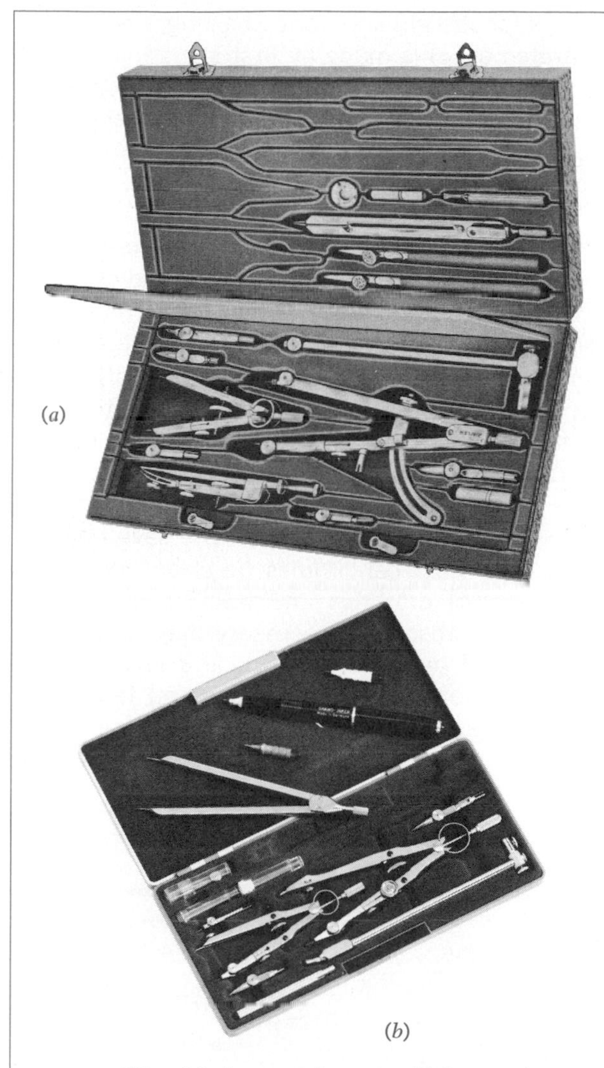

FIG. 22. (*a*) A special case which can be adapted to many combinations by removing fillers. (*b*) A set of instruments in plastic case. This set is supplied with technical fountain pen.

## 22. The Use of Instruments.

In beginning to use drawing instruments, it is important to learn to handle them correctly. Carefully read the instructions and observe strictly all details of technique.

Facility will come with continued practice, but it is essential to adhere to good form from the outset. Bad form in drawing can be traced in every instance to the formation of bad habits in the early stages of learning. Once formed, these habits are difficult to overcome.

It is best to make a few drawings solely to become familiar with the handling and feel of the instruments so that later, in working a drawing problem, you will not lose time because of faulty manipulation. Practice accurate penciling first, and do not attempt inking until you have become really proficient in penciling. With practice, the correct, skillful use of drawing instruments will become a subconscious habit.

For competence in drawing, *accuracy* and *speed* are essential, and in commercial work neither is worth much without the other. It is well to learn as a beginner that a *good* drawing can be made as quickly as a *poor* one. Erasing is expensive and most of it can be avoided. The draftsman of course erases occasionally, and a student must learn to make corrections, but in beginning to use instruments, strive for sheets without blemish or inaccuracy.

## 23. Preparation for Drawing.

The drawing table should be set so that the light comes from the left, and it should be adjusted to a convent height, that is, 36 to 40 in., for use while sitting on a standard drafting stool or while standing. There is more freedom in drawing standing, especially when working on large drawings. The board, for use in this manner, should be inclined at a slope of about 1 to 8. Since it is more tiring to draw standing, many modern drafting rooms use tables so made that the board can be used in an almost vertical position and can be raised or lowered so that the draftsman can use a lower stool with swivel seat and backrest, thus working with comfort and even greater freedom than when an almost horizontal board is used.

The instruments should be placed within easy reach, on the table or on a special tray or stand which is located beside the table. The table, the board, and the instruments should be wiped with a dustcloth before starting to draw.

**24. The Pencil and Its Use.**
The grade of pencil must be selected carefully, with reference to the surface of the paper as well as to the line quality desired. For a pencil layout on detail paper of good texture, a pencil as hard as 5H or 6H may be used, while for finished pencil drawings on the same paper, 2H, 3H, or 4H pencils give the blacker line needed. For finished pencil drawings or tracings on vellum, softer pencils, H to 3H, are employed to get printable lines. The F pencil is much used for technical sketching, and the H is popular for lettering. In every case the pencil must be hard enough not to blur or smudge but not so hard as to cut grooves in the paper under reasonable pressure.

*To sharpen a pencil,* cut away the wood from the unlettered end with a pen-knife or mechanical sharpener, as shown in Fig. 23*a*, and then sharpen the lead to make a long, conic point, as at (*b*), by twirling the pencil as the lead is rubbed with long even strokes against the sandpaper pad or file or placed in a special lead sharpener.

A flat or wedge point will not wear away in use so fast as a conic point, and on that account some prefer it for straight-line work. The long, wedge point illustrated at (*c*) is made by first sharpening, as at (*a*), then making the two long cuts on opposite sides, as shown, then flattening the lead on the sandpaper pad or file, and finishing by touching the corners to make the wedge point narrower than the diameter of the lead.

Have the sandpaper pad within easy reach, and *keep the pencils sharp*. Some hang the pad or file on a cord attached to the drawing table. The professional draftsman sharpens his pencil every few minutes. After sharpening the lead, wipe off excess graphite dust before using the pencil. Form the habit of sharpening the lead as often as you might dip a writing pen into the inkwell. Most commercial and many college drafting rooms are equipped with Dexter or other pencil sharpeners to save time.

Not only must pencil lines be clean and sharp, but for pencil drawings and tracings to be blueprinted, it is absolutely necessary that all the lines of each kind be uniform, firm, and opaque. This means a careful choice of pencils and the proper use of them. The attempt to make a dark line with too hard a pencil results in cutting deep grooves in

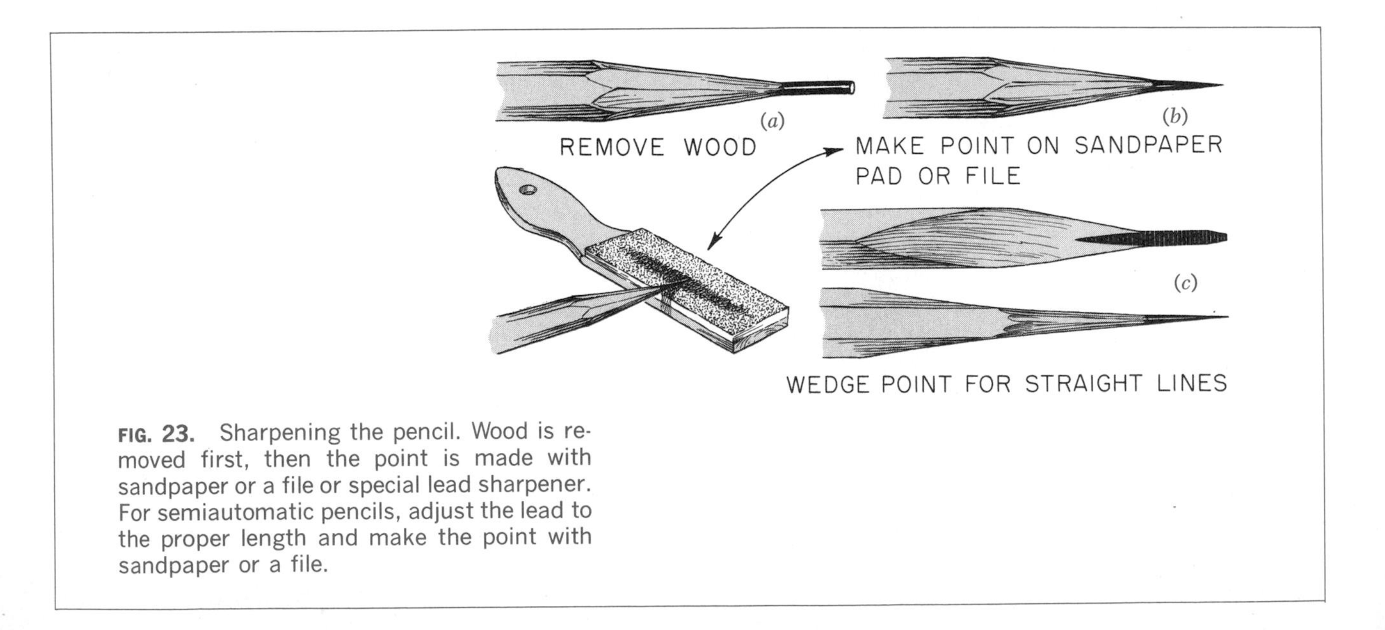


**FIG. 23.** Sharpening the pencil. Wood is removed first, then the point is made with sandpaper or a file or special lead sharpener. For semiautomatic pencils, adjust the lead to the proper length and make the point with sandpaper or a file.

the paper. Hold the pencil firmly, yet with as much ease and freedom as possible.

Keep an even constant pressure on the pencil, and when using a conic point, rotate the pencil as the line is drawn so as to keep both the line and pencil sharp. Use a draftsman's brush or soft cloth occasionally to dust off excess graphite from the drawing.

Too much emphasis cannot be given to the importance of clean, careful, accurate penciling. Never entertain the thought that poor penciling can be corrected in tracing.

### 25. Placing the Paper.

Since the T-square blade is more rigid near the head than toward the outer end, the paper, if much smaller than the size of the board, should be placed close to the left edge of the board (within an inch or so) with its lower edge several inches from the bottom of the board. With the T square against the left edge of the board, square the top of the paper; hold it in this position, slipping the T square down from the edge, and put a thumbtack in each upper corner, pushing it in up to the head so that the head aids in holding the paper. Then move the T square down over the paper to smooth out possible wrinkles, and put thumbtacks in the other two corners. Drafting tape may be used instead of thumbtacks.

### 26. Use of the T Square.

The T square and the triangles have straight edges and are used for drawing straight lines. Horizontal lines are drawn with the T square, which is used with its head against the left edge of the drawing board and manipulated as follows: Holding the head of the tool, as shown in Fig. 24*a*, slide it along the edge of the board to a spot very near the position desired. Then, for closer adjustment, change your hold either to that shown at (*b*), in which the thumb remains on top of the T-square head and the other fingers press against the underside of the board, or, as is more usual, to that shown at (*c*), in which the fingers remain on the T square and the thumb is placed on the board.

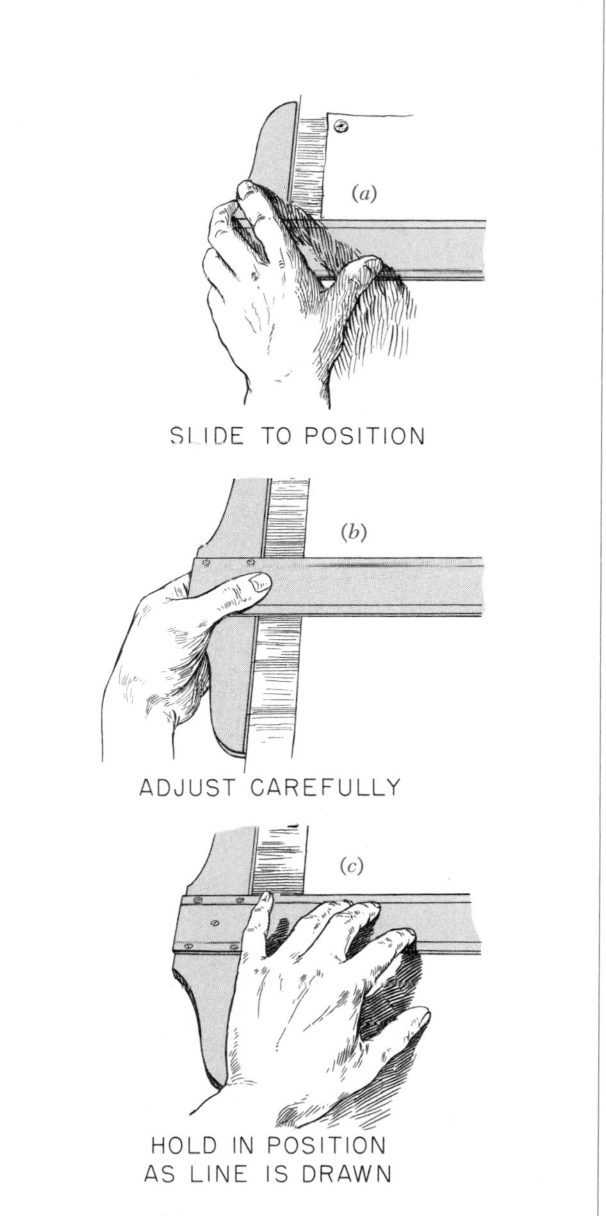


**FIG. 24.** Manipulating the T square. The head must be firmly against the straight left edge of the board.

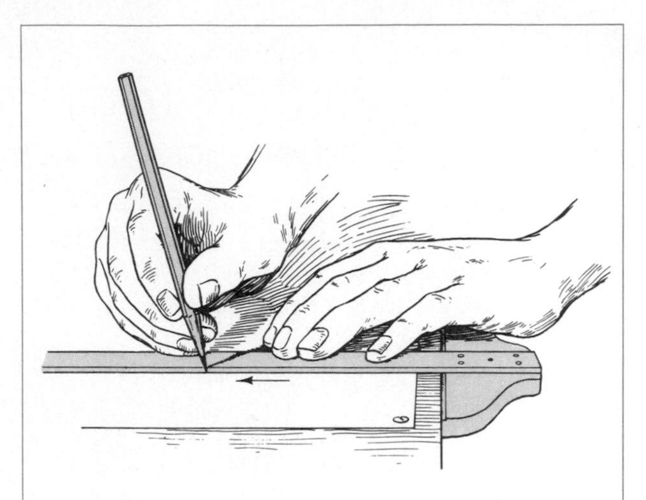

**FIG. 25.** Drawing a horizontal line. Hold the T square with the left hand; draw the line from left to right; incline the pencil in the direction of stroke, so that the pencil "slides" over the paper.

Figure 25 shows the position of the hand and pencil for drawing horizontal lines. Note that the pencil is inclined in the direction the line is drawn, that is, toward the right, and also slightly away from the body so that the pencil point is as close as possible to the T-square blade.

In drawing lines, take great care to keep them accurately parallel to the guiding edge of the T square. The pencil should be held lightly, but close against the edge, and the angle should not vary during the progress of the line. Horizontal lines should always be drawn from left to right. A T-square blade can be tested for straightness by drawing a sharp line through two points and then turning the square over and with the same edge drawing another line through the points, as shown in Fig. 26.

### 27. Use of the Triangles.

Vertical lines are drawn with the triangle, which is set against the T square with the perpendicular edge nearest the head of the square and thus toward the light (Fig. 27). These lines are always drawn upward, from bottom to top.

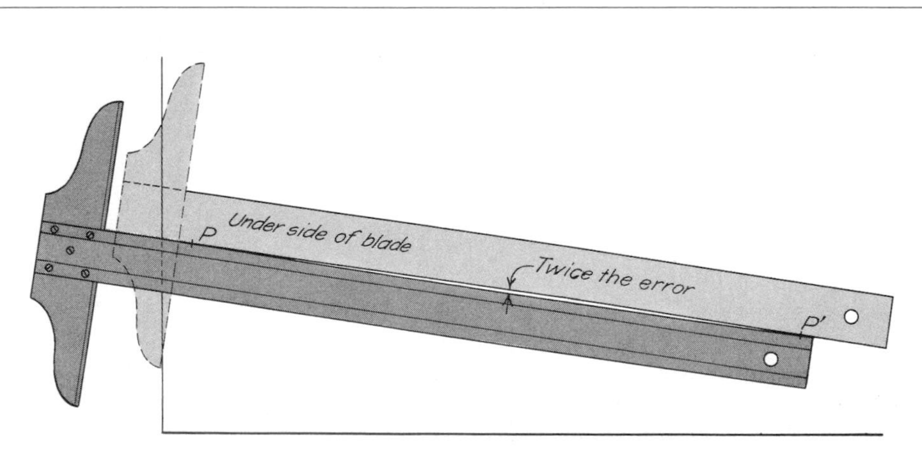


**FIG. 26.** To test a T square. Turn the T square upside down; draw the line; turn it right side up and align it with the original line; draw the second line and compare it with the first.

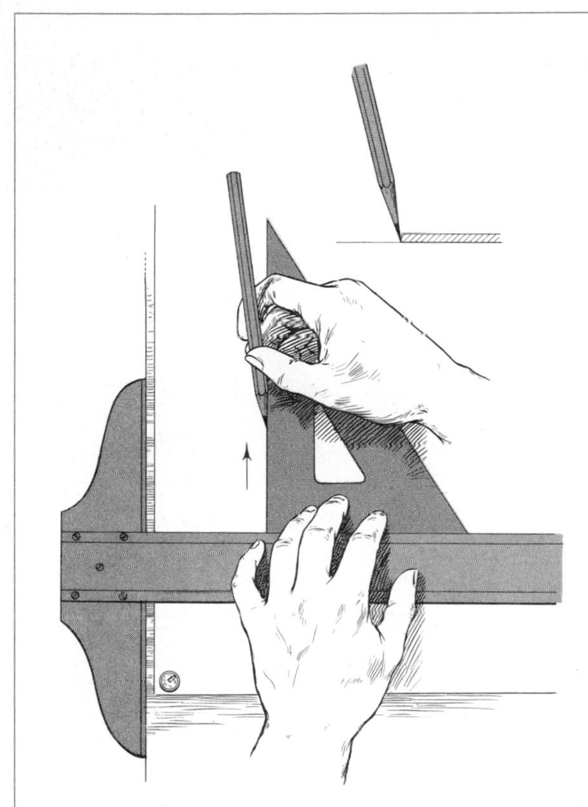

**FIG. 27.** Drawing a vertical line. With the T square and triangle in position, draw the line from bottom to top—always away from the body.

In drawing vertical lines, the T square is held in position against the left edge of the board by the thumb and little finger of the left hand while the other fingers of this hand adjust and hold the triangle. You can be sure that the T square is in contact with the board when you hear the little double click as the two come together, and slight pressure of the thumb and little finger toward the right will maintain the position. As the line is drawn, pressure of all the fingers against the board will hold the T square and triangle firmly in position.

As in using the T square, care must be taken to keep the line accurately parallel to the guiding edge. Note the position of the pencil in Fig. 27.

In both penciling and inking, the triangles must always be used in contact with a guiding straightedge. To ensure accuracy, never work to the extreme corner of a triangle; to avoid having to do so, keep the T square below the lower end of the line to be drawn.

With the T square against the edge of the board, lines at 45° are drawn with the standard 45° triangle, and lines at 30° and 60° with the 30–60° triangle, as shown in Fig. 28. With vertical and horizontal lines included, lines at increments of 45° are drawn with the 45° triangle as at (*b*), and lines at 30° increments with the 30–60° triangle as at (*a*). The two triangles are used in combination for angles of 15, 75, 105°, etc. (Fig. 29). Thus any multiple of 15° is drawn directly; and a circle is divided with the 45° triangle into 8 parts, with the 30–60° triangle into 12 parts, and with both into 24 parts.

*To Draw One Line Parallel to Another (Fig. 30).* Adjust to the given line a triangle held against a straightedge, hold the guiding edge in position, and slide the triangle on it to the required position.

*To Draw a Perpendicular to Any Line (Fig. 31).* Place a triangle with one edge against the T square (or another triangle), and move the two until the hypotenuse of the triangle is coincident with the line, as at position *a*; hold the T square in position and turn the triangle, as shown, until its other side is against the T square; the hypotenuse will then be perpendicular to the original line. Move the triangle to the required position. A quicker method is to set the triangle with its hypotenuse against the guiding edge, fit one side to the line, slide the triangle to the required point and draw the perpendicular, as shown at position *b*.

Never attempt to draw a perpendicular to a line with only one triangle by placing one leg of the triangle along the line.

Through internal strains, triangles sometimes lose their accuracy. They may be tested by drawing a perpendicular and then reversing the triangle, as shown in Fig. 32.

## 28. The Left-handed Draftsman.

If you are left-handed, reverse the T square and

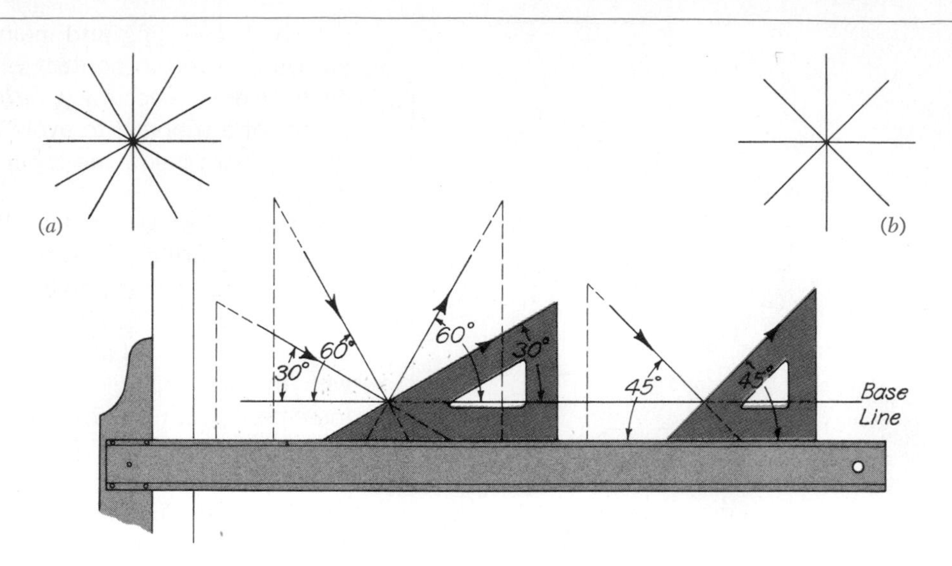


**FIG. 28.** To draw angles of 30°, 45°, 60°. Multiples of 30° are drawn with the 30–60° triangle, multiples of 45° with the 45° triangle.

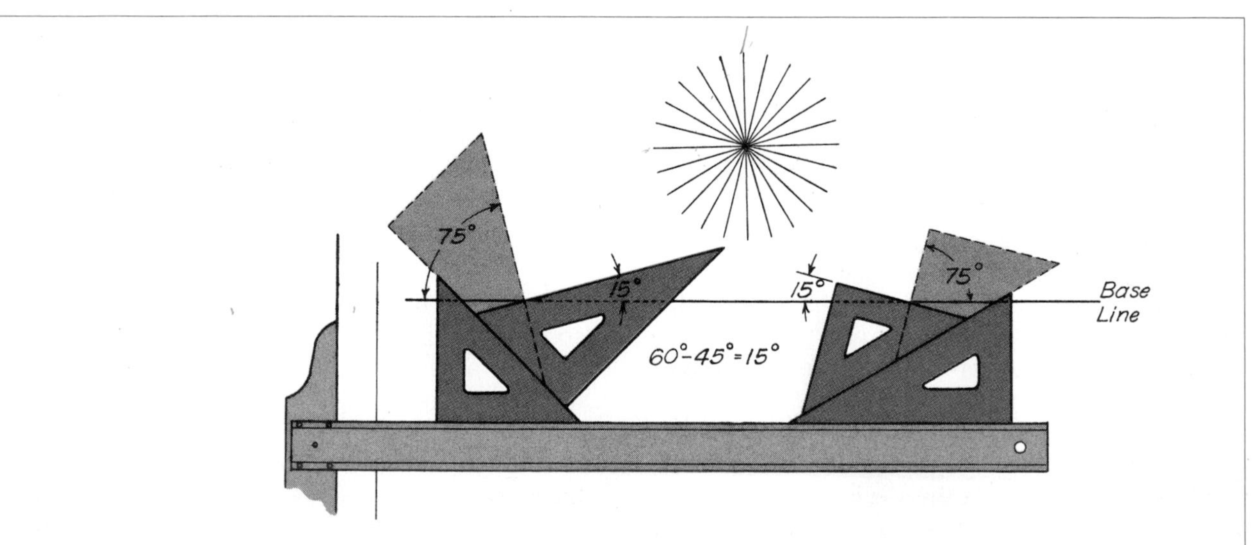


**FIG. 29.** To draw angles of 15° and 75°. Angles in increments of 15° are obtained with the two triangles in combination.

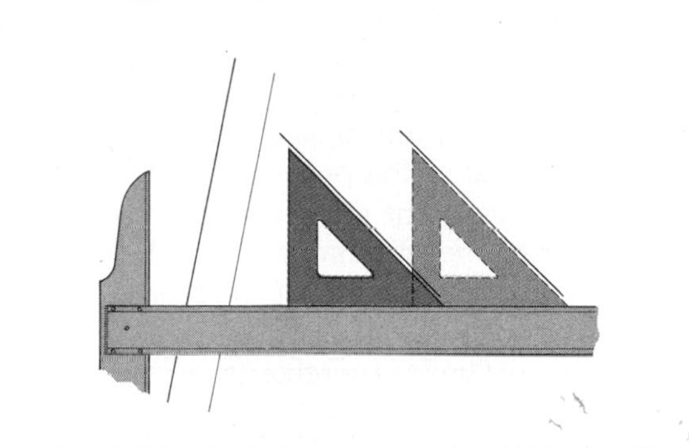

**FIG. 30.** To draw parallel lines. With the T square as a base, the triangle is aligned and then moved to the required position.

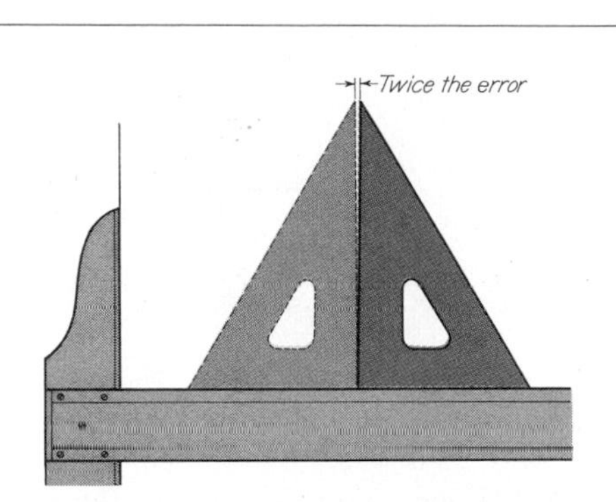


**FIG. 32.** To test a triangle for right angle, draw a line with the triangle in each position; twice the error is produced.

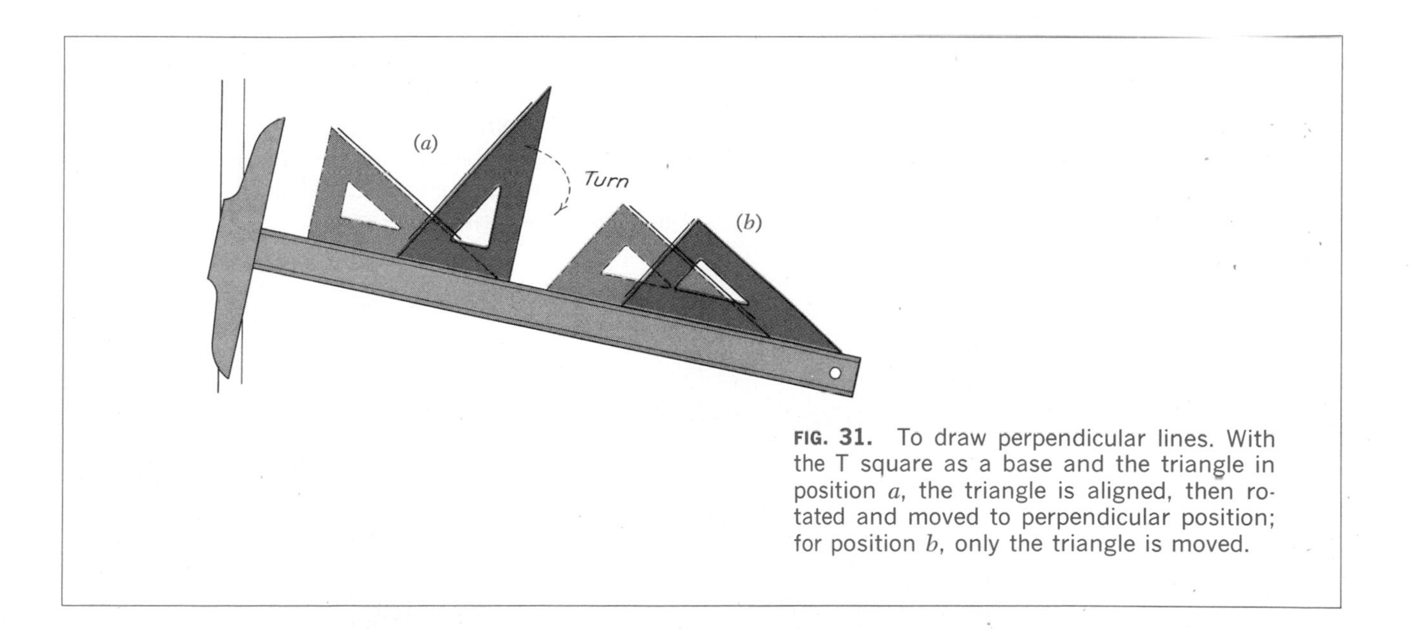


**FIG. 31.** To draw perpendicular lines. With the T square as a base and the triangle in position $a$, the triangle is aligned, then rotated and moved to perpendicular position; for position $b$, only the triangle is moved.

triangles left for right as compared with the regular right-handed position. Use the head of the T square along the right edge of the board, and draw horizontal lines from right to left. Place the triangle with its vertical edge to the right, and draw vertical lines from bottom to top. The drawing table should be placed with the light coming from the right.

**29. Use of the Scale.**
Scale technique is governed largely by the requirements of accuracy and speed. Before a line can be drawn, its relative position must be found by scaling, and the speed with which scale measurement can be made will greatly affect the total drawing time.

Precise layouts and developments, made to scale

**FIG. 33.** Making a measurement. Place the scale in position; the distance is marked on paper by short, light *lines*.

for the workmen, must be very accurately drawn, at the expense of speed; conversely, drawings with figured dimensions need not be quite so carefully scaled, and better speed may be attained.

To make a measurement, place the scale on the drawing where the distance is to be laid off, align the scale in the direction of the measurement, and make a *light* short dash with a sharp pencil at the proper graduation mark (Fig. 33). In layout work where extreme accuracy is required, a "pricker," or needle point set in a wood handle, may be substituted for the pencil, and a *small* hole pricked into the paper in place of the pencil mark. It is best to start with the "zero" of the scale when setting off lengths or when measuring distances. In using an open-divided scale, inches (or fractions) are accounted for in one direction from the zero graduation while feet (or units) are recorded in the opposite direction.

Measurements should not be made on a drawing by taking distances off the scale with dividers, as this method is time-consuming and no more accurate than the regular methods.

To avoid cumulative errors, successive measurements on the same line should, if possible, be made without shifting the scale. In representing objects that are larger than can be drawn to their natural or full size, it is necessary to reduce the size of the drawing in some regular proportion, and for this purpose one of the standard mechanical engineer's, civil engineer's, or architect's scales is used. Standard scales are given in Fig. 34.

**SCALES**

**MECHANICAL ENGINEER'S**

| | |
|---|---|
| 1″ = 1″ (full size) | ½″ = 1″ (½ size) |
| ¼″ = 1″ (¼ size) | ⅛″ = 1″ (⅛ size) |

**ARCHITECT'S OR MECHANICAL ENGINEER'S**

| | | |
|---|---|---|
| 12″ = 1′-0″ (full size) | 1″ = 1′-0″ (1/12 size) | ¼″ = 1′-0″ (1/48 size) |
| 6″ = 1′-0″ (½ size) | ¾″ = 1′-0″ (1/16 size) | 3/16″ = 1′-0″ (1/64 size) |
| 3″ = 1′-0″ (¼ size) | ½″ = 1′-0″ (1/24 size) | ⅛″ = 1′-0″ (1/96 size) |
| 1½″ = 1′-0″ (⅛ size) | ⅜″ = 1′-0″ (1/32 size) | 3/32″ = 1′-0″ (1/128 size) |

**CIVIL ENGINEER'S**

10, 20, 30, 40, 50, 60, or 80 divisions to the inch representing feet, 10 ft, 100 ft, rods, miles, or any other necessary unit

**FIG. 34.** Standard scales. Special scales are available. See manufacturers' catalogues.

The first reduction is to *half size,* or to the scale of 6″ = 1′-0″. In other words, ½ in. on the drawing represents a distance of 1 in. on the object. Stated in terms used for the architect's scales, a distance of 6 in. on the drawing represents 1 ft on the object. This scale is used even if the object is only slightly larger than could be drawn full size. If this reduction is not sufficient, the drawing is made to *quarter size,* or to the scale of 3″ = 1′-0″. If the quarter-size scale is too large, the next reduction is *eighth size,* or 1½″ = 1′-0″, the smallest proportion usually supplied on standard mechanical engineer's scales; but the architect's scales are used down to 3/32″ = 1′-0″, as shown by the listings in Fig. 34.

In stating the scale used on a drawing, the information should be given in accordance with the scale used to make the drawing. If a standard mechanical engineer's scale is employed, the statement may read that the scale is (1) *full size,* (2) *half size,* (3) *quarter size,* or (4) *eighth size.* These scales may also be given as (1) 1″ = 1″, (2) ½″ = 1″, (3) ¼″ = 1″, or (4) ⅛″ = 1″. If a standard architect's scale is used, the statement is given in terms of inches to the foot. Examples are (1) 3″ = 1′-0″, (2) 1½″ = 1′-0″, or (3) 1″ = 1′-0″. In stating the scale, the first figure always refers to the drawing and the second to the object. Thus 3″ = 1′-0″ means that 3 in. on the *drawing* represents 1 ft on the *object.*

Drawings to odd proportions, such as 9″ = 1′-0″, 4″ = 1′-0″, 5″ = 1′-0″, are used only in rare cases when drawings are made for reduction and the conditions of size demand a special scale.

The terms "scale" and "size" have different meanings. The scale ¼″ = 1′-0″ is the usual one for ordinary house plans and is often called by architects the "quarter scale." This term should not be confused with term "quarter size," as the former means ¼ in. to 1 ft and the latter ¼ in. to 1 in.

The size of a circle is generally stated by giving its diameter, while to draw it the radius is necessary. Two scales are usually supplied together on the same body, for example, half and quarter size. Therefore, in drawing to half size, it is often convenient to lay off the amount of the diameter with the quarter-size scale and use this distance as the radius.

Small pieces are often made "double size," and very small mechanisms, such as watch parts, are drawn to greatly enlarged sizes: 10 to 1, 20 to 1, 40 to 1, and 50 to 1, using special enlarging scales.

For plotting and map drawing, the civil engineer's scales of decimal parts, with 10, 20, 30, 40, 50, 60, and 80 divisions to the inch, are used. These scales are not used for machine or structural work but in certain aircraft drawings.

The important thing in drawing to scale is to think and speak of each dimension in its full size and not in the reduced (or enlarged) size it happens to be on the paper. This practice prevents confusion between *actual* and *represented* size.

### 30. Reading the Scale.

Reading the standard mechanical engineer's scales is rather simple, because the scale is plainly marked in inches, and the smaller graduations are easily recognized as the regular divisions of the inch into ½, ¼, ⅛, and 1/16. Thus the scales for half size, quarter size, and eighth size are employed in exactly the same manner as a full-size scale.

The architect's scales, being open-divided and to stated reductions, such as 3″ = 1′-0″, may require some study by the beginner in order to prevent confusion and mistakes. As an example, consider the scale of 3″ = 1′-0″. This is the first reduction scale of the usual triangular scale; on it the distance of 3 in. is divided into 12 equal parts, and each of these is subdivided into eighths. This distance should be thought of not as 3 in. but as a foot divided into inches and eighths of an inch. Notice that the divisions start with the *zero* on the inside, the inches of the divided foot running to the *left* and the open divisions of feet to the *right,* so that dimensions given in feet and in inches may be read directly, as 1′-0½″ (Fig. 35). On the other end will be found the scale of 1½″ = 1′-0″, or eighth size, with the distance of 1½ in. divided on the right of the zero into 12 parts and subdivided into quarter inches, with the foot divisions to the left of the zero coinciding with the marks of the 3-in. scale. Note again that in reading a distance in feet and inches,

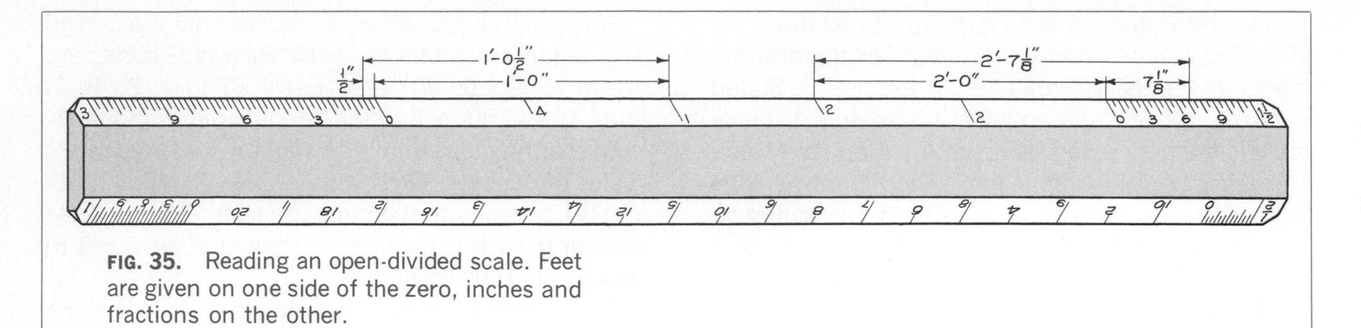


**FIG. 35.** Reading an open-divided scale. Feet are given on one side of the zero, inches and fractions on the other.

for example, the 2′-7⅛″ distance in Fig. 35, feet are determined to the left of the zero and inches to the right of it. The other scales, such as ¾″ = 1′-0″ and ¼″ = 1′-0″, are divided in a similar way, the only difference being in the value of the smallest graduations. The scale of 3/32″ = 1′-0″, for example, can be read only to the nearest 2 in.

### 31. "Laying out" the Sheet.

The paper is usually cut somewhat larger than the desired size of the drawing and trimmed to size after the work is finished. Suppose the finished size is to be 11 by 17 in. with a ½-in. border inside. Lay the scale down on the paper close to the lower edge and measure 17 in., marking the distance with the pencil; at the same time mark ½ in. inside at each end for the border line. Use a short dash forming a continuation of the division line on the scale in laying off a dimension. Do not bore a hole with the pencil. Near the left edge mark 11- and ½-in. borderline points. Through these four marks on the left edge, draw horizontal lines with the T square; and through the points on the lower edge, draw vertical lines, using the triangle against the T square.

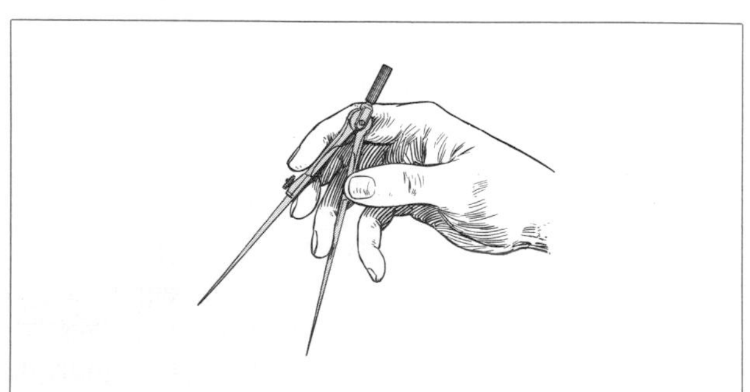

**FIG. 36.** Handling the dividers. The instrument is opened and adjusted with one hand.

### 32. Use of Dividers.

Dividers are used for transferring measurements and for dividing lines into any number of equal parts. Facility in their use is essential, and quick and absolute control of their manipulation must be gained. The instrument should be opened with one hand by pinching the chamfer with the thumb and second finger. This will throw it into correct position with thumb and forefinger outside the legs and the second and third fingers inside, with the head resting just above the second joint of the forefinger (Fig. 36). It is thus under perfect control, with the thumb and forefinger to close it and the other two to open it. Practice this motion until you can adjust the dividers to the smallest fraction. In coming down to small divisions, the second and third fingers must be gradually slipped out from between the legs as they are closed down upon them. Notice that the little finger is not used in manipulating the dividers.

### 33. To Divide a Line by Trial.

In bisecting a line, the dividers are opened at a guess to roughly half the length. This distance is stepped off on the line, holding the instrument by the handle with the thumb and forefinger. If the division is short, the leg should be thrown out to half the remainder (estimated by eye), without removing the other leg from the paper, and the line spaced again with this new setting (Fig. 37). If the result does not come out exactly, the operation can be repeated. With a little experience, a line can be divided rapidly in this way. Similarly, a line, either straight or curved, can be divided into any number of equal parts, say, five, by estimating the first division, stepping this lightly along the line, with the dividers held vertically by the handle, turning the instrument first in one direction and then in the other. If the last division falls short, one-fifth of the remainder should be added by opening the dividers, keeping one point on the paper. If the last division is over, one-fifth of the excess should be taken off and the line re-spaced. If it is found difficult to make this small adjustment accurately with the fingers, the hair-spring may be used. You will find the bow spacers more convenient than the dividers for small or numerous divisions. Avoid pricking unsightly holes in the paper. The position of a small prick point may be preserved, if necessary, by drawing a small circle around it with the pencil.

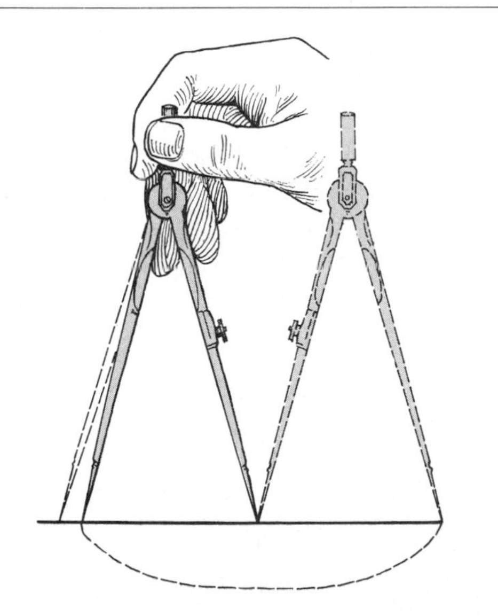

**FIG. 37.** Bisecting a line. Half is estimated; then the dividers are readjusted by estimating half the original error.

### 34. Use of the Compasses.

The compasses have the same general shape as the dividers and are manipulated in a similar way. First of all, the needle should be permanently adjusted. Insert the pen in place of the pencil leg, turn the needle with the shoulder point out, and set it a trifle longer than the pen, as in Fig. 38; replace the pencil leg, sharpen the lead to a long bevel, as in Fig. 39, and adjust it to the needle point. All this is done so that the needle point will be in perfect position for using the pen; the pencil, which must be sharpened frequently, can be adjusted each time to mate in length with the needle point.

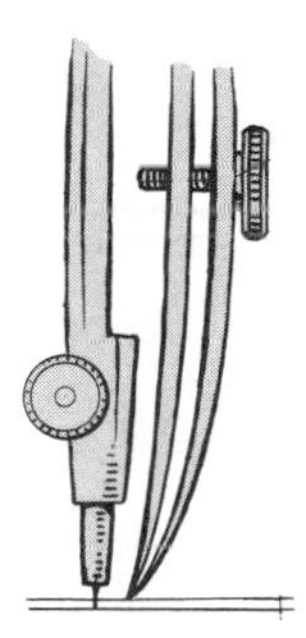

**FIG. 38.** Adjusting the needle point of a large compass. The point is adjusted to the pen; the pen is then replaced by the pencil leg and the pencil adjusted to the point.

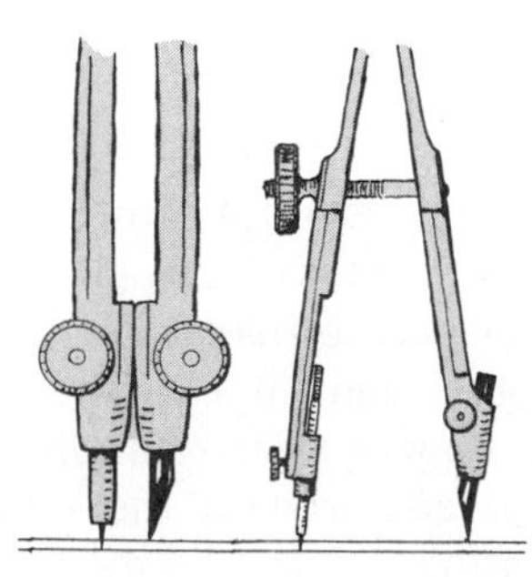

**FIG. 39.** Adjusting the pencil lead. The length is adjusted so that the instrument will be vertically centered.

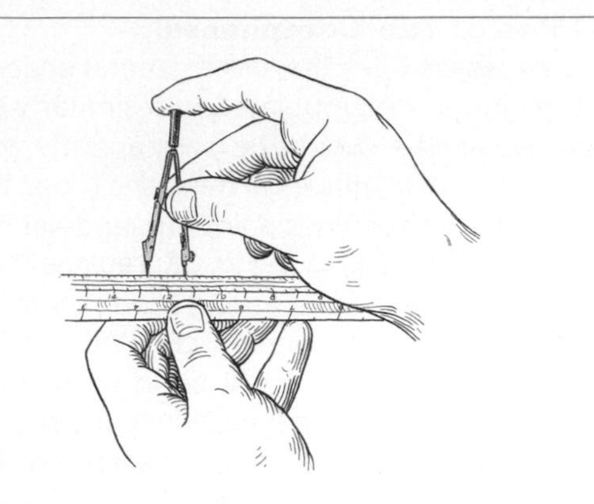

**FIG. 40.** Setting the compass to radius size. Speed and accuracy are obtained by adjusting directly on the scale.

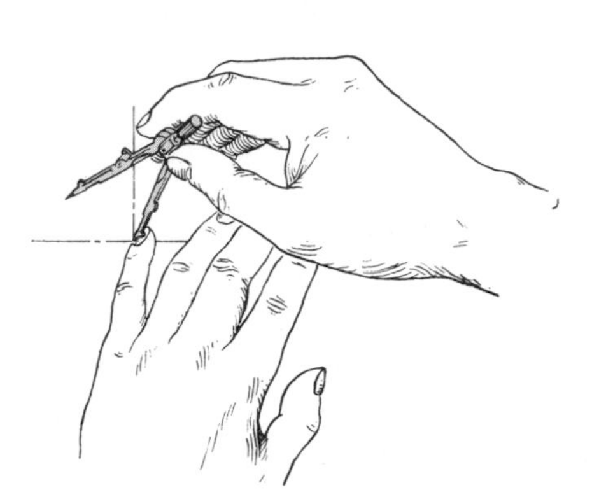

**FIG. 41.** Guiding the needle point. For accuracy of placement, guide with the little finger.

*To Draw a Circle.* Set the compass on the scale, as shown in Fig. 40, and adjust it to the radius needed; then place the needle point at the center on the drawing, guiding it with the left hand (Fig. 41). Raise the fingers to the handle and draw the circle in one sweep, rolling the handle with the thumb and forefinger, inclining the compass slightly in the direction of the line (Fig. 42).

The position of the fingers after the rotation is shown in Fig. 43. The pencil line can be brightened, if necessary, by making additional turns. Circles up to perhaps 3 in. in diameter can be drawn with the legs of the compass straight, but for larger sizes, both the needlepoint leg and the pencil or pen leg

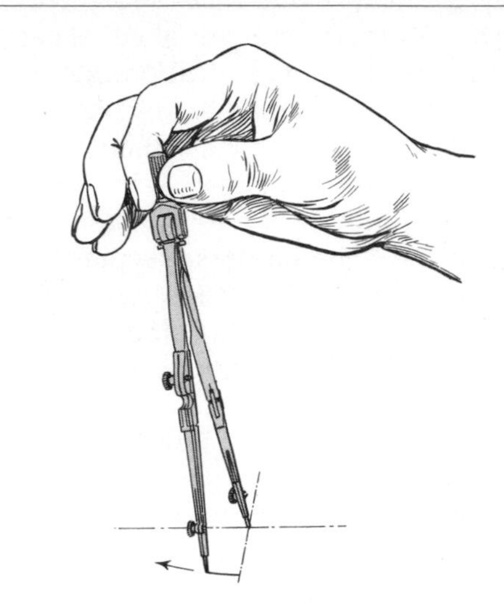

**FIG. 42.** Starting a circle. The compass is inclined in the direction of the stroke.

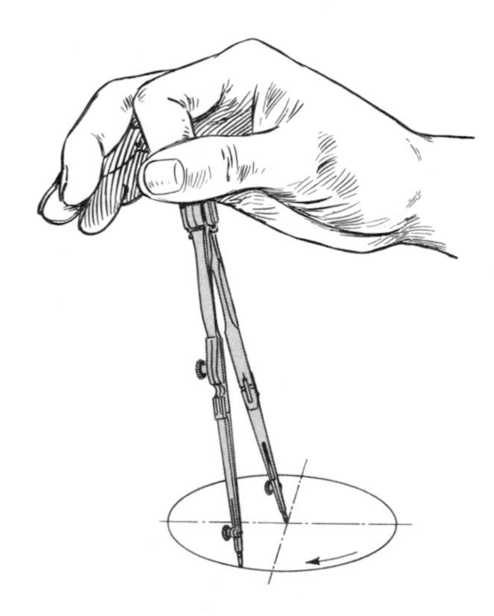

**FIG. 43.** Completing a circle. The stroke is completed by twisting the knurled handle in the fingers.

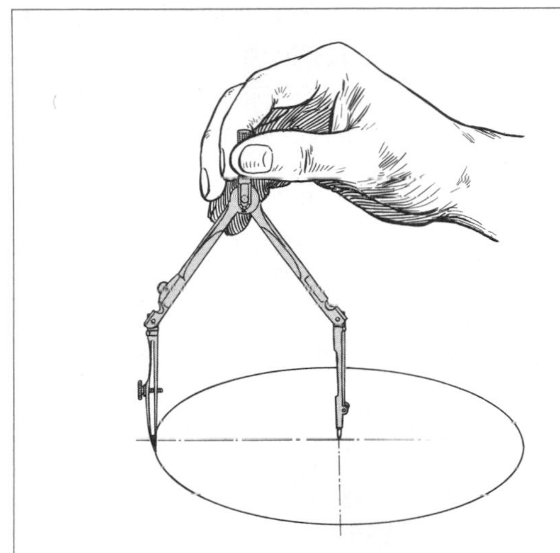

FIG. 44. Drawing a large circle. Knuckle joints are bent to make the legs perpendicular to the paper.

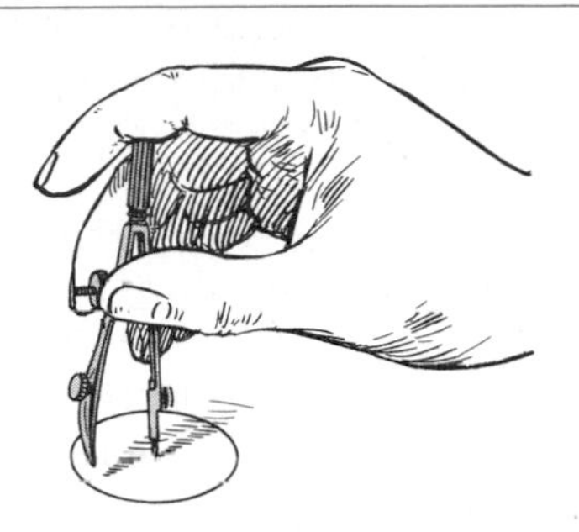

FIG. 46. Adjusting a bow instrument. One hand only is needed. Center-wheel bows are held similarly.

should be bent at the knuckle joints so as to be perpendicular to the paper (Fig. 44). The 6-in. compass may be used in this way for circles up to perhaps 10 in. in diameter; larger circles are made by using the lengthening bar, as illustrated in Fig. 45, or the beam compass (Fig. 18). In drawing concentric circles, the *smallest* should always be drawn first, before the center hole has become worn.

The bow instruments are used for small circles, particularly when a number are to be made of the same diameter. To avoid wear (on side-wheel instruments), the pressure of the spring against the nut can be relieved in changing the setting by holding the points in the left hand and spinning the nut in or out with the finger. Small adjustments should be made with one hand with the needle point in position on the paper (Fig. 46).

When several concentric circles are drawn, time may be saved by marking off the several radii on the paper from the scale and then setting the compass to each mark as the circles are made. In some cases it may be advantageous to measure and mark

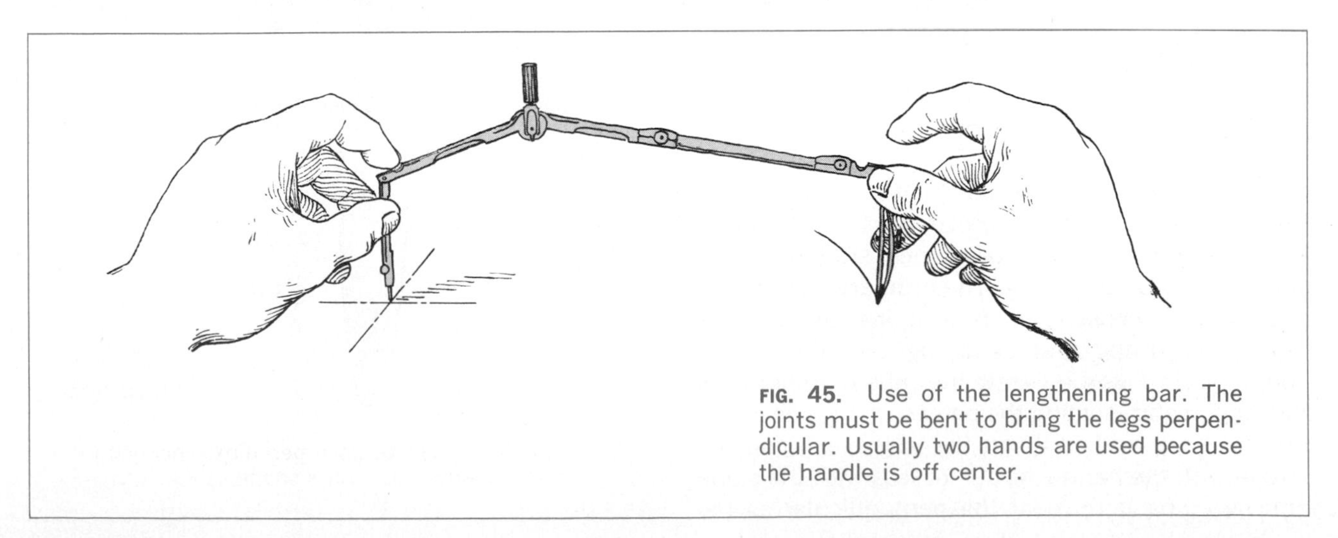

FIG. 45. Use of the lengthening bar. The joints must be bent to bring the legs perpendicular. Usually two hands are used because the handle is off center.

the radius on the paper instead of setting the compass directly on the scale. This method must be used whenever the radius is greater than the length of the scale.

When *extreme accuracy* is required, the compass is set, a light circle is drawn on the paper, and the diameter is checked with the scale; if the size is not satisfactory, the compass is adjusted and the operation is repeated until the size needed is obtained.

### 35. The Ruling Pen.

The ruling pen is for inking straight lines and noncircular curves. Several types are illustrated in Fig. 19. The important feature is the shape of the blades; they should have a well-designed ink space between them, and their points should be rounded (actually elliptical in form) equally, as in Fig. 47. If pointed, as in Fig. 48, the ink will arch up as shown and will be provokingly hard to start. If rounded to a blunt point, as in Fig. 49, the ink will flow too freely, forming blobs and overruns at the ends of the lines. Pens in constant use become dull and worn, as illustrated in Fig. 50. It is easy to tell whether a pen is dull by looking for the reflection of light that travels from the side and over the end of the point when the pen is turned in the hand. If the reflection can be seen all the way, the pen is too dull. A pen in poor condition is an abomination, but a well-sharpened one is a delight to use. Every draftsman should be able to keep his pens in fine condition.

High-grade pens usually come from the makers well sharpened. Cheaper ones often need sharpening before they can be used.

### 36. To Sharpen a Pen.

The best stone for the purpose is a hard Arkansas knife piece. It is well to soak a new stone in oil for several days before using. The ordinary carpenter's oilstone is too coarse for drawing instruments.

The nibs must first be brought to the correct shape, as in Fig. 47. Screw the nibs together until they touch and, holding the pen as in drawing a line, draw it back and forth on the stone, starting the stroke with the handle at 30° or less with the stone and swinging it up past the perpendicular as the

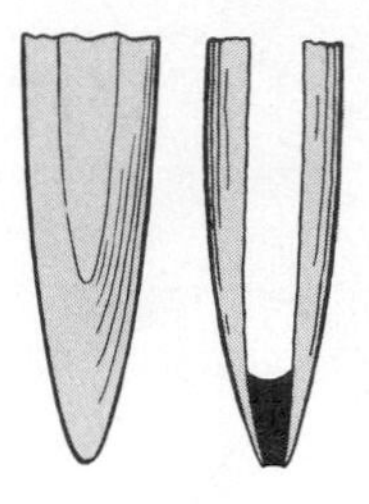

**FIG. 47.** Correct shape of pen nibs. A nicely uniform elliptical shape is best.

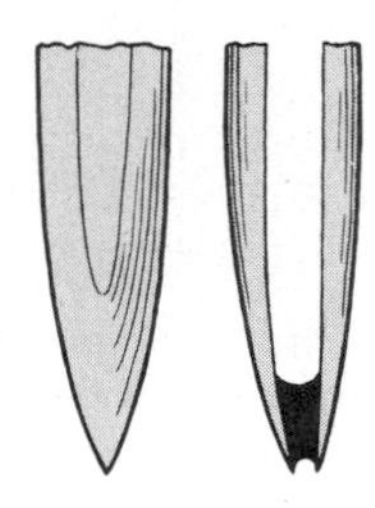

**FIG. 48.** Incorrect shape of pen nibs. This point is much too sharp. Ink will not flow well.

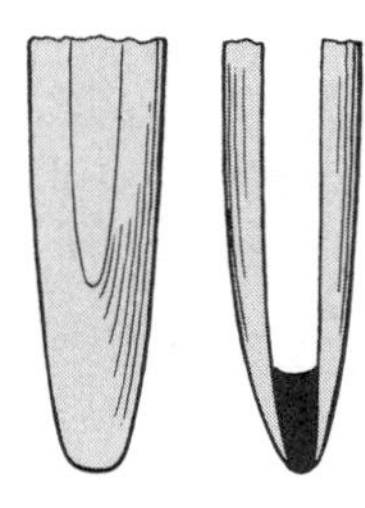

**FIG. 49.** Incorrect shape of pen nibs. This point is too flat. Ink will blob at the beginning and end of the line.

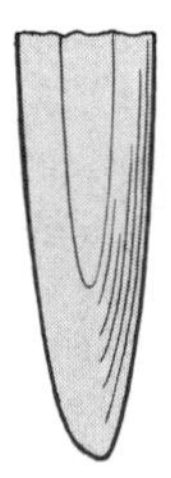

**FIG. 50.** Shape of worn pen nibs. This point needs sharpening to the shape of Fig. 47.

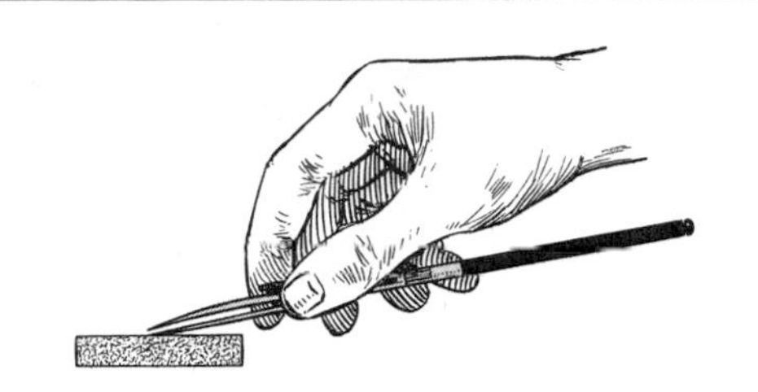

FIG. 51. Sharpening a pen. After bringing the point to the shape of Fig. 47, work down the sides by a rocking motion to conform to blade contour.

line across the stone progresses. This will bring the nibs to exactly the same shape and length, leaving them very dull. Then open them slightly, and sharpen each blade in turn, on the outside only, until the bright spot on the end has just disappeared. Hold the pen, as in Fig. 51, at a small angle with the stone and rub it back and forth with a slight oscillating or rocking motion to conform to the shape of the blade. A stone 3 or 4 in. long held in the left hand with the thumb and fingers gives better control than one laid on the table. Silicon carbide cloth or paper can be substituted for the stone, and for a fine job, crocus cloth may be used for finishing. A pocket magnifying glass may be helpful in examining the points. The blades should not be sharp enough to cut the paper when tested by drawing a line across it without ink. If oversharpened, the blades should again be brought to touch and a line swung very lightly across the stone as in the first operation. When tested with ink, the pen should be capable of drawing clean sharp lines down to the finest hairline. If these finest lines are ragged or broken, the pen is not perfectly sharpened. It should not be necessary to touch the inside of the blades unless a burr has been formed, which might occur if the metal is very soft, the stone too coarse, or the pressure too heavy. To remove such a burr, or wire edge, draw a strip of detail paper between the nibs, or open the pen wide and lay the entire inner surface of the blade flat on the stone and move it with a very light touch.

### 37. Use of the Ruling Pen.

The ruling pen is always used in connection with a guiding edge—T square, triangle, or curve. The T square and triangle should be held in the same positions as for penciling.

To fill the pen, take it to the bottle and touch the quill filler between the nibs. Be careful not to get any ink on the outside of the blades. If the newer plastic squeeze bottle is used, place the small spout against the sides of the nibs and carefully squeeze a drop of ink *between* the nibs. Not more than $\frac{3}{16}$ to $\frac{1}{4}$ in. of ink should be put in; otherwise the weight of the ink will cause it to drop out in a blot. The pen should be held in the fingertips, as illustrated in Fig. 52, with the thumb and second finger against the sides of the nibs and the handle resting on the forefinger. Observe this hold carefully, as the tendency will be to bend the second finger to the position used when a pencil or writing pen is held. The position illustrated aids in keeping the pen at the proper angle and the nibs aligned with the ruling edge.

The pen should be held against the straightedge or guide with the blades parallel to it, the screw on the outside and the handle inclined slightly to the right and always kept in a plane passing through the line and perpendicular to the paper. The pen is thus directed by the upper edge of the guide, as

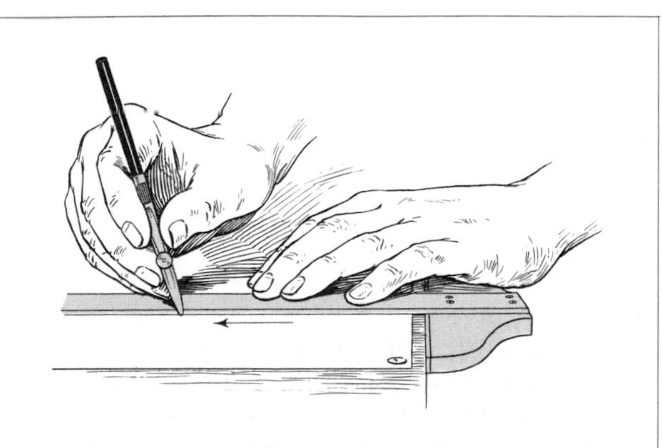

FIG. 52. Correct position of ruling pen. The pen is inclined in the direction of the stroke but held perpendicular to the paper, as in Fig. 53.

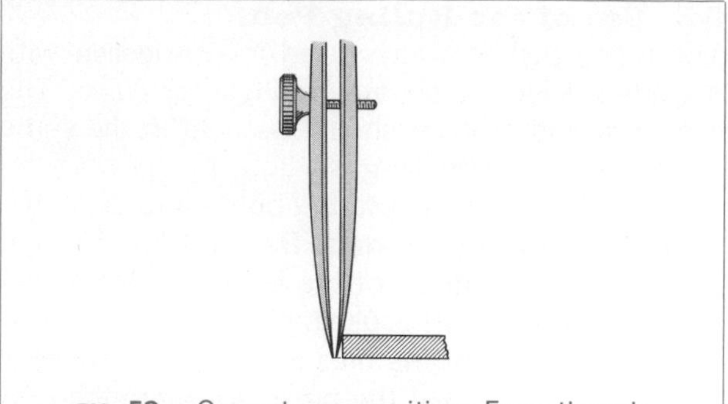

FIG. 53. Correct pen position. Even though the pen is inclined in the direction of the stroke, both nibs must touch the paper equally.

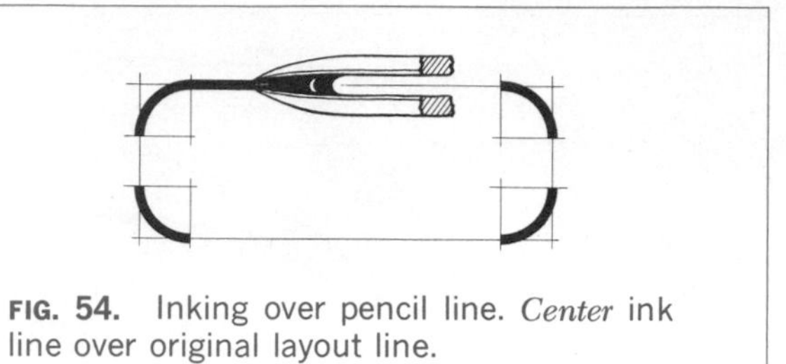

FIG. 54. Inking over pencil line. *Center* ink line over original layout line.

illustrated in actual size in Fig. 53. If the pen point is thrown out from the perpendicular, it will run on one blade and make a line that is ragged on one side. If the pen is turned in from the perpendicular, the ink is likely to run under the edge of the guide and cause a blot.

A line is drawn with a steady, even arm movement, the tips of the third and fourth fingers resting on, and sliding along, the straightedge, keeping the angle of inclination constant. Just before the end of the line is reached, the two guiding fingers on the straightedge should be stopped and, without stopping the motion of the pen, the line finished with a finger movement. Short lines are drawn with this finger movement alone. When the end of the line is reached, the pen is lifted quickly and the straightedge moved away from the line. The pressure on the paper should be light but sufficient to give a clean-cut line, and it will vary with the kind of paper and the sharpness of the pen. The pressure against the T square, however, should be only enough to guide the direction.

If the ink refuses to flow, it may be because it has dried in the extreme point of the pen. If pinching the blades slightly or touching the pen on the finger does not start it, the pen should immediately be wiped out and fresh ink supplied. Pens must be wiped clean after using.

In inking on either paper or cloth, the full lines will be much wider than the pencil lines. You must be careful to have the center of the ink line cover the pencil line, as illustrated in Fig. 54.

Instructions in regard to the ruling pen apply also to the compass. The compass should be slightly inclined in the direction of the line and both nibs of the pen kept on the paper, bending the knuckle joints, if necessary, to effect this.

It is a universal rule in inking that *circles and circle arcs must be inked first.* It is much easier to connect a straight line to a curve than a curve to a straight line.

### 38. Tangents.

It should be noted particularly that two lines are tangent to each other when the center lines of the lines are tangent and not simply when the lines touch each other; thus at the point of tangency, the width will be equal to the width of a single line (Fig. 55). Before inking tangent lines, the point of tangency should be marked in pencil. For an arc tangent to a straight line, this point will be on a line

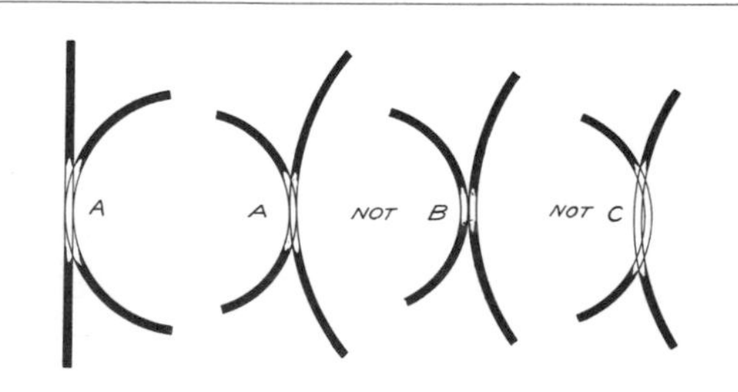


FIG. 55. Correct and incorrect tangents. Lines must be the width of *one* line at tangent point.

through the center of the arc and perpendicular to the straight line, and for two circle arcs it will be on the line joining their centers, as described in Secs. 5 to 16, Chap. 3.

### 39. The "Alphabet of Lines."

As the basis of drawing is the line, a set of conventional symbols covering all the lines needed for different purposes may properly be called an alphabet of lines. Figures 56 and 57 show the alphabet of lines adopted by the ASA as applied to the following:

1. Drawings made directly or traced in pencil on tracing paper or pencil cloth, from which blueprints or other reproductions are to be made (Fig. 56).
2. Tracings in ink on tracing cloth or tracing paper and inked drawings on white paper for display or photoreproductions (Fig. 57).

The ASA recommends three widths of lines for finished drawings: *thick* for visible outlines and cutting-plane and short-break lines; *medium* for hidden outlines; and *thin* for section, center, extension, dimension, long-break, adjacent-part, alternate-position, and repeat lines. The actual

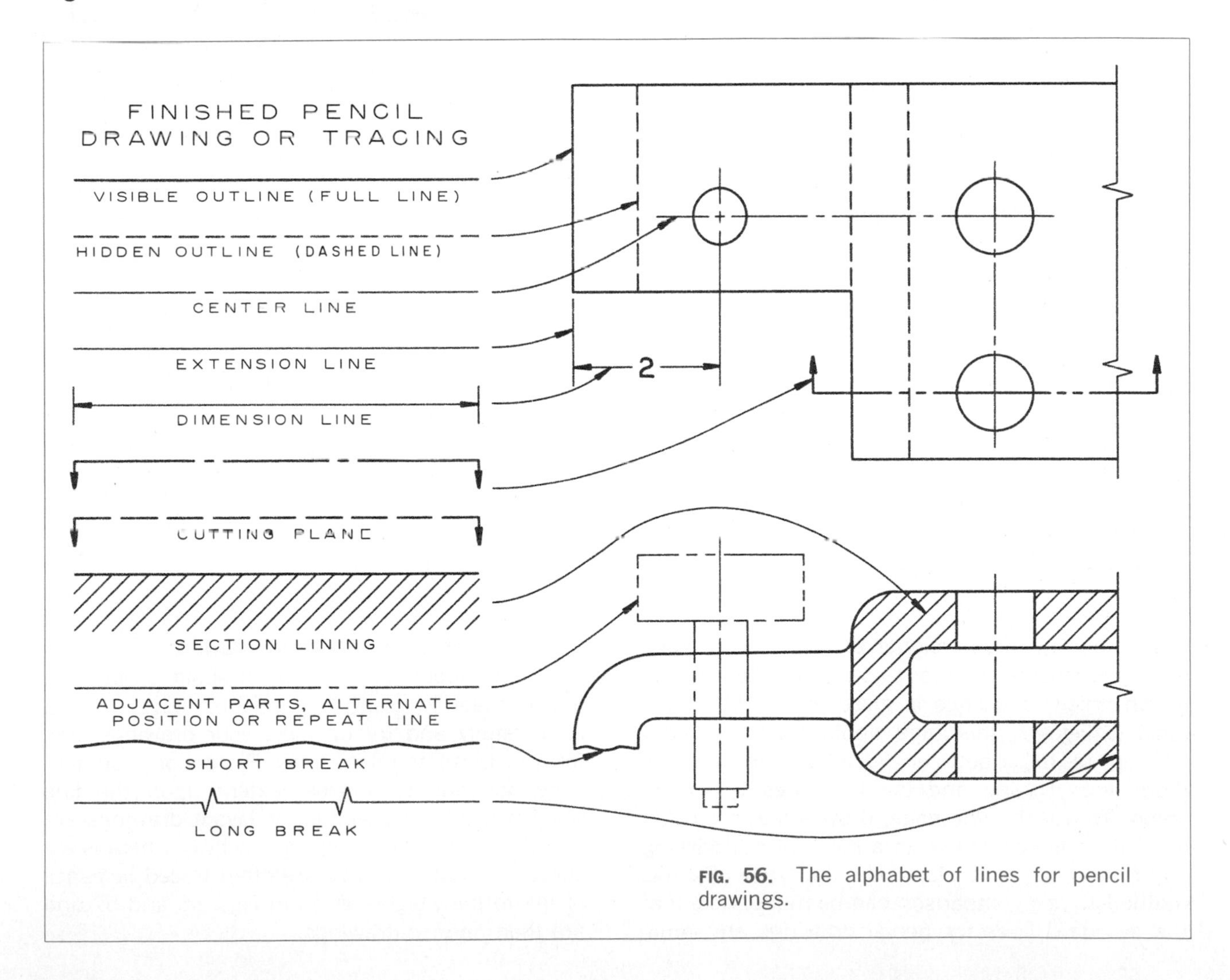


FIG. 56. The alphabet of lines for pencil drawings.

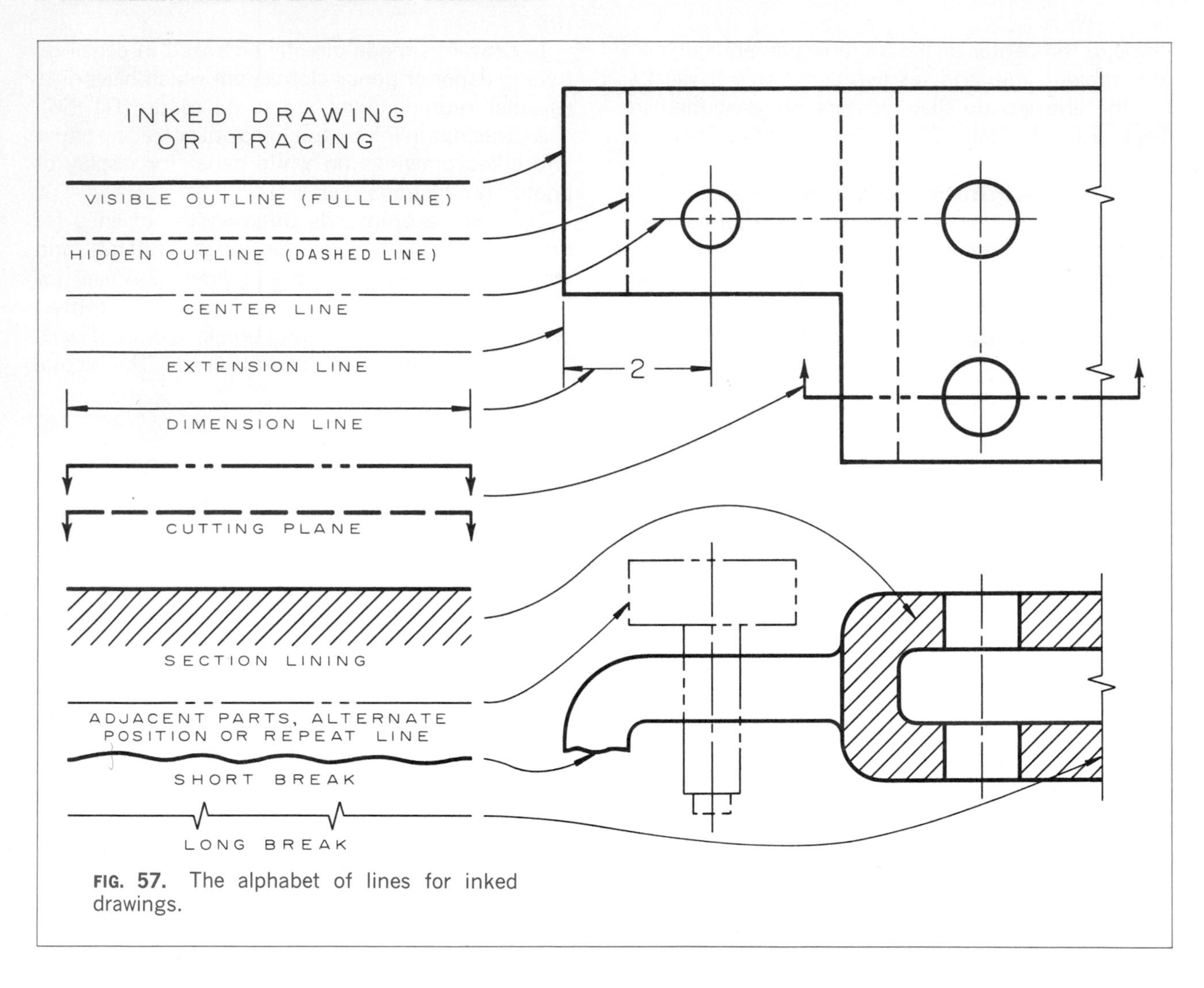


**FIG. 57.** The alphabet of lines for inked drawings.

widths of the three weights of lines, on average drawings, should be about as in Figs. 56 and 57. A convenient line gage is given in Fig. 58. If applied to Fig. 57, this gage would show the heavy lines in ink to be between $\frac{1}{30}$ and $\frac{1}{40}$ in., the medium lines $\frac{1}{60}$ in., and the fine lines $\frac{1}{100}$ in. in width. To use the line gage, draw a line about $1\frac{1}{2}$ in. long in pencil or ink on a piece of the drawing paper and apply it alongside the gage. By this method a good comparison can be made. Note that the standard lines for pencil drawings are somewhat thinner than for inked drawings, the thick line being about $\frac{1}{60}$ in., medium about $\frac{1}{80}$ in., and thin between $\frac{1}{100}$ and $\frac{1}{150}$ in. Study Figs. 56 and 57 carefully and try to make your drawings conform to these standard lines. Professional appearance depends to a great extent upon the line weights used. Line widths for layout drawings are *thin* throughout because the watchword here is accuracy. Layout drawings are often traced in pencil or ink to the weights given in Figs. 56 and 57 and are then *finished* drawings.

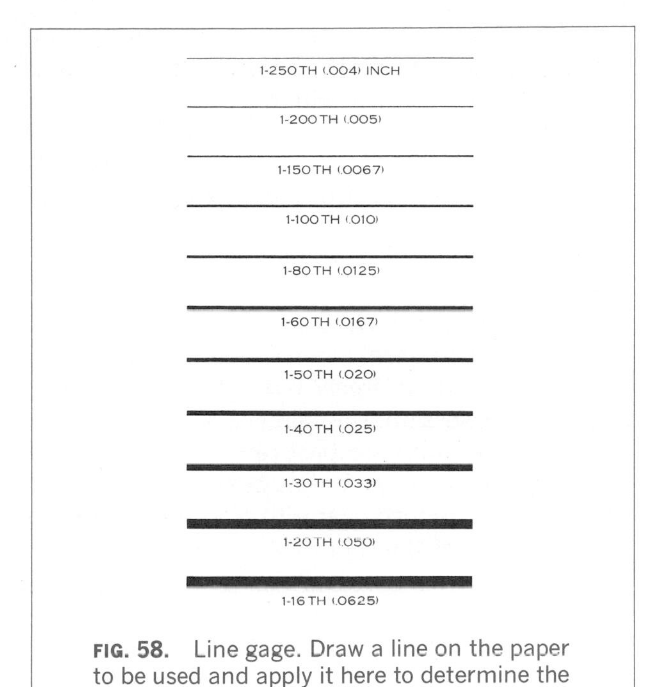


**FIG. 58.** Line gage. Draw a line on the paper to be used and apply it here to determine the width.

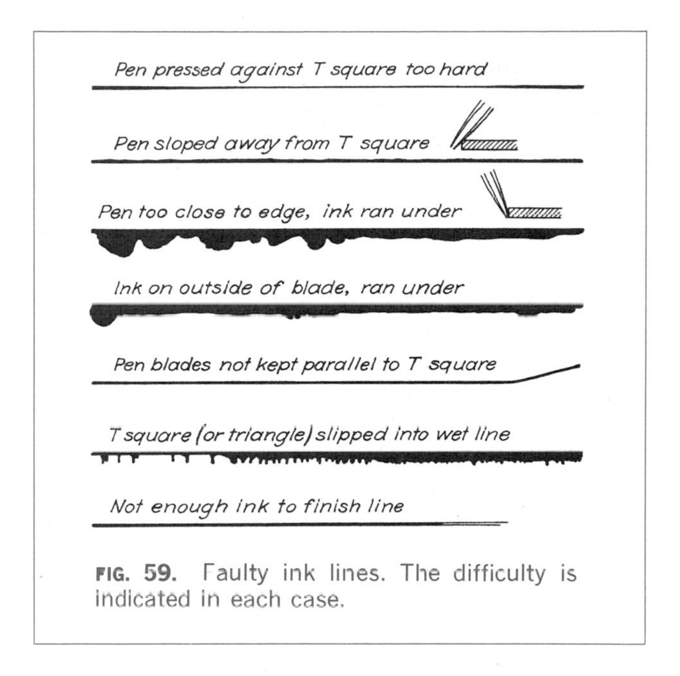


**FIG. 59.** Faulty ink lines. The difficulty is indicated in each case.

## 40. Line Practice.

After reading the preceding paragraphs, take a blank sheet of paper and practice making straight lines and circles in all the forms—full, dashed, etc.—shown in Figs. 56 and 57. Include starting and stopping lines, with special attention to tangents and corners.

*In pencil,* try to get all the lines uniform in width and color for each type. Circle arcs and straight lines should match exactly at tangent points.

*In ink,* proceed as for pencil practice and pay particular attention to the weight of lines and to the spacing of dashed lines and center lines.

If the inked lines appear imperfect in any way, ascertain the reason immediately. It may be the fault of the pen, the ink, the paper, or the draftsman; the probabilities are greatly in favor of the last if the faults resemble those in Fig. 59, which illustrates the characteristic appearance of several kinds of poor line. The correction in each case will suggest itself.

## 41. Use of the French Curve.

The French curve is a guiding edge for noncircular curves. When sufficient points have been determined, it is best to sketch in the line lightly in pencil, freehand and without losing the points, until it is clean, smooth, continuous, and satisfactory to the eye. Then apply the curve to it, selecting a part that will fit a portion of the line most nearly and seeing to it, particularly, that the curve is so placed that the direction in which its curvature increases is the direction in which the curvature of the line increases (Fig. 60). In drawing the part of the line matched by the curve, *always* stop a little short of the distance in which the guide and the line seem to coincide. After drawing this portion, shift the curve to find another place that will coincide with the continuation of the line. In shifting the curve, take care to preserve smoothness and continuity and to avoid breaks or cusps. Do this by seeing that in its successive positions the curve is always adjusted so that it coincides for a short distance with the part of the line already drawn. Thus at each junction the tangents will coincide.

If the curved line is symmetrical about an axis, marks locating this axis, after it has been matched accurately on one side, may be made in pencil on the curve and the curve then reversed. In such a case take exceptional care to avoid a "hump" at the joint.

It is often better to stop a line short of the axis on each side and close the gap afterward with another setting of the curve.

When using the curve in inking, the pen should be held perpendicular and the blades kept parallel to the edge. The inking of curves is excellent practice.

Sometimes, particularly at sharp turns, a combination of circle arcs and curves may be used: In inking a long, narrow ellipse, for example, the sharp curves may be inked by selecting a center on the major diameter by trial, drawing as much arc as will practically coincide with the ends of the ellipse, and then finishing the ellipse with the curve. The experienced draftsman will sometimes ink a curve that cannot be matched accurately by varying the distance of the pen point from the ruling edge as the line progresses.

## 42. Erasing.

The erasing of pencil lines and ink lines is a necessary technique to learn. When changing some detail, a designer, working freely but lightly, uses a soft pencil eraser so as not to damage the finish of the paper. Heavier lines are best removed with a Ruby pencil eraser. If the paper has been grooved by the line, it may be rubbed over with a burnisher or even with the back of the thumbnail. In erasing an ink

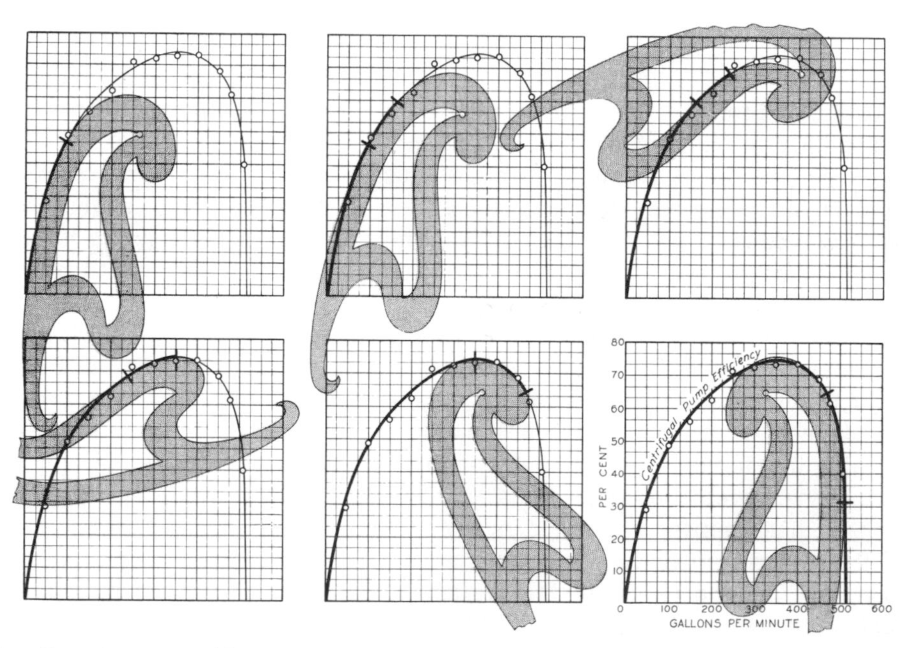


**FIG. 60.** Use of the French curve. The changing curvature of line and curve must match.

line, hold the paper down firmly and rub lightly and patiently, with a Ruby pencil eraser, first along the line and then across it, until the ink is removed. A triangle slipped under the paper or cloth gives a good backing surface.

When an erasure is made close to other lines, select an opening of the best shape on the erasing shield (Fig. 3*d*) and rub through it, holding the shield down firmly, first seeing that both of its sides are clean. Wipe the eraser crumbs off the paper with a dustcloth or brush. Never scratch out a line or blot with a knife or razor blade, and use so-called ink erasers sparingly, if at all. A skilled draftsman sometimes uses a sharp blade to trim a thickened spot or overrunning end on a line.

For extensive erasing, an electric erasing machine is a great convenience. Several successful models are on the market.

**43. Cautions in the Use of Instruments.** To complete this discussion of instruments, here are a few points worth noting:

**Never** use the scale as a ruler for drawing lines.

**Never** draw horizontal lines with the lower edge of the T square.

**Never** use the lower edge of the T square as a horizontal base for the triangles.

**Never** cut paper with a knife and the edge of the T square as a guide.

**Never** use the T square as a hammer.

**Never** put either end of a pencil into the mouth.

**Never** work with a dull pencil.

**Never** sharpen a pencil over the drawing board.

**Never** jab the dividers into the drawing board.

**Never** oil the joints of compasses.

**Never** use the dividers as reamers, pincers, or picks.

**Never** use a blotter on inked lines.

**Never** screw the pen adjustment past the contact point of the nibs.

**Never** leave the ink bottle uncorked.

**Never** hold the pen over the drawing while filling.

**Never** put into the drawing-ink bottle a writing pen that has been used in ordinary writing ink.

**Never** try to use the same thumbtack holes in either paper or board when putting paper down a second time.

**Never** scrub a drawing all over with an eraser after finishing. It takes the life out of the lines.

**Never** begin work without wiping off the table and instruments.

**Never** put instruments away without cleaning them. This applies with particular force to pens.

**Never** put bow instruments away without opening to relieve the spring.

**Never** work on a table cluttered with unneeded instruments or equipment.

**Never** fold a drawing or tracing.

**44. Exercises in the Use of Instruments.** The following problems can be used as progressive exercises for practice in using the instruments. Do them as finished pencil drawings or in pencil layout to be inked. Line work should conform to that given in the alphabet of lines (Figs. 56 and 57).

The problems in Chap. 3 afford excellent additional practice in accurate penciling.

# PROBLEMS

## Group 1. Straight Lines.

**1.** An exercise for the T square, triangle, and scale. Through the center of the space draw a horizontal and a vertical line. Measuring on these lines as diameters, lay off a 4-in. square. Along the lower side and upper half of the left side measure $\frac{1}{2}$-in. spaces with the scale. Draw all horizontal lines with the T square and all vertical lines with the T square and triangle.

**2.** An interlacement. For T square, triangle, and dividers. Draw a 4-in. square. Divide the left side and lower side into seven equal parts with dividers. Draw horizontal and vertical lines across the square through these points. Erase the parts not needed.

**3.** A street-paving intersection. For 45° triangle and scale. An exercise in starting and stopping short lines. Draw a 4-in. square. Draw its diagonals with 45° triangle. With the scale, lay off $\frac{1}{2}$-in. spaces along the diagonals from their intersection. With 45° triangle, complete the figure, finishing one quarter at a time.

**4.** A square pattern. For 45° triangle, dividers, and scale. Draw a 4-in. square and divide its sides into three equal parts with dividers. With 45° triangle, draw diagonal lines connecting these points. Measure $\frac{3}{8}$ in. on each side of these lines, and finish the pattern as shown in the drawing at the bottom of the page.

**5.** An acoustic pattern. For 45° triangle, T square, and scale. Draw two intersecting 45° diagonals 4 in. long, to form a field. With the scale lay off $\frac{1}{2}$-in. spaces from their intersection. Add the narrow border $\frac{3}{16}$ in. wide. Add a second border $\frac{1}{2}$ in. wide. The length of the border blocks is projected from the corners of the field blocks.

**6.** Five cards. Visible and hidden lines. Five cards $1\frac{3}{4}$ by 3 in. are arranged with the bottom card in the center, the other four overlapping each other and placed so that their outside edges form a 4-in. square. Hidden lines indicate edges covered.

**7.** A Maltese cross. For T square, spacers, and 45° and 30–60° triangles. Draw a 4-in. square and a $1\frac{3}{8}$-in. square. From the corners of the inner square, draw lines to the outer square at 15° and 75°, with the two triangles in combination. Mark points with spacers $\frac{1}{4}$ in. inside each line of this outside cross, and complete the figure with triangles in combination.

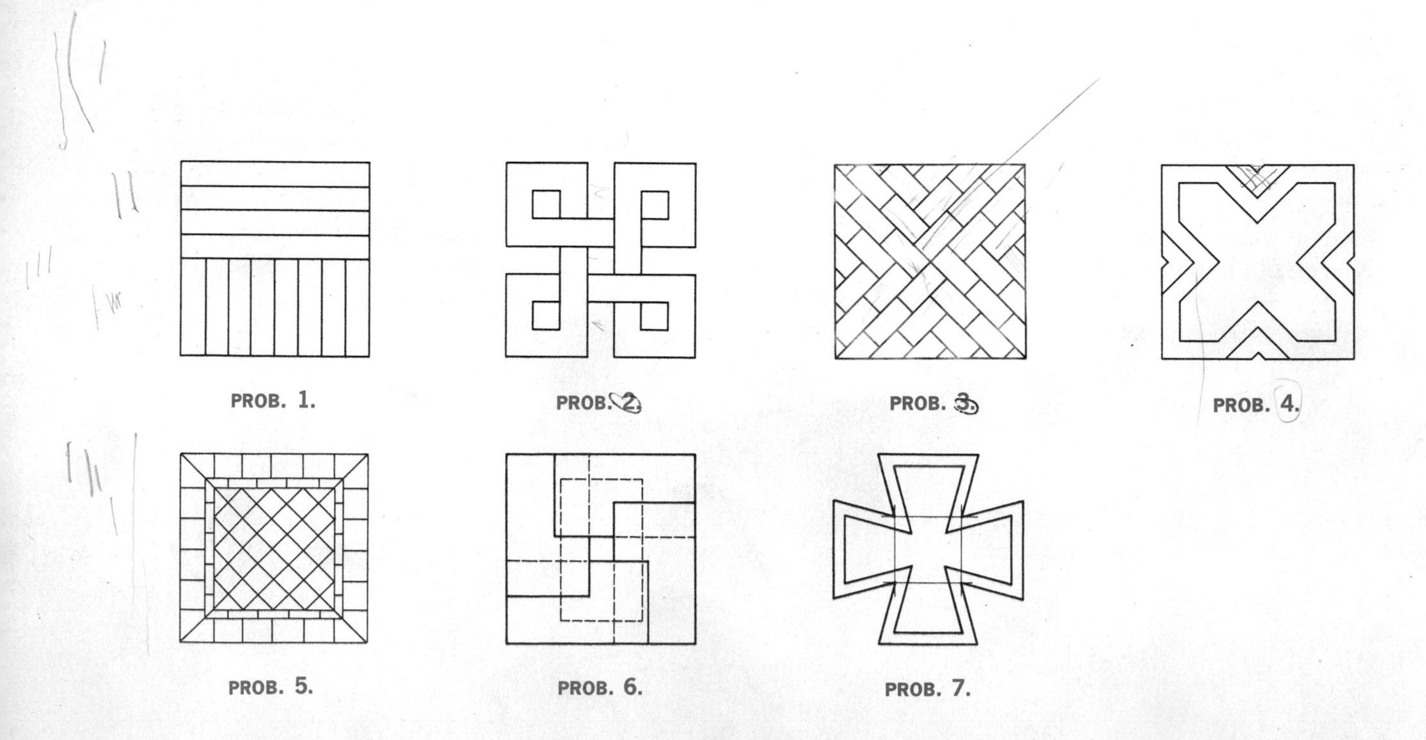

## Group 2. Straight Lines and Circles.

**8.** Insignia. For T square, triangles, scale, and compasses. Draw the 45° diagonals and the vertical and horizontal center lines of a 4-in. square. With compass, draw a $\frac{3}{4}$-in.-diameter construction circle, a $2\frac{3}{4}$-in. circle, and a $3\frac{1}{4}$-in. circle. Complete the design by adding a square and pointed star as shown.

**9.** A six-point star. For compass and 30–60° triangle. Draw a 4-in. construction circle and inscribe the six-point star with the T square and 30–60° triangle. Accomplish this with 4 successive changes of position of the triangle.

**10.** A stamping. For T square, 30–60° triangle, and compasses. In a 4-in. circle draw six diameters 30° apart. Draw a 3-in. construction circle to locate the centers of $\frac{5}{16}$-in.-radius circle arcs. Complete the stamping with perpendiculars to the six diameters as shown.

**11.** Insignia. The device shown is a white star with a red center on a blue background. Draw a 4-in. circle and a $1\frac{1}{4}$-in. circle. Divide the large circle into five equal parts with the dividers and construct the star by connecting alternate points as shown. Red is indicated by vertical lines and blue by horizontal lines. Space these by eye approximately $\frac{1}{16}$-in. apart.

**12.** A 24-point star. For T square and triangles in combination. In a 4-in. circle draw 12 diameters 15° apart, using T square and triangles singly and in combination. With same combinations, finish the figure as shown.

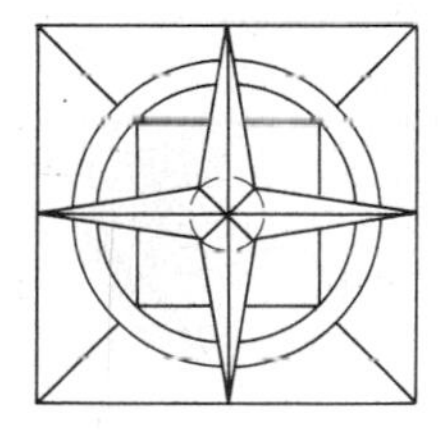

PROB. 8.

PROB. 9.

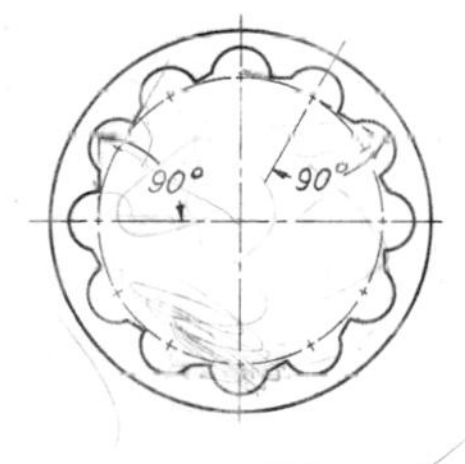


PROB. 10.

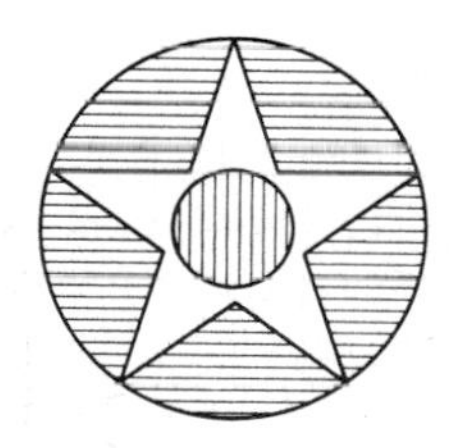

PROB. 11.

PROB. 12.

## Group 3. Circles and Tangents.

**13.** Concentric circles. For compass (legs straight) and scale. Draw a horizontal line through the center of a space. On it mark off radii for eight concentric circles $\frac{1}{4}$ in. apart. In drawing concentric circles, always draw the smallest first.

**14.** A four-centered spiral. For accurate tangents. Draw a $\frac{1}{8}$-in. square and extend its sides as shown. With the upper-right corner as center, draw quadrants with $\frac{1}{8}$- and $\frac{1}{4}$-in. radii. Continue with quadrants from each corner in order until four turns have been drawn.

**15.** A loop ornament. For bow compass. Draw a 2-in. square, about center of space. Divide $AE$ into four $\frac{1}{4}$-in. spaces with scale. With bow pencil and centers $A$, $B$, $C$, and $D$, draw four semicircles with $\frac{1}{4}$-in. radius, and so on. Complete the figure by drawing the horizontal and vertical tangents as shown.

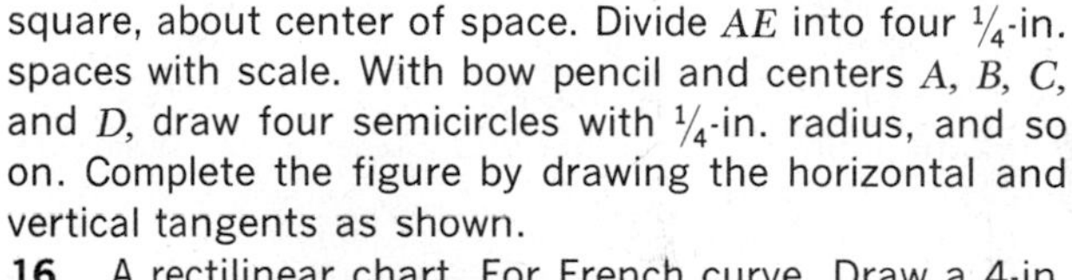

**16.** A rectilinear chart. For French curve. Draw a 4-in. field with $\frac{1}{2}$-in. coordinate divisions. Plot points at the intersections shown, and through them sketch a smooth curve very lightly in pencil. Finish by marking each point with a $\frac{1}{16}$-in. circle and drawing a smooth line with the French curve.

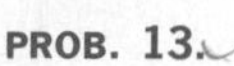

PROB. 13.

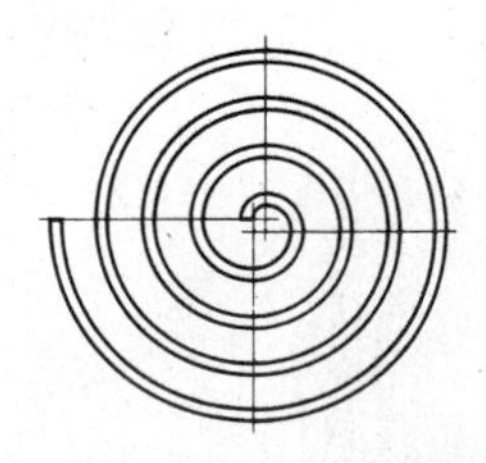

PROB. 14.

PROB. 15.

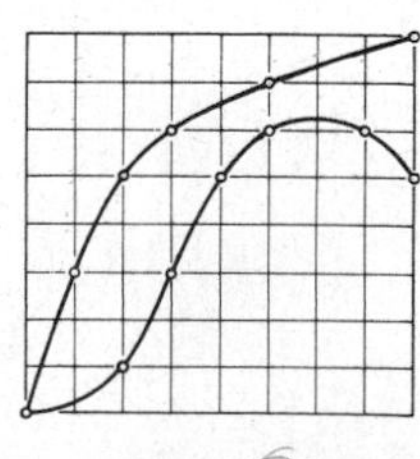

PROB. 16.

## Group 4. Scales.

**17.** Scale practice.

(*a*) Measure lines *A* to *G* to the following scales: *A*, full size; *B*, $\frac{1}{2}$ size; *C*, 3″ = 1′-0″; *D*, 1″ = 1′-0″; *E*, $\frac{3}{4}$″ = 1′-0″; *F*, $\frac{1}{4}$″ = 1′-0″; *G*, $\frac{3}{16}$″ = 1′-0″.

(*b*) Lay off distances on lines *H* to *N* as follows: *H*, 3$\frac{3}{16}$″, full size; *I*, 7″, $\frac{1}{2}$ size; *J*, 2′-6″, 1$\frac{1}{2}$″ = 1′-0″; *K*, 7′-5$\frac{1}{2}$″, $\frac{1}{2}$″ = 1′-0″; *L*, 10′-11″, $\frac{3}{8}$″ = 1′-0″; *M*, 28′-4″, $\frac{1}{8}$″ = 1′-0″; *N*, 40′-10″, $\frac{3}{32}$″ = 1′-0″.

(*c*) For engineer's scale. Lay off distances on lines *H* to *N* as follows: *H*, 3.2″, full size; *I*, 27′-0″, 1″ = 10′-0″; *J*, 66′-0″, 1″ = 20′-0″; *K*, 105′-0″, 1″ = 30′-0″; *L*, 156′-0″, 1″ = 40′-0″; *M*, 183′-0″, 1″ = 50′-0″; *N*, 214′-0″, 1″ = 60′-0″.

A, B, C, D, E, F, G

H, I, J, K, L, M, N

PROB. 17.

## Group 5. Combinations.

**18.** A telephone dial plate. Draw double size.

**19.** A film-reel stamping. Draw to scale of 6″ = 1′-0″.

**20.** Box cover. Make a one-view drawing for rectangular stamping 3 by 4 in., corners rounded with $\frac{1}{2}$-in. radius. Four holes, one in each corner, $\frac{3}{16}$-in. diameter, 3 and 2 in. center to center, for fasteners. Rectangular hole in center, $\frac{3}{8}$ by 1 in., with 1-in. side parallel to 4-in. side. Two slots $\frac{1}{4}$ in. wide, 2 in. long with semicircular ends, midway between center and 4-in. edges, with 2-in. side parallel to 4-in. side and centered between 3-in. edges.

**21.** Spacer. Make a one-view drawing for circular stamping 4 in. OD (outside diameter), 2 in. ID (inside diameter). Six $\frac{1}{4}$-in.-diameter holes equally spaced on 3-in.-diameter circle, with two holes on vertical center line. Two semicircular notches 180° apart made with $\frac{3}{8}$-in. radius centered at intersections of horizontal center line and 4-in.-OD circle.

**22.** Blank for wheel. Make a one-view drawing for stamping 5 in. OD; center hole $\frac{1}{2}$ in. in diameter; eight spokes $\frac{3}{8}$ in. wide connecting 1$\frac{1}{2}$-in.-diameter center portion with $\frac{1}{2}$-in. rim. Eight $\frac{1}{4}$-in.-diameter holes with centers at intersection of center lines of spokes and 4$\frac{1}{2}$-in. circle; $\frac{1}{8}$-in. fillets throughout to break sharp corners.

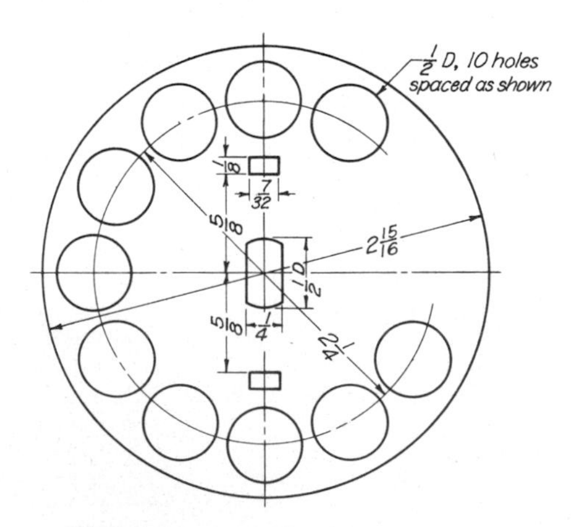


PROB. 18. A telephone dial plate.

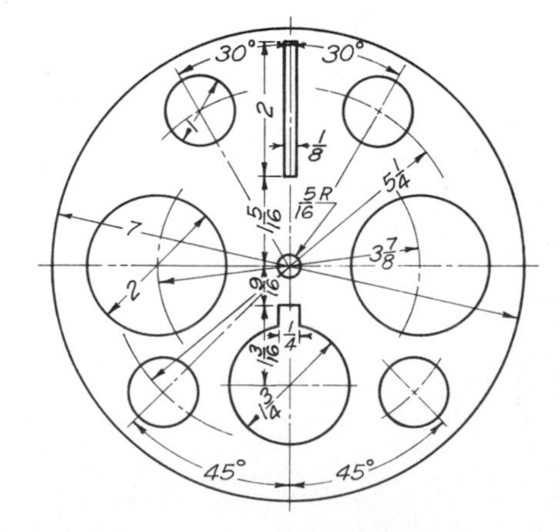


PROB. 19. A film-reel stamping.

**23.** Cover plate. Make a one-view drawing for rectangular stamping 3 by 4 in., corners beveled $\frac{1}{2}$ in. each way. Four holes, one in each corner, $\frac{1}{4}$-in. diameter, 3 and 2 in. center to center, for fasteners. Rectangular hole in center, $\frac{1}{2}$ by 1 in. with 1-in. side parallel to 4-in. side. Two holes, $\frac{3}{4}$-in. diameter, located midway between the slot and short side of rectangle on center line through slot.

**24.** Drawing of fixture base. Full size. Drill sizes specify the diameter (see Glossary).

**25.** Drawing of gage plate. Scale, twice size. Drill sizes specify the diameter (see Glossary).

**26.** Drawing of milling fixture plate. Scale, twice size. Drill size specifies diameter (see Glossary).

**27.** Drawing of dial shaft. Use decimal scale and draw 10 times size.

**28.** Drawing of inner toggle for temperature control. Use decimal scale and draw 10 times size.

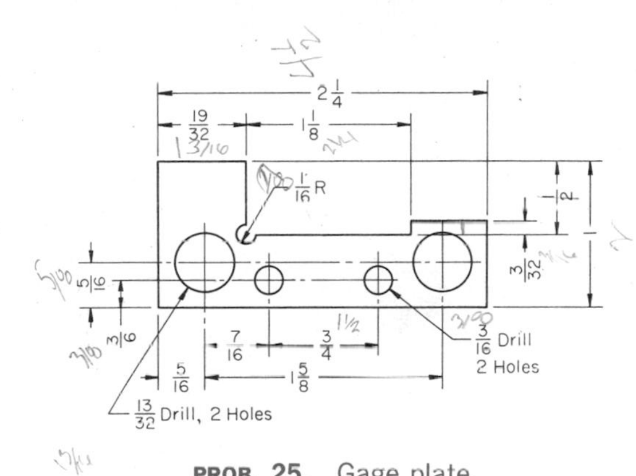


**PROB. 25.** Gage plate.

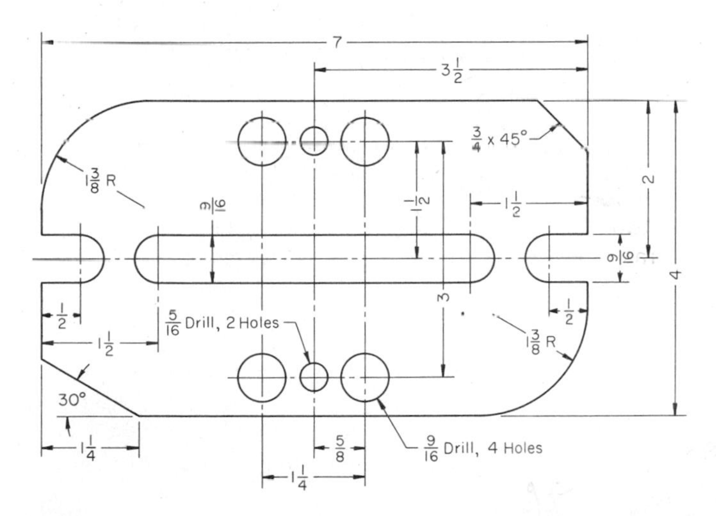


**PROB. 24.** Fixture base.

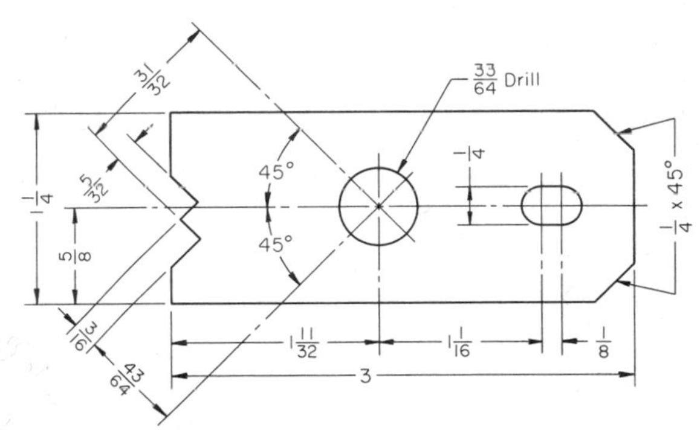


**PROB. 26.** Milling fixture plate.

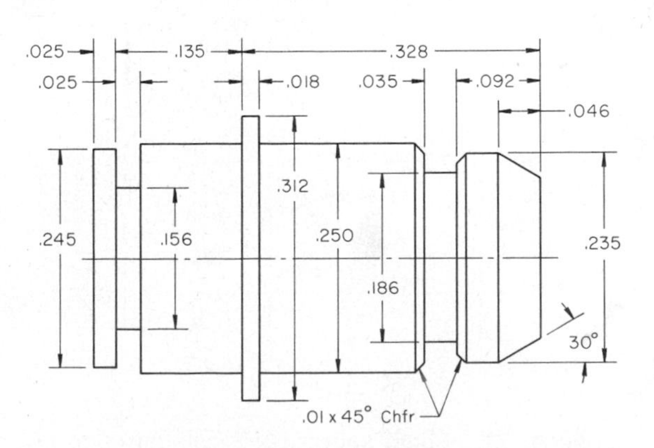


**PROB. 27.** Dial shaft.

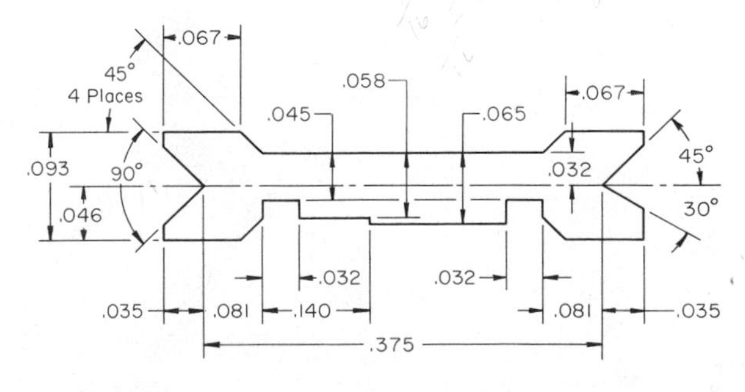


**PROB. 28.** Inner toggle for temperature control.

**29.** Drawing of mounting surface—O control. Scale, twice size.
**30.** Drawing of mounting leg—O control. Scale, full size.
**31.** Drawing of cooling fin and tube support. Full size.
**32.** Drawing of cone, sphere, and cylinder combinations. Scale, half size.
**33.** Drawing of cone-and-ball check. Scale, four times size.
**34.** Drawing of bell crank. Stamped steel. Scale, full size.
**35.** Drawing of control plate (aircraft hydraulic system). Stamped aluminum. Scale, full size.
**36.** Drawing of torque disk (aircraft brake). Stamped steel. Scale, half size.
**37.** Drawing of brake shoe. Stamped steel with molded asbestos composition wear surface. Scale, full size.

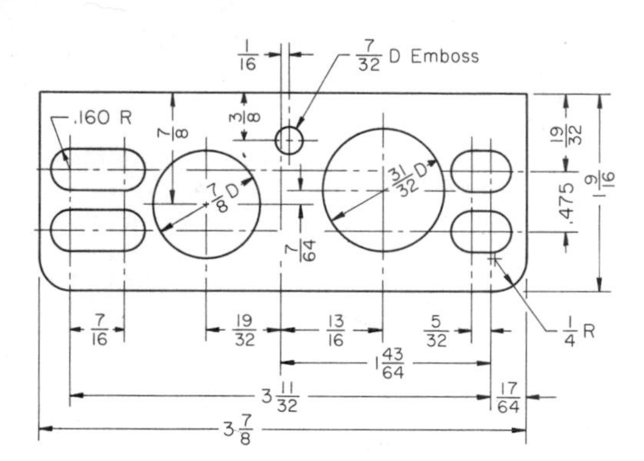

**PROB. 29.** Mounting surface: O control.

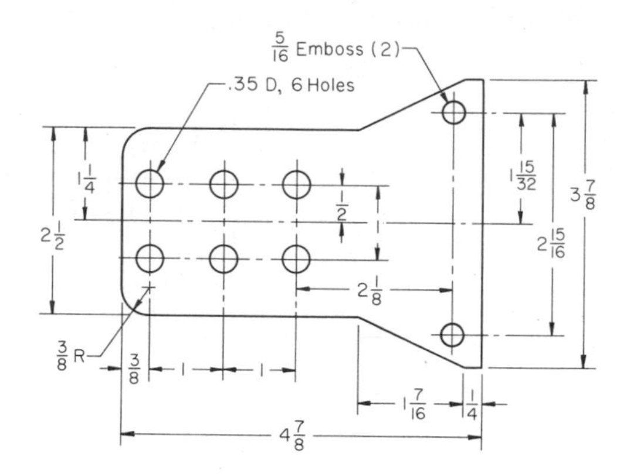

**PROB. 30.** Mounting leg: O control.

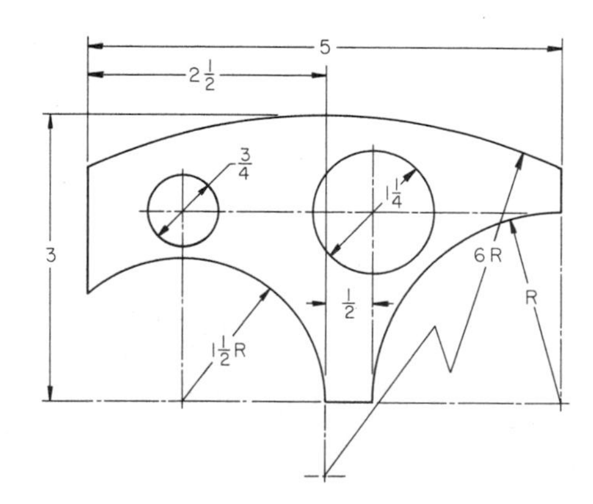

**PROB. 31.** Cooling fin and tube support.

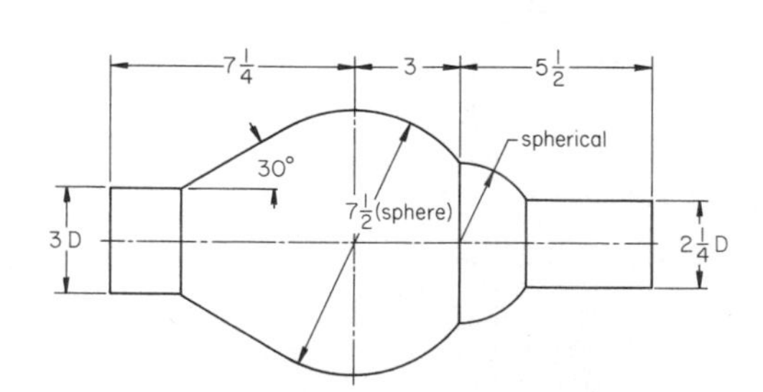

**PROB. 32.** Cone, sphere, and cylinder combinations.

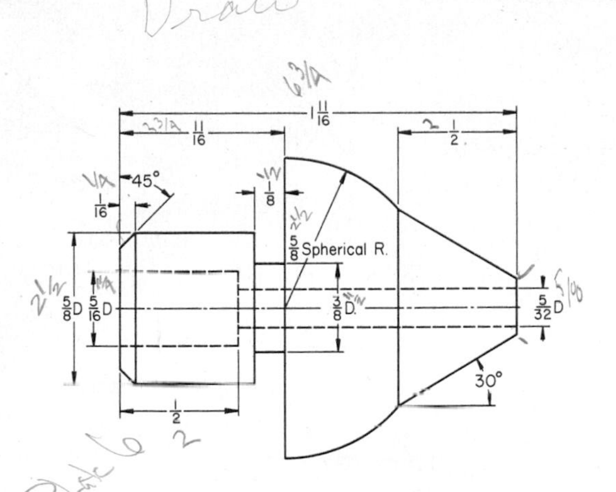


PROB. 33. Cone and ball check.

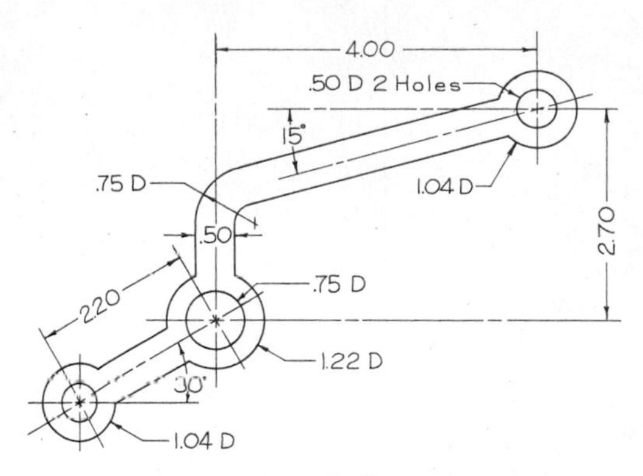


PROB. 34. Bell crank.

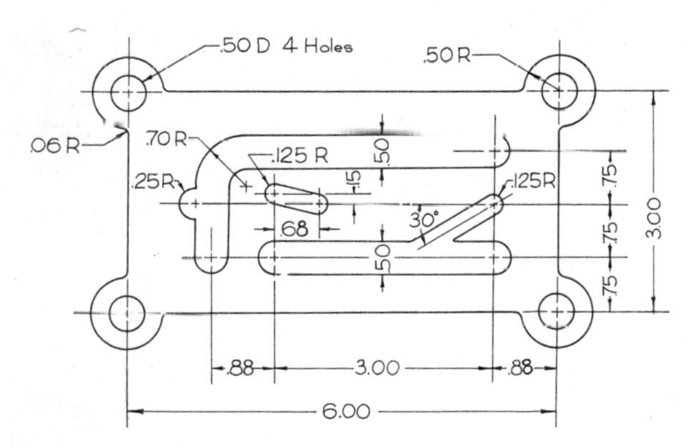


PROB. 35. Control plate.

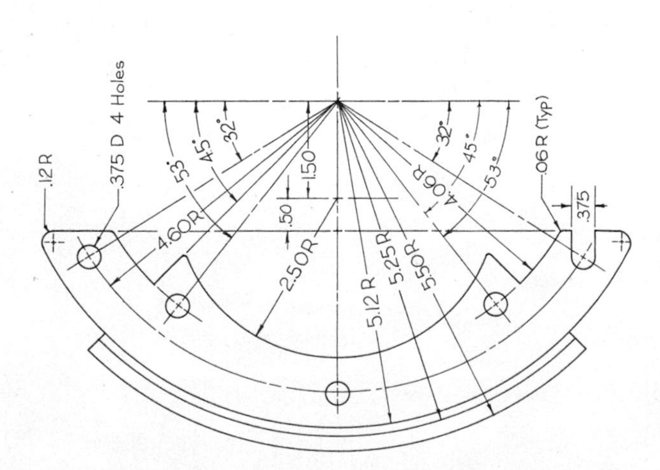


PROB. 36. Torque disk.

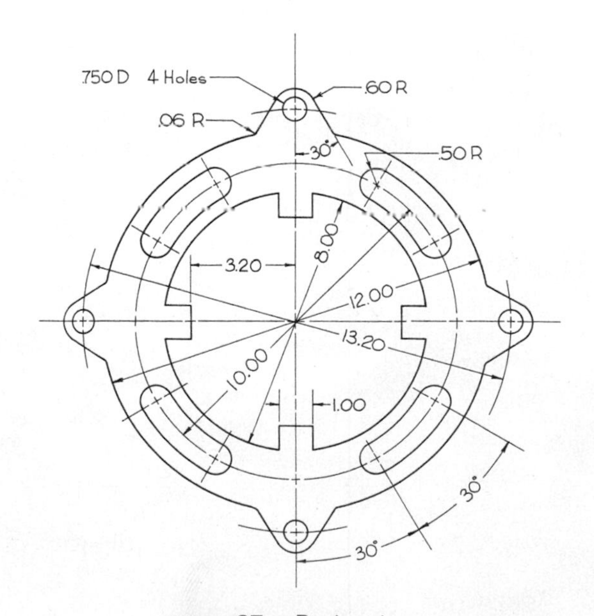


PROB. 37. Brake shoe.

This Chapter Presents the Geometry of:

- Straight Lines, Parallels, Perpendiculars
- Tangents, Tangent Points
- Tangent Lines, Circles, Curves
- Bisections, Trisections, Divisions
- Angles
- Geometric Shapes
- The Ellipse
- Noncircular Curves
- The Parabola
- The Hyperbola
- Cycloids, Involute, Spirals, Helices
- Tangents to Curves
- Problems

# GRAPHIC GEOMETRY

**Accurate, Concise Representation Requires a Knowledge of Geometric Constructions**

## 1. Graphic Geometry— Comparison and Classification.

Strict interpretation of constructional geometry allows use of only the compass and an instrument for drawing straight lines, and with these the geometer, following mathematical theory, accomplishes his solutions. In graphics the principles of geometry are employed constantly, but instruments are not limited to the basic two as T square, triangles, scales, curves, etc., are used to make constructions with speed and accuracy. Since there is continual application of geometric principles, the methods given in this chapter should be mastered thoroughly. It is assumed that students using this book understand the elements of plane geometry and will be able to apply their knowledge.

The constructions given here afford excellent practice in the use of instruments. Remember that the results you obtain will be only as accurate as your skill makes them. Take care in measuring and in drawing so that your work will be accurate and professional in appearance.

For easy reference, the various geometric figures are given in Fig. 101 at the end of this chapter.

This chapter is divided, for convenience and logical arrangement, into four parts: Line Relationships

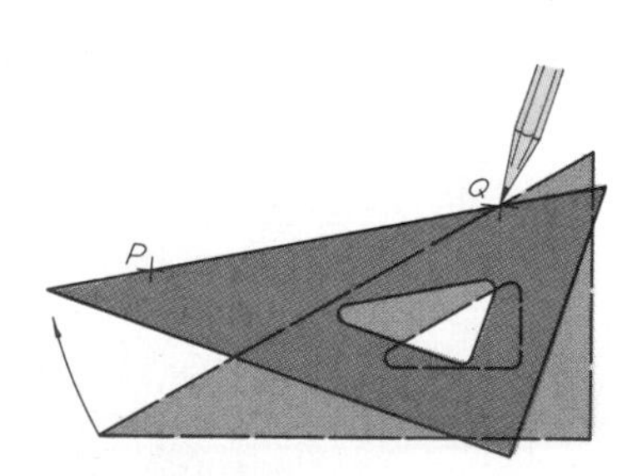


**FIG. 1.** To draw a line through two points. Use the pencil as a pivot and align the triangle or T square with the second point.

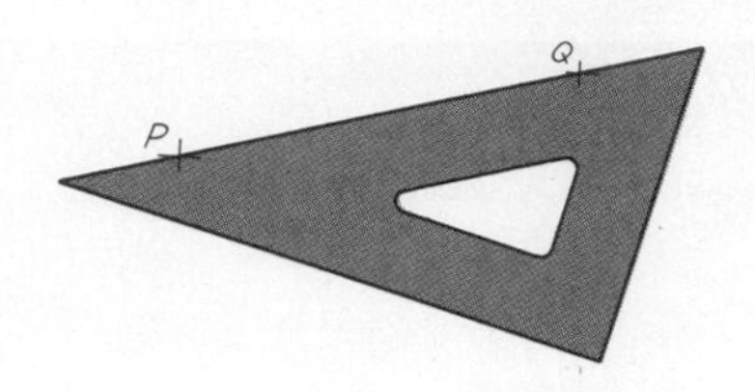


**FIG. 2.** To draw a line through two points (*alternate method*). Carefully align the triangle or T square with the points, and draw the required line.

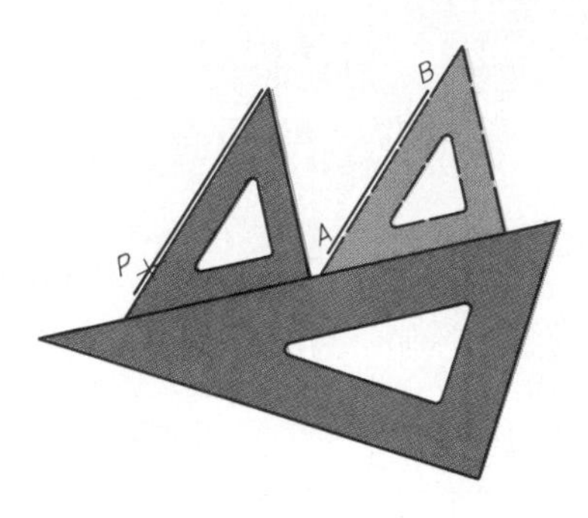


**FIG. 3.** To draw a line parallel to another. Align a triangle with the given line $AB$ using a base as shown; move it to position through the given point $P$ and draw the required line.

**FIG. 4.** To draw a line parallel to and a given distance from a line. Space the distance with circle arc $R$; then align a triangle on a base as shown, move to position tangent to the circle arc, and draw the required line.

and Connections, which represent the bulk of the geometry needed in everyday work; Geometry of Straight-line Figures; Geometry of Curved Lines; and Constructions for Lofting and Large Layouts.

## LINE RELATIONSHIPS AND CONNECTIONS

### 2. To Draw Straight Lines (Figs. 1 and 2).

Straight lines are drawn by using the straight edge of the T square or one of the triangles. For short lines a triangle is more convenient. Observe the directions for technique in Secs. 2, 3, and 4.

To draw a straight line through two points (Fig. 1), place the point of the pencil at $Q$ and bring the triangle ( or T square) against the point of the pencil. Then, using this point as a pivot, swing the triangle until its edge is in alignment with point $P$, and draw the line.

To draw a straight line through two points, *alternate method* (Fig. 2), align the triangle or T square with points $P$ and $Q$, and draw the line.

### 3. To Draw Parallel Lines (Figs. 3 and 4).

Parallel lines may be required in any position. Parallel horizontals or verticals are most common. Horizontals are drawn with T square alone, verticals with T square and triangle. The general cases (odd angles) are shown in Figs. 3 and 4.

To draw a straight line through a point, parallel to another line (Fig. 3), adjust a triangle to the given line $AB$, with a second triangle as a base. Slide the aligned triangle to its position at point $P$ and draw the required line.

To draw a straight line at a given distance from and parallel to another line (Fig. 4), draw an arc with the given distance $R$ as radius and any point on the given line $AB$ as center. Then adjust a triangle to line $AB$, with a second triangle as a base. Slide the aligned triangle to position tangent to the circle arc and draw the required line.

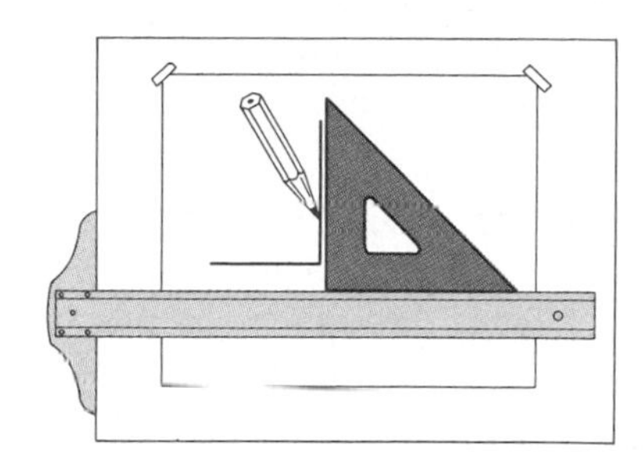

FIG. 5. To draw a line perpendicular to another (when the *given line* is *horizontal*). Place the triangle on the T square and draw the required line.

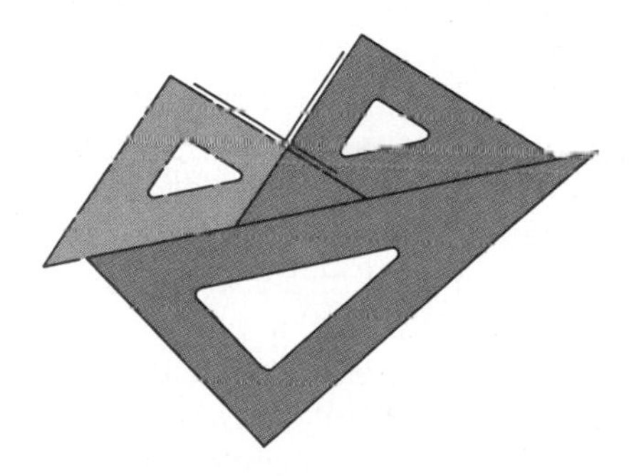

FIG. 6. To draw a line perpendicular to another (general position). Align a triangle with the given line as shown; slide it (on another triangle as a base) to position of perpendicular and draw the required line.

### 4. To Draw Perpendicular Lines (Figs. 5 and 6).

Perpendiculars occur frequently as horizontal-to-vertical (Fig. 5) but also often in other positions (Fig. 6). Note that the construction of a perpendicular utilizes the 90° angle of a triangle.

To erect a perpendicular to a given straight line (*when the given line is horizontal, Fig. 5*), place a triangle on the T square as shown and draw the required perpendicular.

To erect a perpendicular to a given straight line (*general position, Fig. 6*), set a triangle with its hypotenuse against a guiding edge and adjust one side to the given line. Then slide the triangle so that the second side is in the position of the perpendicular and draw the required line.

### 5. Tangents (Figs. 7 to 20).

A *tangent* to a curve is a line, either straight or curved, that passes through two points on the curve infinitely close together. One of the most frequent geometric operations in drafting is the drawing of tangents to circle arcs and the drawing of circle arcs tangent to straight lines or other circles. These should be constructed accurately, and on pencil drawings that are to be inked or traced the points of tangency should be located by short cross marks to show the stopping points for the ink lines. The method of finding these points is indicated in the following constructions. Note in all the following tangent constructions that the location of tangent points is based on one of these geometric facts: (1) the tangent point of a straight line and circle will lie at the intersection of a perpendicular to the straight line that passes through the circle center, and (2) the tangent point of two circles will lie on the circumferences of both circles and on a straight line connecting the circle centers. See Fig. 7.

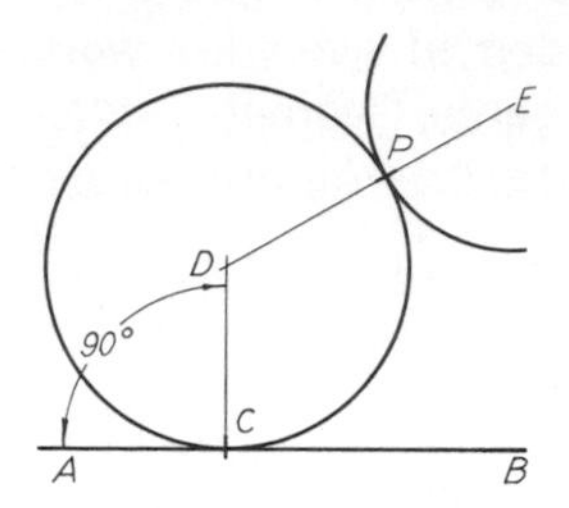


FIG. 7. Tangent points. The tangent point of a straight line and circle is on the perpendicular from the circle center. The tangent point of two circles is on a line connecting their centers.

### 6. Tangent Points (Fig. 7).

To find the point of tangency for line $AB$ and a circle with center $D$, draw $DC$ perpendicular to line $AB$. Point $C$ is the tangent point.

To find the tangent point for two circles (Fig. 7) with centers at $D$ and $E$, draw $DE$, joining the centers. Point $P$ is the tangent point.

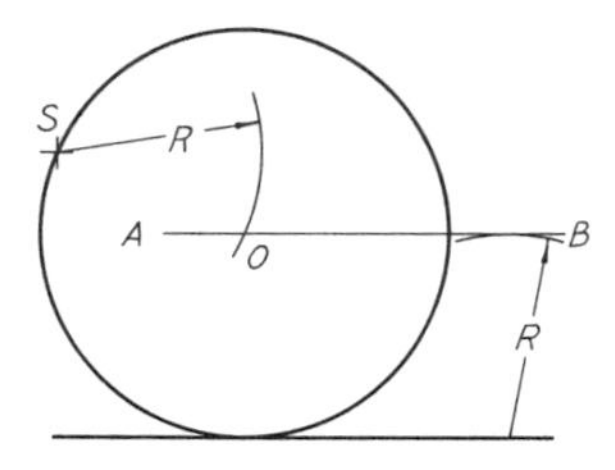


FIG. 8. A circle tangent to a line and passing through a point. The circle center must be equidistant from the line and the point.

### 7. To Draw a Circle of Given Size Tangent to a Line and Passing through a Point (Fig. 8).

Draw a line $AB$, the given radius distance $R$ away from and parallel to the given line. Using the given point $S$ as center, cut line $AB$ at $O$ with the given radius. $O$ is the center of the circle. Note that there are two possible positions for the circle.

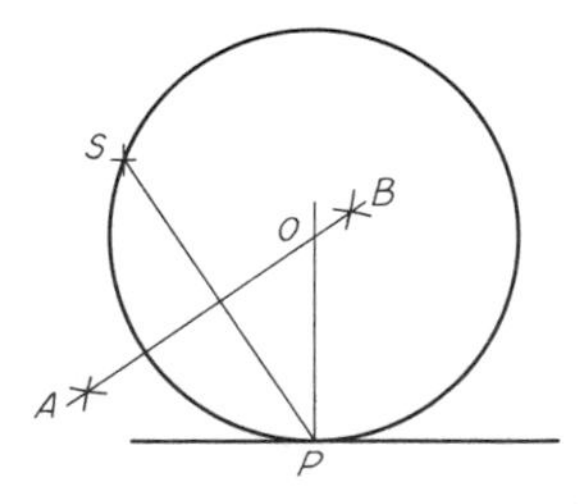


FIG. 9. A circle tangent to a line at a point and passing through a second point. The circle center is at the intersection of the bisector of the chord connecting the points and the perpendicular from the point on the line tangent.

### 8. To Draw a Circle Tangent to a Line at a Point and Passing through a Second Point (Fig. 9).

Connect the two points $P$ and $S$ and draw the perpendicular bisector $AB$ (see Sec. 38). Draw a perpendicular to the given line at $P$. The point where this perpendicular intersects the line $AB$ is the center $O$ of the required circle.

### 9. To Draw a Tangent to a Circle at a Point on the Circle (Fig. 10).

Given the arc $ABC$, draw a tangent at the point $C$. Arrange a triangle in combination with the T square (or another triangle) so that its hypotenuse passes through center $O$ and point $C$. Holding the T square firmly in place, turn the triangle about its square corner and move it until the hypotenuse passes through $C$. The required tangent then lies along the hypotenuse.

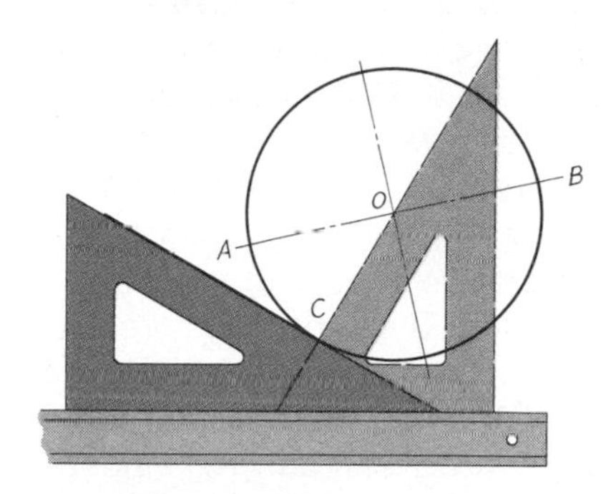


**FIG. 10.** A tangent at a point on a circle. The tangent line must be perpendicular to the line from the point to the center of the circle.

### 10. To Draw a Tangent to a Circle from a Point Outside (Fig. 11).

Given the arc $ACB$ and point $P$, arrange a triangle in combination with another triangle (or T square) so that one side passes through point $P$ and is tangent to the circle arc. Then slide the triangle until the right-angle side passes through the center of the circle and mark lightly the tangent point $C$. Bring the triangle back to its original position and draw the tangent line.

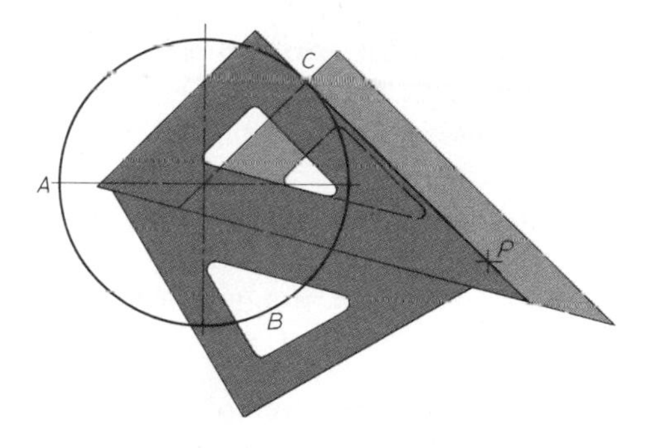


**FIG. 11.** A tangent to a circle from a point outside. The line is drawn by alignment through the point and tangent to the circle; the tangent point is then located on the perpendicular from the circle center.

### 11. To Draw a Circle Arc of Given Radius Tangent to Two Lines at Right Angles to Each Other (Fig. 12).

Draw an arc of radius $R$, with center at corner $A$, cutting the lines $AB$ and $AC$ at $T$ and $T_1$. Then with $T$ and $T_1$ as centers and with the same radius $R$, draw arcs intersecting at $O$, the center of the required arc.

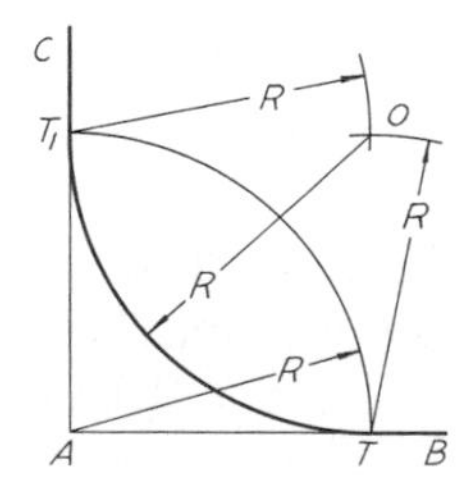


**FIG. 12.** An arc tangent at right-angle corner. The arc center must be equidistant from both lines.

### 12. To Draw an Arc of Given Radius Tangent to Two Straight Lines (Fig. 13).

Given the lines *AB* and *CD*, set the compass to radius *R*, and at any convenient point on the given lines draw the arcs *R* and $R_1$. With the method of Fig. 4, draw parallels to the given lines through the limits of the arcs. These parallels are the loci of the centers of all circles of radius *R* tangent to lines *AB* and *CD*, and their intersection at point *O* will be the center of the required arc. Find the tangent points by erecting perpendiculars, as in Fig. 11, to the given lines through the center *O*. The figure above shows the method for an obtuse angle, below for an acute angle.

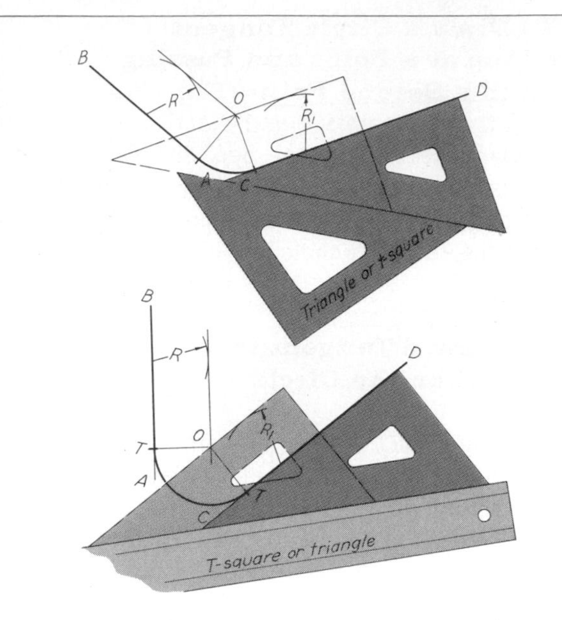


**FIG. 13.** An arc tangent to two straight lines. The arc center must be equidistant from both straight lines.

### 13. To Draw a Tangent to Two Circles (Fig. 14, Open Belt).

Arrange a triangle in combination with a T square or triangle so that one side is in the tangent position. Move to positions 2 and 3, marking lightly the tangent points $T_1$ and $T_2$. Return to the original position and draw the tangent line. Repeat for the other side.

### 14. To Draw a Tangent to Two Circles (Fig. 15, Crossed Belt).

Arrange a triangle in combination with a T square or triangle so that one side is in the tangent position. Move to positions 2 and 3, marking lightly the tangent points $T_1$ and $T_2$. Return to the original position and draw the tangent line. Repeat for the other side.

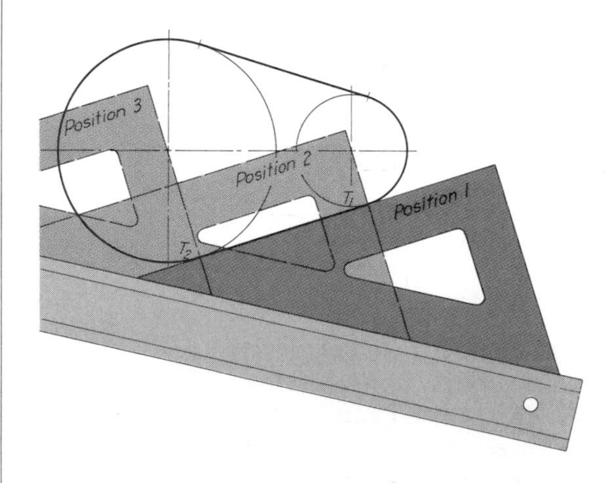


**FIG. 14.** Tangents to two circles (open belt). Tangent lines are drawn by alignment with both circles. Tangent points lie on perpendicular lines from circle centers.

### 15. To Draw a Circle of Radius *R* Tangent to a Given Circle and a Straight Line (Fig. 16).

Let *AB* be the given line and $R_1$ the radius of the given circle. Draw a line *CD* parallel to *AB* at a distance *R* from it. With *O* as center and radius $R + R_1$, swing an arc intersecting *CD* at *X*, the desired center. The tangent point for *AB* will be on a perpendicular to *AB* from *X*; the tangent point for the two circles will be on a line joining their centers *X* and *O*.

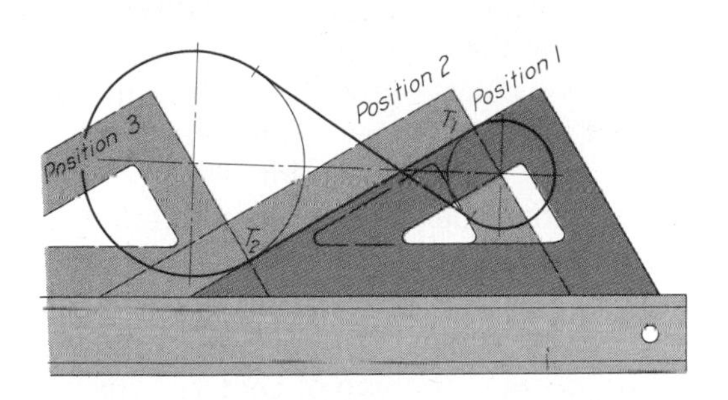


FIG. 15. Tangents to two circles (crossed belt). Tangent lines are drawn by alignment with both circles. Tangent points lie on perpendicular lines from circle centers.

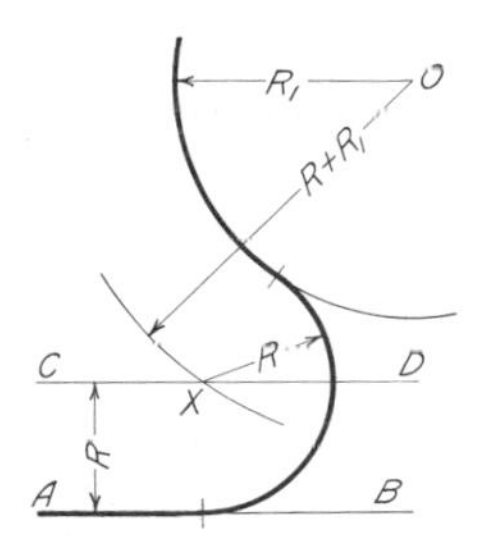

FIG. 16. An arc tangent to a straight line and a circle. The arc center must be equidistant from the line and the circle.

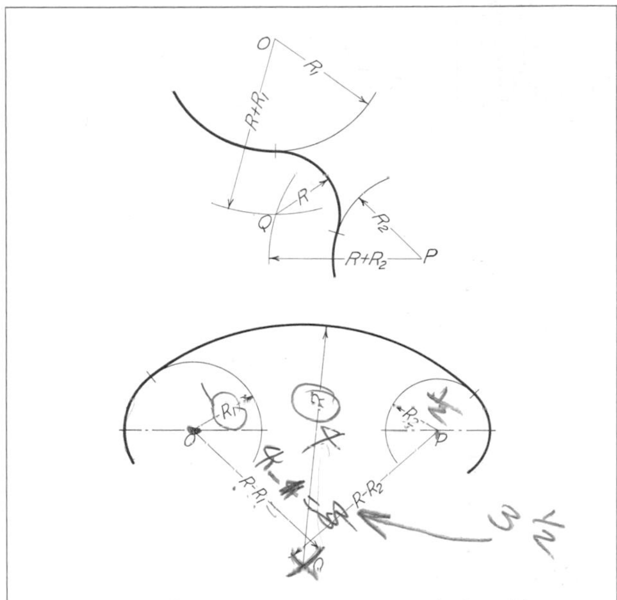

FIG. 17. An arc tangent to two circles. The arc center must be equidistant from both circles.

## 16. To Draw a Circle of Radius *R* Tangent to Two Given Circles.

*First Case* (*Fig. 17, Top*). The centers of the given circles are outside the required circle. Let $R_1$ and $R_2$ be the radii of the given circles and $O$ and $P$ their centers. With $O$ as center and radius $R + R_1$, describe an arc. With $P$ as center and radius $R + R_2$, swing another arc intersecting the first arc at $Q$, which is the center sought. Mark the tangent points in line with $OQ$ and $QP$.

*Second Case* (*Fig. 17, Bottom*). The centers of the given circles are inside the required circle. With $O$ and $P$ as centers and radii $R - R_1$ and $R - R_2$, describe arcs intersecting at the required center $Q$.

## 17. To Draw a Reverse, or Ogee, Curve (Fig. 18).

Given two parallel lines $AB$ and $CD$, join $B$ and $C$ by a straight line. Erect perpendiculars at $B$ and $C$. Any arcs tangent to lines $AB$ and $CD$ at $B$ and $C$ must have their centers on these perpendiculars. On

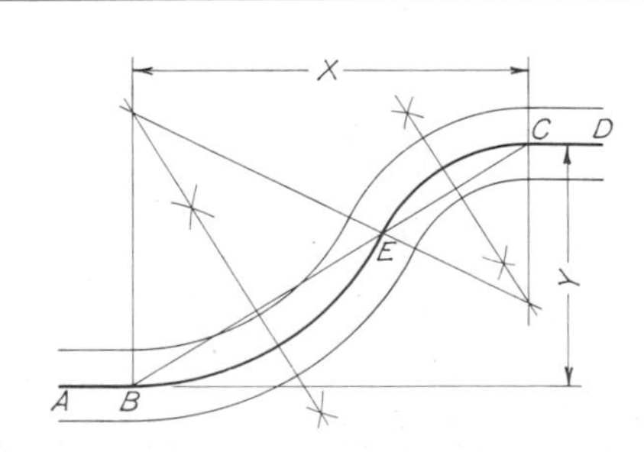

FIG. 18. An ogee curve. It is made of circle arcs tangent to each other and to straight lines.

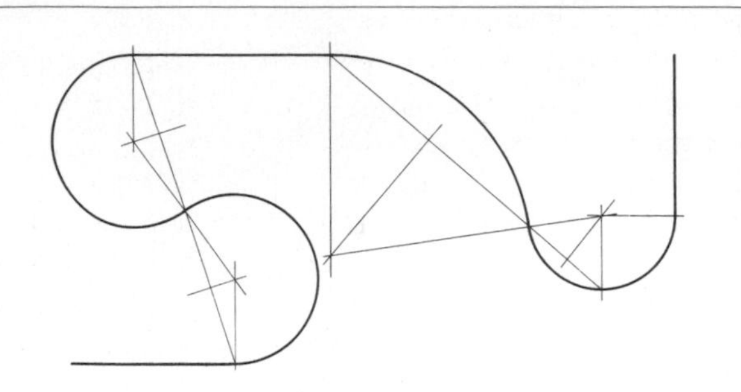

**FIG. 19.** Ogee applications. Note that the circle arcs are tangent on a line connecting the centers.

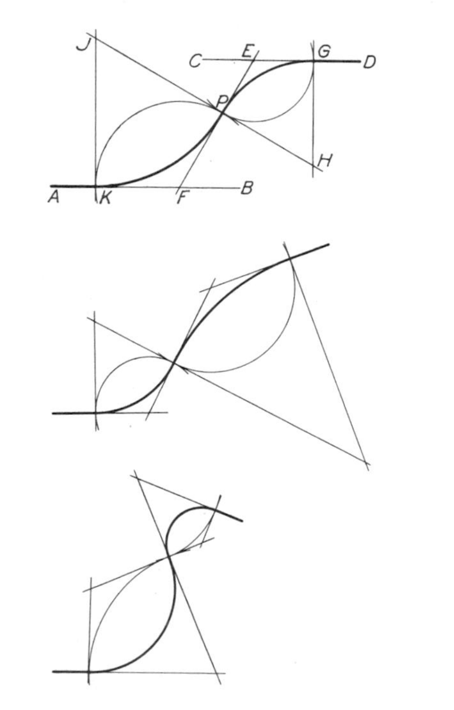


**FIG. 20.** Reverse curve tangent to three lines. Note that the arc centers must lie on a line perpendicular to the tangent line at the point of tangency.

the line *BC* assume point *E*, the point through which it is desired that the curve shall pass. Bisect *BE* and *EC* by perpendiculars. Any arc to pass through *B* and *E* must have its center somewhere on the perpendicular from the middle point. The intersection, therefore, of these perpendicular bisectors with the first two perpendiculars will be the centers for arcs *BE* and *EC*. This line might be the center line for a curved road or pipe. The construction may be checked by drawing the line of centers, which *must* pass through *E*. Figure 19 illustrates the principle of reverse-curve construction in various combinations.

### 18. To Draw a Reverse Curve Tangent to Two Lines and to a Third Secant Line at a Given Point (Fig. 20).

Given two lines *AB* and *CD* cut by the line *EF* at points *E* and *F*, draw a perpendicular *JH* to *EF* through a given point *P* on *EF*. With *E* as center and radius *EP*, intersect *CD* at *G*. Draw a perpendicular from *G* intersecting *JH* at *H*. With *F* as center and radius *FP*, intersect *AB* at *K*. Draw a perpendicular to *AB* from *K* intersecting *JH* at *J*. *H* and *J* will be the centers for arcs tangent to the three lines.

## GEOMETRY OF STRAIGHT-LINE FIGURES

### 19. To Bisect a Line (Fig. 21).

*With Compass.* From the two ends of the line, swing arcs of the same radius, greater than one-half the length of the line, and draw a line through the arc intersections. This line bisects the given line and is also the *perpendicular* bisector. Many geometric problems depend upon this construction.

*With T Square and Triangles* (*Fig. 22*). At two points, *A* and *B*, on the line, draw lines *AC* and *BC* at equal angles with *AB*. A perpendicular *CD* to *AB* then cuts *AB* at *D*, the midpoint, and *CD* is the perpendicular bisector of *AB*.

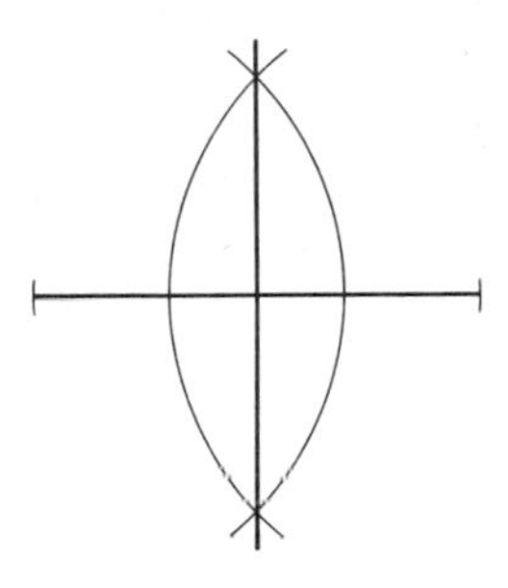

**FIG. 21.** To bisect a line (with compass). The intersections of equal arcs locate two points on the perpendicular bisector.

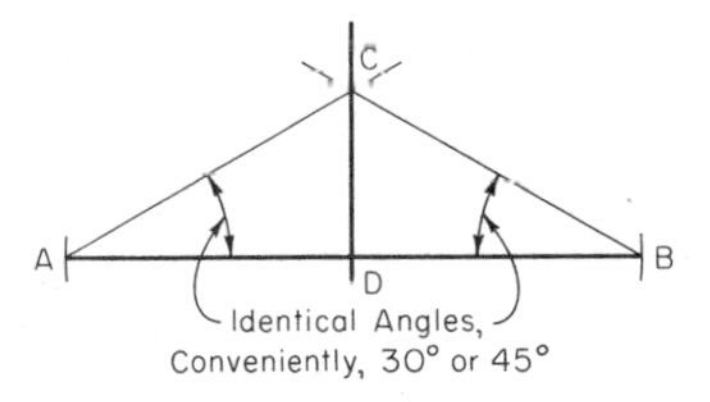


**FIG. 22.** To bisect a line (with T square and triangle). Equal angles locate one point on the perpendicular bisector.

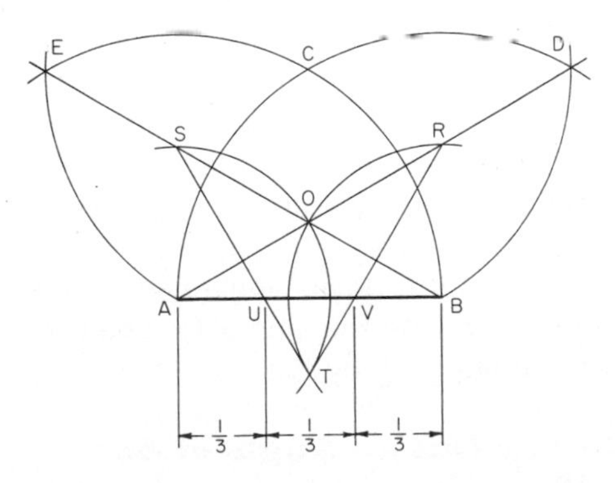


**FIG. 23.** To trisect a line (with compass). Construction is based on the geometry of an equilateral triangle.

## 20. To Trisect a Line.

Trisection of a line is not nearly so often needed as bisection, but will occasionally be required.

*With Compasses and Straightedge* (*Fig. 23*). On given line $AB$ and with radius $AB$, draw two arcs of somewhat more than quarter circles, using $A$ and $B$ as centers. These arcs will intersect at $C$. Using the same radius $AB$ and with $C$ as center cut the first arcs at $D$ and $E$. Then draw $DA$ and $EB$, which intersect at $O$. Using $OA$ (or $OB$) as radius and $A$ and $B$ as centers, cut $AD$ and $BE$ at $R$ and $S$, extend these arcs to intersect at $T$. Then draw $RT$ and $ST$, which will intersect $AB$ at third points $U$ and $V$. You will recognize, upon analysis, that this construction is based on the geometry of an equilateral triangle inscribed in a circle, where the diameter of the circle is equal to one side of the triangle.

*With T Square and Triangles* (*Fig. 24*). The above construction can be accomplished with less detail by using a 30-60° triangle to obtain the needed angles. From $A$ and $B$ draw lines at 30° to intersect at $C$. Then at 60° to $AB$ draw lines from $C$ that intersect $AB$ at $U$ and $V$, the third points.

## 21. To Divide a Line into 2, 3, 4, 6, 8, 9, 12, or 16 Parts (Figs. 25 and 26).

Divisions into successive halves, 2, 4, 8, etc., are the most common. The principle of Fig. 22 can easily be applied to accomplish the division. As shown in Fig. 25, equal angles from $A$ and $B$ locate $C$, and the perpendicular from $C$ to $AB$ then gives the mid-point. Successive operations will give 4, 8, 16, etc., parts.

The principles of bisection (Fig. 22) and trisection (Fig. 24) can be combined to get 6, 9, or 12 parts. As indicated in Fig. 26, lines at 30° to $AB$ from $A$ and $B$ locate $C$, and lines at 60° to $AB$ from $C$ locate $D$ and $E$, the third points. Then 30° lines from $A$ and $D$ will bisect $AD$ at $F$, giving sixth divisions; or trisecting by the principle of Fig. 24 as from $D$ and $E$, giving $G$ and $H$, will produce ninth points; or as from $E$ and $B$, the half point located at $J$, and again the half points of $EJ$ and $BJ$ located at $K$ and $L$, give twelfth divisions of $AB$.

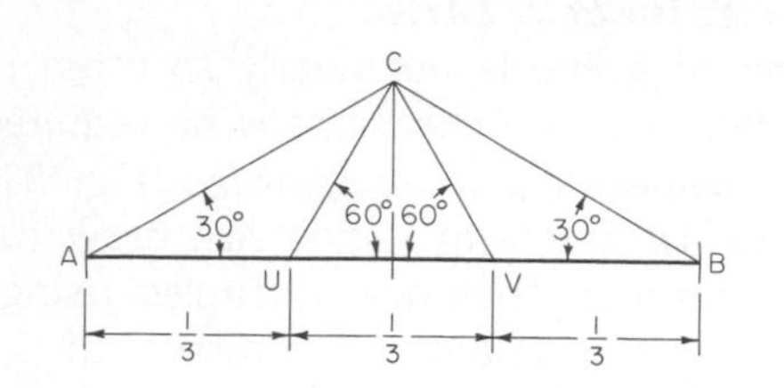


**FIG. 24.** To trisect a line (with T square and triangle). Third points are located by 30° and 60° angles.

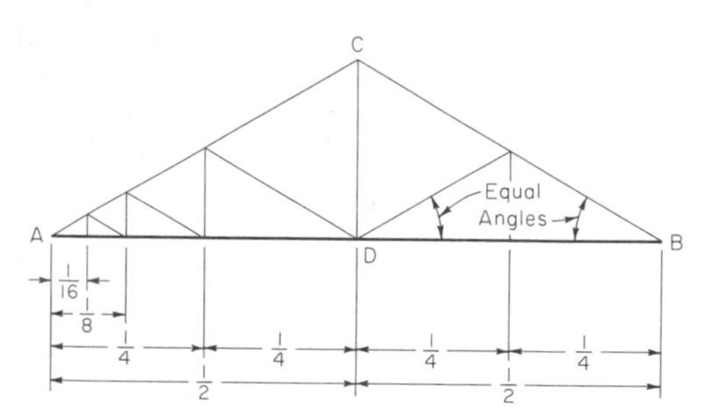


**FIG. 25.** To divide a line into 2, 4, 8, or 16 parts. Equal angles locate a point for bisection.

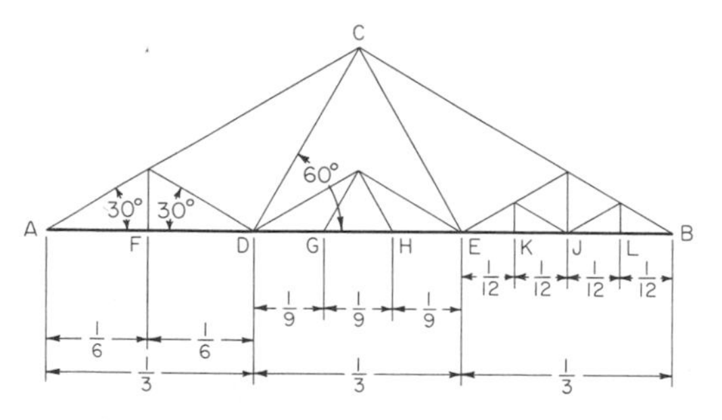


**FIG. 26.** To divide a line into 3, 6, 9, or 12 parts. The 30-60° triangle is used to obtain third points.

## 22. To Divide a Line into Any Number of Parts.

*First Method (Fig. 27).* To divide a line $AB$ into, say, five equal parts, draw any line $BC$ of indefinite length. On it measure, or step off, five divisions of convenient length. Connect the last point with $A$, and using two triangles as shown in Fig. 3, draw lines through the points parallel to $CA$ intersecting $AB$.

*Second Method (Figs. 28 and 29).* Draw a perpendicular $AC$ from $A$. Then place a scale so that five convenient equal divisions are included between $B$ and the perpendicular, as in Fig. 28. With a triangle and T square draw perpendiculars through the points marked, dividing the line $AB$ as required. Figure 29 illustrates an application in laying off stair risers. This method can be used for dividing a line into any series of proportional parts.

## 23. To Lay Out a Given Angle.

*Tangent Method (Fig. 30).* The trigonometric tangent of an angle of a triangle is the ratio of the length of the side opposite the angle divided by the length of the adjacent side. Thus, tan $A = Y/X$, or $X$ tan $A = Y$. To lay out a given angle, obtain the value of the tangent from a table of natural tangents (see the Appendix), assume any convenient distance $X$, and multiply $X$ by the tangent to get distance $Y$. Note that the angle between the sides $X$ and $Y$ must be a right angle.

## 24. To Lay Out a Given Angle.

*Chordal Method (Fig. 31).* If the length of a chord is known for an arc of given radius and included angle, the angle can be accurately laid out. Given an angle in degrees, to lay out the angle, obtain the chord length for a l-in. circle arc from the table in the Appendix. Select any convenient arc length $R$ and multiply the chord length for a 1-in. arc by this distance, thus obtaining the chord length $C$ for the radius distance selected. Lay out the chord length on the arc with compass or dividers and complete the sides of the angle.

The chord length for an angle can be had from a sine table by taking the sine of one-half the given angle and multiplying by two.

## 25. To Lay Out an Angle of 45° (Fig. 32).

Angles of 45° occur often, and are normally drawn with the 45° triangle. However, for large constructions or when great accuracy is needed, the method of equal legs is valuable (tangent 45° =

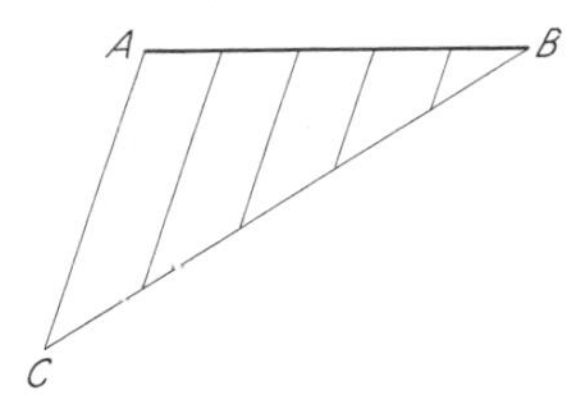


**FIG. 27.** To divide a line. Equal divisions on any line *BC* are transferred to *AB* by parallels to *AC*.

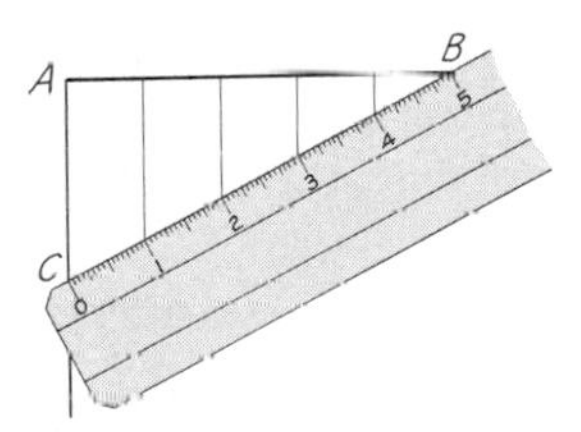


**FIG. 28.** To divide a line. Scale divisions are transferred to given line *AB*.

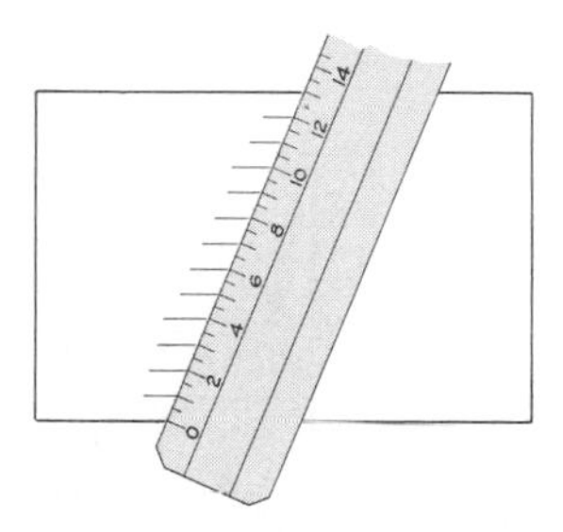

**FIG. 29.** To divide a line. Scale divisions divide (in this case) the vertical space.

1.0). With any distance *AB* on the given line, with center *B* and radius *AB* draw more than a quarter circle. Erect a perpendicular at *B* (to *AB*) which intersects the arc at point *C*. Then *CA* makes 45° with *AB*.

## 26. To Lay Out Angles of 30°, 60°, 90°, 120°, etc. (Fig. 33).

As with 45° angles, angles in multiples of 30° may also be needed for large constructions or when great accuracy is required. On a given line, with any convenient radius *AB*, swing an arc with *A* as center.

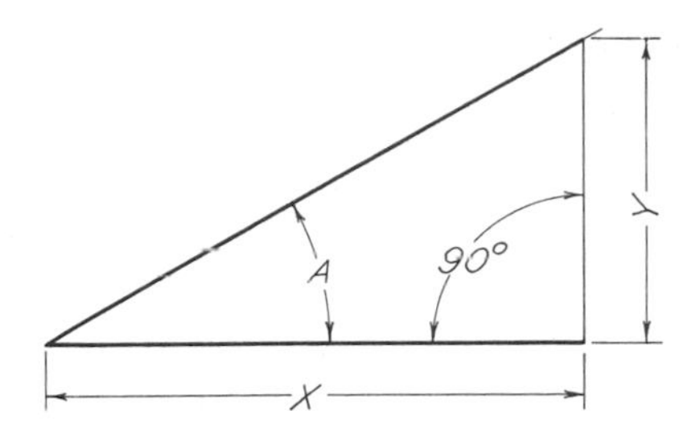


**FIG. 30.** Angle by tangent. The proportion of *Y* to *X* is obtained from a table of natural tangents.

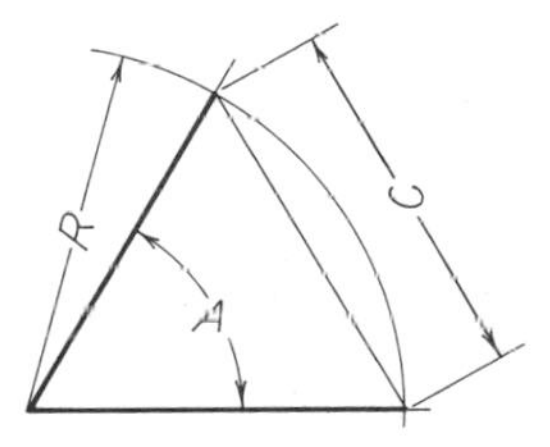


**FIG. 31.** Angle by chord. The proportion of *C* to *R* is obtained from a table of chords.

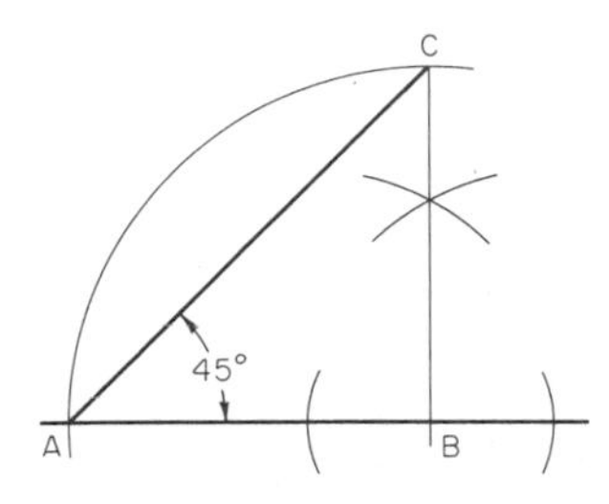


**FIG. 32.** To lay out a 45° angle. The two legs, *AB* and *BC*, are equal and perpendicular.

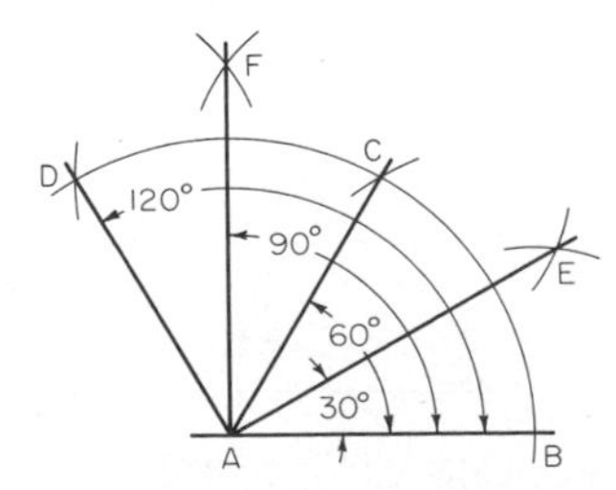


**FIG. 33.** To construct angles in multiples of 30°. Chords equal to the radius will divide a circle into six equal parts.

With the same radius and *B* as center cut the original arc at *C*. The included angle *CAB* is 60° and by bisection (equal arcs from *C* and *B* to locate *E*) 30° is obtained at *CAE* and *EAB*. For 120°, *AB* laid off again from *C* to *D* gives *DAB* as 120°, and by bisecting *DAC*, 90° is given at *FAB*.

### 27. To Bisect an Angle (Fig. 34).

Given angle *SOR*, using any convenient radius shorter than *OR* or *OS*, swing an arc with *O* as center, locating *A* and *B*. Then with the same radius or another radius longer than one-half the distance from *A* to *B*, swing arcs with *A* and *B* as centers, locating *C*. The bisector is *CO*. With repeated bisection, the angle can be divided into 4, 8, 16, 32, 64, etc., parts.

### 28. To Divide an Angle (Fig. 35).

Sometimes an angle must be divided into some number of parts, say 3, 5, 6, 7, or 9, that cannot be obtained by bisection. It is possible to trisect an angle by employing a movable scale, but this method is more complicated and not as accurate as dividing an arc with the dividers and is of no use for odd parts such as 5 and 7.

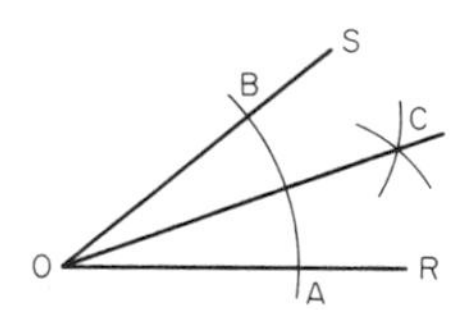


**FIG. 34.** To bisect an angle. Equal arcs from *A* and *B* locate *C* on the bisector.

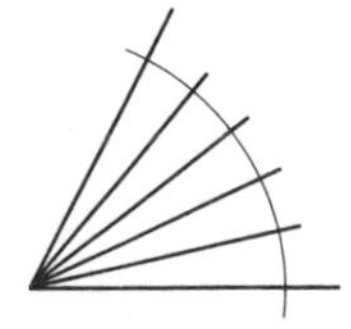

**FIG. 35.** To divide an angle into equal parts. Equal divisions (chords) are stepped off on an arc of the angle.

To divide an angle into, for example, 5 parts (Fig. 35), draw any arc across the angle as shown, and use the bow dividers to divide the arc into equal parts. Lines through these points to the apex are the required divisions.

### 29. To Construct a Triangle Having Given the Three Sides (Fig. 36).

Given the lengths *A*, *B*, and *C*. Draw one side *A* in the desired position. With its ends as centers and radii *B* and *C*, draw two intersecting arcs as shown. This construction is used extensively in developments by triangulation.

### 30. To Locate the Geometric Center of an Equilateral Triangle (Fig. 37).

With *A* and *B* as centers and radius *AB*, draw arcs as shown. These will intersect at *C*, the third corner.

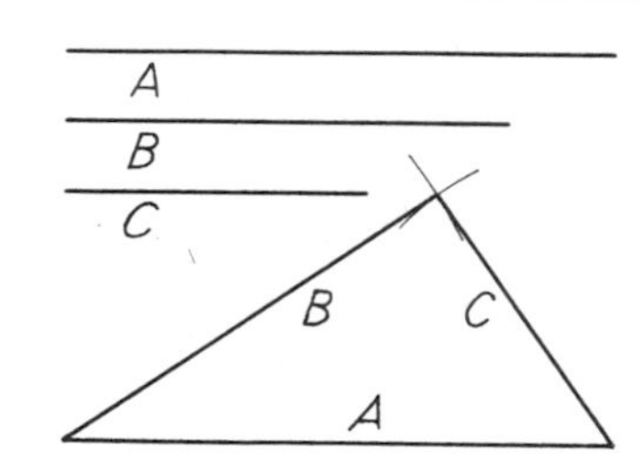


**FIG. 36.** To construct a triangle. The legs are laid off with a compass.

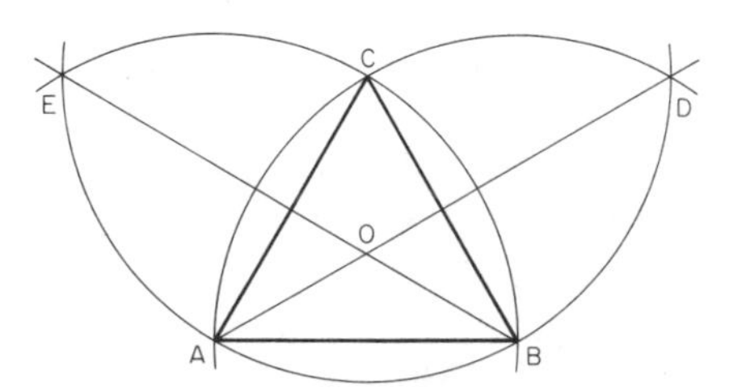


**FIG. 37.** To locate the geometric center of an equilateral triangle. Perpendicular bisectors of the sides intersect at the center.

With center $C$ and radius $AB$ cut the original arcs at $D$ and $E$. Then $AD$ and $BE$ intersect at $O$, the geometric center.

## 31. To Transfer a Polygon to a New Base.

*By Triangulation (Fig. 38).* Given polygon $ABCDEF$ and a new position of base $A'B'$, consider each point as the vertex of a triangle whose base is $AB$. With centers $A'$ and $B'$ and radii $AC$ and $BC$, describe intersecting arcs locating the point $C'$. Similarly, with radii $AD$ and $BD$ locate $D'$. Connect $B'C'$ and $C'D'$ and continue the operation always using $A$ and $B$ as centers.

*Box or Offset Method (Fig. 39).* Enclose the polygon in a rectangular "box." Draw the box on the new base and locate the points $ABCEF$ on this box. Then set point $D$ by rectangular coordinates as shown.

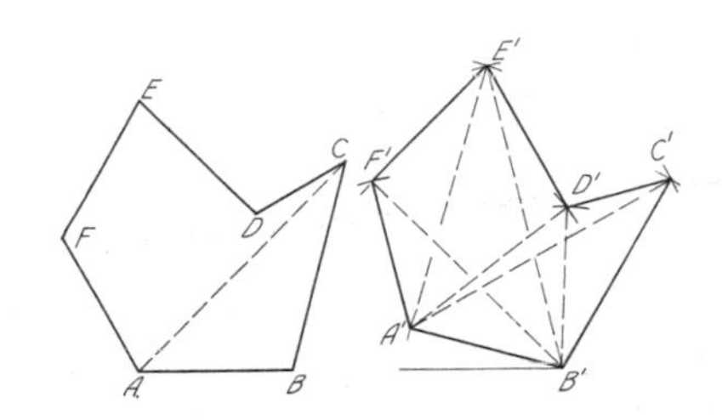


**FIG. 38.** To transfer a polygon: by triangulation. All corners are located by triangles having a common base.

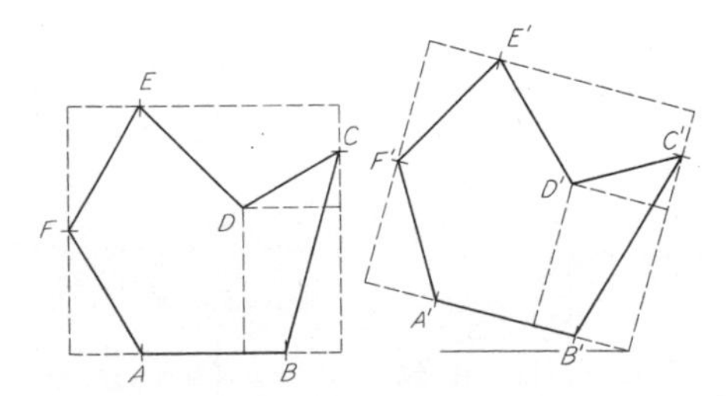


**FIG. 39.** To transfer a polygon: by "boxing." Each corner is located by its position on a rectangle.

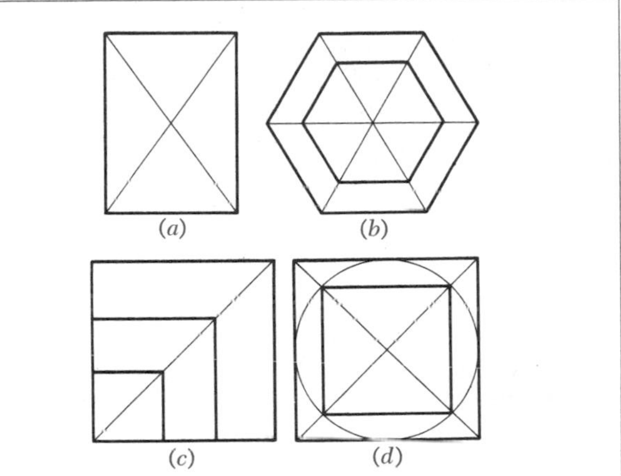


**FIG. 40.** Uses of the diagonal. The diagonals will locate the center of any regular or symmetrical geometric shape that has an even number of sides.

## 32. Uses of the Diagonal.

The diagonal is used in many ways to simplify construction and save drafting time. Figure 40 illustrates the diagonal used at ($a$) for locating the center of a rectangle, at ($b$) for enlarging or reducing a geometric shape, at ($c$) for producing similar figures having the same base, and at ($d$) for drawing inscribed or circumscribed figures.

## 33. To Construct a Regular Hexagon, Given the Distance across Corners.

*First Method (Fig. 41).* Draw a circle with $AB$ as a diameter. With the same radius and $A$ and $B$ as centers, draw arcs intersecting the circle and connect the points.

*Second Method (without Compass).* Draw lines with the 30-60° triangle in the order shown in Fig. 42.

### Given the Distance across Flats.

The distance across flats is the diameter of the inscribed circle. Draw this circle, and with the 30-60° triangle draw tangents to it as in Fig. 43.

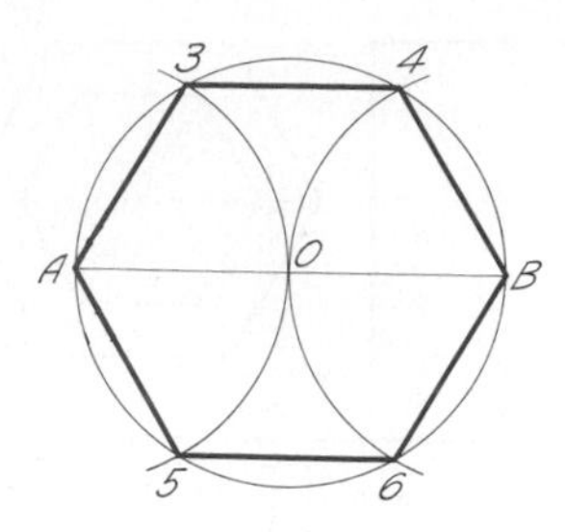


**FIG. 41.** A hexagon. Arcs equal to the radius locate all points. (Corner distance *AB* is known.)

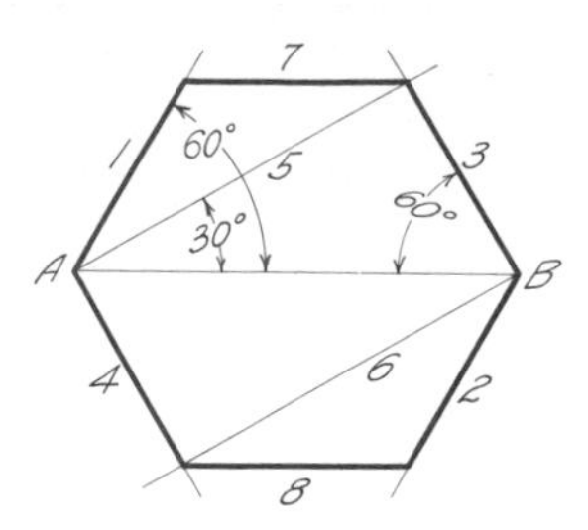


**FIG. 42.** A hexagon. The 30-60° triangle gives construction for all points. (Corner distance *AB* is known.)

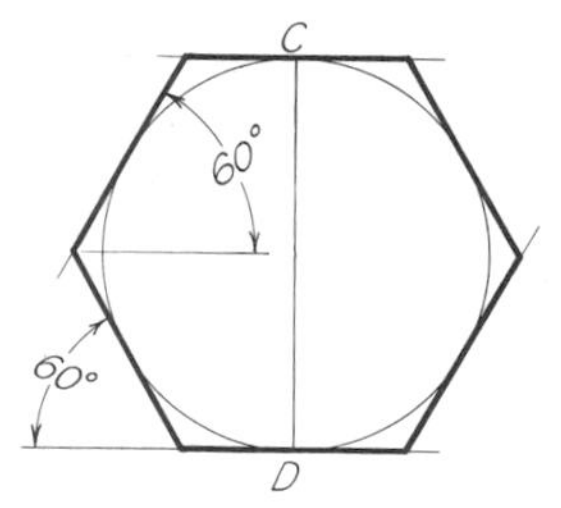


**FIG. 43.** A hexagon. Tangents drawn with the 30-60° triangle locate all points. (Across-flats distance is known.)

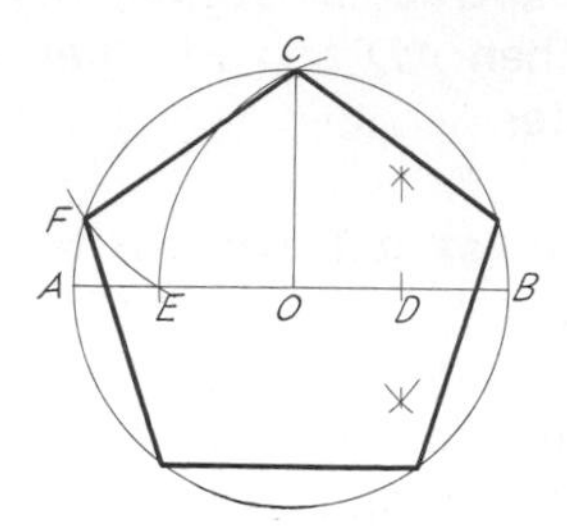


**FIG. 44.** To inscribe a regular pentagon in a circle. Arc construction locates one side; others are stepped off.

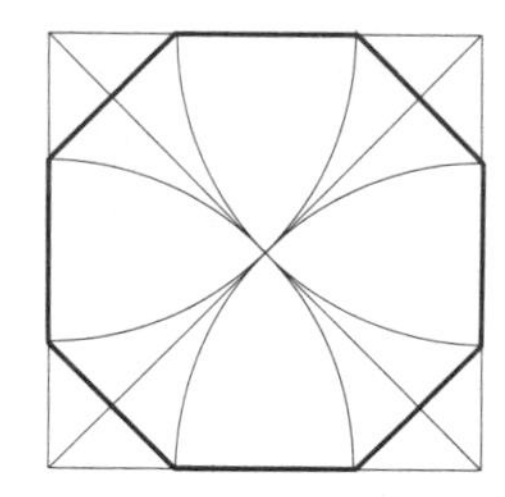

**FIG. 45.** To inscribe a regular octagon in a square. Arcs from the corners of the enclosing square locate all points.

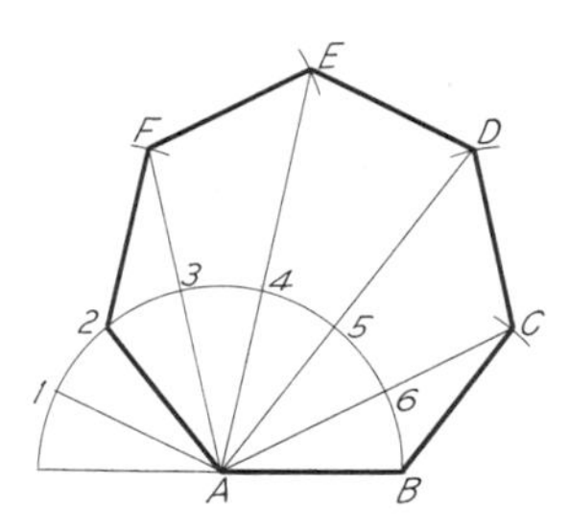


**FIG. 46.** To construct a regular polygon. Given one side, a polygon of any number of sides can be drawn by this method.

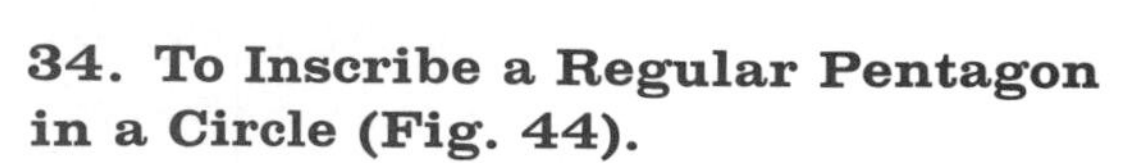

### 34. To Inscribe a Regular Pentagon in a Circle (Fig. 44).

Draw a diameter *AB* and a radius *OC* perpendicular to it. Bisect *OB*. With this point *D* as center and radius *DC*, draw arc *CE*. With center *C* and radius *CE*, draw arc *EF*. *CF* is a side of the pentagon. Step off this distance around the circle with dividers.

### 35. To Inscribe a Regular Octagon in a Square (Fig. 45).

Draw the diagonals of the square. With the corners of the square as centers and a radius of half the diagonal, draw arcs intersecting the sides and connect these points.

**36. To Construct a Regular Polygon, Given One Side (Fig. 46).** Let the polygon have seven sides. With the side *AB* as radius and *A* as center, draw a semicircle and divide it into seven equal parts with dividers. Through the second division from the left draw radial line *A-2*. Through points, *3*, *4*, *5*, and *6* extend radial lines as shown. With *AB* as radius and *B* as center, cut line *A-6* at *C*. With *C* as center and the same radius, cut *A-5* at *D*, and so on at *E* and *F*. Connect the points *or*, after *A-2* is found, draw the circumscribing circle. Another method is to guess at *CF* and divide the circle by trial; this is more common than the geometric method because the intersections of arcs are sometimes difficult.

Application of circles and arcs

## GEOMETRY OF CURVED LINES

**37. Definitions.** A *curved line* is generated by a point moving in a constantly changing direction, according to some mathematical or graphic law. Curved lines are classified as single-curved or double-curved.

A *single-curved line* is a curved line having all points of the line in a plane. Single-curved lines are often called "plane curves."

A *double-curved line* is a curved line having no four consecutive points in the same plane. Double-curved lines are also known as "space curves."

We have defined (Sec. 5) a *tangent* to a curved line as a line, either straight or curved, that passes through two points on the curve that are infinitely close together. Note that this definition of tangency places the tangent in the plane of the curve for single-curved lines and in an instantaneous plane of the curve for double-curved lines.

A *normal* is a line (or plane) perpendicular to a tangent line of a curve at the point of tangency. A normal to a single-curved line will be a line in the plane of the curve and perpendicular to a straight line connecting two consecutive points on the curve.

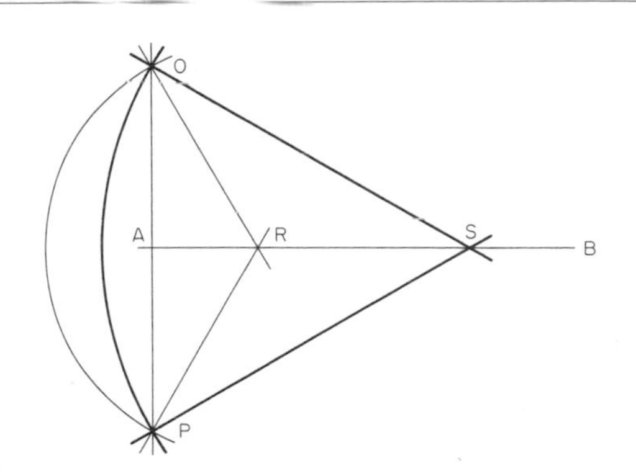


FIG. 47. Arc centers. The center of any arc lies on the perpendicular bisector of any chord of the arc.

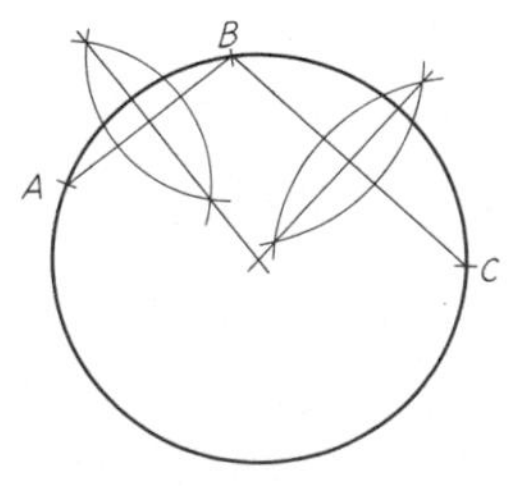


FIG. 48. Circle through three points. Perpendicular bisectors of chords locate the center.

**38. Arc and Circle Centers (Figs. 47 to 50).** The center of any arc must lie on the perpendicular bisector of any and all chords of the arc. Thus, in

Fig. 47, through two points *O* and *P*, an infinite number of arcs may be drawn but all will have centers such as *R* and *S* which lie on the perpendicular bisector *AB* of chord *OP*. This construction is used in Fig. 9 where the circle is required to pass through points *S* and *P*.

Using the above principle, only one circle can be drawn through three points, as illustrated in Fig. 48. The center must lie on the perpendicular bisectors of both chords, *AB* and *BC*. Incidentally, the center will also lie on the perpendicular bisector of chord *AC*. This construction can be used as a check on accuracy.

Any diameter of a circle and a third point on the circumference, connected to form a triangle, will produce two chords of the circle that are perpendicular to each other. Therefore, the center of a circle may be found, as in Fig. 49, by selecting any two points such as *A* and *B*, drawing the perpendicular chords *AC* and *BD* and then the two diameters *CB* and *DA*, which cross at center *O*. Note in Fig. 50 that all chords connecting the diameter *AB*—*AD* and *DB*, *AC* and *CB*, *AE* and *EB*, *AF* and *FB*—form a right angle between the chords.

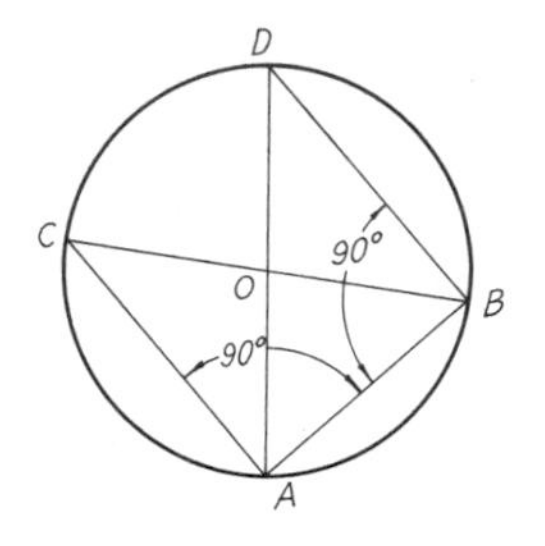


**FIG. 49.** To locate the center of a circle. Perpendicular chords to chord *AB* produce diameters *BC* and *AD*.

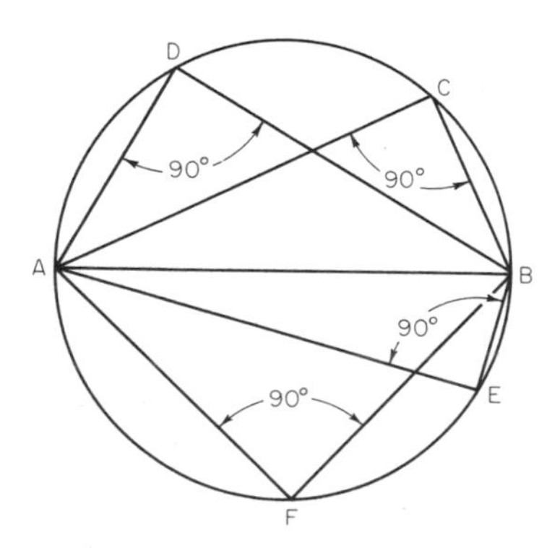


**FIG. 50.** Chords connecting a diameter. Chords from a point, connected to a diameter, form a right angle between the chords.

## 39. To Inscribe a Circle in an Equilateral Triangle (Fig. 51).

With center *A* and radius *AB*, draw an arc as shown. With *B* and *C* as centers and radius *AB*, cut this arc at *D* and *E*. Then *EC* and *DB* intersect at *O*, the geometric center, and the radius of the subscribing circle is *OG* and/or *OF*.

## 40. To Inscribe an Equilateral Triangle in a Circle (Fig. 52).

Using the radius of the circle and starting at *A*, the known point of orientation, step off distances *AB*, *BC*, *CD*, and *DE*. The equilateral triangle is *ACE*. Note that six chords equal in length to the radius will give six equally spaced points on the circumference.

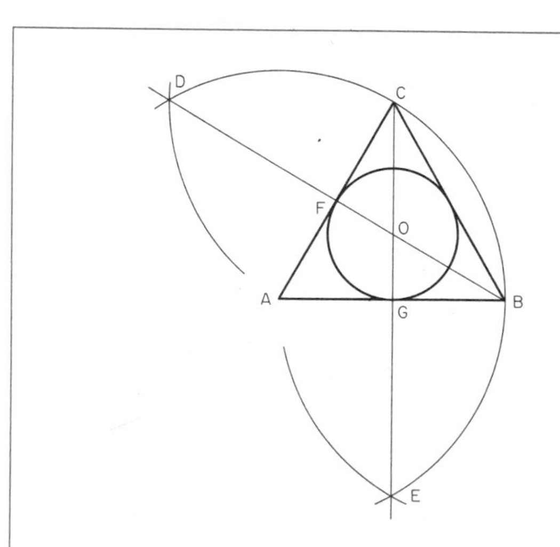


**FIG. 51.** To inscribe a circle in an equilateral triangle. Construction (to get perpendicular bisectors of two sides) locates the circle center.

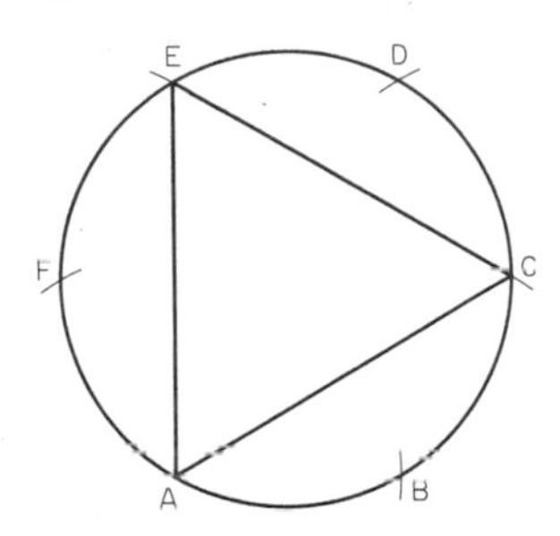


**FIG. 52.** To inscribe an equilateral triangle in a circle. By stepping off the radius twice, corners are located. Compare with Fig. 41.

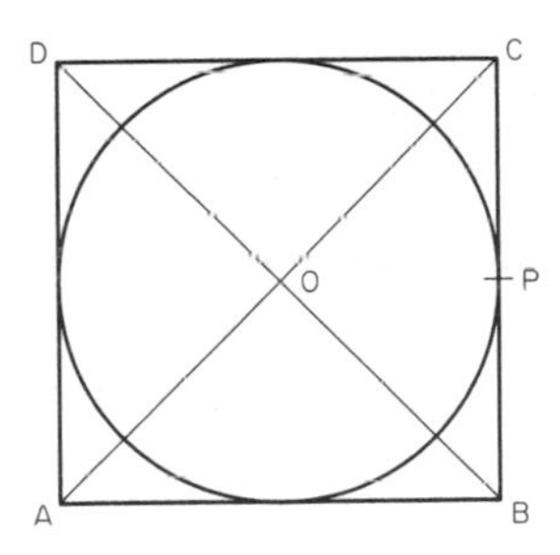


**FIG. 53.** To inscribe a circle in a square. Diagonals locate the circle center.

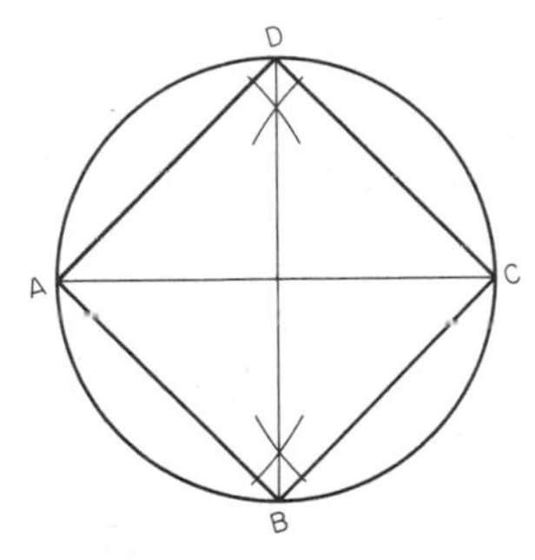


**FIG. 54.** To inscribe a square in a circle. The perpendicular bisector of one diameter locates the two corners to be found.

## 41. To Inscribe a Circle in a Square (Fig. 53).

Given the circle *ABCD*, draw diagonals *AC* and *BD*, which intersect at point *O*, the center. *OP* is the radius of the circle.

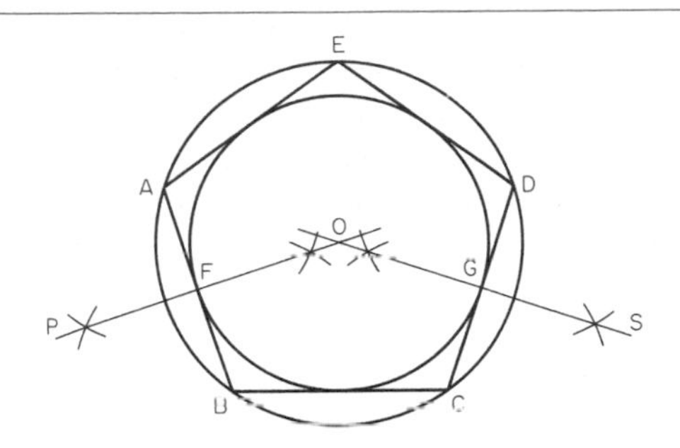


**FIG. 55.** To draw a circle on a regular polygon. Perpendicular bisectors of two sides locate the center.

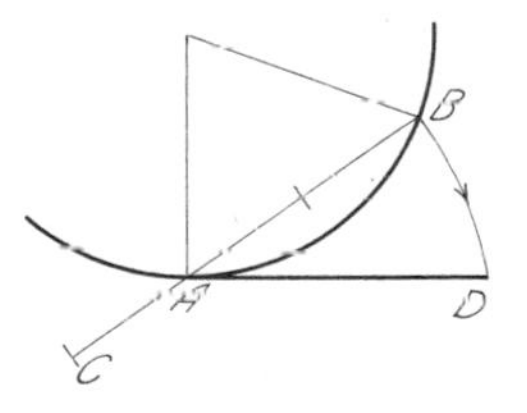


**FIG. 56.** To approximate the length of an arc. See footnote 1 for accuracy.

## 42. To Inscribe a Square in a Circle (Fig. 54).

From *A*, the known point of orientation, draw diameter *AC* and then erect *BD*, the perpendicular bisector of *AC*. The square is *ABCD*.

## 43. To Scribe a Circle in or on a Regular Polygon (Fig. 55).

Draw perpendicular bisectors of any two sides, for example *PO* and *SO* of *AB* and *CD*, giving *O*, the center. The inner-circle radius is *OF* or *OG* and the external-circle radius is *OA, OB,* etc.

## 44. To Lay off on a Straight Line the Approximate Length of a Circle Arc (Fig. 56).

Given the arc *AB*. At *A* draw the tangent *AD* and the chord produced, *BA*. Lay off *AC* equal to half the chord *AB*. With center *C* and radius *CB*, draw an arc intersecting *AD* at *D*; then *AD* will be equal in length to arc *AB* (very nearly).[1] If the given arc

[1] In this (Professor Rankine's) solution, the error varies as the fourth power of the subtended angle. For 60° the line will be $\frac{1}{900}$ part short, while at 30° it will be only $\frac{1}{14,400}$ part short.

is between 45° and 90°, a closer approximation will result by making *AC* equal to the chord of half the arc instead of half the chord of the arc.

The usual way of rectifying an arc is to set the dividers to a space small enough to be practically equal in length to a corresponding part of the arc. Starting at *B*, step along the arc to the point nearest *A* and without lifting the dividers step off the same number of spaces on the tangent, as shown in Fig. 57.

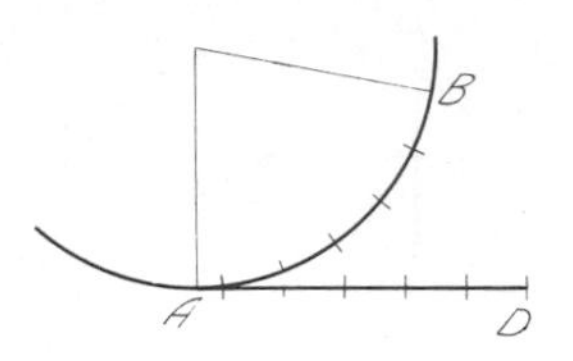


**FIG. 57.** To approximate the length of an arc. The sum of short chords closely approaches the arc length.

### 45. To Lay Off on a Given Circle the Approximate Length of a Straight Line (Fig. 58).

Given the line *AB* tangent to the circle at *A*. Lay off *AC* equal to one-fourth *AB*. With *C* as center and radius *CB*, draw an arc intersecting the circle *D*. The line *AD* is equal in length to *AB* (very nearly).[1] If arc *AD* is greater than 60°, solve for one-half *AB*.

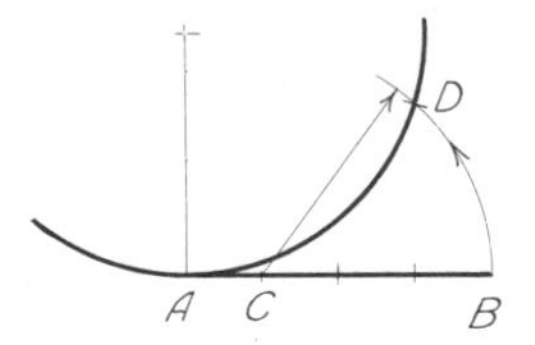


**FIG. 58.** To lay off, on an arc, a specified distance. See footnote 1 for accuracy.

### 46. Plane Curves: the Conic Sections.

In cutting a right-circular cone (a cone of revolution) by planes at different angles, we obtain four curves called "conic sections" (Fig. 59). These are the *circle*, cut by a plane perpendicular to the axis; the *ellipse*, cut by a plane making a greater angle with the axis than do the elements; the *parabola*, cut by a plane making the same angle with the axis as do the elements; the *hyperbola*, cut by a plane making a smaller angle than do the elements.

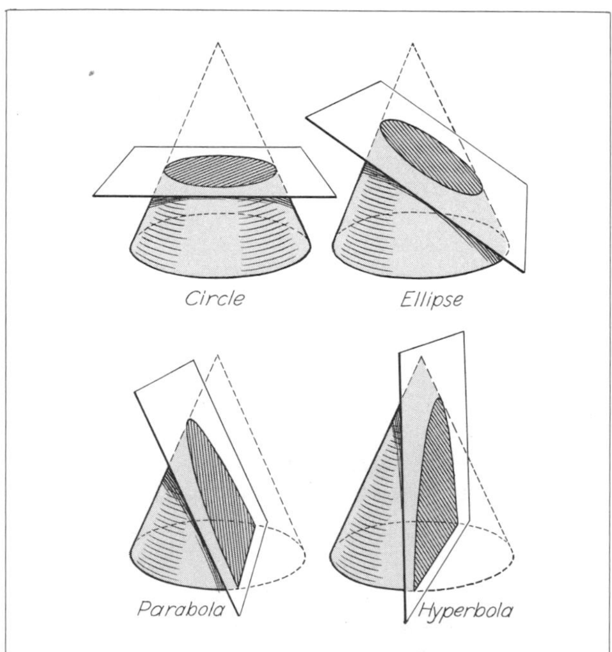


**FIG. 59.** The conic sections. Four plane curves are produced by "cutting" a cone.

### 47. The Ellipse: Major and Minor Diameters (Fig. 60).

An ellipse is the plane curve generated by a point moving so that the sum of its distances from two fixed points ($F_1$ and $F_2$), called "focuses," is a constant equal to the major axis,[2] or *major diameter, AB*.

The minor axis, or *minor diameter, DE*, is the line through the center perpendicular to the major diam-

[2] "Major axis" and "minor axis" are the traditional terms. However, because of the confusion between these and the mathematical *X* and *Y* axes and the central axis (axis of rotation), the terms "major diameter" and "minor diameter" will be used in this discussion.

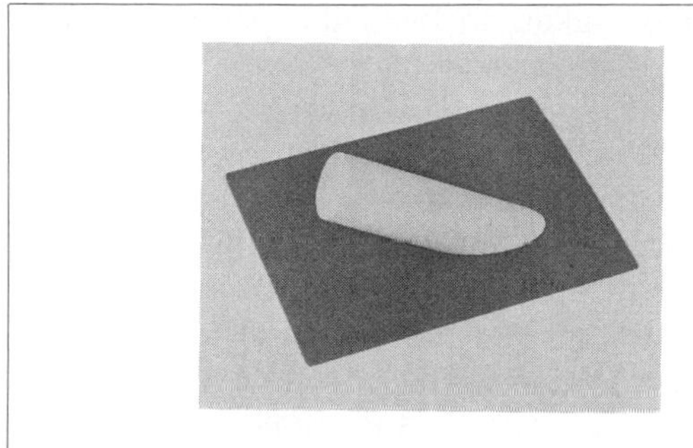

Application of the ellipse (intersection)

eter. The focuses may be determined by cutting the major diameter with an arc having its center at an end of the minor diameter and a radius equal to one-half the major diameter.

Aside from the circle, the ellipse is met with in practice much more often than any of the other conics, so it is important to be able to construct it readily. Several methods are given for its construction, both as a true ellipse and as an approximate curve made by circle arcs. In the great majority of cases when the curve is required, its major and minor diameters are known.

The mathematical equation of an ellipse with the origin of rectilinear coordinates at the center of the ellipse is $x^2/a^2 + y^2/b^2 = 1$, where $a$ is the intercept on the $X$ axis and $b$ is the intercept on the $Y$ axis.

## 48. The Ellipse: Conjugate Diameters (Fig. 61).

Any line through the center of an ellipse may serve as *one* of a pair of conjugate diameters. Each of a pair of conjugate diameters is always parallel to the tangents to the curve at the extremities of the other. For example, $AB$ and $CD$ are a pair of conjugate diameters. $AB$ is parallel to the tangents $MN$ and $PQ$; and $CD$ is parallel to the tangents $MP$ and $NQ$. Also, each of a pair of conjugate diameters bisects all the chords parallel to the other. A given ellipse may have an unlimited number of pairs of conjugate diameters.

*To Determine the Major and Minor Diameters, Given the Ellipse and a Pair of Conjugate Diameters. First Method* (*Fig. 62*). The conjugate diameters are $CN$ and $JG$. With center $O$ and radius $OJ$, draw a semicircle intersecting the ellipse at $P$. The major and minor diameters will be parallel to the chords $GP$ and $JP$, respectively.

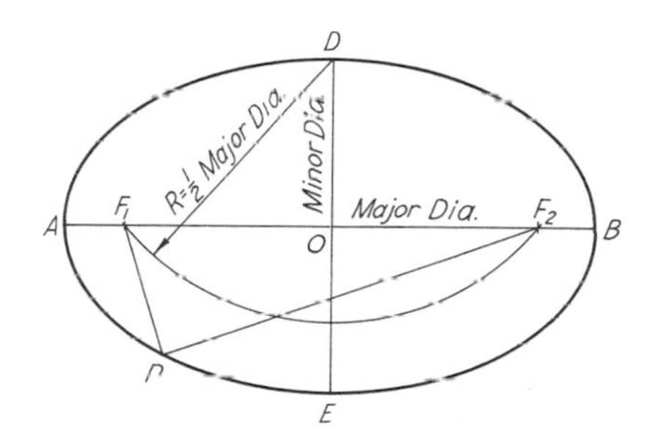


**FIG. 60.** The ellipse: major and minor diameters. These are perpendicular to each other.

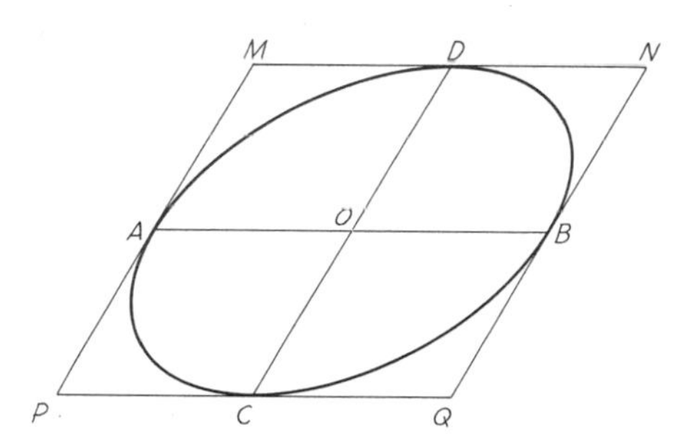


**FIG. 61.** The ellipse: conjugate diameters. Each is parallel to the tangent at the end of the other.

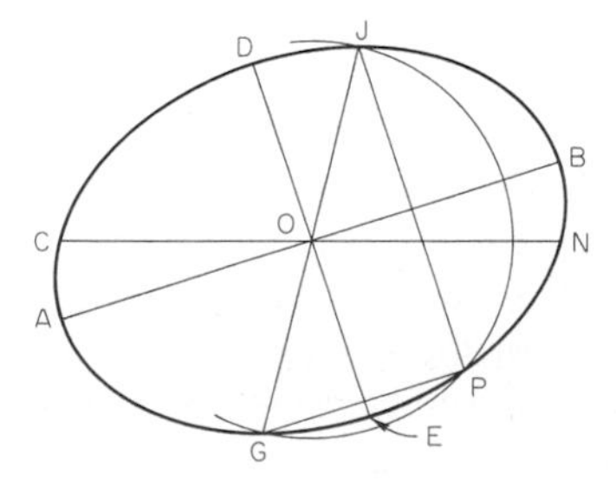


**FIG. 62.** Determination of major and minor diameters from conjugate diameters. Curve is given.

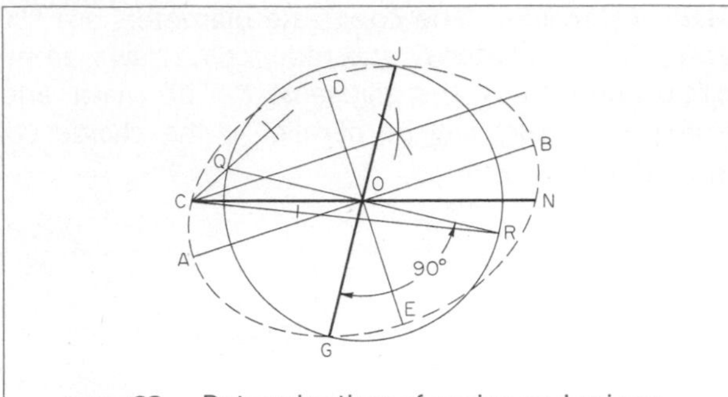


**FIG. 63.** Determination of major and minor diameters from conjugate diameters. Curve is not given.

*Second Method: When the Curve Is Not Given* (*Fig. 63*). The conjugate diameters *CN* and *JG* are given. With center *O* and radius *OJ*, describe a circle and draw the diameter *QR* at right angles to *JG*. Bisect the angle *QCR*. The major diameter will be parallel to this bisector and equal in length to $CR + CQ$; that of the minor diameter will be $CR - CQ$.

### 49. Ellipse Construction: Pin-and-string Method.

This well-known method, sometimes called the "gardener's ellipse," is often used for large work and is based on the definition of the ellipse. Drive pins at the points $D$, $F_1$, and $F_2$ (Fig. 60), and tie an inelastic thread or cord tightly around the three pins. If the pin *D* is removed and a marking point moved in the loop, keeping the cord taut, it will describe a true ellipse.

### 50. Ellipse Construction: Trammel Method for Major and Minor Diameters.

*First Method* (*Fig. 64*). On the straight edge of a strip of paper, thin cardboard, or sheet of celluloid, mark the distance *ao* equal to one-half the major diameter and *do* equal to one-half the minor diameter. If the strip is moved, keeping *a* on the minor diameter and *d* on the major diameter, *o* will give points on the ellipse. This method is convenient as no construction is required, but for accurate results great care must be taken to keep the points *a* and *d* exactly on the major and minor diameters.

*Second Method* (*Fig. 65*). On a strip—as used in the first method—mark the distance *do* equal to one-half the minor diameter and *oa* equal to one-half the major diameter. If this strip is moved, keeping *a* on the minor diameter and *d* on the major diameter, *o* will give points on the ellipse. This arrangement is preferred where the ratio between the major and minor diameters is small.

### 51. Ellipse Construction: Triangle Trammel for Conjugate Diameters (Fig. 66).

The conjugate diameters *AB* and *DE* are given. Erect the perpendicular *AC* to the diameter *ED* and lay off distance *AC* from *E* to locate point *G*. Erect the perpendicular *EF* to the diameter *AB*. Transfer to a piece of paper, thin cardboard, or sheet of celluloid, and cut out the triangle *EFG*. If this triangle is moved, keeping *f* on *AB* and *g* on *ED*, *e* will give points on the ellipse. Extreme care must be taken to keep points *f* and *g* on the conjugate diameters.

### 52. Ellipse Construction: Concentric-circle Method for Major and Minor Diameters (Fig. 67).

This is perhaps the most accurate method for determining points on the curve. On the two principal diameters, which intersect at *O*, describe circles. From a number of points on the outer circle, as *P* and *Q*, draw radii *OP*, *OQ*, etc., intersecting the inner circle at $P'$, $Q'$, etc. From *P* and *Q* draw lines parallel to *OD*, and from $P'$ and $Q'$ draw lines parallel to *OB*. The intersection of the lines through *P* and $P'$ gives one point on the ellipse, the intersection of the lines through *Q* and $Q'$ another point, and so on. For accuracy, the points should be taken closer together toward the major diameter. The process may be repeated in each of the four quadrants and the curve sketched in lightly freehand; or one quadrant only may be constructed and repeated in the remaining three by marking the French curve.

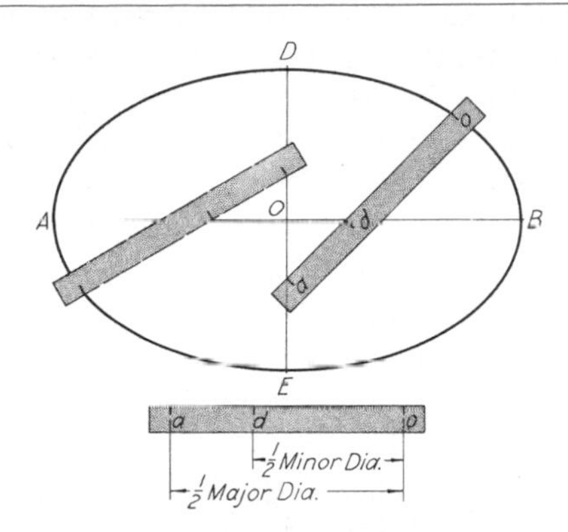


**FIG. 64.** An ellipse by trammel (first method). Points on the curve are plotted.

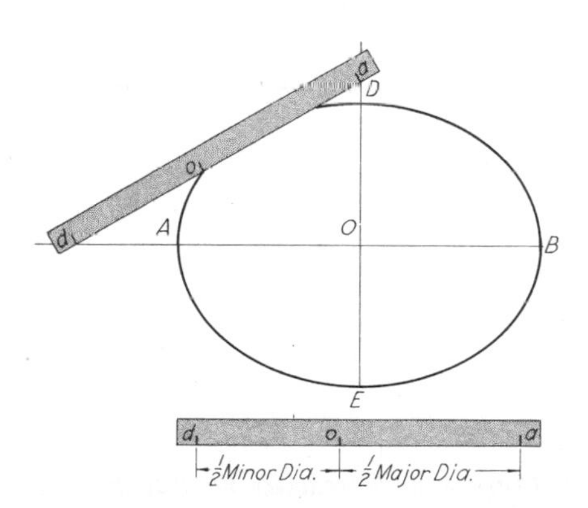


**FIG. 65.** An ellipse by trammel (second method). Points on the curve are plotted.

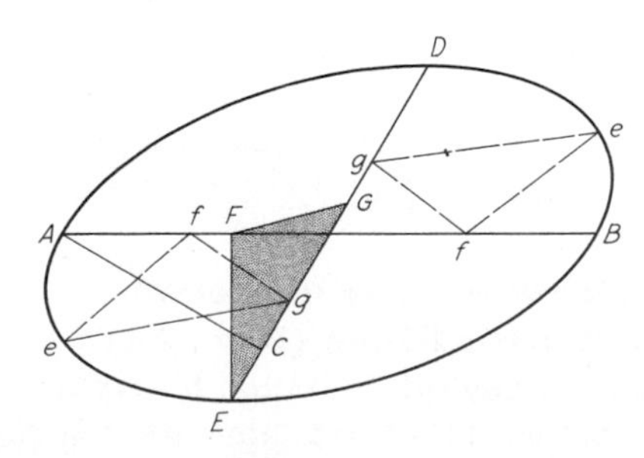


**FIG. 66.** An ellipse by triangle trammel (conjugate diameters). Points are plotted at the apex of the triangle.

## 53. Ellipse Construction: Circle Method for Conjugate Diameters (Fig. 68).

The conjugate diameters $AB$ and $DE$ are given. On the conjugate diameter $AB$, describe a circle; then from a number of points, as $P$, $Q$, and $S$, draw perpendiculars as $PP'$, $QO$, and $SS'$ to the diameter $AB$. From $S$ and $P$, etc., draw lines parallel to $QD$, and from $S'$ and $P'$ draw lines parallel to $OD$. The intersection of the lines through $P$ and $P''$ gives one

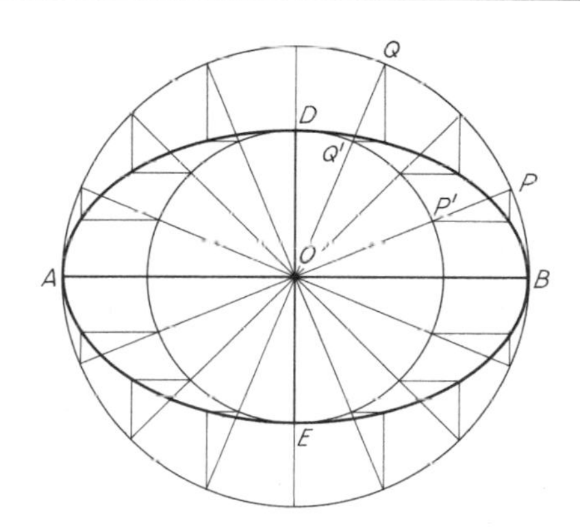


**FIG. 67.** An ellipse by concentric-circle method. Major- and minor-diameter circles and construction give points on the curve.

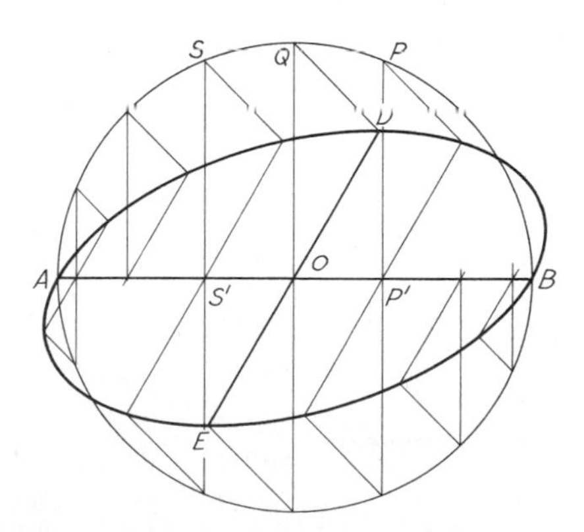


**FIG. 68.** An ellipse by circle method (conjugate diameters). After first construction, parallels plot points on the curve.

point on the ellipse, the intersection of the lines through $S$ and $S'$ another point, and so on.

### 54. Ellipse Construction: Parallelogram Method (Figs. 69 and 70).

This method can be used either with the major and minor diameters or with any pair of conjugate diameters. On the given diameters construct a parallelogram. Divide $AO$ into any number of equal parts and $AG$ into the same number of equal parts, numbering points from $A$. Through these points draw lines from $D$ and $E$, as shown. Their intersections will be points on the curve.

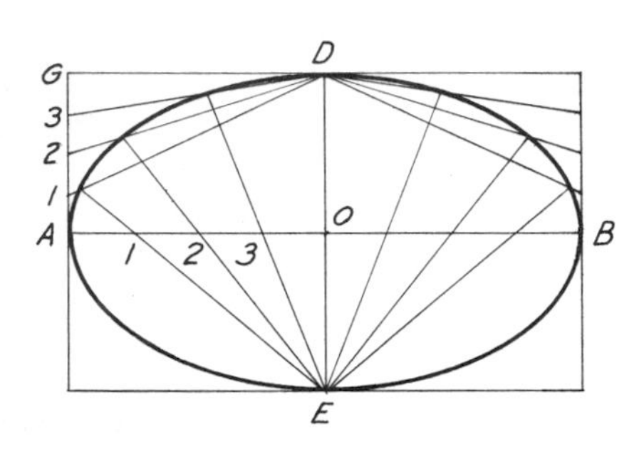


**FIG. 69.** An ellipse by parallelogram method. Points are plotted by construction through equal divisions of $AO$ and $AG$.

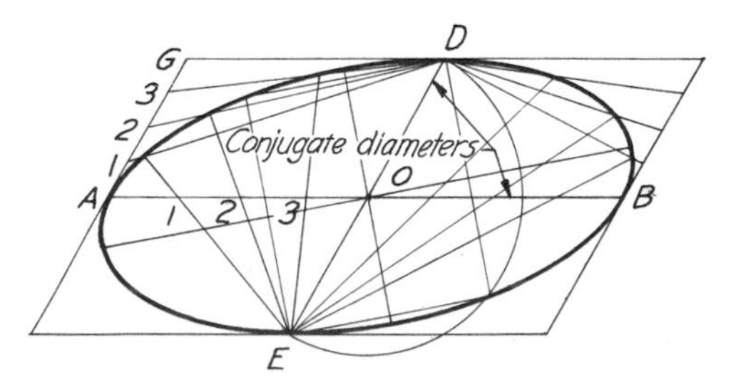


**FIG. 70.** An ellipse by parallelogram method (conjugate diameters). The constructive method of Fig. 69 is used, but here applied to conjugate diameters.

### 55. To Draw a Tangent to an Ellipse.

*At a Given Point on the Curve (Fig. 71).* Draw lines from the given point $P$ to the focuses. The line bisecting the exterior angle of these focal radii is the required tangent.

*Parallel to a Given Line (Fig. 71).* Draw $F_1E$ perpendicular to the given line $GH$. With $F_2$ as center and radius $AB$, draw an arc cutting $F_1E$ at $K$. The line $F_1K$ cuts the ellipse at the required point of tangency $T$, and the required tangent passes through $T$ parallel to $GH$.

*From a Point Outside (Fig. 72).* Find the focuses $F_1$ and $F_2$. With the given point $P$ and radius $PF_2$, draw the arc $RF_2Q$. With $F_1$ as center and radius $AB$, strike an arc cutting this arc at $Q$ and $R$. Connect $QF_1$ and $RF_1$. The intersections of these lines with the ellipse at $T_1$ and $T_2$ will be the tangent points of tangents to the ellipse from $P$.

*Concentric-circle Method (Fig. 73).* When the ellipse has been constructed by the concentric-circle method (Fig. 67), a tangent at any point $H$ can be drawn by dropping a perpendicular to $AB$ from the point to the outer circle at $K$ and drawing the auxiliary tangent $KL$ to the outer circle, cutting the major diameter at $L$. From $L$ draw the required tangent $LH$.

### 56. To Draw a Normal to an Ellipse (Fig. 73).

From point $P$ on the curve, project a parallel to the minor diameter to intersect the major diameter circle at $Q$. Draw $OQ$ extended to intersect (at $N$) an arc with center at $O$ and radius $AO + OE$. $NP$ is the required normal.

*Or,* normals may be drawn perpendicular to the tangents of Figs. 71 and 72.

### 57. To Draw a Four-centered Approximate Ellipse (Fig. 74).

Join $A$ and $D$. Lay off $DF$ equal to $AO - DO$. This is done graphically as indicated on the figure by swinging $A$ around to $A'$ with $O$ as center where now $DO$ from $OA'$ is $DA'$, the required distance. With $D$ as center, an arc from $A'$ to the diagonal $AD$ locates $F$. Bisect $AF$ by a perpendicular crossing $AO$

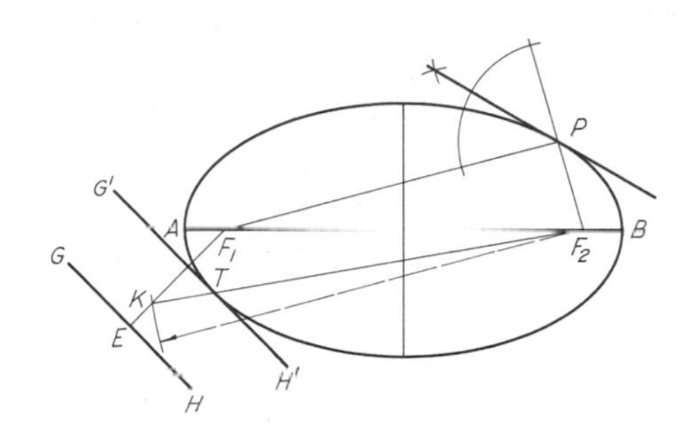


**FIG. 71.** Tangents to an ellipse at a point and parallel to a given line. Focuses are used in construction.

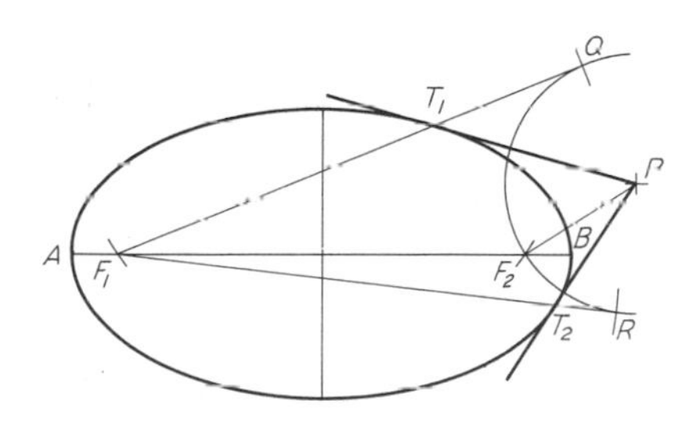


**FIG. 72.** Tangents to an ellipse from an outside point. Tangent points $T_1$ and $T_2$ are accurately located.

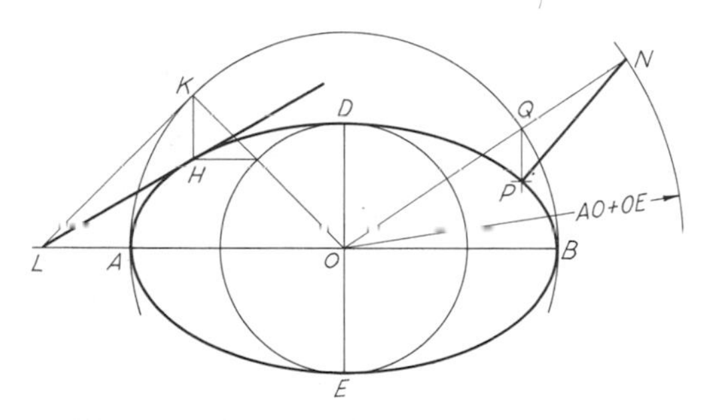


**FIG. 73.** Tangent and normal to an ellipse. The concentric-circle method is used.

at $G$ and intersecting $DE$ produced (if necessary) at $H$. Make $OG'$ equal to $OG$, and $OH'$ equal to $OH$. Then $G$, $G'$, $H$, and $H'$ will be centers for four tangent circle arcs forming a curve *approximating* the shape of an ellipse.

Another mehtod is shown in Fig. 75. This should be used only when the minor diameter is at least two-thirds the length of the major diameter.

### 58. To Draw an Eight-centered Approximate Ellipse (Fig. 76).

When a closer approximation is desired, the eight-centered ellipse, the upper half of which is known in masonry as the "five-centered arch," may be constructed. Draw the rectangle $AFDO$. Draw the diagonal $AD$ and a line from $F$ perpendicular to it, intersecting the extension of the minor diameter at $H$. Lay off $OK$ equal to $OD$, and, on $AK$ as a diameter, draw a semicircle intersecting the extension of the minor diameter at $L$. Make $OM$ equal to $LD$. With center $H$ and radius $HM$, draw the arc $MN$. From $A$, along $AB$, lay off $AQ$ equal to $OL$. With $P$ as center and radius $PQ$, draw an arc intersecting $MN$ at $N$; then $P$, $N$, and $H$ are centers for one-quarter of the eight-centered approximate ellipse.

It should be noted that an ellipse changes its radius of curvature at every successive point and that these approximations are therefore not ellipses, but simply curves of the same general shape and, incidentally, not nearly so pleasing in appearance.

### 59. To Draw Any Noncircular Curve (Fig. 77).

This may be approximated by drawing tangent circle arcs: Select a center by trial, draw as much of an arc as will practically coincide with the curve, and then, changing the center and radius, draw the next portion, remembering always that *if arcs are to be tangent, their centers must lie on the common normal at the point of tangency.*

Curves are sometimes inked in this way in preference to using irregular curves.

### 60. The Parabola.

The parabola is a plane curve generated by a point so moving that its distance from a fixed point, called the "focus," is always equal to its distance from a straight line, called the "directrix." Among its practical applications are searchlights, parabolic reflec-

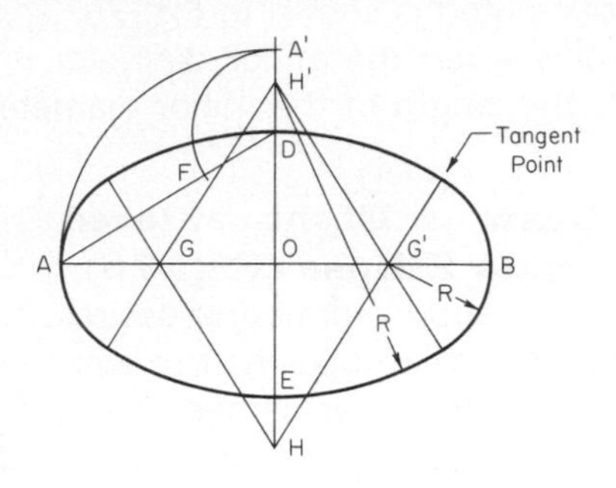


**FIG. 74.** A four-centered approximate ellipse. Four tangent circle arcs give elliptical shape.

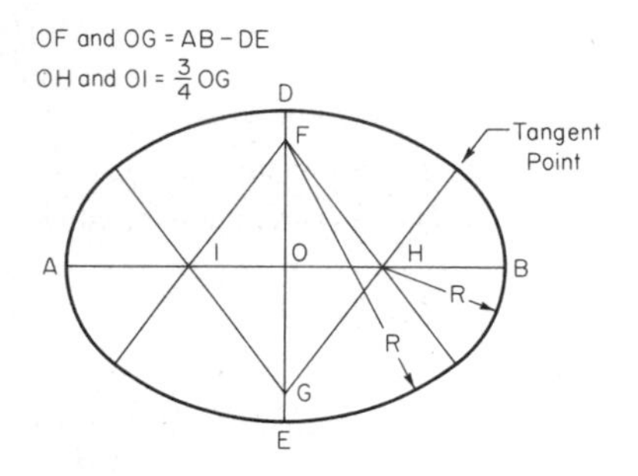


**FIG. 75.** A four-centered approximate ellipse. This method works best when major and minor diameters are nearly equal.

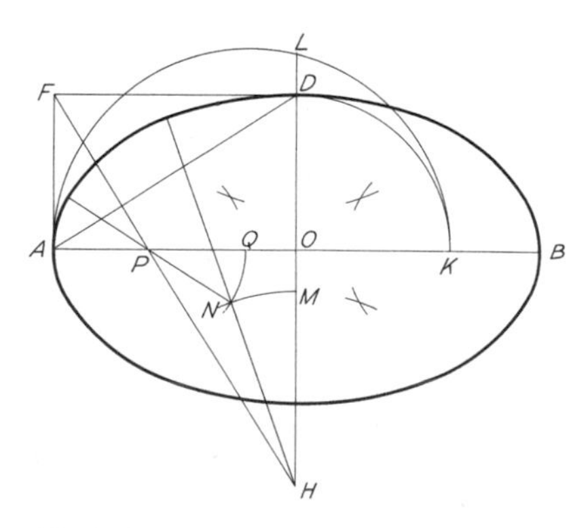


**FIG. 76.** An eight-centered approximate ellipse. This gives a much better approximation than the four-center methods but requires more construction.

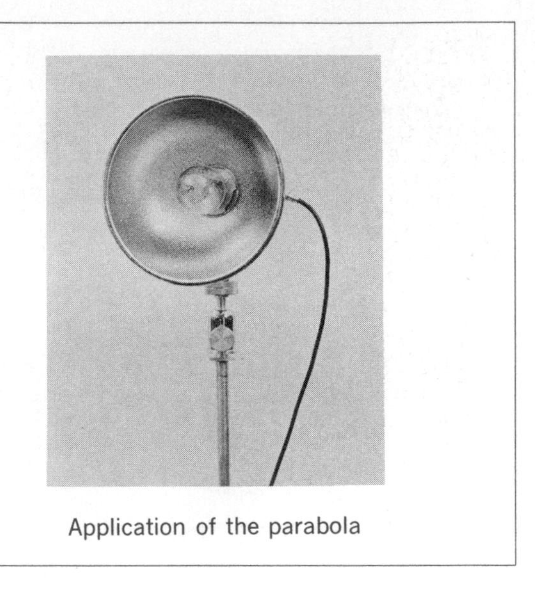

Application of the parabola

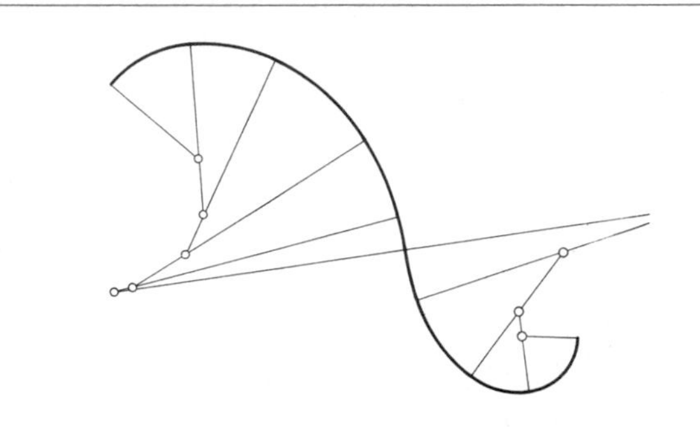

**FIG. 77.** A curve constructed with circle arcs. Note that lines through pairs of centers locate the tangent points of arcs.

tors, some loud-speakers, road sections, and certain bridge arches.

The mathematical equation for a parabola with the origin of rectilinear coordinates at the intercept of the curve with the $X$ axis and the focus on the axis is $y^2 = 2\,px$, where $p$ is twice the distance from the origin to the focus.

To draw a parabola when the focus $F$ and the directrix $AB$ are given (Fig. 78), draw the axis through $F$ perpendicular to $AB$. Through any point $D$ on the axis, draw a line parallel to $AB$. With the

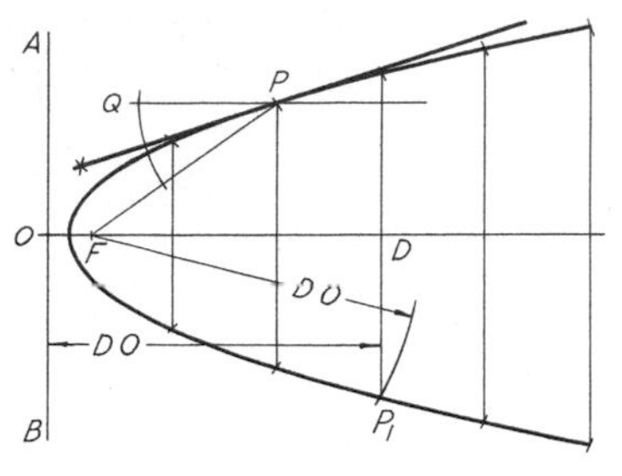

**FIG. 78.** A parabola. Points on the curve are equidistant from the focus $F$ and directrix $AB$.

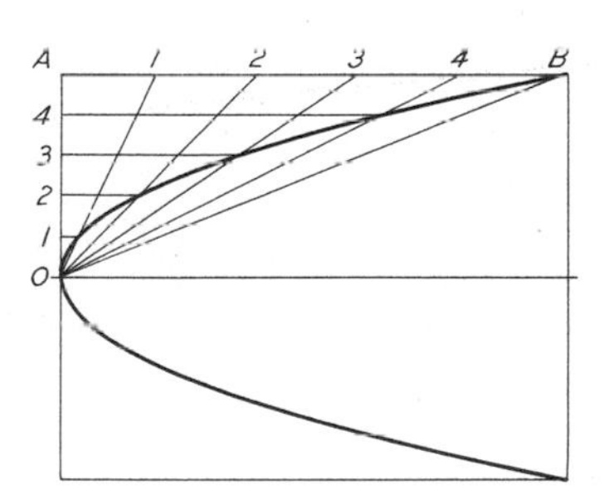


**FIG. 79.** A parabola by parallelogram method. Points are plotted by lines through equal-numbered divisions of $OA$ and $AB$.

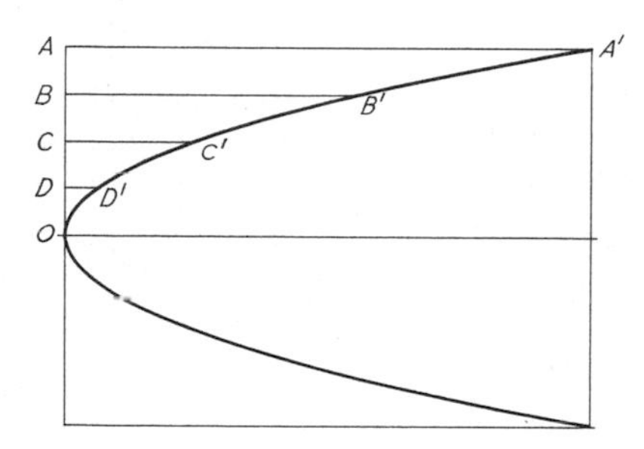


**FIG. 80.** A parabola by offset method. Offsets are proportionate to squares of divisions of $OA$.

distance $DO$ as radius and $F$ as center, draw an arc intersecting the line, thus locating a point $P$ on the curve. Repeat the operation as many times as needed.

*To Draw a Tangent at Any Point P.* Draw $PQ$ parallel to the axis and bisect the angle $FPQ$.

## 61. Parabola Construction: Parallelogram Method.

Usually when a parabola is required, the dimensions of the enclosing rectangle, that is, the width and depth of the parabola (or span and rise), are given, as in Fig. 79. Divide $OA$ and $AB$ into the same number of equal parts. From the divisions on $AB$, draw lines converging at $O$. From the divisions on $OA$, draw lines parallel to the axis. The intersections of these with the lines from the corresponding divisions on $AB$ will be points on the curve.

## 62. Parabola Construction: Offset Method (Fig. 80).

Given the enclosing rectangle, the parabola can be plotted by computing the offsets from the line $OA$. These offsets vary in length as the square of their distances from $O$. Thus if $OA$ is divided into four parts, $DD'$ will be $1/16$ of $AA'$; $CC'$, since it is twice as far from $O$ as $DD'$, will be $4/16$ of $AA'$; and $BB'$, $9/16$. If $OA$ had been divided into five parts, the relations would be $1/25$, $4/25$, $9/25$, and $16/25$, the denominator in each case being the square of the number of divisions. This method is the one generally used by civil engineers in drawing parabolic arches.

## 63. Parabola Construction: Parabolic Envelope (Fig. 81).

This method of drawing a pleasing curve is often used in machine design. Divide $OA$ and $OB$ into the

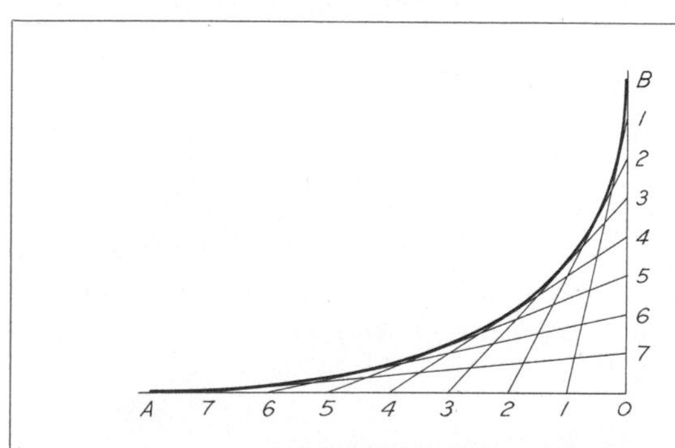


**FIG. 81.** A parabola by envelope method. The curve is drawn tangent to straight-line construction.

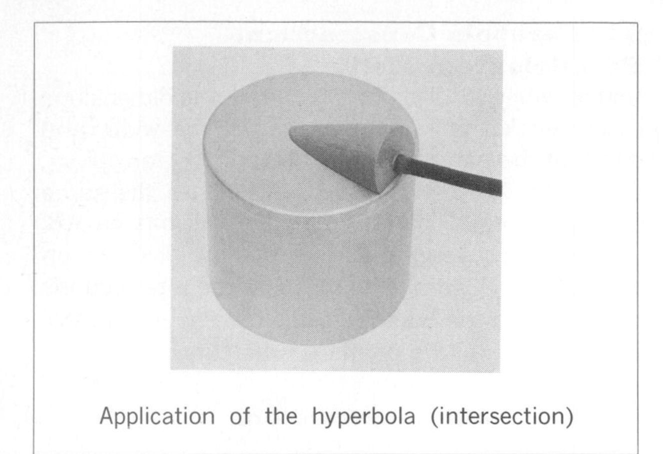

Application of the hyperbola (intersection)

same number of equal parts. Number the divisions from $O$ and $B$ and connect the corresponding numbers. The tangent curve will be a portion of a parabola—but a parabola whose axis is not parallel to either coordinate.

## 64. The Hyperbola.

The hyperbola is a plane curve generated by a point moving so that the difference of its distances from two fixed points, called the "focuses," is a constant. (Compare this definition with that of the ellipse.)

The mathematical equation for a hyperbola with the center at the origin of rectilinear coordinates and the focuses on the $X$ axis is $x^2/a^2 - y^2/b^2 = 1$, where $a$ is the distance from the center to the $X$ intercept and $b$ is the corresponding $Y$ value of the asymptotes, lines that the tangents to the curve meet at infinity.

*To Draw a Hyperbola When the Focuses $F_1$ and $F_2$ and the Transverse Axis AB (Constant Difference) Are Given (Fig. 82).* With $F_1$ and $F_2$ as centers and any radius greater than $F_1B$, as $F_1P$, draw arcs. With the same centers and any radius $F_1P - AB$, strike arcs intersecting these arcs, giving points on the curve.

*To Draw a Tangent at Any Point P.* Bisect the angle $F_1PF_2$.

## 65. Equilateral Hyperbola.

The case of the hyperbola of commonest practical interest to the engineer is the equilateral, or rectangular, hyperbola referred to its asymptotes. With it, the law $PV = c$, connecting the varying pressure and volume of a portion of steam or gas, can be graphically presented.

*To Draw an Equilateral Hyperbola (Fig. 83).* Let $OA$ and $OB$ be the asymptotes of the curve and $P$ any point on it (this might be the point of cutoff on an indicator diagram). Draw $PC$ and $PD$. Mark any points 1, 2, 3, etc., on $PC$, and through these points draw a system of lines parallel to $OA$ and a second system through the same points converg-

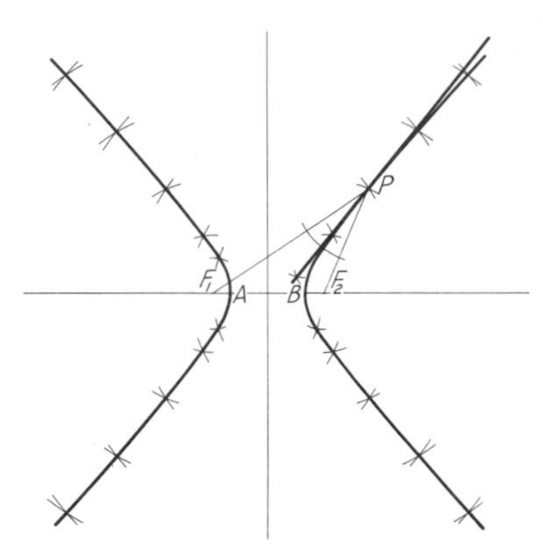


**FIG. 82.** A hyperbola. From any point, the difference in distance to the focuses is a constant.

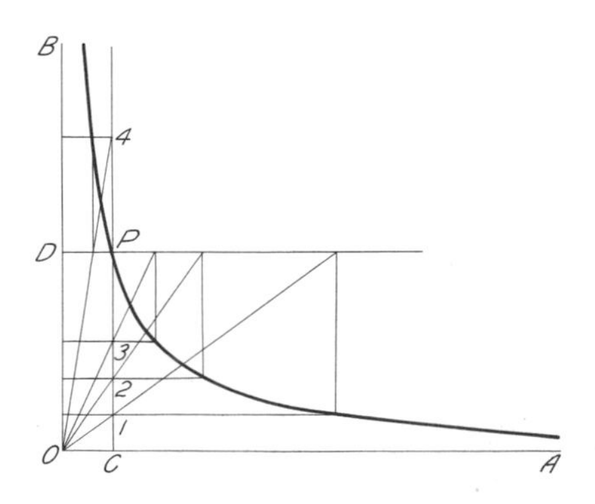


**FIG. 83.** An equilateral hyperbola. The asymptotes are perpendicular to each other.

ing at $O$. From the intersections of the lines of the second system with $PD$ extended, draw perpendiculars to $OA$. The intersections of these perpendiculars with the corresponding lines of the first system give points on the curve.

### 66. Cycloid Curves.

A cycloid is the curve generated by the motion of a point on the circumference of a circle rolled in a plane along a straight line. If the circle is rolled on the outside of another circle, the curve generated is called an "epicycloid"; if rolled on the inside, it is called a "hypocycloid." These curves are used in drawing the cycloid system of gear teeth.

The mathematical equation for a cycloid (parametric form) is $x = r\theta - r \sin \theta$, $y = r - r \cos \theta$, where $r$ is the radius of a moving point and $\theta$ is the turned angle, in radians, about the center from zero position.

*To Draw a Cycloid (Fig. 84).* Divide the rolling circle into a convenient number of parts (say, eight), and, using these divisions, lay off on the tangent $AB$ the rectified length of the circumference. Draw through $C$ the line of centers $CD$, and project the division points up to this line by perpendiculars to $AB$. Using these points as centers, draw circles representing different positions of the rolling circle, and project, in order, the division points of the original circle across to these circles. The intersections thus determined will be points on the curve. The epicycloid and hypocycloid are drawn similarly, as shown in Fig. 85.

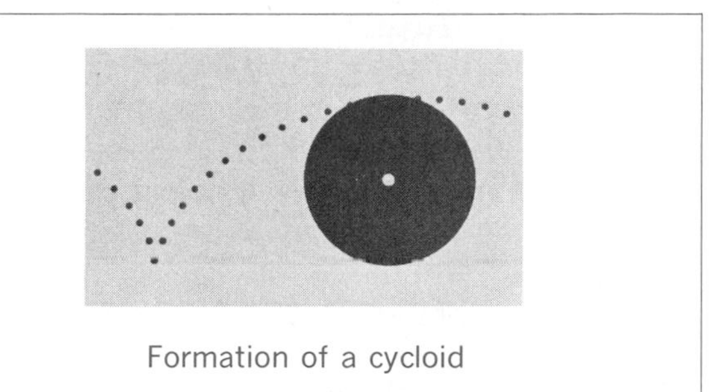

Formation of a cycloid

Application of the involute

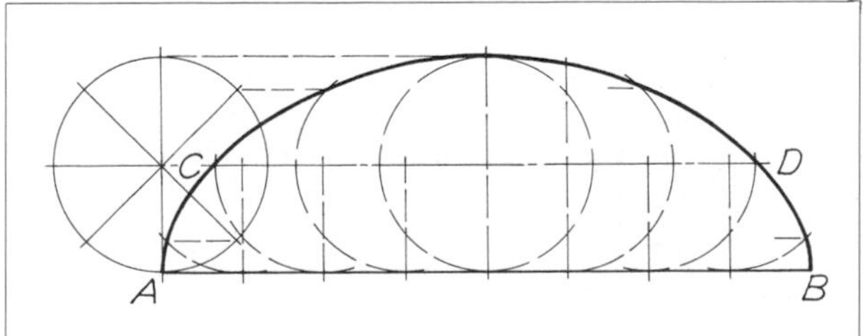


**FIG. 84.** A cycloid. A point on the circumference of a rolling wheel describes this curve.

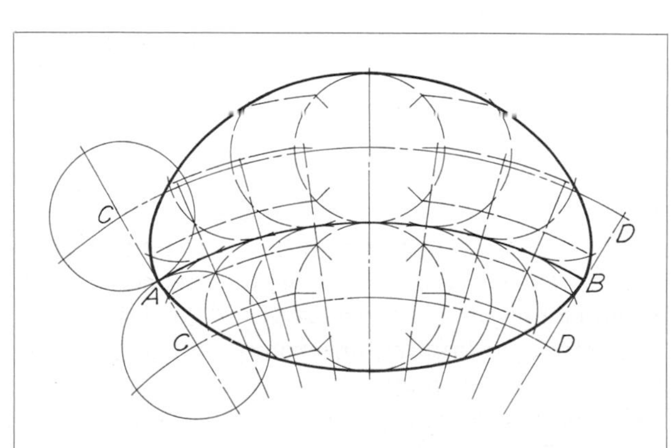


**FIG. 85.** An epicycloid and hypocycloid. Both are formed by a circle rolling on another circle.

## 67. The Involute.

An involute is the spiral curve traced by a point on a taut cord unwinding from around a polygon or circle. Thus the involute of any polygon can be drawn by extending its sides, as in Fig. 86, and, with the corners of the polygon as successive centers, drawing arcs terminating on the extended sides.

The equation of the involute of a circle (parametric form) is $x = r\ (\sin\theta - \theta\cos\theta)$, $y = r\ (\cos\theta + \theta\sin\theta)$, where $r$ is the radius of the circle and $\theta$ is the turned angle for the tangent point.

In drawing a spiral in design, as for example of bent ironwork, the easiest way is to draw it as the involute of a square.

A circle may be conceived of as a polygon of an infinite number of sides. Thus to draw the involute of a circle (Fig. 87), divide it into a convenient number of parts, draw tangents at these points, lay off on these tangents the rectified lengths of the arcs from the point of tangency to the starting point, and connect the points by a smooth curve. The involute of the circle is the basis for the involute system of spur gearing.

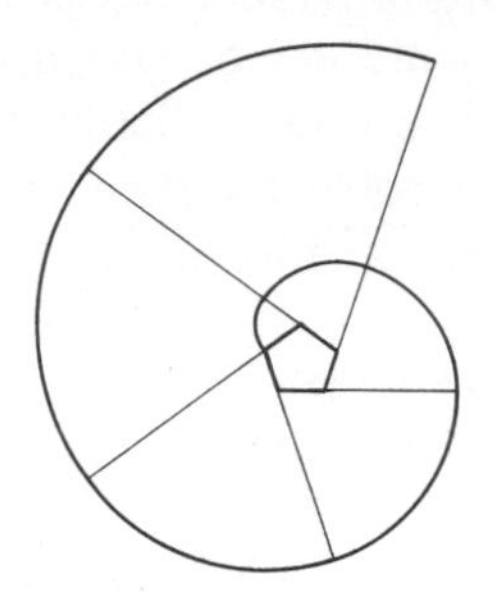

**FIG. 86.** An involute of a pentagon. A point on a cord unwound from the pentagon describes this curve.

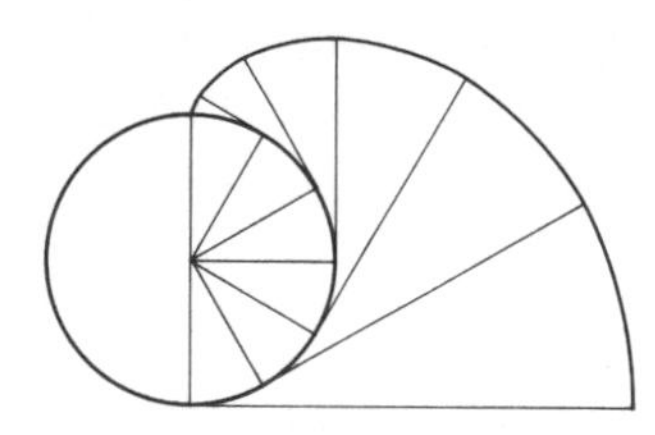

**FIG. 87.** An involute of a circle. As a taut cord is unwound from the circle, it forms a series of tangents to the circle.

## 68. The Spiral of Archimedes.

The spiral of Archimedes is the plane curve generated by a point moving uniformly along a straight line while the line revolves about a fixed point with uniform angular velocity.

*To Draw a Spiral of Archimedes that Makes One Turn in a Given Circle (Fig. 88).* Divide the circle into a number of equal parts, drawing the radii and numbering them. Divide the radius $O$-8 into the same number of equal parts, numbering from the center. With $O$ as center, draw concentric arcs intersecting the radii of corresponding numbers, and draw a smooth curve through these intersections. The Archimedean spiral is the curve of the heart cam used for converting uniform rotary motion into uniform reciprocal motion.

The mathematical equation is $p = a\theta$ (polar form).

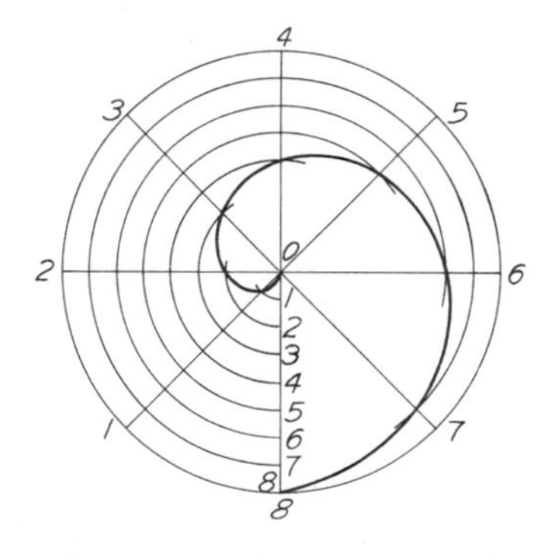


**FIG. 88.** The spiral of Archimedes. This curve increases uniformly in distance from the center as rotation (angular velocity) is constant.

## 69. Other Plane Curves.

The foregoing paragraphs discuss common plane curves that occur frequently in scientific work. Other curves, whenever needed, can be found in any good textbook of analytic geometry and calculus. Typical of the curves that may be met with are the catenary, cardioid, sine curve, cosine curve, logarithmic spiral, reciprocal (hyperbolic) spiral, parabolic spiral, logarithmic curve, exponential curve, and curves of velocity and acceleration.

### 70. Double-curved Lines.

The scarcity of geometric double-curved lines, by comparison with the numerous single-curved lines, is surprising. There are only two double-curved lines, the cylindrical and the conic helix, much used in engineering work. Double-curved lines will, however, often occur as the lines of intersection between two curved solids or surfaces. These lines are not geometric but are double-curved lines of general form.

### 71. The Helix.

The helix is a space curve generated by a point moving uniformly along a straight line while the line revolves uniformly about another line as an axis. If the moving line is parallel to the axis, it will generate a cylinder. The word "helix" alone always means a cylindrical helix. If the moving line intersects the axis at an angle less than 90°, it will generate a cone, and the curve made by the point moving on it will be a "conic helix." The distance parallel to the axis through which the point advances in one revolution is called the "lead." When the angle becomes 90°, the helix degenerates into the Archimedean spiral.

Application of the helix

### 72. To Draw a Cylindrical Helix (Fig. 89).

Draw the two views of the cylinder (see Chap. 5), and then measure the lead along one of the contour elements. Divide this lead into a number of equal parts (say, 12) and the circle of the front view into the same number. Number the divisions on the top view starting at point 1 and the divisions on the front view starting at the front view of point 1. When the generating point has moved one-twelfth of the distance around the cylinder, it has also advanced one-twelfth of the lead; when halfway around the cylinder, it will have advanced one-half the lead. Thus points on the top view of the helix can be found

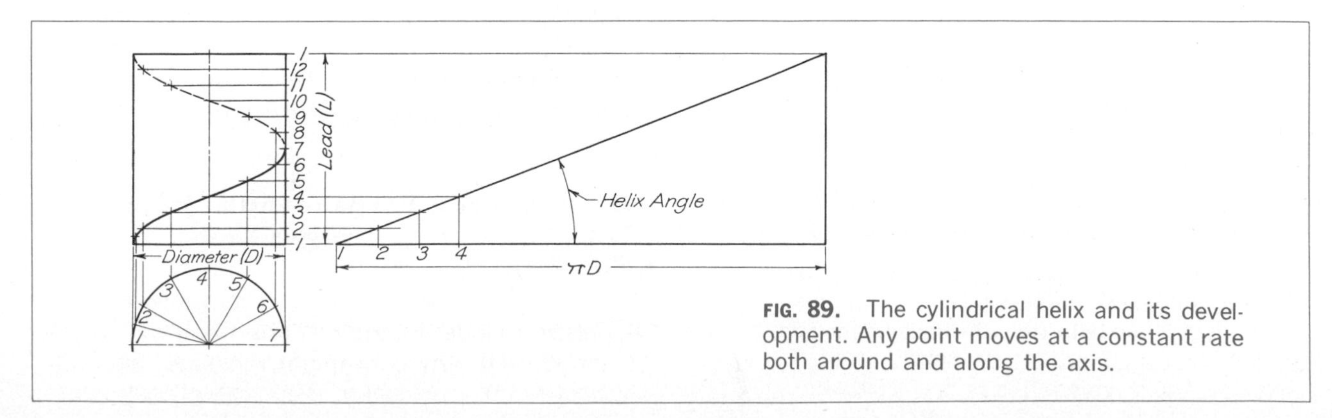


FIG. 89. The cylindrical helix and its development. Any point moves at a constant rate both around and along the axis.

by projecting the front views of the elements, which are points on the circular front view of the helix, to intersect lines drawn across from the corresponding divisions of the lead. If the cylinder is developed, the helix will appear on the development as a straight line inclined to the base at an angle, called the "helix angle," whose tangent is $L/\pi D$, where $L$ is the lead and $D$ the diameter.

### 73. To Draw a Conic Helix (Fig. 90).

First make two views of the right-circular cone (see Chap. 5) on which the helix will be generated. Then lay out uniform angular divisions in the view showing the end view of the axis (in Fig. 90, the top view) and divide the lead into the same number of parts. Points can now be plotted on the curve. Each plotted point will lie on a circle cut from the cone by a plane dividing the lead and will also lie on the angular-division line. Thus, for example, to plot point 14, draw the circle diameter obtained from the front view, as shown in the top view. Point 14 in the top view then lies at the intersection of this circle and radial-division line 14. Then locate the front view by projection from the top view to plane 14 in the front view.

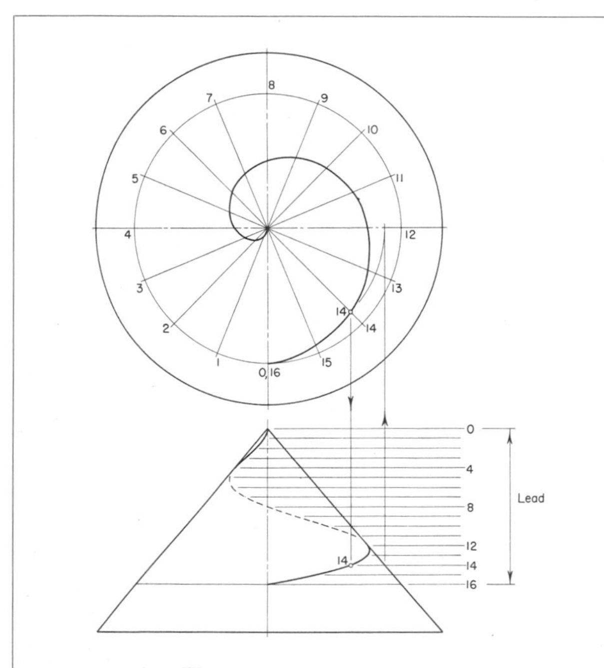


**FIG. 90.** The conic helix. A point traverses the surface of the cone while moving at a constant rate both around and along the axis.

## CONSTRUCTIONS FOR LOFTING AND LARGE LAYOUTS

### 74.

There are cases when the regular drafting instruments are impractical, either because of size limitations or when extreme accuracy is necessary. The geometric methods for common cases of parallelism, perpendicularity, and tangency are given in the following paragraphs.

### 75. To Draw a Line through a Point and Parallel to a Given Line (Fig. 91).

With the given point $P$ as center and a radius of sufficient length, draw an arc $CE$ intersecting the given line $AB$ at $C$. With $C$ as center and the same radius, draw the arc $PD$. With $C$ as center and radius $DP$, draw an arc intersecting $CE$ at $E$. Then $EP$ is the required line.

### 76. To Draw a Line Parallel to Another at a Given Distance.

*For Straight Lines (Fig. 92).* With the given distance as radius and two points on the given line as centers (as far apart as convenient), draw two arcs. A line tangent to these arcs will be the required line.

*For Curved Lines (Fig. 93).* Draw a series of arcs with centers along the line. Draw tangents to these arcs with a French curve (see Fig. 60, Chap. 2).

### 77. To Erect a Perpendicular from a Point to a Given Straight Line (Fig. 94).

With point $P$ as center and any convenient radius $R_1$, draw a circle arc intersecting the given line at $A$ and $B$. With any convenient radius $R_2$ and with

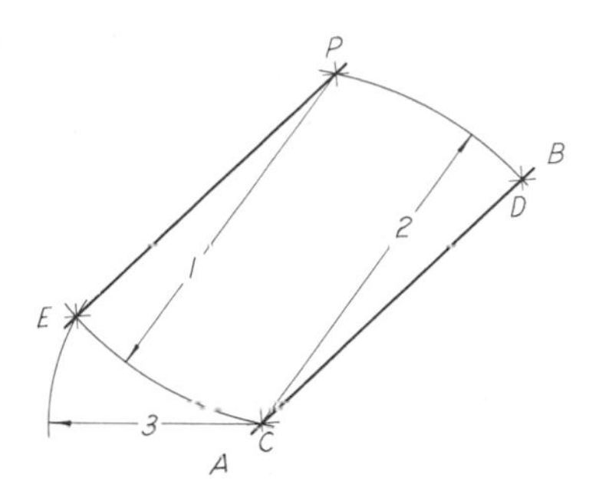


FIG. 91. Parallel lines. Two points determine the parallel.

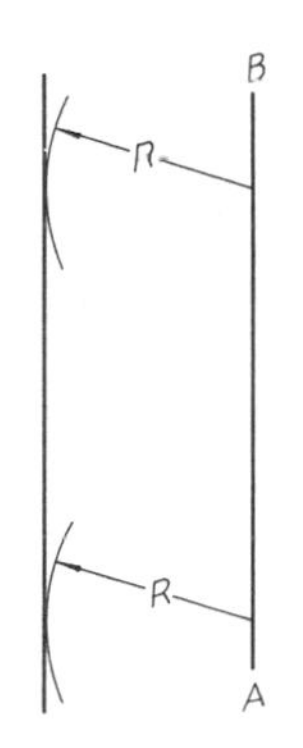


FIG. 92. Parallel lines. The parallel is drawn tangent to a pair of arcs.

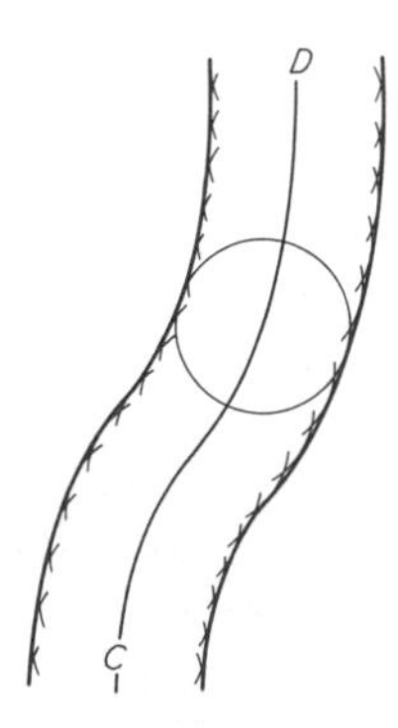


FIG. 93. Curved parallel lines. The curves are drawn tangent to circle arcs.

centers at $A$ and $B$, draw intersecting arcs locating $Q$. The required perpendicular is $PQ$, with $S$ the intersection of the perpendicular and the given line.

**78. To Erect a Perpendicular from a Point on a Given Straight Line.**
*First Method* (*Fig.* 95). With point $P$ on the line as center and any convenient radius $R_1$, draw circle arcs to locate points $A$ and $B$ equidistant from $P$. With any convenient radius $R_2$ longer than $R_1$ and with centers at $A$ and $B$, draw intersecting arcs locating $Q$. $PQ$ is the required perpendicular.

*Second Method* (*Fig.* 96). With any convenient center $C$ and radius $CP$, draw somewhat more than a semicircle from the intersection of the circle arc with the given line at $A$. Draw $AC$ extended to meet the circle arc at $Q$. $PQ$ is the required perpendicular.

**79. To Draw a Tangent to a Circle at a Point on the Circle (Fig. 97).**
Given the arc $ABC$ and to draw a tangent at point $C$, draw the extended diameter $CD$ and locate point $M$. Then with any convenient radius $R_1$, locate point $P$ equidistant from $C$ and $M$. With $P$ as center and the same radius $R_1$, draw somewhat more than a semicircle and draw the line $MPQ$. The line $QC$ is a tangent to the circle at point $C$.

**80. To Draw a Tangent to a Circle from a Point Outside (Fig. 98).**
Connect the point $P$ with the center of the circle $O$. Then draw the perpendicular bisector of $OP$, and with the intersection at $D$ as center, draw a semicircle. Its intersection with the given circle is the point of tangency. Draw the tangent line from $P$.

**81. To Draw a Tangent to Two Circles.**
*First Case: Open Belt* (*Fig.* 99). At center $O$ draw a circle with radius $R_1 - R_2$. From $P$ draw a tangent to this circle by the method of Fig. 98. Extend $OT$ to $T_1$, and draw $PT_2$ parallel to $OT_1$. Join $T_1$ and $T_2$, giving the required tangent.

*Second Case: Crossed Belt* (*Fig.* 100). Draw $OA$ and $O_1B$ perpendicular to $OO_1$. From $P$, where $AB$ crosses $OO_1$, locate tangents as in Fig. 98.

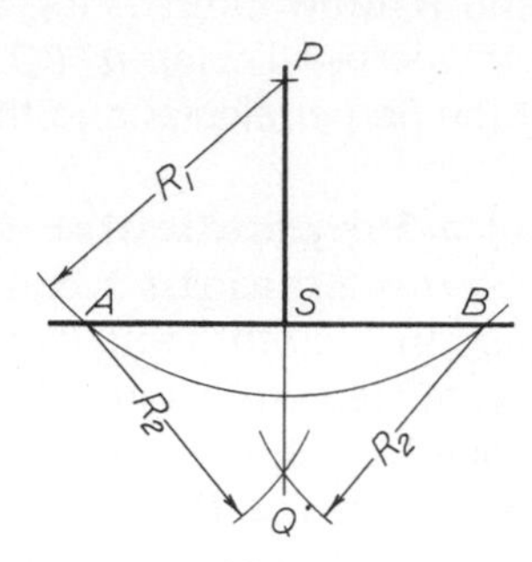


**FIG. 94.** A perpendicular from a point outside. Equal radii determine the perpendicular.

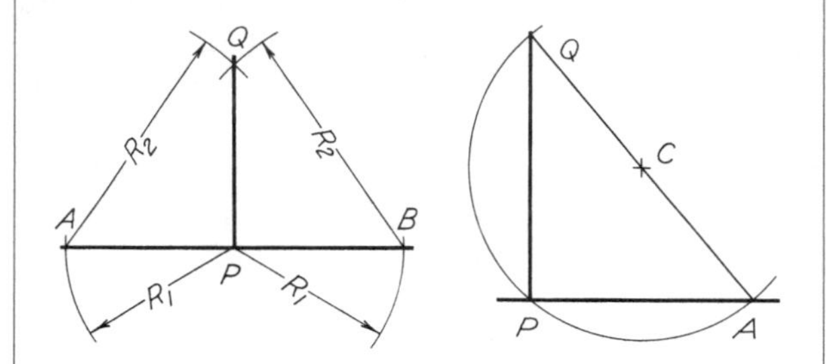


**FIG. 95 (LEFT).** A perpendicular from a point on a line. Equal radii determine a point on the perpendicular.

**FIG. 96 (RIGHT).** A perpendicular from a point on a line. Geometrically, perpendicular chords of a circle produce this solution.

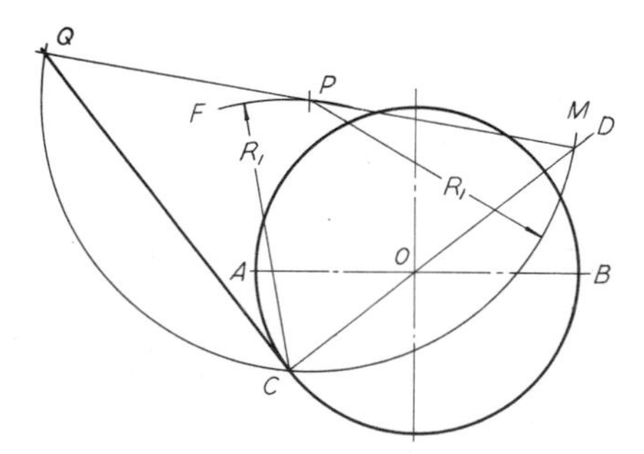


**FIG. 97.** To draw a tangent at a point on a circle. This solution sets up two right-angle chords of a circle. One chord passes through the center of the given circle.

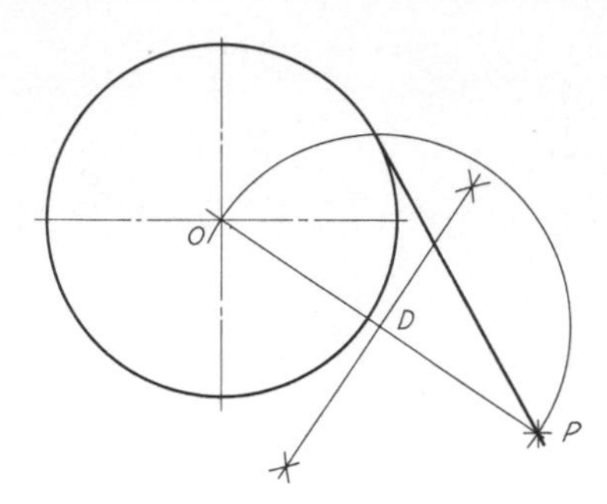


**FIG. 98.** To draw a tangent from an outside point. Here, as in Fig. 97, perpendicular chords determine the solution.

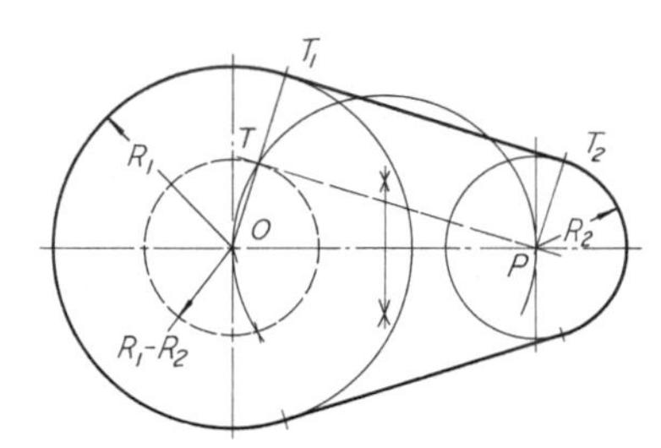


**FIG. 99.** Tangent lines (open belt). The difference in radius of the two circles is the geometric basis. Then perpendicular chords are used.

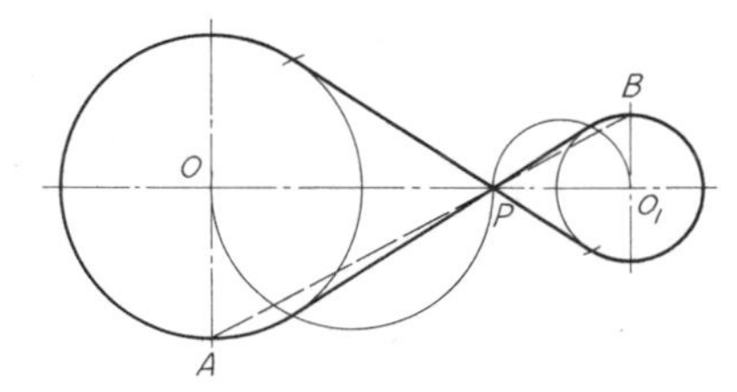


**FIG. 100.** Tangent lines (crossed belt). A duplicate of Fig. 99, except for position of the tangent lines.

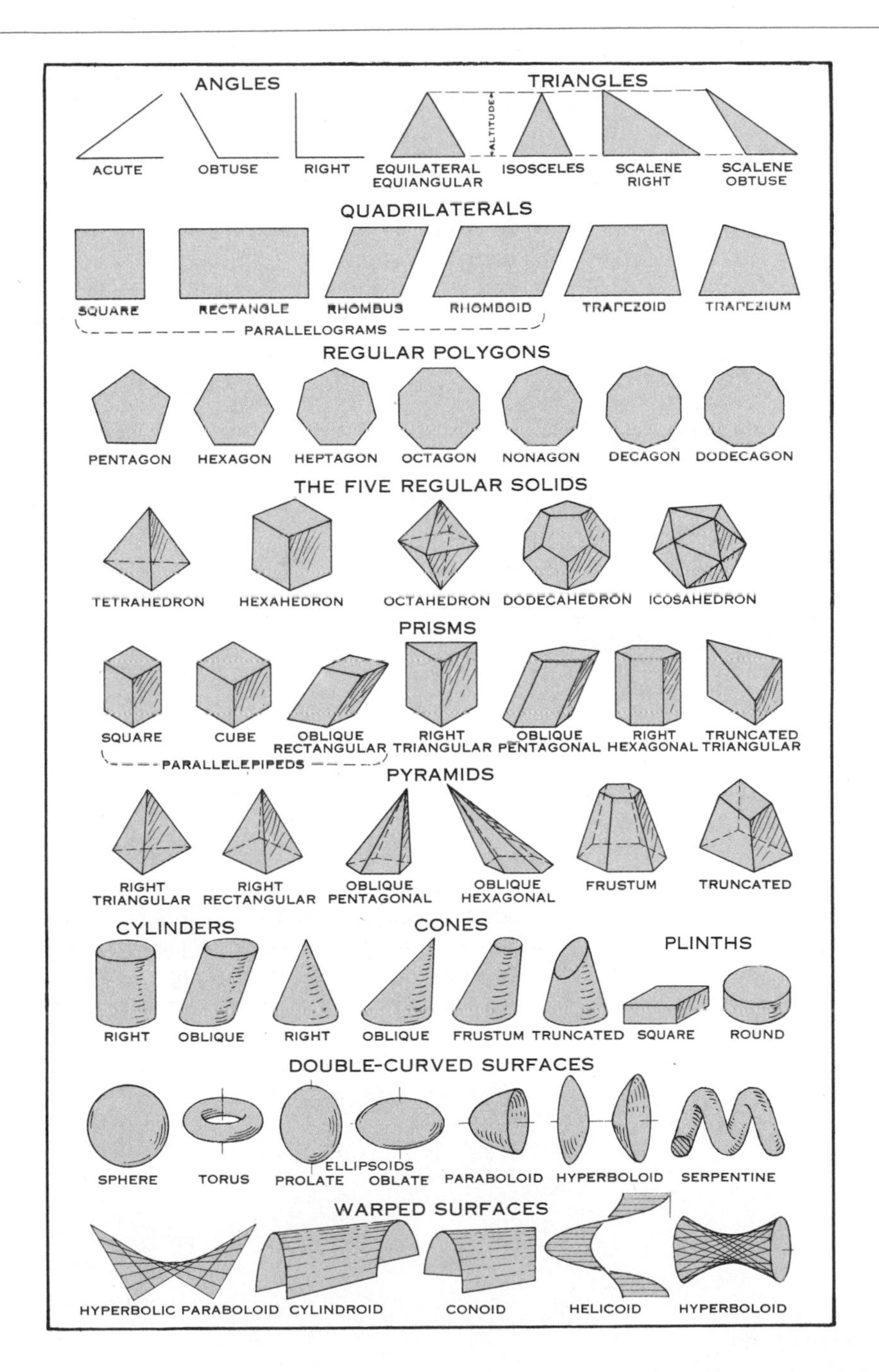


**FIG. 101.** Geometric shapes. These plane figures, solids, and surfaces should be studied and remembered.

# PROBLEMS

To be of value both as drawing exercises and as solutions, geometric problems must be worked accurately. Keep your pencil sharp, and use comparatively light lines. Locate a point by two intersecting lines; indicate the length of a line by short dashes across it.

## Group 1. Lines and Plane Figures.

**1.** Near the center of the working space, draw a horizontal line $4\frac{1}{2}$ in. long. Divide it into seven equal parts by the method of Fig. 27.

**2.** Draw a vertical line 1 in. from the left edge of the space and $3\frac{7}{8}$ in. long. Divide it into parts proportional to 1, 3, 5, and 7.

**3.** Construct a polygon as shown in the problem illustration, drawing the horizontal line $AK$ (of indefinite length) $\frac{3}{8}$ in. above the bottom of the space. From $A$ draw and measure $AB$. Proceed in the same way for the remaining sides. The angles can be obtained by proper combinations of the two triangles (see Figs. 28 and 29, Chap. 2).

**4.** Draw a line $AK$ making an angle of 15° with the horizontal. With this line as base, transfer the polygon of Prob. 3.

**5.** Draw a regular hexagon having a distance across corners of 4 in.

**6.** Draw a regular hexagon, distance across flats $3\frac{3}{8}$ in.

**7.** Draw a regular dodecagon, distance across flats $3\frac{3}{8}$ in.

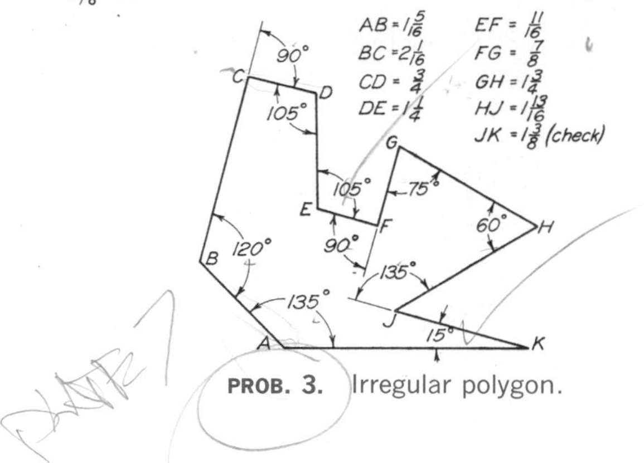


**PROB. 3.** Irregular polygon.

## Group 2. Problems in Accurate Joining of Tangent Lines.

**8.** Draw the offset swivel plate.

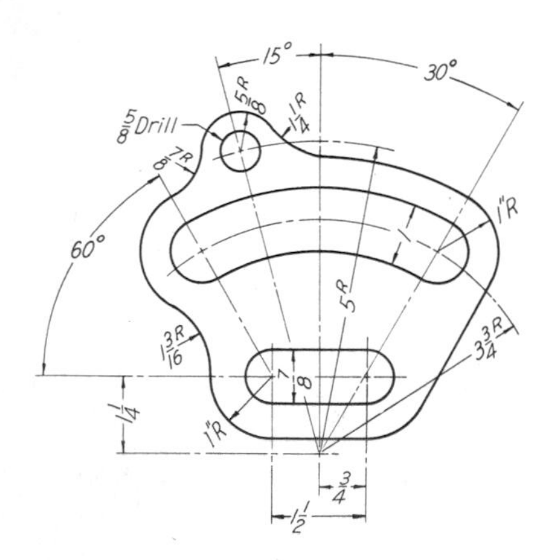

**PROB. 8.** Offset swivel plate.

**9.** Draw two lines $AB$ and $AC$ making an included angle of 30°. Locate point $P$, 4 in. from $A$ and $\frac{1}{2}$ in. from line $AB$. Draw two circle arcs centered at point $P$, one tangent to line $AB$, the other to $AC$. Then draw two lines tangent to the opposite sides of the arcs and passing through point $A$. Locate all tangent points by construction.

**10.** Construct an ogee curve joining two parallel lines $AB$ and $CD$ as in Fig. 18, making $X = 4$ in., $Y = 2\frac{1}{2}$ in., and $BE = 3$ in. Consider this as the center line for a rod $1\frac{1}{4}$ in. in diameter, and draw the rod.

**11.** Make contour view of the bracket. In the upper ogee curve, the radii $R_1$ and $R_2$ are equal. In the lower one, $R_3$ is twice $R_4$.

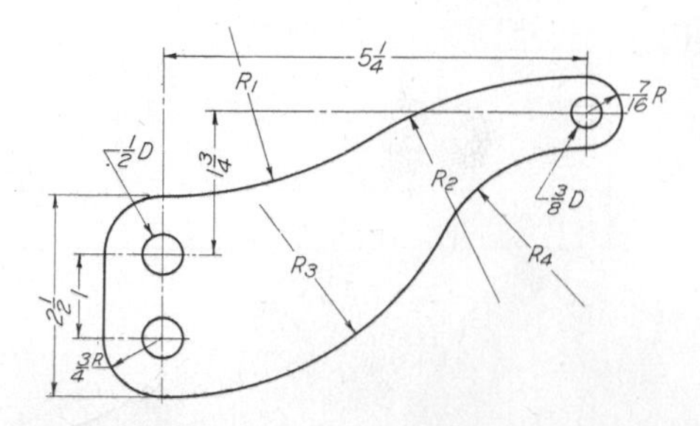

**PROB. 11.** Bracket.

**12.** Draw an arc of a circle having a radius of $3\frac{13}{16}$ in., with its center $\frac{1}{2}$ in. from the top of the space and $1\frac{1}{4}$ in. from the left edge. Find the length of an arc of 60° by construction; compute the length arithmetically, and check the result.
**13.** Front view of washer. Draw half size.
**14.** Front view of shim. Draw full size.
**15.** Front view of rod guide. Draw full size.
**16.** Front view of a star knob. Radius of circumscribing circle, $2\frac{3}{8}$ in. Diameter of hub, $2\frac{1}{2}$ in. Diameter of hole, $\frac{3}{4}$ in. Radius at points, $\frac{3}{8}$ in. Radius of fillets, $\frac{3}{8}$ in. Mark tangent points in pencil.
**17.** Front view of sprocket. OD, $4\frac{3}{4}$ in. Pitch diameter, 4 in. Root diameter, $3\frac{1}{4}$ in. Bore, $1\frac{1}{4}$ in. Thickness of tooth at the pitch line is $\frac{9}{16}$ in. Splines, $\frac{1}{4}$ in. wide by $\frac{1}{8}$ in. deep. Mark tangent points in pencil. See Glossary for terms.
**18.** Front view of a fan. Draw full size.
**19.** Front view of a level plate. Draw full size.
**20.** Front view of an eyelet. Draw full size.
**21.** Front view of a stamping. Draw full size.
**22.** Front view of spline lock. Draw full size.
**23.** Front view of gage cover plate. Draw full size.
**24.** Drawing of heater tube. Draw twice size.
**25.** Drawing of exhaust-port contour. Draw twice size.
**26.** Drawing of toggle spring for leaf switch. Use decimal scale and draw 20 times size.

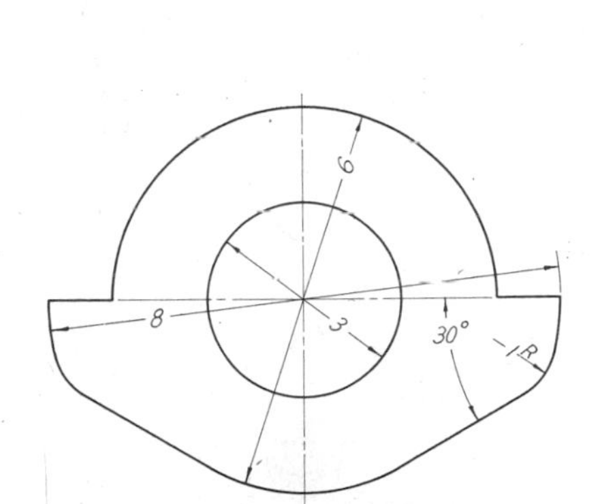

**PROB. 13.** Washer.

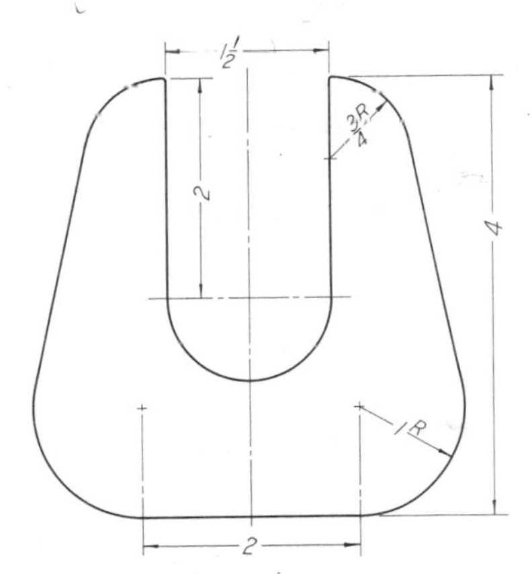

**PROB. 14.** Shim.

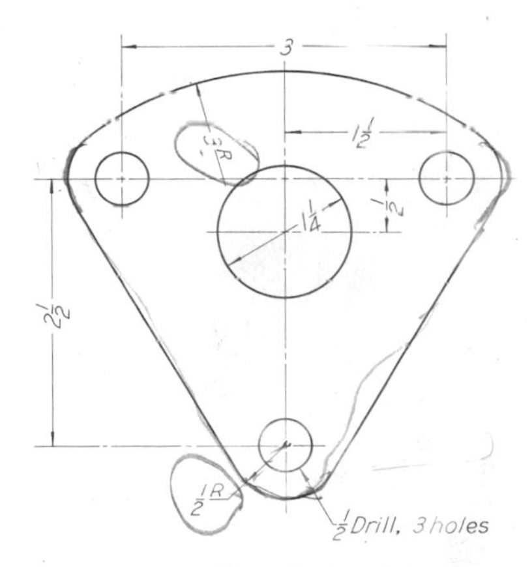

**PROB. 15.** Rod guide.

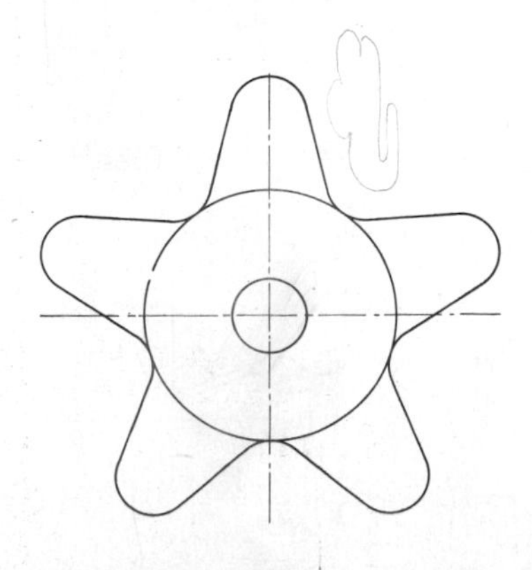
**PROB. 16.** Star knob.

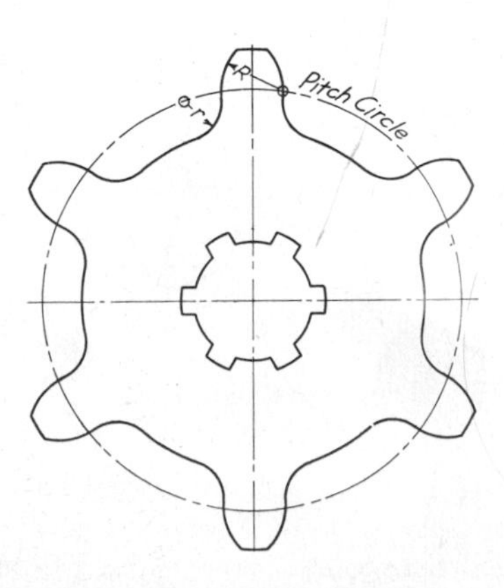

**PROB. 17.** Sprocket.

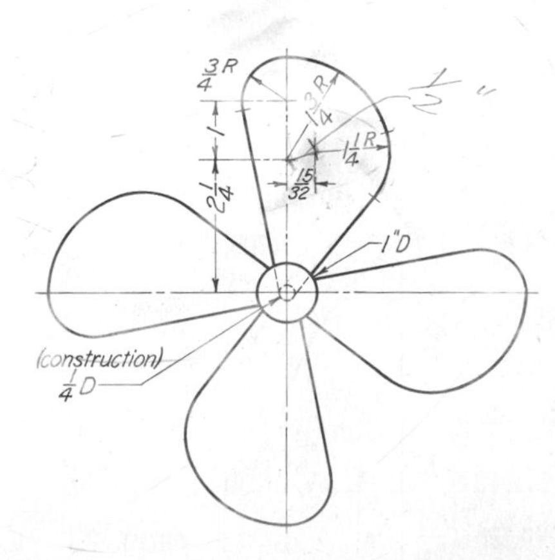

**PROB. 18.** Fan.

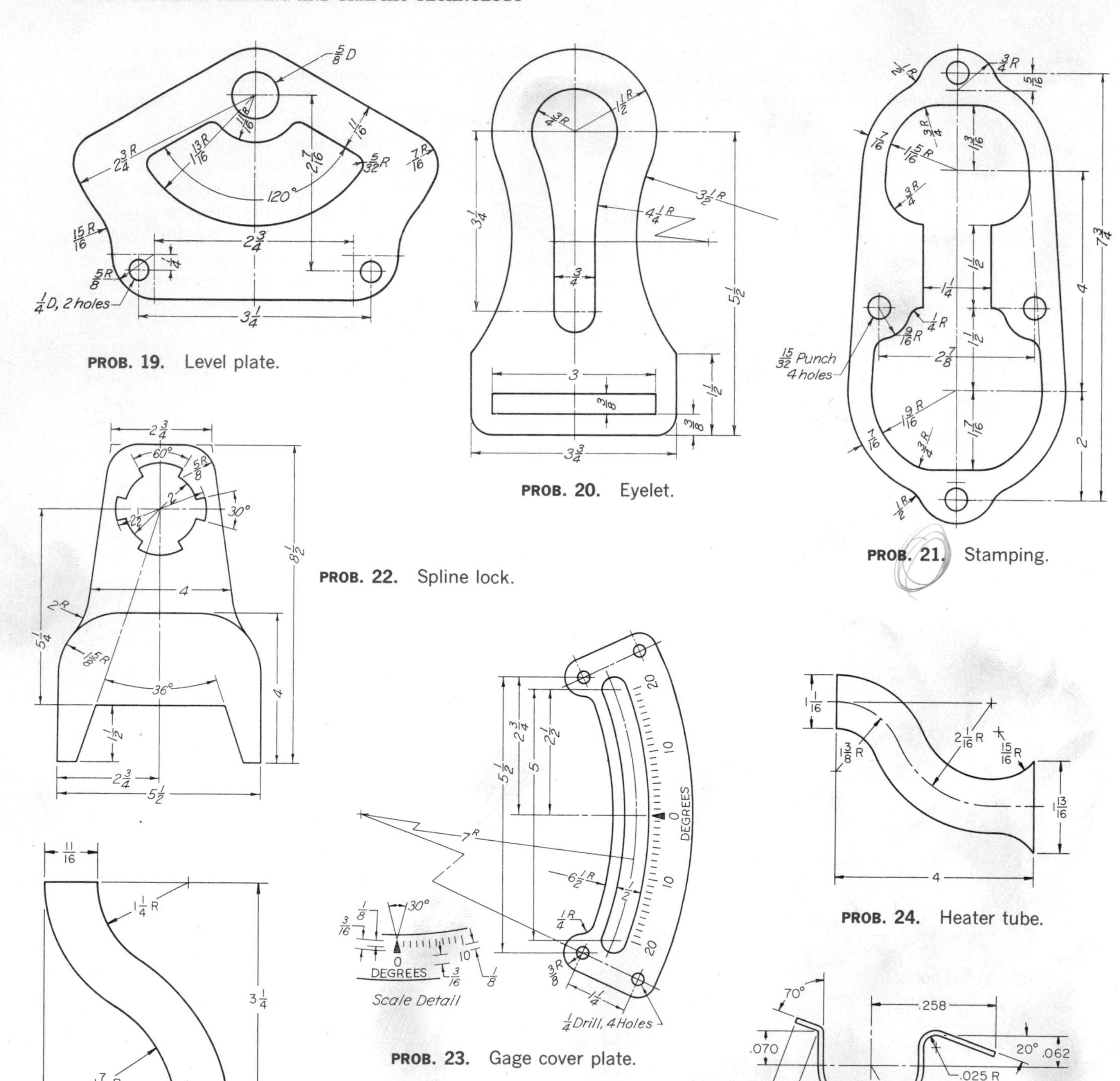


**PROB. 19.** Level plate.

**PROB. 20.** Eyelet.

**PROB. 21.** Stamping.

**PROB. 22.** Spline lock.

**PROB. 23.** Gage cover plate.

**PROB. 24.** Heater tube.

**PROB. 25.** Exhaust-port contour.

**PROB. 26.** Toggle spring for leaf switch.

**27.** Drawing of pulley shaft. Draw full size.
**28.** Drawing of cam for type A control. Plot cam contour from the polar coordinates given. Use decimal scale and draw five times size.

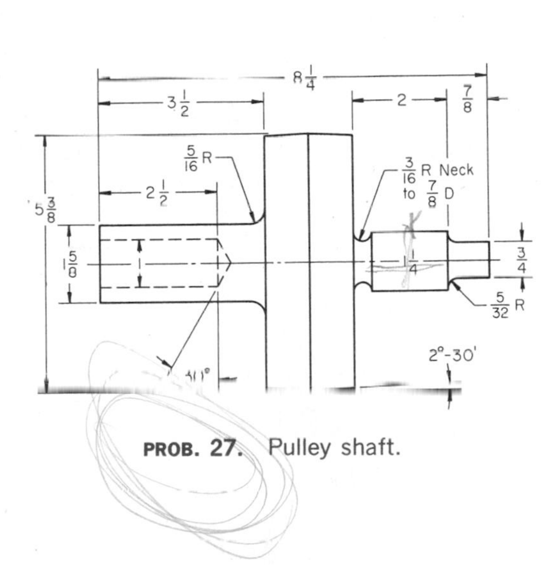

PROB. 27. Pulley shaft.

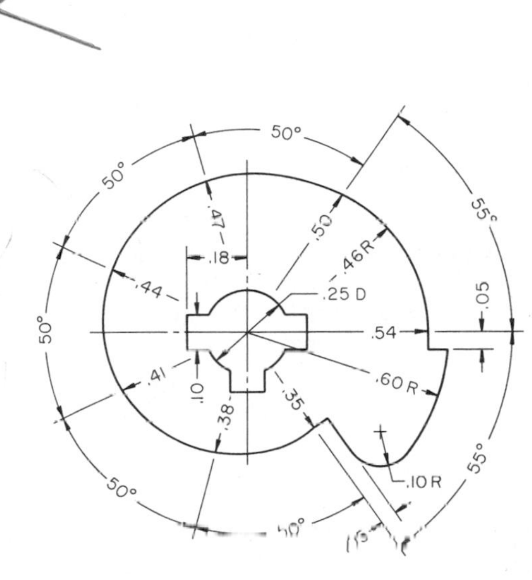

PROB. 28. Cam for type A control.

## Group 3. Plane Curves.

**29.** The conjugate diameters of an ellipse measure 3 and 4 in., the angle between them being 60°. Construct the major and minor diameters of this ellipse and draw the ellipse.
**30.** Using the pin-and-string method, draw an ellipse having a major diameter of 6 in. and a minor diameter of $4\frac{1}{4}$ in.
**31.** Using the trammel method, draw an ellipse having a major diameter of $4\frac{1}{2}$ in. and a minor diameter of 3 in.
**32.** Using the trammel method, draw an ellipse having a major diameter of $4\frac{1}{2}$ in. and a minor diameter of 4 in.
**33.** Using the concentric-circle method, draw an ellipse having a major diameter of $4\frac{5}{8}$ in. and a minor diameter of $1\frac{1}{2}$ in.
**34.** Draw an ellipse on a major diameter of 4 in. One point on the ellipse is $1\frac{1}{2}$ in. to the left of the minor diameter and $\frac{7}{8}$ in. above the major diameter.
**35.** Draw an ellipse having a minor diameter of $2\frac{3}{16}$ in. and a distance of $3\frac{1}{4}$ in. between focuses. Draw a tangent at a point $1\frac{3}{8}$ in. to the right of the minor diameter.
**36.** Draw an ellipse whose major diameter is 4 in. A tangent to the ellipse intersects the minor diameter $1\frac{3}{4}$ in. from the center, at an angle of 60°.
**37.** Draw a five-centered arch with a span of 5 in. and a rise of 2 in.
**38.** Draw an ellipse having conjugate diameters of $4\frac{3}{4}$ in. and $2\frac{3}{4}$ in., making an angle of 75° with each other. Determine the major and minor diameters.
**39.** Draw the major and minor diameters for an ellipse having a pair of conjugate diameters 60° apart, one horizontal and $6\frac{1}{4}$ in. long, the other $3\frac{1}{4}$ in. long.
**40.** Using the circle method, draw an ellipse having a pair of conjugate diameters, one making 15° with the horizontal and 6 in. long, the other making 60° with the first diameter and $2\frac{1}{2}$ in. long.
**41.** Using the triangle-trammel method, draw an ellipse having a pair of conjugate diameters 60° apart, one horizontal and 6 in. long, the other 4 in. long.
**42.** Using the triangle-trammel method, draw an ellipse having a pair of conjugate diameters 45° apart, one making 15° with the horizontal and 6 in. long, the other 3 in. long.

**43.** Draw a parabola, axis vertical, in a rectangle 4 by 2 in.

**44.** Draw a parabolic arch, with 6-in. span and a $2\frac{1}{2}$-in. rise, by the offset method, dividing the half span into eight equal parts.

**45.** Draw an equilateral hyperbola passing through a point $P$, $\frac{1}{2}$ in. from $OB$ and $2\frac{1}{2}$ in. from $OA$. (Reference letters correspond to Fig. 83.)

**46.** Draw two turns of the involute of a pentagon whose circumscribed circle is $\frac{1}{2}$ in. in diameter.

**47.** Draw one-half turn of the involute of a circle $3\frac{1}{4}$ in. in diameter whose center is 1 in. from the left edge of the space. Compute the length of the last tangent and compare with the measured length.

**48.** Draw a spiral of Archimedes making one turn in a circle 4 in. in diameter.

**49.** Draw the cycloid formed by a rolling circle 2 in. in diameter. Use 12 divisions.

**50.** Draw the epicycloid formed by a 2-in.-diameter circle rolling on a 15-in.-diameter directing circle. Use 12 divisions.

**51.** Draw the hypocycloid formed by a 2-in.-diameter circle rolling inside a 15-in.-diameter directing circle. Use 12 divisions.

**52.** Front view of cam. Draw full size.

**53.** Front view of fan base. Draw full size.

**54.** Front view of trip lever. Draw half size.

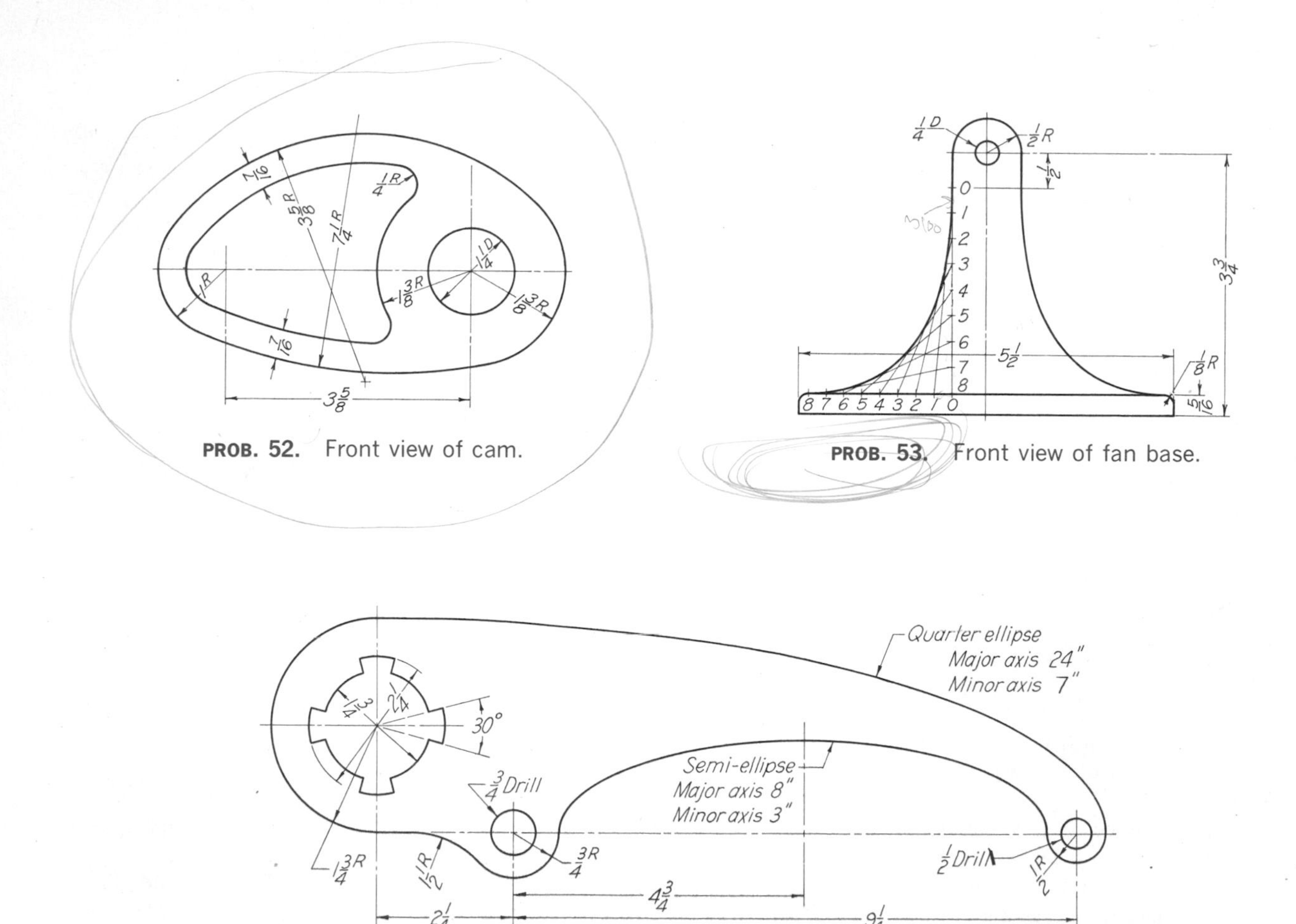


**PROB. 52.** Front view of cam.

**PROB. 53.** Front view of fan base.

**PROB. 54.** Front view of trip lever.

## Group 4. Problems for Slide Rule and Mathematical Tables.

**55.** Find the altitude of an equilateral triangle, each side of which is 4 in.

**56.** Find the value of one-third of an angle of 231°4′2″.

**57.** A tank is 31 in. in diameter and 4 ft $7\frac{1}{2}$ in. long. Find the capacity in liters.

**58.** A 4-in.-diameter cold-rolled steel bar is 96 cm long. What is the weight in pounds? In grams?

**59.** A circle of $12\frac{5}{8}$-in. diameter used as a target for optical devices is to be compared in area with a lens of 80 mm diameter. What percentage is the lens area to the target area?

**60.** A guided missile is fired, hits a target 841 miles away, and during flight rises to a height of 91 miles. What are these distances in kilometers?

**61.** Find one side of a square whose equivalent area is a circle of $6\frac{3}{8}$-in. diameter. Answer in centimeters.

**62.** An ellipse has a major axis of 14 in. and a minor axis of $8\frac{1}{2}$ in. How far from the geometric center is each focus? What is the equivalent distance in centimeters?

**63.** A certain pressure is recorded by a column of water 2 ft $4\frac{1}{2}$ in. high. What is the equivalent pressure in inches of mercury?

**64.** Find the area of a tract of land designated by a traverse that is an equilateral triangle each side of which is 228 ft 4 in. long. Answer in square miles. Also calculate the area in square kilometers.

**65.** A square is inscribed in a circle of 4-in. diameter. What is the area of the square?

**66.** An equilateral triangle is inscribed in a circle of $9\frac{3}{8}$-in. diameter. What is the sum of the sides of the triangle? What is the circumference of the circle? What is the area of the triangle and of the circle?

**67.** A round copper bar 2 in. in diameter and 4 ft long is to be balanced in weight by a cube of aluminum. What is the size of the cube?

**68.** If a 2-in. round bar of steel 2 ft 6 in. long is to be replaced by a bar of Bakelite, what will be the difference in weight?

**69.** A circular opening 6 in. in diameter is to be replaced by openings 0.70 cm in diameter. How many openings must be used to give an equivalent area?

**70.** A $3\frac{1}{8}$-in. cube of aluminum is to be replaced by steel. What is the percentage difference in weight?

**71.** A vertical water tank 4 ft 0 in. square holds 980 gallons. What is the pressure, per square inch, on the bottom of the water tank?

Discussed and Illustrated in this Chapter:

Single-stroke Lettering
Proportions
Guide Lines
Lettering in Pencil
Vertical Capital Letters
Order of Strokes
Vertical Lower-case Letters
Inclined Letters
Composition
Titles
Pen-and-ink Lettering
Supplemental Letter Styles
Problems

# 4

# LETTERING

## Complete Graphic Depiction and Specification Requires Word Supplements

## 1. Lettering—Word Supplements on Drawings.

Graphic representation of the shape of a part, machine, or structure gives one aspect of the information needed for its construction. To this must be added, to complete the description, figured dimensions, notes on material and finish, and a descriptive title—all lettered, freehand, in a style that is perfectly legible, uniform, and capable of rapid execution. As far as the appearance of a drawing is concerned, the lettering is the most important part. But the usefulness of a drawing, too, can be ruined by lettering done ignorantly or carelessly, because illegible figures are apt to cause mistakes in the work.

In a broad sense, lettering is a branch of design. Students of lettering fall into two general classes: those who will use letters and words to convey information on drawings, and those who will use lettering in applied design, for example, art students, artists, and craftsmen. The first group is concerned mainly with legibility and speed, the second with beauty of form and composition. In our study of engineering graphics we are concerned only with the problems of the first group. The engineering student takes up lettering as an early part of his work in graphics and continues its practice throughout his course, becoming more and more skillful and proficient.

## 2. Single-stroke Lettering.

By far the greatest amount of lettering on drawings is done in a rapid single-stroke letter, either vertical or inclined, and every engineer must have absolute command of these styles. The ability to letter well can be acquired only by continued and careful practice, but it can be acquired by anyone with normal muscular control of his fingers who will practice faithfully and intelligently and take the trouble to observe carefully the shapes of the letters, the sequence of strokes in making them, and the rules for their composition. It is not a matter of artistic talent or even of dexterity in handwriting. Many persons who write poorly letter very well.

The term "single-stroke," or "one-stroke," does not mean that the entire letter is made without lifting the pencil or pen but that the width of the stroke of the pencil or pen is the width of the stem of the letter.

## 3. General Proportions.

There is no standard for the proportions of letters, but there are certain fundamental points in design and certain characteristics of individual letters that must be learned by study and observation before composition into words and sentences should be attempted. Not only do the widths of letters in any alphabet vary, from *I*, the narrowest, to *W*, the widest, but different alphabets vary as a whole. Styles narrow in their proportion of width to height are called "**COMPRESSED**," or "**CONDENSED**," and are used when space is limited. Styles wider than the normal are called "**EXTENDED**."

The proportion of the thickness of stem to the height varies widely, ranging from $\frac{1}{3}$ to $\frac{1}{20}$. Letters with heavy stems are called "**BOLDFACE**," or "**BLACKFACE**," those with thin stems "LIGHTFACE."

## 4. The Rule of Stability.

In the construction of letters, the well-known optical illusion in which a horizontal line drawn across the middle of a rectangle appears to be below the middle must be provided for. In order to give the appearance of stability, such letters as *B*, *E*, *K*, *S*, *X*, and *Z* and the figures *3* and *8* must be drawn smaller at the top than at the bottom. To see the effect of this illusion, turn a printed page upside down and notice the appearance of the letters mentioned.

## 5. Guide Lines.

Always draw light guide lines for both tops and bottoms of letters, using a sharp pencil. Figure 1 shows a method of laying off a number of equally spaced lines. Draw the first base line; then set the bow spacers to the distance wanted between base lines and step off the required number of base lines. Above the last line mark the desired height of the letters. With the same setting, step down from this upper point, thus obtaining points for the top of each line of letters.

The Braddock-Rowe triangle (Fig. 2*a*) and the Ames

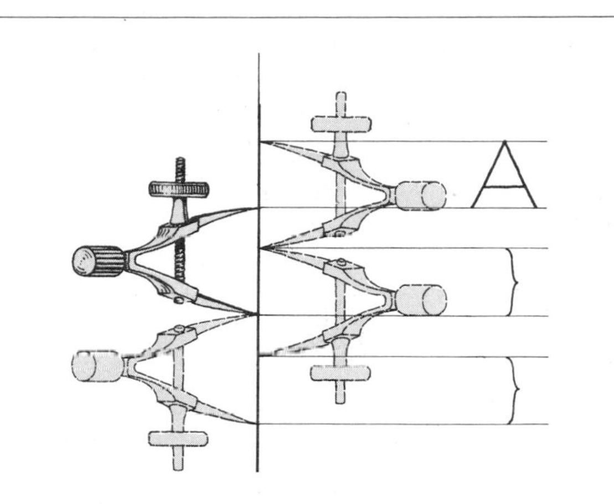


FIG. 1. To space guide lines. Bow dividers are spaced for the distance between base lines; this distance stepped off from capital and base lines locates successive lines of lettering.

lettering instrument (Fig. 2*b*) are convenient devices for spacing lines of letters. In using these instruments, a sharp pencil is inserted in the proper row of countersunk holes, and the instrument, guided by a T-square blade, is drawn back and forth by the pencil, as indicated by Fig. 3. The holes are grouped for capitals and lower case, the numbers indicating the height of capitals in thirty-seconds of an inch; thus no. 6 spacing of the instrument means that the capitals will be $^6/_{32}$, or $^3/_{16}$, in. high.

## 6. Lettering in Pencil.

Good technique is as essential in lettering as in drawing. The quality of the lettering is important whether it appears on finished work to be reproduced by one of the printing processes or as part of a pencil drawing to be inked. In the first case, the penciling must be clean, firm, and opaque; in the second, it may be lighter. The lettering pencil should be selected carefully by trial on the paper. In one instance, the same grade may be chosen as

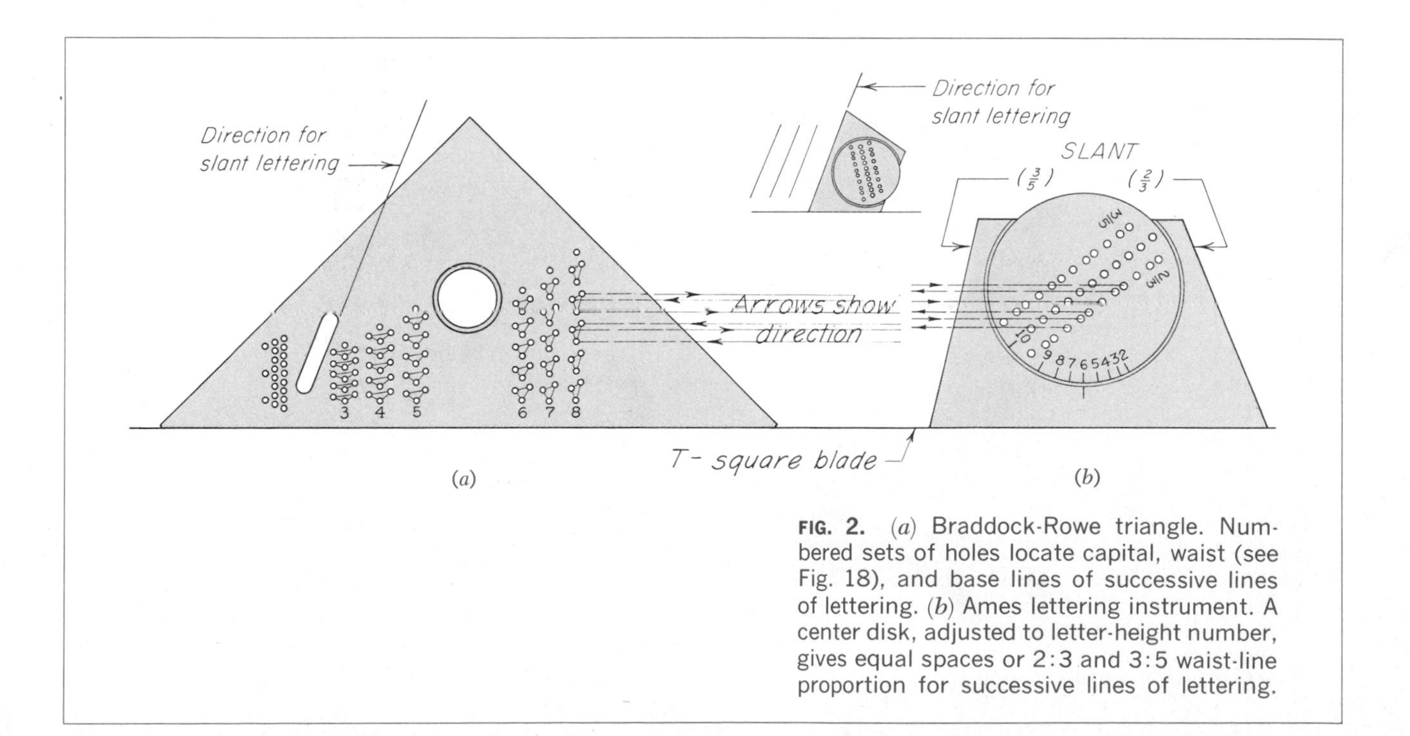


FIG. 2. (*a*) Braddock-Rowe triangle. Numbered sets of holes locate capital, waist (see Fig. 18), and base lines of successive lines of lettering. (*b*) Ames lettering instrument. A center disk, adjusted to letter-height number, gives equal spaces or 2:3 and 3:5 waist-line proportion for successive lines of lettering.

that used for the drawing; in another, a grade or two softer may be preferred. Sharpen the pencil to a long, conic point, and then round the lead slightly on the end so that it is not so sharp as a point used for drawing.

The first requirement in lettering is the correct holding of the pencil or pen. Figure 4 shows the pencil held comfortably with the thumb, forefinger, and second finger on alternate flat sides and the third and fourth fingers on the paper. Vertical, slanting, and curved strokes are drawn with a steady, even, *finger* movement; horizontal strokes are made similarly but with some pivoting of the hand at the wrist (Fig. 5). Exert pressure that is firm and uniform but not so heavy as to cut grooves in the paper. To keep the point symmetrical, form the habit of rotating the pencil after every few strokes.

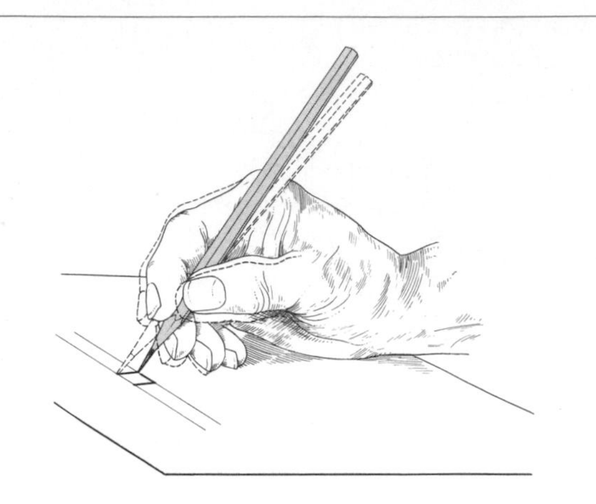

**FIG. 4.** Vertical strokes. These are made entirely by finger movement.

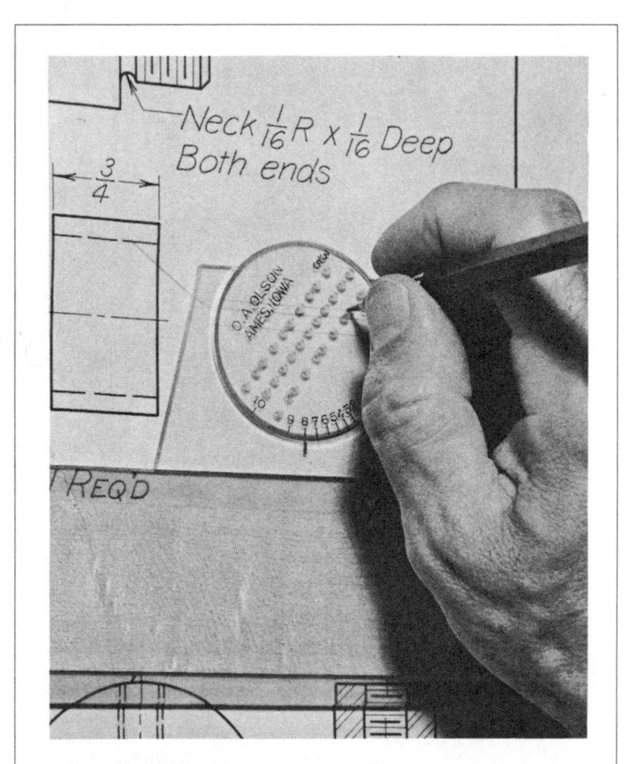


**FIG. 3.** Using the Ames lettering instrument. Lines are drawn as the pencil moves the instrument along the T square (or triangle).

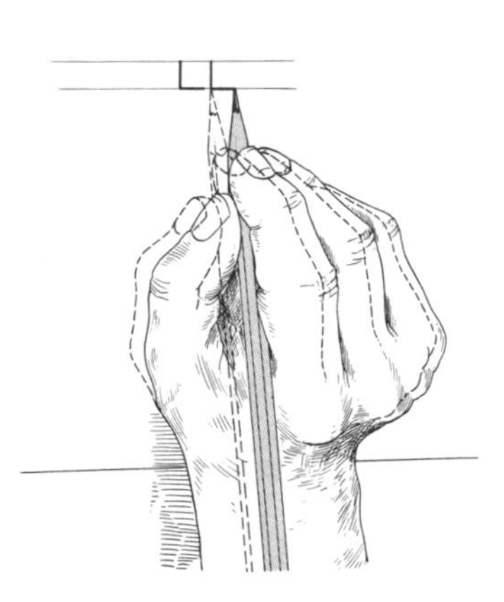

**FIG. 5.** Horizontal strokes. These are made by pivoting the whole hand at the wrist; fingers move slightly to keep the stroke perfectly horizontal.

### 7. Single-stroke Vertical Capitals.

The vertical, single-stroke, commercial Gothic letter is a standard for titles, reference letters, etc. As for the proportion of width to height, the general rule is that the smaller the letters are, the more extended they should be in width. A low extended letter is more legible than a high compressed one and at the same time makes a better appearance.

For proficiency in lettering it is essential to learn the form and peculiarity of each of the letters. Although lettering must be based on a careful regard for the fundamental letter forms, this is not to say that it will be without character. Individuality in lettering is often nearly as marked as in handwriting.

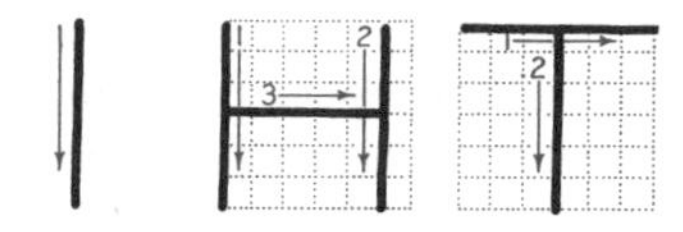

**FIG. 6.** The *I-H-T* group. Note the direction of fundamental horizontal and vertical strokes.

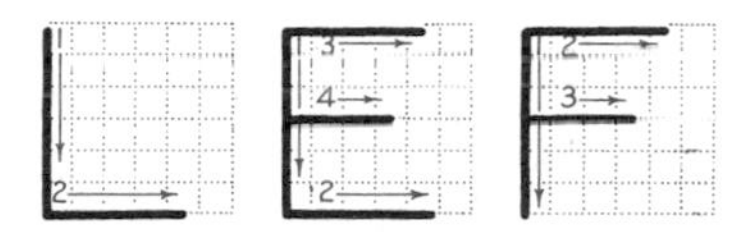

**FIG. 7.** The *L-E-F* group. Note the successive order of strokes.

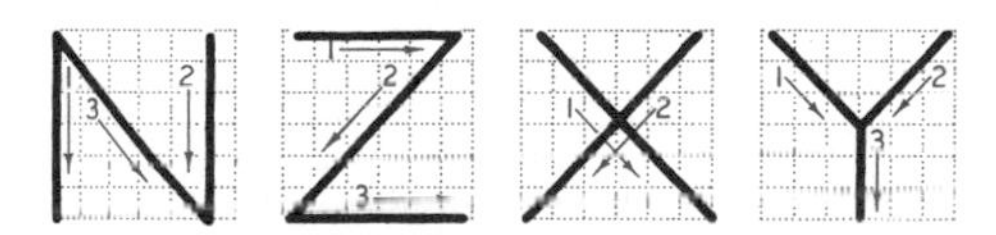

**FIG. 8.** The *N-Z-X-Y* group. Note that *Z* and *X* are smaller at the top than at the bottom, in accordance with the rule of stability.

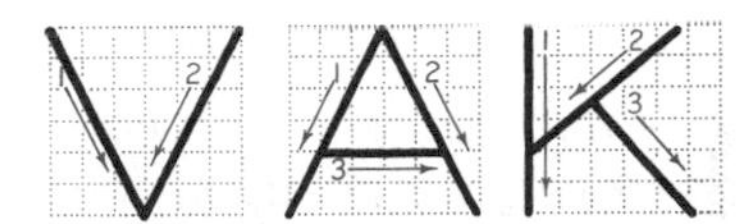

**FIG. 9.** The *V-A-K* group. The horizontal of *A* is one-third from the bottom; the second and third strokes of *K* are perpendicular to each other.

### 8. Order of Strokes.

In the following figures an alphabet of slightly extended vertical capitals has been arranged in family groups. Study the shape of each letter, with the order and direction of the strokes forming it, and practice it until its form and construction are perfectly familiar. Practice it first in pencil to large size, perhaps $\frac{3}{8}$ in. high, then to smaller size, and finally directly in ink.

To bring out the proportions of widths to heights and the subtleties in the shapes of the letters, they are shown against a square background with its sides divided into sixths. Several of the letters in this alphabet, such as *A* and *T*, fill the square; that is, they are as wide as they are high. Others, such as *H* and *D*, are approximately five spaces wide, or their width is five-sixths of their height. *These proportions must be learned visually* so well that letters of various heights can be drawn in correct proportion without hesitation.

*The I-H-T Group (Fig. 6).* The letter *I* is the foundation stroke. You may find it difficult to keep the stems vertical. If so, draw direction lines lightly an inch or so apart to aid the eye. The *H* is nearly square (five-sixths wide), and in accordance with the rule of stability, the cross-bar is just above the center. The top of the *T* is drawn first to the full width of the square, and the stem is started accurately as its middle point.

*The L-E-F Group (Fig. 7).* The *L* is made in two strokes. The first two strokes of the *E* are the same as for the *L*; the third, or upper, stroke is slightly shorter than the lower; and the last stroke is two-thirds as long as the lower and just above the middle. *F* has the same proportions as *E*.

*The N-Z-X-Y Group (Fig. 8).* The parallel sides of *N* are generally drawn first, but some prefer to make the strokes in consecutive order. *Z* and *X* are both started inside the width of the square on top and run to full width at the bottom. This throws the crossing point of the *X* slightly above the center. The junction of the *Y* strokes is at the center.

*The V-A-K Group (Fig. 9).* *V* is the same width as *A*, the full breadth of the square. The *A* bridge is one-third up from the bottom. The second stroke of *K* strikes the stem one-third up from the bottom; the third stroke branches from it in a direction starting from the top of the stem.

*The M-W Group* (*Fig. 10*). These are the widest letters. *M* may be made in consecutive strokes or by drawing the two vertical strokes first, as with the *N*. *W* is formed of two narrow *V*'s, each two-thirds of the square in width. Note that with all the pointed letters the width at the point is the width of the stroke.

*The O-Q-C-G Group* (*Fig. 11*). In this extended alphabet the letters of the *O* family are made as full circles. The *O* is made in two strokes, the left side a longer arc than the right, as the right side is harder to draw. Make the kern of the *Q* straight. A large-size *C* and *G* can be made more accurately with an extra stroke at the top, whereas in smaller letters the curve is made in one stroke (Fig. 19). Note that the bar on the *G* is halfway up and does not extend past the vertical stroke.

*The D-U-J Group* (*Fig. 12*). The top and bottom strokes of *D* must be horizontal. Failure to observe this is a common fault with beginners. In large letters *U* is formed by two parallel strokes, to which the bottom stroke is added; in smaller letters, it may be made in two strokes curved to meet at the bottom. *J* has the same construction as *U*, with the first stroke omitted.

*The P-R-B Group* (*Fig. 13*). With *P*, *R*, and *B*, the number of strokes depends upon the size of the letter. For large letters the horizontal lines are started and the curves added, but for smaller letters only one stroke for each lobe is needed. The middle lines of *P* and *R* are on the center line; that of *B* observes the rule of stability.

*The S-8-3 Group* (*Fig. 14*). The *S*, *8*, and *3* are closely related in form, and the rule of stability must be observed carefully. For a large *S*, three strokes are used; for a smaller one, two strokes; and for a very small size, one stroke only is best. The *8* may be made on the *S* construction in three strokes, or in "head and body" in four strokes. A perfect *3* can be finished into an *8*.

*The 0-6-9 Group* (*Fig. 15*). The cipher is an ellipse five-sixths the width of the letter *O*. The backbones of the *6* and *9* have the same curve as the cipher, and the lobes are slightly less than two-thirds the height of the figure.

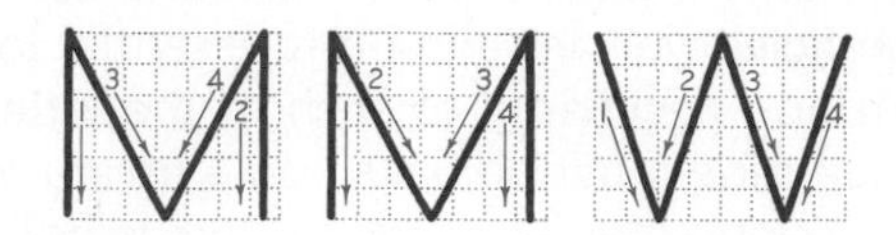

**FIG. 10.** The *M-W* group. *M* is one-twelfth wider than it is high; *W* is one-third wider than it is high.

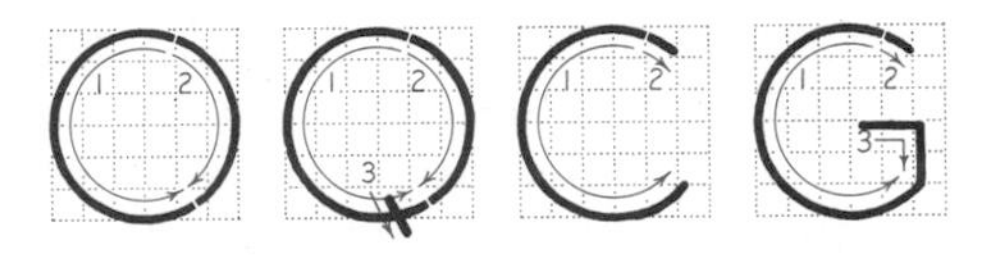

**FIG. 11.** The *O-Q-C-G* group. All are based on the circle.

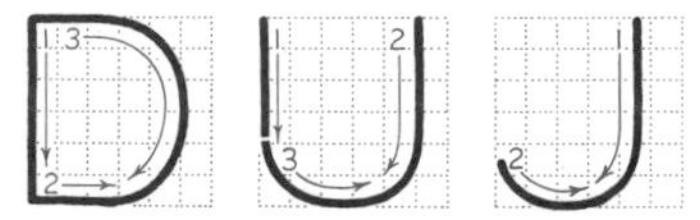

**FIG. 12.** The *D-U-J* group. These are made with combinations of straight and curved strokes.

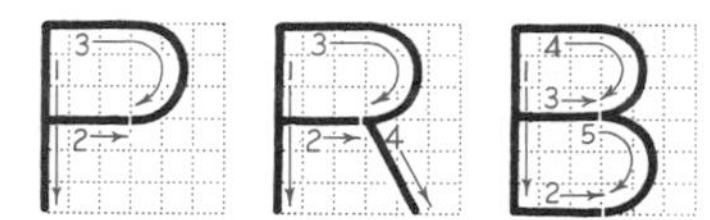

**FIG. 13.** The *P-R-B* group. Note the rule of stability with regard to *R* and *B*.

*The 2-5-7-& Group* (*Fig. 16*). The secret in making the *2* lies in getting the reverse curve to cross the center of the space. The bottom of *2* and the tops of *5* and *7* should be horizontal straight lines. The second stroke of *7* terminates directly below the middle of the top stroke. Its stiffness is relieved by curving it slightly at the lower end. The ampersand (&) is made in three strokes for large letters and two for smaller ones and must be carefully balanced.

*The Fraction Group* (*Fig. 17*). Fractions are always made with horizontal bar. Integers are the same

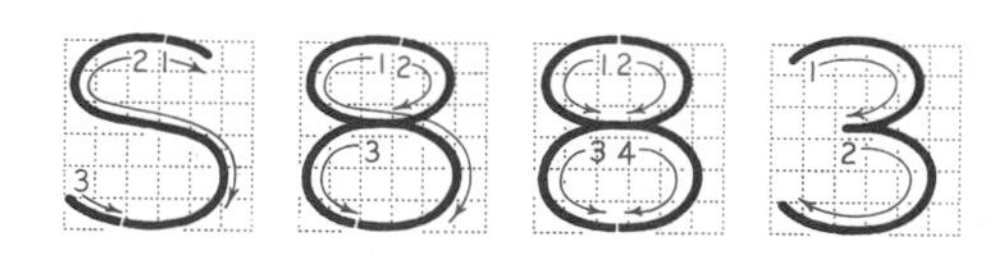

**FIG. 14.** The *S-8-3* group. A perfect *S* and *3* can be completed to a perfect *8*.

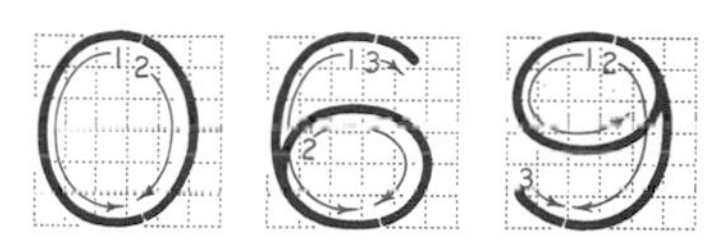

**FIG. 15.** The *0-6-9* group. The width is five-sixths of the height.

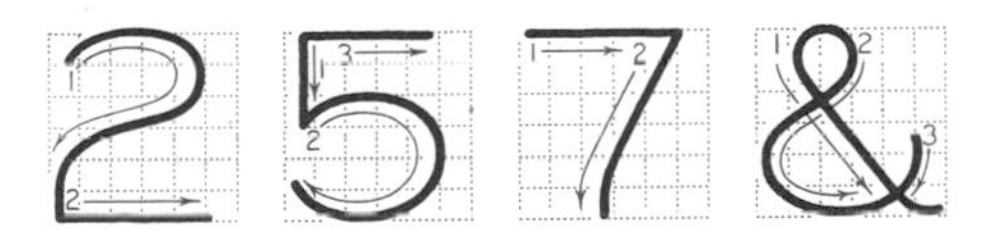

**FIG. 16.** The *2-5-7-&* group. Note the rule of stability. The width is five-sixths of the height.

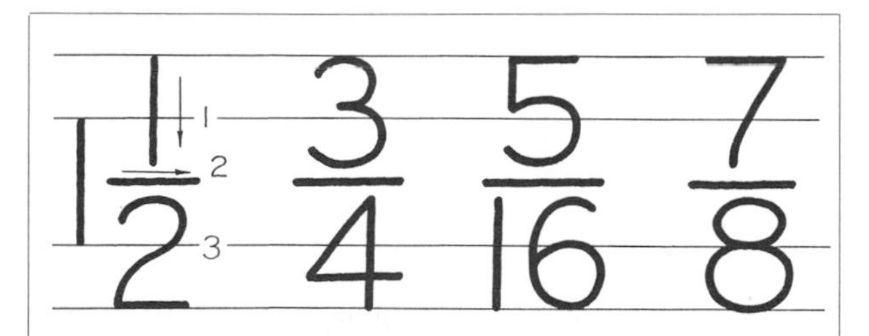

**FIG. 17.** Fractions. The total height of a fraction is twice that of the integer.

height as capitals. The total fraction height is best made twice the height of the integer. The numerator and denominator will be about three-fourths the height of the integer. Be careful to leave a clear space above and below the horizontal bar. Guide lines for fractions are easily obtained with lettering instruments by using the set of uniformly spaced holes or by drawing the integer height above and below the center, the position of the horizontal bar.

**FIG. 18.** Basic forms for lower-case letters. For standard letters, the waist-line height is two-thirds of capital height; capital line and drop line are therefore one-third above and one-third below the body of the letter.

## 9. Vertical Lower-case Letters.

The single-stroke, vertical lower-case letter is not commonly used on machine drawings but is used extensively in map drawing. It is the standard letter for hypsography in government topographic drawing. The bodies are made two-thirds the height of the capitals with the ascenders extending to the capital line and the descenders dropping the same distance below. The basic form of the letter is the combination of a circle and a straight line (Fig. 18). The alphabet, with some alternate shapes, is shown in Fig. 19, which also gives the capitals in alphabetic order.

## 10. Single-stroke Inclined Capitals.

Many draftsmen use the inclined, or slant, letter in preference to the upright. The order and direction of strokes are the same as in the vertical form.

After ruling the guide lines, draw slanting "direction lines" across the lettering area to aid the eye in keeping the slope uniform. These lines may be drawn with a special lettering triangle of about $67\frac{1}{2}°$, or the slope of 2 to 5 may be fixed on the paper by marking two units on a horizontal line and five on a vertical line and using T square and triangle as shown in Fig. 20. The Braddock-Rowe triangle and the Ames instrument (Fig. 2*a* and *b*) both provide for the drawing of slope lines. The form taken by the rounded letters when inclined is illustrated in Fig. 21, which shows that curves are sharp in all upper right-hand and lower left-hand corners and flattened in the other two corners. Take particular care with the letters that have sloping sides, such

**FIG. 19.** Single-stroke vertical capitals and lower case. Note the alternate strokes for small-size capitals and the alternate shapes for lower-case *a*, *g*, and *y*.

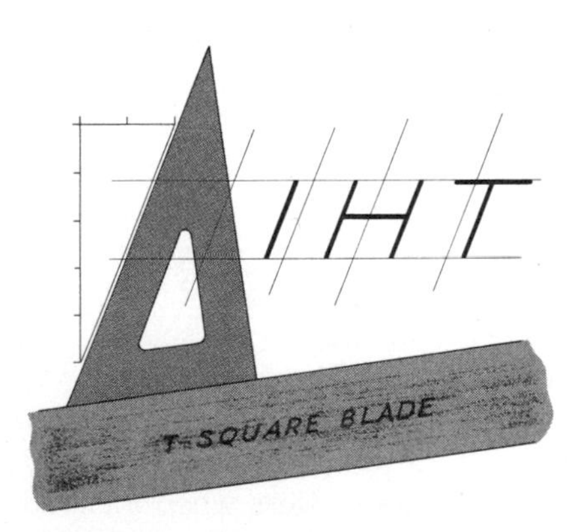


**FIG. 20.** Slope guide lines. The standard slope angle of $67\frac{1}{2}°$ (see Fig. 2*a* and *b*) can be approximated by a slope of 2 to 5.

**FIG. 21.** Form of curved-stroke inclined capitals. The basic shape is elliptical.

**FIG. 22.** Form of straight-stroke inclined capitals. The "center line" must be a slope line.

FIG. 23. Single-stroke inclined capitals and lower case. Note the alternate strokes for small-size capitals and the alternate shapes for the lower case *a*, *g*, and *y*.

as *A*, *V*, and *W*. The sloping sides of these letters must be drawn so that they appear to balance about a slope guide line passing through each vertex, as in Fig. 22. The alphabet is given in Fig. 23. Study the shape of each letter carefully.

Professional appearance in lettering is due to three things: (1) keeping to a uniform slope, (2) having the letters full and well shaped, and (3) keeping them close together. The beginner invariably cramps the individual letters and spaces them too far apart.

**11. Single-stroke, Inclined Lower-case Letters.**
The inclined lower-case letters (Fig. 23) have bodies two-thirds the height of the capitals with the ascenders extending to the capital line and the descenders dropping the same distance below the base line. Among older engineers, particularly civil engineers, this letter is generally known as the "Reinhardt letter," in honor of Charles W. Reinhardt, who first systematized its construction. It is legible and effective and, after its swing has been mastered, can be made rapidly. The lower-case letter is suitable for notes and statements on drawings because it is much more easily read than all capitals, since we read words by the word shapes, and it can be done much faster.

All the letters of the Reinhardt alphabet are based on two elements, the straight line and the ellipse, and have no unnecessary hooks or appendages. They may be divided into four groups, as shown in

Figs. 24 to 27. The dots of *i* and *j* and the top of the *t* are on the "*t* line," halfway between the waist line and the capital line. The loop letters are made with an ellipse whose long axis is inclined about 45° in combination with a straight line. In lettering rapidly, the ellipse tends to assume a pumpkin-seed form; guard against this.

The *c*, *e*, and *o* are based on an ellipse of the shape of the capitals but not inclined quite so much as the loop-letter ellipse. In rapid small work, the *o* is often made in one stroke, as are the *e*, *v*, and *w*. The *s* is similar to the capital but, except in letters more than ⅛ in. high, is made in one stroke. In the hook-letter group, note particularly the shape of the hook.

The single-stroke letter may, if necessary, be much compressed and still be clear and legible (Fig. 28). It is also used sometimes in extended form.

### 12. For Left-handers Only.

The order and direction of strokes in the preceding alphabets have been designed for right-handed persons. The principal reason that left-handers sometimes find lettering difficult is that whereas the right-hander progresses away from the body, the left-hander progresses toward the body; consequently his pencil and hand partially hide the work he has done, making it harder to join strokes and

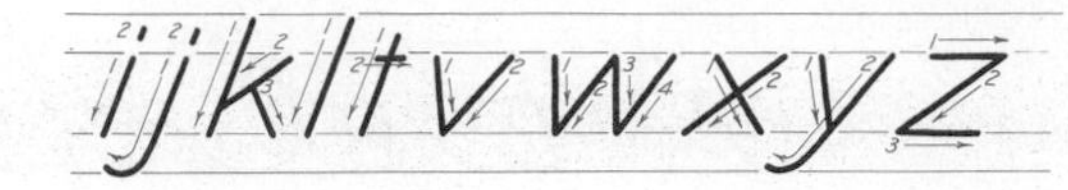


**FIG. 24.** The straight-line inclined lower-case letters. Note that the center lines of the letters follow the slope angle.

**FIG. 25.** The loop letters. Note the graceful combination of elliptical body, ascenders, and descenders.

**FIG. 26.** The ellipse letters. Their formation is basically elliptical.

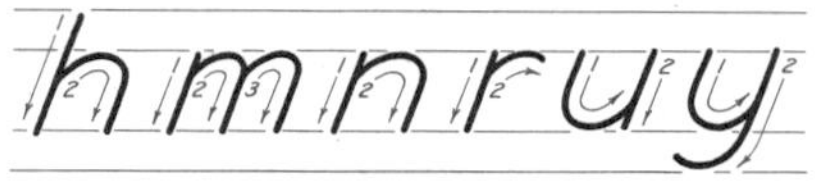

**FIG. 27.** The hook letters. They are combinations of ellipses and straight lines.

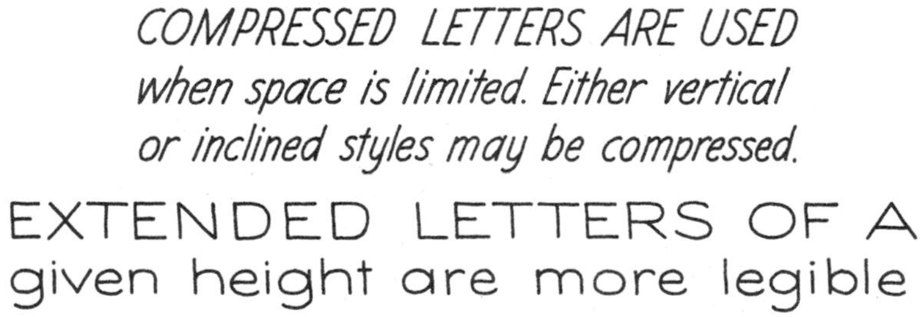

COMPRESSED LETTERS ARE USED
when space is limited. Either vertical
or inclined styles may be compressed.

EXTENDED LETTERS OF A
given height are more legible

**FIG. 28.** Compressed and extended letters. Normal width has *O* the same in width as in height and the other letters in proportion. Wider than high for *O* is extended, narrower is compressed.

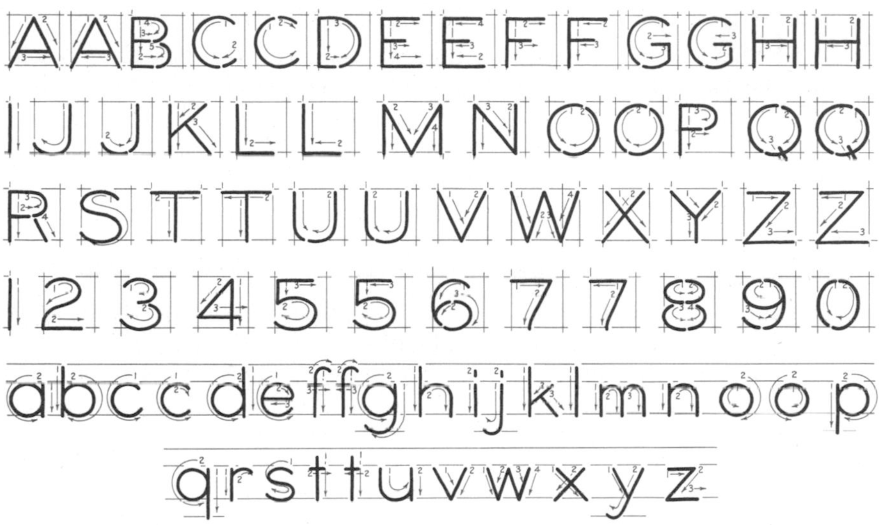

FIG. 29. Strokes for left-handers. Several alternate strokes are given. Choose the one that is most effective for you.

to preserve uniformity. Also, in the case of inclined lettering, the slope direction, instead of running toward his eye, runs off into space to the left of his body, making this style so much harder for him that he is strongly advised to *use vertical letters exclusively*.

For the natural left-hander, whose writing position is the same as a right-hander except reversed left for right, a change in the sequence of strokes of some of the letters will obviate part of the difficulty caused by interference with the line of sight. Figure 29 gives an analyzed alphabet with an alternate for some letters. In *E* the top bar is made before the bottom bar, and *M* is drawn from left to right to avoid having strokes hidden by the pencil or pen. Horizontal portions of curves are easier to make from right to left; hence the starting points for *O*, *Q*, *C*, *G*, and *U* differ from the standard right-hand stroking. *S* is the perfect letter for the left-hander and is best made in a single smooth stroke. The figures *6* and *9* are difficult and require extra practice. In the lower-case letters *a*, *d*, *g*, and *q*, it is better to draw the straight line before the curve even though it makes spacing a little harder.

The hook-wrist left-handed writer, who pushes his strokes from top to bottom, finds vertical lettering more difficult than does the natural left-hander. In Fig. 29, where alternate strokes are given for some of the letters, the hook-wrist writer will probably find the second stroking easier than the first. Some prefer to reverse *all* the strokes, drawing vertical strokes from bottom to top and horizontal strokes from right to left.

By way of encouragement it may be said that many left-handed draftsmen letter beautifully.

## 13. Composition.

Composition in lettering has to do with the selection, arrangement, and spacing of appropriate styles and sizes of letters. On engineering drawings the selection of the style is practically limited to vertical or inclined single-stroke lettering, so composition here means arrangement into pleasing and legible form. After the shapes and strokes of the individual letters

LETTERING
LETTERING

FIG. 30. Background areas. Equal areas between letters produce spacing that is visually uniform.

have been learned, the entire practice should be on composition into words and sentences, since proper spacing of letters and words does more for the appearance of a block of lettering than the forms of the letters themselves. Letters in words are not spaced at a uniform distance from each other but are arranged so that the areas of white space (the irregular backgrounds between the letters) are approximately equal, making the spacing *appear* approximately uniform. Figure 30 illustrates these background shapes. Each letter is spaced with reference to its shape and the shape of the letter preceding it. Thus adjacent letters with straight sides would be spaced farther apart than those with curved sides. Sometimes combinations such as *LT* or *AV* may even overlap. Definite rules for spacing are not successful; it is a matter for the draftsman's judgment and sense of design. Figure 31 illustrates word composition. The sizes of letters to use in any particular case can be determined better by sketching them lightly than by judging from the guide lines alone. A finished line of letters always looks larger than the guide lines indicate. Avoid the use of a coarse pen for small sizes and one that makes thin wiry lines for large sizes. When capitals and small capitals are used, the height of the small capitals should be about four-fifths that of the capitals.

In spacing words, a good principle is to leave the space that would be taken by an assumed letter *I* connecting the two words into one, as in Fig. 32. The space would never be more than the height of the letters.

The clear distance between lines may vary from $\frac{1}{2}$ to $1\frac{1}{2}$ times the height of the letter but for the sake of appearance should not be exactly the same as the letter height. The instruments in Figs. 2 and 3 provide spacing that is two-thirds of the letter height. Paragraphs should always be indented.

### 14. Titles.

The most important problem in lettering composition is the design of titles. Every drawing has a descriptive title that is either all hand-lettered or filled in on a printed form. It gives necessary information concerning the drawing, and the information that is needed will vary with the different kinds of drawings.

The usual form of lettered title is the *symmetrical title,* which is balanced or "justified" on a vertical center line and designed with an elliptical or oval outline. Sometimes the wording necessitates a pyramid or inverted-pyramid ("bag") form. Figure 33 illustrates several shapes in which titles can be composed. The lower right-hand corner of the sheet is, from long custom and because of convenience in filing, the usual location for the title, and in laying

COMPOSITION IN LETTERING
REQUIRES CAREFUL SPACING, NOT ONLY
OF LETTERS BUT OF WORDS AND LINES

FIG. 31. Word composition. Careful spacing of letters and words and the proper emphasis of size and weight are important.

WORDS I SPACED I BY I SKETCHING I AN I I I BETWEEN

WORDS SPACED BY SKETCHING AN I BETWEEN

**FIG. 32.** Word spacing. Space words so that they read naturally and do not run together (too close) or appear as separate units (too far apart).

out a drawing, this corner is reserved for it. The space allowed depends on the size and purpose of the drawing. On an 11- by 17-in. working drawing, the title may be about 3 in. long.

### 15. To Draw a Title.

When the wording has been determined, write out the arrangement on a separate piece of paper as in Fig. 34 (or, better, typewrite it). Count the letters, including the word spaces, and make a mark across the middle letter or space of each line. The lines must be displayed for prominence according to their relative importance as judged from the point of view of the persons who will use the drawing. Titles are usually made in all capitals. Draw the base line for the most important line of the title and mark on it the approximate length desired. To get the letter height, divide this length by the number of letters in the line, and draw the capital line. Start at the center line, and sketch lightly the last half of the line, drawing only enough of the letters to show the space each will occupy. Lay off the length of this right half on the other side, and sketch that half, working forward or backward. When this line is satisfactory in size and spacing, draw the remainder in the same way. Study the effect, shift letters or lines if necessary, and complete in pencil. Use punctuation marks only for abbreviations.

*Scratch-paper Methods.* Sketch each line of the title separately on a piece of scratch paper, using guide lines of determined height. Find the middle point of each of these lines, fold the paper along the base line of the letters, fit the middle point to

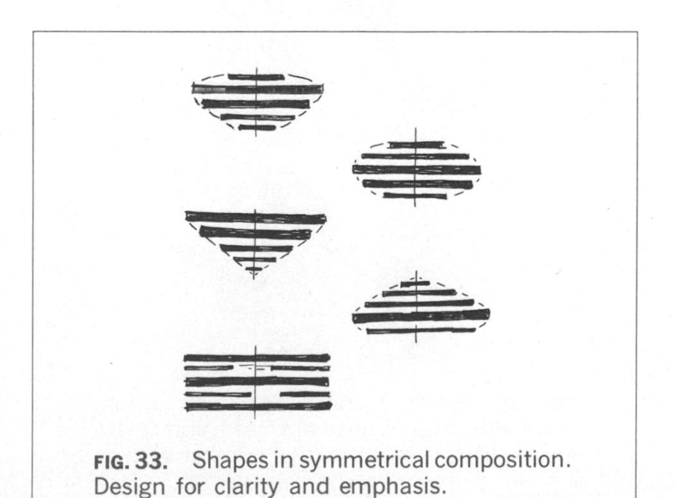

**FIG. 33.** Shapes in symmetrical composition. Design for clarity and emphasis.

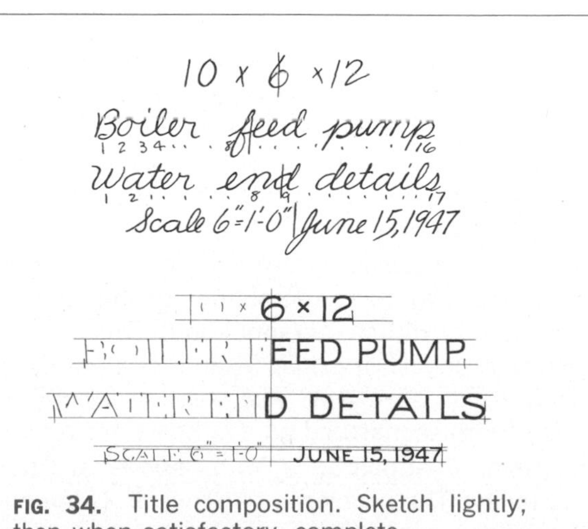


**FIG. 34.** Title composition. Sketch lightly; then when satisfactory, complete.

LEONARDT 516 F: 506 F
HUNT 512: ESTERBROOK 968
Esterbrook 1000 Spencerian No. 1
Gillott 404: Gillott 303
For very fine lines Gillott 170 and 290 or Esterbrook 356 and 355

**FIG. 35.** Pen strokes, full size. These are used principally for average-size lettering on working drawings.

the center line on the drawing, and draw the final letters directly below the sketches. *Or* draw the letters along the edge of the scratch paper, using the upper or lower edge as one of the guide lines. *Or* letter the title on scratch paper, cut it apart and adjust until satisfactory, and then trace it.

### 16. Notes and Filled-in Titles.

By far the principal use of lettering on engineering drawings occurs on working drawings, where the dimensions, explanatory notes, record of drawing changes, and title supplement the graphic description of the object.

Styles vary, but the trend is toward the use of capital letters exclusively. Most draftsmen are able to letter in capital letters with greater readability and fewer mistakes than in lower case. However, for some structural drawings and others where space may be at a premium, lower case is used.

On any working drawing the accuracy of the information conveyed is paramount. For this reason it is important to make the shapes of the letters precisely correct with no personal flourishes, embellishments, or peculiarities. For example, a poorly made *3* may read as an *8* or vice versa. A poorly made *5* may be misconstrued as a *6*. Strive for a smooth, professional appearance with, above all, perfect readability. The problems of Group 5 at the end of this chapter give practice in the lettering of notes.

### 17. Lettering Pens.

There are many steel writing pens that are adaptable to or made especially for lettering. The size of the strokes of a few popular ones is shown in full size in Fig. 35. Several special pens made in sets of graded sizes have been designed for single-stroke lettering; among them are those illustrated in Fig. 36. These are particularly useful for large work. The ink-holding reservoir of the Henry tank pen (Fig. 37) assists materially in maintaining uniform weight of line. A similar device can be made by bending a

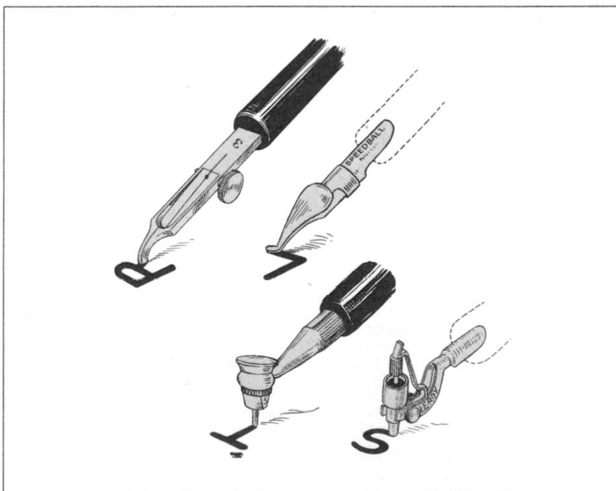

**FIG. 36.** Barch-Payzant, Speedball, Edco, and Leroy pens. These are used principally for displays, large titles, and number blocks.

brass strip from a paper fastener, a piece of annealed watch spring, or—perhaps best—a strip cut from a piece of shim brass into the shape shown in Fig. 38 and inserting it in the penholder so that the curved end just touches the pen nib.

To remove the oil film, always wet a new pen and wipe it thoroughly before using. A lettering pen well broken in by use is worth much more than a new one. It should be kept with care and never lent. A pen that has been dipped into writing ink should never be put into drawing ink. When in use, a pen should be wiped clean frequently with a cloth penwiper.

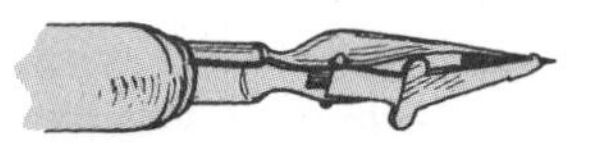

**FIG. 37.** Henry tank pen. The reservoir is designed to prevent spreading of the nibs.

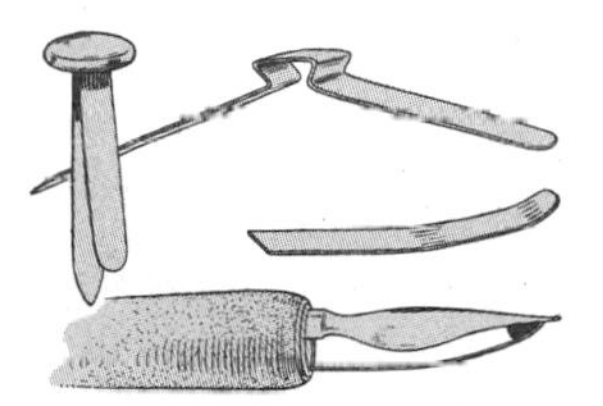

**FIG. 38.** Ink holder. This helps to avoid heavy flow of ink when the pen is full.

EHMNWTZ

**FIG. 39.** Too much ink. Fill the pen sparingly and often.

**FIG. 40.** Holding the pen. Note that the thumb, forefinger, and second finger make a three-point support.

### 18. Using the Pen.

Use a penholder with cork grip (the small size) and set the pen in it firmly. Many prefer to ink a pen with the quill filler, touching the quill to the underside of the pen point, rather than dipping the pen in the ink bottle. If the pen is dipped, the surplus ink should be shaken back into the bottle or the pen touched against the neck of the bottle as it is withdrawn. Lettering with too much ink on the pen gives the results shown in Fig. 39.

In lettering, the penholder is held in the fingers firmly but without pinching, in the position shown in Fig. 40. The strokes of the letters are made with a steady, even motion and a slight, uniform pressure on the paper that will not spread the nibs of the pen.

### 19. Other Lettering Styles.

This chapter has been devoted entirely to the commercial Gothic letter because this style predominates in all types of commercial drawings. However, other letter styles are used extensively on architectural drawings and on maps, in design and for display, and for decorative letters on commercial products. Some traditional letter styles and their uses are given in the Appendix.

# PROBLEMS

## Group 1. Single-stroke Vertical Capitals.

**1.** Large letters in pencil for careful study of the shapes of the individual letters. Starting $\frac{9}{16}$ in. from the top border, draw guide lines for five lines of $\frac{3}{8}$-in. letters. Draw each of the straight-line letters *I, H, T, L, E, F, N, Z, Y, V, A, M, W,* and *X,* four times in pencil only, studying carefully Figs. 6 to 10. The illustration is a full-size reproduction of a corner of this exercise.

**2.** Same as Prob. 1 for the curved-line letters *O, Q, C, G, D, U, J, B, P, R,* and *S.* Study Figs. 11 to 14.

**3.** Same as Prob. 1 for the figures *3, 8, 6, 9, 2, 5,* $\frac{1}{2}$, $\frac{3}{4}$, $\frac{5}{8}$, $\frac{7}{16}$, and $\frac{9}{32}$. Study Figs. 14 to 17.

**4.** Composition. Same layout as for Prob. 1. Read Sec. 13 on composition; then letter the following five lines in pencil: *(a)* WORD COMPOSITION, *(b)* TOPOGRAPHIC SURVEY, *(c)* TOOLS AND EQUIPMENT, *(d)* BRONZE BUSHING, *(e)* JACK-RAFTER DETAIL.

**5.** Quarter-inch vertical letters in pencil and ink. Starting $\frac{1}{4}$ in. from the top, draw guide lines for nine lines of $\frac{1}{4}$-in. letters. In the group order given, draw each letter four times in pencil and then four times directly in ink, as shown in the illustration.

**6.** Composition. Make a three-line design of the quotation from Benjamin Lamme on the Lamme medals: "THE ENGINEER VIEWS HOPEFULLY THE HITHERTO UNATTAINABLE."

**7.** One-eighth-inch vertical letters. Starting $\frac{1}{4}$ in. from the top, draw guide lines for 18 lines of $\frac{1}{8}$-in. letters. Make each letter and numeral eight times directly in ink. Fill the remaining lines with a portion of Sec. 13 on composition.

**8.** Composition. Letter the following definition: "Engineering is the art and science of directing and controlling the forces and utilizing the materials of nature for the benefit of man. All engineering involves the organization of human effort to attain these ends. It also involves an appraisal of the social and economic benefits of these activities."

**PROB. 1.**

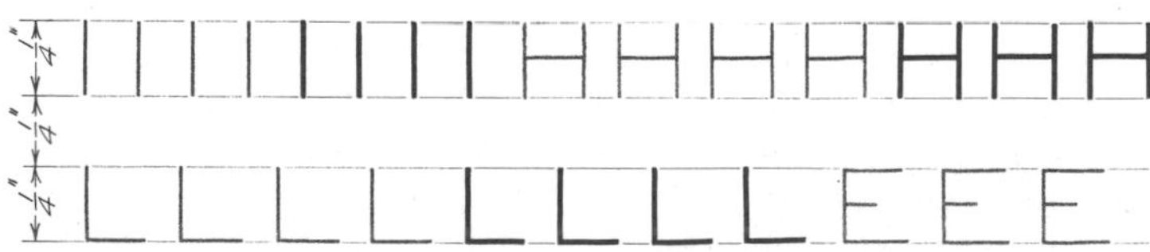

**PROB. 5.**

## Group 2. Single-stroke Inclined Capitals.

**9 to 16.** Same spacing and specifications as for Probs. 1 to 8, but for inclined letters. Study Sec. 10 and Figs. 20 to 23.

## Group 3. Single-stroke Inclined Lower Case.

**17.** Large letters in pencil for use with $\frac{3}{8}$-in. capitals. The bodies are $\frac{1}{4}$ in., the ascenders $\frac{1}{8}$ in. above, and the descenders $\frac{1}{8}$ in. below. Starting $\frac{3}{8}$ in. from the top, draw guide lines for seven lines of letters. This can be done quickly by spacing $\frac{1}{8}$ in. uniformly down the sheet and bracketing capital and base lines. Make each letter of the alphabet four times in pencil only. Study Figs. 23 to 27.

**18.** Lower case for $\frac{3}{16}$-in. capitals. Starting $\frac{1}{2}$ in. from the top, draw capital, waist, and base lines for 13 lines of letters (Braddock or Ames no. 6 spacing). Make each letter six times in pencil and then six times in ink.

**19.** Composition. Same spacing as Prob. 18. Letter the opening paragraph of this chapter.

**20.** Letter Sec. 4.

## Group 4. Titles.

**21.** Design a title for the assembly drawing of a rear axle, drawn to the scale of 6 in. = 1 ft, as made by the Chevrolet Motor Co., Detroit. The number of the drawing is C82746. Space allowed is 3 by 5 in.

**22.** Design a title for the front elevation of a powerhouse, drawn to $\frac{1}{4}$-in. scale by Burton Grant, Architect, for the Citizens Power and Light Company of Punxsutawney, Pennsylvania.

## Group 5. Notes.

**23 to 25.** Vertical capitals. Copy each note in no. 6 (Ames or Braddock) spacing; then in no. 4 spacing.

**26 to 28.** Inclined capitals. Copy in no. 5 or no. 4 spacing.

**29 to 31.** Vertical lower case. Copy in no. 5 spacing.

**32 to 34.** Inclined lower case. Copy in no. 4 spacing.

**35.** For extra practice, select any of Probs. 23 to 34 and letter in a chosen style and size.

PAINT WITH METALLIC SEALER AND TWO COATS LACQUER AS PER CLIENT COLOR ORDER.

PROB. 23.

THIS PRINT IS AMERICAN THIRD-ANGLE PROJECTION

PROB. 24.

TO BE REMOVED AFTER MACHINING AND BEFORE ASSEMBLY.

PROB. 25.

*CUTOFF BURR MUST NOT PROJECT BEYOND THIS SURFACE.*

PROB. 26.

*ALTERNATE MATERIAL:*
*1ST. RED BRASS 85 % CU.*
*2ND. COMM. BRASS 90 % CU.*
*3RD. COMM. BRASS 95 % CU.*

PROB. 27.

*THIS HOLE IN PIECE NO. 821 ONLY. REMOVE BURR ON UPPER SIDE.*

PROB. 28.

Flatten ear on this side to make piece No. 51367. See detail.

PROB. 29.

Deburring slot must be centered on 0.625 hole within ±0.001.

PROB. 30.

Pivot point for high pressure bellows.

PROB. 31.

*Material: #19 Ga (0.024) C R Steel. Temper to Rockwell B-40 to 65.*

PROB. 32.

*This length varies from 0.245 to 0.627. See table (A) below.*

PROB. 33.

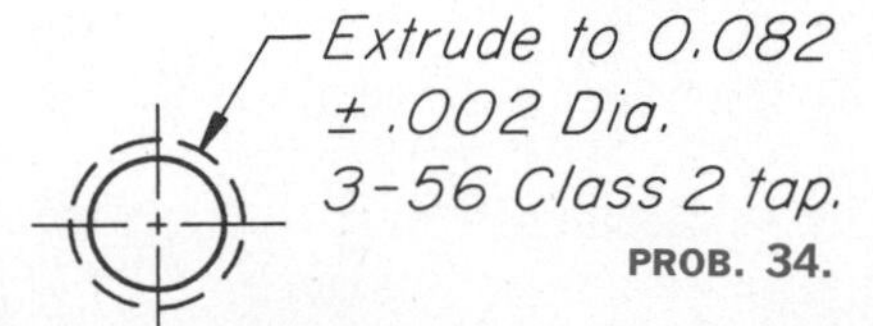

*Extrude to 0.082 ± .002 Dia. 3-56 Class 2 tap.*

PROB. 34.

This Chapter Presents:

Theory
Orthographic Views
The Six Principal Views
Combinations of Views
"Alternate-position" Views
The Three Space Dimensions
Relationship of Planes, View Directions, and Space Dimensions
Classification of Surfaces and Lines
Representation of Lines
Hidden Features
Center Lines
Precedence of Lines
Exercises in Projection
Selection of Views
Projection of Views
Order of Drawing
Freehand Drawing
Freehand Techniques
Reading Orthographic Drawings
Reading Lines, Areas
Learning to Read by Sketching and Modeling
Exercises
Problems

# 5

# ORTHOGRAPHIC DRAWING AND SKETCHING

**Orthographic Drawing and Sketching Is the Basic Graphic Form of Representation for Design and Construction Drawings**

# THEORY

## 1. Methods of Projection—Classification.

The previous chapters have been preparatory to the real subject of graphics as a language. Typically, engineers design and develop machines and structures and direct their construction. Furthermore, to design and then *communicate* every detail to manufacturing groups, descriptions must be prepared that show every aspect of the *shape* and *size* of each part and of the complete machine or structure. Because of this necessity, graphics is the fundamental method of communication. Only as a supplement, for notes and specifications, is the word language used.

In this chapter and Chap. 6 we are concerned with the methods of describing shape. Chapter 12 discusses size description.

Shape is described by projection, that is, by the process of causing an image to be formed by rays of sight taken in a particular direction from an object to a picture plane.[1] Methods of projection vary according to the direction in which the rays of sight are taken to the plane. When the rays are perpendicular to the plane, the projective method is *orthographic.* If the rays are at an angle to the plane, the projective method is called *oblique.* Rays taken to a particular station point result in *perspective projection.* By the methods of perspective the object is represented as it would appear to the eye.

Projective theory is the basis of background information necessary for shape representation. In graphics, two fundamental methods of *shape representation* are used:

(1). *Orthographic views,* consisting of a set of two or more separate views of an object taken from different directions, generally at right angles to each other and arranged relative to each other in a definite way. Each of the views shows the shape of the object for a particular view direction and collectively the views describe the object completely. Orthographic projection *only* is used.

[1]Only in *projective geometry,* a highly theoretical graphic subject, are surfaces other than a plane used.

(2). *Pictorial views,* in which the object is oriented behind and projected upon a single plane. Either orthographic, oblique, or perspective projection is used.

Since orthographic views provide a means of describing the *exact shape* of any material object, they are used for the great bulk of engineering work.

## 2. Theory of Orthographic Projection.

Let us suppose that a transparent plane has been set up between an object and the station point of an observer's eye (Fig. 1). The intersection of this plane with the rays formed by lines of sight from the eye to all points of the object would give a picture that is practically the same as the image formed in the eye of the observer. This is perspective projection.

If the observer would then walk backward from the station point until he reached a theoretically *infinite* distance, the rays formed by lines of sight from his eye to the object would grow longer and finally become infinite in length, parallel to each other, and perpendicular to the picture plane. The image so formed on the picture plane is what is known as "orthographic projection." See Fig. 2.

## 3. Definition.

Basically, orthographic[2] projection could be defined as any single projection made by dropping perpendiculars to a plane. However, it has been accepted through long usage to mean the combination of two or more such views, hence the following definition: *Orthographic projection is the method of representing the exact shape of an object by dropping perpendiculars from two or more sides of the object to planes, generally at right angles to each other; collectively, the views on these planes describe the object completely.* (The term "orthogonal"[3] is sometimes used for this system of drawing.)

## 4. Orthographic Views.

The rays from the picture plane to infinity may be discarded and the picture, or "view," thought of as

[2]Literally, "right writing."
[3]Meaning right-angled.

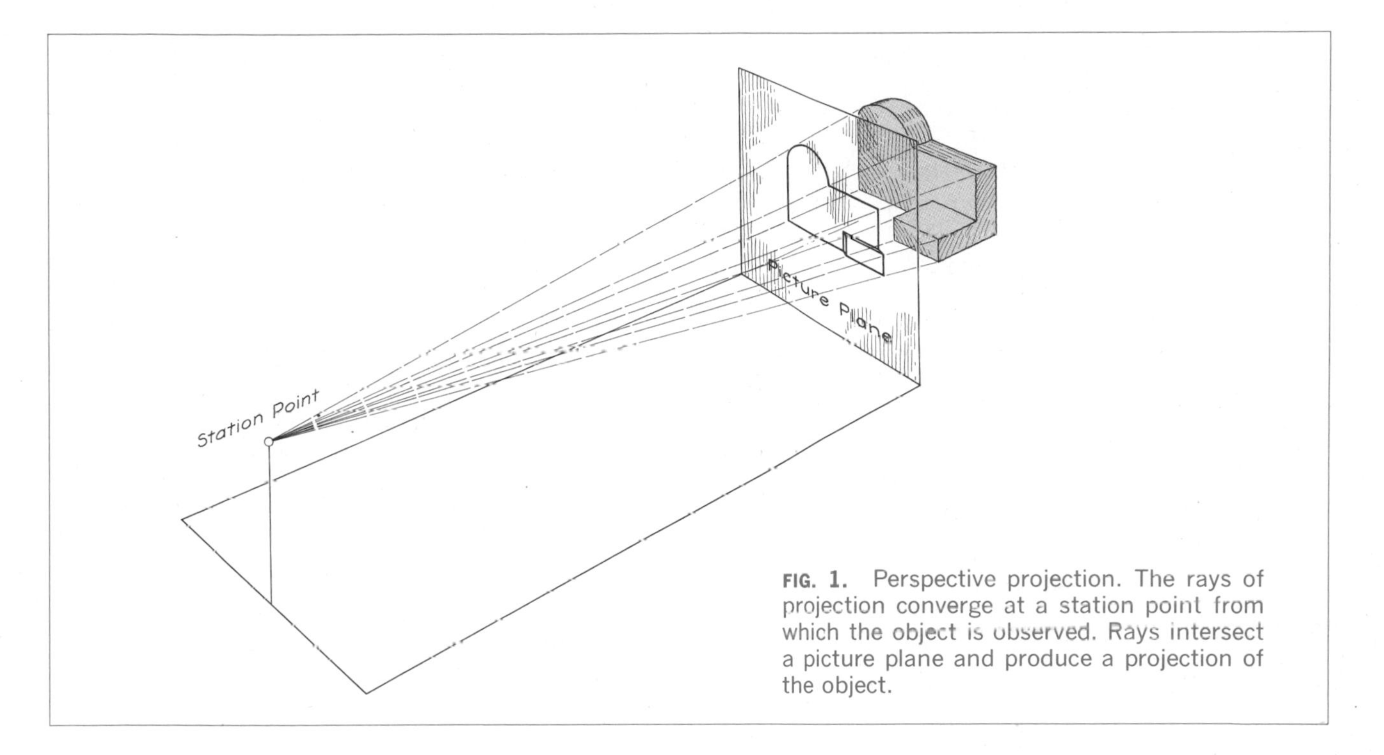


**FIG. 1.** Perspective projection. The rays of projection converge at a station point from which the object is observed. Rays intersect a picture plane and produce a projection of the object.

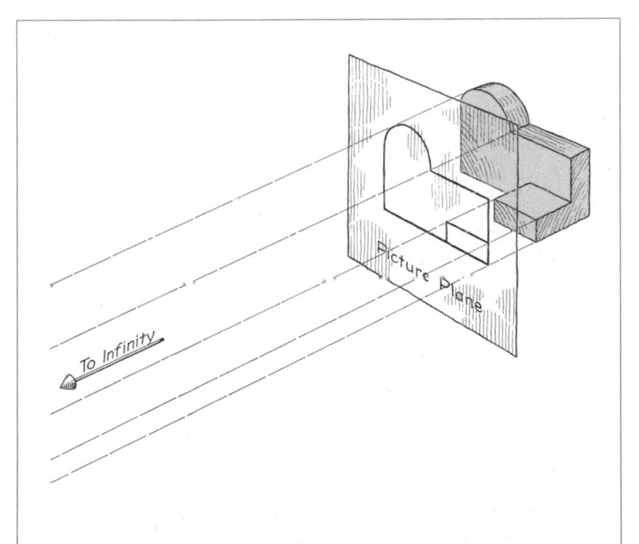


**FIG. 2.** Orthographic projection. The station point is at infinity, making the rays parallel to each other. The rays are perpendicular to the picture plane.

being found by extending perpendiculars to the plane from all points of the object, as in Fig. 3. This picture, or projection on a frontal plane, shows the shape of the object when viewed from the front, but it does not tell the shape or distance from front to rear. Accordingly, more than one projection is required to describe the object.

In addition to the frontal plane, imagine another transparent plane placed horizontally above the object, as in Fig. 4. The projection on this plane, found by extending perpendiculars to it from the object, will give the appearance of the object as if viewed from directly above and will show the distance from front to rear. If this horizontal plane is now rotated into coincidence with the frontal plane, as in Fig. 5, the two views of the object will be in the same plane, as if on a sheet of paper. Now imagine a third plane, perpendicular to the first two (Fig. 6). This plane is called a "profile plane," and a third view can be projected on it. This view shows the shape of the object when viewed from the side and the

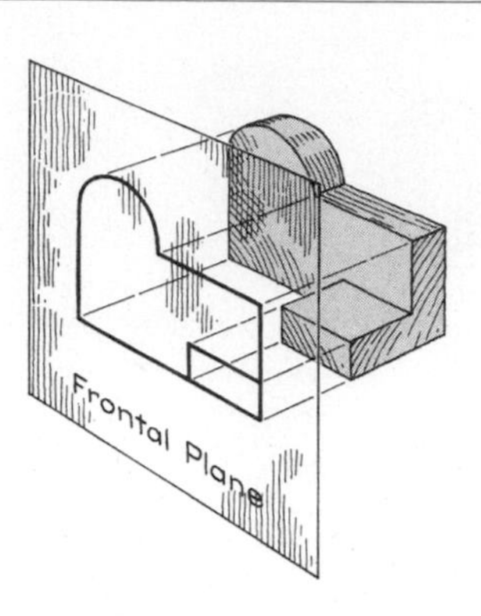


FIG. 3. The frontal plane of projection. This produces the front view of the object.

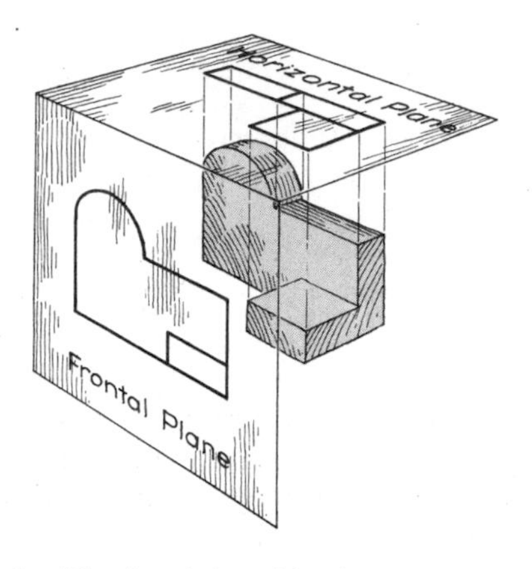


FIG. 4. The frontal and horizontal planes of projection. Projection on the horizontal plane produces the top view of the object. Frontal and horizontal planes are perpendicular to each other.

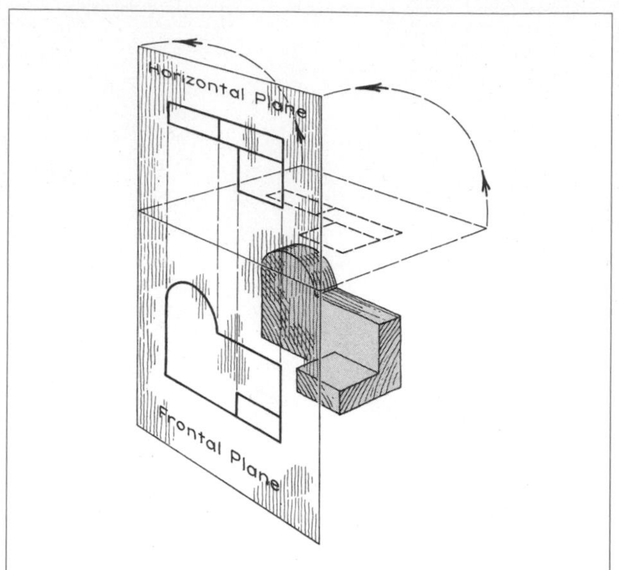


FIG. 5. The horizontal plane rotated into the same plane as the frontal plane. This makes it possible to draw two views of the object on a plane, the drawing paper.

distance from bottom to top and front to rear. The horizontal and profile planes are shown rotated into the same plane as the frontal plane (again thought of as the plane of the drawing paper) in Fig. 7. Thus related in the same plane, they give correctly the three-dimensional shape of the object.

In orthographic projection the picture planes are called "planes of projection"; and the perpendiculars, "projecting lines" or "projectors."

In looking at these theoretical projections, or views, do not think of the views as flat surfaces on the transparent planes, but try to imagine that you are looking *through* the transparent planes at the object itself.

### 5. The Six Principal Views.

Considering the matter further, we find that the object can be entirely surrounded by a set of six planes, each at right angles to the four adjacent to it, as in Fig. 8. On these planes, views can be obtained of the object as it is seen from the top, front, right side, left side, bottom, and rear.

Think now of the six sides, or planes, of the box as being opened up, as in Fig. 9, into one plane, the plane of the paper. The front is already in the plane of the paper, and the other sides are, as it were, hinged and rotated into position as shown. The projection on the frontal plane is the *front view, vertical projection,* or *front elevation;* that on the horizontal plane, the *top view, horizontal projection,* or *plan;* that on the side, or "profile," plane, the

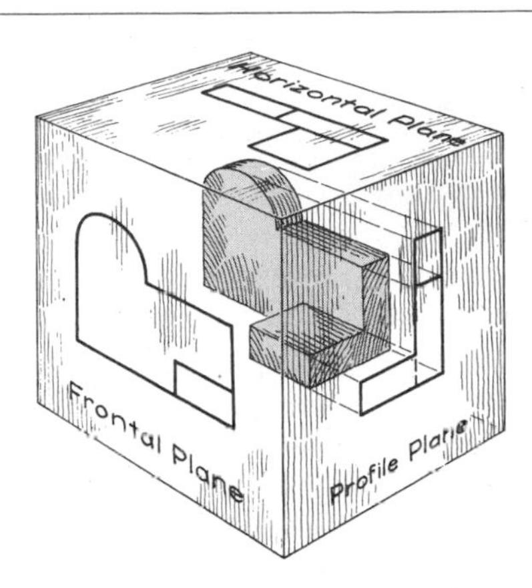


**FIG. 6.** The three planes of projection: frontal, horizontal, and profile. Each is perpendicular to the other two.

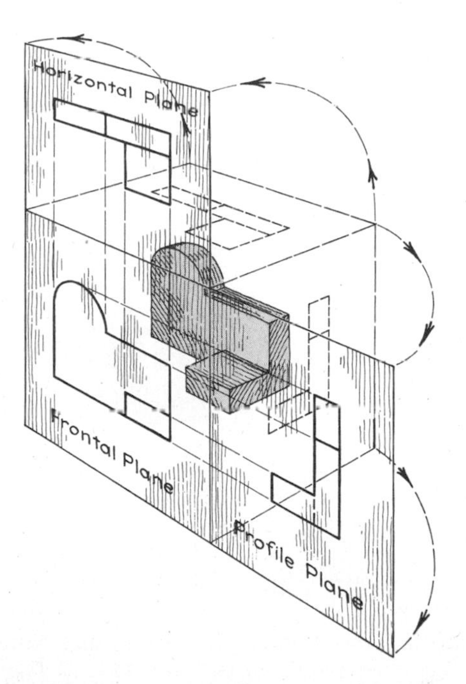


**FIG. 7.** The horizontal and profile planes rotated into the same plane as the frontal plane. This makes it possible to draw three views of the object on a plane, the drawing paper.

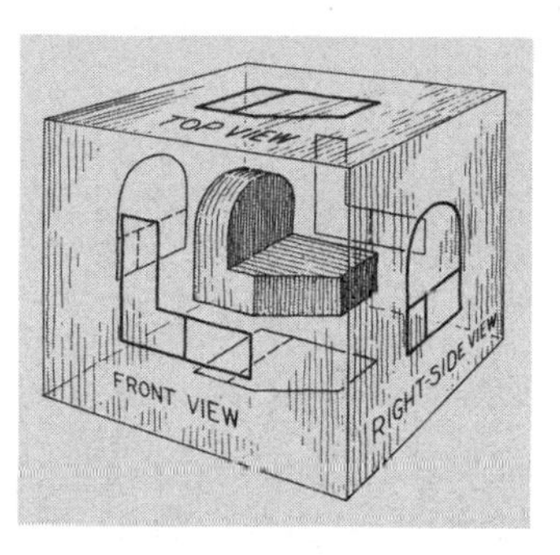


**FIG. 8.** "The transparent box." This encloses the object with another frontal plane behind, another horizontal plane below, and another profile plane to the left of the object.

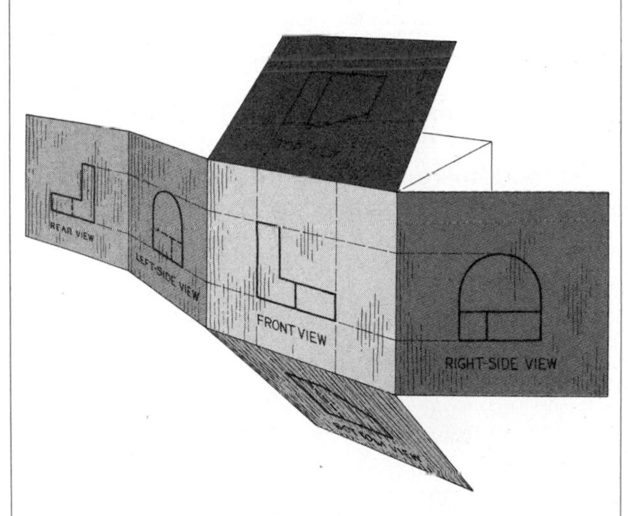


**FIG. 9.** The transparent box as it opens and all planes rotate to the plane of the frontal plane. Note that horizontal and profile planes are hinged to the frontal plane and that the "rear-view" plane is hinged to the left profile plane.

*side view, profile projection, side elevation,* or sometimes *end view* or *end elevation.* By reversing the direction of sight, a *bottom view* is obtained instead of a *top view,* or a *rear view* instead of a *front view.* In comparatively rare cases a bottom view or rear view or both may be required to show some detail

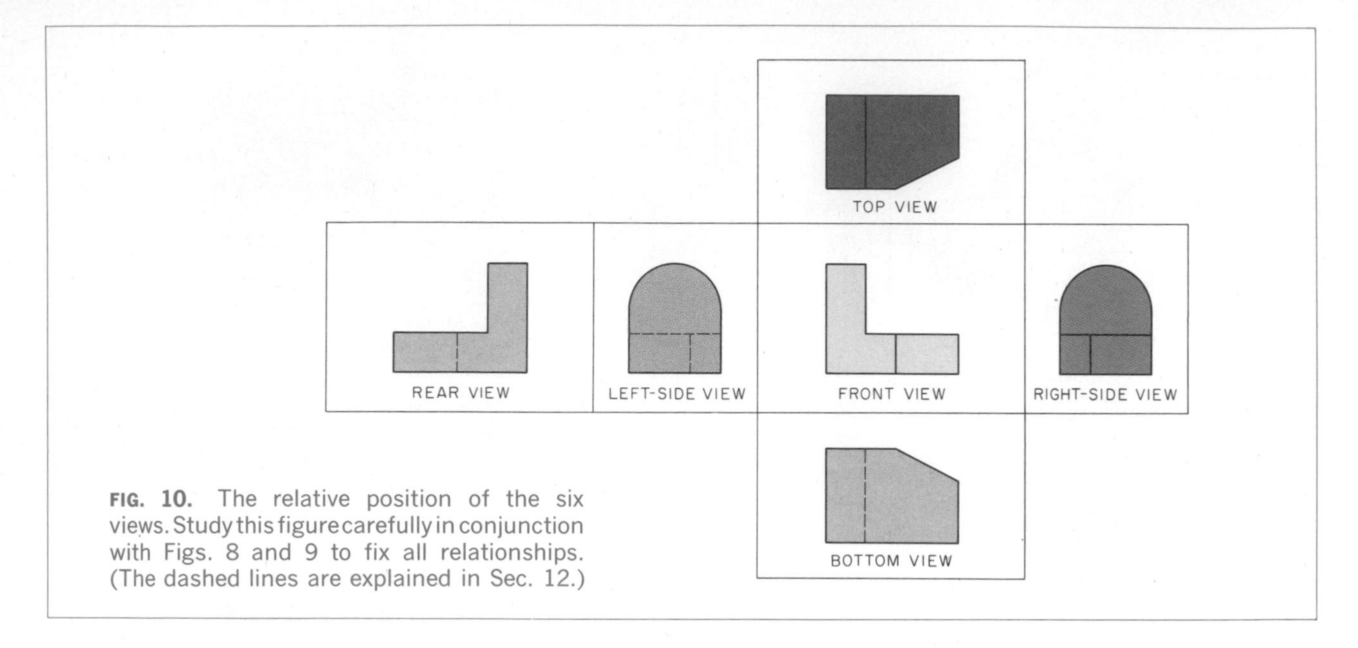


**FIG. 10.** The relative position of the six views. Study this figure carefully in conjunction with Figs. 8 and 9 to fix all relationships. (The dashed lines are explained in Sec. 12.)

of shape or construction. Figure 10 shows the relative position of the six views as set by the ASA. In actual work there is rarely an occasion when all six principal views are needed on one drawing, but no matter how many are required, their positions relative to one another are given in Fig. 10 (except as noted in Sec. 7). All these views are principal views. Each of the six views shows two of the three dimensions of height, width, and depth.

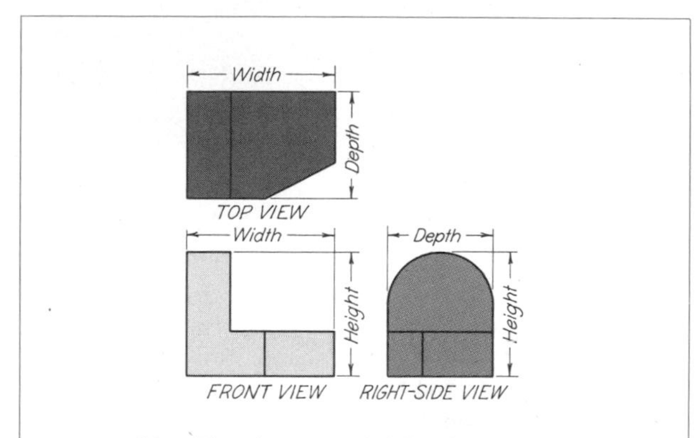


**FIG. 11.** Top, front, and right-side views. This is the most common combination. Note that the top view is directly above and in projection (alignment) with the front view; and that the right-side view is to the right of and in projection with the front view. Observe also that *two* (and remember which *two*) space dimensions of height, width, and depth are represented in each view.

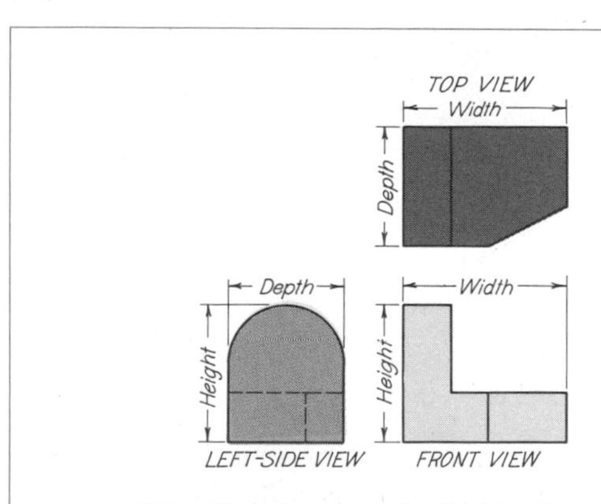


**FIG. 12.** Top, front, and left-side views. Note that the left-side view is drawn to the left of and in projection with the front view. The left-side view is preferred only when, because of the shape of the object, representation is clearer with the left-side view than with the right-side view.

### 6. Combination of Views.

The most usual combination selected from the six possible views consists of the *top, front,* and *right-side* views, as shown in Fig. 11, which, in this case, best describes the shape of the given block. Sometimes the left-side view helps to describe an object more clearly than the right-side view. Figure 12 shows the arrangement of *top, front,* and *left-side* views for the same block. In this case the right-side view would be preferred, as it shows no hidden edges (see Sec. 12 on hidden features). Note that the *side view of the front face of the object is adjacent to the front view* and that the side view of any point will be the same distance from the front surface as is its distance from the front surface on the top view. The combination of *front, right-side,* and *bottom* views is shown in Fig. 13 and of *front, top, left-side,* and *rear* views in Fig. 14.

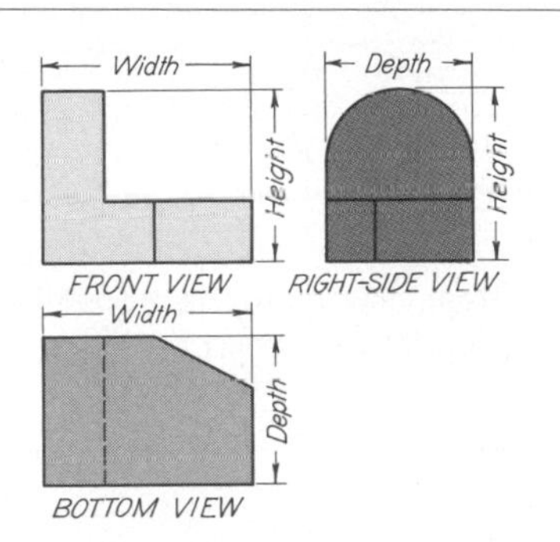


**FIG. 13.** Front, bottom, and right-side views. The bottom view is used instead of the top view only when its use gives clearer representation.

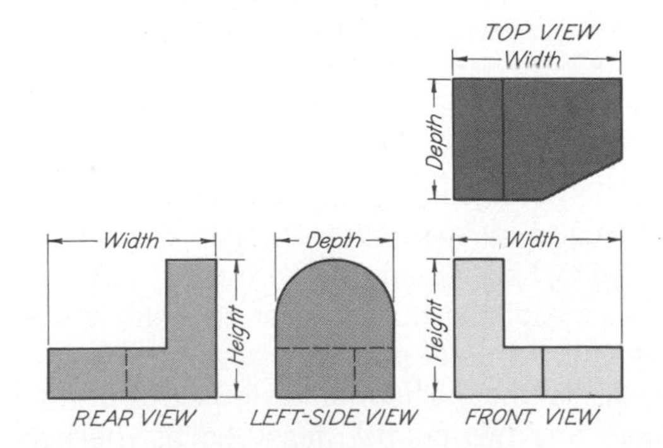


**FIG. 14.** Top, front, left-side, and rear views. The rear view is added *only* when some detail on the rear of the object is important and representation can be improved by its use.

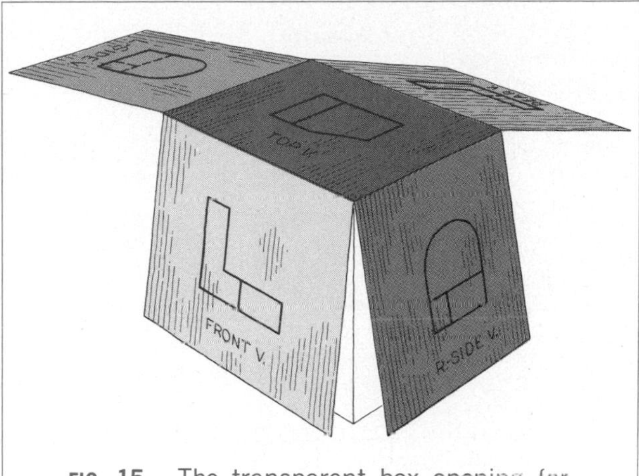


**FIG. 15.** The transparent box opening for alternate-position views. Note that frontal and profile planes are hinged to the horizontal plane.

### 7. "Alternate-position" Views.

The top of the enclosing transparent box may be thought of as in a fixed position with the front, rear, and sides hinged, as in Fig. 15, thus bringing the sides in line with the top view and the rear view above the top view, Fig. 16. This alternate-position arrangement is of occasional use to save space on the paper in drawing a broad, flat object (Fig. 17). The alternate position for the rear view may be used if this arrangement makes the drawing easier to read (Fig. 18).

### 8. The Three Space Dimensions.

As all material objects, from single pieces to complicated structures, have distinct limits and are measurable by three space dimensions,[4] it is desirable

[4]*Space dimensions* and *dimensions of the object* should not be confused. The primary function of orthographic projection is to show the shape of the object. Size is not established until the figured dimensions and/or the scale are placed on the drawing. Space dimensions are *only* the measure of three-dimensional space.

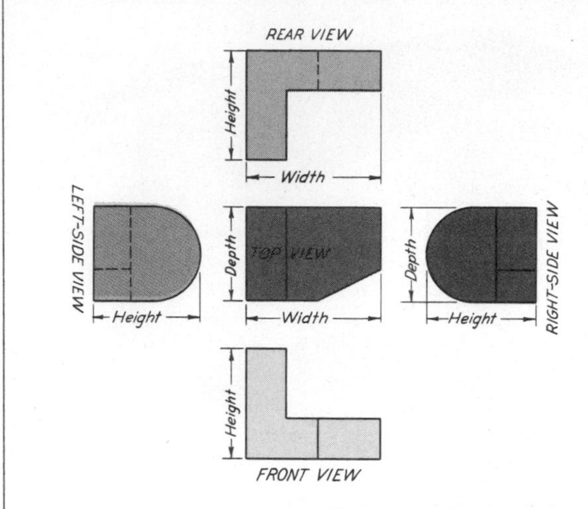


**FIG. 16.** Alternate-position views. Study this figure carefully in conjunction with Fig. 15.

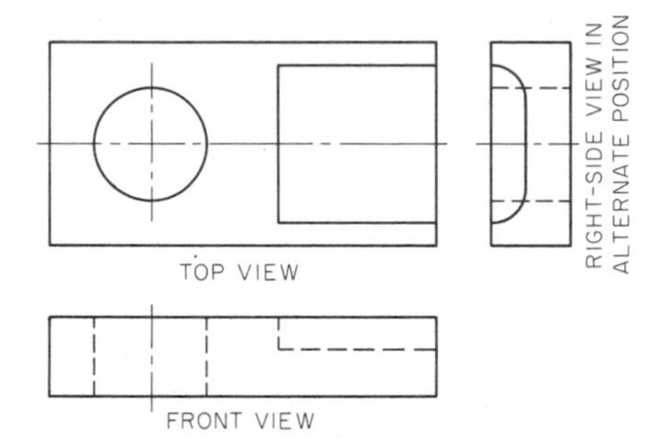


**FIG. 17.** Right-side view in alternate position. Note the saving in paper area (compared with regular position) for this broad, flat object. Compare with Fig. 11.

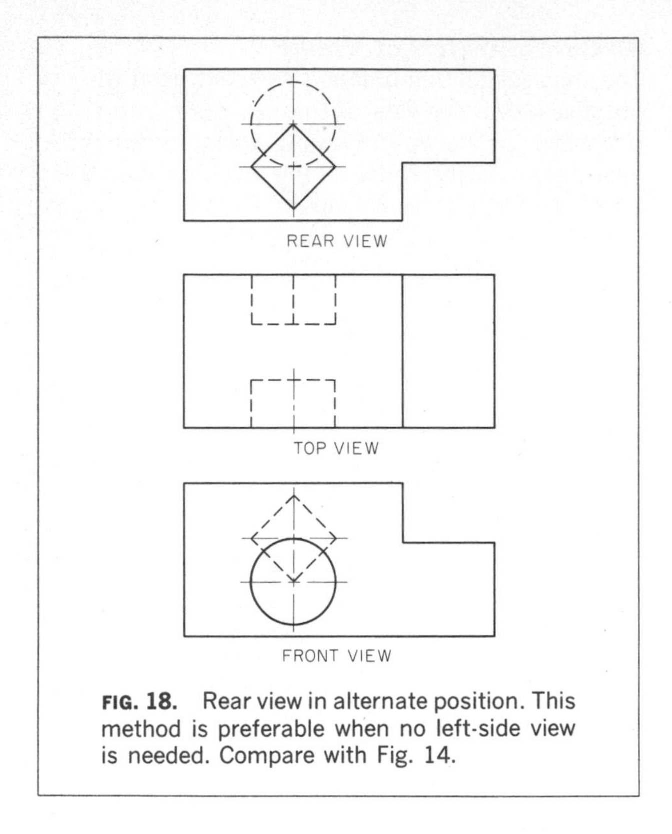


**FIG. 18.** Rear view in alternate position. This method is preferable when no left-side view is needed. Compare with Fig. 14.

for drawing purposes to define these dimensions and to fix their direction.

*Height* is the difference in elevation between any two points, measured as the perpendicular distance between a pair of horizontal planes that contain the points, as shown in Fig. 19. Edges of the object may or may not correspond with the height dimensions. Edge $AB$ corresponds with the height dimension, while edge $CD$ does not, but the space heights of $A$ and $C$ are the same, as are $B$ and $D$. Height is always measured in a vertical direction and has no relationship whatever to the shape of the object.

*Width* is the positional distance left to right between any two points measured as the perpendicular distance between a pair of profile planes containing the points. In Fig. 20 the relative width between points $E$ and $G$ on the left and $H$ and $F$ on the right of an object is shown by the dimension marked "width." The object edge $EF$ is parallel to the width direction and corresponds with the width dimension, but edge $GH$ slopes downward from $G$ to $H$, so this actual edge of the object is longer than the width separating points $G$ and $H$.

*Depth*[5] is the positional distance front to rear between any two points measured as the perpendicular distance between two frontal planes containing the points. Figure 21 shows two frontal

[5] As in the civil engineering sense.

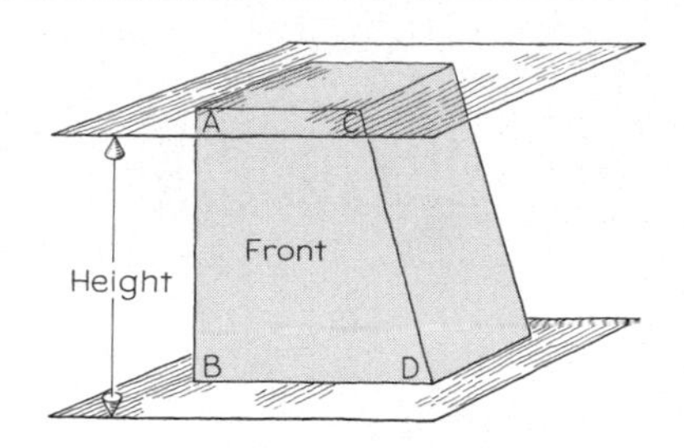


**FIG. 19.** Definition of height. This is the difference in elevation between two points.

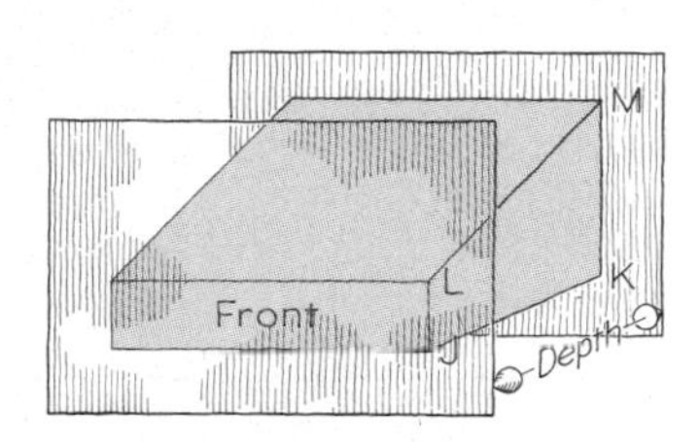


**FIG. 21.** Definition of depth. This is the difference from front to rear between two points.

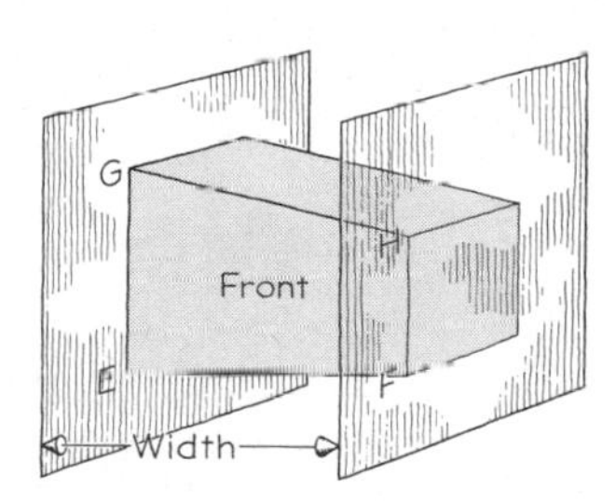


**FIG. 20.** Definition of width. This is the difference from left to right between two points.

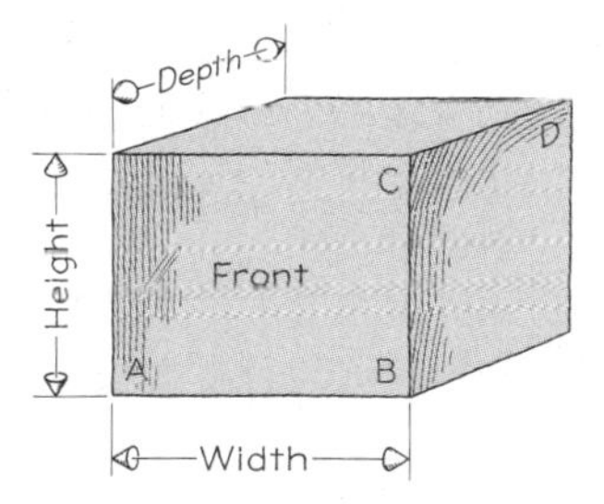


**FIG. 22.** Location of points in space. Height, width, and depth must be designated.

planes, one at the front of the object containing points $J$ and $L$, the other at the rear containing points $K$ and $M$. The relative depth separating the front and rear of the object is the perpendicular distance between the planes as shown.

Any point can be located in space by giving its height, width, and depth relative to some other known point. Figure 22 shows a cube with four identified corners $A$, $B$, $C$, and $D$. Assuming that the plane containing points $A$ and $B$ is the front of the object, height, width, and depth would be as marked. Assuming also that point $A$ is fixed in space, point $B$ could be located from point $A$ by giving the width dimension, including the statement that height and depth measurements are zero. $C$ could be located from $A$ by giving width, height, and zero depth. $D$ could be located from $A$ by giving width, height, and depth measurements.

## 9. The Relationship of Planes, View Directions, and Space Dimensions.

As explained in Secs. 4 and 5, the object to be drawn may be thought of as surrounded by transparent planes upon which the actual views are projected. The three space dimensions—height, width, and depth—and the planes of projection are unchangeably oriented and connected with each other and with the view directions (Fig. 23). Each of the planes of projection is perpendicular, respectively, to its own view direction. Thus the frontal plane is perpendicular to the front-view direction, the horizontal plane is perpendicular to the top-view direction, and the profile plane is perpendicular to the side-view direction. The two space measurements for a view are parallel to the plane of that view and perpendicular to the view direction. Therefore height and width are parallel to the frontal plane and perpen-

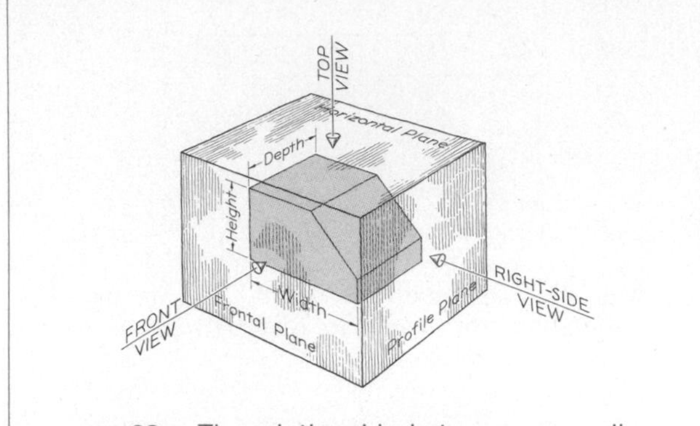


**FIG. 23.** The relationship between space directions and the planes of projection. View directions are perpendicular to their planes of projection. Height is parallel to frontal and profile planes, width is parallel to frontal and horizontal planes, and depth is parallel to horizontal and profile planes.

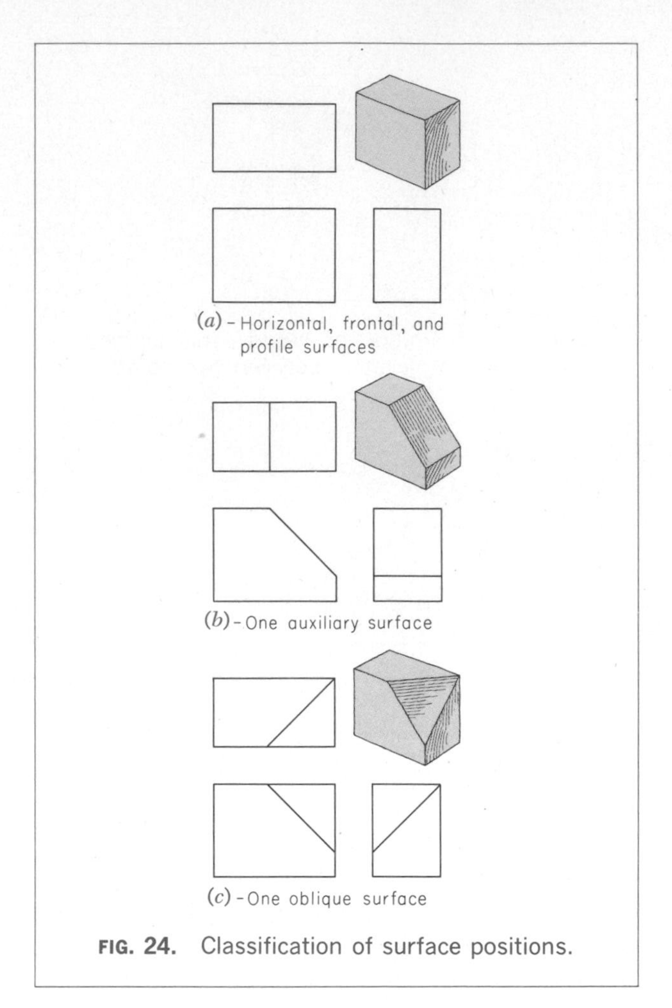


**FIG. 24.** Classification of surface positions.

dicular to the front-view direction; width and depth are parallel to the horizontal plane and perpendicular to the top-view direction; height and depth are parallel to the profile plane and perpendicular to the side-view direction. Note that the three planes of projection are *mutually* perpendicular, as are the three space measurements and the three view directions. Carefully study the views in Figs. 11 to 17 and note the space dimensions marked on each figure.

## 10. Classification of Surfaces and Lines.

Any object, depending upon its shape and space position, may or may not have some surfaces parallel or perpendicular to the planes of projection.

Surfaces are classified according to their space relationship with the planes of projection (Fig. 24). *Horizontal, frontal,* and *profile* surfaces are shown at (*a*). When a surface is inclined to two of the planes of projection (but perpendicular to the third), as at (*b*), the surface is said to be *auxiliary* or *inclined.* If the surface is at an angle to all three planes, as at (*c*), the term *oblique* or *skew* is used.

The edges (represented by lines) bounding a surface may, because of the shape or position of the object, also be in a simple position or inclined to the planes of projection. A line in, or parallel to, a plane of projection takes its name from the plane. Thus a *horizontal line* is a line in a horizontal plane, a *frontal line* is a line in a frontal plane, and a *profile line* is a line in a profile plane. When a line is parallel to two planes, the line takes the name of both planes, as *horizontal-frontal, horizontal-profile,* or *frontal-profile.* A line not parallel to any plane of projection is called an *oblique* or *skew line.* Figure 25 shows various positions of lines.

An edge appears in true length when it is parallel to the plane of projection, as a point when it is perpendicular to the plane, and shorter than true length when it is inclined to the plane. Similarly, a

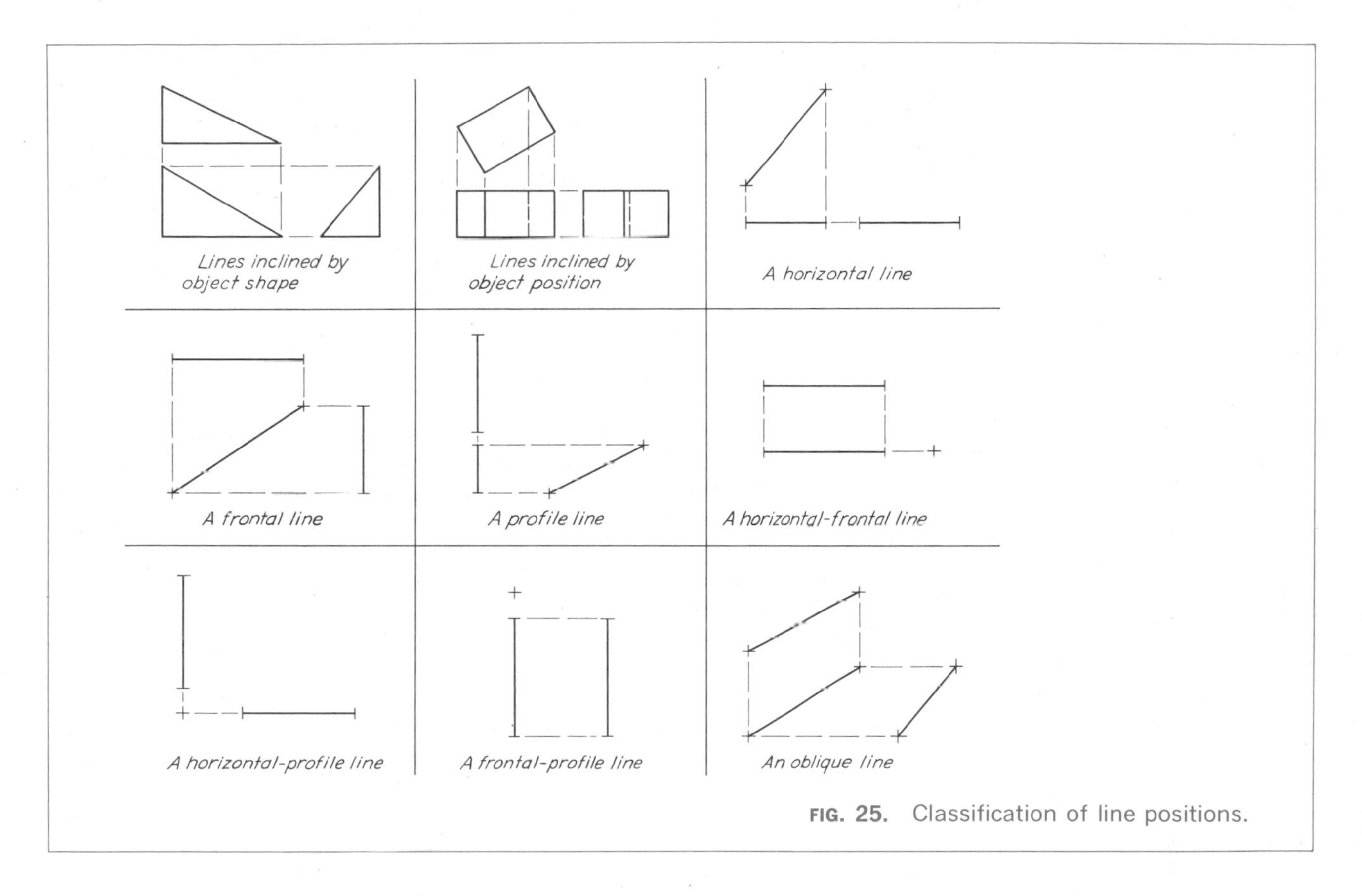


**FIG. 25.** Classification of line positions.

surface appears in true shape when it is parallel to the plane of projection, as a line when it is perpendicular to the plane, and foreshortened when it is inclined to the plane. As an example, Fig. 24*a* shows an object with its faces parallel to the planes of projection; top, front, and right-side surfaces are shown in true shape; and the object edges appear either in true length or as points. The inclined surface of the object at (*b*) does not show in true shape in any of the views but appears as an edge in the front view. The front and rear edges of the inclined surface are in true length in the front view and foreshortened in the top and side views. The top and bottom edges of the inclined surface appear in true length in top and side views and as points in the front view. The oblique (skew) surface of the object at (*c*) does not show in true shape in any of the views, but each of the bounding edges shows in true length in one view and is foreshortened in the other two views.

### 11. Representation of Lines.

Although uniform in appearance, the lines on a drawing may indicate three different types of directional change on the object. An *edge* view is a line showing the edge of a receding surface that is perpendicular to the plane of projection. An *intersection* is a line formed by the meeting of two surfaces when either one surface is parallel and one at an angle or both are at an angle to the plane of projection. A *surface limit* is a line that indicates the reversal of direction of a curved surface (or the series of points of reversal on a warped surface). Figure 26 illustrates the different line meanings, and these are further explained in Sec. 45.

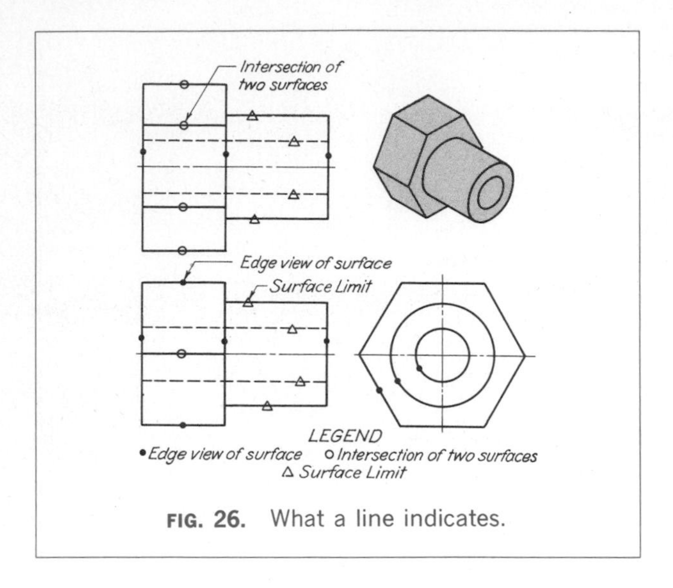


**FIG. 26.** What a line indicates.

## 12. Hidden Features.

To describe an object completely, a drawing should contain lines representing all the edges, intersections, and surface limits of the object. *In any view there will be some parts of the object that cannot be seen from the position of the observer, as they will be covered by portions of the object closer to the observer's eye.* The edges, intersections, and surface limits of these hidden parts are indicated by a discontinuous line called a *dashed line*.[6] See the alphabet of lines (Figs. 56 and 57). In Fig. 27 the drilled hole[7] that is visible in the right-side view is hidden in the top and front views, and therefore it is indicated in these views by a dashed line showing the hole and the shape as left by the drill point. The milled slot (see Glossary) is visible in the front and side views but is hidden in the top view.

The beginner must pay particular attention to the execution of these dashed lines. If carelessly drawn, they ruin the appearance of a drawing and make it harder to read. Dashed lines are drawn lighter than full lines, of short dashes uniform in length with the space between them very short, about one-fourth the length of the dash. It is important that they start and stop correctly. A dashed line always starts with a dash except when the dash would form a continu-

[6]The line indicating hidden features has been traditionally known as a "dotted" line. However, in this treatise the term "dashed" line will be used because it accurately describes the appearance of the line. The term "hidden line," also sometimes used, is completely inaccurate because there are no *lines* on the object, and the line indicating hidden features *is* visible on the drawing.

[7]See Glossary and Index.

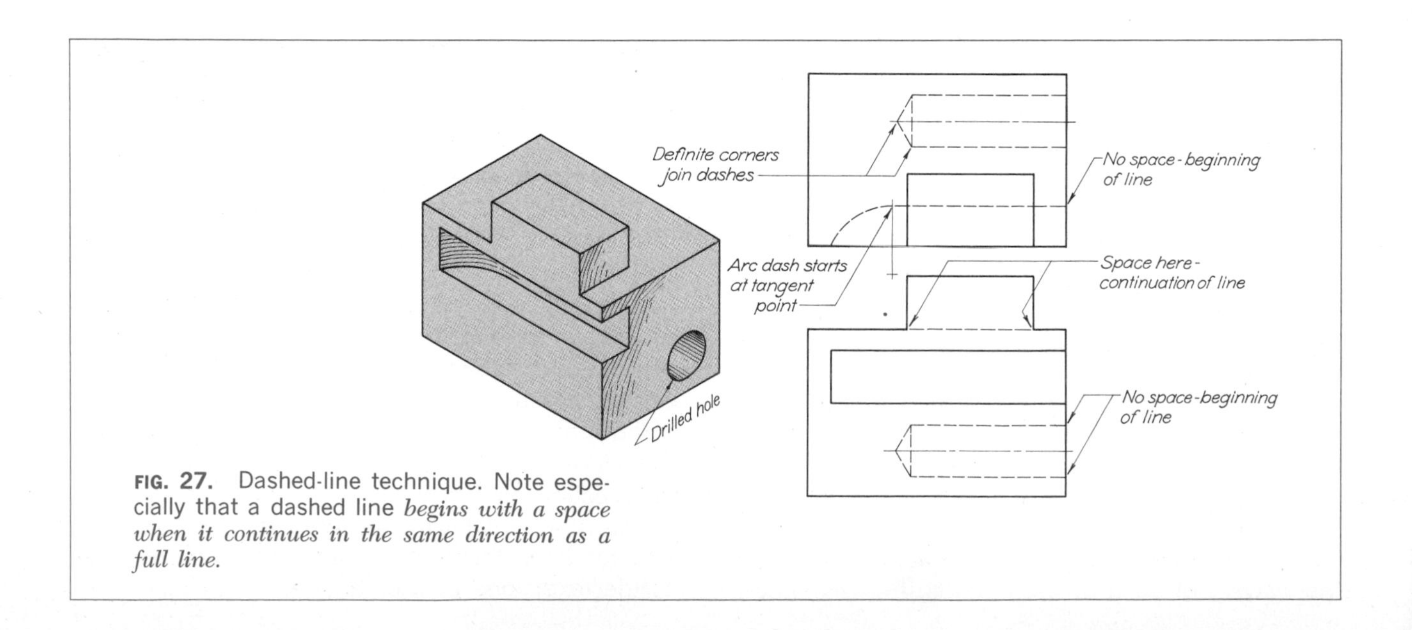


**FIG. 27.** Dashed-line technique. Note especially that a dashed line *begins with a space when it continues in the same direction as a full line.*

ation of a full line; in that case a space is left, as shown in Fig. 27. Dashes always meet at corners. An arc must start with a dash at the tangent point except when the dash would form a continuation of a straight or curved full line. The number of dashes used in a tangent arc should be carefully judged to maintain a uniform appearance (Fig. 28). Study carefully all dashed lines in Figs. 27 and 29.

## 13. Center Lines.

In general, the first lines drawn in the layout of an engineering drawing are the center lines, which are the axes of symmetry for all symmetrical views or portions of views: (1) Every part with an axis, such as a cylinder or a cone, will have the axis drawn as a center line before the part is drawn. (2) Every circle will have its center at the intersection of two mutually perpendicular center lines.

The standard symbol for center lines on finished drawings is a fine line made up of alternate long and short dashes, as shown in the alphabet of lines (Figs. 56 and 57). Center lines are always extended slightly beyond the outline of the view or portion of the view to which they apply. They form the skeleton construction of the drawing; the important measurements are made and dimensions given to and from these lines. Study the center lines in Probs. 90 to 110.

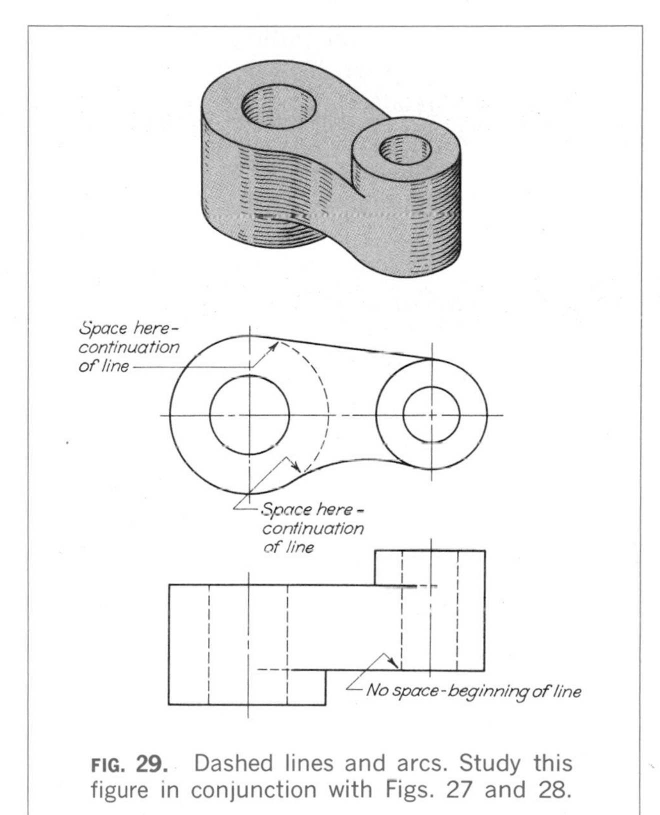


**FIG. 29.** Dashed lines and arcs. Study this figure in conjunction with Figs. 27 and 28.

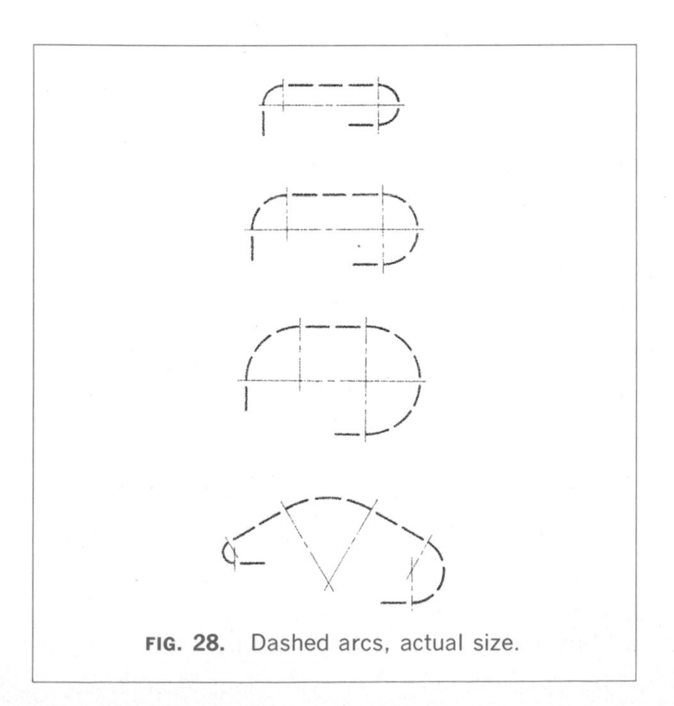

**FIG. 28.** Dashed arcs, actual size.

## 14. Precedence of Lines.

In any view there is likely to be a coincidence of lines. Hidden portions of the object may project to coincide with visible portions. Center lines may occur where there is a visible or hidden outline of some part of the object.

Since the physical features of the object must be represented, full and dashed lines take precedence over all other lines. Since the visible outline is more prominent by space position, full lines take precedence over dashed lines. A full line could cover a dashed line, but a dashed line could not cover a full line. It is evident also that a dashed line could not occur as one of the boundary lines of a view.

When a center line and cutting-plane (explained in Chap. 8) line coincide, the one that is more important for the readability of the drawing takes precedence over the other.

Break lines (explained in Chap. 8) should be placed so that they do not spoil the readability of the over-all view.

Dimension and extension lines must always be placed so as not to coincide with other lines of the drawing.

The following list gives the order of precedence of lines:

1. Full line
2. Dashed line
3. Center line or cutting-plane line
4. Break lines
5. Dimension and extension lines
6. Crosshatch lines

Note the coincident lines in Fig. 30.

### 15. Exercises in Projection.

The principal task in learning orthographic projection is to become thoroughly familiar with the theory and then to practice this theory by translating from a picture of the object to the orthographic views. Figures 31 and 32 contain a variety of objects shown by a pictorial sketch and translated into orthographic views. Study the objects and note (1) how the object is oriented in space, (2) why the orthographic views given were chosen, (3) the projection of visible features, (4) the projection of hidden features, and (5) center lines.

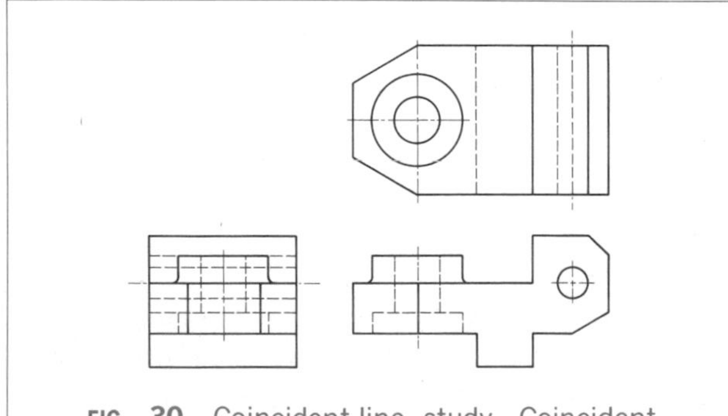

**FIG. 30.** Coincident-line study. Coincident lines are caused by the existence of features of identical size or position, one behind the other.

## WRITING THE GRAPHIC LANGUAGE

### 16. Objectives.

The major objective of a student of the graphic language is to learn orthographic drawing. In addition to an understanding of the theory of orthographic projection, several aspects of drawing are necessary preliminaries to its study. These have been discussed in the preceding chapters and include skill and facility in the use of instruments (Chap. 2), a knowledge of applied geometry (Chap. 3), and fluency in lettering (Chap. 4). In writing the graphic language, a topic we will now take up, always pay careful attention to accuracy and neatness.

### 17. Object Orientation.

An object can, of course, be drawn in any of several possible positions. *The simplest position should be used,* with the object oriented so that the principal faces are perpendicular to the sight directions for the views and parallel to the planes of projection, as shown in Fig. 33. Any other position of the object, with its faces at some angle to the planes of projection, would complicate the drawing, foreshorten the object faces, and make the drawing difficult to make and to read.

### 18. Selection of Views.

In practical work it is important to choose the combination of views that will describe the shape of an object in the best and most economical way. Often only two views are necessary. For example, a cylindrical shape, if on a vertical axis, would require only a front and top view; if on a horizontal axis, only a front and side view. Conic and pyramidal shapes

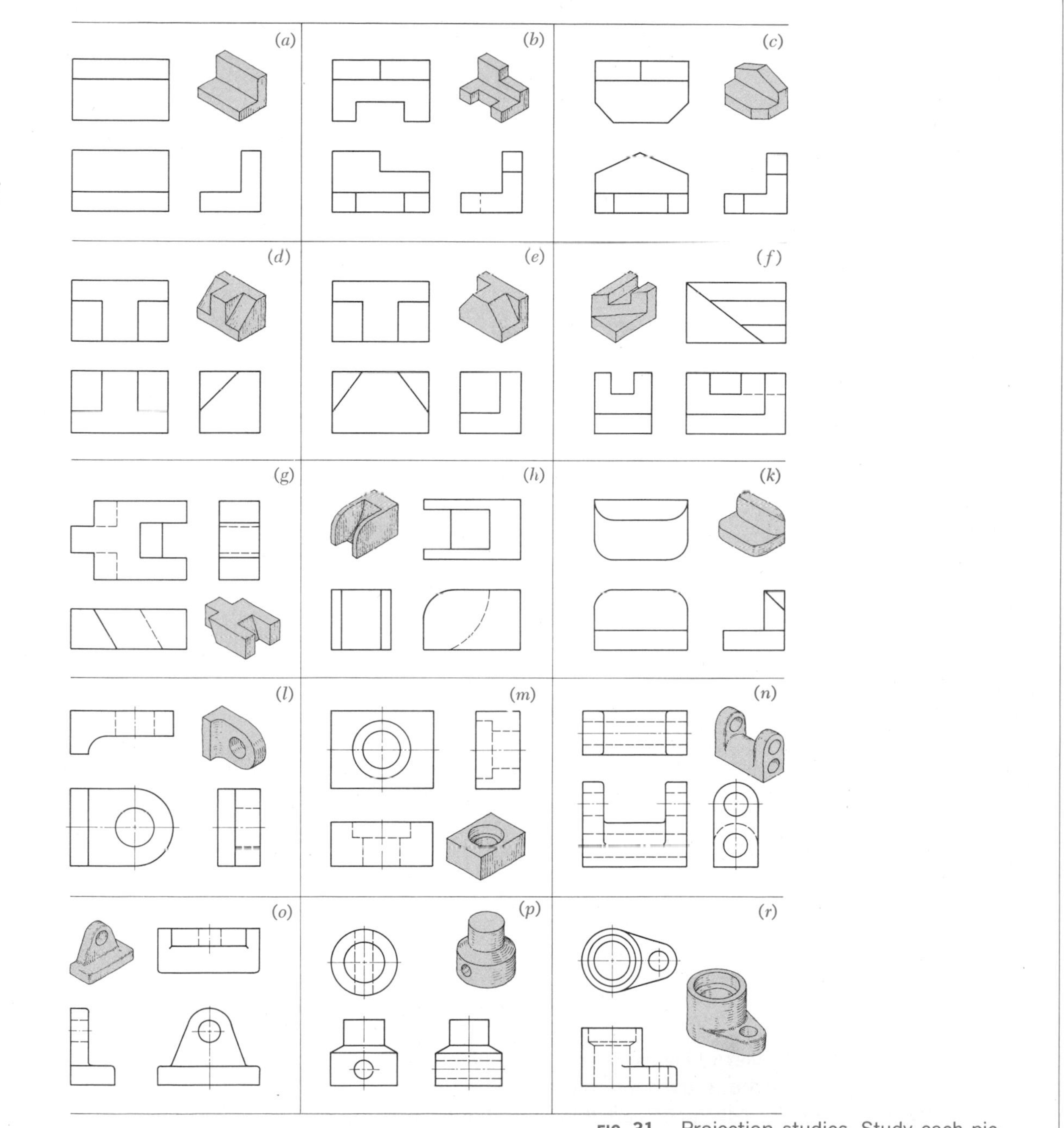


**FIG. 31.** Projection studies. Study each picture and the accompanying orthographic views and note the projection of all features.

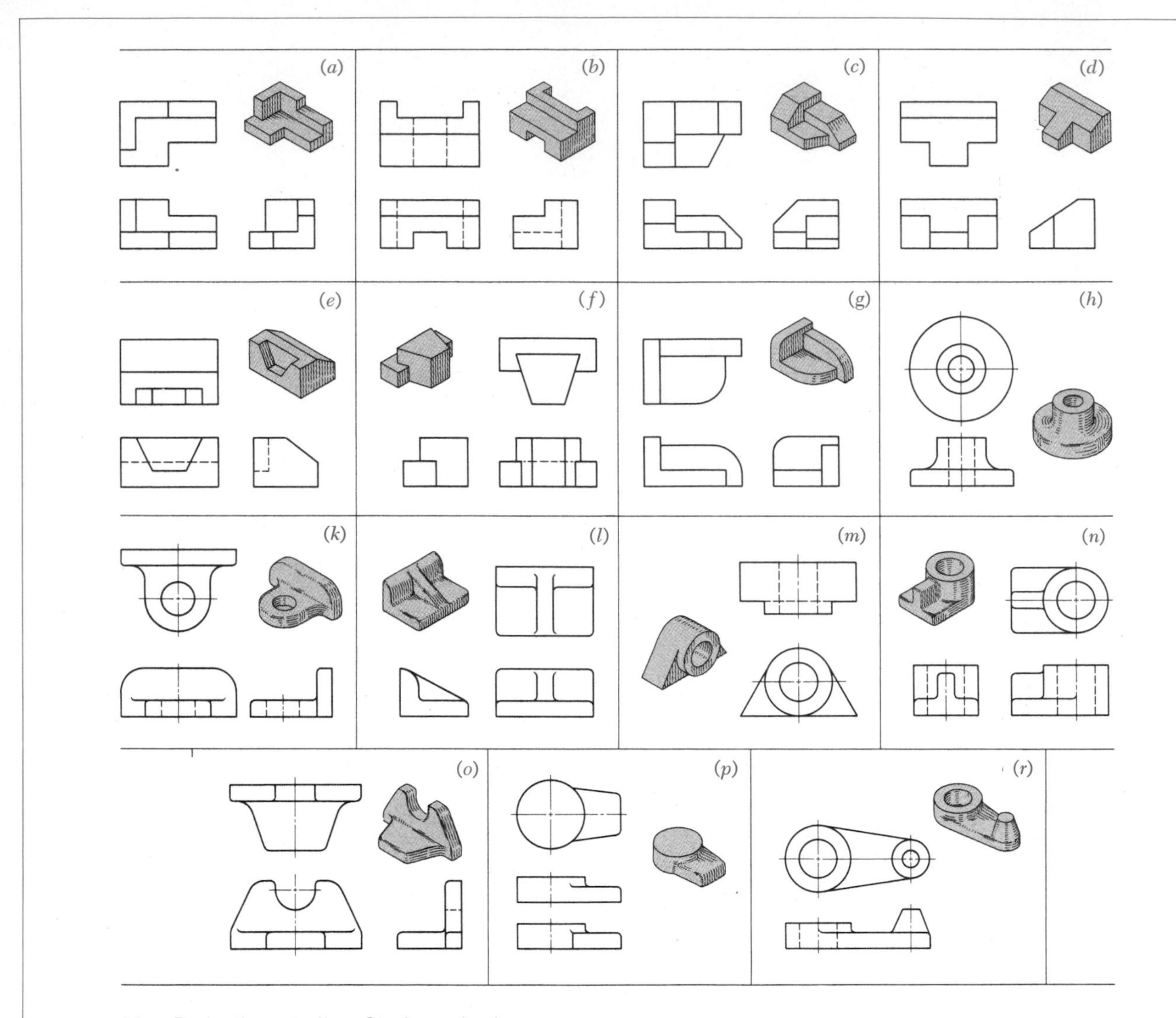


**FIG. 32.** Projection studies. Study each picture and the accompanying orthographic views and note the projection of all features.

can also be described in two views. Figure 34 illustrates two-view drawings. Some shapes will need more than the three regular views for adequate description.

Objects can be thought of as being made up of combinations of simple geometric solids, principally cylinders and rectangular prisms, and the views necessary to describe any object would be determined by the directions from which it would have to be viewed to see the characteristic contour shapes of these parts. Figure 35, for example, is made up of several prisms and cylinders. If each of these

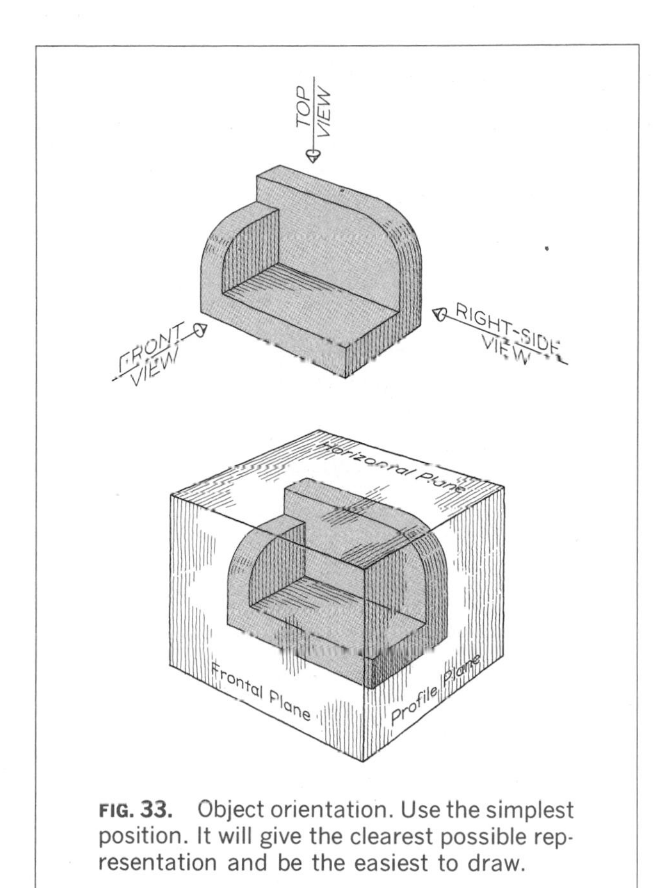


**FIG. 33.** Object orientation. Use the simplest position. It will give the clearest possible representation and be the easiest to draw.

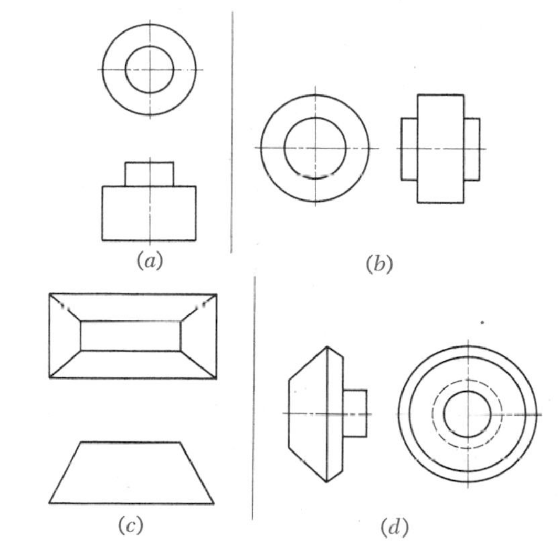


**FIG. 34.** Two-view drawings. These are sufficient for any object having a third view identical with or similar to one of the two views given.

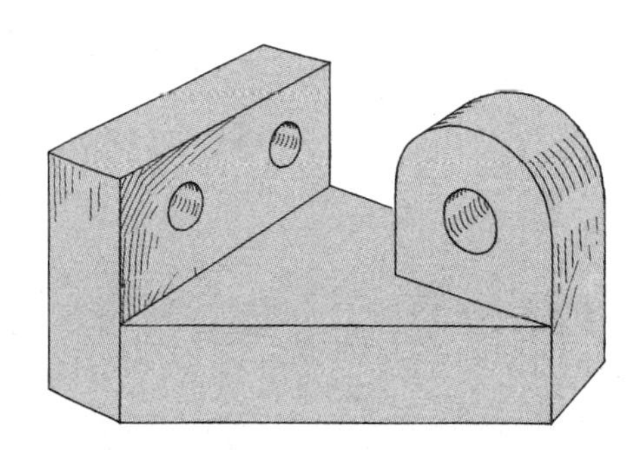

**FIG. 35.** Geometric shapes combined. Even the most complicated objects can be analyzed as combined geometric shapes.

simple shapes is described and its relation to the others is shown, the object will be fully represented. In the majority of cases the three regular views—top, front, and side—are sufficient to do this.

Sometimes two views are proposed as sufficient for an object on the assumption that the contour in the third direction is of the shape that would naturally be expected. In Fig. 36, for example, the figure at (*a*) would be assumed to have a uniform cross section and be a square prism. But the two views *might* be the top and front views of a wedge, as shown in three views at (*b*). Two views of an object, as drawn at (*c*), do not describe the piece at all. The object at (*a*) might be assumed to be square in section, but it could as easily be round, triangular, quarter-round, or of another shape, which should have been indicated by a side view. Sketch several different front views for each top view, Fig. 37*a* to *c*.

With the object preferably in its functioning position and *with its principal surfaces parallel to the planes of projection,* visualize the object, mentally

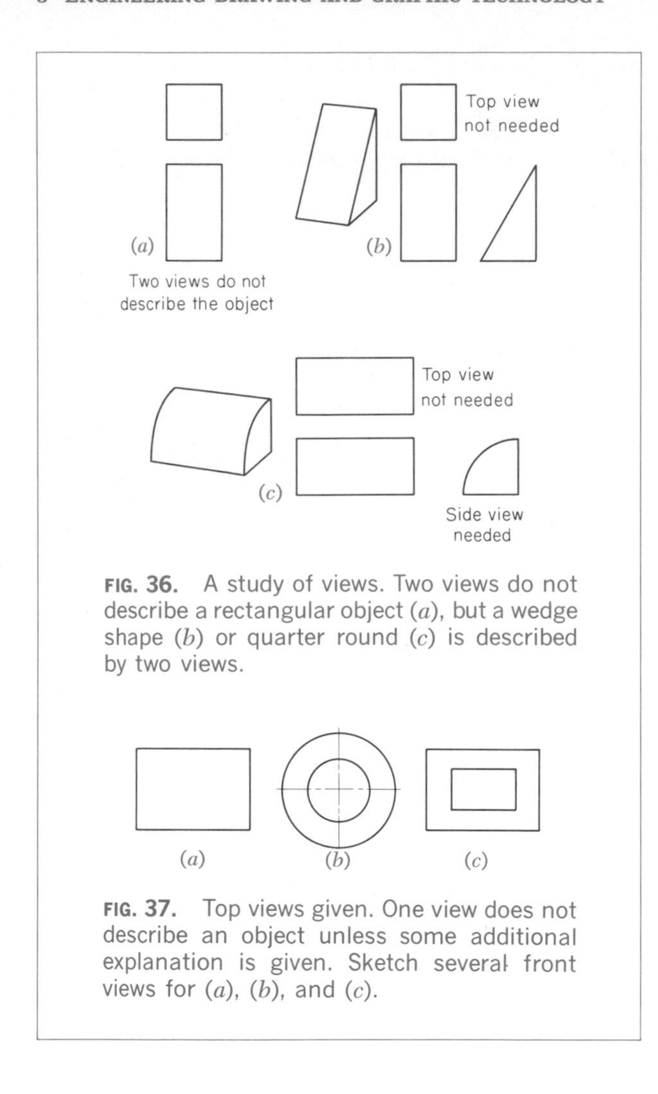


**FIG. 36.** A study of views. Two views do not describe a rectangular object (*a*), but a wedge shape (*b*) or quarter round (*c*) is described by two views.

**FIG. 37.** Top views given. One view does not describe an object unless some additional explanation is given. Sketch several front views for (*a*), (*b*), and (*c*).

**FIG. 38.** Selection of views. A view must be drawn in each direction (top, front, side) needed to conclusively designate every feature of the object, but unnecessary views must *not* be drawn.

picturing the orthographic views one at a time to decide on the best combination. In Fig. 38, the arrows show the direction of observation for the six principal views of an object, and indicate the mental process of the person making the selection. He notes that the front view would show the two horizontal holes as well as the width and height of the piece, that a top view is needed to show the contour of the vertical cylinder, and that the cutout corner calls for a side view to show its shape. He notes further that the right-side view would show this cut in full lines, while the left-side view would give it in dashed lines. He observes also that neither a bottom view nor a rear view would be of any value in describing this object. Thus he has correctly chosen the front, top, and right-side views as the best combination for describing this piece. As a rule, the side view containing the fewer dashed lines is preferred. If the side views do not differ in this respect, the right-side view is preferred in standard practice.

In inventive and design work, any simple object should be visualized mentally and the view selected without a picture sketch. In complicated work, a pictorial or orthographic sketch may be used to advantage, but it should not be necessary, in any case, to sketch all possible views in order to make a selection.

Study the drawings in Fig. 39 and determine why each view was chosen.

## 19. Drawing Sizes.

Standard sizes for sheets of drawing paper, based on multiples of $8\frac{1}{2}$ by 11 in. and 9 by 12 in., are specified for drawings by the ASA. Trimmed sizes of drawing paper and cloth, with suitable border and title dimensions in the final chapter.

## 20. Spacing the Views.

View spacing is necessary so that the drawing will be balanced within the space provided. A little preliminary measuring is necessary to locate the views. The following example describes the procedure: Suppose the piece illustrated in Fig. 40 is to be drawn full size on an 11- by 17-in sheet. With an end-title strip, the working space inside the border will be $10\frac{1}{2}$ by 15 in. The front view will require $7\frac{11}{16}$ in., and the side view $2\frac{1}{4}$ in. This leaves $5\frac{1}{16}$ in. to be distributed between the views and at the ends.

This preliminary planning need not be to exact

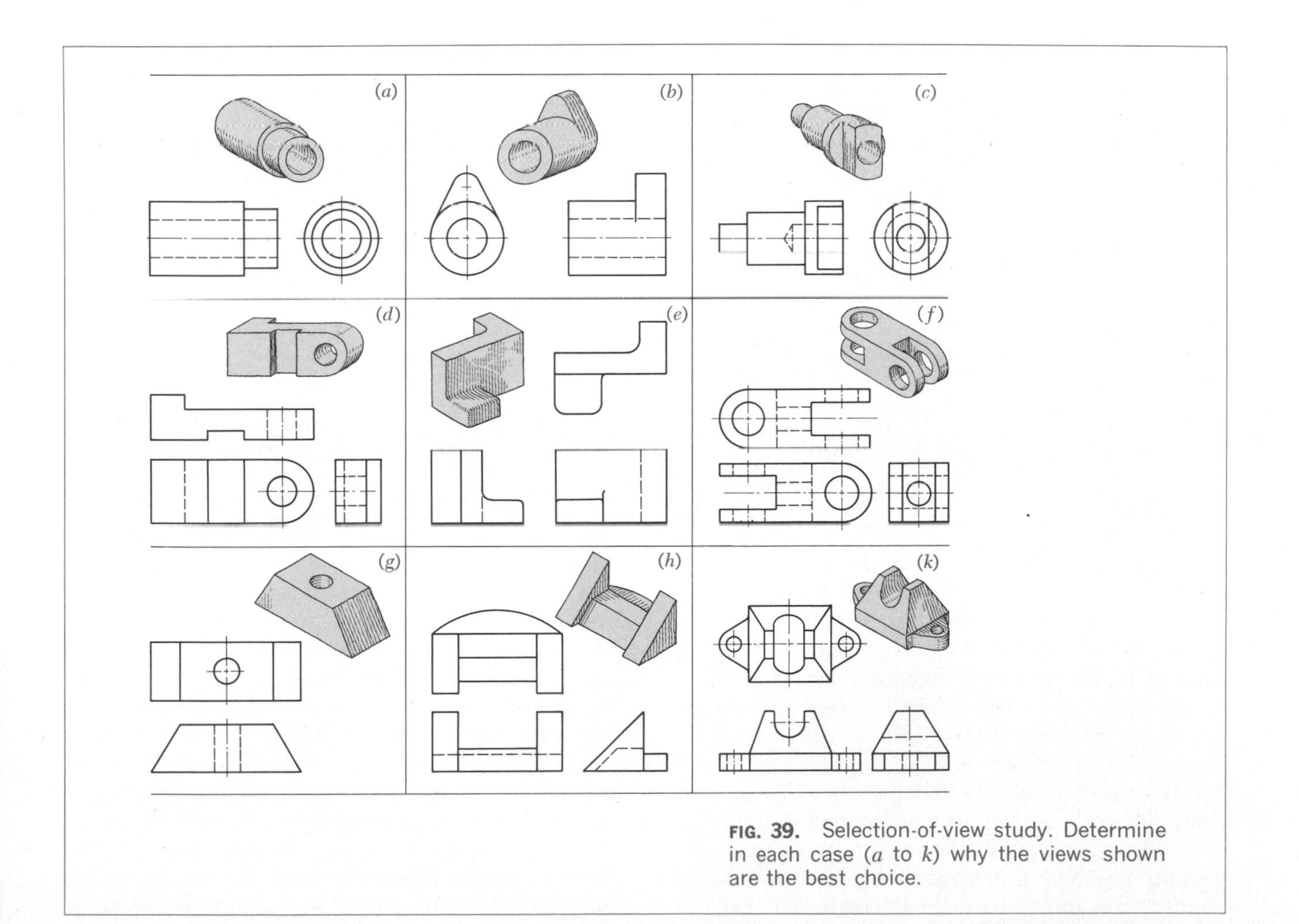


**FIG. 39.** Selection-of-view study. Determine in each case (*a* to *k*) why the views shown are the best choice.

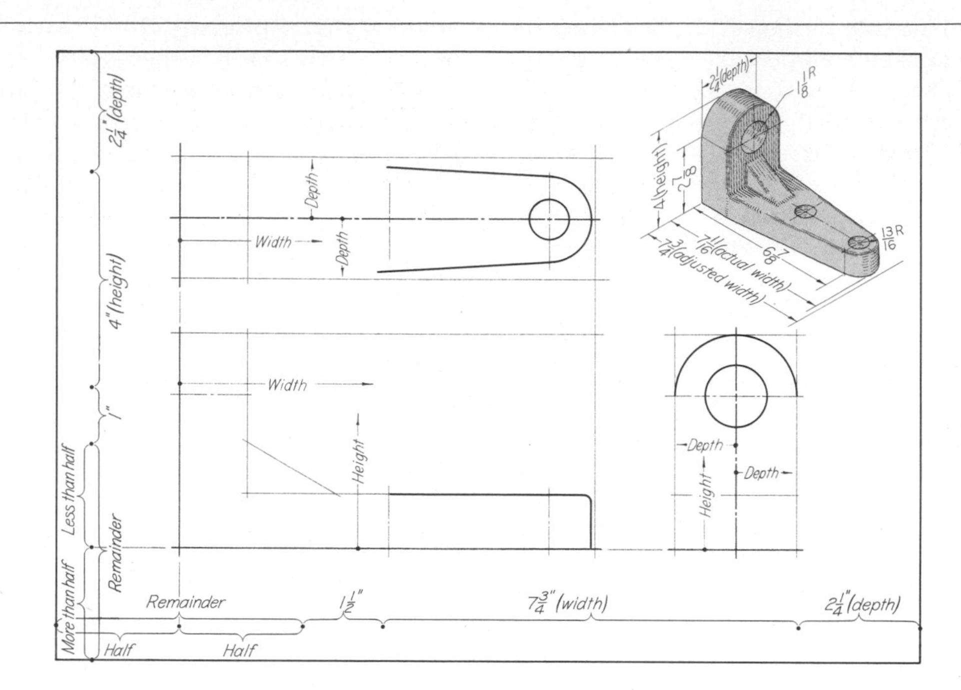


**FIG. 40.** Spacing the views on the paper. This is done graphically. Study the text carefully while referring to this figure and go through the steps by laying out the given object on a standard 11- × 17-in. sheet.

dimensions, that is, small fractional values, such as $^{15}/_{64}$ in. or $^{31}/_{32}$ in., can be adjusted to $^1/_4$ and 1 in., respectively, to speed up the planning. In this case the $7^{11}/_{16}$-in. dimension can be adjusted to $7^3/_4$ in.

Locate the views graphically and quickly by measuring with your scale along the bottom border line. Starting at the lower right corner, lay off first $2^1/_4$ in. and then $7^3/_4$ in. The distance between views can now be decided upon. It is chosen by eye to separate the views without crowding, yet placing them sufficiently close together so that the drawing will read easily (in this case $1^1/_2$ in.). Measure the distance; half the remaining distance to the left corner is the starting point of the front view. For the vertical location: the front view is 4 in. high, and the top view $2^1/_4$ in. deep. Starting at the upper-left corner, lay off first $2^1/_4$ in. and then 4 in.; judge the distance between views (in this case 1 in.), and lay it off; then a point marked at less than half the remaining space will locate the front view, allowing more space at the bottom than at the top for appearance.

Block out lightly the spaces for the views, and study the over-all arrangement, because changes can easily be made at this stage. If it is satisfactory, select reference lines in each view from which the space measurements of height, width, and depth that appear in the view can be measured. The reference line may be an edge or a center line through some dominant feature, as indicated on Fig. 40 by the center lines in the top and side views and the medium-weight lines in all the views. The directions for height, width, and depth measurements for the views are also shown.

## 21. Projecting the Views.

After laying out the views locate and draw the various features of the object. In doing this, carry the views *along together,* that is, *do not* attempt to complete one view before proceeding to another. Draw first the most characteristic view of a feature and then project it and draw it in the other views before going on to a second feature. As an example, the vertical hole of Fig. 41 should be drawn first in the top view, and then the dashed lines representing the limiting elements or portions should be projected and drawn in the front and side views.

In some cases, one view cannot be completed before a feature has been located and drawn in another view. Study the pictorial drawing in Fig. 41, and note from the orthographic views that the horizontal slot must be drawn on the front view before the edge $AB$ on the slanting surface can be found in the top view.

Projections (horizontal) between the front and side views are made by employing the T square to draw the required horizontal line (or to locate a required point), as in Fig. 42.

Projections (vertical) between the front and top views are made by using the T square and a triangle as in Fig. 43.

Projections between the top and side views cannot be projected directly but must be measured and transferred or found by special construction. In carrying the top and side views along together, it is usual to transfer the depth measurement from one to the other with dividers, as in Fig. 44*a*, or with a scale, as at (*b*). Another method, used for an irregular figure, is to "miter" the points around, using a 45° line drawn through the point of intersection of the top and side views of the front face, extended as shown in Fig. 45. The method of Fig. 45, however, requires more time and care than the methods of Fig. 44 and is, therefore, not recommended.

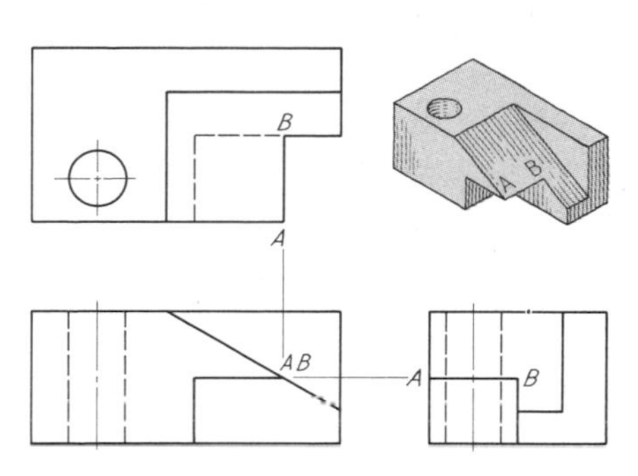


**FIG. 41.** Projection of lines. Carry all views along together. The greatest mistake possible is to try to complete one view before starting another.

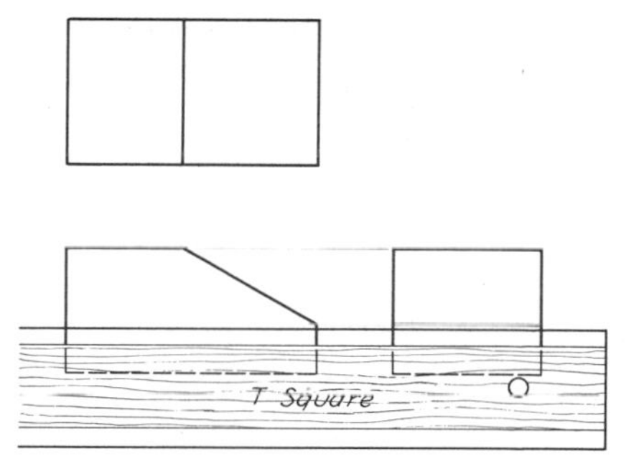


**FIG. 42.** Making a horizontal projection. This is the simplest operation in drawing. The T square provides *all* horizontal lines.

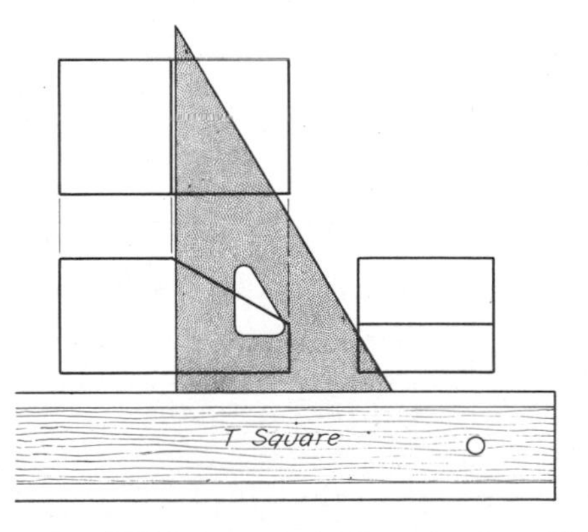


**FIG. 43.** Making a vertical projection. The 90° angle of a triangle with one leg on the horizontal T square produces the vertical.

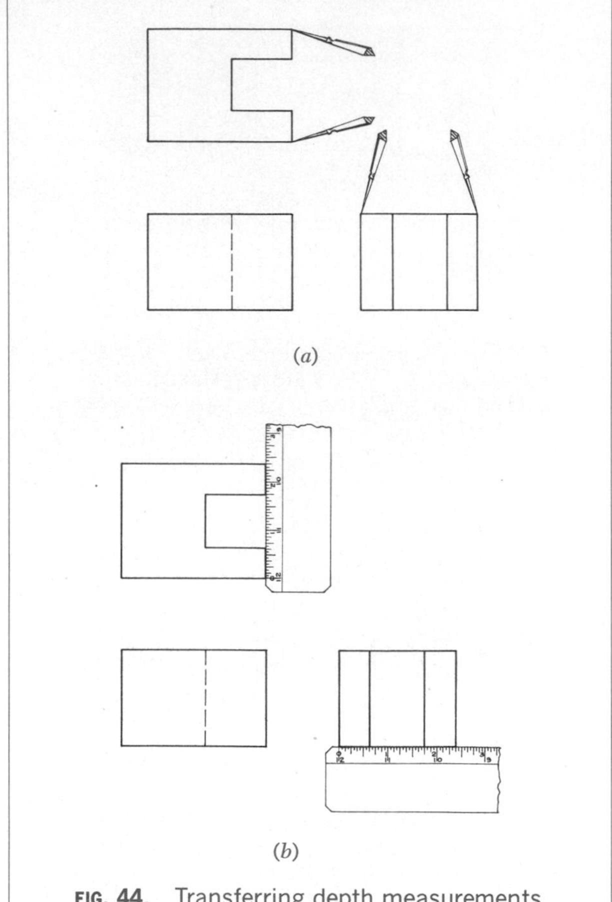


**FIG. 44.** Transferring depth measurements. Depth cannot be projected. Transfer the necessary distances with dividers, as at (*a*), or with scale, as at (*b*).

## 22. Projections of Surfaces Bounded by Linear Edges.

In drawing projections of inclined surfaces, in some cases the corners of the bounding edges may be used, and in other cases the bounding edges themselves may be projected. In illustration of these methods, Fig. 46*a* shows a vertical hexagonal hole that is laid out from specifications in the top view. Then the front view is drawn by projecting from the six corners of the hexagon and drawing the four

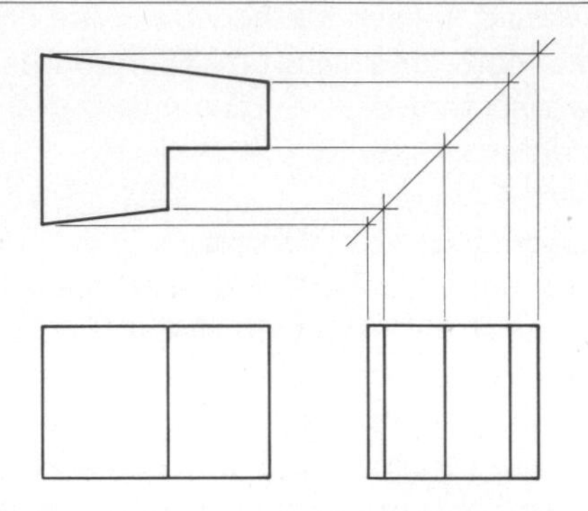

**FIG. 45.** Projecting depth measurements. A "miter line" at 45°, with horizontal and vertical projectors, transfers the depth from top to side view (or vice versa).

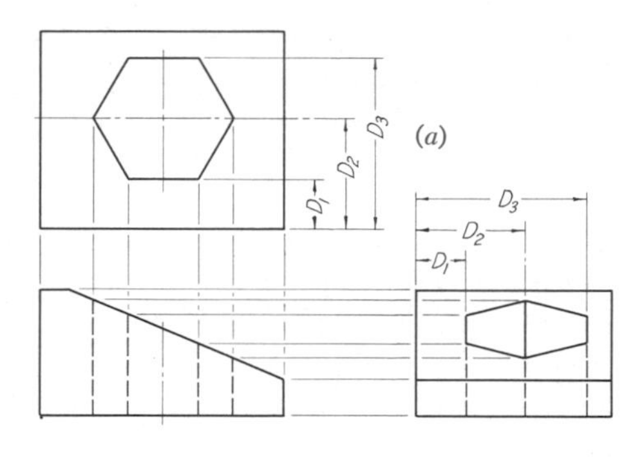


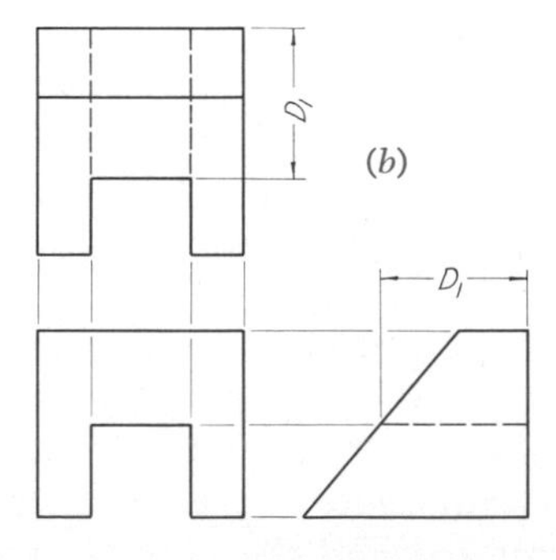


**FIG. 46.** Projections of surfaces bounded by linear edges. Depth measurements cannot be projected directly with T square and triangle. They must be transferred.

dashed lines to complete the front view. To get the side view, a horizontal projection is made from each corner on the front view to the side view, thus locating the height of the points needed on the side view. Then measurements $D_1$, $D_2$, and $D_3$ taken from the top view and transferred to the side view locate all six corners in the side view. The view is completed by connecting these corners and drawing the three vertical dashed lines. The object in Fig. 46*b* shows a horizontal slot running out on an inclined surface. In the front view the true width and height of this slot is laid out from specifications. The projection to the side view is a simple horizontal projection for the dashed line, indicating the top surface of the slot. To get the top view, the width is projected from the front view and then the position of the runout line is measured (distance $D_1$) and transferred to the top view.

In summary, it may be stated that, if a line appears at some angle on a view, its two ends must be projected; if a line appears parallel to its path of projection, the complete line can be projected.

## 23. Projections of an Elliptical Boundary.

The intersection of a cylindrical hole (or cylinder) with a slanting (inclined or skew) surface, as shown in Fig. 47, will be an ellipse, and some projections of this elliptical edge will appear as another ellipse. The projection can be made as shown in Fig. 47*a* by assuming a number of points on the circular view and projecting them to the edge view (front) and then to an adjacent view (side). Thus points 1 to 4 are located in the top view and projected to the front view, and the projectors are then drawn to the side view. Measurements of depth taken from the top view (as $D_1$) will locate the points in the side view. Draw a smooth curve through the points, using a French curve.

For an ellipse on an inclined surface, the projection can also be made by establishing the major and minor diameters of the ellipse, as shown in Fig. 47*b*. A pair of diameters positioned so as to give the largest and smallest extent of the curve will give the required major and minor diameters. Thus, *AB* will

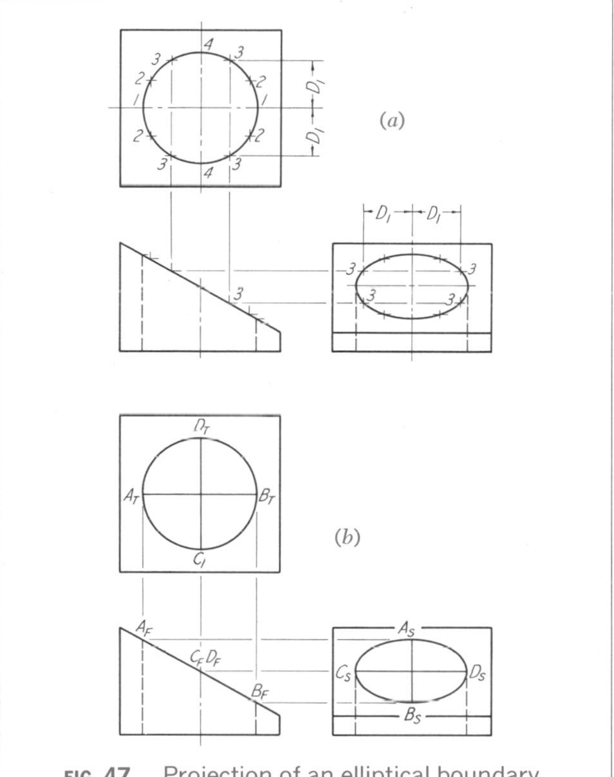


**FIG. 47.** Projection of an elliptical boundary. Points on the curve are projected to determine the curve (*a*), or the curve is determined by major and minor diameters (*b*).

project to the side view as the smaller, or minor, diameter $A_S B_S$, and *CD* will project as the larger, or major, diameter $C_S D_S$. The ellipse can then be drawn by one of the methods of Secs. 50, 52, and 54 in Chap. 3.

If the surface intersected by the cylinder is skew, as shown in Fig. 48, a pair of perpendicular diameters located in the circular view will give a pair of conjugate diameters in an adjacent view. Therefore, $A_T B_T$ and $C_T D_T$ projected to the front view will give conjugate diameters, which can be employed as explained in Secs. 51, 53, and 54 in Chap. 3 to draw the required ellipse.

In projecting the axes, they can be extended to

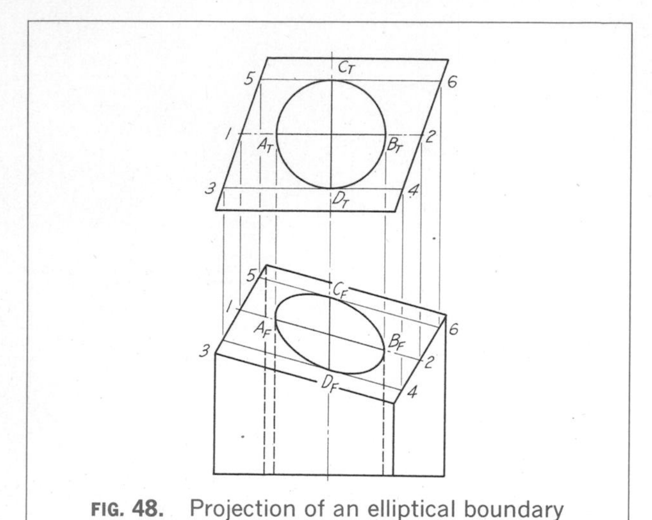


**FIG. 48.** Projection of an elliptical boundary by employing conjugate diameters. Refer to Sec. 53 and Fig. 68, both in Chap. 3.

the straight-line boundary of the skew surface. Thus the line 1–2 located in the front view and intersected by projection of $A_TB_T$ from the top view locates $A_FB_F$. Similarly, lines 3–4 and 5–6 at the ends of the axis $CD$ locate $C_FD_F$.

## 24. Projections of a Curved Boundary.

Any nongeometric curve (or a geometric curve not having established axes) must be projected by locating points on the curve. If the surface is in an inclined position, as in Fig. 49*a*, points may be assumed on the curve laid out from data (assumed in this case to be the top view) and projected first to the edge view (front) and then to an adjacent view (side). Measurements, such as 1, 2, etc., from the top view transferred to the side view complete the projection. A smooth curve is then drawn through the points.

If the surface is skew, as in Fig. 49*b*, elements of the skew surface, such as 1′-1, 2′-2, etc., located in an adjacent view (by drawing the elements parallel to some known line of the skew surface, such as $AB$) make it possible to project points on the curve 1, 2, etc., to the adjacent view, as shown.

## 25. Projections by Identifying Corners.

In projecting orthographic views or in comparing the views with a picture, it is helpful in some cases to letter (or number) the corners of the object and, with these identifying marks, to letter the corresponding

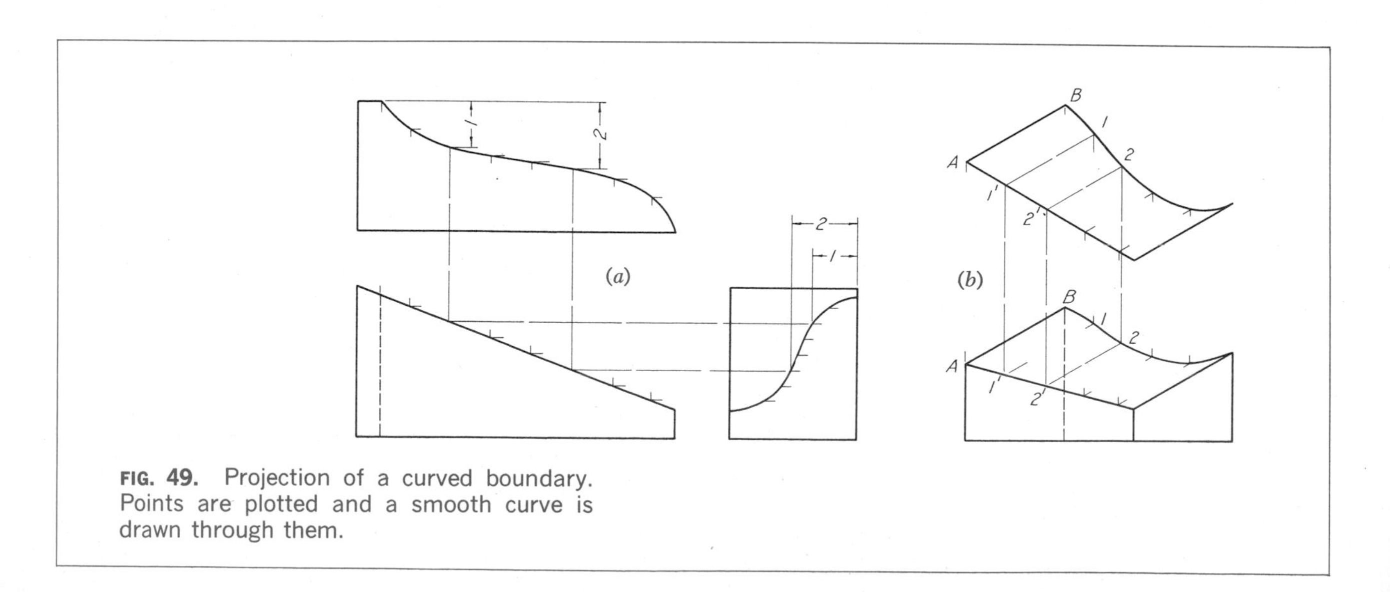


**FIG. 49.** Projection of a curved boundary. Points are plotted and a smooth curve is drawn through them.

points on each of the views, as in Fig. 50. Hidden points directly behind visible points are lettered to the right of the letter of the visible point, and in this figure, they have been further differentiated by the use of "phantom," or dotted, letters. Study Fig. 51, and number or letter the corners of the three views to correspond with the pictorial view.

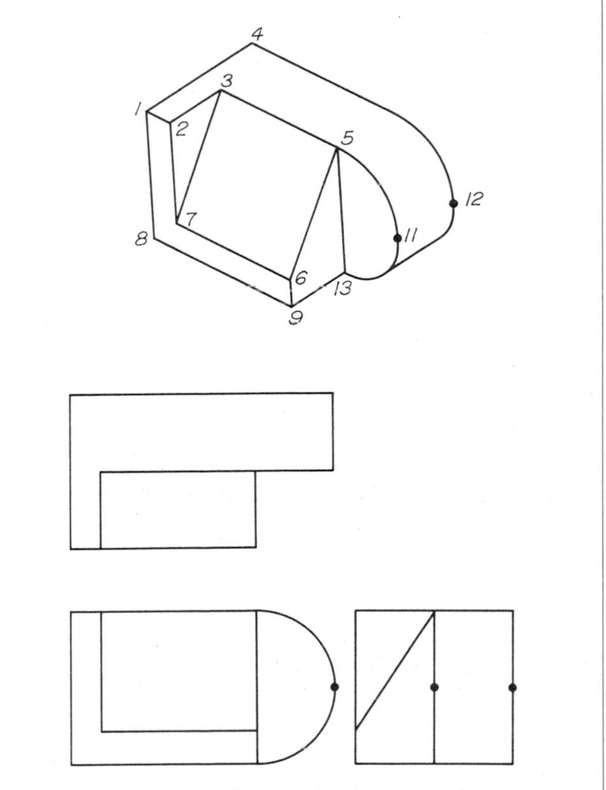


**FIG. 51.** Projection study. Number the corners of the orthographic views to correspond with the numbers on the picture.

## 26. Order of Drawing.

The order of working is important, as speed and accuracy depend largely upon the methods used in laying down lines. Avoid duplications of the same measurement and keep to a minimum changing from one instrument to another. Naturally, *all* measurements cannot be made with the scale at one time or *all* circles and arcs drawn without laying down the compass, but as much work as possible should be done with one instrument before shifting to another. An orderly placement of working tools on the drawing table will save time when changing instruments. The usual order of working is shown in Fig. 52.

1. Decide what combination of views will best describe the object. A freehand sketch will aid in choosing the views and in planning the general arrangement of the sheet.

2. Decide what scale to use, and by calculation or measurement find a suitable standard sheet size;

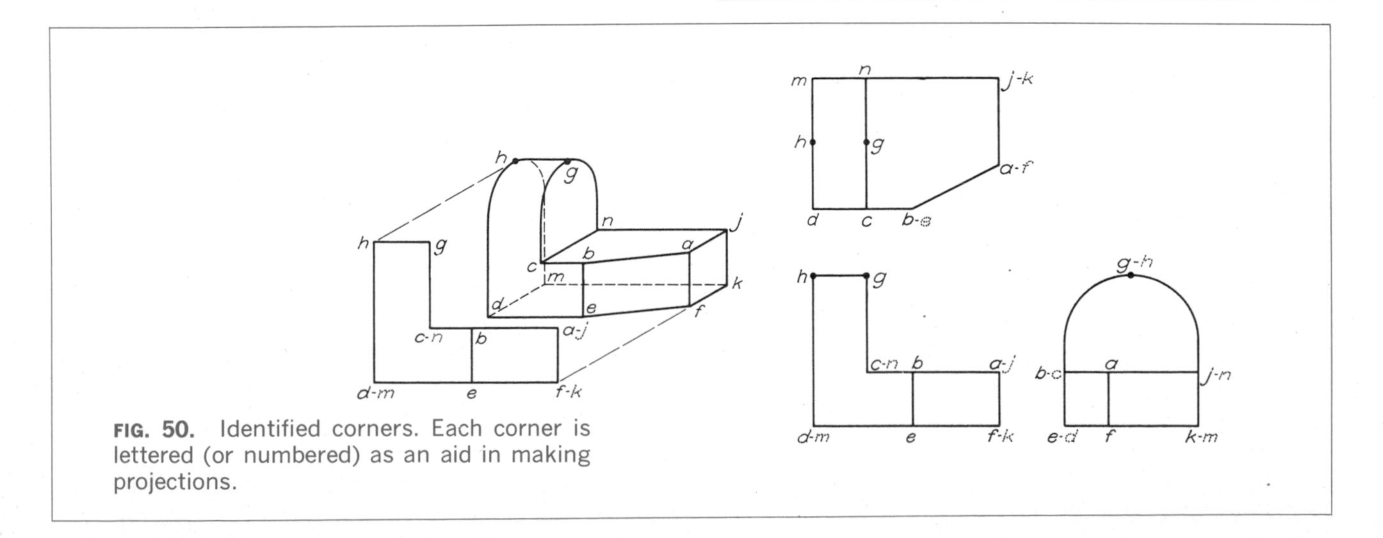


**FIG. 50.** Identified corners. Each corner is lettered (or numbered) as an aid in making projections.

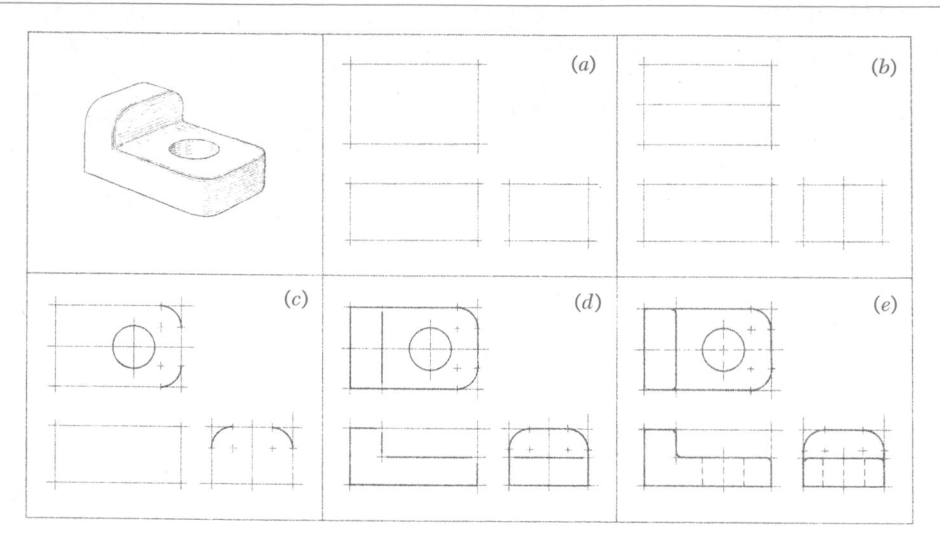


**FIG. 52.** Stages in penciling. (*a*) block out the views; (*b*) locate center lines; (*c*) start details, drawing arcs first; (*d*) draw dominant details; (*e*) finish. See text for explanation.

or pick one of the standard drawing-sheet sizes and find a suitable scale.

3. Space the views on the sheet, as described in Sec. 20.

4. Lay off the principal dimensions, and then block in the views with light, sharp, accurate outline and center lines. Draw center lines for the axes of all symmetrical views or parts of views. Every cylindrical part will have a center line—the projection of the axis of the piece. Every circle will have two center lines intersecting at its center.

5. Draw in the details of the part, beginning with the dominant characteristic shape and progressing to the minor details, such as fillets and rounds. Carry the different views along together, projecting a characteristic shape, as shown in one view to the other views, instead of finishing one view before starting another. Use a minimum of construction and draw the lines to finished weight, if possible, as the views are carried along. *Do not make the drawing lightly and then "heavy" the lines later.*

6. Lay out and letter the title.

7. Check the drawing carefully.

## 27. Order of Tracing.

If the drawing is to be traced in ink as an exercise in the use of instruments or for a finished orthographic drawing without dimensions, the order of working is as follows:

1. Place the pencil drawing to be traced on the drawing board, carefully align it with the T square, and put thumbtacks in the two upper corners. Then place the tracing paper or cloth (dull side up) over the drawing. Holding the cloth in position, lift the tacks one at a time and replace them to hold both sheets. Then put tacks in the two lower corners.

2. To remove any oily film, prepare the surface of the cloth or paper by dusting it lightly with prepared pounce or soft white chalk. *Then wipe the surface perfectly clean with a soft cloth.*

3. Carefully set the pen of the compass to the correct line width, and ink all full-line circles and circle arcs, beginning with the smallest. Correct line weights are given in Fig. 57, Chap. 2.

4. Ink dashed circles and arcs in the same order as full-line circles.

5. Carefully set the ruling pen to draw a line

exactly the same width as the line in the full-line circles. The best way to match the straight lines to the circles is to draw with the compass and ruling pen outside the trim line of the sheet or on another sheet of the same kind of paper and adjust the ruling pen until the lines match.

6. Ink irregular curved lines.
7. Ink straight full lines in this order: horizontal (begin at the top of the sheet and work down), vertical (from the left side of the sheet to right), and inclined (uppermost first).
8. Ink straight dashed lines in the same order. Be careful to match these lines with the lines in the dashed circles.
9. Ink center lines.
10. Crosshatch all areas representing cut surfaces.
11. Draw pencil guide lines and letter the title.
12. Ink the border.
13. Check the tracing for errors and omissions.

**28. Orthographic Freehand Drawing.** Facility in making freehand orthographic drawings is an essential part of the equipment of every engineer, and since ability in sketching presupposes some mastery of other skills (as we have seen in Chaps. 1 to 4), practice should be started early. Although full proficiency in freehand drawing is synonymous with mastery of the graphic language and is gained only after acquiring a background of knowledge and skill in drawing with instruments, sketching is an excellent method for learning the fundamentals of orthographic projection and can be used by the beginner even before he has had much practice with instruments. In training, as in professional work, time can be saved by working freehand instead of with instruments, as with this method more problems can be solved in an allotted amount of time.

Although some experienced teachers advocate the making of freehand sketches before practice in the use of instruments, some knowledge of the use of instruments and especially of applied geometry is a great help because the essentials of line tangents, connections, and intersections as well as the basic geometry of the part should be well defined on a freehand drawing. Drawing freehand is, of course, an excellent exercise in accuracy of observation. Figure 53 is an example of a good freehand drawing.

**29. Line Quality for Freehand Work.** Freehand drawings are made on a wide variety of papers, ranging from inexpensive notebook or writing grades to the finer drawing and tracing papers and even pencil cloth. The surface texture—smooth, medium, or rough—combined with the grade of pencil and pressure used will govern the final result. If a bold rough effect is wanted for a scheming or idea sketch, a soft pencil and possibly rough paper would be employed. For a working sketch or for representation of an object with much small or intricate detail, a harder pencil and smoother paper would help to produce the necessary line quality.

Figure 54 is a photograph (reproduced about half size) showing lines made with pencils of various grades with medium pressure on paper of medium texture. Note that the 6B pencil gives a rather wide and rough line. As the hardness increases from 5B to 4B, etc., up to 2H, the line becomes progressively narrower and lighter in color. This does not mean that a wide line cannot be made with a fairly hard pencil, but with ordinary sharpening and with uniform pressure the softer grades wear down much faster than the harder grades. Unless a soft pencil such as 6B, 5B, or 4B is sharpened after each short stroke, fine lines are impossible. Also, the harder grades such as F, H, and 2H, once sharpened, will hold their point for some time, and fairly fine lines can be obtained without much attention to the point. With ordinary pressure and normal use the soft grades give bold rough results, and the harder grades give light smooth lines.

**30. Range of Pencil Grades.** The 6B grade is the softest pencil made and gives black, rough lines. With normal pressure, the line erases easily but is likely to leave a slight smear. The 6B, 5B, and 4B pencils should be used when a rather rough line is wanted, for example, for scheming or idea sketches, architectural renderings, and illustrations. The range from 3B to HB, inclusive, is usually employed for engineering sketches on medium-textured paper. For example, a sketch

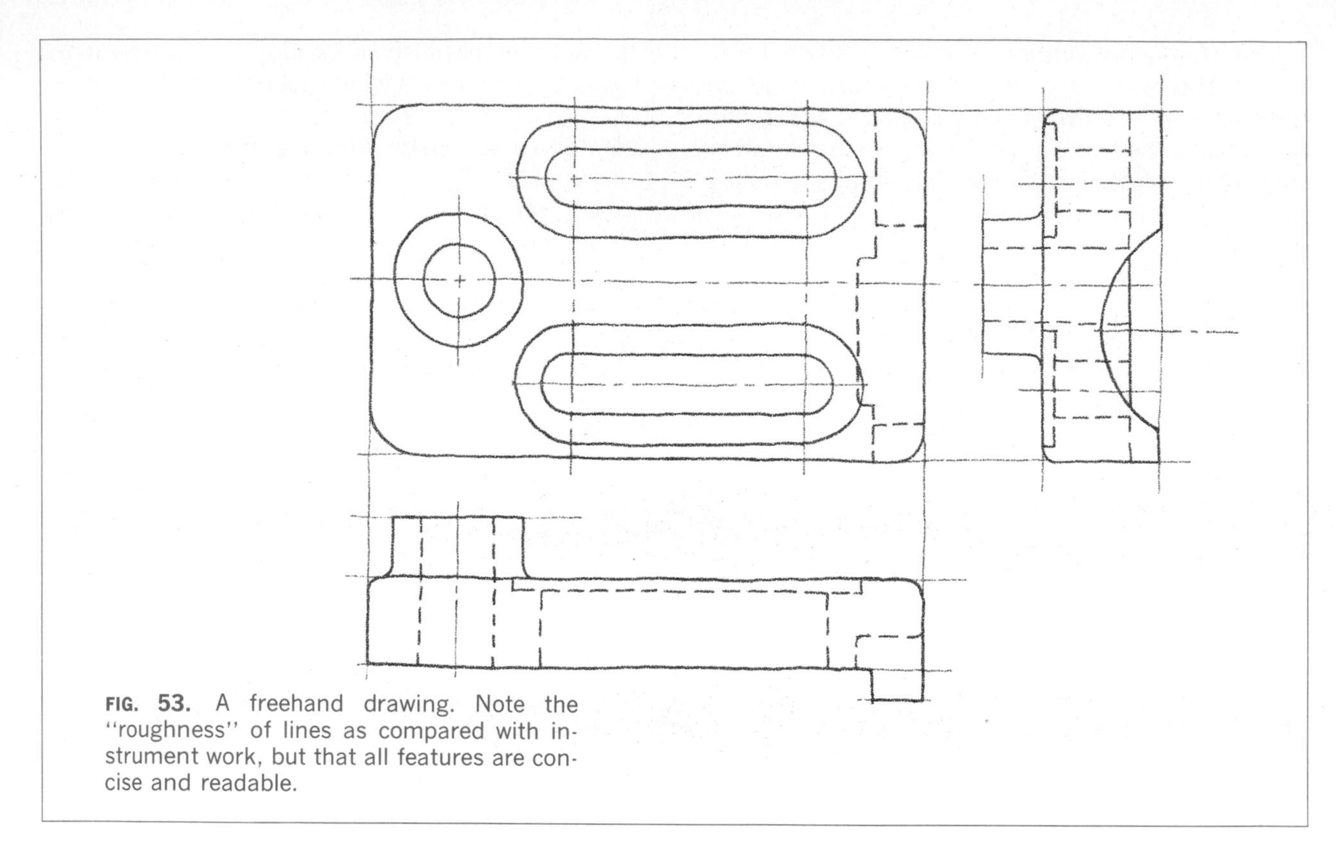

FIG. 53. A freehand drawing. Note the "roughness" of lines as compared with instrument work, but that all features are concise and readable.

of a machine part with a normal amount of small detail can be made effectively with a 2B or B pencil. The smoothness and easy response to variations in pressure make these grades stand out as the preferred grades for a wide variety of work. However, for more critical work where there is much detail and also where the smearing of the soft grades is objectionable, the grades from F to 2H are used. For a sketch on fairly smooth paper, F or H are quite satisfactory. Grades harder than 2H are rarely used for freehand work. Incidentally, a fine pencil for ordinary writing is the 2B grade.

### 31. Sharpening the Pencil.

The pencil should be sharpened to a fairly long, conic point, as explained in Sec. 24, Chap. 2. However, for freehand work a point not quite so sharp as for instrument drawing gives the desired line width without too much pressure. If after sharpening in the regular way the point is too fine, it can be rounded off slightly on a piece of scratch paper before use on the drawing.

### 32. Pencil Pressure and Paper Texture.

The pencil grade, pressure, and paper texture all have an effect on the final result. Figure 54 shows the various pencil grades with medium pressure on paper of medium texture. To mark the difference in line quality obtainable by increasing the pressure, Fig. 55 shows the same paper but with firm pressure on the pencil. Note that the lines in Fig. 55 are much blacker than those in Fig. 54. The line quality of Fig. 55 is about right for most engineering sketches. The rather firm opaque lines are preferred to the type in Fig. 54, especially if reproductions, either photographic or by transparency process, are to be made from the sketch.

To show the difference in line quality produced by paper texture, Fig. 56 shows the same firm pressure used in Fig. 55, but this time on smooth paper. In Fig. 57 the same pressure has been used on rough paper. Figure 58 is given to aid in comparing the effect of pressure and texture. On the left is the medium pressure on medium-textured paper; at (*a*), (*b*), and (*c*) firm pressure has been used but at (*a*) on medium-, at (*b*) on smooth-, and at (*c*) on rough-textured paper.

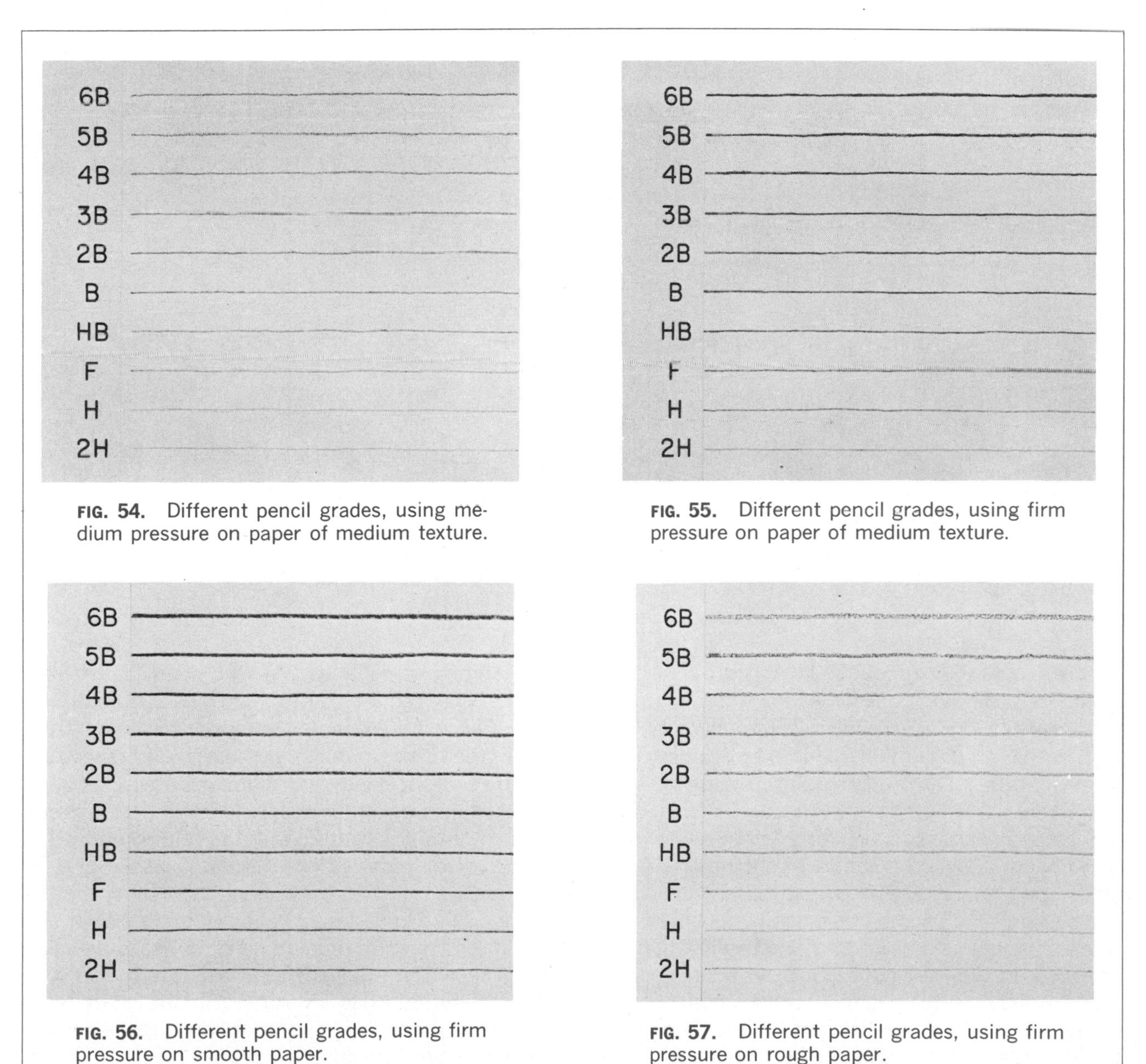


**FIG. 54.** Different pencil grades, using medium pressure on paper of medium texture.

**FIG. 55.** Different pencil grades, using firm pressure on paper of medium texture.

**FIG. 56.** Different pencil grades, using firm pressure on smooth paper.

**FIG. 57.** Different pencil grades, using firm pressure on rough paper.

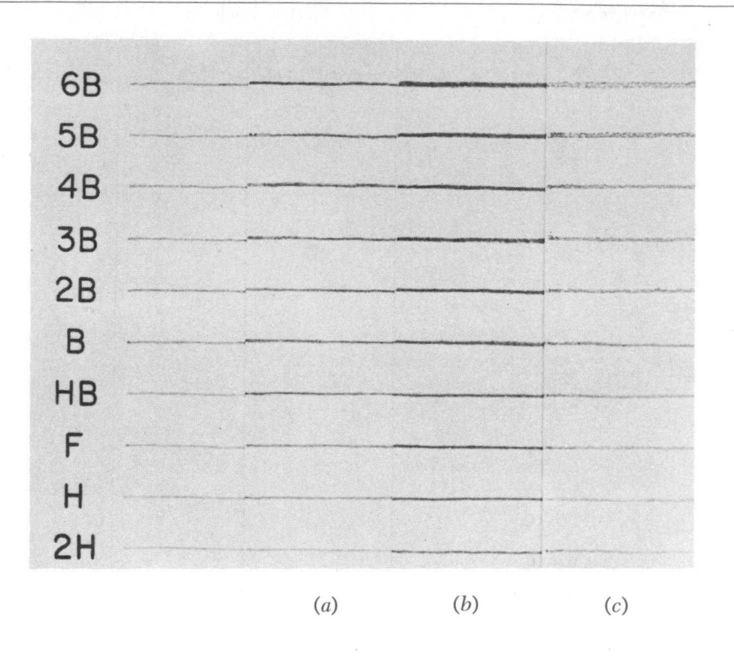


FIG. 58. Comparison of different pencil pressures and paper surfaces. Medium pressure on medium paper—left column. Firm pressure on medium paper at (*a*), on smooth at (*b*), and on rough at (*c*).

## 33. Kinds of Paper.

*Plain and Coordinate.* Sketches are made for many purposes and under a variety of circumstances, and as a result on a number of different paper types and surfaces. A field engineer in reporting information to the central office may include a sketch made on notebook paper or a standard letterhead. On the other hand, a sketch made in the home office may be as important as any instrument drawing and for this reason may be made on good-quality drawing or tracing paper and filed and preserved with other drawings in a set. Figure 59 is an example of a sketch made on plain paper, which might be the letterhead paper of the field engineer or a piece of fine drawing or tracing paper. The principal difficulty in using plain paper is that proportions and projections must be estimated by eye. A good sketch on plain paper requires better-than-average ability and experience. Use of some variety of coordinate paper is a great aid in producing good results. There are many kinds of paper and coordinate divisions available, from smooth to medium texture and coordinate divisions of $\frac{1}{8}$ or $\frac{1}{10}$ to $\frac{1}{2}$ in., printed on tracing paper or various weights of drawing paper. Usually one coordinate size on tracing paper and another (or the same) on drawing paper will supply the needs of an engineering office. Figure 60 is an example of a sketch on paper with coordinate divisions of $\frac{1}{8}$ in. Figure 61 has divisions of $\frac{1}{4}$ in. Figure 62 is a sketch made on $\frac{1}{4}$-in. coordinate paper, actual size.

The paper type, tracing or regular, is another

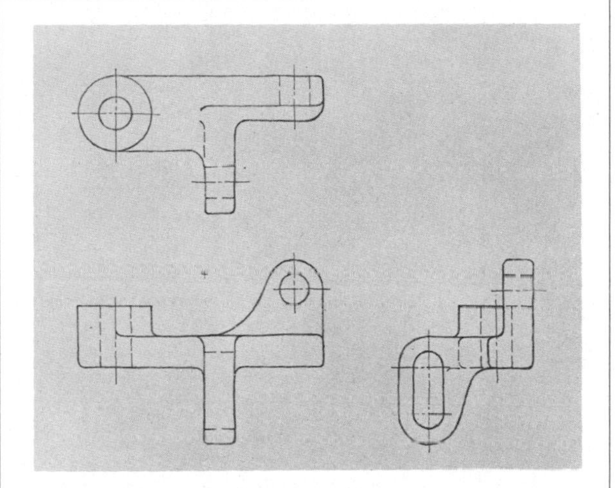

**FIG. 59.** A freehand drawing on plain paper. This drawing was made by an expert of long experience. More "roughness" or "waviness" of lines is permissible on freehand than on instrument drawings, but care should be taken to keep all details concise and readable.

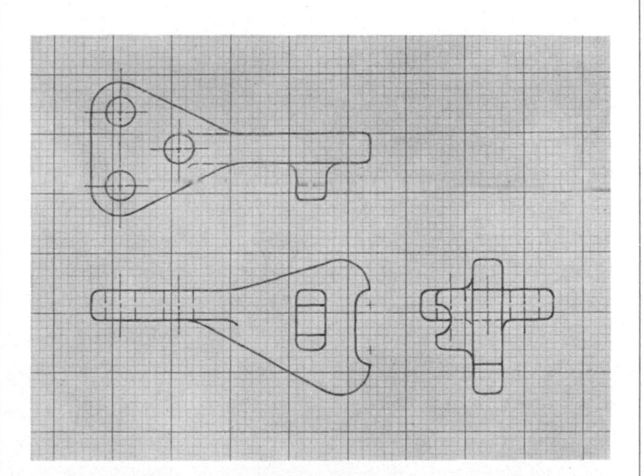

**FIG. 60.** A freehand drawing on coordinate tracing paper. The coordinates, on the back of the paper, aid greatly in making projections, in keeping the right proportions, and in drawing straight and accurate lines.

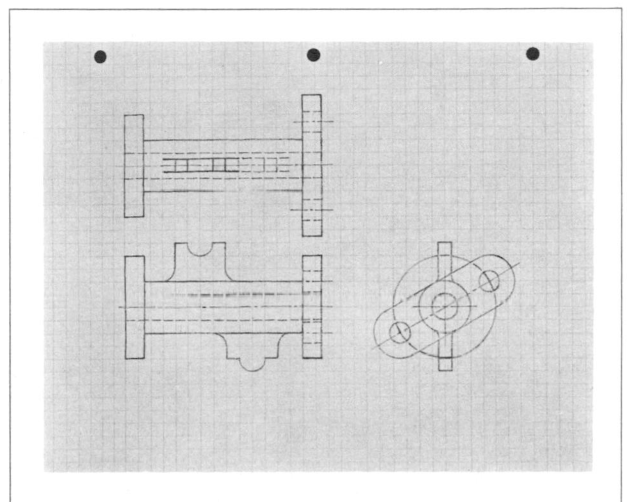

**FIG. 61.** A freehand drawing on coordinate notebook paper. Coordinates are $\frac{1}{4}$ in. apart.

factor to be considered. Reproduction by any of the transparency methods demands the use of tracing paper. If coordinate paper is used, it may be desirable to obtain prints on which the coordinate divisions do not show. Figure 60 is an example of a sketch on tracing paper with the coordinate divisions printed on the back of the paper in faint purplish-blue ink. Since the divisions are on the back, erasures and corrections can be made without erasing the coordinate divisions. Normally the divisions will not reproduce, so prints give the appearance of a sketch made on plain paper. Figure 61 is a sketch on standard, three-hole notebook paper with $\frac{1}{4}$-in. divisions in pale blue ink.

The use of coordinate paper is a great aid in freehand drawing, and speeds up the work considerably. Projections are much easier to make on it than on plain paper, and to transfer distances from top to side view, the divisions can be counted.

### 34. Technique.

The pencil is held with freedom and not close to the point. Vertical lines are drawn downward with a finger movement in a series of overlapping strokes, the hand somewhat in the position of Fig.

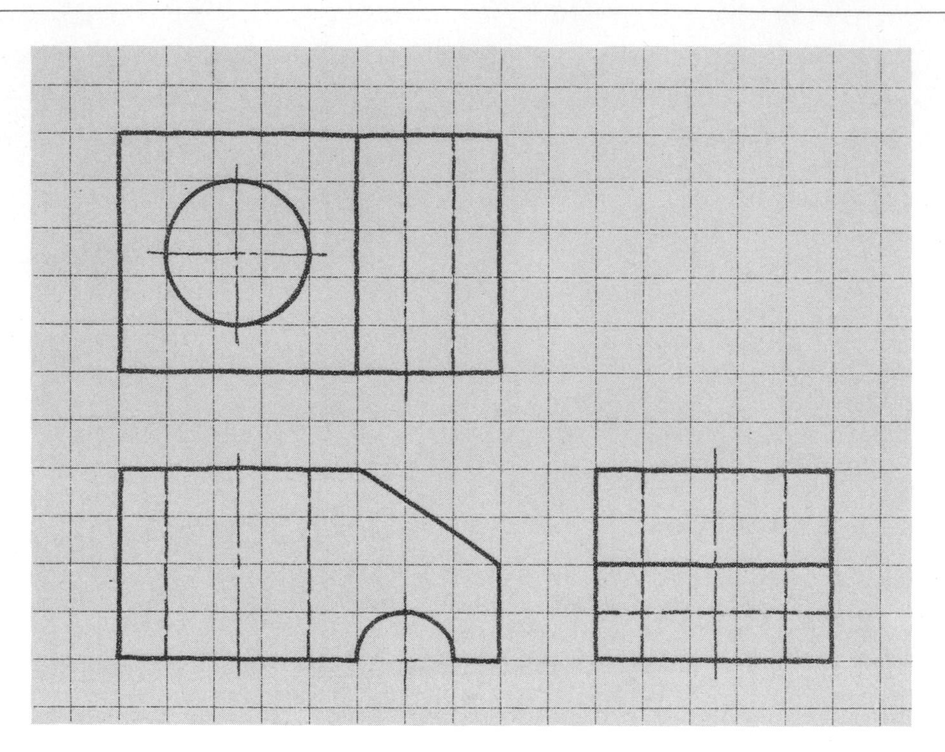

**FIG. 62.** A freehand drawing on coordinate paper (actual size). Note the bold but concise technique.

63. Horizontal lines are drawn with the hand shifted to the position of Fig. 64, using a wrist motion for short lines and a forearm motion for longer ones. In drawing any straight line between two points, *keep your eyes on the point to which the line is to go rather than on the point of the pencil.* Do not try to draw the whole length of a line in a single stroke. It may be helpful to draw a very light line first, as in Fig. 65*A*, and then to sketch the finished line, correcting the direction of the light line and bringing the line to final width and blackness by using strokes of convenient length, as at (*B*). The finished line is shown at (*C*). Do not be disturbed by any nervous waviness. Accuracy of direction is more important than smoothness of line.

### 35. Straight Lines.

Horizontal lines are drawn from left to right as in Fig. 66, vertical lines from top to bottom as in Fig. 67.

Inclined lines running downward from right to left (Fig. 68) are drawn with approximately the same movement as vertical lines, but the paper may be turned and the line drawn as a vertical (Fig. 69).

Inclined lines running downward from left to right (Fig. 70) are the hardest to draw because the hand is in a somewhat awkward position; for this reason, the paper should be turned and the line drawn as a horizontal, as in Fig. 71.

The sketch paper can easily be turned in any direction to facilitate drawing the lines because there is no necessity to fasten the paper to a drawing-table top. The paper may, of course, be taped to a drawing board or attached to a clip board.

It is legitimate in freehand drawing to make long vertical or horizontal lines using the little finger as

**FIG. 63.** Sketching a vertical line. Draw downward with finger movement, overlapping the strokes for long lines.

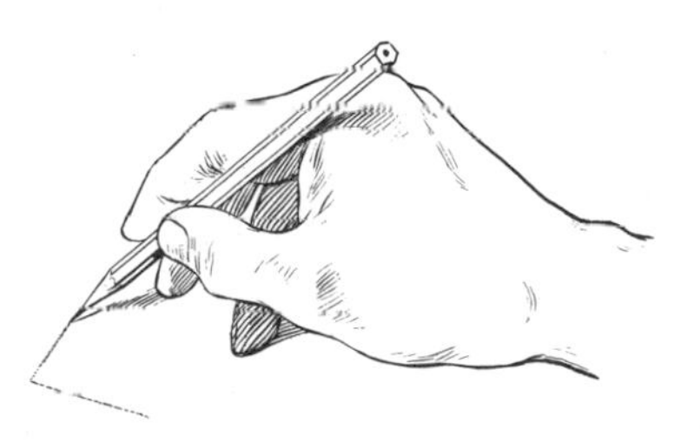

**FIG. 64.** Sketching a horizontal line. Draw from left to right with wrist pivot for short lines and forearm movement for long lines. Overlap strokes if necessary.

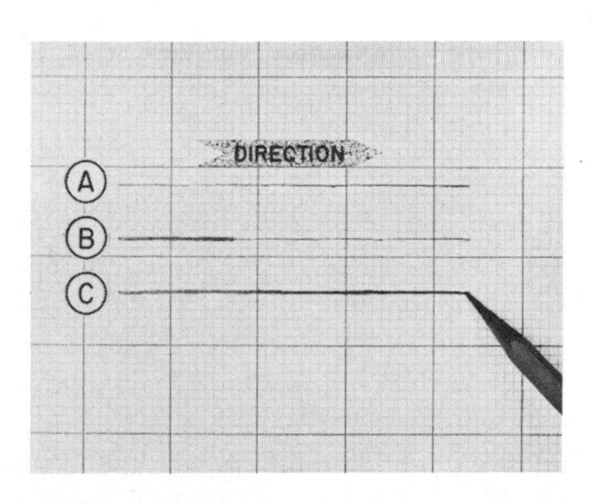


**FIG. 65.** Technique of sketching lines. (*a*) set direction with a *light* construction line; (*b*) first stroke; (*c*) complete line with a series of overlapping strokes.

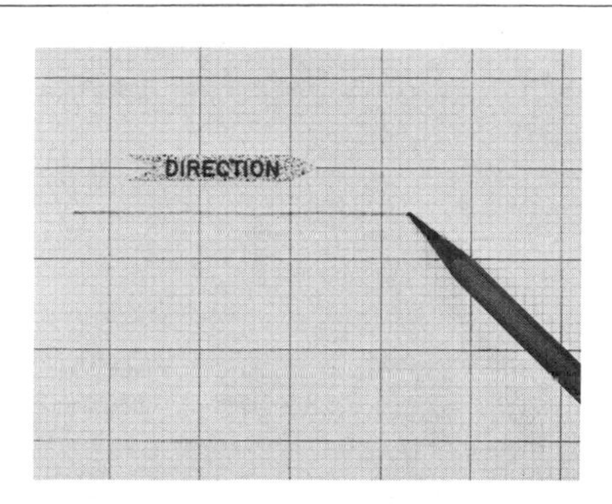


**FIG. 66.** Sketching a horizontal line. Draw from left to right.

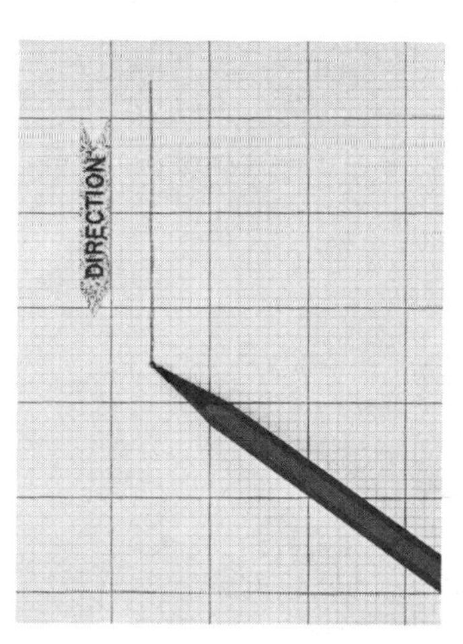


**FIG. 67.** Sketching a vertical line. Draw from top to bottom.

a guide along the edge of the pad or clip board. The three important things about a straight line are that it (1) be essentially straight, (2) be the right length, and (3) go in the right direction.

### 36. Circles.

Circles can be drawn by marking the radius on each side of the center lines. A more accurate method is to draw two diagonals in addition to the center lines and mark points equidistant from the center of the eight radii; at these points, draw short arcs

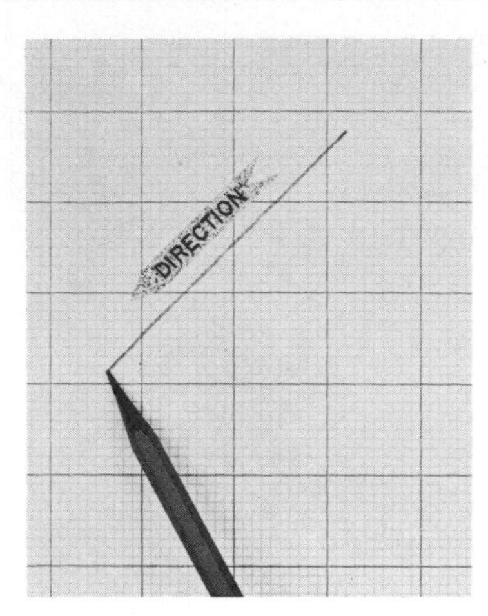


FIG. 68. Sketching an inclined line sloping downward from right to left. This line may be drawn in either direction, whichever is more convenient by personal preference.

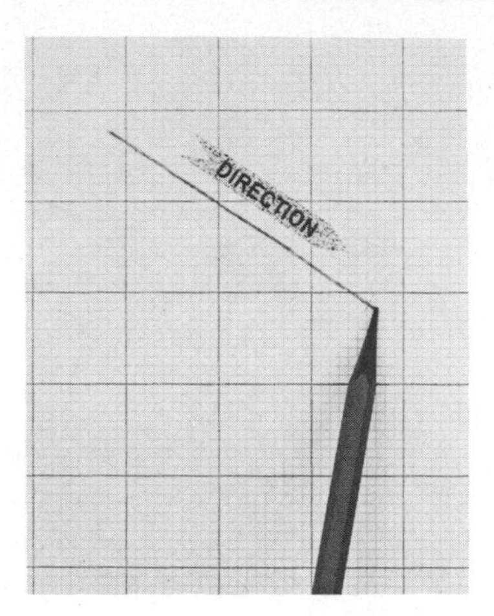


FIG. 70. Sketching an inclined line sloping downward from left to right. This is the most awkward position of any line. The paper should be turned as in Fig. 71 to aid in obtaining a smooth, accurate line.

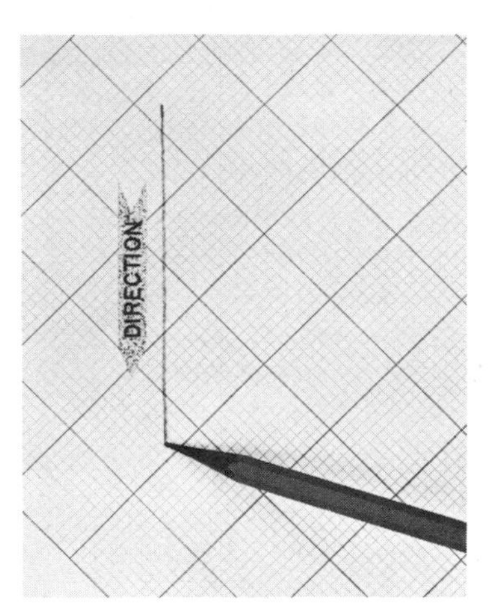


FIG. 69. Turning the paper to sketch an inclined line as a vertical line. This is often a great help because of the awkward position of the inclined line. An alternate turn is to the position of Fig. 71.

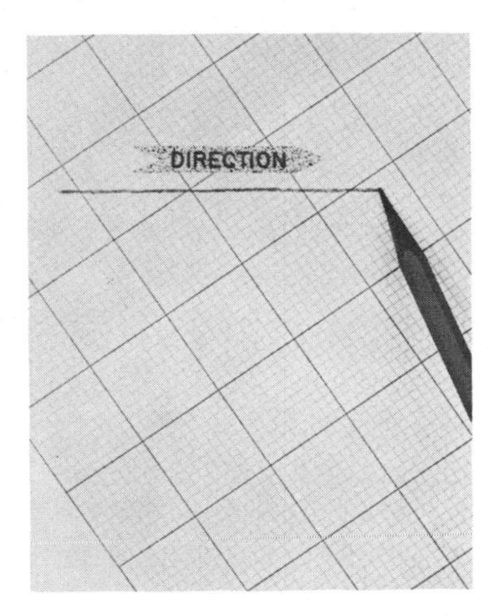


FIG. 71. Paper turned to sketch an inclined line as a horizontal line. This should be done especially for the type of line shown in Fig. 70.

perpendicular to the radii, and then complete the circle as shown in Fig. 72. A modification is to use a slip of paper as a trammel. Large circles can be done smoothly, after a little practice, by using the third or fourth finger as a pivot, holding the pencil stationary and rotating the paper under it, or by holding two pencils and using one as a pivot about which to rotate the paper. Another way of drawing a circle is to sketch it in its circumscribing square.

## 37. Projection.

In making an orthographic sketch, remember and apply the principles of projection and applied geometry. Sketches are *not* made to scale but are made to show fair proportions of objects sketched. It is legitimate, however, when coordinate paper is used, to count the spaces or rulings as a means of proportioning the views and as an aid in making projections. Take particular care to have the various details of the views in good projection from view to view. It is an inexcusable mistake to have a detail sketched to a different size on one view from that on another.

When working on plain paper, projections between the top and front views or between the front and side views are easily made by simply "sighting" between the views or using *very* light construction lines, as in Fig. 73. Projections between the top and side views are laid off by judging the distance by eye, by measuring the distance by holding the finger at the correct distance from the end of the pencil and transferring to the view, or by marking the distance on a small piece of paper and transferring to the view. Note in Fig. 73 that distances *A*, *B*, and others could be transferred from the top to the side view by the methods just mentioned.

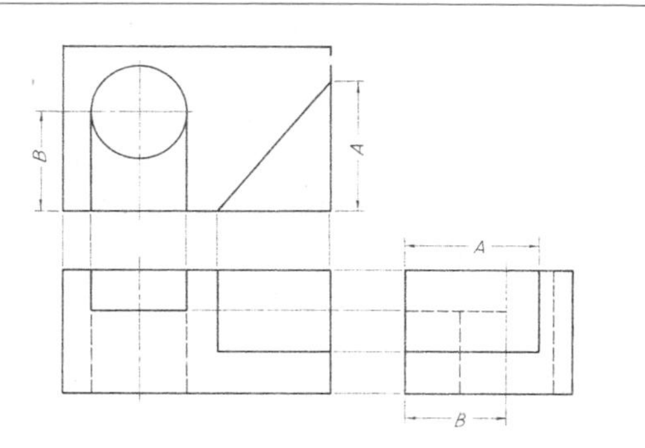


**FIG. 73.** Freehand projections. Judge the projections by eye (aided by coordinate lines if lined paper is used). Judge the measurements (*A* and *B*) by eye; *or* measure the distances with a pencil or by marking on a strip of paper.

Even though freehand lines are somewhat "wavy" and not so accurate in position as ruled lines, a good freehand drawing should present the same clean appearance as a good instrument drawing.

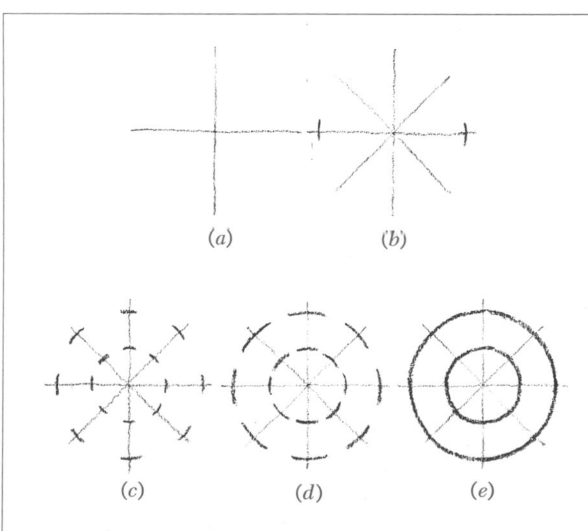


**FIG. 72.** Method of drawing freehand circles. (*a*) draw center lines; (*b*) draw diagonals; (*c*) space points on the circle with *light*, short lines (by eye); (*d*) correct and begin filling in; (*e*) finish.

## 38. Method.

Practice in orthographic freehand drawing should be started by drawing the three views of a number of simple pieces, developing the technique and the ability to "write" the orthographic language, while exercising the constructive imagination in visualizing the object by looking at the three projections. Observe the following order of working:

1. Study the pictorial sketch and decide what combination of views will best describe the shape of the piece.

2. Block in the views, as in Fig. 74*a*, using a very light stroke of a soft pencil (2B, B, HB, or F) and spacing the views so as to give a well-balanced appearance to the drawing.

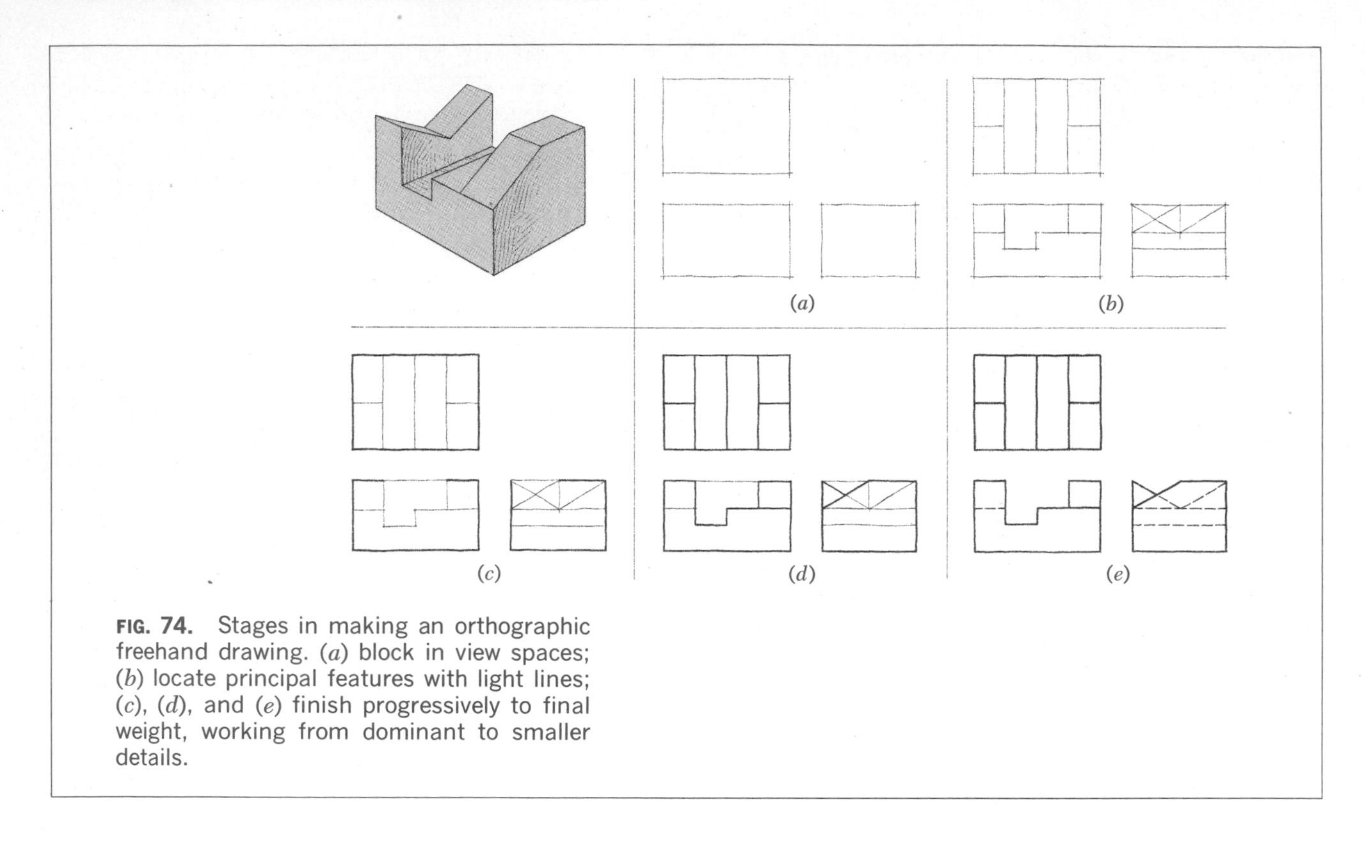


**FIG. 74.** Stages in making an orthographic freehand drawing. (*a*) block in view spaces; (*b*) locate principal features with light lines; (*c*), (*d*), and (*e*) finish progressively to final weight, working from dominant to smaller details.

3. Build up the detail in each view, carrying the three views along together as at (*b*).

4. Brighten the outline of each view with bold strokes as at (*c*).

5. Brighten the detail with bold strokes, thus completing the full lines of the sketch as at (*d*).

6. Sketch in all dashed lines, using a stroke of medium weight and making them lighter than the full lines, as at (*e*), thus completing the shape description of the object.

7. Check the drawing carefully. Then cover the pictorial sketch and visualize the object from the three views.

After drawing a number of simple pieces freehand, try more complicated problems such as Probs. 27 to 55. Faintly ruled coordinate paper (illustrated in Fig. 60) may be used if desired.

### 39. Shop Processes.

Shop processes are properly a part of dimensioning and specification for working drawings and in this text are given in Chap. 11, preceding dimensioning and tolerancing, screw threads and fasteners, and working drawings. However, in order to read the pictorial drawings (problems) to be drawn in orthographic projection, some knowledge of processing fundamentals is necessary. Therefore, Chap. 11, Methods Used in Manufacture, should be studied, especially for a knowledge of hole processing (drilling, reaming, etc.) and for information on fillets, rounds, finished surfaces, and methods of part manufacture. Consult the Glossary for unfamiliar terms.

### 40. Dimensioning Systems.

For the same reasons given in the previous paragraph, some knowledge of dimensioning systems is necessary at this time. Basically, two methods are used: fractional and decimal. These are explained in Sec. 12, Chap. 12. Note especially that a decimally dimensioned part is laid out with a decimal scale.

# READING THE GRAPHIC LANGUAGE

## 41. Orthographic Reading.

The engineer must be able to *read* and *write* the orthographic language. The necessity of learning to read is absolute because everyone connected with technical industry must be able to read a drawing without hesitation or concede technical illiteracy.

Reading the orthographic language is a mental process; a drawing is not read aloud. To describe even a simple object with words is almost impossible. Reading proficiency develops with experience, as similar conditions and shapes occur so often that a person in the field gradually acquires a background of knowledge that enables him to visualize readily the shapes shown. Experienced readers read quickly because they can draw upon their knowledge and recognize familiar shapes and combinations without hesitation. However, reading a drawing should always be done carefully and deliberately, as a whole drawing cannot be read at a glance any more than a whole page of print.

## 42. Prerequisites and Definition.

Before attempting to read a drawing, familiarize yourself with the principles of orthographic projection, as explained in Secs. 1 to 14. Keep constantly in mind the arrangement of views and their projection, the space measurements of height, width, and depth, what each line represents, etc.

*Visualization* is the medium through which the shape information on a drawing is translated to give the reader an understanding of the object represented. The *ability to visualize* is often thought to be a "gift" that some people possess and others do not. This, however, is not true. Any person of reasonable intelligence has a visual memory, as can be seen from his ability to recall and describe scenes at home, actions at sporting events, and even details of acting and facial expression in a play or motion picture.

The ability to visualize a shape shown on a drawing is almost completely governed by a person's knowledge of the principles of orthographic projection. The common adage that "the best way to learn to read a drawing is to learn how to make one" is quite correct, because in learning to make a drawing you are forced to study and apply the principles of orthographic projection.

*Reading a drawing* can be defined as *the process of recognizing and applying the principles of orthographic projection to interpret the shape of an object from the orthographic views.*

## 43. Method of Reading.

A drawing is read by visualizing units or details one at a time from the orthographic projection and mentally orienting and combining these details to interpret the whole object finally. The form taken in this visualization, however, may not be the same for all readers or for all drawings. Reading is primarily a reversal of the process of making drawings; and inasmuch as drawings are usually first made from a picture of the object, the beginner often attempts to carry the reversal too completely back to the pictorial. The result is that the orthographic views of an object like those shown in Fig. 75 are translated to the accompanying picture, with the thought of the object as positioned in space or placed on a table or similar surface. Another will need only to recognize in the drawing the geometry of the solid, which in the case of Fig. 75 would be a rectangular prism so high, so wide, and so deep with a hole passing vertically through the center of it. This second reader will have read the views just as completely as the first but with much less mental effort.

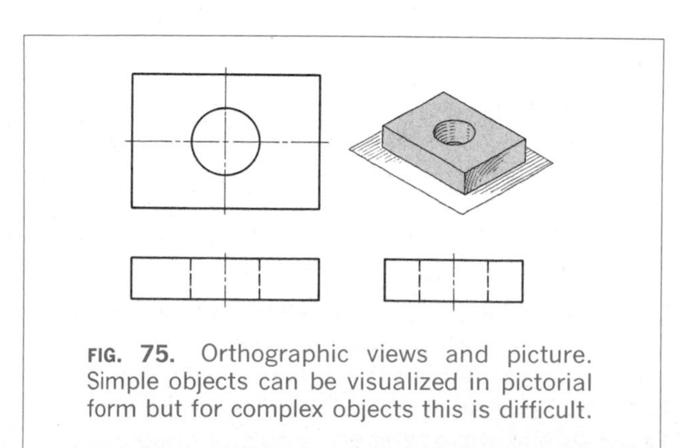

FIG. 75. Orthographic views and picture. Simple objects can be visualized in pictorial form but for complex objects this is difficult.

To most, it is a mental impossibility (and surely unnecessary) to translate more than the simplest set of orthographic views into a complete pictorial form that can be pictured in its entirety. Actually, the reader goes through a routine pattern of procedure (listed in Sec. 44). Much of this is done subconsciously. For example, consider the object in Fig. 76. A visible circle is seen in the top view. Memory of previous projection experience indicates that this must be a hole or the end of a cylinder. The eyes rapidly shift back and forth from the top view to the front view, aligning features of the same size ("in projection"), with the mind assuming the several possibilities and finally accepting the fact that, because of the dashed lines and their extent in the front view, the circle represents a hole that extends through the prism. Following a similar pattern of analysis, the reader will find that Fig. 77 represents a rectangular prism surmounted by a cylinder. This thinking is done so rapidly that the reader is scarcely aware of the steps and processes involved.

The foregoing is the usual method of reading; but how does the beginner develop this ability?

*First,* as stated in Sec. 42, he must have a reasonable knowledge of the principles of orthographic projection.

*Second,* as described in Secs. 45 and 46 he must acquire a complete understanding of the principles behind the meaning of lines, areas, etc., and the mental process involved in interpreting them, as these principles are applied in reading.

There is very little additional learning required. Careful study of all these items plus practice will develop the ability and confidence needed.

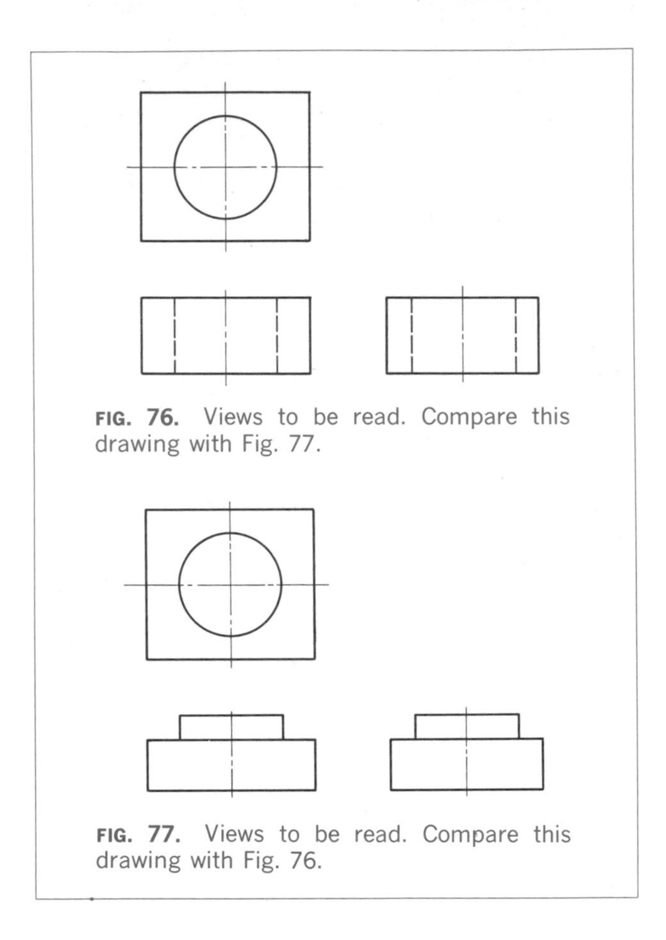

**FIG. 76.** Views to be read. Compare this drawing with Fig. 77.

**FIG. 77.** Views to be read. Compare this drawing with Fig. 76.

## 44. Procedure for Reading.

The actual steps in reading are not always identical because of the wide variety of subject matter (drawings). Nevertheless, the following outline gives the basic procedure and will serve as a guide:

*First,* orient yourself with the views given.

*Second,* obtain a general idea of the over-all shape of the object. Think of each view as the object itself, visualizing yourself in front, above, and at the side, as is done in making the views. Study the dominant features and their relation to one another.

*Third,* start reading the simpler individual features, beginning with the most dominant and progressing to the subordinate. Look for familiar shapes or conditions that your memory retains from previous experience. Read all views of these familiar features to note the extent of holes, thickness of ribs and lugs, etc.

*Fourth,* read the unfamiliar or complicated features. Remember that every point, line, surface, and solid appears in every view and that you must find the projection of every detail in the given views to learn the shape.

*Fifth,* as the reading proceeds, note the relationship between the various portions or elements of the object. Such items as the number and spacing of holes, placement of ribs, tangency of surfaces, and the proportions of hubs, etc., should be noted and remembered.

*Sixth,* reread any detail or relationship not clear at the first reading.

## 45. The Meaning of Lines.

As explained in Sec. 11, a line on a drawing indicates (1) the *edge of a surface,* (2) an *intersection of two surfaces,* or (3) a *surface limit.* Because a line on a view may mean any one of these three conditions, the corresponding part of another view must be consulted to determine the meaning. For example, the meaning of line *AB* on the front view of Fig. 78 cannot be determined until the side view is consulted. The line is then found to be the edge view of the horizontal surface of the cutout corner. Similarly, line *CD* on the top view cannot be fully understood without consulting the side view, where it is identified as the edge view of the vertical surface of the cutout corner. Lines *EF* on the top view and *GH* on the front view are identical in appearance. However, the side view shows that line *EF* represents the edge view of the rear surface of the triangular block and that line *GH* is the intersection of the front and rear surfaces of the triangular block.

The top and front views of the objects shown in Figs. 78 and 79 are identical. Nevertheless, lines *AB* and *CD* in Fig. 79 do not represent what they represent in Fig. 78 but are in Fig. 79 the intersection of two surfaces. Also, lines *EF* and *GH* in Fig. 79 are identical in appearance with those in Fig. 78, but in Fig. 79 they represent the surface limits of the circular boss.

From Figs. 78 and 79 it is readily seen that a drawing cannot be read by looking at a single view. Two views are not always enough to describe an object completely, and when three or more views are given, all must be consulted to be sure that the shape has been read correctly. To illustrate with Fig. 80, the front and top views show what appears to be a rectangular projection on the front of the object, but the side view shows this projection to be quarter-round. Similarly, in the front and side views the rear portion of the object appears to be a rectangular prism, but the top view shows that the two vertical rear edges are rounded.

A shape cannot be assumed from one or two views—*all the views must be read carefully.*

The several *lines* representing one feature must be read in all the views. As an exercise in reading the lines on an orthographic drawing, find *all* the lines representing the hole, triangular prism, slot, and cutoff corner in Fig. 81.

## 46. The Meaning of Areas.

The term "area" as used here means the contour

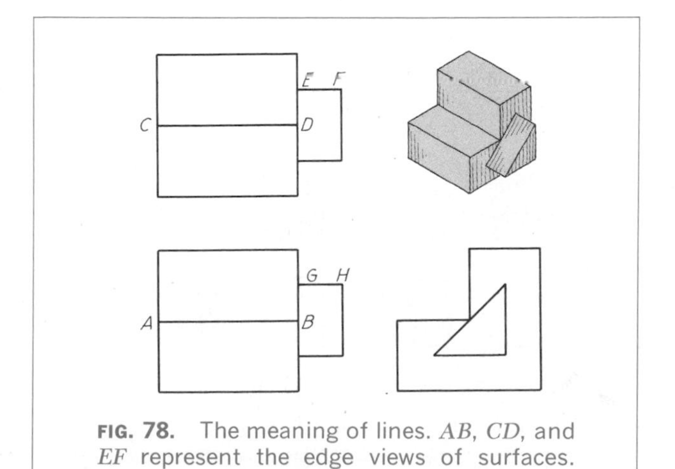


**FIG. 78.** The meaning of lines. *AB, CD,* and *EF* represent the edge views of surfaces. *GH* represents an edge. Study carefully, reading *all* views.

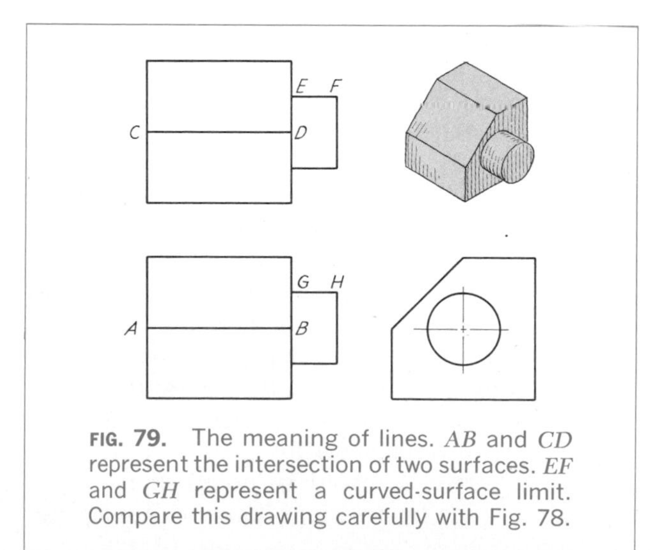


**FIG. 79.** The meaning of lines. *AB* and *CD* represent the intersection of two surfaces. *EF* and *GH* represent a curved-surface limit. Compare this drawing carefully with Fig. 78.

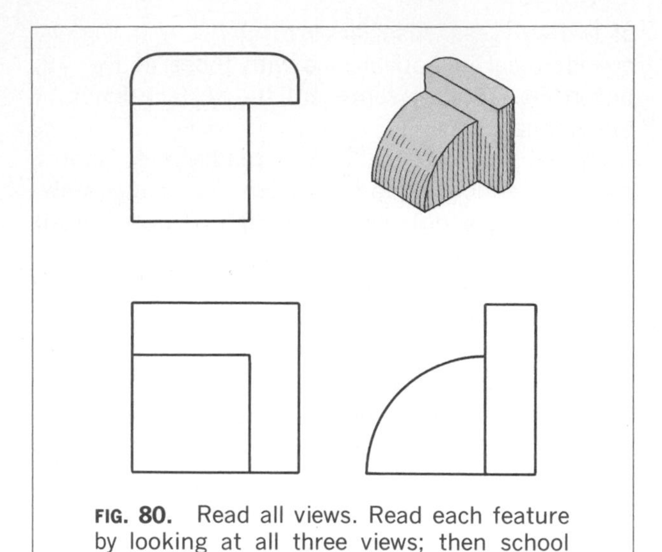

**FIG. 80.** Read all views. Read each feature by looking at all three views; then school yourself to remember all features.

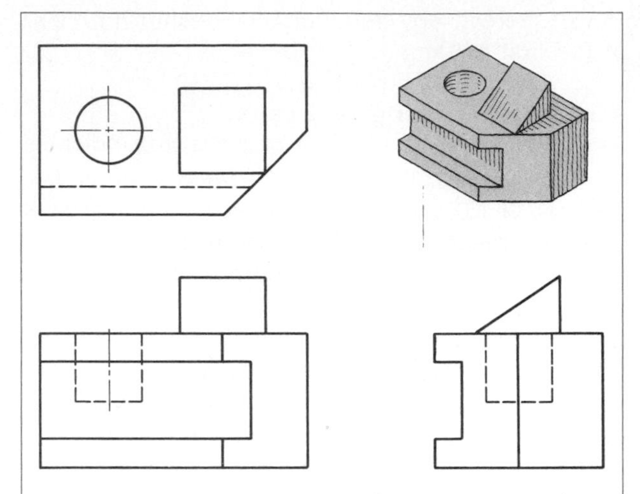

**FIG. 81.** Read all views, features, and lines. This drawing is more complex than Fig. 80. Read all features in all views, by reading all lines, then school yourself to remember all features.

limits of a surface or combination of tangent surfaces as seen in the different orthographic views. To illustrate, an area of a view as shown in Fig. 82 may represent (1) a surface in true shape as at *A*, (2) a foreshortened surface as at *B*, (3) a curved surface as at *C*, or (4) a combination of tangent surfaces as at *D*.

When a surface is in an oblique position, as surface *E* of Fig. 83, it will appear as an area in all principal views of the surface. A study of the surfaces in Figs. 82, 83, and others will establish with the force of a rule that *a plane surface, whether it is positioned in a horizontal, frontal, profile, or an inclined or skew position, will always appear in a principal orthographic view as a line or an area.* Principal views that show a skew surface as an area in more than one view will always show it in like shape. As an example, surface *A* of Fig. 84 appears as a triangular area in all the principal views; the length of the edges and the angles between the edges may change, but all views show the area with the same number of sides. It should be noted that a plane surface bounded by a certain number of sides can

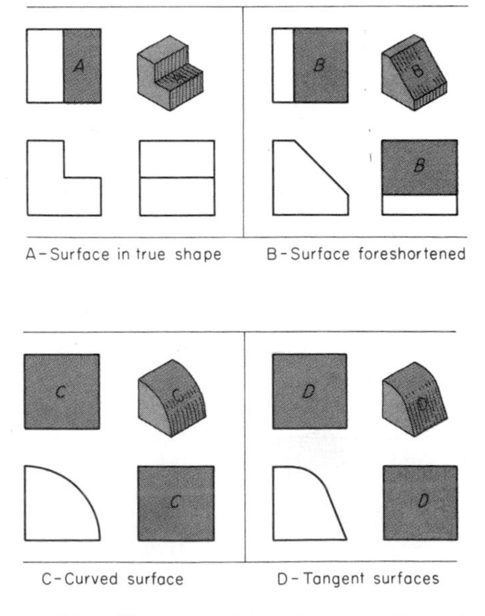


**FIG. 82.** The meaning of areas. Two views must be read to determine what an area means. Compare *A* with *B* and *C* with *D*.

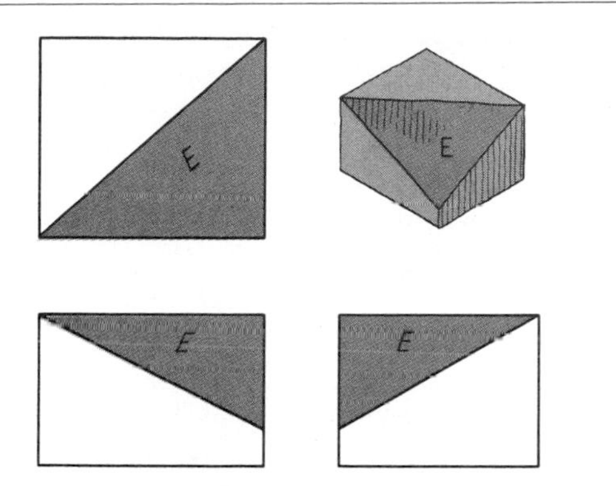


**FIG. 83.** Oblique surface. This will appear as an area in all three principal views.

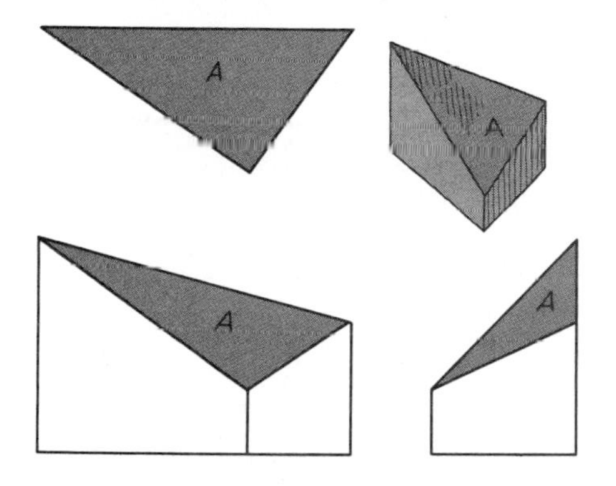


**FIG. 84.** Oblique surface. Note that area $A$, representing this surface, appears in all views as an area of similar shape formed by lines connected in the same order.

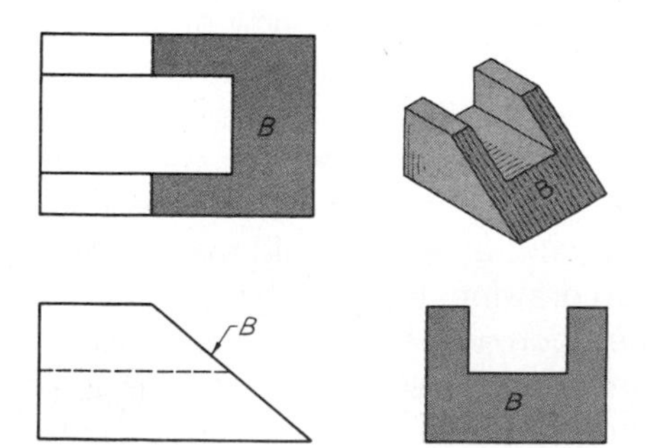


**FIG. 85.** Auxiliary surface. Note that the surface appears as an edge in the front view and as an area of similar shape in top and side views.

never appear to have more or fewer sides except when the surface appears as an edge. Moreover, the sides in any view will always connect in the same sequence. For example, in Fig. 85 the front view shows surface $B$ as an edge; the top and side views show the surface as an area having a similar shape, the same number of sides, and with the corners in the same sequence.

## 47. Adjacent Areas.

No two adjacent areas can lie in the same plane. It is simple logic that, if two adjacent areas *did* lie in the same plane, there would be no boundary between the areas, and therefore, orthographically, the two adjacent areas would not exist. As an illustration, note that in Fig. 86 areas $A$, $B$, $C$, and $D$ are shown in the front and side views to lie in different planes.

Further proof of these principles is given by Fig. 87, in which two top views are shown. By analysis of the projection between top and front views, it is seen that areas $G$ and $H$ shown in the top views must lie in planes $G$ and $H$, respectively, shown in the front view. Also, by projection, it is seen that area $J$ of top view $A$ must lie in plane $H$ and that area $K$ must lie in plane $G$. Because areas $H$ and $J$ lie in plane $H$, and areas $G$ and $K$ lie in plane $G$, the correct top view, therefore, is top view $B$.

*Hidden areas* may sometimes be confusing to read because the areas may overlap or even coincide with each other. For example, areas $A$, $B$, and $C$ in Fig. 88 are not separate areas because they are all formed by the slot on the rear of the object. The apparent separation into separate areas is caused by the dashed lines from the rectangular hole, which is not connected with the slot in any way.

## 48. Reading Lines and Areas.

The foregoing principles regarding the meaning of lines and areas must be used to analyze any given set of views by correlating a surface appearing in one view as a line or an area with its representation in the other views, in which it may appear as a line or an area. Study, for example, Fig. 89, first orienting yourself with the given views. From their arrange-

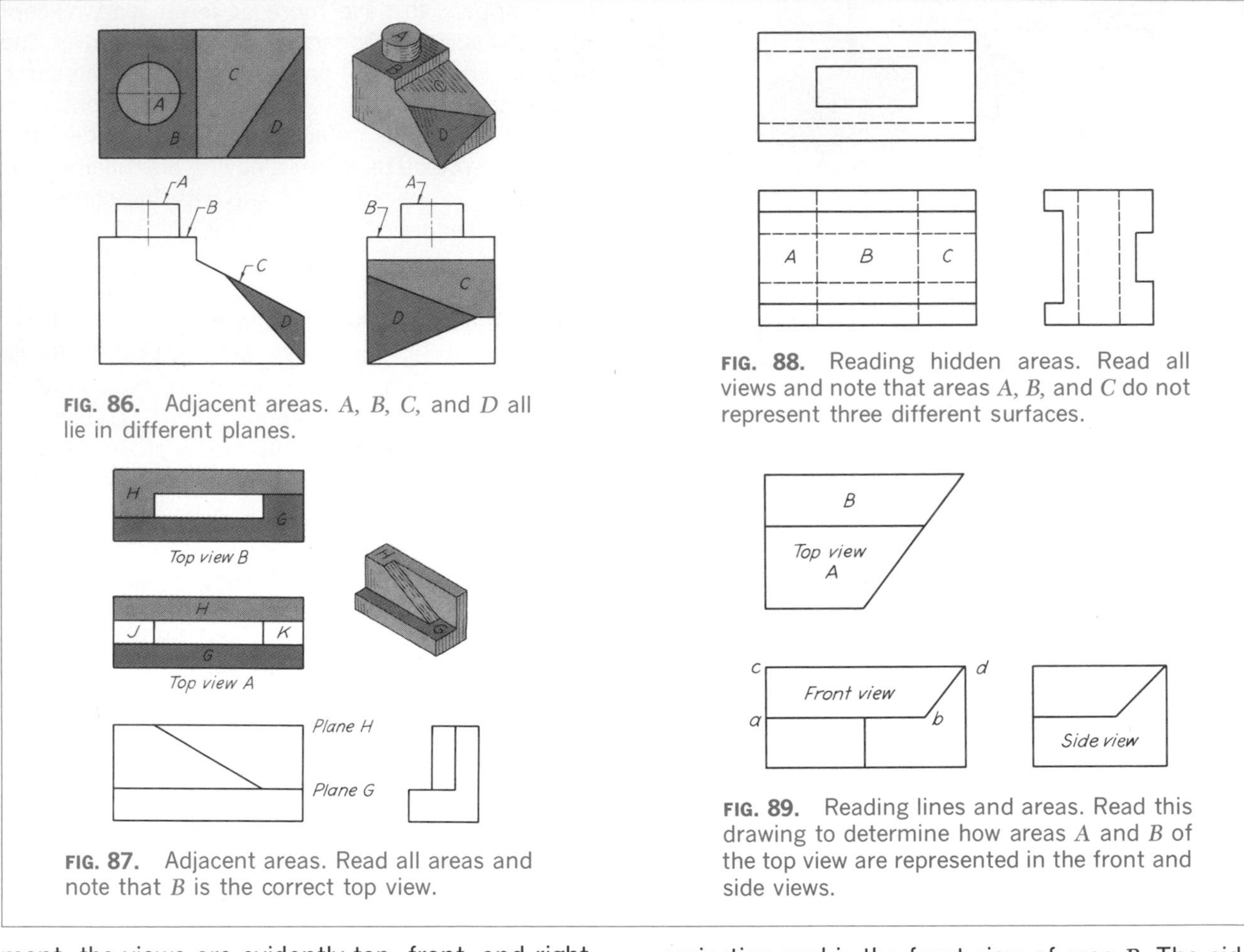


**FIG. 86.** Adjacent areas. *A, B, C,* and *D* all lie in different planes.

**FIG. 87.** Adjacent areas. Read all areas and note that *B* is the correct top view.

**FIG. 88.** Reading hidden areas. Read all views and note that areas *A, B,* and *C* do not represent three different surfaces.

**FIG. 89.** Reading lines and areas. Read this drawing to determine how areas *A* and *B* of the top view are represented in the front and side views.

ment, the views are evidently top, front, and right-side. An over-all inspection of the views does not reveal a familiar geometric shape, such as a hole or boss, so an analysis of the surfaces is necessary. Beginning with the trapezoidal area *A* in the top view and then moving to the front view, note that a similar-shaped area of the same width is not shown; therefore, the front view of area *A* must appear as an edge, the line *ab*. Next, consider area *B* in the top view. It is shown as a trapezoidal area (four sides) the full width of the view. Again, going to the front view for a mating area or line, the area *abcd* is of similar shape and has the same number of sides with the corners in projection. Area *abcd*, therefore, satisfies the requirements of orthographic projection and is the front view of area *B*. The side view should be checked along with the other views to see if it agrees. Proceed in this way with additional areas, correlating them one with another and visualizing the shape of the complete object.

Memory and experience aid materially in reading any given drawing. However, every new set of views must be approached with an open mind because sometimes a shape that looks like a previously known condition will crowd the correct interpretation from the mind of the reader. For example, area *E* in Fig. 90 is in a vertical position. The front view in Fig. 91 is identical with the front view in Fig. 90, but in Fig. 91 the surface *F* is inclined to the rear and is not vertical.

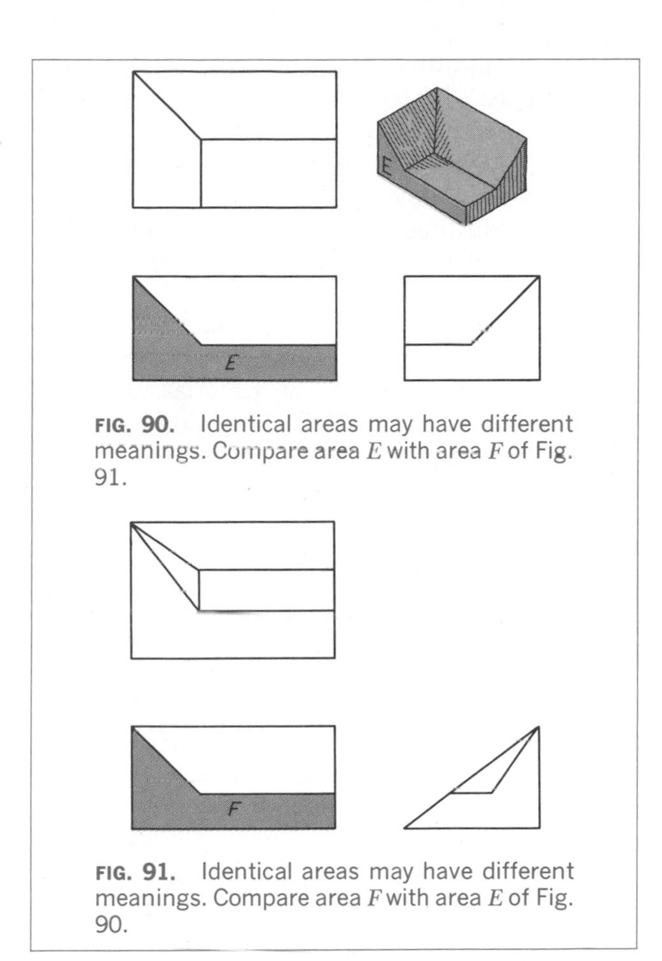


FIG. 90. Identical areas may have different meanings. Compare area $E$ with area $F$ of Fig. 91.

FIG. 91. Identical areas may have different meanings. Compare area $F$ with area $E$ of Fig. 90.

### 49. Reading Corners and Edges.

The corners and edges of areas may be numbered or lettered to identify them in making additional views or as an aid in reading some complicated shape. If there are no coincident conditions, the corners and edges are easily named by projection; that is, the top view is directly over the front view, and the side view lies on a horizontal projector to the front view. When coincident conditions are present, it may be necessary to coordinate a point with an adjacent point as shown in Fig. 92. Corner $c$ in the front view ($c_F$) may be in projection with $c_T$ at any one of the three positions marked 1, 2, 3. However, the point $c$ is one end of an edge $dc$, and the front view of $c$ must therefore be at position 2. An

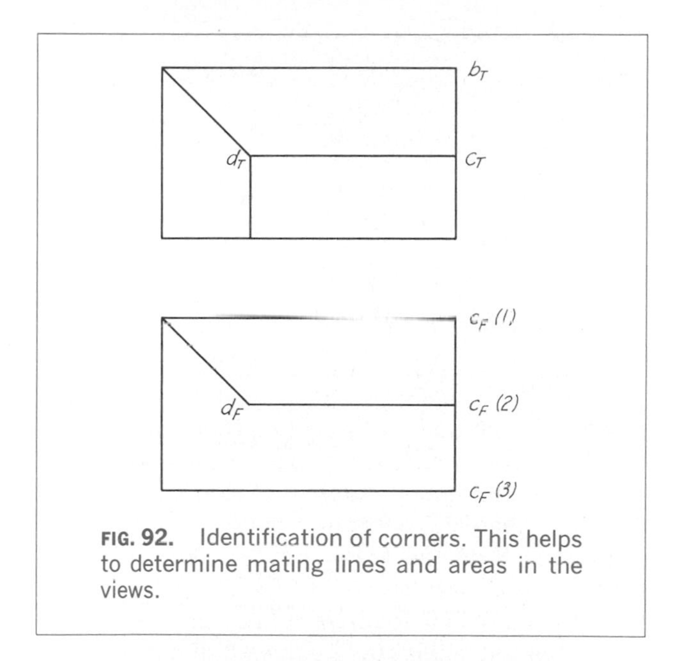


FIG. 92. Identification of corners. This helps to determine mating lines and areas in the views.

experienced reader could probably make the above observations without marking the points, but a beginner can in many cases gain much valuable experience by marking corners and edges, especially if the object he is studying has an unusual combination of surfaces.

### 50. Learning to Read by Sketching.

A drawing is interpreted by mentally understanding the shape of the object represented. You can prove that you have read and understood a drawing by making the object in wood or metal, by modeling it in clay, or by making a pictorial sketch of it. Sketching is the usual method. Before attempting to make a pictorial sketch, make a preliminary study of the method of procedure. Pictorial sketching may be based on a skeleton of three axes, one vertical and two at 30°,[8] representing three mutually perpendicular lines (Fig. 93). On these axes are marked the proportionate width, depth, and height of any rectangular figure. Circles are drawn in their circumscribing squares.

Study the views given in Fig. 94, following the

[8]Isometric position. Oblique or other pictorial methods may also be used. See Chap. 6 for pictorial methods.

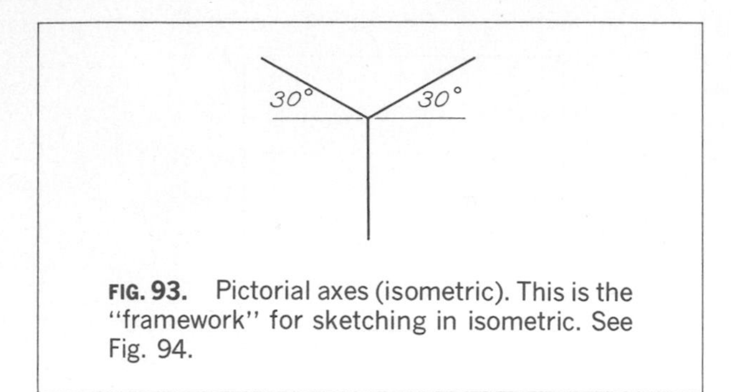


**FIG. 93.** Pictorial axes (isometric). This is the "framework" for sketching in isometric. See Fig. 94.

procedure outlined in Sec. 44. Then with a soft pencil (F) and notebook paper make a *very light* pictorial construction sketch of the object, estimating its height, width, and depth and laying the distances off on the axes as at (*a*); sketch the rectangular box that would enclose the piece, or the block from which it could be cut (*b*). On the top face of this box sketch lightly the lines that occur on the top view of the orthographic drawing (*c*). Note that some of the lines in top views may not be in the top plane. Next sketch lightly the lines of the front view on the front face of the box or block, and if a side view is given, outline it similarly (*d*). Now begin to cut the figure from the block, strengthening the visible edges and adding the lines of intersection where faces of the object meet (*e*). Omit edges that do not appear as visible lines unless they are necessary to describe the piece. Finish the sketch, checking back to the three-view drawing. The construction lines need not be erased unless they confuse the sketch.

## 51. Pictorial Drawing and Sketching.

The foregoing discussion of pictorial sketching should suffice as a guide for making rough sketches as an aid in reading a drawing. However, many experienced teachers like to correlate orthographic drawing, sketching, and reading, with pictorial drawing (which we will take up in Chap. 6). This is valuable training in understanding all methods of graphic expression but also because the making of a pictorial drawing from an orthographic drawing forces the student to read the orthographic drawing.

## 52. Learning to Read by Modeling.

Modeling the object in clay or modeling wax is another interesting, and effective, aid in learning to read a drawing. It is done in much the same way

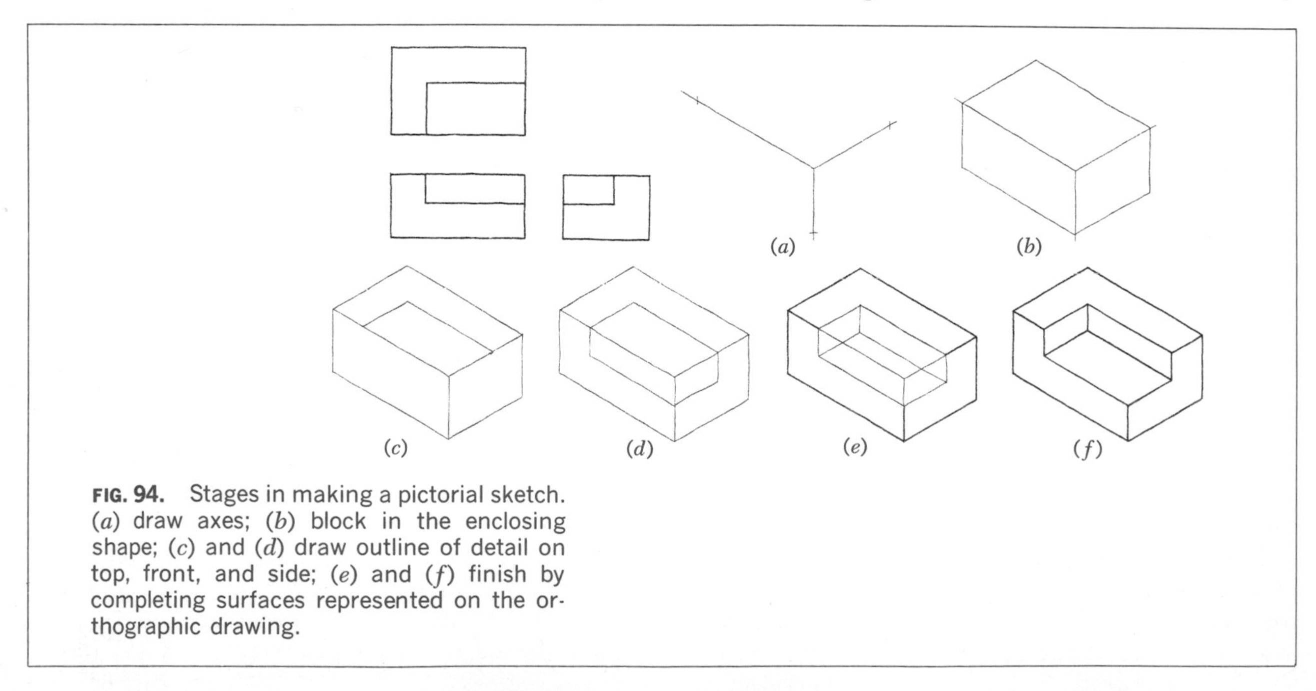


**FIG. 94.** Stages in making a pictorial sketch. (*a*) draw axes; (*b*) block in the enclosing shape; (*c*) and (*d*) draw outline of detail on top, front, and side; (*e*) and (*f*) finish by completing surfaces represented on the orthographic drawing.

as reading by pictorial sketching. Some shapes are easily modeled by cutting out from the enclosing block; others, by first analyzing and dividing the object into its basic geometric shapes and then combining these shapes.

Starting with a rectangular block of clay, perhaps 1 in. square and 2 in. long, read Fig. 95 by cutting the figure from the solid. With the point of the knife or a scriber, scribe lightly the lines of the three views on the three corresponding faces of the block (Fig. 96*a*). The first cut could be as shown at (*b*) and the second as at (*c*). Successive cuts are indicated at (*d*) and (*e*), and the finished model is shown at (*f*).

Figure 97 illustrates the type of model that can be made by building up the geometric shapes of which the object is composed.

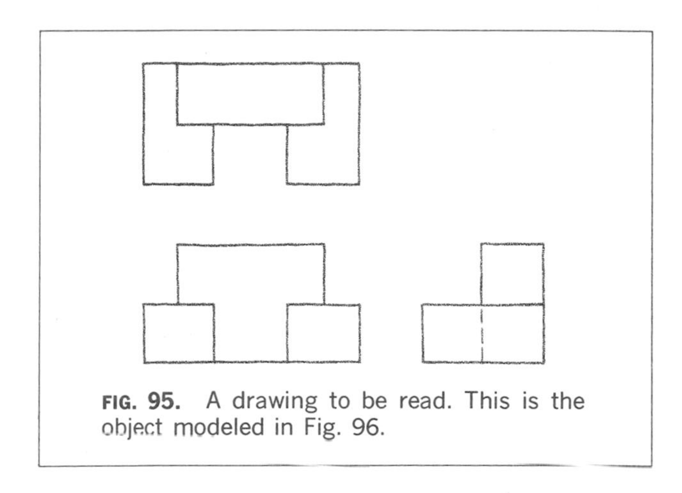

**FIG. 95.** A drawing to be read. This is the object modeled in Fig. 96.

## 53. Calculation of Volume as an Aid in Reading.

To calculate the volume of an object, it must be broken down into its simple geometric elements and the shape of each element must be carefully analyzed before beginning computation. Thus the calculation of volume is primarily an exercise in reading a drawing. Before the computations are completed, the object has usually been visualized, but the mathematical record of the volume of each portion and the correct total volume and weight are proof that the drawing has been read and understood.

The procedure closely follows the usual steps in reading a drawing. Figure 98 illustrates the method.

1. Study the orthographic drawing and pick out the principal masses (*A*, *B*, and *C* on the breakdown and in the pictorial drawing). Pay no attention in the beginning to holes, rounds, etc., but study each principal over-all shape and its relation to the other masses of the object. Record the dimensions of each of these principal portions and indicate plus volume by placing a check mark in the plus-volume column.

2. Examine each principal mass and find the secondary masses (*D* and *E*) that must be added to or subtracted from the principal portions. Bosses, lugs, etc., must be added; cutout portions, holes, etc., subtracted. Record the dimensions of these secondary masses, being careful to indicate plus or minus volume.

3. Further limit the object to its actual shape by locating smaller details, such as holes, fillets, rounds, etc. (parts *F*, *G*, and *H*). Record the dimensions of these parts.

4. Compute the volume of each portion. This may be done by longhand multiplication or, more conveniently, with a slide rule. Record each volume in the proper column, plus or minus. When all unit volumes are completed, find the net volume by subtracting total minus volume from total plus volume.

5. Multiply net volume by the weight per cubic inch of metal to compute the total weight.

The calculations are simplified if all fractional dimensions are converted to decimal form. When a slide rule is used, fractions *must* be converted to decimals. A partial conversion table is given in Fig. 98 and a more complete one in the Appendix.

Complete volume and weight calculation not only gives training in recognition of the fundamental geometric portions of an object but also serves to teach neat and concise working methods in the recording of engineering data.

## 54. Exercises in Reading.

Figures 99 and 100 contain a number of three-view drawings of block shapes made for exercises in reading orthographic projection and translating into

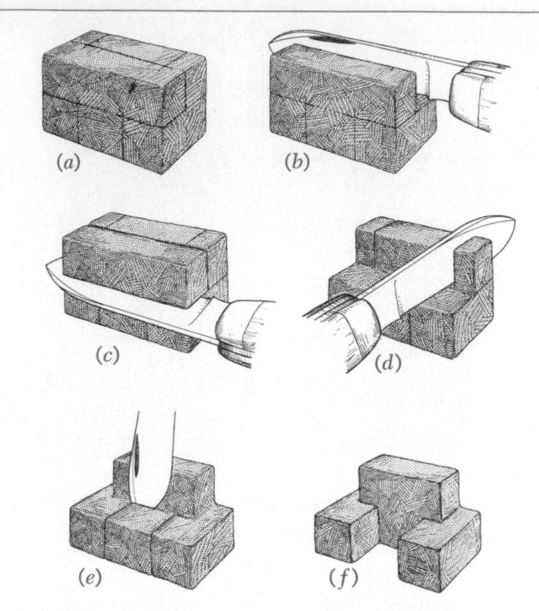


**FIG. 96.** Stages in modeling. (*a*) score details on top, front, and side; then make cuts at (*b*), (*c*), (*d*), and (*e*) to finish as at (*f*).

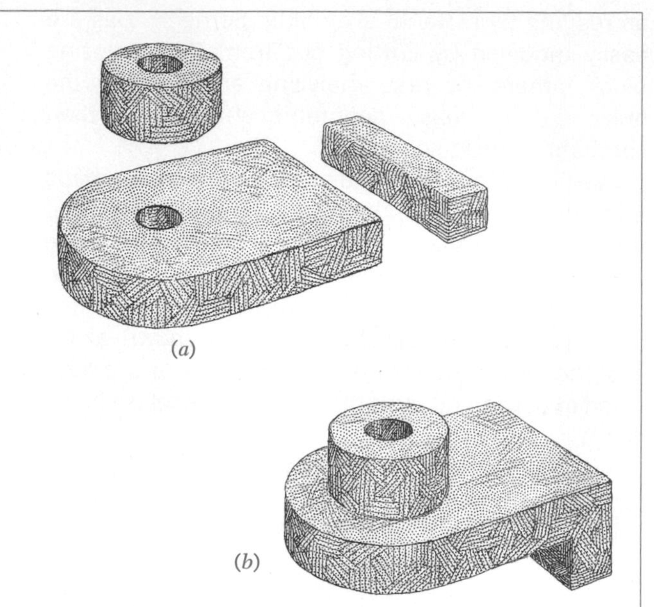


**FIG. 97.** A built-up model. Separate pieces (*a*) are combined to make the finished model (*b*).

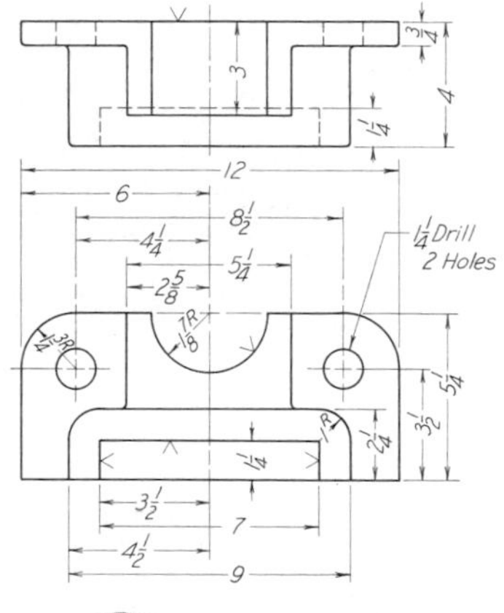


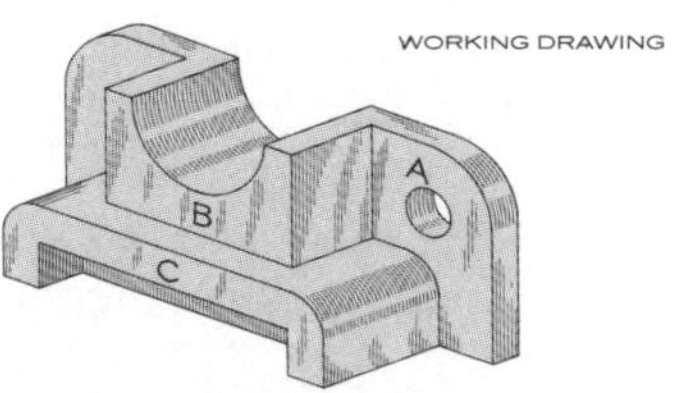


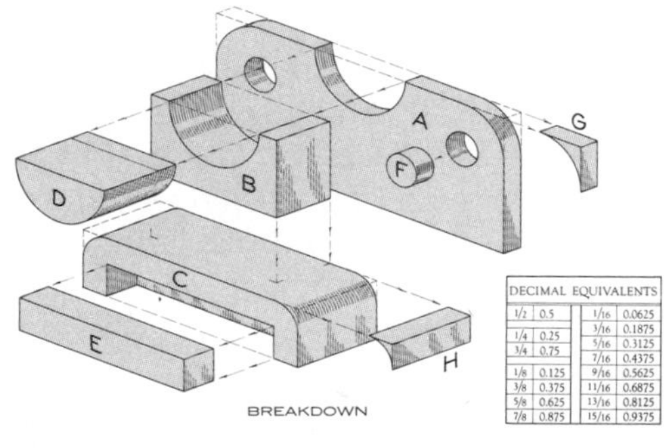


| DECIMAL EQUIVALENTS | | | |
|---|---|---|---|
| 1/2 | 0.5 | 1/16 | 0.0625 |
| | | 3/16 | 0.1875 |
| 1/4 | 0.25 | 5/16 | 0.3125 |
| 3/4 | 0.75 | 7/16 | 0.4375 |
| 1/8 | 0.125 | 9/16 | 0.5625 |
| 3/8 | 0.375 | 11/16 | 0.6875 |
| 5/8 | 0.625 | 13/16 | 0.8125 |
| 7/8 | 0.875 | 15/16 | 0.9375 |

| BEARING REST | MATERIAL | WT. – LB. PER CU. IN. | |
|---|---|---|---|
| PART | DIMENSIONS | PLUS VOLUME | MINUS VOLUME |
| A | | | |
| B | | | |
| C | | | |
| D | | | |
| E | | | |
| F | | | |
| G | | | |
| H | | | |
| | TOTALS | | |
| WEIGHT PER CUBIC INCH<br>GRAY CAST IRON 0.260 LB.<br>TOBIN BRONZE 0.304 LB.<br>CAST STEEL 0.282 LB. | TOTAL NET VOLUME IN CUBIC INCHES | | |
| | WEIGHT IN POUNDS | | |

**FIG. 98.** Shape breakdown for volume and weight calculations. Each feature of the object is analyzed individually.

**FIG. 99.** Reading exercises. Read each drawing, (*a*) to (*r*). Make pictorial sketches or models if necessary for understanding.

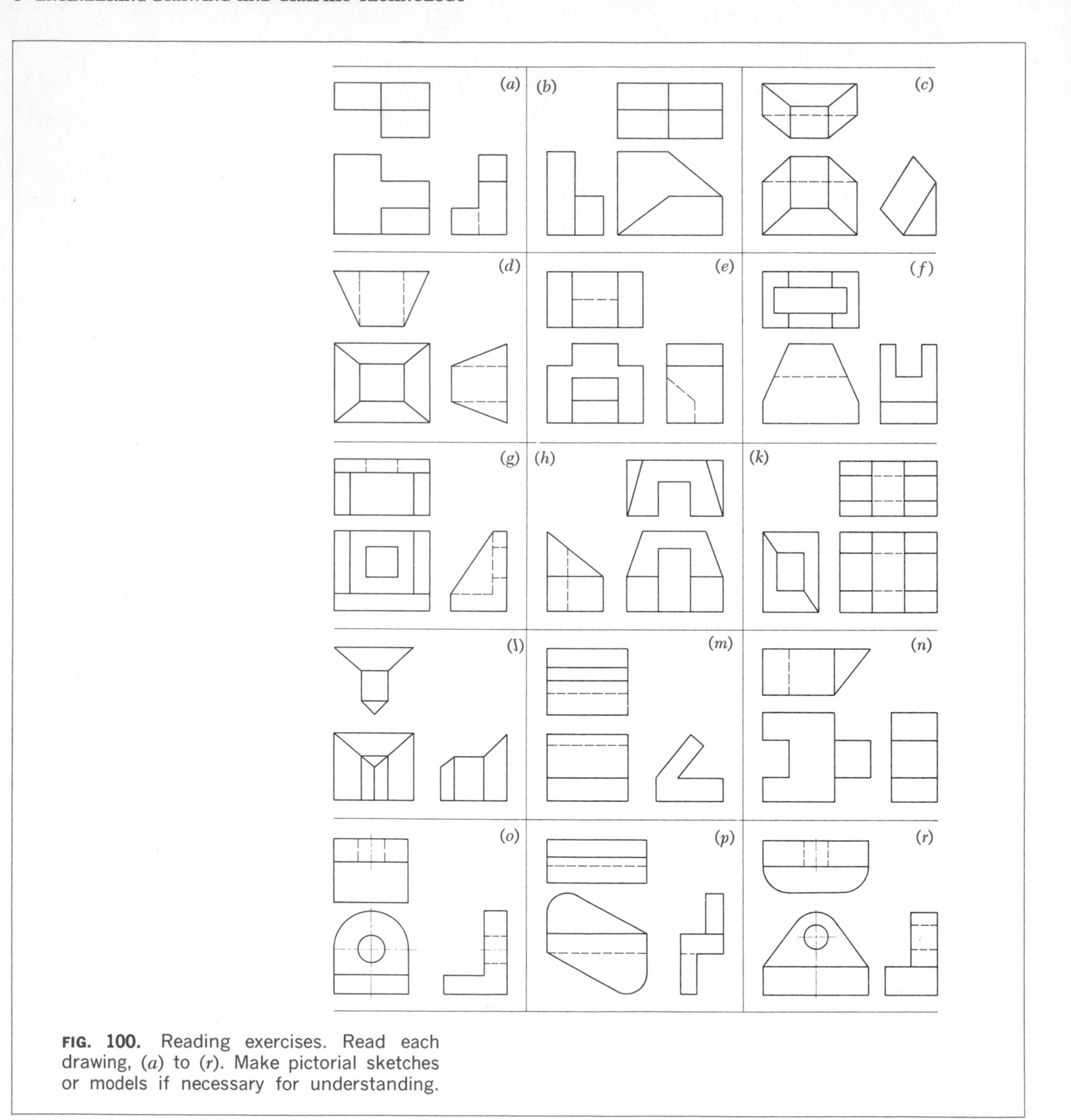


**FIG. 100.** Reading exercises. Read each drawing, (*a*) to (*r*). Make pictorial sketches or models if necessary for understanding.

pictorial sketches or models. Proceed as described in the previous paragraphs, making sketches not less than 4 in. over-all. Check each sketch to be sure that all intersections are shown and that the original

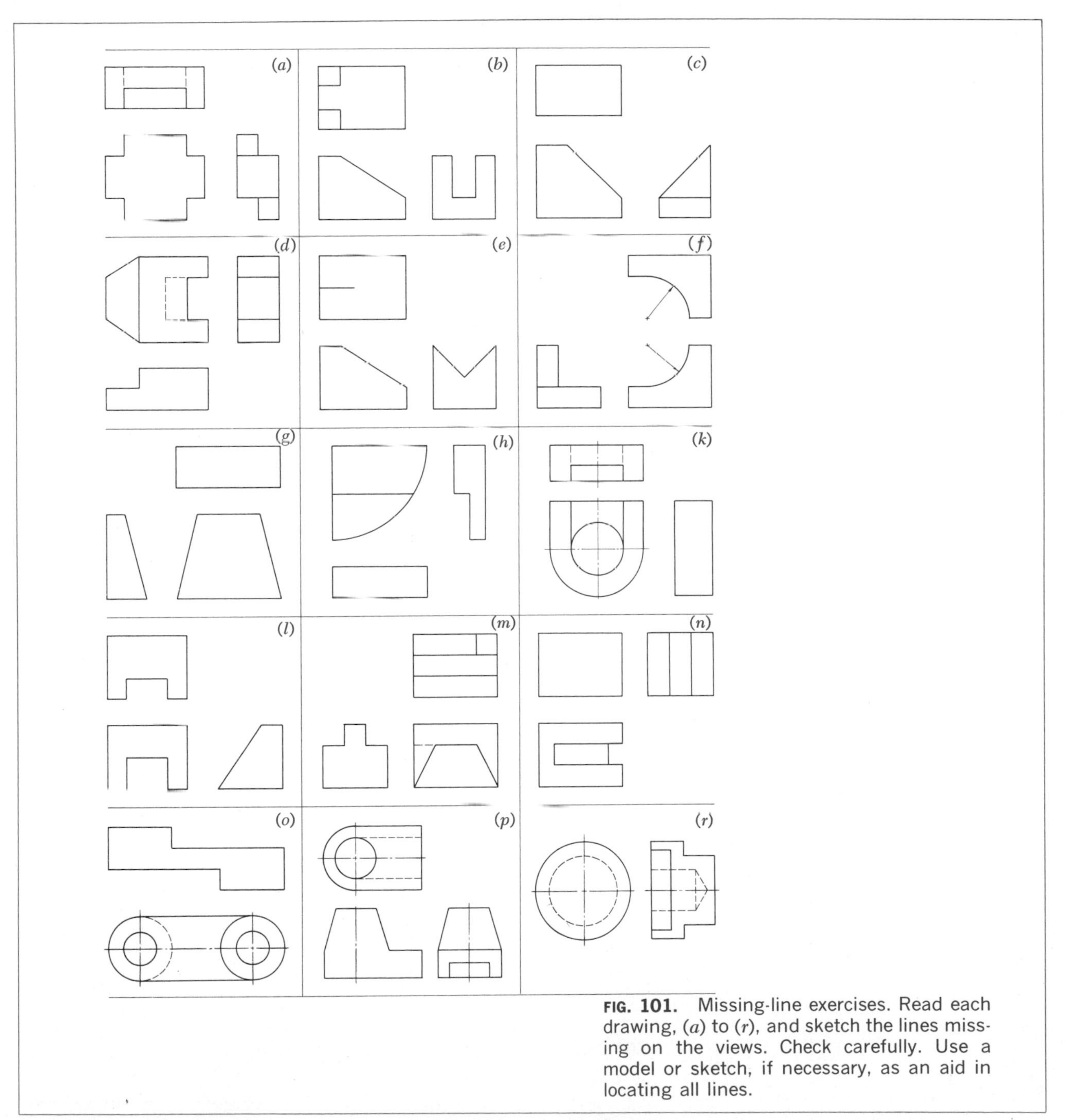


**FIG. 101.** Missing-line exercises. Read each drawing, (*a*) to (*r*), and sketch the lines missing on the views. Check carefully. Use a model or sketch, if necessary, as an aid in locating all lines.

three-view drawing could be made from the sketch. In each drawing in Fig. 101 some lines have been intentionally omitted. Read the drawings and supply the missing lines.

# PROBLEMS

For practice in orthographic freehand drawing, select problems from the following group.

## Group 1. Freehand Projections from Pictorial Views.

The figures for Probs. 1 to 16 contain a number of pictorial sketches of pieces of various shapes which are to be translated into three-view orthographic freehand drawings. Make the drawings of fairly large size, the front view, say, 2 to $2\frac{1}{2}$ in. in length, and estimate the proportions of the different parts by eye or from the proportionate marks shown but without measuring. The problems are graduated in difficulty for selection depending on ability and experience.

Problem 16 gives a series that can be used for advanced work in freehand drawing or that can be used later on, by adding dimensions, as dimensioning studies or freehand working-drawing problems.

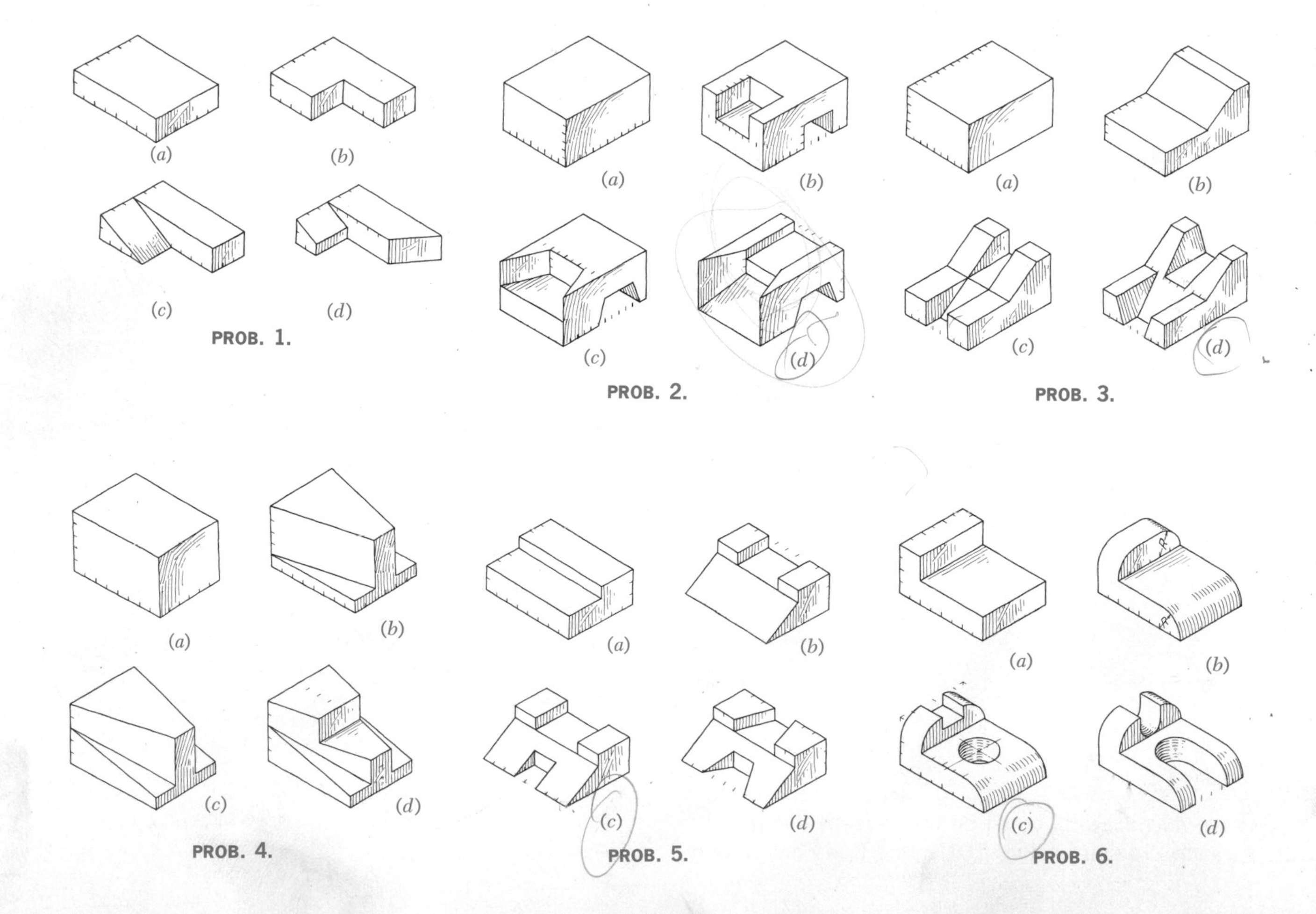

PROB. 7.

PROB. 8.

PROB. 9.

PROB. 10.

PROB. 11.

PROB. 12.

PROB. 13.

PROB. 14.

PROB. 15.

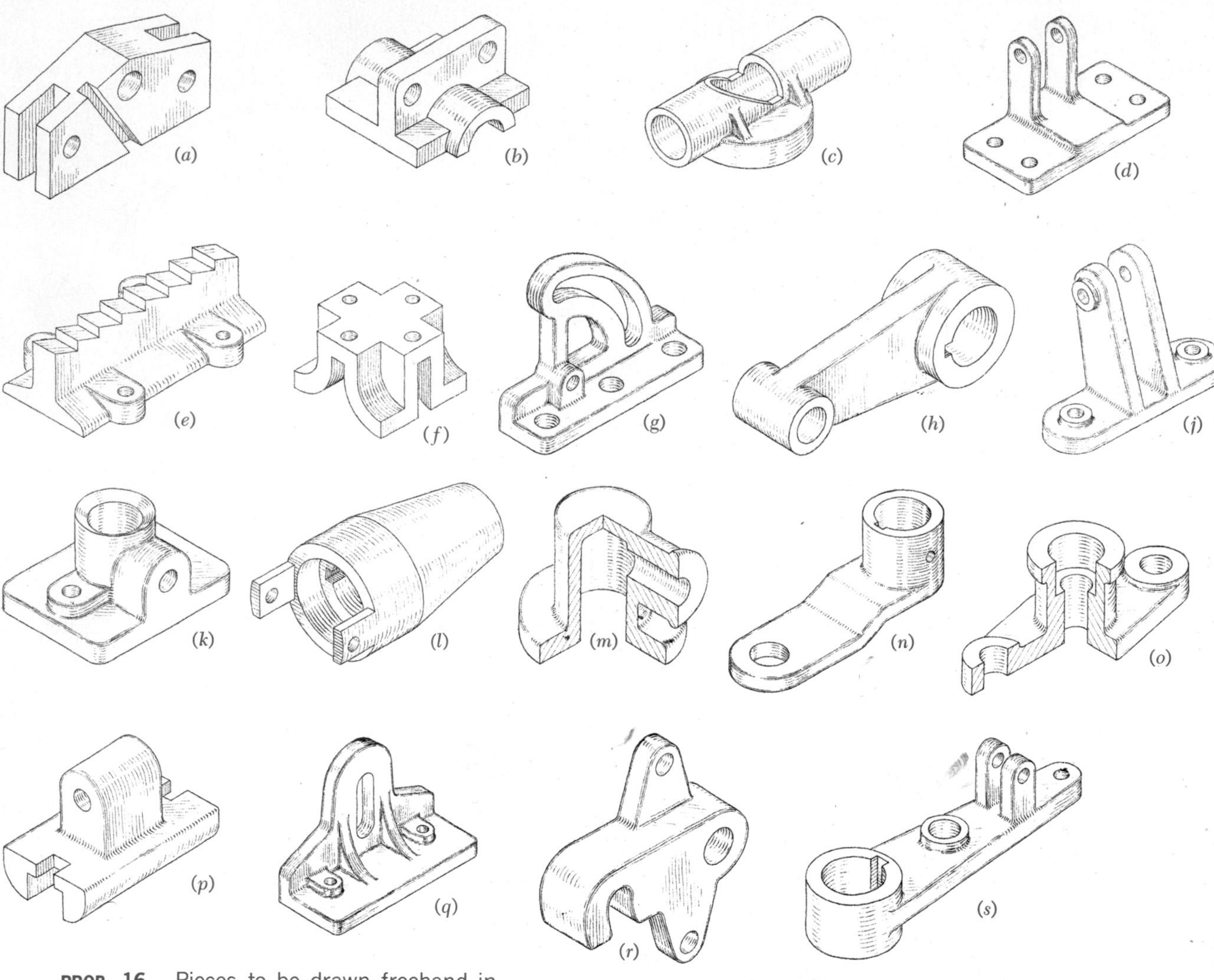


**PROB. 16.** Pieces to be drawn freehand in orthographic projection.

**17.** Make a freehand drawing of the end bracket.
**18.** Make a freehand drawing of the radar wave guide.

Select problems from the following groups for practice in projection drawing. Most of the problems are intended to be drawn with instruments but will give valuable training done freehand, on plain or coordinate paper.

The groups are as follows:

2. Projections from pictorial views

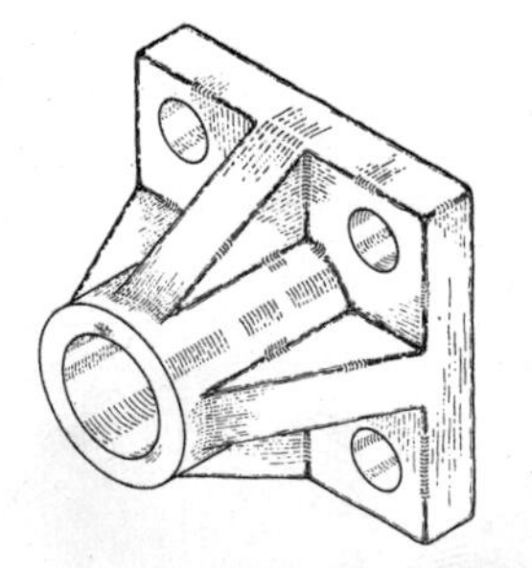

**PROB. 17.** End bracket.

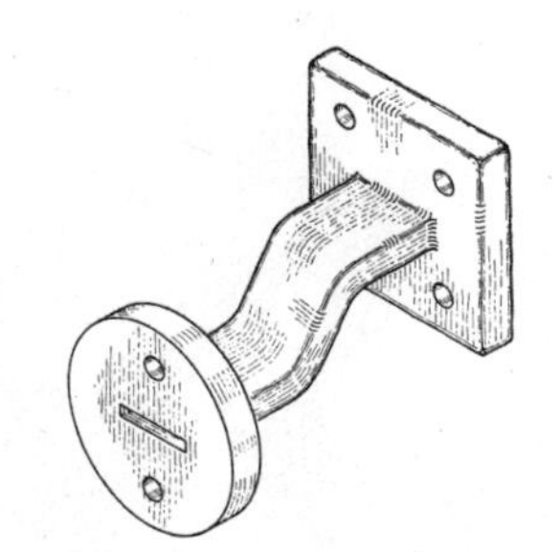

**PROB. 18.** Radar wave guide.

3. Special scales, decimal sizes, projections from photo drawings, references to Appendix material
4. Views to be supplied, freehand
5. Views to be supplied
6. Views to be changed
7. Drawing from memory
8. Volume and weight calculations with slide rule

**Group 2. Projections from Pictorial Views.**

**19.** Draw the top, front, and right-side views of the beam support.

**20.** Draw the top, front, and right-side views of the vee rest.

**21.** Draw three views of the saddle bracket.

**22.** Draw three views of the wedge block.

**23.** Draw three views of the slotted wedge.

**24.** Draw three views of the pivot block.

**25.** Draw three views of the inclined support.

**26.** Draw three views of the corner stop.

**27.** Draw three views of the switch base.

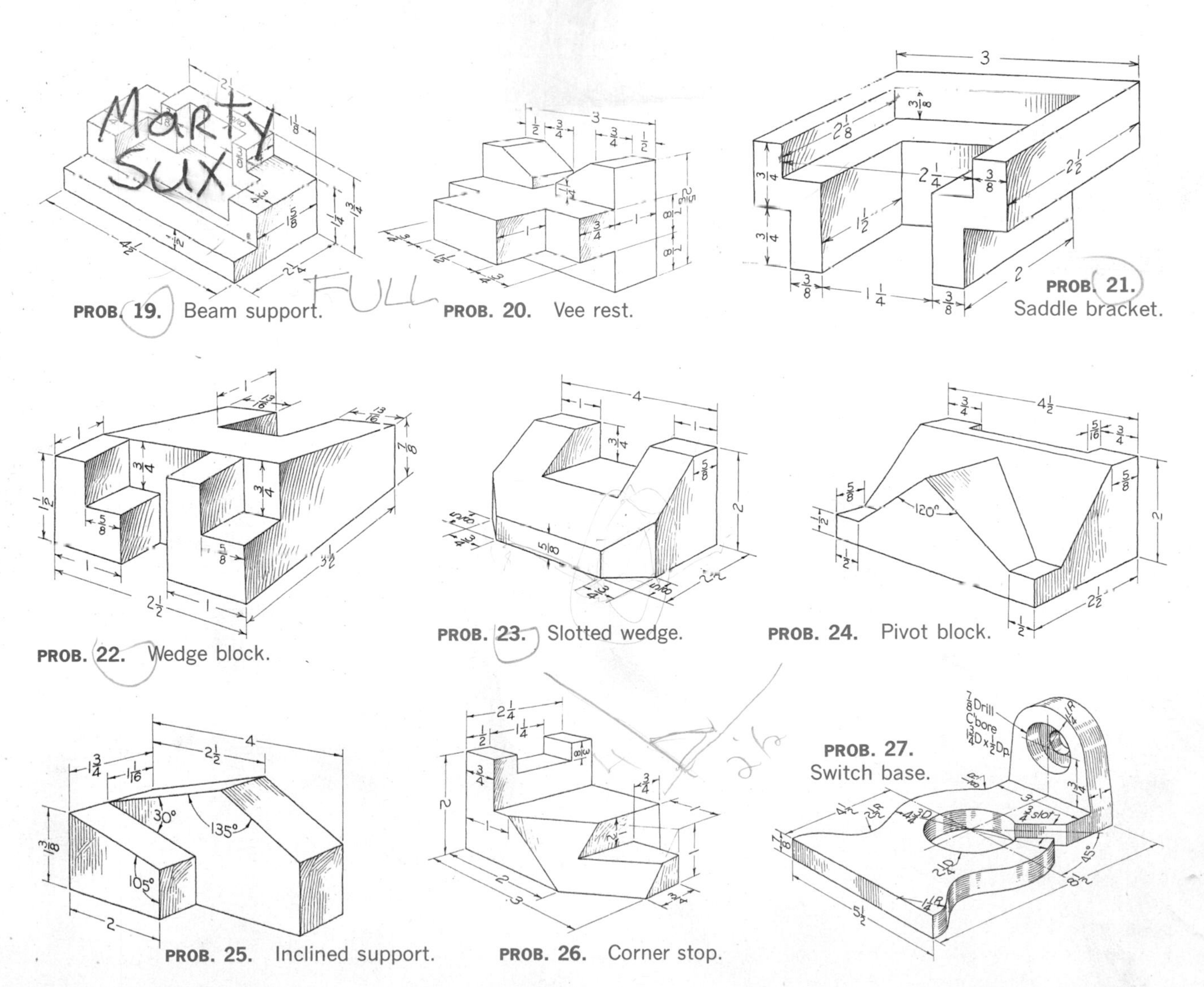


PROB. 19. Beam support.

PROB. 20. Vee rest.

PROB. 21. Saddle bracket.

PROB. 22. Wedge block.

PROB. 23. Slotted wedge.

PROB. 24. Pivot block.

PROB. 25. Inclined support.

PROB. 26. Corner stop.

PROB. 27. Switch base.

**28.** Draw three views of the adjusting bracket.
**29.** Draw three views of the guide base.
**30.** Draw three views of the bearing rest.
**31.** Draw three views of the swivel yoke.
**32.** Draw three views of the truss bearing.
**33.** Draw three views of the sliding-pin hanger.
**34.** Draw two views of the wire thimble.
**35.** Draw three views of the hanger jaw.
**36.** Draw three views of the adjustable jaw.
**37.** Draw two views of the shifter fork.

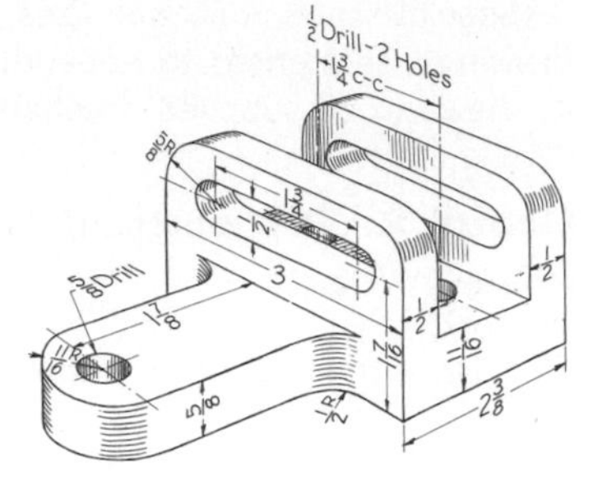


**PROB. 28.** Adjusting bracket.

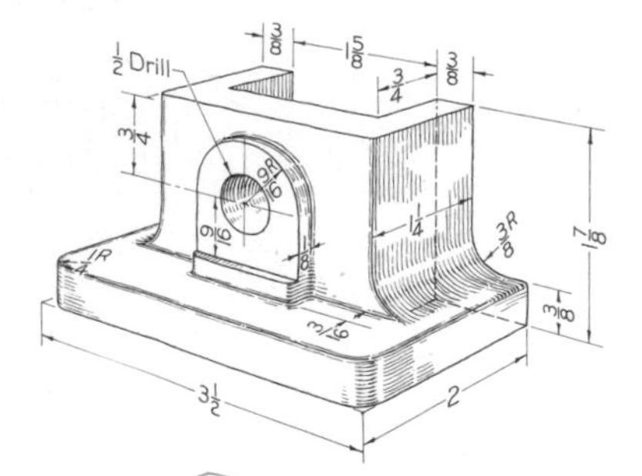


**PROB. 29.** Guide base.

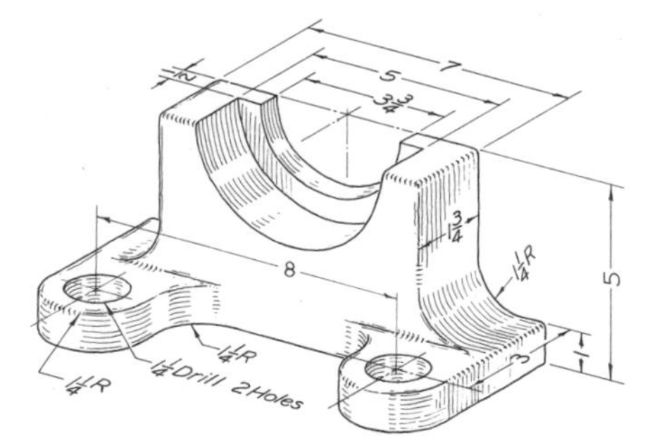


**PROB. 30.** Bearing rest.

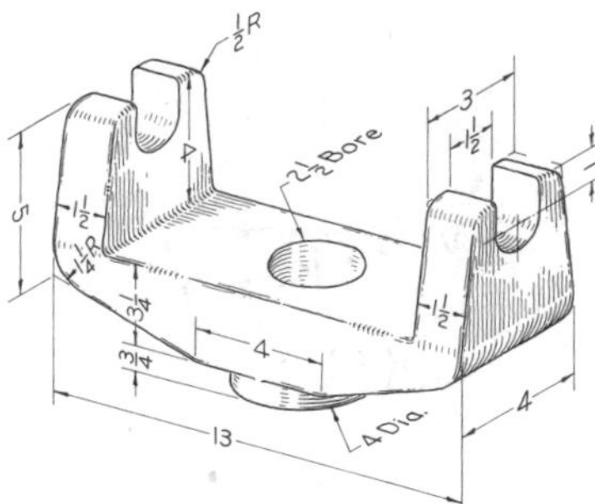


**PROB. 31.** Swivel yoke.

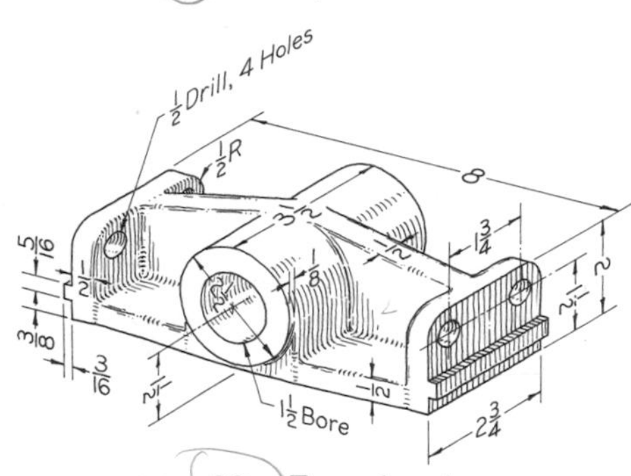


**PROB. 32.** Truss bearing.

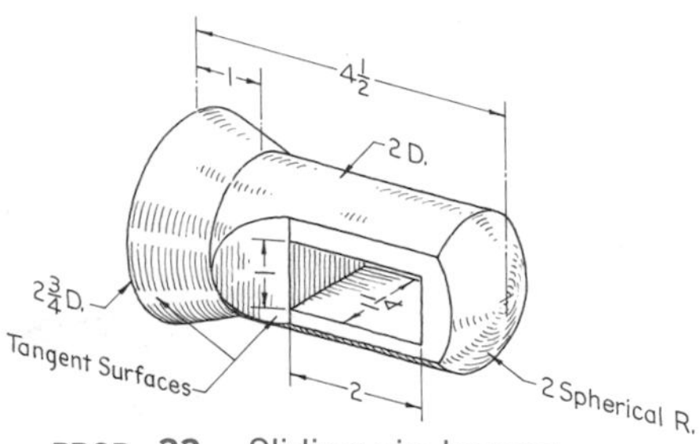


**PROB. 33.** Sliding-pin hanger.

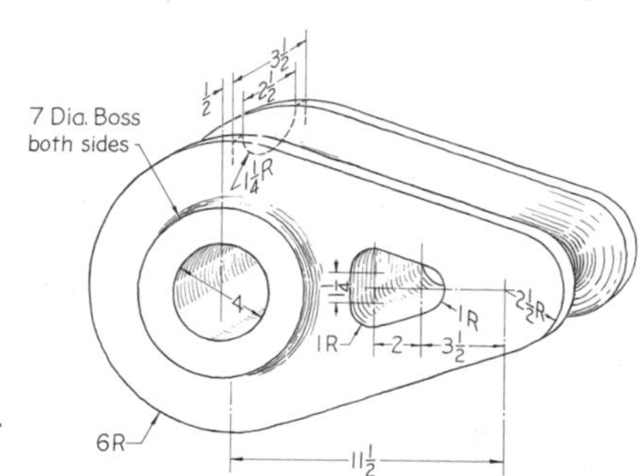


**PROB. 34.** Wire thimble.

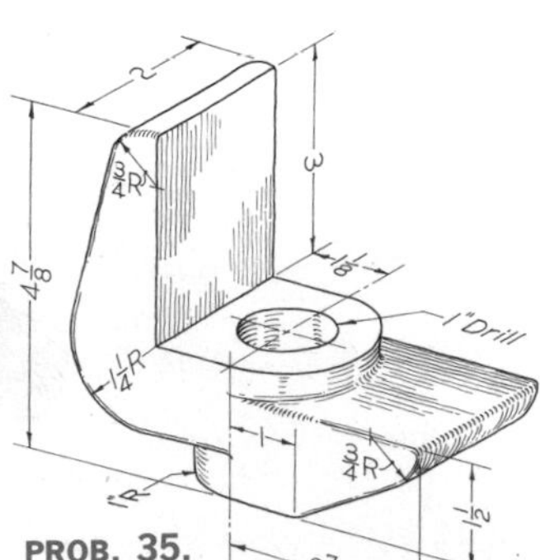


**PROB. 35.** Hanger jaw.

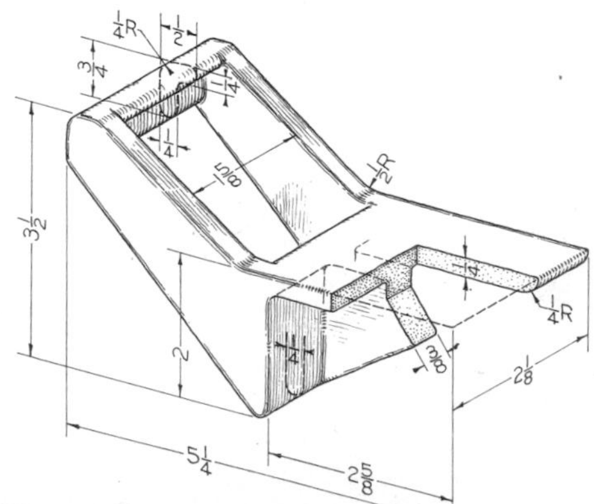

**PROB. 36.** Adjustable jaw.

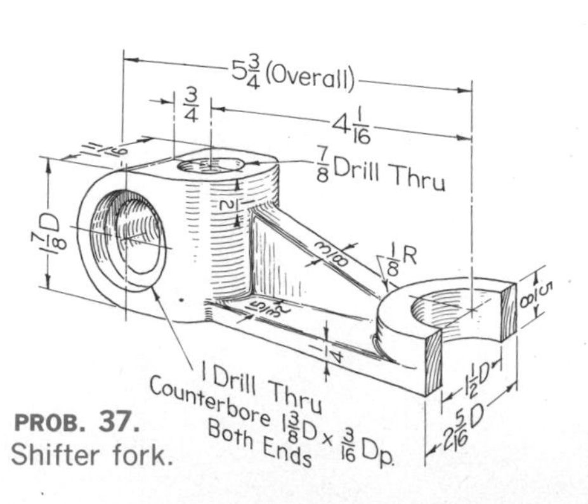


**PROB. 37.** Shifter fork.

38. Draw three views of the mounting bracket.
39. Draw three views of the hinged bearing.
40. Draw two views of the clamp lever.
41. Draw three views of the bedplate stop.
42. Draw top, front, and partial side views of the spanner bracket.
43. Draw two views of the sliding stop.
44. Draw three views of the clamp bracket.
45. Draw three views of the tube hanger.
46. Draw three views of the gage holder.
47. Draw three views of the shaft guide.

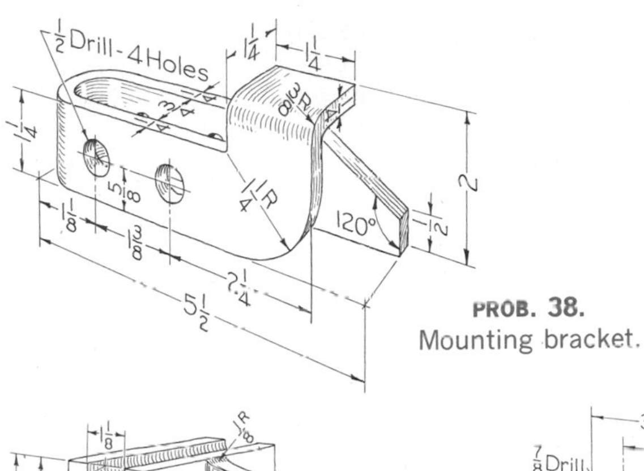


PROB. 38. Mounting bracket.

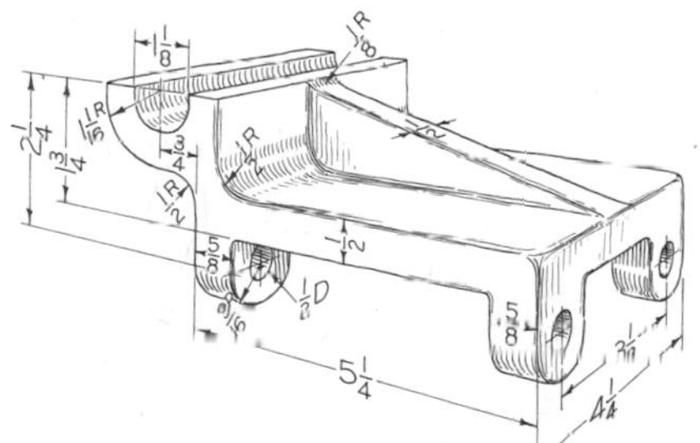

PROB. 39. Hinged bearing.

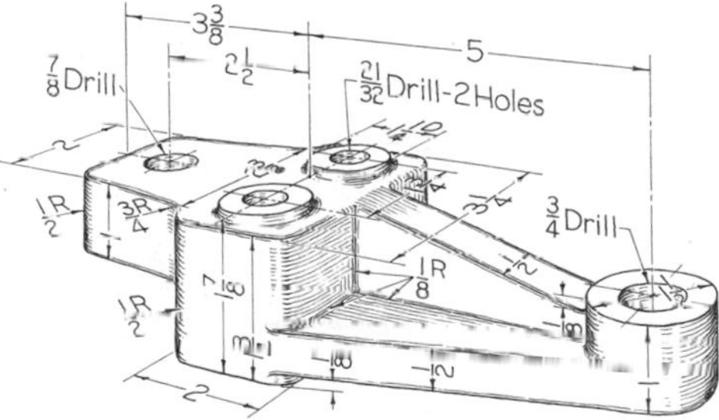


PROB. 40. Clamp lever.

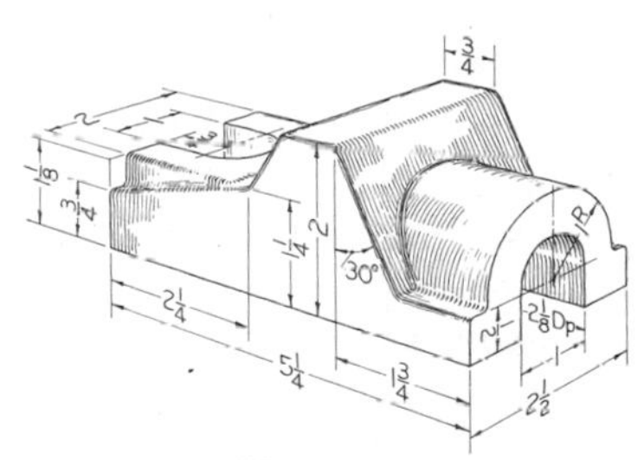


PROB. 41. Bedplate stop.

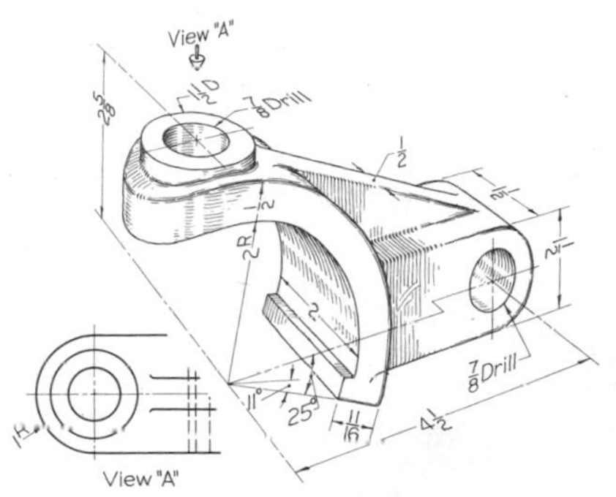


PROB. 42. Spanner bracket.

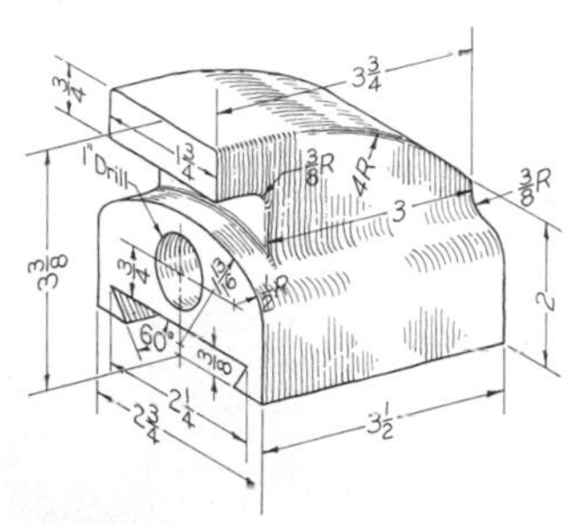


PROB. 43. Sliding stop.

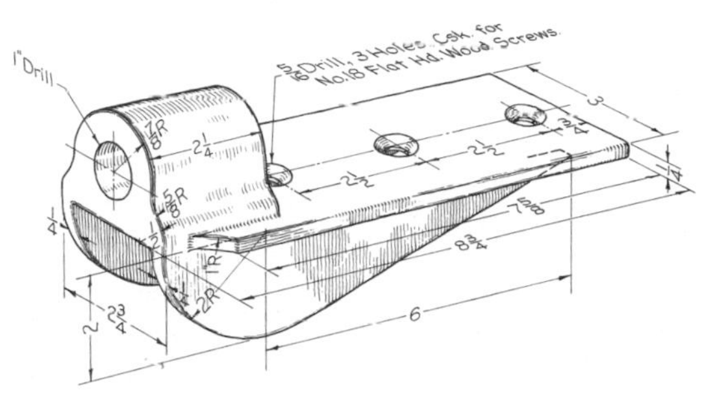


PROB. 44. Clamp bracket.

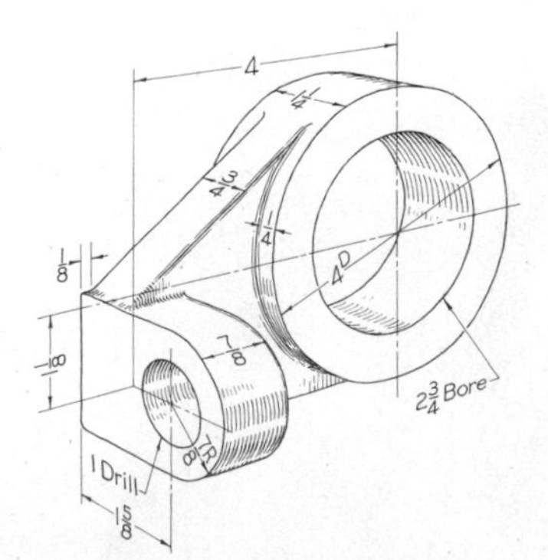


PROB. 45. Tube hanger.

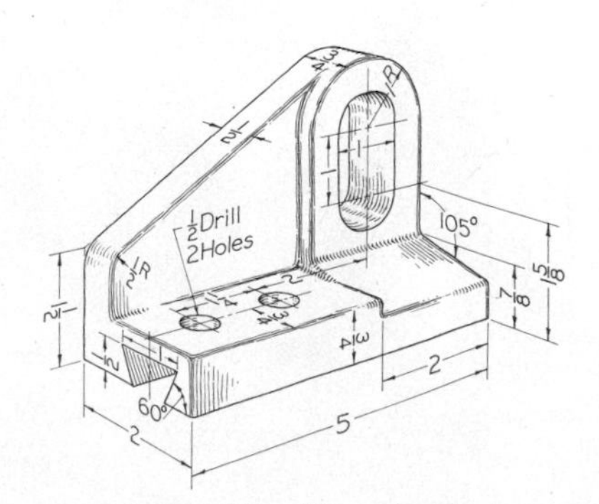


PROB. 46. Gage holder.

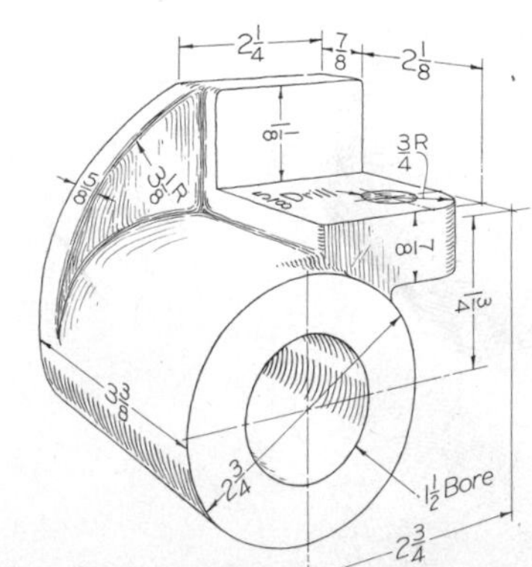


PROB. 47. Shaft guide.

**48.** Draw three views of the clamp block.
**49.** Draw three views of the offset yoke.
**50.** Draw three views of the angle connector.
**51.** Draw three views of the buckstay clamp.
**52.** Draw three views of the stop base.
**53.** Draw three views of the sliding buttress.
**54.** Draw two views of the end plate.
**55.** Draw three views of the plastic switch base.
**56.** Draw two views of the pawl hook.
**57.** Draw three views of the step-pulley frame.

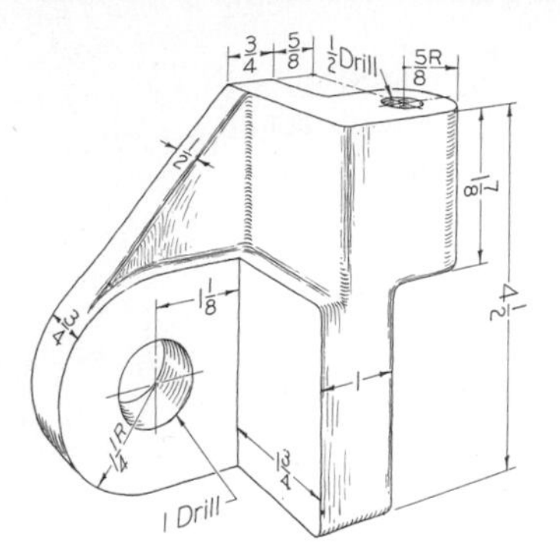


**PROB. 48.** Clamp block.

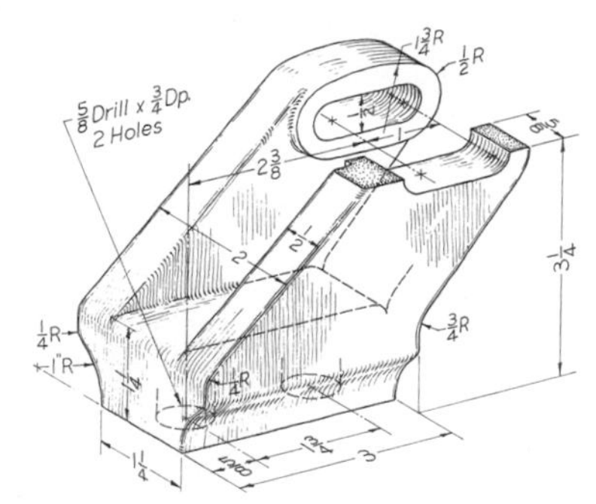


**PROB. 49.** Offset yoke.

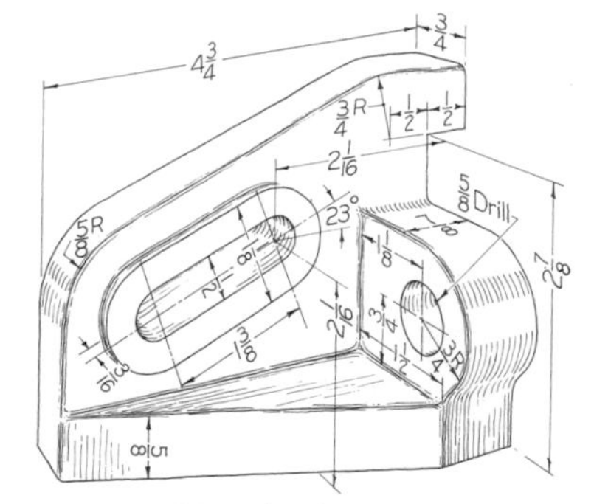


**PROB. 50.** Angle connector.

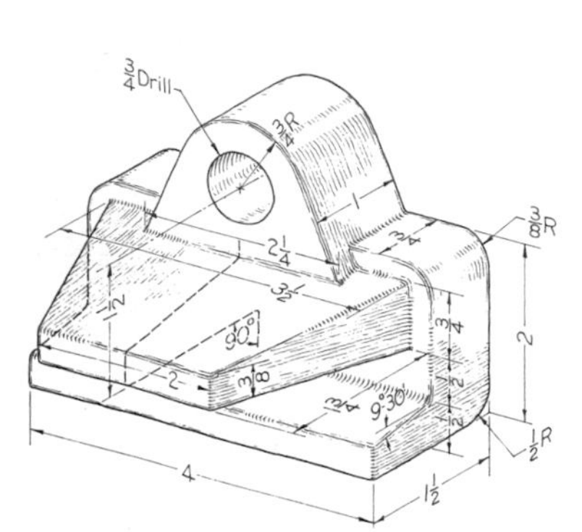


**PROB. 51.** Buckstay clamp.

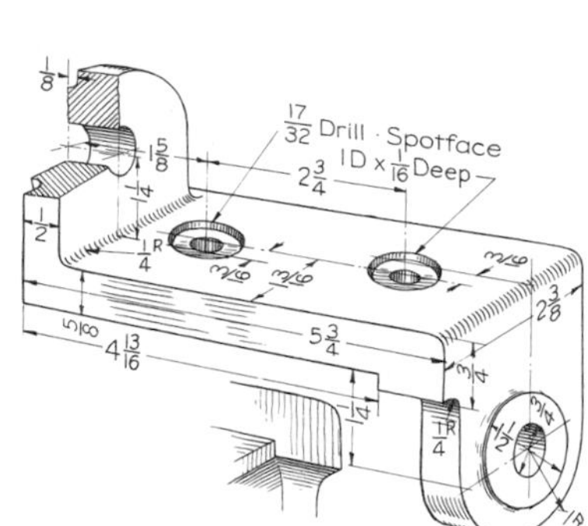


**PROB. 52.** Stop base.

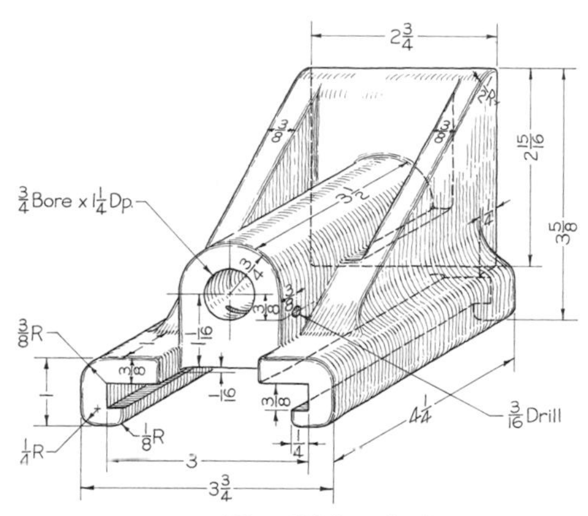


**PROB. 53.** Sliding buttress.

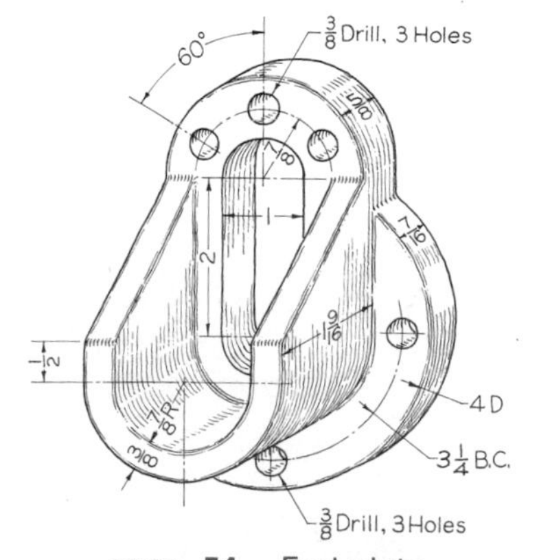


**PROB. 54.** End plate.

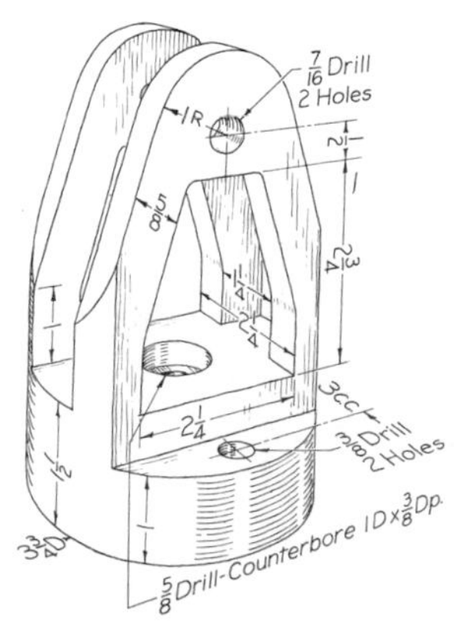


**PROB. 55.** Switch base.

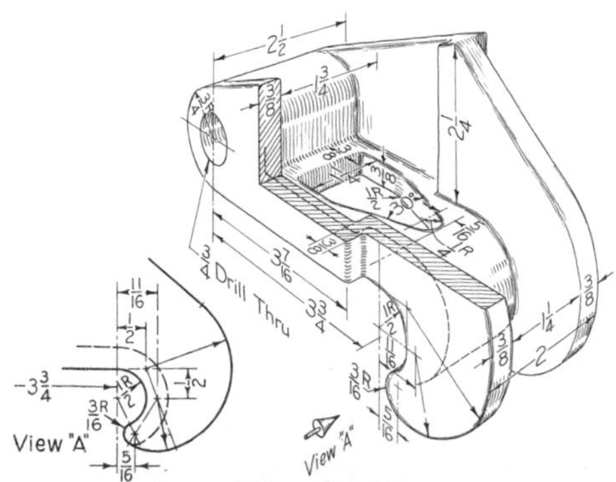


**PROB. 56.** Pawl hook.

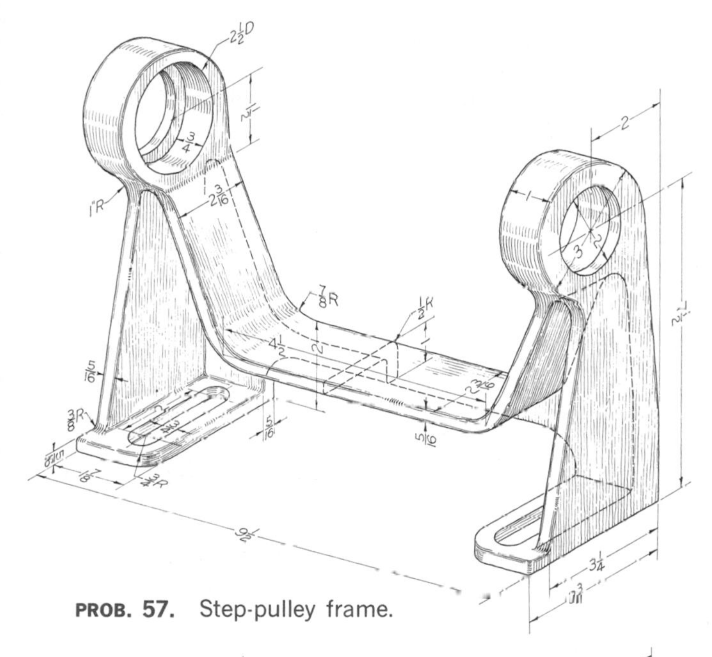

**PROB. 57.** Step-pulley frame.

## Group 3. Special Scales, Decimal Sizes, Projections from Photo Drawings, References to Appendix Material.

The problems in this group (58 to 73) will give practice in the use of a decimal scale for layout. The projections from the photo drawings of the final chapter are given here not only to serve as problem material, but to acquaint the student with this form of pictorial representation. Details of photo-drawing methods are also given in the final chapter. Wood-screw and bolt sizes are given in the Appendix.

**58.** Draw top, front, and right-side views.
**59.** Draw top, front, and right-side views.
**60.** Draw top, front, and right-side views.
**61.** Draw top, front, and left-side views.
**62.** Draw top, front, and right-side views. Bend radii and setbacks are 0.10 in.
**63.** Draw top, front, and right-side views. Screw and bolt sizes are given in the Appendix. Clearance over bolt and screw diameters are necessary in dimensioning but not in drawing the views.

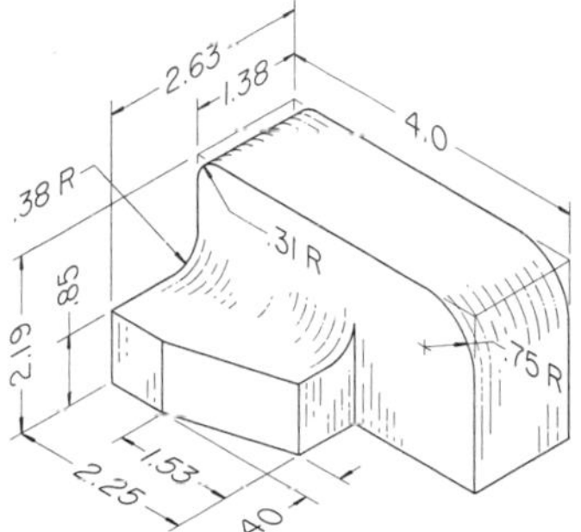

**PROB. 58.** Motor mount.

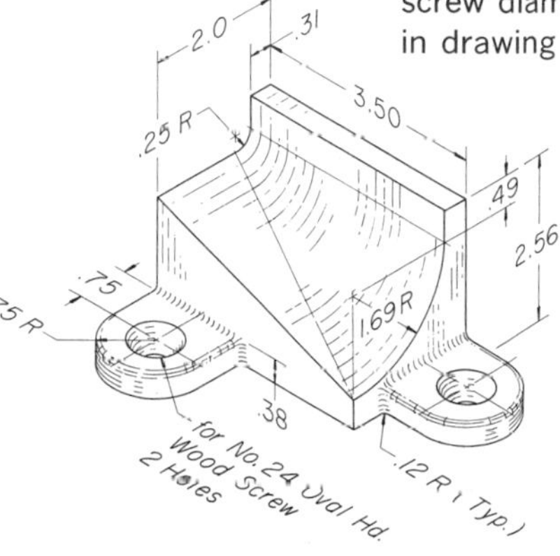

**PROB. 59.** Assembly-jig base.

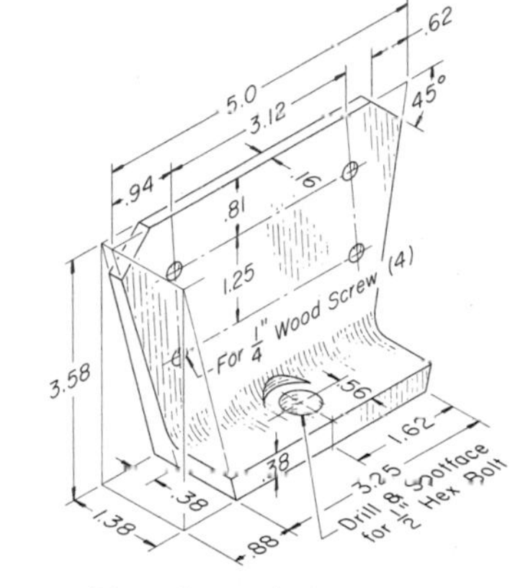

**PROB. 60.** Cargo-hoist tie-down.

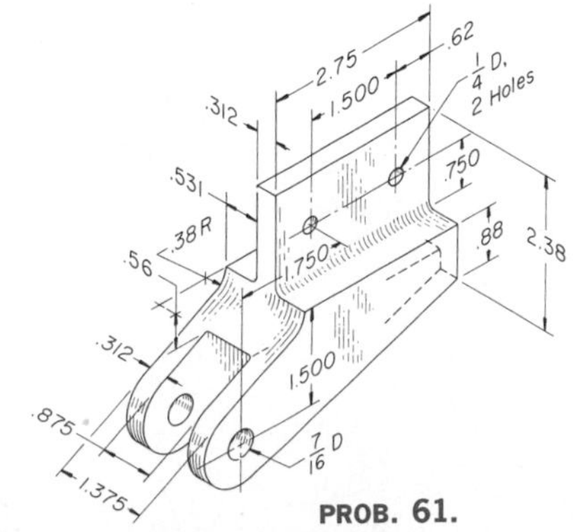

**PROB. 61.** Aileron tab-rod servo fitting.

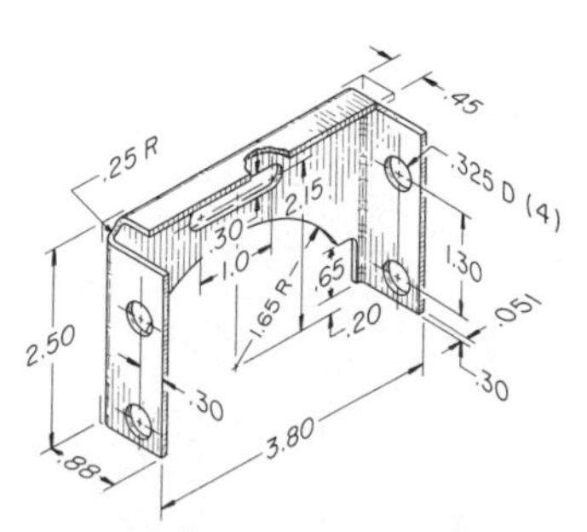

**PROB. 62.** Cover bracket.

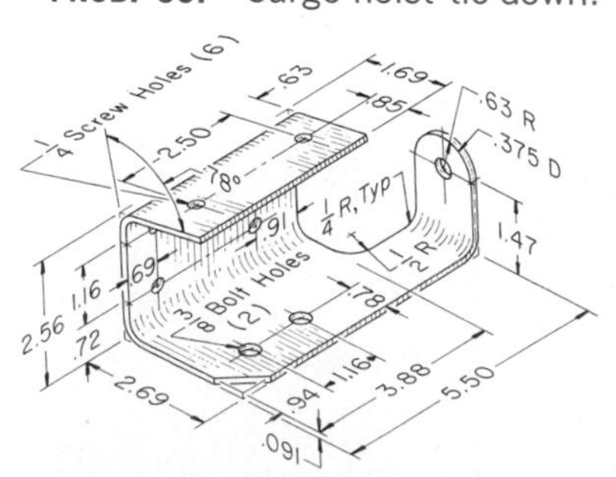

**PROB. 63.** Seat-latch support.

**64.** Draw top, front, left-side, and right-side views. Show only *necessary* hidden detail. The $0.44D$ hole extends through the piece. The limit-dimensioned holes

$$\left(.75 \begin{array}{l} +0.005 \\ -0.002 \end{array} \text{ and } \frac{.86}{.89}\right)$$

are given here as an introduction to precise methods, to be given in detail later. Regardless of dimensional accuracy, features are drawn to basic size with no more scaled accuracy than for other less accurate features.

**65.** Draw top and front views. See Prob. 64 for comment on limit-dimensioned holes.

**66.** Draw top, front, and left-side views. Bolts should have clearance. Undimensioned radii are $\frac{1}{4}R$.

**67.** Draw top and front views. See Prob. 64 for comment on limit-dimensioned holes.

**68.** Draw two views of the conveyor link, 71, final chapter.

**69.** Draw necessary views of lift-strut pivot, 73, final chapter.

**70.** Draw necessary views of hydro-cylinder support (right- and left-hand), Prob. 72, final chapter.

**71.** Draw necessary views of length-adjuster tube, Prob. 74, Chap. 23.

**72.** Draw necessary views of third terminal, Prob. 76, final chapter.

**73.** Select a part from the conveyor link assembly, Prob. 77, final chapter, and make an orthographic drawing.

**74.** Make an orthographic drawing of the jet-engine bracket.

**75.** Make an orthographic drawing of the missile gyro support.

**76.** Make an orthographic drawing of the reclining-seat ratchet plate.

**77.** Make an orthographic drawing of the rigging yoke.

**78.** Make an orthographic drawing of the door bracket.

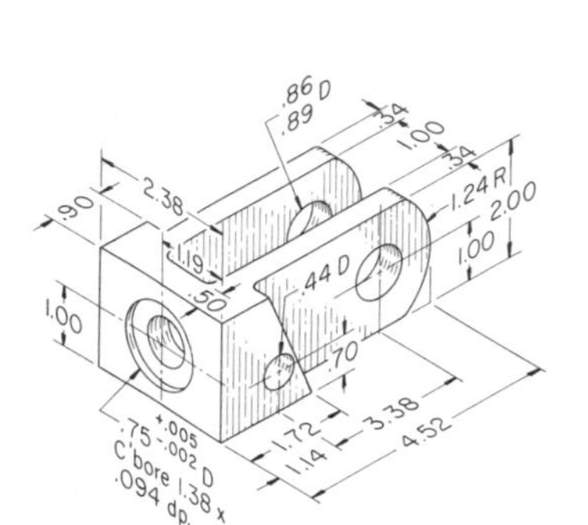


**PROB. 64.** Latch bracket.

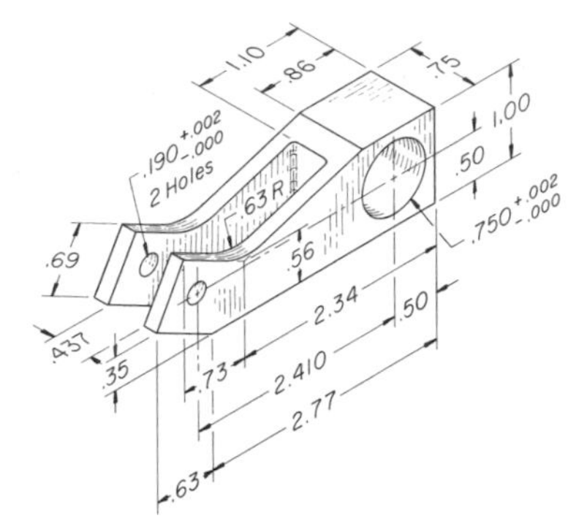


**PROB. 65.** Control crank.

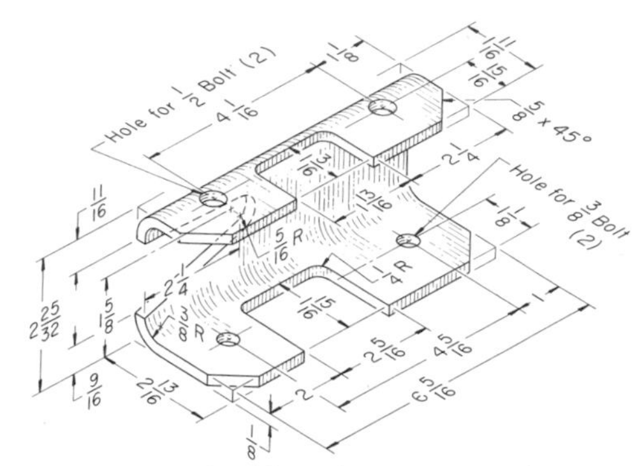


**PROB. 66.** Transformer mounting.

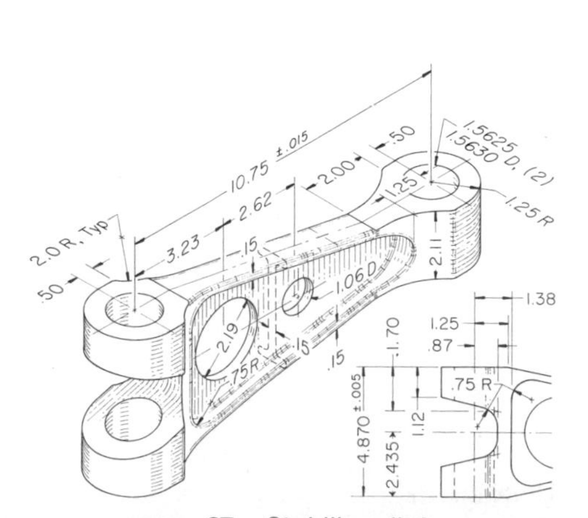


**PROB. 67.** Stabilizer link.

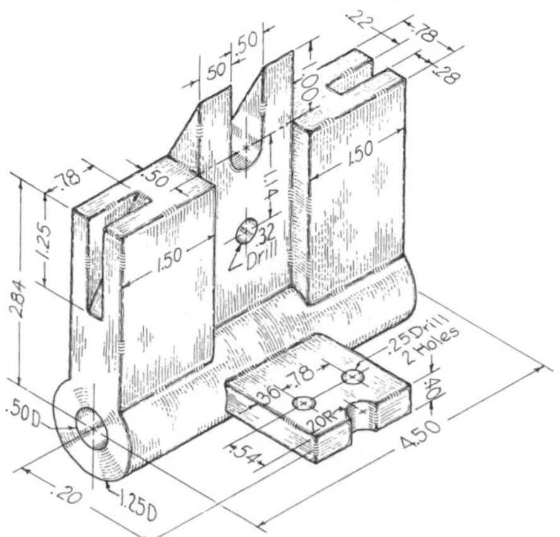


**PROB. 74.** Jet-engine bracket.

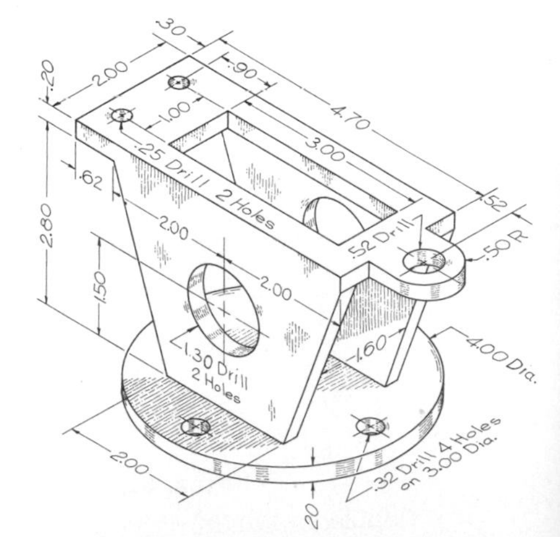


**PROB. 75.** Missile gyro support.

**79.** Make an orthographic drawing of the missile release pawl.

**80.** Make an orthographic drawing of the jet-engine inner-strut bracket.

**81.** Make an orthographic drawing of the transmission transfer fork.

**82.** Make an orthographic drawing of the reversing fork.

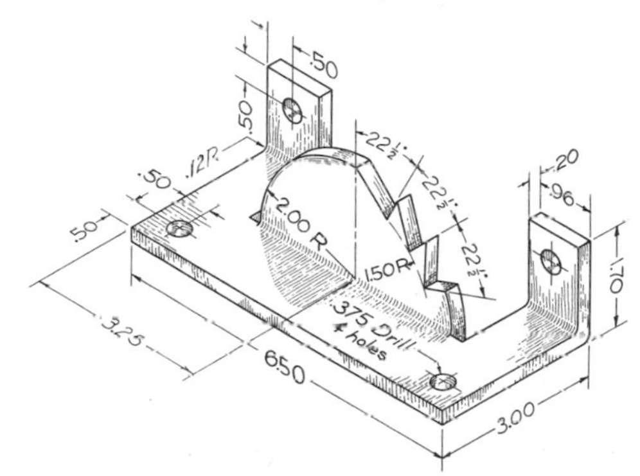


**PROB. 76.** Reclining-seat ratchet plate.

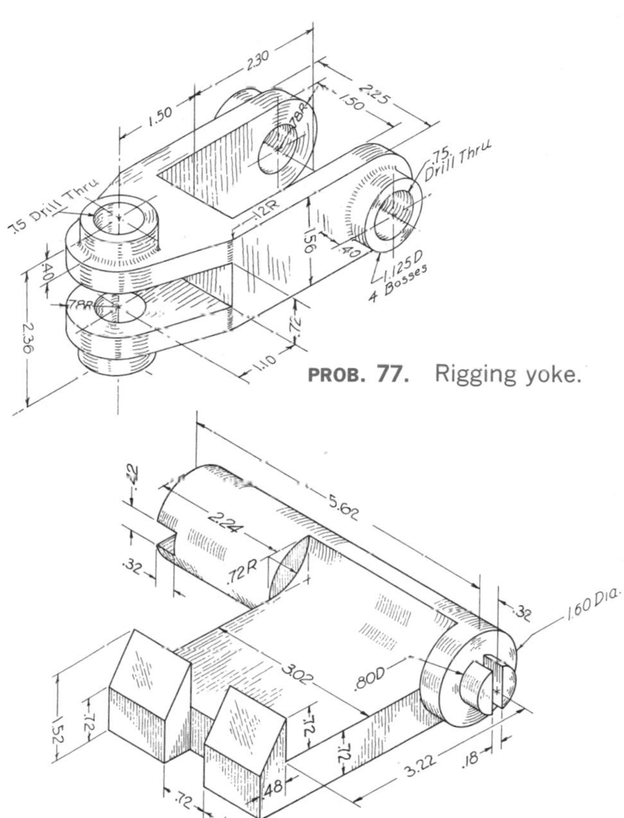


**PROB. 77.** Rigging yoke.

**PROB. 79.** Missile release pawl.

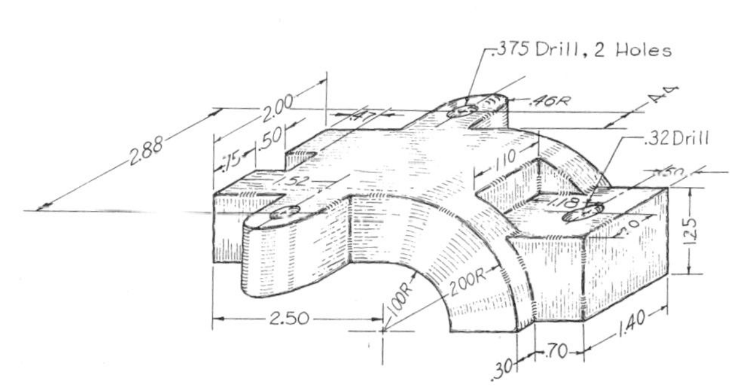


**PROB. 78.** Door bracket.

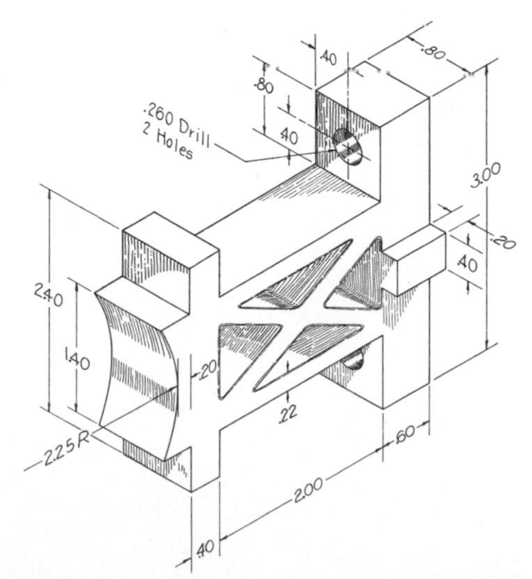


**PROB. 80.** Jet-engine inner-strut bracket.

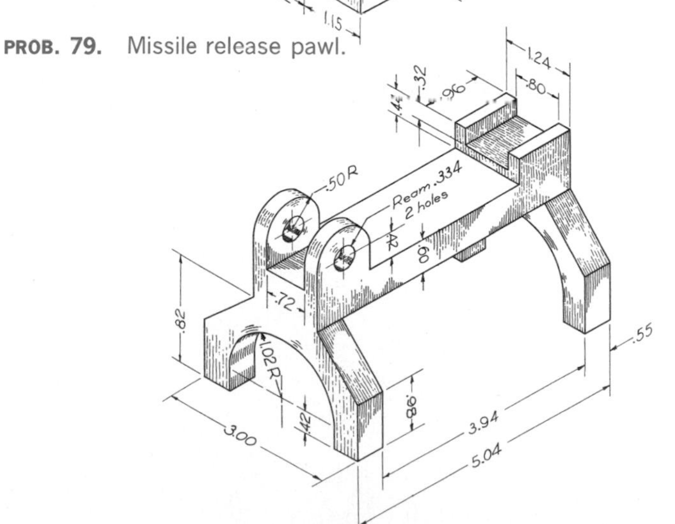


**PROB. 81.** Transmission transfer fork.

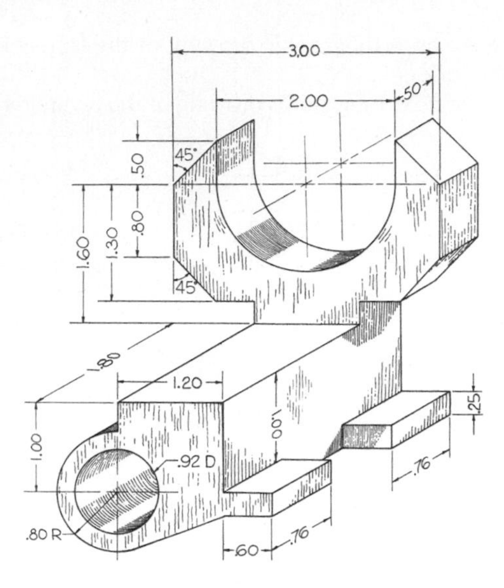


**PROB. 82.** Reversing fork.

## Group 4. Views to be Supplied Freehand.

These problems (83 to 85) will give valuable training in reading orthographic views, as well as further practice in applying the principles of orthographic projection.

Study the meaning of lines, areas, and adjacent areas. Corners or edges of the object may be numbered or lettered to aid in the reading or to aid later in the projection.

A pictorial sketch may be used, if desired, as an aid in reading the views. This sketch may be made before the views are drawn and completed or at any time during the making of the drawing. For some of the simpler objects, a clay model may be of assistance.

After the views given have been read and drawn, project the third view or complete the views as specified in each case.

Remember that every line representing an edge view of a surface, an intersection of two surfaces, or a surface limit will have a mating projection in the other views. Be careful to represent all hidden features and pay attention to the precedence of lines.

The figures in this group contain a number of objects with two views drawn and the third to be supplied. In addition to helping to develop the ability to draw freehand, this exercise will give valuable practice in reading. These problems may be worked directly in the book or on plain or coordinate paper.

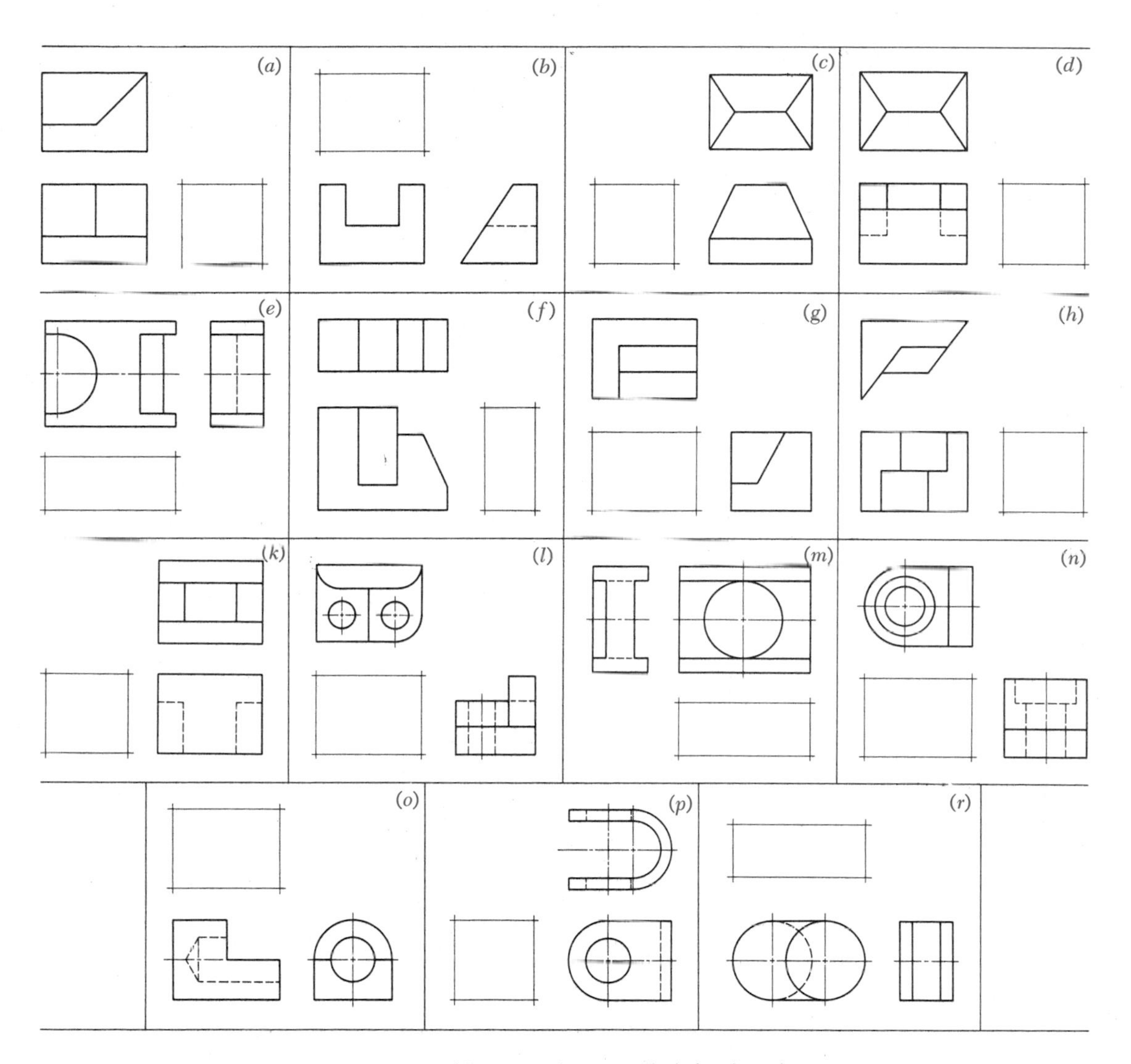


**PROB. 83.** Views to be supplied freehand.

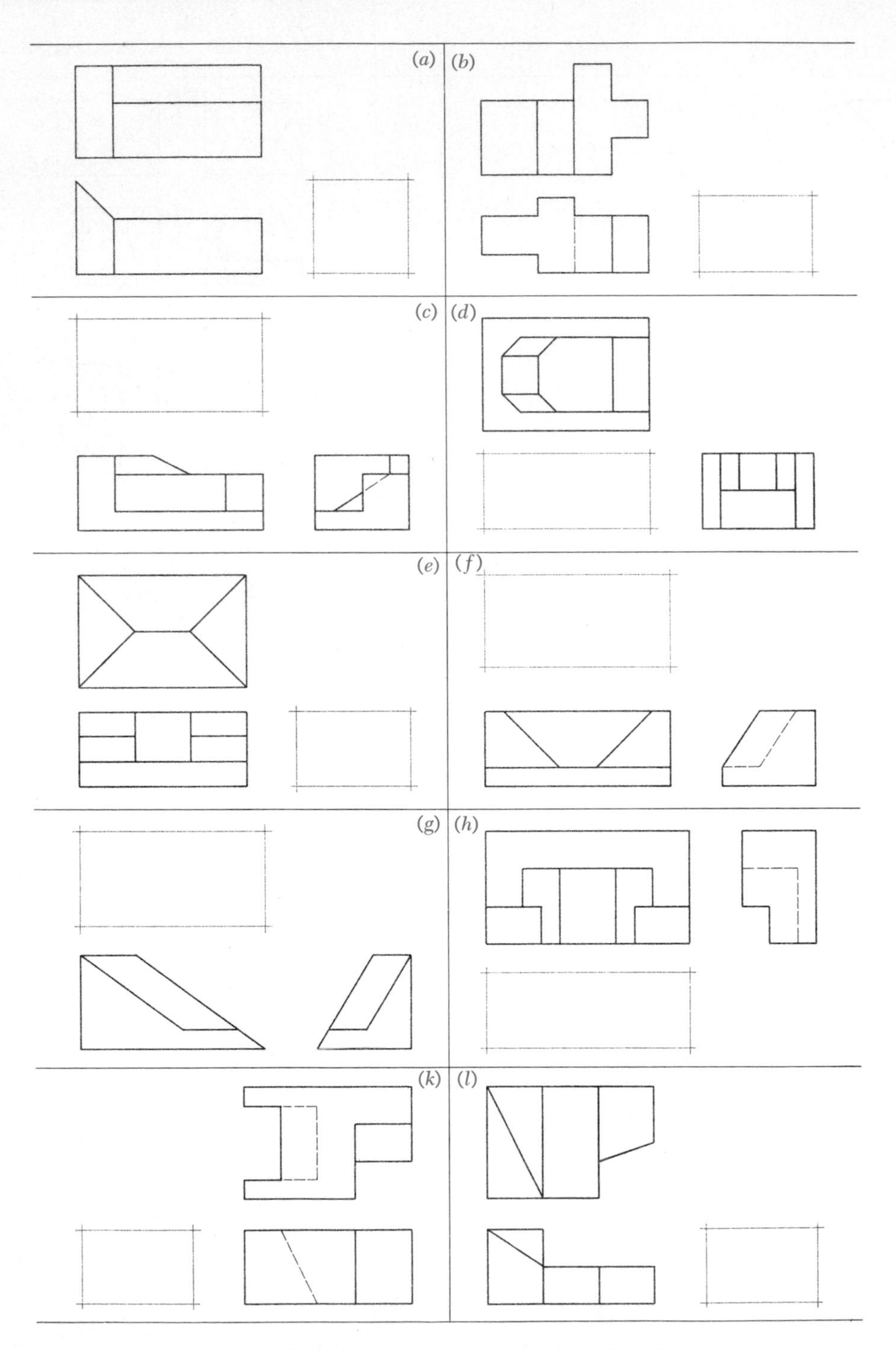


PROB. 84. Views to be supplied freehand.

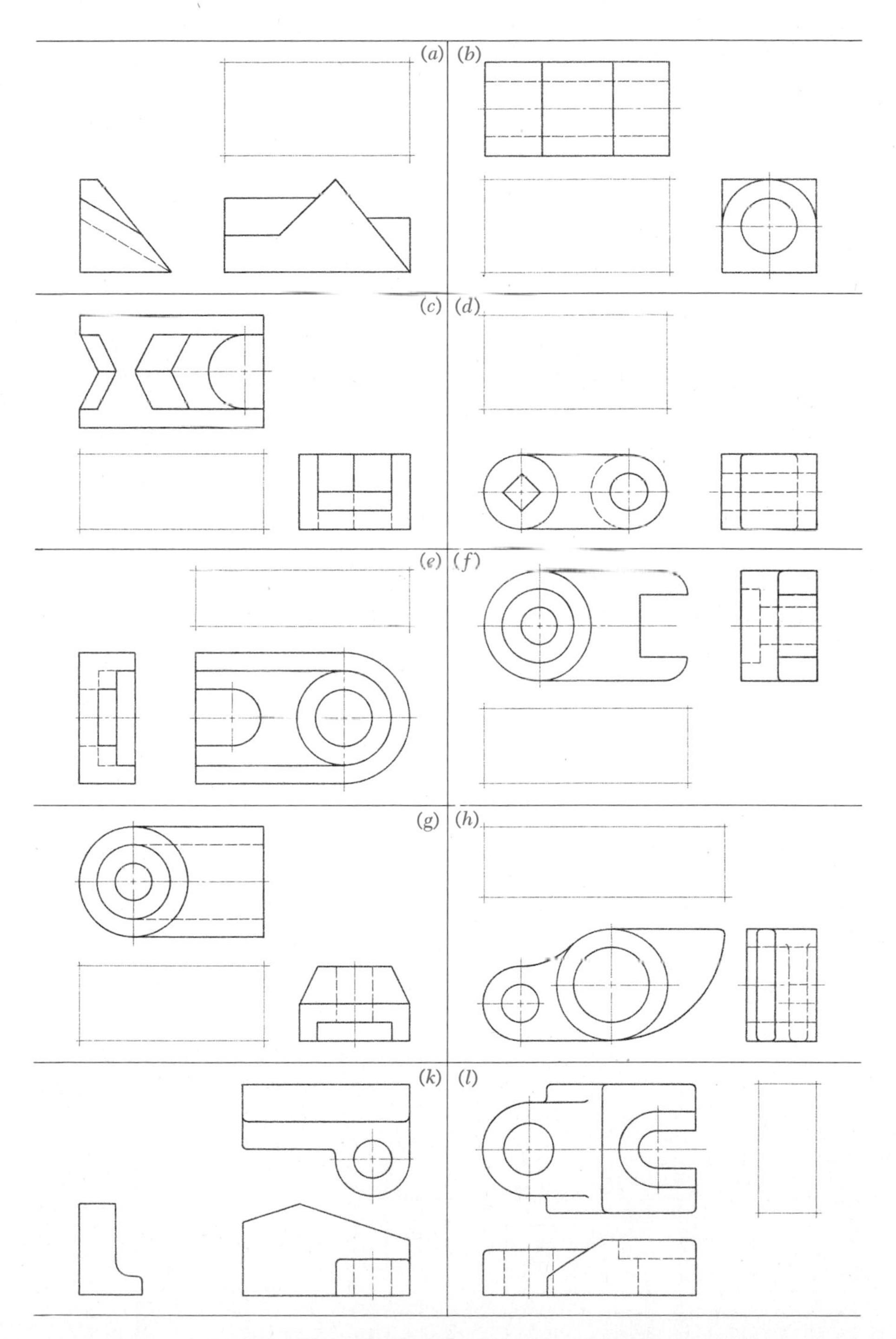


**PROB. 85.** Views to be supplied freehand.

**86.** Given top and front views, add side view. Find at least three solutions. Use tracing paper for the second and third solutions.
**87.** Given front and side views, add top view. Find at least three solutions. Use tracing paper for the second and third solutions.
**88.** Given top and front views, add side view. Find two solutions. Use tracing paper for the second solution.
**89.** Given top and front views, add side view. Find at least three solutions. Use tracing paper for the second and third solutions.

## Group 5. Views to be Supplied.

**90.** Draw the views given, completing the top view from information given on the front and side views. Carry the views along together.
**91.** Given top and front views of the block, add side view. See that dashed lines start and stop correctly.
**92.** Given front and right-side views, add top view.
**93.** Given front and right-side views, add top view.
**94.** Given front and top views, add right-side view.
**95.** Given top and front views, add right-side view.
**96.** Complete the three views given.
**97.** Given front and left-side views, add top view.
**98.** Given front and right-side views, add top view.
**99.** Given front and top views, add right-side view.
**100.** Assume this to be the right-hand part. Draw three views of the left-hand part.
**101.** Given front and top views, add side view.
**102.** Given front and top views, add side view.
**103.** Given top and front views, add left-side view.
**104.** Given front and top views, add side view.
**105.** Given front and top views, add side view.
**106.** Given front and top views, add side view.
**107.** Given top and front views, add side view.
**108.** Given top and front views, add left-side view.
**109.** Given top and front views, add side view.
**110.** Given front and top views, add side view.
**111.** Given front and right-side views of electric-motor support. Add top view.
**112.** Given front and left-side views of master brake cylinder. Add top view.
**113.** Given front and right-side views of end frame for engine starter. Add top view.

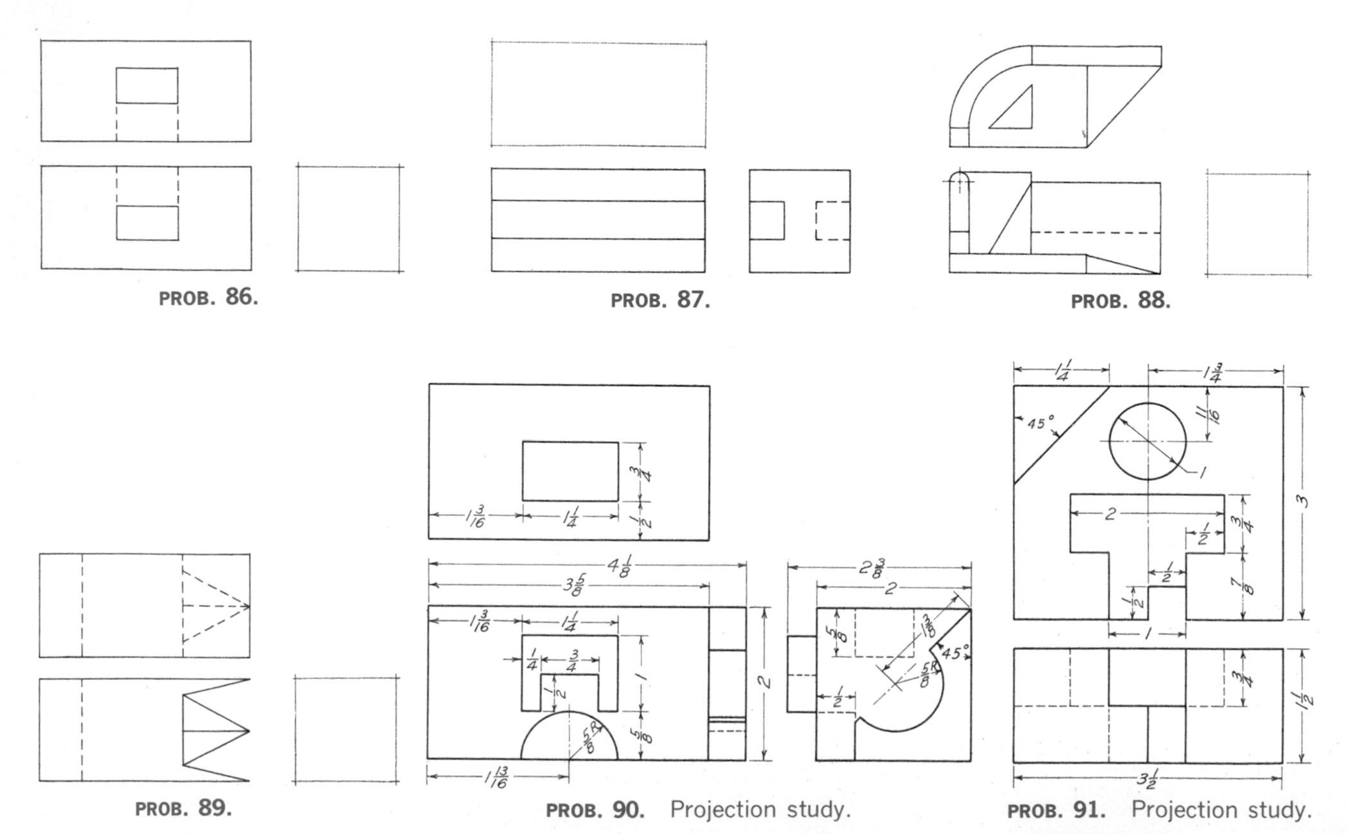

PROB. 86.

PROB. 87.

PROB. 88.

PROB. 89.

PROB. 90. Projection study.

PROB. 91. Projection study.

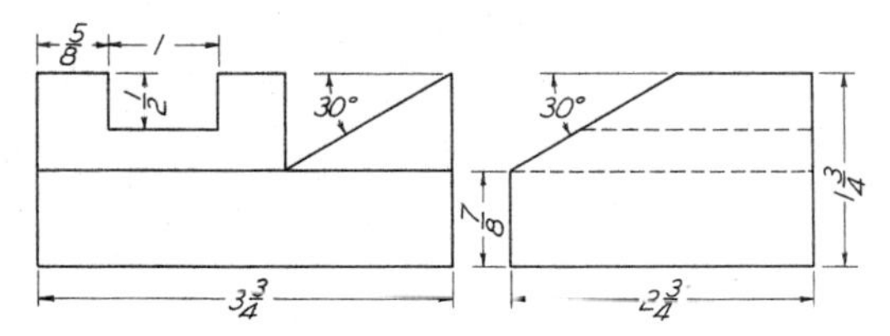

**PROB. 92.** Projection study.

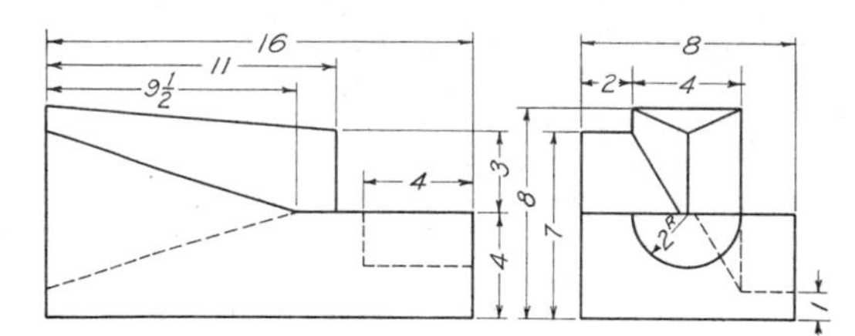

**PROB. 93.** Bit-point forming die.

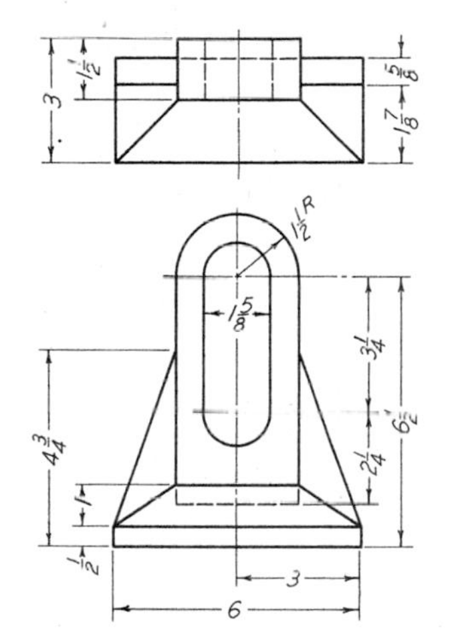

**PROB. 94.** Rabbeting-plane guide.

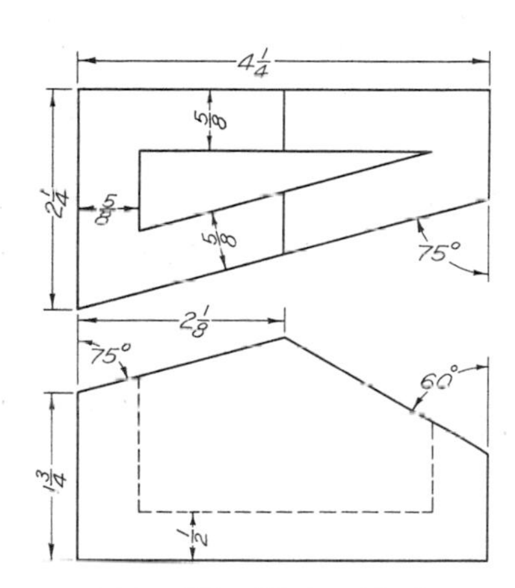

**PROB. 95.** Wedge block.

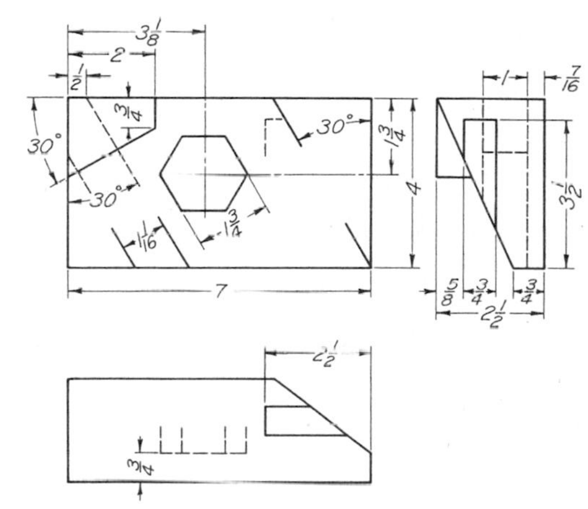

**PROB. 96.** Projection study.

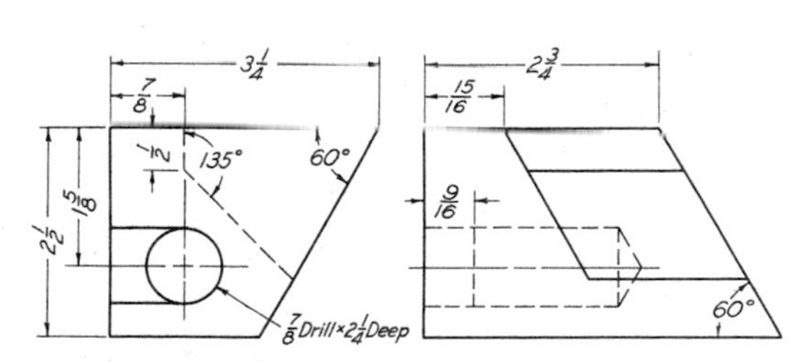

**PROB. 97.** Burner-support key.

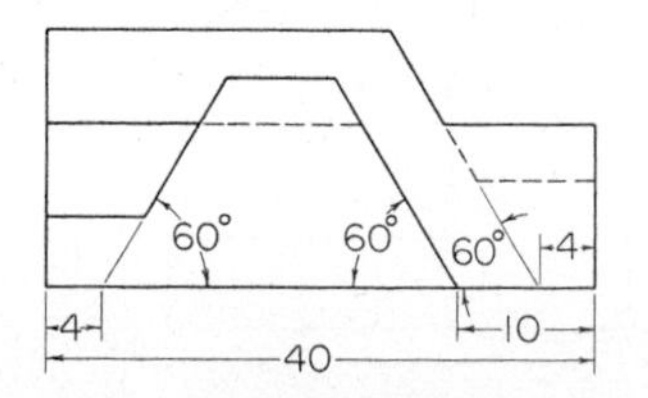

**PROB. 98.** Abutment block.

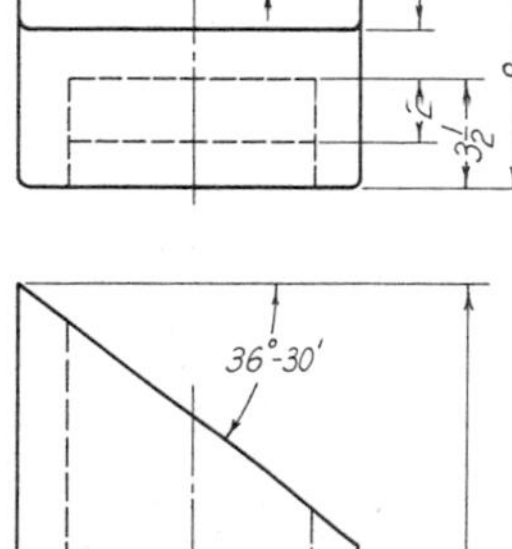

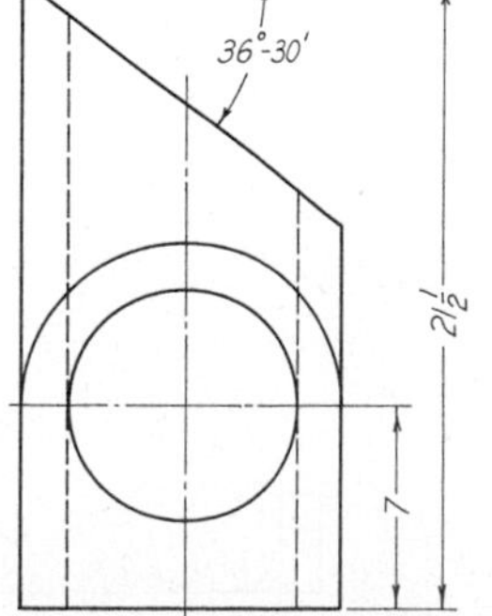

**PROB. 99.** Sliding port.

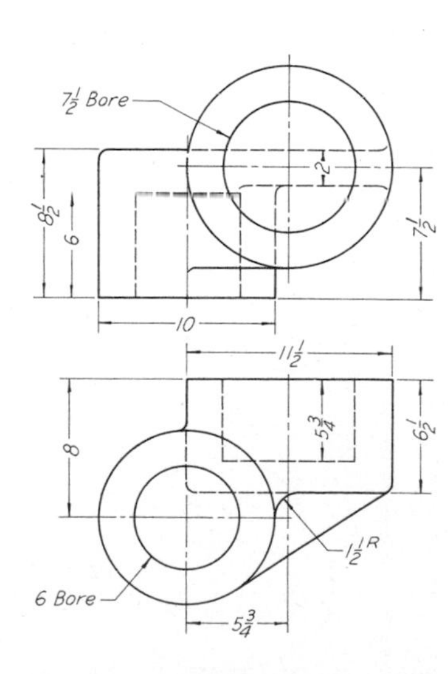

**PROB. 100**
Bumper support and post cap.

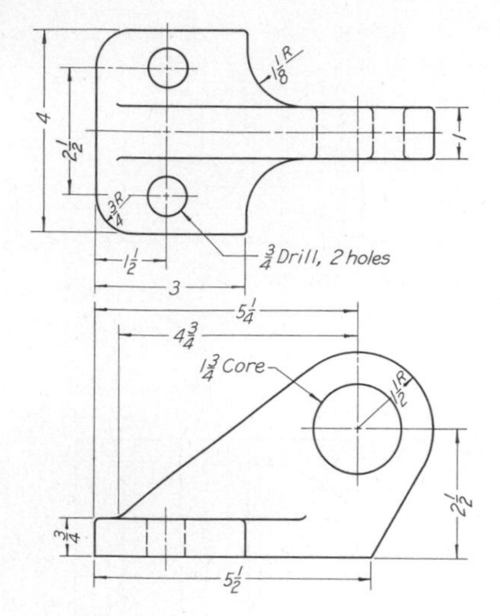


**PROB. 101.** Anchor bracket.

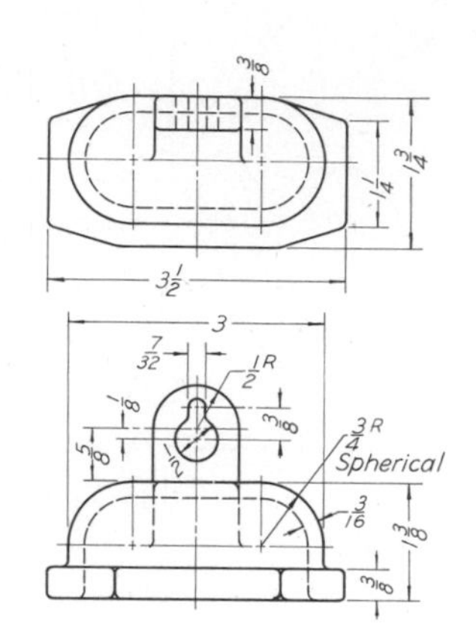


**PROB. 102.** Entrance head.

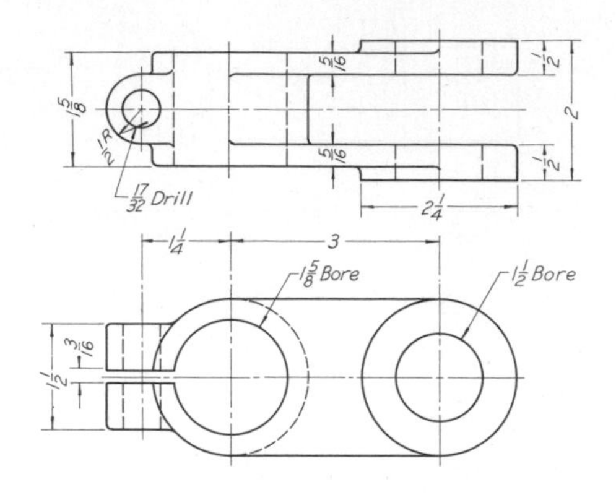


**PROB. 103.** Yoked link.

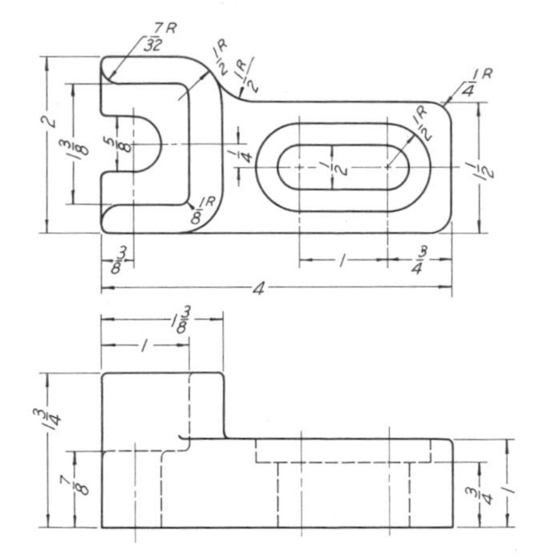

**PROB. 106.** Tool holder.

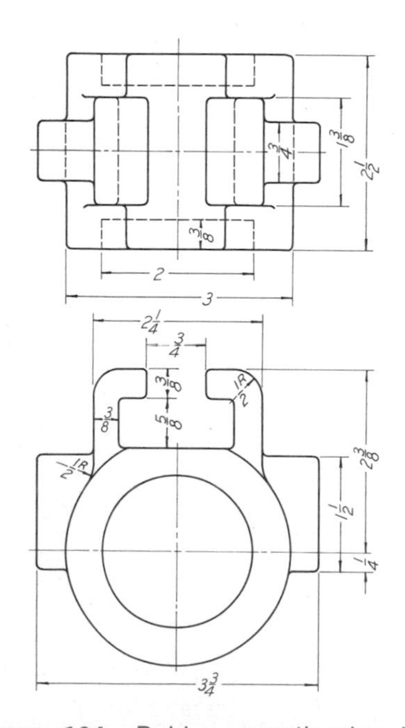

**PROB. 104.** Rubber-mounting bracket.

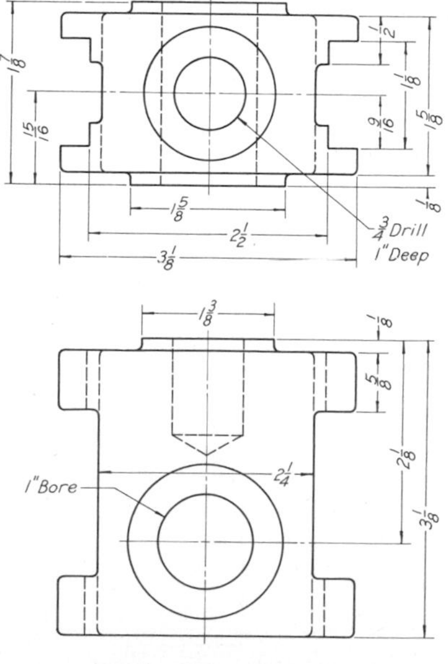


**PROB. 105.** Crosshead.

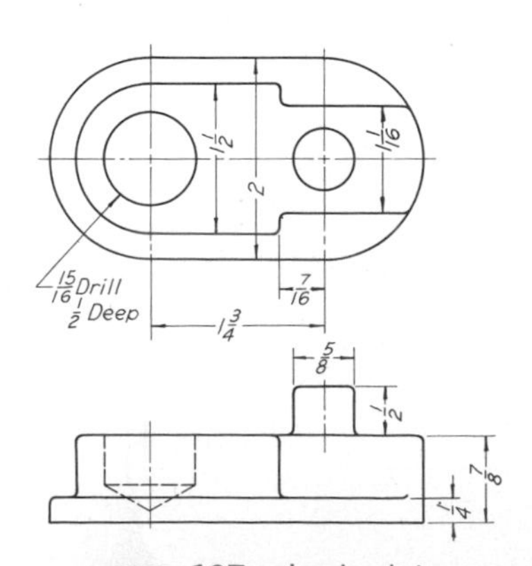


**PROB. 107.** Lock plate.

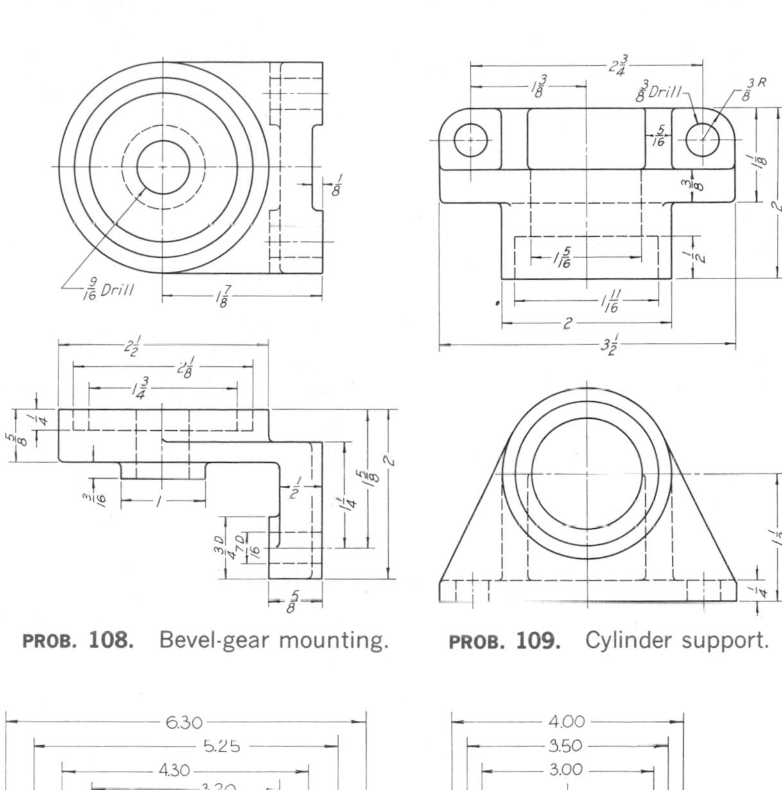

PROB. 108. Bevel-gear mounting.

PROB. 109. Cylinder support.

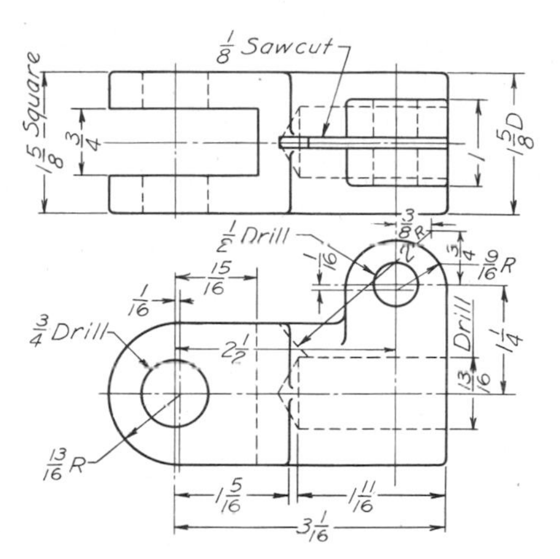

PROB. 110. Rod yoke.

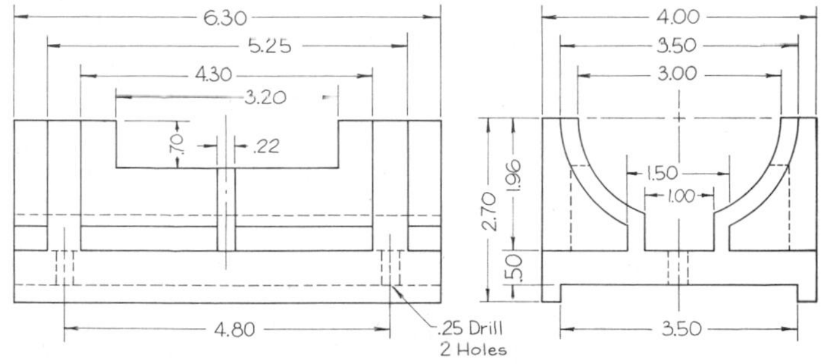

PROB. 111. Electric-motor support.

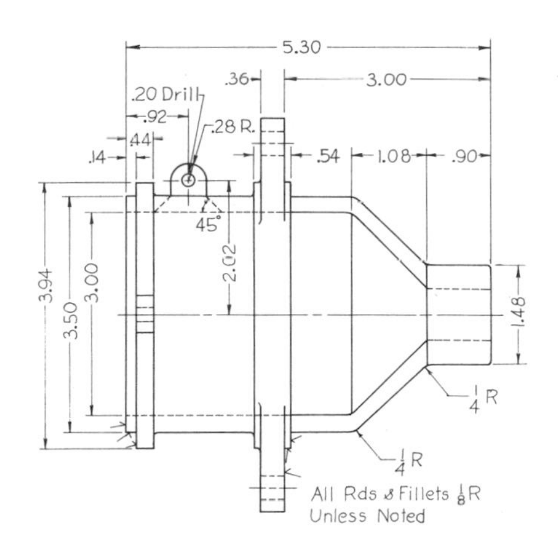


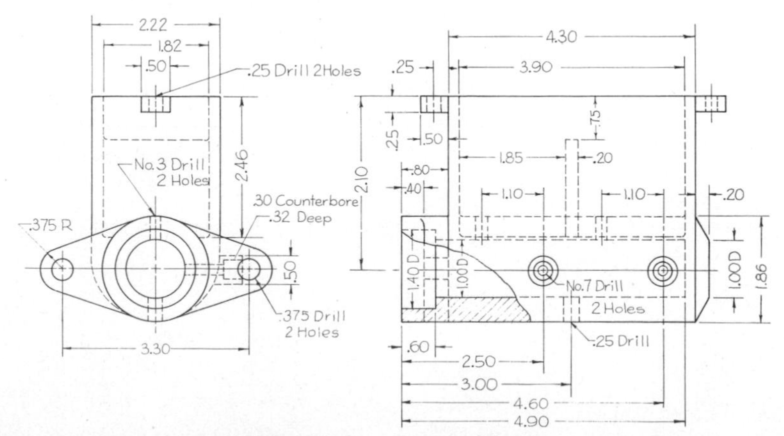

PROB. 112. Master brake cylinder.

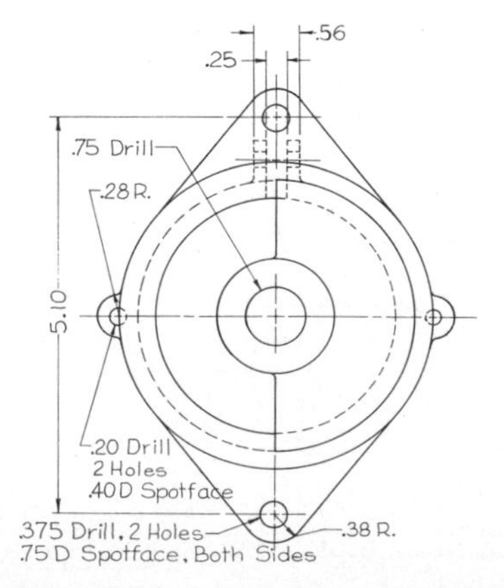

PROB. 113. End frame for engine starter.

## Group 6. Views to be Changed.

These problems (114 to 120) are given to develop the ability to visualize the actual piece in space and from this mental picture to draw the required views as they would appear if the object were looked at in the directions specified.

In addition to providing training in reading orthographic views and in orthographic projection, these problems are valuable exercises in developing graphic technique. Note that all the problems given are castings containing the usual features found on such parts, that is, fillets, rounds, runouts, etc., on unfinished surfaces. Note also that sharp corners are formed by the intersection of an unfinished and finished surface or by two finished surfaces. After finishing one of these problems, check the drawing carefully to make sure that all details of construction have been represented correctly.

**114.** Given front and top views, new front, top, and side views are required, turning the block so that the back becomes the front and the top the bottom. The rib contour is straight.

**115.** Given front, left-side, and bottom views, draw front, top, and right-side views.

**116.** Given front, right-side, and bottom views, draw front, top, and left-side views.

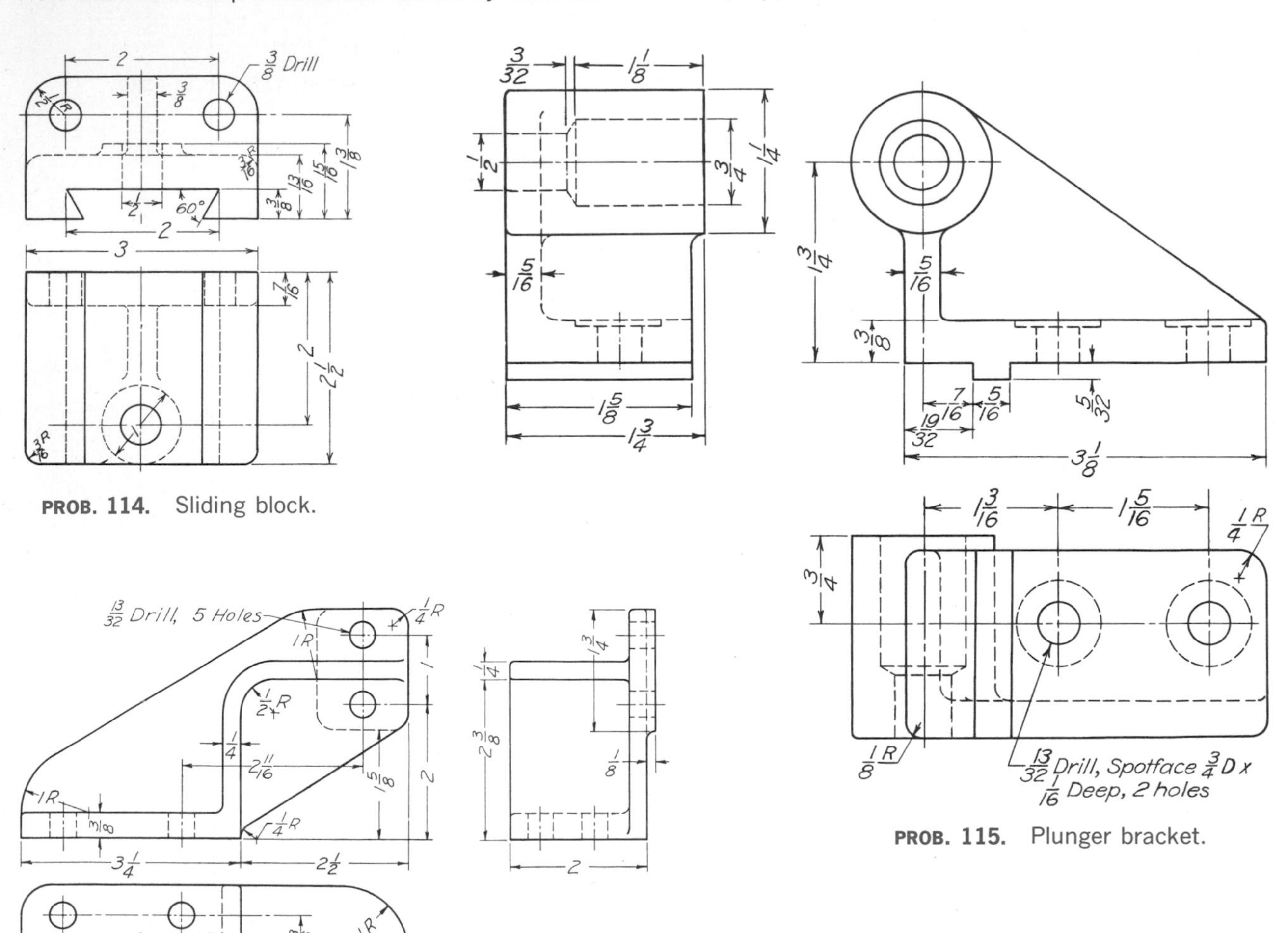


**PROB. 114.** Sliding block.

**PROB. 115.** Plunger bracket.

**PROB. 116.** Offset bracket.

**117.** Given front, right-side, and bottom views, draw new front, top, and right-side views, turning the support so that the back becomes the front.

**118.** Given front and left-side views of the left-hand part, draw the right-hand part.

**119.** Given front, left-side, and bottom views, draw front, top, and right-side views.

**120.** Given top and front views of jet-engine hinge plate. Add right- and left-side views.

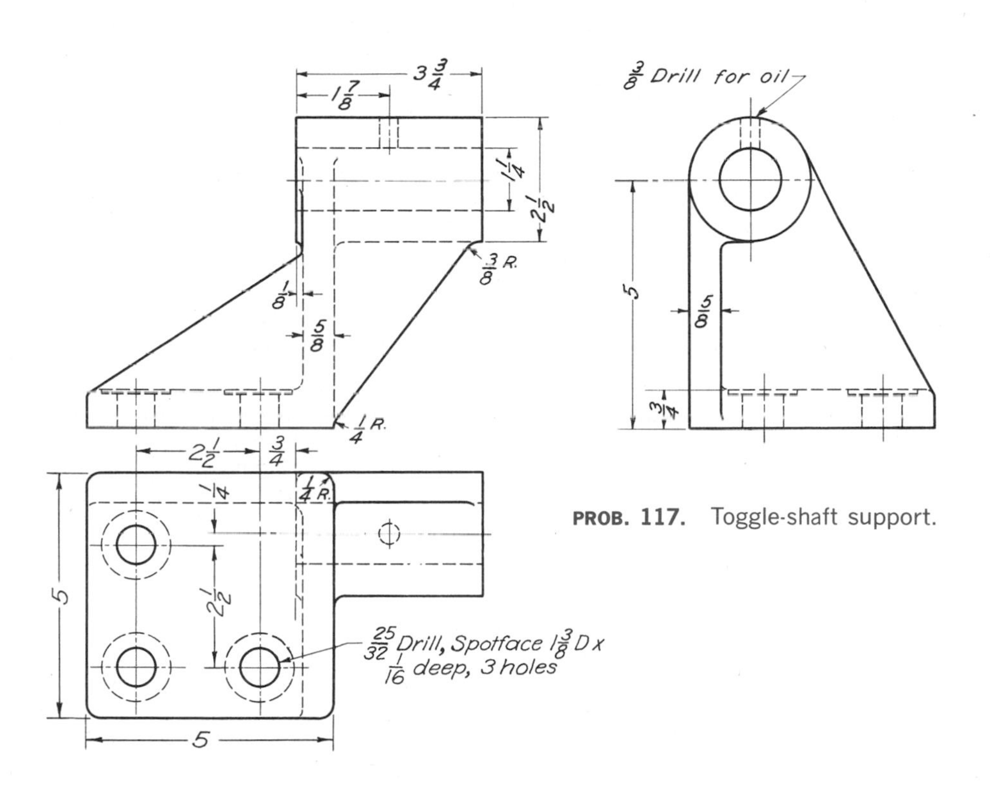


**PROB. 117.** Toggle-shaft support.

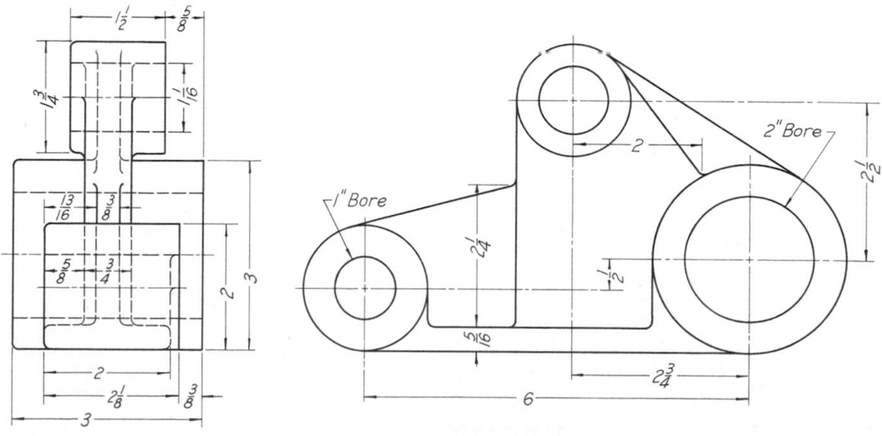


**PROB. 118.** Compound link.

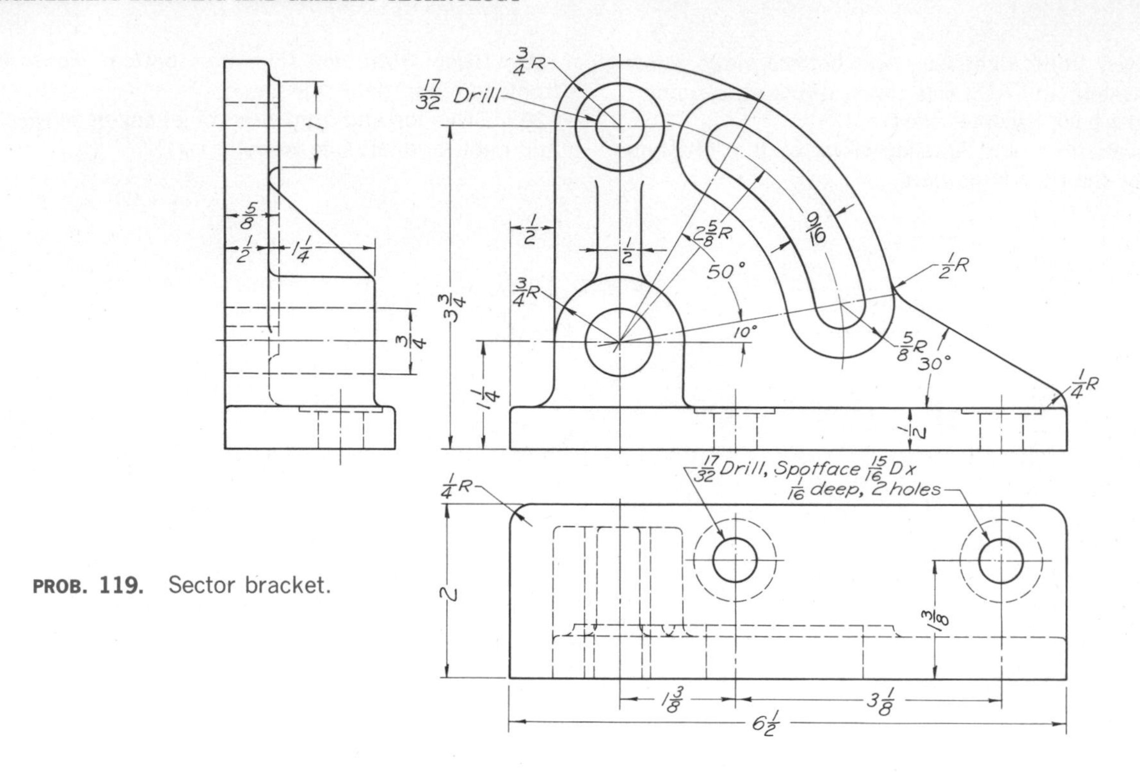


**PROB. 119.** Sector bracket.

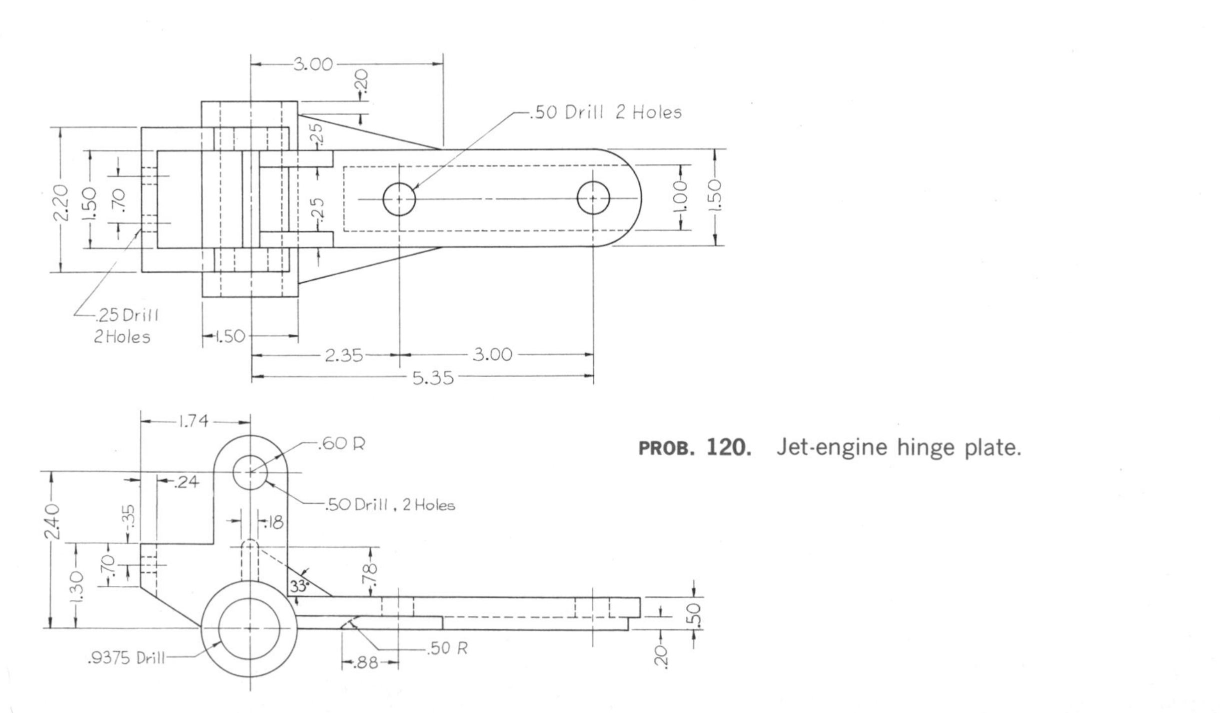


**PROB. 120.** Jet-engine hinge plate.

## Group 7. Drawing from Memory.

One of the valuable assets of an engineer is a trained memory for form and proportion. A graphic memory can be developed to a surprising degree in accuracy and power by systematic exercises in drawing from memory. It is well to begin this training as soon as you have a knowledge of orthographic projection.

Select an object not previously used; look at it with concentration for a certain time (from 5 sec to ½ min or more), close the book, and make an accurate orthographic sketch. Check with the original and correct any mistakes or omissions. Follow with several different figures. The next day, allow a 2-sec view of one of the objects, and repeat the orthographic views of the previous day.

## Group 8. Volume and Weight Calculations with Slide Rule.

In calculating the weight of a piece from a drawing, the object is broken up into the geometric solids (prisms, cylinders, pyramids, or cones) of which it is composed. The volume of each of these shapes is calculated and the individual volumes are added, or subtracted, to find the total volume. The total volume multiplied by the weight of the material per unit of volume gives the weight of the object.

A table of weights of materials is given in the Appendix.

**121.** Find the weight of the cast-iron anchor bracket, Prob. 101.

**122.** Find the weight of the cast-iron bracket, Prob. 104.

**123.** Find the weight of the wrought-iron tool holder, Prob. 106.

**124.** Find the weight of the cast-steel cylinder support, Prob. 109.

**125.** Find the weight of the malleable-iron sliding block, Prob. 114.

This Chapter Offers:

Pictorial Methods Classified

Isometric, Dimetric, Trimetric, and Perspective Drawing and Sketching

Layout of Circles, Arcs, Curves

Object Orientation

Starting Planes and Picture Planes

Projection from Orthographic Views

Graticulation

Pictorial Sketching

Stages in Sketching

Light and Shade

Rendering

Problems

# 6

# PICTORIAL DRAWING AND SKETCHING

**Pictorial Methods Are Used Either as a Basic Form of Shape Description or as a Supplement to Orthographic Depiction**

**1. Pictorial Method–Comparison with Orthographic.**
In discussing the theory of projection in Chap. 5, we noted that in perspective projection the object is represented as it appears to the eye. However, its lines cannot be measured directly for accurate description of the object; and in orthographic projection the object is shown, in two or more views, as it really is in form and dimensions but interpretation requires experience to visualize the object from the views. To provide a system of drawing that represents the object pictorially and in such a way that its principal lines can be measured directly, several forms of one-plane conventional or projectional picture methods have been devised in which the third dimension is taken care of either by turning the object so that its three dimensions are visible or by employing oblique projection. A knowledge of these picture methods and of perspective projection is extremely desirable as they can all be used to great advantage.

Mechanical or structural details not clear in orthographic views can be drawn pictorially or illustrated by supplementary pictorial views. Pictorial views are used advantageously in technical illustrations, Patent Office drawings, layouts, piping plans, and the like. Pictorial methods are useful also in making freehand sketches, and this is one of the most important reasons for learning them.

**2. Pictorial Methods.**
There are three main divisions of pictorial drawing: (1) axonometric, with its divisions into trimetric, dimetric, and isometric; (2) oblique, with several variations; and (3) perspective. These methods are illustrated in Fig. 1.

The trimetric form gives an effect more pleasing to the eye than the other axonometric and oblique methods and allows almost unlimited freedom in orienting the object, but is difficult to draw. With the dimetric method the result is less pleasing and there is less freedom in orienting the object, but execution is easier than with trimetric. The isometric method gives a result less pleasing than dimetric or trimetric, but it is easier to draw and has the distinct advantage that it is easier to dimension. The oblique method is used principally for objects with circular or curved features only on one face or on parallel faces, and for objects of this type the oblique is easy to draw and dimension. Perspective drawing gives a result most pleasing to the eye, but it is of limited usefulness because many lines are unequally foreshortened; isometric and oblique are the forms most commonly used.

**3. Axonometric Projection.**
Axonometric projection is theoretically orthographic projection in which only one plane is used, the object being turned so that three faces show. Imagine a transparent vertical plane with a cube behind it, one face of the cube being parallel to the plane. The projection on the plane, that is, the front view of the cube, will be a square (Fig. 2). Rotate the cube about a vertical axis through any angle less than

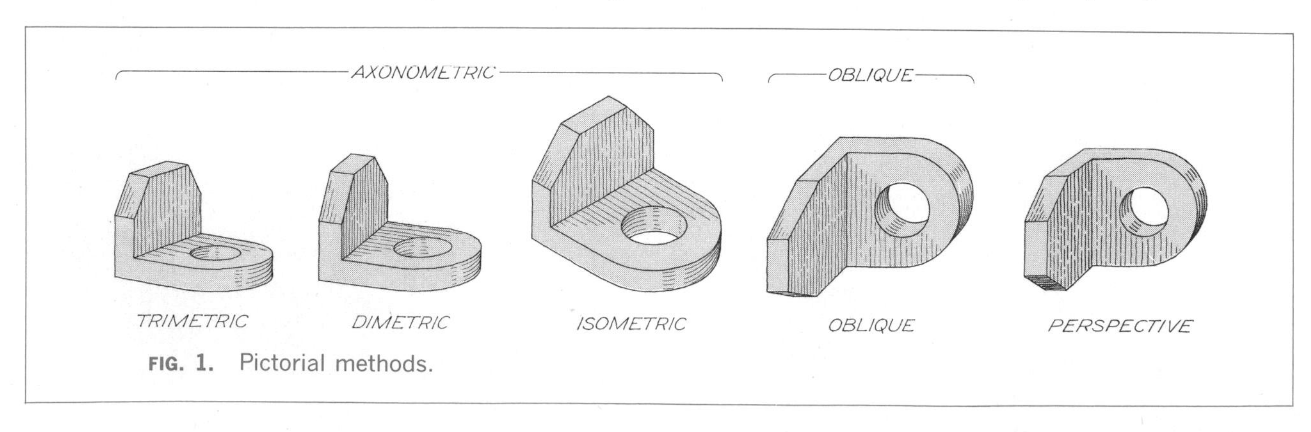


**FIG. 1.** Pictorial methods.

90°, and the front view will now show two faces, both foreshortened (Fig. 3). From this position, tilt the cube forward (rotation axis perpendicular to profile) any amount less than 90°. Three faces will now be visible on the front view (Fig. 4). There can be an infinite number of axonometric positions, depending upon the angles through which the cube is rotated. Only a few of these positions are ever used for drawing. The simplest is the isometric (equal-measure) position, in which the three faces are foreshortened equally.

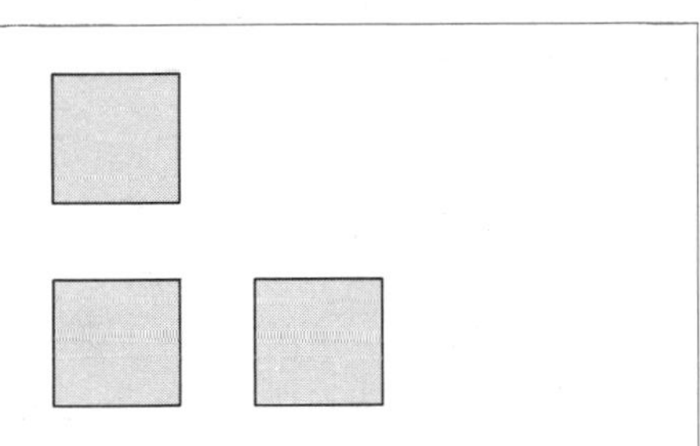

**FIG. 2.** The object face is parallel to the picture plane. One face only is seen in the front view.

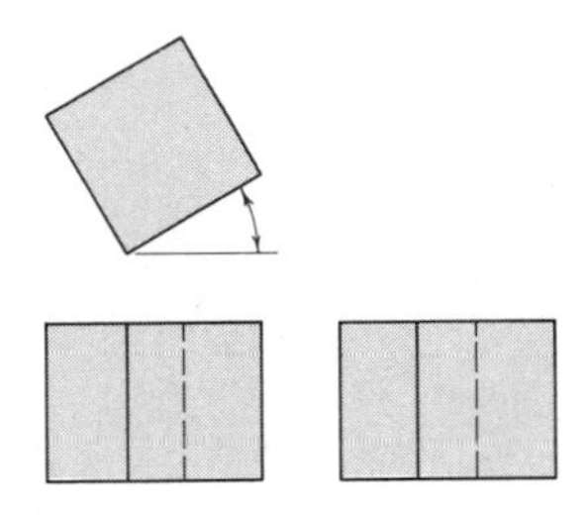

**FIG. 3.** The object is rotated about a vertical axis. Two faces are seen.

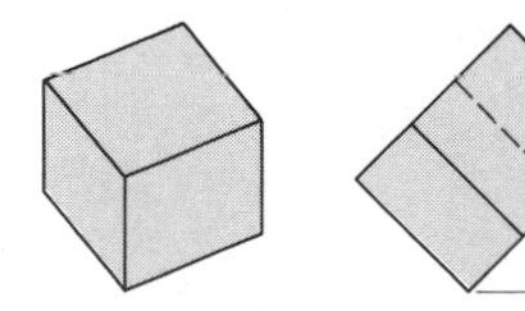

**FIG. 4.** The object is rotated about both a vertical and a profile axis. Three faces are seen.

## 4. Isometric Projection.

If the cube in Fig. 5*a* is rotated about a vertical axis through 45°, as shown at (*b*), and then tilted forward, as at (*c*), until the edge *RU* is foreshortened equally with *RS* and *RT*, the front view of the cube in this position is said to be an "isometric projection." (The cube has been tilted forward until the body diagonal through *R* is perpendicular to the front plane. This makes the top face slope approximately 35°16'.[1]) The projections of the three mutu-

[1]The only difference between rotation and auxiliary projection is that in the former the object is moved and in the latter the plane is moved or the observer is considered to have changed his viewing position. Thus an auxiliary view on a plane perpendicular to a body diagonal of the cube in position (*b*) would be an isometric projection, as illustrated by the dotted view.

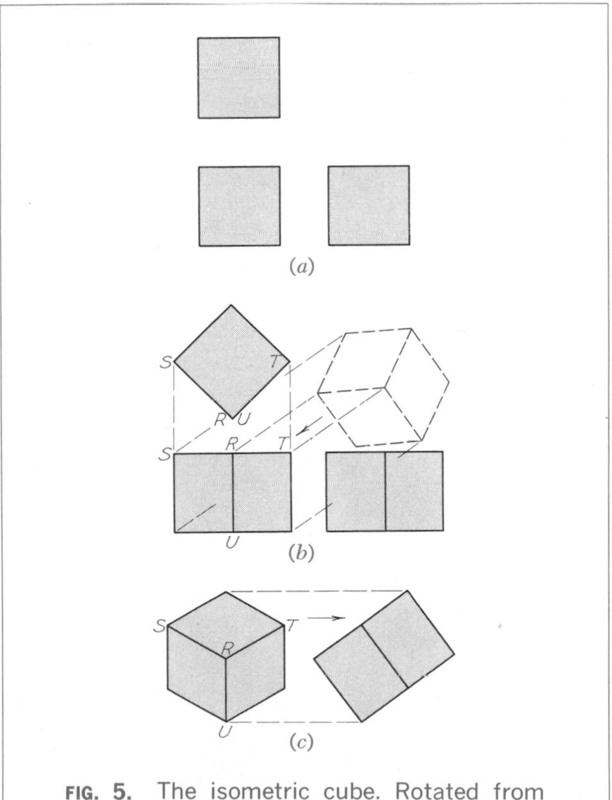


**FIG. 5.** The isometric cube. Rotated from position (*a*) to (*b*) then to (*c*), the three perpendicular edges are now equally foreshortened.

ally perpendicular edges *RS*, *RT*, and *RU* meeting at the front corner *R* make equal angles, 120°, with each other and are called "isometric axes." Since the projections of parallel lines are parallel, the projections of the other edges of the cube will be, respectively, parallel to these axes. Any line parallel to the edge of the cube, whose projection is thus parallel to an isometric axis, is called an "isometric line." The planes of the faces of the cube and all planes parallel to them are called "isometric planes."

The isometric axes *RS*, *RT*, and *RU* are all foreshortened equally because they are at the same angle to the picture plane.

## 5. Isometric Drawing.

In nearly all practical use of the isometric system, this foreshortening of the lines is disregarded, and *their full lengths are laid off on the axes,* as explained in Sec. 6. This gives a figure of exactly the same shape but larger in the proportion of 1.23 to 1, linear, or in optical effect $1.23^3$ to $1.00^3$ (Fig. 6). Except when drawn beside the same piece in orthographic projection, the effect of increased size is usually of no consequence, and since the advantage of measuring the lines directly is of great convenience, isometric drawing is used almost exclusively rather than isometric projection.

In isometric projection the isometric lines have been foreshortened to approximately $^{81}/_{100}$ of their length, and an isometric scale to this proportion can be made graphically as shown in Fig. 7 if it is necessary to make an isometric projection by the method of isometric drawing.

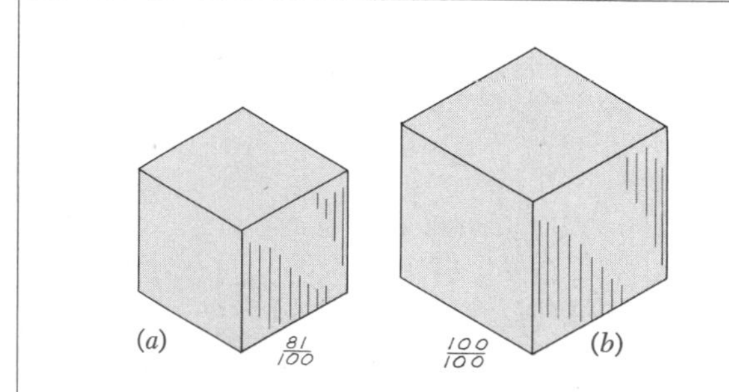


**FIG. 6.** (*a*) isometric projection; (*b*) isometric drawing.

## 6. To Make an Isometric Drawing.

If the object is rectangular (Fig. 8), start with a point

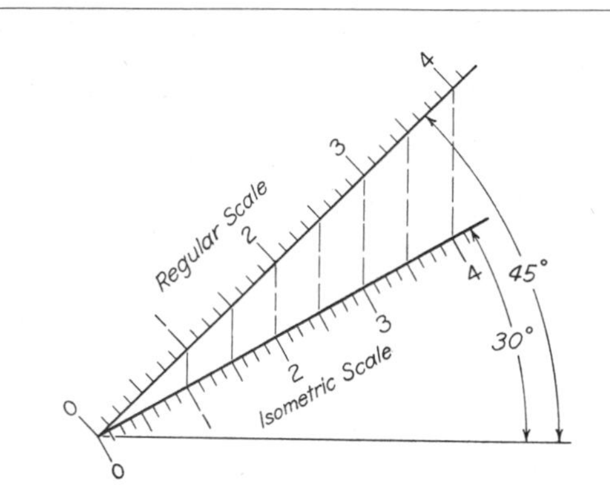


**FIG. 7.** To make an isometric scale.

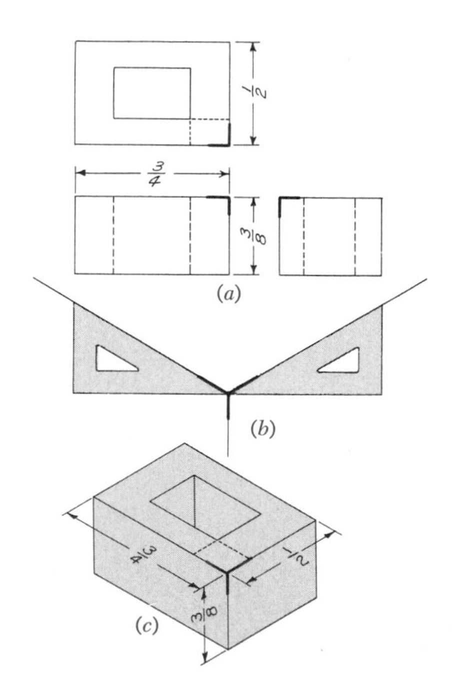


**FIG. 8.** Isometric axes, first position. The starting point is the upper front corner.

representing a front corner, shown at (*a*) with heavy lines, and draw from it the three isometric axes 120° apart, one vertical (*b*), the other two with the 30° triangle. On these three lines measure the height, width, and depth of the object, as indicated at (*c*); through the points so determined draw lines parallel to the axes, completing the figure. When drawing in isometric, remember the direction of the three principal isometric planes. Hidden lines are omitted except when they are needed to describe the piece.

It is often convenient to build up an isometric drawing from the lower front corner, as illustrated in Fig. 9, starting from axes in what may be called the "second position." The location of the starting corner is again shown by heavy lines at (*a*), (*b*), and (*c*).

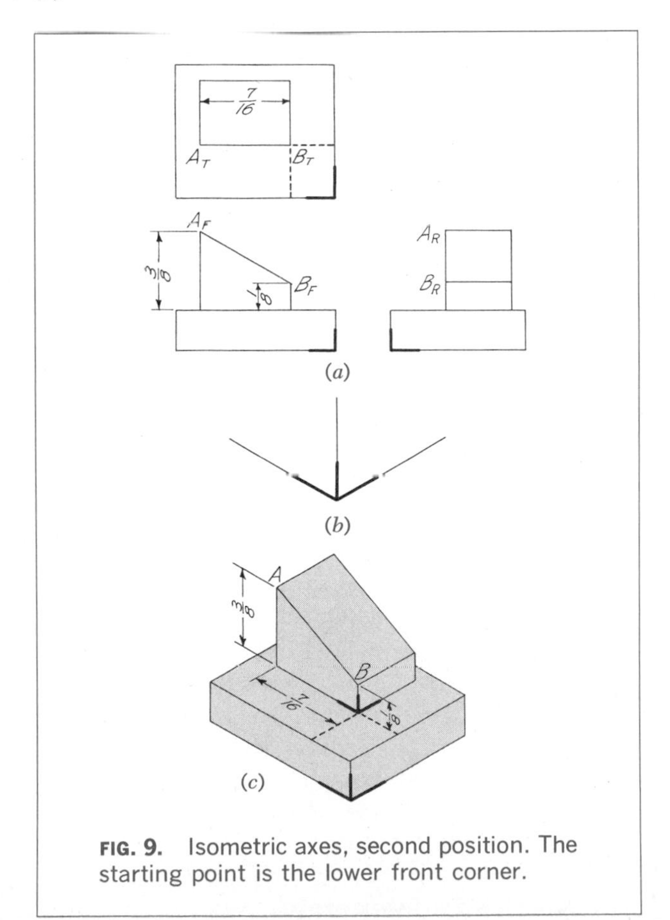


**FIG. 9.** Isometric axes, second position. The starting point is the lower front corner.

### 7. Nonisometric Lines.

Edges whose projections or drawings are not parallel to one of the isometric axes are called "nonisometric lines." The one important rule is that *measurements can be made only on the drawings of isometric lines;* conversely, measurements *cannot* be made on the drawings of *nonisometric* lines. For example, the diagonals of the face of a cube are nonisometric lines; although equal in length, their isometric drawings will not be at all of equal length on the isometric drawing of the cube. Compare the length of the diagonals on the cube in Fig. 6. Since a nonisometric line does not appear in the isometric drawing in its true length, the isometric view of each end of the line must be located and the isometric view of the line found by joining these two points. In Fig. 9*a* and *c*, line *AB* is a nonisometric edge whose true length cannot be measured on the isometric drawing. However, the vertical distances above the base to points *A* and *B* are parallel to the vertical isometric axis. These lines can, therefore, be laid off, as shown at (*c*), to give the isometric view of line *AB*.

### 8. Nonisometric Lines: Boxing Method.

When an object contains many nonisometric lines, it is drawn by the "boxing method" or the "offset method." When the boxing method is used, the object is enclosed in a rectangular box, which is drawn around it in orthographic projection. The box is then drawn in isometric and the object located in it by its points of contact, as in Figs. 10 and 12. It should be noted that the isometric views of lines that are parallel on the object are parallel. This knowledge can often be used to save a large amount of construction, as well as to test for accuracy. Figure 10 might be drawn by putting the top face into isometric and drawing vertical lines equal in length to the edges downward from each corner. It is not always necessary to enclose the whole object in a rectangular "crate." The pyramid (Fig. 11) would have its base enclosed in a rectangle and the apex located by erecting a vertical axis from the center.

The object shown in Fig. 12 is composed almost entirely of nonisometric lines. In such cases the isometric drawing cannot be made without first

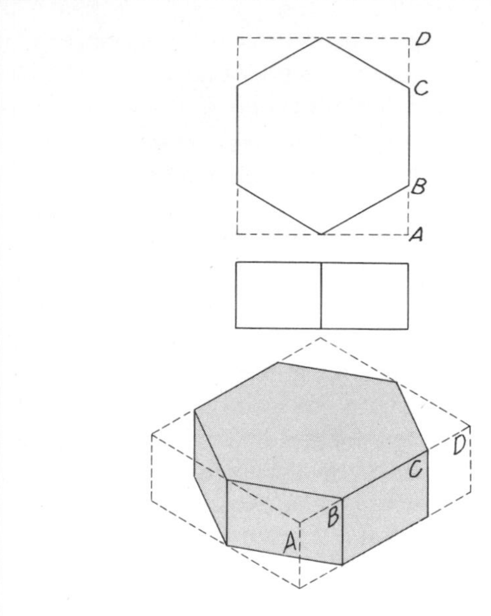


FIG. 10. Box construction. Points on the orthographic box are transferred to the pictorial box. Identical scale must be used.

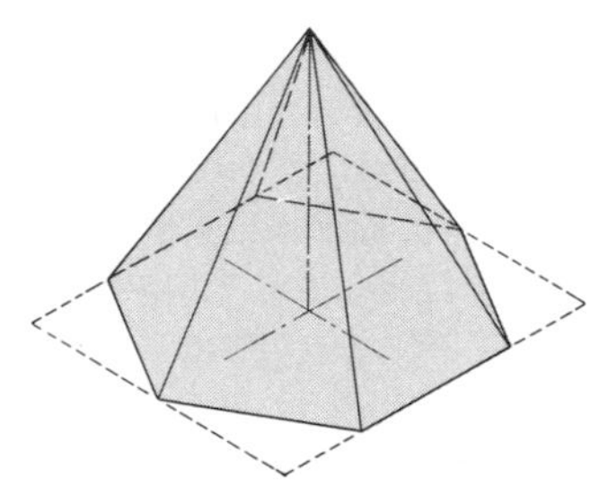

FIG. 11. Semibox construction. Points on the base are transferred by boxing. The altitude is located by a vertical from the base center. Identical scale must be used.

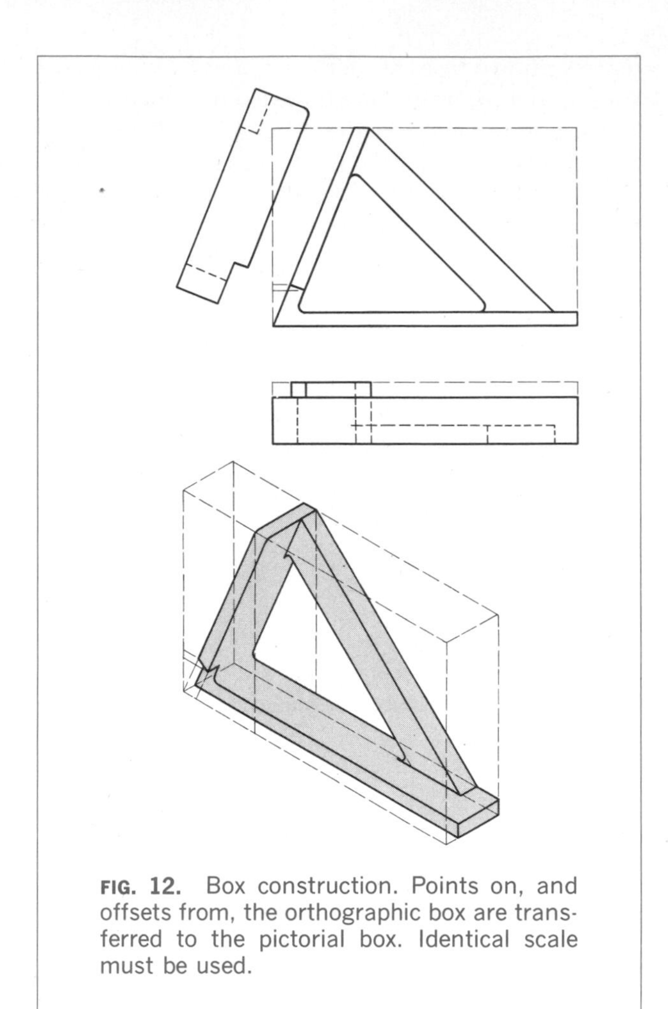

FIG. 12. Box construction. Points on, and offsets from, the orthographic box are transferred to the pictorial box. Identical scale must be used.

making the orthographic views necessary for boxing. In general, the boxing method is adapted to objects that have the nonisometric lines in isometric planes.

**9. Nonisometric Lines: Offset Method.** When an object is made up of planes at different angles, it is better to locate the ends of the edges by the offset method rather than by boxing. When the offset method is used, perpendiculars are extended from each point to an isometric reference plane. These perpendiculars, which are isometric lines, are located on the drawing by isometric coordinates, the dimensions being taken from the orthographic views. In Fig. 13, line $AB$ is used as a base line and measurements are made from it as shown, first to locate points on the base; then verticals from these points locate $e$, $f$, and $g$. Figure 14 is another example of offset construction. Here a vertical plane is used as a reference plane. Note that, as in Fig. 13, the *base* of the offset is located first; then the offset distance is measured.

**10. Angles in Isometric.** The three isometric axes, referred back to the iso-

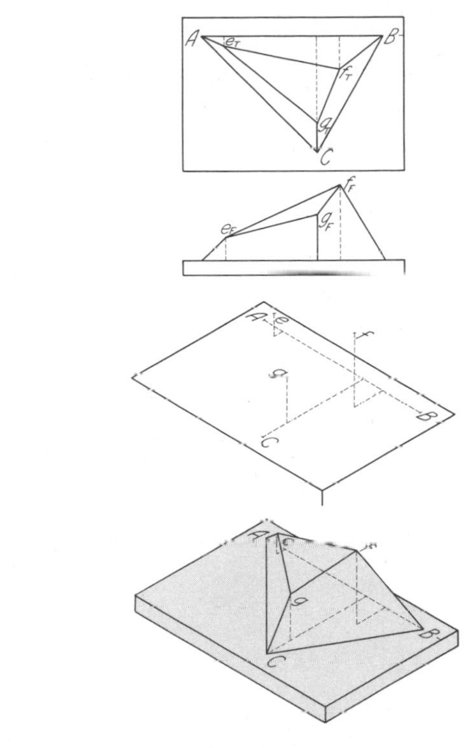


**FIG. 13.** Offset construction. A base (*AB*) is assumed; then all points are located from it.

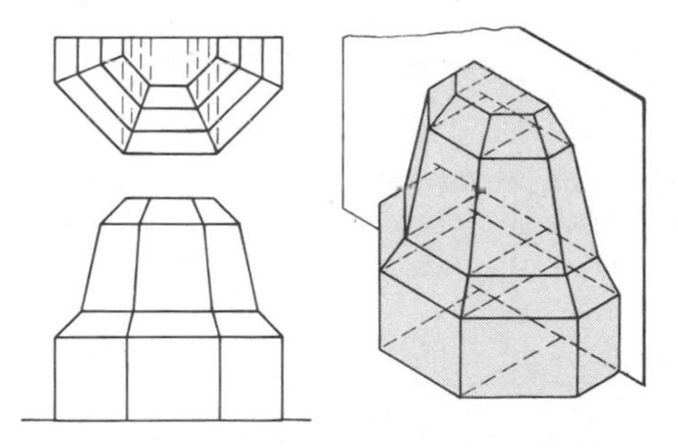

**FIG. 14.** Offset construction. All points are located on a plane or by offsets from the plane. Identical scale must be used.

metric cube, are mutually perpendicular but in an isometric drawing appear at 120° to each other. For this reason, angles specified in degrees do not appear in their true size on an isometric drawing and must be laid off by coordinates that will be parallel to the isometric axes. Thus if an orthographic drawing has edges specified by angular dimensions, as in Fig. 15A, *a view to the same scale as the isometric drawing* is made as at (*B*); from this view the coordinate dimensions *a*, *b*, and *c* are transferred with dividers or scale to the isometric drawing.

## 11. Curves in Isometric.

For the reasons given in Secs. 7 and 10, a circle or any other curve will not show in its true shape when drawn in isometric. A circle on any isometric plane will be an ellipse, and a curve will be shown as the isometric projection of the true curve.

Any curve can be drawn by plotting points on it from isometric reference lines (coordinates) that are parallel to the isometric axes, as shown in Fig. 16.

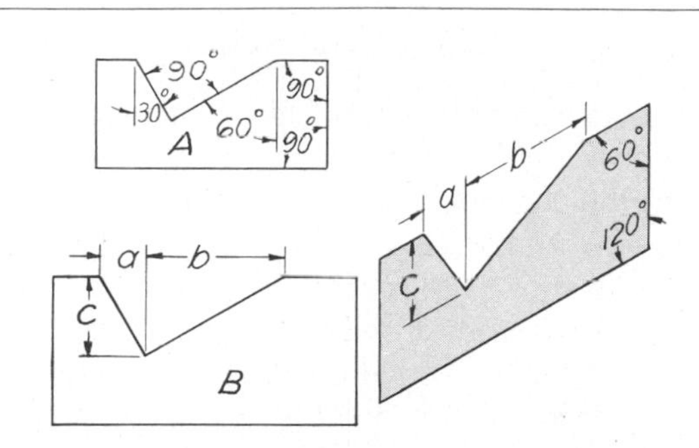


**FIG. 15.** Angles in isometric. These must be laid out by offsets from an orthographic view to the same scale.

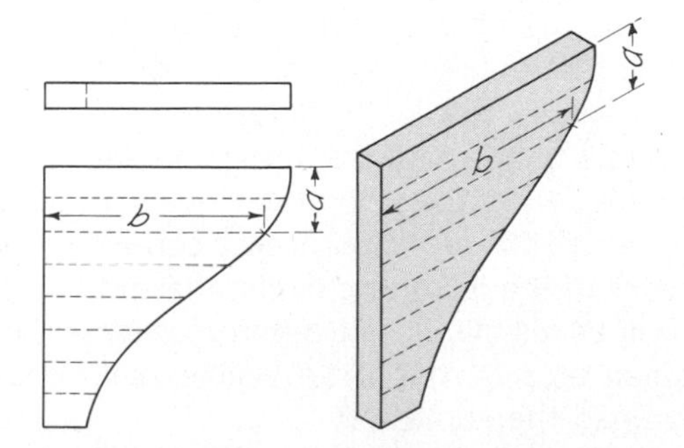


**FIG. 16.** Curves in isometric. Points are transferred from the orthographic view to the pictorial by offsets. Identical scale must be used.

A circle plotted in this way is shown in Fig. 17. Note that in both these figures coordinates *a* and *b* are parallel to the isometric axes and the coordinate distances must be obtained from an orthographic view drawn to the same scale as the isometric.

## 12. Isometric Circles.

Circles occur so frequently that they are usually drawn by a four-centered approximation, which is sufficiently accurate for ordinary work. Geometrically, the center for any arc tangent to a straight line lies on a *perpendicular from the point of tangency* (Fig. 18*a*). In isometric, if perpendiculars are drawn from the middle point of each side of the circumscribing square, the intersections of these perpendiculars will be centers for arcs tangent to two sides (*b*). Two of these intersections will evidently fall at the corners *A* and *C* of the isometric square, as the perpendiculars are altitudes of equilateral triangles. Thus the construction at (*b*) to (*d*) can be made by simply drawing 60° lines (horizontals also at *C* and *D*) from the corners *A* and *C* and then drawing arcs with radii $R$ and $R_1$, as shown.

Figure 19 shows the method of locating and laying out a hole in isometric from the given orthographic views. First locate and then draw the center lines for the hole by laying out the distances *X* and *Y*, as shown. On these lines construct an isometric square with sides equal to the diameter of the hole by laying out the radius *R* in each direction from the intersection of the center lines. Then use the four-center method, as shown in Fig. 18. Should the piece be thin enough, a portion of the backside of the hole will be visible. To determine this, drop the thickness *T* back on an isometric line and swing the large radius $R_1$ of the isometric circle with this point as center. If the arc thus drawn comes within the boundary of the isometric circle, that portion of the back will be visible. In extra-thin pieces, portions of the small arcs $R_2$ might be visible. This would be determined the same way.

If a true ellipse is plotted by the method of Sec. 11 in the same square, it will be a little longer and narrower and of much more pleasing shape than this four-center approximation, but in most drawings the difference is not sufficient to warrant the extra

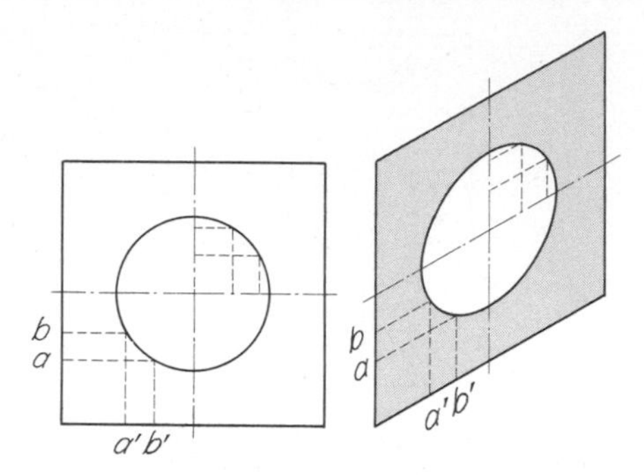


**FIG. 17.** Isometric circle, points plotted. Points are transferred from the orthographic view to the pictorial by offsets. Identical scale must be used.

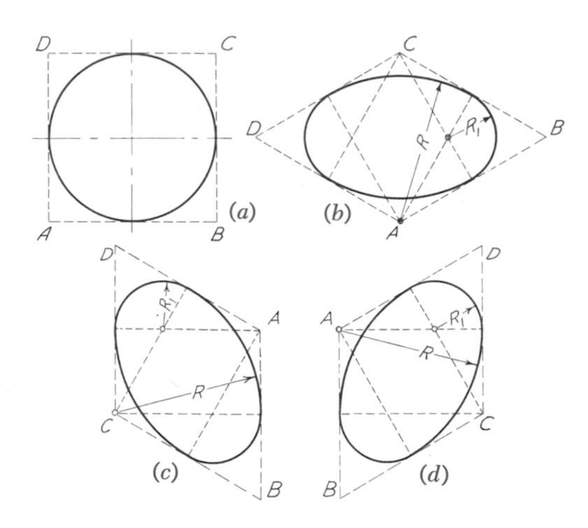


**FIG. 18.** Isometric circles, four-centered method. The ellipse is approximated by circle arcs.

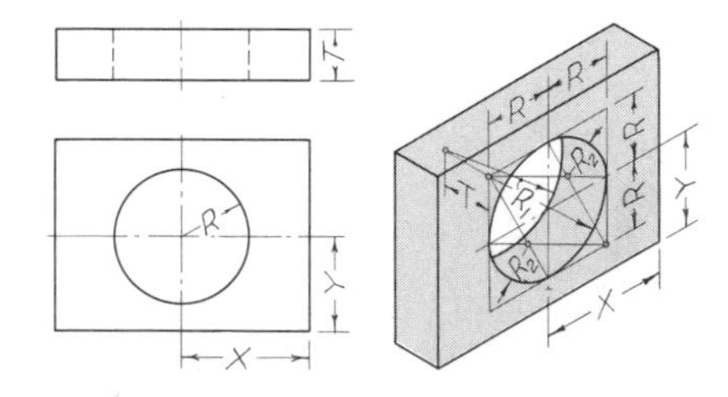


**FIG. 19.** Locating and laying out a hole in isometric. Locate the center, draw the enclosing isometric square, and then draw the circle by the method of Fig. 18.

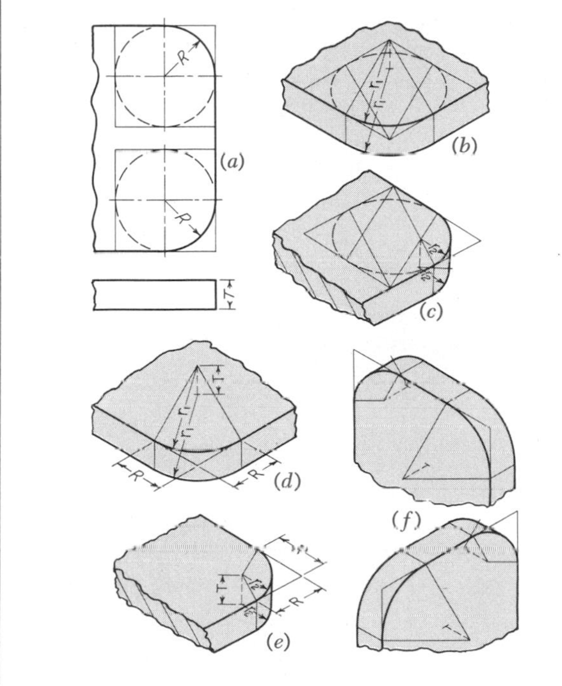


**FIG. 20.** Isometric quarter circles. The radius center lies on perpendiculars from tangent points that are the radius distance from the corner.

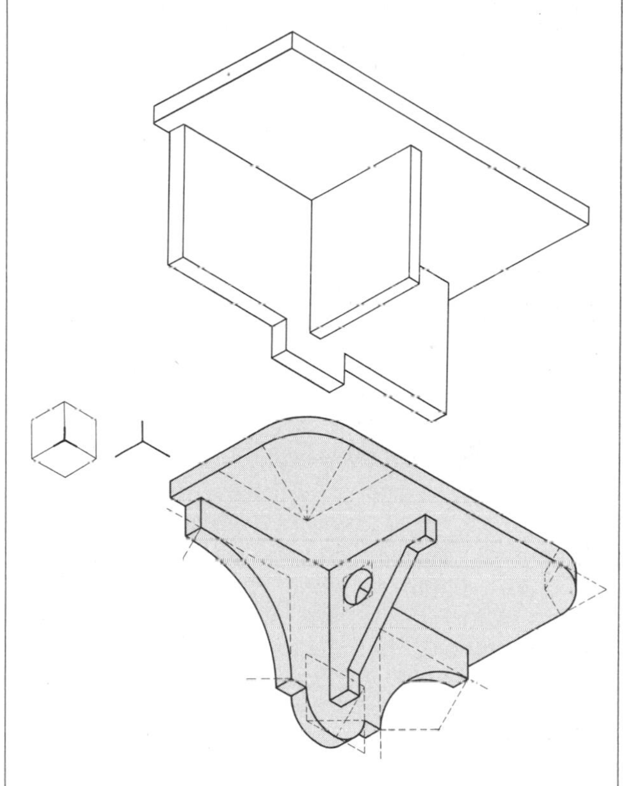

**FIG. 21.** Isometric with reversed axes. The bottom and two sides are shown. Construction methods are the same as for regular position.

expenditure of time required in execution.

The isometric drawing of a *sphere* is a circle with its diameter equal to the long axis of the ellipse that is inscribed in the isometric square of a great circle of the sphere. It would thus be 1.23/1.00 of the actual diameter (the isometric *projection* of a sphere would be a circle of the actual diameter of the sphere).

## 13. Isometric Circle Arcs.

To draw any circle arc, draw the isometric square of its diameter in the plane of its face, with as much of the four-center construction as is necessary to find centers for the part of the circle needed, as illustrated in Fig. 20. The arc occurring most frequently is the quarter circle. Note that in illustrations (*d*) and (*e*) only two construction lines are needed to find the center of a quarter circle in an isometric plane. Measure the true radius $R$ of the circle from the corner on the two isometric lines as shown, and draw *actual* perpendiculars from these points. Their intersection will be the required center for radius $R_1$ or $R_2$ of the isometric quadrant. (*f*) illustrates the construction for the two vertical isometric planes.

## 14. Reversed Isometric.

It is often desirable to show the lower face of an object by tilting it *back* instead of *forward,* thus reversing the usual position so as to show the underside. The construction is the same as when the top is shown, but the directions of the principal isometric planes must be kept clearly in mind. Figure 21 shows the reference cube and the position of the axes, as well as the application of reversed-isometric construction to circle arcs. A practical use of this construction is in the representation of such architectural features as are naturally viewed from below. Figure 22 is an example of this use.

Sometimes a piece is shown to better advantage with the main axis horizontal, as in Fig. 23.

## 15. Isometric Sections.

Isometric drawings are, from their pictorial nature, usually outside views, but sometimes a sectional view (see Chap. 8) is used to good advantage to show a detail of shape or interior construction. The cutting planes are taken as isometric planes, and the section lining is done in the direction that gives the best effect; this is, in almost all cases, the direction of the long diagonal of a square drawn on the surface. As a general rule, a half section is made by outlining the figure in full and then cutting out the front quarter, as in Fig. 24; for a full section, the cut face is drawn first and then the part of the object behind it is added (Fig. 25).

## 16. Dimetric Projection.

The reference cube can be rotated into any number of positions in which two edges are equally foreshortened, and the direction of axes and ratio of foreshortening for any one of these positions might be taken as the basis for a system of dimetric draw-

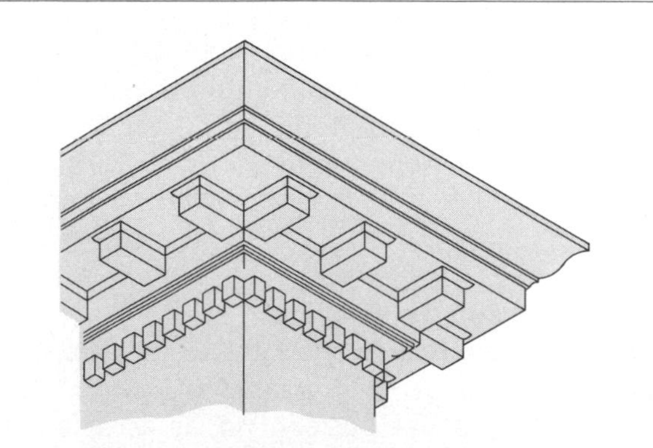

**FIG. 22.** An architectural detail on reversed axes.

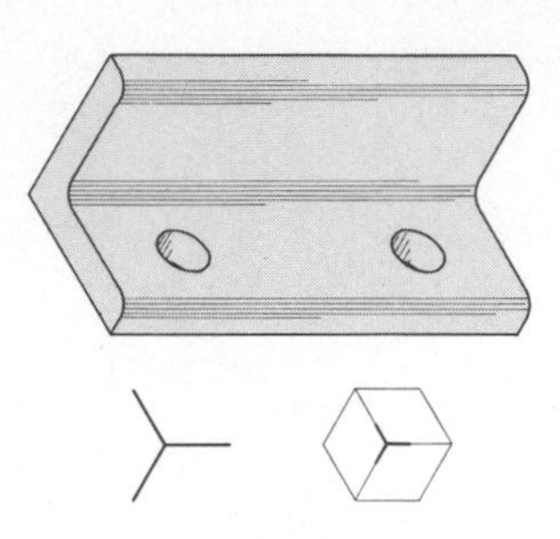

**FIG. 23.** Isometric with the main axis horizontal. It is used when the object looks more natural in this position.

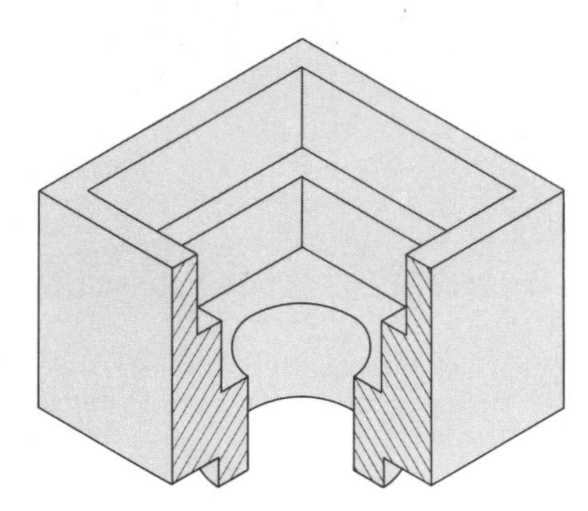

**FIG. 24.** Isometric half section. One-fourth of the object is removed to reveal interior construction.

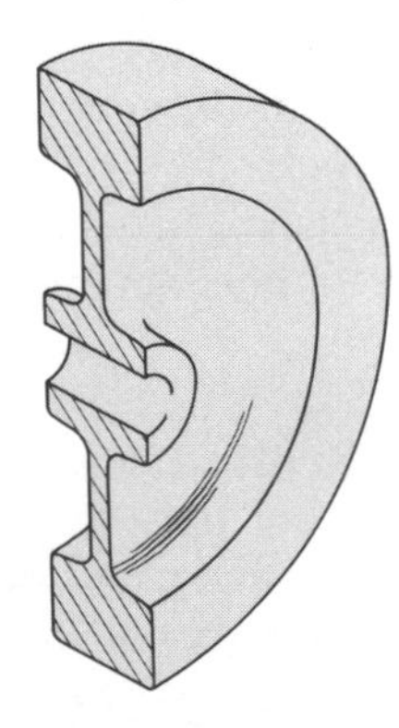

**FIG. 25.** Isometric full section. Half of object is removed to reveal object shape.

ing. A simple dimetric position is one with the ratios 1 to 1 to ½. In this position the tangents of the angles are ⅛ and ⅞, making the angles approximately 7° and 41°. Figure 26 shows a drawing in this system. Dimetric is seldom used because of the difficulty of drawing circles in this projection.

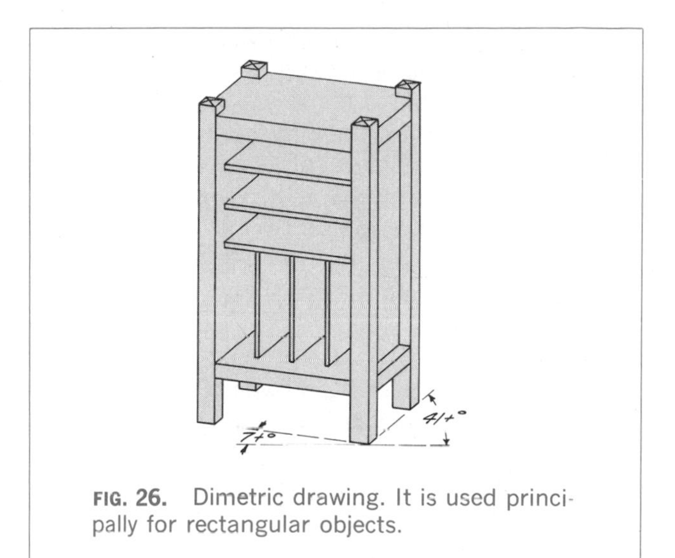


**FIG. 26.** Dimetric drawing. It is used principally for rectangular objects.

### 17. Trimetric Projection.

Any position in which all three axes are unequally foreshortened is called "trimetric." Compared with isometric and dimetric, distortion is reduced in trimetric projection, and even this effect can be lessened with some positions. However, because it is slower to execute than isometric or dimetric, it is seldom used except when done by projection. Axonometric projection from orthographic views is given in Sec. 26.

### 18. Oblique Projection.

When the projectors make an angle other than 90° with the picture plane, the resulting projection is called "oblique." The name "cavalier projection" is given to the special and most used type of oblique projection in which the projectors make an angle of 45° with the plane of projection. Cavalier projection is often called by the general name "oblique projection," or "oblique drawing." The principle is as follows: Imagine a vertical plane with a rectangular block behind it, having its long edges parallel to the plane. Assume a system of parallel projecting lines in any direction making an angle of 45° with the picture plane (they could be parallel to any one of the elements of a 45° cone with its base in the picture plane). Then that face of the block which is parallel to the plane is projected in its true size, and the edges perpendicular to the plane are projected in their true length. Figure 27 illustrates the principle. The first panel shows the regular orthographic projection of a rectangular block with its front face in the frontal plane. An oblique projector

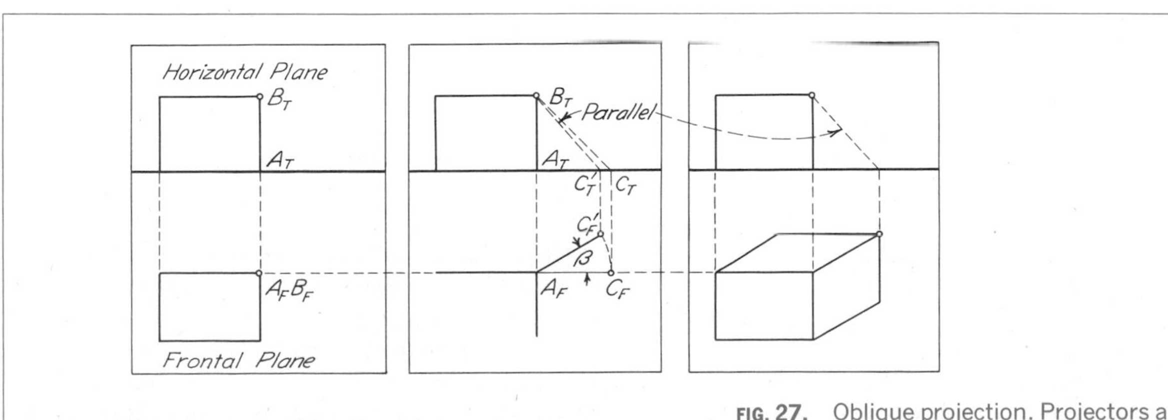


**FIG. 27.** Oblique projection. Projectors are at an oblique angle to the picture plane.

from the back corner $B$ is the hypotenuse of a 45° right triangle of which $AB$ is one side and the projection of $AB$ on the plane is the other side. When this triangle is horizontal, the projection on the plane will be $AC$. If the triangle is rotated about $AB$ through any angle $\beta$, $C$ will revolve to $C'$ and $A_FC_F'$ will be the oblique projection of $AB$.

### 19. To Make an Oblique Drawing.

Oblique drawing is similar to isometric drawing in that it has three axes that represent three mutually perpendicular edges and upon which measurements can be made. Two of the axes are always at right angles to each other, as they are in a plane parallel to the picture plane. The third, or depth, axis may be at any angle to the horizontal, 30° or 45° being generally used (Fig. 28). Oblique drawing is thus more flexible than isometric drawing. To draw a rectangular object (Fig. 29) start with a point representing a front corner ($A$) and draw from it the three oblique axes, one vertical, one horizontal, and one at an angle. On these three axes measure the height, width, and depth of the object. In this case the width is made up of the $2\frac{1}{2}$-in. distance and the $1\frac{5}{16}$-in. radius. Locate the center of the arc, and draw it as shown. The center for the arc of the hole in the figure will be at the same point as the center for the outside arc on the front face. The center for the rear arc of the hole will be $1\frac{1}{8}$-in. rearward on a depth-axis line through the front center.

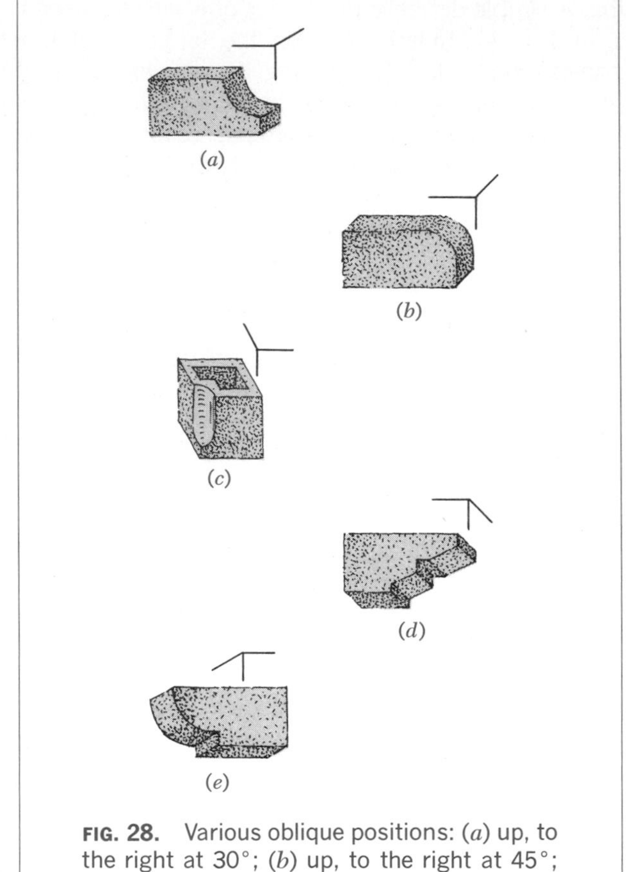


**FIG. 28.** Various oblique positions: (*a*) up, to the right at 30°; (*b*) up, to the right at 45°; (*c*) up, to the left at 45°; (*d*) down, to the right at 30°; (*e*) down, to the left at 30°.

### 20. Object Orientation for Oblique.

Any face parallel to the picture plane will evidently be projected without distortion. In this, oblique projection has an advantage over isometric that is of particular value in representing objects with circular or irregular outline.

The *first rule* for oblique projection is to *place the object with the irregular outline or contour parallel to the picture plane.* Note in Fig. 30 the greater distortion at (*b*) and (*c*) than at (*a*).

One of the greatest disadvantages in the use of isometric or oblique drawing is the effect of distortion produced by the lack of convergence in the receding lines—a violation of perspective. In some cases, particularly with large objects, this becomes so painful as practically to preclude the use of these methods. This is perhaps even more noticeable in oblique than in isometric and of course increases with the length of the depth dimension.

Hence the *second rule: preferably, the longest dimension should be parallel to the picture plane.* In Fig. 31, (*a*) is preferable to (*b*).

In case of conflict between these two rules, *the first always takes precedence,* as the advantage of having the irregular face without distortion is greater than that gained by the second rule, as illustrated in Fig. 32. The first rule should be given precedence

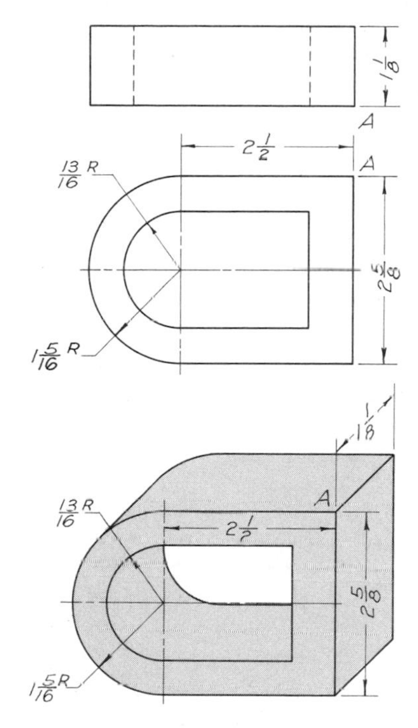


**FIG. 29.** Oblique drawing. The front face, parallel to the picture plane, is identical with an orthographic view.

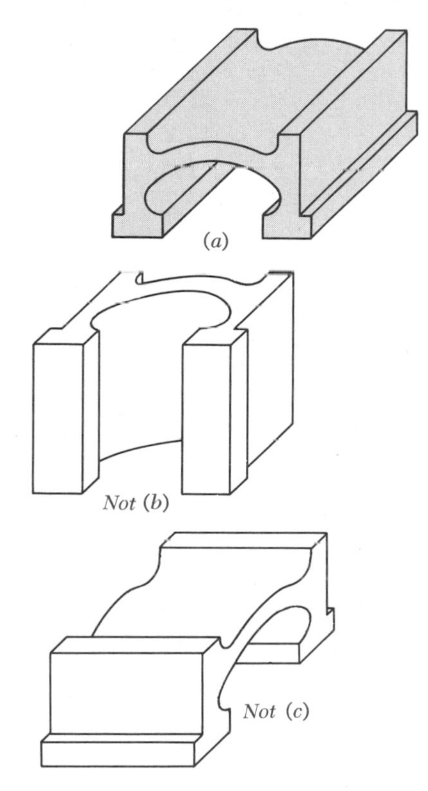


**FIG. 30.** Illustration of the first rule. Note the distortion at (*b*) and (*c*).

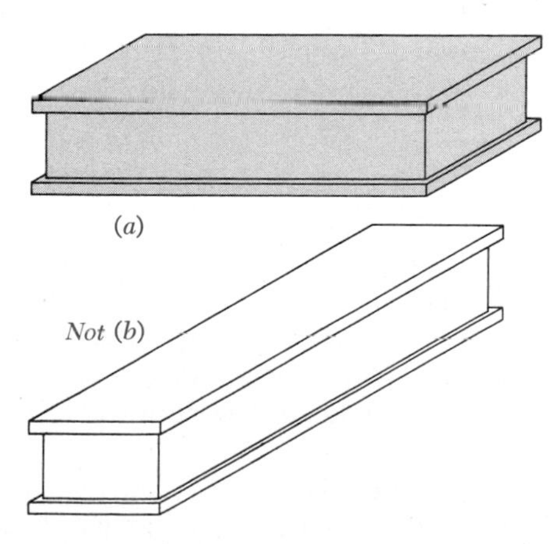


**FIG. 31.** Illustration of the second rule. Note the exaggerated depth at (*b*).

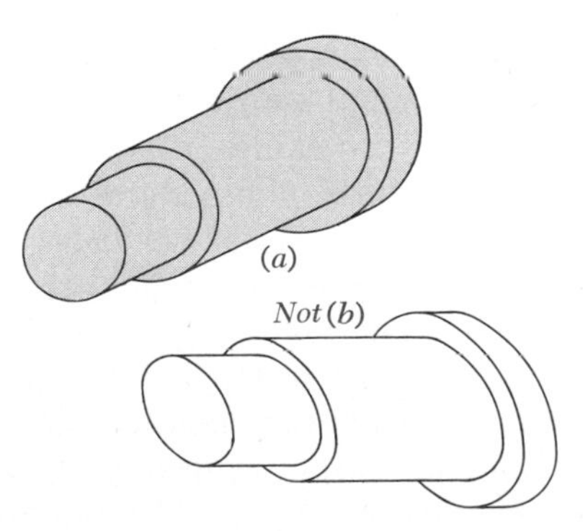


**FIG. 32.** Precedence of the first rule. (*a*), following the first rule, is easier to draw and also shows less distortion than (*b*).

even with shapes that are not irregular if, in the draftsman's judgment, the distortion can be lessened, as in Fig. 33, where (*b*) is perhaps preferable to (*a*).

## 21. Starting Plane.

Note that as long as the front of the object is in one plane parallel to the plane of projection, the front face of the oblique projection is *exactly the same as in the orthographic front view.* When the front is made up of more than one plane, take care to preserve the relationship between the planes by selecting one as the starting plane and working from it. In a piece such as the link in Fig. 34, the front bosses can be imagined as cut off on the plane *A-A*, and the front view, that is, the section on *A-A*, drawn as the front of the oblique projection. Then lay off depth axes through the centers *C* and *D*, the distances, for example, *CE* behind and *CF* in front of the plane *A-A*.

When an object has no face perpendicular to its base, it can be drawn in a similar way by cutting a right section and measuring offsets from it, as in Fig. 35. This offset method, previously illustrated in the isometric drawings in Figs. 13, 14, and 16, is a rapid and convenient way of drawing almost any figure, and it should be studied carefully.

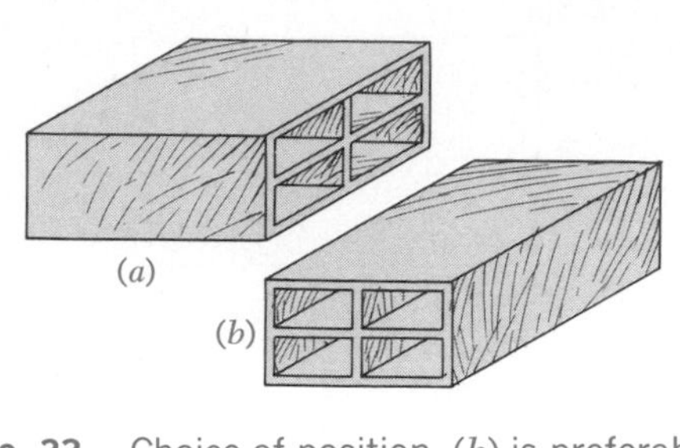


FIG. 33. Choice of position. (*b*) is preferable to (*a*).

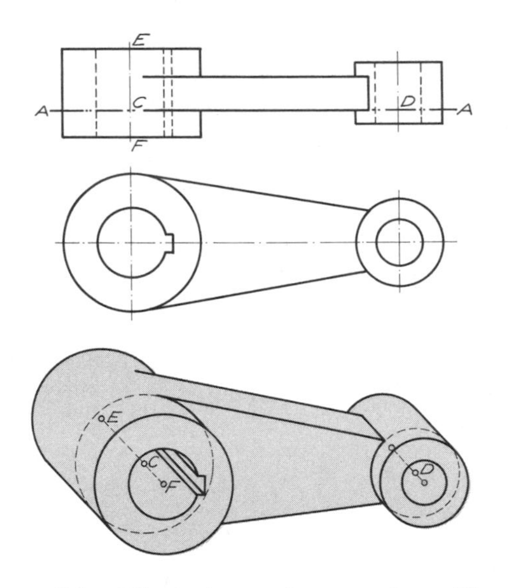


FIG. 34. Offsets from reference plane. Distances forward and rearward are measured from the frontal plane.

## 22. Circles in Oblique.

When it is necessary to draw circles that lie on oblique faces, they can be drawn as circle arcs, with the compasses, on the same principle as the four-center isometric approximation shown in Fig. 18. In isometric it happens that *two of the four intersections of the perpendiculars from the middle points* of the containing square fall at the corner of the square, and advantage is taken of the fact. In oblique, the position of the corresponding points depends on the angle of the depth axis. Figure 36 shows three squares in oblique positions at different angles and the construction of their inscribed circles. The important point to remember is that the circle arcs *must* be tangent at the mid-points of the sides of the oblique square.

## 23. Arcs in Oblique.

Circle arcs representing rounded corners, etc., are drawn in oblique by the same method given for isometric arcs in Sec. 13. The only difference is that the angle of the sides tangent to the arc will vary according to the angle of the depth axis chosen.

## 24. Cabinet Drawing.

This is a type of oblique projection in which the parallel projectors make an angle with the picture plane of such a value that distances measured parallel to the depth axis are reduced one-half that of

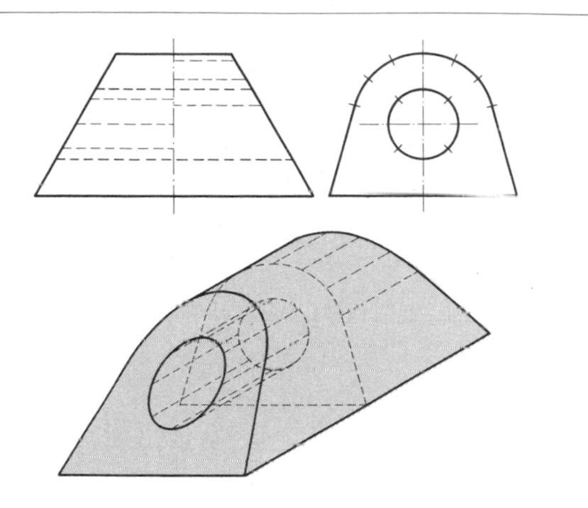

FIG. 35. Offsets from right section. Measurements forward and rearward are made from the frontal plane.

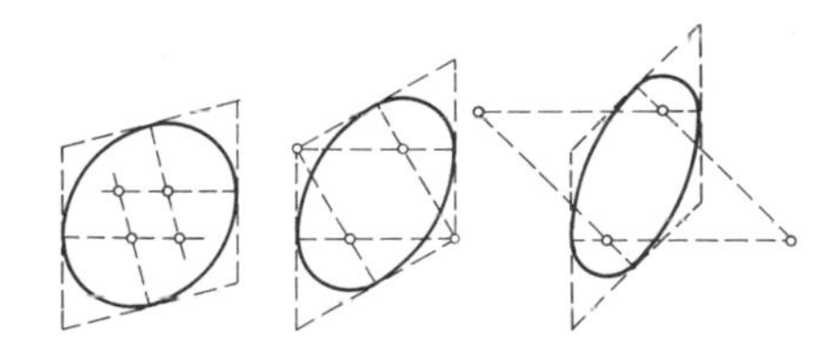

FIG. 36. Oblique circle construction. Note that the tangent points of arcs must be at the midpoints of the enclosing oblique square.

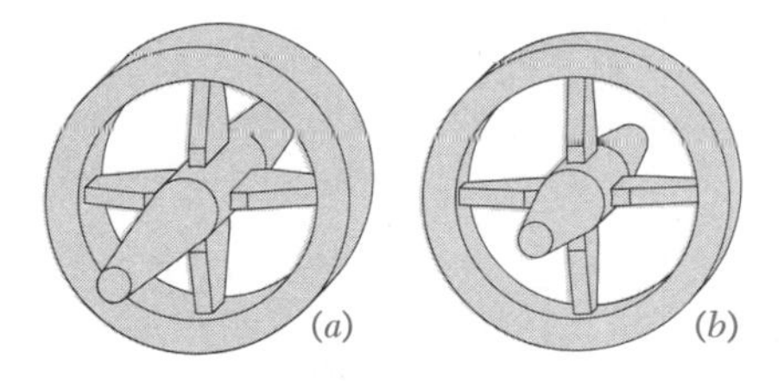


FIG. 37. Oblique (*a*) and cabinet drawing (*b*). Note less exaggeration of depth at (*b*).

cavalier projection. The appearance of excessive thickness that is so disagreeable in cavalier projection is entirely overcome in cabinet projection. The depth axis may be at any angle with the horizontal but is usually taken at 30° or 45°. The appearance of cavalier and cabinet drawing is shown in Fig. 37.

## 25. Other Forms.

Cabinet drawing is popular because of the easy ratio, but the effect is often too thin. Other oblique drawing ratios, such as 2 to 3 or 3 to 4, may be used with pleasing effect.

## 26. Axonometric Projection from Orthographic Views.

In making pictorial drawings of complicated parts, especially whenever curves are plotted, axonometric projection from orthographic views may give an advantage in speed and ease of drawing over axonometric projection made directly from the object. Any position—isometric, dimetric, or trimetric—may be used.

The three axes of an axonometric drawing are *three mutually perpendicular edges* in space. If the angle of rotation and the angle of tilt of the object are known or decided upon, the three axes for the pictorial drawing and the location of the orthographic views for projection to the pictorial can easily be found. Figure 38 illustrates the procedure. The three orthographic views of a cube are shown at (*g*). The three mutually perpendicular edges $OA$, $OB$, and $OC$ will be foreshortened differently when the cube is rotated in space for some axonometric position, but the ends of the axes $A$, $B$, and $C$ will always lie on the surface of a sphere whose radius is $OA = OB = OC$, as illustrated by ($G'$). At any particular angle of tilt of the cube, the axis ends $A$ and $B$ will describe an ellipse, as shown, if the cube is rotated about the axis $OC$. The axis $OC$ will appear foreshortened at $oc'$. Thus for any particular position of the cube in space, representing some desired axonometric position, the axes can be located and their relative amounts of foreshortening found.

Moreover, if a face of the cube is rotated about a *frontal axis perpendicular to the axis that is at right angles to the face*, an orthographic view of the face, in projection with the axonometric view, will result. Thus, the top and right-side views may be located as at (*j*) and projected as at (*k*) to give the axonometric drawing.

The drawings at (*h*), (*j*), and (*k*) illustrate the practical use of the theory of rotation just described. The

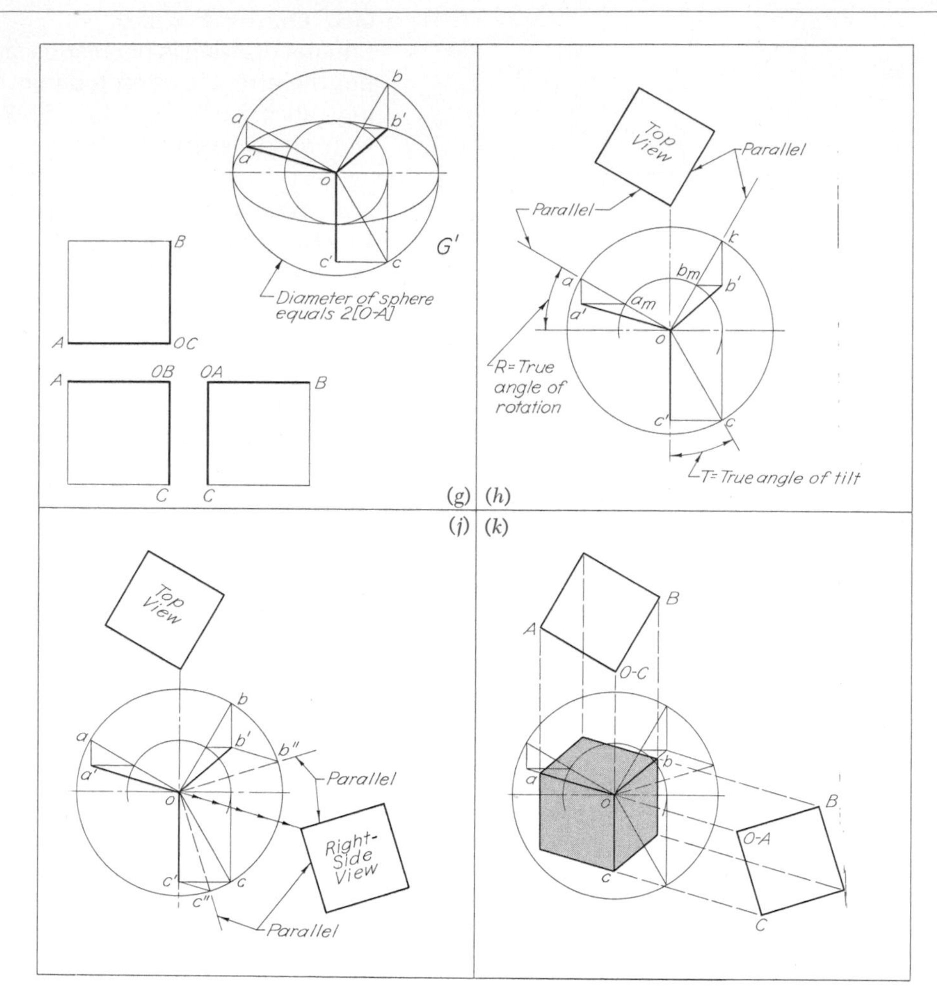


**FIG. 38.** Axonometric projection from orthographic views. (*g*) Construction to locate axes; (*h*) location of top view; (*j*) location of side view; (*k*) projection.

actual size of the sphere is unimportant, as it is used only to establish the direction of the axes. First, the desired angle of rotation $R$ and the angle of tilt $T$ are decided upon and laid out as at (*h*). The minor diameter for the ellipse upon which $A$ and $B$ will lie is found by projecting vertically from $c$ and drawing the circle as shown. $A$ and $B$ on the major-diameter circle of the ellipse will be at $a$ and $b$; on the minor-diameter circle, they will be at $a_m$ and $b_m$; and they are found in the axonometric position by projecting, as in the concentric-circle ellipse method, to $a'$ and $b'$. The foreshortened position of $C$ is found by projecting horizontally across from $c$ to $c'$.

The top orthographic view of the cube (or object) will be parallel to $oa$ and $ob$, and projection from the orthographic view to the axonometric will be vertical (parallel to $aa'$ and $bb'$).

Projection from an orthographic right-side view would be as shown at (*j*). The right side of the cube, containing axes $OC$ and $OB$, is found by projecting

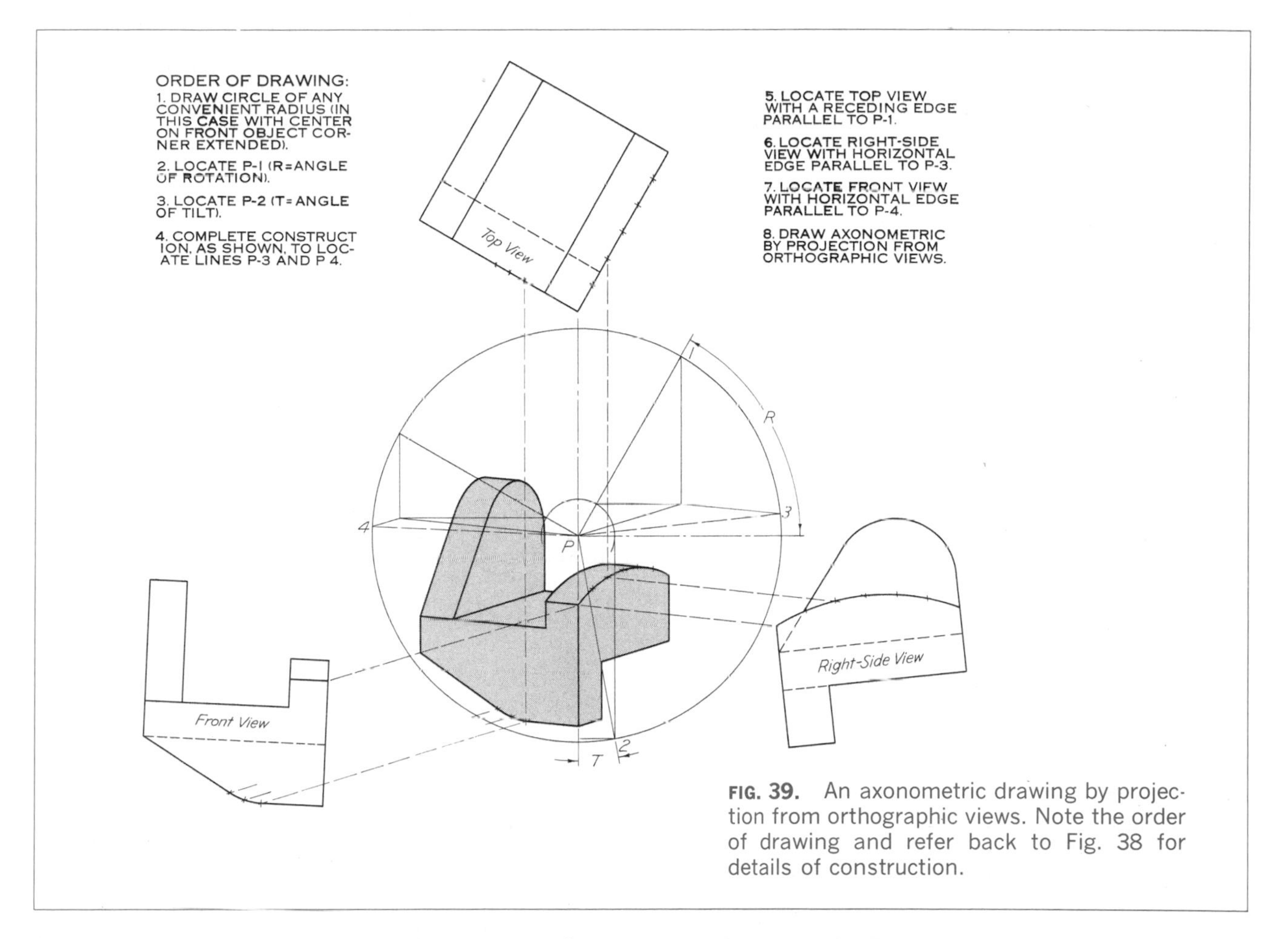

ORDER OF DRAWING:
1. DRAW CIRCLE OF ANY CONVENIENT RADIUS (IN THIS CASE WITH CENTER ON FRONT OBJECT CORNER EXTENDED).
2. LOCATE P-I (R=ANGLE OF ROTATION).
3. LOCATE P-2 (T=ANGLE OF TILT).
4. COMPLETE CONSTRUCTION, AS SHOWN, TO LOCATE LINES P-3 AND P 4.
5. LOCATE TOP VIEW WITH A RECEDING EDGE PARALLEL TO P-1.
6. LOCATE RIGHT-SIDE VIEW WITH HORIZONTAL EDGE PARALLEL TO P-3.
7. LOCATE FRONT VIEW WITH HORIZONTAL EDGE PARALLEL TO P-4.
8. DRAW AXONOMETRIC BY PROJECTION FROM ORTHOGRAPHIC VIEWS.
Top View
Right-Side View
Front View
R
T
P
1
2
3
4


**FIG. 39.** An axonometric drawing by projection from orthographic views. Note the order of drawing and refer back to Fig. 38 for details of construction.

from $b'$ and $c'$, parallel to $oa'$, to locate $b''$ and $c''$ on the circle representing the sphere. The sides of the cube (or object) are parallel to $ob''$ and $oc''$, as shown at ($j$). Projection from the right-side view to the axonometric view is in the direction of $oa'$, as indicated.

The axonometric drawing is shown projected at ($k$). The dashed lines indicate the actual projectors, and the light solid lines and circles show the necessary construction just described.

One advantage of this method is that the angle of rotation and tilt can be decided upon so that the object will be shown in the best position. Figure 39 is an example of an axonometric drawing made by projection from orthographic views. The curved faces are plotted by projecting points as shown.

## 27. Isometric Projection From Orthographic Views.

Isometric is, of course, a special type of axonometric projection in which all three axes are foreshortened equally. The work of finding the axes for isometric projection from orthographic views is reduced if the views are located by angle, as illustrated in Fig. 40.

## 28. Oblique Projection From Orthographic Views.

In oblique projection the projectors make some oblique angle with the picture plane. It should be noted that the actual angle of the projectors (with horizontal and frontal planes) is not critical, so that a variety of angles may therefore be used. The

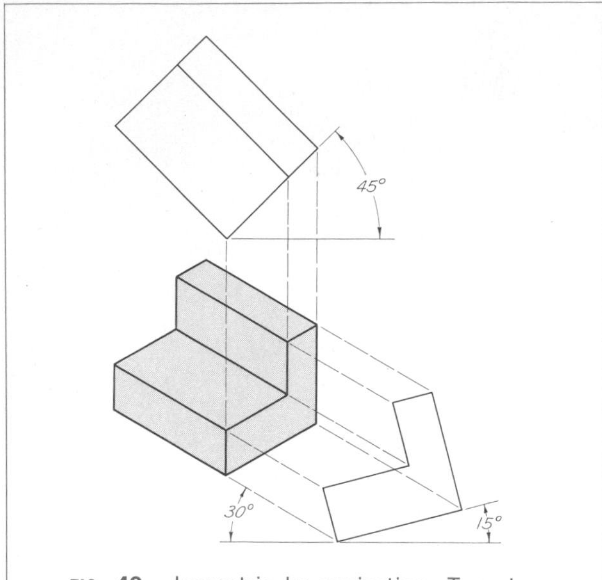


**FIG. 40.** Isometric by projection. Top view turned 45° and side view turned 15° locates the views for projection to the isometric.

making of an oblique drawing by projection from orthographic views is simple, as illustrated by Fig. 41. The picture plane is located, and one face of the object is made coincident with the picture plane. The front view is located at a convenient place on the paper. The angle of the projectors in the top view may be assumed (in this case 45°) and projections made to the picture plane as shown. The angle of the projectors in the front view can than be assumed (in this case 30°). Projection from the front view at the assumed angle and vertically from the picture plane, as shown, will locate the necessary lines and points for the oblique view.

Reversed axes can be obtained by projecting downward from the front view. An axis to the left can be located by changing the direction of the projectors in the top view. Any desired oblique axes can be located by altering the angles (top and front) for the projectors.

## 29. Perspective Drawing.

Perspective drawing represents an object as it appears to an observer stationed at a particular position relative to it. The object is seen as the figure resulting when visual rays from the eye to the object are cut by a picture plane. There is a difference between an artist's use of perspective and geometric perspective. The artist often disregards true perspective since he draws the object as he sees it through his creative imagination, while geometric perspective is projected instrumentally on a plane from views or measurements of the object represented. Projected geometric perspective is, theoretically, very similar to the optical system in photography.

In a technical way, perspective is used more in architecture and in illustration than in other fields, but every engineer will find it useful to know the principles of the subject.

## 30. Fundamental Concepts.

Imagine an observer standing on the sidewalk of a city street, as in Fig. 42, with the picture plane erected between him and the street scene ahead. Visual rays from his eye to the ends of lamppost $A$ intercept a distance $aa'$ on the picture plane. Similarly, rays from post $B$ intercept $bb'$, a smaller distance than $aa'$. This apparent diminution in the size of like objects as the distance from the objects to the eye increases agrees with our everyday experience and is the keynote of perspective drawing. It is evident from the figure that succeeding lampposts will intercept shorter distances on the picture plane than the preceding ones, and that a post at infinity would show only as a point $o$ at the level of the observer's eye.

In Fig. 43 the plane of the paper is the picture plane, and the intercepts $aa'$, $bb'$, etc., show as the heights of the respective lampposts as they diminish in their projected size and finally disappear at a point on the horizon. In a similar way the curbings and balustrade appear to converge at the same point $O$. Thus a system of parallel horizontal lines will vanish at a single point on the horizon, and all horizontal planes will vanish on the horizon. Verticals such as the lampposts and the edges of the buildings, being parallel to the picture plane, pierce the picture plane at an infinite distance and therefore show as vertical lines in the picture.

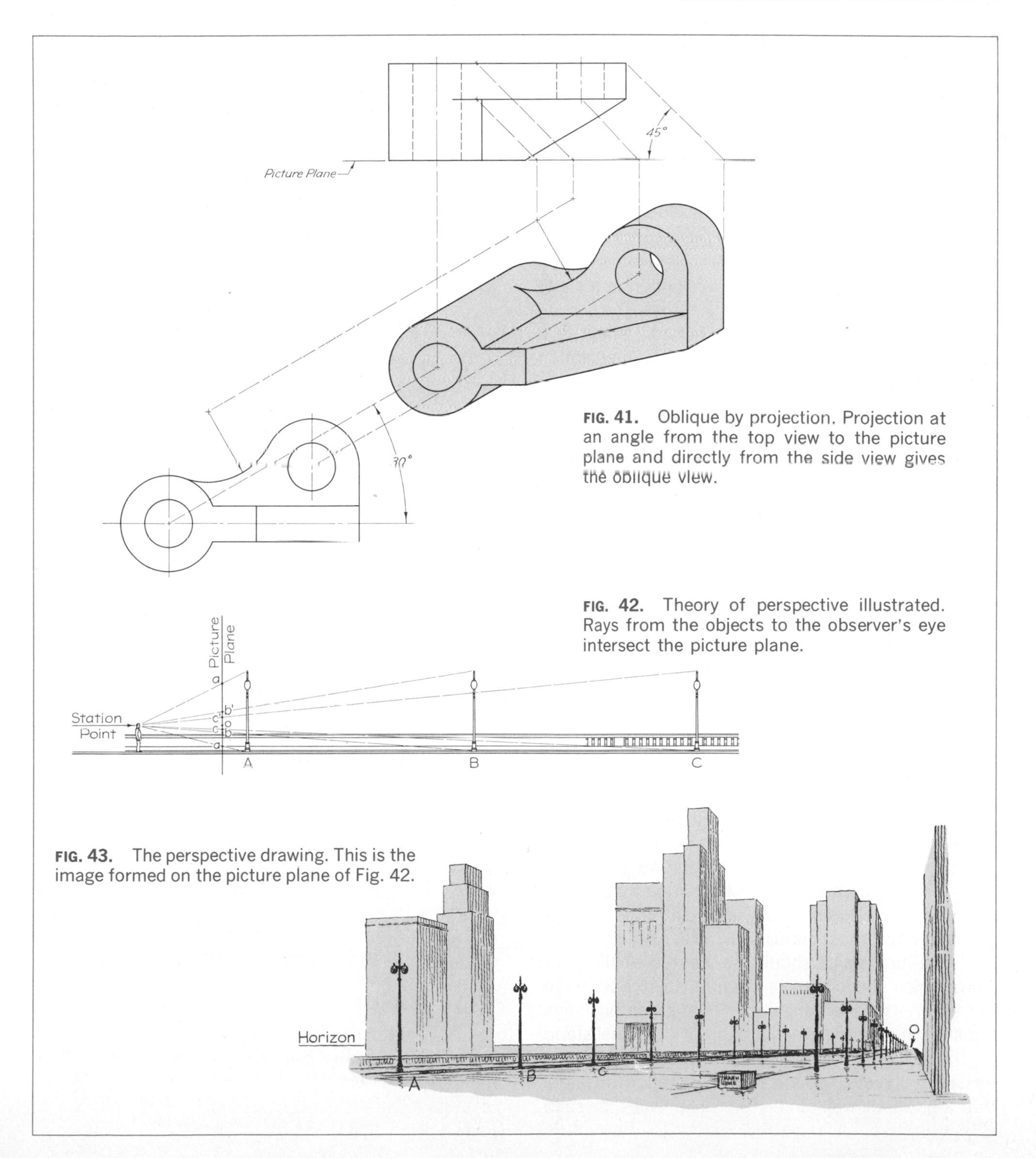


**FIG. 41.** Oblique by projection. Projection at an angle from the top view to the picture plane and directly from the side view gives the oblique view.

**FIG. 42.** Theory of perspective illustrated. Rays from the objects to the observer's eye intersect the picture plane.

**FIG. 43.** The perspective drawing. This is the image formed on the picture plane of Fig. 42.

## 31. Definitions and Nomenclature.

Figure 44 illustrates perspective theory and names the points, lines, and planes used. An observer in viewing an object selects his *station point* and thereby determines the *horizon plane,* as the horizontal plane is at eye level. This horizon plane is normally above the horizontal *ground plane* upon which the object is assumed to rest. The *picture plane* is usually located between the station point and the object being viewed and is ordinarily a vertical plane perpendicular to the horizontal projection of the line of sight to the object's center of interest. The *horizon line* is the intersection of the horizon plane and picture plane, and the *ground line* is the intersection of the ground plane and picture plane. The *axis of vision* is the line through the station point which is perpendicular to the picture plane. The piercing point of the axis of vision with the picture plane is the *center of vision.*

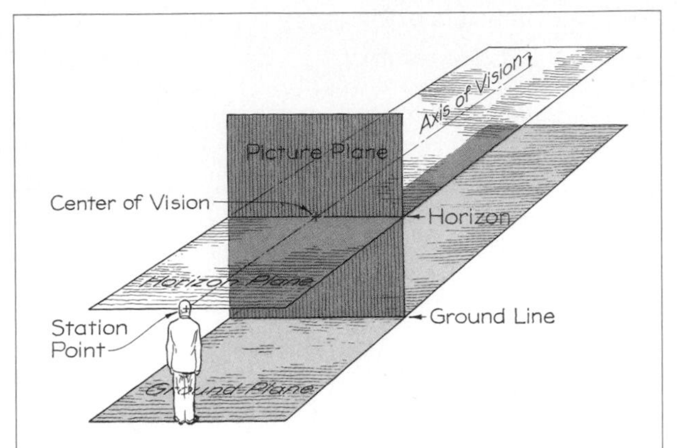


**FIG. 44.** Perspective nomenclature.

## 32. Selection of the Station Point.

In beginning a perspective drawing, take care in selecting the station point, as an indiscriminate choice may result in a distorted drawing. If the station point is placed to one side of the drawing, the same effect is obtained as when a theater screen is viewed from a position close to the front and well off to one side: heights are seen properly but not horizontal distances. Therefore, *the center of vision should be somewhere near the picture's center of interest.*

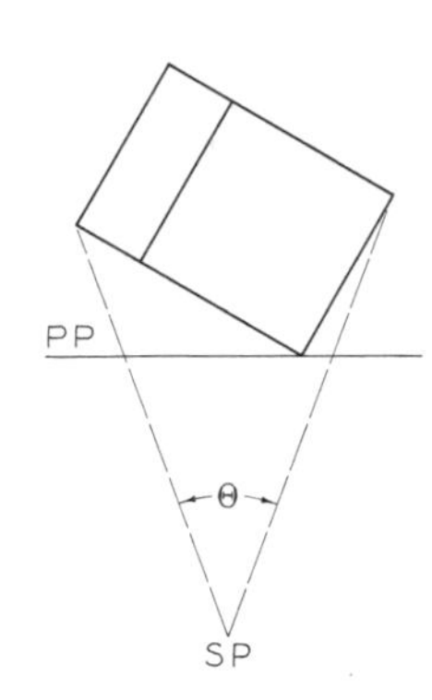


**FIG. 45.** Lateral angle of view.

Wide angles of view result in a violent convergence of horizontal lines and so should be avoided. The angle of view is the included angle $\theta$ between the widest visual rays (Fig. 45). Figure 46 shows the difference in perspective foreshortening for different lateral angles of view. In general, an angle of about 20° gives the most natural picture.

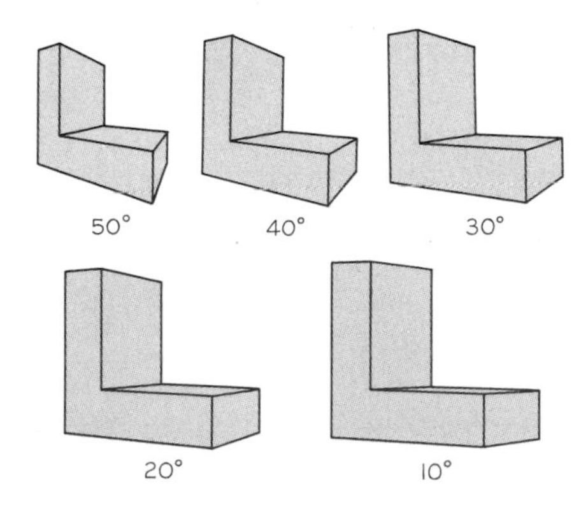


**FIG. 46.** Comparative lateral angles of view. Angles greater than 30° give an unpleasing perspective.

The station point should be located at the point from which the object is seen to best advantage. For this reason, for large objects such as buildings, the station point is usually taken at a normal standing height of about 5 ft above the ground plane; for small objects, the best representation demands that the top, as well as the lateral surfaces, be seen, and the station point must be elevated accordingly. Figure 47 shows the angle of elevation Ω between the horizon plane and the extreme visual ray. By illustrating several different angles of elevation (Ω), Fig. 48 shows the effect of elevation of the station point. In general, the best picturization is obtained at an angle of about 20° to 30°.

Accordingly, *the visual rays to the object should be kept within a right circular cone whose elements make an angle of not more than* 15° *with the cone axis* (total included angle of 30°).

In choosing the station point, see that its position is always offset to one side and also that it is offset vertically from the exact middle of the object, or a rather stiff and awkward perspective will result. Similarly, in locating the object with reference to the picture plane, avoid having the faces make identical angles with the picture plane, or the same stiffness will appear.

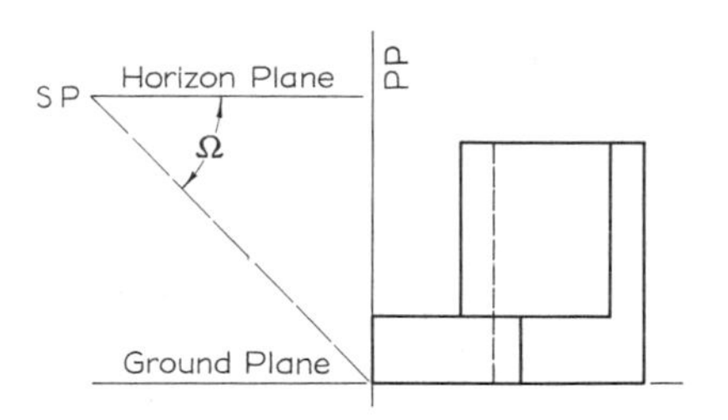


**FIG. 47.** Elevation angle of view.

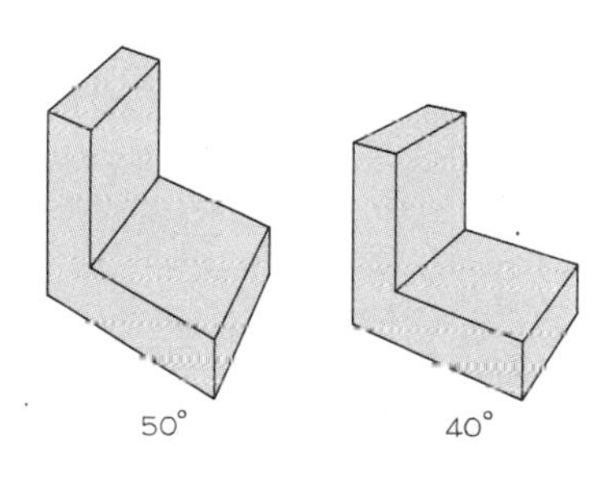


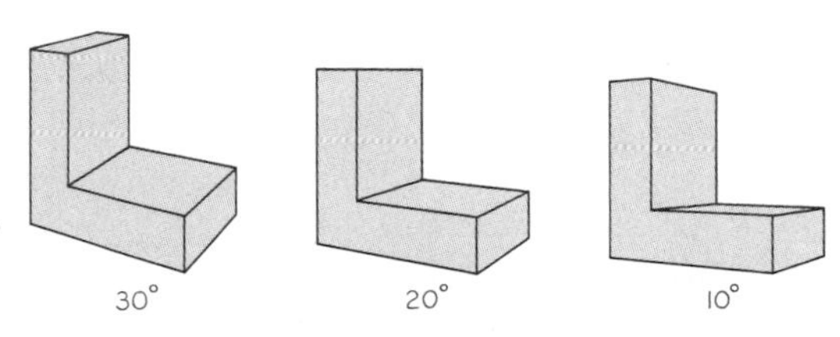


**FIG. 48.** Comparative elevation angles of view. Angles greater than 30° give an unpleasing perspective.

### 33. To Draw a Perspective.

Perspective projection is based on the theory that visual rays from the object to the eye pierce the picture plane and form an image of the object on the plane. Thus in Fig. 49, the image of line *YZ* is formed by the piercing points *y* and *z* of the rays. Several projective methods may be used. The simplest method, basically, but the most laborious to draw is illustrated by the purely orthographic method of Fig. 50, in which the top and side views are drawn in orthographic. The picture plane (edge view) and the station point are located in each view. Assuming that the line *YZ* in Fig. 49 is one edge of the L-shaped block in Fig. 50, visual rays from *Y* and *Z* will intersect the picture plane in the top view, thus locating the perspective of the points laterally. Similarly, the intersections of the rays in the side view give the perspective heights of *Y* and *Z*. Projection from the top and side views of the picture plane gives the perspective of *YZ*, and a repetition of the process for the other lines will complete the drawing. Note that *any* point such as *Y* or *Z* can be located on the perspective, and thus the perspective is actually plotted, by projection, point after point.

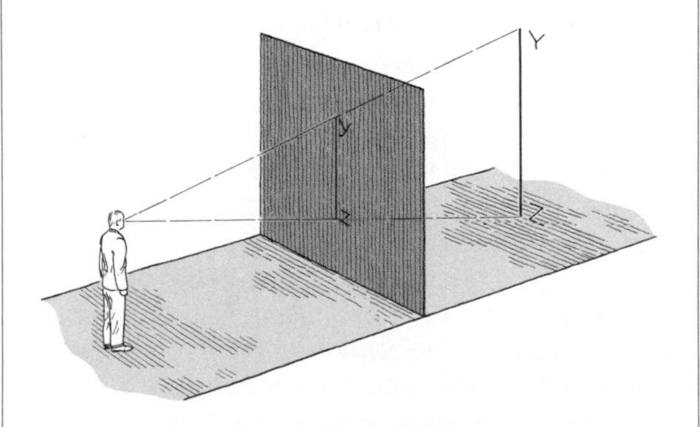


**FIG. 49.** Perspective of a line.

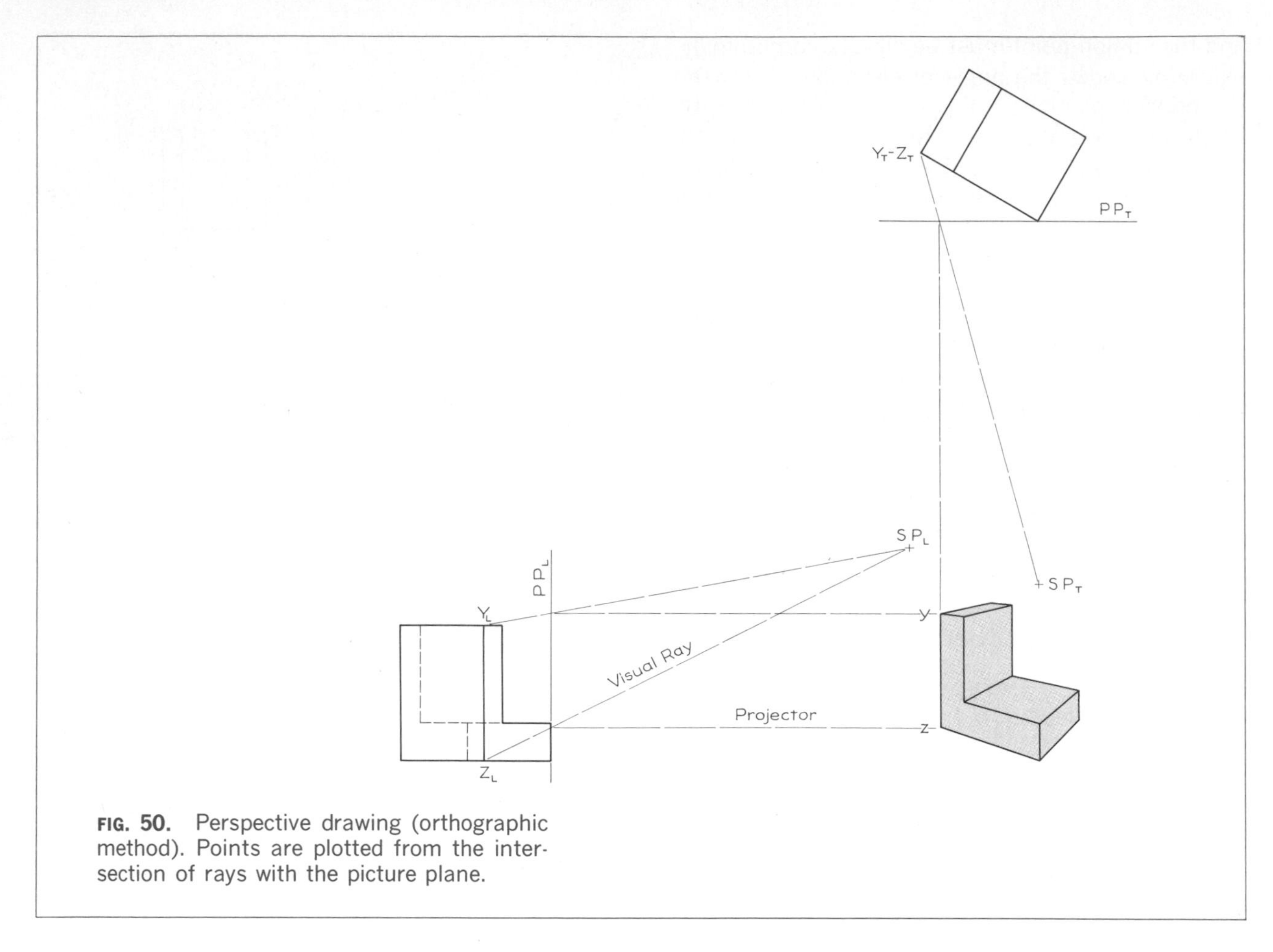


**FIG. 50.** Perspective drawing (orthographic method). Points are plotted from the intersection of rays with the picture plane.

## 34. The Use of Vanishing Points and Measuring Lines.

These facilitate the projections. Let it be required to make a perspective of the sliding block in Fig. 51. The edge view of the picture plane (plan view) is drawn (Fig. 52), and behind it the top view of the object is located and drawn. In this case, one side of the object is oriented at 30° to the picture plane in order to emphasize the L shape more than the end of the block. The station point is located a little to the left of center and far enough in front of the picture plane to give a good angle of view. The ground line is then drawn, and on it is placed the front view of the block from Fig. 51. The height of the station point is then decided—in this case, well

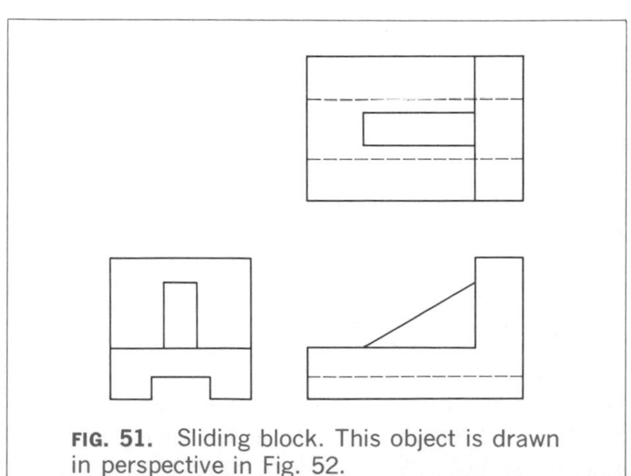

**FIG. 51.** Sliding block. This object is drawn in perspective in Fig. 52.

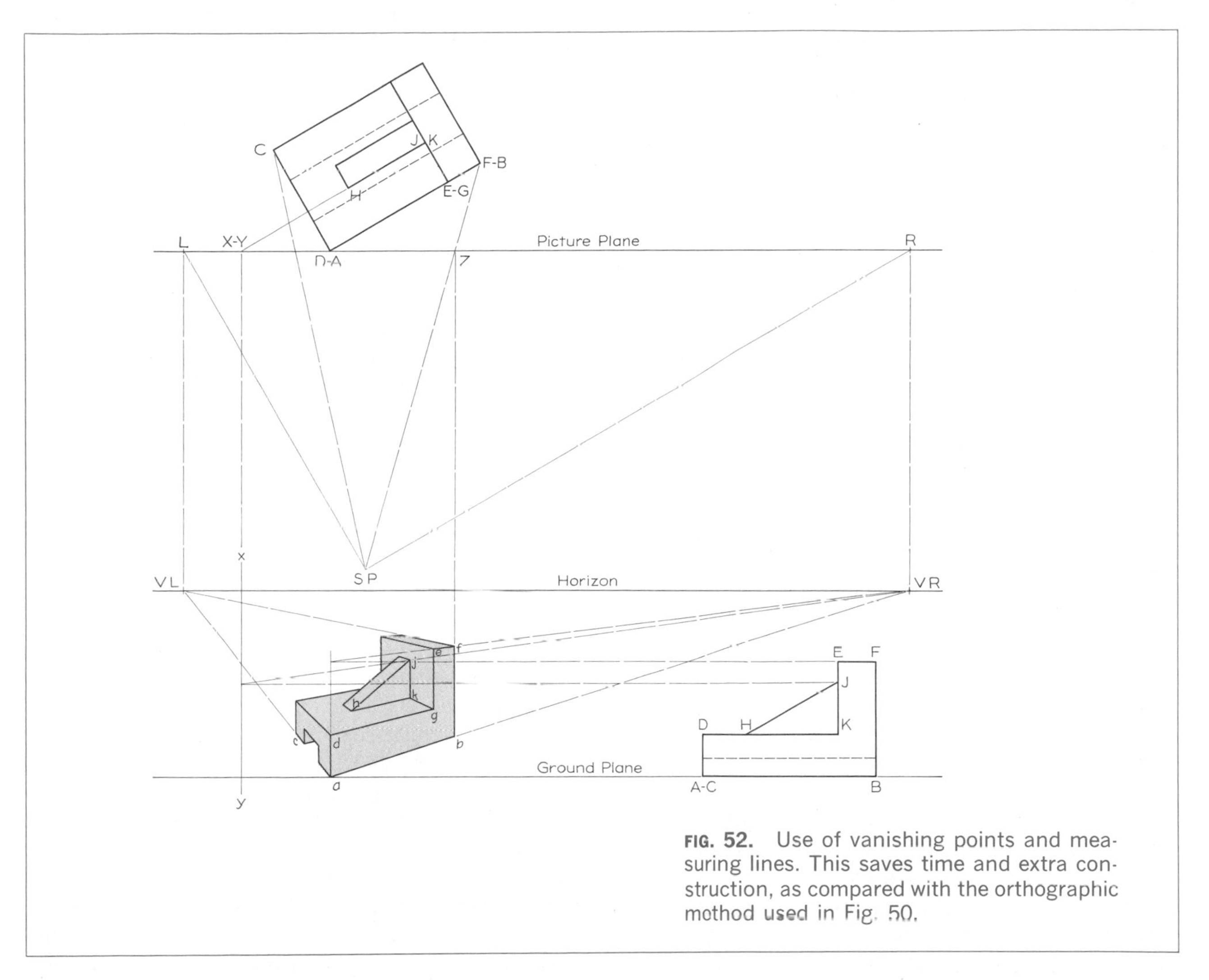


**FIG. 52.** Use of vanishing points and measuring lines. This saves time and extra construction, as compared with the orthographic method used in Fig. 50.

above the block so that the top surfaces will be seen—and the horizon line is drawn at the station-point height.

To avoid the labor of redrawing the top and front views in the positions just described, the views can be cut from the orthographic drawing, oriented in position, and fastened with tacks or tape.

The *vanishing point* for any horizontal line can be found by drawing a visual ray from the station point *parallel* to the horizontal line and finding the piercing point of this visual ray with the picture plane. Thus, in Fig. 52, the line *SP* to *R* is parallel to the edge *AB* of the object, and *R* is the piercing point. Point *R* is then projected to the horizon line, locating *VR*, the vanishing point for *AB* and all edges parallel to *AB*. The vanishing point *VL* for *AC* and edges parallel to *AC* is found similarly, as shown.

In visualizing the location of a vanishing point, imagine that the edge, for example, *AB*, is moved to the right along the ground, still making the same angle with the picture plane; the intercept of *AB* will become less and less until, when *A* is in coincidence with *R*, the intercept will be zero. *R* then must be the top view of the vanishing point for all lines parallel to *AB*.

Point *A* lies in both the picture plane and the

ground plane and will therefore be shown in the perspective at *a*, on the ground line, and in direct projection with the top view. The perspective of *AB* is determined by drawing a line from *a* to *VR* (the perspective *direction* of *AB*) and then projecting the intercept *Z* (of the visual rays *SP* to *B*) to the line, thus locating *b*.

All lines behind the picture plane are foreshortened in the picture, and only those lying in the picture plane will appear in their true length. For this reason, *all measurements must be made in the picture plane*. Since *AD* is in the picture plane, it will show in its actual height as *ad*.

A *measuring line* will be needed for any verticals such as *BF* that do not lie in the picture plane. If a vertical is brought forward to the picture plane along some established line, the true height can be measured in the picture plane. If, in Fig. 52, *BF* is imagined as moved forward along *ab* until *b* is in coincidence with *a*, the true height can be measured vertically from *a*. This vertical line at *a* is then the measuring line for all heights in the vertical plane containing *a* and *b*. The height of *f* is measured from *a*, and from this height point, a vanishing line is drawn to *VR*; then from *Z* (the piercing point in the picture plane of the visual ray to *F*), *f* can be projected to the perspective.

The measuring line can also be thought of as the intersection of the picture plane with a vertical plane that contains the distance to be found. Thus *ad*, extended, is the measuring line for all heights in surface *ABFEGD*. The triangular rib in Fig. 52 is located by continuing surface *HJK* until it intersects the picture plane at *XY*, thereby establishing *xy* as the measuring line for all heights in *HJK*. In the figure, the height of *J* is measured on the measuring line *xy*, and *j* is found as described for *f*.

Note that heights can be measured with a scale on the measuring line or they can be projected from the front view, as indicated in Fig. 52.

*To Make a Perspective Drawing:*

1. Draw the top view (edge of the picture plane).
2. Orient the object relative to the picture plane so that the object will appear to advantage, and draw the top view of the object.
3. Select a station point that will best show the shape of the object.
4. Draw the horizon and ground line.
5. Find the top view of the vanishing points for the principal horizontal edges by drawing lines parallel to the edges, through the station point, and to the picture plane.
6. Project from the top views of the vanishing points to the horizon line, thus locating the vanishing points for the perspective.
7. Draw the visual rays from the station point to the corners of the object in the top view, locating the piercing point of each ray with the picture plane.
8. Start the picture, building from the ground up and from the nearest corner to the more distant ones.

### 35. Planes Parallel to the Picture Plane.

Objects with circles or other curves in a vertical plane can be oriented with their curved faces parallel to the picture plane. The curves will then appear in true shape. This method, often called "parallel perspective," is also suitable for interiors and for street vistas and similar scenes where considerable depth is to be represented.

The object in Fig. 53 has been placed so that the planes containing the circular contours are parallel to the picture plane. The horizontal edges parallel to the picture plane will appear horizontal in the picture and will have no vanishing point. Horizontals perpendicular to the picture plane are parallel to the axis of vision and will vanish at the center of vision *CV*. Except for architectural interiors, the station point is usually located above the object and either to the right or left, yet not so far in any direction as to cause unpleasant distortion. For convenience, one face of the object is usually placed in the picture plane and is therefore not reduced in size in the perspective.

In Fig. 53, the end of the hub is in the picture plane; thus the center *o* is projected from *O* in the top view, and the circular edges are drawn in their true size. The center line *ox* is vanished from *o* to *CV*. To find the perspective of center line *MN*, a

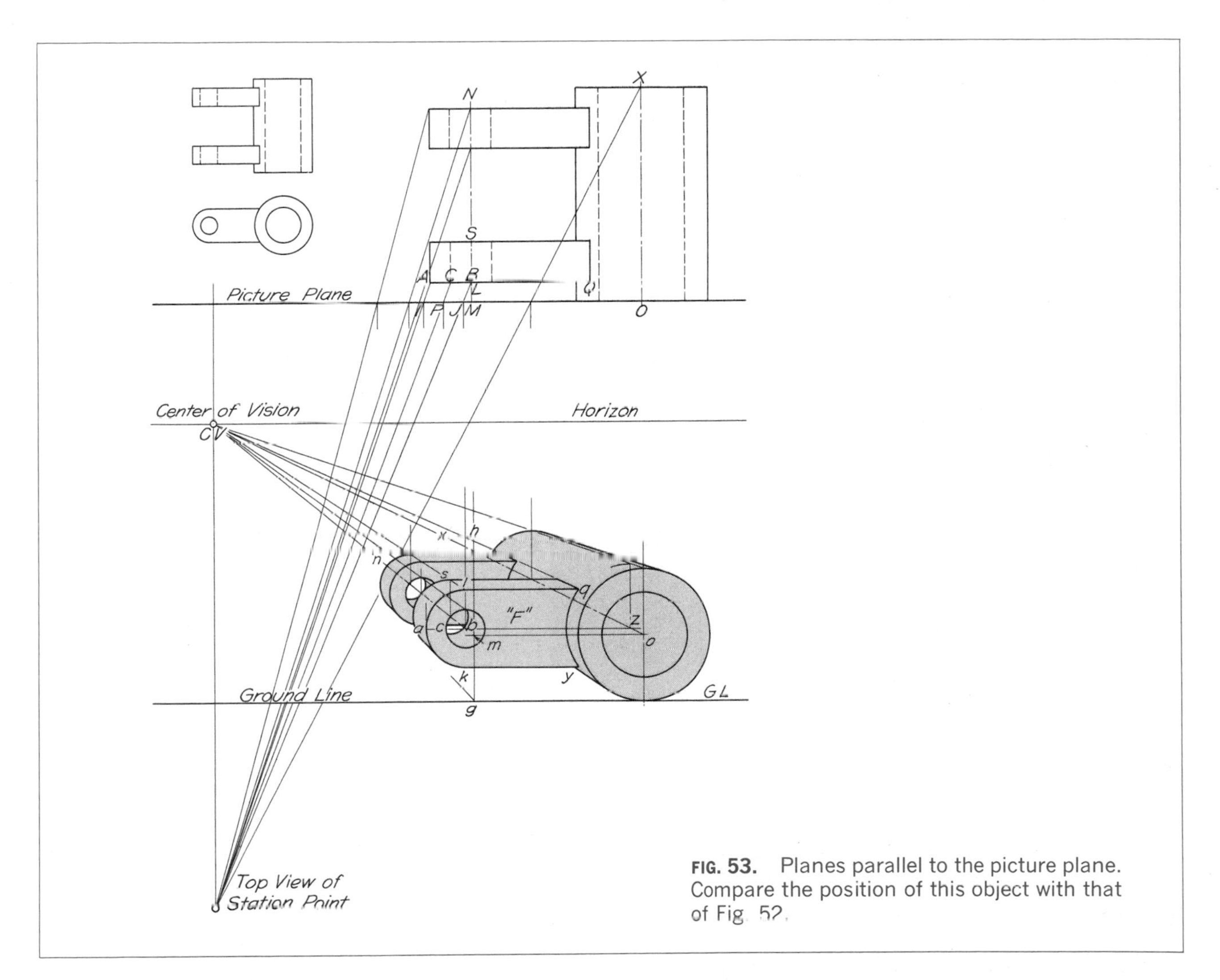


**FIG. 53.** Planes parallel to the picture plane. Compare the position of this object with that of Fig. 52.

vertical plane is passed through $MN$ intersecting the picture plane in measuring line $gh$. A horizontal line from $o$ intersecting $gh$ locates $m$, and $m$ vanished to $CV$ is the required line.

By using the two center lines from $o$ and $m$ as a framework, the remaining construction is simplified. A ray from the station point to $B$ pierces the picture plane at $J$, which, projected to $mn$, locates $b$. The horizontal line $bz$ is the center line of the front face of the nearer arm, and the intercept $IJ$ gives the perspective radius $ab$. The circular hole having a radius $CB$ has an intercept $PJ$, giving $cb$ as the perspective radius. The arc $qy$ has its center on $ox$ at $z$. On drawing the tangents $lq$ and $ky$, the face "$F$" is completed.

The remaining construction for the arms is exactly the same as that for "$F$." The centers are moved back on the center lines, and the radii are found from their corresponding intercepts on the picture plane.

## 36. Circles in Perspective.

The perspective of a circle is a circle only when its plane is parallel to the picture plane; the circle appears as a straight line when its plane is receding from the station point. In all other positions the circle projects as an ellipse whose major and minor diameters are not readily determinable. The major diameter of the ellipse will be at some odd angle except when a vertical circle has its center on the horizon plane; then the major diameter will be vertical. Also, when a horizontal circle has its center directly above, below, or on the center of vision, the major diameter will be horizontal. It should be noted that in all cases the center of the circle is not coincident with the center of the ellipse representing the circle and that concentric circles are not represented by concentric ellipses. The major and minor diameters of the ellipses for concentric circles are not even parallel except in special cases.

The perspective of a circle can be plotted point by point, but the most rapid solution is had by enclosing the circle in a square, as shown in Fig. 54, and plotting points at the tangent points and at the intersections of the diagonals. The eight points thus determined are usually sufficient to give an accurate curve. The square, with its diagonals, is first drawn in the perspective. From the intersection of the diagonals, the vertical and horizontal center lines of the circle are established; where these center lines cross the sides of the square are four points on the curve. In the orthographic view, the measurement $X$ is made, then laid out *in the picture plane* and vanished, crossing the diagonals at four additional points.

Note that the curve is tangent to the lines enclosing it and that the *direction* of the curve is established by these tangent lines; if the lines completing the circumscribing octagon are projected and drawn, the direction of the curve is established at eight points.

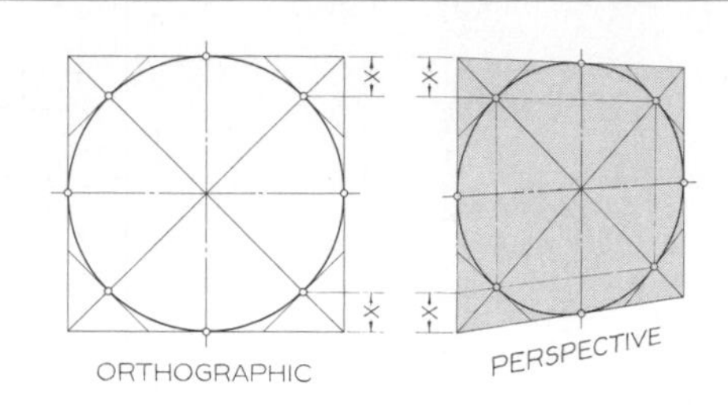


**FIG. 54.** Perspective of a circle. Points are plotted.

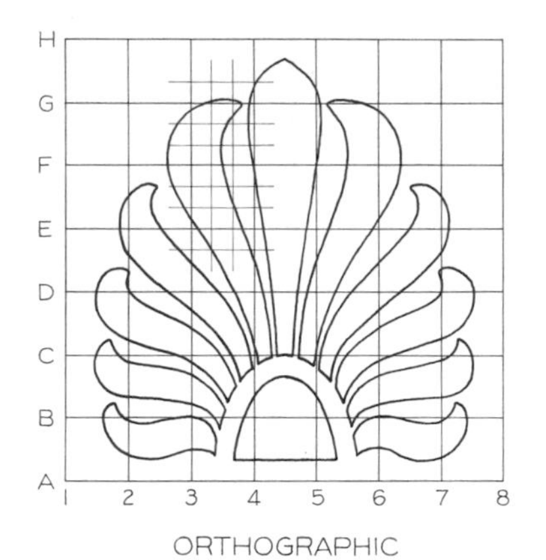


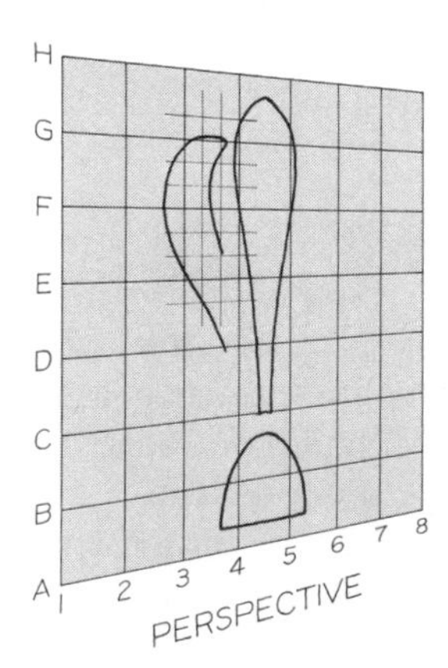


**FIG. 55.** Graticulation. Points are plotted.

## 37. Graticulation.

The perspectives of irregular curves can be drawn by projecting a sufficient number of points to establish the curve, but if the curve is complicated, the method of graticulation may be used to advantage. A square grid is overlaid on the orthographic view as shown in Fig. 55; then the grid is drawn in perspective and the outlines of the curve are transferred by inspection from the orthographic view.

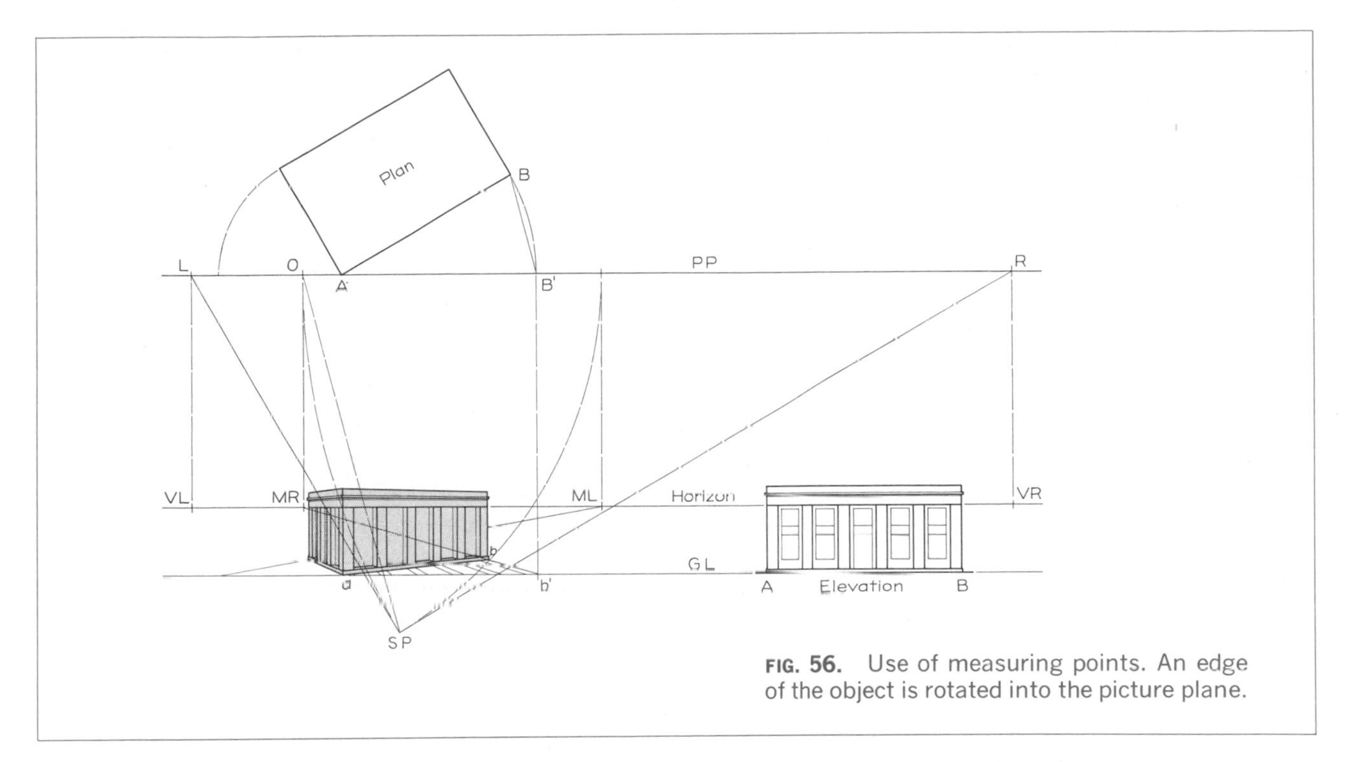


FIG. 56. Use of measuring points. An edge of the object is rotated into the picture plane.

## 38. Measuring Points.

It has been shown that all lines lying in the picture plane will be their own perspectives and can be scaled directly on the perspective drawing. The adaptation of this principle has an advantage in laying off a series of measurements, such as a row of pilasters, because it avoids a confusion of intercepts on the picture plane and the inaccuracies due to long projection lines.

In the measuring-points method, a surface, such as the wall between *A* and *B* in Fig. 56, is rotated into the picture plane for the purpose of making measurements, as shown at *AB′*. While in the picture plane, the entire surface can be laid out directly to the same scale as the top view; therefore, *ab′* and other horizontal dimensions of the surface are established along the ground line as shown. The counterrotation of the wall to its actual position on the building and the necessary projections in the perspective are based on the principle that the rotation has been made about a vertical axis and that any point has traveled in a horizontal plane. By drawing, as usual, a line parallel to *BB′*, from the station point to the picture plane, and then projecting to the horizon, the vanishing point *MR* is found. This vanishing point is termed a *measuring point* and may be defined as the vanishing point for lines joining corresponding points of the actual and rotated positions of the face considered. The divisions on *ab′* are therefore vanished to *MR*; where this construction intersects *ab* (the perspective of *AB*), the lateral position of the pilasters, in the perspective, is determined. Heights are scaled on the vertical edge through *a*, as this edge lies in the picture plane. The perspective of the wall between *A* and *B* is completed by the regular methods previously described. For work on the end of the building, the end wall is rotated as indicated, measuring point *ML* is found, and the projections are continued as described for the front wall.

Measuring points can be more readily located if the draftsman recognizes that the triangles *ABB′* and *R O SP* are similar. Therefore, a measuring point is as far from its corresponding vanishing point as the station point is from the picture plane, measuring the latter parallel to the face concerned. *MR* can then be found by measuring the distance from the station point to *R* and laying off *RO* equal to the measurement, or by swinging an arc, with *R* as center, from the station point to *O*, as shown. The measuring point *MR* is then projected from *O*.

### 39. Inclined Lines.

Any line that is not parallel or perpendicular to either the picture plane or the horizon plane is termed an inclined line. Any line may have a vertical plane passed through it, and if the vanishing line of the plane is found, a line in the plane will vanish at some point on the vanishing line of the plane. Vertical planes will vanish on vertical lines, just as horizontal planes vanish on a horizontal line, the horizon. In Fig. 57, the points *a* to *e* have all been found by regular methods previously described. The vanishing point of the horizontal *ab* is *VR*. The vertical line through *VR* is the vanishing line of the plane of *abc* and all planes *parallel to abc*. This vanishing line is intersected by the extension of *de* at *UR*, thereby determining the vanishing point for *de* and all edges *parallel to de*.

The vanishing point for inclined lines can also be located on the theory that the vanishing point for any line can be determined by moving the line until it appears as a point, while still retaining its original angle with the picture plane. The vanishing point of *de* can therefore be located by drawing a line through the station point parallel to *DE* and finding its piercing point with the picture plane. This is done by laying out *SP T* at the angle $\beta$ to *SP R* and erecting *RT* perpendicular to *SP R*. Then *RT* is the height of the vanishing point *UR* above *VR*.

If measuring points are used for the initial work on the perspective, it will be an advantage to recog-

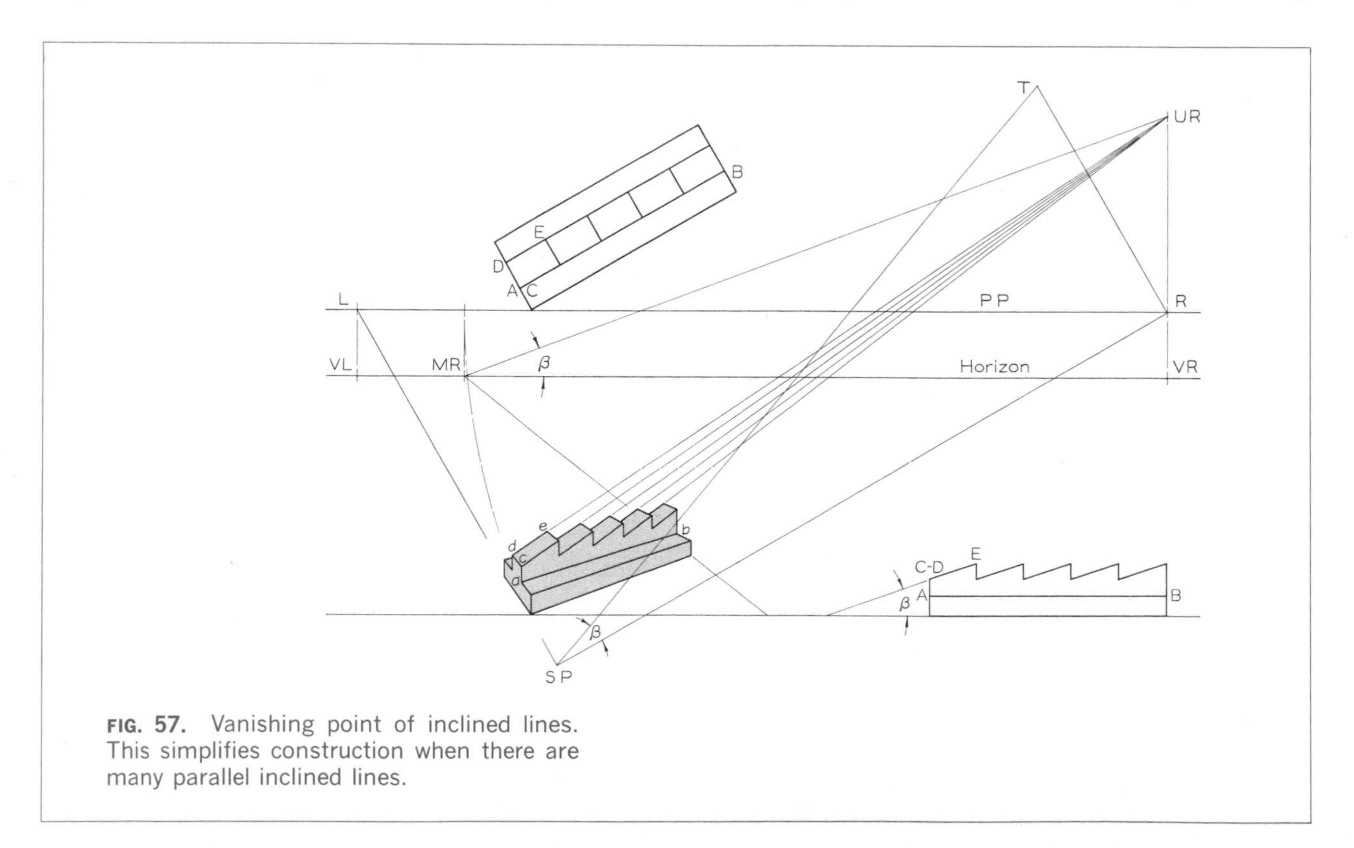


**FIG. 57.** Vanishing point of inclined lines. This simplifies construction when there are many parallel inclined lines.

nize which one of the measuring points was used for determining horizontal measurements in the parallel vertical planes containing the inclined lines; at that measuring point, the angle $\beta$ is laid out, above or below the horizon depending upon whether the lines slope up or down as they go into the distance. Where this construction intersects the vanishing line for the vertical planes containing the inclined lines, the vanishing point is located.

### 40. Inclined Planes.

An inclined plane is any plane not parallel or perpendicular to either the picture plane or the horizon plane. The vanishing line for an inclined plane can be found by locating the vanishing points for any two systems of parallel lines in the inclined plane. To determine the vanishing line of plane *ABCD* in Fig. 58, the vanishing point *VL* of the horizontal edges *AD* and *BC* is one point, and the vanishing point *UR* for the inclined edges *AB* and *DC* gives a second point on the vanishing line *VL UR* for plane *ABCD*.

It is often necessary to draw the line of intersection of two inclined planes. The intersection will vanish at the point of intersection of the vanishing lines of both planes. The intersection *J* of the two vanishing lines of the roof planes in Fig. 58 is the vanishing point of the line of intersection of the two planes.

### 41. Pictorial Sketching.

The need for the engineer to be trained in freehand sketching was emphasized in Chap. 5, where the discussion referred particularly to sketching in orthographic projection. Before he can be said to have a command of the graphic language, his training in freehand drawing must include also acquiring the ability to sketch *pictorially* with skill and facility.

In designing and inventing, the first ideas come into the mind in pictorial form, and sketches made in this form preserve the ideas as visualized. From this record the preliminary orthographic design sketches are made. A pictorial sketch of an object or of some detail of construction can often be used

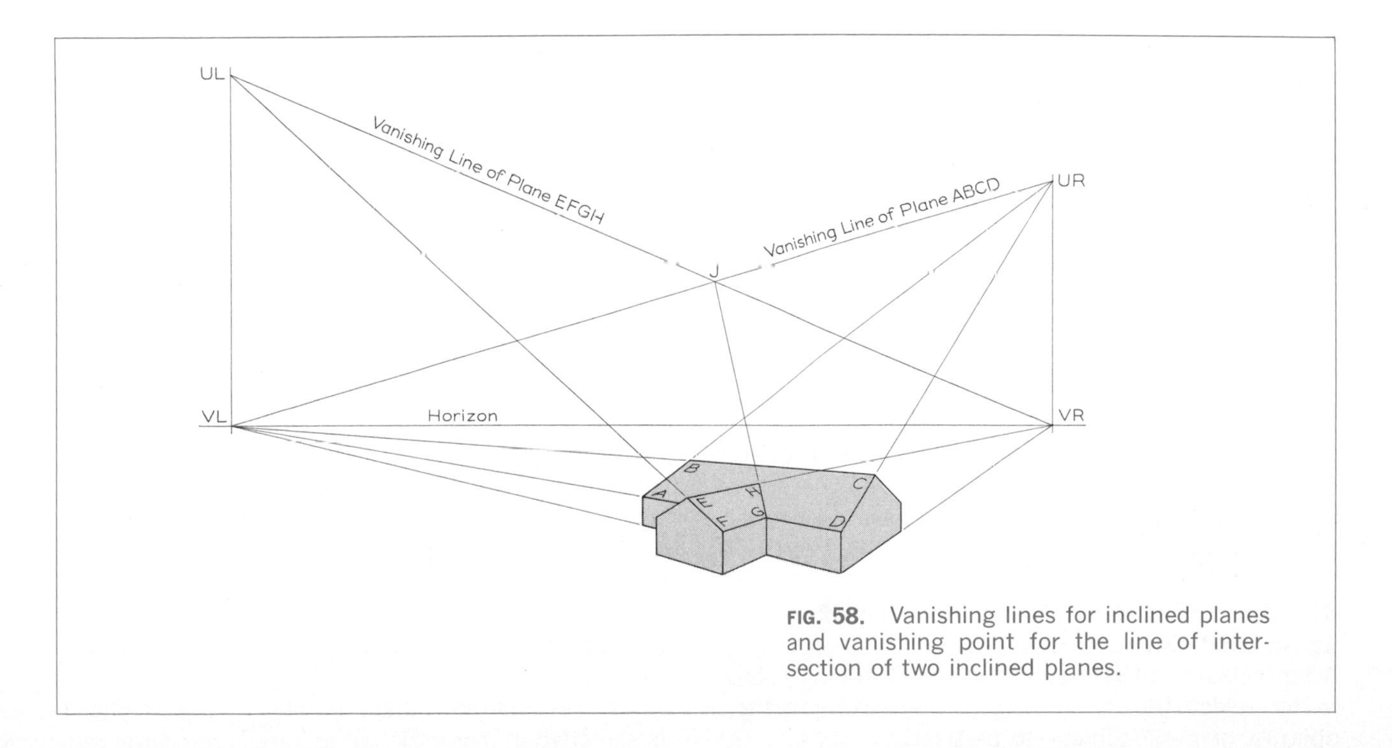


FIG. 58. Vanishing lines for inclined planes and vanishing point for the line of intersection of two inclined planes.

to explain it to a client or workman who cannot read the orthographic projection intelligently. One of the best ways of reading a working drawing that is difficult to understand is to start a pictorial sketch of it. Usually before the sketch is finished, the orthographic drawing is perfectly clear. Often a pictorial sketch can be made more quickly and serve as a better record than orthographic views of the same piece. A young engineer should not be deterred by any fancied lack of "artistic ability." An engineer's sketch is a record of information, not a work of art. The one requirement is *good proportion*.

## 42. Methods.

Although this is not a complete classification, there may be said to be three pictorial methods of sketching: axonometric, oblique, and perspective. The mechanical construction has been explained in detail

## 43. Prerequisites.

It should be clearly understood at the outset that pictorial sketching means the making of a pictorial drawing *freehand*. The same construction that is used for locating points and lines and for drawing circles and arcs with instruments will be used in pictorial sketching. From this standpoint, a knowledge of the constructions already given is necessary before attempting pictorial sketching. Note in Figs. 69 to 75 that the ellipses representing holes and rounded contours have, before being drawn, been boxed in with construction lines representing the enclosing square in exactly the same manner as for an instrument drawing.

## 44. Materials and Technique.

The same materials, pencil grades, etc., used for orthographic freehand drawing, described in Chap. 5, are employed for pictorial sketching. The directions given there for drawing straight lines, circles, and arcs will apply here also.

## 45. Pictorial Sketching: Choice of Type and Direction of View.

After a clear visualization of the object, the first step is to select the type of pictorial—axonometric, oblique, or perspective—to be used.

Isometric is the simplest axonometric position, and it will serve admirably for representing most objects. Although dimetric or trimetric may be definitely advantageous for an object with some feature that is obscured or misleading in isometric, it is best to try isometric first, especially if there is doubt that another form will be superior. This is principally because proportions are easier to judge in isometric.

Oblique forms (cavalier or cabinet) may be used to advantage for cylindrical objects or for objects with a number of circular features in parallel planes. Nevertheless, a true circle, representing a circular feature parallel to the picture plane in oblique, is much harder to sketch than an ellipse, representing the same feature in isometric, as the slightest deviation from a circle is evident, while the same deviation in an ellipse is unnoticed. Also, inherently, there is more distortion in the oblique than in axonometric forms. Therefore, especially for the sake of professional appearance, an axonometric form has the advantage.

Perspective is the best form for pictorial sketching because it is free from any distortion. A perspective is not much more difficult to sketch than an axonometric or an oblique, but attention must be paid to the convergence of the lines and to keeping good proportion. However, do not discard axonometric and oblique from consideration. As will be seen in Secs. 46 and 47, these forms can be handled as successfully as perspective by some flattening of the axes and by converging the lines properly.

Choose carefully the *direction* in which the object is to be viewed. There are many possibilities. The object may be turned so that any lateral face will be represented on the right or the left side of the pictorial. Orient the object so that the two *principal* faces will show to advantage. Use reversed axes if necessary. Do this by mentally visualizing and turning the object in every possible position in order to arrive at the best representation for all features. Be alert to see that some feature will not be hidden by a portion in front of it. The proper choice of direction is an important factor in pictorial sketching.

Fig. 59 illustrates these points. At (*a*) an object is sketched in isometric; all features are clearly rep-

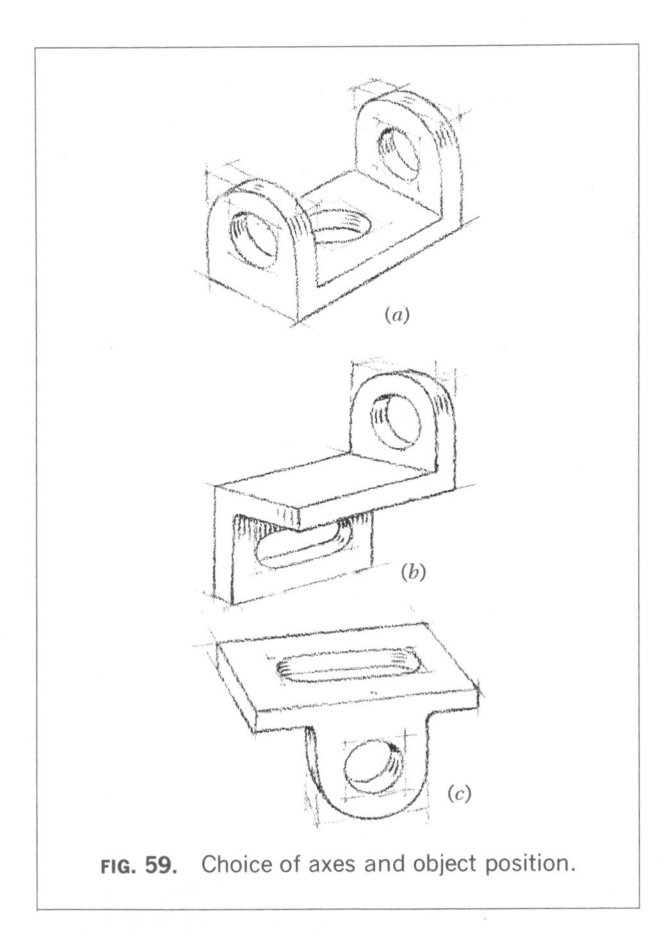


FIG. 59. Choice of axes and object position.

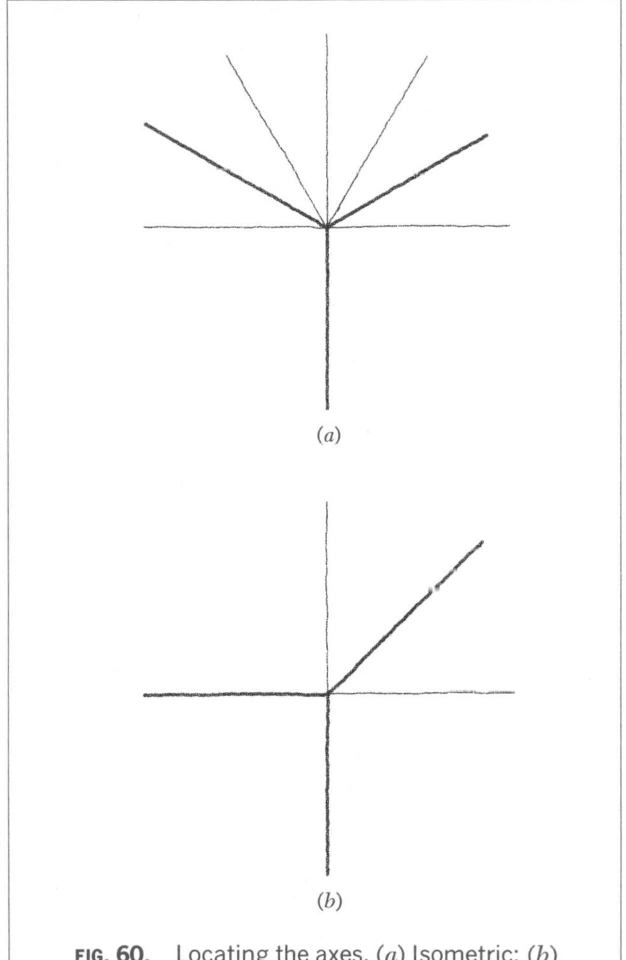


FIG. 60. Locating the axes. (*a*) Isometric; (*b*) oblique at 45°.

resented and the object appears natural in this form. At (*b*) trimetric has been used so that the slot in the lower portion is not obscured by the horizontal middle portion. At (*c*) another trimetric position has been chosen to present the semicircular ear as the definite front face of the object. For further illustration of the possibilities, study the pictorial drawings in Chap. 5.

## 46. Sketching the Axes.

After the type of pictorial and the position of the object have been decided upon, the first step in making the sketch is to draw the axes.

*In isometric,* the three axes should be located as nearly as possible 120° from one another (one vertical and two at 30° with the horizontal). Because no triangles are used in sketching, the angles must be located by judgment. Figure 60*a* shows a satisfactory method of judging the position of the lines. First draw a *light* horizontal and vertical and then divide both upper quadrants into thirds. The lines at the top of the lower thirds are then the two axes at 30° to the horizontal, and the third axis is the vertical. This method is simple and accurate because it is easy to estimate equal thirds of a quadrant.

*In dimetric,* the standard angles are 7° and 41°. For the 7° axis, again referring to Fig. 60*a*, draw the bisector of the 30° axis for isometric to get 15°, and then bisect again to get $7\frac{1}{2}$°, which will be quite satisfactory and can be done fairly accurately. For the 41° axis, take the mid-line of the second 30°

section, which is 45°, and shade it a little to approximate 41°.

*In trimetric,* almost any combination representing three mutually perpendicular lines is possible. However, a pronounced distortion will occur if the two transverse axes are more than 30° from the horizontal. Read the cautions in Sec. 47.

*In oblique,* the common angles are 30° and 45° although theoretically any angle is possible. To prevent violent distortion, never make the depth axis greater than 45°. To locate a 45° axis, sketch a vertical and horizontal as in Fig. 60*b*, and then sketch the bisector of the quadrant (see oblique positions) where the axis is wanted. For an axis at 30°, proceed as in Fig. 60*a* for isometric.

*For perspective,* as explained in Sec. 32, take care to have a reasonable included angle of view; otherwise a violent convergence will occur. In order to prevent difficulty, first locate two vanishing points, as in Fig. 61, as widely separated as the paper will allow (attach extra paper with scotch tape if necessary). Then sketch the *bottom* edges of the object (the heavier lines in Fig. 61) and on these sketch a rectangular shape, as shown. It will be immediately evident that the arrangement is satisfactory or that the vanishing points must be moved *or* that the base lines must be altered. Remember that the two vanishing points *must* be at the same level, the horizon.

### 47. Sketching the Principal Lines.

Almost without exception, the first lines sketched should be those that box in the whole object, or at least its major portion. These first lines are all-important to the success of the sketch, for mistakes at the outset are difficult to correct later. Observe carefully the following three points:

1. *Verticals must be parallel to the vertical axis.* In Fig. 62 a cube is sketched correctly at (*a*). At (*b*) and (*c*) the same cube is sketched but at (*b*) the verticals converge downward and at (*c*), upward. It is evident that (*b*) and (*c*) do not look cubical at all, but like frustums of pyramids. This is proof that verticals *must be kept accurately vertical* Be critical of the verticals throughout the construction. Accurate verticals add a stability and crispness not attained in any other way.

2. *Transverse lines must be parallel or converging.* In Fig. 63 a cube is sketched at (*a*) with the transverse lines (receding right and left edges) made accurately parallel. At (*b*) these lines are made to converge as they recede. Note that (*b*) looks more natural than (*a*) because of the effect of perspective foreshortening. The monstrosity at (*c*) is produced by the separating of the receding lines as they recede. In his attempt to get the lines parallel, the beginner often makes a mistake like that at (*c*). Converge the lines deliberately, as in (*b*), to avoid the results of (*c*)!

3. *Axes must be kept flat to avoid distortion.* In Fig. 64 an accurate isometric sketch is shown at (*a*). At (*b*) the axes have been flattened to less than 30° with the horizontal. (*b*) possibly looks more natural than (*a*). At (*c*), however, the axes are somewhat more than 30° to the horizontal. Note the definite

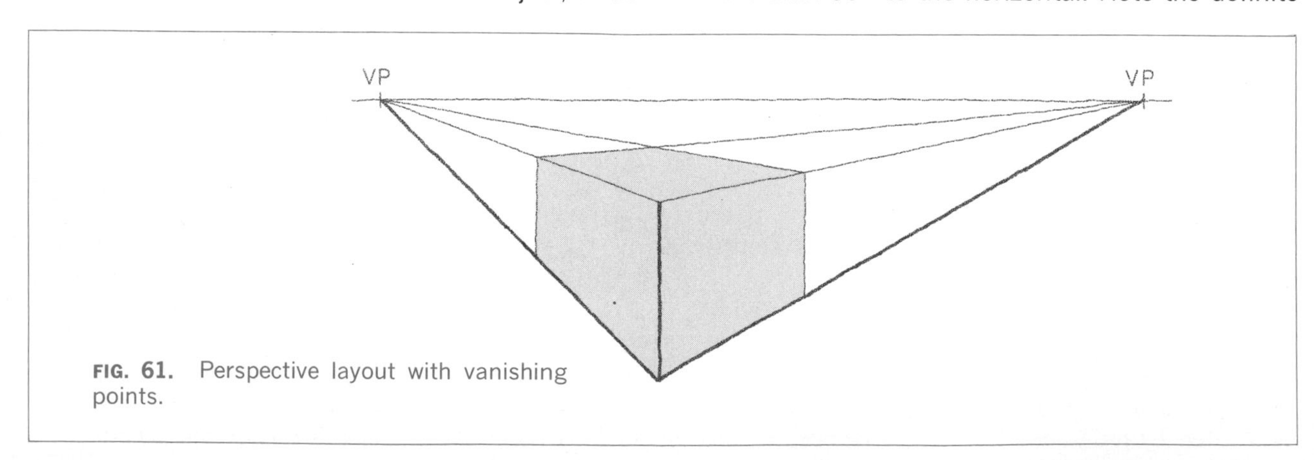


**FIG. 61.** Perspective layout with vanishing points.

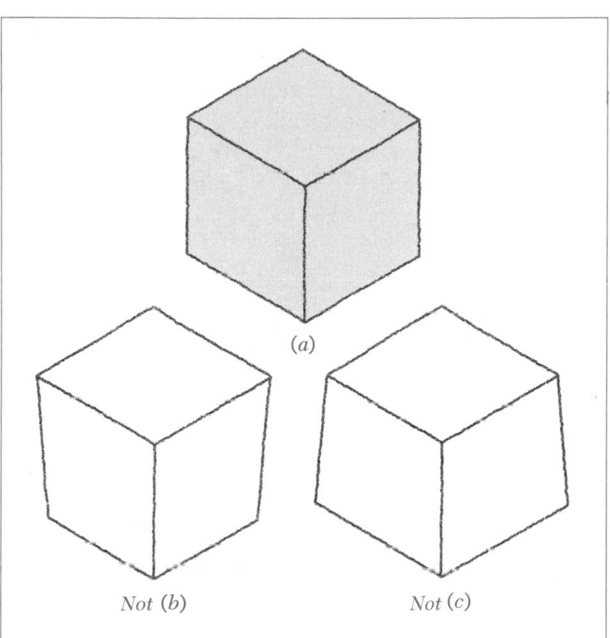


**FIG. 62.** Sketching vertical lines. These must be accurately vertical to define object shape.

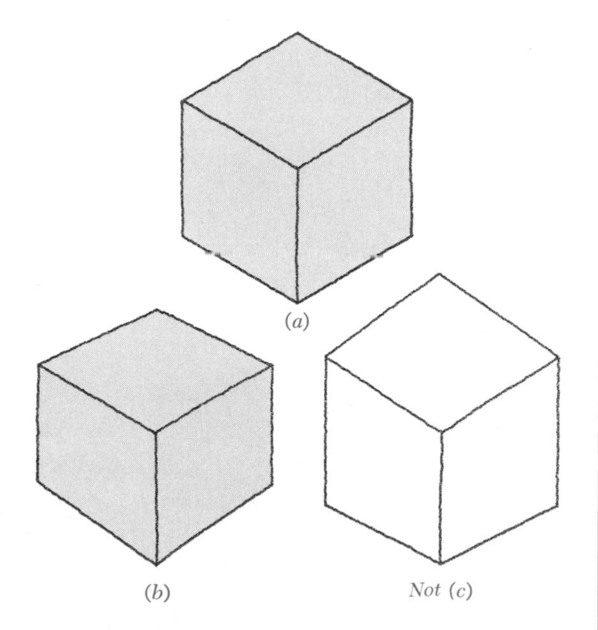


**FIG. 63.** Sketching the receding edges. These must be parallel (*a*); or converging (*b*); *never* separating (*c*).

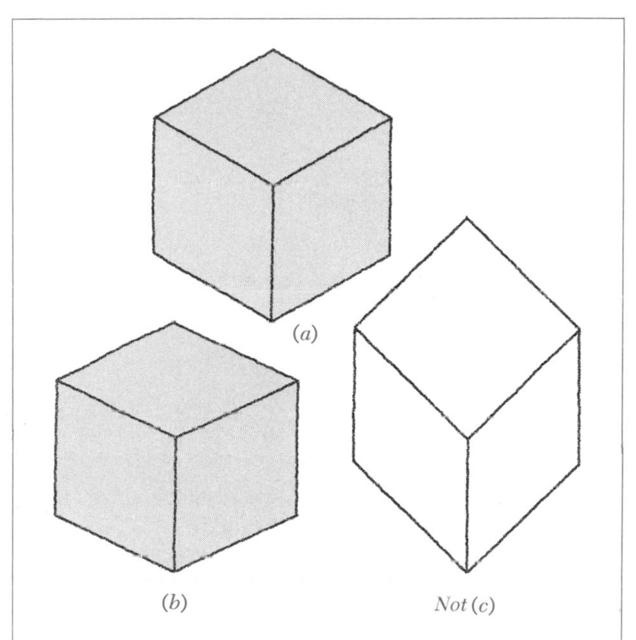


**FIG. 64.** Angle of axes. Isometric position (*a*) or flattened (*b*) gives a natural appearance. Distortion is inherent in steep axes as at (*c*).

distortion and awkward appearance of (*c*). Therefore, especially in isometric, but also in other forms, keep the axes at their correct angle or flatter than the normal angle.

The foregoing three points must be kept constantly in mind. Hold your sketch at arm's length often during the work, so that you will see errors that are not so evident in the normal working position. Become critical of your own work and you will soon develop confidence and a good sense of line direction.

## 48. Divisions for Symmetry.

Continually in sketching, centers must be located and divisions of a face must be made into thirds, fourths, fifths, etc. Division into halves to locate a center line is the simplest and most common, and is easily accomplished by judging the mid-point along one of the sides, as indicated on the top face of the rectangular shape in Fig. 65. Also shown on this top face are additional divisions into quarters and eights. Practice this a few times to develop your

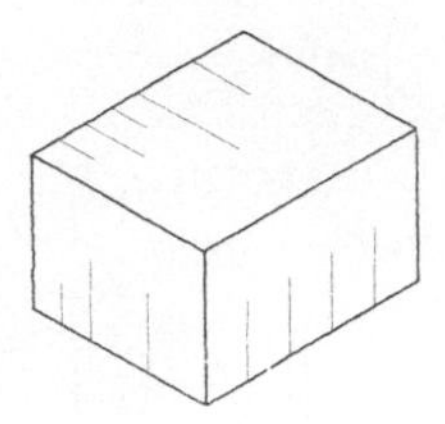

**FIG. 65.** Judging equal spaces. This can be done satisfactorily by eye.

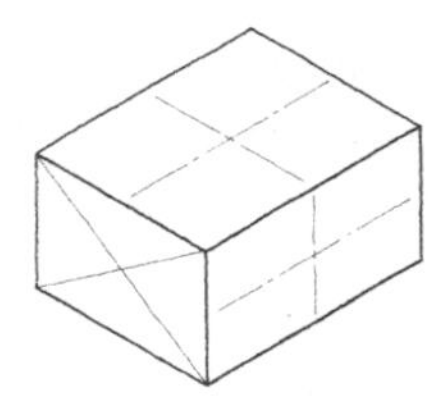

**FIG. 66.** Location of centers. Use diagonals or judge the position of center lines.

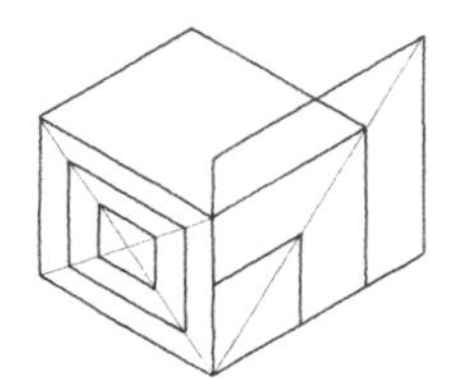

**FIG. 67.** Use of diagonals for increasing or decreasing rectangular shapes.

judgment of equal spacings. It is really quite easy to do.

Division into thirds is a little harder but after short practice is readily done. The left face of Fig. 65 shows third divisions and one space divided in half to give a sixth point. The right face of Fig. 65 shows division into fifths.

Practice dividing lines or spaces into various equal units. The experience will be valuable in later work.

## 49. Uses of the Diagonal.

The two diagonals of a square or rectangle will locate its geometric center, as indicated on the left face of Fig. 66. The center is also readily located by drawing two center lines, estimating the middle of the space as shown on the top and right faces.

The diagonals of a rectangular face can also be used to increase or decrease the rectangle symmetrically about the same center and in proportion as shown on the left face of Fig. 67, or with two sides coincident as shown on the right face. To increase or decrease by equal units, the distance (or space) between lines must be judged as explained in Sec. 48.

## 50. Sketching Circular Features.

A circle in pictorial is an ellipse whose major diameter is always perpendicular to the *rotation axis*. Thus its minor diameter coincides on the drawing with the rotation axis (Fig. 68). These facts can be used to advantage when drawing an object principally made up of cylinders on the same axis. Note particularly from the above that *all* circles on horizontal planes are drawn as ellipses *with the major diameter horizontal*, as shown in Fig. 69.

Most objects, however, are made up of combina-

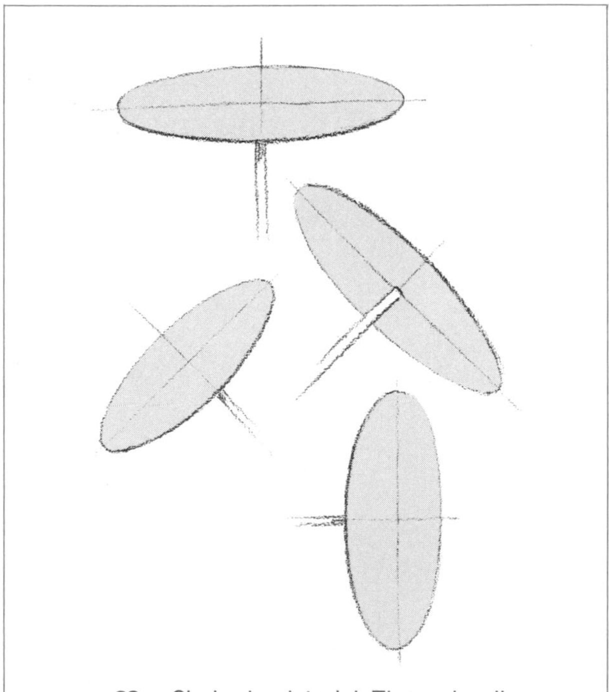

**FIG. 68.** Circles in pictorial. The major diameter of the ellipse is perpendicular to the axis of rotation.

tions of rectangular and circular features, and for this reason it is best to draw the enclosing pictorial square for all circular features. Figure 70 shows circles on all three axonometric planes. Note particularly that the ellipses must be tangent to the sides of the pictorial square at the mid-points of the sides; accordingly, it is best always to draw center lines, also shown. Always sketch the enclosing pictorial square for *all* circular features because by this method the size of the ellipse and the thickness of the cylindrical portion are easily judged. Figure 71 illustrates the boxing of circular features. This object would be difficult to sketch without first boxing in the circular portions.

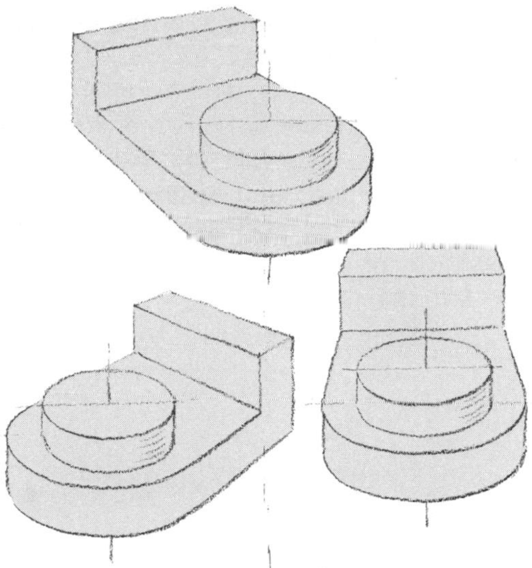

**FIG. 69.** Circular features on horizontal planes. The major diameter of the ellipse is horizontal.

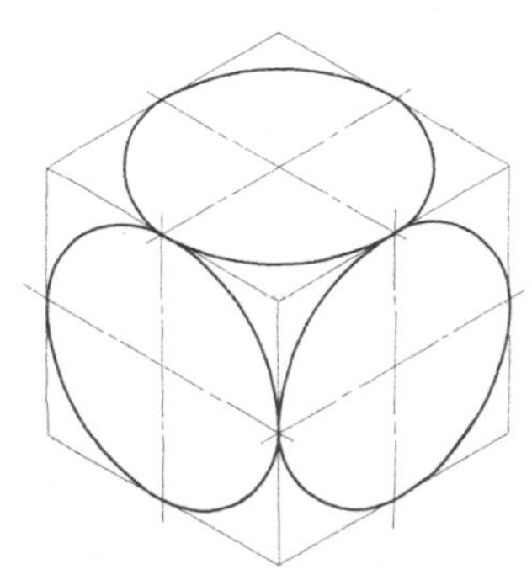

**FIG. 70.** Circles in isometric. Ellipses are tangent to the enclosing isometric squares at the mid-points of the sides.

## 51. Proportioning the Distances.

The ability to make divisions into equal units, discussed in Sec. 48, is needed in proportioning distances on a sketch. The average object does not have distances that are easily divisible into even *inches, half inches,* etc., but as sketches are *not* made to scale, only to good proportions, great accuracy is not necessary. Also, most objects, dimensioned as they invariably are in various odd distances, are difficult to judge from the standpoint of one distance being $\frac{1}{2}$, $\frac{1}{3}$, $\frac{1}{4}$, etc., of another. Nevertheless, the proportioning can be done easily and quickly by the method we are about to discuss.

Figure 72 shows a simple rectangular object, but the distances are not multiples of any simple unit. The best way to proportion this object (and others)

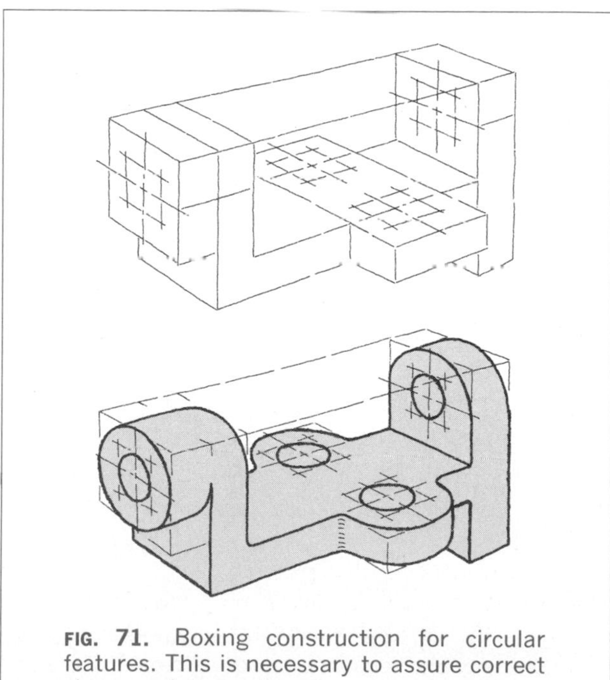

**FIG. 71.** Boxing construction for circular features. This is necessary to assure correct shape and proportion.

is to lay off on *one* axis a distance that is to represent one side of the object. This *sets the size* of the sketch. Then divide this first side into some unit that can be used easily to proportion other distances. At (*a*) the *left* side has been laid off and divided into three parts. Each of these parts now represents *approximately* a half inch. To get the distance for the vertical axis, the last third (at the rear) has been divided in half, and again in half, so that the dimension shown is approximately $1\frac{1}{8}$ in. Transfer this distance to the vertical axis (1) by judging by eye, (2) by measuring with the finger on the pencil, or (3) by marking the distance on a piece of scratch paper. The method to be used will suggest itself according to the relative accuracy needed. The right-axis distance is obtained similarly, by first transferring the whole left distance (representing approximately $1\frac{1}{2}$ in.) and then adding two-thirds of the left distance (approx. 1 in.), which gives a total of $2\frac{1}{2}$ in., close enough to the actual $2\frac{5}{8}$ in. dimension of the object. Any side may be chosen to start the sketch. At (*b*) the right side has been laid off and divided, this time, into five parts so that again each unit is approximately $\frac{1}{2}$ in. The procedure is then similar to that described for (*a*).

Remember that great accuracy is unnecessary. Do

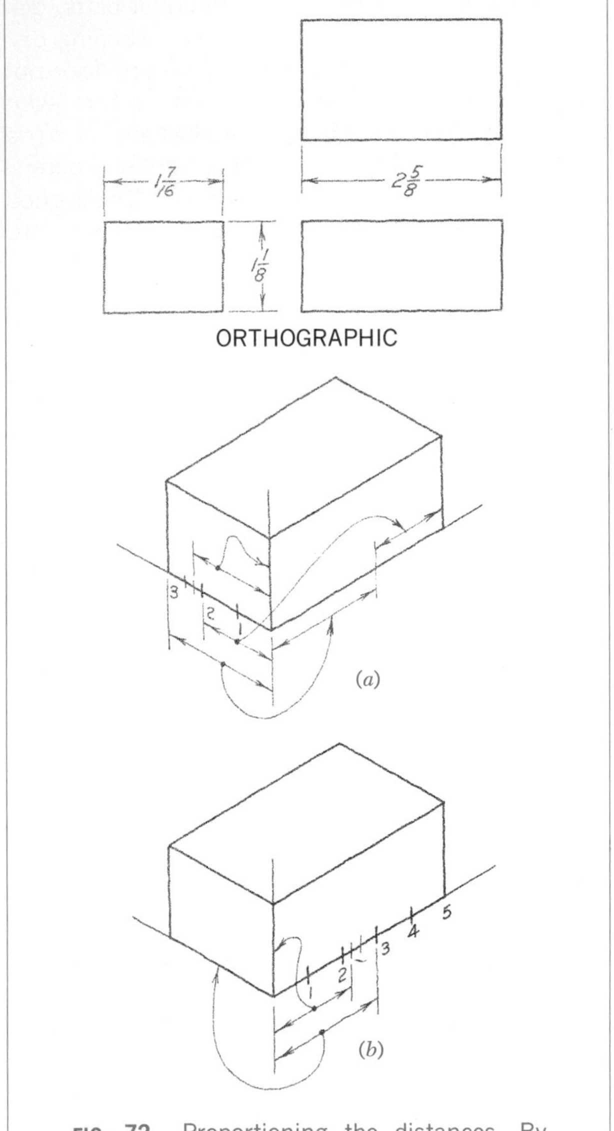


**FIG. 72.** Proportioning the distances. By dividing one side into units, all other distances are proportioned.

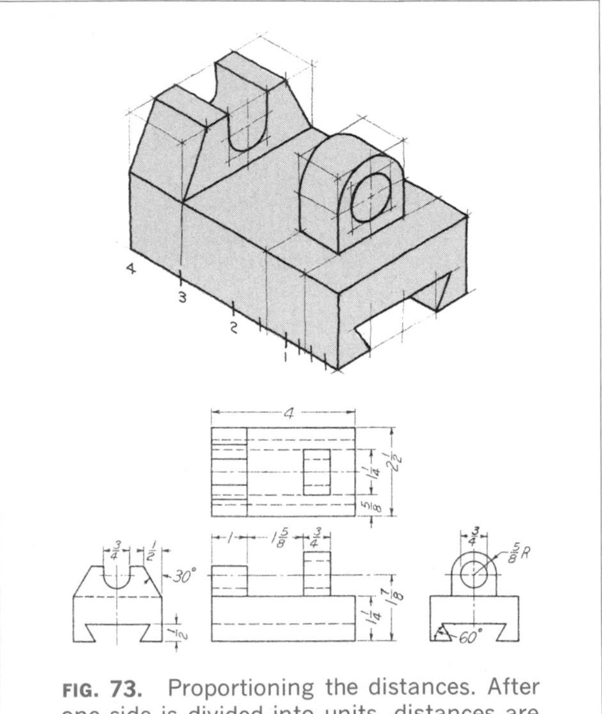


**FIG. 73.** Proportioning the distances. After one side is divided into units, distances are projected or transferred.

not make the proportioning a burden. It is simply a method of getting distances reasonably close to the actual distance and eliminating the need for wild guesses or extra construction on the sketch.

Figure 73 shows the method applied to a more complex object. The left axis is divided into four parts. In this case the upper details have been located by projection upward from the left axis divisions, a method that is often used. Note the boxing of circular features, as described in Sec. 50. Study this figure carefully, with particular attention to the proportioning and construction.

### 52. Steps in Making a Pictorial Sketch.

Because a variety of objects are sketched, the order of procedure will not always be the same, but the following will serve as a guide:

*A*. Visualize the shape and proportions of the object from the orthographic views, a model, or other source.
*B*. Mentally picture the object in space and decide the pictorial position that will best describe its shape.
*C*. Decide on the type of pictorial to use—axonometric, oblique, or perspective.
*D*. Pick a suitable paper size.
*E*. Then proceed as shown in Fig. 74.

The numbers that follow refer to the figure; 1 through 6 shows light construction, 7, 8, and 9 completion to final weight.

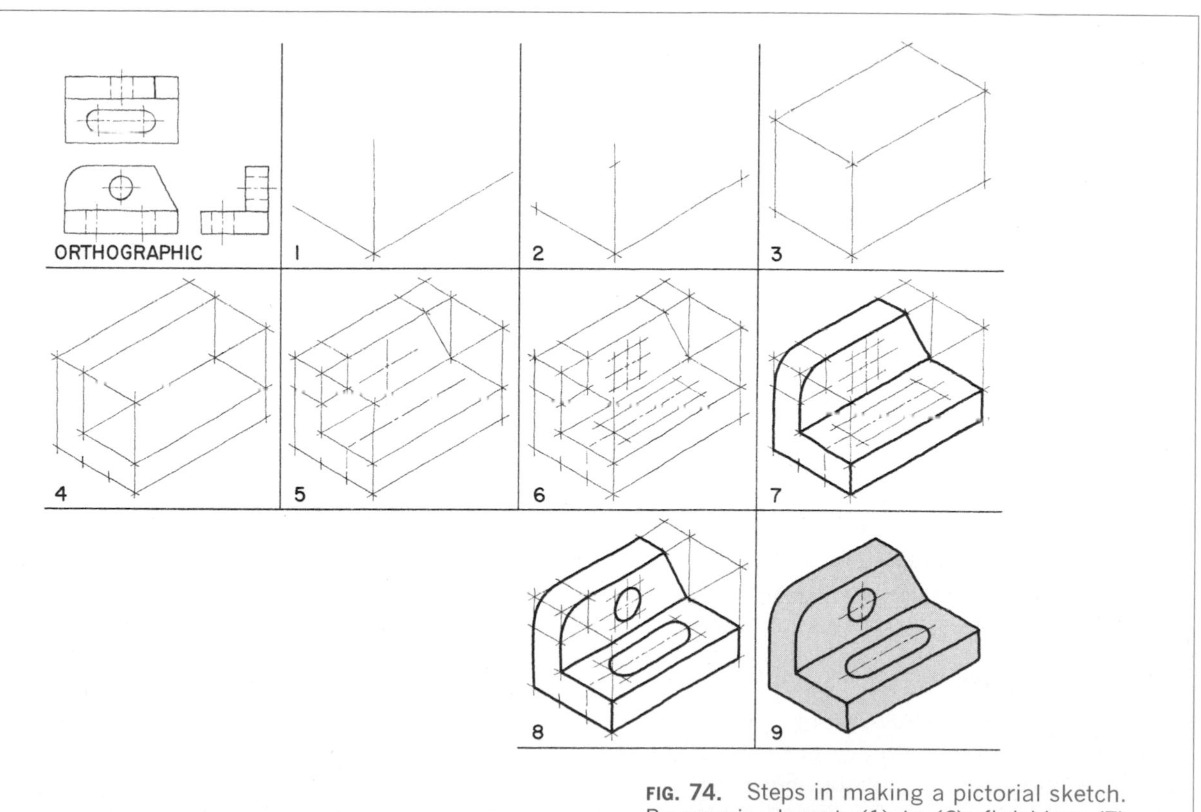


FIG. 74. Steps in making a pictorial sketch. Progressive layout, (1) to (6); finishing, (7) to (9).

1. Sketch the axes.
2. Lay off the proportions of an enclosing rectangular box for the whole object or a principal portion of it.
3. Sketch the enclosing box.
4. Divide one axis for proportioning distances and sketch the most dominant detail of the object.
5. Proportion smaller details by reference to the divided axis and sketch the enclosing boxes, center lines, or outlines.
6. Complete the boxes for circular features. Check to be sure that all features are in good proportion and that there are no errors in representation (see Sec. 47). In working entirely with light lines, corrections are easily made.
7. Start sketching to the final line width. Begin with the most dominant feature.
8. Sketch the smaller details.
9. Remove construction.

### 53. Axonometric Sketching.

The methods presented thus far have been directed toward the making of axonometric sketches because they are the type most used. However, the practices given apply to oblique and perspective sketching, for which additional help is given in the two sections that follow.

### 54. Oblique Sketching.

The advantage of oblique projection in preserving one face without distortion is of particular value in sketching, as illustrated by Fig. 75. The painful effect of distortion in oblique drawing that is done instrumentally can be greatly lessened in sketching by foreshortening the depth axis to a pleasing proportion. The effect of perspective is obtained by converging the lines parallel to the depth axis. This converging in axonometric or oblique is sometimes called "fake perspective."

### 55. Perspective Sketching.

A sketch made in perspective gives a better effect than in axonometric or oblique. For constructing a perspective drawing of a proposed structure from its plans and elevations, a knowledge of the principles of perspective drawing is required, but for making a perspective sketch from the object, you can get along by observing the ordinary phenomena of perspective which affect everything we see: the fact that objects appear proportionately smaller as their distance from the eye increases, that parallel lines appear to converge as they recede, and that horizontal lines and planes appear to "vanish" on the horizon.

In perspective sketching from the model, make the drawing simply by observation, estimating the directions and proportionate lengths of lines by sighting and measuring on the pencil held at arm's length and use your knowledge of perspective phenomena as a check. With the drawing board or sketch pad held in a comfortable drawing position perpendicular to the line of sight from the eye to the object, test the direction of a line by holding the pencil at arm's length parallel to the board, rotating the arm until the pencil appears to coincide with the line on the model, and then moving it parallel to this position back to the board. Estimate the apparent lengths of lines in the same way; holding the pencil in a plane perpendicular to the line of sight, mark with the thumb the length of pencil which covers the line of the model, rotate the arm with the thumb held in position until the pencil coincides with another line, and then estimate the proportion of this measurement to the second line (Fig. 76).

Make the sketch lightly, with free sketchy lines, and do not erase any lines until the whole sketch has been blocked in. *Do not make the mistake of getting the sketch too small.*

In starting a sketch from the object, set it in a position to give the most advantageous view, and sketch the directions of the principal lines, running them past the limits of the figure toward their vanishing points. Block in the enclosing squares for all circles and circle arcs and proceed with the figure, drawing the main outlines first and adding details later; then brighten the sketch with heavier lines.

The drawing in Fig. 77 shows the general appearance of a "one-point" perspective sketch before the construction lines have been erased. Figure 78 is an example showing the object turned at an angle to the picture plane.

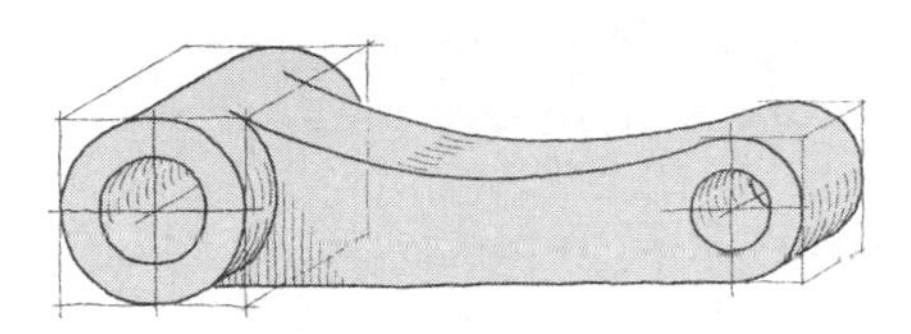

**FIG. 75.** An oblique sketch. Note boxing of circular features.

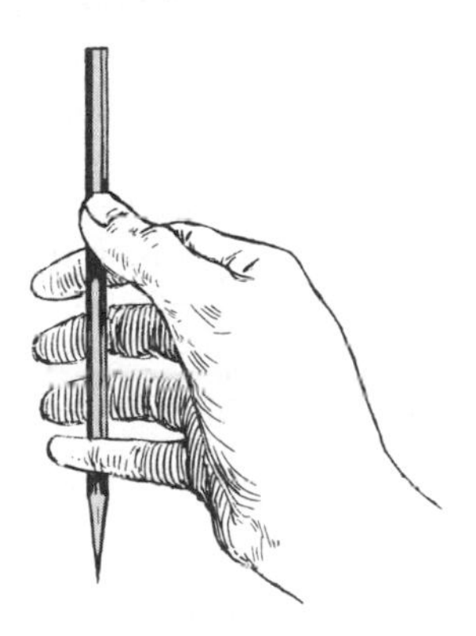

**FIG. 76.** Estimating distances and proportion.

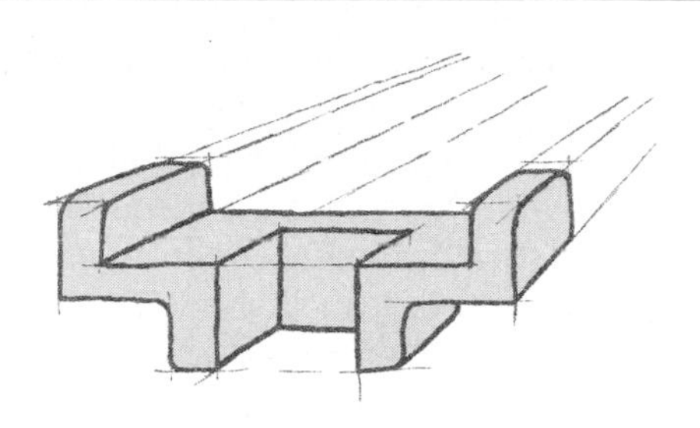

**FIG. 77.** A perspective sketch. The front face of the object is parallel to the picture plane.

## 56. Pictorial Illustration.

Pictorial illustration combines any one of the regular pictorial methods with some method of shading or "rendering." In considering a specific problem, decide upon the pictorial form—axonometric, oblique, or perspective—and then choose a method of shading that is suited to the method of reproduction and the general effect desired.

## 57. Light and Shade.

The conventional position of the light in light-and-shade drawing is the same as that used for orthographic line shading, that is, a position to the left, in front of, and above the object. Any surface

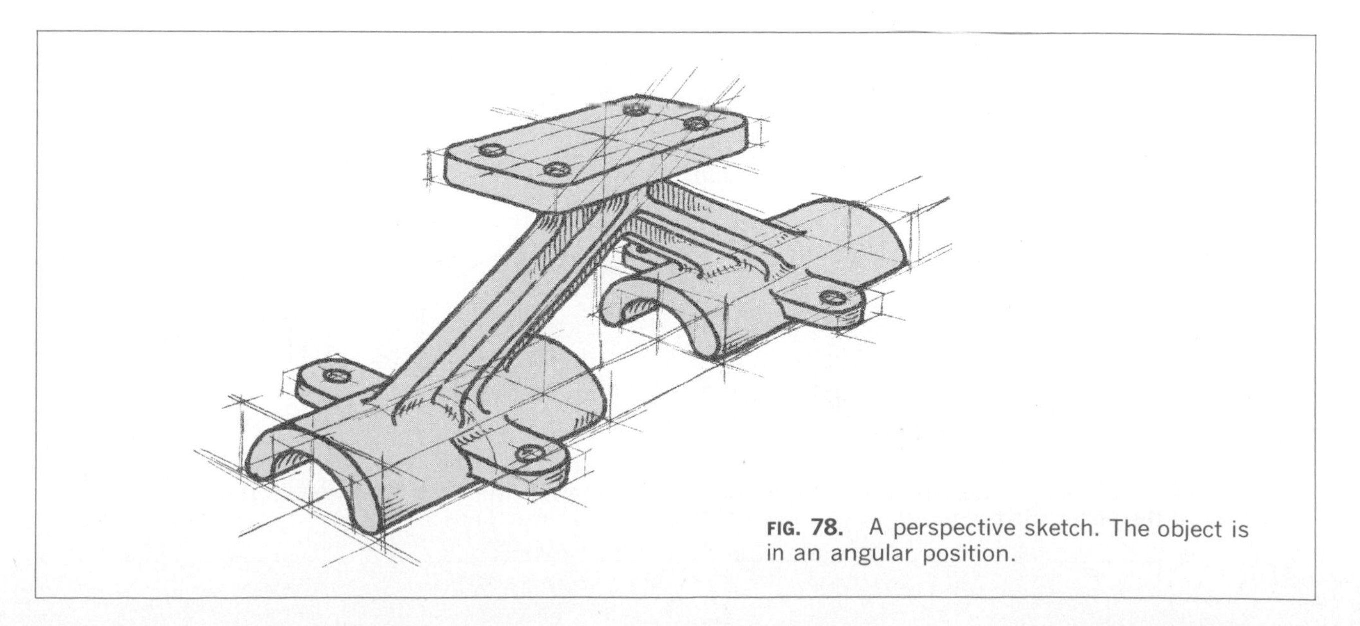

**FIG. 78.** A perspective sketch. The object is in an angular position.

or portion of a surface perpendicular to the light direction and directly illuminated by the light would receive the greatest amount of light and be lightest in tone on the drawing; any face not illuminated by the light would be "in shade" and darkest on the drawing. Other surfaces, receiving less light than the "high" light but more than a shade portion, would be intermediate in tone.

An understanding of the simple one-light method of illumination is needed at the outset, as well as some artistic appreciation for the illumination on various surfaces of the object. Figure 79 shows a sphere, cylinder, cone, and cube illuminated as described and shaded accordingly. Study the tone values in this illustration.

## 58. Shade Lines.

Shade lines, by their contrast with other lines, add some effect of light and shade to the drawing. These lines used alone, without other shading, give the simplest possible shading method. Usually the best effect is obtained by using heavy lines only for the left vertical and upper horizontal edges of the dark faces (Fig. 80). Holes and other circular features are drawn with heavy lines on the shade side. Shade lines should be used sparingly as the inclusion of too many heavy lines simply adds weight to the drawing and does not give the best effect.

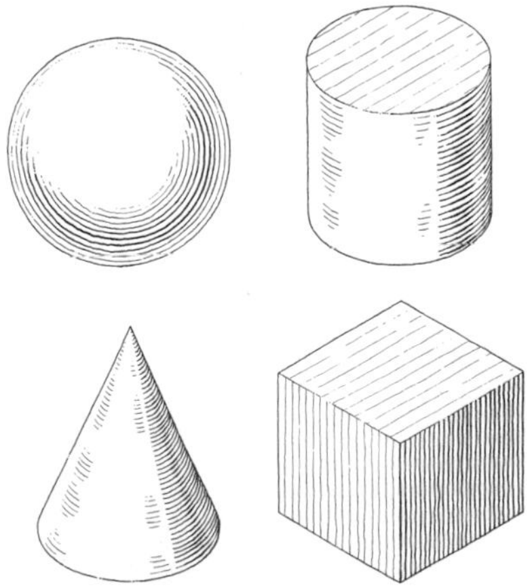

FIG. 79. Light and shade. The light source is from the upper left front.

## 59. Pencil Rendering.

There are two general methods of pencil shading—continuous tone and line tone. Continuous-tone shading is done with a fairly soft pencil with its point flattened. A medium-rough paper is best for the purpose. Start with a light, over-all tone and then build the middle tones and shade portions gradually. Figure 81 is an example. Clean high lights with an eraser.

Line-tone shading requires a little more skill, as the tones are produced by line spacing and weight. Light lines at wide spacing produce the lightest tone, and heavy lines at close spacing make the darkest shade. Leave high lights perfectly white. Pure black may be used sparingly for deep shade or shadow. Figure 82 is an example, drawn with only a light outline.

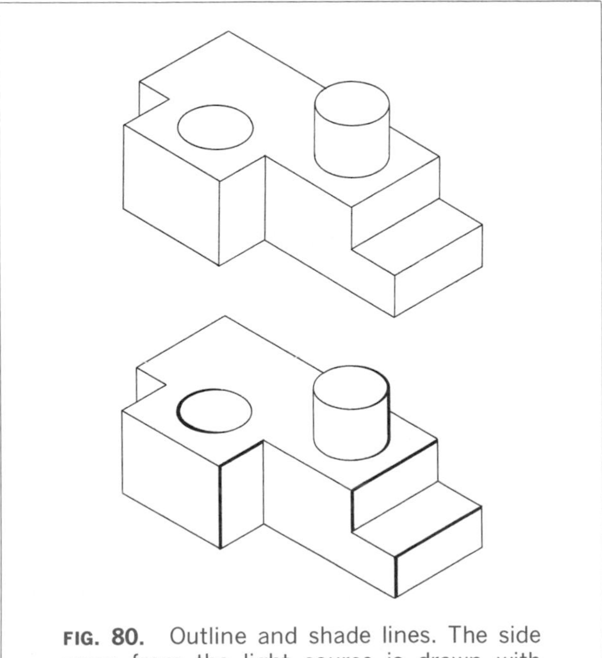

FIG. 80. Outline and shade lines. The side away from the light source is drawn with heavier lines.

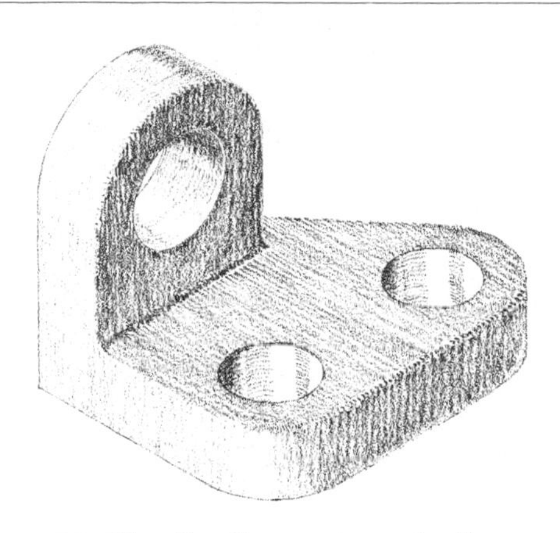

**FIG. 81.** Continuous-tone shading.

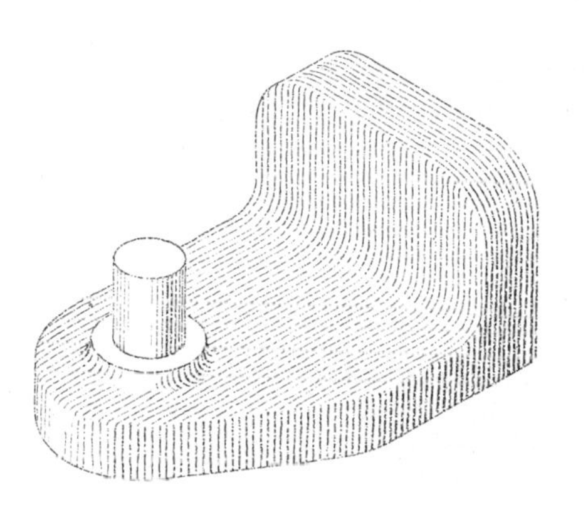

**FIG. 82.** Line-tone shading.

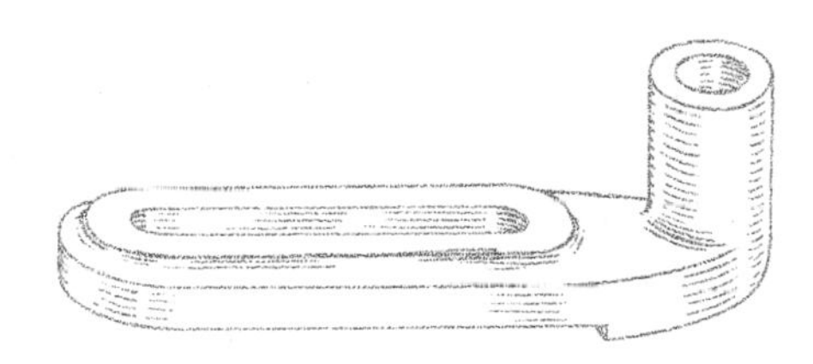

**FIG. 83.** Line-shading technique in pencil.

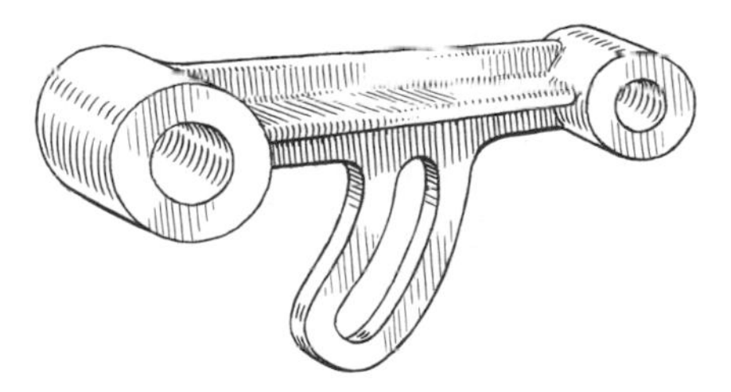

**FIG. 84.** Line-shading technique in ink.

Complete over-all shading is somewhat heavy, and a lighter, more open treatment is usually desired. To achieve this, leave light portions of the object with little or no shading, and line middle tones and shade sparingly. The few lines used strongly suggest light, shade, and surface finish (Fig. 83). There are many variations which can be made in this type of rendering.

## 60. Pen-and-ink Rendering.

Pen-and-ink methods follow the same general pattern as work in pencil, with the exception that no continuous tone is possible. However, there are some variations not ordinarily used in pencil work. Figure 84 shows line techniques. As in pencil work, the common and usually the most pleasing method is the partially shaded, suggestive system.

# PROBLEMS

The following problems are intended to furnish practice (1) in the various methods of pictorial representation and (2) in reading and translating orthographic projections.

In reading a drawing, remember that a line on any view always means an edge or a change in direction of the surface of the object and always look at another view to interpret the meaning of the line.

## Group 1. Isometric Drawings.

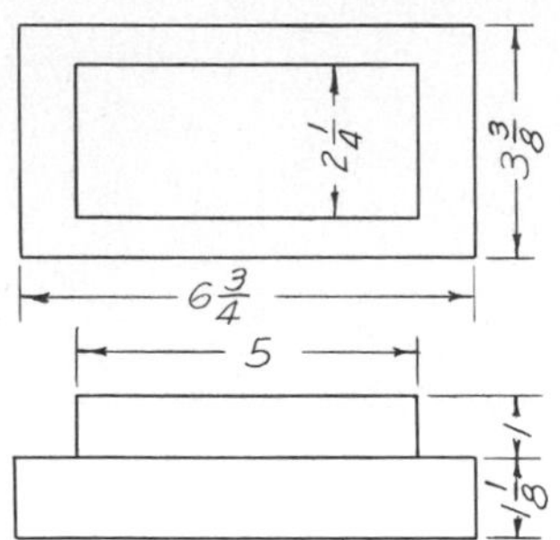

**PROB. 1.** Jig block.

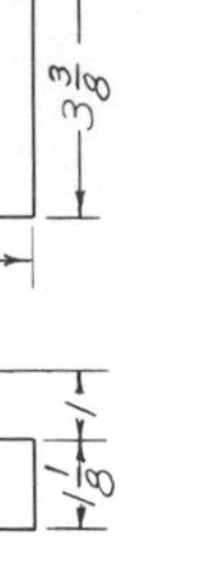

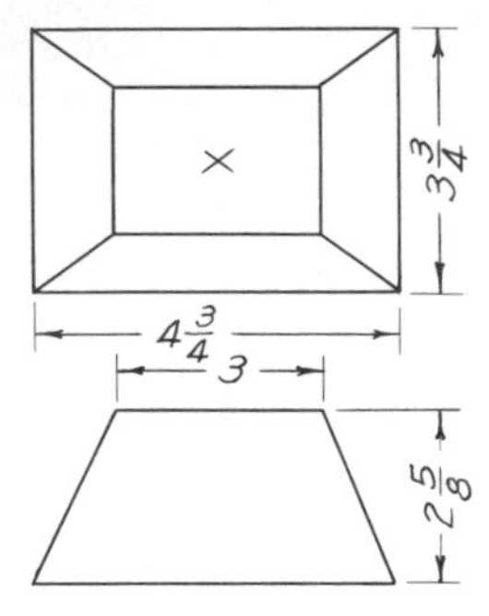

**PROB. 2.** Frustum of pyramid.

**PROB. 3.** Hopper.

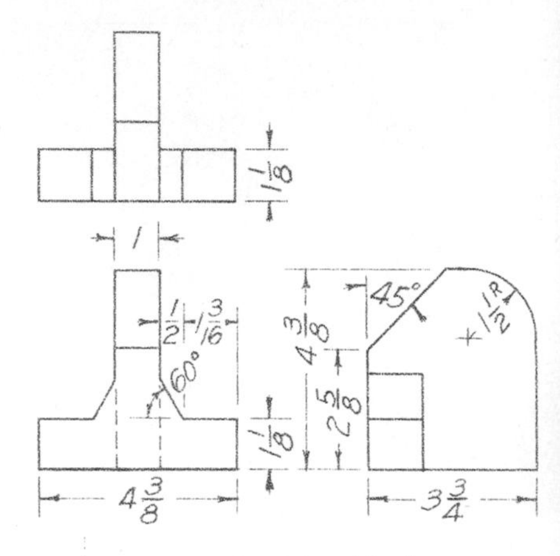

**PROB. 4.** Stop block.

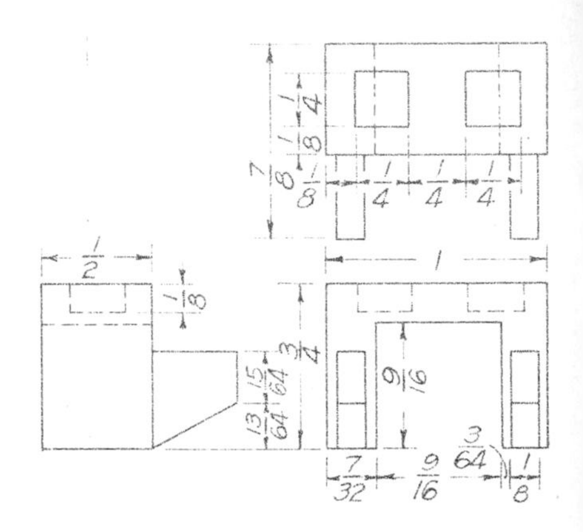

**PROB. 5.** Skid mount.

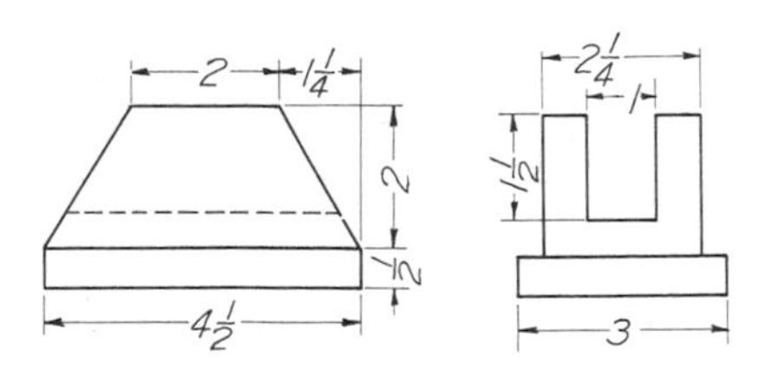

**PROB. 6.** Guide block.

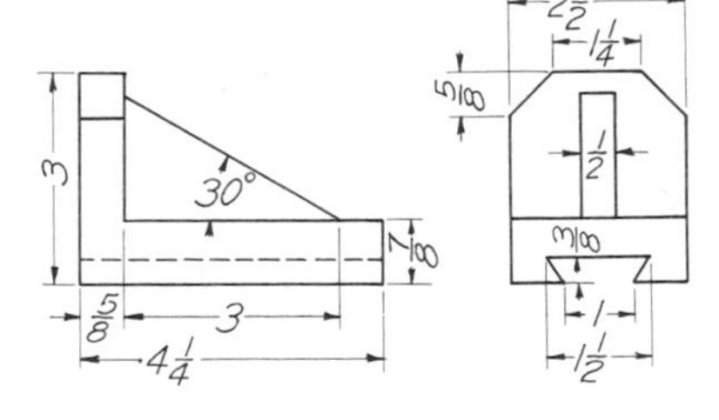

**PROB. 7.** Dovetail stop.

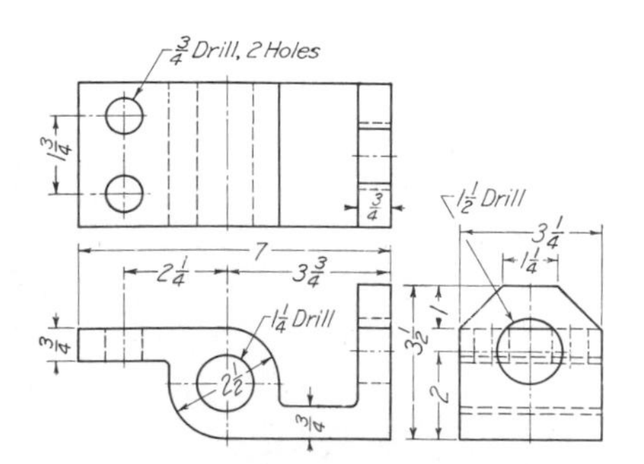


**PROB. 8.** Bracket.

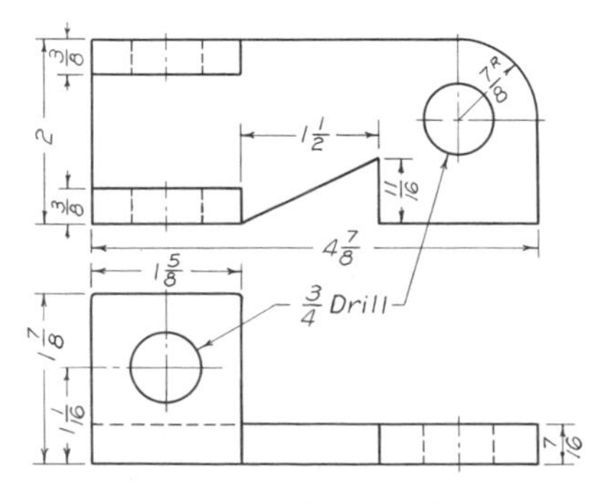


**PROB. 9.** Hinged catch.

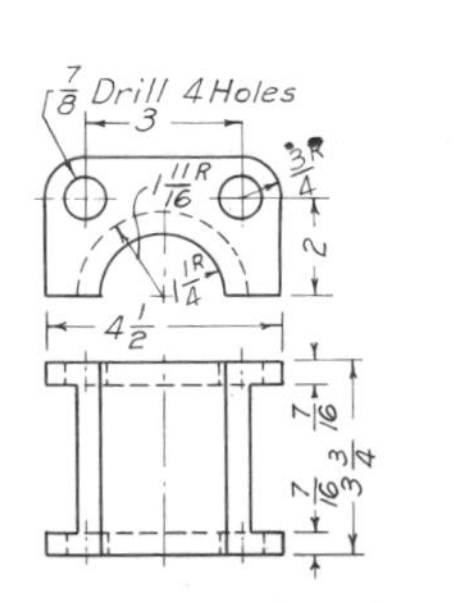

**PROB. 10.** Bearing.

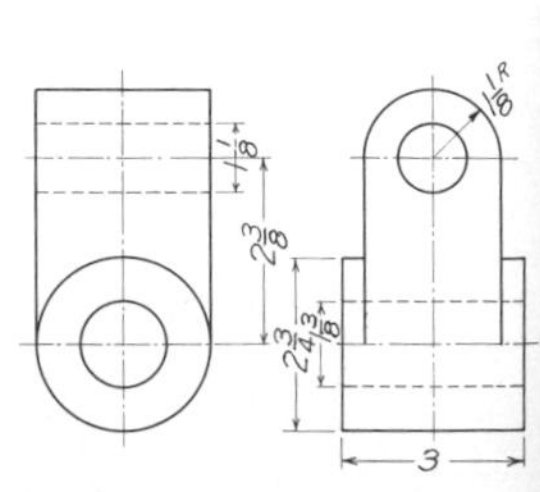

**PROB. 11.** Cross link.

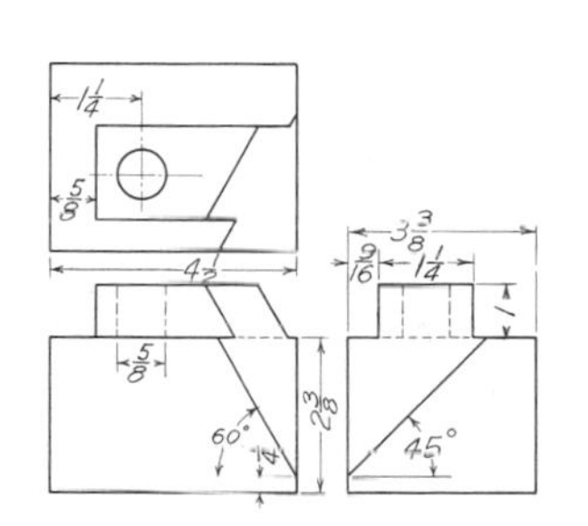

**PROB. 12.** Wedge block.

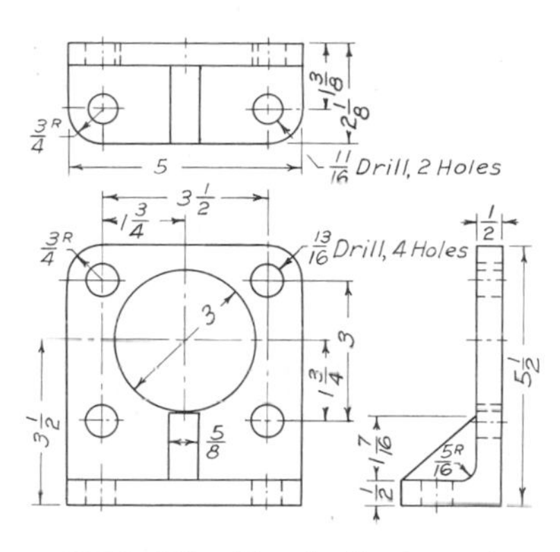


**PROB. 13.** Head attachment.

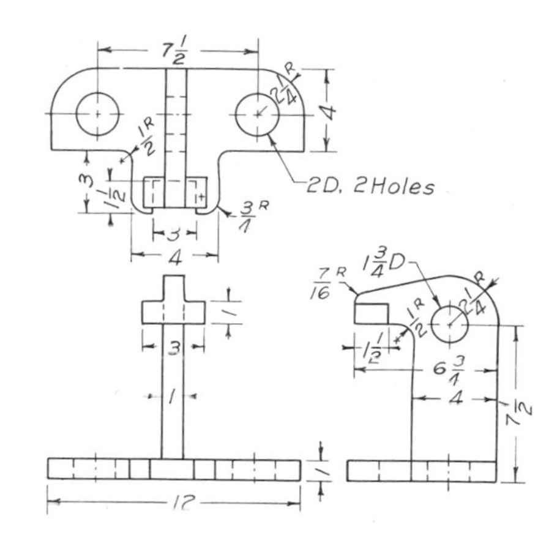


**PROB. 14.** Slide stop.

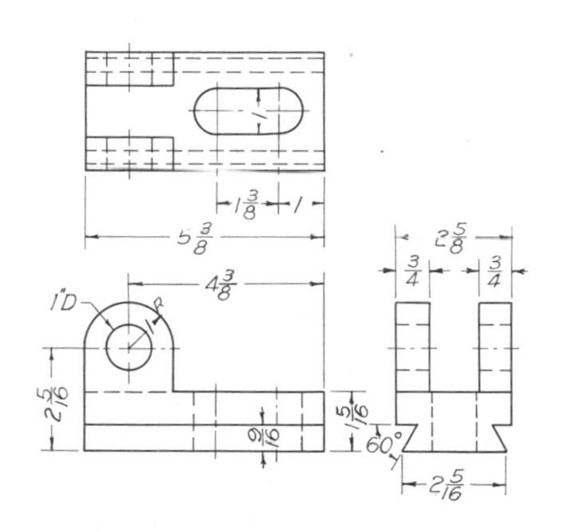

**PROB. 15.** Dovetail bracket.

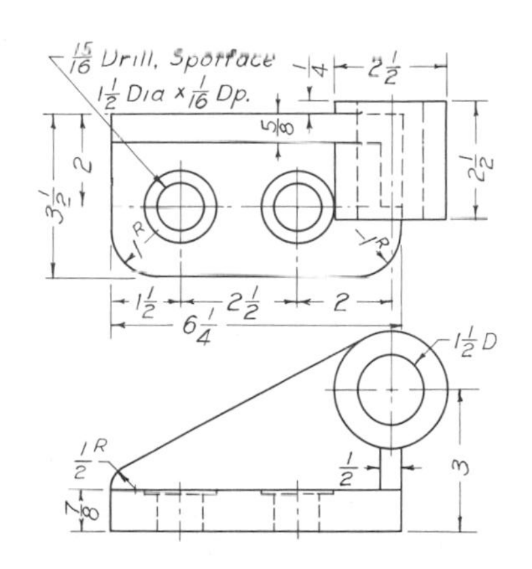


**PROB. 16.** Offset bracket.

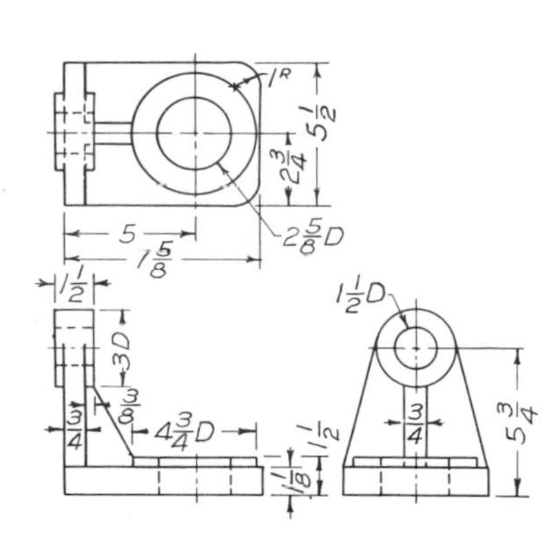

**PROB. 17.** Cradle bracket.

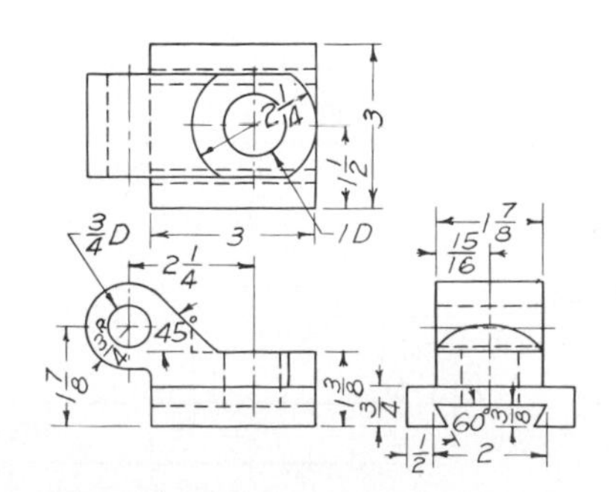

**PROB. 18.** Dovetail hinge.

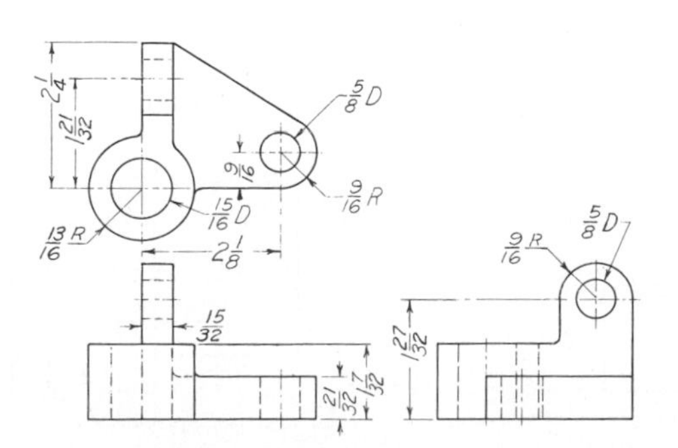

**PROB. 19.** Cable clip.

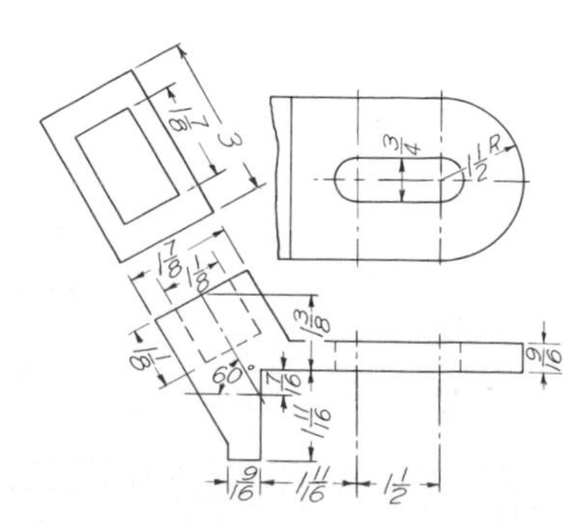

**PROB. 20.** Strut anchor.

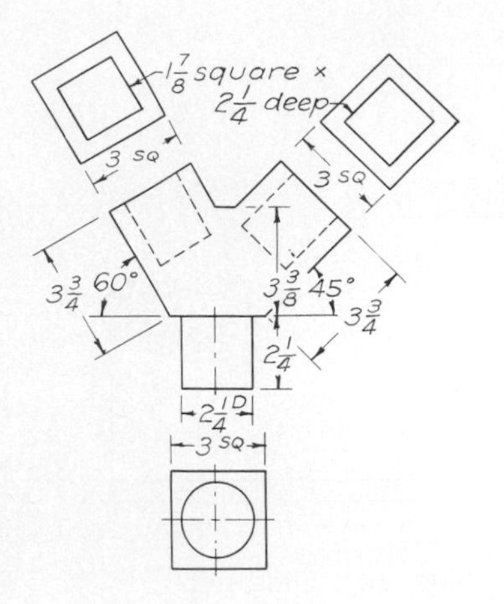


**PROB. 21.** Strut swivel.

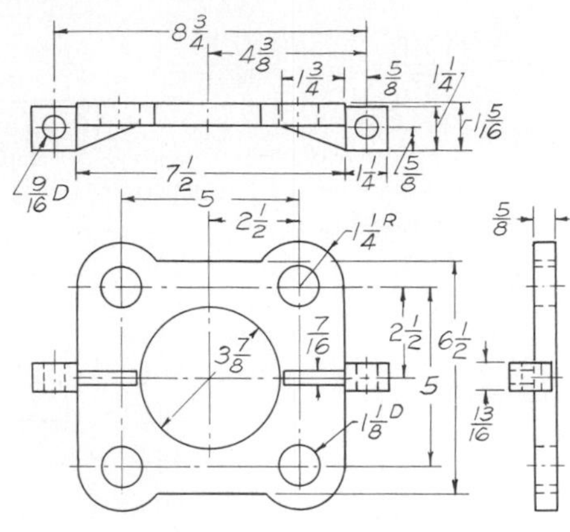

**PROB. 22.** Tie plate.

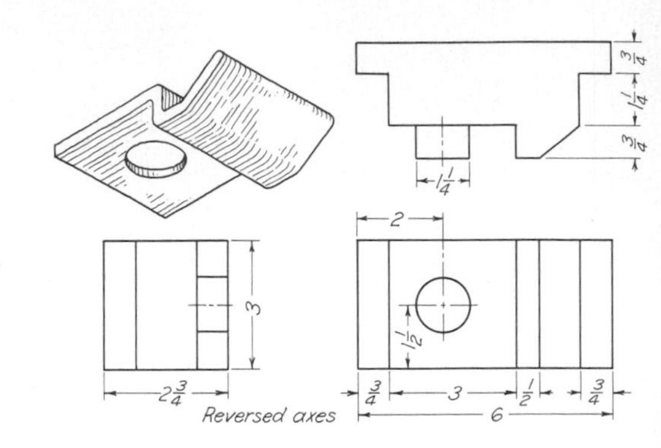


**PROB. 23.** Forming punch.

## Group 2. Isometric Sections.

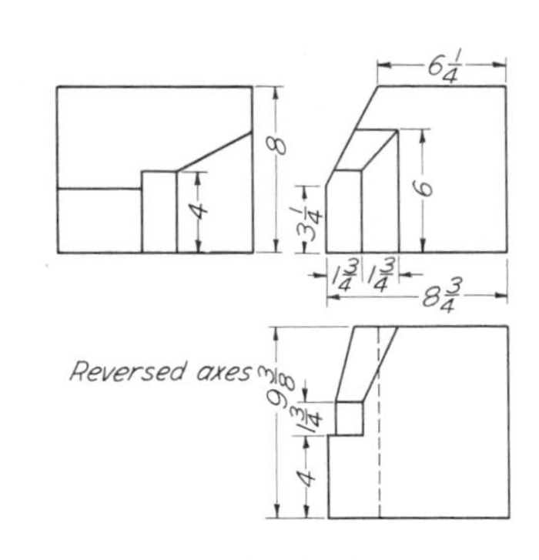


**PROB. 24.** Springing stone.

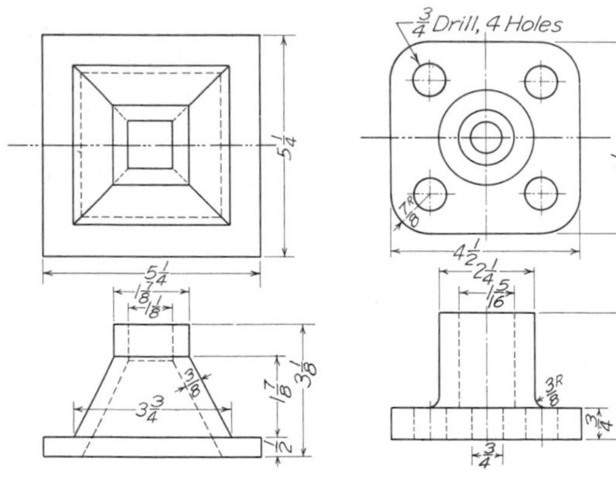


**PROB. 25.** Column base. **PROB. 26.** Base plate.

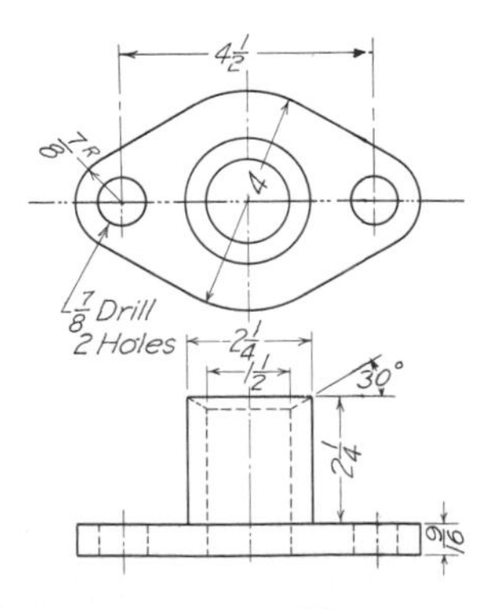


**PROB. 27.** Gland.

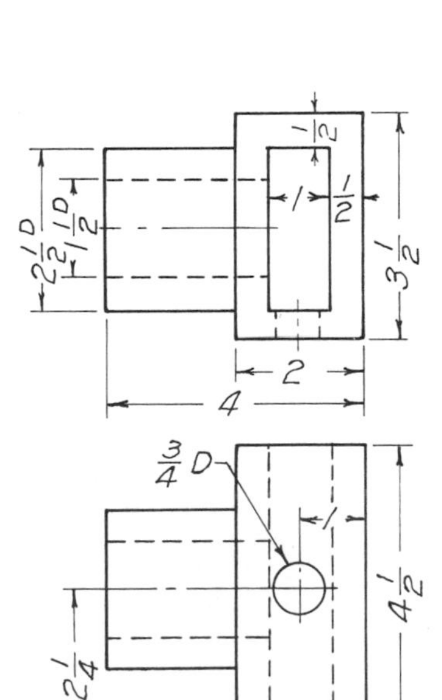

**PROB. 28.** Squared collar.

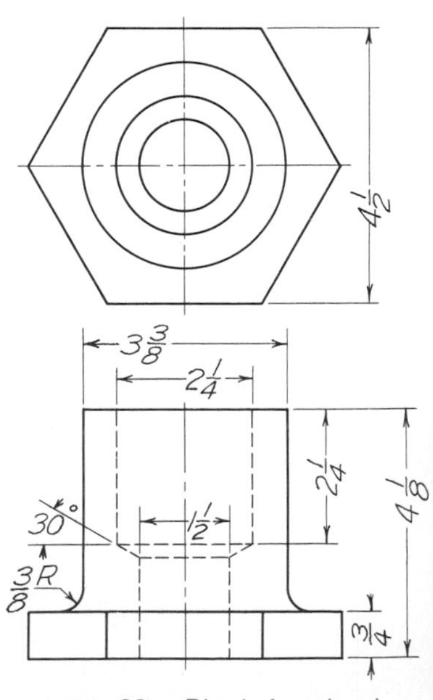

**PROB. 29.** Blank for gland.

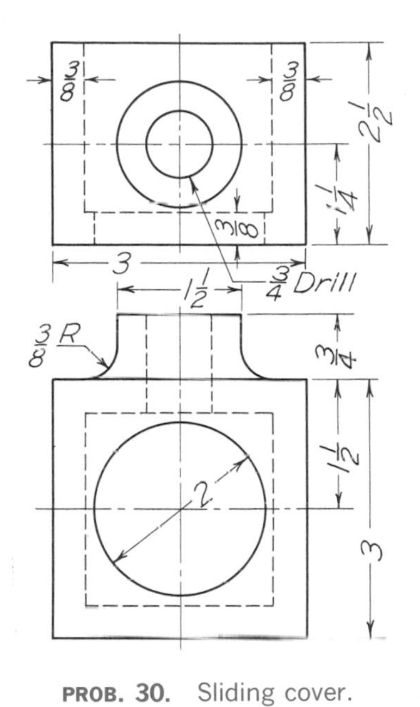


**PROB. 30.** Sliding cover.

**PROB. 31.** Rod support.

**PROB. 32.** Side-beam bracket.

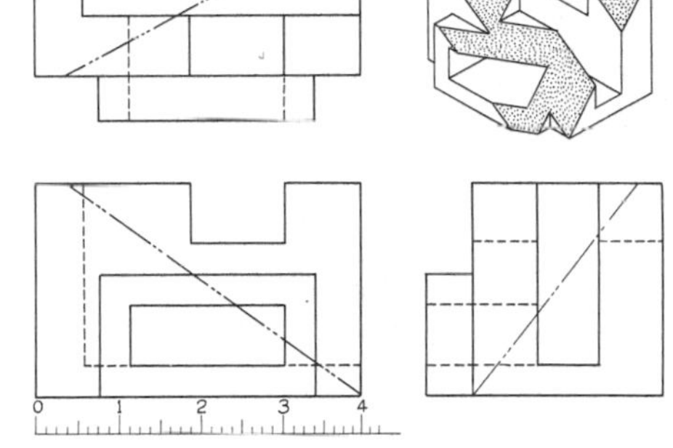

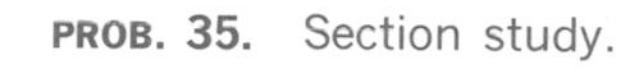


**PROB. 35.** Section study.

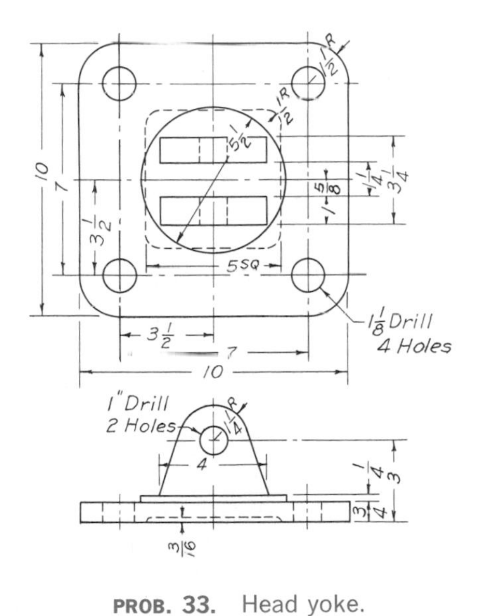


**PROB. 33.** Head yoke.

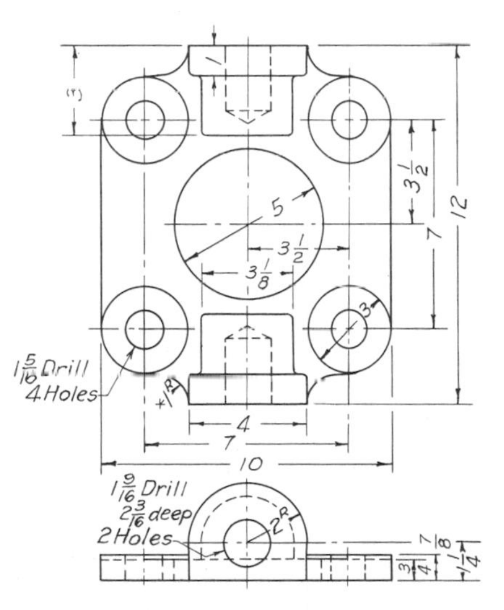


**PROB. 34.** Trunnion plate.

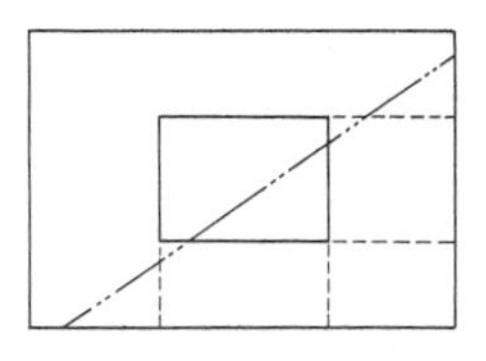

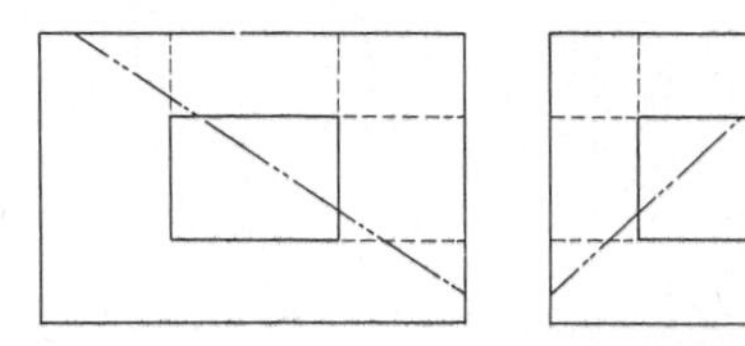

**PROB. 36.** Section study.

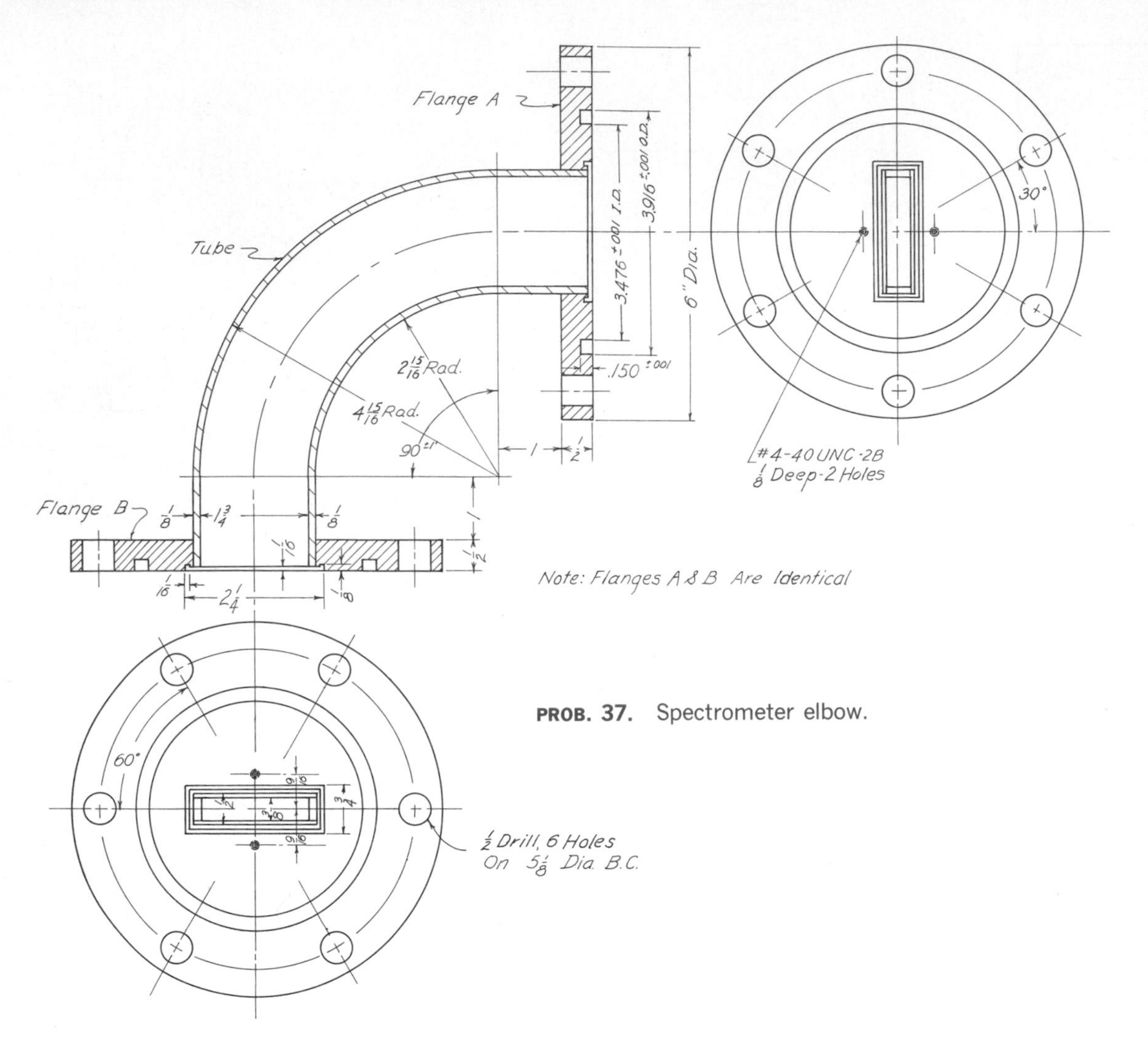


**PROB. 37.** Spectrometer elbow.

## Group 3. Oblique Drawings.

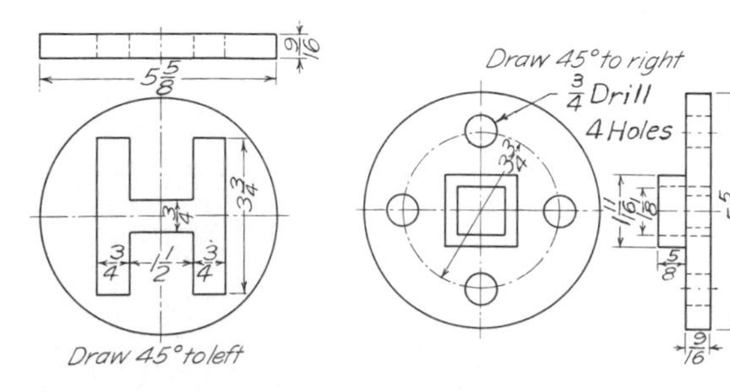


**PROB. 38.** Letter die.

**PROB. 39.** Guide plate.

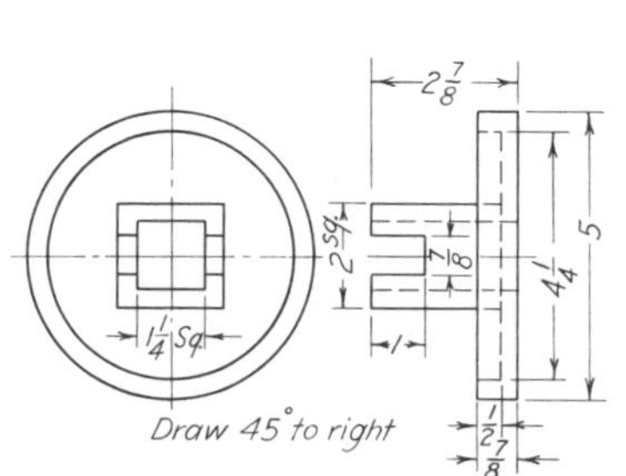


**PROB. 40.** Brace base.

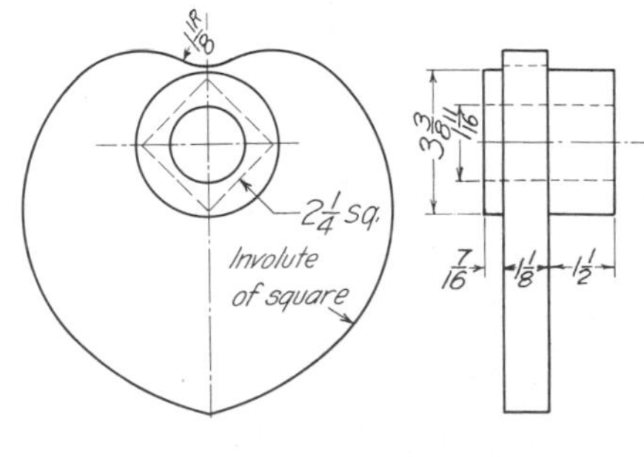


**PROB. 41.** Heart cam.

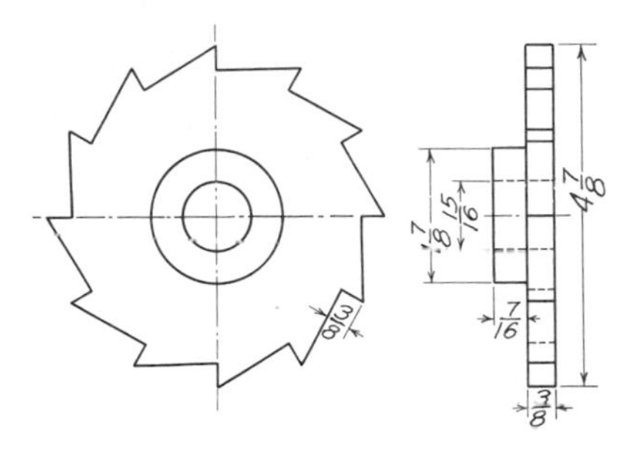


**PROB. 42.** Ratchet wheel.

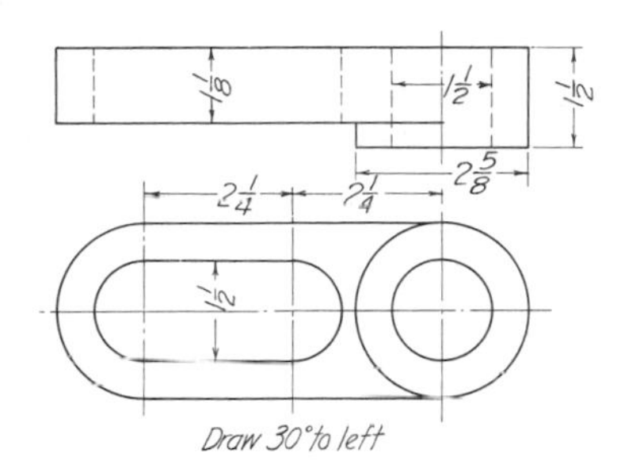


**PROB. 43.** Slotted link.

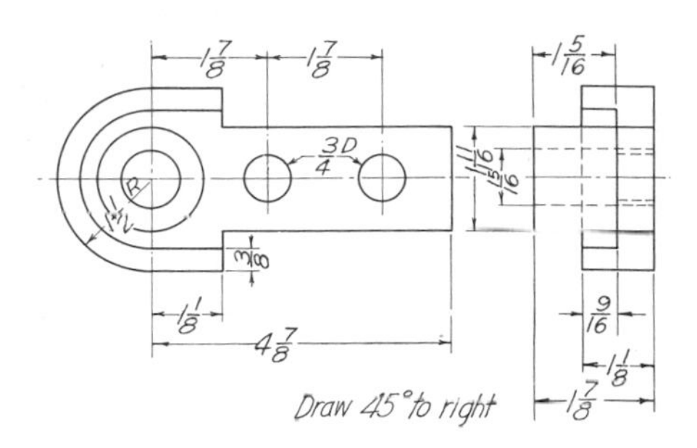


**PROB. 44.** Swivel plate.

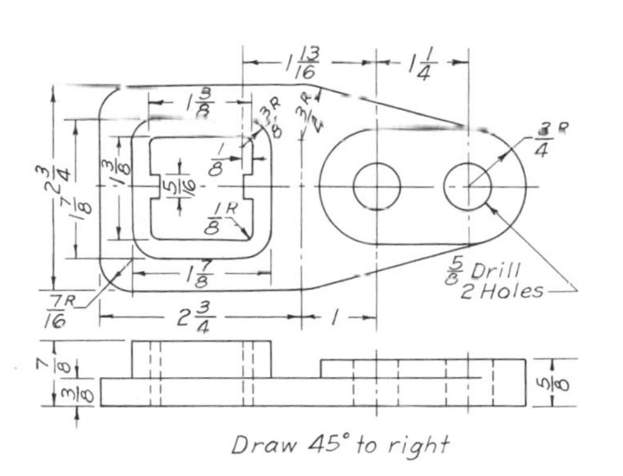


**PROB. 45.** Slide bracket.

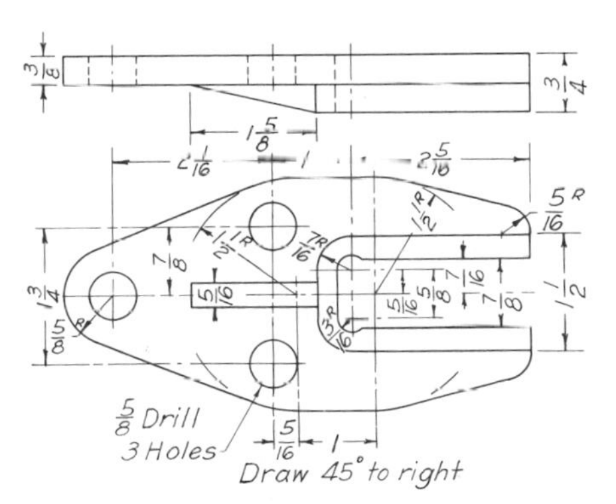


**PROB. 46.** Jaw bracket.

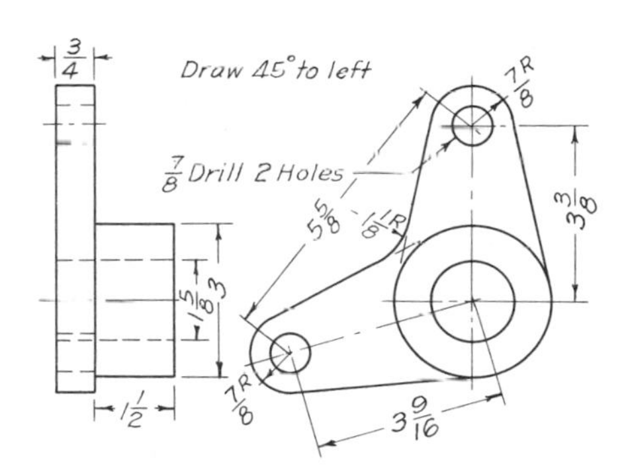


**PROB. 47.** Bell crank.

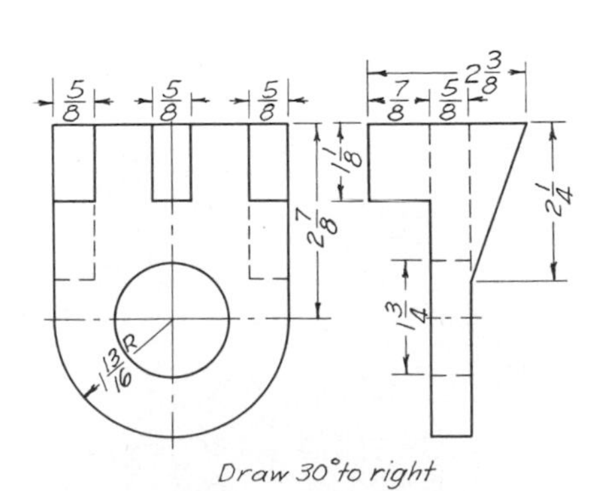


**PROB. 48.** Stop plate.

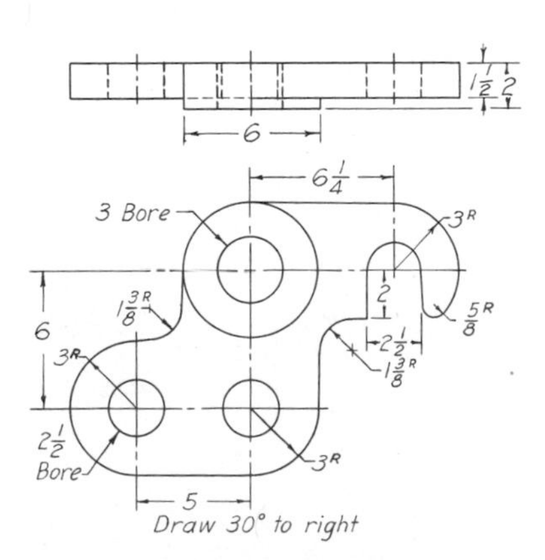


**PROB. 49.** Hook brace.

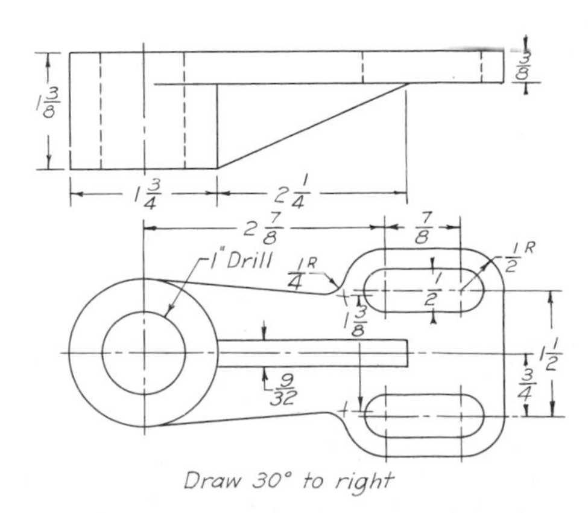


**PROB. 50.** Adjusting-rod support.

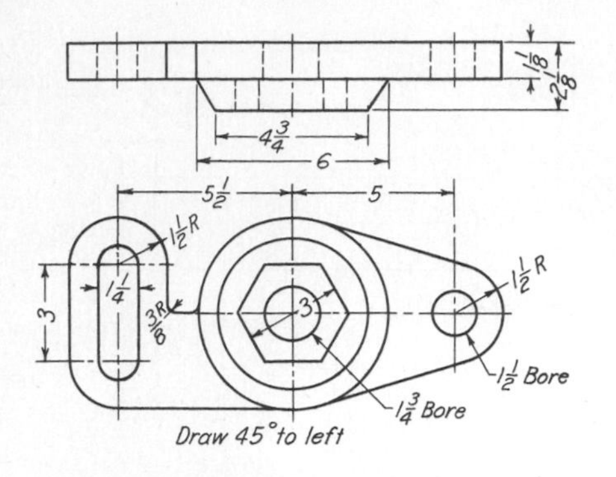

PROB. 51. Link.

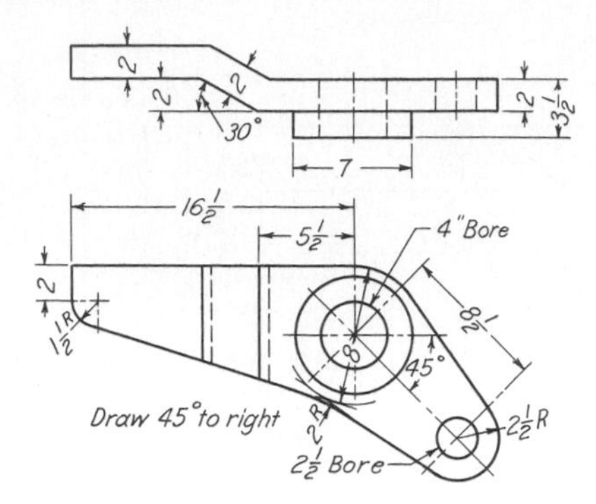

PROB. 52. Pawl.

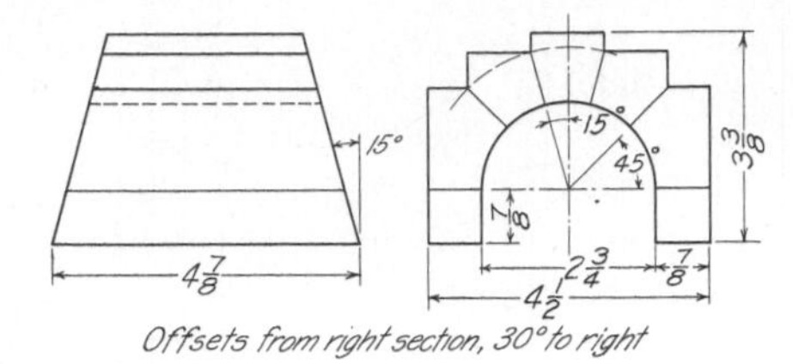

PROB. 53. Culvert model.

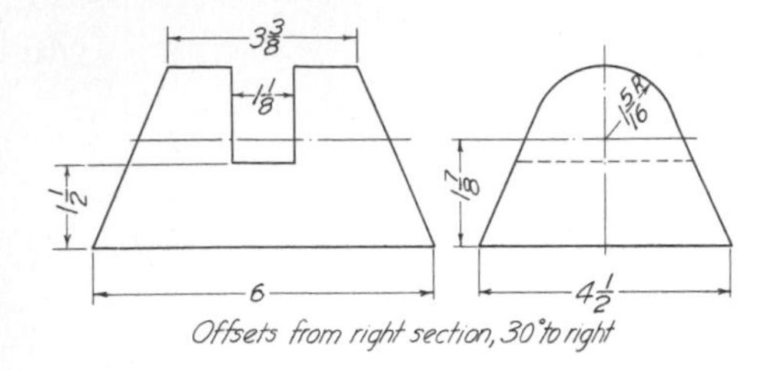

PROB. 54. Slotted guide.

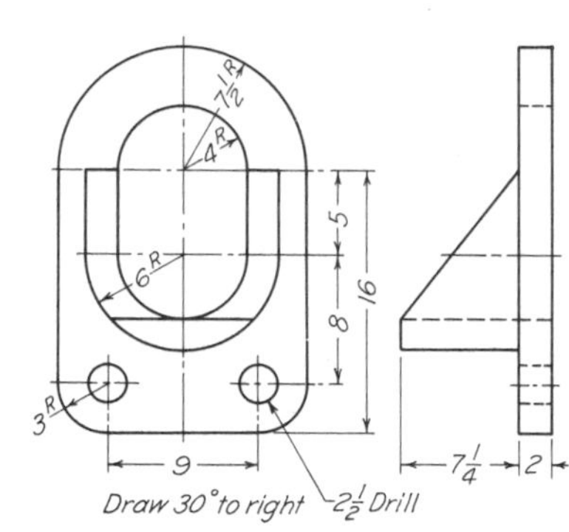

PROB. 55. Support bracket.

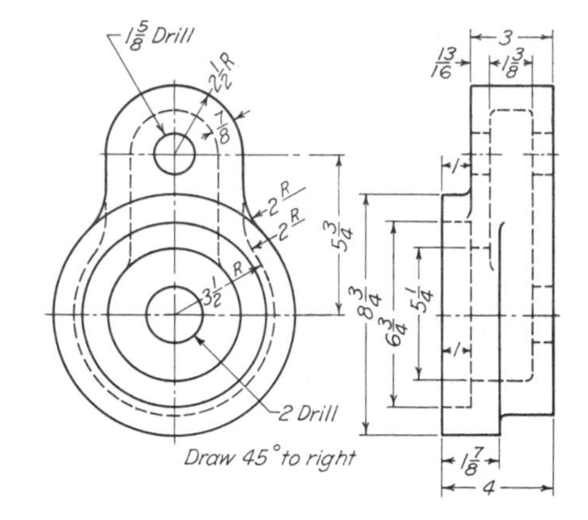

PROB. 56. Port cover.

## Group 4. Oblique Sections.

**57.** Oblique full section of sliding cone.
**57A.** Oblique half section of sliding cone.
**58.** Oblique full section of conveyor-trough end.
**58A.** Oblique half section of conveyor-trough end.
**59.** Oblique full section of ceiling flange.
**59A.** Oblique half section of ceiling flange.

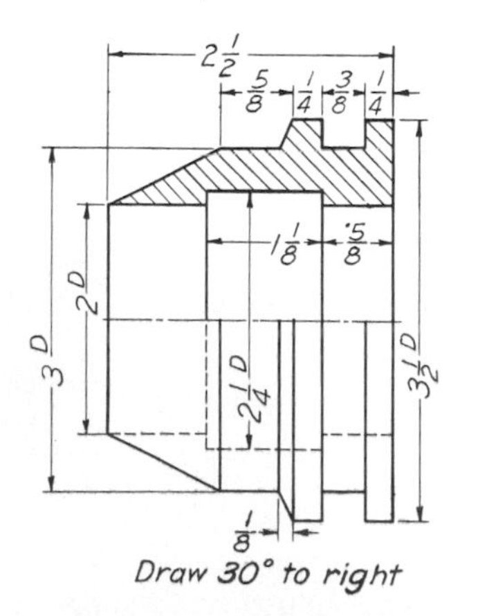

PROB. 57. Sliding cone.

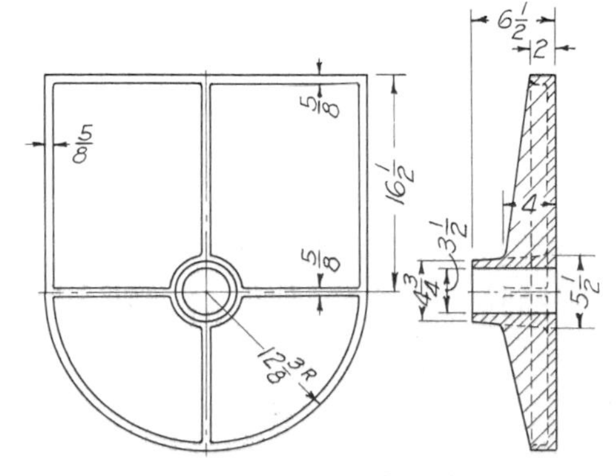

PROB. 58. Conveyor-trough end.

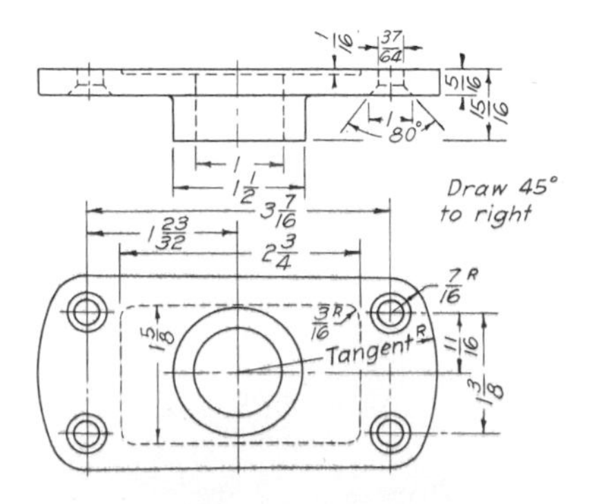

PROB. 59. Ceiling flange.

**60.** Oblique full section of hanger flange.
**60A.** Oblique half section of hanger flange.
**61.** Oblique half section of the base plate of Prob. 26.
**62.** Oblique half section of the gland of Prob. 27.
**63.** Oblique half section of regulator level.
**64.** Oblique full section of anchor plate.

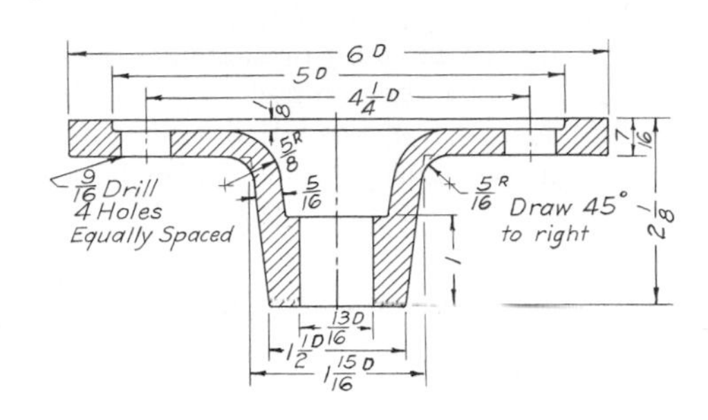

**PROB. 60.** Hanger flange.

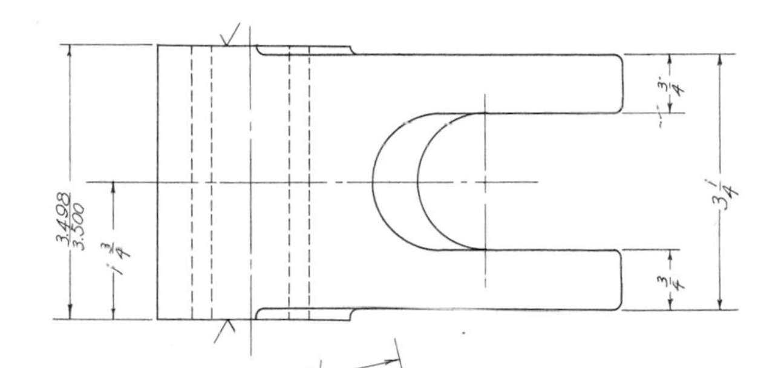

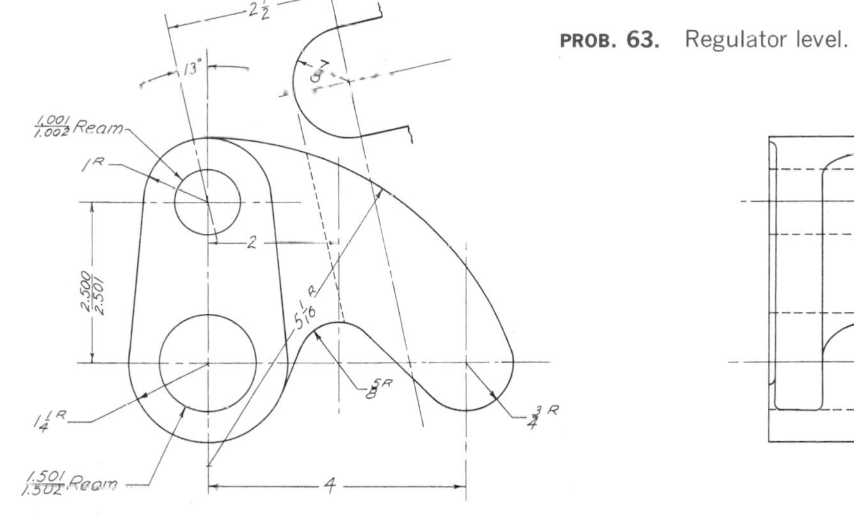

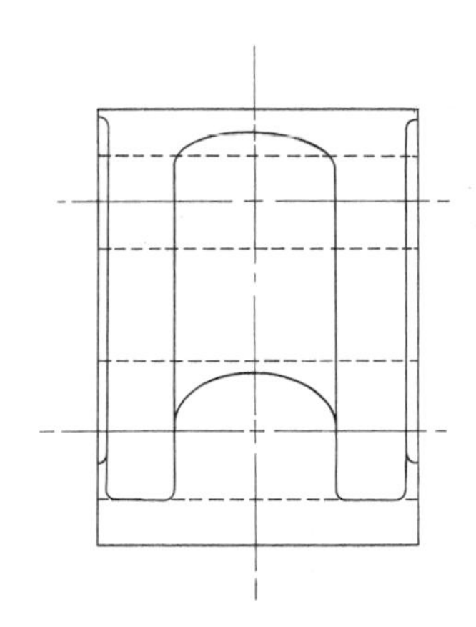

**PROB. 63.** Regulator level.

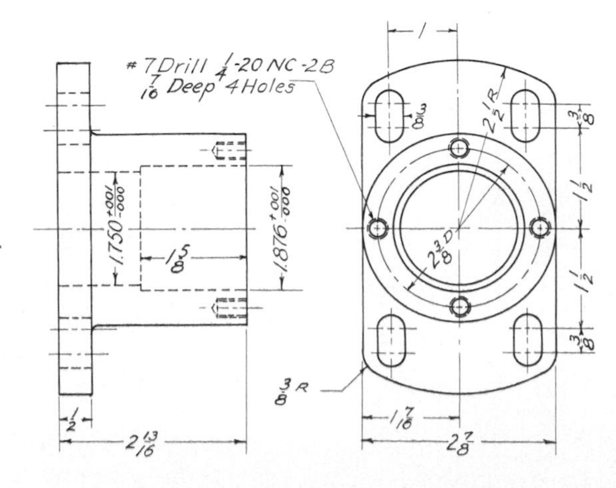

**PROB. 64.** Anchor plate.

**65.** Oblique drawing of shaft guide. Use sections, as needed, to describe the part.

**66.** Oblique half section of rod support. Partial sections and phantom lines can also be used to describe the part.

## Group 5. Dimetric and Cabinet Drawing.

**67.** Make a dimetric drawing of the jig block, Prob. 1.

**68.** Make a dimetric drawing of the guide block, Prob. 6.

**69.** Make a cabinet drawing of the gland, Prob. 27.

**70.** Make a cabinet drawing of the ceiling flange as shown in Prob. 59.

## Group 6. Perspective Drawings.

The following are a variety of different objects to be drawn in perspective. A further selection can be made from the orthographic drawings in other chapters.

**71.** Double wedge block.
**72.** Notched holder.
**73.** Crank.
**74.** Corner lug.
**75.** House.
**76.** Church.

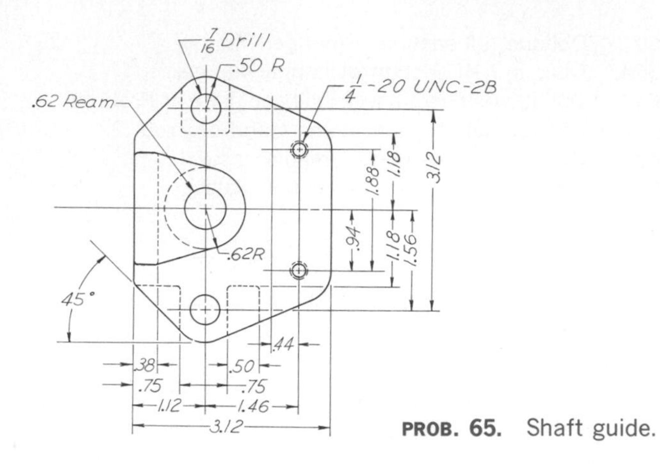


**PROB. 65.** Shaft guide.

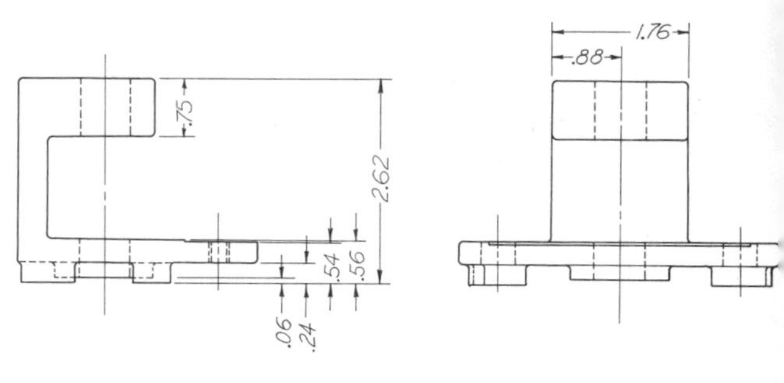


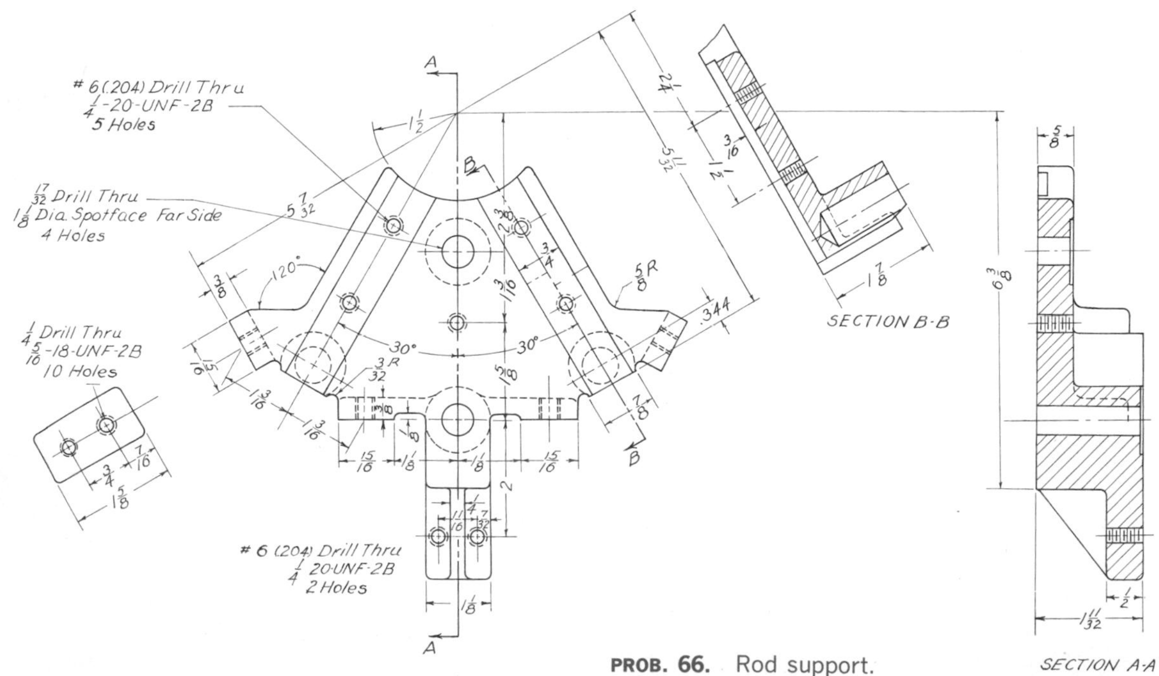


**PROB. 66.** Rod support.

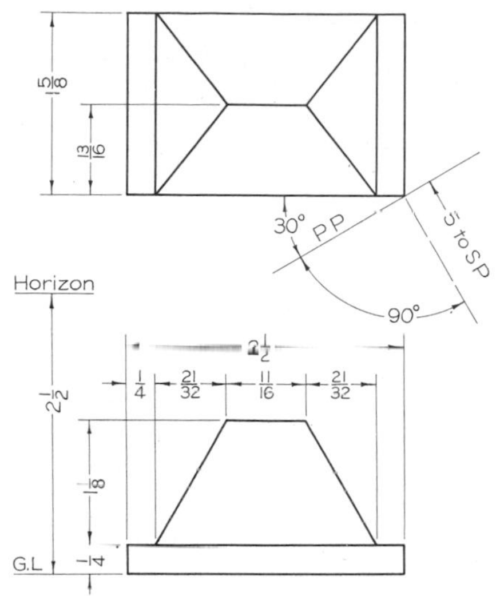


**PROB. 71.** Double wedge block.

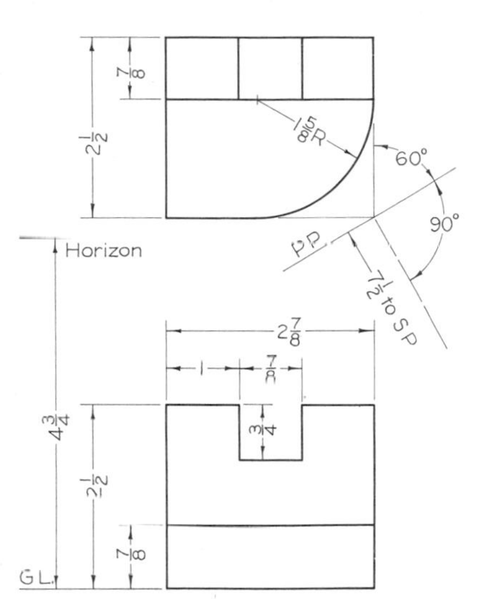


**PROB. 72.** Notched holder.

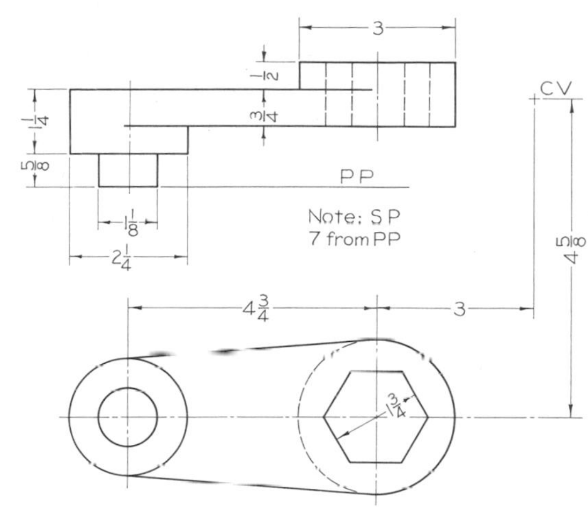


**PROB. 73.** Crank.

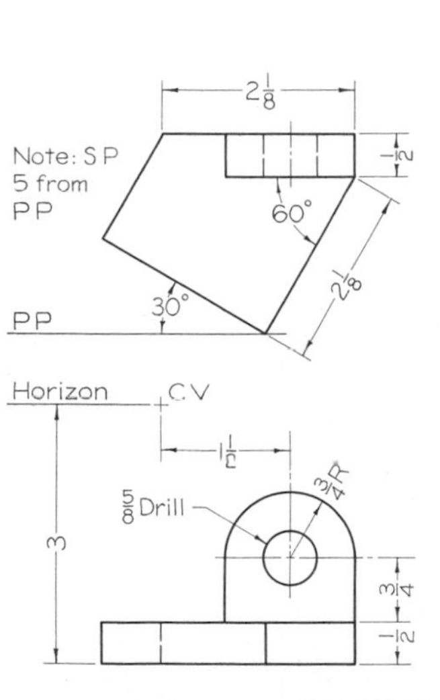


**PROB. 74.** Corner lug.

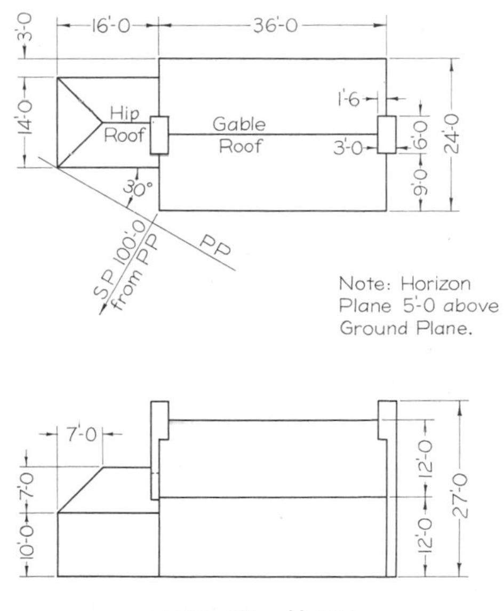


**PROB. 75.** House.

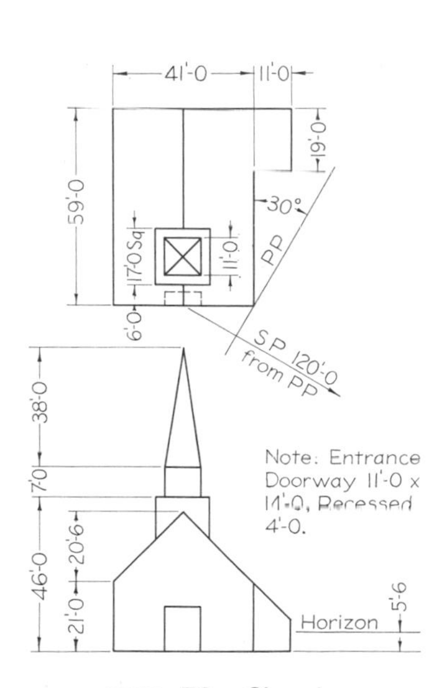


**PROB. 76.** Church.

## Group 7. Pictorial Sketching.

The following problems are planned to develop skill not only in pictorial sketching but also in reading orthographic drawings. Make the sketches to suitable size on $8\frac{1}{2}$- by 11-in. paper, choosing the most appropriate form of representation—axonometric, oblique, or perspective—with partial, full, or half sections as needed. Small fillets and rounds may be ignored in these problems and shown as sharp corners.

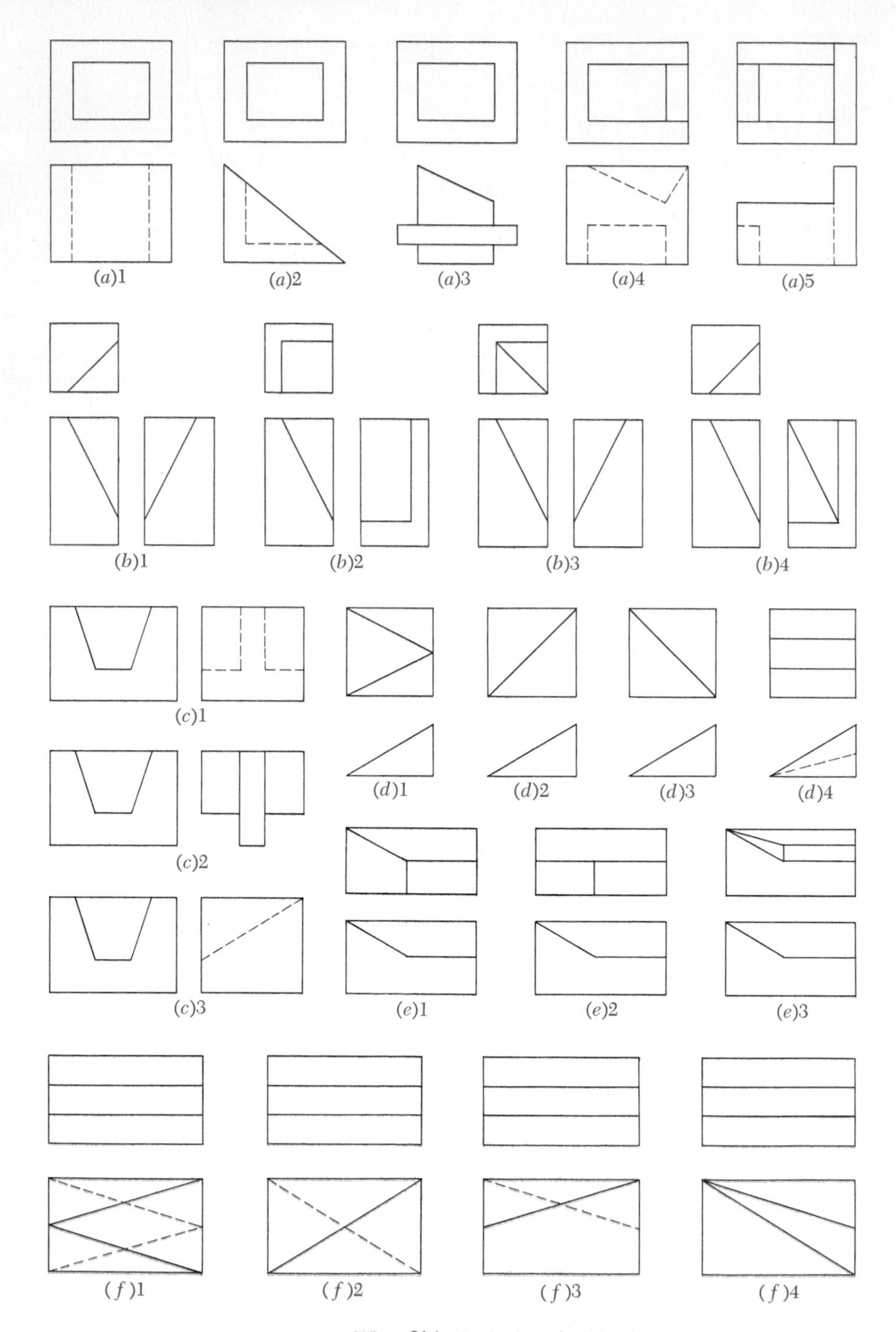


**PROB. 77.** Objects to be sketched.

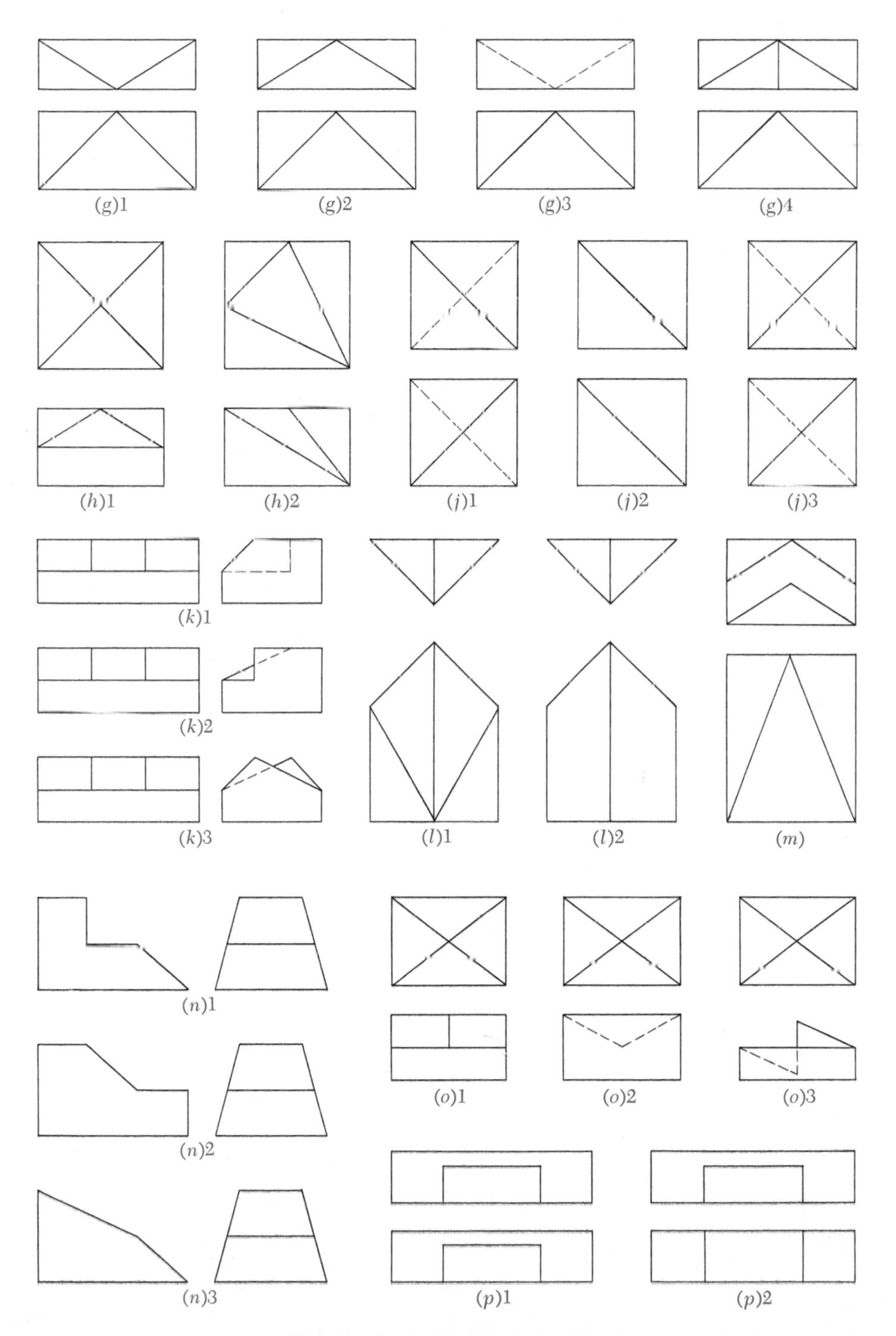


**PROB. 78.** Objects to be sketched.

**PROB. 79.** Objects to be sketched.

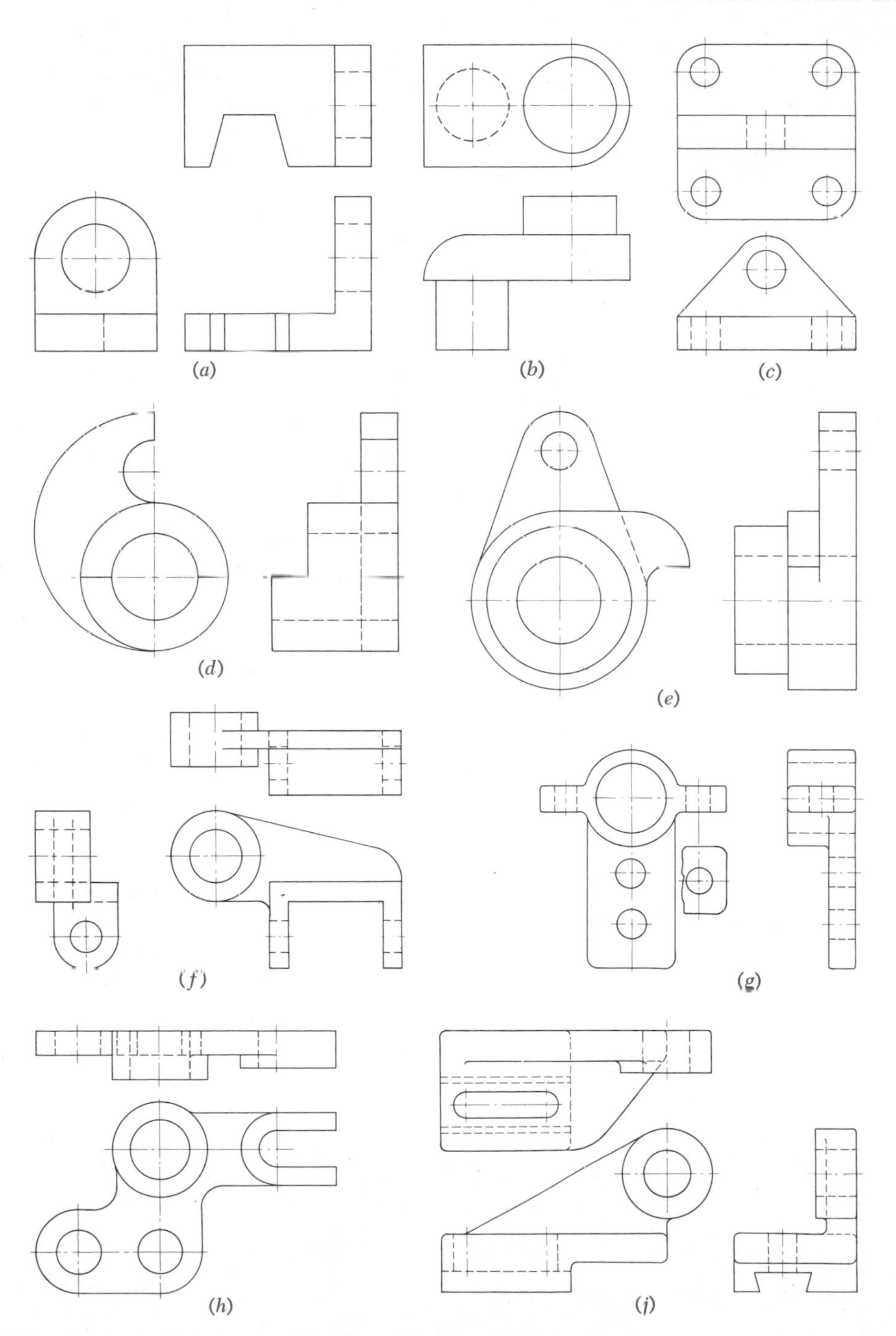


**PROB. 80.** Objects to be sketched.

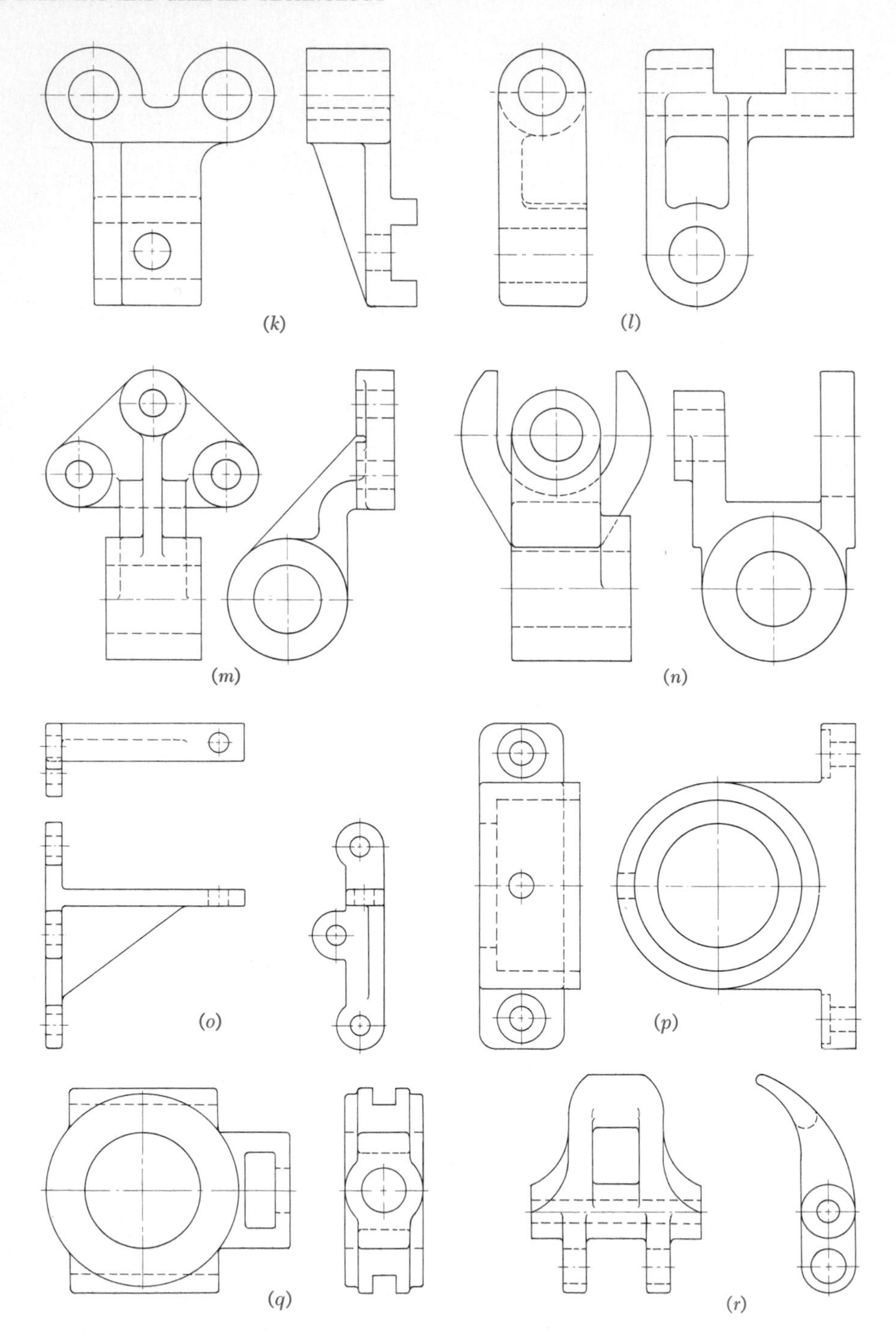


**PROB. 81.** Objects to be sketched.

### Group 8. Oblique Sketching.

Select an object from Group 3 not previously drawn with instruments, and make an oblique sketch.

### Group 9. Perspective Sketching.

Select an object not previously drawn with instruments, and make a perspective sketch.

### Group 10. Axonometric and Oblique Projection from Orthographic Views.

Any of the problems given in this chapter may be used for making axonometric or oblique projections from the orthographic views by first drawing the orthographic views to suitable scale and then employing these views as described in Secs. 26 to 28 to obtain the pictorial projection.

### Group 11. Pictorial Drawings from Machine Parts.

Machine parts, either rough castings and forgings or finished parts, offer valuable practice in making pictorial drawings. Choose pieces to give practice in isometric and oblique drawing. Use the most appropriate form of representation, and employ section and half-section treatments where necessary to give clearer description.

### Group 12. Pictorial Working Drawings.

Any of the problems in this chapter offer practice in making complete pictorial working drawings. Follow the principles of dimensioning in Chap. 11. The form and placement of the dimension figures are given in Sec. 37, Chap. 11.

**82.** Pictorial working drawing of dovetail stop, Prob. 7.
**83.** Pictorial working drawing of hinged catch, Prob. 9.
**84.** Pictorial working drawing of tie plate, Prob. 22.
**85.** Pictorial working drawing of head yoke, Prob. 33.
**86.** Pictorial working drawing of jaw bracket, Prob. 46.
**87.** Pictorial working drawing of adjusting-rod support, Prob. 50.
**88.** Pictorial working drawing of port cover, Prob. 56.
**89.** Pictorial working drawing of ceiling flange, Prob. 59.
**90.** Make a pictorial working drawing of the extension block. Use any reasonable technique such as exploding the part (with note to explain), partial sections, extra views, or phantom lines to completely describe the part.

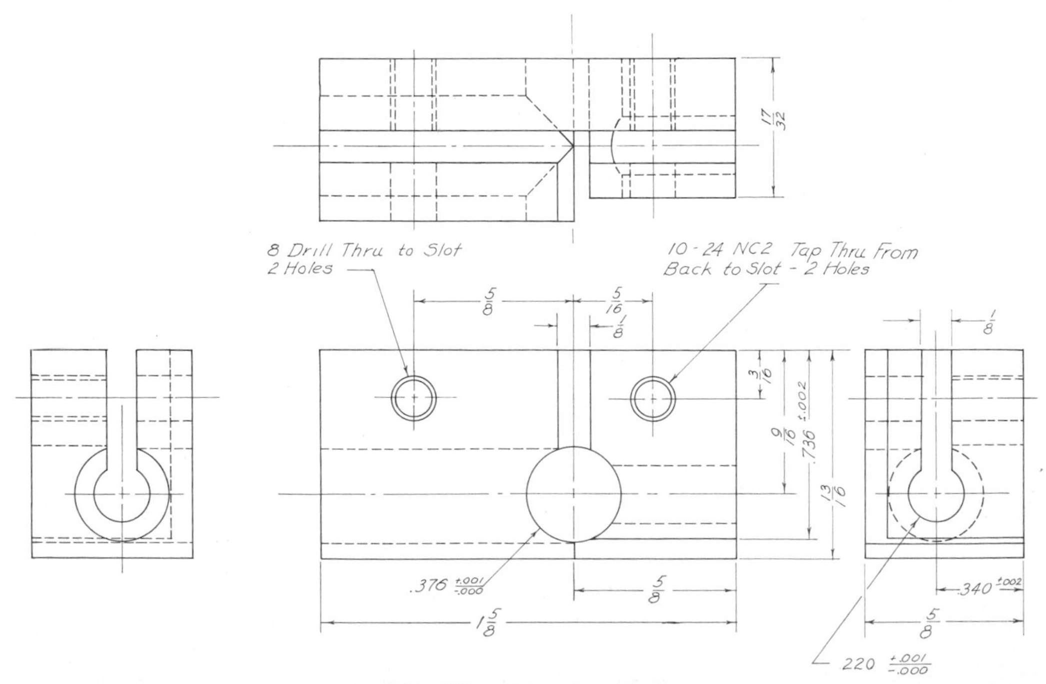


**PROB. 90.** Extension block.

**91.** Make a pictorial working drawing of the double sleeve clamp. Show the two halves as an exploded assembly.

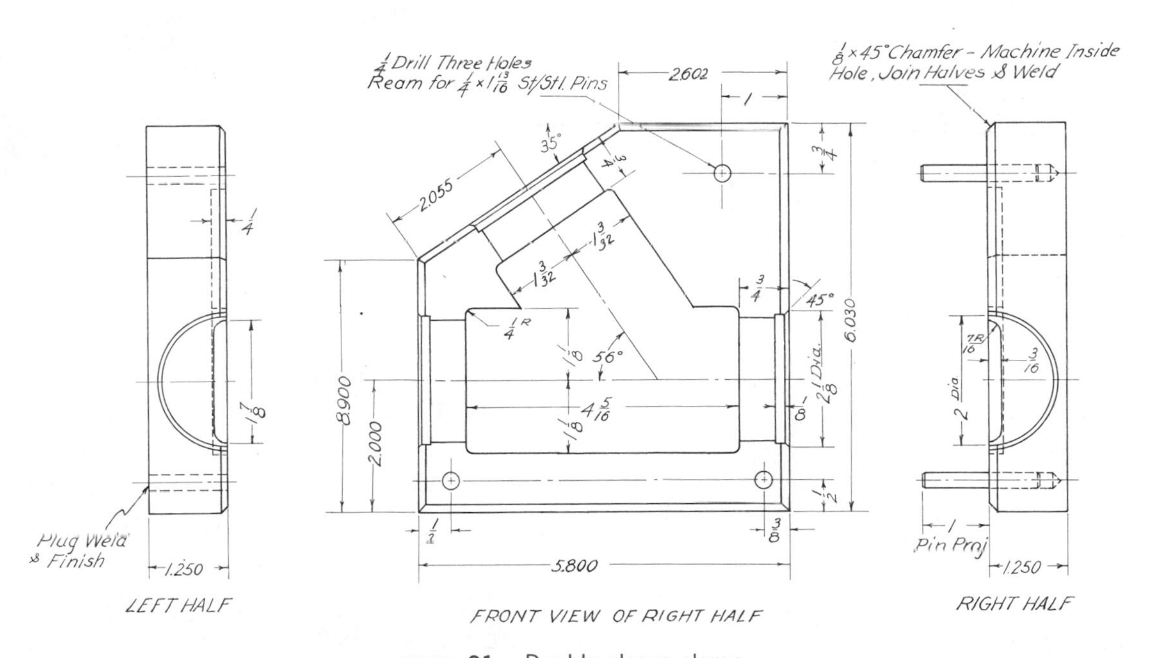


**PROB. 91.** Double sleeve clamp.

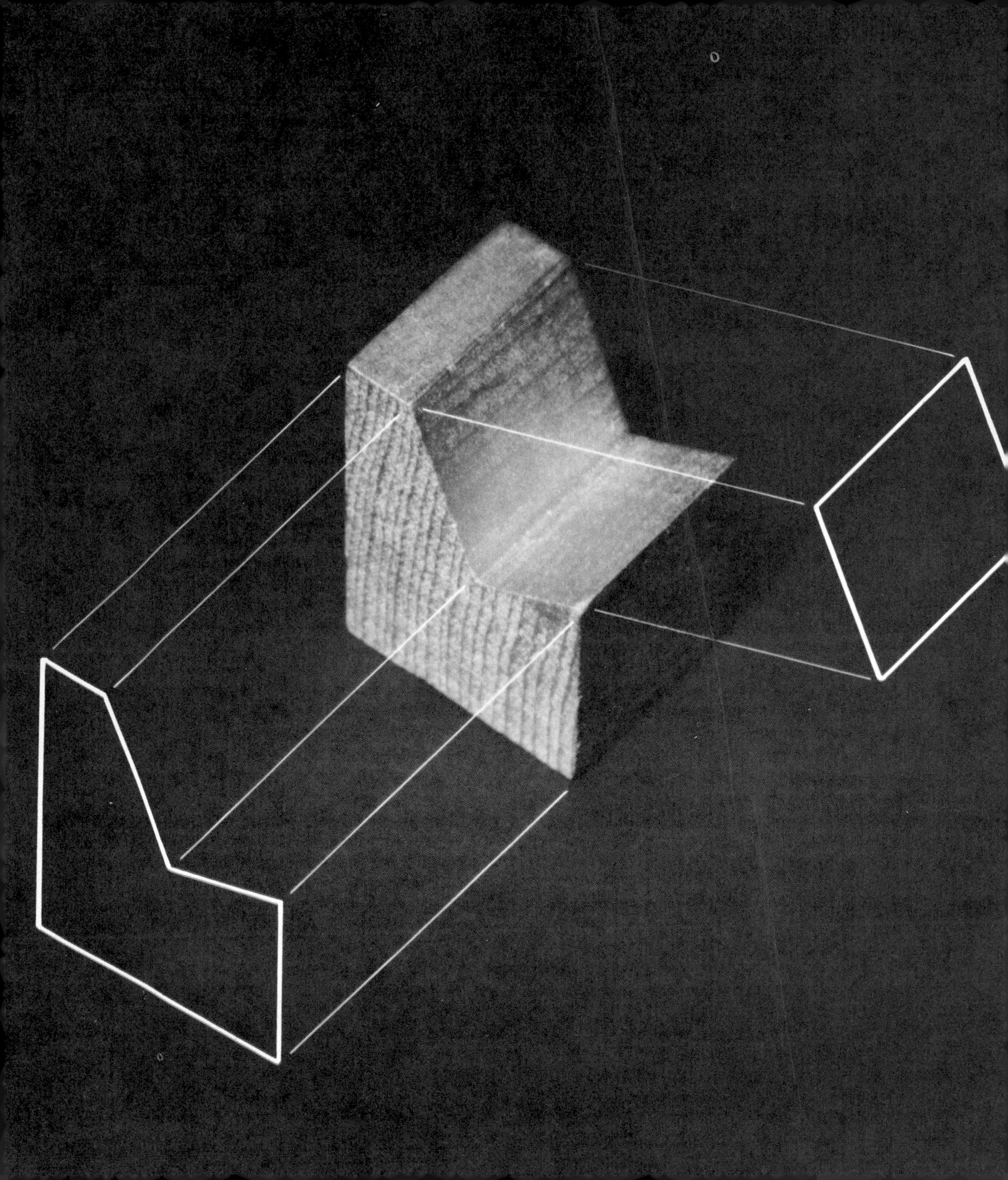

Items Given in this Chapter:

# 7 AUXILIARY VIEWS

Auxiliary Views Are Special Projections Used to Clarify and Complete Orthographic Descriptions of Shape

**1. Auxiliary Views–Basic Concepts.** A plane surface is shown in true shape when the direction of view is perpendicular to the surface; for example, rectangular objects can be placed with their faces parallel to the principal planes of projection and be fully described by the principal views. In Fig. 1 the top, front, and right-side views show the true shape, respectively, of the top, front, and right side of the object. Note especially that the planes of projection are *parallel* to the top, front, and right side of the object and that the directions of observation are *perpendicular* to the object faces and to the planes of projection. Figure 2 is a pictorial of Fig. 1, given here to aid in visualizing the relationship of object faces, planes of projection, and view directions.

Note in Figs. 1 and 2 that each view also shows the *edge* of certain surfaces of the object. For example, the front view shows the edge of the top, bottom, and both sides of the object. For a surface to appear as an edge it must be perpendicular to the plane of projection for the view.

Sometimes an object will have one or more *inclined* surfaces whose true shape it is necessary to show, especially if they are irregular in outline. Figure 3 shows an object with one inclined face, *ABDC*. This face is inclined to the horizontal and profile planes and perpendicular to the frontal plane. The face *ABDC* therefore appears as an edge in the front view, but none of the principal views shows the true size and shape of the surface. To show the true size and shape of *ABDC*, a view is needed that has a direction of observation perpendicular to *ABDC* and is projected on a plane parallel to *ABDC*, as shown in Fig. 4. This view is known as a *normal* view. The top, front, and side views of Fig. 1 are normal views of the top, front, and side surfaces of the object, because they *all* show the true size and shape of a surface by having the direction of observation at right angles to the surface. The dictionary defines a *normal*, in geometry, as "any perpendicular." In graphics, a plane of projection is involved as well as a direction of observation; hence, *a normal view is a projection that has the viewing direction perpendicular to, and made on a plane parallel to, the object face.*

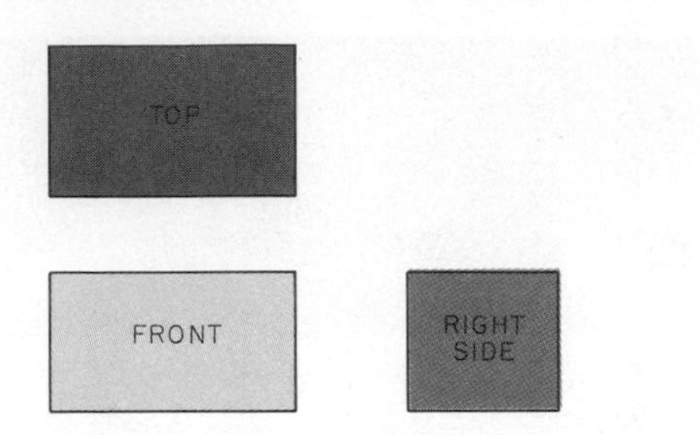


**FIG. 1.** Object faces parallel to the principal projection planes. Top, front, and side views are normal views, respectively, of top, front, and side of the object.

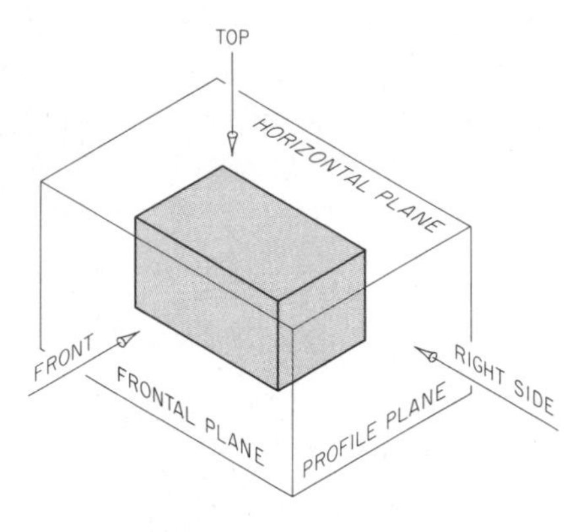


**FIG. 2.** Pictorial of Fig. 1. This is given to aid in visualizing the relationship of principal planes, view directions, and object position.

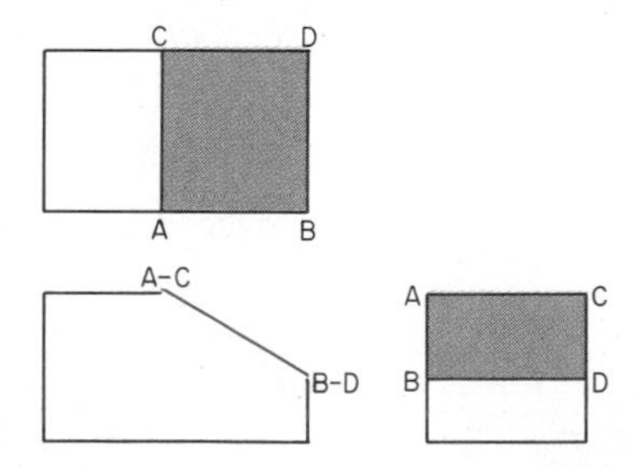


**FIG. 3.** One object face inclined to two principal planes. The inclined face does not appear in its true size and proportions in any view.

Figure 5 is the orthographic counterpart of Fig. 4. To get the normal view of surface *ABDC*, a projec-

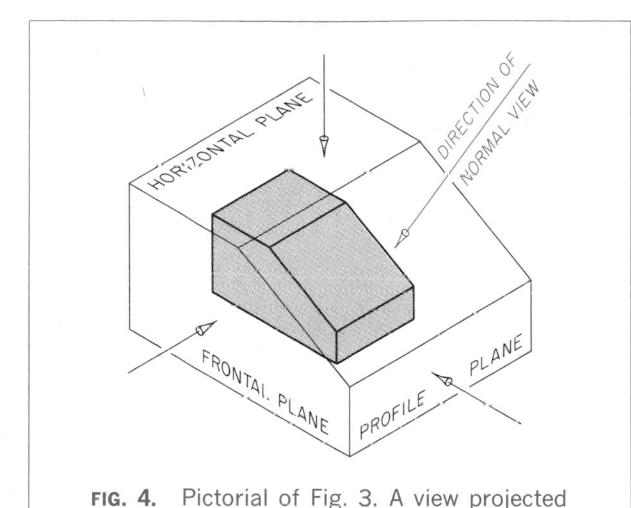


**FIG. 4.** Pictorial of Fig. 3. A view projected perpendicular to the inclined face will show it in true size and proportions.

tion is made perpendicular to $ABDC$. This projection is made *from* the view where the surface shows as an edge, in this case, the front view, and thus perpendicularity from the surface is seen in true relationship. Extra views such as the normal view of Fig. 5 are known as *auxiliary views* to distinguish them from the principal (top, front, side, etc.) views. However, since an auxiliary is made for the purpose of showing the true configuration of a surface, the terms *normal view* or *edge view* state positively what the view is and what it shows.

Detailed instructions for making a normal view of any surface in *any possible position* are given in the sections that follow. Edge views are discussed in Sec. 9.

**2. Classification of Surfaces.**
Surfaces may occur in any of the positions shown in Fig. 6. At (*a*) all the surfaces are aligned with the principal planes of projection, and thus each of the principal views is a normal view. At (*b*) the shaded surface is at an angle to *two* of the principal planes but perpendicular to one plane and is called an *inclined* surface. At (*c*) the shaded surface is at an angle to *all three* principal planes of projection and is known as a *skew*, or *oblique*, surface (that is, one that takes a slanting or oblique course or direction).

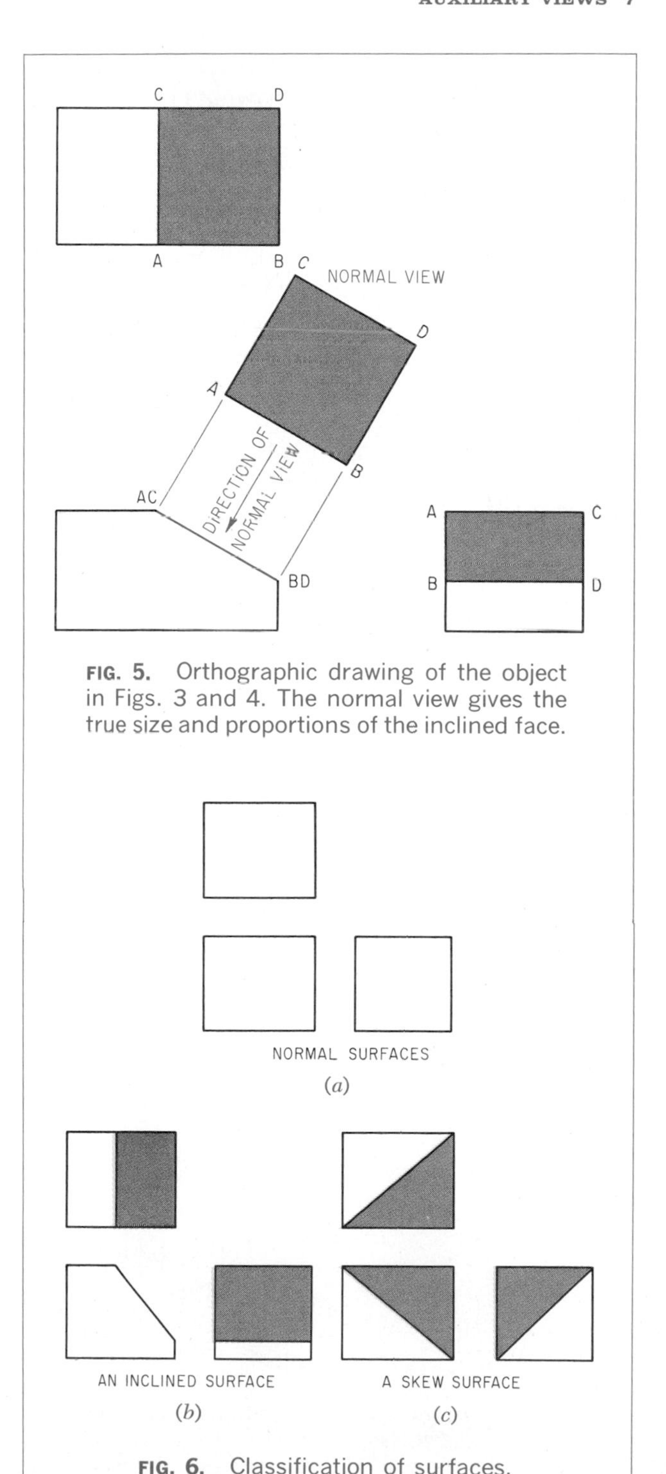


**FIG. 5.** Orthographic drawing of the object in Figs. 3 and 4. The normal view gives the true size and proportions of the inclined face.



**FIG. 6.** Classification of surfaces.

### 3. Directions of Inclined Surfaces.

Inclined surfaces may occur anywhere on an object, and because other features of the object must also be represented, the inclined surface may be in any one of the twelve positions shown in Fig. 7. The first column, (*b*) to (*e*), shows surfaces inclined to the front and side so that the surface may be on the (*b*) right front, (*c*) left front, (*d*) left rear, or (*e*) right rear. The second column, (*f*) to (*j*), shows surfaces inclined to the top and side, so that the surface may be on the (*f*) upper right, (*g*) lower right, (*h*) lower left, or (*j*) upper left. The third column, (*k*) to (*n*), shows surfaces inclined to the top and front, so that the surface may be on the (*k*) upper front, (*l*) upper rear, (*m*) lower rear, or (*n*) lower front.

Inclined surfaces occur at *angles* of inclination differing from those shown in Fig. 7 but, for general position, no other locations are possible.

### 4. The Normal View of an Inclined Surface.

No matter what the position of an inclined surface may be, the fundamentals of projecting a normal view of the surface are the same, as will be seen from Figs. 8 to 21.

Figure 8 is the orthographic drawing of the object in position (*b*) of Fig. 7. The inclined surface, identified by the letter *A*, is on the right front of the object and shows, in Fig. 8, as an *edge* in the top view. Because surface *A* appears as an edge in the top view, the direction of observation for the normal view is established perpendicular to the edge view, as shown. We are looking, in this case, in a horizontal direction, *directly at* surface *A*. Projectors parallel to the viewing direction (also, perpendicular to surface *A*) establish one dimension of the surface needed for the normal view. This is the *horizontal*

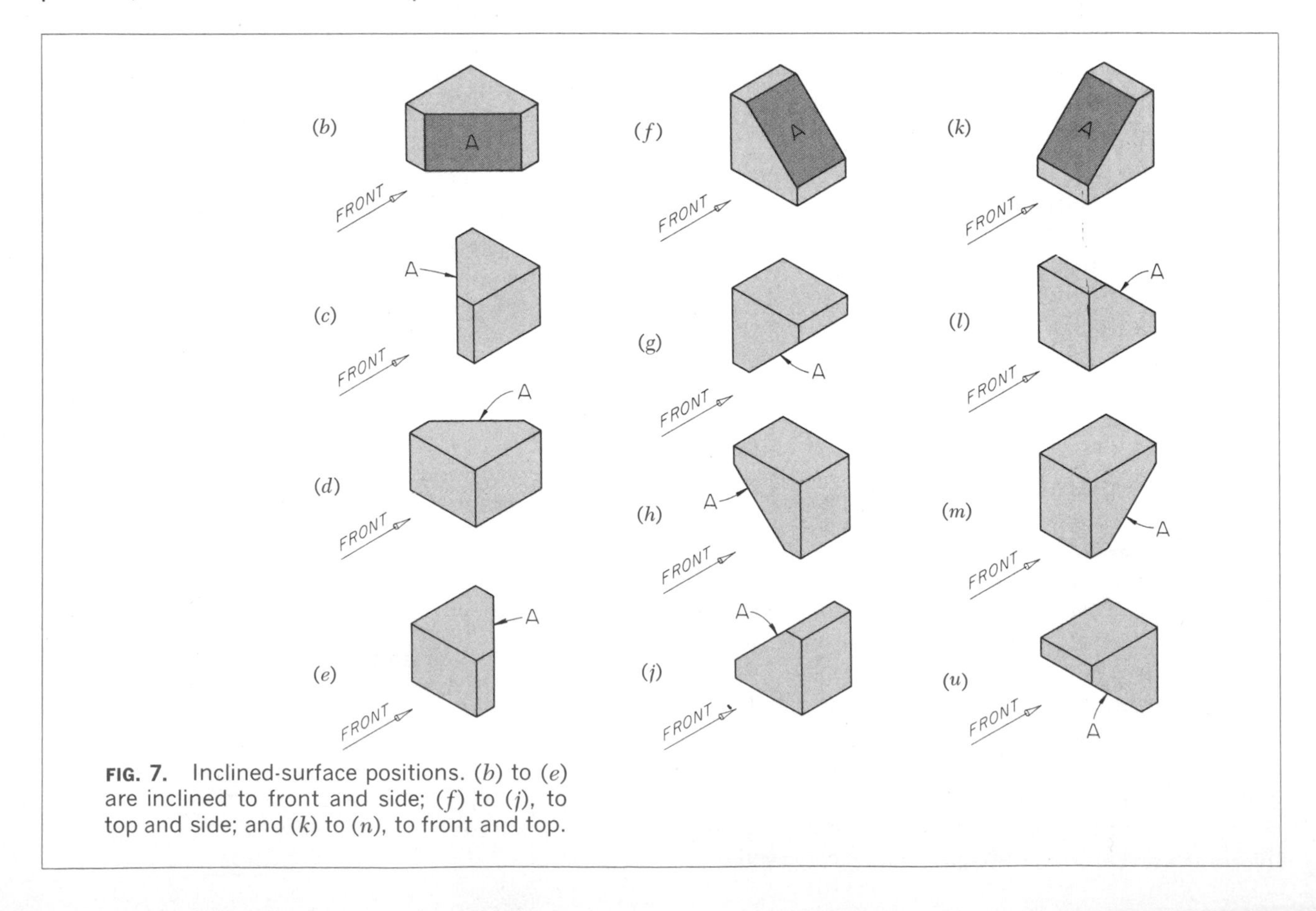


**FIG. 7.** Inclined-surface positions. (*b*) to (*e*) are inclined to front and side; (*f*) to (*j*), to top and side; and (*k*) to (*n*), to front and top.

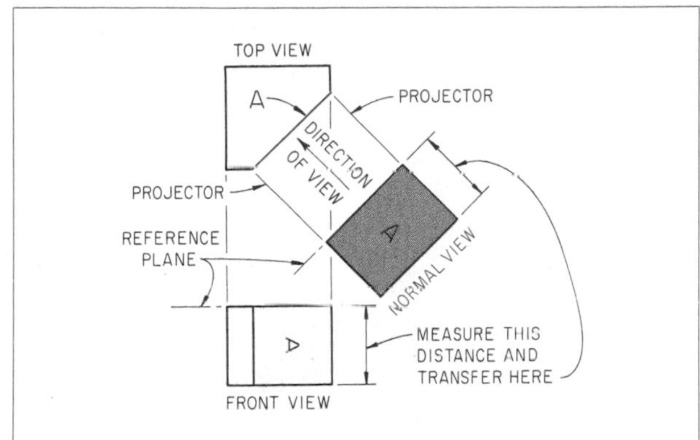


**FIG. 8.** Normal view of an inclined surface. The object is in position (*b*) of Fig. 7.

view, projected directly out from its original coincidence on the object, occurs in the position first shown in Fig. 8. To get the view into coincidence with the plane of the drawing paper, the view is rotated to the position shown in Fig. 10. Carefully study Figs. 8 to 10 and reread the text to fix all relationships.

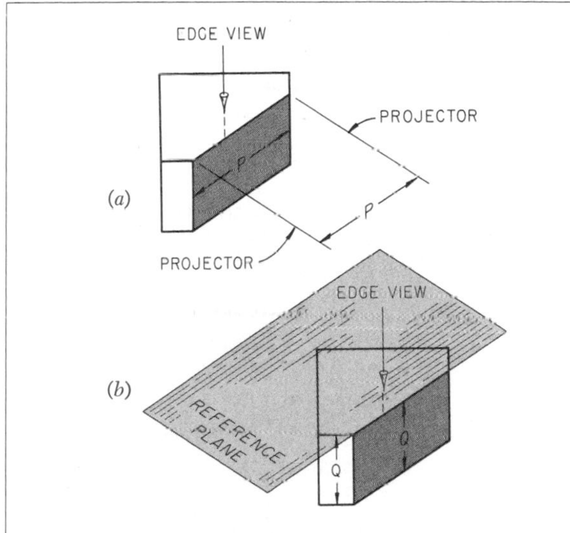


**FIG. 9.** Relationship of projectors, reference plane, and surface dimensions.

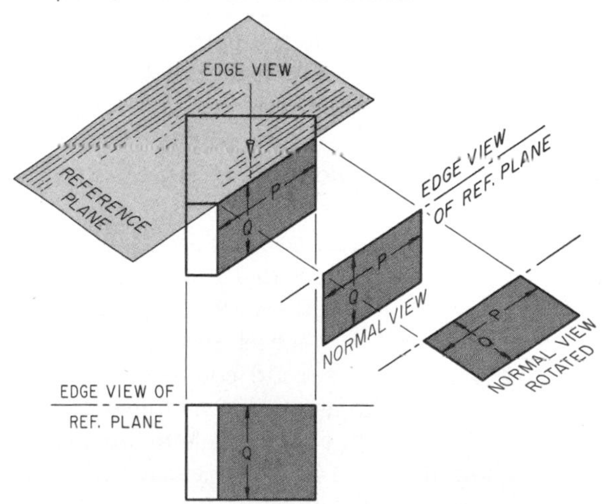


**FIG. 10.** Relationship of projectors, reference plane, surface dimensions, normal view, and rotated position of normal view.

distance from the left-front vertical edge to the right-rear vertical edge of surface $A$. Figure 9*a* shows that the distance *between* the projectors (dimension $P$) is equal to dimension $P$ on the object.

All orthographic views have two dimensions, and to complete the normal view of surface $A$, we now need its second dimension. That dimension is, in this case, the vertical distance $Q$, shown at (*b*), which is the vertical height of surface $A$ and appears in true length *on the front view* (Fig. 8). This distance cannot be projected but must be transferred from the front view to the normal view as shown in Fig. 8. In order to facilitate transferring the distance (this will apply particularly in later problems where the surface may not be rectangular), a *reference plane* (Figs. 9*b* and 10) is established on both the normal view and front view. This reference plane *must* be perpendicular to the distance to be measured and transferred; it is therefore perpendicular to the projectors between the top and front views *and in the front view,* and also perpendicular to the projectors from the top view to the normal view *and in* the normal view, as shown in Fig. 8. The reference plane, here, is at the *top edge* of surface $A$ in both views. This can be visualized in the normal view by considering the reference plane as a hinge that will rotate the normal view 90° downward to its original position in coincidence with the top view. This is shown pictorially in Fig. 10 to aid in visualizing the relationship. It must be understood that the normal

The foregoing constitutes all the geometry and projection needed for drawing *any* normal view of an inclined surface. The projection will in some cases be made from a different view and the measurement transferred will be width or depth instead of height, but the fundamental relationships of the views will *be the same in every case.*

Consider the other positions of the object, (*c*) to (*e*), of Fig. 7. In every case the inclined surface is similarly oriented in space to that of Fig. 8, but turned to a different position. Figure 11 [position (*c*) of Fig. 7] has the inclined surface on the left front of the object. As in the first case, the direction of view must be perpendicular to the inclined face, and is seen in true relationship in the top view, where surface $A$ appears as an edge. The projectors for the normal view are parallel to the viewing direction and of course perpendicular to surface $A$. The reference plane is established in the normal view perpendicular to the projectors for the normal view and in the front view perpendicular to the projectors between top and front views. The measurement of height, transferred from the front view to the normal view, completes the normal view.

Note that the reference plane will *always* appear in the two views that are in projection *with* the view that shows the inclined surface as an edge, and that the reference plane is always perpendicular to the projectors between these pairs of views.

Figure 12 shows position (*d*) of Fig. 7. This time the inclined surface is at the left rear of the object. Figure 13 shows position (*e*) of Fig. 7. Read the explanation given for Figs. 8, 9, and 10 while studying these two figures (12 and 13) and note that the directions given for Figs. 8 to 10 apply equally well for Figs. 12 and 13. This shows that the actual position of the inclined surface is simply a variation.

Turning now to the second column of positions (*f*) to (*j*) of Fig. 7, note again that all these positions have the inclined surface at an angle to top and side. Then, looking at Figs. 14 to 17 inclusive (these are all similar because the inclined surface appears in every case as an edge on the front view), observe that Fig. 14 is position (*f*), Fig. 15 position (*g*), Fig. 16 position (*h*), and Fig. 17 position (*j*).

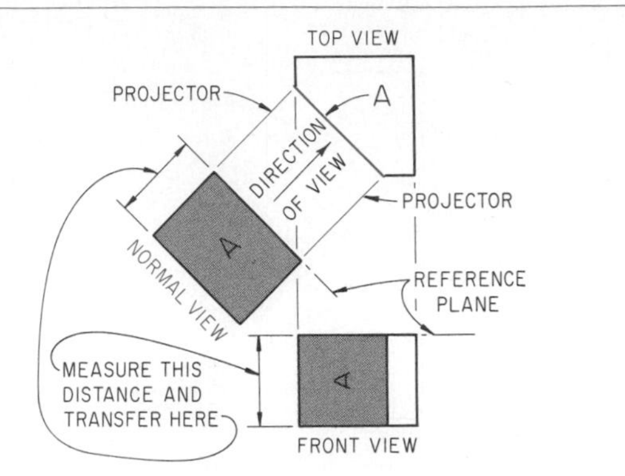


**FIG. 11.** Normal view of an inclined surface. The object is in position (*c*) of Fig. 7.

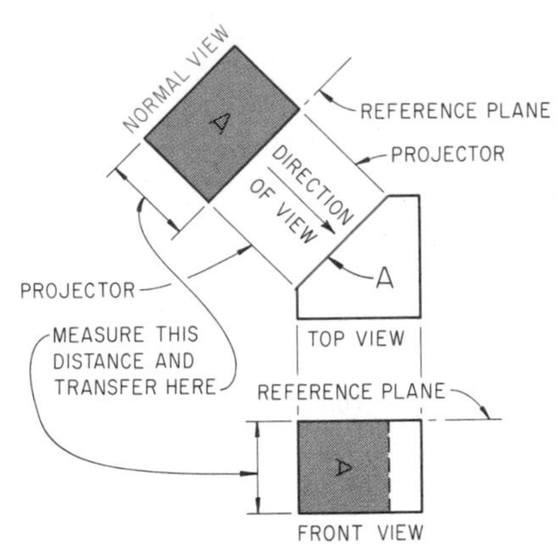


**FIG. 12.** Normal view of an inclined surface. The object is in position (*d*) of Fig. 7.

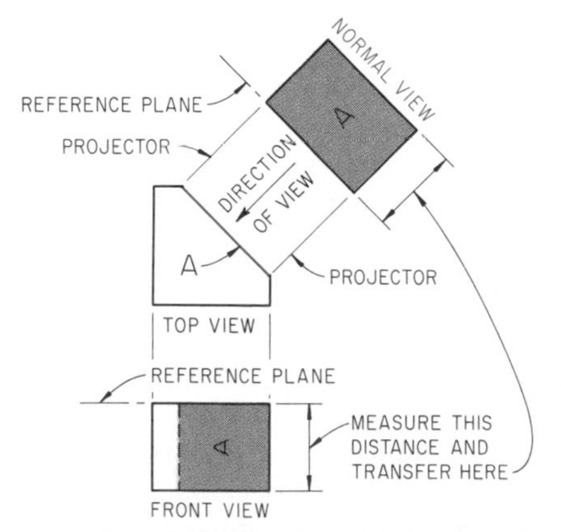


**FIG. 13.** Normal view of an inclined surface. The object is in position (*e*) of Fig. 7.

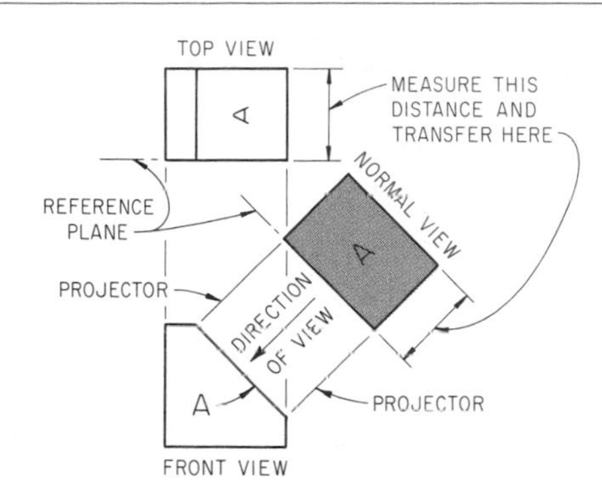


**FIG. 14.** Normal view of an inclined surface. The object is in position (*f*) of Fig. 7.

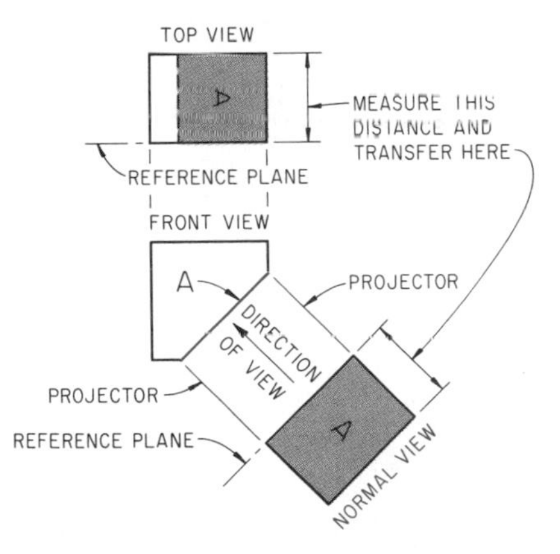


**FIG. 15.** Normal view of an inclined surface. The object is in position (*g*) of Fig. 7.

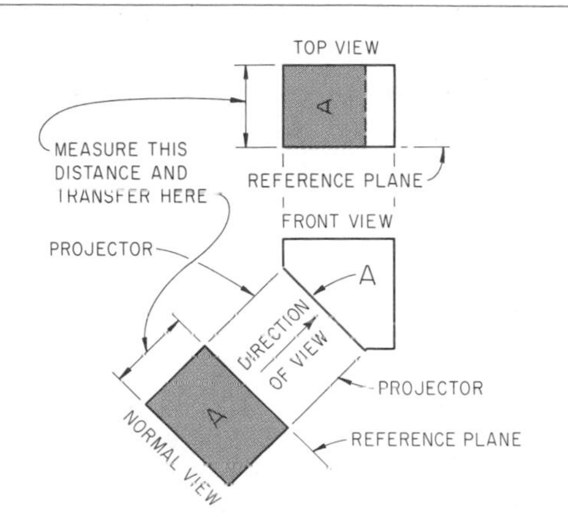


**FIG. 16.** Normal view of an inclined surface. The object is in position (*h*) of Fig. 7.

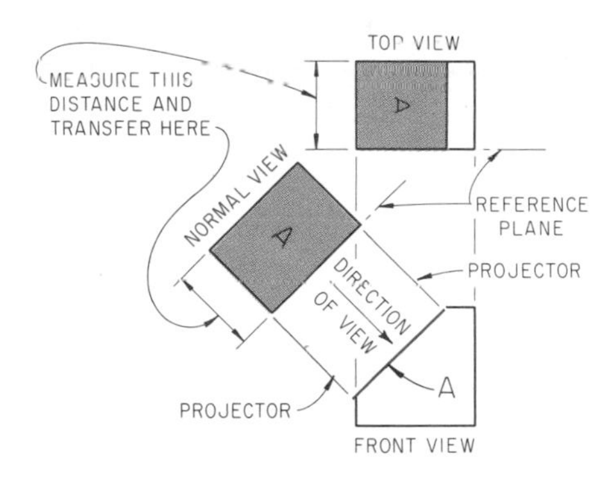


**FIG. 17.** Normal view of an inclined surface. The object is in position (*j*) of Fig. 7.

The projection of the normal view in Figs. 14 to 17 is unchanged geometrically and projectively as compared with Figs. 8 to 13. The *only* difference is that the projectors for the normal view this time emanate from the front view, where the inclined surface appears as an edge. While studying Figs. 14 to 17, consider again the principles of projection involved and the procedure followed: (1) The direction of observation for the normal view is perpendicular to the inclined surface and, on the drawing, perpendicular to the edge view of the surface. (2) The projectors for the normal view are parallel to the viewing direction, perpendicular to the inclined surface and, on the drawing, perpendicular to the edge view of the surface. (3) The projectors (or the spacing between them) determine one dimension of the normal view. (4) The second dimension for the normal view is now needed. (5) The reference plane for making the needed measurement will appear in the normal perpendicular to the projectors between edge view and normal view, and in the principal view (in projection with the edge view) perpendicular to the projectors between edge view and principal view. (6) The measurement perpendicular to the reference plane in the principal view is then transferred to the normal view, perpendicular to the reference plane, thus completing the layout of the normal view.

Figures 18 to 21 show the inclined surface positions of the third column of Fig. 7 [(*k*) to (*n*)], in the same order. Note that the projective system is the same as before. The only difference now is that the edge view of the inclined surface appears in the *side* view, and because *either* side view *might* be drawn, the projectors from the edge view to the normal view will have to be drawn from whichever side view, right or left, is used. To illustrate this variation, Figs. 18 and 19 have been drawn with the right-side view and Figs. 20 and 21 with the left-side view. Note also that the reference plane, with measurements taken from it, will in these cases be in the normal view and the front view, since the front view is the view in *direct* projection with the edge view of the inclined surface.

Study Figs 7 to 21 carefully, note the recurrence of certain principles of projection, and study the relationships of views, projectors, directions, and measurements.

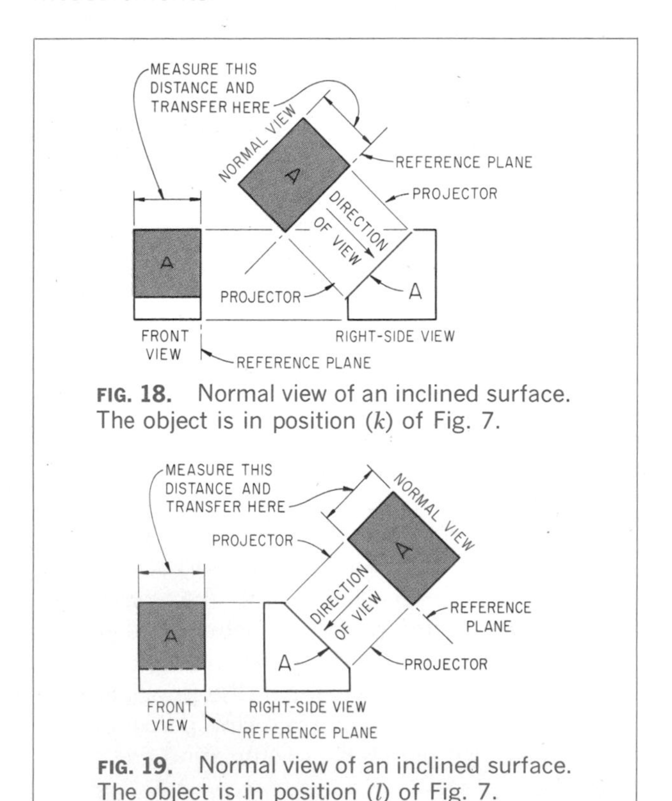


**FIG. 18.** Normal view of an inclined surface. The object is in position (*k*) of Fig. 7.

**FIG. 19.** Normal view of an inclined surface. The object is in position (*l*) of Fig. 7.

## 5. Normal Views of Inclined Surfaces on Practical Objects.

So far we have presented only the pertinent principles of projection and the procedure for drawing a normal view, using a rectangular object for illustration. Practical objects are, of course, made up of rectangular, conic, cylindrical, and other shapes, and this is sometimes considered by the uninitiated to be a complication. However, this is not the case; a machine part is often easier to draw than a purely geometric shape because it is more readily visualized.

Figure 22 shows a part with a surface inclined to front and side, about in the same position as the inclined surface in Fig. 8. At (*a*) the object is shown pictorially, surrounded by the planes of projection, and at (*b*) the projection planes are opened up into

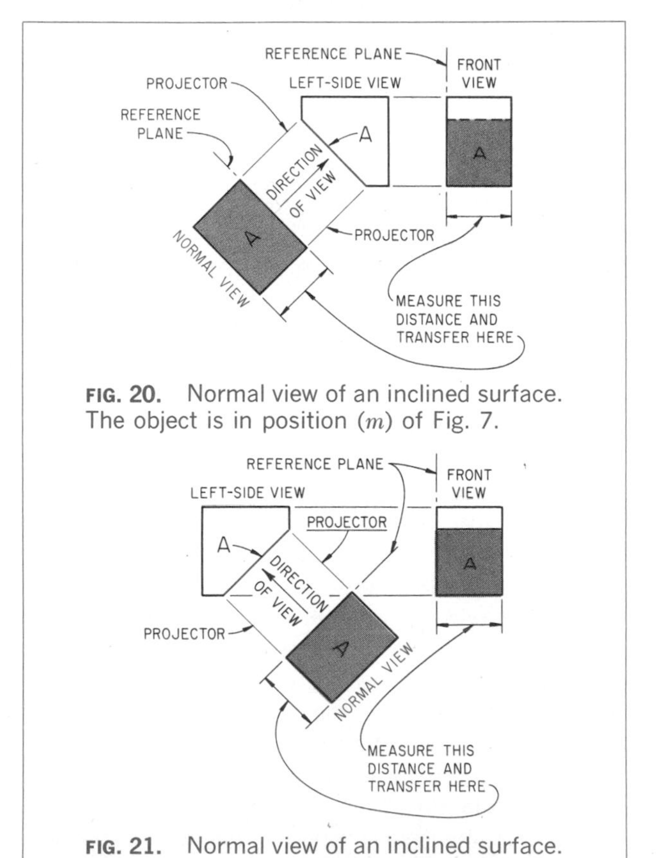


**FIG. 20.** Normal view of an inclined surface. The object is in position (*m*) of Fig. 7.

**FIG. 21.** Normal view of an inclined surface. The object is in position (*n*) of Fig. 7.

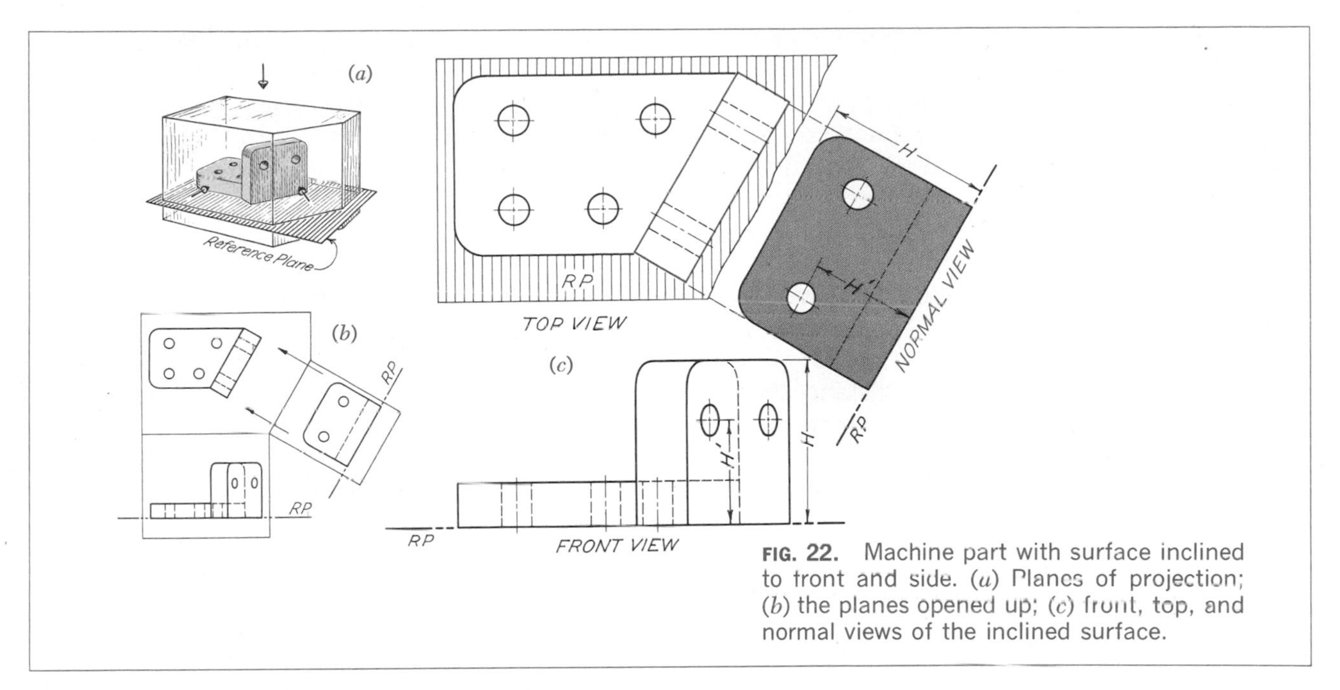


**FIG. 22.** Machine part with surface inclined to front and side. (*a*) Planes of projection; (*b*) the planes opened up; (*c*) front, top, and normal views of the inclined surface.

the plane of the paper. Note that the reference plane is placed, in this case, at the *base* of the object because this is a natural reference surface. Thus at (*c*) the reference plane is drawn at the base in the front and normal views, and measurements are made *upward* (dimensions $H$ and $H'$) to needed points. Note that the top view is the normal view of the reference plane, which appears as an edge in the front and normal views.

Figure 23 illustrates the procedure for drawing the normal view of an inclined surface on a machine part. Steps are as follows:

1. Draw the partial top and front views, as at (*b*), and locate the view direction by drawing projectors

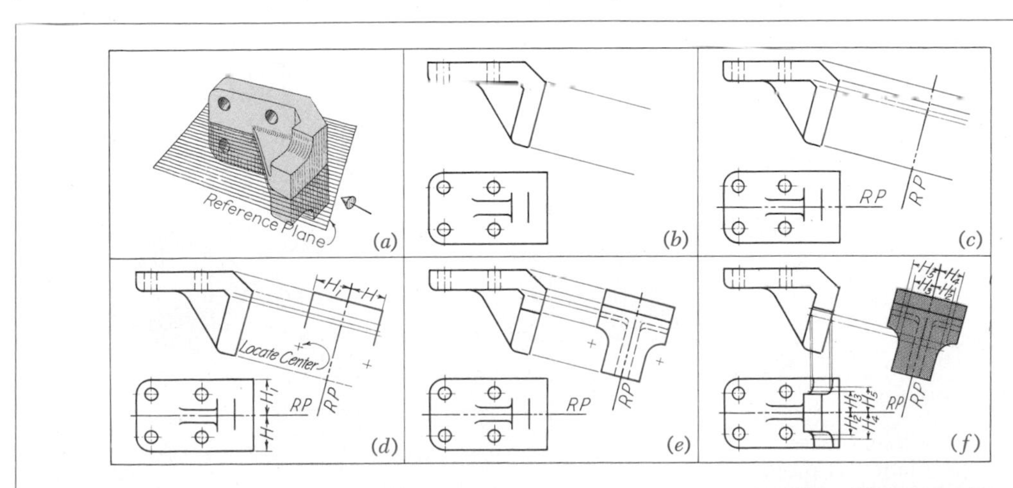


**FIG. 23.** Stages in drawing a machine part with a surface inclined to front and side.

perpendicular to the edge view of the inclined surface, as shown.

2. Locate the reference plane (*RP*) in the front view. The reference plane may be taken above, below, or through the view, and is chosen for convenience in measuring. In this case it is taken through the natural center line of the front view. The reference plane in the normal view will be perpendicular to the projectors already drawn, and is located at a convenient distance from the top view, as shown at (*c*).

3. As shown at (*d*), measure the distance (height) from the reference plane of various points needed, as, for example, $H$ and $H_1$, and transfer these measurements with dividers or scale to the normal view, measuring from the normal-view reference plane.

4. Complete the normal view from specifications of the rounds, etc., as shown at (*e*). Note that any measurement in the front view, made *toward* the top view, is transferred to the *normal view, toward* the top view. Note also that the front view could not be completed without using the normal view.

5. To get the front view of the circular portions, the true shape of which shows only in the normal view as circle arcs, select points in the normal view, project them back to the top view, and then to the front view. On these projectors transfer the heights $H_2$ and $H_3$ from the normal view to find the corresponding points in the front view. $H_4$, $H_5$, and others will complete the curve in the front view. Note that this procedure is exactly reversed from the operation of measuring from the front view to locate a distance in the normal view.

Figure 24 shows an object with *two* inclined surfaces. Note that the normal views are only partial views because the base of the object is fully described in top and front views. The reference plane is taken through the center because the object is symmetrical.

Figure 25 illustrates the steps in drawing the normal view of a face inclined to top and side:

1. Draw the partial top and front views, as at (*b*), and locate the view direction by drawing projectors perpendicular to the inclined surface, as shown.

2. Locate the reference plane in the top view. The reference plane may be taken in front of, through, or to the rear of the view but is here located at the rear flat surface of the object because of convenience in measuring. The reference plane in the normal view will be perpendicular to the projectors already drawn, and is located at a convenient distance from the front view, as shown at (*c*).

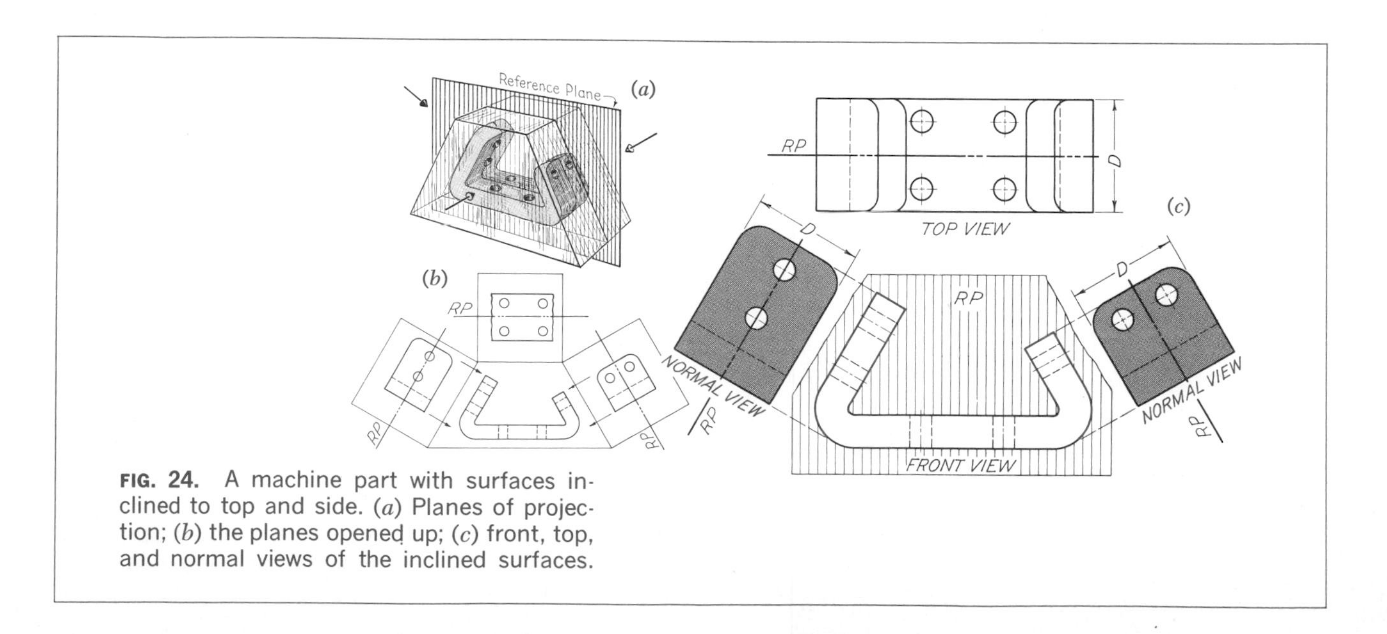


**FIG. 24.** A machine part with surfaces inclined to top and side. (*a*) Planes of projection; (*b*) the planes opened up; (*c*) front, top, and normal views of the inclined surfaces.

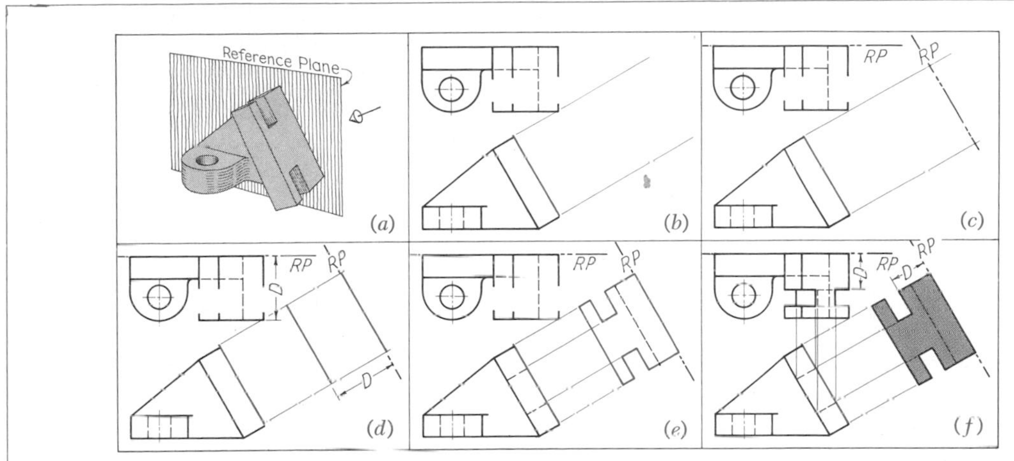


**FIG. 25.** Stages in drawing a machine part with a surface inclined to top and side.

3. As shown at (*d*), measure the distance (depths) from the reference plane (of various points needed), and transfer these measurements with dividers or scale to the normal view, measuring from the reference plane in the normal view. Note that the points are in front of the reference plane in the top view and are therefore measured toward the front in the normal view.

4. From specifications, complete the normal view, as shown at (*e*).

5. Complete the drawing, as shown at (*f*). In this case the top view could have been completed before the normal view was drawn. However, it is considered better practice to lay out the normal view before completing the view that will show the surface foreshortened.

A part might also be drawn with the inclined surface on the lower or upper front or rear. Orientation of this type is shown in Fig. 26, where the inclined surface is on the upper front. The steps in drawing an object of this type are given in Fig. 27 as follows:

1. Draw the partial front, top, and right-side views, as at (*b*), and locate the normal view direction by drawing projectors perpendicular to the inclined surface, as shown.

2. Locate the reference plane in the front view. This reference plane is taken at the left side of the object, because both the vertical and inclined portions have a left surface in the same profile plane. The reference plane in the normal view will be perpendicular to the projectors already drawn, and is located at a convenient distance from the right-side view, as shown at (*c*).

3. As shown at (*d*), measure the distances (widths) from the reference plane (of various points needed), and transfer these measurements with dividers or scale to the normal view, measuring from the reference plane in the normal view. Note that points to the right of the reference plane in the front view will be measured in a direction *toward* the right-side view in the normal view.

4. From specifications of the surface contour complete the normal view (*e*).

5. Complete the right-side and front views by projecting and measuring from the normal view. As an example, one intersection of the cut corner is projected to the right-side view and from there to the front view; the other intersection is measured (distance $W$) from the normal view and then laid off in the front view.

### 6. Purposes of Normal Views.

In practical work the chief reason for using a normal view is to show the true shape of an inclined surface.

In a normal view, the inclined surface will be shown in its true shape, but the other faces of the

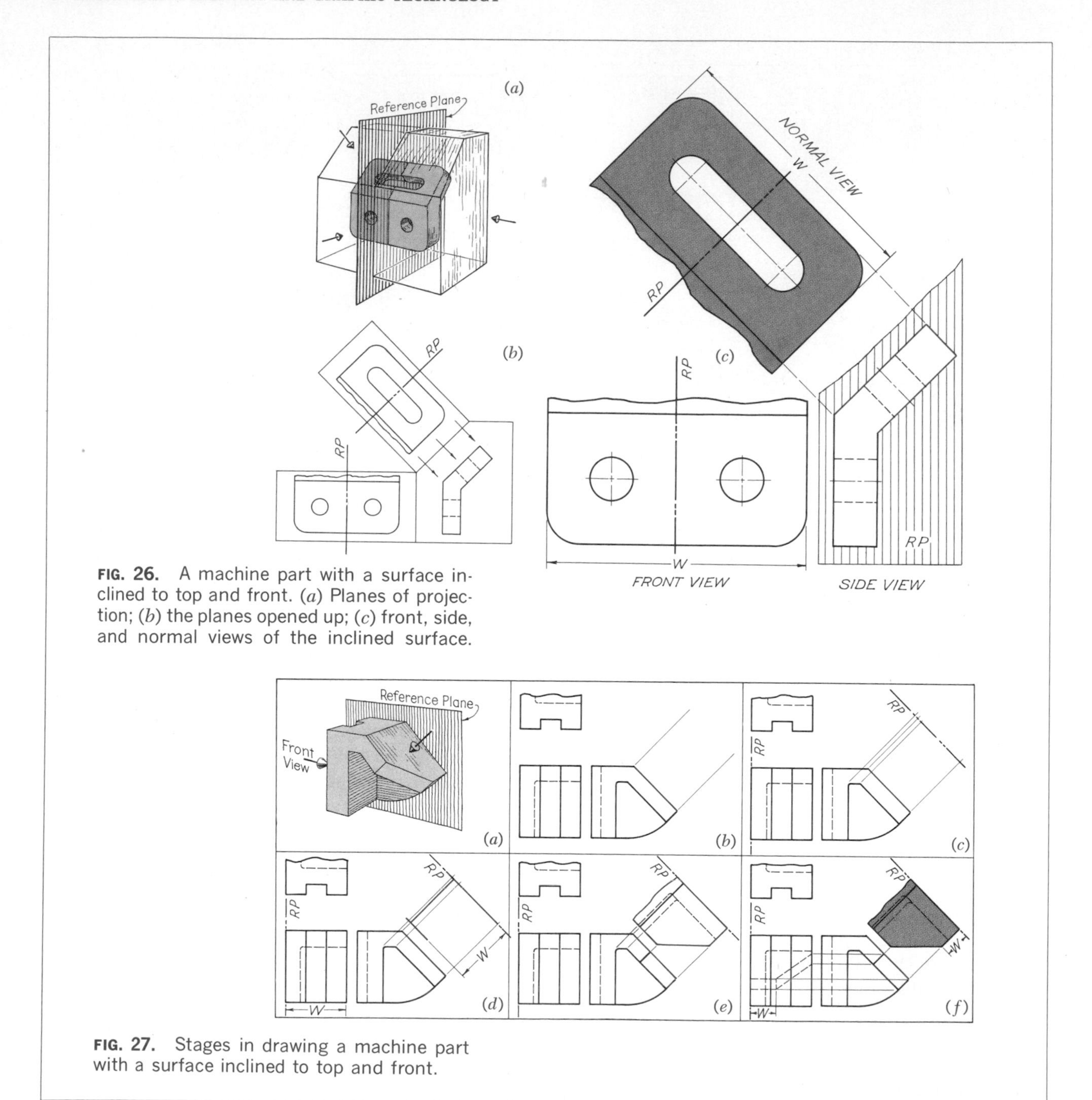


**FIG. 26.** A machine part with a surface inclined to top and front. (*a*) Planes of projection; (*b*) the planes opened up; (*c*) front, side, and normal views of the inclined surface.

**FIG. 27.** Stages in drawing a machine part with a surface inclined to top and front.

object appearing in the view will be foreshortened. In practical work these foreshortened parts are usually omitted, as in Fig. 28. Views thus drawn are called *partial views*. The exercise of drawing the complete view, however, may aid the student in understanding the subject.

Another important use of a normal view is in the case where a principal view has a part in a foreshortened position which cannot be drawn without first constructing a normal view in its true shape from which the part can be projected back to the principal view. Figure 29 illustrates this procedure.

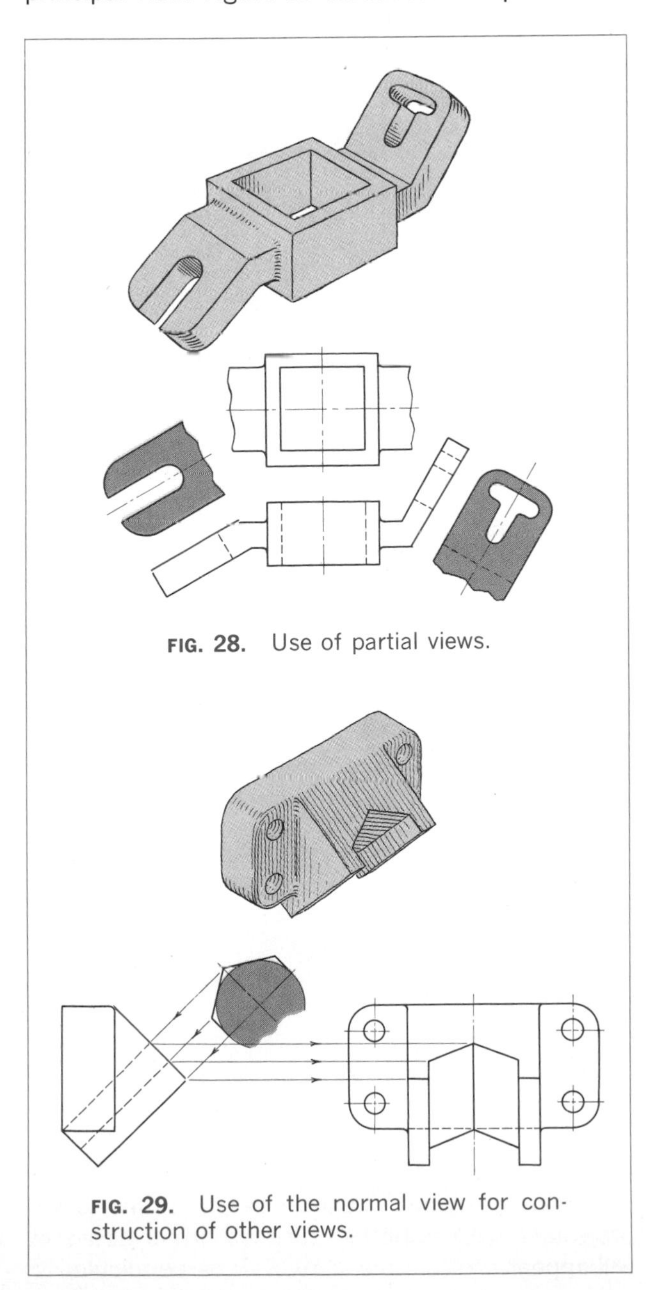

FIG. 28. Use of partial views.

FIG. 29. Use of the normal view for construction of other views.

Note in this figure that the view direction is set up looking along the semihexagonal slot and perpendicular to the face that is at a right angle to the slot. From the normal view showing the true semihexagonal shape, the side view and then the front view can be completed.

In most cases the normal view cannot be projected from the principal views but *must be drawn from dimensional specifications of the surface shape.*

Another practical example is the flanged 45° elbow of Fig. 30, a casting with an irregular inclined face, which not only cannot be shown in true shape in any of the principal views but also is difficult to draw in its foreshortened position. An easier and more practical selection of views for this piece is shown in Fig. 31, where normal views looking in directions perpendicular to the inclined faces show the true shape of the surfaces and allow for simplification. This is because each view can be laid out independently from specifications, and there is not even need for a reference plan, measurements, etc.

### 7. Normal View of a Line.

The normal view of a surface shows the true length of *all* lines in the surface. To prove this, go back to Fig. 8 and visualize *any* line on surface $A$ of the figure. You will find that the line shows in true length in the normal view. Often, however, especially in solving geometric space problems, a line is not a part of a surface but a separate line located in space by a pair of orthographic views. Any line that is parallel to one of the planes of projection will have a principal view that is the normal view. Thus in Fig. 32 where the lines are horizontal (*a*), frontal (*b*), and profile (*c*), the normal views of the lines are at (*a*) the top view, at (*b*) the front view, and at (*c*) the side view.

Often a line is skew to *all* the planes of projection and an extra view is required to give the normal view. To find the normal view of a surface, the direction of observation must be *perpendicular* to the surface; the same direction of observation is needed in finding the normal view of a line—the direction of observation *must* be perpendicular to the line. Thus at (*d*) projectors are drawn in the top view *perpendicular* to the top view of the line. The direction of

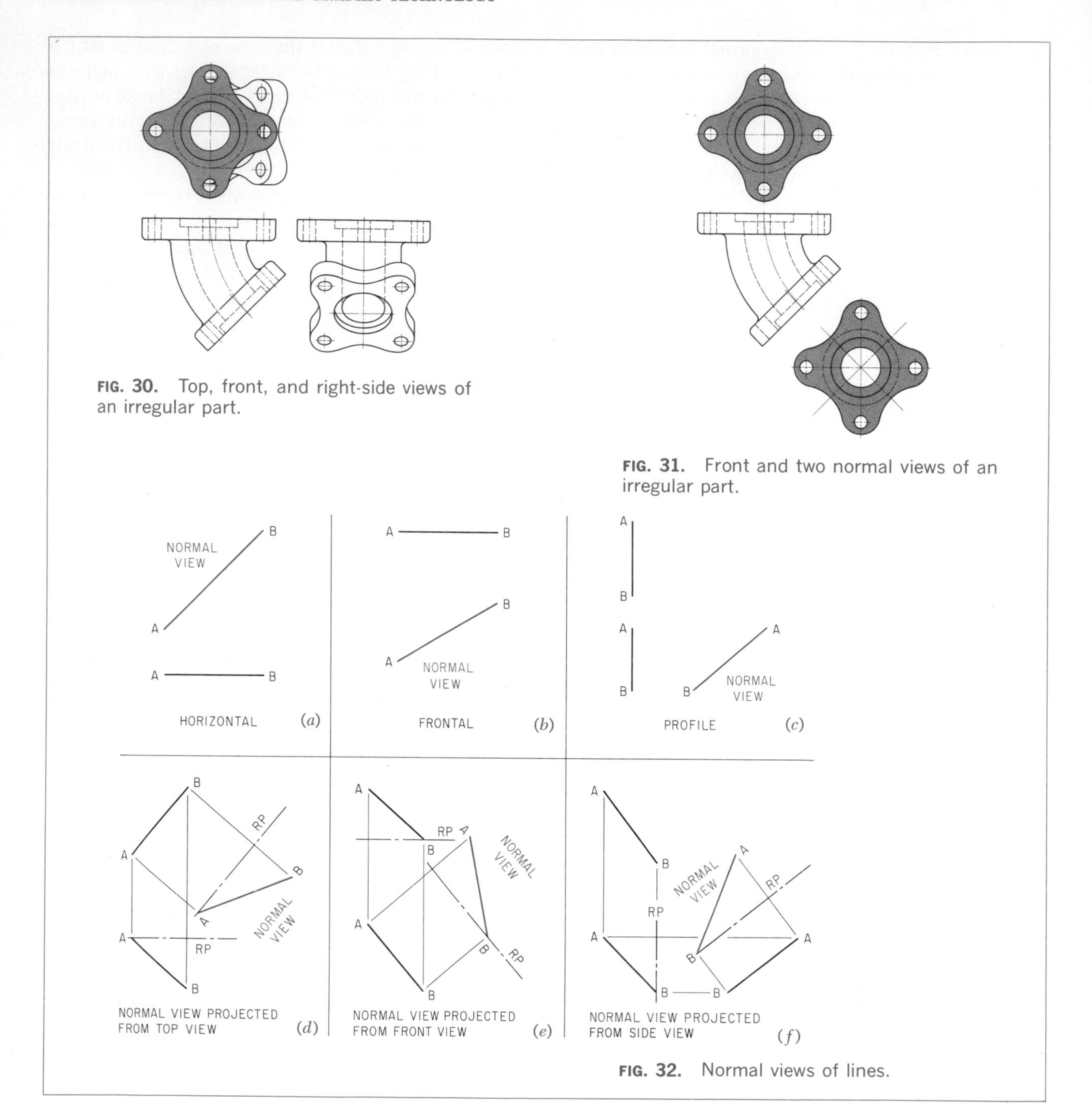


**FIG. 30.** Top, front, and right-side views of an irregular part.

**FIG. 31.** Front and two normal views of an irregular part.

**FIG. 32.** Normal views of lines.

view is now established perpendicular to the line, since the line could be on a surface, the top view being the *edge* view of the assumed surface. Next, the reference plane will appear perpendicular to the projectors just established for the normal view and will appear again in the front view perpendicular to the projectors between top and front views. Finally, point $A$ is *on* the reference plane and on the projector from the top view and is thus located in the normal view. Point $B$, then measured from the reference plane in the front view, transferred to the normal view and laid off from the reference plane *on* the projector for point $B$, completes the normal view. A similar normal view can be projected from the front view as at $E$, or from the side view as at $F$. Note in both cases that the projectors for the normal view *must* be perpendicular to the line in the view the normal is projected from.

## 8. Skew Surfaces.

In Sec. 2, in the classification of surfaces, the skew surface was described and illustrated. For convenience, Fig. 33 again shows this type of surface both pictorially and orthographically. Remember that a skew surface is one that occurs at an angle to *all* the principal planes of projection.

The normal view of a skew surface cannot (without resorting to special projective methods) be projected directly from the principal views, but must be projected from an *edge* view of the surface. Also, the edge view is almost always needed to show other features, such as the angle a surface makes with its base. Thus, *two* views are required and we must first learn how to get the edge view.

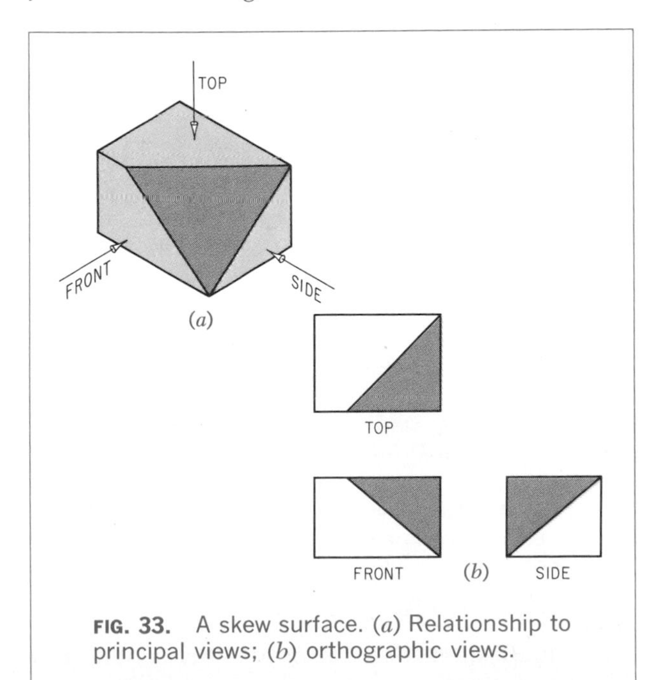


**FIG. 33.** A skew surface. (*a*) Relationship to principal views; (*b*) orthographic views.

## 9. Edge Views of Skew Surfaces.

The key to determining the direction of observation needed to obtain the edge view of a surface is to locate a direction that will give the *end view of one line* of the surface. This is easily done if the surface contains a line that is horizontal, frontal, or profile. In Fig. 34*a* the line is horizontal and a view made looking in a horizontal direction that is *aligned with the line* will give the end view. Note in Fig. 34*a* that points $A$ and $B$ (the two ends of the line) are both on the same projector and that both lie on the reference plane thus giving coincidence of the two ends of the line in the end view. Also note that the end view must be projected from the normal view of the line, in this case the top view. The end view of a frontal line is shown at (*b*); it is projected from the front (normal) view. The end view of a profile line is shown at (*c*); it is projected from the side (normal) view.

As stated before, the edge view of a surface is made by first obtaining the end view of any line of the surface. In Fig. 35 surface $ABDC$ is a skew surface (at an angle to *all* principal planes). However, lines $AB$ and $DC$ are horizontal lines of the surface. Selecting $DC$ as the key line, a projector is drawn aligned with $DC$ and another projector (parallel) is drawn for $AB$. This establishes the direction of view for the end view of both $DC$ and $AB$. The reference plane for the edge view will appear as an edge perpendicular to these projectors, and will appear again as an edge in the front view perpendicular to the projectors between top and front views. For convenience, the reference plane is placed through points $D$ and $C$. Then $D$ and $C$ both project to the edge view on the same projector and both lie on the reference plane, giving $DC$ in the edge view. $A$ and $C$ both project to the edge view on the same projector. The distance from the reference plane in the front view to points $A$ and $C$ is measured and transferred to the edge view, thus locating $AC$ in the edge view. The line connecting

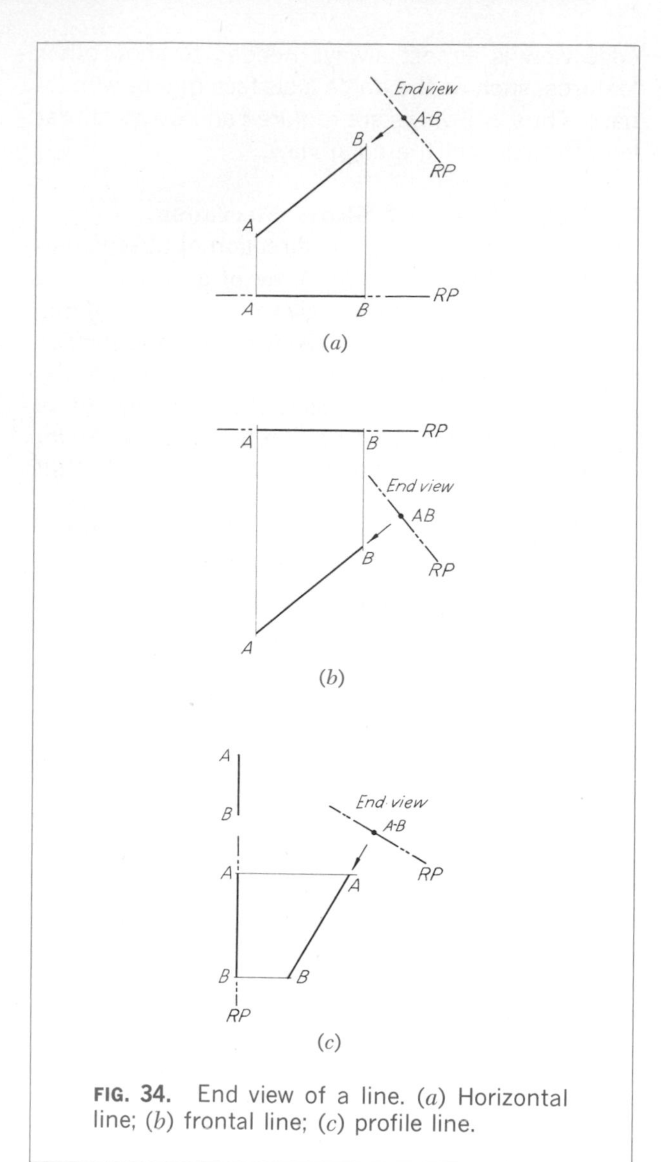


FIG. 34. End view of a line. (*a*) Horizontal line; (*b*) frontal line; (*c*) profile line.

*DC* and *AB* is the edge view of surface *ABDC*.

Figure 36 shows the identical surface *ABDC* of Fig. 35, but this time the edge view has been obtained by looking along the two frontal lines *AC* and *BD*. The edge view is projected in the direction of *AC* and *BD* from the front view because *AC* and *BD* are normal in the front view. The reference plane then located in the edge view and top view, and *AC* and *BD* located with reference to it, complete the edge view.

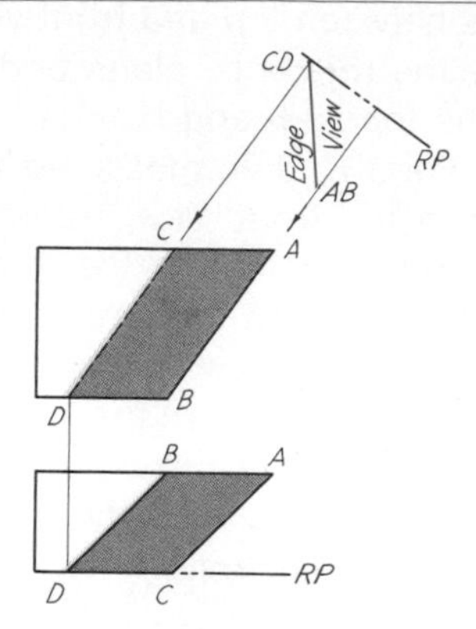


FIG. 35. Edge view of a skew surface. Projection is in the direction of the horizontal line *DC*.

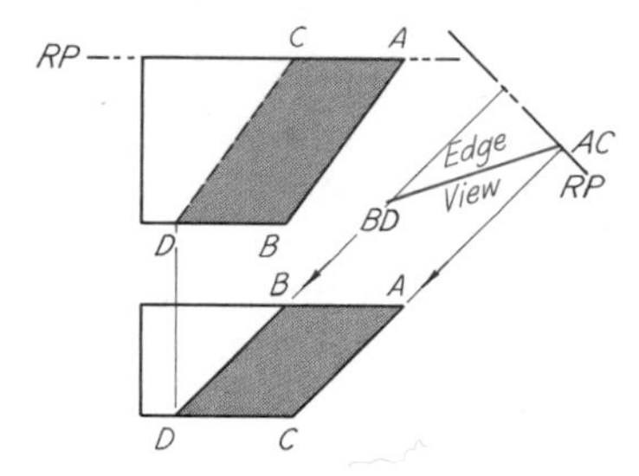


FIG. 36. Edge view of a skew surface. Projection is in the direction of frontal line *AC*.

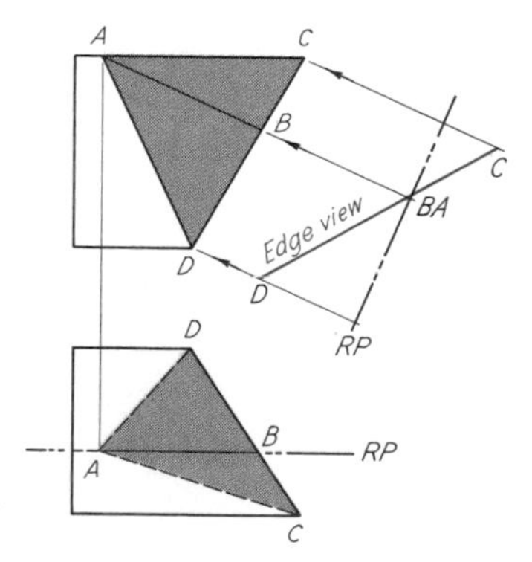


FIG. 37. Edge view of a skew surface. The direction of the edge view is located by the horizontal line *AB*.

In Fig. 37 the surface *ACD* does not contain any horizontal, frontal, or profile line, and so to get an

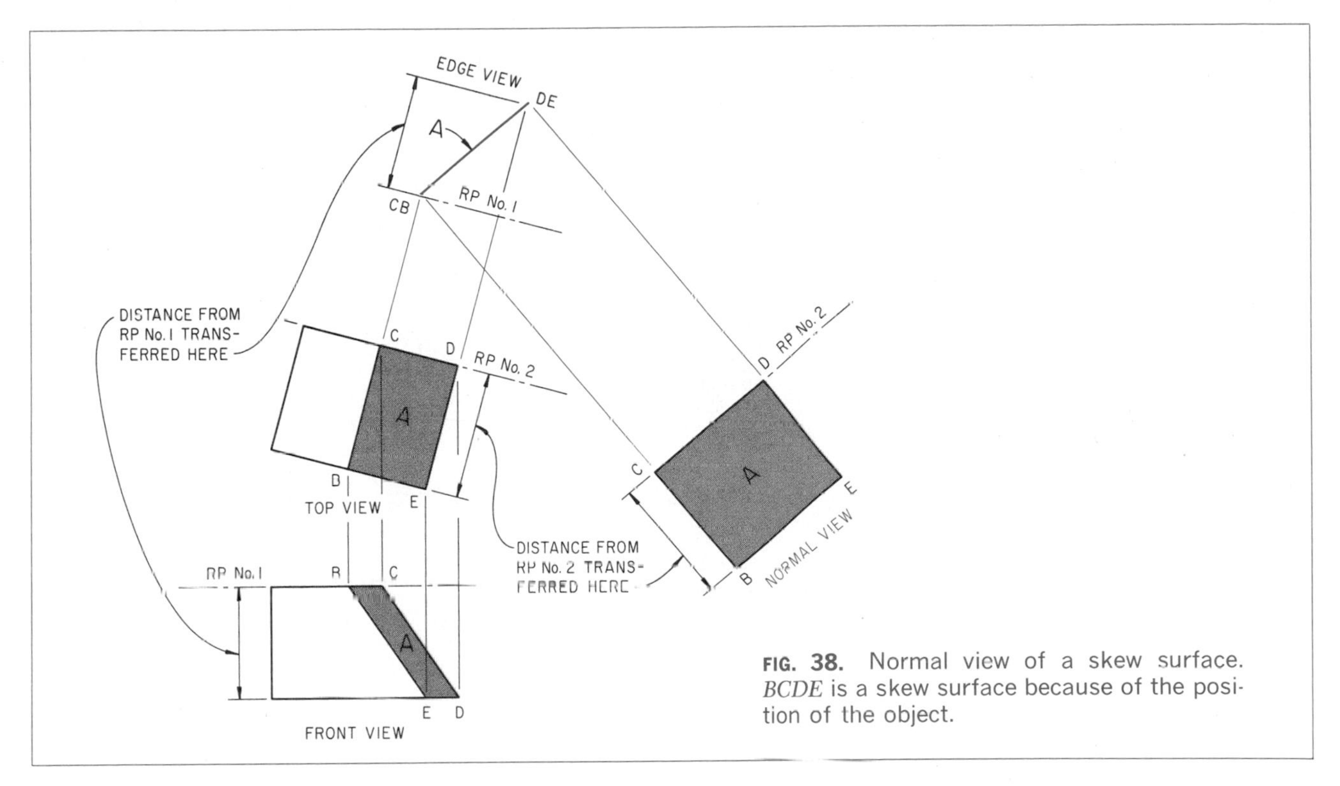


**FIG. 38.** Normal view of a skew surface. *BCDE* is a skew surface because of the position of the object.

end view of one line of the surface, a horizontal, frontal, or profile line must be laid out on the surface. A horizontal line has been selected and is drawn at *AB* in the front view and then projected to the top view. This line (*AB*) is then normal in the top view and projectors parallel to it from *AB*, *C*, and *D* will establish the direction for the edge view of *ACD*. The reference plane is conveniently located through *AB* in front and edge views; then *D* and *C* are measured from the reference plane in the front view and transferred to the edge view. Note that *D* lies *above* the reference plane and *C below* and that they are transferred accordingly to the edge view. To avoid reversing a normal or edge view, remember this simple rule: if a point is on the side of the reference plane *toward* the view projected from in making an edge or normal view (the top view, in this case), the point will be transferred *toward* that same view. As an example of the rule, note in Fig. 37 that *D* is on the side *toward* the top view from the reference plane in both front and edge views while *C* is on the side *away* from the reference plane.

## 10. Normal Views of Skew Surfaces.

A skew surface is produced by an object with an inclined surface that is turned at an angle to the principal planes, as in Fig. 38, or by an object with a face at an angle to three object faces, as in Figs. 39 and 40. Figure 38 is easier to visualize, so we will discuss it first.

Surface *BCDE* of Fig. 38 is a skew surface. *CB* and *DE* are horizontal lines of the surface. The edge view is obtained by looking in the direction of *CB* and *DE*, as shown in Fig. 38 and described in Sec. 9.

The normal view is made with the direction of observation perpendicular to the edge view. To draw the normal view, first draw projectors perpendicular to the edge view (Fig. 38), establishing the direction for the normal view. As in all earlier examples, the reference plane for the normal view (*RP* No. 2 in Fig. 38) is perpendicular to the projectors. The reference plane appears normal in the edge view and appears again as an edge in *any* other view in direct

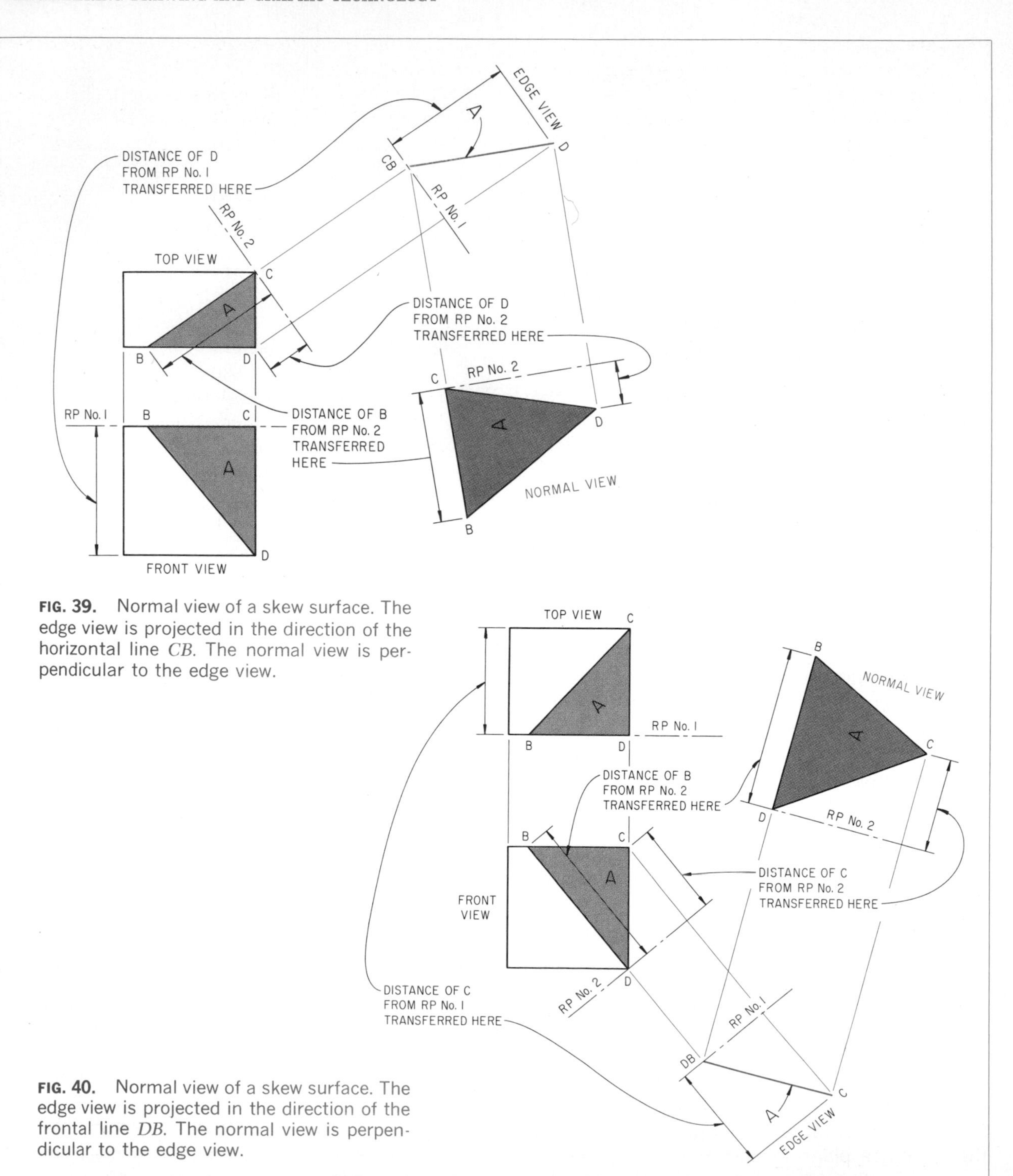


**FIG. 39.** Normal view of a skew surface. The edge view is projected in the direction of the horizontal line *CB*. The normal view is perpendicular to the edge view.

**FIG. 40.** Normal view of a skew surface. The edge view is projected in the direction of the frontal line *DB*. The normal view is perpendicular to the edge view.

projection with the edge view. In this case therefore, the reference plane shows as an edge in the top view, and is perpendicular to the projectors between top and edge views. To prove this relationship to yourself, note that here the top view, edge view, and normal view have *exactly* the same relationship as the front view, top view, and normal view of Fig. 8. Finally, transfer the distance of *B* and *E* from the reference plane (No. 2) in the top view to the normal view, as shown.

Figure 39 is similar to Fig. 38, but this time the object faces are parallel to the principal planes, and surface *BCD* is skew to object faces and principal planes. Again, one line (*BC*) of the skew surface is horizontal and gives the direction for the edge view. In the edge view, *B* and *C* are on the reference plane. Point *D*, measured from reference plane No. 1 in the front view, is transferred to the edge view, as shown. Then, projectors perpendicular to the edge view give the direction for the normal view, and reference plane No. 2 is drawn in normal and top views perpendicular to the projectors connecting each with the edge view. Finally the distances of *B* and *D* from reference plane No. 2 are transferred to the normal view.

The object of Fig. 40 is similar to that of Fig. 39, but in Fig. 40 the frontal line (*BD*) of surface *BCD* has been used for the direction of the edge view. The edge view is therefore projected from the front view. All other constructions are the same as before and are evident from the figure.

### 11. Normal Views of Skew Surfaces on Practical Objects.

Figure 41 illustrates the successive steps in drawing a normal view. The pictorial illustration (*a*) shows a typical object with a skew surface. The line of intersection between the skew portion and the horizontal base is line *AB*. In order to get an edge view of the skew surface, a view may be taken looking in the direction of line *AB*, thus giving an end view of *AB*. Because *AB* is a line of the skew surface, the edge view will result. The reference plane (*RP* No. 1) for this view will be horizontal. The direction of observation for the normal view will be perpendicular to the skew surface. The reference plane (*RP*

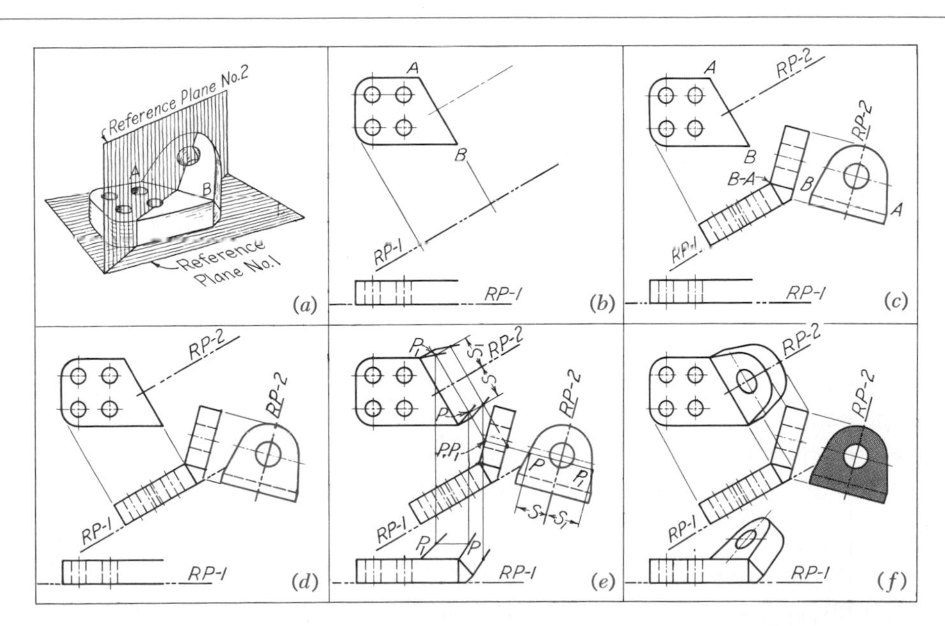


**FIG. 41.** Stages in drawing a machine part with a skew surface.

No. 2) for the normal view will be perpendicular to the edge-view direction and thus perpendicular to edge $AB$, as shown at ($a$).

At ($b$), parital top and front views are shown. The projectors and reference plane for the required edge view are also shown.

At ($c$), the edge view has been drawn. Note that line $AB$ appears as a point in the edge view. The angle that the skew surface makes with the base is laid out in this view from specifications.

At ($d$), the normal view is added. The projectors for the view are perpendicular to the edge view. The reference plane is drawn perpendicular to the projectors for the normal view and at a convenient distance from the edge view. The reference plane in the top view is drawn midway between points $A$ and $B$ in the top view because the skew surface is symmetrical about this reference plane. The normal view is drawn from specifications of the shape. The projection back to the edge view can then be made.

The views thus completed at ($d$) describe the object, but the top and front views may be completed for illustrative purposes or as an exercise in projection. The method is illustrated at ($e$) and ($f$). Any point, say $P$, may be selected and projected back to the edge view. From this view a projector is drawn back to the top view. Then the distance $S$ from the normal view is transferred to the reference plane in the top view. A number of points so located will complete the top view of the circular portion, and the straight-line portion can be projected in similar manner. The front view is found by drawing projectors to the front view for the points needed, measuring the heights from the reference plane in the edge view, and transferring these distances to the front view. Note that this procedure for completing the top and front views is the same as for drawing the views originally but in reverse order.

## 12. Terms of Reference for Auxiliary Views.

In this text, emphasis is given to the *purpose* for which auxiliary views are made; namely, to show the *edge view* or *normal view* of a surface. This practice is basic to the representation of solid objects, and, fundamentally, an auxiliary view used to depict a surface needs no further identification than the designation *edge view* or *normal view*. However, auxiliaries are named in various ways in other texts on engineering drawing and graphics, and for this reason the following terms of reference are given:

*Reference Group 1*

*Elevation auxiliary* or *auxiliary elevation.* Any view made by looking in a horizontal direction but inclined to frontal and profile planes. These are the views illustrated in Fig. 7*b* to *e*.

*Right auxiliary.* Any view made by looking from the right side in a frontal direction but inclined to horizontal and profile planes. See Fig. 7*f* and *g*.

*Left auxiliary.* Any view made by looking from the left side in a frontal direction but inclined to horizontal and profile planes. See Fig. 7*h* and *j*.

*Front auxiliary.* Any view made by looking from the front in a profile direction but inclined to horizontal and frontal planes. See Fig. 7*k* and *l*.

*Rear auxiliary.* Any view made by looking from the rear in a profile direction but inclined to horizontal and frontal planes. See Fig. 7*m* and *n*.

*Reference Group 2*

*Top-adjacent auxiliary.* Any auxiliary view projected from the top view. Such auxiliaries are the same as elevation auxiliaries (group 1). See Fig. 7*b* to *e*.

*Front-adjacent auxiliary.* Any auxiliary view projected from a front view. Such auxiliaries are the same as right-and-left auxiliaries. See Fig. 7*f* to *j*.

*Side-adjacent auxiliary.* Any auxiliary projected from a side view. Such auxiliaries are the same as front-and-rear auxiliaries. See Fig. 7*k* to *n*.

*Reference Group 3*

*Oblique view.* This is the normal view of a skew (oblique) surface and is projected from an edge view. See Figs. 38 to 40.

*Reference Group 4*

*Auxiliary-adjacent auxiliary view.* This is the normal view of a skew surface, projected from an edge view. See Figs. 38 to 40.

These terms of reference are used in books on descriptive geometry. However, since the book on descriptive geometry you are using with this text may give other terms, the space below is provided for listing them.

## PROBLEMS

### Group 1. Normal Views of Inclined Surfaces.

**1 to 7.** Draw given views and add normal views of the inclined surface using the reference plane indicated.

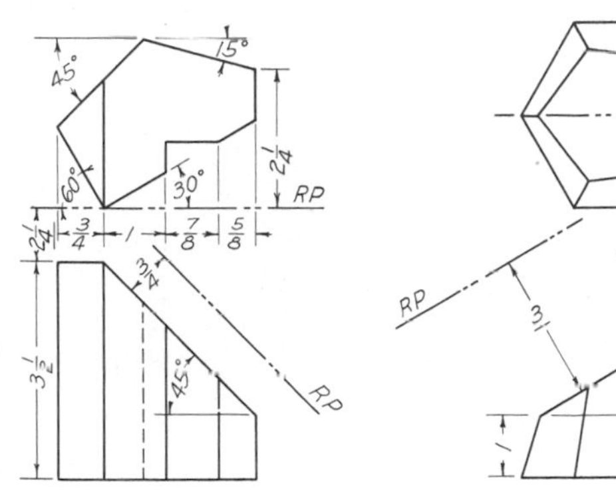


**PROB. 1.** Locking wedge.

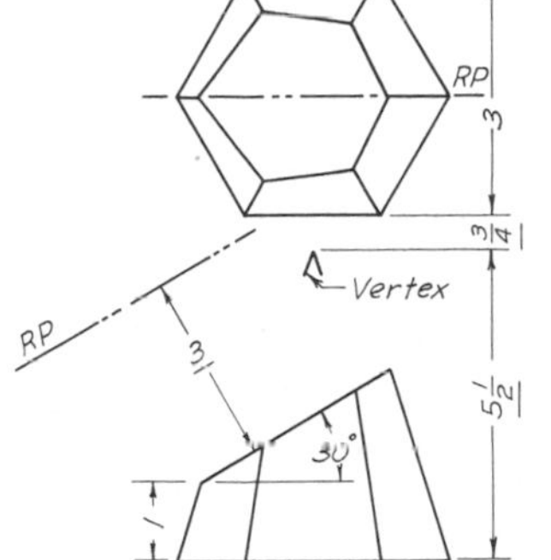


**PROB. 2.** Statue base.

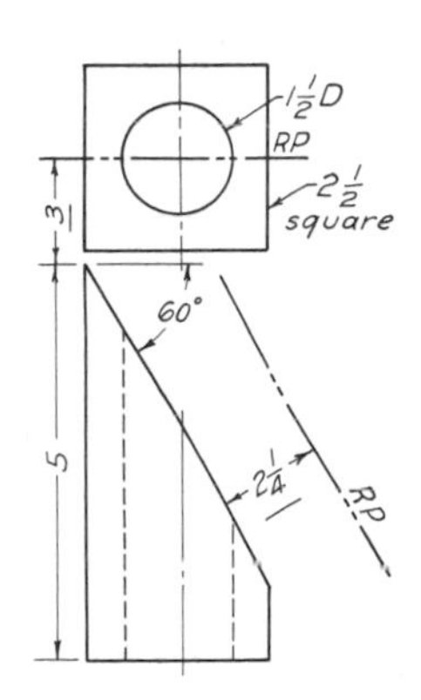


**PROB. 3.** Rod slide.

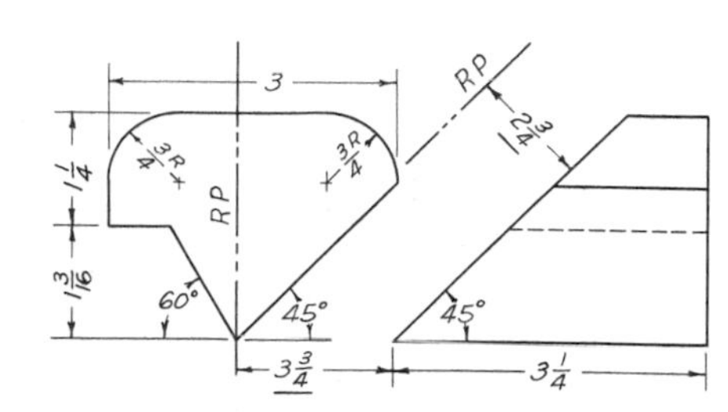


**PROB. 4.** Fulcrum.

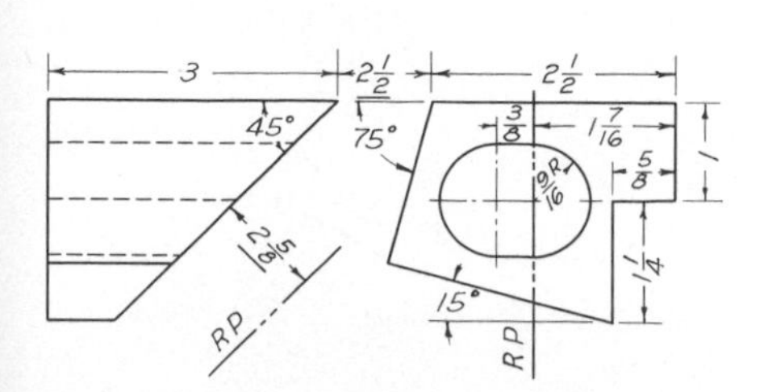


**PROB. 5.** Adjustable pawl.

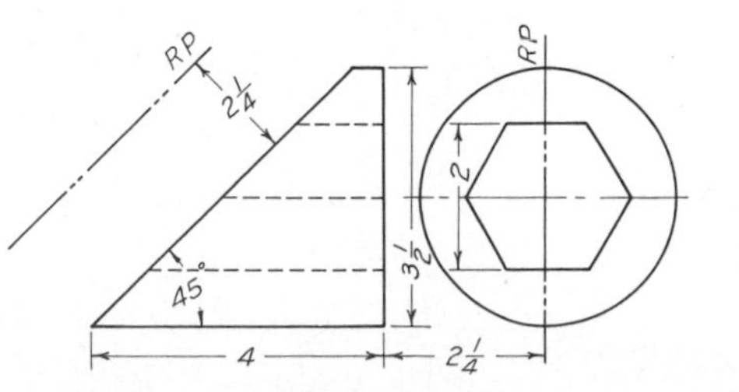


**PROB. 6.** Hexagonal-shaft lock.

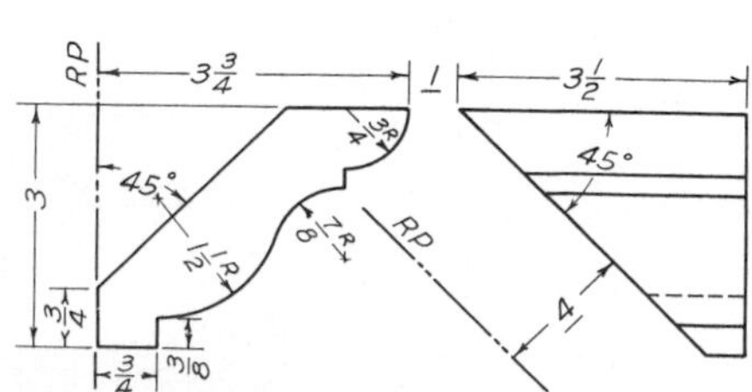


**PROB. 7.** Molding.

**8.** Draw the front view, partial top view, and normal view of the inclined surface.
**9.** Draw the partial front view, right-side view, partial top view, and normal view of the inclined surface.
**10.** Draw the front view, partial top view, and normal view of the inclined surfaces.
**11.** Draw the top view, partial front view, and normal view of the inclined surfaces.
**12.** Draw the front view, partial top view, and normal view of the inclined surface.
**13.** Draw the front view, partial right-side view, and normal view of the inclined surface. Draw the normal view before completing the front view.
**14.** Draw the front view, parital top view, and normal view of the inclined surface.
**15.** Draw the front view, partial top view, and normal view of the inclined surface.
**16.** Draw the front view, partial top and right-side views, and normal view of the inclined surface.

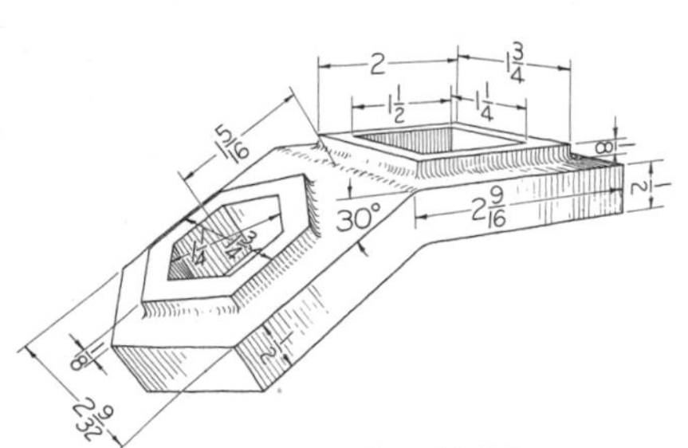


**PROB. 8.** Holder.

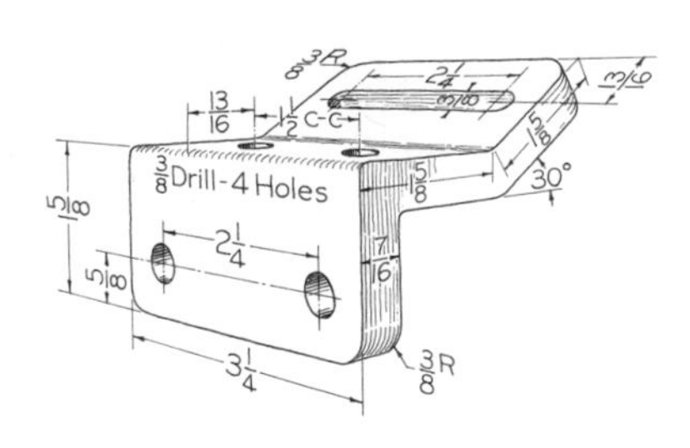


**PROB. 9.** Slotted anchor.

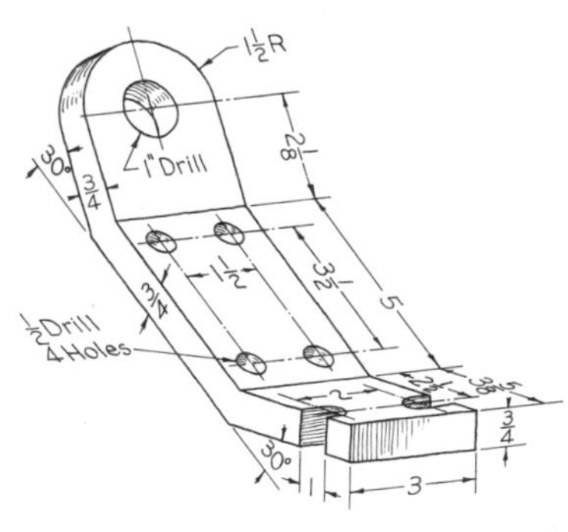


**PROB. 10.** Connector strip.

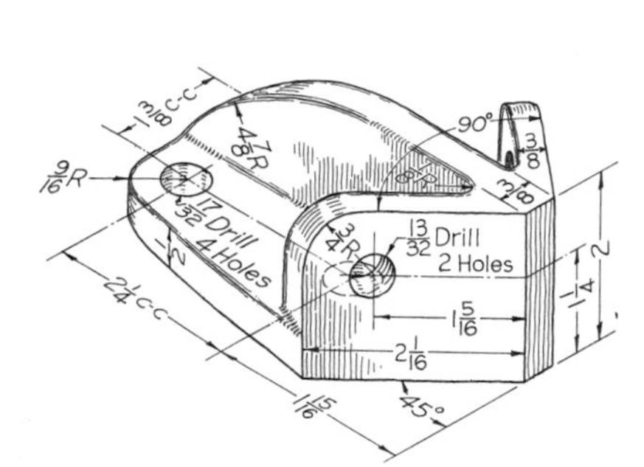


**PROB. 11.** Push plate.

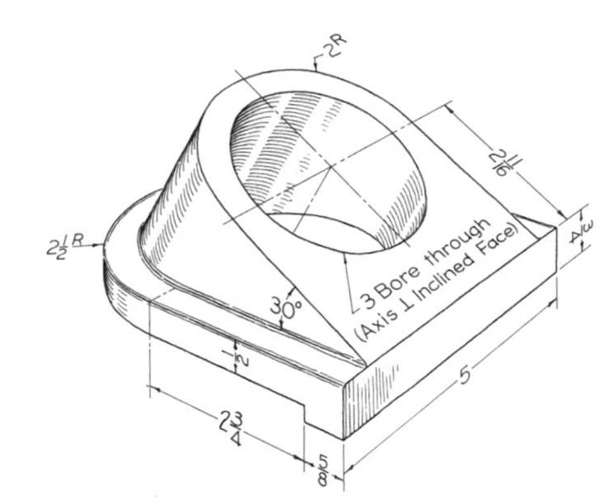


**PROB. 12.** Bevel washer.

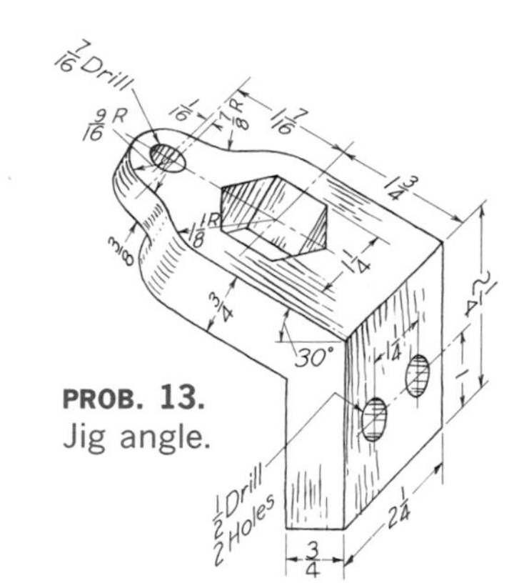


**PROB. 13.** Jig angle.

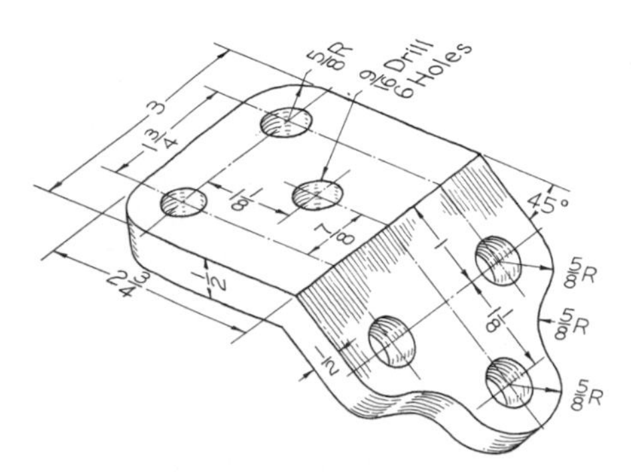


**PROB. 14.** Angle clip.

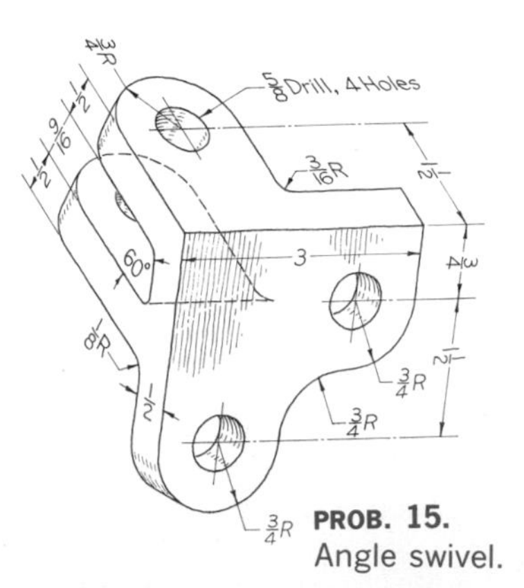


**PROB. 15.** Angle swivel.

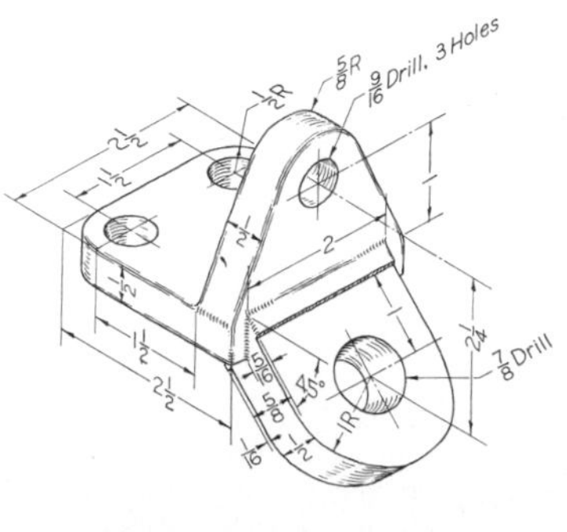


**PROB. 16.** Corner tie.

**17.** Draw the front view, partial top and left-side views, and normal view of the inclined surface.

**18 and 19.** Determine what views and partial views will best describe the part. Sketch the proposed views before making the drawing with instruments.

**20.** Draw the front view, partial bottom view, and normal view of the inclined surface.

**21.** Draw the front view, partial left-side and bottom views, and normal view of the inclined surface.

**22 and 23.** Determine what views and partial views will best describe the part.

**24.** Draw the given front view and add the views necessary to describe the part.

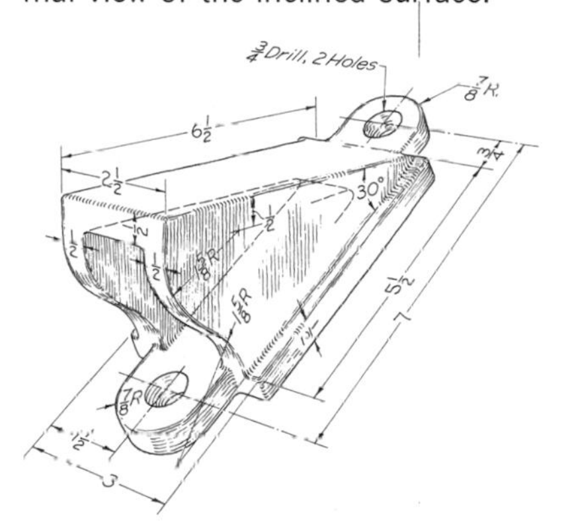

**PROB. 17.** Channel support.

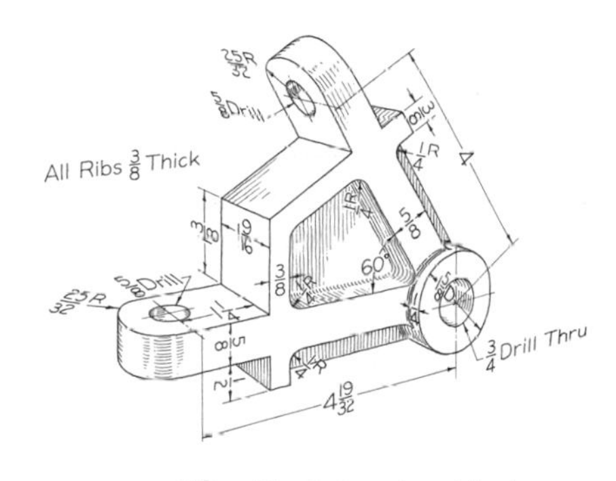


**PROB. 18.** Radial swing block.

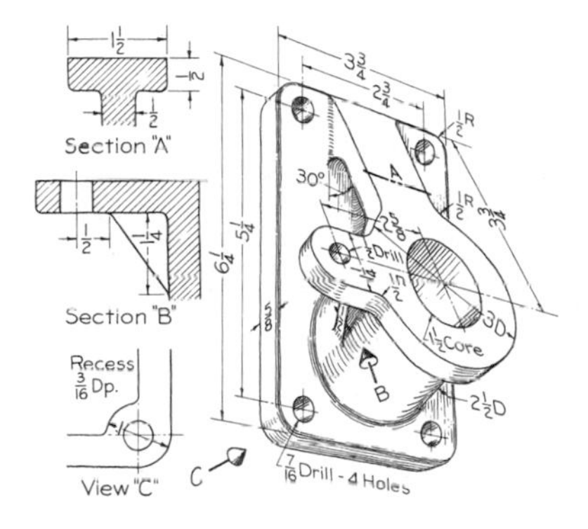


**PROB. 19.** Angle-shaft base.

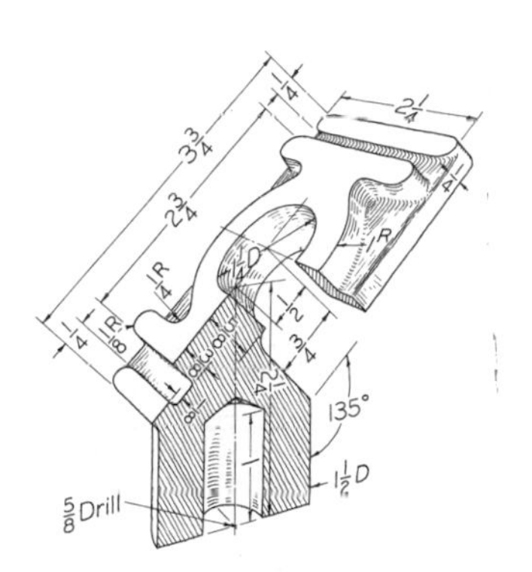


**PROB. 20.** Catenary clip.

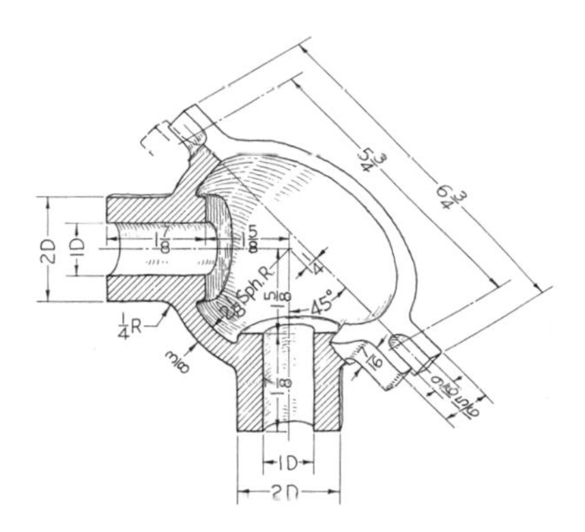

**PROB. 21.** Bevel-gear housing.

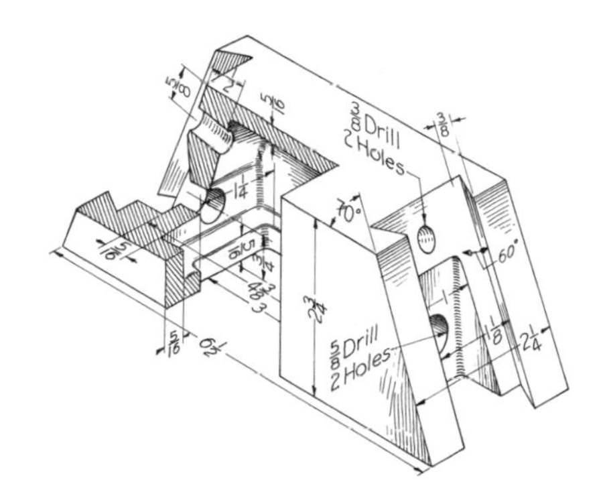

**PROB. 22.** Slide base.

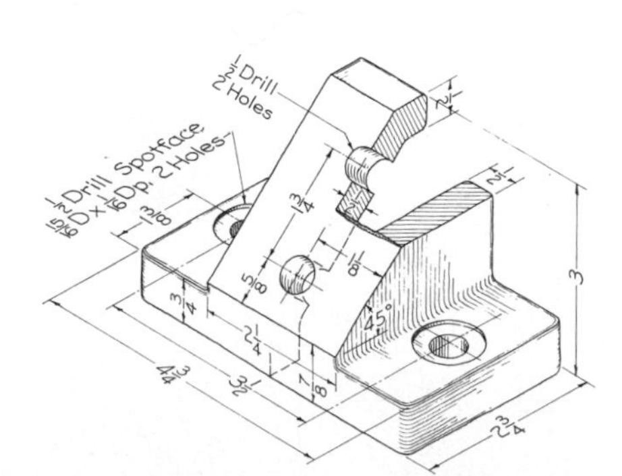

**PROB. 23.** Idler bracket.

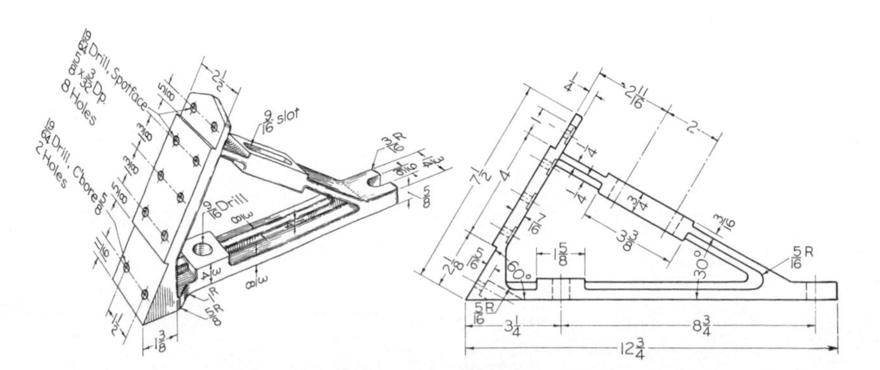

**PROB. 24.** Corner brace.

**25.** Draw the front view, partial left-side view, and normal view of the inclined surface.

**26 and 27.** This pair of similar objects has the upper lug in two different positions. Layouts are for 11- × 17-in. paper. Draw the views and partial views as indicated on the layouts.

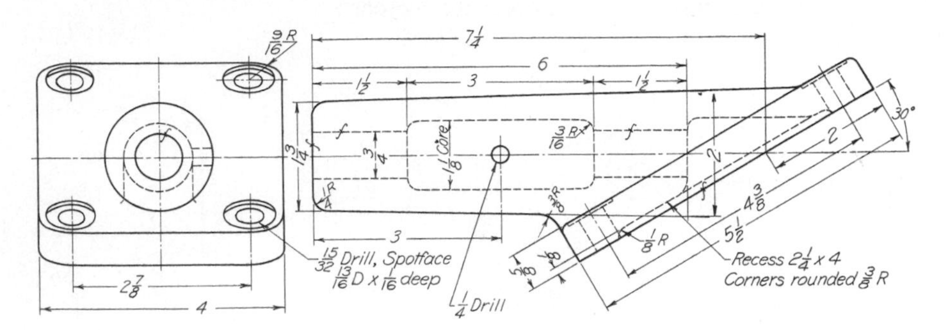


**PROB. 25.** Spindle support.

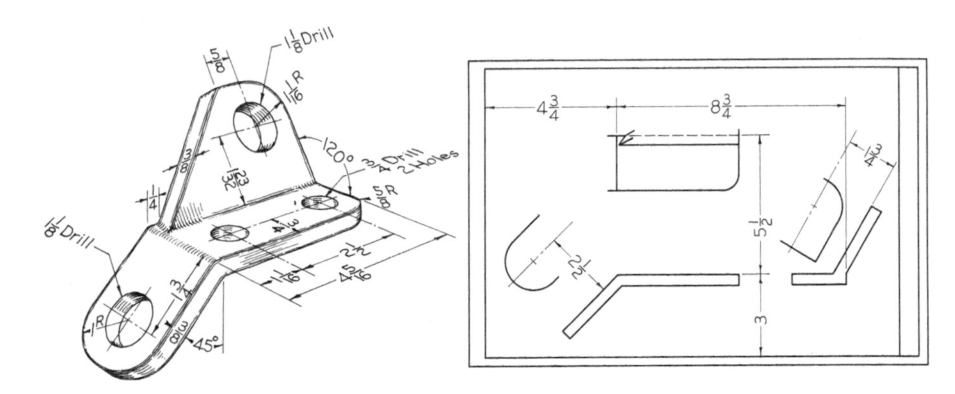


**PROB. 26.** Spar clip, 90°.

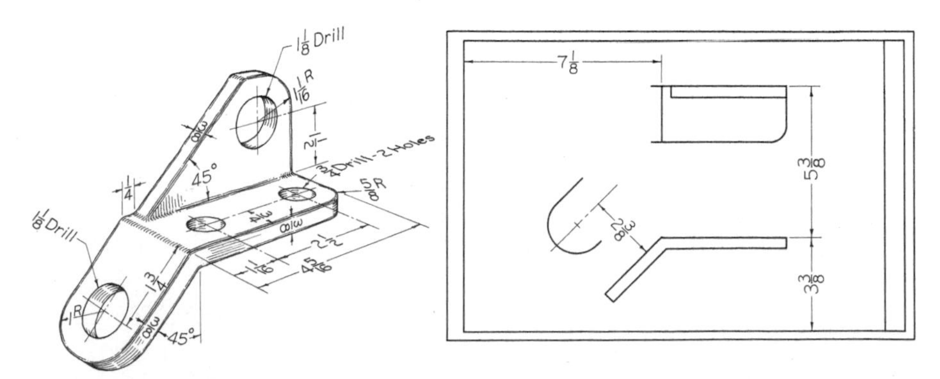


**PROB. 27.** Spar clip, 120°.

**28.** Draw the front view; partial top, right-side, and left-side views; and normal view of the inclined surface. Use decimal scale for layout.

**29.** Draw the views and partial views that will best describe the part.

**30.** Draw top and front views and a normal view of the inclined face. Will the normal view of the inclined face show the true cross section of the square hole?

**31.** Draw front, top, and side views and normal views of the inclined surfaces.

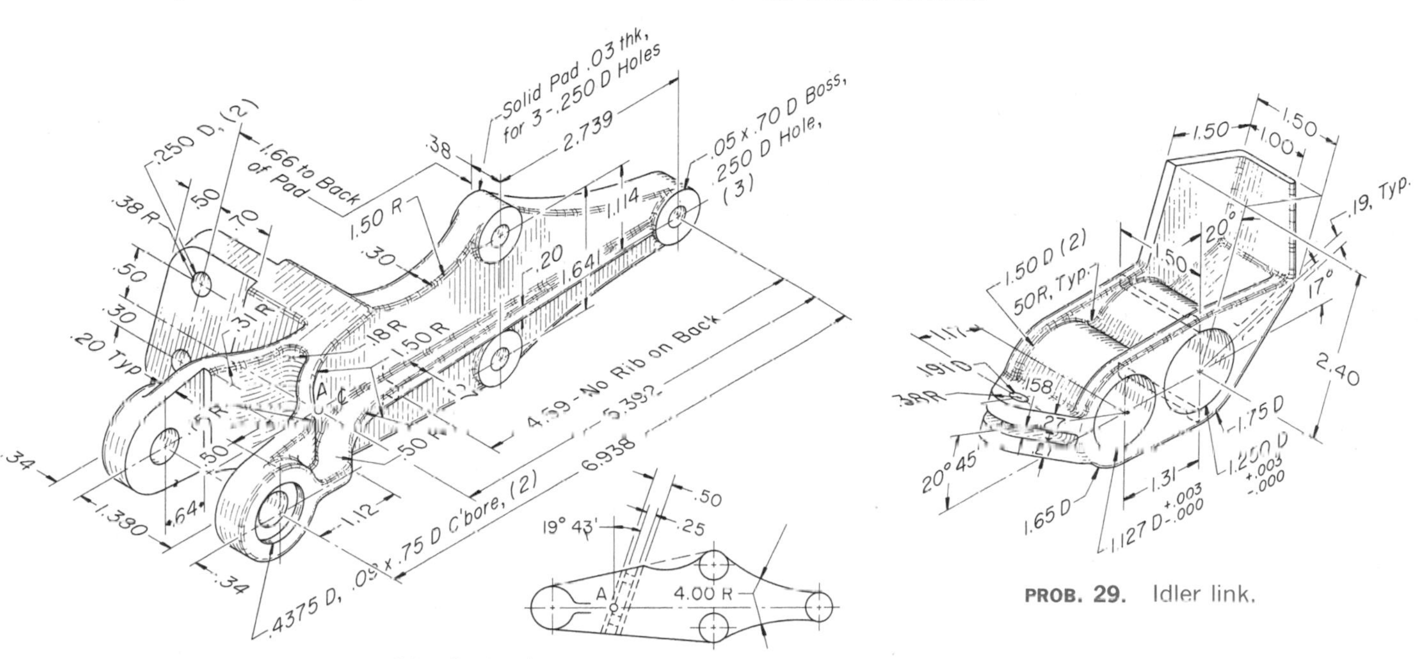


**PROB. 28.** Seat release.

**PROB. 29.** Idler link.

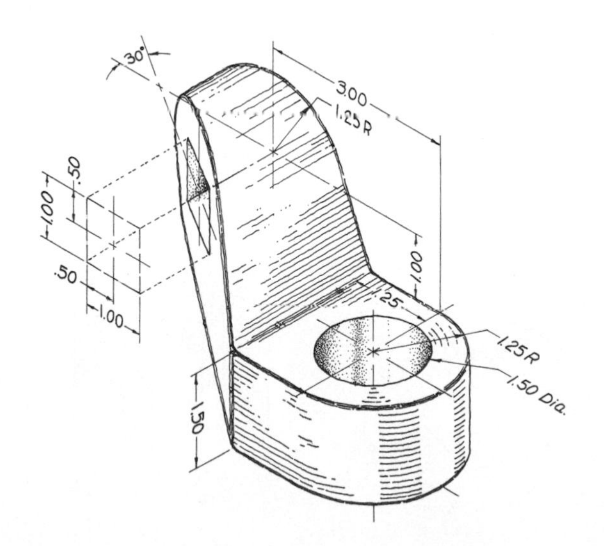


**PROB. 30.** Actuator bracket.

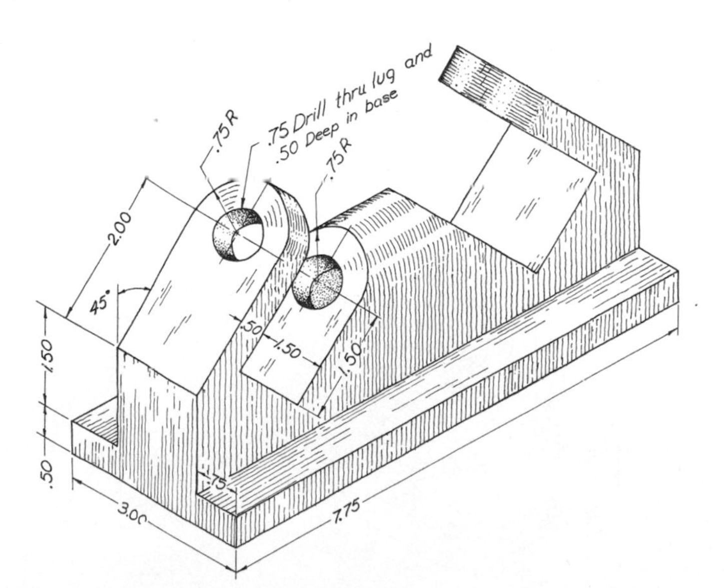


**PROB. 31.** Assembly-fixture base.

**32.** Front and right-side views of the shaft-locator wedge are shown on a layout for 11- × 17-in. paper. Add normal view of the inclined faces at centerline position shown.

## Group 2. Normal Views of Skew Surfaces.

**33.** Draw the partial front and top views, edge view showing the contour of the slot, and normal view of the skew surface.

**34.** Draw the partial front and top views, and edge and normal views of the skew surface. Draw the normal view before completing the edge view.

**35.** Draw the partial front and top views, and edge and normal views of the skew surface.

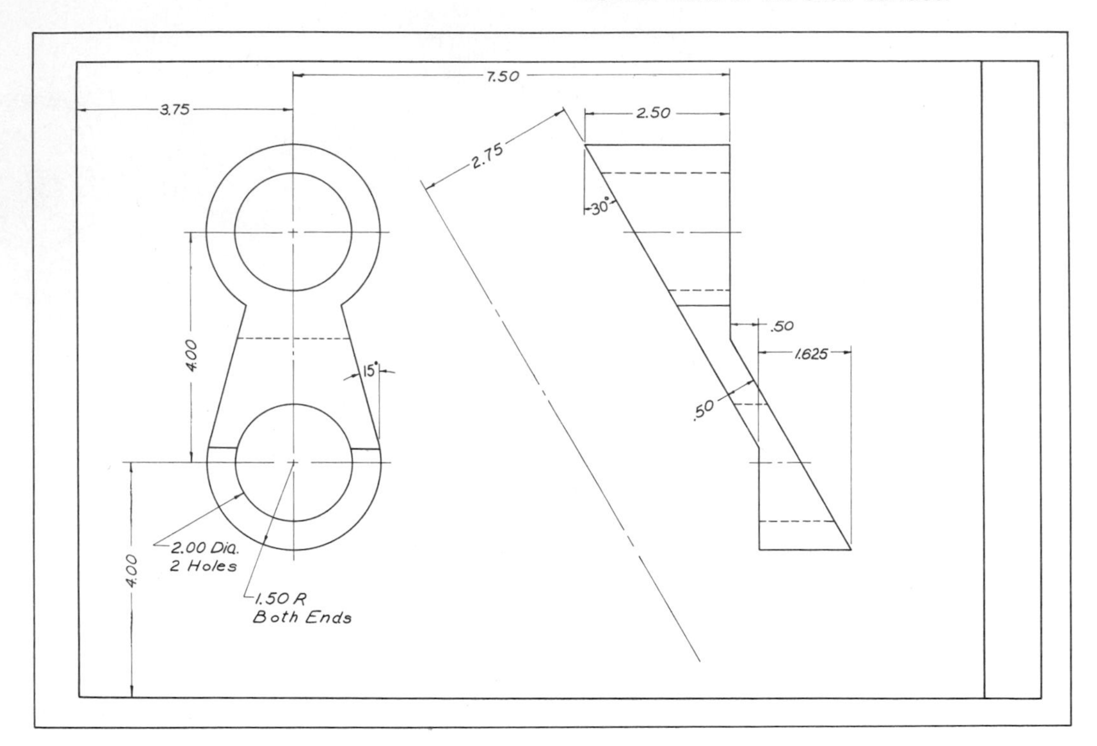


**PROB. 32.** Shaft-locator wedge.

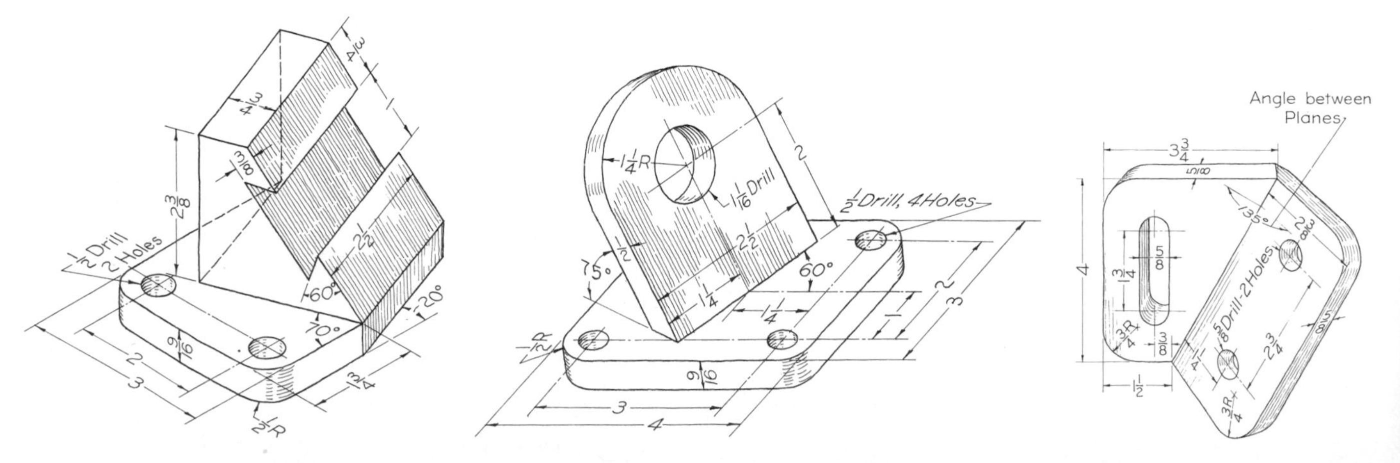


**PROB. 33.** Dovetail clip. **PROB. 34.** Anchor base. **PROB. 35.** Adjusting clip.

**36.** Draw the views given, omitting the lugs in the top view. Add normal views to describe the lugs.
**37.** Draw the views given, using edge and normal views to obtain the shape of the lugs.
**38.** Draw the partial top, front, and side views. Add edge and normal views to describe the lugs.
**39.** Draw the top and front views and use edge and normal views to describe the slots and skew surfaces. The part is symmetrical about the main axis.
**40.** Draw the spar clip, using the layout shown for 11- × 17-in. paper. Note that an edge and two normal views are required.
**41.** Draw the views given, using edge and normal views to describe the lugs.
**42.** Draw top and front views and auxiliary views that will describe the skew surface.

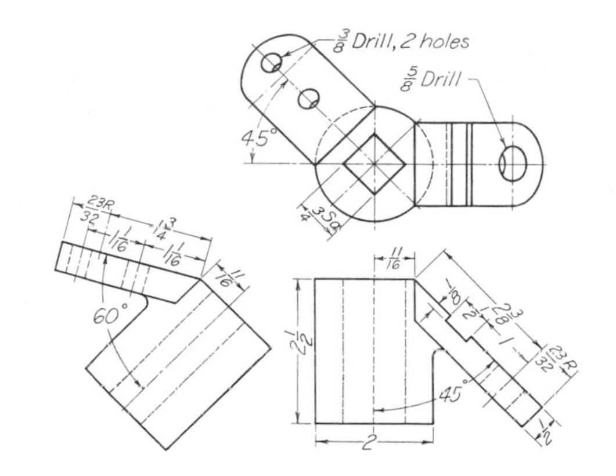


**PROB. 36.** Bar-strut anchor.

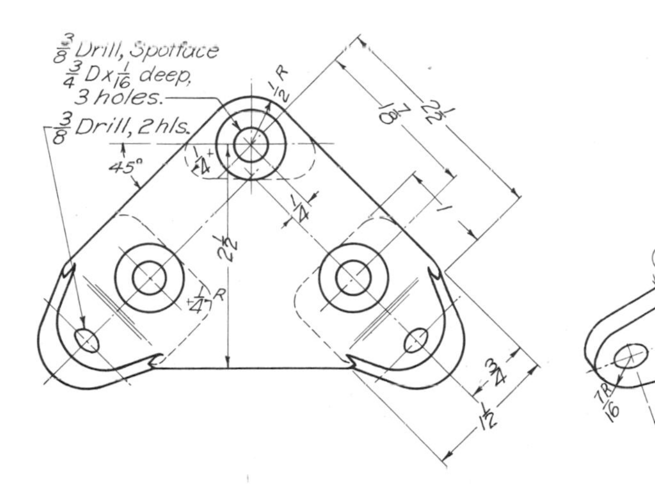


**PROB. 37.** Cable anchor.

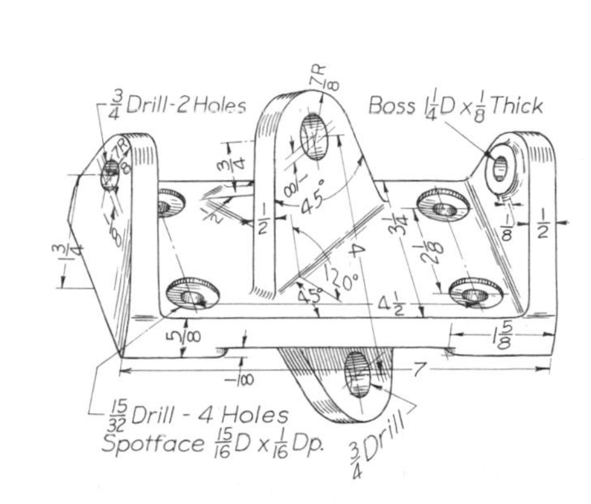


**PROB. 38.** Transverse connection.

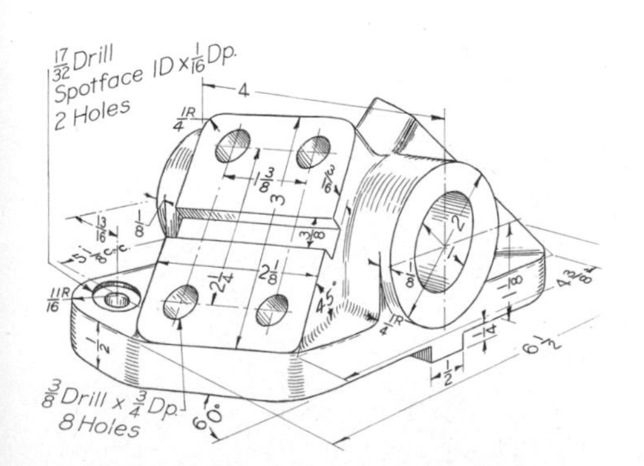


**PROB. 39.** Chamfer-tool base.

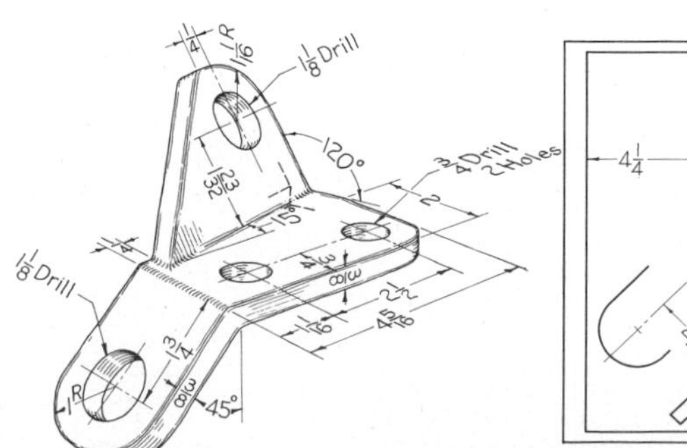


**PROB. 40.** Spar clip.

**43.** Shown on layout for 11- × 17-in. paper are partial front and right-side views and partial auxiliary views that describe the position and shape of the skew surface. Complete front, right-side, and first auxiliary views.

**44.** Shown on layout for 11- × 17-in. paper are partial top and front views and auxiliaries that describe the position and shape of the clamp portion of the part. Complete the top and front views.

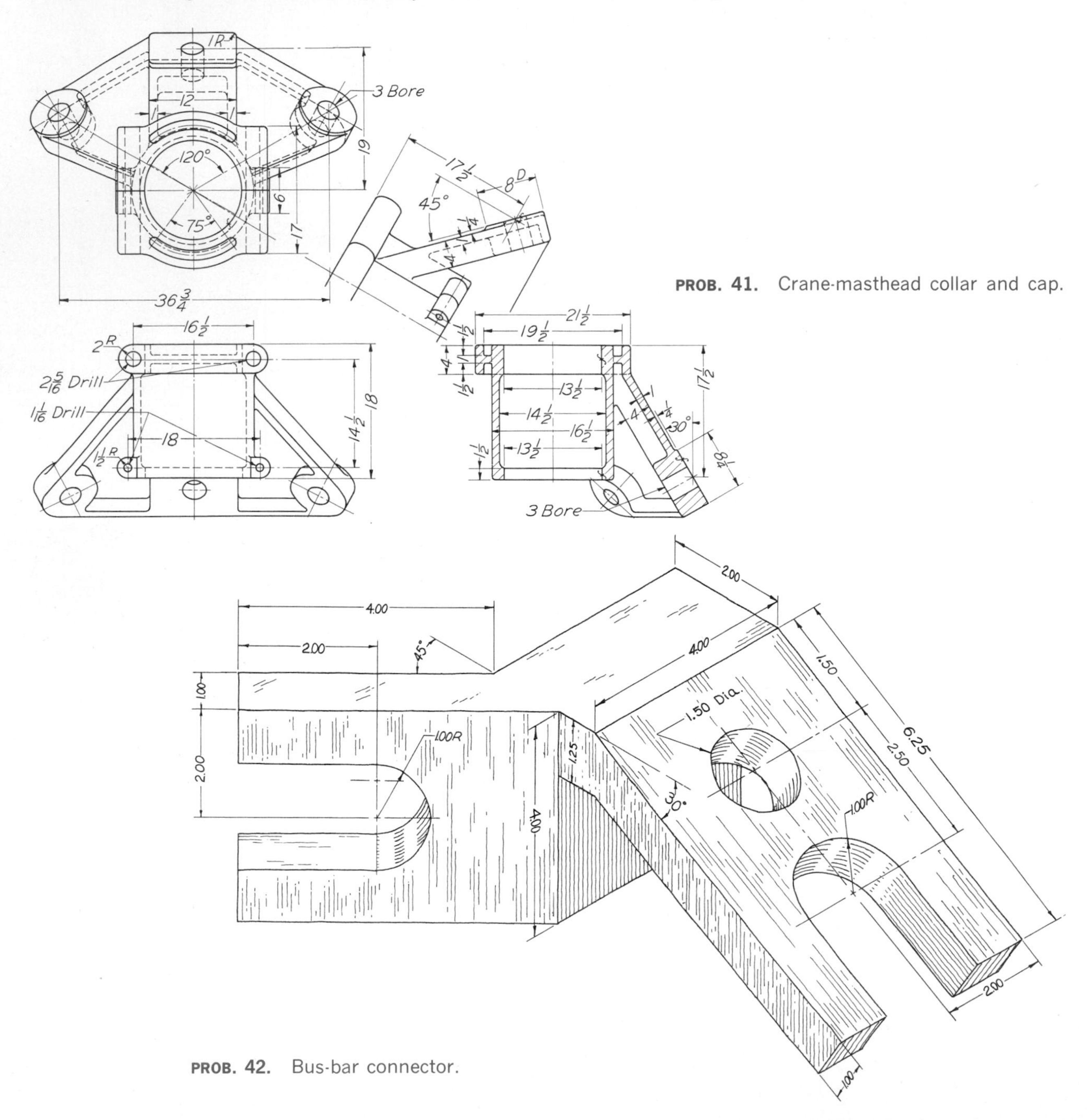


**PROB. 41.** Crane-masthead collar and cap.

**PROB. 42.** Bus-bar connector.

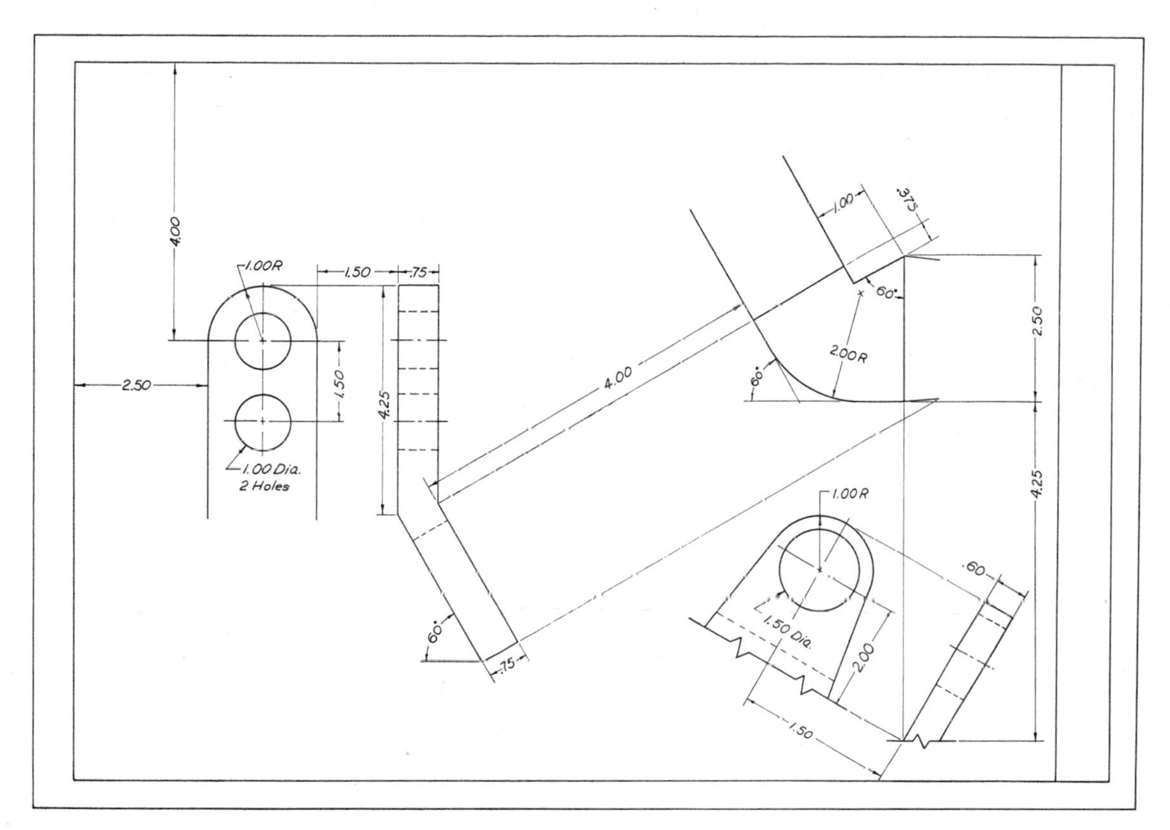


**PROB. 43.** Valve control-shaft bracket.

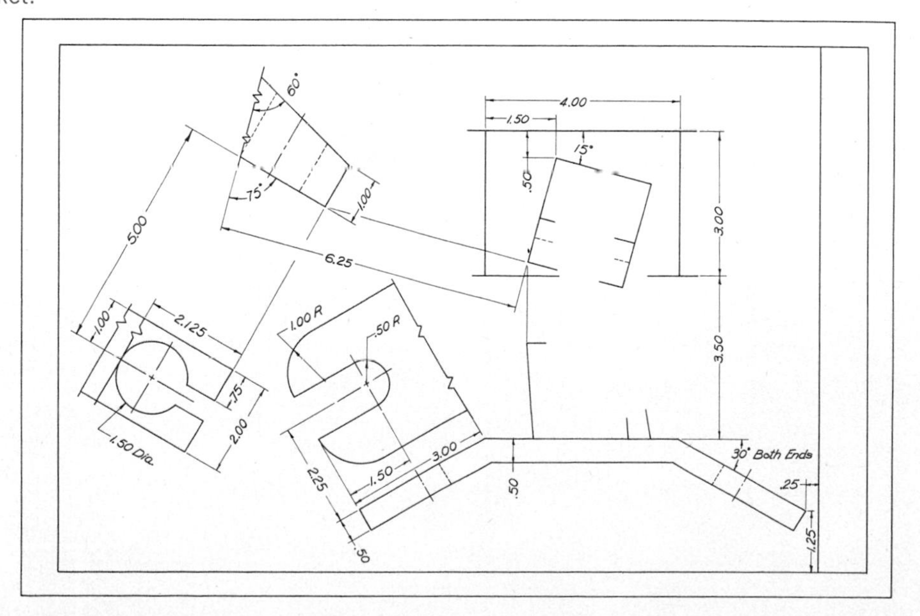


**PROB. 44.** Bipod shaft clamp.

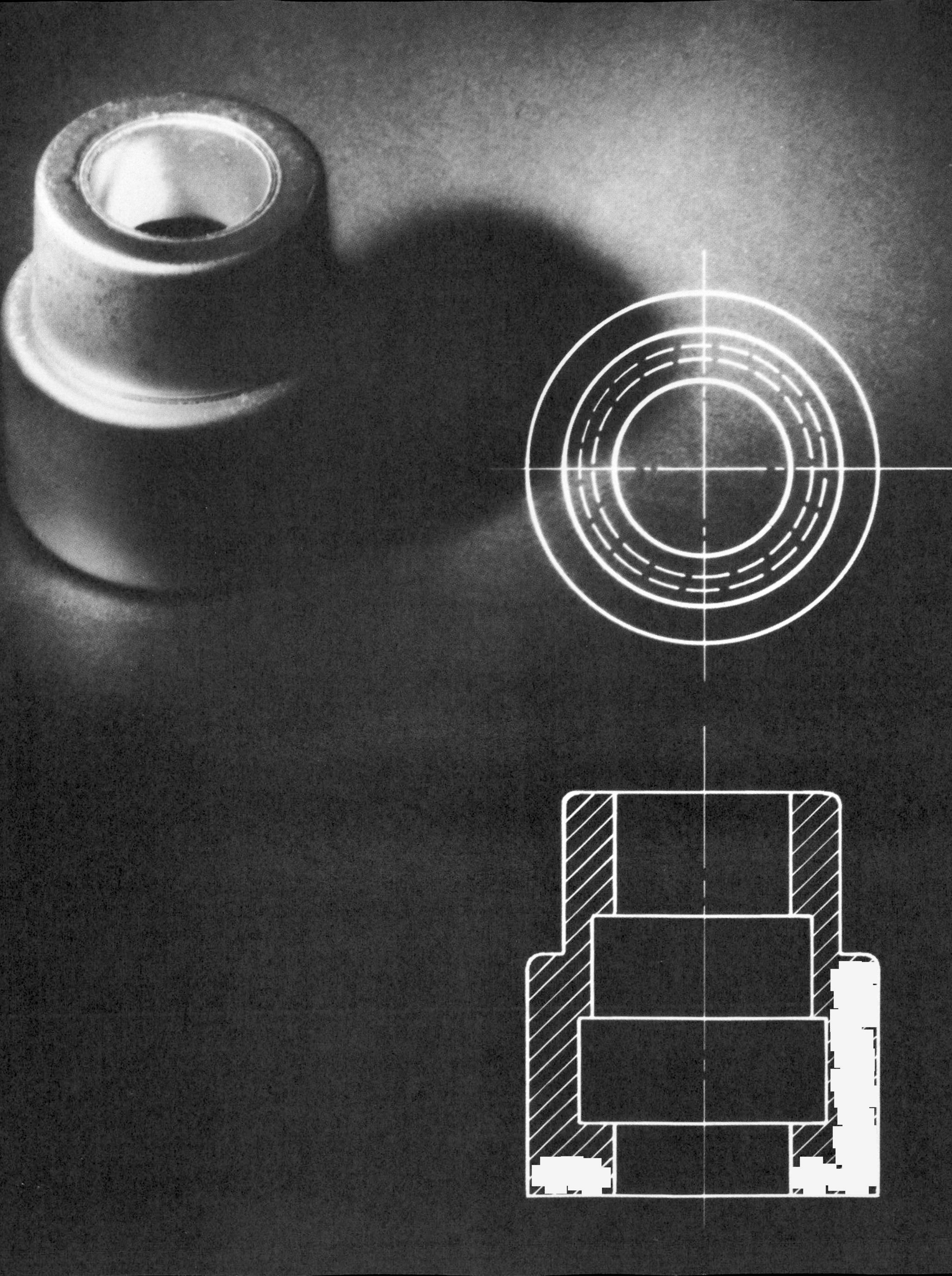

**Shown in this Chapter:**



# 8

# SECTIONAL VIEWS AND CONVENTIONS

**These Special Views and Practices Are Specific Aids to Complete and Accurate Orthographic Representation**

## 1. Sectional Views Defined.

When the interior of an object is complicated or when the component parts of a machine are drawn assembled, an attempt to show hidden portions by the customary dashed lines in regular orthographic views often results in a confusing network, as shown in Fig. 1 at (*a*), which is difficult to draw and almost impossible to read clearly.

In cases of this kind, to aid in describing the object, one or more views are drawn to show the object as if a portion had been cut away to reveal the interior, as at (*b*). Also, if some detail of the shape of an object is not clear, a cut taken through the portion and then turned up, or turned up and removed, as at (*c*), will describe the shape concisely and often eliminate the need for an extra complete view.

Either of these conventions is called a *section,* which is defined as an imaginary cut made through an object to expose the interior or to reveal the shape of a portion. A view in which all or a substantial portion of the view is sectioned is known as a *sectional view.*

For some simple objects where the orthographic, *unsectioned* views can be easily read, sectional views are often preferable because they show clearly and emphasize the solid portions, the voids, and the shape.

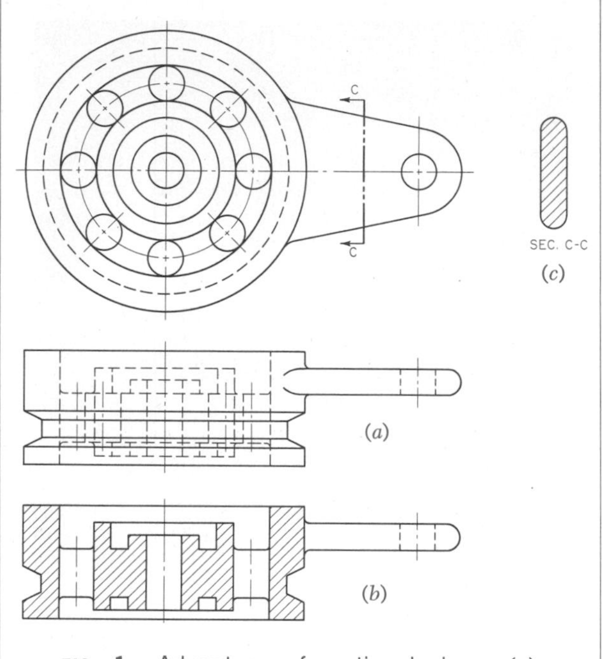


**FIG. 1.** Advantage of sectional views. (*a*) Orthographic view with hidden edges indicated by dashed lines; (*b*) the same view but made as a section to clarify the shape; (*c*) cross-sectional shape of lug shown by removed section.

## 2. How Sections are Shown.

The place from which the section is taken must be identifiable on the drawing, and the solid portions and voids must be distinguished on the sectional view. The place from which the section is taken is in many cases obvious, as it is for the sectional view (*b*) in Fig. 1; the section is quite evidently taken through the center (at the center line of the top view). In such cases no further description is needed. If the place from which a section is taken is not obvious, as at (*c*), a *cutting plane,* directional arrows, and identification letters are used to identify it. Whenever there is any doubt of the position where the section is taken, the cutting plane (see alphabet of lines, Sec. 39, Chap. 2) should be shown. A cutting plane is the imaginary medium used to show the path of cutting an object to make a section. Cutting planes for each kind of section will be discussed in the following sections.

A sectional view must show which portions of the object are solid material and which are spaces. This is done by section lining, sometimes called "cross-hatching," the solid parts with lines, as shown at (*b*) and (*c*). Section-lining practice is given in Sec. 11, where codes for materials are also discussed.

## 3. Types of Sections.

Although the different sections and sectional views have been named for identification and for speci-

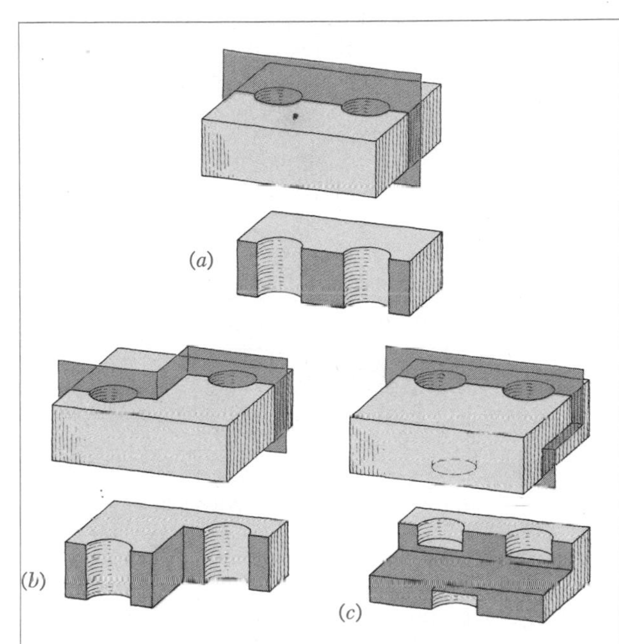


**FIG. 2.** Cutting planes for a full section. The plane may cut straight across (*a*) or change direction (*b* and *c*) to pass through features to be shown.

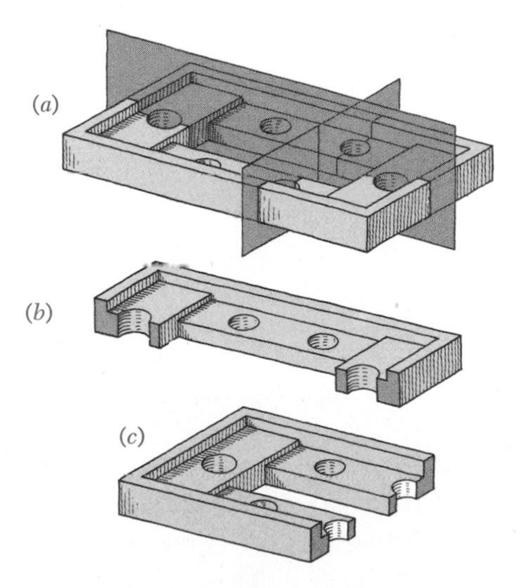


**FIG. 3.** Two cutting planes. The two planes at (*a*) will produce sections (*b*) and (*c*). Each section is considered separately, without reference to what has been removed for the other view.

fying the type of view required in a drawing, the names are not shown on the drawing for the same reason that a top, front, or side view would not be so labeled—the views are easily interpreted and a workman does not require the name to read the drawing. The names are assigned by the character of the section or the amount of the *view* in section, *not* by the amount of the object removed.

## 4. Full Section.

A full section is one in which the cutting plane passes entirely across the object, as in Fig. 2, so that the resulting view is completely "in section." The cutting plane may pass straight through, as at (*a*), or be offset, changing direction forward or backward, to pass through features it would otherwise have missed, as at (*b*) and (*c*). Sometimes *two* views are drawn in section on a pair of cutting planes, as at (*a*) in Fig. 3. In such cases each view is considered separately, without reference to what has been removed for another view. Thus (*b*) shows the portion remaining and the cut surface for one sectional view, and (*c*) for the other sectional view. Figure 4 is the orthographic drawing of the object, with indication for sectioning, as shown in Fig. 3. Both the front and side views are full-sectional views.

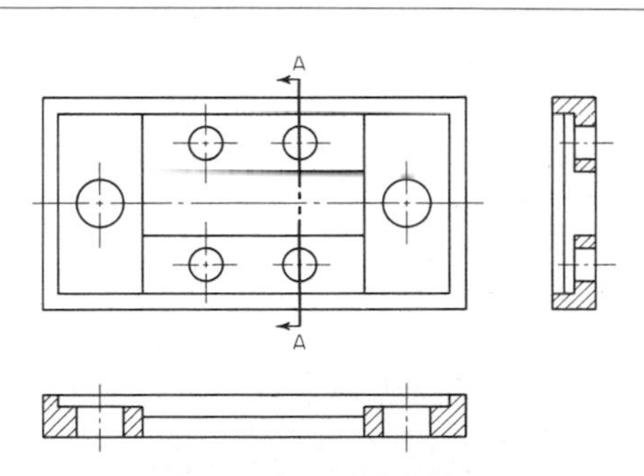


**FIG. 4.** Full sections. This is the orthographic drawing of the object in Fig. 3. The cutting-plane position is obvious for the front-view section and is not identified in the top view. Section *A-A* (side view) has its cutting plane identified because the section might be made elsewhere.

Figure 5 shows a full section made on an offset cutting plane. Note that the *change in plane direction* is *not* shown on the sectional view, for the cut is purely imaginary and *no edge* is present on the object at this position.

## 5. Half Section.

This is a view sometimes used for symmetrical objects in which one half is drawn in section and the other half as a regular exterior view. The cutting plane is imagined to extend halfway across, then forward, as in Fig. 6. A half section has the advantage of showing both the interior and exterior of the object on one view without using dashed lines, as at (*a*) in Fig. 7. However, a half section thus made is difficult to dimension without ambiguity, and so, if needed for clarity, dashed lines may be added, as at (*b*).

Note particularly that a *center line* separates the exterior and interior portions on the sectional view. This is for the same reason that the change in plane direction for the offset of the cutting plane of Fig. 5 is not shown—no edge exists *on the object* at the center.

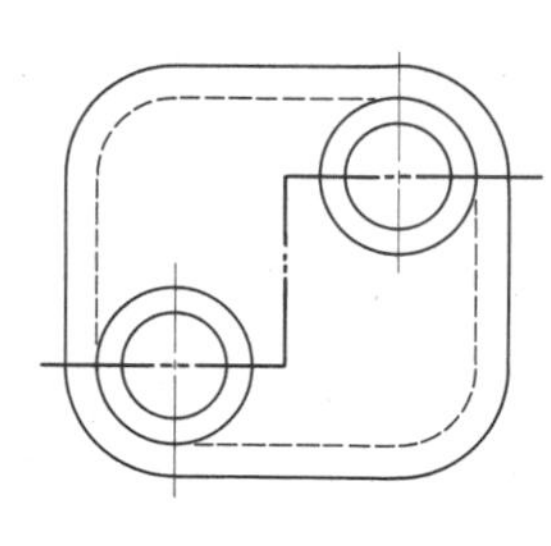

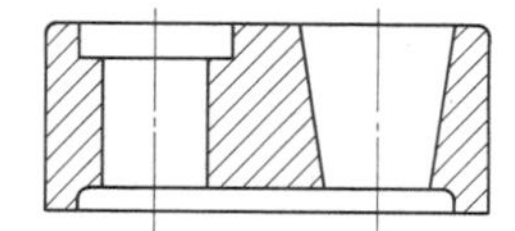

**FIG. 5.** A full section. The cutting plane is offset to pass through both principal features of the object.

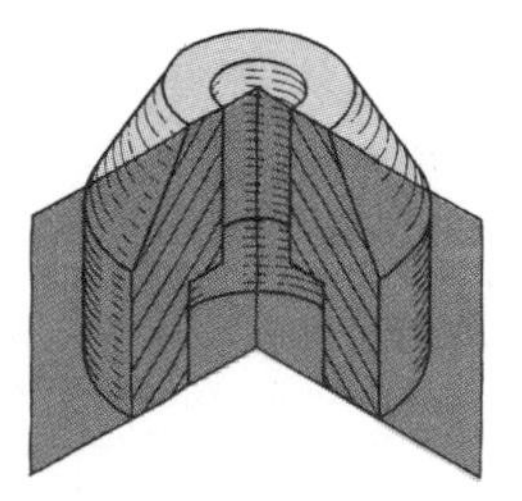

**FIG. 6.** The cutting plane for a half section. The resulting sectional view will be half in section and half an external view.

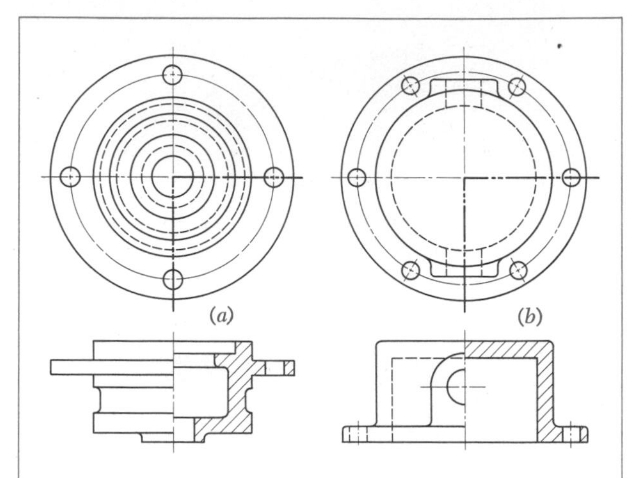


**FIG. 7.** Half sections. Dashed lines are rarely necessary (*a*), but may be used if needed for clarity or to aid in dimensioning (*b*).

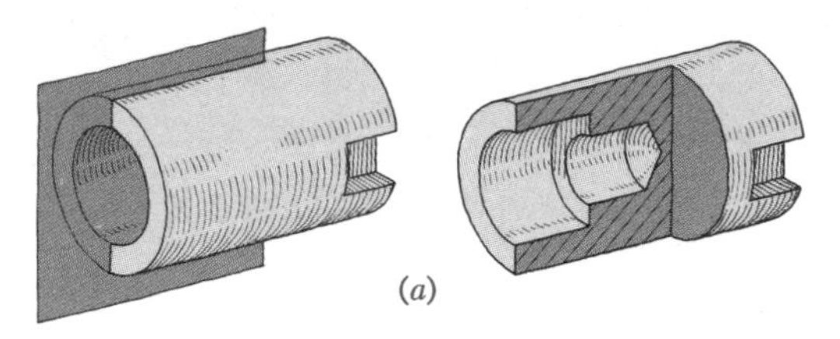


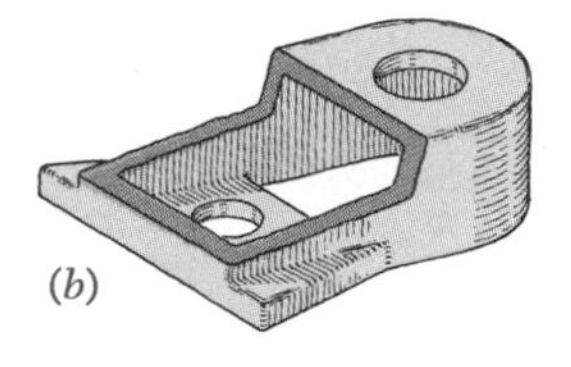


**FIG. 8.** Cutting planes for broken-out sections. The cutting plane extends only slightly beyond the features to be shown in section.

## 6. Broken-out Section.

Often an interior portion must be shown but a full or half section cannot be used because the cutting plane would remove some feature that must be included. For this condition the cutting plane is extended only so far as needed, as illustrated by Fig. 8*a* and *b*, and then is thought of as "broken out." Figure 9 is an example. Note the irregular break line that limits the extent of the section.

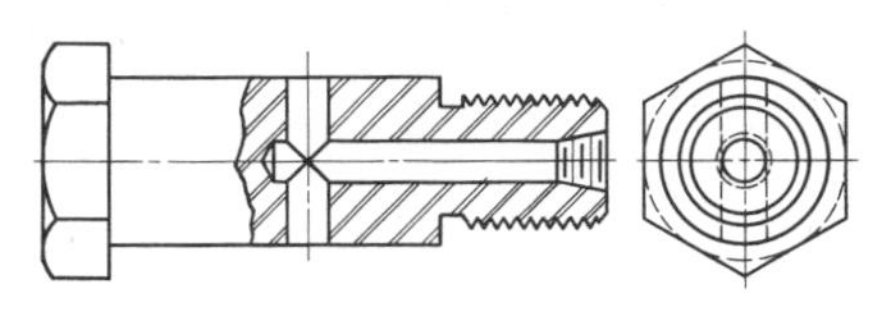

**FIG. 9.** A broken-out section. Note that a standard break line terminates the sectional portion.

## 7. Rotated Section.

As indicated in Sec. 1, a section may be a slice of negligible thickness used to show a shape that would otherwise be difficult to see or describe. The cutting plane for such a section is shown in Fig. 10. If the resulting section is then rotated 90° *onto* the view as in Fig. 11, the section is called a rotated section. Whenever the view outline interferes with the section, the view is broken, as in Fig. 12.

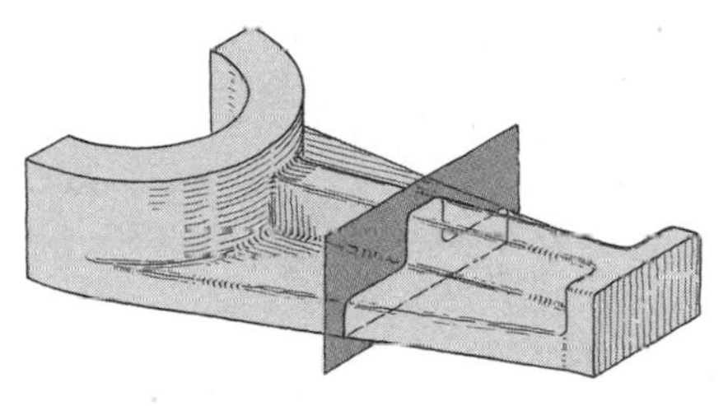

**FIG. 10.** The cutting plane for a rotated or a removed section. A slice of negligible thickness is taken.

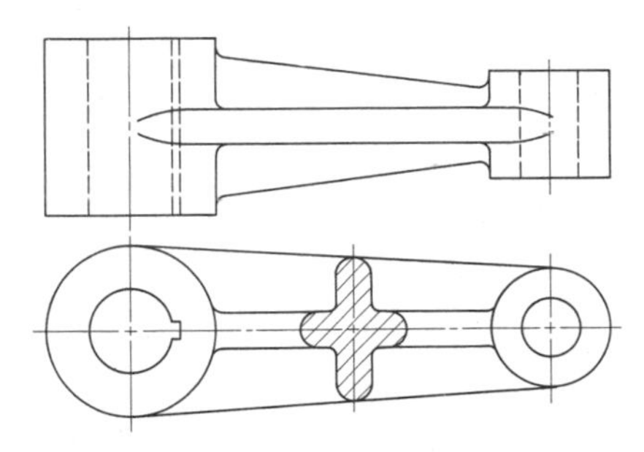

**FIG. 11.** A rotated section. The section is rotated 90° to bring it into the plane of the view.

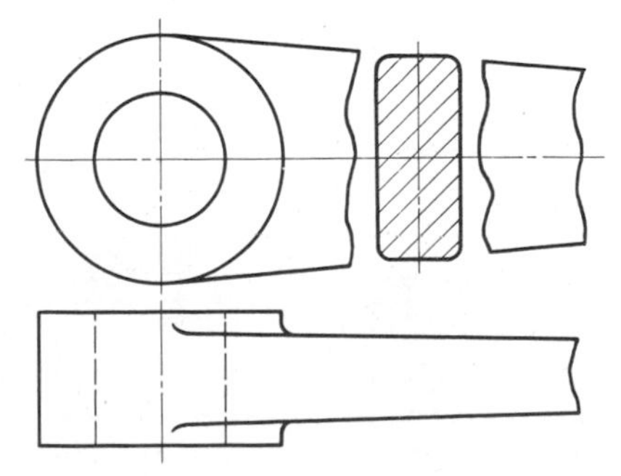

**FIG. 12.** A rotated section with broken view. The view is broken thus whenever the view outline interferes with the section.

## 8. Removed Sections.

These are used for the same purpose as rotated sections, but instead of being drawn *on* the view, they are removed to some adjacent place on the paper, as in Fig. 13.

The cutting plane with reference letters should always be indicated unless the place from which the section has been taken is obvious. Removed sections are used whenever restricted space for the section

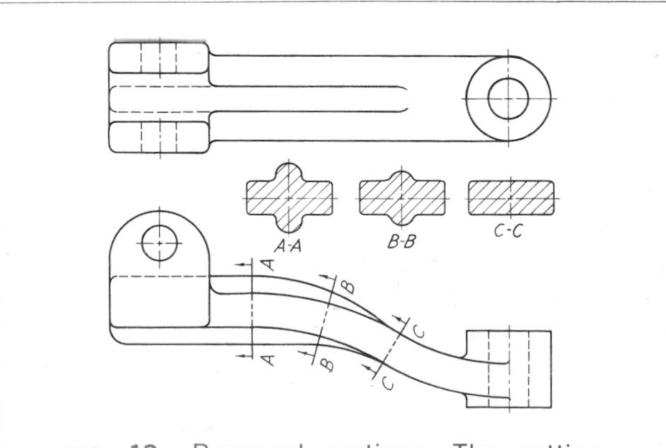


**FIG. 13.** Removed sections. The cutting planes and the mating sections must be identified.

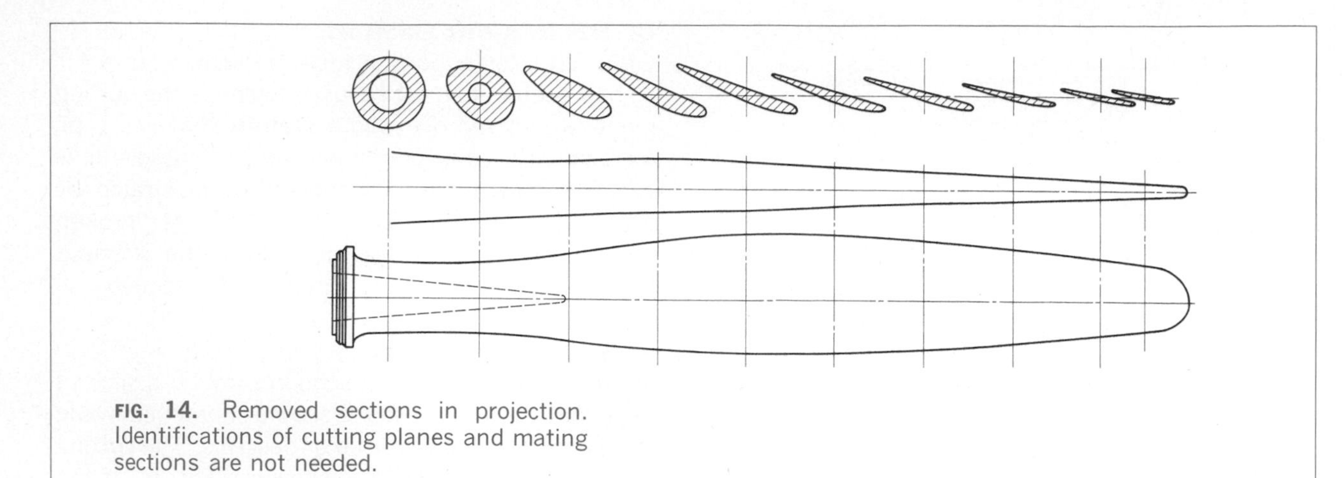

FIG. 14. Removed sections in projection. Identifications of cutting planes and mating sections are not needed.

or the dimensioning of it prevents the use of an ordinary rotated section. When the shape of a piece changes gradually or is not uniform, several sections may be required (Fig. 14). It is often an advantage to draw the sections to larger scale than that of the main drawing in order to show dimensions more clearly. Sometimes sections are removed to a separate drawing sheet. When this is done, the section must be carefully shown on the main drawing with cutting plane and identifying letters. Often these identifying letters are made as a fraction in a circle, with the numerator a letter identifying the section and the denominator a number identifying the sheet. The sectional view is then marked with the same letters and numbers. The ASA recommends that, whenever possible, a removed section be drawn in its natural projection position. Note in Fig. 14 that this practice is followed.

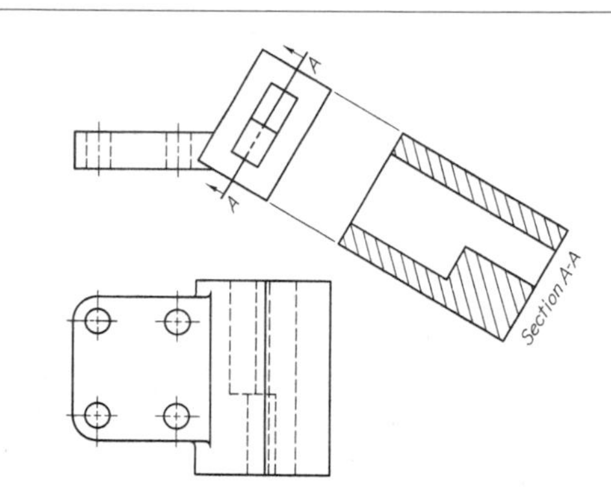


FIG. 15. Auxiliary section. The section is a normal view of the cutting plane.

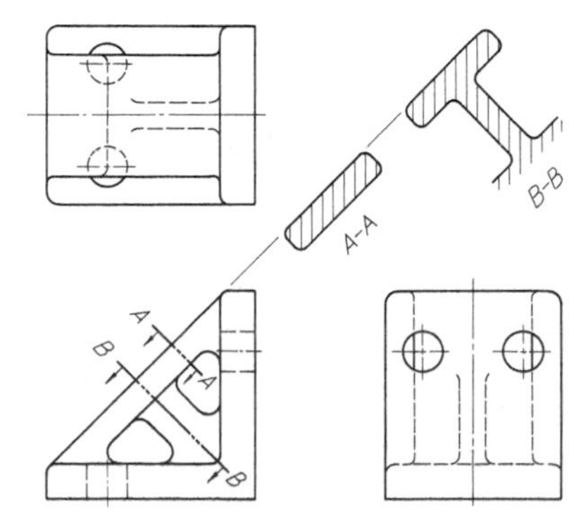


FIG. 16. Auxiliary sections (partial). Identifications of the cutting planes and mating sections are necessary.

## 9. Auxiliary Sections.

These are sectional views conforming to all the principles of edge and normal views given in Chap. 7. The section shows the *normal* view of a cutting plane that is in a position on an inclined feature so as to reveal the interior, as shown in Fig. 15. The edge view of any perpendicular faces will also be seen in the normal view, as in Fig. 15. All types of sections—full, half, broken-out, rotated, and removed—are used on auxiliaries. Figure 16 shows auxiliary partial sections, which are also properly called re-

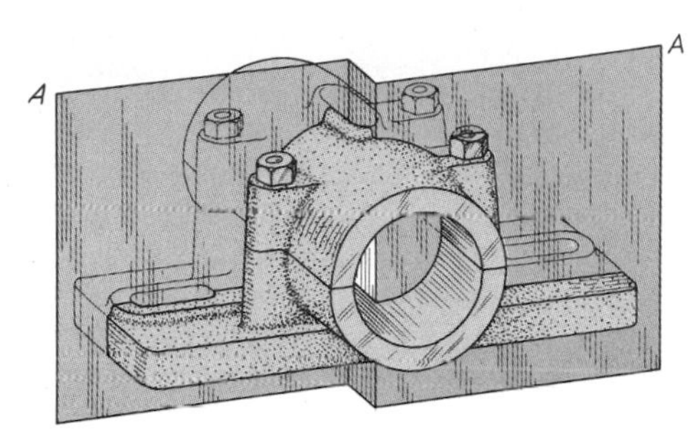


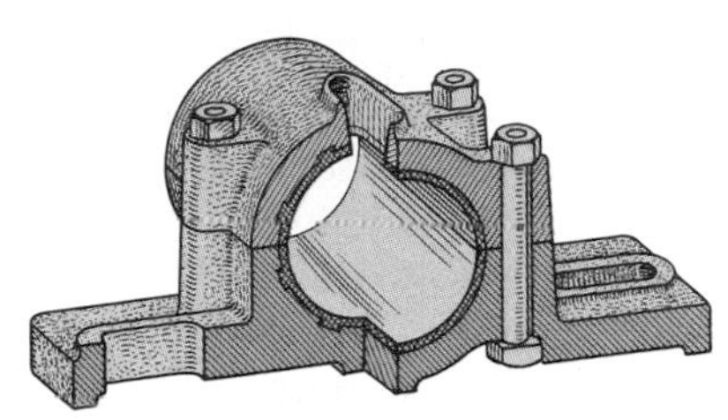

**FIG. 17.** The cutting plane for an assembly section. It is often offset to pass through features to be shown in the section.

moved sections in auxiliary position. Note in Fig. 16 that the two sections are normal views of their cutting planes.

## 10. Assembly Sections.

As the name implies, an assembly section is made up of a combination of parts. All the previously mentioned types of sections may be used to increase the clarity and readability of assembly drawings. The cutting plane for an assembly section is often offset, as in Fig. 17, to reveal the separate parts of a machine or structure.

The purpose of an assembly section is to reveal the interior of a machine or structure so that the separate parts can be clearly shown and identified, but the separate parts do not need to be completely described. Thus only such hidden details as are needed for part identification or dimensioning are shown. Also, the small amount of clearance between mating or moving parts is not shown because, if shown, the clearance would have to be greatly exaggerated, thus confusing the drawing. Even the clearance between a bolt and its hole, which may be as much as $^{1}/_{16}$ in., is rarely shown. Figure 18 is an example of an assembly section in tabular form. The component is similar to that in Fig. 17. Note that a half section is used in Fig. 18 on both views.

Crosshatching practice for assembly sections is explained in Sec. 11.

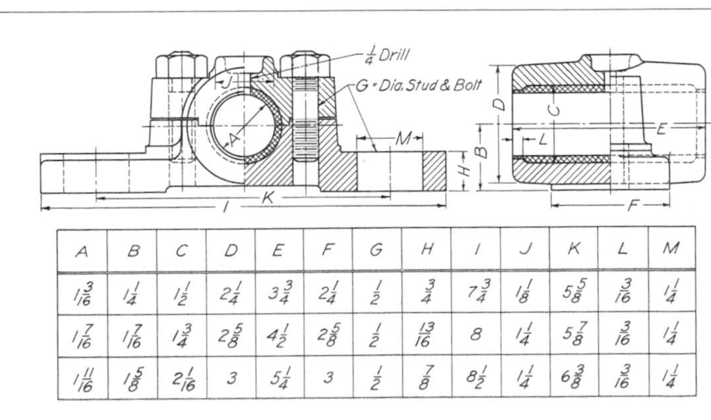


| A | B | C | D | E | F | G | H | I | J | K | L | M |
|---|---|---|---|---|---|---|---|---|---|---|---|---|
| $1\frac{3}{16}$ | $1\frac{1}{4}$ | $1\frac{1}{2}$ | $2\frac{1}{4}$ | $3\frac{3}{4}$ | $2\frac{1}{4}$ | $\frac{1}{2}$ | $\frac{3}{4}$ | $7\frac{3}{4}$ | $1\frac{1}{8}$ | $5\frac{5}{8}$ | $\frac{3}{16}$ | $1\frac{1}{4}$ |
| $1\frac{7}{16}$ | $1\frac{7}{16}$ | $1\frac{3}{4}$ | $2\frac{5}{8}$ | $4\frac{1}{2}$ | $2\frac{5}{8}$ | $\frac{1}{2}$ | $\frac{13}{16}$ | 8 | $1\frac{1}{4}$ | $5\frac{7}{8}$ | $\frac{3}{16}$ | $1\frac{1}{4}$ |
| $1\frac{11}{16}$ | $1\frac{5}{8}$ | $2\frac{1}{16}$ | 3 | $5\frac{1}{4}$ | 3 | $\frac{1}{2}$ | $\frac{7}{8}$ | $8\frac{1}{2}$ | $1\frac{1}{4}$ | $6\frac{3}{8}$ | $\frac{3}{16}$ | $1\frac{1}{4}$ |

**FIG. 18.** A sectional assembly drawing. The sectional views give emphasis to construction and separate parts.

## 11. Drawing Practices for Sectional Views.

In general, the rules of projection are followed in making sectional views. Figure 19 shows the picture of a casting intersected by a cutting plane, giving the appearance that the casting has been cut through by the plane *A-A* and the front part removed, exposing the interior. Figure 20 shows the drawing of the casting with the front view in section. The edge of the cutting plane is shown in the top view by the cutting-plane symbol, with reference letters and arrows to show the direction in which the view is taken. It must be understood that in thus removing the nearer portion of the object to make the sectional view, the portion assumed to be removed is not omitted in making other views. Therefore, the top and right-side views of the object in Fig. 20 are full and complete, and only in the front view has part of the object been represented as removed.

The practices recommended by the ASA for inclusion of the cutting-plane symbol, of visible and hidden edges, and for crosshatching are as follows:

*The Cutting-plane Symbol.* It may be shown on the orthographic view where the cutting plane appears as an edge and may be more completely identified with reference letters along with arrows to show the direction in which the view is taken. The cutting-plane line symbol is shown in the alphabet of lines, Figs. 56 and 57, Chap. 2. Use of the symbol is illustrated in Figs. 1 and 20. Often when the position of the section is evident, the cutting-plane symbol is omitted (Fig. 21). It is not always desirable to show the symbol through its entire length; so in such cases the beginning and ending of the plane is shown, as in sections *A-A* and *B-B* in Prob. 47 in the final chapter. Removed sections usually need the cutting-plane symbol with arrows to show the direction of sight and with letters to name the resulting sectional view (Fig. 16).

*Unnecessary Hidden Detail.* Hidden edges and surfaces are not shown unless they are needed to describe the object. Much confusion may result if all detail behind the cutting plane is drawn. In Fig. 21, (*a*) shows a sectional view with all the hidden edges and surfaces shown by dashed lines. These

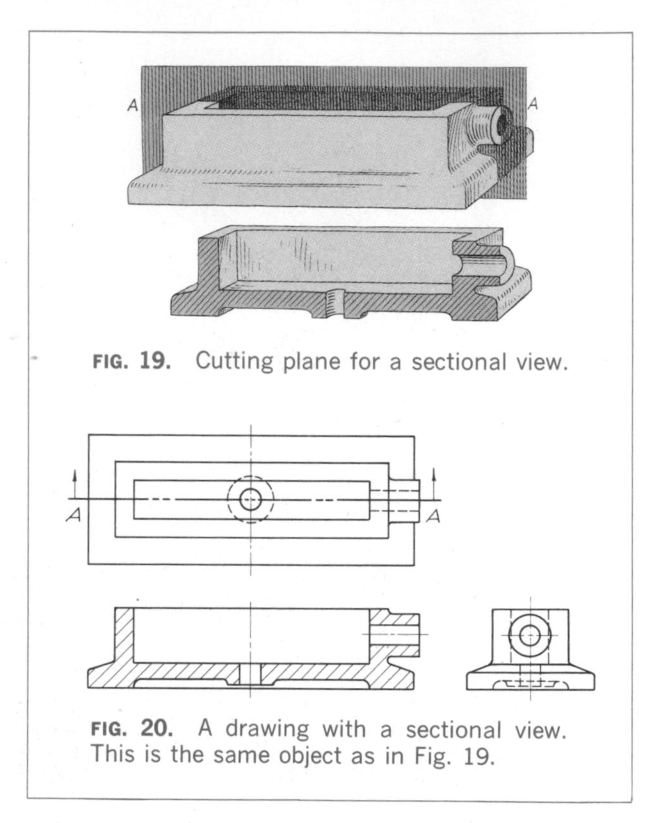


**FIG. 19.** Cutting plane for a sectional view.

**FIG. 20.** A drawing with a sectional view. This is the same object as in Fig. 19.

lines complicate the view and do not add any information. The view at (*b*) is preferred because it is simpler, less time-consuming to draw, and more easily read than the view at (*a*). The holes lie on a circular center line; and where similar details repeat, all may be assumed to be alike.

*Necessary Hidden Detail.* Hidden edges and surfaces are shown if necessary for the description of the object. In Fig. 22 view (*a*) is inadequate since it does not show the thickness of the lugs. The correct treatment is in view (*b*), where the lugs are shown by dashed lines.

*Visible Detail Shown in Sectional Views.* Figure 23 shows an object pictorially with the front half removed, thus exposing edges and surfaces behind the cutting plane. At (*a*) a sectional view of the cut surface only is shown, with the visible elements omitted. This treatment should *never* be used. The view should be drawn as at (*b*) with the visible edges and surfaces behind the cutting plane included in the sectional view.

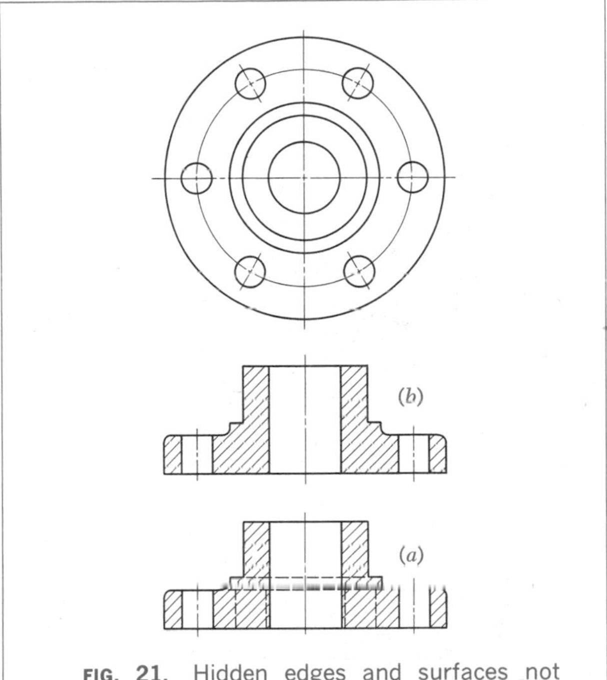


FIG. 21. Hidden edges and surfaces not shown; (*b*) is preferable to (*a*) because the dashed lines are not necessary for description of the part.

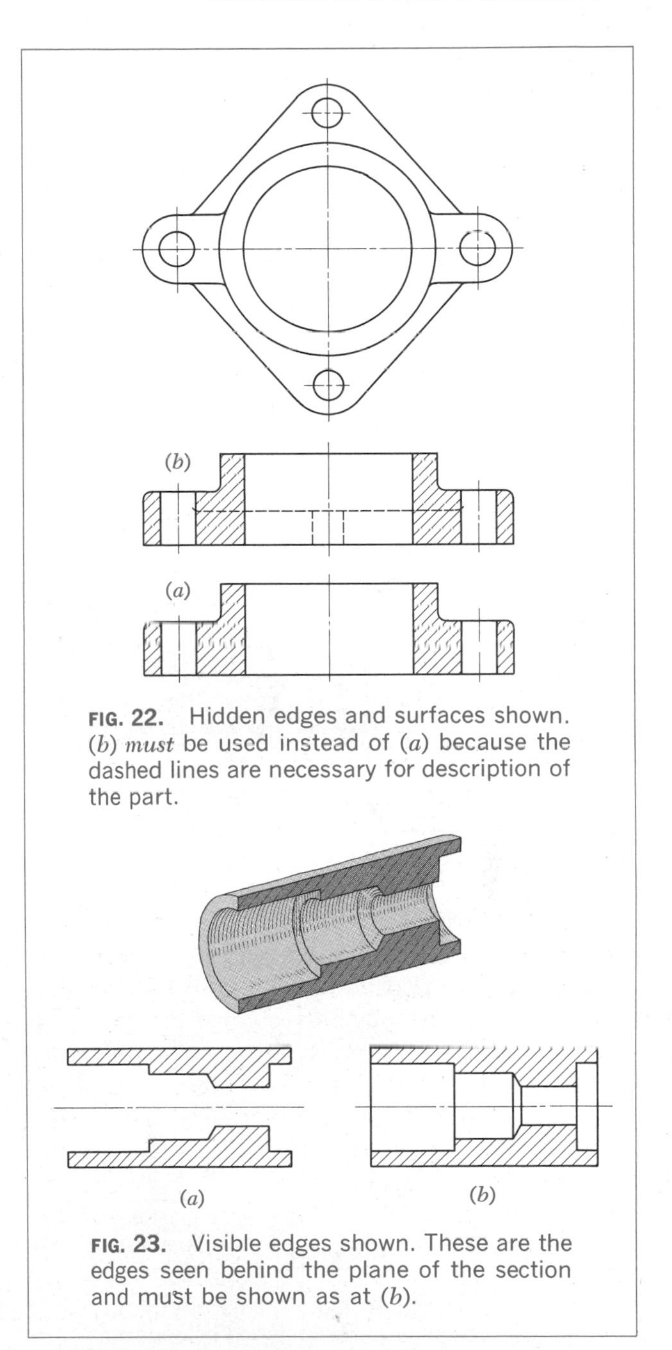


FIG. 22. Hidden edges and surfaces shown. (*b*) *must* be used instead of (*a*) because the dashed lines are necessary for description of the part.

FIG. 23. Visible edges shown. These are the edges seen behind the plane of the section and must be shown as at (*b*).

*Visible Detail Not Shown in Sectional Views.* Sometimes confusion results if all visible detail behind the cutting plane is drawn, and it may be omitted if it does not aid in readability. Omission of detail should be carefully considered and may be justified as time saved in drawing. This applies mainly to assembly drawings for showing how the pieces fit together rather than for giving complete information for making the parts (see Fig. 24).

*Section Lining.* Wherever material has been cut by the section plane, the cut surface is indicated by section lining done with fine lines generally at 45° with the principal lines in the view and spaced uniformly to give an even tint. These lines are spaced entirely by eye except when some form of mechanical section liner is used. The pitch, or distance between lines, is governed by the size of the surface. For ordinary working drawings, it will not be much less then $\frac{1}{16}$ in. and rarely more than $\frac{1}{8}$ in. *Very* small pieces may require a spacing closer than $\frac{1}{16}$ in. Take care in setting the pitch by the first two or three lines, and glance back at the first lines often to see that the pitch does not gradually

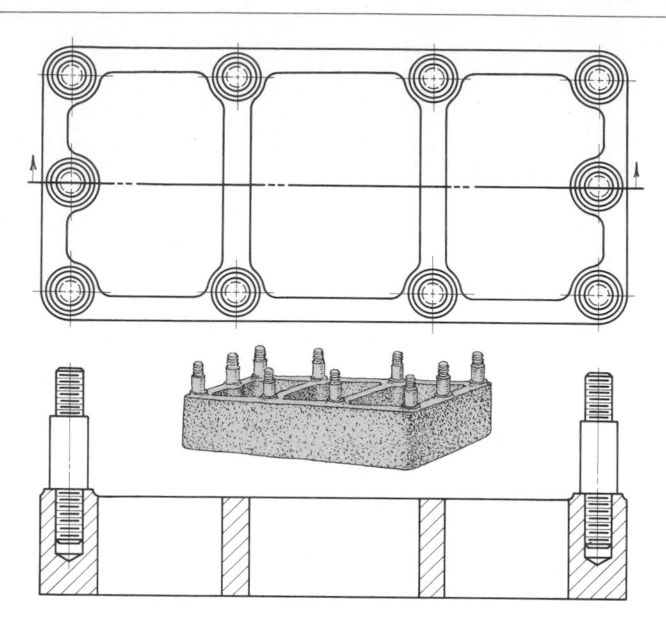

**FIG. 24.** Omission of detail. Detail behind the plane of the section may be omitted when it does not aid in readability or when it might cause ambiguity.

increase or decrease. Nothing mars the appearance of a drawing more than poor section lining. The alphabet of lines, Figs. 56 and 57, Chap. 2, gives the weight of crosshatch lines.

Two adjacent pieces in an assembly drawing are crosshatched in opposite directions. If three pieces adjoin, one of them may be sectioned at other than 45° (usually 30° or 60°, Fig. 25), or all pieces may be crosshatched at 45° by using a different pitch for each piece. If a part is so shaped that 45° sectioning runs parallel, or nearly so, to its principal outlines, another direction should be chosen (Fig. 26).

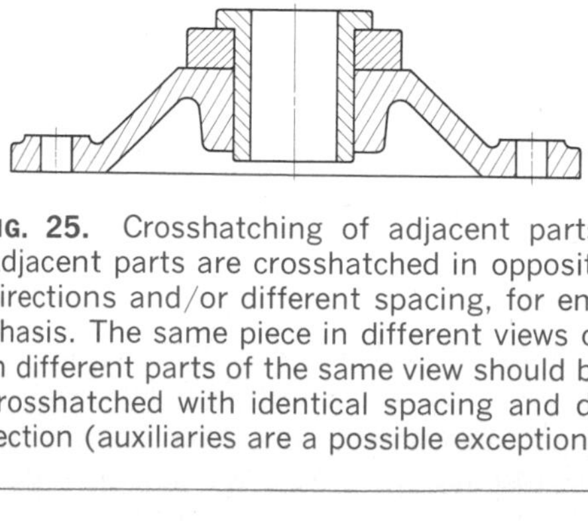

**FIG. 25.** Crosshatching of adjacent parts. Adjacent parts are crosshatched in opposite directions and/or different spacing, for emphasis. The same piece in different views or in different parts of the same view should be crosshatched with identical spacing and direction (auxiliaries are a possible exception).

Large surfaces are sometimes sectioned around the edge only, as illustrated in Fig. 27.

Very thin sections, as of gaskets, sheet metal, or structural-steel shapes to small scale, may be shown in solid black, with white spaces between the parts where thin pieces are adjacent (Fig. 28).

Section lining for the same piece in different views or for the same piece in different parts of the same view should be identical in spacing and direction[1] (Figs. 18 and 24).

Adjacent pieces are section-lined in opposite directions and are often distinguished more clearly by varying the pitch of the section lines for each piece, using closer spacing for the smaller pieces (Figs. 25 and 29).

[1] An exception to this rule is made for the crosshatching of an auxiliary view in order to avoid crosshatch lines that are parallel or perpendicular, or nearly so, to the outlines of the view.

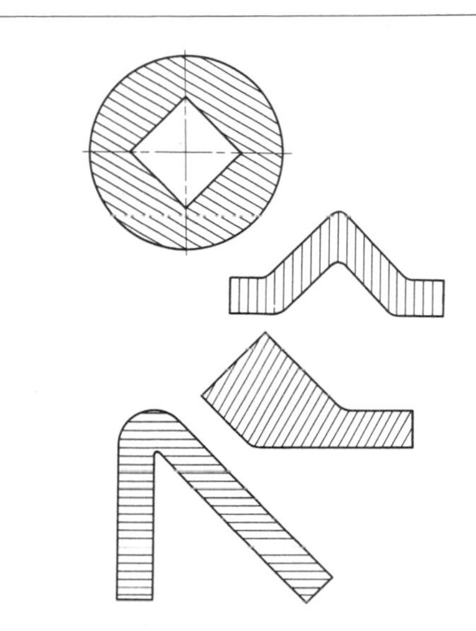

FIG. 26. Section-line directions for unusual shapes. Avoid crosshatch directions parallel to the view outlines.

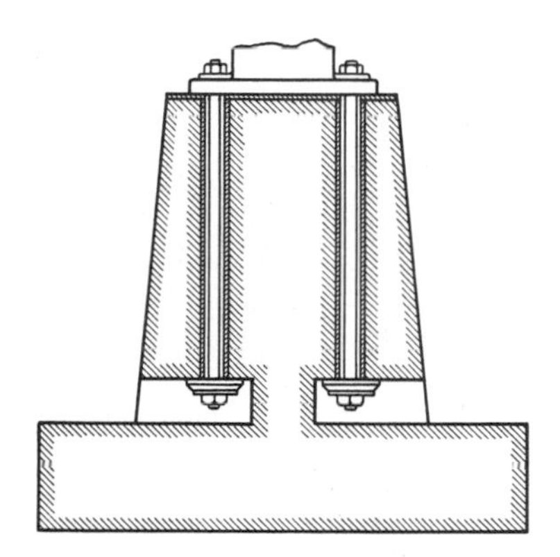

FIG. 27. Outline sectioning. This saves drafting time on large views or on large drawings.

*Code for Materials in Section.* Symbolic section lining is not commonly used on ordinary working drawings, but in an assembly section it is sometimes useful to show a distinction between materials, and a recognized standard code is an obvious advantage. The ASA symbols for indicating different materials is given in the Appendix. Code section lining is used only as an aid in reading a drawing and is not to be taken as the official specification of the materials. Exact specifications of the material for each piece are always given on the detail drawing.

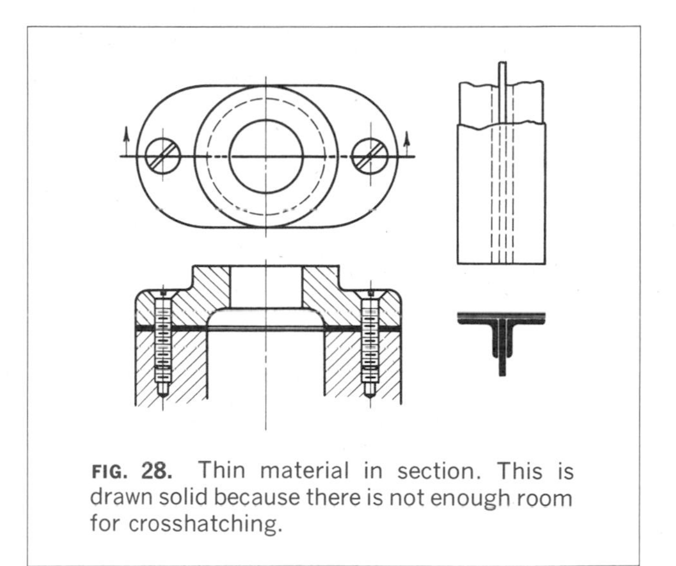

FIG. 28. Thin material in section. This is drawn solid because there is not enough room for crosshatching.

**12. Conventional Practices.**
All sections are conventions in that they represent an assumed imaginary cut from which, following the theory of projection rather closely, the sectional views are made. However, the strict rules of projection may be disregarded if by so doing a type of view can be drawn which more accurately depicts the shape of the object. It is impossible to illustrate all the conditions that might occur, but the following discussion shows the principles that are recognized as good practice, since observing them results in added clearness and readability.

**13. Parts Not Sectioned.**
Many machine elements, such as fasteners, pins, and shafts, have no internal construction and, in addition, are more easily recognized by their exterior views. These parts often lie in the path of the section plane, but if they are sectioned (and crosshatched), they are more difficult to read because their typical identifying features (boltheads, rivet heads, chamfers on shafts, etc.) are removed. Thus features of this kind should be left in full view and *not sectioned.* To justify this treatment, the assembly is thought of as being sectioned on a particular plane, with the

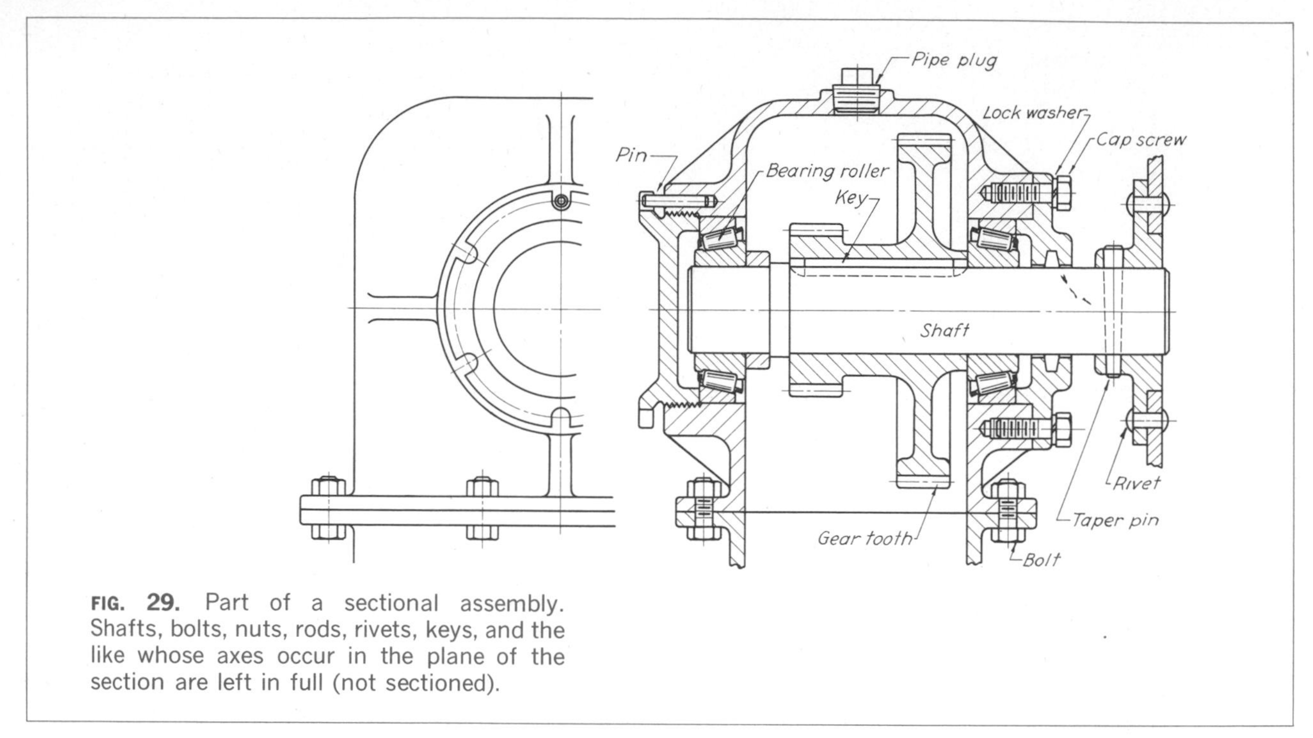


**FIG. 29.** Part of a sectional assembly. Shafts, bolts, nuts, rods, rivets, keys, and the like whose axes occur in the plane of the section are left in full (not sectioned).

nonsectioned parts placed in the half holes remaining after the section is made. Figure 17 shows a full section made on an offset cutting plane that passes through a bolt, with the bolt in full view. Figure 29 shows several nonsectioned parts; it is evident that if the shaft, bolts, nuts, rivets, etc., were sectioned, the drawing would be confusing and difficult to read.

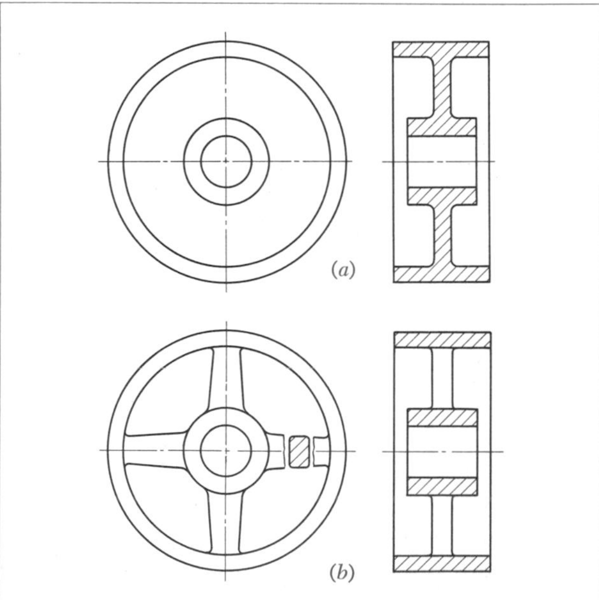


**FIG. 30.** Spokes in section. The wheel with spokes (*b*) is treated as though the cutting plane were in front of the spokes, to avoid misreading the section as a solid web line (*a*).

## 14. Spoke and Arms in Section.

A basic principle for sectioning circular parts is that any element not continuous (not solid) around the axis of the part should be drawn without crosshatching in order to avoid a misleading effect. For example, consider the two pulleys in Fig. 30. Pulley (*a*) has a solid web connecting the hub and rim. Pulley (*b*) has four spokes. Even though the cutting plane passes through two of the spokes, the sectional view of (*b*) must be made without crosshatching the spokes in order to avoid the appearance of a solid web, as in pulley (*a*).

Other machine elements treated in this manner are teeth of gears and sprockets, vanes and sup-

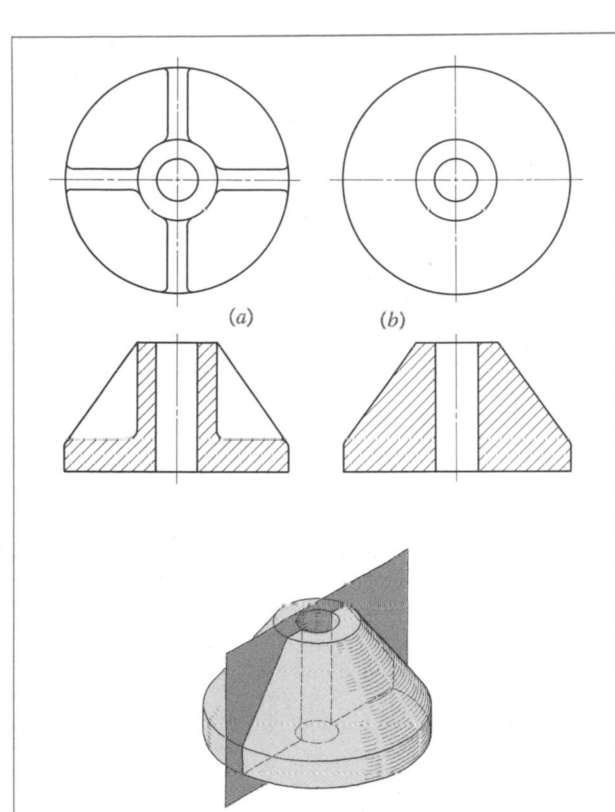


**FIG. 31.** Ribs in section. Ribs at (*a*) are treated as though the cutting plane were in front of them, to avoid misreading the section as a solid (*b*).

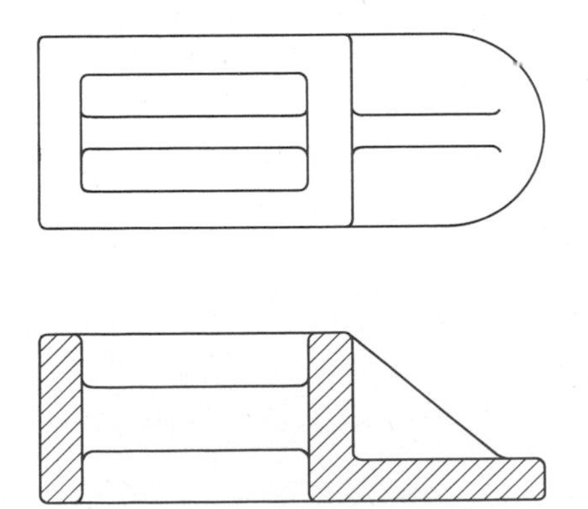

**FIG. 32.** Ribs in section. These are treated as though the cutting plane were in front of the ribs.

porting ribs of cylindrical parts, equally spaced lugs, and similar parts.

### 15. Ribs in Section.

For reasons identical with those given in Sec. 14, when the cutting plane passes longitudinally through the center of a rib or web, as in Fig. 31*a*, the crosshatching is eliminated from the ribs as if the cutting plane were just in front of them or as if they had been temporarily removed and replaced after the section was made. A true sectional view with the ribs crosshatched gives a heavy, misleading effect suggesting a cone shape, as shown at (*b*). The same principle applies to ribs cut longitudinally on rectangular parts (Fig. 32). When the cutting plane cuts a rib transversely, that is, at right angles to its length or axis direction (the direction that shows its thickness), it is always crosshatched (Fig. 24).

### 16. Lugs in Section.

For the same reasons given in Secs. 14 and 15, a lug or projecting ear (Fig. 33*a*), usually of *rectangular* cross section, is not crosshatched; note that crosshatching either of the lugs would suggest a circular flange. However, the somewhat similar condition at (*b*) should have the projecting ears crosshatched as shown because these ears *are* the base of the part.

### 17. Alternate Crosshatching.

In some cases omitting the crosshatching of ribs or similar parts gives an inadequate and sometimes ambiguous treatment. To illustrate, Fig. 34*a* shows a full section of an idler pulley. At (*b*) four ribs have been added. Note that the top surfaces of the ribs are flush with the top of the pulley. Without crosshatching, the section at (*b*) is identical with (*a*) and the ribs of (*b*) are not identified at all on the sectional view. A better treatment in this case is to use alternate crosshatching for the ribs, as at (*c*), where half (alternating) the crosshatch lines are carried through the ribs. Note that the line of demarcation between rib and solid portions is a *dashed* line.

### 18. Aligned Spokes and Arms.

Any part with an odd number (3, 5, 7, etc.) of spokes or ribs will give an unsymmetrical and misleading

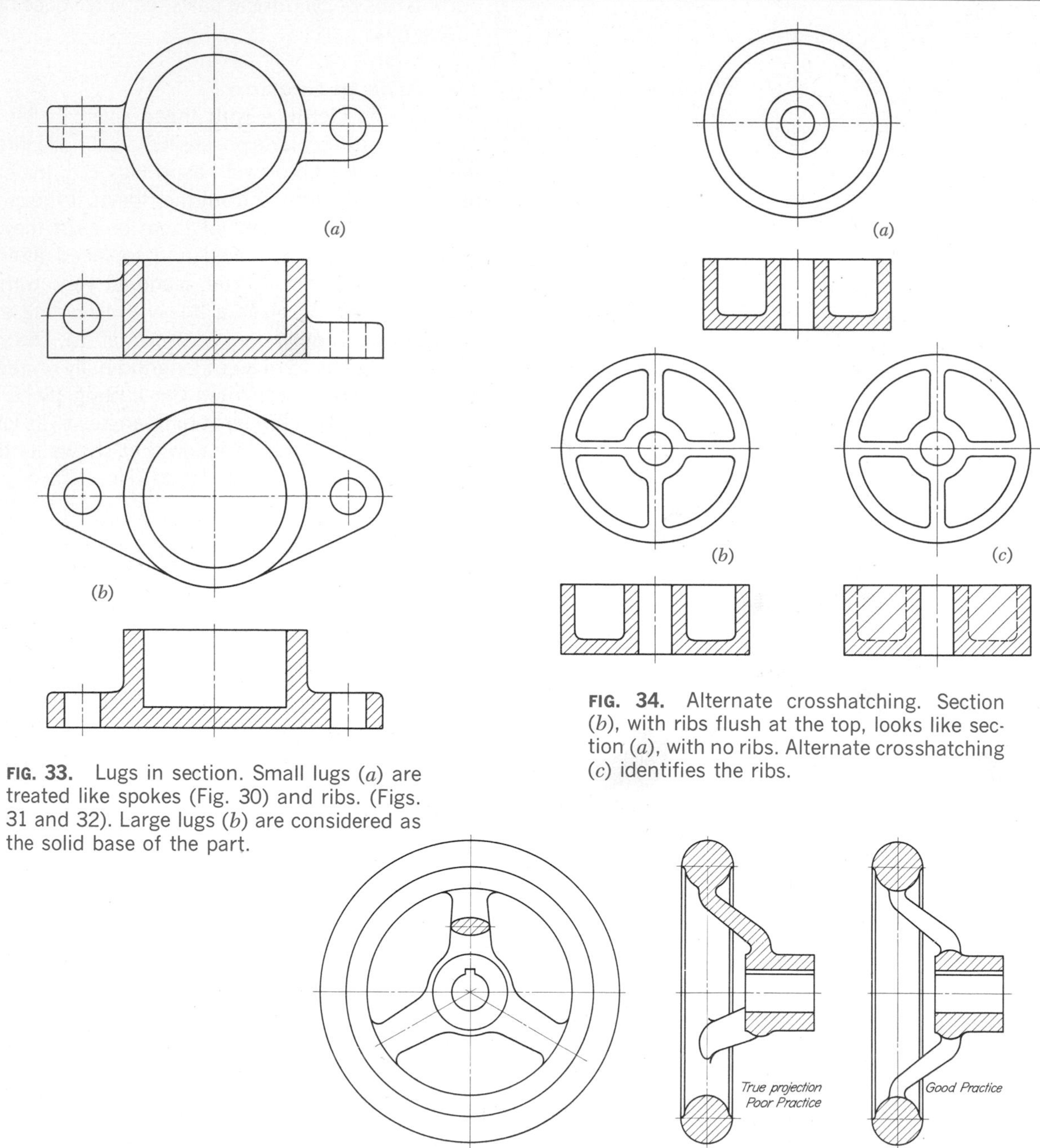


**FIG. 33.** Lugs in section. Small lugs (*a*) are treated like spokes (Fig. 30) and ribs. (Figs. 31 and 32). Large lugs (*b*) are considered as the solid base of the part.

**FIG. 34.** Alternate crosshatching. Section (*b*), with ribs flush at the top, looks like section (*a*), with no ribs. Alternate crosshatching (*c*) identifies the ribs.

**FIG. 35.** Aligned spokes. True projection is misleading and difficult to draw. Alignment gives a symmetrical section for a symmetrical part.

section if the principles of true projection are strictly adhered to, as illustrated by the drawing of a handwheel in Fig. 35. The preferred projection is shown in the second sectional view, where one arm is drawn as if aligned, or in other words, the arm is rotated to the path of the vertical cutting plane and then projected to the side view. Note that neither arm should be sectioned, for the reasons given in Sec. 14.

This practice of alignment is well justified logically because a part with an odd number of equally spaced elements is just as symmetrical as a part with an even number and, therefore, should be shown by a symmetrical view. Moreover, the symmetrical view shows the true *relationship* of the elements, while the true projection does not.

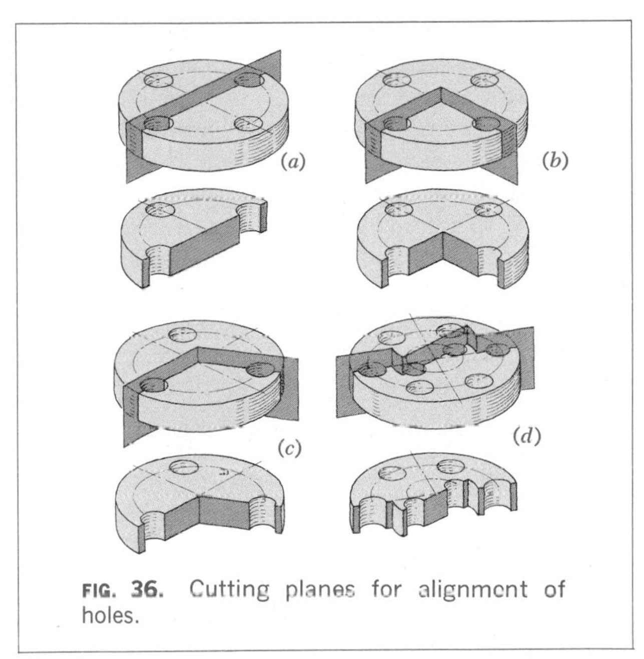


**FIG. 36.** Cutting planes for alignment of holes.

### 19. Aligned Ribs, Lugs, and Holes.

Ribs, lugs, and holes often occur in odd numbers and following the principles given in Sec. 18, should be aligned to show the true relationship of the elements. Figure 36 shows several examples of how the cutting plane may pass through a symmetrical object, permitting the removal of a portion of the object so as to describe the shape better. Note how the cutting planes may change direction so as to pass through the holes. In Fig. 37, true projection

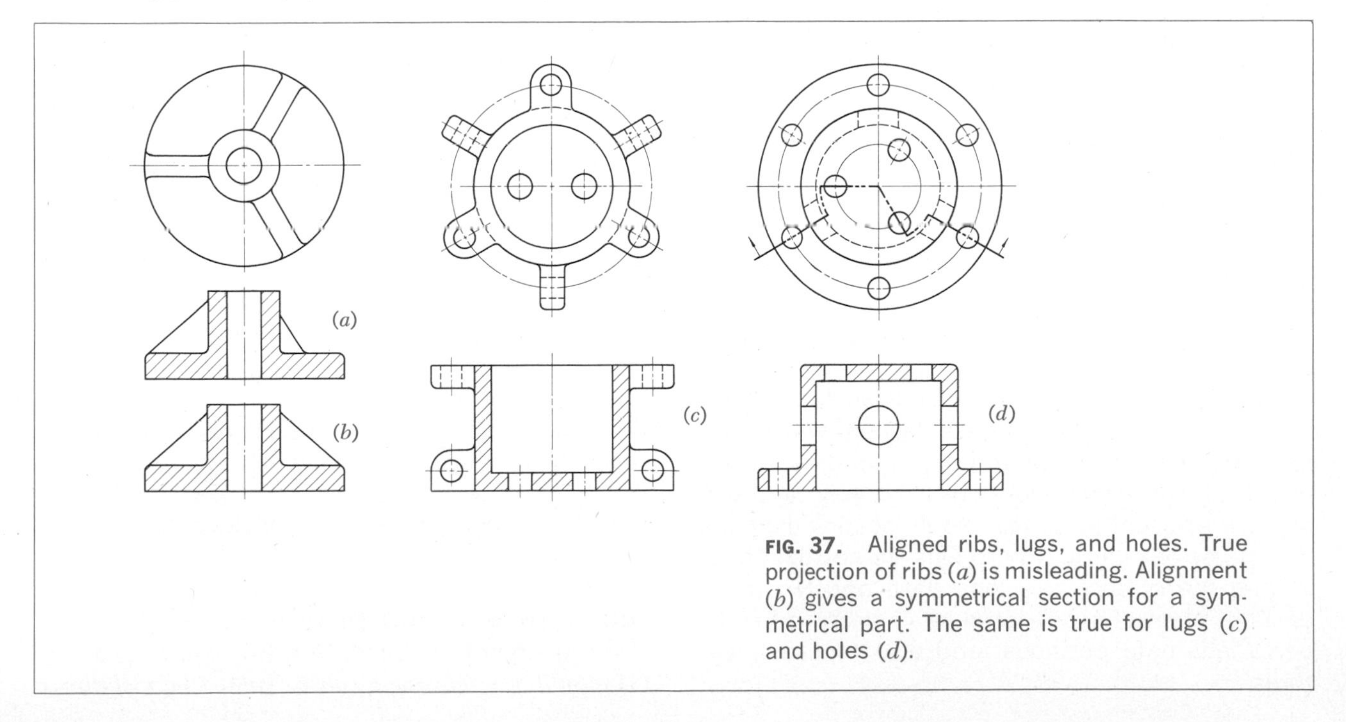


**FIG. 37.** Aligned ribs, lugs, and holes. True projection of ribs (*a*) is misleading. Alignment (*b*) gives a symmetrical section for a symmetrical part. The same is true for lugs (*c*) and holes (*d*).

of the ribs would show the pair on the right foreshortened, as at (*a*), suggesting in the sectional view that they would not extend to the outer edge of the base. Here, again, the alignment shown at (*b*) gives a symmetrical section of a symmetrical part and shows the ribs in their true relationship to the basic part. To illustrate further, at (*c*) and (*d*) the lugs and holes are aligned, thus showing the holes at their true radial distance from the axis, and, incidently, eliminating some difficult projections.

In all cases of alignment, the element can be thought of as being swung around to a common cutting plane and then projected to the sectional view. Note at (*c*) that because an offset cutting plane is used, each hole is brought separately into position on a common cutting plane before projection to the sectional view. The cutting plane used here is similar to that of (*d*) in Fig. 36.

### 20. Alignment of Elements in Full Views.

In full views, as well as in sectional views, certain violations of the rules of true projection are recognized as good practice because they add to the clearness of the drawing. For example, if a front view shows a hexagonal bolthead "across corners," the theoretical projection of the side view would be "across flats"; but in a working drawing, boltheads are drawn across corners in both views to show better the shape and the space needed. As another example, the slots of screw heads are always drawn at an angle of 45° in all views so that the closely spaced lines will not be confused with horizontal and vertical center lines.

Lugs or parts cast on for holding purposes and to be machined off are shown by "adjacent-part" lines (Figs. 56 and 57, Chap. 2). If such parts are in section, the section lines are dashed. "Alternate-position" lines (Figs. 56 and 57, Chap. 2) are used to indicate the limiting positions of moving parts and to show adjacent parts that aid in locating the position or use of a piece. Note in Figs. 56 and 57, Chap. 2, that the symbol of a long dash and two short dashes, alternating, is used to indicate adjacent parts, alternate positions, and also repeated features.

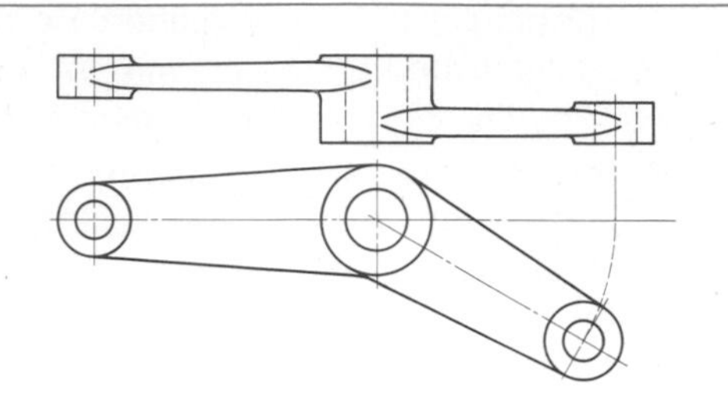

**FIG. 38.** Aligned view. This is to avoid drawing in the foreshortened position.

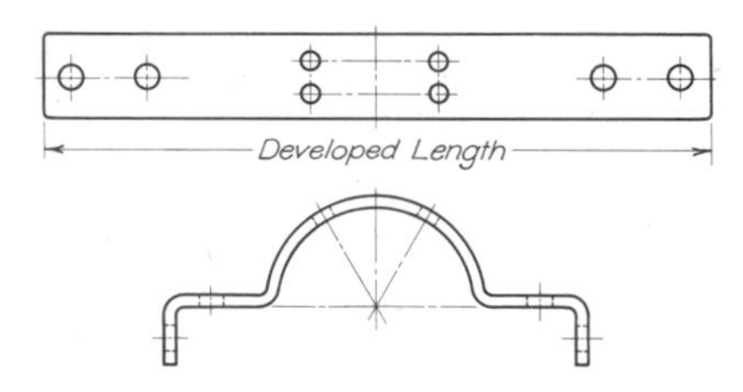


**FIG. 39.** Developed view. This part is drawn as though it were straightened out in one plane.

### 21. Aligned and Developed Views.

Pieces that have elements at an angle to one another, as the lever of Fig. 38, may be shown straightened out or aligned in one view. Similarly, bent pieces of the type of Fig. 39 should have one view made as a developed view of the *blank* to be punched and formed. Extra metal must be allowed for bends.

### 22. Half Views.

When space is limited, it is allowable to make the top or side view of a symmetrical piece as a half view. If the front is an exterior view, the *front* half of the top or side view would be used, as in Fig. 40; but if the front view is a sectional view, the *rear* half would be used, as in Fig. 41. Figure 42 shows another space-saving combination of a half view with a half section. Examples of half views occur in Probs. 44 to 46, Chap. 23.

### 23. Conventional Practices.

One statement can be made with the force of a rule: *If anything in clearness can be gained by violating a*

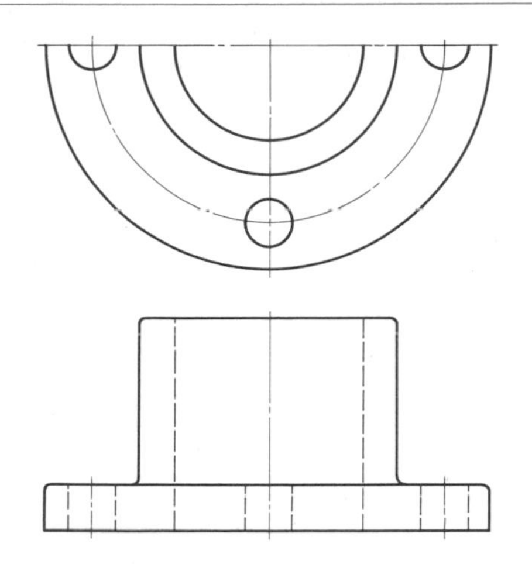

**FIG. 40.** Half view. The front half is drawn when the mating view is external.

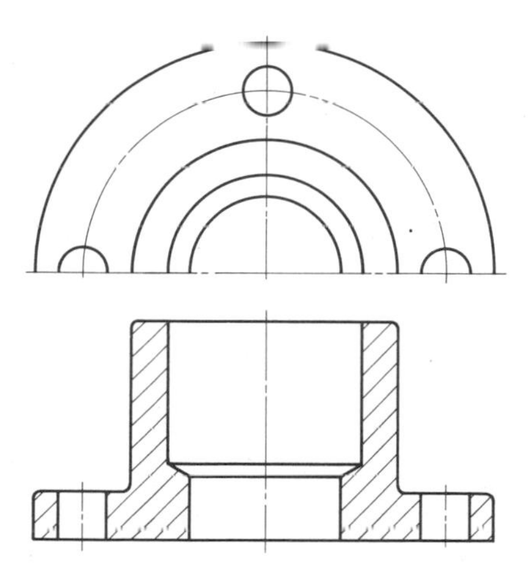

**FIG. 41.** Half view. The rear half is drawn when the mating view is a full section.

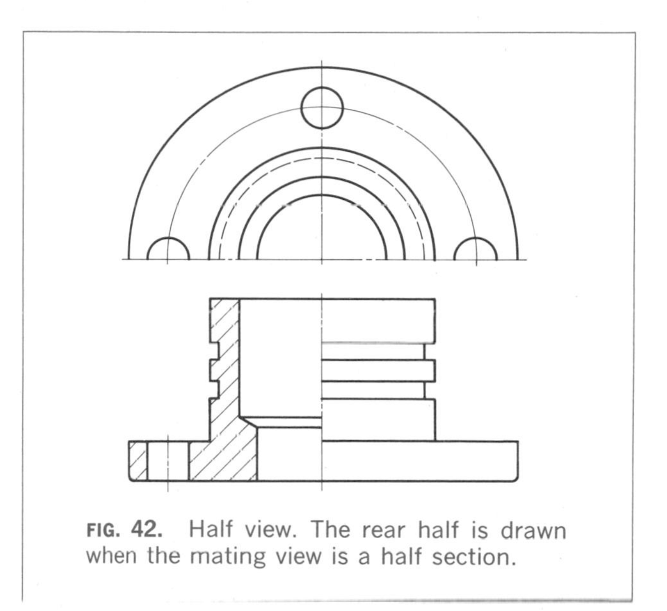

**FIG. 42.** Half view. The rear half is drawn when the mating view is a half section.

*principle of projection, violate it.* This applies to full as well as sectional views. Permissible violations are not readily apparent to the reader, since when they occur the actual conditions are described in a better and usually simpler form than if they were shown in true projection.

However, some care and judgment must be used in applying conventional treatments. Persons trained in their use ordinarily understand their meaning, but workmen often do not. Some unusual convention may be more confusing than helpful, and result in greater ambiguity than true projection.

There are occasions when the true lines of intersection are of no value as aids in reading and should be ignored. Some typical examples are shown in Fig. 43. It must be noted, however, that in certain cases that are similar but where there is a major difference in line position when the true projection is given, as compared with conventional treatment, the true line of intersection should be shown. Compare the treatment of the similar objects in Figs. 43 and 44. It would not be good practice to conventionalize the intersections on the objects in Fig. 44 because the difference between true projection and the convention is too great. To develop your judgment of the use of conventional intersections, carefully study Figs. 43 and 44 and observe the difference between true projection and conventional treatment in each case.

## 24. Fillets and Rounds.

In designing a casting, never leave sharp internal angles because of the liability of fracture at those points. The radius of the fillet depends on the thickness of the metal and other design conditions. When not dimensioned, it is left to the patternmaker. External angles may be rounded for appearance or

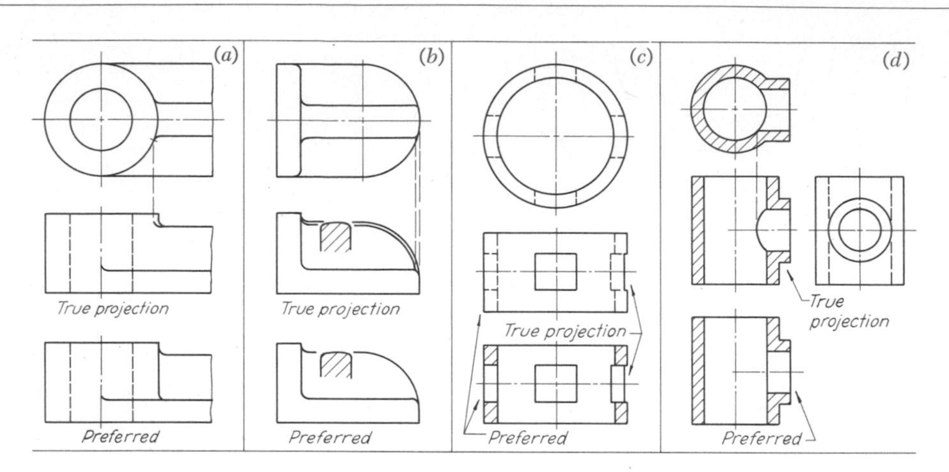


**FIG. 43.** Conventional intersections. These are used when there is a small difference as compared with true projection.

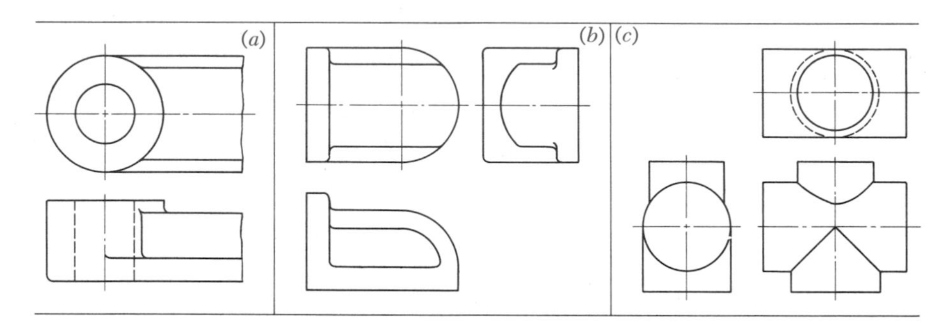


**FIG. 44.** True intersections. These are used when there is a major difference as compared with conventional treatment.

comfort, with radii ranging from enough merely to remove the sharp edges to an amount nearly equal to the thickness of the piece. An edge made by the intersection of two unfinished surfaces of a casting should always be "broken" by a very small round. A sharp corner on a drawing indicates that one or both of the intersecting surfaces are machined. Small fillets, rounds, and "runouts" are best put in freehand, both in pencil and ink. Runouts, or "die-outs," as they are sometimes called, are conventional indications of filleted intersections where, theoretically, there would be no line because there is no abrupt change in direction. Figure 45 shows some conventional representations of fillets and rounds with runouts of arms and ribs intersecting other surfaces.

## 25. Conventional Breaks.

In making the detail of a long bar or piece with a uniform cross section, it is rarely necessary to draw its whole length. It may be shown to a larger and therefore better scale by breaking out a piece, moving the ends together, and giving the true length by a dimension, as in Fig. 46. The shape of the cross

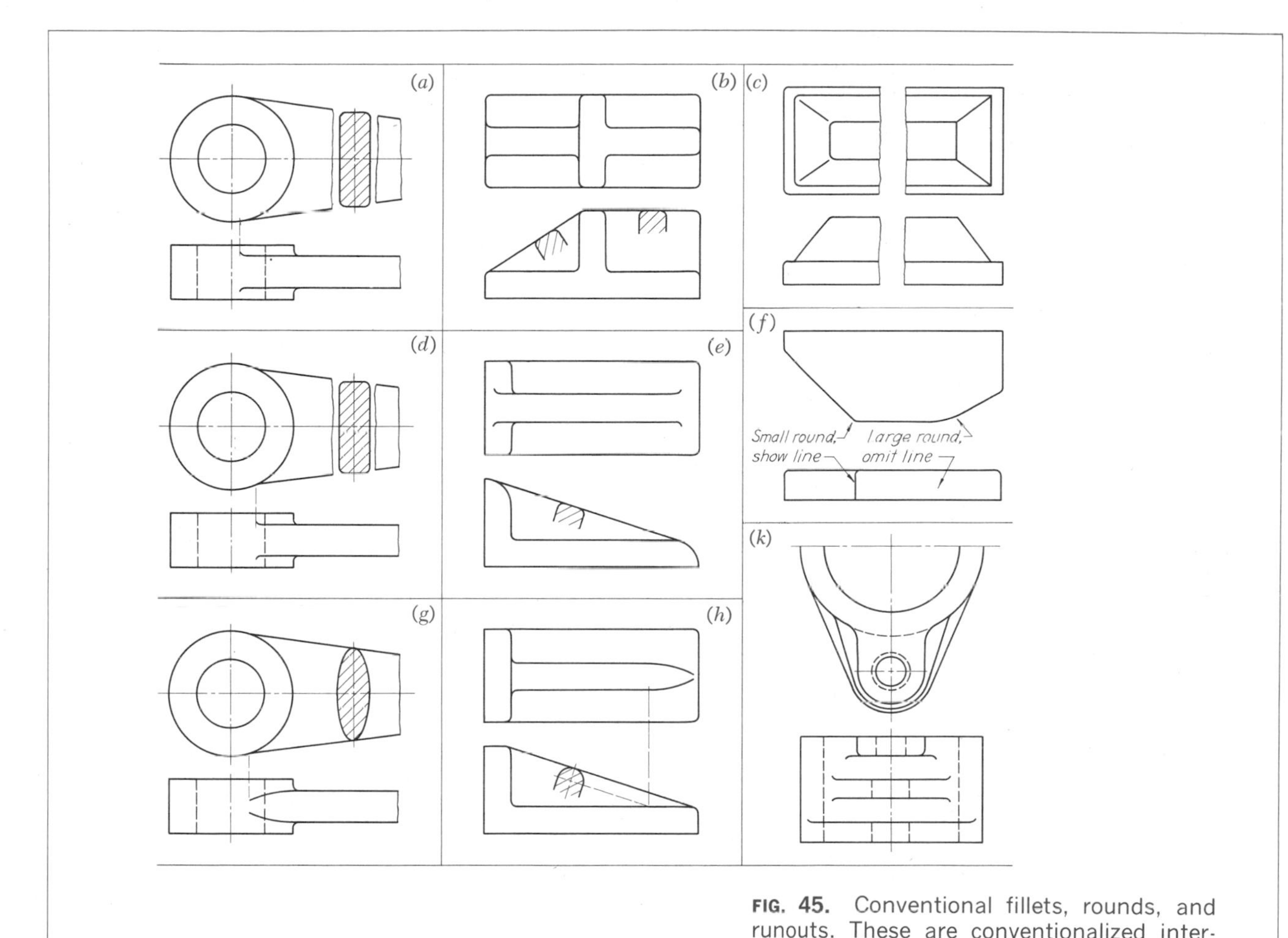


**FIG. 45.** Conventional fillets, rounds, and runouts. These are conventionalized intersections.

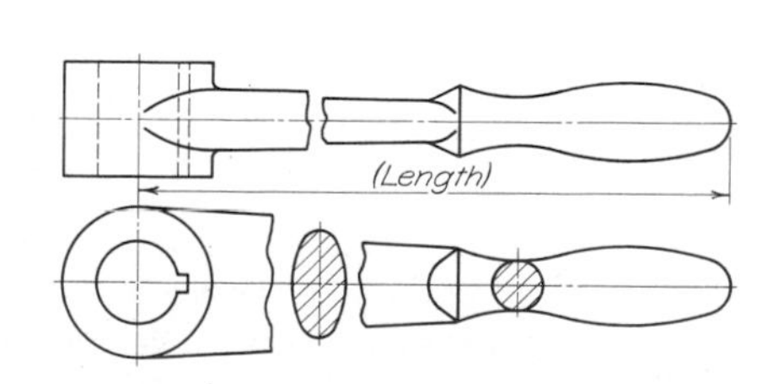


**FIG. 46.** Broken view with rotated sections. A broken view saves drawing the whole length of the part.

section is indicated by a rotated section or more often by a semipictorial break line, as in Fig. 47.

## 26. Conventional Symbols.

Engineers and draftsmen use conventional representation to indicate many details, such as screw threads, springs, pipe fittings, and electrical apparatus. These have been standardized by the ASA, whose code for materials in section has already been referred to in Sec. 11.

The symbol of two crossed diagonals is used for two distinct purposes: (1) to indicate on a shaft the

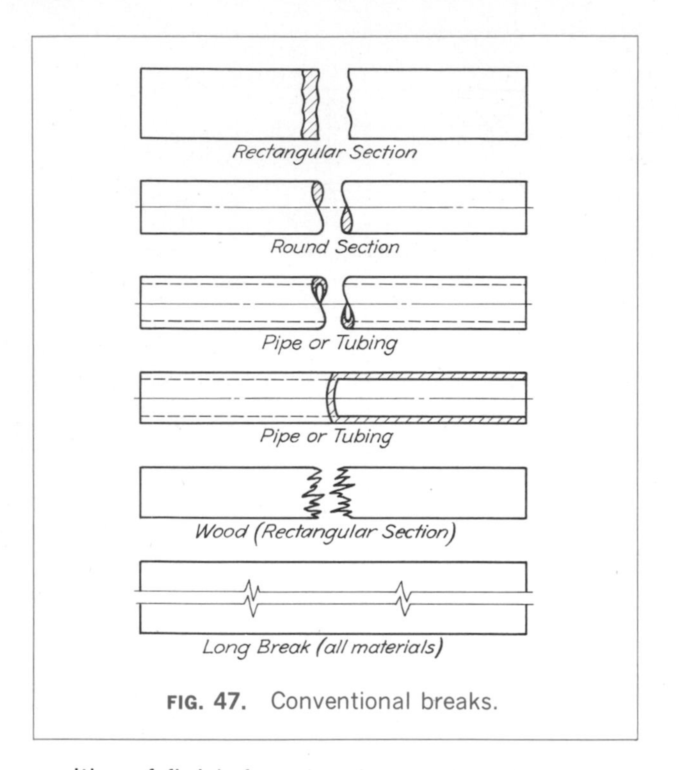


**FIG. 47.** Conventional breaks.

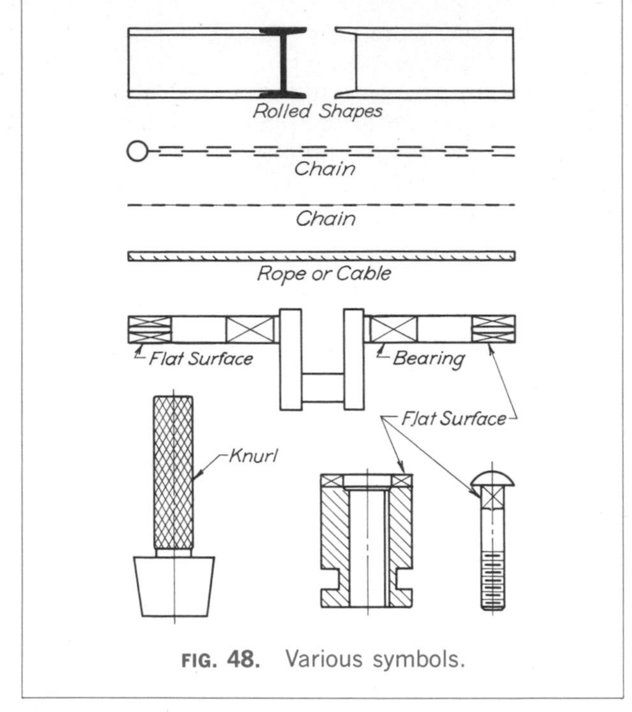


**FIG. 48.** Various symbols.

position of finish for a bearing, and (2) to indicate that a certain surface (usually parallel to the picture plane) is flat. These two uses are not apt to be confused (Fig. 48).

Because screw threads recur constantly, the designation for them is one of the most important items under conventional symbols. Up to the time of official standardization by the ASA, there were a dozen different thread symbols in use. Now a regular symbol and a simplified one have been adopted for American drawings, and both are understood internationally. The symbols for indicating thread on bolts, screws, and tapped holes will be given in a later chapter.

The conventional symbols mentioned are used principally on machine drawings. Architectural drawing, because of the small scales employed, uses many conventional symbols, and topographic drawing is made up almost entirely of symbols. Electrical *diagrams* are completely symbolic.

# PROBLEMS

## Group 1. Single Pieces.

The following problems can be used for practice in shape description only or, by adding dimensions, in making working drawings.

**1.** Draw the top view, and change the front and side views to sectional views as indicated.

**2.** Draw the top view, and make the front and two side views in section on cutting planes as indicated. Scale to suit.

**3 to 5.** Given the side view, draw full front and side views in section. Scale to suit.

**6 and 7.** Change the right-side view to a full section.

**8 and 9.** Change the right-side view to a full section.

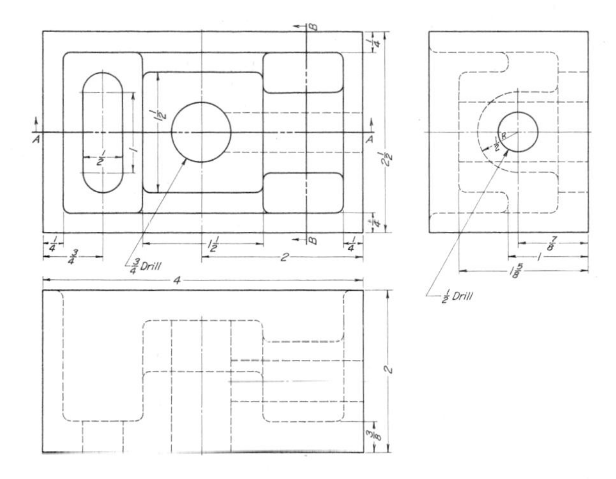

**PROB. 1.** Section study.

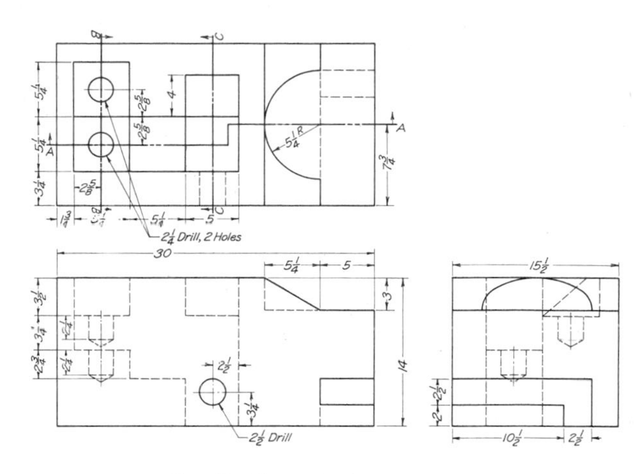

**PROB. 2.** Section study.

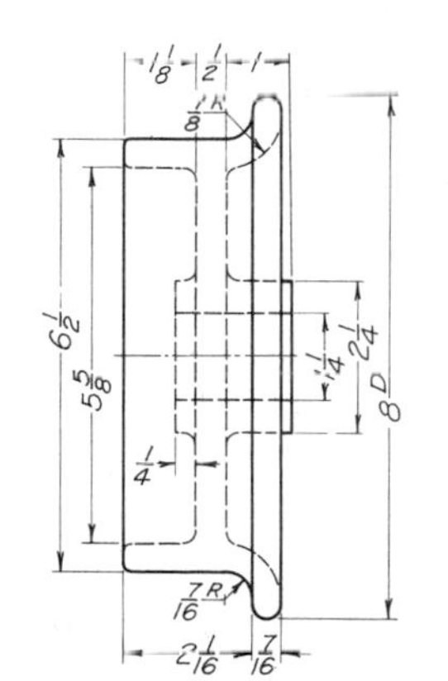

**PROB. 3.** Flanged wheel.

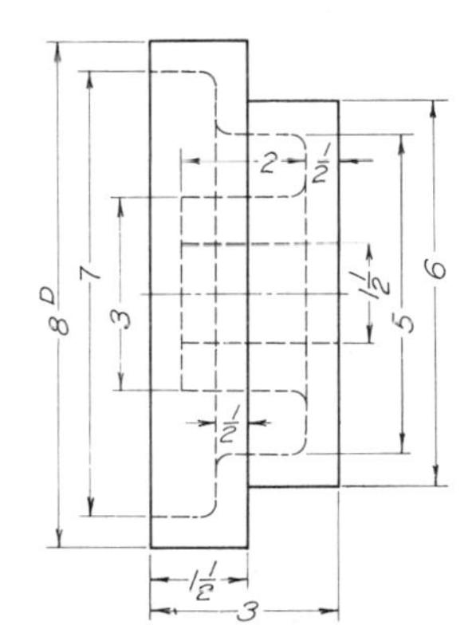

**PROB. 4.** Step pulley.

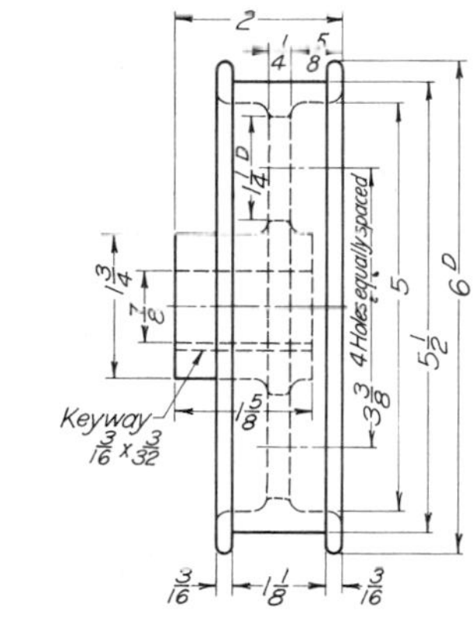

**PROB. 5.** Double-flanged wheel.

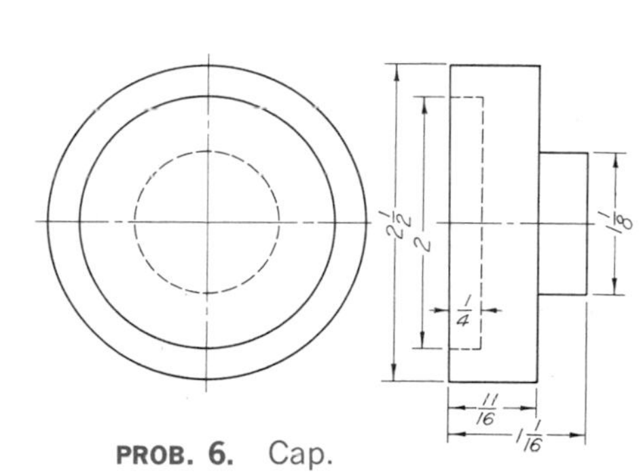

**PROB. 6.** Cap.

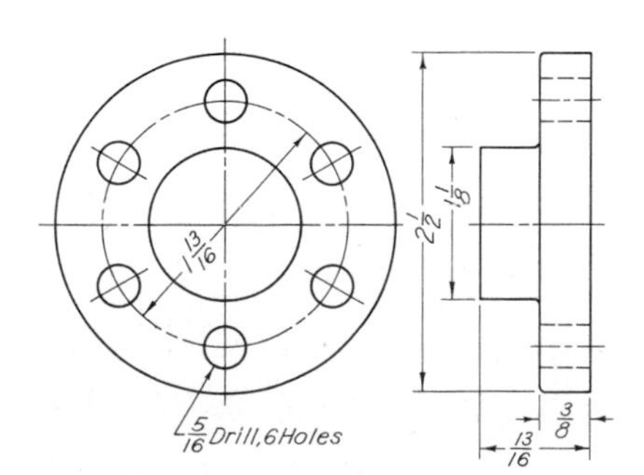

**PROB. 7.** Flanged cap.

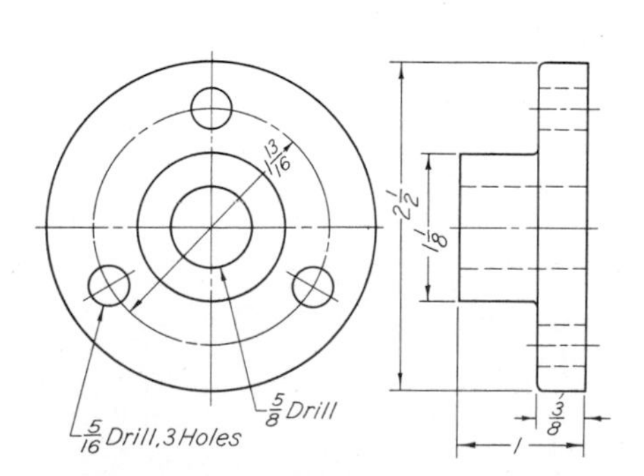

**PROB. 8.** Pump-rod guide.

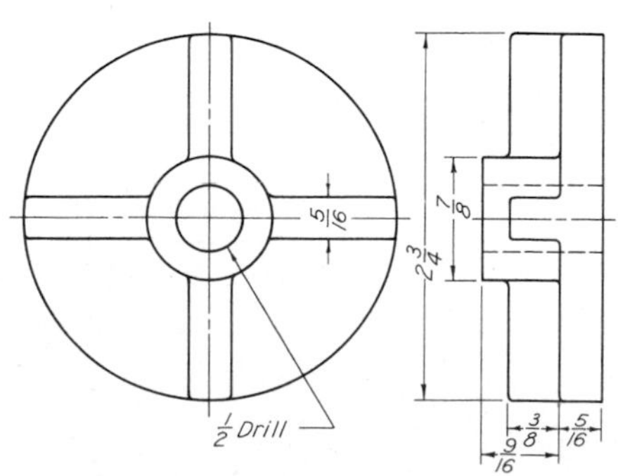

**PROB. 9.** Face plate.

**10.** Change the right-side view to a full section.
**11.** Change the right-side view to a sectional view as indicated.
**12 and 13.** Change the front view to a full section.
**14 and 15.** Change the front view to a full section.
**16 and 17.** Select views that will best describe the piece.
**18.** Draw the top view as shown in the illustration and the front view as a full section.
**19.** Draw the top view as illustrated and the front view in half section on *A-A*.
**20.** Draw the top view as shown, and change the front and side views to sections as indicated.
**21.** Draw the top view and sectional view (or views) to best describe the object.
**22.** Turn the object through 90°, and draw the given front view as the new top view; then make the new front view as section *B-B* and auxiliary section *A-A*. Refer to Sec. 9, for instructions on making auxiliary sections. See also Chap. 7 for the method of projection for the auxiliary section. Note in this case that the new top view, front-view section *B-B*, and auxiliary section *A-A* will completely describe the object. However, if desired, the side view may also be drawn, as shown or as an aligned view (described in Sec. 21).

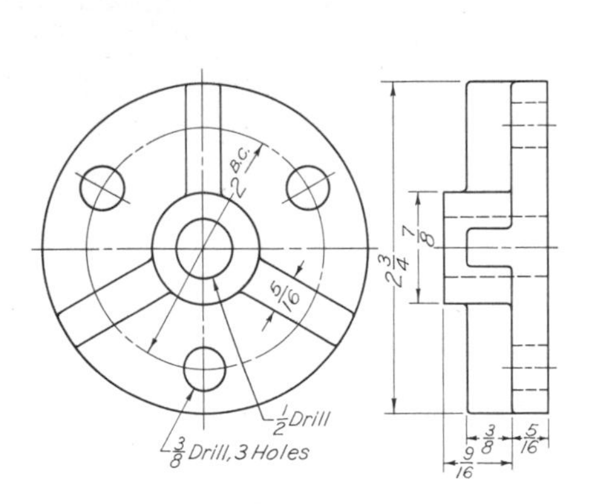


**PROB. 10.** Ribbed support.

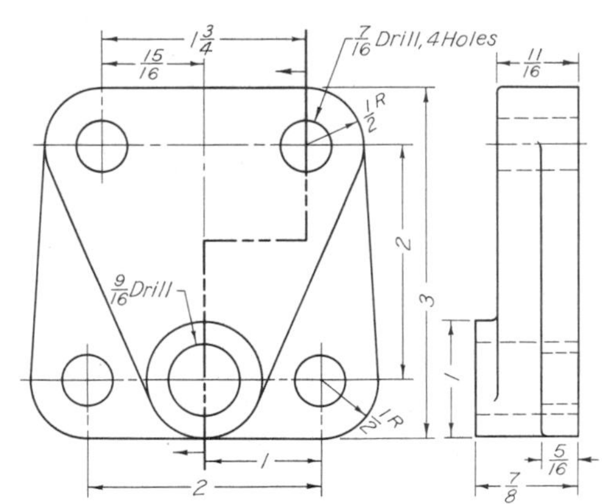


**PROB. 11.** Housing cover.

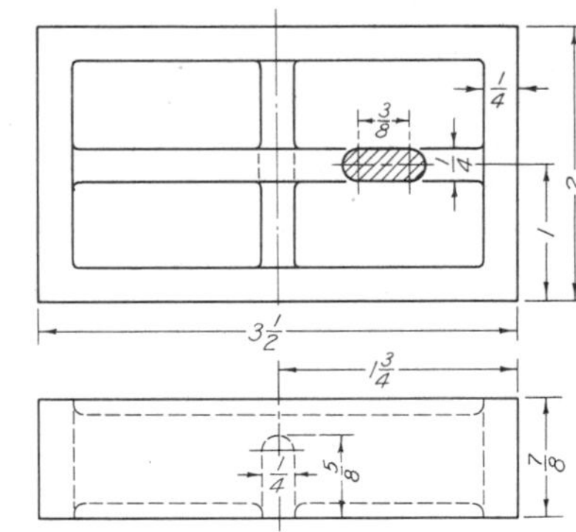

**PROB. 12.** Filler block.

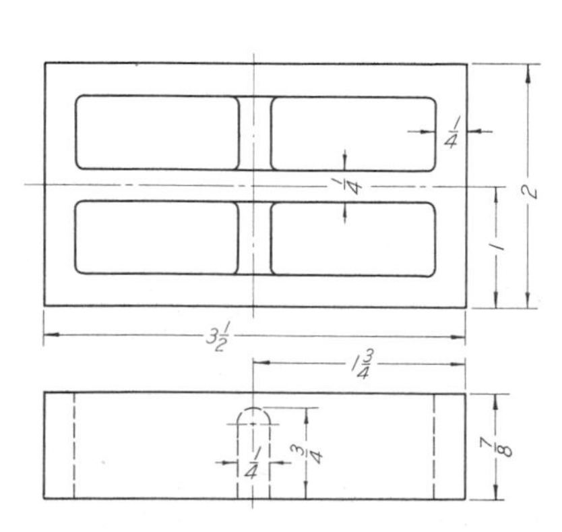

**PROB. 13.** Filler block.

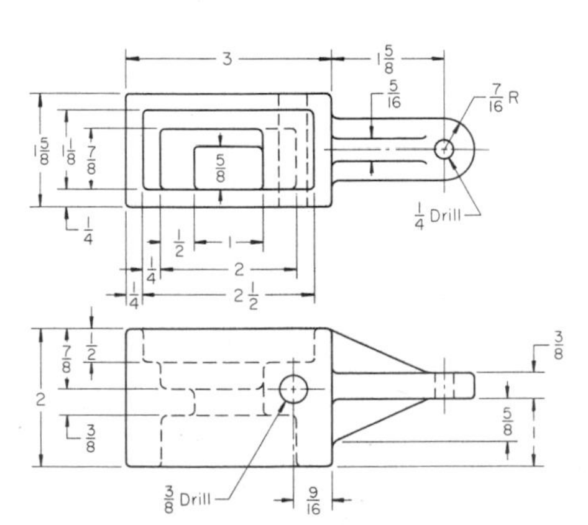


**PROB. 14.** Bumper body.

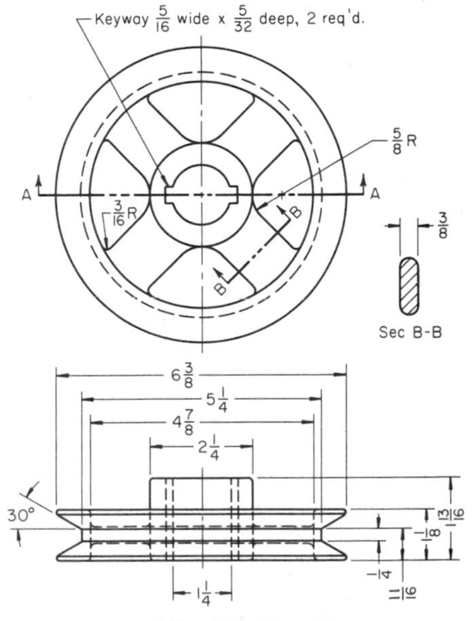


**PROB. 15.** V-belt pulley.

**22***A*. As an alternate for Prob. 22, draw views as follows: with the object in the position shown, draw the front view as shown, draw the left-side view as section *B-B*, and draw the new top view as an aligned view.

**23.** Draw the top view and front view in section.

**24 and 25.** Draw the views and add the sectional views indicated.

**26 and 27.** Draw a view and sectional view to best describe the piece.

**28.** Draw the top view and necessary sectional view (or views) to best describe the object.

**29.** Draw three views, making the side view as a section on *B-B*.

**29***A*. Draw three views, making the top view as a half section on *A-A*.

**30.** Select views that will best describe the piece.

**31.** Top view and front view in section, of rigging yoke, Prob. 77, Chap. 5.

**32.** Three views with front view in section, of missile gyro support, Prob. 75, Chap. 5.

**33.** Three views with front view in section, of end frame for engine starter, Prob. 113, Chap. 5.

**34.** Three views with front view in section, of master brake cylinder, Prob. 112, Chap. 5.

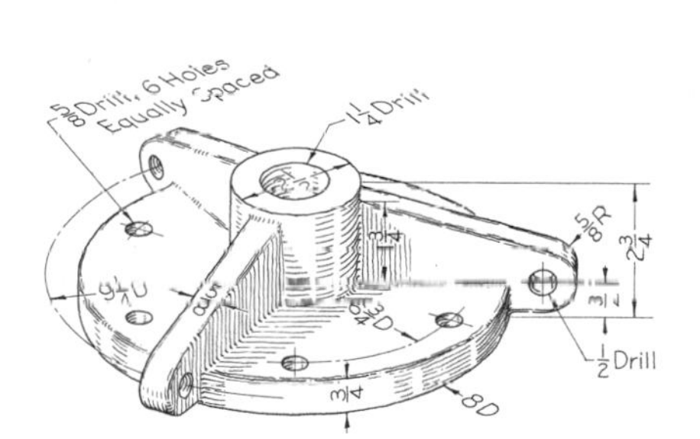


**PROB. 16.** End plate.

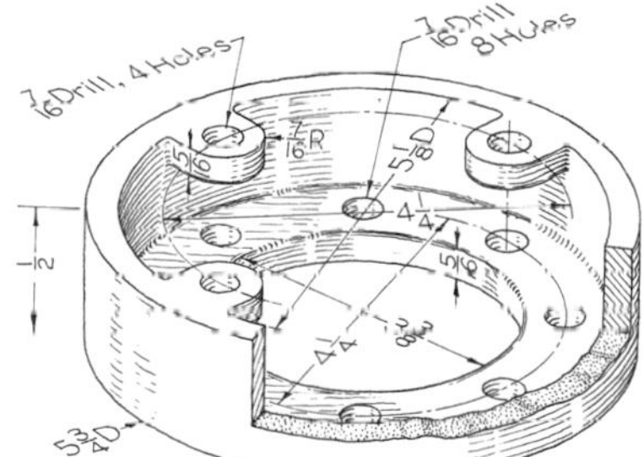


**PROB. 17.** Piston cap.

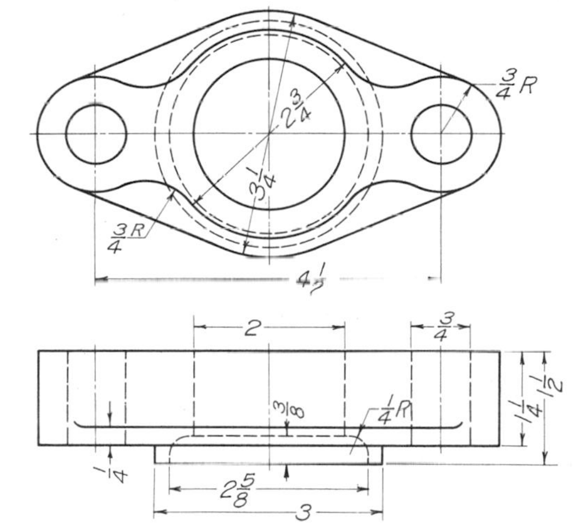

**PROB. 18.** Pump flange.

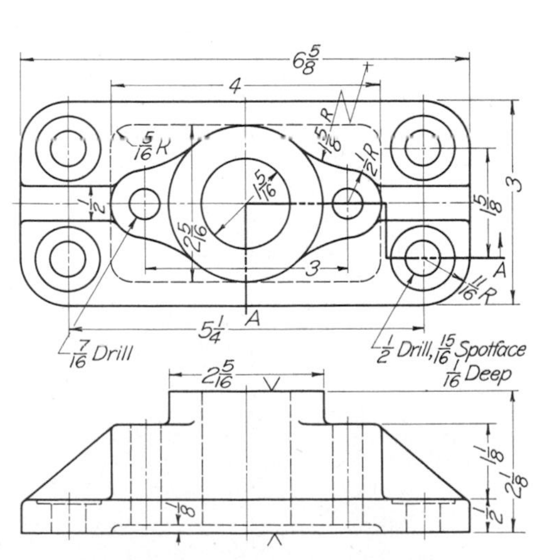


**PROB. 19.** Brake-rod bracket.

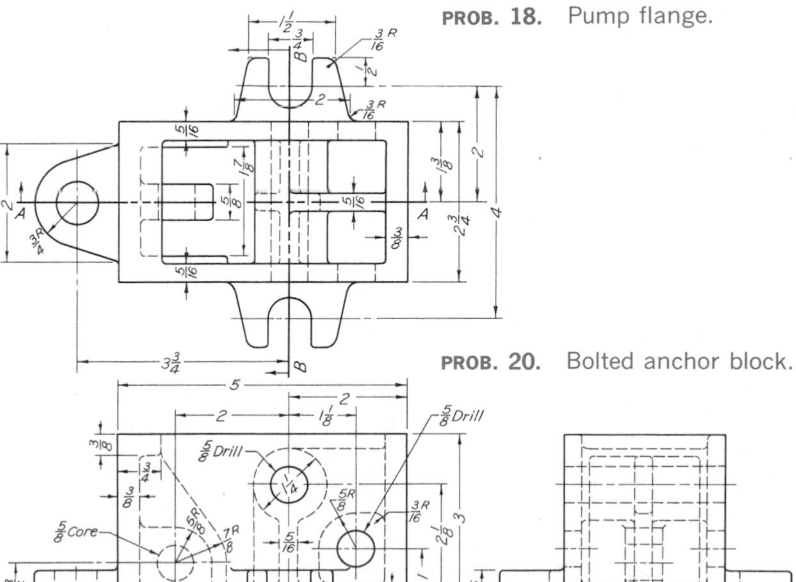


**PROB. 20.** Bolted anchor block.

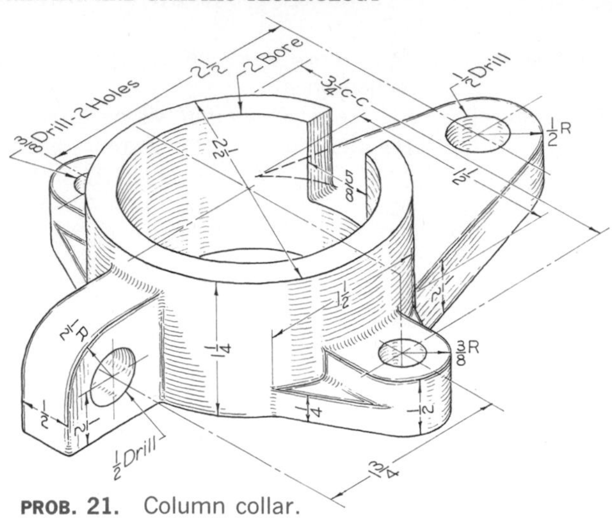


**PROB. 21.** Column collar.

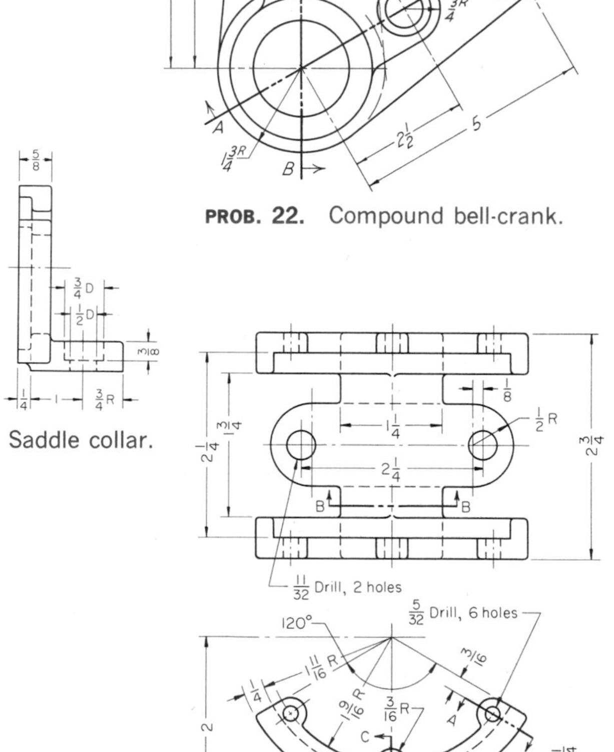


**PROB. 22.** Compound bell-crank.

**PROB. 24.** Spool base.

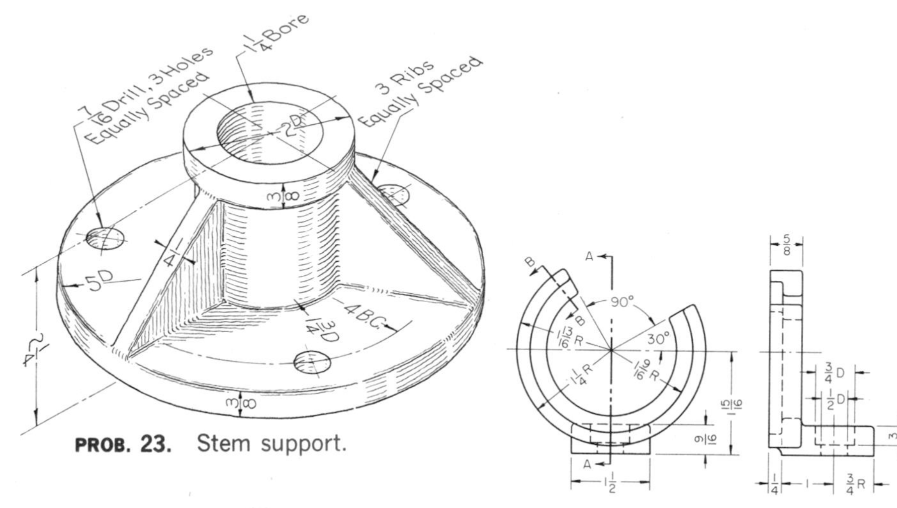


**PROB. 23.** Stem support.

**PROB. 25.** Saddle collar.

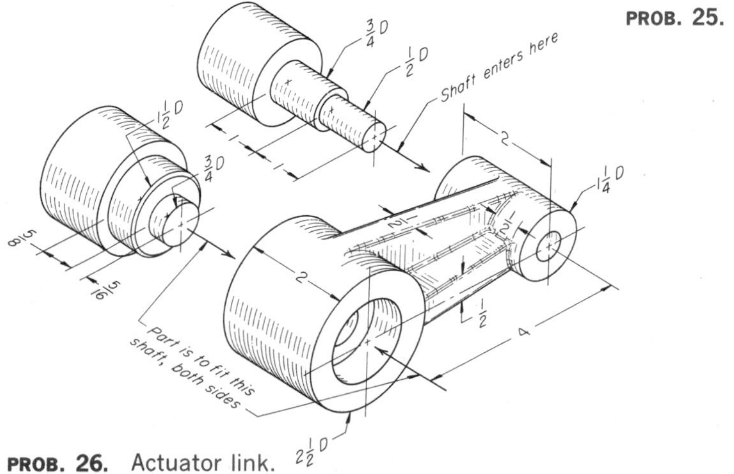


**PROB. 26.** Actuator link.

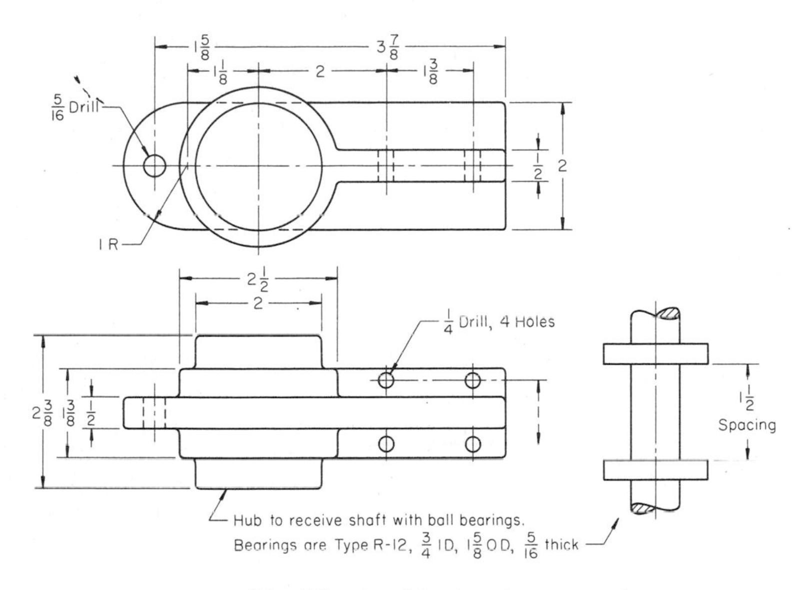


**PROB. 27.** Vibrator-drive bearing support.

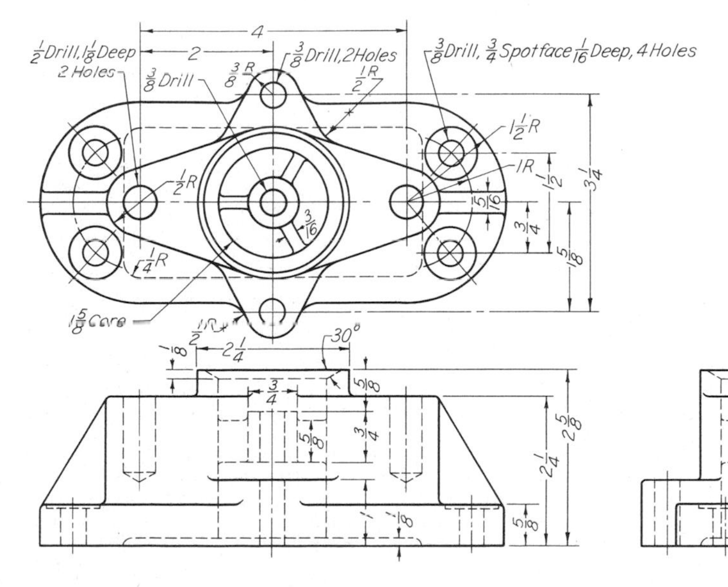


**PROB. 28.** Cover and valve body.

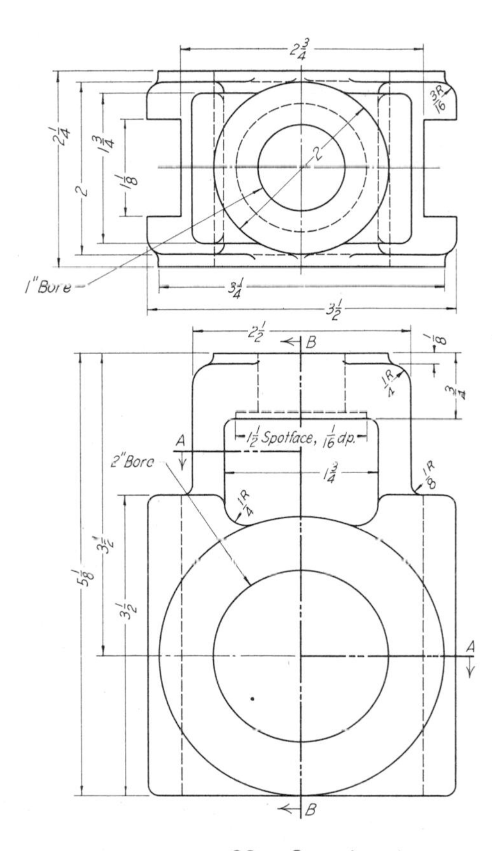


**PROB. 29.** Crosshead.

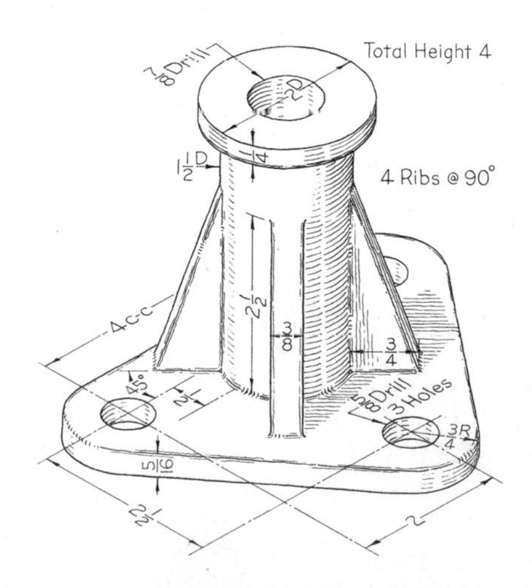


**PROB. 30.** Spindle support.

## Group 2. Assemblies.

**35.** Choose views and sectional views that will best describe the electric-motor support, Prob. 111, Chap. 5.

**36.** Choose views and sectional views that will best describe the jet-engine hinge plate, Prob. 120, Chap. 5.

**37.** Two views with sectional treatment that will best describe the anchor plate, Prob. 64, Chap. 6.

**38.** Draw a half end view and a longitudinal view as full section.

**39.** Draw the top view as shown in the figure and new front view in section. Show the shape (right section) of the link with rotated or removed section. The assembly comprises a cast-steel link, two bronze bushings, steel toggle pin, steel collar, steel taper pin, and part of the cast-steel supporting lug.

**40.** Draw the front view and longitudinal section. The assembly comprises a cast-iron base, a bronze bushing, a bronze disk, and two steel dowel pins.

**41.** Draw two half end views and a longitudinal section. The assembly consists of cast-iron body, two bronze bushings, steel shaft, cast-iron pulley, and steel taper pin.

**42.** Make an assembly drawing in section. The bracket is cast iron, the wheel is cast steel, the bushing is bronze, and the pin and taper pin are steel. Scale: full size.

**42*A*.** Make a drawing of the bracket with one view in section. Material is cast iron. Scale: full size.

**42*B*.** Make a drawing of the wheel with one view in section. Material is cast steel. Scale: full size.

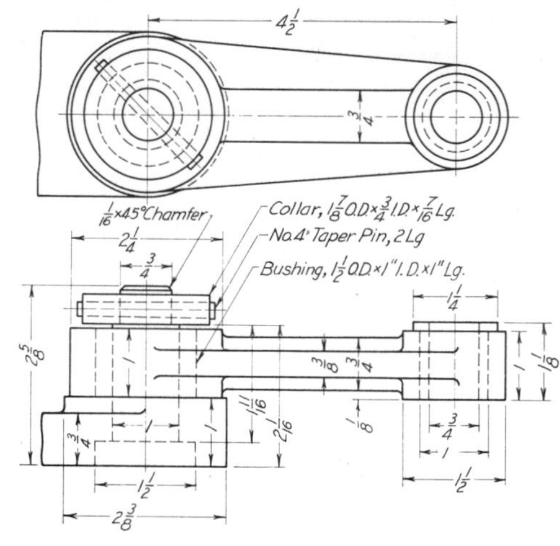


**PROB. 39.** Link assembly.

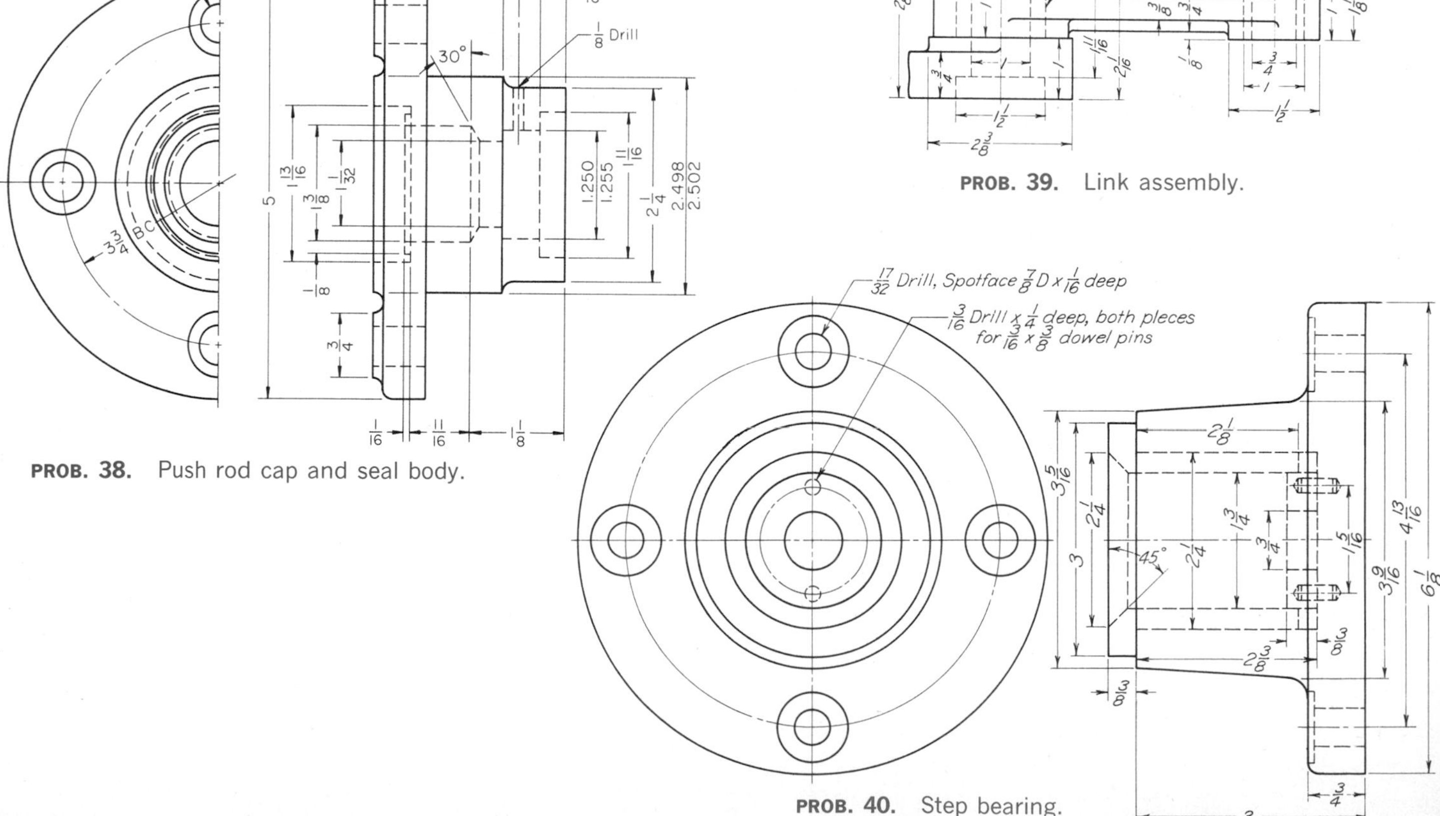


**PROB. 38.** Push rod cap and seal body.

**PROB. 40.** Step bearing.

**43.** Make an assembly drawing in section. The assembly comprises two cast-iron brackets, two bronze bushings, steel shaft, cast-steel roller, and cast-iron base. The bushings are pressed into the roller, and the shaft is drilled for lubrication. Scale: full size.

**44A.** Make a drawing of the roller and bushing assembly that gives one view in section. See Prob. 43 for materials. Scale: full size.

**45.** Sectional assembly of sealed shaft unit, Prob. 34 of the final chapter.

**46.** Sectional assembly of crane hook, Prob. 35, final chapter.

**47.** Sectional assembly of caster, Prob. 36, final chapter.

**48.** Sectional assembly of antivibration mount, Prob. 53, of the final chapter.

**49.** Sectional assembly of pump valve, Prob. 55, of the final chapter.

**50.** Sectional assembly of boring-bar holder, Prob. 58, final chapter.

**51.** Sectional assembly of conveyor link unit, Prob. 77, final chapter.

**52.** Sectional assembly of pressure cell piston, Prob. 62, final chapter.

**53.** Sectional assembly of pivot nut and adjustable screw, Prob. 63, final chapter.

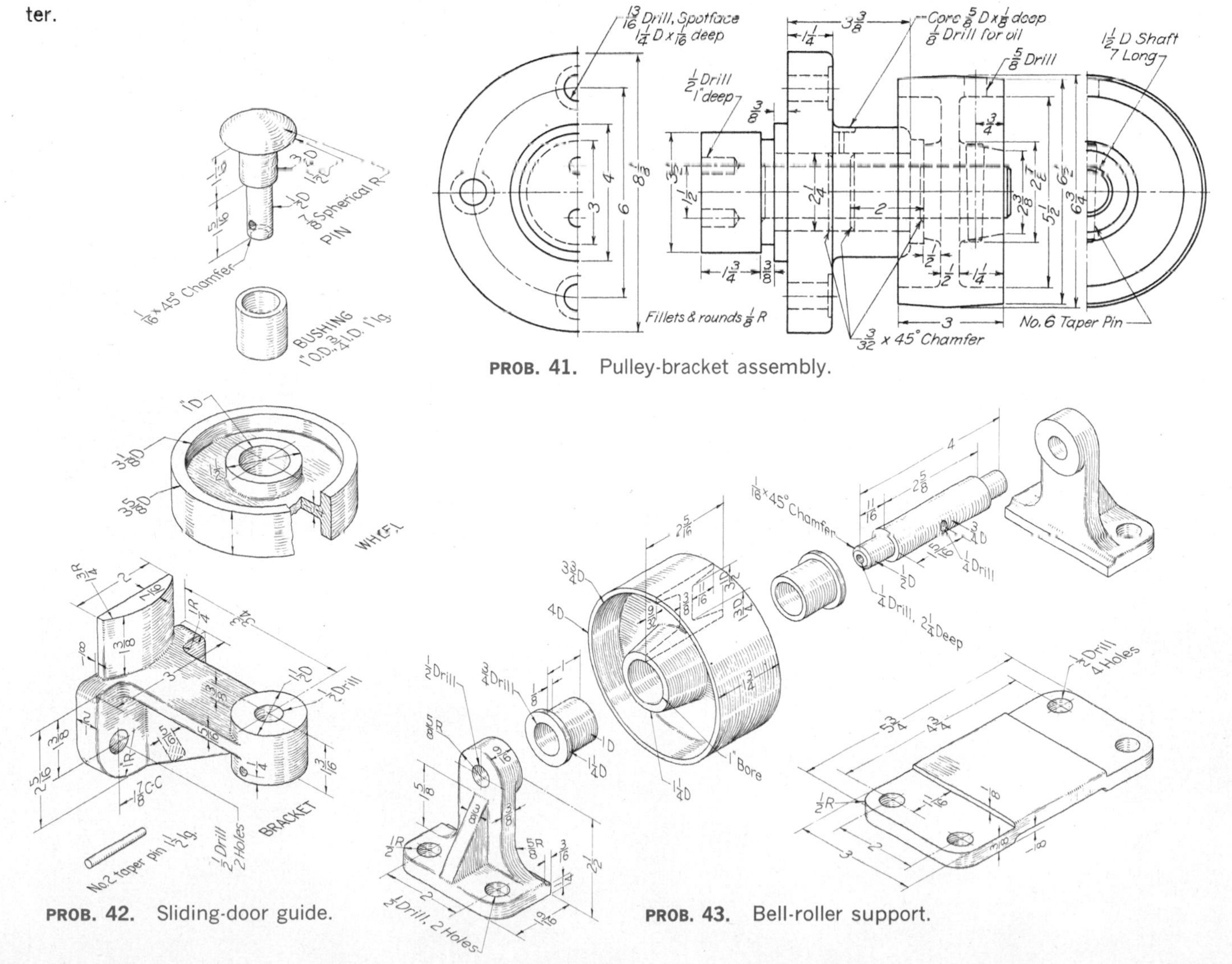

**PROB. 41.** Pulley-bracket assembly.

**PROB. 42.** Sliding-door guide.

**PROB. 43.** Bell-roller support.

Included Here:

Definitions and Classifications
Intersections of Lines and Planes
Intersections of Prisms
Intersections of Pyramids
Intersections of Curved Surfaces
Intersections of Cylinders and Cones with Lines and Planes
Intersections of Cylinders and Cones
Problems

# 9

# SURFACE INTERSECTIONS

Geometric Components of an Object or Assembly Meet in Intersections Which Must Be Shown in order to Complete the Graphic Description

**1. Definition and Classifications.**
In making orthographic drawings, there is the repeated necessity to represent the *lines of intersection* between the various surfaces of a wide variety of objects. Nearly every line on a drawing is a line of intersection, generally the intersection of two planes, giving a straight line, or of a cylinder and a plane, giving a circle or an ellipse. The term "intersection of surfaces" refers, however, to the more complicated lines that occur when geometric surfaces such as planes, cylinders, and cones intersect one another. These lines of intersection are shown by one of two basic methods: (1) *conventional intersections,* ordinarily used to represent a fillet, round, or runout, as explained in Sec. 23, Chap. 8, and shown in Fig. 43 of that chapter, or (2) *plotted intersections,* used when an intersection must be located accurately for purposes of dimensioning or for development of the surfaces. In sheet-metal combinations the intersection *must* be found before the piece can be developed. In this chapter we are concerned solely with the methods of projecting plotted intersections.

*Classification of Surfaces.* A surface may be considered to be generated by the motion of a line: the generatrix. Surfaces are thus divided into two general classes: (1) those that can be generated by a moving *straight* line, and (2) those that can be generated by a moving *curved* line. The first are called *ruled surfaces;* the second, *double-curved surfaces.* Any position of the generatrix is called an *element* of the surface.

*Ruled surfaces* are divided into (*a*) *the plane,* (*b*) *single-curved surfaces,* and (*c*) *warped surfaces.*

The plane is generated by a straight line moving so as to touch two other intersecting or parallel straight lines or a plane curve.

*Single-curved surfaces* have their elements parallel or intersecting. In this class are the cylinder and the cone and also a third surface, which we shall not consider, known as the "convolute," in which only consecutive elements intersect.

*Warped surfaces* have no two consecutive elements that are parallel or intersecting. There are a great variety of warped surfaces. The surface of a screw thread and that of an airplane wing are two examples.

*Double-curved surfaces* are generated by a curved line moving according to some law. The commonest forms are *surfaces of revolution,* made by revolving a curve about an axis in the same plane, such as the sphere, torus or ring, ellipsoid, paraboloid, hyperboloid. Illustrations of various surfaces may be found in Fig. 101, Chap. 3.

*Definitions and Details of Common Surfaces.* A *prism* is a polyhedron whose bases or ends are equal parallel polygons and whose lateral faces are parallelograms. A right prism is one whose lateral faces are rectangles; all others are called oblique prisms. The axis of a prism is a straight line connecting the centers of the bases. A truncated prism is that portion of a prism lying between one of its bases and a plane which cuts all its lateral edges.

A *pyramid* is a polyhedron whose base is a polygonal plane and whose other surfaces are triangular planes meeting at a point called the "vertex." The axis is a line passing through the vertex and the mid-point of the base. The altitude is a perpendicular from the vertex to the base. A pyramid is *right* if the altitude coincides with the axis; it is *oblique* if they do not coincide. A truncated pyramid is that portion of a pyramid lying between the base and a cutting plane which cuts all the lateral edges. The frustum of a pyramid is that portion of a pyramid lying between the base and a cutting plane parallel to the base which cuts all the lateral edges.

A *cylinder* is a single-curved surface generated by the motion of a straight-line generatrix remaining parallel to itself and constantly intersecting a curved directrix. The various positions of the generatrix are elements of the surface. It is a *right* cylinder when the elements are perpendicular to the bases, an *oblique* cylinder when they are not. A truncated cylinder is that portion of a cylinder which lies between one of its bases and a cutting plane which cuts all the elements. The axis is the line joining the centers of the bases.

A *cone* is a single-curved surface generated by the movement, along a curved directrix, of a straight-line generatrix, one point of which is fixed. The directrix is the base, and the fixed point is the vertex of the cone. Each position of the generatrix is an element of the surface. The axis is a line connecting the vertex and the center of the base. The altitude is a perpendicular dropped from the vertex to the base. A cone is *right* if the axis and altitude coincide; it is *oblique* is they do not coincide. A truncated cone is that portion of a cone lying between the base and a cutting plane which cuts all the elements. The frustum of a cone is that portion of a cone lying between the base and a cutting plane parallel to the base which cuts all the elements.

### 2. Intersections of Plane Surfaces.

The intersection of a line and a plane is a *point* common to both. The intersection of two planes is a *line* common to both. After mastering the graphic methods of locating intersections (Secs. 3 to 10), apply the principles you have learned to finding the line of intersection between objects made up of plane surfaces (prisms and pyramids, Secs. 11 to 13). The method of solution for the intersection of other polyhedrons should follow logically from the examples given.

### 3. Intersection of a Line and a Plane, Both in Principal Positions.

Principal positions of planes are horizontal, frontal, or profile. Principal positions of lines occur when the lines are *not* inclined to any principal plane; thus a line is (1) horizontal-frontal, (2) horizontal-profile, or (3) frontal-profile. Therefore, if the line is parallel to, or lies in, the principal plane, no single point of intersection is possible, but in other positions when the line is perpendicular to the plane, a single point exists. To illustrate, Fig. 1 shows a rectangular object made up of horizontal, frontal, and profile planes, with an accompanying horizontal-profile line $AB$. Because the line is parallel to horizontal and profile planes, the line and these planes do not intersect except at infinity. However, the line is perpendicular to frontal planes and there is a single point of intersection, observed in the top view at $P_T$ and in the side view at $P_R$, where the front of the object appears as an edge. The front view of the intersection is coincident with $A_F B_F$ at $P_F$ because this is the end view of line $AB$.

Thus it is established that *the intersection of a line and plane is a point on the line coincident with the edge view of the plane.* The point of intersection of a line and plane is often called a "piercing point."

### 4. Intersection of an Inclined Line with Planes in Principal Positions.

Two points of intersection are possible, one with each plane to which the line is not parallel. For example, Fig. 2 shows a frontal line $AB$ and an object made up of horizontal, frontal, and profile planes. The line will not intersect any frontal plane (except

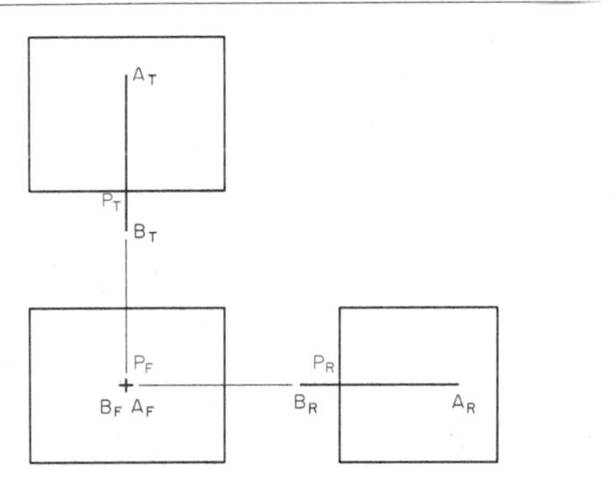


**FIG. 1.** Intersection of a line and plane. The line is perpendicular to the plane.

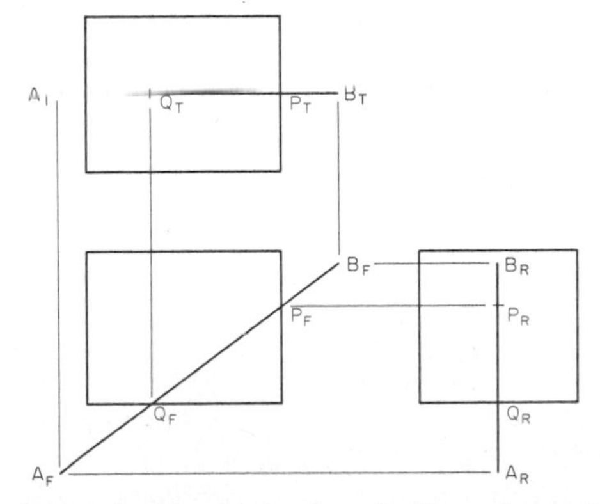


**FIG. 2.** Intersection of an inclined line with planes. In this case the line intersects the horizontal and profile sides of an object.

at infinity). The intersection with the (profile) right side is seen at $P_T$ and $P_F$, where the plane appears as an edge, and is then easily projected to $P_R$. The intersection with the horizontal bottom is observed at $Q_F$ and $Q_R$, where the plane appears as an edge, and is projected from $Q_F$ to $Q_T$.

Observe that line $AB$ does not intersect the top or left side of the *object*, but if the top and side planes were extended, a point of intersection would exist.

## 5. Intersection of a Skew Line with Planes in Principal Positions.

Theoretically, a skew line (a line inclined to *all* principal planes) will intersect all three principal planes. However, on a *rectangular object*, depending upon the length and position of the line, only one or two intersections exist. Figure 3 illustrates all possibilities. At (*a*), the skew line from point $A$ inside the object emanates upward, backward, and to the right, and obviously intersects the top (horizontal) surface of the object at $P$, observed first at $P_F$ and $P_R$, where the top surface appears as an edge, and then projected to $P_T$. But line $AB$ does not intersect any other surface of the object.

At (*b*), line $AB$ is seen to intersect two surfaces of the object, the right side at $Q$ and the front at $P$. Note that in each case the points are found first where the surface appears as an edge, $Q_T$ and $Q_F$ for point $Q$, and $P_T$ and $P_R$ for point $P$.

At (*c*), the top surface of the object is extended so that three intersections are possible. Physically this occurs on an object when two rectangular shapes are offset or when a base or lug extends. The line $AB$ intersects the right side at $P$ and the front at $Q$. The top surface and the line give an intersection at $R$, shown on the figure with dashed lines for the extensions.

Note in every case, (*a*), (*b*), and (*c*), of Fig. 3, that the intersection is found by observing where the line intersects the surface *in the edge view* of the surface. For horizontal, frontal, and profile surfaces, two views always show the surface as an edge. Thus it is easily seen that, for example, at (*b*) the line crosses the *edge* view of the *right side* of the object in the top view at $Q_T$ and in the front view at $Q_F$,

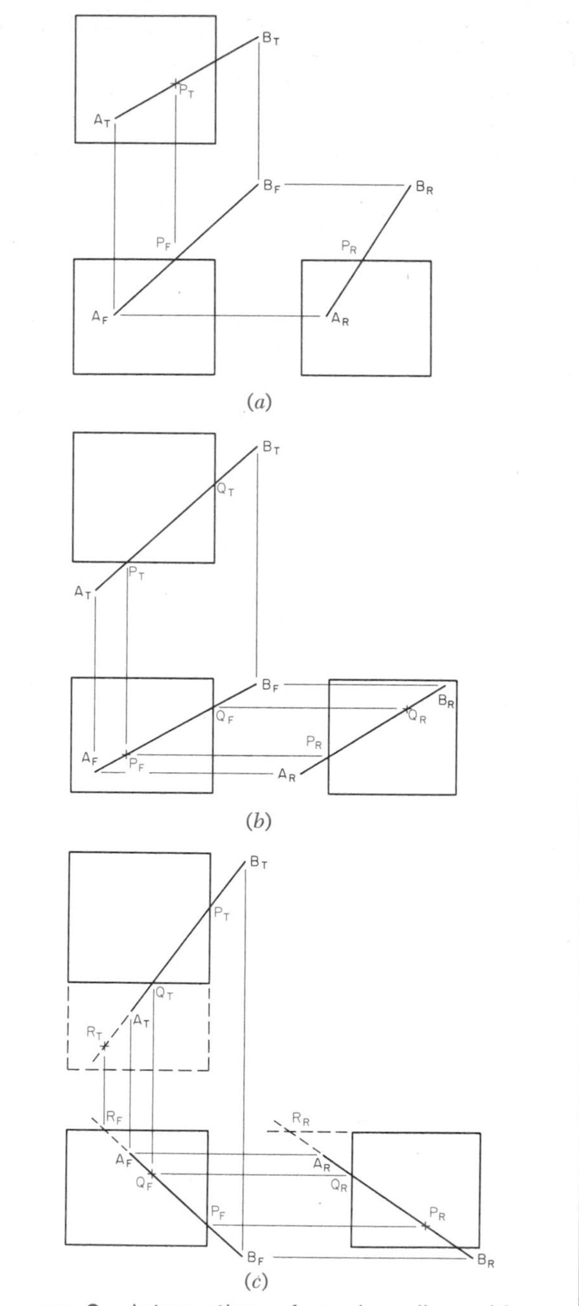


**FIG. 3.** Intersection of a skew line with planes. (*a*) Intersection with horizontal; (*b*) intersection with frontal and profile; (*c*) intersection with horizontal, frontal, and profile.

and because point $Q$ falls *within the confines* of the surface, a real point of intersection has been found. Also observe at (*a*) that line $AB$ *apparently* crosses the edge view of the right side of the object *in the top view*. Nevertheless, reading the front view, line $AB$ is seen to miss the right side and there is therefore no intersection of $AB$ with the right side. To illustrate further, line $AB$ of Fig. 4*a* apparently intersects the edge view of the top surface, as observed in the front view. But projecting $Q_F$ to the top view at $Q_T$ on $AB$, we see that $Q_T$ falls within the confines of the top surface of the object and is therefore an intersection. However, at (*b*) when $Q_F$ is projected to the top view, $Q_T$ is outside the confines of the top surface and there is no intersection. From this it is clear that to determine a real point of intersection, (1) *two views must be consulted*, and (2) *the apparent point of intersection in one view must project within the confines of the surface in another view.*

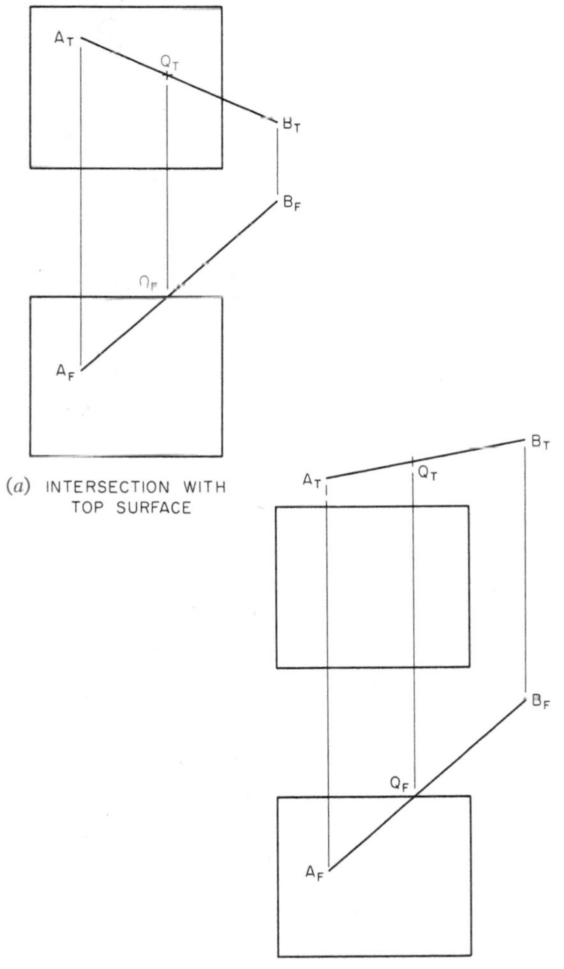

(*a*) INTERSECTION WITH TOP SURFACE

(*b*) NO INTERSECTION WITH TOP SURFACE

**FIG. 4.** To determine an intersection. The intersection (piercing point) must fall within the surface area in two (or more) views.

### 6. Intersection of a Skew Line with Inclined Planes.

An inclined plane appears as an edge in one view and, because of this, its intersection with any line is easily found. In Fig. 5*a* the inclined surface appears as an edge in the front view. The apparent intersection $P_F$ is projected to $P_T$ on $A_TB_T$, where it is seen that $P_T$ falls within the confines of the inclined surface and therefore is a real point of intersection. Figure 5 shows the principle applied at (*b*) to a surface inclined to front and side and at (*c*) to a surface inclined to top and front. Note that at (*b*) the intersection will be located first in the top view, and at (*c*) it will be found first in the side view.

### 7. Intersection of a Skew Line with a Skew Surface.

A skew surface does not appear as an edge in any principal view, so if the methods of Secs. 3 to 6 are used, an edge view will have to be made. This is illustrated in Fig. 6 where the edge view has been made by looking in the direction of $SR$, a horizontal line of the skew surface. Line $AB$ is now projected into the edge view, and the intersection $P$ is located and then projected back to top, front, and side views.

It is not necessary, however, to employ an extra (edge) view. Figure 7 shows pictorially a line and skew plane. If any edge-view plane is passed through the line, the intersection of this plane with the skew plane will contain the point of intersection between the line and the skew plane. The intersection of the edge-view plane and the skew plane is easily located,

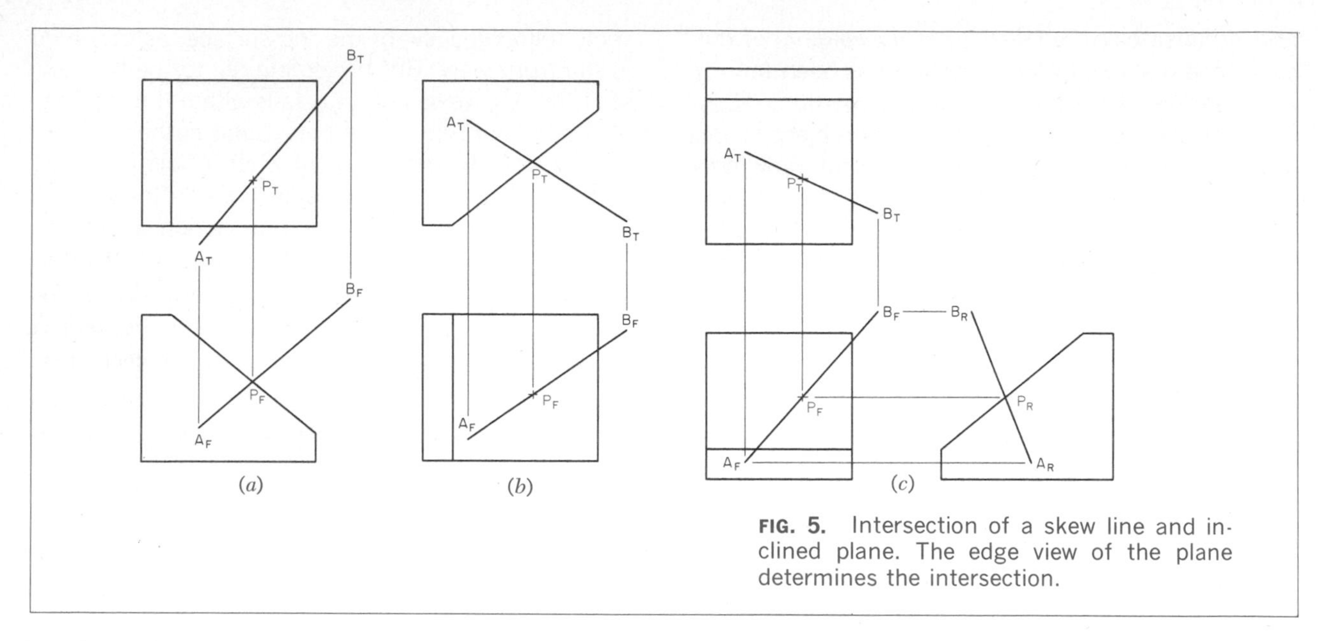


**FIG. 5.** Intersection of a skew line and inclined plane. The edge view of the plane determines the intersection.

as shown in Fig. 8, where $AB$ is a line and $RST$ a skew plane. A plane *appearing as an edge in the top view*, passed through $AB$, will intersect $RS$ at $Z_T$ and $RT$ at $X_T$, and these points projected to the front view at $Z_F$ and $X_F$ establish $ZX$ as the line of intersection between the edge-view plane and $RST$, the skew plane. The front view shows the intersection of $AB$ with $ZX$ at $P_F$, the point common to line, skew plane, and edge-view plane. To complete, $P$ is projected to the top view at $P_T$.

It often happens that one edge of a skew surface is profile, as is $ST$ in Figs. 8 and 9; and when the line, in this case $AB$, is in a position like that of Fig.

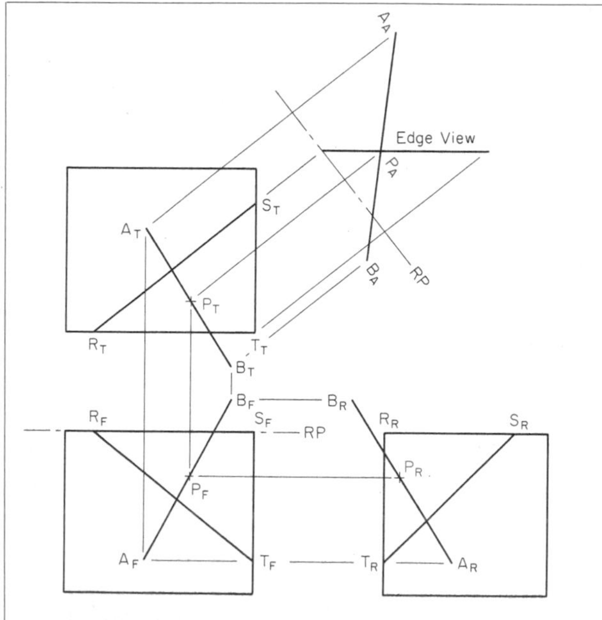


**FIG. 6.** Intersection of a skew line and skew plane. An auxiliary (edge view) determines the intersection.

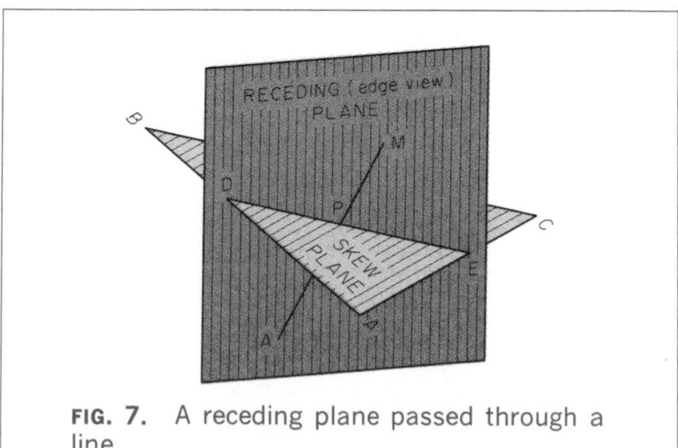


**FIG. 7.** A receding plane passed through a line.

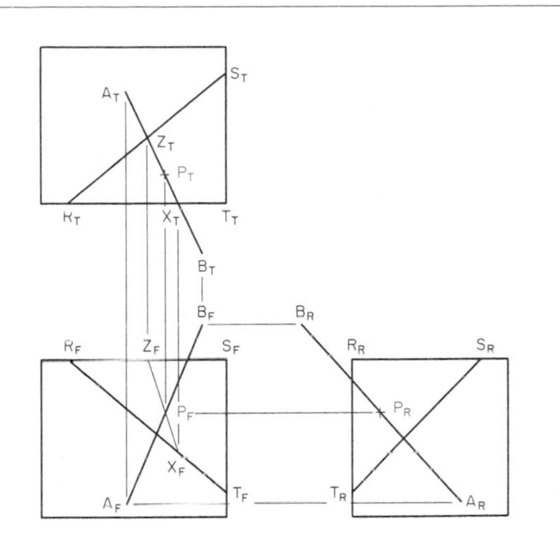


**FIG. 8.** Intersection of a skew line and skew plane. The intersection is determined by a receding plane (edge view plane in top view) passed through line $AB$.

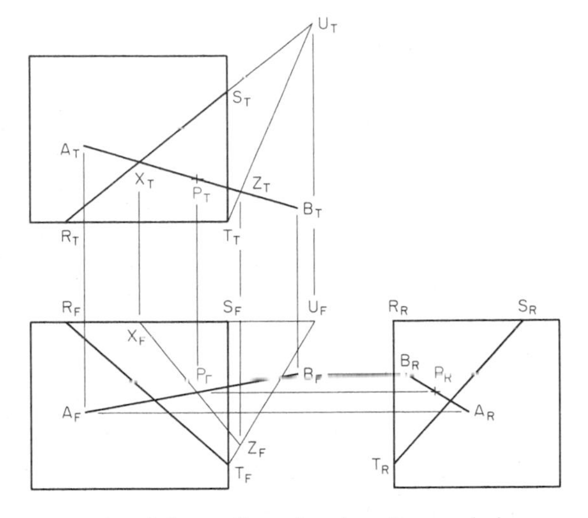


**FIG. 9.** Intersection of a skew line and skew plane. This illustrates enlargement of the skew plane to eliminate difficult projection.

9, the intersection (with the skew plane) of an edge-view plane through the line is not easily projected. An accepted method in such cases is to *extend* the skew plane to eliminate the profile edge. In Fig. 9, $RS$ has been extended to $U$, making the skew surface $RUT$, still the *same* surface but larger. An edge-view plane (in the top view) passed as before gives $XZ$ as its line of intersection with $RUT$. Point $P$, the point common to edge-view plane, skew plane, and line, is then located first in the front view at $P_F$ and then projected to top and side views.

## 8. Intersection of a Plane with a Plane in a Principal Position.

Any line of a given plane will intersect a second plane *on the line of intersection* between two planes. Therefore, to establish the line of intersection between two planes, the intersection of either of two lines of one plane with the other plane (or one line of each plane with the opposite plane) will establish two points on the line of intersection of the planes. The simplest case is when one of the planes is in a principal position as is the top surface of the object in Fig. 10. Lines $A$-$A'$ and $B$-$B'$ are edges of a skew plane, which, extended, are seen to intersect the top surface of the rectangular object at $P$ and $Q$, as explained in Sec. 5.

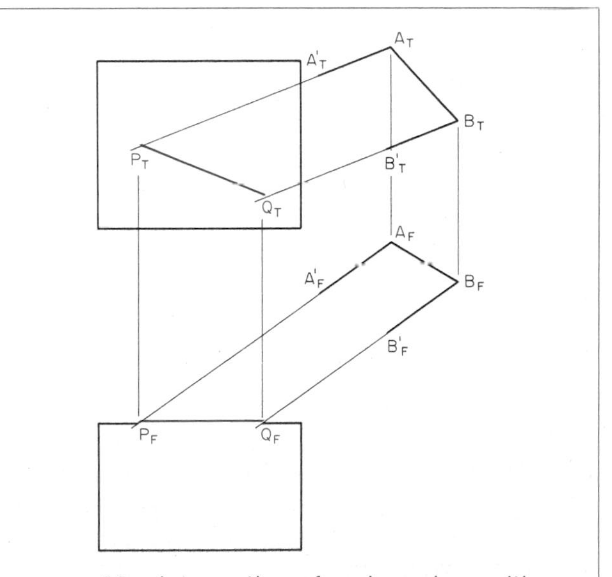


**FIG. 10.** Intersection of a skew plane with a plane in principal position. The edge view of the principal plane determines the intersection.

### 9. Intersection of a Plane with an Inclined Plane.

Figure 11 is similar to Fig. 10 but the intersection this time is with the inclined surface of the object. The intersection is again found by extending *A-A′* and *B-B′* to the edge view of the inclined surface. Refer to Sec. 6.

### 10. Intersection of Two Skew Planes.

As before, the points of intersection of two lines determine the intersection of the planes. In Fig. 12 two edge-view (in the front view) planes have been passed through *A-A′* and *B-B′* extended. The edge-view planes then intersect the skew plane *RSTU* in lines *WX* and *YZ*. Then $P_T$ and $Q_T$, the intersections of *A-A′* and *B-B′* extended, in the top view determine the top view of the line of intersection, which, projected to the front view at $P_F$ and $Q_F$, completes the solution. Compare Fig. 12 (edge-view plane in front view is used) with Fig. 8 (edge-view plane in top view is used). An edge-view plane can also be passed in a side view whenever it is convenient to do so.

### 11. To Find the Intersection of Two Prisms (Fig. 13).

In general, find the line of intersection of a surface on one prism with all surfaces on the other. Then take a surface adjacent to the first surface, and find its intersection with the other prism. Continue in this manner until the complete line of intersection of the prisms is determined.

The method of locating end points on the line of intersection of two surfaces depends upon the position of the surfaces, as follows:

*Both Surfaces Receding (Edge View).* Their intersection appears as a point in the view in which they recede. Project the intersection to an adjacent view, locating the two ends of the intersection on the edges of one or both intersecting surfaces so that they will lie within the boundaries of the other surface. The intersection 4-5 of surfaces *QRST* and *EF*-3 was obtained in this manner.

*One Surface Receding, the Other Skew.* An edge of the oblique surface may appear to pierce the receding surface in a view in which these conditions exist. If, in an adjacent view, the piercing point lies on the edge of the oblique surface and within the

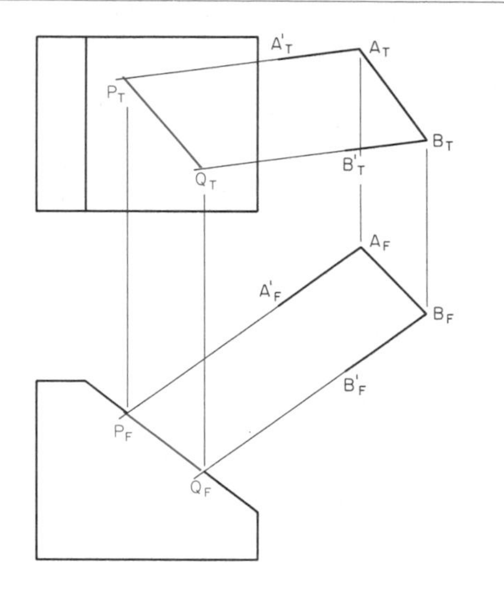


**FIG. 11.** Intersection of a skew plane with an inclined plane. The edge view of the inclined plane determines the intersection.

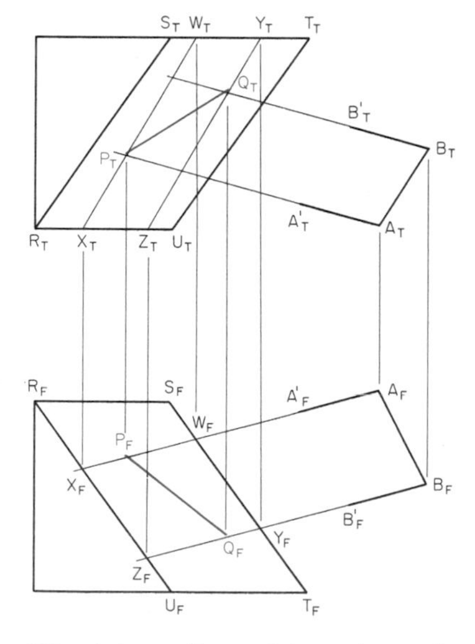


**FIG. 12.** Intersection of two skew planes. Receding planes, passed through two lines of one plane, determine the intersection.

boundaries of the other surface, it is an end point on the intersection of the surfaces. Point 5, lying on edge *F*-5 of the oblique surface *FG*-1-5 and the surface *QRST*, is located in the top view in this manner. Point 1 was similarly established. Point 6, lying on edge *ST*, is found by passing a vertical plane *AA* through edge *ST*. Plane *AA* cuts line 1-*c* from plane *GF*-5-1, giving point 6 where line 1-*c* crosses *ST*.

*Both Surfaces Skew.* Find the piercing point of an edge of one surface with the other surface, as follows: Pass a receding plane through an edge of one surface. Find the line of intersection of the receding plane and the other surface, as explained above. The piercing point of the edge and surface is located where the line of intersection, just found, and the edge intersect. Repeat this operation to establish the other end of the line of intersection of the surfaces. Point 3, on the line of intersection 2-3 of the oblique surfaces *NORS* and *EG*-1-3, was found in this manner by passing the receding plane *BB* through edge *E*-3, finding the intersection *b*-4 of the surfaces, and then locating point 3 at the intersection of *b*-4 and *E*-3.

## 12. To Find the Intersection of Two Pyramids.

In general, find the point where one edge on one pyramid pierces a surface of the other pyramid. Then find where a second edge pierces, and so on. To complete the line of intersection, the piercing points of the edges of the second pyramid with surfaces of the first will probably also have to be found. Figure 14 illustrates the method. Find where edge *AD* pierces plane *EHG*, by assuming a vertical cutting plane through edge *AD*. This plane cuts line 1-2 from plane *EHG*, and the piercing point is point *P*, located first on the front view and then projected to the top view. Next, find where *AD* pierces plane *EFG*, by using a vertical cutting plane through *AD*. This plane cuts line 3-4 from plane *EFG*, and the piercing point is point *Q*.

After finding point *P* on plane *EHG* and point *Q* on plane *EFG*, locate the piercing point of edge *EG* with plane *ABD* in order to draw lines of intersection.

A vertical plane through *EG* cuts line 5-6 from plane *ABD*, and the intersection is point *R* on edge *EG*.

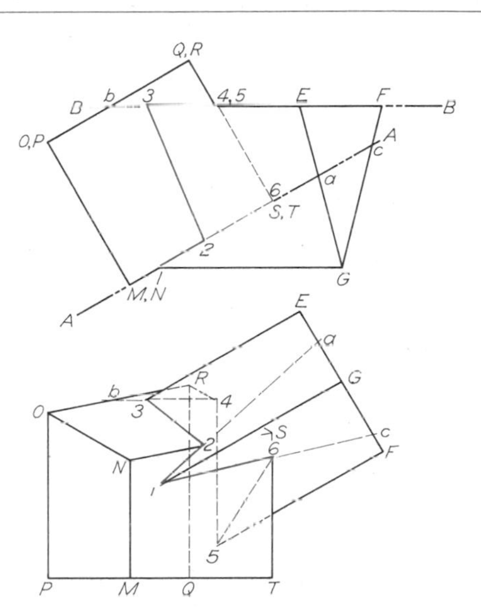


**FIG. 13.** Intersection of two prisms. The edges of one, intersecting the faces of the other, determine the complete "line" of intersection.

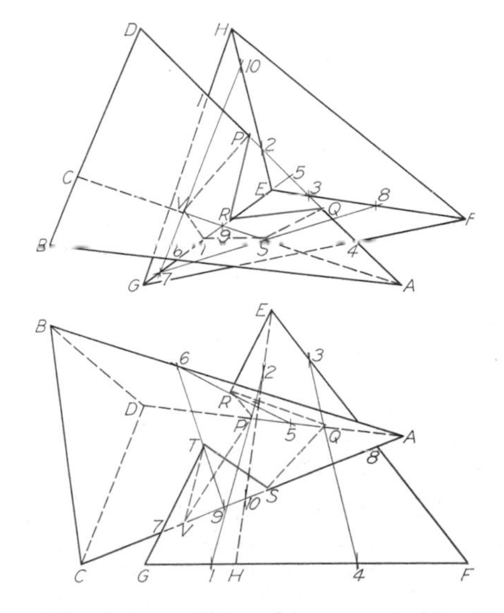


**FIG. 14.** Intersection of two pyramids. The edges of one, intersecting the faces of the other, determine the intersection.

Thus edges of the "first" pyramid pierce surfaces of the "second," and edges of the second pierce surfaces of the first. Continue in this manner until the complete line of intersection *PRQSTV* has been found.

The use of a vertical cutting plane to obtain the piercing points is perhaps the simplest method and the easiest to visualize. Nevertheless, it should be noted that a plane receding from the frontal or profile plane could also be used. As an example of the use of a plane receding from the frontal, consider that such a plane has been passed through line *CA* in the front view. This plane cuts line 7-8 from plane *EFG*, and the point of intersection is *S* on line *CA*. The use of a plane receding from the profile plane would be basically the same but would require a side view.

### 13. To Find the Line of Intersection between a Prism and a Pyramid (Fig. 15).

The method is basically the same as for two pyramids. Thus a vertical plane through edge *G* cuts line 1-2 from surface *AED*, and a vertical plane through edge *K* cuts line 3-4 from surface *AED*, giving the two piercing points *P* and *Q* on surface *AED*. A vertical plane through edge *AE* cuts elements 7 and 8 from the prism and gives piercing points *R* and *T*. Continue in this manner until the complete line of intersection *PQSTVR* is found.

### 14. Intersections of Curved Surfaces.

The intersections of single- and double-curved surfaces with lines and planes and with each other, especially when one or both of the surfaces are in a skew position, are properly a part of the study of engineering geometry and therefore not included here. However, the intersections of cylinders and cones in simple positions occur more or less frequently on machine parts, and so are discussed in the paragraphs that follow.

### 15. Intersection of a Cylinder and Line.

The common case occurs when the cylinder is in a simple position, as in Fig. 16, with one view the

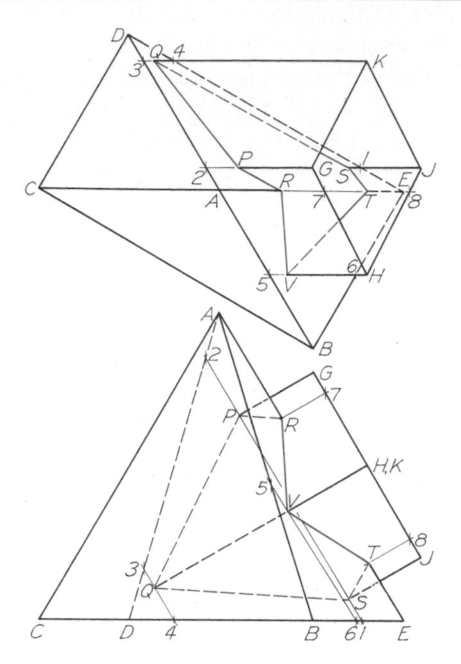


**FIG. 15.** Intersection of a prism and pyramid. The edges of one, intersecting the faces of the other, determine the intersection.

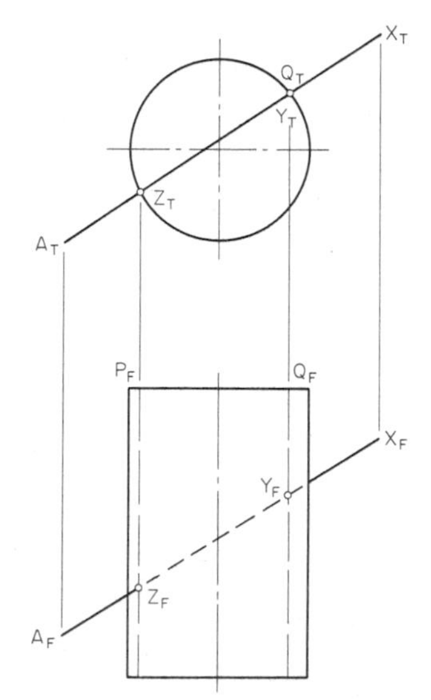


**FIG. 16.** Intersection of a line and cylinder. The edge view of the cylinder determines the two piercing points.

end view—the *edge* view of the surface and the other view showing the normal view of all the elements. Any line such as $AX$ is therefore seen to intersect the cylinder where the surface appears as an edge, in this case the top view at $Y_T$ and $Z_T$. These two points lie on elements of the cylinder $P$ and $Q$, which are projected to the front view thus locating $Y_F$ and $Z_F$.

### 16. Intersection of a Cylinder and Plane.

Planes theoretically contain an infinite number of lines, and so, by the method of Sec. 15, the intersection of a number of *lines* of a plane will determine the intersection of a plane with a cylinder. In Fig. 17 the intersection of $ABC$ with the cylinder is to be located. Any line $AX$ is drawn on the plane and the intersections $Z$ and $Y$ found, giving two points on the curved line of intersection. The process repeated with a number of lines similar to $AX$ will give points to complete the curve.

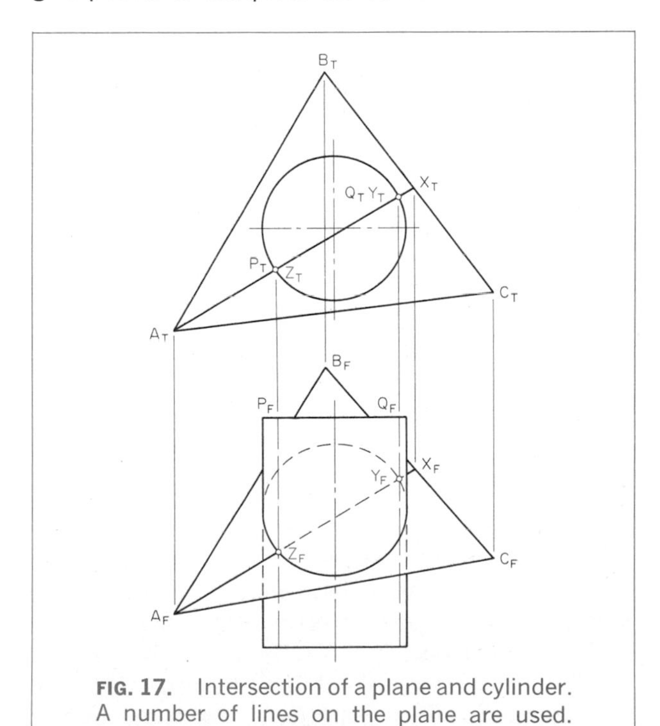


**FIG. 17.** Intersection of a plane and cylinder. A number of lines on the plane are used.

### 17. Intersection of a Cone and a Line.

Figure 18*a* illustrates pictorially the problem of finding the points where a line pierces a cone. Select a cutting plane, $VXY$, containing the given line $AB$ and passing through the apex, $V$, of the cone. A plane passing through the apex of a cone will intersect the cone in straight lines. The lines of intersection of the cutting plane and the cone are lines $V$-1 and $V$-2. These lines are determined by finding the line of intersection, $RS$, between the cutting plane and the base plane of the cone and then finding the points 1 and 2 where this line crosses

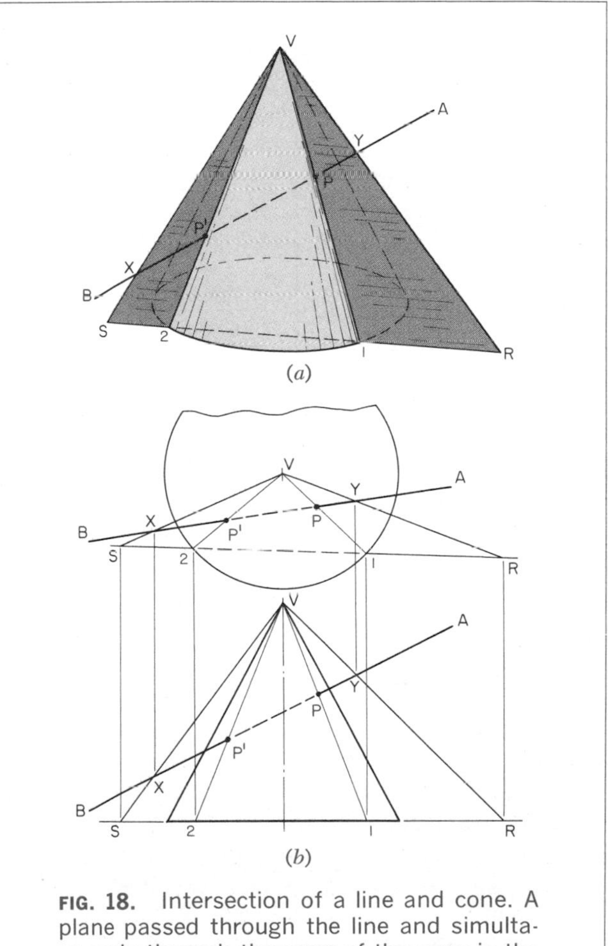


**FIG. 18.** Intersection of a line and cone. A plane passed through the line and simultaneously through the apex of the cone is the basis for determination.

the base curve. Line $AB$ intersects both lines $V$-1 and $V$-2, establishing the points $P$ and $P'$ where line $AB$ pierces the cone.

The orthographic solution is illustrated at (*b*). Convenient points $X$ and $Y$ on line $AB$ are selected so that plane $VXY$ (a plane containing the apex of the cone and line $AB$) extended will intersect the base plane of the cone in the line $RS$. In the top view $RS$ cuts the base curve at 1 and 2, thus locating elements $V$-1 and $V$-2 *in* the cutting plane and *on* the surface of the cone. $V$-1 and $V$-2 then intersect $BA$ at $P$ and $P'$, the piercing points of line $AB$ and the cone.

**18. Intersection of a Cone and a Plane.** Figure 19*a* illustrates pictorially the determination of the line of intersection between a cone and a plane. Lines (elements) of the cone, such as $V$-1, $V$-2, etc., will intersect the plane at $a$, $b$, etc., respectively. If the edge view of the plane appears in one of the views, as at (*b*), the solution is quite simple. Selected elements of the cone, $V$-1, $V$-2, etc., are drawn in both views. These elements are seen to intersect the plane in the front view at $a$, $b$, etc. Projection of these points to the top view then gives points through which a smooth curve is drawn to complete the solution. The edge view of the plane (within the confines of the cone) is, of course, the intersection in the front view.

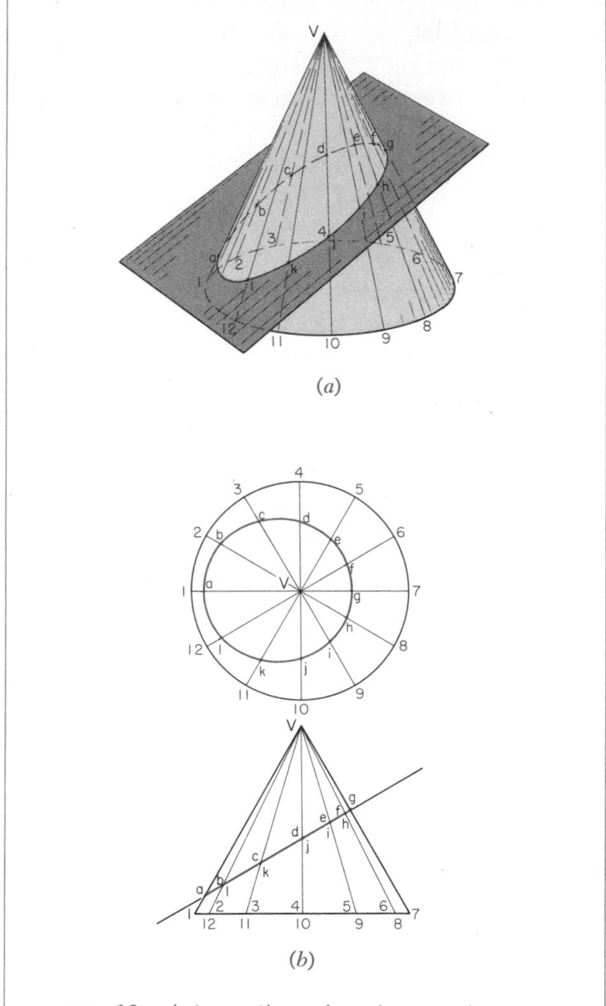


**FIG. 19.** Intersection of a plane and cone. The edge view of the plane determines the intersection.

If the plane intersecting the cone is skew, as in Fig. 20, the cutting-plane method is probably preferable because of the simplicity of the solution. Any plane passed through the apex of the cone will cut straight lines from the cone. Therefore, as shown at (*a*), if vertical cutting planes are used, all passing through $V$, the apex, these planes will cut lines such as $V$-2 and $V$-8 from the cone. At the same time, a vertical cutting plane will cut $XZ$ from the plane. $XZ$ then intersects $V$-2 at $b$ and $V$-8 at $h$, giving two points on the line of intersection. A series of planes thus passed and points found will complete the solution.

Even though the above plane method is simple and requires only the given views for solution, it *might* be advantageous to draw an extra view (auxiliary), as at (*b*), where, by projecting parallel to a horizontal line of the plane, *the plane appears as an edge*. In this event, selected lines of the cone are seen to intersect the plane in the auxiliary view (where the plane appears as an edge), and the solution becomes basically the same as in Fig. 19*b*. In addition, points located first in the auxiliary view and then in the top view will have to be found in the front view by projection from the top view and

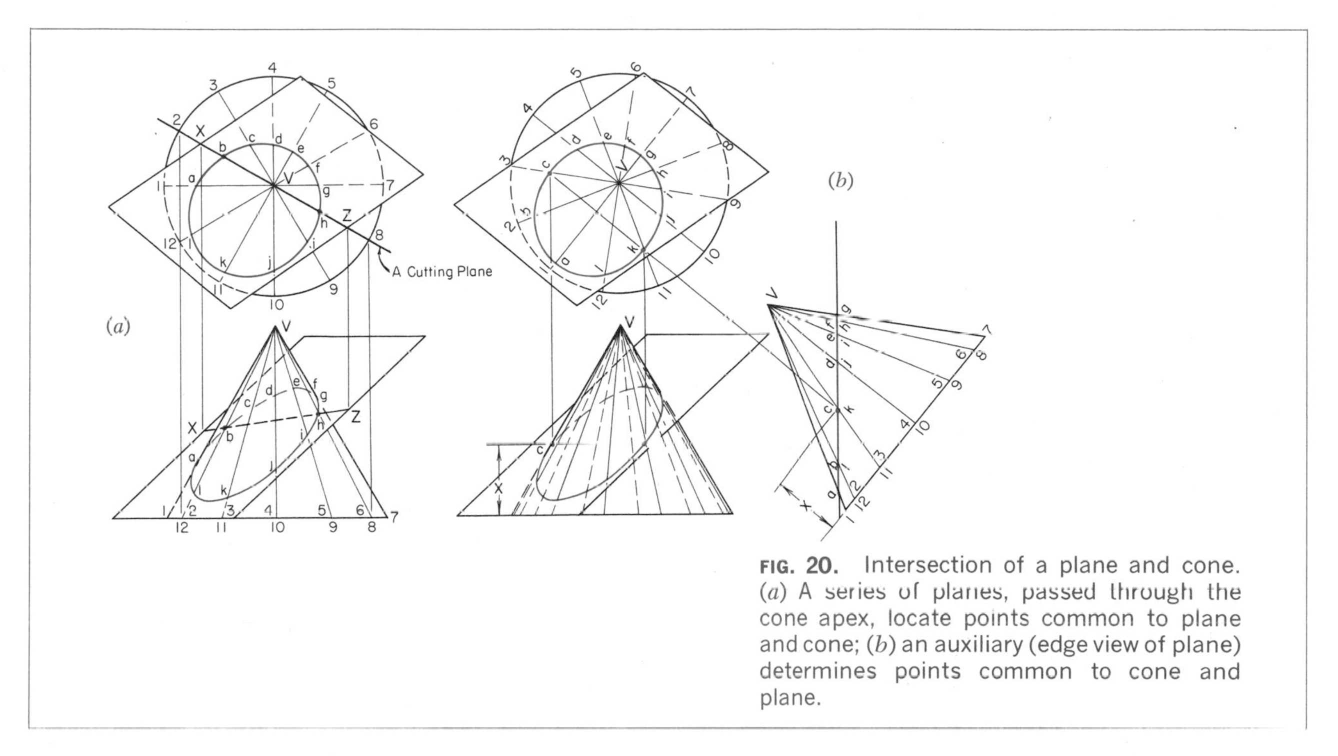


**FIG. 20.** Intersection of a plane and cone. (*a*) A series of planes, passed through the cone apex, locate points common to plane and cone; (*b*) an auxiliary (edge view of plane) determines points common to cone and plane.

measurement from the auxiliary, indicated by distance *X* on the figure.

## 19. Intersection of Two Cylinders (Fig. 21).

Cutting planes parallel to the axis of a cylinder will cut straight-line elements from the cylinder. The frontal cutting planes *A*, *B*, *C*, and *D*, *parallel to the axis of each cylinder,* cut elements from each cylinder, the intersections of which are points on the curve. The pictorial sketch shows a slice cut by a plane from the object, which has been treated as a solid in order to illustrate the method more easily.

When the axes of the cylinders do not intersect, as in Fig. 22, the same method is used. Certain "critical planes" give the limits and turning points of the curve. Such planes should always be taken through the contour elements. For the position shown, planes *A* and *D* give the depth of the curve, the plane *B* the extreme height, and the plane *C* the tangent or turning points on the contour element of the vertical cylinder. After the critical points have been determined, a sufficient number of other cutting planes are used to give an accurate curve.

## 20. Intersection of a Cylinder and a Cone (Fig. 23).

Cutting planes may be taken, as at (*a*), so as to pass through the vertex of the cone and parallel to the axis of the cylinder, thus cutting the straight-line elements from both cylinder and cone; or, as at (*b*), with a right-circular cone, when the cylinder's axis is parallel or perpendicular to the cone's axis, cutting planes may be taken parallel to the base so as to cut circles from the cone. Both systems of planes are illustrated in the figure. The pictorial sketches show slices taken by each plane through the objects, which have been treated as solids in order to illustrate the method more easily. Some judgment is necessary in the selection of both the direction and the number of cutting planes. More points need to be found at the places of sudden curvature or change of direction of the projections of the line of intersections.

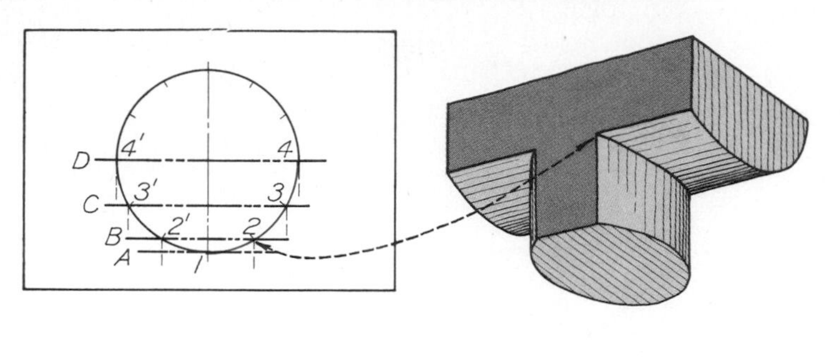


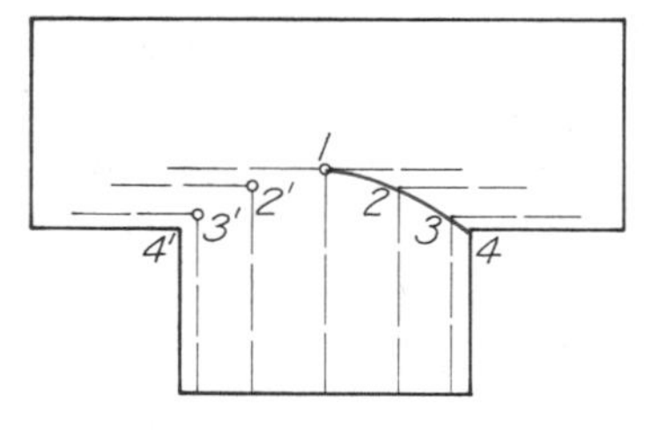


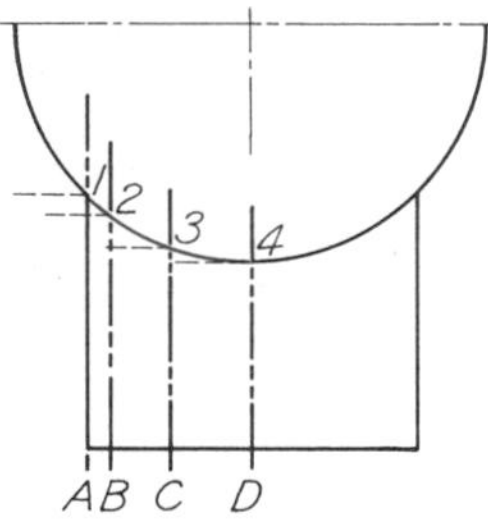


**FIG. 21.** Intersection of two cylinders. Planes cut straight-line elements from both.

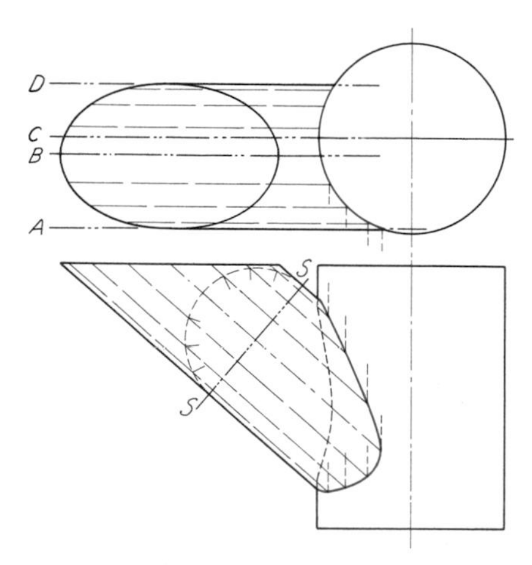


**FIG. 22.** Intersection of two cylinders, axes not intersecting. Planes, parallel to each of the axes, cut straight-line elements from both surfaces.

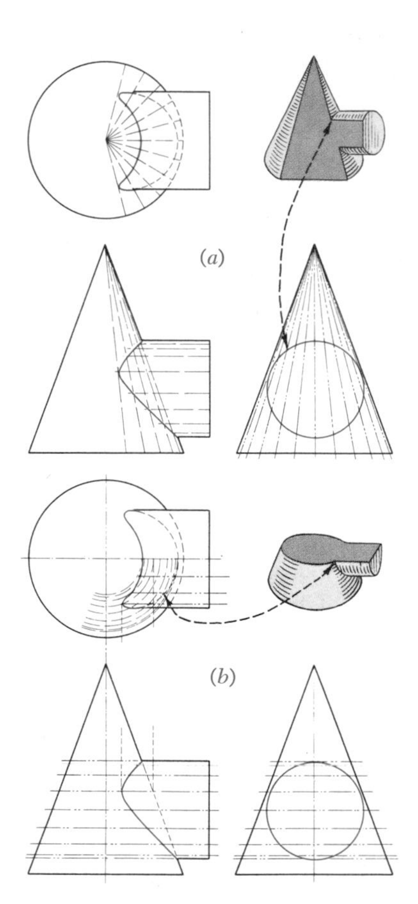


**FIG. 23.** Intersection of a cylinder and cone. (*a*) Elements of the cone are observed to intersect the cylinder in the side view (edge view of cylinder); or (*b*) horizontal planes cut circles from both surfaces and locate points common to both.

In Fig. 23*a* the cutting planes appear as edges in the right-side view, where the cylinder surface also appears as an edge. The observed intersections in the side view are then projected to the cone elements formed by the cutting planes, in the top and front views, to complete the solution.

At (*b*), horizontal cutting planes cut circles from the cone and straight-line elements from the cylinder. The top view then reveals intersections of cone circles and cylinder elements. Projection to the front view will complete the solution.

## PROBLEMS

Selections may be made from the following problems. Construct the figures accurately in pencil without inking. Any practical problem can be resolved into some combination of the "type solids," and the exercises given illustrate the principles involved in the various combinations.

The following problems may be drawn on $8\frac{1}{2}$- × 11-in. or 11- × 17-in. sheets. Assume the objects to be made of thin metal with open ends unless otherwise specified.

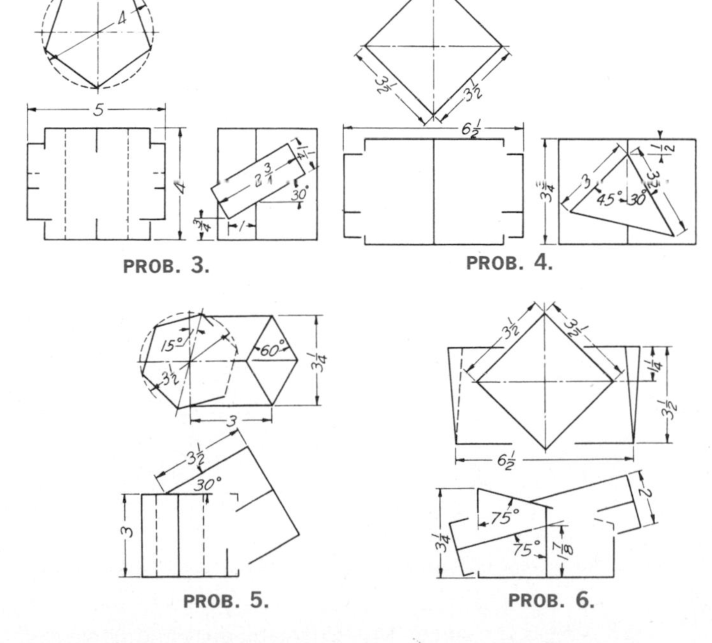

PROB. 1. PROB. 2. PROB. 3. PROB. 4. PROB. 5. PROB. 6.

### Group 1. Intersections of Prismatic Ducts.

**1. to 3.** Find the line of intersection, considering the prisms as pipes opening into each other. Use care in indicating visible and invisible parts of the line of intersection.

**4 to 6.** Find the line of intersection, indicating visible and invisible parts and considering prisms as pipes opening into each other.

**7.** Find the intersection between the two prismatic ducts.

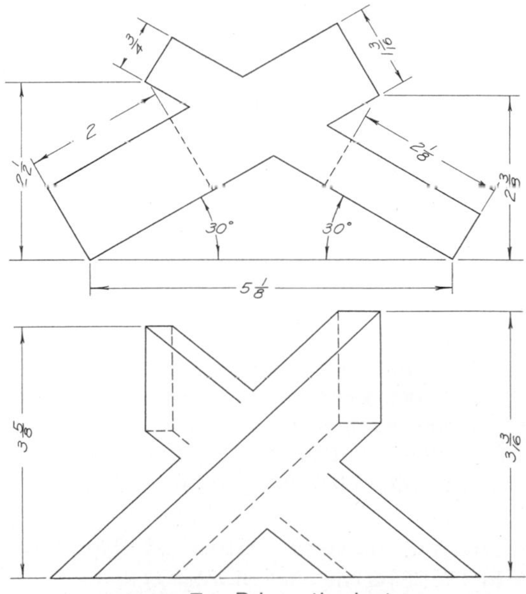

PROB. 7. Prismatic ducts.

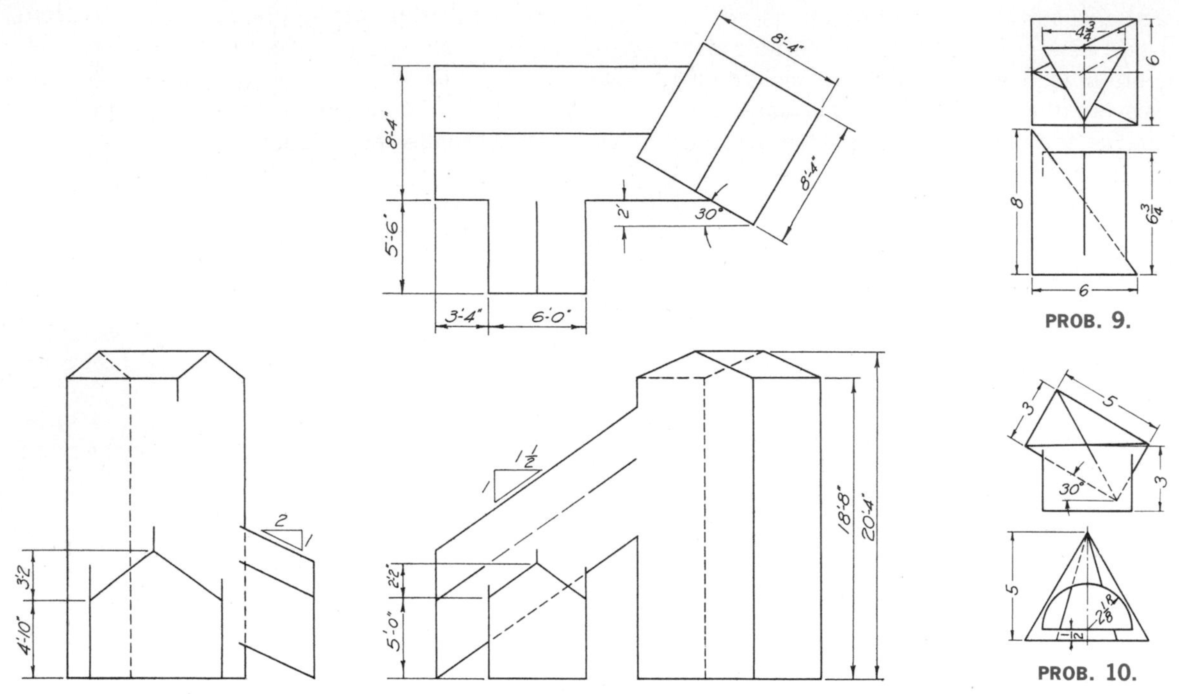


PROB. 9.

PROB. 10.

**PROB. 8.** Tower and conveyor-belt galleys.

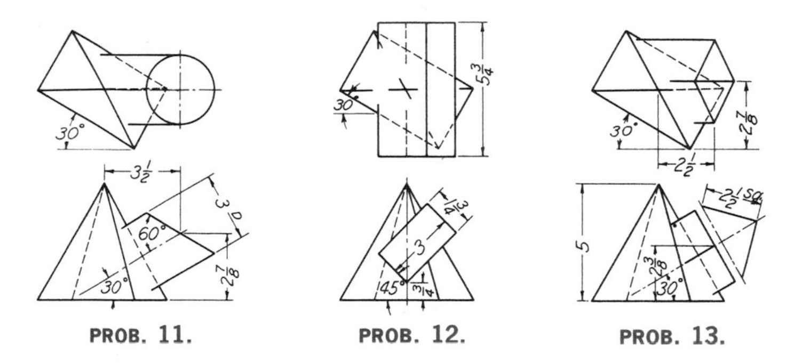


**PROB. 11.** **PROB. 12.** **PROB. 13.**

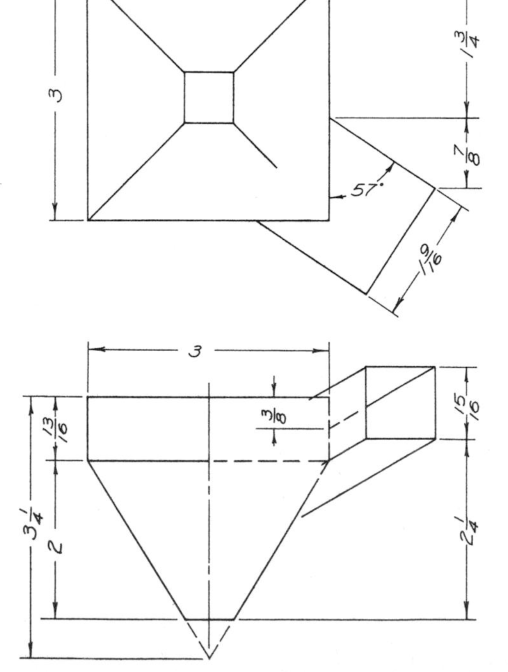


**PROB. 14.** Prismatic duct and hopper.

**8.** The layout as illustrated shows a tower and proposed conveyor-belt galleys. Find the intersection between the tower and the main supply galley.

**8*A*.** Same layout as Prob. 8. After solving Prob. 8., find the intersection between the two galleys.

## Group 2. Intersections of Pyramidal Objects.

**9 to 13.** Find the lines of intersection.

**14.** Find the line of intersection between the prismatic duct and hopper.

## Group 3. Intersections of Cylindrical Ducts.

**15 to 17.** Find the line of intersection, indicating visible and invisible portions and considering cylinders as pipes opening into each other.

**18.** The layout shows a pump casing composed of a cylinder and elbow. Find the intersection between the cylinder and elbow.

**19.** The layout shows a portion of a liquid oxygen pumping system consisting of pipes $A$, $B$, $C$, and $D$. Find the intersection of the pipes.

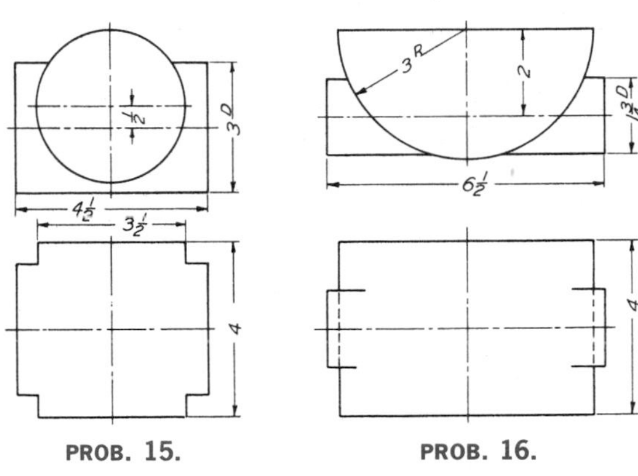


**PROB. 15.** **PROB. 16.**

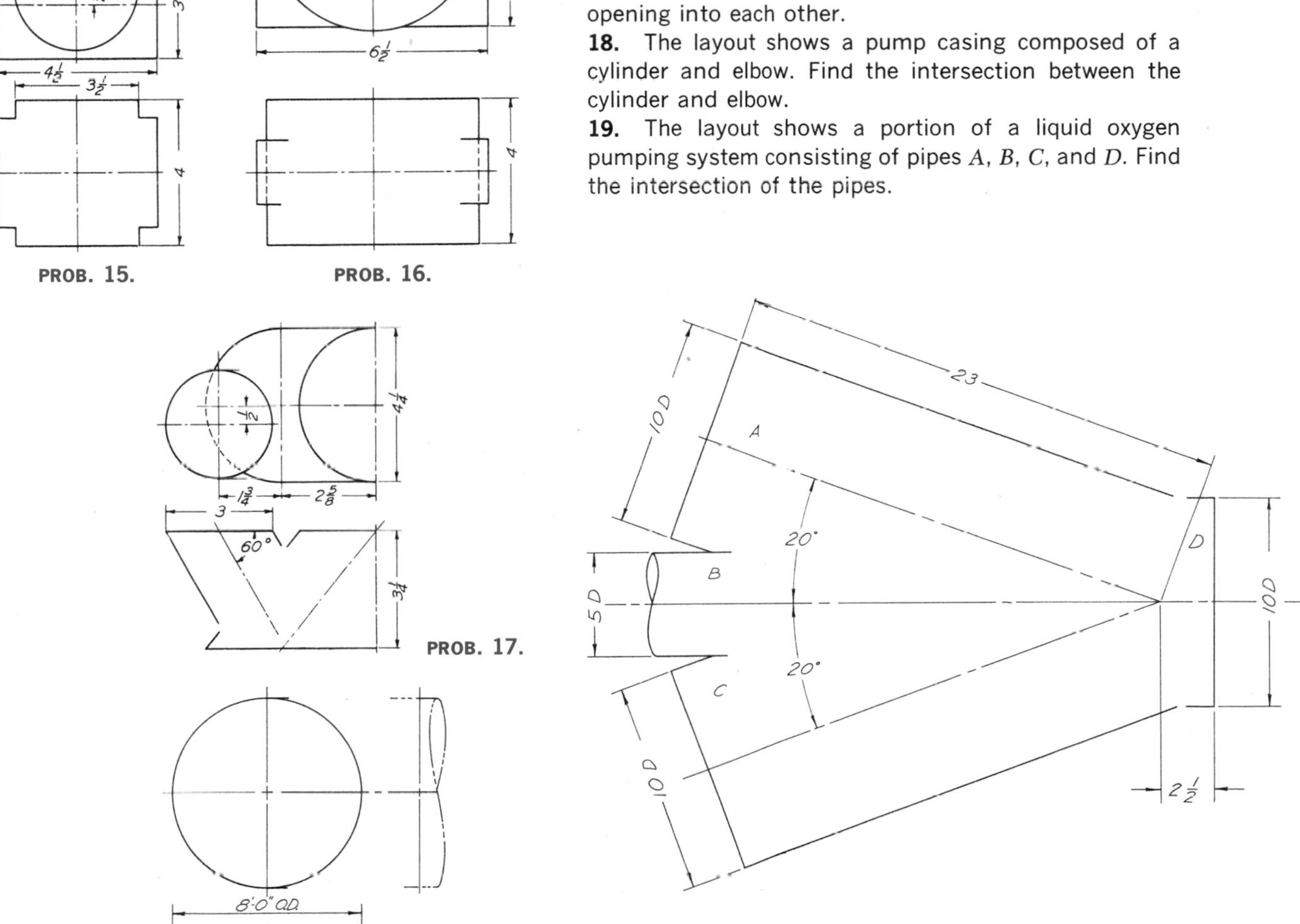


**PROB. 17.**

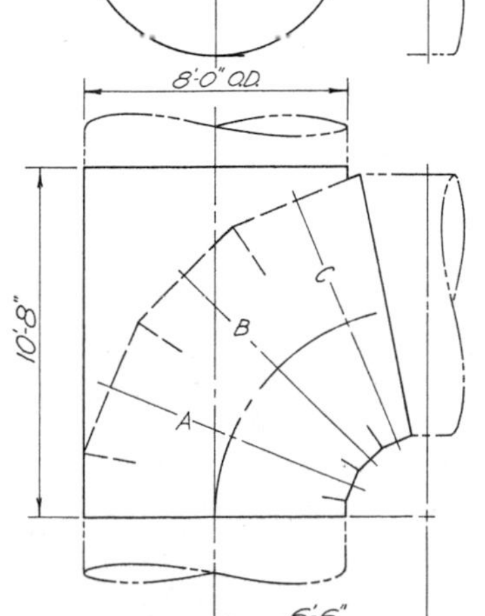


**PROB. 18.** Pump casing.

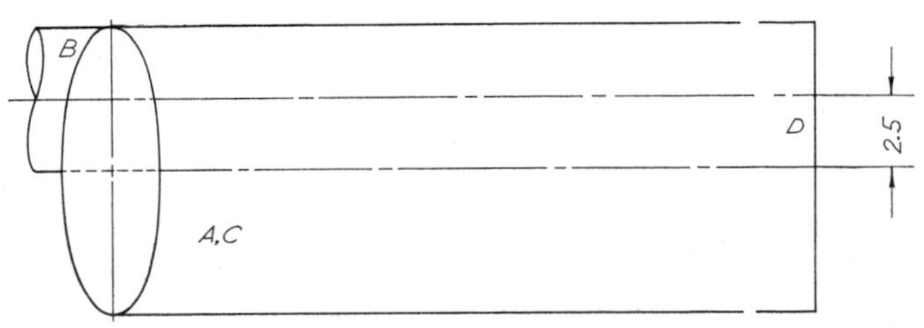


**PROB. 19.** Lox pump outlet.

**20.** The layout shows an oblique pipe meeting a cylindrical inlet in a power-station piping system. Find the intersection between the inlet and pipe.

## Group 4. Intersections of Conic Objects and Ducts.

**21 to 31.** Find the line of intersection.

**32.** The layout shows the plan view of an intake manifold consisting of a conical connector intersected by two cylindrical pipes. Find the intersection of the pipe that

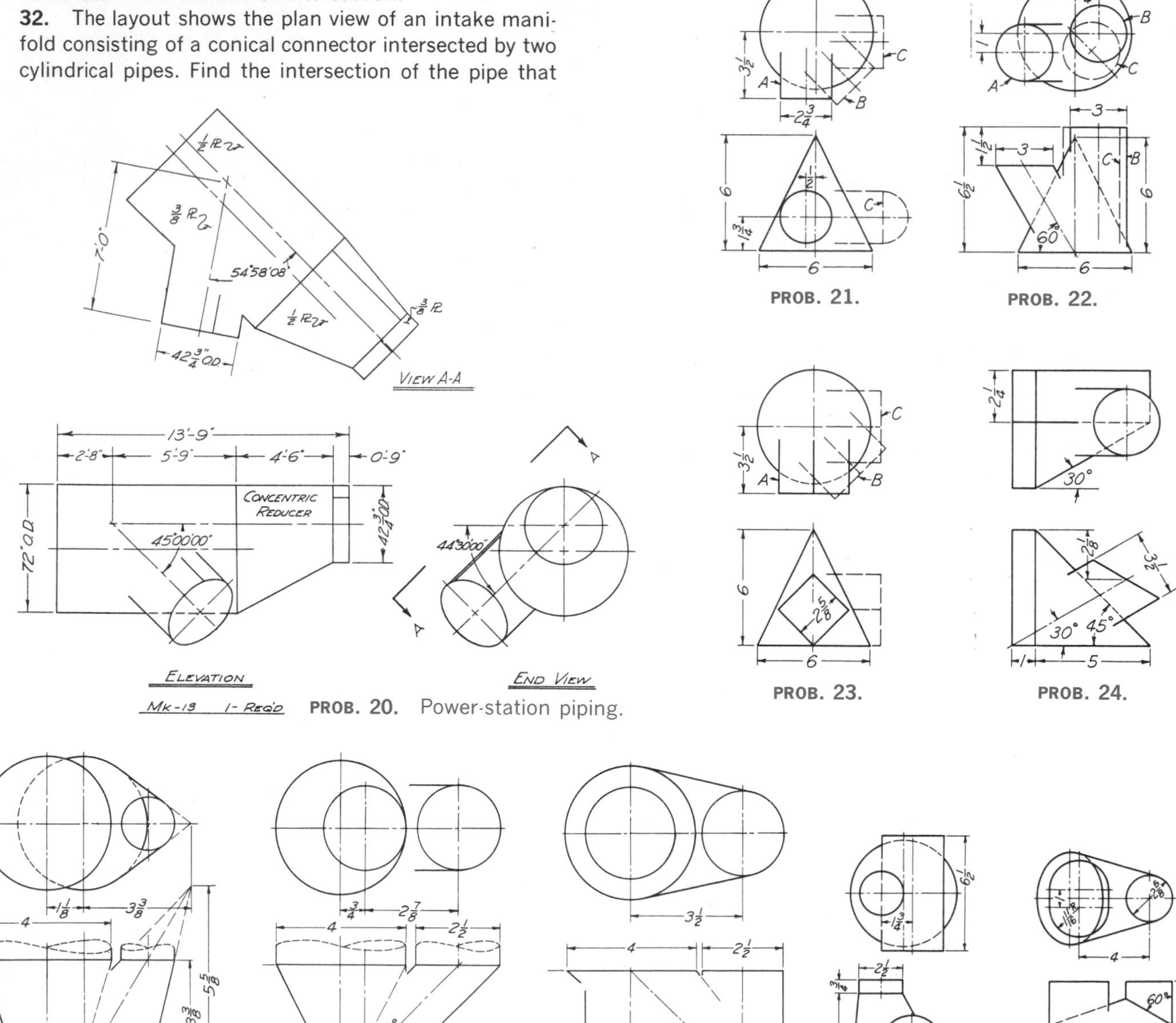


PROB. 20. Power-station piping.

PROB. 21.

PROB. 22.

PROB. 23.

PROB. 24.

PROB. 25.

PROB. 26.

PROB. 27.

PROB. 28.

PROB. 29.

enters the side of the conical connector. The axes of the two shapes intersect. Solve this problem using only the plan view.

**33.** The layout shows the cylindrical outlet pipe and conical inlet pipe of a "Kaplan-system" outlet. Find the intersection of the two shapes, using the cutting-sphere method.

**34.** The layout shows a right circular cylinder and an elliptical-base cone. Find the intersection of the two surfaces.

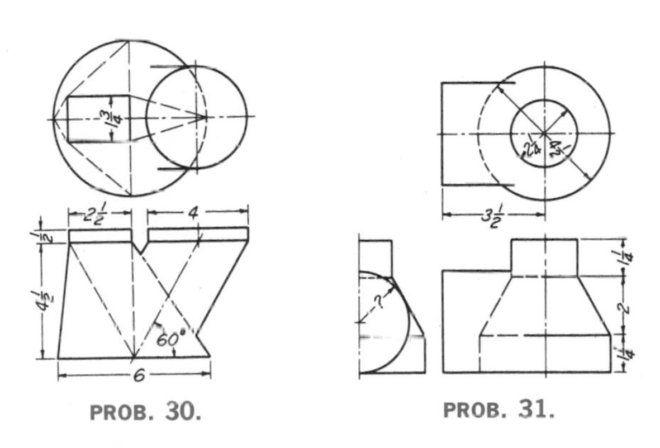

**PROB. 30.**

**PROB. 31.**

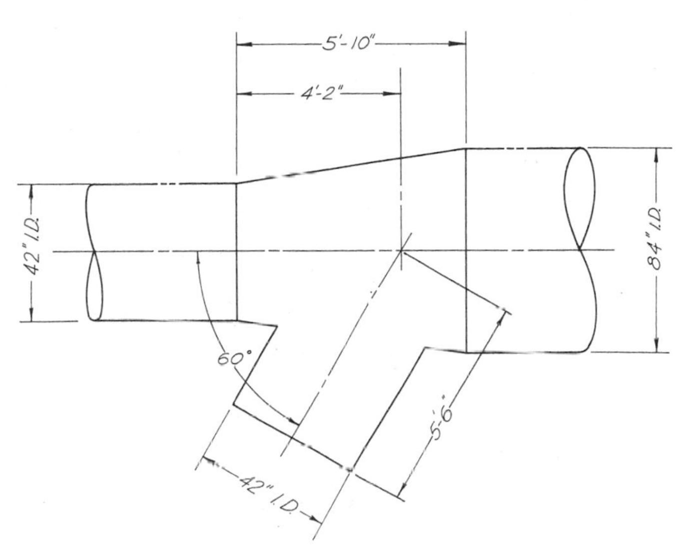


**PROB. 32.** Discharge manifold.

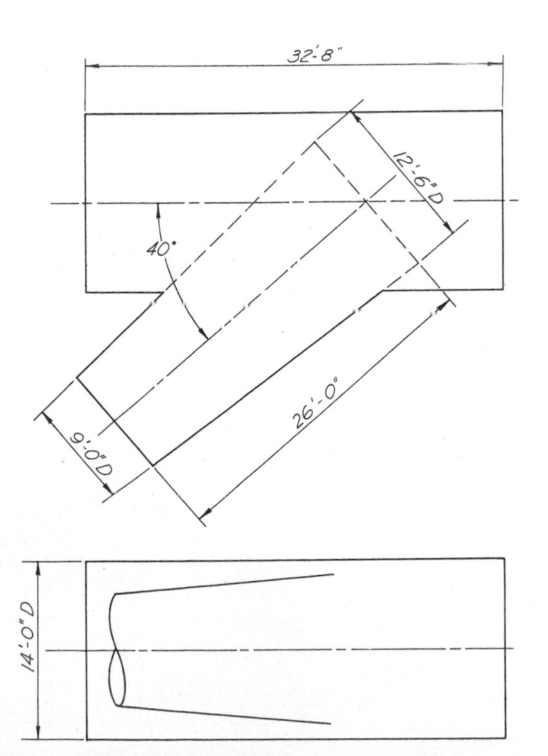


**PROB. 33.** "Kaplan-system" outlet.

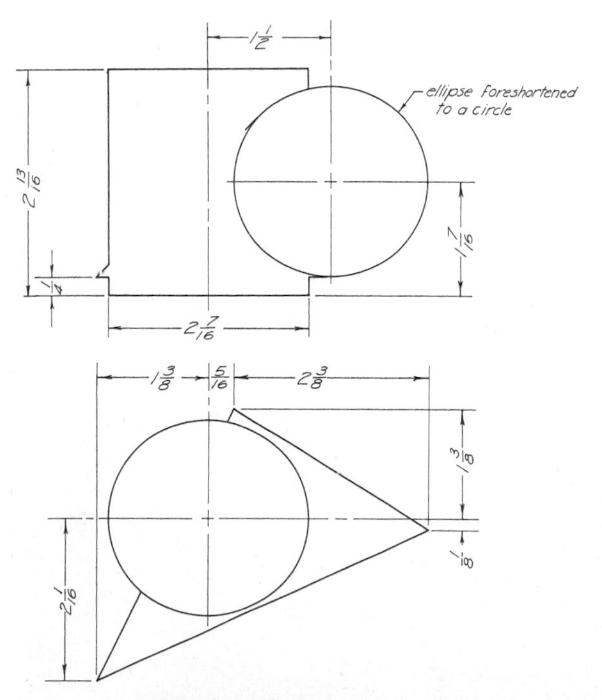


**PROB. 34.** Cylinder and cone.

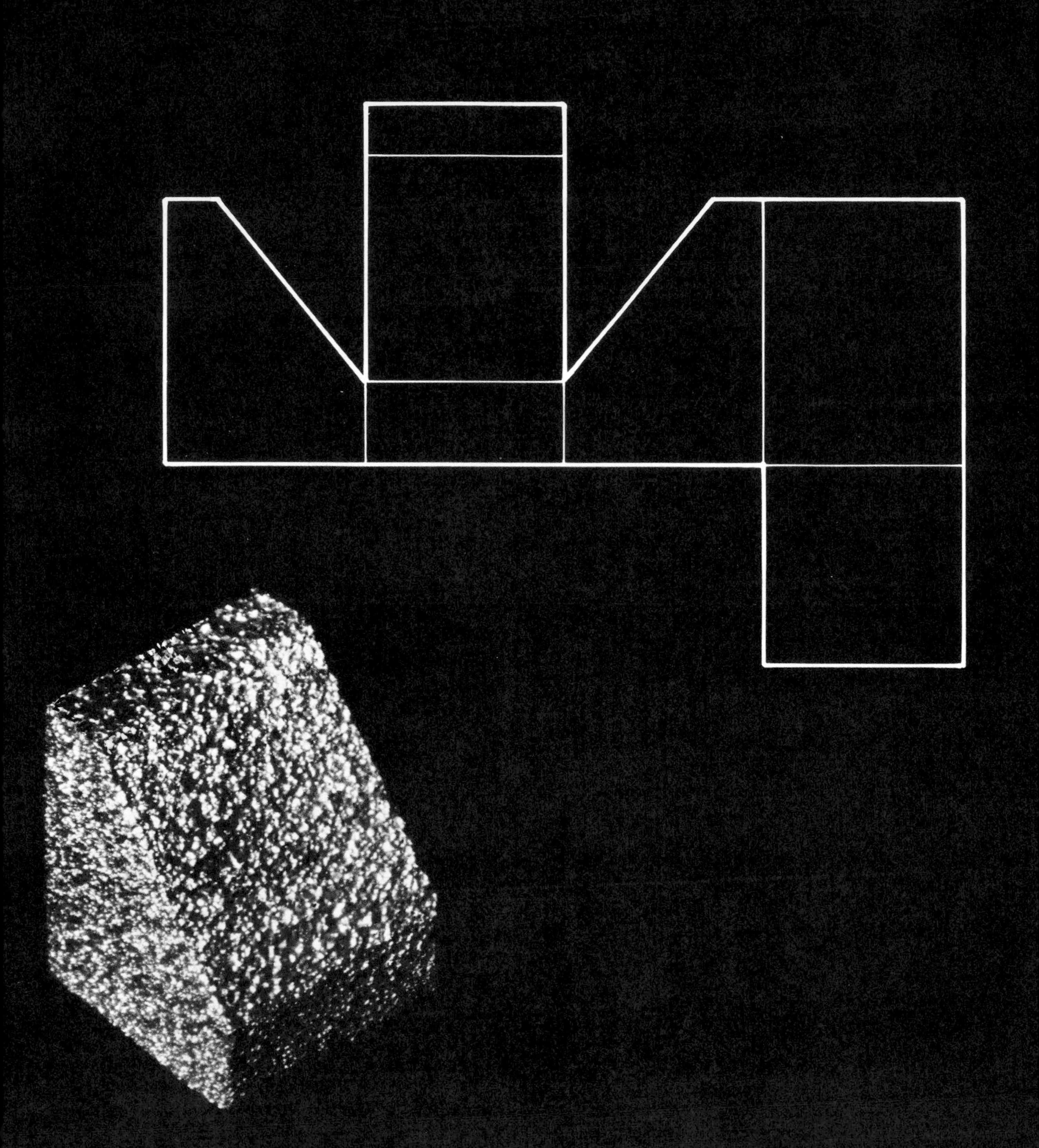

This is a Discussion and Illustration of:



# 10

# DEVELOPED VIEWS

These Show a Representation of the Shape and Size of Thin Material Used to Make an Object by Folding, Rolling, or Forming

## 1. Developments, Defined.

In many different kinds of construction full-size patterns of some or all of the faces of an object are required; for example, in stonecutting, a template or pattern giving the shape of an irregular face, or in sheet-metal work, a pattern to which a sheet may be cut so that when rolled, folded, or formed it will make the object.

The complete surface laid out in a plane is called the "development" of the surface.

Surfaces about which a thin sheet of flexible material (such as paper or tin) can be wrapped smoothly are said to be developable; these include objects made up of planes and single-curved surfaces only. Warped and double-curved surfaces are nondevelopable; and when patterns are required for their construction, they can be made only by methods that are approximate, but, assisted by the ductility or pliability of the material, they give the required form. Thus while a ball cannot be wrapped smoothly, a two-piece pattern developed approximately and cut from leather can be stretched and sewed on in a smooth cover, or a flat disk of metal can be die-stamped, formed, or spun to a hemispherical or other shape.

## 2. Basic Considerations.

We have learned the method of finding the true size of a plane surface by projecting its normal view. If the true size of all the plane faces of an object are found and joined in order at their common edges so that all faces lie in a common plane, the result will be the developed surface. Usually this may be done to the best advantage by finding the true length of the edges.

The development of a right cylinder is evidently a rectangle whose width is the altitude and length the rectified circumference (Fig. 1); and the development of a right-circular cone is a circular sector with a radius equal to the slant height of the cone and an arc equal in length to the circumference of its base (Fig. 1).

As illustrated in Fig. 1, developments are drawn with the inside face up. This is primarily the result of working to inside rather than outside dimensions of ducts. This procedure also facilitates the use of fold lines, identified by punch marks at each end, along which the metal is folded in forming the object.

In laying out real sheet-metal designs, an allowance must be made for seams and lap and, in heavy sheets, for the thickness and crowding of the metal; there is also the consideration of the commercial sizes of material as well as the question of economy in cutting. In all of this some practical shop knowledge is necessary. Figure 17 and Sec. 14 indicate the usage of some of the more common joints, although the developments given in this chapter will be confined to the principles.

## 3. To Develop a Truncated Hexagonal Prism (Fig. 2).

First draw two projections of the prism: (1) a normal view of a right section (a section or cut obtained by a plane perpendicular to the axis) and (2) a normal view of the lateral edges. The base *ABCDEF* is a right section shown in true size in the bottom view. Lay off the perimeter of the base on line *AA* of the development. This line is called by sheet-metal workers the "stretchout" or "girth" line. At points *A*, *B*, *C*, etc., erect perpendiculars called "measuring lines" or "bend lines," representing the lateral edges along which the pattern is folded to form the prism. Lay off on each of these its length *A*-1, *B*-2, *C*-3, etc., as given on the front view. Connect the points 1, 2, 3, etc., in succession, to complete the development of the lateral surfaces. Note on the pattern that the inside of the lateral faces is toward the observer. For the development of the entire surface in one piece, attach the true sizes of the upper end and the base as shown, finding the true size of the upper end by an auxiliary (normal) view as described in Sec. 4, Chap. 7. For economy of solder or rivets and time, it is customary to make the seam on the shortest edge or surface. In seaming along the intersection of surfaces whose dihedral angle is other than 90°, as is the case here, the lap seam lends itself to convenient assembling. The flat lock could be used if the seam were made on one of the lateral faces.

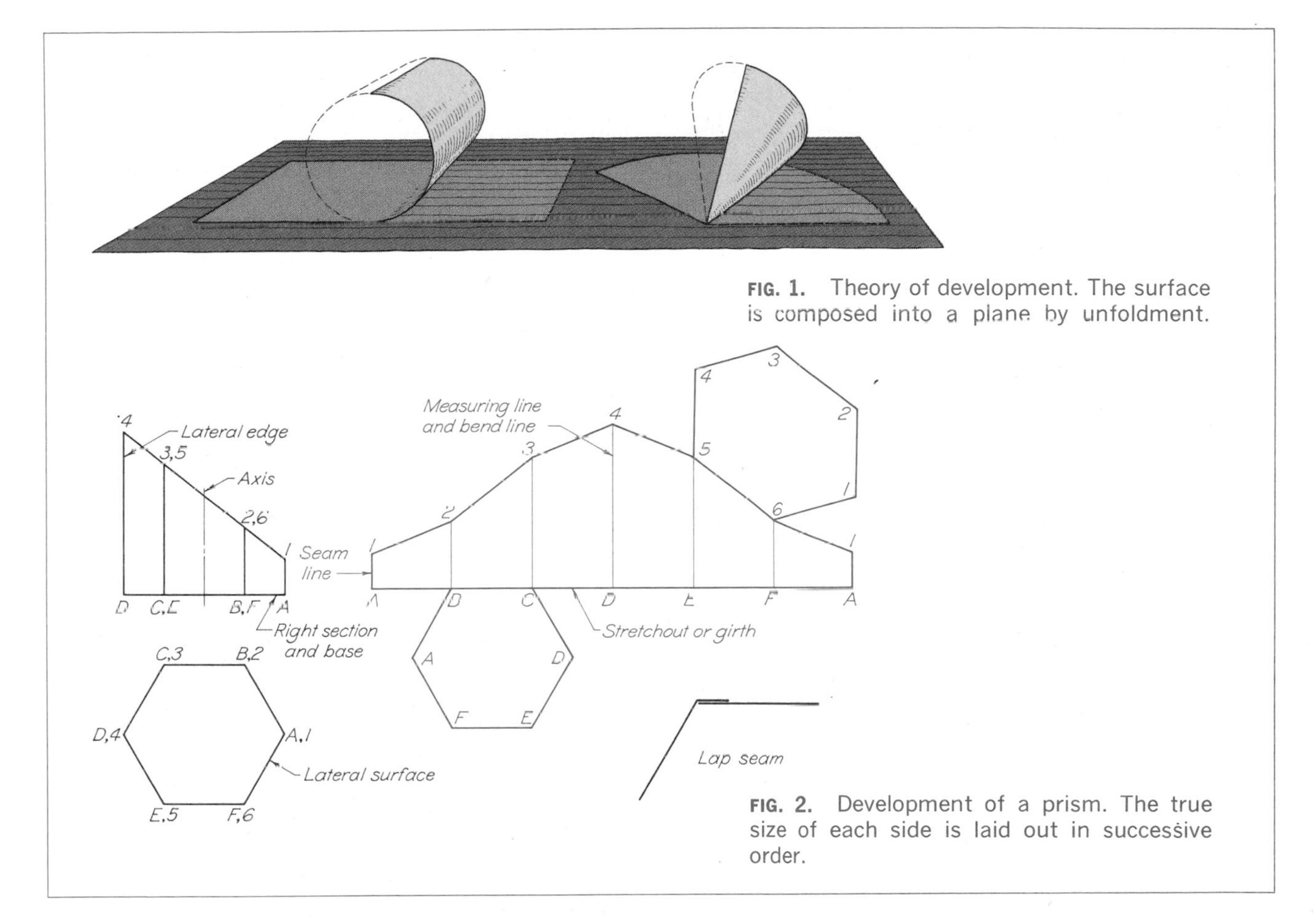


**FIG. 1.** Theory of development. The surface is composed into a plane by unfoldment.

**FIG. 2.** Development of a prism. The true size of each side is laid out in successive order.

## 4. To Develop a Truncated Right Pyramid (Fig. 3).

Draw the projections of the pyramid which show (1) a normal view of the base or right section and (2) a normal view of the axis. Lay out the pattern for the pyramid, and then superimpose the pattern of the truncation.

Since this is a portion of a right regular pyramid, the lateral edges are all of equal length. The lateral edges $OA$ and $OD$ are parallel to the frontal plane and consequently show in their true length on the front view. With center $O_1$, taken at any convenient place, and a radius $O_FA_F$, draw an arc that is the stretchout of the pattern. On it step off the six equal sides of the hexagonal base, obtained from the top view, and connect these points successively with each other and with the vertex $O_1$, thus forming the pattern for the pyramid.

The intersection of the cutting plane and lateral surfaces is developed by laying off the true length of the intercept of each lateral edge on the corresponding line of the development. The true length of each of these intercepts, such as $OH$ and $OJ$, is found by rotating it about the axis of the pyramid until they coincide with $O_FA_F$. The path of any point, as $H$, will be projected on the front view as a horizontal line. To obtain the development of the entire surface of the truncated pyramid, attach the base; also find the true size of the cut face and attach it on a common line.

The lap seam is suggested for use here also for convenient assembling.

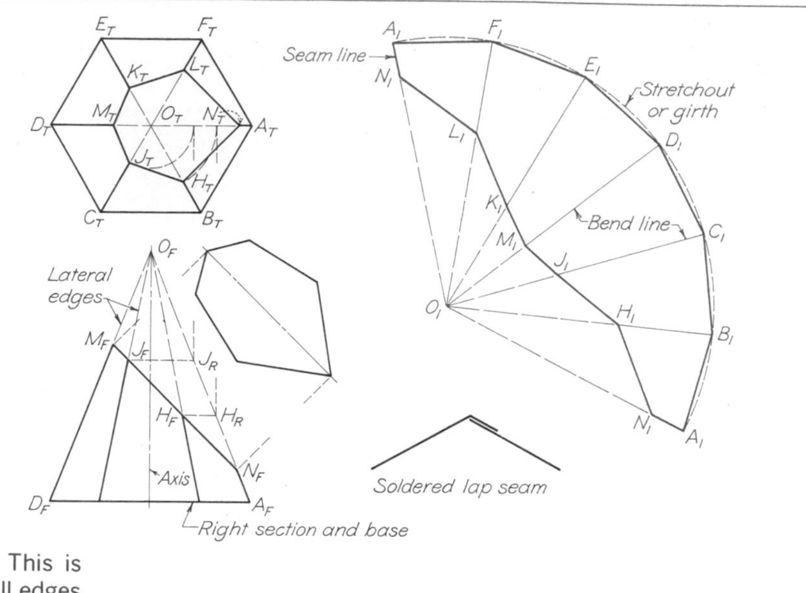


**FIG. 3.** Development of a pyramid. This is a right pyramid. The true lengths of all edges (from the vertex to the base) are equal.

*The right-rectangular pyramid,* Fig. 4, is developed in a similar way, but as the edge $OA$ is not parallel to the plane of projection, it must be rotated to $O_FA_R$ to obtain its true length.

## 5. To Develop an Oblique Pyramid (Fig. 5).

Since the lateral edges are unequal in length, the true length of each must be found separately by rotating it parallel to the frontal plane. With $O_1$ taken at any convenient place, lay off the seam line $O_1A_1$ equal to $O_FA_R$. With $A_1$ as center and radius $A_1B_1$ equal to the true length of $AB$, describe an arc. With $O_1$ as center and radius $O_1B_1$ equal to $O_FB_R$, describe a second arc intersecting the first in vertex $B_1$. Connect the vertices $O_1$, $A_1$, and $B_1$, thus forming the pattern for the lateral surface $OAB$. Similarly, lay out the patterns for the remaining three lateral surfaces, joining them on their common edges. The stretchout is equal to the summation of the base edges. If the complete development is required, attach the base on a common line. The lap seam is suggested as the most suitable for the given conditions.

## 6. To Develop a Truncated Right Cylinder (Fig. 6).

The development of a cylinder is similar to the development of a prism. Draw two projections of the cylinder: (1) a normal view of a right section and (2) a normal view of the elements. In rolling the cylinder out on a tangent plane, the base or right section, being perpendicular to the axis, will develop

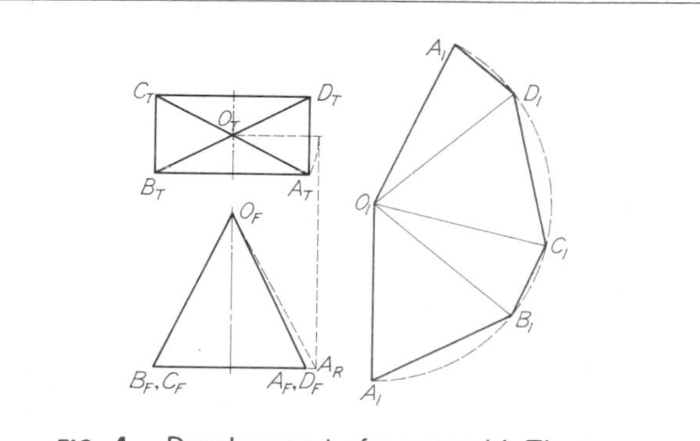


**FIG. 4.** Development of a pyramid. The true length from vertex to base is found by rotating one edge until frontal.

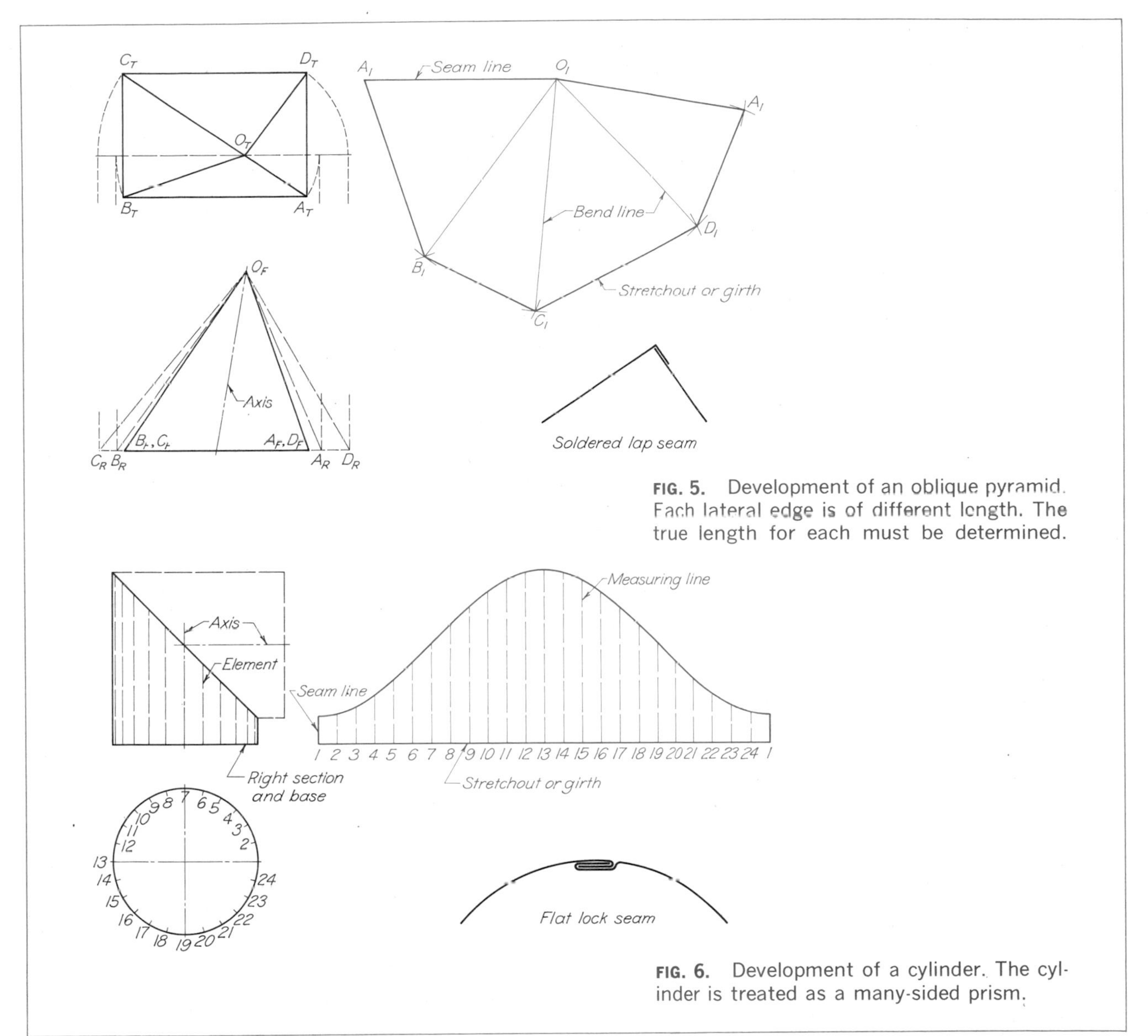


**FIG. 5.** Development of an oblique pyramid. Each lateral edge is of different length. The true length for each must be determined.

**FIG. 6.** Development of a cylinder. The cylinder is treated as a many-sided prism.

into a straight line. For convenience in drawing, divide the normal view of the base, here shown in the bottom view, into a number of equal parts by points that represent elements. These divisions should be spaced so that the chordal distances closely enough approximate the arc to make the stretchout practically equal to the periphery of the base or right section. Project these elements to the front view. Draw the stretchout and measuring lines as in Fig. 2, the cylinder now being treated as a many-sided prism. Transfer the lengths of the elements in order, by projection or with dividers, and join the points thus found by a smooth curve, sketching it in freehand very lightly before fitting the French curve to it. This development might be the pattern of one-half of a two-piece elbow. Three-piece, four-piece, or five-piece elbows can be drawn similarly, as illustrated in Fig. 7. As the base is symmetrical, only one-half of it need be drawn. In these cases, the intermediate pieces, as $B$, $C$, and $D$, are developed on a stretchout line formed by laying off the perimeter of a right section. If the right section is taken through the middle of the piece, the stretchout line becomes the center line of the development.

Evidently any elbow could be cut from a single sheet without waste if the seams were made alternately on the long and short sides. The flat lock seam is recommended for Figs. 6 and 7, although other types could be used.

*The octagonal dome* (Fig. 8) illustrates an application of the development of cylinders. Each piece is a portion of a cylinder. The elements are parallel to the base of the dome and show in their true lengths in the top view. The true length of the stretchout line for sections $A$ and $A'$ shows in the

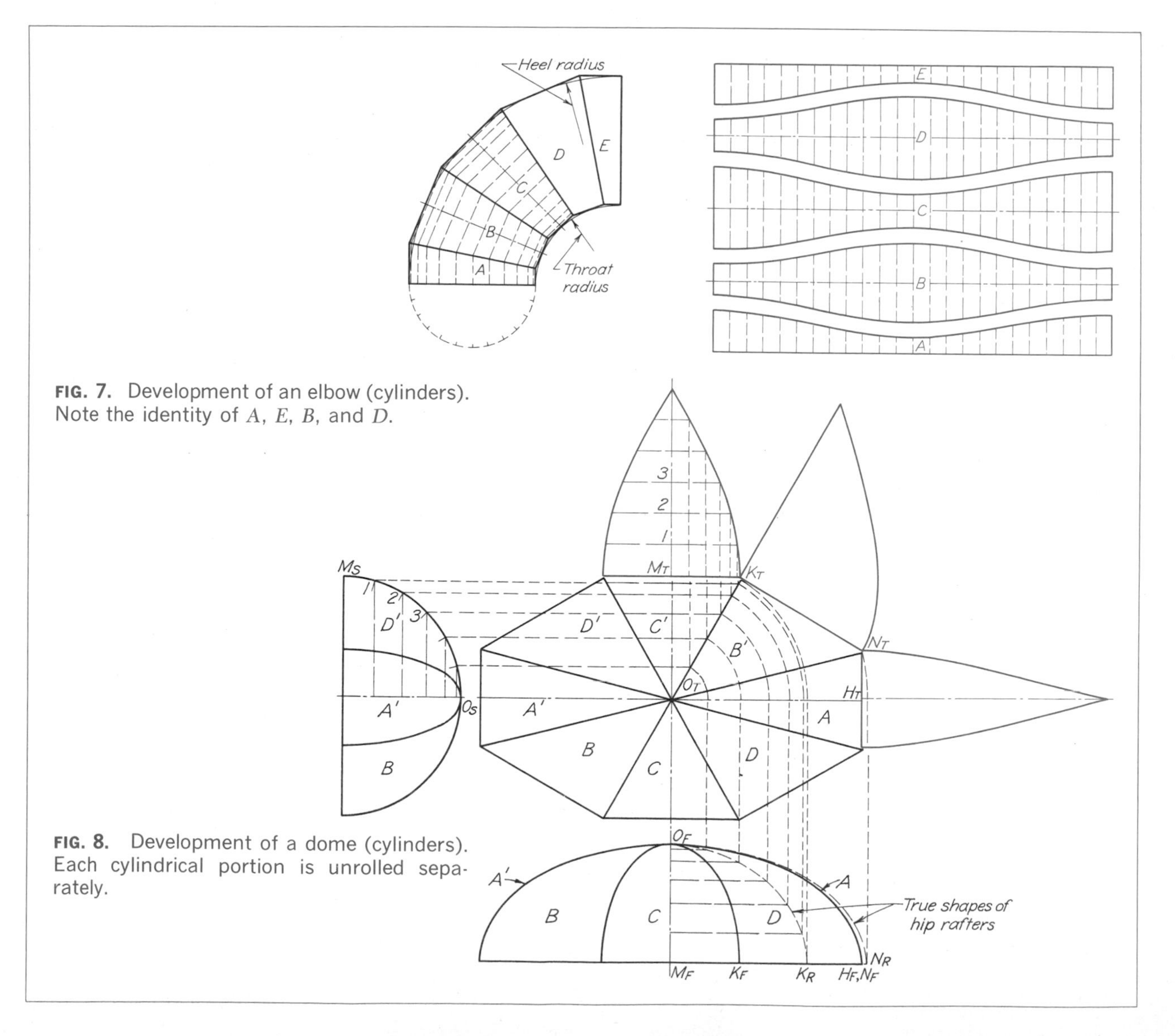


**FIG. 7.** Development of an elbow (cylinders). Note the identity of $A$, $E$, $B$, and $D$.

**FIG. 8.** Development of a dome (cylinders). Each cylindrical portion is unrolled separately.

front view at $O_F H_F$. By considering $O_T H_T$ as the edge of a plane cutting a right section, the problem is identical with the preceding problem.

Similarly, the stretchout line for sections $B$, $B'$, $D$, and $D'$ shows in true length at $O_F K_R$ in the front view, and for section $C$ and $C'$ at $O_S M_S$ in the side view.

The true shape of hip rafter $ON$ is found by rotating it until it is parallel to the frontal plane, as at $O_F N_R$, in the same manner as in finding the true length of any line. A sufficient number of points should be taken to give a smooth curve.

### 7. To Develop a Truncated Right-Circular Cone (Fig. 9).

Draw the projections of the cone which will show (1) a normal view of the base or right section and (2) a normal view of the axis. First develop the surface of the complete cone and then superimpose the pattern for the truncation.

Divide the top view of the base into a sufficient number of equal parts so that the sum of the resulting chordal distances will closely approximate the periphery of the base. Project these points to the front view, and draw front views of the elements through them. With center $A_1$ and a radius equal to the slant height $A_F$-$1_F$, which is the true length of all the elements, draw an arc, which is the stretchout, and lay off on it the chordal divisions of the base, obtained from the top view. Connect these points $1_1$, 2, 3, etc., with $A$-1, thus forming the pattern for the cone. Find the true length of each element from vertex to cutting plane by rotating it to coincide with the contour element $A$-1, and lay off this distance on the corresponding line of the development. Draw a smooth curve through these points. The flat lock seam along element $S$-1 is recommended, although other types could be employed. The pattern for the inclined surface is obtained from the auxiliary (normal) view.

### 8. Triangulation.

Nondevelopable surfaces are developed approximately by assuming them to be made of narrow sections of developable surfaces. The commonest and best method for approximate development is that of triangulation, that is, the surface is assumed to be made up of a large number of triangular strips,

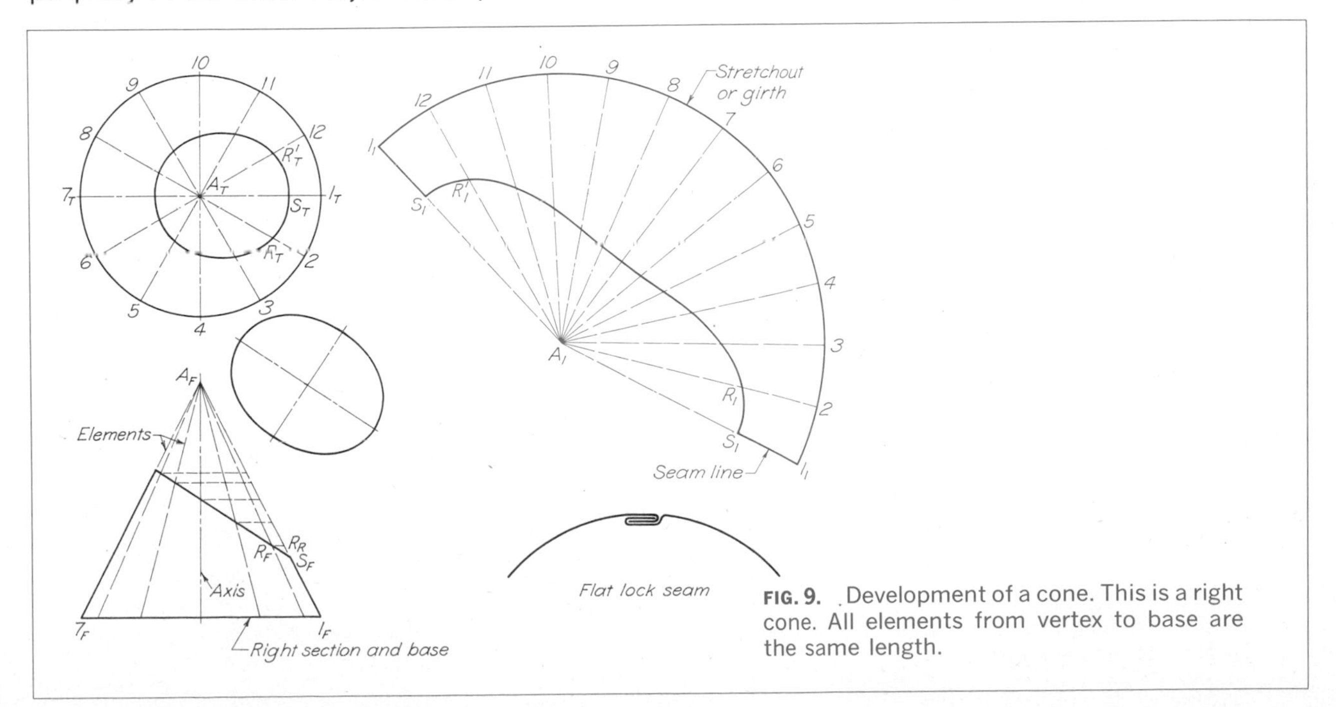


**FIG. 9.** Development of a cone. This is a right cone. All elements from vertex to base are the same length.

or plane triangles with very short bases. This method is used for all warped surfaces and also for oblique cones. Oblique cones are single-curved surfaces and thus are theoretically capable of true development, but they can be developed much more easily and accurately by triangulation, a simple method which consists merely of dividing the surface into triangles, finding the true lengths of the sides of each, and constructing them one at a time, joining these triangles on their common sides.

**9. To Develop an Oblique Cone (Fig. 10).** An oblique cone differs from a cone of revolution in that the elements have different lengths. The development of the right-circular cone is, practically, made up of a number of equal triangles which meet at the vertex and whose sides are elements and whose bases are the chords of short arcs of the base of the cone. In the oblique cone each triangle must be found separately.

Draw two views of the cone showing (1) a normal view of the base and (2) a normal view of the altitude. Divide the true size of the base, here shown in the top view, into a sufficient number of equal parts so that the sum of the chordal distances will closely approximate the length of the base curve. Project these points to the front view of the base. Through these points and the vertex, draw the elements in each view. Since this cone is symmetrical about a frontal plane through the vertex, the elements are shown only on the front half of it. Also, only one-half of the development is drawn. With the seam on the shortest element, the element $OC$ will be the center line of the development and can be drawn directly at $O_1C_1$, as its true length is given at $O_FC_F$. Find the true length of the elements by rotating them until parallel to the frontal plane or by constructing a "true-length diagram." The true length of any element would be the hypotenuse of a triangle, one leg being the length of the projected element as seen in the top view and the other leg being equal to the altitude of the cone. Thus to make the diagram, draw the leg $OD$ coinciding with or parallel to $O_FD_F$. At $D$ and perpendicular to $OD$, draw the other leg, on which lay off the lengths $D$-1, $D$-2, etc., equal to $D_T$-$1_T$, $D_T$-$2_T$, etc., respectively. Distances from $O$ to points on the base of the diagram are the true lengths of the elements.

Construct the pattern for the front half of the cone as follows: With $O_1$ as center and radius $O$-1, draw an arc. With $C_1$ as center and radius $C_T$-$1_T$, draw a second arc intersecting the first at $1_1$; then $O_1$-$1_1$ will be the developed position of the element $O$-1. With $1_1$ as center and radius $1_T$-$2_T$, draw an arc

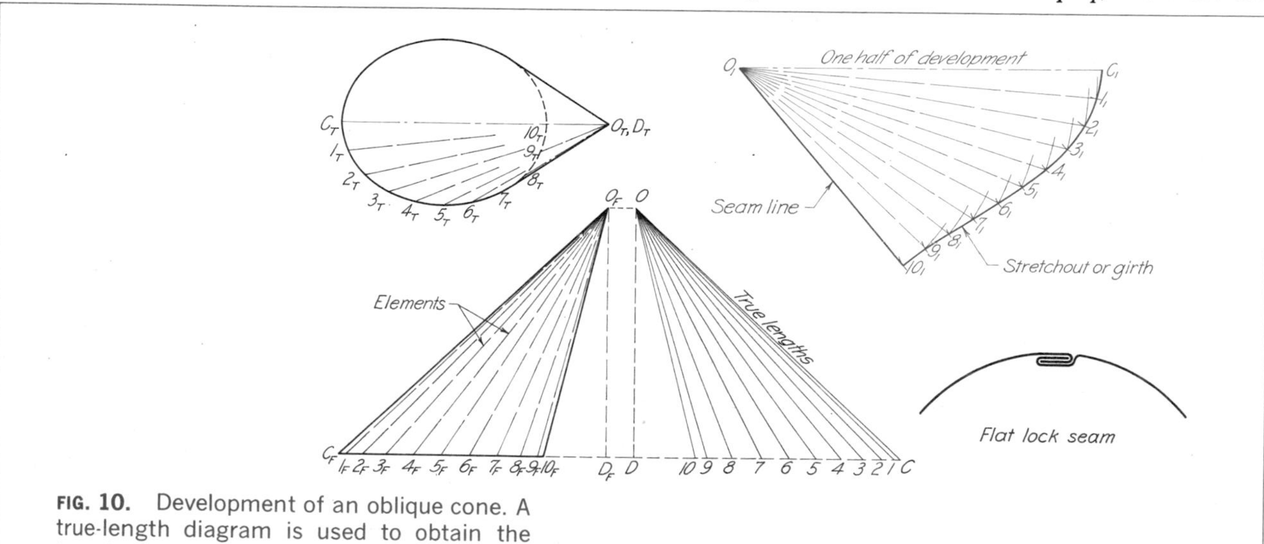


**FIG. 10.** Development of an oblique cone. A true-length diagram is used to obtain the lengths of the elements.

intersecting a second arc with $O_1$ as center and radius $O$-2, thus locating $2_1$. Continue this procedure until all the elements have been transferred to the development. Connect the points $C_1$, $1_1$, $2_1$, etc., with a smooth curve, the stretchout line, to complete the development. The flat lock seam is recommended for joining the ends to form the cone.

### 10. A Conic Connection between Two Parallel Cylindrical Pipes of Different Diameters.

This is shown in Fig. 11. The method used in drawing the pattern is an application of the development of an oblique cone. One-half of the elliptical base is shown in true size in an auxiliary view, here attached to the front view. Find the true size of the base from its major and minor diameters, divide it into a number of equal parts so that the sum of these chordal distances closely approximates the periphery of the curve, and project these points to the front and top views. Draw the elements in each view through these points, and find the vertex $O$ by extending the contour elements until they intersect. The true length of each element is found by using the vertical distance between its ends as the vertical leg of the diagram and its horizontal projection as the other leg. As each true length from vertex to base is found, project the upper end of the intercept horizontally across from the front view to the true length of the corresponding element to find the true length of the intercept. The development is drawn by laying out each triangle in turn, from vertex to base, as in Sec. 9, starting on the center line $O_1C_1$ and then measuring on each element its intercept length. Draw smooth curves through these points to complete the pattern. Join the ends with a flat lock seam.

### 11. Transition Pieces.

These are used to connect pipes or openings of different shapes of cross section. Figure 12, showing a transition piece for connecting a round pipe and a rectangular pipe, is typical. Transition pieces are always developed by triangulation. The piece shown in Fig. 12. is made up of four triangular planes, whose bases are the sides of the rectangle, and four parts of oblique cones, whose common bases are arcs of the circle and whose vertices are at the corners of the rectangle. To develop it, make a true-length diagram as in Fig. 10. When the true length of $O$-1 is found, all the sides of triangle $A$ will be known. Attach the development of cones $B$ and $B'$, then those of triangles $C$ and $C'$, and so on.

Figure 13 is another transition piece joining a rectangular to a circular pipe whose axes are nonparallel. By using a partial right-side view of the round opening, the divisions of the bases of the oblique cones can be found (as the object is sym-

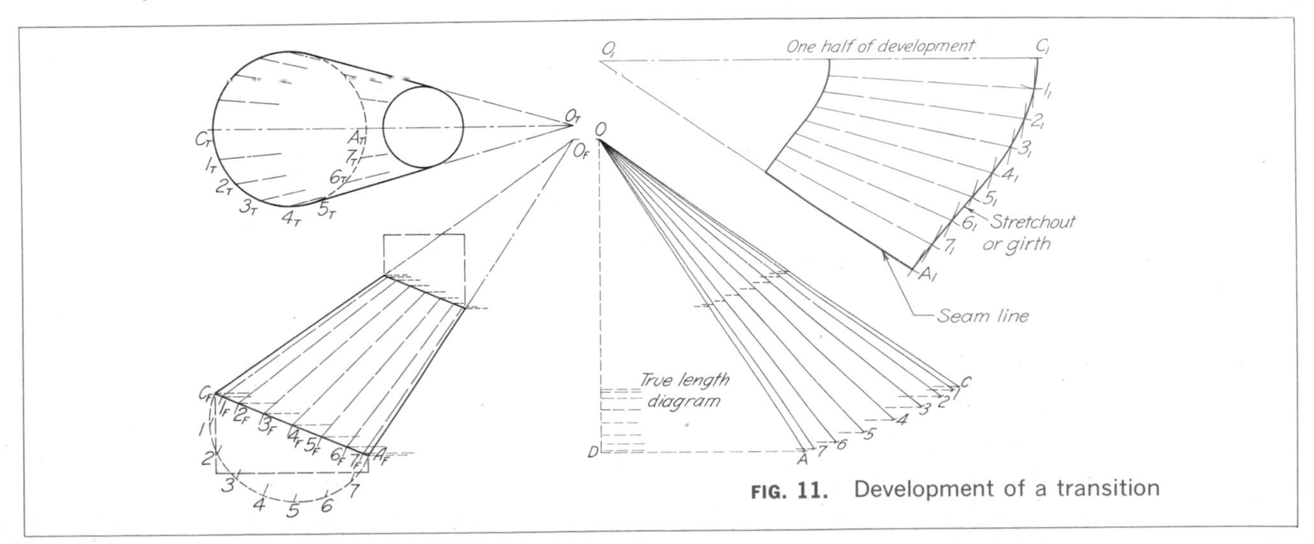


**FIG. 11.** Development of a transition

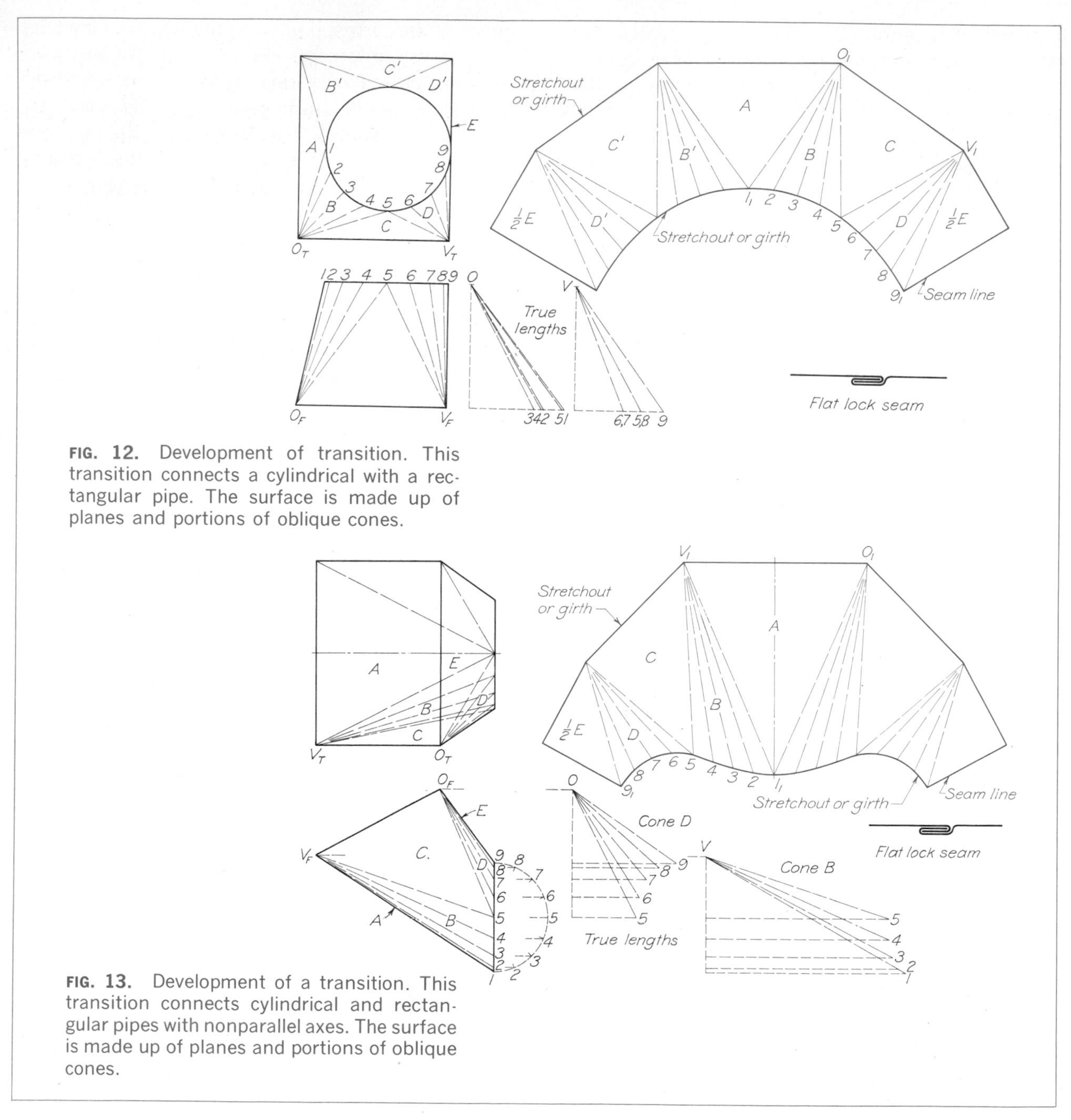


**FIG. 12.** Development of transition. This transition connects a cylindrical with a rectangular pipe. The surface is made up of planes and portions of oblique cones.

**FIG. 13.** Development of a transition. This transition connects cylindrical and rectangular pipes with nonparallel axes. The surface is made up of planes and portions of oblique cones.

metrical, one-half only of the opening need be divided). The true lengths of the elements are obtained as in Fig. 11.

With the seam line the center line of the plane $E$ in Figs. 12 and 13, the flat lock is recommended for joining the ends of the development.

**12. Triangulation of Warped Surfaces.** The approximate development of a warped surface is made by dividing it into a number of narrow quadrilaterals and then splitting each of these quadrilaterals into two triangles by a diagonal, which is assumed to be a straight line, although really a curve. Figure 14 shows a warped transition piece to connect an ovular (upper) pipe with a right-circular cylindrical pipe (lower). Find the true size of one-half the elliptical base by rotating it until horizontal about an axis through 1, when its true shape appears on the top view. The major diameter is $1\text{-}7_R$, and the minor diameter through $4_R$ will equal the diameter of the lower pipe. Divide the semiellipse into a sufficient number of equal parts, and project these to the top and front views. Divide the top semicircle into the same number of equal parts, and connect similar points on each end, thus dividing the surface into approximate quadrilaterals. Cut each into two triangles by a diagonal. On true-length diagrams find the lengths of the elements and the diagonals, and draw the development by constructing the true sizes of the triangles in regualr order. The flat-lock seam is recommended for joining the ends of the development.

**13. To Develop a Sphere.**
The sphere may be taken as typical of double-curved surfaces, which can be developed only approximately. It may be cut into a number of equal meridian sections, or lunes, as in Fig. 15, and these may be considered to be sections of cylinders. One of these sections, developed as the cylinder in Fig. 15, will give a pattern for the others.

Another method is to cut the sphere into horizontal sections, or zones, each of which may be

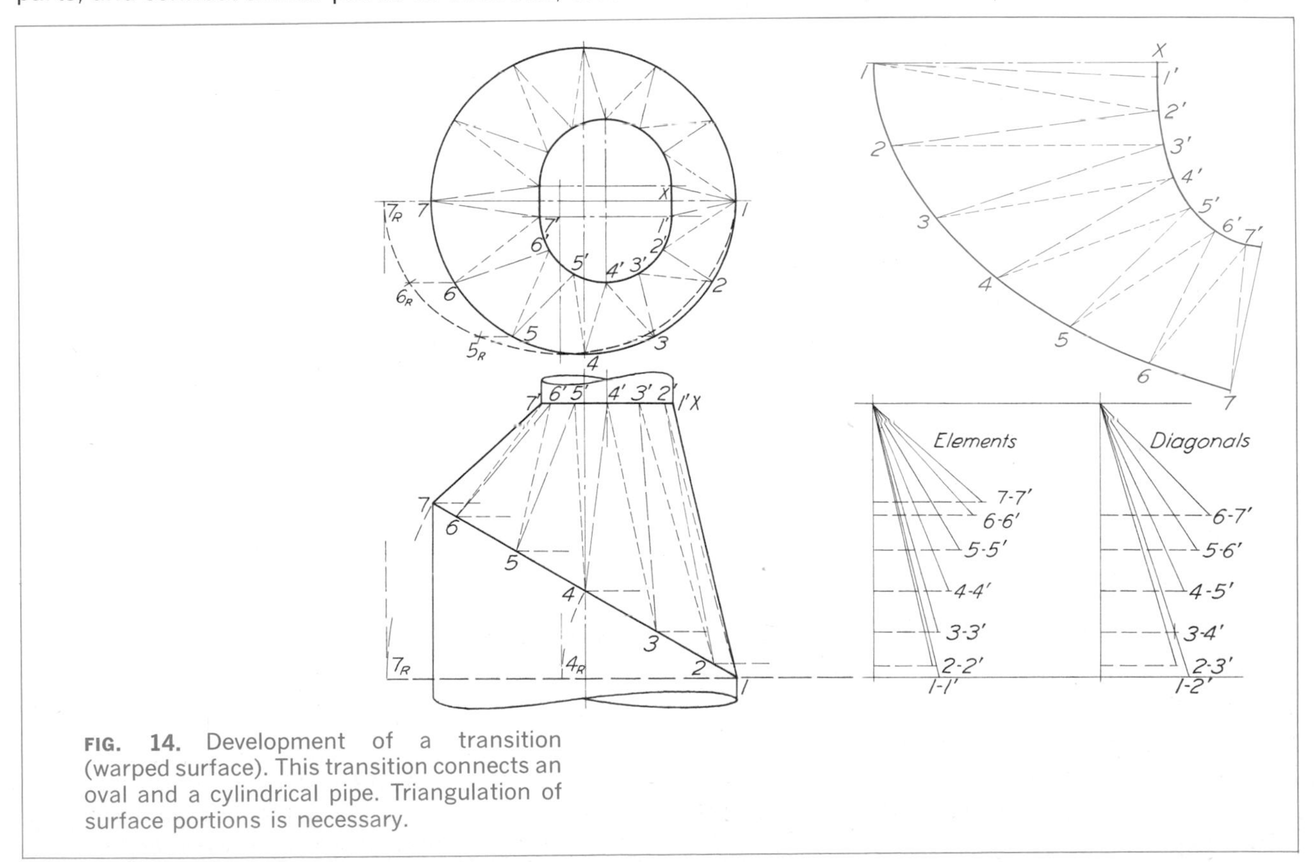


**FIG. 14.** Development of a transition (warped surface). This transition connects an oval and a cylindrical pipe. Triangulation of surface portions is necessary.

taken as the frustum of a cone whose vertex is at the intersection of the extended chords (Fig. 16).

**14. Joints, Connectors, and Hems.** There are numerous joints used in seaming sheet-metal ducts and in connecting one duct to another. Figure 17 illustrates some of the more common types, which may be formed by hand on a break or by special seaming machines. No attempt to dimension the various seams and connections has been made here because of the variation in sizes for different gages of metal and in the forming machines of manufacturers.

Hemming is used in finishing the raw edges of the end of the duct. In wire hemming, an extra allowance of about $2\frac{1}{2}$ times the diameter of the wire is made for wrapping around the wire. In flat hemming, the end of the duct is bent over once or twice to relieve the sharp edge of the metal.

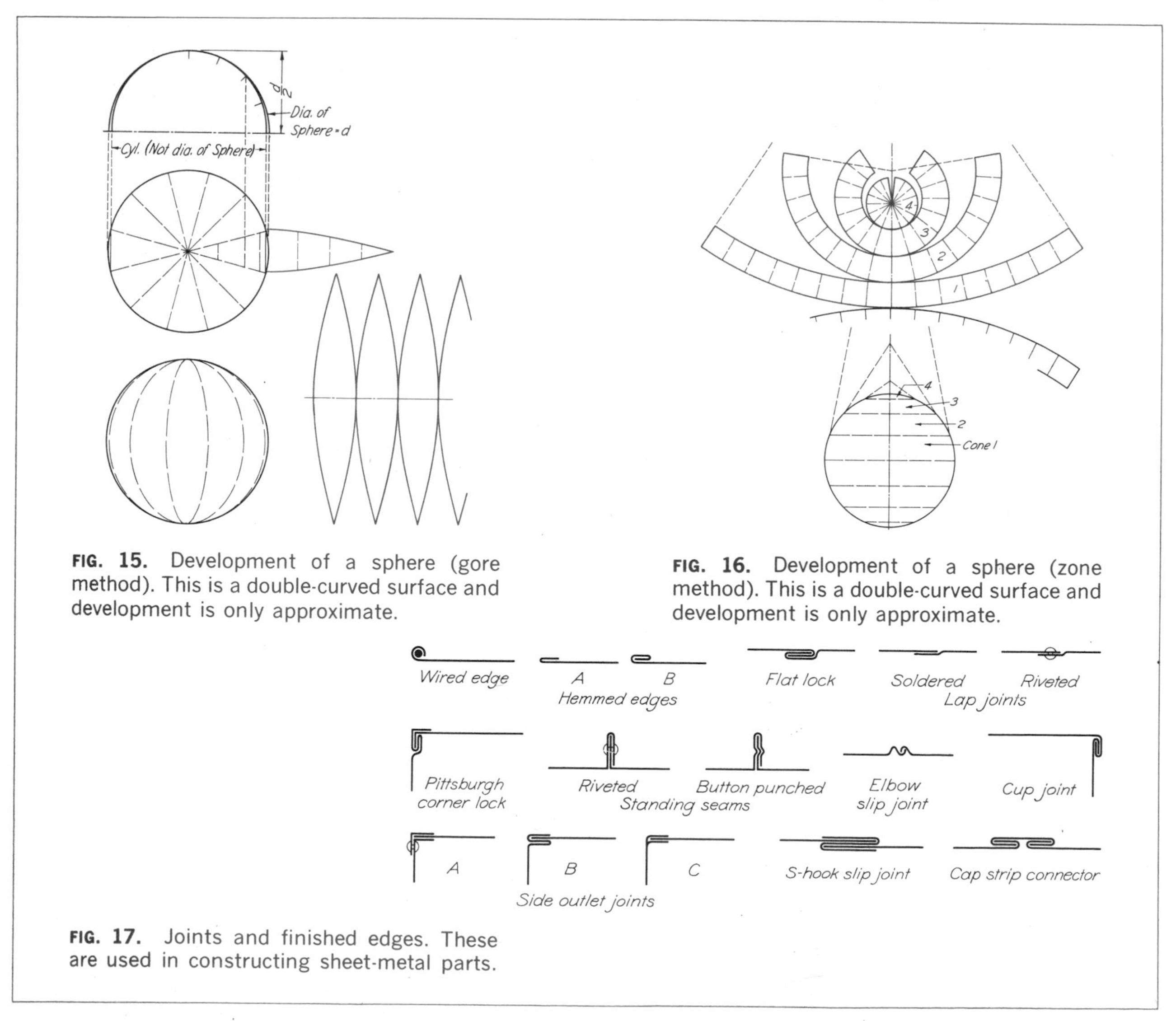


**FIG. 15.** Development of a sphere (gore method). This is a double-curved surface and development is only approximate.

**FIG. 16.** Development of a sphere (zone method). This is a double-curved surface and development is only approximate.

**FIG. 17.** Joints and finished edges. These are used in constructing sheet-metal parts.

## PROBLEMS

Selections may be made from the following problems. Construct the figures accurately in pencil without inking. Any practical problem can be resolved into some combination of the "type solids," and the exercises given illustrate the principles involved in the various combinations.

An added interest in developments may be found by working problems on suitable paper, allowing for fastenings and lap, and cutting them out. It is recommended that at least one or two models be constructed in this manner.

In sheet-metal shops, development problems, unless very complicated, are usually laid out directly on the metal.

The following problems may be drawn on $8\frac{1}{2}$- × 11-in. or 11- × 17-in. sheets. Assume the objects to be made of thin metal with open ends unless otherwise specified.

### Group 1. Developments of Prisms.

**1 to 6.** Develop lateral surfaces of the prisms.

### Group 2. Developments of Pyramids.

**7 to 9.** Develop lateral surfaces of the hoppers.

**10 and 11.** Develop lateral surfaces of the pyramids.

### Group 3. Developments of Cylinders.

**12 to 18.** Develop lateral surfaces of the cylinders.

**19.** Develop the oblique pipe section of the power-station piping of Prob. 20, Chap. 9. Also develop the cylindrical inlet, showing the opening to be made for the oblique pipe.

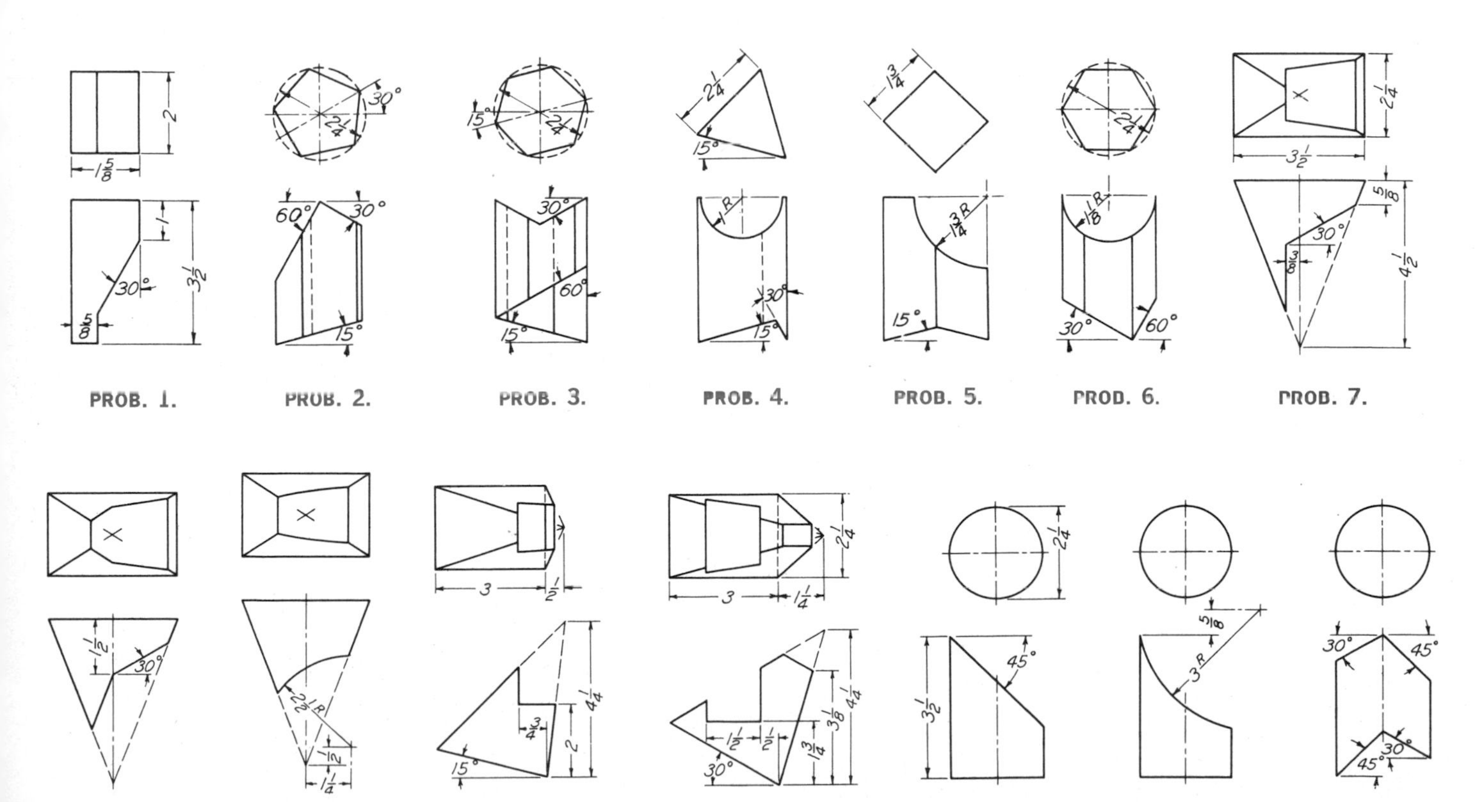

PROB. 1. PROB. 2. PROB. 3. PROB. 4. PROB. 5. PROB. 6. PROB. 7.

PROB. 8. PROB. 9. PROB. 10. PROB. 11. PROB. 12. PROB. 13. PROB. 14.

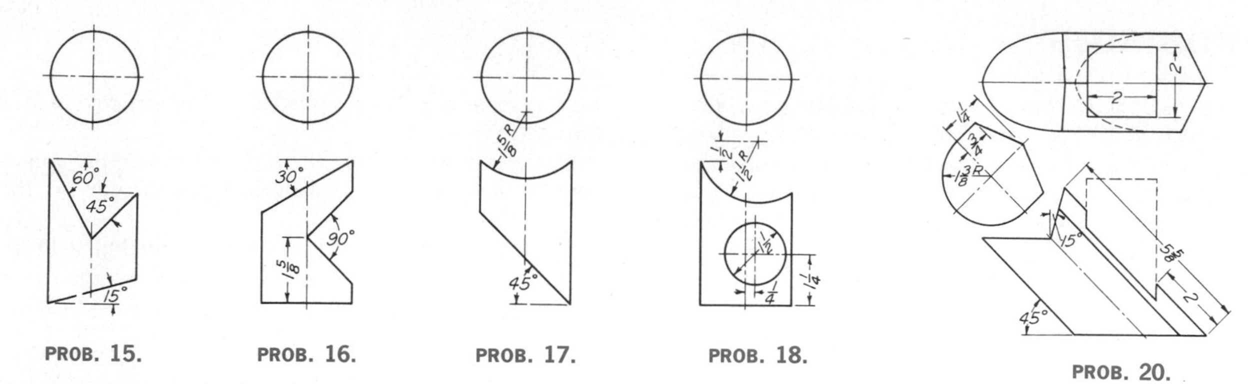

PROB. 15. PROB. 16. PROB. 17. PROB. 18. PROB. 20.

## Group 4. Developments of Combinations of Prisms and Cylinders.

**20 to 22.** Develop lateral surfaces.

## Group 5. Developments of Cones.

**23 to 27.** Develop lateral surfaces.

**28.** Develop the concentric reducer for the power-station piping of Prob. 20, Chap. 9.

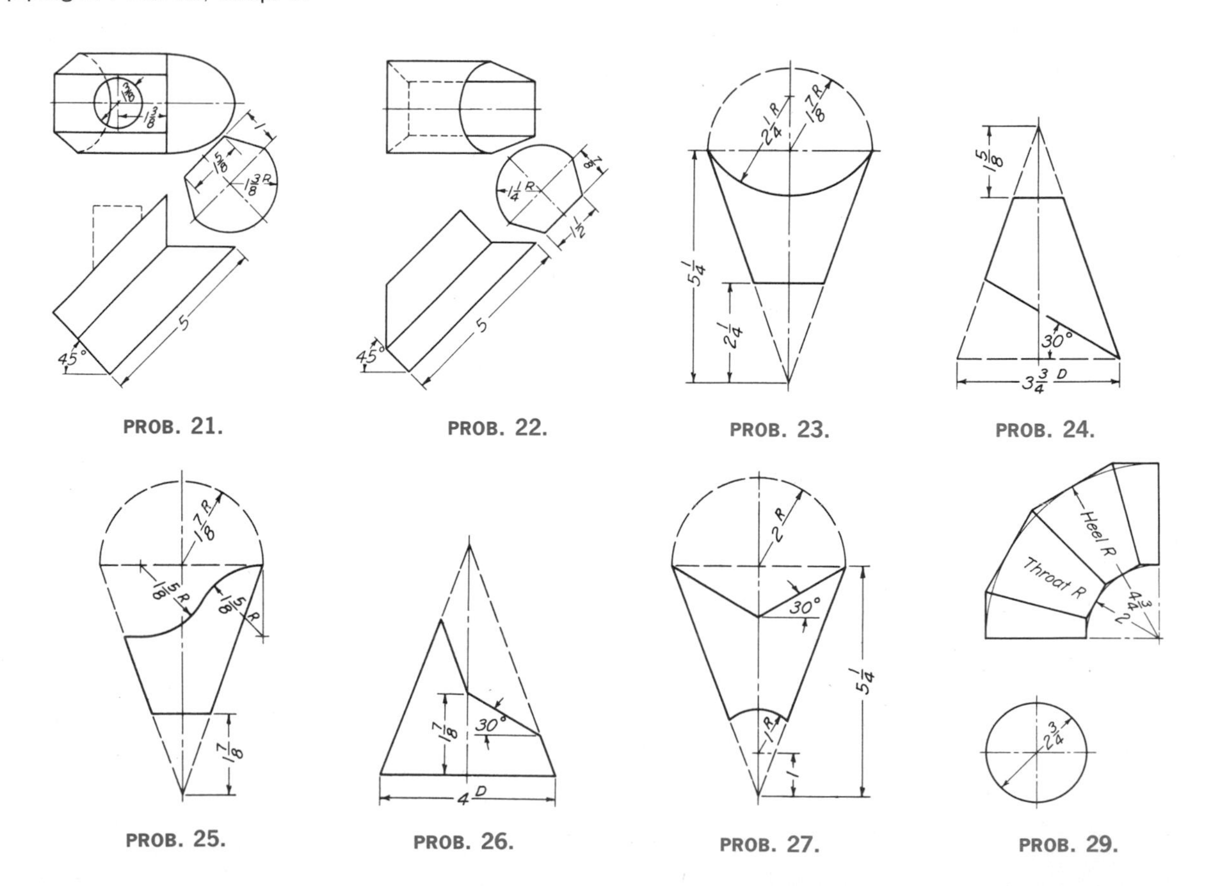


PROB. 21. PROB. 22. PROB. 23. PROB. 24.

PROB. 25. PROB. 26. PROB. 27. PROB. 29.

## Group 6. Developments of Combinations of Surfaces.

**29 to 32.** Develop lateral surfaces of the objects.

**33.** Develop the cylindrical pipe for the pump casing of Prob. 18, Chap. 9.

**34.** Develop the elbow sections, *A*, *B*, and *C*, of the pump casing of Prob. 18, Chap. 9.

## Group 7. Developments of Cones and Transition Pieces.

**35 to 42.** Develop lateral surfaces of the objects (one-half in Probs. 35, 36, 38, 39, and 41).

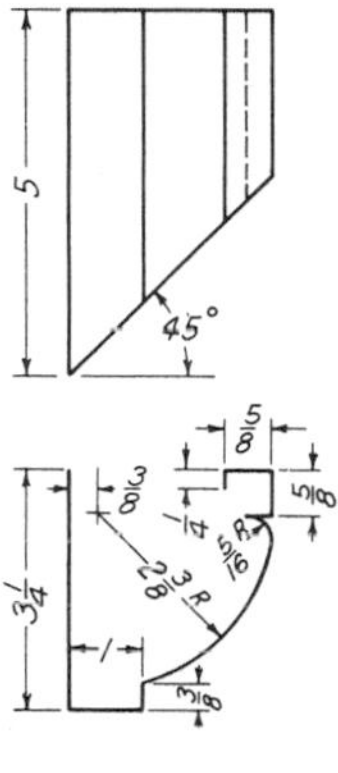


PROB. 30.

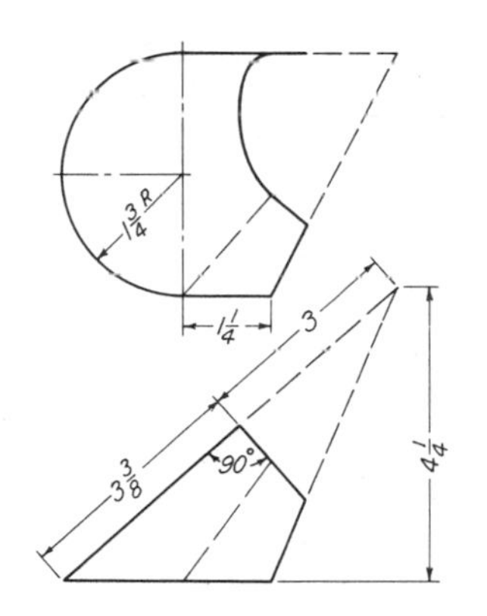


PROB. 31.

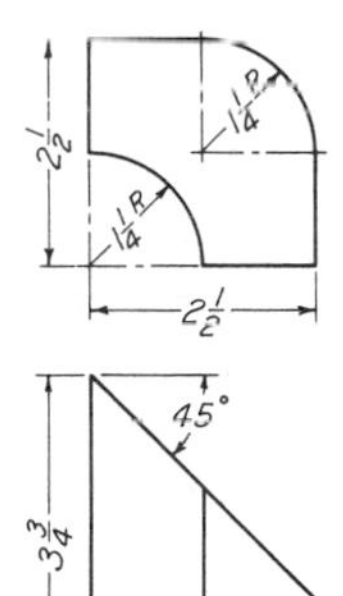


PROB. 32.

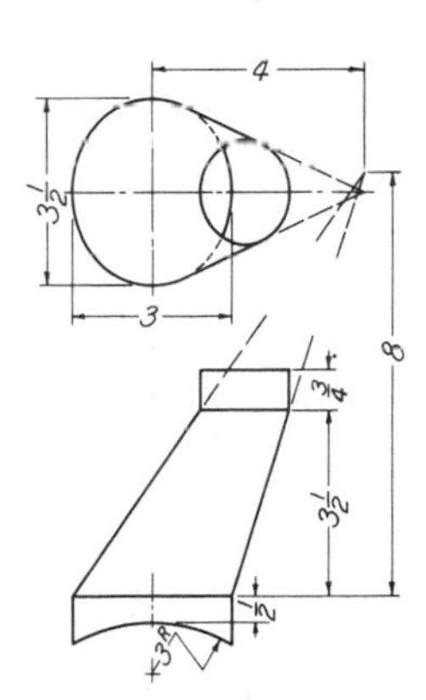


PROB. 35.

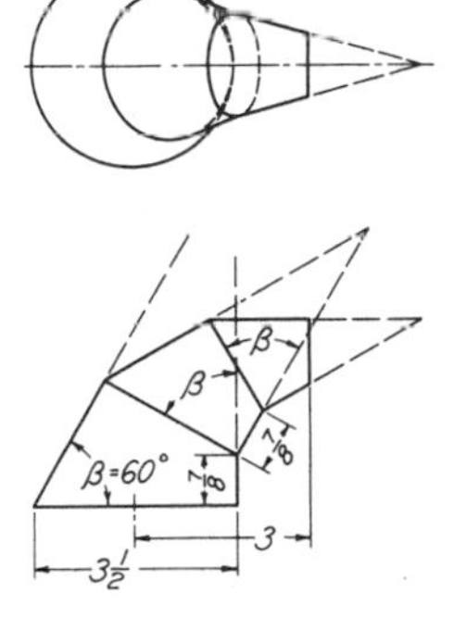


PROB. 36.

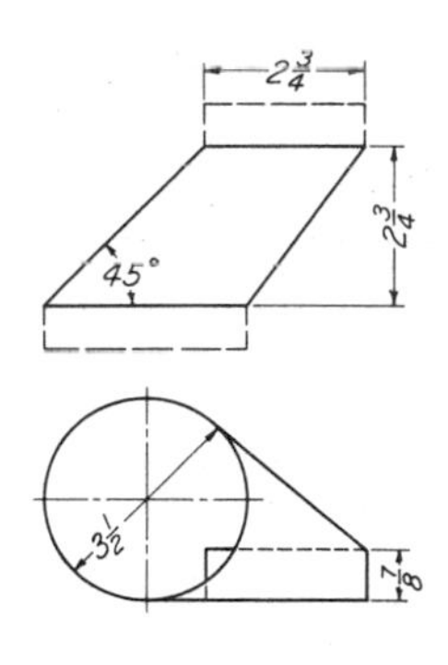


PROB. 37.

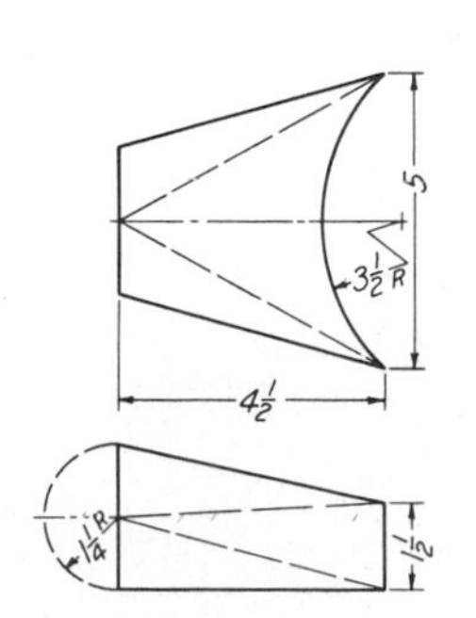


PROB. 38.

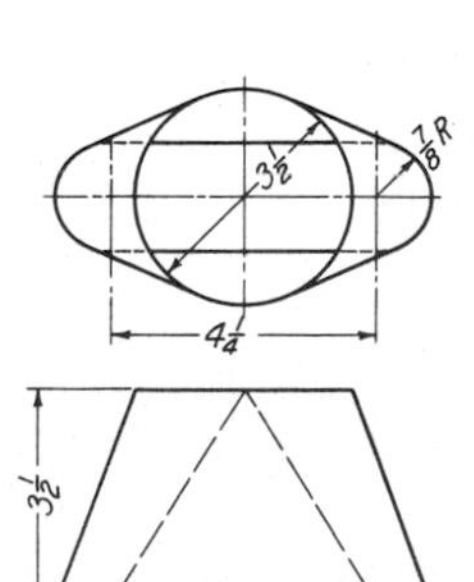

PROB. 39.

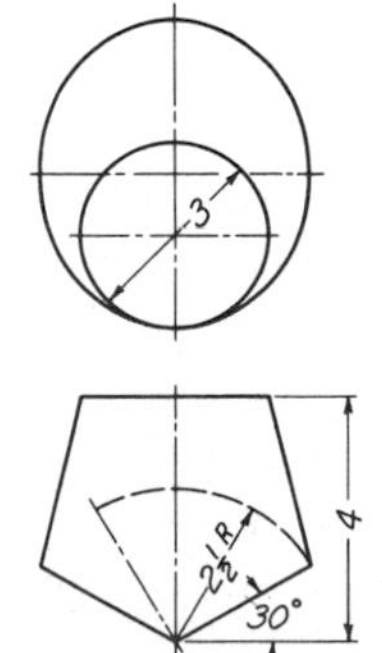


PROB. 40.

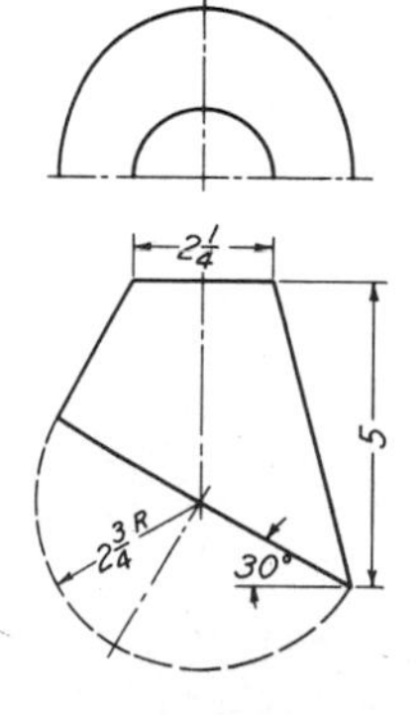


PROB. 41.

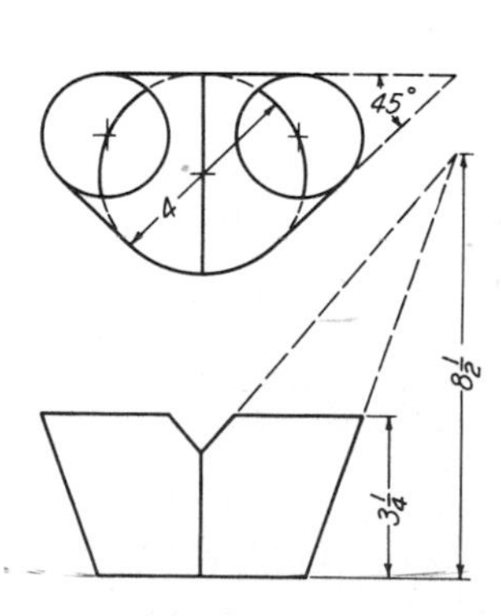


PROB. 42.

## Group 8. Developments of Furnace-pipe Fittings.

**43 to 50.** Develop surfaces and make paper models.

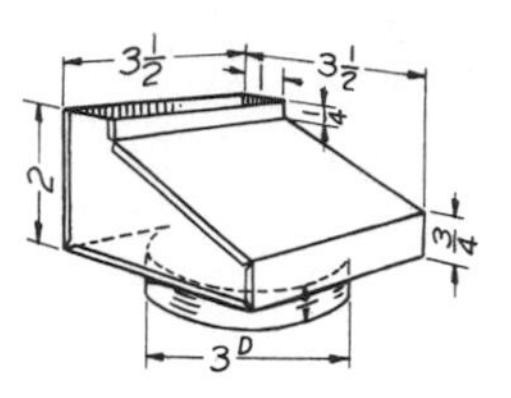

PROB. 43.

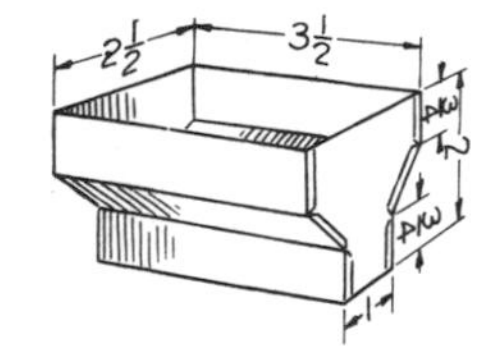

PROB. 44.

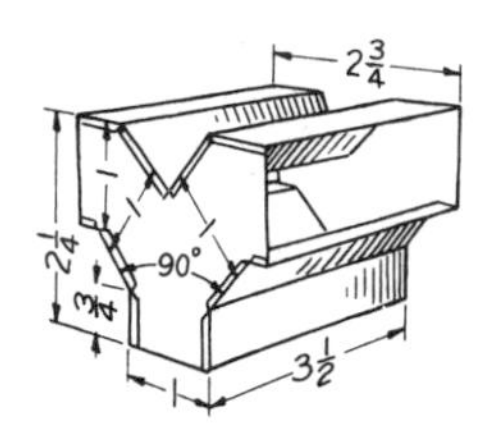


PROB. 45.

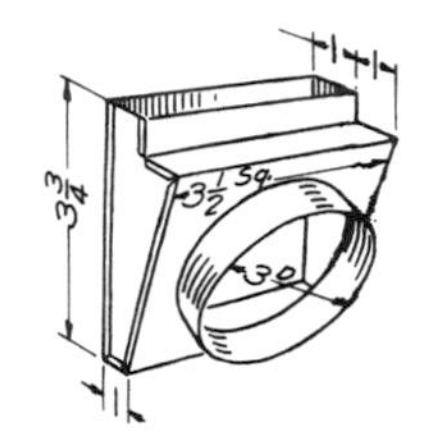

PROB. 46.

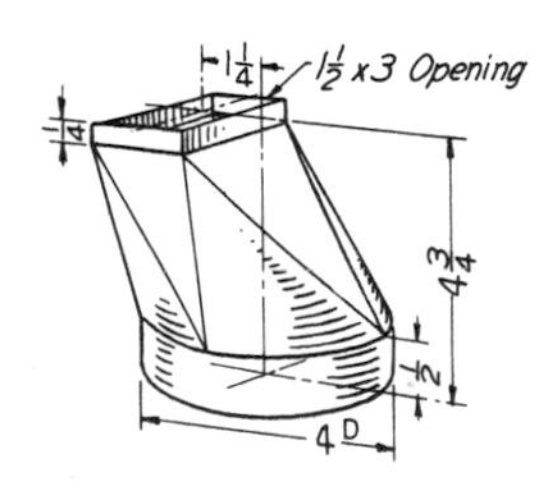


PROB. 47.

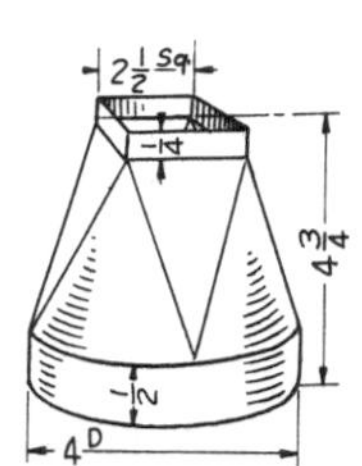

PROB. 48.

## Group 9. Specialities.

There are innumerable combinations of geometric shapes and, sometimes, nongeometric forms used in modern engineering work that, fundamentally, are problems of layout and intersection and then either development or representation. The following three problems are typical examples.

**51.** The diverter-duct system for a VTOL jet consists essentially of a formed elbow and a cylinder. A typical layout is shown in Prob. 51. Find the intersection between the cylinder and elbow. Develop the cylinder and make an accurate representational drawing of the elbow.

**52.** The evaporator shown in the layout consists of a combustion sphere and cone. Connected to this combination is a steam outlet $A$ and an exhaust pipe $B$. Find all intersections and then develop the parts that can be developed.

**53.** The microwave horn reflector antenna consists essentially of three surfaces: a cone, a cylinder, and a paraboloid of revolution, which, if chosen correctly, should intersect one another on the same curve. To verify this construction, lay out the complete horn according to the dimensions shown in the illustration. Then, find the intersection between the cone and cylinder and between the cylinder and paraboloid. Develop all parts that can be developed.

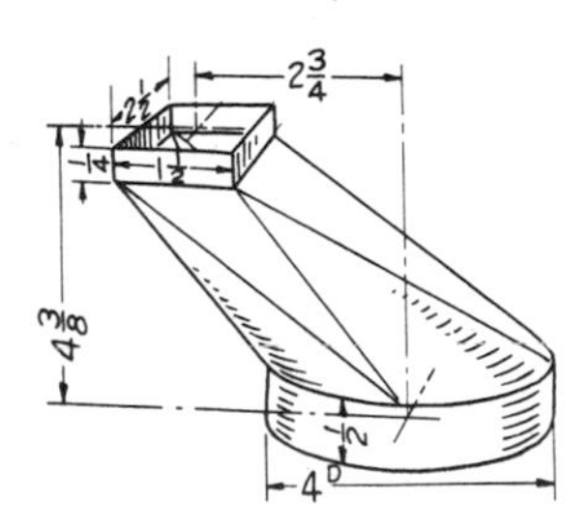

PROB. 49.

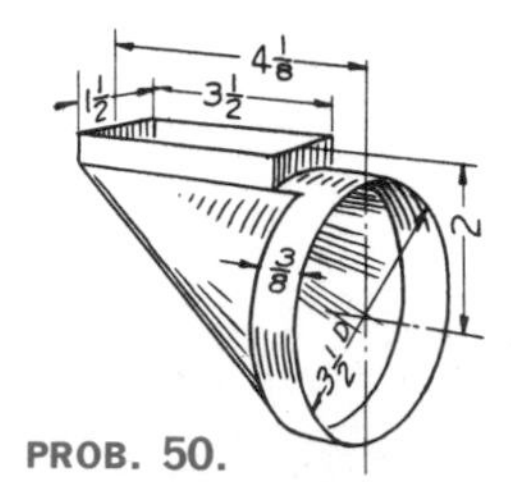

PROB. 50.

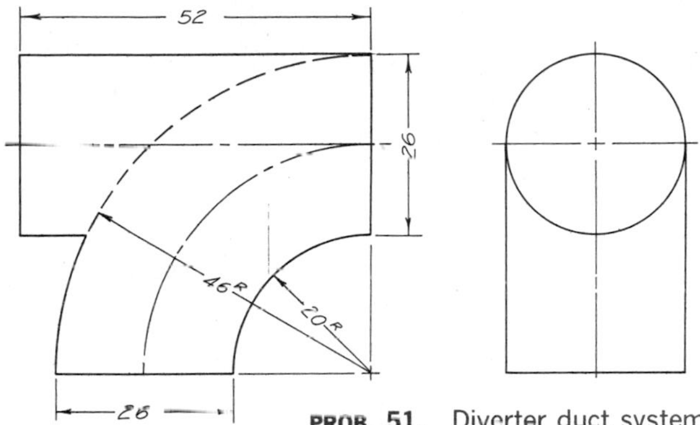


**PROB. 51.** Diverter duct system.

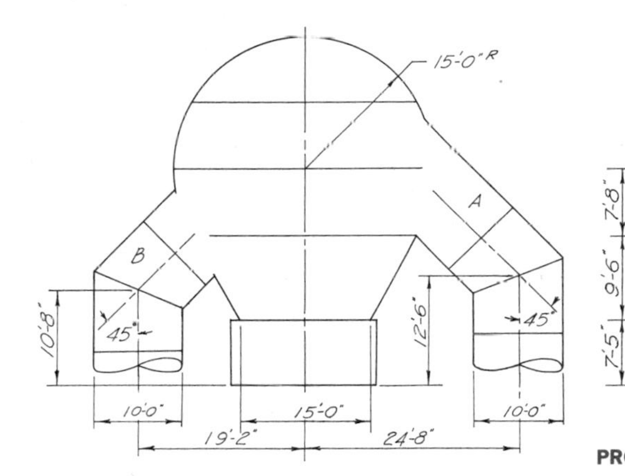


**PROB. 52.** Evaporator.

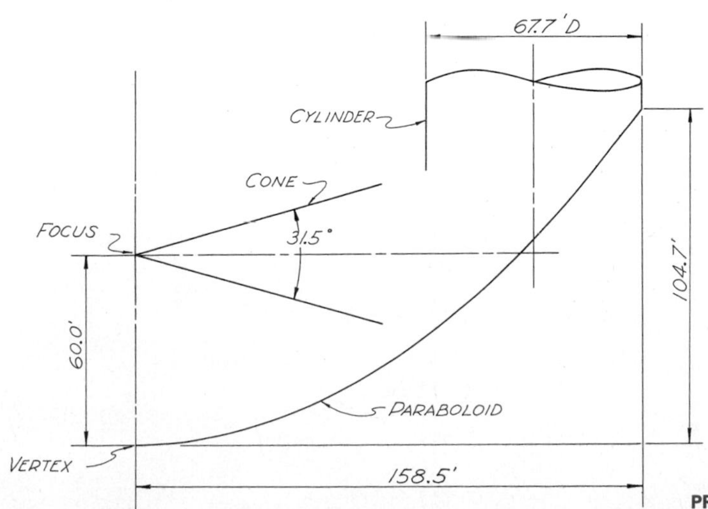


**PROB. 53.** Reflector horn.

In this Chapter:

The Effect of the Basic Manufacturing Method on the Drawing

Drawings for Sand Castings, Pattern Shop, Foundry

Forgings and Drawings of Forgings

Fundamentals of Machining

The Lathe, Drill Press, Shaper, Planer

Parts Machined from Stock

The Turret Lathe, Milling Machine, Grinder

Lapping, Honing, and Superfinishing

Broaching, Small Tools, Welded Parts, Parts Made from Sheet Stock, Plastics

Heat Treatment, Tools for Mass Production, Jigs and Fixtures, Inspection, Assembly

# METHODS USED IN MANUFACTURE

**Concise and Complete Specification of Size Requires a Knowledge of Manufacturing Procedures**

## 1. Drawings—Their Relationship to Manufacturing.

The test of any working drawing for legibility, completeness, and accuracy is the production of the object or assembly by the shop without further information than that given on the drawing. A knowledge of shop methods will, to a great extent, govern the effectiveness and completeness of the drawing. Study the glossary of shop terms to become familiar with the terms and the form of designation in notes. This chapter is given as an introduction to those on dimensioning (Chap. 12) and working drawings (final chapter).

The relation of drawings and the prints made from them to the operations of production is illustrated in the accompanying chart (Fig. 1). This chart shows in diagrammatic form the different steps in the development of drawings and their distribution and use in connection with shop operations from the time the order is received in the plant until the finished machine is delivered to the shipping room.

## 2. Effect of the Basic Manufacturing Method on the Drawing.

In drawing any machine part, consider first the manufacturing process to be used, as on this depends the representation of the detailed features of the part and, to some extent, the choice of dimensions. Special or unusual methods may occasionally be used, but most machine parts are produced by (1) casting, (2) forging, (3) machining from standard stock, (4) welding, or (5) forming from sheet stock.

Each of the different methods produces a characteristic detailed shape and appearance of the parts, and these features must be shown on the drawing. Figure 2 shows and lists typical features of each method and indicates the differences in drawing practice.

## 3. The Drawings.

For the production of any part, a detail working drawing is necessary, complete with shape and size description and giving, where needed, the opera-

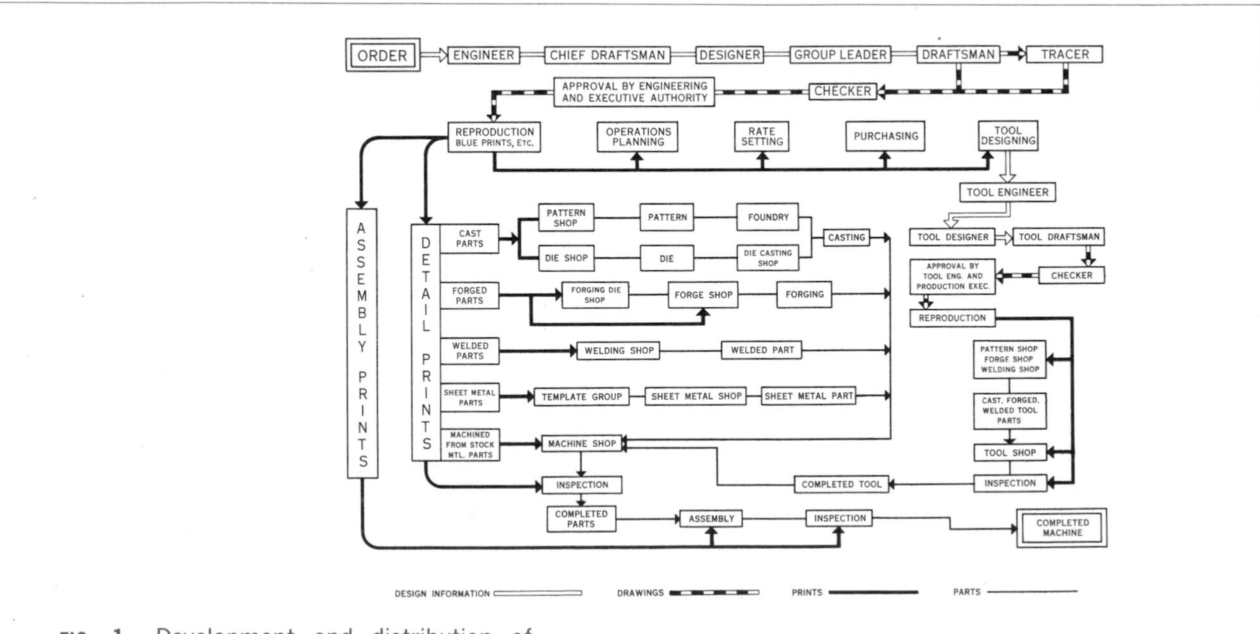


**FIG. 1.** Development and distribution of drawings. This shows the routing of drawings, prints, and parts through all design and manufacturing units.

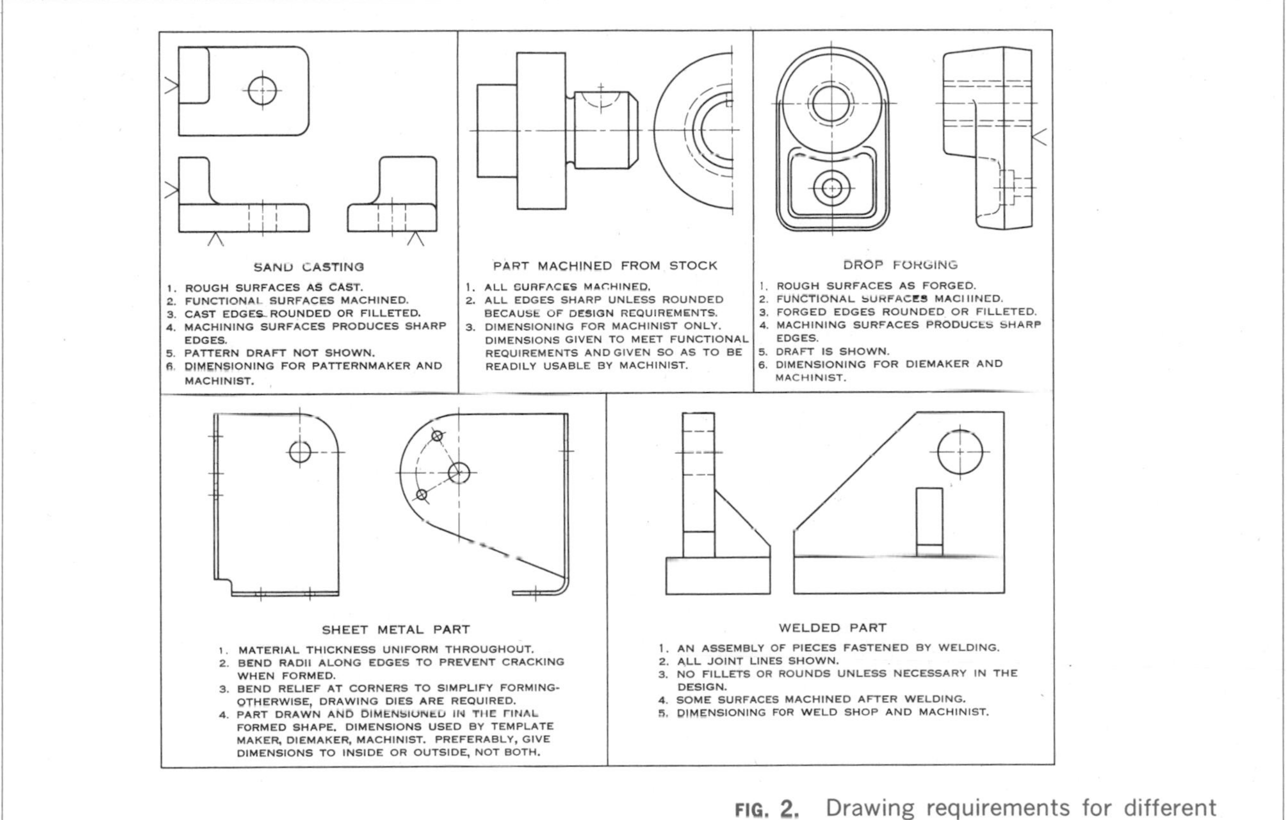


**FIG. 2.** Drawing requirements for different manufacturing methods.

tions that are to be performed by the shop. Machined surfaces must be clearly indicated, with dimensions chosen and placed so as to be useful to the various shops without the necessity of adding or subtracting dimensions or scaling the drawing.

Two general practices are followed: (1) the "single-drawing" system, in which only one drawing, showing the finished part, is made to be used by all the shops involved in producing the part; and (2) the "multiple-drawing" system, in which different drawings are prepared, one for each shop, giving only the information required by the shop for which the drawing is made.

The second practice is recommended, as the drawings are easier to dimension without ambiguity, somewhat simpler and more direct, and therefore easier for the shop to use. Figure 3 is a single drawing, to be used by the pattern shop and the machine shop. Figures 107 and 108, Chap. 12, are multiple drawings, Fig. 107 for the patternmaker and Fig. 108 for the machine shop.

## 4. Sand Castings.

Figure 3 shows (in the title strip) that the material to be used is cast iron (*C.I.*), indicating that the part will be formed by pouring molten iron into a mold (in this case a "sand mold"), resulting in a sand casting.

After casting, subsequent operations produce the finished part. Figure 4 illustrates the shop interpretation of the casting drawing in Fig. 3 and indicates the order of operations to be performed.

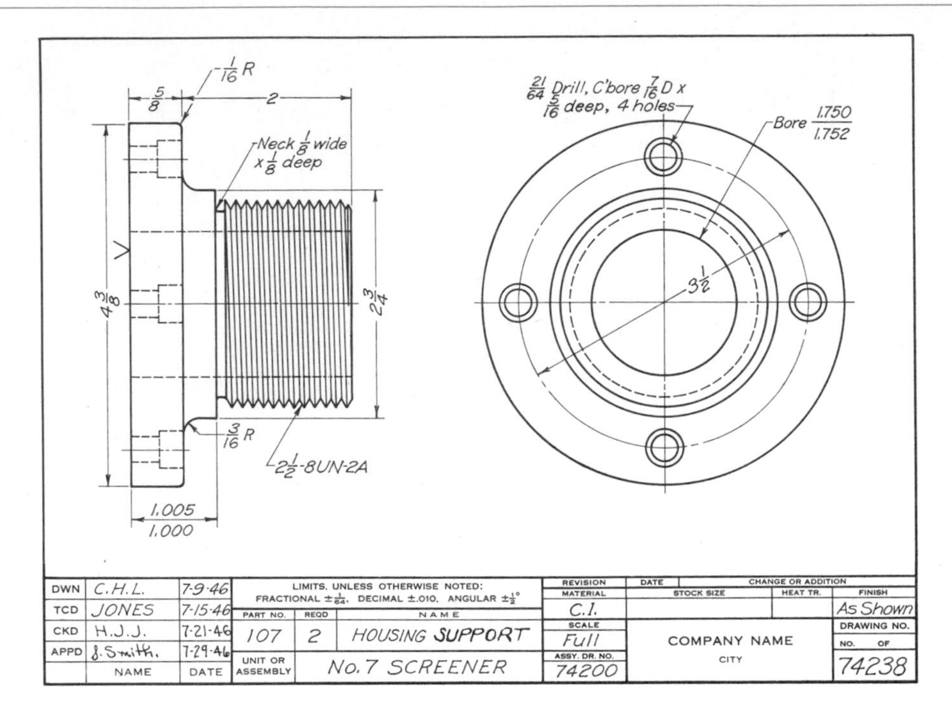


**FIG. 3.** A working drawing of a cast part. Complete information is given for all shops and all operations.

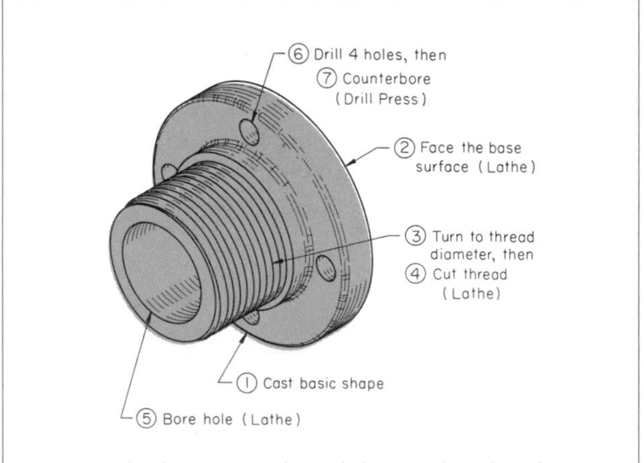


**FIG. 4.** Interpretation of the casting drawing in Fig. 3. These are the operations that must be performed to produce the part.

## 5. The Pattern Shop.

The drawing is first used by the patternmaker, who makes a pattern, or "model," of the part in wood. From this, if a large quantity of castings is required, a metal pattern, often of aluminum, is made. The patternmaker provides for the shrinkage of the casting by making the pattern oversize, using a "shrink rule" for his measurements. He also provides additional metal (machining allowance) for the machined surfaces, indicated on the drawing by (1) finish marks, (2) dimensions indicating a degree of precision attainable only by machining, or (3) notes giving machining operations. The patternmaker also provides the "draft," or slight taper, not shown on the drawing, so that the pattern can be withdrawn easily from the sand. A "core box," for making sand cores for the hollow parts of the casting, is also made in the pattern shop. A knowledge of patternmaking

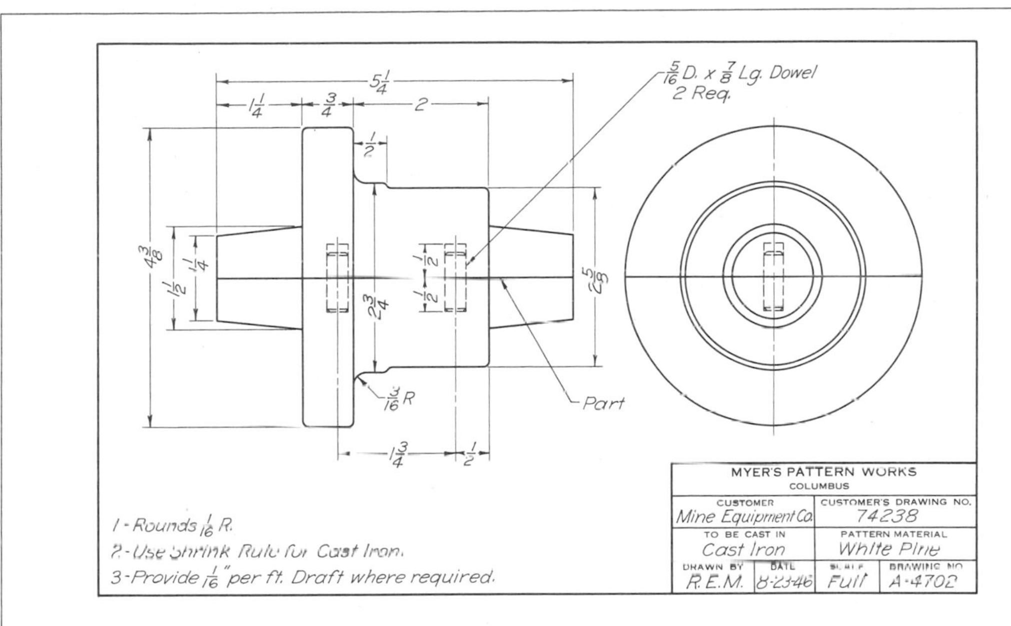


FIG. 5. A pattern drawing. This is a drawing of the pattern for the casting in Fig. 3.

is a great aid in dimensioning, as almost all the dimensions are used by the patternmaker, while only the dimensions for finished features are used by the machine shop.

## 6. Drawings of Castings.

A casting drawing is usually made as a single drawing of the machined casting, with dimensions for both the patternmaker and the machinist (Fig. 3). If the multiple-drawing system is followed, a drawing of the unmachined casting, with allowances for machining accounted for and with no finish marks or finish dimensions, is made for the patternmaker; then a second drawing for the machinist shows the finished shape and gives machining dimensions.

For complicated or difficult castings, a special "pattern drawing" may be made (Fig. 5), showing every detail of the pattern, including the amount of draft, the parting line, "core prints" for supporting the cores in the mold, and the pattern material. Similar detail drawings may also be made for the core boxes.

## 7. The Foundry.

The pattern and core box or boxes are sent to the foundry, and sand molds are made so that molten metal can be poured into the molds and allowed to cool, forming the completed rough casting. Figure 6 is a cross section of a two-part mold, showing the space left by the pattern and the core in place. Only

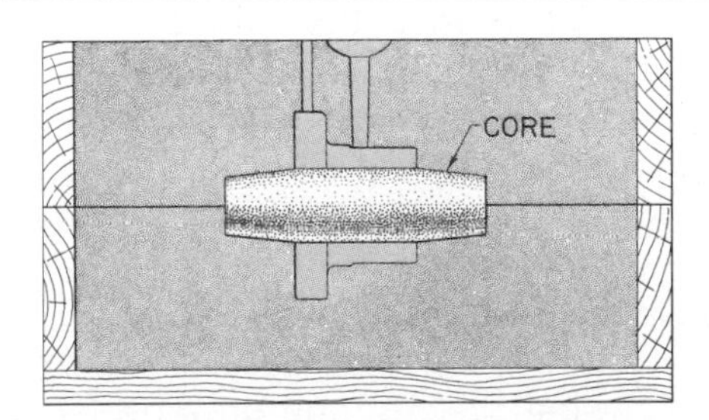


FIG. 6. Cross section of a two-part mold. This shows the cavity produced by the pattern in Fig. 5.

in occasional instances does the foundryman call for assistance from the drawing, as his job is simply to reproduce the pattern in metal.

*Permanent molds,* made of cast iron coated on the molding surfaces with a refractory material, are sometimes an advantage in that the mold can be used over and over again, thus saving the time to make an individual sand mold for each casting. This method is usually limited to small castings.

*Die castings* are made by forcing molten metal under pressure into a steel die mounted in a special die-casting machine. Alloys with a low melting point are used in order to avoid damaging the die. Because of the accuracy possible in making a die, a fine finish and accurate dimensions of the part can be obtained; thus machining may be unnecessary.

## 8. Forgings.

Forgings are made by heating metal to make it plastic and then forming it to shape on a power hammer with or without the aid of special steel dies. Large parts are often hammered with dies of generalized all-purpose shape. Smaller parts in quantity may warrant the expense of making special dies. Some small forgings are made with the metal cold.

*Drop forgings* are the most common and are made in dies of the kind shown in Fig. 7. The lower die is held on the bed of the drop hammer, and the upper die is raised by the hammer mechanism. The hot metal is placed between the dies, and the upper die is dropped several times, causing the metal to flow into the cavity of the dies. The slight excess of material will form a thin fin, or "flash," surrounding the forging at the parting plane of the dies (Fig. 7). This flash is then removed in a "trimming" die made for the purpose. Considerable draft must be provided for release of the forging from the dies.

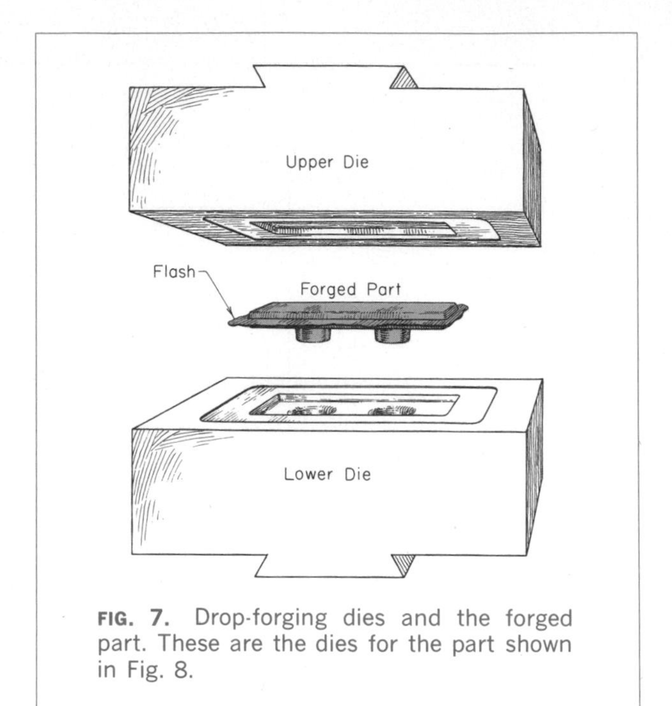


FIG. 7. Drop-forging dies and the forged part. These are the dies for the part shown in Fig. 8.

## 9. Drawings of Forgings.

Forging drawings are prepared according to the multiple-drawing system, one drawing for the diemaker and one for the machinist (Fig. 110, Chap. 12); or the single-drawing system, one drawing for both (Fig. 8). In either case the parting line and draft should be shown and the amount of draft specified.

On the single drawing (Fig. 8), the shape of the finished forging is shown in full outline, and the machining allowance is indicated by "alternate-position" lines, thus completing the shape of the rough forging. This single drawing combines two drawings in one, with complete dimensions for both diemaker and machinist.

Figure 9 illustrates the shop interpretation of the forging drawing in Fig. 8 and indicates the order of operations to be performed.

## 10. The Machine Shop.

The machine shop produces parts machined from stock material and finishes castings, forgings, etc., requiring machined surfaces. Cylindrical and conic surfaces are machined on a lathe. Flat or plane surfaces are machined on a planer, shaper, milling machine, broaching machine, or in some cases (facing) a lathe. Holes are drilled, reamed, counterbored, and countersunk on a drill press or lathe; holes are bored on a boring mill or lathe. For exact work, grinding machines with wheels of abrasive

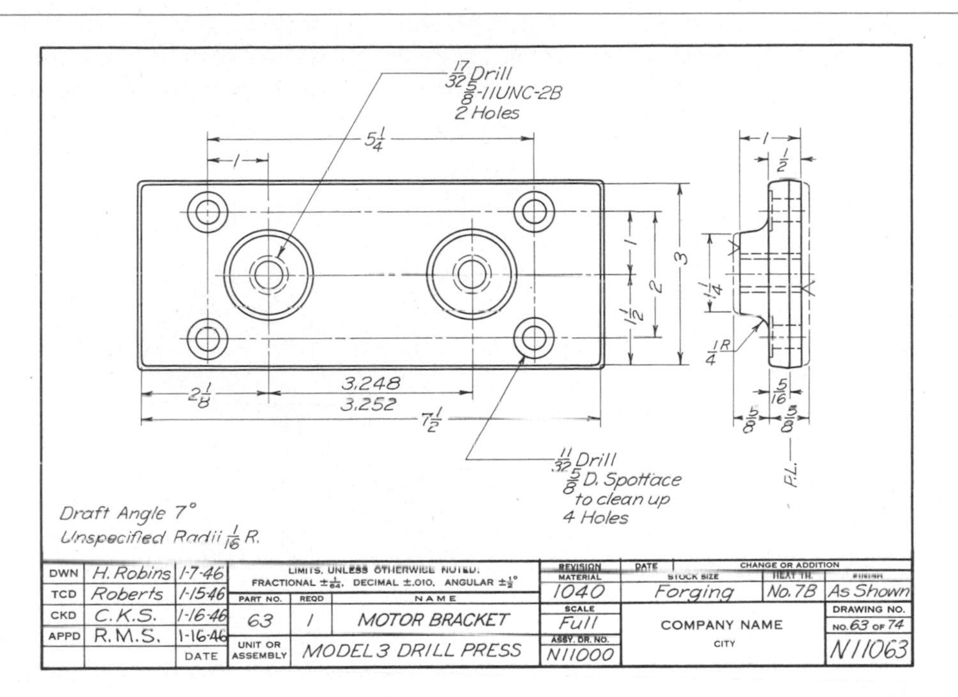


**FIG. 8.** A working drawing of a forged part. Note the indication of parting line and extra material for machining.

material are used. Grinders are also coming into greatly increased use for operations formerly made with cutting tools. In quantity production many special machine tools and automatic machines are in use. The special tools, jigs, and fixtures made for the machine parts are held in the toolroom ready for the machine shop.

## 11. Fundamentals of Machining.

All machining operations remove metal, either to make a smoother and more accurate surface, as by planing, facing, milling, etc., or to produce a surface not previously existing, as by drilling, punching, etc. The metal is removed by a hardened steel, carbide, or diamond cutting tool (machining) or an abrasive wheel (grinding); the product, or "work piece," as well as the tool or wheel, being held and guided by the machine. When steel cutting tools are used, the product must remain relatively soft until after all

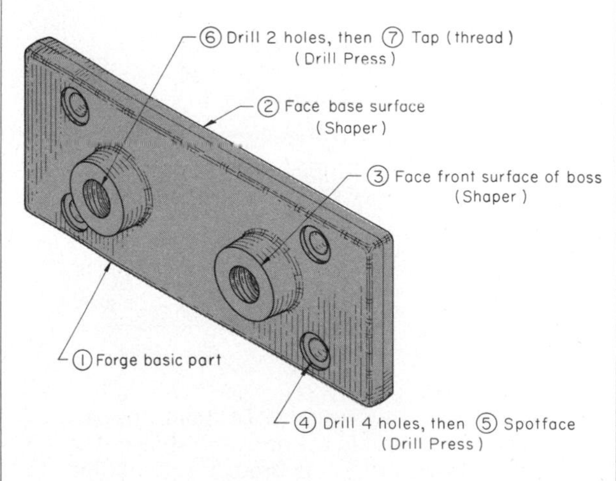


**FIG. 9.** Interpretation of the forging drawing in Fig. 8. These are the separate manufacturing steps.

machining has been performed upon it, but if diamond-tipped tools are used or if grinding wheels are employed, the product may be hardened by heat-treatment before finishing.

All machining methods are classified according to the operating principle of the machine performing the work:

1. The surface may be *generated* by moving the work with respect to a cutting tool or the tool with respect to the work, following the geometric laws for producing the surface.

2. The surface may be *formed* with a specially shaped cutting tool, moving either work or tool while the other is stationary.

The forming method is, in general, less accurate than the generating method, as any irregularities in the cutter are reproduced on the work. In some cases a combination of the two methods is used.

**12. The Lathe.**

Called the "king of machine tools," the lathe is said to be capable of producing all other machine tools. Its primary function is machining cylindrical, conic, and other surfaces of revolution, but with special attachments it can perform a great variety of operations. Figure 10 shows the casting made from the drawing of Fig. 3 held in the lathe chuck. As the work revolves, the cutting tool is moved across per-

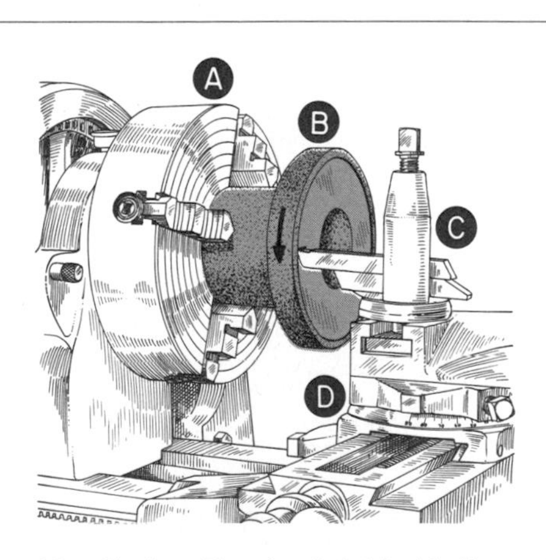


**FIG. 10.** Facing. The chuck (*A*) holds the part (*B*) and revolves it in the direction shown. The tool in the holder (*C*) is brought against the work surface by the lathe carriage, and the motion of the tool across the face is controlled by the cross slide (*D*).

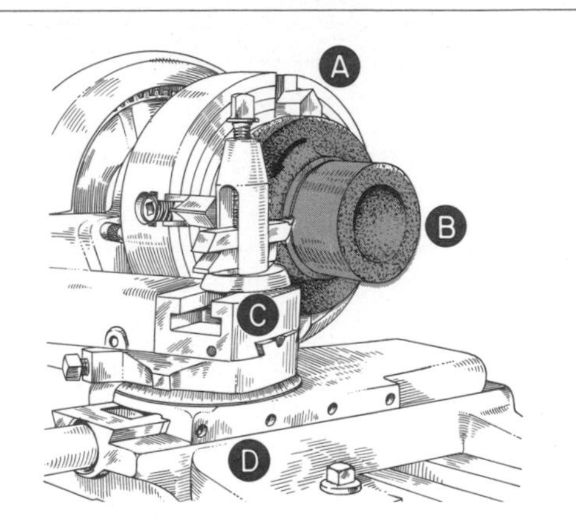


**FIG. 11.** Turning. The chuck (*A*) holds the part (*B*) and revolves it in the direction shown. The tool in the holder (*C*) is brought against the work surface by the cross slide, and the motion of the tool is controlled by the lathe carriage (*D*).

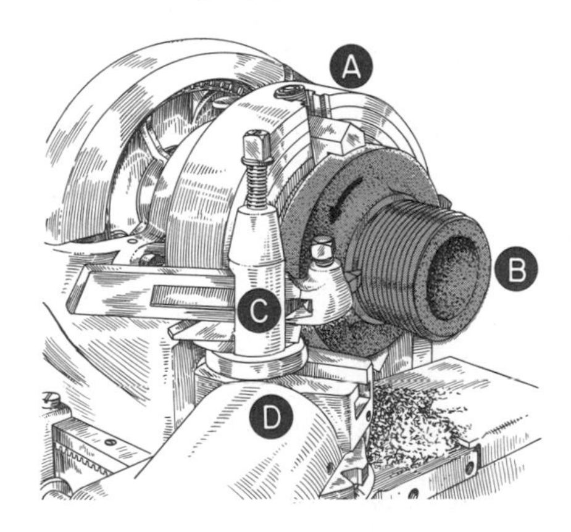


**FIG. 12.** Threading. The chuck (*A*) holds the part (*B*) and revolves it in the direction shown. The tool, ground to the profile of the thread space, in the holder (*C*) is brought into the work surface by the angled cross slide (*D*), and the rate of travel is controlled by gears and a lead screw that moves the lathe carriage.

pendicular to the axis of revolution, removing metal from the base and producing a plane surface by generation. This operation is called *facing*. After being faced, the casting is turned around, and the finished base is aligned against the face of the chuck, bringing the cylindrical surface into position for *turning* to the diameter indicated in the thread note on the drawing. The neck shown at the intersection of the base with the body is turned first, running the tool into the casting to a depth slightly greater than the depth of the thread. The cylindrical surface is then turned (generated) by moving the tool parallel to the axis of revolution (Fig. 11). Figure 12 shows the thread being cut on the finished cylinder. The tool is ground to the profile of the thread space, carefully lined up to the work, and moved parallel to the axis of revolution by the lead screw of the lathe. This operation is a combination of the fundamental processes, the thread profile being formed while the helix is generated.

The hole through the center of the casting, originally cored, is now finished by *boring*, as the cutting of an interior surface is called (Fig. 13). The tool is held in a boring bar and moved parallel to the axis of revolution, thus generating an internal cylinder.

Note that in these operations the dimensions used by the machinist have been (1) the finish mark on the base and thickness of the base, (2) the thread note and outside diameter of the thread, (3) the dimensions of the neck, (4) the distance from the base to the shoulder, and (5) the diameter of the bored hole.

Long cylindrical pieces to be turned in the lathe are supported by conic centers, one at each end. Figure 22 illustrates the principle.

## 13. The Drill Press.

The partially finished piece of Fig. 3 is now taken to the drill press for drilling and counterboring the holes in the base according to the dimensions on the drawing. These dimensions give the diameter of the drill, the diameter and depth of the counterbore, and the location of the holes. The casting is clamped to the drill-press table (Fig. 14) and the rotating drill brought into the work by a lever operating a rack and pinion in the head of the machine. The cutting is done by two ground lips on the end of the drill (Fig. 24). Drilling can also be done in a lathe, the work revolving while the drill is held in

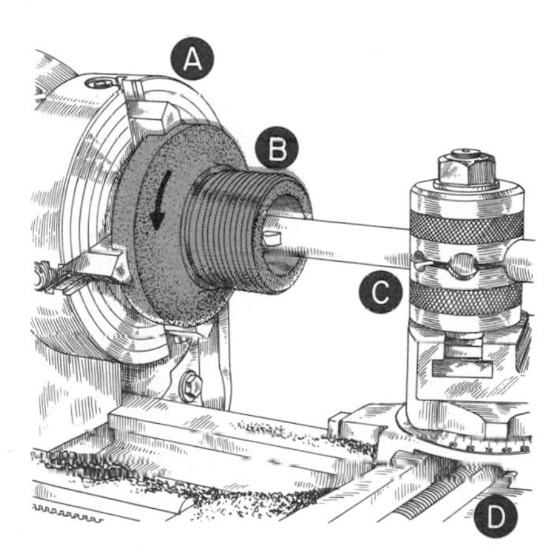


**FIG. 13.** Boring. The chuck ($A$) holds the part ($B$) and revolves it in the direction shown. The boring bar with the tool in the holder ($C$) is brought against the work surface by the cross slide, and advance is controlled by the lathe carriage ($D$).

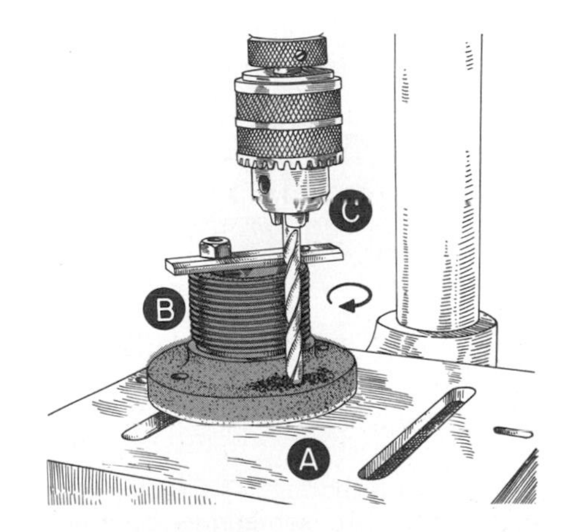


**FIG. 14.** Drilling. The table ($A$) supports the part ($B$), sometimes clamped as shown. The drill in the chuck ($C$) revolves in the direction shown, and is forced downward by a gear and rack in the drill-press head, to make the hole.

and moved by the tailstock. In Fig. 15 the drill has been replaced by a counterboring tool (Fig. 24) of which the diameter is the size specified on the drawing and which has a cylindrical pilot on the end to fit into the drilled hole, thus ensuring concentricity. This tool is fed in to the depth shown on the drawing.

Study the drawing of Fig. 3 with the illustrations of the operations, and check, first, the dimensions that would be used by the patternmaker and, second, those required by the machinist.

### 14. The Shaper and the Planer.

The drop forging of Fig. 8 requires machining on the base and boss surfaces.

Flat surfaces of this type are machined on a shaper or a planer. In this case the shaper (Fig. 16) is used because of the relatively small size of the part. The tool is held in a ram that moves back and forth across the work, taking a cut at each pass forward. Between the cuts, the table moves laterally so that closely spaced parallel cuts are made until the surface is completely machined.

The planer differs from the shaper in that its bed, carrying the work, moves back and forth under a stationary tool. It is generally used for a larger and heavier type of work than that done on a shaper.

### 15. Parts Machined from Standard Stock.

The shape of a part will often lend itself to machining directly from standard stock, such as bars, rods, tubing, plates, and blocks, or from extrusions and rolled shapes, such as angles and channels. Hot-rolled (HR) and cold-rolled (CR) steel are common materials.

Parts produced from stock are usually finished on all surfaces, and the general note "Finish all over" on the drawing eliminates the use of finish marks. Figure 17 is the drawing of a part to be made from bar stock. Note the specification of material, stock size, etc., in the title.

Figure 18 illustrates the shop interpretation in Fig. 17 and indicates the order of machining operations to be performed.

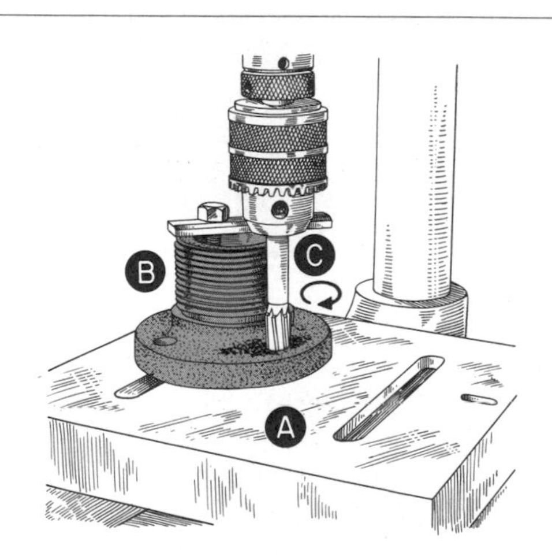


**FIG. 15.** Counterboring. The table (*A*) supports the part (*B*), sometimes clamped as shown. The counterboring tool with piloted end to fit the previously drilled hole is held in the chuck (*C*) and revolves in the direction shown. The gear and rack in the drill-press head bring the tool into the work.

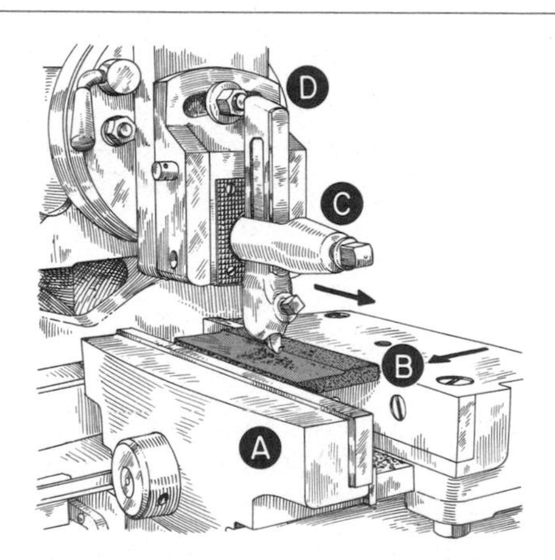


**FIG. 16.** Shaping. The vise (*A*) holds the part (*B*). The tool in the holder (*C*) is forced across the work surface by the ram (*D*) in the forward-stroke direction shown. The tool lifts on the return stroke by the pivot head on the ram. To make successive cuts, the table holding (*A*) and (*B*) moves in the direction shown.

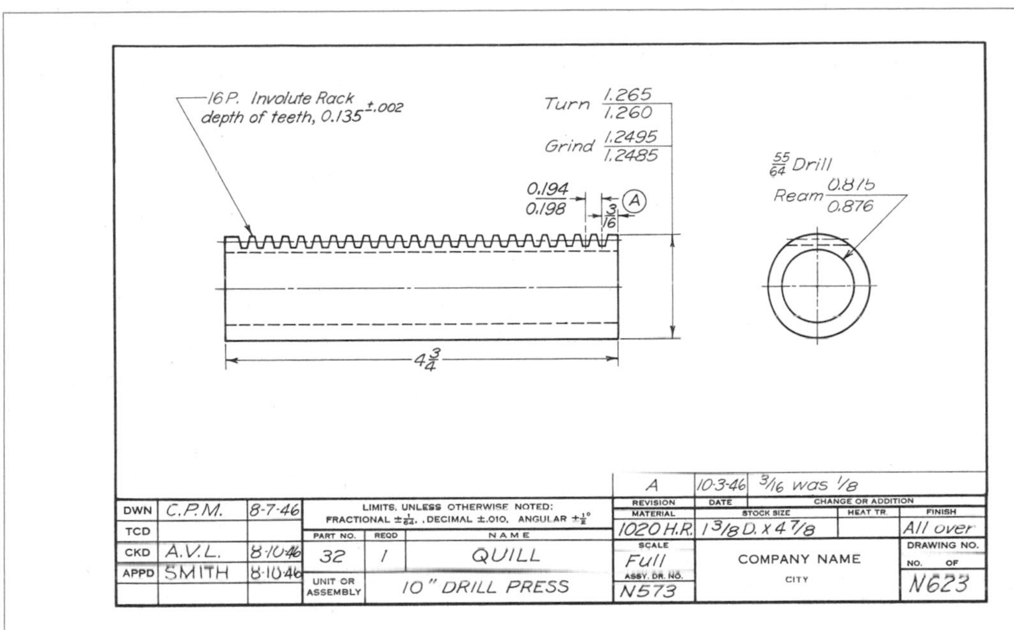


FIG. 17. Working drawing of a part machined from stock. Note the specification of material and stock size in the title block.

## 16. The Turret Lathe.

The *quill* of Fig. 17, produced in quantity, may be made on a turret lathe, except for the rack teeth and the outside-diameter grinding. The stock is held in the collet chuck of the lathe. First the end surface is faced, and then the cylindrical surface (OD) is turned. The work piece is then ready for drilling and reaming. The turret holds the various tools and swings them around into position as needed. A center drill starts a small hole to align the larger drill, and then the drill and reamer are brought successively into position. The drill provides a hold slightly undersize, and then the reamer, cutting with its fluted sides, cleans out the hole and gives a smooth surface finished to a size within the dimensional limits on the drawing. Figure 19 shows the turret indexed so that the drill is out of the way and the reamer in position. At the right is seen the cutoff tool ready to cut the piece to the length shown on the drawing.

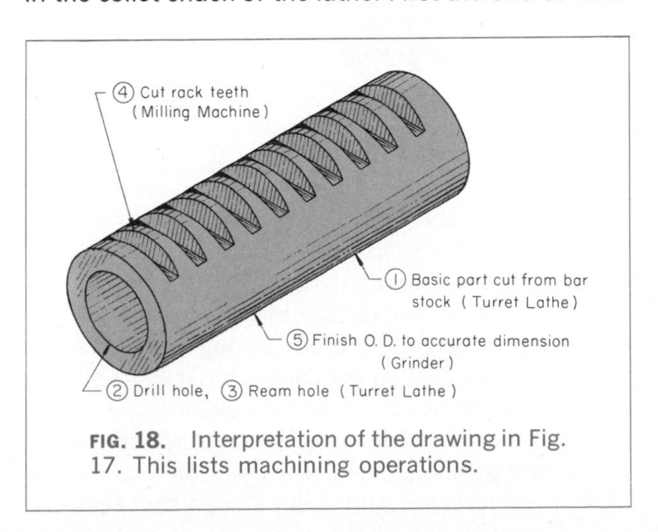


FIG. 18. Interpretation of the drawing in Fig. 17. This lists machining operations.

## 17. The Milling Machine.

The dimensions of the rack teeth (Fig. 17) give the depth and spacing of the cuts and the specifications for the cutter to be used. This type of work may be done on a milling machine. The work piece is

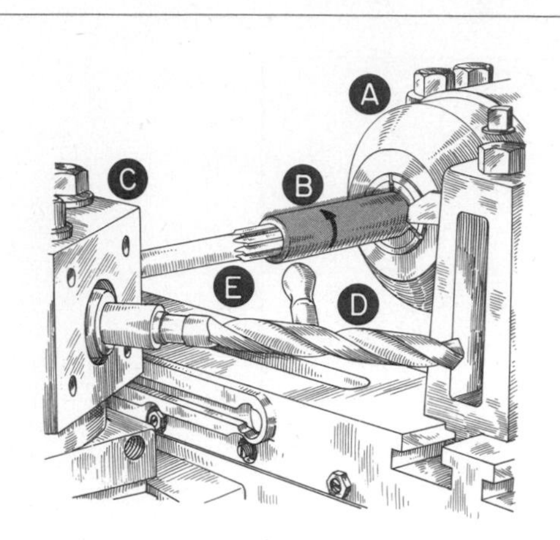


**FIG. 19.** Turret-lathe operations (drilling and reaming). The chuck (*A*) holds, indexes (for length), and rotates the stock (*B*) in the direction shown. The turret (*C*) swings the drill (*D*) and reamer (*E*) successively into position.

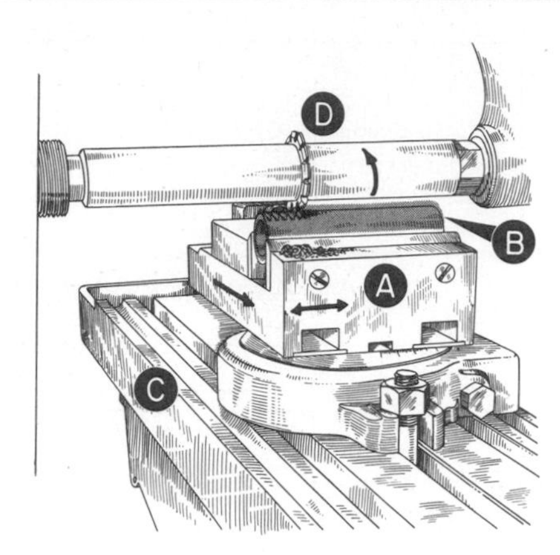


**FIG. 20.** Milling. The vise (*A*) holds the part (*B*) to the table (*C*), which moves laterally to index cuts and longitudinally, in the direction shown, for individual cuts. The milling cutter (*D*) revolves in the direction shown.

**FIG. 21.** Milling cutters.

held in a vise and moved horizontally into the rotating milling cutter, which, in profile, is the shape of the space between the teeth (Fig. 20). The cuts are spaced by moving the table of the machine to correspond with the distance shown on the drawing. Note that this operation is a forming process, as the shape depends upon the contour of the cutter. With several cutters mounted together (gang milling), a number of teeth can be cut at the same time.

There are many types of milling cutters made to cut on their peripheries, their sides, or their ends, for forming flat, curved, or special surfaces. Three milling cutters are shown in Fig. 21.

## 18. The Grinder.

The general purpose of grinding is to make a smoother and more accurate surface than can be obtained by turning, planing, milling, etc. In many cases, pieces hardened by heat-treatment will warp slightly; and as ordinary machining methods are impractical with hardened materials, such parts are finish-ground after hardening.

The limit dimensions for the outside diameter of the quill (Fig. 17) indicate a grinding operation on a cylindrical grinder (Fig. 22). The abrasive wheel rotates at high speed, while the work piece, mounted on a mandrel between conic centers, rotates slowly in the opposite direction. The wheel usually moves laterally to cover the surface of the work piece. The work piece is gaged carefully during the operation to bring the size within the dimensional limits shown on the drawing and to check for a cylindrical surface without taper. The machine for flat surfaces, called a "surface grinder," holds the work piece on a flat table moving back and forth under the abrasive

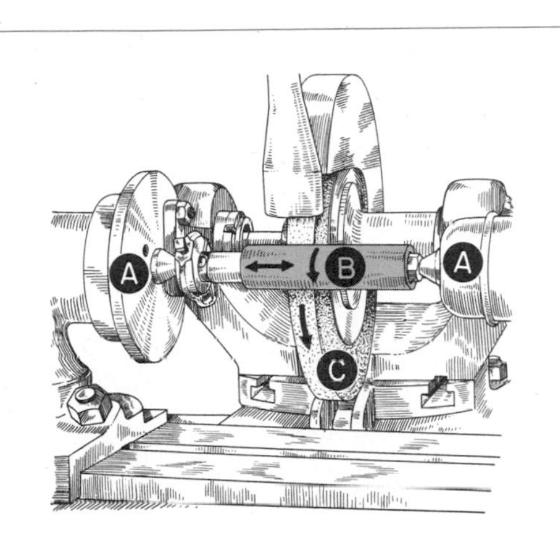


**FIG. 22.** Grinding. The cone centers (*A*) hold the mandrel, which mounts and rotates the part (*B*) in the direction shown. The wheel (*C*) moves laterally to traverse the entire surface of (*B*).

wheel. The table "indexes" laterally after each pass under the work.

### 19. Lapping, Honing, and Superfinishing.

These are methods of producing smooth, accurate, mirrorlike surfaces after grinding. All three methods use fine abrasives (1) powdered and carried in oil on a piece of formed soft metal (lapping) or (2) in the form of fine-grained compact stones (honing and superfinishing) to rub against the surface to be finished and reduce scratches and waviness.

### 20. The Broaching Machine.

A broach is a long, tapered bar with a series of cutting edges (teeth), each successively removing a small amount of material until the last edge forms the shape desired. For flat or irregular external surfaces, the broach and work piece are held by the broaching machine, and the broach is passed across the surface of the work piece. For internal surfaces, the broach is pulled or pushed through a hole to give the finished size and shape.

Some machined shapes can be more economically produced by broaching than by any other method. Figure 23 shows several forms of broaches and the shapes they produce.

### 21. Small Tools.

The shop uses a variety of small tools, both in powered machines and as hand tools. Figure 24 shows a *twist drill,* available in a variety of sizes (numbered, lettered, fractional, and metric) for producing holes in almost any material; a *reamer,* used to enlarge and smooth a previously existing hole and to give greater accuracy than is possible by drilling alone; a *counterbore* and a *countersink,* both used to enlarge and alter the end of a hole (usually for screwheads). A *spot-facing tool* is similar to a counterbore. *Taper, plug,* and *bottoming taps* for cutting the thread of a tapped hole and a die for threading a rod or shaft are also shown.

### 22. Welded Parts.

Simple shapes cut from standard rod, bar, or plate stock can be combined by welding to form a finished part. Some machining after welding is frequently necessary.

### 23. Parts from Standard Sheet.

A relatively thin sheet or strip of standard thickness may first be cut to size "in the flat" and then bent, formed, punched, etc., to form the final required part. The drawing should be made so as to give information for the "template maker" and the information required for bending and forming the sheet. Sometimes separate developments (Chap. 10) are made. The thickness of sheet stock is specified by giving (1) the gage (see table, Appendix) and the equivalent thickness in decimals of an inch or (2) only the decimal thickness (the practice followed in specifying aluminum sheet). Figure 111, Chap. 12, shows a working drawing of a sheet-metal part.

### 24. Plastics.

Plastics are available either in standard bar, rod, tubing, sheet, etc., from which parts can be made by machining, or in granular form to be used in "molding," a process similar to die-casting, in which the material is heated to a plastic state and com-

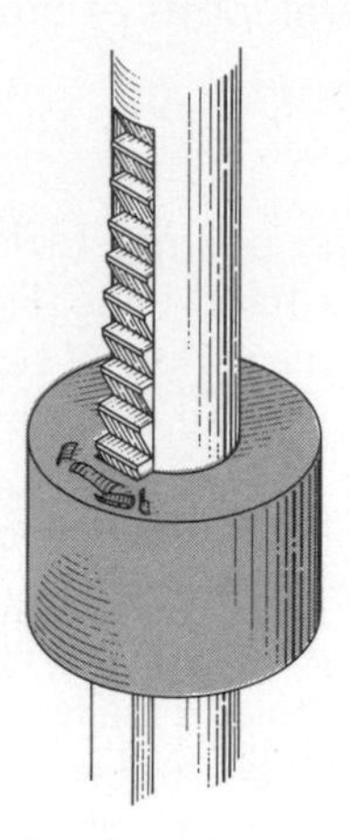


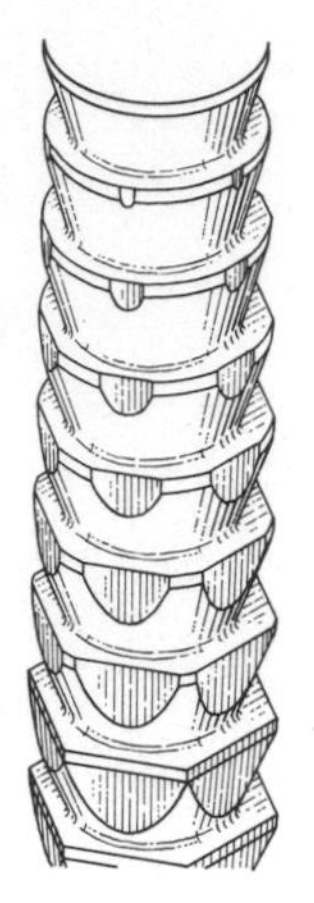


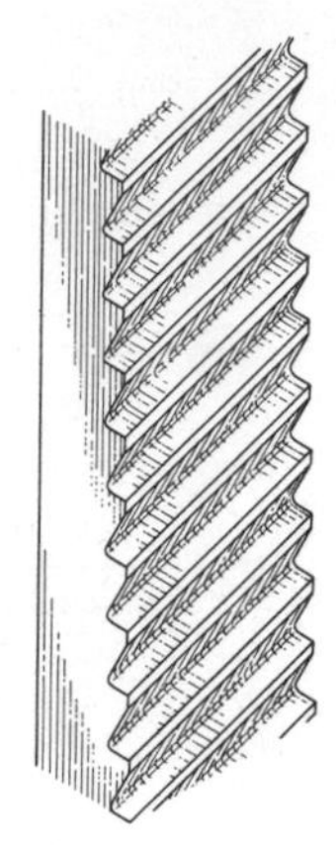


**FIG. 23.** Broaches. Each tooth takes its small cut as the broach is forced across or through the work piece.

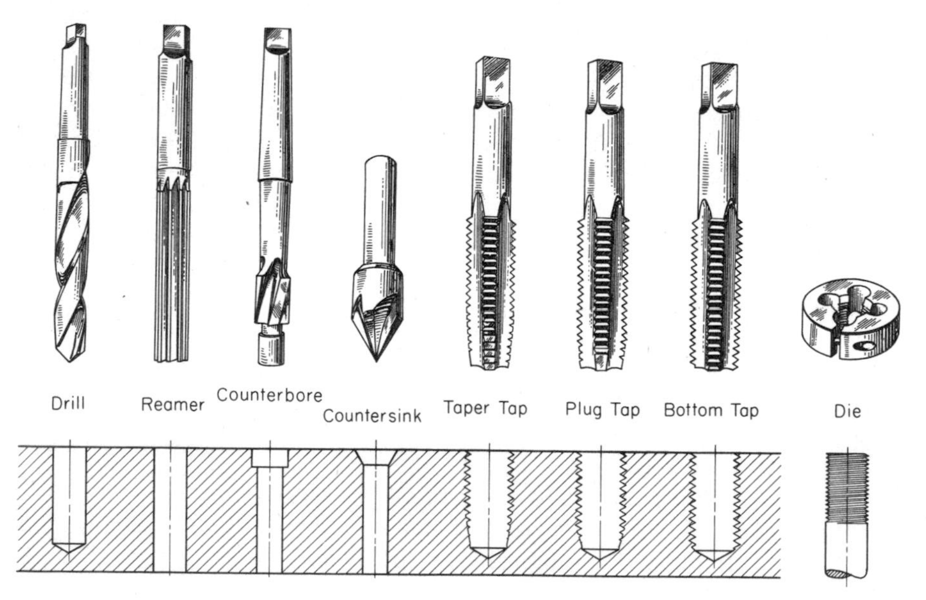


**FIG. 24.** Small tools.

pressed by a die (compression molding) or injected under pressure into a die (injection molding). Metal inserts for threads, wear bushings, etc., are sometimes cast into the part. Consideration should be given the diemaker when dimensioning the drawing.

### 25. Heat-treatment.

This is a general term applied to the processing of metals by heat and chemicals to change the physical properties of the material.

The glossary of shop terms gives definitions of such heat-treatment processes as annealing, carburizing, casehardening, hardening, normalizing, and tempering.

The specification of heat-treatment may be given on the drawing in several ways: (1) by a general note listing the steps, temperatures, and baths to be used, (2) by a standard heat-treatment number (SAE or company standard) in the space provided in the title block, (3) by giving the Brinell or Rockwell hardness number to be attained, or (4) by giving the tensile strength, in pounds per square inch, to be attained through heat-treatment.

Figure 8, and Figs. 67, 106, and 110 in Chap. 12, illustrate these methods.

### 26. Tools for Mass Production.

Many special machine tools, both semi-automatic and fully automatic, are used in modern factories. These machines are basically the same as ordinary lathes, grinders, etc., but contain mechanisms to control the movements of cutting tools and produce identical parts with little attention from the operator once the machine has been "tooled up." Automatic screw machines and centerless grinders are examples.

### 27. Jigs and Fixtures.

Jigs for holding the work and guiding the tool and fixtures for holding the work greatly extend the production rate for general-purpose machine tools.

### 28. Inspection.

Careful inspection is an important feature of modern production. Good practice requires inspection after each operation. For production in quantity, special gages are usually employed, but in small-quantity production, the usual measuring instruments, calipers and scale, micrometers, dial gages, etc., are used. For greater precision in gaging, electrical, air, or optical gages, etc., are often employed.

### 29. Assembly.

The finished separate pieces come to the assembly department to be put together according to the assembly drawings. Sometimes it is desirable or necessary to perform some small machining operation during assembly, often drilling, reaming, or hand-finishing. In such cases the assembly drawing should carry a note explaining the required operation and give dimensions for the alignment or location of the pieces. If some parts are to be combined before final assembly, a subassembly drawing or the detail drawings of each piece will give the required information; "$\frac{1}{8}$ drill in assembly with piece No. 107" is a typical note form for an assembly machining operation.

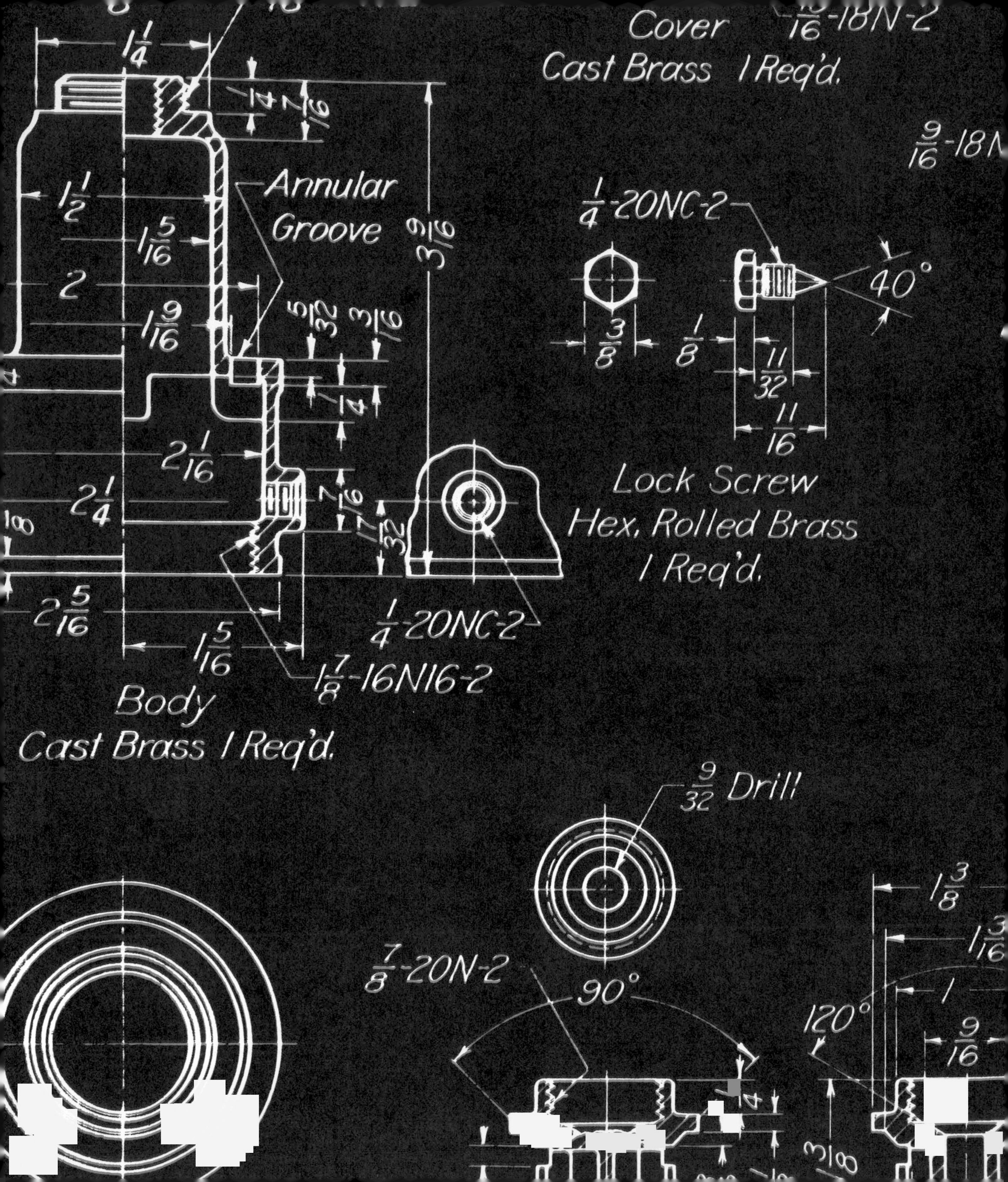

Cover
Cast Brass I Req'd.
Annular Groove
Lock Screw
Hex. Rolled Brass
I Req'd.
Body
Cast Brass I Req'd.
Drill

Size Description Requires a Knowledge of:

Lines and Symbols

Selection of Distances

Placement of Dimensions

The Dimensioning of Standard Features

Precision and Tolerance

Production Methods

Problems

# 12

# DIMENSIONS, NOTES, LIMITS, AND PRECISION

**These Are the Elements Used To Describe Size, which along with a Description of Shape Comprise a Complete Graphic Representation**

**1. Size Description—
First Considerations.**
After the shape of an object has been described by orthographic (or pictorial) views, the value of the drawing for the construction of the object depends upon dimensions and notes that describe the *size*. In general, the description of shape and size together gives complete information for producing the object represented.

The dimensions put on the drawing are *not necessarily* those used in making the drawing but are those required for the proper functioning of the part after assembly, selected so as to be readily usable by the workers who are to make the piece. Before dimensioning the drawing, study the machine and understand its functional requirements; then put yourself in the place of the patternmaker, diemaker, machinist, etc., and mentally construct the object to discover which dimensions would best give the information.

**2. Method.**
The basic factors in dimensioning practice are:

1. *Lines and Symbols.* The first requisite is a thorough knowledge of the elements used for dimensions and notes and of the weight and spacing of the lines on the drawing. These lines, symbols, and techniques are the "tools" for clear, concise representation of size.

2. *Selection of Distances.* The most important consideration for the ultimate operation of a machine and the proper working of the individual parts is the selection of distances to be given. This selection is based upon the functional requirements, the "breakdown" of the part into its geometric elements, and the requirements of the shop for production.

3. *Placement of Dimensions.* After the distances to be given have been selected, the next step is the actual placement of the dimensions showing these distances on the drawing. The dimensions should be placed in an orderly arrangement that is easy to read and in positions where they can be readily found.

4. *Dimensioning Standard Features.* These include angles, chamfers, standard notes, specifications of holes, spherical shapes, round-end shapes, tapers, and others for which, through long usage and study, dimensioning practice has been standardized.

5. *Precision and Tolerance.* The ultimate operation of any device depends upon the proper interrelationship of the various parts so that they operate as planned. In quantity production, each part must meet standards of size and position to assure assembly and proper functioning. Through the dimensioning of the individual parts, the limits of size are controlled.

6. *Production Methods.* The method of manufacturing (casting, forging, etc.) affects the detailed information given on the drawing proper, and in notes and specifications. The operations of various shops must be known, in order to give concise information.

## LINES AND SYMBOLS

**3. Dimension Forms.**
Two basic methods are used to give a distance on a drawing: a *dimension* (Fig. 1) or a *note* (Fig. 2).

A dimension is used to give the distance between two points, lines, or planes, or between some combination of points, lines, and planes. The numerical value gives the actual distance, the dimension line indicates the direction in which the value applies, and the arrowheads indicate the points between which the value applies. Extension lines refer the dimension to the view when the dimension is placed outside the view.

A note provides a means of giving explanatory information along with a size. The leader and arrowhead refer the word statement of the note to the proper place on the drawing. Notes applying to the object as a whole are given without a leader in some convenient place on the drawing.

The lines and symbols used in dimensioning are dimension lines, arrowheads, extension lines, leaders, numerical values, notes, finish marks, etc.

**4. Line Weights.**
Dimension lines, extension lines, and leaders are made with fine full lines the same width as center

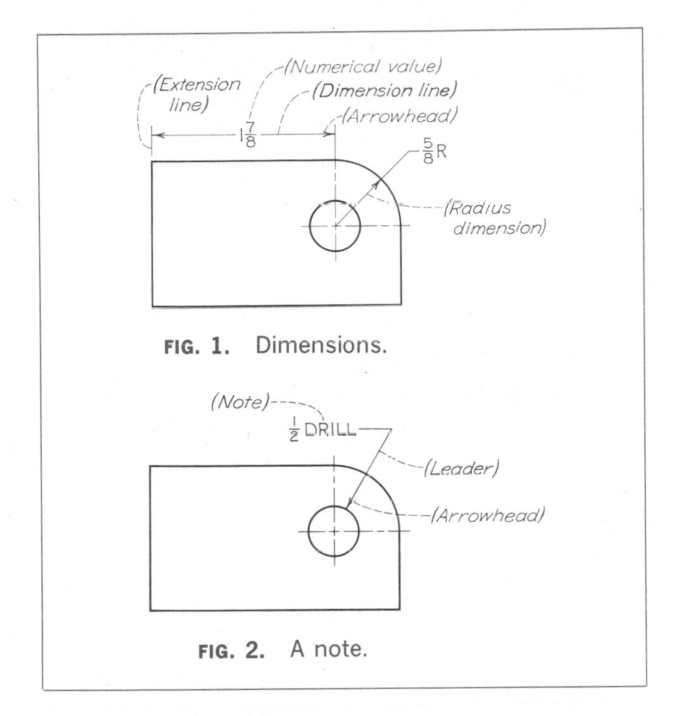


FIG. 1. Dimensions.

FIG. 2. A note.

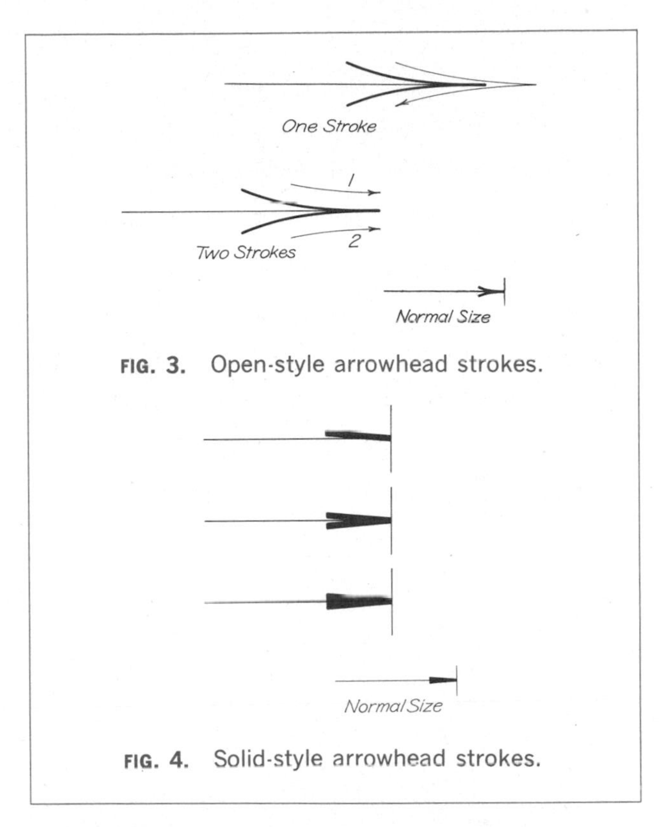


FIG. 3. Open-style arrowhead strokes.

FIG. 4. Solid-style arrowhead strokes.

lines so as to contrast with the heavier outlines of the views. Note the line widths given in the alphabet of lines, Figs. 56 and 57, Chap. 2.

## 5. Arrowheads.

These are carefully drawn freehand. The sides of the arrowhead are made either in one stroke, toward the point and then away from it, or in two strokes toward the point, as shown in enlarged form in Fig. 3. The general preference is for the solid head, as in Fig. 4. The solid head is usually made narrower and slightly longer than the open head and has practically no curvature to the sides. It is made in one stroke and then filled, if necessary, without lifting the pen or pencil; a rather blunt pencil or pen is required for this style of head. The bases of arrowheads should not be made wider than one-third the length. All arrowheads on the same drawing should be the same type, open or solid, and the same size, except in restricted spaces. Arrowhead lengths vary somewhat depending upon the size of the drawing. One-eighth inch is a good general length for small drawings and $^3/_{16}$ in. for larger drawings.

Poor arrowheads ruin the appearance of an otherwise carefully made drawing. Avoid the incorrect shapes and placements shown in Fig. 5.

## 6. Extension Lines.

These extend from the view to a dimension placed outside the view. They should not touch the outline of the view but should start about $^1/_{16}$ in. from it and extend about $^1/_8$ in. beyond the last dimension line (Fig. 6*a*). This example is printed approximately one-half size.

Dimensions may also terminate at *center lines* or *visible outlines of the view*. Where a measurement between centers is to be shown, as at (*b*), the center lines are continued to serve as extension lines, extending about $^1/_8$ in. beyond the last dimension line. Usually the outline of the view becomes the terminal for arrowheads, as at (*c*), when a dimension must be placed inside the view. This might occur because of limited space, when extension lines crossing parts

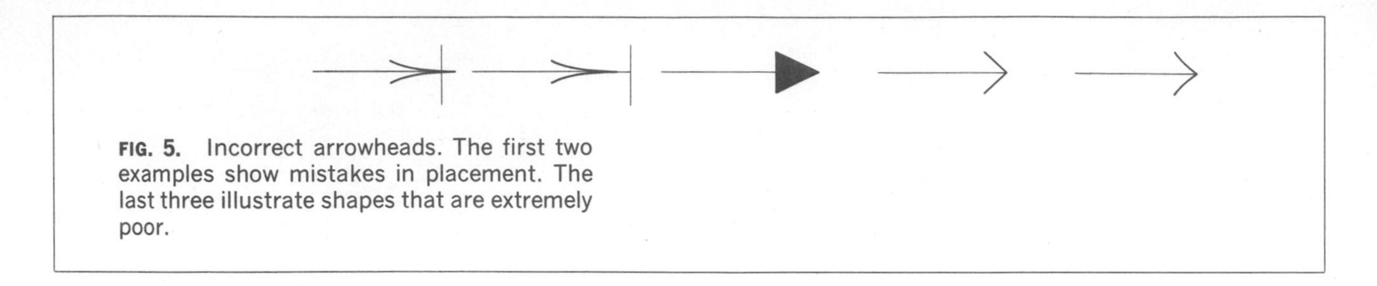

**FIG. 5.** Incorrect arrowheads. The first two examples show mistakes in placement. The last three illustrate shapes that are extremely poor.

of the view would cause confusion, or when long extension lines would make the dimension difficult to read.

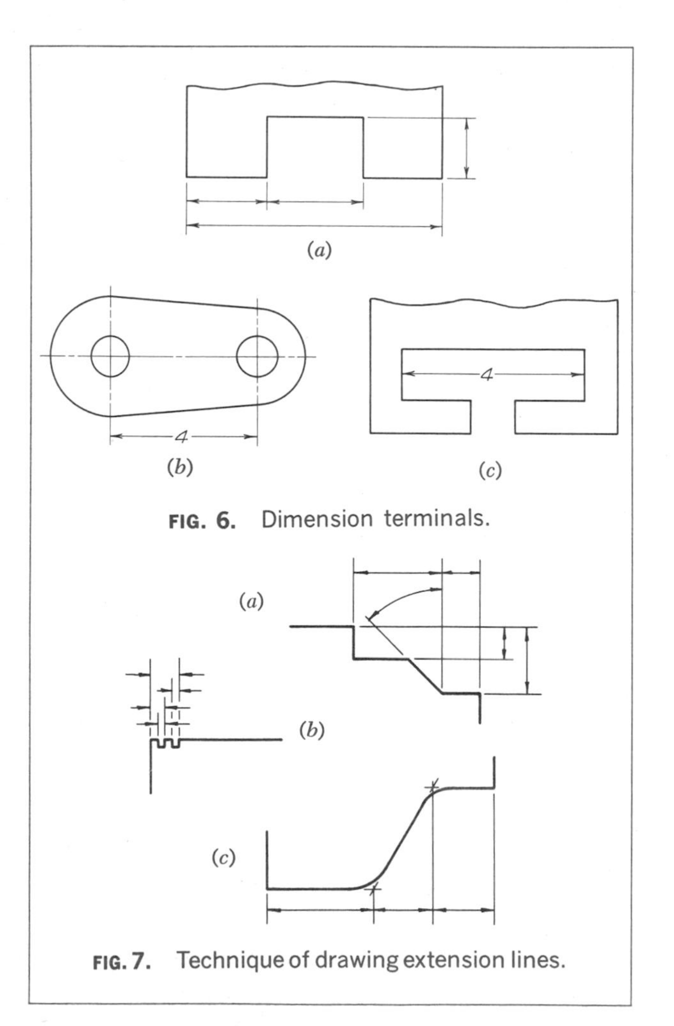


**FIG. 6.** Dimension terminals.



**FIG. 7.** Technique of drawing extension lines.

Extension lines for an angular dimension are shown in Fig. 7*a*, which also shows one of the extension lines used for a linear dimension, a common occurrence.

Extension lines should not be broken where they cross each other or an outline of a view, as shown in Figs. 6*a* and 7*a*. However, when space is restricted and the extension lines come close to arrowheads, the extension lines may be broken for clarity, as in Fig. 7*b*.

Where a point is located by extension lines alone, the extension lines should pass through the point, as at (*c*).

## 7. Leaders.

These are *straight* (not curved) lines leading from a dimension value or an explanatory note to the feature on the drawing to which the note applies (Fig. 8). An arrowhead is used at the pointing end of the leader but never at the note end. The note end of the leader should terminate with a short horizontal bar at the mid-height of the lettering and should run to the beginning or the end of the note, never to the middle.

Leaders should be drawn at an angle to contrast with the principal lines of the drawing, which are mainly horizontal and vertical. Thus leaders are usually drawn at 30°, 45°, or 60° to the horizontal; 60° looks best. When several leaders are used, the appearance of the drawing is improved if the leaders can be kept parallel.

The ASA allows the use of a "dot" for termination of a leader, as in Fig. 9*a*, when the dot is considered to be a clearer representation than an arrowhead. The dot should fall within the outline of the object.

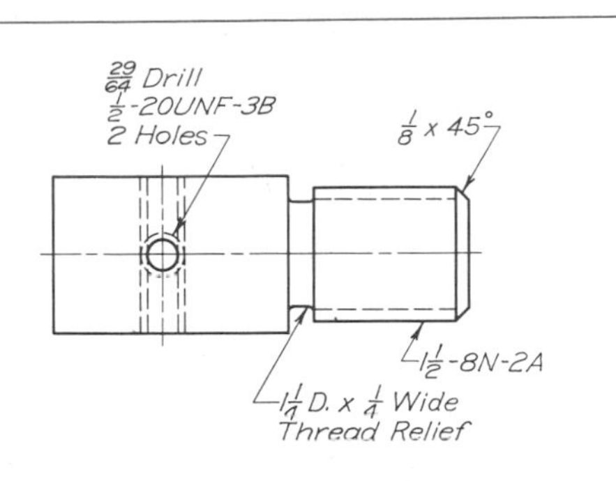


**FIG. 8.** Leaders for notes. Observe that the leader emanates from the *beginning* or the *end* of the lettered information.

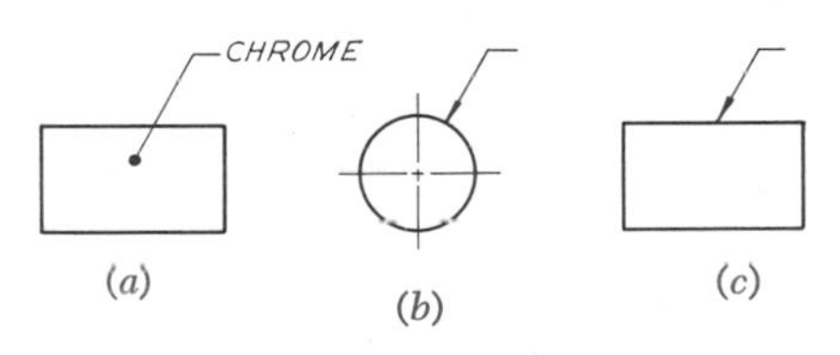


**FIG. 9.** Leaders. (*a*) Special terminal; (*b*) and (*c*) proper angle.

When dimensioning a circular feature with a note, the leader should be *radial* as shown at (*b*). A leader directed to a flat surface should meet the surface at an angle of *at least* 30°; an angle of 45° or 60° as at (*c*) is best.

If possible *avoid* crossing leaders; *avoid* long leaders; *avoid* leaders in a horizontal or vertical direction; *avoid* leaders parallel to adjacent dimension lines, extension lines, or crosshatching; and *avoid* small angles between leaders and the lines on which they terminate.

## 8. Figures.

For dimension values, figures must be carefully lettered in vertical or inclined style. In an effort to achieve neatness, the beginner often gets them too small. One-eighth inch for small drawings and $\frac{5}{32}$ in. for larger drawings are good general heights.

The general practice is to leave a space in the dimension line for the dimension value (Fig. 10). It is universal in structural practice and common in architectural practice to place the value above a continuous dimension line (Fig. 10).

## 9. Common Fractions.

These should be made with the fraction bar parallel to the guide lines for making the figure and with the numerator and denominator each somewhat smaller than the height of the whole number so that the *total* fraction height is twice that of the integer (Fig. 11). Avoid the incorrect forms shown. The figures should not touch the fraction bar.

## 10. Feet and Inches.

Indicate these thus: 9′-6″. When there are no inches, it should be so indicated, as 9′-0″, 9′-0½″. When dimensions are all in inches, the inch mark is preferably omitted from all the dimensions and notes unless there is some possibility of misunderstanding; if that is the case, "1 bore," for example, should be given for clarity as "1″ bore."

In some machine industries, all dimensions are given in inches. In others where feet and inches are used, the ASA recommends that dimensions up to and including 72 in. be given in inches and greater lengths in feet and inches.

In structural drawing, length dimensions should be given in feet and inches. Plate widths, beam sizes, etc., are given in inches. Inch marks are omitted, even though the dimension is in feet and inches (Fig. 10).

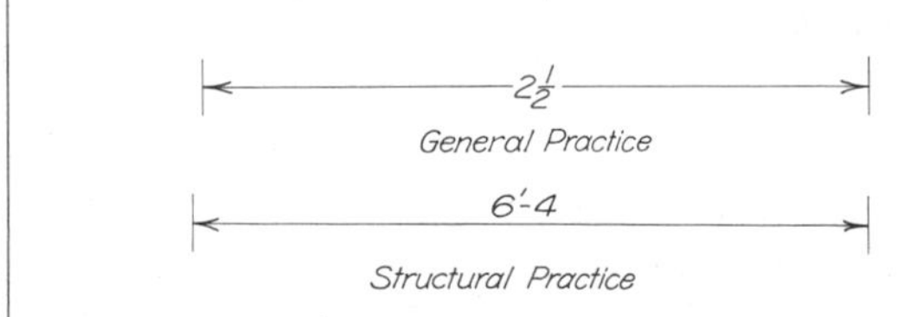


**FIG. 10.** Placement of dimension values.

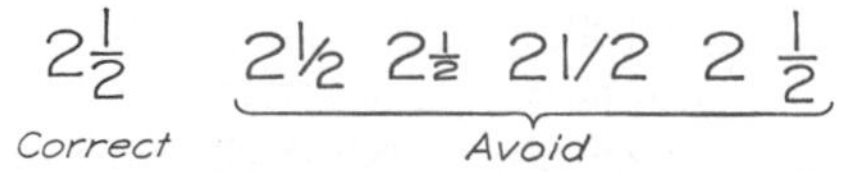

**FIG. 11.** Technique for lettering values with common fractions.

In the United States, if no foot or inch marks appear on a drawing, the dimension values indicate inches unless a different unit of measurement is indicated by a general note. Drawings made in foreign countries employing the metric system are commonly dimensioned in millimeters.

## 11. Reading Direction of Figures.

This is arranged according to the aligned system or the unidirectional system.

The *aligned system* is the older of the two methods. The figures are oriented to be read from a position *perpendicular* to the dimension line; thus the guide lines for the figures will be parallel to the dimension line, and the fraction bar in line with the dimension line (Fig. 12). The figures should be arranged so as to be read from the *bottom* or *right side* of the drawing. Avoid running dimensions in the directions included in the shaded area of Fig. 12; if this is unavoidable, they should read downward with the line.

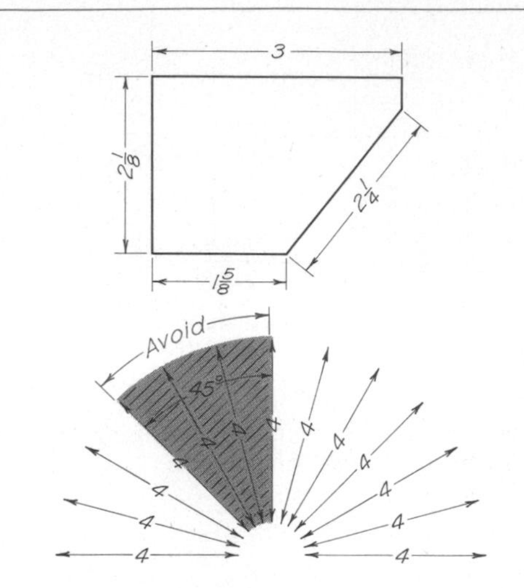


**FIG. 12.** Reading direction of values (aligned system). The drawing is read from the bottom and right side.

The *unidirectional system* originated in the automotive and aircraft industries, and is sometimes called the "horizontal system." All figures are oriented to read from the bottom of the drawing. Thus the guide lines and fraction bars are horizontal regardless of the direction of the dimension (Fig. 13). The "avoid" zone of Fig. 12 has no significance with this system.

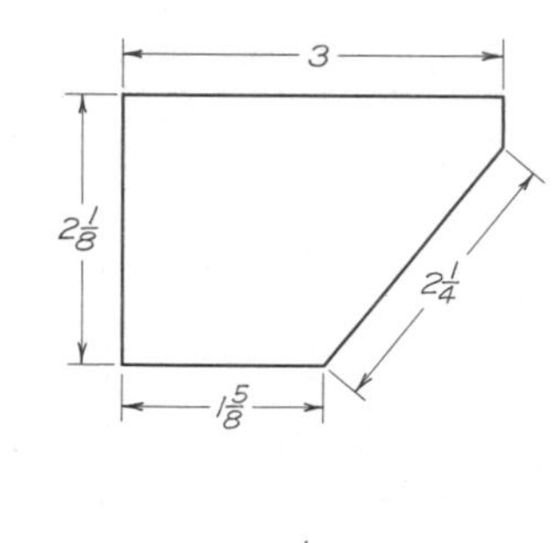


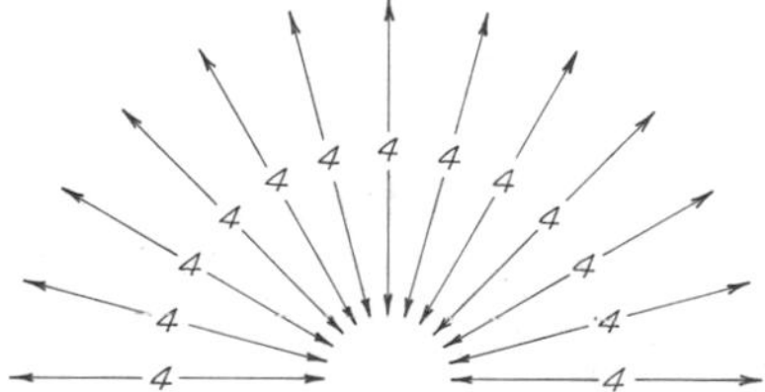


**FIG. 13.** Reading direction of values (unidirectional system). The drawing is read from the bottom.

*Notes must be lettered horizontally and read from the bottom of the drawing in either system.*

## 12. Systems of Writing Dimension Values.

Dimension values may be given as common fractions, $\frac{1}{4}$, $\frac{3}{8}$, etc., or as decimal fractions, 0.25, 0.375, etc.; and from these, three systems are evolved.

The *common-fraction system,* used in general drawing practice, including architectural and structural work, has all dimension values written as units and common fractions, as $3\frac{1}{2}$, $1\frac{1}{4}$, $\frac{3}{8}$, $\frac{1}{16}$, $\frac{3}{32}$, $\frac{1}{64}$. Values thus written can be laid out with a steel tape or scale graduated in sixty-fourths of an inch.

The *common-fraction, decimal-fraction system* is used principally in machine drawing whenever the degree of precision required calls for fractions of an inch smaller than those on the ordinary steel scale. To continue the use of common fractions below $\frac{1}{64}$, such as $\frac{1}{128}$ or $\frac{1}{256}$, is considered impractical. The method followed is to give values: (1) in units and common fractions for distances not requiring an accuracy closer than 1/64 in.; and (2) in units and decimal fractions, as 2.375, 1.250, 0.1875, etc., for distances requiring greater precision. The decimal fractions are given to as many decimal places as needed for the degree of precision required.

The *complete decimal system* uses decimal fractions exclusively for all dimension values. This system has the advantages of the metric system but uses the inch as its basis, thus making it possible to use present measuring equipment.

The ASA complete decimal system[1] uses a two-place decimal for all values where common fractions would ordinarily be used. The digits after the decimal point are preferably written to even fiftieths, .02, .10, .36, etc., so that when halved, as for radii, etc., two-place decimals will result. Writing the values in even fiftieths allows the use of scales divided in

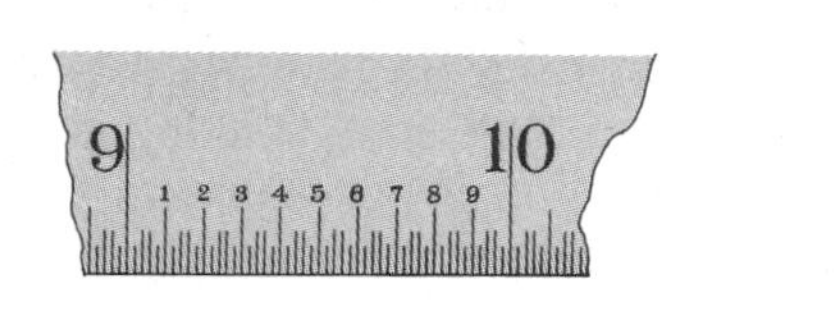


**FIG. 14.** Decimal scale. Graduations are in fiftieths.

fiftieths (Fig. 14), which are much easier to read than scales divided in hundredths.

Dimension values for distances requiring greater precision than that expressed by the two-place decimal are written to three, four, or more decimal places as needed for precision.

Figure 15 is a detail drawing dimensioned according to the ASA decimal system. The advantage of this system in calculating, adding, and checking and

[1]Y14.5—1964.

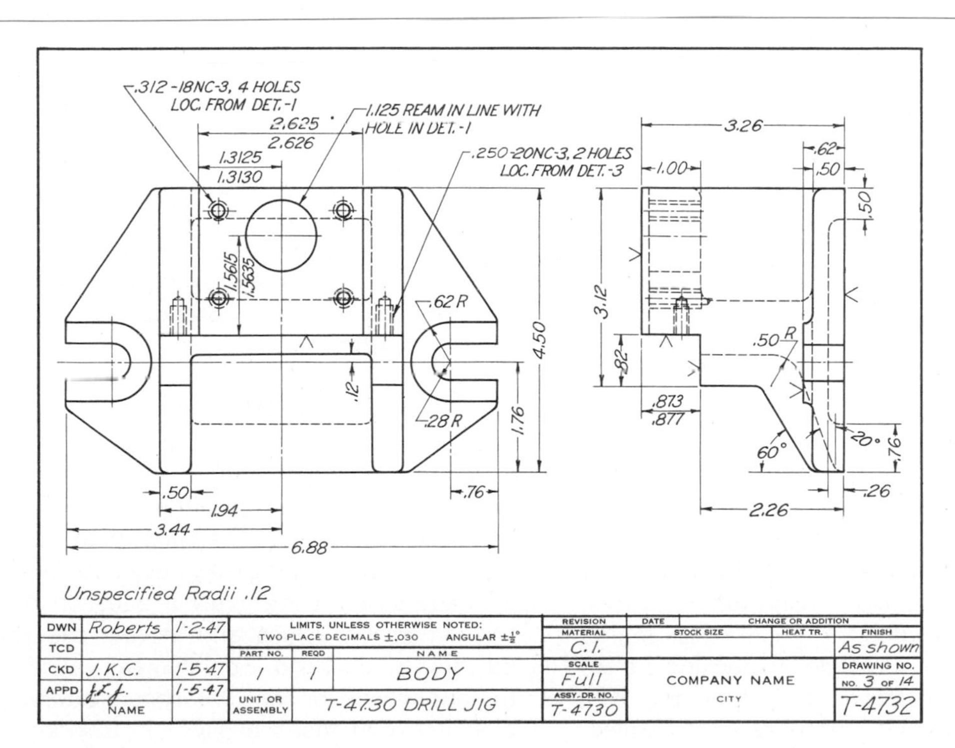


**FIG. 15.** A drawing dimensioned in the decimal system.

in doing away with all conversion tables, as well as in lessening chances for error, is apparent.

Designers and draftsmen working in the complete decimal system find it necessary to think in terms of tenths and hundredths of inches instead of common fractions. New designs must be made in decimal sizes without reference to common fractional sizes. However, until standard-stock materials, tools, and commercial parts are available in decimal sizes, some dimensions will have to be given as the decimal equivalent of a common fraction. Thus, for example, a standard $\frac{3}{8}$-16UNC-2A thread would be given as 0.375-16UNC-2A.

*Decimal equivalents* of some common fractions come out to a greater number of decimal places (significant digits) than is necessary or desirable for use as a dimension value, and in such cases the decimal should be adjusted, or "rounded off," to a smaller number of decimal places. The following procedure from the SAE Aerospace–Automotive Drafting Standard[2] is recommended:

When the figure beyond the last figure to be retained is less than 5, the last figure retained should not be changed. Example: 3.46325, if cut off to three places, should be 3.463.

When the figure beyond the last figure to be retained is more than 5, the last figure retained should be increased by 1. Example: 8.37652, if cut off to three places, should be 8.377.

When the figure beyond the last place to be retained is exactly 5 with only zeros following, the preceding number, if even, should be unchanged; if odd, should be increased by 1. Example: 4.365 becomes 4.36 when cut off to two places. Also, 4.355 becomes 4.36 when cut off to two places.

## 13. Finish Marks.

These are used to indicate that certain surfaces of metal parts are to be machined and that allowance must therefore be provided for finish. Finish marks need not be used for parts made by machining from rolled stock, as the surfaces are necessarily machined. Neither are they necessary on drilled, reamed, or counterbored holes or on similar machined features when the machining operation is specified by note. Also, when limit dimensions are given, the accuracy required is thereby stated, and a finish mark on the surface is unnecessary.

The standard finish mark recommended by the ASA is a 60° **V** with its point touching the line representing the edge view of the surface to be machined. The **V** is placed on the "air side" of the surface. Figure 16 shows the normal size of the **V** and its position for lines in various directions as given on a drawing.

Finish marks should be placed on all views in which the surface to be machined appears as a line, including dashed lines. If the part is to be machined on all surfaces, the note "Finish all over," or "FAO," is used, and the marks on the views are omitted.

In addition to using the finish mark to indicate a machined surface, it may be necessary in some cases to indicate the degree of smoothness of the surface. The ASA gives a set of symbols to indicate the degree of smoothness of a surface, the various conditions of *surface quality*. These symbols are explained and illustrated in Sec. 79.

## 14. Scale of the Drawing.

Even though a workman is never expected to scale a distance on a drawing to obtain a dimension value, the scale to which the drawing is made should be stated in the title block. Standard scales are listed in Fig. 34, Chap. 2.

## 15. Revision of Dimensions.

As a project is being developed, changes in design, in engineering methods, etc., may make it necessary to change some drawings either before or after they have been released to the shop. If the change is a major one, the drawing may have to be remade. But in many cases it is merely a question of altering the dimension values and leaving the shape description unchanged. For changes in dimensions, the out-of-scale dimensions should be indicated by one of the methods in Fig. 17. Drawing changes should be listed in tabular form, in connection with the title block, or in the upper right corner, with

[2]Section A.6 (1963).

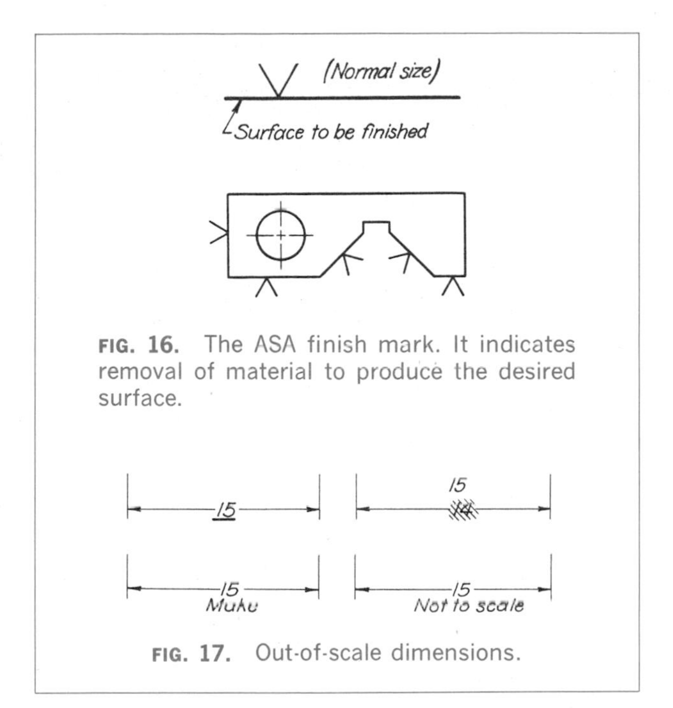


**FIG. 16.** The ASA finish mark. It indicates removal of material to produce the desired surface.

**FIG. 17.** Out-of-scale dimensions.

reference letters and the date, as explained in Sec. 19.

### 16. Abbreviations.

Many abbreviations are used on drawings to save time and space. The common ones such as DIA, THD, MAX, MIN, ID, and OD are universally understood. Uncommon abbreviations such as PH BRZ (Phosphor bronze) should be used cautiously because of the possibility of misinterpretation. Abbreviations should conform to the SAE Aerospace–Automotive Drafting Standard, Sec. Z.1.

### 17. Decimal Points.

These should be made distinctly, in a full letter space, and should be aligned with the bottom edges of digits and letters.

### 18. Dashes.

These are used in many standard expressions such as ½—20 and should be made clearly, one letter space in length, at the mid-height of letters or digits and parallel to the direction of the expression.

## SELECTION OF DISTANCES

### 19. Theory of Dimensioning.

Any object can be broken down into a combination of basic geometric shapes, principally prisms and cylinders. Occasionally, however, there will be parts of pyramids and cones, now and then a double-curved surface, and rarely, except for surfaces of screw threads, a warped surface. Any of the basic shapes may be positive or negative, in the sense that a hole is a negative cylinder. Figure 98, Chap. 5, illustrates a machine part broken down into its fundamental shapes.

If the *size* of each of these elementary shapes is dimensioned and the relative position of each is given, measuring from center to center, from base lines, or from surfaces, the dimensioning of any piece can be done systematically. Dimensions can thus be classified as dimensions of *size* and dimensions of *position*.

### 20. Dimensions of Size.

Since every solid has three dimensions, each of the geometric shapes making up the object must have its height, width, and depth indicated in the dimensioning.

The *prism,* often in plinth or flat form, is the most common shape and requires three dimensions for square, rectangular, or triangular (Fig. 18*a*). For regular hexagonal or octagonal types, usually only two dimensions are given, either the distance "across corners" and the length or the distance "across flats" and the length.

The *cylinder,* found on nearly all mechanical pieces as a shaft, boss, or hole, is the second most common shape. A cylinder obviously requires only two dimensions, diameter and length (*b*). Partial cylinders, such as fillets and rounds, are dimensioned by radius instead of diameter. A good general rule is to dimension complete circles with the diameter and circle arcs (partial circles) with the radius.

*Right cones* can be dimensioned by giving the altitude and the diameter of the base. They usually occur as frustums, however, and require the diameters of the ends and the length (*c*). Sometimes it

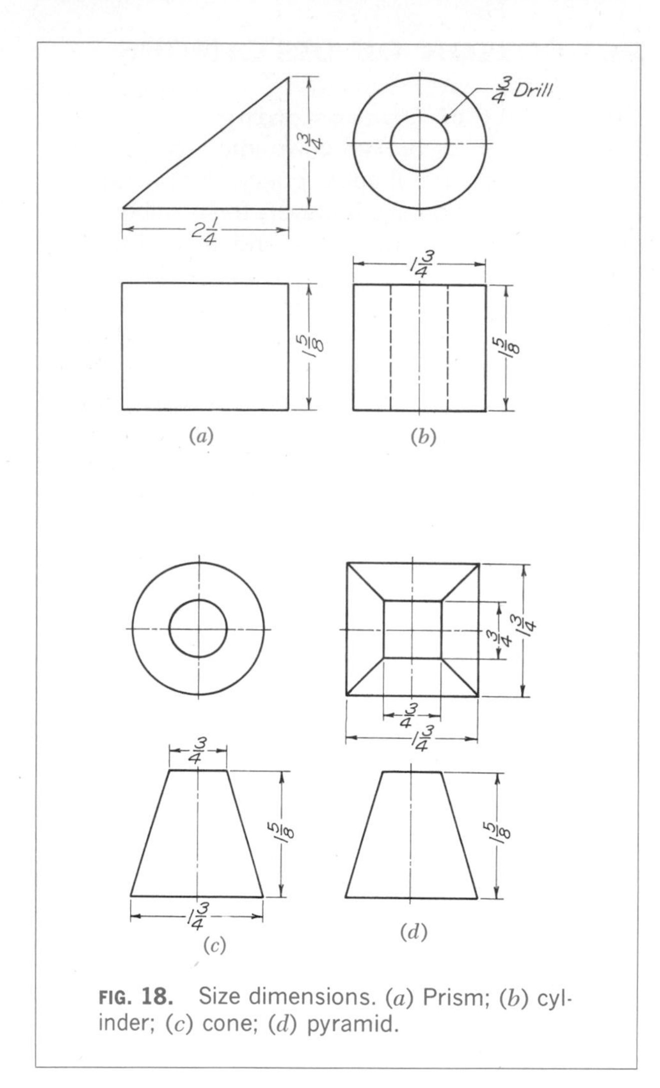


FIG. 18. Size dimensions. (*a*) Prism; (*b*) cylinder; (*c*) cone; (*d*) pyramid.

is desirable to dimension cone frustums as *tapers* or with an angular dimension, as described in Sec. 41.

*Right pyramids* are dimensioned by giving the dimensions of the base and the altitude. Right pyramids are often frustums, requiring dimensions of both bases (*d*).

*Oblique cones* and *pyramids* are dimensioned in the same way as right cones and pyramids but with an additional dimension parallel to the base to give the offset of the vertex.

*Spheres* are dimensioned by giving the diameter and other surfaces of revolution by dimensioning the generating curve.

*Warped surfaces* are dimensioned according to their method of generation; and as their representation requires numerous sections, each of these must be fully dimensioned by ordinate and abscissa dimensions.

## 21. Dimensions of Position.

After the basic geometric shapes have been dimensioned for size, the position of each relative to the others must be given. *Position must be established in height, width, and depth directions.* Rectangular shapes are positioned with reference to their faces, cylindrical and conic shapes with reference to their center lines and their ends.

One basic shape will often coincide or align with another on one or more of its faces. In such cases the alignment serves partially to locate the parts and eliminates the need of a dimension of position in a direction perpendicular to the line of coincidence. Thus in Fig. 19, prism *A* requires only one dimension for complete positioning with respect to prism *B*, as two surfaces are in alignment and two in contact.

Coincident center lines often eliminate the need for position dimensions. In the cylinder in Fig. 18*b* the center lines of the hole and of the cylinder coincide, and no dimensions of position are needed. The two holes of Fig. 19 are on the same center line, and the dimension perpendicular to the common center line positions both holes in that direction.

## 22. Selection of Dimensions.

The dimensions arrived at by reducing the part to its basic geometric shapes will, in general, fulfill the requirements of practical dimensioning. However, sometimes other dimensions are required to ensure satisfactory functioning of the part and to give the information in the best way from the standpoint of production.

*The draftsman must therefore correlate the dimensions on drawings of mating parts to ensure satisfactory functioning and, at the same time, select dimensions convenient for the workmen to use.*

Here our study of graphics as a language must be supplemented by a knowledge of shop methods.

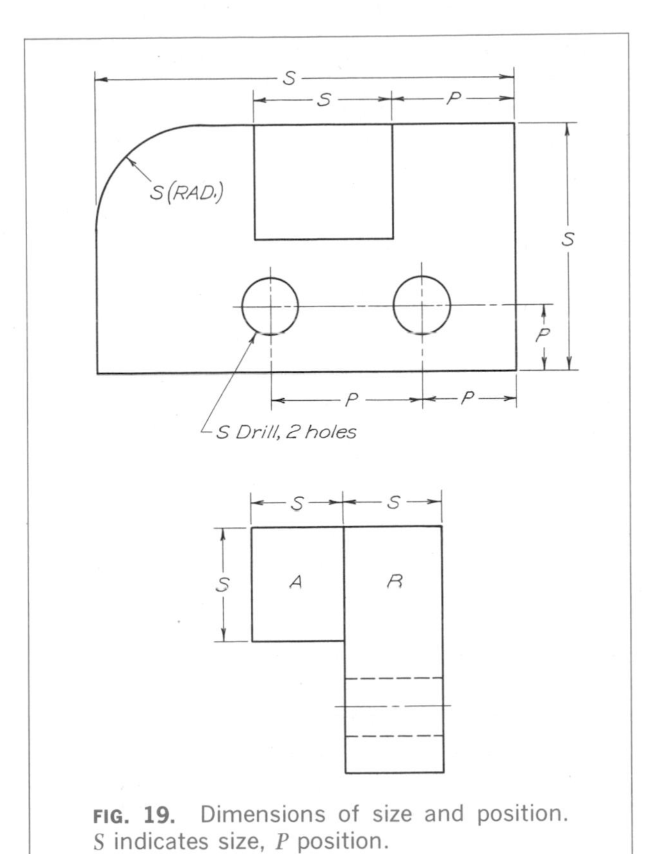


FIG. 19. Dimensions of size and position. *S* indicates size, *P* position.

To be successful, the machine draftsman must have an intimate knowledge of patternmaking, foundry practice, forging, and machine-shop practice, as well as, in some cases, sheet-metal working, metal and plastic die casting, welding, and structural-steel fabrication.

The beginning student who is without this knowledge should not depend upon his instructor alone but should learn by observing work going through the shops and reading books and periodicals on methods used in modern production work.

The *selection of dimensions of size* arrived at by shape breakdown will usually meet the requirements of the shop since the basic shapes result from the fundamental shop operations. However, a shop often prefers to receive dimensions of size in note form rather than as regular dimensions when a shop process is involved, such as drilling, reaming, counterboring, and punching.

*Selecting dimensions of position* ordinarily requires more consideration than selecting dimensions of size because there are usually several ways in which a position might be given. In general, positional dimensions are given between finished surfaces, center lines, or a combination thereof (Fig. 22). Remember that rough castings or forgings vary in size; so do not position machined surfaces from unfinished surfaces—except in one situation: A machined surface may be positioned from an unmachined surface only when the *initial*, or *starting*, dimension gives the position for the first surface to be machined—a position from which the other machined surfaces will in turn be positioned. *Coinciding center lines of unfinished and finished surfaces often take the place of a starting dimension.*

The position of a point or center by offset dimensions from two center lines or surfaces (Fig. 20) is preferable to angular dimensions (Fig. 21) unless

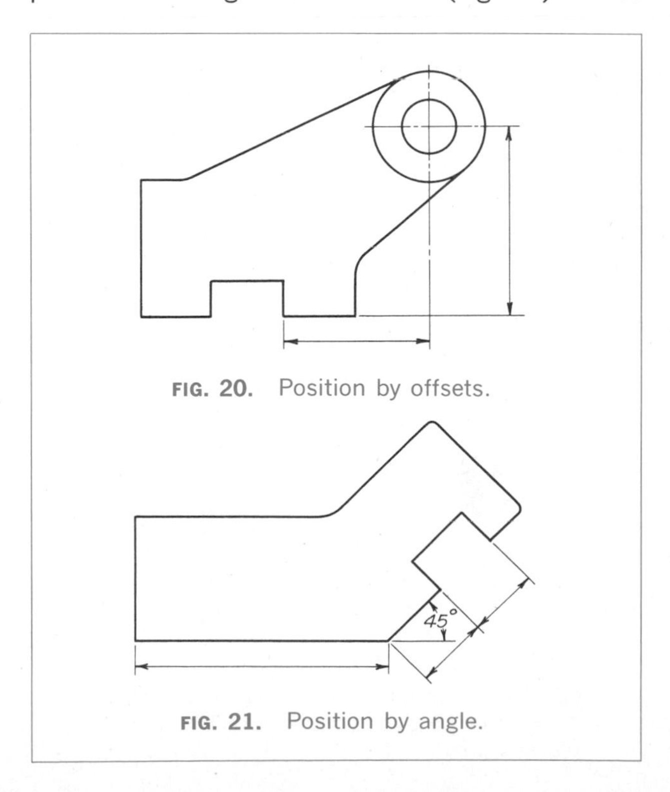


FIG. 20. Position by offsets.

FIG. 21. Position by angle.

the angular dimension is more practical from the standpoint of construction.

**23. Correlation of Dimensions.**
Mating parts must have their dimensions correlated so that the two parts will fit and function as intended. Figure 22 illustrates this principle. Note that the tongue of the bracket is to fit the groove in the body, and that the drilled holes in both pieces must align. Study the dimensioning of both pieces and observe that the dimensions of position (*P*) are correlated so that the intended alignment and fitting of the parts will be accomplished.

Not only must dimensions be correlated with the dimensions of the mating part, but the accuracy to which these distances are produced must meet certain requirements, or the parts still may not fit and function properly. Distances between the surfaces or center lines of finished features of an object must usually be more accurately made than distances between unfinished features. In Fig. 23, note that the position dimensions between center lines or surfaces of finished features are given as three-place decimals, as dimension *A*. Dimensions of position for unfinished features are given as common fractions, as dimension *B*. The decimal dimensions call for greater precision in manufacture than do the common fractions. Dimension *B* is in this case used by the patternmaker to position cylinder *C* from the right end of the piece. The machinist will first locate the finished hole in this cylinder, making it concentric with the cylinder; then all other machined surfaces will be positioned from this hole, as, for example, the spline position by dimension *A*. The four spot-faced holes are positioned with reference to each other with fractional dimensions since the holes are oversize for the fastenings used, allowing enough shifting of the fastenings in the holes so that great accuracy in location is not necessary. The mating part, with its holes to receive the screws, would be similarly dimensioned.

Study Fig. 23 and note the classification, size, and position of each dimension.

**24. Superfluous Dimensions.**
*Duplicate* or *unnecessary* dimensions are to be avoided because they may cause confusion and delay. When a drawing is changed or revised, a duplicate dimension may not be noticed and changed along with its counterpart; hence a distance

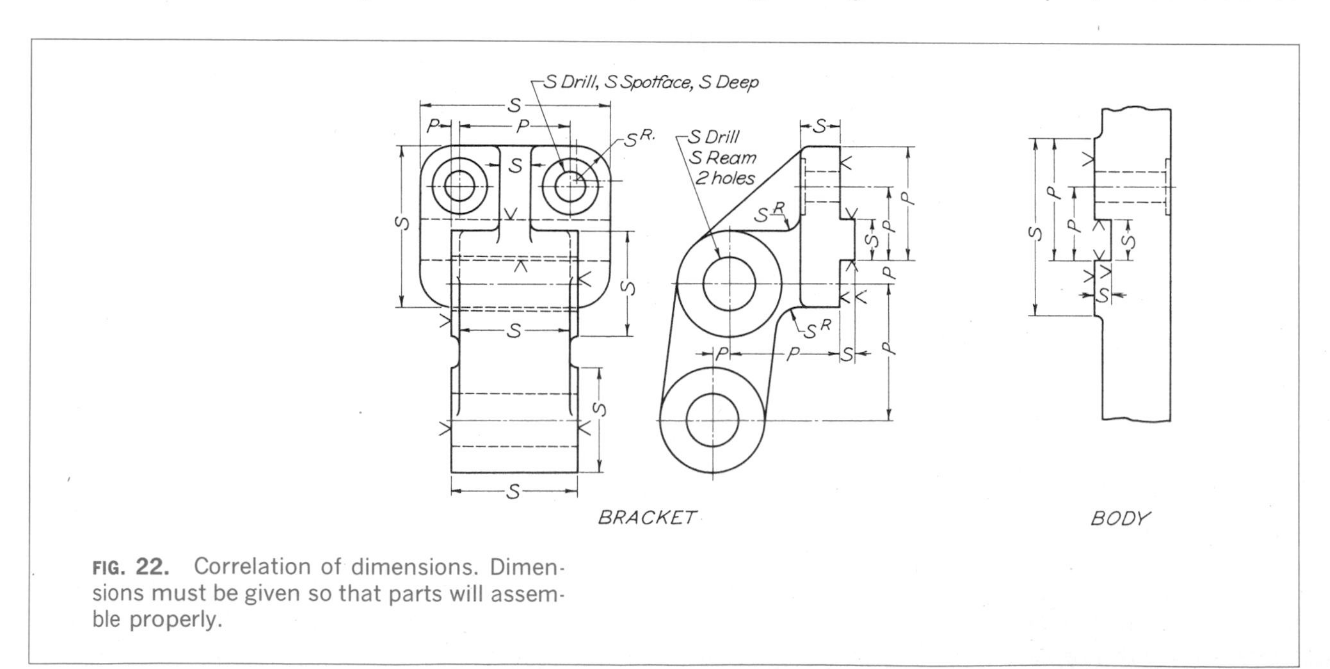


**FIG. 22.** Correlation of dimensions. Dimensions must be given so that parts will assemble properly.

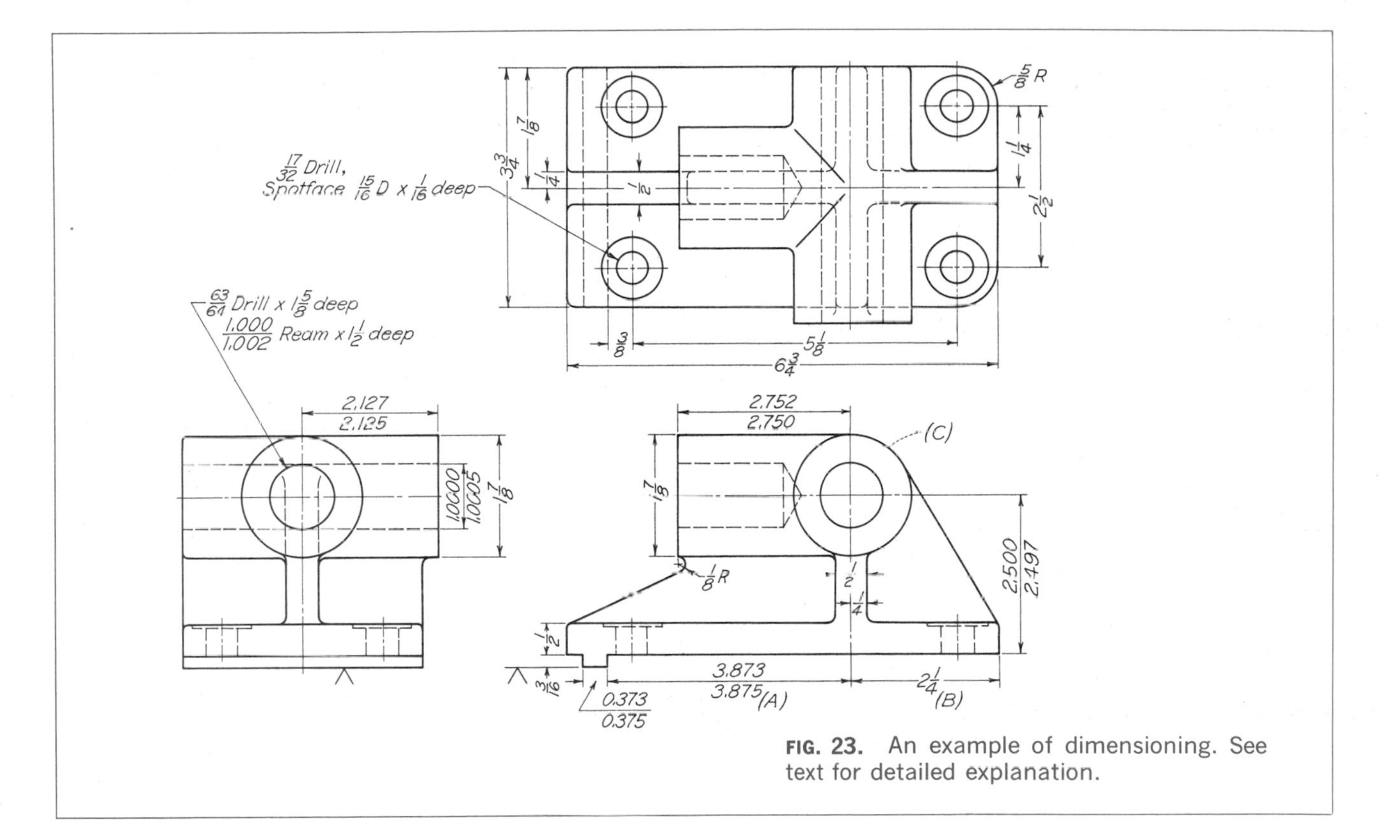


FIG. 23. An example of dimensioning. See text for detailed explanation.

will have two different values, one incorrect. An unnecessary dimension is any dimension, other than a duplicate, that is not essential in making the piece. Because of the allowable variation permitted the manufacturer on each dimension (see Sec. 54), difficulties will be encountered if unnecessary dimensions occur when parts are to be interchangeable. Actually, if the proper dimensions have been selected, it will be possible to establish a point on the object in any given direction with only one dimension. Unnecessary dimensions always occur when all the individual dimensions are given, in addition to the over-all dimension (Fig. 24). One dimension of the series must be omitted if the over-all dimension is used, thus allowing only one possible positioning from each dimension (Fig. 25).

In architectural and structural work, where the interchangeability of parts is usually irrelevant, unnecessary dimensions cause no difficulty and all dimensions are given.

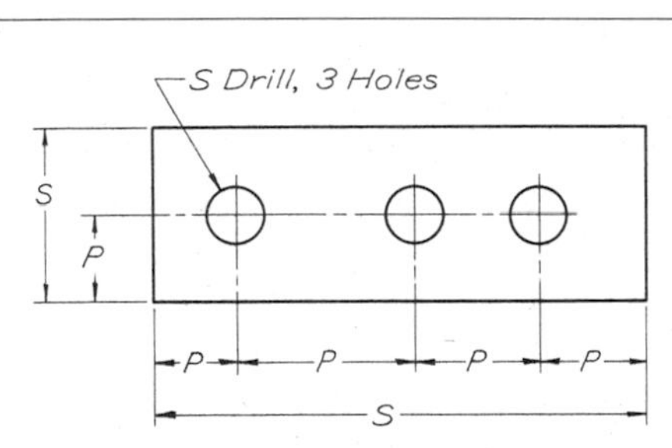


FIG. 24. One unnecessary dimension.

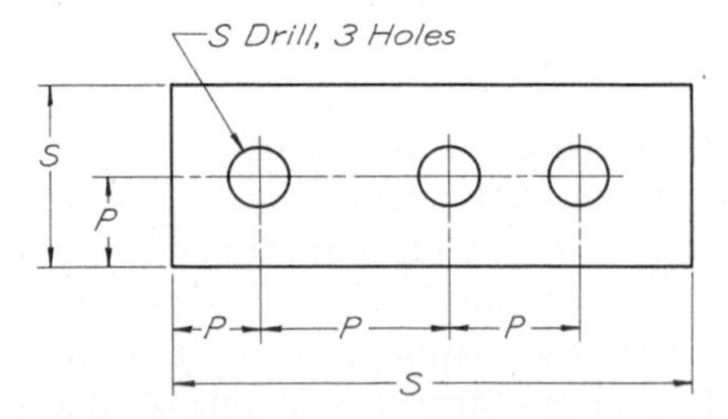


FIG. 25. Unnecessary dimension omitted.

Although it is important not to "over-dimension" a part, it is equally important to give all the dimensions that are needed to locate every point, line, or surface of the object. Every dimension that the workman will require in making the part must be given.

Dimensions for similar features, such as the thickness of several ribs obviously of the same size, need not be repeated (Fig. 26). Also, such details as the size of fillets and rounds can be provided for with a general note. Any superfluous or omitted dimensions can be easily discovered by mentally going through the manufacture, or even the drawing, of the part, checking each dimension as it is needed.

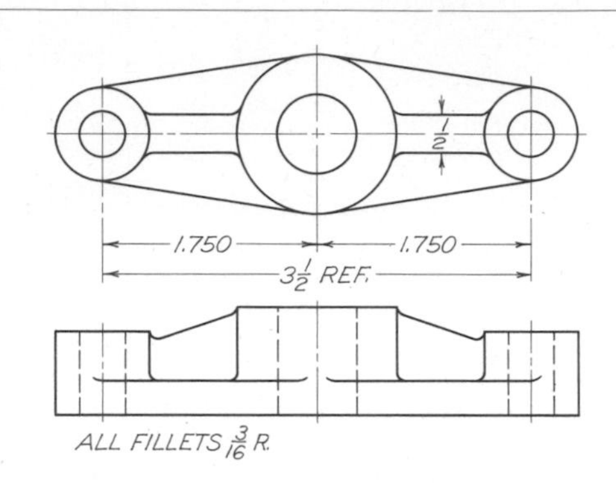


FIG. 26. One reference dimension.

## 25. Reference Dimensions.

Occasionally it is useful, for reference and checking purposes, to give all dimensions in a series as well as the over-all dimension. In such cases one dimension is marked with the abbreviation "REF" as shown in Fig. 26. According to the American Standard definition, a reference dimension is "a dimension without tolerance, used for informational purposes only, and does not govern machining or inspection operations."

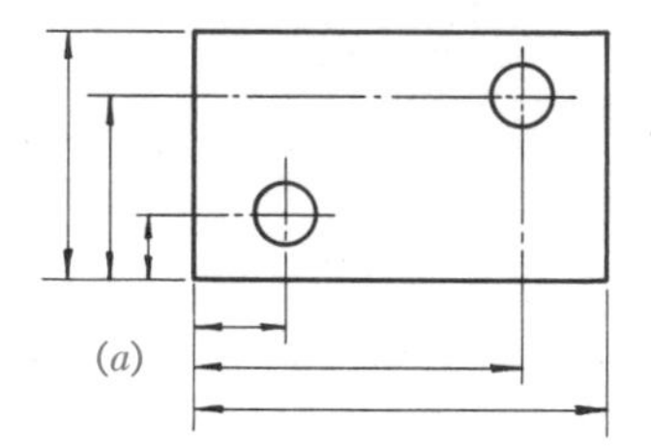


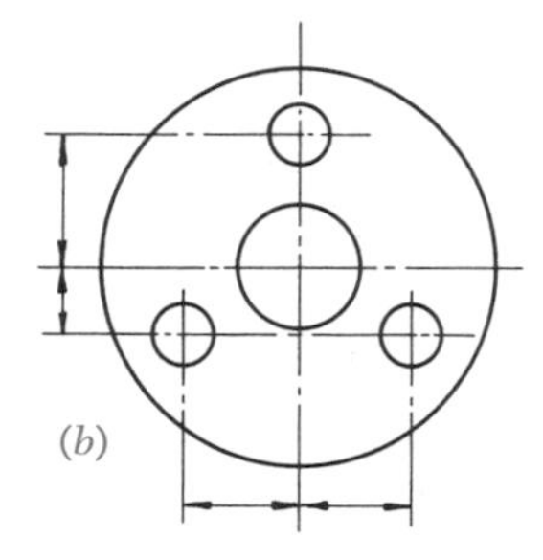


FIG. 27. Dimensions from datum. (*a*) Datum surfaces; (*b*) datum center.

## 26. Dimensions from Datum.

Datum points, lines, and edges of surfaces of a part are features that are assumed to be exact for purposes of computation or reference, and *from* which the position of other features is established. In Fig. 27*a* the left side and bottom surfaces of the part are the datum surfaces, and at (*b*) the center lines of the central hole in the part are datum lines. Where positions are specified by dimensions from a datum, different features are always positioned from this datum and *not* with respect to one another.

## 27. Selection of a Datum.

A feature selected to serve as a datum must be clearly identified and readily recognizable. Note in Fig. 27 that the datum lines and edges are obvious. On an actual part, the datum features must be accessible during manufacture so that there will be no difficulty in making measurements. In addition, *corresponding* features on mating parts must be employed as datum features to assure assembly and proper functioning.

A datum surface (on a physical part) must be more accurate than the allowable variation on any dimension of position which is referred to the datum. Thus, it may be necessary to specify the perfection of datum surfaces for flatness, straightness, roundness, etc. See Secs. 77 and 78.

## 28. Basic Dimensions.

Any dimension on a drawing specified as BASIC is a *theoretical* value used to describe the exact size, shape, or position of a feature. It is used as a reference from which permissible variations are established.

**29. Maximum and Minimum Sizes.**
In some cases a maximum or a minimum size represents a limit beyond which any variation in size cannot be permitted, but in the *opposite* direction there is no difficulty. To illustrate, Fig. 28*a* shows a shaft and hub assembly. If the change in diameter of the shaft (rounded) is too large or the hub edge (chamfer) too small, the hub will not seat against the shaft shoulder; but if the shaft radius is somewhat smaller or the hub edge somewhat larger, no difficulty exists. Therefore, in the dimensioning of both, at (*b*) and (*c*), MAX and MIN are applied to the dimensions. This method is often used in connection with depths of holes, lengths of threads, chamfers, and radii.

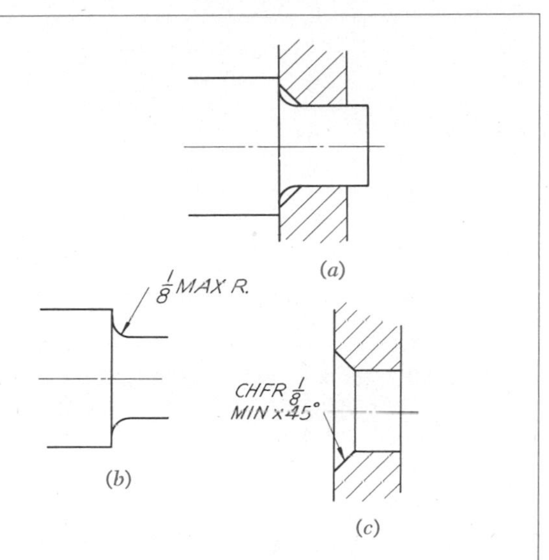


**FIG. 28.** Maximum and minimum sizes. Mating edges at (*a*) are dimensioned as at (*b*) and (*c*) so that there will be no assembly interference.

## PLACEMENT OF DIMENSIONS

**30. Placement of Dimensions.**
After the distances have been selected, it is possible to decide (1) the *view* on which the distance will be indicated, (2) the particular *place* on that view, and (3) the *form* of the dimension itself. Numerous principles, some with the force of a rule, can be given, but in all cases the important consideration is *clarity*.

**31. Views.**
*The Contour Principle.* One of the views of an object will usually describe the shape of some detailed feature better than will the other view or views, and the feature is then said to be "characteristic" in that particular view. In reading a drawing, it is natural to look for the dimensions of a given feature wherever that feature appears most characteristic, and an advantage in clarity and in ease of reading will certainly result if the dimension is placed there. In Fig. 29 the rounded corner, the drilled hole, and the lower notched corner are all characteristic in, and dimensioned on, the front view. The projecting shape on the front of the object is more characteristic in the top view and is dimensioned there.

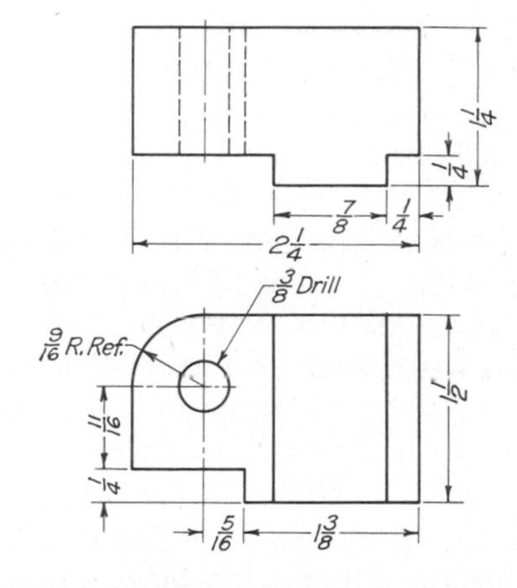


**FIG. 29.** The contour principle. Dimensions are placed where the feature dimensioned is most easily recognized.

*Dimensions for prisms* should be placed so that two of the three dimensions are on the view showing the contour shape and the third on one of the other views (Figs. 18 and 29).

*Dimensions for cylinders,* the diameter and length, are usually best placed on the noncircular view (Fig.

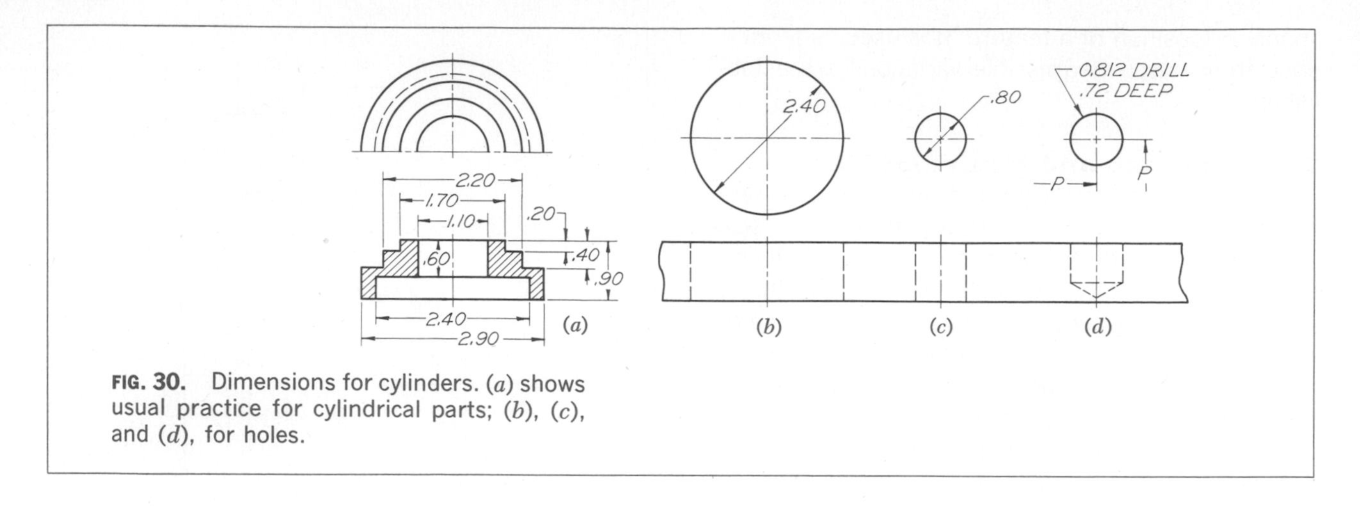


**FIG. 30.** Dimensions for cylinders. (*a*) shows usual practice for cylindrical parts; (*b*), (*c*), and (*d*), for holes.

30*a*). This practice keeps the dimensions on one view, a convenience for the workman. Occasionally a cylindrical hole is dimensioned with the diameter at an angle on the circular view, as indicated at (*b*). This practice should never be used unless there is a clear space for the dimension value. In some cases, however, the value can be carried outside the view, as at (*c*). When a round hole is specified by a note, as at (*d*), the leader should point to the circular view if possible. The note has an advantage in that the diameter, operation, and depth can all be given together. Giving the diameter on the circular view as at (*b*), (*c*), and (*d*) may make for ease of reading, as the dimensions of position will probably be given there also, as indicated at (*d*). When it is not obvious from the drawing, a dimension may be indicated as a diameter by following the value with the letter *D*, as shown in Fig. 31.

*Principles for the Placement of Dimensions*:

1. Dimensions outside the view are preferred, unless added clearness, simplicity, and ease of reading will result from placing some of them inside. For good appearance, dimensions should be kept off the cut surfaces of sections. When it is not possible to do this, the section lining is omitted around the numbers, as shown in Fig. 31 (see Sec. 35).

2. Dimensions between the views are preferred unless there is some reason for placing them elsewhere, as in Fig. 29 where the dimensions for the lower notched corner and the position of the hole must come at the bottom of the front view.

3. Dimensions should be applied to one view only; that is, with dimensions between views, the extension lines should be drawn from one view, not from both views (Fig. 32).

4. Dimensions should be placed on the view that shows the distance in its true length (Fig. 33).

5. Dimension lines should be spaced, in general, $\frac{1}{2}$ in. away from the outlines of the view. This applies to a single dimension or to the first dimension of several in a series.

6. Parallel dimension lines should be spaced uniformly with at least $\frac{3}{8}$ in. between lines.

7. Values should be midway between the arrowheads, except when a center line interferes (Fig. 34) or when the values of several parallel dimensions are staggered (Fig. 35).

8. Continuous or staggered dimension lines may be used, depending upon convenience and readability. Continuous dimension lines are preferred where possible (Figs. 36 and 37).

9. Always place a longer dimension line outside a shorter one to avoid crossing dimension lines with the extension lines of other dimensions. Thus an over-all dimension (maximum size of piece in a given direction) will be outside all other dimensions.

10. Dimensions should never be crowded. If the space is small, follow one of the methods given in Sec. 32.

11. Center lines are used to indicate the sym-

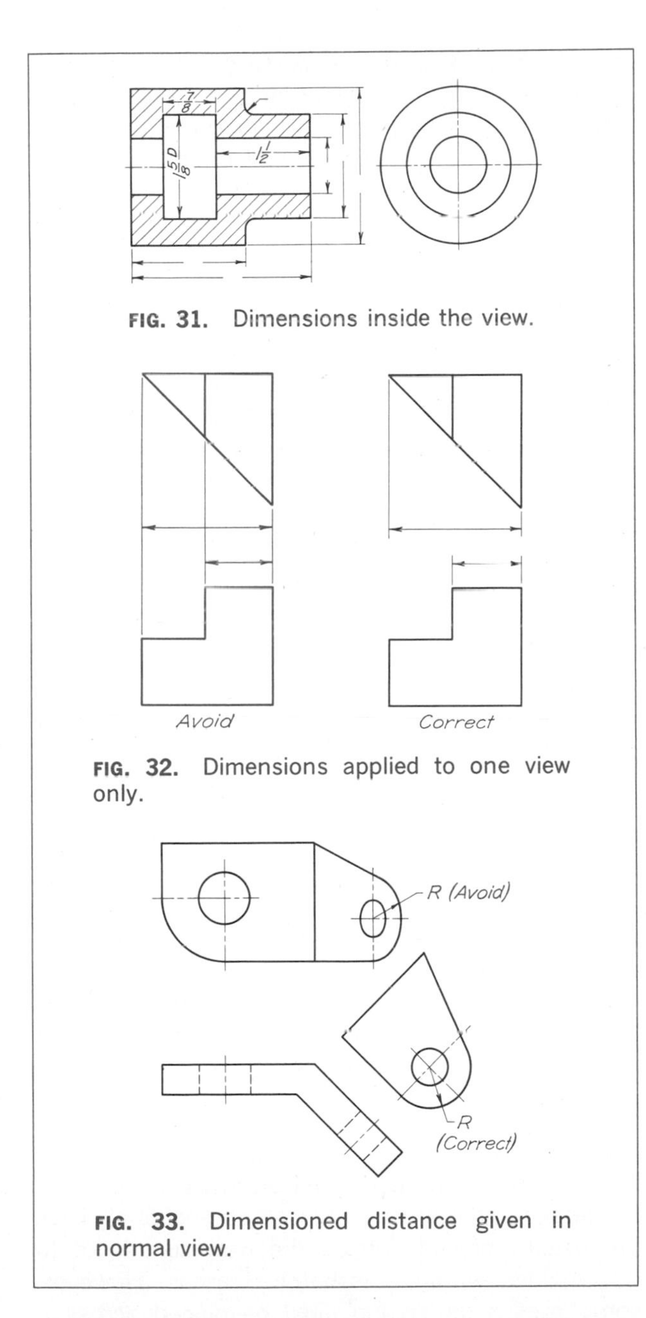


**FIG. 31.** Dimensions inside the view.

**FIG. 32.** Dimensions applied to one view only.

**FIG. 33.** Dimensioned distance given in normal view.

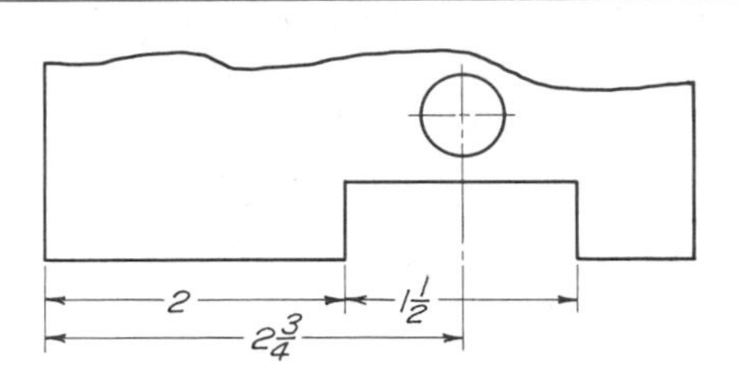


**FIG. 34.** Values midway between arrowheads.

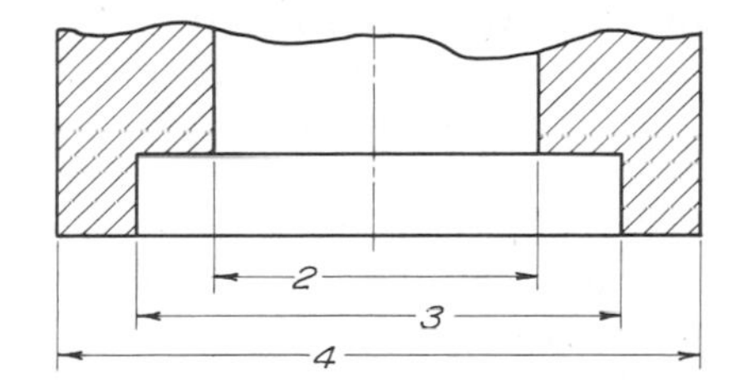


**FIG. 35.** Values staggered for clarity.

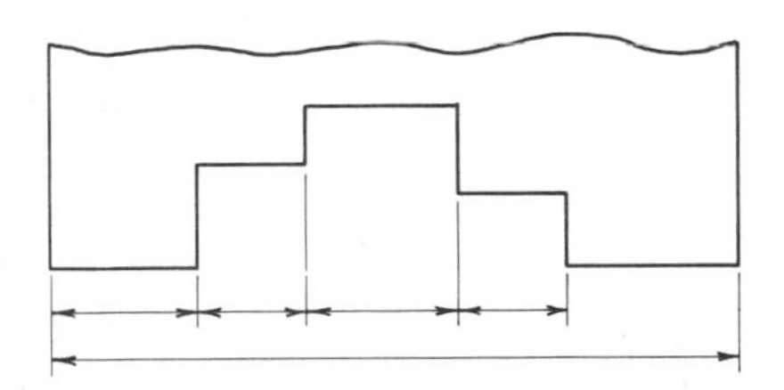

**FIG. 36.** Dimensions arranged in continuous form.

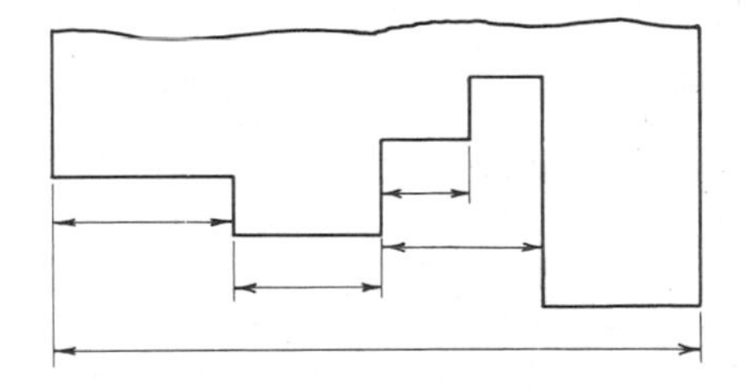

**FIG. 37.** Dimensions staggered.

metry of shapes, and frequently eliminate the need for a positioning dimension. They should be considered as part of the dimensioning and drawn in finished form at the time of dimensioning. They should extend about $\frac{1}{8}$ in. beyond the shape for which they indicate symmetry unless they are carried further to serve as extension lines. Center lines should not be continued between views.

12. All notes must read horizontally (from the bottom of the drawing).

*Cautions:*

1. Never use a center line, a line of a view, or an

extension line as a dimension line.

2. Never place a dimension line on a center line or place a dimension line where a center line should properly be.

3. Never allow a line of any kind to pass through a dimension figure.

4. Never allow the crossing of two dimension lines or an extension line and a dimension line.

5. Avoid dimensioning to dashed lines if possible.

### 32. Dimensioning in Limited Space.

Dimensions should never be crowded into a space too small to contain them. One of the methods in Fig. 38 can be used where space is limited. Sometimes a note is appropriate. If the space is small and crowded, an enlarged removed section or part view can be used (Fig. 39).

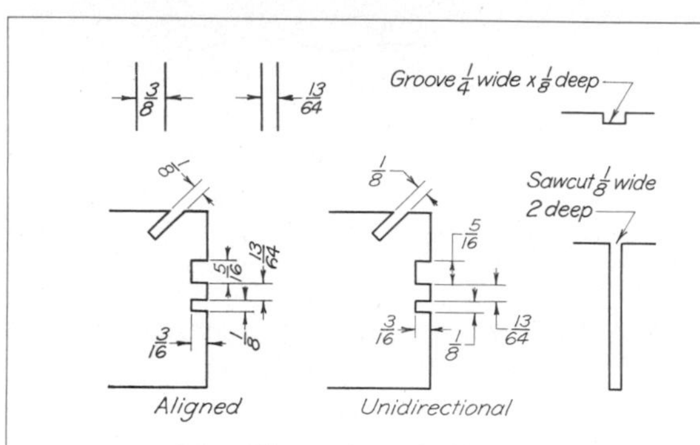


**FIG. 38.** Dimensions in limited space.

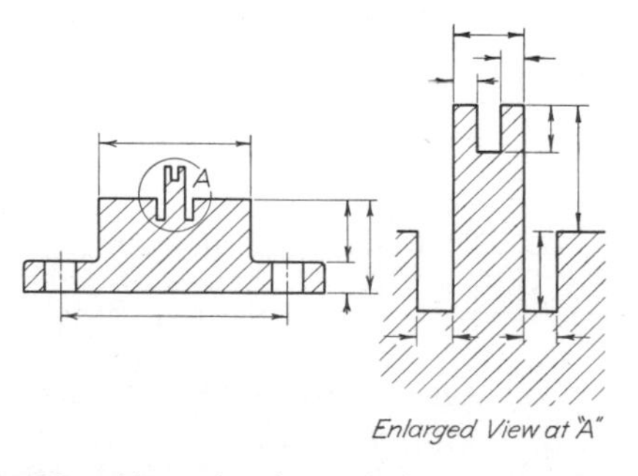


**FIG. 39.** Use of enlarged view to clarify dimensions.

### 33. Order of Dimensioning.

A systematic order of working is a great help in placing dimensions. Figure 40 illustrates the procedure. First complete the shape description (*a*). Then place the extension lines and extend the center lines where necessary (*b*), thus planning for the location of both size and position dimensions; study the placement of each dimension and make alterations if desirable or necessary. Add the dimension lines (*c*). Draw arrowheads and leaders for notes (*d*). Then add values and letter notes (*e* and *f*).

It is desirable to add the notes *after* the dimensions have been placed. If the notes are placed first, they may occupy a space needed for a dimension. Because of the freedom allowed in the use of leaders, notes may be given in almost any available space.

### 34. Dimensioning of Auxiliary Views.

In placing dimensions on an auxiliary view, the same principles of dimensioning apply as for any other drawing, but special attention is paid to the contour principle given in Sec. 31. An auxiliary view is made for the purpose of showing the normal view (size) of some inclined or skew face, and for this reason the dimensioning of the face should be placed where it is easiest to read, *which will be on the normal view*. Note in Fig. 41 that the spacing and size of holes as well as the size of the inclined face are dimensioned on the auxiliary view. Note further that the angle and dimension of position tying the inclined face to the rest of the object could not be placed on the auxiliary view.

### 35. Dimensioning of Sectional Views.

Dimensions that must be placed on sectional views are usually placed outside the view so as not to be crowded within crosshatched areas. However, sometimes a dimension *must* be placed across a crosshatched area. When this is the case, the crosshatching is left out around the dimension figures, as illustrated in Fig. 31. Examples showing dimensioning practice on sectional views are given in Probs. 36, 37, and 49 of the final chapter.

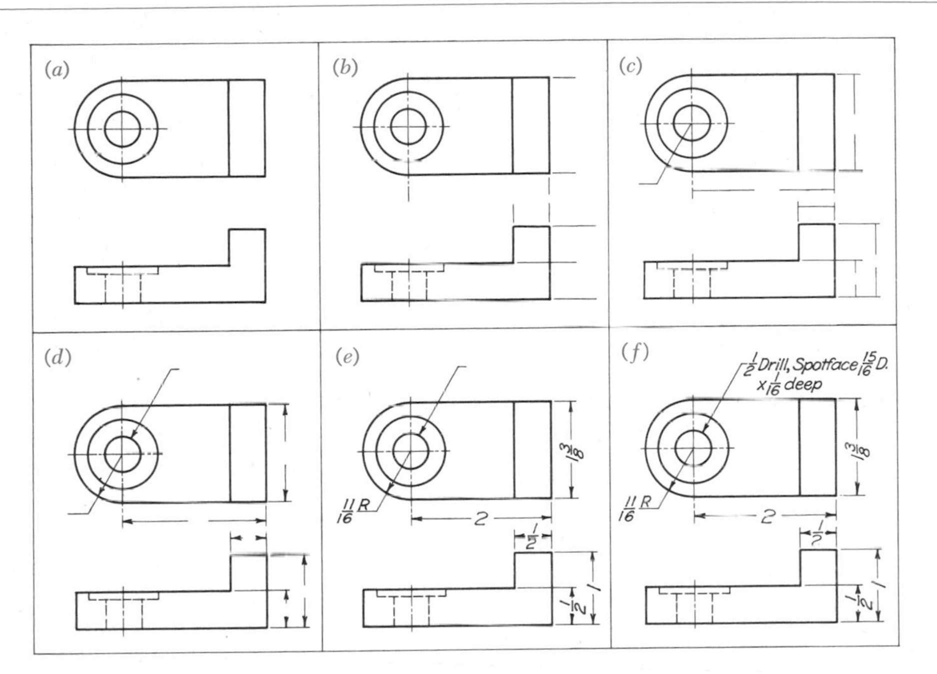


FIG. 40. Order of dimensioning. See text for details.

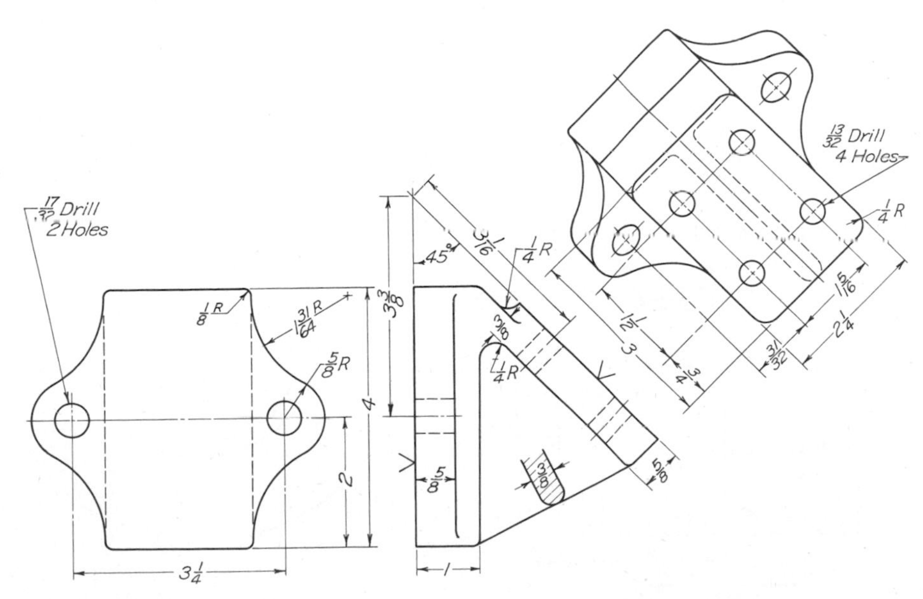


FIG. 41. Dimensioning on an auxiliary (normal) view.

**36. Dimensioning a Half Section.** In general, the half section is difficult to dimension clearly without some possibility of crowding and giving misleading or ambiguous information. Generous use of notes and careful placement of dimension lines, leaders, and figures will in most cases make the dimensioning clear; but if a half section cannot be clearly dimensioned, an extra view or part view should be added on which to describe the size.

Inside diameters should be followed by the letter *D* and the dimension line carried over the center line, as in Fig. 42, to prevent the possibility of reading the dimension as a radius. Sometimes the view and the dimensioning can both be clarified by showing the dashed lines on the unsectioned side. Dimensions of internal parts, if placed inside the view, will prevent confusion between extension lines and the outline of external portions.

**37. Dimensioning of Pictorial Drawings.** Pictorial drawings are often more difficult to dimension than orthographic drawings because there is one view instead of several, and the dimensioning may become crowded unless the placement is carefully planned. In general, the principles of dimensioning for orthographic drawings should be followed whenever possible. The following rules should be observed:

1. Dimension and extension lines should be placed so as to lie *in* or *perpendicular* to the face on which the dimension applies. See Probs. 4 and 5, Chap. 23.

2. Dimension numerals should be placed so as to lie in the plane in which the dimension and extension lines lie. See Probs. 8 and 14, final chapter.

3. Leaders for notes and the lettered note should be placed so as to lie in a plane parallel or perpendicular to the face on which the note applies. See Probs. 18 and 59 in the final chapter.

4. Finish may be indicated by the standard finish symbol (**V**). The symbol is applied *perpendicular to* the face with its point touching a short line lying in the surface. The symbol and line should be parallel to one of the principal axes. If the symbol cannot be applied in the above manner, the finish symbol may be attached to a leader pointing to the face. See Probs. 5 and 18 in the final chapter.

5. Lettering of dimension values and of notes

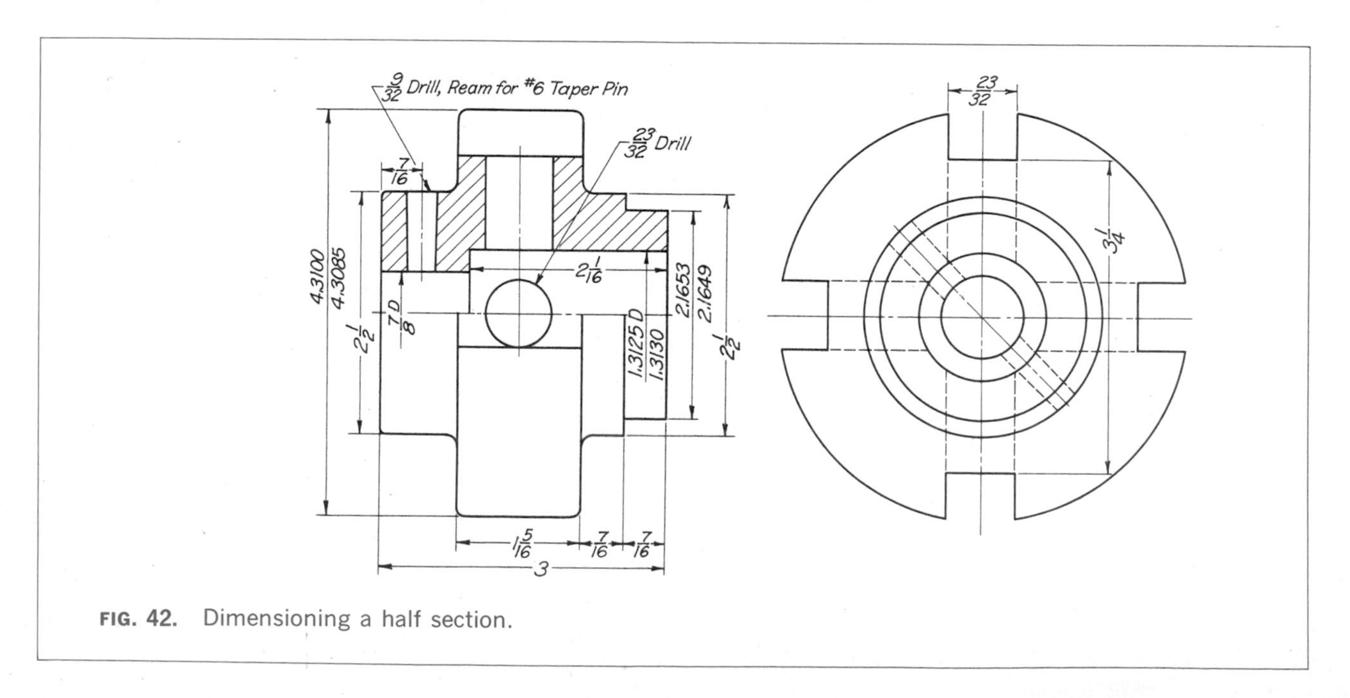


**FIG. 42.** Dimensioning a half section.

should be made so that the lettering appears to lie in or parallel to one of the principal faces of the pictorial drawing. To do this, the lettering must be the *pictorial representation of vertical figures*. Note the placement and lettering of the dimension values and the notes in Probs. 54 and 56, Chap. 23.

The American Standard permits lettering of pictorial drawings according to the unidirectional system, with either vertical or slant lettering. This is principally to make possible the use of mechanical lettering devices. If the unidirectional system is employed,

1. Dimension values are lettered to read from the bottom of the sheet.
2. Notes are lettered so that they lie *in the picture plane*, to read from the bottom of the sheet. Notes should be kept off the view, if possible.

## DIMENSIONING STANDARD FEATURES

### 38. Notes.

These are word statements giving information that cannot be given by the views and dimensions. They almost always specify some standard shape, operation, or material, and are classified as *general* or *specific*. A general note applies to the entire part, and a specific note applies to an individual feature. Occasionally a note will save making an additional view or even an entire drawing, for example, by indicating right- and left-hand parts.

Do not be afraid to put notes on drawings. Supplement the graphic language with the English language whenever added information can be conveyed by so doing, but be careful to word the note so clearly that the meaning cannot possibly be misunderstood.

*General notes* do not require the use of a leader and should be grouped together above the title block. Examples are "Finish all over," "Fillets $\frac{1}{4}$R, rounds $\frac{1}{8}$R, unless otherwise specified," "All draft angles 7°," "Remove burs."

Much of the information provided in the title strip of a machine drawing is a grouping of general notes. Stock size, material, heat-treatment, etc., are general notes in the title of the drawing of Fig. 106.

*Specific notes* almost always require a leader and should therefore be placed fairly close to the feature to which they apply. Most common are notes giving an operation with a size, as "$\frac{1}{2}$ Drill, 4 holes."

Recommended wordings for notes occurring more or less frequently are given in Fig. 43.

When lower-case lettering is used, capitalization of words in notes depends largely on company policy. One common practice is to capitalize all important words. However, for long notes, as on civil engineering or architectural drawings, the grammatical rules for capitalization usually prevail.

### 39. Angles.

The dimension line for an angle is a circle arc with its center at the intersection of the sides of the angle (Fig. 44). The value is placed to read horizontally, with the exception that in the aligned system large arcs have the value aligned with the dimension arc. Angular values should be written in the form 35°7′ with no dash between the degrees and the minutes.

### 40. Chamfers.

Chamfers may be dimensioned by note, as in Fig. 45*a*, if the angle is 45°. The linear size is understood to be a short side of the chamfer triangle; the dimensioning without a note shown at (*b*) is in conformity with this. If the chamfer angle is other than 45°, it is dimensioned as at (*c*).

### 41. Tapers.

The term "taper" as used in machine work usually means the surface of a cone frustum. The dimensioning will depend on the method of manufacture and the accuracy required. If a standardized taper (see Appendix) is used, the specification should be accompanied by one diameter and the length as shown at (*a*) in Fig. 46. The general method of giving the diameters of both ends and the taper per foot is illustrated at (*b*). An alternate method is to give one diameter, the length, and the taper per foot. Taper per foot is defined as the difference in diameter in inches for 1 ft of length. The method of dimensioning for precision work, where a close fit between the parts as well as a control of entry dis-

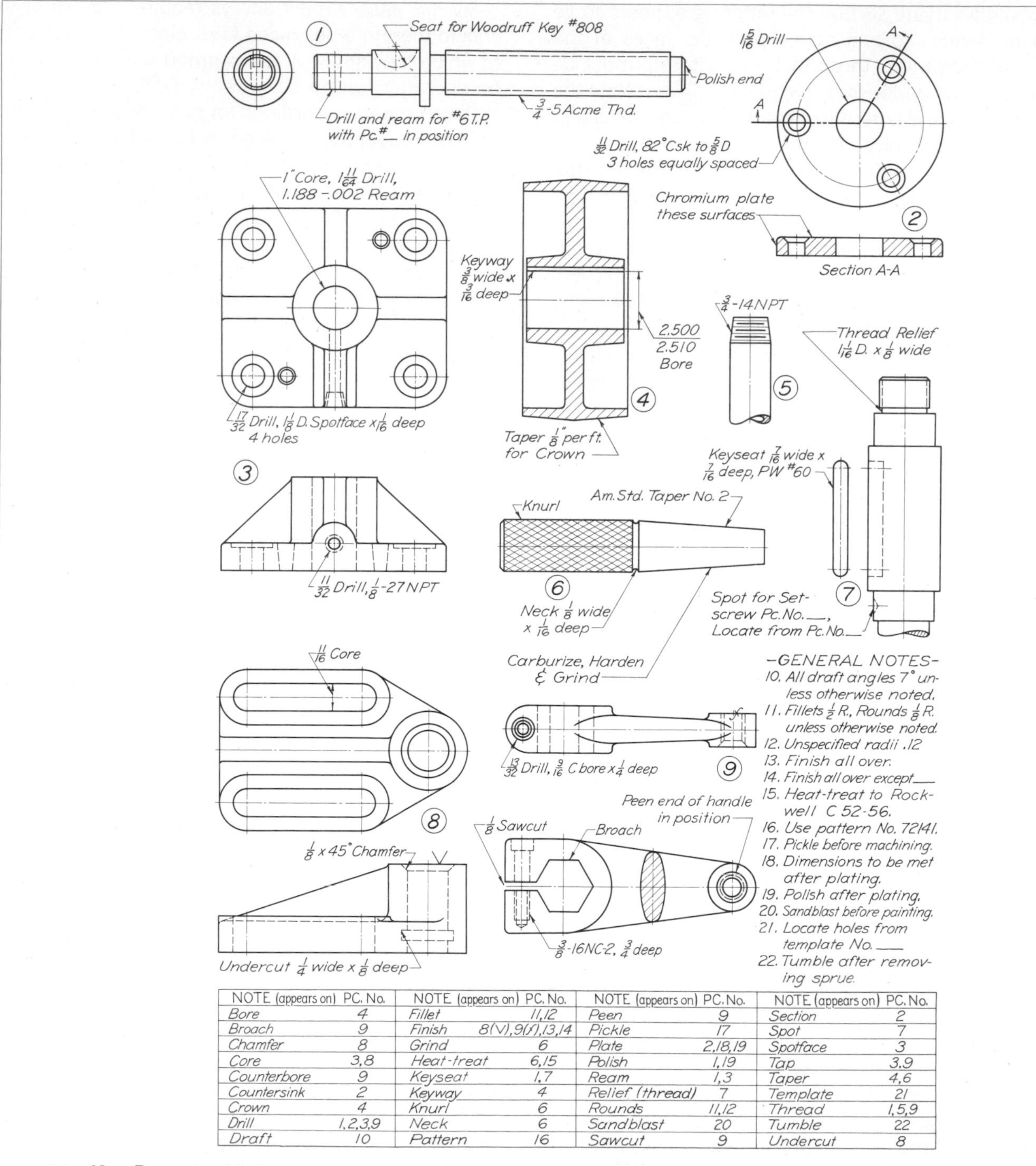


| NOTE (appears on) | PC. No. | NOTE (appears on) | PC. No. | NOTE (appears on) | PC. No. | NOTE (appears on) | PC. No. |
|---|---|---|---|---|---|---|---|
| Bore | 4 | Fillet | 11,12 | Peen | 9 | Section | 2 |
| Broach | 9 | Finish | 8(V),9(f),13,14 | Pickle | 17 | Spot | 7 |
| Chamfer | 8 | Grind | 6 | Plate | 2,18,19 | Spotface | 3 |
| Core | 3,8 | Heat-treat | 6,15 | Polish | 1,19 | Tap | 3,9 |
| Counterbore | 9 | Keyseat | 1,7 | Ream | 1,3 | Taper | 4,6 |
| Countersink | 2 | Keyway | 4 | Relief (thread) | 7 | Template | 21 |
| Crown | 4 | Knurl | 6 | Rounds | 11,12 | Thread | 1,5,9 |
| Drill | 1,2,3,9 | Neck | 6 | Sandblast | 20 | Tumble | 22 |
| Draft | 10 | Pattern | 16 | Sawcut | 9 | Undercut | 8 |

**FIG. 43.** Recommended wording of notes.

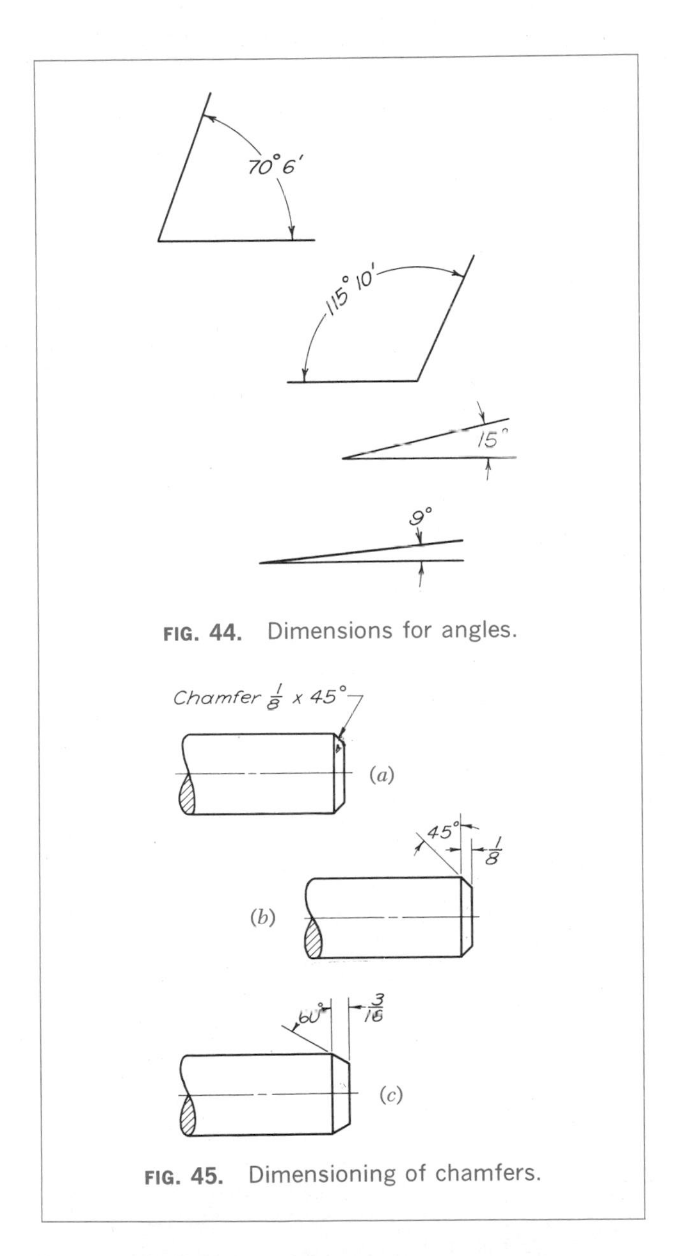


FIG. 44. Dimensions for angles.

FIG. 45. Dimensioning of chamfers.

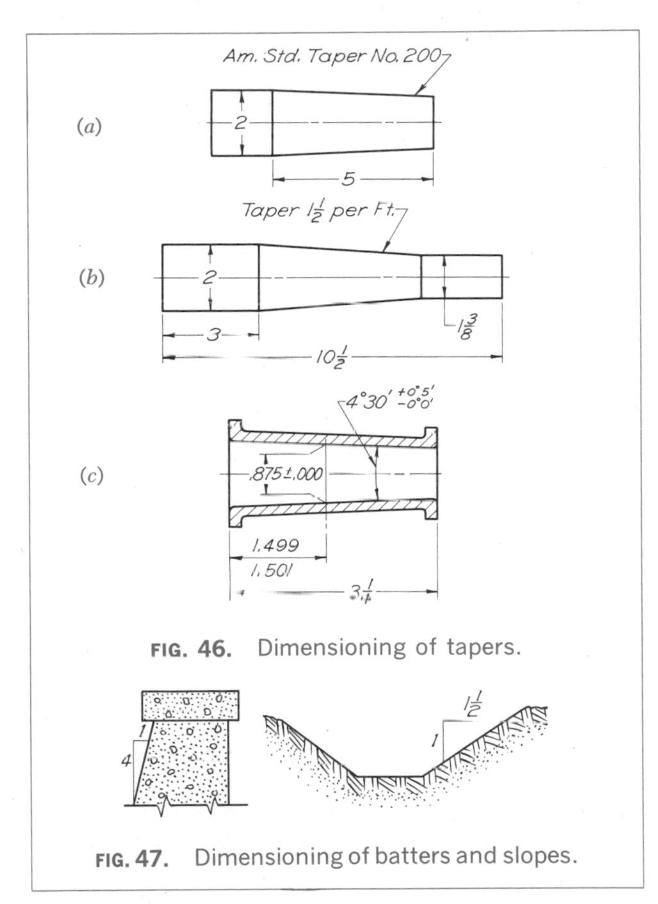


FIG. 46. Dimensioning of tapers.

FIG. 47. Dimensioning of batters and slopes.

tance is required, is shown at (*c*). Because of the inaccuracy resulting from measuring at one of the ends, a gage line is established where the diameter is to be measured. The entry distance is controlled through the allowable variation in the location of the gage line, and the fit of the taper is controlled by the accuracy called for in the specification of the angle.

## 42. Batters, Slopes, and Grade.

*Batter* is a deviation from the vertical, such as is found on the sides of retaining walls, piers, etc., and *slope* is a deviation from the horizontal. Both are expressed as a ratio with one factor equal to unity, as illustrated in Fig. 47. *Grade* is identical with slope but is expressed in percentage, the inclination in feet per hundred feet. In structural work angular measurements are shown by giving the ratio of run to rise with the larger side 12 in.

## 43. Arcs.

Arcs should be dimensioned by giving the radius on the view that shows the true shape of the curve. The dimension line for a radius should always be drawn as a radial line at an angle (Fig. 48), never horizontal or vertical; and only one arrowhead is used. There is no arrowhead at the arc center. The numerical

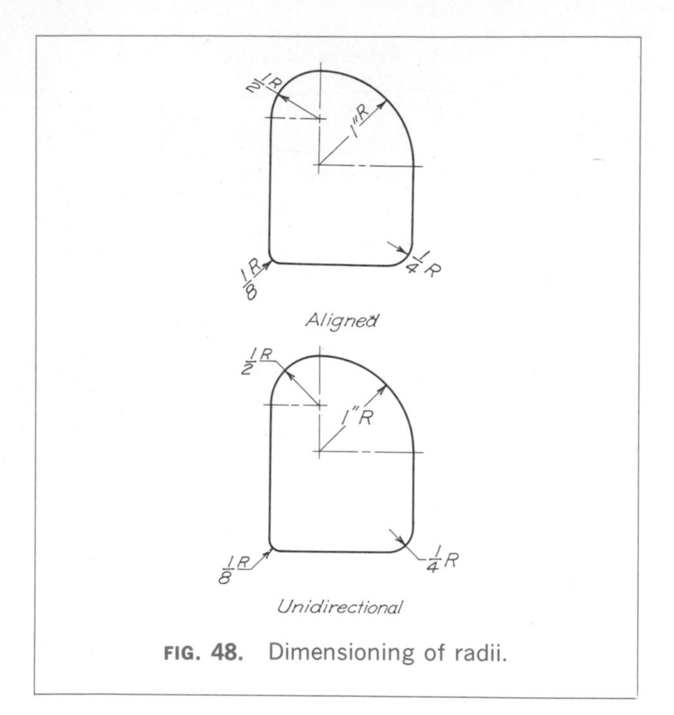


FIG. 48. Dimensioning of radii.

value should be followed by the letter $R$. Depending upon the size of the radius and the available space for the value, the dimension line and value are both inside the arc, or the line is inside and the value outside, or, for small arcs, both are outside, as shown in the illustration.

When the center of an arc lies outside the limits of the drawing, the center is moved closer along a center line of the arc and the dimension line is jogged to meet the new center (Fig. 49). The portion of the dimension line adjacent to the arc is a radial line of the true center.

A curved line made up of circle arcs is dimensioned by radii with the centers located, as in Fig. 50.

## 44. Curves.

Curves for which great accuracy is not required are dimensioned by offsets, as in Fig. 51. For greater accuracy, dimensions from datum features, as in Fig. 52, are recommended. Note in Fig. 52 that *any* pair of dimensions (indicated at $x$) could be given to greater accuracy than the others, in order to position a point for which greater accuracy is required.

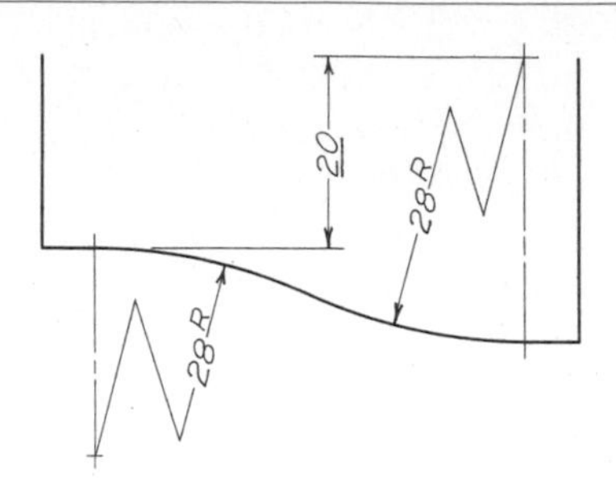


FIG. 49. Dimensioning of radii having inaccessible centers. Note that the positioning dimension is marked out of scale.

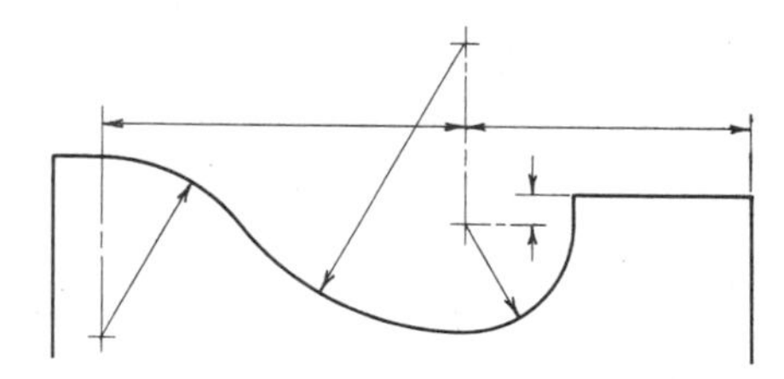

FIG. 50. Dimensioning a curve made up of radii. The position of centers must be given.

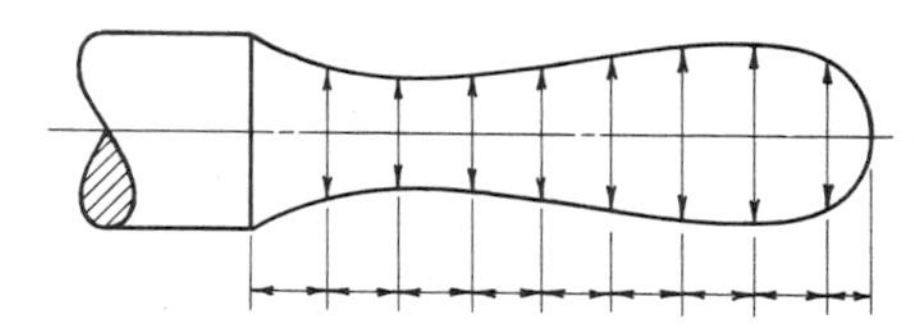

FIG. 51. A curve dimensioned by offsets.

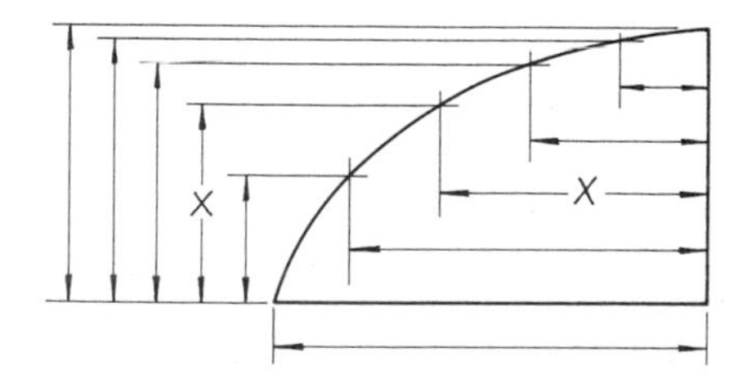


FIG. 52. A curve dimensioned from datum edges.

## 45. Shapes with Rounded Ends.

These should be dimensioned according to their method of manufacture. Figure 53 shows several similar contours and the typical dimensioning for each. The link (*a*), to be cut from thin material, has the radius of the ends and the center distance given as it would be laid out. At (*b*) is shown a cast pad dimensioned as at (*a*), with the dimensions most usable for the patternmaker. The drawing at (*c*) shows a slot machined from solid stock with an end-milling cutter. The dimensions give the diameter of the cutter and the travel of the milling-machine table. The slot at (*d*) is similar to that at (*c*) but it is dimensioned for quantity production, where, instead of the table travel, the over-all length is wanted for gaging purposes. Pratt and Whitney keys and key seats are dimensioned by the method shown at (*d*).

## 46. Dimensions and Specifications for Holes.

Drilled, reamed, bored, punched, or cored holes are usually specified by note giving the diameter, operation, and depth if required. If there is more than one hole of the same kind, the leader needs to point to but one hole, and the number of holes is stated in the note (Fig. 54). Several operations involving one hole may be grouped in a common note. Figures 54 to 57 show typical dimensioning practice for drilled and reamed, counterbored, countersunk, and spot-faced holes.

The ASA specifies that standard drill sizes be given as decimal fractions, such as 0.250, 0.375, 0.750, and 1.500. If the size is given as a common fraction, the decimal equivalent should be added.

The leader to a hole should point to the *circular* view if possible. The pointing direction is toward the center (Fig. 9*b*). With concentric circles, the arrowhead should touch the inner circle (usually the first operation) unless an outer circle would pass through the arrowhead. In such a case, the arrow should be drawn to touch the outer circle.

Holes made up of several diameters and involving several stages of manufacture may be dimensioned

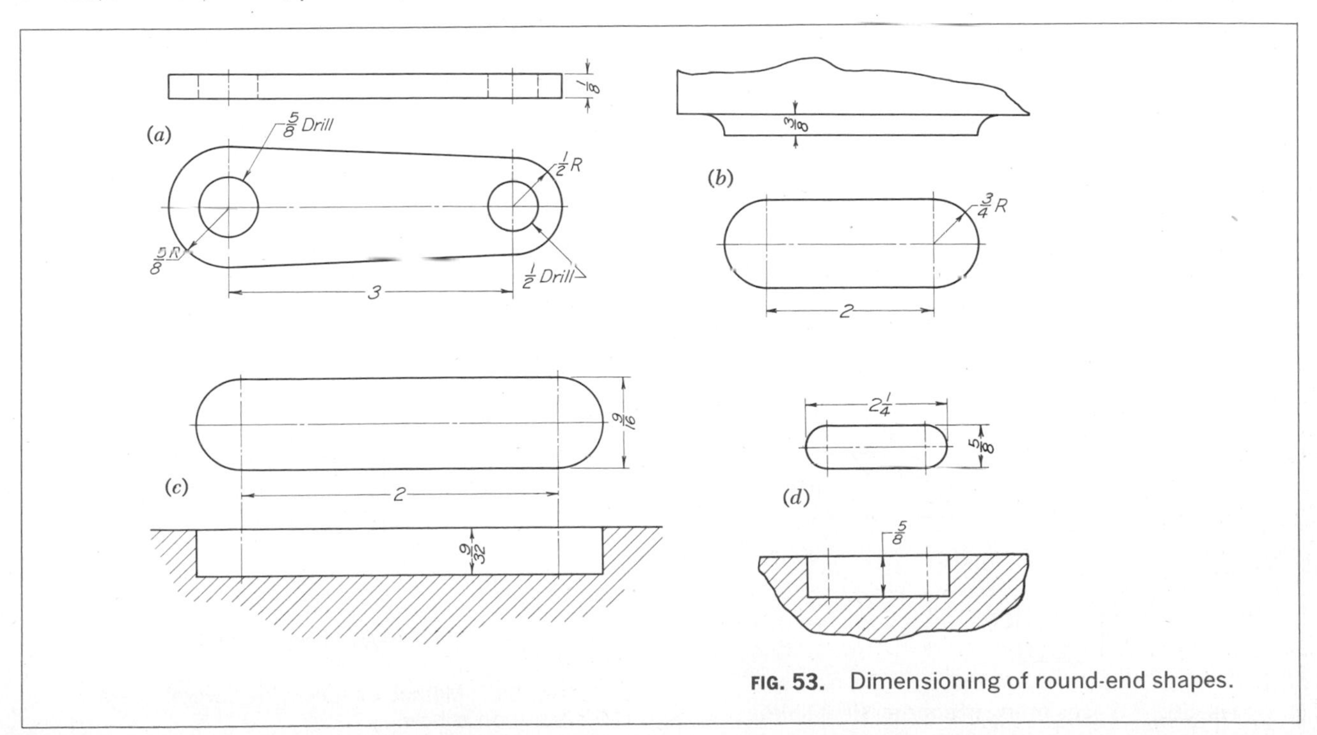


**FIG. 53.** Dimensioning of round-end shapes.

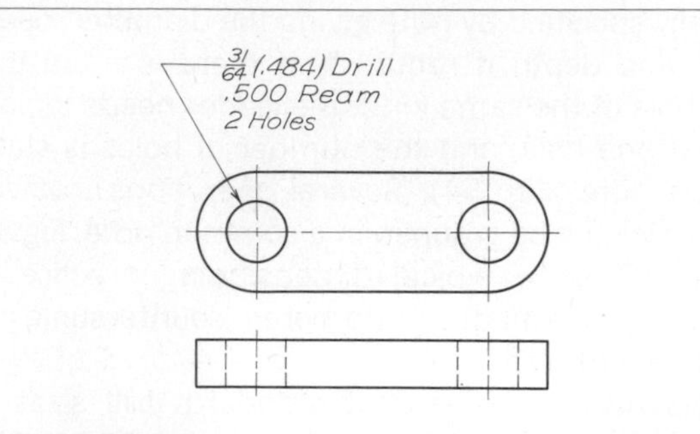


**FIG. 54.** Dimensioning of drilled and reamed holes.

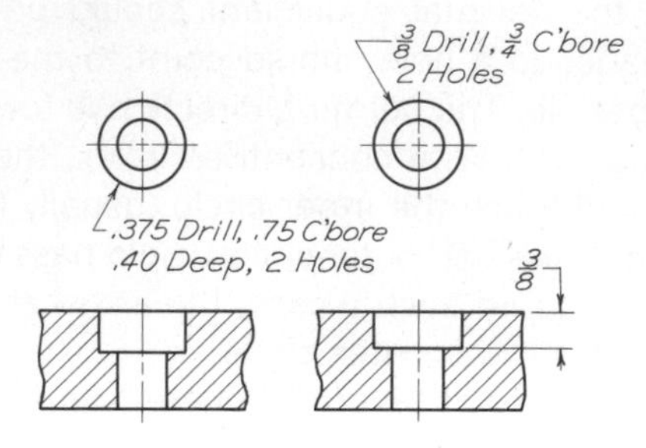


**FIG. 55.** Dimensioning of counterbored holes.

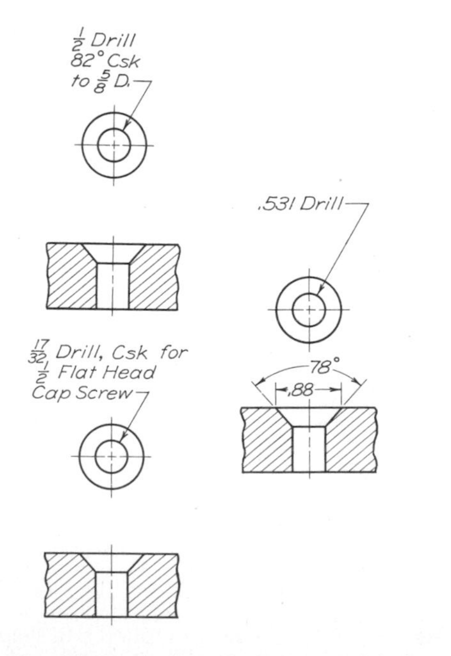


**FIG. 56.** Dimensioning of countersunk holes.

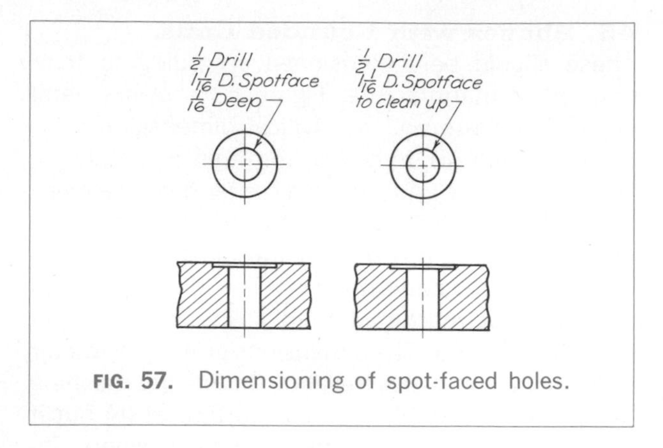


**FIG. 57.** Dimensioning of spot-faced holes.

as shown in Fig. 58. This method of dimensioning combines notes with the regular dimensions.

## 47. Positioning of Holes.

Mating parts held together by bolts, screws, rivets, etc., must have holes for fastenings positioned from common datum surfaces or lines in order to assure matching of the holes. When two or more holes are on an established center line, the holes will require a dimension of position in one direction only (Fig. 59). If the holes are not on a common center line, they will require positioning in two directions, as in Fig. 60. The method at (*b*) is preferred when it is

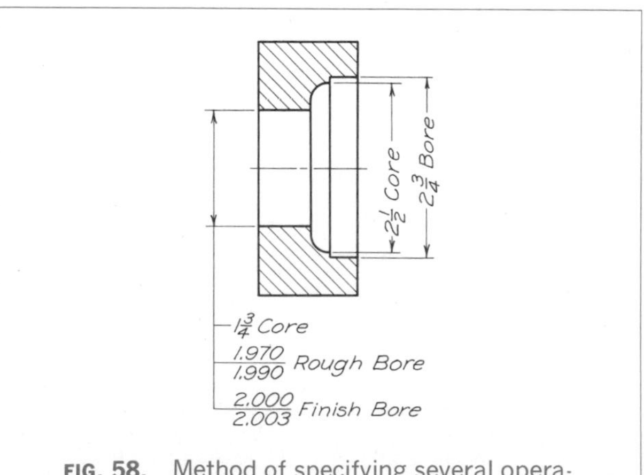


**FIG. 58.** Method of specifying several operations on a hole.

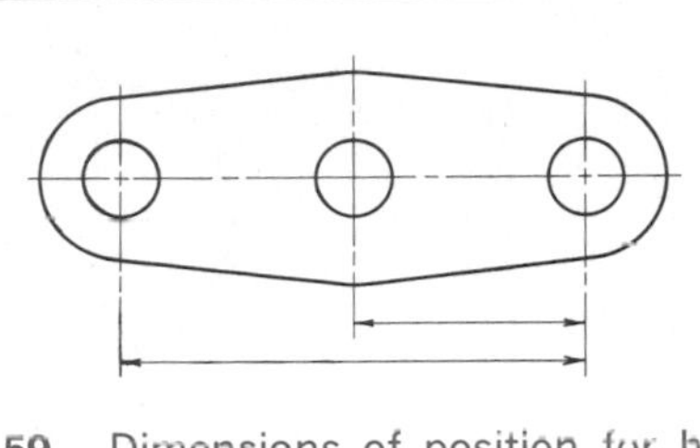

FIG. 59. Dimensions of position for holes.

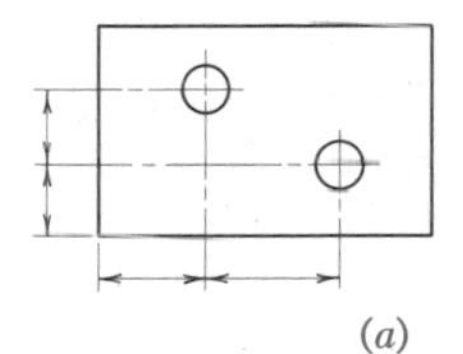


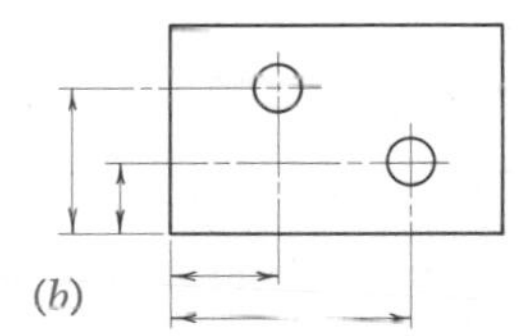


FIG. 60. Dimensions of position for holes.

important to have the positions of both holes established from datum features (left and lower sides of the part).

The coordinate method for the positioning of holes (Fig. 61*a*) is preferred in precision work. The hole circle is often drawn and its diameter given for reference purposes, as in the figure. The diameter of a hole circle is invariably given on the circular view. The datum lines in this case are the center lines of the part.

*Hole circles* are circular center lines, often called "bolt circles," on which the centers of a number of holes are located.

One practice is to give the diameter of the hole circle and a note specifying the size of the holes, the number required, and the spacing, as in Fig. 61*b*. If one or more holes are not in the regular equally spaced position, their location may be given by an offset dimension, as shown at (*c*).

Figure 62 shows holes located by polar coordinates. This method should be used only when modern, accurate shop equipment is available for locating holes by angle.

## 48. Cylindrical Surfaces.

Cylindrical surfaces on a drawing having an end view may be dimensioned as in Fig. 63*a*; if there is no end view, "DIA" should be placed after each value as in Fig. 63*b*.

## 49. Spherical Surfaces.

Spherical surfaces should be dimensioned as in Fig. 64, that is, by placing the abbreviation "SPHER" after the dimension value.

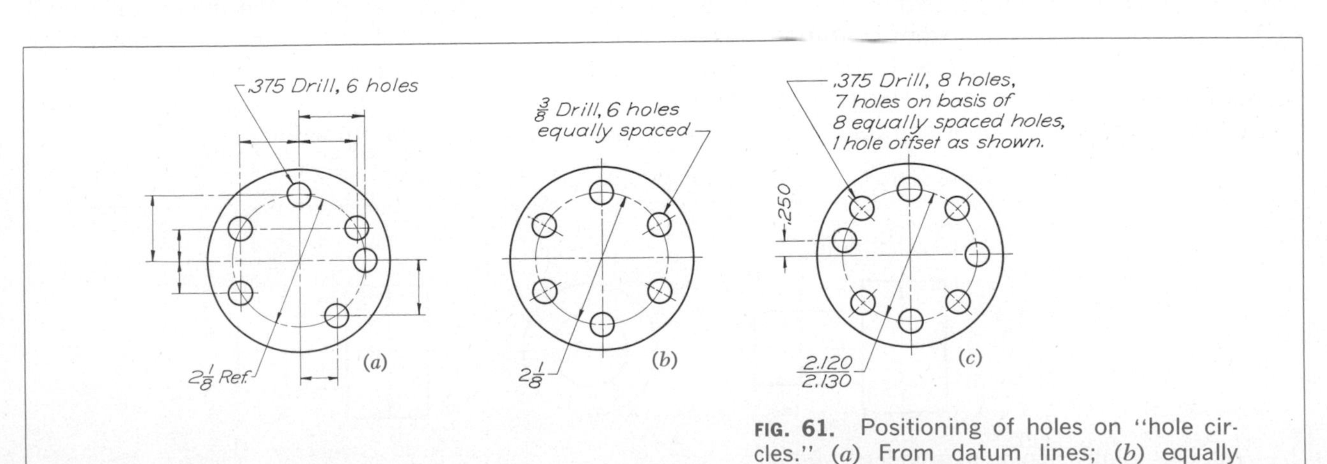


FIG. 61. Positioning of holes on "hole circles." (*a*) From datum lines; (*b*) equally spaced; (*c*) one hole offset.

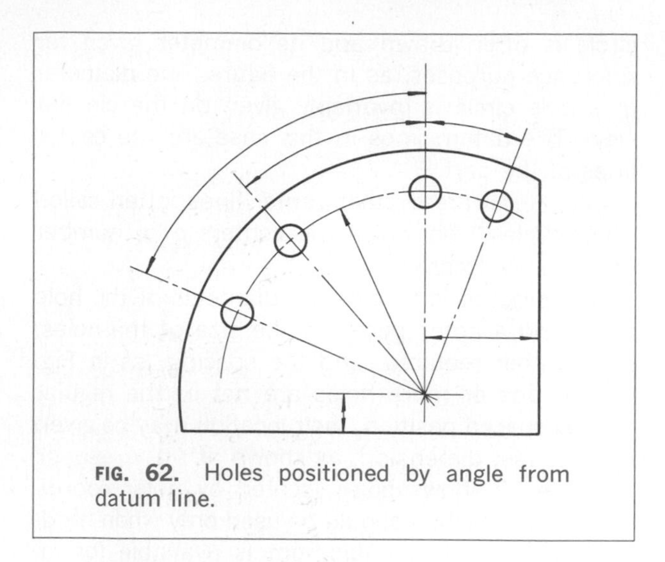

**FIG. 62.** Holes positioned by angle from datum line.

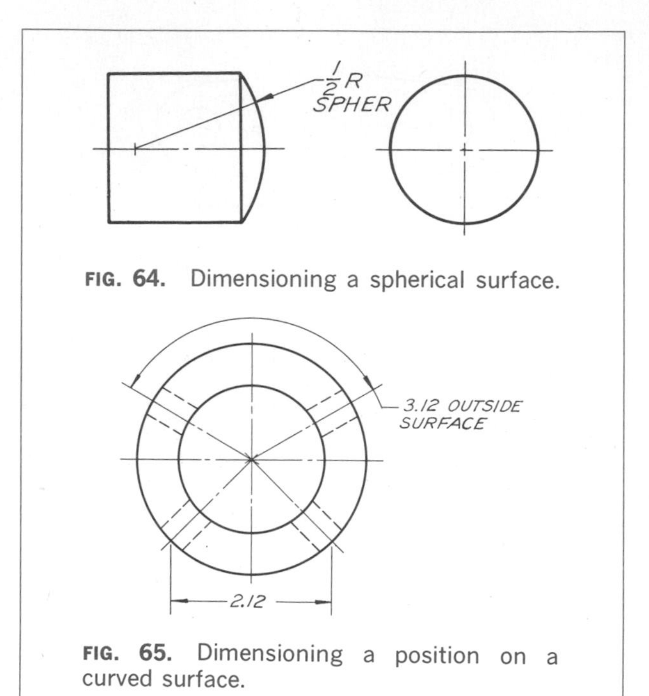


**FIG. 64.** Dimensioning a spherical surface.



**FIG. 65.** Dimensioning a position on a curved surface.

## 50. Curved Surfaces.

A position on a curved part may be misconstrued unless there is a specification showing the *surface* of the part to which the dimension applies, as in Fig. 65.

## 51. Threads, Fasteners, Keyways, and Keyseats.

The dimensioning of these will be given in a later section.

## 52. The Metric System.

A knowledge of the metric system is advantageous, as it is used on all drawings from countries where this system is standard and with increasing frequency on drawings made in the United States. The first instance of international standardization of a mechanical device is that of ball bearings, which have been standardized in the metric system.

Scale drawings in the metric system are not made to English or American scales, but are based on divisions of 10 as full size and then 1 to 2, 1 to $2\frac{1}{2}$, 1 to 5, 1 to 10, 1 to 20, 1 to 50, and 1 to 100. The unit of measurement is the millimeter (mm), and the figures are all understood to be millimeters, without any indicating marks. Figure 66 is an example of metric dimensioning. A table of metric equivalents is given in the Appendix.

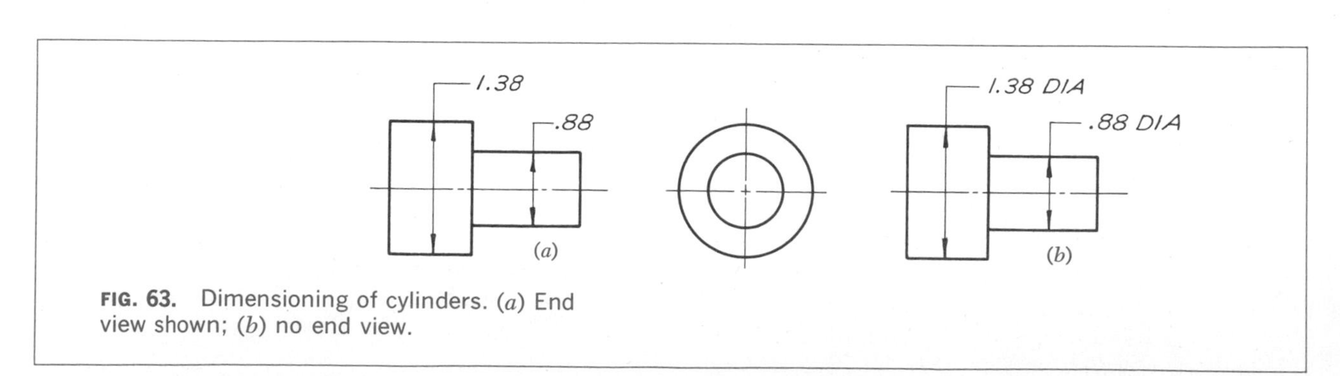


**FIG. 63.** Dimensioning of cylinders. (*a*) End view shown; (*b*) no end view.

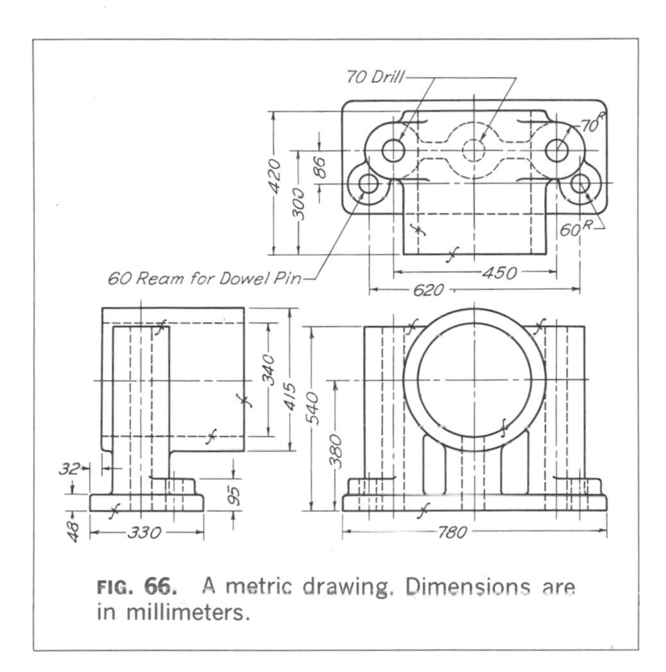


FIG. 66. A metric drawing. Dimensions are in millimeters.

**53. Standard Sizes, Parts, and Tools.** In dimensioning any machine part, it is often necessary to specify some standard thickness or diameter or the size produced by some standard tool. The American Standard, prevailing company standard, or manufacturer's standard should be consulted in order to assure giving correct information.

*Wire and sheet-metal gages* are given by number and are followed by the equivalent thickness or diameter in decimal form.

*Bolts and screws* are supplied in fractional and numbered sizes.

*Keys* are available in manufacturer's numbered sizes or, for square and flat keys, in fractional sizes.

*Rivets*, depending upon the variety, are supplied in fractional or numbered sizes.

*Drills* are available in numbered, lettered, fractional, and metric sizes.

*Reamers, milling cutters*, and other standard tools are available in a variety of standard sizes.

The Appendix gives tables of standard wire and metal gages, bolt and screw sizes, key sizes, etc. ASA or manufacturer's standards will give further information required.

# PRECISION AND TOLERANCE

**54. Precision and Tolerance.** In the manufacture of any machine or structure, quality is a primary consideration. The manufacturing care put into the product determines its quality in relation to competitive products on the market and, in part, its accompanying relative cost and selling price.

*Precision* is the degree of accuracy necessary to ensure the functioning of a part as intended. As an example, a cast part usually has two types of surfaces: mating surfaces and nonmating surfaces. The mating surfaces are machined to the proper smoothness and to be at the correct distance from each other. The nonmating surfaces, exposed to the air and with no important relationship to other parts or surfaces, are left in their original rough-cast form. Thus mating surfaces ordinarily require much greater manufacturing precision than nonmating surfaces. The dimensions on a drawing must indicate which surfaces are to be finished and the degree of precision required in finishing. However, because it is impossible to produce any distance to an absolute size, some variation must be allowed in manufacture.

*Tolerance* is the allowable variation for any given size and provides a practical means of achieving the precision required. The tolerance on any given dimension varies according to the degree of precision necessary for the particular surface. For nonmating surfaces, the tolerance may vary from 0.01 in. for small parts to as much as 1 in. on very large parts. For mating surfaces, tolerances as small as a few millionths of an inch are sometimes necessary (for extremely close-fitting surfaces), but usually surfaces are finished to an accuracy of 0.001 to 0.010, depending upon the function of the part. Figure 67 shows variously toleranced dimensions on a machine drawing. Methods of expressing tolerances are given in Sec. 61.

In some cases, particularly in structural and architectural work, tolerances are not stated on the drawing but are given in a set of specifications or are understood to be of an order standard for the industry.

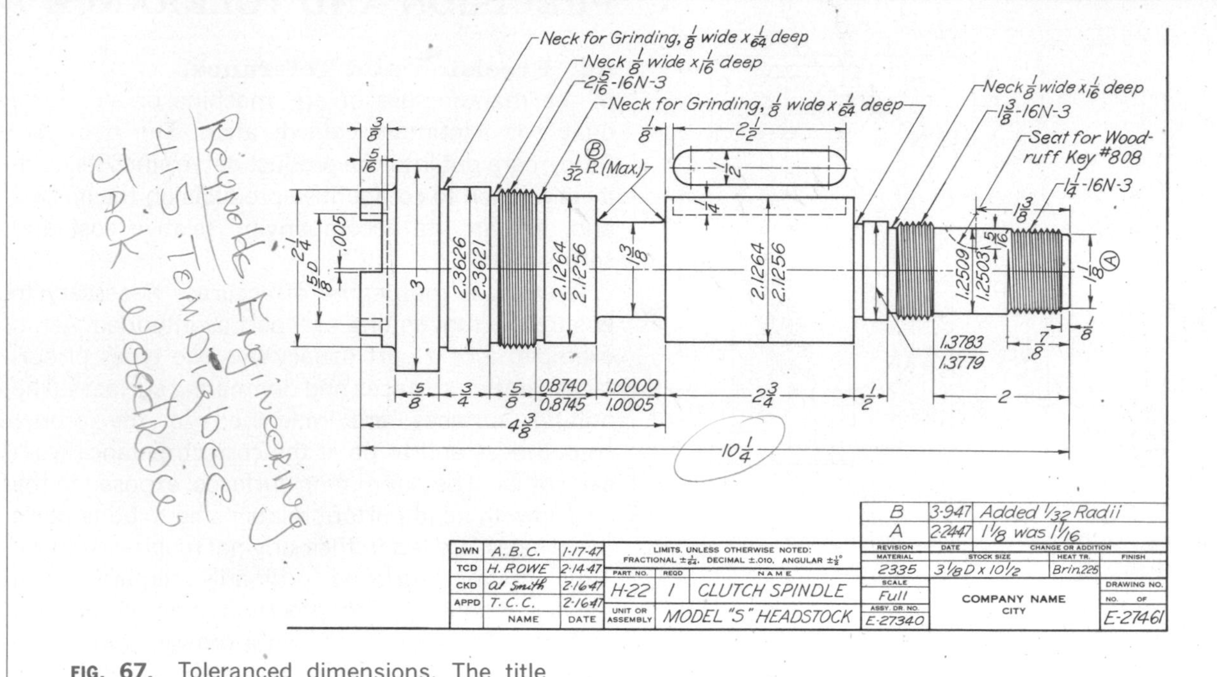


**FIG. 67.** Toleranced dimensions. The title strip gives tolerances on fractional, decimal, and angular dimensions. All other tolerances are specified directly as limits.

## 55. Fits of Mating Parts.

The working parts of any machine have some definite relationship to their mating parts in order to achieve a particular function, such as free rotation, free longitudinal movement, clamping action, permanent fixed position, etc. To ensure the proper relationship, the old practice was to mark the drawings of both parts with the same fractional dimension and add a note such as "running fit" or "drive fit," leaving the difference in size required (the allowance) to the experience and judgment of the machinist.

The tongue of Fig. 68 is to slide longitudinally in the slot. Thus, if the slot is machined first and measures 1.499 in. and the machinist, from his experience, assumes an allowance of 0.004 in., he will carefully machine the tongue to 1.495 in.; the parts will fit and function as desired. In making up a second machine, if the slot measured, say, 1.504 in. after machining, the tongue would be made 1.500 in. and an identical fit obtained; but the tongue of the first machine would be much too loose in the slot of the second machine, and the tongue of the second would not enter the slot of the first. The parts would, therefore, not be interchangeable.

Since it is not possible to work to absolute sizes, it is necessary where interchangeable assembly is required to give the dimensions of mating parts with "limits," that is, the maximum and minimum sizes within which the actual measurements must fall in order for the part to be accepted. The dimensions for each piece are given by three- or four-place decimals, the engineering department taking all the responsibility for the correctness of fit required.

Figure 69 shows the same tongue and slot as in Fig. 68 but dimensioned for interchangeability of parts. In this case, for satisfactory functioning, it has been decided that the tongue must be at least

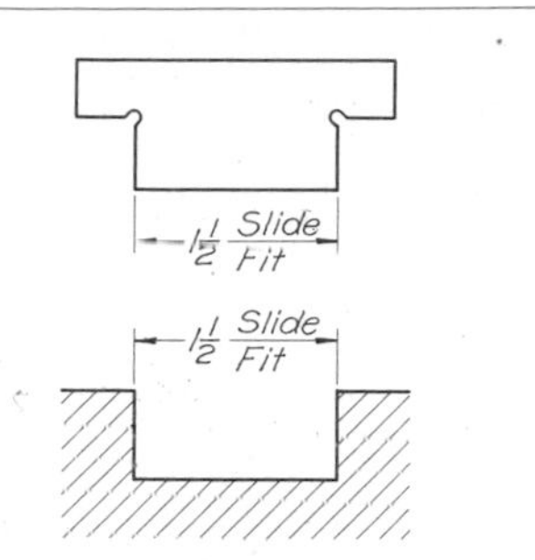


**FIG. 68.** Dimensioning a fit. According to this practice the machinist fits one part to the other.

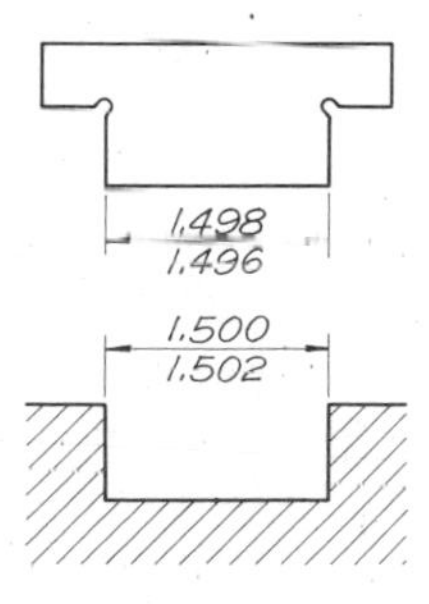


**FIG. 69.** Dimensioning a fit with limits. Tolerance on tongue, 0.002 in.; tolerance on groove, 0.002 in.; allowance, 0.002 in.

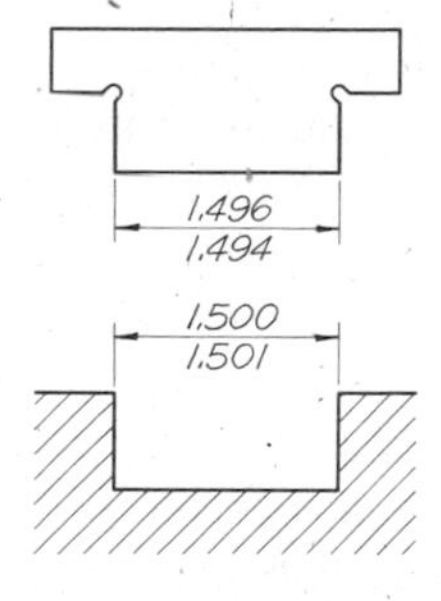


**FIG. 70.** An example of limit dimensioning. Tolerance on tongue, 0.002 in.; tolerance on groove, 0.001 in.; allowance, 0.004 in.

0.002 in. smaller than the slot but not more than 0.006 in. smaller. This would provide an average fit similar to that used in the previous example. The maximum and minimum sizes acceptable for each part are then figured.

The value 1.500 in. has been assigned as the size of the minimum acceptable slot. This value minus the minimum clearance, 0.002 in., gives a size for the maximum tongue of 1.498. The maximum allowable clearance, 0.006, minus the minimum allowable clearance, 0.002, gives the amount, 0.004, available as the total manufacturing tolerance for both parts. This has been evenly divided and applied as 0.002 to the slot and 0.002 to the tongue. Thus the size of the maximum slot will be the size of the minimum slot *plus* the slot tolerance, or $1.500 + 0.002 = 1.502$. The size of the minimum tongue will be the size of the maximum tongue *minus* the tongue tolerance, or $1.498 - 0.002 = 1.496$.

A study of Fig. 69 will show that, made in any quantity, the two parts will allow interchangeable assembly and that any pair will fit approximately as any other pair, as planned. This system is essential in modern quantity-production manufacture. Quantity-production and unit-production methods are discussed in Sec. 80.

## 56. Nomenclature.

The terms used in limit dimensioning are so interconnected that their meaning should be clearly understood before a detailed study of the method is attempted.

The following are adapted from ASA definitions:

*Nominal Size.* The nominal size is the designation that is used for the purpose of general identification.

*Dimension.* A dimension is a geometric characteristic such as diameter, length, angle, or center distance.

*Size.* Size is a designation of magnitude. When a value is assigned to a dimension, it is referred to as the size of that dimension.

*Allowance.* An allowance is an intentional difference between the maximum material limits of mating parts. (See definition of "Fit.") It is a minimum clearance (positive allowance) or maximum interference (negative allowance) between mating parts.

*Tolerance.* A tolerance is the total permissible variation of a size. The tolerance is the difference between the limits of size.

*Basic Size.* The basic size is that size from which the limits of size are derived by the application of allowances and tolerances.

*Design Size.* The design size is that size from which the limits of size are derived by the application of tolerances. When there is no allowance, the design size is the same as the basic size.

*Actual Size.* An actual size is a measured size.

*Limits of Size.* The limits of size are the applicable maximum and minimum sizes.

*Maximum Material Limit.* A maximum material limit is the maximum limit of size of an external dimension or the minimum limit of size of an internal dimension.

*Minimum Material Limit.* A minimum material limit is the minimum limit of size of an external dimension or the maximum limit of size of an internal dimension.

*Tolerance Limit.* A tolerance limit is the variation, positive or negative, by which a size is permitted to depart from the design size.

*Unilateral Tolerance.* A unilateral tolerance is a tolerance in which variation is permitted only in one direction from the design size.

*Bilateral Tolerance.* A bilateral tolerance is a tolerance in which variation is permitted in both directions from the design size.

*Unilateral Tolerance System.* A design plan that uses only unilateral tolerance is known as a Unilateral Tolerance System.

*Bilateral Tolerance System.* A design plan that uses only bilateral tolerances is known as a Bilateral Tolerance System.

*Fit.* Fit is the general term used to signify the range of tightness which may result from the application of a specific combination of allowances and tolerances in the design of mating parts.

*Actual Fit.* The actual fit between two mating parts is the relation existing between them with respect to the amount of clearance or interference that is present when they are assembled.

*Clearance Fit.* A clearance fit is one having limits of size so prescribed that a clearance always results when mating parts are assembled.

*Interference Fit.* An interference fit is one having limits of size so prescribed that an interference always results when mating parts are assembled.

*Transition Fit.* A transition fit is one having limits of size so prescribed that either a clearance or an interference may result when mating parts are assembled.

*Basic-hole System.* A basic-hole system is a system of fits in which the design size of the hole is the basic size and the allowance is applied to the shaft.

*Basic-shaft System.* A basic-shaft system is a system of fits in which the design size of the shaft is the basic size and the allowance is applied to the hole.

In illustration of some of these terms, a pair of mating parts is dimensioned in Fig. 70. In this example the *nominal size* is $1\frac{1}{2}$ in. The *basic size* is 1.500. The *allowance* is 0.004. The *tolerance* on the tongue is 0.002, and on the slot it is 0.001. The *limits* are, for the tongue, 1.496 (maximum) and 1.494 (minimum) and, for the slot, 1.501 (maximum) and 1.500 (minimum).

## 57. General Fit Classes.

The fits established on machine parts are classified as follows:

A *clearance fit* is the condition in which the internal part is smaller than the external part, as illustrated by the dimensioning in Fig. 71. In this case the largest shaft is 1.495 in. and the smallest hole 1.500 in., leaving a clearance of 0.005 for the tightest possible fit.

An *interference fit* is the opposite of a clearance fit, having a definite interference of metal for all possible conditions. The parts must be assembled by pressure or by heat expansion of the external member. Figure 72 is an illustration. The shaft is 0.001 in. larger than the hole for the loosest possible fit. The allowance in this case is 0.003 in. interference.

A *transition fit* is the condition in which either a clearance fit or an interference fit may be had. Figure 73 illustrates a transition fit; the smallest shaft in the largest hole results in 0.0003 in. clearance and the largest shaft in the smallest hole results in 0.0007 in. interference.

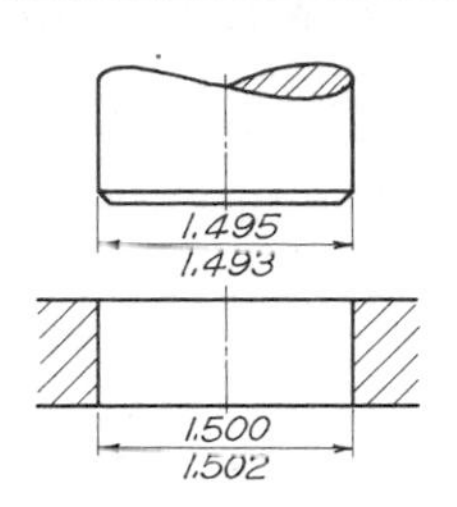


**FIG. 71.** A clearance fit. The tightest fit is 0.005 in. clearance; the loosest, 0.009 in. clearance.

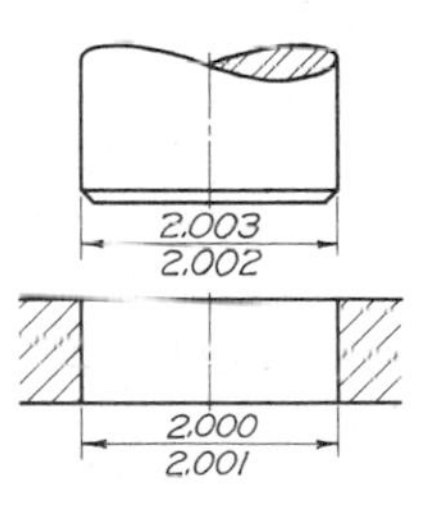


**FIG. 72.** An interference fit. The loosest fit is 0.001 in. interference; the tightest, 0.003 in. interference.

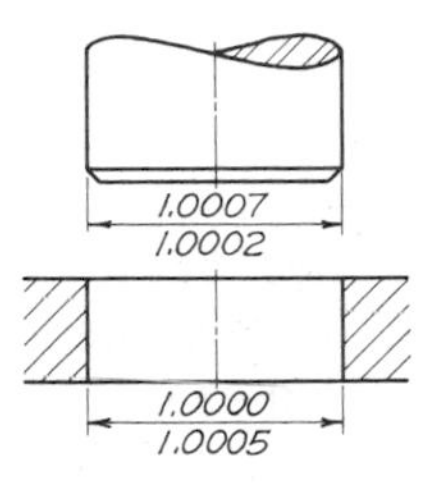


**FIG. 73.** A transition fit. The loosest fit is 0.0003 in. clearance; the tightest, 0.0007 in. interference.

**58. Selective Assembly.**
Sometimes the fit desired may be so close and the tolerances so small that the cost of producing interchangeable parts is prohibitive. In such cases tolerances as small as practical are established; then the parts are gaged and graded as, say, *small, medium,* and *large.* A small shaft in a small hole, medium in medium, or large in large will produce approximately the same fit allowance. Transition and interference fits often require a selection of parts in order to get the amount of clearance or interference desired. Antifriction bearings are usually assembled selectively.

**59. Basic-hole and Basic-shaft Systems.**
Production economy depends to some extent upon which mating part is taken as a standard size. In the *basic-hole system,* the minimum size of the hole is taken as a base from which all variations are made; the hole can often be made with a standard tool.

Where a number of different fits of the same nominal size are required on one shaft, as, for example, when bearings are fitted to line shafting, the *basic-shaft system* is employed, in which the maximum shaft size is taken as the basic size.

**60. Unilateral and Bilateral Tolerances.**
A unilateral tolerance is one in which the total allowable variation is in *one* direction, plus or minus (not both) from the basic value. A bilateral tolerance is one in which the tolerance is divided, with part plus and the remainder minus from the basic value.

**61. Methods of Expressing Tolerances.**
Tolerances may be *specific,* given with the dimension value; or *general,* given as a note in the title block. The general tolerances apply to all dimensions not carrying a specific tolerance. The general tolerance should be allowed to apply whenever possible, using specific tolerances only when necessary. If no tolerances are specified, the value usually assumed for fractional dimensions is $\pm\frac{1}{64}$ in.; for angular dimensions $\pm\frac{1}{2}°$; and for decimal dimensions plus or minus the nearest significant figure, as, for example, $\pm0.01$ in. for a two-place decimal and $\pm0.001$ in. for a three-place decimal.

There are several methods of expressing tolerances. The method preferred in quantity-production work, where gages are employed extensively, is to write the two limits representing the maximum and minimum acceptable sizes as in Fig. 74. An internal dimension has the *minimum* size above the line, and an external dimension has the *maximum* size above

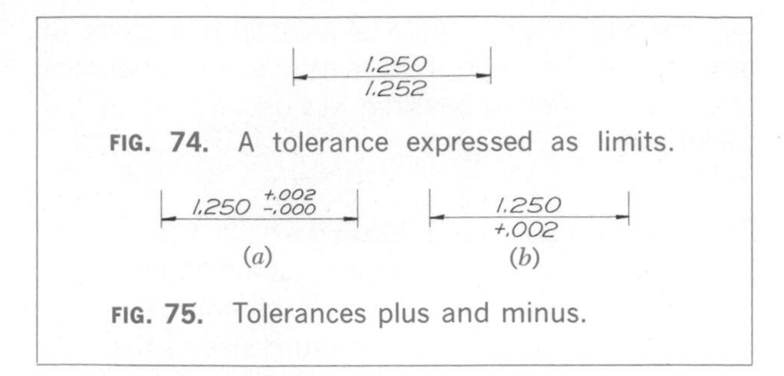


FIG. 74. A tolerance expressed as limits.

FIG. 75. Tolerances plus and minus.

the line. This arrangement is for convenience in machining.

Another method is to give the basic size followed by the tolerance, plus and minus (with the plus above the minus), as in Fig. 75. If only one tolerance value is given, as at (*b*), the other value is assumed to be zero.

*Unilateral* tolerances are expressed by giving the two limits, as in Fig. 74, or by giving one limiting size and the tolerance, as 2.750 + 0.005 or 2.750 $\begin{smallmatrix}+0.005\\-0.000\end{smallmatrix}$; for fractional dimensions, $\frac{1}{2} - \frac{1}{32}$ or $\frac{1}{2}$ $\begin{smallmatrix}+0\\-\frac{1}{32}\end{smallmatrix}$; for angular dimensions, 64°15′30″ + 0°45′0″ or 64°15′30″ $\begin{smallmatrix}+0°45'0''\\-0°0'0''\end{smallmatrix}$

*Bilateral* tolerances are expressed by giving the basic value followed by the divided tolerance, both plus and minus (commonly equal in amount), as 1.500 $\begin{smallmatrix}+0.002\\-0.002\end{smallmatrix}$ or 1.500 ± 0.002; for fractional dimensions, $1\frac{1}{2}$ $\begin{smallmatrix}+\frac{1}{64}\\-\frac{1}{64}\end{smallmatrix}$ or $1\frac{1}{2} \pm \frac{1}{64}$; for angular dimensions, 30°0′ $\begin{smallmatrix}+0°10'\\-0°10'\end{smallmatrix}$ or 30°0′ ± 0°10′.

## 62. Decimal Places.

A dimension value should be carried to the same number of decimal places as the tolerance. For example, with a tolerance of 0.0005 on a nominal dimension of $1\frac{1}{2}$ in., the basic value should be written 1.5000. Tolerances for common fractional values should be given as common fractions, as $\frac{7}{8} \pm \frac{1}{64}$. Tolerances for decimal values should be given as decimal fractions, as 0.750 ± 0.010.

## 63. Fundamentals for Tolerance Selection.

Before the engineer can decide on the precision necessary for a particular part and specify the proper fits and tolerances, he must have experience in the manufacturing process used and understand the particular mechanism involved. The following quotation from the ASA Standard is pertinent: "Many factors, such as length of engagement, bearing load, speed, lubrication, temperature, humidity, and materials, must be taken into consideration in the selection of fits for a particular application."

A table of fits, such as the ASA table of cylindrical fits (Appendix) explained in Sec. 67, may be taken as a guide for ordinary work.

In many cases practical experience is necessary in determining the fit conditions guaranteeing proper performance. Often it is difficult to determine the definite size at which performance fails, and critical tolerances are sometimes determined through exhaustive testing of experimental models.

It is essential to know the precision attainable with various machine tools and machining methods. As an example, holes to be produced by drilling must not be specified to a smaller tolerance than can be attained by drilling. Attainable manufacturing precision is discussed in Sec. 64. A knowledge of kinds and types of equipment is needed to assure that the tolerances specified can be attained.

## 64. Manufacturing Precision.

The different manufacturing processes all have inherent minimum possible accuracies, depending upon the size of the work, the condition of the equipment, and, to some extent, the skill of the workmen. The following *minimum* tolerances are given as a guide and are based on the assumption that the work is to be done on a quantity-production basis with equipment in good condition. Greater precision can be attained by highly skilled workmen on a unit-production basis.

In general, the following are recommended as tolerances for dimensions having *no effect on the function of the part;* for sizes of 0 to 6 in., $\pm\frac{1}{64}$; 6 to 18 in., $\pm\frac{1}{32}$; 18 in. and larger, $\pm\frac{1}{16}$ (or more).

*Sand Castings.* For unmachined surfaces, a tolerance of $\pm 1/32$ is recommended for small castings and a tolerance of $\pm 1/16$ for medium-sized castings. On larger castings the tolerance should be increased to suit the size. Small- and medium-sized castings are rarely below the nominal size since the pattern is "rapped" for easy removal from the sand, and this tends to increase the size.

*Die Castings and Plastic Moulding.* A tolerance of $\pm 1/64$ or less can easily be held with small and medium-sized parts; for large parts, the tolerance should be increased slightly. Hole-center distances can be maintained within 0.005 to 0.010, depending on the distance of separation. Certain alloys can be die-cast to tolerances of 0.001 or less.

*Forgings.* The rough surfaces of drop forgings weighing 1 lb or less can be held to $\pm 1/32$; for weights up to 10 lb, $\pm 1/16$; for weights up to 60 lb, $\pm 1/8$. Because of die wear, drop forgings tend to increase in size as production from the die increases.

*Drilling.* For drills from no. 60 to no. 30, allow a tolerance of +0.002 − 0.000; from no. 29 to no. 1, +0.0004 − 0.000; from ¼ to ½ in., +0.005 − 0.000; from ½ to ¾ in., +0.008 − 0.000; from ¾ to 1 in., +0.010 − 0.000; from 1 to 2 in., a tolerance of +0.015 − 0.000.

*Reaming.* In general, a tolerance of +0.0005 − 0.0000 can be held with diameters up to ½ in. For diameters from ½ to 1 in., +0.001 − 0.000; from 1 in. and larger, +0.0015 − 0.0000.

*Lathe Turning: Rough Work.* For diameters of ¼ to ½ in., allow a total tolerance of 0.005; for diameters of ½ to 1 in., 0.007; for diameters of 1 to 2 in., 0.010; for diameters of 2 in. and larger, 0.015.

*Finish Turning.* For diameters of ¼ to ½ in., allow a total tolerance of 0.002; for diameters of ½ to 1 in., 0.003; for diameters of 1 to 2 in., 0.005; for diameters of 2 in. and larger, 0.007.

*Milling.* When single surfaces are to be milled, tolerances of 0.002 to 0.003 can be maintained. When two or more surfaces are to be milled, the most important can be toleranced to 0.002 and the remainder to 0.005. In general, 0.005 is a good value to use with most milling work.

*Planing and Shaping.* These operations are not commonly used with small parts in quantity-production work. For larger parts, tolerances of 0.005 to 0.010 can be maintained.

*Broaching.* Diameters up to 1 in. can be held within 0.001; diameters of 1 to 2 in., 0.002; diameters of 2 to 4 in., 0.003. Surfaces up to 1 in. apart can be held within 0.002; 1 to 4 in. apart, 0.003; 4 in. apart and over, 0.004.

*Threads.* Tolerances for ASA threads are provided on the pitch diameter through the thread class given with the specification. For a given class, the tolerances increase as the size of the thread increases.

*Grinding.* For both cylindrical and surface grinding, a tolerance of 0.0005 can be maintained.

## 65. Selection of Tolerances.

A common method of determining and applying tolerances is to determine at the outset how much clearance or interference there can be between the mating parts *without impeding their proper functioning.* The difference between the tightest and loosest conditions will be the *sum* of the tolerances of both parts. To obtain tolerances for the individual parts, take half of this value; or if it seems desirable because of easier machining on one part, use slightly less tolerance for that part, with a proportionately larger tolerance for the part more difficult to machine. The following example will illustrate the procedure:

Assume that a running fit is to be arranged between a 2-in. shaft and bearing. It has been determined that in order to provide clearance for a film of oil, the parts cannot fit closer than 0.002 in., and in order to prevent excessive looseness and radial movement of the shaft, the parts cannot be looser than 0.007 in. Then we have the following calculations:

| | | |
|---|---|---|
| Loosest fit | = 0.007 | (maximum clearance) |
| Tightest fit | = 0.002 | (minimum clearance—allowance) |
| Difference | = 0.005 | (sum of tolerances) |

$\frac{1}{2}$ of difference = 0.0025 (possible value for each tolerance)

Assuming that the shaft will be ground and the bearing reamed, 0.002 can be used for the shaft tolerance and 0.003 for the bearing tolerance since these values conform better to the precision attainable by these methods of production.

Figure 76 illustrates the completed dimensions. Note that the minimum hole is taken as the basic size of 2.000 and the tolerance of 0.003 applied. Then the largest shaft size will be the basic size minus the value for the tightest fit (the allowance); the shaft tolerance then subtracted from the maximum shaft size gives the minimum shaft size. From the figure, the 2.003 bearing minus the 1.996 shaft gives 0.007, the loosest fit; the 2.000 bearing minus the 1.998 shaft gives 0.002, the tightest fit.

## 66. ASA Preferred Limits and Fits for Cylindrical Parts.[3]

This standard conforms with the recommendations of American-British-Canadian conferences. Agreement has been reached for diameters up to 20 in., and larger diameters are under study.

## 67. Designation of ASA Fits.

The standard ASA fits are designated by symbols that facilitate reference for educational purposes.

[3]Excerpted from American Standard preferred limits and fits for cylindrical parts (ASA B4.1—1955) with the permission of the publisher, the American Society of Mechanical Engineers, 345 East 47th Street, New York, N.Y. 10017.

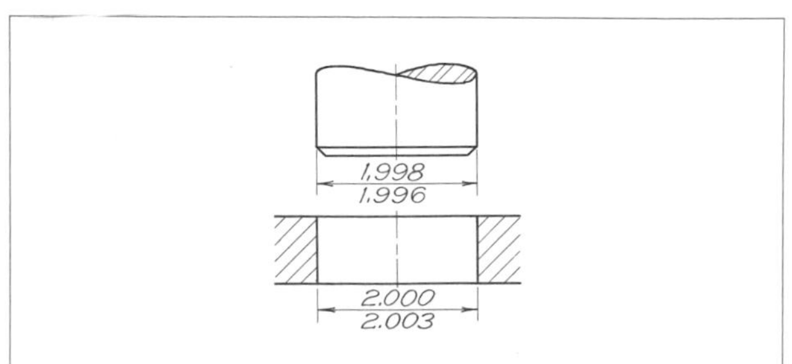


**FIG. 76.** Limits calculated from maximum and minimum clearances. See text for details.

These symbols are not to be shown on manufacturing drawings; instead, sizes should be specified. The letter symbols used are as follows:

- RC, running or sliding fit
- LC, locational clearance fit
- LT, transition fit
- LN, locational interference fit
- FN, force or shrink fit

These letter symbols are used in conjunction with numbers for the class of fit; thus FN4 represents a class 4 force fit. Each symbol (two letters and a number) represents a complete fit, for which the minimum and maximum clearance or interference and the limits of size for the mating parts are given in the Appendix.

## 68. Description of Fits.

The following is a description of each class of fit, with a reference in each case to the tables in the Appendix.

*Running and Sliding Fits (Table 1).* These fits provide a similar running performance, with suitable lubrication allowance, throughout the range of sizes. The clearances for the first two classes, used chiefly as slide fits, increase more slowly than for the other classes, so that accurate location is maintained, even when this is at the expense of free relative motion.

RC 1, *close sliding fits* accurately locate parts that must assemble without perceptible play.

RC 2, *sliding fits* are for accurate location, but with greater maximum clearance than RC 1. Parts move and turn easily but do not run freely, and in the larger sizes may seize with small temperature changes.

RC 3, *precision running fits* are about the closest fits expected to run freely, and are for precision work at slow speeds and light journal pressures. They are not suitable under appreciable temperature differences.

RC 4, *close running fits* are chiefly for running fits on accurate machinery with moderate surface speeds and journal pressures, where accurate location and minimum play are desired.

RC 5 and RC 6, *medium running fits* are for higher running speeds, heavy journal pressures, or both.

RC 7, *free running fits* are for use where accuracy is not essential, where large temperature variations are likely, or under both these conditions.

RC 8 and RC 9, *loose running fits* are for materials such as cold-rolled shafting and tubing, made to commercial tolerances.

*Locational Fits (Tables 2, 3, and 4).* These fits determine only the location of mating parts and may provide rigid or accurate location—as in interference fits—or some freedom of location—as in clearance fits. They fall into three groups:

LC, *locational clearance fits* are for normally stationary parts that can be freely assembled or disassembled. They run from snug fits for parts requiring accuracy of location, through the medium clearance fits for parts such as spigots, to the looser fastener fits where freedom of assembly is of prime importance.

LT, *transition locational fits* fall between clearance and interference fits for application where accuracy of location is important, but a small amount of clearance or interference is permissible.

LN, *locational interference fits* are used where accuracy of location is of prime importance, and for parts needing rigidity and alignment with no special requirements for bore pressure. Such fits are not for parts that transmit frictional loads from one part to another by virtue of the tightness of fit; these conditions are met by force fits.

*Force Fits (Table 5).* A force fit is a special type of interference fit, normally characterized by maintenance of constant bore pressures throughout the range of sizes. Thus the interference varies almost directly with diameter, and to maintain the resulting pressures within reasonable limits, the difference between its minimum and maximum value is small.

FN 1, *light drive fits* require light assembly pressures and produce more or less permanent assemblies. They are suitable for thin sections, long fits, or cast-iron external members.

FN 2, *medium drive fits* are for ordinary steel parts or for shrink fits on light sections. They are about the tightest fits that can be used with high-grade cast-iron external members.

FN 3, *heavy drive fits* are suitable for heavier steel parts or for shrink fits in medium sections.

FN 4 and FN 5, *force fits* are for parts that can be highly stressed, or for shrink fits where heavy pressing forces are impractical.

## 69. Examples of Dimensioning Standard ASA Fits.

Assume that a 1-in. shaft is to run with moderate speed but with a fairly heavy journal pressure. The fit class chosen is RC 6. From Table 1 (Appendix) the limits given are:

Hole: $\begin{matrix} +1.2 \\ -0 \end{matrix}$

Shaft: $\begin{matrix} -1.6 \\ -2.8 \end{matrix}$

These values are thousandths of an inch. The basic size is 1.0000. Therefore, for the hole, 1.0000 + 0.0012 gives 1.0012 as the maximum limit, and 1.0000 + 0.0000 gives 1.0000 for the minimum limit. For the shaft, 1.0000 − 0.0016 gives 0.9984 for the maximum limit, and 1.0000 − 0.0028 gives 0.9972 for the minimum limit. The dimensioning is shown in Fig. 77. To analyze the fit, the tightest condition (largest shaft in smallest hole) is 1.0000 − 0.9984, or 0.0016, clearance. This is the *allowance*. The loosest condition (smallest shaft in largest hole) is 1.0012 − 0.9972, or 0.0040, clearance. The val-

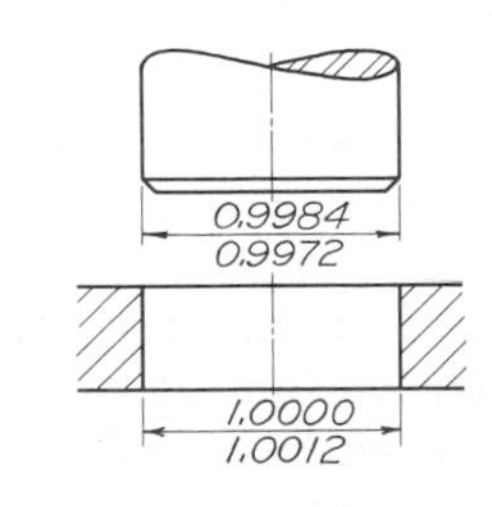


FIG. 77. An ASA clearance fit. This one is class RC 6. Tolerance on shaft, 0.0012 in.; tolerance on hole, 0.0012 in.; allowance, 0.0016 in. clearance.

ues of these two limits are given in the table under "limits of clearance."

To illustrate further, suppose a 2-in. shaft and hub are to be fastened permanently with a drive fit. The hub is high-grade steel. Fit FN 4 has been chosen. Table 5 gives:

Hole: $\begin{matrix} +1.2 \\ -0 \end{matrix}$

Shaft: $\begin{matrix} +4.2 \\ +3.5 \end{matrix}$

These values are in thousandths of an inch. The basic size is 2.0000. Thus, for the hole 2.0000 + 0.0012 gives 2.0012 for the maximum limit, and 2.0000 − 0.0000 gives 2.0000 as the minimum limit. For the shaft, 2.0000 + 0.0042 gives 2.0042 for the maximum limit, and 2.0000 + 0.0035 gives 2.0035 for the minimum limit. The limits of fit are 0.0023 minimum interference and 0.0042 maximum interference. The allowance is 0.0042. The dimensioning of this fit is shown in Fig. 78.

Note in both the above cases that, in dimensioning, the maximum limit of the hole and the minimum limit of the shaft are placed below the dimension line. This is for convenience in reading the drawing and to help prevent mistakes by the machinist.

The ASA fits are based on standard hole practice. Note in the tables (1 to 5) that the minimum hole size is always the basic size.

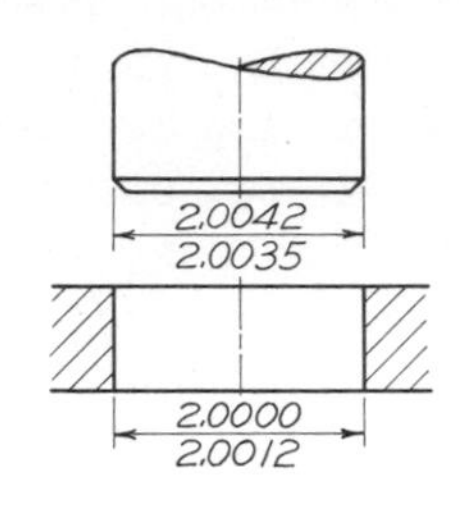


**FIG. 78.** An ASA interference fit. This one is class FN 4. Tolerance on shaft, 0.0007 in.; tolerance on hole, 0.0012 in.; allowance, 0.0042 in. interference.

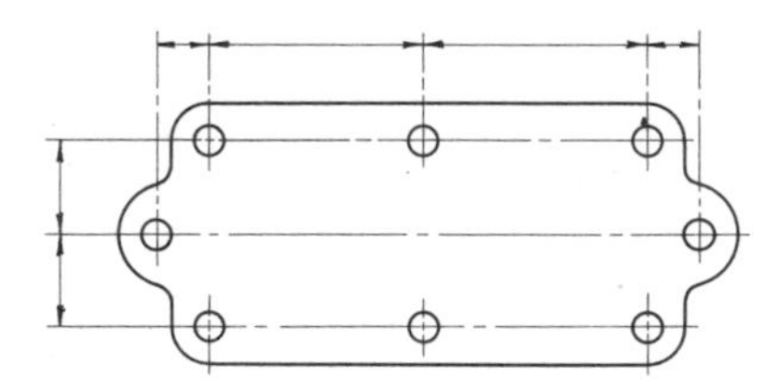

**FIG. 79.** Successive dimensioning. Tolerances accumulate.

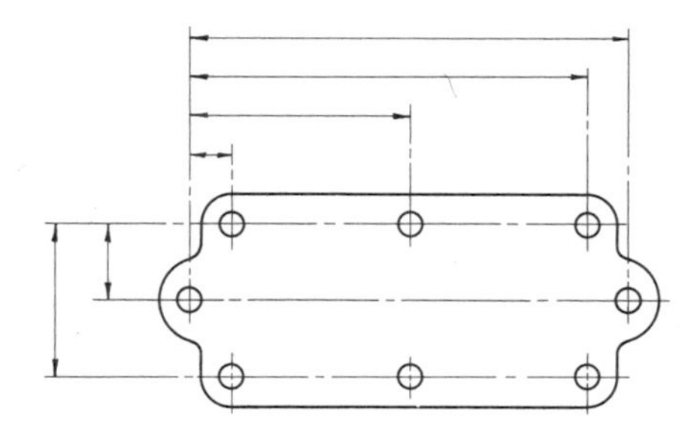

**FIG. 80.** Dimensions from datum. Position from datum is subject to only one tolerance, but the center distance between two holes positioned from datum is subject to variation of two tolerances.

### 70. Cumulative Tolerances.

Tolerances are said to be cumulative when a position in a given direction is controlled by more than one tolerance. In Fig. 79 the holes are positioned one from another. Thus, the distance between two holes separated by two, three, or four dimensions will vary in position by the sum of the tolerances on all the dimensions. This difficulty can be eliminated by dimensioning from *one* position, which is used as a datum for all dimensions, as shown in Fig. 80. This system is commonly called base-line dimensioning.

Figure 81 is a further example of the effect of cumulative tolerances. The position of surface $Y$ with respect to surface $W$ is controlled by the additive tolerances on dimensions $A$ and $B$. If it is important, functionally, to hold surface $Y$ with respect to surface $X$, the dimensioning used is good. If, however, it is more important to hold surface $Y$ with respect to surface $W$, the harmful effect of cumulative tolerances can be avoided by dimensioning as in Fig. 82. Cumulative tolerance, however, is always present; in Fig. 82 the position of surface $Y$ with respect

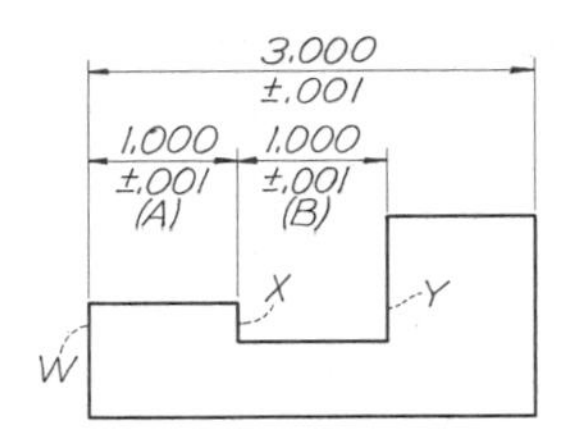


**FIG. 81.** Control of surface position through different toleranced dimensions. Here one dimension is successive.

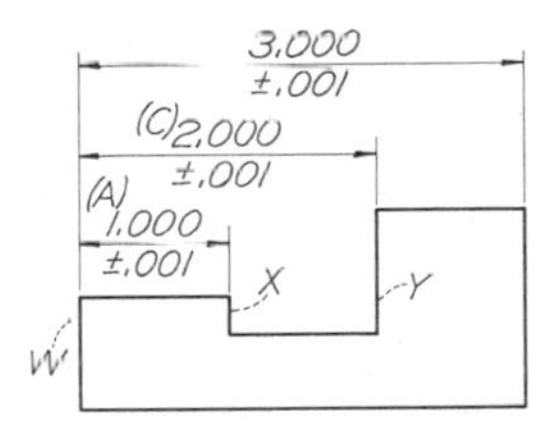


**FIG. 82.** Control of surface position through different toleranced dimensions. All dimensions are from datum.

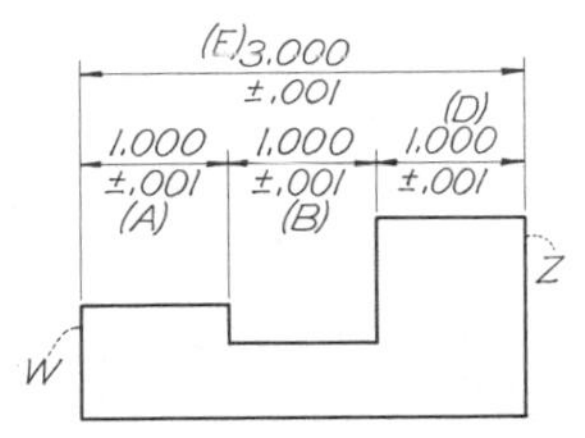


**FIG. 83.** Control of surface position through different toleranced dimensions. This drawing is overdimensioned.

to surface $X$ is now subject to the cumulative tolerances of dimensions $A$ and $C$.

In machine drawing, overdimensioning a drawing may cause confusion in the shop due to cumulative tolerances. This is illustrated in Fig. 83, where one of the surfaces will be positioned by two dimensions, both of which are subject to a tolerance. Thus surface $Z$ may be positioned with respect to surface $W$ by means of dimensions $A$, $B$, and $D$ and be within $\pm 0.003$ in. of the basic position; this variation is inconsistent with the tolerance on dimension $E$. The situation can be clarified by assigning smaller tolerances to dimensions $A$, $B$, and $D$ so that, cumulatively, they will be equal to $\pm 0.001$ or less. This is poor practice, however, since it will probably increase the production cost. Another solution is to increase the tolerance on dimension $E$ to $\pm 0.003$ if the function of the part will permit. The best solution, however, is to eliminate one of the four dimensions, since one dimension is superfluous. If all four dimensions are given, one should be marked "REF," and its tolerance thus eliminated.

### 71. Tolerance between Centers.

In all cases where centers are arranged for interchangeable assembly, the tolerance on shafts, pins, etc., and the tolerance on bearings or holes in the mating pieces will affect the possible tolerance between centers. In Fig. 84, observe that smaller tolerances on the holes would necessitate a smaller tolerance on the center distances. A smaller allowance for the fit of the pins would make a tighter fit and reduce the possible tolerances for the center-to-center dimensions. Study carefully the dimensions of both pieces.

### 72. Tolerance of Concentricity.

Tolerance of concentricity is a special case of tolerance in which there is a coincidence of centers. In most cases, concentric cylinders, cones, etc., gener-

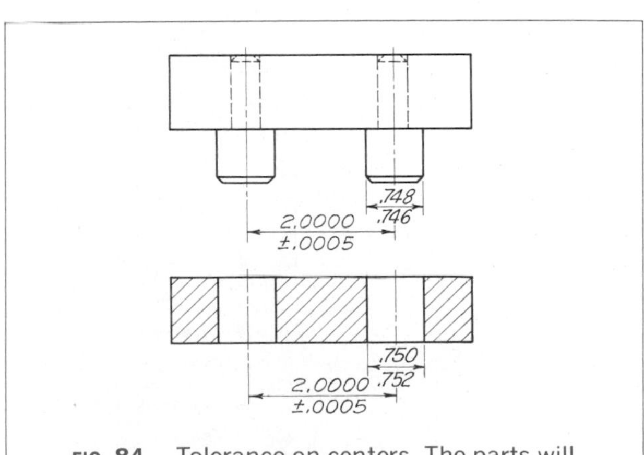


**FIG. 84.** Tolerance on centers. The parts will not assemble unless all features are held within the limits shown.

ated about common axes in manufacture, will be concentric to a degree of precision more than adequate for functional requirements, and no statement is required on the drawing concerning the allowable variation. However, mating pairs of two (or more) precise, close-fitting, machined cylindrical surfaces must have the axes of adjoined cylinders closely coinciding in order to permit assembly of the parts; thus it is sometimes necessary to give the permissible deviation from concentricity. Since the center lines of adjoining cylinders coincide on a drawing, the tolerance cannot be given as a dimension. One method of indicating it is to mark the diameters with reference letters and give the tolerance in note form as in Fig. 85*a*. The reference letters can be dispensed with if the note is applied directly to the surfaces, as at (*b*).

### 73. Tolerance for Angular Dimensions.

When it is necessary to give the limits of an angular dimension, the tolerance is generally bilateral, as $32 \pm \frac{1}{2}°$. When the tolerance is given in minutes, it is written $\pm 0°10'$; and when it is given in seconds, it is written $\pm 0'30''$. Where the location of a hole or other feature depends upon an angular dimension, the length along the leg of the angle governs the angular tolerance permitted. A tolerance of $\pm 1°$ gives a variation of 0.035 in. for a length of 1 in. and can be used as a basis for computing the tolerance in any given problem.

As an example, assume an allowable variation of 0.007 in.; then $(0.007/0.035) \times 1° = \frac{1}{5}°$ is the angular tolerance at 1 in. If the length is assumed as 2 in., the tolerance would be one-half the tolerance computed for 1 in., or $\frac{1}{5}° \times \frac{1}{2} = \frac{1}{10}°$, or $0°6'$.

### 74. Coinciding Center Lines and Dimensions.

In many cases the center lines for two different features of a part will coincide. Often one center line

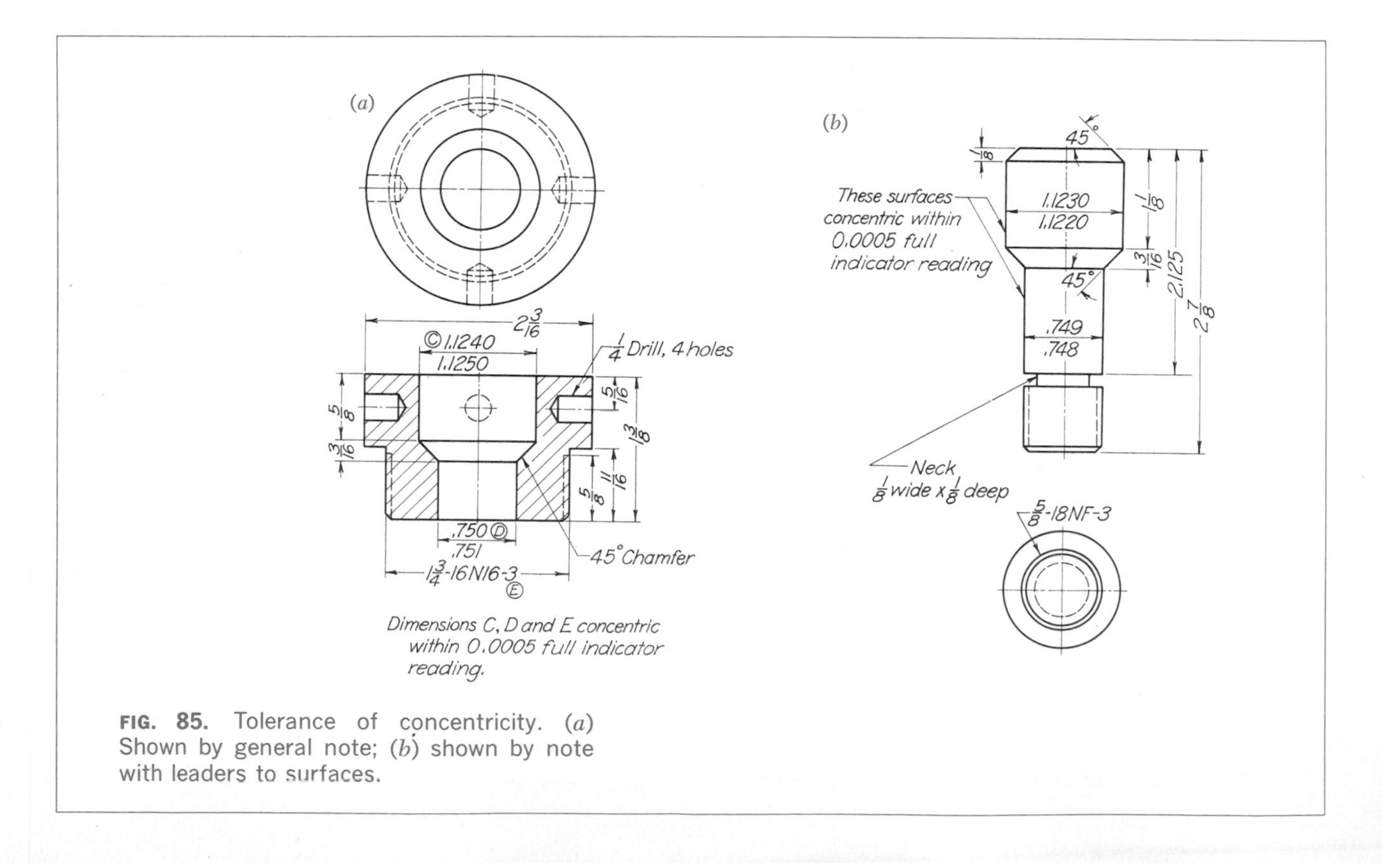


**FIG. 85.** Tolerance of concentricity. (*a*) Shown by general note; (*b*) shown by note with leaders to surfaces.

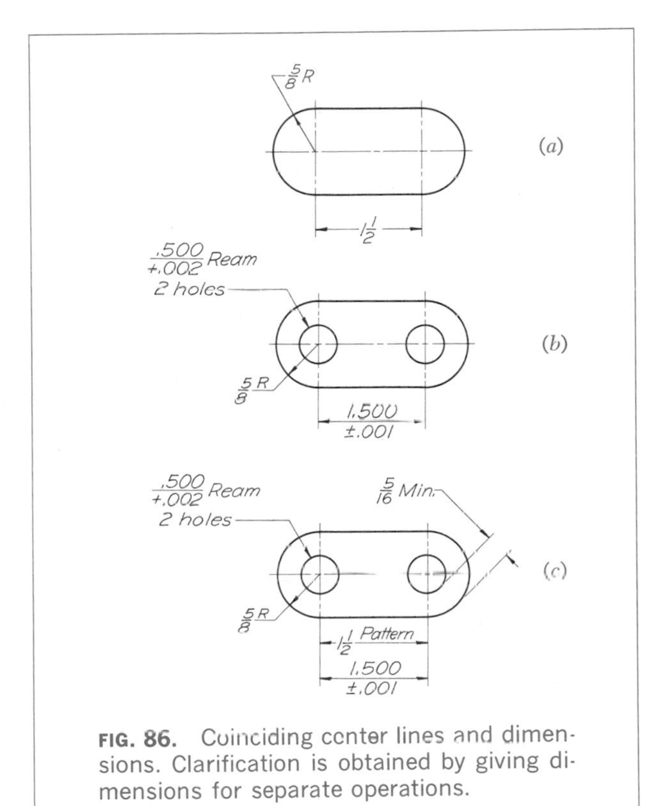


**FIG. 86.** Coinciding center lines and dimensions. Clarification is obtained by giving dimensions for separate operations.

is for an unfinished feature and the other (and coincident) center line for a finished feature. Figure 86*a* shows the drawing of a link dimensioned for the patternmaker. If holes are to be machined in this link, the drawing would be as at (*b*); the patternmaker would not use the dimension between centers, as shown, but would assume the nominal dimension of $1\frac{1}{2}$ in. with the usual pattern tolerance of $\pm\frac{1}{32}$ in. The clearest dimensioning in this case would be as at (*c*).

In any instance where there is a coincidence of centers, it may be difficult to indicate the limits within which the coincidence must be maintained. At (*c*) there are actually two horizontal center lines, one for the cast link and another for the hole centers. One method of controlling the deviation from coincidence is to give the wall thickness as a minimum, which is understood to apply in all radial directions.

In cases where the coincident center lines are both for finished features with differing tolerances, there may be a serious ambiguity on the drawing unless the dimensioning is specially arranged. Figure 87*a* shows a milled slot with nominal dimensions and, on the coincident center lines, two accurate holes with a closely toleranced center distance. Unless the dimensioning is cleared by two separate dimensions, as shown, the machinist would not know the difference in tolerance. A somewhat more difficult case is shown at (*b*) where pairing holes are diagonally opposite. Unless all the holes are to be toleranced the same on their center distance, the dimensioning must be made clear with notes, as shown.

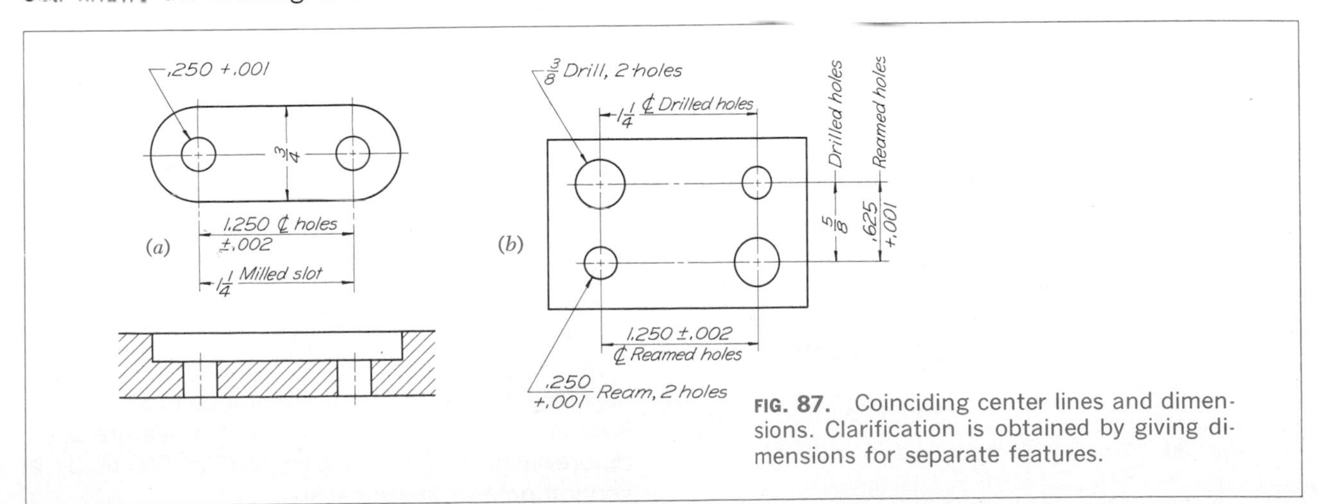


**FIG. 87.** Coinciding center lines and dimensions. Clarification is obtained by giving dimensions for separate features.

## 75. Positional Tolerances.

Figures 61 and 62 show holes positioned by rectangular or polar coordinates. This method of positioning by tolerances on two dimensions produces a square tolerance zone for the common case where the positioning dimensions are at right angles to each other. The engineering intent can often be specified more accurately by giving information that states the *true position*, with tolerances to indicate how far actual position on the part can vary from true position. True position denotes the basic or theoretically exact position of a feature. This method results in a circular tolerance zone when the tolerance applies in all directions from true position.

Features such as holes may be allowed to vary in any direction from the specified true position. Features such as slots may be allowed to vary from the specified true position on either side of the true-position plane. Thus there are two methods of applying true-position tolerances. Both methods are shown in Fig. 88.

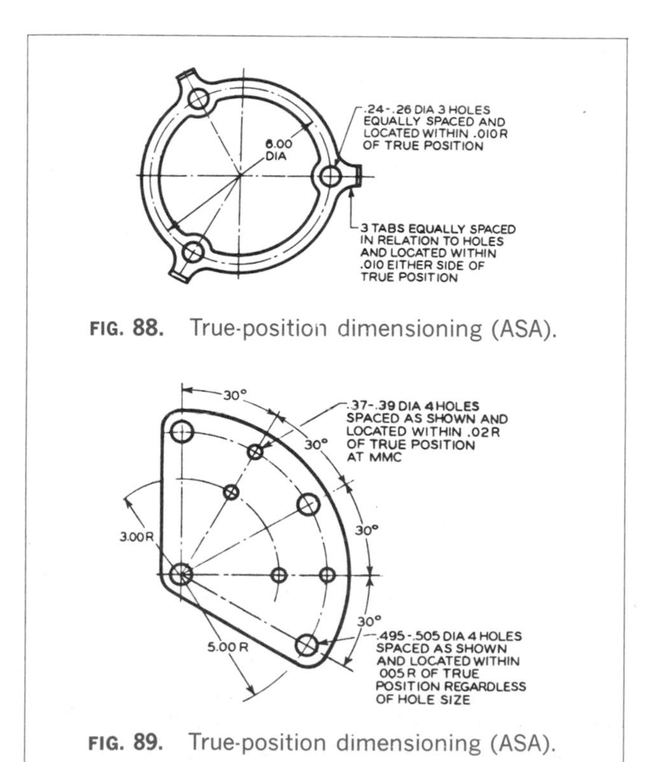


**FIG. 88.** True-position dimensioning (ASA).

**FIG. 89.** True-position dimensioning (ASA).

When a feature is allowed to vary in any direction, a note in one of the following forms should be used:

(*a*) 6 holes located at true position within 0.010 dia.

(*b*) 6 holes located within 0.005R of true position.

When features are allowed to vary from a true-position plane, a note in one of the following forms should be used:

(*a*) 6 slots located at true position within 0.010 wide zone.

(*b*) 6 slots located within 0.005 either side of true position.

When a feature is allowed to vary in any direction from true position, the position of the feature is given by untoleranced dimensions, as shown in Figs. 88 and 89. The fact that there is no allowable variation on the positioning dimensions should be indicated either on the drawing, in the title block, or in a separate specification, by a note—"Dimensions locating true position are basic." If this is not done, the word "basic" should be shown on each dimension subject to true-position tolerance.

## 76. Maximum Material Condition.

Since all features of a part have allowable variations in size, for *mating* parts the least favorable assembly condition exists when the mating parts are both at their maximum material condition. This means that a hole is at its minimum size and a shaft at its maximum. In terms of the cylindrical surface of a hole, it means that no point on the surface will be inside a cylinder having a diameter equal to the *actual* diameter of the hole, minus the true-position tolerance (diameter or twice the radius of the tolerance circle), the axis of the cylinder being at true position.

Where the maximum material condition applies, it is stated by the addition of "maximum material condition" to the true-position note, or by using the abbreviation "MMC," as shown in Fig. 89. Also, MMC may be stated in a general note or a specification.

In some cases it may be necessary to state a positional tolerance without reference to MMC. This is done by the reference "regardless of feature size" (abbreviated RFS), as shown on Fig. 89, or by a general note or specification.

In most cases the datum for true position is obvious from the dimensioning itself, but where there may be any doubt, the positional tolerance note should read "XX holes located within .xxxR of true position in relation to datum *A*." Of course, datum *A* must be clearly marked on the drawing.

For a complete discussion of true-position tolerancing and related details, see ASA Y14.5—1965 (tentative) or any subsequent revisions thereof.

## 77. Zone Tolerances.

A zone tolerance may be specified where a uniform variation can be permitted along a contour, as in Fig. 90. For complete coverage, see ASA Y14.5—1965 (tentative) or any subsequent revisions thereof.

## 78. Form Tolerances.

Tolerances of form state how far actual surfaces may vary from the perfect geometry implied by the drawing. The methods of indicating straightness, flatness, and parallelism are shown in Fig. 91; squareness in Fig. 92; angularity, symmetry, concentricity, and roundness in Fig. 93; and parallelism with a surface and another hole in Fig. 94. For complete details on form tolerances, see ASA Y14.5—1965 (tentative).

## 79. Surface Quality.

The proper functioning and wear life of a part frequently depend upon the smoothness quality of its surfaces. American Standard B46.1—1962 defines the factors of surface quality and describes the meaning and use of symbols on drawings. Any surface, despite its apparent smoothness, has minute peaks and valleys, the height of which is termed "surface roughness" and which may or may not be superimposed on a more general "*waviness*." The most prominent direction of tool marks and minute scratches is called "lay."

*Roughness,* produced principally by cutting edges and tool feed, is expressed as the arithmetical average from the mean in microinches (Fig. 95).

*Roughness width* is rated in inches as the maximum permissible spacing between repetitive units of the surface pattern (Fig. 95).

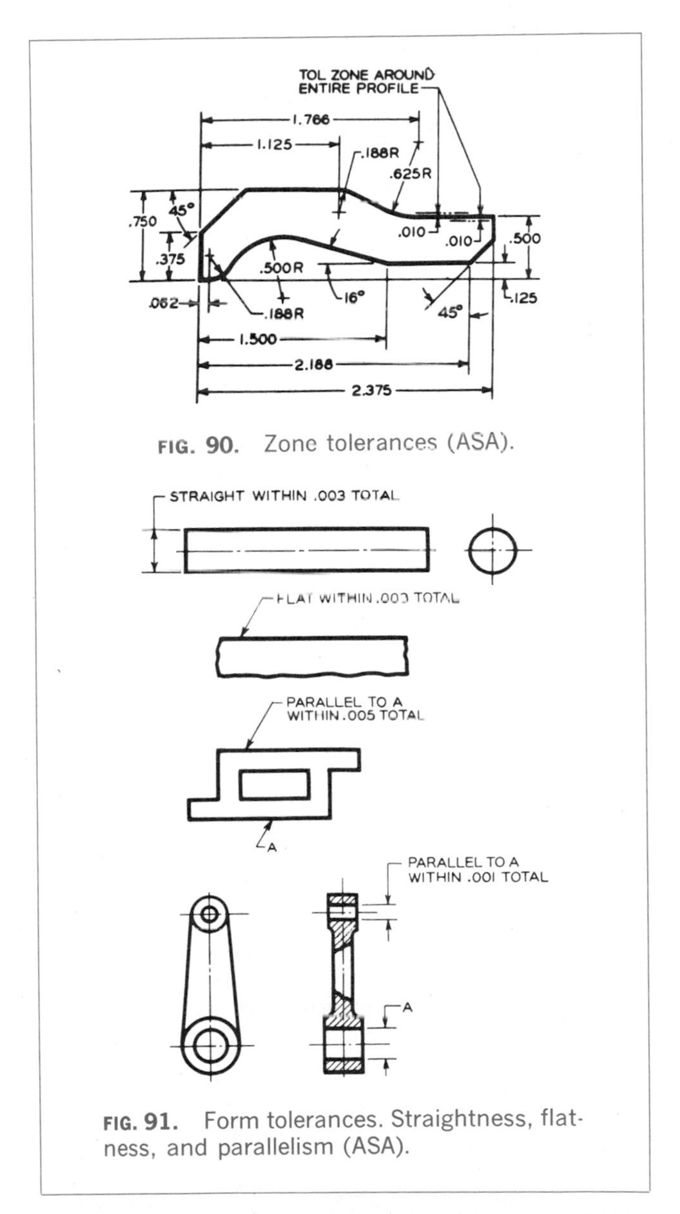


FIG. 90. Zone tolerances (ASA).

FIG. 91. Form tolerances. Straightness, flatness, and parallelism (ASA).

*Roughness-width cutoff* is the maximum width in inches of surface irregularities to be included in the measurement of roughness height (Fig. 95).

*Waviness* designates irregularities of greater spacing than the roughness, resulting from factors such as deflection and vibration (Fig. 95). The height is rated in inches as peak-to-valley height. The width is rated in inches as the spacing of adjacent waves.

*Lay* is the direction of the predominant surface

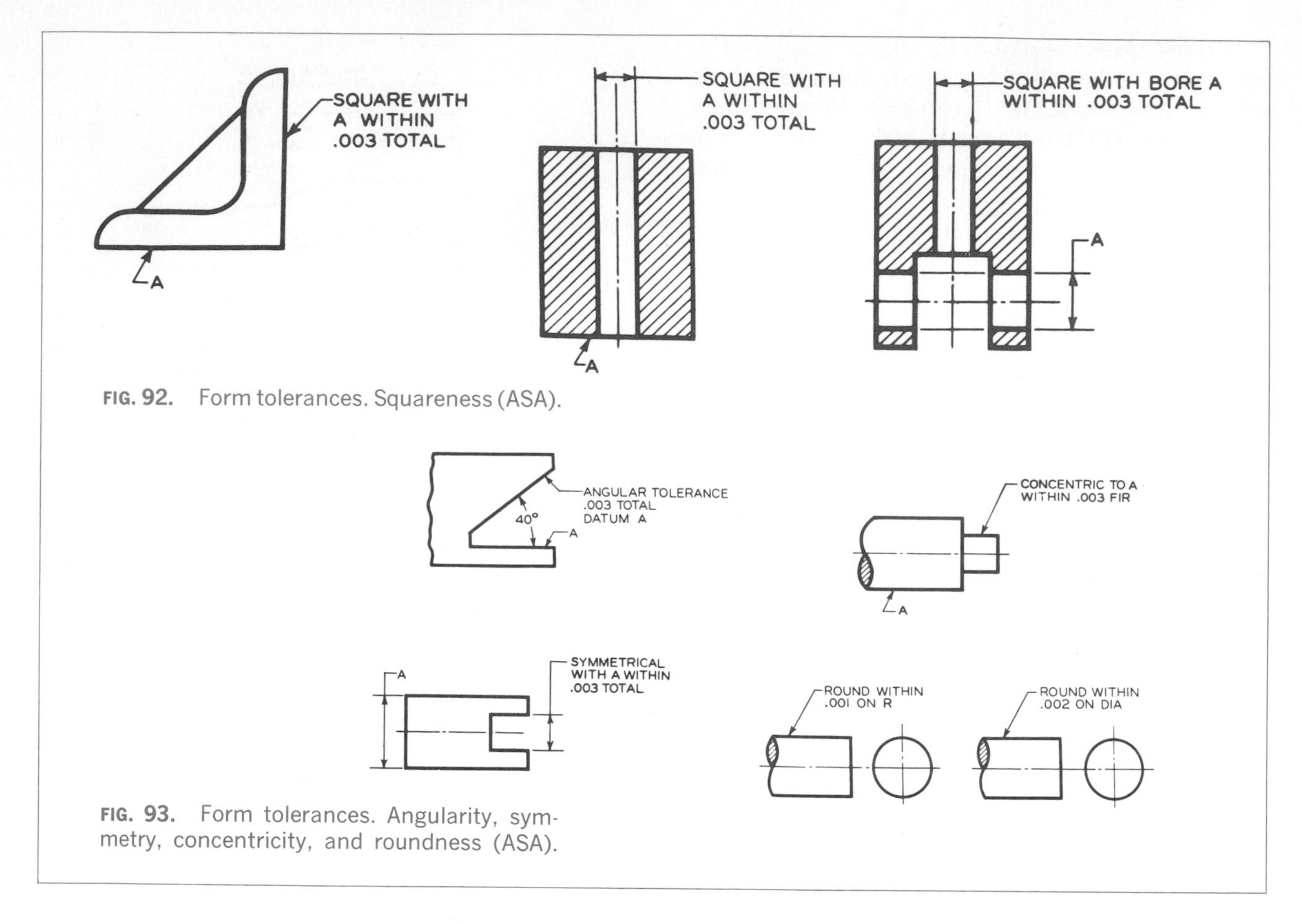


FIG. 92. Form tolerances. Squareness (ASA).

FIG. 93. Form tolerances. Angularity, symmetry, concentricity, and roundness (ASA).

pattern, produced by tool marks or grains of the surface ordinarily determined by the production method used (Fig. 95).

Although other instruments are used, the common method of measuring surface roughness employs electrical amplification of the motion of a stylus over the surface.

*The symbol* for indicating surface irregularities is a "check mark" with the horizontal extension as shown in Figs. 96 to 103. Symbols for lay and the application of the symbols on a drawing are shown in Fig. 104. The caption explains each case.

*Values* for roughness height, waviness height, and roughness-width cutoff are shown in Table 1.

Figure 105 is a chart adapted from several sources, giving the range of surface roughness for various methods.

**Table 1**

**Recommended roughness-height values (microinches)**

| | | | | |
|---|---|---|---|---|
| 0.25 | 5 | 20 | 80 | 320 |
| 0.5 | 6 | 25 | 100 | 400 |
| 1.0 | 8 | 32 | 125 | 500 |
| 2.0 | 10 | 40 | 160 | 600 |
| 3.0 | 13 | 50 | 200 | 800 |
| 4.0 | 16 | 63 | 250 | 1000 |

**Recommended waviness-height values (inches)**

| | | | | | |
|---|---|---|---|---|---|
| 0.00002 | 0.00008 | 0.0003 | 0.001 | 0.005 | 0.015 |
| 0.00003 | 0.0001 | 0.0005 | 0.002 | 0.008 | 0.020 |
| 0.00005 | 0.0002 | 0.0008 | 0.003 | 0.010 | |

**Recommended roughness-width-cutoff values (inches)**

| | | | | | |
|---|---|---|---|---|---|
| 0.003 | 0.010 | 0.030 | 0.100 | 0.300 | 1.0000 |

# PRODUCTION METHODS

## 80. Production Methods and Dimensioning Practice.

Production methods can be classified as (1) *unit production*, when one or only a few devices or structures are to be built, and (2) *quantity* or *mass production*, when a large number of practically identical machines or devices are to be made with the parts interchangeable from one machine to another.

*Unit-production* methods almost always apply to large machines and structures, especially if they are custom-made. The large size to some extent eliminates the need for great accuracy. Each individual part is produced to fit or is fitted to the adjacent parts, frequently on the job, by experienced workmen in accordance with common fractional dimensions and directions given on the drawings. Since interchangeability of parts is no object, tolerances are not ordinarily used.

Similar methods are employed for unit production of smaller machines and mechanical devices. The drawings may have common fractional dimensions exclusively, on the assumption that the parts will be individually fitted in the shop. If this practice is followed, the manufacturing group accepts the re-

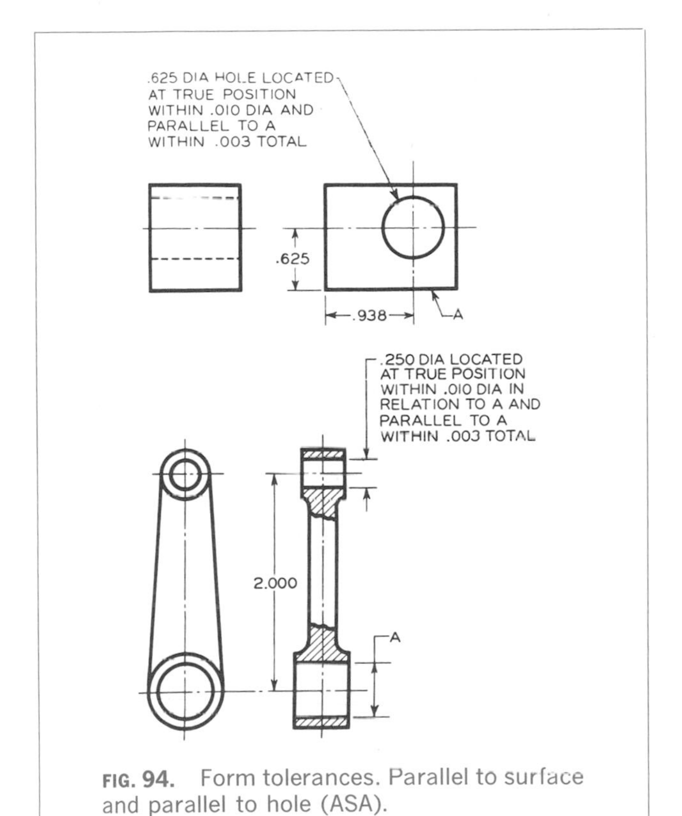


**FIG. 94.** Form tolerances. Parallel to surface and parallel to hole (ASA).

**FIG. 95.** Surface roughness (ASA).

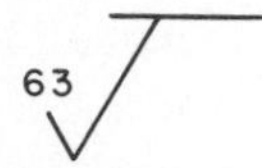


**FIG. 96.** Symbol. The roughness height rating is placed at the left of the long leg.

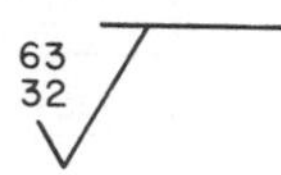


**FIG. 97.** Symbol. The specification of maximum and minimum roughness height indicates the allowable range.

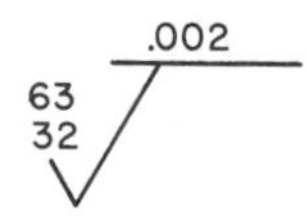


**FIG. 98.** Symbol. The maximum waviness height rating is placed above the horizontal extension.

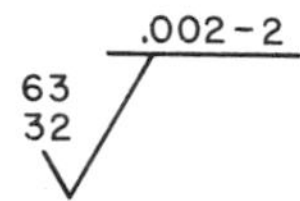


**FIG. 99.** Symbol. The maximum waviness width rating is placed to the right of the waviness height rating.

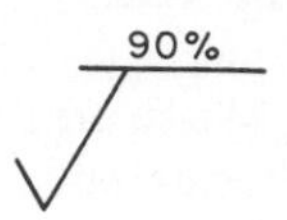


**FIG. 100.** Symbol. To specify the contact area, when required, the percentage value is placed above the extension line.

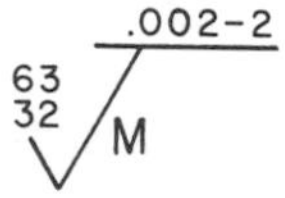

**FIG. 101.** Symbol. Lay designation is given by the lay symbol placed to the right of the long leg.

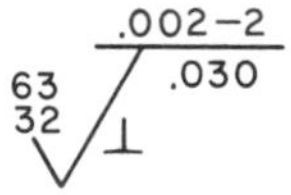

**FIG. 102.** Symbol. The roughness width cutoff rating is placed below the horizontal extension.

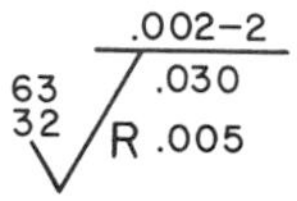

**FIG. 103.** Symbol. When it is required, the maximum roughness width rating is placed at the right of the lay symbol.

sponsibility for the proper functioning of the machine, and in some cases even for designing some of the parts. Skilled workmen are employed for this work. Usually one machine is completed before another is started, and the parts are not interchangeable.

*Quantity-production* methods are used whenever a great many identical products are made. After a part has been detailed, the operations-planning group of the engineering department plans the shop operations step by step. Then special tools are designed by the tool-design group so that, in production, semiskilled workmen can perform operations that would otherwise require skilled workmen. These tools, built by a highly skilled toolmaker, simplify and greatly increased the rate of production.

One workman performs a single operation on a part; then it is passed to a second workman, who performs another operation; and so on until the part has been completed. Specially designed tools and equipment make it possible to produce economical parts that have dimensional exactness consistent with the requirements for interchangeability. The assembly may also be made by semiskilled workers using special assembly fixtures and tools.

With this system nothing can be left to the judgment of the workman. In preparing drawings intended for quantity production, the engineering department must assume full responsibility for the success of the machine by making the drawings so exact and complete that, if followed to the letter, the resulting parts cannot fail to be satisfactory. The engineering department alone is in a position to correlate corresponding dimensions of mating parts, establish dimensional tolerances, and give complete directions for the entire manufacturing job.

It is sometimes expedient for concerns doing unit or small-production work to follow the methods of the quantity-production system. The advantage they gain is interchangeability of parts, which can be produced without reference or fitting to mating parts.

## 81. Principles for the Selection of Dimensions.

Systematic selection of dimensions demands consideration of the *use* or *function* of the part and the *manufacturing process* to be used in producing it.

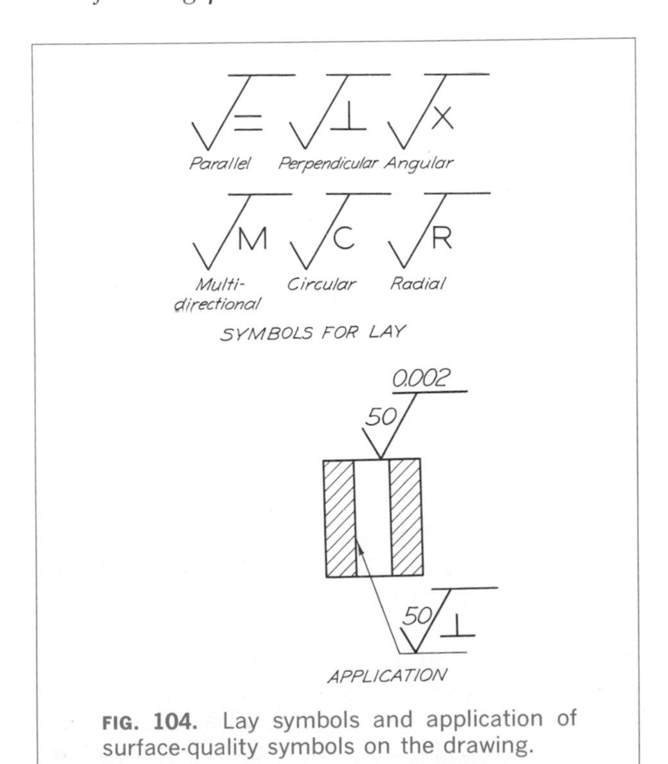


FIG. 104. Lay symbols and application of surface-quality symbols on the drawing.

The *functional principle* recognizes that it is essential to dimension between points or surfaces associated through their functional relationship with points or surfaces of mating parts. This is accomplished by correlating the dimensions on a drawing of one part with the mating dimensions on the drawing of a mating part and arranging the tolerances of these dimensions to ensure interchangeability and proper functioning.

The *process principle*, or "workman's rule," as it is sometimes called, recognizes that the work of manufacture can be made easier by giving directly the dimensions the shop will find most convenient to "work to" in producing the part. Here a knowledge of manufacturing processes and procedure is necessary, as explained in Chap. 11.

In some cases there may be a conflict between these two principles. Whenever there is, the functional principle must take precedence, as any attempt to satisfy both principles would result in overdimensioning, as described in Sec. 24, causing confusion for the workmen and possible malfunctioning of the part. With few exceptions, however, dimensions can be chosen to satisfy both principles.

## 82. Procedure for the Selection of Dimensions.

A systematic procedure for selecting dimensions is of course desirable. The following steps can serve as a guide:

1. Carefully study the part along with its mating part or parts. Pay particular attention to the mating and controlling surfaces. Plan dimensions meeting functional requirements before placing any dimensions on the drawing so that you can correlate dimensions of mating parts.

2. Study the part to determine whether or not the manufacturing processes can be simplified by an alteration of any of the functional dimensions. Do not make changes if the functioning of the part would be impaired in any way.

3. Select the nonfunctional dimensions, being guided by the process principle, so that the dimensions will be readily usable by the workmen. Avoid overdimensioning and duplication.

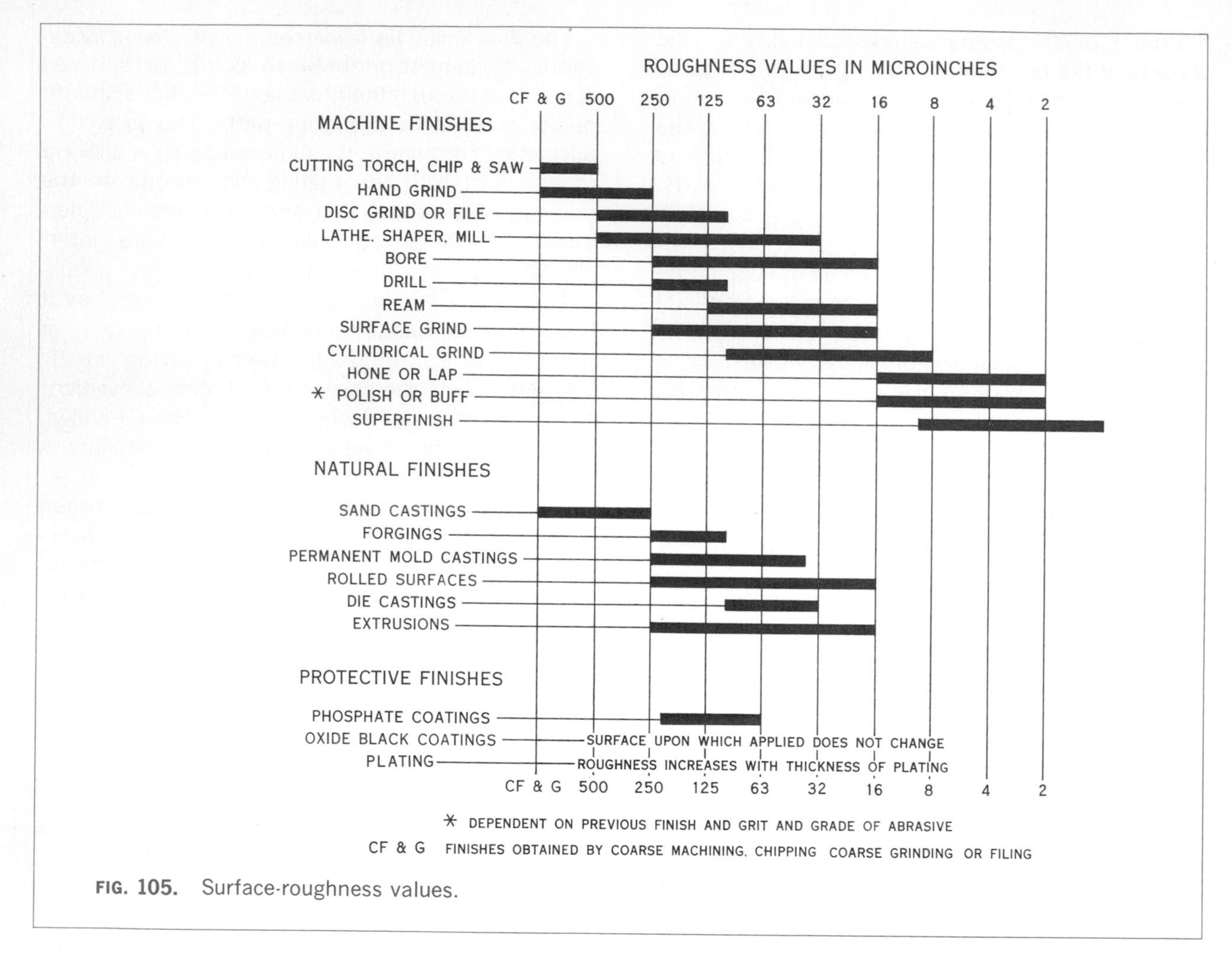


**FIG. 105.** Surface-roughness values.

In general, dimensions for mating surfaces are governed by the functional and process principles, and the dimensions for nonmating surfaces are governed by the process principle only.

Occasionally the manufacturing process will not be known at the time of dimensioning. This may happen when there are optional methods of manufacture, all equally good, or when the details of the manufacturing equipment of a contracting firm are not known. When this is the case, the dimensions should be selected and toleranced in a logical manner so that the part, regardless of the method by which it is produced, cannot fail to be satisfactory. The size and location dimensions arrived at by the shape-breakdown system described in Sec. 22 apply here to a great extent since these dimensions fulfill most production requirements. Contracting firms often redraw incoming part drawings, dimensioning them for the most economical production with their own shop equipment.

## 83. Methods of Part Production.

In following the process principle, the basic methods of part production—casting, forging, etc.—as described in Chap. 11, must be known. The manufacturing procedures followed with the particular type of part involved are then considered in selecting the dimensions. The only workman to be considered in dimensioning a part cut from solid stock is the machinist. For parts produced by casting, the workmen to be considered are the patternmaker (for sand castings) or the diemaker (for die castings) and, for finishing, the machinist. Forged parts subject to quantity production are dimensioned for the diemaker and machinist. For parts produced from sheet stock, the template maker, the diemaker, and the machinist must be considered; information for making the template and for forming the blank is obtained from a detail drawing showing the part as is should be when completed. In every case one drawing, appropriately dimensioned, must show the finished part.

The sections that follow give examples of the dimensioning of machine parts for quantity production.

### 84. Dimensioning a Part Machined from Stock.

Figure 106 is a detail drawing of the stud from the rail-transport hanger in Prob. 51, final chapter. Study the drawing to determine the function of the stud. The stud is produced by machining on a lathe. Cold-rolled steel stock, $1\frac{1}{4}$ in. in diameter, is used. The stock diameter is the same as the large end of the stud, thus eliminating one machining operation.

Shape breakdown of the part results in a series of cylinders each requiring two dimensions—diameter and length. The important functional dimensions have been marked (on Fig. 106) with the

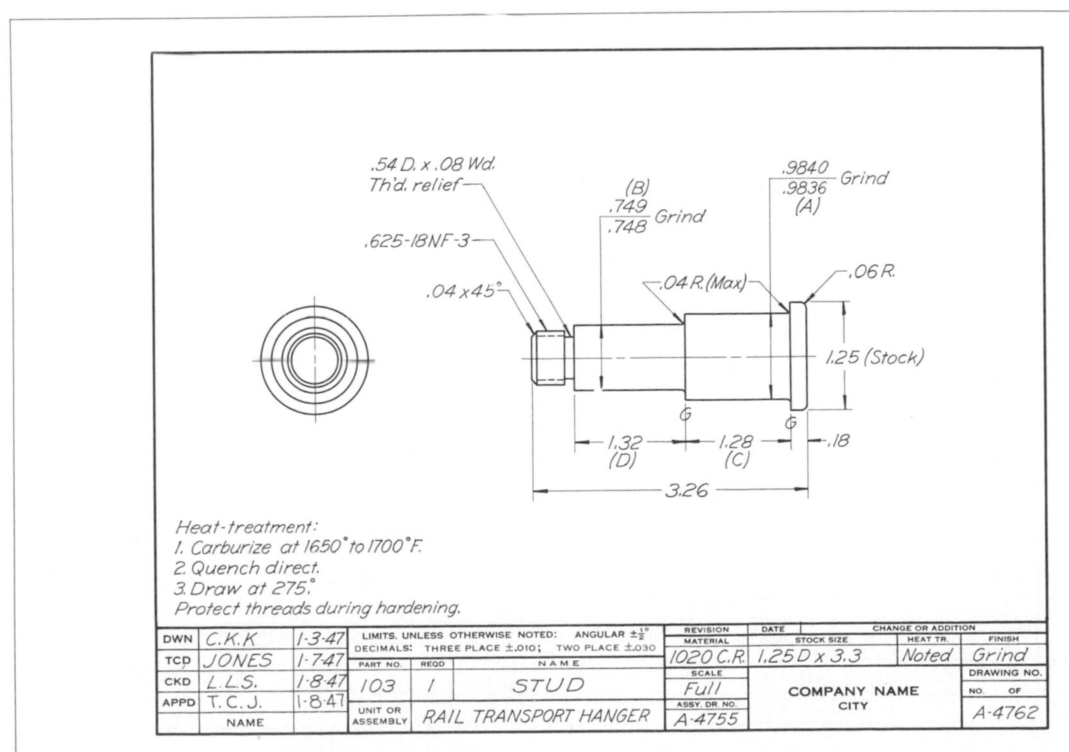


FIG. 106. Dimensioning a part machined from stock.

letters *A*, *B*, *C*, and *D*. Diameter *A* is given to correlate with the bore of the bearings; a four-place decimal limit provides for the desired fit. Dimension *B* is a three-place decimal limit to correlate with a similar dimension for the hole in the hanger. Dimension *C* is made 0.03 in. larger than the combined width of the two bearings in order to allow the inner races of the bearings to "creep." Dimension *C*, a two-place decimal, can vary ±0.030 in., but clearance for the bearings is assured under all conditions. Dimension *D* is made approximately 0.05 in. less than the length of the hanger hub to ensure that the nut will bear against the hanger rather than on the shoulder of the stud.

Functional dimensions need not always be extremely accurate. Note that dimensions *C* and *D*, with the relatively broad tolerance of two-place decimals, will allow the part to function as intended.

The thread specification can be considered as a functional dimension with the tolerance provided through the thread class.

The remainder of the dimensions selected are those that best suit shop requirements. Note that the thread length and over-all dimension cannot both be given, or the part would be overdimensioned.

## 85. Dimensioning a Casting.

The dimensions required for sand castings can be classified as those used by the patternmaker and those used by the machinist. Since a cast part has two distinct phases in its manufacture, we will discuss a particular case first with the drawings made according to the multiple system explained in Chap. 11, one for the patternmaker (Fig. 107) and one for the machinist (Fig. 108).

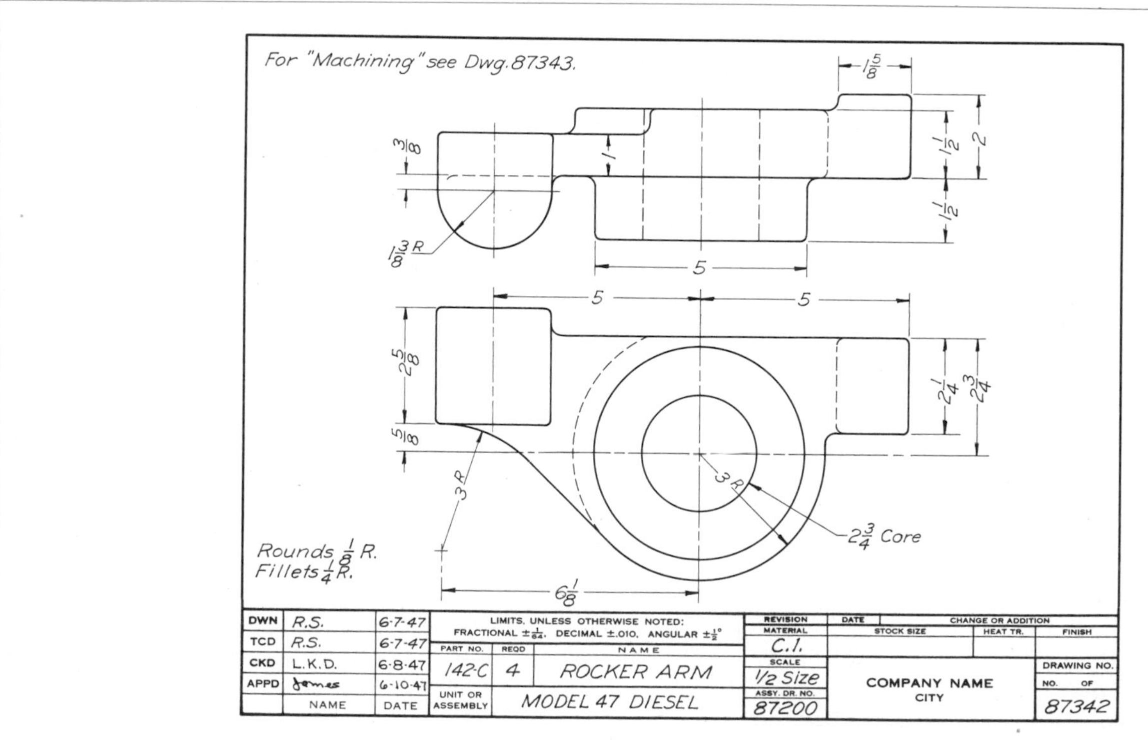


**FIG. 107.** Dimensioning an unmachined casting.

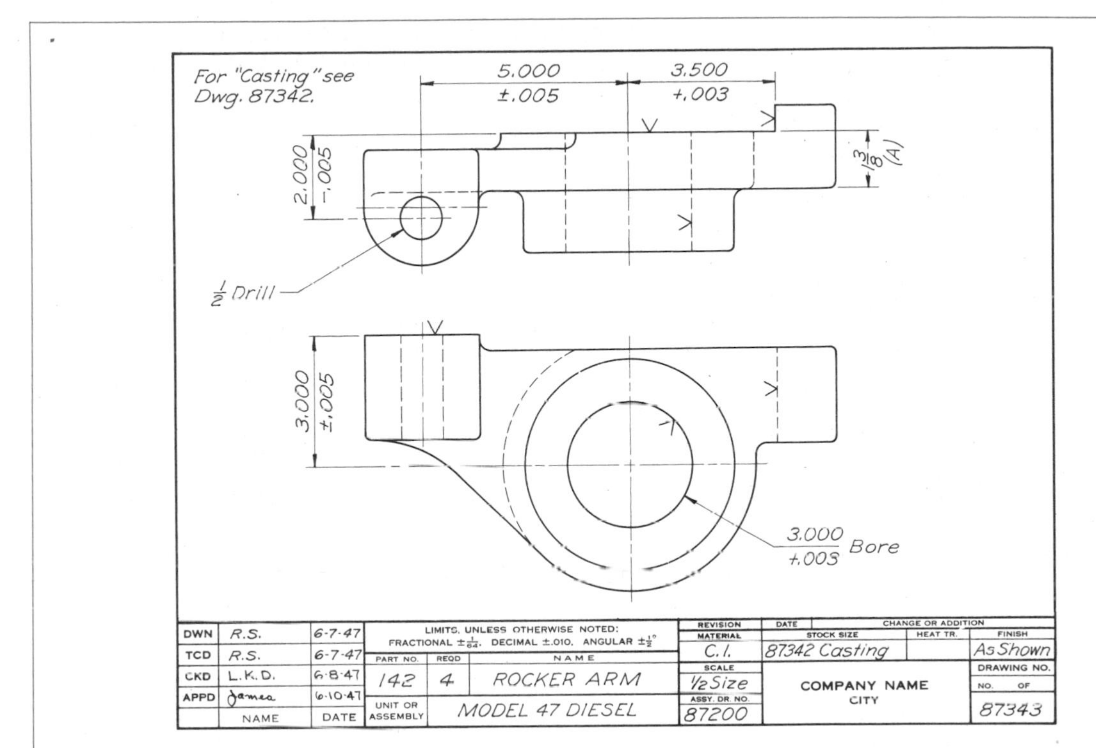


**FIG. 108.** Dimensions for machining a casting.

The *casting drawing* gives the shape of the unmachined casting and carries dimensions for the patternmaker only. Shape breakdown will show that each geometric shape has been dimensioned for size and then positioned, thus providing dimensions easily usable by the workman. Some of the dimensions might be altered, depending upon how the pattern is made; the most logical and easily usable combination should be given. Note that the main central shape is dimensioned as it would be laid out on a board. Note also that several of the dimensions have been selected to agree with required functional dimensions of the machined part although the dimensions employed have been selected so as to be directly usable by the patternmaker; they also achieve the *main objective,* which is to state *the sizes that the unmachined casting must fulfill when produced.*

The *machining drawing* shows only the dimensions required by the machinist. These are almost all functional dimensions and have been selected to correlate with mating parts. It is important to note that a starting point must be established in each of the three principal directions for machining the casting. In this case a starting point is provided by (1) the coincidence of the center lines of the large hole and cylinder (positioning in two directions) and (2) dimension $A$ to position the machined surface on the back, from which is positioned the drilled hole. Dimension $A$ is a common fraction carrying the broad tolerance of $\pm 1/64$ in., as there is no functional reason for working to greater precision.

Figure 109 is a drawing of the same part used in Figs. 107 and 108 but with the casting drawing dispensed with and the patternmaker's dimensions incorporated in the drawing of the finished part. In combining the two drawings, some dimensions have been eliminated, as the inclusion of all the dimensions of both drawings would result in overdimensioning; thus the patternmaker must make use of certain machining dimensions in his work. In working from the drawing in Fig. 109, the patternmaker provides for machining allowance, being guided by the finish marks. In the drawing in Fig. 107 the engineering department provides for the machining allowance by showing and dimensioning the rough casting oversize where necessary for machining, and no finish marks are used.

**86. Dimensioning a Drop Forging.** Figure 110 is a drawing of a drop forging showing, at the left, the unmachined forging and, at the right, the machined forging. The drawing of the unmachined forging carries the dimensions it must fulfill when produced; these dimensions have been selected so as to be most useful to the diemaker for producing the forging dies. As the draft on drop forgings is considerable, it is shown on the drawing and dimensioned (usually) by a note. If the draft varies for different portions of the part, the angles may be given on the views. The dimensions parallel to the horizontal surfaces of the die are usually given so as to specify the size at the *bottom of the die cavity*. Thus, in dimensioning, visualize the draft as stripped off; then its apparent complication will no longer be a difficulty.

The machining drawing shows the dimensions for finishing. These are all functional, selected from the standpoint of the required function of the part. Study the illustration carefully.

**87. Dimensioning a Sheet-metal Part.** Parts to be made of thin materials are usually drawn showing the part in its finished form, as in Fig. 111.

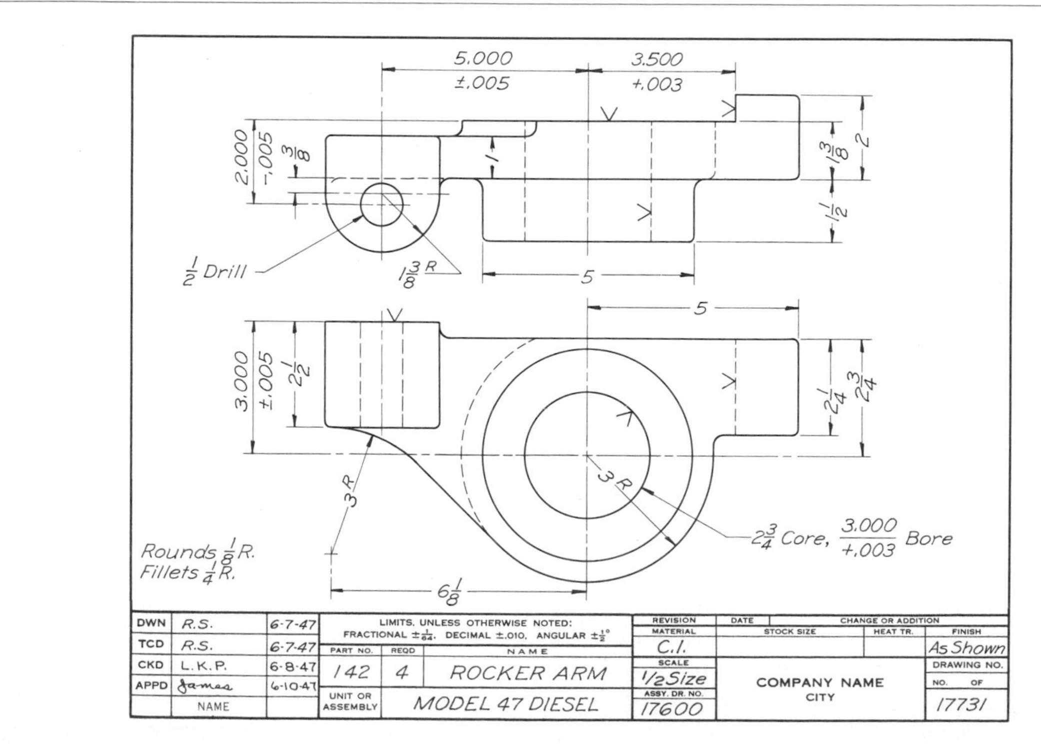


**FIG. 109.** All dimensions for a casting.

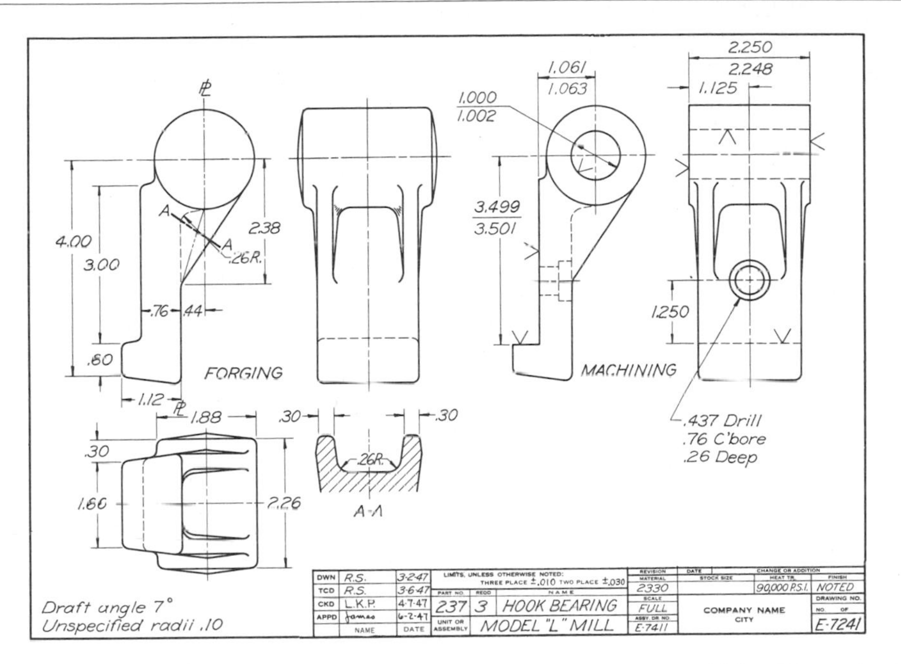


FIG. 110. Dimensioning a drop forging.

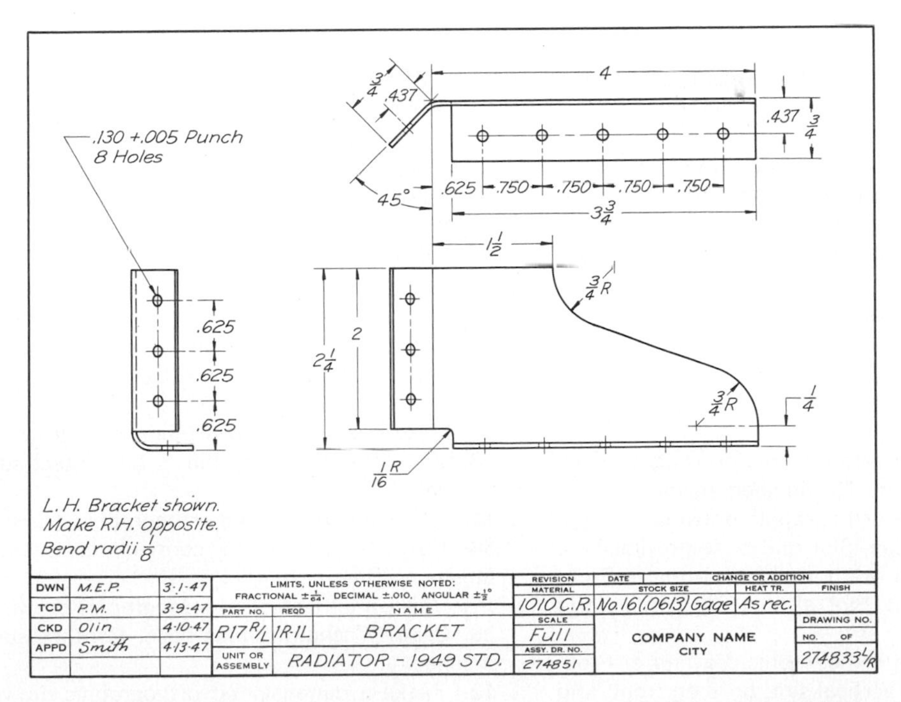


FIG. 111. Dimensioning a sheet-metal part.

The template maker first uses the drawing to lay out a flat pattern of the part. If only a few parts are to be made, this template will serve as a pattern for cutting the blanks. Then the part will be formed and completed by hand. If a large number of parts are to be made, the diemaker will use the template and drawing in making up the necessary dies for blanking, punching, and forming. The work of both template maker and diemaker is simplified by giving the dimensions to the same side of the material inside or outside, whichever is more important from the functional standpoint, as shown in Fig. 111. Dimensions to rounded edges (bends) are given to the theoretical sharp edges, which are called *mold lines*. The thickness of the material is given in the "stock" block of the title strip. Note in the figure that the holes are positioned in groups (because of functional requirements) and that important functional dimensions are three-place decimals.

## PROBLEMS

The following problems are given as studies in dimensioning in which the principles presented in this chapter are to be applied. Attention should be given to the methods of manufacture, as described in Chap. 11. Assume a function for the part in order to fix the position of finished surfaces and limit possibilities in the selection of dimensions.

### Group 1. Dimensioned Drawings from Pictorial Views.

The problems that are presented in pictorial form in Chaps. 5 to 8 can be used as dimensioning problems. Either dimension a drawing you have already made as an exercise in shape description or, for variety, select another. Because of the difference in method of representation, the dimensions on a pictorial drawing and those on an orthographic drawing of the same object will not necessarily correspond; therefore, pay no attention to the placement of dimensions on the pictorial drawings except for obtaining the sizes needed. A selection of 12 problems, graded in order of difficulty, is given below.

**1.** Make a dimensioned orthographic drawing of Prob. 19, Chap. 5, beam support. No finished surfaces.
**2.** Make a dimensioned orthographic drawing of Prob. 23, Chap. 5, slotted wedge. Slot and base are finished.
**3.** Make a dimensioned orthographic drawing of Prob. 26, Chap. 5, corner stop. Slot at top, cut corner, and base are finished.
**4.** Make a dimensioned orthographic drawing of Prob. 29, Chap. 5, guide base. Vertical slot, boss on front, and base are finished.
**5.** Make a dimensioned orthographic drawing of Prob. 37, Chap. 5, shifter fork. All contact surfaces are finished.
**6.** Make a dimensioned orthographic drawing of Prob. 47, Chap. 5, shaft guide. L-shaped pad and end of hub are finished.
**7.** Make a dimensioned orthographic drawing of Prob. 13, Chap. 7, jig angle. Finished all over.
**8.** Make a dimensioned orthographic drawing of Prob. 19, Chap. 7, angle-shaft base. Base and slanting surface are finished.
**9.** Make a dimensioned orthographic drawing of Prob. 18, Chap. 7, radial swing block. All contact surfaces are finished.
**10.** Make a dimensioned orthographic drawing of Prob. 38, Chap. 7, transverse connection. Base pads are finished.
**11.** Make a dimensioned orthographic drawing of Prob. 39, Chap. 7, chamfer-tool base. Contact surfaces are finished.
**12.** Make a dimensioned orthographic drawing of the desurger case.

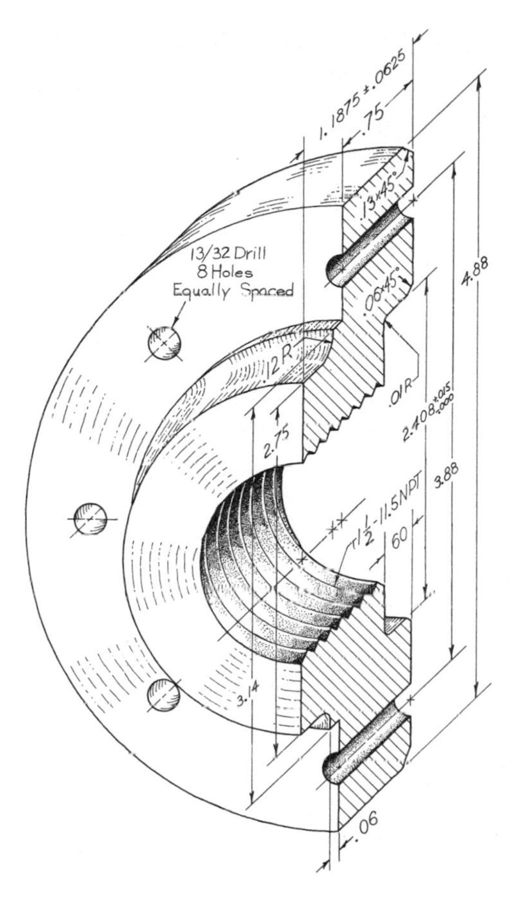


PROB. 12. Desurger case. (*Courtesy of Westinghouse Air Brake Co.*)

**Group 2. Dimensioned Drawings from Models.** Excellent practice in dimensioning is afforded by making a detail drawing from a pattern, casting, or forging or from a model made for the purpose. Old or obsolete patterns can often be obtained from companies manufacturing a variety of small parts, and "throw-out" castings or forgings are occasionally available. In taking measurements from a pattern, a shrink rule should be used; allowance must be made for finished surfaces.

**Group 3. Pieces to be Drawn and Dimensioned.** The problem illustrations are printed to scale, as indicated in each problem. Transfer distances with dividers or by scaling, and draw the objects to a convenient scale on paper of the size to suit. For proper placement of dimensions, more space should be provided between the views than is shown in the illustrations.

Use the aligned or horizontal dimensioning system, as desired. It is suggested that some problems be dimensioned in the complete decimal system.

**13.** Stud shaft, shown half size. Machined from steel-bar stock.

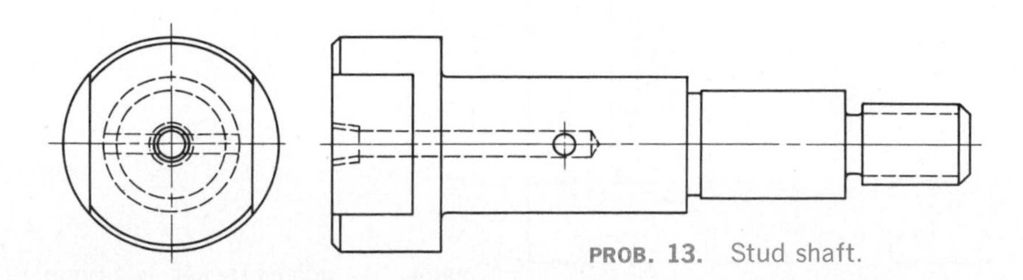

PROB. 13. Stud shaft.

**14.** Shaft bracket, shown half size. Malleable iron. Hole in base is drilled and counterbored for a socket-head cap screw. Base slot and front surface of hub are finished. Hole in hub is bored and reamed. The function of this part is to support a shaft at a fixed distance from a machine bed, as indicated by the small pictorial view.

**15.** Idler bracket, shown half size. Cast iron. Hole is bored and reamed. Slot is milled.

**15*A*.** See Prob. 15. Draw and dimension the right-hand part.

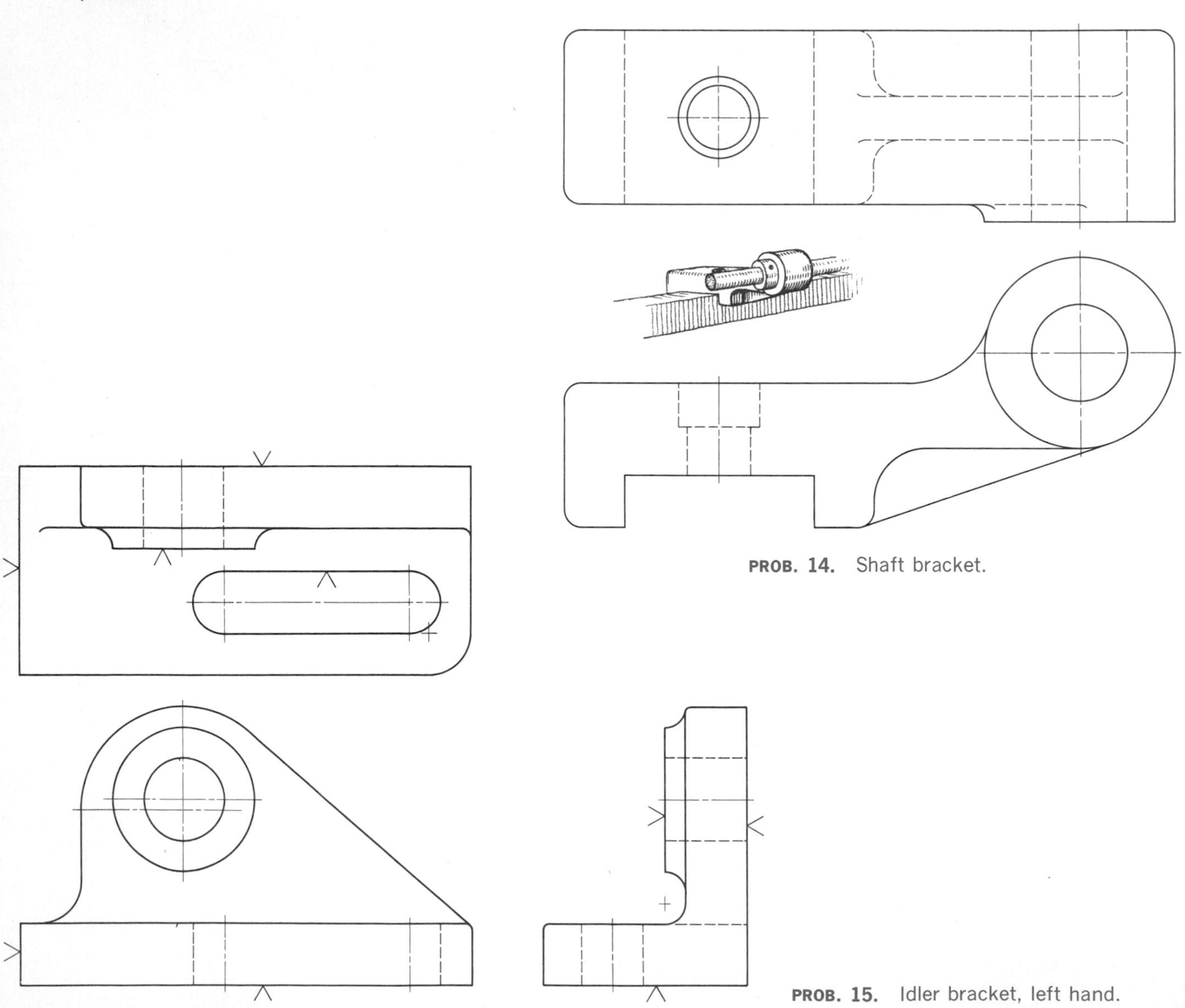

**PROB. 14.** Shaft bracket.

**PROB. 15.** Idler bracket, left hand.

**16.** Filter flange, shown half size. Cast aluminum. The small holes are drilled. Add spot faces.

**17.** Boom-pin rest. Steel drop forging. Shown half size; draw half size or full size. Add top view if desired. Show machining allowance with alternate position lines, and dimension as in Fig. 8, Chap. 11. All draft angles 7°. Holes are drilled, corner notches milled.

**17A.** Same as Prob. 17, but make two drawings; (*a*) the unmachined forging dimensioned for the diemaker and (*b*) the machined forging dimensioned for the machinist. Reference: Fig. 110.

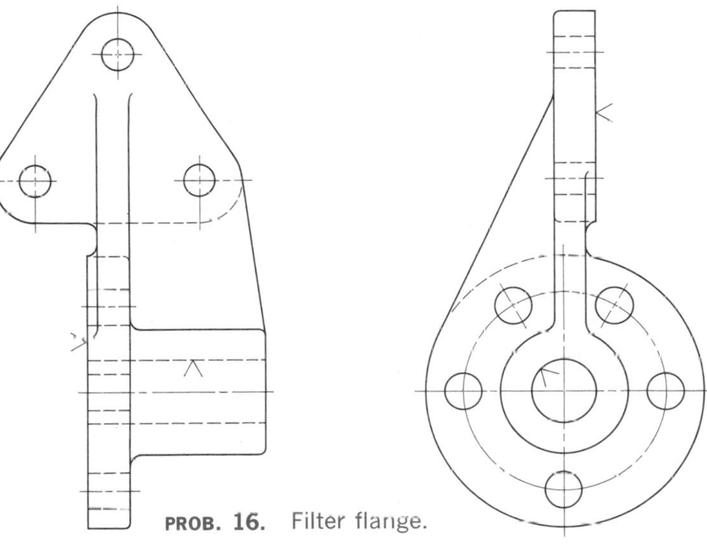

**PROB. 16.** Filter flange.

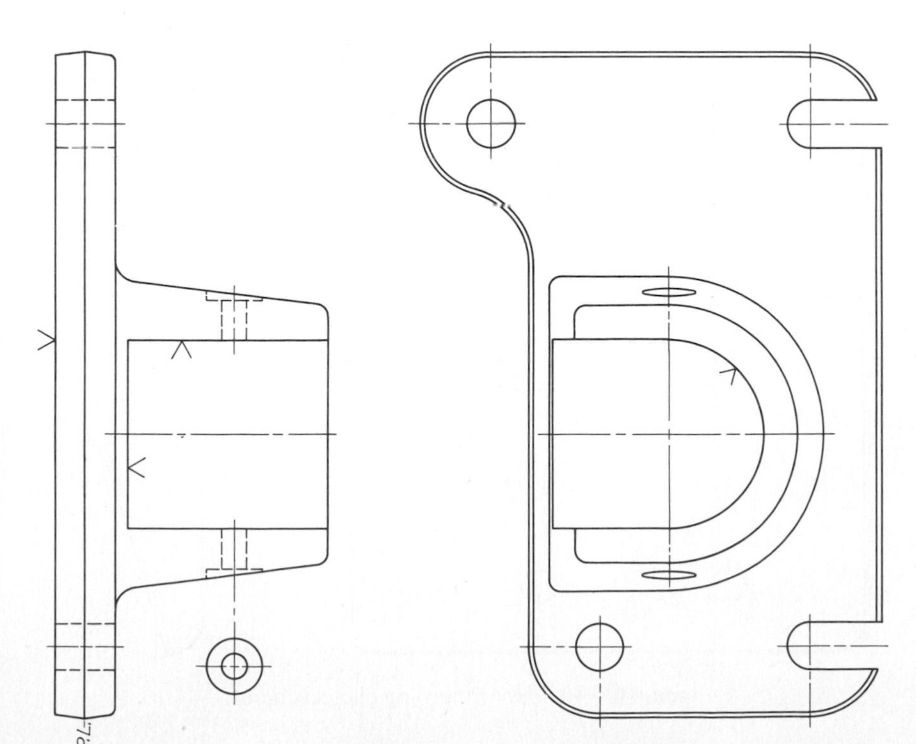


**PROB. 17.** Boom-pin rest.

**18.** Clutch lever. Aluminum drop forging. Shown half size, draw full size or twice size. Add top view if desired. Holes are drilled and reamed; ends of hub are finished; left-end lug is straddle milled; slot in lower lug is milled. Show machining allowance with alternate position lines, and dimension as in Fig. 8, Chap. 11. Draft angles 7°.
**18*A*.** Same as Prob. 18, but make two drawings: (*a*) the unmachined forging dimensioned for the diemaker and (*b*) the machined forging dimensioned for the machinist. Reference: Fig. 110.
**19.** Radiator mounting clip, LH, no. 16 (0.0625) steel sheet. Shown half size. Holes and slot are punched. Reference: Fig. 111.

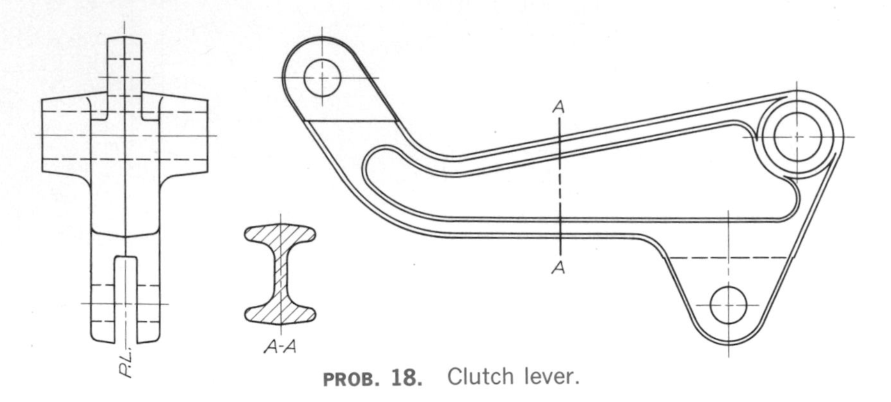


**PROB. 18.** Clutch lever.

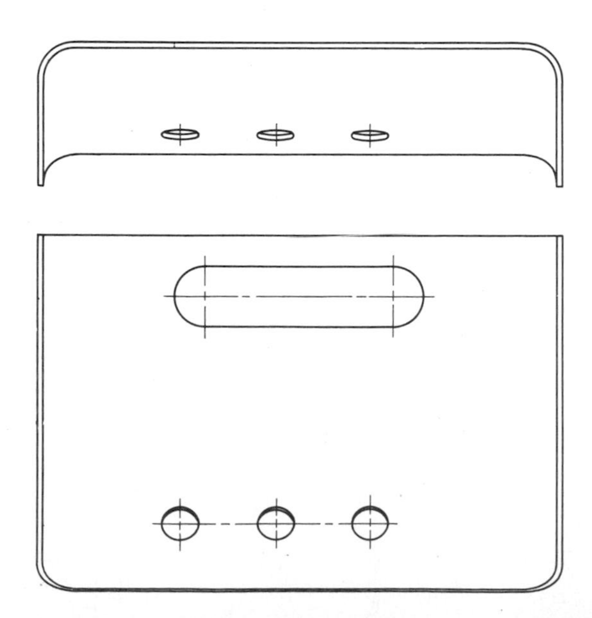

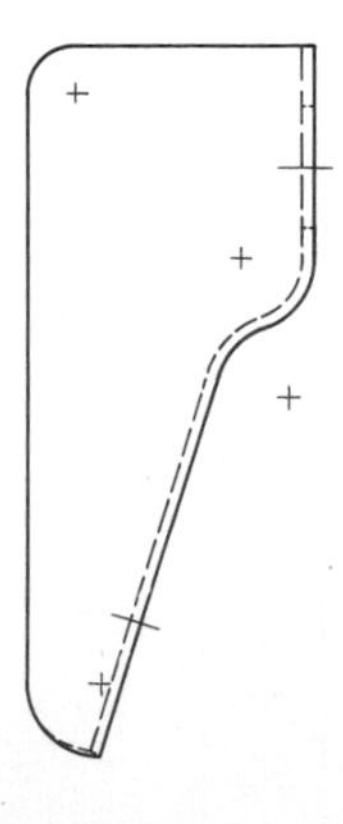

**PROB. 19.** Radiator mounting clip, left hand.

**20.** Pulley bracket. Shown half size. Aluminum sheet, 24-ST, 0.032 in. thick. Reference: Fig. 111.

**21.** Using the scale shown, draw and dimension the check-valve body.

**22.** Using the scale shown, draw and dimension the hydrostatic pressure housing. Draw to convenient scale on paper of a size adequate for clear representation. Use either the fractional or decimal system.

**23.** Using the scale shown, draw and dimension the universal-joint housing. Use a scale and paper size adequate for clear representation. Either the fractional or decimal system may be used.

**24.** Using dividers on the scale shown, draw and dimension the universal joint. Note that the flexing-rod portions are helical. Use fractional or decimal system.

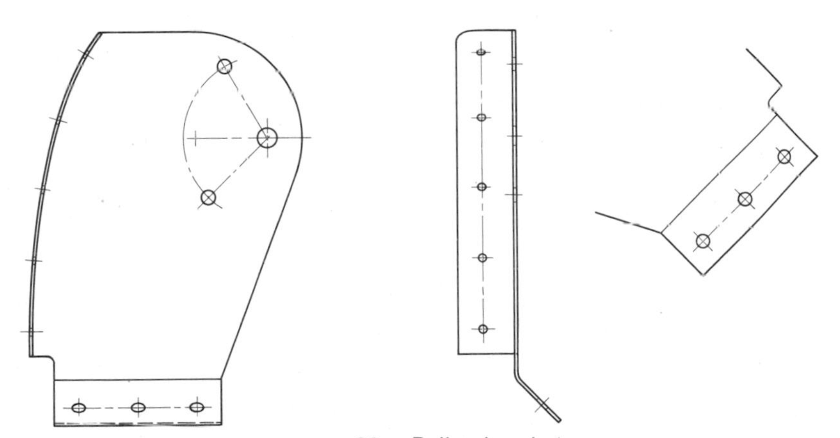

**PROB. 20.** Pulley bracket.

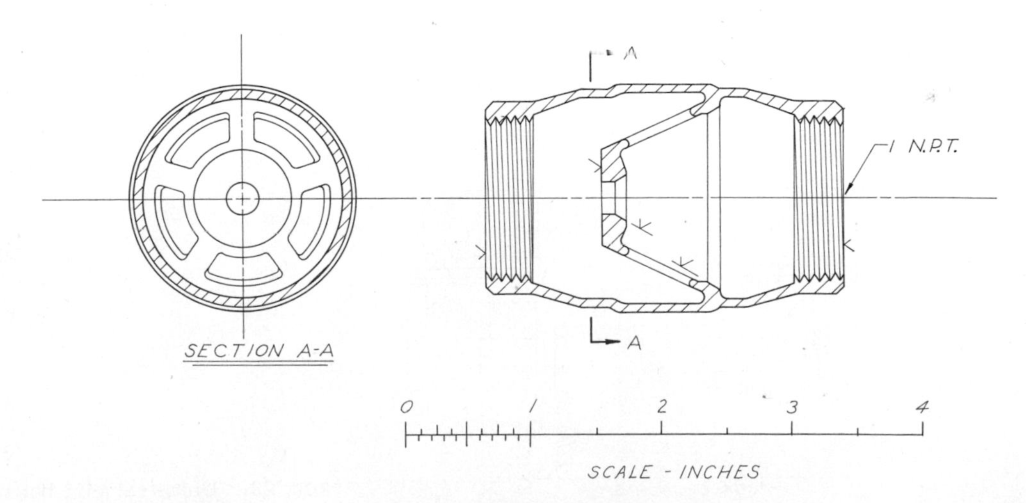


**PROB. 21.** Check-valve body.

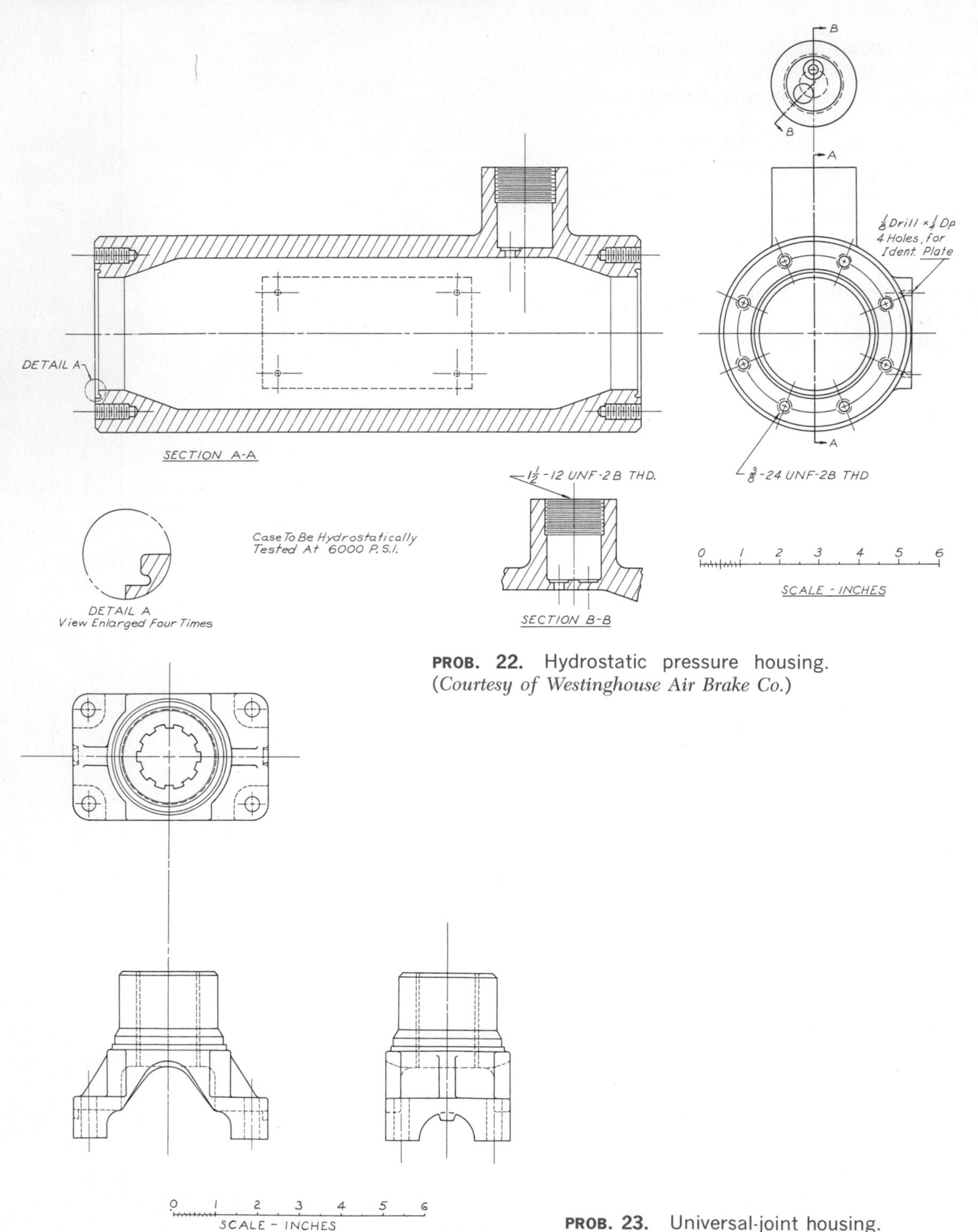


**PROB. 22.** Hydrostatic pressure housing. (*Courtesy of Westinghouse Air Brake Co.*)

**PROB. 23.** Universal-joint housing.

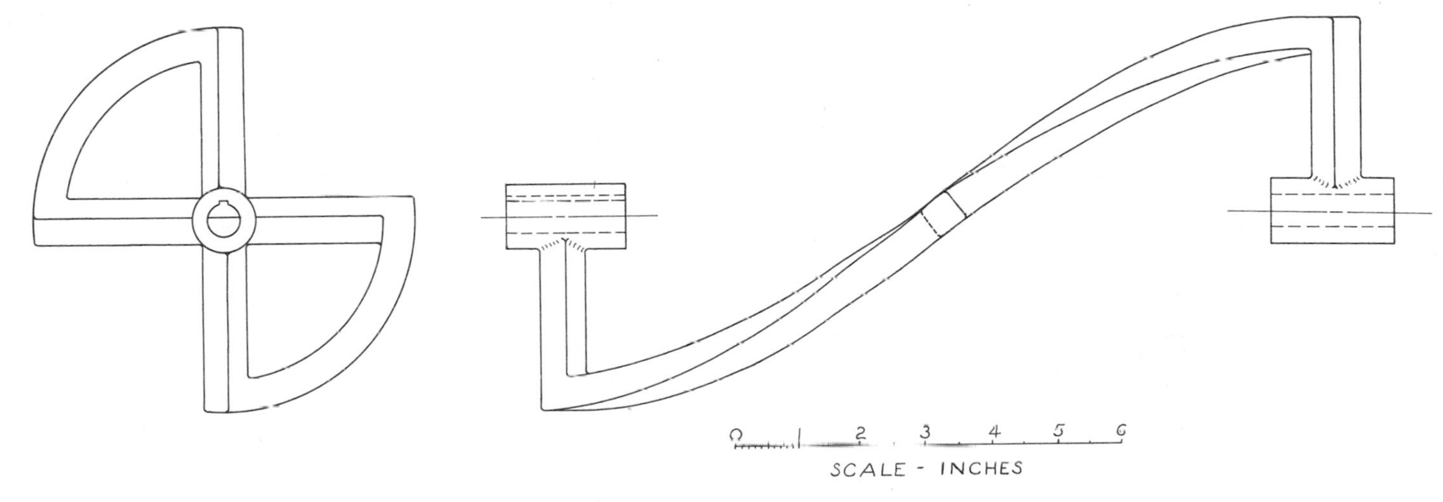


**PROB. 24.** Universal joint.

## Group 4. Dimensioned Drawings from an Assembly or Design Drawing.

The assembly drawings given in the problem section of the final chapter are well suited for exercises in dimensioning detail working drawings. The assembly shows the position of each part, and the function can be understood by a study of the motion, relationship, etc., of the different parts. Note particularly the mating and controlling surfaces and the logical reference surfaces for dimensions of position. Inasmuch as the dimensioning of an assembly drawing is crowded, and it will probably not even have the same views as the detail drawing of one of the parts, disregard the position and selection of the assembly dimensions, and use them only to obtain sizes. The following are suggested. Note that the problem numbers given refer to those in the final chapter.

**25.** Detail drawing of shaft, Prob. 41.
**26.** Detail drawing of bushing, Prob. 41.
**27.** Detail drawing of bracket, Prob. 41.
**28.** Detail drawing of body, Prob. 45.
**29.** Detail drawing of hanger, Prob. 51.
**30.** Detail drawing of rack, Prob. 47.
**31.** Detail drawing of rack housing, Prob. 47.
**32.** Detail drawing of cover, Prob. 47.
**33.** Detail drawing of base, Prob. 39.
**34.** Detail drawing of base, Prob. 80.
**35.** Detail drawing of jaw, Prob. 80.
**36.** Detail drawing of screw, Prob. 80.
**37.** Detail drawing of screw bushing, Prob. 80.
**38.** Detail drawing of body (drop forging), Prob. 59.
**39.** Detail drawing of base, Prob. 78.
**40.** Detail drawing of frame, Prob. 78.
**41.** Detail drawing of frame, Prob. 79.
**42.** Detail drawing of ram, Prob. 79.
**43.** Detail drawing of pinion shaft, Prob. 79.
**44.** Detail drawing of base, Prob. 79.
**45.** Detail drawing of cover, Prob. 79.
**46.** Detail drawing of sleeve ball, Prob. 79.
**47.** Detail drawing of stud ball, Prob. 79.
**48.** Detail drawing of body, Prob. 57.
**49.** Detail drawing of spring, Prob. 57.

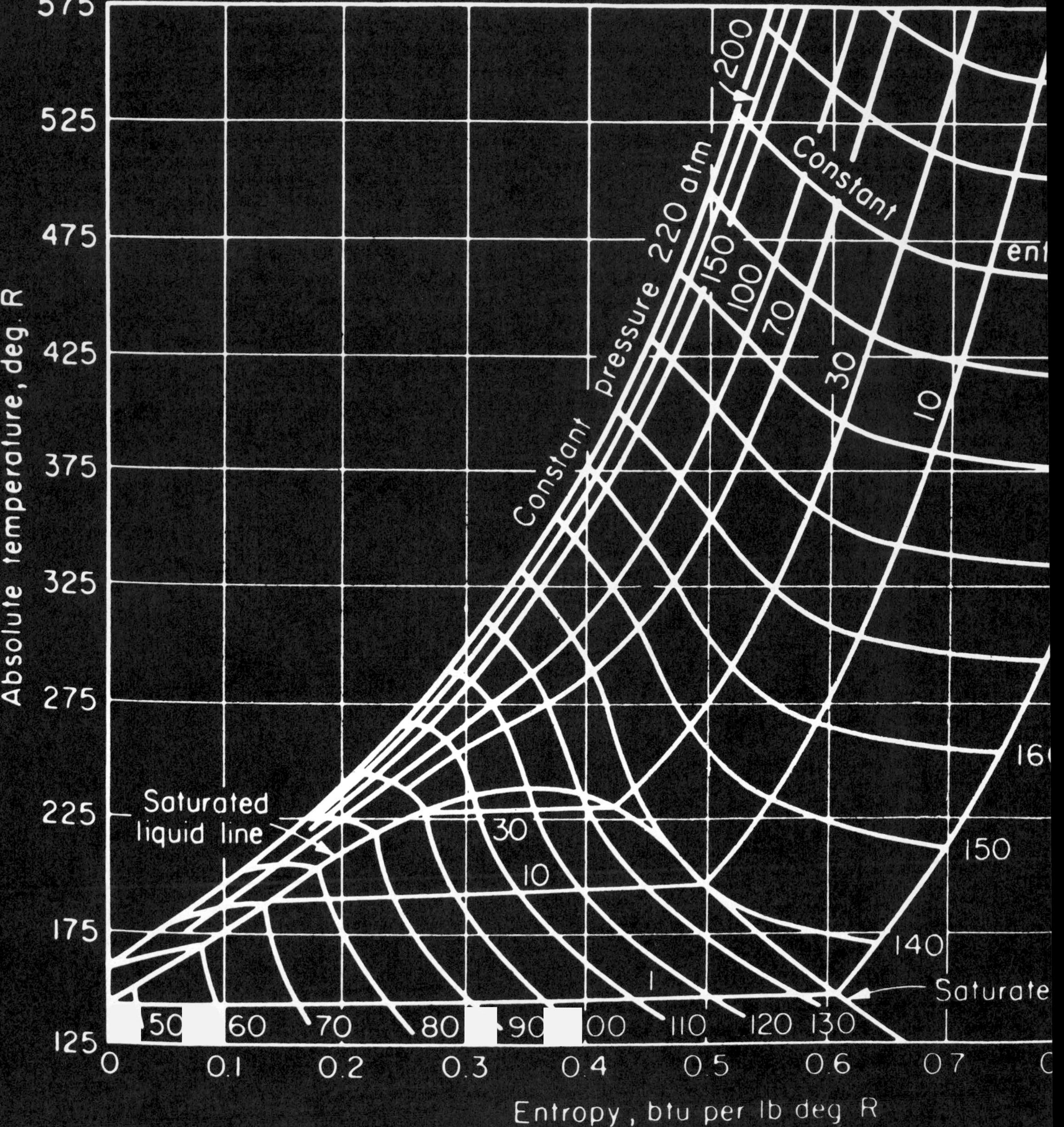

575
525
475
425
375
325
275
225
175
125
Absolute temperature, deg. R
0
0.1
0.2
0.3
0.4
0.5
0.6
0.7
Entropy, btu per lb deg R
Constant pressure 220 atm
200
150
100
70
30
10
30
10
1
Constant
ent
Saturated liquid line
Saturate
50
60
70
80
90
00
110
120
130
140
150
160

Graphic Presentation of Facts:



# 13

# CHARTS, GRAPHS, AND DIAGRAMS

These Are Graphic Representations of Data and Are Fundamental to All of Science and Engineering

## 1. Graphic Presentations—Their Importance.

This chapter is given as an introduction to the use of graphic methods in tabulating data for analysis, solving problems, and presenting facts. We will discuss the value of this application of graphics in engineering and suggest ways of studying the subject further.

The graphic chart is an excellent method for presenting a series of quantitative facts quickly. When properly constructed, charts, graphs, and diagrams constitute a powerful tool for computation, analysis of engineering data, and the presentation of statistics for comparison or prediction.

## 2. Classification.

Charts, graphs, and diagrams fall roughly into two classes: (1) those used for purely technical purposes and (2) those used in advertising or in presenting information in a way that will have popular appeal. The engineer is concerned mainly with those of the first class, but he should be acquainted also with the preparation of those of the second and understand their potential influence. The aim here is to give a short study of the types of charts, graphs, and diagrams with which engineers and those in allied professions should be familiar.

It is assumed that the student is familiar with the use of rectangular coordinates and that such terms as "axes," "ordinates," "abscissas," "coordinates," and "variables," are understood.

## 3. Representation of Data: Rectilinear Charts.

Because the preliminary chart work in experimental engineering is done on rectilinear graph paper, the student should become familiar with this form of chart early in his course. The rectilinear chart is made on a sheet ruled with equispaced horizontal lines crossing equispaced vertical lines. The spacing is optional. One commercial graph paper is divided into squares of $\frac{1}{20}$ in., with every fifth line heavier, to aid in plotting and reading. Sheets are available with various other rulings, such as 4, 6, 8, 10, 12, and 16 divisions per inch.

It is universal practice to use the upper right-hand quadrant for plotting experimental-data curves, making the lower left-hand corner the origin. In case both positive and negative values of a function are to be plotted, as occurs with many mathematical curves, the origin must be placed so as to include all desired values.

Figure 1 shows a usual form of rectilinear chart, such as might be made for inclusion in a report.

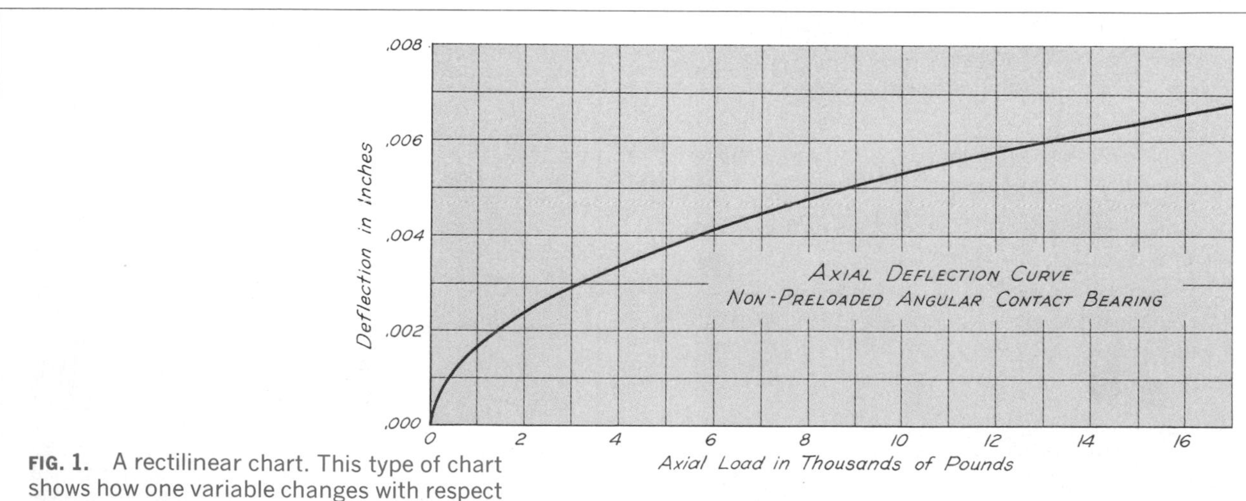


**FIG. 1.** A rectilinear chart. This type of chart shows how one variable changes with respect to another. Corresponding values can be determined.

## 4. Drawing the Curve.

In drawing graphs from experimental data, it is often a question whether the curve should pass through all the points plotted or strike a mean between them. In general, observed data not backed up by definite theory or mathematical law are shown by connecting the points plotted with straight lines, as in Fig. 2*a*. An empirical relationship between curve and plotted points may be used, as at (*b*), when, in the opinion of the engineer, the curve should exactly follow some points and go to one side of others. Consistency of observation is indicated at (*c*), in which case the curve should closely follow a true theoretical curve.

## 5. Titles and Notation.

The title is an important part of a chart, and its wording should be clear and concise. In every case, it should contain sufficient description to tell what the chart is, the source or authority, the name of the observer, and the date. Approved practice places the title at the top of the sheet, arranged in phrases symmetrically about a center line. If it is placed within the ruled space, a border line or box should set it apart from the sheet. Each sheet of curves should have a title; and when more than one curve is shown on a sheet, the different curves should be drawn so as to be easily distinguishable. This can be done by varying the character of the lines, using full, dashed, and dot-and-dash lines, with a tabular key for identification, or by lettering the names of the curves directly along them. When the charts are not intended for reproduction, inks of different colors can be used.

## 6. To Draw a Chart.

In drawing a coordinate chart, the general order is: (1) compute and assemble all data; (2) determine the size and kind of chart best adapted and whether to use printed or plain paper; (3) determine, from the limits of the data, the scales for abscissas and ordinates to give the best effect to the resulting curve; (4) lay off the independent variable (often *time*) on the horizontal, or $X$, axis and the dependent variable on the vertical, or $Y$, axis; (5) plot points from the data and pencil the curves; (6) ink the curve; (7) compose and letter the title and coordinates.

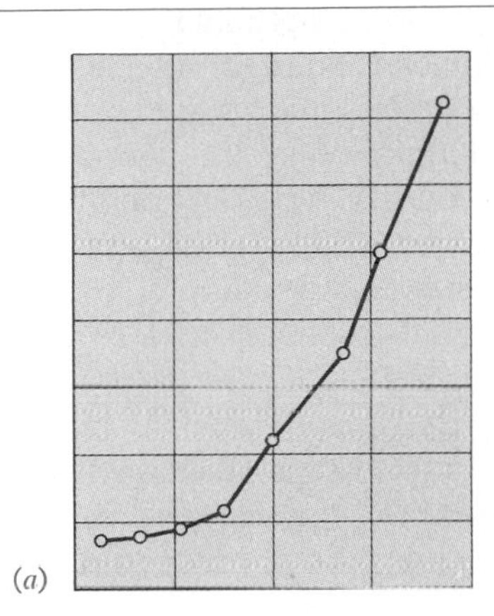

(*a*)

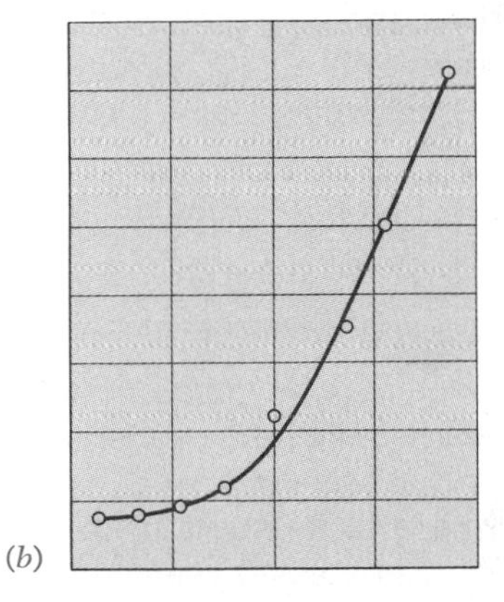

(*b*)

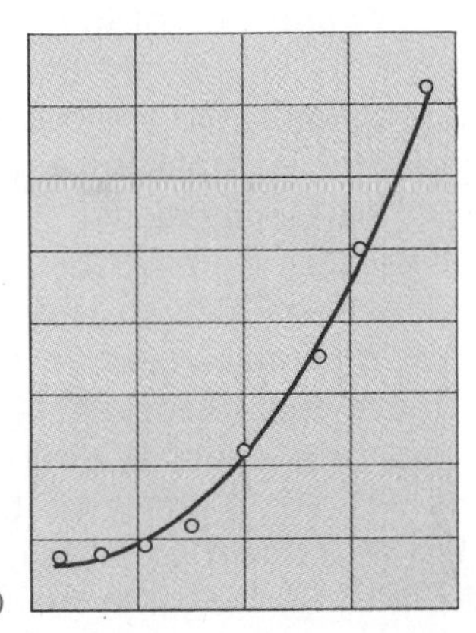

(*c*)

FIG. 2. Methods of drawing curves. (*a*) "Curve" for data not supported by theory and therefore not necessarily producing a smooth curve; (*b*) and (*c*) curves for data known to produce smooth curves.

The construction of a graphic chart requires good draftsmanship, especially for the lettering, but in engineering and scientific work the primary considerations are judgment in the proper selection of coordinates, accuracy in plotting points and drawing the graph, and an understanding of the functions and limitations of the resulting chart.

When the chart is drawn on a printed form, to be blueprinted, the curve can be drawn on the reverse side of the paper, enabling erasures to be made without injuring the ruled surface.

Green is becoming the standard color for printed forms. Blue will not print or photograph; red is trying on the eyes.

If the curve is for purposes of computation, it should be drawn with a fine, accurate line; if for demonstration, it should be fairly heavy, for contrast and effect.

The following rules are adapted from ASA Y15 (formerly Z15):

*Standards for Graphic Presentations*

1. A graph should be free of all lines and lettering that are not essential to the reader's clear understanding of its message.

2. All lettering and numbers on a graph should be placed so as to be easily read from the bottom and from the right-hand side of the graph, not the left-hand side.

3. Standard abbreviations should be used where space is limited as, for example, in denoting the unit of measurement in scale captions.

4. The range of scales should be chosen so as to ensure effective and efficient use of the coordinate area in attaining the objective of the chart.

5. The zero line should be included if visual comparison of plotted magnitudes is desired.

6. When it is desired to show whether the rate of change of the dependent variable is increasing, constant, or decreasing, a logarithmic vertical scale should be used in conjunction with an arithmetical horizontal scale.

7. The horizontal (independent variable) scale values should usually increase from left to right and the vertical (dependent variable) scale values from bottom to top.

8. Scale values and scale captions should be placed outside the grid area, normally at the bottom for the horizontal scale and at the left side for the vertical scale. On wide graphs it may be desirable to repeat the vertical scale at the right side.

9. For arithmetical scales, the scale numbers shown on the graph and the space between coordinate rulings should preferably correspond to 1, 2, or 5 units of measurement, multiplied or divided by 1, 10, 100, etc.

10. The use of many digits in scale numbers should be avoided.

11. The scale caption should indicate both the variable measured and the unit of measurement. For example: EXPOSURE TIME IN DAYS.

12. Coordinate rulings should be limited in number to those necessary to guide the eye in making a reading to the desired degree of approximation. Closely spaced coordinate rulings are appropriate for computation charts but not for graphs intended primarily to show relationship.

13. Curves should preferably be represented by solid lines.

14. When more than one curve is presented on a graph, relative emphasis or differentiation of the curves may be secured by using different types of line, that is, solid, dashed, dotted, etc., or by different widths of line. A solid line is recommended for the most important curve.

15. The observed points should preferably be designated by circles.

16. Circles, squares, and triangles should be used rather than crosses or filled-in symbols to differentiate observed points of several curves on a graph.

17. Curves should, if practicable, be designated by brief labels placed close to the curves (horizontally or along the curves) rather than by letters, numbers, or other devices requiring a key.

18. If a key is used, it should preferably be placed within the grid in an isolated position, and enclosed by a light line border—grid lines, if convenient.

19. The title should be as clear and concise as possible. Explanatory material, if necessary to ensure clearness, should be added as a subtitle.

20. Scale captions, designations, curves, and blank spaces should, so far as practicable, be ar-

ranged to give a sense of balance around vertical and horizontal axes.

21. The appearance and the effectiveness of a graph depend in large measure on the relative widths of line used for its component parts. The widest line should be used for the principal curve. If several curves are presented on the same graph, the line width used for the curves should be less than that used when a single curve is presented.

22. A simple style of lettering, such as Gothic with its uniform line width and no serifs, should in general be used.

### 7. Logarithmic Scales.

An important type of chart is that in which the divisions are made proportional to the logarithms of the numbers instead of to the numbers themselves, and accordingly are not equally spaced. When ruled logarithmically in one direction with equal spacing at right angles, the spacing is called "semilogarithmic."

Logarithmic spacing may be done directly from the graduations on one of the scales of a slide rule. Logarithmic paper is sold in various combinations of ruling. It is available in one, two, three, or more cycles; in multiples of 10; and also in part-cycle and split-cycle form. In using logarithmic paper, interpolations should be made logarithmically; arithmetical interpolation with coarse divisions might lead to considerable error.

### 8. Semilogarithmic Charts.

These charts have equal spacing on one axis, usually the $X$ axis, and logarithmic spacing on the other axis. Owing to a property by virtue of which the slope of the curve at any point is an exact measure of the rate of increase or decrease in the data plotted, the semilogarithmic chart is frequently called a "ratio chart." Often called the "rate-of-change" chart as distinguished from the rectilinear, or "amount-of-change," chart, it is extremely useful in statistical work as it shows at a glance the rate at which a variable changes. By the use of this chart it is possible to predict a trend, such as the future increase of a business, growth of population, etc.

In choosing between rectilinear ruling and semilogarithmic ruling, the important point to consider is whether the chart is to represent *numerical* increases and decreases or *percentage* increases and decreases. In many cases it is desirable to emphasize the percentage, or rate, change, not the numerical change; for these a semilogarithmic chart should be used.

An example of a semilogarithmic chart is given in Fig. 3. This curve was drawn from data compiled for the *World Almanac*. The dashed line shows the actual production by years, and the full line is the trend curve, the extension of which predicts future production.

Exponential equations of the form $y = ae^{bx}$ plot on semilogarithmic coordinates as a straight line. This represents the type of relationship in which a quantity increases or decreases at a rate proportional to the amount present at any time. For example, the passage of light through a translucent substance varies (in intensity) exponentially with the thickness of material.

### 9. Logarithmic Charts.

These are charts in which both abscissas and ordinates are spaced logarithmically. Any equation of the form $y = ax^b$, in which one quantity varies directly as some power of another, will plot as a straight line on logarithmic coordinates. Thus, multiplication, division, powers, and roots are examples. Figure 4 shows a logarithmic plot of sound intensity (and the power required to produce it) which varies as a power of the distance from the source. Other examples are the distance-time relationship of a falling body and the period of a simple pendulum.

### 10. Polar Charts.

The use of polar coordinate paper for representing intensity of illumination, intensity of heat, polar forms of curves, etc., is common. Figure 5 shows a candle-power distribution curve for an ordinary Mazda B lamp, and Fig. 6 the curve for a certain type of reflector. The candle power in any given direction is determined by reading off the distance from the origin to the curve. Use of polar forms of curves enables the determination of the foot-candle intensity at any point.

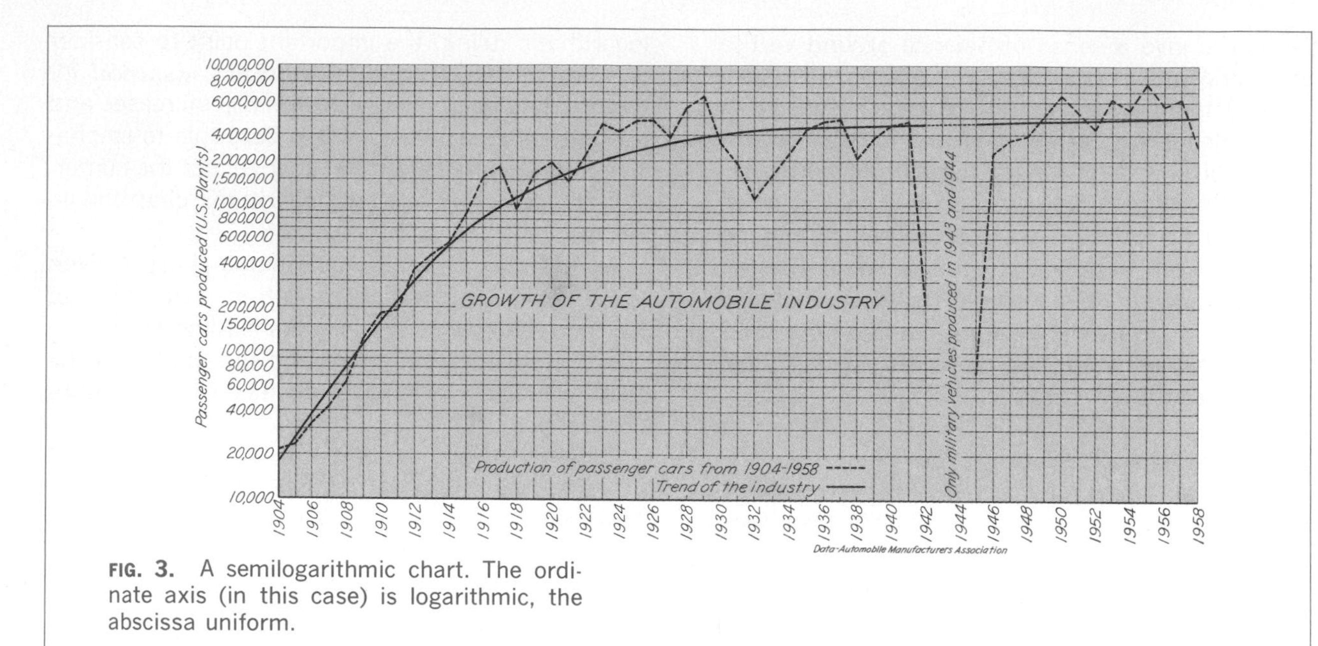


**FIG. 3.** A semilogarithmic chart. The ordinate axis (in this case) is logarithmic, the abscissa uniform.

## 11. Trilinear Charts.

The trilinear chart, or "triaxial diagram," as it is sometimes called, affords a valuable means of studying the properties of chemical compounds consisting of three elements, alloys of three metals or compounds, and mixtures containing three variables. The chart has the form of an equilateral triangle, the altitude of which represents 100 per cent of each of the three constituents. Figure 7, showing the ultimate tensile strength of copper-tin-zinc alloys, is a typical example of its application. The usefulness of such diagrams depends upon the geometric principle that the sum of the perpendiculars to the sides from any point within an equilateral triangle is a constant and is equal to the altitude.

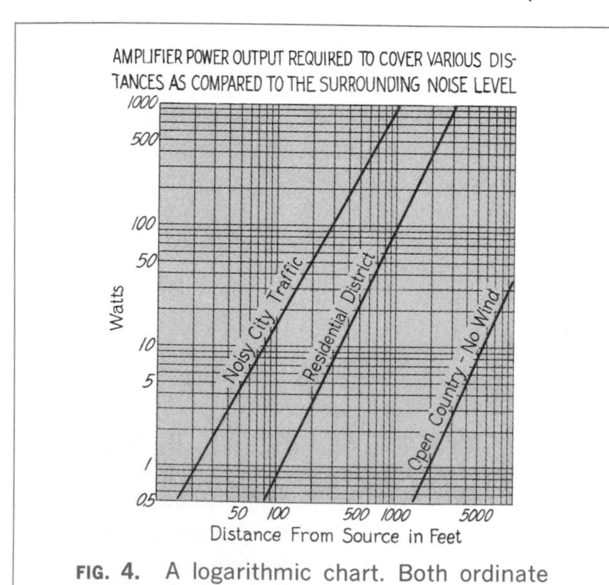


**FIG. 4.** A logarithmic chart. Both ordinate and abscissa are divided logarithmically.

## 12. Choice of Type and Presentation.

The function of a chart is to reveal facts. It may be entirely misleading if wrong paper or coordinates are chosen. The growth of an operation plotted on a rectilinear chart might, for example, entirely mislead an owner analyzing the trend of his business, while if plotted on a semilogarithmic chart, it would give a true picture of conditions. Intentionally misleading charts have been used many times in advertising; the commonest form is the chart with a greatly exaggerated vertical scale. In engineering

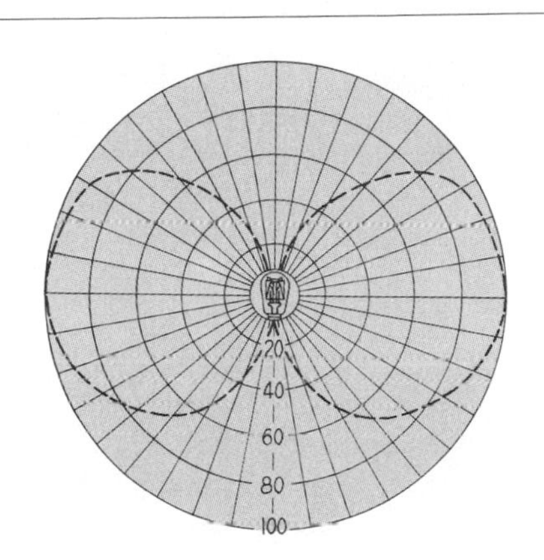


FIG. 5. A polar chart. This type is used when direction from a point (pole) must be shown.

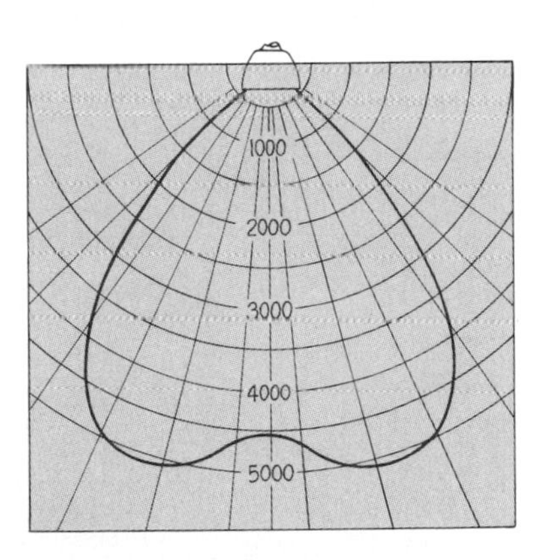


FIG. 6. A polar chart. This one shows the area of illumination from a reflector.

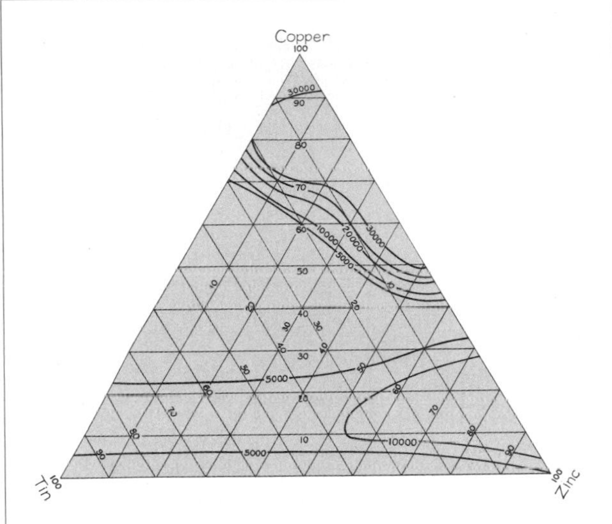


FIG. 7. A trilinear chart. This type is used to represent properties of three combined elements.

work it is essential to present the facts honestly and with scientific accuracy.

## 13. Analysis of Experimental Data.

So far we have discussed only the representation of data. In many cases representation alone on suitable coordinates gives an accurate description—for, from the plot, the *value* of one variable relative to a second variable can be ascertained. Nevertheless, the equation of a relationship can often be obtained from the data. To do this, the data must be plotted on coordinates that will "justify" the curve to a straight line. The equation of a straight line on any system of coordinates is easily determined. Equations thus resolved are known as *empirical* equations.

The standard procedure is first to plot the data on rectangular coordinates. If a straight line results, it is known that the equation is of the first degree (linear). If a curve results from the plot, the equation is of a higher form, and semilogarithmic coordinates or logarithmic coordinates should be tried; if the data justifies to a straight line on either system, the equation can be determined.

## 14. Rectangular Plots: Linear Equations.

If a plot of data results in a straight line on rectilinear coordinates, the equation is of the first degree. Figure 8 shows a plot that justifies to a straight line on rectilinear coordinates. To determine the equation of the line, select two points, such as $y_1x_1$ and $y_2x_2$ shown in Fig. 8. Then by simple proportion,

$$\frac{y - y_1}{x - x_1} = \frac{y_2 - y_1}{x_2 - x_1}$$

and from Fig. 8,

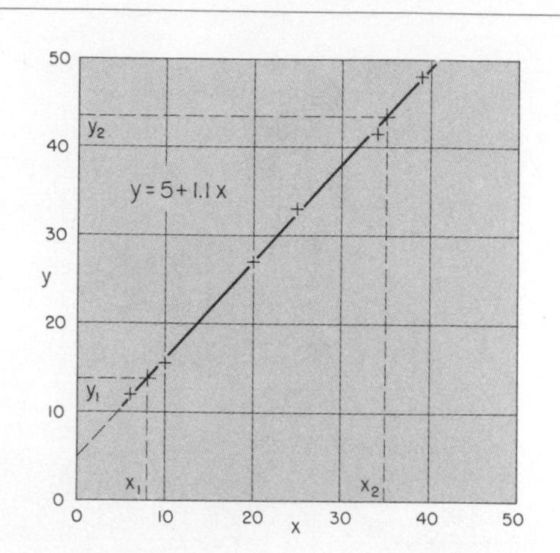


**FIG. 8.** A straight-line plot on rectilinear coordinates. The equation of the line is linear, $y = a + bx$.

$y_1 = 13.8$, $y_2 = 43.5$; $x_1 = 8$, $x_2 = 35.0$

substituting,

$$\frac{y - 13.8}{x - 8.0} = \frac{43.5 - 13.8}{35.0 - 8.0}$$

whence,

$$\frac{y - 13.8}{x - 8.0} = 1.1$$

and $y = 5.0 + 1.1x$

Thus, this equation is of a straight line, $y = a + bx$, where $a = 5.0$ and $b = 1.1$.

The coefficient 1.1 is the slope of the line and the constant 5.0 is the ordinate-axis intercept.

**15. Logarithmic Plots: Power Equations.** If the data justifies to a straight line on logarithmic coordinates, the equation is a power relationship of the form $y = ax^b$. Figure 9 shows a plot that justifies to a straight line on logarithmic coordinates. Two selected points give

$x_1 = 2.5$ $\quad y_1 = 3.8$
$x_2 = 55.0$ $\quad y_2 = 26.5$

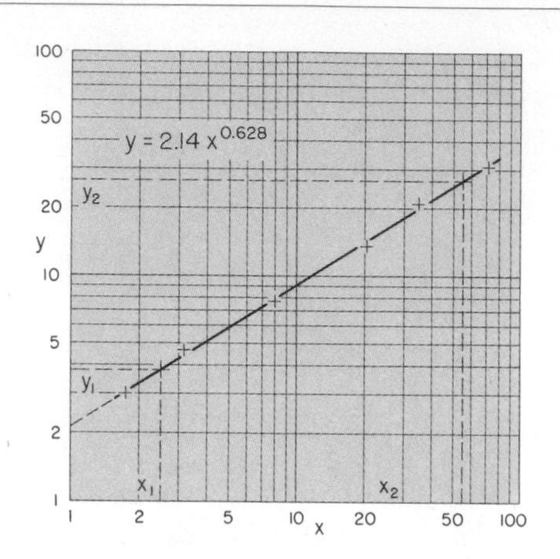


**FIG. 9.** A straight-line plot on logarithmic coordinates. The equation of the line is a power form, $y = ax^b$.

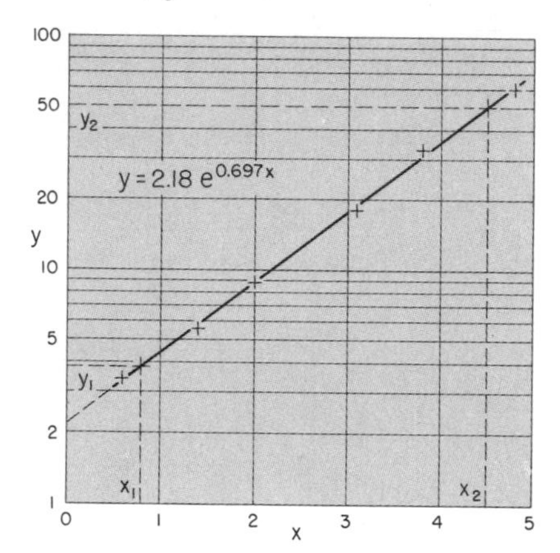


**FIG. 10.** A straight-line plot on semilogarithmic coordinates. The equation of the line is an exponential form, $y = ae^{bx}$.

Because the plot is logarithmic,

$$\frac{\log y - \log 3.8}{\log x - \log 2.5} = \frac{\log 26.5 - \log 3.8}{\log 55.0 - \log 2.5}$$

whence,

$$\log y = 0.330 + 0.628 \log x$$

and

$$y = 2.14x^{0.628}$$

## 16. Semilogarithmic Plots: Exponential Equations.

If plotted data justifies to a straight line on semilogarithmic coordinates, the equation is an exponential form, $y = ae^{bx}$ or $y = a10^{bx}$ (for base 10 logarithms). Figure 10 shows a typical plot. Two selected points give

$x_1 = 0.8 \quad y_1 = 3.8$
$x_2 = 4.5 \quad y_2 = 50.0$

Writing the equation of the straight line,

$$\frac{\ln y - \ln 3.8}{x - 0.8} = \frac{\ln 50.0 - \ln 3.8}{4.5 - 0.8}$$

whence,

$$\ln y = 0.777 + 0.697x$$

and

$$y = 2.18e^{0.697x}$$

## 17. Nomography.

After the equation of a relationship has been determined, specific values of the variables can be found by substitution in the equation. To facilitate such calculations, a chart called a nomogram can be made in which the variables are represented by plotted scales arranged so that a straightedge placed across the scales will give corresponding values of the variables. A simple example of a nomogram (conversion chart) is shown in Fig. 11.

The construction of a nomogram requires a knowledge of functional scales, chart forms, and the necessary mathematics. Thus, nomography is a special graphic field and is beyond the scope of this book. For a discussion of it, see "Graphic Science" by Thomas E. French and Charles J. Vierck.[1]

## 18. Graphic Calculus.

Considered in the same light as nomography, this is a specialized field of chart construction in which the rates of change of variables are determined graphically. This field is also beyond the scope of this book. For a discussion of graphic calculus, see "Graphic Science" by Thomas E. French and Charles J. Vierck.[1]

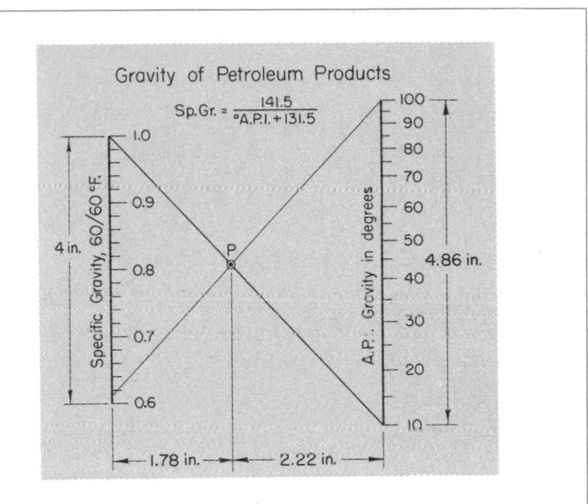


FIG. 11. A nomogram. This one is a conversion chart for specific gravity to degrees API.

## 19. Classification Charts, Route Charts, and Flow Sheets.

The uses to which these three classes of charts can be put are widely different; but their underlying principles are similar, and so they have been grouped together for convenience.

A *classification chart,* illustrated in Fig. 12, is intended to show the subdivisions of a whole and the interrelation of its parts with one another. Such a chart often takes the place of a written outline as it gives a better visualization of the facts than can be achieved with words alone. A common application is an organization chart of a corporation or business. It is customary to enclose the names of the divisions in rectangles, although circles or other shapes may be used. The rectangle has the advantage of being convenient for lettering, while the circle can be drawn more quickly and possesses greater popular appeal. Often a combination of both is used.

The *route chart* is used mainly for the purpose of showing the various steps in a process, either of manufacturing or other business. The *flow sheet,* illustrated in Fig. 13, is an example of a route chart applied to a chemical process. Charts of this type show in a dynamic way facts that might require

[1] McGraw-Hill, New York, 2d ed., 1963.

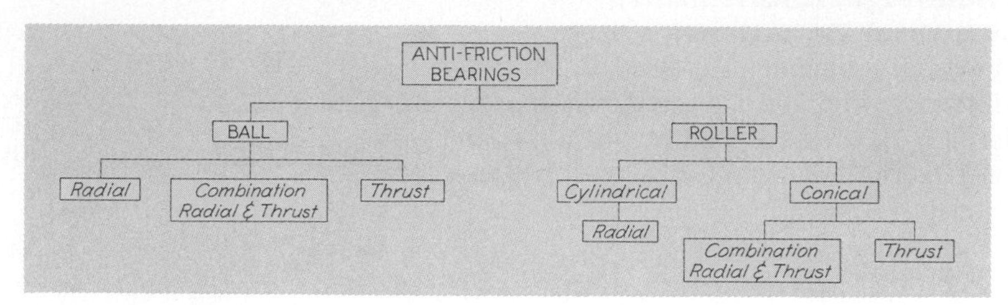


FIG. 12. A classification chart. This type of chart shows the relationship of parts of a whole.

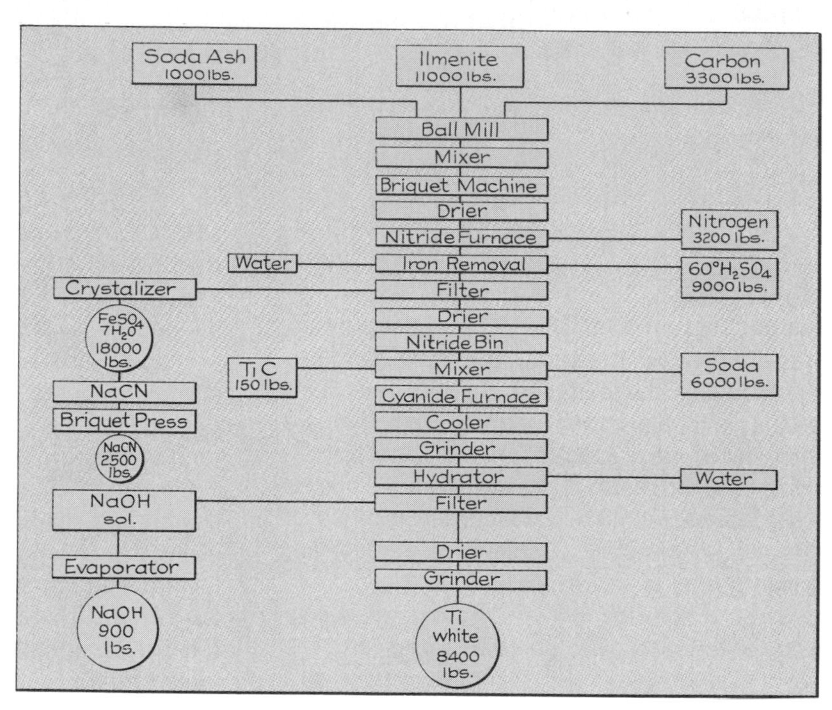


FIG. 13. A flow sheet. This is an effective way to show the fundamentals of a process or method.

considerable study to be understood from a written description. Figure 1, Chap. 11, showing the course of a drawing through the shops, illustrates a different form of route chart.

## 20. Popular Charts.

Engineers and draftsmen are frequently called upon to prepare charts and diagrams that will be understood by nontechnical readers. For such charts it is often advisable to present the facts, not by means of curves on coordinate paper but in a way that is more easily read, even though the representation may be somewhat less accurate. Charts for popular use must be made so that the impression they produce is both quick and accurate. They are seldom studied critically but are taken in at a glance; hence the method of presentation requires the exercise of careful judgment and the application of a certain amount of psychology.

## 21. Bar Charts.

The bar chart is a type of chart easily understood by nontechnical readers. One of its simplest forms is the *100 per cent bar* for showing the relations of the constituents to a given total. Figure 14 is an example. The different segments should be crosshatched, shaded, or distinguished in some other effective manner, the percentage represented should be placed on the related portion of the diagram or directly opposite, and the meaning of each segment should be clearly stated. Bars may be placed vertically or horizontally, the vertical position giving an advantage for lettering and the horizontal an advantage in readability, as the eye judges horizontal distances readily.

Figure 15 is an example of a *multiple-bar chart,* in which the length of each bar is proportional to the magnitude of the quantity represented. Means should be provided for reading numerical values represented by the bars. If it is necessary to give the exact value represented by the individual bars, the values should not be lettered at the ends of the bars as this would increase their apparent length. This type of chart is made horizontally, with the description at the base, and vertically. The vertical form is sometimes called the "pipe-organ chart." When vertical bars are drawn close together, touching along the sides, the diagram is called a "staircase chart." This chart is more often made as the "staircase curve," a line plotted on coordinate paper representing the profile of the tops of the bars.

A *compound-bar chart* is made when it is desirable to show two or more components in each bar. It is really a set of 100 per cent bars of different lengths set together in pipe-organ or horizontal form.

**FIG. 14.** A 100 per cent bar chart. This is a simple and impressive method of showing percentages of a whole.

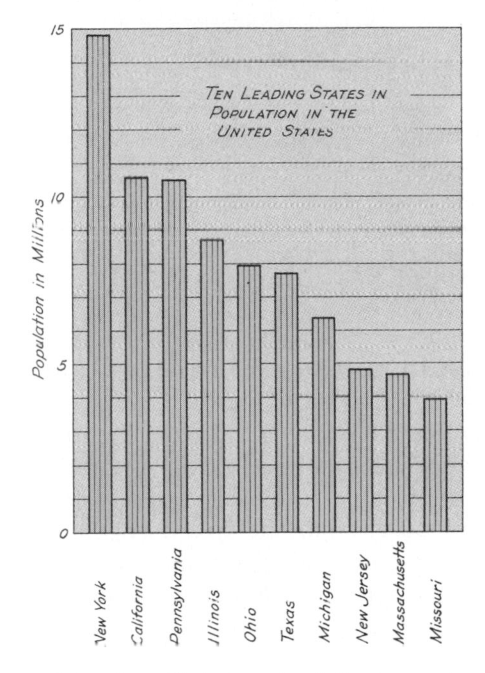


**FIG. 15.** A multiple-bar chart. Each bar gives (graphically) the magnitude of the item represented.

## 22. Pie Charts.

The "pie diagram," or 100 per cent circle (Fig. 16) is much inferior to the bar chart but is used constantly because of its popular appeal. It is a simple form of chart and, except for the lettering, is easily constructed. It can be regarded as a 100 per cent bar bent into circular form. The circumference of the circle is divided into 100 parts, and sectors are used to represent percentages of the total.

To be effective, this diagram must be carefully lettered and the percentages marked on the sectors or at the circumference opposite the sectors. For contrast, it is best to crosshatch or shade the individual sectors. If the original drawing is to be displayed, the sectors may be colored and the diagram supplied with a key showing the meaning of each

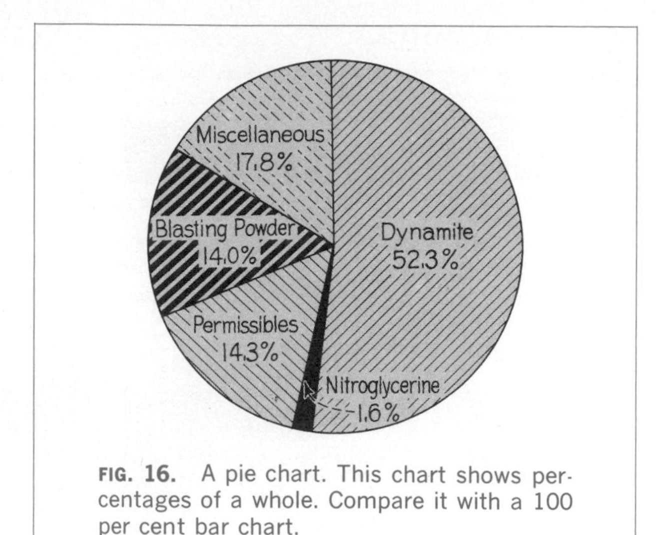


**FIG. 16.** A pie chart. This chart shows percentages of a whole. Compare it with a 100 per cent bar chart.

color. The percentage notation should always be placed where it can be read without removing the eyes from the diagram.

## 23. Strata and Volume Diagrams.

The use of strata and volume diagrams has been common although they are usually the most deceptive of the graphic methods of representation. Figure 17 is a strata chart showing a change in amounts over a period of time. The area between the curves is shaded to emphasize each variable represented. However, in reading it should be remembered that the areas *between* curves have no significance.

## 24. Pictorial Charts.

Pictorial charts were formerly much used for comparisons, for example of costs, populations, standing armies, livestock, and various products. It was customary to represent the data by human or other figures whose heights were proportional to numerical values or by silhouettes of the animals or products concerned whose heights or sometimes areas were proportional. Since volumes vary as the cubes of the linear dimensions, such charts are grossly misleading. Bar charts or charts such as the one shown in Fig. 18, where the diameter of the column is constant, should be used for the type of comparisons that are shown in pictorial charts.

## 25. Charts for Reproduction.

Charts for reproduction by the zinc-etching process should be carefully penciled to about twice the size of the required cut. Observe the following order in inking: First, ink the circles around plotted points; second, ink the curves with strong lines. A border pen is useful for heavy lines, and a Leroy pen (in socket holder) may be used to advantage, particularly for dashed lines. Third, ink the title box and

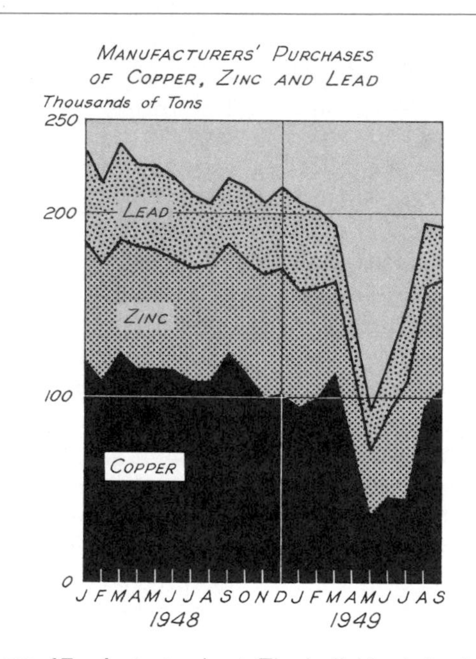


**FIG. 17.** A strata chart. The individual charts are superimposed.

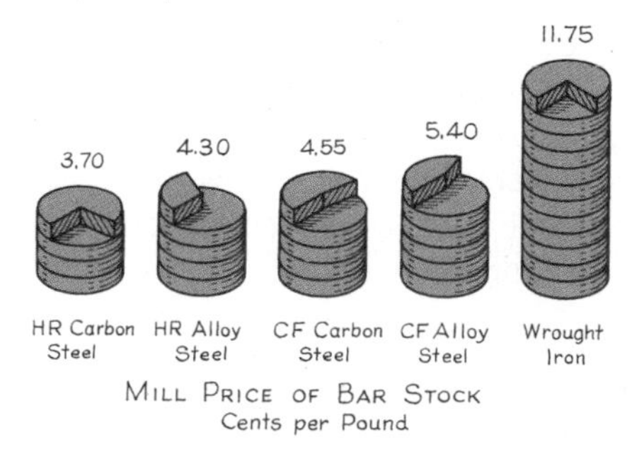


**FIG. 18.** A pictorial chart. Pictorialism adds to popular interest but detracts from accuracy of representation.

all lettering; fourth, ink the coordinates with fine black lines, putting in only as many as are necessary for easy reading and breaking them wherever they interfere with title or lettering or where they cross plotted points.

### 26. Charts for Display.

Large charts for demonstration purposes are sometimes required. These can be drawn on sheets 22 by 28 in. or 28 by 44 in. known as "printer's blanks." The quickest way to make them is with show-card colors and single-stroke sign-writer's brushes. Large bar charts can be made with strips of black adhesive tape. Lettering can be done with the brush or with gummed letters.

### 27. Standards.

Many standards are available for graphic presentations of data. Of particular interest are the ASA standards for graphic symbols (Y32), letter symbols (Y10), and time-series charts (Y15.2—1960), and ASA Y15.1 "Guide for Preparing Technical Illustrations for Publications and Projections."

### 28. The Advantage of Color.

Even though "black-and-white" charts, graphs, and diagrams are satisfactory for statistical and scientific purposes, the use of color will add greatly to readability, separation, and emphasis. In black and white, line codes (dashes, etc.) must be used and areas can be emphasized only by crosshatching. In color, lines stand out prominently and areas are easily distinguished. Especially for popular charts, color is almost mandatory. Today, many of the charts in good periodicals, corporate reports, promotional charts, and many others are being done in color. The use of color is increasing rapidly. Note the use of color techniques in all of the illustrations in this chapter, especially Figs. 12, 13, 15, and Figs. 19 to 23.[2] Also, compare Fig. 19 with Fig. 20.

### 29. Principles for the Use of Color.

The use of color always requires some artistic judgment and consequently it is difficult to give definitive rules. Nevertheless, the following principles will serve as a guide.

*Backgrounds.* A pure white background for a chart is not nearly so effective as a background made of a light tint of color. This is because a white background, even if it is outlined, is the same value as the page itself. A background made as a light tint adds cohesiveness and vigor to the charts. Notice the use of tinted backgrounds in all the charts in this chapter.

*White lines and areas.* White lines and areas can be produced only if a tinted background is employed. This adds one more color to any series used and is quite effective. Notice the white area in Fig. 23 and the white line in Fig. 22.

*Values of tints.* Tints of a color are made either by diluting the color with white or, in printing, by the use of screens which reduce the color intensity. In printing, the screens are rated in percentage, which means that a 75% screen will print (in actual minute areas) 75% of the color and 25% white (blank area). Thus a 30% screen would print 30% of the color and 70% as blank area. For the dilution of a color with white when applying color by hand, it is advantageous to think of the color *value* also in percentage because it is easy to judge color intensity, for example, as approximately $\frac{3}{4}$ the intensity of the pure color. The actual dilution *will not* be the same as the value of the intensity; the dilution must be made by carefully adding white and then judging the intensity. Note the various intensities of colors used in the illustrations of this chapter.

*Contrast.* Contrast may be defined as the degree of dissimilarity. Thus, high contrast is had when a color of strong intensity is placed adjacent to a color of very weak intensity. Low contrast is had when two colors of about the same intensity are placed together. High contrast will accentuate the areas and low contrast will diminish the impact of the areas. Notice the gradual (low-contrast) changes in both color and value of the areas in Fig. 19 and also the high contrast and accentuation of the darkest and white areas of Fig. 20. If emphasis is needed, use high contrast. If a subtle change is desirable, use low contrast.

[2] Figures 19 to 23 are supplied through the courtesy of Mr. W. Anthony Dowell of the Palm Beach office to Merrill, Lynch, Pierce, Fenner, and Smith and are adapted from the U.S. Industrial Outlook, 1965, by the U.S. Department of Commerce.

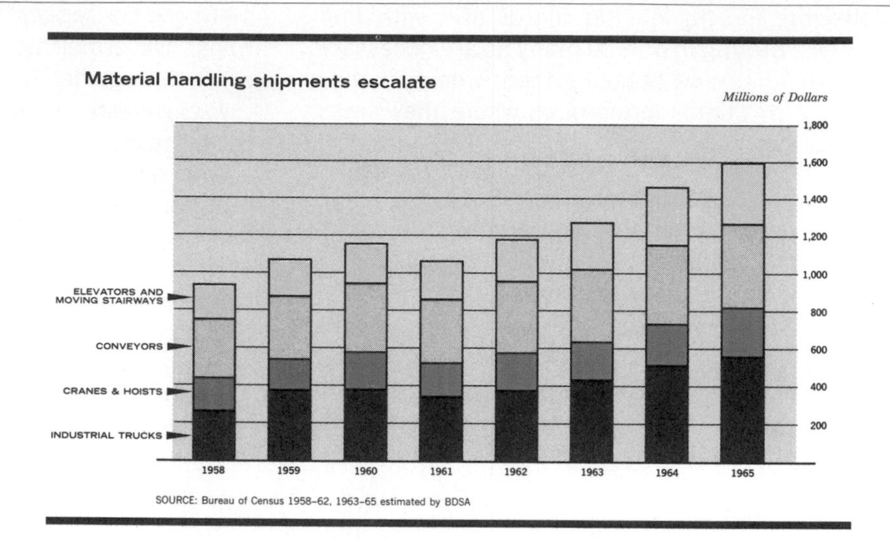


**FIG. 19.** A multiple-bar chart in color. Two basic colors and tints of each are used. Compare with Fig. 20.

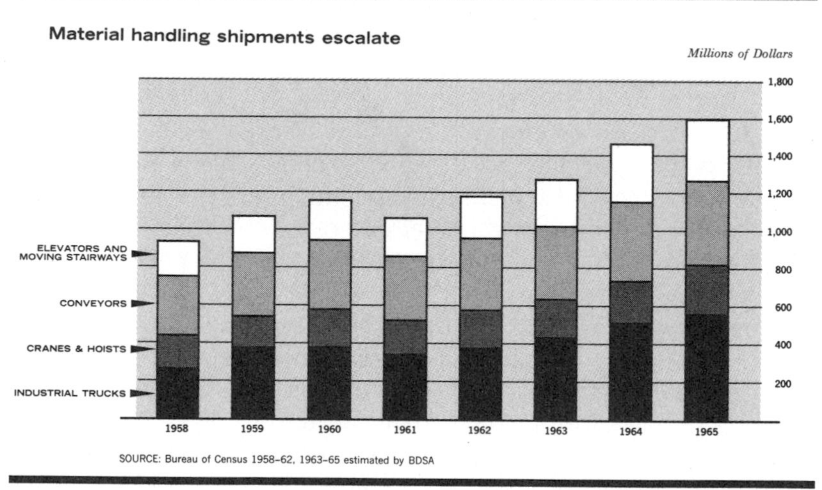


**FIG. 20.** The chart of Fig. 19 in "black and white" and tints. Compare with Fig. 19.

*Adjacent areas.* Adjacent areas are well defined and easily read if they are made of different colors. Notice in Fig. 21 that adjacent areas are different colors and also that different values of the colors have been used. Maximum separation of areas occurs when a strong color and value is placed next to a weak color and value. In some cases maximum separation is needed, in others a more gradual change is desirable. For example, in Fig. 20 the stronger colors are used at the bottom of the bars and the weaker colors at the top. This is done not only for the sake of stability of the bars, but because, if adjacent areas of the bars are made as high-contrast areas, the bars become "spotty" in ap-

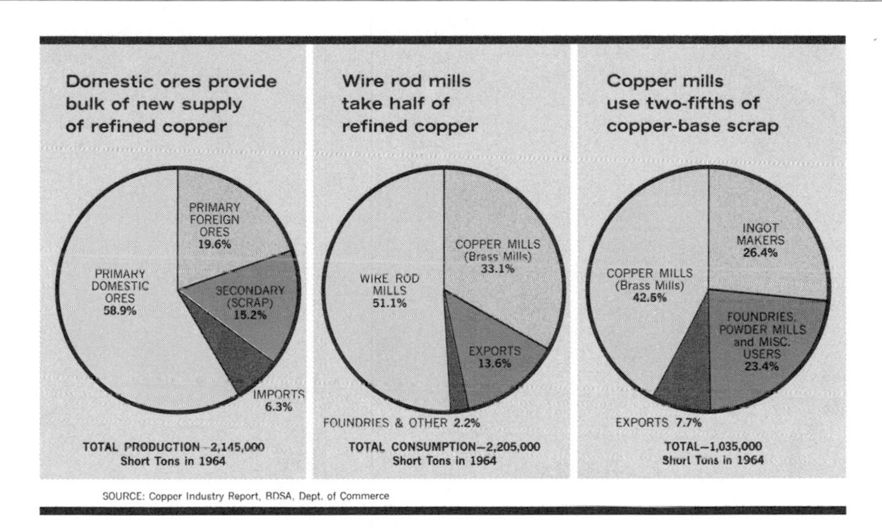


**FIG. 21.** A series of pie charts in color. Readability and appeal is superior to single-color rendition.

pearance and the value of a continuous bar is lost.

*Adjacent lines.* The best separation of lines occurs when adjacent lines are contrasty, either in color or value, or both. Note the use of alternate brown and green lines in Fig. 22 and the contrasting white line at the top.

*Choice of Colors.* The actual choice of colors is almost entirely one of artistic judgment, but this is often tempered by other factors. Colors at the high (wave-length) end of the spectrum are known as "cold" colors. Colors at the low end of the spectrum are known as "warm" colors. The colors of nature are the blue of the sky; the greens of trees, grass, and shrubbery; and the browns of tree trunks, soil, and rock. Note that these colors comprise a cold color, a medium color, and a warm color. However, even though we view the colors of nature daily and are not disturbed by them, such colors are not commonly put together in a chart. Also, strange as it may seem to the layman, a study of the colors of the spectrum and other scientific aspects of color transmission and reflectance will be of little help in the choice of colors to be used for a chart, graph, or diagram. This is because the *only* important considerations are those of pleasing and visually effective combinations, contrasts, and color values. Colors that are not pleasingly compatible should not be used together unless a startling or garish effect is wanted. For example, violet, purple, and yellow are almost never used for charts. The colors used most are blues, greens, reds, browns, white, and black. The symbolic colors of a company or organization may also play a part in the choice, as for example, the blue and white of the Pure Oil Company or the red, white, and blue of The Standard Oil Company. Of interest is the fact that blue has predominated recently in the annual reports of many large companies. In addition, many of the charts and graphs in such magazines as *Scientific American* have been printed in black, shades of black (gray), red, blue, and green with occasionally some yellow, but black, gray, and red are predominant.

The best practice is to lay out the chart and then try several colors by holding colored sheets or strips over the chart to judge finally the most pleasing and effective combination.

### 30. Color for Primary Charts.

By the term "primary charts" is meant the chart that is made by hand for use directly. This excludes

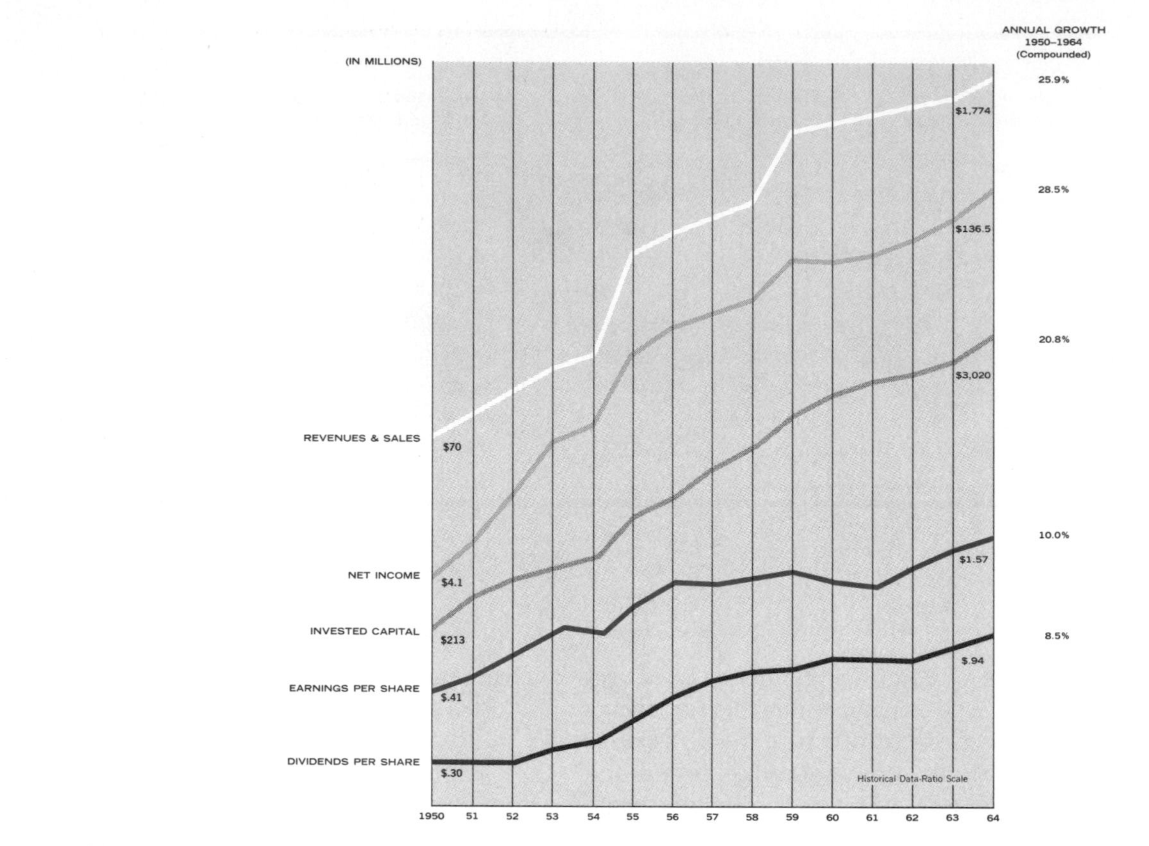


**FIG. 22.** A graph in color. Note that one more colored line is produced by employing a tinted background with the white line on it.

reproduction. Color may be applied in the form of colored inks, water color, or tempera. All of these may be used in a ruling pen for making lines or applied with a brush for areas. Chart-Pak (Chart-Pak Inc., River Road, Leeds, Mass.) colored acetate fiber strip, available in many colors and widths, is a convenient method of coloring for line graphs and bar charts. Colored paper can be used for backgrounds, drawing the chart *on* the colored paper.

**31. Color for Reproduced Charts.** When the chart is made exclusively for reproduction, the basic chart is outlined and drawn in india ink as would be done for a black-and-white chart, with the exception that areas are not blacked-in. Then, an overlay sheet of tracing paper or drafting film such as Mylar is placed over the chart and fastened with drafting (or transparent) tape. On this overlay sheet directions are lettered or written, designating the colors to be used for lines and areas. The engraver, following these directions, will then make plates for the printing of each color. In giving the directions for color and color value, either an approximate value must be given, such as *light blue,* or an exact value must be designated, such as 30% blue.

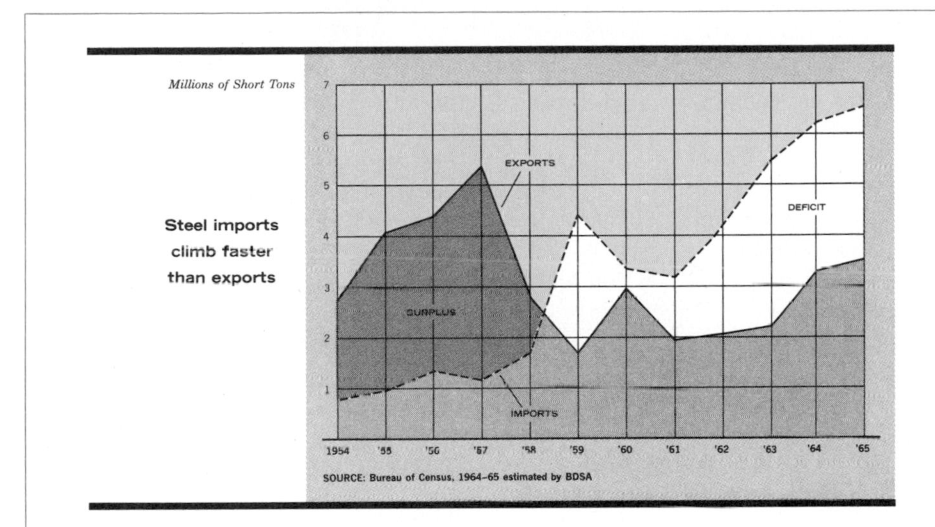


FIG. 23. A strata chart in color. Note that the overlap of colored strata produces readability for the *surplus* area and the separation of strata emphasizes the *deficit* area.

## PROBLEMS

These problems are given for practice in preparing various types of charts, graphs, and diagrams for technical or popular presentation.

**1.** The data given below were obtained in a tension test of a machine-steel bar. Plot the data on rectangular coordinates, using the elongation as the independent variable, the applied load as the dependent variable.

| Applied load, lb per sq in. | Elongation per in. of length |
|---|---|
| 0 | 0 |
| 3,000 | 0.00011 |
| 5,000 | 0.00018 |
| 10,000 | 0.00033 |
| 15,000 | 0.00051 |
| 20,000 | 0.00067 |
| 25,000 | 0.00083 |
| 30,000 | 0.00099 |
| 35,000 | 0.00115 |
| 40,000 | 0.00134 |
| 42,000 | 0.00142 |

**2.** A test of the corrosive effect of 5 per cent sulfuric acid, both air-free and air-saturated, on 70 per cent nickel, 30 per cent copper alloy (Monel) over a temperature range from 20 to 120°C resulted in the data tabulated below. Plot this data on rectangular coordinates with corrosion rate as ordinate versus temperature as abscissa.

| | Corrosion rate, MDD (mg per sq dm per day) | |
|---|---|---|
| Temp., °C | Acid sat. with air | Acid air-free, sat. with $N_2$ |
| 20 | 195 | 35 |
| 30 | 240 | 45 |
| 40 | 315 | 50 |
| 50 | 425 | 63 |
| 60 | 565 | 70 |
| 70 | 670 | 74 |
| 80 | 725 | 72 |
| 83 | 715 | 70 |
| 85 | 700 | 68 |
| 92 | 580 | 57 |
| 95 | 470 | 42 |
| 101 | 60 | 12 |

**3.** In testing a small 1-kw transformer for efficiency at various loads, the following data were obtained: Watts delivered: 948, 728, 458, 252, 000. Losses: 73, 62, 53, 49, 47.

Plot curves on rectangular coordinate paper showing the relation between percentage of load and efficiency, using watts delivered as the independent variable and remembering that efficiency = output ÷ (output + losses).

**4.** The following data were obtained from a test of an automobile engine:

| Rpm | Length of run, min | Fuel per run, lb | Bhp |
|---|---|---|---|
| 1,006 | 11.08 | 1.0 | 5.5 |
| 1,001 | 4.25 | 0.5 | 8.5 |
| 997 | 7.53 | 1.0 | 13.0 |
| 1,000 | 5.77 | 1.0 | 16.3 |
| 1,002 | 2.38 | 0.5 | 21.1 |

Plot curves on rectangular coordinate paper showing the relation between percentage of load and efficiency, using watts delivered as the independent variable between thermal efficiency and brake horsepower developed, assuming the heat value of the gasoline to be 19,000 Btu per lb.

**5.** During the year 1950 the consumption of wood pulp by various grades in the United States was as shown in the table. Show these facts by means of a 100 per cent bar, a pie diagram, and a multiple-bar chart.

| Grade of pulp | Consumption in paper and board manufacture, tons |
|---|---|
| Sulfate | 8,380,864 |
| Sulfite | 3,331,668 |
| Groundwood | 2,483,980 |
| Defibrated, exploded, etc. | 971,912 |
| Soda | 564,355 |
| All other | 754,331 |
| Total (all grades) | 16,487,110 |

**6.** Put the data given in Fig. 16 into 100 per cent bar form.

**7.** Put the data of Fig. 14 into pie-chart form.

**8.** A test of the resistance of alloy steels to high-temperature steam resulted in the data tabulated below. Represent the results of the test graphically by means of a multiple-bar chart.

**Corrosion of Steel Bars in Contact with Steam at 1100°F for 2,000 Hr**

| Steel | Average penetration, in. |
|---|---|
| SAE 1010 | 0.001700 |
| 1.25 Cr-Mo | 0.001169 |
| 4.6 Cr-Mo | 0.000956 |
| 9 Cr 1.22 Mo | 0.000694 |
| 12 Cr | 0.000045 |
| 18-8-Cb | 0.000012 |

**9.** From the data below, plot curves showing the "thinking distance" and "braking distance." From these curves, plot the sum curve "total distance." Title "Automobile Minimum Travel Distances when Stopping—Average Driver."

| Mph | Ft per sec | Thinking distance, ft | Braking distance, ft |
|---|---|---|---|
| 20 | 29 | 22 | 18 |
| 30 | 44 | 33 | 40 |
| 40 | 59 | 44 | 71 |
| 50 | 74 | 55 | 111 |
| 60 | 88 | 66 | 160 |
| 70 | 103 | 77 | 218 |

**10.** Make a semilogarithmic chart showing the comparative rate of growth of the five largest American cities from 1880 to 1950. Data for this chart are in the table in the next column.

**Population**

| City | 1880 | 1890 | 1900 | 1910 |
|---|---|---|---|---|
| New York, N.Y. | 1,911,698 | 2,507,414 | 3,437,202 | 4,766,883 |
| Chicago, Ill. | 503,185 | 1,099,850 | 1,698,757 | 2,185,283 |
| Philadelphia, Pa. | 847,170 | 1,046,964 | 1,293,697 | 1,549,008 |
| Los Angeles, Calif. | 11,181 | 50,395 | 102,479 | 319,198 |
| Detroit, Mich. | 116,340 | 205,876 | 285,704 | 465,766 |

| City | 1920 | 1930 | 1940 | 1950 |
|---|---|---|---|---|
| New York, N.Y. | 5,620,048 | 6,930,446 | 7,454,995 | 7,835,099 |
| Chicago, Ill. | 2,701,705 | 3,376,348 | 3,396,808 | 3,606,436 |
| Philadelphia, Pa. | 1,823,779 | 1,950,961 | 1,931,334 | 2,064,794 |
| Los Angeles, Calif. | 576,673 | 1,238,048 | 1,504,277 | 1,957,692 |
| Detroit, Mich. | 993,678 | 1,568,662 | 1,623,452 | 1,838,517 |

| Angle, ° | (1) Mazda lamp and porcelain enameled reflector, candle power | (2) Type S-1 sun lamp, candle power |
|---|---|---|
| 0 | 600 | 1,000 |
| 5 | 655 | 925 |
| 10 | 695 | 790 |
| 15 | 725 | 640 |
| 20 | 760 | 460 |
| 25 | 800 | 350 |
| 30 | 815 | 260 |
| 35 | 825 | 205 |
| 40 | 830 | 170 |
| 45 | 815 | 150 |
| 50 | 785 | 130 |
| 55 | 740 | 120 |
| 60 | 640 | 0 |
| 65 | 485 | |
| 70 | 330 | |
| 75 | 200 | |
| 80 | 90 | |
| 85 | 25 | |
| 90 | 0 | |

**11.** On polar coordinate paper, plot a curve for the first set of data given below as a solid line (Fig. 6) and a curve for the second set as a broken line (Fig. 5). The lower end of the vertical center line is to be taken as the zero-degree line. The curves will be symmetrical about this line, two points being plotted for each angle, one to the left and one to the right. Mark the candle power along this center line also.

**12.** On trilinear coordinate paper plot the data given below. Complete the chart, identifying the curves and lettering the title below the coordinate lines (Fig. 7).

**Freezing Points of Solutions of Glycerol and Methanol in Water**

| Weight, per cent | | | Freezing points, °F |
|---|---|---|---|
| Water | Methanol | Glycerol | |
| 86 | 14 | 0 | +14 |
| 82 | 8 | 10 | +14 |
| 78 | 6 | 16 | +14 |
| 76 | 24 | ... | −4 |
| 74 | 2 | 24 | +14 |
| 71 | 17 | 12 | −4 |
| 70 | 0 | 30 | +14 |
| 68 | 32 | 0 | −22 |
| 65 | 10 | 25 | −4 |
| 62 | 23 | 15 | −22 |
| 60 | 40 | 0 | −40 |
| 58 | 5 | 37 | −4 |
| 57 | 17 | 26 | −22 |
| 55 | 0 | 45 | −4 |
| 54 | 28 | 18 | −40 |
| 53 | 47 | 0 | −58 |
| 52 | 10 | 38 | −22 |
| 50 | 20 | 30 | −40 |
| 47 | 31 | 22 | −58 |
| 45 | 0 | 55 | −22 |
| 42 | 7 | 51 | −40 |
| 39 | 12 | 49 | −58 |
| 37 | 0 | 63 | −40 |
| 36 | 8 | 56 | −58 |
| 33 | 0 | 67 | −58 |

**13.** Thermal conductivities of a bonded-asbestos-fiber insulating material at various mean temperatures are tabulated below:

| Mean temperature (hot to cold surface) $T$, °F | Thermal conductivity $K$, Btu per hr-sq ft-°F temp diff-in. thickness |
|---|---|
| 100 | 0.365 |
| 200 | 0.415 |
| 320 | 0.470 |
| 400 | 0.520 |
| 460 | 0.563 |
| 600 | 0.628 |
| 730 | 0.688 |
| 900 | 0.780 |

Plotting the data on uniform rectangular coordinates with thermal conductivity as ordinate and mean temperature as abscissa, evaluate the constants $a$ and $b$ in the linear equation $K = a + bT$ relating the variables.

**14.** Corrosion tests on specimens of pure magnesium resulted in values for weight increase in pure oxygen at 525°C tabulated below:

| Elapsed time $T$, hr | Weight increase $W$, mg per sq cm |
|---|---|
| 0 | 0 |
| 2.0 | 0.17 |
| 4.0 | 0.36 |
| 8.2 | 0.65 |
| 11.5 | 1.00 |
| 20.0 | 1.68 |
| 24.1 | 2.09 |
| 30.0 | 2.57 |

Plotting the data on uniform rectangular coordinates with weight increase as ordinate and elapsed time as abscissa, evaluate the constant $a$ in the linear equation $W = aT$ relating the variables.

**15.** Approximate rates of discharge of water under various heads of fall through a 1,000-ft length of 4-in. pipe having an average number of bends and fittings are tabulated on the next page:

| Head of fall $H$, ft | Discharge $Q$, gal per min |
|---|---|
| 1 | 35.8 |
| 2 | 50.6 |
| 4 | 71.6 |
| 6 | 87.7 |
| 9 | 107.5 |
| 12 | 123.7 |
| 16 | 142.9 |
| 20 | 159.7 |
| 25 | 178.9 |
| 30 | 195.8 |
| 40 | 225.8 |
| 50 | 252.2 |
| 75 | 309.8 |
| 100 | 357.9 |

Plotting the data on logarithmic coordinates with discharge as ordinate and head as abscissa, evaluate the constants $a$ and $b$ in the power equation $Q = aH^b$ relating the variables.

**16.** Creep-strength tests of a high-chromium (23 to 27 per cent) ferritic steel used in high-temperature service resulted in the stress values, to produce a 1 per cent deformation in 10,000 hr at various temperatures, tabulated below:

| Temperature $T$, °F | Stress $S$, lb per sq in. |
|---|---|
| 1,000 | 6,450 |
| 1,100 | 2,700 |
| 1,200 | 1,250 |
| 1,300 | 570 |
| 1,400 | 270 |

Plotting the data on semilogarithmic coordinates with stress as ordinate on a logarithmic scale and temperature as abscissa on a uniform scale, evaluate the constants $a$ and $b$ in the exponential equation $S = a10^{bT}$ relating the variables.

**17. COLORED CHARTS.** Select any of Probs. 1 to 16, and draw in color, following the suggestions given in Sec. 30.

This Chapter Includes:



# 14

# FUNDAMENTALS OF DESIGN

**Design Is the Preface to All of Engineering. Only after Design Can Manufacture, Production, Construction, etc., Be Approached**

**1. Introductory—Design Catagories.** The word "design" has many meanings. A digest of various dictionary definitions is: to plan, conceive, invent, and to designate so as to transmit the plan to others. *Design* has many purely artistic connotations. For example, the design of fabrics, clothing, furniture, etc. In engineering, design has come to mean that broad category of invention leading to the production of useful devices.

*Design,* from the latin "designare" (to mark out), is the process of developing plans, schemes, directions, and specifications for something new. Thus it is within context to speak of Hitler's designs for world conquest; the design (conception) of a book, play, or motion picture; or the design of fabric, clothing, furniture, appliances, or other completely physical objects. Design is distinguished from production and craftsmanship: design is the creative original plan, and the production and craftsmanship are a part of the execution of the plan.

Design means creation in the purest sense. Specifically, design does not go beyond creation, and it logically follows that the execution of a design, that is, the carrying out of the plan by presentation, action, production, manufacture, craftsmanship, and use are not design at all but are simply and positively the products of the design. Also, when examining a finished product, it is proper to speak of the *design* of it, and by this term of reference we mean the original plan or scheme and not the product itself.

A design may be presented by means of drawings, models, patterns, specifications, or other similar methods of communication. By whatever means the design is made known, every detail important to consummation must be given. This will include such items as materials and their capabilities, the methods of adapting the materials to their purpose or work, the relationship of parts within the whole, and the effect of the finished product upon those who may see it, use it, or become involved with it.

Design is a word used more or less loosely in all the arts in referring to composition, style, decoration, or any relationship of the parts of a complete entity. In some areas, notably in architecture and in product design, art and engineering affect each other, so that complete freedom is often somewhat restricted. Usually, painters, poets, musicians, and some others can design with great freedom.

Logically, one who designs is called a designer. All designers must be experienced and educationally organized and oriented. In other words a designer must know a great deal about what he is attempting to design or he will fail miserably. As an example, suppose that a person who has never fished and knows nothing about the sport, attempts to design a fishing reel. Because of his ignorance of such aspects as weight, balance, line capacity, drag characteristics, and over-all performance such a person is completely incapable of producing a good design. Nevertheless, most good engineering designers are capable of designing a wide variety of devices because of their knowledge of materials, processes, production methods, and other related aspects. Designers may be likened to executives in large companies, especially such people as editors and directors. The executive will decide policies and business methods and then transmit his ideas to colleagues who carry out the executive orders. A designer conceives his design and then transmits his plans to others who produce the product. This does not mean, however, that a designer never is involved in production. Especially in the fine arts, a designer may actually produce the product himself.

Design in a broad sense can be, and often is, classified according to its relationship to practicality. Thus, *abstract design* has no relationship whatever to useful or physical objects and is intended only to create a visual interest or impact. The frontispiece of Chapter 1 is an example of abstract design. Much of so-called "modern" art is abstract design. *Aesthetic design* is a design applied to some useful object. The design is intended for decorative purposes alone, and has nothing whatever to do with the usefulness of the object. Figure 1 shows the design of a lace mantilla, a traditional head covering for women. The design is aesthetic only and performs no practical purpose in effectively covering the head. Other examples of aesthetic design are found in furniture, architecture, motorcars, appliances, floor coverings, and other useful objects where the design per se has no function but to

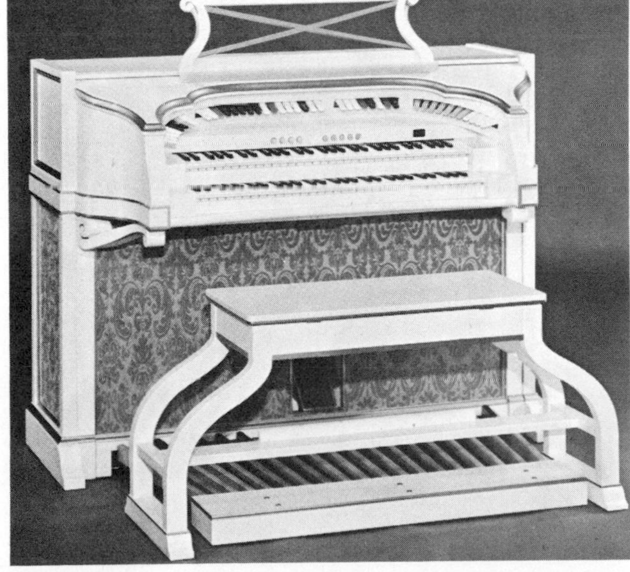

**FIG. 1.** Aesthetic design. The design has no function other than to decorate.

**FIG. 2.** Aesthetic-functional design. Aesthetics and functionalism are affected by each other.

decorate and create an aesthetic impact. *Aesthetic functional* design is that category of design where aesthetic and functional aspects are closely allied. An example is shown in Fig. 2 above. This is a

Baldwin theatre organ, a fine musical instrument as well as an outstanding aesthetic and traditional design. To see how function affects aesthetics note that the seat must be wide enough at the base to cover the width of the 32-note pedal board. Thus, the aesthetic design of the seat must be accommodated to functional requirements. The curved console is not only traditional and aesthetically interesting but also has a function in that the tabs controlling the voices on the two manuals are more convenient to the organist. It is interesting to note here that the Baldwin theatre organ contains completely transistorized tone generators and amplifiers and the most modern of speaker systems, and the cabinet proper must be sized and designed for acoustics and to accommodate the fundamental parts of the instrument. *Purely functional design* is any design where function is completely dominant with aesthetics not considered at all. An example is shown in Fig. 3, a picture of a 3KVAR, 250-volt single-phase 60-cycle capacitor used on electric transmission lines. This piece of equipment is completely functional and aesthetics plays no part in the design. Other examples are machines such as lathes, boring machines, motors, power tools, conveyors, material-handling equipment, and the great bulk of manufacturing and production equipment.

Even though aesthetic considerations may be present in engineering design, the emphasis in this discussion must necessarily be restricted to good functional engineering design. However, never entertain the thought that pure function always prevails as the governing factor. As an example, in early automotive design the machine proper was designed, and then the body was designed to cover the machine. Recent automotive design shows that because of body shape, many components have been redesigned and moved from former positions in order to accommodate to body design.

Some of the best examples of design in all aspects are to be found in automotive and commercial aircraft design. Purely aesthetic features are evidenced in colors used; elegance of fabric, plastic, or leather in upholstery; finish and appearance of appointments, body lines, and artistic configuration.

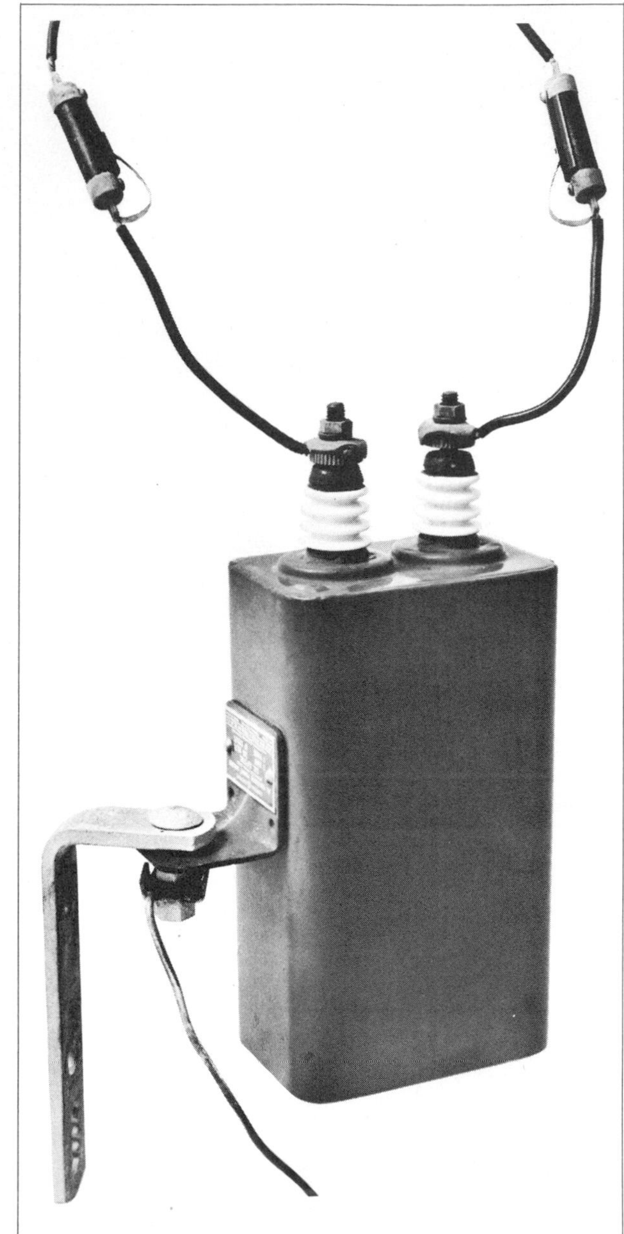

**FIG. 3.** Purely functional design. Aesthetics plays no part, and is not considered important in any way.

Functional-aesthetic features are obvious in such aspects as artful and functional controls, dials, instruments, glass areas, seat comfort, and safety features. Purely functional design is clearly manifested by examination of the power supply and all of its auxiliaries and components.

There are many categories of engineering design, such as machine design and structural design. Paralleling this, the designer working in a particular field is designated by his field or subclassification in it: for example, machine designer, appliance designer, structural designer, automotive parts designer, etc.

Even though the field of design is broad, all designers think and produce in much the same manner. Simply put, the designer draws upon his background of knowledge and experience in producing a new entity. Thus, every designer must have (1) knowledge in his field, (2) experience, (3) inventive ability, (4) a knowledge of materials and processes, and (5) the ability to represent (draw) so as to transmit his designs to others.

All this is not quite as difficult as one might think. The real key lies mostly in the ability to *draw*, both freehand and with instruments; it is then the creation develops. At this time all the designer's knowledge, experience, and skill are brought to fruition. As he thinks of ways to solve the problem—considering methods, materials, combinations, and arrangement of components—he records his thoughts and develops his design. Because of the creative aspect, designing is personally very interesting and satisfying.

It is our purpose in this chapter to describe the processes of creation and development and to give advice on what authorities consider to be good design.

## 2. The Necessary Background of Knowledge and Experience.

No designing is possible for people unfamiliar with the problem at hand and with the possible methods of solution. Thus, first of all, a designer must have a thorough knowledge of all the elements involved. This means also that for every particular "field of endeavor," the background will be dictated by that field. Nevertheless, no field today is quite "pure." As an example, in mechanical engineering there will most certainly be many cases where an electrical application is an important part of the mechanical device, and vice versa. Therefore, a diversified background is a distinct advantage. Furthermore, the necessary background will vary considerably, depending upon the extent of scientific education and training required. For example, a designer working in one of the aerospace fields must have extensive study in physics, chemistry, mathematics, etc., while a designer working for a company which manufactures small home appliances would probably not need anything beyond basic courses. Further, it should be noted that in conference with many top executives and designers in the aerospace as well as in highly technical electronics fields, it has been learned that a number of men having a terminal technical training of only two years have, nevertheless, through experience of a few years, become valuable to the company as designers of fairly complex machines. It is the judgement of this author, however, that the extent of technical training dictates the extent to which a designer can expect to work on highly technical equipment. But it may also be said that experience in the field will make up to some extent for a lack of formal education.

Every designer, no matter what the field or the product, *must* have a thorough training in graphics. Without it, a designer would fail completely because as the design conception proceeds, the designer's own thinking must be recorded in the form of sketches and drawings. Furthermore, as the design is developed it must be discussed with and approved by such people as the chief designer, chief engineer, and management executives. This means that clear and concise communication is necessary, which is accomplished through the sketches and drawings made by the designer. The design drawings will often be augmented and supported by mathematical data and diagrams, sometimes including computer data, but the designer's sketches and drawings and his discussion and explanation are the most significant aspect of communication. As has often been said, although it is quite impossible to describe even a simple component in words, communication is

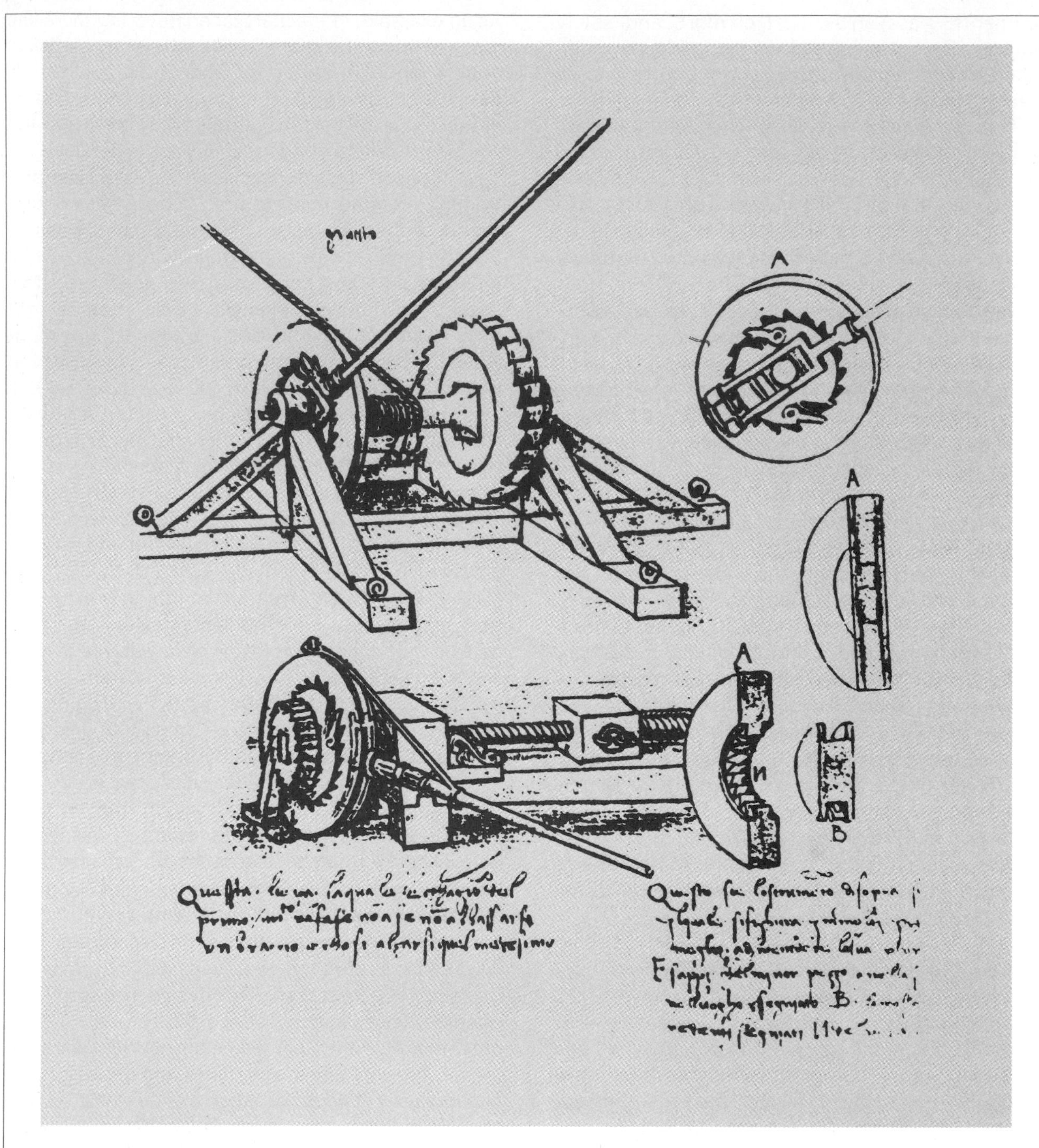


**FIG. 4.** A historical design drawing. Leonardo da Vinci (about A.D. 1500).

very simply and directly accomplished with a drawing.

Figure 4 is an example of a very early design drawing by Leonardo da Vinci (dating from about A.D. 1500). Because of his great ability to represent graphically, da Vinci was able to convey a design with great clarity in a single drawing.

In addition to the ability to express himself graphically, a designer must have a complete knowledge of processes, materials, methods of fastening and of assembly, types of finishes, and economical methods of production, as well as a knowledge of aesthetic values.

The Sections following give information on how a designer carries his work forward to produce the good design of a complete entity made with good components, and the thought he must give to drawings, materials, processes, fastening, assembly, finishes, economics, and aesthetics.

### 3. Procedure—How a Designer Thinks.

*First,* in approaching a new project, the designer must recognize and thoroughly understand every phase of the problem. This will include all information necessary to state what the device is expected to accomplish and the pertinent data such as speeds, pressures, temperatures, and operating conditions. Also, statements of relative size and appearance are often included.

*Second,* the search for solutions is started, and here the real ability to create shows itself. In the beginning, conventional solutions will come to mind, and there may be several of them. Then, as the investigation continues, newer, more modern, and previously unheard-of answers may appear. At this stage the designer should let his imagination really "run rampant." Every unusual physical, chemical, and electrical application, use of material, combination of elements, and their relationships, should be carefully studied. A number of possible solutions will probably emerge. Sensitivity to the problem, the ability and desire to create, coupled with originality and the worker's ability to analyze and synthesize, characterizes this stage of the work. Perseverance and persistence offer the key to success here.

While the solutions are being composed, the designer must make sketches for his own use. To be sure, these may be very "rough" and may lack detailed information, but they will probably include many written notes giving information for later use. This is where real ability and facility in the use of the graphic language play a very important role. An active mind and fluency at recording the mind's products of originality on paper combine to produce successful solutions.

*Third,* all of the solutions should be evaluated. This analysis must include feasibility of the design from every standpoint—from every engineering detail to economics and aesthetics. This must be done carefully and honestly because it is principally self-evaluation. Personal idiosyncrasies and preferences should be subservient to a completely "open mind."

*Fourth,* either decide on the best solution or continue the search for a better solution.

*Fifth,* obtain approval of the design. The practice here varies with the size of the organization. Company practice prevails.

*Sixth,* refine and correct the original design and make more complete sketches as a guide before starting a formal design drawing. Some alterations may be a result of the approval conference on the design.

*Seventh,* make a formal design drawing. This drawing must show all information that will be needed by detailers who will make the individual part drawings, subassemblies, and assemblies.

*Eighth,* obtain final approval of the design.

### 4. Drawings.

Design drawings are not like other drawings (assembly and detail drawings) previously discussed in this text. The difference is that design drawings supply the information *from which* assembly and detail drawings will be made. Design drawings may be divided into two classes, *preliminary* and *final.*

*Preliminary design drawings,* for the most part, are sketches, although some (usually in the later stages of design) may be instrument drawings. Figures 5 to 8 illustrate preliminary design drawings. These are made freehand. While working up the sketches, the

## DESIGN PROCEDURE

**PHASE ONE—THE PROJECT**

A. Discussion with management, engineering, client.
B. Statements and specification of the design problem.
C. Collection of all pertinent information.

**PHASE TWO—FORMULATION**

A. Recognition of requirements.
B. Definition of requirements.
C. Consideration of previous designs.
D. Assembly of all original data needed—mathematical, graphical, mechanical, electrical, etc.

**PHASE THREE—CONCEPTS**

A. Preliminary design sketches.
B. Preliminary design data giving materials, methods, construction details, and projected characteristics.

**PHASE FOUR—ANALYSIS**

A. Critical analysis of all design concepts.
B. Selection of most promising design or designs.

**PHASE FIVE—DESIGN CONFERENCE**

A. Discussion of preliminary designs with engineering, management, client.
B. Approval of design or designs.

**PHASE SIX—REFINEMENT**

A. More complete drawings and specifications of selected design or designs.
B. More complete data supporting projected design.

**PHASE SEVEN—DESIGN CONFERENCE**

A. Discussion of refined design or designs.
B. Approval of most promising design.

**PHASE EIGHT—SYNTHESIS**

A. Projected design supported by mathematical, graphical, and computer-aided and combined systems data.
B. Investigation of all physical aspects and proof of soundness of the design.

**PHASE NINE—MODELS**

A. Components.
B. Mock-ups.
C. Models of critical features.

**PHASE TEN—TESTING**

A. Proof of operating characteristics of components.
B. Proof of soundness of complete entity.

**PHASE ELEVEN—CONFERENCE**

A. Final discussion with originating authority.
B. Approval of final design.

**PHASE TWELVE—FINAL PREPARATION**

A. Final design drawings.
B. Final specifications.

**PHASE THIRTEEN—TRANSMITTAL**

A. Transmittal of final design drawings and specifications to originating authority.

NOTE: In the outline above three conferences (Phases 5, 7, and 11) are scheduled with the originating authority, engineering, and management. This is usual, but it does *not* mean that these are the only conferences during the development of the project. The designer confers frequently with colleagues, and with engineers, components suppliers, materials experts, and others, to obtain information and confirm design features.

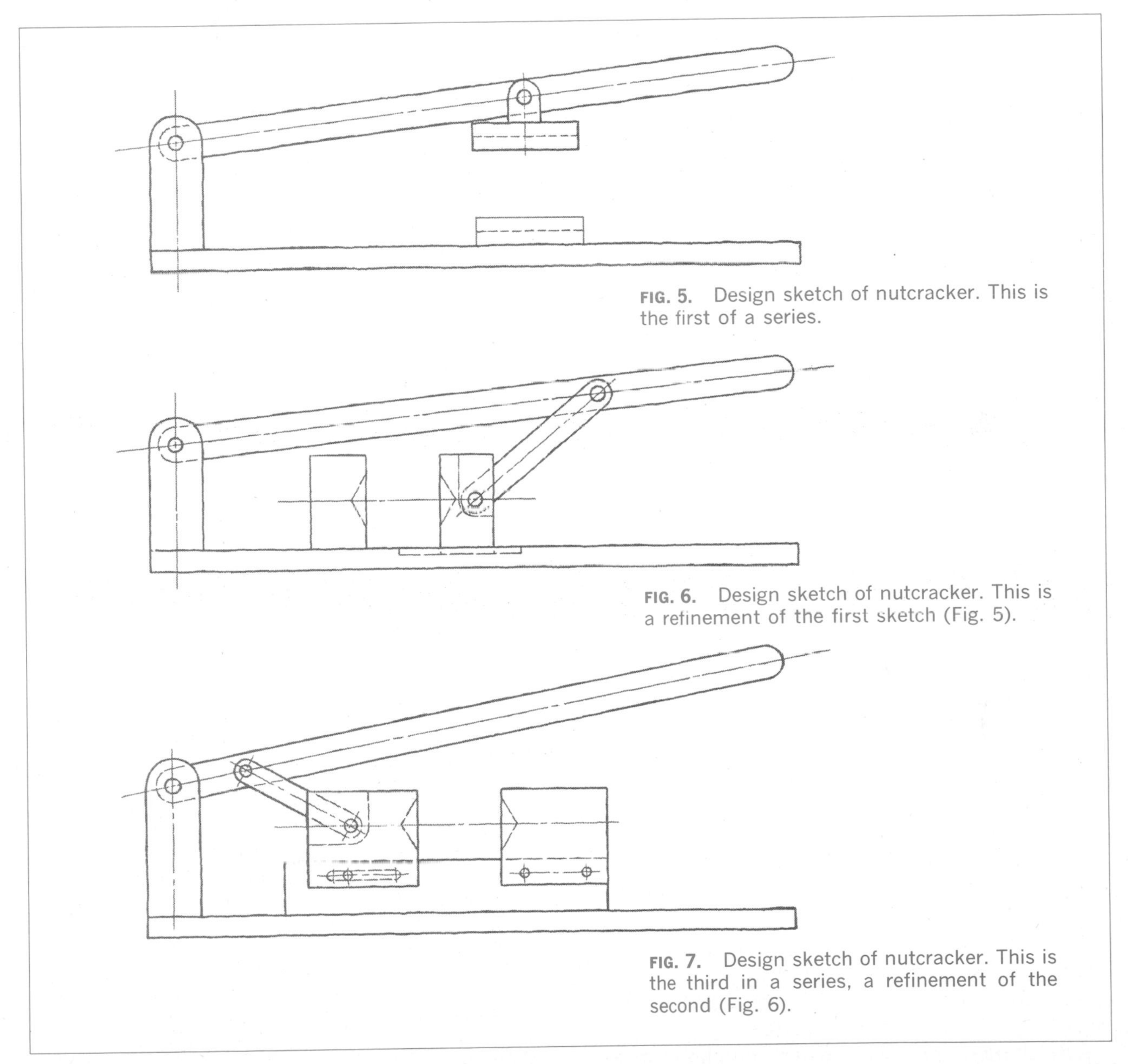

FIG. 5. Design sketch of nutcracker. This is the first of a series.

FIG. 6. Design sketch of nutcracker. This is a refinement of the first sketch (Fig. 5).

FIG. 7. Design sketch of nutcracker. This is the third in a series, a refinement of the second (Fig. 6).

designer is keenly attentive to the solution of the problem, as described in Sec. 3. As the creative process goes forward every aspect of the final design must be considered and recorded.

As an example of how the design thinking and the preliminary drawings are brought forward together, Figs. 5 to 8 illustrate a series of design drawings made of a nutcracker. The first drawing, Fig. 5, shows that the designer conceived a lever and base with fixed and pivoted pads between which the nut would be placed. It is immediately obvious that when the shell breaks, the release of counter

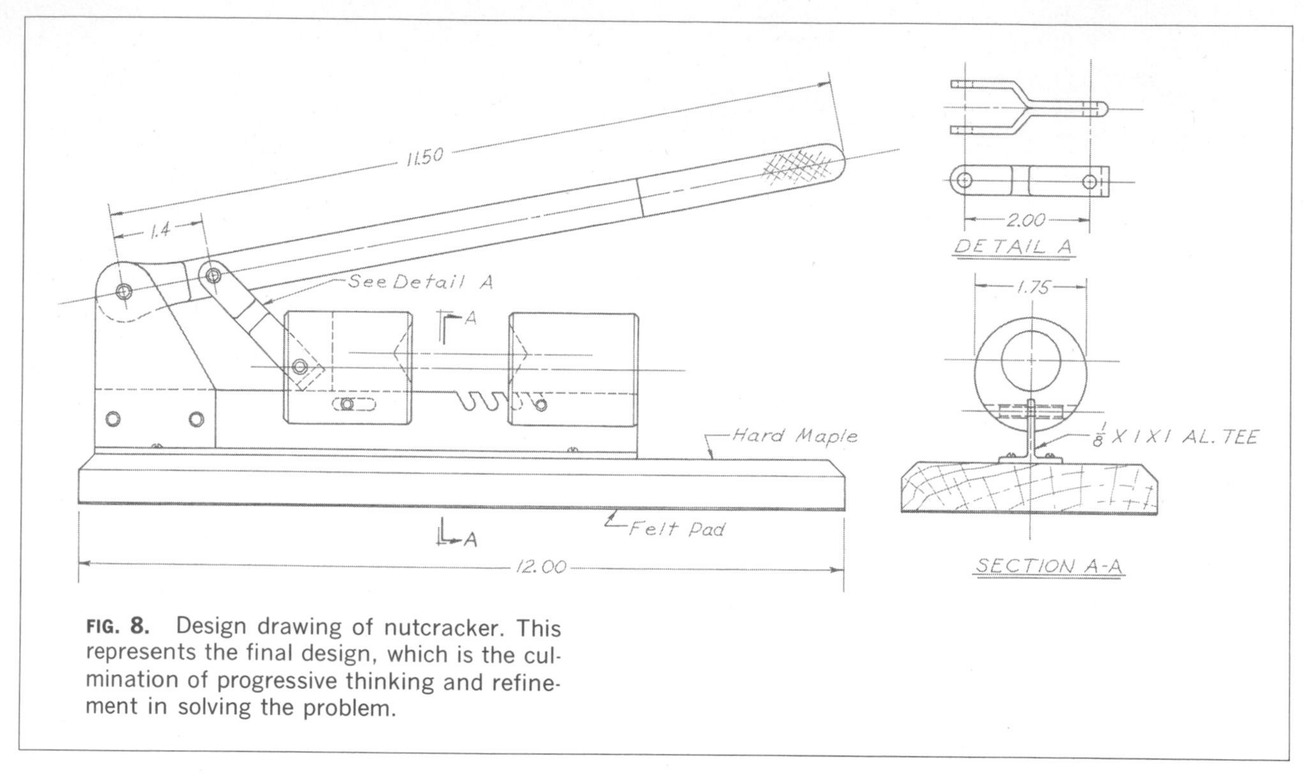


**FIG. 8.** Design drawing of nutcracker. This represents the final design, which is the culmination of progressive thinking and refinement in solving the problem.

force will allow the lever to descend at once and crush the nut meat. The second design, Fig. 6, is a refinement over the first design with the linkage arranged so that when the actuating lever descends to its maximum, the holding pads will come to a minimum separation, but no closer together. A further refinement is seen in Fig. 7, where the linkage is reversed to place it away from the hand position on the operating lever. Finally, in Fig. 8, the basic arrangement of Fig. 7 is refined to include an adjustable distance between the holding pads. Also at this time a more accurate and complete drawing is made, and some basic dimensions and details, including material types, are added.

A much more sophisticated and complex example of design thinking is described below. The accompanying drawings are shown in Figs. 9 to 14. The final design drawing, prepared for presentation at an engineering and management conference, is shown in Fig. 15. For purposes of authenticity the following is an excerpt from the designer's file of notes and comments on the design through all stages of development.

*Problem*

Design a device to collect seawater samples at any desired depth. This device, on one trip to sea bottom and back, must collect at least six samples and preferably two or three times that number.

*Operating Conditions*

Salt water, depth nearly seven miles, maximum pressure about six tons per square inch, water significantly compressible at this pressure.

*Preliminary Solutions and Comments* (see sketches, Figs. 9 to 14)

*Design 1* (*Fig.* 9). A pull on the actuator cable will swing arm in and up, tripping lever and opening port. Slacking off on cable permits return spring to move arm out again. The next pull will open the next

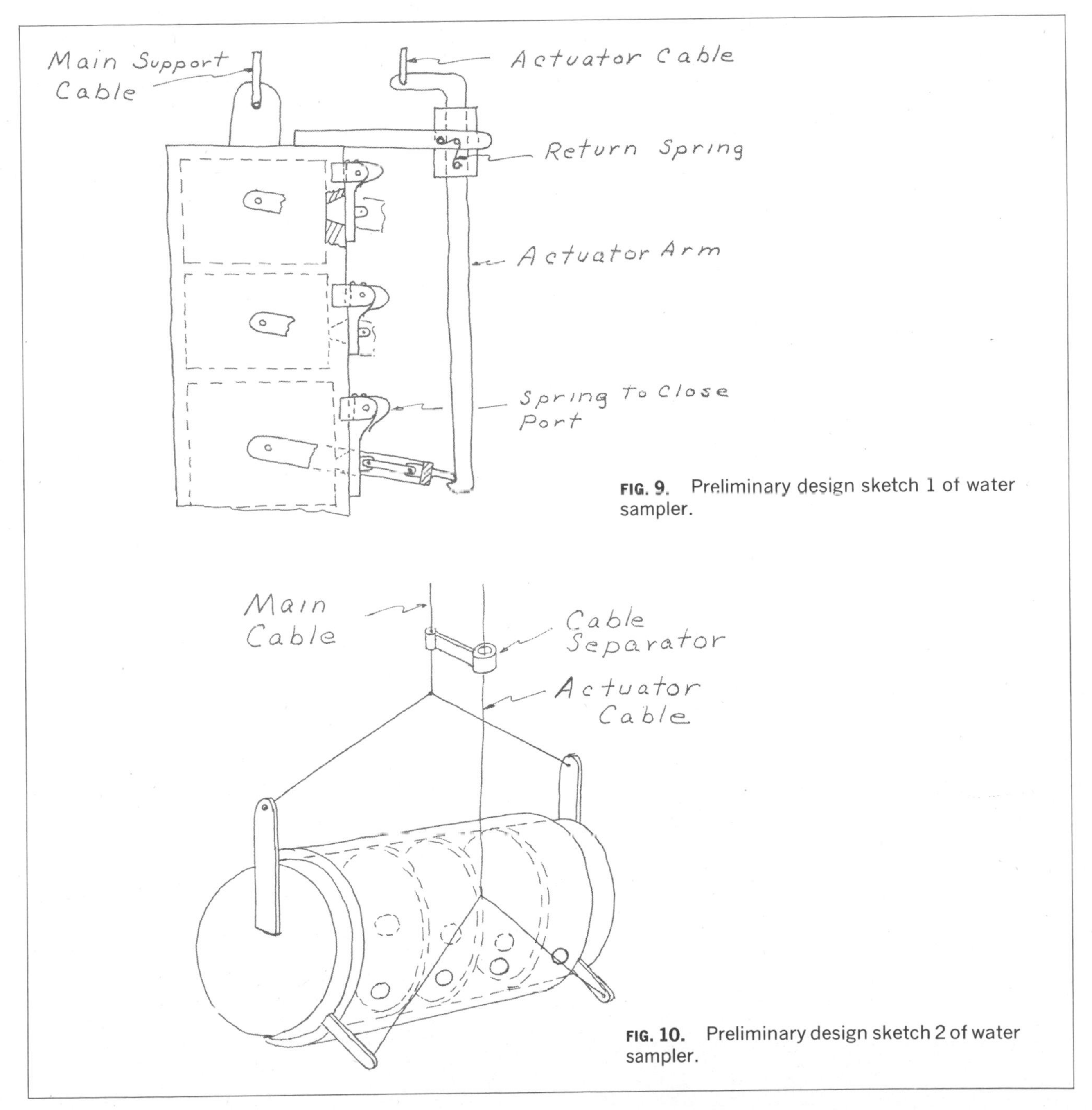


**FIG. 9.** Preliminary design sketch 1 of water sampler.

**FIG. 10.** Preliminary design sketch 2 of water sampler.

chamber above, etc. A spring to close each port after its chamber fills will be required.

*Disadvantages:* Too complicated. Cables will twist together and tangle. Leverage is good, but may not be sufficient to open ports at high pressures. Operation will always be uncertain.

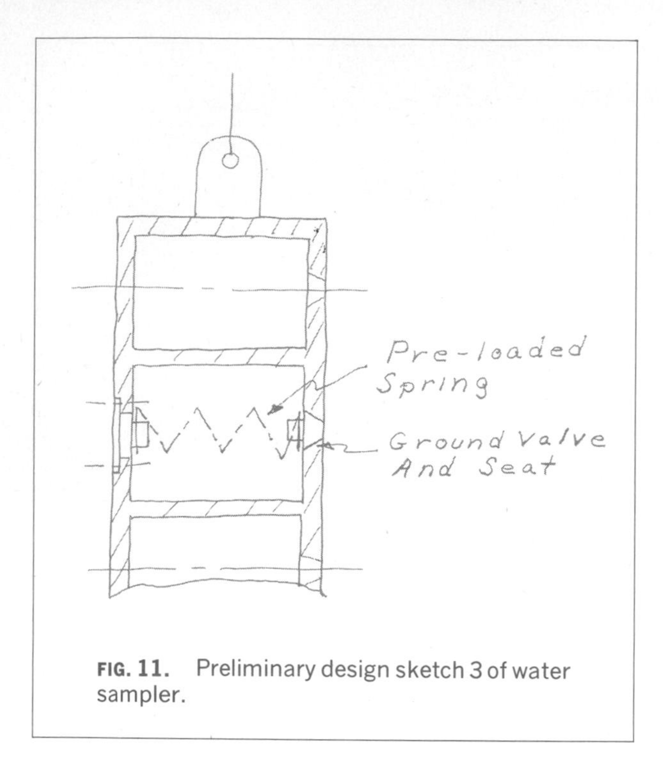


**FIG. 11.** Preliminary design sketch 3 of water sampler.

*Design 2* (*Fig. 10*). A pull on the actuator cable will index the external cylinder relative to the internal cylinder containing the collecting chambers. Each index will line up a pair of holes between the two cylinders permitting each chamber to fill in turn.

*Disadvantages:* Sealing off chambers will be difficult. Cable stretch and tangling will make indexing inaccurate and uncertain. Cylinders may stick together, making indexing uncertain.

*Design 3* (*Fig. 11*). Each chamber contains a preloaded spring pressing a ground valve into its seat. Each spring is carefully loaded to permit the valve to open at a definite pressure.

*Disadvantages:* Spring calibration will change with corrosion and use. Temperature and salinity variations will prevent pressure from varying as a straight-line function of depth, i.e., depth at which each chamber opens will be uncertain.

*Note:* Electrical contacts could be designed to send signal to surface indicating when chamber opened. A premeasured suspension cable will then indicate depth.

*Design 4* (*Fig. 12*). This design is basically the same as design 3. A solenoid plunger engages an extension on the valve so that it will stay closed until an electrical impulse from the surface retracts the plunger. Spring for each depth will exert less force than comparable spring from design 3.

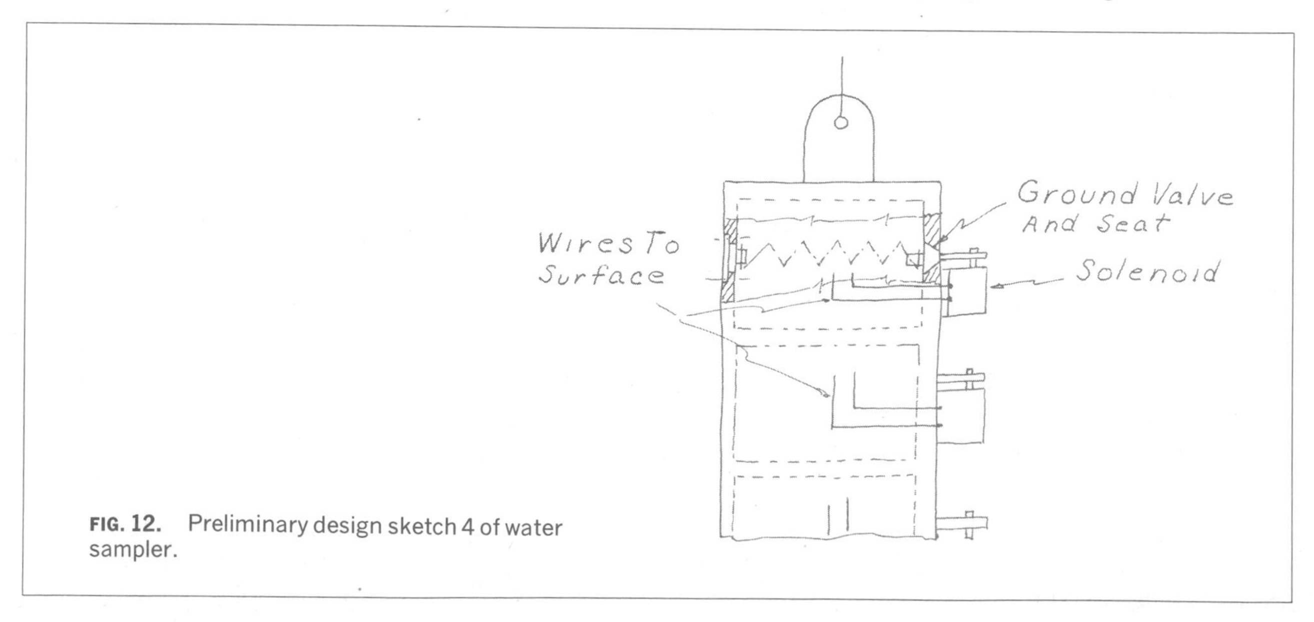


**FIG. 12.** Preliminary design sketch 4 of water sampler.

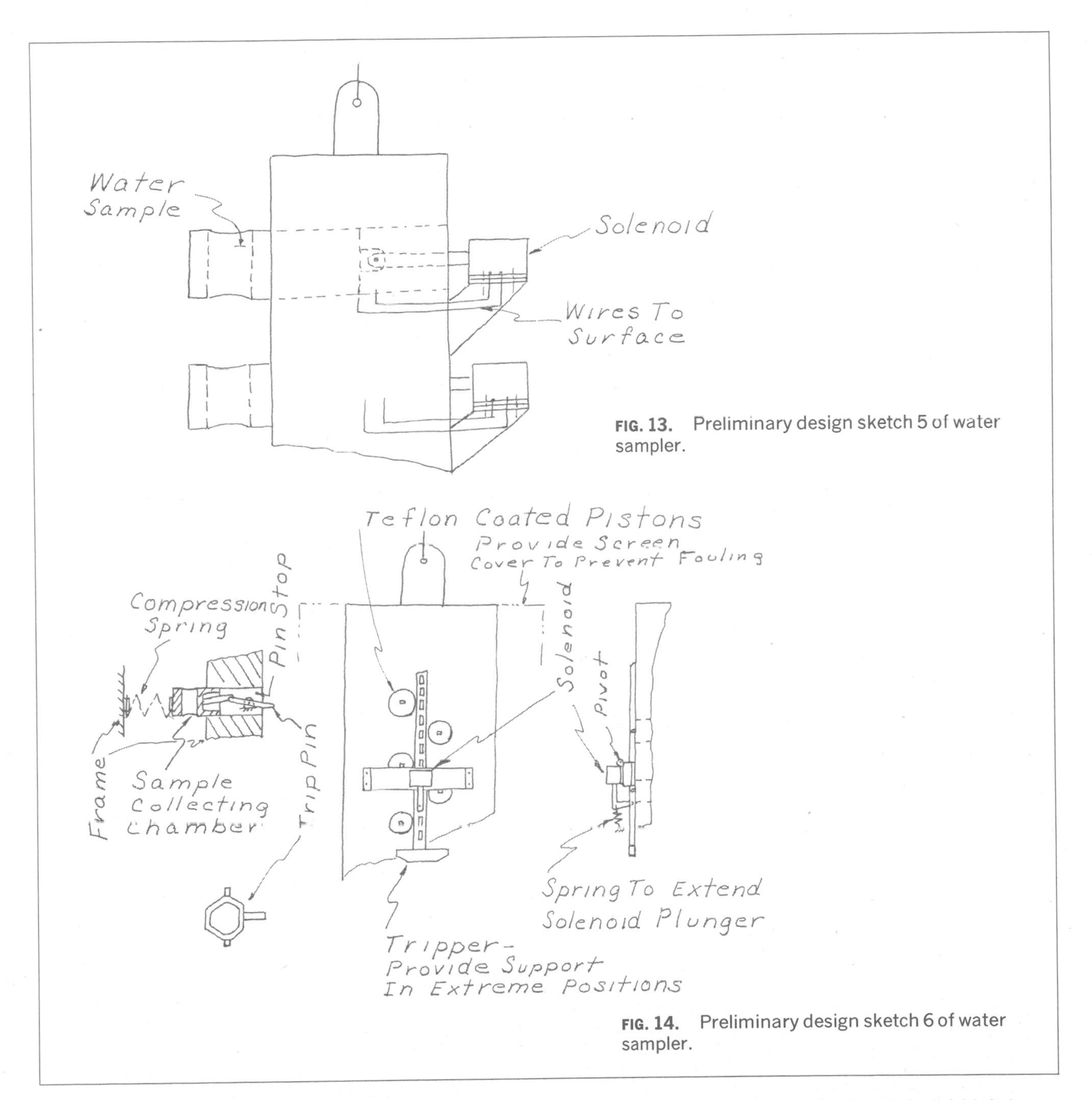


**FIG. 13.** Preliminary design sketch 5 of water sampler.

**FIG. 14.** Preliminary design sketch 6 of water sampler.

*Disadvantages:* Too many wires to run to surface. Trapped air may not allow cylinders to fill completely. Trapped air and slight expansion of water brought up from great depth will build high internal cylinder pressure at surface. Both high external and internal pressures (at different times)

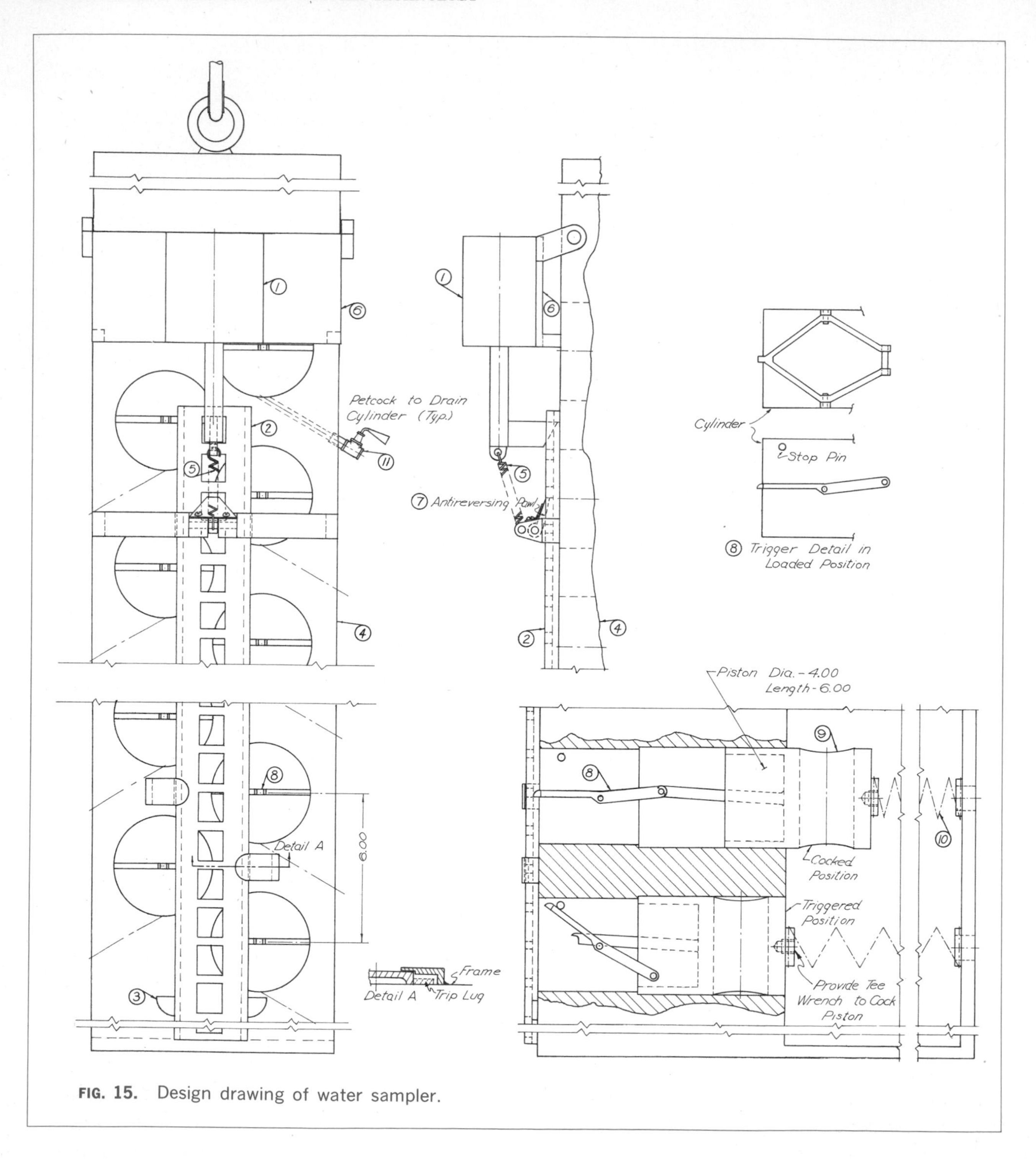


**FIG. 15.** Design drawing of water sampler.

make cylinder design difficult. Removing water samples will be a problem.

*Design 5 (Fig. 13).* The chambers used to collect samples are cut into close-fitting Teflon-coated pistons. Water flows continually through chambers until solenoids retract them into housing. Samples are collected from bottom, coming up. As internal pressure builds up, it will tend to bleed off around piston. External water will not contaminate sample as long as some positive pressure remains in chamber.

*Disadvantages:* Too many solenoids and too many wires. All electrical systems will have to be specially designed to withstand the pressure. Solenoid will have to stay energized to hold sample in place. It may be impossible to get a reasonable-sized solenoid to give a stroke that is long enough.

*Design 6 (Fig. 14).* Each chamber remains in open water as in design 5 until it is pushed into the housing. A single solenoid operated repeatedly indexes the tripper vertically up the housing. After the solenoid indexes the proper number of times, the tripper engages the trip pin holding one of the cylinders in place. As the trip pin is lifted, the compression spring pushes the cylinder into the housing, collecting a sample. The solenoid continues indexing the tripper vertically, with each current impulse from surface tripping each of the cylinders in turn.

*Disadvantages:* Complicated. Electrical components must be specially designed

*Notes:* This looks good enough to make a design layout and to contact manufacturers of solenoids about the electrical design.

The electrical current can be monitored at the surface with an oscilloscope to tell if solenoid has indexed properly with each current impulse sent to it.

This explanation accompanies the design drawing, Fig. 15:

The solenoid (1) is energized, thus moving the trip arm (2) with trip lugs (3) up the housing (4). When the solenoid is deenergized, the return spring (5) pulls the solenoid arm out and over the trip arm and positions the solenoid arm in the next slot on the trip arm. This action is possible because of the hinging of the solenoid mount plate (6). The trip arm is not pushed down by this action due to the spring-loaded antireversing pawl (7), which locks into a slot in the trip arm. As the trip arm moves up, the trip lugs trip the trigger mechanism (8) built within the cylinder.

The piston (9) is then forced into the housing by the spring force of the load spring (10). This operation is continued, triggering cylinders at required depths. The cylinders are machined into the lower portion of the housing, with the upper portion of the housing not bored, but used to serve as additional weight and support for the trip arm during its upward motion.

When this device is brought to the surface, water samples are obtained from each cylinder by opening the individual petcock (11) for each cylinder in turn.

To reload this device a threaded tee wrench is passed through the hole in the housing support for the load spring and screwed into the piston. The piston is then retracted, and the trigger mechanism is reset manually.

This entire device should be surrounded by a wire cage to prevent fouling.

*Final design drawings.* Figure 15 is an example of a typical final design drawing. It represents the culmination of previous design sketches into a final, complete and accurate drawing, accompanied by all necessary data and specifications. From this drawing, detail and assembly drawings will be made. See also Fig. 1, Chap. 23.

**5. Models Used as a Design Tool.** Three-dimensional models serve a purpose in design that normally cannot be achieved by two-dimensional drawings. Whereas almost anyone can look at a model for a short period of time and grasp the design intent, only those trained in reading and interpreting drawings can visualize what is portrayed by plans, elevations, sections, and details shown on drawings.

Models are used:

1. To provide study tools for engineers and designers

2. As a method of presentation and review
3. To communicate the designer's ideas to the construction forces, thus reducing the number of drawings required for some phases of construction

**6. Model Types.**
There are many different types of models which an engineer can use. They can be grouped into four general types: (1) architectural presentation, (2) equipment prototype, (3) equipment arrangement, and (4) process. The utilization of these types of models varies widely from one company to another, but there is increasingly wider acceptance of modeling as an aid in engineering.

*Architectural Presentation.* A model of this type is built strictly for approval of a proposed design. Models for presentation are usually built to a small scale but with considerable detail and elaboration to achieve the realism desired. Some even have simulated lighting and working parts. All have miniature trees, bushes, sidewalks, and other landscaping details, in addition to detailed representation of the architectural treatment. A typical architectural model, which is relatively simple compared with many, is shown in Fig. 16.

*Equipment Prototype.* Prototype models are built to check the appearance or feasibility of a design and to assist in sales efforts. They are usually built to a large scale. Often they are full size, and for large pieces of equipment they are one-quarter or one-half size. Appliance manufacturers always build prototype models before setting up a production run. Several alternative designs for radios, clocks, etc., frequently are modeled and distributed to large buyers to check their preferences before a final design is settled upon. Car makers always make a full-size model of a new car to check the appearance before they proceed with the expensive tooling that leads to the new product. A typical prototype model is shown in Fig. 17.

*Equipment Arrangement.* Note that the equipment-arrangement model has perhaps had the widest acceptance and application. This is due to the ease with which it is constructed and straightforward methods of utilization. The modeling techniques vary slightly with the type of facility under study. There are three general applications of this type of model: (1) small-scale plot plan, (2) machine shop or factory, (3) process equipment.

The scale and the amount of detail on these models vary to suit individual requirements, but in general the scale ranges from ⅛ in. to the foot on plot-plan models to ¾ in. to the foot on some process-equipment models. Scale is an important consideration in planning a model. The size depends on how much detail must be shown for visualization of the problems to be solved.

Plot-plan models are usually built to a small scale for two reasons: (1) not much detail is needed to show the relationships between buildings and/or

**FIG. 16.** Typical architectural model. This type of model is made for presentation of a design to the client. (*Courtesy of Procter & Gamble.*)

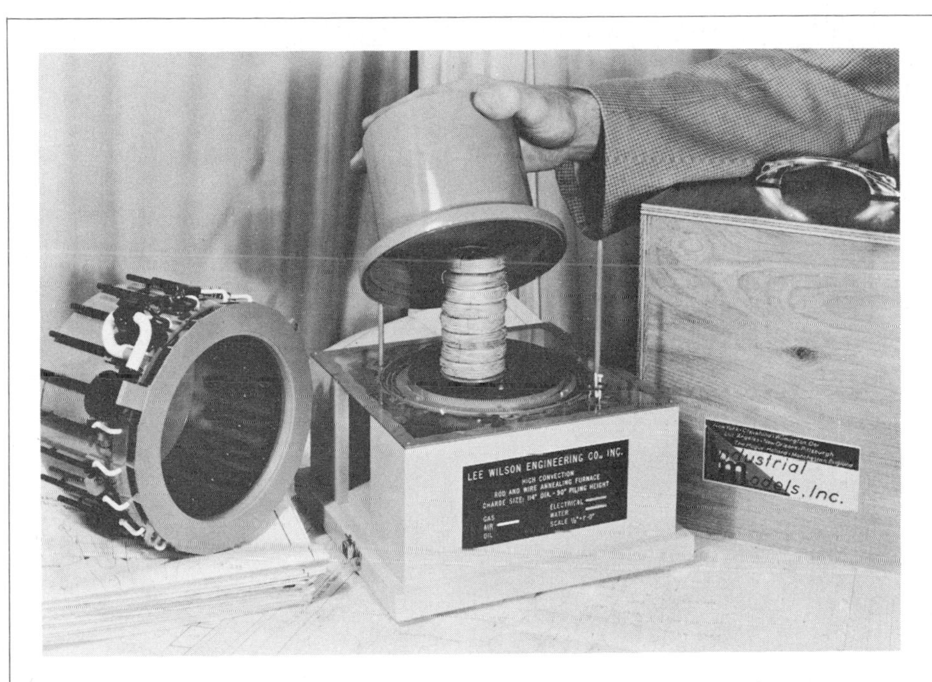


**FIG. 17.** (Above) Typical prototype model of equipment. This kind of model is used to demonstrate a new device or design. (*Courtesy of Industrial Models.*)

**FIG. 18.** (Below) Typical plot-plan model. This model, built to a scale of $\frac{1}{8}'' = 1'\text{-}0''$, is used to establish the over-all layout of a factory. (*Courtesy of Procter & Gamble.*)

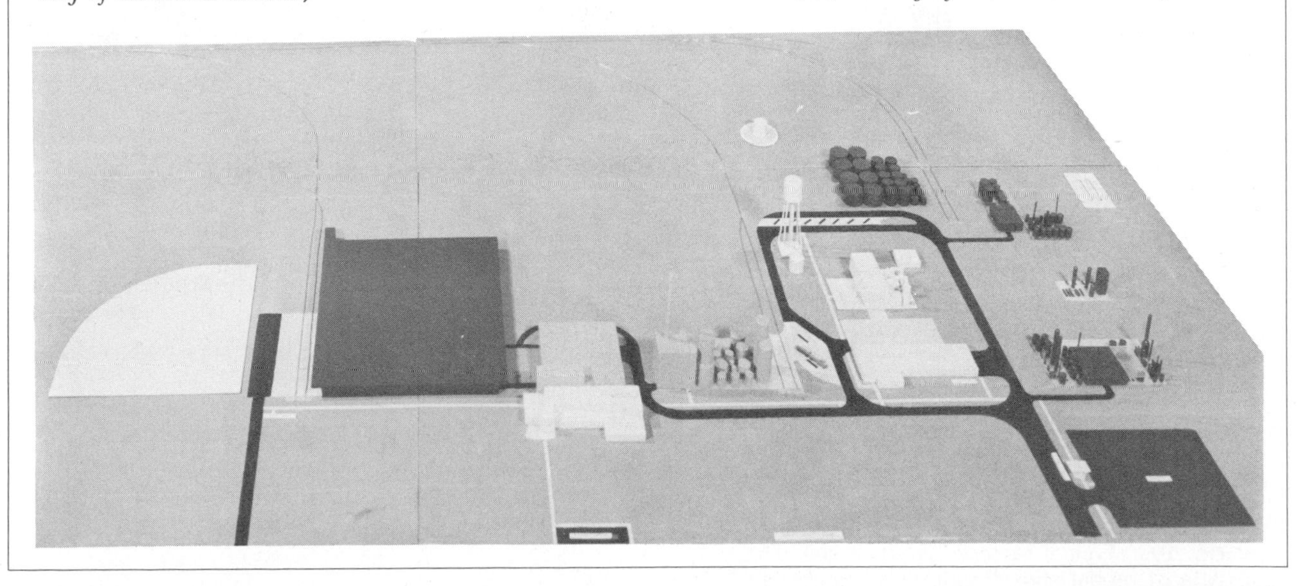

units of equipment; (2) the plot under study is frequently so big that a large scale would make the size of the model inconvenient to work with. A common scale is $\frac{1}{8}$ in. to the foot. The model in Fig. 18 is built to this scale. As can be seen, the buildings and equipment are simple blocks. This type of model is used instead of drawing several alternative plot plans on paper. To study alternatives, an engineer or designer simply moves the blocks around on the study board until he finds the optimum arrangement, taking into account future planning as well as the current project. After he has studied the arrangement, he invites other interested persons to review and approve the layout. With modeling, much more can be contributed by all levels of management than would normally be anticipated with drawings, and this usually results in the best over-all job for a company at the lowest total cost.

Figure 19 shows a typical machine-shop or factory layout model. This type of model is widely used to determine the most efficient manufacturing operations. It is commonly constructed to a scale of $\frac{1}{4}$ in. to the foot, and many companies have models of all the tools in their factory so that each time a manufacturing change is contemplated, its effect can be studied by modeling. This type of effort has become so common that die-cast and injection-molded models of the more common machine tools and factory accessories are available at low cost. On mass-produced models it is practical to have more

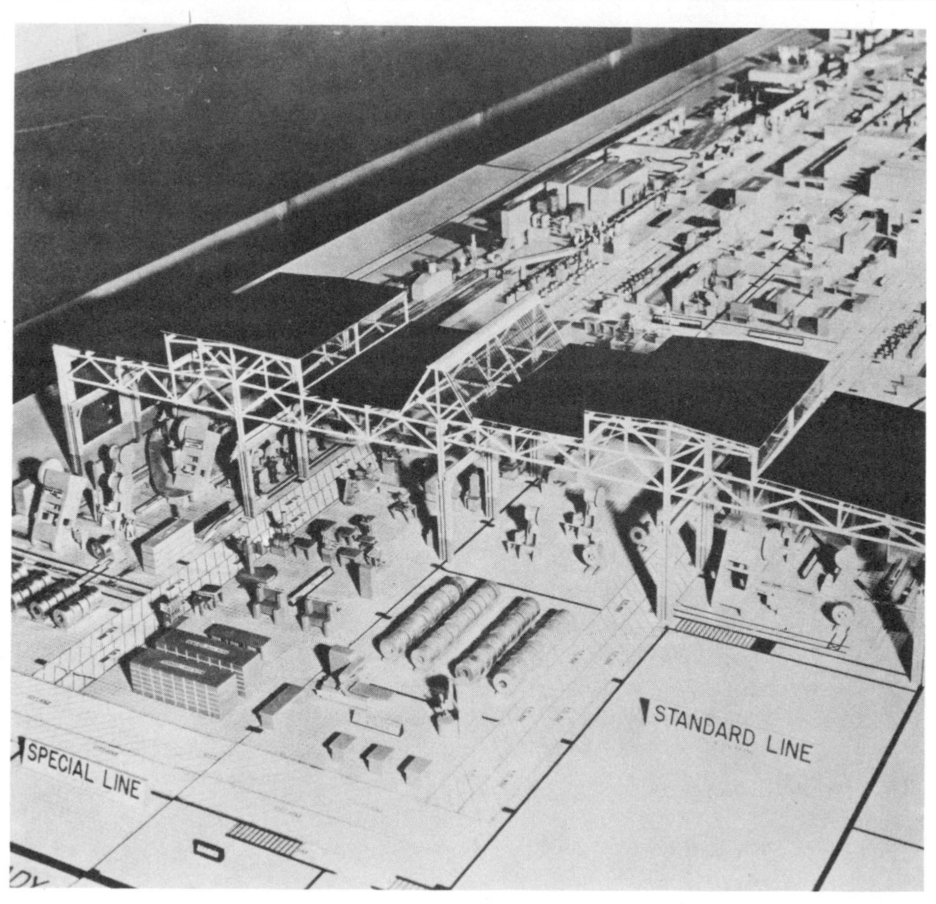


**FIG. 19.** Typical factory layout model. This type of model shows equipment, route and assembly lines, raw and finished material, etc. (*Courtesy of Visual Plant Layouts.*)

detail than would normally be shown if a few units were made for a specific project. This increased detail adds to the realism without running up the cost of the modeling effort. Usually a grid background is used to facilitate the alternative placing of the model components.

Chemical process industries also use equipment-arrangement models to determine the most economical placing of equipment from a first-cost and operating standpoint. Whereas machine-shop or factory models are built to a $\frac{1}{4}$ in. to 1′-0″ scale, the scale of process-equipment layout models varies widely with the physical size of the equipment and the size of the area under study. Models of large areas with large equipment, such as a petroleum refinery, commonly are made to $\frac{1}{4}$ in. to the foot. Small processing units frequently are made to $\frac{3}{4}$ in. to the foot and in special cases even larger. The equipment is represented by wooden dowels or blocks on the smaller-scale models and by styrofoam or some other light-weight plastic on those of larger scales. A typical process-equipment layout model is shown in Fig. 20. Many modeling companies now supply special kits designed to facilitate layout study work. These kits include an assortment of equipment and basic material and usually provide some patented method of building a model with multi-storied units. The application of such a kit is shown in Fig. 21. As with other layout models, the object is to find an arrangement that will give a unit of the lowest over-all cost, by having as many people as practicable make contributions.

One reason why layout models have been so effective is that once the model pieces are built, the

**FIG. 20.** Typical equipment-arrangement model. This one is for processing units in a chemical plant. (*Courtesy of Procter & Gamble.*)

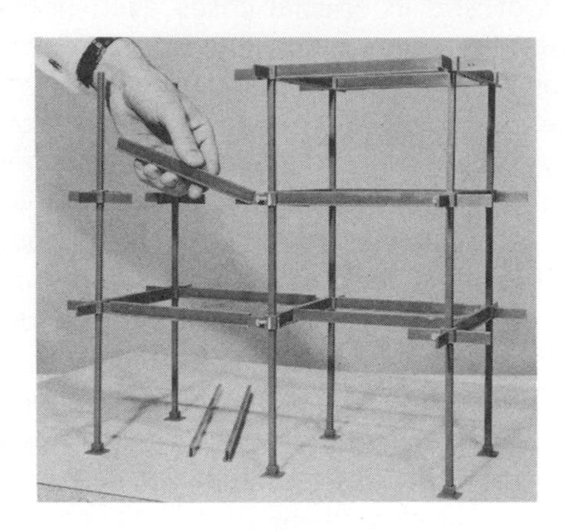

FIG. 21. Structural components of a modeling kit. These are used to facilitate the erection and use of a preliminary process-equipment model. (*Courtesy of Industrial Models.*)

engineer moves them around and studies the layout himself, instead of trying to solve the problem on paper.

*Process Models.* One important use of models is in portraying complete process systems. Full sets of drawings of modern piping systems which represent every detail of design and its relationship to equipment, structural steel, electrical conduit, etc., are time-consuming and expensive to prepare. Many manhours are required to make the necessary checks between all phases—electrical, mechanical, piping—to assure no interferences and to show the piping in the most economical manner. After the drawings are completed, only trained people can read and visualize the piping systems. In view of this, the modern construction practice is to make sketches of each pipe line, and carry out fabrication from these.

Even with the best designers and checkers, it is difficult to find all errors and interferences and to arrive at the minimum cost when stainless steel and other expensive materials are used. Costly mistakes in construction have led many companies to use three-dimensional models of their piping systems. Use of these models varies from company to company, and in most instances the application of modeling as an engineering technique has evolved because of well-established habits and methods of operation. The application of piping models ranges from building a model *after* the design has been worked out on paper, thus making the model an instrument for confirmation and approval; to building a model without assembly piping drawings, thus using the model to communicate design information to the construction forces.

Piping models normally are built to a scale that varies from $\frac{1}{4}$ in. to the foot up to $\frac{3}{4}$ in. to the foot and occasionally to 1 in. to the foot. In general, the scale used depends upon the intended use of the model. If conventional drawings are prepared along with the model, a scale of $\frac{1}{4}$ in. or $\frac{3}{8}$ in. to the foot is large enough. In this case the model is used for confirmation and approval, and detail and accuracy are not required since information is given on the drawings. If the model is used to the utmost—if the designs are worked out on it and it is then used by the construction forces without the aid of drawings—the scale should be $\frac{1}{2}$ in. or $\frac{3}{4}$ in. to the foot. If the piping is large-size (6 in., 10 in., or 12 in., in diameter), such as in a paper mill, the $\frac{1}{2}$-in. scale is adequate. If the average line size is 2 in. in diameter or less, the $\frac{3}{4}$-in. scale should be used. The $\frac{3}{4}$-in. scale has an added advantage in being easier to work, because with this scale $\frac{1}{16}$ in. on the model represents 1 in. actual, and this is the tolerance to which models can be made economically.

Regardless of scale and use, methods of building piping models are similar but of course vary between companies. The first step in building a model is to construct what is called the basic model. The basic model consists of a base board, models of all the equipment, and a replica of the structural design. All equipment is mounted in its appropriate location on the base or structure and forms the background on which the piping and other systems are installed. It is interesting to note that this is the same way the actual plant will be constructed. Also, this method is similar to that used in making conventional piping assembly drawings—the equipment and structural background is drawn on the paper before the piping design is started. In modern practice plastics are used almost exclusively in building the basic model. Several different kinds of plastic meet the requirements for this purpose. They are easy to work, light in weight, not subject to deterioration from moisture and moderate temperature changes, easy to join with solvent cements, and easy to paint. Figure 22 illustrates some unpainted basic-equipment items made of plastic. Figure 23 shows a typical basic model that is ready for installation of the piping.

The piping may be installed on the model from conventional drawings or the design may be worked out on the model directly without making conventional piping assembly drawings. In either case the piping itself can be represented by using the "fine-wire" or the "true-scale" technique. The fine-wire technique normally uses a $\frac{1}{16}$-in.-diameter brass wire, which represents the pipe center line, with fiber disks or rubber sleeves of the appropriate diameter spaced along its length to indicate the outside diameter of the pipe. The true-scale technique uses plastic rod or tubing whose outside diameter is equal to the scale diameter of the various standard pipe sizes. A typical fine-wire model is shown in Fig. 24. Fine-wire models are sometimes unsuccessful because inaccuracies and faulty representation on the model result in a piping system that cannot be installed as shown on the model; this is especially true at the smaller scales. As a result many companies now build true-scale models. Several modeling companies market a complete line of true-scale piping components made from various types of plastic. These are quickly and easily assembled. It is anticipated that in the near future almost all piping models will be constructed by the true-scale technique because this method gives superior representation and better visualization by all concerned. Figure 25 shows a typical true-scale modeling system developed for a scale of $\frac{3}{4}$ in. to the foot.

**FIG. 22.** Examples of equipment models. These are made of plastic and are shown here prior to painting. (*Courtesy of Proctor & Gamble.*)

The installation of the piping on the basic model is straightforward when drawings have been prepared. When the model alone is used, the basic procedure is to design the piping on the model directly from the flow diagrams without any intermediate studies. Since there are no drawings, every pipe line, regardless of size, and each instrument and piping accessory (such as valves, steam traps, and strainers) is installed in its proper location. Before installation of the piping is begun, some guiding rules are usually established by a model specification. One of these rules will establish the piping elevations. For example, it might be specified that all pipes running east to west will run at 8-ft elevation and those running north to south will run at 10-ft elevation. With a few guiding rules such as this, it is practical to install the piping directly on the model from the flow diagram. Figure 26 shows a typical true-scale model with most of the piping installed on it. When the model is nearly completed, it is carefully examined in a "model-review conference" by all parties who have an interest in the project. At this review each pipe is traced out and examined critically to make sure that it fulfills the needs of the process in addition to representing the

**FIG. 23.** Completed basic model. At this stage the model is ready for installation of the piping. (*Courtesy of Procter & Gamble.*)

most economical design. After the review the model is revised to reflect the contribution of the reviewers and then color-coded, tagged, and shipped to the field-construction site.

Normally, model color coding can be carried out in one of two ways. If the model is a confirmation or approval model, the pipe lines are coded according to the system of which they are a part; for example, all water lines might be light blue. If the model illustrates a piping design for which there are no conventional drawings, each pipe line is coded to indicate its construction specification; for example,

FIG. 24. Typical fine-wire model. This one is for a process unit. (*Courtesy of Procter & Gamble.*)

yellow might mean that a pipe is to be constructed of Sched. 10, type 304 stainless steel.

Piping is not the only phase that is shown on a process model. Since one of the difficult things to achieve in working out a design for a complicated piping system is the coordination with other phases, all other phases should be shown on the model with the piping. If all the phases go together on the model, they will fit in the actual construction. How much, and to what detail, other phases are shown varies widely. Many companies now expand the model approach used for piping to some of the other phases, such as electrical conduit routing, lighting, chutes, ducts, and instrument location and mounting. Some companies show designs of minor structural items, such as ladders, handrails, small-equipment supports, and platforms on the model without making additional drawings. The construction forces then install these items directly from the design intent shown on the model. Figure 27 shows a completed true-scale piping model that has been color-coded and tagged, and is ready for shipment to the construction site. It is typical of a final piping model on which all the piping, electrical con-

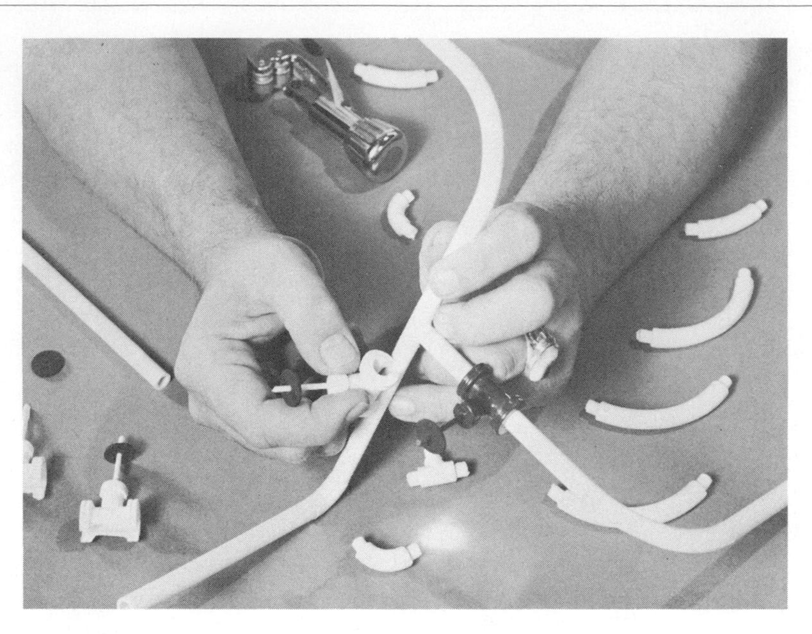

FIG. 25. True-scale modeling components. This illustration shows one of several systems that have been developed. (*Courtesy of Procter & Gamble.*)

duit, lights, chutes, ducts, instruments, and minor structural items are shown. From this model, the unit will be constructed without conventional assembly drawings for the various phases.

All the modeling work mentioned in this chapter has developed in answer to the problems inherent in visualizing a design when the information is presented solely by drawings. Models give a realistic rendition that is not achieved in drawings. However, drawings will always be required because some details that can *not* be shown on models *can* be shown on drawings.

### 7. Mandatory Relationships of Design, Materials, and Construction.

In any design, a prime factor is consideration of materials to be used. This is because, to some extent, the fastening of parts, their relationship, and often their detailed shape and the aesthetic appearance of the complete entity are affected by the materials used. More important reasons, however, are the conditions under which the machine is expected to operate. For example, a machine to be used in both tropical and extremely cold climates will have to be made of materials not adversely affected by either climate. In the hot and humid climate, parts deteriorate because of heat, corrosion due to humidity, and damage by fungous growths. On the other hand, in an extremely cold climate, contraction causes reduction of clearances and parts may "freeze" in position. Also, in extreme cold some materials become brittle and crack or break, operational difficulties are caused by cold "gummy" lubricants, and hazards are created by frost and ice. The same conditions prevail for many aircraft parts. From summertime ground temperatures of plus 90°F or so, in a very short time, an airplane may be flying at 30,000 ft where the temperature may be of the order of minus 40°F. Even if a device will still operate at two temperature extremes, the oper-

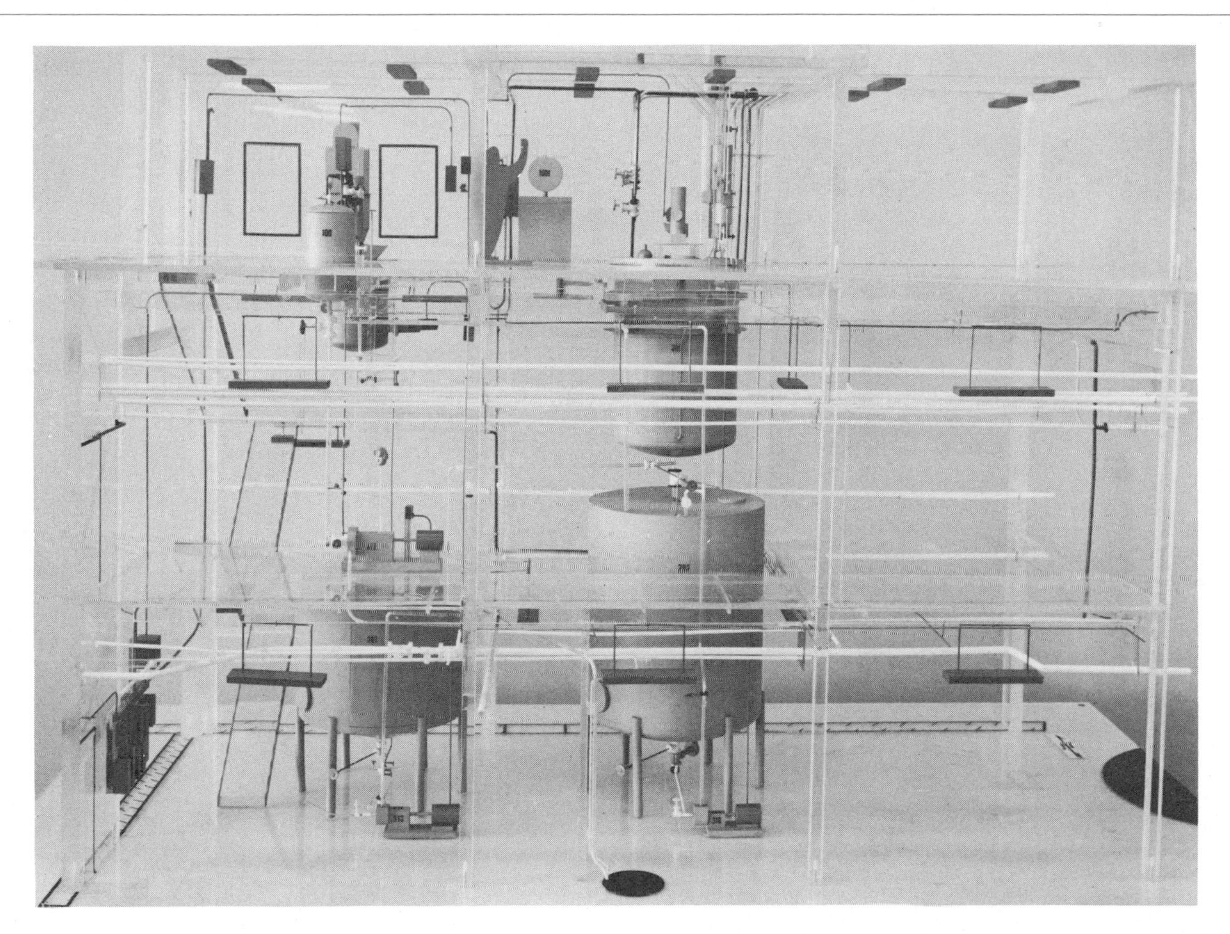

**FIG. 26.** True-scale model under construction. This illustration shows the model at the stage of construction before color-coding and pipe-line tagging. (*Courtesy of Procter & Gamble.*)

ational characteristics, especially in delicate instruments, may change radically, making either insulation or maintenance at a constant temperature (or both) a necessary part of the design.

The actual details of construction also have much to do with the successful operation of a machine. Inadequate bearing surfaces or insufficient lubrication will cause the machine to have a short operating life before repair or replacement. Protective seals against contaminants, abrasive materials, or corrosive liquids and gases, are frequently necessary. Even "foolproof" design and construction against mishandling is important. The designer himself, or another engineer familiar with the equipment, might operate a particular machine with a sensitive regard for its lacks or inadequacy, but especially when home appliances or other devices operated by laymen are designed, precautions against mishandling must be provided. Control knobs that come off and get lost; levers that break; screws and nuts that

**FIG. 27.** Completed true-scale model, showing all phases—piping, electrical and mechanical—and is now ready for use by the construction forces. (*Courtesy of Procter & Gamble.*)

readily loosen and drop off; multiple ("stacked") controls so made that the careless person can inadvertently move two or more while adjusting one; finishes that rust, corrode, pit, and wear off; parts that are difficult to keep clean, and many other similar factors are all abominations that will cause the purchaser and user to take a very dim view of the relative quality of the device. The cardinal rule is to design and dictate construction details to provide as many guards against damage to the device as is consistent with the price range of the product.

To summarize the foregoing discussion, it may be said that the design, the materials used, and the details of construction are all so closely interrelated for ultimate success of the device that they must be considered simultaneously.

Through consultations and discussions with designers in many different fields, the author has formed the opinion that all designers do not think in exactly the same way. Indeed, the same designer may not approach any particular problem according to a fixed pattern. Nevertheless, most designers consider the material first and then, closely following, will think of the design of the part and how it is attached to mating parts. This is because the material used more or less determines the detailed shape of a part. For example, a cast part will have a somewhat different configuration than that of a sheet-metal part or a part machined from solid material. If the part is designed before the material and method of production is determined, some adjustments in the design will probably have to be made.

Following this philosophy, the next four sections give (8) characteristics of materials, (9) methods of parts production, (10) good design and proportioning of parts, and (11) assembly—fastening and joining of parts.

## 8. Characteristics of Materials.

It is impossible, in the space available in this text, to give all available information on materials. Complete listings would include all detailed characteristics of strength, heat-treatment, resistance to corrosion, hardness, toughness, wear resistance, machinability, ductility, and so many others that the descriptions would fill a large book. The summarized material following, however, will serve as a guide to the principal characteristics and uses of most common materials. Detailed information will be found as needed in the standards of the ASA, ASME, ASTM, ASM, AISE, and SAE, and in such publications as *Machinery's Handbook and Materials in Design Engineering*, a periodical published by the Reinhold Publishing Corporation, 600 Summer Street, Samford, Conn. Also see Glossary. It is not considered to be within our premise to engage here in a discussion of the strengths of materials.

### Ferrous Metals.

**Cast Iron—Gray.** Widely used because of the ease with which a great variety of parts can be made. Good machining properties; can be joined by welding; can be brazed; fairly resistant to corrosion.

**Cast Iron—Nodular and Ductile Forms.** Similar to ordinary gray irons, but with controlled alloy composition for ductility and strength. Good machinability; can be welded by most fusion processes; fairly resistant to corrosion.

**Cast Iron—Malleable.** The malleableizing process produces a structure similar to steel. Good machining qualities; can be joined by welding, brazing, and soldering. Resistant to corrosion in rural, industrial, and marine atmospheres, both fresh and salt water.

**Cast Iron—White and Alloy.** These irons are compounded to obtain special properties. Nickel, chromium, and molybdenum are added for wear and abrasion resistance. Large amounts of silicon, nickel, and aluminum produce irons resistant to heat and corrosion.

**Iron—Ingot.** Available as sheets, strips, wire, and rail sections. Readily joined by resistance, arc, and gas welding. Corrosion resistance is poor—all types are rusted by oxygen and water at room temperature.

**Iron—Wrought.** Available as tubular products, plates, sheets, bars, structural shapes, wire, chain; fair corrosion resistance (better than carbon steels).

**Steel, Carbon—Hardening Grades.** Available as hot- and cold-rolled strips, bars, rods, and mill shapes, these steels are almost always heat-treated to give hardness, strength, and resistance to wear. Depending upon composition, machining qualities are fair to excellent. Forging qualities are good; welding requires special practices. Corrosion resistance is poor—all types are rusted by oxygen and water at room temperature.

**Steel, Carbon—Carburizing Grades.** Available in all mill forms, these steels are for case-hardening by carburizing, giving a strong, tough core and a very hard "skin." They can be welded, but some grades require special practice. Machining qualities vary from poor to excellent; forging qualities are generally good. Corrosion resistance is poor—all types are rusted by oxygen and water at room temperature.

**Steel, Nitriding—Wrought.** These steels are treated by nitriding to give an extremely hard outer surface and are heat-treated for toughness and strength. Can be welded by the atomic hydrogen process. Most uses of these steels are based on resistance to wear. If the nitrided outer layer is not removed, the case is quite corrosion-resistant to alkalis, oils, combustion products, and to fresh and salt water.

**Steel, Alloy—Wrought.** This class of steels in both carburizing and hardening grades (AISI 1340 to 9255) offers a variety for various purposes where different degrees of hardness and strength are required. In general, corrosion resistance is fair (better than that of plain, or non-alloy, carbon steels).

**Steel, High-strength—Wrought.** Available in most standard mill forms and structural shapes, these niobium- (columbium-) and vanadium-bearing carbon steels are used principally for structural purposes.

**Steel, Ultra-high-strength—Wrought.** Available in most standard wrought forms, these steels are used primarily where a high strength-to-weight ratio is required. Readily welded.

**Steel, Free-cutting—Wrought.** Available in cold-drawn shapes, these steels are used for a variety of parts such as studs, nuts, fasteners, collars, dowels.

**Steel, High-temperature—Wrought.** Available in sheets, strips, plates, bars, forgings, tubing, these steels are used for such parts as turbine buckets, jet-engine compressor blades, etc., where high strength at high temperature is required. Good resistance to atmospheric corrosion.

**Steel, Stainless Austenitic—Wrought.** Obtainable in sheets, strips, plates, bars, wire, and tubing, these steels, as the name implies, are for corrosion-resistant applications. These stainless steels have a fair index of machinability and are readily welded. Depending upon composition, grades vary from general purpose to types which resist both corrosion and heat.

**Steels, Stainless Ferritic—Wrought.** These stainless steels are general-purpose grades having fair to excellent machinability and fair weldability, which are used in such applications as automobile trim, kitchen equipment, and chemical equipment. Some grades also resist corrosion at high temperature.

**Steels, Martenistic Stainless — Wrought.** These stainless steels have good corrosive resistance to weather and water and to some chemicals; used for instruments, cutlery, springs, and applications requiring good mechanical properties at high temperatures.

**Steels, Stainless Age-hardenable—Wrought, Cast.** These stainless steels are used for structural parts where a high strength-weight ratio is required or for parts requiring high strength, good resistance to corrosion, and ease of fabrication.

**Iron-base Superalloys (Cr-Ni) — Wrought.** These alloys are used for highly stressed parts at elevated temperatures; corrosion resistance excellent; machinability and weldability only fair.

**Iron-base Superalloys (Cr-Ni-Co)—Cast, Wrought.** These alloys have good resistance to corrosion at elevated temperatures; used for such parts as gas-turbine blades, vanes, and nozzles.

**Steel, Carbon—Cast.** These steels have good machinability and weldability; poor corrosion resistance; used for applications requiring good to high strength, good machinability, toughness, wear, and fatigue resistance.

**Steels, Alloy—Cast.** Properties are similar to those of cast carbon steels except that high-temperature properties are better. Some of these steels are intended for applications requiring high strength, wear resistance, high hardness, and high fatigue resistance.

**Steels, Stainless—Cast.** These stainless steels, in a variety of compositions, are easily welded by metal arc. Corrosion resistance is excellent, including strong reducing and oxidizing media. Tensile strengths from about 70,000 to 140,000 psi.

**Alloys, Heat-resistant—Cast.** These metals have good machinability, can be welded by all common methods, and maintain good mechanical properties at high temperatures.

**Steels, Tool—Wrought.** These steels, in a variety of compositions, are used principally for machine tools and dies where strength, toughness, and resistance to wear are required.

## Nonferrous Metals.

**Aluminum and its Alloys—Wrought and Cast.** Most of the alloys for wrought forms are available in sheets, plates, rods, bars, tubes, extrusions, and structural shapes. This class of metals offers a wide variety of compositions all having high resistance to rural atmospheres. Some have good resistance to fresh and sea water and also to organic acids, anhydrides, aldehydes, esters, ketones, petroleum derivatives, ammonium compounds, and other commercial chemicals. Machinability and weldability are good. The uses are many and varied—electrical conductors, chemical equipment, cooking utensils, architectural applications, pressure tubes, bus and truck bodies, transportation equipment, auto and appliance trim, screw machine products, pipe and hardware, structural parts for aircraft, transmission cases, among others.

**Cobalt-base Alloys—Cast, Wrought.** Wrought forms are sheets, bars, billets, forgings, and wire. These metals are used for high-temperature applications requiring strength and corrosion resistance.

**Tungsten, Tantalum, Molybdenum.** These metals are used for a variety of electrical applications such as filaments, contacts, capacitors and for such aerospace applications as rocket nozzles, heat shields, radiation shields, and structural parts.

**Niobium (Columbium) and its Alloys.** These metals are used principally for parts requiring good strength at elevated temperatures, especially in missile and space-vehicle structures and components.

**Copper and its Alloys—Wrought, Cast.** Available in pure form and in combination with many other metals. Alloys include phosphorus, tellurium, sulfur, zirconium, beryllium, chromium. Commercial bronze is 90% copper, 10% zinc. Red brass is 85% copper, 15% zinc. Other brasses have more zinc and less copper down to yellow brass, which is 65% copper and 35% zinc. Muntz metal is 60% copper and 40% zinc. When the copper-zinc combination is altered by the addition of lead, so-called leaded bronzes and brasses are produced. All alloys have good resistance to rural, marine, and industrial atmospheres. With some exceptions, joining by soldering and welding is excellent. Wrought forms are sheets, rods, strips, plates, bars, wire, foil, tubes, and pipe. Uses include a wide variety of electrical, architectural, and mechanical applications; copper is outstanding for thermal and electrical conductivity; copper, bronze, and brass are also used extensively for artistic and decorative purposes.

**Lead and its Alloys.** Lead is used in pure form (chemical lead) and combined with tellurium for chemical apparatus. Common soft lead (Pb 99.73+) is used commercially for such applications as storage batteries, cable sheathing, alloying, and coatings. The wrought forms (1% to 9% Sb) are commonly used for roofing and flashing, extruded pipe, and corrosion-resistant applications requiring more strength than soft lead. All forms are resistant to sulfuric, sulfurous, phosphoric, and chromic acids, but are attacked by acetic, formic, and nitric acids. Very resistant to all atmospheres and to fresh and salt water.

**Magnesium Alloys.** Magnesium alloys are available in wrought sheets, rods, bars, and extrusions and also for sand and die castings. Because of their lightness and strength, these alloys are particularly useful in aircraft, missile, and aerospace applications, and for the same reasons are also used extensively in such applications as electronic housings, office machines, appliances, luggage, cameras and optical equipment, and sporting goods, among others. Magnesium has good resistance to atmospheres, but it is attacked by salt water unless its surface is finished.

**Nickel and its Alloys.** Many different alloys are available. Nickel 200 and 201, Duranickel, and Monel 400 and K-500 are commonly used for parts requiring a combination of strength, ductility, and corrosion resistance. Inconel is used extensively for applications where oxidizing and carburizing atmospheres at elevated temperatures are present. Monel 411 and 505 are used extensively for food-handling equipment, tanks, boilers, valve seats, liners, bushings. Low-expansion nickel alloys are valuable for length-standard bars, bimetal thermostats and similar low-expansion applications. Nickel-base superalloys are much used in jet-engine, missile, and furnace applications where high-temperature strength and resistance to corrosion are important factors.

**Precious Metals.** The precious metals are, of course, much used for jewelry and other art applications, but are also valuable for engineering purposes. The following entries indicate typical uses.

*Gold.* Lining of chemical equipment, high-melting solder alloys, dentistry. Gold does not oxidize when heated in air.

*Silver.* Electrical contacts, corrosion-resistant equipment, photographic equipment, brazing alloys. Does not oxidize when heated in air. Resists most dilute mineral acids and alkalis.

*Platinum.* Chemical equipment, electrical contacts, catalysts, laboratory equipment, thermocouples. Very good oxidation resistance, but is dissolved by aqua regia.

*Palladium.* Electrical contacts, catalysts, hydrogen production, dental alloys. Resists hydrofluoric, acetic, and phosphoric acids.

*Rhodium.* Electrodeposits for nontarnishing finishes, mirrors, thermocouples, catalysts. Resistant to most acids.

*Ruthenium.* Hardener for platinum and palladium. Resistant to most acids.

*Osmium.* Pen tips, small wear-resistant parts, electrical contacts, instrument pivots. Resistant to common acids.

*Iridium.* Alloys with platinum for electrical contacts, hypodermic needles, thermocouples. Resistant to common acids.

**Tin and its Alloys.** Many alloys are available. Soft tin is used for food-can linings, beverage-can linings, pipe for handling beverages, liners for food-handling equipment. Resists atmospheres and waters, but is attacked by strong acids and alkalis. Hard tin is used for collapsible tubes and foil. White metal and pewter are used for jewelry and ornamental holloware for table service. Other alloys (Sn, Sb, Cu, Pb) are used for bearings and die-cast parts. Tin-lead-antimony alloys are used for bearing liners under light, medium, and heavy loads and speeds.

**Titanium and its Alloys.** Titanium is characterized by its superior resistance to corrosive attacks by sea water and most chloride salt solutions. Titanium is mostly used for parts requiring a high-strength formability and weldability. Also for parts requiring a high strength-to-weight ratio.

**Zinc and its Alloys.** Zinc and zinc alloys are available in rolled plates, strips, and sheets; also in extruded rods and shapes and drawn rods and wire. Zinc is an important metal in the alloying of copper-base metals and is much used for coating of steel (galvanizing), but is also used in pure form and as a base metal for alloys. Zinc may be fabricated by drawing, bending, roll-forming, stamping, swaging, coining, and extruding and is easily soldered and welded. Common uses are for dry-battery cases, flashing, weatherstrip, lithoplates, automotive parts, household items, office equipment, hardware. Zinc is very resistant to atmospheres and to moisture- and acid-free hydrocarbons, but it is attacked by strong acids and bases.

**Zirconium and its Alloys.** Available as ingots, billets, sheets, strips, bars, rods, tubes, and wire, zirconium is very ductile and workable and is very easily fabricated. It may be welded under inert atmosphere and can be brazed and soldered. Principal uses are for chemical equipment, flash-bulb filler, fuel-cladding in water, steam-, or gas-cooled, nuclear reactors. Resistant to hydrochloric, nitric, and sulfuric acid and to alkalis at all concentrations and temperatures, zirconium is attacked by hydrofluoric acid and aqua regia.

## Plastics and Rubber.

**ABS Resins—Molded, Extruded.** These plastics in grades designed for medium to very high impact. low-temperature impact, and heat-resistant service are all highly resistant to aqueous acids, alkalis, salts, and are also resistant to phosphoric and hydrochloric acids and alcohols. Disintegrated by sulfuric and nitric acids; soluble in esters and ketones. Typical uses are for piping, appliance housings and wheels, housewares, garden equipment, office equipment, cases, luggage.

**Acrylics—Cast, Molded, Extruded.** These plastics in general-purpose and high-impact grades are all resistant to weak acids, alkalis, and some hydrocarbons, but are attacked by esters, ketones, aromatic and chlorinated hydrocarbons, and concentrated acids. Typical uses are for transparent enclosures, electronic parts; drafting equipment; decorative parts; household, office, and automotive parts; knobs and keys for office, musical, and electronic equipment.

**Alkyds—Molded.** These plastics in granular, putty, and glass-reinforced grades, are all resistant to weak acids but are attacked by alkalis. They are almost unattacked by organic alcohols, hydrocarbons, and fatty acids. Typical uses are for ignition, switch, and circuit-breaker parts; resistors; capacitors; and other electronic parts.

**Cellulose Acetate—Molded, Extruded.** These plastics are all resistant to salt and fresh water and to dilute acetic and sulfuric acid, but are decomposed by strong acids, and dissolved by acetone and ethyl acetate. Typical uses are for photographic film base, tape, appliance housings, optical parts, handles and knobs, toys, buttons.

**Cellulose Acetate, Butyrate, and Propionate—Molded, Extruded.** These plastics are all unaffected by fresh and salt water, 3% hydrogen peroxide, weak acids, and white gasoline, but are softened or dissolved by ethyl alcohol, acetone, ethyl acetate, ethylene dichloride, carbon tetrachloride, and toluene. Typical uses are for decorative parts, knobs and handles, pipe, pens, telephones, steering and control wheels, tooth brushes, optical parts.

**Cellulose Nitrate and Ethyl Cellulose—Molded, Extruded.** These plastics are resistant to weak sulfuric, acetic, and hydrochloric acids and to weak sodium hydroxide and sodium carbonate. Attacked by strong acids, ethyl alcohol, acetone, toluene, and gasoline. Typical uses are for fountain pens, spectacle frames, drawing instruments, radio housings, handles, and knobs.

**Chlorinated Polyether, Phenoxy, and Polyallomer.** Chlorinated polyether is unaffected by both organic and inorganic chemicals except fuming nitric and sulfuric. Phenoxy is very resistant to waters, acids, alkalis, aliphatic hydrocarbons, oils, and greases, but is attacked by many organic solvents. Polyallomer is unaffected by strong acids and bases and only slightly affected by chlorinated compounds and aliphatic hydrocarbons. Typical uses are for valves, pipes, lining, coatings, housings, electrical conduits, film.

**Diallyl Phthalates—Molded.** These plastics are all unaffected by weak acids, alkalis, and organic solvents, and only slightly affected by strong acids and alkalis. Typical uses are for connectors, plugboards, housings, appliance parts, insulators, resistors, aircraft leading edges, nose cones, ducts, decorative sheets for laminating wood and fabric.

**Epoxies—Cast, Molded.** These plastics are highly resistant to water and strong alkalis, and are somewhat less resistant to sulfuric and acetic acid and oxidizing agents. Typical uses are for coatings for electrical components, patching and repair compounds, glues, electrical moldings, coatings for marine hulls and parts, paints, and protective finishes.

**Melamines—Molded.** There are many forms. All are resistant to weak acids and alkalis, organic solvents, greases, and oils, but they are attacked by strong acids and alkalis. Unfilled melamine is used for decorative buttons, moldings, and general ornamental applications. Cellulose-electrical is used for general mechanical and electrical applications at elevated temperatures. Glass fiber is used for applications requiring high shock resistance, good electrical properties, and resistance to burning. Alpha-cellulose–mineral forms are used for applications requiring low shrinkage, good dimensional stability, and good molding characteristics. The alpha-cellulose form is a general-purpose class for electrical and mechanical applications such as tableware, kitchenware, and lighting fixtures. Mineral-electrical is for elevated temperature and electrical applications such as ignition parts and terminal blocks. Fabric forms are for applications requiring improved impact strength.

**Phenolics—Molded.** These plastics are resistant to waters and atmospheres, but are severely attacked by strong acids and alkalis. Resistance to weak acids, alkalis, and organic solvents varies with the reagent and with the material formulation. Several types are available: mechanical and chemical, general-purpose, transparent, arc-resistant (mineral), and rubber phenolic (wood flour, flock, fabric, asbestos). Typical uses are for instrument panels, knobs, coil forms, cases, fuse blocks, handles, connector plugs, furniture hardware, instrument casings, molds, and dies.

**Polyamides (Nylons)—Molded, Extruded.** Many forms are available. In general, polyamides are inert to organic chemicals, esters, ketones, alcohols, hydrocarbons, weak acids, common solvents, and petroleum products. They also resist alkalis and salt solutions, but are attacked by phenols, strong acids, and strong oxidizing agents. Typical uses are for bearings, bushings, coil forms, abrasion-resistant sheathing and coating, electrical insulation, lines, cords, and ropes, seals, tubing, rods, sheets, laminations.

**Polyesters—Cast.** These plastics are attacked by strong acids, alkalis, ketones, and chlorinated solvents. Typical uses are for castings for electrical components and high-temperature applications.

**Polystyrenes—Molded, Extruded.** These plastics are resistant to alkalis, salts, lower alcohols, glycols, and waters, and are somewhat resistant to mineral and vegetable oils. They are attacked by higher alcohols, gasoline, and strong oxidizing agents; soluble in aromatic and chlorinated hydrocarbons. Typical uses are for electrical parts, insulators, coil forms, rigid containers, instrument panels, housewares, storage-battery cases, drafting instruments, instrument housings, knobs, panels for electric parts and refrigerators.

**Polyethylenes—Molded, Extruded.** These plastics are resistant to atmospheres and waters, and to acids and alkalis at normal temperature. They are attacked by oxidizing acids such as nitric and fuming sulfuric. Typical uses are for containers for drugs, squeeze bottles for drugs and toiletries, kitchen utility ware, tank liners, sealing rings, battery parts, wrapping films for food and clothing, high-frequency insulation, pipe, tubing.

**Polypropylenes—Molded, Extruded.** These plastics are resistant to most acids, alkalis, and saline solutions and to organic solvents. Typical uses are for appliances, television and radio housings, housewares, wire coatings, luggage, automotive parts, containers, packaging.

**Polycarbonates—Molded, Extruded.** These plastics are resistant to most organic solvents. They are attacked by strong acids and alkalis. Typical uses are for appliance parts, gears, bearings, bushings, electrical parts, housings, electronic components, aircraft and automotive parts, portable tool housings.

**Polyvinyl Chloride—Molded, Extruded.** Polyvinyl chloride and polyvinyl chloride acetate in nonrigid and rigid forms are resistant to alkalis and weak acids, but are not resistant to strong acids, ketones, and esters. Vinylidene chloride is resistant to all common acids and alkalis. Typical uses are for nonrigid general-purpose garden hose, protective garments, small solid tires, flexible grips, tubes; nonrigid electrical low-tension power cable and wiring insulation, appliance cords; rigid hoods, ducts, tanks, chemical piping.

**Polyvinyl Alcohol—Molded, Extruded.** This plastic resists organic solvents and petroleum products, but is attacked by strong acids. Typical uses are for adhesives, sizing and coatings for fabric and paper, chemical tubing, gaskets, diaphragms, packages for chemicals.

**Polyvinyl Butyral—Molded, Extruded.** This plastic is resistant to alkalis, aliphatics, and hydrocarbons, but is attacked by strong acids. Typical uses are as laminating sheet for safety glass, textile waterproofing, structural adhesive, coatings, and primers.

**Polyvinyl Formal—Molded, Extruded.** This plastic is resistant to alcohols, esters, and ketones, except those having a high acetate content. It is attacked by strong acids. The principal use is for wire and cable coatings requiring toughness and adhesion.

**Fluorocarbons—Molded, Extruded.** These plastics are resistant to most chemicals and solvents with (in some forms) the exception of alkali metals.

Typical uses are for chemical piping, gaskets, diaphragms, electrical insulation, connectors, coil forms, anti-adhesive coatings, bearings, bushings, wear surfaces, molded components, laminates.

**Silicones—Molded.** These plastics are resistant to gasoline and petroleum oils, and to sulfuric and hydrochloric acids. Typical uses are for connector plugs, switch parts, insulators, jet-engine parts, ignition systems, guided-missile parts.

**Ureas—Molded.** These plastics have high resistance to organic solvents, oils, and greases; not resistant to acids and alkalis. Typical uses are for housings for radios, business machines, and food equipment; toilet seats; electric switches and plugs; cosmetic containers.

**Plastic Films.** Plastic films are available in the following materials: cellophane, fluorocarbon, polypropylene, nylon, polycarbonate, polyester, polystyrene, polyethylene, polyvinyl chloride, and rubber hydrochloride.

**Plastic Laminates, Glass Reinforced—Molded.** These are available in the following: phenolic, polyester (rigid) silicone, and epoxy. Typical uses are: air ducts, airframe components, missile nose cones, fuel tanks, printed circuits, roto blades, luggage, boats, car bodies, machine housings, chemical tanks, high-frequency equipment, high-strength components, pressure tanks, high-strength tubing and pipe.

**Plastic and Rubber Foams—Flexible.** Polyethylene, silicone, urethane, vinyl, natural rubber, neoprene, butadiene-acrylonitrile, and butadiene-styrene are typical flexible foams. Uses include a variety of applications for thermal insulation, cushioning, vibration isolation, flotation, packaging, shock absorption, filtering.

**Plastic Foams—Rigid.** Cellulose acetate, epoxy, urethane, silicone, phenolic, and polystyrene are typical rigid foams. Uses include vibration insulation, cushioning, thermal insulation, packaging, void filling, buoyancy components, and pipe coverings.

**Rubber, Hard—Molded, Extruded.** Hard rubber is used for molded parts and is obtainable in sheets, rods, and tubes. Typical uses are for chemical tanks and apparatus and for electrical and electronic components.

**Rubber, Natural and Synthetic—Molded, Extruded.** The following lists typical materials and uses:

*Natural rubber and butadiene-styrene.* Used for pneumatic tires and tubes, power transmission belts, gaskets, hose, tank linings, sound and shock absorption, seals.

*Butyl.* Used for pneumatic tires and tubes, steam hose, diaphragms, flexible electrical insulation.

*Butadiene-acrylonitrile.* Used for diaphragms, self-sealing tank applications, aircraft hose, gaskets, shock mountings, gasoline and oil hose.

*Chloroprene.* Used for petroleum tubes, hoses, tank linings, electric insulation, special tires.

*Polysulfide.* Used for gaskets, diaphragms, seals, shock mountings, hose, chemical applications for solvents.

*Silicone.* Used for high- and low-temperature electric insulation, gaskets, seals, O-rings, diaphragms.

*Urethane.* Used for special aircraft tires, commercial tires, shoe heels, and allied products.

*Polyacrylate.* Used for petroleum hose, gaskets, seals, and O-rings for resistance to high pressures in lubricating systems, especially where sulfur is present.

*Polybutadiene.* Used for seals, gaskets, belting, and pneumatic tires, and in combination with other rubbers to increase resistance to abrasion and to produce greater resilience.

*Vinylidene-fluoride-hexafluoropropylene.* Used for seals, gaskets, diaphragms, flexible mounts, coated fabrics.

*Fluorosilicone.* Used for seals, gaskets, O-rings, etc., where resistance to high-temperature solvents or oils is necessary.

*Ethylene propylene.* Used for weather stripping, hose, tubing, belts, automotive and appliance parts, electric insulation, footwear.

*Chlorosulfonated polyethylene.* Used for chemical and petroleum hose and tubing, tank linings, wire and cable sheathing, footwear, flooring, building products.

*Polytrifluorochloroethylene.* Used for seals, gaskets, O-rings, tubing and diaphragms.

## Ceramics, Glass, Carbon, and Mica.

**Ceramics—Fired Parts.** The following lists materials and typical uses:

*Polycrystalline glass.* Used for heatproof cookware and parts requiring stability under temperature changes, also for high-temperature, high-frequency applications in electronics.

*Cordierite.* Used for insulators, terminal blocks, heater-coil supports, foundry parts, fuel-burner tips, and similar high-temperature applications.

*Forsterite.* Used principally for low-loss insulators.

*Standard electrical ceramic.* Used for low- and high-voltage insulators, high-temperature wire supports, lightning-arrester parts, suspension (strain) insulators, x-ray equipment (rods and tubes).

*Refractory mullite.* Used for high-temperature insulators, spark-plug cores, high-temperature laboratory ware.

*Steatite.* Used for electronic tube sockets, electric line insulators, electrical component spacers and bushings, thermostat components.

*Zicron.* Used for electronic tube sockets, coil forms, component spacers, printed circuit plates.

**Refractory Ceramics and Cermets—Fired or Sintered Parts.** Refractory ceramics and cermets (composite materials consisting of an intimate mixture of ceramic and metallic components) are used for a wide variety of mechanical and electrical applications. Cermets are usually sintered, that is, powdered elements are combined, pressed to form the part, and then heated to about half the melting temperature (on the absolute scale) so that the elements fuse together with increased contact area to form a homogeneous crystalline structure. Refractory ceramics are used for pots, dishes, bowls, tableware, insulators, switch parts, connector blocks, spacers, and for magnetic, piezoelectric, and high-dielectric-constant parts. Sintered cermets are important because of their qualities of high strength, high thermal-shock resistance, high corrosion resistance, high wear resistance, high-temperature resistance, and hardness. Depending upon composition, sintered cermets are used successfully for bearings, bushings, seal parts, gears, friction parts, cutting and drilling tools, etc., and in such electric applications as insulators, capacitors, dielectrics, microwave parts, nuclear-reactor elements, filaments.

**Industrial Glass.** Many forms are commercially available: soda-lime glass is used for general plate applications and lamp bulbs; alumina-silicate for blown ware; soda-zinc for pressed ware; high-lead for radiation shielding; potash-lead for capacitors; borosilicate for electric insulation, industrial glassware, ultraviolet-light transmission; high-silica for high-temperature applications.

**Carbon and Graphite—Molded, Extruded.** Carbon, graphite, and carbon-graphite combinations are used for a variety of mechanical and electrical applications. Typical uses are for bearings, seals, pistons, wear surfaces, pump and valve parts, battery carbons, electric contacts, welding carbons, brushes for electric motors. Chemical uses include pumps, valves, fittings, pipe, and filtering media.

**Mica-Molded Sheet.** Natural muscovite is used for furnace peepholes, gage glass, capacitors, tube spacers, high-temperature insulators, and spacers. Glass-bonded and ceramoplastic mica are used for high-temperature insulators, computer parts, and electromechanical devices requiring high stability.

## Fibers—Natural, Synthetic, and Inorganic; Felts, Woods, Papers.

**Fibers—Natural.** Vegetable fibers, specifically, cotton, flax, jute, hemp, rami, manila, sisal, and hennaquin, and zoological fibers of wool, horsehair, and silk are used extensively for thread, cord, and rope which are combined to form mats and fabric. Loose or cut fibers are formed into mats, pads, and felts.

**Fibers—Synthetic.** Regenerated cellulose, cellulose esters, fluorocarbon, acrylics, polyamides, polyesters, vinly, and polyethylene are used for monofilament and multifilament fibers. These are used principally in thread, cord, pad, and rope applications and for the making of fabric.

**Fibers—Inorganic.** Glass, asbestos, and aluminum silicate in crude fiber or chopped, are used for insulation, pads, additives to plastics, reinforcement. Asbestos is made into rolls, sheets, and plates for thermal insulation. Glass fibers are made into threads and woven into cloth both for decorative purposes and for engineering applications such as thermal insulation, filtering, wicks, air filters.

**Felts—Natural.** Natural fibers, principally cotton and wool, are made into felts, usually supplied in sheets and rolls. These are used in many ways by cutting, laminating, coating, impregnating, cementing, molding and shaping to form a variety of components. Commonly used for bearing seals, vibration mounts, lubricators and wicks, grease retainers, weatherstrips, dust shields, noise and vibration dampers.

**Felts—Synthetic.** Synthetic fibers of polyester, polypropylene, rayon, nylon, and acrylics are made into felts. Uses are similar to those of natural-fiber felts, but particularly for cases where the chemical resistive qualities of the synthetics are needed.

**Woods.** Natural wood (cut boarding) is a structural material familiar to almost everyone. Its use should not be overlooked in selecting materials. Hardboard (fibrous), particle board, softboard, and plywood panels are widely used for structural, insulation, and decorative purposes.

**Papers.** Papers of wood, cotton, and linen are used extensively for electric insulation, spacers, gaskets, washers, and seals. Papers are treated with oils, plastics, and resins to extend useful applications.

## Coatings and Finishes.

**Coatings—Sprayed Metal.** Sprayed metal coatings are used for several purposes, for example, to increase corrosion resistance, to increase hardness and wear resistance, or to provide better bearing surfaces. Aluminum is used for corrosion and heat resistance; babbitt for bearing properties; brass and bronze for appearance and resistance to corrosion and wear; copper for electrical conductivity; lead, tin, zinc, Monel, nickel, and stainless steel for corrosion resistance; steel for hardness.

**Coatings—Electrodeposited.** Coatings deposited electrically are used to improve appearance, to increase electrical qualities, and to increase resistance to wear, to corrosion, or to specific environments. Not all metals can be satisfactorily deposited on all base metals.[1]

**Coatings—Ceramic, Cermet, and Refractory.** Fired porcelain enamel frits and refractory materials are used as corrosion-resistant coatings and also for color appeal and decorative effect. Examples are refrigerators, ranges, food-handling equipment, photographic trays and chemical laboratory equipment, and parts subjected to oxidation and high temperatures. These coatings are especially valuable for use on ferrous metals. Flame-applied ceramics and cermets, depending upon the material used, give coatings that are resistant to wear, abrasion, heat, oxidation, and to thermal and mechanical shock.

**Coatings—Hot-dip.** These coatings, used principally on steel, cast iron, and copper, provide corrosion resistance at low cost. The materials used are aluminum, zinc, lead, tin, and lead-tin. Examples of use are for roofing, nails, wire, outdoor and indoor hardware, barrels, cans, food cans, kitchen utensils, electronic parts, printed circuits.

**Coatings—Immersion.** These coatings can be applied to most ferrous and nonferrous metals, with a few exceptions. Materials used are nickel, tin, copper, gold, silver, and platinum. Examples of use are for decorative finishes, corrosion protection, increased conductivity (electronic parts and printed circuits), facilitation of soldering and brazing.

**Coatings—Diffusion.** Diffusion coatings are produced by the application of heat while the base material is in contact with a powder or solution. Common diffusion coatings on steel are carburizing, chromizing, cyaniding, nitriding, sherardizing, calorizing. Most diffusion coatings are intended to obtain hard and wear-resistant surfaces and to increase resistance to corrosion.

**Coatings—Vapor-Deposited.** Vapor-deposited metalizing consists of vaporization or evaporation of a metal in a vacuum chamber, where it then condenses on all cool surfaces. Most metals and nonmetals can be used as base materials to be coated. Examples are mirrors and optical reflectors, metalized plastics, lens coatings, costume jewelry, instrument parts.

**Coatings—Organic.** These consist of alkyds, celluloses, epoxies, phenolics, silicones, vinyls, rubbers, and others. They are used for color, decoration, and resistance to corrosion.

**Coatings—Chemical Conversion.** These are chemical coatings which react with the base metal to produce a surface structure that will improve paint bonding, corrosion resistance, decorative properties, and wear resistance. Phosphate, chromate, anodic, and oxide coatings are common.

**Coatings—Rust-Prevention.** These are oils, petroleum derivatives, and waxes that form a film which will resist attack, principally from industrial and rural atmospheres.

**Finishes—Mechanical.** Sandblasting, peening, hammering, wire brushing, wheel or belt polishing, burnishing, and buffing are used to produce a decorative finish. Surfaces obtained vary from dull (sandblasting) to very bright (buffing).

**Facings.** Facings are overlays applied by welding operations. The purpose is to produce a hard surface designed for special service conditions, usually resistance to wear and abrasion. Typical materials used for facing are carbon steels, chromium and chromium-nickel steels, and cobalt, nickel, and tungsten alloys.

## Composites.

**Laminates—Plastic-Metal.** These consist of a metal (steel, aluminum, or magnesium) precoated with a plastic film. At the present time, vinyl and polyester predominate. They are available in a wide variety of textures and colors.

**Laminates—Metal-Wood.** The most common are wood cores with steel or aluminum faces. These are used extensively in automotive, appliance, and architectural applications.

[1]Consult detailed listings given by suppliers, standards, or such sources as the supplements of *Materials in Design Engineering* magazine.

**Laminates—Bonded Combinations.** Practically any material can be combined with plastics by adhesive bonding. Common varieties are plastic laminates faced with copper, silver, gold, aluminum, vulcanized fiber, asbestos.

**Laminates—Bimetallic.** These are composite material wherein aluminum or magnesium alloys are molecularly bonded to a ferrous metal. The advantages are that the strength and other characteristics of ferrous metals are retained along with the high heat conductivity and other characteristics of light metals.

**Laminates—Honeycomb.** These include a wide variety of materials and structures. Their greatest advantage lies in a structural core separating two facings of high-strength material, thus combining light weight and strength because the high-strength material is some distance from the neutral axis. Common facing materials are steel, aluminum, stainless steel, superalloys, plastics, and plywood. Core materials include paper (impregnated), aluminum, steel, stainless steel, superalloys, and plastics.

**Laminates, General-purpose—Sheet, Rod, Tube.** These laminates include phenolics with canvas, asbestos, paper, asbestos cloth, glass cloth; also, melamine, epoxy, and polyester with glass cloth. Typical uses are for switch panels, electronic insulation, printed circuits, terminal blocks, motor parts, electric appliance insulation.

**Laminates, Mechanical—Sheet, Rod, Tube.** These laminates of phenolic with cotton canvas or cotton-linen canvas and melamine with cotton canvas are for mechanical purposes. Typical uses are for gears, pinions, pulleys, knobs, bushings, cams, threaded and tapped parts.

**Laminates, Electrical—Sheet, Rod, Tube.** These laminates of phenolic with paper, cotton, linen, glass cloth, or nylon and silicone with glass cloth are used for a variety of electrical applications. Typical uses are for panels for electrical and electronic purposes, terminal blocks, coil forms, switch rotors and stators, wave-change switch parts and electronic applications where low-loss insulation is needed.

**Metals—Preclad and Precoated.** Base metals, principally steel, aluminum, copper, bronze, and brass, precoated with such metals as aluminum, copper, lead, silver, gold, nickel, platinum, and stainless steel are available as sheets, rods, strips, tubing, and wire. Also available are base metals with coatings of alkyd, polyvinyl chloride, vinyl lacquer, or enamel. Preplated or precoated metals often have an economic advantage because they can be fabricated without further finishing.

## 9. Methods of Parts Production.

As in the case of the listing of materials in Sec. 8, it is impossible to give here all details of parts production. The following descriptions are only a guide; detailed information may be found in textbooks on parts production and processing. See Glossary.

### Casting.

**Sand Castings.** Moist sand is packed around a pattern and the pattern is then removed. Molten metal is poured into the cavity left by the pattern. When the metal solidifies, the sand mold is broken away and the casting removed.

**Shell-mold Castings.** Sand mixed with a thermosetting plastic is placed in a heated metal pattern in which the sand mixture solidifies, forming a mold or "shell." Molten metal poured into the shell solidifies, and the shell is then broken away.

**Permanent-mold Castings.** Mold cavities are machined into two or more metal blocks which are hinged or clamped together. After molten metal is poured into the mold cavity and allowed to solidify, the mold is removed.

**Plaster-mold Castings.** A slurry of gypsum and other ingredients is poured over a pattern and allowed to set. The pattern is then removed and the mold is baked. Next, molten metal is poured into the mold and allowed to solidify, and the mold is then broken away.

**Investment Castings.** A refractory slurry is formed around a pattern of wax, meltable plastic, or frozen mercury. After the slurry sets up, the pattern is melted out, and the mold is baked. Molten metal is then poured into the mold and allowed to solidify, after which the mold is broken away.

**Die Castings.** Molten metal is forced under pressure into a closed steel die. When the metal solidifies, the die is opened and the casting ejected.

**Centrifugal Castings.** A mold of sand, graphite, or metal is rotated; molten metal introduced into the mold is thrown to the mold wall, where it remains by centrifugal force until solidified.

**Continuous Casting.** Molten metal (usually a copper alloy such as bronze) is fed into a mold, rapidly cooled and withdrawn, and then cut to the desired length.

### Forging.

**Open-die Forgings.** Mechanical hammering is applied to metal heated to the plastic state. The desired shape is achieved by turning and manipulating the workpiece between hammer blows.

**Closed-die Forgings.** Compressive forces produced by a mechanical or hydraulic hammer are applied to metal heated to the plastic state and placed between dies having cavities which will form the desired shape. In some cases several successive dies are required.

**Upset Forgings.** Metal heated to the plastic state is gripped by dies which also compress to form the desired shape.

### Other Forming Operations.

**Cold Heading.** This is similar to upset forging, except that the metal is worked cold. Common practice is to feed wire (up to 1-in. diameter) to a gripping die with a portion protruding. The force of a punch against the protruding end forms a "head" on the part. Head shape is determined by the shape of the punch die.

**Impact Extrusions.** *Forward* extrusions are produced by placing a preformed thick metal blank in a die and striking the blank with a high-velocity punch. The metal flows forward through an opening in the die. *Rearward* extrusions are made by placing a thick metal blank in the bottom of a die cavity. The blank is struck by a high-velocity punch and the metal flows upward around the punch.

**Die Extrusions.** Metal heated to the plastic state is forced through a die having an opening of the desired cross-sectional shape. The metal emerges from the die in a continuous piece, which is then cut to the desired length.

**Stampings.** Parts made of plate, strip, and sheet stock are produced by one or more of these operations: cutting, forming, drawing. *Cutting* operations include blanking, punching, shearing, piercing, notching, etc.; the metal is parted by stressing beyond the ultimate strength. *Forming* includes bending, stretch forming, coining, embossing; the metal is stressed beyond the yield point and permanently deformed. *Drawing* is an operation in which a flat blank is pressed into the desired shape.

**Spun Parts.** A preformed or flat metal blank is drawn over a male spinning (rotating) form by the application of pressure from either a round-ended or roller-ended tool. The method is restricted to surfaces of revolution.

**Screw-machine Parts.** Bar or tube stock is fed through the hollow spindles of an automatic lathe. The stock is indexed and clamped in a collet chuck, and is then cut by various tools carried on a turret and/or tool slides. Operations performed include turning, facing, threading, tapping, drilling, boring, reaming, and other similar cutting operations.

**Powdered Metal and Cermet Parts.** Powdered metal combinations are compressed in a die to the desired finished form. The part is then sintered (heated) in a furnace at a temperature of one-half to two-tnirds of the melting point of the base constituent, which produces a homogeneous structure.

**Electroforming.** This is a process in which a mandrel, having a shape which is the reverse mating form of the part wanted, is placed in an electroplating bath and plated until the desired thickness is obtained. The electro-deposited part is then separated from the mandrel.

**Tubing and Formed Tube Parts.** Tubing is produced by extruding, by piercing or drawing of rod, or by bending and welding of sheet, strip, or plate. Tube parts are produced by forming a tube by methods such as bending, swaging, upsetting, flaring, spinning, and machining.

**Plastics.** Plastics, basically, are molded. There are many ways in which the molding can be accomplished, but fundamentally the process consists of placing or forcing the material in a plastic, powdered, or liquid state into a mold where the material solidifies and then is ejected or removed. The details of all the processes are too extensive to be given here, but injection moldings, extrusions, sheet moldings, blow moldings, slush and dip moldings, compression moldings, transfer moldings, and reinforced moldings are common methods. Plastics can also be cast. Sheets, tubes, etc., can be cut and machined. Inserts for fastening and for wear surfaces can be molded in place.

**Rubber.** Rubber is molded in a manner similar to the molding of plastic. Uncured rubber is forced or placed in a mold where under heat and pressure the rubber vulcanizes. Extrusions and die-cut parts from sheets and other shapes are also common.

**Ceramics, Glass, Carbon, Graphite.** Ceramics are formed by extruding, injection molding, pressing and casting. Glass is formed by blowing, pressing, drawing, and sintering. Carbon and graphite are formed by molding and extruding.

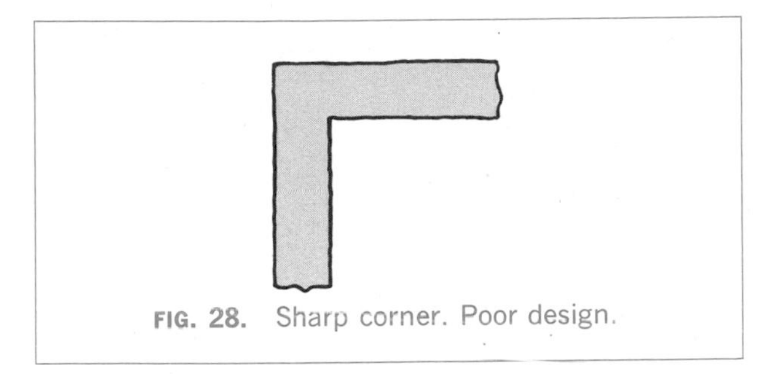

FIG. 28. Sharp corner. Poor design.

## 10. Good Design and Proportioning of Parts.

There are principles of design that govern the shape and proportion of machine parts. These principles have been determined by geometry, by experience in part-production methods, by calculations of strength, and by testing procedures. Study the descriptions and accompanying illustrations carefully, remember all the details you can, and refer back to the discussion when necessary.

*Corners.* Sharp corners are dangerous. In cast parts cracks may occur at a sharp corner, due to cooling strains. In other parts (forged, molded, etc.) breaks caused by overstressing will always start at a sharp corner. A corner like that shown in Fig. 28 is very poor design. Design with a rounded corner as shown in Fig. 29 if at all possible. The minimum inside radius is usually about one-half the metal thickness, making the minimum outer radius about one and one-half times the metal thickness.

*Tee sections.* For the same reasons given for outside corners, sharp corners should never be used on tee sections. Figure 30 represents a dangerous condition and is poor design. The rounded inside corners in Fig. 31 show good design. The radius is usually a minimum of about one-half the metal thickness.

*Differences in Metal Thickness.* When joined sections vary greatly in metal thickness, especially in castings and forgings, cooling strains may produce cracks and warping. Figure 32 is very poor design because of the wide difference in metal thickness and also because of the sharp corners. Keep joined sections as uniform as possible and avoid sharp corners, as in Fig. 33.

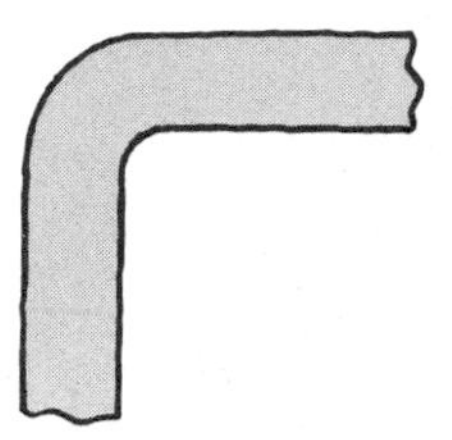

FIG. 29. Rounded corner. Good design.

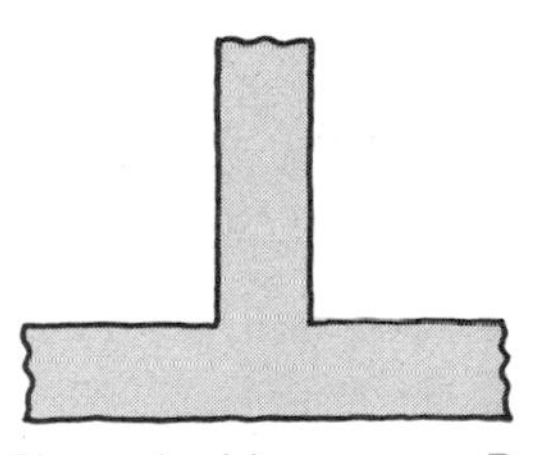

FIG. 30. Sharp inside corner. Poor design.

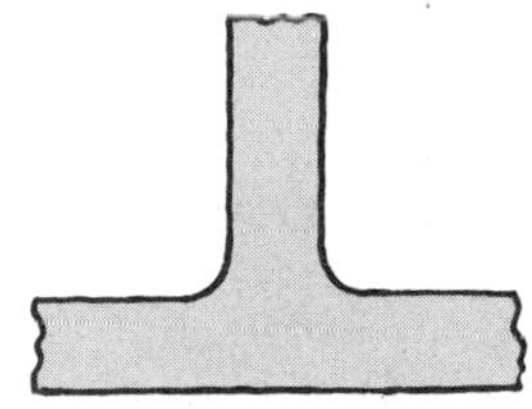

FIG. 31. Rounded inside corner. Good design.

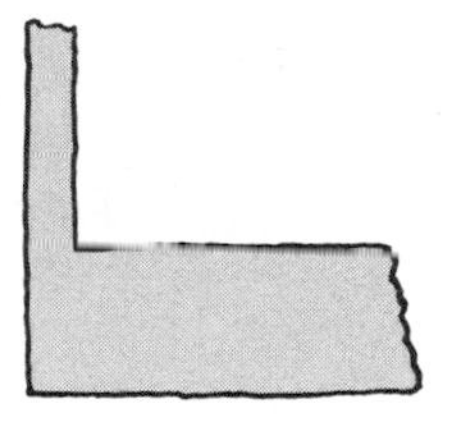

FIG. 32. Wide difference in metal thickness. Poor design.

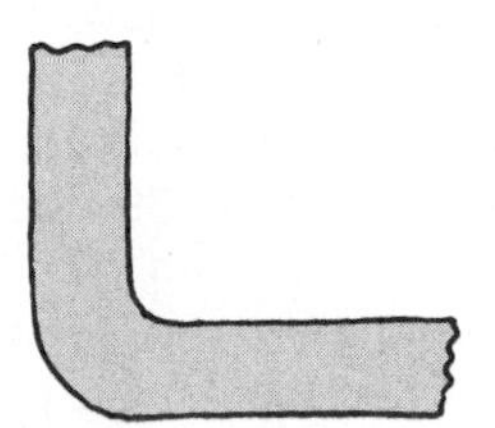

FIG. 33. Uniform section. Good design.

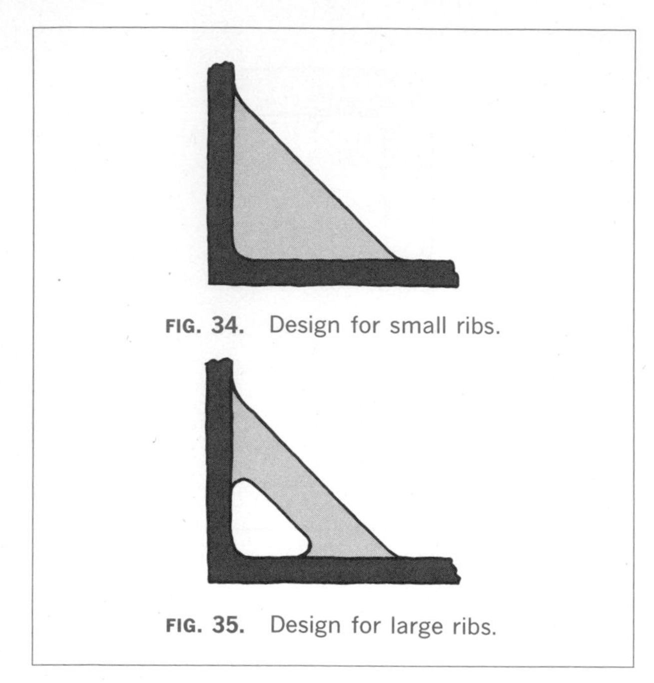

**FIG. 34.** Design for small ribs.

**FIG. 35.** Design for large ribs.

*Ribs.* Small ribs can be designed as solid ribs, as in Fig. 34. However, for large ribs it is best to open the inside of the rib, as in Fig. 35. The opened inside design is especially good for light-weight parts because the inside portion of the rib, structurally, adds very little strength.

*Bracing Ribs or Supports.* Internal bracing should be designed according to geometrical principles. The ribs of Fig. 36 lend very little stiffness to the part because the geometry of the part is based on the square, which is an unstable shape. Stiffening ribs based on the triangle, a stable shape, should be employed if possible, as shown in Figs. 37 and 38. These designs are for torsioned stiffening only and are not applied to beam stiffening.

*Size of Corner Radii.* We have seen before that corners should be rounded for several reasons. However, if the corner radius is too large for intersecting members, as in Fig. 39, too much material is left at the intersection and the design becomes poor because of wide differences in metal thickness. Keep the radius to a reasonable (minimum) value of about half the metal thickness, as in Fig. 40.

When several (five or more) members intersect, as in Fig. 41, even a small corner radius will not

**FIG. 36.** Stiffening ribs. Poor design because of an unstable geometry.

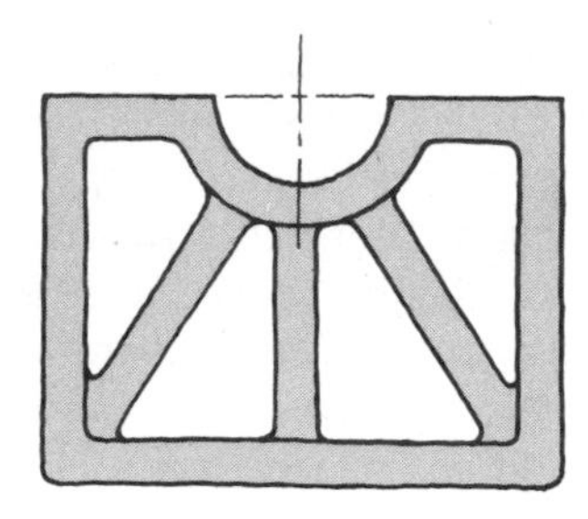

**FIG. 37.** Stiffening ribs. Good design because of stable geometric shapes.

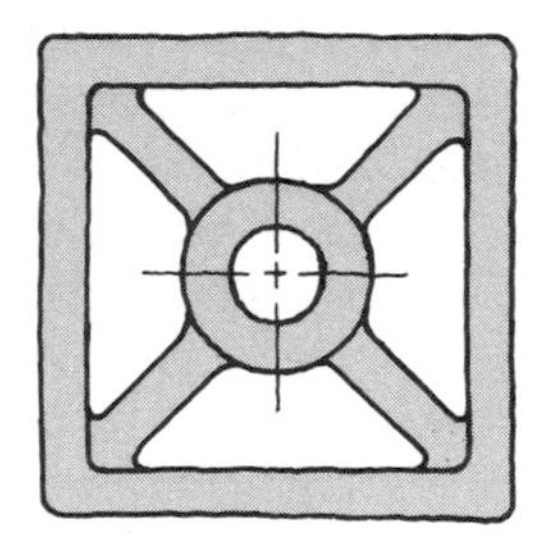

**FIG. 38.** Stiffening ribs. Good design because of stable geometric shapes.

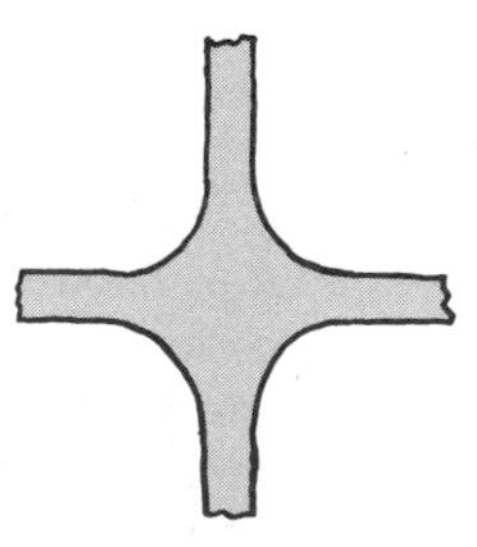

**FIG. 39.** Corner radius too large. Poor design.

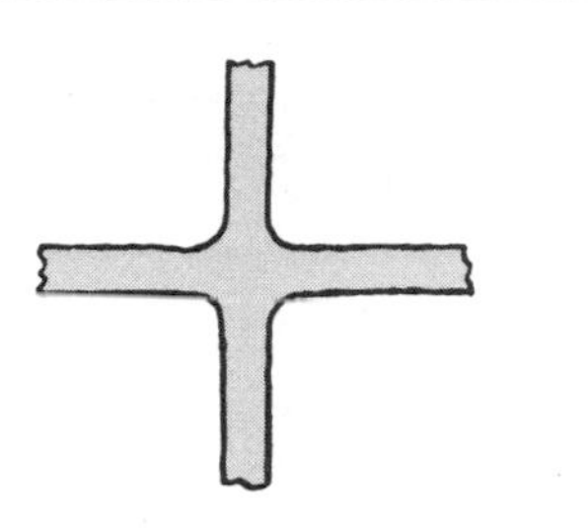

FIG. 40. Proper corner radius. Good design.

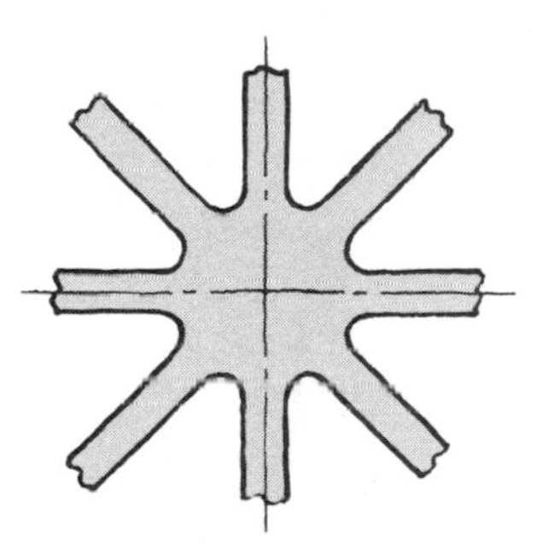

FIG. 41. Several intersecting members produce a great difference in metal thickness. Poor design.

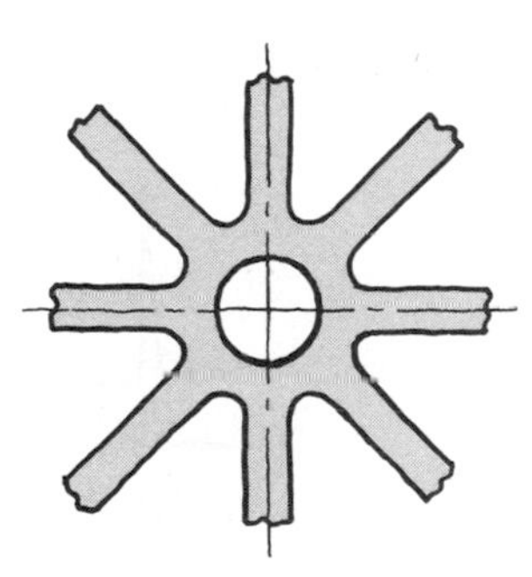

FIG. 42. Great difference in metal thickness relieved with a hole. Good design. Compare with Fig. 41.

solve the problem and the difference in metal thickness becomes critical. The solution of the problem is to relieve the center of the part with a hole, as in Fig. 42, regardless of whether the hole has any function for the part or not.

*Good Proportions.* Many combinations of hubs, ribs, holes, corner sections, flanges, and other elements occur in machine parts. It is impossible to describe all combinations, but the following examples will cover the bulk of the elements usually combined. Please note that the values given for corner radii are *maximum* values. Also note that the proportioning is based on the theory that differences in metal thickness should be "blended" into each other. This produces the best insurance against distortion and cracking during cooling and develops the maximum strength capability of the part.

Figure 43 shows good proportioning for a corner rib.

Figure 44 shows good proportioning for a centralized hub.

Figure 45 shows good proportions for a flared section (hub) on one side only of the section.

Figure 46 shows good proportions for a support rib. Note that the tapered section of the rib provides a good blend of metal thickness from the body section.

Figure 47 shows good proportions for a ribbed tee section.

Figure 48 shows good proportions for a support rib on a section having two different thicknesses. Note that the thinner section ($S$) is any thickness less than $R = 3T$ and that the radius values of $3T$ are maximum values. The tapered rib gives a good blend between the rib and body section.

Figure 49 shows good proportions for an end rib. The tapered form provides a good blend between the rib and body section.

Figure 50 shows good proportions for a flanged rib. The flanged section provides extra strength.

Figure 51 shows good proportions for a flange at the end of a tube.

Figure 52 shows good proportions for extra material needed for a tapped hole in a thin wall section.

Figure 53 shows good proportions for a hub. The sizes given are based on the diameter of the hole needed.

Figure 54 shows good proportions for a symmetrical enlarged section necessary to provide a hole equal to, or larger than, a wall section.

Figure 55 shows an enlarged section similar to Fig. 54, but for a nonsymmetrical section.

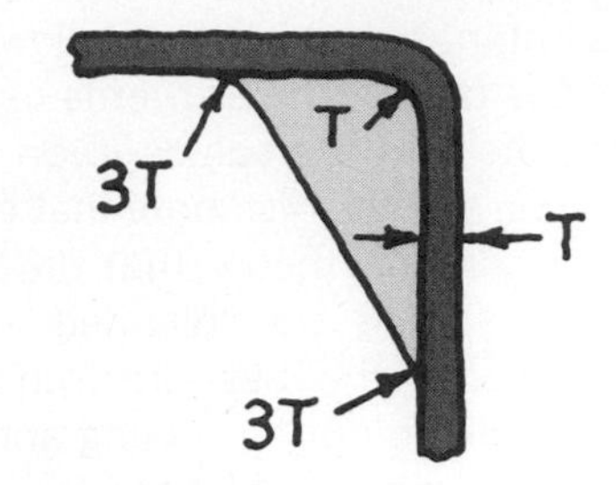


**FIG. 43.** Good proportioning for a corner rib. Radius values are maximum.

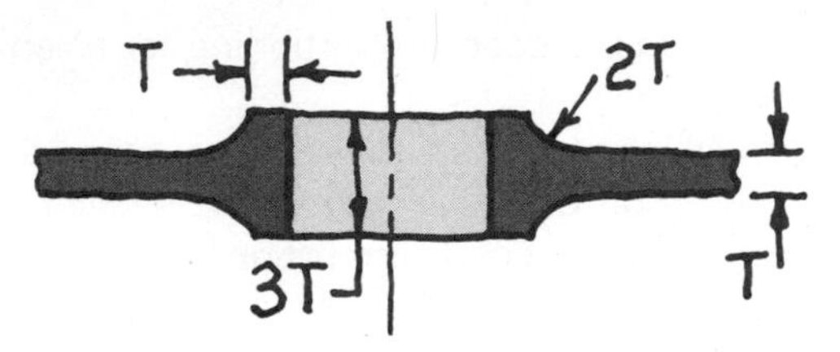


**FIG. 44.** Good proportioning for a centralized hub. Radius value is maximum.

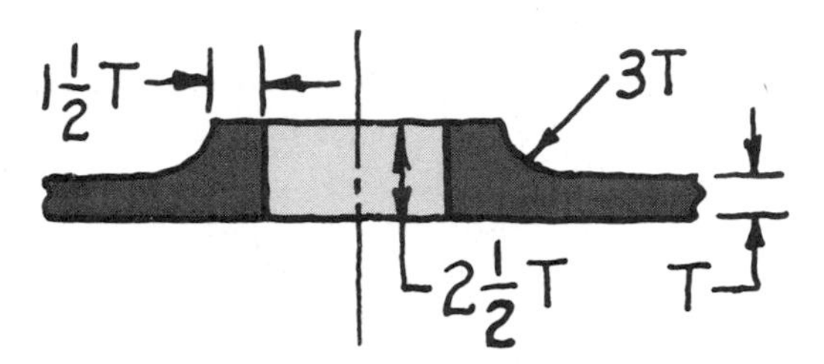


**FIG. 45.** Good proportioning for a flared hub. Radius value is maximum.

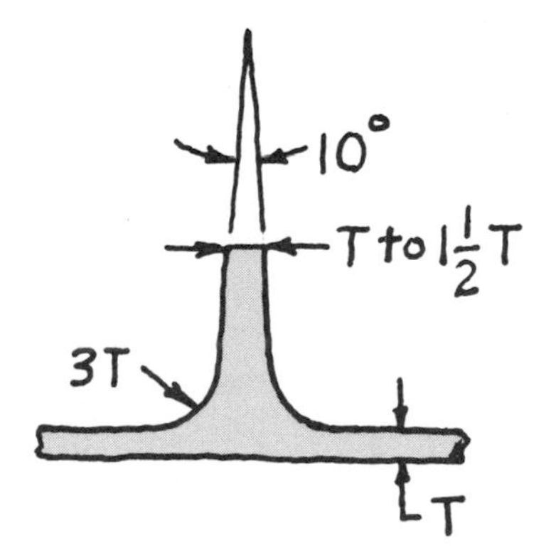


**FIG. 46.** Good proportioning for a support rib. Radius value is maximum.

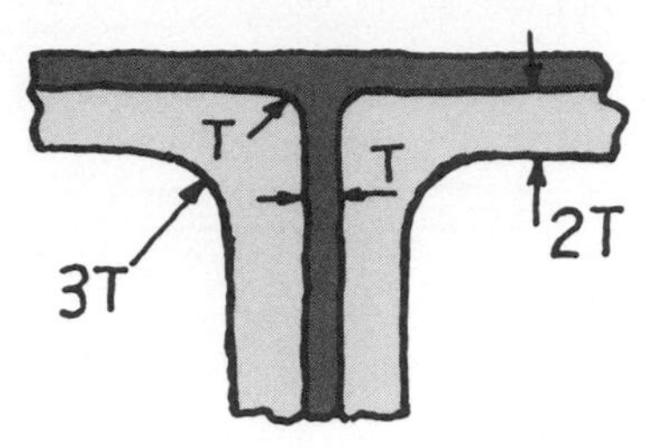


**FIG. 47.** Good proportioning for a ribbed tee section. Radius values are maximum.

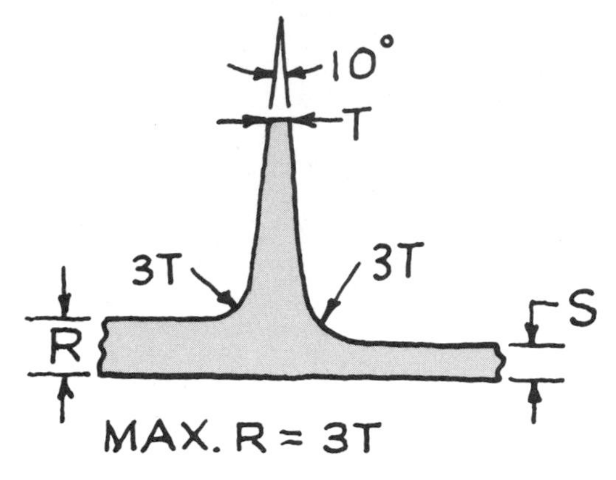


**FIG. 48.** Good proportioning for a support rib for unequal body sections. Radius values are maximum.

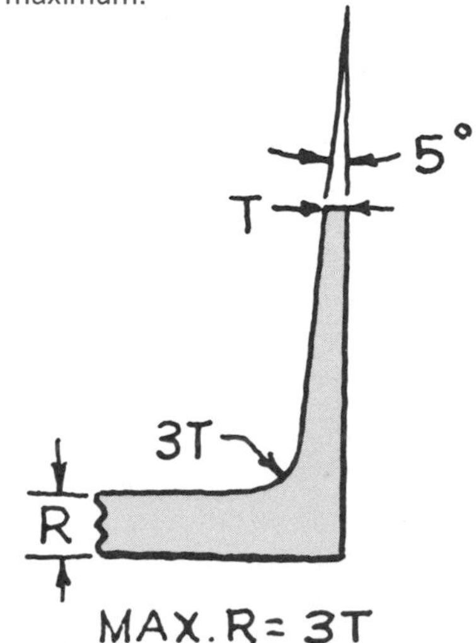


**FIG. 49.** Good proportioning for an end rib. The radius value is maximum.

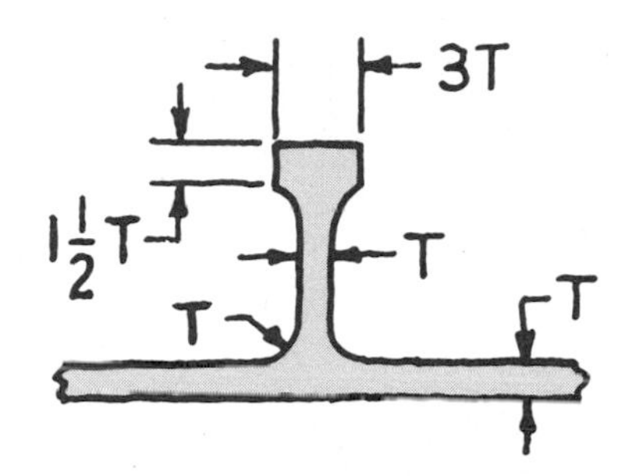


FIG. 50. Good proportioning for a flanged rib. Radius value is maximum.

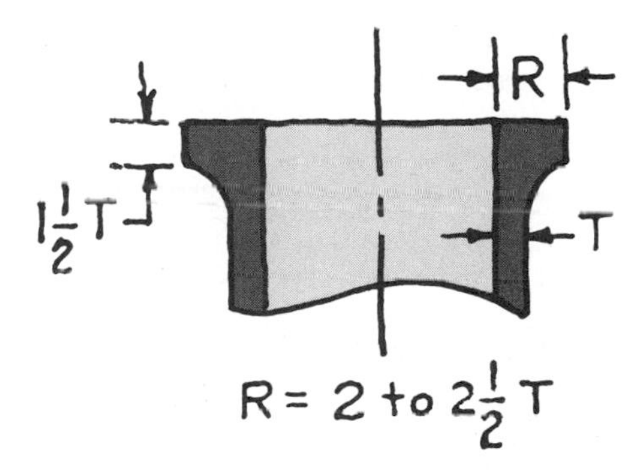


FIG. 51. Good proportioning for a flange at the end of a tube.

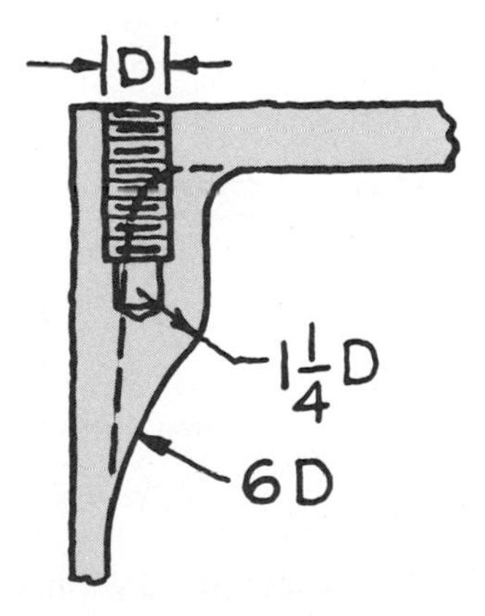


FIG. 52. Good proportioning for extra material needed for a tapped hole in a thin wall.

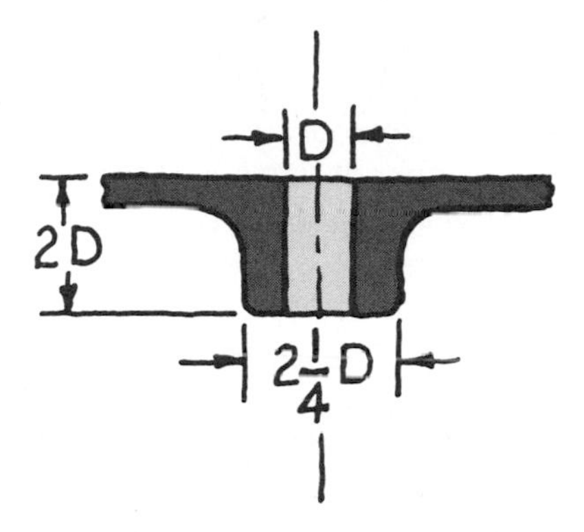


FIG. 53. Good proportioning based on the diameter of hole needed.

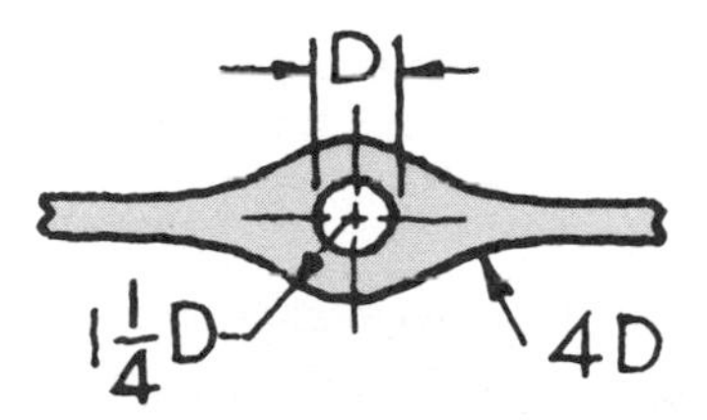


FIG. 54. Good proportioning for an enlarged portion.

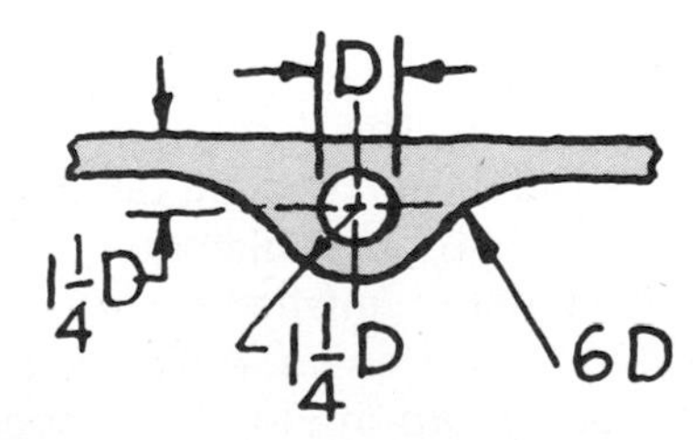


FIG. 55. Good proportioning for an enlarged section.

Figure 56 shows a simple cylindrical part machined from solid bar stock. The sharp edges weaken the part and even though it may not be highly stressed, they may cause difficulties in assembly and malfunctioning later. A better design is shown in Fig. 57, where the edges are chamfered.

An even better design is shown in Fig. 58, where the edges are rounded.

Figure 59 shows a keyseat designed so that its bottom surface corresponds with the smaller shaft diameter. This is poor design because the machining of the keyseat will probably cut into (or score) the shaft. The part should be designed for clearance between the bottom of the seat and the shaft, as in Fig. 60.

It is not impossible to cut a completely sharp corner at the change in diameter of a headed shaft, bolt, nut, or similar part, but it is difficult and expensive, and the part may be unreliable. When such a part must seat on a mating part, the certain interference because of an imperfect sharp corner can be relieved by necking the shaft as in Fig. 61. However, the neck requires an extra machining operation and also weakens the shaft. A better method, especially for highly stressed parts and for threaded parts, is to round the intersection of head and body and chamfer the mating part as in Fig. 62. The round radius must be less than the right-angle legs of the chamfer.

Parts can be severely weakened by poorly chosen placement of keyseats, grooves, and other similar elements. Notice in Fig. 63 that the keyway severely weakens the rounded end portion of the part. The difficulty is easily eliminated by placing the keyway as shown in Fig. 64.

In round parts such as hubs and collars there is no position for a keyway that will not weaken the part. An example is shown in Fig. 65. One solution of the problem is to add extra material by flaring the hub, as in Fig. 66. The flared portion is usually designed with a radius and straight tangent portions connecting to the hub outer diameter.

In good machinery, finished-surface seats are always provided for fasteners. A simple unfinished surface, like that shown in Fig. 67, is poor design. Surfaces are finished either by adding a boss with machined surface as in Fig. 68 or by spot-facing the surface as in Fig. 69.

It is not necessary to finish the complete base surface of a hub or similar part unless for some functional reason a completely flat surface is needed. Figure 70 shows a part with the complete base surface machined. When the complete flat surface is not necessary, the center portion can be relieved, as in Fig. 71. This saves machining time and produces a good seating surface.

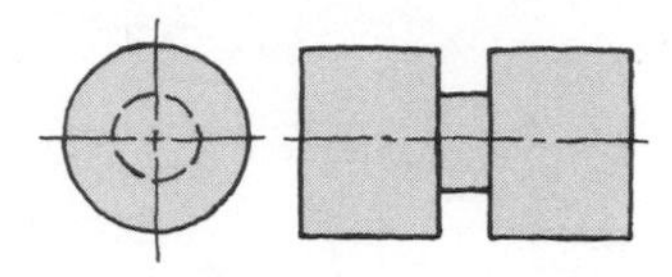

**FIG. 56.** A machined part with sharp edges. Poor design.

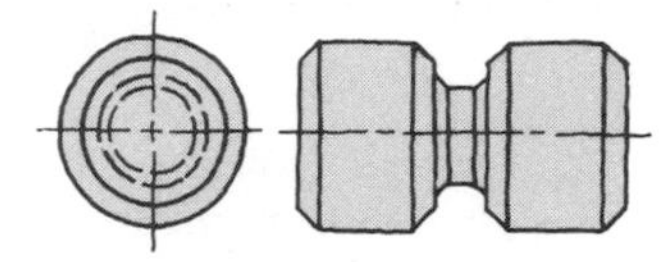

**FIG. 57.** A machined part with chamfered edges. Good design.

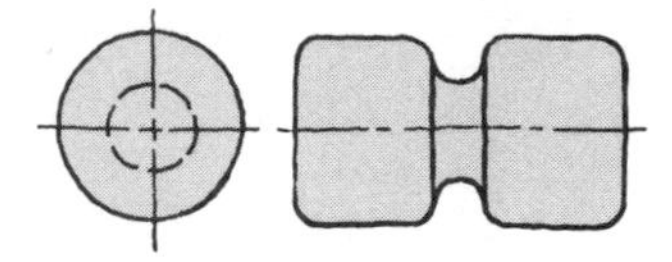

**FIG. 58.** A machined part with rounded edges. Good design.

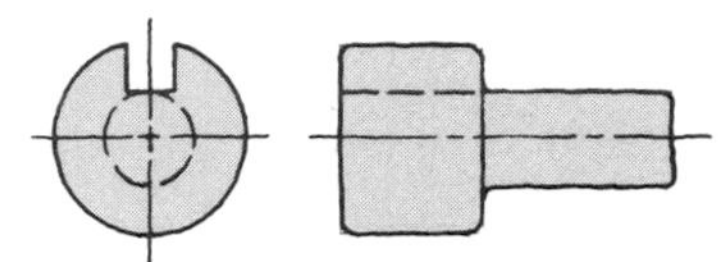

**FIG. 59.** Keyseat bottom and shaft correspond. Poor design.

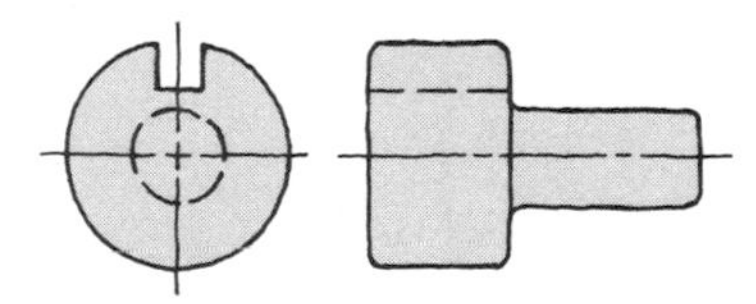

**FIG. 60.** Clearance between keyseat bottom and shaft. Good design.

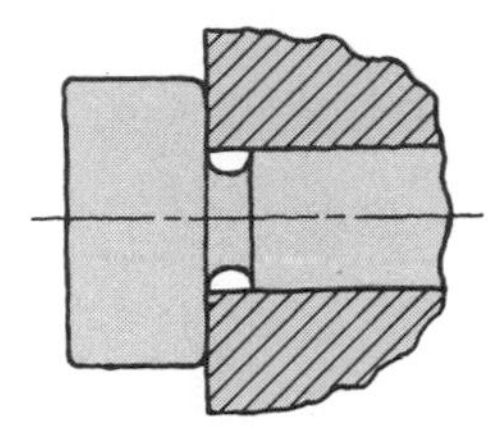

FIG. 61. Necking of a shaft to eliminate interference. Allowable design for lightly stressed parts.

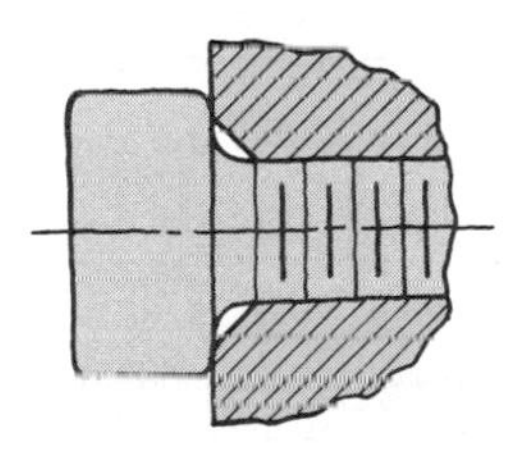

FIG. 62. Round on shaft and chamfer on mating part to eliminate interference. Good design.

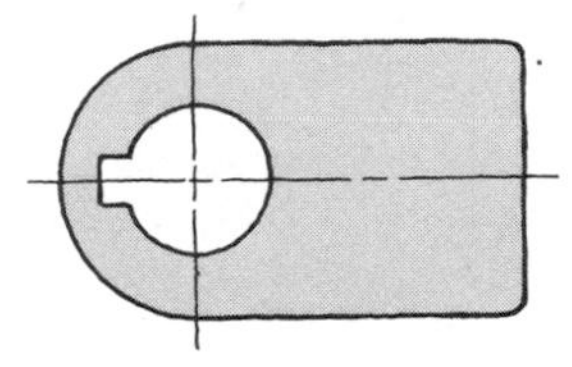

FIG. 63. Keyway in a position that severely weakens the part. Poor design.

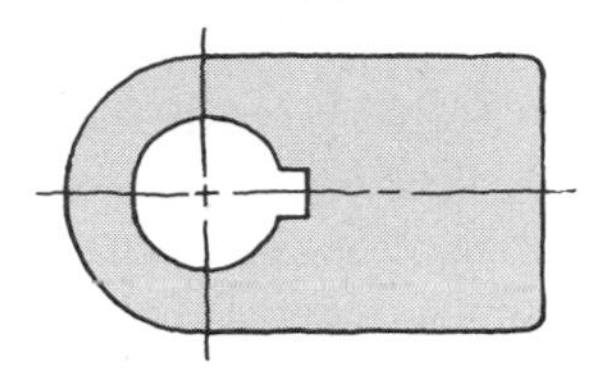

FIG. 64. Keyway position does not weaken the part. Good design.

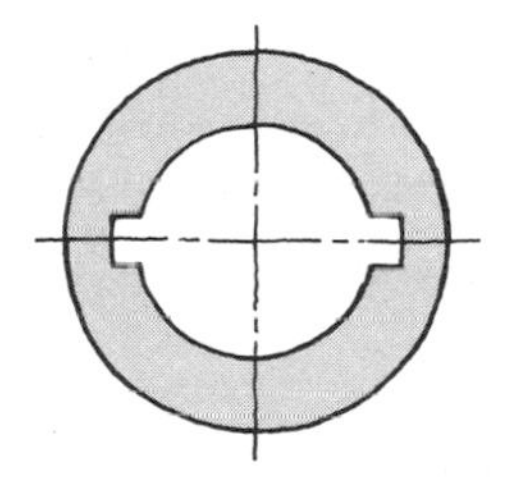

FIG. 65. Hub is weakened by keyways. Poor design.

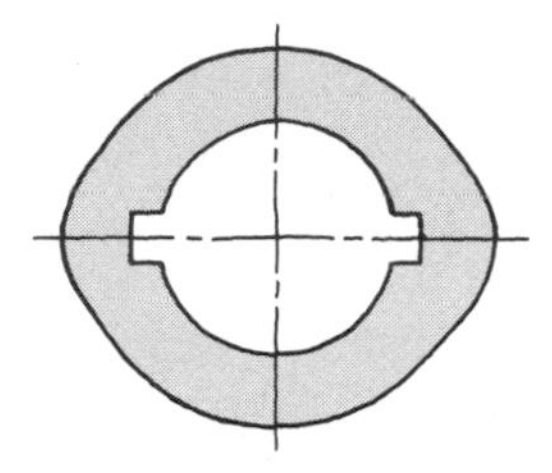

FIG. 66. Extra material added to hub so that keyways will not weaken the part. Good design.

Parts should never be designed so that a hole must be drilled at an angle to a surface, as in Fig. 72. Even though a drill jig is used, the design will cause drill breakage, inaccuracy, and severe wear on the drill bushing. Provide a normal surface for the hole by building up the piece as in Fig. 73.

In parts formed from sheet metal two practices should be avoided if possible. First, full 90-degree bends will cause undue stretching of the material by the die and the part may be difficult to strip from the die. Second, small radii less than the metal thickness (internal) may cause cracking of the material at the bend. Such a poorly designed part is shown in Fig. 74. Unless full 90-degree bends are needed for some functional reason, a far better design is to use bends of less than 90 degrees and radii equal to the metal thickness (internal) or larger, as shown in Fig. 75.

Sharp pointed notches in sheet metal parts should be avoided because stress breaks will start at the sharp point. For this reason Fig. 76 is poor design. It costs no more and is much safer to provide a round bottom on the notch as in Fig. 77, or, better still, to provide also rounds at the upper ends of

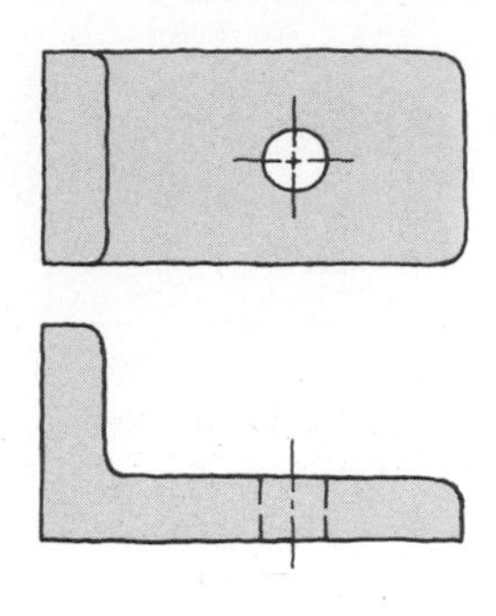

**FIG. 67.** No finished surface for bolt or screw seat. Poor design.

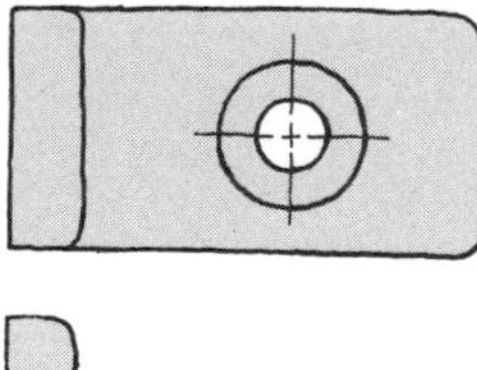

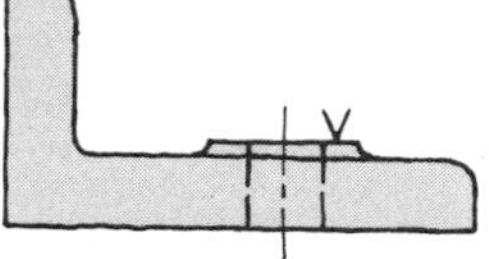

**FIG. 68.** Seat for bolt or screw provided by a finished boss. Good design.

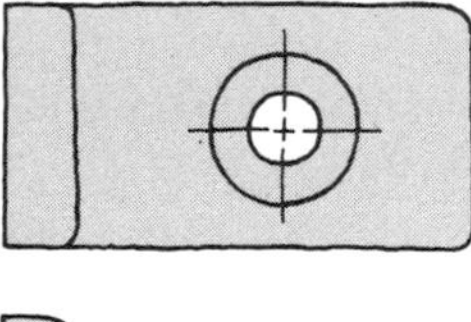

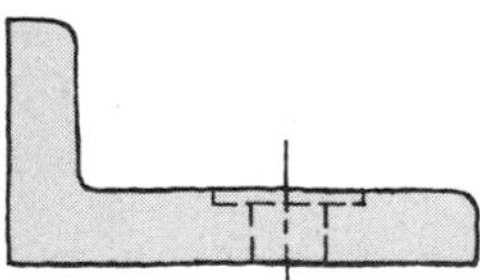

**FIG. 69.** Seat for bolt or screw provided by spotface. Good design.

the notch as in Fig. 78.

When bent lugs are needed on a sheet-metal part for fastening or a similar purpose, notches should be provided at the sides of the lug. In Fig. 79 notches

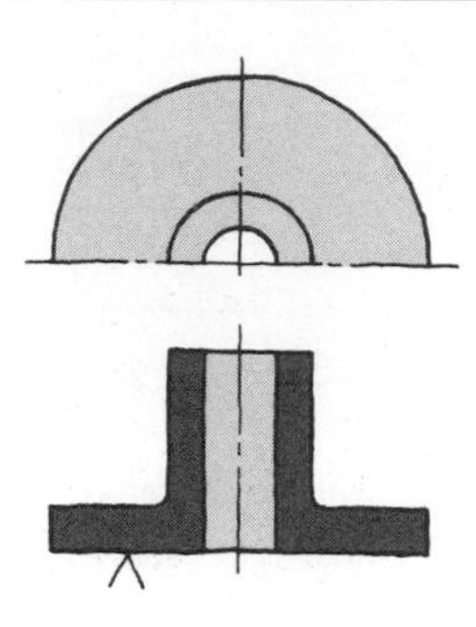

**FIG. 70.** Full finished surface on base of hub. Good design when a complete flat surface is needed.

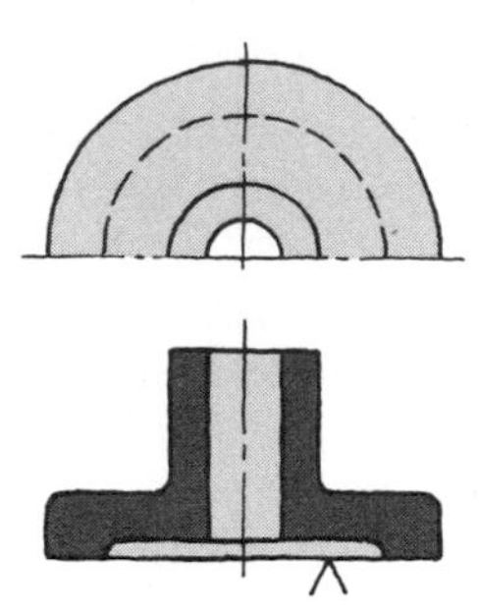

**FIG. 71.** Center portion of hub base relieved to save machining. Good design.

are not provided and because of the bend radius required, undue stretching and distortion will occur at the sides of the lug. Side notches, such as shown in Fig. 80, will eliminate the difficulty. Note also in Fig. 80 that the outer surface can be brought flush with the side surface of the part proper.

Bent stiffening flanges on round corners of sheet-metal parts must be specially designed or crowding and crinkling of the metal will occur, as shown in Fig. 81. A design with notches to relieve the crowding of the material will solve the problem, as indicated in Fig. 82.

Formed flanges on the outside rounded corners of sheet-metal parts are almost impossible to make without undue stretching, weakening, and cracking of the material, as indicated in Fig. 83. The solution

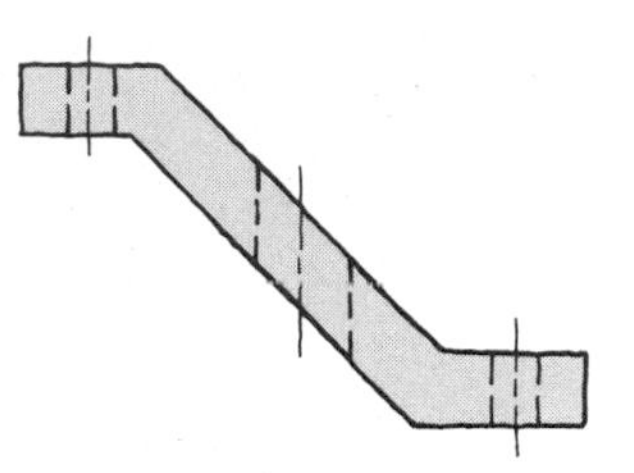

FIG. 72. Drilled hole at an angle to the part surface. Poor design.

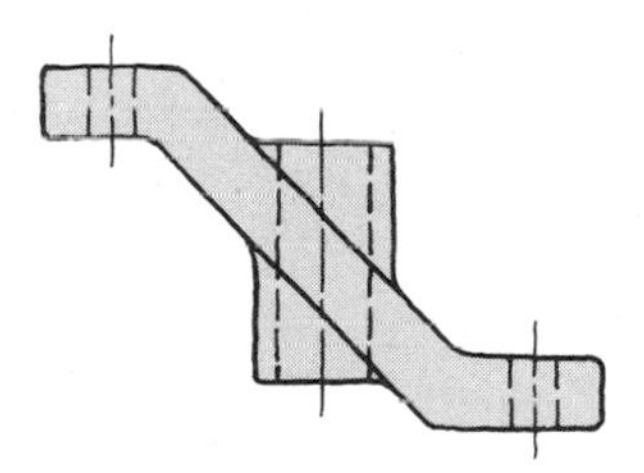

FIG. 73. Normal surface provided for drilled hole. Good design.

FIG. 74. Formed sheet-metal part with 90° bends and small bend radii. Poor design.

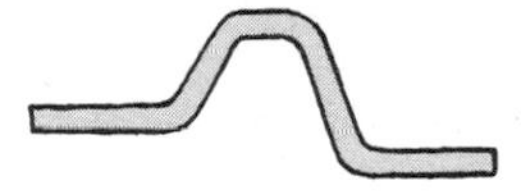

FIG. 75. Formed sheet-metal part with bends less than 90° and bend radii (internal) greater than the metal thickness. Good design.

of the problem is to provide notches, as in Fig. 84.

Formed parts of spherical or similar shape should not be made, as in Fig. 85, without a flange for stiffening. A flange, such as shown in Fig. 86, makes for a much more stable part and also provides a seating surface, often needed.

When a portion of a sheet-metal part must be made by shearing or cutting and bending over as

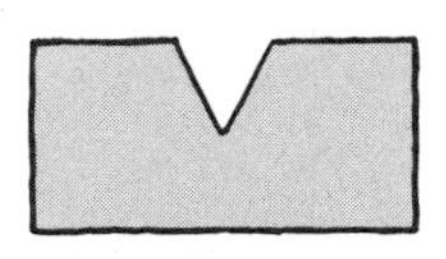

FIG. 76. Sharp-pointed notch. Poor design.

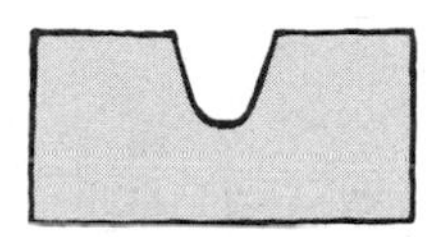

FIG. 77. Round-bottomed notch. Good design.

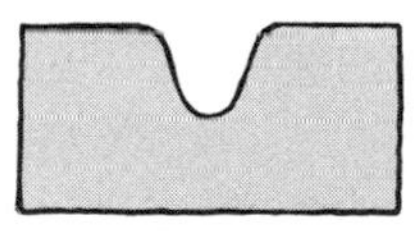

FIG. 78. All corners of notch rounded. Good design.

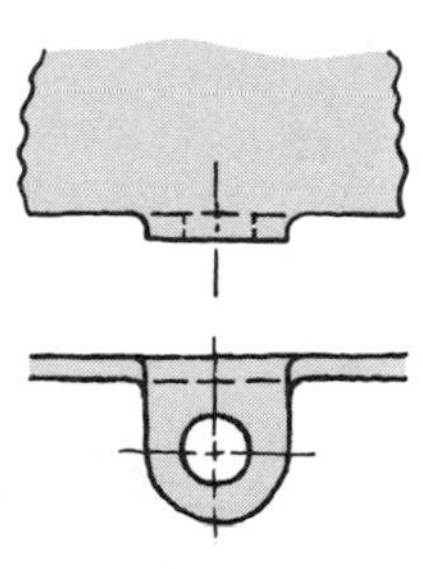

FIG. 79. No corner notches for bend relief. Poor design.

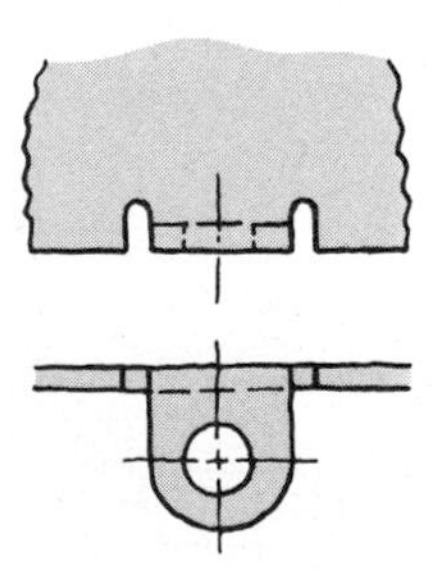

FIG. 80. Corner notches provided for bend relief. Good design.

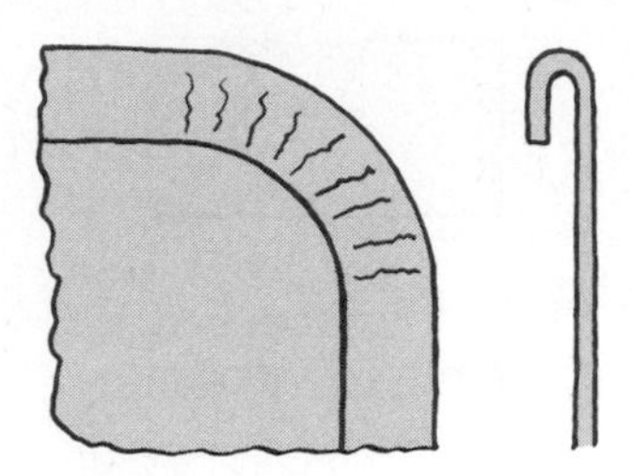

FIG. 81. Bent flange on round corner causes crowding of the metal. Poor design.

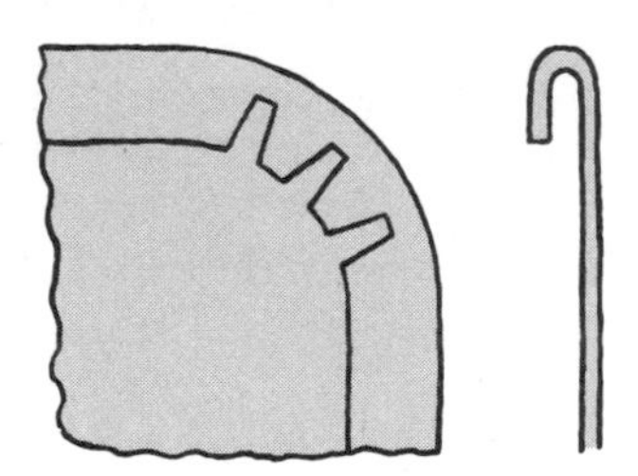

FIG. 82. Notches are used to relieve the crowding of metal on a bent corner flange. Good design.

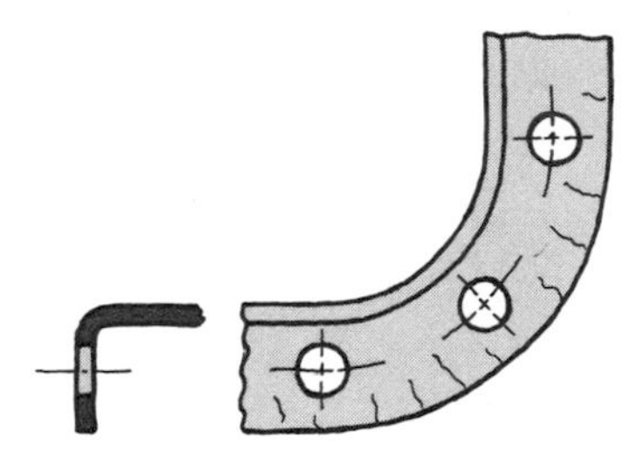

FIG. 83. Formed outside flange on rounded corner produces breaks in the material. Poor design.

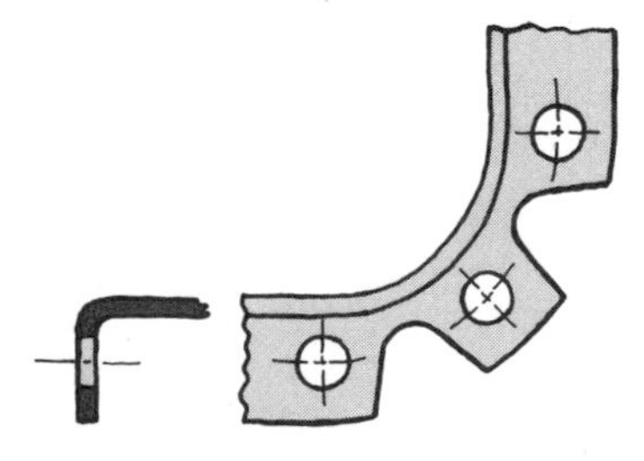

FIG. 84. Notches on the outside formed flange on a rounded corner prevent material stretching and breakage. Good design.

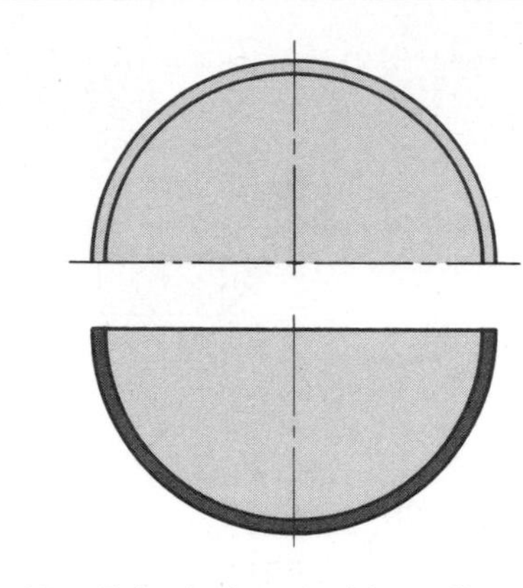

FIG. 85. Spherical part without flange. Poor design.

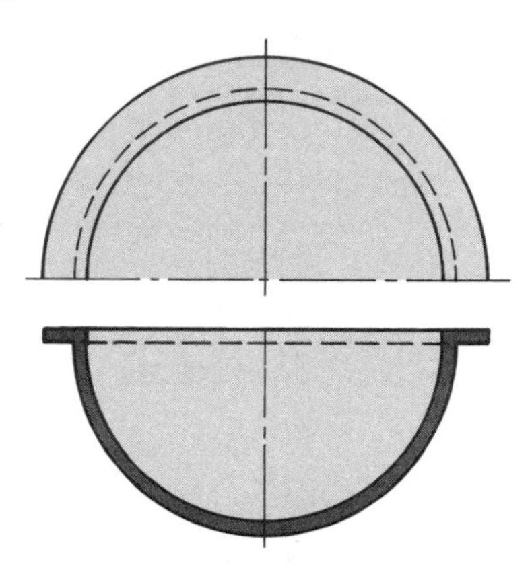

FIG. 86. Spherical part with flange. Good design.

in Fig. 87, breakage because of overstressing or fatigue will start from the end of the sheared or cut surface. The sheared or cut line should be terminated by a hole (drilled or punched), as in Fig. 88. The hole contour will distribute stresses and prevent breakage.

Holes in sheet metal made by punching or drilling are satisfactory when a completely flat surface is needed, as in Fig. 89. However, a stronger hole can be made by punching or drilling and forming, as in Fig. 90. Another advantage of a hole so made is that it may be tapped to receive a screw.

Punched holes too close to the edge of a sheet-metal part may cause distortion and possible breakage of the material, as indicated in Fig. 91. The hole should be set back with the minimum edge distance shown in Fig. 92.

If the corner of a sheet-metal pan does not have to be closed for functional purposes, an economical part can be made by relieving the corner as shown in Fig. 93.

If possible, sheet metal parts should be designed so that maximum use can be made of the stock without excessive scrap, as indicated in Fig. 94.

Sheet-metal parts should be designed so as to simplify the shape of the blank unless there is some functional reason for not doing so. The design of the part shown in Fig. 95 produces an irregular blank and excessive scrap from the blanking operation.

A more economical design is shown in Fig. 96 where the blank has straight sides. By reversal in the blanking operation excessive scrap is eliminated and maximum use of the stock is attained.

### 11. Assembly—Fastening and Joining of Parts.

There are many methods of joining parts. The designer should be familiar with all possible methods

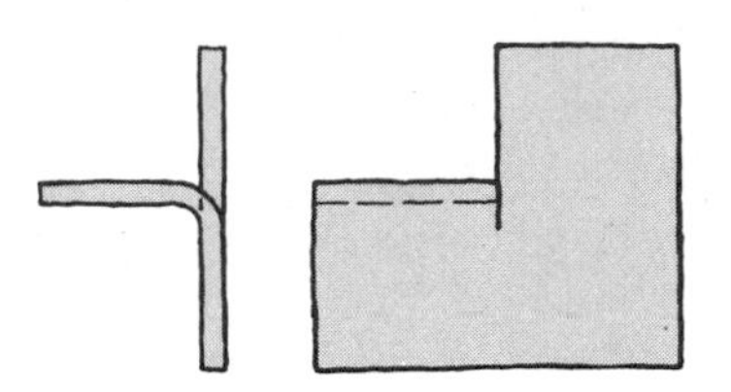

FIG. 87. Bent portion of part formed by shearing and bending. Poor design.

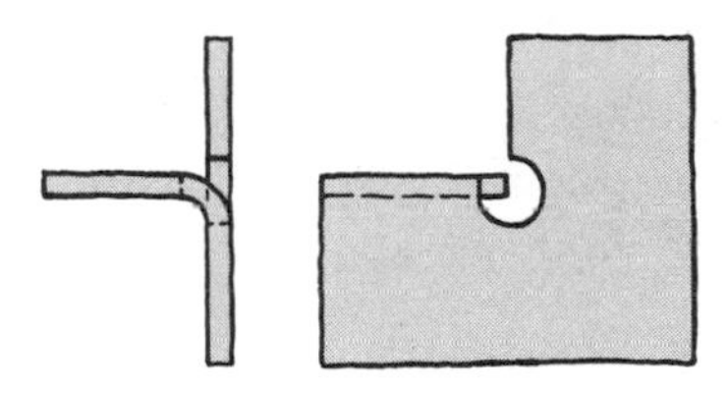

FIG. 88. Bent portion of part formed by shearing to a terminal hole and bending. Good design.

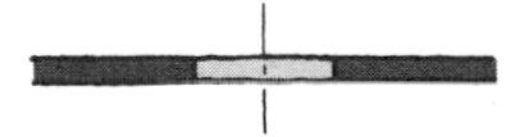

FIG. 89. Drilled hole in sheet metal.

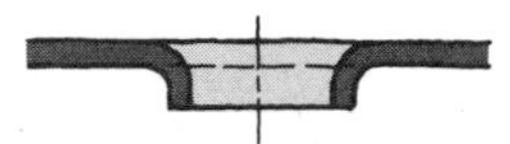

FIG. 90. Punched (or drilled) and formed hole in sheet metal.

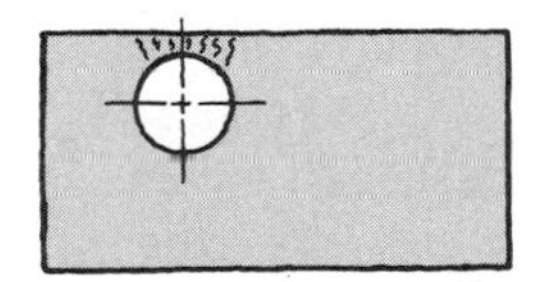

FIG. 91. Hole too close to the edge causes distortion and possible breakage. Poor design.

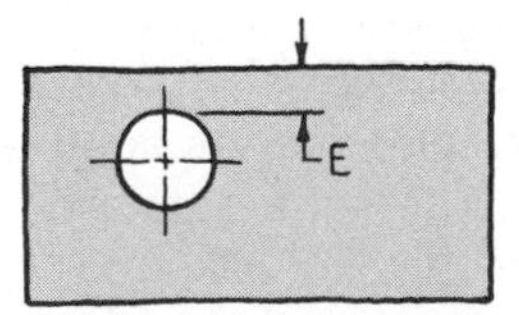


$E = 1\frac{1}{2}$ METAL THICKNESS (MIN.)

FIG. 92. Proper setback for a punched hole in sheet metal. Good design.

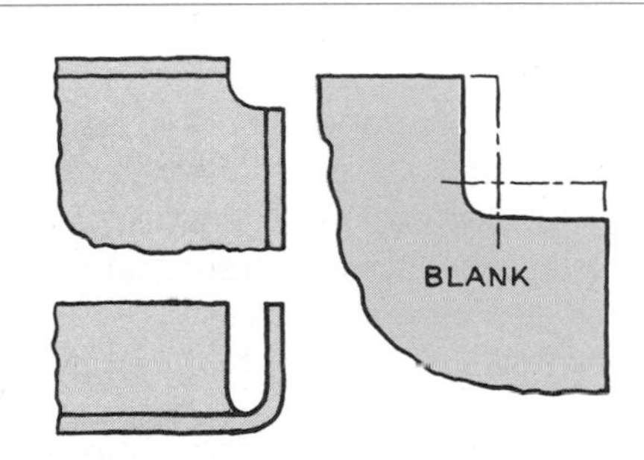


FIG. 93. Corner relief for a formed sheet-metal pan. Economical design.

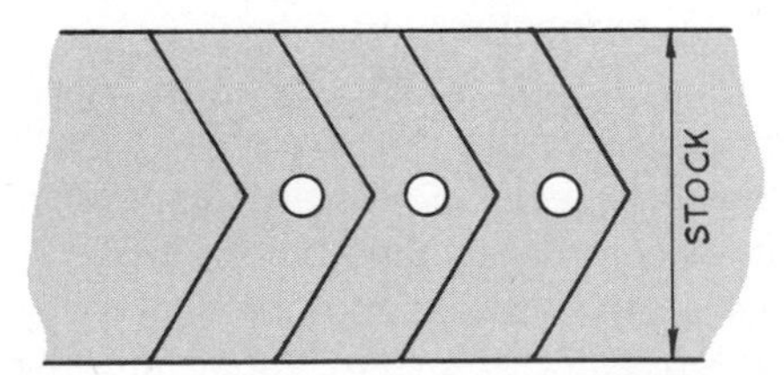


FIG. 94. Maximum use of stock. Good design.

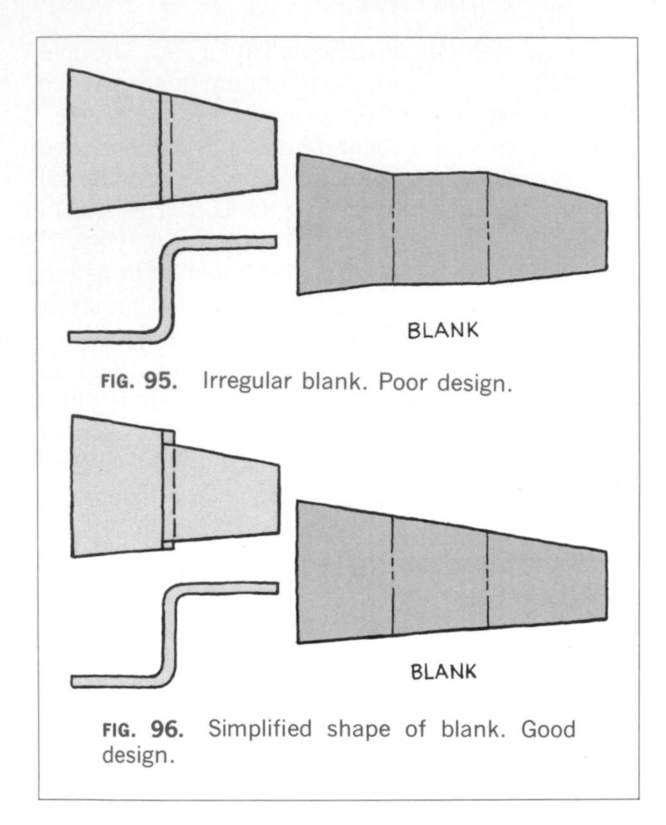


FIG. 95. Irregular blank. Poor design.

FIG. 96. Simplified shape of blank. Good design.

and should select the method (in any particular case) that will satisfactorily hold the parts together and also be economical from the standpoints of cost, repair and replacement, and general maintenance. In any particular machine several different methods may be used depending somewhat on whether the fastening is to be permanent or removable. The following lists describe the most common methods.

*Mechanical Fasteners.* These include bolts and screws, rivets, special springs, snap rings, O-rings, rollpins, and keys. For details, see Chap. 15.

*Adhesives.* Adhesives are classified by *bonding type* (heat, pressure, time, catalyst, vulcanizing); *form* (liquid, paste, powder, mastic); *flow* (flowable or nonflowable); *vehicle* (solvent dispersion, water emulsion, or complete solids); or chemical composition, which is the most significant.

*Natural adhesives* of casein, blood albumin, hide, bone, fish, starch, rosin, shellac, asphalt, sodium silicate, or glycerin-litharge are used for wood, paper, cork, packaging, textiles, some metals, and plastics.

*Thermoplastic adhesives* of polyvinyl acetate, polyvinyl alcohol, acrylic, cellulose, nitrate, and asphalt are used for practically all materials, but principally for nonmetallics such as wood, leather, and cork.

*Thermosetting adhesives* of phenolic, resorcinol, phenolresorcinol, epoxy, urea, melamine, and alkyd are used for many metals and nonmetals. Specifically, epoxies are used to join dissimilar materials, metals and plastics; phenolics for metal, glass, and wood; ureas and melamines for wood; alkyds for metal laminations.

*Elastomeric adhesives* of natural rubber, butadiene-styrene, neoprene, and silicone are used for wood, rubber, fabric, foil, paper, leather, plastic film, and for tape.

Almost all adhesives are strongest in shear, somewhat weaker in tension, and weakest in peel strength. Therefore, in the design of the joint strength is an important consideration.

A joint like that shown in Fig. 97 is weak in tension because, as the parts joined are stressed, the metal will distort and will be peeled away from the adhesive bond.

A simple butt joint like that shown in Fig. 98 is not strong in tension because of the weakness of the bond itself, and also because in most joints of this type the actual bond area is limited.

A better joint may be made by lapping the parts as in Fig. 99. This puts the bond in shear, but as the parts are stressed, some distortion due to the offset will occur and the adhesive may peel.

Probably the best joint in tension is that shown in Fig. 100, where tension on the joint will not distort the members and the adhesive remains in shear.

When an adhesive bond is used for a right-angle member as in Fig. 101, the member should be flared as shown to get a greater bonding area.

When a joint is designed as in Fig. 102 for a right-angle corner, the adhesive bond will be in peel and will be weak. The joint should be designed as in Fig. 103, where the adhesive bond is in shear.

Adhesive-bonded joints are not used for highly stressed parts because of the relatively low strength of adhesives. However, adhesive joints and connections may be very satisfactory and economical for lightly stressed parts and for attaching coverings,

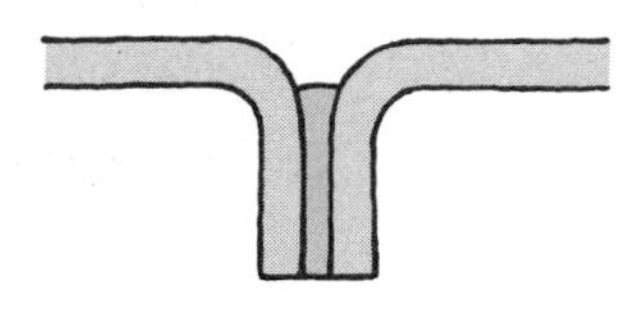

FIG. 97. Adhesive-bonded joint. Weak in tension because of peel.

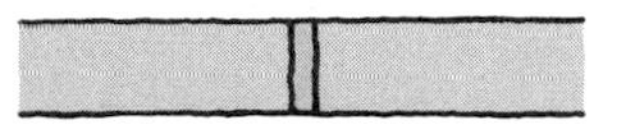

FIG. 98. Adhesive-bonded butt joint. Weak in tension because of bond strength and lack of bond area.

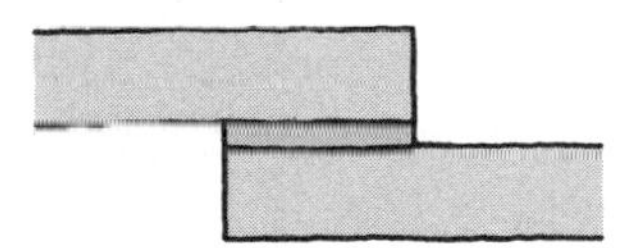

FIG. 99. Adhesive lap joint. Better in tension than the joints of Figs. 97 and 98.

FIG. 100. Adhesive-bonded joint. Better in tension than the joints of Figs. 97, 98, or 99.

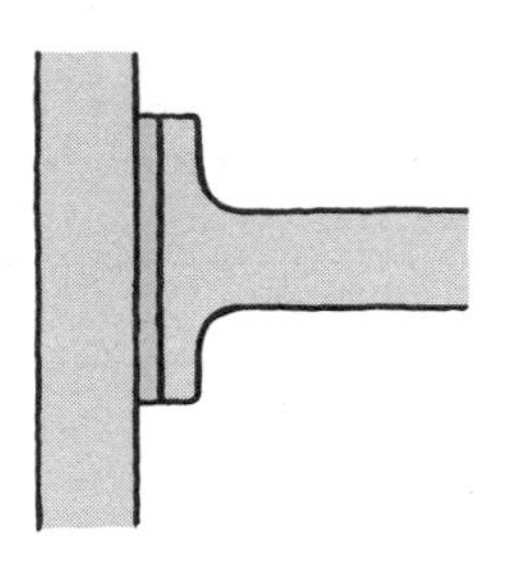

FIG. 101. Right-angle member flared for greater area of bonded joint.

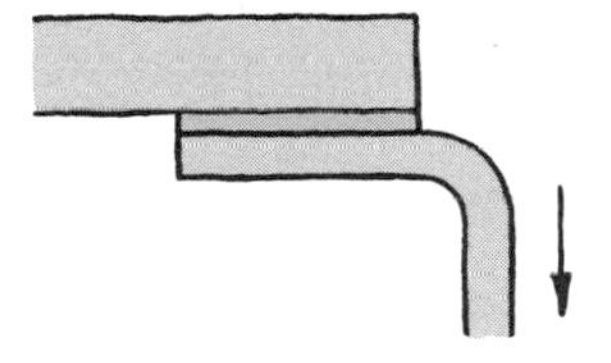

FIG. 102. Adhesive-bonded joint for right angle. Poor design.

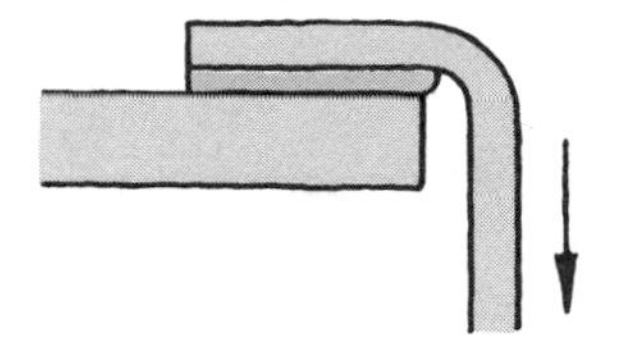

FIG. 103. Adhesive-bonded joint for right angle. Good design.

medallions, name plates and similar parts which are not stressed at all.

*Solders.* Many combinations of solder metals are commercially available. *Tin-lead* solders are used for general-purpose applications on ferrous and nonferrous metals for joining and coating. *Tin-lead-antimony* solders are used for machine and torch soldering and coating of metals, except galvanized iron. *Tin-antimony* solders are used for joints on copper in electrical, plumbing, and heating applications. *Tin-silver* solders are used principally for work on fine instruments. *Tin-zinc* solders are used to join aluminum. *Lead-silver* solders are used on copper, brass, and similar metals (with torch heating). *Cadmium-zinc* solders are used for soldering aluminum. *Cadmium-silver* solders are used for joining aluminum to aluminum and to dissimilar metals. *Zinc-aluminum* solders are used for high-strength aluminum joints.

*Brazing.* Brazing is the operation of joining two similar or dissimilar metals by partial fusion with a metal or alloy applied under very high temperature. Brazing alloys include aluminum-silicon, copper-phosphorus, silver, gold, copper, copper-zinc, magnesium, and nickel.

*Welding.* Welding is the operation of joining two similar metals with a fused joint of the same or very similar metal. Steel, cast iron, copper and copper alloys, nickel alloys, and aluminum can be welded. See Chap. 16 for welding practices and symbols.

The accompanying table lists joinability of common materials.

**Joinability of Materials***

| | Arc welding | Oxyacetylene welding | Resistance welding | Brazing | Soldering | Adhesive bonding | Threaded fasteners | Riveting and stitching |
|---|---|---|---|---|---|---|---|---|
| Cast iron | × | × | | | | × | × | × |
| Carbon and low-alloy steels | × | × | × | × | | × | × | × |
| Stainless steels | × | × | × | × | | × | × | × |
| Aluminum magnesium | × | × | × | × | | × | × | × |
| Copper and its alloys | × | × | × | × | × | × | × | × |
| Nickel and its alloys | × | × | × | × | × | × | × | × |
| Titanium | × | | × | | | × | × | |
| Lead, zinc | × | × | | | × | × | × | × |
| Thermoplastics | × | × | × | | | × | × | × |
| Thermosets | | | | | | × | × | × |
| Elastomers | | | | | | × | | × |
| Ceramics | | | | | | × | × | |
| Glass | | × | | | | × | × | |
| Wood | | | | | | × | × | × |
| Leather | | | | | | × | | × |
| Fabric | | | | | | × | | × |
| Dissimilar metals | | | | × | × | × | × | × |
| Metals to nonmetals | | | | | | × | × | × |
| Dissimilar nonmetals | | | | | | × | | × |

*The crossmark (×) in the table indicates common practice. Difficult and unreliable methods are not given.

**12. Aesthetics.**
The aesthetics of any completely new and original creation can usually be designed with freedom because there is no other similar entity to compare or compete with. The choice of configuration, colors, contrasts, materials, and other similar factors offers

extensive opportunity for experiment. The designer may try a number of combinations, build mock-ups, and then choose the most appealing.

The aesthetics or "styling" of products built previously, but redesigned for a new model, are affected somewhat by the former model and also by what most industries call styling trends. As examples, note the trends in each year's models of automobiles, refrigerators, radios, television receivers, and other well-known products. Each year, new ways are found to combine colored finishes, chrome or other metal trim, plastics, leather, wood, and many other materials. Configurations are improved and "cleaned up." Operative elements such as knobs and levers are altered to give interesting and pleasing appearances.

The aesthetics of any product, new or redesigned, presents no simple problem. The project requires artistic ability, a keen color sense, good judgment of configurations and combinations, and an instinct for supplying the factors that appeal strongly to most buyers and users. This may sound difficult to a person who thinks that he has no artistic ability. A measure of success can be attained, however, by diligent and purposeful study, observation, and work.

Consider the following list of factors.

*Configuration.* The "over-all shape" of any product gives the first and most important impact. As an example, automobiles are made in many colors and color combinations, but the particular model is at once recognized by its distinctive shape. In designing the configuration of a product, try to find a combined shape that is recognizable, distinctive, and functionally reasonable. Attempts to cover function with a shape that has no relationship to function usually fail. Experiment with every possible combination of shapes, blended curves, and surfaces. If the product is redesigned (new model), attempt to find distinctive forms that follow styling trends.

*Colors.* After configuration, color probably has the most important impact. Choose colors or combinations that are traditionally accepted, but also try new shades, tints, and combinations.

*Contrasts.* Contrasts can be obtained in many ways. Examples are combining a dark color with a light color; combining a rough texture with a smooth texture; combining definitely different materials (leather and gold, wood and metal, fabric and wood or metal, etc.); combining materials with differences in reflectance (bright finish and dull finish); combinations of color, material, texture, brightness.

*Details.* Small details should not be overlooked. Quite often the success of a design lies in attention to the apparently insignificant parts. The color, material, texture, or relative brightness of a small part may add a definite distinction to the product. Small details to be considered are such items as knobs, handles, lights and lighting accessories, indicators, meters, dials, and other similar appurtenances.

*Accents.* Accents are the smaller but very important features of aesthetic design exemplified by a decorative line of contrasting color, texture, or material on an otherwise plain surface; a decorative medallion or name plate so placed as to add to the over-all composition; any small detail that adds distinction to an otherwise plain, uninteresting configuration.

*Models.* The best way to study the aesthetics of a product is to make several models of the design and then use a number of different combinations of colors, textures, and materials for later appraisal of the over-all impact.

*Appraisal.* In a company of any significant size the appraisal of styling is usually made by a group. This is a very sound method, because the appraisal will represent the combined thinking of several people. If, however, the judgment must be made by one or two persons only, the cardinal rule is to make every attempt to be as objective as possible.

### 13. Summary of Design Procedure.

The accompanying diagrammatic outline has been assembled from recorded standard practices in a number of large engineering companies and from discussion with a number of designers who are personal friends. Thus this outline represents an authentic cross section of usual procedure in this country.

## DESIGN PROCEDURE

**PHASE ONE—THE PROJECT**

A. Discussion with management, engineering, client.
B. Statements and specification of the design problem.
C. Collection of all pertinent information.

**PHASE TWO—FORMULATION**

A. Recognition of requirements.
B. Definition of requirements.
C. Consideration of previous designs.
D. Assembly of all original data needed—mathematical, graphical, mechanical, electrical, etc.

**PHASE THREE—CONCEPTS**

A. Preliminary design sketches.
B. Preliminary design data giving materials, methods, construction details, and projected characteristics.

**PHASE FOUR—ANALYSIS**

A. Critical analysis of all design concepts.
B. Selection of most promising design or designs.

**PHASE FIVE—DESIGN CONFERENCE**

A. Discussion of preliminary designs with engineering, management, client.
B. Approval of design or designs.

**PHASE SIX—REFINEMENT**

A. More complete drawings and specifications of selected design or designs.
B. More complete data supporting projected design.

**PHASE SEVEN—DESIGN CONFERENCE**

A. Discussion of refined design or designs.
B. Approval of most promising design.

**PHASE EIGHT—SYNTHESIS**

A. Projected design supported by mathematical, graphical, and computer-aided and combined systems data.
B. Investigation of all physical aspects and proof of soundness of the design.

**PHASE NINE—MODELS**

A. Components.
B. Mock-ups.
C. Models of critical features.

**PHASE TEN—TESTING**

A. Proof of operating characteristics of components.
B. Proof of soundness of complete entity.

**PHASE ELEVEN—CONFERENCE**

A. Final discussion with originating authority.
B. Approval of final design.

**PHASE TWELVE—FINAL PREPARATION**

A. Final design drawings.
B. Final specifications.

**PHASE THIRTEEN—TRANSMITTAL**

A. Transmittal of final design drawings and specifications to originating authority.

NOTE: In the outline above three conferences (Phases 5, 7, and 11) are scheduled with the originating authority, engineering, and management. This is usual, but it does *not* mean that these are the only conferences during the development of the project. The designer confers frequently with colleagues, and with engineers, components suppliers, materials experts, and others, to obtain information and confirm design features.

## DESIGN PROJECTS

A series of design projects is presented on the following pages. The projects are designed to tax the ingenuity of the student without requiring any special knowledge of engineering mechanics or higher mathematics.

Each project has an accompanying suggested solution. These solutions are deliberately left undimensioned and incomplete so that the minimum requirement for the student is to complete the design and make a set of working drawings. They do not necessarily represent the best possible design for the project. The student is urged to consider as many other ideas as possible, weighing all such factors as cost and reliability before making a final selection.

Use of standard purchased parts such as rollpins, retainer rings, screws, etc., is encouraged.

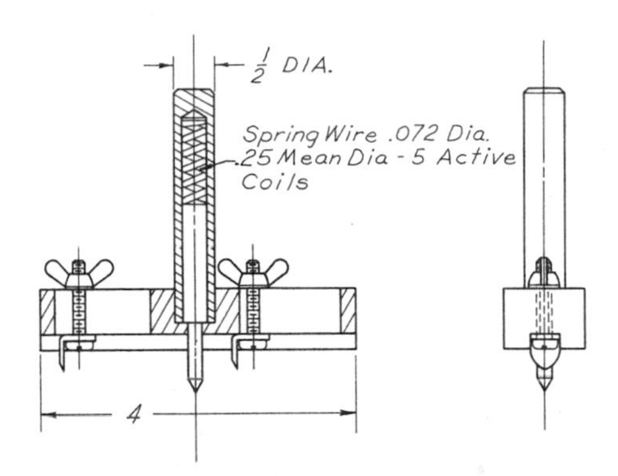


DESIGN PROJECT 1.

**1.** Design a device to be used in a drill press to cut circular gaskets from sheet cork, rubber, etc. It must cut gaskets to varying sizes, as large as 4 in. outside diameter, with the interior hole adjusting down as small as practicable.

The design indicated makes use of a spring-loaded centering pin to keep the material from slipping during the cutting. Torque must be transmitted from the spring tube to the cutter bar by some fastening device not shown. If this design is used, the means incorporated to make this connection must permit the parts to be disassembled for spring replacement and sharpening of the center pin.

**2.** An industrial dolly made from steel angle is to be rebuilt to carry delicate electronics instruments. The dolly is equipped with four casters similar to that shown in Prob. 36, Chap. 23.

Design a new caster mount to absorb shock incurred when the dolly is used on rough floors. A design load of about 100 lb per caster (including dolly) is to be carried.

The suggested design makes use of a compression spring having a spring constant of 100 psi, with a total deflection of 2 in.

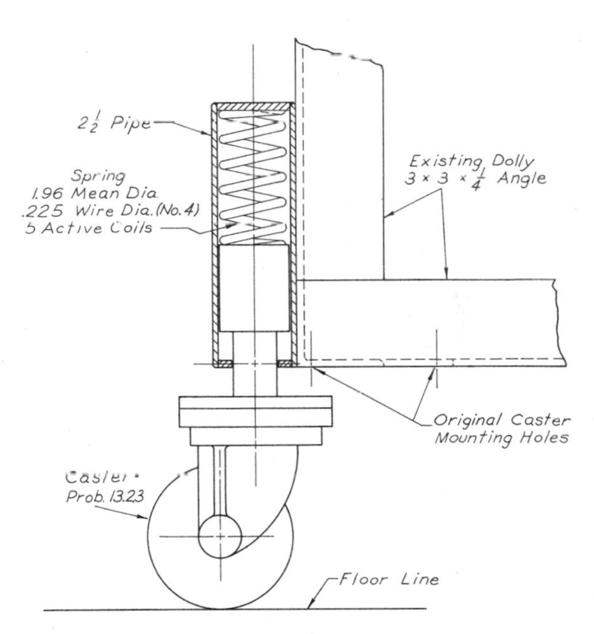


DESIGN PROJECT 2.

**3.** Design a rust-resistant portable stand suitable for holding serving trays, campers' cookstoves, laundry baskets, etc. This stand is to be folded into as small and neat a package as possible.

The suggested solution makes use of aluminum angles and flat aluminum bar stock. The fasteners are threadless pins held in place by pushnuts (see Appendix).

**4.** Design a portable solar stove suitable for survey parties and campers, for use in relatively sunny climates. It must be lightweight and disassemble to pack into a small box. A parabolic reflector at least 3 ft in diameter must be used, and its focal point must not be too precise or it will tend to burn a hole in a small container. The reflector must be on an adjustable mounting so that it

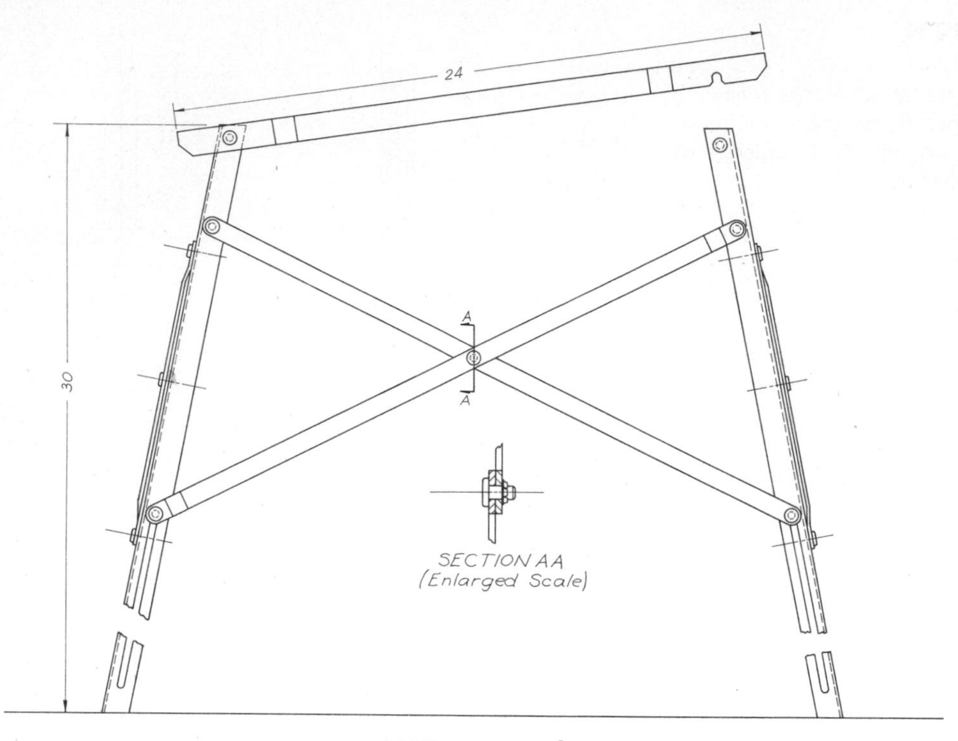


DESIGN PROJECT 3.

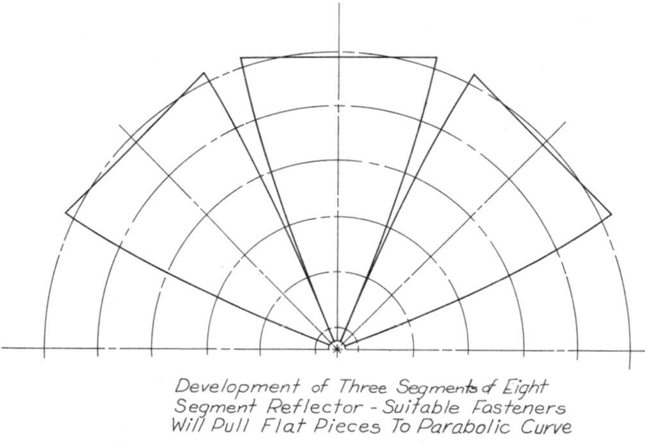

Development of Three Segments of Eight
Segment Reflector - Suitable Fasteners
Will Pull Flat Pieces To Parabolic Curve


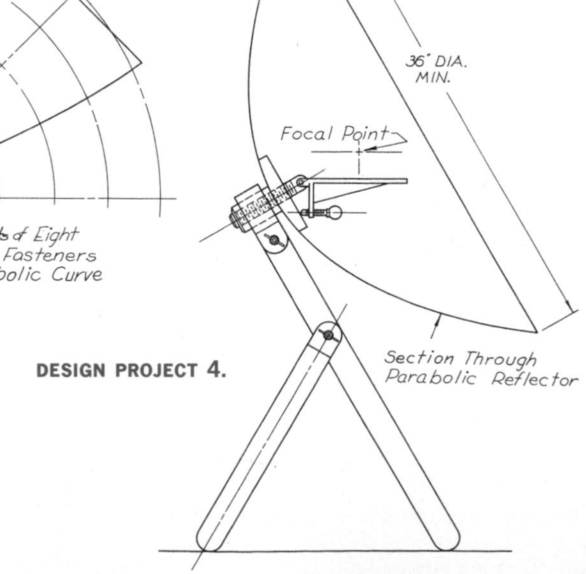


DESIGN PROJECT 4.

can be focused on the sun at different times of the day and year, and at different latitudes.

The suggested solution consists of a reflector made of eight properly developed segments of polished sheet-metal. These segments lie flat when not in use, and are bent to the proper shape by fasteners holding the segment edges together.

**5.** A small brass shell is fed from a hopper into the existing supply tube shown, that in turn supplies a press which further processes the part. All parts must have the closed end down.

Design a device to turn those pieces which have the open end down. The hopper feeds the pieces at such a rate that they do not touch each other.

One possible solution is shown; it utilizes an adjustable pin which enters the open end of those pieces that need turning. Gravity will then flip the pieces as indicated. The closed ends of the other pieces will hit the pin, the pieces will bounce up and fall without turning.

The design must be mounted on the existing 6-in. I-section column shown.

**6.** An industrial drying process is supplied with warm air through an 8-in.-diameter round duct. A variable-diameter orifice is to be installed in the duct to control the airflow. To aid in maintaining laminar flow the opening is to be kept approximately round and concentric in the duct. The diameter must reduce down to approximately one-half of the original diameter.

Design a suitable device for this service.

The solution indicated is based on the idea of six curved blades pivoted outside the periphery of the duct and moved in and out of the airstream by a control ring which is concentric with the duct.

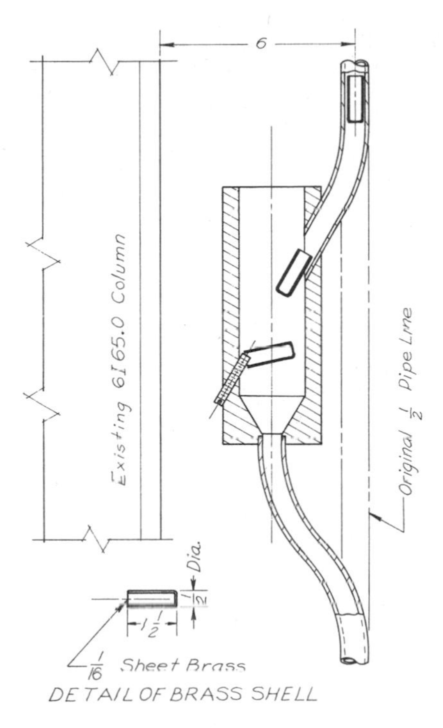


DESIGN PROJECT 5.

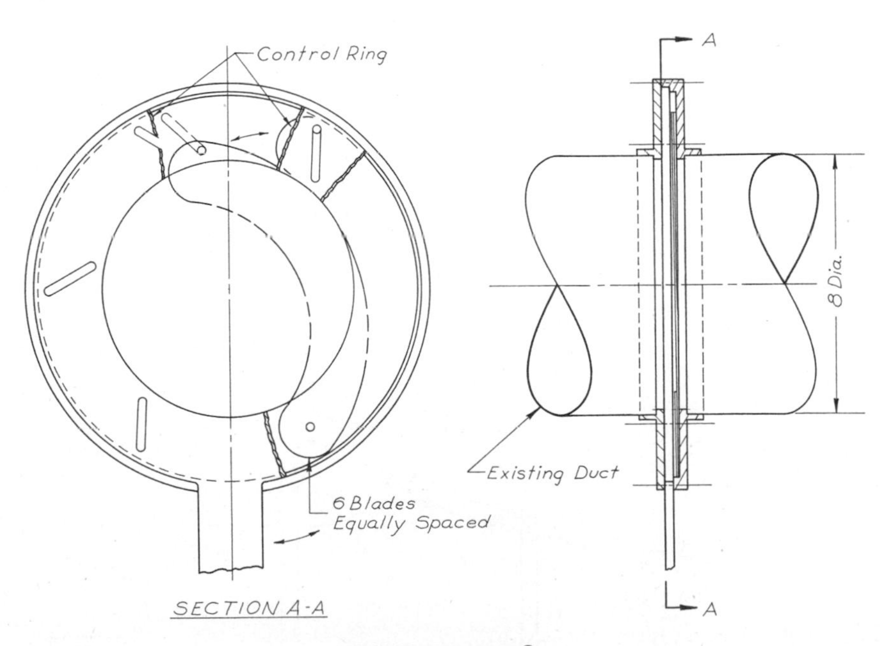


DESIGN PROJECT 6.

**7.** Many mechanical devices are lubricated for life at the time of assembly. This requires that a metered amount of oil be placed in the machine.

Design a device that will meter out 2 fluid ounces of oil with a single "push-pull" motion of the operator (1 U.S. fluid ounce = 1.805 cubic inches).

Some adjustment is desirable so that precise control is possible. Oil must not flow when the device is not in use.

The partial solution indicated uses two pistons, mounted on a single rod, sliding back and forth in a carefully finished cylinder.

Air trapped in the cylinder may impede movement of either the oil or the piston. The plastic float valve indicated will permit air to enter and leave the cylinder, but will trap the oil.

**8.** A "tin"-can food-container manufacturing plant is supplying cans directly to a food processer. The flow of cans must be equally divided and supplied to the two conveyor belts which will be installed as shown. Design a device and make the necessary assembly, working, and installation drawings for this.

The solution indicated consists of a "tee" member pivoted as shown. Each can will rotate the member in one direction or the other, evenly dividing the flow.

The cans are 4 in. in diameter and $4\frac{3}{4}$ in. high. They are supplied spaced out so that they do not touch each other.

**9.** Design a device to check the compression-type coil spring specified below for the following:

Spring rate (pounds to deflect spring 1 inch)
Maximum free length
Maximum length when compressed solid

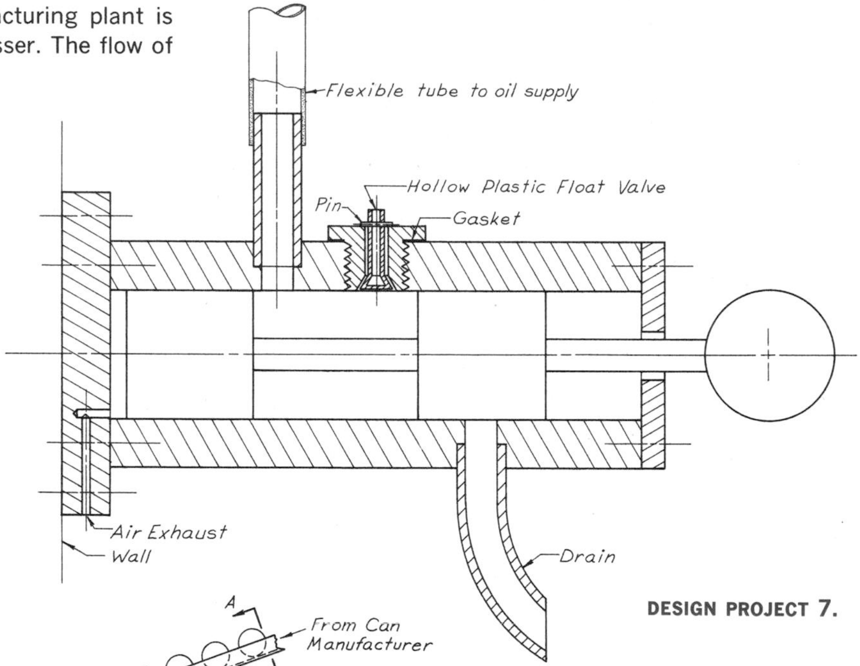

**DESIGN PROJECT 7.**

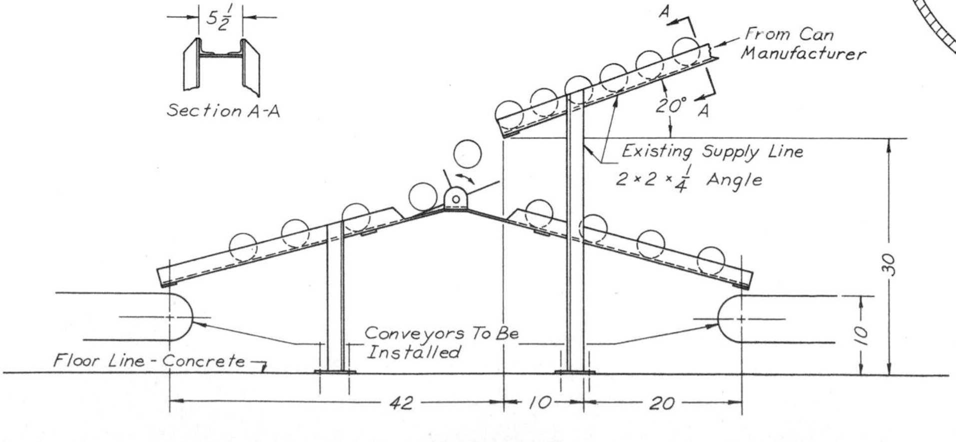

**DESIGN PROJECT 8.**

Specification of spring: 1 in. mean diameter, .0625-in.-diameter wire, five active coils, ends squared and ground, 2-in. maximum free length, 0.44-in. maximum solid length, and 4.4 psi spring rate. A load of 4.4 lb must deflect spring from 0.9 to 1.1 in.

The device indicated is one possible solution. A cam turned by hand deflects the master spring 1 in. (It is carefully set to give a force of 4.4 lb at 1-in. deflection.) If the two shoes are equidistant from the center pivot, the spring under test will carry a 4.4-lb load. Lines scribed on the base then will show, if the spring is within tolerance. The other lines scribed on the base will show, if spring meets length-tolerance specifications.

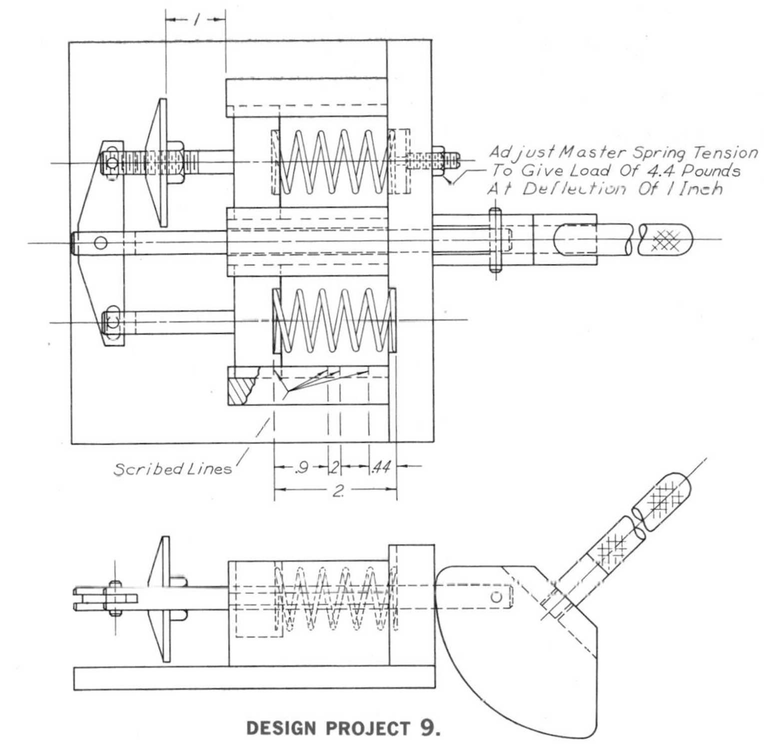


DESIGN PROJECT 9.

**10.** An electrical manufacturer makes transformer cores by welding or riveting stacks of thin steel laminations together. Because of the variation in thickness of the laminations due to the rolling tolerance (both total thickness variation and variation from side to side on the same sheet) neither the number of laminations per stack nor the height of the stack are good measures of the ideal stack. As the electrical properties are primarily a function of the mass of steel in the core, the cores are best gaged by weight.

The lamination shown is stamped from a 22-gage sheet steel and is to be used in a core weighing 1.4 lb with as little variation as possible.

The ultimate goal is an automatic machine to receive the laminations directly from the punch press, select the stock, and feed it to an automatic welder. However, because of the complexity of the problem, a hand-powered, hand-loaded, hand-unloaded experimental machine is to be designed and tried first.

A partial solution is shown. Complete this design or prepare an alternate design to handle this problem.

DESIGN PROJECT 10.

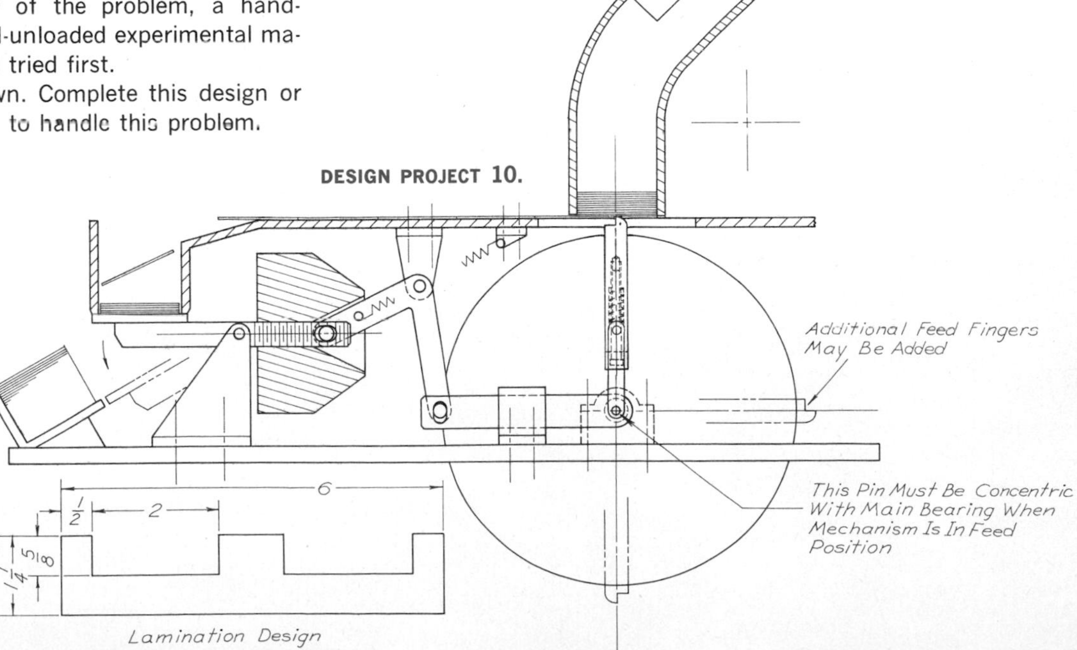

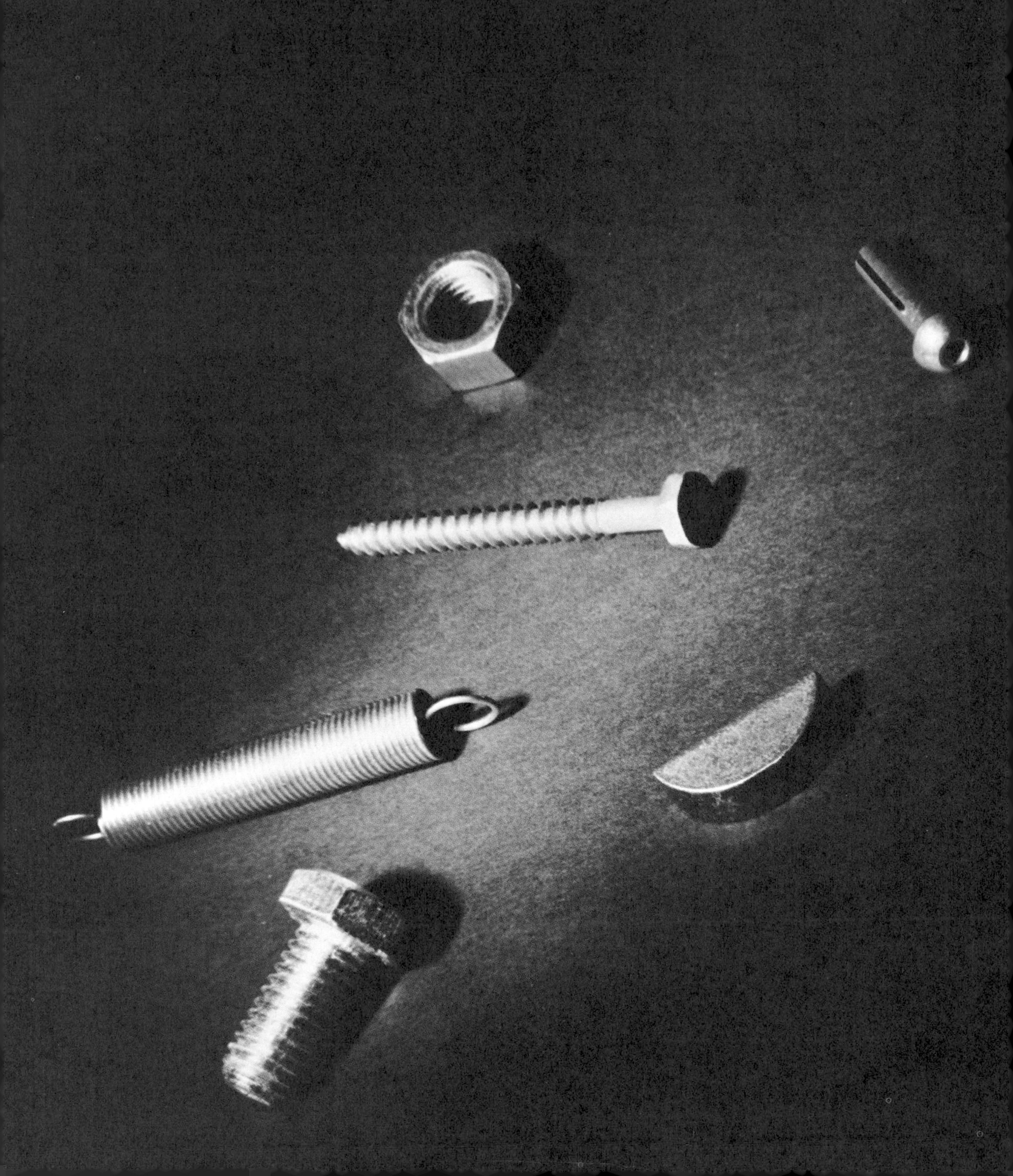

# 15

# SCREW THREADS, FASTENERS, KEYS, AND SPRINGS

Practically All Designs of Devices, Machines, Structures, etc., Require Fasteners and Related Elements to Accomplish their Function

This Chapter Discusses:

## 1. Screw Threads—Their Importance.

A screw thread is the functional element used on bolts, nuts, cap screws, wood screws, and the like and on shafts or similar parts employed for transmitting power or for adjustment. Screw threads occur in one form or another on practically all engineering products. Consequently, in making working drawings, there is the repeated necessity to *represent* and *specify* screw threads.

## 2. History.

The earliest records of the screw are found in the writings of Archimedes (278 to 212 B.C.). Although specimens of ancient Greek and Roman screws are so rare as to indicate that they were seldom used, there are many from the later Middle Ages; and it is known that crude lathes and dies were used to cut threads in the latter period. Most early screws were made by hand, by forging the head, cutting the slot with a saw, and fashioning the screw with a file. In America in colonial times wood screws were blunt on the ends; the gimlet point did not appear until 1846. Iron screws were made for each threaded hole. There was no interchanging of parts, and nuts had to be tied to their own bolts. In England, Sir Joseph Whitworth made the first attempt to set up a uniform standard in 1841. His system was generally adopted there but not in the United States.

## 3. Standardization.

The initial attempt to standardize screw threads in the United States came in 1864 with the adoption of a report prepared by a committee appointed by the Franklin Institute. The system, designed by William Sellers, came into general use and was known as the "Franklin Institute thread," the "Sellers thread," or the "United States thread." It fulfilled the need for a general-purpose thread at that period; but with the coming of the automobile, the airplane, and other modern equipment, it became inadequate. Through the efforts of the various engineering societies, the Bureau of Standards, and others, the National Screw Thread Commission was authorized by act of Congress in 1918 and inaugurated the present standards. The work has been carried on by the ASA and by the Interdepartmental Screw Thread Committee of the U.S. Departments of Defense (Army, Navy, Air Force) and Commerce. Later, these organizations, working in cooperation with representatives of the British and Canadian governments and standards associations, developed an agreement covering a general-purpose thread that fulfills the basic requirements for interchangeability of threaded products produced in the three countries. The "Declaration of Accord" establishing the Unified Screw Thread was signed in Washington, D.C., on November 18, 1948.

Essential features of the Unified and other threads are given in this chapter, while standards covering them are listed in the Appendix.

## 4. Screw-thread Terminology.

*Screw Thread (Thread).* A ridge of uniform section in the form of a helix on the external or internal surface of a cylinder or cone.

*Straight Thread.* A thread formed on a cylinder (Fig. 1).

*Taper Thread.* A thread formed on a cone.

*External Thread (Screw).* A thread on the external surface of a cylinder or cone (Fig. 1).

*Internal Thread (Nut).* A thread on the internal surface of a cylinder or cone (Fig. 1).

*Right-hand Thread.* A thread which, when viewed axially, winds in a clockwise and receding direction. Threads are always right-hand unless otherwise specified.

*Left-hand Thread.* A thread which, when viewed axially, winds in a counterclockwise and receding direction. All left-hand threads are designated "LH."

*Form.* The profile (cross section) of the thread. Figure 3 shows various forms.

*Crest.* The edge or surface that joins the sides of a thread and is farthest from the cylinder or cone from which the thread projects (Fig. 1).

*Root.* The edge or surface that joins the sides of adjacent thread forms and coincides with the cylinder or cone from which the thread projects (Fig. 1).

*Pitch.* The distance between corresponding points on adjacent thread forms measured parallel to the axis (Fig. 1). This distance is a measure of the size of the thread form used.

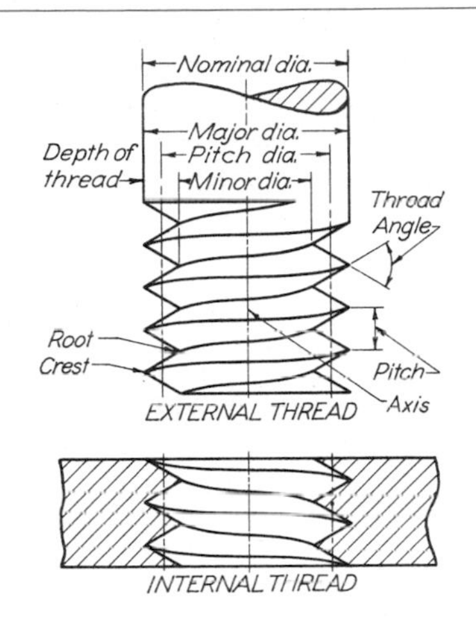


**FIG. 1.** Screw-thread terminology. See Sec. 4 for details.

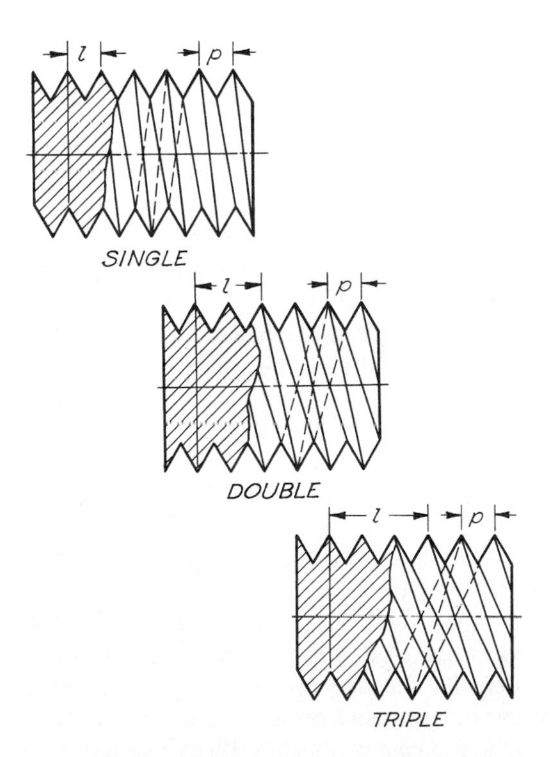


**FIG. 2.** Multiple threads. Note that, for a single thread, lead and pitch are identical values; for a double thread, lead is twice the pitch; and for a triple thread, lead is three times the pitch.

*Lead.* The distance a threaded part moves axially, with respect to a fixed mating part, in one complete revolution. See multiple thread (below) and Fig. 2.

*Threads per Inch.* The reciprocal of the pitch and the value specified to govern the size of the thread form.

*Major Diameter.* The largest diameter of a screw thread (Fig. 1).

*Minor Diameter.* The smallest diameter of a screw thread (Fig. 1).

*Pitch Diameter.* On a straight thread, the diameter of an imaginary cylinder, the surface of which cuts the thread forms where the width of the thread and groove are equal (Fig. 1). The clearance between two mating threads is controlled largely by closely toleranced pitch diameters.

*Depth of Thread.* The distance between crest and root measured normal to the axis (Fig. 1).

*Single Thread.* A thread having the thread form produced on but one helix of the cylinder (Fig. 2). See multiple thread (below). On a single thread, the lead and pitch are equivalent. Threads are always single unless otherwise specified.

*Multiple Thread.* A thread combination having the same form produced on two or more helices of the cylinder (Fig. 2). For a multiple thread, the lead is an integral multiple of the pitch, that is, on a *double thread* the lead is twice the pitch, on a *triple thread,* three times the pitch, etc. A multiple thread permits a more rapid advance without a coarser (larger) thread form. Note that the helices of a double thread start 180° apart; those of a triple thread, 120° apart; and those of a quadruple thread, 90° apart.

## 5. Thread Forms.

Screw threads are used on fasteners, on devices for making adjustments, and for the transmission of power and motion. For these different purposes, a number of thread forms are in use (Fig. 3). The dimensions given in the figure are those of the basic thread forms. In practical usage, clearance must be provided between the external and internal threads.

The *sharp V,* formerly used to a limited extent, is rarely employed now; it is difficult to maintain sharp roots in quantity production. The form is of interest, however, as the basis of more practical

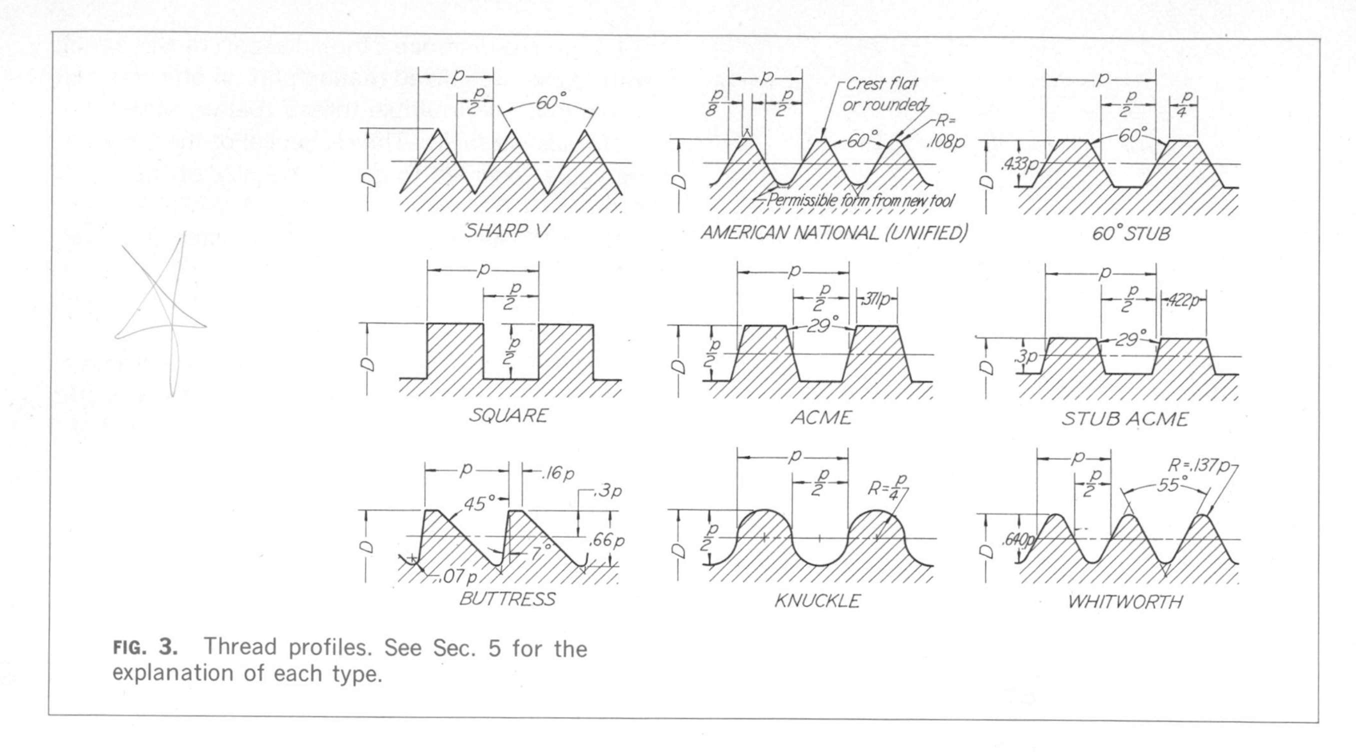


**FIG. 3.** Thread profiles. See Sec. 5 for the explanation of each type.

V-type threads; also, because of its simplicity, it is used on drawings as a conventional representation for other (V-type) threads.

V-type threads are employed primarily on fasteners and for making adjustments. For these purposes, the standard thread form in the United States is the *American National*. This form is also the basis for the Unified Screw Thread standard of the United States, Canada, and Great Britain and as such is known as the *Unified* thread. As illustrated in Fig. 3, the form is that of the maximum external thread. Observe that while the crest may be flat or rounded, the root is rounded by design or as a result of tool wear. The American National thread is by far the most commonly used thread in this country.

The 60° *stub* form is sometimes preferred when, instead of multiple threads, a single National-form thread would be too deep.

The former British standard was the *Whitworth*, at 55°, with crests and roots rounded. The *British Association Standard*, at 47½°, measured in the metric system, is used for small threads. The *French* and the *International Metric Standards* have a form similar to the American National but are dimensioned in the metric system.

The V shapes are not desirable for transmitting power since part of the thrust tends to burst the nut. This does not happen when a *square thread* is used as it transmits all the forces nearly parallel to the axis of the screw. The square thread can have, evidently, only half the number of threads in the same axial space as a V thread of the same pitch, and thus in shear it is only half as strong. Because of manufacturing difficulties, the square-thread form is sometimes modified by providing a slight (5°) taper to the sides.

The *Acme* has generally replaced the square thread in use. It is stronger, more easily produced, and permits the use of a disengaging or split nut that cannot be used on a square thread.

The *Stub Acme* is a strong thread suited to power applications where space limitation makes it desirable.

The *buttress*, for transmitting power in only one direction, has the efficiency of the square and the strength of the V thread. It was formerly produced with a perpendicular pressure flank (face); the newer 7° slope is more easily manufactured. Sometimes called the "breechblock" thread, it is used to withhold the pressure on breech blocks of large guns.

The *knuckle* thread is especially suitable when threads are to be molded or rolled in sheet metal. It can be observed on glass jars and in a shallow form on bases of ordinary incandescent lamps.

Internal screw threads are produced by cutting, while external threads are made by cutting or rolling. Tests show that rolled threads are considerably stronger than cut threads. Through cold forging, rolling adds toughness and strength to the threaded portion.

### 6. Thread Representation.

The true representation of a screw thread is almost never used in making working drawings. In true representation, the crest and root lines (Fig. 4) appear as the projections of helices (see Helix, Sec. 71, Chap. 3), which are extremely laborious to draw.

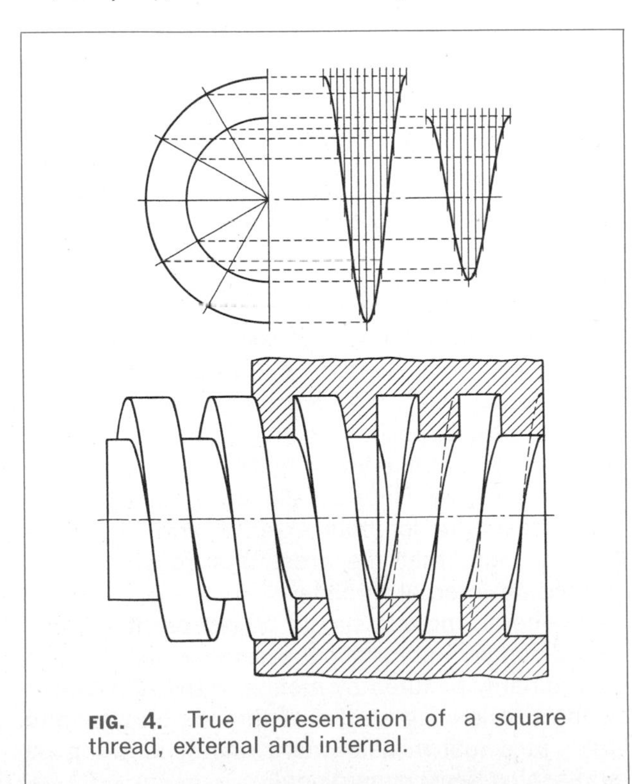

**FIG. 4.** True representation of a square thread, external and internal.

For instances where true representation is desirable, that is, on advertising, elaborate display drawings, etc., templates can be made of cardboard or celluloid to assist in drawing the helices.

On practical working drawings, threads are given a semiconventional or symbolic representation, which provides thread pictures adequate for manufacturing purposes. The methods of representing threads are discussed in the following paragraphs.

### 7. Semiconventional Representation.

This simplifies the drawing of the thread principally by conventionalizing the projection of the helix into a straight line. Where applicable with certain thread forms, further simplifications are made. For example, the 29° angle of the Acme and 29° stub forms are drawn at 30°, and the American National form is represented by the sharp V. In general, true pitch should be shown, although a small increase or decrease in pitch is permissible so as to have even units of measure in making the drawing. Thus seven threads per inch may be increased to eight, or four and one-half may be decreased to four. Remember this is only to simplify the drawing—the actual threads per inch must be specified in the dimensioning.

To draw a thread semiconventionally, you must know whether it is external or internal, the form, the major diameter, the pitch, its multiplicity, and whether it is right- or left-hand. Figure 5 illustrates the stages in drawing an American National, or sharp V; Fig. 6, an Acme; and Fig. 7, a square thread. Figure 5, showing the V thread, illustrates the method of drawing a thread semiconventionally. At (*a*) the diameter is laid out, and on it the pitch is measured on the upper line. This thread is a single thread; therefore, the pitch is equal to the lead, and the helix will advance $p/2 = l/2$ in 180°. This distance is laid off on the bottom diameter line, and the crest lines are drawn in. One side of the V form is drawn at (*b*), and it is completed at (*c*). At (*d*), the root lines are added. As can be seen from Figs. 5 to 7, the stages in drawing any thread semiconventionally are similar; the principal difference is in the thread form. The square thread in Fig. 7 is double, while that in Fig. 8 is single and left-hand.

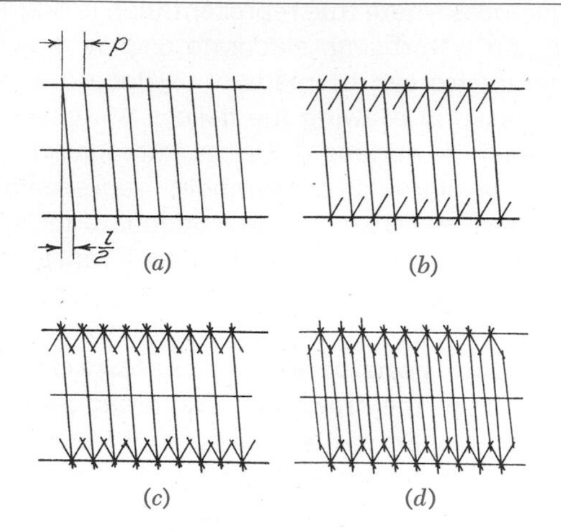


**FIG. 5.** Stages in construction for drawing a single V thread. Lay out pitch and half-lead and draw crest lines (*a*), start thread contour (*b*), complete thread contour (*c*), and draw root lines (*d*).

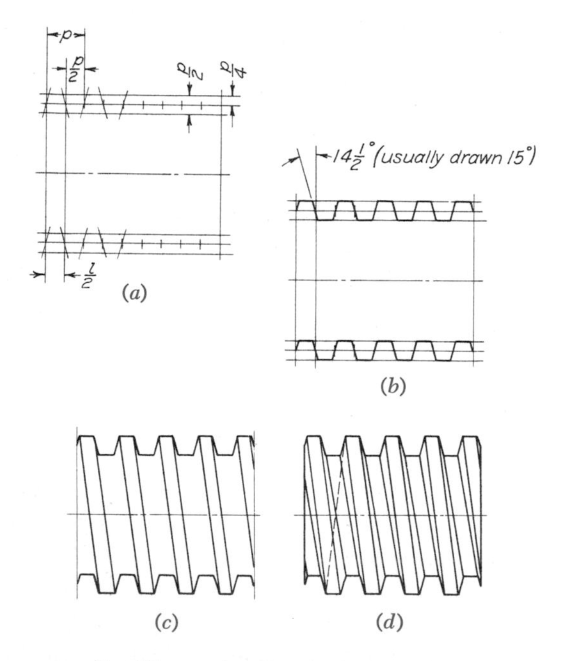


**FIG. 6.** Stages in drawing a single Acme thread. Because of the shape of the thread, the pitch diameter must first be located (*a*); then the thread form is drawn (*b*) and completed (*c*) and (*d*).

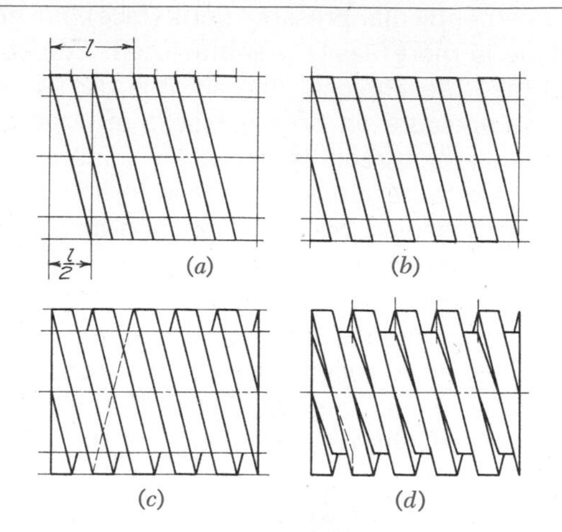


**FIG. 7.** Stages in drawing a double square thread. The lead is twice the pitch. Note that after spacing the lead and pitch (*a*), crest and root lines are drawn (*b*) and (*c*); then the base line of the thread is added (*d*).

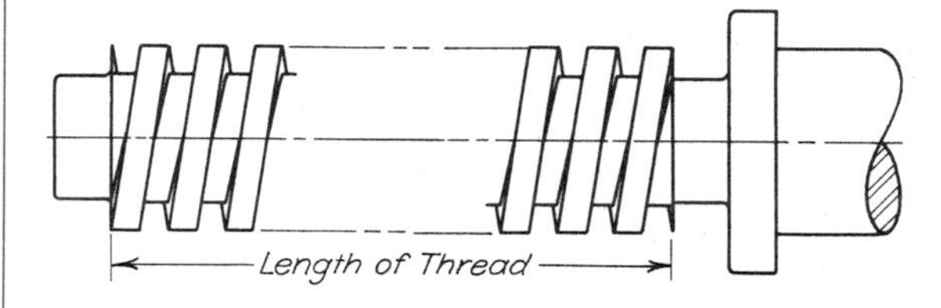


**FIG. 8.** Thread representation on a long screw. The "repeat" lines save much drawing time.

Observe in Fig. 8 that it is unnecessary to draw the threads the whole length of a long screw. If the thread is left-hand, the crest and root lines are slanted in the opposite direction from those shown in Figs. 5 to 7, as illustrated by the left-hand square thread in Fig. 8. Figure 9 illustrates semiconventional treatment for both external and internal V threads. Note that the crest and root lines are omitted on internal threads.

In general, threads should be represented semiconventionally except for the smaller sizes, which are ordinarily pictured by means of the ASA thread symbols. It is suggested that threads of 1 in. and larger in actual measurement on the drawing be represented semiconventionally.

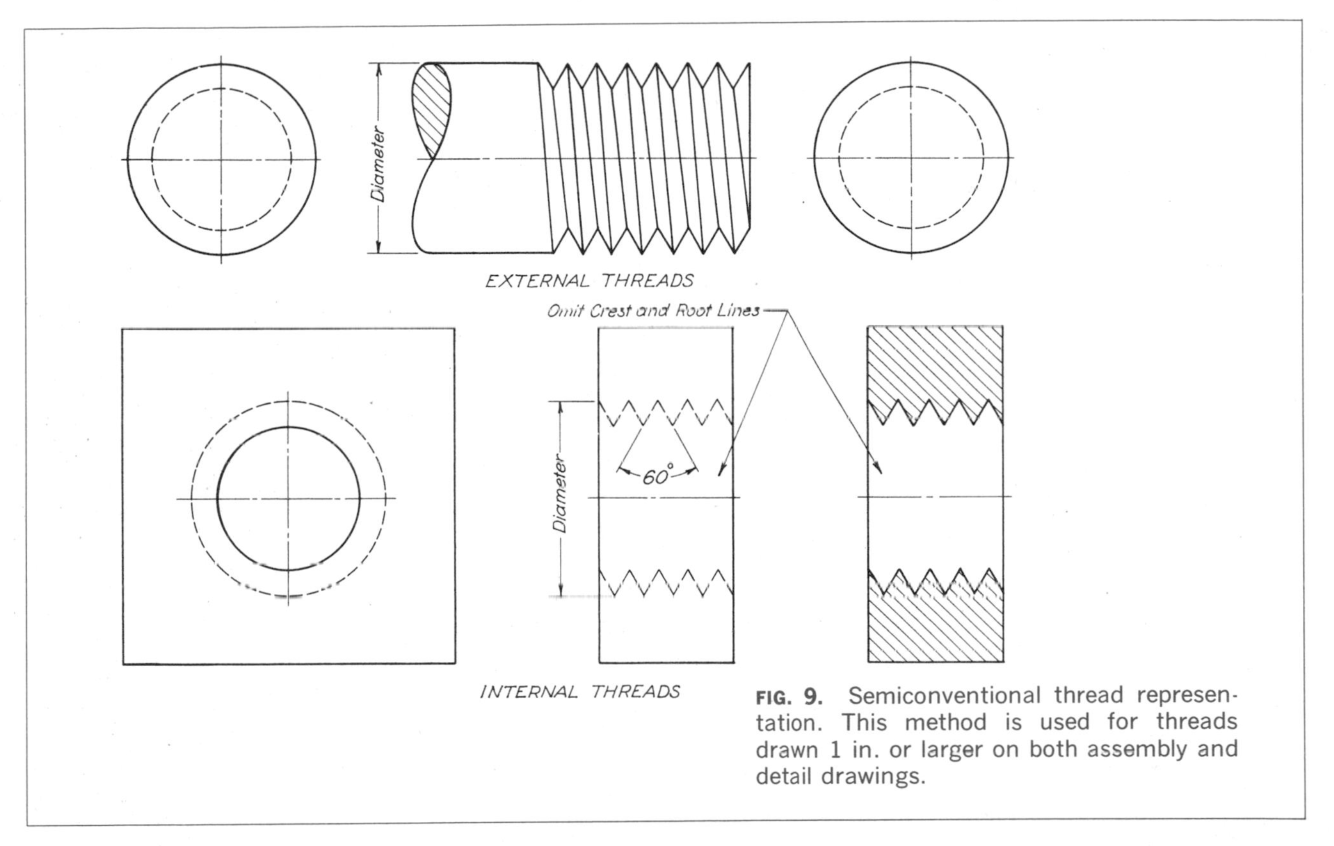


FIG. 9. Semiconventional thread representation. This method is used for threads drawn 1 in. or larger on both assembly and detail drawings.

**8. Thread Symbols.**
The ASA provides two types of thread symbols: "regular" and "simplified." It is recommended that the ASA symbols be used for indicating the smaller threads, those drawn under 1 in., and that the regular symbols be used on assembly drawings and the simplified symbols on detail drawings. The following paragraphs describe these symbols in detail.

**9. ASA Regular Thread Symbols (Fig. 10).**
The regular symbols omit the profile on longitudinal views and indicate the crests and roots by lines perpendicular to the axis. However, exceptions are made for internal threads not drawn in section and external threads drawn in section as shown in the figure.

**10. ASA Simplified Symbols (Fig. 11).**
The simplified symbols omit both form and crest lines and indicate the threaded portion by dashed lines parallel to the axis at the approximate depth of thread. The simplified symbols are less descriptive than the regular symbols, but they are quicker to draw and for this reason are preferred on detail drawings.

**11. To Draw The ASA Symbols.**
The two sets of symbols should be carefully studied and compared. Note that the regular and simplified symbols are identical for hidden threads. Note also that the end view of an external thread differs from the end view of an internal thread but that regular and simplified end-view symbols are identical.

No attempt need be made to show the actual pitch of the threads or their depth by the spacing of lines in the symbol. Identical symbols may be used for several threads of the same diameter but of different pitch. Only in the larger sizes for which the symbols are used could the actual pitch and the true depth

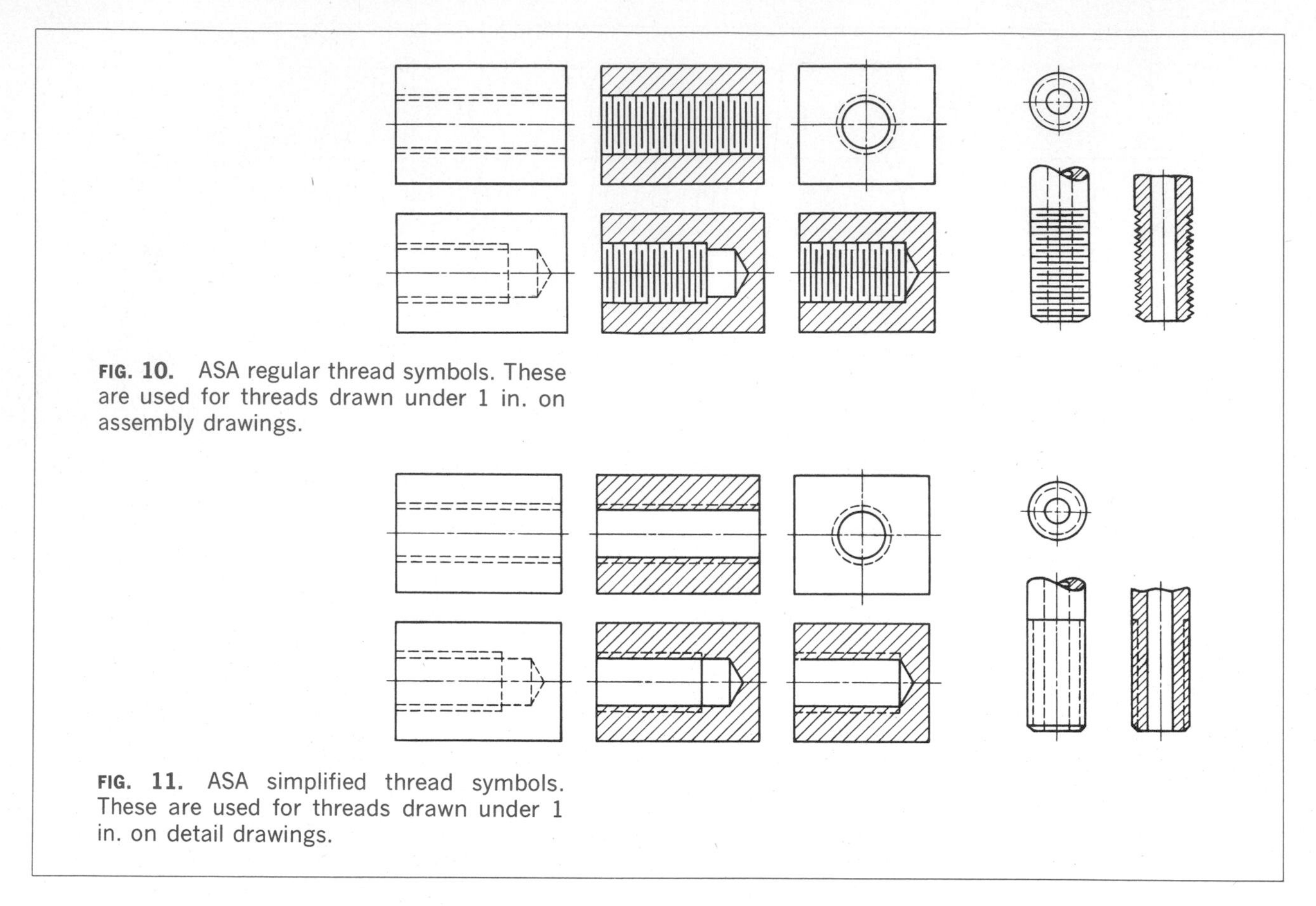

**FIG. 10.** ASA regular thread symbols. These are used for threads drawn under 1 in. on assembly drawings.

**FIG. 11.** ASA simplified thread symbols. These are used for threads drawn under 1 in. on detail drawings.

of thread be shown without a confusion of lines that would defeat the purpose of the symbol. The symbols should therefore be made so as to read clearly and look well on the drawing, without other considerations.

*To draw* a symbol for any given thread, only the major diameter and length of thread must be known, and for a blind tapped hole, the depth of the tap drill is also needed.

A *regular or simplified symbol for a tapped hole* is drawn in the stages shown in Fig. 12. The lines representing depth of thread are not drawn to actual scale but are spaced so as to look well on the drawing and to avoid crowding.

A *regular symbol for a tapped hole in section* is drawn in the stages shown in Fig. 13. The lines representing the crests are spaced by eye or scale to look well and need not conform to the actual thread pitch. The lines representing the roots of the thread are equally spaced by eye between the crest lines and are usually drawn heavier. Their length need not indicate the actual depth of thread but should be kept uniform by using light guidelines.

A *simplified symbol for an external thread* is drawn in the stages shown in Fig. 14. The 45° chamfer extends to the root line of the thread. Note that in the end view the chamfer line is shown.

A *regular symbol for external threads* is drawn in the stages shown in Fig. 15. The chamfer is 45° and to the depth of thread. Crest lines are spaced by eye or scale. Root lines are spaced by eye, are usually drawn heavier, and need not conform to actual thread depth.

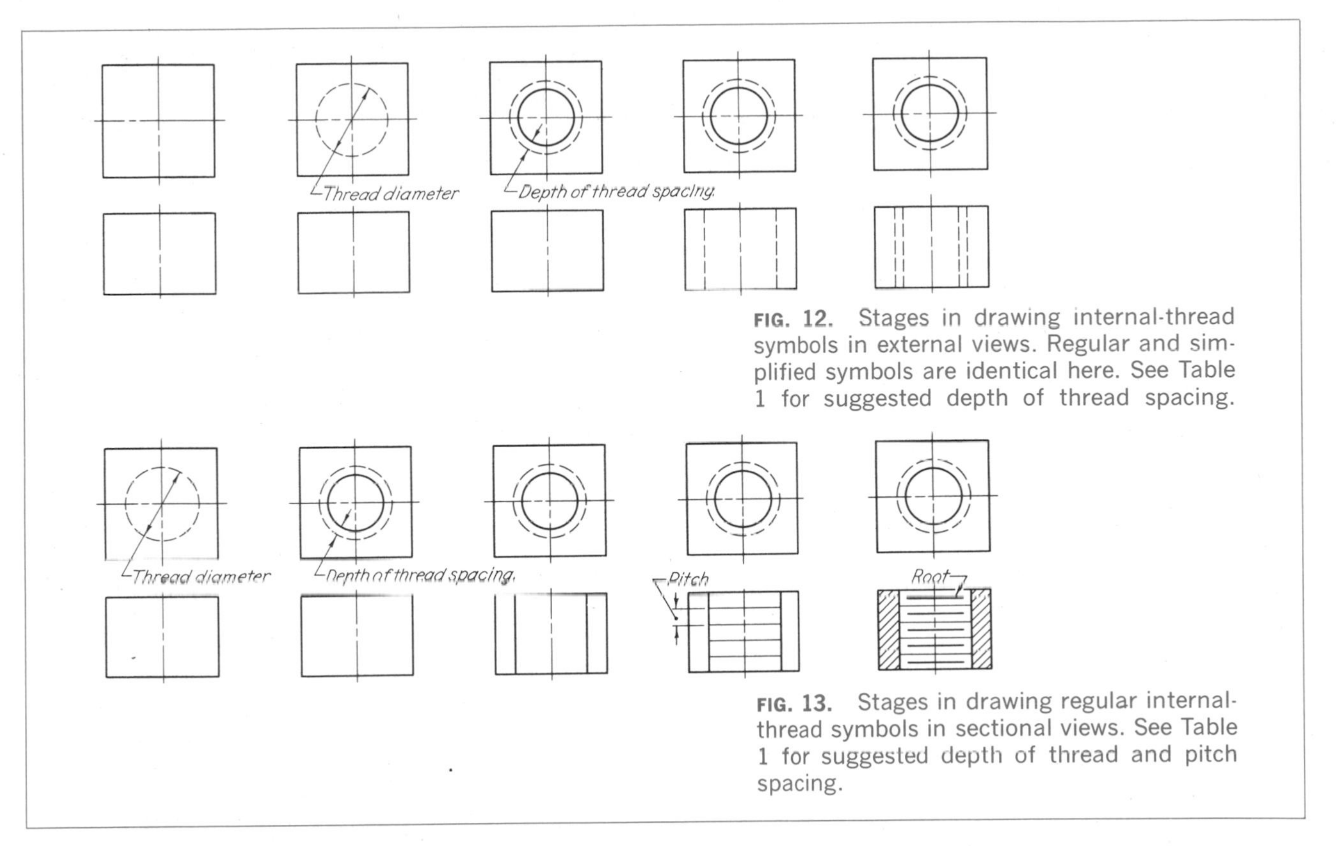


FIG. 12. Stages in drawing internal-thread symbols in external views. Regular and simplified symbols are identical here. See Table 1 for suggested depth of thread spacing.

FIG. 13. Stages in drawing regular internal-thread symbols in sectional views. See Table 1 for suggested depth of thread and pitch spacing.

*Line Spacings.* Table 1 gives suggested values of "pitch" and "depth of thread" for purposes of drawing the ASA symbols. Figure 16 shows both regular and simplified symbols, full size, drawn according to the values given in Table 1. No distinction is made in the symbol between coarse and fine threads.

**Table 1. Suggested Values for Drawing Thread Symbols.**

| Major diam. of thread $D$ | Pitch $p$ for dwg. purposes | Depth of thread $p/2$ for dwg. purposes |
|---|---|---|
| 1/8 and 3/16 | 1/16 scant | 1/32 scant |
| 1/4 and 3/8 | 1/16 | 1/32 |
| 1/2 and 5/8 | 1/8 | 1/16 |
| 3/4 and 7/8 | 3/16 | 3/32 |

## 12. Threads in Section.

Figure 4 shows the true form of an internal square thread in section. Observe that the far side of the thread is visible, causing the root and crest lines to slope in the opposite direction from those on the external thread. Figure 9 shows the semiconventional treatment for V threads 1 in. or larger in diameter. Note that the crest and root lines are omitted. The regular and simplified symbols for threads in section are shown in Figs. 10 and 11. When two pieces screwed together are shown in section, the thread form should be drawn to aid in reading (Fig. 17). With small diameters, it is desirable to decrease the number of threads per inch, thus eliminating monotonous detail and greatly improving the readability of the drawing.

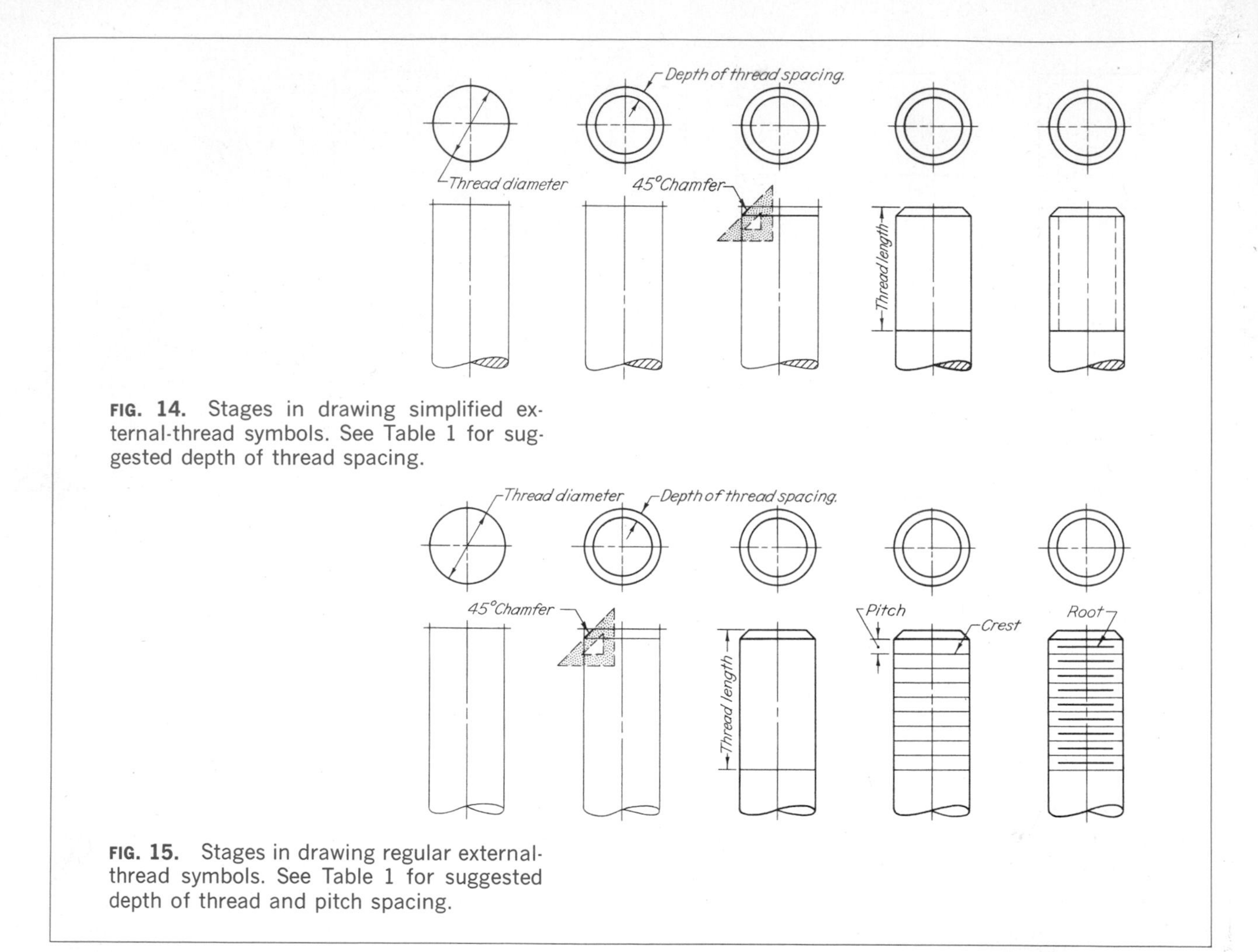


**FIG. 14.** Stages in drawing simplified external-thread symbols. See Table 1 for suggested depth of thread spacing.

**FIG. 15.** Stages in drawing regular external-thread symbols. See Table 1 for suggested depth of thread and pitch spacing.

## 13. Unified and American (National) Screw Threads.

The Unified thread standards adopted by the United States, Canada, and Great Britain for the bulk of threaded products basically constitutes the American Standard "Unified and American Screw Threads" (ASA B1.1—1960). The form of thread employed has been described in Sec. 5 and is essentially that of the former (1935) American Standard. Threads produced according to the former and present American Standards will interchange. Important differences between the two standards are in the liberalization of manufacturing tolerances, the provision of allowances for most classes of threads, and the changes in thread designations. The new American Standard contains, in addition to the Unified diameter-pitch combinations and thread classes (adopted in common by the three countries), additional diameter-pitch combinations and two thread classes retained from the 1935 standard.

## 14. Thread Series.

Threads are classified in "series" according to the number of threads per inch used with a specific diameter. For example, an American (Unified) thread having 20 threads per inch applied to a $\frac{1}{4}$-in. diameter results in a thread belonging to the coarse-thread series, while one with 28 threads per

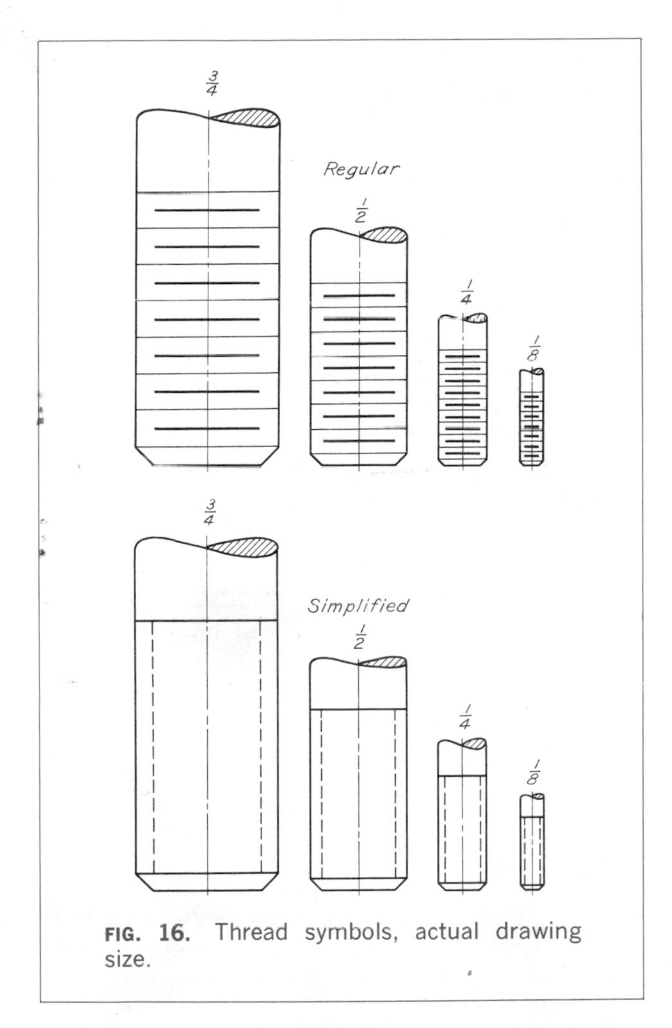


FIG. 16. Thread symbols, actual drawing size.

inch on the same diameter gives a thread belonging to the fine-thread series.

In the United States the thread forms that have been subjected to series standardization by the ASA include the Unified or American National, the Acme, Stub Acme, pipe threads, buttress, and the knuckle thread as used on electric sockets and lamp bases. Except for the Unified, or American National, only one series is provided for each of the thread forms so standardized (see Appendix for Acme and Stub Acme threads and for pipe threads).

The American Standard "Unified and American (National) Screw Threads" covers six series of screw threads and, in addition, certain other preferred special diameter-pitch combinations. In the descrip-

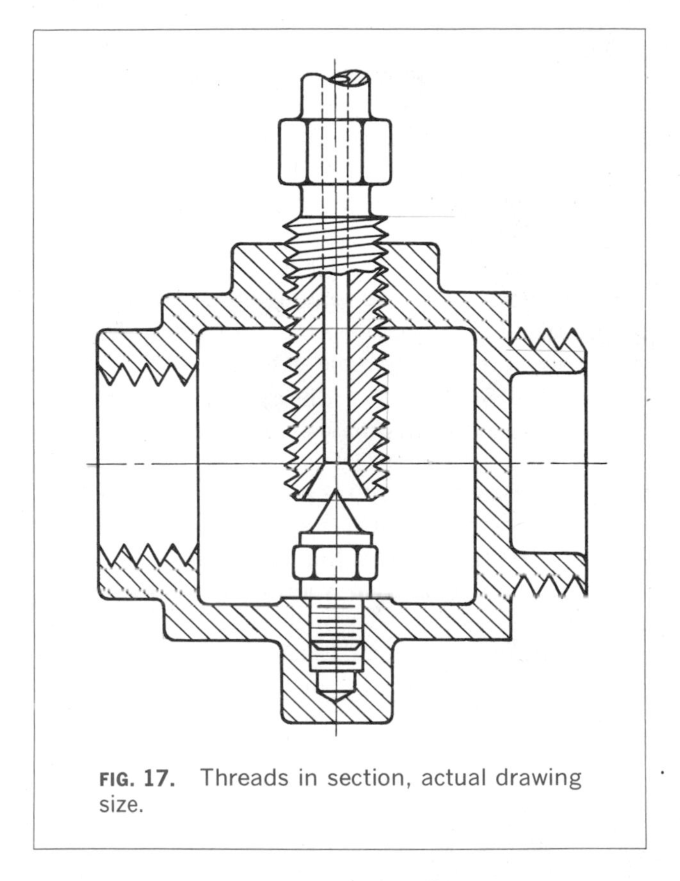

FIG. 17. Threads in section, actual drawing size.

tions of the series which follow, the letters "U" and "N" used in the series designations stand for the words "Unified" and "National (form)," respectively.

The *coarse-thread series,* designated "UNC" or "NC," is recommended for general use where conditions do not require a fine thread.

The *fine-thread series,* designated "UNF" or "NF," is recommended for general use in automotive and aircraft work and where special conditions require a fine thread.

The *extra-fine-thread series,* designated "UNEF" or "NEF," is used particularly in aircraft work, where an extremely shallow thread or a maximum number of threads within a given length is required.

The *8-thread series,* designated "8N," is a uniform-pitch series using eight threads per inch for all diameters concerned. Bolts for high-pressure pipe flanges, cylinder-head studs, and similar fasteners against pressure require that an initial tension be set up by elastic deformation of the fastener and that the components be held together so that the joint will not open when steam or other pressure is applied. To secure a proper initial tension, it is not practicable that the pitch should increase with the diameter of the thread, as the torque required to assemble would be excessive. Accordingly, for such purposes, the 8-thread series has come into general use in many classes of engineering work and as a substitute for the coarse-thread series.

The *12-thread series,* designated "12UN" or "12N," is a uniform-pitch series using 12 threads per inch for all diameters concerned. Sizes of 12-pitch threads from $\frac{1}{2}$ to $1\frac{3}{4}$ in. in diameter are used in boiler practice, which requires that worn stud holes be retapped with the next larger size. The 12-thread series is also widely used in machine construction for thin nuts on shafts and sleeves and provides continuation of the fine-thread series for diameters larger than $1\frac{1}{2}$ in.

The *16-thread series,* designated "16UN" or "16N," is a uniform-pitch series using 16 threads per inch for all diameters concerned. This series is intended for applications requiring a very fine thread, such as threaded adjusting collars and bearing retaining nuts. It also provides continuation of the extra-fine-thread series for diameters larger than 2 in.

*Special threads,* designated "UN," "UNS," or "NS," as covered in the standards include nonstandard or special combinations of diameter, pitch, and length of engagement.

The diameter-pitch combinations of the above series will be found in the Appendix, where the Unified combinations (combinations common to the standards of the United States, Canada, and Great Britain) are printed in bold type.

## 15. Unified and American Screw-thread Classes.

A class of thread is distinguished by the tolerance and allowance specified for the member threads and is therefore a control of the looseness or tightness of the fit between mating screws and nuts. The Unified and American (National) and Acme are at present the only threads in this country standardized to the extent of providing several thread classes to control the fit.

The classes provided by the American Standard "Unified and American Screw Threads" are classes 1A, 2A, and 3A, applied to *external threads only;* classes 1B, 2B, and 3B, applied to *internal threads only;* and classes 2 and 3, applied to *both external and internal threads.* These classes are achieved through toleranced thread dimensions given in the standards.

*Classes 1A and 1B* are intended for ordnance and other special uses where free assembly and easy production are important. Tolerances and allowance are largest with this class.

*Classes 2A and 2B* are the recognized standards for the bulk of screws, bolts, and nuts produced and are suitable for a wide variety of applications. A moderate allowance provides a minimum clearance between mating threads to minimize galling and seizure.

*Classes 3A and 3B* provide a class where accuracy and closeness of fit are important. No allowance is provided.

*Classes 2 and 3,* each applying to both external and internal threads, have been retained from the former (1935) American Standard. No allowance is provided with either class, and tolerances are in general closer than with the corresponding new classes.

## 16. Unified Threads.

Not all the diameter-pitch combinations listed in the American Standard "Unified and American Standard Screw Threads" appear in the British and Canadian standards. Combinations used in common by the three countries are called Unified threads and are identified by the letter U in the series designation; they are printed in bold type in the Appendix table. Unified threads may employ classes 1A and 1B, 2A and 2B, and 3A and 3B only. When one of these classes is used and the U does not appear in the designation, the thread conforms to the principles on which Unified threads are based.

The Unified and American National screw-thread table (Appendix) indicates the classes for which each series has data tabulated in the standards. Threads of standard diameter-pitch combinations but of a class for which data are not tabulated in the standard are designated UNS if a Unified combination and NS if not.

**17. Acme and Stub Acme Threads.** Acme and Stub Acme threads have been standardized by the ASA in one series of diameter-pitch combinations (see Appendix). In addition, the standard provides for two general applications of Acme threads: general purpose and centralizing. The three thread classes 2G, 3G, and 4G, standardized for general-purpose applications, have clearances on all diameters for free movement. Centralizing Acme threads have a close fit on the major diameter to maintain alignment of the screw and nut and are standardized in five thread classes, 2C, 3C, 4C, 5C, and 6C. The Stub Acme has only one thread class for general usage.

**18. Buttress Threads.** Buttress threads have been standardized in a different fashion from the other types. Thread form and class of fit is completely defined. Thread series is not defined, but a table of recommended thread diameters and associated thread pitches is provided. Since the number of associated thread pitches varies from three to eleven (depending upon the thread diameter), the designation of a thread series would merely add complication. Three classes of fit are standard: class 1 (free), class 2 (medium), and class 3 (close).

**19. Thread Specification.** The orthographic views of a thread are necessary in order to locate the position of the thread dimensionally on the part. In addition, the views describe whether the thread is external or internal. All other information, called the "specification," is normally conveyed by means of a note or dimensions and a note. In addition to appearing on drawings, the specification may be needed in correspondence, on stock and parts lists, etc.

Features of a thread on which information is essential for manufacture are form, nominal (major) diameter, threads per inch, and thread class or toleranced dimensions. In addition, if the thread is left-hand, the letters LH must be included in the specification; also, if the thread is other than single, its multiplicity must be indicated.

In general, threads other than the Unified and American National, Acme, Stub Acme, and buttress require toleranced dimensions to control the fit (Fig. 18).

*Unified and American National threads* are specified completely by note. The form of the specification always follows the same order: The nominal size is given first, then the number of threads per inch and the series designation (UNC, NC, etc.), and then the thread class (Fig. 18*a* to *c*). If the thread is left-hand, the letters LH follow the class (Fig. 18*a*). Examples:

| | |
|---|---|
| ¼-20UNC-3A | ¼-20NC-2 |
| 1-12UNF-2B-LH | 1-20NEF-3 |
| 2-8N-2 | 2-12UN-2A |
| 2-16UN-2B | 2-6NS-2A |

Note that, when the Unified-thread classes are employed, the letters A and B indicate whether the thread is external or internal.

*Acme and Stub Acme threads* are also specified by note. The form of the specification follows that of the Unified (Fig. 18*d*). Examples:

1¾-4Acme-2G (general-purpose class 2 Acme; 1¾ in. major diameter, 0.25 in. pitch; single, right-hand)

1-5Acme-4C-LH (centralizing class 4 Acme; 1 in. major diameter, 0.2 in. pitch; single, left-hand)

2½-0.333p-0.666L-Acme-3G (general-purpose class 3 Acme; 2½ in. major diameter, 0.333 in. pitch, 0.666 in. lead; double, right-hand)

¾-6Stub Acme (Stub Acme; ¾ in. major diameter, 0.1667 in. pitch; right-hand)

*Buttress threads* can be completely specified by note but require one additional item of information in addition to diameter, threads per inch, type of thread, and class of fit: the direction of pressure as exerted by the internal member (screw). The

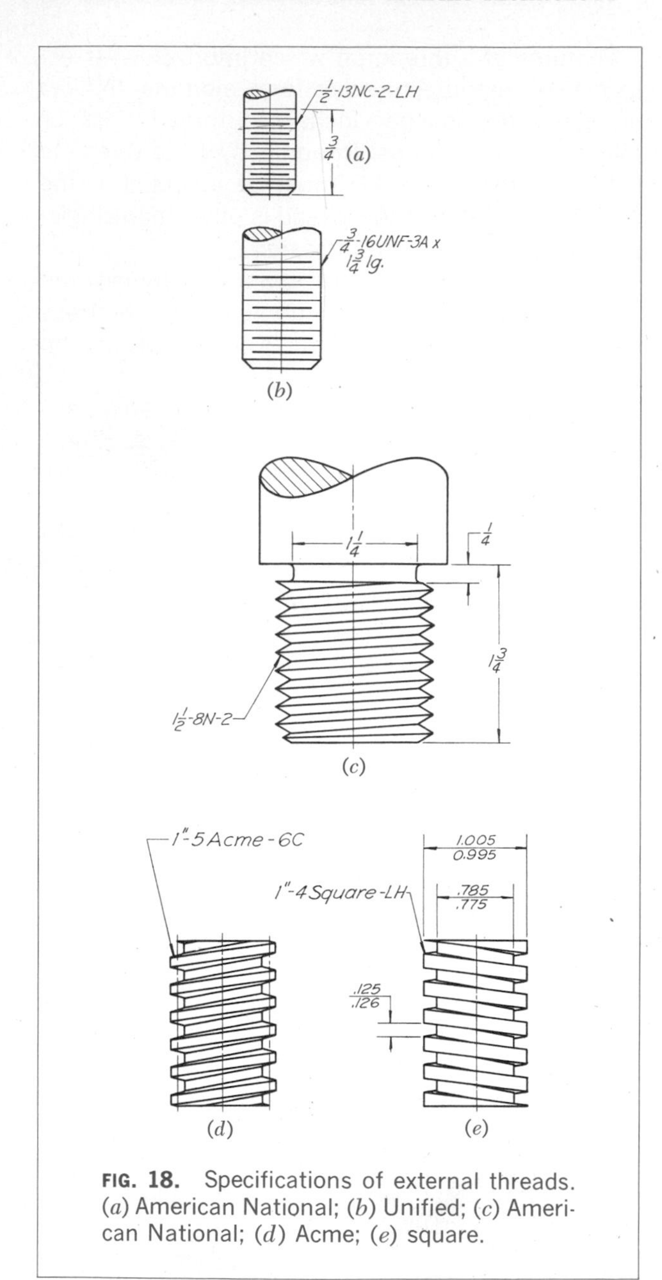


**FIG. 18.** Specifications of external threads. (*a*) American National; (*b*) Unified; (*c*) American National; (*d*) Acme; (*e*) square.

following symbols are recommended for this purpose:

(← screw *pushing* to left (screw pressure flank facing left)
←( screw *pulling* to right (screw pressure flank facing right

Note that the arrow points left but that the parenthesis mark indicates the direction of pressure flank. Where the buttress thread is multiple, the pitch and lead distances should be given instead of threads per inch. Examples:

5/8-20(←N. Butt.-1
4-8←(N. Butt.-2LH
10-0.1p-0.2L(←N. Butt.-1

## 20. Tapped-hole Specifications.

Always specify by note, giving the tapdrill diameter and depth of hole followed by the thread specification and length of thread (Fig. 19). For tapdrill sizes, see the Appendix. It is general commercial practice to use 75 per cent of the theoretical depth of thread for tapped holes. This gives about 95 per cent of the strength of a full thread and is much easier to cut.

## 21. Depth of Tapped Holes and Entrance Length.

For threaded rods, studs, cap screws, machine screws, and similar fasteners, the depth of tapped holes and entrance length may be found by using an empirical formula based on the diameter of the fastener and the material tapped. See Fig. 20 and Table 2.

## 22. Threaded Fasteners.

All engineering products, structures, etc., are composed of separate parts that must be held together by some means of fastening. Compared with permanent methods of fastening, such as welding and riveting, threaded fasteners provide an advantage in that they can be removed, thus allowing disassembly of the parts. As distinguished from other fastening devices such as pins, rivets, and keys, a *threaded* fastener is a cylinder of metal with a screw thread on one end and, usually, a head on the other.

The quantity of threaded fasteners used each year is tremendous. Many varieties are obtainable, some standardized and others special. The standardization of such widely used products results in uniform and interchangeable parts obtainable without complicated and detailed specification and at

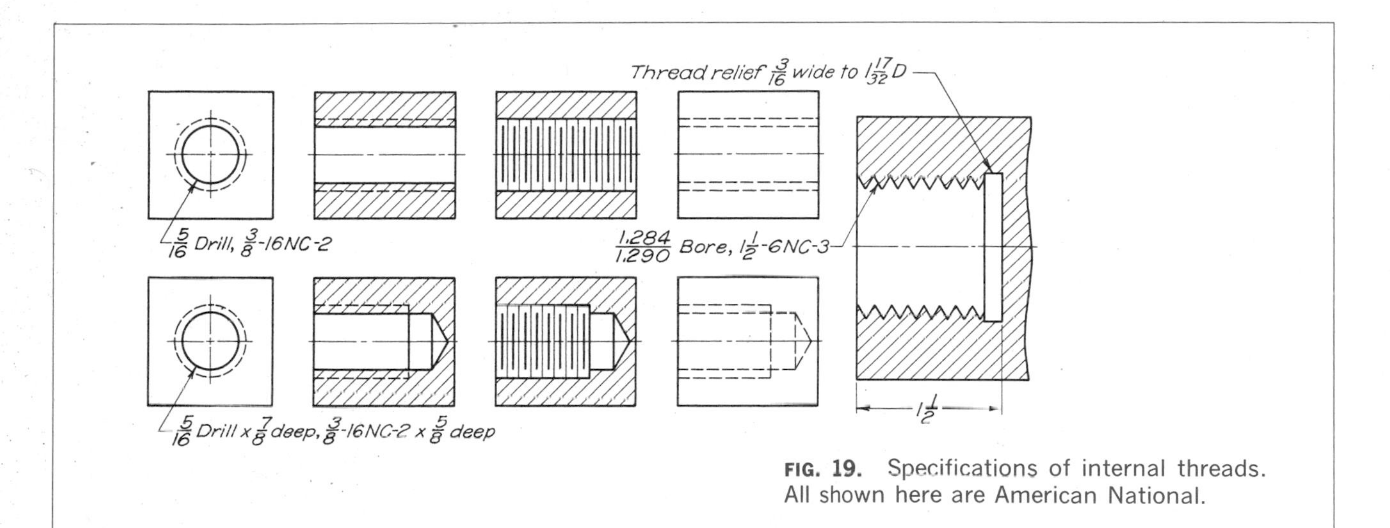


FIG. 19. Specifications of internal threads. All shown here are American National.

**Table 2. Detailed Depths for Drilling and Tapping Holes in Common Materials**

| Material | Entrance length for cap screws, etc., $A$ | Thread clearance at bottom of hole $B$ | Thread depth $C$ | Unthreaded portion at bottom of hole $E$ | Depth of drilled hole $F$ |
|---|---|---|---|---|---|
| Aluminum | $2D$ | $4/n$ | $2D + 4/n$ | $4/n$ | $C + E$ |
| Cast iron | $1\frac{1}{2}D$ | $4/n$ | $1\frac{1}{2}D + 4/n$ | $4/n$ | $C + E$ |
| Brass | $1\frac{1}{2}D$ | $4/n$ | $1\frac{1}{2}D + 4/n$ | $4/n$ | $C + E$ |
| Bronze | $1\frac{1}{2}D$ | $4/n$ | $1\frac{1}{2}D + 4/n$ | $4/n$ | $C + E$ |
| Steel | $D$ | $4/n$ | $D + 4/n$ | $4/n$ | $C + E$ |

$A$ = entrance length for fastener.
$B$ = thread clearance at bottom of hole.
$C$ = total thread depth.
$D$ = diameter of fastener.
$E$ = unthreaded portion at bottom of hole.
$F$ = depth of tap-drill hole.
$n$ = threads per inch.

low cost. Standardized fasteners should be employed wherever possible.

Most fasteners have descriptive names, as the "setscrew," which holds a part in a set, or fixed, position. The bolt derives its name from an early-English use; it was employed as a fastener or pin for bolting a door. Five types—the bolt, stud, cap screw, machine screw, and setscrew—represent the bulk of threaded fasteners.

**23. American Standard Bolts and Nuts.** A *bolt* (Fig. 21*a*), having an integral head on one end and a thread on the other end, is passed through clearance holes in two parts and draws them together by means of a nut screwed on the threaded end.

Two major groups of bolts have been standardized: round-head bolts and wrench-head bolts (sometimes called "machine bolts").

*Round-head bolts* are used as through fasteners with a nut, usually square. Eleven head types have standard proportions and include carriage bolts, step bolts, elevator bolts, and spline bolts. Several head types are intended for wood construction and have square sections, ribs, or fins under the head to prevent the bolts from turning. These bolts are hot or cold formed, with no machining except threading; hence they present a somewhat coarse

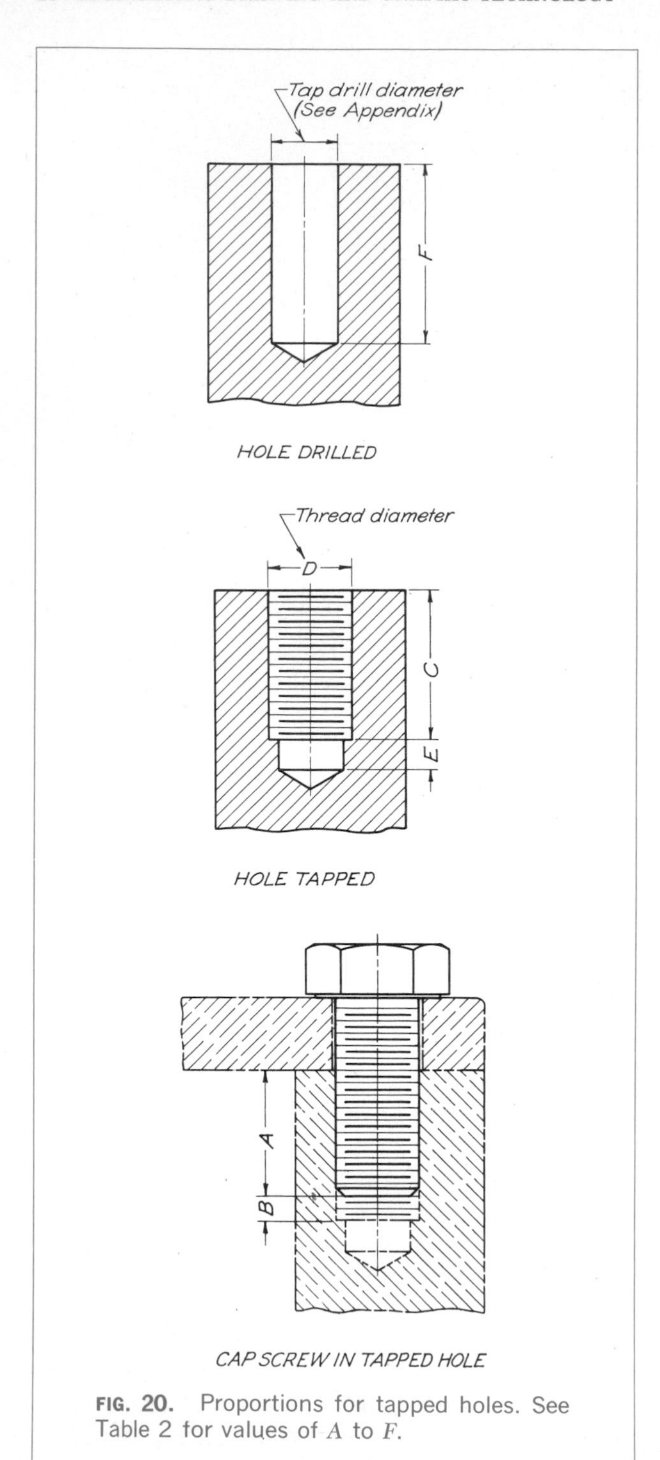


**FIG. 20.** Proportions for tapped holes. See Table 2 for values of $A$ to $F$.

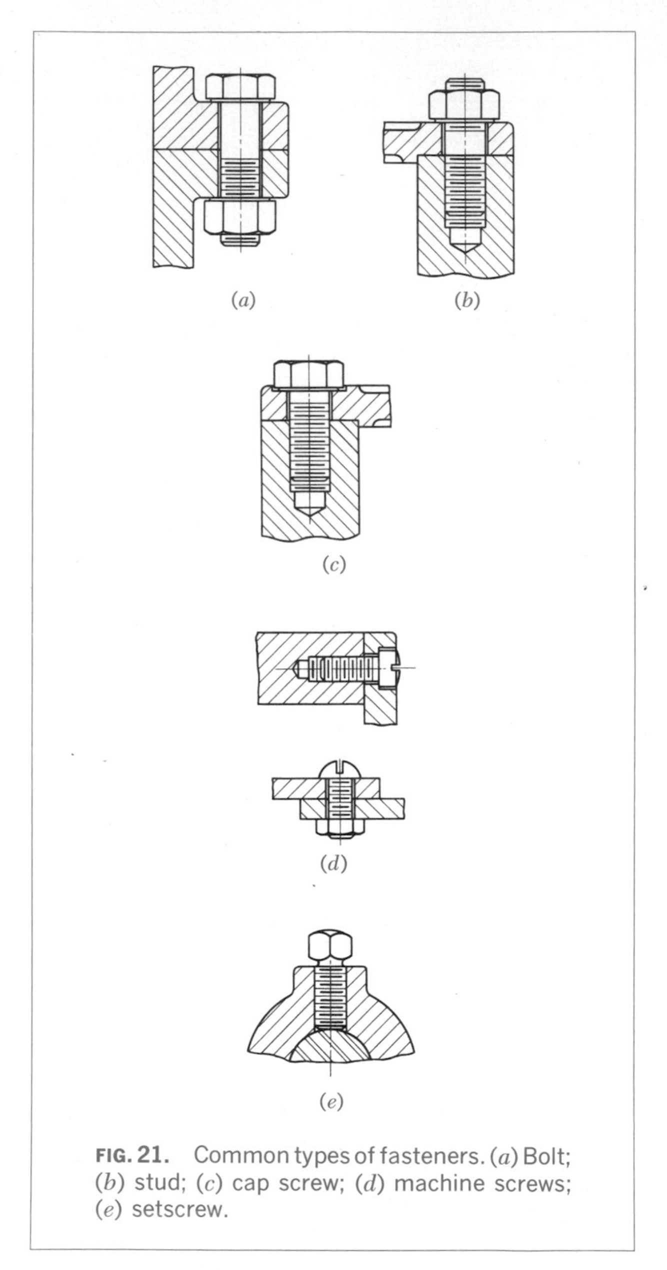


**FIG. 21.** Common types of fasteners. (*a*) Bolt; (*b*) stud; (*c*) cap screw; (*d*) machine screws; (*e*) setscrew.

and irregular appearance. A table in the Appendix shows the various head forms and gives nominal dimensions suitable for drawing purposes.

*Wrench-head (machine) bolts* have two standard head forms: square and hexagonal. Although intended for use as a through fastener with a nut, machine bolts are sometimes used as cap screws. Nuts to match the bolthead form and grade are available, but any nut of correct thread will fit a bolt. Machine bolts vary in grade from coarsely finished products resembling round-head bolts to a well-finished product matching a hexagon cap screw in appearance. Dimensions for drawing purposes are given in the Appendix.

## 24. Studs.

A *stud* (Fig. 21*b*) is a rod threaded on each end. As used normally, the fastener passes through a clearance hole in one piece and screws permanently into a tapped hole in the other. A nut then draws the parts together. The stud (Fig. 22) is used when through bolts are not suitable for parts that must be removed frequently, such as cylinder heads and chest covers. One end is screwed tightly into a tapped hole, and the projecting stud guides the removable piece to position. The end to be screwed permanently into position is called the "stud end"; and the opposite end, the "nut end." The nut end is sometimes identified by rounding instead of chamfering. Studs have not been standardized by the ASA. The length of thread on the stud end is governed by the material tapped, as indicated in Sec. 20. The threads should jam at the top of the hole to prevent the stud from turning out when the nut is removed. The fit of the thread between the stud and tapped hole should be tight.

The length of thread on the nut end should be such that there is no danger of the nut binding before the parts are drawn together. The name "stud bolt" is often applied to a stud used as a through fastener with a nut on each end. The stud, a nonstandard part, is usually described on a detail drawing (Fig. 22*b*). The nut, being a standard part, is described by note on the assembly drawing (Fig. 22*a*) or on the parts list.

## 25. Cap Screws

A *cap screw* (Fig. 21*c*) passes through a clearance hole in one piece and screws into a tapped hole in the other. The head, an integral part of the screw, draws the parts together as the screw enters the tapped hole. Cap screws are used on machine tools

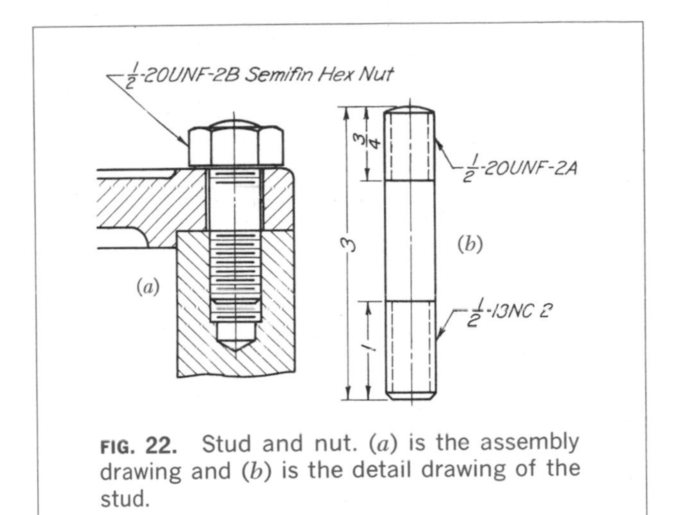


FIG. 22. Stud and nut. (*a*) is the assembly drawing and (*b*) is the detail drawing of the stud.

and other products requiring close dimensions and finished appearances. They are well-finished products; for example, the heads of the slotted- and socket-head screws are machined, and all have chamfered points. The five types of heads shown in Fig. 23 are standard. Detail dimensions and length increments are given in the Appendix.

## 26. Machine Screws.

A *machine screw* (Fig. 21*d*) is a small fastener used with a nut to function in the same manner as a bolt; or without a nut, to function as a cap screw. Machine screws are small fasteners used principally in numbered diameter sizes. The finish is regularly bright, and the material used is commonly steel or brass. The nine standardized head shapes (Fig. 24), except for the hexagon, are available in slotted form as shown or with cross recesses (Fig. 25). The size of the recess varies with the size of the screw. Two recess types occur: intersecting slots with parallel sides converging to a sharp apex at the bottom of the recess (the same driver is used for all size recesses) and large center opening, tapered wings, and a blunt bottom (five sizes of drivers needed).

The hexagon machine screw is not made with a cross recess but may be optionally slotted. Dimensions for drawing machine screw heads will be found in the Appendix.

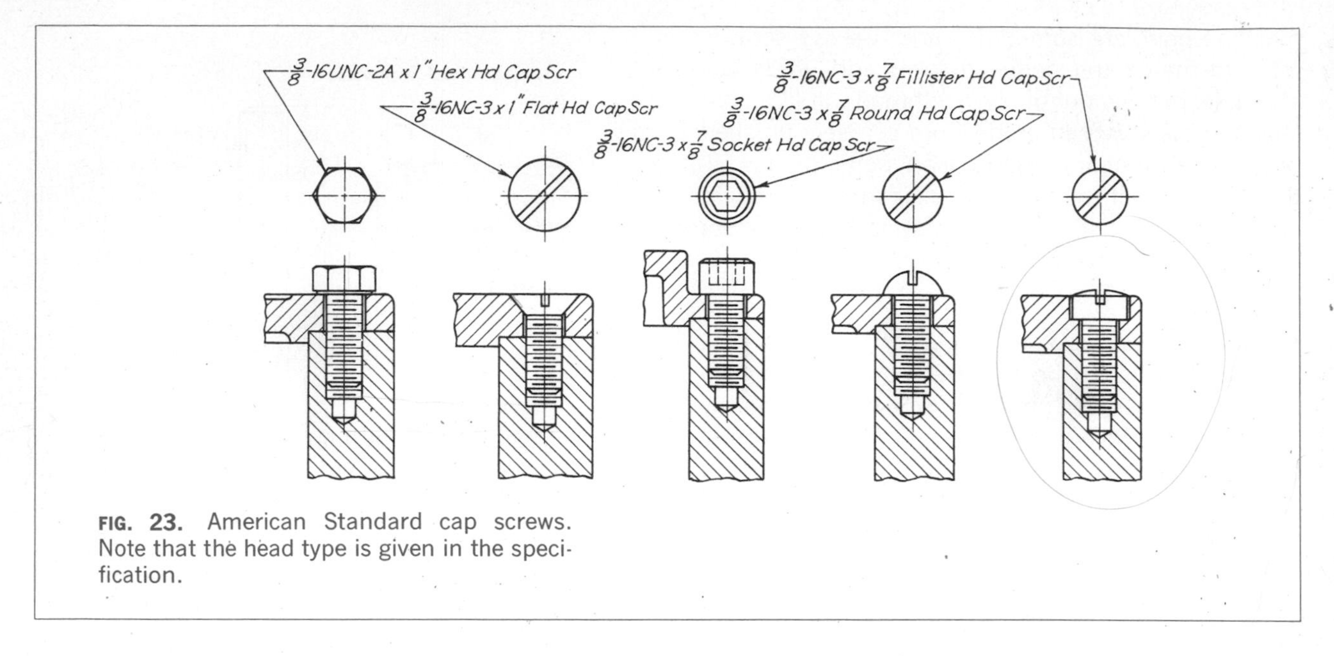


**FIG. 23.** American Standard cap screws. Note that the head type is given in the specification.

## 27. Setscrews.

A *setscrew* (Fig. 21*e*) screws into a tapped hole in an outer part, often a hub, and bears with its point against an inner part, usually a shaft. Setscrews are made of hardened steel and hold two parts in relative position by having the point set against the inner part. The American Standard square-head and headless screws are shown in Fig. 26. Types of points are shown in Fig. 27. Headless setscrews are made to comply with the safety code of factory-inspection laws, which are strict regarding the use of projecting screws on moving parts. Square-head setscrews have head proportions following the formulas in Fig. 26 and can be drawn by using the radii suggested there. A neck or a radius may be used under the head, and the points have the same dimensions as headless screws. Dimensions for headless setscrews are given in the Appendix.

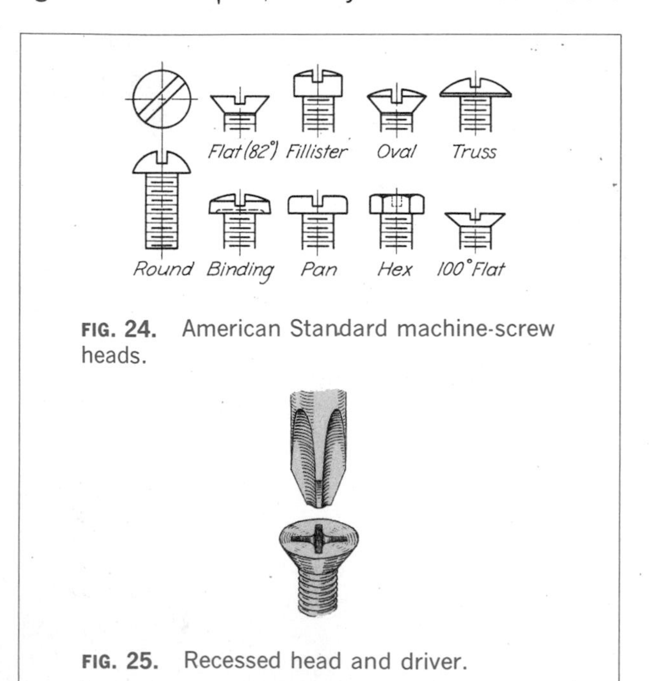


**FIG. 24.** American Standard machine-screw heads.

**FIG. 25.** Recessed head and driver.

## 28. American Standard Nuts.

These are available in two wrench-type styles: square and hexagonal. In addition to the plain form usually associated with bolts, several special-purpose styles are available. Jam, hexagonal slotted, and hexagonal castle nuts are shown in Fig. 28. Machine screw and stove-bolt nuts are made in small sizes only.

## 29. Standard Fasteners.

The ASA Standards provide uniform dimensions and proportions for all fasteners (except studs) listed in the preceding paragraphs. With the exception of

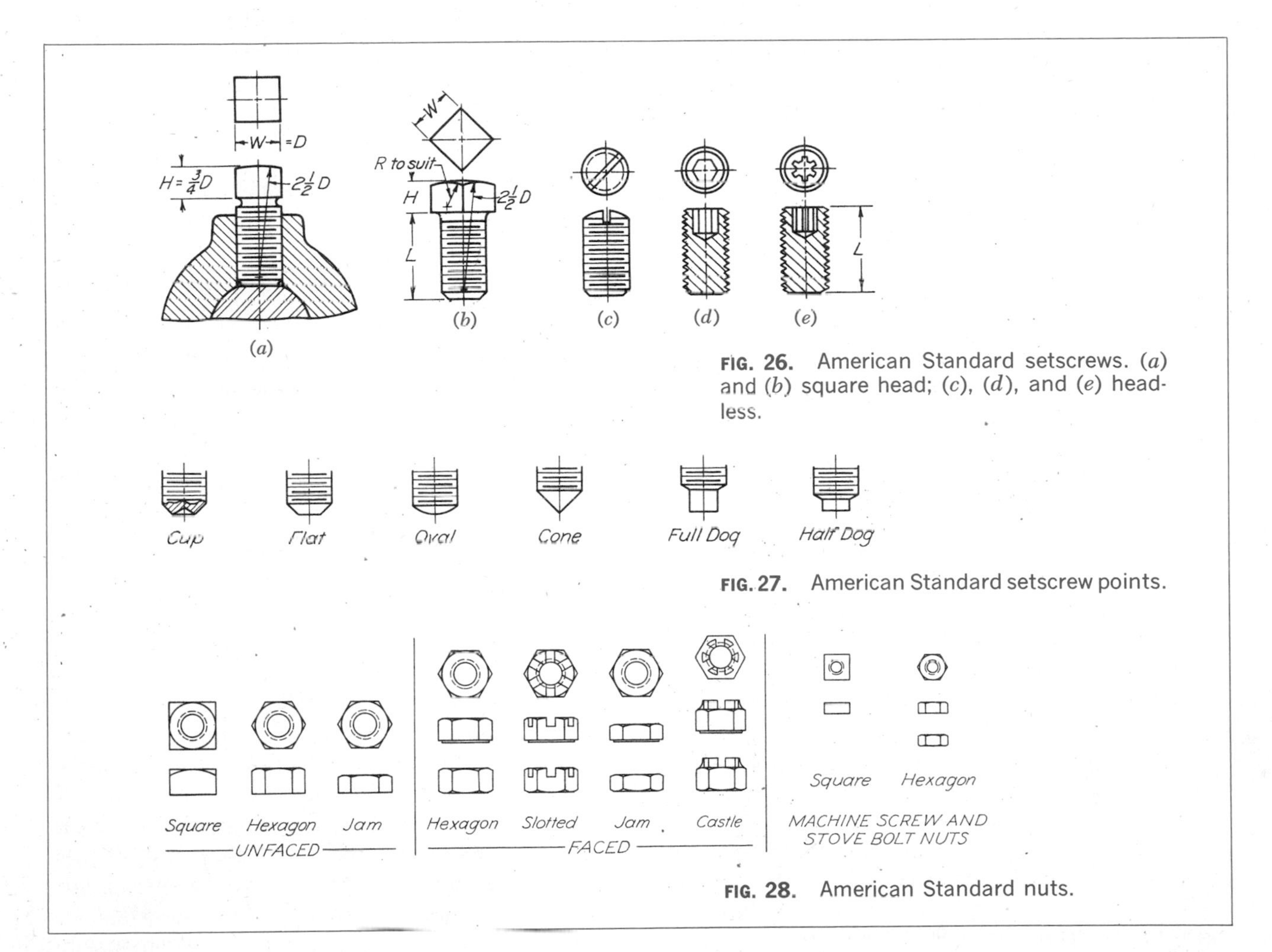


**FIG. 26.** American Standard setscrews. (*a*) and (*b*) square head; (*c*), (*d*), and (*e*) headless.

**FIG. 27.** American Standard setscrew points.

**FIG. 28.** American Standard nuts.

machine bolts, the only options for any of the preceding fasteners are the head style and thread series. Wrench head bolts are available in two head styles, two head weights, three finishes, and three thread series. Table 3 lists the available standard options, and Sec. 30 describes the differences.

**Table 3. Available Types and Thread Details of Standard Fasteners**

| Fastener group | Head styles | Thread series | Thread class |
|---|---|---|---|
| Round-head bolts | All | UNC | 2A |
| Cap screws | Hexagonal | UNC, UNF, 8UN | 2A |
| | Slotted | | 2A |
| | Socket | | 3A |
| Machine screws | All | NC, NF | 2 |
| Setscrews | Socket | UNC, UNF, 8UN, UNC, UNF | 3A |
| | Square | | 2A |
| | Slotted | | 2A |

**Table 3. Available Types and Thread Details of Standard Fasteners (Continued)**

**Wrench-head (Machine) Bolts**

| Finish class | Square | | Hexagonal | | Thread | |
|---|---|---|---|---|---|---|
| | Regular | Heavy | Regular | Heavy | Series | Class |
| Unfaced | X | . . . | X | X | UNC | 2A |
| Semifinished | . . . | . . . | X | X | UNC | 2A |
| Finished | . . . | . . . | X | X | UNC, UNF, 8UN | 2A |

**American Standard Nuts**

| Finish class | Square | | Hexagonal | | | Hexagonal slotted | | |
|---|---|---|---|---|---|---|---|---|
| | Regular | Heavy | Regular | Heavy | Thick | Regular | Heavy | Thick |
| Unfaced | X | X | X | X | | | | |
| Semifinished | . . . | . . . | X | X | . . . | X | X | |
| Finished | . . . | . . . | X | . . . | X | X | . . . | X |

## 30. Fastener Terms.

*Nominal Diameter.* The basic major diameter of the thread.

*Width across Flats—W.* The distance separating parallel sides of the square or hexagonal head or nut, corresponding with the nominal size of the wrench. See table in Appendix for dimension *W*.

*Tops of Boltheads and Nuts.* The tops of heads and nuts are flat with a chamfer to remove the sharp corners. The angle of chamfer with the top surface is 25° for the square form and 30° for the hexagonal form; both are drawn at 30°. The diameter of the top circle is equal to the width across flats.

*Washer Face.* The washer face is a circular boss turned or otherwise formed on the bearing surface of a bolthead or nut to make a smooth surface. The diameter is equal to the width across flats. The thickness is $\frac{1}{64}$ in. for all fasteners. A circular bearing surface can be obtained on a nut by chamfering the corners. The angle of chamfer with the bearing face is 30°; the diameter of the circle is equal to the width across flats.

*Fastener Length.* The nominal length is the distance from the bearing surface to the point. For flat-head fasteners and for headless setscrews it is the over-all length.

*Regular Series.* Regular boltheads and nuts are for general use. The dimensions and the resulting strengths are based on the theoretical analysis of the stresses and on results of numerous tests.

*Heavy Series.* Boltheads and nuts in this series are for use where greater bearing surface is necessary. Therefore, for the same nominal size, they are larger in over-all dimensions than regular heads and nuts. They are used where a large clearance between the bolt and hole or a greater wrench-bearing surface is considered essential.

*Thick Nuts.* These have the same dimensions as regular nuts, except that they are higher.

*Class of Finish. Unfaced bolts and nuts* are not machined on any surface except the threads. The bearing surface is plain. Dimensional tolerances are as large as practicable.

*Semifinished Boltheads and Nuts.* These have a smooth bearing surface machined or formed at right angles to the axis (see Fig. 29). For boltheads, this is a washer face; and for nuts, a washer face or a circular bearing surface produced by chamfering the corners. Dimensional tolerances of the fastener are otherwise the same as for the unfaced group.

*Finished Bolts and Nuts.* These differ from the semifinished in two ways: The bearing surface may

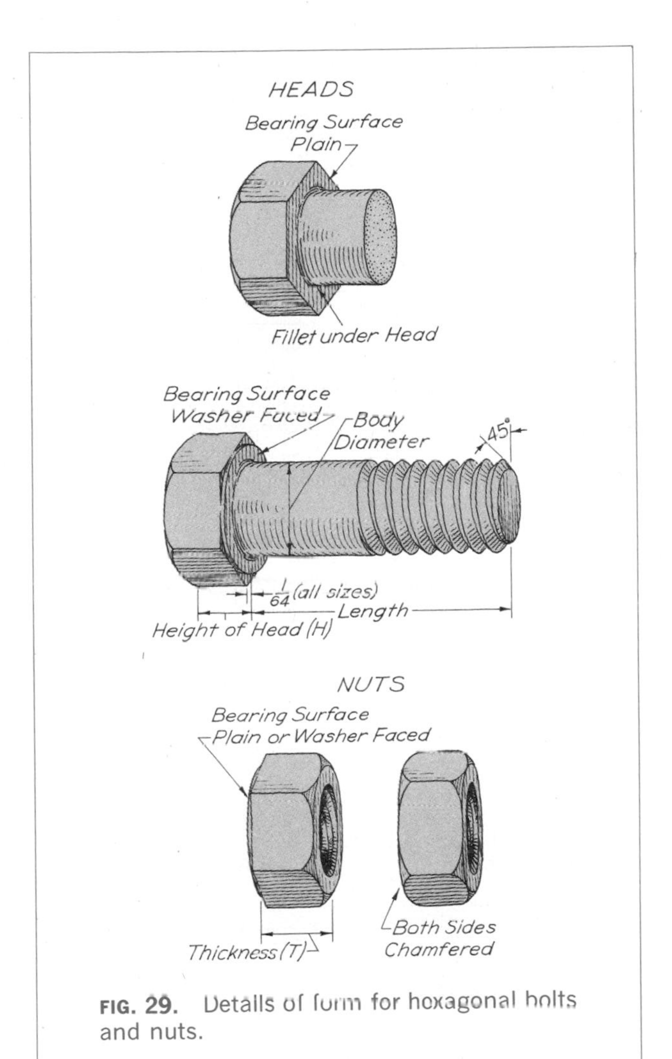


**FIG. 29.** Details of form for hexagonal bolts and nuts.

be washer-faced or produced by double chamfering (see Fig. 29). Dimensional tolerances are smaller than for the semifinished form. Finished hexagonal bolts have dimensional tolerances and workmanship similar to hexagonal cap screws.

## 31. Standard Fastener Specifications.

All standard fasteners may be specified on drawings or parts lists by notes. The following items may be included in a fastener specification in the order listed. Obviously, many of the items will not apply to some fastener types.

1. Thread specification (without length)
2. Fastener length
3. Fastener series (bolts only)
4. Class of finish (bolts only)
5. Material (if other than steel)
6. Head style or form
7. Point type (setscrews only)
8. Fastener group name

Sample specifications follow. Note carefully the machine-bolt notes.

½-13UNC-2A × 3
Reg Sq Bolt
¼-20UNC-2A × 4
Reg Semifin Hex Bolt
⅜-16UNC-2A × 3½
Hvy Semifin Hex Bolt
¾-16UNF-2A × 2½
Finished Hex Bolt (omit Reg)
⅝-11UNC-2A × 4¼
Hvy Finished Hex Bolt
#10-24NC-2 × 2
Brass Slotted Flat Hd Mach Scr
#6-32NC-2 × ½
Recess Fil Hd Mach Scr
¼-20UNC-2A × ¾
Sq Hd Cone Pt Setscrew
½-13UNC-3A × 2
Soc, Flt Pt Setscrew

See Fig. 23 for cap-screw notes.

## 32. The Drawing of Fasteners.

Before drawing a fastener, its *type, nominal diameter,* and *length,* if a bolt or screw, must be known. Knowing the type and diameter, other dimensions can be found in the tables (see Appendix).

## 33. To Draw Square and Hexagon-form Fasteners.

Square and hexagonal heads and nuts are drawn "across corners" in all views showing the faces unless a special reason exists for drawing them "across flats." Figure 30 shows stages in drawing square heads both across corners and across flats, and Fig. 31 shows the same for hexagonal heads.

## End View of Square Bolthead

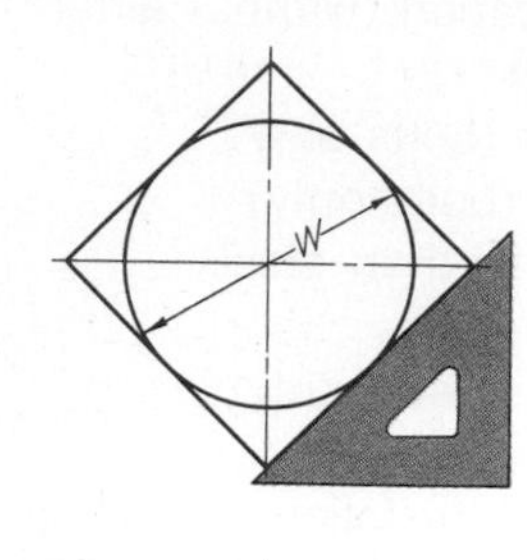


Draw a circle of diameter $W$, and then draw the square with T square and 45° triangle.

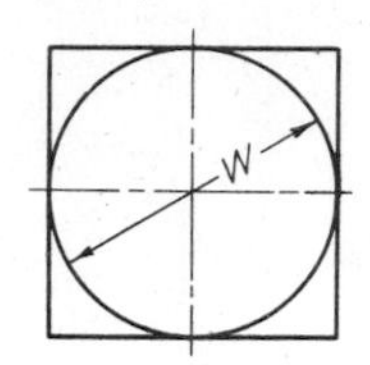


## Face View of Square Bolthead

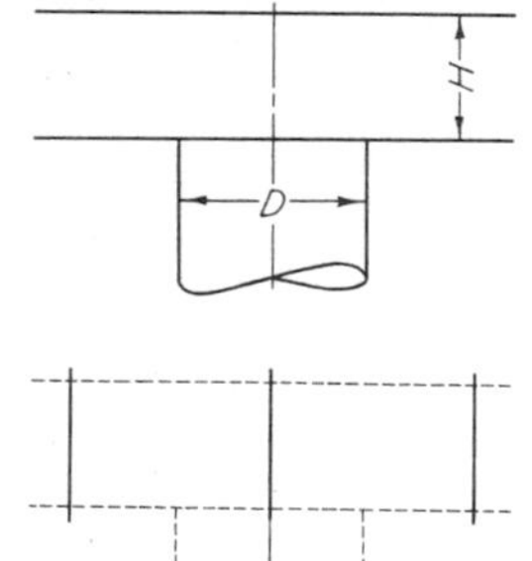


1. Establish the diameter and the height of head.

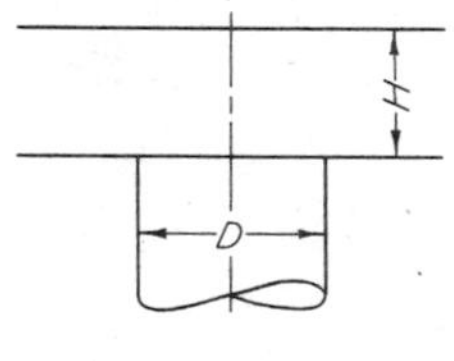


2. Draw, lightly, the vertical edges of the faces, projecting from the end view.

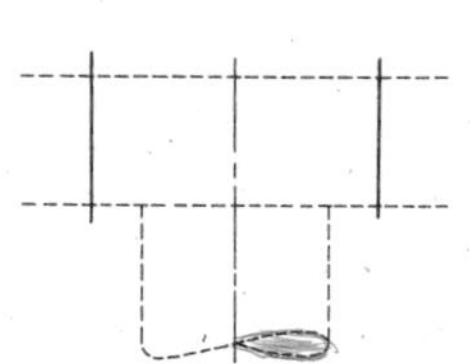

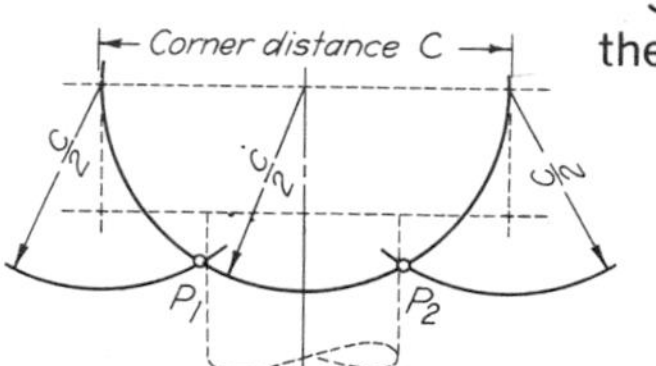


3. Set compass to radius of $C/2$, and draw the circle arcs locating centers $P_1$ and $P_2$.

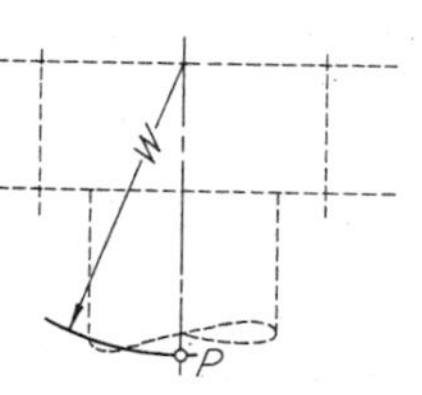


3. Set compass to radius of $W$, and draw the circle arc locating the center $P$.

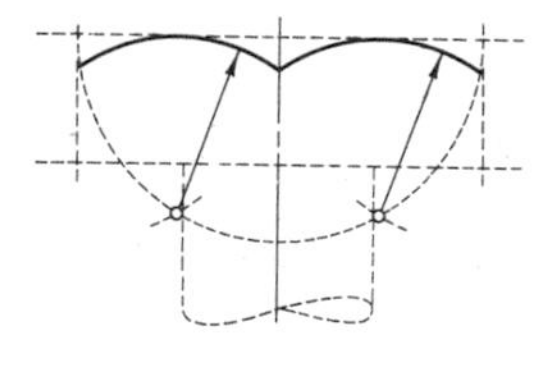

4. Draw chamfer arcs, using radii and centers shown.

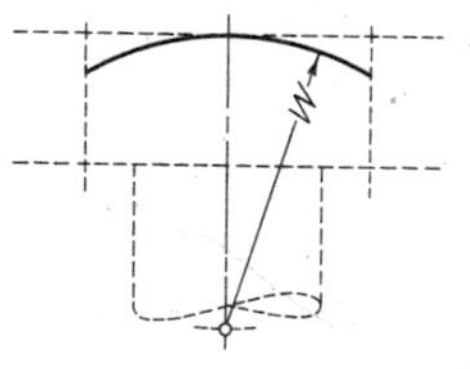


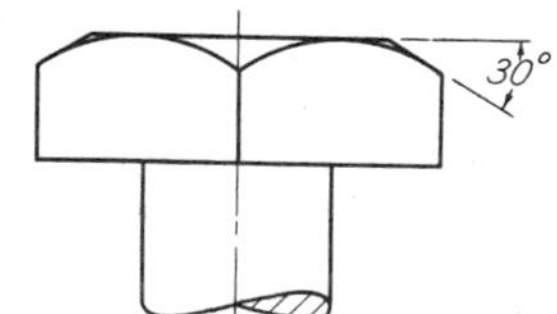


5. Complete the views. Show 30° chamfer on across-corners view.

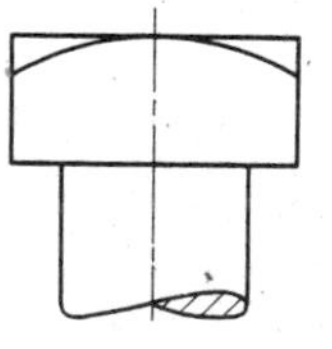

**FIG. 30.** Stages in drawing a square head.

The principles apply equally to the drawing of nuts. The following information must be known: (1) type of head or nut, (2) whether regular or heavy, (3) whether unfinished or semifinished, and (4) the nominal diameter. Using this information, additional data on $W$ (width across flats), $H$ (height of head),

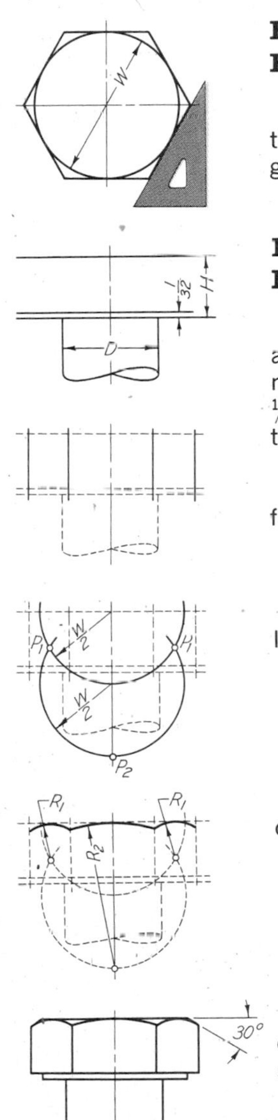

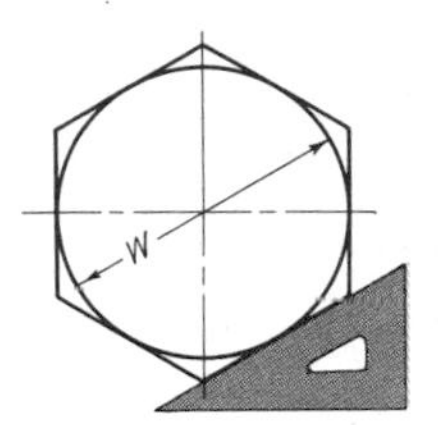

## End View of Hexagonal Bolthead

Draw a circle of diameter $W$, and then draw the hexagon with T square and 30–60° triangle.

## Face View of Hexagonal Bolthead

1. Establish the diameter, height of head, and washer face thickness. The actual thickness of the washer face for all fasteners is $\frac{1}{64}$ in. but may be increased up to $\frac{1}{32}$ in. for the drawing.

2. Draw, lightly, the vertical edges of the faces, projecting from the end view.

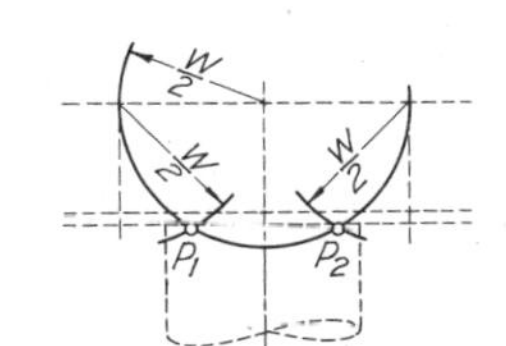

3. With radius of $W/2$ draw the circle arcs locating centers $P_1$ and $P_2$.

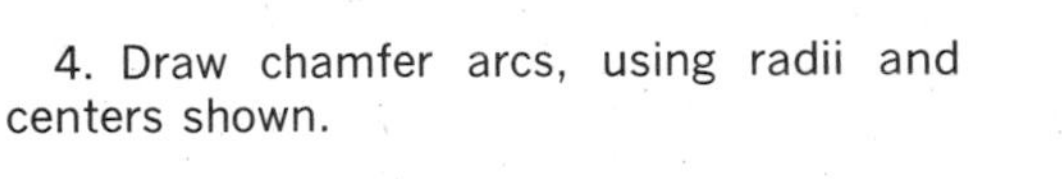

4. Draw chamfer arcs, using radii and centers shown.

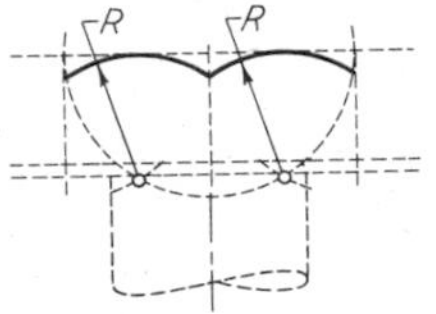

5. Complete the views. Washer-face diameter is equal to $W$. For across-corners view, show 30° chamfer.

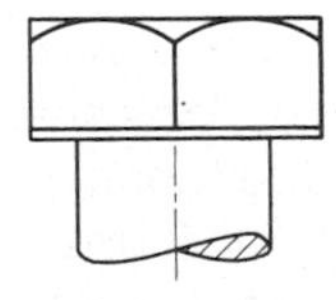

**FIG. 31.** Stages in drawing a hexagonal head.

and $T$ (thickness of nut) are obtained from the tables (see Appendix).

Figure 32 shows a regular semifinished hexagonal bolt and nut, drawn by the method of Fig. 31, showing the head across flats and the nut across corners. The length is selected from the bolt-length tables and the length of thread determined from the footnote to the bolt table (see Appendix). Observe that the washer face shown in Fig. 31 will occur only with semifinished hexagonal fasteners and is sometimes

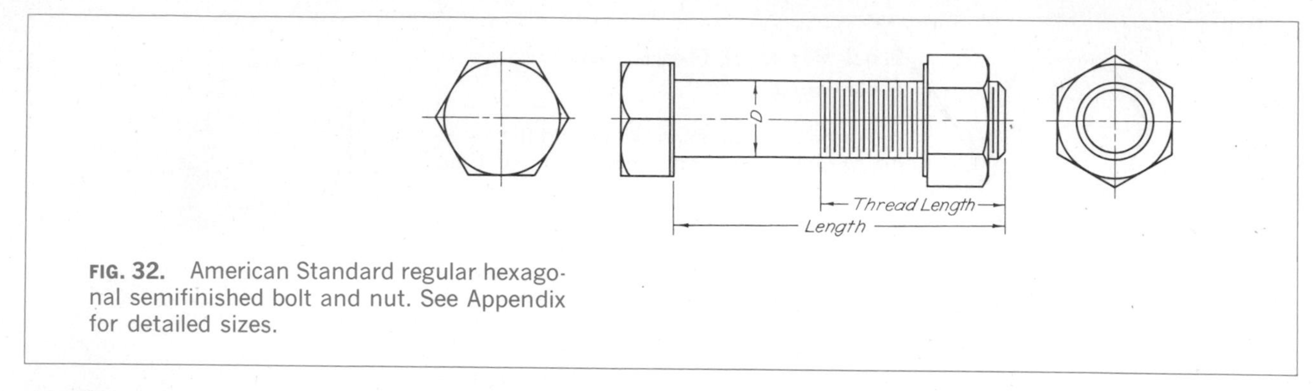


FIG. 32. American Standard regular hexagonal semifinished bolt and nut. See Appendix for detailed sizes.

omitted from the drawings of these. Other views of bolts and nuts are shown in Figs. 33 to 36.

The drawing may require considerable time when, for example, clearance conditions necessitate an accurately drawn fastener, using exact dimensions from tables. Often, however, the representation of fasteners may be approximate or even symbolic because the note specifications invariably accompany and exactly specify them. If an accurate drawing of the fastener is not essential, the $W$, $H$, and $T$ dimensions of hexagonal and square heads and nuts may be obtained from the nominal diameter and the following formulas; the resulting values will be quite close to the actual dimensions. For the regular series, $W = 1\frac{1}{2}D$, $H = \frac{2}{3}D$, and $T = \frac{7}{8}D$; for the heavy series, $W = 1\frac{1}{2}D + \frac{1}{8}$ in., $H = \frac{3}{4}D$, and $T = D$.

A wide variety of *templates* to facilitate the drawing of fasteners is available.

### 34. Specialized Fasteners.

There are a great many special fasteners, a number of which are the developments of the companies that supply a variety of forms. The common types are discussed in the following sections.

### 35. Shoulder Screws (Fig. 37).

These are widely used for holding machine parts together and providing pivots, such as with cams, linkages, etc. They are also used with punch and die sets for attaching stripper plates and are then commonly called "stripper bolts." Threads are coarse series, class 3. Detail dimensions are given in the Appendix.

### 36. Plow Bolts (Fig. 38).

These are used principally in agricultural equipment. The No. 3 and No. 7 heads are recommended for most new work. Threads are coarse series, class 2A. For particulars and proportions, see ASA B18.9—1958.

### 37. American Standard Tapping Screws.

Tapping screws are hardened fasteners that form their own mating internal threads when driven into a hole of the proper size. For certain conditions and materials, these screws give a combination of speed and low production cost, which makes them preferred. Many special types are available. The ASA has standardized head types conforming with all machine-screw heads except the binding and 100° flat heads. For drawing purposes, dimensions of machine-screw heads may be used. Sizes conform, in general, with machine-screw sizes. The ASA provides three types of thread and point combinations (Fig. 39). The threads are 60° with flattened crest and root; types $A$ and $B$ are interrupted threads, and all fasteners are threaded to the head. Full details are given in ASA B18.6.1—1961.

### 38. Fasteners for Wood.

Many forms of threaded fasteners for use in wood are available. Some are illustrated in Fig. 40. Threads are interrupted, 60° form, with a gimlet point. The ASA has standardized lag bolts with square heads and also wood screws with flat, round, and oval heads. Lag boltheads follow the same dimensions and, in general, the same nominal sizes as regular square bolts. For specific information see

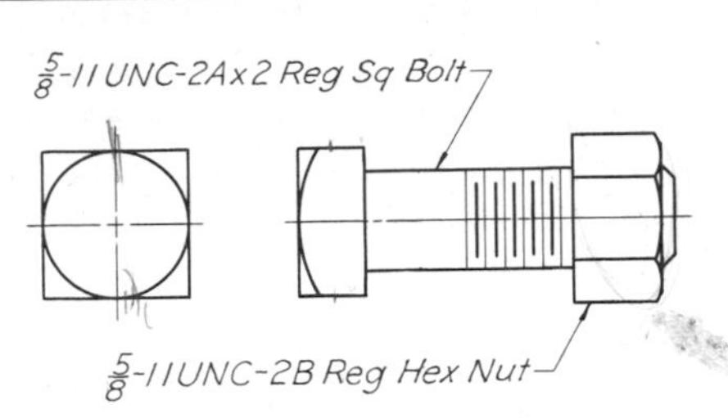


**FIG. 33.** ASA regular square bolt and regular unfaced hexagonal nut. See Appendix for detailed sizes.

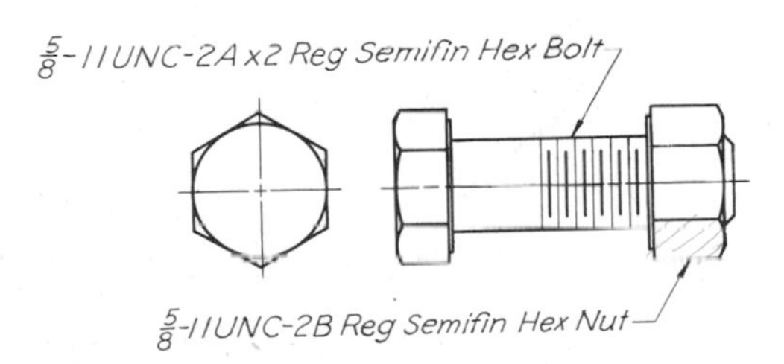


**FIG. 34.** ASA regular semifinished hexagonal bolt and regular semifinished hexagonal nut. See Appendix for detailed sizes.

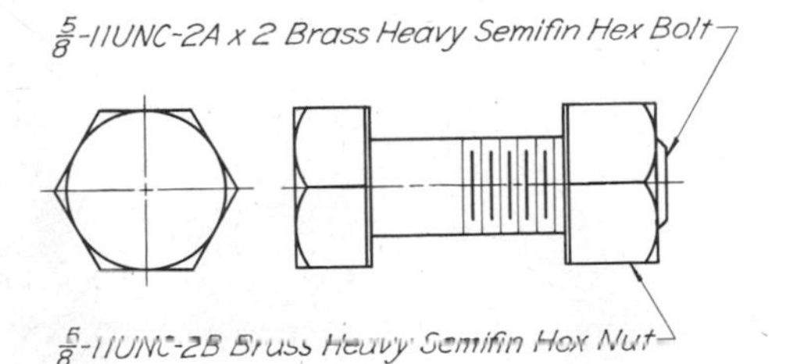


**FIG. 35.** ASA heavy semifinished hexagonal bolt and heavy semifinished hexagonal nut. See Appendix for detailed sizes.

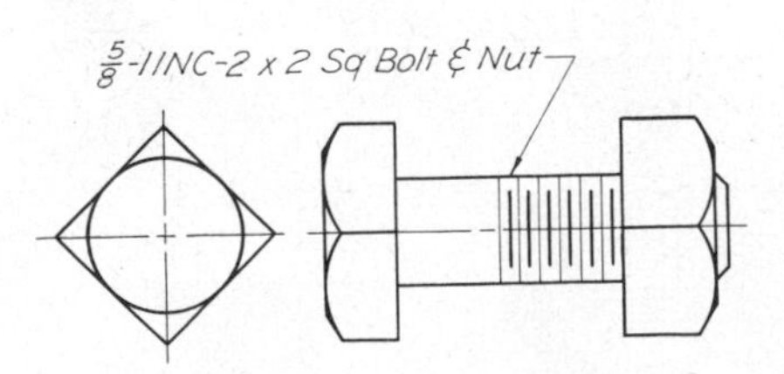


**FIG. 36.** ASA regular unfaced square bolt and nut. See Appendix for detailed sizes.

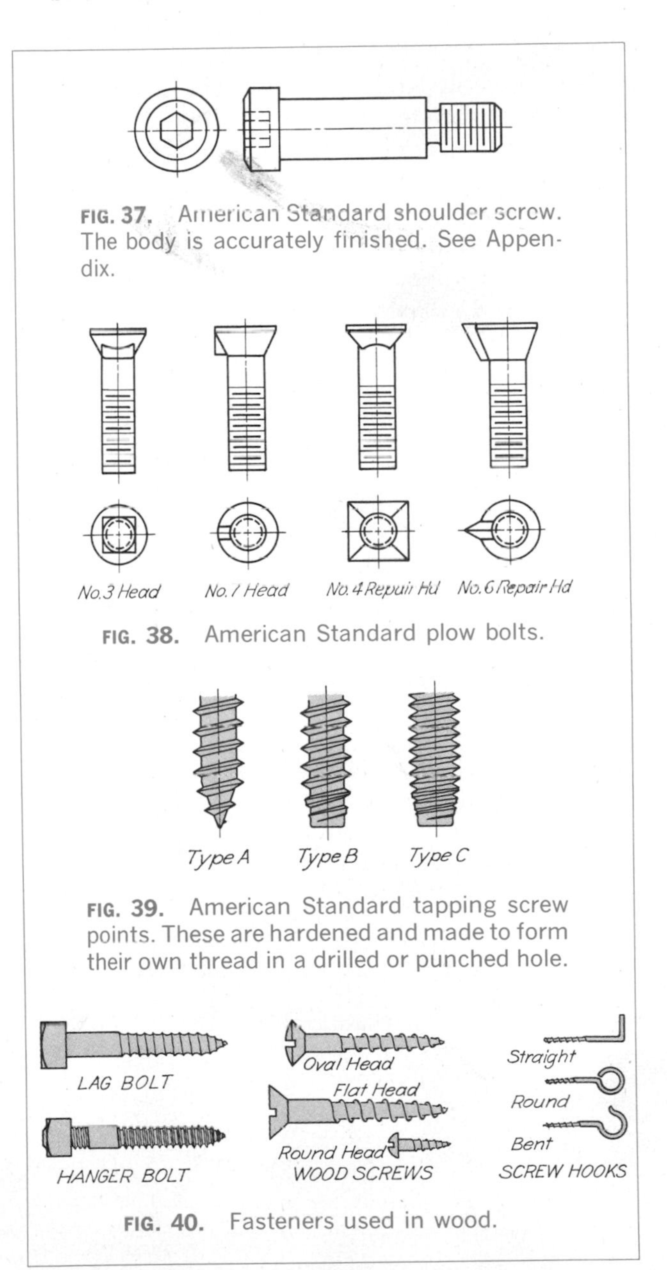


**FIG. 37.** American Standard shoulder screw. The body is accurately finished. See Appendix.

**FIG. 38.** American Standard plow bolts.

**FIG. 39.** American Standard tapping screw points. These are hardened and made to form their own thread in a drilled or punched hole.

**FIG. 40.** Fasteners used in wood.

ASA B18.10—1963. Wood screws follow the same head dimensions and, in general, the same nominal sizes as the corresponding flat-, round-, and oval-head machine screws. The nominal sizes, however, are carried to higher numbers. See table in the Appendix. Like machine screws, wood-screw heads may be plain slotted or have either style of cross recess. Several kinds of finish are available, for example, bright steel, blued, or nickel-plated. Material is usually steel or brass.

**39. Other Forms of Threaded Fasteners.** Many other forms of threaded fasteners, most of which have not been standardized, are in common use. Figure 41 illustrates some of these.

### 40. Lock Nuts and Locking Devices (Fig. 42).

Many different locking devices are used to prevent nuts from working loose. A screw thread holds securely unless the parts are subject to impact and vibration, as in a railroad-track joint or an automobile engine. A common device is the *jam nut* shown at (*A*). *Slotted nuts* (*L*) and *castle nuts* (*M*), to be held with a cotter or wire, are commonly used in auto-

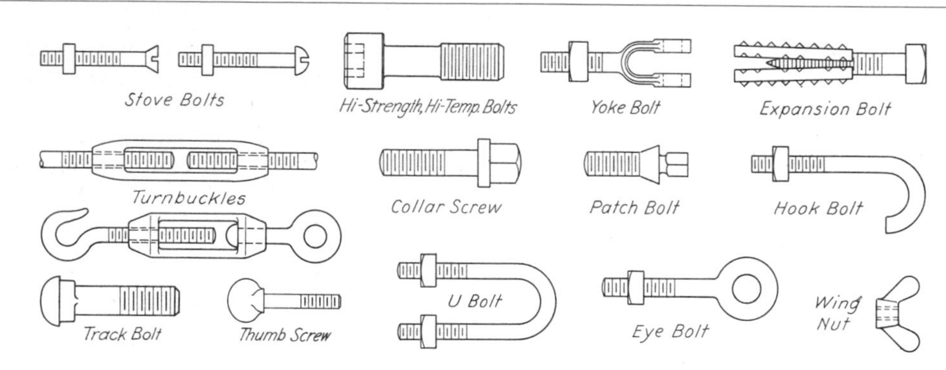


**FIG. 41.** Miscellaneous threaded fasteners. Many special fasteners are available.

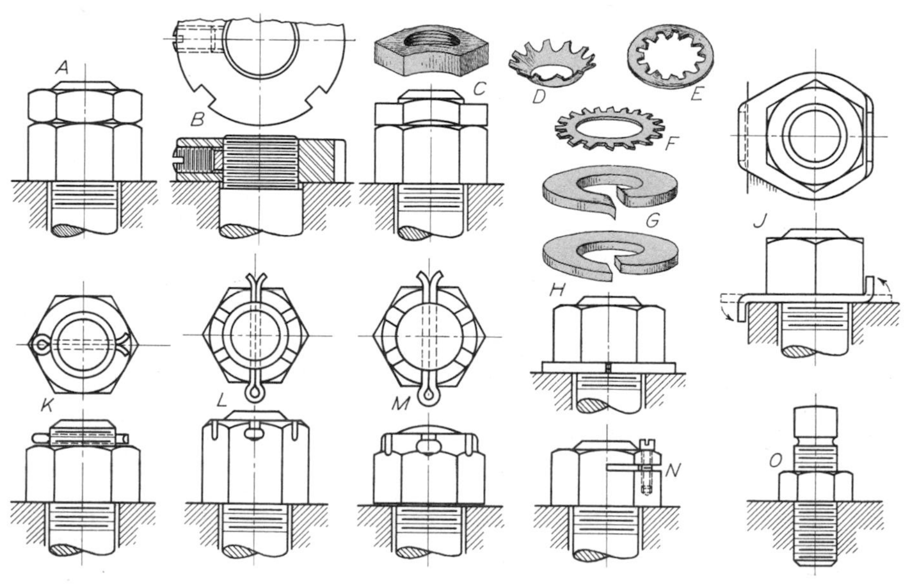


**FIG. 42.** Various locking devices. See Sec. 40.

motive and allied work. For additional information on jam, slotted, and castle nuts, see Sec. 28.

At (*B*) is shown a *round nut* locked by means of a setscrew. A brass plug is placed under the setscrew to prevent damage to the thread. This is a common type of adjusting nut used in machine-tool practice. (*C*) is a *lock nut*, in which the threads are deformed after cutting. Patented *spring washers*, such as are shown at (*D*), (*E*), and (*F*), are common devices. Special patented nuts with plastic or fiber inserts or with distorted threads are in common use as locking devices. The locking action of (*J*), (*K*), (*N*), and (*O*) should be evident from the figure.

### 41. ASA Standard Plain and Lock Washers.

There are four standard ASA spring lock washers: light, medium, heavy, and extra heavy. These are shown in Fig. 42*G* and *H* and are specified by giving nominal diameter and series, for example,

½ Heavy Lock Washer

ASA plain washers are also standardized in four series—light, medium, heavy, and extra heavy—and are specified by giving nominal diameter and series, for example,

$^7/_{16}$ Light Plain Washer

Dimensions of ASA plain and lock washers are given in the Appendix.

### 42. Keys.

In making machine drawings there is frequent occasion for representing key fasteners, used to prevent the rotation of wheels, gears, etc., on their shafts. A key is a piece of metal (Fig. 43) placed so that part of it lies in a groove, called the "key seat," cut in a shaft. The key then extends somewhat above the shaft and fits into a "keyway" cut in a hub. Thus, after assembly, the key is partly in the shaft and partly in the hub, locking the two together so that one cannot rotate without the other.

### 43. Key Types.

The simplest key, geometrically, is the square key, placed *half* in the shaft and *half* in the hub (Fig. 44). A flat key is rectangular in cross section and is used in the same manner as the square key. The gib-head key (Fig. 45) is tapered on its upper surface and is driven in to form a very secure fastening. Both square and flat (parallel and tapered stock) keys have been standardized by the ASA. Tables of standard sizes are given in the Appendix.

The Pratt and Whitney key (Fig. 46) is a variation on the square key. It is rectangular in cross section and has rounded ends. It is placed two-thirds in the shaft and one-third in the hub. The key is proportioned so that the key seat is square and the keyway is half as deep as it is wide. Sizes are given in the Appendix.

Perhaps the most common key is the Woodruff (Fig. 47). This key is a flat segmental disk with a flat (*A*) or round (*B*) bottom. The key seat is semicylindrical and cut to a depth so that *half* the *width* of the key extends above the shaft and into the hub. Tables of dimensions are given in the Appendix. A good basic rule for proportioning a Woodruff key to a given shaft is to have the width of the key one-fourth the diameter of the shaft and its radius equal to the radius of the shaft, selecting the standard key that comes nearest to these proportions. In drawing Woodruff keys, take care to place the center for the arc above the top of the key to a distance equal to one-half the thickness of the saw used in splitting the blank. This amount is given in column *E* of the table in the Appendix.

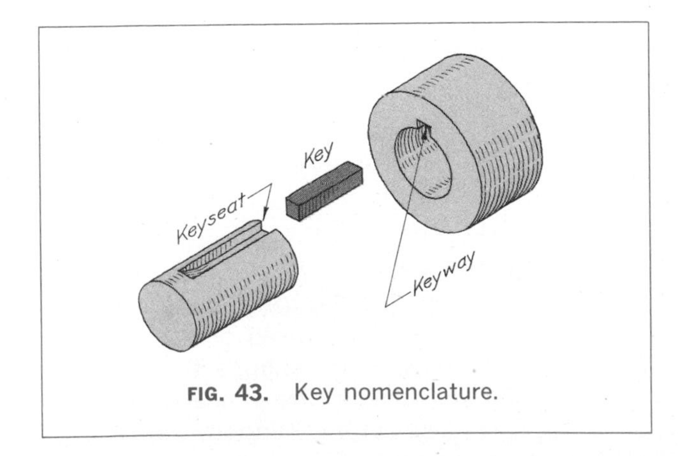


**FIG. 43.** Key nomenclature.

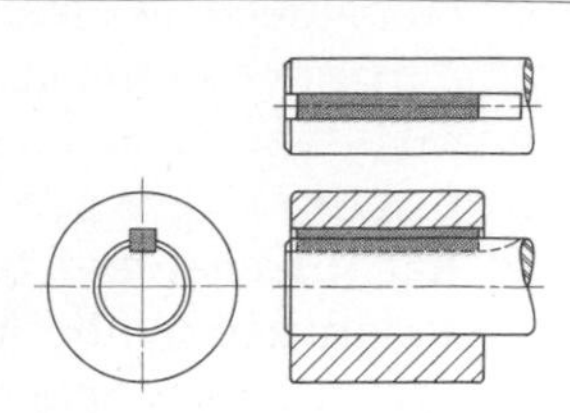

**FIG. 44.** Square (or flat) key. One-half of the key is in the keyway, one-half in the key seat.

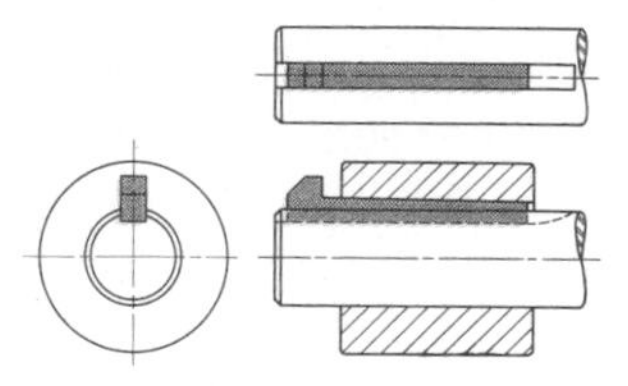

**FIG. 45.** Gib-head key. The head shape provides removal.

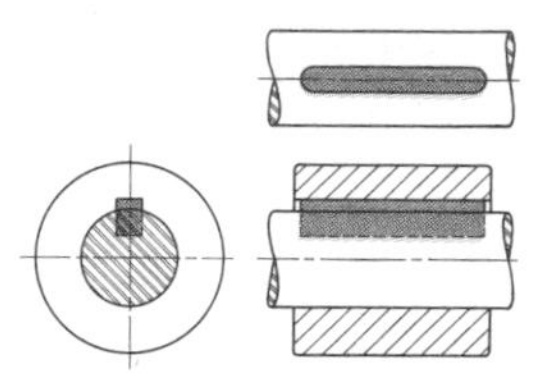

**FIG. 46.** Pratt and Whitney key. The ends of the key are rounded.

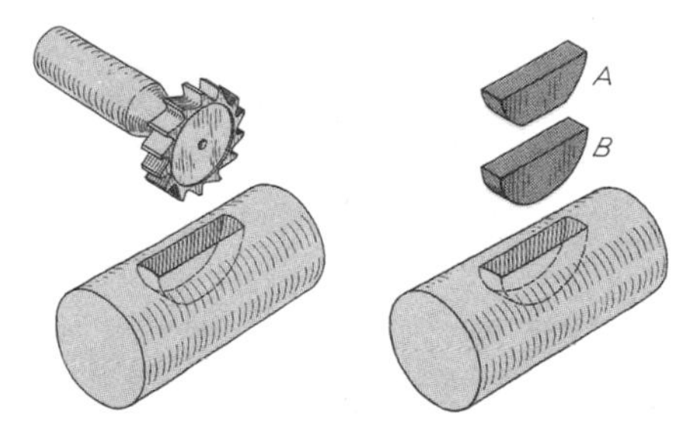


**FIG. 47.** Woodruff keys, cutter, and key seat. This is most widely used.

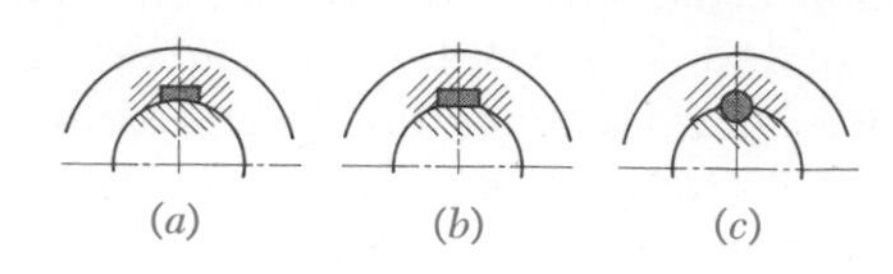


**FIG. 48.** Keys for light duty. (*a*) Saddle; (*b*) flat; (*c*) Nordberg.

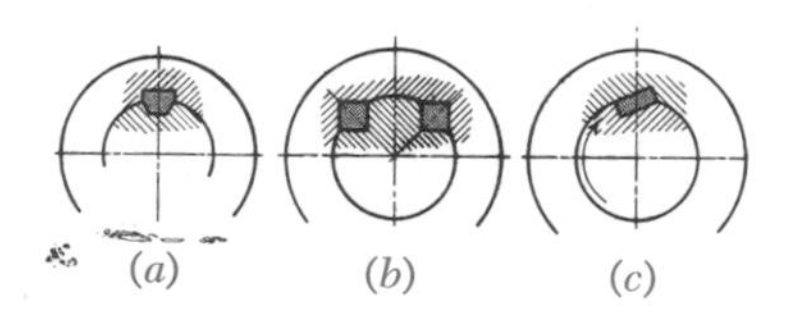


**FIG. 49.** Keys for heavy duty. (*a*) Barth; (*b*) Kennedy; (*c*) Lewis.

Figure 48 shows three keys for light duty: the saddle key (*a*), the flat key (*b*), and the pin, or Nordberg, key (*c*), which is used at the end of a shaft, as, for example, in fastening a handwheel.

Figure 49 shows some forms of heavy-duty keys. (*a*) is the Barth key, (*b*) the Kennedy key, and (*c*) the Lewis key for driving in one direction. In the latter two, the line of shear is on the diagonal.

For very heavy duty, keys are not sufficiently strong and splines (grooves) are cut in both shaft and hub, arranged so that they fit one within the other (Fig. 50). (*a*) and (*b*) are two forms of splines widely used instead of keys; (*b*) is the newer ASA involute spline (B5.15—1960).

## 44. Specification of Keys.

Keys are specified by note or number, depending upon the type.

*Square and flat keys* are specified by a note giving the width, height, and length, for example,

$\frac{1}{2}$ Square Key $2\frac{1}{2}$ Lg
$\frac{1}{2} \times \frac{3}{8}$ Flat Key $2\frac{1}{2}$ Lg

*Plain taper stock keys* are specified by giving the width, the height at the large end, and the length. The height at the large end is measured at the distance *W* (width) from the large end. The taper is 1 to 96 (see Appendix). For example,

$\frac{3}{8} \times \frac{3}{8} \times 1\frac{1}{2}$ Square Plain Taper Key
$\frac{1}{2} \times \frac{3}{8} \times 1\frac{1}{4}$ Flat Plain Taper Key

*Gib-head taper stock keys* are specified by giving, except for name, the same information as for square or flat taper keys (see Appendix), for example,

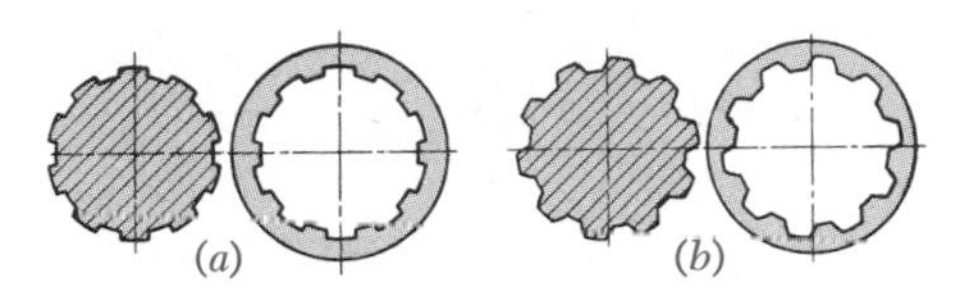


**FIG. 50.** Splined shafts and hubs. (*a*) Straight-radial; (*b*) involute. Both provide strong resistance to torque forces.

$\frac{3}{4} \times \frac{3}{4} \times 2\frac{1}{4}$ Square Gib-head Taper Key
$\frac{7}{8} \times \frac{5}{8} \times 2\frac{1}{2}$ Flat Gib-head Taper Key

*Pratt and Whitney keys* are specified by number or letter (see Appendix), for example,

Pratt and Whitney Key No. 6

*Woodruff keys* are specified by number (see Appendix).

Dimensions and specifications of other key types may be found in handbooks or manufacturers' catalogues.

## 45. Dimensioning Key Seats and Keyways.

The dimensioning of the *seat* and *way* for keys depends upon the purpose for which the drawing is intended. For unit production, when the keys are expected to be fitted by the machinist, nominal dimensions may be given, as in Fig. 51. For quantity production, the limits of width (and depth, if necessary) should be given as in Fig. 52. Pratt and Whitney key seats and keyways are dimensioned as in Fig. 53. Note that the *length* of the key seat is given to correspond with the specification of the key. If interchangeability is important and when careful gaging is necessary, the dimensions should be given as in Fig. 54 and the values expressed with limits.

## 46. Springs.

A spring can be defined as an elastic body designed to store energy when deflected. Springs are classified according to their geometric form: helical or flat.

## 47. Helical Springs.

These are further classified as (1) compression, (2) extension, or (3) torsion, according to the intended action. On working drawings, helical springs are drawn as a single-line convention, as in Fig. 55; or semiconventionally, as in Figs. 56 to 58, by laying

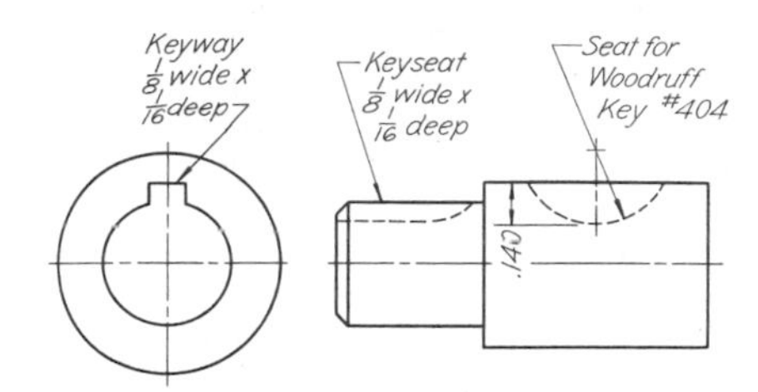


**FIG. 51.** Nominal dimensions of key seats and keyways (square and Woodruff). See Sec. 45.

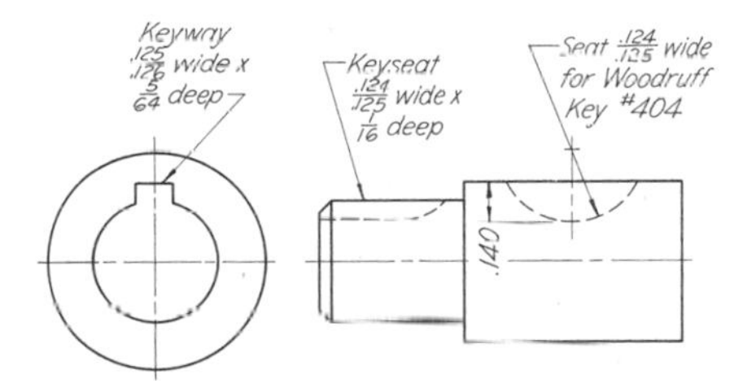


**FIG. 52.** Limit dimensions of key seats and keyways (square and Woodruff). See Sec. 45.

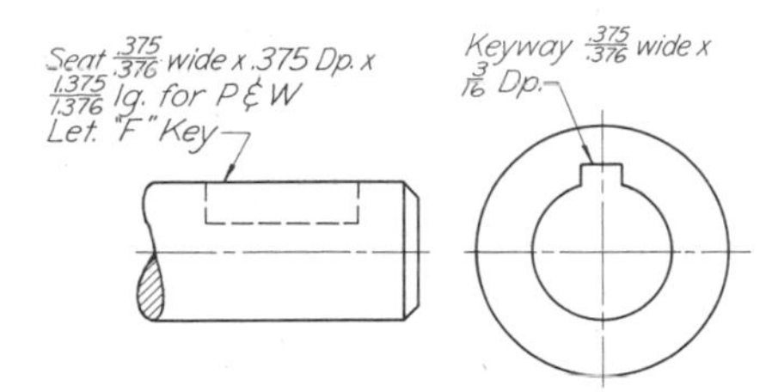


**FIG. 53.** Dimensioning for a Pratt and Whitney key seat and keyway.

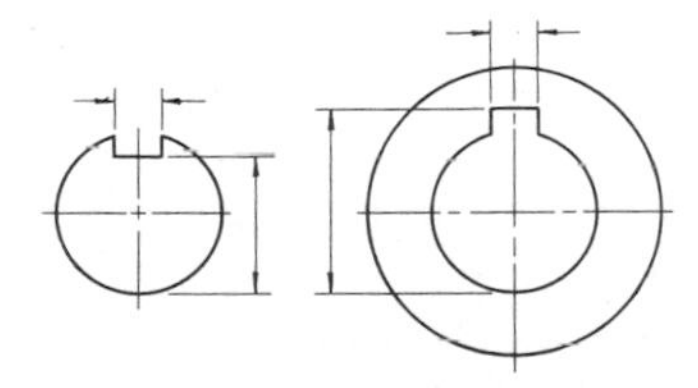

**FIG. 54.** Dimensions of keyway and key seat for interchangeable assembly. These correspond to sizes on gages used.

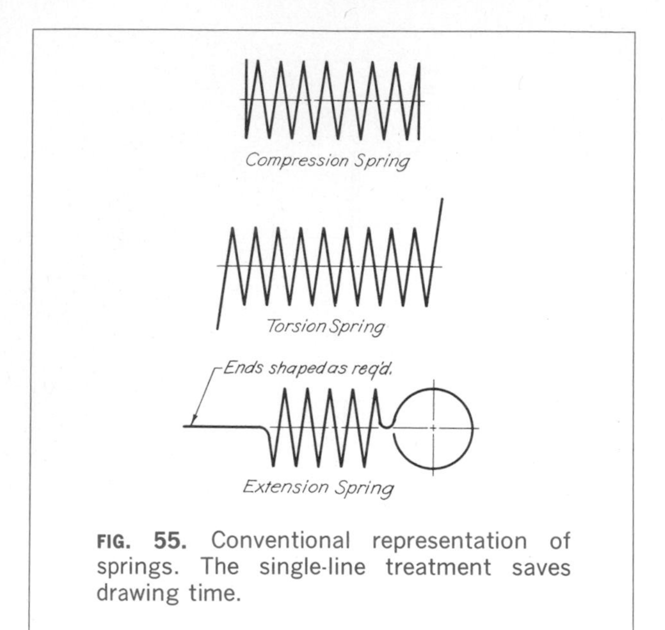


**FIG. 55.** Conventional representation of springs. The single-line treatment saves drawing time.

out the diameter ($D$) and pitch ($P$) of coils and then drawing a construction circle for the wire size at the limiting positions and conventionalizing the helix with straight lines. Helical springs may be wound of round-, square-, or special-section wire.

*Compression springs* are wound with the coils separated so that the unit can be compressed, and the ends may be open or closed and may be left plain or ground, as shown in Fig. 56. The information that must be given for a compression spring is as follows:

1. Controlling diameter: (*a*) outside, (*b*) inside, (*c*) operates inside a tube, or (*d*) operates over a rod
2. Wire or bar size
3. Material (kind and grade)
4. Coils: (*a*) total number and (*b*) right- or left-hand
5. Style of ends
6. Load at deflected length of ___
7. Load rate between ___ inches and ___ inches
8. Maximum solid height
9. Minimum compressed height in use

*Extension springs* are wound with the loops in contact so that the unit can be extended, and the ends are usually made as a loop, as shown in Fig. 57. Special ends are sometimes required and are described by the ASA. The information that must be given for an extension spring is as follows:

1. Free length: (*a*) over-all, (*b*) over coil, or (*c*) inside of hooks
2. Controlling diameter: (*a*) outside diameter, (*b*) inside diameter, or (*c*) operates inside a tube
3. Wire size
4. Material (kind and grade)
5. Coils: (*a*) total number and (*b*) right- or left-hand
6. Style of ends
7. Load at inside hooks
8. Load rate, pounds per 1-in. deflection
9. Maximum extended length

*Torsion springs* are wound with closed or open coils, and the load is applied torsionally (at right angles to the spring axis). The ends may be shaped as hooks or as straight torsion arms, as indicated in Fig. 58. The information that must be given for a torsion spring is as follows:

1. Free length (dimension $A$, Fig. 58)
2. Controlling diameter: (*a*) outside diameter, (*b*) inside diameter, (*c*) operates inside a hole, or (*d*) operates over a rod
3. Wire size
4. Material (kind and grade)
5. Coils: (*a*) total number and (*b*) right- or left-hand
6. Torque, pounds at ___ degrees of deflection
7. Maximum deflection (degrees from free position)
8. Style of ends

## 49. Flat springs.

A flat spring can be defined as any spring made of flat or strip material. Flat springs (Fig. 59) are classified as (1) simple flat springs, formed so that the desired force will be applied when the spring is deflected in a direction opposite to the force; (2) power springs, made as a straight piece and then coiled inside an enclosing case; (3) Belleville springs, stamped of thin material and shaped so as to store energy when deflected; and (4) leaf springs, in elliptic or semielliptic form, made of several pieces of varying length, shaped so as to straighten when a

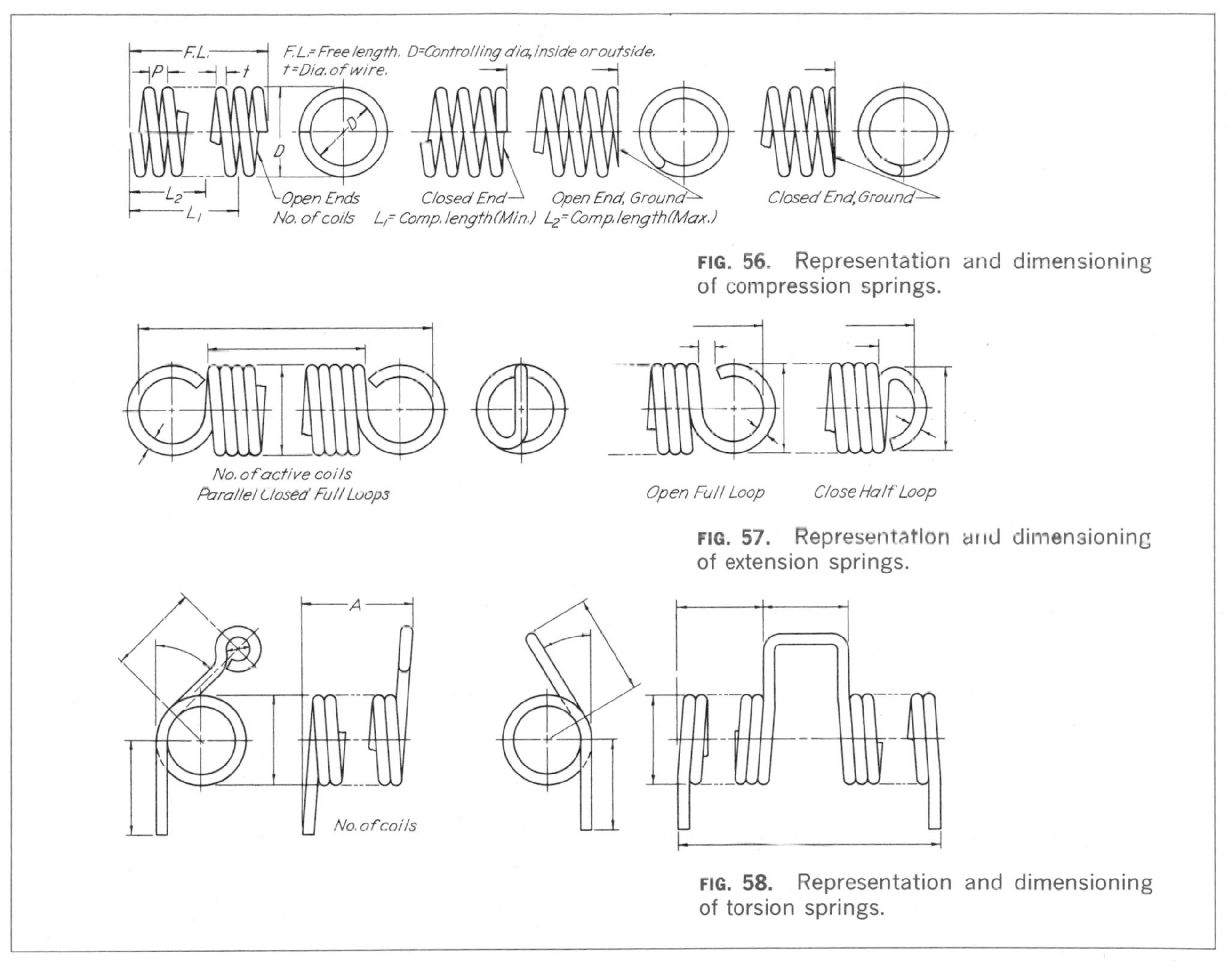


**FIG. 56.** Representation and dimensioning of compression springs.

**FIG. 57.** Representation and dimensioning of extension springs.

**FIG. 58.** Representation and dimensioning of torsion springs.

load is applied. The information that must be given for a flat spring is as follows:

1. Detailed shape and dimensions of the spring shown in a drawing
2. Material and heat-treatment
3. Finish

## 50. Manufacturers' Specialties.

A common problem in all commercial drafting rooms is the design and representation of parts to accommodate manufacturers' specialties such as fasteners, keys, rivets, springs, and ball bearings. Also necessary is the specification of the specialty, including the trade name or other pertinent information such as the size and company number. These are the principal reasons why the following descriptions and accompanying problems are given. There are many industrial companies that manufacture fastening specialities. Many of these are made specifically for the purpose of supplying a low-cost product. Others are designed for speed of assembly. Still others are devised for special uses or to replace standard fasteners, keys, or springs. Most of these designs are not standardized and possibly never will be, either because the manufacturing companies are continually improving the component characteristics

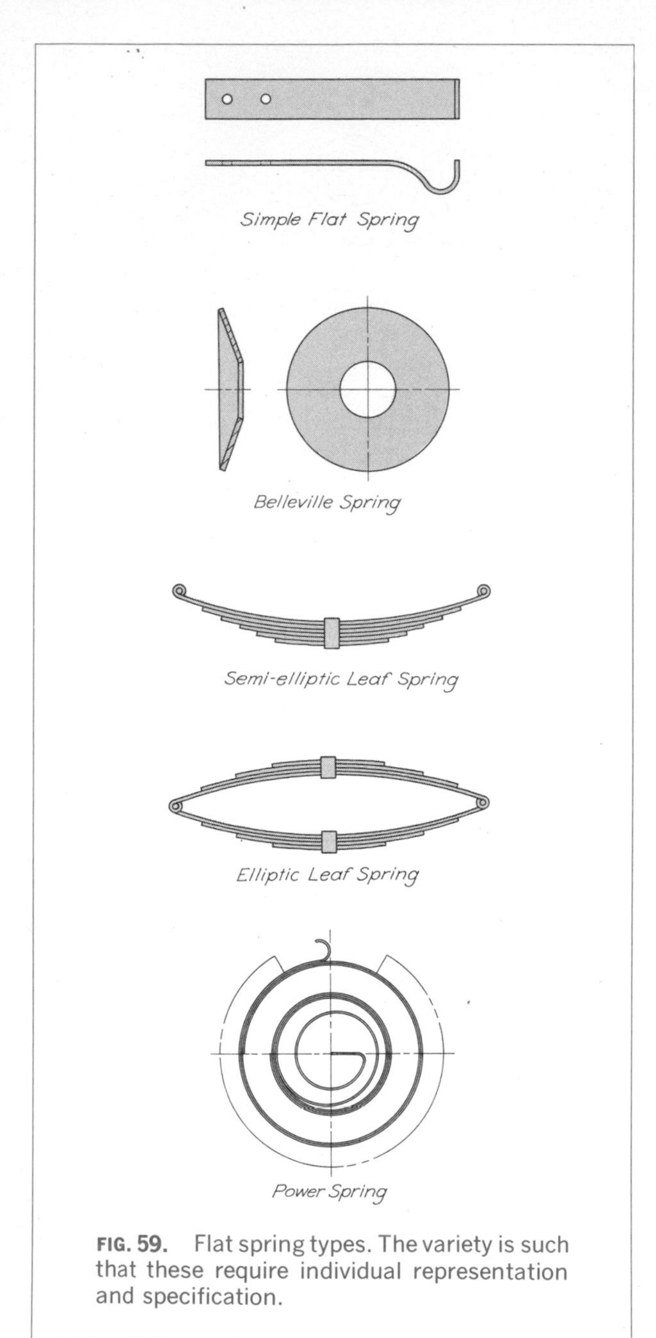


**FIG. 59.** Flat spring types. The variety is such that these require individual representation and specification.

and forms or because they are patented devices.

It is impossible to give complete coverage here to all obtainable products or even to list all varieties made by a particular manufacturer. However, the following paragraphs describe some well-known and widely used parts. National organizations such as the ASME supply comprehensive lists of manufacturers' products, and the advertising in current periodicals is a good source of information. Descriptive literature is obtainable from the manufacturers.

*Self-threading nuts* are inexpensive heat-treated spring-steel parts that are easily assembled on an unthreaded stud, shaft, rod, wire, pin, or tube of suitable dimensions and material. Details of *PALNUTS,* made by The Palnut Company, Mountainside, N.J., are given in the Appendix.

*Push-on fasteners* are spring-steel washerlike parts that are pushed onto an unthreaded stud, shaft, rod, wire, pin, or tube and make a permanent, quick, and inexpensive fastening by means of the gripping action of the specially designed hole in the part. *PUSHNUTS,* also made by The Palnut Company, Mountainside, N.J., are given in the Appendix.

*ROLLPINS* are parts similar to standard straight or tapered pins but are tubular in form with a slotted and chamfered longitudinal opening. They are made of heat-treated spring steel and can be used to replace a grooved pin, key, rivet shaft, cotter pin, setscrew, clevis pin, hinge pin, dowel pin, bolt and nut, rivet, or taper pin. *ROLLPINS,* made by the Elastic Stop Nut Corporation of America, Union, N.J., are given in the Appendix.

*Retaining rings* are tempered spring-steel rings designed as either an external ring to fit into a groove on a shaft or as an internal ring to be fitted into a groove in a hub. Their principal use is to locate or prevent axial movement of a shaft in a hub. They are made in a great variety of forms and sizes. *Waldes Truarc* retaining rings, made by Waldes Kohinoor, Inc., Long Island City, New York, N.Y., are given in the Appendix.

# PROBLEMS

## Group 1. Helices.

**1.** Draw three complete turns of a helix; diameter, 3 in.; pitch, $1\frac{1}{4}$ in. See Chap. 3.

**2.** Draw three complete turns of a conic helix, end and front views, with $1\frac{1}{2}$-in. pitch, whose large diameter is 4 in. and small diameter is $1\frac{1}{2}$ in. See Chap. 3.

## Group 2. Screw Threads.

**3.** Draw in section the following screw-thread forms, 1 in. pitch: American National, Acme, stub Acme, square.

**4.** Draw two views of a square-thread screw and a section of the nut separated; diameter, $2\frac{1}{2}$ in.; pitch, $\frac{3}{4}$ in.; length of screw, 3 in.

**5.** Same as Prob. 4 but for V thread with $\frac{1}{2}$-in. pitch.

**6.** Draw screws 2 in. in diameter and $3\frac{1}{2}$ in. long: single square thread, pitch $\frac{1}{2}$ in.; single V thread, pitch $\frac{1}{4}$ in.; double V thread, pitch $\frac{1}{2}$ in.; left-hand double square thread, pitch $\frac{1}{2}$ in.

**7.** Working space, 7 by $10\frac{1}{2}$ in. Divide space as shown and in left space draw and label thread profiles, as follows: at *A*, sharp V, $\frac{1}{2}$-in. pitch; at *B*, American National (Unified), $\frac{1}{2}$-in. pitch; at *C*, square, 1-in. pitch; at *D*, Acme, 1-in. pitch. Show five threads each at *A* and *B* and three at *C* and *D*.

In right space, complete the views of the object by showing at *E* a $3\frac{1}{2}$-4UNC-2B threaded hole in section. The thread runs out at a $3\frac{17}{32}$-in.-diameter by $\frac{1}{4}$-in.-wide thread relief. The lower line for the thread relief is shown. At *F*, show a $1\frac{3}{4}$-5UNC-3A thread. The thread is 1 in. long and runs out at a neck that is $1\frac{3}{8}$ in. in diameter by $\frac{1}{4}$ in. wide. The free end of the thread should be chamfered at 45°. At *G*, show the shank threaded $\frac{5}{8}$-18UNF-3A by 1 in. long; at *H*, show $\frac{3}{8}$-16UNC-2B through tapped holes in section, four required. Completely specify all threads.

**8.** Working space, 7 by $10\frac{1}{2}$ in. Complete the views and show threaded features as follows: at *A*, a $\frac{1}{2}$-20UNF-3B through tapped hole; at *B*, a $\frac{5}{8}$-11UNC-2B by $\frac{3}{4}$-in.-deep tapped hole with tap drill $1\frac{1}{8}$ in. deep; at *C*, a $\frac{3}{8}$-16UNC-2B by $\frac{1}{4}$-in.-deep tapped hole with the tap drill through to the 1-in. hole; at *D*, a $1\frac{1}{2}$-8N-2 threaded hole running out at the large cored hole. Completely specify all threads.

**9.** Complete the offset support, showing threaded holes as follows: at *A*, $1\frac{3}{8}$-6UNC-2B; at *B*, $\frac{1}{2}$-13NC-2; at *C*, $\frac{5}{8}$-18UNF-3B. *A* and *C* are through holes. *B* is a blind hole to receive a stud. Material is cast iron. Specify the threaded holes.

**10.** Complete the views of the cast-steel rocker, showing threads as follows: on center line *A-A*, 2-8N-2; on center line *B-B*, a $\frac{1}{2}$-20UNF-2B tapped hole, $\frac{5}{8}$ in. deep, with tap drill $\frac{7}{8}$ in. deep; on center line *C-C*, a $2\frac{3}{4}$-12UN-2A external thread; on center line *D-D*, a $\frac{7}{8}$-14UNF-3A external thread.

**11.** Minimum working space, $10\frac{1}{2}$ by 15 in. Complete the views of the objects, and show threads and other details as follows: *upper left*, on center line *A-A*, show in section a $\frac{3}{4}$-10UNC-2B tapped hole, $1\frac{1}{2}$ in. deep, with tap-drill hole $2\frac{1}{2}$ in. deep. At *B*, show a $1\frac{3}{4}$-8N-2 external thread in section. At *C*, show a $\frac{3}{4}$-10UNC-2A thread in section. From *D* on center line *A-A*, show a $\frac{1}{4}$-in. drilled hole extending to the tap-drill hole. At *E*, show six $\frac{1}{4}$-in.

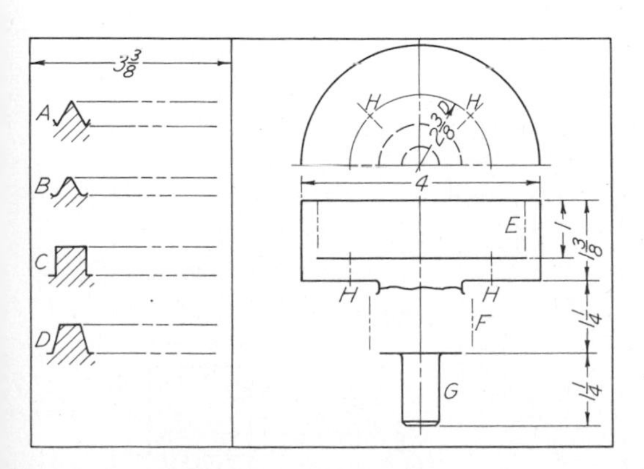


**PROB. 7.** Screw threads.

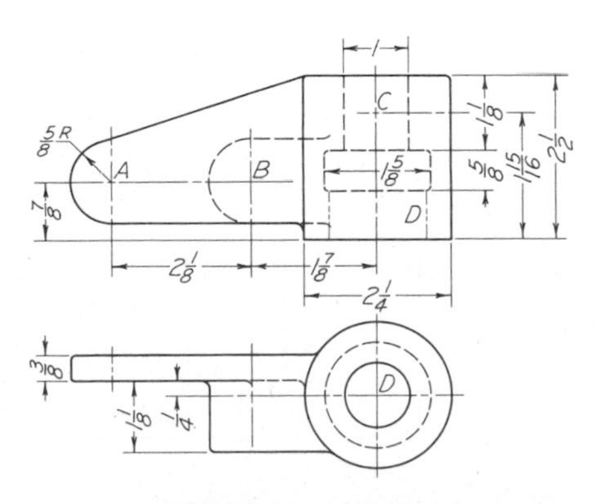


**PROB. 8.** Intermediate lever.

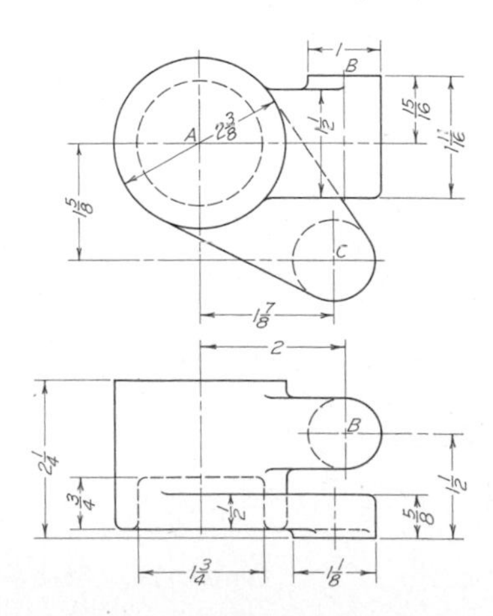


**PROB. 9.** Offset support.

drilled holes, $\frac{1}{4}$ in. deep, equally spaced for spanner wrench. *Upper right,* at *F*, show a $1\frac{1}{2}$-6UNC-2A thread. *Lower left,* at *G*, show (three) $\frac{3}{8}$-16UNC-2B through holes. At *H*, show a $\frac{7}{8}$-9UNC-2B tapped hole, $1\frac{1}{8}$ in. deep, with the tap-drill hole going through the piece. On center line *K-K*, show a $\frac{7}{8}$-14UNF-3B through tapped hole. *Lower center,* at *M*, show a $\frac{5}{8}$-18UNF-3A thread $1\frac{3}{4}$ in. long. *Lower right,* on center line *P*, show a 2-$4\frac{1}{2}$UNC-2B hole in section.

Completely specify all threads.

## Group 3. Threaded Fasteners.

**12.** Draw one view of a regular semifinished hexagonal bolt and nut, across corners; the diameter is 1 in.; the length is 5 in.

**13.** Same as Prob. 12 for a heavy unfinished bolt and nut.

**14.** Same as Prob. 12 for a square bolt and nut.

**15.** Draw four $\frac{1}{2}$- by $1\frac{1}{2}$-in. cap screws, each with a different kind of head. Specify each.

**16.** Show the pieces fastened together on center line *E-E* with a $\frac{3}{4}$-in. hexagonal-head cap screw and light lock washer. On center line *A-A* show a $\frac{5}{8}$- by $1\frac{1}{2}$-in. shoulder screw.

**17.** Show pieces fastened together with a $\frac{3}{4}$-in. square bolt and heavy square nut. Place nut at bottom.

**18.** Fasten pieces together with a $\frac{3}{4}$-in. stud and regular semifinished hexagonal nut.

**19.** Fasten pieces together with a $\frac{3}{4}$-in. fillister-head cap screw.

**20.** Draw one view of a round-head square-neck (carriage) bolt; diameter, 1 in.; length, 5 in.; with regular square nut.

**21.** Draw one view of a round-head rib-neck (carriage) bolt; diameter, $\frac{3}{4}$ in.; length, 4 in.; with regular square nut.

**22.** Draw two views of a shoulder screw; diameter, $\frac{3}{4}$ in.; shoulder length, 5 in.

**23.** Draw two views of a socket-head cap screw; diameter, $1\frac{1}{4}$ in.; length, 6 in.

**24.** Draw the stuffing box and gland, showing the required fasteners. At *A*, show $\frac{1}{2}$-in. hexagonal-head cap screws (six required). At *B*, show $\frac{1}{2}$-in. studs and regular semifinished hexagonal nuts. Specify fasteners.

**25.** Draw the bearing plate, showing the required fasteners. At *C*, show $\frac{1}{2}$-in. regular semifinished hexagonal bolts and nuts (four required). At *D*, show $\frac{1}{2}$-in. socket setscrew. At *E*, show $\frac{1}{2}$-in. square-head setscrew. Setscrews are to have cone points. Specify fasteners.

**26.** Draw the ball-bearing head, showing the required fasteners. At *A*, show $\frac{1}{2}$- by $1\frac{3}{4}$-in. regular semifinished hexagonal bolts and nuts (six required), with heads to

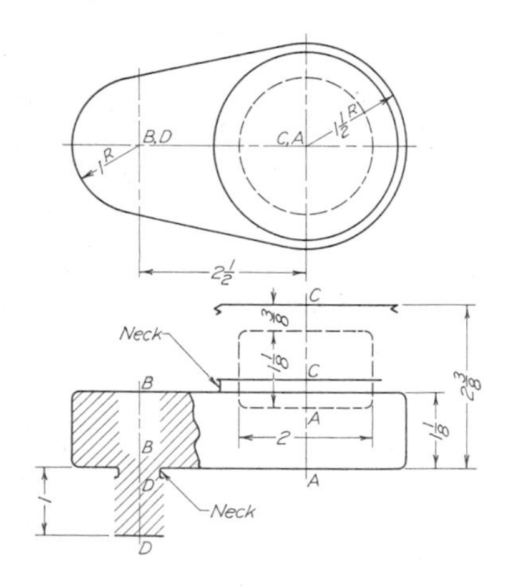


**PROB. 10.** Rocker.

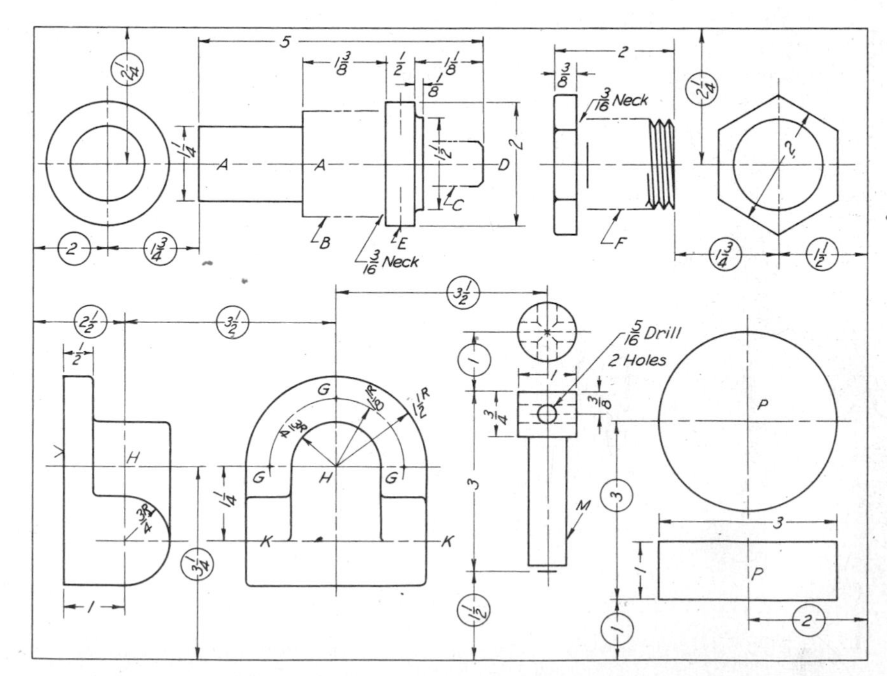


**PROB. 11.** Screw threads.

left and across flats. Note that this design prevents the heads from turning. At *B*, show 5/16- by 3/4-in. fillister-head cap screws (four required). At *C*, show a 3/8-by 1/2-in. slotted flat-point setscrew with fiber disk to protect threads of spindle. Specify fasteners.

**27.** Draw the plain bearing head, showing the required fasteners. At *D*, show 1/2- by 2-in. studs and regular semifinished hexagonal nuts (six required); spot-face 1 in. diameter by 1/16 in. deep. At *E*, show 3/8- by 1-in. hexagonal-head cap screws (four required). At *F*, show a 7/16- by 7/8-in. square-head cup-point setscrew. At *C*, show a 1/8-27NPT hole with a pipe plug. Show the 11/32-in. tap drill through for gun packing the gland. Specify fasteners.

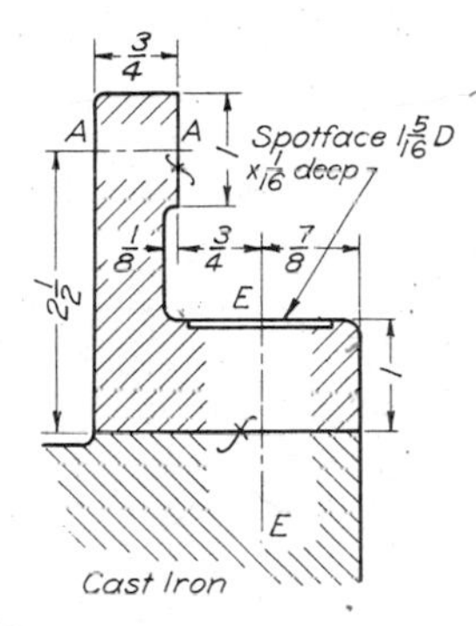


**PROB. 16.** L-lug.

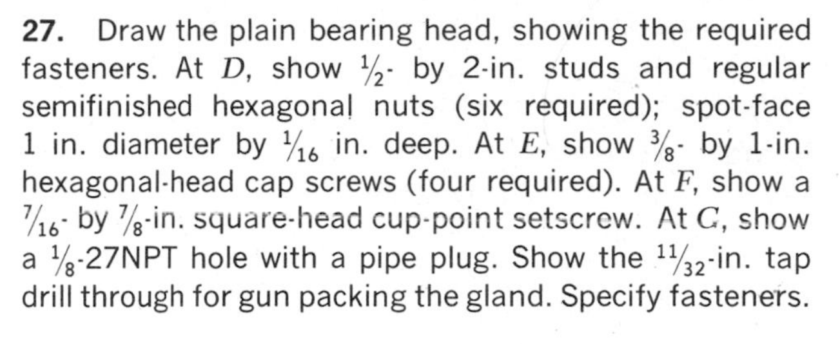

**PROB. 17.** Double cover.

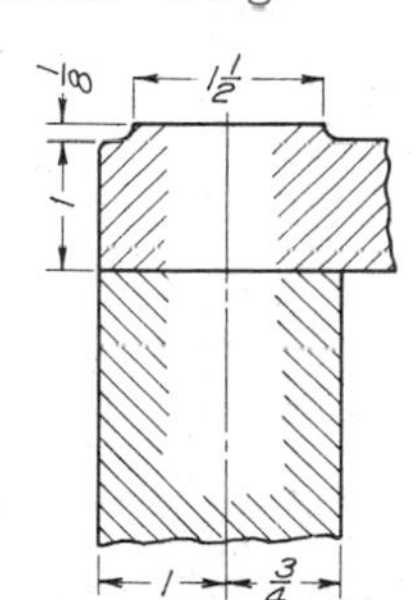

**PROB. 18.** Bracket and support.

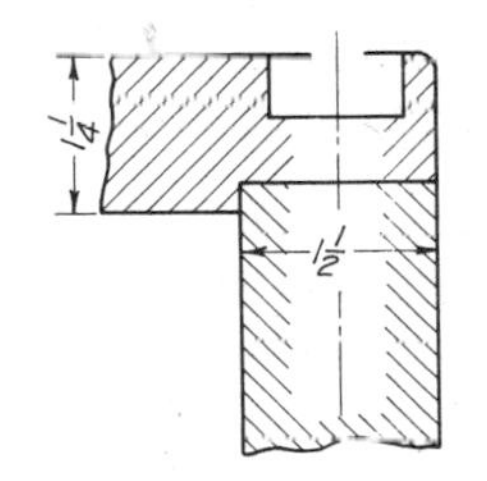

**PROB. 19.** Centered closure.

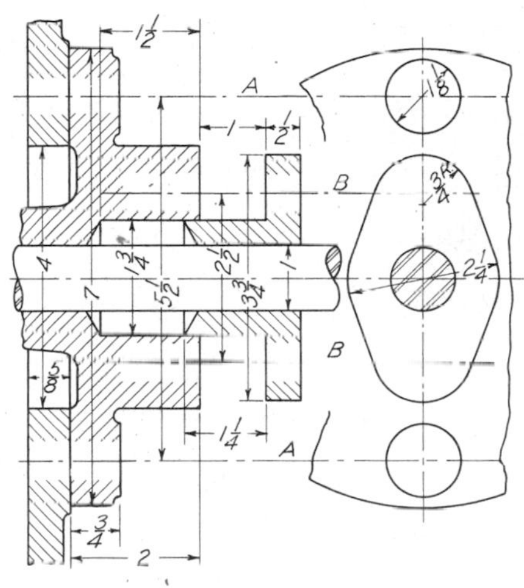

**PROB. 24.** Stuffing box and gland.

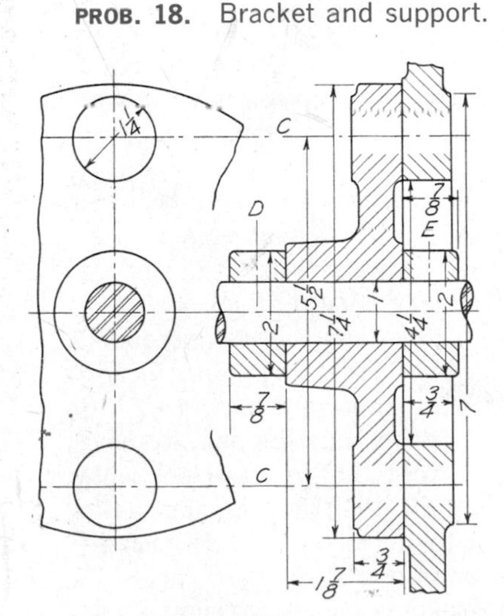

**PROB. 25.** Bearing plate.

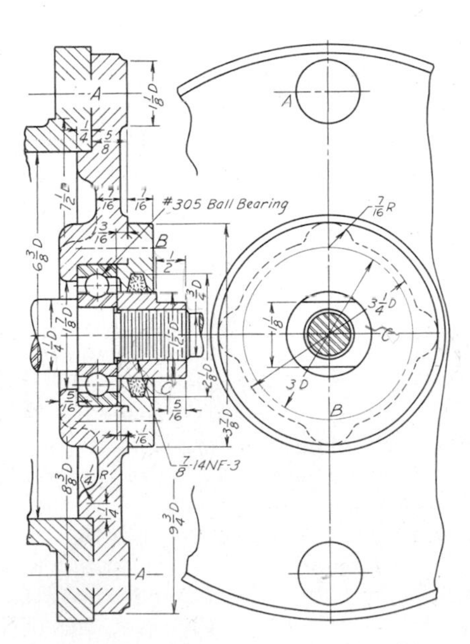


**PROB. 26.** Ball-bearing head.

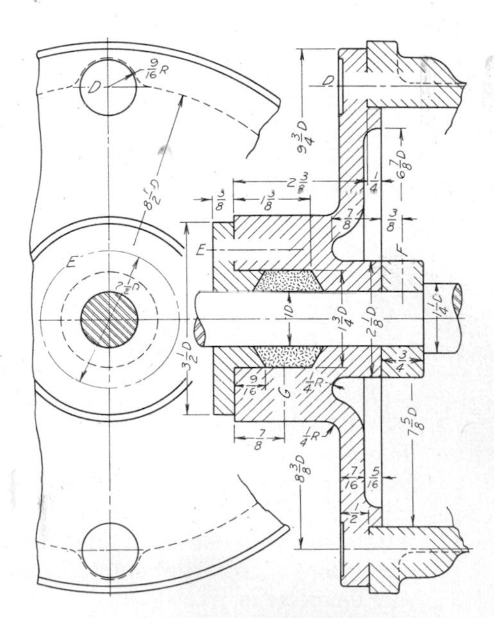

**PROB. 27.** Plain bearing head.

Problems 24 and 25 together and 26 and 27 together may be drawn on an 11- by 17-in. sheet or on separate sheets, showing full diameter of flanges.

## Group 4. Keys.

Key sizes are given in the Appendix.

**28.** Draw hub and shaft as shown, with a Woodruff key in position.

**29.** Draw hub and shaft as shown, with a square key 2 in. long in position.

**30.** Draw hub and shaft as shown, with a gib-head key in position.

**31.** Draw hub and shaft as shown, with a Pratt and Whitney key in position.

## Group 5. Springs.

**32.** Draw a compression spring as follows: inside diameter, $\frac{3}{4}$ in.; wire size, $\frac{1}{8}$ in. diameter; coils 14, right-hand; squared and ground ends; free length, $3\frac{1}{2}$ in.

**33.** Draw a compression spring as follows: outside diameter, 1 in.; wire size, $\frac{3}{32}$ in. diameter; coils 12, left-hand; open ends, not ground; free length, 4 in.

**34.** Draw an extension spring as follows: free length over coils, 2 in.; outside diameter, $1\frac{1}{8}$ in.; wire size, $\frac{1}{8}$ in. diameter; coils 11, right-hand; the ends parallel, closed loops.

**35.** Draw an extension spring as follows: free length inside hooks, $2\frac{3}{4}$ in.; inside diameter, $\frac{3}{4}$ in.; wire size, $\frac{1}{8}$ in. diameter; coils 11, left-hand; the ends parallel, closed half loops.

**36.** Draw a torsion spring as follows: free length over coils, $\frac{3}{4}$ in.; inside diameter, $1\frac{1}{8}$ in.; wire size, $\frac{1}{8}$ in. diameter; coils 5, right-hand; ends straight and turned to follow radial lines to center of spring and extend $\frac{1}{2}$ in. from outside diameter of spring.

**37.** A compression spring made of 26-gage steel wire with an outside diameter of 2 in. has 8 active coils wound to a pitch of $\frac{1}{2}$ in. Determine the free length if:

(*a*) The spring has open ends.
(*b*) The spring has closed ends.
(*c*) The spring has open ends ground.
(*d*) The spring has closed ends ground.

**38.** If the above spring is compressed solid, determine its outside diameter.

**38***A*. If the above spring has a pitch tolerance of $\pm 0.015$ in. and an outside-diameter tolerance of $\pm 0.015$-in. dia., determine the maximum outside diameter. (*Note:* The wire manufacturers' tolerance on the wire diameter is $\pm 0.001$ in.)

**39.** Redesign the ball-bearing heads of Prob. 26 using a lubricated and sealed-for-life type of bearing held in place with a retainer ring.

## Group 6. Manufacturers' Specialties.

**40.** Draw and dimension the limits of the drilled hole in the external collar necessary to hold a Rollpin in place. Show and specify the Rollpin in position on center line and draw three times actual size.

**41.** Draw and dimension the limits of the drilled hole in the outer member necessary to hold a $\frac{3}{8}$-in.-dia. Rollpin. The inner member is to have a minimum clearance of 0.003 in. Dimension the maximum and minimum diameter of the drilled hole in the inner member. Show and specify the Rollpin in position on the center line. Draw three times actual size.

**42.** Show and specify a Waldes Truarc 5000 series retaining ring in position to limit the axial travel of the shaft, until it reaches a minimum distance $X$, as shown in the Appendix on retaining rings. Determine and show the dimension $X$ and the complete groove dimensions as well as the maximum 45° chamfer dimension. Draw three times actual size.

**43.** Show and specify a Waldes Truarc 5100 series retaining ring in position to limit axial travel of the shaft. Determine and show the groove dimensions and the maximum 45° chamfer allowable on the external member. Draw three times actual size.

**44.** Same as Prob. 41, but use series 5133 E ring.

## Group 7. Design Problems.

**45.** The shaft and bar shown are to be joined so that a light torque can be transmitted from one member to the other. The parts must be readily disassembled. Show at least three ways of solving this problem. In each case give the instructions and dimensions required to complete the design.

**46.** A shaft passing through a fixed panel is to have a plastic knob fastened to it. Its angular motion is to be limited to 45° of travel. Select appropriate standard fasteners, and make a drawing showing them installed. Give whatever instructions and dimensions are required to complete the design. (*Note:* This is a large production design and time and cost are important factors.)

**47.** The $\frac{1}{2}$-in.-dia. shaft shown must be able to move axially $\frac{3}{4}$ in. in the frame without rotating. The block must be rigidly attached to this shaft. The two $\frac{3}{8}$-in.-dia. shafts must be free to rotate in the block with no appreciable axial movement. Select appropriate standard fasteners to complete this design. Make necessary drawings showing any additional machining operations, and an assembly drawing. Give complete instructions and all necessary dimensions to complete this design.

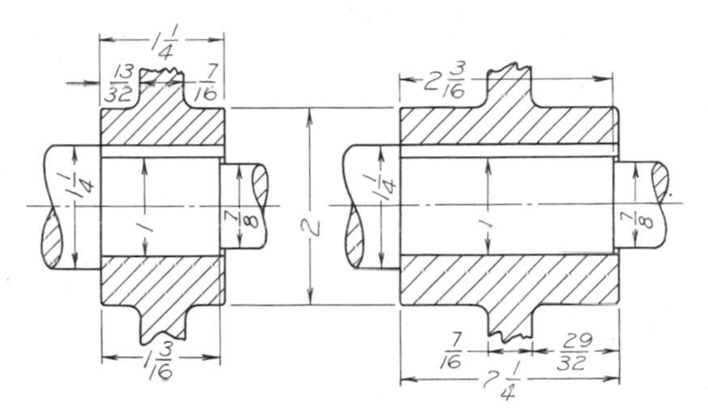

PROB. 28. Hub and shaft. PROB. 29. Hub and shaft.

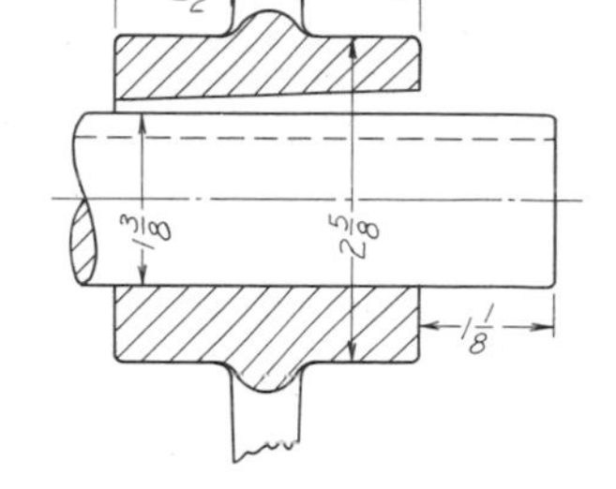

PROB. 30. Layout.

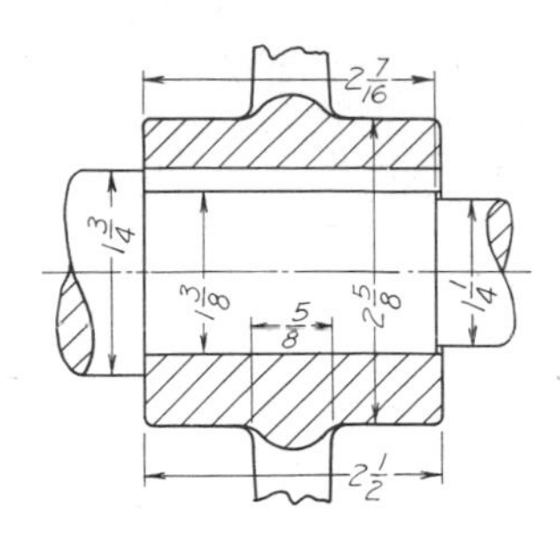

PROB. 31. Layout.

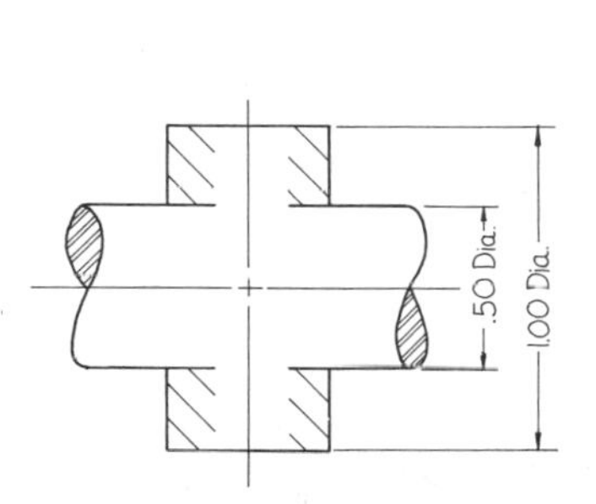


PROB. 40. Layout.

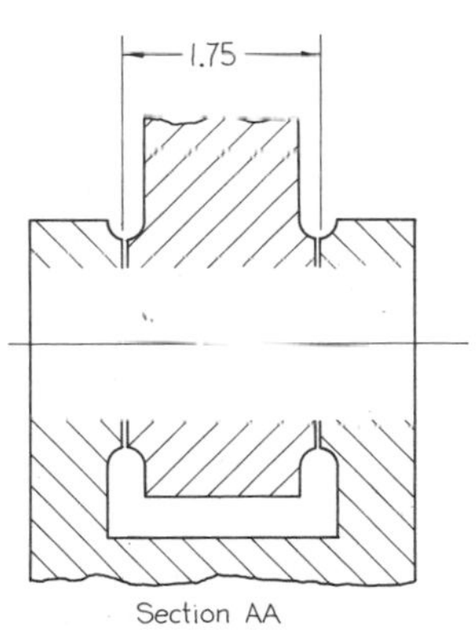


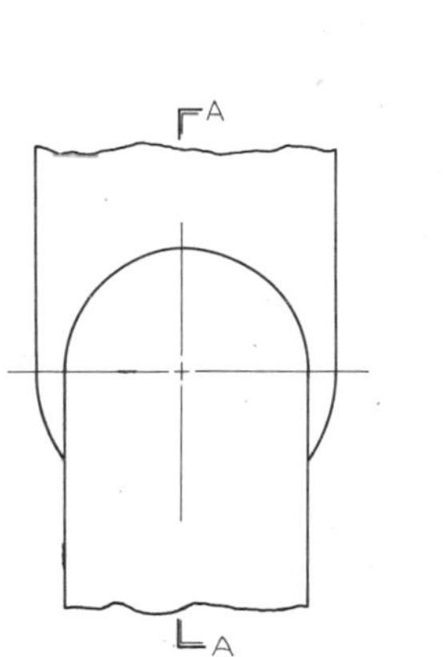


PROB. 41. Layout.

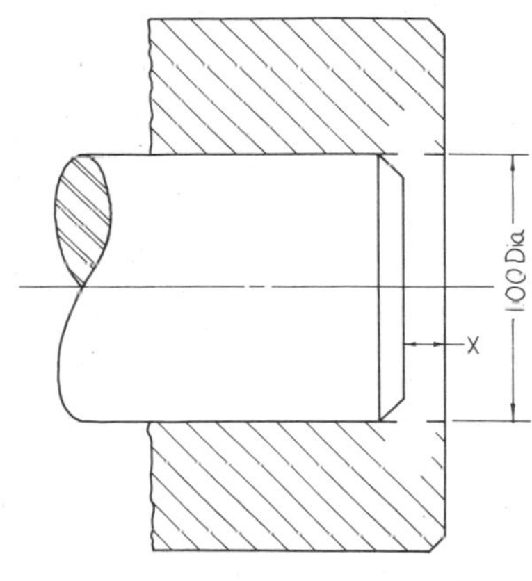


PROB. 42. Layout.

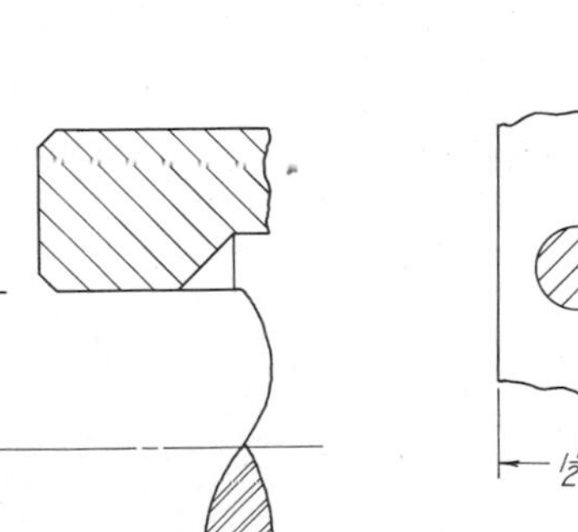


PROB. 43. Layout

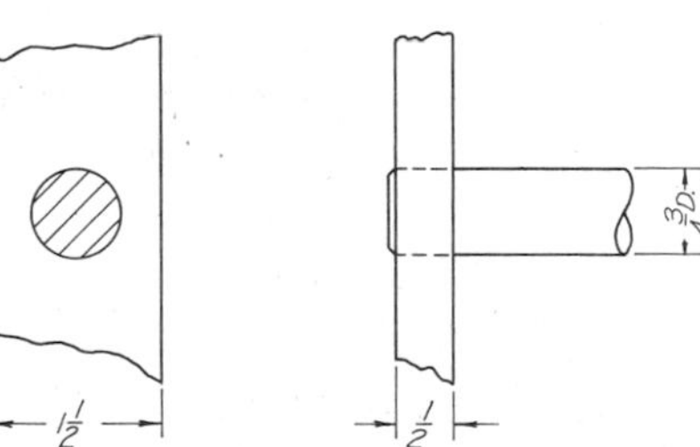

PROB. 45. Layout.

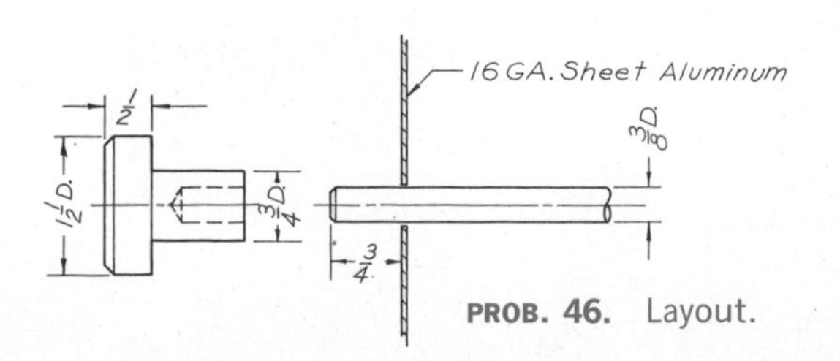


PROB. 46. Layout.



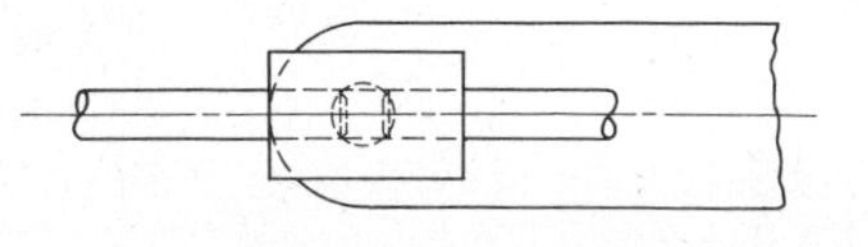

PROB. 47. Layout.

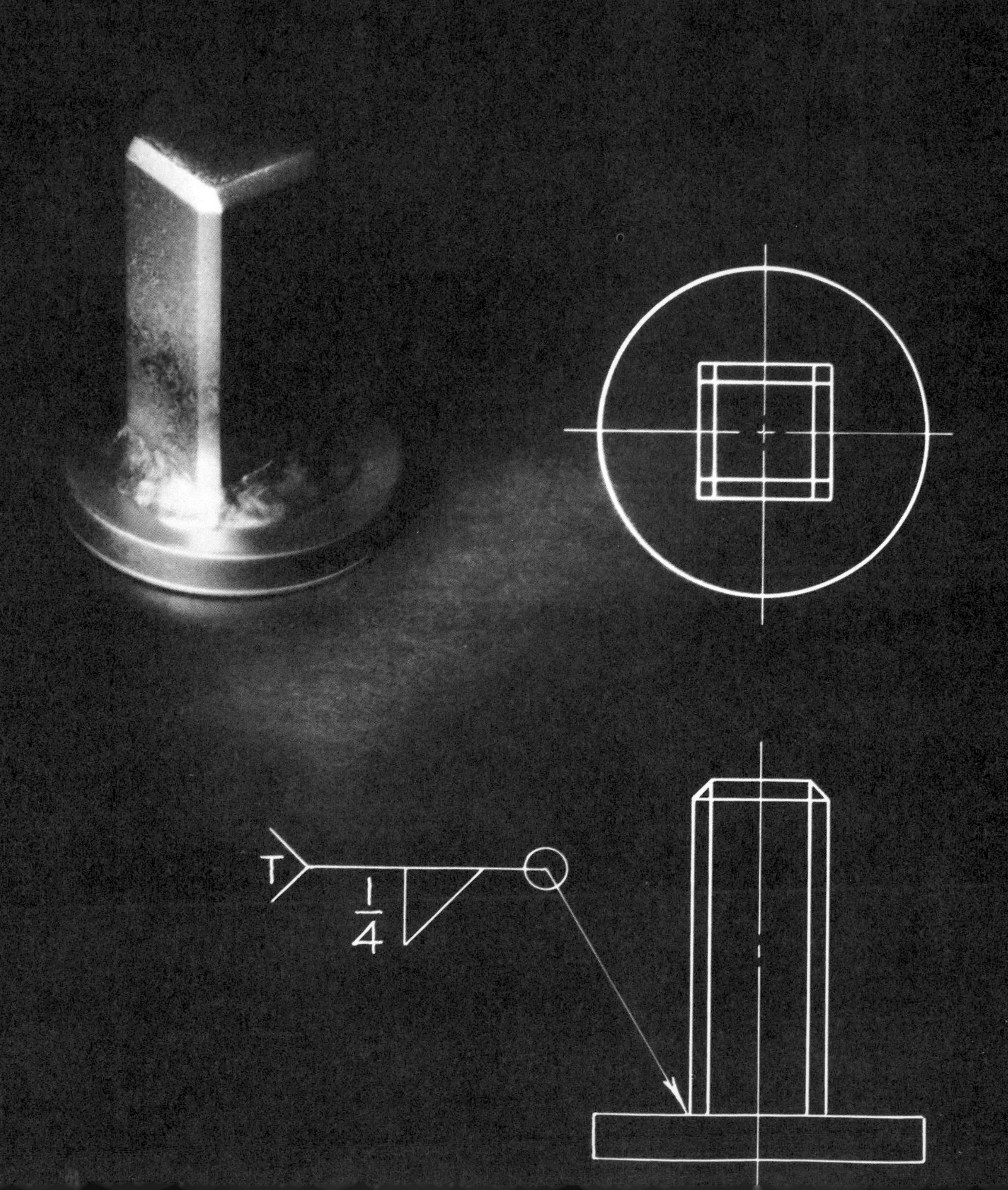
T
1
4

This Chapter Shows and Describes:

Welding Processes
Types of Welds
Welding Symbols
Sizes of Welds
Welding Drawings
Use of Welding Symbols
Rivets
Rivet Types
Manufacturers' Specialties
Problems

# 16

# WELDING AND RIVETING

## These Methods Are Used to Permanently Attach Components to Each Other

## 1. Welding Classified.

The subject of welding is of particular interest to the engineer and draftsman for two reasons. First, welding is used extensively for permanent fastenings of the components of machines and structures. Second, the production of machine parts by fabricating and welding instead of by casting or forging is a method that, for many purposes, forms a lighter and stronger part, often at an economy of cost.

## 2. Welding Processes.

These are classified according to the manner in which the welded joint is completed: (1) pressure welding (forging) and (2) nonpressure welding (fusion and brazing). Actually all welding is a fusion process, but by long usage, fusion welding is understood to include the arc, gas, and thermit processes.

In arc welding, pieces of metal to be welded are brought to the proper welding temperature at point of contact by the heat liberated at the arc terminals so that the metals are completely fused into each other.

In gas or oxyacetylene welding a high-temperature flame is produced by igniting a mixture of two gases, usually oxygen and acetylene.

Resistance welding is a heat and squeeze process. The parts that are to be welded, while being forced together by mechanical pressure, are raised to the temperature of fusion by the passage of a heavy electrical current through the junction.

## 3. Classification of Welded Joints.

Figure 1 shows the basic types of joints, which are classified by the method of assembly of the parts. The applicable welds for each type are listed under each joint.

## 4. Types of Welds.

Figure 2 shows in cross section the fundamental types of welds. For bead and fillet welds, the pieces are not prepared by cutting before making the weld, and the essential difference between V, bevel, U, and J welds is in the preparation of the parts joined. Pairs of fundamental welds—double V, double bevel, etc.—make a further variety. Almost any combination is possible for complicated connections.

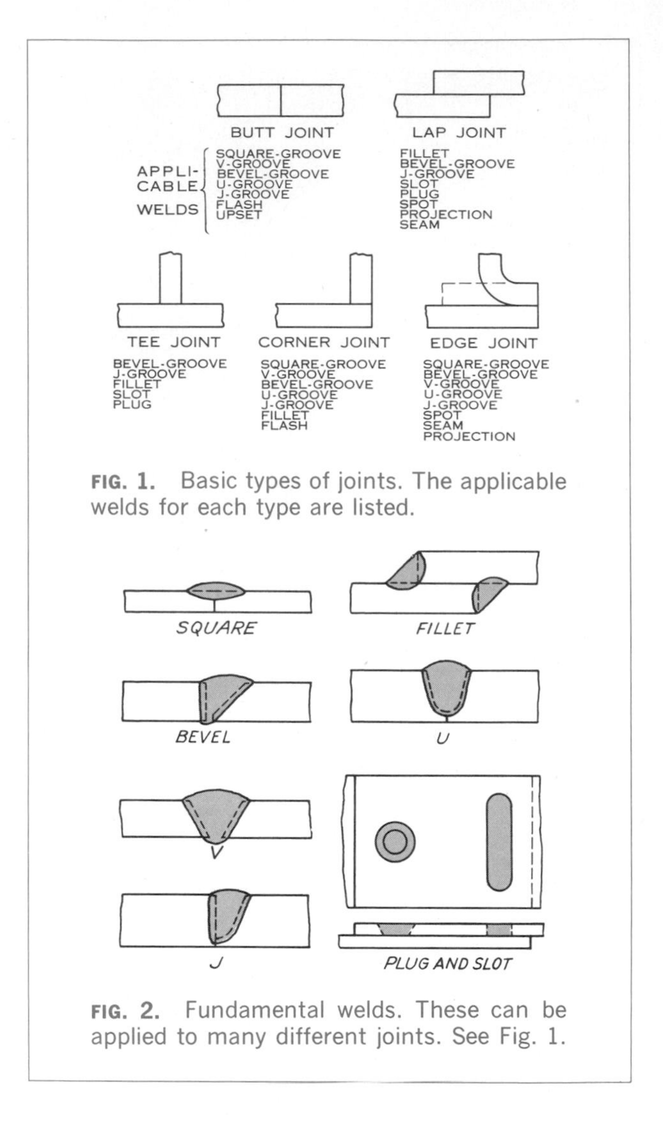


FIG. 1. Basic types of joints. The applicable welds for each type are listed.



FIG. 2. Fundamental welds. These can be applied to many different joints. See Fig. 1.

## 5. Individual Basic Weld Symbols.

These are derived from the preparation of the pieces making up the joint or, where no preparation is necessary, from the section shape of the weld. Figure 3 shows the fundamental welds and the basic symbols that are used in specifying them.

Figure 4 shows the American Standard basic arc, gas, and resistance weld symbols, including supplementary symbols.

## 6. Sizes of Welds.

In addition to specifying the type of weld to be made,

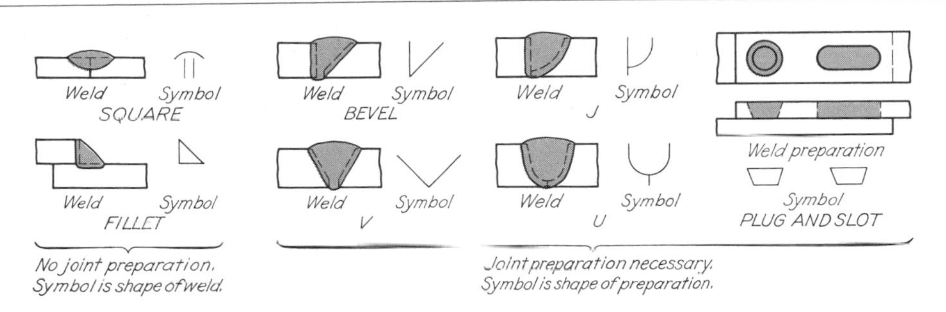


FIG. 3. Fundamental welds and individual basic symbols. Note that the symbols are pictorially descriptive of weld shapes.

| ARC AND GAS WELD SYMBOLS | | | | | | | | | | | | |
|---|---|---|---|---|---|---|---|---|---|---|---|---|
| TYPE OF WELD | | | | | | | | | SUPPLEMENTARY | | | |
| BEAD | FILLET | PLUG OR SLOT | GROOVE | | | | | WELD ALL AROUND | FIELD WELD | CONTOUR | |
| | | | SQUARE | V | BEVEL | U | J | | | FLUSH | CONVEX |
| | | | | | | | | | | — | |

| RESISTANCE WELD SYMBOLS | | | | | | | |
|---|---|---|---|---|---|---|---|
| TYPE OF WELD | | | | SUPPLEMENTARY | | | |
| SPOT | PROJECTION | SEAM | FLASH OR UPSET | WELD ALL AROUND | FIELD WELD | CONTOUR | |
| | | | | | | FLUSH | CONVEX |
| | | | \| | | | — | |

FIG. 4. American Standard weld symbols.

a description of the size is also necessary. Thus, the *basic weld symbol* plus *size specifications* are the elements included in a complete welding symbol. Figure 5 shows weld sizes. The dimensions of root opening, depth of preparation, and included angle are the important sizes to specify for grooved welds. The size of a 45° fillet weld is the dimension shown. Unequal-leg fillet welds are specified by giving the size of both legs, as described in Sec. 9, Art. B-2*b*; sizes of plug and slot welds are given as specified in Arts. E. and F.

(A) Included angle
(Size) S
Depth of weld (Size) S
Root opening (0)

FIG. 5. Weld sizes. These are fundamental dimensions to be given on weld symbols. See Fig. 6.

## 7. The Complete Welding Symbol.

Figure 6 shows the basic form of the welding symbol and gives the position of the various marks and dimensions. There is a distinction between the terms *weld symbol* (Sec. 5) and *welding symbol.*

The *weld symbol* is the ideograph (Fig. 4) used to indicate the desired type of weld. The assembled *welding symbol* consists of the following eight elements, or such of these elements as are necessary:

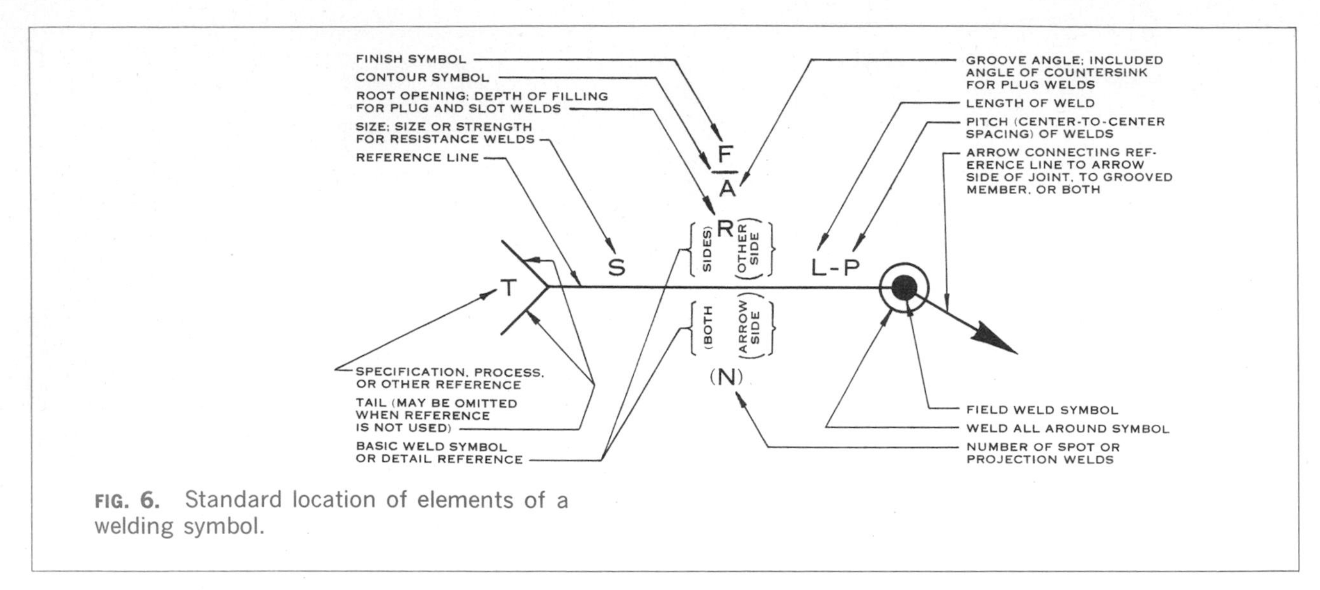


FIG. 6. Standard location of elements of a welding symbol.

(1) reference line, (2) arrow, (3) basic weld symbols, (4) dimensions and other data, (5) supplementary symbols, (6) finish symbols, (7) tail, (8) specification process or other references.

Figure 7 summarizes the important points dealing with the specification of welds and the location of weld symbols. It also gives some general directions for drawing the symbols and shows the completed symbols.

Figure 8 classifies welded joints and shows the symbols for each. The arrow points to the grooved member at a point near the weld. The side of the weld pointed to is always called the arrow side. For nonsymmetrical welds (bevel and J), in order to show *which* piece is to be prepared, the leader must be made with a definite break pointing toward the piece to be prepared, as shown by the bevel and J welds in Fig. 8 and described in Art. A-15. The tail of the arrow is used to hold a reference only when specification of strength, type of rod, etc., is to be given. The individual basic symbols are placed on the basic form to describe any possible combination of welds for a complete joint. Every simple weld that is a part of the complete joint must be specified.

## 8. Welding Drawings.

A welding drawing shows a unit or part made of several pieces of metal, with each welded joint described and specified. Figure 9 shows the detail drawing of a part made of cast iron, and Fig. 10 shows a part identical in function but made up by welding. Comparison of the two drawings shows the essential differences both in construction and in drawing technique. Note the absence of fillets and rounds in the welding drawing. Note also that all the pieces making up the welded part are dimensioned so that they can be cut easily from standard stock.

All joints between the individual pieces of the welded part must be shown, even though the joint would not appear as a line on the completed part. The lines marked $A$ in Fig. 11 illustrate this principle. Each individual piece should be identified by number (Fig. 10).

## 9. Use of the Symbols.

The following instructions for the placement and form of welding symbols are adapted from ASA Y32.3—1959.

### A. General Provisions.

*1. Location Significance of Arrow*

*a.* In the case of groove, fillet, and flash or upset welding symbols, the

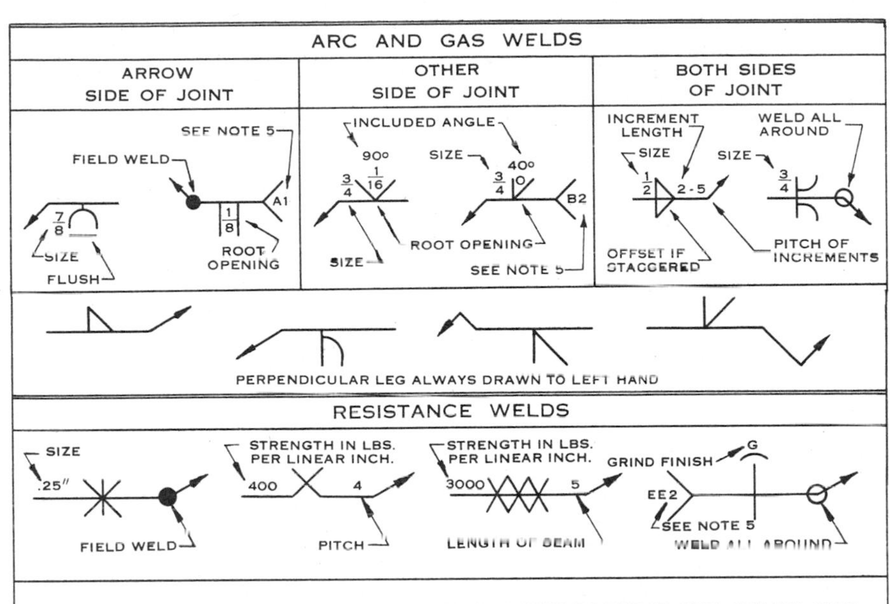


1. THE SIDE OF THE JOINT TO WHICH THE ARROW POINTS IS THE ARROW SIDE.
2. BOTH-SIDES WELDS OF SAME TYPE ARE OF SAME SIZE UNLESS OTHERWISE SHOWN.
3. SYMBOLS APPLY BETWEEN ABRUPT CHANGES IN DIRECTION OF JOINT OR AS DIMENSIONED (EXCEPT WHERE ALL AROUND SYMBOL IS USED).
4. ALL WELDS ARE CONTINUOUS AND OF USER'S STANDARD PROPORTIONS. UNLESS OTHERWISE SHOWN.
5. TAIL OF ARROW USED FOR SPECIFICATION REFERENCE (TAIL MAY BE OMITTED WHEN REFERENCE NOT USED).
6. DIMENSIONS OF WELD SIZE. INCREMENT LENGTHS AND SPACINGS IN INCHES.

**FIG. 7.** Location of weld symbols and specifications.

arrow connects the welding symbol reference line to one side of the joint, and this side is considered the *arrow side* of the joint. The side opposite the arrow side of the joint is considered the *other side* of the joint (Fig. 6).

*b*. In the case of plug, slot, spot, seam, and projection welding symbols, the arrow connects the welding symbol reference line to the outer surface of one of the members of the joint at the center line of the desired weld. The member to which the arrow points is considered the *arrow-side* member; the other member, the *other-side* member (Art. E-1*a*).

*c*. When a joint is depicted by a single line on the drawing and the arrow of a welding symbol is directed to this line, the arrow side of the joint is considered the near side of the joint in accordance with the usual conventions of drafting (Fig. 8).

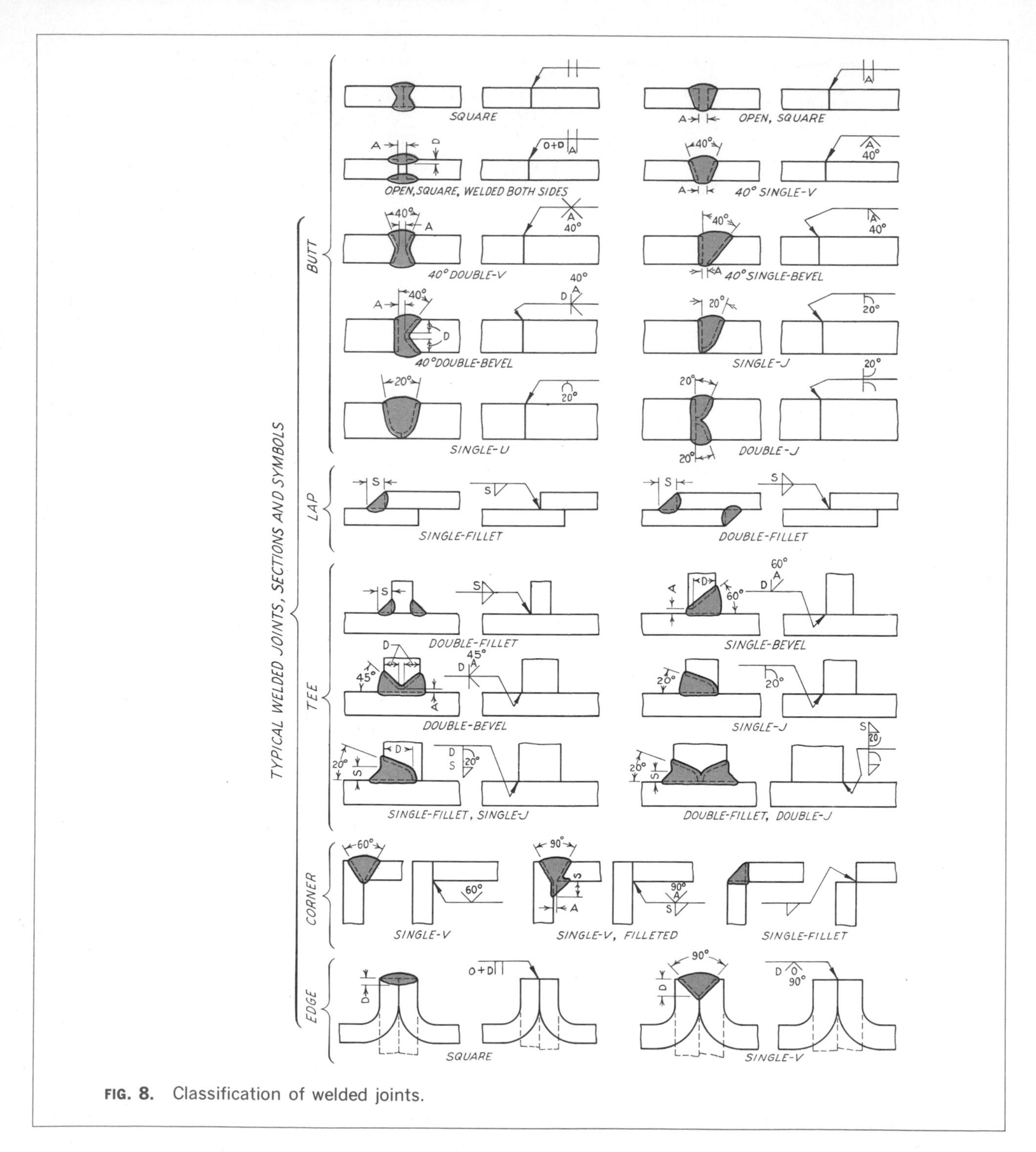


**FIG. 8.** Classification of welded joints.

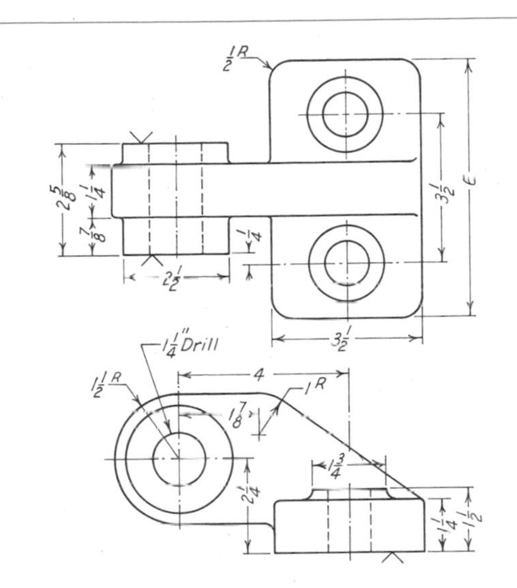


**FIG. 9.** Detail drawing of a casting. Compare with welded part, Fig. 10.

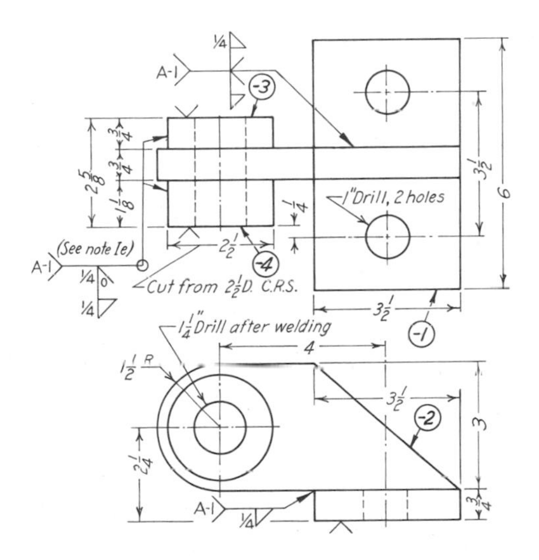


**FIG. 10.** Detail drawing of a welded part. Compare with cast part, Fig. 9.

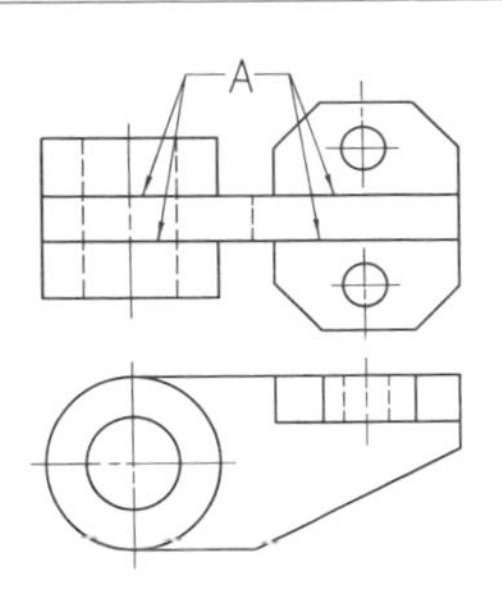


**FIG. 11.** Joint lines. These are shown on drawings of welded parts.

*d.* When a joint is depicted as an area parallel to the plane of projection in a drawing and the arrow of a welding symbol is directed to near area, the *arrow-side* member of the joint is considered the near side of the joint in accordance with the usual conventions of drafting (Fig. 8).

2. *Location of Weld with Respect to Joint*

*a.* Welds on the arrow side of the joint are shown by placing the weld symbol on the side of the reference line toward the reader.

*b.* Welds on the other side of the joint are shown by placing the weld symbol on the side of the reference line away from the reader.

*c.* Welds on both sides of the joint are shown by placing weld symbols on both sides of the reference line, toward and away from the reader.

*d.* Spot, seam, flash, and upset weld symbols have no arrow-side or other-side significance in themselves, although supplementary symbols used in conjunction with them may have such significance. Spot, seam, flash, and upset weld symbols are centered on the reference line.

3. *Method of Drawing Symbols*

*a.* Symbols may be drawn mechanically or freehand.

*4. Use of Inch, Degree, and Pound Marks*

*a.* Inch, degree, and pound marks may be used on welding symbols or not, as desired, except that inch marks must be used for indicating the diameter of spot and projection welds and the width of seam welds when such welds are specified by linear dimension.

*5. Location of Specification, Process, or Other References*

*a.* When a specification, process, or other reference is used with a welding symbol, the reference is placed in the tail.

*6. Use of Symbols without References*

*a.* When desired, weld symbols may be used without specification, process, or other references in the following instances:

(1) When a note such as the following is included on the drawing: "Unless otherwise designated, all welds are to be made in accordance with Specification No.__."

(2) When the welding procedure to be used is prescribed elsewhere.

*7. Use of General Notes*

*a.* When desired, general notes such as the following may be placed on a drawing to provide detailed information pertaining to the predominating welds, and this information need not be repeated on the symbols.

"Unless otherwise indicated, all fillet welds are $5/16$ inch size."

"Unless otherwise indicated, root openings for all groove welds are $3/16$ inch."

*8. Use of Weld-all-around Symbol*

*a.* Welds extending completely around a joint are indicated on the drawing by means of the weld-all-around symbol.

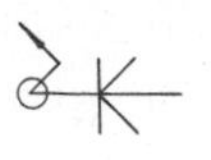

*9. Use of Field-weld Symbol*

*a.* Field welds (welds not made in a shop or at the place of initial construction) are indicated by means of the field-weld symbol.

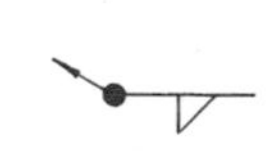

*10. Extent of Welding Denoted by Symbols*

*a.* Symbols apply between abrupt changes in the direction of the welding or to the extent of hatching or dimension lines, except when the weld is indicated by means of the weld-all-around symbol.

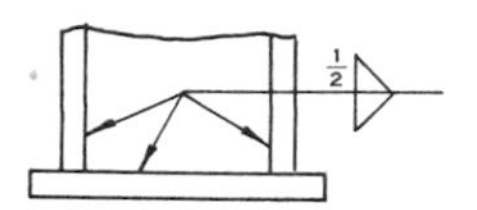

*11. Weld Proportions*

*a.* All welds are continuous and of user's standard proportions unless otherwise indicated.

*12. Finishing of Welds*

*a.* Finishing of welds, other than cleaning, is indicated by suitable contour and finish symbols (Art. B-9).

*13. Location of Weld Symbols*

*a.* Weld symbols, except spot and seam, are shown only on the welding symbol reference line and not on the lines of the drawing.

*b.* Spot and seam weld symbols may be placed directly on drawings at the locations of the desired welds.

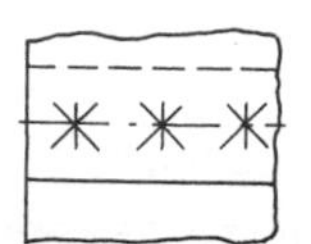

*14. Construction of Fillet, and Bevel- and J-groove Welding Symbols*

*a.* Fillet and bevel- and J-groove weld symbols are shown with the perpendicular leg *always* to the left.

*15. Use of Break in Arrow of Bevel- and J-groove Welding Symbols*

*a.* When a bevel- or J-groove weld symbol is used, the arrow points with a definite break toward the member that is to be chamfered, as shown. In cases where the member to be chamfered is obvious, the break in the arrow may be omitted.

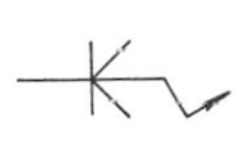

*16. Reading of Information on Welding Symbols*

*a.* Information on welding symbols is placed to read from left to right along the reference line in accordance with the usual conventions of drafting.

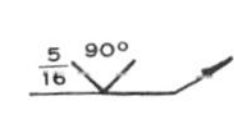


*17. Combined Weld Symbols*

*a.* For joints having more than one weld, a symbol is shown for each weld.

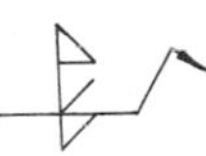

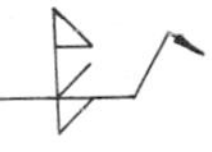

*18. Designation of Special Types of Welds*

*a.* When the basic weld symbols are inadequate to indicate the desired weld, the weld is shown by a cross section, detail, or other data, with a reference to it on the welding symbol, observing the usual location significance.

## B. Fillet Welds.

*1. General*

*a.* Dimensions of fillet welds are shown on the same side of the reference line as the weld symbol.

*b.* When no general note on the drawing governs the dimensions of fillet welds, the dimensions of fillet welds on both sides of the joint are shown.

(1) When both welds have the same dimensions, one or both may be dimensioned.

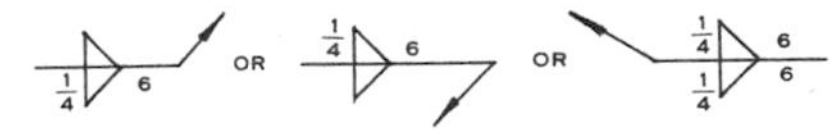


(2) When the welds differ in dimensions, both are dimensioned.

*c.* When there appears on the drawing a general note governing the dimensions of fillet welds, such as "All fillet welds 5/16 in. size unless otherwise noted," the dimensions of fillet welds on both sides of the joint are indicated as follows:

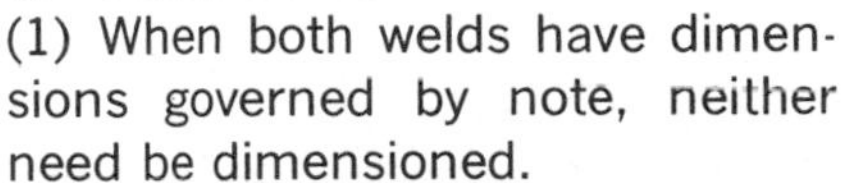

(1) When both welds have dimensions governed by note, neither need be dimensioned.

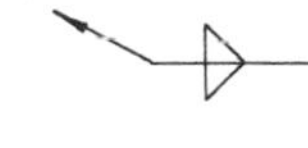

(2) When the dimensions of one or both welds differ from the dimensions given in the general note, both welds are dimensioned.

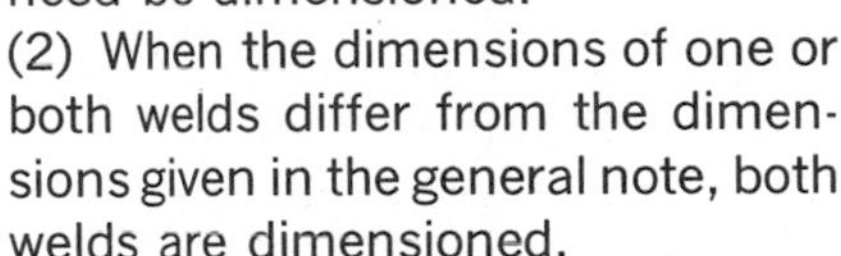

*2. Size of Fillet Welds*

*a.* The size of a fillet weld is shown to the left of the weld symbol.

*b.* The size of a fillet weld with unequal legs is shown in parentheses to the left of the weld symbol, as shown. Weld orientation is not shown by the symbol and is shown on the drawing when necessary.

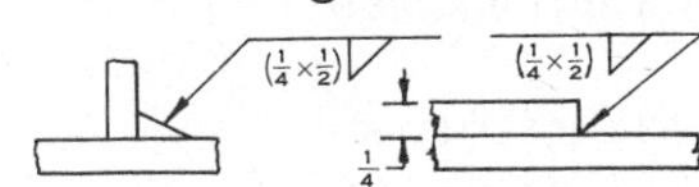


*3. Length of Fillet Welds*

*a.* The length of a fillet weld, when indicated on the welding symbol, is shown to the right of the weld symbol.

*b*. When fillet welding extends for the full distance between abrupt changes in the direction of the welding, no length dimension need be shown on the welding symbol.

*c*. Specific lengths of fillet welding may be indicated by symbols in conjunction with dimension lines.

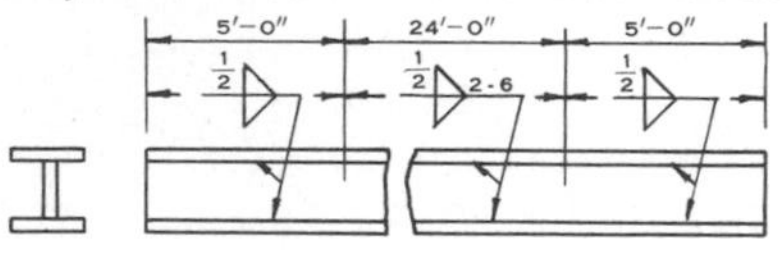


*4. Extent of Fillet Welding*

*a*. When it is desired to show the extent of fillet welding graphically, one type of hatching with definite end lines is used.

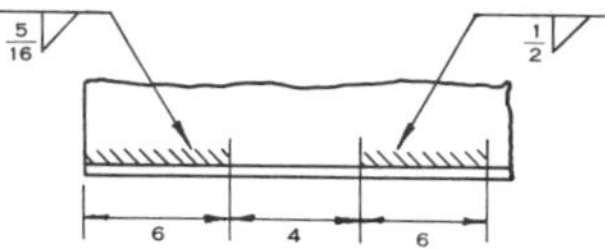


*b*. Fillet welding extending beyond abrupt changes in the direction of the welding is indicated by means of additional arrows pointing to each section of the joint to be welded, as shown in Art. A-10, except when the weld is indicated by means of the weld-all-around symbol.

*5. Dimensioning of Intermittent Fillet Welding*

*a*. The pitch (center-to-center spacing) of intermittent fillet welding is indicated as the distance between centers of increments on *one* side of the joint.

*b*. The pitch (center-to-center spacing) of intermittent fillet welding is shown to the right of the length dimension.

3-5

*c*. Chain intermittent fillet welding is shown thus:

4-6

*d*. Staggered intermittent fillet welding is shown thus:

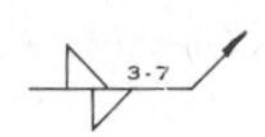


*6. Termination of Intermittent Fillet Welding*

*a*. When intermittent fillet welding is used by itself, the symbol indicates that increments are located at the ends of the dimensioned length.

*b*. When intermittent fillet welding is used between continuous fillet welding, the symbol indicates that spaces equal to the pitch minus the length of one increment are to be left at the ends of the dimensioned length (Art B-3).

*7. Combination of Intermittent and Continuous Fillet Welding*

*a*. Separate symbols are used for intermittent and continuous fillet welding when the two are used in combination (Art B-3*c*).

*8. Fillet Welds in Holes and Slots*

*a*. Fillet welds in holes and slots are indicated by means of fillet-weld symbols.

*9. Surface Contour of Fillet Welds*

*a*. Fillet welds that are to be welded approximately flat-faced without recourse to any method of finishing are shown by adding the flush-contour symbol to the weld symbol, observing the usual location significance.

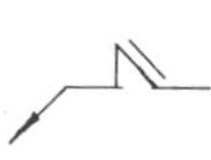

*b*. Fillet welds that are to be made flat-faced by mechanical means are shown by adding both the flush-contour symbol and the user's standard finish symbol[1] to the weld

[1] Finish symbols used here indicate the method of finishing (C, chipping; G, grinding; M, machining) and not the degree of finish.

symbol, observing the usual location significance.

*c.* Fillet welds that are to be mechanically finished to a convex contour are shown by adding both the convex-contour symbol and the user's standard finish symbol to the weld symbol, observing the usual location significance.

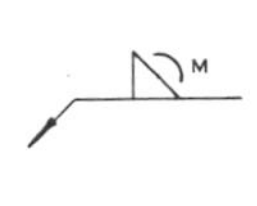


## C. Groove Welds.

*1. General*

*a.* Dimensions of groove welds are shown on the same side of the reference line as the weld symbol.

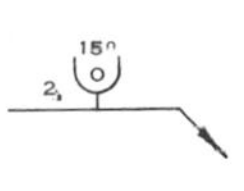


*b.* When no general note governing the dimensions of groove welds appears on the drawing, the dimensions of double-groove welds are shown as follows:

(1) When both welds have the same dimensions, one or both may be dimensioned.

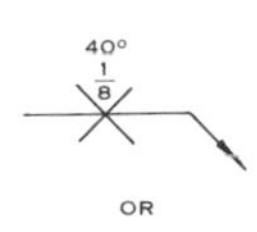


(2) When the welds differ in dimensions, both are dimensioned.

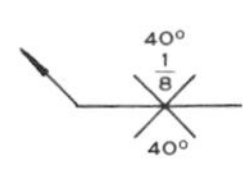


*c.* When there appears on the drawing a general note governing the dimensions of groove welds, such as "All V-groove welds shall have a 60° groove angle unless otherwise noted," the dimensions of double-groove welds are indicated as follows:

(1) When both welds have dimensions governed by the note, neither need be dimensioned.

(2) When the dimensions of one or both welds differ from the dimensions given in the general note, both welds are dimensioned.

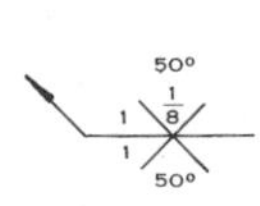


*2. Size of Groove Welds*

*a.* Size of groove welds is shown left of the weld symbol (groove depth plus root penetration).

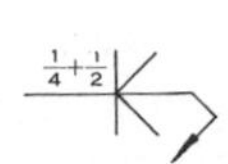


*b.* The size of groove welds with no specified root penetration is shown as follows:

(1) The size of single-groove and symmetrical double-groove welds that extend completely through the member or members being joined need not be shown on the welding symbol.

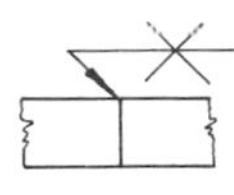

(2) The size of groove welds that extend only partly through the member or members being joined is shown on the welding symbol.

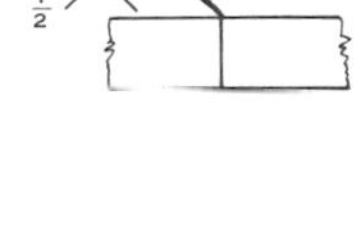


*c.* The size of groove welds with specified root penetration is indicated by showing both the depth of chamfering and the root penetration, separated by a plus mark and placed to the left of the weld symbol. The depth of chamfering and the root penetration is read in that order from left to right along the reference line.

*3. Groove Dimensions*

*a.* Root opening of groove welds is the user's standard unless otherwise indicated. Root opening of groove welds, when not the user's standard, is shown inside the weld symbol.

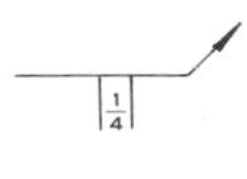


*b.* Groove angle of groove welds is the user's standard, unless otherwise indicated. Groove angle of groove welds, when not the user's standard is shown thus:

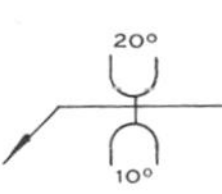


*c.* Groove radii and root faces of U- and J-groove welds are the user's standard unless otherwise indicated. When groove radii and root faces of U- and J-groove welds are not the user's standard, the weld is shown by a cross section, detail, or other data, with a reference to it on the welding symbol, observing the usual location significance.

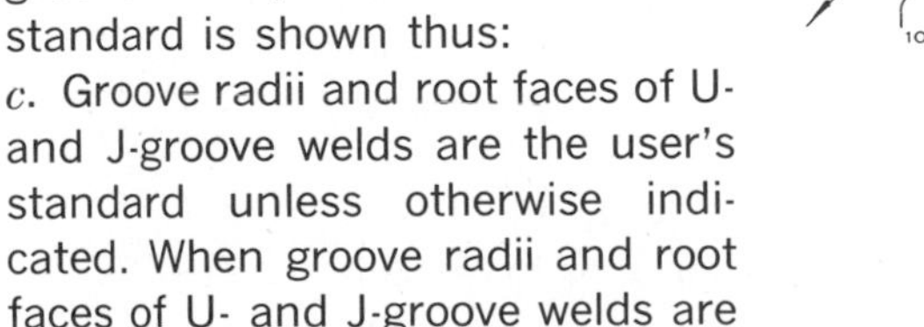

*4. Designation of Back and Backing Welds*

Bead-type back and backing welds of single-groove welds are shown by means of the bead weld symbol (Art. D-2).

*5. Surface Contour of Groove Welds*

*a.* Groove welds that are to be welded approximately flush without recourse to any method of finishing are shown by adding the flush-contour symbol to the weld symbol, observing the usual location significance.

*b.* Groove welds that are to be made flush by mechanical means are shown by adding both the flush-contour symbol and the user's standard finish symbol to the weld symbol, observing the usual location significance.

*c.* Groove welds that are to be mechanically finished to a convex contour are shown by adding both the convex-contour symbol and the user's standard finish symbol to the weld symbol, observing the usual location significance.

## D. Bead Welds.

*1. General*

*a.* The single bead weld symbol is used to indicate bead-type back or backing welds of single-groove welds.

*b.* The dual bead weld symbol is used to indicate surfaces to be built up by welding.

*2. Use of Bead Weld Symbol to Indicate Bead-type Back or Backing Welds*

*a.* Bead welds used as back or

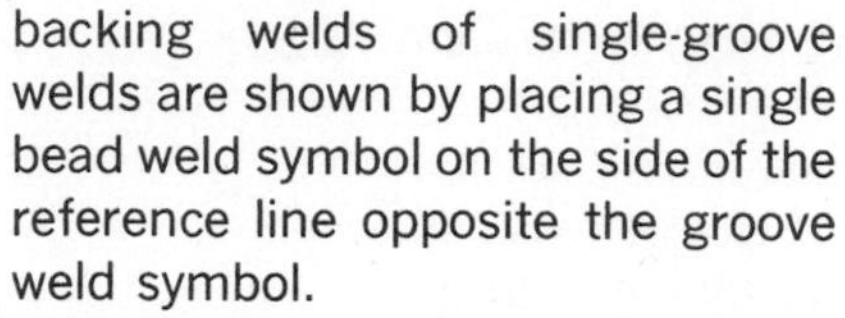

backing welds of single-groove welds are shown by placing a single bead weld symbol on the side of the reference line opposite the groove weld symbol.

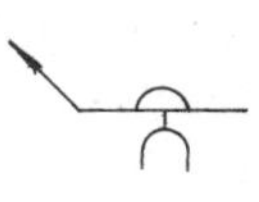

*b.* Dimensions of bead welds used as back or backing welds are not shown on the welding symbol. If it is desired to specify these dimensions, they are shown on the drawing.

*3. Surface Contour of Back or Backing Welds*

*a.* Back or backing welds that are to be welded approximately flush without recourse to any method of finishing are shown by adding the flush-contour symbol to the bead weld symbol.

*b.* Back or backing welds that are to be made flush by mechanical means are shown by adding the flush-contour symbol and the user's standard finish symbol (see n. 1, p. 552) to the bead weld symbol.

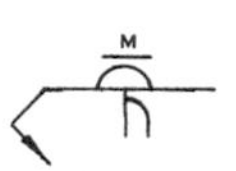


*c.* Back or backing welds that are to be mechanically finished to a convex contour are shown by adding the convex-contour symbol and the user's standard finish symbol (see n. 1, p. 552) to the bead weld symbol.

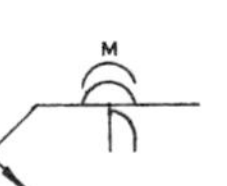


*4. Use of Bead Weld Symbol to Indicate Surfaces Built up by Welding*

*a.* Surfaces built up by welding, whether by single- or multiple-pass bead welds, are shown by the dual bead weld symbol.

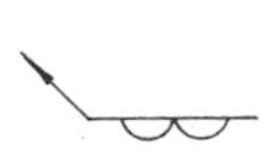

*b.* The dual bead weld symbol does not indicate the welding of a joint, and hence has no arrow- or other-side significance. This symbol is drawn on the side of the reference line toward the reader, and the arrow must point clearly to the surface on which the weld is to be deposited.

*c.* Dimensions used in conjunction with the dual bead weld symbol are shown on the same side of the reference line as the weld symbol.

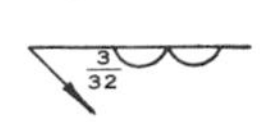


*5. Size (Height) of Surfaces Built up by Welding*

*a.* The size of a surface built up by welding is indicated by showing the minimum height of the weld deposit to the left of the weld symbol.

*b.* When no specific height of weld deposit is desired, no size dimension need be shown on the welding symbol.

*6. Extent, Location, and Orientation of Surfaces Built up by Welding*

*a.* When the entire area of a plane or curved surface is to be built up by welding, no dimension other than size (height of deposit) need be shown on the welding symbol.

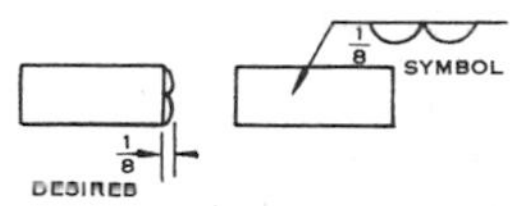


*b.* When a portion of the area of a plane or curved surface is to be built up by welding, the extent, location and orientation of the area to be built up is indicated on the drawing.

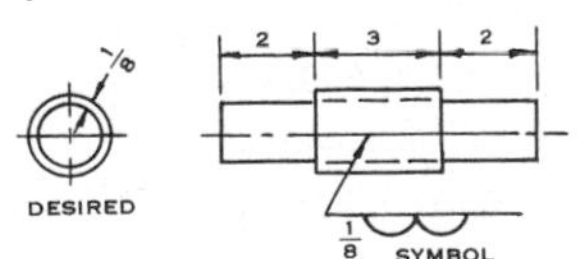


## E. Plug Welds.

*1. General*

*a.* Holes in the arrow-side member of a joint for plug welding are indicated by placing the weld symbol on the side of the reference line toward the reader.

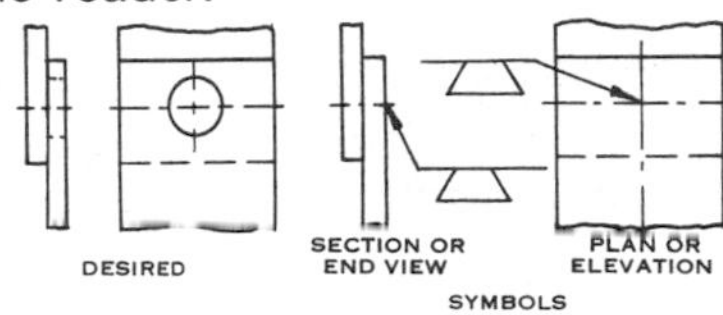


*b.* Holes in the other-side member of a joint for plug welding are indicated by placing the weld symbol on the side of the reference line away from the reader.

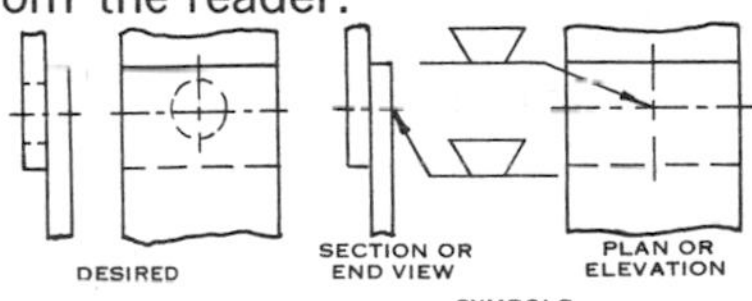


*c.* Dimensions of plug welds are shown on the same side of the reference line as the weld symbol.

*d.* The plug weld symbol is not used to designate fillet welds in holes (Art. B-8).

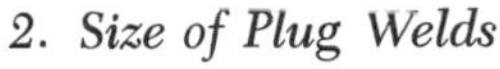

*2. Size of Plug Welds*

*a.* The size of a plug weld is shown to the left of the weld symbol.

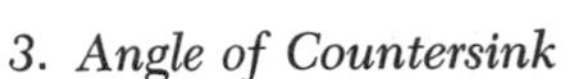

*3. Angle of Countersink*

*a.* Included angle of countersink of plug welds is the user's standard unless otherwise indicated. Included angle of countersink, when not the user's standard, is shown thus:

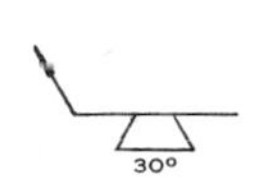


*4. Depth of Filling*

*a.* Depth of filling of plug welds is complete unless otherwise indicated. When the depth of filling is less than complete, the depth of filling, in inches, is shown inside the weld symbol.

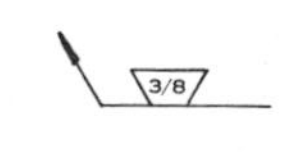


*5. Spacing of Plug Welds*

*a.* Pitch (center-to-center spacing) of plug welds is shown to the right of the weld symbol.

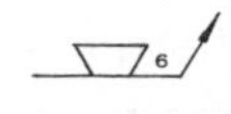

*6. Surface Contour of Plug Welds*

*a.* Plug welds that are to be welded approximately flush without recourse to any method of finishing are shown by adding the flush-contour symbol to the weld symbol.

*b.* Plug welds that are to be made flush by mechanical means are shown by adding both the flush-contour symbol and the user's standard finish symbol (see n. 1, p. 552) to the weld symbol.

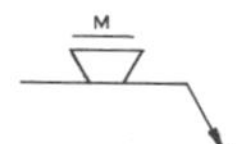


## F. Slot Welds.

*1. General*

*a.* Slots in the arrow-side member of a joint for slot welding are indicated by placing the weld symbol on the side of the reference line toward the reader.

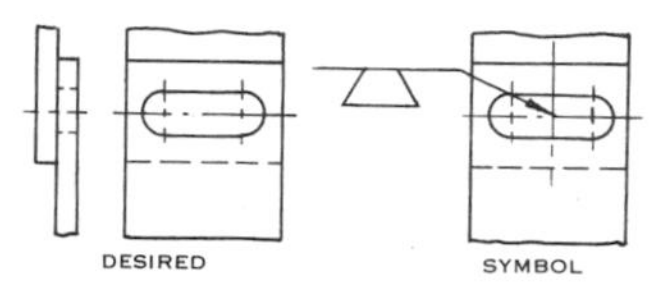


*b.* Slots in the other-side member of a joint for slot welding are indicated by placing the weld symbol on the side of the reference line away from the reader.

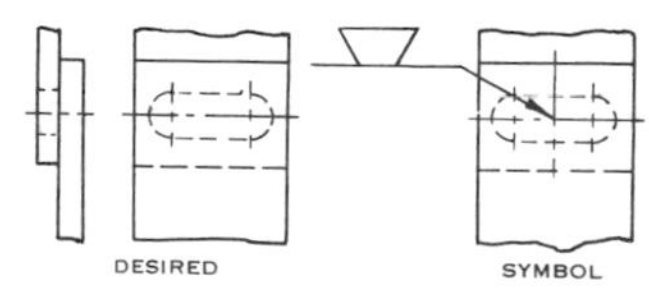


*c.* Dimensions of slot welds are shown on the same side of the reference line as the weld symbol.

*d.* The slot weld symbol is not used to designate fillet welds in slots (Art. B-8).

*2. Depth of Filling*

*a.* Depth of filling of slot welds is complete unless otherwise indicated. When the depth of filling is less than complete, the depth of filling, in inches, is shown inside the weld symbol.

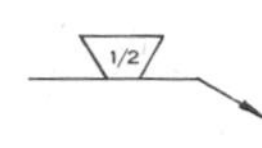


*3. Details of Slot Welds*

*a.* The length, width, spacing, included angle of countersink, orientation, and location of slot welds cannot be shown on the welding symbol. These data are shown on the drawing or by a detail with a reference to it on the welding symbol, observing the usual location significance.

*4. Surface Contour of Slot Welds*

*a.* Slot welds that are to be welded approximately flush without any method of finishing are shown by adding the flush-contour symbol to the weld symbol.

*b.* Slot welds that are to be made flush by mechanical means are shown by adding both the flush-contour symbol and the user's standard finish symbol (see n. 1, p. 552) to the weld symbol.

## G. Spot Welds.

*1. General*

*a.* Spot weld symbols have no arrow- or other-side significance in themselves, although supplementary symbols used with them may have such significance (Art. G-6). Spot weld symbols are centered on the reference line (Art. G-4).

*b.* Dimensions of spot welds may be shown on either side of the reference line.

*2. Size of Spot Welds*

*a.* Spot welds are dimensioned by size or strength, as follows:

(1) The size of spot welds is designated as the diameter of the weld expressed decimally in hundredths of an inch, and is shown, with inch marks, to the left of the weld symbol.

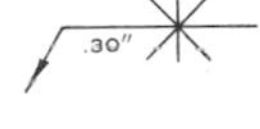


(2) The strength of spot welds is designated as the minimum acceptable shear strength in pounds per spot, and is shown to the left of the weld symbol.

*3. Spacing of Spot Welds*

*a.* The pitch (center-to-center spacing) of spot welds is shown to the right of the weld symbol.

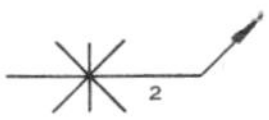

*b.* When spot weld symbols are shown directly on the drawing, spacing is shown by dimensions.

*4. Extent of Spot Welding*

*a.* When spot welding extends less than the distance between abrupt changes in the direction of the welding, or less than the full length of the joint (Art. A-10), the extent is dimensioned, thus:

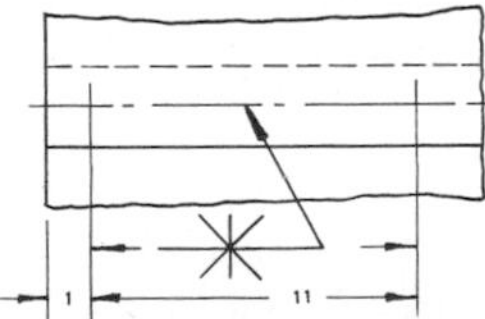


*5. Number of Spot Welds*

*a.* When a definite number of spot welds is desired in a certain joint, the number is shown in parentheses above or below the weld symbol.

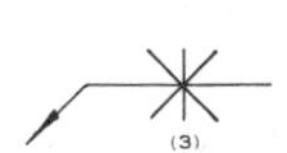


*6. Flush Spot-welded Joints*

*a.* When the exposed surface of one member of a spot-welded joint is to be flush, that surface is indicated by adding the flush-contour symbol to the weld symbol, observing the usual location significance.

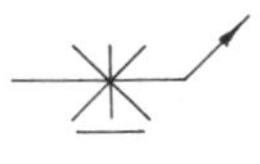

## H. Seam Welds.

*1. General*

*a.* Seam weld symbols have no arrow- or other-side significance in themselves, although the supplementary symbols used in conjunction with them may have such significance (Art. H-7). Seam weld symbols are centered on the reference line.

*b.* Dimensions of seam welds may be shown on either side of the reference line.

*2. Size of Seam Welds*

*a.* Seam welds are dimensioned by size or strength as follows:

(1) The size of seam welds is designated as the width of the weld expressed decimally in hundredths of an inch, and is shown, with inch marks, to the left of the weld symbol.

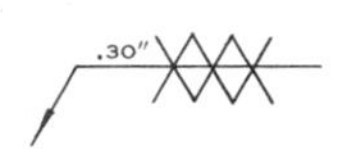


(2) The strength of seam welds is designated as the minimum acceptable shear strength in pounds per linear inch, and is shown to the left of the weld symbol.

*3. Length of Seam Welds*

*a.* The length of a seam weld, when indicated on the welding symbol, is shown to the right of the weld symbol.

*b.* When seam welding extends for the full distance between abrupt changes in the direction of the welding (Art. A-10), no length dimension need be shown on the welding symbol.

*c.* When seam welding extends less than the distance between abrupt changes in the direction of the welding, or less than the full length of the joint (Art. A-10), the extent is dimensioned in the same manner as for spot welds (Art. G-4).

*4. Dimensioning of Intermittent Seam Welding*

*a.* The pitch (center-to-center spacing) of intermittent seam welding is shown as the distance between centers of the weld increments.
*b.* The pitch (center-to-center spacing) of intermittent seam welding is shown to the right of the length dimension.

*5. Termination of Intermittent Seam Welding*

*a.* When intermittent seam welding is used by itself, the symbol indicates that increments are located at the ends of the dimensioned length.
*b.* When intermittent seam welding is used between continuous seam welding, the symbol indicates that spaces equal to the pitch minus the length of one increment are to be left at the ends of the dimensioned length.

*6. Combination of Intermittent and Continuous Seam Welding*

*a.* Separate symbols are used for intermittent and continuous seam welding when the two are used in combination.

*7. Flush Seam-welded Joints*

*a.* When the exposed surface of one member of a seam-welded joint is to be flush, that surface is indicated by adding the flush-contour symbol to the weld symbol, observing the usual location significance.

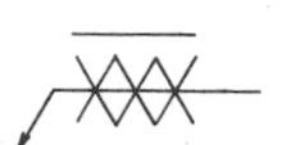

**I. Projection Welds.**

*1. General*

*a.* Embossments on the arrow-side member of a joint for projection welding is indicated by placing the weld symbol on the side of the reference line toward the reader.

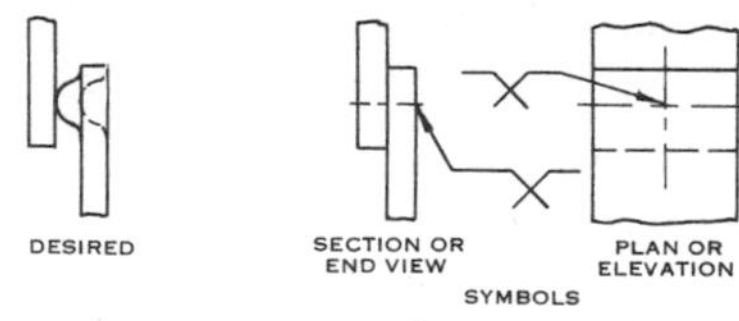


*b.* Embossments on the other-side member of a joint for projection welding is indicated by placing the weld symbol on the side of the reference line away from the reader.

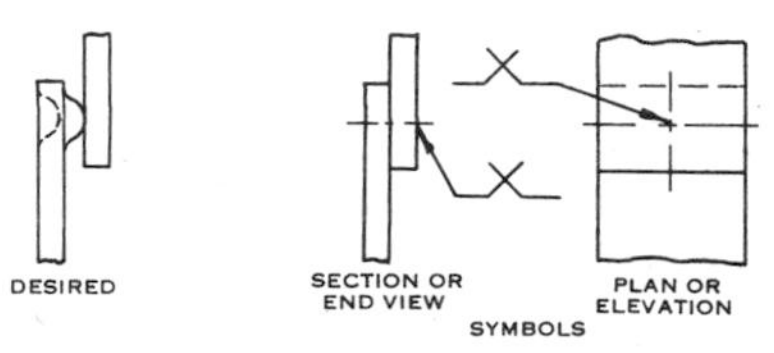


*c.* Proportions of projections are shown by a detail or other suitable means.
*d.* Dimensions of projection welds are shown on the same side of the reference line as the weld symbol.

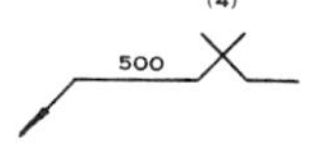


*2. Size of Projection Welds*

*a.* Projection welds are dimensioned by size or strength, as follows:
(1) The size of projection welds is designated as the diameter of the weld expressed decimally in hundredths of an inch, and is shown, with inch marks, to the left of the weld symbol.

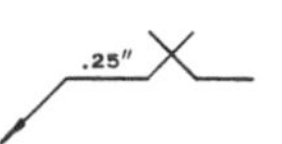


(2) The strength of projection welds is designated as the minimum ac-

ceptable shear strength in pounds per weld, and is shown to the left of the weld symbol.

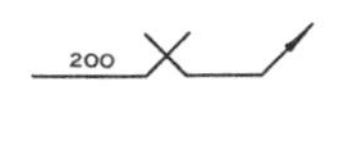


3. *Spacing of Projection Welds*

*a.* The pitch (center-to-center spacing) of projection welds is shown to the right of the weld symbol.

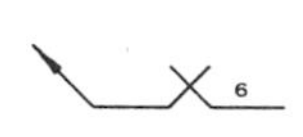


4. *Extent of Projection Welding*

*a.* When projection welding extends less than the distance between abrupt changes in the direction of the welding, or less than the full length of the joint, the extent is dimensioned in the same manner as for spot welds (Art. G-4).

5. *Number of Projection Welds*

*a.* When a definite number of projection welds is desired in a certain joint, the number is shown in parentheses.

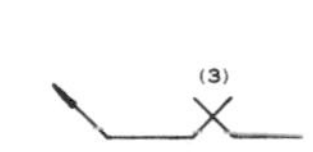


6. *Flush Projection-welded Joints*

*a.* When the exposed surface of one member of a projection-welded joint is to be made flush, that surface is indicated by adding the flush-contour symbol to the weld symbol, observing the usual location significance.

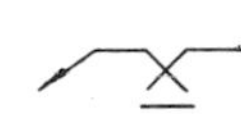

## J. Flash and Upset Welds.

1. *General*

*a.* Flash and upset weld symbols have no arrow-side or other-side significance in themselves although supplementary symbols used in conjunction with them may have such significance (Art. J-2). Flash or upset weld symbols are centered in the reference line.

*b.* The dimensions of flash and upset welds are not shown on the welding symbol.

2. *Surface Contour of Flash and Upset Welds*

*a.* Flash and upset welds that are to be made flush by mechanical means are shown by adding both the flush-contour symbol and the user's standard finish symbol to the weld symbol, observing the usual location significance.

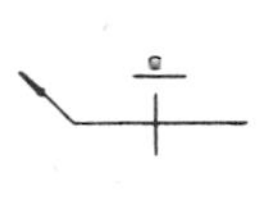


*b.* Flash and upset welds that are to be mechanically finished to a convex contour are shown by adding both the convex-contour symbol and the user's standard finish symbol to the weld symbol, observing the usual location significance.

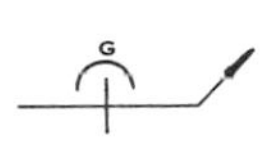


## 10. Rivets.

Rivets are used for making permanent fastenings, generally between pieces of sheet or rolled metal. They are round bars of steel or wrought iron with a head formed on one end and are often put in place red hot so that a head can be formed on the other end by pressing or hammering. Rivet holes are punched, punched and reamed, or drilled larger than the diameter of the rivet, and the shank of the rivet is made just long enough to give sufficient metal to fill the hole completely and make the head.

Large rivets are used in structural-steel construction and in boiler and tank work. In structural work, only a few kinds of heads are normally needed: the button, high button, and flat-top countersunk heads (Fig. 12).

For boiler and tank work, the button, cone, round-top countersunk, and pan heads are used. Plates are connected by either lap or butt joints. Figure 13*a* is a single-riveted lap joint; (*b*) is a double-riveted lap joint; (*c*) is a single-strap butt joint; and (*d*) a double-strap butt joint.

Large rivets are available in diameters of $\frac{1}{2}$ to $1\frac{3}{4}$ in., by increments of even $\frac{1}{8}$ in. The length needed is governed by the "grip," as shown in Fig. 13*d*, plus the length needed to form the head. Length of rivets for various grip distances may be found in the handbook "Steel Construction" published by the

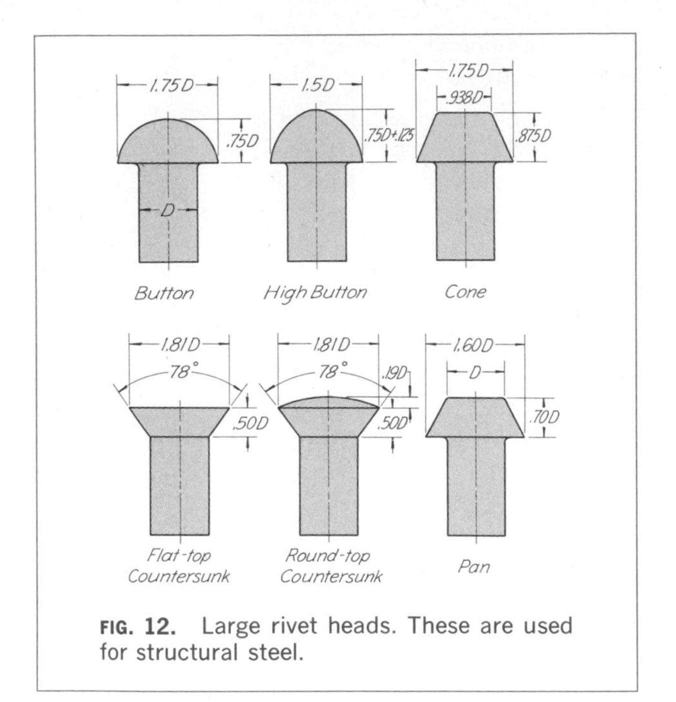


**FIG. 12.** Large rivet heads. These are used for structural steel.

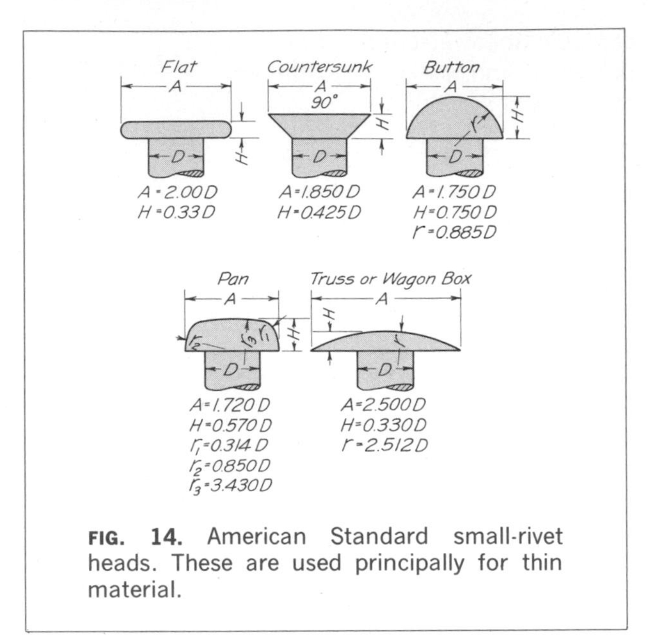


**FIG. 14.** American Standard small-rivet heads. These are used principally for thin material.

American Institute of Steel Construction.

Small rivets are used for fabricating light structural shapes and sheet metal. ASA small-rivet heads are shown in Fig. 14. Small rivets are available in diameters of $^3/_{32}$ to $^7/_{16}$ in., by increments of even $^1/_{32}$ in. to $^3/_8$ in. diameter.

Tinners', coopers', and belt rivets are used for fastening thin sheet metal, wood, leather, rubber, etc. Standard heads and proportions are given in the Appendix.

For large structures *rivet symbols* are used for representation because it is impossible to show the details of head form on a small-scale drawing. The ASA symbols for indicating various kinds of heads are given in Fig. 15.

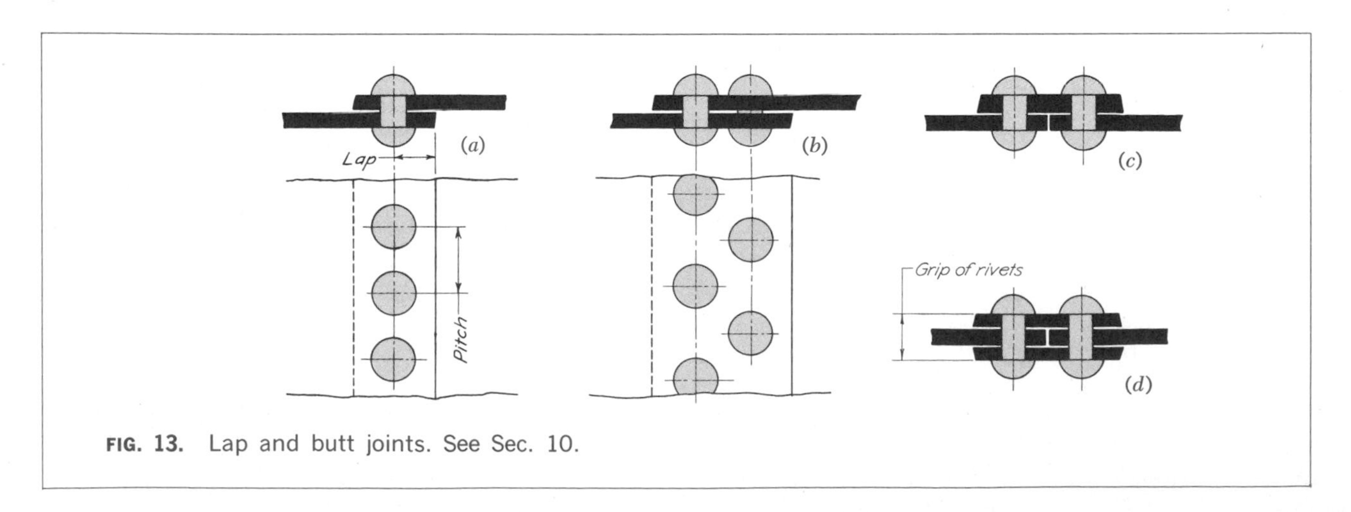


**FIG. 13.** Lap and butt joints. See Sec. 10.

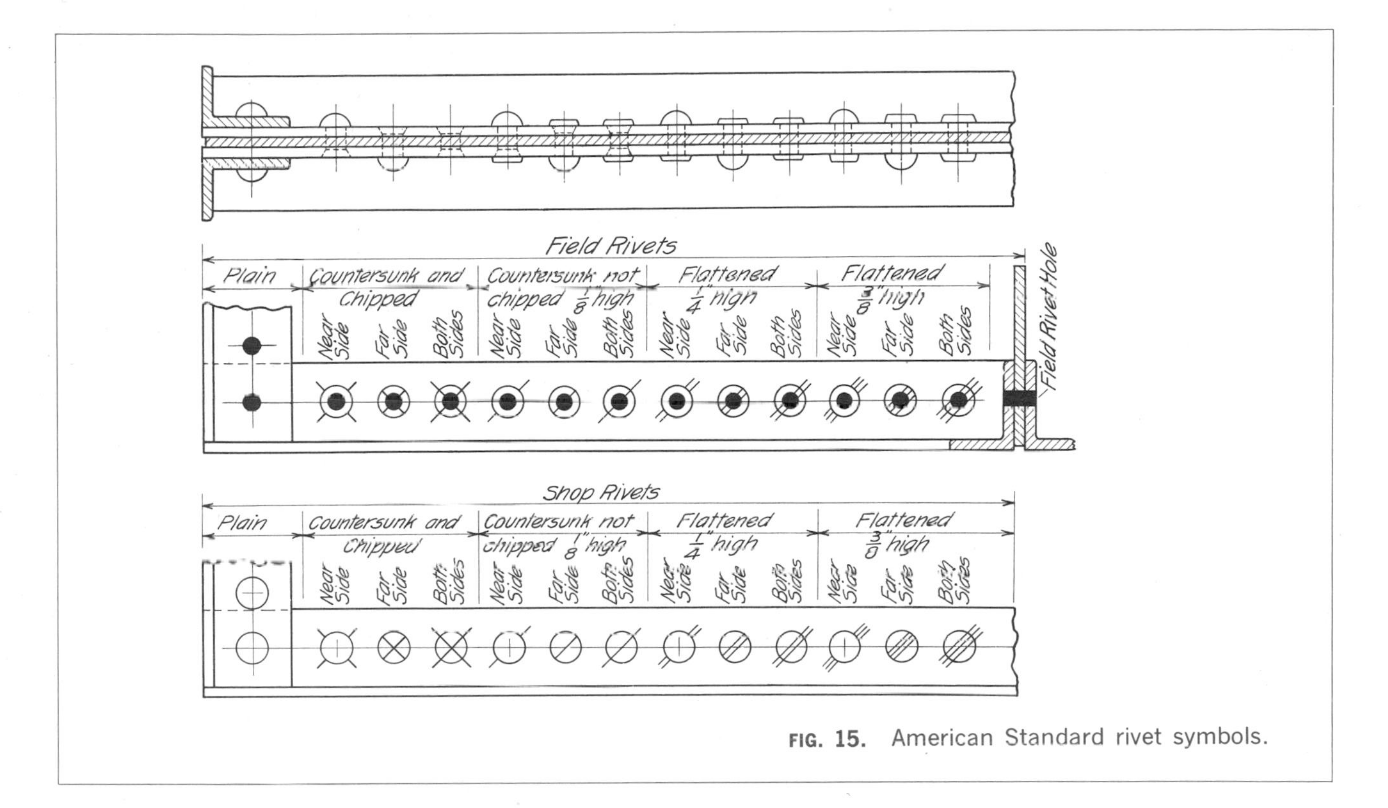


**FIG. 15.** American Standard rivet symbols.

## 11. Manufacturers' Specialties.

A common problem in all commercial drafting rooms is the design and representation of parts to accommodate manufacturers' specialties. Also necessary is the specification of the specialty, including the trade name or other pertinent information such as the size and company number. These are the principal reasons why the following descriptions and accompanying problems are given. There are many industrial companies that manufacture rivet specialities. Many of these are made specifically for the purpose of supplying a low-cost product. Others are designed for speed of assembly. Still others are devised for special uses.

It is impossible to give complete coverage here to all obtainable products or even to list all varieties made by a particular manufacturer. However, the following paragraphs describe some well-known and widely used parts. National organizations such as the ASME supply comprehensive lists of manufacturers' products, and the advertising in current periodicals is a good source of information. Descriptive literature is obtainable from the manufacturers.

*Rivet-nuts* are, as the name implies, a combination of a rivet and nut. They are tubular rivets with internal threads and have a body so designed that a hand or power assembly tool, which engages the thread, causes the shank to expand and fill the prepared hole in the part to be attached. *RIVNUTS*, made by the B. F. Goodrich Company, Akron, Ohio, are given in the Appendix.

*Drive rivets* are "blind"-type rivets consisting of a slotted and cored body and a steel grooved pin. In the installation the body is inserted in prepared holes in the parts to be fastened and then the pin is driven into the body. There are many sizes and lengths. *SOUTHCO* aluminum drive rivets and steel drive rivets are given in the Appendix. These rivets are made by the Southco Division of the South Chester Corporation, Lester, Penna.

# PROBLEMS

## Group 1. Welds.

Study the welding symbols so that you will be able to write and read them without hesitation. Problems 1 and 2 give practice in reading, Probs. 3 and 4 in writing. Problems 5 to 9 give practice in use of the symbols on working drawings.

**1 and 2.** Make full-size cross-sectional sketches of the joints indicated. Dimension each sketch.

**3 and 4.** Sketch members and show welding symbol for each complete joint. Estimate weld size from plate thickness.

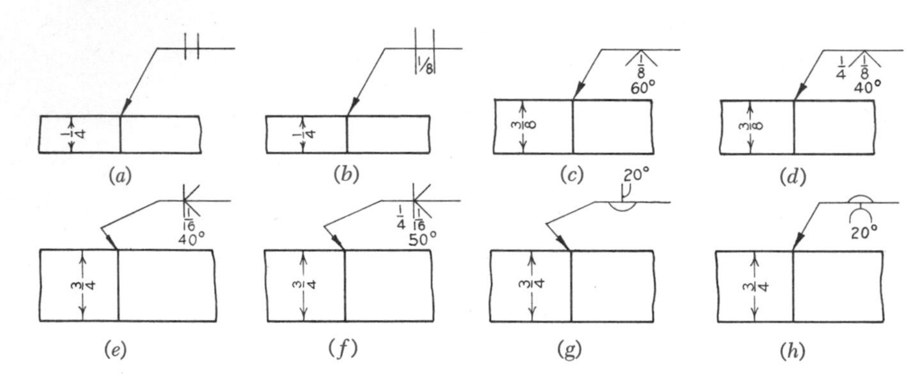


**PROB. 1.** Butt joints.

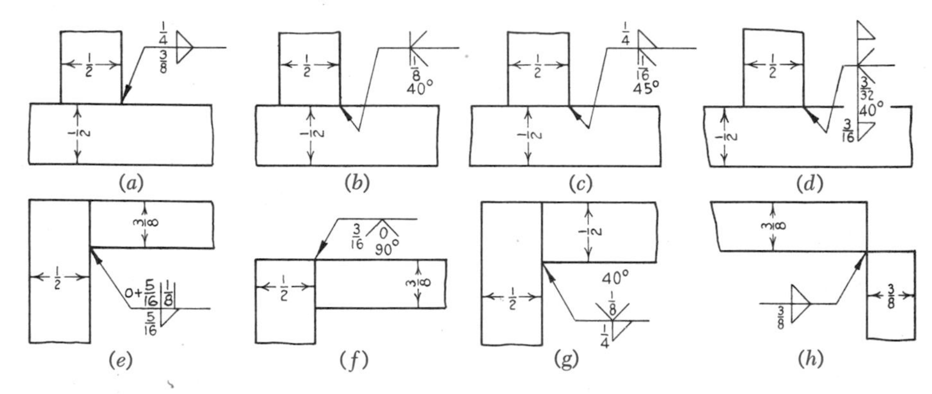


**PROB. 2.** Tee and corner joints.

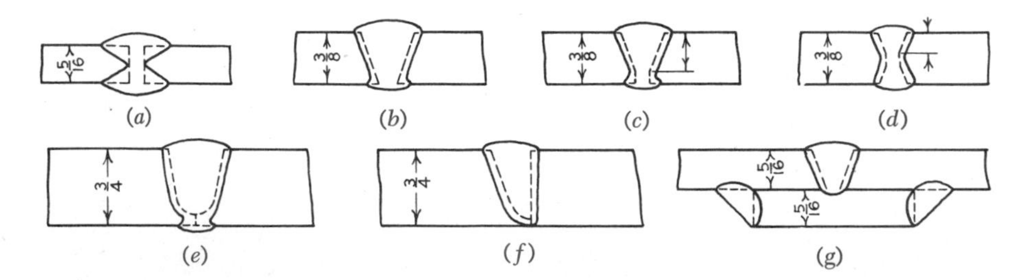


**PROB. 3.** Butt joints.

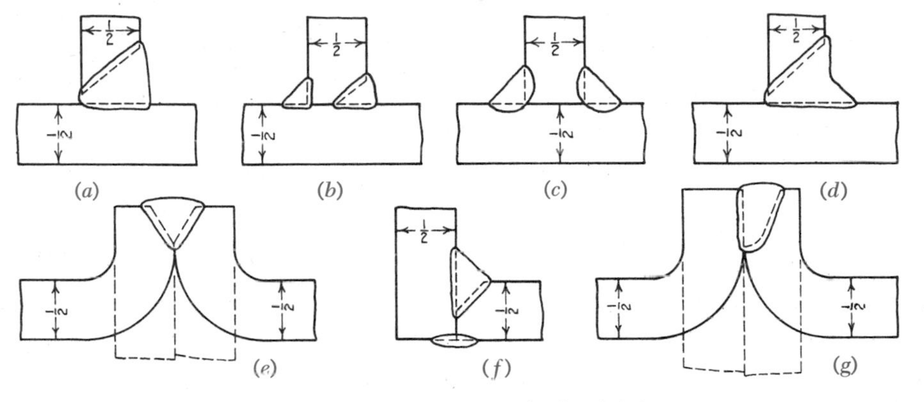


**PROB. 4.** Tee, corner, and edge joints.

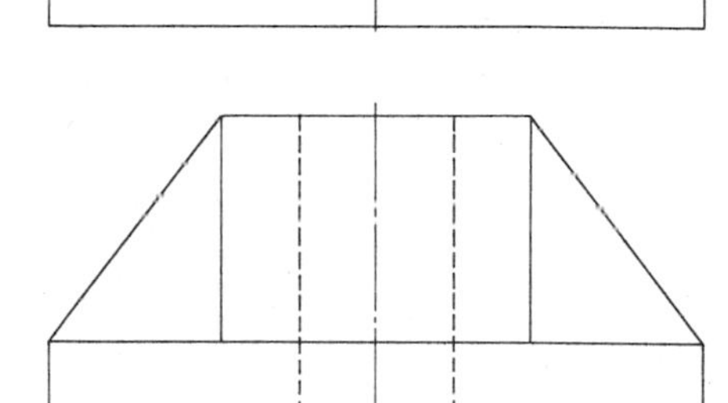

**PROB. 5.** Base.

## Group 2. Welded Parts.

**5 and 6.** Make a complete welding drawing for each object. These problems are printed quarter size. Draw full size by scaling or transferring with dividers.

**7.** Draw the views given, add welding symbols, dashed numbers for identification of the individual pieces, and complete the materials list.

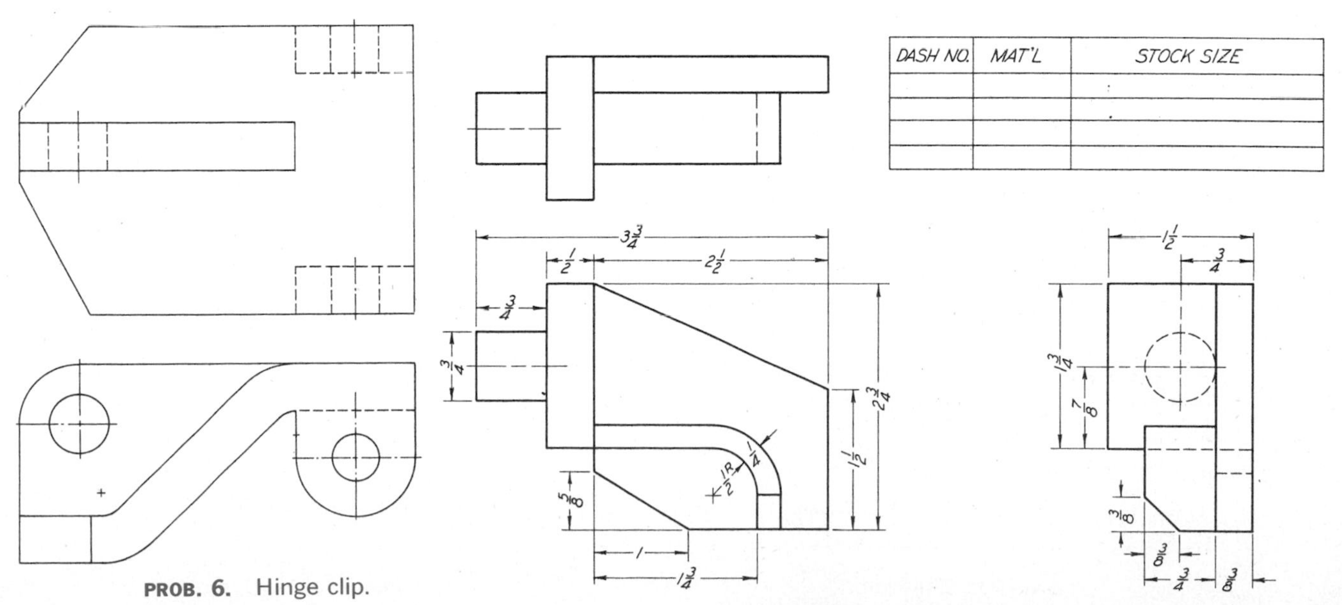

| DASH NO. | MAT'L | STOCK SIZE |
|---|---|---|
| | | |
| | | |
| | | |
| | | |

**PROB. 6.** Hinge clip.

**PROB. 7.** Pivoted spacer.

**8.** Draw the views given and add welding symbols and dashed numbers. Make a materials list similar to the one used in Prob. 7.

**9.** Draw the views given and add welding symbols and dashed numbers. Make a materials list. Determine length of material for rim.

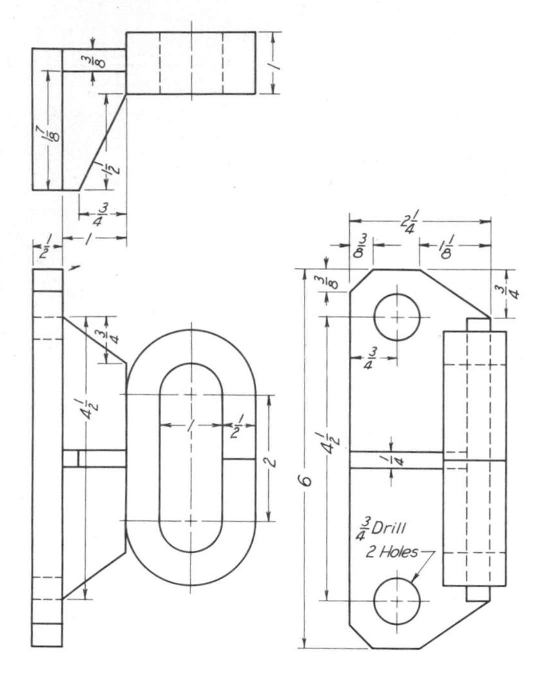


**PROB. 8.** Belt-tightener bracket.

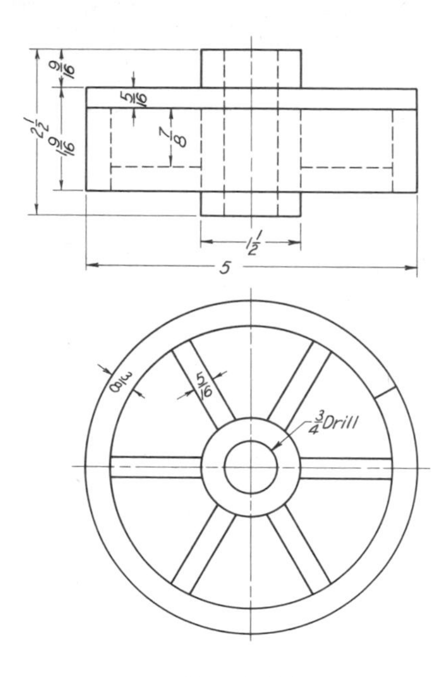


**PROB. 9.** Ribbed-disk wheel.

## Group 3. Rivets.

**10.** Draw top view and section of a single-riveted butt joint $10\frac{5}{8}$ in. long. Pitch of rivets is $1\frac{3}{4}$ in. Use cone-head rivets.

**11.** Draw a column section made of 15-in. by 33.9-lb channels with cover plates as shown, using $\frac{7}{8}$-in. rivets (dimensions from the handbook of the American Institute of Steel Construction). Use button-head rivets on left side and flat-top countersunk-head rivets on right side so that the outside surface is flush.

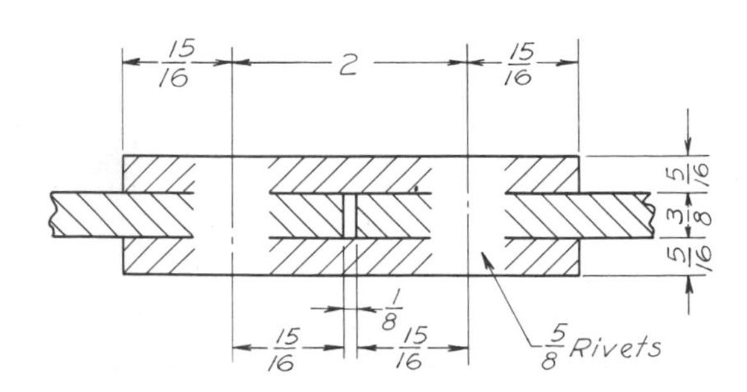


**PROB. 10.** Butt joint.

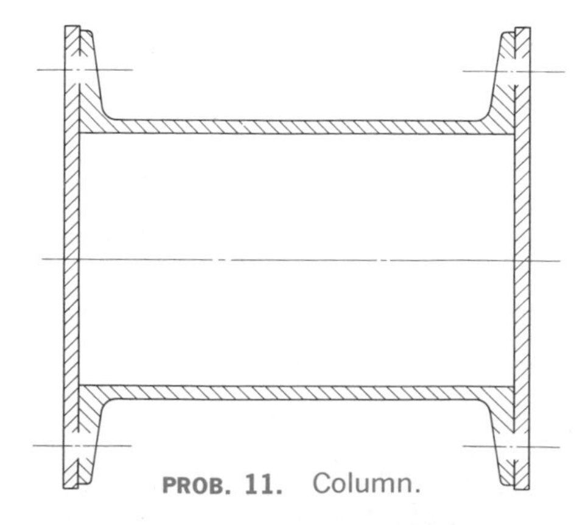

**PROB. 11.** Column.

## Group 4. Manufacturers' Specialties.

**12.** Two sheets of 16-gage steel are to be fastened together with a Rivnut. This Rivnut is to be $\frac{1}{4}$-in. dia., flathead, open end, keyless, aluminum. Specify Rivnut and draw inserted on center line, but not upset. Draw nine times actual size.

**13.** Same as Prob. 12. Two sheets of 16-gage steel are to be fastened together with a Rivnut. This Rivnut is to be 0.221-in. dia., flathead, closed end, keyed, steel. Specify Rivnut and draw inserted on center line, but not upset. Draw nine times actual size.

**14.** A $\frac{1}{4}$-in. steel plate is to be fastened to a sheet of 16-gage steel by a Rivnut. This is to be a countersunk head, $\frac{1}{4}$-in. dia., closed end, keyless, steel Rivnut.

(*a*) Draw and specify Rivnut with fastener inserted on center line, but not upset. Draw three times size.

(*b*) Completely specify the flathead machine screw of maximum length that will fasten a $\frac{1}{8}$-in. plate to the above after upsetting.

**15.** Same as Prob. 14. A $\frac{1}{4}$-in. steel plate is to be fastened to a sheet of 16-gage sheet metal by a Rivnut. This is to be a countersunk head, 0.490-in. dia., open end, keyless, corrosion-resistant steel Rivnut. Specify Rivnut and draw on center line with Rivnut inserted, but not upset. Draw three times size.

**16.** Two sheets of 16-gage sheet metal are to be fastened together with a $\frac{1}{8}$-in.-dia. aluminum drive rivet made by the Southco Division of the South Chester Corporation. Show a universal head drive rivet in position on center line but not upset. Specify the rivet and draw nine times actual size.

**17.** Same as Prob. 16, but full brazier head instead of universl head.

**18.** Two $\frac{1}{4}$-in. plates are to be fastened together with a Southco $\frac{3}{16}$-in.-dia. aluminum drive rivet. Show an all-purpose liner head drive rivet in position on center line, but not upset. Specify the rivet and draw three times actual size.

**19.** Same as Prob. 18, but for 100° countersunk head.

**20.** Same as Prob. 18, but for full brazier head.

**21.** Same as Prob. 18, but for universal head.

**22.** Same as Prob. 18, but for round ply-head.

**23.** Same as Prob. 18, but for flat ply-head.

**24.** A $\frac{1}{4}$-in. plate is to be fastened to a 1-in.-thick plywood member. The fastening device is to be a Southco $\frac{1}{4}$-in.-dia. aluminum drive rivet. The plywood is to have a $\frac{3}{4}$-in.-deep hole. Show a 100° countersunk head drive rivet in place on center line, but not upset. Specify the rivet and draw three times actual size.

**25.** Same as Prob. 24, but for universal head.

**26.** Same as Prob. 24, but for full brazier head.

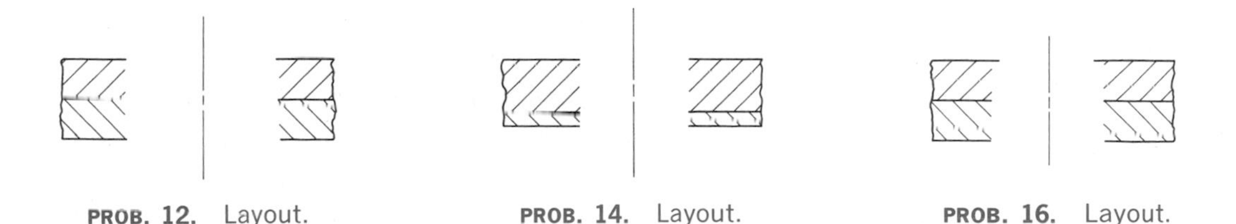

**PROB. 12.** Layout. **PROB. 14.** Layout. **PROB. 16.** Layout.

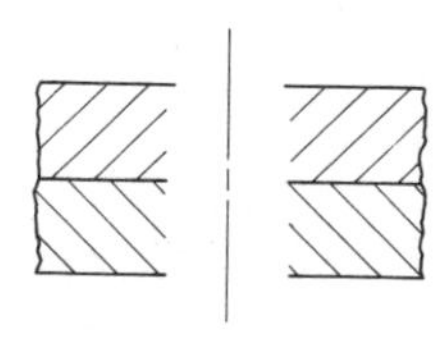

**PROB. 18.** Layout.

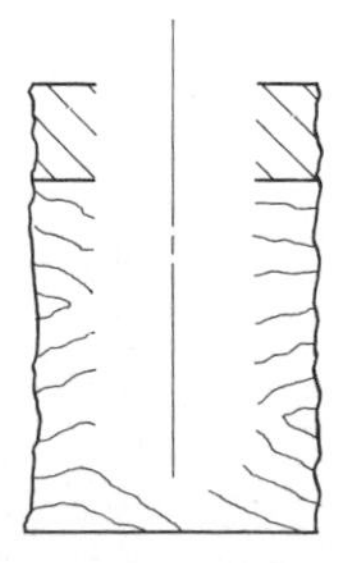

**PROB. 24.** Layout.

# 17

# JIGS AND FIXTURES

**In Quantity Production, Jigs and/or Fixtures Are Used in Manufacture in order to Assure Consistently Accurate, Precise Parts**

**Included Here:**

### 1. Production in Quantity—The Need for Consistency.

Manufacturing methods for producing a single unit or very few units require individual measurements for locations, careful and accurate gaging during machining, and meticulous care to bring parts to required specifications. All this must be done by trained and experienced workers, an expensive procedure because the parts are virtually handmade.

Many products—for example, automobiles, home appliances, and components such as valves, motors—are manufactured by the thousands, even millions. Manufacturing parts in such quantities by ordinary unit methods would be impractical not only because of the time and care required but, more important, because of parts thus produced being costly and not consistently interchangeable.

Even when producing only a few parts, say up to 100, the use of devices known as jigs and fixtures assists greatly in obtaining uniform parts at reduced cost.

### 2. Jig and Fixture Drawings.

Jig and fixture drawings are essentially the same as the average working drawing except for the complication, on an *assembly* drawing, of the necessity to show both the jig or fixture and the production piece which the jig or fixture is manufactured to produce. Thus, some means of identification, especially on and for the drawing proper and also for identification on a print of the drawing, is needed. The jig or fixture is drawn in black (pencil or ink) and the production piece is drawn in a color. This practice results in easy identification on the original drawing; and because colored pencil lines and also colored inks print lighter than black lines, the production piece is also identifiable on the print. Traditionally, the production piece has been drawn in red. However, since the development of new printing methods, more spectrally sensitive than blueprinting (which prints red lines almost as densely as black), other colors such as blue and green which are on the opposite end of the spectrum (from red) are preferred. Remember that the real purpose of using color for the production piece is for identification purposes only.

When choosing colored pencils for showing the production piece on an assembly drawing, use the *hardest* obtainable. This is because some colored pencils, such as the type that will mark practically any material—paper, glass, plastic, and metal—will smear easily and are also very difficult to erase from paper or from some plastic sheets which are "grained" to receive pencil lines.

Further, it is universal practice to draw the jig or fixture as it would appear if the production piece were not in place. Thus, an assembly jig or fixture drawing is actually *two* drawings, superimposed.

Detail drawings of the individual components of a jig or fixture are quite naturally drawn as any detail drawing would be, in black, because the superimposition of the production piece is not considered or shown.

### 3. Jigs.

A *jig* is a device that holds a production and guides a rotating tool that performs some machining operation. Thus a jig is used for drilling, reaming, counterboring, countersinking, tapping, or combinations of these operations. The part being processed is often called "the production" or "the subject." Figure 1 is a picture of a small jig for performing several drilling operations on a part. The jig is placed on the table of a drill press, and the drill is passed through a "drill bushing" in the jig. The drill is thus accurately located (positioned) and guided by the bushing to drill a hole in the production. The production is accurately located and clamped tightly in the jig so that each part produced will be essentially identical. The accuracy built into the jig controls the accuracy of the finished product and eliminates the necessity for measurements on each (individual) part in order to locate the holes.

### 4. Principles of Design.

To illustrate some of the principles of jig design, a simple jig for drilling two $\frac{3}{4}$-in. holes and reaming one 1-in. hole in a pawl carrier is shown in Fig. 2.

**FIG. 1.** A jig. The product to be processed ($A$) is placed in the jig ($B$) where it is located against finished surfaces and held in position by clamps ($C$). The drill then passes through a bushing at ($D$) to make the hole in part ($A$).

The principles that should be followed as far as possible in designing a jig are:

1. The production must go into the jig easily and quickly.
2. The production must be located accurately.
3. The bushings must be accessible to the operator.
4. The production must be securely clamped in the jig.
5. The production must be removable easily and quickly.

It is universal practice in jig drawing to show the production in *red;* the jig itself is drawn in *black*. Visible features are drawn as they would be if the production were not in place. On blueprints and direct-process prints, the red lines that show the production print *lighter* than the black lines. Thus, also on the print, the production is distinguishable

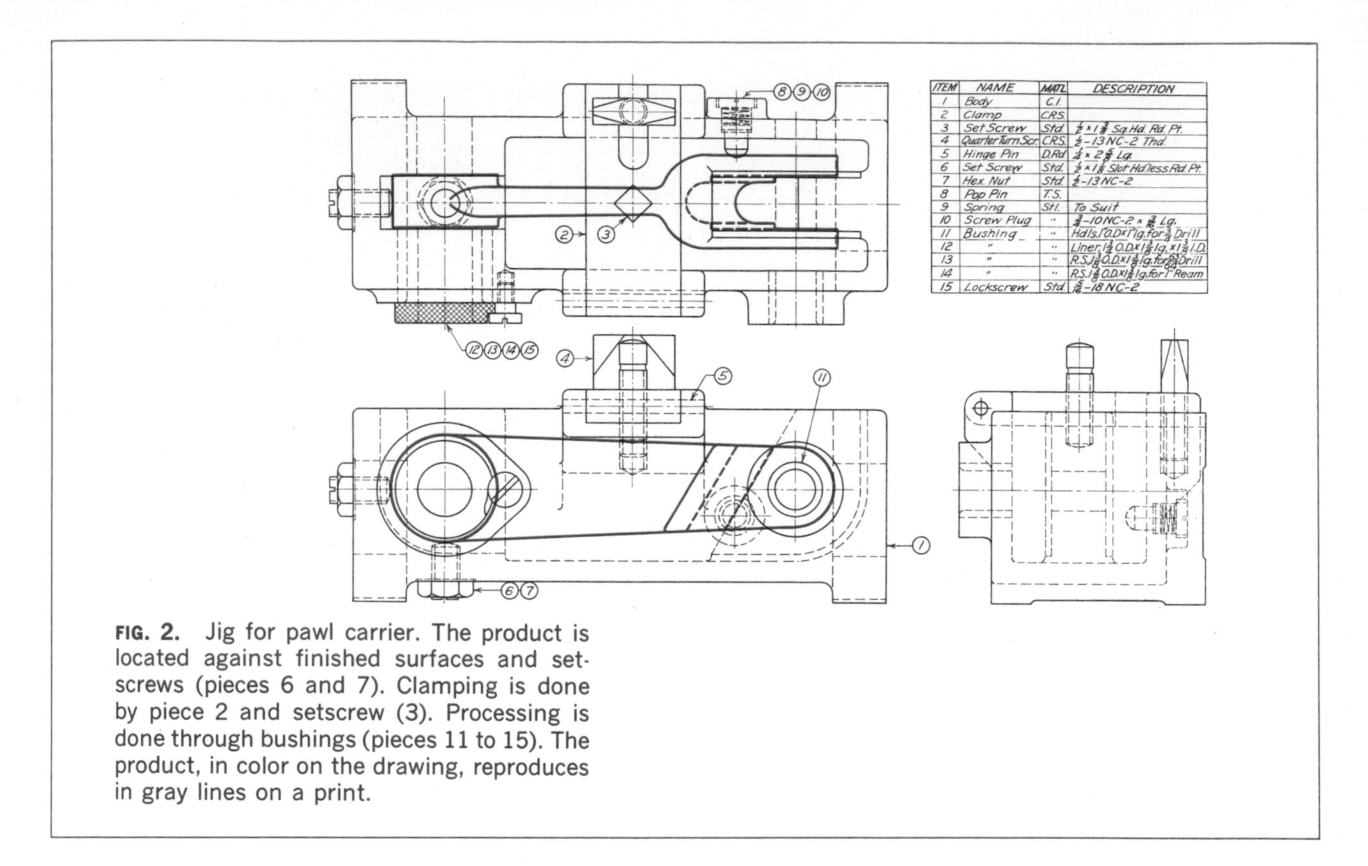

| ITEM | NAME | MATL | DESCRIPTION |
|---|---|---|---|
| 1 | Body | C.I. | |
| 2 | Clamp | C.R.S. | |
| 3 | Set Screw | Std. | $\frac{1}{2} \times 1\frac{3}{8}$ Sq. Hd. Rd. Pt. |
| 4 | Quarter Turn Scr. | C.R.S. | $\frac{1}{2}$-13 NC-2 Thd. |
| 5 | Hinge Pin | D. Rd. | $\frac{1}{4} \times 2\frac{5}{8}$ Lg. |
| 6 | Set Screw | Std. | $\frac{1}{2} \times 1\frac{1}{8}$ Slot Hd'less Rd. Pt. |
| 7 | Hex. Nut | Std. | $\frac{1}{2}$-13 NC-2 |
| 8 | Pop Pin | T.S. | |
| 9 | Spring | Stl. | To Suit |
| 10 | Screw Plug | " | $\frac{3}{4}$-10 NC-2 $\times \frac{3}{16}$ Lg. |
| 11 | Bushing | " | Hdls. 1" O.D. × 1" lg. for $\frac{3}{4}$ Drill |
| 12 | " | " | Liner, $1\frac{3}{4}$ O.D. × $1\frac{3}{8}$ lg. × $1\frac{3}{8}$ I.D. |
| 13 | " | " | R.S. $1\frac{3}{8}$ O.D. × $1\frac{3}{8}$ lg. for $\frac{63}{64}$ Drill |
| 14 | " | " | R.S. $1\frac{3}{8}$ O.D. × $1\frac{3}{8}$ lg. for 1" Ream |
| 15 | Lockscrew | Std. | $\frac{5}{16}$-18 NC-2 |

**FIG. 2.** Jig for pawl carrier. The product is located against finished surfaces and setscrews (pieces 6 and 7). Clamping is done by piece 2 and setscrew (3). Processing is done through bushings (pieces 11 to 15). The product, in color on the drawing, reproduces in gray lines on a print.

from the jig or fixture. All the jig and fixture drawings in this chapter illustrate shop drawings; therefore the production shows in red lines.

In all jig designs four main points of design related to the above principles must be considered. Briefly they are:

1. Locating the production accurately
2. Clamping the production securely
3. Selecting bushings of correct style and size
4. Designing the jig body to accommodate the production and satisfy the principles

By observing these four cardinal points, a drill jig can be designed that will satisfy the requirements of commercial production. Study each of the four aspects of the design carefully to determine the best solution. There are many different methods of providing for each. Select the one that will be most suitable in combination with the method of providing for the other three. We will discuss several methods of meeting each of the four requirements, but selection is not confined to these in designing a particular piece of work.

## 5. Locating the Production.

The shape of the production, previous finishing before drilling, and other points of design will influence the type of locator best suited for the production. Location must be thoroughly considered, as it is perhaps the most important of the four points for accuracy in the jig.

*Finished surfaces* are often used to locate—a finished surface of the production is placed against a finished surface of the jig; or, when necessary, an unfinished surface of the production is placed against a finished surface of the jig. Location surfaces may take the form of pads, counterbores, or two finished surfaces at right angles to each other.

*Pins* give an easy and relatively inexpensive

method of location and at the same time a very accurate one. A finished or an unfinished surface of the production is held against a pin and a finished surface, or against two or three pins, by a clamp or screws (Fig. 3).

In the jig in Fig. 6 two pins, one circular and the other flattened, are used to locate. Both pins are accurate to the size necessary to fit into two previously drilled or reamed holes in the production. (One pin must be flattened because the center-to-center distances of the holes may vary enough to make a fit impossible with two round pins.) The round pin locates along the line of centers, and both pins locate at right angles to the line of centers, since the flats of the flattened pin are always placed perpendicular to the line of centers. Note that the pins are made so that they can be pressed into the fixtures only up to the shoulder. The ends are chamfered, preferably at 30°, to allow easy entry into the production.

Small pins are usually made of tool steel, hardened and ground. Large pins may be made of cold-rolled steel, pack-hardened and ground. The shoulder of the pin is pressed against a finished surface or into a light spot face or counterbore to a suitable depth.

*Bushings* serve as locators in certain designs, as illustrated in Fig. 4. This jig is to serve in drilling and reaming holes in a gear case, on which the out-to-out distance between bosses is an accurate dimension with limits; hence the surfaces of these bosses provide good points for location, and the accurately located shoulder bushings are designed to come to contact with them. Location of the subject in the other directions is by using finished surfaces, against which it is clamped by screws and a bar clamp.

*V blocks* are often used in jig design, both as locators and as clamps. An example was seen in Fig. 3. The jig in Fig. 5 employs a V whose purpose is mainly for location, though it also serves as a "backstop" in clamping, while the setscrews are the clamps proper. A V block is more easily made as a separate piece, secured to the body of the jig by screws and dowel pins. Fastening the V block without dowel pins, by using slotted holes for the cap

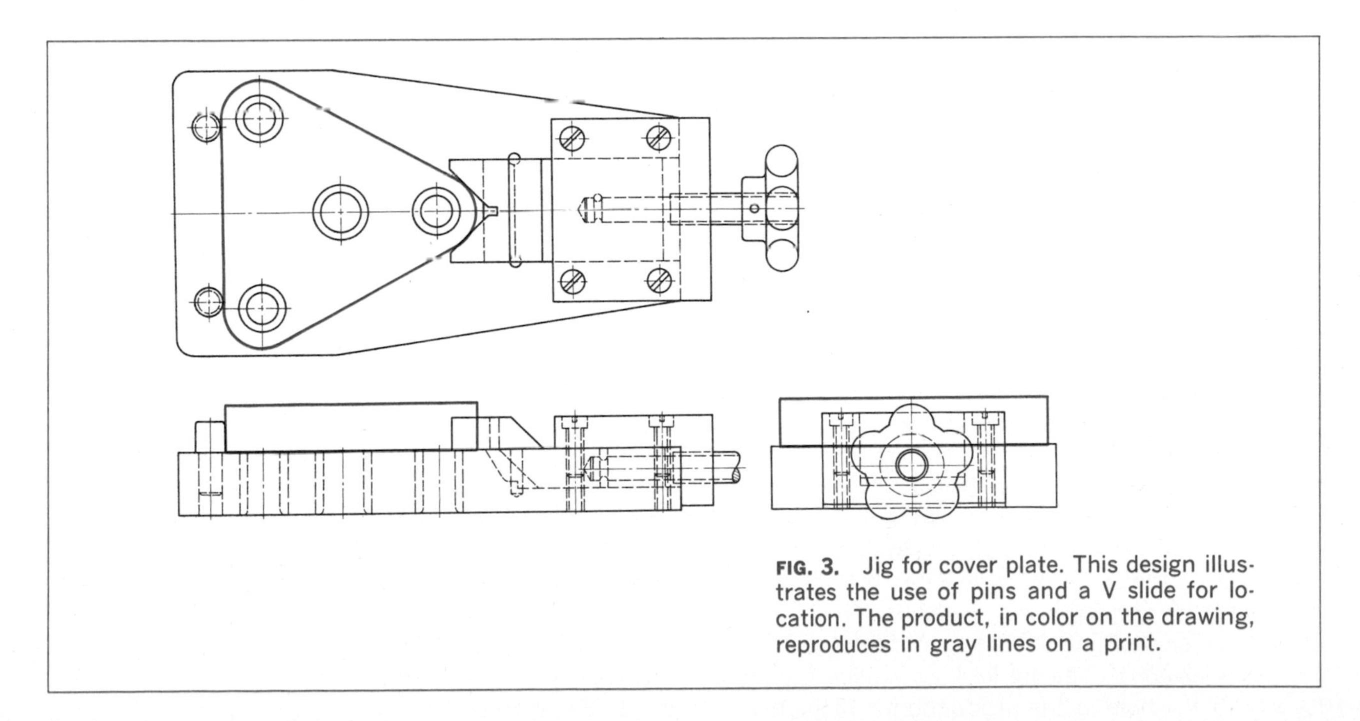

**FIG. 3.** Jig for cover plate. This design illustrates the use of pins and a V slide for location. The product, in color on the drawing, reproduces in gray lines on a print.

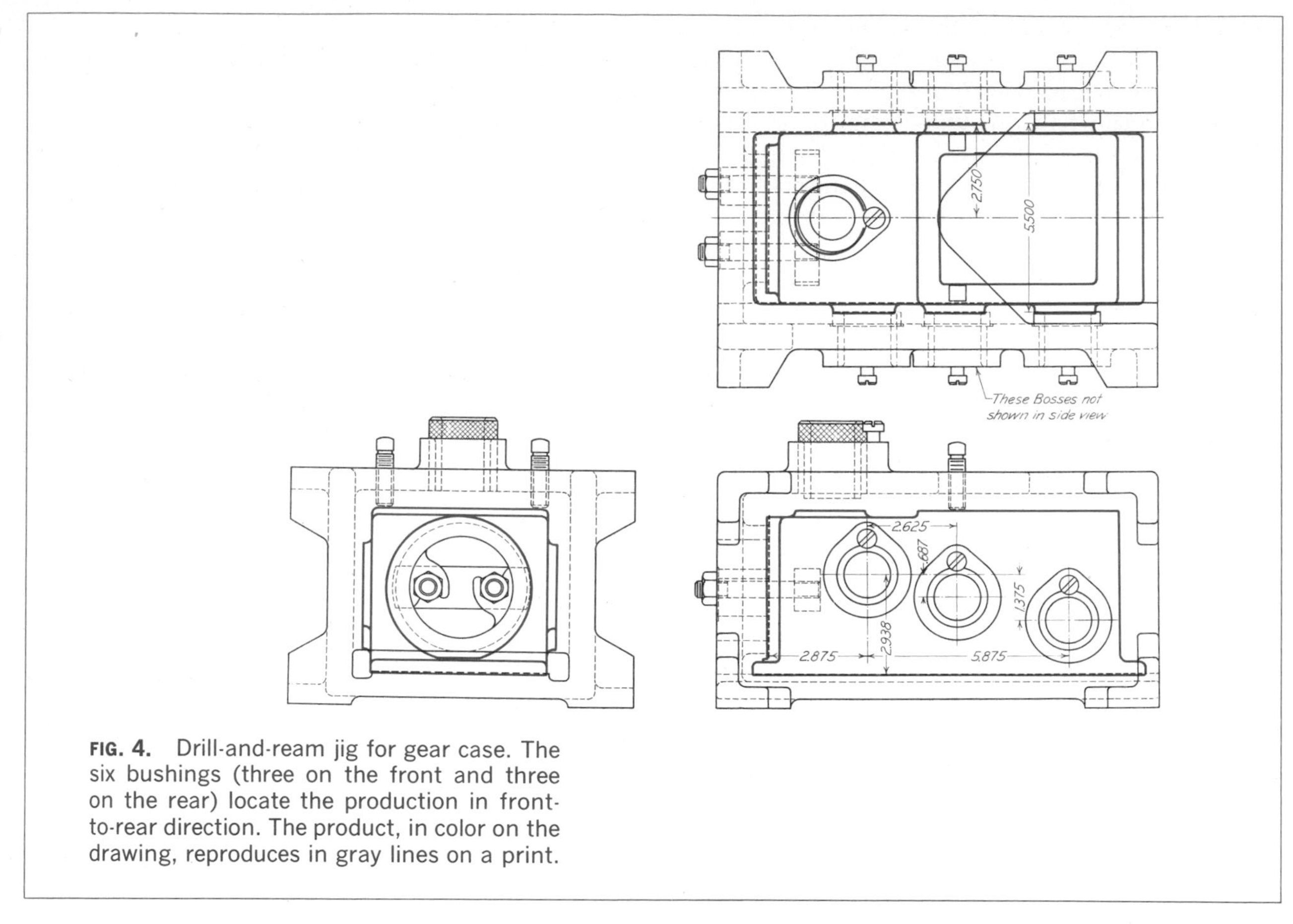


**FIG. 4.** Drill-and-ream jig for gear case. The six bushings (three on the front and three on the rear) locate the production in front-to-rear direction. The product, in color on the drawing, reproduces in gray lines on a print.

screws, might be necessary in case the circular boss on the production varies considerably in diameter with different castings.

*Accurate holes for pins.* A small plate jig is often used to drill a recurring series of holes in making a large jig, as shown in Fig. 6, or in a production itself. The small jig in the figure carries two locating pins, one round and the other flattened (for reasons already mentioned in Sec. 5). On the large jig the holes marked $A$ have been previously drilled and reamed, with the jig borer or on a radial boring machine, to match the pins of the small jig; by using the small jig, the series of holes $B$ can then be drilled with far more accuracy and speed than if each hole had to be located individually.

*Center locators.* The jig in Fig. 7 uses a center locator. If the hole in the production is of such size that it has been bored on a lathe and the piece faced at the same time, this method of locating is indicated. The shank for locating is of such diameter that the hole is a slip fit over it.

*Keyways* serve as locators in cases where it is necessary to drill holes that must be in a particular position. In a jig designed for a coupling where the holes must be in the center of the bosses and at the same time located with respect to the keyway, the center locator would include a key to slip into the keyway of the subject.

## 6. Clamps and Clamping.

Some of the more commonly used clamps for fastening the subject securely in the jig are the following:

1. Bar clamp

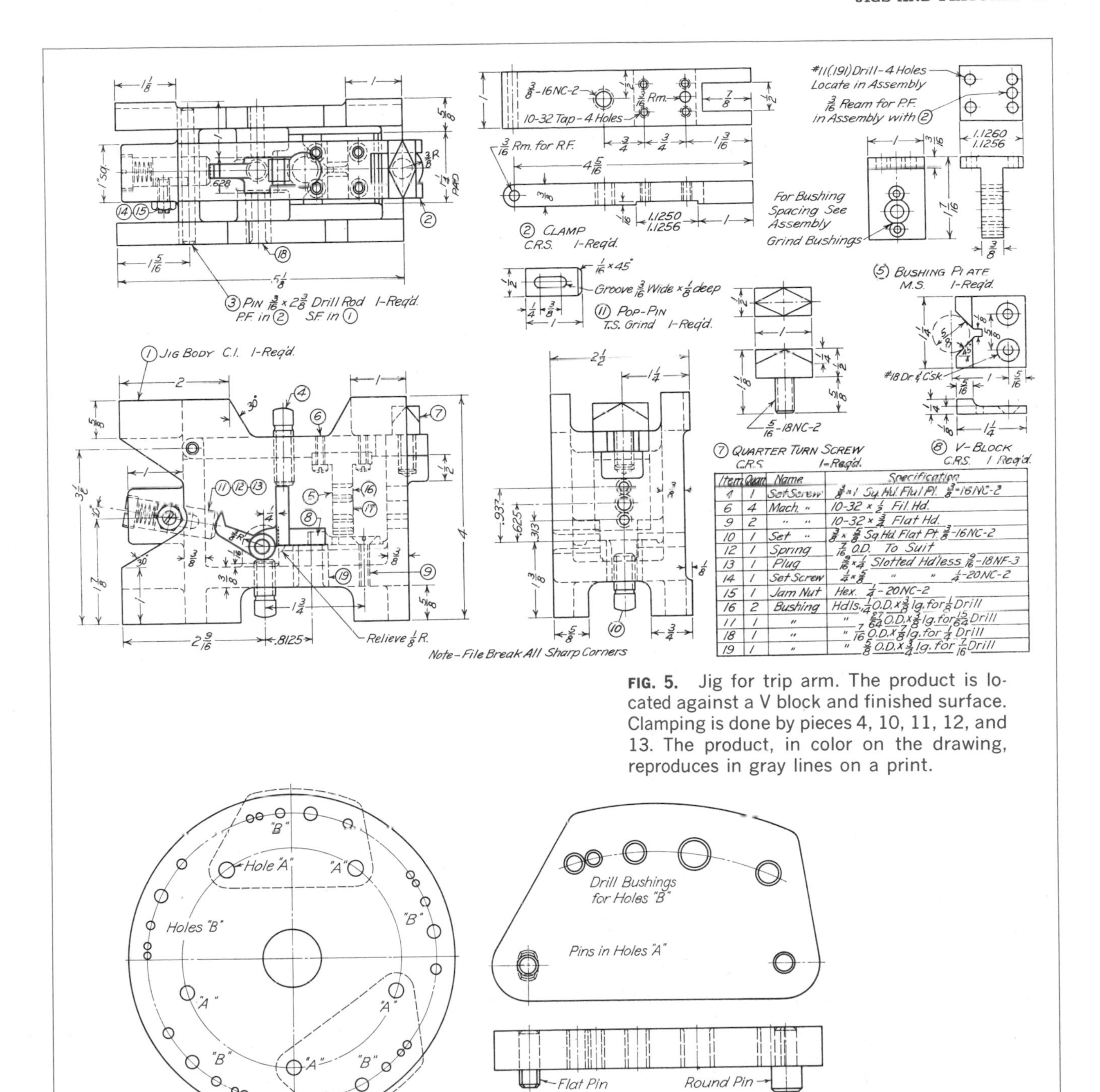


| Item | Quan | Name | Specification |
|---|---|---|---|
| 4 | 1 | Set Screw | $\frac{3}{8}$ × 1 Sq. Hd. Flat Pt. $\frac{3}{8}$-16NC-2 |
| 6 | 4 | Mach. " | 10-32 × $\frac{1}{2}$ Fil. Hd. |
| 9 | 2 | " " | 10-32 × $\frac{3}{4}$ Flat Hd. |
| 10 | 1 | Set " | $\frac{3}{8}$ × $\frac{5}{8}$ Sq. Hd. Flat Pt. $\frac{3}{8}$-16NC-2 |
| 12 | 1 | Spring | $\frac{7}{16}$ O.D. To Suit |
| 13 | 1 | Plug | $\frac{9}{16}$ × $\frac{1}{4}$ Slotted Hdless $\frac{9}{16}$-18NF-3 |
| 14 | 1 | Set Screw | $\frac{1}{4}$ × $\frac{5}{8}$ " " $\frac{1}{4}$-20NC-2 |
| 15 | 1 | Jam Nut | Hex. $\frac{1}{4}$-20NC-2 |
| 16 | 2 | Bushing | Hdls. $\frac{1}{4}$ O.D. × $\frac{3}{8}$ lg. for $\frac{1}{8}$ Drill |
| 17 | 1 | " | " $\frac{27}{64}$ O.D. × $\frac{3}{8}$ lg. for $\frac{15}{64}$ Drill |
| 18 | 1 | " | " $\frac{7}{16}$ O.D. × $\frac{7}{8}$ lg. for $\frac{1}{4}$ Drill |
| 19 | 1 | " | " $\frac{5}{8}$ O.D. × $\frac{3}{4}$ lg. for $\frac{7}{16}$ Drill |

**FIG. 5.** Jig for trip arm. The product is located against a V block and finished surface. Clamping is done by pieces 4, 10, 11, 12, and 13. The product, in color on the drawing, reproduces in gray lines on a print.



**FIG. 6.** Jig for making a jig. This type of jig is often used to obtain accuracy for identical sets of holes.

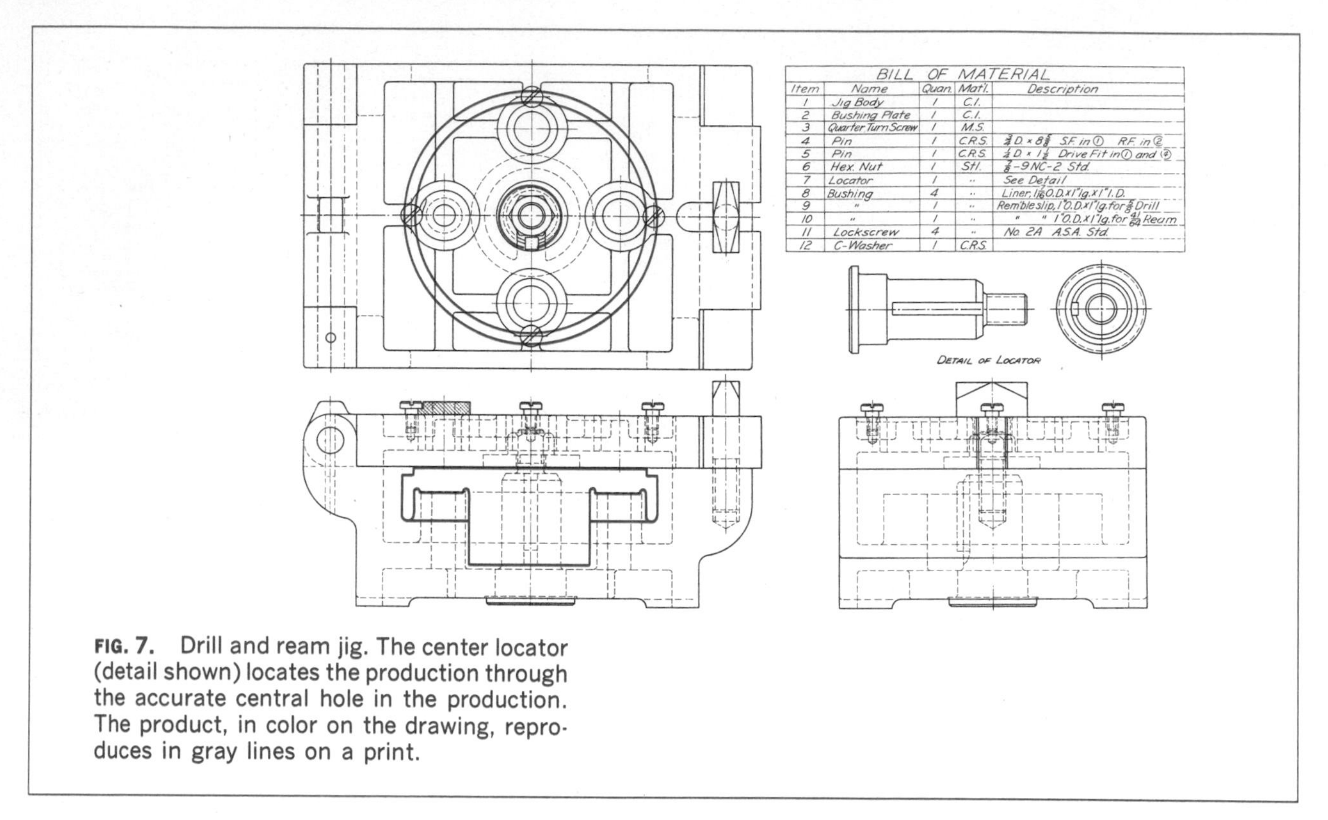

BILL OF MATERIAL

| Item | Name | Quan. | Mat'l. | Description |
|---|---|---|---|---|
| 1 | Jig Body | 1 | C.I. | |
| 2 | Bushing Plate | 1 | C.I. | |
| 3 | Quarter Turn Screw | 1 | M.S. | |
| 4 | Pin | 1 | C.R.S. | $\frac{3}{4}$ D. × $8\frac{5}{8}$ S.F. in ① R.F. in ② |
| 5 | Pin | 1 | C.R.S. | $\frac{1}{4}$ D. × $1\frac{1}{2}$ Drive Fit in ① and ④ |
| 6 | Hex. Nut | 1 | Stl. | $\frac{7}{8}$-9 NC-2 Std. |
| 7 | Locator | 1 | " | See Detail |
| 8 | Bushing | 4 | " | Liner, $1\frac{7}{16}$" O.D. × 1" lg. × 1" I.D. |
| 9 | " | 1 | " | Rem'ble slip, 1" O.D. × 1" lg. for $\frac{5}{8}$ Drill |
| 10 | " | 1 | " | " " 1" O.D. × 1" lg. for $\frac{41}{64}$ Ream |
| 11 | Lockscrew | 4 | " | No. 2A A.S.A. Std. |
| 12 | C-Washer | 1 | C.R.S. | |



**FIG. 7.** Drill and ream jig. The center locator (detail shown) locates the production through the accurate central hole in the production. The product, in color on the drawing, reproduces in gray lines on a print.

2. Slotted clamp
3. Setscrews and studs
4. C washer
5. V slide
6. Spiral-rise cam
7. Star knob and stud
8. Adjustable pins
9. Hydraulic piston

The jigs illustrated in this chapter show several of these methods of clamping. The important point to observe is that *the clamp must not distort the production, as such distortion introduces inaccuracy in the drilled holes upon release of the clamping pressure. Clamping must be applied at a point on the production that will withstand the strain introduced* and as close as possible to the point drilled. It is not always possible to apply clamping close to the point drilled, but this is an important consideration.

Clamps suffer the most wear of any part of a jig, and cyanide hardening is advisable. When there is a possibility of marring a finished surface, a soft-nosed clamp should be employed.

*Slotted clamps* are widely used. To get proper action, they should have the stud at the center or closer to the production than to the tail of the clamp. Studs and nuts are preferred to cap screws, as they do not wear out the body of the jig and are cheap to replace.

*Setscrews* are cheap and highly efficient as a means of holding the production securely (Fig. 5). In clamping four sides of an object, the setscrews on two sides will have lock nuts, and those on the other two sides will be used for locking and unlocking the piece.

Use may be made of a stud and nut in combination with a C washer in cases where the production slips over the stud or a locating pin (Fig. 7).

*V slide.* In using the V slide in clamping, take care to see that its length is at least equal to its width, to avoid having an unstable action. Its thickness depends upon the production and whether the V is fixed or movable. In the latter case, the size of the control screw will have a bearing on the thickness.

*Spiral-rise cam clamps* are useful in quantity production where quick clamping and unclamping with little thought required from the operator are the prime requisites. The maximum variation in the production pieces at the point of clamping must be known—the rise of the cam is computed from this point—in order that the cam face will clamp the production with a 90° turn of the handle. Figure 8 shows two types of locking cams that give locking action in either of two directions from the axis.

*Star knobs* and studs are an adaptation of the setscrew principle for hand operation.

*Adjustable pins* are used to support fragile sections of the production. Correct designs are shown in Fig. 9. They should be locked into position by a setscrew.

*Hydraulic pistons* are for heavy work and work requiring special clamping. Their design is of too specialized a nature to include here.

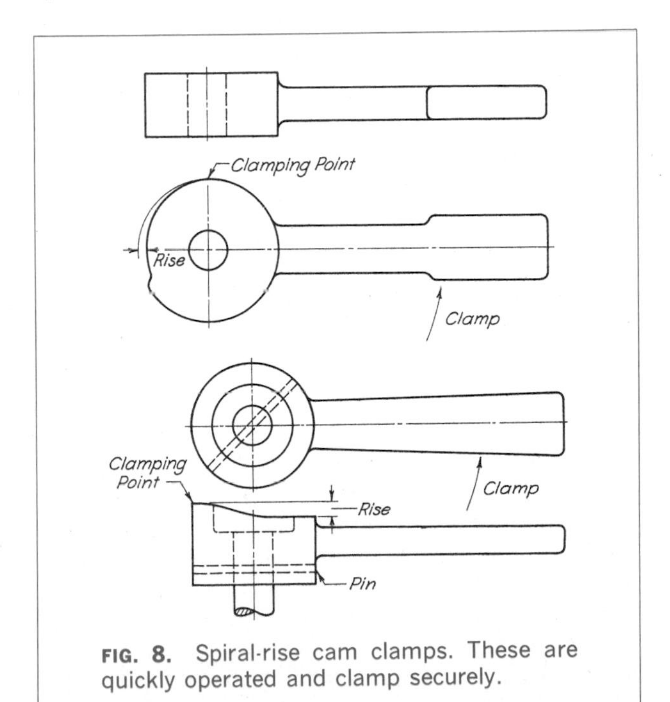


**FIG. 8.** Spiral-rise cam clamps. These are quickly operated and clamp securely.

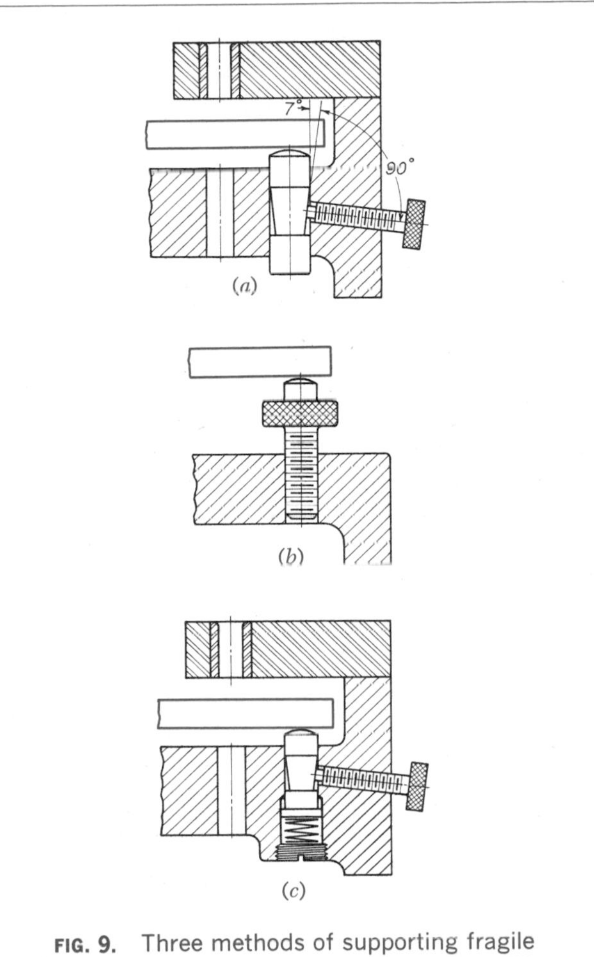


**FIG. 9.** Three methods of supporting fragile sections. (*a*) Adjustable pin and clamp; (*b*) adjustable screw; (*c*) spring-backed pin and clamp. The product, in color on the drawing, reproduces in gray lines on a print.

## 7. American Standard Bushings.

Drill bushings are standardized items, made in five different styles (Fig. 10). They are available in from six to eight lengths in each style for use with all the numbered, lettered, and fractional drill sizes up to 2 inches.

*Plain, stationary press-fit bushings,* used when the bushing is expected to last during the life of the jig, are made in two types: the *headless* and the *head* types. The headless type is used when the center

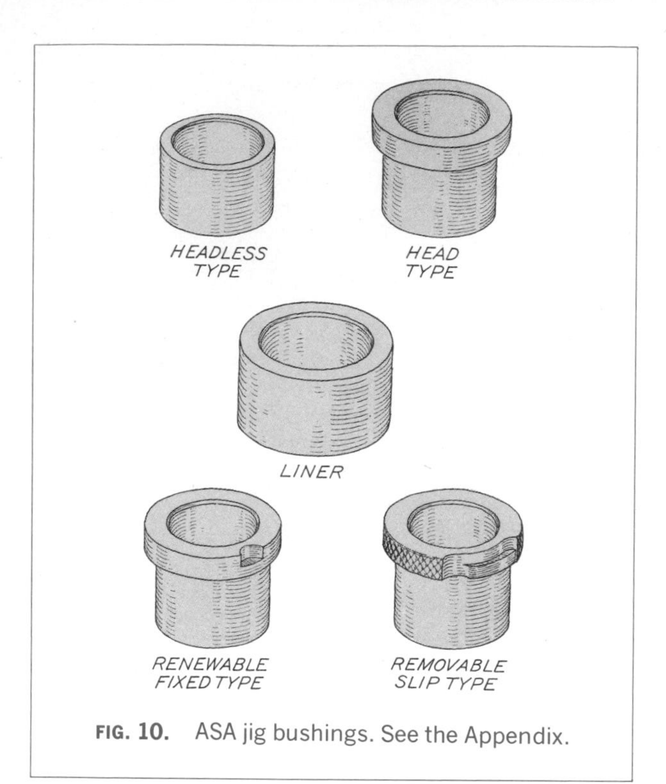


**FIG. 10.** ASA jig bushings. See the Appendix.

distance of holes is too close for a bushing with a head and when it is desirable to have the top of the bushing flush with the top of the jig plate. Both types are used without liners.

*Renewable bushings,* with liners, are used in cases where the bushing will wear enough for replacement or where it is necessary to interchange bushings in one hole. They are of two types: the *renewable fixed* and the *renewable-slip*[1] types. Renewable fixed bushings are used for one operation only. Renewable slip bushings are used where two or more operations, requiring different inside diameters, are performed in a single jig without removing the production from the jig, for example, when drilling is followed by reaming, spot-facing, counterboring, tapping, etc. They should be used in combination with a liner and lock screw unless the design cannot possibly allow the additional space required by the liner.

[1]Also called *removable-slip* type.

Bushings are specified by giving, in order, (1) type, (2) outside diameter, (3) length, (4) drill size.

For correct installation, design drawings should show clearly the bell-mouthed end of the bushing as the entry end for the drill. For accurate drilling, the other end should be not more than one drill diameter from the production. The thickness of the production, type of material being drilled, and design of the jig will all influence the minimum distance between the end of the bushing and the production. Chip clearance must be considered to avoid drill binding and the creation of unusual pressures. Sometimes the bushing is designed to touch the production, and the chips are carried up and out at the top.

Some shops prefer to use the type of bushings that have $\frac{1}{64}$-in. grinding stock on the outside diameter, for fitting the bushing to the hole.

Bushings for special work should be described by following the tables of standard wall thicknesses, size of head, etc., and specifying the proper finish and heat-treatment.

Standard drill bushings are listed in the Appendix.

## 8. The Jig Body.

Jig bodies are of two general classes: the open body and the closed or box type. In general, open jigs have drill bushings in the same plane, parallel to one another. Closed or box-shaped jigs are for drilling holes from various planes and directions. Occasionally there may be an overlap in the nomenclature of the two general types.

Because of the rigidity required, cast iron has been the usual material for jig bodies, but welded steel is now being used successfully.

The weight of the body should be considered. For ease of handling, it should have no excess weight, but it must not be lightened at the expense of the stiffness and rigidity needed for accuracy. It is often possible, however, to core out metal in various places without decreasing strength. For the comfort and safety of the operator, corners should be rounded and all burrs and sharp edges removed by filing. For convenience in moving, small jigs can be equipped with handles, and large ones with hooks for handling with a crane.

Finished feet should be provided on the sides opposite the drill bushings. For proper machining, small lugs are often placed on other sides to act as stops. The jig feet are generally part of the casting but in some cases are inserts. Four should always be used in preference to three, because with four feet any unevenness in setting, such as a chip under one foot, will at once, by rocking, draw the attention of the operator.

On the inside of the jig and at other places where machining is to be done, particular care should be taken to allow proper clearance for the machining tools. Points of location should, if possible, be visible to the operator.

Small jigs do not need to be clamped to the table, but large jigs and all fixtures (see Sec. 11) should be provided with means of clamping securely to the machine on which they are used.

## 9. Summary—Points in Jig Design.

1. Provide the best method of locating.
2. Provide the best method of clamping.
3. Select correct types of bushings.
4. Have bushings accessible to the operator.
5. Design for quick loading and unloading.
6. Design for ability to withstand abuse without affecting accuracy.
7. Keep in mind the safety of the operator.
8. Provide clearance for drills after passing through the work.
9. Provide for chip clearance and easy removal of chips.
10. Design so that cheaper parts wear out first.
11. See that finished surfaces will not be marred by clamping devices.
12. Provide means for lifting heavy jigs.
13. If loose parts are unavoidable, chain them to the body of the jig.
14. Consider the cost of materials and labor, but do not attempt to cut the cost of the jig at the expense of efficiency.

## 10. Making a Jig Drawing.

The procedure in designing a jig or fixture should follow approximately this order:

1. Sketch the design freehand to get the proper choice of views and an idea of space requirements. This original sketch will take into account previously finished surfaces of the production.
2. Allowing ample space between views, carefully draw the production in *red* in its several views. When the jig drawing is reproduced, the red lines print gray, so the production remains distinguishable from the jig.
3. Build the jig around the production, following the correct principles of location, clamping, bushings, and body design. With respect to visibility, one should draw the jig as though the production were not in place.
4. Dimension the drawing of the jig, using decimal dimensions for all locators and bushings, following the system of base-line dimensioning, from zero coordinate axes.
5. Give each part an item number.
6. Prepare a bill of material of all items in sequence.
7. Check the drawing.

## 11. Fixtures.

A fixture is a device that holds and locates a production so that some machining operation can be performed. Fixtures are used on machine tools that have their own (built-in) guiding and controlling mechanism for the tool. Thus, on a lathe the fixture may be for facing, turning, or boring; on a milling machine, for milling one or more surfaces; on a grinder, for grinding cylindrical or flat surfaces.

Figure 11 is a photograph of a turning fixture; Fig. 12, of a milling fixture.

Two examples are given in Figs. 13 and 14 to illustrate the principles of fixture construction. Figure 13 is a lathe fixture to aid in boring the hole and facing the projection and bottom of the flange of Prob. 18, Chap. 8. The flange locates over the pins. The center clamp is removed, and slotted clamps are used while the hole and projection are being machined. To complete the finishing of the bottom, the clamps are slid back and the center clamp is put in place.

Figure 14 is a fixture for holding the toggle-shaft support of Prob. 117, Chap. 5, in boring the hole

**FIG. 11.** A turning fixture. The production (*A*) is located over pins on the fixture (*B*) and is held in place by the clamp (*C*). The tool (*D*) then makes the required cut.

and facing the end. The bracket locates over two pins and is held in place by the clamp.

Both these fixtures clamp to the faceplate of the lathe, the entire fixture and production rotating. Being unbalanced, the offset bracket fixture requires a counterbalance to reduce vibration and aid in accuracy of work. To compute the size of the counterbalance, the center of gravity and the moment of the fixture and production together about the working center must be found. From this, the area and thickness of counterbalance and its working distance from the center may be calculated to find an equivalent moment to balance that on the opposite side of center. Common practice is to provide a slightly oversize counterbalance thickness, which allows the shop to complete the balance by removal of metal. This is a time-saver and permits slight changes that may take place in some parts used in the fixture.

In designing fixtures for milling, slotting, saw cutting, and similar operations, the same principles of location and clamping as those given for jig de-

**FIG. 12.** A milling fixture. The production (*A*) is located in the fixture (*B*) against the finished surfaces and over a pin. The clamp (*C*) holds the production in position as the milling cutter (*D*) makes the required cut.

signing are followed. To obtain secure clamping, finished bases should be provided with two square keys for aligning the fixture with the T slots in the milling table, as well as with slots at each end for T bolts.

If a previous operation has been performed on the subject, a gaging surface should be provided, if at all possible, to which the cutter may be set, thus obtaining accurately the required distance between the two finished surfaces.

**12. Assembly Jigs and Fixtures.** Sometimes it is necessary to perform some machining operation at the time of assembly of parts. Thus, an assembly jig locates two (or more) parts and provides bushings for drilling, etc. Or an assembly fixture may locate two (or more) parts for a machining operation. In some cases a device is used *only* to *locate* one part, relative to another, at assembly. This is termed (1) an assembly fixture, (2) a locating fixture, or (3) an adjusting fixture.

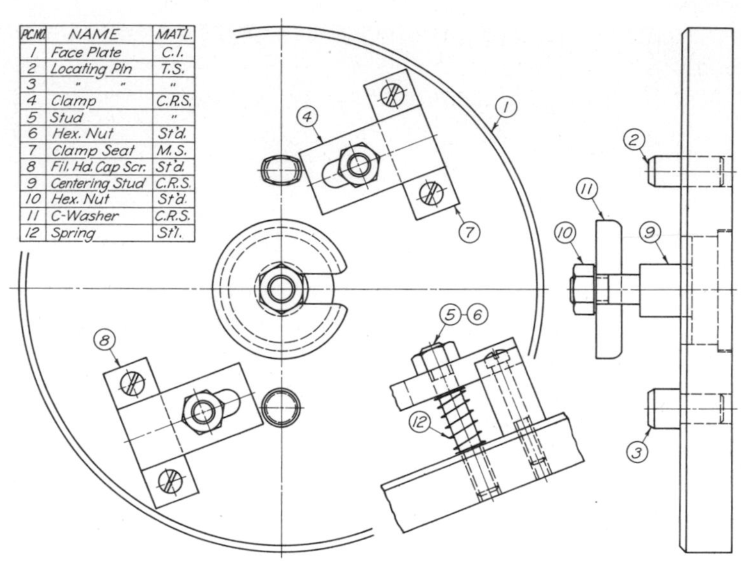

| PC.NO. | NAME | MAT'L. |
|---|---|---|
| 1 | Face Plate | C.I. |
| 2 | Locating Pin | T.S. |
| 3 | " " | " |
| 4 | Clamp | C.R.S. |
| 5 | Stud | " |
| 6 | Hex. Nut | St'd. |
| 7 | Clamp Seat | M.S. |
| 8 | Fil. Hd. Cap Scr. | St'd. |
| 9 | Centering Stud | C.R.S. |
| 10 | Hex. Nut | St'd. |
| 11 | C-Washer | C.R.S. |
| 12 | Spring | St'l. |

**FIG. 13.** Lathe fixture for boring and facing. This fixture is for Prob. 18, Chap. 8.

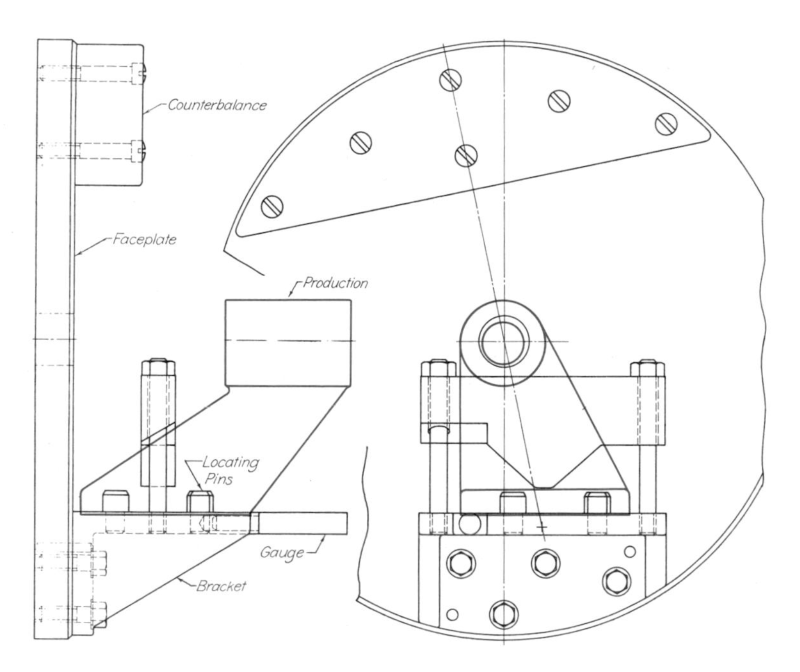


**FIG. 14.** Counterbalanced boring-and-facing fixture. This fixture is for Prob. 117, Chap. 5. The product, in color on the drawing, reproduces in gray lines on a print.

### 13. Standards.

The scope of this book allows only an introduction to the very broad field of tooling and toolmaking. For further information, consult the standards of the ASA and SAE. Textbooks are listed in the bibliography. Especially pertinent are the standards on small tools and machine-tool elements.

## PROBLEMS

**1.** Design a drill-and-ream jig for Fig. 23, Chap. 12. Locate on a finished pad the lug in a finished slot and the finished end of the cylinder, 4 in. long, against a finished boss of the jig. Clamp from the top and back with one setscrew each. Provide plain bushings for the base holes in the pad and two liners with removable drill-and-ream bushings for the larger holes. Two lengths of removable bushings will be required, the longer being to reach into within a drill diameter of the cylinder $2\frac{3}{4}$ in. long. Design a box-type body with feet opposite the sides in which bushings are placed. The right side of the body will be open to permit loading. Spot-face the holes outside the jig.

**2.** Design a drill jig for special nut (Fig. 85*a*, Chap. 12). Locate over a center locator on a finished pad through the $1\frac{1}{8}$-in. hole, the locator reducing in size sufficiently to use a nut and C washer for clamping on the $1\frac{3}{4}$-in.-diameter end. Nut should be small enough to clear the hole when loading and unloading. Provide bushings in an open-type body.

*Alternative method.* Use the same method of locating but provide only one drill bushing and an index pin 90° from the bushing. Drill one hole, index this hole to the pin, drill the next hole, etc.

**3.** Design a drill jig for the $\frac{1}{4}$-in. oil hole in the bracket of Prob. 41, Chap. 23. Locate on a horizontal locator through the large center hole of the production, using a nut and C washer for removal. Index the production with a pin through one of the capscrew holes, so that the drill will enter the oil reservoir in the casting at the proper place.

**4.** Design a boring-and-facing fixture for flange, Prob. 18, Chap. 8. Refer to the design shown in Fig. 13. Assume that the top flat surface has been finished and the $\frac{3}{4}$-in. holes reamed prior to the operation of boring and facing in this fixture. For boring the $2\frac{5}{8}$- and 2-in.-diameter holes, use the two clamps, the center stud and washer being removed. For facing, remove or slip the clamps out of place and use the centering stud with washer and nut. On a production line, two fixtures could be made using the parts necessary for successive operations in each fixture.

**5.** Design a boring-and-facing fixture for plunger bracket, Prob. 115, Chap. 5. Refer to the design shown in Fig. 14. Assume that the base is finished and that the base holes are reamed to size. Use these for locating over two pins, one flat and one round, and clamp down to the projecting shelf with a clamp as shown, or equivalent. This fixture should be counterbalanced to reduce vibration and obtain greater accuracy.

In this Chapter:



# 18

# GEARS AND CAMS

Many Machines Require Some Form of Mechanism to Change Speed, Alter Relative Movement, Produce a Required Design Characteristic, or Provide a Mechanical Advantage

## 1. Gears—Classification.

The representation and specification of gears are of such common occurrence that it is important to become familiar with the nomenclature, basic proportions, and formulas for calculation.

Gears are an adaptation of rolling cylinders and cones, designed to ensure positive motion. There are numerous variations, but the basic forms (Fig. 1) are *spur gears,* for transmitting power from one shaft to a parallel shaft; *spur gear* and *rack,* for changing rotary motion to linear motion; *bevel gears,* for shafts whose axes intersect; and *worm gears,* for nonintersecting shafts at right angles to each other.

## 2. Gear Teeth.

The teeth of gears are projections designed to fit into the tooth spaces of the mating gear and contact mating teeth along a common line known as the "pressure line" (Fig. 2). The most common form for the tooth flank is the *involute,* and when it is made in this form the gears are known as involute gears. The angle of the pressure line determines the particular involute the flank will have. The ASA has standardized two pressure angles, $14\frac{1}{2}°$ and 20°. A composite $14\frac{1}{2}°$ tooth and a 20° stub tooth are also used (see ASA B6.1).

## 3. Letter Symbols and Formulas for Spur Gearing.

The names of the various portions of a spur gear and the teeth are given in Fig. 3. The ASA standard letter symbols and formulas for calculation are as follows:

$N$ = number of teeth $= P_d \times D$
$P_d$ = diametral pitch = number of teeth on the gear for each inch of pitch diameter $= N/D$
$D$ = pitch diameter $= N/P_d$
$p$ = circular pitch = length of the arc of the pitch-diameter circle subtended by a tooth and a tooth space $= \pi D/N = \pi/P_d$
$t$ = circular (tooth) thickness = length of the arc of the pitch-diameter circle subtended by a tooth $= p/2 = \pi D/2N = \pi/2P_d$
$t_c$ = chordal thickness = length of the chord subtended by the circular thickness arc $= D \sin (90°/N)$
$a$ = addendum = radial distance between the pitch-diameter circle and the top of a

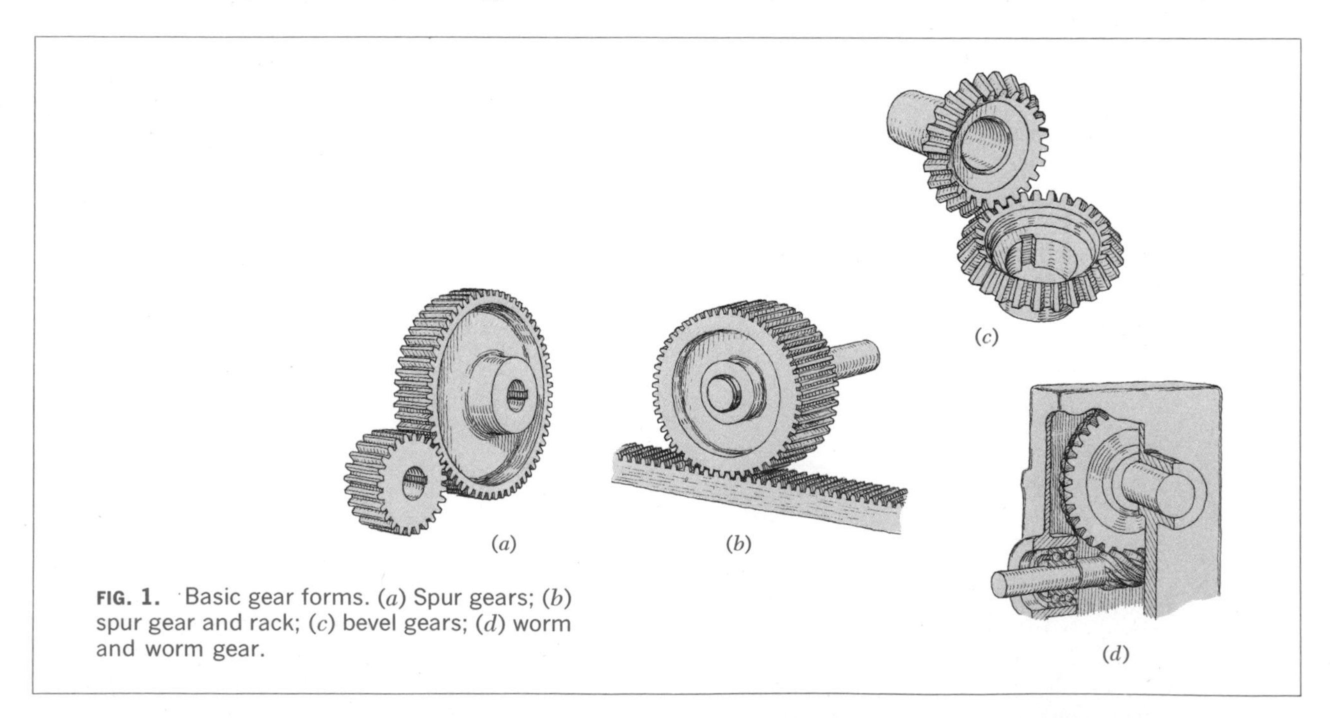


FIG. 1. Basic gear forms. (*a*) Spur gears; (*b*) spur gear and rack; (*c*) bevel gears; (*d*) worm and worm gear.

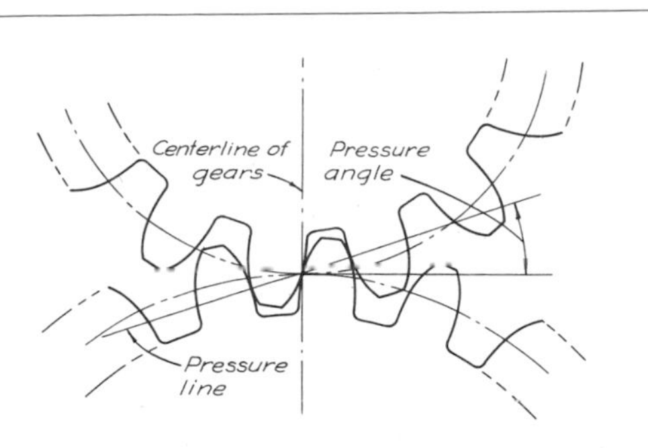


FIG. 2. Mating gear teeth. The tooth faces are involute surfaces that "roll" on each other.

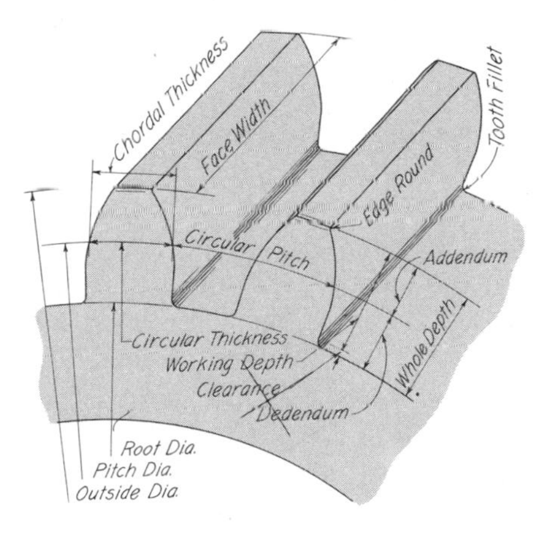


FIG. 3. Gear nomenclature.

tooth = a constant/$P_d$ = for standard $14\frac{1}{2}°$ or 20° involute teeth, $1/P_d$

$b$ = dedendum = radial distance between the pitch-diameter circle and the bottom of a tooth space = a constant/$P_d$ = for standard $14\frac{1}{2}°$ or 20° involute teeth, $1.157/P_d$

$c$ = clearance = radial distance between the top of a tooth and the bottom of a mating tooth space = a constant/$P_d$ = for standard $14\frac{1}{2}°$ or 20° involute teeth, $0.157/P_d$

$h_t$ = whole depth = radial distance between the top and bottom of a tooth = $a + b$ = for standard $14\frac{1}{2}°$ or 20° involute teeth, $2.157/P_d$

$h_k$ = working depth = greatest depth a tooth of one gear extends into a tooth space of a mating gear = $2a$ = for standard $14\frac{1}{2}°$ or 20° involute teeth, $2/P_d$

$D_O$ = outside diameter = diameter of the circle containing the top surfaces of the teeth = $D + 2a = (N + 2)/P_d$

$D_R$ = root diameter = diameter of the circle containing the bottom surfaces of the tooth spaces = $D - 2b = (N - 2.314)/P_d$

$F$ = face width = width of the tooth flank

$f$ = tooth fillet = fillet joining the tooth flank and the bottom of the tooth space = $0.157/P_d$ max

$r$ = edge round = radius of the circumferential edge of a gear tooth (to break the sharp corner)

$n$ = revolutions per unit of time

In addition to the above basic letter symbols, subscripts $G$ and $P$ are used to denote gear and pinion, respectively. (When one of two mating gears is smaller than the other, it is known as the pinion.)

$m$ = gear ratio = $m_G$ for the gear = $N_G/N_P = n_P/n_G = D_G/D_P$

$m$ = gear ratio = $m_P$ for the pinion = $N_P/N_G = n_G/n_P = D_P/D_G$

(The pitch diameter and number of teeth are inversely proportional to speed.)

### 4. Spur-gear Calculations.

There are many different combinations producing various individual problems, but the following is typical and will illustrate the procedure:

Assume that the center distance and speeds are known.

Center distance = 7 in.

Speed of gear 500 rpm, speed of pinion 1,500 rpm

Then $m_G = n_P/n_G = 1{,}500/500 = 3/1 = D_G/D_P$

Therefore, $3D_P = D_G$

From the center distance, $D_P + D_G = 14$ in.

Substituting for $D_G$ in terms of $D_P$,

$D_P + 3D_P = 14$. Solving, $D_P = 3\frac{1}{2}$ in.

Solving for $D_G = 3D_P = 3 \times 3\frac{1}{2} = 10\frac{1}{2}$ in.

At this time, the diametral pitch will have to be assumed.

Assume $P_d = 5$.

Then, $N_P = P_d \times D_P = 5 \times 3\frac{1}{2} = 17\frac{1}{2}$, which is obviously impossible since a gear could not have a half tooth.

Assuming $P_d = 6$,
$N_P = 6 \times 3\frac{1}{2} = 21$
$N_G = 6 \times 10\frac{1}{2} = 63$

Now all the values such as addendum, dedendum, outside diameter, can be calculated from the equations in Sec. 3. Note that if, in a given problem, the two pitch diameters are known, the center distance must be found; if one pitch diameter is known and the center distance given, the second pitch diameter must be found, etc.

### 5. To Draw a Spur Gear (Fig. 4).

To draw the teeth of a standard involute-toothed spur gear by an approximate circle-arc method, lay off the pitch circle, root circle, and outside circle. Start with the pitch point and divide the pitch circle into distances equal to the circular thickness. Through the pitch point draw a line of $75\frac{1}{2}°$ with the center line for a $14\frac{1}{2}°$ involute tooth (for convenience, draftsmen use 75°), or 70° for a 20° tooth. Draw the basic circle tangent to the pressure line. With compass set to a radius equal to one-fourth the radius of the pitch circle, describe arcs through the division points on the pitch circle, keeping the needle point on the base circle. Darken the arcs for the tops of the teeth and bottoms of the spaces and add the tooth fillets. For 16 or fewer teeth, the radius value of one-fourth pitch radius *must be increased* to suit, in order to avoid the appearance of excessive undercut. For a small number of teeth, the radius may be as large as or equal to the pitch radius.

This method of drawing gear teeth is useful on display drawings, but on working drawings the tooth outlines are not drawn. Figure 5 illustrates the method of indicating the teeth and dimensioning a spur gear on a working drawing.

Gears must be precisely made in order to prevent noisy and faulty running characteristics. Thus, the dimensioning almost always includes surface roughness and form-tolerance specifications as shown on Fig. 5. The tabular information given is for shop use in cutting and checking the teeth and as a record of tooth dimensions for checking with the mating gear.

### 6. To Draw a Rack (Fig. 6).

A rack is, theoretically, a spur gear having an infinite pitch diameter. Therefore, compared with the mating gear, all circular dimensions become linear. To draw the teeth of a standard involute rack by an approximate method, draw the pitch line and lay off the addendum and dedendum distances. Divide the pitch line into spaces equal to the circular thickness of the mating gear. Through these points of division draw the tooth faces at $14\frac{1}{2}°$ (15° is used by draftsmen). Draw tops and bottoms and add the tooth fillets. For standard 20° full-depth or stub teeth, use 20° instead of $14\frac{1}{2}°$. Specifications of rack teeth (to be given on a detail drawing) are axial

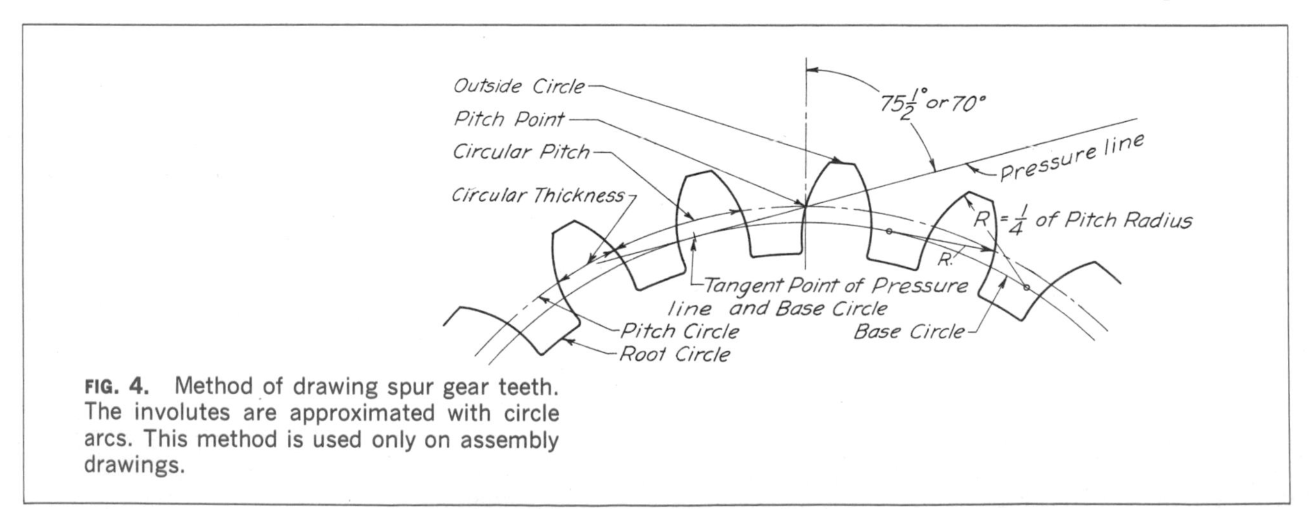


**FIG. 4.** Method of drawing spur gear teeth. The involutes are approximated with circle arcs. This method is used only on assembly drawings.

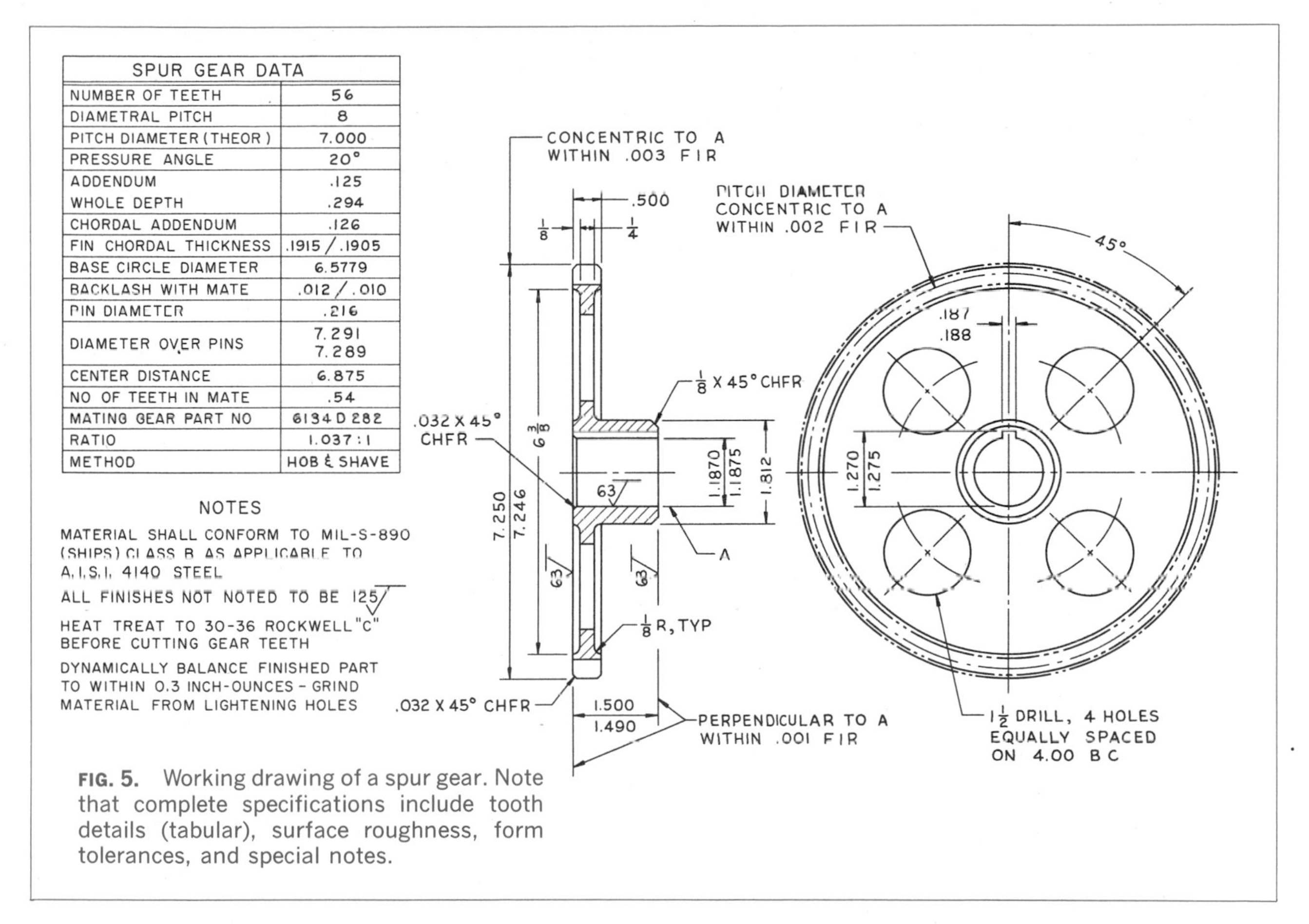

| SPUR GEAR DATA | |
|---|---|
| NUMBER OF TEETH | 56 |
| DIAMETRAL PITCH | 8 |
| PITCH DIAMETER (THEOR) | 7.000 |
| PRESSURE ANGLE | 20° |
| ADDENDUM | .125 |
| WHOLE DEPTH | .294 |
| CHORDAL ADDENDUM | .126 |
| FIN CHORDAL THICKNESS | .1915 / .1905 |
| BASE CIRCLE DIAMETER | 6.5779 |
| BACKLASH WITH MATE | .012 / .010 |
| PIN DIAMETER | .216 |
| DIAMETER OVER PINS | 7.291 7.289 |
| CENTER DISTANCE | 6.875 |
| NO OF TEETH IN MATE | .54 |
| MATING GEAR PART NO | 6134 D 282 |
| RATIO | 1.037 : 1 |
| METHOD | HOB & SHAVE |

NOTES

MATERIAL SHALL CONFORM TO MIL-S-890 (SHIPS) CLASS B AS APPLICABLE TO A.I.S.I. 4140 STEEL

ALL FINISHES NOT NOTED TO BE 125

HEAT TREAT TO 30-36 ROCKWELL "C" BEFORE CUTTING GEAR TEETH

DYNAMICALLY BALANCE FINISHED PART TO WITHIN 0.3 INCH-OUNCES - GRIND MATERIAL FROM LIGHTENING HOLES



**FIG. 5.** Working drawing of a spur gear. Note that complete specifications include tooth details (tabular), surface roughness, form tolerances, and special notes.

(linear) pitch (equal to circular pitch of the mating gear), number of teeth, diametral pitch, whole depth.

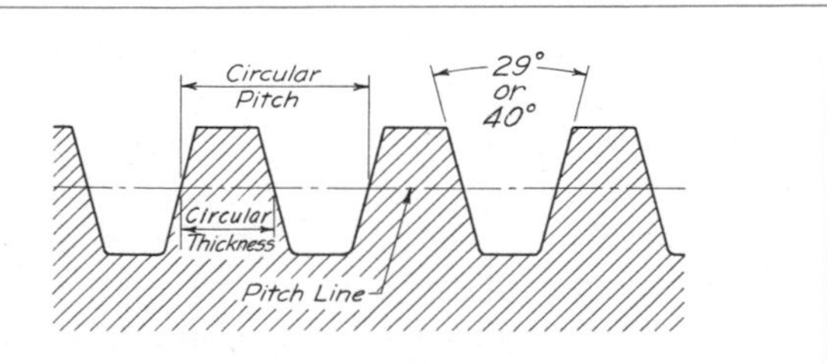


**FIG. 6.** Method of drawing an involute rack. Circular pitch and thickness are linear because the pitch diameter is infinite.

## 7. Bevel Gears.

For bevel gears the theoretical rolling surfaces become cones. The pitch diameter $D$ of the gear, as shown in Fig. 7*a*, is the base diameter of the cone. The addendum and dedendum are calculated in the same way as for a spur gear and are measured on a cone, called the "back cone," whose elements are normal to the face-cone elements (Fig. 7*a*). The diametral pitch, circular pitch, etc., are the same as for a spur gear.

In addition to the letter symbols and calculations for spur gears, the dimensions shown in Fig. 7*b*, with formulas for calculating, are needed:

$\Gamma$ = pitch-cone angle = angle between a pitch-cone element and the cone axis. $\tan \Gamma = D_G/D_P = N_G/N_P$ (gear) $= D_P/D_G = N_P/N_G$ (pinion)

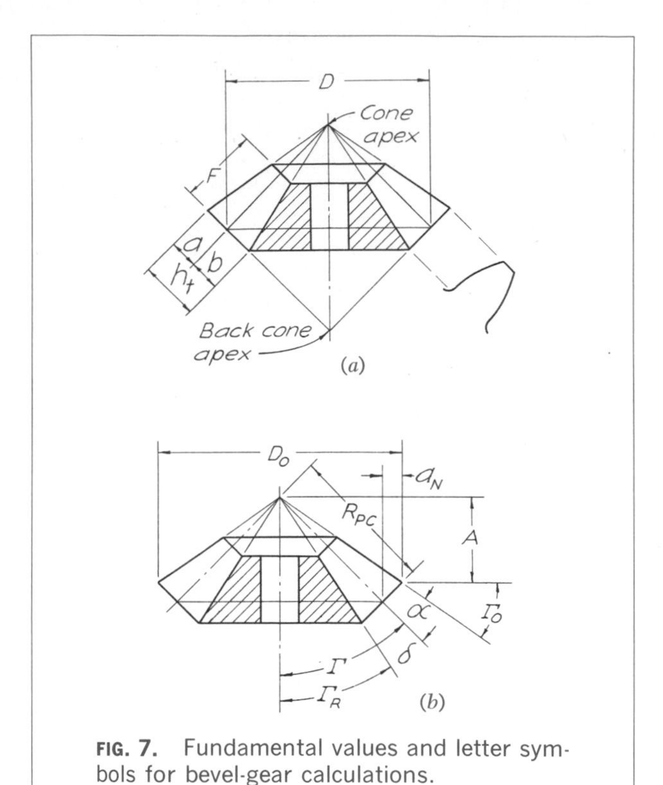


**FIG. 7.** Fundamental values and letter symbols for bevel-gear calculations.

$R_{PC}$ = pitch-cone radius = length of an element of the pitch cone. $R_{PC} = D/2 \sin \Gamma$

$\alpha$ = addendum angle = angle between the pitch cone and the outside of a tooth. $\tan \alpha = a/R_{PC}$

$\delta$ = dedendum angle = angle between the pitch cone and the bottom of a tooth space. $\tan \delta = b/R_{PC}$

$\Gamma_O$ = face angle = angle between a line *normal* to the gear axis and the outside (face) of the teeth. $\Gamma_O = 90° - (\Gamma + \alpha)$. (Note that this angle is the complement of the angle an element of the outside cone makes with the axis.)

$\Gamma_R$ = root-cone angle = cutting angle = angle between the axis and an element of the root cone. $\Gamma_R = \Gamma - \delta$

$a_N$ = angular addendum = addendum distance measured *normal* to gear axis. $a_N = a \cos \Gamma$

$D_O$ = outside diameter = outside diameter at the base of the teeth. $D_O = D + 2a_N$

$A$ = apex distance = altitude of the outside-diameter cone. $A = (D_o/2) \tan \Gamma$

**8. To Draw a Bevel Gear (Fig. 8).** To draw the teeth of an involute-toothed bevel gear by an approximate method (the Tredgold method), draw the center lines, intersecting at $O$. Across the center lines lay off the pitch diameters, and project them parallel to the center lines until the projectors intersect at the pitch point $P$. From the pitch point, draw the pitch-circle diameters for each gear, and from their extremities, draw the "pitch cones" to the vertex or "cone center" $O$. Lay off the addendum and dedendum distances for each gear on lines through the pitch points perpendicular to the cone elements. Extend one of these normals for each gear to intersect the axis, as at $B$ and $C$, making the "back cones." With $B$ as center, swing arcs 1, 2, and 3 for the top, pitch line, and bottom, respectively, of a developed tooth. On a radial center line $AB$, draw a tooth, by the method of Fig. 4. Start the plan view of the gear by projecting points 1, 2, and 3 across to its vertical center line and drawing circles through the points. Lay off the radial center lines for each tooth. With dividers take the circular thickness distances from $A$ and transfer them to each tooth. This will give three points on each side of each tooth through which a circle arc, found by trial, will pass, giving the foreshortened contour of the large end of the teeth in this view. From this point the drawing becomes a problem in projection drawing. In every view the lines converge at the cone center $O$ and by finding three points on the contour of each tooth, circle arcs can be found which will approximate the tooth shape.

This method is used for finished display drawings. Working drawings for cut bevel gears are drawn without tooth outlines and dimensioned as in Fig. 9, from calculations in Sec. 7. Information on the drawing gives details for cutting and checking the teeth. The drawing is dimensioned with limits where required. Surface roughness and form-tolerance specifications are sometimes necessary.

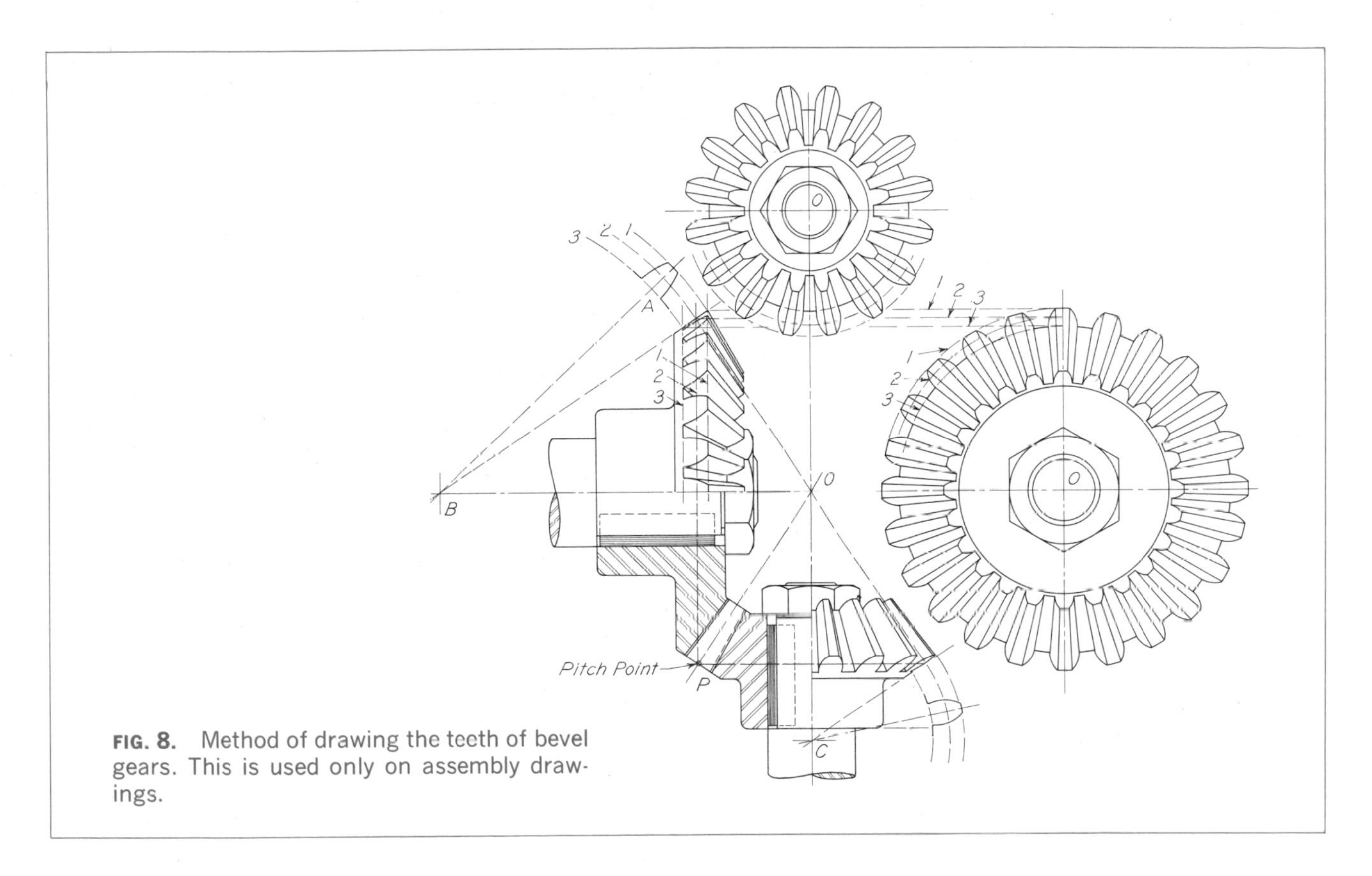


**FIG. 8.** Method of drawing the teeth of bevel gears. This is used only on assembly drawings.

## 9. Worm Gears.

Worm gears are used primarily to get great reductions in relative speed and to obtain a large increase in effective power. The worm is similar to a screw thread, and the computing of pitch diameter, etc., is also similar. On a section taken through the axis of the worm, the worm gear and worm have the same relationship as a spur gear and rack. Therefore, the tooth shape, addendum, dedendum, etc., will be the same as for a spur gear and rack.

In addition to the calculations necessary for a spur gear and rack (addendum, dedendum, etc.), the following are needed:

*Worm*

$l$ = lead of worm threads
$N_W$ = multiplicity of worm threads
$p_{XW}$ = axial pitch = $l/N_W$
$D_W$ = pitch diameter of worm
$\lambda$ = lead angle. $\tan \lambda = l/\pi D$

*Worm Gear*

$C$ = center distance between worm and wheel
$D_G$ = pitch diameter of worm gear
$D_{tp}$ = pitch throat diameter = pitch diameter of wheel = $D_G$
$D_{to}$ = outer throat diameter = pitch diameter plus twice the addendum = $D_G + 2a$
$D_{ti}$ = inner throat diameter = pitch diameter minus twice the dedendum = $D_G - 2b$
$R_{tp}$ = pitch throat radius = pitch radius of worm = $D_W/2$
$R_{to}$ = outer throat radius = $(D_W/2) - a$
$R_{ti}$ = inner throat radius = $(D_W/2) + b$
$C_{Rt}$ = center of throat radius = $C = (D_G/2) + (D_W/2)$

## 10. To Draw Worm Gears.

In assembly, a worm and worm gear are generally

| BEVEL GEAR DATA | |
|---|---|
| NUMBER OF TEETH | 81 |
| DIAMETRAL PITCH | 10.4516 |
| PITCH DIAMETER (THEOR) | 7.750 |
| PITCH ANGLE | 82° 58' |
| ROOT ANGLE (BASIC) | 80° 56' |
| PRESSURE ANGLE | 20° |
| ADDENDUM (THEOR) | .042 |
| WHOLE DEPTH (APPROX) | .181 |
| CONE CENTER TO CROWN | .437 |
| CONE DISTANCE | 3.9044 |
| BACKLASH WITH MATE | .010/.014 |
| NO OF TEETH IN MATE | 10 |
| MATING GEAR PART NO | 1620 D 10 |
| DRIVING MEMBER | PINION |

NOTES

(B) CARBURIZE AND HARDEN ENCLOSED ZONE TO ROCKWELL "C" 58 MIN-FINISHED CASE DEPTH .020-.025 DP-CORE AND UNCARBURIZED AREAS TO BE ROCKWELL "C" 27 MIN

FINISH UNLESS NOTED TO 100 ALL OVER

GEAR AND PINION TO BE MATCHED AND LAPPED AS A SET-ETCH GEAR WITH SET NO, MOUNTING DIST, AND BACKLASH

MAGNETIC INSPECT PER MIL-I-6868

STATICALLY BALANCE TO WITHIN 0.06 INCH-OUNCES

PITCH DIAMETER OF GEAR CONCENTRIC TO A WITHIN .002 FIR

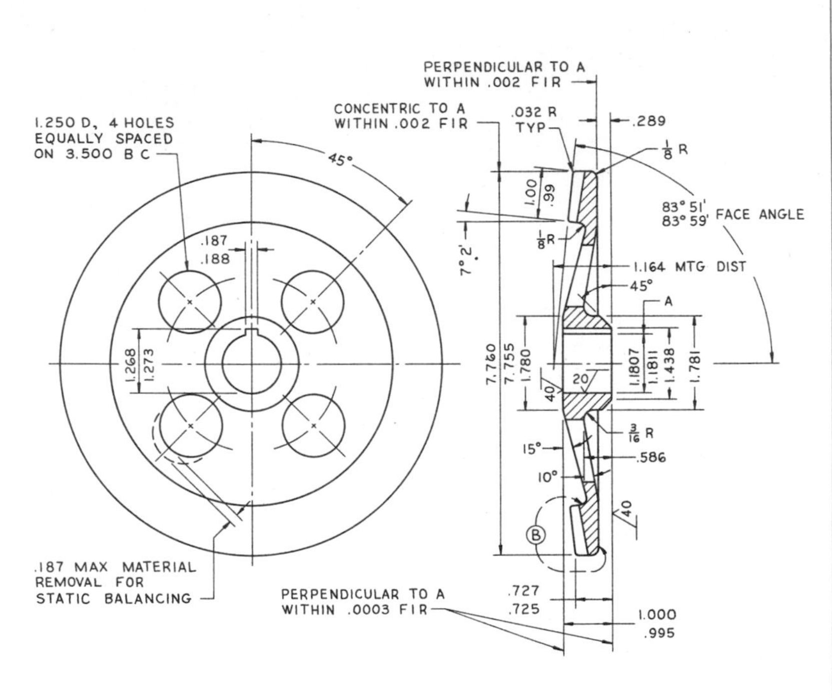


**FIG. 9.** Working drawing of a bevel gear. Note that complete specifications include tooth details (tabular), surface roughness, form tolerances, and special notes.

shown with the worm in section and the worm gear drawn as a conventional end view like the end view in Fig. 11. Detail drawings are made up as in Figs. 10 and 11. The calculated dimensions described in Sec. 9 are given as shown in the figures.

When tooth information is tabulated and given on drawings, as in Figs. 5 and 9, it should include:

*For the Worm Wheel*

1. Number of teeth
2. Addendum
3. Whole depth
4. Pitch diameter
5. Center distance between worm and wheel
6. Helix angle
7. Radius of throat
8. Face angle

*For the Worm*

1. Linear pitch
2. Addendum
3. Whole depth
4. Pitch diameter
5. Outside diameter
6. Root diameter
7. Helix angle
8. Minimum length of worm for complete action
9. Center distance between worm and wheel

**11. Cams.**

A cam is a machine element with surface or groove formed to produce special or irregular motion in a second part, called a "follower." The shape of the

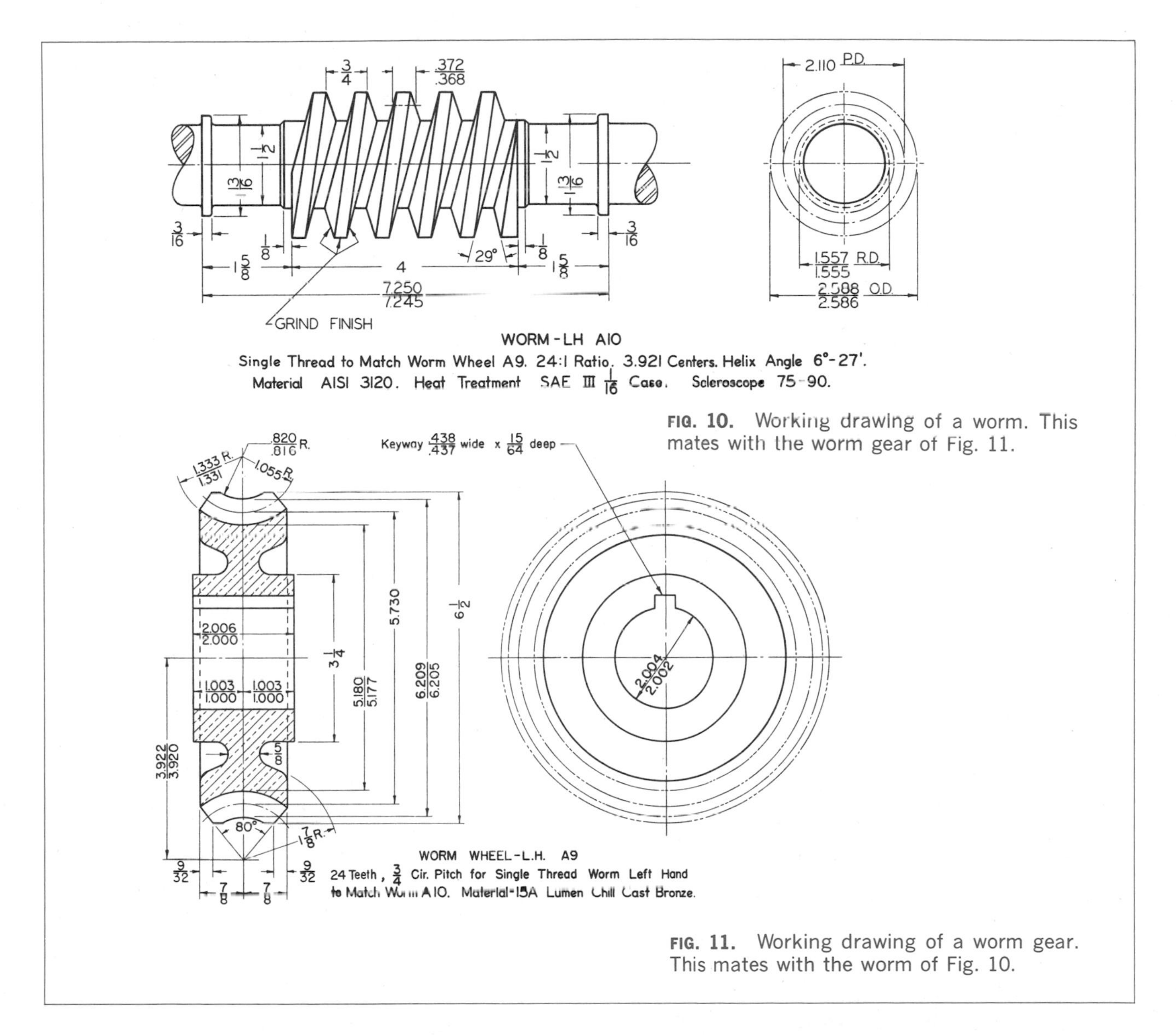


**FIG. 10.** Working drawing of a worm. This mates with the worm gear of Fig. 11.

**FIG. 11.** Working drawing of a worm gear. This mates with the worm of Fig. 10.

cam is dependent upon the motion that is required and the type of follower that is used. The type of cam is dictated by the required relationship of the parts and the motions of both.

## 12. Types of Cams.

The direction of motion of the follower with respect to the cam axis determines two general types of cams: (1) radial or disk cams, in which the follower moves in a direction perpendicular to the cam axis; and (2) cylindrical or end cams, in which the follower moves in a direction parallel to the cam axis.

Figure 12 shows at (*a*) a *radial cam,* with a roller follower held against the cam by gravity or by a spring. As the cam revolves, the follower is raised and lowered. Followers are also made with pointed

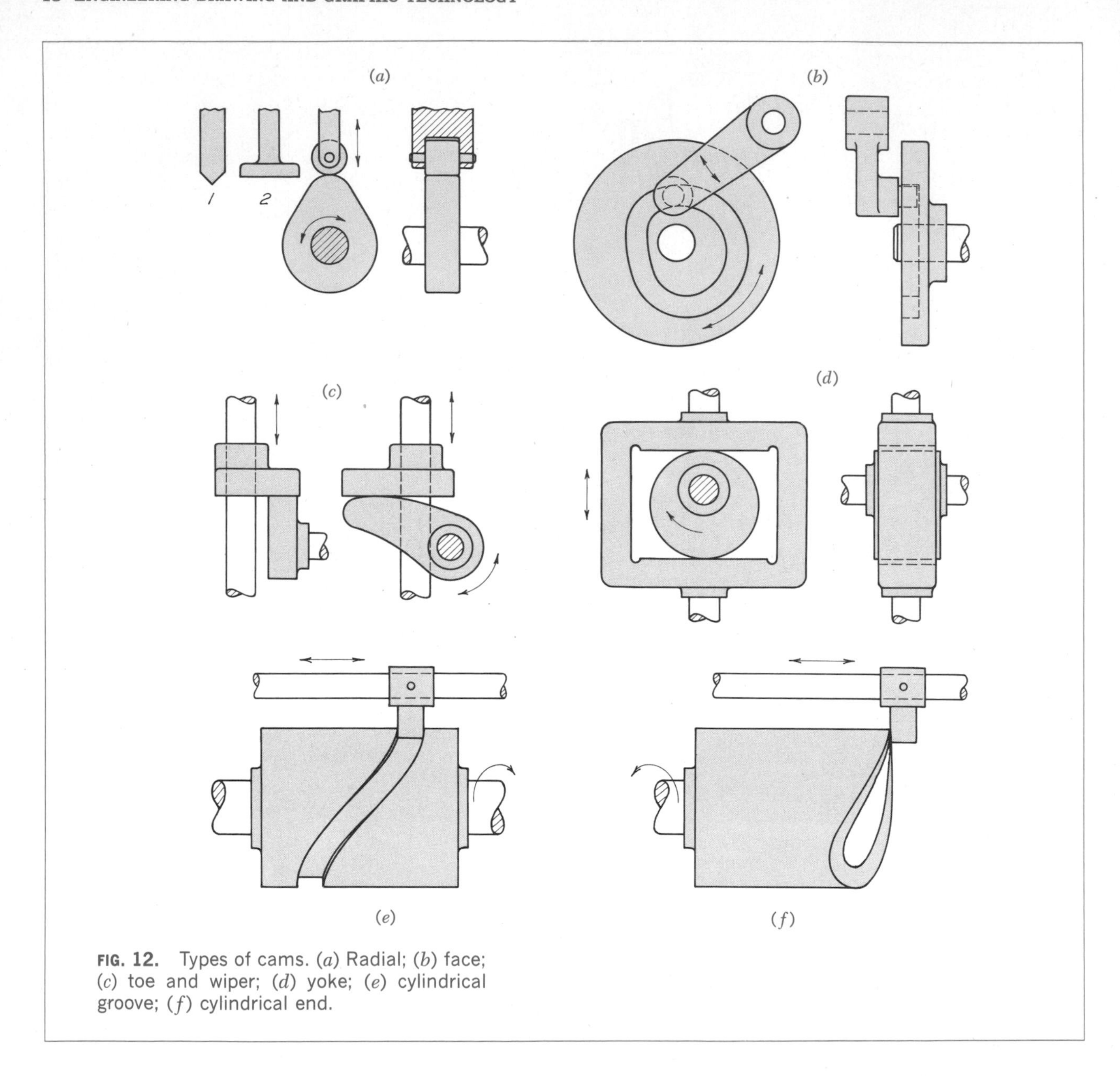


**FIG. 12.** Types of cams. (*a*) Radial; (*b*) face; (*c*) toe and wiper; (*d*) yoke; (*e*) cylindrical groove; (*f*) cylindrical end.

ends and with flat ends. At (*b*) is shown a *face cam*, with a roller follower at the end of an arm or link, the follower oscillating as the cam revolves. When the cam itself oscillates, the *toe* and *wiper* are used, as at (*c*). The toe, or follower, may also be made in the form of a swinging arm. A *yoke* or *positive-motion cam* is shown at (*d*), the enclosed follower making possible the application of force in either direction. The sum of the two distances from the center of the cam to the points of contact must always be equal to the distance between the follower surfaces. The cylindrical *groove cam* at (*e*) and the *end cam* at (*f*) both move the follower parallel to the cam axis, force being applied to the follower in both directions with the groove cam, and in only one direction with the end cam.

## 13. Kinds of Motion.

Cams can be designed to move the follower with constant velocity, acceleration, or harmonic motion. In many cases combinations of these motions, together with surfaces arranged for sudden rise or fall, or to hold the follower stationary, make up the complete cam surface.

## 14. Cam Diagrams.

In studying the motion of the follower, a diagram showing the height of the follower for successive cam positions is useful and is frequently employed. The cam position is shown on the abscissa, the full 360° rotation of the cam being divided, generally, every 30° (intermediate points may be used if necessary). The follower positions are shown on the ordinate, divided into the same number of parts as the abscissa. These diagrams are generally made to actual size.

Constant velocity gives a uniform rise and fall and can be plotted, as at (*a*) in Fig. 13, by laying off the cam positions on the abscissa, measuring the total follower movement on the ordinate and dividing it into the same number of parts as the abscissa. As the cam moves one unit of its rotation, the follower likewise moves one unit, producing the straight line of motion shown.

With constant acceleration, the distance traveled is proportional to the square of the time, or the total distance traveled is proportional to 1, 4, 9, 16, 25, etc.; and if the increments of follower distance are made proportional to 1, 3, 5, 7, etc., the curve can be plotted as shown at (*b*). Using a scale, divide the follower rise into the same number of parts as the abscissa, making the first part one unit, the second three units, and so on. Plot points at the intersection of the coordinate lines, as shown. The curve at (*b*) accelerates and then decelerates to slow up the follower at the top of its rise.

Harmonic motion (sine curve) can be plotted as at (*c*) by measuring the rise and drawing a semicircle, dividing it into the same number of parts as the abscissa, and projecting the points on the semicircle as ordinate lines. Points are plotted at the intersection of the coordinate lines, as shown.

Figure 14 is the cam diagram for the cam in Fig. 16. The follower rises with harmonic motion in 180°, drops halfway down instantly, and then returns with uniform motion to the point of beginning.

## 15. Timing Diagrams.

When two or more cams are used on the same machine and their functions are dependent on each other, the "timing" and relative motions of each can be studied by means of a diagram showing each follower curve. The curves can be superimposed, but it is better to place one above the other as in Fig. 15.

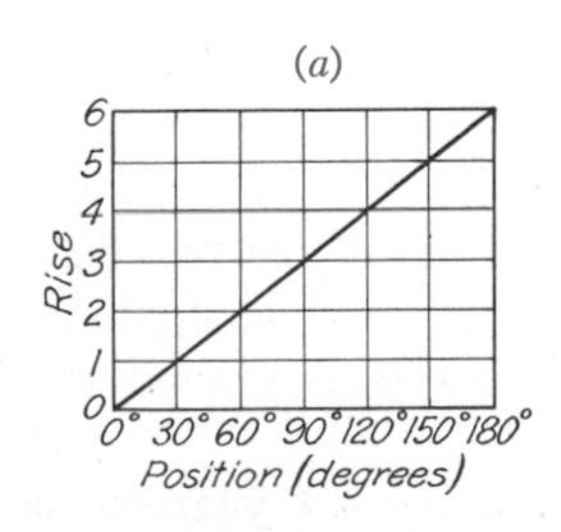


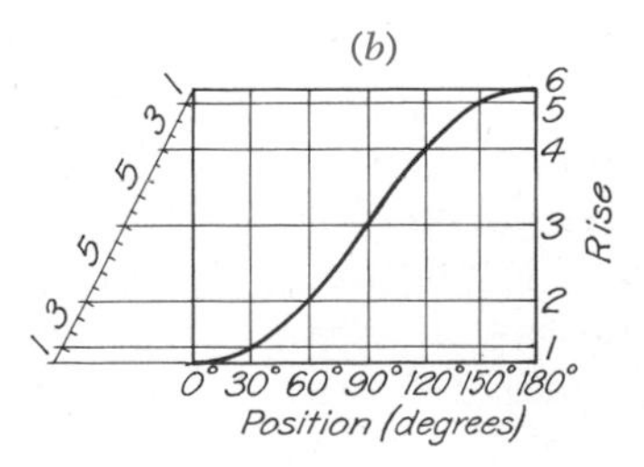


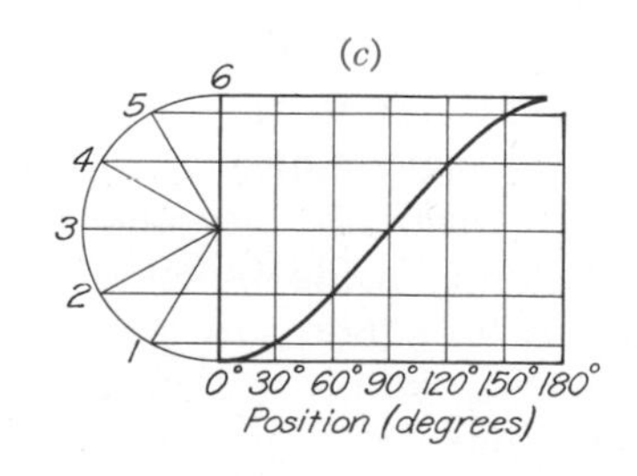


**FIG. 13.** Methods of plotting cam diagrams. (*a*) Constant velocity; (*b*) constant acceleration; (*c*) harmonic motion.

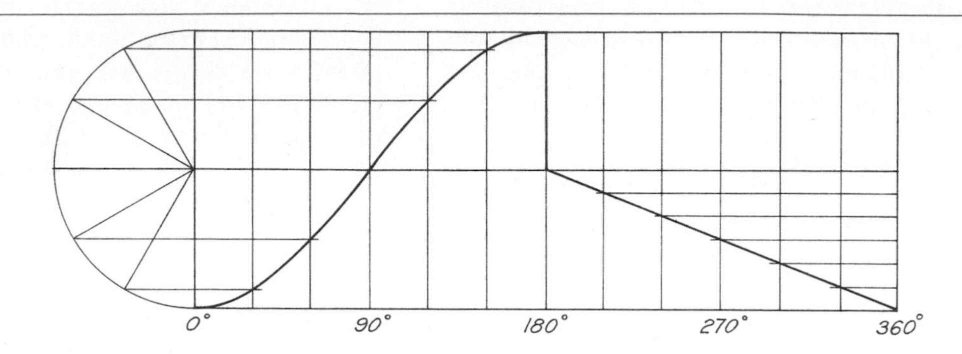


FIG. 14. A cam diagram. This is for the cam in Fig. 16.

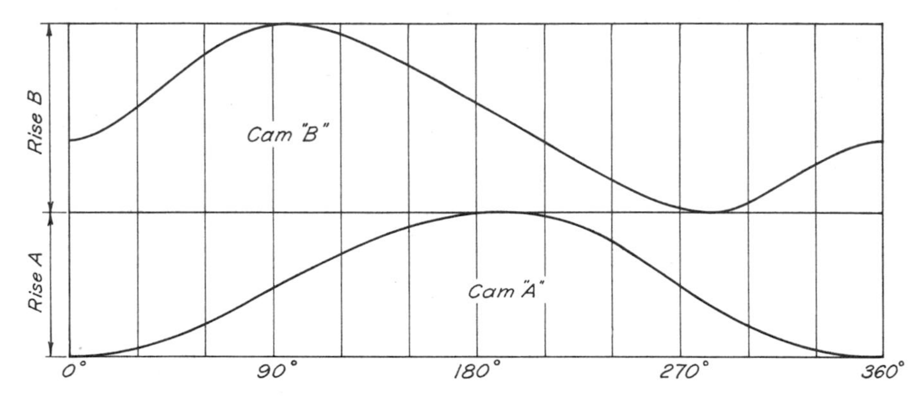


FIG. 15. A timing diagram. This is made to study the characteristics of two separate but related motions.

## 16. To Draw a Plate Cam.

The same principle is involved in drawing all types of cams. The cam in Fig. 16, for which the diagram in Fig. 14 was made, will serve as an example. The point $C$ is the center of the shaft, and $A$ is the lowest and $B$ the highest position of the center of the roller follower. Divide the rise into six parts harmonically proportional. Divide the semicircle $ADE$ into as many equal parts as there are spaces in the rise, and draw radial lines. With $C$ as center and radius $C1$, draw an arc intersecting the first radial line at $1'$. In the same way locate points $2'$, $3'$, etc., and draw a smooth curve through them. If the cam is revolved in the direction of the arrow, it will raise the follower with the desired harmonic motion. Draw $B'F$ equal to one-half $AB$. Divide $A3$ into six equal parts and the arc $EGA$ into six equal parts. Then for equal angles the follower must fall equal distances. Circle arcs drawn as indicated will locate the required points on the cam outline.

This outline is for the center of the roller; allowance for the roller size can be made by drawing the roller in its successive positions and then drawing a tangent curve as shown in the auxiliary figure.

## 17. To Draw a Cylindrical Cam

A drawing of a cylindrical cam differs somewhat from that of a plate cam, as in addition to the regular views, it generally includes a developed view, from which a template is made. Assume that the follower is to move upward $1\frac{1}{2}$ in. with harmonic motion in 180°, and then return with uniform acceleration. Top and front views of the cylinder are drawn (Fig. 17) and the development of the surface is laid out. Divide the surface as shown, also the top view to show the positions of points plotted. Divide the rise for harmonic motion by drawing the semicircle and projecting the points (refer to Fig. 13*c*). Divide the

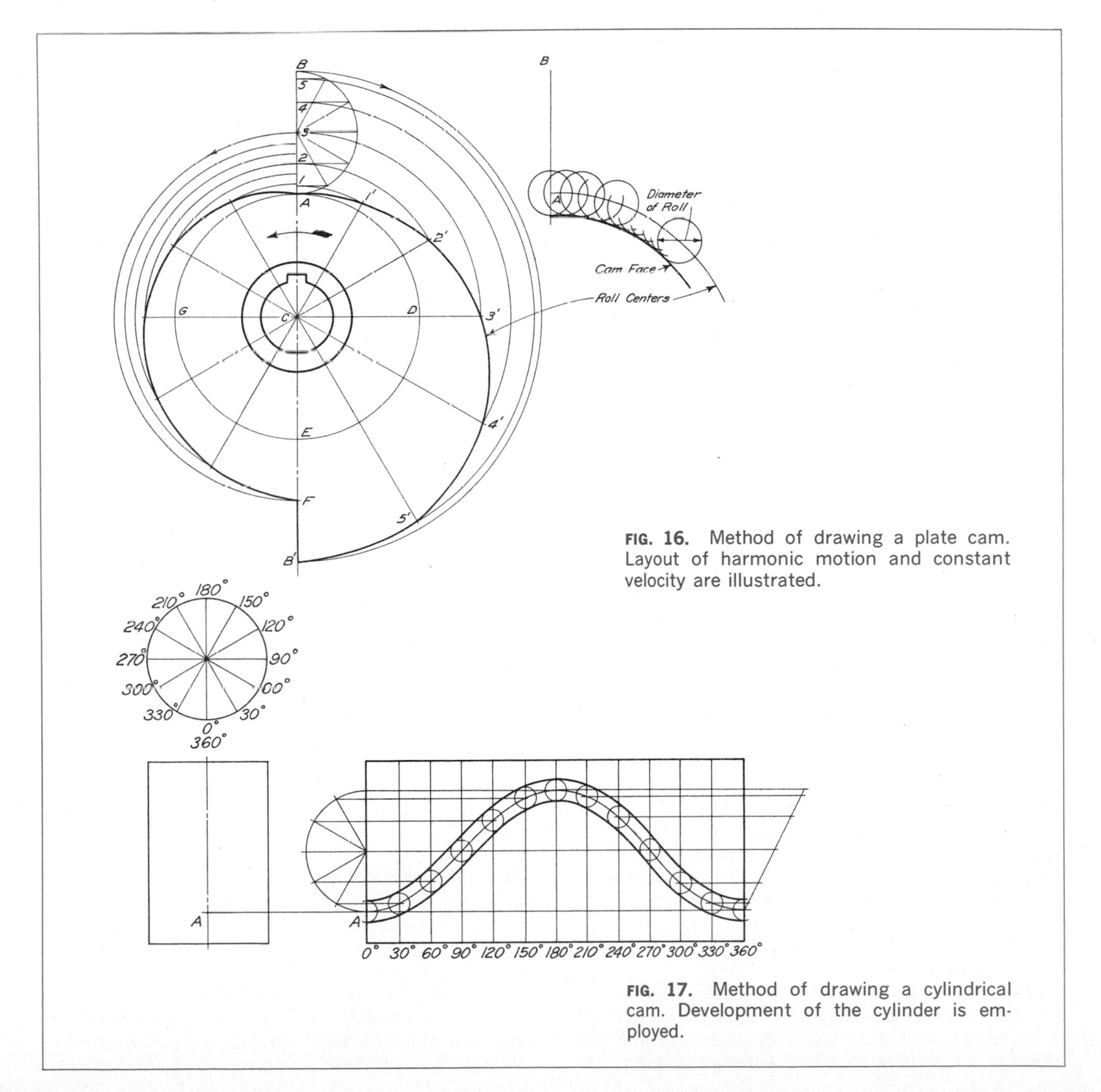


**FIG. 16.** Method of drawing a plate cam. Layout of harmonic motion and constant velocity are illustrated.

**FIG. 17.** Method of drawing a cylindrical cam. Development of the cylinder is employed.

return for acceleration as shown (refer to Fig. 13*b*). The curve thus obtained is for the center of the follower. Curves drawn tangent to circles representing positions of the follower will locate the working surfaces of the cam. The development made as described is the drawing used to make the cam.

### 18. Dimensioning a Cam Drawing.

In general, two practices are followed, depending upon the function of the cam: (1) For cams where follower position is not critical, full-size layouts, as in Fig. 16, with necessary fractional or decimal dimensions suffice. Such cams can be made by profiling the surface by sawing, nibbling and finishing, and torch cutting, etc. (2) For cams where the follower position and relative motion is critical, limit dimensioning from datum features is necessary. The dimensioning in Prob. 28, Chap. 3, is an example. Where the degree of precision requires, surface roughness and form tolerances are needed.

## PROBLEMS

### Group 1. Spur Gears.

**1.** Make an assembly drawing of a pair of spur gears, from the following information: On an 11- by 17-in. sheet locate centers for front view of gear $B$ $4\frac{1}{2}$ in. from right border and $3\frac{1}{2}$ in. from bottom border. Gear $A$ is to the left of gear $B$. Center distances between gears are 5.250 in. Gear $A$ revolves at 300 rpm and has four spokes, elliptical in cross section, 1-in. major and $\frac{1}{2}$-in. minor axes; inside flange diameter, $4\frac{3}{8}$ in.; hub, 2 in. diameter, $1\frac{1}{2}$ in. long. Gear $B$ revolves at 400 rpm and is disk-type with $\frac{1}{2}$-in. web; inside flange diameter, $3\frac{1}{4}$ in.; hub, 2 in. diameter, $1\frac{1}{2}$ in. long. Material is cast steel; face width, 1 in.; diametral pitch, 4; shaft diameters, 1 in.; $\frac{1}{4}$-in. Woodruff keys. Draw front view and sectional top view.

**2.** The figure shows a gear box for a reducing mechanism using spur gears. Using the scale shown on the drawing, transfer the necessary distances to obtain the net available space for the gears and design gears as follows: center distance, 4 in.; gear $A$ revolves at 900 rpm, gear $B$ at 1,500 rpm; diametral pitch, 4; face width, 1 in. Use standard $14\frac{1}{2}°$ involute teeth. Calculate the necessary values (shown in table in figure for Prob. 2). Draw the top view in section. Show four teeth on each gear on front view and complete the view with conventional lines.

**3.** Same as Prob. 2 but for $A = 525$ rpm, $B = 675$ rpm. Use 20° involute teeth.

**4.** Make working drawings for the gears in Prob. 2.

**5.** Make working drawings for the gears in Prob. 3.

**6.** A broken spur gear has been measured and the following information obtained: number of teeth, 33; outside diameter, $4\frac{3}{8}$ in.; width of face, 1 in.; diameter of shaft, $\frac{7}{8}$ in.; length of hub, $1\frac{1}{4}$ in. Make a drawing of a gear blank with all dimensions and information necessary for making a new gear. The dimensions that are not specified here may be made to suit as the drawing is developed.

**7.** Make a drawing of a spur gear. The only information available is as follows: root diameter, 7.3372 in.; outside diameter, 8.200 in.; width of face, $1\frac{7}{8}$ in.; diameter of shaft, $1\frac{3}{8}$ in.; length of hub, 2 in.

### Group 2. Racks.

**8.** Draw a spur gear and rack as follows: gear, 4 in. pitch diameter, 20 teeth, face width, 1 in.; standard $14\frac{1}{2}°$ involute teeth. Rack is to move laterally 5 in. Compute axial pitch, addendum, and dedendum for rack, and draw the rack and gear in assembly showing all teeth on rack and four teeth on gear.

**9.** Same as Prob. 8 but gear is 5 in. pitch diameter, has 25 teeth and $\frac{7}{8}$ in. face width, with 20° standard involute teeth.

### Group 3. Bevel Gears.

**10.** The figure shows a gear box for a reducing mechanism using bevel gears. Using the scale shown on the drawing, transfer the necessary distances to obtain net available space for the gears and design gears as follows: gear $A$ has 6 in. pitch diameter, diametral pitch, 4; speed, 350 rpm; face width, 1 in. Gear $B$ has a speed of 600 rpm. Use standard $14\frac{1}{2}°$ involute teeth. Compute the necessary values (shown in table in figure for Prob. 10).

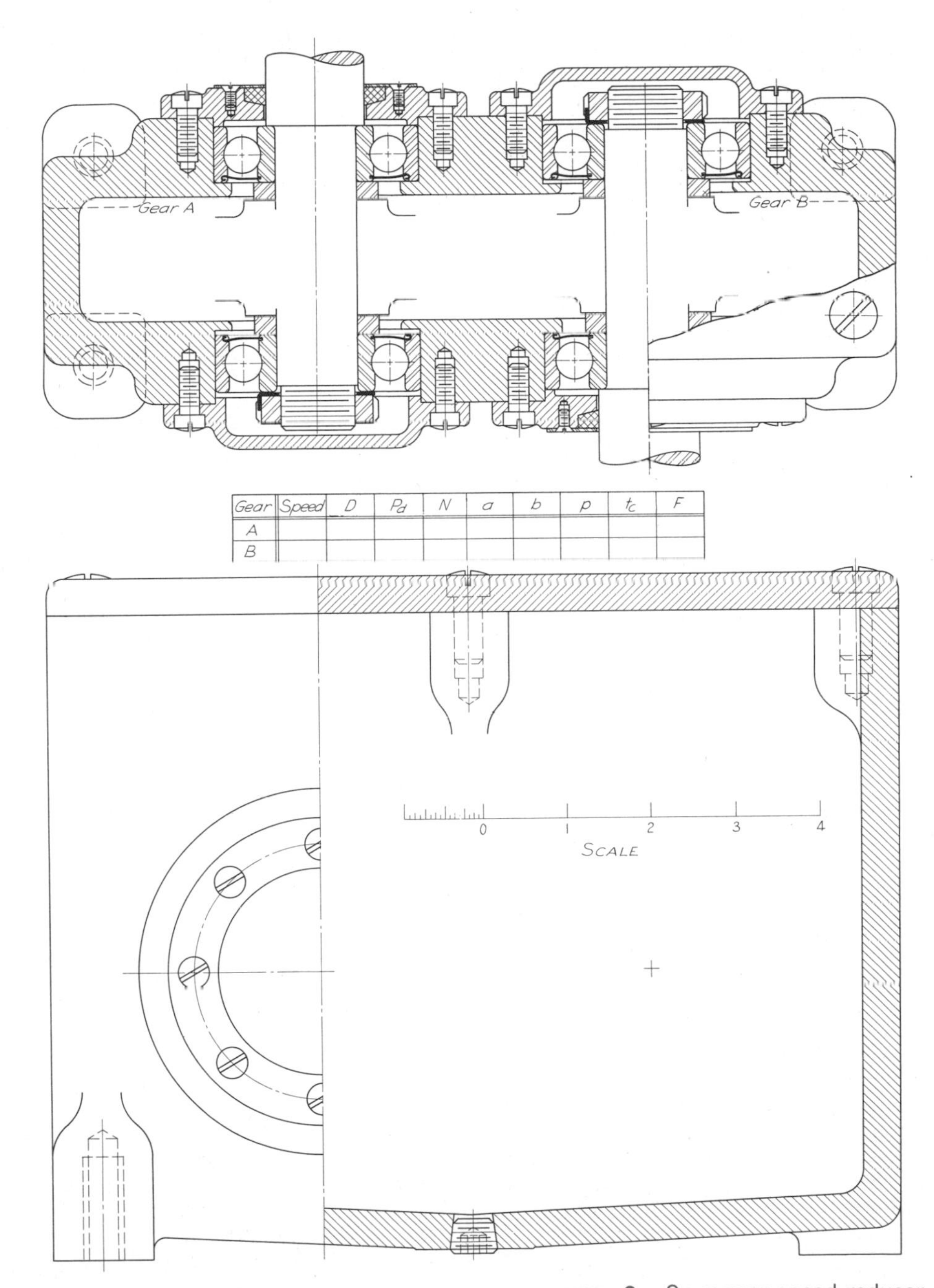


**PROB. 2.** Spur-gear speed reducer.

Draw top view in section and front view as an external view which shows all teeth.

**11.** Same as Prob. 10 but gear *A* has $5\frac{1}{2}$ in. pitch diameter and 22 teeth; speed, 750 rpm. Gear *B* has a speed of 1,100 rpm. Use standard 20° involute teeth. Face width is 1 in.

**12.** Make working drawings for the bevel gears in Prob. 10; refer to Sec. 7 to compute values needed.

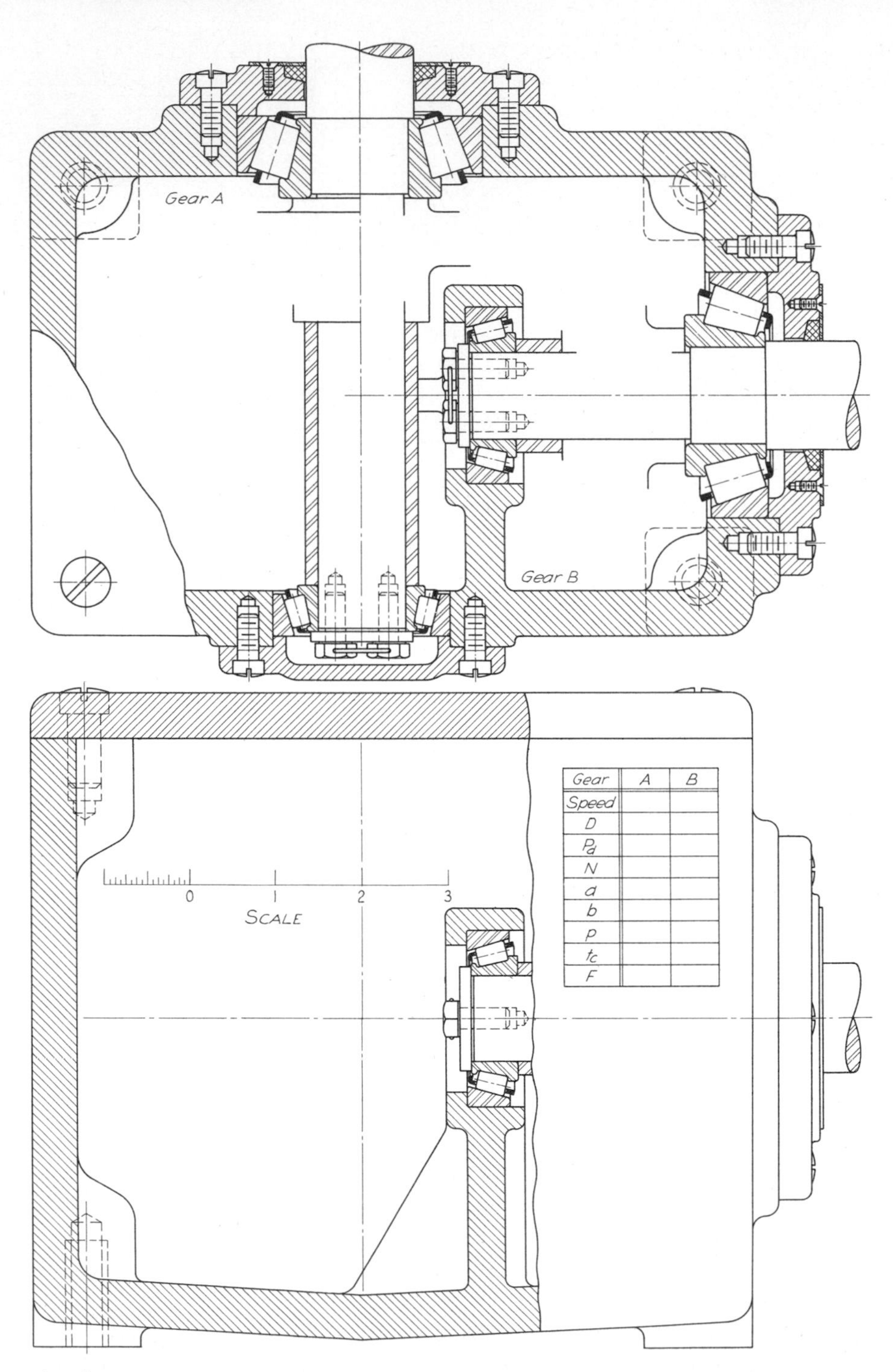


**PROB. 10.** Bevel-gear speed reducer.

**13.** Make working drawings for the bevel gears in Prob. 11; refer to Sec. 7 to compute values needed.

## Group 4. Spur- and Bevel-gear Trains.

**14.** Make an assembly drawing of gear train, as follows: $A$ and $B$ are bevel gears, with $\frac{7}{8}$ in. face width and diametral pitch of 6. $A$ is 3 in. pitch diameter, revolves at 150 rpm. $B$ revolves at 100 rpm. $C$ and $D$ are spur gears with diametral pitch of 8, 1 in. face width. $C$ engages $D$, which revolves at 40 rpm. All shafts are 1 in. Draw $A$ in full section, $B$ with lower half in section, $C$ and $D$ in full section, quarter-end view of gear $B$ in space indicated, and end views of $C$ and $D$.

**15.** A 3-in-diameter, 3-diametral-pitch bevel gear $R$ on shaft $AB$ running 1,120 rpm drives another bevel gear $S$ on shaft $AC$ at 840 rpm. On shaft $AC$ centered at $P$, an 8-in.-diameter, 4-diametral-pitch spur gear $T$ drives a pinion $U$ at 1,680 rpm. All shaft diameters are 1 in.; face widths, 1 in. Hub diameters of $R$ and $S$ are $1\frac{3}{4}$ in. Gear $C$ has four spokes, elliptical, $\frac{5}{8}$ by 1 in.; hub, $1\frac{7}{8}$ in. diameter; thickness of flange, $\frac{1}{2}$ in. Draw gear $R$ with upper half in section; $S$, $T$, and $U$ in full section. Put quarter-end view of $R$ in space indicated and end views of $T$ and $U$ on center line $MN$.

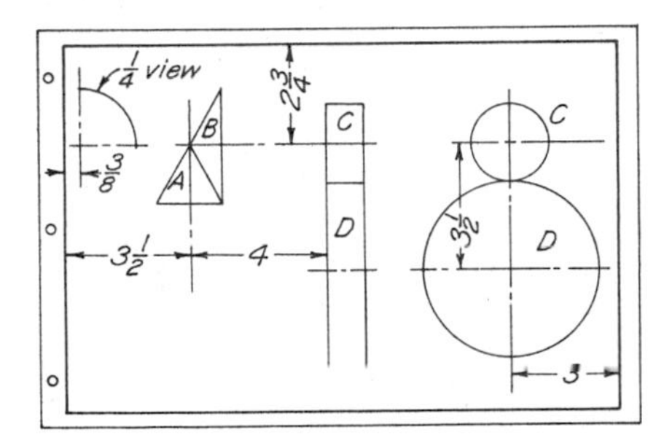


**PROB. 14.** Spur-and-bevel-gear train.

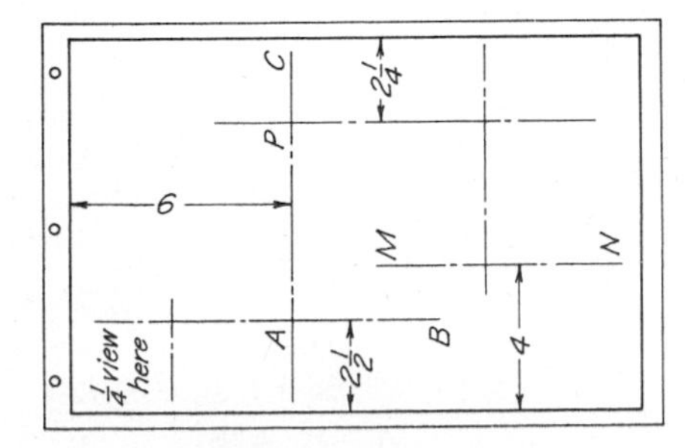


**PROB. 15.** Bevel-and-spur-gear train.

## Group 5. Worm Gears.

**16.** Make an assembly drawing of a pair of worm gears, as follows: center distance, 6 in.; pitch diameter of worm gear, $8\frac{1}{2}$ in.; face width of worm gear, $1\frac{7}{8}$ in.; pitch diameter of worm, $3\frac{1}{2}$ in.; standard $14\frac{1}{2}°$ involute worm teeth. Single-threaded worm, left hand 4 in. long, 1 in. lead. Show appropriate spindle ends for worm and suitable hub for worm gear. Make assembly drawing as a half section through axis of worm.

**17.** Make detail drawings of the worm gear and worm of Prob. 16. Refer to Sec. 9 and Figs. 10 and 11 for calculation of necessary dimensions.

## Group 6. Cams.

**18.** Make a drawing for a plate cam to satisfy the following conditions: On a vertical center line, a point $A$ is $\frac{7}{8}$ in. above a point $O$, and a point $B$ is $1\frac{3}{4}$ in. above $A$. With center at $O$, revolution clockwise, the follower starts at $A$ and rises to $B$ with uniform motion during one-third revolution, remains at rest one-third revolution, and drops with uniform motion the last one-third revolution to the starting point. Diameter of shaft, $\frac{3}{4}$ in.; diameter of hub, $1\frac{1}{4}$ in.; thickness of plate, $\frac{1}{2}$ in.; length of hub, $1\frac{1}{4}$ in.; diameter of roller, $\frac{1}{2}$ in.

**19.** Make a drawing for a face cam using the data of Prob. 18.

**20.** Make a drawing for a toe-and-wiper cam. The toe shaft is vertical, $\frac{3}{4}$ in. in diameter. Starting at a point 1 in. directly above center of wiper shaft, the toe is to move upward 2 in. with simple harmonic motion, with 135° turn of the shaft. Wiper has $1\frac{1}{4}$-in.-diameter hub; is $1\frac{1}{4}$ in. long; has $\frac{3}{4}$-in.-diameter shaft. Design toe to suit.

**21.** Make a drawing for a positive-motion cam. Starting at a point 1 in. above center of cam shaft, upper follower surface is to move upward 1 in. with simple harmonic motion in 180° turn of cam. Return is governed by necessary shape of cam. Follower is $\frac{1}{2}$ in. thick on $\frac{1}{2}$-in. vertical shaft. Cam is $\frac{1}{2}$ in. thick on $\frac{3}{4}$-in.-diameter shaft; hub, $1\frac{1}{4}$ in. diameter, $1\frac{1}{4}$ in. long.

**22.** Make a drawing, with development, for a cylindrical cam. The $\frac{1}{2}$-in.-diameter roller follower is to move 2 in. leftward with constant velocity in 180° turn of cylinder and return with simple harmonic motion. Cam axis is horizontal; cylinder, 4 in. diameter, 4 in. long on 1-in. shaft. Follower is pinned to $\frac{5}{8}$-in. shaft 3 in. center to center from cylinder.

**Described in this Chapter:**



# 19

# PIPING

**Pipes and Tubes in Great Variety Are Available for the Transmission of Gasses and Liquids to Components of Machines or from a Source of Supply to a Machine**

### 1. Piping—Fundamental Considerations.

A knowledge of pipe and pipe fittings is necessary not only for making piping drawings but because pipe is often used as a construction material. An understanding of pipe threads is also essential, because in making machine drawings it is frequently necessary to represent and specify tapped holes to receive pipe for liquid or gas supply lines.

### 2. Metal Pipe.

Standard pipe of steel or wrought iron up to 12 in. in diameter is designated by its nominal inside diameter, which differs somewhat from the actual inside diameter. Early pipe manufacturers made the walls in the smaller sizes much too thick and in correcting this error in design took the excess from the inside to avoid changing the sizes of fittings. Three weights of pipe—standard, extra strong, and double extra strong—are in common use. In the same nominal size all three have the same outside diameter, that of standard-weight pipe, the added thickness for the extra and double extra (XX) strong being on the inside. Thus the outside diameter of 1-in. pipe in all three weights is 1.315 in. The inside diameter of standard 1-in. pipe is 1.05 in.; of 1-in. extra strong, 0.951 in.; and of XX, 0.587 in. The ASA in B36.10—1959 gives a means of specifying wall thicknesses by a series of schedule numbers which indicate approximate values for the expression 1,000 $\times$ $(P/S)$, where $P$ is pressure and $S$ the allowable stress. Recommended values for $S$ can be obtained from the ASME Boiler Code, the American Standard Code for Pressure Piping (ASA, B31.1), etc. The designer computes the exact value of wall thickness as required for a given condition and selects from the schedule numbers the one nearest to the computed values. In the ASA system, pipe is designated by giving nominal pipe size and wall thickness, or nominal pipe size and weight per foot.

All pipe over 12 in. in diameter is designated as OD (outside-diameter) pipe and is specified by its outside diameter and thickness of metal. Boiler tubes in all sizes are known by their outside diameters.

Brass, copper, stainless steel, and aluminum pipe has the same nominal diameters as iron pipe but some has thinner wall sections. There are two standard weights: regular and extra strong. Commercial lengths are 12 ft, with longer lengths made to order.

Lead pipe and lead-lined pipe are used in chemical work. Cast-iron pipe is used for water and gas in underground mains and for drains in buildings.

Many other kinds of pipe are in more or less general use and are known by trade names, such as hydraulic pipe, merchant casing, API (American Petroleum Institute) pipe, etc. Details are given in manufacturers' catalogues.

Most small-line plumbing in homes, buildings, and industries, for hot and cold water installations, employ copper pipe with soldered-joint fittings.

### 3. Tubing.

Seamless flexible metal tubing is used for conveying steam, gases, and liquids in all types of equipment such as locomotives, diesel engines, hydraulic presses, etc., where vibration is present, where outlets are not in alignment, and where there are moving parts.

Copper tubing is available in nominal diameters of $\frac{1}{8}$ to 12 in. and in four weights known as classes K, L, M, and O. Class K is extra-heavy hard; class L is heavy hard; class M is standard hard; and class O is light hard. Boiler tubes in all sizes are designated by their outside diameters.

Tubing is made in a variety of materials—glass, steel, aluminum, copper, brass, aluminum bronze, asbestos, fiber, lead, and others. The mechanical catalogue of the American Society of Mechanical Engineers, New York, N.Y., lists manufacturers.

### 4. Plastic Pipe.

Since plastic pipe does not corrode and has high resistance to a broad group of industrial chemicals, it is used extensively in place of metal pipe. Polyvinyl chloride, polyethelyne, and styrene are basic plastic materials. Of these, polyvinyl chloride (PVC) is the most widely used. It does not support combustion, is nonmagnetic and nonsparking, imparts no odor or taste to contents, is light ($\frac{1}{2}$ the weight of aluminum), has low flow resistance, resists weathering, and is easily bent and fabricated by solvent cementing or, in heavy weights, by threading. Its chief limitations are higher cost (partially offset by lower installation cost), low temperature limit (150°F in continuous service), and low pressure limits. Also, it does not resist all solvents; and it requires more supports, and contracts and expands more (about 5 times) than steel.

Metal pipe lined with plastic has the advantage of combining the strength of metal with the chemical resistance of plastic. Seran rubber is also used for lining metal pipe.

The mechanical catalogue of the American Society of Mechanical Engineers, New York, N.Y., lists manufacturers from whom catalogues and special information are available.

## 5. Pipe Joints.

Pipe is connected by methods dependent upon the material and the demands of service. Steel, wrought-iron, brass, or bronze pipe is normally threaded and screwed into a coupling (or a fitting), as shown in Fig. 1 at (*a*). At (*b*) a screwed flange is illustrated; this joint is easily disassembled for cleaning or repair. At (*c*) a permanent welded joint is shown. When welded pipe must be disassembled periodically, ring joints (*d*) are used, where necessary, in the system; these joints are bolted together. Cast-iron pipe cannot be successfully welded or threaded; thus, a bell-and-spigot joint, caulked and leaded, as at (*e*) is used.

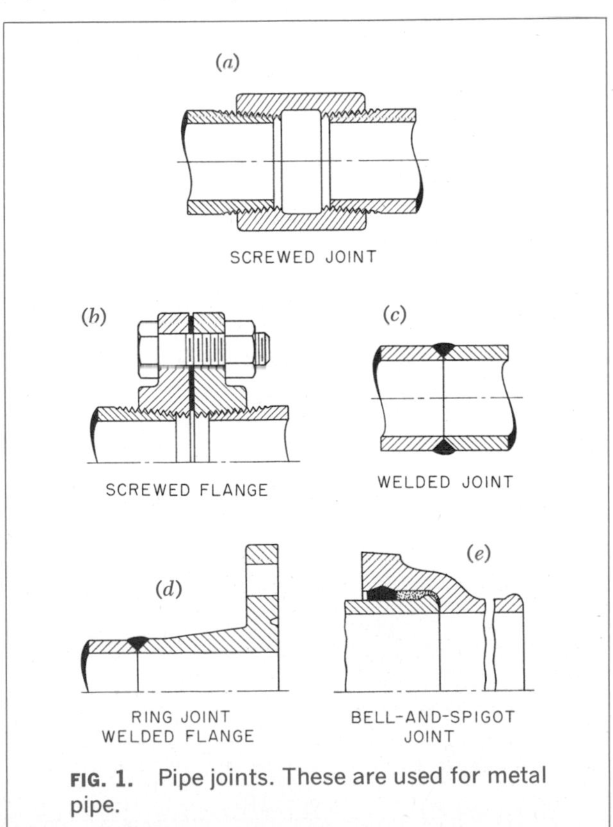


**FIG. 1.** Pipe joints. These are used for metal pipe.

## 6. Tube Joints.

Tubing is commonly employed to connect small components for liquid or gas service. Figure 2 illus-

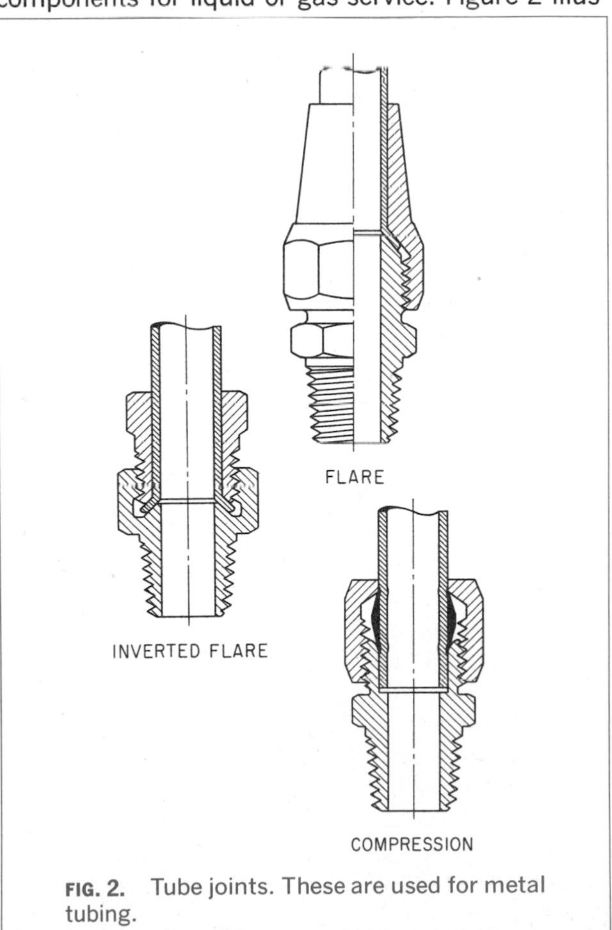


**FIG. 2.** Tube joints. These are used for metal tubing.

trates three common methods of connection. The *flare* and *inverted flare* can both be disassembled without serious injury to the joint, and can be used for fairly high pressures. The compression joint is used for lower pressures and when the joint is not expected to be taken apart and reassembled.

## 7. Pipe Fittings.

Pipe fittings are the components used in connecting and "making up" pipe systems. Cast-iron or malleable-iron fittings (Fig. 3) are commonly used with threaded wrought-iron pipe. Brass and other alloys are employed for special purposes. Steel butt-welding fittings (Fig. 4) are used with steel pipe. Soldered-joint fittings (Fig. 5) are used with copper pipe. Cast-iron bell-and-spigot fittings are used with cast-iron pipe.

*Elbows* are used to change the direction of a pipe line either 90° or 45°. The *street elbow* has male threads on one end, thus eliminating one pipe joint if used at a fitting. *Tees* connect three pipes, and *crosses* connect four. *Laterals* are made with the third opening at 45° or 60° to the straight run.

Straight sections of pipe are made in 12- to 20-ft lengths and are connected by *couplings*. These are short cylinders, threaded on the inside. A right-hand coupling has right-hand threads at both ends. To close a system of piping, although a union is preferable, a *right-and-left coupling* is sometimes used. A *reducer* is similar to a coupling but has the two ends threaded for different-sized pipe. Pipes are also connected by screwing them into cast-iron flanges and bolting the flanges together. Unless the pressures are very low, flanged fittings are recommended for all systems requiring pipe over 4 in. in diameter.

*Nipples* are short pieces of pipe threaded on both ends. If the threaded portions meet, the fitting is a *close nipple;* if there is a short unthreaded portion, it is a *short nipple*. Long and extra-long nipples range in length up to 24 in.

A *cap* is used to close the end of a pipe. A *plug* is used to close an opening in a fitting. A *bushing* is used to reduce the size of an opening. *Unions* are used to close systems and to connect pipes that are to be disassembled occasionally. A screwed

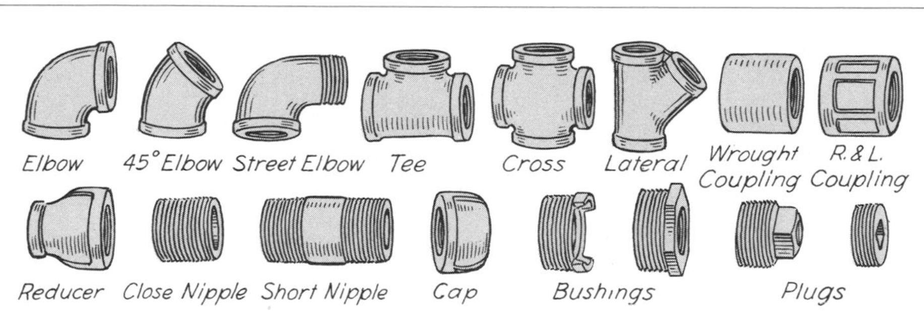


**FIG. 3.** Screwed fittings. See the Appendix for detailed dimensions of malleable and cast-iron fittings.

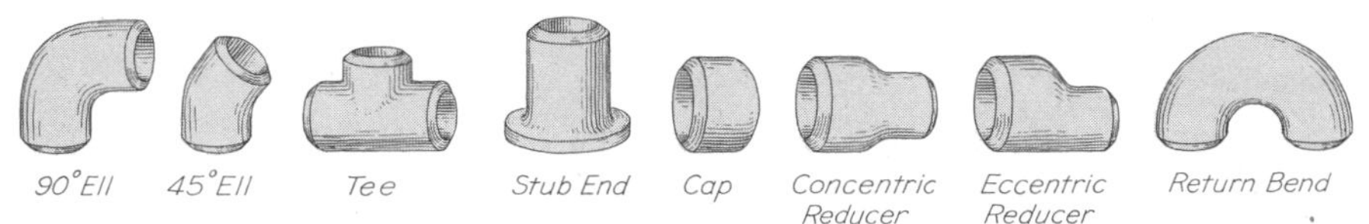


**FIG. 4.** Butt-welding fittings. See the Appendix for detailed dimensions.

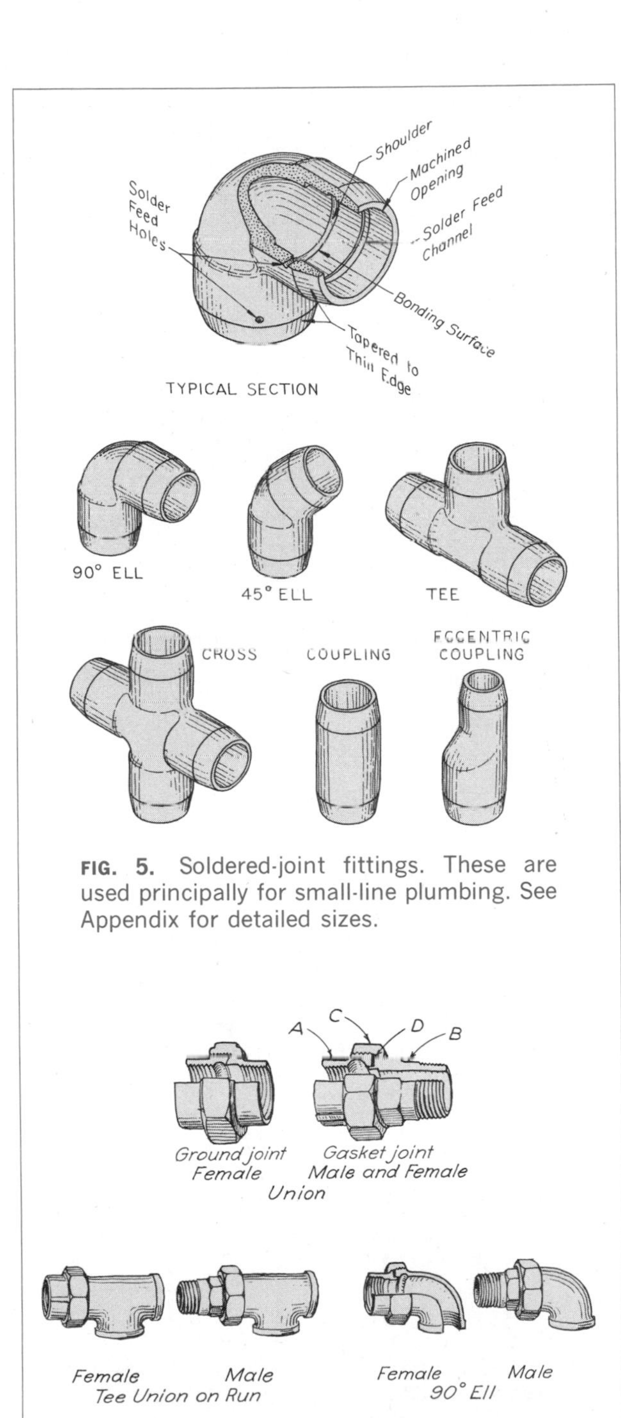


**FIG. 5.** Soldered-joint fittings. These are used principally for small-line plumbing. See Appendix for detailed sizes.



**FIG. 6.** Screwed unions and union fittings. These are used to "close" a pipe system.

union (Fig. 6) is composed of three pieces, two of which, $A$ and $B$, are screwed firmly on the ends of the pipes to be connected. The third piece $C$ draws them together, the gasket $D$ forming a tight joint. Unions are also made with ground joints or with special metallic joints instead of gaskets. Several forms of screwed unions and union fittings are shown in Fig. 6. Flange unions in a variety of forms are used for large sizes of pipe.

## 8. Valves.

Figure 7 shows a few types of valves used in piping. (*a*) is a gate valve, used for water and other liquids, as it allows a straight flow. (*b*) is a plug valve, opened and closed with a quarter turn; (*c*), a ball-check valve; and (*e*), a swing-check valve permitting flow in one direction. For heavy liquids the ball-check valve is preferred. (*d*) is a globe valve, used for throttling steam or fluids; (*f*) is a butterfly valve, opened and closed with a quarter turn, but not steamtight, and used only as a check or damper.

## 9. Specification of Fittings.

Fittings are specified by giving the nominal pipe size, material, and name, for example,

2″ M.I. Elbow    1½″ Brass Tee

When a fitting connects more than one size of pipe, the size of the largest run opening is given first, followed by the size at the opposite end of the run. Figure 8 shows the order of specifying reducing fittings. The word "male" must follow the size of the opening if an external thread is wanted, for example,

2 × 1 (male) × ¾ M.I. Tee

Valves are specified by giving the nominal size, material, and type, for example,

1″ Iron Body Brass-mounted Globe Valve. (If a particular valve is required it is best to give, in addition, "Manufacturer's No.___ or equal.")

## 10. Pipe Threads.

When screwed fittings are used or when a connec-

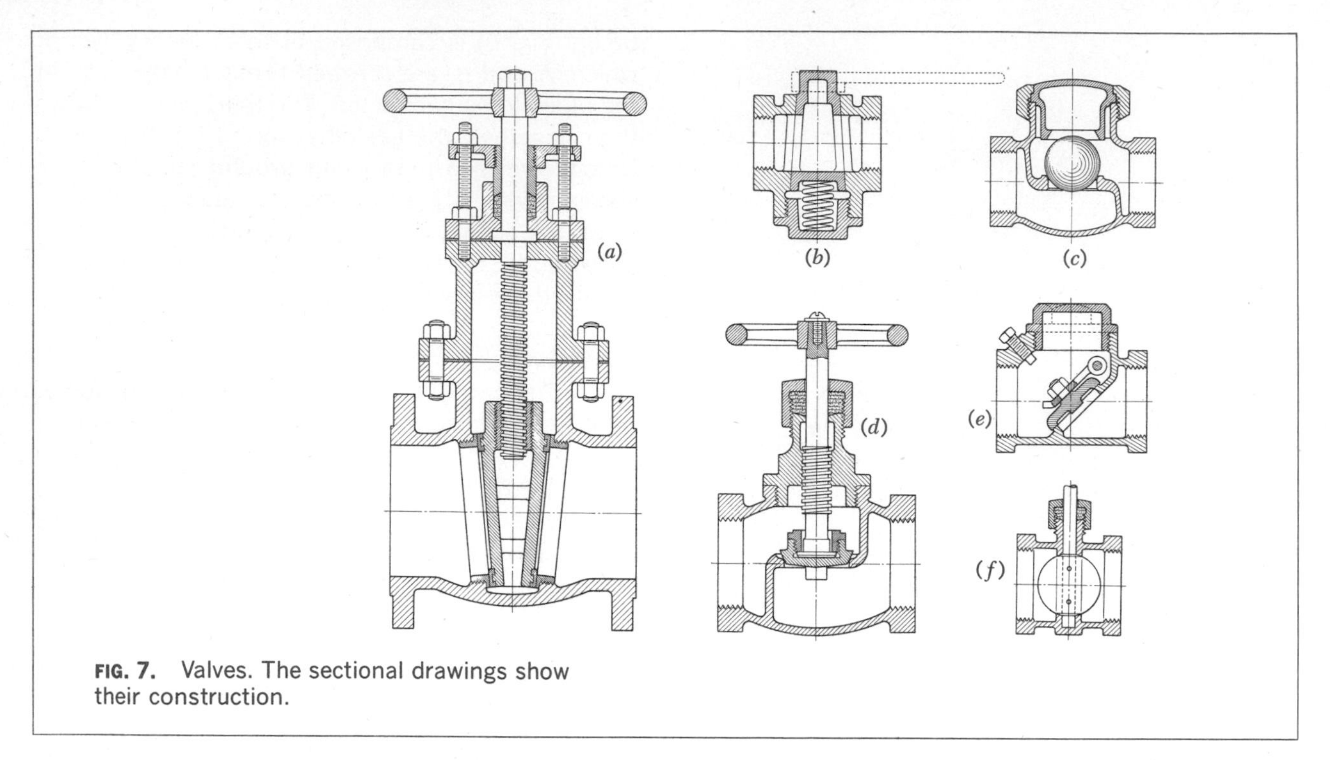


**FIG. 7.** Valves. The sectional drawings show their construction.

tion must be made to a tapped hole, pipe is threaded on the ends for the purpose. The ASA provides two types of pipe thread: tapered and straight. The normal type employs a taper internal and taper external thread. This thread (originated in 1882 as the Briggs Standard) is illustrated in Fig. 9. The threads are cut on a taper of $\frac{1}{16}$ in. per inch, measured on the diameter, thus fixing the distance a pipe enters a fitting and ensuring a tight joint. Taper threads are recommended by the ASA for all uses with the exception of the following five types of joints: type 1, pressure-tight joints for pipe couplings; type 2, pressure-tight joints for grease-cup, fuel, and oil fittings; type 3, free-fitting mechanical joints for fixtures; type 4, loose-fitting mechanical joints with lock nuts; type 5, loose-fitting mechanical joints for hose coupling. For these joints straight pipe threads can be used. The number of threads per inch is the same in taper and straight pipe threads. Actual diameters vary for the different types of joints. When needed they can be obtained from the ASA bulletins. A common practice is to use a taper external thread with a straight internal thread, on the assumption that the materials are sufficiently ductile to allow the threads to adjust themselves to the taper thread.

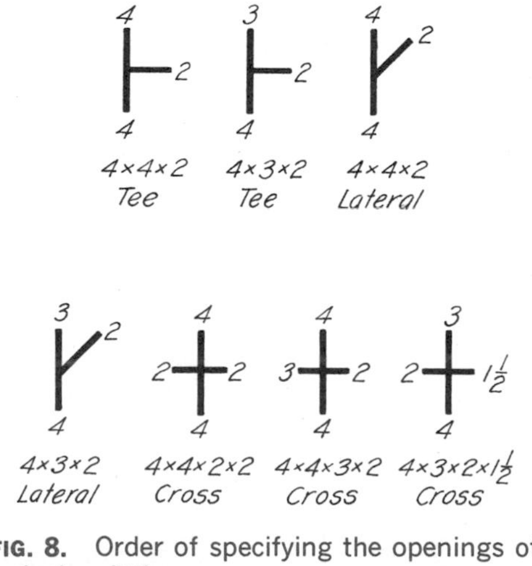

4×4×2 Tee
4×3×2 Tee
4×4×2 Lateral
4×3×2 Lateral
4×4×2×2 Cross
4×4×3×2 Cross
4×3×2×1½ Cross

**FIG. 8.** Order of specifying the openings of reducing fittings.

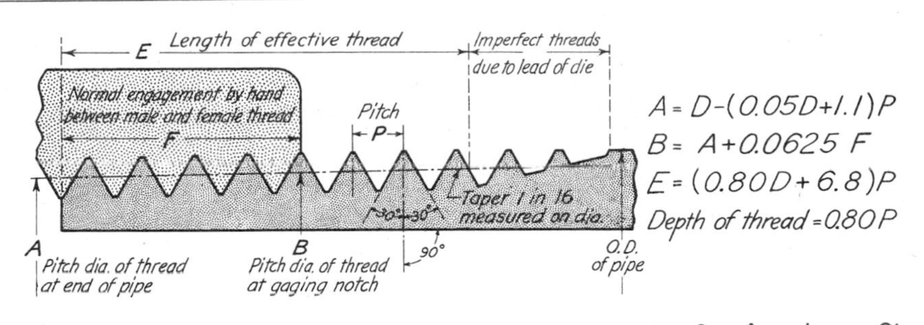


**FIG. 9.** American Standard taper pipe thread. "Wedging" action of the taper produces a tight seal.

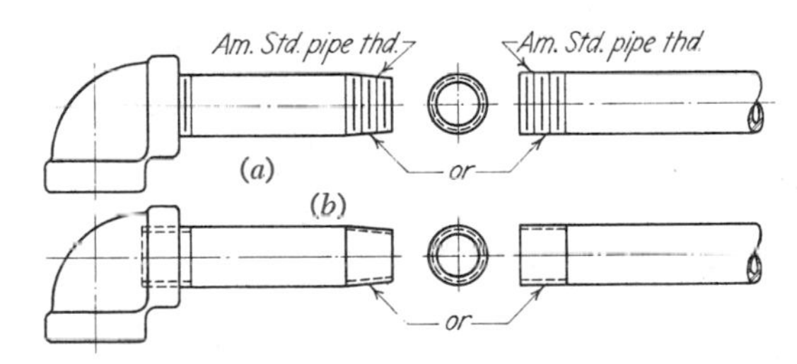


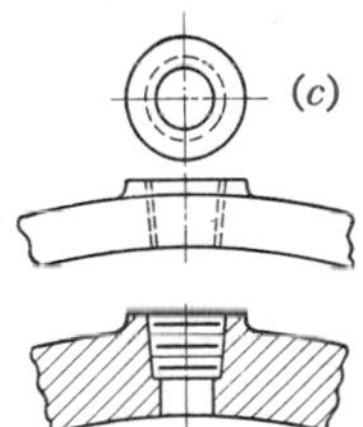


**FIG. 10.** Conventional methods of drawing pipe threads. (*a*) Regular method; (*b*) simplified method; (*c*) hole tapped for pipe.

All pipe threads are assumed to be tapered unless otherwise specified.

Pipe threads are represented by the same conventional symbols as bolt threads. The taper is so slight that it shows only if exaggerated. It need not be indicated unless it is desired to call attention to it, as in Fig. 10. In plan view (*c*), the dashed circle should be the actual outside diameter of the pipe specified. The length of effective thread is $E = (0.80D + 6.8)P$ (Fig. 9).

### 11. Specification of Threads.

Pipe threads are specified by giving the nominal pipe diameter, number of threads per inch, and the standard letter symbol to denote the type of thread. The following ASA symbols are used:

NPT = taper pipe thread
NPTF = taper pipe thread (dryseal)
NPS = straight pipe thread
NPSC = straight pipe thread in couplings
NPSI = intermediate internal straight pipe thread (dryseal)
NPSF = internal straight pipe thread (dryseal)
NPSM = straight pipe thread for mechanical joints
NPSL = straight pipe thread for locknuts and locknut pipe threads
NPSH = straight pipe thread for hose couplings and nipples
NPTR = taper pipe thread for railing fittings

Examples: $\frac{1}{2}$-14NPT   $2\frac{1}{2}$-8NPTR

The specification for a tapped (pipe-thread) hole must include the tap drill size, for example,

$\frac{59}{64}$ Drill, $\frac{3}{4}$-14NPT

Dimensions of ASA taper pipe threads (NPT) are given in the Appendix. Dimensions of other pipe threads are given in ASA B2.1 and in manufacturers' catalogues.

## 12. Piping Drawings.

Two general systems are used: (1) *scale layout* and (2) *diagrammatic*. Scale layouts are used principally for large pipe (usually flanged), as in boiler and power-plant work where lengths are critical and especially when the pipe is not cut and fitted in the field. Also, smaller pipe can be thus detailed when the parts are cut and threaded and then shipped to the job. Figure 11 is an example of a scale layout. The fittings can be specified on the drawing, as in Fig. 11, or on a bill of material. On small-scale drawings such as architectural plans, plant layouts, etc., or on sketches, the diagrammatic system is used. According to this system the fittings are shown by symbols (see Appendix) and the runs of pipe are shown by a single line, regardless of the pipe diameters, as shown in Fig. 12. When lines carry different liquids or different states of a liquid, they are identi-

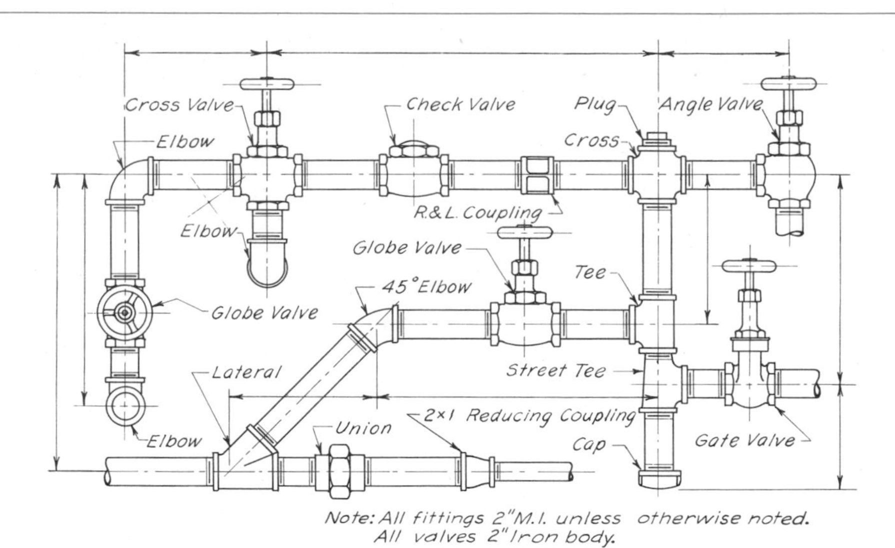


**FIG. 11.** Scale layout of piping. Pipe and fittings are drawn from dimensional specifications (see Appendix).

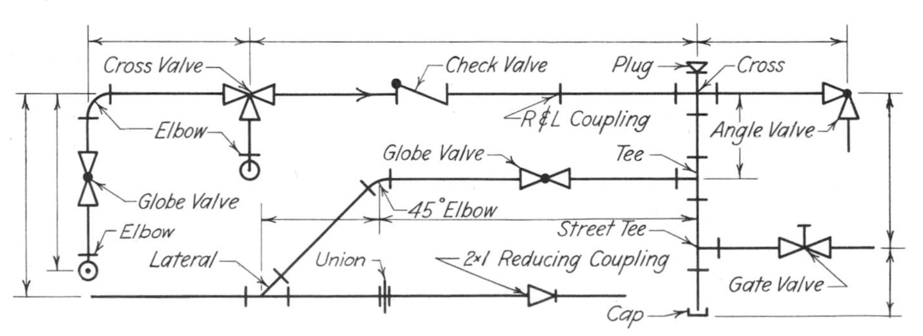


**FIG. 12.** Diagrammatic drawing of piping. Fittings are indicated by standard symbols (see Appendix).

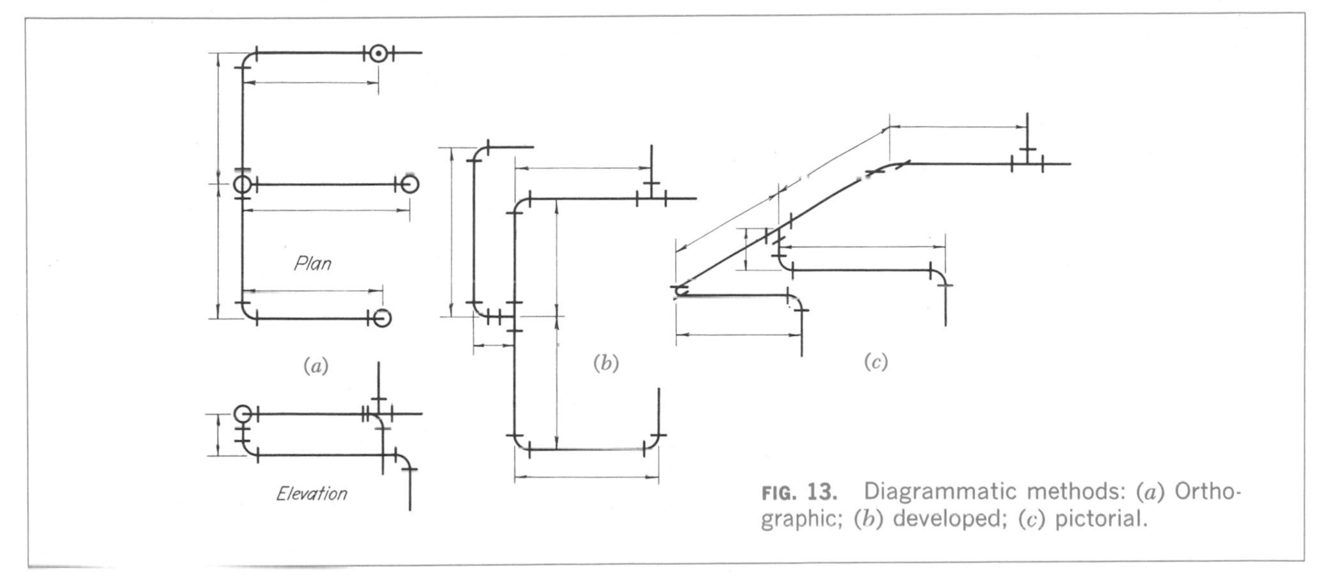


FIG. 13. Diagrammatic methods: (*a*) Orthographic; (*b*) developed; (*c*) pictorial.

fied by coded line symbols. The standard code for hot water, steam, cold water, etc., is given in the Appendix. The single line should be made heavier than the other lines of the drawing.

The arrangement of views is generally in orthographic projection (Fig. 13*a*). Sometimes, however, it is clearer to rotate all the piping into one plane and make only one "developed view" as at (*b*). Isometric and oblique diagrams, used alone or in conjunction with orthographic or developed drawings, are often employed in representing piping. The representation at (*c*) is drawn in oblique. Problem 14 is an example of an isometric layout.

## 13. Dimensioning a Pipe Drawing.

The dimensions on a piping drawing are principally *location* dimensions, all of which are made to center lines, both in single-line diagrams and in scale layouts, as shown in Figs. 11 and 12. Valves and fittings are located by measurements to their centers, and the allowances for make-up are left to the pipe fitter. In designing a piping layout, take care to locate valves so that they are easily accessible and have ample clearance at the handwheels. The *sizes* of pipe should be specified by notes telling the nominal diameters, never by dimension lines on the drawing of the pipes. The fittings are specified by note, as described in Sec. 9. Very complete notes are an essential part of all piping drawings and sketches.

When it is necessary to dimension the actual length of a piece of pipe, the distance can be calculated by using the over-all fitting dimensions and accounting for the entrance length of the pipe threads.

Dimensions for standard pipe and for various fittings are given in the Appendix.

## 14. Pipe Hangers and Supports.

Small pipe and light tubing in short runs can be supported by connections to various machines or fittings. Straps of various types are used to tie pipe to posts, columns, walls, ceilings, etc. Pipe hangers and supports are available for almost any size and type of installation. In accordance with ASA B31.1—1955, Code for Pressure Piping, all piping systems require sway braces, guides, and supports. In Figure 14 a few commonly used hangers and supports are shown. A split ring (*a*) is used with a threaded rod attached to the building proper. The locking device prevents change of adjustment due to vibration and assures proper pitch of the line. A double bolt clamp, shown at (*b*), is for service where it is desirable to have the yoke extend outside of pipe covering. The I-beam clamp, two styles shown

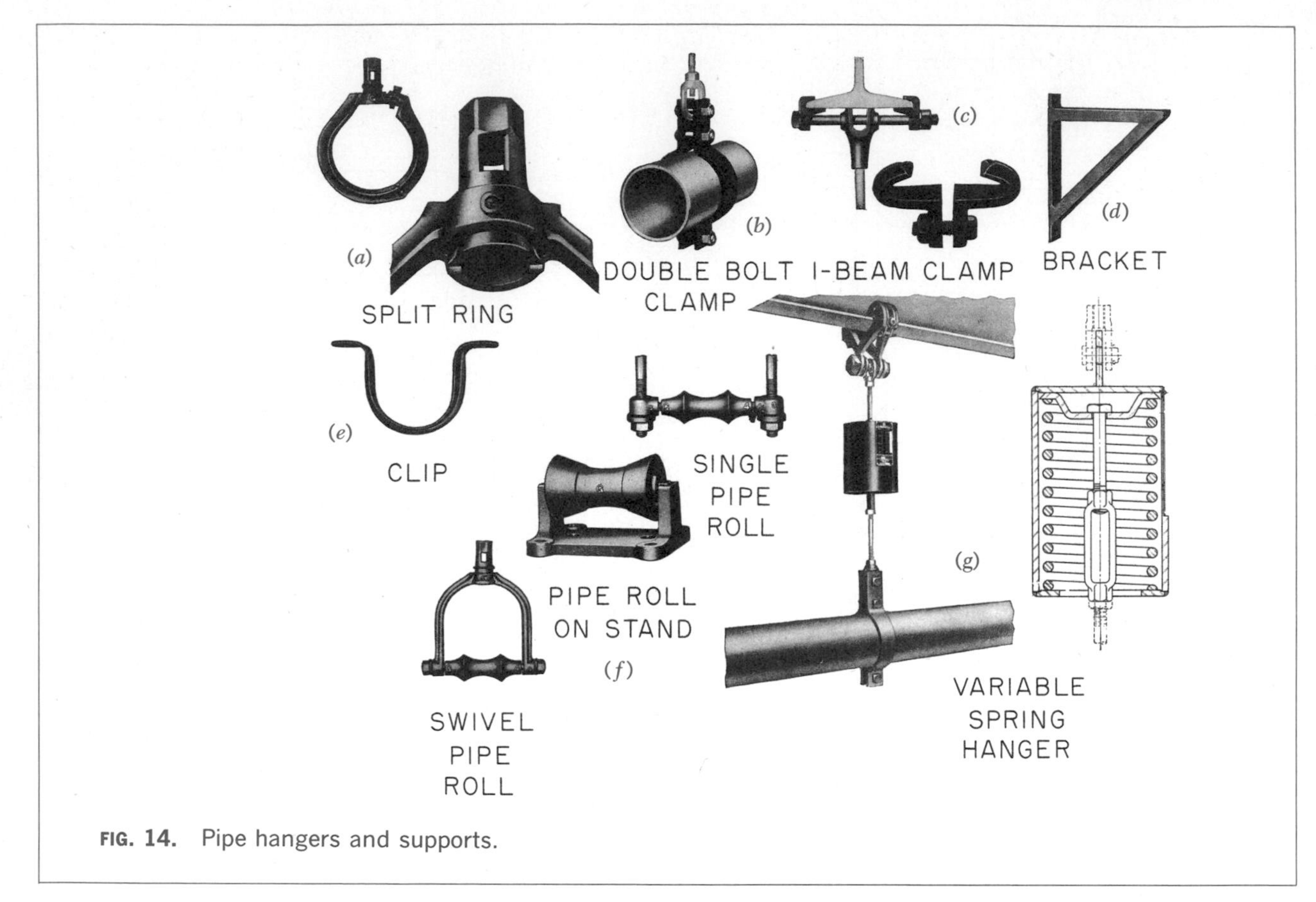


**FIG. 14.** Pipe hangers and supports.

at (*c*), is suitable for clamping on flanges ranging in width from 2 to $6\frac{1}{2}$ inches. The steel bracket (*d*) can be bolted to a wall and pipes can be supported on, as well as hung from, the horizontal member. The strap or clip (*e*) can be used for small pipe; this type is used when pipe is to be flush with a ceiling or wall. Pipe rolls (*f*) are designed to support piping so that longitudinal movement resulting from expansion and contraction can take place. Three types are shown for varying support conditions. The variable spring hanger (*g*) can be obtained in several different sizes and arrangements.

# PROBLEMS

## Group 1. Pipe Fittings.

**1.** In the upper left-hand corner of one sheet draw a 2-in. tee (full size). Plug one outlet. In the second outlet place a 2- by $1\frac{1}{2}$-in. bushing; in the remaining outlet use a 2-in. close nipple and on it screw a 2- by $1\frac{1}{2}$-in. reducing coupling. Lay out the remainder of the sheet so as to include the following $1\frac{1}{2}$-in. fittings: coupling, globe valve, R&L coupling, angle valve, 45° ell, 90° ell, 45° Y, cross, cap, three-part union, flange union. Add extra pipe, nipples, and fittings so that the system will close at the reducing fitting first drawn.

2. Make a one-view drawing of a $1\frac{1}{2}$-in. globe valve. Use an $8\frac{1}{2}$- by 11-in. sheet. See Appendix for proportions.
3. Same as Prob. 2 for a $1\frac{1}{2}$-in. angle globe valve.
4. Same as Prob. 2 for a $1\frac{1}{4}$-in. gate valve.

## Group 2. Piping Layouts to Scale.

5. (*a*) Make a scale layout of the sodium hydroxide, $NaOH + H_2O$, lines of the hydrogen generator. Show a cross section of the headers, one cylinder of the pump, a portion of the sodium hydroxide tank, and a portion of the top of one generator. (*b*) Make a scale layout of the hydrogen ($H_2$) line of the hydrogen generator. Show a portion of each generator with the lines in place. Use 11- by 17-in. paper. Scale, 3″ = 1′-0″. Use standard pipe, malleable fittings, brass valves. Make a bill of material listing all pipe, valves, and fittings needed.

The operation of the unit is as follows: After the generators have been charged with ferrosilicon ($Fe + Si$), a mixture of sodium hydroxide ($NaOH$) and water ($H_2O$) is pumped into the generators. When the ferrosilicon, sodium hydroxide, and water are together in the generator, they react to form hydrogen ($H_2$) and a sludge ($Fe$, $Na_2SiO_2$, and $H_2O$). As the hydrogen gas forms, the internal pressure of the generator is built up. To keep the reaction going, additional water and sodium hydroxide solution has to be pumped into the generators against this pressure. As the hydrogen gas is formed, it is piped to a storage tank. The sludge is siphoned off periodically to prevent excessive accumulation that might retard the reaction.

A four-cylinder pump is used to provide a more uniform flow of fluid than would be obtained if a large single-cylinder pump were used. The four cylinders of the pump are connected to one intake line and to one exhaust line through headers (manifolds). Instead of building an intake and an exhaust valve into the cylinders of the pump, standard ball-type check valves are used and are connected as near as possible to the pump cylinders by means of standard pipe and fittings.

6. Make a scale layout of the gas-burner installation. Specify fittings and give center-line dimensions. Use 11- by 17-in. paper. Scale, 3″ = 1′-0″.

7. Make a scale layout of a Grinnell industrial heating unit, closed return, gravity system. Use center-line distances and placement of fittings as shown in the diagram. Use 3-in. supply main, 2-in. pipe and fittings to unit, $\frac{3}{4}$-in. pipe and fittings from unit to 2-in. return main. Add all necessary notes and dimensions. Use an 11- by 17-in. sheet. Scale, 3″ = 1′-0″.

## Group 3. Piping Layouts, Diagrammatic.

8. *A* is a storage tank for supplying the mixing tanks *B*, *C*, and *D* and is located directly above them. The capacities of the mixers are in the ratios of 1, 2, and

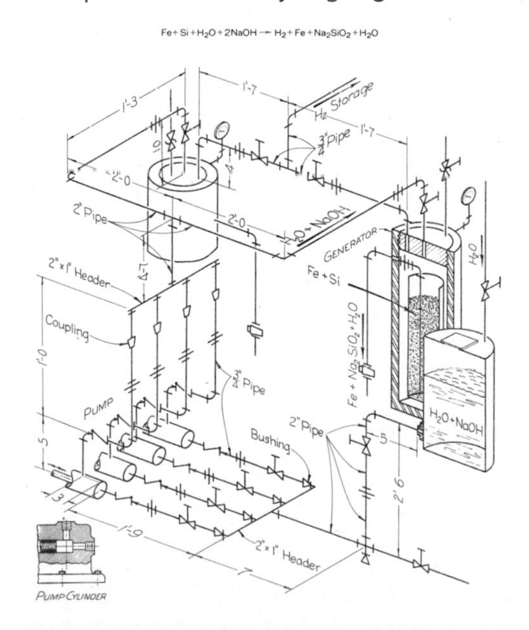


PROB. 5. Hydrogen generator.

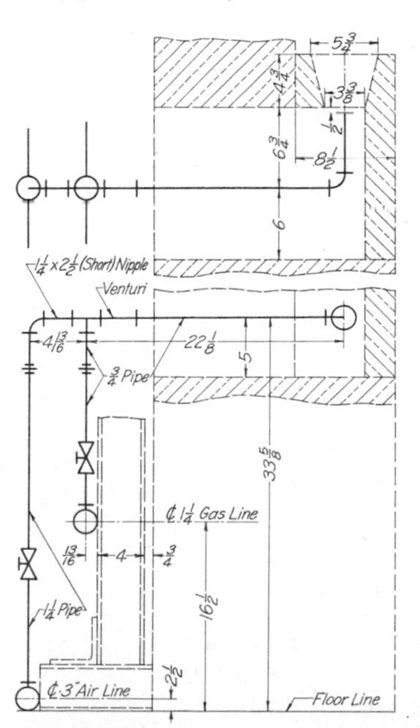


PROB. 6. Gas-burner installation.

3" Supply main
Pitch down
3x3x2 Red. Tee
¾" Air Valve as high as possible
2x2x¾ Red. Tee
1"x¾ Red. Bush.
Pitch down
2" Return

PROB. 7. Grinell industrial heating unit.

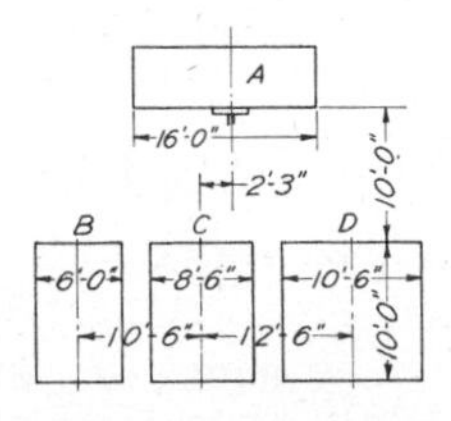


PROB. 8. Storage and mixing tanks.

3. Design (in one view) a piping system with sizes such that, neglecting frictional losses, the three tanks will fill in approximately the same time. So arrange the piping that any one of the tanks can be cut out or removed for repairs without disturbing the others. Use single-line conventional representation. Dimension to center lines and specify the fittings.

**9.** The figure shows the arrangement of a set of mixing tanks. Make an isometric drawing of an overhead piping system to supply water to each tank. Water supply enters the building through a $2\frac{1}{2}$-in. main at point $A$, 3 ft below floor level. Place all pipe 10 ft above floor level, except riser from water main and drops to tanks, which are to end with globe valves 5 ft above floor level. Arrange the system to use as little pipe and as few fittings as possible. Neglecting frictional losses, sizes of pipe used should be such that they will deliver approximately an equal volume of water to each tank if all were being filled at the same time. The pipe size at the tank should not be less than $\frac{3}{4}$ in. Dimension and specify all pipe and fittings.

**10.** Make a drawing of the system of Prob. 9. Show the layout in a developed view. Dimension from center to center and specify all pipe and fittings.

**11.** Make a list of the pipe and fittings to be ordered for the system of Prob. 9. Arrange the list in a table, heading the columns as below.

**12.** Make an oblique drawing of a system of piping to supply the tanks in Prob. 9. All piping except risers is to be in a trench 1 ft below floor level. Risers should not run higher than 6 ft above floor level. Other conditions as in Prob. 9.

**13.** The figure shows the outline of the right-hand half of a bank of eight heat-treating furnaces. $X$ and $Y$ are the lead-ins from the compressed-air and fuel mains. Draw the piping layout, using single-line representation, to distribute the air and fuel to the furnaces. The pipe sizes

| Size | Pipe lengths | Valves | | Fittings | | Material | Remarks (make, kind of threads, etc.) |
|---|---|---|---|---|---|---|---|
| | | Number | Kind | Number | Kind | | |

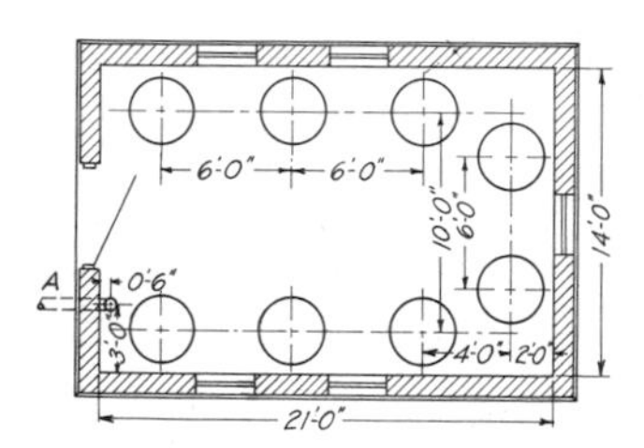


PROB. 9. Mixing tanks.

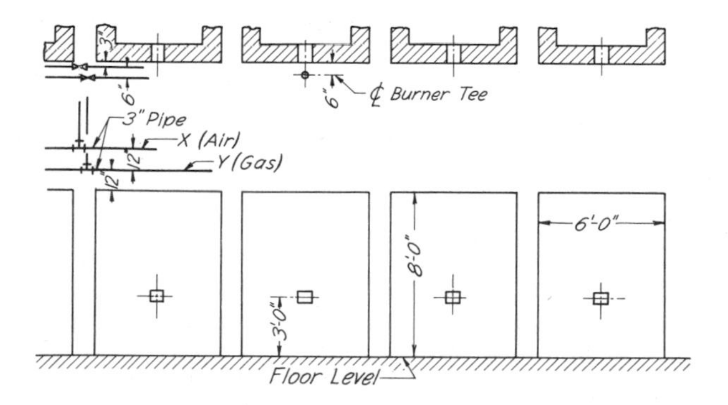


PROB. 13. Heat-treating furnaces.

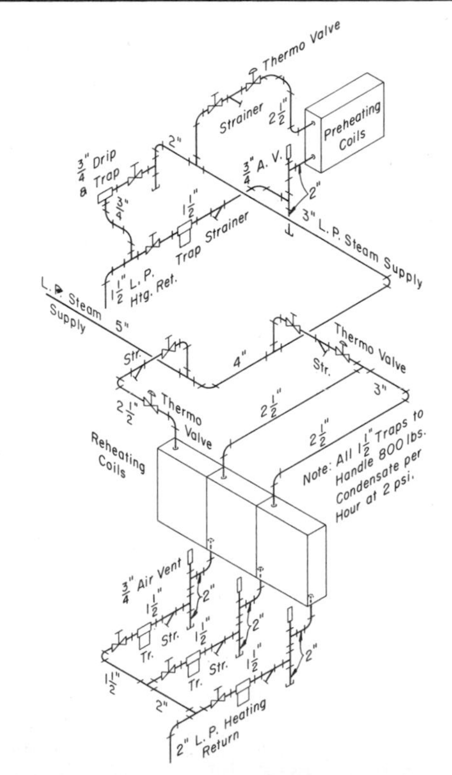


PROB. 14. Industrial heater system.

should be reduced proportionately as the oven leads are taken off. Each tail pipe should be removable without disturbing the other leads or closing down the other furnaces. Dimension the piping layout and make a bill of material for the pipe and fittings.

**14.** The figure is the diagrammatic isometric layout of an industrial heater system. Using the basic layout as shown, assume positions for the preheating and reheating coils and make a complete, dimensioned, isometric diagrammatic layout.

**15.** Make an orthographic diagrammatic layout for Prob. 14.

**16.** Make a developed diagrammatic layout for Prob. 14.

**17.** Shown is a scale layout of a meter and valve enclosure for an industrial plant. Make a scale drawing of the enclosure and show all piping diagrammatically.

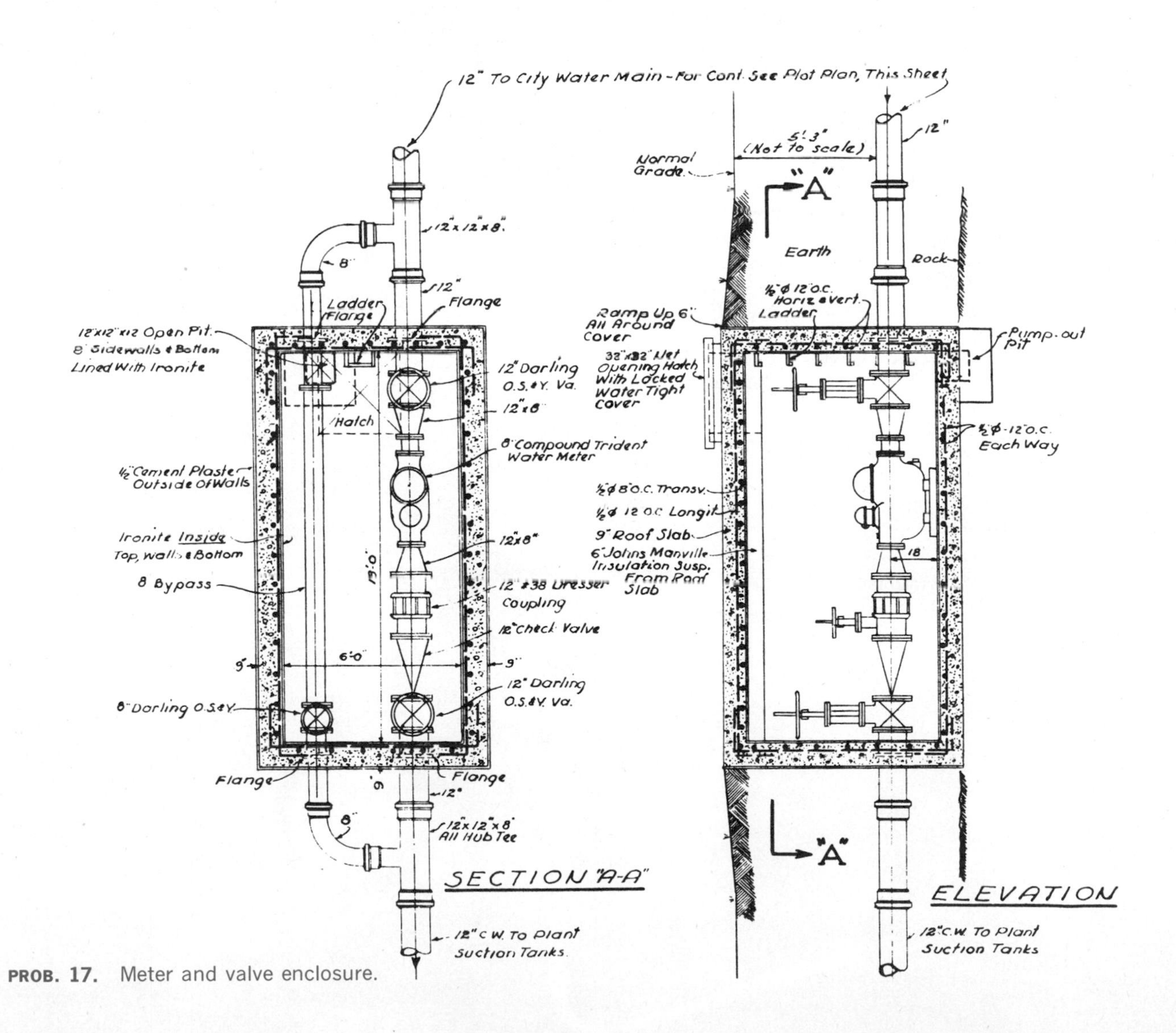


**PROB. 17.** Meter and valve enclosure.

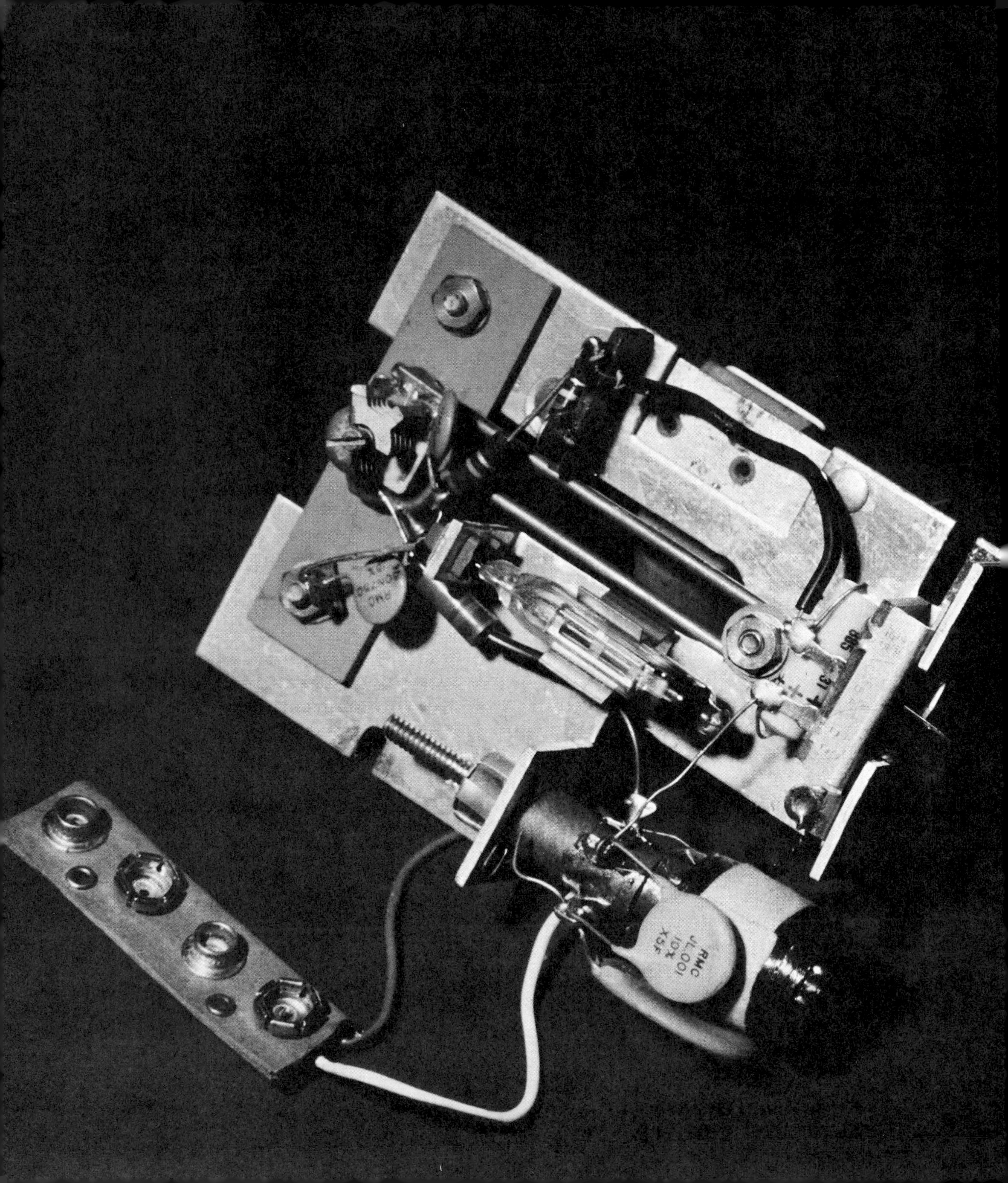
RMC
JL.001
10X
X5F

Discussed Here:



# 20

# ELECTRIC SYSTEMS

Representation of Electric or Electronic Systems Is Accomplished Predominantly by the Use of Orthographic Descriptions Combined with Symbolic Diagrams

## 1. Electric Systems— Important Considerations.

Almost all engineering projects include mechanical and electrical components. The electrical components consist of items such as power supply, motors, generators, controls, heating and cooling devices, and lighting for industry; energy conversion, control apparatus, guidance systems and other devices for transportation in the air, on land, and in water; wired and wireless systems for communication; and home electrical and electronic appliances for everyday comfort and convenience. Thus, a knowledge of electrical systems is important for every engineer. Our purpose here is to show the kinds of drawings required for electrical systems, give the special drafting practices in electrical engineering, and point out the importance of graphics in this field.

Drafting for electrical equipment such as machinery, switchgear, and component circuitry is based upon the same principles as that for mechanical equipment. Because of the complex devices used in electrical circuits, graphic symbols are necessary to simplify representation. Graphic symbols for electrical diagrams have been standardized by the ASA in bulletin Y32.2—1962. "American Drafting Standards Manual," Y14.15—1960, gives drafting details of electrical diagrams.

## 2. Diagrams.

Many different types of diagrams are employed to show arrangements of components and wiring connections. Figure 1 is an example.

The following definitions for electrical diagrams adopted as standard by the ASA are quoted from Y14.15—1960

*Single-line or One-line Diagram.* "A diagram which shows, by means of single lines and graphical symbols, the course of an electric circuit or system of circuits and the component devices or parts used therein." See Sec. 21 and Fig. 20.

*Schematic or Elementary Diagram.* "A diagram which shows, by means of graphical symbols, the electrical connections and functions of a specific circuit arrangement. The schematic diagram facili-

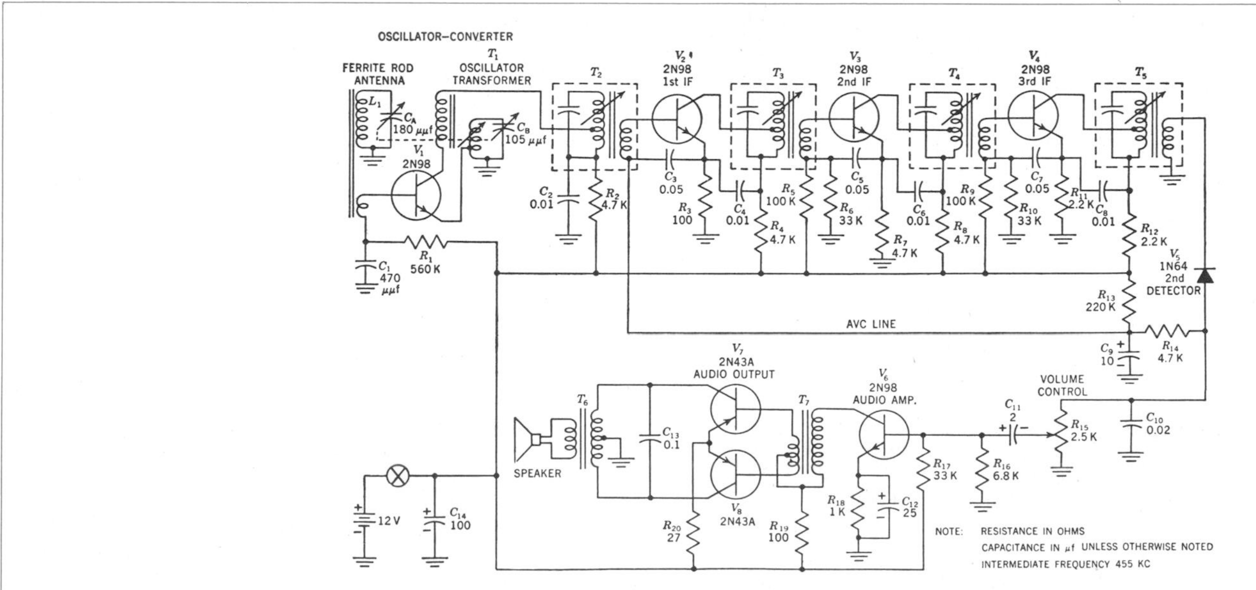


**FIG. 1.** A schematic diagram. This is for a small transistorized radio receiver. The symbols show the components; the notes give values and identifications.

tates tracing the circuit and its functions without regard to the actual physical size, shape, or location of the component device or parts." See Sec. 22 and Figs. 1 and 21.

*Connection or Wiring Diagram.* "A diagram which shows the connections of an installation or its component devices or parts. It may cover internal or external connections, or both, and contains such detail as is needed to make or trace connections that are involved. The connection diagram usually shows general physical arrangement of the component devices or parts."

*Interconnection Diagram.* "A form of connection or wiring diagram which shows only external connections between unit assemblies or equipment. The internal connections of the unit assemblies or equipment are usually omitted."

The text material and illustrations in Secs. 3 to 22 are adapted from ASA Y14.15—1960, Section 15, Electrical Diagrams, by special permission of the ASME. The material given here is intended for educational purposes only. For commercial use, follow the latest issue of ASA or other national-level standards.

### 3. Diagram Titles.

The title of a diagram is given to state the type of diagram and the system or component for which the diagram is made. Examples are: single-line diagram—audio circuit; schematic diagram—vertical oscillator.

### 4. Combined Diagrams.

Diagrams that combine features of more than one type are often more useful than those in the pure forms, defined in Sec. 2. For example, a schematic diagram may include wiring information or interconnection details, especially when the diagram is made for only a portion of a complete circuit. Note in Fig. 49 that connections to other parts of the circuit are shown and identified. The title for a combined diagram conforms to the *principal* purpose of the diagram.

### 5. Drawing Size and Format.

Drawing sizes and formats for electrical drawings conform to the standards of the ASA. See Sec. 12 and Fig. 8, both in Chap. 23.

### 6. Line Conventions and Lettering.

A line of medium thickness, as shown in Fig. 2, is recommended for general use on electrical diagrams. Thin lines are used for brackets and leaders. A heavy line is used to emphasize a special feature. However, the thickness of a line has no significance in meaning—the line thickness is chosen solely for readability of the diagram. When a drawing for reproduction is to be reduced in size, for example, for a manual, the line weights are made heavier to give good readability. See Sec. 47, Chap. 23, on drawing for reproduction.

Lettering on electrical drawings and diagrams should be done in capital letters. For drawings made for direct reproduction (same size), the smallest letters should be not less than $\frac{1}{8}$ in. high. Drawings reduced for manuals should be lettered in a size that will be not less than $\frac{1}{16}$ in. high when reduced.

### 7. Graphic Symbols.

Graphic symbols on electrical diagrams conform to "American Standard Graphical Symbols for Electri-

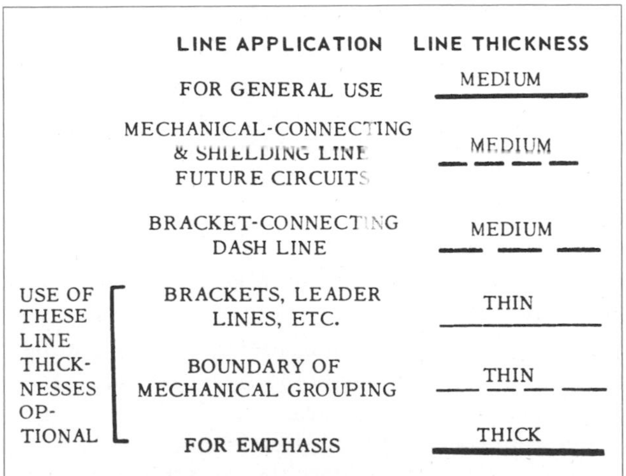

LINE APPLICATION
LINE THICKNESS
FOR GENERAL USE
MEDIUM
MECHANICAL-CONNECTING
& SHIELDING LINE
FUTURE CIRCUITS
MEDIUM
BRACKET-CONNECTING
DASH LINE
MEDIUM
USE OF THESE LINE THICKNESSES OPTIONAL
BRACKETS, LEADER LINES, ETC.
THIN
BOUNDARY OF MECHANICAL GROUPING
THIN
FOR EMPHASIS
THICK


FIG. 2. Line conventions for electrical diagrams. Medium and thin lines are used for all lines normally needed on electrical diagrams. Thick lines are used only for emphasis.

cal Diagrams," Y32.2, or other national-level standards if the symbol is not covered in the American Standard. A selection of symbols, adapted from Y32.2, is given in the Appendix. For special components, if no suitable symbol exists, an appropriate symbol may be made but must be explained in a note. Symbols that can be reversed or rotated without affecting their meaning may be so drawn to simplify the circuit layout. Symbols may be drawn to any convenient size that suits the diagram and produces good readability.

## 8. Layout of Electrical Diagrams.

The best way to start an electrical diagram is to make a freehand sketch of the proposed arrangement. While doing this, place the symbols—remembered or obtained from the Standard (Appendix)—in the best position to eliminate long connecting lines and require as few crossovers as possible. In making the drawing proper, use the sketch as a guide. The layout should prominently show the main features. Space the symbols and lines so that there is sufficient area for notes and reference information. However, avoid large blank spaces except when an area is needed for circuits or information to be added later. In general, electrical diagrams are drawn to follow the circuit, signal, or transmission path, from input to output, source to load, or in the order of functional sequence.

*Connecting lines* are drawn horizontally or vertically and with as few bends and crossovers as possible. No more than three lines should be shown connected at one point when an alternate arrangement is possible. Parallel lines are arranged in groups, preferably three to a group, with double space between groups. In grouping parallel lines, lines representing related functions are grouped together. Parallel lines should never be closer than $\frac{1}{16}$ in. in final reduced size.

*Interrupted paths* are used as needed in a diagram, but each path must be identified and the destination shown. Letters, numbers, or abbreviations used for identification should be placed as close as possible to the point of interruption. Figure 3 shows two examples of interrupted paths, multiple paths on the left and single paths on the right.

*Interrupted grouped lines* are employed in a diagram as needed. The destination of the group is shown by bracketing and identifying as in Fig. 4,

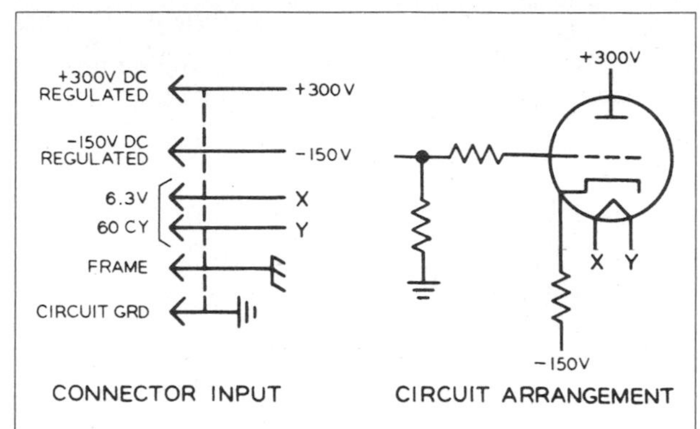


**FIG. 3.** Identification of interrupted lines. Left side of figure: a group of lines interrupted on the diagram. Right side of figure: single lines interrupted on the diagram.

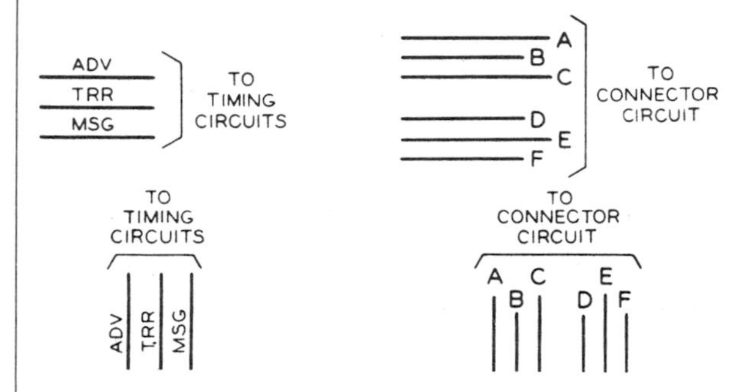


**FIG. 4.** Typical arrangement of line identifications and destinations.

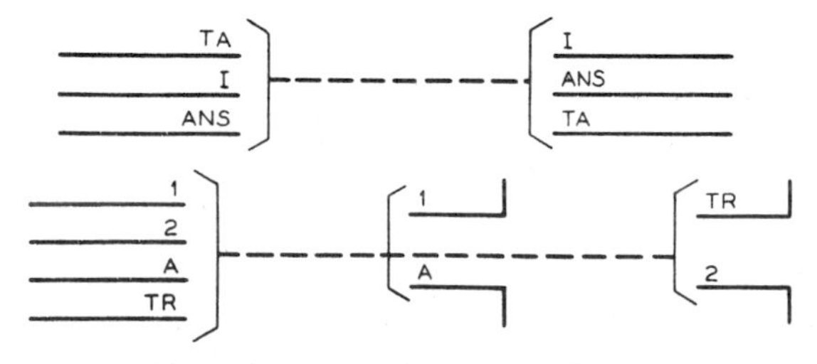


**FIG. 5.** Typical interrupted lines interconnected by dashed lines. The dashed line shows the interrupted paths that are to be connected. Individual line identifications indicate matching connections.

or by using a dashed line connecting the brackets as in Fig. 5. The dashed line is drawn so that it will not form a continuation of one of the lines of the interrupted group.

*Abbreviations* used on electrical diagrams should conform to "American Standard Abbreviations for Use on Drawings," Z32.13, or other national-level standard if the abbreviation is not covered in the American Standard. If no suitable abbreviation exists, a special abbreviation may be used but must be explained by a note on the diagram.

**9. Representation of Electrical Contacts.** Switch symbols (see Appendix and ASA Y32.2) should be shown in the position of no applied operating force. For switches that may be in any of two or more positions with no operating force applied, or for switches actuated by mechanical or electrical means, a drawing note should identify the functional phase shown on the diagram. Relay contacts should be shown in the unenergized position.

**10. Identification of Terminals.** Terminal identifications are added to graphic symbols to show actual physical markings on or near part terminations; also, when no markings exist on a part, arbitrary number or letter designations may be assigned. Figures 6 and 7 show terminal identifications on the symbols of schematic diagrams and on orientation diagrams for the actual part. When terminal designations are arbitrarily assigned, the orientation diagram *must* be included in order to relate the designation to actual physical locations on the part. Pictorial drawings of a part, such as Fig. 8, are used when identification on a symbol and

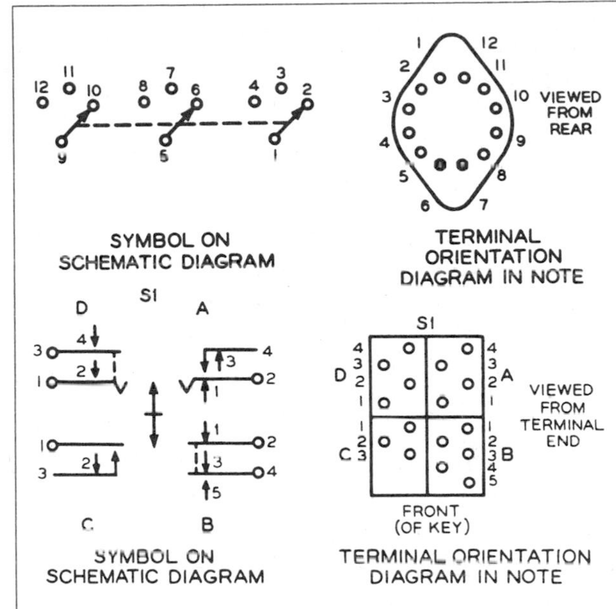


FIG. 7. Terminal identification. Upper half: rotary switch. Lower half: multicontact lever-type key.

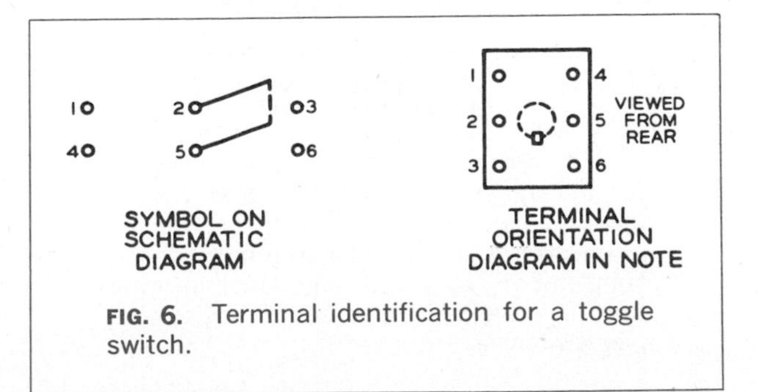


FIG. 6. Terminal identification for a toggle switch.

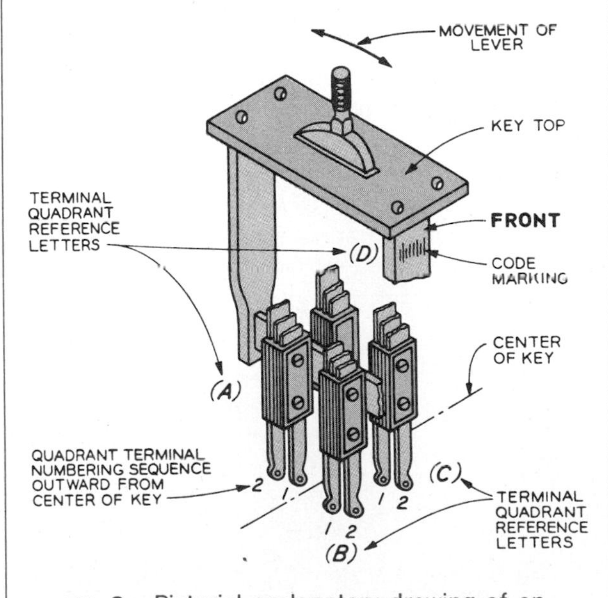


FIG. 8. Pictorial explanatory drawing of an electrical part. This shows the relationship and orientation of elements and gives the identification sequence to be used on a diagram.

a simple terminal orientation note are insufficient. Pictorial drawings are often used in service manuals.

When terminals or leads are identified on the part by color code, letter, number, or geometric symbol, this identification is shown on or near the connecting line adjacent to the symbol.

## 11. Terminal Identification of Adjustable Resistors.

If it is necessary to indicate direction of rotation, it is customary to refer to rotary motion as clockwise or counterclockwise when rotation is viewed from the knob or actuator end of the control. The preferred method of terminal identification is to designate with the letters CW the terminal adjacent to the movable contact when it is in the extreme clockwise position, as shown in Fig. 9*a*. Numbers may also be used with the resistor symbol in which no. 2 is assigned to the adjustable contact as shown at (*b*). Additional fixed taps are numbered sequentially, 4, 5, as at (*c*).

## 12. Switch Terminals and Circuit Functions.

The relation of switch position to circuit function is shown on a schematic diagram by notation near the terminals as at the left in Fig. 10, or in tabular form as shown at the right. For simple toggle switches, the notations ON and OFF are usually sufficient.

Rotary switches that perform involved functions, as illustrated in Fig. 11, should be explained by using the tabular form shown. In tabular listings, dashes link the terminals that are connected.

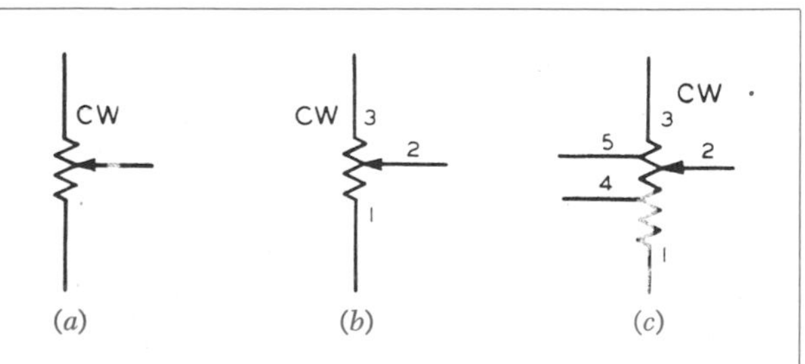


**FIG. 9.** Terminal identification. This example illustrates identifications on a variable resistor for (*a*) tap (CW) nearest variable contact in extreme clockwise position; (*b*) all taps (no. 2 assigned to variable contact); (*c*) additional fixed taps (4 and 5).

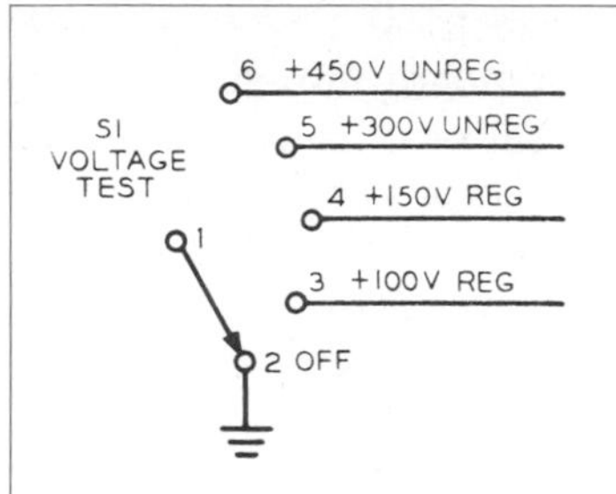


| S1 VOLTAGE TEST | |
|---|---|
| FUNCTION | TERM. |
| OFF | 1-2 |
| +100V REG | 1-3 |
| +150V REG | 1-4 |
| +300V UNREG | 1-5 |
| +450V UNREG | 1-6 |
| | |

FUNCTIONS SHOWN IN TABULAR FORM

**FIG. 10.** Position and function identification for a rotary switch. The choice of method—diagram (left) or table (right)—is optional These methods are recommended for simple switches.

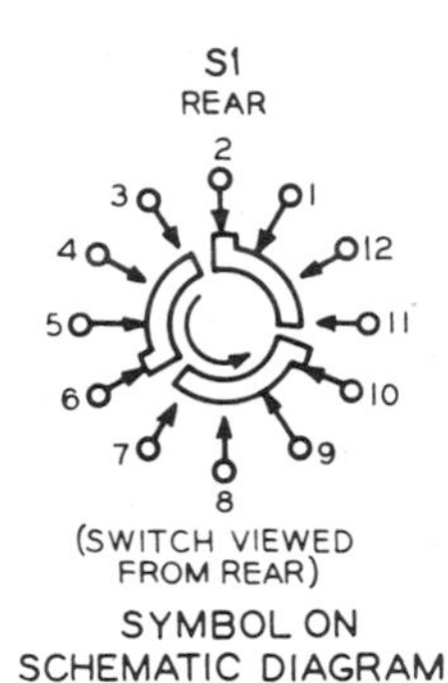


| S1 | | |
|---|---|---|
| POS | FUNCTION | TERM. |
| 1 | OFF(SHOWN) | 1-2,5-6,9-10 |
| 2 | STANDBY | 1-3,5-7,9-11 |
| 3 | OPERATE | 1-4,5-8,9-12 |
| | | |

FUNCTIONS SHOWN IN TABULAR FORM

**FIG. 11.** Position and function identification for a rotary switch. Position and contact identification (only) are shown on the diagram proper. The tabular information then supplements by giving functional information. This method is recommended for complex switches.

**13. Electron Tube-pin Identification.** Tube-pin numbers are shown outside the tube envelope and directly above or at one side of the connecting line, as in Fig. 12*a*. An alternative method is shown at (*b*), where the circles correspond to approximate physical locations of the pins when the tube is viewed from the bottom. Starting with the pin adjacent to the tube-base key or similar point of reference, the circles are numbered consecutively in clockwise order.

**14. Identification of Parts by Suffix Number.** Subdivisions of parts are indicated by adding a suffix letter to the reference designation. For example, C1A and C1B will designate electrically separate sections of a dual capacitor C1.

Suffix letters are also used to identify subdivisions of a complete part when the individual parts are shown enclosed or associated in a unit, as shown in Fig. 13.

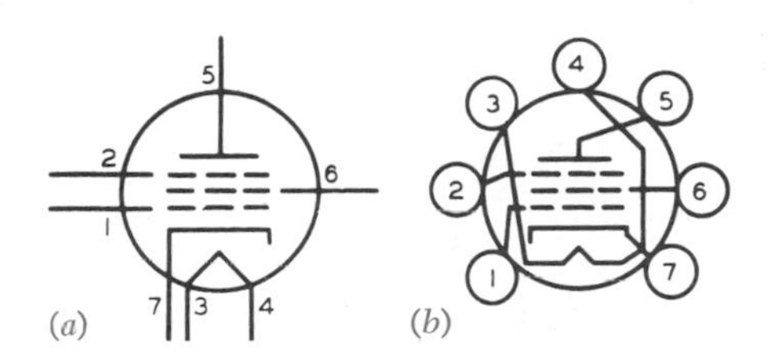


FIG. 12. Terminal identification for electron tube pins. (*a*) Arbitrary numbers are assigned; (*b*) physical location is shown when the tube is viewed from the bottom.

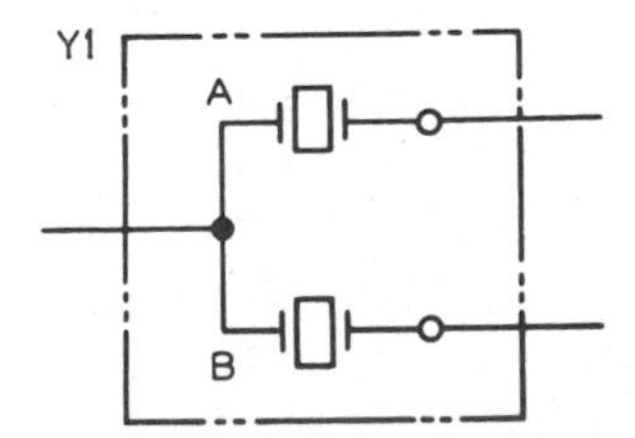


FIG. 13. Identification of parts by suffix letters. A and B are subdivisions of part Y1. Identifications are Y1A and Y1B.

**15. Identification of Rotary-switch Parts.** When parts or rotary switches are designated S1A, S1B, S1C, etc., the suffix letters start with *A* at the knob or actuator end and are assigned sequentially away from this position. Each section of the switch is shown viewed from the same end, as in Fig. 14*a*. When both sides of a rotary switch are used for separate functions, the front (actuator end) and rear symbols should be differentiated by modifying the reference designation, for example, S1A FRONT and S1A REAR as at (*b*).

**16. Identification of Parts by "Part of" Prefix.** When parts of connectors, terminal boards, or rotary-switch sections are functionally separated on

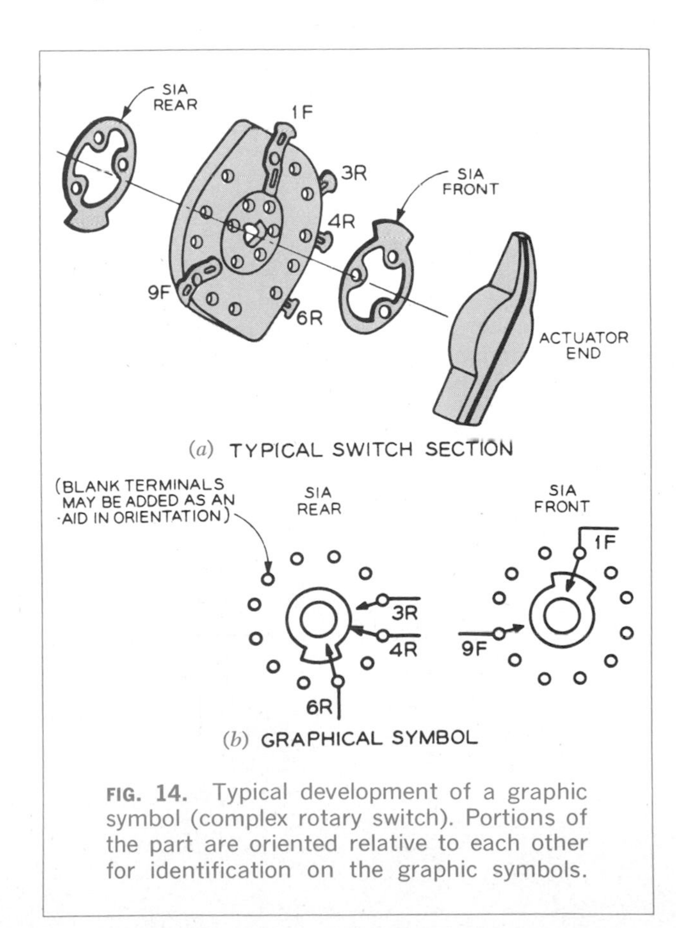


FIG. 14. Typical development of a graphic symbol (complex rotary switch). Portions of the part are oriented relative to each other for identification on the graphic symbols.

the diagram, the words PART OF are used to precede the reference designation as in Fig. 15.

### 17. Identification of Parts as Individual Terminals.

When the separation of parts of connectors or terminal boards on the same drawing becomes extensive, the separated parts may be identified as individual terminals, as shown in Fig. 16.

### 18. Reference Designations.

These are combinations of letters and numbers which identify parts shown on single-line and schematic diagrams and also on related drawings such as the assembly drawing and connection diagram.

The number portion of the designation follows the letter or letters without a space or hyphen and is of the same size as the letter(s), for example, S4, MG5, T6A, etc.

The assignment of numbers should start with the lowest numbers in the upper left-hand corner of the diagram and proceed consecutively from left to right throughout the drawing. When parts are eliminated as a result of a drawing revision, the remaining parts should not be renumbered. As an aid in accounting for all reference designations, a table, as in Fig. 17, may be included showing the highest reference designations and the reference designations not used.

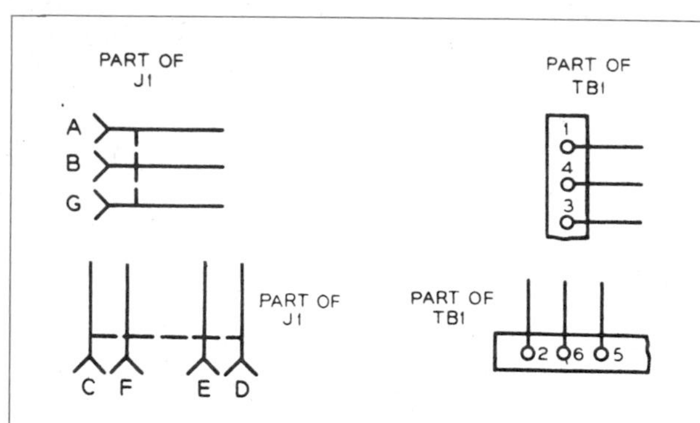


**FIG. 15.** Identification of parts by "part of" prefix. All "part of" portions (separated on the diagram) form a single unit on the actual connector or terminal board.

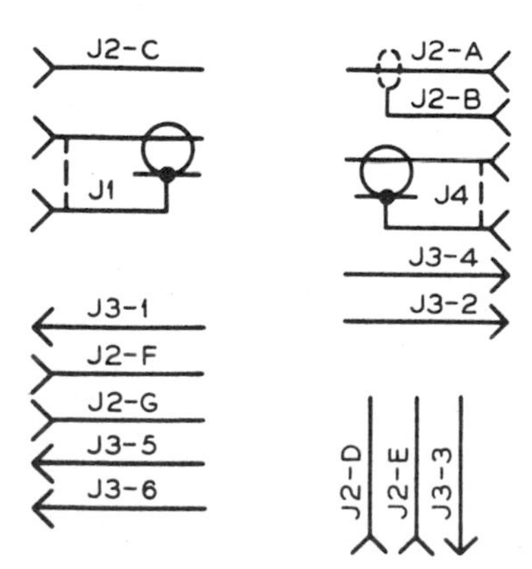


**FIG. 16.** Identification of parts as individual terminals.

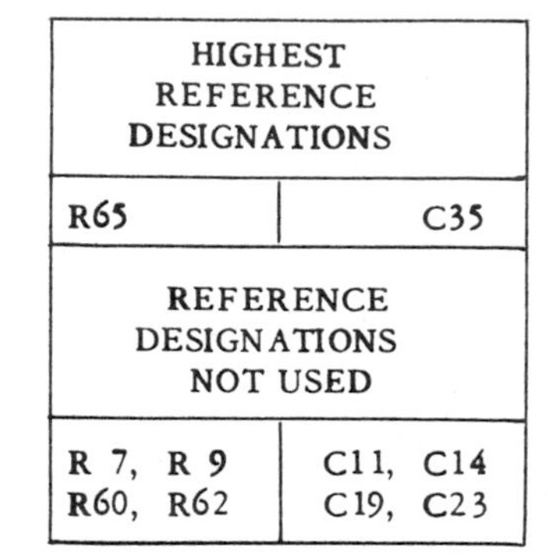

| HIGHEST REFERENCE DESIGNATIONS | |
|---|---|
| R65 | C35 |
| REFERENCE DESIGNATIONS NOT USED | |
| R 7, R 9<br>R60, R62 | C11, C14<br>C19, C23 |

**FIG. 17.** Typical table indicating highest and omitted numerical reference designations.

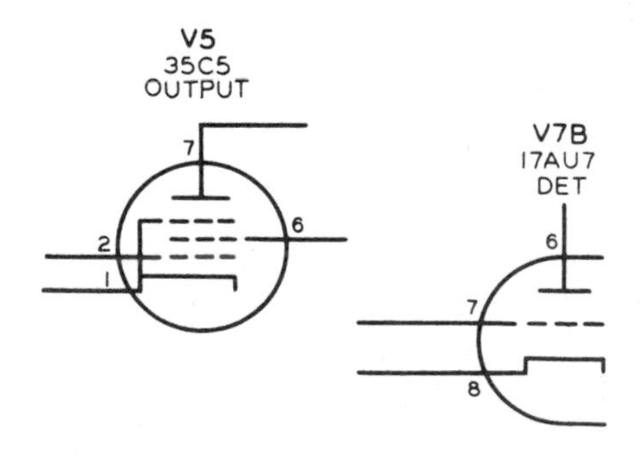


**FIG. 18.** Reference designations for electron tubes. The reference designation is given on the top, the tube type number next, and then function.

Electron tubes are identified by tube type number in addition to the reference designation. The circuit function may also be given below the tube type number, as in Fig. 18.

## 19. Numerical Values for Resistance, Capacitance, Inductance.

The guiding principle is to use expressions that require the fewest ciphers.

*Resistance.* In expressing resistances, alternative methods of expression may be used depending on the range of values expressed.

(*a*) For the entire range of values of 1 to and including 99,999, *ohms* may be used. For example, 15,000Ω. For the range of 100,000 ohms and above, the expression *megohm* (equivalent to 1,000,000 ohms) may be used. For example, 200,000 ohms may be expressed as .2 MEG (megohms).

(*b*) For the range of values from 1,000 to 999,000, the abbreviation K (*kilohm*) may be used instead of ohms. For example, 56K instead of 56,000. This method of expression is preferable to the method described in (*a*) when the last three places of the value are ciphers.

(*c*) For resistance values less than 1 ohm, the value of resistance is expressed as a decimal. For example, 0.031Ω.

*Capacitance.* Capacitance up to and including 9,999 *micromicrofarads* should be expressed in micromicrofarads. Capacitance of 10,000 micromicrofarads and above should be expressed as microfarads (equivalent to 1,000,000 micromicrofarads). For example, 92,000 UUF (micromicrofarads) should be expressed as 0.092 UF (microfarads).

*Inductance.* Inductance may be expressed in *henries* (H), *millihenries* (MH), or *microhenries* (UH). For example, 2UH instead of 0.002MH, and 5MH instead of either 0.005H or 5000UH.

*Commas.* In numerical values of four digits, the comma is omitted.

*Value Placement.* Numerical values of components are located as near as practical to the symbol. Preferred arrangements are shown in Fig. 19.

## 20. Notes.

To eliminate giving units of measurement which are applicable throughout the diagram, a general note can be used and only the numerical value given on the diagram. A recommended form is: "Unless otherwise specified: resistance values are in ohms. Capacitance values are in microfarads."

## 21. Single-line Diagrams.

A single-line diagram conveys basic information about the operation of a circuit or system of circuits. Much of the detailed information such as reference designations, tube numbers, and part values is omitted. Also, a single-line diagram is a simplification of multiconductor circuits, such as communication and power systems, in which a single line represents a multiconductor group. Figure 20 is an example of a single-line diagram.

## 22. Schematic Diagrams.

A schematic diagram shows the electrical functions and connections of a circuit. Electrical components

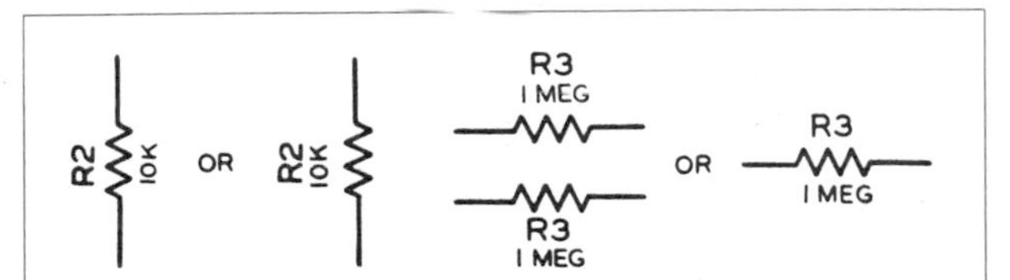


**FIG. 19.** Recommended placement of numerical values.

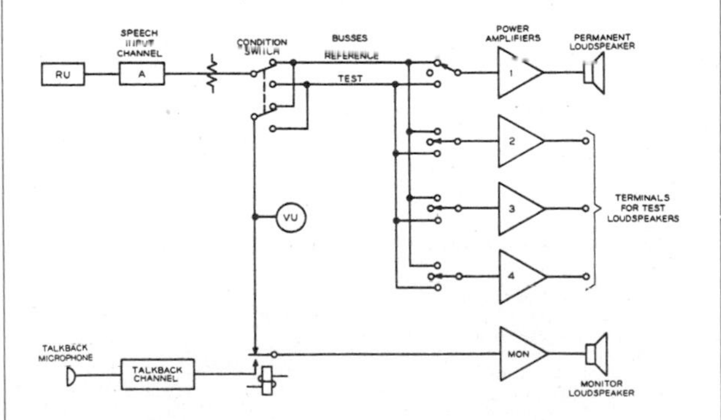


**FIG. 20.** A typical single-line diagram. This type of diagram conveys basic functional information but does not give detailed information. The one illustrated is for a complex audio system.

are shown by symbols (Appendix and ASA Y32.2). Components are identified by reference designations (Sec. 18). Values of components are given by numerical value of resistance, capacitance, inductance (Sec. 19). Paragraphs 3 to 20 give details for the preparation of schematic diagrams. Figure 21 is an example.

## 23. Special Drafting Practices.

Most of the commercial practices and economies described in Chap. 23 are applicable to the making of electrical diagrams and drawings. Of particular value in electrical drafting are templates and overlays (Secs. 38 and 39, Chap. 23), special uses of reproduction (Sec. 40, Chap. 23), photo drawing (Sec. 41, Chap. 23), and the use of illustrated drawings (Sec. 44, Chap. 23). An example of a gummed overlay is shown in Fig. 22. Many standard circuits, portions of circuits, and graphic symbols are available. The Stanpat Co. manufactures a variety printed on transparent material. The backing is peeled off and the stanpat stuck in place on the drawing or sketch. Templates such as those shown in Fig. 32, Chap. 23, save much time in drawing symbols. These are available from drawing-instrument manufacturers (see Bibliography).

When the accuracy of an instrument drawing is not necessary, drafting time can be saved by making electrical diagrams freehand. Coordinate paper with divisions printed on the back (see Sec. 33, Chap. 5) promotes economy. The coordinate lines aid in

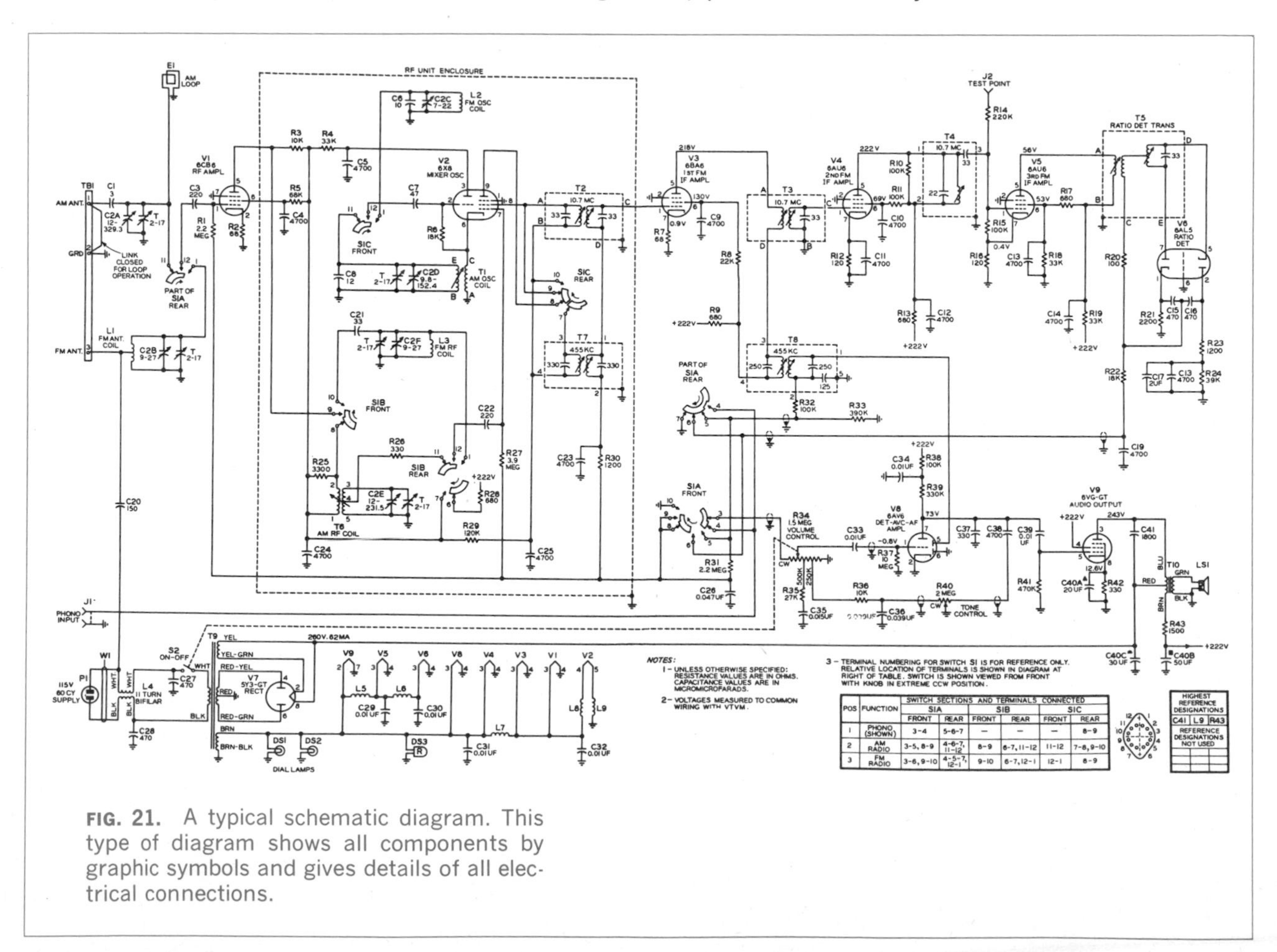


| POS | FUNCTION | SWITCH SECTIONS AND TERMINALS CONNECTED | | | | | |
|---|---|---|---|---|---|---|---|
| | | S1A | | S1B | | S1C | |
| | | FRONT | REAR | FRONT | REAR | FRONT | REAR |
| 1 | PHONO (SHOWN) | 3-4 | 5-6-7 | — | — | — | 8-9 |
| 2 | AM RADIO | 3-5, 8-9 | 4-6-7, 11-12 | 8-9 | 6-7, 11-12 | 11-12 | 7-8, 9-10 |
| 3 | FM RADIO | 3-6, 9-10 | 4-5-7, 12-1 | 9-10 | 6-7, 12-1 | 12-1 | 8-9 |

**FIG. 21.** A typical schematic diagram. This type of diagram shows all components by graphic symbols and gives details of all electrical connections.

keeping lines straight—particularly the long parallel connecting lines. The symbol sizes are not important but should be kept to proportions that look well and read easily on the diagram.

## 24. Models.

When models of plants are used, the conduit is included with other details in a full representation, to scale, of all conductors. These are added to the model after machines, process piping, plumbing, special equipment, etc., are in place. Figure 27, Chap. 14, shows a plant model in which the conduit for lighting, power, and control circuits is included. Giving the conduit on the model has the advantage that it simplifies the arrangement, develops a true scale for installation, allows tagging of all lines for size and circuits included, and reduces the number of drawings required. The tagging of power and control lines on models is sometimes supplemented or replaced by a conduit schedule.

To speed up connection and tagging of wiring, or for service use, photo drawings such as Fig. 23 are sometimes prepared. A photograph of the actual equipment is made, and the connections, instructions, or identifications are added to the screened positive.

## 25. Drawings of Printed Circuits.

Printed circuits can be obtained from very accurately drawn layouts. The method employs efficient lami-

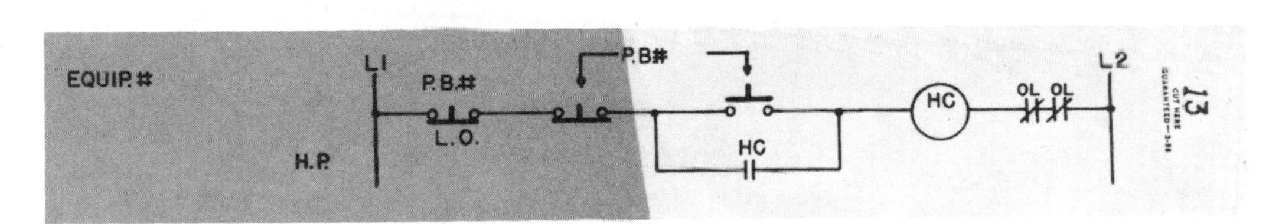


**FIG. 22.** A gummed overlay. The overlays are stripped from the backing paper and applied to the drawing. Many standard portions of circuits are available. (*Courtesy of Stanpat Co.*)

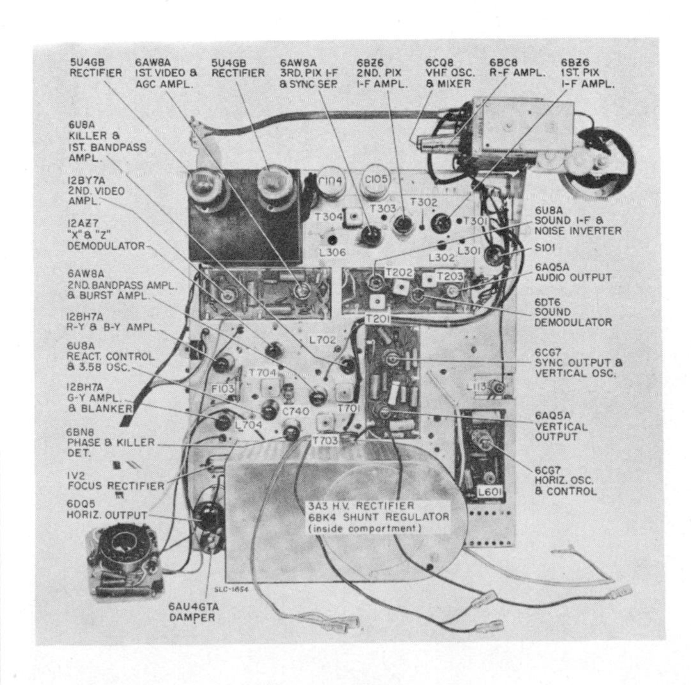


**FIG. 23.** A photo drawing. Photo drawings are used principally for service manuals. This one shows component and tube locations in a color-television chassis.

nated wiring on a wide variety of permanently bonded insulating plastics, from low cost vinyls and polyethylenes to fluorocarbons. Printed circuits allow miniaturization and the elimination of circuit errors—advantages that cannot be obtained by other methods. Once a pattern or suitable design is established, preparation of a black and white drawing can start. Scales for reduction, for example, 4 to 1, 3 to 1, or 2 to 1, are used. To ensure sufficient bonding area of the metal laminate during soldering operations, lines should not be less than $\frac{1}{32}$ inch in width when reduced. Line separation should never be closer than $\frac{1}{32}$ inch on the final circuit. Figure 24 illustrates the drawing of printed circuits.

To produce the circuit, a sheet of copper foil is bonded to a plastic board. The copper foil is processed by printing photographically from the layout and then etching to leave the circuits as originally drawn. Holes are punched (or drilled) in the board (Fig. 25) and various components inserted through

**FIG. 24.** Drawing a printed circuit. The drawing is carefully planned and executed for photographic reproduction on the circuit board.

**FIG. 25.** Punching holes in a circuit board. The holes are to receive component wires.

them. The leads of the various components are cut and bent over the copper-foil wiring. The wiring side of the board is then dipped in molten solder to make all solder connections at once, and all copper-foil wiring becomes covered, thus increasing its ability to carry current. A coat of silicone resin varnish is applied to the wiring side of the board; this prevents short circuits caused by dust or moisture. The method is known as the etched-wiring method, and results in uniformity of wiring, compactness, and freedom from wiring errors. It is used more widely than other methods, such as embossed wiring, stamped wiring, and pressed powdered wiring, because of its reliability, its great flexibility, and low set-up cost. Figure 26 is a photographic reproduction of a printed circuit board; the caption explains several details of printed circuitry.

## 26. Installation Drawings and Diagrams.

The electrical drawings required for an industrial plant are: (1) contractors' drawings, (2) system-design drawings, (3) component and special-equipment drawings. These vary considerably depending upon the industry because of the diversity of processes and equipment.

## 27. Contractors' Drawings.

To illustrate the types of drawings needed, we will describe briefly several contractors' electrical draw-

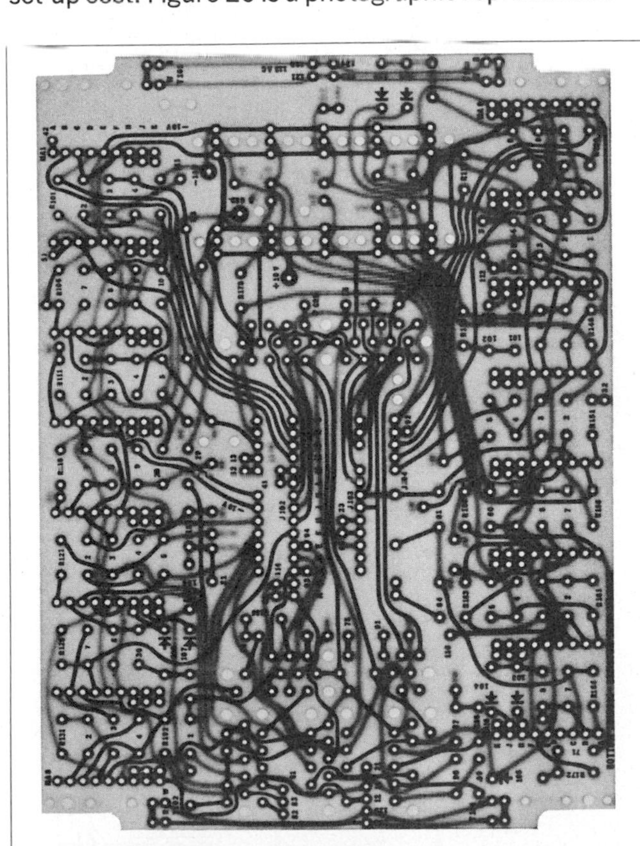

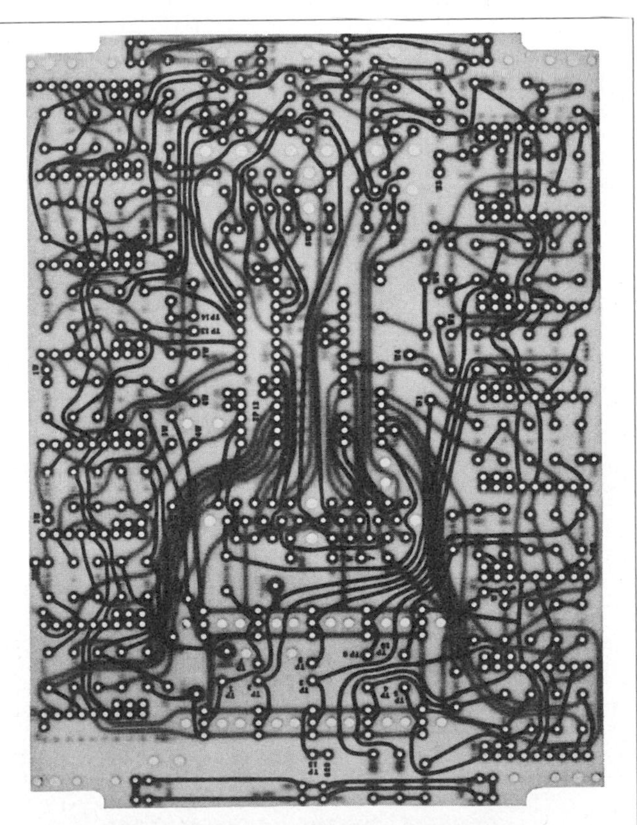

FIG. 26. A "double"printed circuit board. This type of circuit board has the circuit integrated and printed on both sides. The left illustration shows the resistor side of the board; the right illustration shows the side used for other components. The opposite side can be seen in both illustrations—the fainter, diffused lines. By printing both sides of the board, circuit crossovers are easily handled. Note that this complicated circuit can be produced in quantity without circuit errors.

ings for an industrial manufacturing plant in which the products are machine parts. The plant consists of several buildings. The greater part of the manufacturing area is shown in the plot plan in Fig. 27. In laying out the underground distribution along with substations, manholes, and duct locations, it is necessary to know all the power and lighting requirements for the entire plant as well as all breakdowns for power and lighting for each area. From these data the power to be supplied is determined and the outdoor substation located. The master substation and the seven other substations are located on the basis of area power requirements. The 34,500-volt overhead transmission line is shown in the figure where it enters the plant. All substations are shown on the plot plan except substation no.

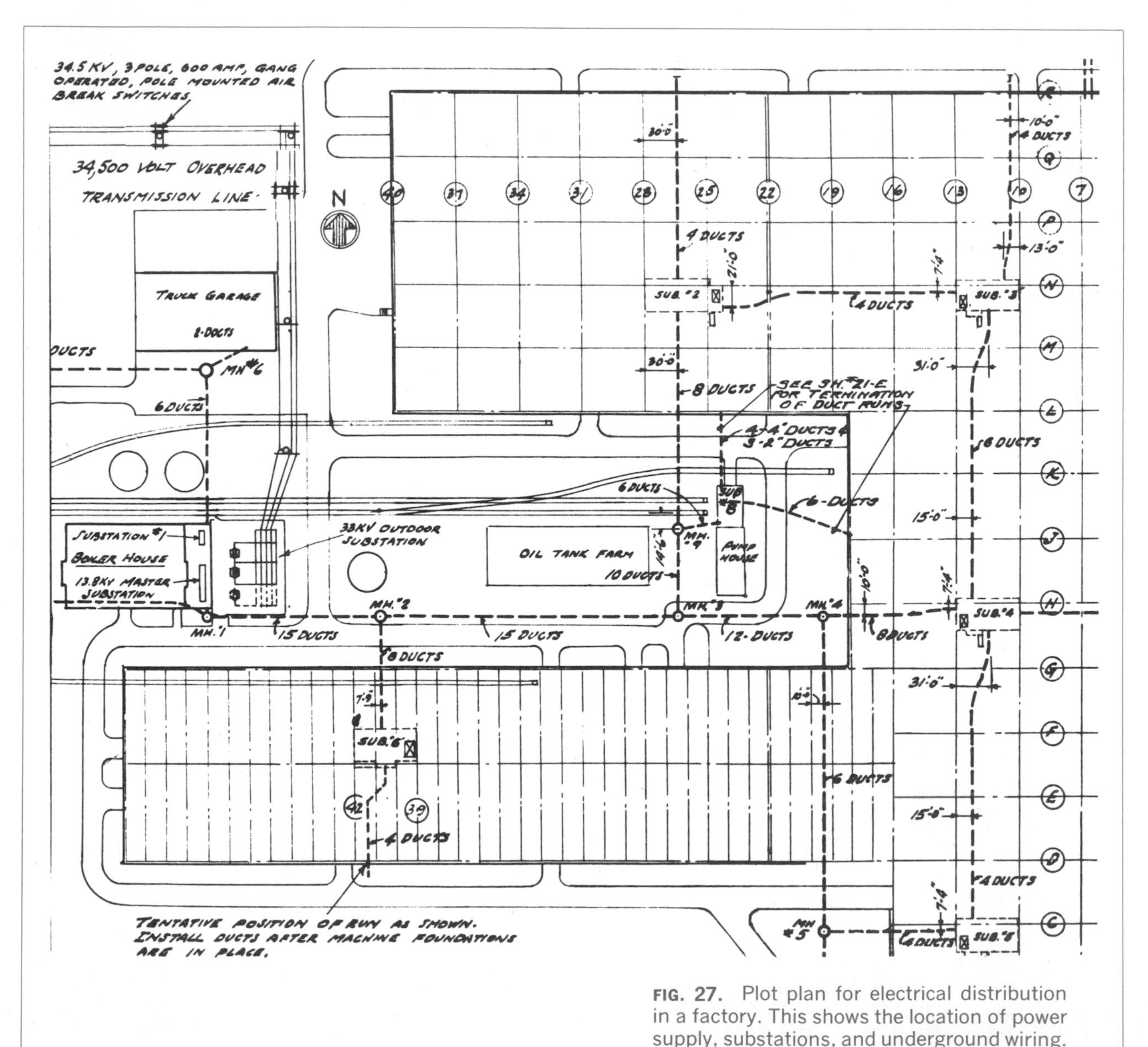


FIG. 27. Plot plan for electrical distribution in a factory. This shows the location of power supply, substations, and underground wiring.

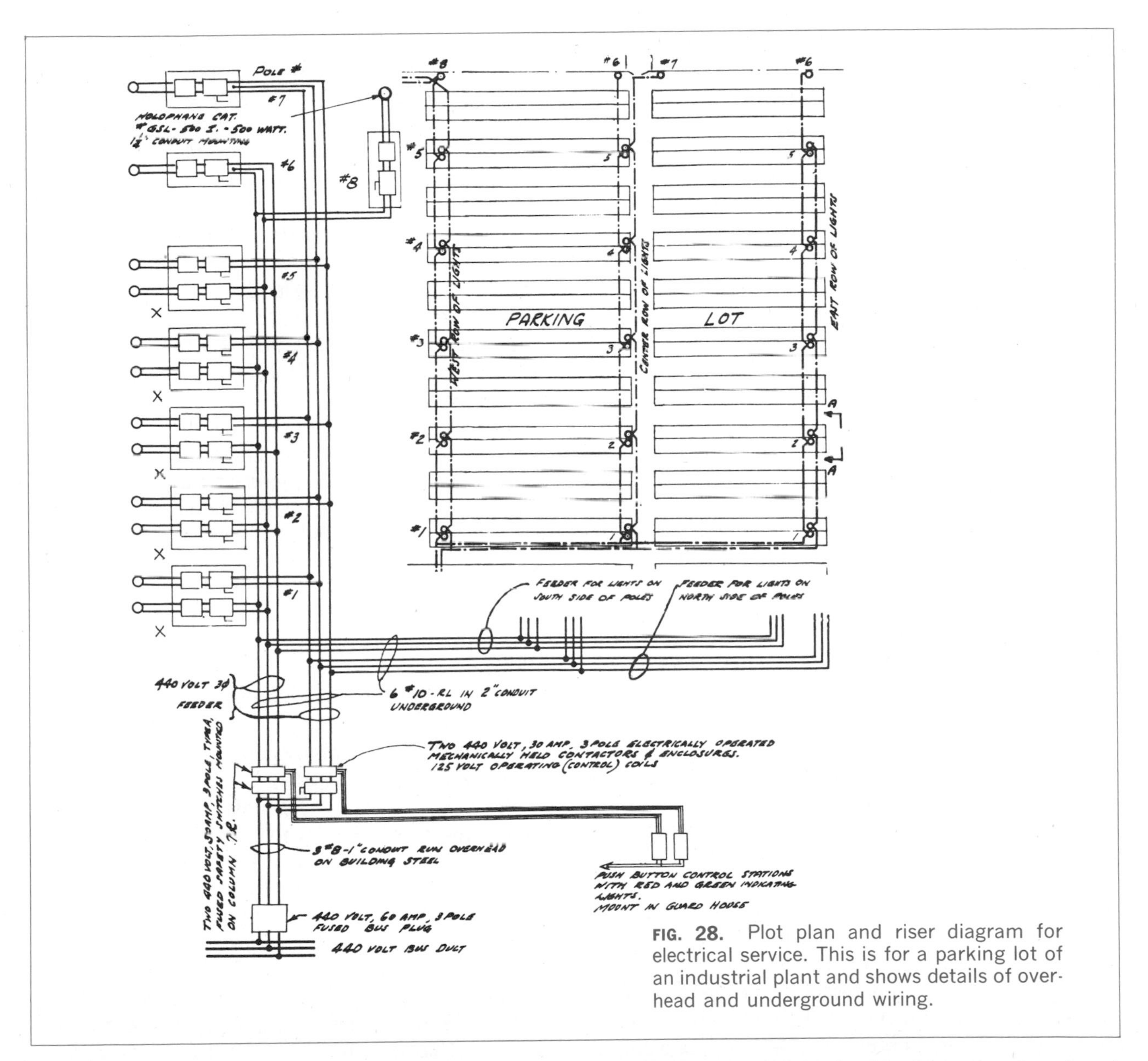


**FIG. 28.** Plot plan and riser diagram for electrical service. This is for a parking lot of an industrial plant and shows details of overhead and underground wiring.

7, which controls the power requirements for the engineering and office area. The underground distribution is shown in the figure by dashed lines.

Figure 28 shows the parking-lot plot plan (upper right) for this plant and a riser diagram (left and lower portion) indicating the power distribution for this system. Each rectangle ($X$) on the riser diagram represents a waterproof cabinet housing two fused safety switches and two single-phase dry-type transformers, shown on the detail of the lighting pole in Fig. 29. The feeder conduits are brought up through the concrete piers into each of these cabinets. As shown on the plot plan, there are three rows of lights. Each pair is mounted on a pole supported

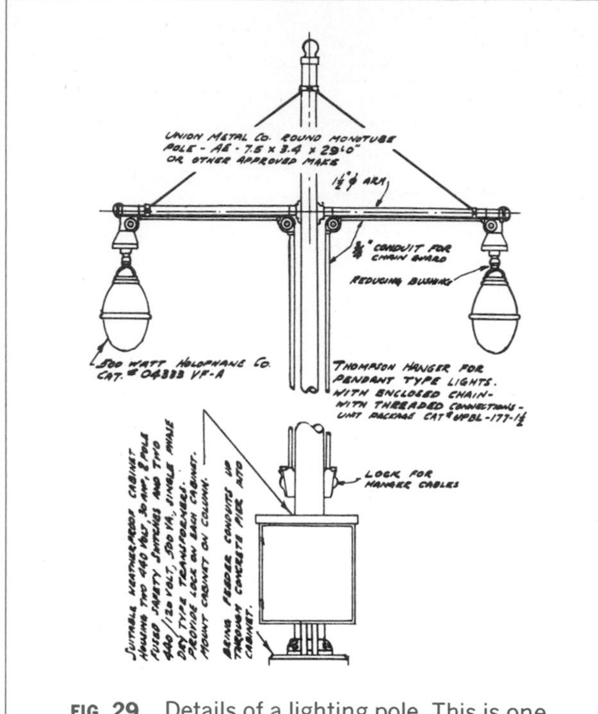


**FIG. 29.** Details of a lighting pole. This is one of the detail drawings to accompany the plot plan in Fig. 28.

from a square reinforced-concrete pier that projects above the ground.

Figure 30 is a plan view of the outdoor substation enlarged from the plot plan (Fig. 27). The incoming power line enters the outdoor substation from the north and attaches to the superstructure with strain insulators and strain clamps. All major transmission lines are $1\frac{1}{4}$-in. copper tubing.

Figure 31 gives three sectional views, *A-A*, *B-B*, and *C-C*, which show how the transformers, lightning arrestors, circuit breakers, metering, and outlet are related to the master substation.

The power ducts run from the master substation to the various substations shown in Fig. 27. Spare ducts are added so that additional circuits may be pulled when expansion or changes require more wiring. All circles in the dashed lines on Fig. 27 marked MH are manholes from which circuits are pulled and cable splices made. A detail of manhole no. 9 is shown in Fig. 32. This is a typical manhole although many are rectangular in shape rather than triangular. Shown in Fig. 32 are ten ducts with cables entering from the south from manhole no. 3. Some of these ducts are for spare cables. The distribution from this area is north with eight ducts to substation no. 2 and six ducts to substation no. 8. Shown projecting from the side walls are steel pulling eyes (used in pulling cables) and necessary cable supports. Other details sometimes necessary for duct layout are profile or elevation sheets showing variations in level to clear all obstructions such as footers, storm sewers, drains.

In addition to the plot-plan distribution shown in Fig. 27 a single-line wiring diagram is drawn for the entire power system. Separate single-line riser diagrams are then drawn to show the blocking distribution of the various circuits by floors and areas. Figure 33 is a portion of one of these drawings for substation no. 7, showing distribution circuits for lighting and power for the office building. The master lighting panel has three mains and three branches. All lighting panels have automatic circuit breakers. All power panels feed circuits in the underfloor ducts.

Figure 34 is an enlarged plan of substation no. 5 with a sectional elevation. It shows cable distribution with switchgear and panels. The lighting layout is also given with lights, fixtures, and switches.

Typical lighting layouts are indicated in Figs. 35 and 36. Figure 35 shows fluorescent fixtures with two tubes. Figure 36 shows holophane lights with switching arrangement. The power requirements are specified on Fig. 36.

Figure 37 gives a detail for mounting lighting fixtures from a messenger wire supported from purlins. This is a typical mounting for holophane lights when a cable hanger is used. The wiring diagram shows the usual circuit through an autotransformer. The lights are mounted so that the bottom of the fixtures are at the same elevation as the bottom chords of the trusses in low bays.

There are many more drawings and detailed specifications necessary in a complete set of electri-

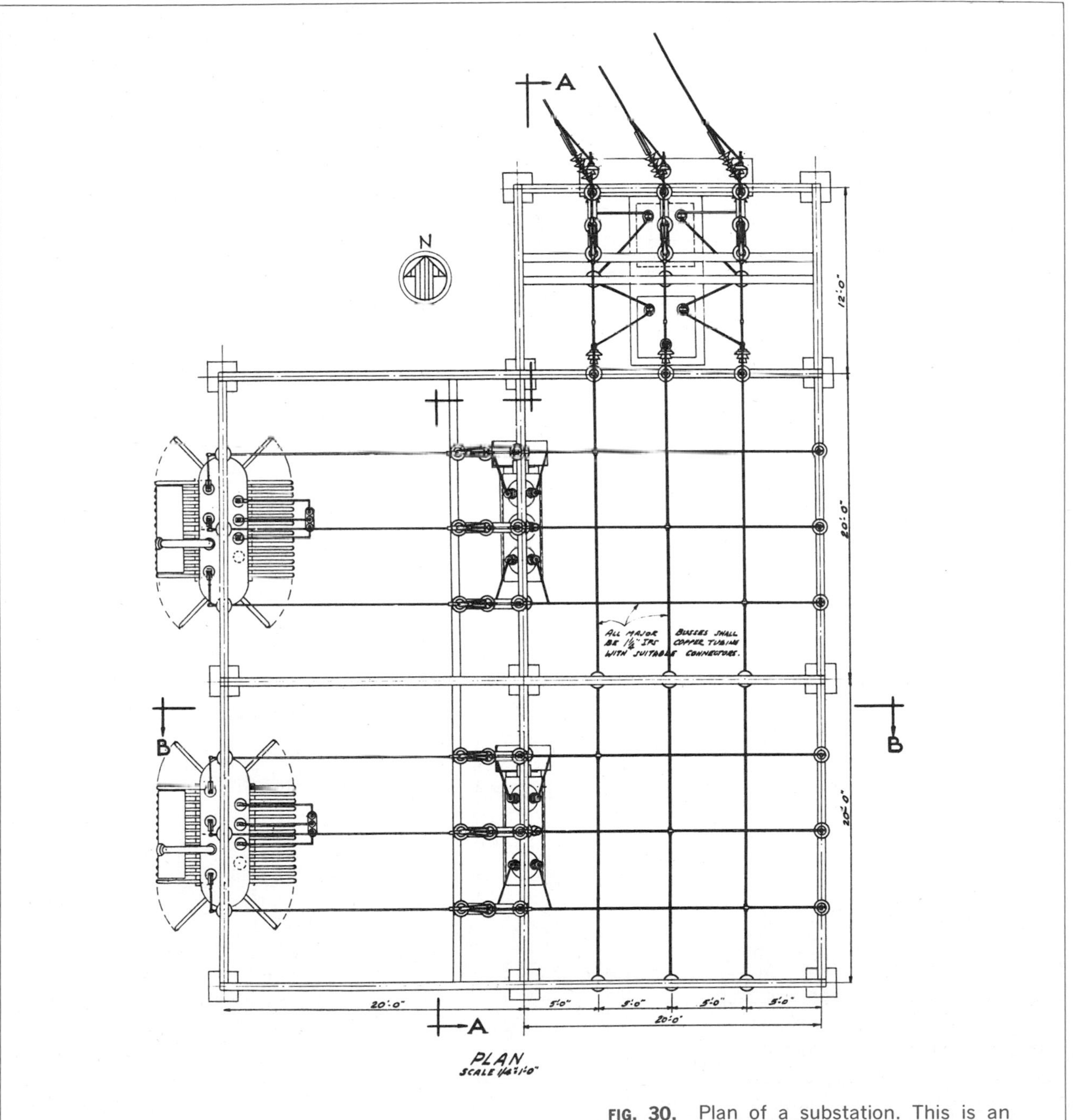


**FIG. 30.** Plan of a substation. This is an enlarged detail to accompany the plot plan in Fig. 27.

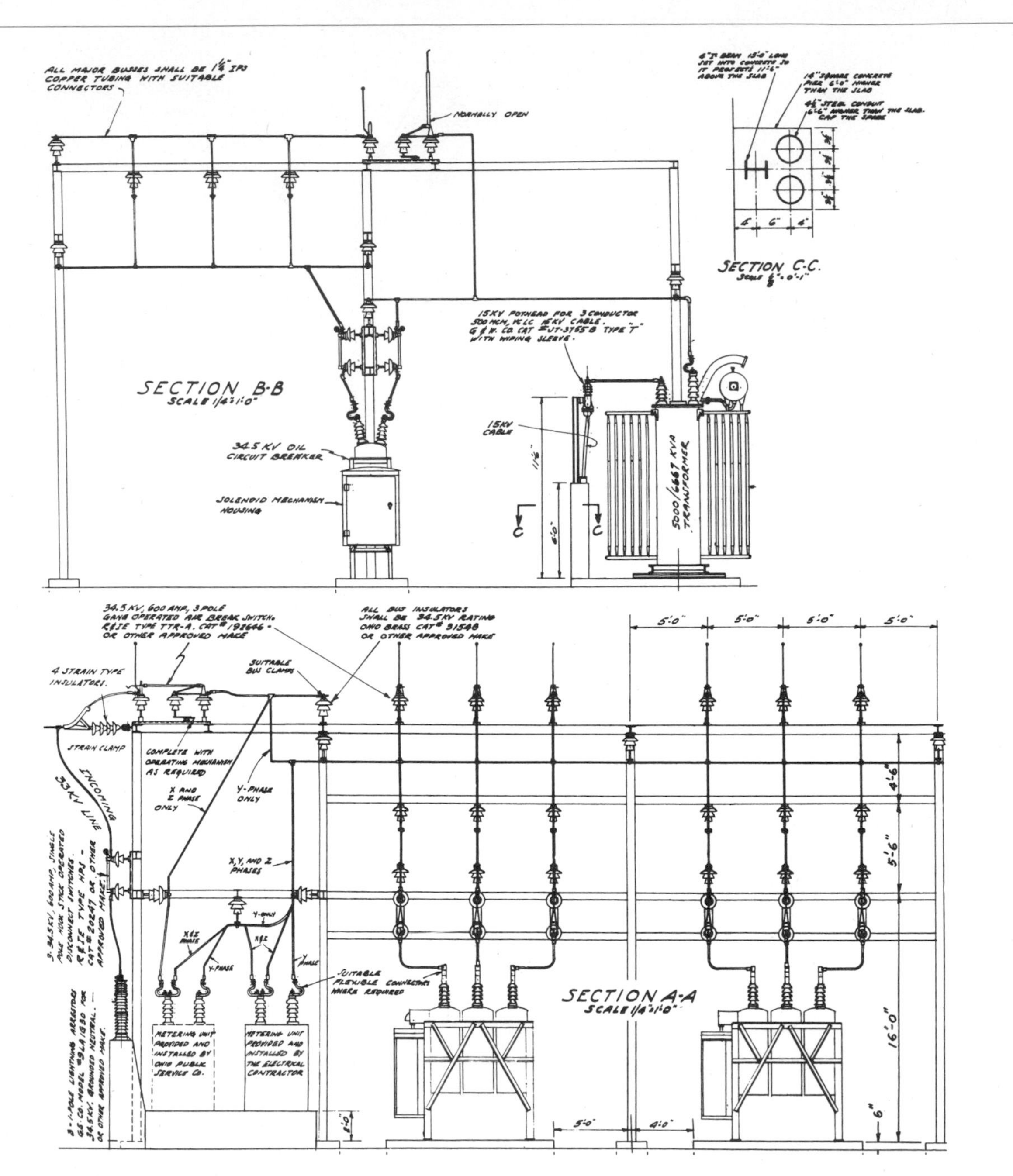


**FIG. 31.** Sections of a substation. These include *A-A* and *B-B*, taken from the plan in Fig. 30, and *C-C* identified on section *B-B*.

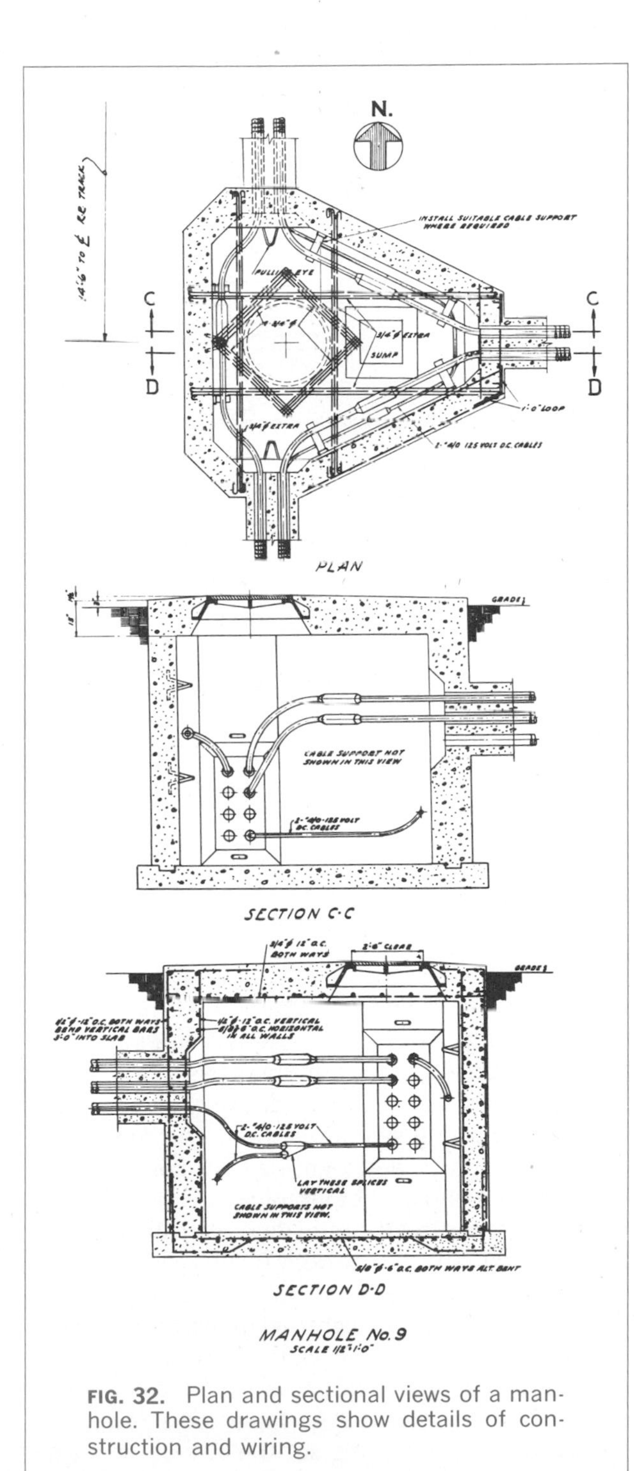


FIG. 32. Plan and sectional views of a manhole. These drawings show details of construction and wiring.

cal drawings for a plant of this size. These include a lighting layout for each factory and office area, an emergency-power layout, riser diagrams and lighting-panel layouts, power-transmission-duct layouts, night- and emergency-lighting layouts, telephone system and details, public-address system, and other miscellaneous services. Figures 27 to 37 show typical drawings from a complete set.

## 28. System-design Drawings.

The starting point in designing an electrical system for an industrial plant is generally the motor list together with devices and interlocks list. From these the number, size, and characteristics of components and the location by buildings and floors may be set forth in a bill of material. (A typed bill of material is shown in Fig. 38.) From it the equipment is procured. If a special item of equipment is needed, for example, the impulse timer in Fig. 39, a picture or pictorial may be shown with the wiring diagram.

The next division of field effort is to install equipment and conduit and pull wire. A conduit schedule such as is shown in Fig. 40 is common. Numbers refer to numbers on the drawings or model. Figure 41 is an example of a conduit layout where approximate physical location of conduit and electrical equipment is shown on a background of the equipment layout. Figure 42 is an example of a conduit block diagram. It shows the equipment as blocks (hence the term) and the connecting conduits but does not indicate the physical routing. Photo drawing lends itself well to this type of job involving long routes in *existing* buildings (see Sec. 41, Chap. 23). Figure 43 shows a single-line power-flow diagram. Figure 44 is a freehand schematic diagram, also known as an elementary wiring diagram. This type of drawing is normally accompanied by a description-of-operation sheet and a devices list, typed either on a separate sheet (Fig. 45) or on stanpat paper and applied directly to the drawing.

## 29. Architectural Layouts.

Figure 12, Chap. 21, is a typical architectural drawing which shows, among other details, the electrical outlets, switches, lights, and wall receptacles. Exact

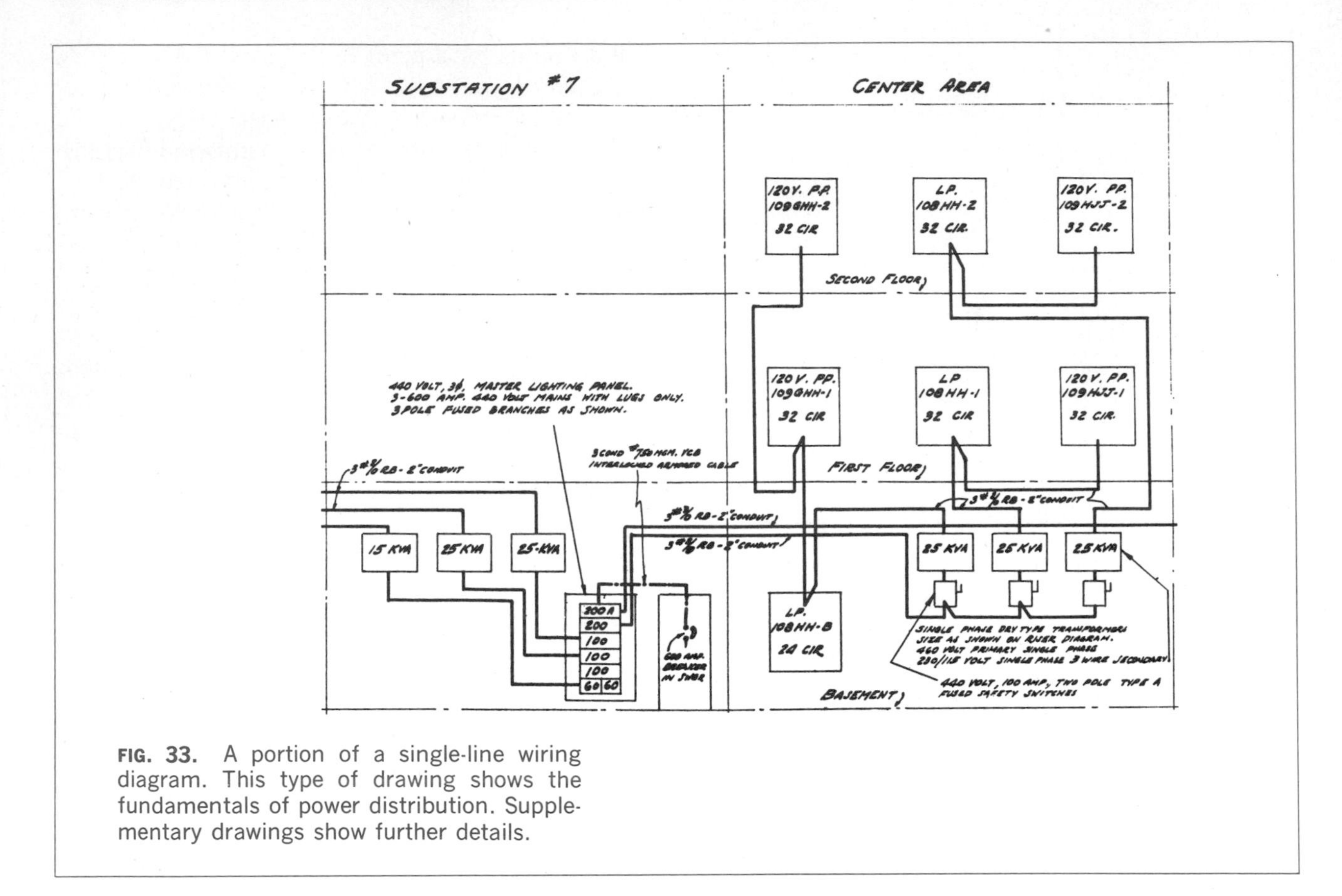


**FIG. 33.** A portion of a single-line wiring diagram. This type of drawing shows the fundamentals of power distribution. Supplementary drawings show further details.

dimensions for location of the components are not given but are maintained within reasonable proximity to the position shown on the drawing. The listing of details for switch boxes, wiring, fixtures, and the procedures and acceptable standards of installation are given in the specifications.

## 30. Special Equipment.

Figures 46 and 47 show a regulated power-supply chassis assembly—a representative example of modular construction of electronic equipment. The schematic wiring diagram is shown in Fig. 46. The signal input is a direct-current supply from an a-c to d-c rectifier circuit. The output is a direct current regulated to within 0.1 per cent and supplied to other components of the system. This unit is a part of an AccuRay Density Measuring System manufactured by Industrial Nucleonics Corporation. The system employs a radioactive source and detector for measuring the density of materials flowing inside a pipe or tube. Parts used in this chassis are shown and identified in Fig. 47, a drawing showing two views of the wiring side of the unit. The connecting wires to all parts are identified by color on this drawing. The three photographs shown in Fig. 48 illustrate the finished production chassis from the top, the side, and the bottom (wiring side).

Transistors, because of their small size, lend themselves to compact light-weight assemblies such as portable radios. As an example of the drawings needed in a service manual, a schematic diagram of a small transistorized portable radio receiver is shown in Fig. 49. In this set the line-up of stages includes a converter, two intermediate-frequency

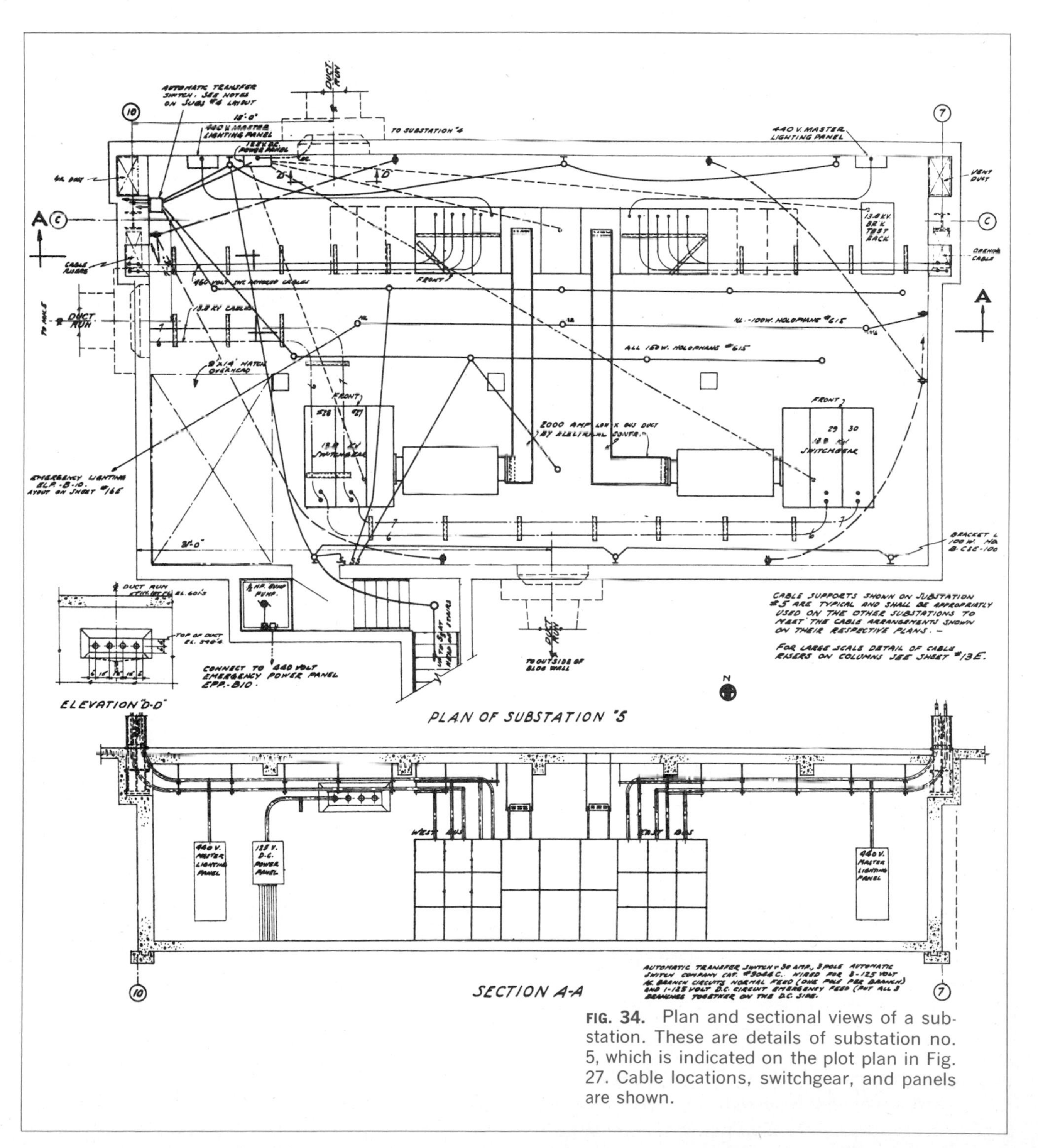


**FIG. 34.** Plan and sectional views of a substation. These are details of substation no. 5, which is indicated on the plot plan in Fig. 27. Cable locations, switchgear, and panels are shown.

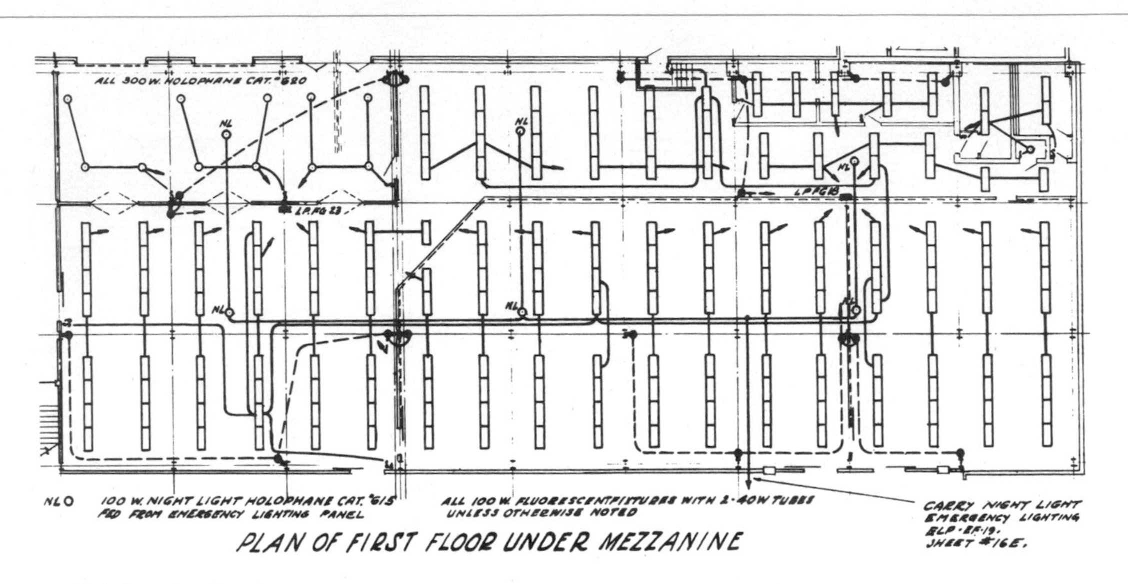


FIG. 35. Lighting details. This drawing gives location of wiring and fixtures.

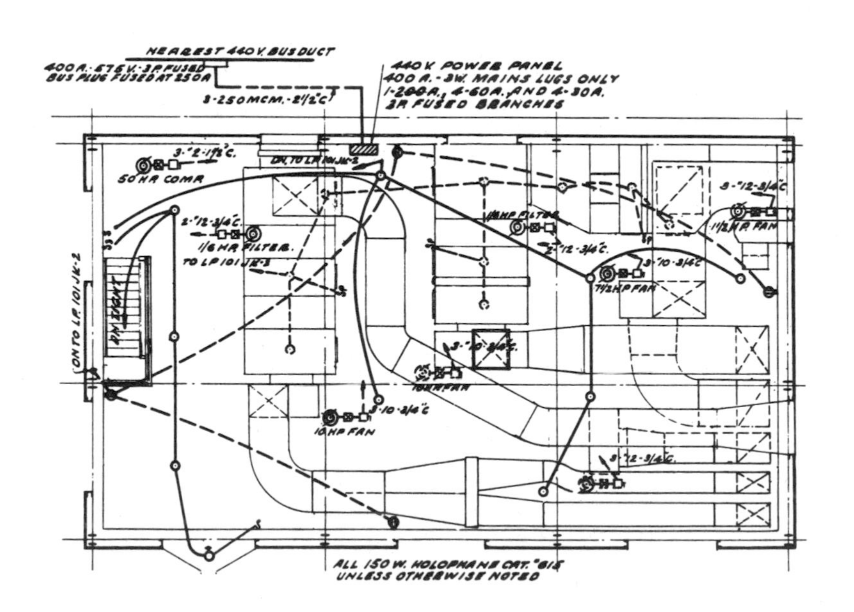


FIG. 36. Lighting details. This drawing shows power panel, wiring, and fixture locations.

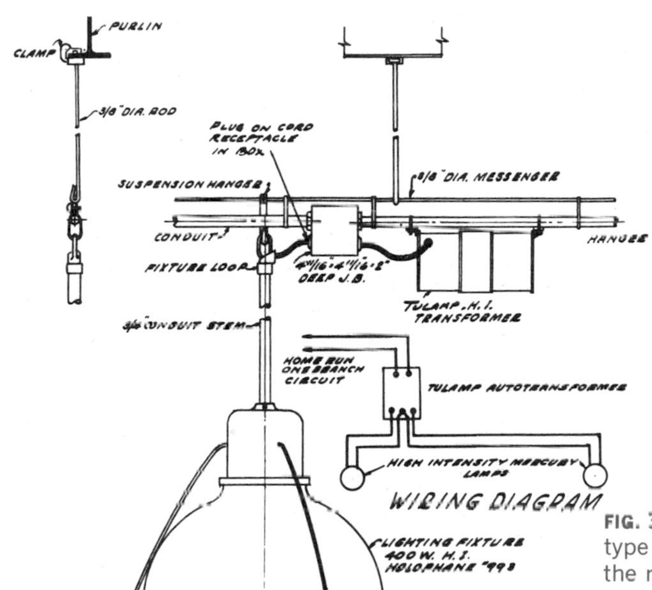


DETAIL OF MOUNTING LIGHTING FIXTURES ON MESSENGER WIRE SUPPORTED FROM PURLINS

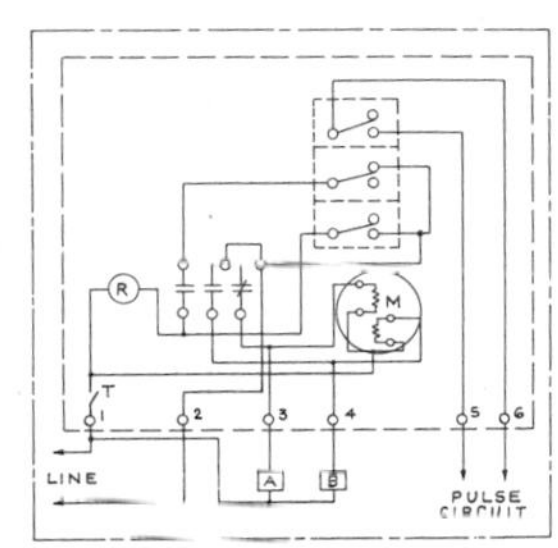


**FIG. 39** (Above). Illustration of special equipment. This type of supplementary drawing is used to specify unusual items.

**FIG. 37** (Left). Fixture-mounting details. This type of drawing specifies fixtures and shows the mounting and wiring.

**FIG. 38** (Below). A bill of material for electrical equipment. This is needed for the purchase of equipment.

| ITEM No. | DRAWING No. | DESCRIPTION | MANUFACTURER & CATALOG NO. | QUAN. | REMARKS |
|---|---|---|---|---|---|
| 1 | C-19678 | Pushbutton with double throw contact block. | Sq. D, Class 9001, Type TR1B marked "UP". | 2 | P.B. Sta. #1 |
| 2 | C-19678 | Pushbutton with double throw contact block. | Sq. D., Class 9001, Type TR2B marked "DOWN" | 2 | P. B. Sta. #1 |
| 3 | C-19678 | ON-OFF Selector Switch | Sq. D., Class 9001, Type TS1A, marked "No. 1 CLAMP-ON OFF". | 1 | SW. 1, P.B. Sta. #1 |
| 4 | C-19678 | ON-OFF Selector Switch | Sq. D., Class 9001, Type TS1A, marked "VIB. #1 - MAN-AUTO. | 1 | SW. 3, P.B. Sta. #2 |
| 5 | C-19678 | Selector switch with double throw contact block. | SQ. D., Class 9001, Type TS1B, marked "DUMP - No. 1, No. 2" | 1 | SW. 5, P.B. Sta. #2 |
| 6 | C19678 | Control relay 2 N. O. contacts, 110 volt, 60 cycle coil, open type. | Allen-Bradley, Bulletin 700, Type No. BX-220, 110 V., 60 cycle coil. | 2 | R1, R2 |
| 7 | C-19678 | Control relay 4 N. O. contacts, 110 volt, 60 cycle coil, open type. | Allen-Bradley, Bulletin 700, Type No. BX-440, 110 V., 60 cycle coil. | 1 | R3 |
| 8 | C-19678 | Timing Relay 1 N.O. & 1 N.C. contact, 110 V. 60 cycle coil, Time delay on energization. | AGASTAT, Type NE-12. | 1 | TR1 |
| 9 | C-19678 | Reversing starter for 1 HP Motor, 440 V., 3 phase, 60 cycle. | Complete combination unit to accommodate 1 HP, 440 V., 60 cycle, 3 phase, reversing induction motor complete with door and external brk'r. operating handle for installation in existing MCC serial No. ________ purchased in ____________. Starter coil voltage 110 V, 60 cycle, single phase with no control transf. required. 1 N. O. aux. contact on the breaker is required. | 2 | HC1U - HC1D HC2U - HC2D |
| 10 | C-19678 | Alarm Horn, Projector-less type with grille front, for 110 V. operation | Benjamin, Type SNP, Cat. No. 8741-115 V. | 1 | |
| 11 | C-19678 | NEMA 12 sheet steel enclosure for 6 units. | Sq. D., Class 9001, Type TYA-6 | 2 | P. B. Sta. #1 & #2 |
| 12 | C-19678 | Closing plate for unused hole in P.B. Sta. #2 | Sq. D., Class 9001, Type TO-1 | 1 | |
| 13 | C-19678 | Thermal O.L. relay elements to accommodate 1 1/2 HP, 440 V., 3 phase, induction motor. Suitable for mounting in existing M.C.C. Serial No.________, Purchased in_______. | Square D, Class 9065, Hand reset type. | 1 | |

| WIRES POWER NO. | POWER SIZE | POWER TYPE | CONTROL NO. | CONTROL SIZE | CONTROL TYPE | CND SIZE (Inches) | NOTES | CONDUIT No. | FROM | TO |
|---|---|---|---|---|---|---|---|---|---|---|
| 3 | 2/10 | RH-RW | | | | 2 | Feeder Tap (FT) | C1 | Junction Box | Feeder Disconnect |
| 3 | 6 | RH-RW | | | | 1 | 6 FT1, 6FT2, 6FT3 | C2 | Feeder Disconnect | Conduit Fitting |
| 3 | 6 | RH-RW | 4 | 14 | TW | 1 1/4 | 44, 45, 46, 47, 6FT1 | C3 | Conduit Fitting | Conduit Fitting |
| | | | 4 | 14 | TW | 1/2 | 44, 45, 46, 47 | C4 | Conduit Fitting | Control Station #737 |
| 3 | 2/0 | RH-RW | | | | 2 | | C5 | Feeder Disconnect | Wiring Trough |
| 3 | 12 | TW | 3 | 12 | TW | 1 | Eq. #733, 35, 65, 67 | C6 | Circuit Brkr. #733 | Alarm Relay Box |
| 3 | 12 | TW | 2 | 12 | TW | 3/4 | Eq. #733A, 61, 62 | C7 | Alarm Relay Box | Starter #733A |
| 3 | 12 | TW | 2 | 12 | TW | 3/4 | Eq. #733B, 63, 64 | C8 | Alarm Relay Box | Starter #733B |
| 3 | 12 | TW | | | | 1 1/2 | EQ. #709-3T1, 3T2, 3T3 | C9 | Wiring Trough | Conduit Fitting |
| 3 | 2 | RH-RW | | | | 1 1/4 | 2T1, 2T2, 2T3 | C10 | Conduit Fitting | Motor #707 |
| 3 | 12 | TW | | | | 1/2 | 3T1, 3T2, 3T3 | C11 | Conduit Fitting | Motor #709 |
| 3 | 8 | RH-RW | 2 | 14 | TW | 1 | 24, 24A, 4T1, 4T2, 4T3 | C12 | C. T. Cabinet | Equipment #731 |
| 3 | 10 | TW | 2 | 14 | TW | 3/4 | 40, 40A, 5T1, 5T2, 5T3 | C13 | Starter #715 | Equipment #715 |
| 3 | 8 | RH-RW | | | | 1 1/4 | Eq. #705, 1T1, 1T2, 1T3 | C14 | Wiring Trough | Conduit Fitting |
| 3 | 12 | TW | 2 | 14 | TW | 3/4 | Eq. #877, 9T1, 9T2, 9T3 | C15 | Conduit Fitting | Conduit Fitting |
| | | | 2 | 14 | TW | 1/2 | 48, 50 | C16 | Conduit Fitting | P.B. #14 |
| | | | 2 | 14 | TW | 1/2 | 49, 50 | C17 | P.B. #14 | Float Switch |
| 3 | 12 | TW | | | | 1/2 | 9T1, 9T2, 9T3 | C18 | Conduit Fitting | Motor #877 |
| 3 | 8 | RH-RW | | | | 3/4 | 1T1, 1T2, 1T3 | C19 | Conduit Fitting | Motor #705 |
| | | | 28 | 14 | TW | 2 | 1, 3, 4, 5, 6, 10, 11, 13, 14 | C20 | Wiring Trough | Control Station #737 |
| 2 | 12 | TW | | | | 1/2 | 31, 35 | C21 | Control Station #737 | Weigh Hopper |
| | | | 3 | 14 | TW | 1/2 | 6, 14, 15 | C22 | Control Station #737 | P.B. #5 |
| | | | 2 | 14 | TW | 1/2 | 14, 15 | C23 | P.B. #5 | Solenoid Valve S-1 |
| 3 | 12 | TW | 8 | 14 | TW | 1 1/4 (#12 are 31, 32, 35) | 7, 8, 17, 18, 25, 26, 31, 32, 35 | C24 | Control Station #737 | Buggy Fill Control Station |
| | | | 3 | 12 | TW | 1/2 | 35, 38, 39 | C25 | Buggy Fill Control Station | Conduit Fitting |
| | | | 2 | 12 | TW | 1/2 | 35, 39 | C26 | Conduit Fitting | Solenoid Valve S-3 |
| | | | 2 | 12 | TW | 1/2 | 35, 38 | C27 | Conduit Fitting | Solenoid Valve S-2 |
| 3 | 12 | TW | 3 | 12 | TW | 1 | Eq. #733, 35, 65, 67 | C28 | Wiring Trough | Circuit Brkr. #733 |
| 6 | 12 | TW | 4 | 14 | TW | 1 1/4 | 16, 20, 22, 23, 3FT1, 3FT2 | C29 | Wiring Trough | Starter #709 |
| 2 | 12 | TW | | | | 1/2 | | C30 | Wiring Trough | Control Transformer |
| 3 | 10 | TW | | | | 3/4 | | C31 | Control Transformer | 110/220V Panel |
| 3 | 12 | TW | | | | 1/2 | 31, 35, 65 | C32 | 110/220V Panel | Wiring Trough |
| 6 | 2 | RH-RW | 4 | 14 | TW | 2 1/2 | 6, 10, 13, 14, 2FT1, 2FT2, | C33 | Wiring Trough | Starter #707 |
| 2 | 2 | RH-RW | | | | 1 1/4 | 2T1, 2T1 | C34 | Starter #707 | C. T. Cabinet |
| 6 | 8 | RH-RW | 4 | 14 | TW | 1 1/4 | 1, 3, 4, 5, 1FT1, 1FT2 | C35 | Wiring Trough | Starter #705 |

**FIG. 40.** A conduit schedule. This is used to specify conduit details. Letters and numbers refer to a master layout.

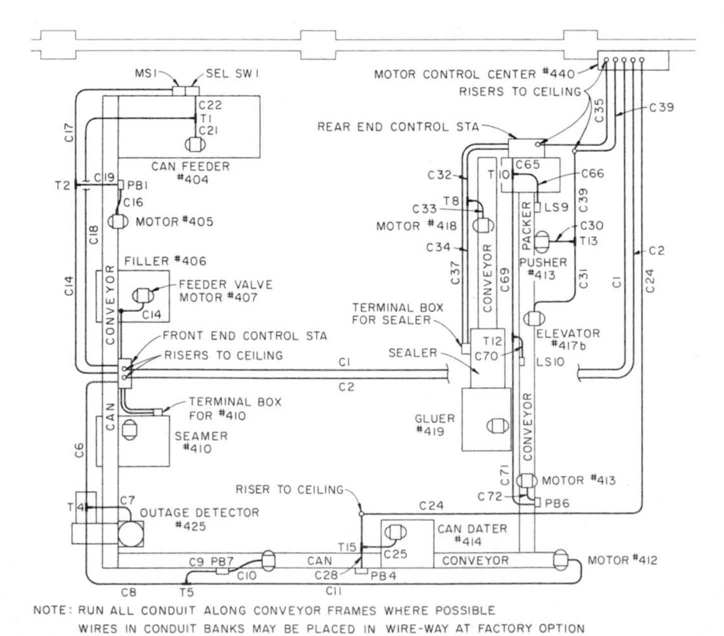


**FIG. 41.** A conduit layout. This shows conduit location on a drawing showing equipment location.

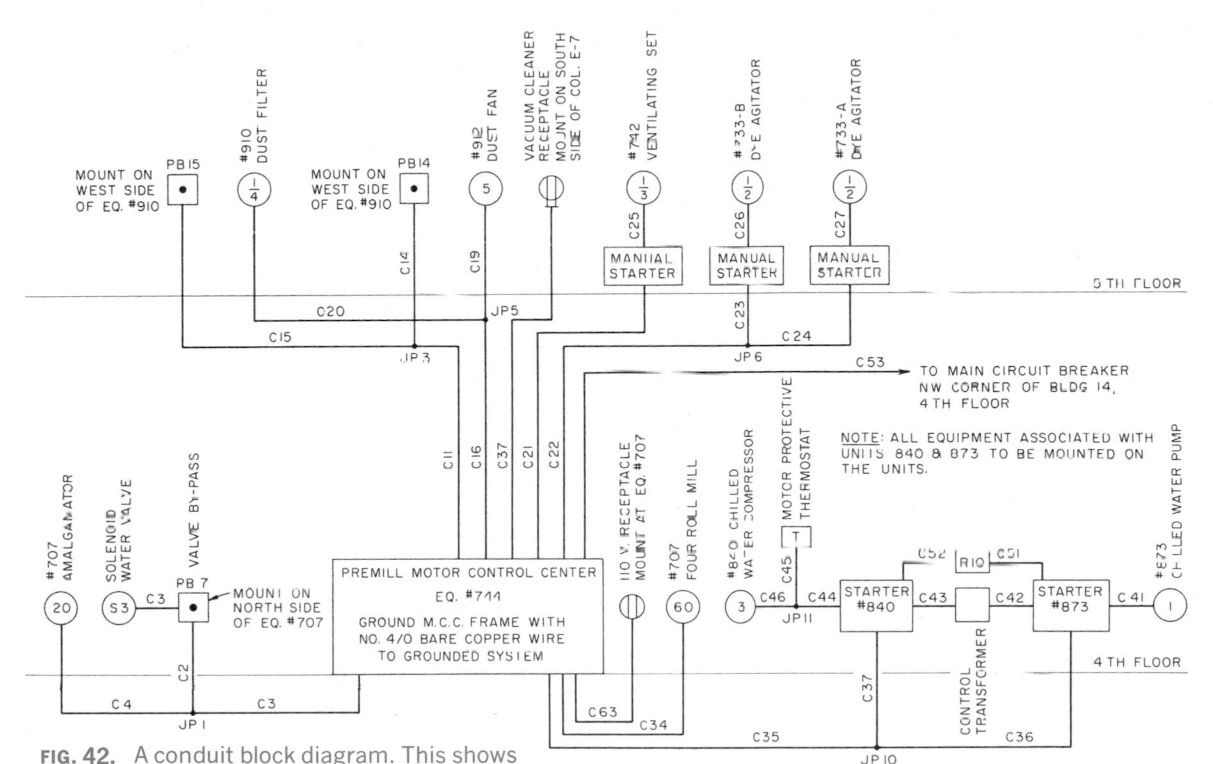


**FIG. 42.** A conduit block diagram. This shows fundamentals of conduit wiring but does not show physical locations.

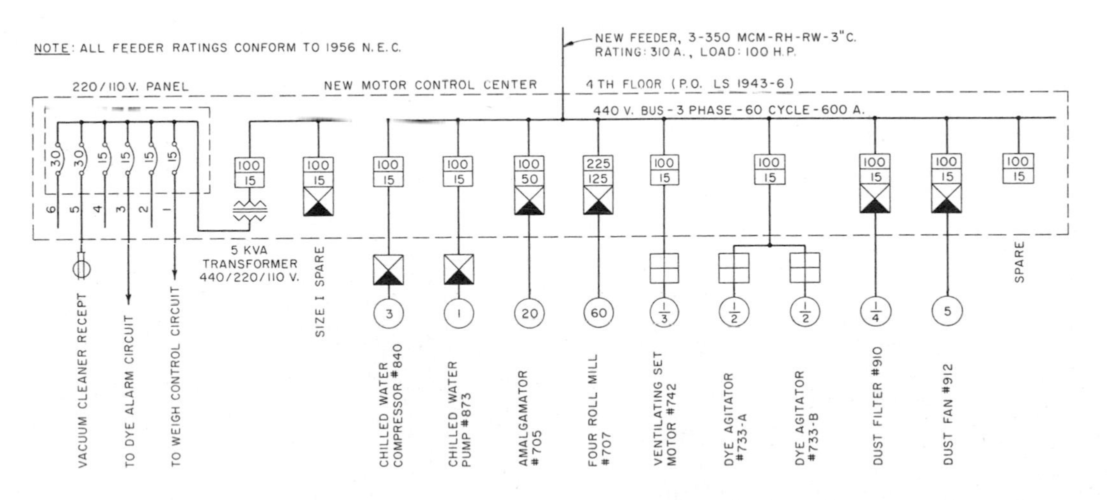


**FIG. 43.** A single-line power-flow diagram. This shows the necessary power supply to equipment but not physical details.

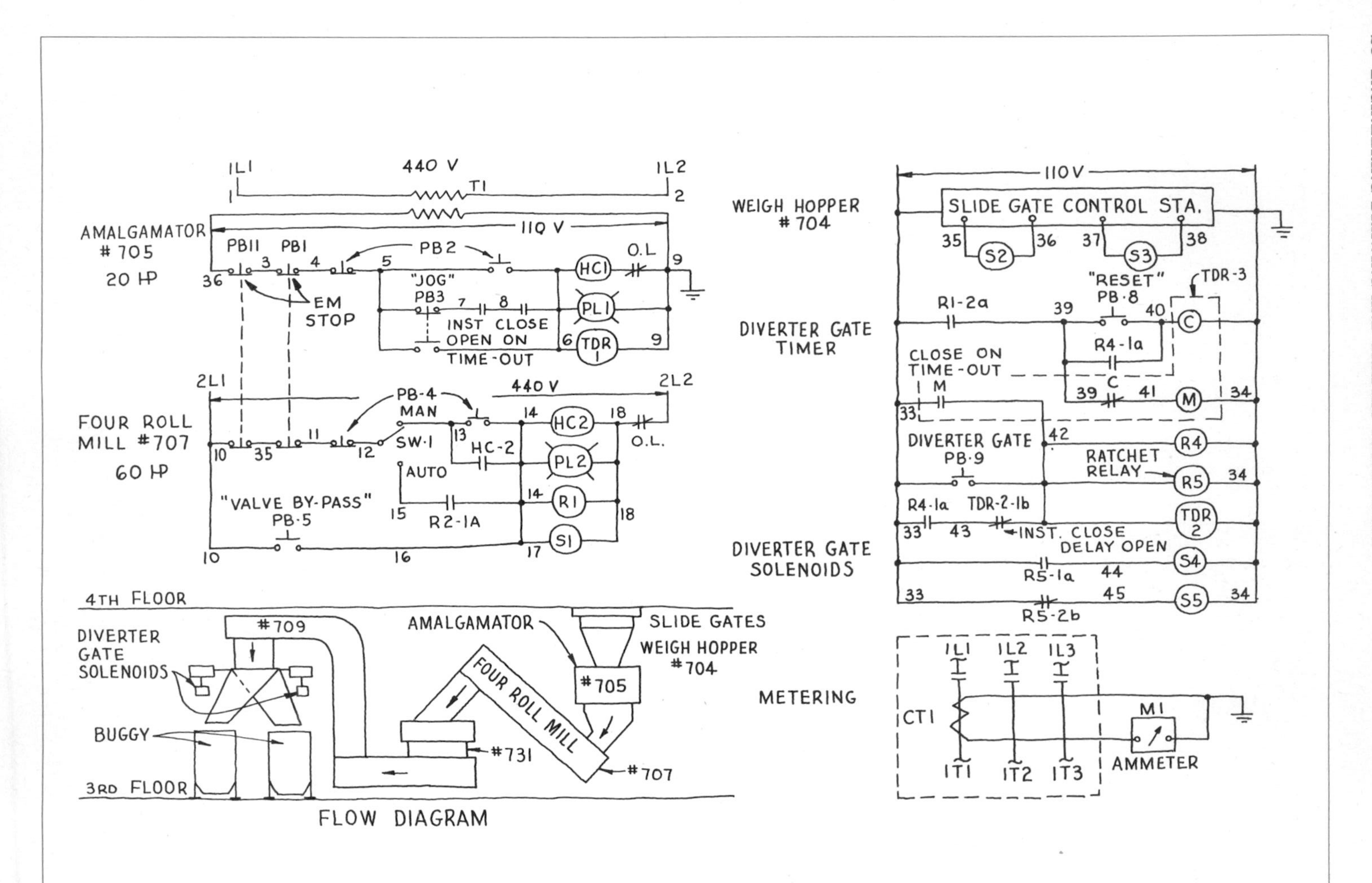


**FIG. 44.** A schematic diagram, also called an elementary wiring diagram. This one is made freehand, a common drawing practice for original planning.

DESCRIPTION OF OPERATION
ZEST PREMILLING

Purpose
To provide necessary electrical control for Zest premilling equipment.

Operation
Amalgamator 705
When start button PB2 is depressed to start motor, timer TDR1 closes contact TDR1-la and starts timing period. At the end of the timing period, contact TDR1-la opens and stops the Amalgamator. When timer is de-energized, it resets to the zero position. Jog button PB3 will also operate the amalgamator.

Four Roll Mill 707, Elevator 709, Premill Plodder 731
Manual Operation (Selector Switch #1 in MANUAL position)
Push button PB4 starts the Four Roll Mill and energizes relay R1. Contact R1-la closes and energizes solenoid S1 which admits water to the mill. Push button PB7 starts the Bucket Elevator. Jog button PB10 will also operate the elevator. Push button PB6 starts the Premill Plodder.

Automatic Operation (Selector Switch #1 in AUTO position)
Elevator must be started before the Premill Plodder and the Four Roll Mill. Depressing start button PB7 starts Elevator and energizes relay R3. Contact R3-la closes in Premill Plodder circuit. Relay R2 is energized and starts the Four Roll Mill. Contact R1-la closes to energize solenoid S1. It the Elevator is stopped, contact R3-la opens to stop Premill Plodder. Contact R2-la opens to stop Four Roll Mill, and R1-la opens to de-energize colenoid S1.

Manual or Automatic Operation (Selector Switch #1 in AUTO or MANUAL position)
Bypass push button PB5 can be used to energize solenoid S1 when the Four Roll Mill is not running. The Four Roll Mill, Elevator and Premill Plodder can be stopped by means of stop buttons. In case of an emergency, all three motors may be stopped simultaneously by means of stop button PB1, or PB11.

Diverter Gats
The purpose of this circuit is to provide a means of shifting a diverter gate after approximately twenty minutes of running time of the Four Roll Mill. Whenever contact R1-2a from relay R1 in the Four Roll Hill circuit is closed, timer TDR-3 will time. At the end of the timing period, contact M closes and energizes relays R4, R5, and TDR2. Contact R4-la closes and resets timer. Ratchet relay contact R5-la closes and contact R5-2b opens. Energization of solenoid S4 shifts the diverter gate. At the end of the next timing period the contacts will return to the position shown and energization of solenoid S5 will cause the diverter gate to return to its initial position. After timer has had sufficient time to reset, contact TDR2-lb opens and de-energizes relays R4, R5, and TDR2. Timer can be reset at any time by means of the reset push button PB8. The diverter gate can be operated manually and the timer reset by means of push button PB9.

Weigh Hopper Slide Gate Control
A 110 volt circuit is supplied for the Weigh Hopper slide gate control.

DEVICES LIST

(Bill of Materials are Drawings #2A-14047 and #2A-14048)

| Notation | Description | Remarks |
|---|---|---|
| PB #1-11 | Push Button | B/M Items #1-6, 16 |
| SW #1 | Selector Switch | B/M Item #7 |
| PL #1-4 | Pilot Light | B/M Items 8, 9 |
| R1-5 | Relay | B/M Items 14, 15, 18 |
| TDR1-3 | Time Delay Relay | B/M Items 11-13 |
| S1 | Solenoid Valve | Asco #81210A47, 2", 440V |
| S2, 3 | Slide Gate Solenoid | Westinghouse Type D Pilotair Valve #PD-4-41-01-14, 110V |
| S4, 5 | Diverter Gate Solenoid | Westinghouse Type D Pilotair Valve #PD-2-41-13-13, 110V |
| M1 | Ammeter | B/M Item #10 |
| CT 1 | Current Transformer | B/M Item #17 |

**FIG. 45.** Description of operation and devices list. These accompany the schematic diagram in Fig. 44. The schematic diagram and the description of operation and devices list, together, give fundamental information for the production of more detailed drawings.

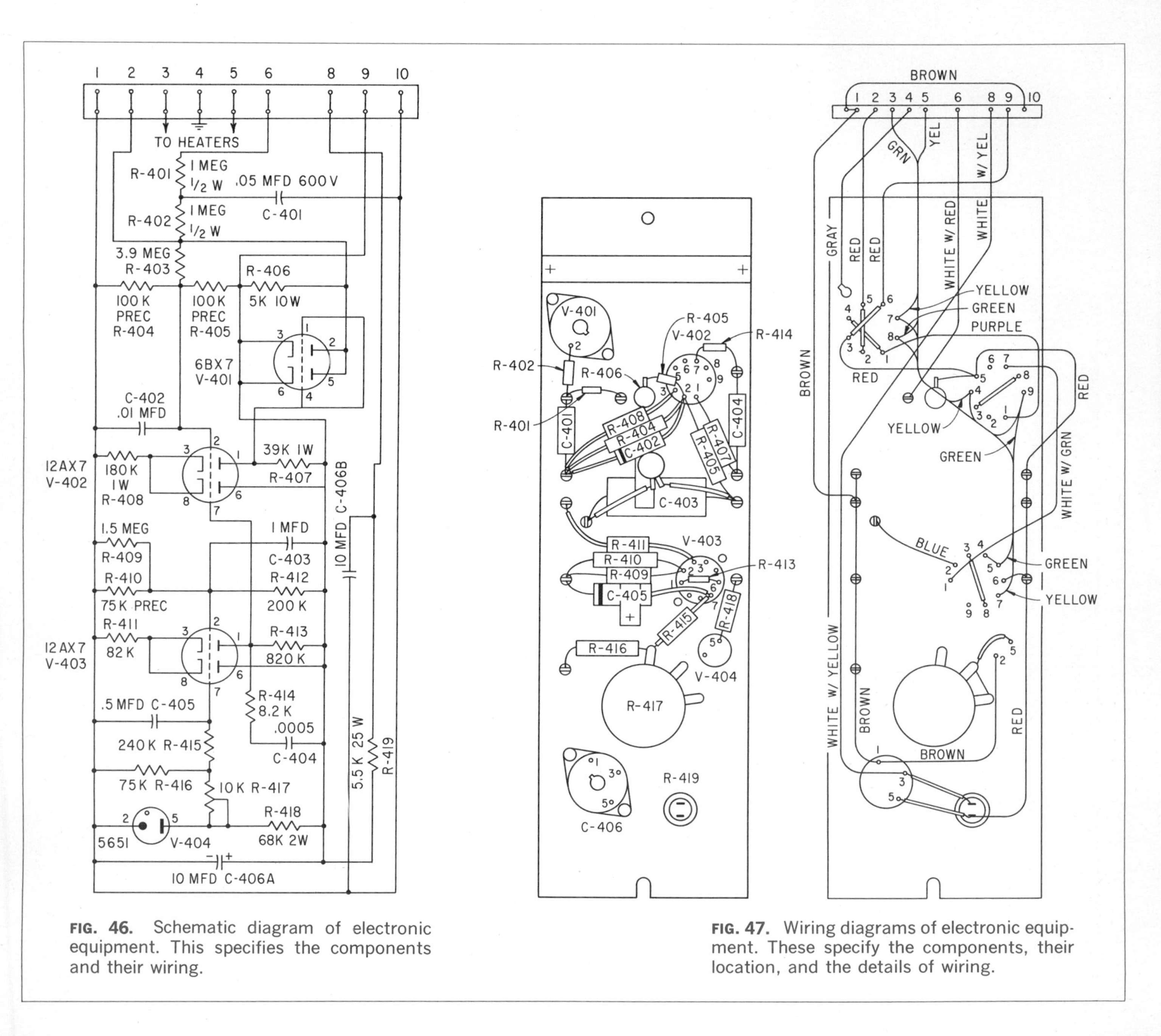


**FIG. 46.** Schematic diagram of electronic equipment. This specifies the components and their wiring.

**FIG. 47.** Wiring diagrams of electronic equipment. These specify the components, their location, and the details of wiring.

amplifier stages, second detector, audio driver, and a matched pair of audio amplifiers. Further examples of the drawings required for this type of equipment are given in Fig. 50, the drawing of the printed circuit

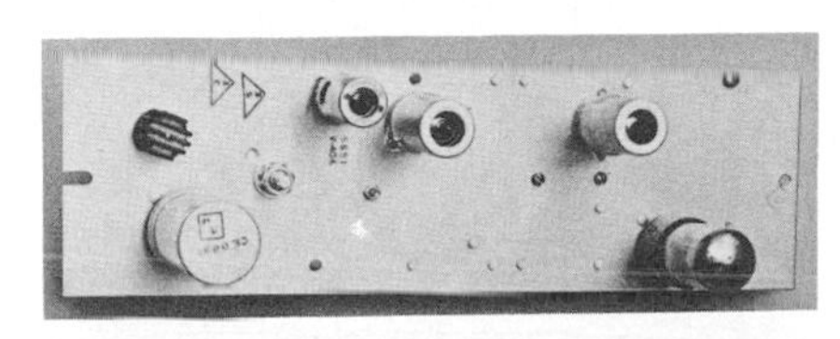

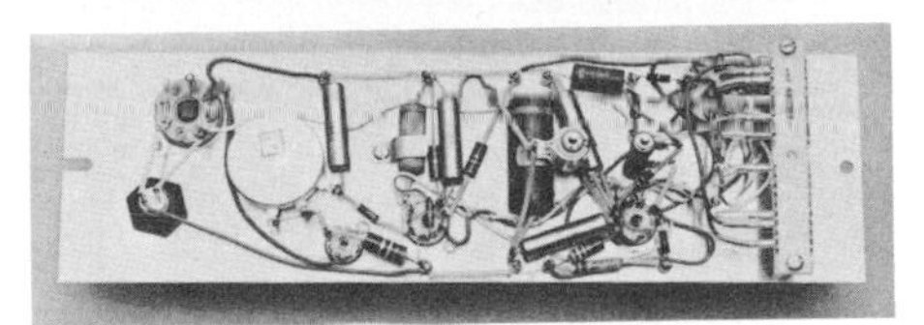

**FIG. 48.** Pictures of electronic equipment that represent electronic devices in Figs. 46 and 47, and may be used for presentation to a customer or service-manual illustration.

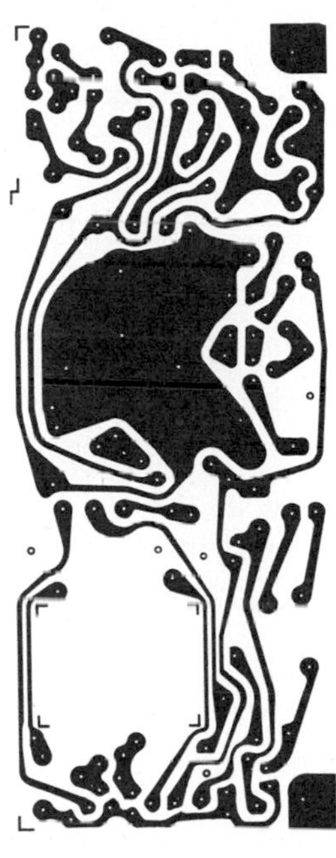

**FIG. 50.** Drawing of a printed circuit board for a service manual. This is produced from the original drawing of the circuit board.

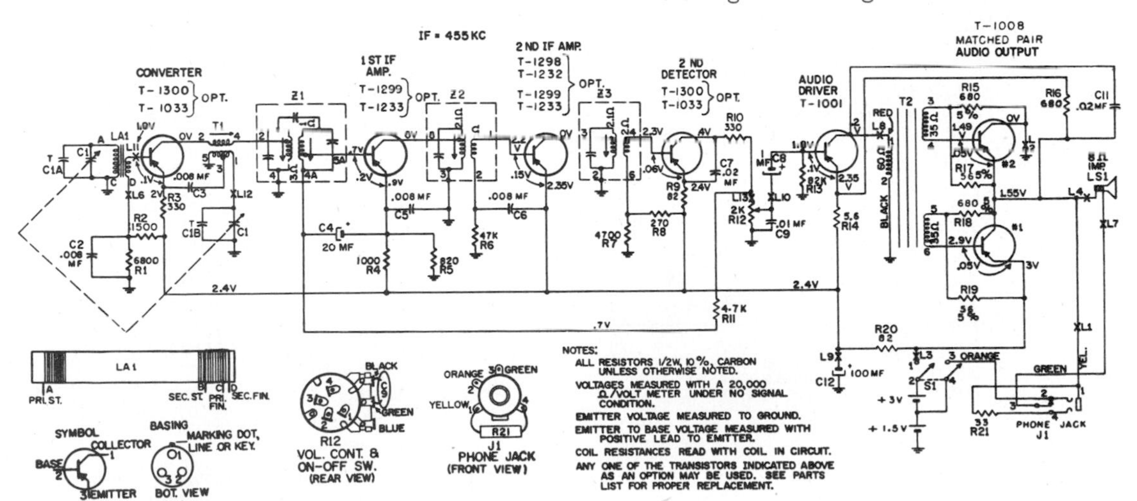


**FIG. 49.** Schematic diagram for a service manual. This gives the circuit and also specifies some details.

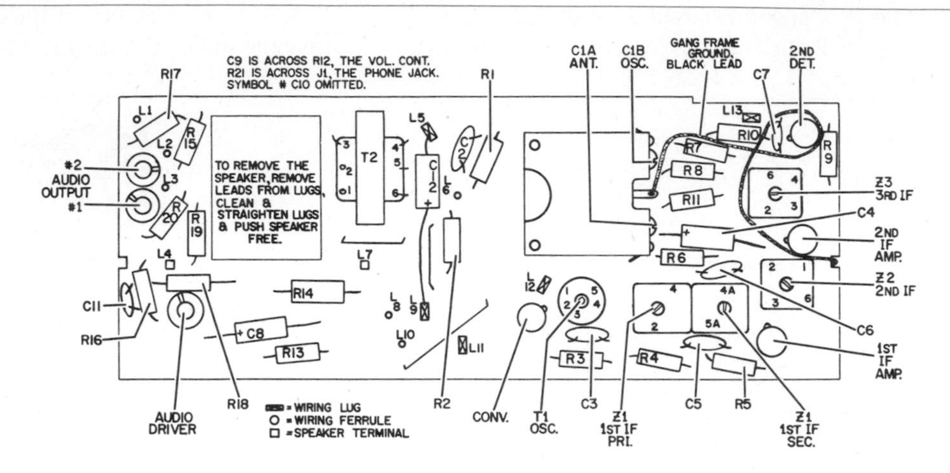


**FIG. 51.** Overlay drawing for a service manual. This mates with the circuit-board drawing in Fig. 50 and shows the components.

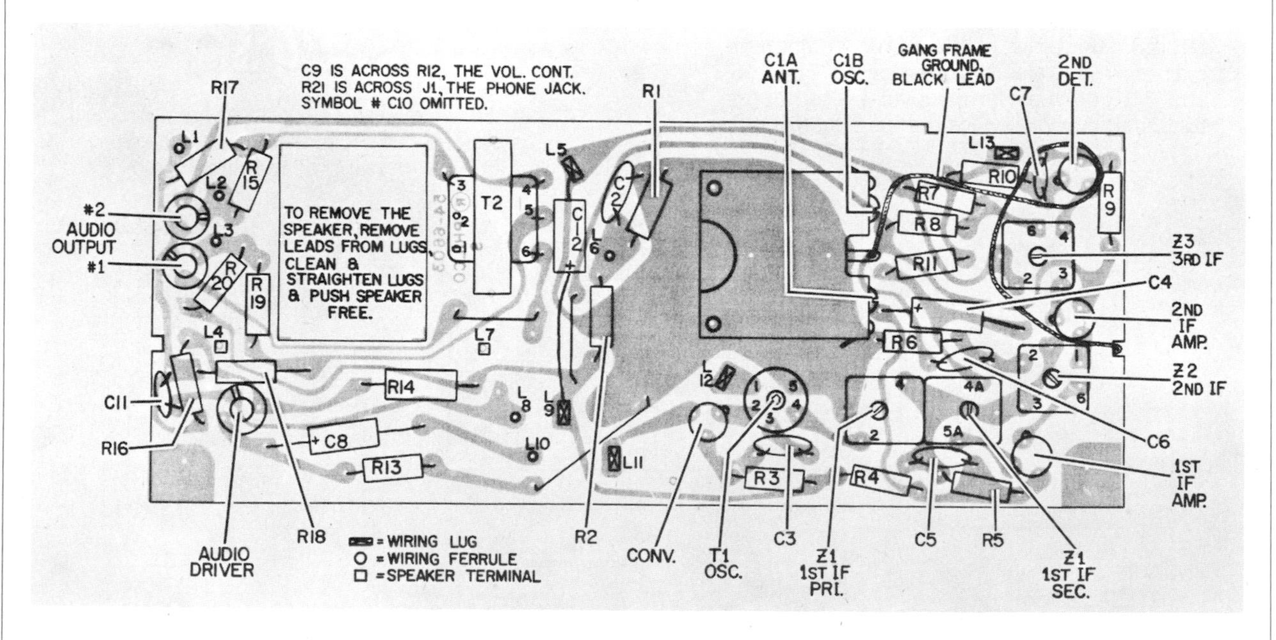


**FIG. 52.** Completed drawing for a service manual. The screened circuit-board drawing (Fig. 50) is printed with the overlay (Fig. 51) to give the final photocopy for reproduction.

board, and Fig. 51, the drawing for identification of components. These two drawings are made to the same scale with register marks so that they can be combined and printed together (photographically) to produce the drawing needed for a service manual (Fig. 52). Figure 53 is a photograph of the finished product.

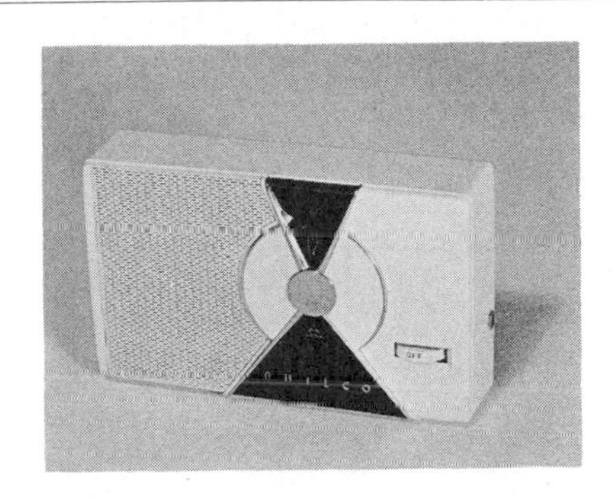

**FIG. 53.** Picture of electronic equipment for a service manual. This accompanies the drawings showing details of wiring and construction.

### 31. Electrical Charts.

Many charts, graphs, and diagrams are used in electrical work to plot and determine the design and operating characteristics of components and of complete electrical devices. Chapter 13 describes and illustrates fundamental charts, graphs, and diagrams. These are used in electrical practice as well as in other fields.

## PROBLEMS

### Group 1. Circuits.

1. The drawing shows the arrangement of a circuit devised by an electrical engineer for testing the diode characteristics of a type-80 electronic tube. Draw the schematic diagram of this circuit. Include a voltmeter in parallel for measuring the voltage across the tube. It should be noted that the oscilloscope is a testing device which will trace a curve proportional to the voltages applied to the vertical and horizontal axes and that the electrical circuit will operate without this piece of equipment.

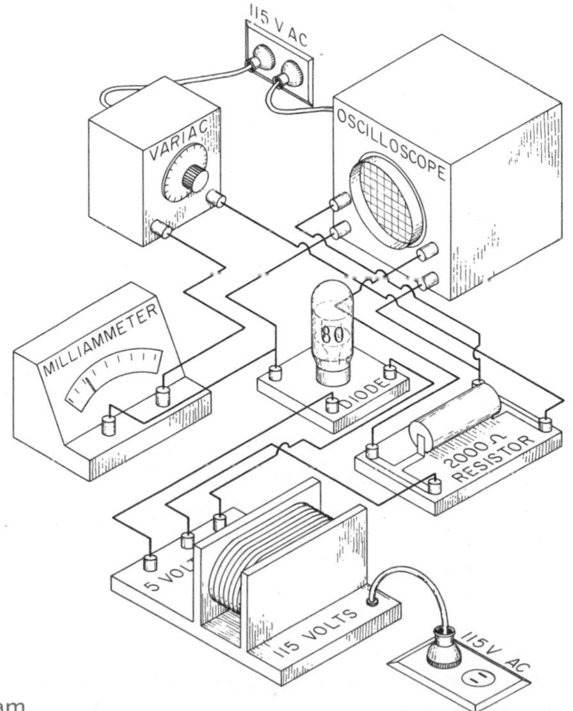


**PROB. 1.** A pictorial schematic diagram.

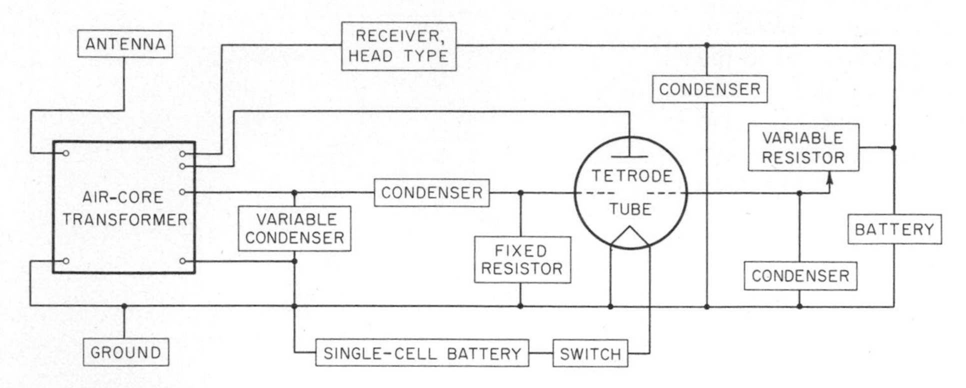


**PROB. 2.** Block diagram of a radio circuit.

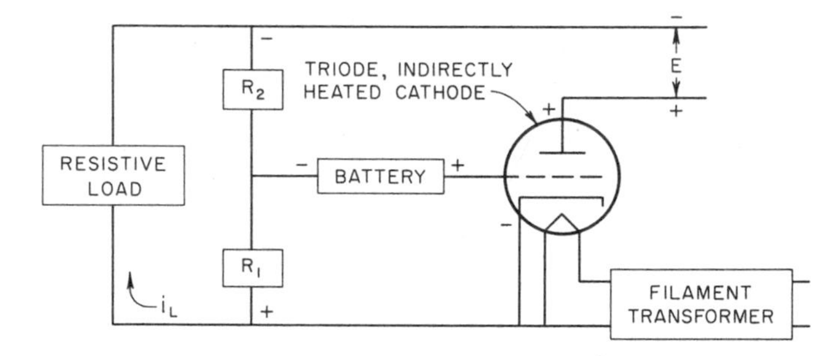


**PROB. 3.** Block diagram of a servo system.

**2.** The drawing is a block diagram of the circuit for a simple regenerative radio. Draw the schematic diagram of this circuit on an $8\frac{1}{2}$- by 11-in. sheet of drawing paper.

**3.** The block diagram of a simple electrical servo system is shown. In this system the voltage across the load is maintained almost constant when the load or the supply voltage is changed. For instance, if $R_L$ or $E$ would decrease, the voltage drop across $R_1$ would decrease. This decrease would in turn make the grid more positive and the current $i_L$ would be increased back to (or near) the original value. Draw a schematic diagram of this circuit, making sure that the polarity is correct.

**4.** The upper graph describes the flow of liquid raw materials necessary for a certain chemical process. These materials are forced to flow through a single solenoid-actuated automatic valve. In order to obtain the correct actuating mechanism, it is necessary to obtain the maximum dynamic requirements of the valve, that is, the maximum change in valve opening per unit of time. A manufacturer of automatic valves has supplied a curve of valve-flow rate capacity versus valve opening for the valve size required (lower graph). Draw the curve of valve opening versus time, and find the point on the curve where maximum change in valve opening occurs (the point where the slope of the curve is a maximum).

**4*A*.** Calculate the maximum change in valve opening per unit of time for Prob. 4. (Draw the slope of the curve and obtain $\frac{dy}{dx\,(\text{max})} = \frac{\text{in}}{\text{sec}}$.) Show graphically how the solution was obtained.

**4*B*.** Draw a block diagram showing the combined mechanical and electrical system for Prob. 4. A pressure-operated metering switch actuates a relay that in turn actuates the valve solenoid.

**4*C*.** Draw a wiring diagram for the electrical system of Prob. 4*B*.

**5.** This problem illustrates combined electrical and mechanical components. An industrial furnace is to be carried through a prescribed temperature cycle. This is to be done by adjusting the gas valve that supplies fuel to the furnace. It has been determined by calculation and tests that a synchronous, electrically controlled, cam-operated valve can be designed to give the desired time-temperature relationship. The cam turns at a constant speed and is so geared that it takes 90 minutes for a complete revolution. This time is equal to the length of the heat-treating cycle. From the curves shown, design the shape of the cam that will give the desired valve openings and thus the desired furnace temperature. See Chap. 18 for drawing the cam.

**5*A*.** Draw a block diagram for the system of Prob. 5. Include a temperature recorder, automatic "closed" and "open" warning lights on the furnace door, and a solenoid-operated door lock to prevent opening the door during the heating cycle.

**5*B*.** Draw a wiring diagram showing operation of the system in Prob. 5. List the required components.

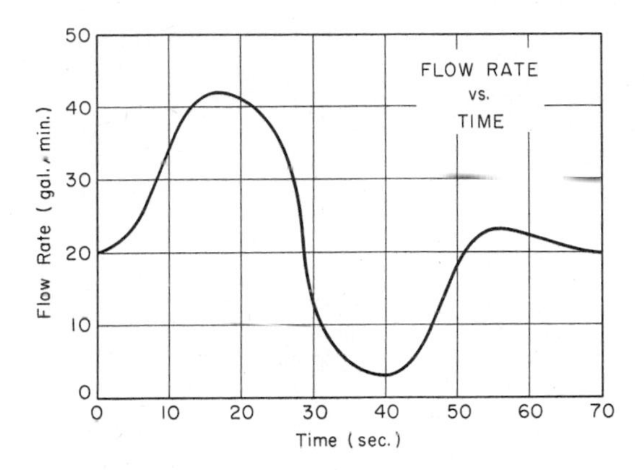


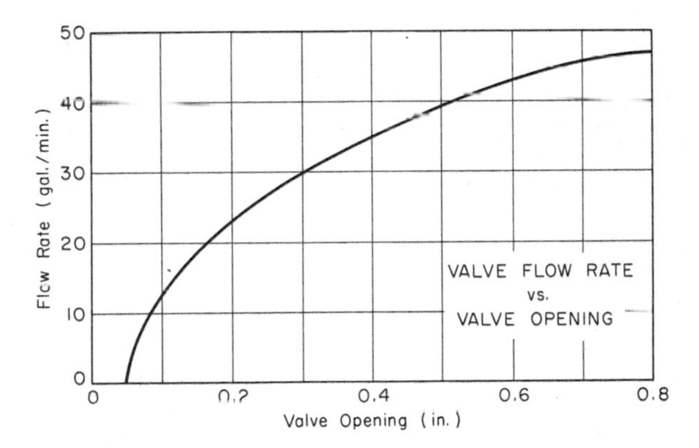


**PROB. 4.** Data for determination of value characteristics.

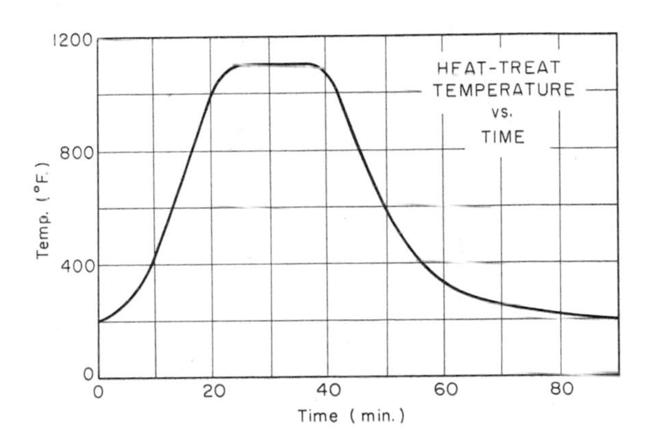


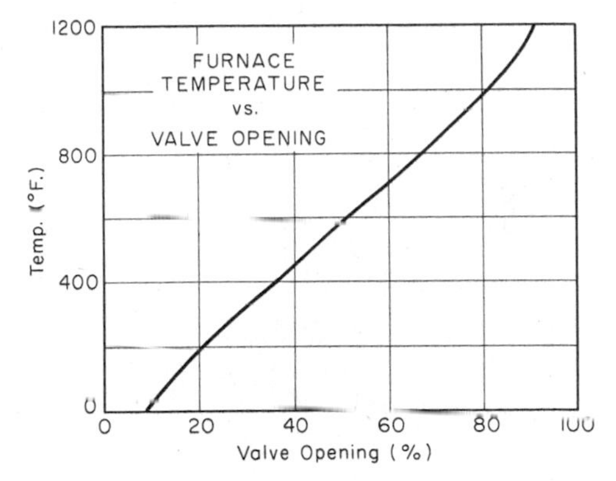


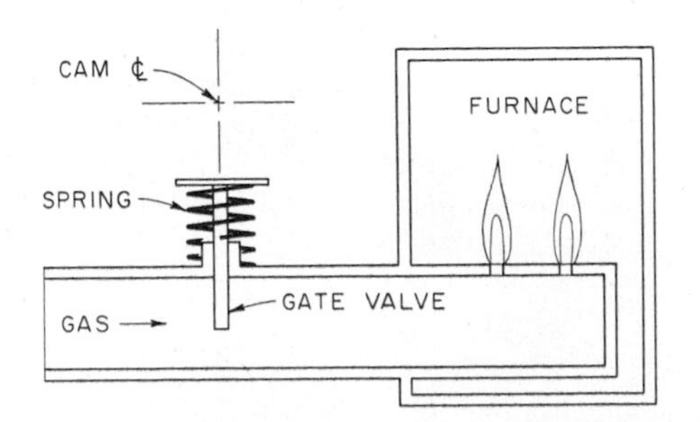


**PROB. 5.** Data for design of heat-treating-furnace control.

Included Here:

Architectural Drawings
Studies, Presentations, Models
Materials
Size Coordination, Grids
Dimensioning, Working Drawings
Plans and Elevations
Specifications
Structural Drawings
Classification, Shapes, Drawings
Detailing Practice
Timber, Masonry, Reinforced Concrete Structures
Problems

# 21

# STRUCTURES

The Chapter Presents Graphic Knowledge Needed in Architectural and Structural Work

### 1. Structures—Art and Science Combined.

The combined art and science of building structures, especially the art of building houses, churches, bridges, and the like, for the purposes of civil life, is properly placed, professionally, between engineering and the fine arts. Architectural designing is that phase of architecture based primarily upon study of the fine arts, but architects must understand structural engineering requirements. Conversely, engineers often work with architects in building structures and should therefore understand some of the procedures followed in an architectural drafting room. Structural engineers design columns, beams, floor slabs, etc.; mechanical engineers design heating and ventilating equipment; electrical engineers determine power requirements and design the electrical system of a structure to meet the requirements of the client; other engineers are often concerned with structures to house manufacturing and processing equipment.

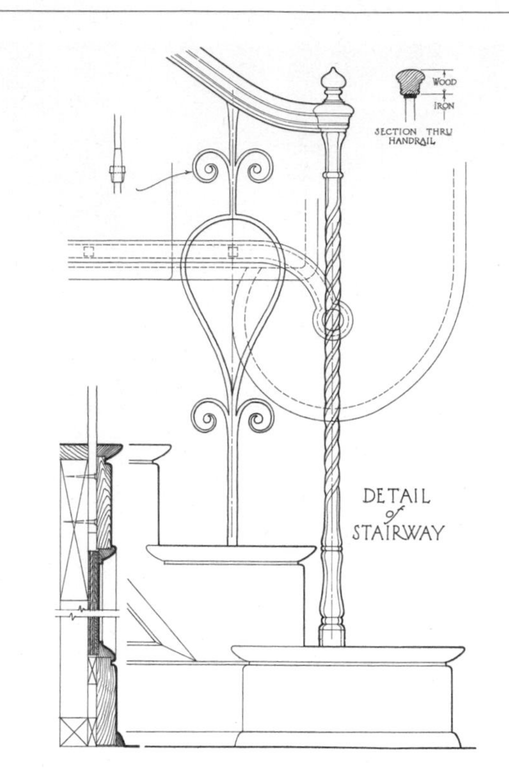


FIG. 1. Special techniques in an architectural detail. Distinctive features of this drawing are a profiled section and a superimposed plan view.

## ARCHITECTURAL DRAWINGS

### 2. Characteristics of Architectural Drawing.

The architect uses certain expressions that are peculiar to the architectural field. Engineers should be familiar with these expressions as they are often required to read and work with architectural drawings. Presentation techniques, as in a profiled section (Fig. 1), or unusual views, such as a reflected plan (Fig. 2), or even the occasional use of first and second angles of projection may be foreign to an engineer whose previous experience has not required collaboration with an architect. To prepare himself fully for such an association, an engineering student should make a careful study of architectural drawings, noting the reason for the differences between them and the more usual engineering drawings.

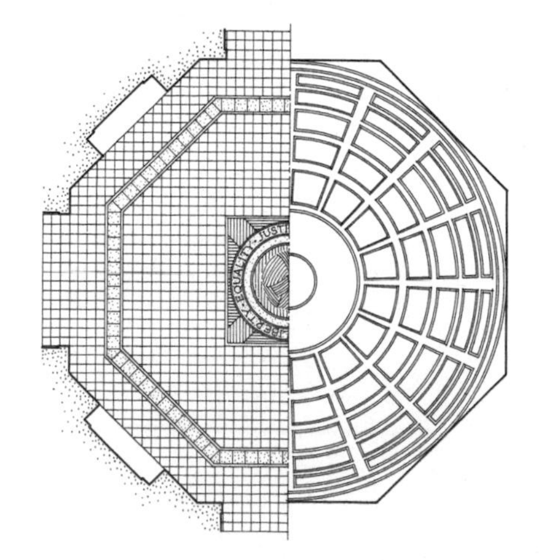

FIG. 2. Special techniques in an architectural detail. This drawing illustrates the use of a floor plan and a reflected plan of the ceiling.

The general principles of drawing are the same for all kinds of technical work. However, architectural drawings usually reveal a more spontaneous approach to drawing problems than that shown by drawings in other fields of engineering. The prepa-

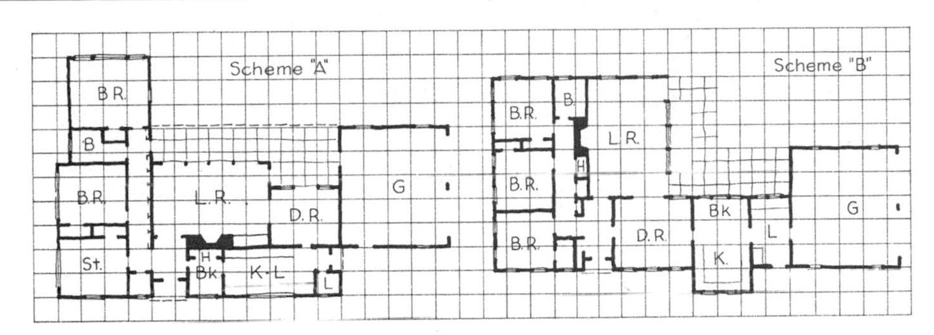


**FIG. 3.** Preliminary-plan studies. These are sketched on coordinate paper and are for studying space and arrangement.

ration of preliminary studies, presentation drawings, building models, and architectural working drawings requires various techniques and drawing skills. Thus, an architect, even more than his fellow engineers, must be able to draw both freehand and with instruments, using all the materials and techniques at a draftsman's disposal.

### 3. Preliminary Studies.

The design for a building is initiated by the client, who presents his requirements to the architect. This may be done in conferences or through the use of written material; at times a combination of both procedures is used. If the building is for industrial purposes, the engineers of the industry present flow sheets of the manufacturing processes to be carried on in the proposed structure. Close cooperation between engineering and architectural personnel is absolutely necessary to ensure that the building will meet the needs of the industry. Similarly, for all types of buildings, a program of minimum requirements, maximum allowable costs, architectural style, building materials, etc., is determined through client-architect contacts.

The architect then establishes a structural grid (see Secs. 8 to 10) in keeping with the function and structural requirements of the building.[1] With this grid as a background several solutions to the problem are developed, usually in rough sketches on tracing paper. Then a selection of the better solutions is made for further study in plan, elevation, and section.

Throughout this phase of the work sketches are kept relatively small, although care is taken to maintain proper proportions. The smallness of these "thumbnail sketches" forces the designer to consider the essential elements only, for a study of detail is impossible at a small scale. Preliminary plans, showing two schemes for a one-floor house, are reproduced in Fig. 3 at approximately one-half the original scale.

### 4. Presentation Drawings.

Presentation drawings are made to provide realistic and effective representations of proposed designs. They are used to illustrate proposed buildings for clients and in submitting designs in competition. They are usually drawn in perspective, then rendered in water color, ink, crayon, or pencil, giving the effect of color or light and shade. Such accessories as human figures, adjacent buildings, and foliage are introduced for scale, to give an idea of relative size. Should the presentation show one or more floor plans, the scale of the plan is indicated by including furniture and floor designs on the interior and walks, drives, and planting on the grounds immediately surrounding the building. Examples of presentation drawings are given in Fig. 4.

[1]Various modular dimensions for the grid may be employed, but in residential work a 4′-0″ square is most adaptable, whereas in commercial work, with reinforced concrete or steel construction, 20′-0″ or larger is advantageous.

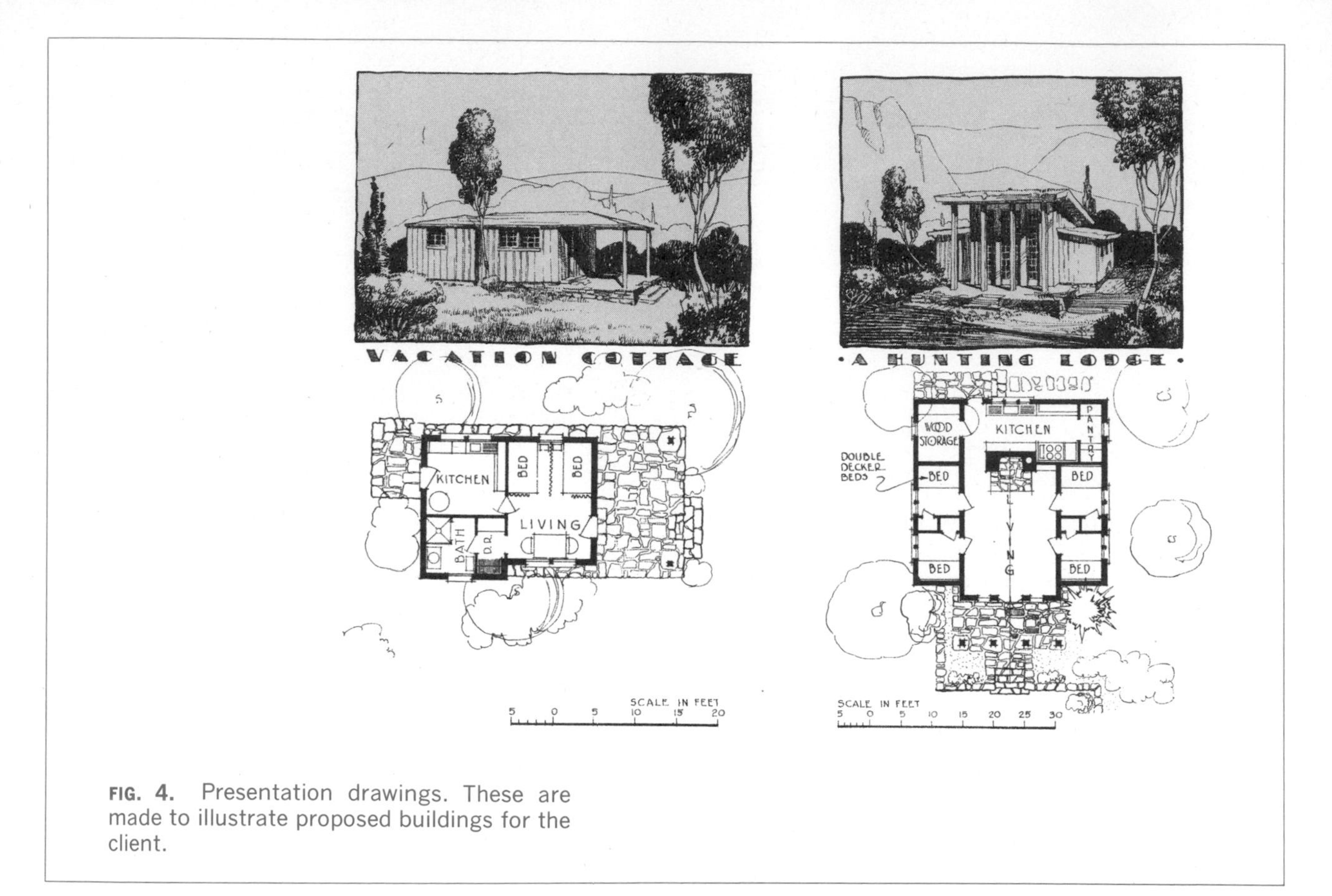


**FIG. 4.** Presentation drawings. These are made to illustrate proposed buildings for the client.

### 5. Models.

The use of building models, for presentation purposes, is widespread throughout the architectural profession. They are made in the drafting room, using drawing paper, illustration board, balsa wood, etc. The different walls and roof surfaces are laid out in developed form, rendered, folded, and then mounted on a rigid base. It is important that all features, such as moldings, railings, and plantings, be kept to scale. The advantage of a building model in showing the appearance of a completed building from any station point is of obvious value, when compared with a presentation drawing, showing it from a single station point.

For reproduction purposes a photograph of the model, superimposed on a photograph of the actual site, makes an effective presentation.

### 6. Materials for Construction.

The selection of building materials must be made at some time prior to the preparation of the working drawings and specifications. Items that influence the choice include appearance, initial cost, erection time, and maintenance. Conventional symbols for the more common building materials are shown in Fig. 5.

### 7. Size Coordination of Building Materials and Equipment.

Size coordination of a building's many parts is accomplished by cutting and fitting during field operations or by shop fabrication. In actual practice a combination of both techniques is used. However, it is apparent that tighter fits and lower costs result when shop-fabrication methods replace the more

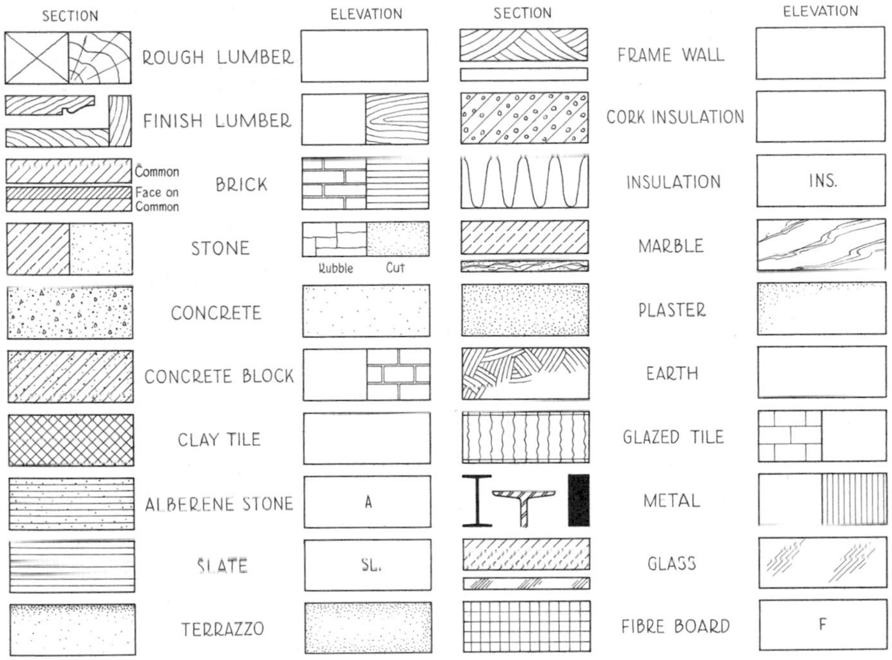

SECTION | ELEVATION | SECTION | ELEVATION

ROUGH LUMBER
FINISH LUMBER
BRICK (Common; Face on Common)
STONE (Rubble; Cut)
CONCRETE
CONCRETE BLOCK
CLAY TILE
ALBERENE STONE (A)
SLATE (SL.)
TERRAZZO
FRAME WALL
CORK INSULATION
INSULATION (INS.)
MARBLE
PLASTER
EARTH
GLAZED TILE
METAL
GLASS
FIBRE BOARD (F)

**FIG. 5.** Symbols for building materials. These indicate types of materials. Complete details are given in notes and specifications.

wasteful procedures inherent in the majority of field operations.

After considerable study and research the ASA approved a 4-in. unit as the standard module for building materials. From a practical standpoint, it must be understood that not all manufacturers are producing materials sized in 4-in. increments, nor is such standardization ever likely to occur. *Standardization, for the purpose of size coordination of building materials and equipment, is achieved by a judicious selection from the available sizes, followed by a tying of the resulting details into the modular grid.*

## 8. Grid Position of Details.

A symmetrical location of walls and openings, with respect to the horizontal grid, is desirable in order to effect standardization. In the instance of wall locations, Fig. 6 shows that only one reference to the grid is possible if the walls are (*a*) centered on

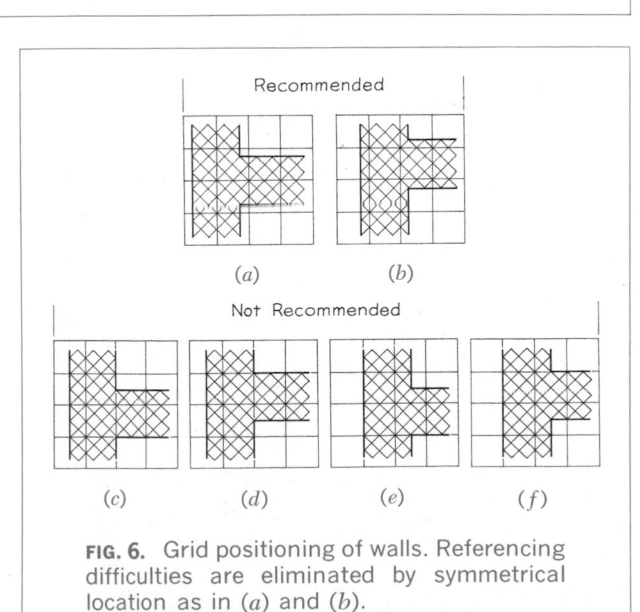


**FIG. 6.** Grid positioning of walls. Referencing difficulties are eliminated by symmetrical location as in (*a*) and (*b*).

the grid lines, or (*b*) centered between grid lines. Any other system of positioning walls results in numerous possibilities of grid referencings. There are four possible grid positions, as shown in (*c*) to (*f*), if one face of each wall is positioned on a grid line.

Vertical referencing is usually restricted to maintaining floor levels on grid lines. Openings need not be symmetrical in their vertical positioning, since head and sill details differ. However, different-sized openings are usually positioned with all heads on the same grid line, for any one story. Necessary adjustments due to variations in floor thicknesses are made at the intersections of the walls and ceilings (Fig. 7).

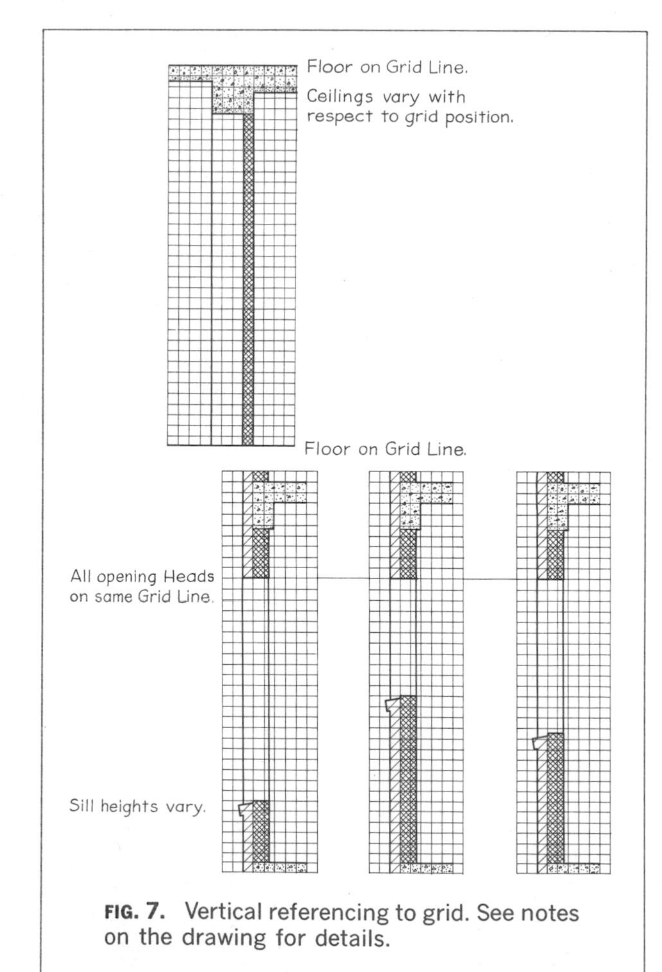


**FIG. 7.** Vertical referencing to grid. See notes on the drawing for details.

## 9. Nominal and Actual Size.

Two systems of dimension notation are currently in use in architectural offices. The majority of offices show actual values in nearly all instances. However, the more modern procedure of grid referencing allows, in fact makes preferable, the use of nominal values on all small-scale layout drawings.

Figure 8 illustrates the difference between nominal and actual sizes. Nominal thicknesses of walls are established by the nominal size of the structural members. Since 2- by 4-in. studding is used in both the exterior and partition frame walls shown, the 4-in. size is assigned as their nominal thickness. The actual size of a 2 by 4 is $1\frac{5}{8}$ by $3\frac{5}{8}$ in. The glazed tile in the masonry partition wall is actually $3\frac{5}{8}$ by $11\frac{5}{8}$ by $3\frac{5}{8}$ in. and nominally 4 by 12 by 4 in. (allowance for a $\frac{3}{8}$-in. mortar joint having been made in each of the three dimensions). A $\frac{3}{8}$-in. mortar joint is again designed for in the sizing of modular brick; actual dimensions are $2\frac{5}{16}$ by $7\frac{5}{8}$ by $3\frac{5}{8}$ in., while nominally they are $2\frac{2}{3}$ by 8 by 4 in. Note that interior and exterior treatment, or finish, is not considered in the assignment of the nominal size to any wall thickness; the structural members alone are considered.

Comparison of the line of actual dimensions with the line of nominal dimensions in Fig. 9 shows the comparative ease in determining total dimensions (such as $X$) when nominal values are used in preference to actual quantities. A detail, dimensioned with actual values and referenced to the grid, completes the information required for the proper use of nominal values, thus simplifying the arithmetic processes ever present in checking, estimating from, and field use of architectural drawings.

## 10. Dimensioning.

The correct dimensioning of an architectural drawing requires a knowledge of building-construction methods. The dimensions are placed so as to be most convenient for the workman, given to and from accessible points, and selected so that commercial variation in the sizes of materials will not affect the principal dimensions. In general, dimensions are given to the faces of masonry walls, the outside face of the studs in an exterior wall, and the center of

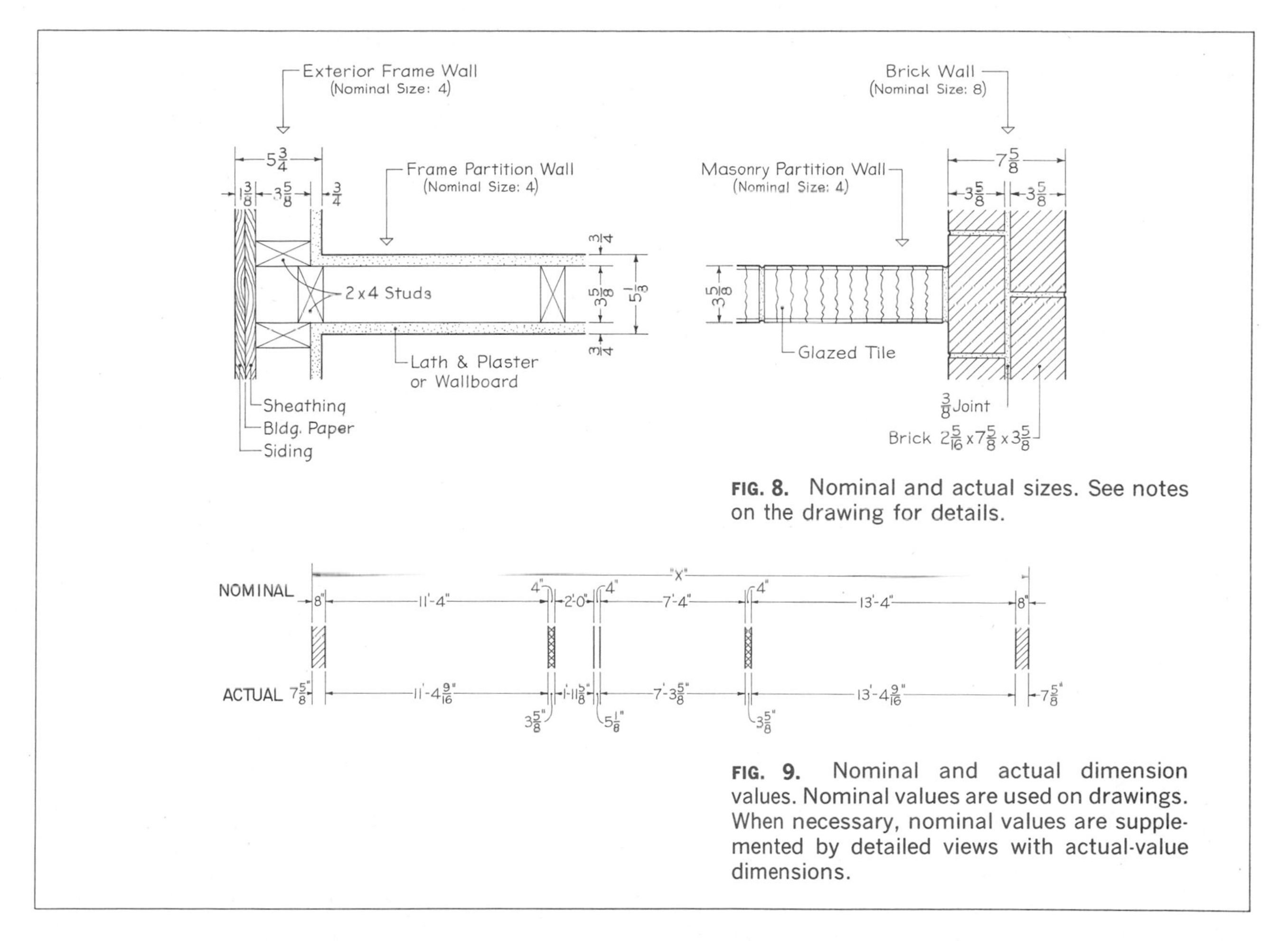


**FIG. 8.** Nominal and actual sizes. See notes on the drawing for details.

**FIG. 9.** Nominal and actual dimension values. Nominal values are used on drawings. When necessary, nominal values are supplemented by detailed views with actual-value dimensions.

the studs in a partition wall. Columns and beams are located by dimensioning to their center lines, as are window and door openings in frame walls (in masonry walls the openings are dimensioned). Compared with drawings dimensioned according to the techniques used in the other engineering fields, architectural working drawings appear overdimensioned. There is no attempt to avoid duplicating a dimension as there is in fields where parts are made for interchangeable assembly. No usable dimension should be omitted on an architectural drawing, and, as in other fields, care must be taken to see that every dimension that is needed is given even though that dimension could be determined by the addition and/or subtraction of other dimensions on the sheet.

Legibility is of prime importance in dimensioning. All numerals must be carefully lettered. If a dimension value is not clearly indicated, a crease in the paper, a smudge, or a poor print of the drawing may cause it to be misread. Watch particularly the numerals 3 and 5.

In dimensioning drawings that have been laid out in the modular system, it is essential that the grid referencings be clearly indicated since the grid will usually be omitted from small-scale drawings. *An arrowhead is used at the end of the dimension line when the dimension is to a point on a grid line, and a dot is used for all points not on grid lines* (Fig. 10). Regardless of whether actual or nominal values are shown, the architect does not find the graceful but slim arrowhead used by other engineers suitable for

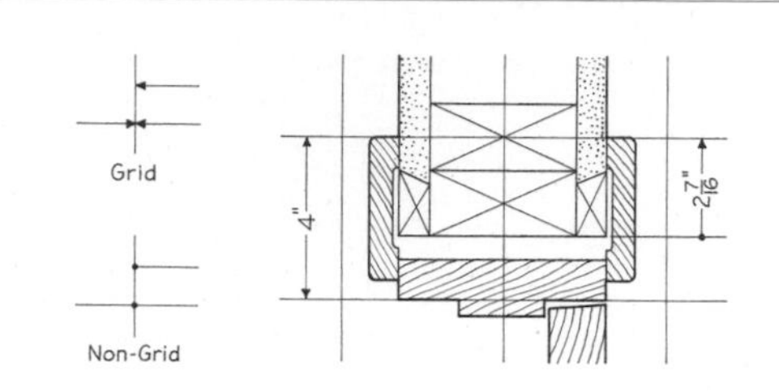


**FIG. 10.** Dimension-line terminals. Arrowheads indicate points on the grid; dots indicate points not on the grid.

**FIG. 11.** Arrowheads. The type shown at (*b*) is used on architectural drawings.

his use. It is clear in Fig. 11*a* that the dimensions might be taken to the solid or the dashed line; in (*b*), where a more bulky arrowhead is used, there is no question regarding the line to which the dimensions refer.

## 11. Working Drawings.

The general principles of Chap. 23 on production and working drawings are applicable to architectural working drawings. As a general rule, details that are related should be shown together, and information should be grouped by craft whenever feasible. Many present-day buildings are so complicated that it is advantageous to draw several sets of plans, a set for each of the several crafts—structural, plumbing, electrical, heating, and ventilating—in addition to the architectural set.

Architectural handbooks, American Institute of Architects' files, Sweet's Catalogues, and other literature should be freely consulted during the preparation of all working drawings. From them, sizes for building materials and equipment are determined, together with other factual data necessary for the proper selection and specification of such items. In preparing working drawings, publications such as local and state building codes must also be studied and their provisions adhered to. Legal requirements as to approval of the plans and specifications, securement of building permits, etc., must also be met.

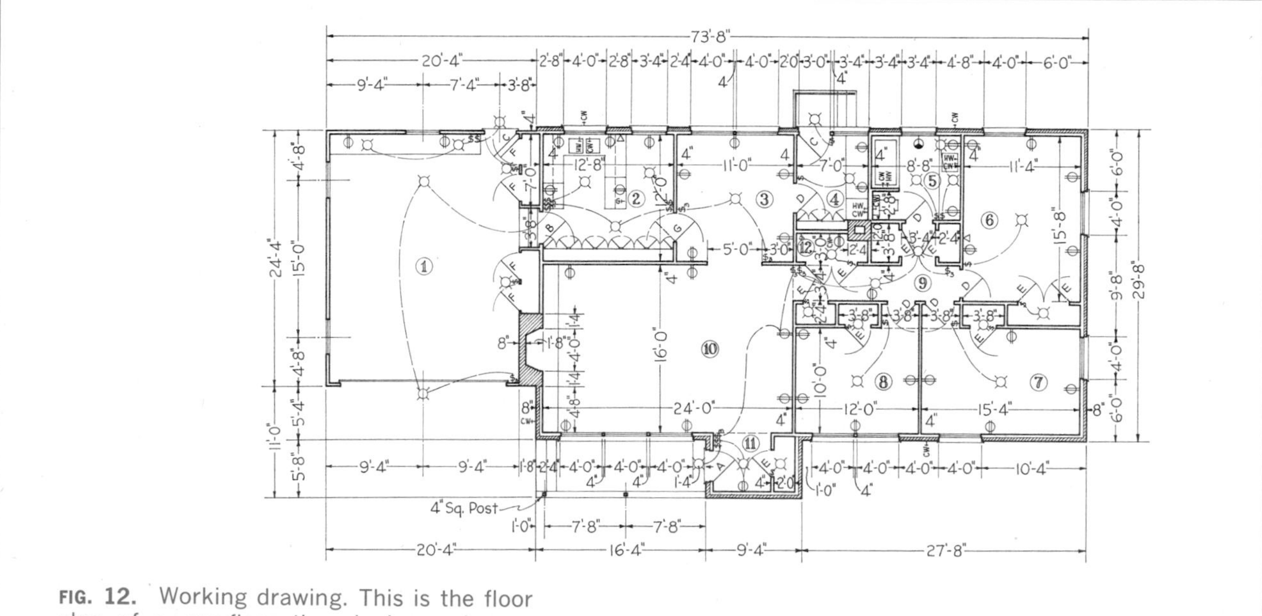


**FIG. 12.** Working drawing. This is the floor plan of a one-floor three-bedroom house.

| ROOM FINISH SCHEDULE | | | | | | |
|---|---|---|---|---|---|---|
| Rm. No. | Floor | Base | Walls | Ceiling | Trim | Remarks |
| 1 | Concrete | Concrete | Sheetrock | Sheetrock | Poplar | Built-in bench w/dwrs. |
| 2 | Linoleum | Cove | Linowall | " | " | See cabinet details. |
| 3 | Concrete | Birch | Sheetrock | " | Birch | See details. |
| 4 | Linoleum | Cove | Linowall | " | Poplar | " |
| 5 | Cer. Tile | Cer. Tile | cer. tile wnsct sheetrock | " | " | " " |
| 6 | Concrete | Birch | Sheetrock | " | Birch | " " |
| 7 | " | " | " | " | " | " " |
| 8 | " | " | " | " | " | " " |
| 9 | " | " | " | " | " | " " |
| 10 | " | " | " | " | " | " " |
| 11 | Slate | " | " | " | " | " " |
| 12 | Concrete | Concrete | " | " | Poplar | " " |

**FIG. 13.** Working drawing. This is the room finish schedule for the house in Fig. 12. Numbers are referenced to the floor plan.

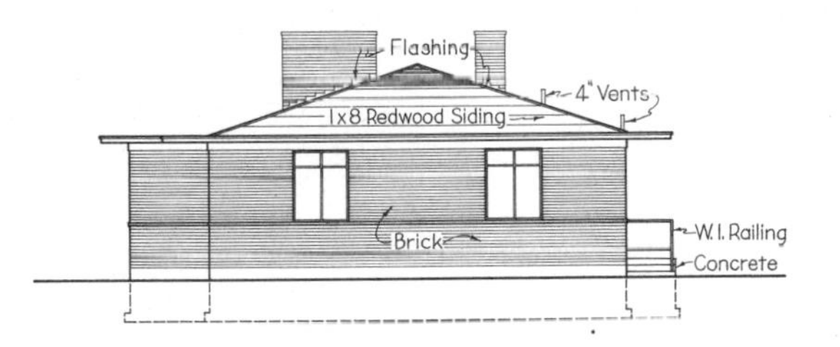


**FIG. 14.** Working drawing. This is the right-side elevation for the house in Fig. 12.

| CUTTING SCHEDULE | | |
|---|---|---|
| Sym. | No./Truss | Diagram |
| A | 2 | $1\frac{5}{16}$", $1\frac{5}{16}$", 2x4, 15'-$3\frac{3}{4}$" |
| B | 2 | 2x6x17'-$6\frac{7}{8}$" Sq. Cut |
| C | ? | $1\frac{5}{16}$", 2x4, 2'-0" |
| D | 2 | Same as "C" except "D" is 3'-11" Long. |
| E | 1 | 10", 10", 3'-$5\frac{1}{2}$" |
| F | ? | $2\frac{5}{8}$", 2x4, 6'-$7\frac{1}{2}$" |
| G | 2 | $3\frac{3}{16}$", $2\frac{3}{16}$", 2x4, 6'-5", 2" |
| H | 4 | 2x4x10" Sq. Cut |
| J | 1 | 2x6x4'-0" Sq. Cut |

**FIG. 16.** Working drawing. This is the roof truss detail and cutting schedule for the house in Fig. 12.

Figures 12 to 18 are selections from a set of drawings for a one-floor three-bedroom house. Omitted from reproduction here are the following drawings: plot plan, left-side elevation, rear elevation, foundation details, radiant-heating layout, roof-framing plan, door and window schedules, doorway details, kitchen and bath elevations, fireplace details, cabinet detail, etc.

## 12. Plot Plan.

Before designing any structure of importance, a plot plan is made giving the property line, contours, available utilities, location of trees, building lines, and other pertinent data. The building is then designed to fit the site. The plot plan is completed as one of the working drawings by locating on it the building, approaches, and finished grade contours (Fig. 19).

## 13. Floor Plans.

A floor plan is a horizontal section taken at distances above the floor that vary so as to show best all the features of the building between that floor and the

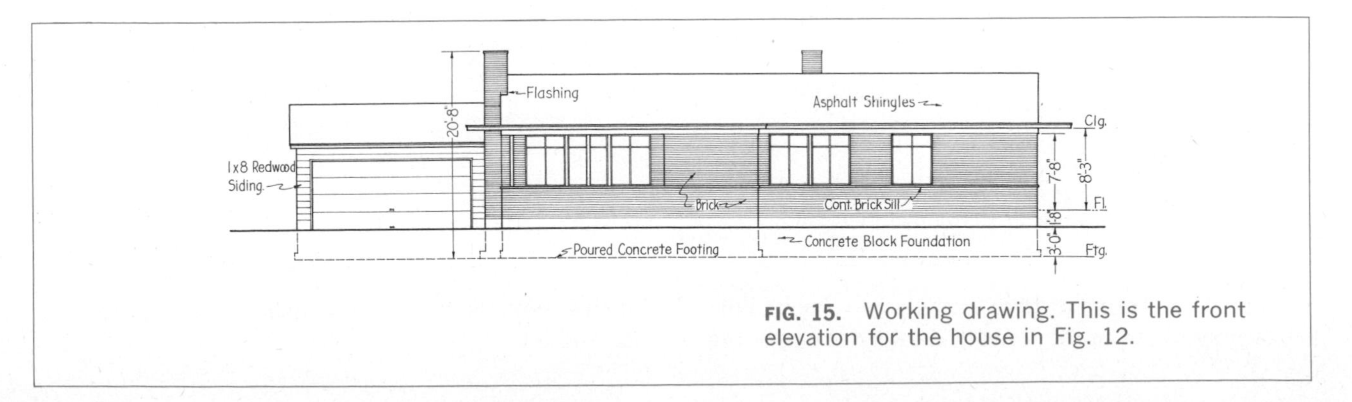


**FIG. 15.** Working drawing. This is the front elevation for the house in Fig. 12.

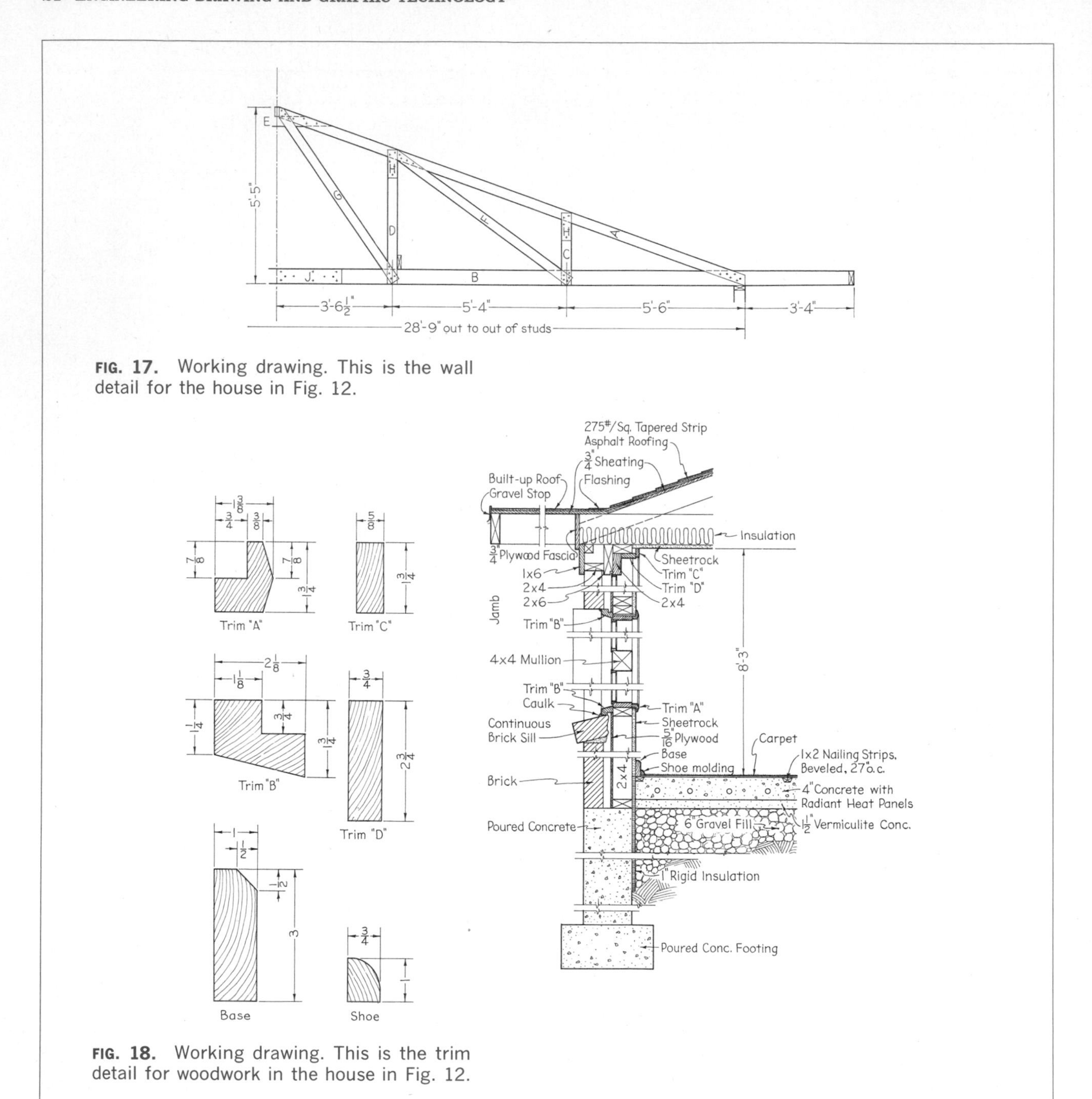


**FIG. 17.** Working drawing. This is the wall detail for the house in Fig. 12.

**FIG. 18.** Working drawing. This is the trim detail for woodwork in the house in Fig. 12.

one above. Because of their small scale, floor plans are largely made up of conventional symbols. Some of the symbols currently in use are shown in Figs. 20 and 21.

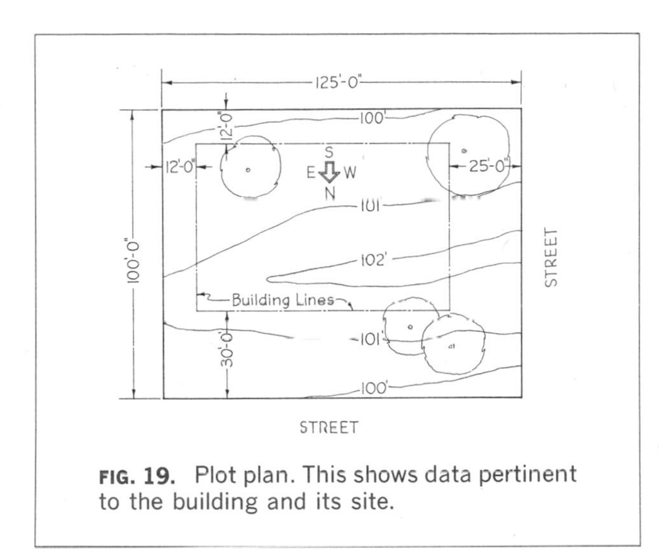


FIG. 19. Plot plan. This shows data pertinent to the building and its site.

In drawing a floor plan, the large structural grid is drawn first to serve as a guide. The entrance front is placed at the bottom of the sheet without regard to compass directions. Columns and walls are then positioned on the grid. Axes of communication, or route lines between the several rooms, are established. Interior doorways, stairways, elevator and service shafts, etc., are located. Exterior wall openings, however, are positioned *only after elevation studies have been made*. A completely dimensioned floor plan of a warehouse is shown in Fig. 22 (because of the simplicity of the design, this drawing gives both architectural and structural information).

## 14. Elevations.

An elevation is a view seen looking in a horizontal direction. For practical purposes, the elevations drawn are those showing normal views of wall surfaces.

In drawing an elevation, a wall section (showing the grade line, floor heights, head, and sill of a

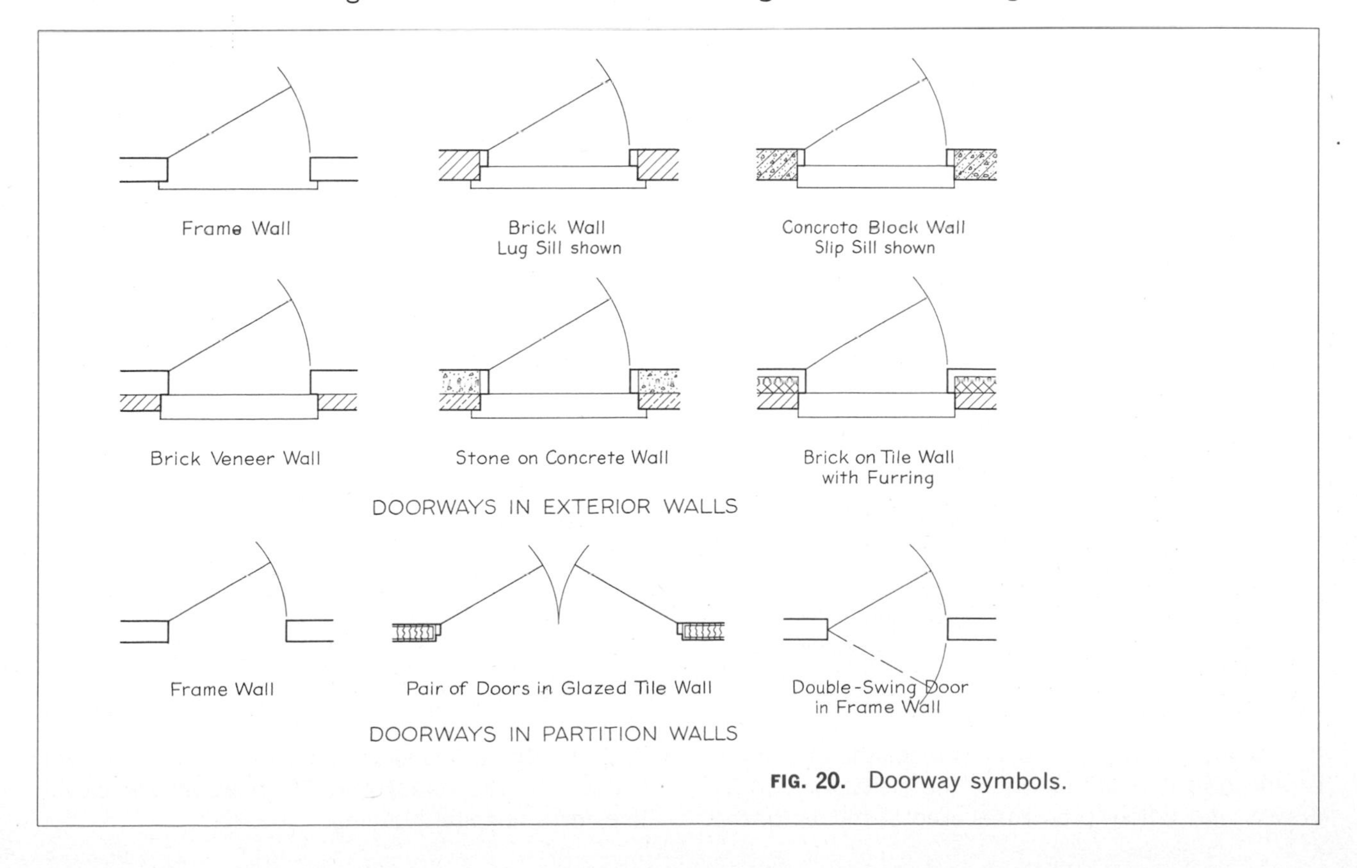


FIG. 20. Doorway symbols.

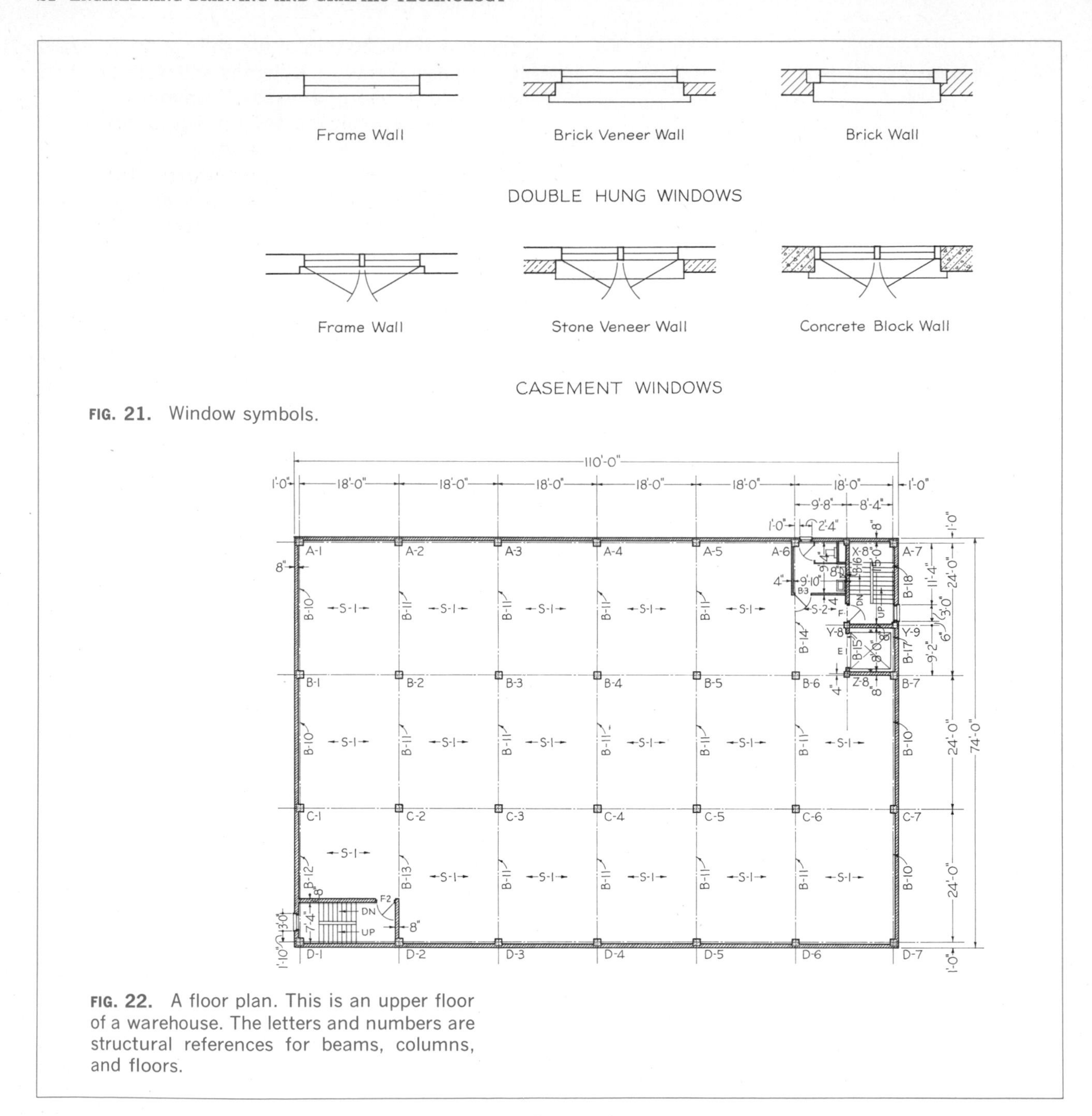


**FIG. 21.** Window symbols.

**FIG. 22.** A floor plan. This is an upper floor of a warehouse. The letters and numbers are structural references for beams, columns, and floors.

typical window in each story, cornice, and roof ridge) is placed to the side of where the elevation is to be drawn, and a floor plan is secured to the top of the drawing board. The elevation is then projected from the section and plan. Figure 23 shows an end elevation for a hospital building.

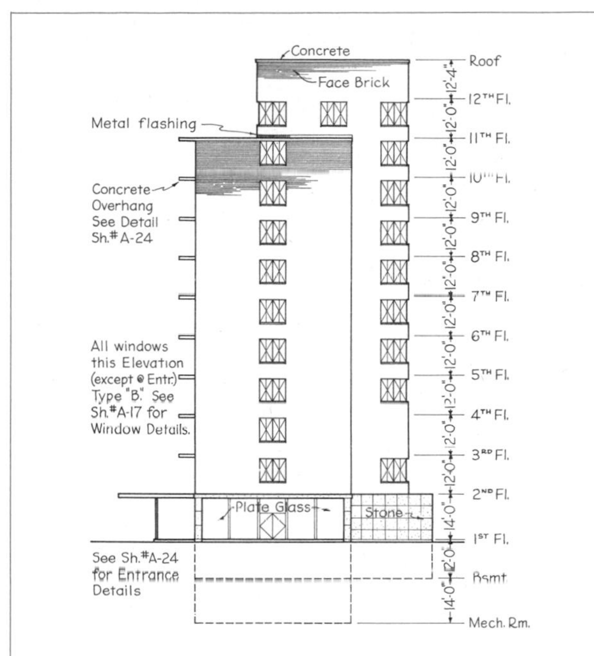


**FIG. 23.** An elevation. This is the end elevation for a hospital building. Note references to other drawings.

## 15. Sectional Elevations (Fig. 24).

Sectional elevations through the entire structure show interior construction and architectural treatment. Two such sections, called the longitudinal and transverse sections, at right angles to each other, are usually included in a set of working drawings. The cutting planes for these sections do not need to be continuous, but may be offset so as to include as much information as possible.

## 16. Detail Drawings.

A set of architectural drawings contains, in addition to plans and elevations, larger-scale drawings of parts not indicated with sufficient definiteness on the small-scale drawings. Details and detail sections of items such as footings, windows, and framing are required to show details of both construction and architectural design. Details are best grouped so that each sheet contains the references that are made on one sheet of the general drawings (the floor plans and elevations).

As the construction progresses, drawings are made of full-size details of moldings, millwork, ornamental iron, etc. Such drawings are made *after*

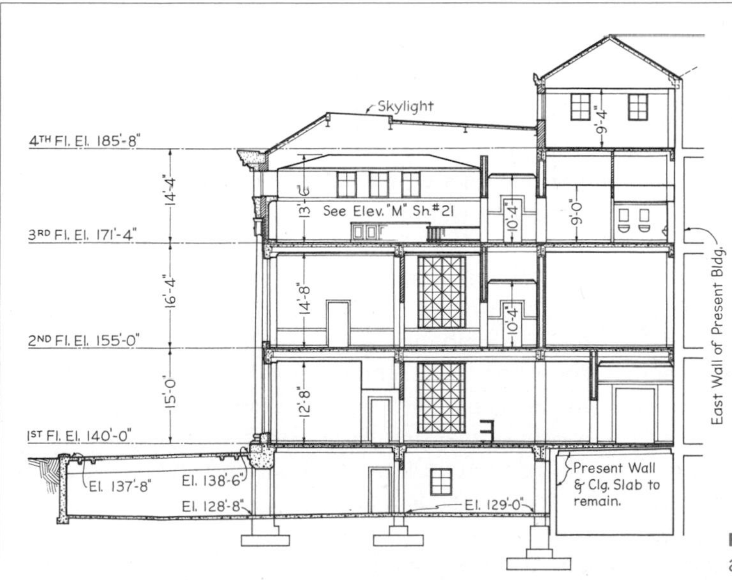


**FIG. 24.** A sectional elevation. This is for an addition to a municipal building. Note inclusion of elevations, notes, and references.

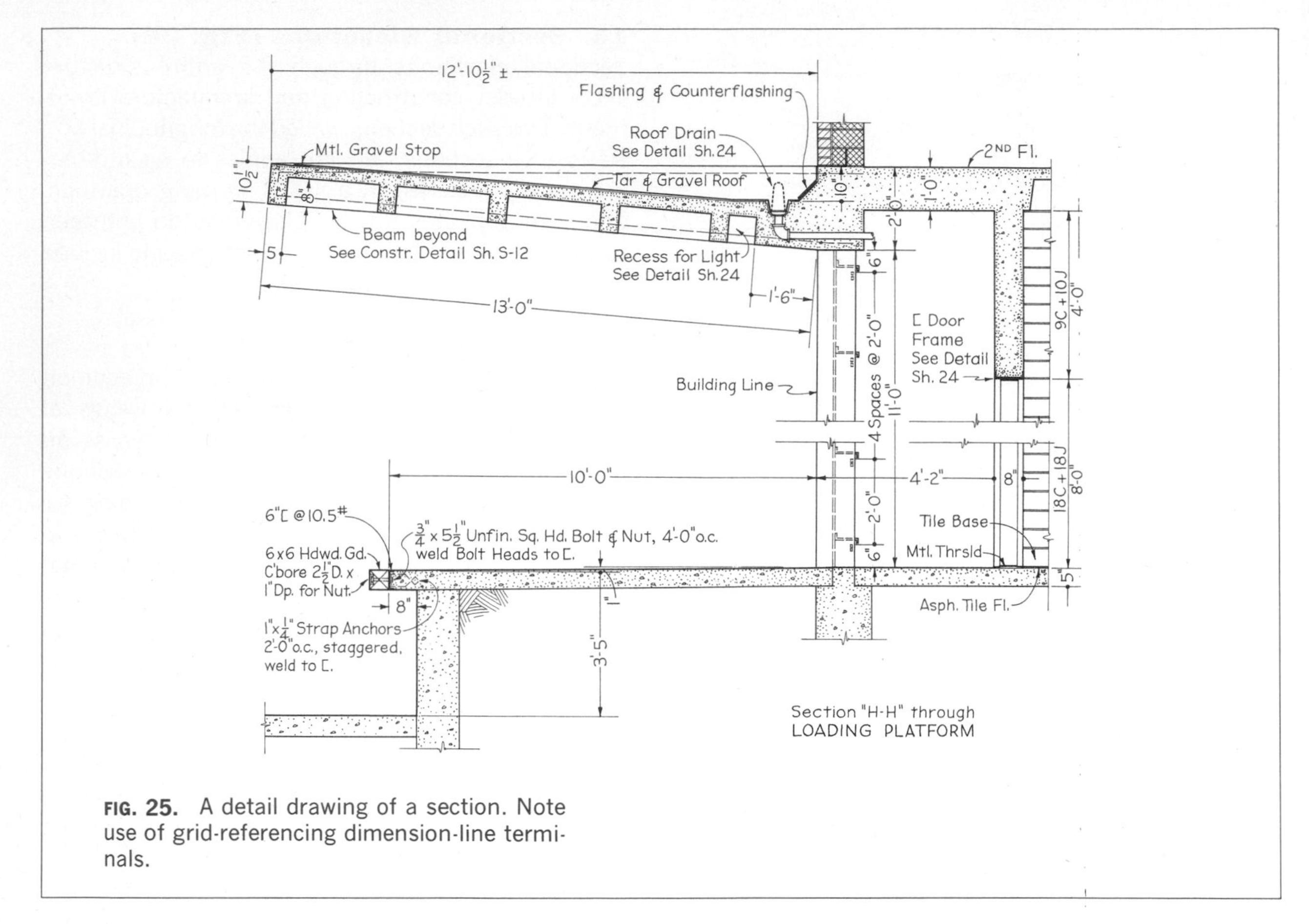


**FIG. 25.** A detail drawing of a section. Note use of grid-referencing dimension-line terminals.

measurements of the building have been taken. Sections are used freely on these drawings.

Figures 25 and 26 illustrate typical detail drawings.

## 17. Lettering.

There are two distinct divisions in the use of lettering by an architect. The first, office lettering, includes all titles and notes put on drawings for information; the second, design lettering, covers the drawing of letters to be executed in stone, bronze, or some other material, in connection with the architectural design. Legibility is the primary consideration in both divisions. In design lettering appearance is as important as legibility; in office lettering speed of execution is the second factor to consider. Pleasant appearance results from uniformity regardless of the letter style; speed in drawing necessarily depends upon simplicity of the letter forms.

The Old Roman alphabet is generally employed in design work. Before attempting to use this letter, the designer should become thoroughly familiar with its construction, character, and beauty. If he realizes that the alphabet was designed for inscription in stone and if he knows how the work is done, he will develop a sympathetic understanding of how to use these classic forms. See the Appendix.

Single-stroke letters, illustrated in the Appendix, are excellent for use in office lettering.

## 18. Titles.

An architectural title may be hand lettered on each sheet, printed, or impressed with a rubber stamp. In the latter two instances some bits of information

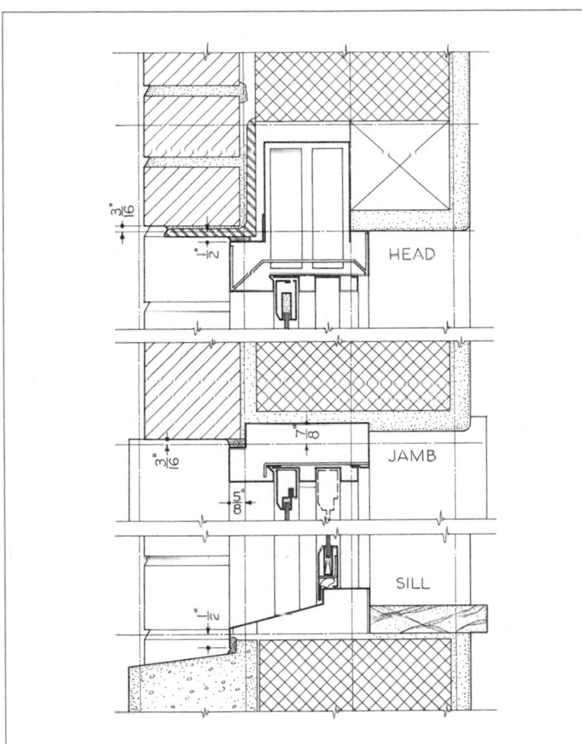


FIG. 26. A detail drawing. This is of a double-hung, spring-balanced metal window. Note 4-in. grid and referencing.

will necessarily be hand lettered to complete the title. Figure 27 shows several examples of architectural titles.

A complete title contains the following information:

1. Name and location of the building
2. Kind of view (as "First Floor Plan") and scale
3. Name and address of the client
4. Name and address of the architect
5. Sheet number
6. Office record, to include draftsman's and checker's initials and dates

## 19. Shop Drawings.

The purchase of special items of equipment for installation in a building requires *shop drawings,* prepared by the equipment manufacturers. Such drawings show over-all sizes, details of construction, methods for securing the equipment to the structure, and any other pertinent data the architect needs for the final positioning of the equipment.

## 20. Specifications, Notes, and Schedules.

Architectural drawings are amplified by a document called the "specifications," a detailed account of

FIG. 27. Architectural titles.

material and workmanship for the many parts of a structure. Since the workmen on a job seldom see these specifications, they should be briefed; and specification should be given, as notes or schedules, on the working drawing sheets. An example of a schedule is given in Fig. 28, a partial door schedule taken from the working drawings for the municipal building whose sectional elevation is shown in Fig. 22.

| DOOR SCHEDULE | | | | | | | |
|---|---|---|---|---|---|---|---|
| | W. | H. | T. | TYPE | MATERIAL | GLASS | REMARKS |
| A | 3'-0" | 8'-4" | Std. | Solid | Steel | — | "Vulcan" Stor. Rm. Door |
| B | 3'-6" | 7'-6" | $1\frac{3}{4}$ | 10 Pan. | Kalamein | — | |
| C | 2'-6" | " | " | " | " | — | |
| D | 3'-6" | " | " | Glaz. | " | Cl. Wr. | |
| E | 2'-2" | 7'-0" | $1\frac{5}{8}$ | " | Hol. Met. | " | 2 Speed Sliding |
| F | 5'-4" | 8'-4" | Std. | Tin Clad | Steel | — | |
| G | 3'-6" | 7'-6" | $1\frac{3}{4}$ | Glaz. | Hol. Met. | Cl. Wr. | |
| H | 2'-8" | 6'-8" | " | 1 Pan. | Kalamein | — | Stock |
| I | 4'-0" | 9'-0" | $2\frac{1}{4}$ | 10 Pan. | Wh. Pine | Plate | 2 Sp. Bi-Part. Sl. |
| J | 2'-10" | 7'-0" | " | Glaz. | " | " | |
| K | 3'-0" | " | $1\frac{3}{4}$ | 10 Pan. | Wood | — | |
| L | " | " | " | Glaz. | " | Tap. | |

**FIG. 28.** A door schedule. Note that sizes and materials are given.

## 21. Checking.

As the architect develops the working drawings, he keeps a constant check for accuracy in his work. A tentatively final set of blueprints is then gone over by a responsible checker, who checks all correct items and marks, with red pencil, all mistakes. The drawings are then corrected, and another check set of blueprints is prepared.

All checking is done in a definite order with each item followed through separately and systematically. The order is dictated by the checker's preference or by conditions of the problem. The following is suggested as a guide:

1. Check over-all dimensions on the plans, seeing that all plans agree.
2. Check structural dimensions on the plans, seeing that column centers line up through the several stories.
3. Check location dimensions on the plans, seeing that openings line up vertically and that axes of communication are maintained by proper horizontal positioning of the openings.
4. See that detail drawings are dimensioned to correspond to the dimensions of the plan and will allow for proper fitting with adjacent features.
5. Check all doors for proper size and swing.
6. Check design and notation for all structural members.
7. Check sizes and locations of all ducts and flues.
8. Check location and notation of electrical outlets and switching for same.
9. See that several items are not indicated for the same location. For example, wiring to a switch should not share wall space with a heat duct.
10. See that all notes are complete and accurate.
11. Check the specifications with the drawings, making sure that no discrepancy exists and that the specifications are complete in every detail.
12. Check for conformity with building codes and laws.

# STRUCTURAL DRAWINGS

## 22. Introduction

The term "structural drawings" means the drawings of steel, masonry, wood, concrete, etc., for structures such as buildings, bridges, and dams. Structural drawings differ from other drawings only in certain details and practices that have developed because of the materials worked with and the method of their fabrication. The differences are so well established that it is essential for every engineer to know something of the methods of representation used in structural work.

## 23. Classification of Structural Drawings.

Structural drawings are classified as follows:[2]

(1) *General plan.* This will include a profile of the ground; location of the structure; elevations of ruling points in the structure; clearances; grades; direction of flow, high water, and low water (for a bridge); and all other data necessary for designing the substructure and superstructure.

(2) *Stress diagram.* This will give the main dimensions of the structure, the loading, stresses in all members for the dead loads, live loads, wind loads, etc., itemized

[2]From Milo S. Ketchum, "Structural Engineers' Handbook," McGraw-Hill, 1924.

separately; total maximum stresses and minimum stresses; sizes of members; typical sections of all built members showing arrangement of material; and all information necessary for detailing the various parts of the structure.

(3) *Shop drawings.* Shop detail drawings should be made of all steel and ironwork, and detail drawings should be made of all timber, masonry, and concrete work.

(4) *Foundation or masonry plan.* The foundation or masonry plan should contain detail drawings of all foundations, walls, piers, etc., that support the structure. The plans should show the loads on the foundations, the depth of footings, the spacing of piles where used, the proportions for the concrete, the quality of masonry and mortar, the allowable bearing on the soil, and all data necessary for accurately locating and constructing the foundations.

(5) *Erection diagram.* The erection diagram should show the relative location of every part of the structure, shipping marks for the various members, all main dimensions, number of pieces in a member, packing of pins, size and grip of pins, and any special feature or information that may assist the erector in the field. The approximate weight of heavy pieces will materially assist the erector in designing his falsework and derricks.

(6) *Falsework plans.* For ordinary structures it is not common to prepare falsework plans in the office, this important detail being left to the erector in the field. For difficult or important work, erection plans should be worked out in the office and should show in detail all members and connections of the falsework and also give instructions for the successive steps in carrying out the work. Falsework plans are especially important for concrete and masonry arches and other concrete structures and for forms for all walls, piers, etc. Detail plans of travelers, derricks, etc., should also be furnished the erector.

(7) *Bills of material.* Complete bills of material showing the different parts of the structure with their marks and shipping weights should be prepared. This is necessary to permit checking of shipping weights and shipment and arrival of materials.

(8) *Rivet list.* The rivet list should show the dimensions and number of all field rivets, field bolts, spikes, etc., used in the erection of the structure.

(9) *List of drawings.* A list should be made showing the contents of all drawings belonging to the structure.

## 24. Structural-steel Shapes.

Steel structures are made up of "rolled shapes" fastened together with rivets, bolts, or welds. The function of structural-steel drawings is to show how these shapes are to be fabricated in the shop and then erected to form the various members of bridges, buildings, etc. Sections of the common shapes are shown in Fig. 29. Dimensions for detail-

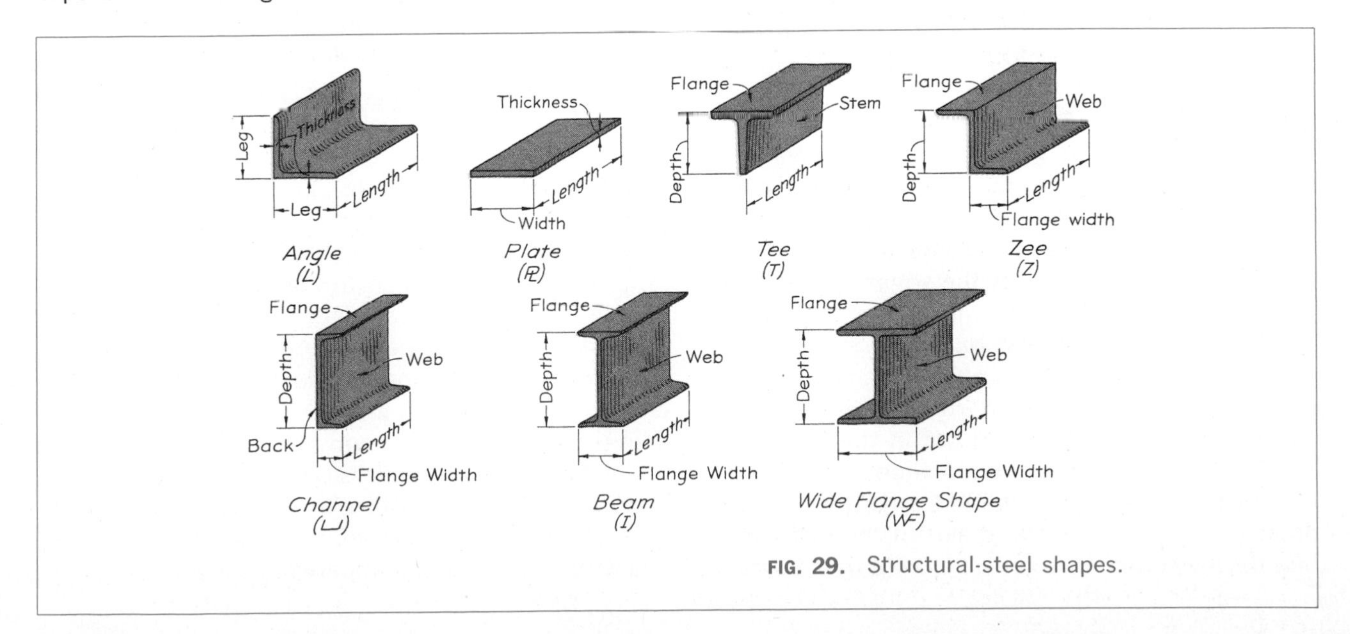


**FIG. 29.** Structural-steel shapes.

ing these and other less common shapes are shown in the American Institute of Steel Construction handbook. Dimensions for detailing common sizes of channels, I beams, wide flange sections, and angles are given in the Appendix.

In designating rolled-steel shapes on drawings, the AISC standard abbreviations and order of specifications are used, as follows:

*Plates.* Pl 18 × ½ × 4′-2 (width × thickness × length)

*Equal angles.* L3 × 3 × ¼ × 7′-6 (size of legs × thickness × length)

*Unequal angles.* L6 × 4 × 5/16 × 10′-0 (size of long leg × size of short leg × thickness × length)

*Standard channels.* 9⊔13.4 × 12′-4 (depth of section × weight per foot × length)

*Standard I beams.* 12 I 31.8 × 14′-6 (depth of section × weight per foot × length)

*Wide flange sections.* 14 WF 48 × 16′-3 (nominal depth of section × weight per foot × length)

*Tees.* T5 × 3 × 11.5 × 8′-0 (width of flange × depth × weight per foot × length)

*Structural tees.* Cut from WF section: ST 5 10.5 × 4′-7 (depth × weight per foot × length)

Cut from I beam: ST 5 I 12.5 × 5′-4 (depth × weight per foot × length)

*Zees.* Z6 × 3½ × 15.7 × 9′-10 (depth × width of flange × weight per foot × length)

## 25. Drawings of Steel Structures.

The general drawings include the general plan, the stress diagram, and the erection diagram and correspond in many respects to the design drawings of the mechanical engineer. In some cases the design is worked out completely by the engineer, who gives the sizes and weights of members and the number and spacing of all rivets; but in most cases the general dimensions, positions, and sizes of the members and the number of rivets are shown, leaving the details to be worked out in the shop or to be given on separate complete shop drawings.

In order to show the details clearly, the structural draftsman often uses two scales in the same view, one for the center lines or skeleton of the structure, showing the shape, and a larger one for the parts composing it. The scale used for the skeleton is determined by the size of the structure as compared with the sheet; ¼″, ⅜″, and ½″ to 1′-0″ are commonly used. Figure 30 is a typical shop drawing of a small roof truss, giving complete details. Such drawings are made about the stress-diagram lines (used in calculating the stresses and sizes of the members), which are then employed as the gravity lines of the members and form the skeleton, as illustrated separately to small scale in the box on the figure. The intersections of these lines are called "working points" and are the points from which all distances are figured. The length of each working line is computed accurately, and from it the intermediate dimensions are obtained.

The design diagram is often put on the same sheet as the shop drawing, as in the drawing of the truss.

## 26. Detailing Practice for Structural Steel.

Separate drawings made to a sufficiently large scale to carry complete information are called "shop detail drawings" (Fig. 31). When possible, the drawings of all members are shown in the same relative position that they will occupy in the completed structure: vertical, horizontal, or inclined. Long vertical or inclined members may be drawn in a horizontal position, a vertical member always having its lower end at the left, and an inclined member drawn in the direction it would fall. Except in plain building work, a diagram to small scale (showing by a heavy line the relative position of the member in the structure) should be drawn on every detail sheet.

In steel construction a member is composed of either a single rolled shape or a combination of two or more rolled shapes. Figure 31 is a shop detail drawing of a member made from a single rolled shape; Figs. 30, 32, and 33 are shop detail drawings of members made up of several rolled shapes. The figures illustrate detailing practice. Note in Fig. 30 that only one-half of the truss is shown. When thus drawn it is always the left end, looking toward the side on which the principal connections are made.

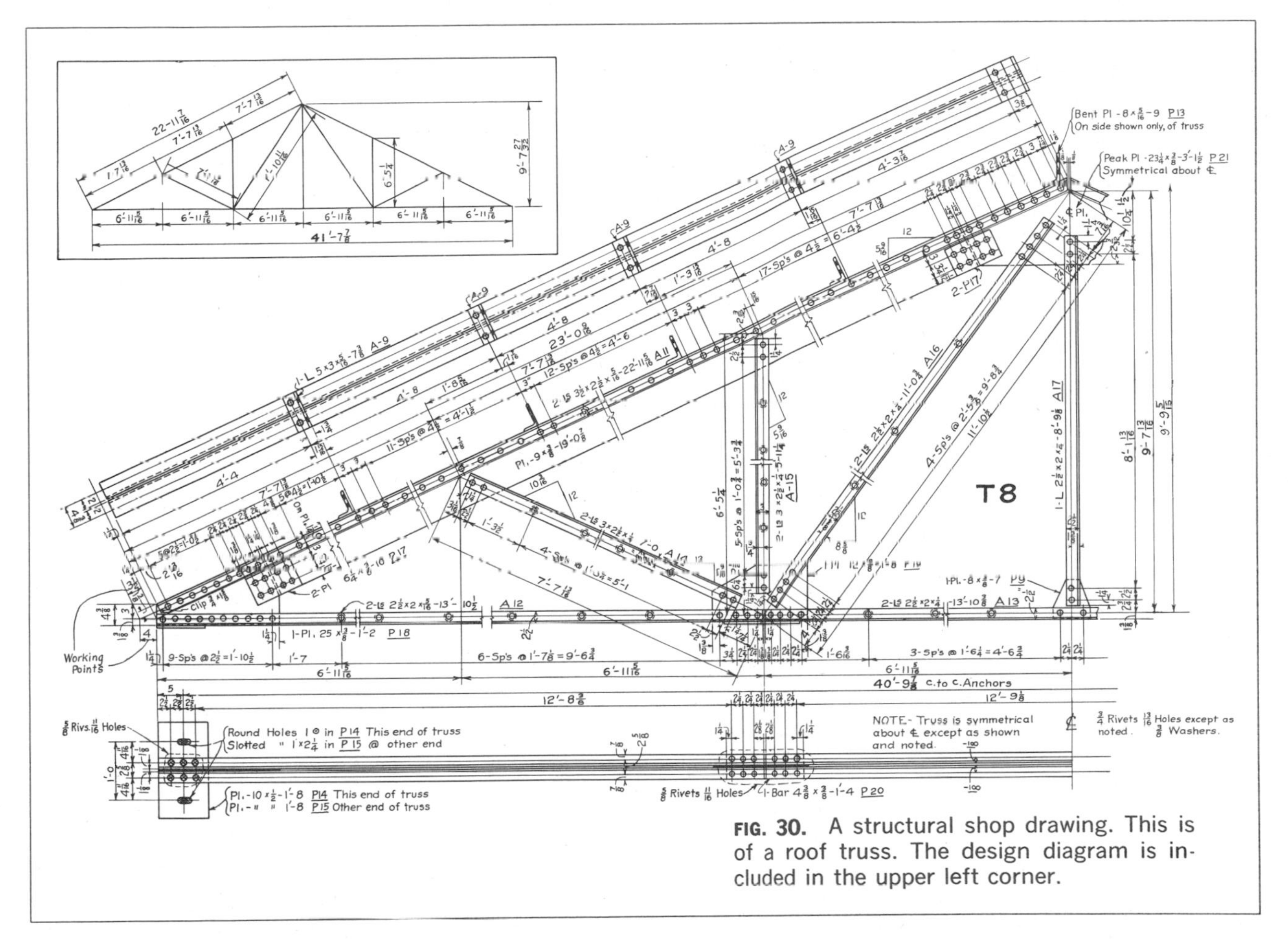


**FIG. 30.** A structural shop drawing. This is of a roof truss. The design diagram is included in the upper left corner.

The scale of shop drawings ranges from $\frac{1}{4}'' = 1'\text{-}0$ to $1'' = 1'\text{-}0$. Often, for long members, the cross section is drawn to a larger scale than the length. Sometimes it is even advantageous to pay no attention to scaled length but to draw the member as though there were breaks in the length (but not shown on the drawing, as in Fig. 33) so that rivet spacings at the ends (and intermediately) can be drawn to the same scale as the cross section.

*Dimensions* are always placed above the dimension line, and the dimension lines are not broken but continuous. Length dimensions are expressed in feet and inches. All inch symbols are omitted unless there is the possibility of misunderstanding, thus 1 bolt should be 1″ bolt to distinguish between size and number. Inch symbols are omitted even though the dimensions are in feet and inches, and dimensions should be hyphenated thus: 7′-0, 7′-0$\frac{1}{2}$, 7′-4. Plate widths and section depths of rolled shapes are given in inches. Dimensions are given to commercial sizes of materials.

*Rolled shapes* are specified by abbreviated notes, as described in paragraph 24. The specification is given either with the length dimension as in Fig. 31 or near and parallel to the shape as in Fig. 32.

*Erection marks* are necessary in order to identify the member. These are indicated on the drawing by letters and numbers in the subtitle. The erection diagram then carries these marking numbers and identifies the position of the member. Problem 17

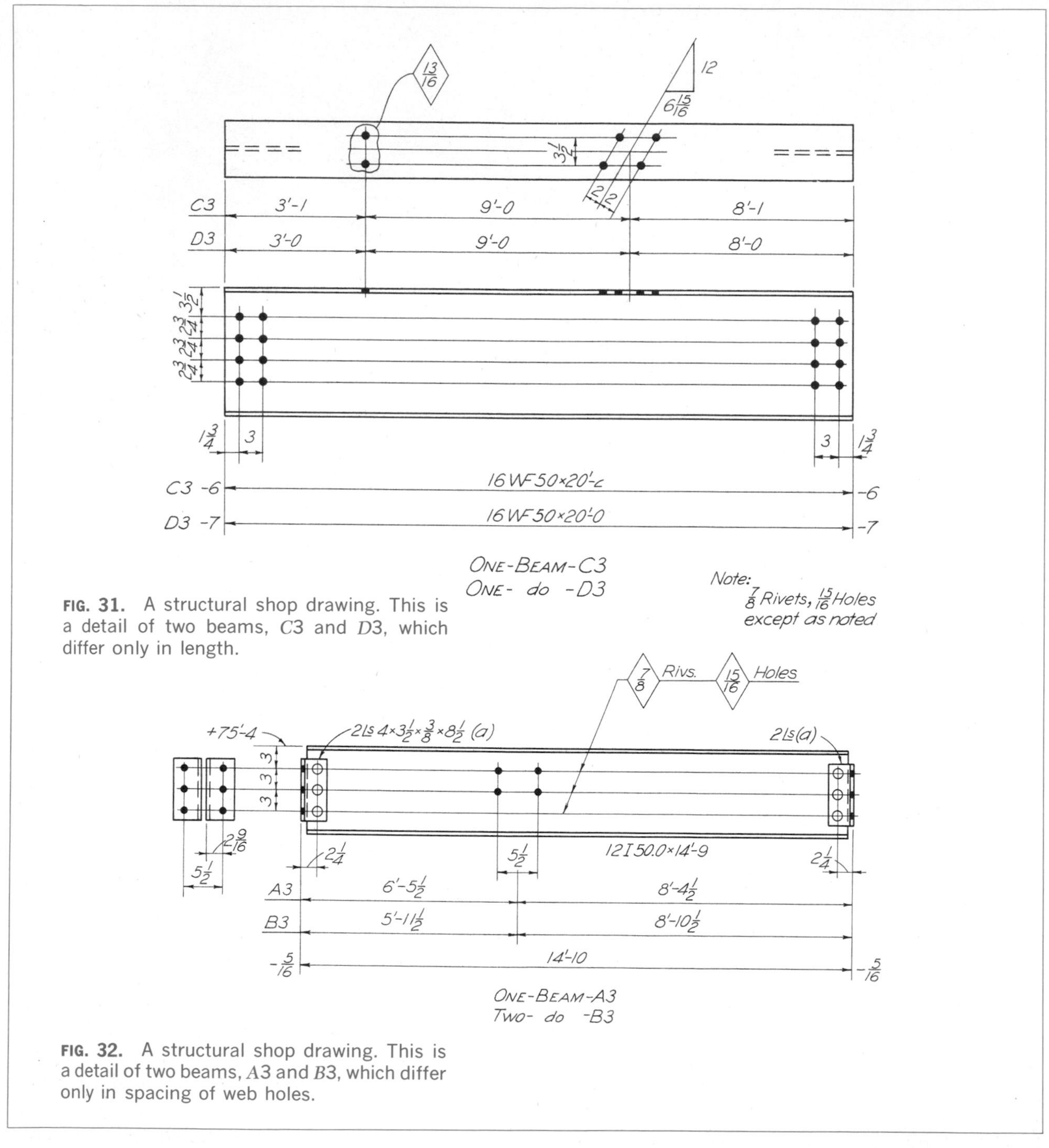


**FIG. 31.** A structural shop drawing. This is a detail of two beams, *C*3 and *D*3, which differ only in length.

**FIG. 32.** A structural shop drawing. This is a detail of two beams, *A*3 and *B*3, which differ only in spacing of web holes.

shows an example of an erection diagram with marking numbers.

The erection mark also identifies the member in the shop and serves as a shipping mark. The mark

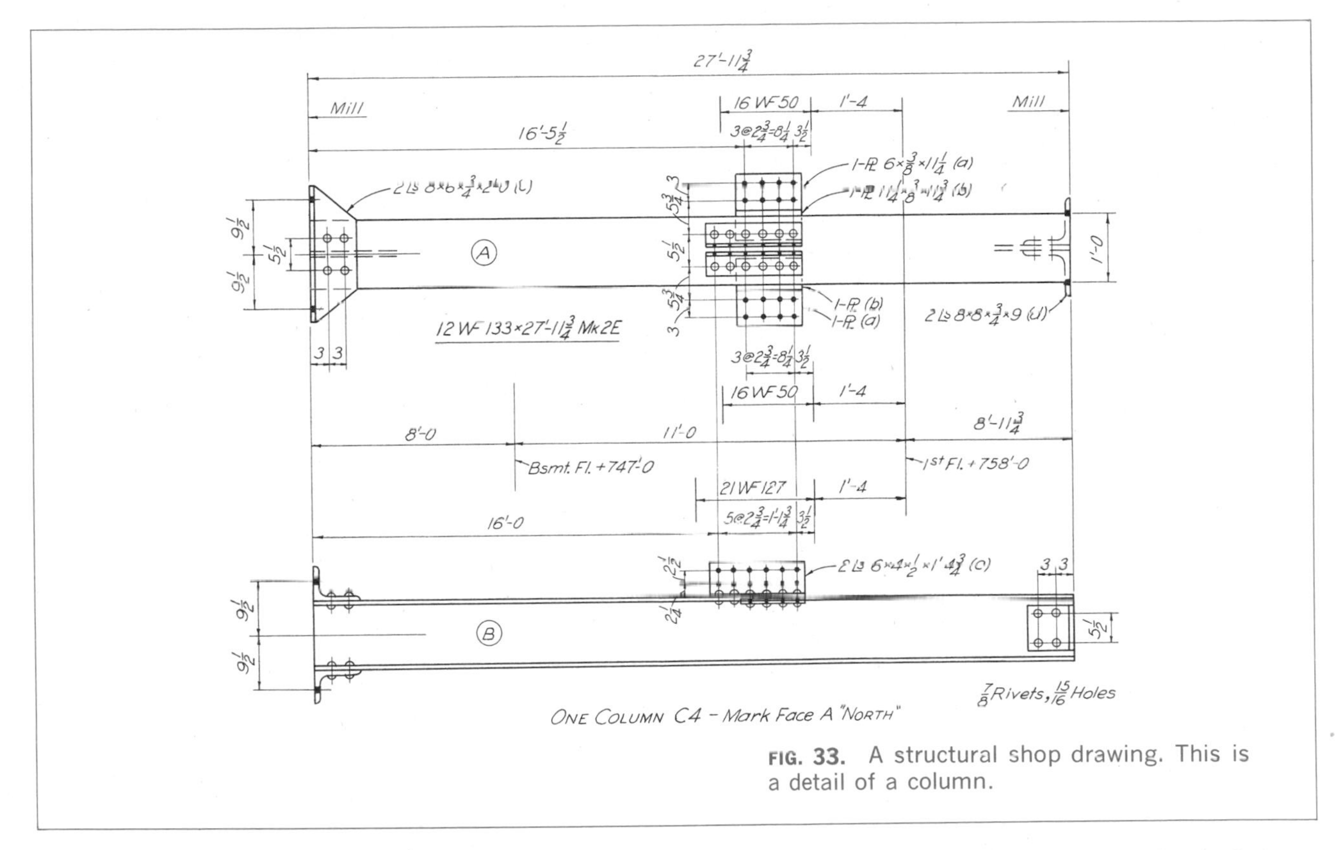


**FIG. 33.** A structural shop drawing. This is a detail of a column.

is composed of a letter and number. Capital letters are used, *B* for beam, *C* for column, *T* for truss, etc. The number gives the specific member in an assembly.

*Assembly marks* are used when the same shape is used in more than one place on a member. The member is completely specified *once* followed by the assembly mark (lower-case letters, to distinguish from erection marks). Then the complete specification is not repeated; only the assembly mark is given. As an example, see the angles "2 ∠ $4 \times 3\frac{1}{2} \times \frac{3}{8} \times 8\frac{1}{2}$(a)" in Fig. 32.

*Different members* may be detailed together on the same drawing when they differ only in length or spacing of rivets or holes, or when one member has special holes or an extra piece added. Figures 31 and 32 both show two beams detailed on the same drawing. Note that the different lengths are given with the erection mark at the end of the dimension line.

When one of two similar members has a hole or other feature on one member only, the special feature is indicated by drawing a freehand circle around the detail with a leader to a note such as, "on A-3 only."

*Distances to center lines* of another connecting member are given, as in Figs. 31 and 32, by placing the distance, preceded by a minus sign ( − ), at the end of the length dimension. These distances to center are a great aid in checking the details with an assembly or layout drawing.

*Inclinations of members* and inclined center lines, cuts, etc., are indicated by their tangents. As in Figs. 30 and 31, a small right triangle is drawn with its hypotenuse parallel to the line or on the line whose inclination is to be specified. The long leg of the triangle is always specified as 12.

*Rivets and holes* are dimensioned in the view that shows them as circles. Dimensions pertaining to a row of rivets are given in a single line as in Fig. 30.

Gage dimensions, even though standard, are always given. The size of rivets and of holes is given in a general note, as in Fig. 31.

The length of rivets is not usually given on the drawing; the workman picks a rivet length long enough to go through the members and protrude far enough so that the head can be formed.

Common practice is to make rivet holes $\frac{1}{16}$ in. larger than the rivets (see Figs. 31 to 33).

If some of the rivets or holes for a member are of a different size, these are indicated by the size in a diamond, as in Figs. 31 and 32.

*Rivet symbols* are used because it is impossible to show the details of head form on a small-scale drawing. The ASA symbols for indicating various kinds of heads are given in Fig. 34.

The size of most structures prevents their being completely fabricated in the shop. Therefore, portions as large as transportation facilities or ease of handling will allow are assembled and then connected to other portions in the field. Shop rivets are indicated by an open circle, the size of the rivet head. Holes for field rivets are indicated by "blacked-in" circles the diameter of the hole (see Figs. 31 to 34). Rivets are drawn to scale. Figure 35 shows the proportions of rivet heads with respect to the diameter.

*Rivet spacing* is an important item in detailing, and there are a number of conditions that control the placement. Rivets spaced too close together or too close to a projecting part cannot be properly driven. Rivets placed too close to the edge or end of a member may weaken the member.

Rivets are spaced along "gage" lines, parallel to the axis of the member and at certain "pitch" distances along the gage line (Fig. 36). The minimum distance between rivets in any direction is three rivet diameters. The minimum edge distance (*e*, Fig. 37), from sheared edges of a member is $1\frac{1}{2}$ rivet diameters. Distance to a rolled edge may be slightly less than $1\frac{1}{2}$ diameters and is usually controlled by the location of the gage line. Whenever possible the same edge distance is used on all members, and the distance is then given in a general note such as "edge distance $1\frac{1}{4}$ except as noted."

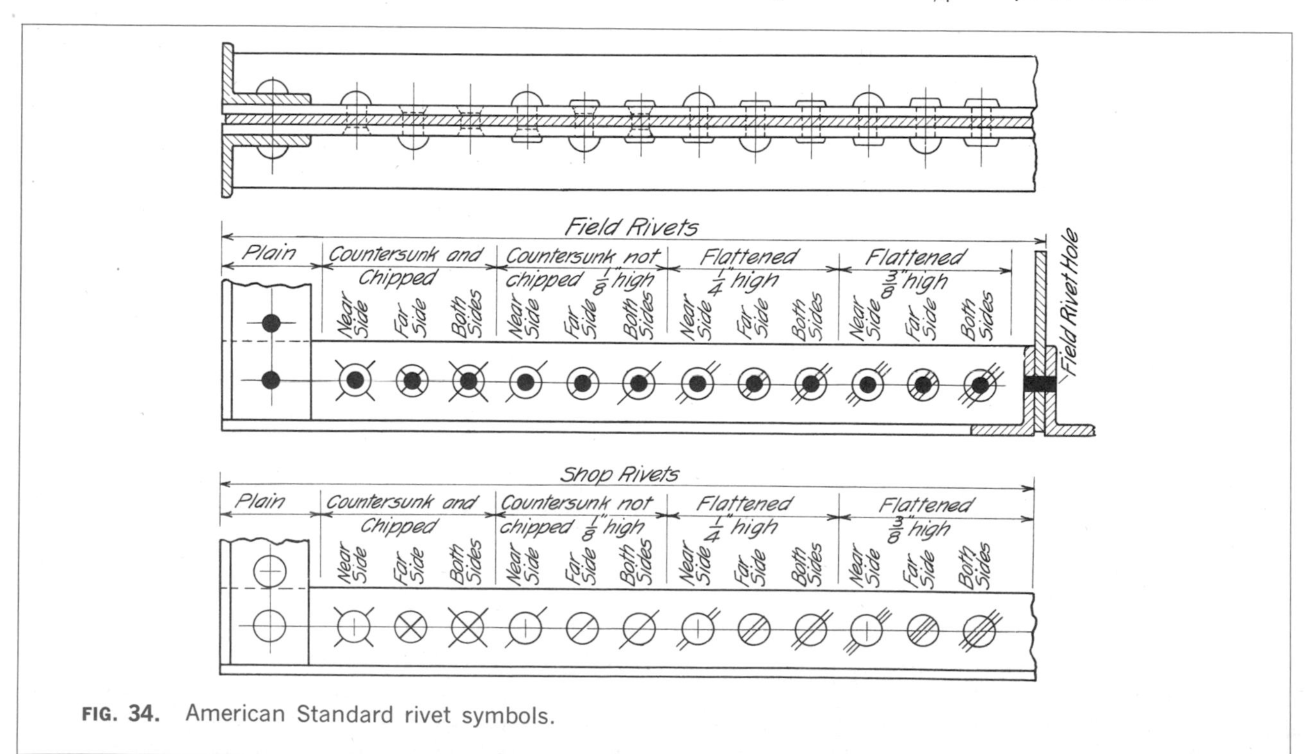


**FIG. 34.** American Standard rivet symbols.

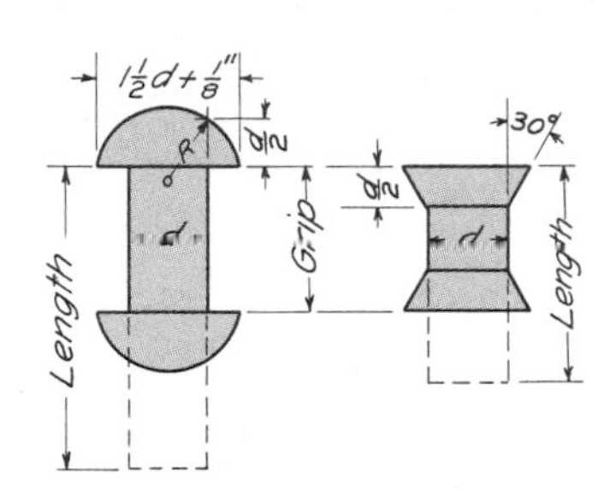


FIG. 35. Proportions of structural rivets.

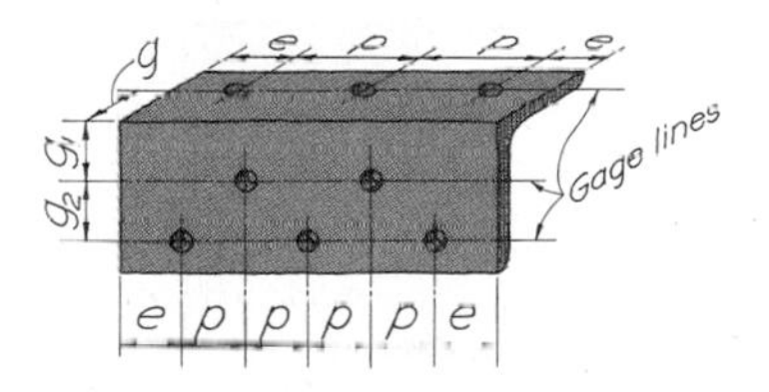


FIG. 36. Hole spacings. These are: gage ($g$), pitch ($p$) and edge ($e$) distances.

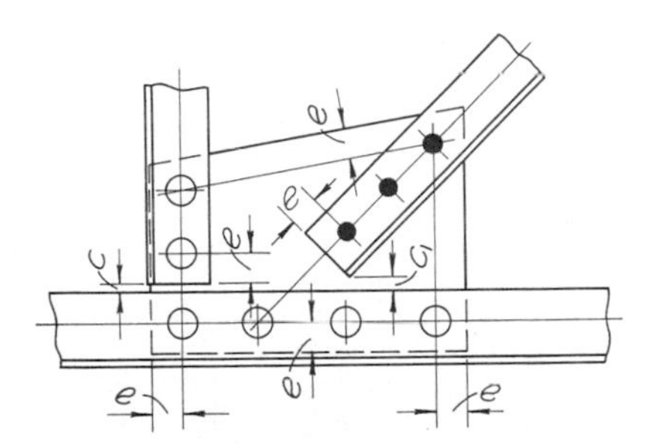

FIG. 37. Spacings on built-up members. These are edge ($e$) and clearance ($c$) distances.

Gage-line distances and minimum driving clearances for common rolled shapes are given in the Appendix.

Standard beam connections are also given in the Appendix.

*Clearances* between various members of a structure are necessary so that there will be no interferences, because of manufacturing inaccuracies. More clearance should be allowed for field erection than for shop fabrication. Field clearance ($C_1$, Fig. 37) should be approximately $\frac{1}{2}$ in., and shop clearance ($C$) should be about $\frac{1}{4}$ in. Note the clearances shown between the angles (fastened with a gusset plate) in Fig. 37.

*Elevations, sections,* and other views are placed in relation to each other by the rules of third-angle projection, except that when a view is given under a front view, as in Fig. 30, it is made as a section taken above the lower flange, looking down, instead of as a regular bottom view looking up. Large sections of materials are shown with uniform cross-hatching. Small-scale sections are blacked-in solid, with white spaces left between adjacent parts (see Sec. 11, Chap. 8).

*Bent plates* should be developed and the "stretch-out" length of bent forged bars given. The length of a bent plate may be taken as the inside length of the bend plus half the thickness of the plate for each bend.

A *bill of material* always accompanies a structural drawing. This can be put on the drawing, but the best practice is to attach it as a separate "bill sheet," generally on $8\frac{1}{2}$- by 11-in. paper.

*Lettering* is done in rapid single stroke, inclined or vertical.

*Checking* is indicated by a red-pencil check mark placed under the dimension, note, or specification.

*General notes* either accompany the general drawing or appear in the title of the shop drawing. Items included are painting, shipping instructions, etc.

## 27. Timber Structures.

The representation of timber-framed structures involves no new principles but requires particular attention to details. Timber members are generally rectangular in section and are specified to nominal sizes in even inches as 8″ × 12″. The general drawing must give center and other important distances accurately. Details drawn to larger scale give specific information as to separate parts. The particulars of joints, splices, methods of fastening, etc., are detailed.

Two scales are sometimes used to advantage on the general drawing, as is done in Fig. 38. Complete notes are an essential part of such drawings, especially when an attempt at dimensioning the smaller details results in confusion.

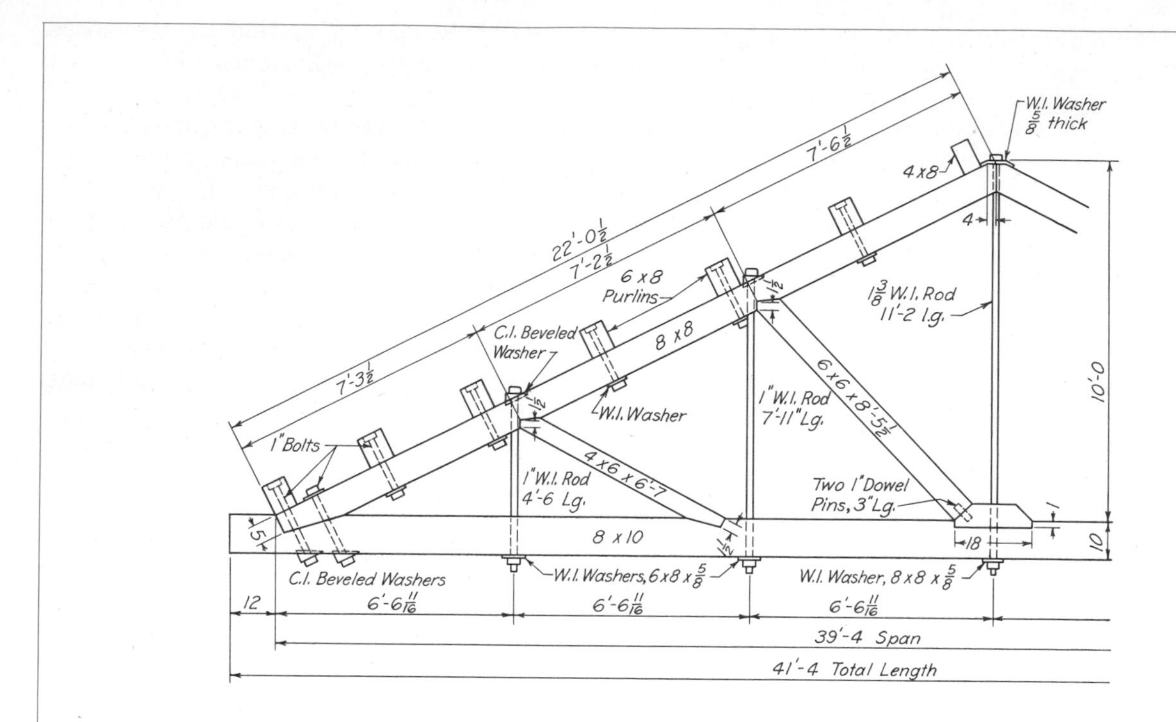


**FIG. 38.** A structural shop drawing. This is a detail of a timber truss.

Joints in timber structures are fastened with nails or spikes, wood screws, bolts, or ring-shaped or flat *connectors*, similar in action to a dowel or key. Some common types are shown in Fig. 39. A split ring, shown at (*a*), is assembled in grooves in each piece and held together with a bolt, as indicated at (*b*). The sharp projections on the alligator connector (*c*) are forced into the members by pressure. The claw-plate connector (*d*) is used either in pairs, back to back, for timber-to-timber connections; or single for timber to metal. A typical assembly is shown at (*e*). The Kubler wood dowel connector (*f*) fits into a bored hole in each timber face, and a bolt holds the parts together.

The Forest Products Laboratory publication, "Wood Handbook," gives basic information on wood as a material of construction, including the connectors, and is available at small cost from the Superintendent of Documents, Washington, D.C.

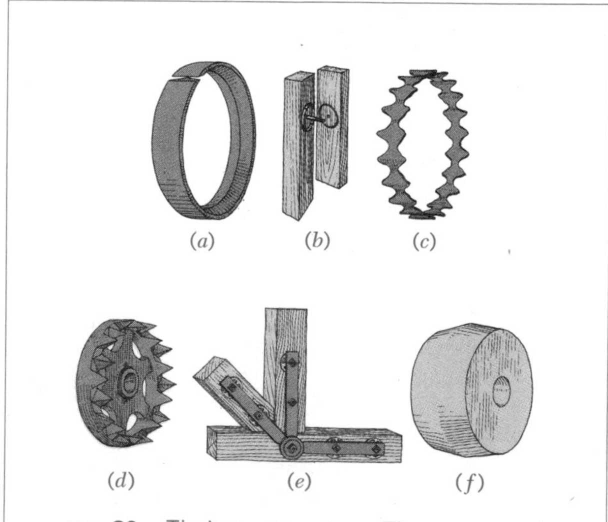


**FIG. 39.** Timber connectors. These are used for joints in wooden structures.

## 28. Masonry Structures.

In drawing masonry, the symbols used bear some resemblance to the material represented. Figure 40 gives those in common use and shows the stages followed to secure uniformity of effect in rendering earth and concrete. Drawings for piers, foundations for machines, and other masonry structures occur in all kinds of engineering work. Grade levels, floor levels, and other fixed heights are always given, together with accurate location dimensions for foundation bolts. All materials are marked plainly with name or notes.

## 29. Reinforced Concrete.

This is an important division of masonry construction and requires care in representation and specification. It is almost impossible to show the shape and location of reinforcing bars in concrete by the usual orthographic views without using a systematic scheme of conventional symbols and marking.

Figure 41 shows a portion of the general structural plan for the first floor of a building. By means of the plan, the tabular data, the general notes, and the specifications, the engineer has completely specified the reinforced concrete floor of the building. The location of the reinforced beams is given on the plan; the size of the beam and reinforcing to be used are indicated in the beam schedule; and the typical beam sections give the basic information for bending the reinforcing bars. The type of floor slab used is shown on the plan and in the slab schedule by the letters $A$, $B$, etc. The slab schedule, in conjunction with the typical slab plan, indicates the direction and spacing of reinforcing.

General notes on the drawing or in the specifications cover items such as maximum strength of concrete, grade of steel, and minimum cover.

The general drawing in Fig. 41, although completely specifying the reinforced concrete construction, does not give the exact details of how the bars are to be bent and placed. The contractor supplying the steel makes another drawing (Fig. 42), showing in detail the bending, spacing, etc., for all the steel.

Walls, columns, and other portions of reinforced structures are represented similarly to the beams and floor slabs illustrated.

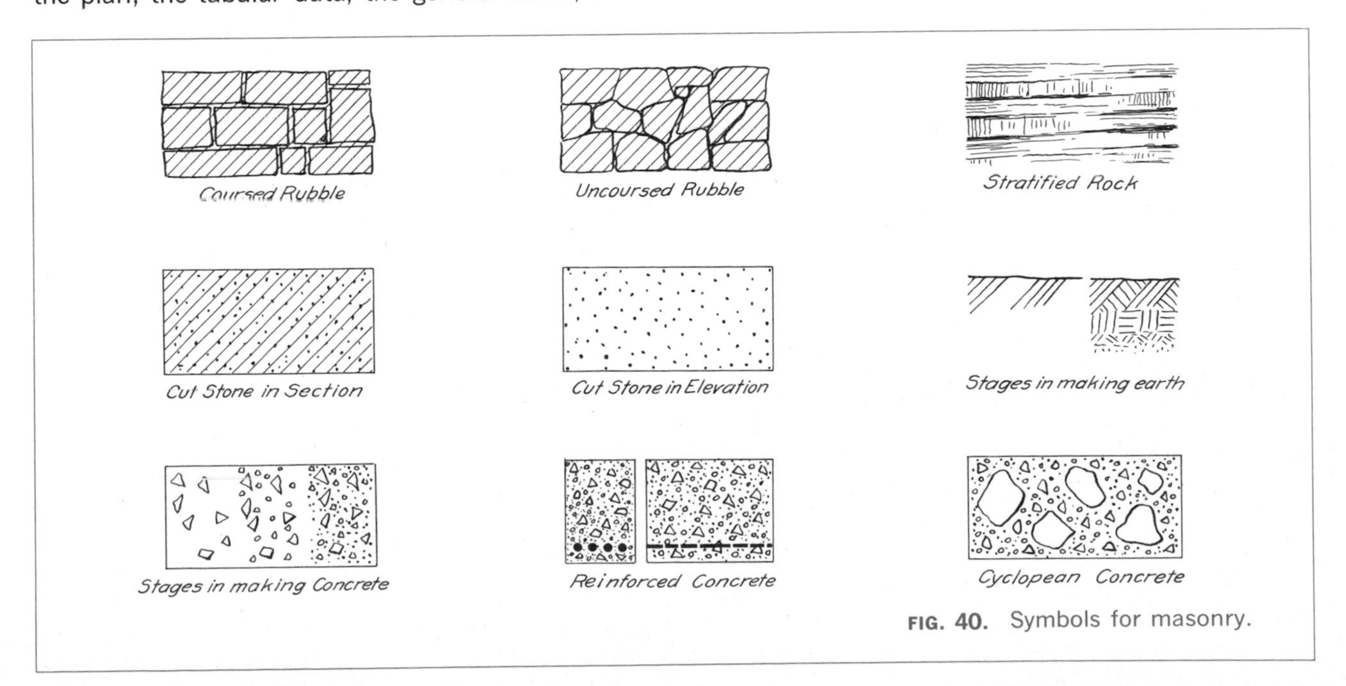


**FIG. 40.** Symbols for masonry.

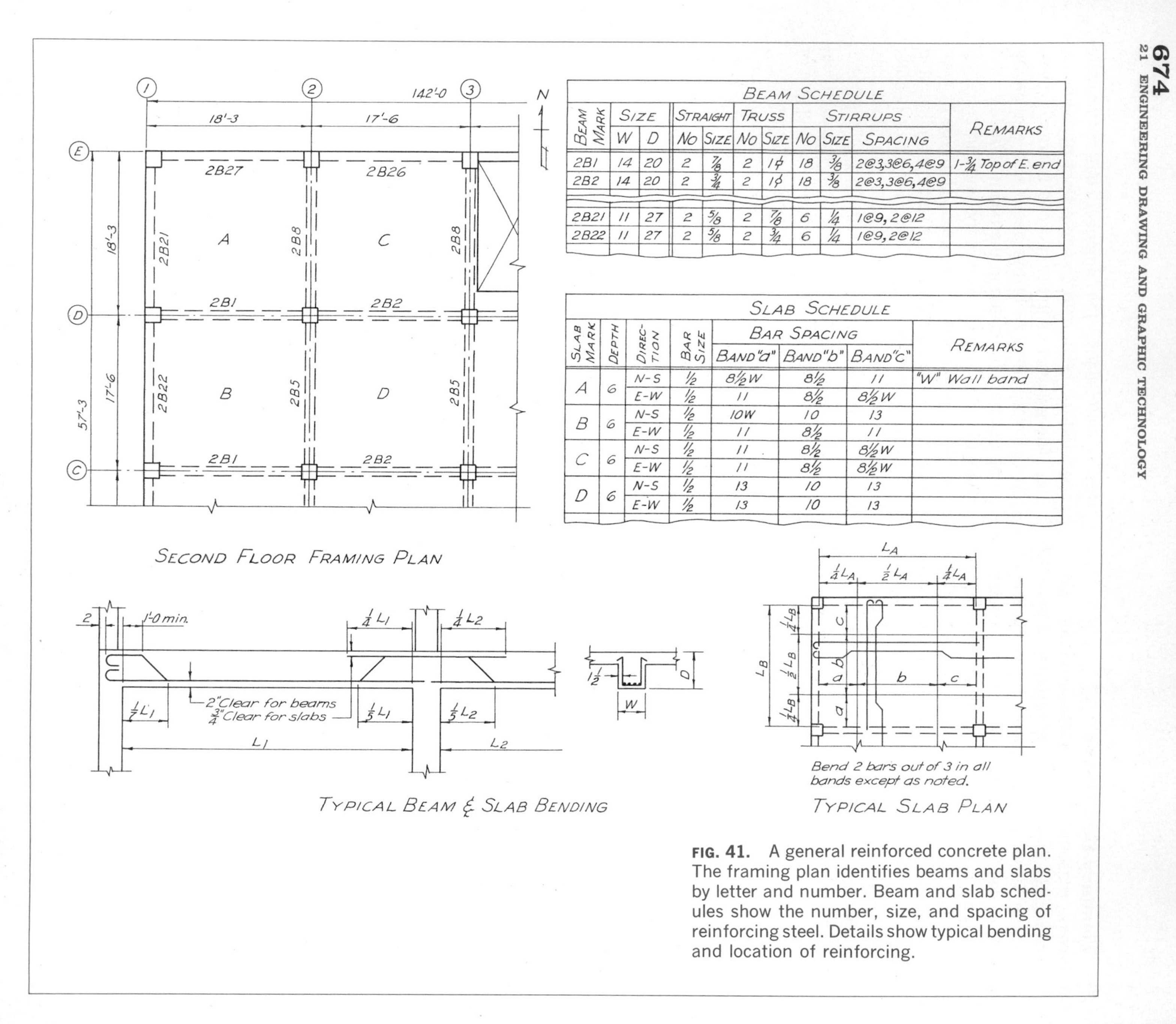


BEAM SCHEDULE

| Beam Mark | Size W | Size D | Straight No | Straight Size | Truss No | Truss Size | Stirrups No | Stirrups Size | Stirrups Spacing | Remarks |
|---|---|---|---|---|---|---|---|---|---|---|
| 2B1 | 14 | 20 | 2 | 7/8 | 2 | 1ϕ | 18 | 3/8 | 2@3,3@6,4@9 | 1-3/4 Top of E. end |
| 2B2 | 14 | 20 | 2 | 3/4 | 2 | 1ϕ | 18 | 3/8 | 2@3,3@6,4@9 | |
| 2B21 | 11 | 27 | 2 | 5/8 | 2 | 7/8 | 6 | 1/4 | 1@9, 2@12 | |
| 2B22 | 11 | 27 | 2 | 5/8 | 2 | 3/4 | 6 | 1/4 | 1@9, 2@12 | |

SLAB SCHEDULE

| Slab Mark | Depth | Direction | Bar Size | Bar Spacing Band "a" | Bar Spacing Band "b" | Bar Spacing Band "c" | Remarks |
|---|---|---|---|---|---|---|---|
| A | 6 | N-S | 1/2 | 8 1/2 W | 8 1/2 | 11 | "W" Wall band |
| | | E-W | 1/2 | 11 | 8 1/2 | 8 1/2 W | |
| B | 6 | N-S | 1/2 | 10W | 10 | 13 | |
| | | E-W | 1/2 | 11 | 8 1/2 | 11 | |
| C | 6 | N-S | 1/2 | 11 | 8 1/2 | 8 1/2 W | |
| | | E-W | 1/2 | 11 | 8 1/2 | 8 1/2 W | |
| D | 6 | N-S | 1/2 | 13 | 10 | 13 | |
| | | E-W | 1/2 | 13 | 10 | 13 | |

**FIG. 41.** A general reinforced concrete plan. The framing plan identifies beams and slabs by letter and number. Beam and slab schedules show the number, size, and spacing of reinforcing steel. Details show typical bending and location of reinforcing.

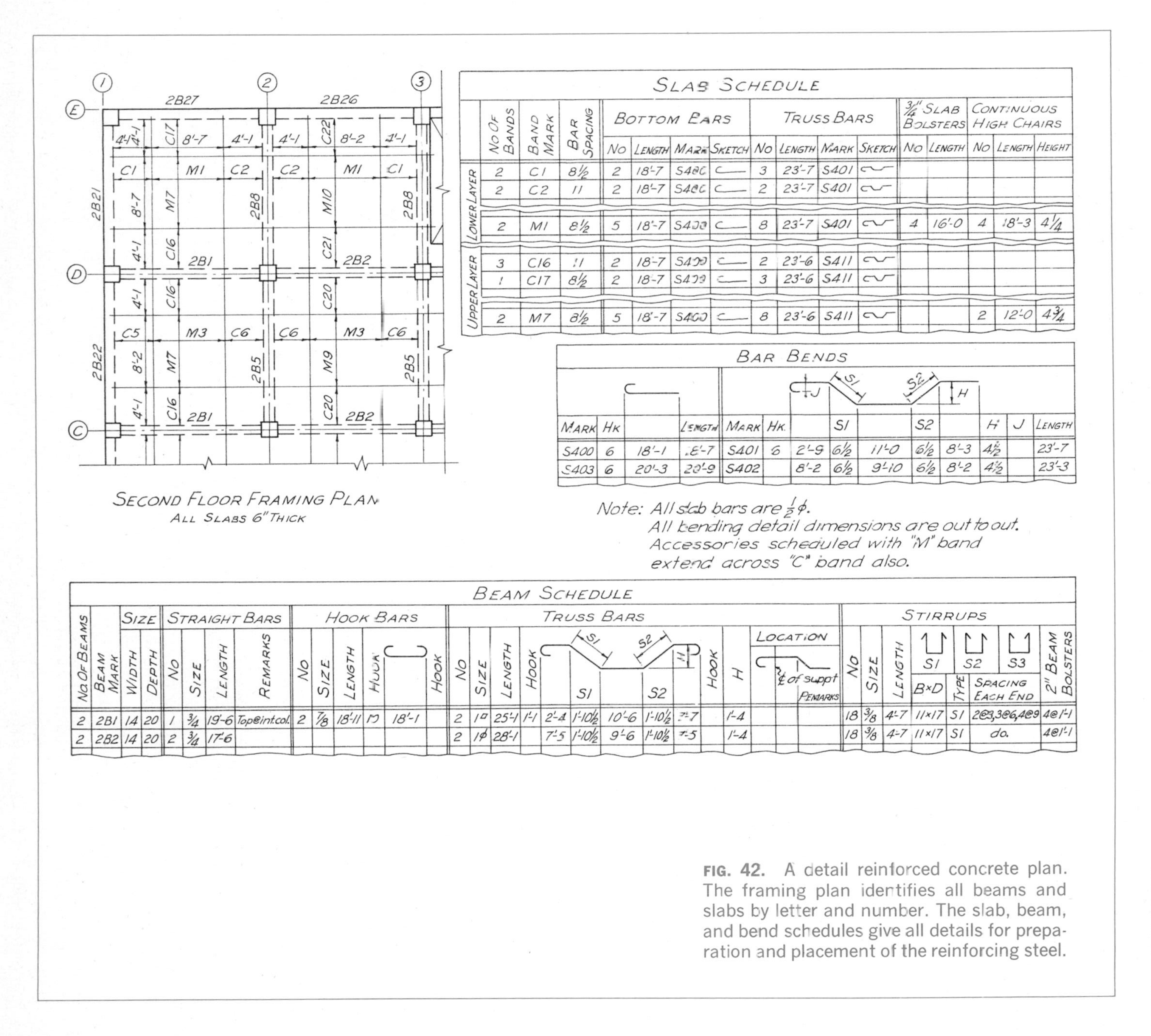


**FIG. 42.** A detail reinforced concrete plan. The framing plan identifies all beams and slabs by letter and number. The slab, beam, and bend schedules give all details for preparation and placement of the reinforcing steel.

# PROBLEMS

## Group 1. Architectural Drawings.

**1.** Prepare a preliminary sketch plan for a vacation cottage to meet the following requirements: kitchen-dining area (approximately 140 sq ft of floor space); living-recreation area (approximately 280 sq ft); two bunk rooms (approximately 70 sq ft each); bathroom; four closets (one for each bunk room, one for outer garments, and one for linens); and a porch. Scale: $\frac{1}{16}$″ = 1′-0″. It is recommended that the plan be developed on $\frac{1}{4}$-in. coordinate paper using sketch technique.

**2.** Prepare a preliminary sketch plan for a gasoline service station to be located on a corner lot (100 × 175 ft). Scale: $\frac{1}{16}$″ = 1′-0″. Requirements: office-salesroom; one lubrication rack; one wash rack; work space for battery charging, tire repair, etc.; two rest rooms (women's rest room to be entered from the outside); two air stations; and two gasoline islands, each with three pumps.

**3.** Prepare presentation perspective drawing of house in Figs. 12 to 18. Scale $\frac{1}{4}$″ = 1′-0″. Picture plane to make an angle of 30° with front wall and to contain left front corner of house. Perspective to show front and left-side walls. Horizon: 5′-6″ above finished grade.

**4.** Make a paper model of the house shown. Scale of model: $\frac{1}{8}$″ = 1′-0″. Mount on cardboard base.

Optional requirement: Draw windows and doors on all walls, after the instructor has approved preliminary first- and second-floor plans.

**5.** Prepare working-drawing details for fireplace with a 40 in. wide by 29 in. high opening. Scale of drawings: $\frac{3}{4}$″ = 1′-0″. Refer to fireplace-equipment manufacturers' catalogues and architectural handbooks for data.

**6.** Make an isometric drawing of the piping for the house shown in Figs. 12 to 18. Scale: $\frac{1}{2}$″ = 1′-0″. Determine pipe sizes from published laws governing plumbing and drainage and from architectural handbooks; note pipe sizes on the isometric. (See Chap. 19, Piping.)

**7.** Draw a completed plot plan showing the house of scheme *A*, Fig. 3, positioned on the lot shown in figure for Prob. 7. Scale $\frac{1}{8}$″ = 1′-0″.

**8.** Complete the working drawings for the garage shown.

**9.** Prepare the working drawings for the weekend cottage shown. Scale of elevation and plan: $\frac{1}{4}$″ = 1′-0″. Scale of wall section and fireplace details: $\frac{3}{4}$″ = 1′-0″. Scale of cornice, door, and window details: $1\frac{1}{2}$″ = 1′-0″.

**10.** (*a*) Prepare preliminary plan and elevations for a shelter with a floor area approximately 24 by 40 ft. One end of the shelter is to have a large fireplace. Materials

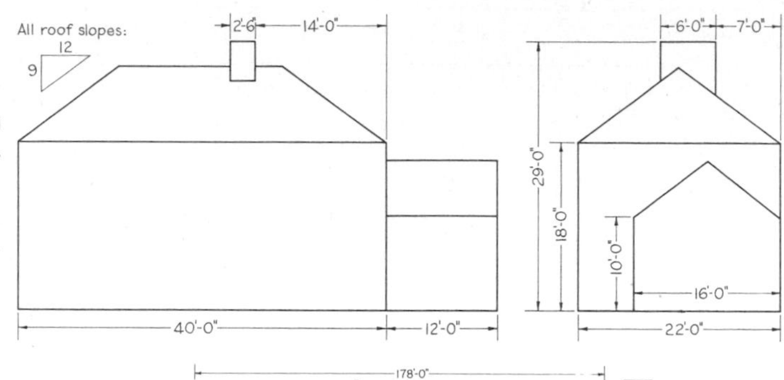


**PROB. 4.** House, in block form.

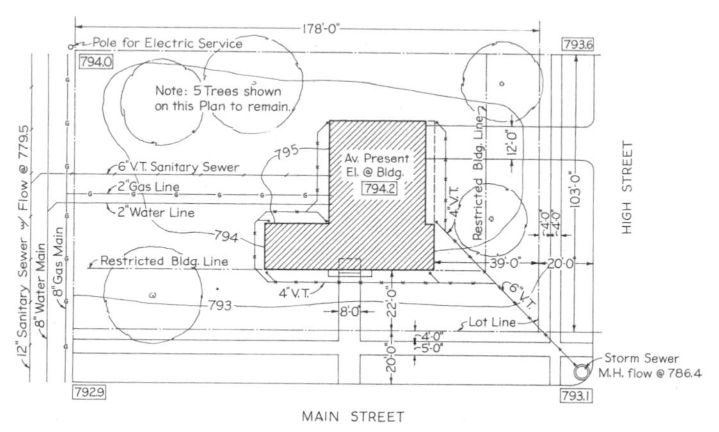


**PROB. 7.** Building lot.

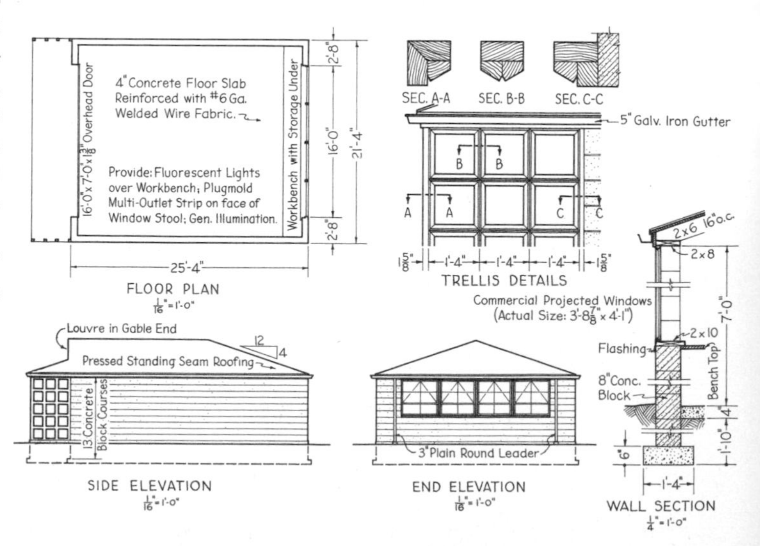


**PROB. 8.** Garage: a partial working drawing.

of construction: stone and heavy timbers. (*b*) After preliminary drawings have been approved by the instructor, prepare the working drawings for the shelter.

**11.** Complete the set of working drawings for the house in Figs. 12 to 18. Required views are listed in Sec. 11.

## Group 2. Structural Drawings.

**12.** Make shop drawing of item shown. Assume complete shop fabrication. Use ⅝-in. rivets.

**12*A*.** Make shop drawing of item shown. Assume shop fabrication of all parts except long channels and 3- × 3- × ¼-in. angle brace; use ⅝-in. shop rivets, ⅝-in. field bolts.

**12*B*.** Redesign Prob. 12*A* for welded construction.

**13.** Make a shop drawing of the column base. Assume

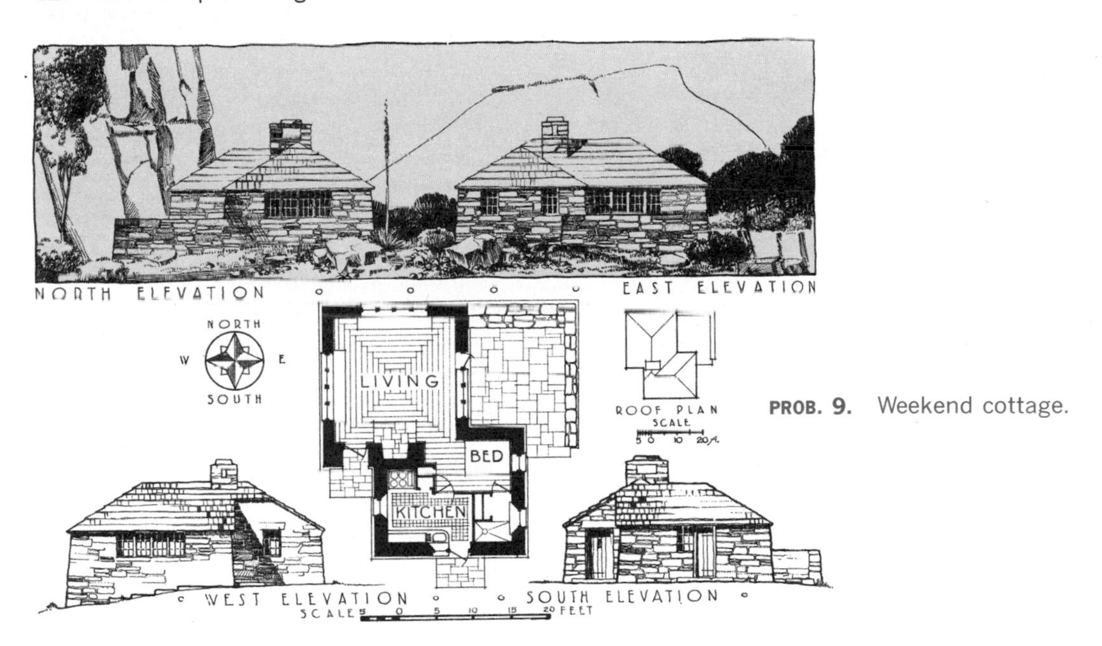


**PROB. 9.** Weekend cottage.

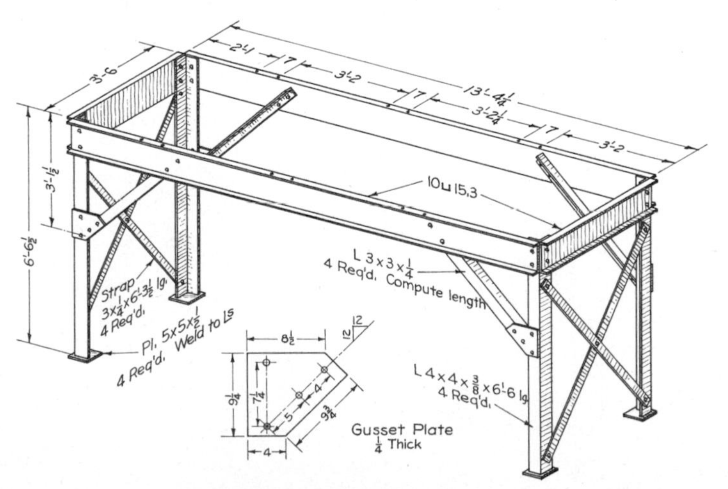


**PROB. 12.** Triple-effect evaporator support.

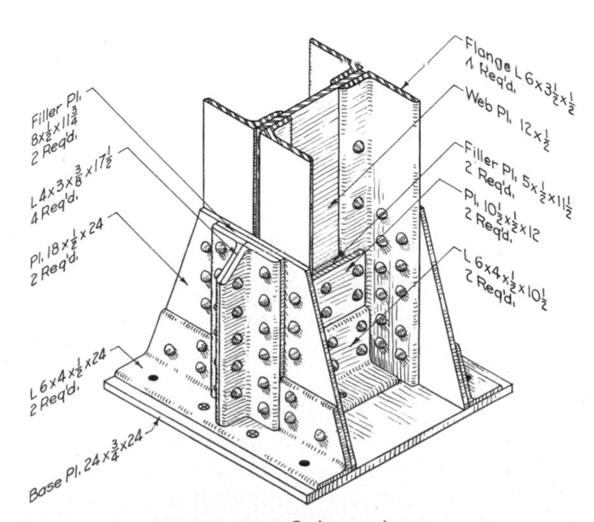


**PROB. 13.** Column base.

complete assembly in the shop. Use 3/4-in. rivets. Open holes are for 1-in. anchor bolts.

**13A.** Make a bill of material of parts used in Prob. 15.

**14.** Make a shop drawing for the crane trolley. Assume fabrication as indicated in the figure. One end is shown; the other end is similar. Use 5/8-in. rivets wherever possible.

**14A.** Make a bill of material of parts used in Prob. 14.

**15.** Redesign Fig. 38, using (wherever possible) fasteners shown in Fig. 39.

**16.** Make a shop drawing for the motor support. Assume shop fabrication as follows: (*a*) 8-in. channels, 4- by 4- by 1/2-in. angles, 1/2-in. plate, and 6- by 4- by 3/8-in. beam connectors (angles), two members required, one right hand, one left hand; (*b*) 21 by 3/4 by 2′-4 plate; (*c*) 12-in. I beam, two required with 4- by 4- by 3/8-in. beam connectors temporarily bolted in place. Assume columns to be part of the existing structure. Use 3/4-in. rivets and 3/4-in. bolts.

**16A.** Make an erection drawing for the motor support. Assume columns to be part of the existing structure.

**17.** Make a shop drawing of roof truss *T*1 for the steel-frame mill building. This drawing is to be similar to the drawing shown in Fig. 30. For truss *T*1, use the same size members and the same roof pitch as is used for truss *T*8, Fig. 30. Detail only the left half of the truss to a scale suitable for 11- by 17-in. paper.

**17A.** Prepare shop drawings for the following steel members used in fabricating the steel mill building: (*a*) bracket *M*1; (*b*) beams *B*1 and *B*3 (one detail for both); (*c*) beam *B*5 (assume 3/8-in. plate to be shop fabricated to column); and (*d*) beam *B*6. Assume shop and field fabrication indicated. Use 3/4-in. rivets. These details, along with the shipping list and general notes, may be

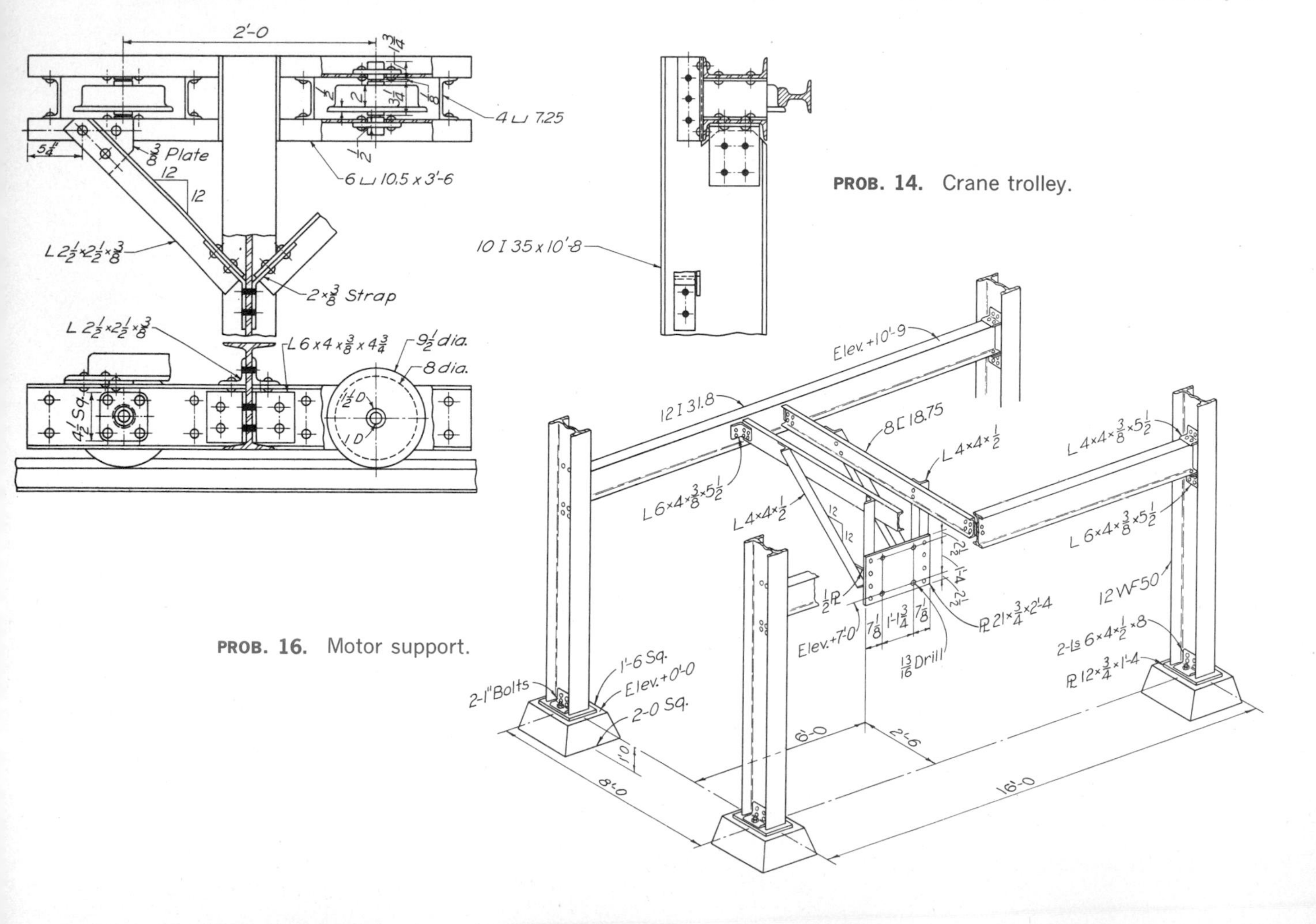


**PROB. 14.** Crane trolley.

**PROB. 16.** Motor support.

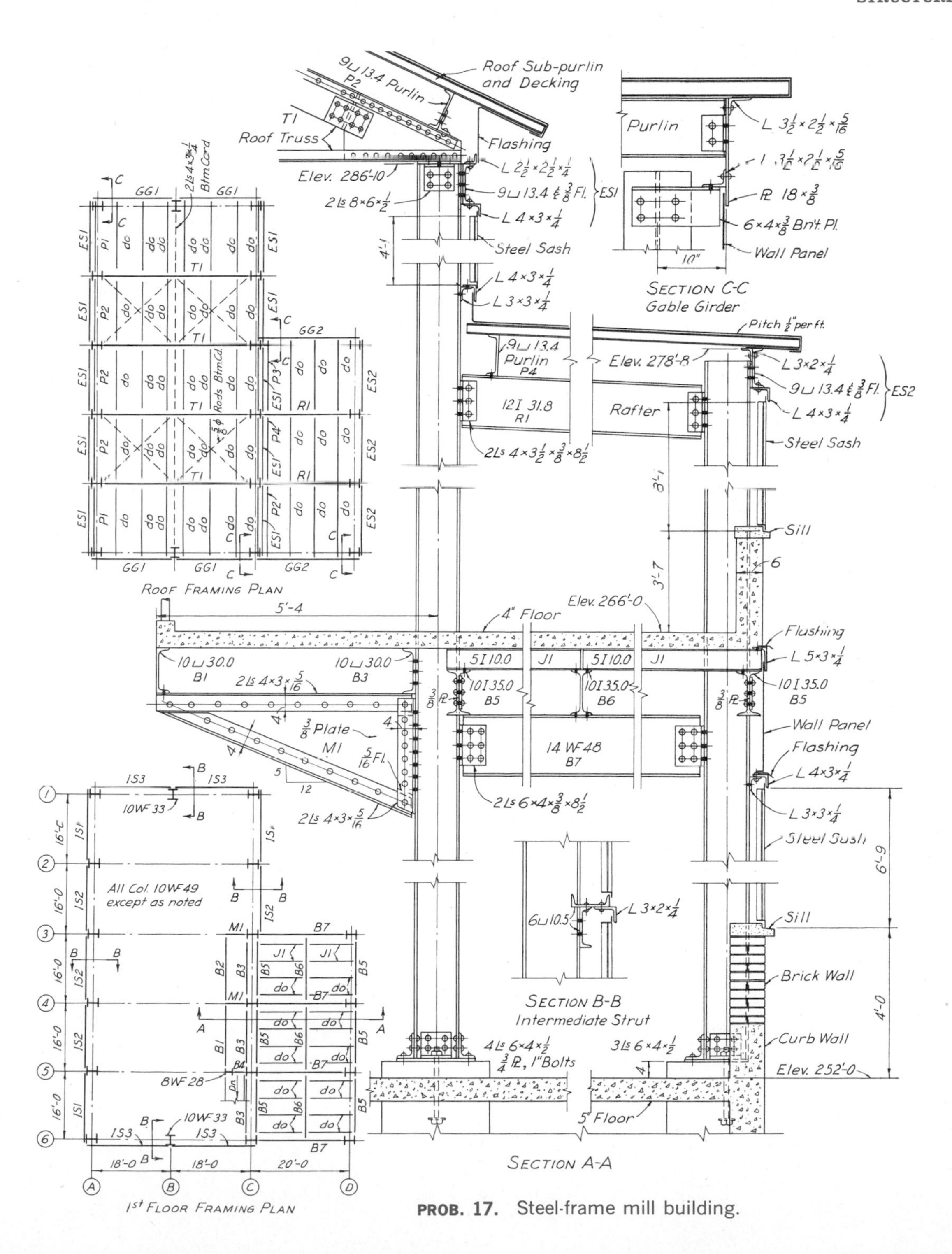


**PROB. 17.** Steel-frame mill building.

drawn on 11- by 17-in. paper to a scale of 1″ = 1′-0″.

**17*B*.** Prepare shop drawings for the following steel members used in fabricating the steel mill building: (*a*) bracket *M*1, and (*b*) column *C*4. Assume shop and field fabrication indicated. Use ¾-in. rivets. Three faces of the column will have to be shown.

**17*C*.** Make a shop drawing of column *D*4. Assume shop and field fabrication as indicated. Use ¾-in. rivets.

**17*D*.** Make a shop drawing of beam *B*7.

**17*E*.** Make a shop drawing of rafter *R*1.

**17*F*.** Make a general structural plan, similar to Fig. 41, of the concrete floor at elevation 266′-0. Use ½-in. round reinforcing bars spaced 9 in. on centers, alternate bars are bent for the floor. In 6-in. wall, use 3½-in. round bars in bottom and ⅜-in. round vertical bars at 18 in. hooked 18 in. into floor slab.

**17*G*.** Make a reinforcing detail drawing, similar to Fig. 42, for Prob. 17*F*.

**18.** Make a detailed shop drawing of the structural members in a section of the pipe bridge. Do not include the bents.

**18*A*.** Make a detailed drawing of the base section of

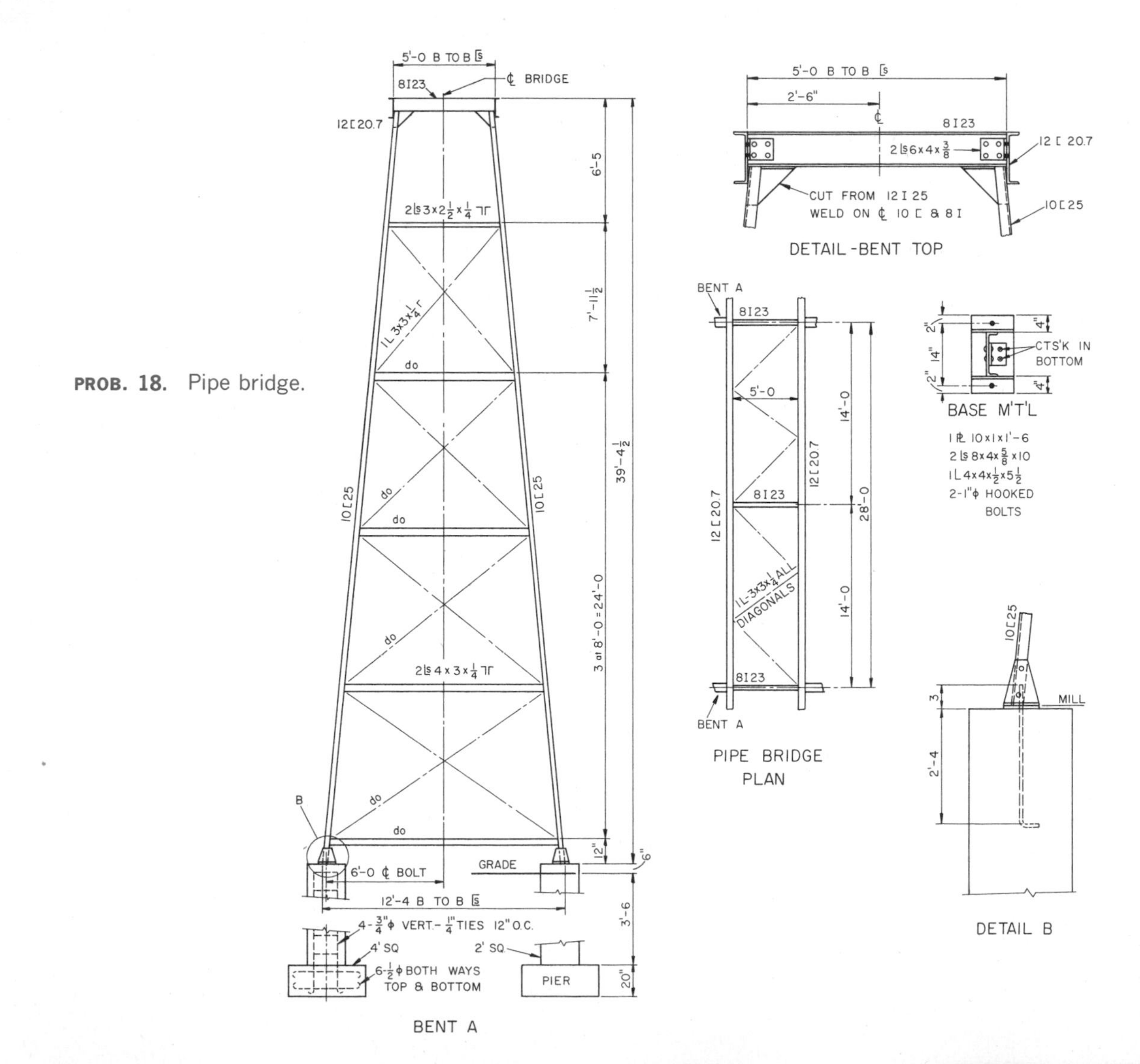


**PROB. 18.** Pipe bridge.

bent *A* including the two bottom diagonal braces.

**18*B*.** Make a detailed shop drawing of the entire bent *A*.

**18*C*.** Make a detailed drawing of the pier support for bent *A*.

**18*D*.** Make a detail of a typical connection to the 10 in. channel using the horizontal members as 2 ∠ $3 \times 2\frac{1}{2} \times \frac{1}{4}$ and each of the two diagonal members coming to this connection as $3 \times 3 \times \frac{1}{4}$ angles.

**18*E*.** The pipe bridge carries 10-in., 7-in., 4-in., and 3-in. insulated pipe lines suspended from the 8 in. 23 lb I beams on each bent. Keep the pipes in the order indicated from left to right and show a detail of the strap supports and their connection to the bent.

**19.** The center lines of a timber truss are shown. The joints are to be made using split-ring connectors—one at each joint—except at the junction of the short compression member and the rafter where a 1- × 4-in. wood scab plate nailed over the joint is all that is needed.

Complete the design for the necessary trusses to roof over the structure indicated. Allow for a 2-ft overhang all the way around. The end trusses must be made flush with the wall, and designed so that some form of sheeting to cover the gabled end can be nailed in place. The trusses are to be placed on 2-ft centers (approximately) and sheeted over with ½-in. plywood. Plan for the transition trusses between the two wings to rest on the sheeting of the 50-ft wing. The design of these trusses can be varied to fit their decreasing length.

These trusses used as specified will carry approximately 35 lb per sq ft dead plus live load of the roof and 10 lb per sq ft dead load on the ceiling.

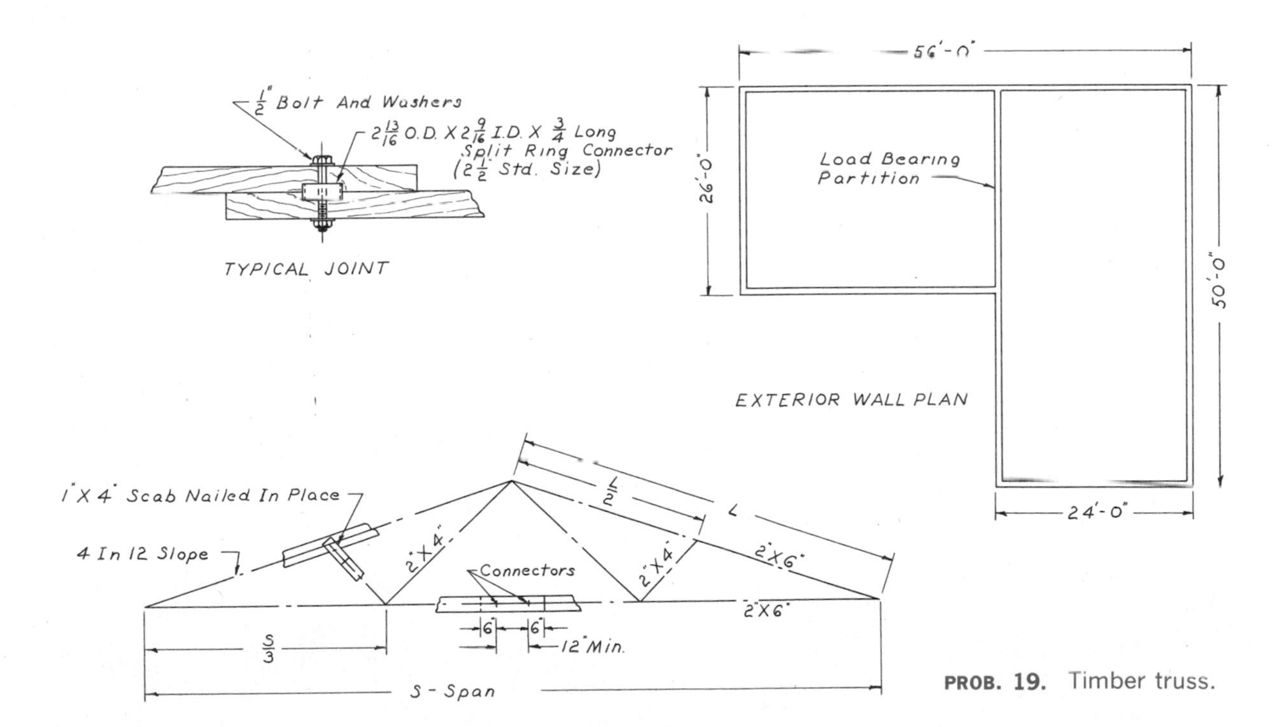


**PROB. 19.** Timber truss.

This Chapter Discusses:



# 22

# MAPS AND TOPOGRAPHY

Many Workers in Fields Such as Geology, Navigation, Exploration, and Archaeology, as well as All Engineers, Should Have a Knowledge of Maps and Topography. This Chapter Gives Graphic Fundamentals

### 1. Maps—One-view Drawings.

Thus far in our consideration of drawing as a graphic language we have had to represent the three dimensions of an object pictorially or, in the usual case, by drawing two or more views of it. In map drawing, that is, in representing features on parts of the earth's surface, there is the distinct difference that the drawing is complete with two dimensions given in one view; the third dimension, the height, is either represented in this view or omitted as not required for the purpose for which the particular map was made.

Surveying and mapping the site is the first preliminary in engineering projects, and all engineers should be familiar with the methods and symbols used in map and topographic drawing. Without considering the practice of surveying and plotting or the various methods used by cartographers in projecting the curved surface of the earth on a plane, we are interested in the use and details of execution of plats and topographic maps.

### 2. Classification.

The content or information on maps is in general classified under three divisions:

1. The representation of boundaries, bearings, and distances, such as divisions between areas subject to different authority or ownership, either public or private; or lines indicating geometric measurements on land, on sea, and in the air. This division includes plats or land maps, farm surveys, city subdivisions, plats of mineral claims, and nautical and aeronautical charts.

2. The representation of real or material features or objects within the limits of the tract, showing their relative location or size and location, depending upon the purpose of the map. When relative location only is required, the scale may be small, and symbols may be employed to represent objects, such as houses, bridges, or even towns. When the size of the objects is an important consideration, the scale must be large and the map becomes a real orthographic top view.

3. The representation of the relative elevations of the surface of the ground. Maps with this feature are called "relief maps" or, if contours are used with elevations marked on them, "contour maps." Hydrographic maps show fathom-line depth curves.

Various combinations of these three divisions are required for different purposes. Maps are classified according to purpose, as follows:

1. *Geographical maps* include large areas and consequently must be to small scale. They show important towns and cities, streams and bodies of water, political boundaries, and relief.

2. *Topographic maps* are complete descriptions of certain areas and show to larger scale the geographical positions of the natural features and the works of man. The relief is usually represented by contours.

3. *Hydrographic maps* deal with information concerning bodies of water, such as shore lines, sounding depths, subaqueous contours, navigation aids, and water control.

4. *Nautical maps* or charts show aids to water navigation, such as buoys, beacons, lighthouses, lanes of traffic, sounding depths, shoals, and radio compass stations.

5. *Aeronautical maps* or charts give prominent landmarks of the terrain and accentuate the relief by layer tints, hachures, and 500- or 1,000-ft contours as aids to air navigation.

6. *Cadastral maps* are very accurate control maps for cities and towns, made to large scale with all features drawn to size. They are used to control city development and operation, particularly taxation.

7. *Engineering maps* are working maps for engineering projects and are designed for specific purposes to aid construction. They provide accurate horizontal and vertical control data and show objects on the site or along the right of way.

8. *Photogrammetric maps* represent features on the earth's surface from terrestrial and aerial photographs. These photographs are perspectives from which orthographic views are obtained by stereoscopic instruments. Ground-control stations are necessary to bring the photographs to a required datum.

9. *Military maps* contain information of military importance in the area represented.

### 3. Plats.

A map plotted from a plane survey and having the third dimension omitted is called a "plat" or "land map." It is used in the description of any tract of land when it is not necessary to show relief, as in a farm survey or a city plat.

The plotting is done from field notes by (1) latitudes and departures, (2) bearings and distances, (3) azimuths and distances, (4) deflection angles and distances, or (5) rectangular coordinates. Or the plotting is done by the total latitude and departure from some fixed origin for each separate point; this method is necessary to distribute plotting errors over the entire survey. Angles are laid off from bearing or azimuth lines by plotting the tangent of the angle or the sine of half the angle, by sine-and-cosine method, or by an accurate protractor.

*Simplicity* is of first importance in the execution of this kind of drawing. Information should be clear, concise, and direct. The lettering should be done in single stroke, and the north point and border should be of the simplest character. The day of the intricate border corner, elaborate north point, and ornamental title is, happily, past, and all such embellishments are rightly considered to be a waste of time and in bad taste.

### 4. Plat of a Survey.

The plat of a survey should give clearly all the information necessary for the legal description of a parcel of land. It should contain:

1. Direction and length of each line
2. Acreage
3. Location and description of monuments found and set
4. Location of highways, streams, rights of way, and any appurtenances required
5. Official division lines within the tract
6. Names of owners of abutting property
7. Title, scale, date
8. North point with certification of horizontal control
9. Plat certification properly executed
10. Reference to state plane-coordinate system

Figure 1 illustrates the general treatment of this kind of drawing. It is almost always traced and blueprinted, and no water-lining of streams or other elaboration should be attempted. The size of the lettering used for the several features must be in proportion to their importance.

### 5. Industrial Plats.

Of the many kinds of plats used in industrial work only one is illustrated here, a portion of a railway situation or station map (Fig. 2). This might represent also a plant-valuation map, a type of plat often required. The information on such maps varies to meet the requirements of particular cases. In addition to the items in the preceding list, the map might include pipe lines, fire hydrants, location and description of buildings, railroads and switch points, outdoor crane runways, etc.

### 6. Plats of Subdivisions.

The plats of subdivisions and allotments in cities are filed with the county recorder for record and must be complete in their information concerning the location and size of the various lots and parcels composing the subdivisions (Fig. 3). All monuments set should be shown and all directions and distances recorded, so that it will be possible to locate any lot with precision.

Sometimes landowners use these maps in display to prospective buyers and often include a blueprint or black-line print bound with the deed. Some degree of embellishment is allowable, but care must be taken not to overdo the ornamentation. Figure 4 is an example showing an acceptable style of execution and finish.

### 7. City Plats.

Under this head are included chiefly maps or plats drawn from subdivision plats or other sources for the record of city improvements. These plats are used to record a variety of information, such as the location of sewers, water mains, gas, power and steam lines, telephone installations, and street improvements.

The records maintained on these maps provide valuable data for assessments and constitute prog-

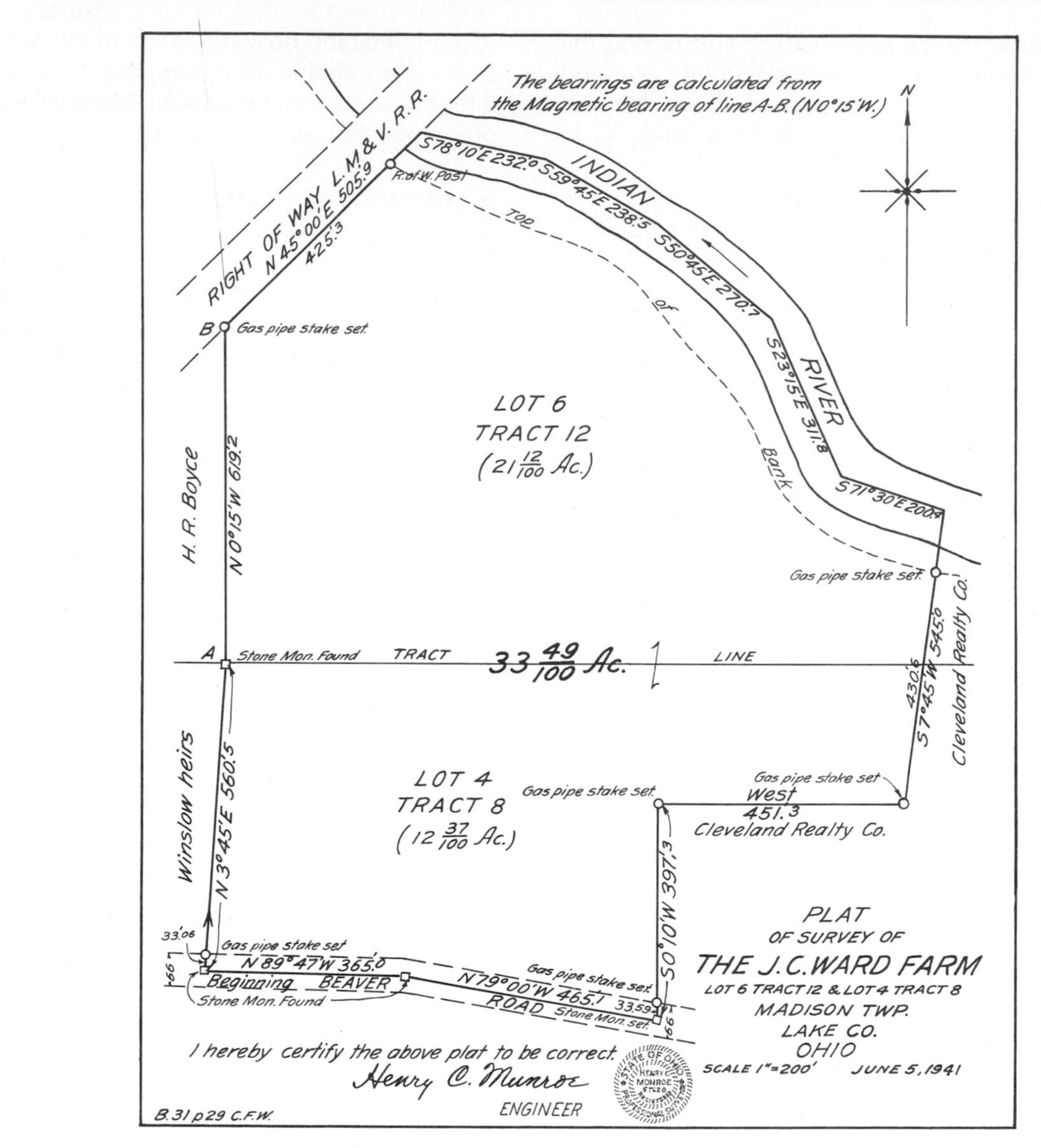


**FIG. 1.** Plat of a survey. This type of map includes the information listed in Sec. 4.

ress reports on the growth of a city. These maps are made for a definite purpose and should contain only the information needed for that purpose; hence they will not include all the details as to sizes of lots, which are given on subdivision plats, but they should carry both horizontal and vertical control points for proper location of utilities. They are usually made on mounted paper and should be to a scale large enough to show clearly the features required; 100 and 200 ft to the inch are common scales, and as large as 50 ft is sometimes used. For small cities the entire area may be covered by one

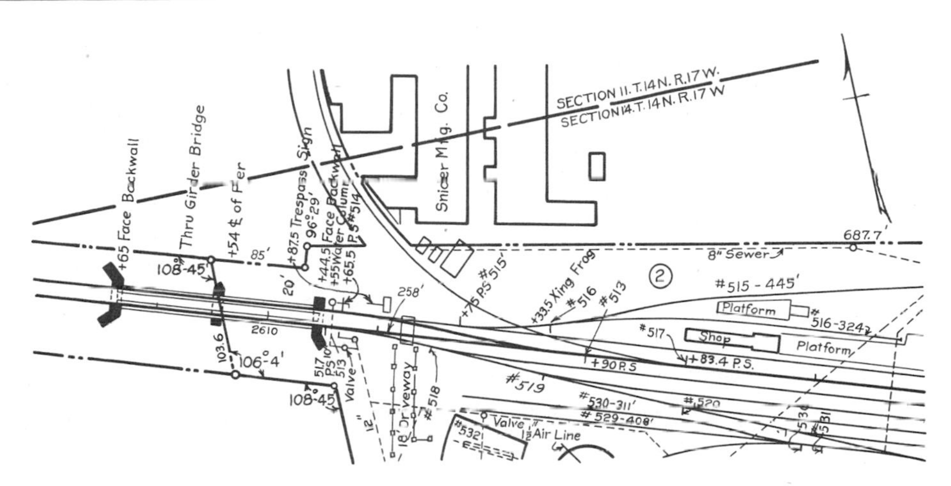


**FIG. 2.** Part of an industrial map. This is a railroad property map.

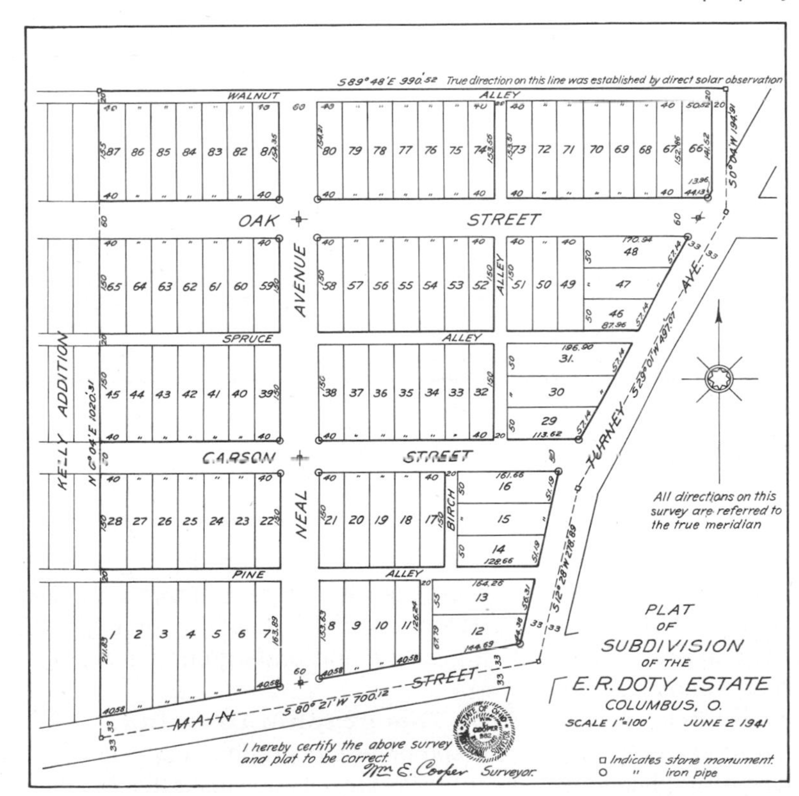


**FIG. 3.** Plat of a city subdivision. This type of map shows the size and location of lots and streets.

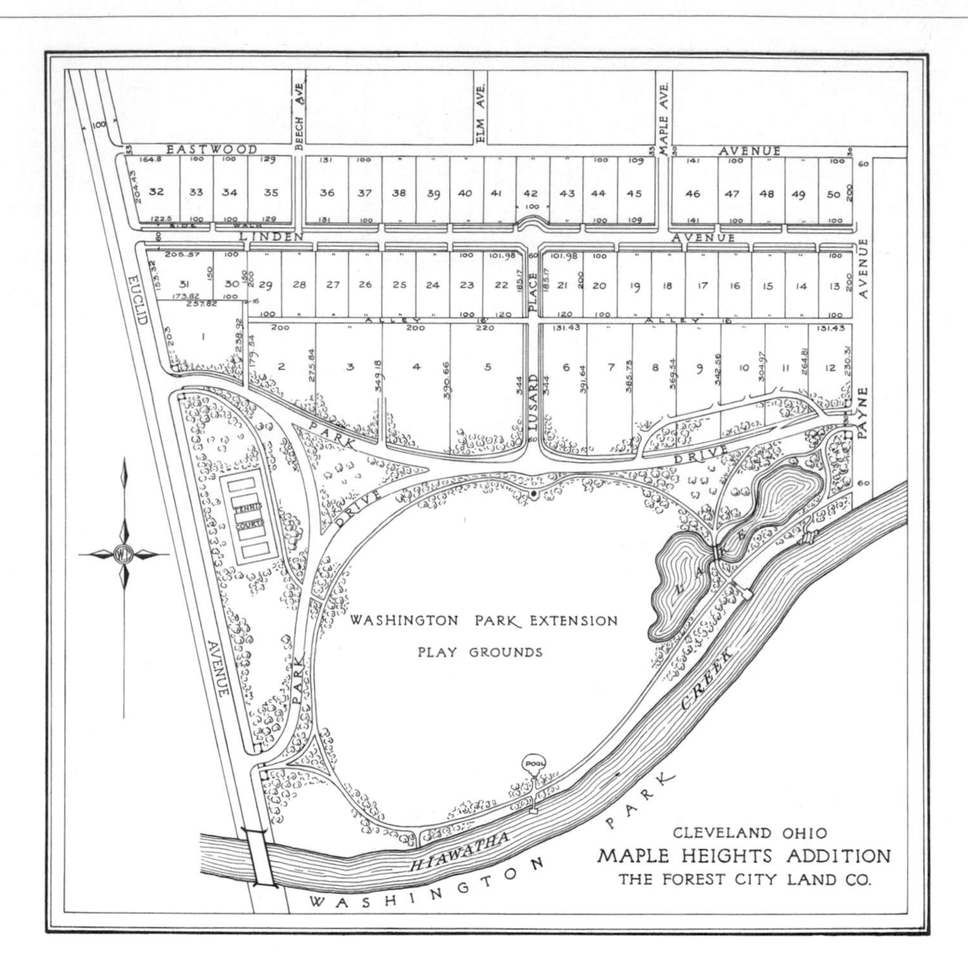


**FIG. 4.** A real-estate display map. This type of map shows the size of lots and streets but does not include the details given in a survey.

map; for large cities the maps are made in sections that can be conveniently filed.

A study of Fig. 5, a sewer map, will show the general treatment of city plats. The appearance of the drawing is improved by adding shade lines on the lower and right-hand side of the blocks, that is, treating the streets and water features as depressions. A few of the more important public buildings are shown, to facilitate reading. The various wards, subdivisions, or districts may be shown by large outline letters or numerals as illustrated in the figure. Contours are often put on these maps in red or brown ink, either on the original or on a positive print from it. Figure 6 shows a modern system of horizontal control used by the city of Cleveland for a geodetic and underground survey.

## 8. Topographic Drawing.

A complete topographic map will contain

1. The lines establishing the divisions of authority or ownership.

2. The geographical position of both the natural features and the works of man. It may also include information on the vegetation.

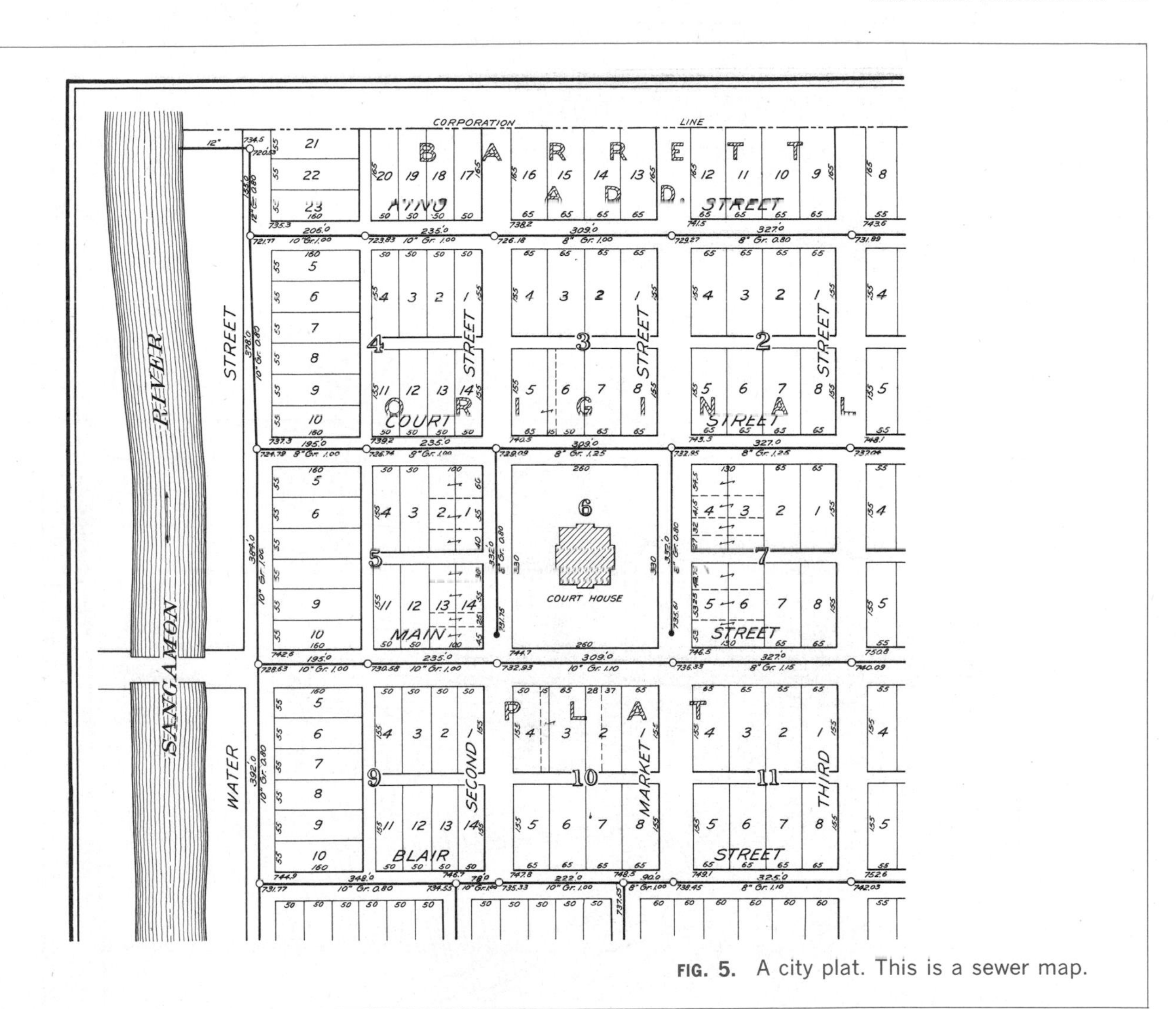


**FIG. 5.** A city plat. This is a sewer map.

3. The relief, or indication of the relative elevations and depressions. The relief, which is the third dimension, is represented in general by contours or by hill shading.

These are represented on the drawing by lines, symbols, and lettered information, as explained in the paragraphs that follow.

## 9. Contours.

A contour is a theoretical line on the surface of the ground which at every point passes through the same elevation; thus the shore line of a body of water represents a contour. If the water should rise 1 ft, the new shore line would be another contour, with 1-ft "contour interval." A series of contours may be illustrated approximately as in Fig. 7.

Figure 8 is a perspective view of a tract of land. Figure 9 is a contour map of the same area, and Fig. 10 shows the same surface with hill shading by hachures. Contours are drawn as fine full lines, with every fifth one of heavier weight and with the elevations in feet marked on them at intervals, usually with the sea level as datum. They may be drawn with a swivel pen, or a fine pen such as Gillott's 170 or Esterbrook's 356. On paper drawings they are usually made in brown.

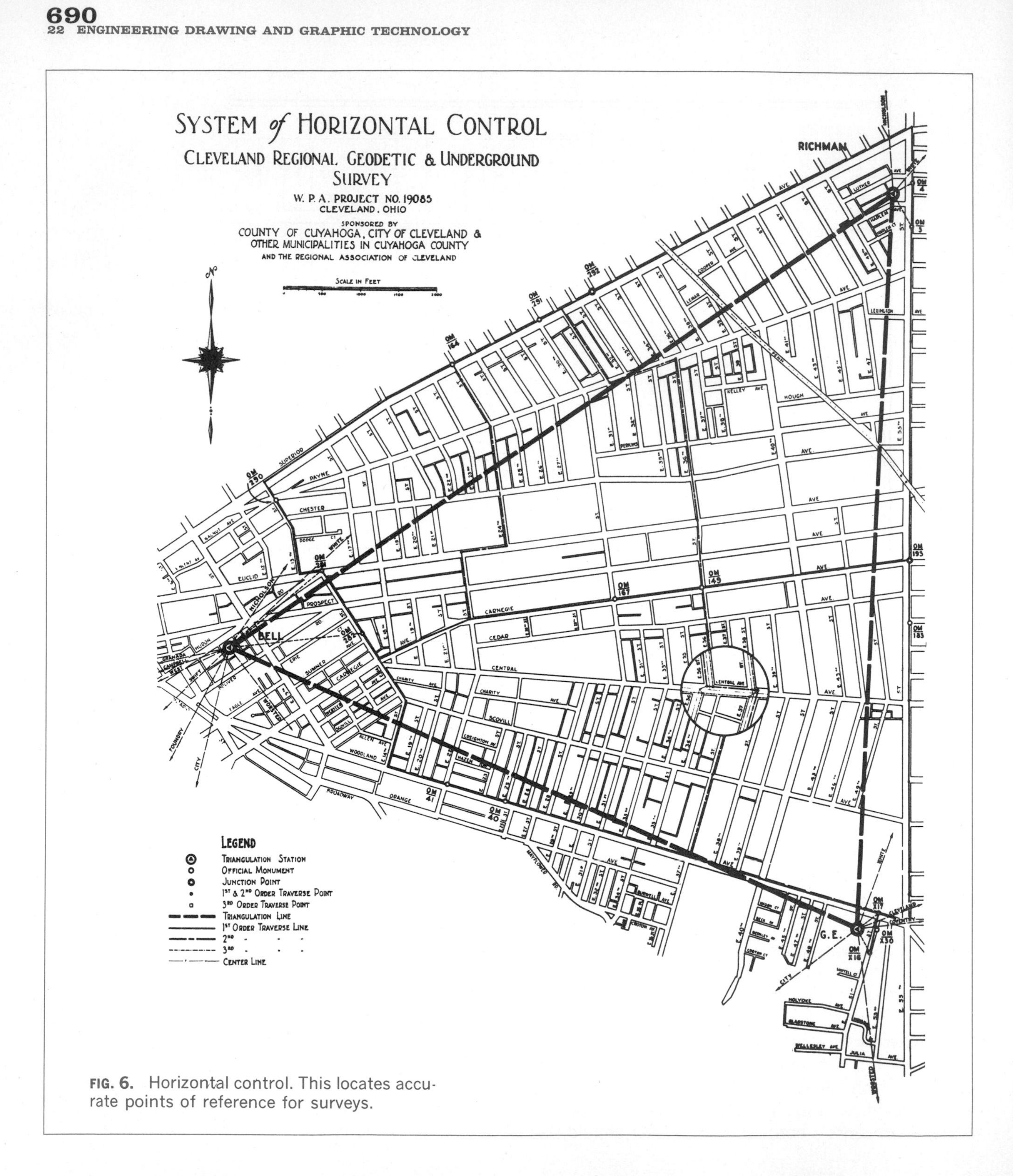


FIG. 6. Horizontal control. This locates accurate points of reference for surveys.

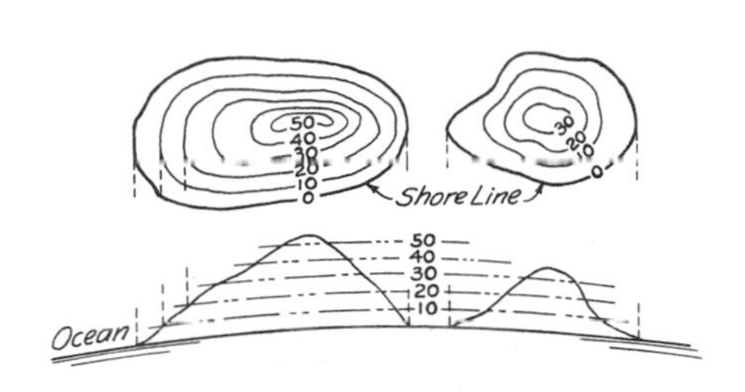


**FIG. 7.** Contours. These are theoretical lines of constant elevation.

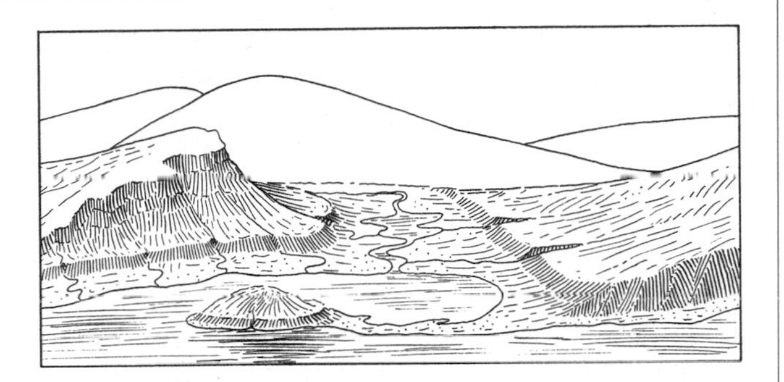

**FIG. 8.** A perspective view. This is a pictorial map in which relief is indicated by line shading. Maps of this type are used only for illustrative purposes.

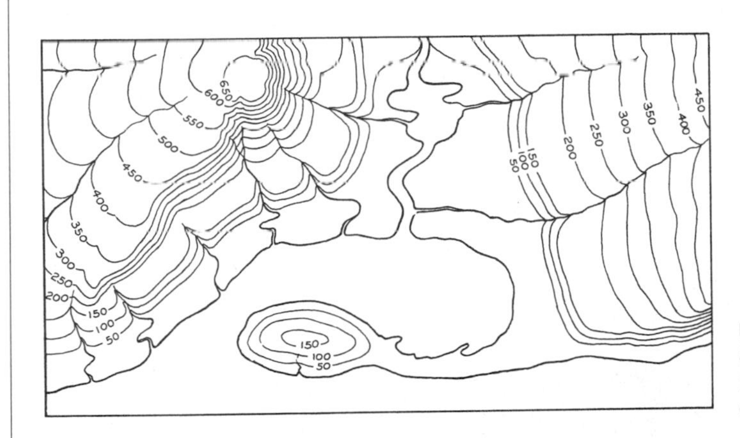


**FIG. 9.** A contour map. This is a map of the area shown in Fig. 8. The relief is read by noting the elevation, position, and spacing of the contours.

**FIG. 10.** A relief map. In this map of the area given in Fig. 8 the relief is shown by hachures. High areas are light, low areas dark. This type of map shows only relative and approximate relief and is therefore not used for scientific purposes.

Figure 11 is a topographic map of the site of a proposed filtration plant and illustrates the use of the contour map as the necessary preliminary drawing for engineering projects. Often the same drawing shows, by lines of different character, both the existing contours and the required finished grades.

## 10. Hill Shading.

Showing relief by means of hill shading gives a pleasing effect but is difficult to execute, does not give exact elevations, and would not be applied on maps to be used for engineering purposes. Hill shading is sometimes used to advantage in reconnaissance maps or in small-scale maps for illustration. There are several systems, of which hachuring, as shown in Fig. 10, is the commonest. This is done by sketching the contours lightly in pencil and drawing the hachures perpendicular to them, starting at the summit and grading the weight of line to the degree of slope. A scale of hachures, graded from black for 45° to white for horizontal, is often made to use for reference. The rows of strokes should touch the pencil line to avoid white streaks along the contours. Two other systems in use are the horizontal, or English, system using graded hachure lines parallel to the contours; and the oblique illumination, or French, system using hachures graded to give sunlight effect as well as the degree of slope.

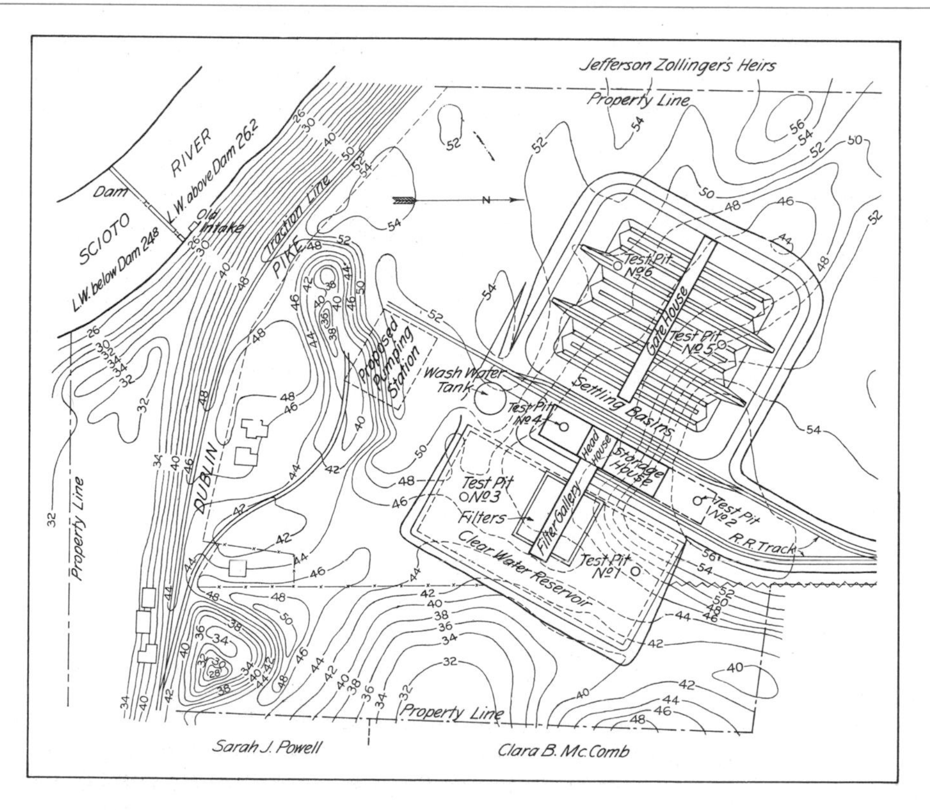


**FIG. 11.** A contour map of an engineering project. This type of map shows relief by contours and includes other physical features such as buildings, roadways, and streams.

## 11. Water Lining.

On topographic maps made for display or reproduction, the water features are usually finished by "water lining," that is, by running a system of fine lines parallel to the shore lines, either in black or in blue (it must be remembered that blue will not photograph for reproduction or print well from a tracing). Poor water lining will ruin the appearance of an otherwise well-executed map, and it is better to omit it rather than do it hastily. The shore line is drawn first, and the water lining done with a fine mapping pen, the draftsman always drawing toward his body, with the preceding line to his left. The first line should follow the shore line closely; the distances between the succeeding lines should be gradually increased and the irregularities lessened. Sometimes the weight of lines is graded as well as the intervals, but this is a difficult operation and not necessary for the effect. A common mistake is to make the lines excessively wavy or rippled.

In water lining a stream of varying width, the lines are not crowded so as to be carried through the narrower portions; corresponding lines are brought together in the middle of the stream, as illustrated in Fig. 10. Avoid spots of sudden increase or decrease in spacing.

## 12. Topographic Symbols.

The various symbols used in topographic drawing can be grouped under four heads:

1. Culture, or the works of man
2. Relief—relative elevations and depressions
3. Water features
4. Vegetation

When color is used the culture is done in black, the relief in brown, the water features in blue, and the vegetation in black or green.

Topographic symbols, used to represent characteristics on the earth's surface, are made, when possible, to resemble somewhat the features or objects represented as they would appear in plan or in elevation. No attempt is made here to give symbols for all the features that might occur in a map; indeed one may have to invent symbols for some particular condition.

Figure 12 illustrates a few of the conventional

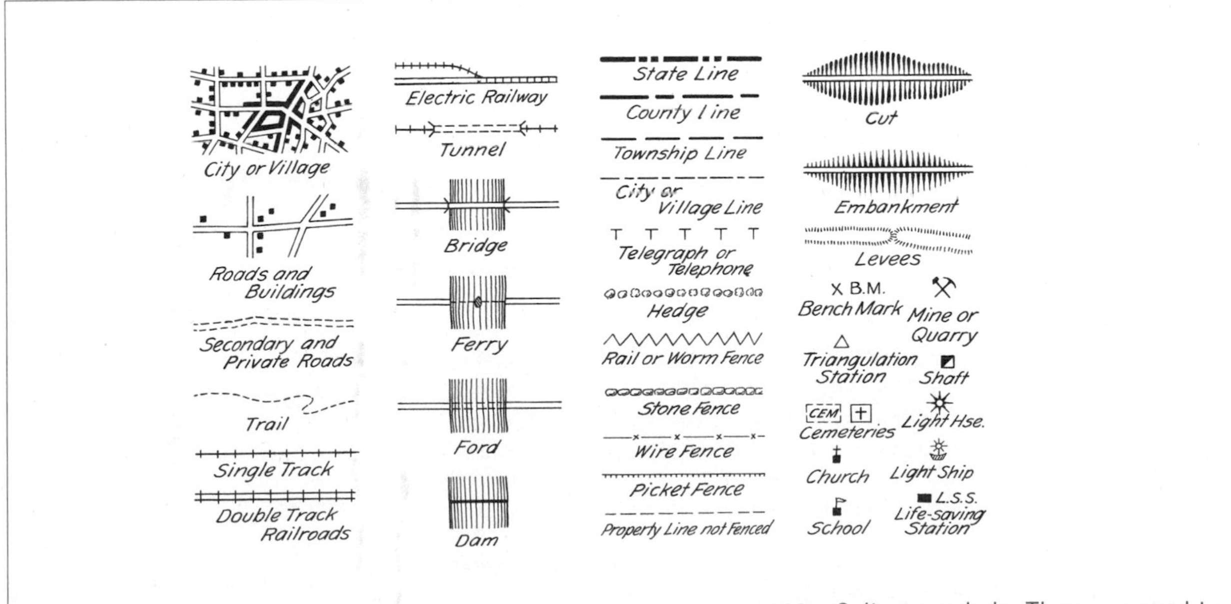


**FIG. 12.** Culture symbols. These are used to indicate the works of man.

symbols used for culture or the works of man. When the scale used is large, houses, bridges, roads, and even tree trunks can be plotted so that their principal dimensions can be scaled. The landscape architect is interested not only in the size of the trunk of a tree but also in the spread of its branches. A small-scale map can give by its symbols only the relative locations.

Some military symbols are shown in Fig. 13, symbols for aerial navigation in Fig. 14, and aids to water navigation in Fig. 15—all as adopted by the U.S. Board of Surveys and Maps. Figure 16 gives the standard symbols used in the development of oil and gas fields, Fig. 17 the symbols used to show relief, Fig. 18 water features, and Fig. 19 some of the common symbols for vegetation and cultivation.

Keep in mind the purpose of the map and in some measure indicate the relative importance of the features, varying their prominence by the weight of lines used or by varying the scale of the symbol.

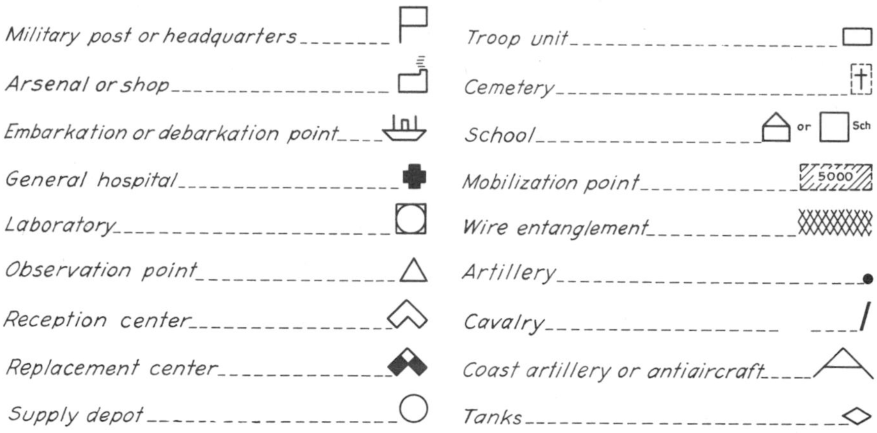

Military post or headquarters
Arsenal or shop
Embarkation or debarkation point
General hospital
Laboratory
Observation point
Reception center
Replacement center
Supply depot
Troop unit
Cemetery
School — or Sch
Mobilization point — 5000
Wire entanglement
Artillery
Cavalry
Coast artillery or antiaircraft
Tanks

**FIG. 13.** Military symbols. These are used on military maps.

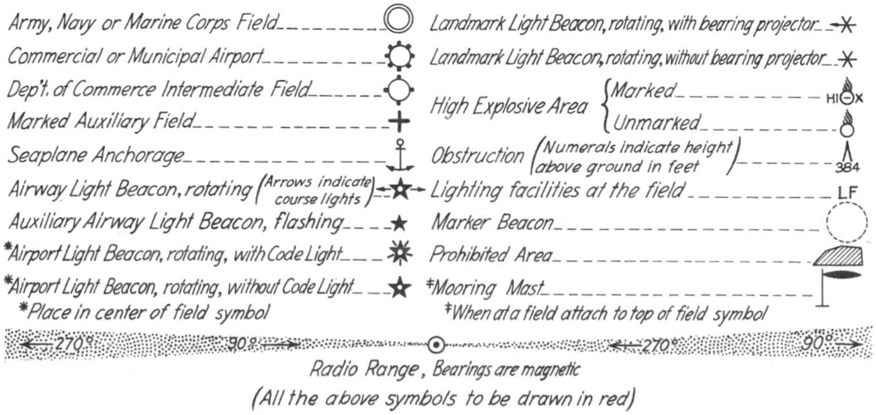

Army, Navy or Marine Corps Field
Commercial or Municipal Airport
Dep't. of Commerce Intermediate Field
Marked Auxiliary Field
Seaplane Anchorage
Airway Light Beacon, rotating (Arrows indicate course lights)
Auxiliary Airway Light Beacon, flashing
*Airport Light Beacon, rotating, with Code Light
*Airport Light Beacon, rotating, without Code Light
*Place in center of field symbol
Landmark Light Beacon, rotating, with bearing projector
Landmark Light Beacon, rotating, without bearing projector
High Explosive Area {Marked — HI X; Unmarked}
Obstruction (Numerals indicate height above ground in feet) — 384
Lighting facilities at the field — LF
Marker Beacon
Prohibited Area
‡Mooring Mast
‡When at a field attach to top of field symbol
270° — 90° — 270° — 90°
Radio Range, Bearings are magnetic
(All the above symbols to be drawn in red)

**FIG. 14.** Aviation symbols. These are used on navigation maps for aircraft.

For instance, in a map made for military maneuvers a cornfield might be an important feature; or in maps made to show the location of special features, such as fire hydrants, these objects would be plainly indicated. The map of an airport or a golf course would emphasize features significant for an airport or golf course. The principle calls for some originality in meeting various cases.

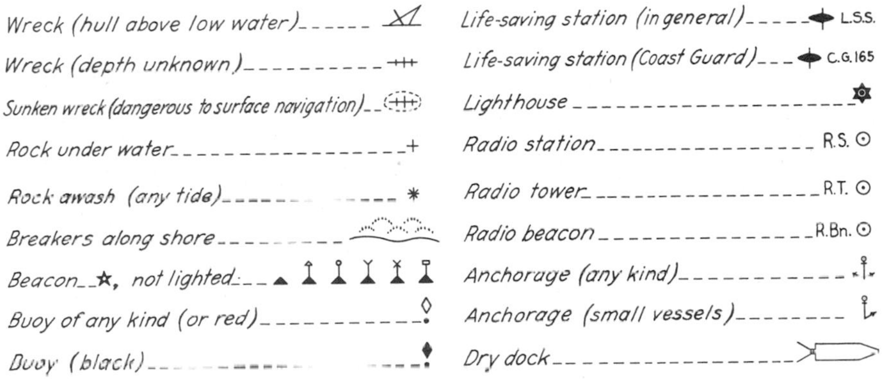

**FIG. 15.** Marine symbols. These are used on navigation maps for water craft.

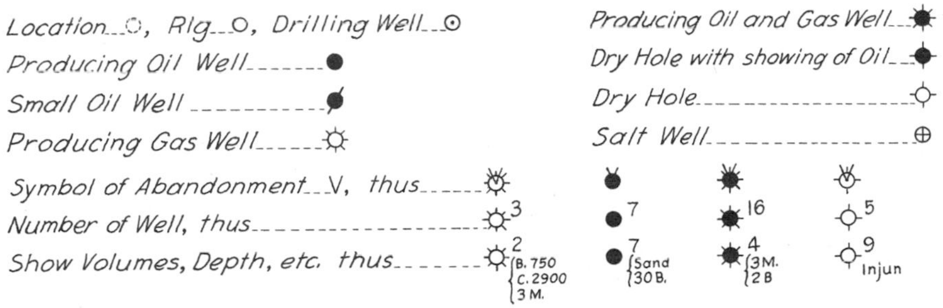

**FIG. 16.** Oil and gas symbols. These are used on contour and relief maps.

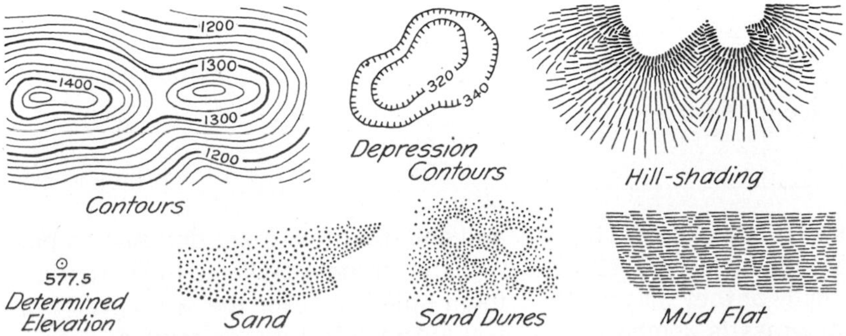

**FIG. 17.** Relief symbols. These are used on contour and relief maps.

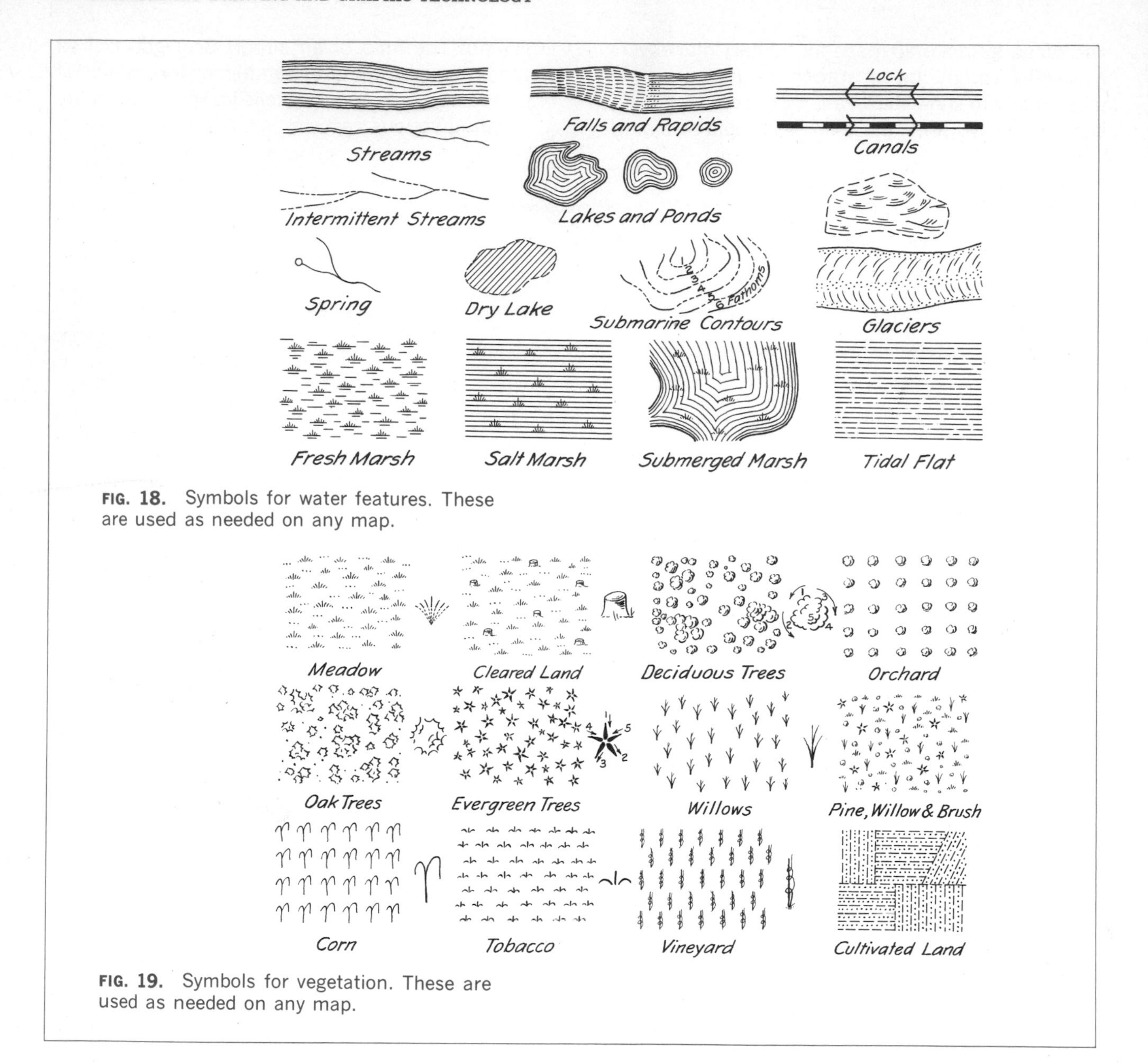


**FIG. 18.** Symbols for water features. These are used as needed on any map.

**FIG. 19.** Symbols for vegetation. These are used as needed on any map.

A beginner will often make symbols too large. The symbols for grass, shown under "meadow" (Fig. 19), must be made and spaced correctly or they will spoil an entire map. The symbol for grass is composed of five to seven short strokes radiating from a common center and starting along a horizontal line, as shown in the enlarged example, each tuft beginning and ending with a mere dot. Always place the tufts with the bottom parallel to the border and distribute them uniformly over the space, but not in rows. A few incomplete tufts or rows of dots improve the appearance. Grass-tuft symbols should never be as heavy as tree symbols. In drawing the symbol for deciduous trees, the sequence of strokes shown should be followed.

Figure 20 is a topographic map, given to illustrate the general execution and placing of symbols.

The well-known maps of the U.S. Coast and Geodetic Survey and the Geological Survey illustrate the application of topographic drawing. The *quadrangle sheets* issued by the topographic branch of the U.S. Geological Survey are excellent examples and so easily available that every engineer should be familiar with them. These sheets represent 15 minutes of latitude and 15 minutes of longitude to the scale of 1:62,500, or approximately 1 in. to the mile. The entire United States is being mapped by the department in cooperation with the different states. This work is now greatly facilitated through the use of aerial photography. Much territory in the West and South has been mapped ½ in. to the mile; and earlier some in the West was mapped ¼ in. to the

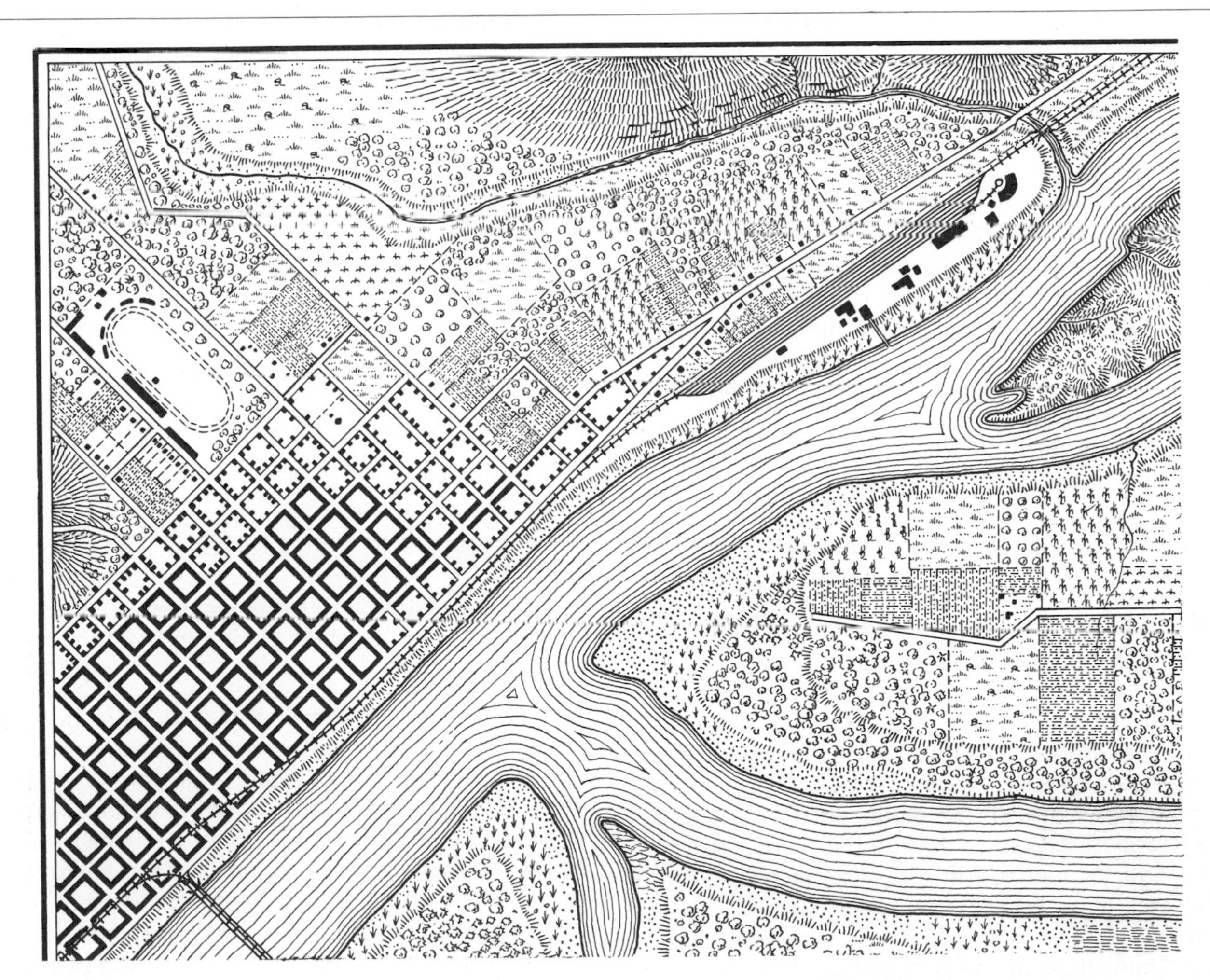

**FIG. 20.** Part of a topographic map. This map shows the application of relief, water, and vegetation symbols.

mile. These maps may be secured from The Director, U.S. Geological Survey, Washington, D.C., who will also give information about the completion of any particular locality or the progress in any state.

### 13. Landscape Maps.

A topographic map made to a relatively large scale and showing all details is called a "landscape map." Such maps are required by architects and landscape gardeners for use in planning buildings to fit the natural topographic features and in landscaping parks, playgrounds, and private estates. These are generally maps of small areas, and a scale of 1″ = 20′ to 1″ = 50′, depending upon the amount of detail, is used.

The contour interval varies from 6 in. to 2 ft according to the ruggedness of the surface. The commonest interval is 1 ft. Landscape maps are often reproduced in black-line prints, upon which contours in different color are drawn to show the landscape treatment proposed. Natural features and culture are added in more detail than on ordinary topographic maps. Trees are designated as to size, species, and sometimes spread of branches and condition. It is often necessary to invent symbols suitable for a particular survey and to include a key or legend on the map. Roads, walks, streams, flower beds, houses, etc., should be plotted carefully to scale, so that measurements can be taken from them.

### 14. Colors.

Instead of using colored inks, which are thin and unsatisfactory to handle in the pen and do not photograph or blueprint well, it is better to use water colors for contours, streams, and other colored features in topographic mapping: for contours, burnt sienna, either straight or darkened with a drop of black, and mixed rather thick; for streams, Prussian blue; and for features in red, alizarin crimson. All work well in crow-quill or contour pens and make good blueprints. Colors in tubes are more convenient than those in cakes or pans.

### 15. Lettering.

The style of lettering on a topographic map will depend upon the purpose for which the map is made. If it is for construction purposes, such as a contour map for the study of municipal problems, street grades, plants, or railroads, the single-stroke Gothic and Reinhardt is preferred. For a finished map, vertical Modern Roman letters should be used, capitals for important land features and lower case for less important features, such as small towns and villages; for water features, inclined Roman and stump letters (see Appendix). The scale should always be drawn as well as stated.

### 16. Titles.

Modern Roman is the standard letter for finished map titles. The design should be symmetrical, with the height of the letters proportioned to the relative importance of the line. A map title should contain as many of the following items as are necessary.

1. Kind—"Map of," etc.
2. Name
3. Location of tract
4. Purpose, if special features are represented
5. For whom made
6. Engineer in charge
7. Date (of survey)
8. Scale—stated and drawn, contour interval, datum
9. Authorities
10. Legend or key to symbols
11. North point, with certification of horizontal control
12. Certification, properly executed
13. Reference to state plane-coordinate system

### 17. Profiles.

Perhaps no kind of drawing is used more by civil engineers than the ordinary profile, which is simply a vertical section taken along a given line, either straight or curved. Such drawings are indispensable in problems of railroad construction, highway and street improvements, sewer construction, and many other problems where a study of the surface of the ground is required. Frequently engineers other than civil engineers are called upon to make profile drawings. Several different types of profile and cross-section paper are in use; they are described in the catalogues of the various firms dealing in drawing materials. One type of profile paper in common use

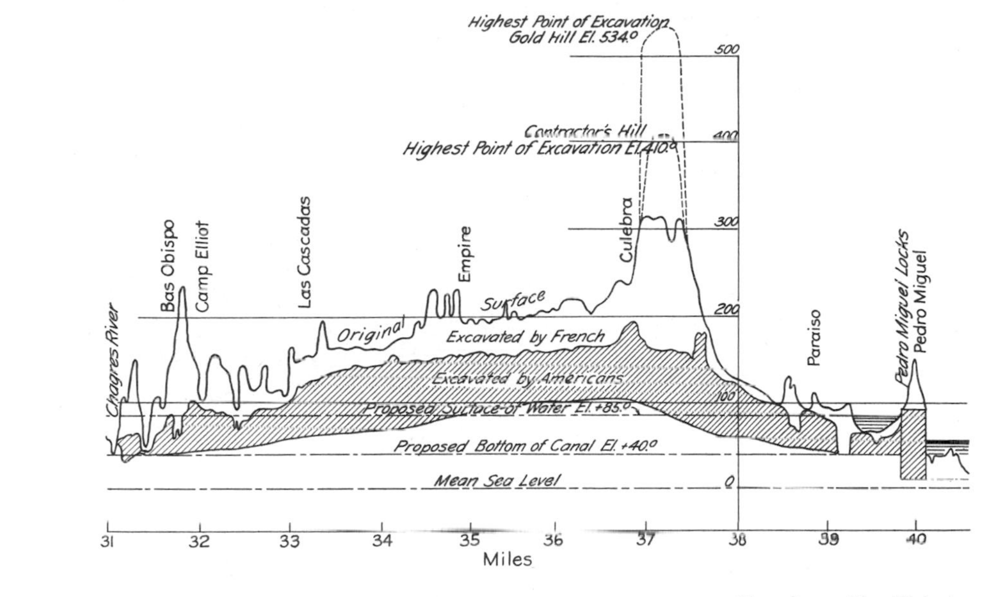


**FIG. 21.** A profile. This is used to study surface variations. In this profile the vertical scale is 50 times the horizontal scale.

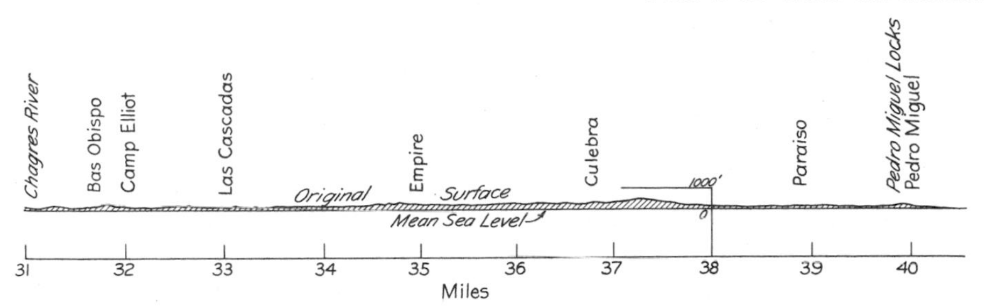


**FIG. 22.** A profile. This is the same surface as in Fig. 21 but plotted on equal horizontal and vertical scales. Note the superiority of Fig. 21 for analysis.

is known as "Plate A" and has 4 divisions to the inch horizontally and 20 to the inch vertically. Other divisions in use are 4 × 30 to the inch and 5 × 25 to the inch. At intervals, both horizontally and vertically, heavier lines are made to facilitate reading.

Horizontal distances are plotted as abscissas and elevations as ordinates. Since the vertical distances represent elevations and are plotted to larger scale, a vertical exaggeration is obtained that is useful in studying profiles that are to be used for establishing grades. The vertical exaggeration is sometimes confusing to the layman or inexperienced engineer, but ordinarily a profile will fail in the purpose for which it was intended if the horizontal and vertical scales are the same. Furthermore, if the profile were not distorted in this way, it would be a long and unwieldy affair, perhaps even impossible to make. The difference between profiles with and without vertical exaggeration is shown in Figs. 21 and 22.

Figure 23 is a portion of a typical state highway

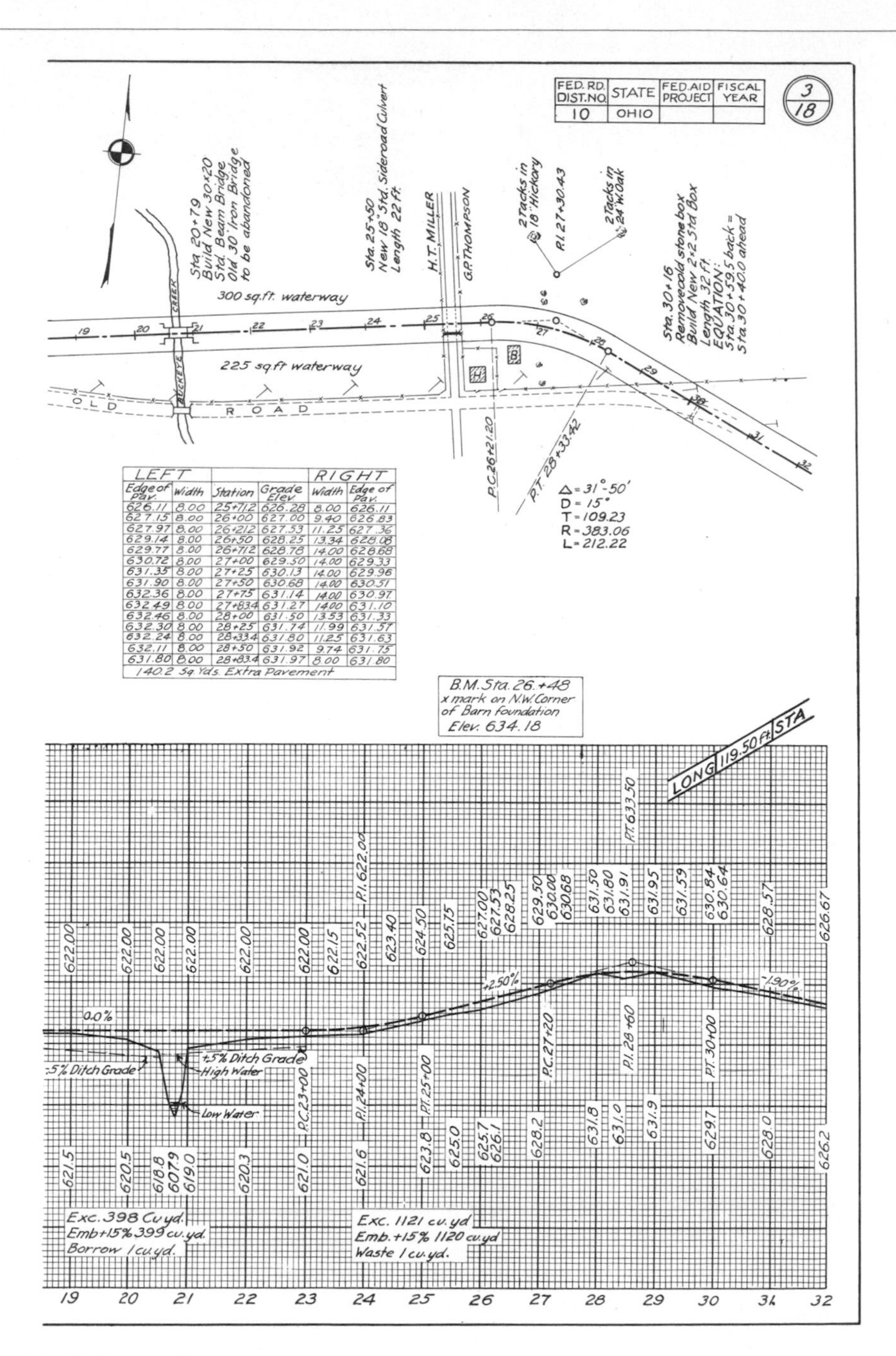


| LEFT | | | | RIGHT | |
|---|---|---|---|---|---|
| Edge of Pav. | Width | Station | Grade Elev. | Width | Edge of Pav. |
| 626.11 | 8.00 | 25+71.2 | 626.28 | 8.00 | 626.11 |
| 627.15 | 8.00 | 26+00 | 627.00 | 9.40 | 626.83 |
| 627.97 | 8.00 | 26+21.2 | 627.53 | 11.25 | 627.36 |
| 629.14 | 8.00 | 26+50 | 628.25 | 13.34 | 628.08 |
| 629.77 | 8.00 | 26+71.2 | 628.78 | 14.00 | 628.68 |
| 630.72 | 8.00 | 27+00 | 629.50 | 14.00 | 629.33 |
| 631.35 | 8.00 | 27+25 | 630.13 | 14.00 | 629.96 |
| 631.90 | 8.00 | 27+50 | 630.68 | 14.00 | 630.51 |
| 632.36 | 8.00 | 27+75 | 631.14 | 14.00 | 630.97 |
| 632.49 | 8.00 | 27+83.4 | 631.27 | 14.00 | 631.10 |
| 632.46 | 8.00 | 28+00 | 631.50 | 13.53 | 631.33 |
| 632.30 | 8.00 | 28+25 | 631.74 | 11.99 | 631.57 |
| 632.24 | 8.00 | 28+33.4 | 631.80 | 11.25 | 631.63 |
| 632.11 | 8.00 | 28+50 | 631.92 | 9.74 | 631.75 |
| 631.80 | 8.00 | 28+83.4 | 631.97 | 8.00 | 631.80 |
| 140.2 Sq. Yds. Extra Pavement | | | | | |

**FIG. 23.** Part of a state highway alignment and profile sheet. This gives fundamental information needed for construction of a highway.

alignment and profile sheet, plotted to a horizontal scale of 1″ = 100′ and a vertical scale of 1″ = 10′. For this type of drawing, tracing cloth is available with the coordinates printed in red on the back so that any changes or erasures on the profile will not damage the coordinate lines. Lettering or other features are sometimes brought out by erasing the lines on the back. The alignment and profile sheet is one of a set of drawings used in the construction of a highway for estimating cost and as a working drawing, by the contractor, during construction. Other drawings in the set would consist of a title sheet showing the location plan with detours provided, a sheet indicating conventional signs, a sheet giving an index to bound sheets, and a sheet with space reserved for declarations of approval and signatures of proper officials.

There would also be sheets of cross sections taken at each 100-ft station and all necessary intermediate stations to estimate earthwork for grading; working drawings for drainage structures; site plans for bridges; specifications for guard rails and other safety devices; standard or typical road sections for cut and fill and various other conditions; and finally summary sheets for separate tables and quantities of materials for roadway, pavement, and structures.

## PROBLEMS

In Probs. 1 and 2, contour maps, which can be obtained from local, county, state, or federal authorities, are needed.

### Group 1. Contour Problems.

**1.** Obtain a contour map showing a river. Plot the watershed of the river. This is done by locating the series of highest points on each side of the river and drawing a watershed line connecting the points. Indicate the watershed area by crosshatching.

**2.** Obtain a contour map showing a river. At a point on the river where the banks are steep, or where there is a maximum rise above the river, locate a dam. Plot the lake formed by the dam. Represent the lake by crosshatching.

### Group 2. Plats of Subdivisions.

**3.** From Fig. 1, lay out the complete tract of land enclosed by the survey. Divide this tract into lots with (approximately) 70-foot frontage. Locate appropriate 30-foot roadways.

**4.** Lay out the tract of land in Fig. 3 and plan streets and lots as follows: Provide a park-type entrance on Main Street, use two curved roadways extending back to the former Walnut Alley; provide shorter streets as needed in the plan; use lots of approximately 100-foot frontage.

**5.** Make a sewer map for Prob. 3.

**6.** Make a gas-supply map for Prob. 4. Scale to fit a 17 × 22 in. sheet.

### Group 3. Industrial Plats.

**7.** Lay out the complete tract of land in Fig. 1 on 17- × 22-in. paper. On the tract, locate an industrial building 175 feet × 100 feet. Locate a spur track from the LM and V railway to one side of the building. Provide a parking area and access street from Beaver Road.

**8.** Convert the city subdivision in Fig. 3 to an industrial site. Specifications of building are the same as in Prob. 7. Provide a large parking lot, with access road from Main Street.

### Group 4. Profiles.

**9.** Plot the profile of Fig. 21 to suitable scale on a 17- × 22-in. sheet.

**10.** Plot the profile of a roadway similar to Fig. 23 but with these elevations:

**Station Point and Elevation**

| Sta. | 19 | 20 | 21 | 22 | 23 | 24 | 25 | 26 | 27 | 28 | 29 | 30 |
|---|---|---|---|---|---|---|---|---|---|---|---|---|
| Elev. | 511.0 | 511.0 | 511.0 | 511.0 | 511.2 | 511.5 | 513.5 | 516.0 | 518.5 | 520.5 | 520.9 | 519.8 |

Add a profile of ditch elevations deviating from the roadway as follows:

**Station Point and Elevation**

| Sta. | 19 | 20 | 21 | 22 | 23 | 24 | 25 | 26 | 27 | 28 | 29 | 30 |
|---|---|---|---|---|---|---|---|---|---|---|---|---|
| Elev. | −1.0 | −1.5 | −1.5 | −1.6 | −1.4 | −1.0 | −0.5 | +1.0 | +1.5 | +2.5 | +3.5 | +4.5 |

In this Chapter:



# 23

# DRAWINGS FOR ENGINEERING DESIGN AND CONSTRUCTION

This Chapter Discusses Professional Practices Employed in the Making of Design, Detail, Assembly, Production, Construction, and Other Drawings

## 1. Drawings—Fundamental Concepts.

A "production," "construction," or "working drawing" is any drawing used to give information for the manufacture or construction of a machine or the erection of a structure. Complete knowledge for the production of a machine or structure is given by a *set* of drawings conveying all the facts fully and explicitly so that further instructions are not required.

The description given by the set of drawings will include:

1. The full graphic representation of the shape of each part (shape description).
2. Dimensions of each part (size description).
3. Explanatory notes, general and specific, on the individual drawings, giving the specifications of material, heat-treatment, finish, etc. Often, particularly in architectural and structural work, the notes of explanation and information concerning details of materials and workmanship are too extensive to be included on the drawings and so are made up separately in typed or printed form and called the "specifications"—thus the term "drawings and specifications."
4. A descriptive title on each drawing.
5. A description of the relationship of each part to the others (assembly).
6. A parts list or bill of material.

A set of drawings will include, in general, two classes of drawings: *detail drawings*, giving the information included in items 1 to 4; and an *assembly drawing*, giving the location and relationship of the parts, item 5.

## 2. Engineering Procedure.

When a new machine or structure is designed, the first drawings are usually in the form of freehand sketches on which the original ideas, scheming, and inventing are worked out. These drawings are accompanied or followed by calculations to prove the suitability of the design. Working from the sketches and calculations, the design department produces a *design assembly* (also called a "design layout" or "design drawing")(Fig. 1). This is a preliminary pencil drawing on which more details of the design are worked out. It is accurately made with instruments, full size if possible, and shows the shape and position of the various parts. Little attempt is made to show all the intricate detail; only the essential dimensions, such as basic calculated sizes, are given. On the drawing, or separately as a set of written notes, will be the designer's general specifications for materials, heat-treatments, finishes, clearances or interferences, etc., and any other information needed by the draftsman in making up the individual drawings of the separate parts.

Working from the design drawing and notes, draftsmen (detailers) then make up the individual detail drawings. Figure 2 shows a detail drawing taken from the design drawing of Fig. 1. On a detail drawing, all the views necessary for complete shape description of a part are provided, and all the necessary dimensions and manufacturing directions are given. Dimension values for nonmating surfaces are obtained by scaling the design drawing, and the more critical values are determined from the design notes and from drafting-room standards. The detailer correlates the dimensions of mating parts and gives all necessary manufacturing information.

The set of drawings is completed with the addition of an assembly drawing and a parts list or bill of material.

If a machine is to be quantity produced, "operation" or "job" sheets will be prepared giving the separate manufacturing steps required and indicating the use and kind of special tools, jigs, fixtures, etc. The tool-design group, working from the detail drawings and the operation sheets, designs and makes the drawings for the special tools needed.

## 3. Assembly Drawings.

An *assembly drawing* is, as its name implies, a drawing of the machine or structure put together, showing the relative positions of the different parts.

The term "assembly drawings" includes preliminary design drawings and layouts, piping plans, unit assembly drawings, installation diagrams, and final complete drawings used for assembling or erecting the machine or structure.

The design drawing is the preliminary layout on which the scheming, inventing, and designing are accurately worked out. The assembly drawing is in

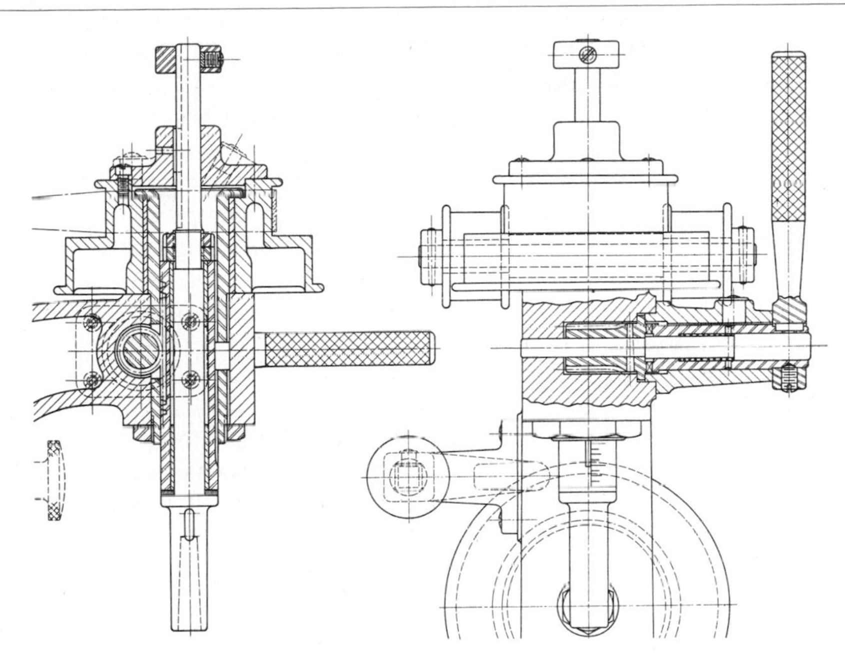

**FIG. 1.** A portion of a design drawing. Notes and specifications accompany the drawing.

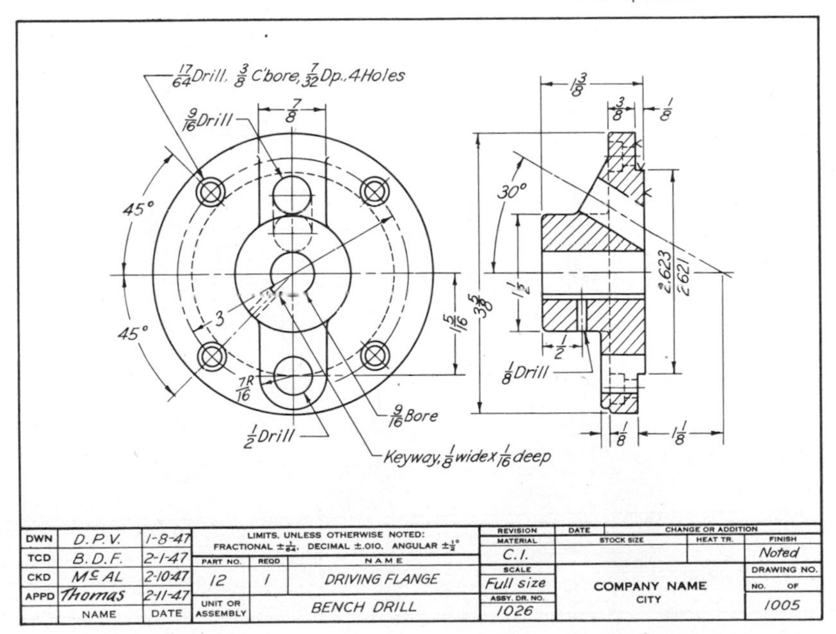


**FIG. 2.** A detail drawing. This gives complete information for the production of a single part.

some cases made by tracing from the design drawing. More often it is drawn from the dimensions of the detail drawings; this provides a valuable check on the correctness of the detail drawings.

The assembly drawing sometimes gives the overall dimensions and the distances between centers or from part to part of the different pieces, thus fixing the relation of the parts to each other and aiding in erecting the machine. However, many assembly drawings need no dimensions. An assembly drawing should not be overloaded with detail, particularly hidden detail. Unnecessary dashed lines should not be used on any drawing, least of all on an assembly drawing.

Assembly drawings usually have reference letters or numbers designating the different parts. These "piece numbers," sometimes enclosed in circles (called "balloons" by draftsmen) with a leader pointing to the piece (Fig. 3), are used in connection with the details and bill of material.

A *unit assembly drawing* or subassembly (Fig. 3) is a drawing of a related group of parts used to show the assembly of complicated machinery where it would be practically impossible to show all the features on one drawing. Thus, in the drawing of a lathe, there would be included unit assemblies of groups of parts such as the headstock, tailstock, gearbox, etc.

An *outline assembly drawing* is used to give a general idea of the exterior shape of a machine or structure and contains only the principal dimensions (Fig. 4). When it is made for catalogues or

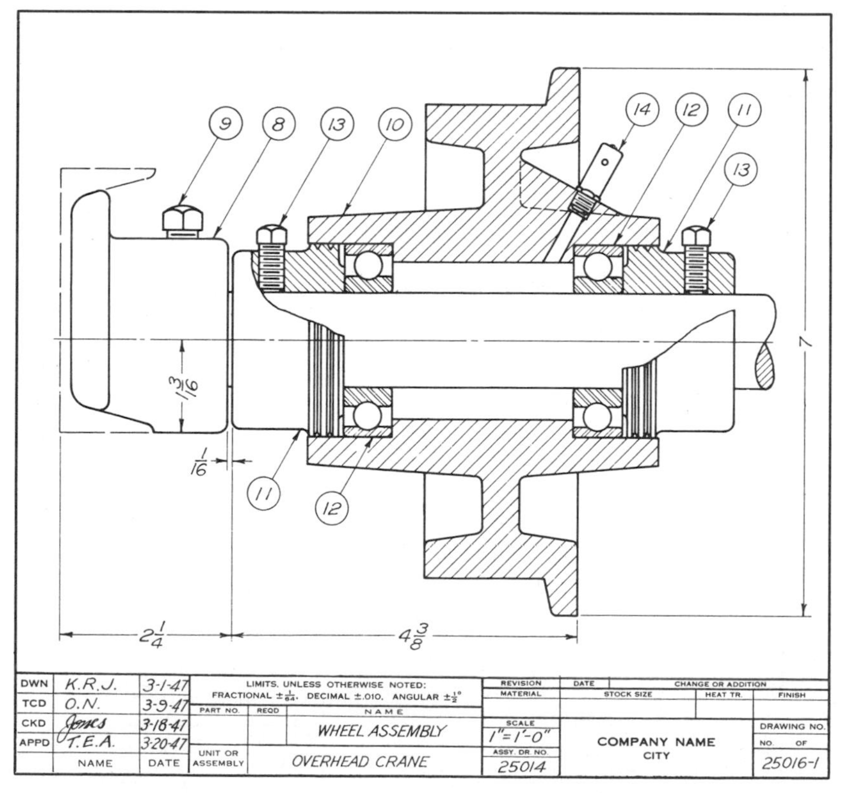


**FIG. 3.** A unit assembly drawing. For complex machines, a number of units may be used in place of one complete assembly drawing.

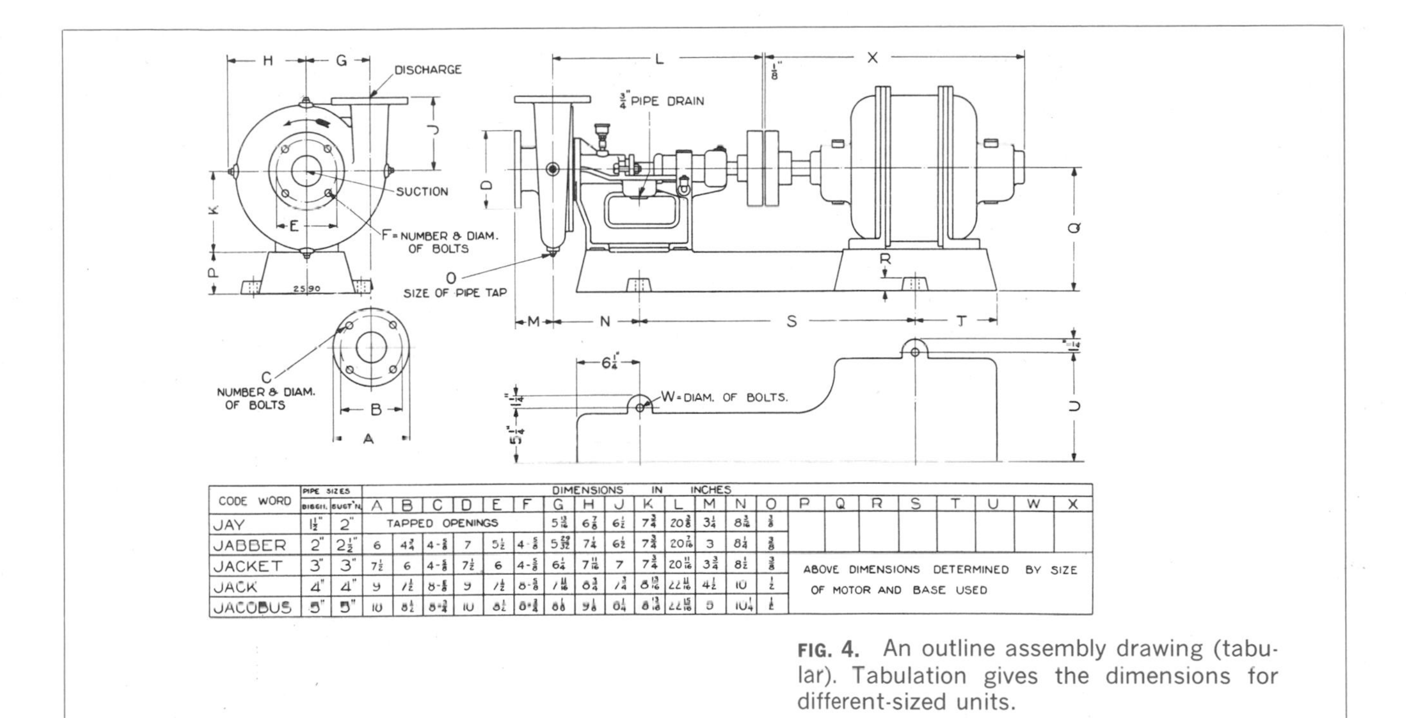


| CODE WORD | PIPE SIZES DISCH. | PIPE SIZES SUCT'N | A | B | C | D | E | F | G | H | J | K | L | M | N | O | P | Q | R | S | T | U | W | X |
|---|---|---|---|---|---|---|---|---|---|---|---|---|---|---|---|---|---|---|---|---|---|---|---|---|
| JAY | 1½" | 2" | TAPPED OPENINGS | | | | | | 5 13/16 | 6 7/8 | 6½ | 7¾ | 20 3/8 | 3¼ | 8 3/16 | 3/8 | | | | | | | | |
| JABBER | 2" | 2½" | 6 | 4¾ | 4-5/8 | 7 | 5½ | 4-5/8 | 5 29/32 | 7¼ | 6½ | 7¾ | 20 7/16 | 3 | 8¼ | 3/8 | | | | | | | | |
| JACKET | 3" | 3" | 7½ | 6 | 4-5/8 | 7½ | 6 | 4-5/8 | 6¼ | 7 11/16 | 7 | 7¾ | 20 11/16 | 3¾ | 8½ | 3/8 | ABOVE DIMENSIONS DETERMINED BY SIZE OF MOTOR AND BASE USED | | | | | | | |
| JACK | 4" | 4" | 9 | 7½ | 8-5/8 | 9 | 7½ | 8-5/8 | 7 11/16 | 8¾ | 7¾ | 8 13/16 | 22 11/16 | 4½ | 10 | ½ | | | | | | | | |
| JACOBUS | 5" | 5" | 10 | 8½ | 8-¾ | 10 | 8½ | 8-¾ | 8⅛ | 9⅛ | 8¼ | 8 13/16 | 22 15/16 | 5 | 10¼ | ½ | | | | | | | | |

**FIG. 4.** An outline assembly drawing (tabular). Tabulation gives the dimensions for different-sized units.

other illustrative purposes, dimensions are often omitted. Outline assembly drawings are frequently used to give the information required for the installation or erection of equipment and are then called *installation drawings*.

An *assembly working drawing* gives complete information for producing a machine or structure on one drawing. This is done by providing adequate orthographic views together with dimensions, notes, and a descriptive title. The figure for Prob. 45 is an example.

A *diagram drawing* is an assembly showing, symbolically, the erection or installation of equipment. Erection and piping and wiring diagrams are examples. Diagram drawings are often made in pictorial form.

## 4. Detail Drawings.

A *detail* drawing is the drawing of a single piece, giving a complete and exact description of its form, dimensions, and construction. A successful detail drawing will tell the workman *simply* and *directly* the shape, size, material, and finish of a part; what shop operations are necessary; what limits of accuracy must be observed; the number of parts wanted; etc. It should be so exact in description that, if followed, a satisfactory part will result. Figure 5 illustrates a commercial detail drawing.

Detailing practice varies somewhat according to the industry and the requirements of the shop system. For example, in structural work, details are often grouped together on one sheet, while in modern mechanical practice a separate sheet is used for each part.

If the parts are grouped on one sheet, the detailed pieces should be set, if possible, in the same relative position as on the assembly and, to facilitate reading, placed as nearly as possible in natural relationship. Parts of the same material or character are usually grouped together, as, for example, forgings on one sheet, castings on another, and parts machined from stock on another. A subtitle must be provided for each part, giving the part number, material, number required for each assembly, etc.

The accepted and best system in mechanical work is to have each piece, no matter how small, on a

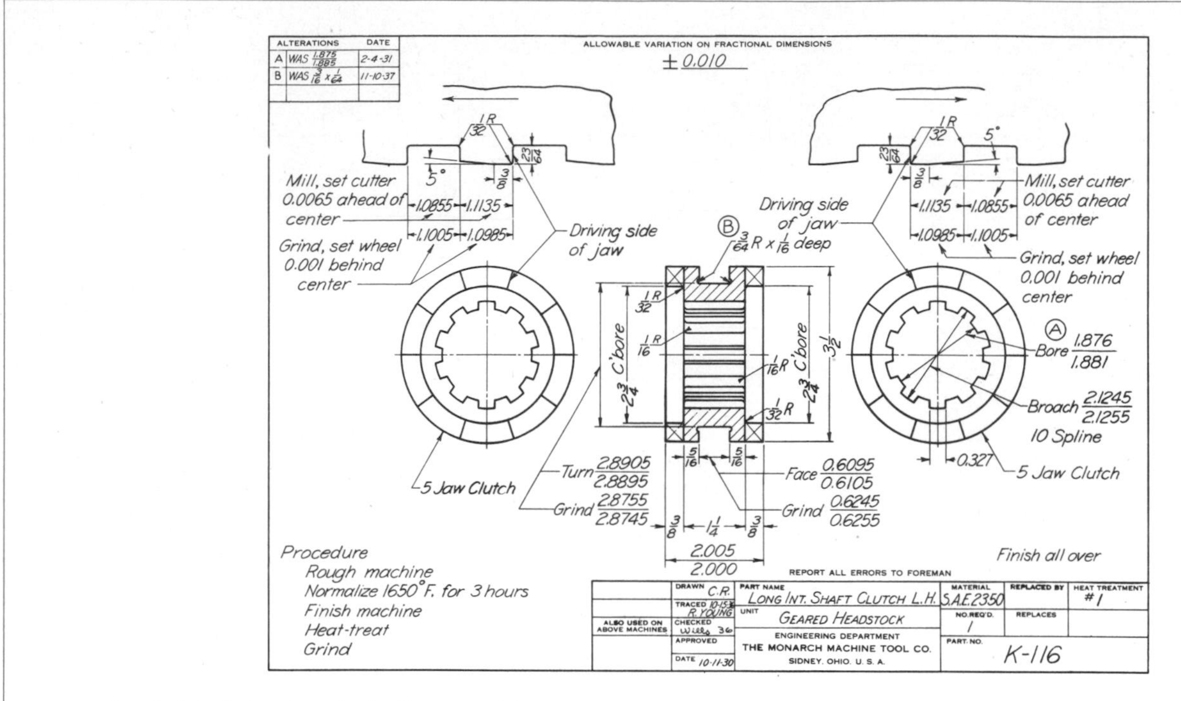


FIG. 5. A detail drawing. Note the completeness of manufacturing information given on the drawing.

separate sheet. As described in Sec. 3, Chap. 11, if the single-drawing system is followed, one drawing will be used by all shops. If the multiple system is used, a separate drawing must be made for each shop; thus there may be a *pattern drawing*, a *casting drawing*, and a *machining drawing*, all for a single cast part. A detail drawing should be a complete unit for the purpose intended and should not be dependent in any way upon any other detail drawing.

## 5. Tabular Drawings.

A tabular drawing, either assembly or detail, is one on which the dimension values are replaced by reference letters, an accompanying table on the drawing listing the corresponding dimensions for a series of sizes of the machine or part, thus making one drawing serve for the range covered. Some companies manufacturing parts in a variety of sizes use this tabular system of size description, but a serious danger with it is the possibility of misreading the table. Figure 4 illustrates a tabular assembly drawing.

## 6. Standard Drawings.

To avoid the difficulties experienced with tabular drawings, some companies make a "standard drawing," complete except for the actual figured dimensions. This drawing is reproduced by offset printing or black-and-white reproduction on vellum paper, and the reproductions are dimensioned separately for the various sizes of parts. This method gives a separate complete drawing for each size of part; and when a new size is needed, the drawing is easily and quickly made. Figure 6 shows a standard drawing and Fig. 7, the drawing filled in, making a completed working drawing.

## 7. Standard Parts.

Purchased or company standard parts are specified by name and size or by number and, consequently,

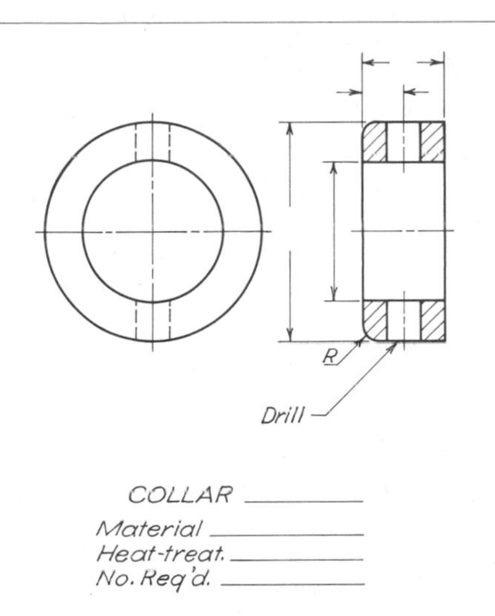


FIG. 6. A standard drawing. This is reproduced on tracing paper to provide the shape description needed for the drawing.

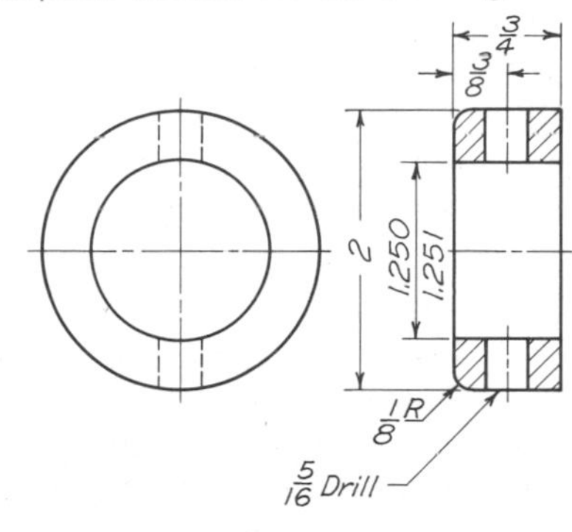


FIG. 7. A filled-in standard drawing. With the addition of dimensions, a standard drawing becomes a completed drawing to be used for manufacturing purposes.

do not need to be detailed. All standard parts, such as bolts, screws, antifriction bearings, etc., are shown on the assembly drawing and are given a part number. The complete specifications for their purchase are given in the parts list.

Sometimes, however, a part is made by *altering* a standard or previously produced part. In this case a detail drawing is made, showing and specifying the original part with changes and dimensions for the alteration.

### 8. The Bill of Material, or Parts List.

This is a tabulated statement, usually placed on a separate sheet in the case of quantity production (as in Prob. 37) or on the assembly drawing in other cases (as in Prob. 50). This table gives the piece number, name, quantity, material, sometimes the stock size of raw material, detail drawing numbers, weight of each piece, etc. A final column is usually left for remarks. The term "bill of material" is ordinarily used in structural and architectural drawing. The term "parts list" applies more accurately in machine-drawing practice. In general, the parts are listed in the order of their importance, with the larger parts first and the standard parts such as screws, pins, etc., at the end.

The blank ruling for a bill of material should not be crowded. Lines should never be spaced closer than $\frac{1}{4}$ in.; $\frac{5}{16}$ or $\frac{3}{8}$ in. spacing is better, with the height of the lettering not more than half the space and centered between the lines. Instead of being lettered, bills of material are frequently typed on forms printed on thin paper. Intensifying the impression by typing with carbon paper on the back increases the opacity of the letters, and a clearer blueprint will result.

### 9. Set of Drawings.

A *complete set* of drawings consists of detail sheets and assembly sheets, the former giving all necessary information for the manufacture of each of the individual parts which cannot be purchased and the latter showing the parts assembled as a finished unit or machine. The set includes the bill of material or parts list and may also contain special drawings for the purchaser, such as foundation plans or oiling diagrams.

### 10. Making a Working Drawing: Basic Concepts.

Although pictorial drawings are used to some extent in special cases, the basis of all working drawings is orthographic projection. Thus, to represent an object completely, at least two views are ordinarily needed, often more. The only general rule is: *make as many views as are necessary to describe the object clearly—and no more.* Instances may occur in which the third dimension is so evident as to make one view sufficient, as, for example, in the drawing of a shaft or bolt. In other cases, perhaps a half-dozen views will be required to describe a piece completely. Sometimes half, partial, or broken views can be used to advantage.

Select for the front view the face showing the largest dimension, preferably the obvious front of the object when in its functioning position, and then decide what other views are necessary. A vertical cylindrical piece, for example, would require only a front and a top view; a horizontal cylindrical piece, only a front and a side view. Determine which side view to use or whether both are needed. The one with the fewest dashed lines should be preferred. In some cases the best representation will be obtained by using *both* side views with all unnecessary dashed lines omitted. See whether an auxiliary view or a note will eliminate one or more other views and whether a section is better than an exterior view. One statement can be made with the force of a rule:

*If anything in clearness can be gained by violating a principle of projection, violate it.*

Secs. 12 to 26, Chap. 8, give a number of examples of conventions that are in violation of theoretical representation but are in the interest of clearness. The draftsman must remember that his responsibility is to the reader of the drawing and that he is not justified in saving himself any time or trouble at the expense of the drawing, by making it less plain or easy to read. The time so saved by the draftsman may be lost to the company a hundredfold in the shop, where the drawing is used not once but repeatedly.

There is a *style* in drawing, just as there is in literature, which indicates itself in one way by ease of reading. Some drawings stand out, while others, which may contain all the information, are difficult to decipher. Although dealing with mechanical thought, there is a place for some artistic sense in mechanical drawing. The number, selection, and disposition of views; the omission of anything unnecessary, ambiguous, or misleading; the size and placement of dimensions and lettering; and the contrast of lines are all elements in the style.

In commercial drafting, *accuracy* and *speed* are the two requirements. The drafting room is an expensive department, and time is an important element. The draftsman must therefore have a ready knowledge not only of the principles of drawing but of the conventional methods and abbreviations and of any device or system that will save time without sacrificing clearness.

The usual criticism of the beginner by the employer is the result of the former's lack of appreciation of the necessity for speed.

### 11. Materials Used for Working Drawings.

Working drawings go to the shop in the form of blueprints, black-line prints, or other similar forms of reproduction, and the drawings must therefore be made on translucent material, either directly or as tracings. Pencil drawings may be made on tracing paper or on pencil cloth; inked drawings, on tracing paper or on tracing cloth.

*Tracing paper* is a thin translucent material, commonly called "vellum." Considerable time and expense may be saved by making the original pencil drawing on this material. Excellent prints can be obtained if the lines are of sufficient blackness and intensity.

*Pencil cloth* is a transparentized fabric with one or both sides of its surface prepared to take pencil so that the original drawing can be made on it and prints made either from the pencil drawing directly or after it has been inked. Some of the newer cloths are moisture resistant, others are really waterproof. Pencil cloth is made for pencil drawings, and perfect blueprints can be made from drawings made on it with sharp, hard pencils. Ink lines, however, do not adhere well and have a tendency to chip or rub off in cleaning.

*Tracing cloth* is a fine-thread fabric sized and transparentized with a starch preparation. The smooth side is considered by the makers as the working side, but most draftsmen prefer to work on the dull side, which will take pencil marks. The cloth should be fastened down smoothly over the pencil drawing and its selvage torn off. To remove the traces of grease that sometimes prevent the flow of ink, dust the tracing cloth with chalk or prepared pounce (a blackboard eraser may be used) and then rub it off with a cloth. Carbon tetrachloride is an effective cleaning agent. Rub a moistened cloth over the surface—any excess will evaporate in a moment.

A *plastic material,* known as "Mylar,"[1] is now available for both pencil and ink drawings. It is evenly translucent, has a fine matte surface, good lasting qualities, and requires no special storage precautions.

[1]DuPont Manufacturing Co.

**12. Drawing Sizes.**
Drawing paper and cloth are available in rolls of various widths and in standard trimmed sizes. Most drafting rooms use standard sheets, printed with border and title block. The recommended sizes shown in Table 1, based on multiples of $8\frac{1}{2}$ by 11 in. and 9 by 12 in., permit the filing of prints in a standard letter file.

Figure 8 shows the most common trimmed sizes. Larger drawings can be made on rolled stock of

**Table 1. Finished Flat-sheet Sizes.**

| Designation | Width | Length | Designation | Width | Length |
|---|---|---|---|---|---|
| *A* | $8\frac{1}{2}$ | 11 | *A* | 9 | 12 |
| *B* | 11 | 17 | *B* | 12 | 18 |
| *C* | 17 | 22 | *C* | 18 | 24 |
| *D* | 22 | 34 | *D* | 24 | 36 |
| *E* | 34 | 44 | *E* | 36 | 48 |
| *F** | 28 | 40 | *F** | 28 | 40 |

*Not a multiple of the basic size. To be used when width of *E* size is not adaptable.

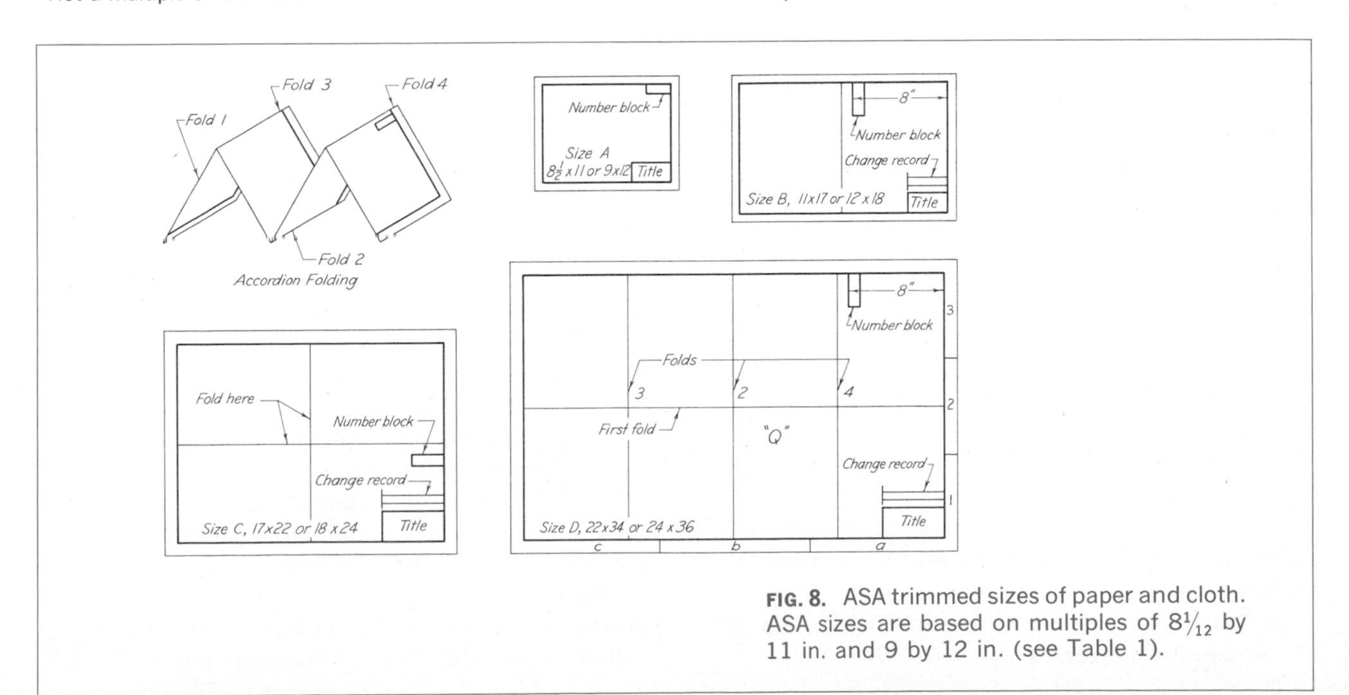


**FIG. 8.** ASA trimmed sizes of paper and cloth. ASA sizes are based on multiples of $8\frac{1}{12}$ by 11 in. and 9 by 12 in. (see Table 1).

standard width, with the length as a multiple of 11 or 12 in., not to exceed 144 in.

### 13. Order of Penciling.

After the scheming, inventing, and calculating have been done and the design drawing has been completed, the order of procedure for making the detail drawings is:

1. Select a suitable standard sheet or lay off a sheet to standard size, with the excess paper to the right, as a convenient space for making sketches and calculations, and block out the space for the title.
2. Decide what scale to use, choosing one large enough to show all dimensions without crowding, and plan the arrangement of the sheet by making a little preliminary freehand sketch, estimating the space each view will occupy and placing the views to the best advantage for preserving, if possible, a balance in the appearance of the sheet. Be sure to leave sufficient space *between* views for the dimensions.
3. Draw the center lines for each view, and on these "block in" the views by laying off the principal dimensions and outlines, using *light, sharp, accurate* pencil lines. Center lines are drawn for the axes of symmetry of all symmetrical views or parts of views. Thus every cylindrical part should have a center line—the projection of the axis of the piece. Every circle should have two center lines intersecting at its center.
4. Draw the views, beginning with the most dominant features and progressing to the subordinate. Carry the different views on together, projecting a characteristic shape as shown on one view to the other views and *not* finishing one view before starting another. Draw the lines to the final finished weight (wherever possible), using a minimum of construction. *Never* make a drawing lightly and "heavy" it later.
5. Finish the projections, putting in last the minor details. Check the projections and make sure that all views are complete and correct.
6. Draw all necessary dimension lines; then put in the dimension values.
7. Draw guide lines for the notes, and then letter them.
8. Lay out the title.
9. Check the drawing carefully.

The overrunning lines of the constructive stage should not be erased before tracing or inking. These extensions are often convenient in showing the stopping points. Avoid unnecessary erasing as it abrades the surface of the paper so that it catches dirt more readily.

As an aid in stopping tangent arcs in inking, it is desirable to mark the tangent point on the pencil drawing with a short piece of the normal to the curve at the point of tangency.

Figure 9 illustrates the stages of penciling.

### 14. Order of Inking.

To ensure good printing, the ink should be perfectly black and the ruling pens in good condition. Red ink should not be used unless it is desired to have some lines inconspicuous on the print. Blue ink will not print well. Sometimes, on maps, diagrams, etc., it is desirable to use colored inks on the tracing to avoid confusion of lines; in such cases, add a little Chinese white to the colored inks and it will render them sufficiently opaque to print.

Ink lines can be removed from tracing cloth by rubbing with a hard pencil eraser, slipping a triangle under the tracing to give a harder surface. The rubbed surface should afterward be burnished with a burnisher or fingernail. In tracing a part that has been section-lined, a piece of white paper can be slipped under the cloth and the section lining done without reference to the section lines underneath.

Tracing cloth is sensitive to atmospheric variations, often changing overnight so as to require restretching. If a large tracing cannot be finished in one day, some views should be finished and no figure left only partly traced.

In making a large tracing, it is well to cut the required piece from the roll and lay it exposed, flat, for a short time before fastening it down.

Water will ruin a tracing on starch-coated cloth, and moist hands or arms should not come in contact with it. Form the habit of keeping the hands off

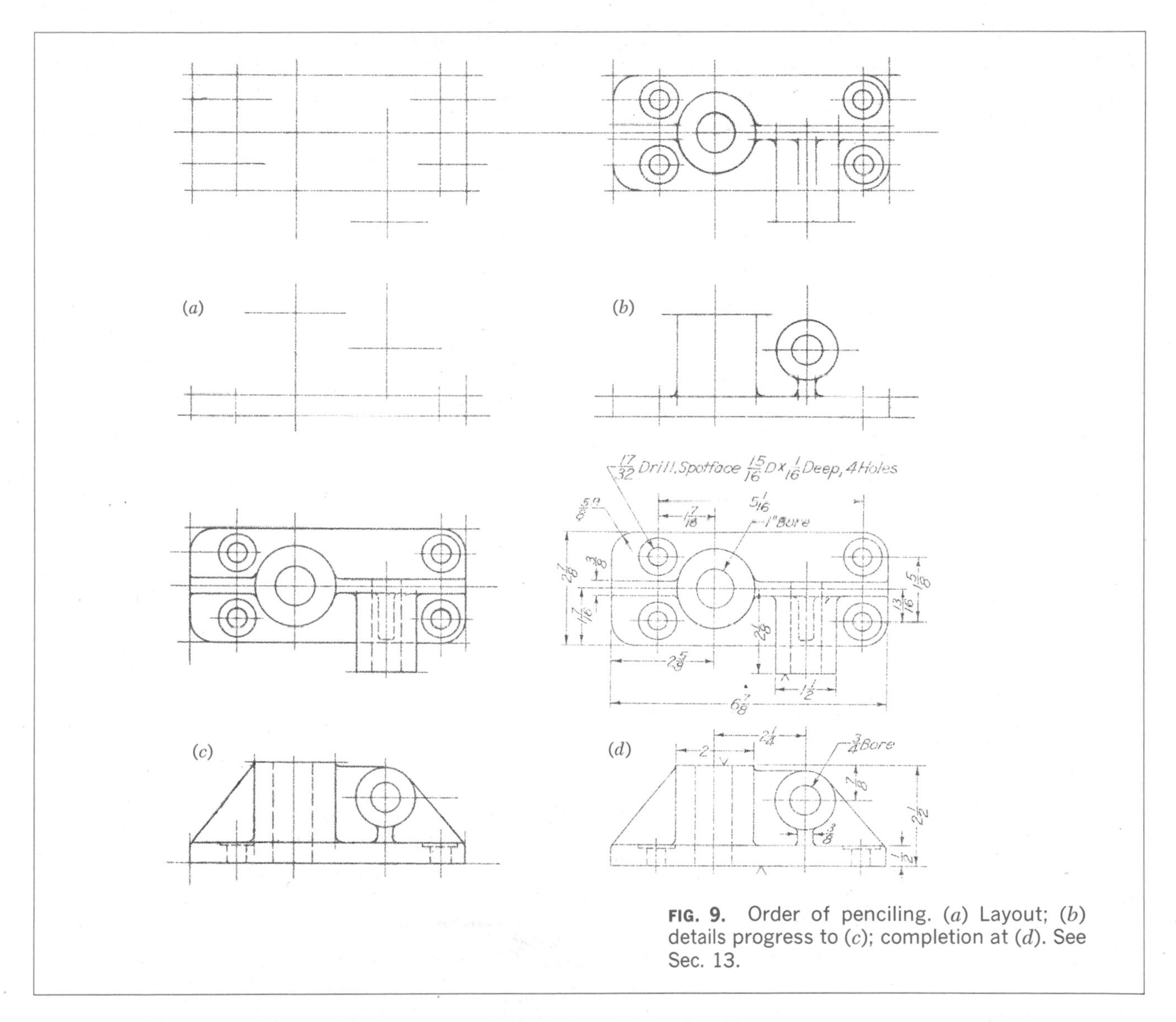


FIG. 9. Order of penciling. (*a*) Layout; (*b*) details progress to (*c*); completion at (*d*). See Sec. 13.

drawings. In both drawing and tracing on large sheets, it is a good plan to cut a mask of drawing paper to cover all but the view being worked on. Unfinished drawings should be covered overnight.

Tracings can be cleaned of pencil marks and dirt by wiping with a cloth moistened with benzine or carbon tetrachloride. To prevent smearing when using this method of cleaning, borders and titles should be printed in an ink not affected by benzine.

*Order of Inking*

1. Ink all full-line circles, beginning with the smallest, and then circle arcs.
2. Ink dashed circles and arcs in the same order as full-line circles.
3. Ink any irregular curved lines.
4. Ink straight full lines in this order: horizontal, vertical, and inclined.
5. Ink straight dashed lines in the same order.

6. Ink center lines.
7. Ink extension and dimension lines.
8. Ink arrowheads and dimensions.
9. Section-line all areas representing cut surfaces.
10. Letter notes and titles. (On tracings, draw pencil guide lines first.)
11. Ink the border.
12. Check the inked drawing.

Figure 10 shows the stages of inking.

## 15. Title Blocks.

The title of a working drawing is usually placed in the lower right-hand corner of the sheet, the size of the space varying with the amount of information to be given. The spacing and arrangement are designed to provide the information most helpful in a particular line of work.

In general, the title of a machine drawing should contain the following information:

1. Name of company and its location

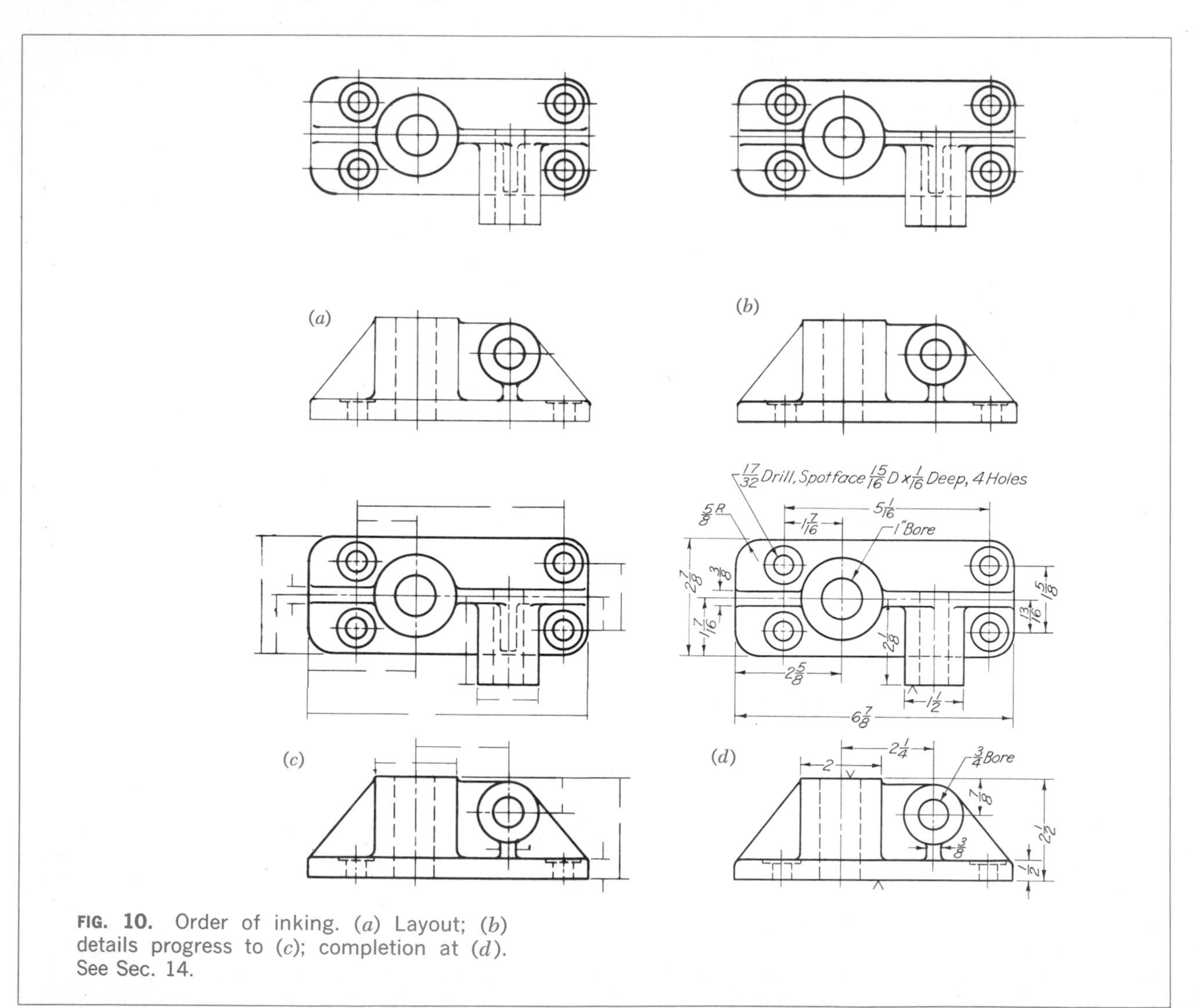


**FIG. 10.** Order of inking. (*a*) Layout; (*b*) details progress to (*c*); completion at (*d*). See Sec. 14.

2. Name of machine or unit
3. Name of part (if a detail drawing)
4. Drawing number
5. Part number (if a detail drawing)
6. Number of parts required (for each assembly)
7. Scale
8. Assembly-drawing number (given on a detail drawing to identify the part in assembly)
9. Drafting-room record: names or initials of draftsman, tracer, checker, approving authority; each with date
10. Material

To these, depending upon the need for the information, may be added:

11. Stock size
12. Heat-treatment
13. Finish
14. Name of purchaser, if special machine
15. The notation that the drawing "supersedes" and is "superseded by"

*Form of Title.* Every drafting room has its own standard form for titles. In large offices the blank form is often printed in type on paper, plastic, film or cloth (see Figs. 11 and 12 for characteristic examples.)

A form of title that is used to some extent is the *record strip*, a strip marked off across the lower part or right end of the sheet, containing the information required in the title and space for the recording of orders, revisions, changes, etc., that should be noted, with the date, as they occur. Figure 13 illustrates one form.

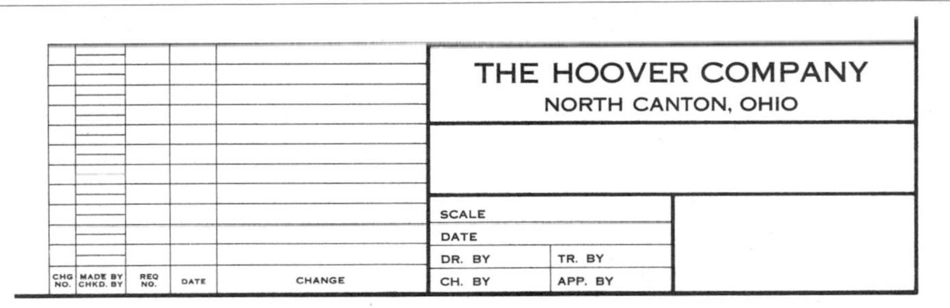

THE HOOVER COMPANY
NORTH CANTON, OHIO

| CHG NO. | MADE BY CHKD. BY | REQ NO. | DATE | CHANGE |
|---|---|---|---|---|

| SCALE | |
|---|---|
| DATE | |
| DR. BY | TR. BY |
| CH. BY | APP. BY |

FIG. 11. A printed title form. This one contains a "change-record" block.

HARRISON COUPLER CO.
ELECTRICAL DEPT. PHILADELPHIA

| PATTERN NO. | MATERIAL | EST. WEIGHT | SUPERSEDES DWG. |
|---|---|---|---|

TITLE:

ELECTRICAL ENGINEER — SUPT. ELECTRICAL DEPARTMENT

| DRAWN BY | DATE | TRACED BY | DATE |
|---|---|---|---|
| | | CHECKED BY | |

NO.

FIG. 12. A printed title form. This one is for an electrical department.

| REVISION | DATE | CHANGE OR ADDITION |
|---|---|---|

UNIT — NAME OF PIECE

| | DATE | SYMBOL OF MACHINES USED ON | SUPERSEDES DRAW. | STOCK CASTING DROP FORGING | |
|---|---|---|---|---|---|
| DR. | | | | | |
| CH. | | | | | |
| TR. | | THE LODGE & SHIPLEY MACHINE TOOL CO. | SUPERSEDED BY DRAW. | MATERIAL | PIECE NO. |
| TR. CH. | | Form 796 CINCINNATI, OHIO. U. S. A. | | | |

FIG. 13. A strip title. This type extends across one side of the drawing.

It is sometimes desired to keep the records of orders and other private information on the tracing but not to have them appear on the print. In such cases a record strip is put outside the border and trimmed off the print before sending it out.

*To Letter a Title.* The title should be lettered freehand in single-stroke capitals, vertical or inclined—not both styles in the same title. Write out the contents on a separate piece of paper; then refer to Sec. 15, Chap. 4, where full instructions have been given.

### 16. Zoning.

As an aid in locating some item on a large drawing, the lower and right borders may be ruled and marked as shown on the *D*-size drawing in Fig. 8. Item *Q* would be located in zone *b*2. A separate column in the change-record block is often used to indicate the position of each drawing change.

### 17. Checking.

When a working drawing is finished, it must be checked by an experienced person, who, in signing his name to it, becomes responsible for any errors. This is the final "proofreading" and cannot be done by the one who has made the drawing nearly so well as by another person. In small offices all the work is checked by the chief draftsman, or sometimes draftsmen check one another's work; in large drafting rooms one or more checkers are employed who devote all their time to this kind of work. All notes, computations, and checking layouts should be preserved for future reference.

Students can gain experience in this work by checking one another's drawings.

To be effective, checking must be done in an absolutely systematic way and with thorough concentration.

### 18. Order of Checking.

Each of the following items should be gone through separately. As each dimension or feature is verified, a check mark in colored pencil should be placed on or above it and corrections indicated with a different-colored pencil.

1. Put yourself in the position of those who are to read the drawing, and find out whether it is easy to read and tells a straight story. Always do this before checking any individual features, in other words, before you have had time to become accustomed to the contents.

2. See that each piece is correctly designed and illustrated and that all necessary views are shown but none that is not necessary.

3. Check all the dimensions by scaling and, where advisable, also by calculation. Preserve the calculations.

4. See that dimensions for the shop are given as required by the shop and that the shop is not left to do any adding or subtracting in order to get a needed dimension.

5. Check for tolerances. See that they are neither too "fine" nor too "coarse" for the particular conditions of the machine, so that they will not, on the one hand, increase unnecessarily the cost of production or, on the other, impair accuracy of operation or duplication.

6. Go over each piece and see that finishes are properly specified.

7. See that every specification of material is correct and that all necessary ones are given.

8. Look out for "interferences." This means that each detail must be checked with the parts that will be adjacent to it in the assembled machine to see that proper clearances have been allowed.

9. When checking for clearances in connection with a mechanical movement, lay out the movement to scale, figure the principal angles of motion, and see that proper clearances are maintained in all positions, drawing small mechanisms to double size or larger.

10. Check to see that all the small details such as screws, bolts, pins, and rivets are standard and that, where it is possible, stock sizes have been employed.

11. Check every feature of the title or record strip and bill of material.

12. Review the drawing in its entirety, adding such explanatory notes as will increase its efficiency.

## 19. Alterations.

Once a drawing has been printed and the prints have been released to the shop, any alterations or changes should be recorded on the drawing and new prints issued. If the changes are extensive, the drawing may be *obsoleted* and a new drawing made which *supersedes* the old drawing. Many drawing rooms have "change-record" blocks printed in conjunction with the title, where minor changes are recorded (Fig. 11). The change is identified in the record and on the face of the drawing by a letter.

New designs may be changed so often that the alterations cannot be made fast enough to reach the shop when needed. In this case sketches showing the changes are rapidly made, reproduced, and sent to the shop, where they are fastened to each print of the drawing. These sketches, commonly known as "engineering orders," are later incorporated on the drawing.

Portions of a drawing can be canceled by drawing closely spaced parallel lines, usually at 45°, over the area to be voided.

## 20. Working Sketches.

Facility in making freehand orthographic drawings is an essential part of the equipment of every engineer. Routine men such as tracers and detailers may get along with skill and speed in mechanical execution, but the designer must be able to record his ideas freehand. In all inventive mechanical thinking, in all preliminary designing, in all explanations and instructions to draftsmen, freehand sketching is the mode of expression. Its mastery means mastery of the language, and it is gained only after full proficiency in drawing with instruments has been acquired. It is mastery which the engineer, inventor, designer, chief draftsman, and contractor, with all of whom time is too valuable to spend in mechanical execution, must have.

Working sketches may be made in orthographic or pictorial form. Chapter 5 gives the fundamentals for orthographic freehand drawings, and Chap. 6 discusses pictorial sketching.

## 21. Kinds of Working Sketches.

Working sketches can be divided into two general classes: (1) those made before the structure is built and (2) those made after the structure is built.

In the first class are included the sketches made in connection with the designing of the structure. These can be classified as (*a*) *scheme* or *idea* sketches, used in studying and developing the arrangement and proportion of parts; (*b*) *computation sketches*, made in connection with the figured calculations for motion and strength; (*c*) *executive sketches*, made by the chief engineer, inventor, or consulting engineer to give instructions for special arrangements or ideas that must be embodied in the design; (*d*) *design sketches*, used in working up the schemes and ideas into suitable form so that the design drawing can be started; and (*e*) *detail sketches*, made as substitutes for detail drawings.

The second class includes (*a*) *detail sketches*, drawn from existing models or parts, with complete notes and dimensions, from which duplicate parts can be constructed directly or from which working drawings can be made (Fig. 14); (*b*) *assembly sketches*, made from an assembled machine to show the relative positions of the various parts, with center and location dimensions, or sometimes for a simple machine, with complete dimensions and specifications; and (*c*) *outline* or *diagrammatic sketches*, generally made for the purpose of location: sometimes, for example, to give the size and location of pulleys and shafting, piping, or wiring, that is, information for use in connection with the setting up of machinery; sometimes to locate a single machine, giving the over-all dimensions, sizes, and center distances for foundation bolts and other necessary information.

## 22. Making a Working Sketch.

In making a working sketch, the principles of projection and the rules of practice for working drawings are to be remembered and applied. A systematic order should be followed for both idea sketches and sketches from objects, as listed below:

1. Visualize the object.

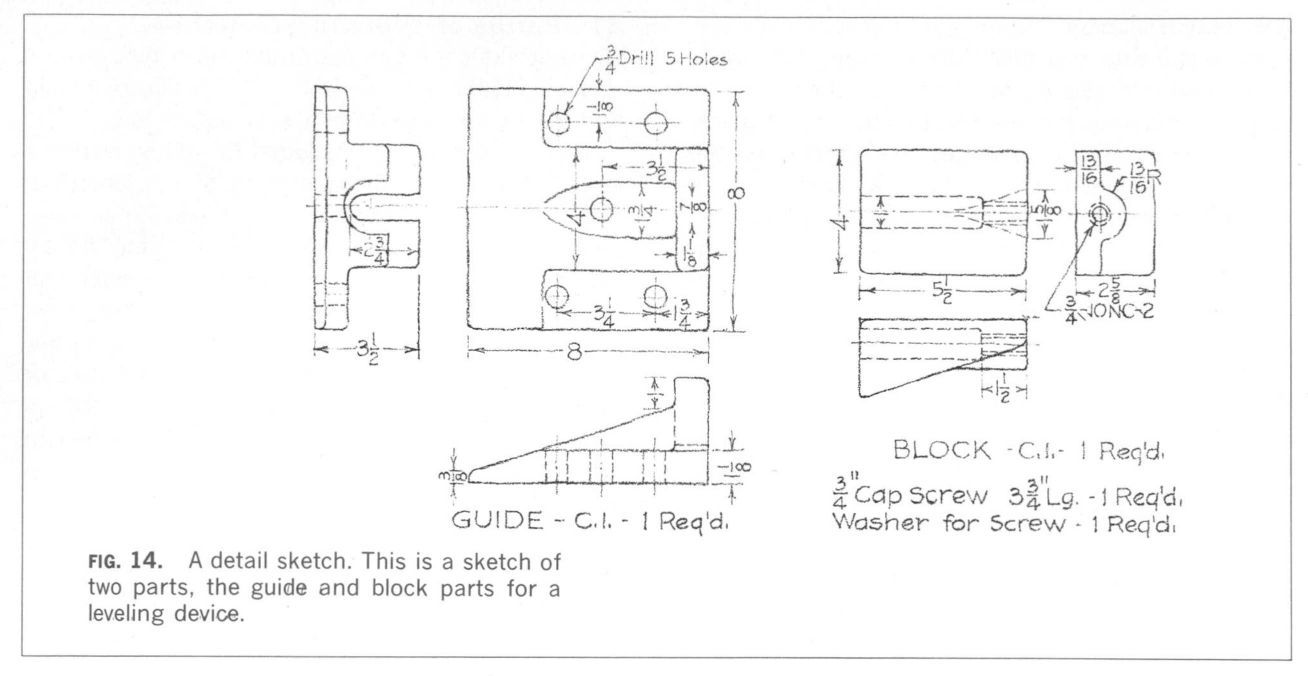


**FIG. 14.** A detail sketch. This is a sketch of two parts, the guide and block parts for a leveling device.

2. Decide on the treatment, orthographic or pictorial.
3. Determine the view or views.
4. Determine the size of the sketch.
5. Locate the center lines.
6. Block in the main outlines.
7. Complete the detail.
8. Add dimension lines and arrowheads.
9. Put on the dimension values.
10. Letter notes and title, with date.
11. Check the drawing.

Before a good graphic description of an object or idea can be developed, it is essential that the mental image of it be definite and clear. The clearness of the sketch is a direct function of this mental picture. Hence the first step is to concentrate on visualization, which leads directly to the second step: determination of the best method of representation.

The method of representation will probably not be just the same as would be used in a scale drawing. For example, a note in regard to thickness or shape of section will often be used in a working drawing to save a view (Fig. 15); thus one view of a piece that is circular in cross section would be sufficient. In other cases additional part views and extra sections may be sketched rather than complicate the regular views with added lines that might confuse the sketch, although the same lines might be perfectly clear in a measured drawing.

The third step is to determine the view (pictorial) or views (orthographic). Draw the object in its functioning position, if possible, but if another position will show the features to better advantage, use it. A machine should, of course, be represented right-side up, in its natural working position. If symmet-

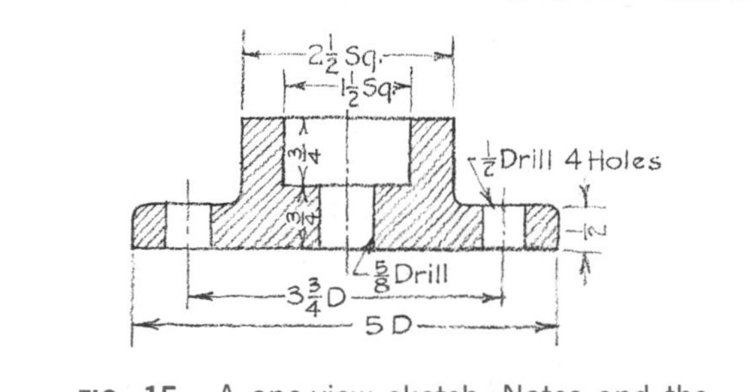


**FIG. 15.** A one-view sketch. Notes and the dimensioning details make another view unnecessary.

rical about an axis, often only one-half need be sketched. If a whole view cannot be made on one sheet, it may be put on two, each part being drawn up to a break line used as a datum line.

The fourth step is to proportion the size of the sketch to the sheet. Have it large enough to show all detail clearly, but allow plenty of room for dimensions, notes, and memorandums. Small parts can be sketched larger than full size. Do not try to crowd all the views on one sheet of paper. Use as many sheets as may be required, but name each view, indicating the direction in which it is taken. Sometimes one view alone will require a whole sheet.

*In drawing on plain paper,* the location of the principal points, centers, etc., should be so judged that the sketches will fit on the sheet, and the whole sketch, with as many views, sections, and auxiliary views as are necessary to describe the piece, will be drawn in as nearly correct proportion as the eye can determine, *without making any measurements.*

*Cross-section Paper.* Sketches are often made on coordinate paper ruled faintly in $\frac{1}{16}$, $\frac{1}{8}$, or $\frac{1}{4}$ in., used simply as an aid in drawing straight lines and judging proportions, or for drawing to approximate scale by assigning suitable values to the unit spaces. The latter use is more applicable to design sketches than to sketches from the object (Fig. 16).

In order to gain skill through practice, sketches should be made entirely freehand. However, in commercial work the engineer often saves time by making a hybrid sketch, drawing circles with the compass or even with a coin from his pocket, ruling some lines with a pocket scale or a triangle and making some freehand but always keeping a workmanlike quality and good proportion.

## 23. Dimensioning and Finishing.

After completing the views of a piece, go over it and add *dimension lines* for all the dimensions needed for its construction, drawing extension lines and arrowheads carefully and checking to see that none is omitted.

The dimension values are now added. If the sketch is made from reference drawings and specifications, these sources give the dimension values.

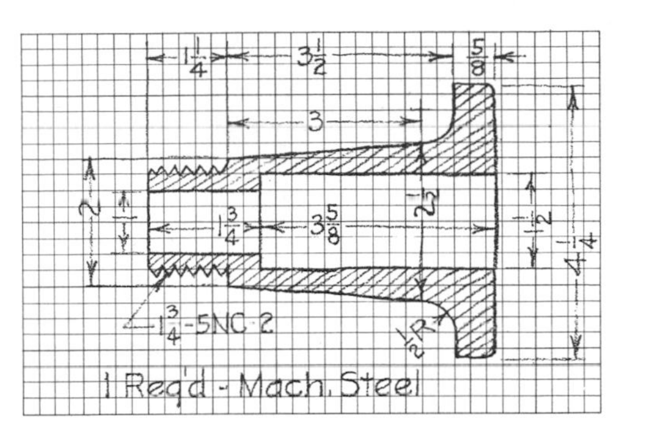


FIG. 16. A sketch on coordinate paper.

If the sketch is of an existing part or model, measurements must be made to determine the dimension values, as explained in Sec. 24.

Add all remarks and notes that seem to be of possible value. The title should be written or lettered on the sketch.

*Always Date a Sketch.* Valuable inventions have been lost through inability to prove priority because the first sketches were not dated. In commercial work the draftsman's notebook with sketches and calculations is preserved as a permanent record, and its sketches should be made so as to stand the test of time and be legible after the details of their making have been forgotten.

## 24. Measuring and Dimensioning.

Before adding values, *if the sketch is of an existing model or part,* the object has not been handled, and so the drawing has been kept clean. The measurements for the dimensions indicated on the drawing are now needed. A flexible rule or steel scale will serve for getting most of the dimensions. Never use a draftsman's scale for measuring castings, as it would become soiled and its edges marred. The diameter of cylindrical shapes or the distance between outside surfaces can be measured by using outside calipers and scale (Fig. 17); and the sizes of holes or internal surfaces, by using inside calipers. Figure 18 illustrates the inside transfer caliper, used when a projecting portion prevents removing

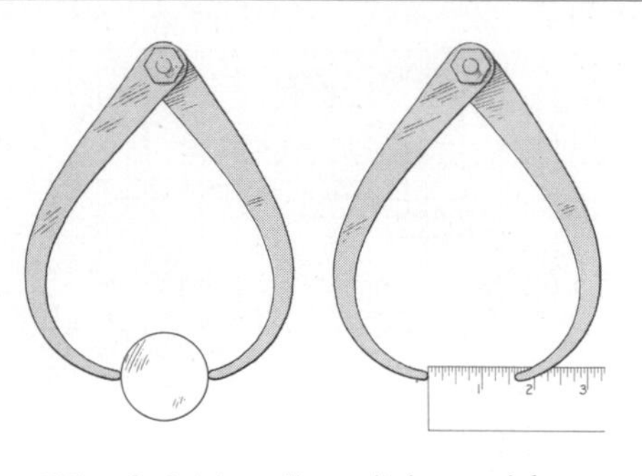

**FIG. 17.** Outside caliper. It is used for measuring diameters or thicknesses.

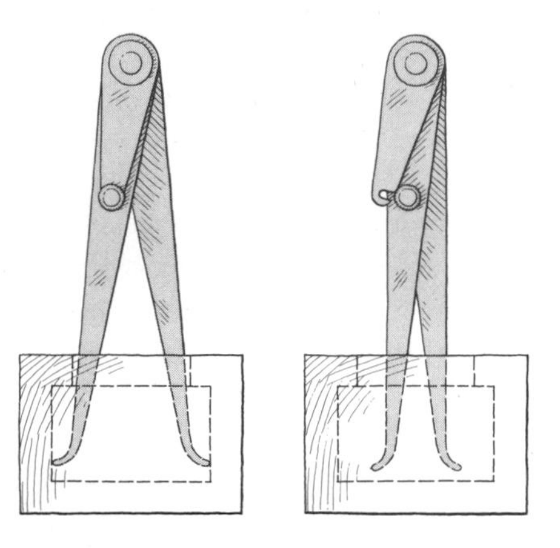

**FIG. 18.** Inside transfer caliper. It is used for measuring internal distances when a shoulder prevents direct removal.

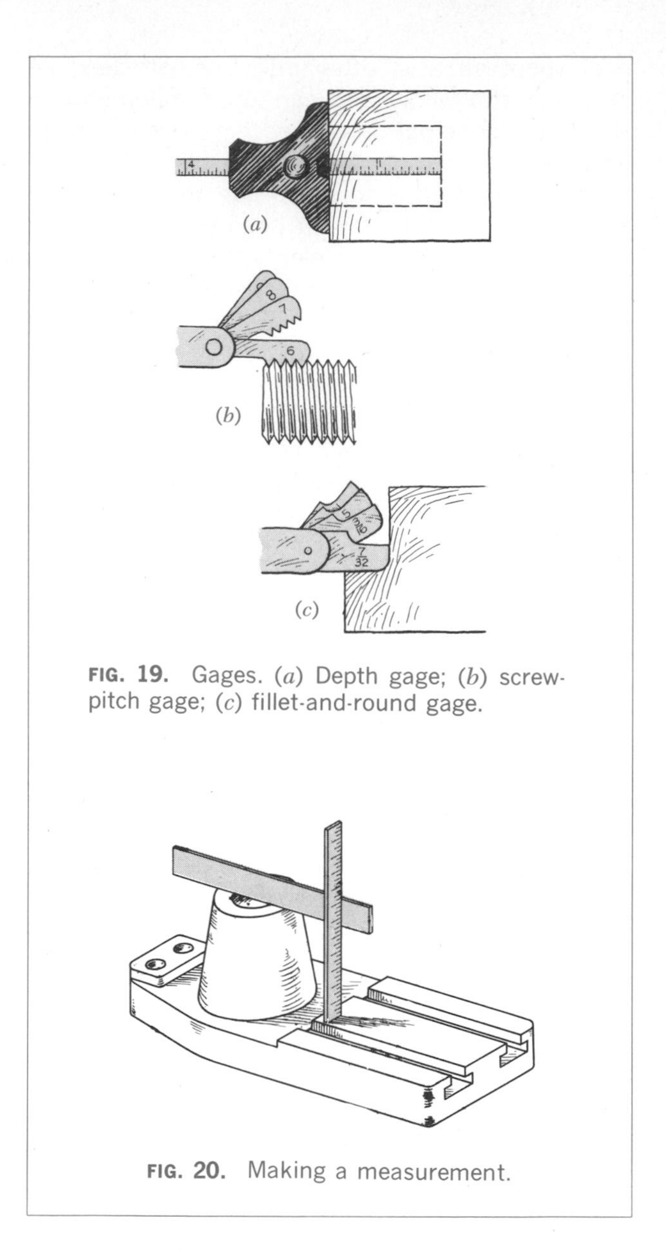

**FIG. 19.** Gages. (*a*) Depth gage; (*b*) screw-pitch gage; (*c*) fillet-and-round gage.

**FIG. 20.** Making a measurement.

the ordinary caliper. The outside transfer caliper is used for a similar condition occurring with an outside measurement. The depth of a hole is easily measured with the depth gage (Fig. 19*a*). Screw threads are measured by calipering the body diameter and either counting the number of threads per inch or gaging with a screw-pitch gage (Fig. 19*b*). A fillet-and-round gage measures radii, the gage fitting to the circular contour (Fig. 19*c*). It is often necessary to lay a straightedge across a surface, as in Fig. 20. This type of measurement could be made conveniently with a combination square or with a surface gage. The combination square has two different heads, the regular 90-45° head (Fig. 21*a*) for a variety of measurements, and the protractor head (Fig. 21*b*) for measuring or laying out angles. For accurate measurements, outside or inside micrometer calipers are necessary. The outside type is illustrated in Fig. 22. Readings to 0.001 in. are easily obtained. Accurate measurements of holes can be made with a telescopic gage in conjunction with an outside micrometer.

A variety of gages made for special purposes, such

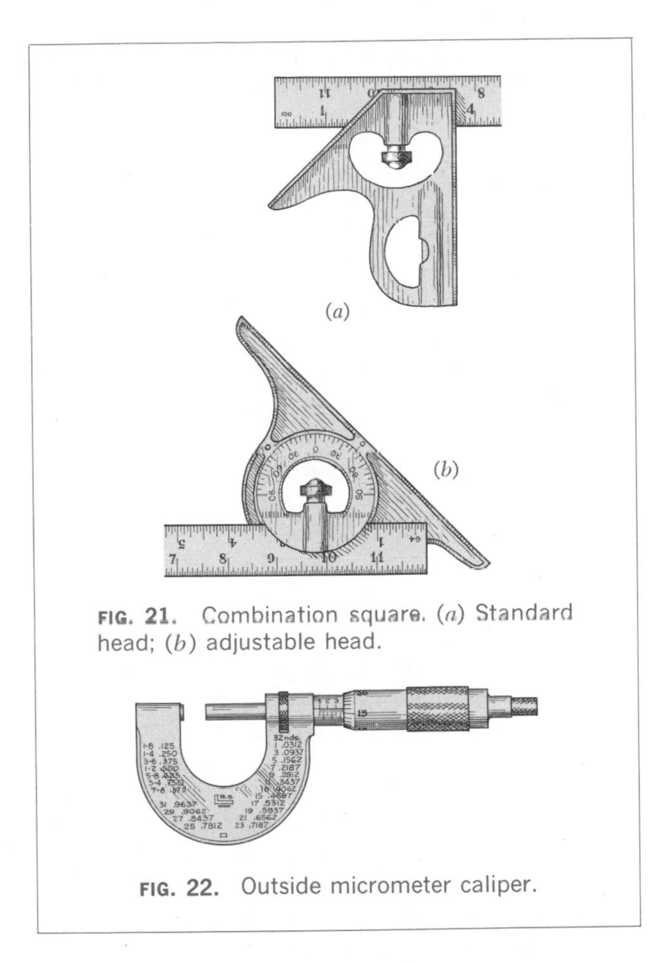


**FIG. 21.** Combination square. (*a*) Standard head; (*b*) adjustable head.

**FIG. 22.** Outside micrometer caliper.

as a wire gage, gage for sheet metal, can be used as occasion demands. With some ingenuity, measurements can often be made with the simpler instruments when special ones are not available.

Always measure from finished surfaces, if possible. Judgment must be exercised in measuring rough castings so as not to record inequalities.

In finding the distance between centers of two holes of the same size, measure from the edge of one to the corresponding edge of the other. Curves are measured by coordinates or offsets, as shown in Figs. 51 and 52, Chap. 12. A curved outline can be recorded by laying a sheet of paper on it and making a rubbing.

## 25. Checking the Sketch.

The final step is to check the sketch. It is a curious fact that when a beginner omits a dimension, it is usually a basic, vital one, such as that of the center height of a machine or an over-all length.

Sketches are not always made on paper with a printed title, but essentially the same information as for an instrument drawing should be recorded in some convenient place on the sheet. All notes and special directions should be checked for accuracy along with the drawing proper. In general, follow the order of checking given in Sec. 17.

## 26. Reproduction of Drawings.

Working drawings go to the shop in the form of prints made from the original drawings. Several different printing processes are in use, all of which give the best results from tracings inked on tracing cloth or paper. However, quite satisfactory prints can be obtained from pencil drawings on translucent paper when the penciling is done skillfully, with uniform, opaque lines. In fact, most of the drawings of industry are not inked; only those of a permanent nature, such as maps, charts, etc., and tracings that must be printed a great many times, are inked.

*Blueprints.* The simplest and most generally used copying process is the blueprinting process, in which the prints are made by exposing a piece of sensitized paper and a tracing in close surface contact with each other to sunlight or electric light in a printing frame or machine made for the purpose. On exposure to the light, a chemical action takes place, which, when fixed by washing in water, gives a strong blue color. The parts protected from the light by the black lines of the tracing wash out, leaving the white paper.

*Vandyke Paper.* This is a thin sensitized paper that turns dark brown when exposed to light and properly "fixed." A reversed negative of a tracing can be made on it by exposing it to light with the inked side of the drawing next to the sensitized side of the paper; then this negative can be printed on blueprint paper, giving a positive print with blue lines on white.

*BW Prints and Directo Prints.* These have black lines on a white ground and are made directly from the original tracing, in a blueprinting machine (and developed by hand) or in a special machine made for the purpose. They are used extensively when positive prints are desired.

*Ozalid Prints.* This process is based on the chemical action of light-sensitive diazo compounds. It is a contact method of reproduction in which the exposure is made in a regular blueprinting machine or an ozalid "whiteprint" machine, and the exposed print is developed dry with ammonia vapors in a developing machine. Standard papers giving black, blue, and maroon lines on a white ground are available. Dry developing has the distinct advantage of giving prints without distortion, and it also makes possible the use of transparent papers, cloth, and foils, which effects savings in drawing time, as these transparent replicas can be changed by additions or erasures and prints made from them without altering the original tracing.

*Photostat Prints.* These are extensively used by large corporations. By this method, a print with white lines on a dark background is made directly from any drawing or tracing, to any desired reduction or enlargement, through the use of a large, specially designed camera. This print can be again photostated, giving a brown-black line on a white ground. This method is extremely useful to engineers for drawings to be included in reports and for matching drawings of different scales which may have to be combined into one.

*Duplicating Tracings.* Tracings with all the qualities of ordinary inked ones are made photographically from pencil drawings by using a sensitized tracing cloth.

*Lithoprinting.* When a number of copies of a drawing (50 or more) are needed, they may be reproduced by lithoprinting, a simplified form of photolithography, at comparatively small cost.

*Copying Methods.* Copying methods such as the mimeograph, ditto machine, and other forms of the hectograph or gelatin pad are often used for small drawings.

*Photo-copying Methods.* Negative copies on high-contrast film are reproduced on paper or cloth either to the same size as the original or to a reduced or increased size. Micromaster copies (Keuffel & Esser Co.) are well known. The reproductions are very good, and when the process is used for reproduction of an old drawing, wrinkles and other blemishes can be eliminated by using opaque on the film negative.

## 27. Filing and Storing Drawings.

Drawings are filed in steel or wooden cabinets made for the purpose. Many engineering offices store their drawings in fireproof vaults and remove them only for making alterations or for printing. Photographic copies are sometimes made as a separate permanent record. Drawings are always filed flat or rolled. Prints, however, are folded for filing or mailing. The usual method is the "accordion" fold illustrated in Fig. 8. To aid in the filing of accordion-folded prints, a supplementary number block may be added, as shown.

## 28. Simplified Practices.

The bulk of commercial design and representation is done in the conventional manner, that is, by drawing orthographic or pictorial views and dimensioning them as described in the preceding chapters. However, there are many modifications of standard practice which result in economies both in making and in using drawings. These economies can be classified as (1) simplified practices, (2) elimination of drafting effort by the use of templates and overlays, (3) special uses of reproduction, (4) photo drawing, (5) model making,[2] (6) illustrated drawings, and (7) use of special methods and equipment. The purpose in this chapter is to describe and illustrate modifications of standard practice and to suggest ways in which they can be applied.

## 29. "Simplified" Drawing: Economical Drafting Practices.

The drafting of any engineering creation is generally considered by management to be an expensive part of the complete operation of producing a finished, salable product. Furthermore, the percentage of total cost relegated to research, development, draft-

[2]The material on photo drawing and model-making in this chapter is adapted from original subject matter and a selection of illustrations by G. P. Hammond of Procter & Gamble.

ing, tooling, factory, labor, and other items of expense can be judged only on the economy of each individual department. Thus the drafting department comes under the same critical scrutiny as any other division of operation. Any saving in time and expense that can be made without injuring the effectiveness and efficiency of the department is true economy.

This question of thrift has been studied by many executives for years, but W. L. Healy and A. H. Rau,[3] of the General Electric Company, are credited with the first publication recommending and describing methods by which economies can be effected in the drafting room. There are a number of economical practices in making drawings and others that are important only to management. The principles on which economical drafting is based are discussed in the paragraphs that follow.

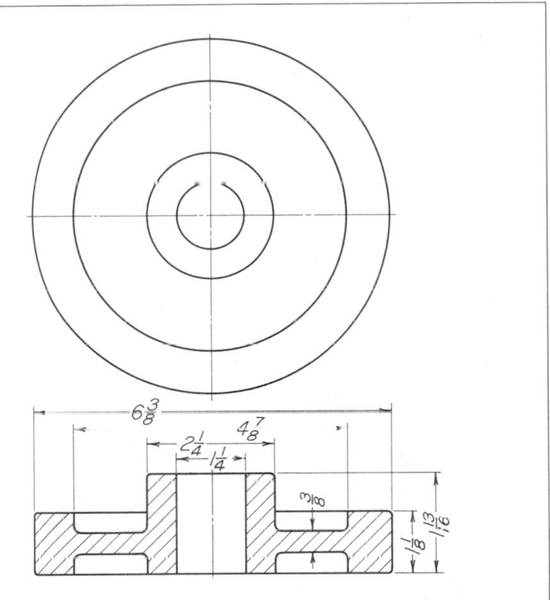


**FIG. 23.** Conventional drawing of a pulley. The two views represent the shape; the dimensions give size description (magnitudes).

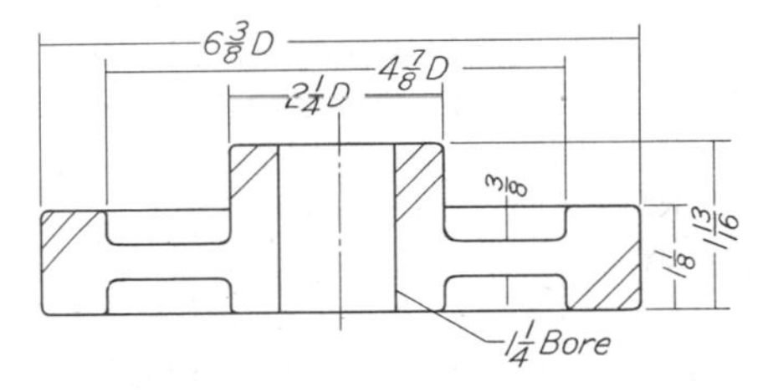


**FIG. 24.** Simplified drawing of a pulley. The single view describes the cross-sectional shape and the letter D on the dimensions specifies that the shape is cylindrical. The figured dimensions give size description (magnitudes).

## 30. Unnecessary Views.

Any view that is not necessary for clear description is a waste of drafting time, paper, reproduction (area and cost), and filing space. Figure 23 is a conventional drawing of a pulley in which the top view is used to describe the cylindrical shapes. This pulley *can* be represented economically, as in Fig. 24, by eliminating the top view and using the letter *D* on the dimensions of the front view to show that the shapes represented are cylinders. Some drafting economists argue that even the letter *D* is not necessary because the title block will tell that the object is a pulley and therefore a combination of cylinders. Each drafting executive must judge the extent to which extra views, letters, etc., can be eliminated without making the interpretation of the drawing obscure or misleading. Nevertheless, the following principles should be observed: (1) Never draw more views than are necessary for clear description; (2) examine the object to see if a view can be eliminated by use of a descriptive title, note, symbols, letters, or numbers; (3) examine the object to determine whether a word statement can eliminate *all* views (this is usually possible only for very simple parts).

[3] "Simplified Drafting Practice" (New York: John Wiley & Sons, Inc., 1953).

## 31. Unnecessary Scale.

The reader of the drawing does not establish distances by scaling the drawing, and most commercial drawings carry a note stating that the drawing is not to be scaled. Drafting time should not be used in drawing to scale *when it is not necessary*. A good example is given in Fig. 25, a structural detail *not to any scale* at all but just as effective and readable as if drawn carefully and meticulously to scale. The

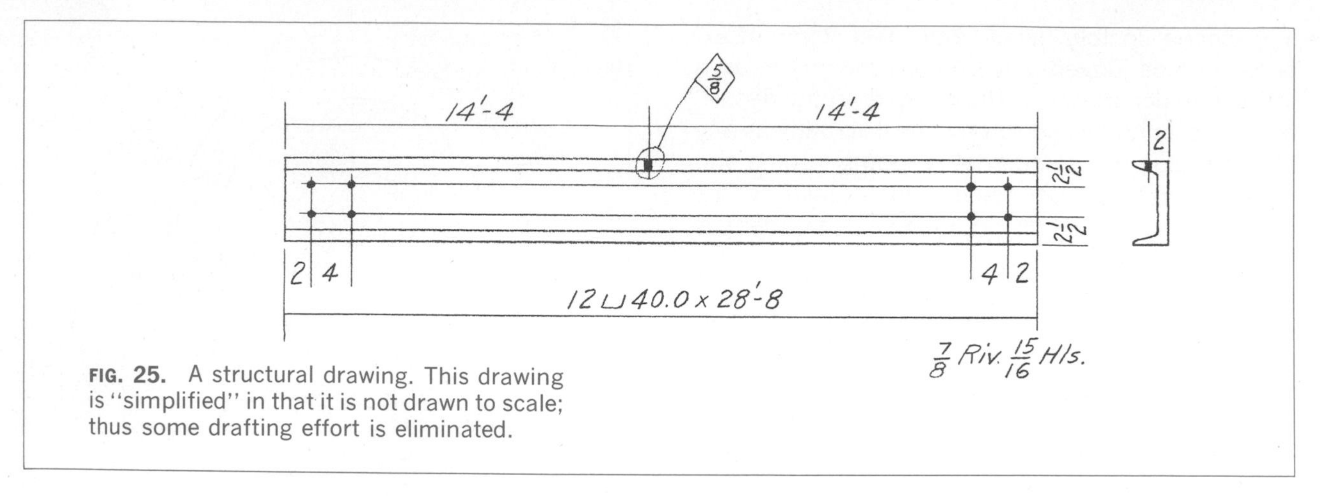


FIG. 25. A structural drawing. This drawing is "simplified" in that it is not drawn to scale; thus some drafting effort is eliminated.

cardinal points are: (1) do not draw to scale unless it is necessary for proper layout and readability; (2) do not cautiously draw to scale when approximations will suffice.

In sketching, *proportions* need not be meticulously maintained. Note in Fig. 26 that the two holes ($\frac{7}{8}$ and $\frac{3}{8}$ in. diameter) are *not* in good proportion and that the two $\frac{1}{2}$ in. radii for the bends are not shown centered according to the dimensioning. Nevertheless, the drawing is readable, and if the dimensions are followed—and of course they *must* be—the correct object will be manufactured. Figure 26 is not an extreme case—proportions could be *much* poorer than in Fig. 26 and the drawing still be perfectly usable.

**32. Unnecessary Lines and Elaboration.** Every line made on a drawing costs in drafting time and the company's money. Therefore, any line that does not accomplish a useful purpose should be eliminated. Such lines will include extra lines (on any view) that are not necessary for describing the object. Meticulous crosshatching and embellishment, especially, can be omitted without seriously affecting readability. Compare the two drawings in Figs. 27 and 28. Painstaking and complete representation of sections such as is given in Fig. 27 is a waste of time and could be advantageously simplified, as in Fig. 28. The drawing in Fig. 28 actually gives more structural information than the detailed representation in Fig. 27.

**33. Arrowless Dimensioning.** Time can be saved by eliminating unnecessary arrows on dimension lines and leaders, as has been done in Figs. 24, 25, 26, 28, and 30. To the uninitiated, these drawings may appear to be unfinished, but careful examination will show that arrows are not really needed.

**34. Simplified Base-line Dimensioning.** If the base-line system of dimensioning (see Sec.

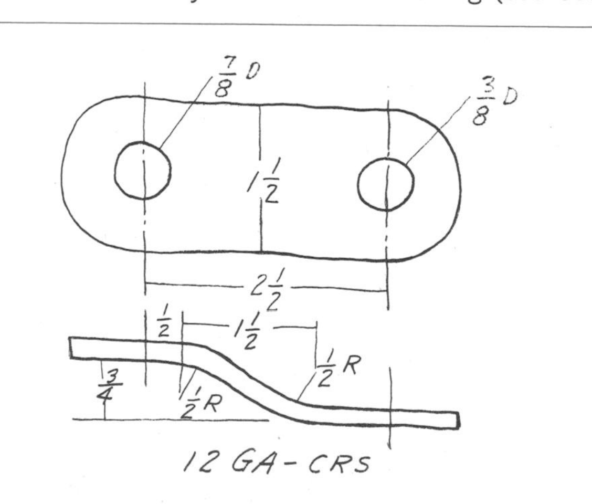


FIG. 26. A simplified sketch. In this example, proportions have not been carefully maintained, thus saving time. The object is built by following the general shape shown and adhering to the dimensions; therefore, careful proportioning is not necessary.

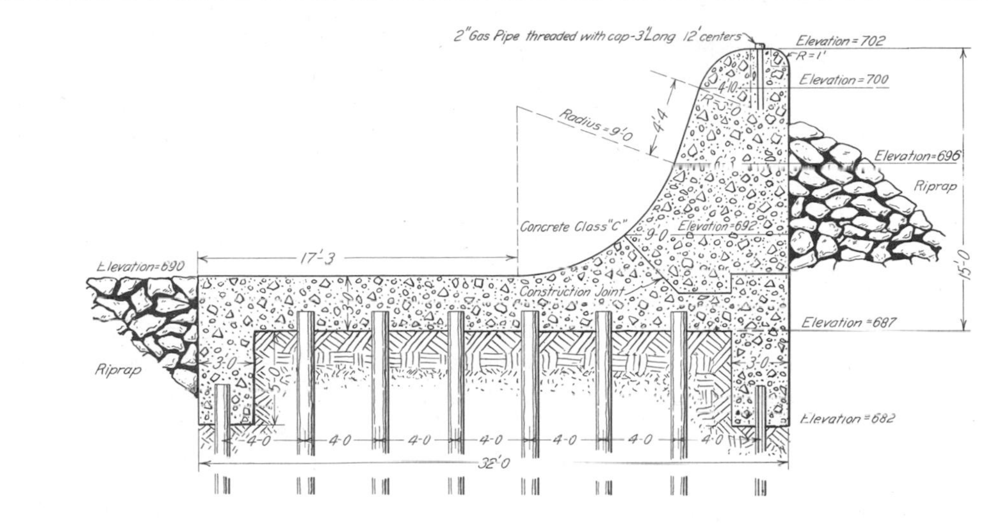


**FIG. 27.** A conventional drawing. Because of the drafting effort required to produce the detail and embellishment, this drawing is expensive.

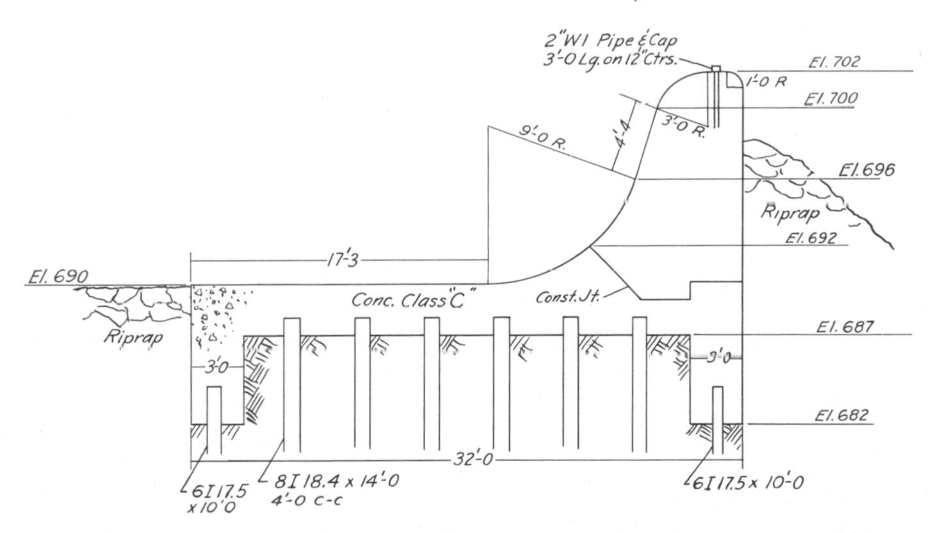


**FIG. 28.** A simplified drawing. This drawing represents the same structure as in Fig. 27. Note the absence of extra detail and embellishment, which reduces drawing cost.

70, Chap. 12) is used, even the dimension *lines* can be eliminated. As shown in Fig. 29, two base lines, one horizontal and the other vertical, are established. The location of base lines may be at the edge of the part, on the center of symmetry, or through some important feature. Distances are then given, as shown in the figure, *from* the base lines (corresponding to any coordinate system) and in-

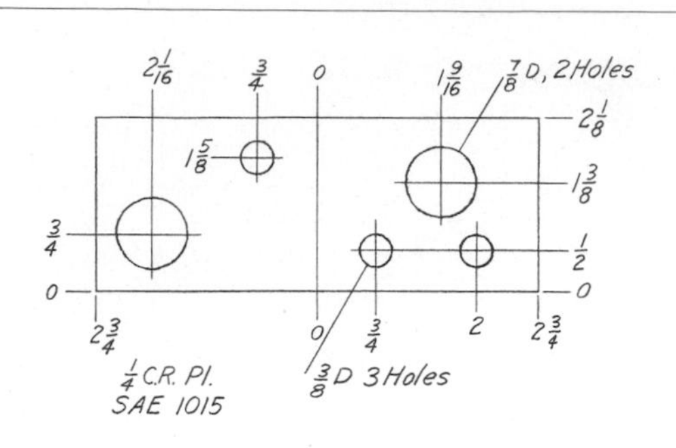


**FIG. 29.** Simplified base-line dimensioning. This system accomplishes the same purpose as standard dimensioning from datum, but eliminates all unnecessary lines and arrowheads, thus reducing drafting time.

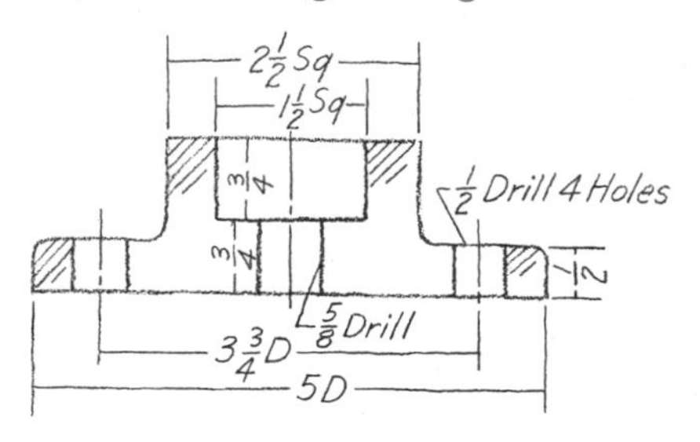


**FIG. 30.** A simplified sketch. This drawing, made freehand, is economical, but economy is carried further by elimination of extra views, embellishment, and arrowheads on the dimensions.

clude all the information necessary for construction.

### 35. Freehand Drawing.

The greatest saving of drafting time can often be realized by using the methods just described and by making the drawing freehand. Compare the simplified sketch in Fig. 30 with the conventional sketch in Fig. 15. See also Sec. 31 and Fig. 26 where proportioning is discussed and illustrated.

### 36. Cautions in the Use of Simplified Practices.

Simplified drafting practices are comparatively new and therefore may not be understood by some engineers and manufacturers. For example, for one unfamiliar with simplified practices, the drawing in Fig. 29 *may* be more difficult to read than a conventional drawing. Further, if mistakes occur in manufacture because of misreading a simplified drawing, the saving in drafting time is offset many times over in costly rejected parts. The common, and safest, use of simplified practices occurs *within an organization* where everyone concerned is thoroughly trained in all company practices. It is potentially dangerous to send a simplified drawing to a subcontractor or supplier *unless it is known* that *all* special drafting practices will be understood.

### 37. Abbreviations.

Another economy results from using abbreviations where this does not detract from the readability of the drawing. American Standard abbreviations are given in the Appendix. Note the use of abbreviations in Figs. 24 to 30.

## TEMPLATES AND OVERLAYS

### 38. Templates.

Use of modern aids such as the drafting machine, pencil sharpener and pointer, and special instruments such as an ellipsograph (Fig. 62) conserves drafting time. For features such as circles, ellipses, boltheads, nuts, and symbols which must be drawn repeatedly, a template greatly shortens the time required for layout. Figure 31 shows several typical templates: (*a*) and (*b*), Rapidesign templates for ellipses and circles; (*c*), a Tilt-hex template for boltheads; (*d*), a Du-all template for a variety of operations; (*e*), a Holometer for boltheads and nuts; (*f*), a K & E thread template; and (*g*), a Rapidesign template for piping symbols.

A number of symbol templates are available for use with Leroy lettering equipment; Fig. 32 shows some of them. Figure 33 is an example of a symbolic drawing.

### 39. Overlays.

An overlay is typed or printed detail reproduced on tracing paper, Mylar, or some other translucent or transparent material. Overlays are fastened to a

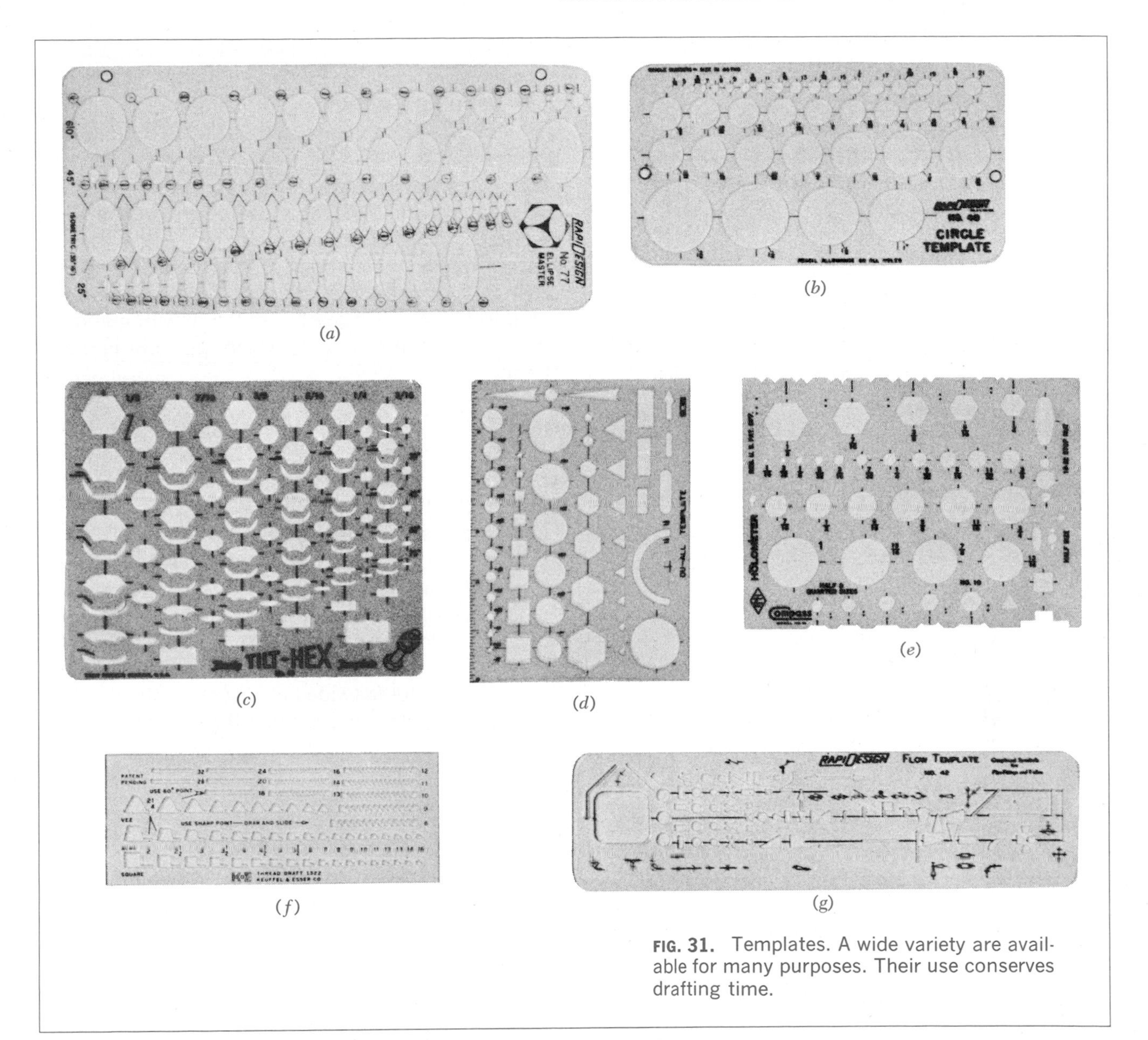


**FIG. 31.** Templates. A wide variety are available for many purposes. Their use conserves drafting time.

drawing by means of transparent tape or the adhesive backing on the overlay.

Most companies have standard items that must be repeated on drawings. The use of overlays eliminates the need for redrawing these items each time they appear. One prominent processing company, for example, is continually building new plants and expanding existing plants, and it must make plant layout drawings showing the location of equipment. The plan views of tanks, pumps, filters, and other items are drawn and identified for size and scale, and then reproduced on a suitable overlay material.

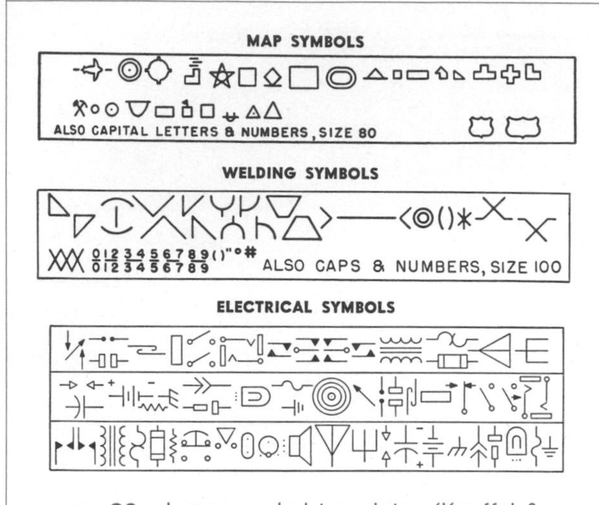


**FIG. 32.** Leroy symbol templates (Keuffel & Esser Co.). A wide variety of these templates, for use with standard Leroy equipment, are available. Special templates of a user's design also can be obtained.

Figure 34 shows the standard overlay sheet. Needed items are cut from this sheet and applied to the drawing with tape. Figure 35 gives the finished result. The outline of the overlays and the tape will show faintly on the reproduced drawing but will not impair its readability. Extra information is lettered on the overlays, as shown in Fig. 35.

Many standard items are available from companies that manufacture a general line of overlays. For details that are time-consuming to draw and that occur again and again, the use of an overlay such as the "north point" illustrated in Fig. 36 is a great economy. This north point is manufactured by the Stanpat Co., which makes a wide variety of overlays, all printed on a transparent material with gummed backing. See also Sec. 23, Chap. 20, and Fig. 22, Chap. 20.

Parts lists, tables, references, and specifications must often appear on drawings, and it is time-consuming to letter these. Laborious lettering can be eliminated by employing a standard overlay such as the one shown in Fig. 37. If a standard overlay cannot be used, information can be typed on blank stanpat or some other overlay material and applied to the drawing.

The use of overlays is not restricted to prepared details and lettering. Black tape is used to block out unwanted areas or in place of drawn material. It is especially useful for wide detail such as the building outline in Fig. 35. Tape can be cut with a sharp blade

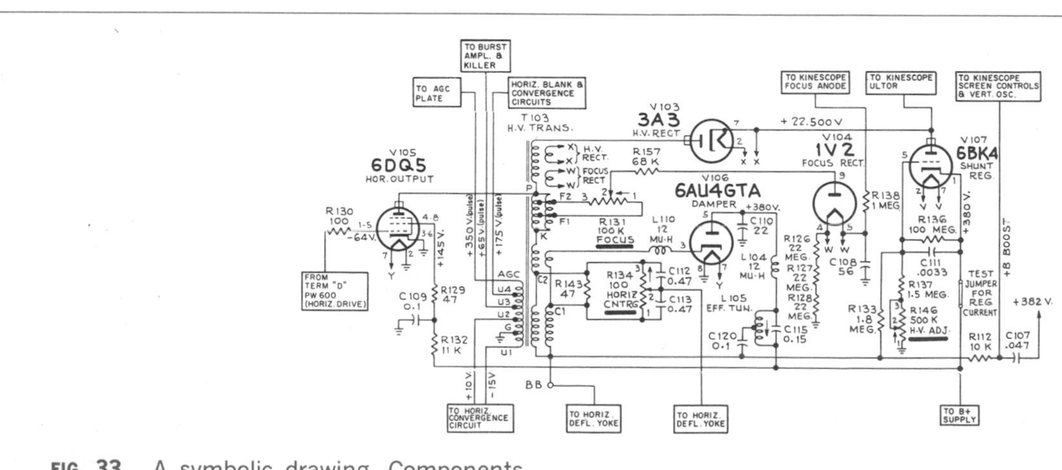


**FIG. 33.** A symbolic drawing. Components are shown by symbols. Magnitudes and other details are given by lettered information. (*Courtesy of Radio Corporation of America.*)

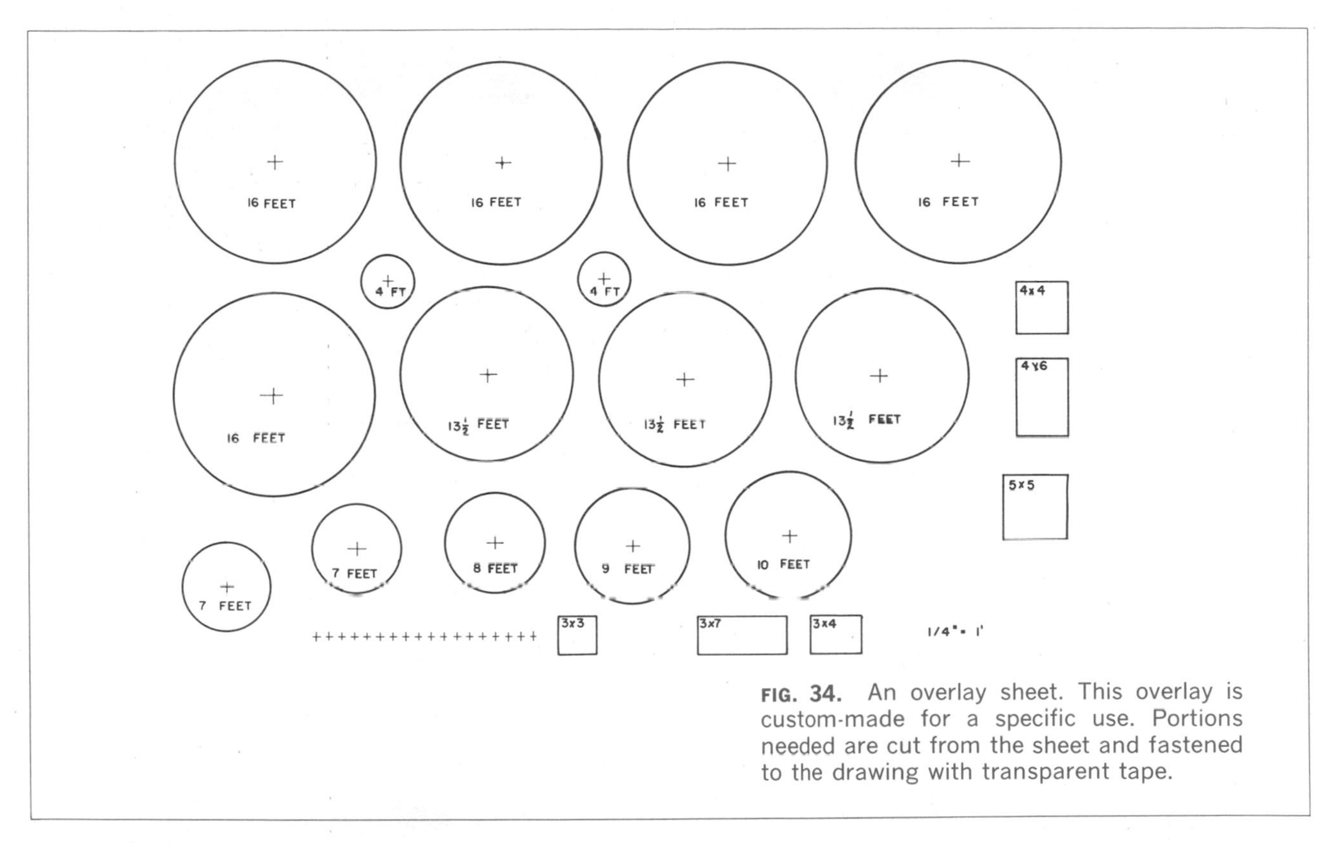


**FIG. 34.** An overlay sheet. This overlay is custom-made for a specific use. Portions needed are cut from the sheet and fastened to the drawing with transparent tape.

to any width needed. Tapes in widths from $\frac{1}{32}$ in. to 1 in. for full, dashed, and special lines and for areas, and a "tape pen" for application are manufactured by Chart-Pac, Inc., Leeds, Massachusetts.

## SPECIAL USES OF REPRODUCTION

### 40. Reproduction Methods.

There are many reproduction methods by which drafting time can be saved. Whole drawings or portions can be reproduced, and alterations can be made by painting out previous detail with an eradicating solution and then drawing in the new feature. Several companies manufacture reproducible materials and machines. The materials of the Eastman Kodak Company are typical. The following information is adapted from their supplement to the Kodak data book on Kodagraph reproduction materials.

An intermediate printed on Kodagraph autopositive paper saves wear and tear on original drawings. Prints are made from the autopositive, thus eliminating handling of the original.

A drawing that is unprintable because of age or hard use can be printed on Kodagraph autopositive paper and then printed from this intermediate. By this method its usefulness is restored without cost in drafting time.

Direct-process prints or white prints are often the only available record. A print of this record made on Kodagraph autopositive paper provides an intermediate from which copies can be made.

When a blueprint is the only available record, additional prints can be obtained by making an intermediate on Kodagraph autopositive paper or Kodak repro-negative paper.

Portions of two or more drawings can be made into a new drawing by printing the portions on separate sheets of Kodagraph autopositive paper and then combining these to make prints.

Kodagraph contact materials, which must be handled

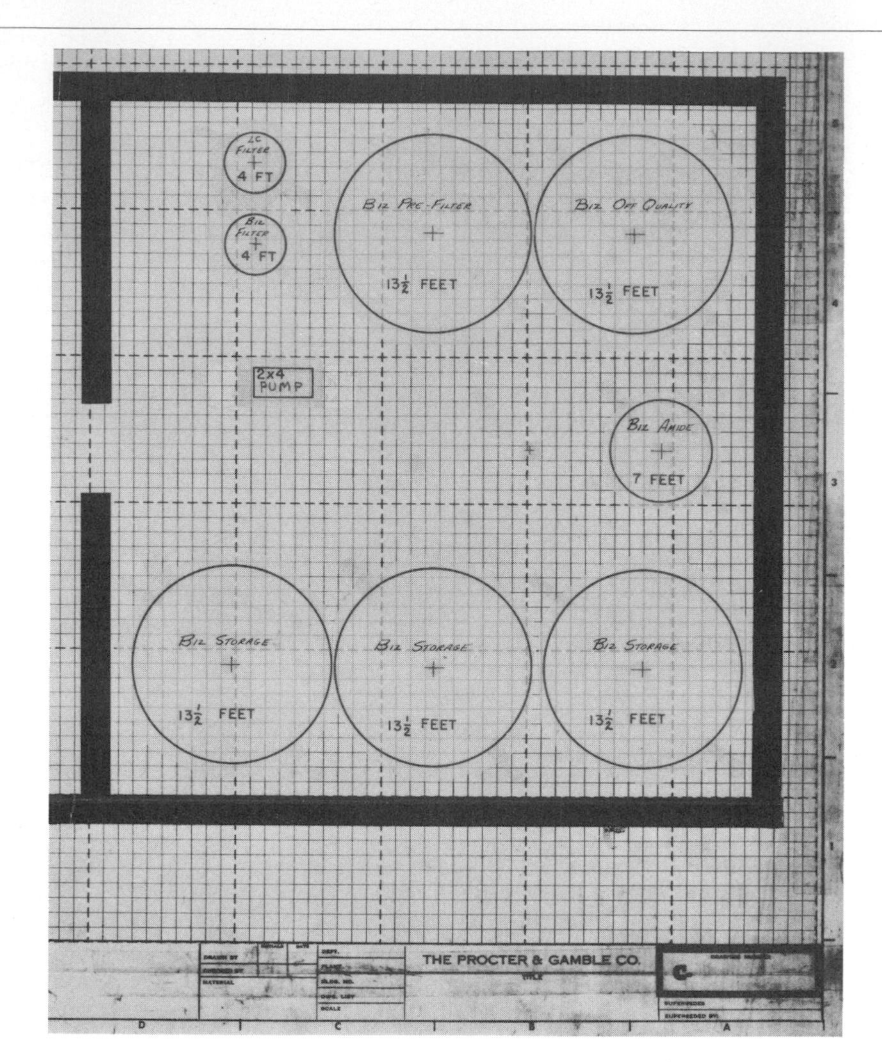


**FIG. 35.** Overlays applied to a drawing. In addition to saving drafting time, overlays can be moved around as the arrangement of elements is planned.

in a safelighted darkroom, are used to produce printable copies of old drawings, blueprints, brownprints, or direct-process prints.

Kodalith film is often used to reproduce details for insertion in other drawings or reproducibles. A reproducible is any medium from which prints can be made.

Kodagraph projection and fast projection materials provide enlarged- or reduced-scale prints from negatives made in a camera.

Reproducibles save much drafting time. The background of Fig. 35 is an intermediate printed on transparent material. Thus the standard grid drawing is available in quantity.

## 41. Photo Drawing.

A recent development in the field of engineering presentation is the use of photography to reduce

**FIG. 36.** A gummed overlay. This "north point" is typical of details that are often repeated. Such details are expensive to draw and an overlay saves practically all drafting time required for the detail. (*Courtesy of Stanpat Co.*)

drafting time or to replace conventional drawings. By recording engineering subjects, the camera has become an important aid.

Individual applications of photo-drawing technique vary widely in industry, depending on the type of business in which the company is engaged. There are two principal areas in which photo drawings are of value: (1) in revising and in adding to existing equipment, (2) in recording and in supplementing model designs. Many process companies constantly change their facilities, to make a new product or to improve and expand; photography can play a major role in such engineering changes. Some companies have manufacturing operations that call for adding completely new units. Photo drawings of models are used to record the design and aid in obtaining the best job on new-unit construction at the lowest cost.

Regardless of how they are used, photo drawings have two main advantages:

1. They lower engineering costs. The camera records information that would have to be drawn if photo drawings were not utilized.

2. They provide better visualization of the design intent than is obtained from a drawing. The camera records an image approximately as the eye sees it—in three dimensions. In contrast, when conventional plan and elevation drawings are made, the reader has to mentally combine the various views to visualize the design intent.

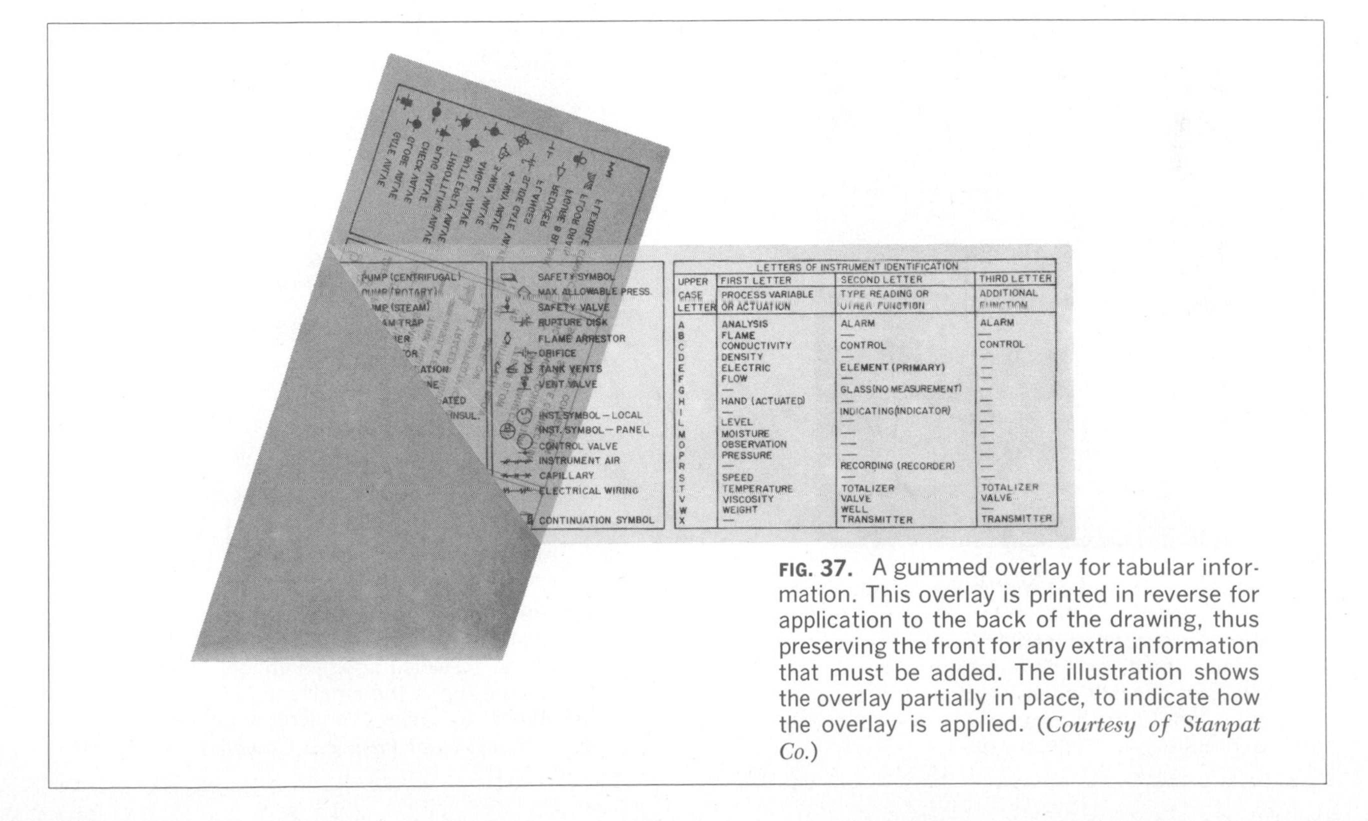


LETTERS OF INSTRUMENT IDENTIFICATION

| UPPER CASE LETTER | FIRST LETTER PROCESS VARIABLE OR ACTUATION | SECOND LETTER TYPE READING OR OTHER FUNCTION | THIRD LETTER ADDITIONAL FUNCTION |
|---|---|---|---|
| A | ANALYSIS | ALARM | ALARM |
| B | FLAME | — | — |
| C | CONDUCTIVITY | CONTROL | CONTROL |
| D | DENSITY | — | — |
| E | ELECTRIC | ELEMENT (PRIMARY) | — |
| F | FLOW | — | — |
| G | — | GLASS (NO MEASUREMENT) | — |
| H | HAND (ACTUATED) | — | — |
| I | — | INDICATING (INDICATOR) | — |
| L | LEVEL | — | — |
| M | MOISTURE | — | — |
| O | OBSERVATION | — | — |
| P | PRESSURE | — | — |
| R | — | RECORDING (RECORDER) | — |
| S | SPEED | — | — |
| T | TEMPERATURE | TOTALIZER | TOTALIZER |
| V | VISCOSITY | VALVE | VALVE |
| W | WEIGHT | WELL | — |
| X | — | TRANSMITTER | TRANSMITTER |

**FIG. 37.** A gummed overlay for tabular information. This overlay is printed in reverse for application to the back of the drawing, thus preserving the front for any extra information that must be added. The illustration shows the overlay partially in place, to indicate how the overlay is applied. (*Courtesy of Stanpat Co.*)

## 42. Photo Drawings for Revisions of Existing Areas.

The cost of engineering revisions of existing facilities can be reduced one-third to one-half through the use of photo drawings. The original and main use of photo drawings is in the field of piping rework, but new applications are being found in the electrical, mechanical, structural, and site-clearance fields.

No matter what phase of rework is being engineered, preparation of a photo drawing is similar and involves four basic steps: (1) taking the photographs, (2) making the reproducible[4] (the transparent base for all drawing effort), (3) drawing on the reproducible to show the design intent of the revision, and (4) making prints from the photo drawing for distribution to the construction forces.

Arrangements for taking the photographs depend on the facilities at the disposal of the engineering personnel. The highest-quality reproducibles, and therefore the best photo drawings, are obtained when the picture is taken by a skilled photographer working in close association with the engineer or designer responsible for the project. The photographer knows the correct camera setting, lighting, etc., that will produce a negative which can be made into a good reproducible; the engineer or designer knows what changes are to be made and can direct the photographer so that the photo drawings will show clearly the work to be done.

[4]See Sec. 40.

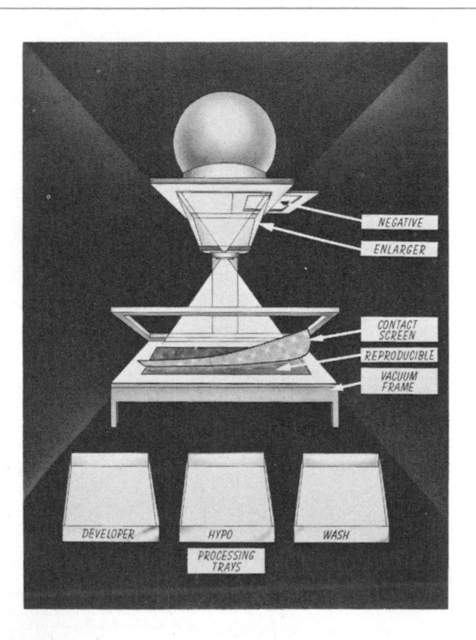


FIG. 38. Method of making a reproducible. A negative in the enlarger is projected through a contact screen onto a sensitized plastic sheet, held in a vacuum frame. The exposed sheet is then developed in conventional photographic trays. (*Courtesy of Procter & Gamble.*)

FIG. 39. Illustration of screening on a reproducible. The drawing at the bottom shows a normal, screened positive image. The portion at the top is the small section enlarged ten times to show the screen pattern of dots. (*Courtesy of Procter & Gamble.*)

A reproducible is made in several ways, depending on the company's facilities. The procedure diagramed in Fig. 38 gives the highest-quality reproducible because it has less photographic steps than other methods. The resulting reproducible is a screened image on which highlights and shadows have been obtained by varying the concentration of white and black dots (see Fig. 39). Screened images, with their tonal characteristics, give much superior and more realistic prints than are available with printing processes that record only in solid black and pure white.

Several materials have been developed and are used as reproducible bases. The best ones are made of polyester resin-base plastics but some acetate-base material is used. The drawing technique is the same no matter what material is used to prepare the reproducible. There are no unusual mechanical problems in drawing on photo drawings. By using a soft pencil, which produces heavy black lines, and by controlling the printing speed, it is possible to make additions stand out in contrast to the existing photographic background, as shown in Fig. 40

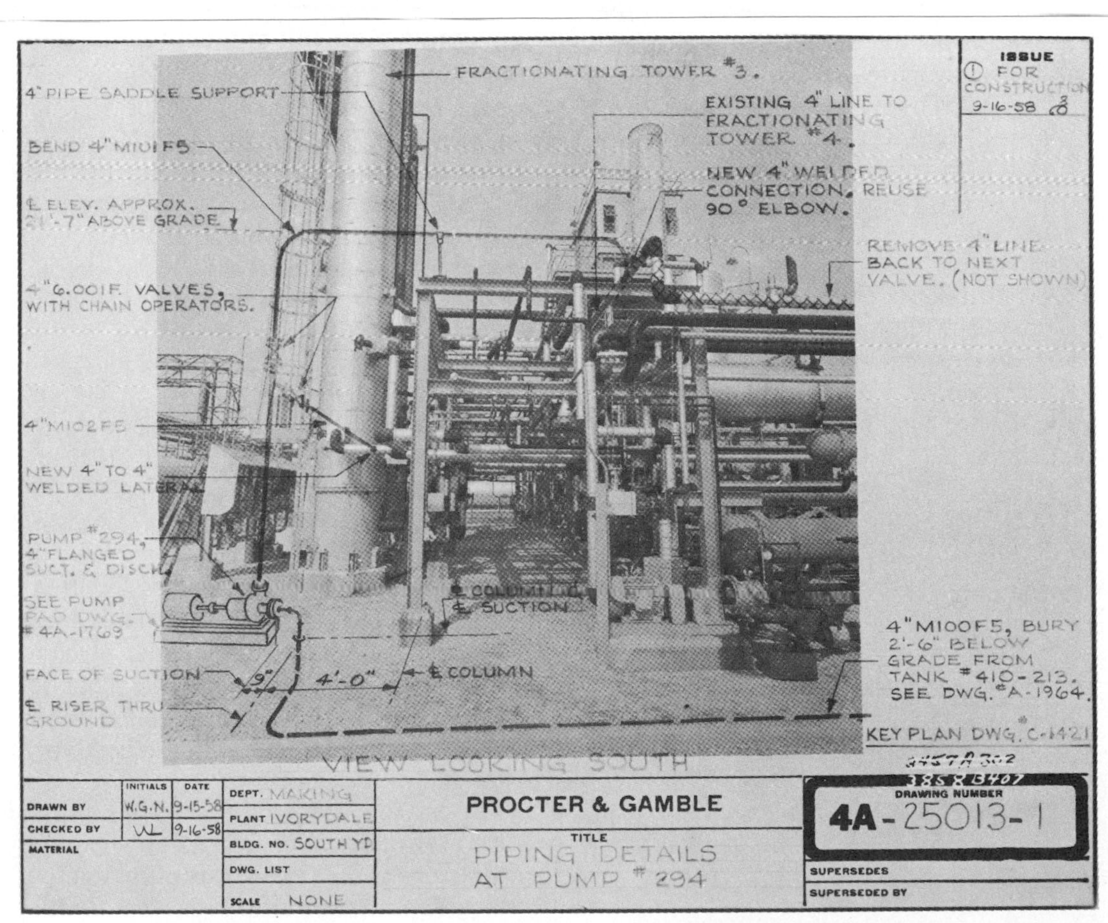


**FIG. 40.** Typical photo drawing of piping for the revision of an existing facility. Observe the extensive notes and directions added to the reproducible in pencil. (*Courtesy of Procter & Gamble.*)

Frequently it is necessary to clear a site before new equipment can be installed. Instructions for site clearance are easily transmitted by crossing out on a photo drawing the items that are to be removed, as indicated in Fig. 41. A second photo drawing of the same area is then made with the existing equipment removed from the picture so that it does not interfere with the drawing of new equipment. Any part of the photographic image can be removed from a photo-drawing reproducible with a wet eraser or with chemicals. Note the removed and drawn-in portions in Fig. 42.

Perspective can cause difficulty when drawn material is added to a photograph. Since each photo drawing has a slightly different station point, it is important that the draftsman determine this point and draw the new information with a similar perspective; otherwise the drawing will have a peculiar appearance and may convey the wrong intent to the construction forces.

Usually photo drawings convey design intent with only the important control dimensions indicated. It would be difficult, but not impossible, to include all of the detailed dimensions normally shown on conventional drawings. Craftsmen are accustomed to revision work and to "field checking" the drawings they receive, so the lack of complete dimensioning is not critical if the design intent is clear. Note the dimensioning and extensive use of notes in Fig. 43.

**FIG. 41.** Typical drafting-room scene when work is being done on a photo drawing.

When more than one photograph is required to portray a change, each photo drawing is referenced to the other. This is done when there are multiple images on one photo drawing as well as when there are several single-image photo drawings. When there are several photo drawings for a project, a key plan is made. The key plan is a simple sketch of the area involved on which are shown the numbers of the various photo drawings with an arrow indicating the direction in which each picture was taken.

### 43. Model Photo Drawings.

The camera can also record a design shown by a model, thus giving the engineer or designer a drawing base to which he can add information.

The basic steps followed in making photo drawings of existing areas (taking the picture, preparing the reproducible, drawing in the engineering information, making the prints) apply also in making model photo drawings. The principal difference between the two uses is in the taking of the pictures.

Model photo drawings are used for (1) equipment layouts, (2) location drawings, (3) information for people remote from the model, and (4) permanent records of the job once the model is destroyed. All these applications require that the photo drawings be shadowless, so that the equipment and piping on the model stand out clearly.

One solution to the shadowless-lighting problem is shown in Fig. 44, where a swinging light is employed. There are other solutions, which vary from a simple "washing with light" technique to building a complete "tent" of lights around the model. A photo drawing produced with the equipment in Fig. 44 is shown in Fig. 45.

The problems of perspective are more difficult to solve. The ultimate in model photo drawings is a complete absence of perspective, that is, purely orthographic photo drawings, representing all dimensions to true scale. Most companies use the more easily attainable minimum perspective, which is obtained through use of a long-focal-length camera lens. Photo drawings in this perspective are not

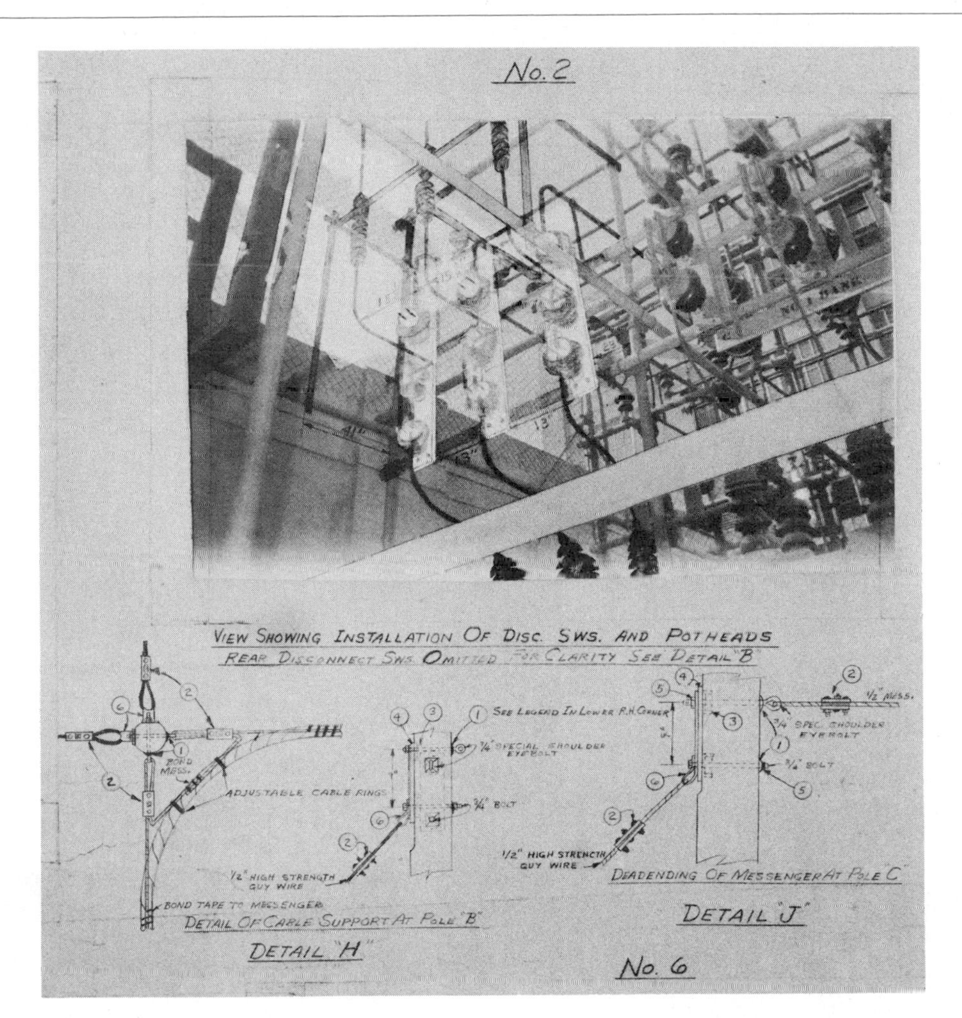


**FIG. 42.** Typical completed photo drawing for piping revision of an existing facility. (Compare with Fig. 40.) In this case the photo portion shows design intent, and the necessary details are given below in conventional orthographic views. (*Courtesy of Procter & Gamble.*)

unreasonably difficult to dimension and understand, but are still not scalable. Development work has been done on equipment that will take orthographic photographs—devices that utilize segments of parabolic mirrors combined to form a telecentric optical system. Such devices can take scalable photographs of a model or other three-dimensional subject. An example is shown in Fig. 46.

Once the picture has been taken, under conditions that will produce either orthographic or minimum-perspective results, the reproducible, which is the base for the model photo drawing, is made in the

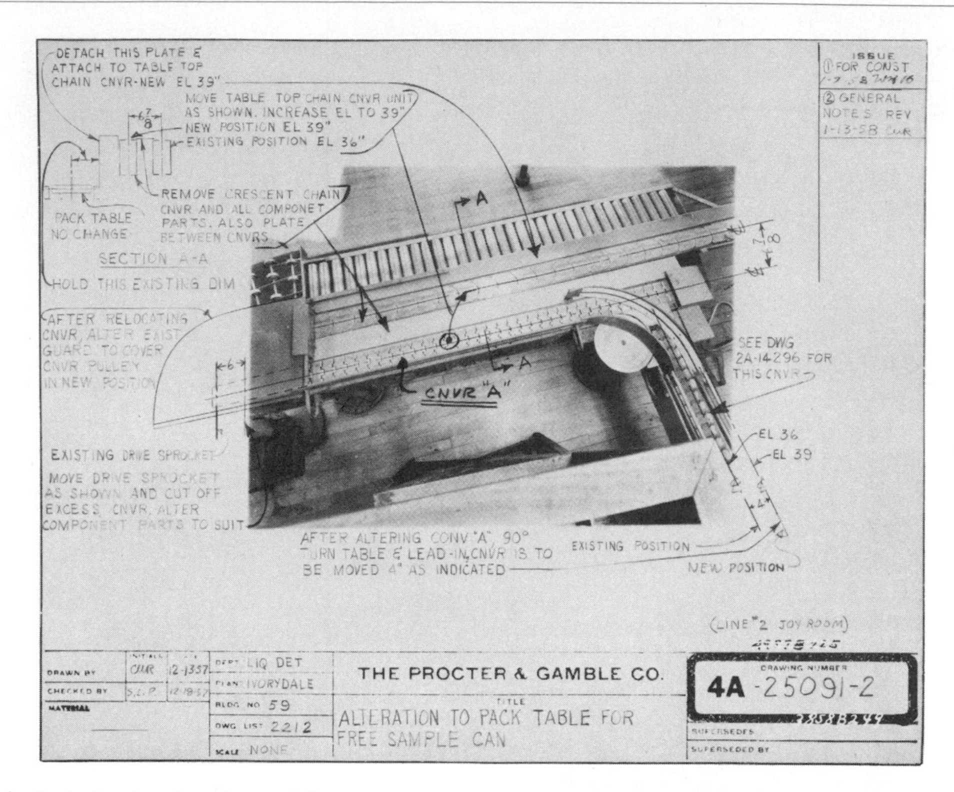


**FIG. 43.** Typical photo drawing for revision of mechanical equipment. Note that this drawing employs demensioning and notes applied to the photo portion, and also uses small explanatory views. (*Courtesy of Procter & Gamble.*)

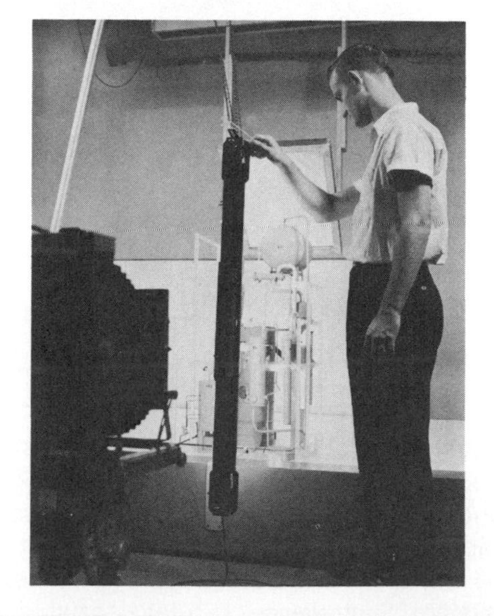

**FIG. 44.** Method for shadowless lighting. The model is placed in front of a light box, which eliminates black shadows. Shadowless front lighting is produced by a fluorescent light. (*Courtesy of Procter & Gamble.*)

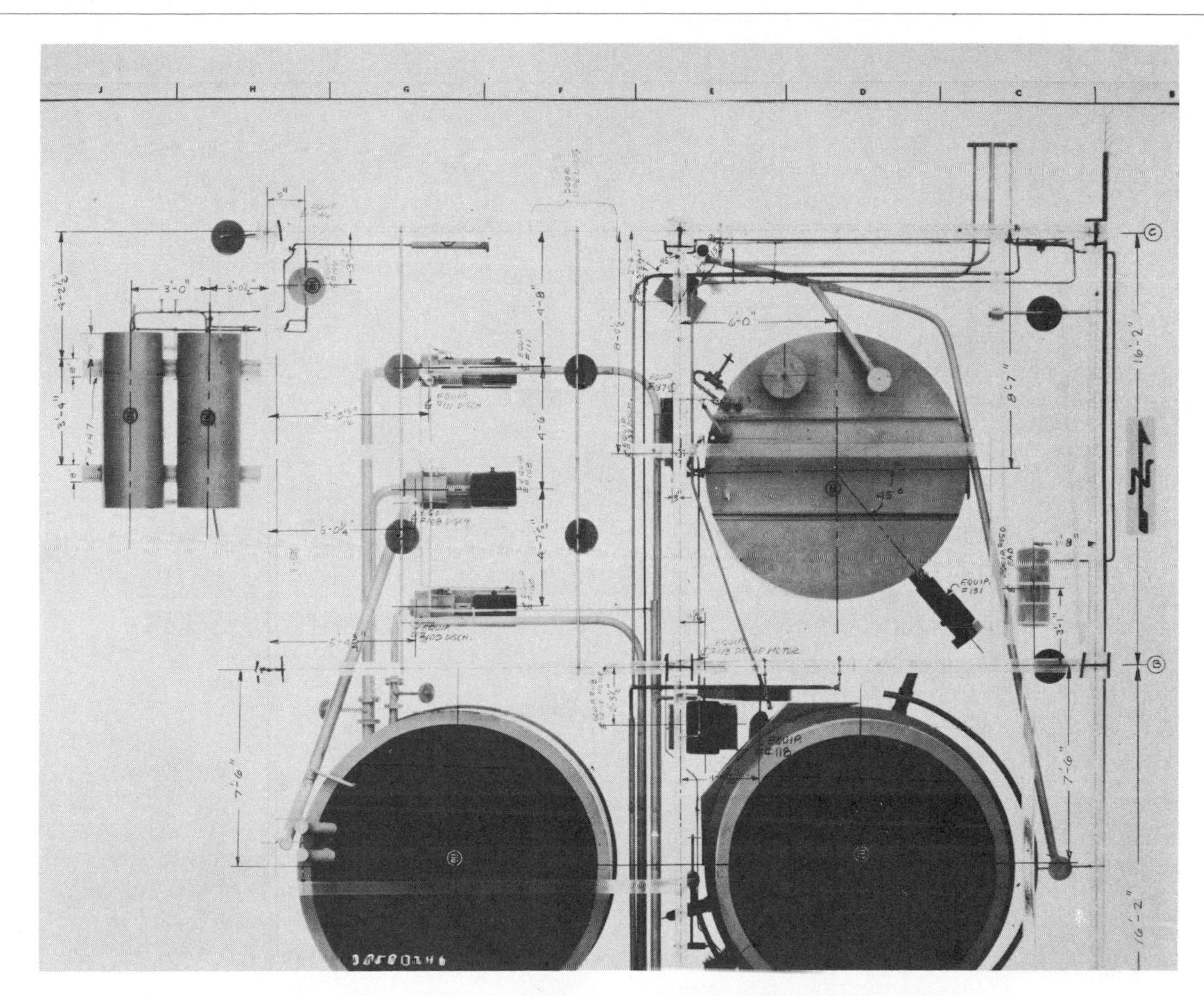

FIG. 45. A portion of a shadowless model photo drawing. The photographic portions were produced by the lighting method shown in Fig. 44. (*Courtesy of Procter & Gamble.*)

same manner as previously outlined for existing-area photo drawings.

## ILLUSTRATED DRAWINGS

### 44. Illustration.

Illustrated drawings are used in commercial practice for many of the same reasons that models are made. As applied to the graphic language, illustration means clarification in the readability of a drawing by the use of shading and special methods of projection and representation. Often a drawing can be made so clear that a person unfamiliar with graphic methods can read it as readily as one experienced in them. Any drawing—orthographic, diagrammatic, or pictorial—made freehand or with instruments, in pencil or ink, may be illustrated. Figure 47 is an example of an illustrated drawing.

**FIG. 46.** Comparison of methods of taking photographs of a model. The lower picture was taken with a conventional camera; the upper picture, with a device employing a telecentric system to produce an orthographic (nonperspective) view. (*Courtesy of Procter & Gamble.*)

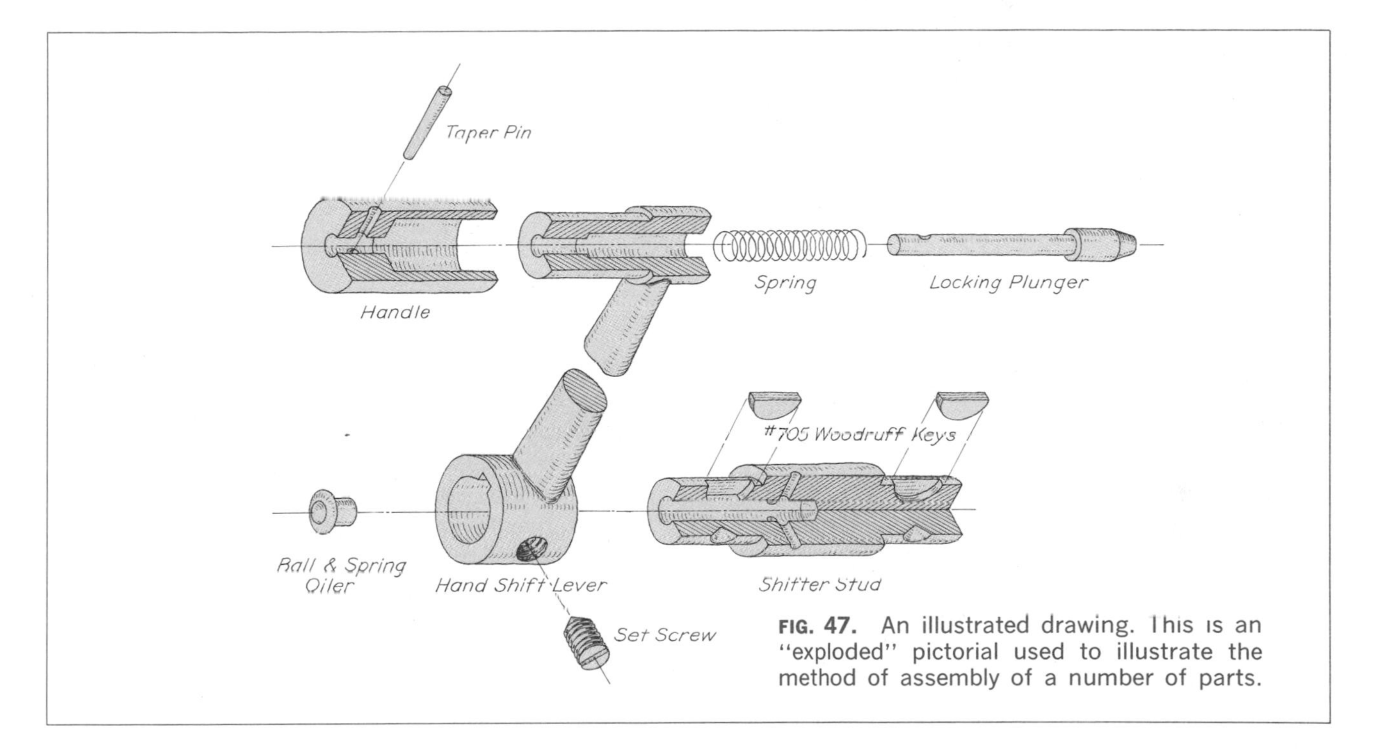


**FIG. 47.** An illustrated drawing. This is an "exploded" pictorial used to illustrate the method of assembly of a number of parts.

## 45. Types of Illustrated Drawings.

Illustrated drawings are used for many purposes and in many different fields. The following examples are typical:

*Advertising illustrations* are usually shaded pictorial drawings, often with color added to make the presentation as forceful as possible.

*Catalogue illustrations,* orthographic or pictorial, are often colored or shaded in pencil or ink, by stippling, air brush, etc.

*Operation, service, and repair charts* are drawings showing the working parts of a machine with appropriate directions for the purpose intended. These drawings are effective in shaded or colored pictorial form.

*Piping, wiring, and installation diagrams* are easy to read when made in shaded or colored pictorial form.

*Architectural and engineering presentation drawings,* sometimes shaded in pencil or ink, often in water color, show the building or structure as it will appear when completed.

*Textbook illustrations* in pictorial form, usually in black and white but sometimes in two colors or full color, add visual clarity and give emphasis not achieved by words alone.

*Patent drawings* are usually shaded to bring out and clarify every feature of the invention.

*Production drawings,* from original design sketches to the final details, subassemblies, and assemblies, are frequently made in pictorial form, often shaded. Such drawings are particularly useful when persons not trained in reading orthographic drawings are employed for manufacturing.

## 46. Illustrated Working Drawings.

Modern mass-production methods demand simplification and breakdown of a multitude of manufacturing operations. Illustrated drawings bring complex and difficult jobs within the grasp of workers not trained in reading orthographic drawings. They are used in every phase of production, from the original design to the final operating instructions.

Actual practice varies somewhat within an industry and widely among industries, but, in general, illustrated drawings can be classified as (1) *design drawings,* (2) *manufacturing illustrations,* and (3) *operation and maintenance illustrations.*

*Design drawings* include a variety of pictorial illustrations that show the machine or structure broken down into small, workable units, and then give details of construction, location of equipment, structural features, function of parts and equipment, tooling methods, etc. These drawings are used to study the complete production job and to plan and correlate the work. As the design progresses they are altered, corrected, or redrawn. The preliminary production breakdown shown in Fig. 48 is an example of the type of illustration used in the design stage of the work.

*Manufacturing illustrations* give detailed information regarding the breakdown of the structure or machine from which a multitude of separate illustrations are made to show the location of subassemblies, parts, and equipment; give directions for performing operations; provide detailed information for the manufacture of parts; and furnish directions for assembly. Figure 49 is an example of a manufacturing illustration taken from the set of details given in Prob. 35.

*Operation and maintenance illustrations* give directions for disassembly, repair and replacement of parts, lubrication, inspection and care of equipment, etc. In many cases the drawings that were originally made for manufacturing can be used in a service manual. Figure 50 is an example.

## 47. Drawing for Reproduction.

Drawings for illustrations in books, periodicals, catalogues, or other printed materials must be made with an understanding of reproduction processes.

*Preparation of the Drawings.* The drawings are usually made larger than the final size of the illustration and reduced photographically in the reproduction process. Consequently the work must be done with visualization of the line weights, contrast, size of lettering, and general effect in reduced size. To preserve the hand-drawn character of the original, the reduction should be slight. For a very smooth effect, the drawing may be as much as three or four times larger than the reproduction. The best general size is one and a half (for one-third reduction) to two times (for one-half reduction) the linear size of the cut.

The reduction is usually an even proportion such as one-fourth, one-third, one-half, etc., although odd reductions may be used. For a drawing marked "Reduce $\frac{1}{3}$," the reproduction will be two-thirds the

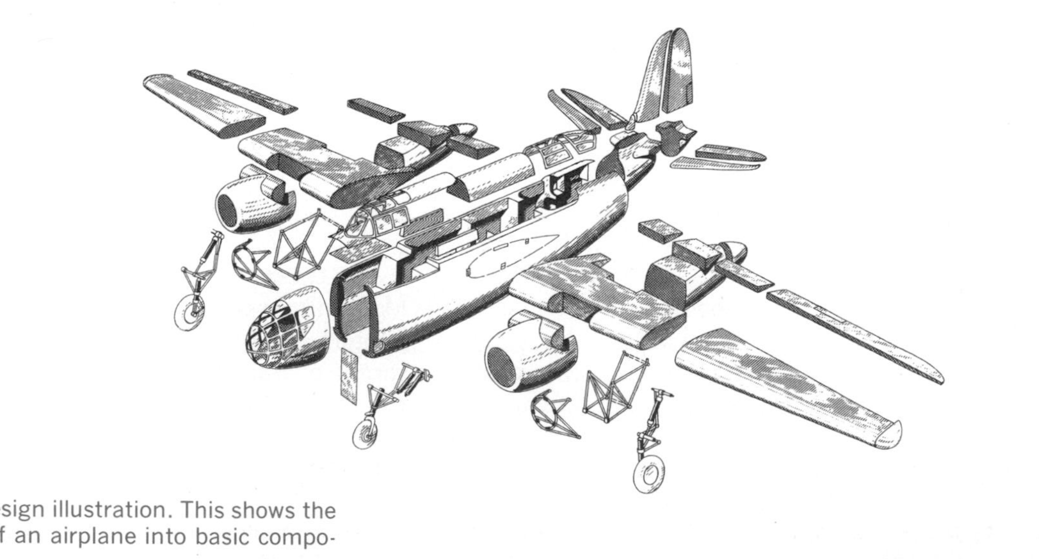

**FIG. 48.** A design illustration. This shows the breakdown of an airplane into basic components.

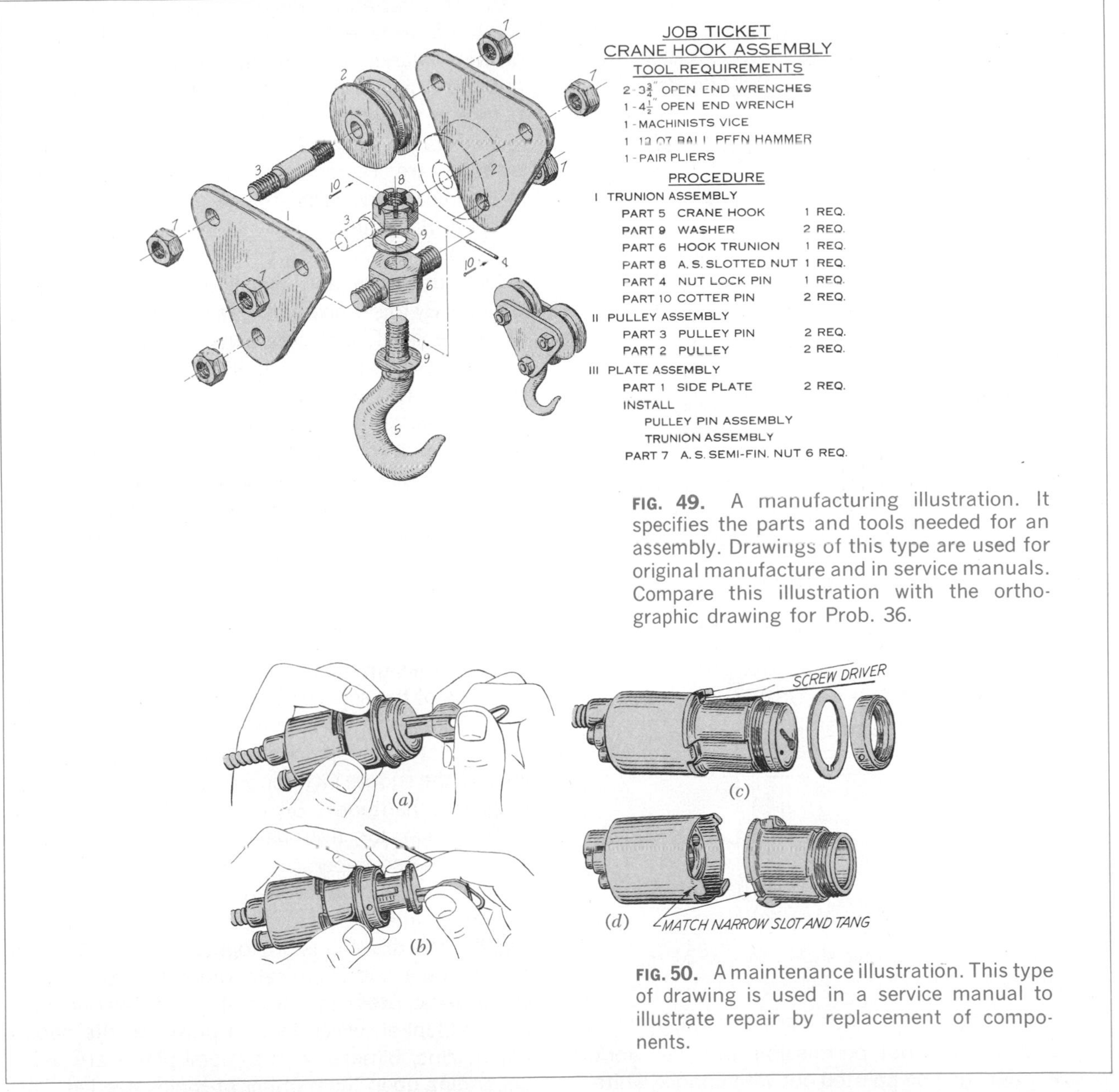

JOB TICKET
CRANE HOOK ASSEMBLY
TOOL REQUIREMENTS
2-3¾" OPEN END WRENCHES
1-4½" OPEN END WRENCH
1-MACHINISTS VICE
1-12 OZ BALL PEEN HAMMER
1-PAIR PLIERS
PROCEDURE
I TRUNION ASSEMBLY
PART 5 CRANE HOOK 1 REQ.
PART 9 WASHER 2 REQ.
PART 6 HOOK TRUNION 1 REQ.
PART 8 A. S. SLOTTED NUT 1 REQ.
PART 4 NUT LOCK PIN 1 REQ.
PART 10 COTTER PIN 2 REQ.
II PULLEY ASSEMBLY
PART 3 PULLEY PIN 2 REQ.
PART 2 PULLEY 2 REQ.
III PLATE ASSEMBLY
PART 1 SIDE PLATE 2 REQ.
INSTALL
PULLEY PIN ASSEMBLY
TRUNION ASSEMBLY
PART 7 A. S. SEMI-FIN. NUT 6 REQ.
SCREW DRIVER
(a)
(b)
(c)
(d)
MATCH NARROW SLOT AND TANG


**FIG. 49.** A manufacturing illustration. It specifies the parts and tools needed for an assembly. Drawings of this type are used for original manufacture and in service manuals. Compare this illustration with the orthographic drawing for Prob. 36.

**FIG. 50.** A maintenance illustration. This type of drawing is used in a service manual to illustrate repair by replacement of components.

linear size of the original. Figure 51 illustrates the appearance of an original drawing, and Fig. 52 shows the same drawing reduced one-half. The coarse appearance, open shading, and lettering size of the original should be noted. The line work must be kept fairly "open," for if lines are drawn close together the space between them may choke in the reproduction and mar the effect. A reducing glass, a concave lens mounted like a reading glass, is sometimes used to aid in judging the appearance of a drawing on reduction.

Drawings for reproduction may be altered and

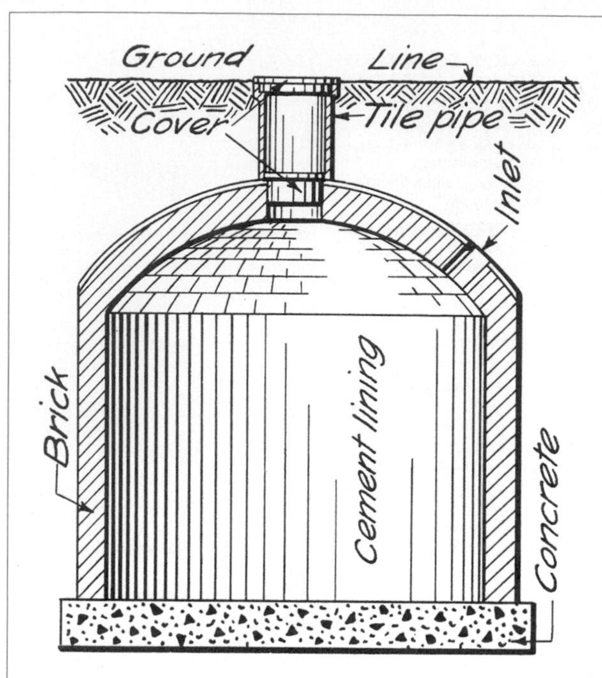


FIG. 51. A drawing for reproduction. This drawing is made for reduction to half size. The reduced reproduction is shown in Fig. 52.

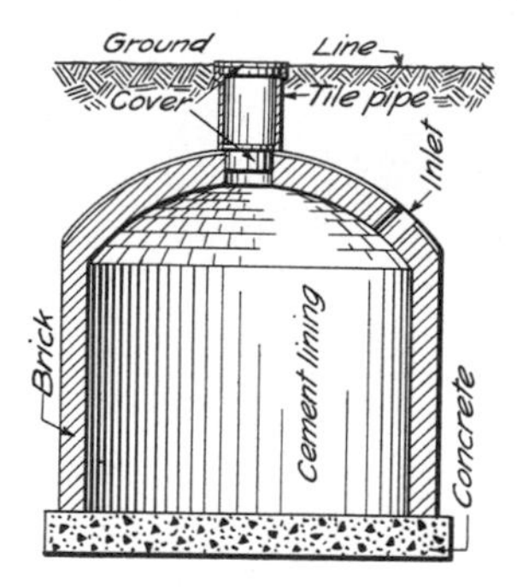


FIG. 52. The drawing shown in Fig. 51 reduced to half size and reproduced.

corrected in ways not permissible in other work. Irregularities may be painted out with opaque white. A sharp blade or "scratcher" may be used to clean off small errors. If a figure that has been inked is to be shifted, it may be cut out and pasted in the required position. A piece of paper may be pasted over a portion of the drawing for blocking out or redrawing. Reference letters and numbers, notes, and other lettering are often cut out of a sheet printed from type of proper size, and pasted on the drawing. Marks made by edges of pasted pieces will not show on the final reproduction; they are eliminated by the engraver when the plate is finished.

*Methods of Reproduction.* Drawings for illustrations are reproduced by one of the photomechanical processes, that is, by zinc or copper etching, halftone, or one of the methods of photolithography.

Line drawings are usually reproduced by zinc etching. By this process the drawing is photographed on a prepared zinc plate (when a particularly fine result is desired, a copper plate is used); then the plate is etched with acid, leaving the lines in relief and giving, when mounted, a block that can be printed along with type in an ordinary printing press. Drawings for zinc etching should be made on smooth white paper or tracing cloth in black drawing ink. The finest lines should be black and definite—weak thin lines will not reproduce well.

Wash drawings and photographs are reproduced in a similar way on copper by the halftone process. A ruled screen is placed in front of the camera lens, and it breaks up the tones into a series of dots. The fineness of the screen varies with the kind of paper used. Screens of 65 to 100 lines to the square inch are used for newspaper halftones, of 110 to 133 lines to the square inch for ordinary commercial and magazine halftones, of 133 to 175 lines to the square inch for fine halftones printed on very smooth coated paper.

The photolithographic processes may be used to reproduce line drawings, wash drawings, or photographs. The drawing is photographed on a thin sensitized sheet and chemically processed to use in a lithographic press in which the plate prints on a rubber blanket, which in turn prints on the paper. Paper, zinc, bimetal, and trimetal plates are used, depending upon the number of copies needed. This book is printed by photolithography.

**48. Special Shading Methods.** There are several methods of representing textures, in addition to the regular methods of shading in pencil and ink, given in Chap. 6.

FIG. 53. Smudge shading. This method gives an appearance similar to airbrush work and photography.

*Smudge shading* is a rapid method often used for smooth surfaces (Fig. 53). Graphite from a soft pencil, powdered graphite, charcoal, or crayon sauce is first rubbed on a piece of paper, then picked up with a piece of cotton or an artist's stump and applied to the drawing. High lights are easily brought out with a sharpened eraser. Be careful in shading an erased portion, however, as the abrading of the paper will cause the shading medium to "take" more heavily.

*Stippling* with pen, pencil, brush, or sponge is an effective method of indicating rough-texture surfaces. In pen and pencil stippling, a good effect of light and shade is achieved with a multitude of dots, widely spaced for light surfaces, more closely for dark surfaces. In brush and sponge stippling, printer's ink or artist's oil color (drier added) is first worked out smoothly on a palette; then the medium is picked up from the palette with a bristle brush or sponge and applied to the drawing with a dabbing motion. Sharp edges of shaded areas can be maintained by the use of masks.

Small areas are easily lightened with a razor blade or "scratcher." High lights are brought out with an eraser after the ink is dry. Figure 54 is an example of brush stippling with the smooth surfaces rendered by smudging.

*Prepared papers* are popular for a variety of commercial drawings. Craftint papers[5] have a shading pattern in the paper that is brought out by a special developer. They are of two types: single-tone and double-tone. When these papers are used, the drawing is penciled and the solid blacks are inked in with waterproof drawing ink. The shading tones are then brought out by brushing on the developer in the areas where shading is wanted. These papers are available in many different shading patterns. Figure 48 is drawn on Craftint paper.

*Shading screens* of clear cellulose with printed line or dot patterns provide a simple and effective shading method. Craftint shading film is an overlay sheet containing a pattern printed in black or white. The shading pattern is easily rubbed off with a smooth

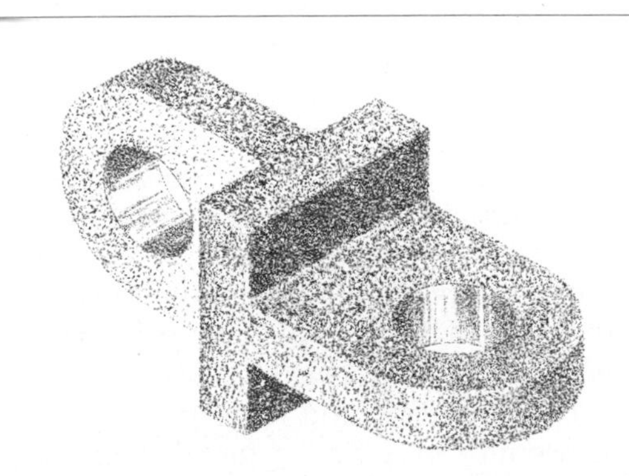

FIG. 54. Brush-stipple shading. This type of shading can also be done with a pen, pencil, or sponge.

[5]Craftint Manufacturing Co., Cleveland, Ohio.

wood stick wherever the shading is not wanted. Any part of the white pattern can be converted to black with a special developer.

Zip-a-tone screens[6] are of clear cellulose printed with a shading pattern and backed with a special adhesive. The screen is applied to the drawing and rubbed down lightly wherever shading is wanted. The shaded sections are then outlined with a cutting needle and the unwanted pieces stripped off. The portions left on the drawing are then rubbed down firmly with a burnisher. High lights can be painted out with opaque white if the area is too small to strip off in the regular way. A variety of shading patterns are available. Figure 60, Chap. 2, is shaded with a Zip-a-tone screen.

*Scratchboard.* Drawing paper with a chalky surface is much used for commercial illustrations because of the ease with which white lines can be produced on a black background, as well as black lines on a white background. Rossboard,[7] available in many surface textures for both pencil and ink drawings, is popular among illustrators.

For ink work, the drawing is penciled on the board in the usual way, and then inked by line-shading the lighter areas, working gradually to the darker areas. The darker areas are painted in with a brush, and when dry, white lines, dots, etc., are produced by scratching off the ink with a sharp-pointed knife, sharp stylus, or a needle point. Corrections are easily made by scratching off unwanted ink. Scratched areas can be reinked if necessary. Figure 55 is an example of scratchboard technique.

### 49. Patent Office Drawings.

In an application for letters patent on an invention or discovery a written description, called the "specification," is required, and for a machine or device a drawing showing every feature of the invention must also be supplied. A high standard of execution and conformity to the rules of the Patent Office must be observed. A pamphlet called "Rules of Practice," giving full information and rules governing Patent Office procedure in applying for a patent, can be obtained from the Commissioner of Patents, Washington, D.C.

[6] Para-tone Company, Chicago, Ill.
[7] Charles J. Ross Co., Philadelphia, Pa.

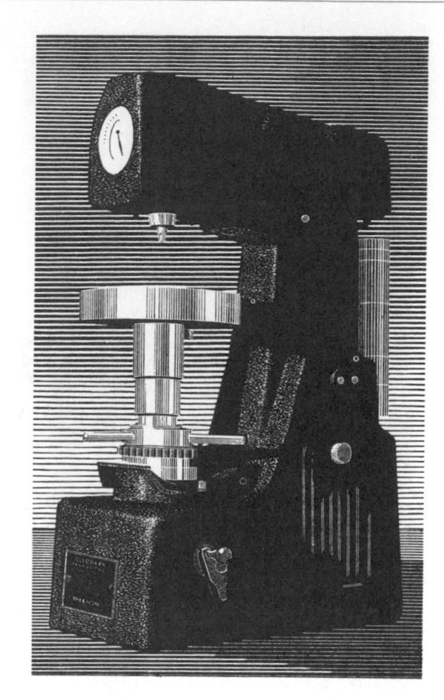

**FIG. 55.** a scratchboard drawing. The effect is similar to a woodcut.

The rules call for drawings made on smooth white paper, 10 by 15 in., with a border line 1 in. from the edges. A space not less than $1\frac{1}{4}$ in. inside the top border must be left blank for the printed title to be added by the Patent Office. Drawings must be made in India ink and drawn for reproduction to a reduced size. As many sheets as necessary may be used.

Patent Office drawings are not working drawings. They are descriptive and pictorial rather than structural, hence they have no center lines, dimensions, notes, or names of views. The views are lettered with figure numbers, and the parts are designated by reference numbers through which the invention is described in the specification.

The drawings may be made in orthographic, axonometric, oblique, or perspective projection. The pictorial system is used extensively, for all or part of the views. Surface shading is used whenever it will aid readability.

Figure 56 is an example of a Patent Office drawing.

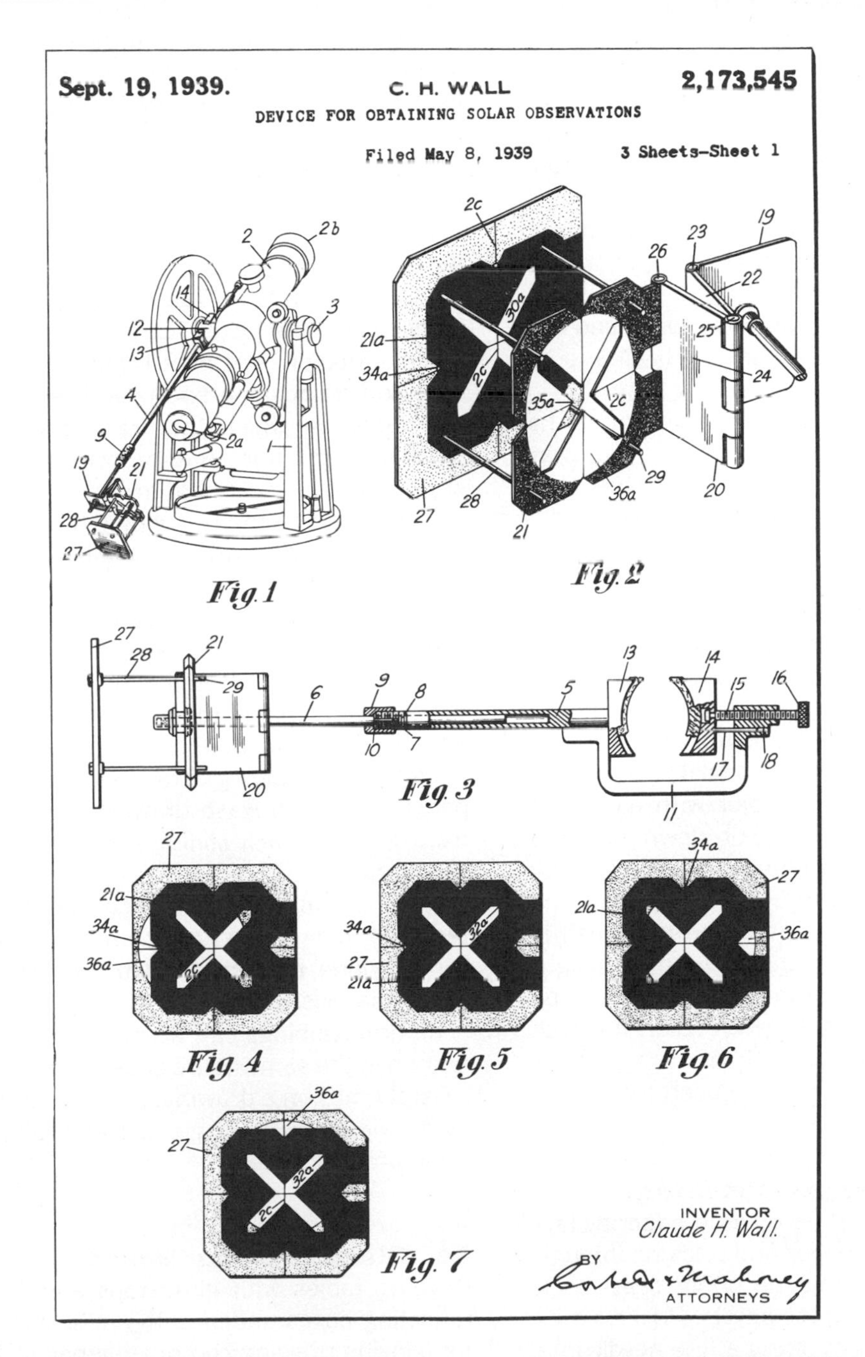

Sept. 19, 1939.
C. H. WALL
2,173,545
DEVICE FOR OBTAINING SOLAR OBSERVATIONS
Filed May 8, 1939
3 Sheets-Sheet 1
Fig. 1
Fig. 2
Fig. 3
Fig. 4
Fig. 5
Fig. 6
Fig. 7
INVENTOR
Claude H. Wall.
BY
ATTORNEYS


**FIG. 56.** A Patent Office drawing. The illustration here is reduced one-half.

## SPECIAL METHODS AND EQUIPMENT

### 50. Stretching Paper.

If a drawing is to be tinted, the paper should be stretched on the board. First dampen it on both sides, with a sponge or under the faucet, until it is limp; then lay it on the drawing board face down, take up the excess water from the edges with a blotter, brush a strip of glue or paste about ½ in. wide around the edge, turn the paper over, and rub its edges down on the board until set. Allow it to dry horizontally.

Drawings or maps on which much work is to be done, even though they are not to be tinted, may be made advantageously on stretched paper; but Bristol or calendered paper should not be stretched.

### 51. Tinting.

This is done with washes made with water colors. The drawing may be inked (with waterproof ink) before or preferably after tinting. It should be cleaned and unnecessary pencil marks removed with a very soft rubber. The tint is mixed in a saucer and applied with a camel's-hair or sable brush. Incline the board and flow the color with horizontal strokes, leading the pool of color down over the surface, taking up the surplus at the bottom by wiping the brush quickly and using it to pick up the excess color. Stir the color each time the brush is dipped into the saucer. Tints should be made in light washes; depth of color is obtained if necessary by repeating the wash. To get an even color, it is well to go over the surface first with a wash of clear water. Diluted colored inks may be used for washes instead of water color.

### 52. Copying Drawings—Pricking.

Drawings are often copied on opaque paper by laying the drawing over the paper and pricking through with a needle point, turning the upper sheet back frequently, and connecting the points. Prickers may be purchased, or made by forcing a fine needle into a softwood handle. They can also be used in accurate drawing to transfer measurements from scale to paper.

### 53. Transfer by Rubbing.

This method, known as *frotté,* is useful, particularly in architectural drawing, in transferring any kind of sketch or design to the paper on which it is to be rendered.

The original is made on any paper and may be worked over, changed, and marked up until the design is satisfactory. To transfer the drawing, lay a piece of tracing paper over the original and trace the outline carefully. Turn the tracing over and retrace the outline just as carefully on the other side, using a medium soft pencil with a *sharp* point. Turn back to first position and tack down smoothly over the paper on which the drawing is to be made, registering the tracing to proper position by center or reference lines on both tracing and drawing. Now transfer the drawing by rubbing the tracing with the rounded edge of a knife handle or other instrument (a smooth-edged coin held between thumb and forefinger and scraped back and forth is commonly used), holding a small piece of tracing cloth with smooth side up between the rubbing instrument and the paper, to protect the paper. Do not rub too hard and be sure that neither the cloth nor the paper moves while rubbing. Transfers in ink instead of pencil, useful on wash drawings, may be made by tracing with *encre à poncer,* a rubbing ink made for this purpose.

If the drawing is symmetrical about any axis, the reversed tracing need not be made, as the rubbing can be done from the first tracing by reversing it about the axis of symmetry.

Several rubbings can be made from one tracing, and when the same figure or detail must be repeated several times on a drawing, much time can be saved by drawing it on tracing paper and rubbing it in the several positions.

### 54. Glass Drawing Board.

Drawing tables with glass tops and with lights in reflecting boxes underneath are successful devices for copying drawings on opaque paper. Even pencil drawings can be copied readily on the heaviest paper or Bristol board by the use of a transparent drawing board.

## 55. Proportional Methods—The Pantograph.

The principle of the pantograph, used for reducing or enlarging drawings in any proportion, is well known. The instrument consists essentially of four bars, which for any setting must form a parallelogram and have the pivot, tracing point, and marking point in a straight line. Any arrangement of four arms conforming to this requirement will work in true proportion. For reduction the tracing point and marking point are interchanged. A suspended pantograph with metal arms, for accurate engineering work, is shown in Fig. 57.

Drawings can be copied to reduced or enlarged scale by using proportional dividers, illustrated in Fig. 58. The divisions marked "lines" are linear proportions; those marked "circles" give the setting for dividing a circle into a desired number of equal parts when the large end is opened to the diameter of the circle.

The well-known method of *proportional squares* is often used for reduction or enlargement. The drawing to be copied is ruled in squares of convenient size, or if it is undesirable to mark the drawing, a sheet of ruled tracing cloth or celluloid is laid over it and the copy made freehand on the paper, which has been ruled in corresponding squares, larger or smaller (Fig. 59).

Reduction or enlargement of a drawing can of course be done photographically. Photographic reproduction eliminates much drafting time, especially for complicated drawings.

## 56. Special Instruments.

There are some instruments not in the usual assortment that are occasionally needed. Beam compasses are used for circles larger than the capacity of ordinary compasses with lengthening bar. A good form is illustrated in Fig. 60. A metal-bar beam compass is shown in Fig. 18, Chap. 2.

With the drop pen or rivet pen (Fig. 61) smaller circles can be made than with the bow pen, and

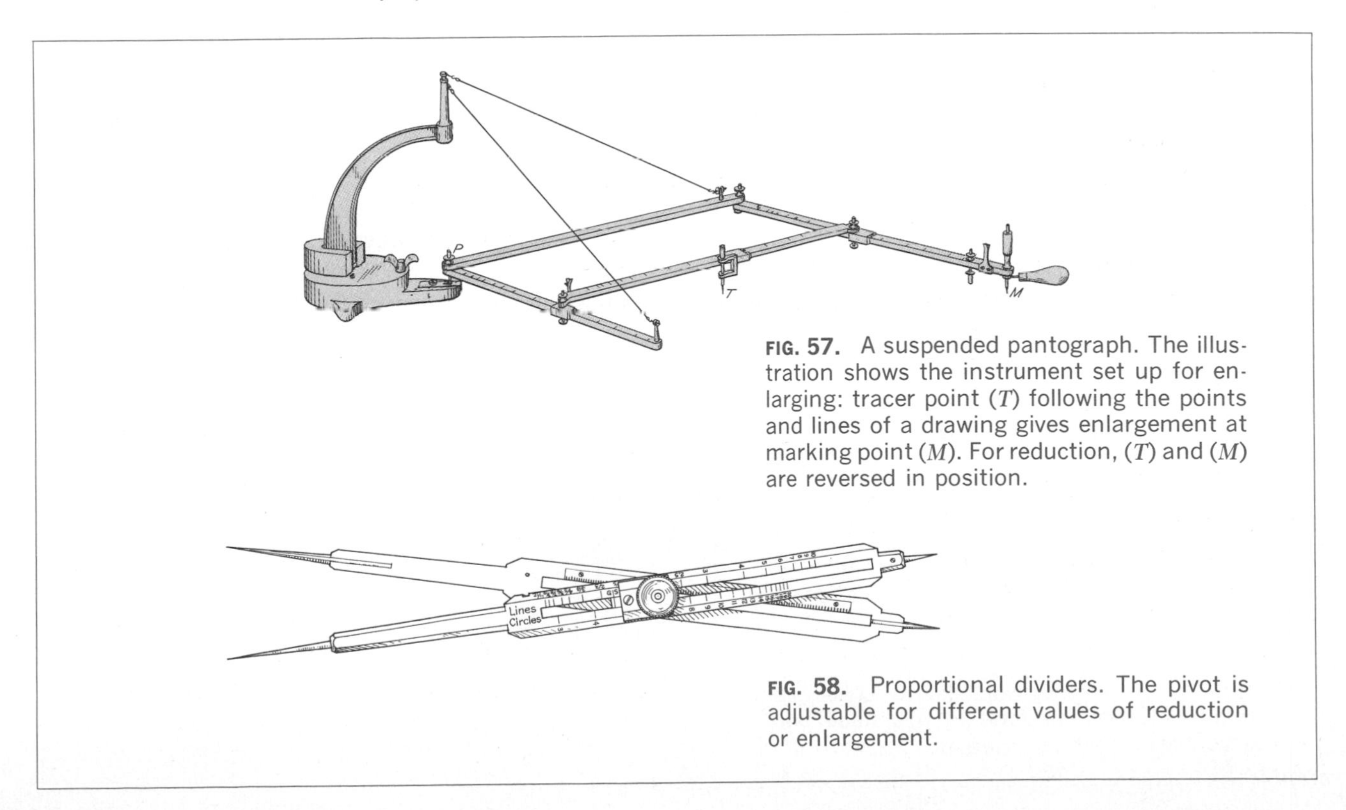


**FIG. 57.** A suspended pantograph. The illustration shows the instrument set up for enlarging: tracer point (*T*) following the points and lines of a drawing gives enlargement at marking point (*M*). For reduction, (*T*) and (*M*) are reversed in position.

**FIG. 58.** Proportional dividers. The pivot is adjustable for different values of reduction or enlargement.

FIG. 59. Enlargement or reduction by squares. Points are plotted by inspection for one square grid to the other. This method is also known as graticulation.

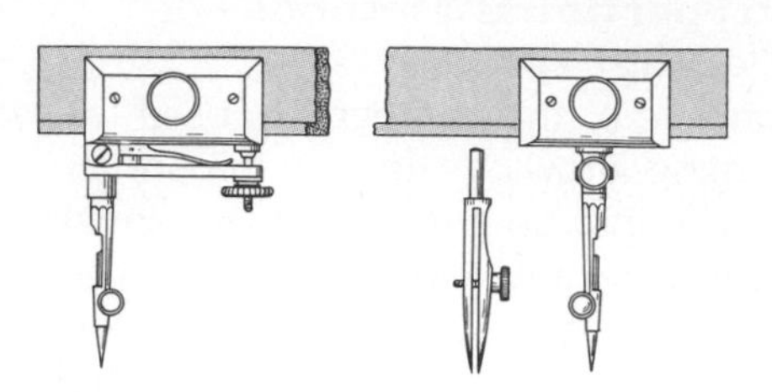

FIG. 60. A beam compass. This one employs a wooden beam, obtainable in various lengths up to 6 feet.

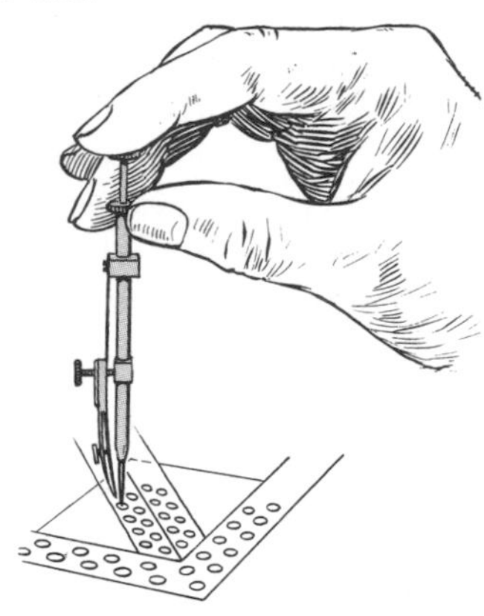

FIG. 61. A drop pen. This type of compass is particularly efficient for drawing a number of small circles of the same diameter. These are much used in structural drawing.

they are made quickly. It is held as shown; the needle point is stationary, and the pen is revolved around it. This pen is particularly convenient for bridge and structural work and topographic drawing.

Several instruments for drawing ellipses have been made. The ellipsograph (Fig. 62) is very satisfactory.

Three special pens are shown in Fig. 63. The *railroad pen* (*a*) is used for double lines. A better pen for double lines up to $\frac{1}{4}$ in. apart is the *border pen* (*b*), as it can be held down to the paper more satisfactorily. It can be used for very wide solid lines by inking the middle space as well as the two pens. The *contour*, or curve, pen (*c*), made with a swivel, is used in map work for freehand curves.

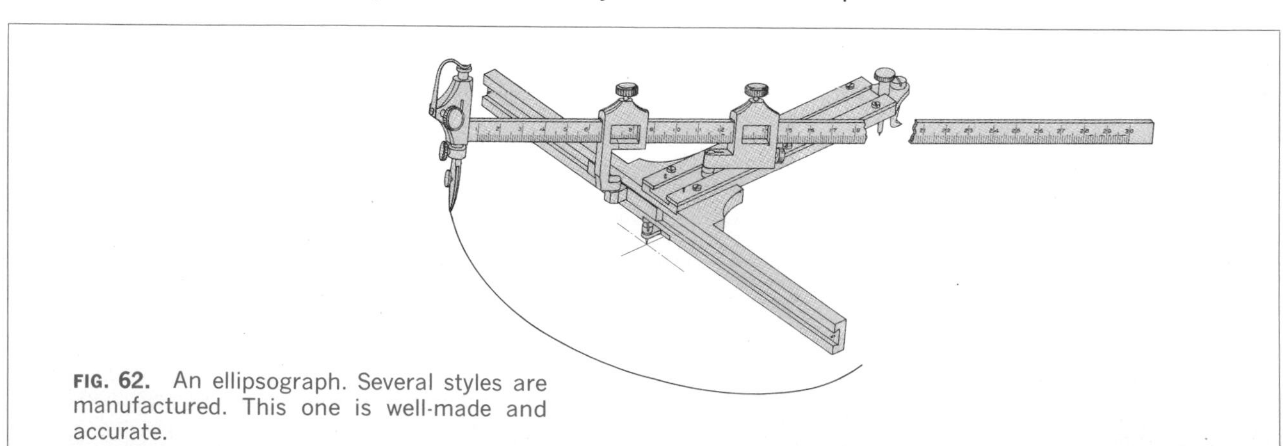

FIG. 62. An ellipsograph. Several styles are manufactured. This one is well-made and accurate.

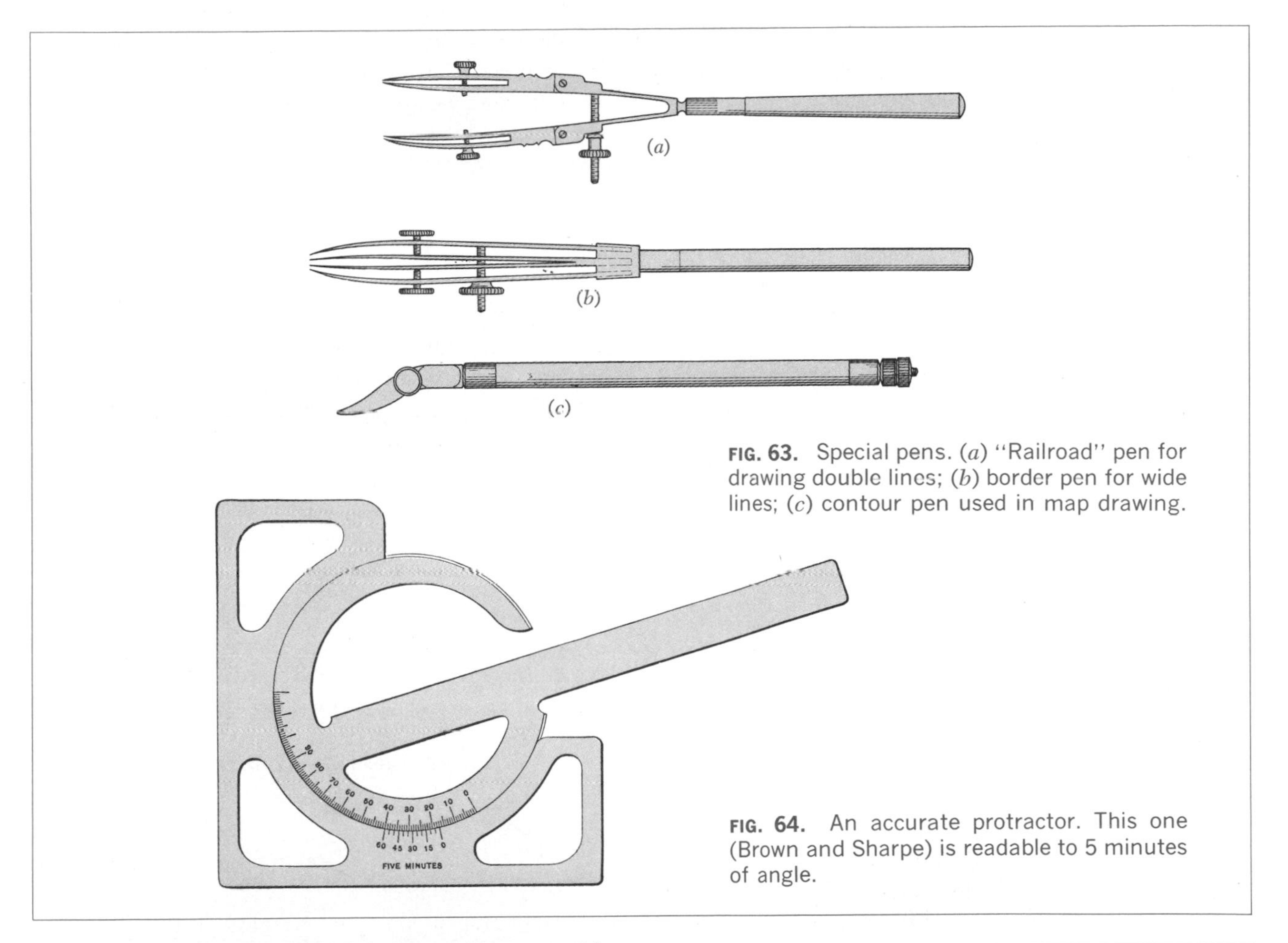


**FIG. 63.** Special pens. (*a*) "Railroad" pen for drawing double lines; (*b*) border pen for wide lines; (*c*) contour pen used in map drawing.

**FIG. 64.** An accurate protractor. This one (Brown and Sharpe) is readable to 5 minutes of angle.

A *protractor* is a necessity in map and topographic work. A semicircular brass or nickel-silver one, 6 in. in diameter, will read to half degrees. Protractors are available with an arm and vernier reading to minutes. Large circular paper protractors 8 and 14 in. in diameter reading to half and quarter degrees are used and preferred by some map draftsmen. Others prefer the Brown and Sharpe protractor (Fig. 64), which reads to 5 minutes.

A combination of triangle and protractor popular with architects and draftsmen is shown in Fig. 65. Numerous different combination "triangles" have been devised; several types are usually carried by dealers.

*Vertical drawing boards* with sliding parallel straightedges are found by some draftsmen to be

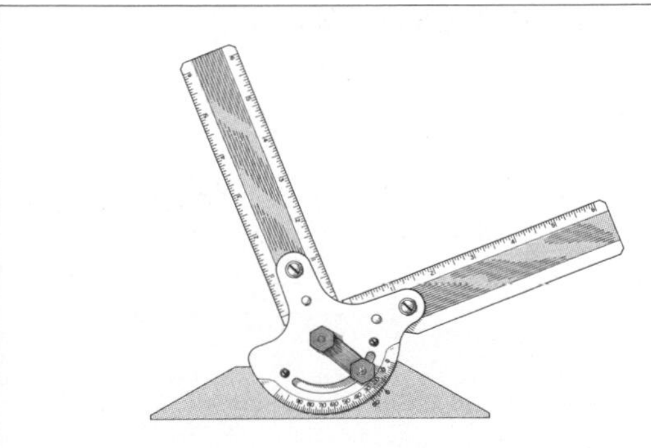

**FIG. 65.** Combined triangle, scale, and protractor. Several different types are made. This one is the Tri-pro-scale.

most suitable for large work.

Figure 66 shows a special *bottleholder* made by the Alteneder Company. The pen can be filled with one hand, saving time.

Some irregular *curves* were illustrated in Fig. 10, Chap. 2. There are also many others on the market. Sometimes for special or recurring curves it is advisable for a draftsman to make his own template. Templates can be cut out of thin holly or basswood, sheet lead, celluloid, or even cardboard or pressboard. To make a paper curve, sketch the desired shape on paper, cut it out with scissors, and sandpaper the edge. For inking, use it over a triangle or another piece of paper. Flexible curves of different kinds are sold. A copper wire or piece of wire solder can be used as a homemade substitute.

The curve shown in Fig. 67 is particularly useful for engineering diagrams, steam curves, etc.

If the glaze is removed from a celluloid irregular curve, by rubbing with fine sandpaper, pencil marks that facilitate the drawing of symmetrical curves can be made on it.

**FIG. 66.** Alteneder bottle-holder. The design makes it possible to fill the pen with one hand.

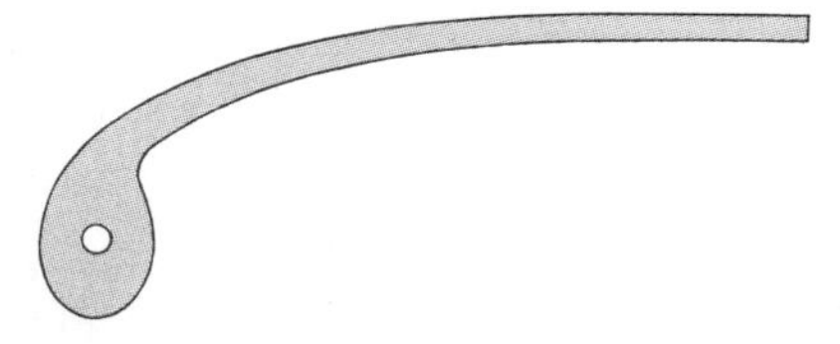

**FIG. 67.** A diagram curve. Many special curves are available from manufacturers.

*Splines* are flexible-curve rulers that are adjusted to the points of the curve to be drawn and held in place by lead weights, called "ducks" by draftsmen. They come in various lengths and are part of the regular equipment of all aircraft drawing rooms (Fig. 68).

## 57. Various Devices.

If a drawing or map is so large that it extends over the bottom edge of the board, a piece of half round should be fastened to the board, as in Fig. 69, to prevent creasing the paper.

A steel edge for a drawing board can be made of an angle iron planed straight and set flush with

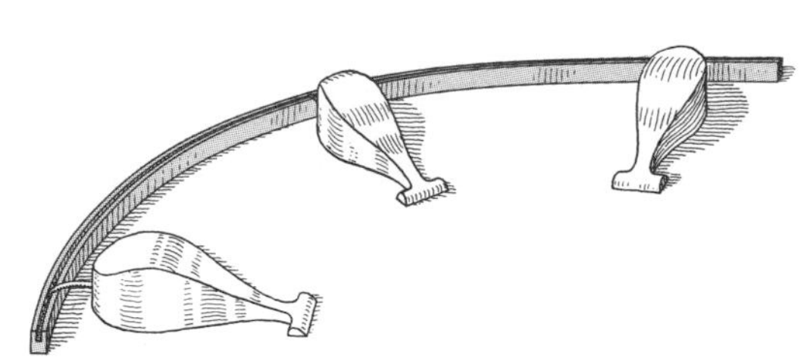

**FIG. 68.** A spline and "ducks." This flexible-curve ruler is used principally for drawing large curves, especially in lofting work.

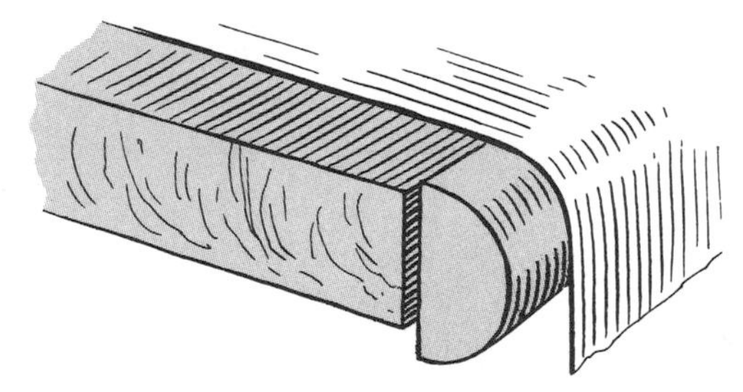

**FIG. 69.** Rounded drawing-board edge. This prevents large drawings from being creased.

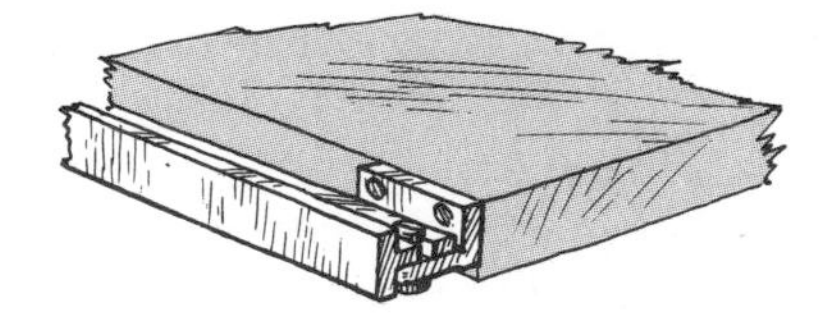

**FIG. 70.** A metal drawing-board edge. Some are fastened permanently to the board. This one (L. S. Starrett & Company) is particularly rugged.

FIG. 71. A section liner. Several designs are available. The spacing and angle of the blade are adjustable.

the edge. A well-liked adjustable metal edge is made by L. S. Starrett & Company (Fig. 70). With a steel edge and steel T square, very accurate plotting can be done. These are often used in bridge offices.

Section lining or crosshatching is a difficult operation for a beginner but is done almost automatically by an experienced draftsman. A number of instruments for mechanical spacing have been devised. *Section liners* are mechanical spacers for use in crosshatching or other evenly spaced line work. One type is shown in Fig. 71.

There are many other devices designed to save labor and for convenience in drafting rooms. The Bostich tacker is used instead of thumbtacks. Many draftsmen like to fasten paper to the board with Scotch tape. For this, *drafting tape* should be used in preference to masking tape or other varieties of Scotch tape. The Dexter "Draftsmen's Special" pencil sharpener removes the wood only, leaving a long exposure of lead. Electric erasing machines are popular.

## 58. Lettering Devices.

The Braddock-Rowe triangle and the Ames lettering instrument (Fig. 2*a* and *b*, Chap. 4) are convenient devices used to draw guide lines for lettering.

Commercial drafting rooms regard mechanical lettering instruments as necessary for producing neat consistent lettering and symbols on inked drawings. The Leroy set (Figs. 72 and 73) is one of the most popular. It is available with a wide variety of letter styles and with many electrical, welding, mapping, geological, and mathematical symbols.

## 59. Drafting Machines.

Since the expiration of the patent on the Universal drafting machine, several makers have come into competition with different designs of this important instrument, which combines the functions of T square, triangles, scale, and protractor and which is used extensively in commercial drafting rooms. It is estimated that 35 per cent of time in machine drawing and over 50 per cent in structural drawing

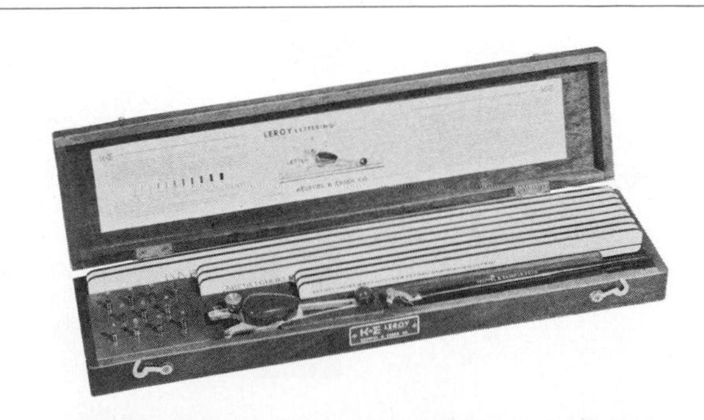

FIG. 72. Leroy lettering set (Keuffel & Esser Co.). Many different templates for lettering symbols, and special designs are available.

**FIG. 73.** The Leroy instrument in use. The scriber is guided by engraved grooves in the template. Many different pen sizes can be inserted in the scriber.

**FIG. 74.** A drafting machine. This combines T square, triangles, scales, and protractor in one instrument.

is saved by the use of drafting machines. Figure 74 shows a band-type drafting machine. Special drafting machines are made for left-handed draftsmen.

### 60. The Slide Rule.

Although not a drawing instrument, the slide rule (see Appendix B) is an engineer's tool, and proficiency in its use is a requirement in every modern drafting room. A good way for a beginner to learn to use a slide rule is in connection with a drawing course. Its use facilitates the rapid calculation of volumes and weights, as an aid in reading drawings or later as an essential part of drafting work. Of the several varieties of slide rule, those recommended for prospective engineers are a Polyphase Duplex Trig,[8] a Polyphase Duplex Decitrig,[8] a Log Log Duplex Trig,[8] a Log Log Duplex Decitrig,[8] or a Versilog[8] in 10-in. size.

[8] Registered trade-mark.

## PROBLEMS

### Group 1. Detail Drawings.

This group of problems gives practice in making drawings with sectional views, auxiliaries, and conventional representation. Several methods of part production are included: casting, forging, forming of sheet metal, and plastic molding.

1. Make complete working drawing with necessary sectional views. Cast iron.
2. Working drawing of friction-shaft bearing. Cast iron.
3. Working drawing of mixing-valve body. Cast brass.

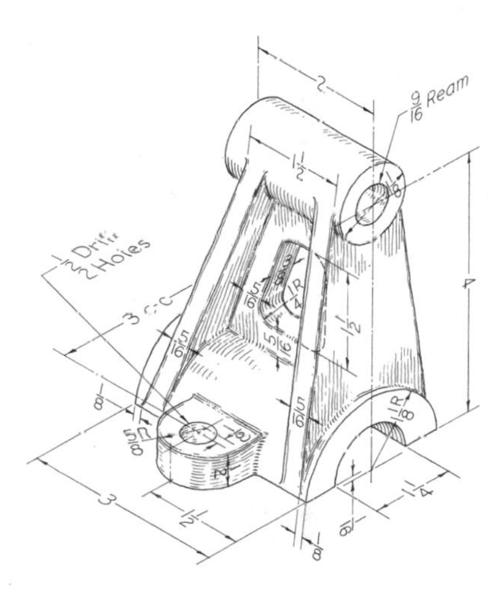


PROB. 2. Friction-shaft bearing.

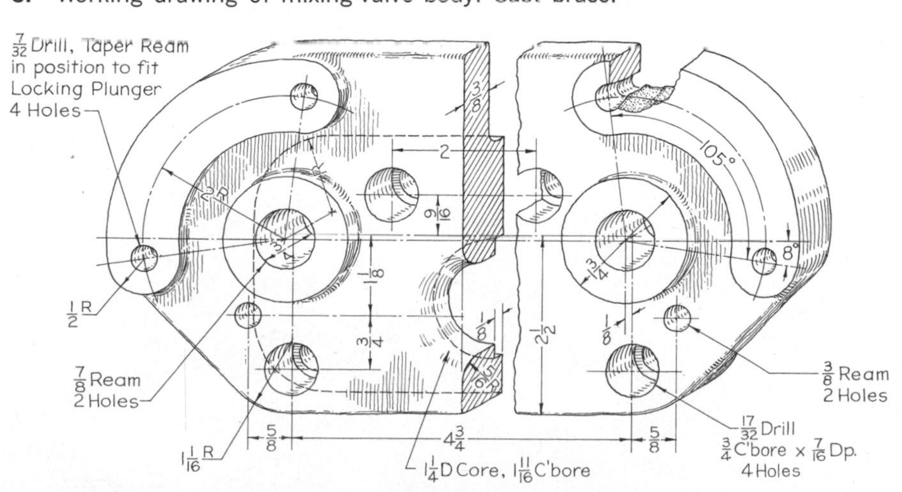


PROB. 1. Gear-shifter bracket.

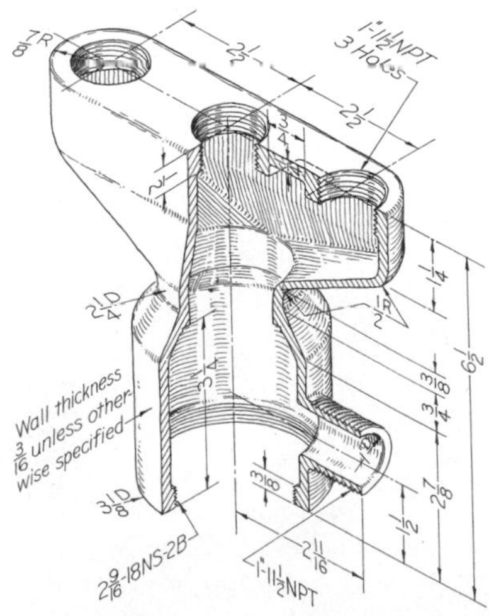


PROB. 3. Mixing-valve body.

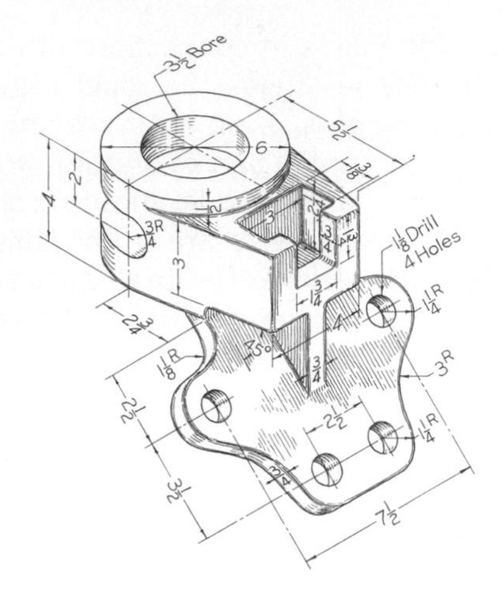


**PROB. 4.** Conveyor hanger.

**4.** Working drawing of conveyor hanger. Determine what views and part views will best describe the piece. Cast steel.

**5.** Working drawing of strut base. Determine what views and part views will best describe the piece. Cast aluminum.

**6.** Draw given front view. Add part views and auxiliaries to describe the piece best. Cast aluminum.

**7.** Working drawing of valve cage. Cast bronze.

**8.** Determine what views and part views will most adequately describe the piece. Malleable iron.

**9.** Working drawing of water-pump cover. Aluminum alloy die casting.

**10.** Working drawing of slotted spider. Malleable iron.

**11.** Make working drawing of relief-valve body. Cast brass.

**12.** Working drawing of breaker. Steel.

**13.** Working drawing of meter case. Molded Bakelite.

**14.** (*a*) Make detail working drawings on same sheet, one for *rough* forging and one for *machining;* or (*b*) make one detail drawing for forging and machining. Alloy steel.

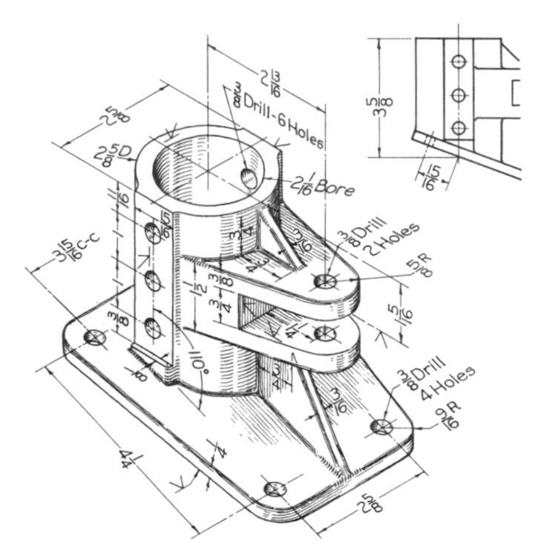


**PROB. 5.** Strut base.

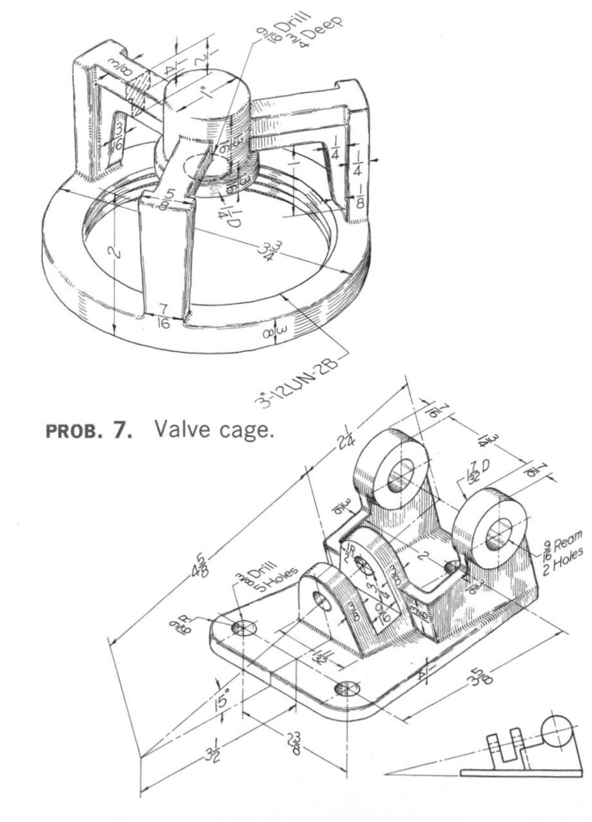


**PROB. 7.** Valve cage.

**PROB. 8.** Brace plate.

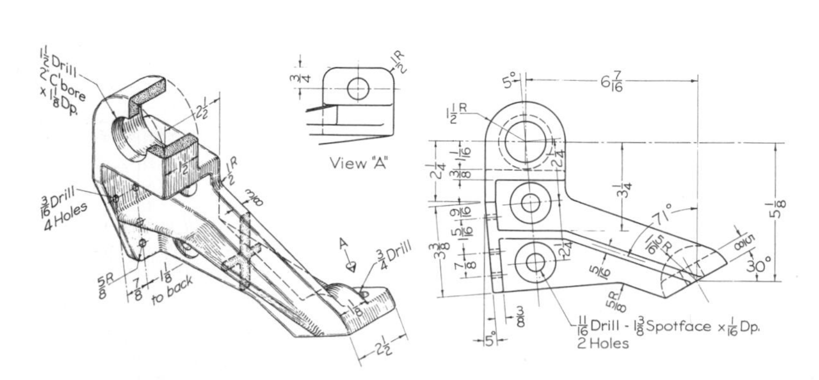


**PROB. 6.** Hinge bracket.

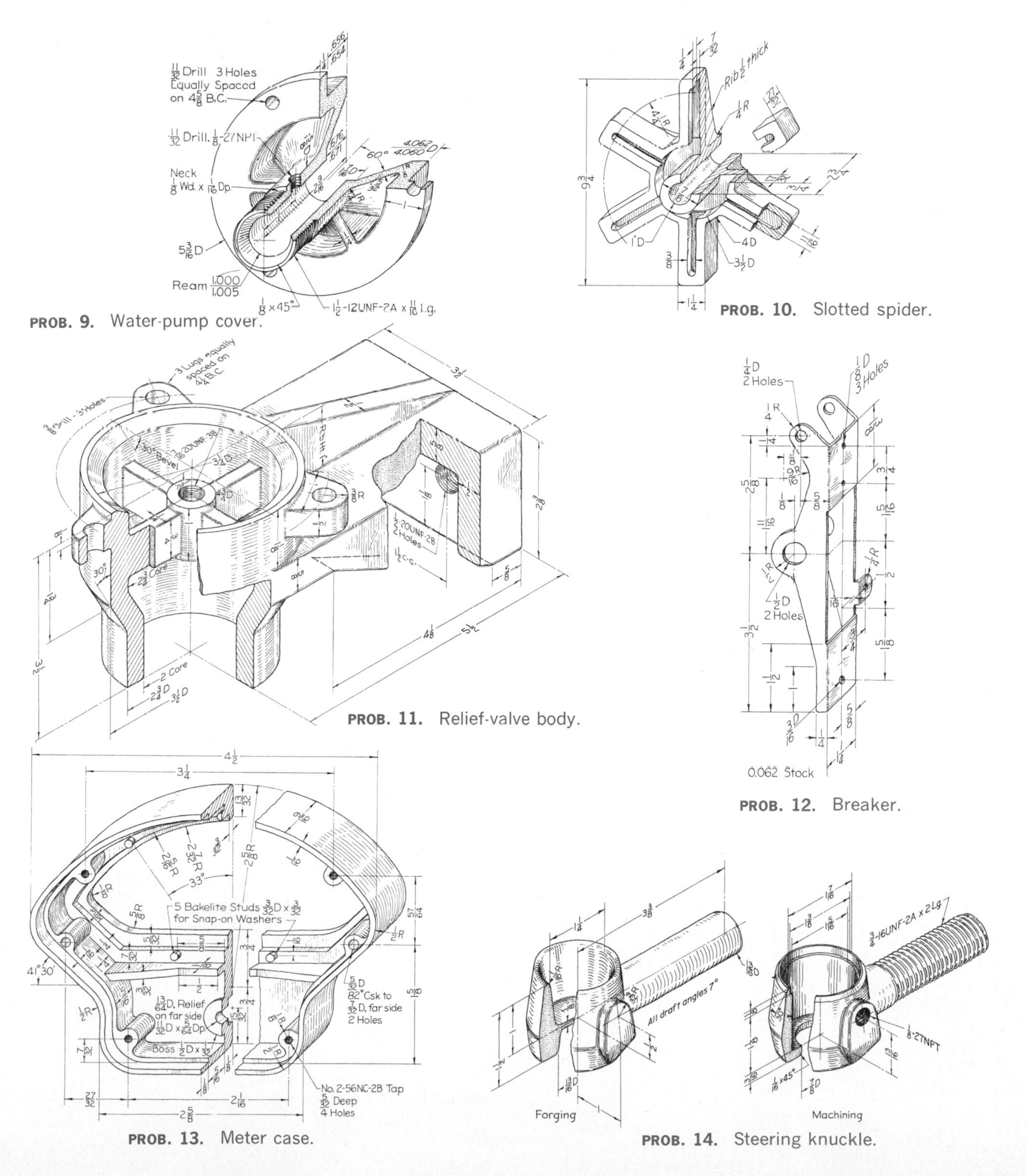


**PROB. 9.** Water-pump cover.

**PROB. 10.** Slotted spider.

**PROB. 11.** Relief-valve body.

**PROB. 12.** Breaker.

**PROB. 13.** Meter case.

**PROB. 14.** Steering knuckle.

**15.** Working drawing of automotive connecting rod. Drop forging, alloy steel. Drawings same as Prob. 14.
**16.** Working drawing of puller body. Press forging, steel. Drawings same as Prob. 14.
**17.** Make working drawing of buffer stand. Steel drop forging. Drawings same as Prob. 14.
**18.** Working drawing of torque-tube support. Drop forging, aluminum alloy. Drawings same as Prob. 14.
**19.** Working drawing of supply head. Cast iron.

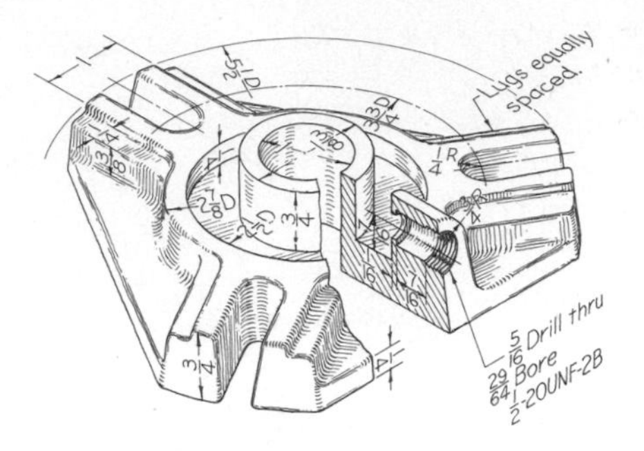


**PROB. 16.** Puller body.

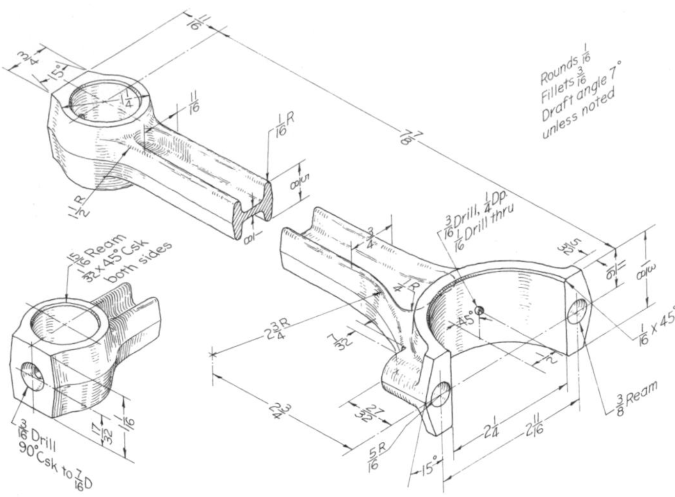


**PROB. 15.** Automotive connecting rod.

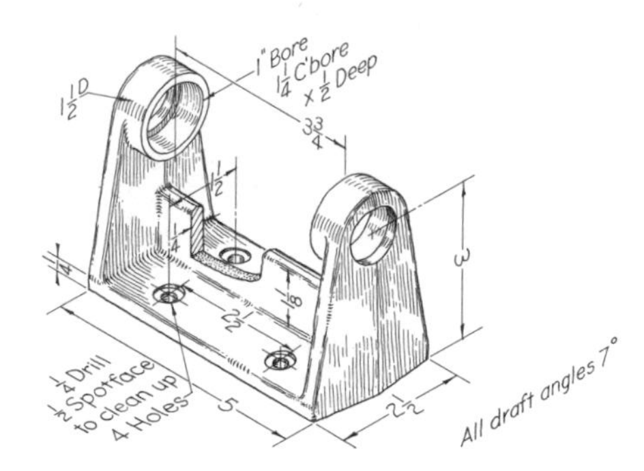


**PROB. 17.** Buffer stand.

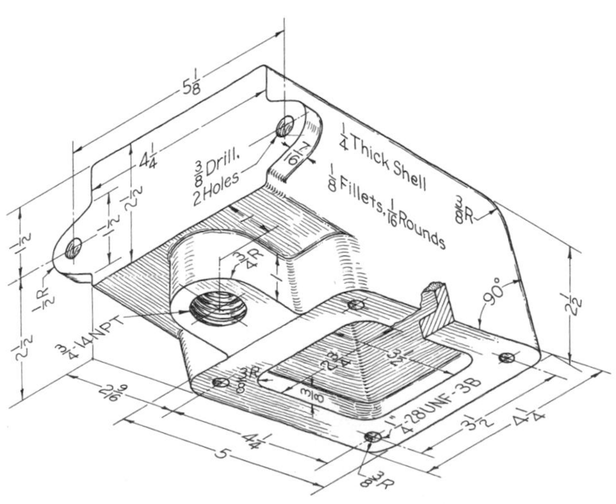


**PROB. 19.** Supply head.

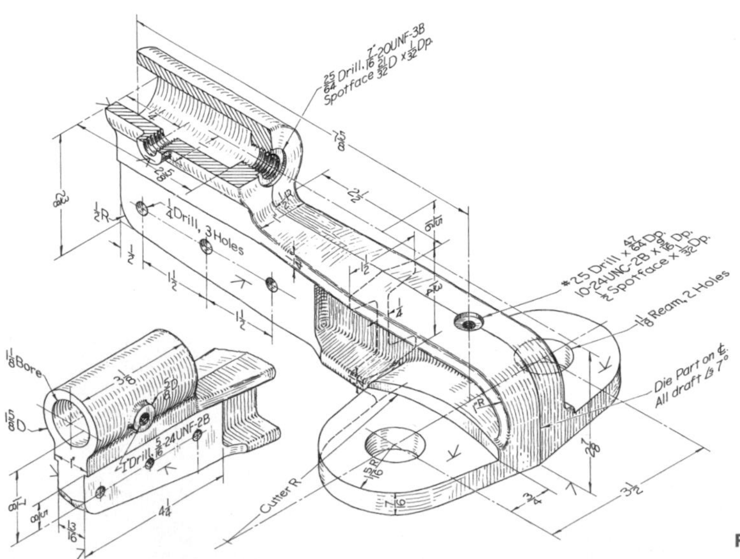


**PROB. 18.** Torque-tube support.

**20.** Working drawing of bearing block, cast iron.
**21.** Working drawing of elevator control bracket. Aluminum alloy, welded.
**22.** Working drawing of flap link. High-strength aluminum alloy.
**23.** Working drawing of wing fitting. High-strength aluminum alloy.
**24.** Working drawing of micro-switch bracket. Aluminum sheet, 0.15 thick.

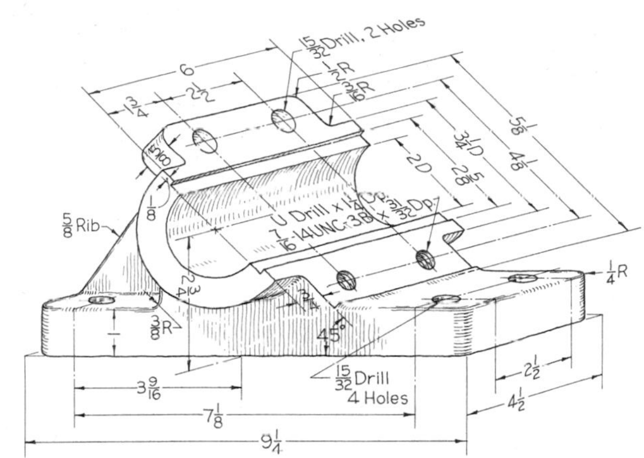


**PROB. 20.** Bearing block.

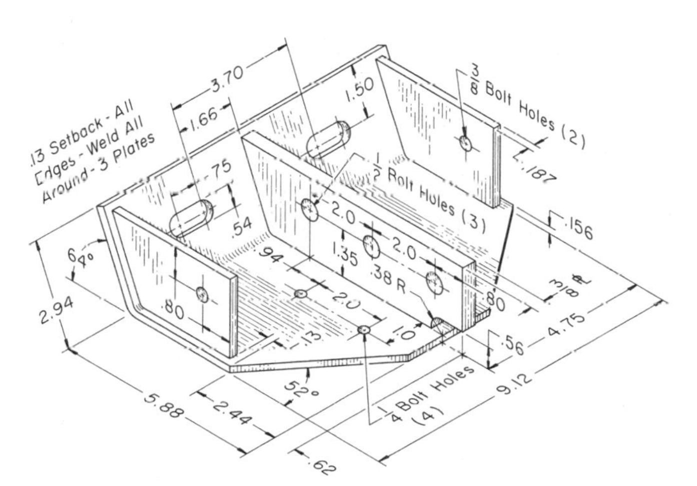


**PROB. 21.** Elevator control bracket.

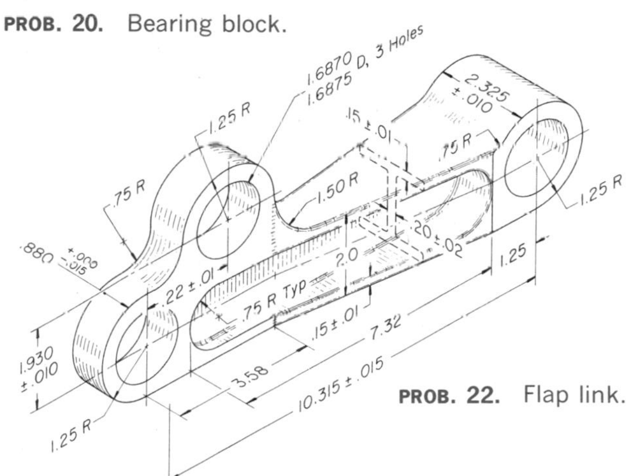


**PROB. 22.** Flap link.

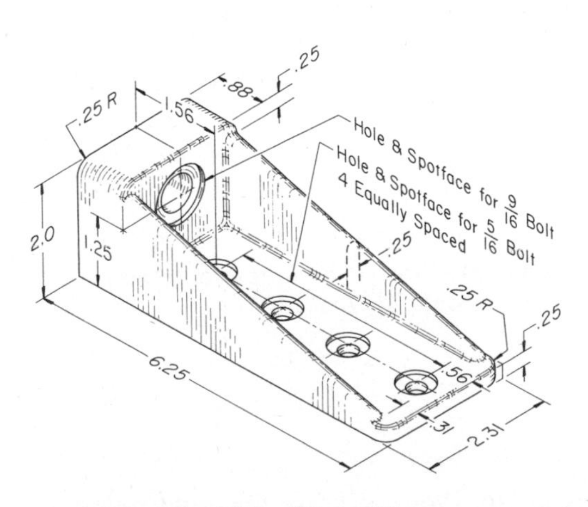


**PROB. 23.** Wing fitting.

**PROB. 24.** Micro-switch bracket.

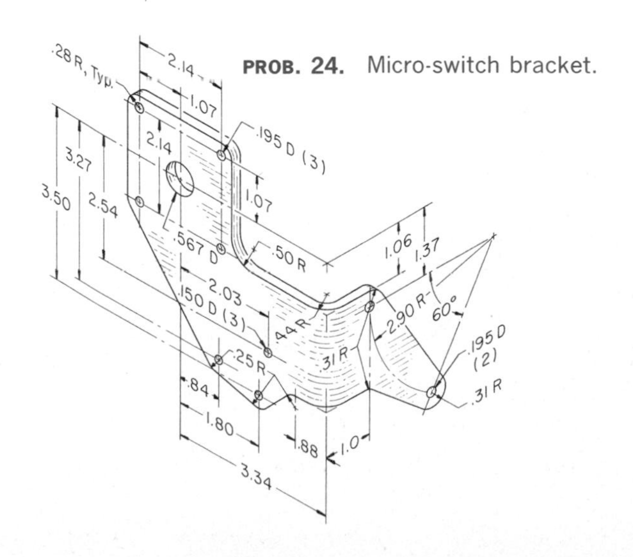

**25.** Working drawing of tab link. High-strength aluminum alloy.

**26.** Working drawing of drag strut. High-strength aluminum alloy.

**27.** Make a complete working drawing of the $1\frac{1}{2}$-in., no. 2000 desurger case head of Prob. 12, Chap. 12. Material is ductile iron.

**28.** Make a complete working drawing of the desurger loading-valve gage connector. This part is made of $\frac{3}{4}$-in. hexagon steel (AISI C1212).

**29.** Make a complete working drawing of the needle-valve needle. Material is stainless steel.

**30.** Make a complete working drawing of the wheel-cylinder boot.

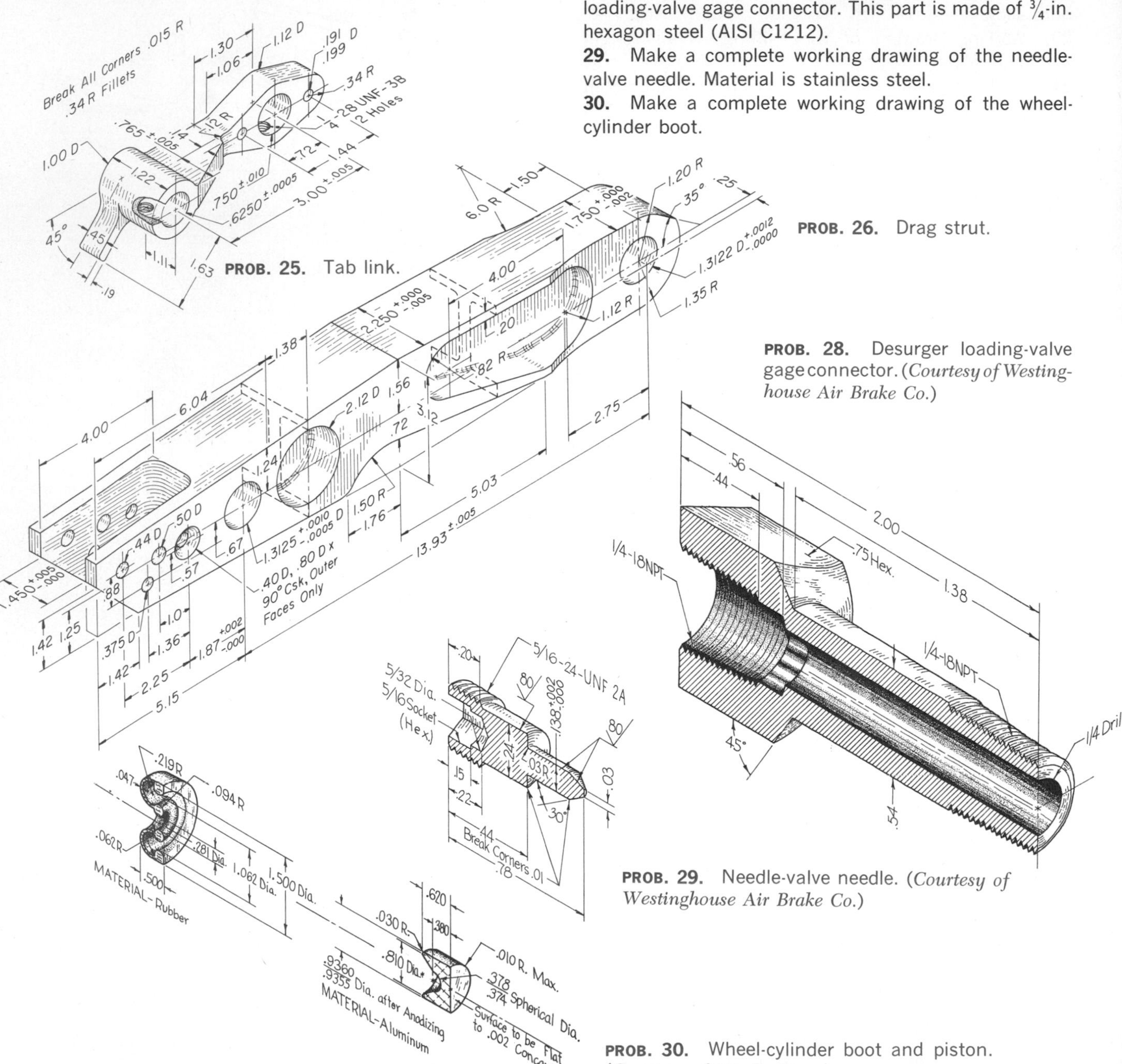


**PROB. 25.** Tab link.

**PROB. 26.** Drag strut.

**PROB. 28.** Desurger loading-valve gage connector. (*Courtesy of Westinghouse Air Brake Co.*)

**PROB. 29.** Needle-valve needle. (*Courtesy of Westinghouse Air Brake Co.*)

**PROB. 30.** Wheel-cylinder boot and piston. (*Courtesy of Bendix Corp.*)

**31.** Make a complete drawing of the wheel-cylinder piston, Prob. 30.

**32.** Make a complete working drawing of the desurger loading-valve body. Material is brass.

**33.** Make a complete working drawing of the wheel-cylinder cup.

## Group 2. An Assembly Drawing from the Details.

**34.** Assembly drawing of sealed shaft unit. Bracket and gland are cast iron. Shaft, collar, and studs are steel. Bushing is bronze. Drill bushing in assembly with bracket.

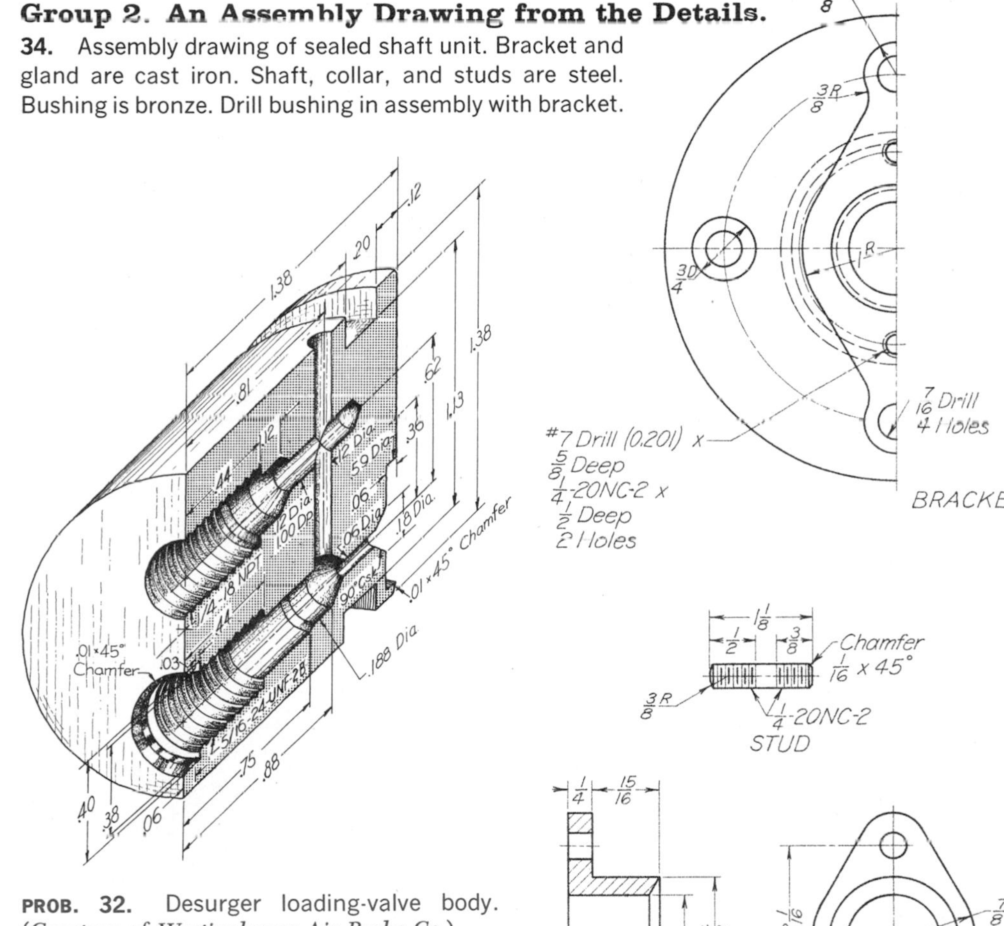


**PROB. 32.** Desurger loading-valve body. (*Courtesy of Westinghouse Air Brake Co.*)

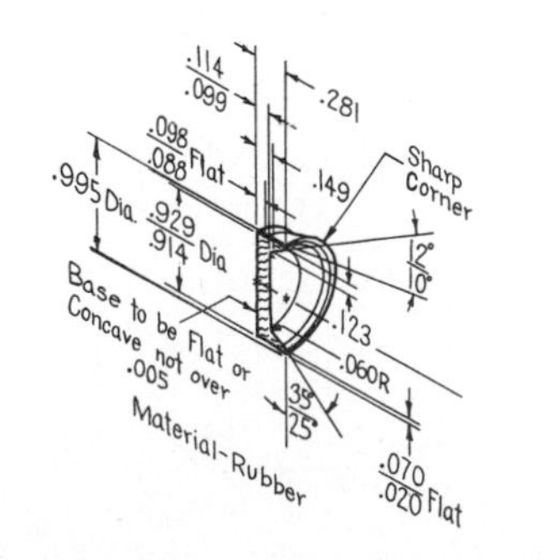


**PROB. 33.** Wheel-cylinder cup. (*Courtesy of Bendix Corp.*)

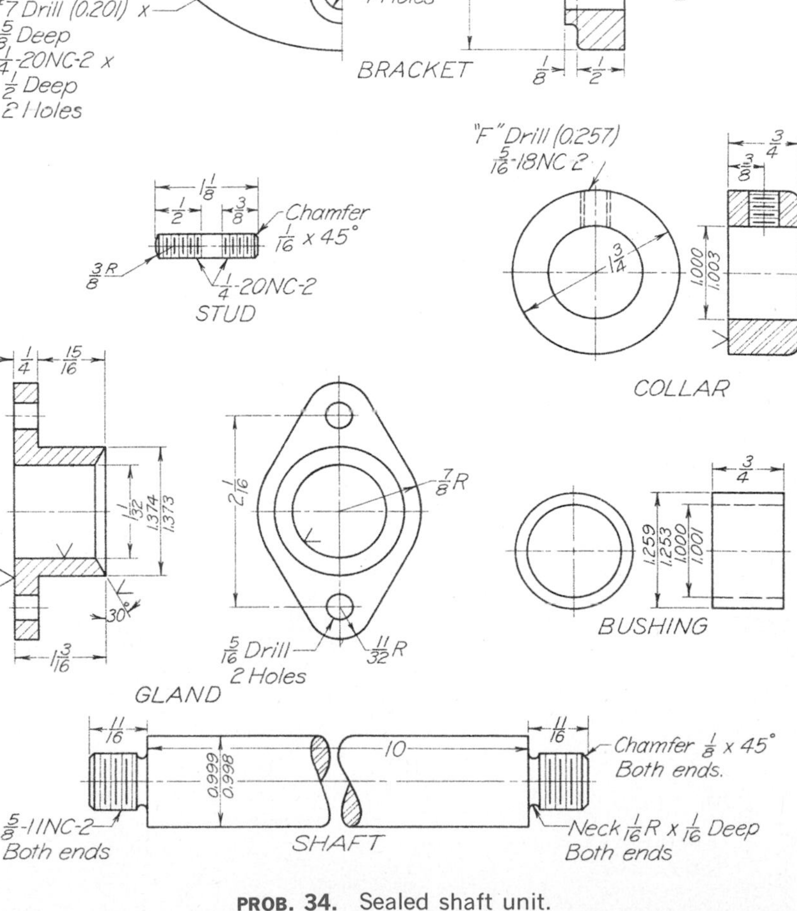


**PROB. 34.** Sealed shaft unit.

**35.** Make an assembly drawing of the crane hook from details given. Standard parts 7 to 10 are not detailed; see Appendix or handbook for sizes.

**36.** Make an assembly drawing, front view in section, of caster.

**36*A*.** Redesign caster for ball-bearing installation.

**37.** Make an assembly drawing of the Brown and Sharpe rotary geared pump, with top view, longitudinal section, and side view. Show direction of rotation of shafts and flow of liquid with arrows. Give dimensions for base holes to be used in setting; also give distance from base to center of driving shaft and size of shaft and key. For the parts that are not detailed, see the parts list.

**38.** Make assembly drawing of parallel-index drive.

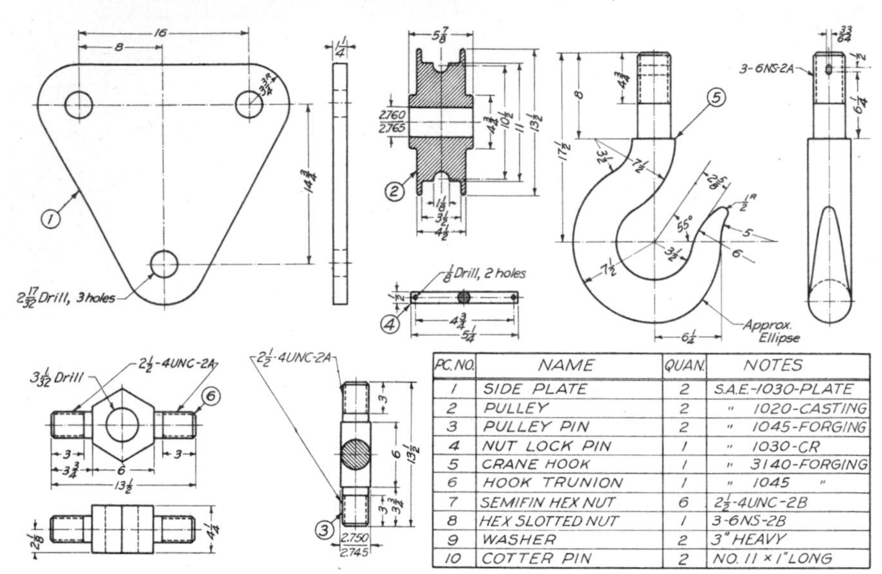

| PC. NO. | NAME | QUAN. | NOTES |
|---|---|---|---|
| 1 | SIDE PLATE | 2 | S.A.E.-1030-PLATE |
| 2 | PULLEY | 2 | " 1020-CASTING |
| 3 | PULLEY PIN | 2 | " 1045-FORGING |
| 4 | NUT LOCK PIN | 1 | " 1030-CR |
| 5 | CRANE HOOK | 1 | " 3140-FORGING |
| 6 | HOOK TRUNION | 1 | " 1045 " |
| 7 | SEMIFIN HEX NUT | 6 | 2½-4UNC-2B |
| 8 | HEX SLOTTED NUT | 1 | 3-6NS-2B |
| 9 | WASHER | 2 | 3" HEAVY |
| 10 | COTTER PIN | 2 | NO. 11 × 1" LONG |

**PROB. 35.** Crane hook.

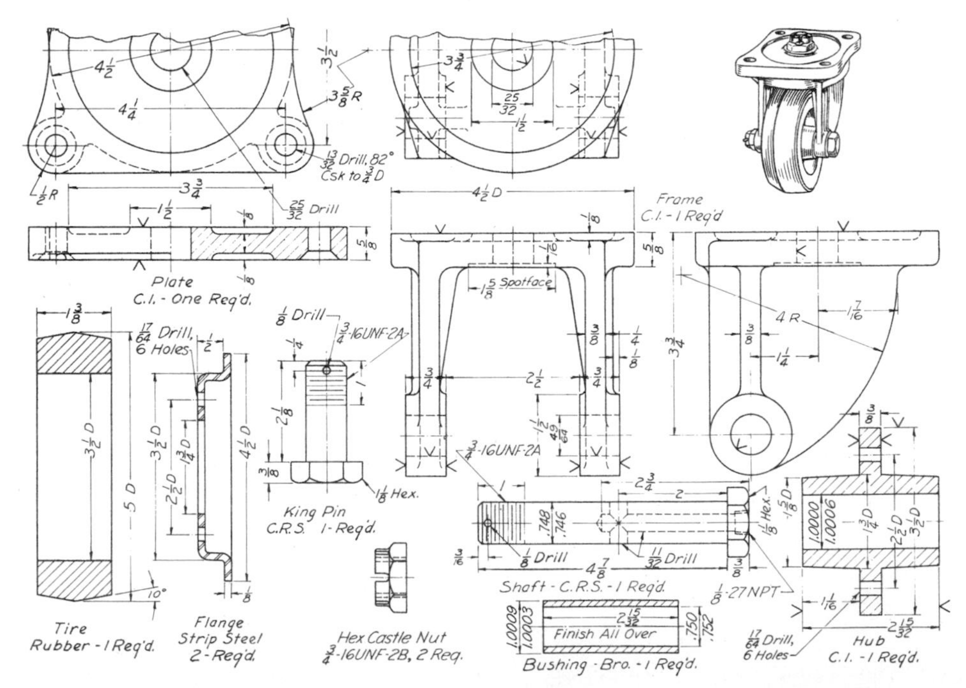

PARTS LIST

BROWN AND SHARPE No. 1 ROTARY GEARED PUMP

| PC. NO. | DRAW SIZE | NAME | QUAN. | MAT. | STOCK DIA. | STOCK LGTH. | USED ON NAME | USED ON PC. NO. | REMARKS |
|---|---|---|---|---|---|---|---|---|---|
| 101 | | Base | 1 | C.I. | | | | | |
| 102 | | Body | 1 | C.I. | | | | | |
| 103 | | Cover | 1 | C.I. | | | | | |
| 104 | | Pulley | 1 | C.I. | | | | | |
| 105 | | Gland | 1 | C.I. | | | | | |
| 106 | | Gland Bushing | 1 | Bro. | | | | | |
| 107 | | Gear Bushing | 4 | Bro. | | | | | |
| 108 | | Driving Gear | 1 | S.A.E. #1045 | 1 9/16 | 5 7/8 | | | |
| 109 | | Driven Gear | 1 | S.A.E. #1045 | 1 9/16 | 2 9/16 | | | |
| 110 | | Gasket | 2 | Sheet Copper | | | Body | 102 | #26 B&S Gage (0.0159) |
| 111 | | #10-32 x 1 5/8 Slotted Hex. Hd. Mach. Scr. & Nut | 4 | | | | Cover | 103 | |
| 112 | | #10-32 x 1 5/8 Slotted Hex. Hd. Cap Scr. | 2 | | | | Cover | 103 | |
| 113 | | #10-32 x 7/8 Slotted Hex. Hd. Cap Scr. | 2 | | | | Gland | 105 | |
| 114 | | Woodruff Key #405 | 1 | | | | Driving Gear | 108 | |
| 115 | | 3/8 x 3/8 Headless Set Scr., 3/8 16NC-2 | 1 | | | | Pulley | 104 | |
| 116 | | 3/16 x 1 7/16 Dowel Pin | 2 | C.R.S. | | | Cover | 103 | |
| | | Packing | To Suit | | | | | | Garlock Rotopac #239 |

**PROB. 37.** Rotary geared pump and parts list.

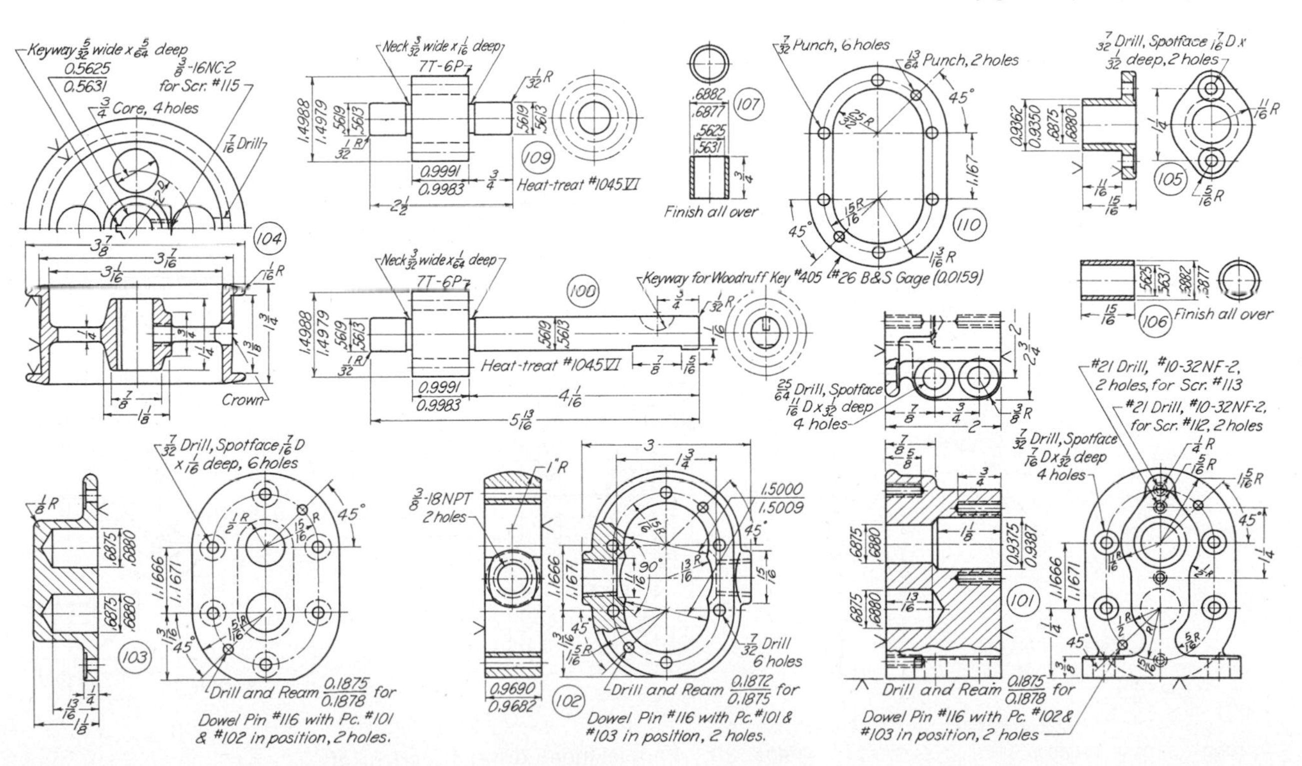

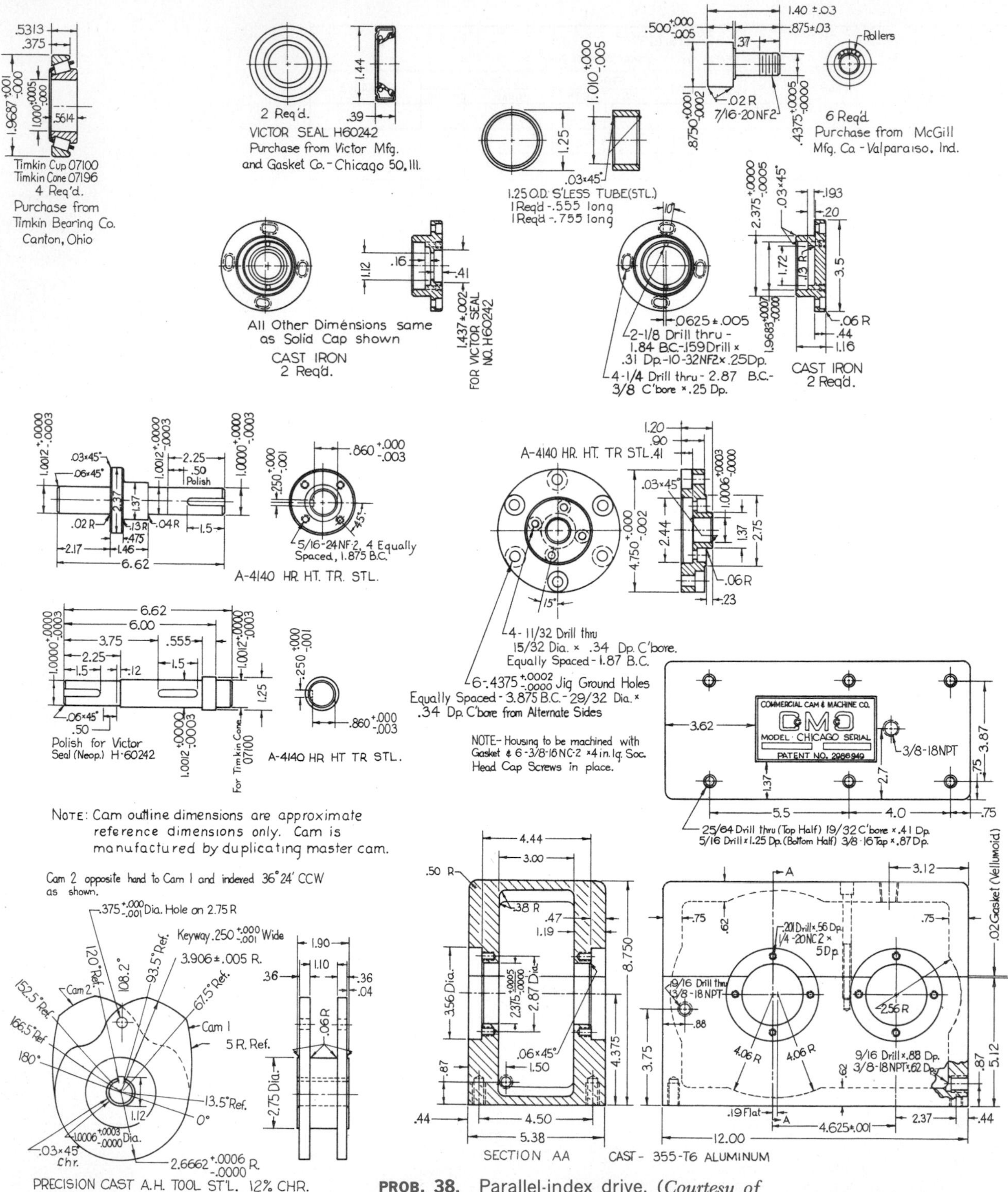


**PROB. 38.** Parallel-index drive. (*Courtesy of Commercial Cam and Machine Co.*)

## Group 3. Detail Drawings from the Assembly.

**39.** Make detail drawings of the jig table. Parts, cast iron.

**40.** Make detail drawings of the door catch.

**41.** Make detail drawings of belt drive. The pulley and bracket are cast iron; the gear and shaft, steel. The bushing is bronze.

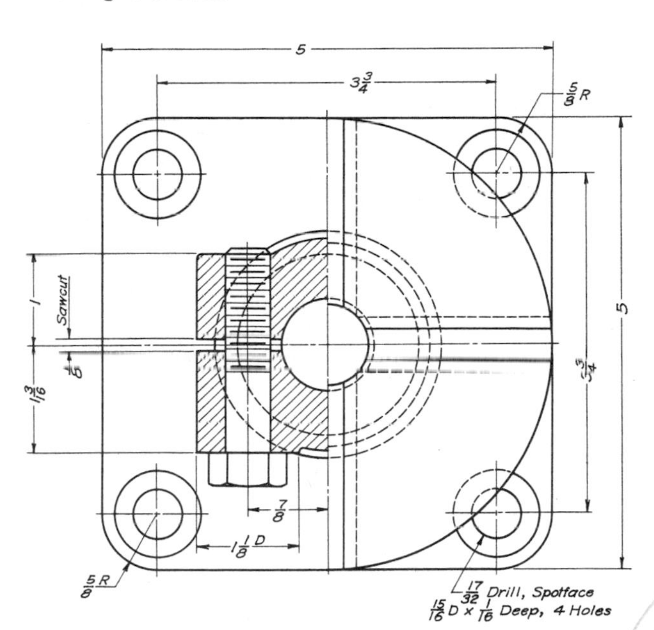


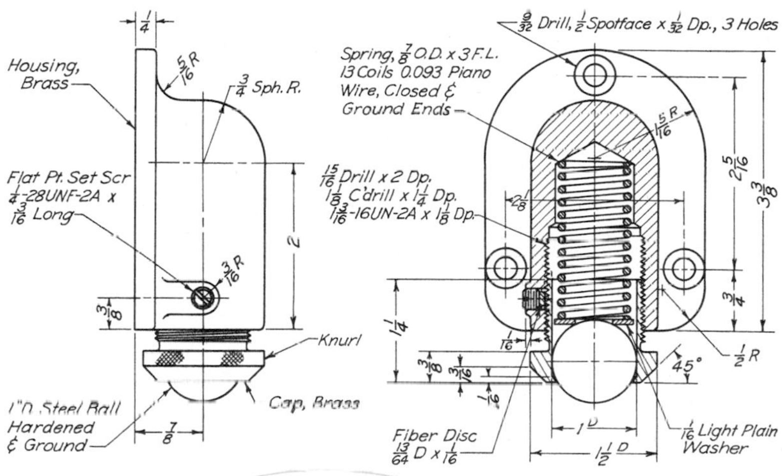


**PROB. 40.** Door catch.

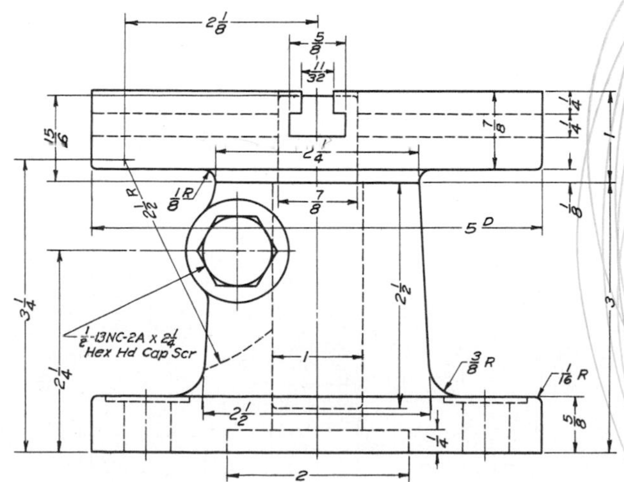


**PROB. 39.** Jig table.

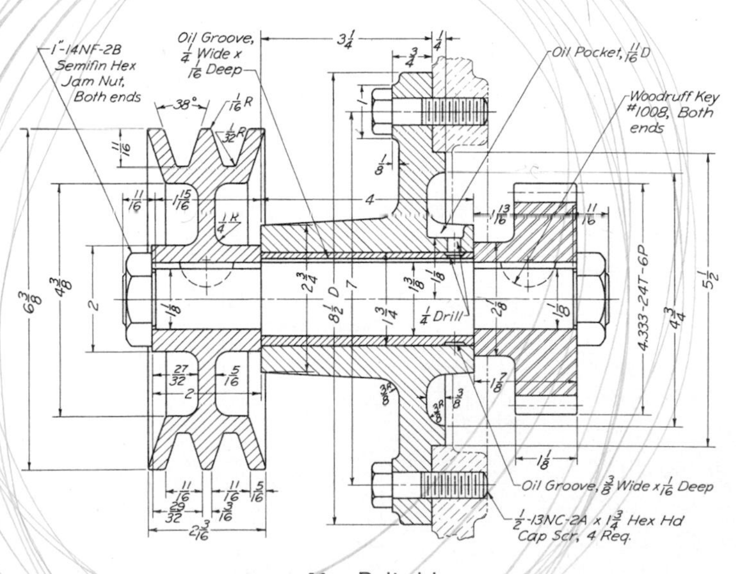


**PROB. 41.** Belt drive.

**42.** Make detail drawings of swing table.

**43.** Make detail drawings of sealed ball joint.

**44.** Make detail drawings of belt tightener. The bracket, pulley, and collar are cast iron. The bushing is bronze; the shaft, steel.

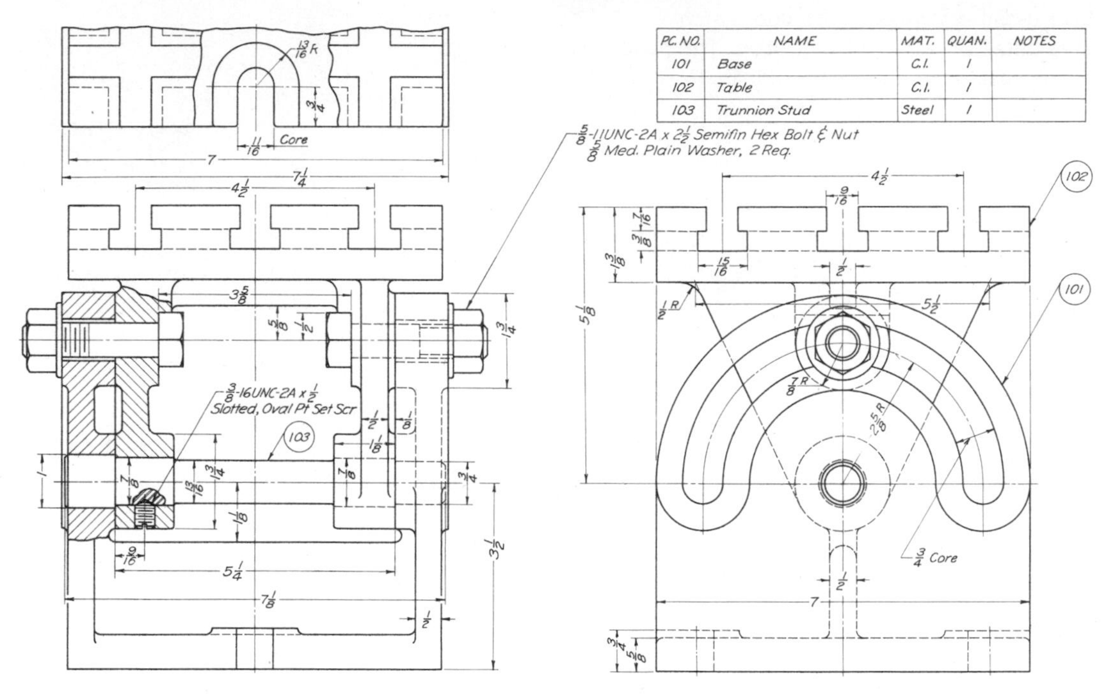


| PC. NO. | NAME | MAT. | QUAN. | NOTES |
|---|---|---|---|---|
| 101 | Base | C.I. | 1 | |
| 102 | Table | C.I. | 1 | |
| 103 | Trunnion Stud | Steel | 1 | |

**PROB. 42.** Swing table.

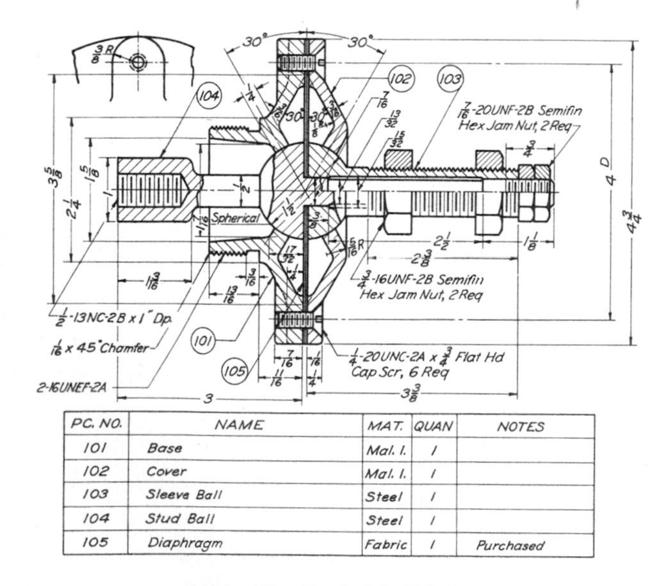


| PC. NO. | NAME | MAT. | QUAN | NOTES |
|---|---|---|---|---|
| 101 | Base | Mal. I. | 1 | |
| 102 | Cover | Mal. I. | 1 | |
| 103 | Sleeve Ball | Steel | 1 | |
| 104 | Stud Ball | Steel | 1 | |
| 105 | Diaphragm | Fabric | 1 | Purchased |

**PROB. 43.** Sealed ball joint.

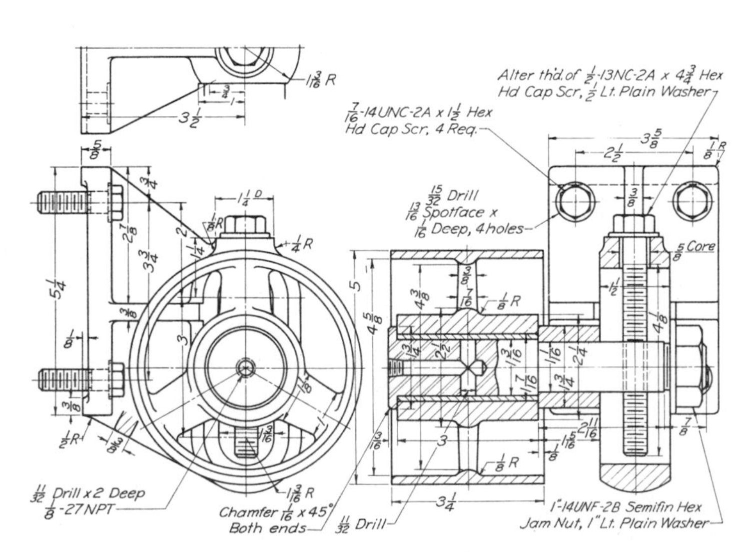


**PROB. 44.** Belt tightener.

**45.** Make detail drawings of rotary pressure joint.
**46.** Make detail drawings of ball-bearing idler pulley.

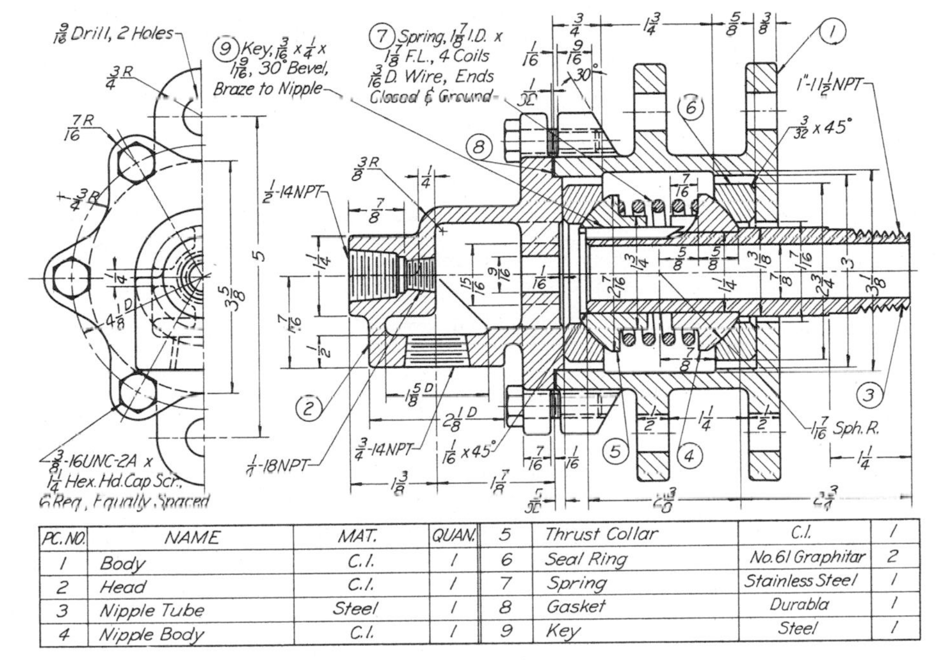


| PC. NO. | NAME | MAT. | QUAN. |
|---|---|---|---|
| 1 | Body | C.I. | 1 |
| 2 | Head | C.I. | 1 |
| 3 | Nipple Tube | Steel | 1 |
| 4 | Nipple Body | C.I. | 1 |
| 5 | Thrust Collar | C.I. | 1 |
| 6 | Seal Ring | No. 61 Graphitar | 2 |
| 7 | Spring | Stainless Steel | 1 |
| 8 | Gasket | Durabla | 1 |
| 9 | Key | Steel | 1 |

**PROB. 45.** Rotary pressure joint.

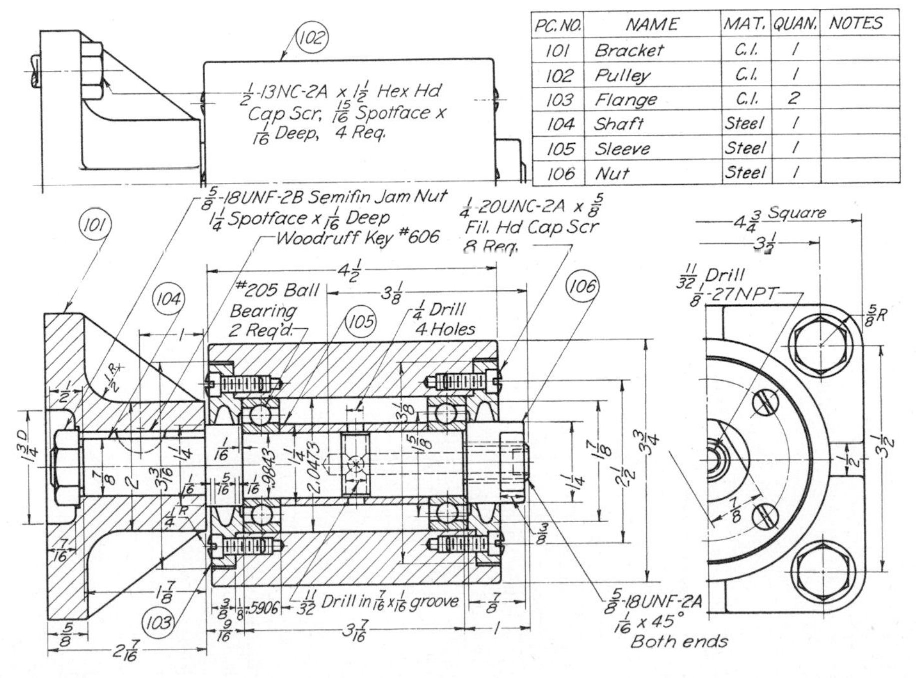


| PC. NO. | NAME | MAT. | QUAN. | NOTES |
|---|---|---|---|---|
| 101 | Bracket | C.I. | 1 | |
| 102 | Pulley | C.I. | 1 | |
| 103 | Flange | C.I. | 2 | |
| 104 | Shaft | Steel | 1 | |
| 105 | Sleeve | Steel | 1 | |
| 106 | Nut | Steel | 1 | |

**PROB. 46.** Ball-bearing idler pulley.

**47.** Make detail drawings of butterfly valve.

**48.** Make detail drawings for V-belt drive.

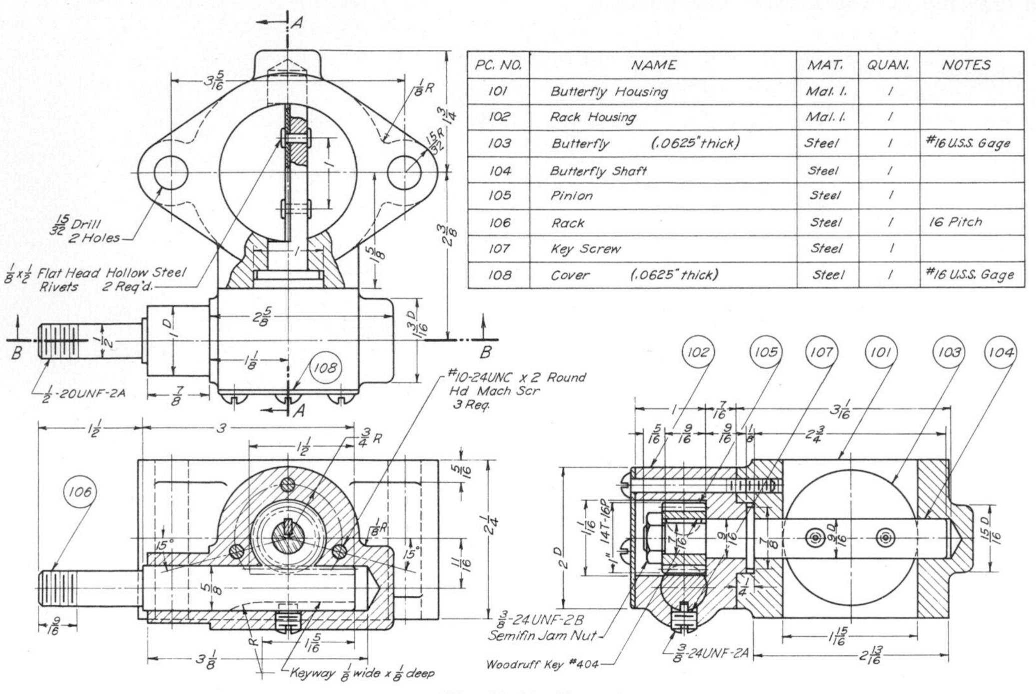

| PC. NO. | NAME | MAT. | QUAN. | NOTES |
|---|---|---|---|---|
| 101 | Butterfly Housing | Mal. I. | 1 | |
| 102 | Rack Housing | Mal. I. | 1 | |
| 103 | Butterfly (.0625" thick) | Steel | 1 | #16 U.S.S. Gage |
| 104 | Butterfly Shaft | Steel | 1 | |
| 105 | Pinion | Steel | 1 | |
| 106 | Rack | Steel | 1 | 16 Pitch |
| 107 | Key Screw | Steel | 1 | |
| 108 | Cover (.0625" thick) | Steel | 1 | #16 U.S.S. Gage |

**PROB. 47.** Butterfly valve.

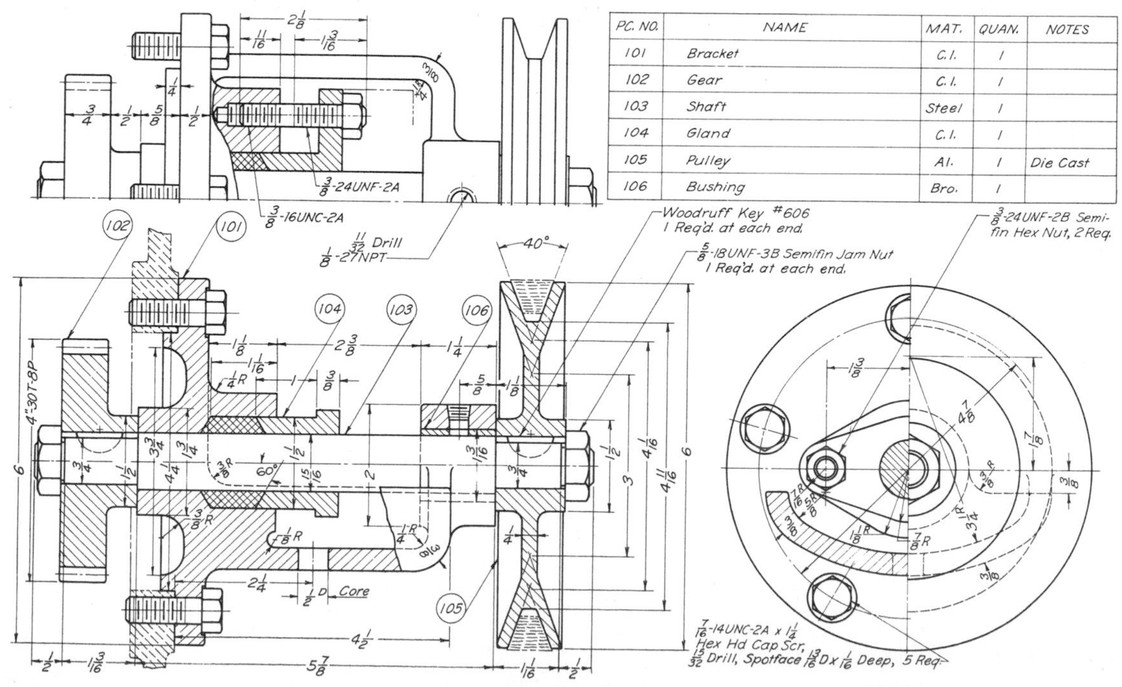

| PC. NO. | NAME | MAT. | QUAN. | NOTES |
|---|---|---|---|---|
| 101 | Bracket | C.I. | 1 | |
| 102 | Gear | C.I. | 1 | |
| 103 | Shaft | Steel | 1 | |
| 104 | Gland | C.I. | 1 | |
| 105 | Pulley | Al. | 1 | Die Cast |
| 106 | Bushing | Bro. | 1 | |

**PROB. 48.** V-belt drive.

**49.** Make detail drawings of hydraulic punch. In action, the punch assembly proper advances until the cap, piece 109, comes against the work. The assembly (piece 106 and attached parts) is then stationary, and the tension of the punch spring (piece 114) holds the work as the punch advances through the work and returns.

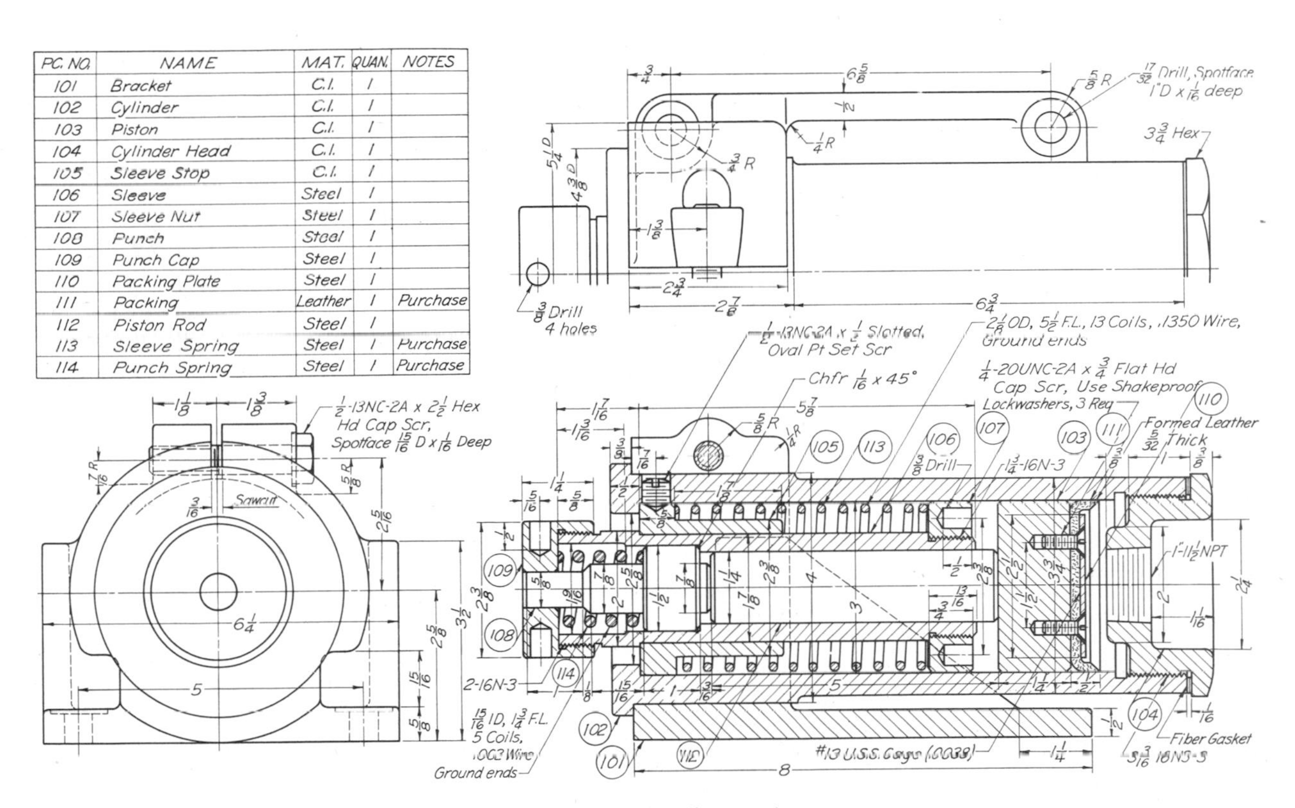

| PC. NO. | NAME | MAT. | QUAN. | NOTES |
|---|---|---|---|---|
| 101 | Bracket | C.I. | 1 | |
| 102 | Cylinder | C.I. | 1 | |
| 103 | Piston | C.I. | 1 | |
| 104 | Cylinder Head | C.I. | 1 | |
| 105 | Sleeve Stop | C.I. | 1 | |
| 106 | Sleeve | Steel | 1 | |
| 107 | Sleeve Nut | Steel | 1 | |
| 108 | Punch | Steel | 1 | |
| 109 | Punch Cap | Steel | 1 | |
| 110 | Packing Plate | Steel | 1 | |
| 111 | Packing | Leather | 1 | Purchase |
| 112 | Piston Rod | Steel | 1 | |
| 113 | Sleeve Spring | Steel | 1 | Purchase |
| 114 | Punch Spring | Steel | 1 | Purchase |



**PROB. 49.** Hydraulic punch.

**50.** Make detail drawings of double-acting air cylinder. Length of stroke to be assigned. Fix length of cylinder to allow for clearance of 1 in. at ends of stroke. Note that pieces 101 and 102 are identical except for the extra machining of the central hole in piece 101 for the shaft, packing, and gland. Make separate drawings for this piece, one for the pattern shop and two for the machine shop.

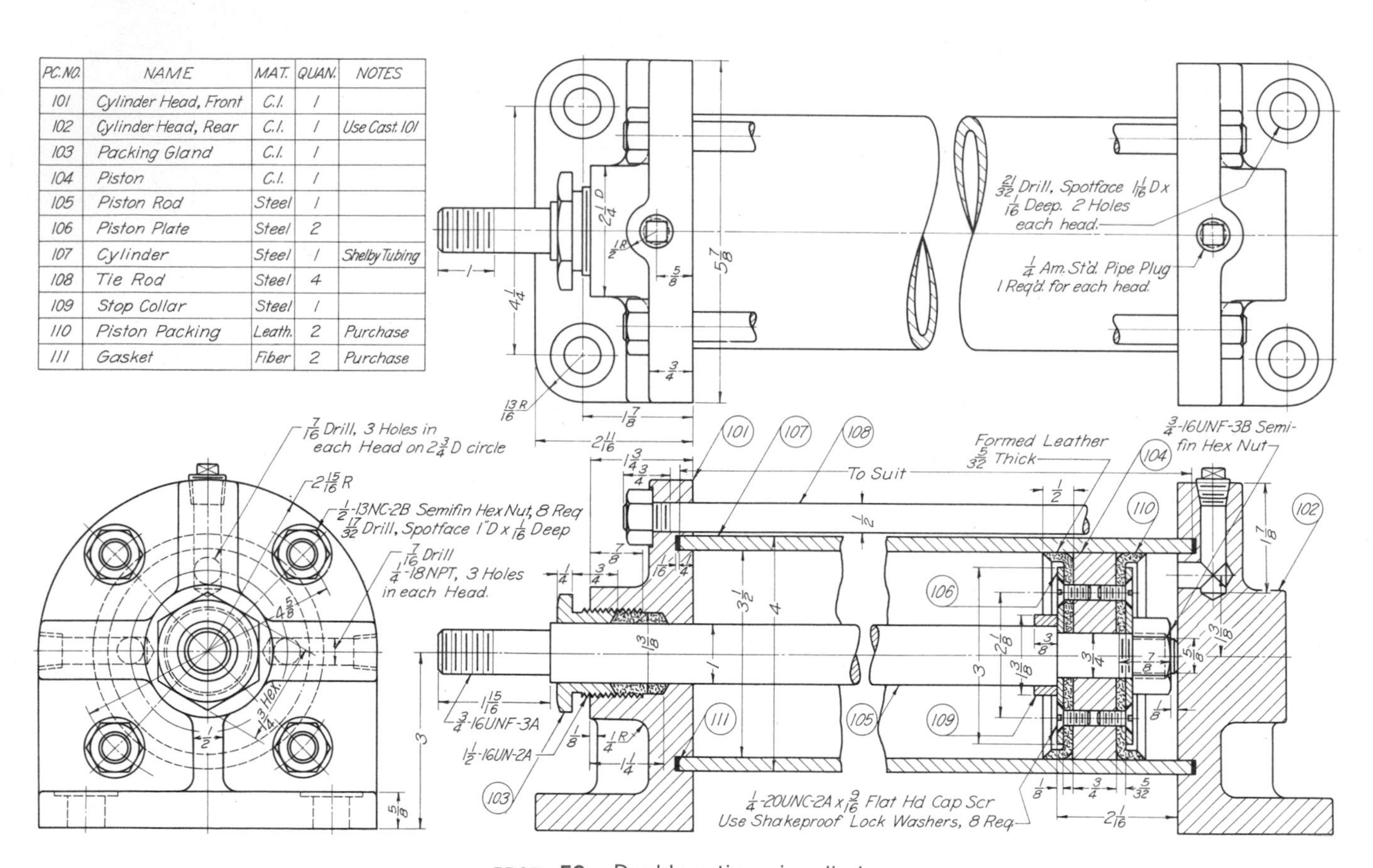

| PC. NO. | NAME | MAT. | QUAN. | NOTES |
|---|---|---|---|---|
| 101 | Cylinder Head, Front | C.I. | 1 | |
| 102 | Cylinder Head, Rear | C.I. | 1 | Use Cast. 101 |
| 103 | Packing Gland | C.I. | 1 | |
| 104 | Piston | C.I. | 1 | |
| 105 | Piston Rod | Steel | 1 | |
| 106 | Piston Plate | Steel | 2 | |
| 107 | Cylinder | Steel | 1 | Shelby Tubing |
| 108 | Tie Rod | Steel | 4 | |
| 109 | Stop Collar | Steel | 1 | |
| 110 | Piston Packing | Leath. | 2 | Purchase |
| 111 | Gasket | Fiber | 2 | Purchase |



**PROB. 50.** Double-acting air cylinder.

**51.** Make detail drawings of rail-transport hanger. Rail is 10-lb ASCE.

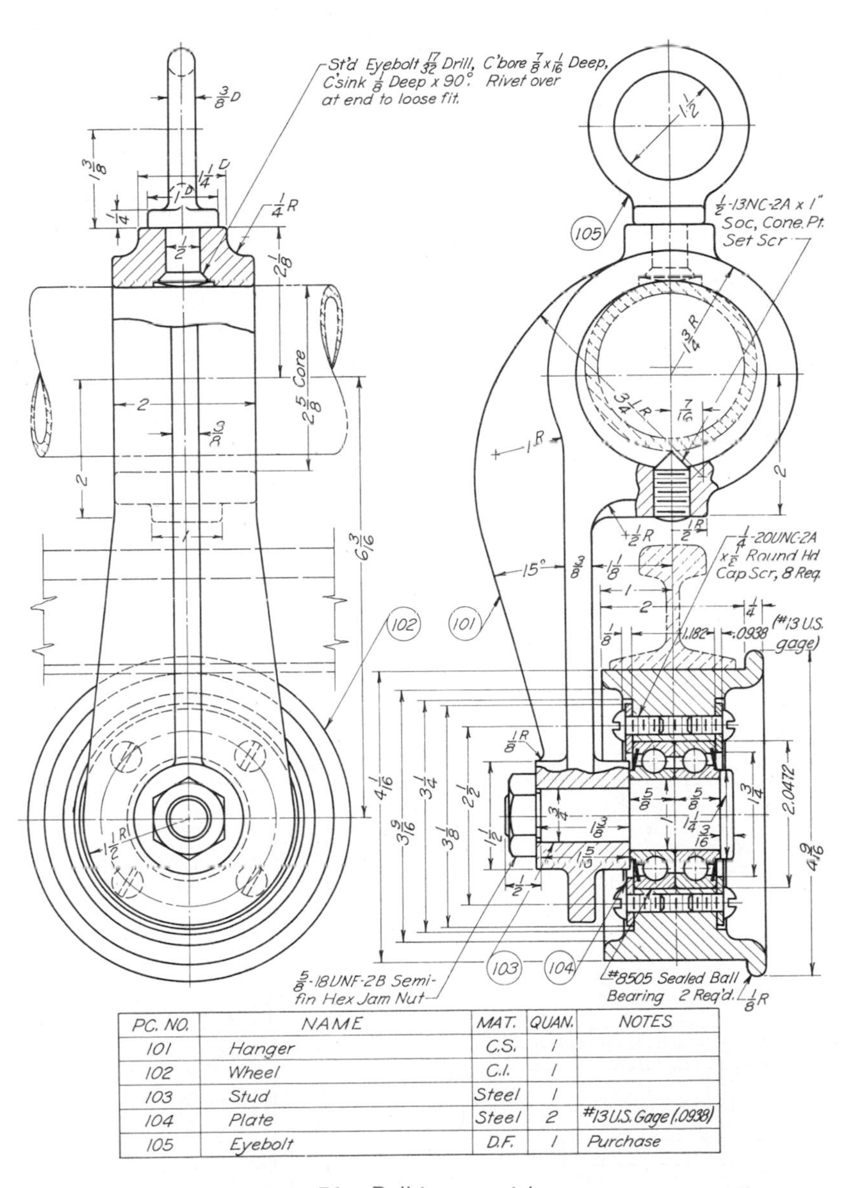


| PC. NO. | NAME | MAT. | QUAN. | NOTES |
|---|---|---|---|---|
| 101 | Hanger | C.S. | 1 | |
| 102 | Wheel | C.I. | 1 | |
| 103 | Stud | Steel | 1 | |
| 104 | Plate | Steel | 2 | #13 U.S. Gage (.0938) |
| 105 | Eyebolt | D.F. | 1 | Purchase |

**PROB. 51.** Rail-transport hanger.

**52.** Make detail drawings of the laboratory pump. Materials are shown on the assembly drawing.

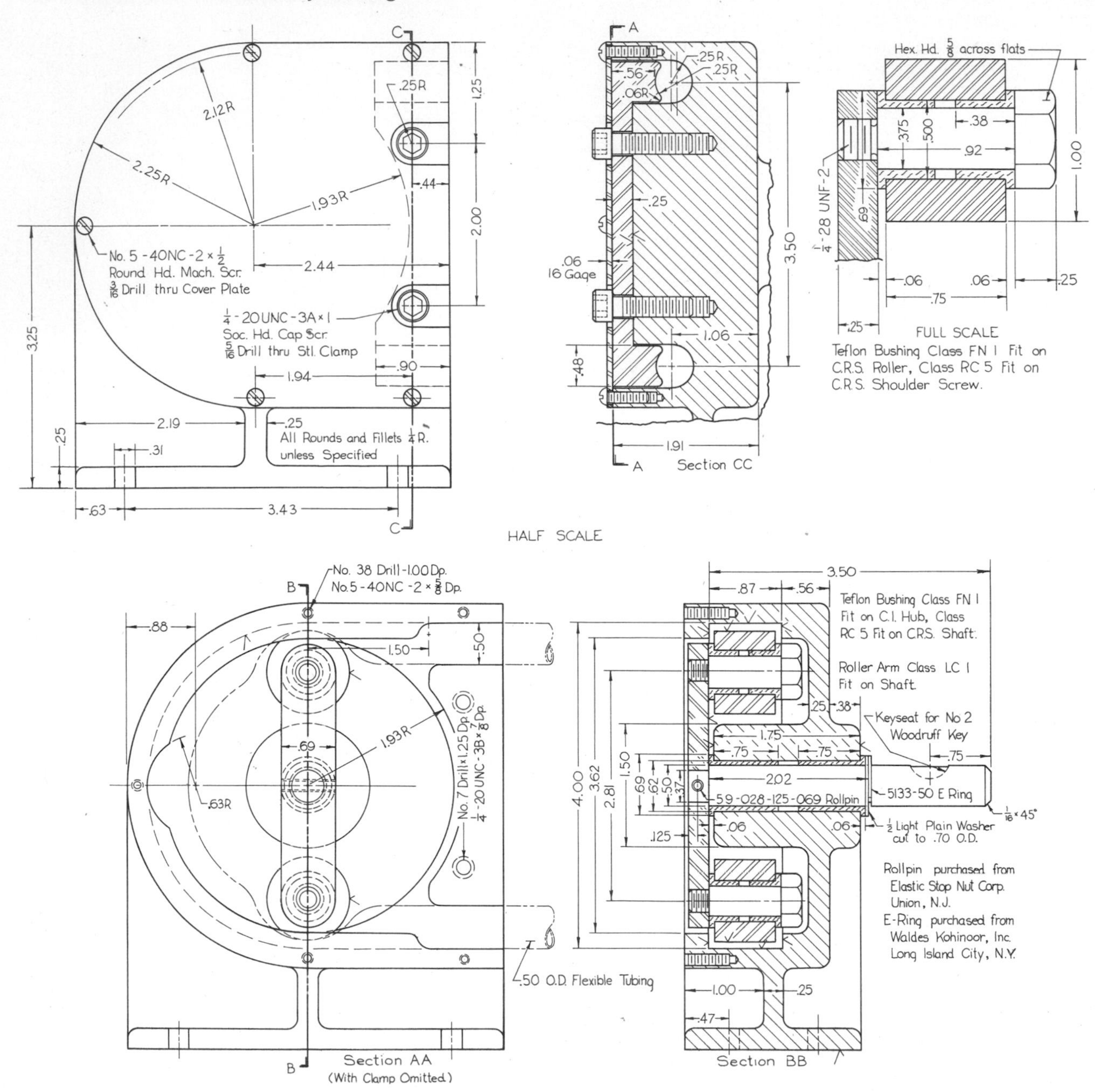


**PROB. 52.** Laboratory pump.

## Group 4. A Set of Drawings from an Exploded Pictorial.

The problems that are given in this group have been arranged as complete exercises in making a set of working drawings. Remember that the dimensions given on the pictorial views are to be used only to obtain distances or information needed. In some cases the data needed for a particular part may have to be obtained from the mating part.

The detail drawings should be made with each part on a separate sheet. Drawings of cast or forged parts may be made in the single-drawing or the multiple-drawing system described in Chap. 12.

The assembly drawing should include any necessary dimensions, such as number, size, and spacing of mounting holes, that might be required by a purchaser or needed for checking with the machine on which a subassembly is used.

For the style and items to be included in the parts list, see Prob. 37.

**53.** Make a complete set of working drawings for the antivibration mount.

**54.** Make a complete set of drawings for pivot hanger, including detail drawings, assembly drawing, and parts list. All parts are steel. This assembly consists of a yoke, base, collar, and standard parts.

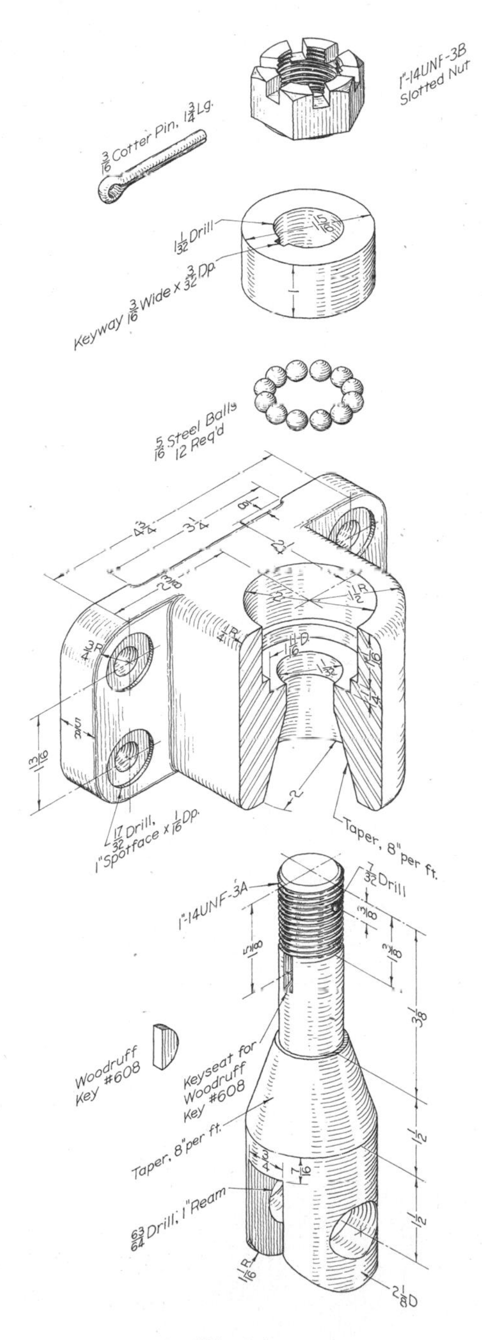


**PROB. 54.** Pivot hanger.

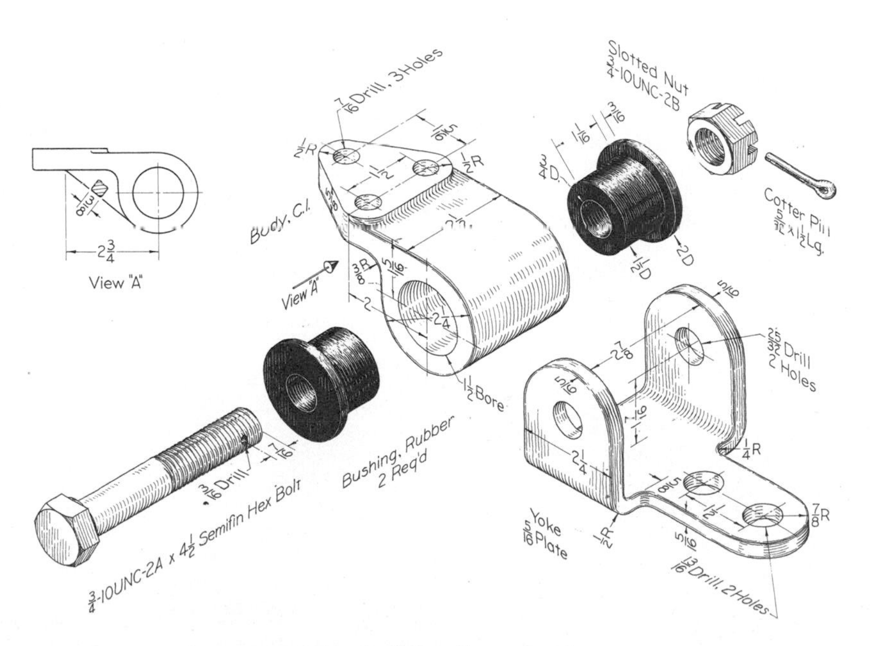


**PROB. 53.** Antivibration mount.

**55.** Make a complete set of drawings for pump valve. The valve seat, stem, and spring are brass; the disk is hard-rubber composition. In operation, pressure of a fluid upward against the disk raises it and allows flow. Pressure downward forces the disk tighter against the seat and prevents flow.

**56.** Make a complete set of working drawings for the cartridge-case trimmer. Note that some of the dimensions will have to be obtained from the mating part. A tabular drawing may be made of the case holder, covering holders for various cartridge cases. The dimensions to be tabulated are (*a*) holder length, (*b*) diameter, and (*c*) taper of hole.

**57.** Make a complete set of working drawings of the hydraulic check valve. Spring is stainless steel; gasket is soft aluminum; all other parts are steel.

**58.** Make a complete set of working drawings of the boring-bar holder. All parts are steel. Note that the *body* is made in one piece, then split with a ⅛-in.-wide cut (exaggerated in the picture). The holder can be seen in use in Fig. 13, Chap. 11.

**58*A*.** Design three boring bars (see Fig. 13, Chap. 11) to fit the boring-bar holder, and make a tabular drawing.

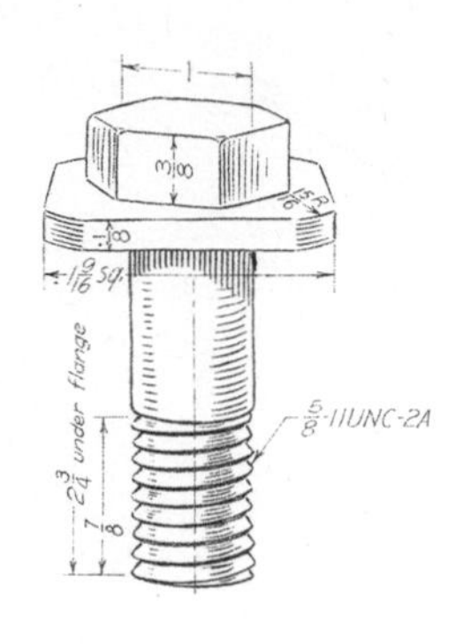


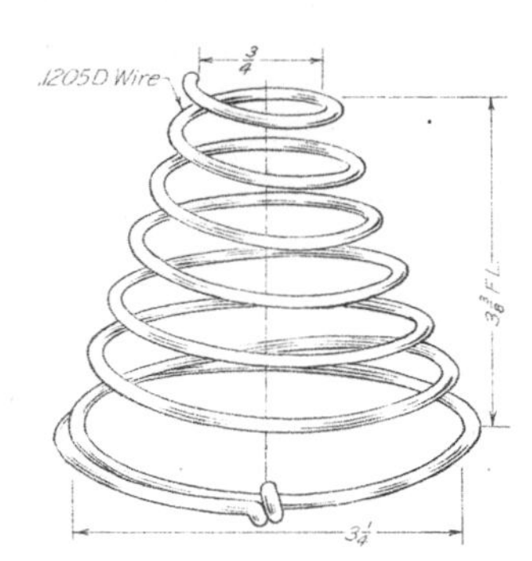


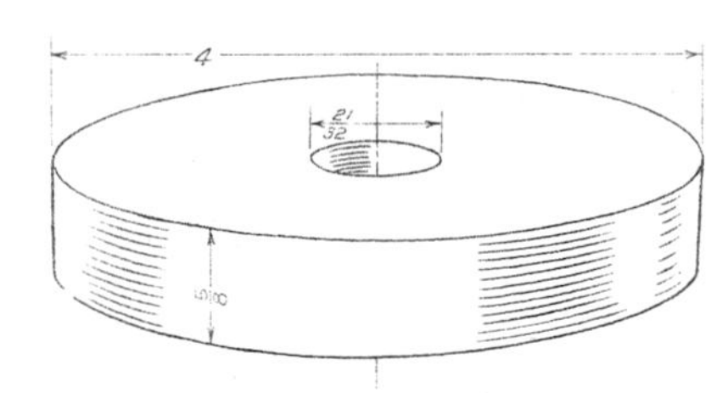

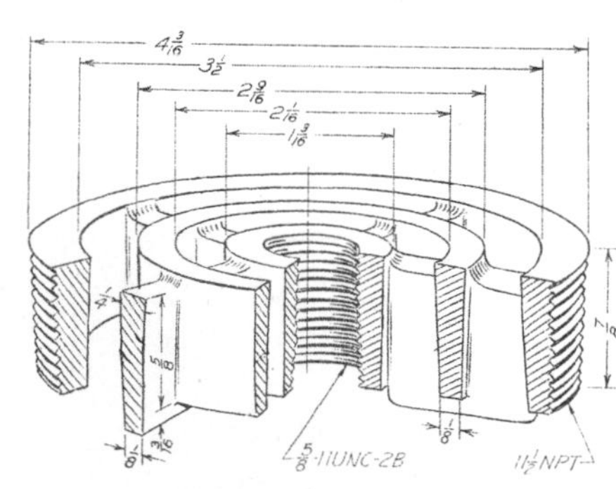


**PROB. 55.** Pump valve.

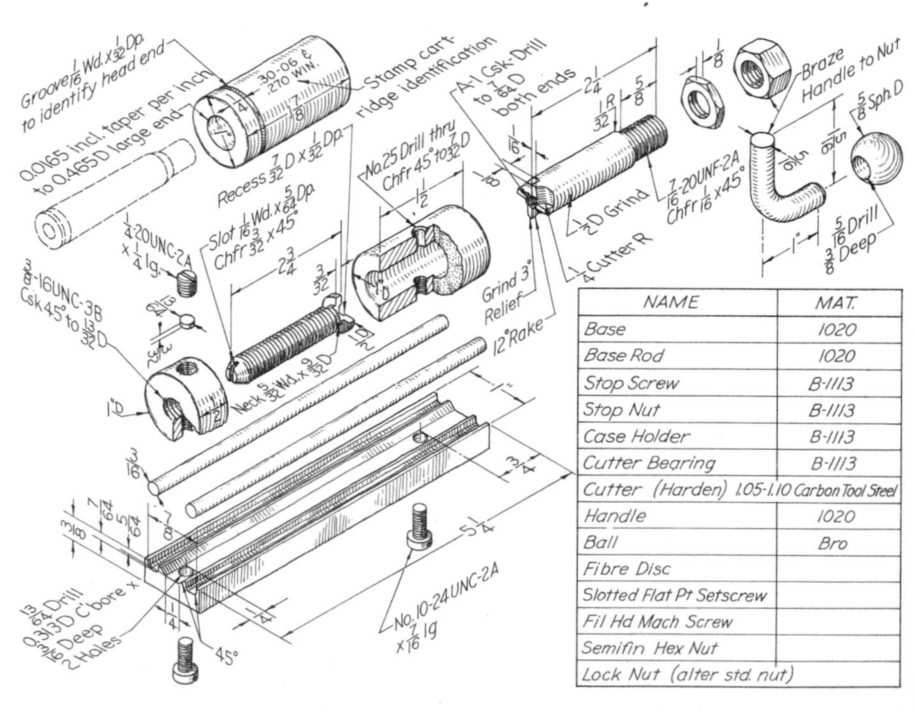


| NAME | MAT. |
|---|---|
| Base | 1020 |
| Base Rod | 1020 |
| Stop Screw | B-1113 |
| Stop Nut | B-1113 |
| Case Holder | B-1113 |
| Cutter Bearing | B-1113 |
| Cutter (Harden) | 1.05-1.10 Carbon Tool Steel |
| Handle | 1020 |
| Ball | Bro |
| Fibre Disc | |
| Slotted Flat Pt Setscrew | |
| Fil Hd Mach Screw | |
| Semifin Hex Nut | |
| Lock Nut (alter std. nut) | |

**PROB. 56.** Cartridge-case trimmer.

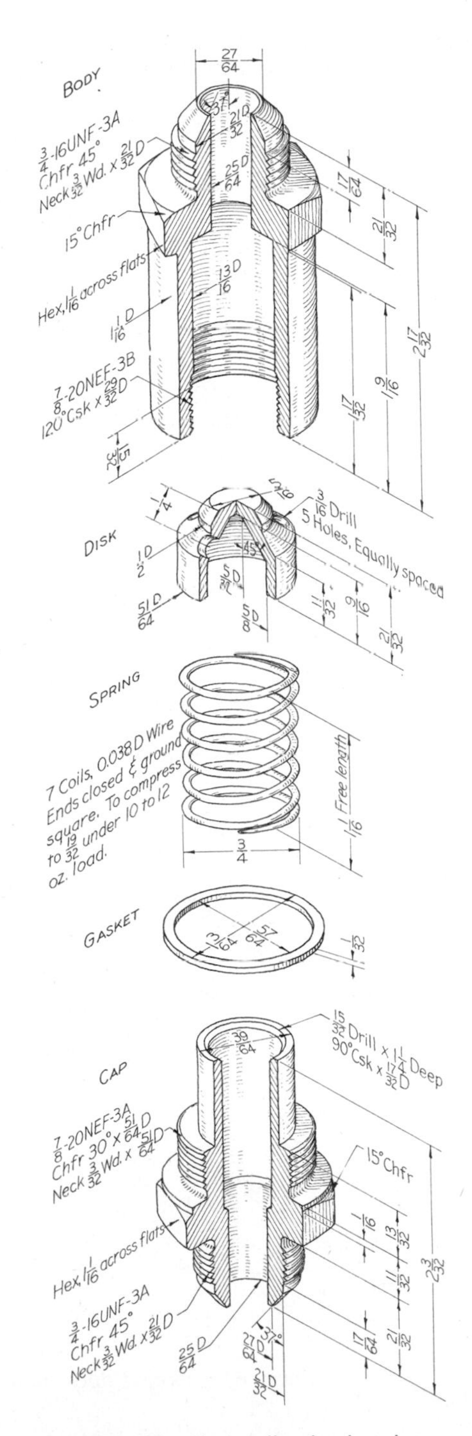

**PROB. 57.** Hydraulic check valve.

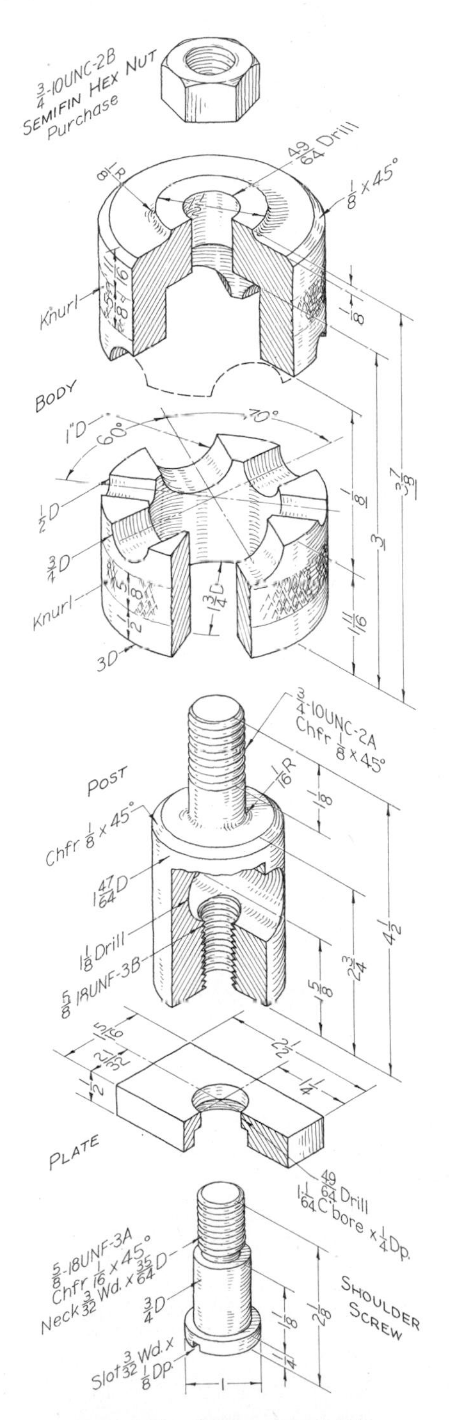

**PROB. 58.** Boring-bar holder.

**59.** Make a complete set of working drawings of the ratchet wrench.
**60.** Make detail and assembly drawings of landing-gear door hinge. High-strength aluminum alloy.
**61.** Make a complete set of working drawings for access-door support.

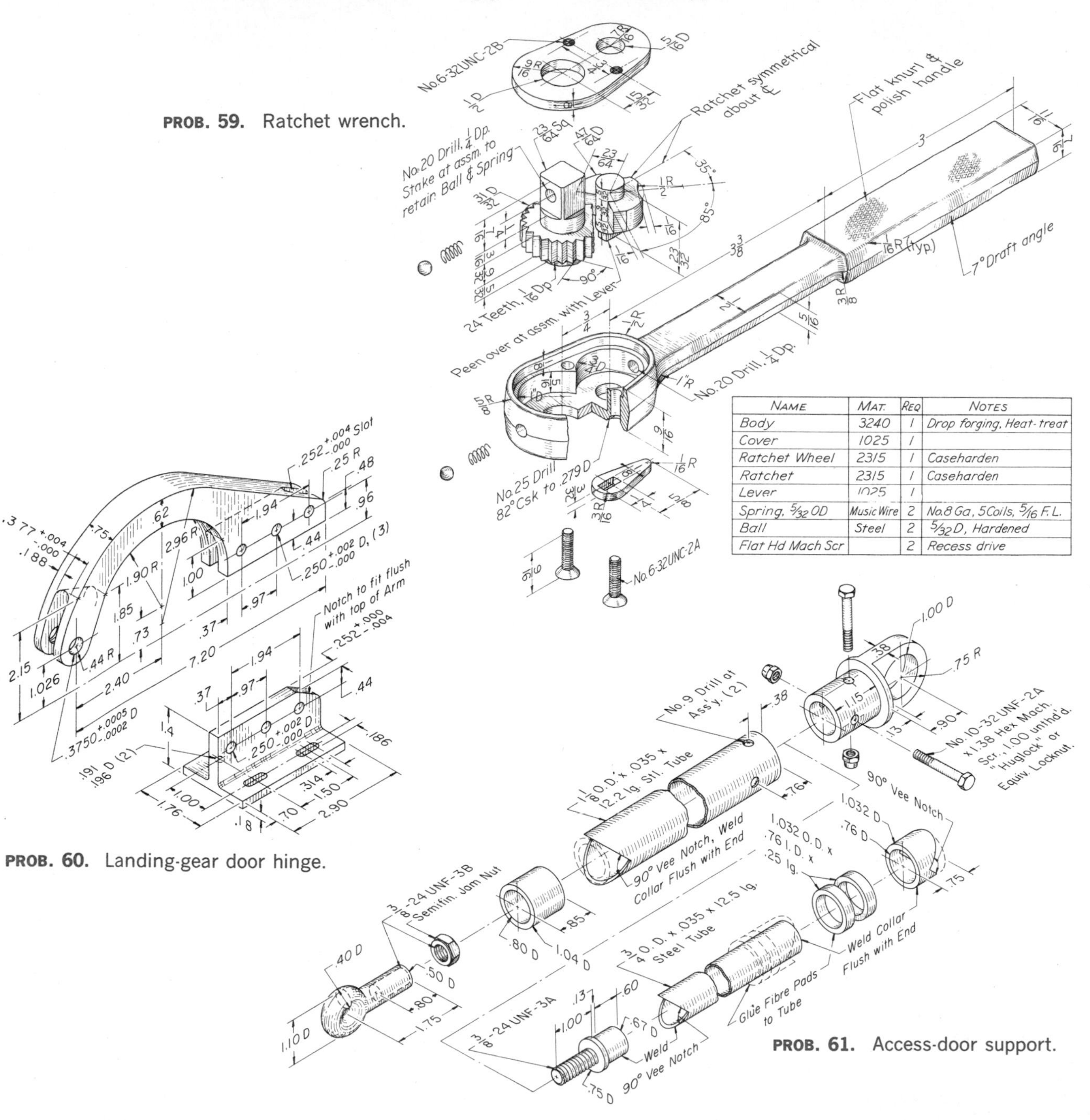


| Name | Mat. | Req | Notes |
|---|---|---|---|
| Body | 3240 | 1 | Drop forging, Heat-treat |
| Cover | 1025 | 1 | |
| Ratchet Wheel | 2315 | 1 | Caseharden |
| Ratchet | 2315 | 1 | Caseharden |
| Lever | 1025 | 1 | |
| Spring, 5/32 OD | Music Wire | 2 | No.8 Ga, 5 Coils, 5/16 F.L. |
| Ball | Steel | 2 | 5/32 D, Hardened |
| Flat Hd Mach Scr | | 2 | Recess drive |

**PROB. 59.** Ratchet wrench.

**PROB. 60.** Landing-gear door hinge.

**PROB. 61.** Access-door support.

**62.** Make detail and assembly drawings of the pressure-cell piston. Body parts are high-strength aluminum alloy. O rings are Teflon.

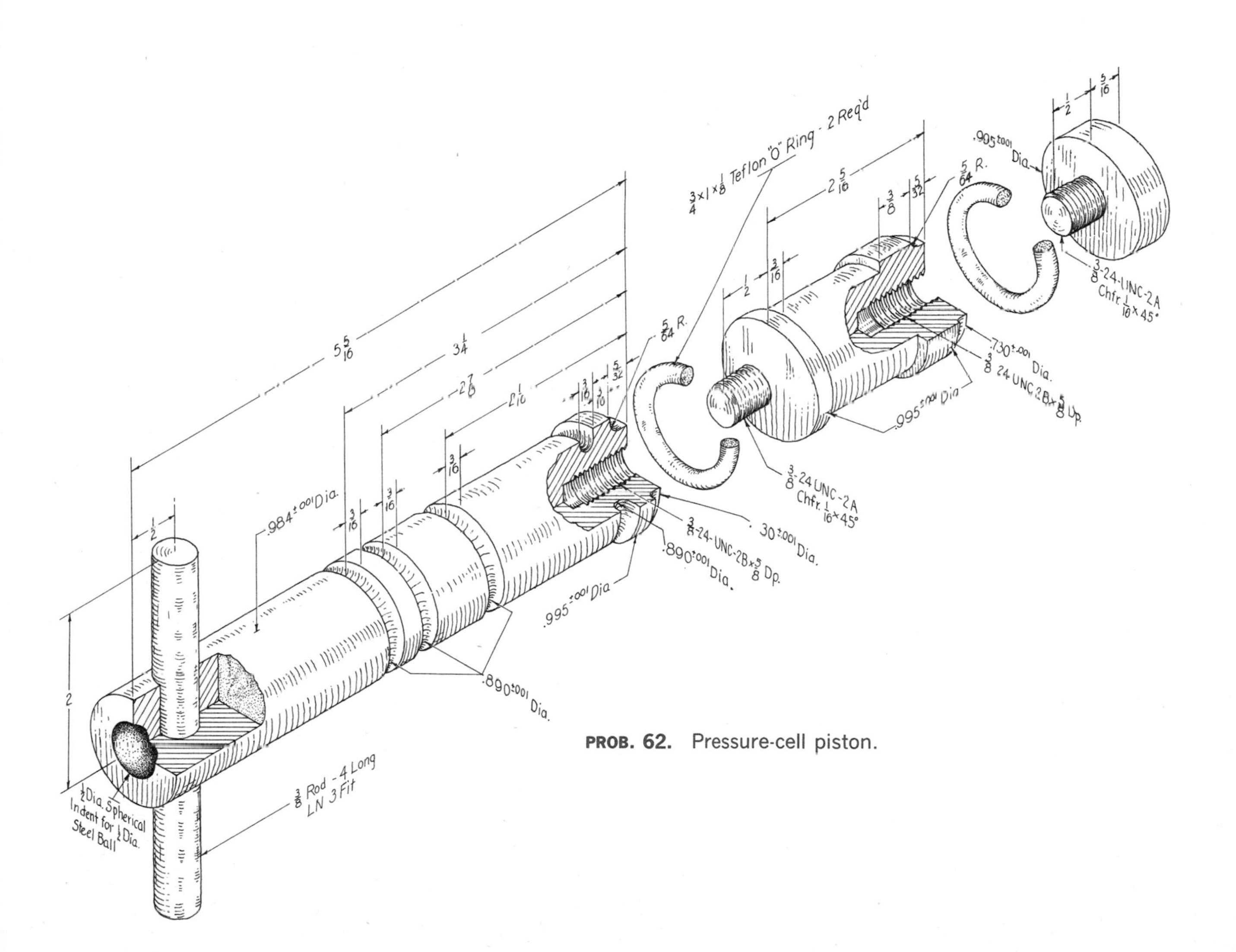


**PROB. 62.** Pressure-cell piston.

**63.** Make a complete set of working drawings of the pivot nut and adjusting screw.

## Group 5. A Set of Drawings from a Pictorial Assembly.

**64.** Make a complete set of working drawings of the high-tension coil mount.

**65.** Make a complete set of working drawings of the pipe clamp. The flange is cast steel.

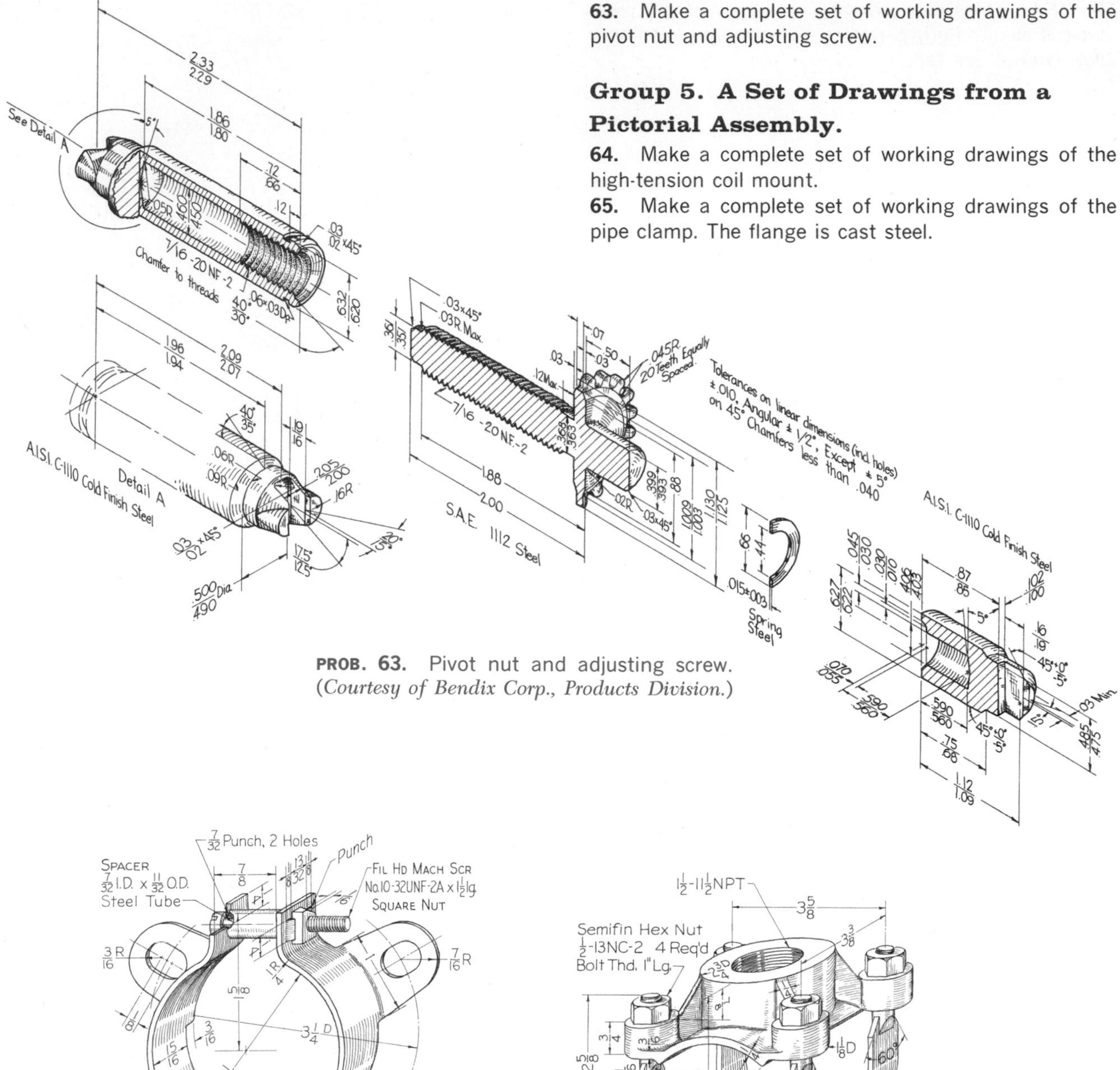


**PROB. 63.** Pivot nut and adjusting screw. *(Courtesy of Bendix Corp., Products Division.)*

**PROB. 64.** High-tension coil mount.

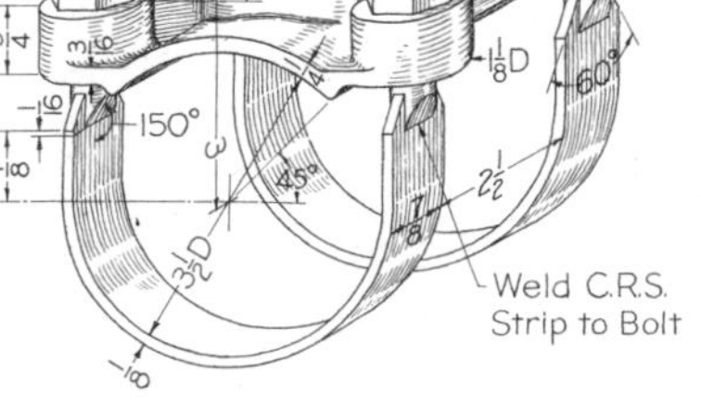


**PROB. 65.** Pipe clamp.

**66.** Make a complete set of working drawings of the stay-rod pivot. Parts are malleable iron.

**67.** Make a complete set of working drawings of the tool post. All parts are steel.

**68.** Make a unit-assembly working drawing of the wing-nose rib.

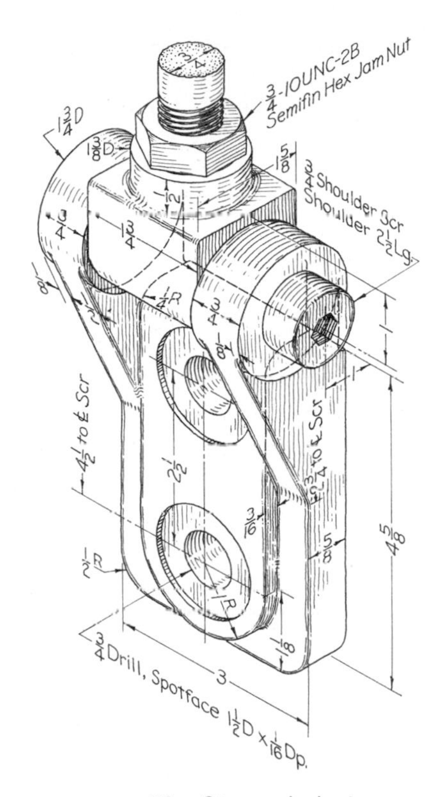

**PROB. 66.** Stay-rod pivot.

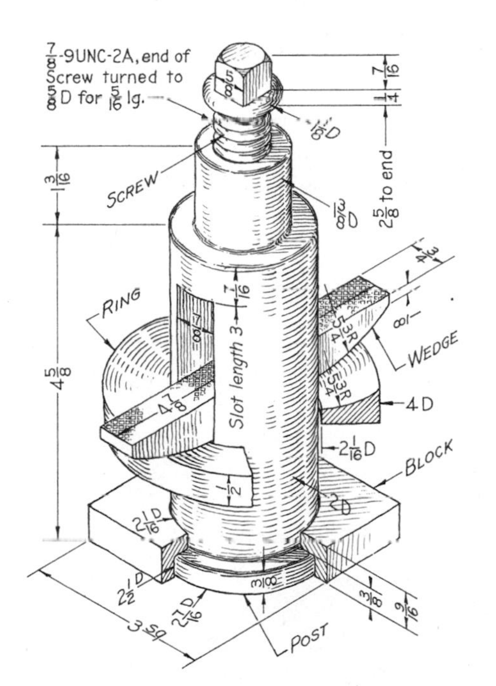

**PROB. 67.** Tool post.

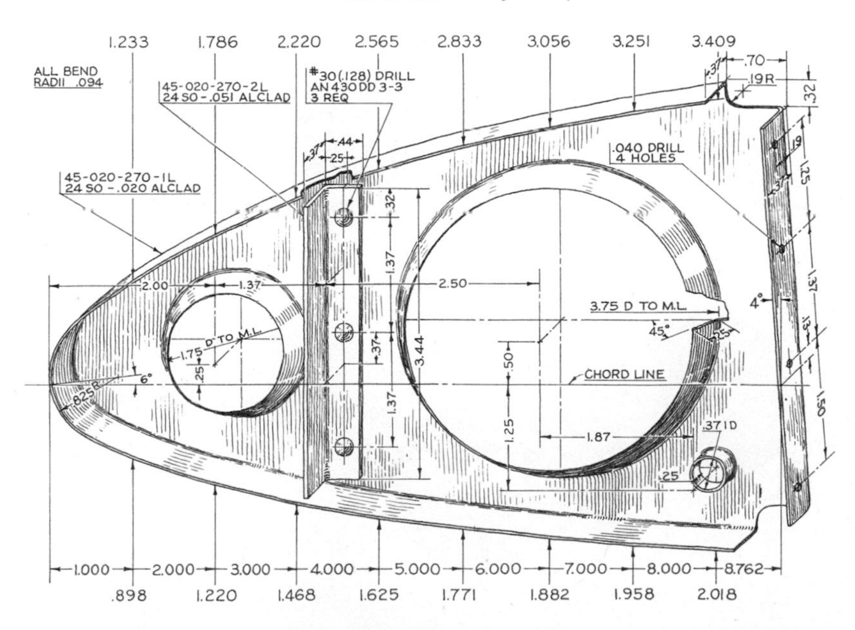

**PROB. 68.** Wing-nose rib.

**69.** Make a complete set of working drawings of the valve-poppet assembly. Flexible washer made of Koroseal 67-77 Durometer (B. F. Goodrich grade 116). Draw ten times actual size.

**70.** Make a complete set of working drawings of the diaphragm assembly. Draw four times actual size.

**71.** Detail drawing of conveyor link. SAE 1040 steel.

**72.** Detail drawing of hydro-cylinder support. Aluminum sheet.

## Group 6. Working Drawings from Photo-drawings.

Photo-drawings are coming into use commercially, especially in the processing industries. These problems give practice in making detail and assembly drawings, and illustrate the possibilities of drawings produced from photographs of models or test parts.

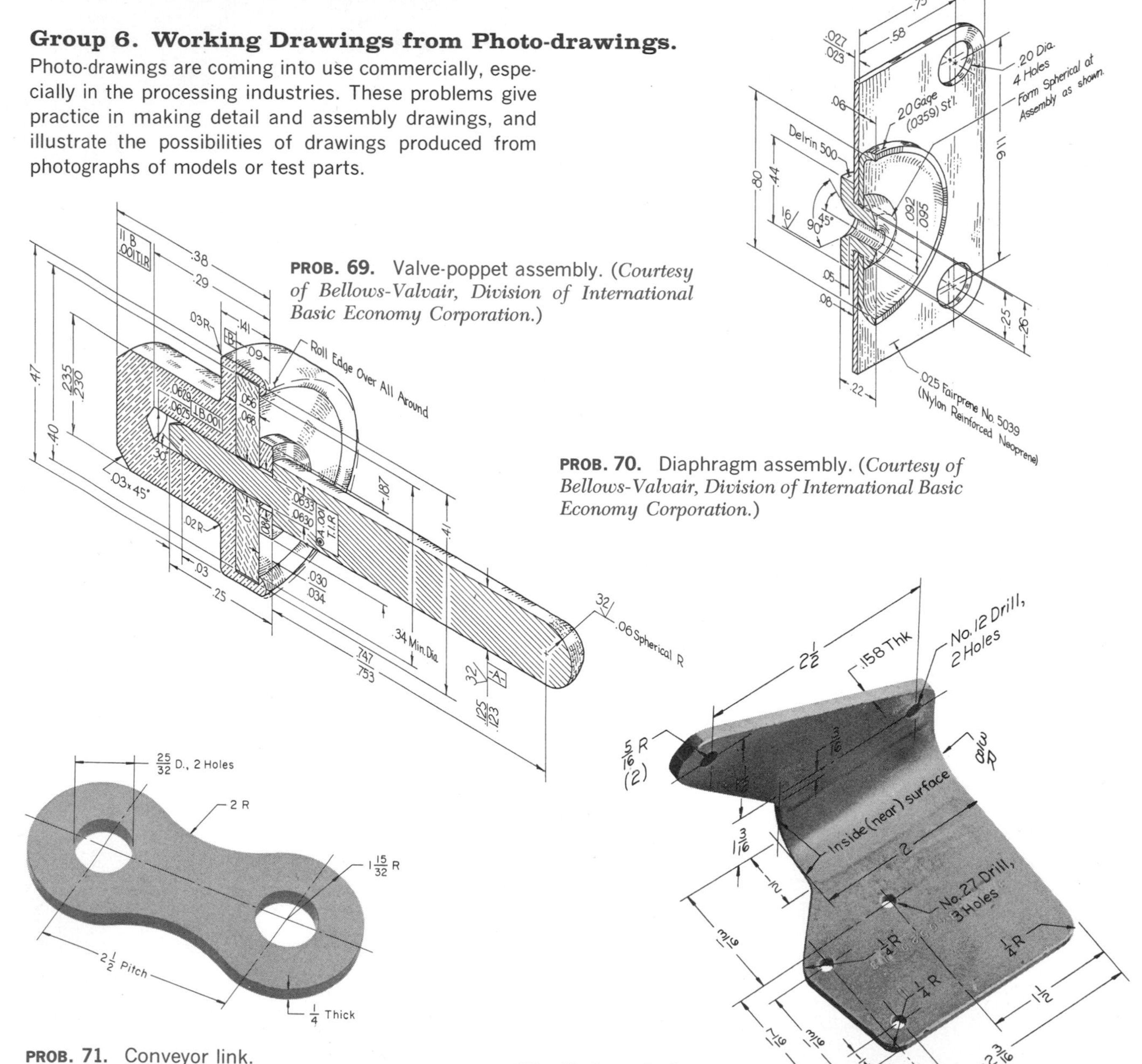


**PROB. 69.** Valve-poppet assembly. (*Courtesy of Bellows-Valvair, Division of International Basic Economy Corporation.*)

**PROB. 70.** Diaphragm assembly. (*Courtesy of Bellows-Valvair, Division of International Basic Economy Corporation.*)

**PROB. 71.** Conveyor link.

**PROB. 72.** Hydro-cylinder support.

**73.** Detail drawing of lift-strut pivot. Forged aluminum.
**74.** Detail drawing of length adjuster-tube. SAE 1220 steel.
**75.** Detail and unit assembly of pivot shaft. Shaft is bronze; nuts are steel.
**76.** Detail drawing of third terminal (temperature control). Material is red brass (85% CU).
**77.** Set of working drawings of conveyor link unit. All parts are steel.

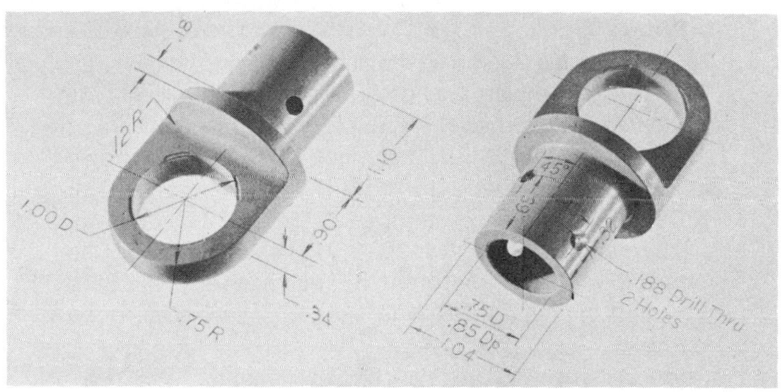


**PROB. 73.** Lift-strut pivot.

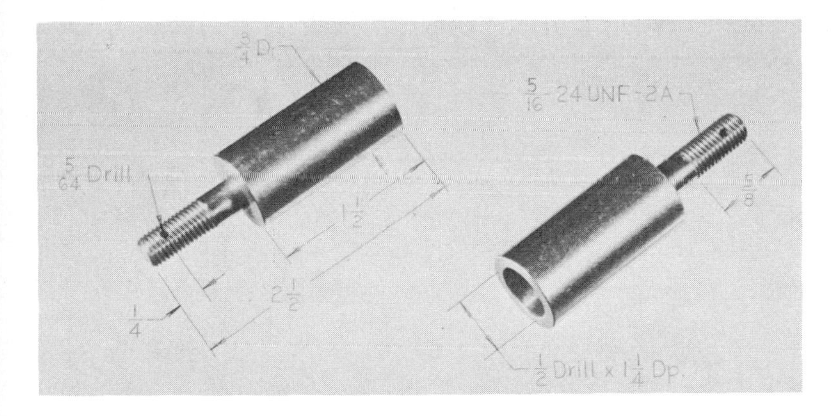


**PROB. 74.** Length adjuster-tube.

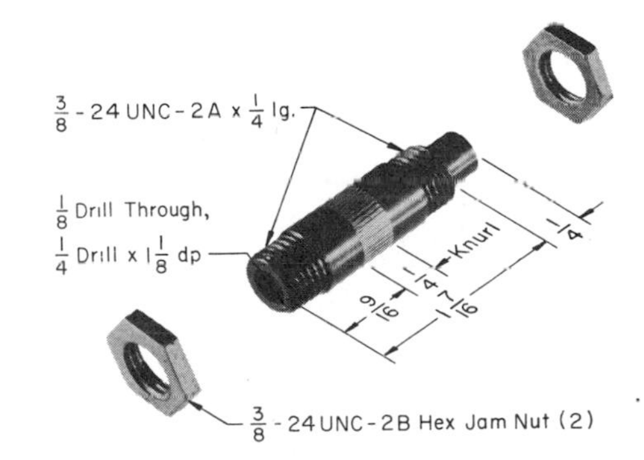


**PROB. 75.** Pivot shaft.

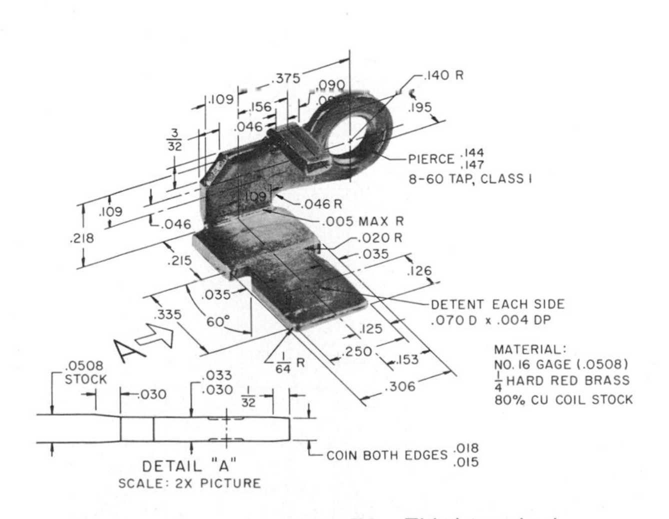


**PROB. 76.** Third terminal.

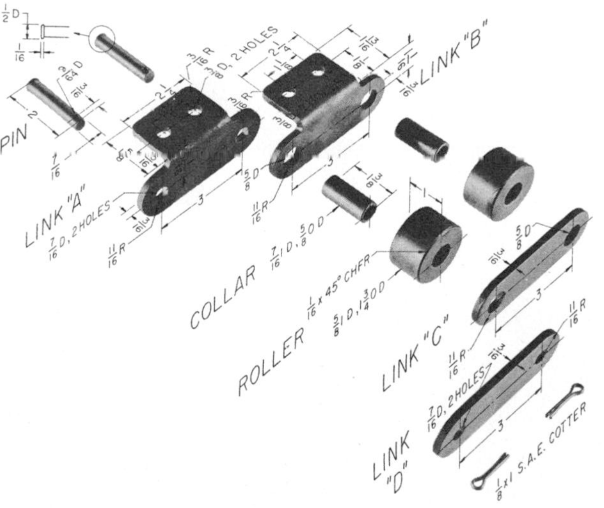


**PROB. 77.** Conveyor link unit.

## Group 7. A Set of Drawings from the Design Drawing.

**78.** Marking machine. From design drawing, shown half size, make complete set of drawings. Base, piece 1, is malleable-iron casting. Frame, piece 2, is cast iron. Ram, piece 3, is 1020 HR; bushing, piece 4, is 1020 cold-drawn tubing; heat-treatment for both is carburize at 1650 to 1700°F, quench direct, temper at 250 to 325°F. Spring, piece 5, is piano wire, No. 20 (0.045) gage, six coils, free length 2 in., heat-treatment is "as received." Marking dies and holders are made up to suit objects to be stamped.

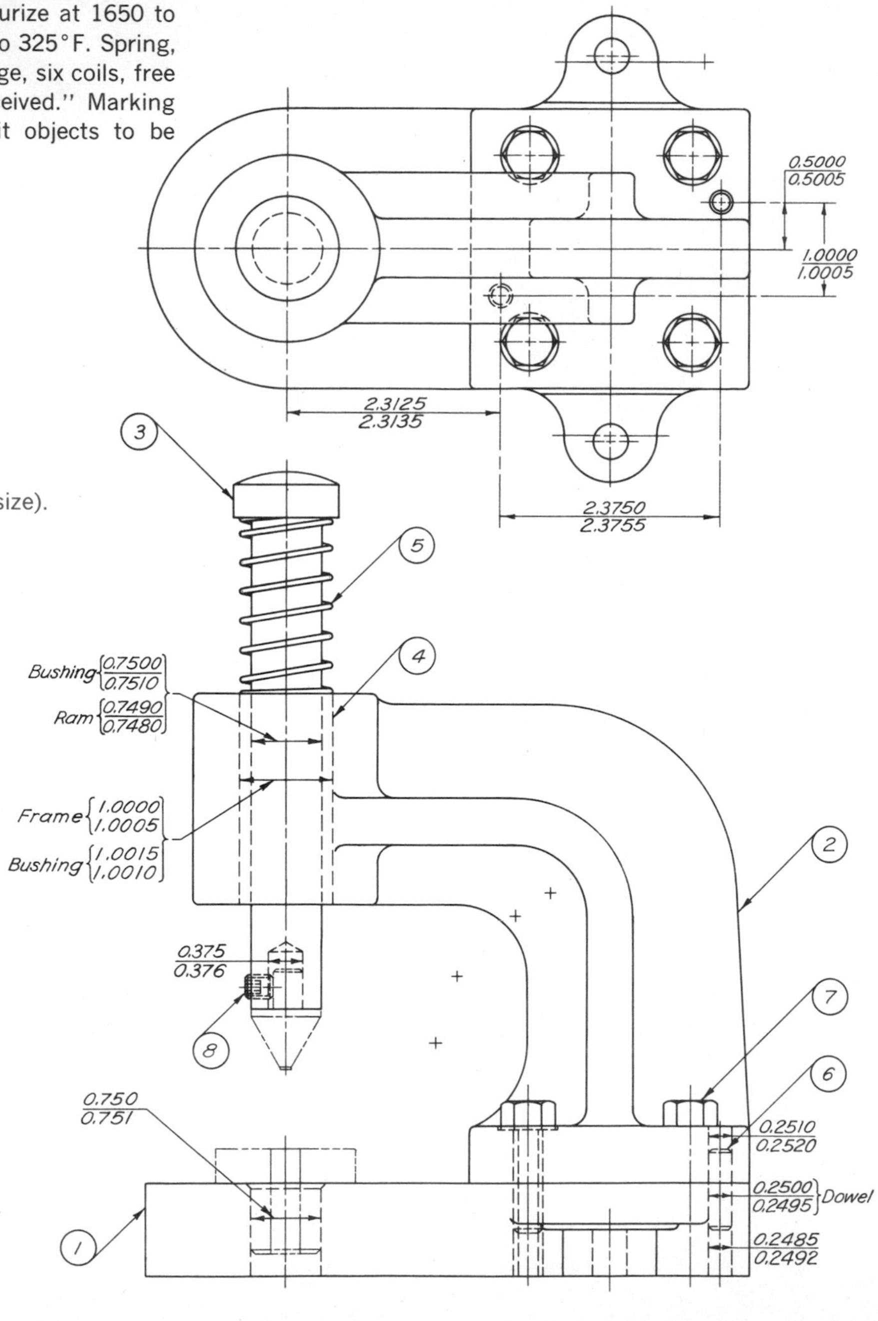

**PROB. 78.** Marking machine (one-half size).

**79.** Arbor press. From design drawing, shown one-quarter size, make complete set of drawings. All necessary information is given on the design drawing.

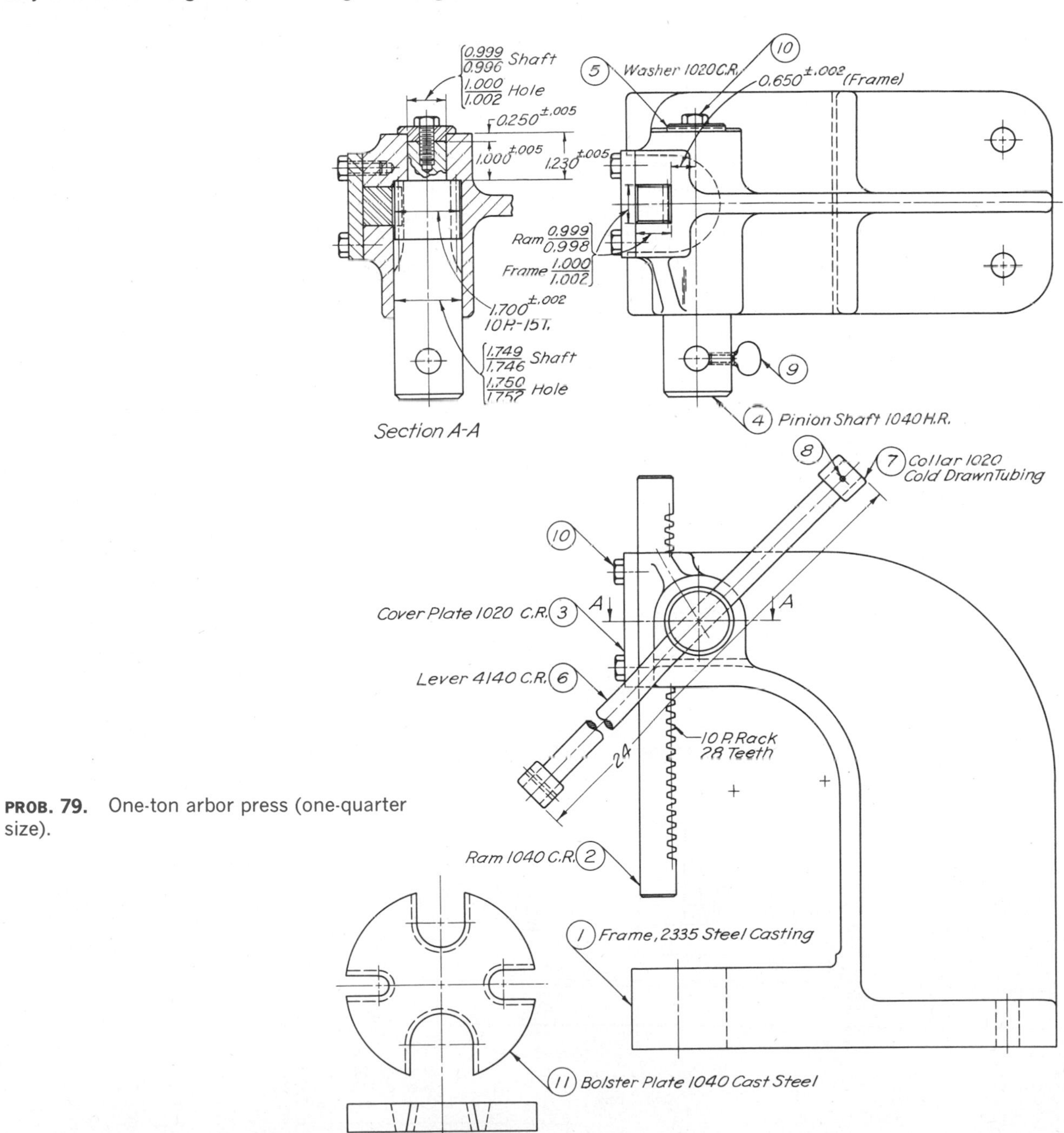


**PROB. 79.** One-ton arbor press (one-quarter size).

**80.** Number 2 flanged vise. From design drawing, make complete set of drawings, including details, parts list, and assembly. Design drawing is shown one-third size. All necessary information is on the design drawing.

**81.** Bench shears. Design drawing is half size. Make a complete set of drawings. For dimensions where close fits are involved, either the decimal limits or the tolerance to be applied to the scaled basic size are given on the design drawing. Heat-treatments should be specified as follows: for base and jaw, "normalize at 1550°F"; for eccentric, "as received"; for blade, "to Rockwell C57-60"; for handle, none. Finish for blades, "grind." Washers are special but may be specified on the parts list by giving inside diameter, outside diameter, and thickness. The key may be specified on the parts list by giving width, thickness, and length.

Following are some specifications for limits and tolerances:

Diameter of shoulder screw and hole in base and jaw:

Screws (as manufactured): $\frac{0.623}{0.621}$

Hole: $\frac{0.624}{0.626}$

Diameter of eccentric and hole in handle:

Eccentric: $\frac{0.999}{0.997}$

Handle: $\frac{0.999}{1.001}$

Diameter of eccentric and hole in base:

Eccentric: $\frac{1.3745}{1.3740}$

Base: $\frac{1.3750}{1.3755}$

Diameter of eccentric and width of slot in jaw:

Eccentric: $\frac{0.874}{0.872}$

Jaw: $\frac{0.875}{0.877}$

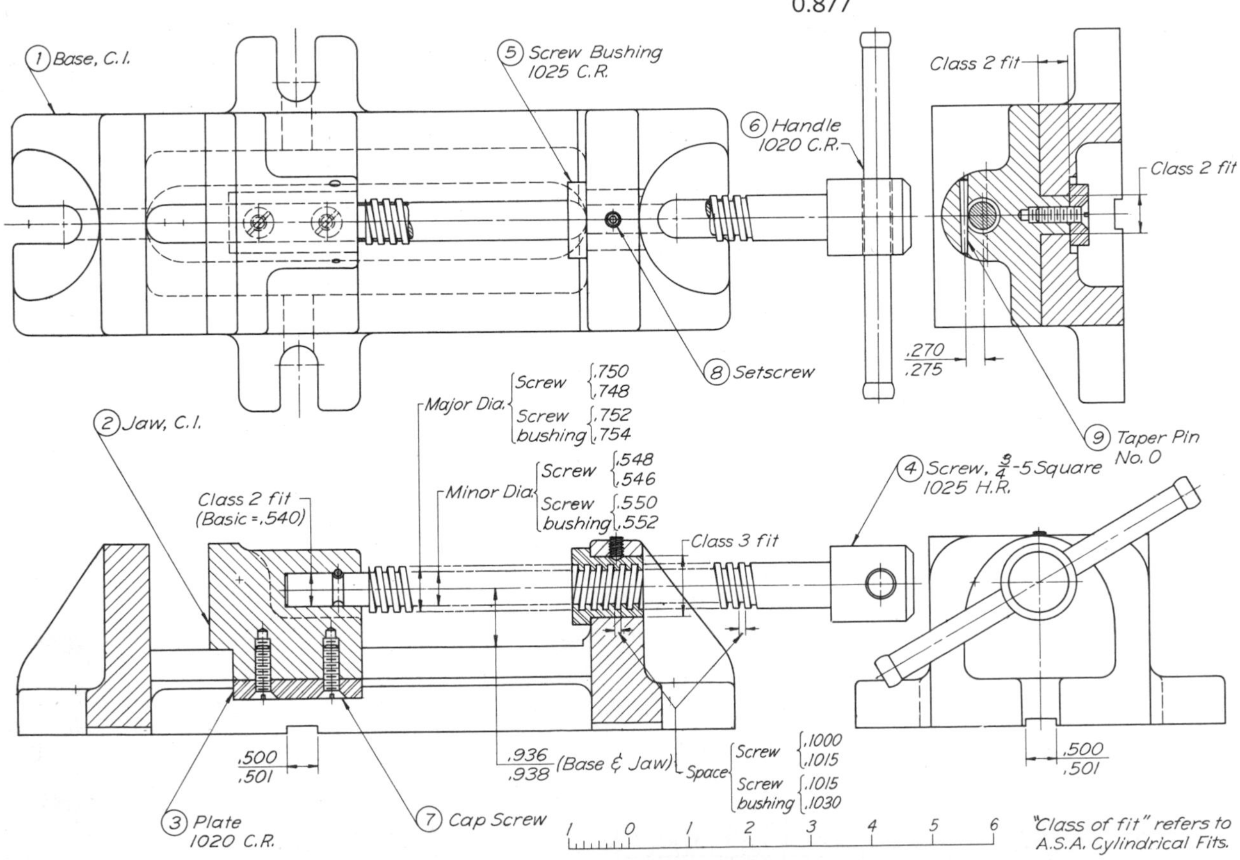


**PROB. 80.** No. 2 flanged vise (one-third size).

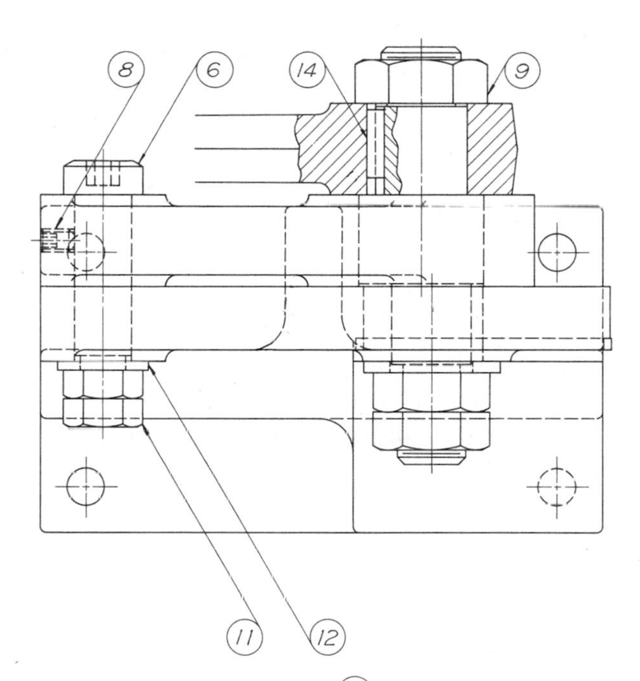


| PC. NO. | NAME | MAT. | REQ | NOTES |
|---|---|---|---|---|
| 1 | Base | 2335 | 1 | Steel Casting |
| 2 | Jaw | 2335 | 1 | Steel Casting |
| 3 | Eccentric | 2340 HR | 1 | |
| 4 | Blade | 1095 HR | 2 | |
| 5 | Handle | 1040 | 1 | Drop Forging |
| 6 | Shoulder Screw | | 1 | |
| 7 | Flat Head Cap Screw | | 4 | |
| 8 | Socket Set Screw | | 1 | Flat Point |
| 9 | Semifin Hex Jam Nut | | 1 | |
| 10 | Semifin Hex Jam Nut | | 2 | |
| 11 | Semifin Hex Jam Nut | | 2 | |
| 12 | Cut Washer | 1112 CR | 1 | |
| 13 | Cut Washer | 1112 CR | 1 | |
| 14 | Key | Key Stock | 1 | |

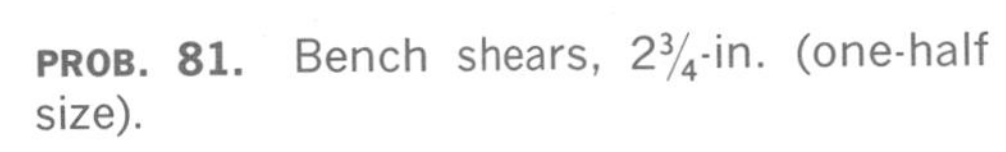
**PROB. 81.** Bench shears, $2\frac{3}{4}$-in. (one-half size).

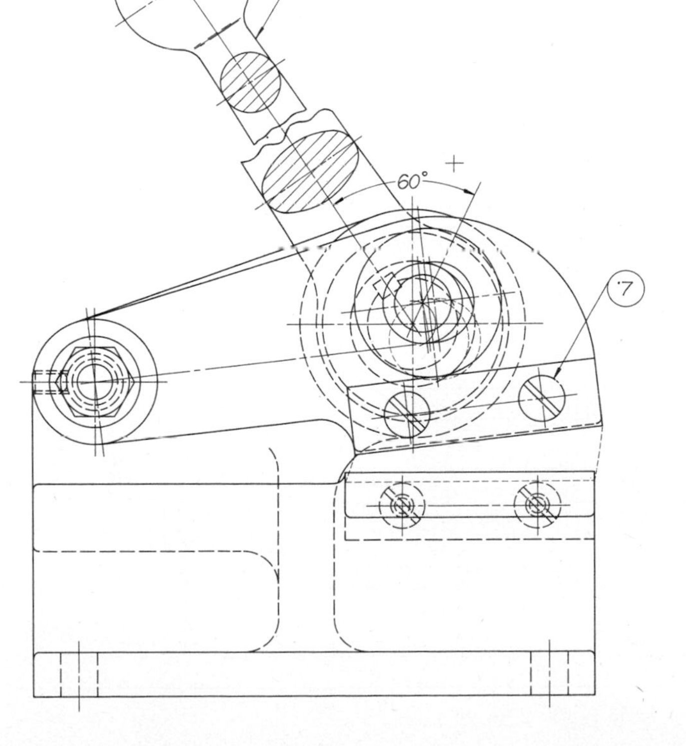


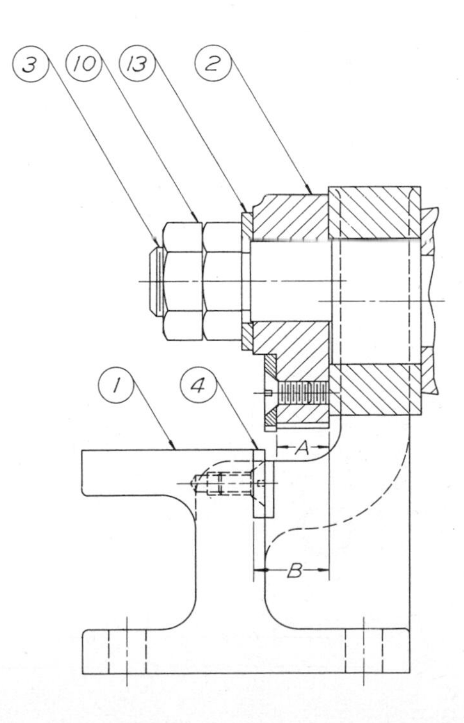

Width of keyway in both handle and eccentric:

Key (as purchased): $\frac{0.250}{0.249}$

Key seat and keyway: $\frac{0.250}{0.251}$

Depth of keyway and key seat in both handle and eccentric:

$\frac{3}{32}'' + \frac{1}{64} - 0$

Control of clearance between blades:

Thickness of blade: $\frac{0.1250}{0.1245}$

Dimension $A$: $\frac{0.562}{0.561}$

Dimension $B$: $\frac{0.812}{0.813}$

Tolerance on shoulder screw and eccentric hole locations: For base and jaw (two dimensions on each part)

$\pm 0.002$

Limits for eccentric offset:

Center to center: $\frac{0.248}{0.252}$

Length of handle:

Center of hub to center of ball: 12 in.

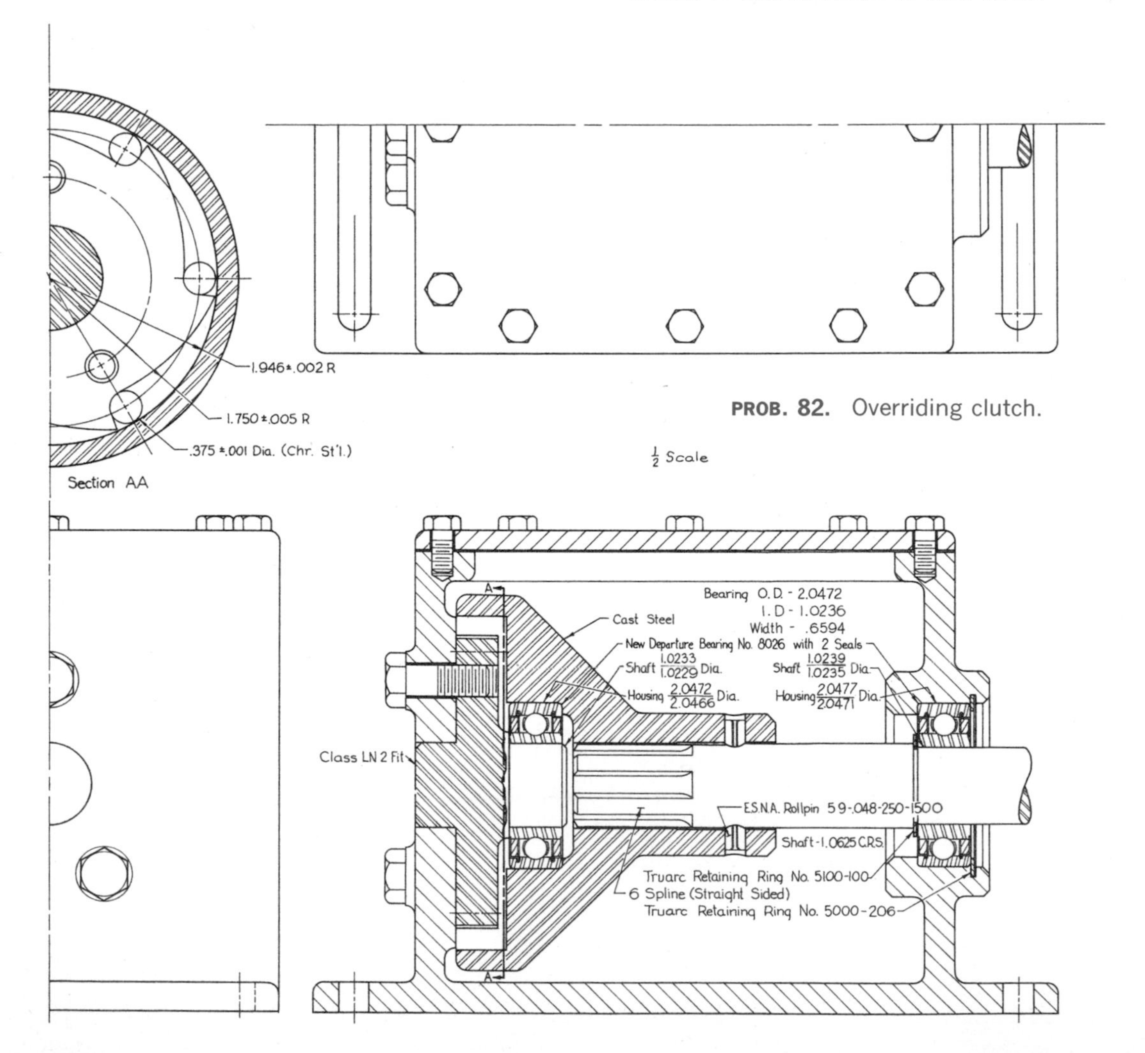


**PROB. 82.** Overriding clutch.

| No. | No. Req'd | Description | Size | Material |
|---|---|---|---|---|
| 1 | 1 | Shaft | 1 Dia. × 13 | Stn. Stl. |
| 2 | 1 | Base | .375 × 1 × 16.5 | Stn. Stl. |
| 3 | 3 | Lug | .375 × 1 × 1 | Stn. Stl. |
| 4 | 2 | Clamp Ring | 1.375 Dia × 2 | Stn. Stl. |
| 5 | 4 | Lug | .188 × .375 × 2 | Stn. Stl. |
| 6 | 1 | Frame | .188 × 2 × 9 | Stn. Stl. |
| 7 | 2 | Clamp Bolt | .375 Dia. × 3.5 | Stn. Stl. |
| 8 | 1 | Spur Gear | 20 Teeth - 20° Press. Angle 10 Pitch - .75 Face Width | Stn. Stl. |
| 9 | 2 | Taper Pin | No. 1 | Stn Stl. |
| 10 | 2 | Bushing | 1 Dia. × .312 | Nylon |
| 11 | 1 | Shaft | .50 Dia. × 2.125 | Stn. Stl. |
| 12 | 1 | Bearing Block | .75 × 1 × 2 | Nylon |
| 13 | 1 | Rack | .75 × 1 × 7.5 | Stn. Stl. |
| 14 | 1 | Handle | .50 Dia. × 8 | Stn. Stl. |
| 15 | 1 | Hub | 1 Dia. × .75 | Stn. Stl. |
| 16 | 4 | Self Tapping Rd. Hd. Screw | No. 8 × .50 Long | Stn. Stl. |
| 17 | 1 | Press Plate | .75 × 3 × 4 | Stn. Stl. |
| 18 | 1 | Bone Juice Drain | .875 Dia × 3 | Stn. Stl. |
| 19 | 1 | Bone Juice Plate | .75 × 5 × 6 | Stn. Stl. |
| 20 | 1 | Web | .50 × 1.50 × 5 | Stn. Stl. |

**82.** The overriding clutch is shown half size. From this design drawing make a complete set of working drawings. The frame is made of cast iron and the shaft is cold-rolled steel. Other parts are as indicated.

**83.** The illustration for this problem is a pictorial design drawing, made for presentation to a medical research group. The device, a bone crusher, is proposed for use by doctors of veterinary medicine in the laboratory analysis of bone disease. Fundamentally, the machine consists of parts welded together as shown by the drawing and indicated on the bill of materials. Make a complete set of drawings, including details, working drawings, subassembly drawings if necessary, and the assembly drawing.

**84.** A common problem in all manufacturing projects is the planning and execution of methods to promote speed, consistent quality of parts, and economy in production. Quite often, parts can be made as a portion or segment of a basic shape. For example, rings can be cut

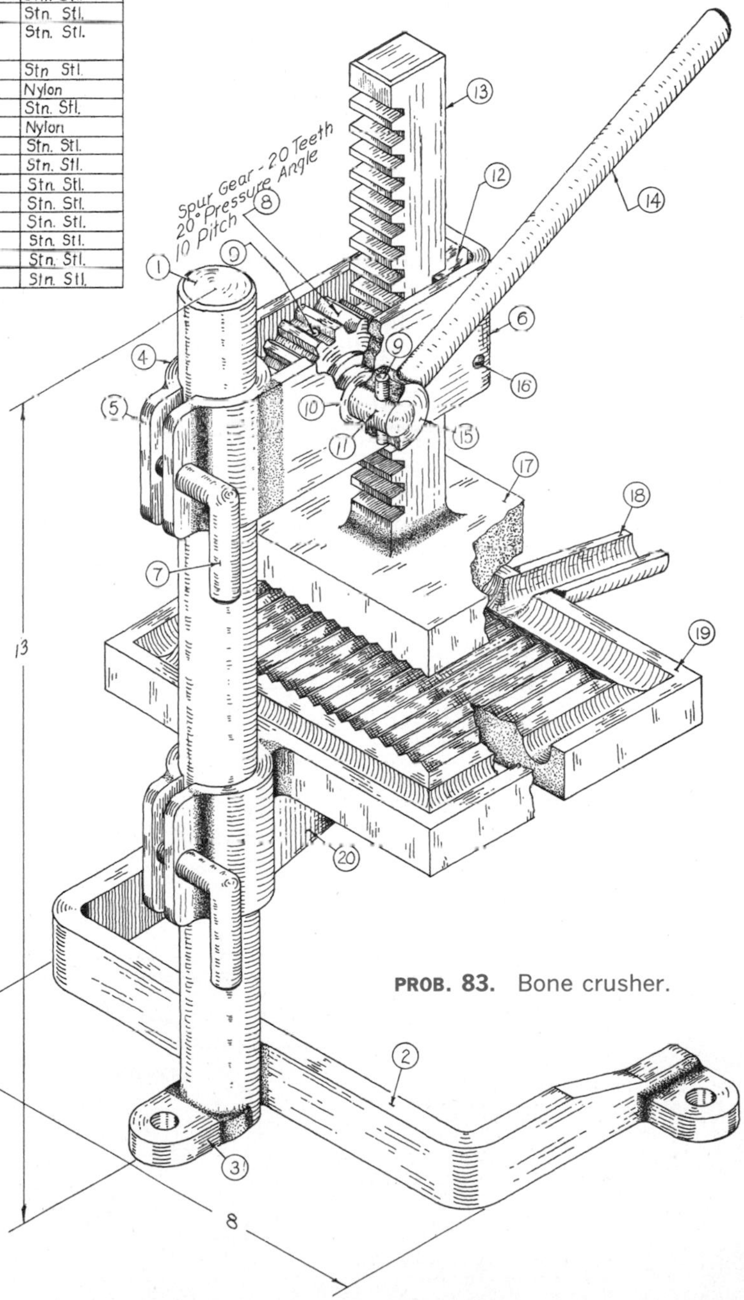

**PROB. 83.** Bone crusher.

into segments and extruded shapes can be cut to length quickly and economically to produce brackets, clips, etc. Bars, tubes and rolled shapes also can be cut to length to produce parts of required shape.

A fine example of intelligent production planning is given in the following problem.

The M-14 rifle has a curved aperture in its sighting mechanism offering a challenge to produce the simple and economical part shown in the illustration. Essentially the part is a serrated sector of a circle having a tapered hole. The aperture is made of SAE 1041 steel heat-treated to Rockwell C40-45 after machining, then phosphate-coated for corrosion protection. A unique process of external and internal broaching is an important phase of the processing to achieve the necessary economy.

Since the body is a portion of a true circle, it was decided to produce the part from steel tubing, broaching where possible. Engineers of National Broach & Machine Co. assisted in making a study to recommend processing sequences with tooling that would produce the part to desired tolerances. Special Red Ring broaches were chosen for three operations.

*Operation 1.* First a ring is turned, bored, cut off and faced on a four-spindle, $4\frac{3}{4}$-in. Conomatic bar machine. The $\frac{1}{8}$-in. width of the serrated tooth section is held to 0.001 in.

*Operation 2.* Next the ring is faced and undercut on both sides in an Ex-Cell-O two-spindle boring machine.

*Operation 3.* The ring is then broached on a 40-ton Colonial horizontal broaching machine. The internal broach shapes the ID of the ring which will later be cut to form eight peephole gun parts. Broaching operation takes 30 sec.

*Operation 4.* The next operation uses two broaches on a 20-ton Oilgear vertical machine that notches and serrates the OD and forms eight segments.

The tool for this sequence is an internal pot-type broach that traverses over the ring mounted on a long fixture. First the rings are finished on the notched OD and serrated with one set of tools. The 2.217-in. OD is held to 0.002 in. while sixteen 0.0628-in. circular pitch 20° P.A., involute serrations are broached on each of the eight sections. Time: 30 sec.

*Operation 5.* After a number of rings have been finished, broach inserts are removed and replaced with another set of inserts that cut the ring into eight aperture parts and finish broach each part to length. Time: another 30 sec.

**PROB. 84.** M-14 rifle aperture.

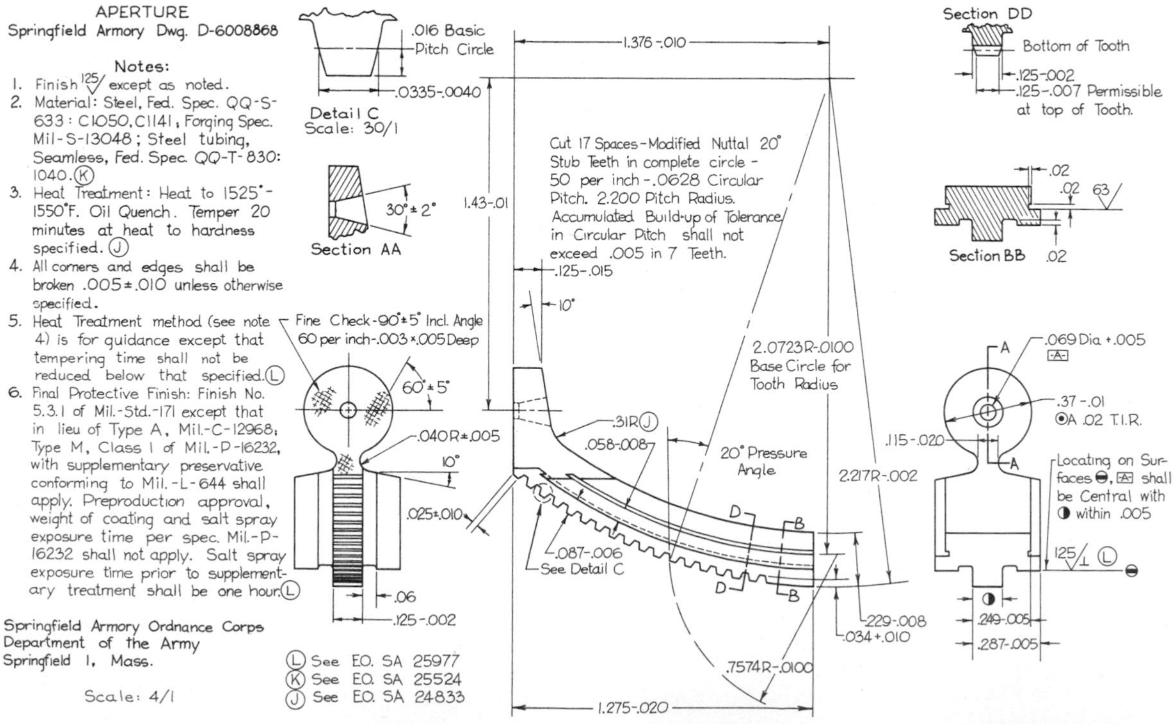

*Operation 6.* Next Red Ring external surface broaches form the $\frac{3}{8}$-in.-dia. circular portion around the peephole on a 5-ton Cincinnati vertical broaching machine. This takes 20 sec.

*Operation 7.* Final machining is the 0.069-in. peephole, which has a 30° included-angle conical section. It is drilled on a Hamilton drill press equipped with a four-station index fixture and special drilling tool. The hole must be accurately positioned within 0.002 in. of the center line of the gear section.

This part is an excellent example of how four broaching operations equipped with properly conceived broaching tools can cut production costs and make a precise component with a total production time of only 110 sec/part.

From the final part drawing of Prob. 84, make separate drawings to show the part at the completion of each stage of manufacture, as shown by the pictures accompanying Prob. 84 and designated as Operations 1 through 7.

This problem and the explanation accompanying it is given through the courtesy of the National Broach & Machine Co. and by Dale O. Miller, Jr., of Rochester Gear, Inc., a personal friend of the author of this text.

## Group 8. Working Sketches.

**85.** Select one of the problems in Group 1, and make a detail working sketch.

**86.** Select one of the problems in Group 2, and make an assembly sketch.

**87.** Select a single part from one of the assemblies of Groups 3 to 5, and make a working sketch.

**88.** Select a single part from one of the Probs. 1 to 59 and make a pictorial working sketch.

**89.** Select one of the problems that are given in Group 3, and make detail working sketches.

**90.** Select one of the problems in Group 4 or 5, and make a complete set of working sketches.

**OPERATION 1:** Machine and cut off.

**OPERATION 2:** Face and undercut.

**OPERATION 3:** Broach inside diameter.

**OPERATION 4:** Broach ouside diameter.

**OPERATION 5:** Broach to separate segments.

**OPERATION 6:** Broach radius.

**OPERATION 7:** Drill peep hole.

# APPENDIXES

Included here:

Glossary
Bibliography of Allied Subjects
Appendix A Lettering
Appendix B The Slide Rule
Appendix C Mathematical Tables
Appendix D Standard Parts, Sizes, Symbols, and Abbreviations
Appendix E Manufacturer's Specialties

# GLOSSARY

## Part A. Shop Terms

**anneal** (*v.*) To soften a metal piece and remove internal stresses by heating to its critical temperature and allowing to cool very slowly.

**arc-weld** (*v.*) To weld by electric-arc process.

**bore** (*v.*) To enlarge a hole with a boring tool, as in a lathe or boring mill. Distinguished from *drill*.

**boss** (*n.*) A projection of circular cross section, as on a casting or forging.

**braze** (*v.*) To join by the use of hard solder.

**broach** (*v.*) To finish the inside of a hole to a shape usually other than round. (*n.*) A tool with serrated edges pushed or pulled through a hole to enlarge it to a required shape.

**buff** (*v.*) To polish with abrasive on a cloth wheel or other soft carrier.

**burnish** (*v.*) To smooth or polish by a rolling or sliding tool under pressure.

**bushing** (*n.*) A removable sleeve or liner for a bearing; also a guide for a tool in a jig or fixture.

**carburize** (*v.*) To prepare a low-carbon steel for heat-treatment by packing in a box with carbonizing material, such as wood charcoal, and heating to about 2000°F for several hours, then allowing to cool slowly.

**caseharden** (*v.*) To harden the surface of carburized steel by heating to critical temperature and quenching, as in an oil or lead bath.

**castellate** (*v.*) To form into a shape resembling a castle battlement, as castellated nut. Often applied to a shaft with multiple integral keys milled on it.

**chamfer** (*v.*) To bevel a sharp external edge. (*n.*) A beveled edge.

**chase** (*v.*) To cut threads in a lathe, as distinguished from cutting threads with a die. (*n.*) A slot or groove.

**chill** (*v.*) To harden the surface of cast iron by sudden cooling against a metal mold.

**chip** (*v.*) To cut or clean with a chisel.

**coin** (*v.*) To stamp and form a metal piece in one operation, usually with a surface design.

Boss

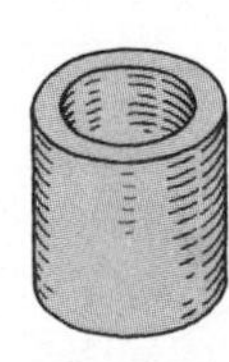
Bushing

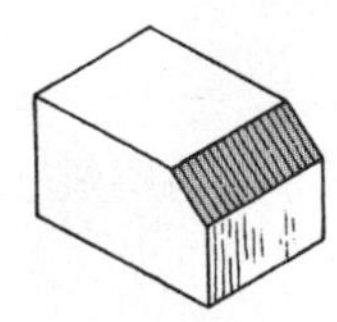
Chamfer

**cold-work** (*v.*) To deform metal stock by hammering, forming, drawing, etc., while the metal is at ordinary room temperature.

**color-harden** (*v.*) To caseharden to a very shallow depth, chiefly for appearance.

**core** (*v.*) To form the hollow part of a casting, using a solid form made of sand, shaped in a core box, baked, and placed in the mold. After cooling, the core is easily broken up, leaving the casting hollow.

**counterbore** (*v.*) To enlarge a hole to a given depth. (*n.*) 1. The cylindrical enlargement of the end of a drilled or bored hole. 2. A cutting tool for counterboring, having a piloted end the size of the drilled hole.

**countersink** (*v.*) To form a depression to fit the conic head of a screw or the thickness of a plate so that the face will be level with the surface. (*n.*) A conic tool for countersinking.

**crown** (*n.*) Angular or rounded contour, as on the face of a pulley.

**die** (*n.*) 1. One of a pair of hardened metal blocks for forming, impressing, or cutting out a desired shape. 2 (thread). A tool for cutting external threads. Opposite of *tap*.

**die casting** (*n.*) A very accurate and smooth casting made by pouring a molten alloy (or composition, as Bakelite) usually under pressure into a metal mold or die. Distinguished from a casting made in sand.

**die stamping** (*n.*) A piece, usually of sheet metal, formed or cut out by a die.

**draw** (*v.*) 1. To form by a distorting or stretching process. 2. To temper steel by gradual or intermittent quenching.

**drill** (*v.*) To sink a hole with a drill, usually a twist drill. (*n.*) A pointed cutting tool rotated under pressure.

**drop forging** (*n.*) A wrought piece formed hot between dies under a drop hammer, or by pressure.

**face** (*v.*) To machine a flat surface perpendicular to the axis of rotation on a lathe. Distinguished from *turn.*

**feather** (*n.*) A flat sliding key, usually fastened to the hub.

**fettle** (*v.*) To remove fins and smooth the corners on unfired ceramic products.

**file** (*v.*) To finish or trim with a file.

**fillet** (*n.*) A rounded filling of the internal angle between two surfaces.

**fin** (*n.*) A thin projecting rib. Also, excess ridge of material.

**fit** (*n.*) The kind of contact between two machined surfaces. 1. *Drive, force,* or *press:* When the shaft is slightly larger than the hole and must be forced in with sledge or power press. 2. *Shrink:* When the shaft is slightly larger than the hole, the piece containing the hole is heated, thereby expanding the hole sufficiently to slip over the shaft. On cooling, the shaft will be seized firmly if the fit allowances have been correctly proportioned. 3. *Running* or *sliding:* When sufficient allowance has been made between sizes of shaft and hole to allow free running without seizing or heating. 4. *Wringing:* When the allowance is smaller than a running fit and the shaft will enter the hole by twisting the shaft by hand.

**flange** (*n.*) A projecting rim or edge for fastening or stiffening.

**forge** (*v.*) To shape metal while hot and plastic by a hammering or forcing process either by hand or by machine.

**galvanize** (*v.*) To treat with a bath of lead and zinc to prevent rusting.

**graduate** (*v.*) To divide a scale or dial into regular spaces.

**grind** (*v.*) To finish or polish a surface by means of an abrasive wheel.

**harden** (*v.*) To heat hardenable steel above critical temperature and quench in bath.

**hot-work** (*v.*) To deform metal stock by hammering, forming, drawing, etc., while the metal is heated to a plastic state.

**kerf** (*n.*) The channel or groove cut by a saw or other tool.

**key** (*n.*) A small block or wedge inserted between shaft and hub to prevent circumferential movement.

Counterbore

Countersink

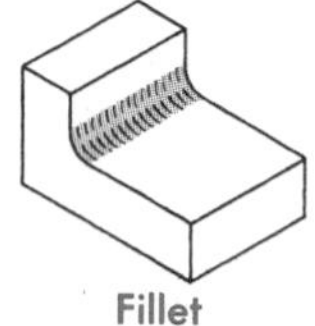
Fillet

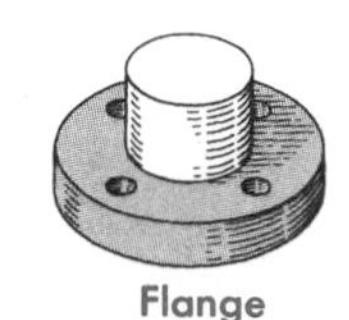
Flange

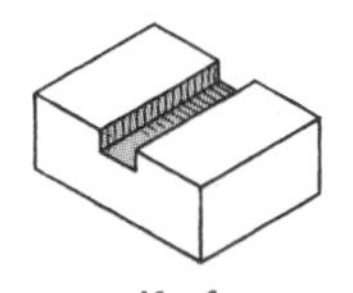
Kerf

**keyway, key seat** (*n.*) A groove or slot cut to fit a key. A key fits into a key seat and slides in a keyway.
**knurl** (*v.*) To roughen or indent a turned surface, as a knob or handle.
**lap** (*n.*) A piece of soft metal, wood, or leather charged with abrasive material, used for obtaining an accurate finish. (*v.*) To finish by lapping.
**lug** (*n.*) A projecting "ear," usually rectangular in cross section. Distinguished from *boss*.
**malleable casting** (*n.*) An ordinary casting toughened by annealing. Applicable to small castings with uniform metal thicknesses.
**mill** (*v.*) To machine with rotating toothed cutters on a milling machine.
**neck** (*v.*) To cut a groove around a shaft, usually near the end or at a change in diameter. (*n.*) A portion reduced in diameter between the ends of a shaft.
**normalize** (*v.*) To remove internal stresses by heating a metal piece to its critical temperature and allowing it to cool very slowly.
**pack-harden** (*v.*) To carburize and caseharden.
**pad** (*n.*) A shallow projection. Distinguished from *boss* by shape or size.
**peen** (*v.*) To stretch, rivet, or clinch over by strokes with the peen of a hammer. (*n.*) The end of a hammer head opposite the face, as *ball peen*.
**pickle** (*v.*) To clean castings or forgings in a hot weak sulfuric acid bath.
**plane** (*v.*) To machine work on a planer having a fixed tool and reciprocating bed.
**planish** (*v.*) To finish sheet metal by hammering with polished-faced hammers.
**plate** (*v.*) The electrochemical coating of a metal piece with a different metal.
**polish** (*v.*) To make smooth or lustrous by friction with a very fine abrasive.
**profile** (*v.*) To machine an outline with a rotary cutter usually controlled by a master cam or die.
**punch** (*v.*) To perforate by pressing a nonrotating tool through the work.
**ream** (*v.*) To finish a drilled or punched hole very accurately with a rotating fluted tool of the required diameter.
**relief** (*n.*) The amount one plane surface of a piece is set below or above another plane, usually for clearance or for economy in machining.
**rivet** (*v.*) 1. To fasten with rivets. 2. To batter the headless end of a pin used as a permanent fastening.
**round** (*n.*) A rounded exterior corner between two surfaces. Compare with *fillet*.
**sandblast** (*v.*) To clean castings or forgings by means of sand driven through a nozzle by compressed air.
**shape** (*v.*) To machine with a shaper, a machine tool differing from a planer in that the work is stationary and the tool reciprocating.
**shear** (*v.*) To cut off sheet or bar metal between two blades.
**sherardize** (*v.*) To galvanize with zinc by a dry heating process.
**shim** (*n.*) A thin spacer of sheet metal used for adjusting.
**shoulder** (*n.*) A plane surface on a shaft, normal to the axis and formed by a difference in diameter.
**spin** (*v.*) To shape sheet metal by forcing it against a form as it revolves.
**spline** (*v.*) A long keyway. Sometimes also a flat key.
**spot-face** (*v.*) To finish a round spot on a rough surface, usually around a drilled hole, to give a good seat to a screw or bolthead, cut, usually $\frac{1}{16}$ in. deep, by a rotating milling cutter.

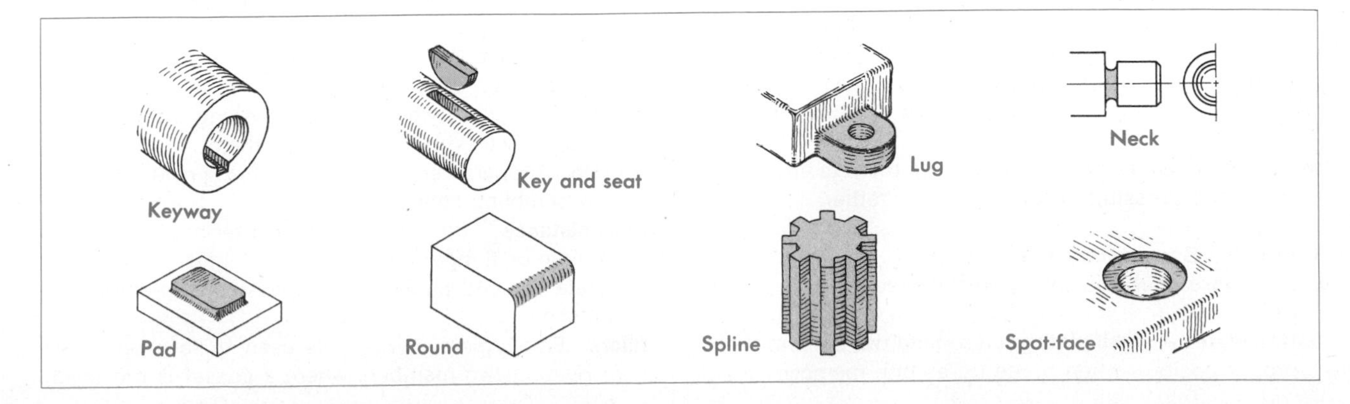

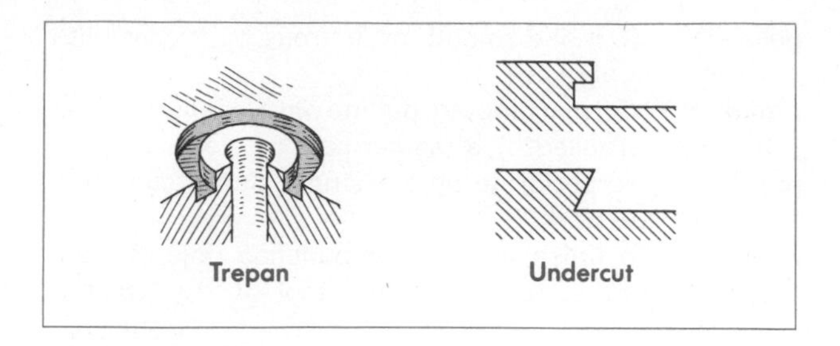


**spot-weld** (*v.*) To weld in spots by means of the heat of resistance to an electric current. Not applicable to sheet copper or brass.
**steel casting** (*n.*) Material used in machine construction. It is ordinary cast iron into which varying amounts of scrap steel have been added in the melting.
**swage** (*v.*) To shape metal by hammering or pressure with the aid of a form or anvil called a "swage block."
**sweat** (*v.*) To join metal pieces by clamping together with solder between and applying heat.
**tack-weld** (*v.*) To join at the edge by welding in short intermittent sections.
**tap** (*v.*) To cut threads in a hole with a rotating tool called a "tap," having threads on it and fluted to give cutting edges.
**temper** (*v.*) To change the physical characteristics of hardened steel by reheating to a temperature below the critical point and allowing to cool.
**template, templet** (*n.*) A flat pattern for laying out shapes, location of holes, etc.
**trepan** (*v.*) To cut an outside annular groove around a hole.
**tumble** (*v.*) To clean, smooth, or polish castings or forgings in a rotating barrel or drum by friction with each other, assisted by added mediums, as scraps, "jacks," balls, sawdust, etc.
**turn** (*v.*) To machine on a lathe. Distinguished from *face.*
**undercut** (*v.*) To cut, leaving an overhanging edge. (*n.*) A cut having inwardly sloping sides.
**upset** (*v.*) To forge a larger diameter or shoulder on a bar.
**weld** (*v.*) To join two pieces by heating them to the fusing point and pressing or hammering together.

## Part B. Structural Terms

**bar** Square or round rod; also flat steel up to 6 in. in width.
**batten plate** A small plate used to hold two parts in their proper position when made up as one member.
**batter** A deviation from the vertical in upright members.
**bay** The distance between two trusses or transverse bents.
**beam** A horizontal member forming part of the frame of a building or structure.
**bearing plate** A steel plate, usually at the base of a column, used to distribute a load over a larger area.
**bent** A vertical framework usually consisting of a truss or beam supported at the ends on columns.
**brace** A diagonal member used to stiffen a framework.
**buckle plate** A flat plate with dished depression pressed into it to give transverse strength.
**built-up member** A member built from standard shapes to give one single stronger member.
**camber** Slight upward curve given to trusses and girders to avoid effect of sag.
**cantilever** A beam, girder, or truss overhanging one or both supports.
**chord** The principal member of a truss on either the top or bottom.
**clearance** Rivet driving clearance is distance from center of rivet to obstruction. Erection clearance is amount of space left between members for ease in assembly.
**clevis** U-shaped shackle for connecting a rod to a pin.
**clip angle** A small angle used for fastening various members together.
**column** A vertical compression member.
**cope** To cut out top or bottom of flanges and web so that one member will frame into another.
**coping** A projecting top course of concrete or stone.
**counters** Diagonal members in a truss to provide for reversal of shear due to live load.
**cover plate** A plate used in building up flanges, in a built-up member, to give greater strength and area or for protection.
**crimp** To offset the end of a stiffener to fit over the leg of an angle.
**diagonals** Diagonal members used for stiffening and wind bracing.
**dowel** An iron or wooden pin extending into, but not through, two timbers to connect them.
**driftpin** A tapered steel pin used to bring rivet holes fair in assembling steel work.
**edge distance** The distance from center of rivet to edge of plate or flange.
**fabricate** To cut, punch, and subassemble members in the shop.
**fillers** Either plate or ring fills used to take up space in riveting two members where a gusset is not used.

**flange** The projecting portion of a beam, channel, or column.

**gage line** The center line for rivet holes.

**gin pole** A guyed mast with block at the top for hoisting.

**girder** A horizontal member, either single or built up, acting as a principal beam.

**girt** A beam usually bolted to columns to support the side covering or serve as window lintels.

**gusset plate** A plate used to connect various members, such as in a truss.

**hip** The intersection between two sloping surfaces forming an exterior angle.

**knee brace** A corner brace used to prevent angular movement.

**lacing or lattice bars** Bars used diagonally to space and stiffen two parallel members, such as in a built-up column.

**laterals** Members used to prevent lateral deflection.

**lintel** A horizontal member used to carry a wall over an opening.

**louvers** Metal slats either movable or fixed, as in a monitor ventilator.

**monitor ventilator** A framework that carries fixed or movable louvers at the top of the roof.

**panel** The space between adjacent floor supports or purlins in a roof.

**pitch** Center distance between rivets parallel to axis of member. Also, for roofs, the ratio of rise to span.

**plate** Flat steel over 6 in. in width and $\frac{1}{4}$ in. or more in thickness.

**purlins** Horizontal members extending between trusses, used as beams for supporting the roof.

**rafters** Beams or truss members supporting the purlins.

**sag ties** Tie rods between purlins in the plane of the roof to carry the component of the roof load parallel to the roof.

**separator** Either a cast-iron spacer or wrought-iron pipe on a bolt for the purpose of holding members a fixed distance apart.

**sheet** Flat steel over 6 in. in width and less than $\frac{1}{4}$ in. in thickness.

**shim** A thin piece of wood or steel placed under a member to bring it to a desired elevation.

**sleeve nut** A long nut with right and left threads for connecting two rods to make an adjustable member.

**span** Distance between centers of supports of a truss, beam, or girder.

**splice** A longitudinal connection between the parts of a continuous member.

**stiffener** Angle, plate, or channel riveted to a member to prevent buckling.

**stringer** A longitudinal member used to support loads directly.

**strut** A compression member in a framework.

**truss** A rigid framework for carrying loads, formed in a series of triangles.

**turnbuckle** A coupling, threaded right and left or swiveled on one end, for adjustably connecting two rods.

**valley** The intersection between two sloping surfaces, forming a reentrant angle.

**web** The part of a channel, I beam, or girder between the flanges.

## Part C. Architectural Terms

**apron** The finished board placed against the wall surface, immediately below a window stool.

**ashlar** Thin, squared, and dressed stone facing of a masonry wall.

**backing** The inner portion of a wall—that which is used behind the facing.

**batten** A strip of wood used for nailing across two other pieces of wood to hold them together and cover a crack.

**batter boards** Boards set up at the corners of a proposed building from which the lines marking off the walls are stretched.

**bearing wall** A wall that supports loads other than its own weight.

**bond** The joining together of building materials to ensure solidity.

**bridging** The braces or system of bracing used between joists or other structural members to stiffen them and to distribute the load.

**centering** A substructure of temporary nature, usually of timber or planks, on which a masonry arch or vault is built.

**coffer** An ornamental panel deeply recessed, usually in a dome or portico ceiling.

**corbel** A bracket formed on a wall by building out successive courses of masonry.

**curtain wall** A wall that carries no building load other than its own weight.

**fenestration** The arrangement and proportioning of window and door openings.

**flashing** The sheet metal built into the joints of a wall, or covering the valleys, ridges, and hips of a roof for the purpose of preventing leakage.

**footing** A course or series of courses projecting at the base of a wall for the purpose of distributing the load from above over a greater area, thereby preventing excessive settlement.

**furring** The application of thin wood, metal, or other building material to a wall, beam, ceiling, or the like to level a surface for lathing, boarding, etc., or to make an air space within a wall.

**glazing** The act of furnishing or fitting with glass.

**ground** Strips of wood, flush with the plastering, to which moldings, etc., are attached. Grounds are usually installed first and the plastering floated flush with them.

**grout** A thin mortar used for filling up spaces where heavier mortar will not penetrate.

**head** The horizontal piece forming the top of a wall opening, as a door or window.

**hip** The intersection of two roof surfaces, which form an external angle on the plan.

**jamb** The vertical piece forming the side of a wall opening.

**lintel** The horizontal structural member that supports the wall over an opening.

**millwork** The finish wordwork, machined and in some cases partly assembled at the mill.

**miter** To match together, as two pieces of molding, on a line bisecting the angle of junction.

**mullion** A vertical division of a window opening.

**muntin** The thin members that separate the individual lights of glass in a window frame.

**party wall** A division wall common to two adjacent pieces of property.

**plate** A horizontal member that carries other structural members; usually the top timber of a wall that carries the roof trusses or rafters directly.

**rail** A horizontal piece in a frame or paneling.

**return** The continuation in a different direction, most often at right angles, of the face of a building or any member, as a colonnade or molding; applied to the shorter in contradistinction to the longer.

**reveal** The side of a wall opening; the whole thickness of the wall; the jamb.

**riser** The upright piece of a step, from tread to tread.

**saddle** A small double-sloping roof to carry water away from behind chimneys, etc.

**scratch coat** The first coat in plastering, roughened by scratching or scoring so that the next coat will firmly adhere to it.

**screeds** A strip of plaster of the thickness proposed for the work, applied to the wall at intervals of 4 or 5 ft, to serve as guides.

**shoring** A prop, as a timber, placed against the side of a structure; a prop placed beneath anything, as a beam, to prevent sinking or sagging.

**sill** The horizontal piece, as a timber, which forms the lowest member of a frame.

**sleepers** The timbers laid on a firm foundation to carry and secure the superstructure.

**soffit** The underside of subordinate parts and members of buildings, such as staircases, beams, arches, etc.

**stile** A vertical piece in a frame or paneling.

**stool** The narrow shelf fitted on the inside of a window against the actual sill.

**threshold** The stone, wood, or metal piece that lies directly under a door.

**trap** A water seal in a sewage system to prevent sewer gas from entering the building.

**tread** The upper horizontal piece of a step, on which the foot is placed.

**valley** The intersection of two roof surfaces which form, on the plan, a reentrant angle.

## Part D. Welding Terms

**air-acetylene welding** A gas-welding process in which coalescence is produced by heating with a gas flame or flames obtained from the combustion of acetylene with air, without the application of pressure and with or without the use of filler metal.

**arc cutting** A group of cutting processes in which the severing of metals is effected by melting with the heat of an arc between an electrode and the base metal.

**arc welding** A group of welding processes in which coalescence is produced by heating with an electric arc or arcs, with or without the application of pressure and with or without the use of filler metal.

**As-welded** The condition of weld metal, welded joints, and weldments after welding prior to any subsequent thermal or mechanical treatment.

**atomic-hydrogen welding** An arc-welding process in which coalescence is produced by heating with an electric arc maintained between two metal electrodes in an atmosphere of hydrogen. Shielding is obtained from the hydrogen. Pressure may or may not be used, and filler metal may or may not be used.

**automatic welding** Welding with equipment which performs the entire welding operation without constant observation and adjustment of the controls by an operator. The equipment may or may not perform the loading and unloading of the work.

**axis of a weld** A line through the length of a weld, perpendicular to the cross section at its center of gravity.

**backing** Material (metal, weld metal, asbestos, carbon, granular flux, etc.) backing up the joint during welding to facilitate obtaining a sound weld at the root.

**bevel** A type of edge preparation.

**braze** A weld in which coalescence is produced by heating to suitable temperatures above 800°F and by using a nonferrous filler metal having a melting point below that of the base metals. The filler metal is distributed between the closely fitted surfaces of the joint by capillary attraction.

**butt joint** A joint between two members lying approximately in the same plane.

**coalesce** To unite or merge into a single body or mass. Fusion.

**die welding** A forge-welding process in which coalescence is produced by heating in a furnace and by applying pressure by means of dies.

**edge joint** A joint between the edges of two or more parallel or nearly parallel members.

**filler metal** Metal to be added in making a weld.

**fillet weld** A weld of approximately triangular cross section joining two surfaces approximately at right angles to each other in a lap joint, tee joint, or corner joint.

**forge welding** A group of welding processes in which coalescence is produced by heating in a forge or other furnace and by applying pressure or blows.

**fusion** The melting together of filler metal and base metal, or of base metal only, which results in coalescence.

**gas welding** A group of welding processes in which coalescence is produced by heating with a gas flame or flames, with or without the application of pressure, and with or without the use of filler metal.

**groove weld** A weld made in the groove between two members to be joined. The standard types of groove welds are: square-groove weld; single-V-groove weld; single-bevel-groove weld; single-U-groove weld; single-J-groove weld; double-V-groove weld; double-bevel-groove weld; double-U-groove weld; double-J-groove weld.

**hammer welding** A forge-welding process in which coalescence is produced by heating in a forge or other furnace and by applying pressure by means of hammer blows.

**intermittent welding** Welding in which the continuity is broken by recurring unwelded spaces.

**joint penetration** The minimum depth a groove weld extends from its face into a joint, exclusive of reinforcement.

**kerf** The space from which metal has been removed by a cutting process.

**lap joint** A joint between two overlapping members.

**oxy-acetylene welding** A gas-welding process in which coalescence is produced by heating with a gas flame or flames obtained from a combustion of acetylene with oxygen, with or without the application of pressure and with or without the use of filler metal.

**peening** The mechanical working of metals by means of hammer blows.

**plug weld** A circular weld made by either arc or gas welding through one member of a lap or tee joint joining that member to the other. The weld may or may not be made through a hole in the first member.

**pressure welding** Any welding process or method in which pressure is used to complete the weld.

**projection welding** A resistance-welding process in which coalescence is produced by the heat obtained from resistance to the flow of electric current through the work parts held together under pressure by electrodes.

**root opening** The separation between the members to be joined, at the root of the joint.

**seam weld** A weld consisting of a series of overlapping spot welds, made by seam welding or spot welding.

**slot weld** A weld made in an elongated hole in one member of a lap or tee joint joining that member to the portion of the surface of the other member which is exposed through the hole.

**spot welding** A resistance-welding process in which coalescence is produced by the heat obtained from resistance to the flow of electric current through the work parts, which are held together under pressure by electrodes.

**tack weld** A weld made to hold parts of a weldment in proper alignment until the final welds are made.

**tee joint** A joint between two members located approximately at right angles to each other in the form of a T.

**upset welding** A resistance-welding process in which coalescence is produced, simultaneously over the entire area of abutting surfaces or progressively along a joint, by the heat obtained from resistance to the flow of electric current through the area of contact of those surfaces. Pressure is applied before heating is started and is maintained throughout the heating period.

# BIBLIOGRAPHY OF ALLIED SUBJECTS

The following classified list is given to supplement this book, whose scope as a general treatise on engineering drawing permits only the mention or brief explanation of some subjects.


### Aeronautical Drafting and Design

Anderson, Newton H.: "Aircraft Layout and Detail Design," 2d ed., McGraw-Hill, New York, 1946.

Apalategui, J. J., and J. J. Adams: "Aircraft Analytic Geometry," McGraw-Hill, New York, 1944.

Davis, D. J., and C. H. Goen: "Aircraft Mechanical Drawing," McGraw-Hill, New York, 1944.

Katz, H. H.: "Aircraft Drafting," Macmillan, New York, 1946.

Leavell, S., and S. Bungay: "Aircraft Production Standards," McGraw-Hill, New York, 1943.

Liston, Joseph: "Powerplants for Aircraft," McGraw-Hill, New York, 1953.

Svensen, C. L.: "A Manual of Aircraft Drafting," Van Nostrand, Princeton, N.J., 1941.

Tharratt, George: "Aircraft Production Illustration," McGraw-Hill, New York, 1944.

### Architectural Drawing

Crane, T.: "Architectural Construction," 2d ed., Wiley, New York, 1956.

Kenney, Joseph E., and John P. McGrail: "Architectural Drawing for the Building Trades," McGraw-Hill, New York, 1949.

Morgan, Sherley W.: "Architectural Drawing," McGraw-Hill, New York, 1950.

Ramsey, C. G., and H. R. Sleeper: "Architectural Graphic Standards," 5th ed., Wiley, New York, 1956.

Saylor, H. H.: "Dictionary of Architecture," Wiley, New York, 1952.

Sleeper, H. R.: "Architectural Specifications," Wiley, New York, 1940.

### Descriptive Geometry

Bradley, H. C., and E. H. Uhler: "Descriptive Geometry for Engineers," International Textbook, Scranton, Pa., 1943.

Grant, Hiram E.: "Practical Descriptive Geometry," 2d ed., McGraw-Hill, New York, 1965.

Higbee, F. G.: "Drawing-board Geometry," Wiley, New York, 1938.

Hood, George J., and Albert S. Palmerlee: "Geometry of Engineering Drawing," 5th ed., McGraw-Hill, New York, 1969.

Johnson, L. O., and I. Wladaver: "Elements of Descriptive Geometry," Prentice-Hall, Englewood Cliffs, N.J., 1953.

Paré, E. G., R. O. Loving, and I. L. Hill: "Descriptive Geometry," 2d ed., Macmillan, New York, 1952.

Rowe, C. E., and James Dorr McFarland: "Engineering Descriptive Geometry," 3d ed., Van Nostrand, Princeton, N.J., 1961.

Shupe, Hollie W., and Paul E. Machovina: "A Manual of Engineering Geometry and Graphics for Students and Draftsmen," McGraw-Hill, New York, 1956.

Street, W. E.: "Technical Descriptive Geometry," Van Nostrand, Princeton, N.J., 1948.

Warner, Frank M., and Matthew McNeary, "Applied Descriptive Geometry," 5th ed., McGraw-Hill, New York, 1959.
Watts, Earle F., and John T. Rule: "Descriptive Geometry," Prentice-Hall, Englewood Cliffs, N.J., 1946.
Wellman, B. Leighton: "Technical Descriptive Geometry," 2d ed., McGraw-Hill, New York, 1957.

### Drawing-instrument Catalogues

Theo. Alteneder and Sons, Philadelphia.
Eugene Dietzgen Co., Chicago.
Gramercy Guild Group, Inc., Denver, Colo.
Keuffel & Esser Co., Hoboken, N.J.
The Frederick Post Co., Chicago.
V and E Manufacturing Co., Pasadena, Calif.

### Electrical Drafting

Baer, C. J.: "Electrical and Electronics Drawing," 2d ed., McGraw-Hill, New York, 1966.
Bishop, Calvin C., C. T. Gilliam, and Associates: "Electrical Drafting and Design," 3d ed., McGraw-Hill, New York, 1952.
Carini, L. F. B.: "Drafting for Electronics," McGraw-Hill, New York, 1946.
Kocher, S. E.: "Electrical Drafting," International Textbook, Scranton, Pa., 1939.
Van Gieson, D. Walter: "Electrical Drafting," McGraw-Hill, New York, 1945.

### Engineering Drawing

French, Thomas E., and Charles J. Vierck: "Fundamentals of Engineering Drawing," 3d ed., McGraw-Hill, New York, 1972.
French, Thomas E., and Charles J. Vierck: "Graphic Science and Design," 3d ed., McGraw-Hill, New York, 1970.
French, Thomas E., and Charles J. Vierck: "A Manual of Engineering Drawing for Students and Draftsmen," 10th ed., McGraw-Hill, New York, 1966.
Giesecke, F. E., A. Mitchell, and H. C. Spencer: "Technical Drawing," 4th ed., Macmillan, New York, 1959.
Healy, W. L., and A. H. Rau: "Simplified Drafting Practice," Wiley, New York, 1953.
Hoelscher, R. P., and C. H. Springer: "Engineering Drawing and Geometry," 2d ed., Wiley, New York, 1961.
Luzadder, W. J.: "Graphics for Engineers," Prentice-Hall, Englewood Cliffs, N.J., 1957.
Zozzora, Frank: "Engineering Drawing," 2d ed., McGraw-Hill, New York, 1958.

### Engineering-drawing Problem Sheets

Cooper, Charles D., and Paul E. Machovina: "Engineering Drawing Problems," Series III, $11 \times 17$, McGraw-Hill, New York, 1960.
Higbee, F. G., and J. M. Russ: "Engineering Drawing Problems," $8\frac{1}{2} \times 11$ in., Wiley, New York, 1955.
Levens, A. S., and A. E. Edstrom: "Problems in Engineering Drawing," Series V, $10 \times 12$ in., McGraw-Hill, New York, 1960.
Vierck, Charles J., and R. I. Hang: "Fundamental Engineering Drawing Problems," $8\frac{1}{2} \times 11$ in., 2d ed., McGraw-Hill, New York, 1968.
Vierck, Charles J., and R. I. Hang: "Engineering Drawing Problems," $8\frac{1}{2} \times 11$ in., 2d ed., McGraw-Hill, New York, 1968.
Vierck, Charles J., and R. I. Hang: "Graphic Science Problems," $8\frac{1}{2} \times 11$ in., 2d ed., McGraw-Hill, New York, 1963.

### Graphic Solutions

Davis, D. S.: "Empirical Equations and Nomography," McGraw-Hill, New York, 1943.
Douglass, Raymond D., and Douglas P. Adams: "Elements of Nomography," McGraw-Hill, New York, 1947.
Hoelscher, R. P., J. Norman Arnold, and S. H. Pierce: "Graphic Aids in Engineering Computation," McGraw-Hill, New York, 1952.
Kulmann, C. Albert: "Nomographic Charts," McGraw-Hill, New York, 1951.
Levens, A. S.: "Nomography," 2d ed., Wiley, New York, 1959.
Levens, A. S.: "Graphics in Engineering and Science," Wiley, New York, 1954.
Lipka, J.: "Graphical and Mechanical Computation," Wiley, New York, 1918.
Rule, John T., and Earle P. Watts: "Engineering Graphics," McGraw-Hill, New York, 1951.

### Handbooks

A great many handbooks, with tables, formulas, and information, are published for the different branches of the engineering profession, and are useful for ready reference. Handbook formulas and figures should of course be used only with an understanding of the principles upon which they are based. Among the best-known handbooks are the following, alphabetized by title:

"Chemical Engineers' Handbook," 4th ed., Robert H. Perry (ed.), McGraw-Hill, New York, 1963.

"Civil Engineering Handbook," 4th ed., by L. C. Urquhart, Porter, Urquhart, McCreary, and O'Brien, McGraw-Hill, New York, 1959.
"Cutting of Metals," American Society of Mechanical Engineers, New York, 1945.
"Definitions of Occupational Specialties in Engineering," American Society of Mechanical Engineers, New York, 1952.
"Design Data and Methods—Applied Mechanics," American Society of Mechanical Engineers, New York, 1953.
"Dynamics of Automatic Controls," American Society of Mechanical Engineers, New York, 1948.
"Engineering Tables," American Society of Mechanical Engineers, New York, 1956.
"Frequency Response," American Society of Mechanical Engineers, New York, 1956.
"General Engineering Handbook," 2d ed., Charles Edward O'Rourke (ed.), McGraw-Hill, New York, 1940.
"Glossary of Terms in Nuclear Science and Technology," American Society of Mechanical Engineers, New York, 1957.
"Machinery's Handbook," 17th ed., The Industrial Press, New York, 1964.
"Manual on Cutting Metals," American Society of Mechanical Engineers, New York, 1952.
"Manual of Standard Practice for Detailing Reinforced Concrete Structures," American Concrete Institute, Detroit, Mich., 1956.
"Mechanical Engineers' Handbook," 6th ed., Lionel S. Marks (ed.), McGraw-Hill, New York, 1958.
"Mechanical Engineers' Handbook," 12th ed., William Kent (ed.), Wiley, New York, 1950.
"Metals Engineering—Design," 2d ed., American Society of Mechanical Engineers, New York, 1965.
"Metals Properties," American Society of Mechanical Engineers, New York, 1954.
"New American Machinists' Handbook," Rupert Le Grand (ed.), McGraw-Hill, New York, 1955.
"Operation and Flow Process Charts," American Society of Mechanical Engineers, New York, 1949.
"Piping Handbook," 5th ed., by Sabin Crocker, McGraw-Hill, New York, 1967.
"Plant Layout Templates and Models," American Society of Mechanical Engineers, New York, 1949.
"Riveted Joints," American Society of Mechanical Engineers, New York, 1945.
"SAE Handbook," Society of Automotive Engineers, New York, yearly.
"Shock and Vibration Instrumentation," American Society of Mechanical Engineers, New York, 1956.
"Standard Handbook for Electrical Engineers," 10th ed., Donald G. Fink and John M. Carroll (eds.), McGraw-Hill, New York, 1957.
"Steel Castings Handbook," Steel Founders Society of America, Cleveland, Ohio, 1968.
"Steel Construction," American Institute of Steel Construction, New York, 1956.
"Structural Shop Drafting," American Institute of Steel Construction, New York, 1950.
"Tool Engineers' Handbook," 2d ed., Frank W. Wilson (ed.), McGraw-Hill, New York, 1959.

### Illustration

Hoelscher, Randolph Philip, Clifford Harry Springer, and Richard F. Pohle: "Industrial Production Illustration," 2d ed., McGraw-Hill, New York, 1946.
Tharratt, George: "Aircraft Production Illustration," McGraw-Hill, New York, 1944.
Treacy, John: "Production Illustration," Wiley, New York, 1945.

### Kinematics and Machine Design

Albert, C. D.: "Machine Design Drawing Room Problems," 4th ed., Wiley, New York, 1951.
Black, Paul H.: "Machine Design," 3d ed., McGraw-Hill, New York, 1968.
Faires, V. M.: "Design of Machine Elements," 3d ed., Macmillan, New York, 1955.
Guillet, G. L.: "Kinematics of Machines," 5th ed., Wiley, New York, 1950.
Ham, C. W., E. J. Crane, and W. L. Rogers: "Mechanics of Machinery," 4th ed., McGraw-Hill, New York, 1958.
Hyland, P. H., and J. B. Kommers: "Machine Design," 3d ed., McGraw-Hill, New York, 1943.
Keown, Robert McArdle, and Virgil Moring Faires: "Mechanism," 5th ed., McGraw-Hill, New York, 1960.
Prageman, I. H.: "Mechanism," International Textbook, Scranton, Pa., 1943.
Sahag, L. M.: "Kinematics of Machines," Ronald, New York, 1948.
Schwamb, Peter, and others: "Elements of Mechanism," Wiley, New York, 1954.
Sloane, A.: "Engineering Kinematics," Macmillan, New York, 1941.
Doughtie, Venton L., and Alex Vallance: "Design of Machine Members," 4th ed., McGraw-Hill, New York, 1964.

### Lettering

French, Thomas E., and W. D. Turnbull: "Lessons in Lettering," McGraw-Hill, New York, 1950.

Svensen, C. L.: "The Art of Lettering," 2d ed., Van Nostrand, Princeton, N.J., 1947.

### Map and Topographic Drawing

Sloane, Roscoe C., and John M. Montz: "Elements of Topographic Drawing," 2d ed., McGraw-Hill, New York, 1943.

### Perspective

Lawson, Philip J.: "Practical Perspective Drawing," McGraw-Hill, New York, 1943.

Lubschez, B.: "Perspective," Van Nostrand, Princeton, N.J., 1926.

### Piping

See Handbooks, as well as the following:

"Catalogue," Crane Co., Chicago.

"Catalogue," Walworth Company, New York.

Crocker, Sabin: "Piping Handbook," 5th ed., McGraw-Hill, New York, 1967.

Plum, S.: "Plumbing Practice and Design," Wiley, New York, 1943.

### Shop Practice and Tools

See Handbooks, as well as the following:

Benedict, Otis J., Jr.: "Manual of Foundry and Pattern Shop Practice," McGraw-Hill, New York, 1947.

Boston, O. W.: "Metal Processing," 2d ed., Wiley, New York, 1951.

Burghardt, Henry D., Aaron Axelrod, and James Anderson: "Machine Tool Operation," pt. I, 5th ed., 1959; pt. II, 4th ed., 1960, McGraw-Hill, New York.

Campbell, James S., Jr.: "Casting and Forming Processes in Manufacturing," McGraw-Hill, New York, 1950.

Colvin, Fred H., and Lucian L. Haas: "Jigs and Fixtures," 5th ed., McGraw-Hill, New York, 1948.

Doe, E. W.: "Foundry Work," Wiley, New York, 1951.

Hine, Charles R.: "Machine Tools for Engineers," 2d ed., McGraw-Hill, New York, 1959.

Schaller, Gilbert S.: "Engineering Manufacturing Methods," 2d ed., McGraw-Hill, New York, 1959.

### Slide Rule

Cajori, F.: "A History of the Logarithmic Slide Rule," Engineering News Publishing Co., New York, 1910.

Machovina, Paul E.: "A Manual for the Slide Rule," McGraw-Hill, New York, 1950.

### Structural Drawing and Design

See Handbooks, as well as the following:

Bishop, C. T.: "Structural Drafting," Wiley, New York, 1941.

Shedd and Vawter: "Theory of Simple Structures," 2d ed., Wiley, New York, 1941.

Urquhart, L. C., and C. E. O'Rourke: "Design of Steel Structures," McGraw-Hill, New York, 1930.

Winter, George, and Leonard C. Urquhart, C. E. O'Rourke, and A. H. Nelson: "Design of Concrete Structures," 7th ed., McGraw-Hill, New York, 1964.


## American Standards

The ASA is working continually on standardization projects. Of its many publications, the following Standards relating to the subjects in this book are now available. A complete list of American Standards will be sent by the Association on application to its offices, 10 East 40th Street, New York 17.


### American Standard Safety Codes

Code for pressure piping, B31.1—1955

Gas-transmission and distribution piping systems, B31.8—1963

Jacks, B30.1—1952

Mechanical power-transmission apparatus, B15.1—1958

Scheme for identification of piping systems, A13.1—1956

### ASME Boiler and Pressure-vessel Codes

Material specifications, 1965

Power boilers, 1965

Welding qualifications, 1965

### Bolts, Nuts, Rivets, and Screws

Hexagonal- and slotted-head cap screws, square-head setscrews, slotted headless set-screws, B18.6.2—1956

High-strength, high-temperature internal wrenching bolts, B18.8—1958

Large rivets, B18.4—1960

Plow bolts, B18.9—1958

Round-head bolts, B18.5—1959

Slotted- and recessed-head wood screws, B18.6.1—1961

Small solid rivets, B18.1—1955

Socket-head cap screws and socket setscrews, B18.3—1961

Square and hexagonal bolts and nuts and lag bolts, B18.2.1—1965

Track bolts and nuts, B18.10—1963

### Drafting, Charts, and Symbols

Abbreviations for scientific and engineering terms, Z10.1—1941—to be revised as Y1
Abbreviations for use on drawings, Z32.13—1950—to be revised as Y1
Drawings and drafting-room practice, Z14.1—1946
Graphical symbols for heating, ventilating, and air conditioning, Z32.2.4—1949—to be revised as Y32
Graphical symbols for heat-power apparatus, Z32.2.6—1950—to be revised as Y32
Graphical symbols for pipe fittings, valves, and piping, Z32.2.3—1953 to be revised as Y32
Graphical symbols for plumbing, Y32.4—1955
Graphical symbols for railway use, Y32.7—1957
Graphical symbols for welding and instructions for their use, Y32.3—1959
A guide for preparing technical illustrations for publications and projections, 1954
Letter symbols for acoustics, Y10.11—1953
Letter symbols for aeronautical sciences, Y10.7—1954
Letter symbols for chemical engineering, Y10.12—1955
Letter symbols for heat and thermodynamics, including heat flow, Y10.4—1957
Letter symbols for hydraulics, Y10.2—1958
Letter symbols for mechanics of solid bodies, Z10.3—1948—to be revised as Y10
Letter symbols for meteorology, Y10.10—1953
Letter symbols for physics, Z10.6—1948—to be revised as Y10
Letter symbols for radio, Y10.9—1953
Letter symbols for structural analysis, Y10.8—1962
Time series charts, Y15.2—1960

### Electrical Drawing

Electrical Diagrams Y14.15—1960
Graphical Electrical Symbols for Architectural Plans, Y32.9—1962
Graphical Symbols for Electrical Diagrams, Y32.2—1962
Letter Symbols for Electrical Quantities, Z10.5—1949—to be revised as Y10

### Gear Design, Dimensions, and Inspection

Design for fine-pitch worm gearing, B6.9—1962
Find-pitch straight bevel gears, B6.8—1950
Gear nomenclature, B6.10—1954
Gear tolerances and inspection, B6.6—1946
Inspection of fine-pitch gears, B6.11—1956
Letter symbols for gear engineering, B6.5—1954
Nomenclature for gear-tooth wear and failure, B6.12—1964
Spur-gear tooth form, B6.1—1932
System for straight bevel gears, B6.13—1955
Twenty-degree involute fine-pitch system for spur and helical gears, B6.7—1956

### Miscellaneous Standards

Indicating pressure and vacuum gages, B40.1—1953
Lock washers, B27.1—1958
Plain washers, B27.2—1958
Preferred thickness for uncoated, thin, flat metals, B32.1—1959
Shaft coupling, B49.1—1947
Surface roughness, waviness, and lay, B46.1—1962
Woodruff keys, keyslots, and cutters, B17f—1955

### Pipe, Pipe Fittings, and Threads

Brass fittings for flared copper tubes, B16.26—1958
Brass or bronze flanges and flanged fittings—150 and 300 lb, B16.24—1962
Brass or bronze screwed fittings—125 lb, B16.15—1964
Brass or bronze screwed fittings—250 lb, B16.15—1964
Butt-welding ends, B16.25—1964
Cast-brass solder-joint drainage fittings, B16.23—1960
Cast-brass solder-joint fittings, B16.18—1963
CI pipe flanges and flanged fittings, 25-psi, B16b2—1952; class 125, B16.1—1960; class 250, B16.2—1960; 800-lb hydraulic pressure, B16b1—1952; class 300-lb refrigerant, B16.16—1952
CI screwed drainage fittings, B16.12—1953
CI screwed fittings, 125- and 250-lb, B16.4—1963
CI soil pipe and fittings, A40.1—1935
Face-to face dimensions of ferrous flanged and welding end valves, B16.10—1957
Ferrous plugs, bushings, and locknuts with pipe threads, B16.14—1953
Malleable-iron screwed fittings, 150-lb, B16.3—1963
Malleable-iron screwed fittings, 300-lb, B16.3—1963
National plumbing code, A40.8—1955
Nonmetallic gaskets for pipe flanges, B16.21—1962
Pipe threads, B2.1—1960
Ring-joint gaskets and grooves for steel pipe flanges, B16.20—1963
Stainless-steel pipe, B36.19—1957
Steel butt-welding fittings, B16.9—1964
Steel pipe flanges and flanged fittings, 150-, 300-, 400-, 600-, 900-, 1,500-, and 2,500-lb, B16.5—1961
Steel-socket welding fittings, B16.11—1952
Threaded cast-iron pipe for drainage, vent, and waste services, A40.5—1943

Wrought-copper and wrought-bronze solder-joint fittings, B16.22—1963
Wrought-steel and wrought-iron pipe, B36.10—1959

**Small Tools and Machine-tool Elements**

Acme screw threads, B1.5—1952
Buttress screw threads, B1.9—1953
Chucks and chuck jaws, B5.8—1959
Hose-coupling screw threads, B33.1—1947
Involute serrations, B5.15—1960
Involute splines, B5.15—1960
Jig bushings, B5.6—1962
Knurling, B5.30—1958
Machine tapers, B5.10—1963
Milling cutters, B5.3—1960
Nomenclature, definitions, and letter symbols for screw threads, B1.7—1953
Preferred limits and fits for cylindrical parts, B4.1—1955
Reamers, B5.14—1959
Screw thread gages and gaging, B1.2—1951
Stub Acme screw threads, B1.8—1952
Taps—cut and ground threads, B5.4—1959
T slots and their bolts, nuts, tongues, and cutters, B5.1—1949
Twist drills, B5.12—1958
Unified and American screw threads standard, B1.1—1960

**Standards under Development**

Abbreviations—for use in text
Drafting standards manual—size and format, line conventions, lettering and sectioning, projections, pictorial drawings, dimensions and notes, castings, die castings, forgings, gears, splines and serrations, helical and flat springs, hydraulic and pneumatic diagrams, metal stampings, plastics, schematic wiring and diagrams, screw threads, structural drafting
Fluid meters—theory and application
Gears—inspection of coarse-pitch spur and helical gears
Screws and screw threads—microscope-objective threads, surveying-instrument mounting threads, national miniature screw threads, "trial" class 5 interference-fit thread
Small tools—inserted-blade milling-cutter bodies, driving and spindle ends for portable electric tools
Symbols—miscellaneous
Washers—precision

## Publications of National Societies

The national engineering organizations publish a wide variety of manuals, standards, handbooks, and pamphlets. Information concerning these publications is available directly from the societies. The following is a selected list of American organizations:

Aerospace Industries Assn. of America, 1725 Desales St. NW, Washington, D.C. 20036
Air Conditioning and Refrigeration Institute, 1815 North Fort Myer Drive, Arlington, Va.
American Association for the Advancement of Science (AAAS), 1515 Massachusetts Ave. NW, Washington 5, D.C.
American Association of Engineers (AAE), 8 South Michigan Ave., Chicago 3, Ill.
American Association of Petroleum Geologists, Inc. (AAPG), P.O. Box 979, Tulsa 1, Okla.
American Association of University Professors, 1785 Massachusetts Ave. NW, Washington 6, D.C.
American Ceramic Society (ACerS), 4055 North High St., Columbus 14, Ohio
American Chemical Society (ACS), ACS Bldg., 1155 16th St. NW, Washington 6, D.C.
American Concrete Institute (ACI), P.O. Box 4754, Redford Station, Detroit 19, Mich.
American Gas Association (AGA), 605 Third Ave., New York, N.Y. 10016
American Inst. of Aeronautics and Astronautics (AIAA) 1290 Ave. of the Americas, New York, N.Y. 10019
American Institute of Chemical Engineers (AIChE), 345 East 47th St., New York, N.Y. 10017
American Institute of Consulting Engineers (AICE), 345 East 47th St., New York, N.Y. 10017
American Institute of Mining and Metallurgical, and Petroleum Engineers, Inc. (AIME), 345 East 47th St., New York, N.Y. 10017
American Institute of Steel Construction, Inc. (AISC), 101 Park Ave., New York 17, N.Y.
American Mining Congress, Ring Bldg., Washington 6, D.C.
American Petroleum Institute (API), 1271 Ave. of the Americas, New York, N.Y. 10020
American Society for Engineering Education (ASEE), University of Illinois, Urbana, Ill.
American Society for Metals (ASM), Metals Park, Ohio
American Society for Quality Control (ASQC), 161 West Wisconsin Ave., Milwaukee 3, Wis.
American Society for Testing and Materials (ASTM), 1916 Race St., Philadelphia 3, Pa.
American Society of Agricultural Engineers (ASAE), 420 Main St., St. Joseph, Mich.

American Society of Civil Engineers (ASCE), 345 East 47th St., New York, N.Y. 10017

American Society of Heating, Refrigerating, and Air-conditioning Engineers (ASHRAE), 345 East 47th St., New York, N.Y. 10017

American Society of Lubrication Engineers, 838 Busse Highway, Park Ridge, Ill. 60068

American Society of Mechanical Engineers (ASME), 345 East 47th St., New York, N.Y. 10017

American Society of Photogrammetry, 44 Leesburg Pike, Falls Church, Va.

American Society of Safety Engineers, 5 N. Wabash Ave., Chicago 2, Ill.

American Society of Tool and Manufacturing Engineers, 10700 Puritan Ave., Detroit 38, Mich.

American Standards Association (ASA), 10 East 40th St., New York, N.Y.

American Welding Society (AWS), 345 East 47th St., New York, N.Y. 10017

Asphalt Institute, Asphalt Institute Bldg., College Park, Md.

Association of Iron and Steel Engineers (AISE), 1010 Empire Bldg., Pittsburgh 22, Pa.

Concrete Pipe Association, 228 North LaSalle St., Chicago 1, Ill.

Electrochemical Society, Inc., The, 30 East 42nd St., New York, N.Y. 10017

Federation of American Scientists (FAS), 1700 K St. NW, Washington, D.C. 20036

Highway Research Board, 2101 Constitution Ave. NW, Washington 25, D.C.

Illuminating Engineering Society, 345 East 47th St., New York, N.Y. 10017

Industrial Management Society (IMS), 330 South Wells St., Chicago 6, Ill.

Industrial Research Institute, Inc., 100 Park Ave., New York 17, N.Y.

Institute of Electrical and Electronics Engineers (IEEE), 345 East 47th St., New York, N.Y. 10017

Mining and Metallurgical Society of America (MMSA), 11 Broadway, New York 4, N.Y.

National Aeronautic Association (NAA), 1025 Connecticut Ave. NW, Washington 6, D.C.

National Institute of Ceramic Engineers, The, 4055 North High St., Columbus 14, Ohio

National Society of Professional Engineers (NSPE), 2029 K St. NW, Washington 6, D.C.

Society of Automotive Engineers (SAE), 485 Lexington Ave., New York, N.Y. 10017

# APPENDIX

# LETTERING

## Outlined Commercial Gothic.

In Chap. 4, the so-called "Gothic" letter was considered only as a single-stroke letter. For sizes larger than say $^5/_8$ in., or for boldface letters, it is drawn in outline and filled in solid. For a given size, this letter is readable at a greater distance than any other style; hence it is used where legibility is the principal requirement. The stems may be one-tenth to one-fifth of the height, and care must be taken to keep them uniform in width at every point on the letter. In inking a penciled outline, keep the *outside* of the ink line on the pencil line (Fig. A-1) or the letter will be heavier than expected.

The general order and direction of penciling large commercial Gothic letters is similar to that for the single-stroke letter but with two strokes made in place of one, as shown in typical examples in Fig. A-2. Free ends, such as on *C*, *G*, and *S*, are cut off perpendicular to the stem. The stiffness of plain letters is sometimes relieved by

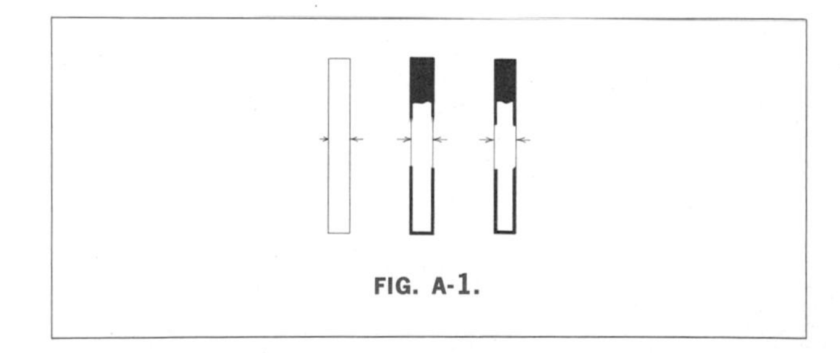

**FIG. A-1.**

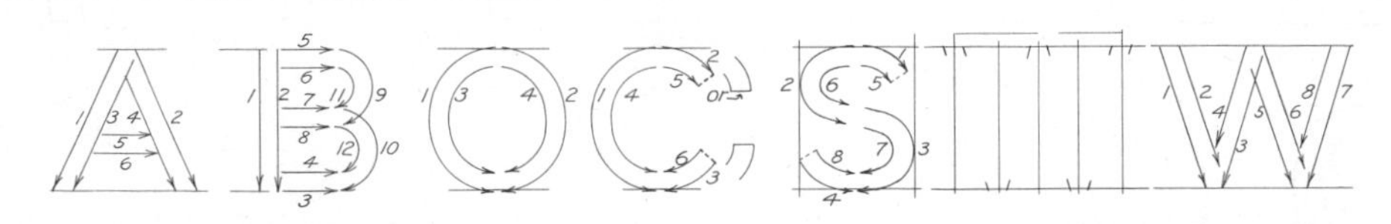

**FIG. A-2.** Typical construction for large commercial Gothic.

ABCDEFGHIJ
KLMNOPQRS
TUVWXYZ &

**FIG. A-3.** Compressed commercial Gothic.

**FIG. A-4.** Large commercial Gothic construction.

finishing the ends with a slight spur, as in Fig. A-3. The complete alphabet in outline, with stems one-sixth of the height, is given in Fig. A-4. The same scale of widths may be used for drawing lighter-face letters. Figure A-3 illustrates a commercial Gothic alphabet compressed to two-thirds the normal width. In this figure the stems are drawn one-seventh of the height, but the scale is given in sixths, as in Fig. A-4.

**The Roman Letter.**

The Roman letter is the parent of all the styles that are in use today. Although there are many variations of it, the three general forms are: (1) early or classic, (2) Renaissance, and (3) Modern. The first two are very similar in effect, and the general term "Old Roman" is used for both.

The Roman letter is composed of two weights of lines, corresponding to the downstroke and the upstroke of a broad reed pen, with which it was originally written. It is an inexcusable fault to shade a Roman letter on the wrong stroke.

*Rule for shading.* All horizontal strokes are light. All vertical strokes are heavy except in *M*, *N*, and *U*. To determine the heavy stroke in letters containing slanting sides, trace the shape of the letter from left to right in one stroke and note which lines were made downward. Figure A-5 is an Old Roman alphabet with the width of the body stroke one-tenth of the height of the letter and

**FIG. A-5.** Old Roman capitals.

the light lines slightly over one-half this width. For inscriptions and titles, capitals are generally used, but sometimes lower case is needed; the examples given in Fig. A-6 are drawn with waist line six-tenths high and the width of the stems one-twelfth of capital height.

Old Roman is the architect's general-purpose letter. A single-stroke adaptation (Fig. A-7) is generally used on architectural drawings.

### Modern Roman.

Civil engineers in particular must be familiar with Modern Roman, as it is the standard letter for titles of finished maps and the names of civil divisions, such as countries and cities. It is a difficult letter to draw and can be mastered only by careful attention to details. The heavy or "body" strokes are one-sixth to one-eighth the height of the letter. Those in Fig. A-8 are one-seventh the height of the letter. A paper scale made by dividing the height into seven parts will aid in penciling.

Modern lower case (Fig. A-9) is used on maps for names of towns and villages. Notice the difference in the serifs of Figs. A-9 and A-6.

The order and direction of strokes used in drawing Modern Roman letters are illustrated in typical letters in Fig. A-10. Serifs on the ends of the strokes extend one space on each side and are joined to the main stroke by small fillets. Curved letters are flattened slightly on their diagonals. A Roman-letter title is shown in Fig. A-11.

abcdefghijklmn
opqrstuvwxyz

**FIG. A-6.** Old Roman lower.

ABCDEFGHIJKLMNOPQRS
TUVWXYZ& 1234567890
abcdefghijklmnopqrstuvwxyz
*Compressed Italic for Limited Space*
Sans-serif for Speed & Simplicity

**FIG. A-7.** Single stroke Roman and italic.

FIG. A-8. Modern Roman capitals.

FIG. A-9. Modern Roman lower case.

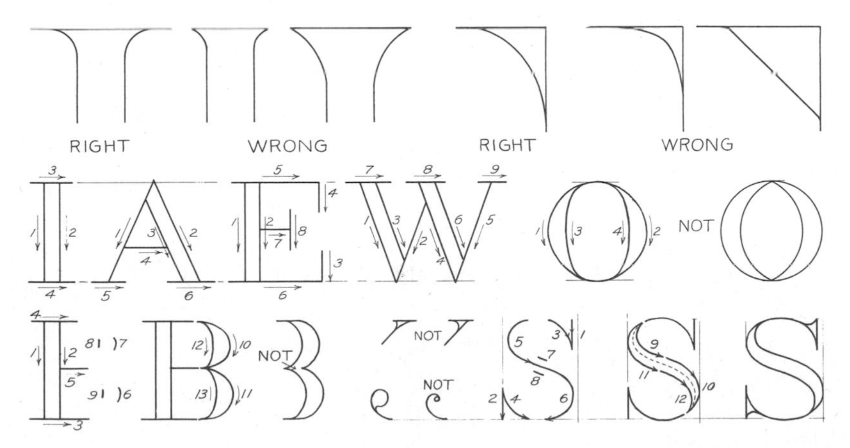


FIG. A-10. Modern Roman construction.

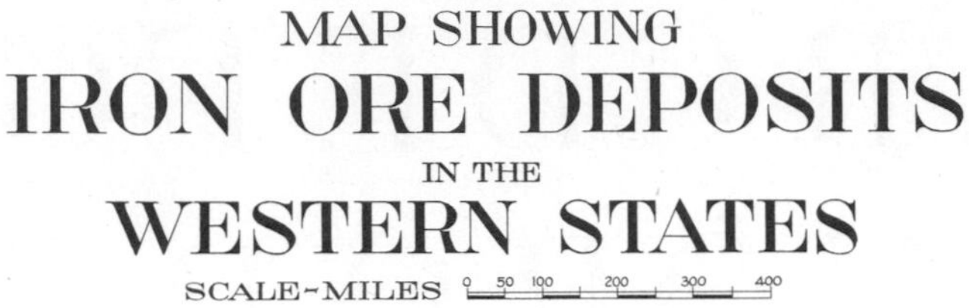

MAP SHOWING
IRON ORE DEPOSITS
IN THE
WESTERN STATES
SCALE–MILES 0 50 100 200 300 400

**FIG. A-11.** A Roman-letter title.

The Roman letter may be extended or compressed, as shown in Fig. A-12. For extended or compressed letters, a scale for widths would be longer or shorter than the normal scale. The compressed letters in Fig. A-12 are made with a scale three-fourths the height divided into sevenths.

### Inclined Roman.

Inclined letters are used for water features on maps. An alphabet of inclined Roman made to the same proportions as the vertical in Fig. A-8 is shown in Fig. A-13. The slope may be 65° to 75°. Those shown are inclined 2 to 5. The lower-case letters in this figure are known as stump letters. For small sizes, lines are made with one stroke of a fine flexible pen; for larger sizes they are drawn and filled in.

### The Proportional Method.

Because of the varying widths of Roman letters, it is sometimes difficult to space a word or line to a given

EXTENDED ROMAN
BCGHJKLPQSUVW

COMPRESSED ROMAN-BHKTWG

**FIG. A-12.** Modern Roman, extended and compressed.

*A B C D E F G H I J K L M N*
*O P Q R S T U V W X Y Z & a b*
*c d e f g h i j k l m n o p q r s t u v or v w or w*
*x y or y z 1 2 3 4 5 6 7 8 9 0*

**FIG. A-13.** Inclined Roman and stump letters.

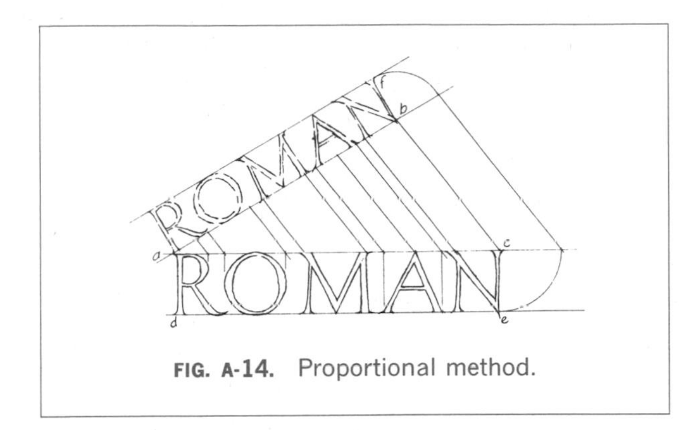


**FIG. A-14.** Proportional method.

length by counting letters. Figure A-14 illustrates the method of spacing by the principle of similar triangles. Suppose it is necessary to put the word "ROMAN" on the line to the length *ab*. Draw line *ac* from *a* at any angle, say 30°, and a second line *de* parallel to it; then sketch the word in this space, starting at *a* and spacing each letter with reference to the one before it, allowing the word to end where it will. Connect the end of the last letter, at *c*, with *b*, and draw lines parallel to *cb* from each letter, thus dividing *ab* proportionately. Obtain the height *bf* from *ce* by the construction shown. Then sketch the word in its final position.

## The Greek Alphabet.

While Greek is not a required subject in most engineering curricula, the engineer often uses letters of the Greek alphabet, both capitals and lower case, as symbols and reference letters. He should therefore be able to draw them readily and to read them without hesitation.

There is variety in Greek alphabets, just as there is in Roman alphabets. The alphabet given in Fig. A-15 is clearly legible; it has accented and unaccented strokes in the capitals which follow closely the rules for shading Roman letters. The lower case has good historical precedent in form, shading, and comparative size.

When Greek letters are used in equations and formulas, the letters are made in a single stroke by simply following the letter shape without any attempt at shading. For display purposes and when shading is desirable for legibility, the letters are outlined and filled in by the method described for outlined commercial Gothic.

**FIG. A-15.** The Greek alphabet.

# APPENDIX B
# THE SLIDE RULE

The slide rule is a graphic instrument for making mathematical calculations. Because the instrument is *graphic*, and also because its use is common in all engineering work, the following pictorial guide to its use is given.

The slide rule is based on the well-known mathematical relationships of logarithms of numbers. For example, the equation for the multiplication of two factors, $A \times B = C$, when written logarithmically becomes: $\log A + \log B = \log C$. Thus, to multiply, logarithms are *added* and, conversely, to divide, logarithms are subtracted. Slide-rule scales are arranged in such a way that the addition and subtraction of logarithms is accomplished *graphically*.

It is beyond our scope here to give a complete discussion of slide-rule use. The pictorial examples are intended only for quick, easy visual reference. Complete discussions on the theory and use of slide rules are given in manuals prepared especially for the slide rule (see Bibliography).

Because slide-rule settings of values and the answers obtained from slide rules depend upon reading a graphic scale, results are not absolutely accurate. However, the instrument is unrivaled as an inexpensive, portable, and efficient device for calculation whenever absolute accuracy is not essential.

### Multiplication of Two Values (C and D Scales)

$A \times B = R$

Set $A$ on D scale with index of C scale (Fig. B-1). Set $B$ on C scale with hairline of glass slide. Read answer $R$ on D scale at hairline. Example shown:

$1.375 \times 1.962 = 2.697$

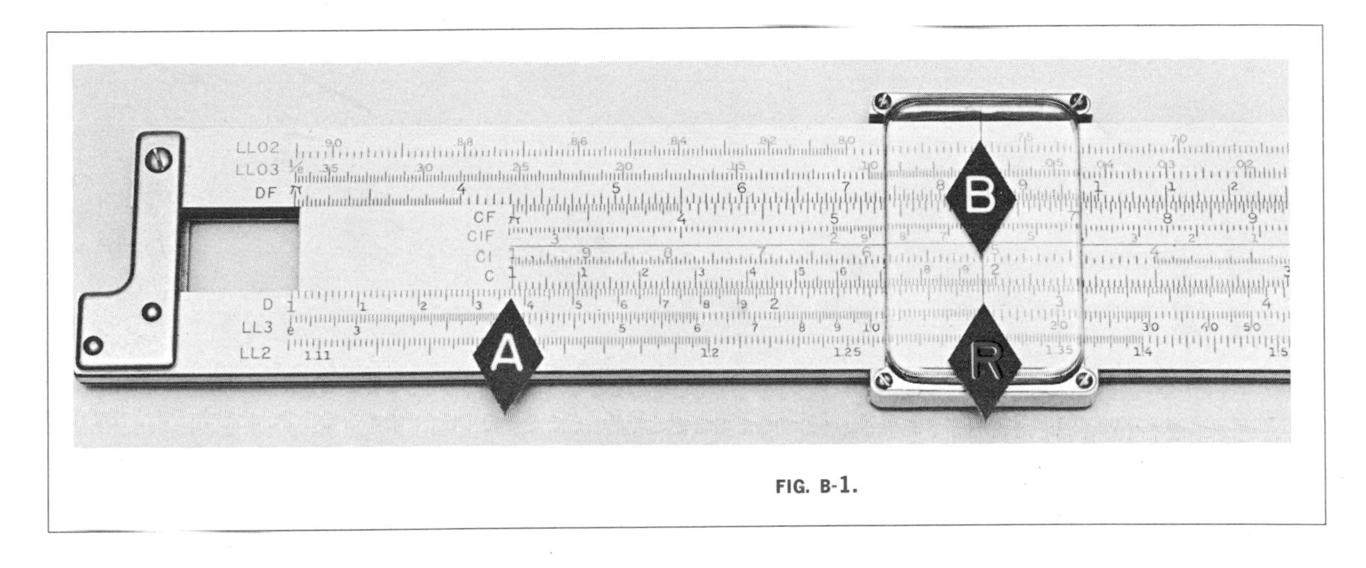


FIG. B-1.

### Division of Two Values (C and D Scales)

$\frac{A}{B} = C$

Set $A$ on D scale with hairline (Fig. B-2). Set $B$ on C scale under hairline. Read answer $R$ on D scale at index of C scale. Example shown:

$\frac{2.70}{1.962} = 1.375$

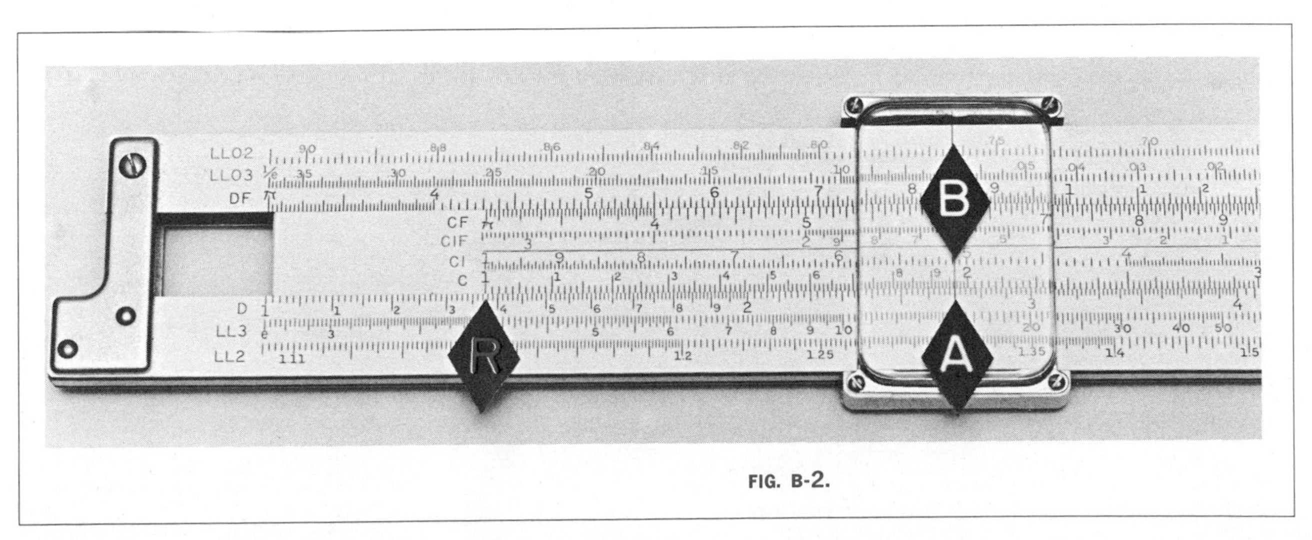


FIG. B-2.

## Proportion (C and D Scales)

$$\frac{A}{B} = \frac{C}{D} = \frac{E}{F}$$

Set $A$ on D scale and $B$ on C scale opposite each other by using the hairline (Fig. B-3). Then any other combination in the same proportion is had by moving the hairline and reading both values on C and D scales. Example shown:

$$\frac{A}{B} = \frac{2.0}{1.5}, \frac{C}{D} = \frac{1.4}{1.05}, \frac{E}{F} = \frac{2.7}{2.025}$$

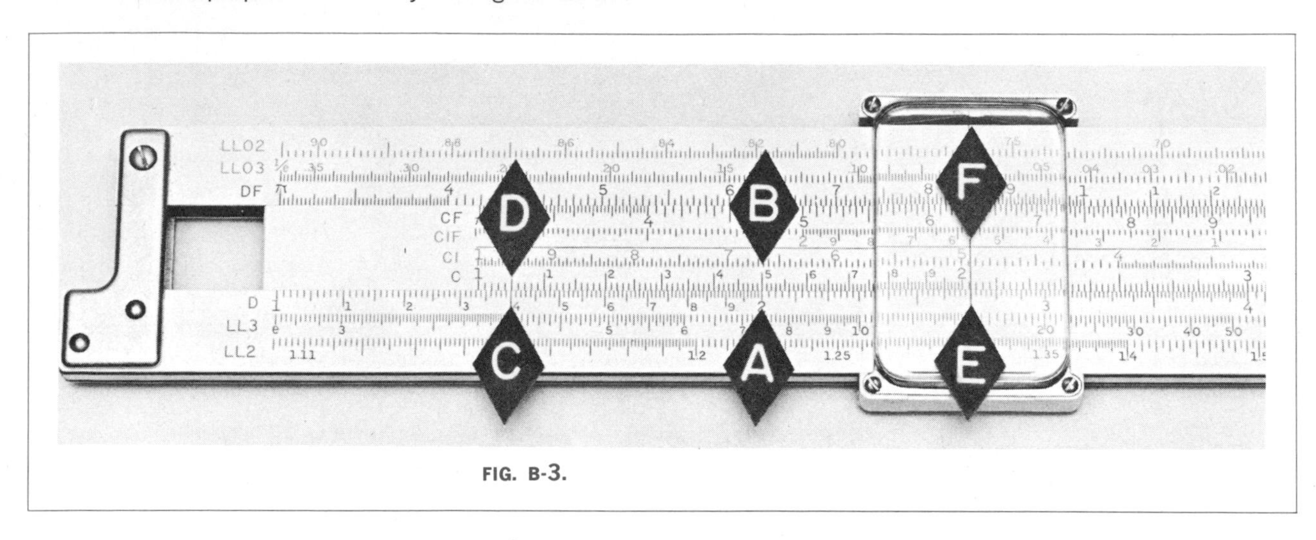


FIG. B-3.

## Successive Multiplication of One Value by Several Others (C and D Scales)

$A \times B = R$
$A \times C = S$
$A \times D = T$

Set $A$ on D scale with index of C scale (Fig. B-4). Then set $B$, $C$, and $D$ with hairline on C scale and read answers on D scale at $R$, $S$, and $T$ under hairline. Example shown:

$2.4 \times 1.8 = 4.32$ $(R)$
$2.4 \times 2.5 = 6.0$ $(S)$
$2.4 \times 3.0 = 7.2$ $(T)$

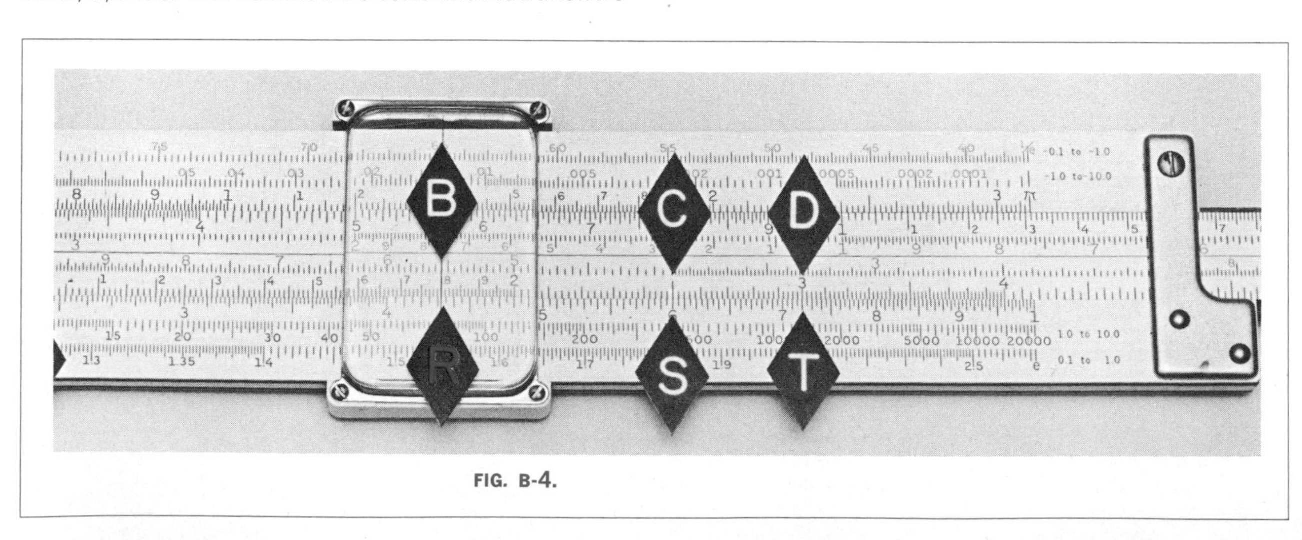


FIG. B-4.

## Successive Division of One Value by Several Others (D and CI Scales)

$$\frac{A}{B} = R$$

$$\frac{A}{C} = S$$

$$\frac{A}{D} = T$$

Set $A$ on D scale with index (right) of CI scale (Fig. B-5). The CI scale is the same as the C scale, but inverted. Then set $B$, $C$, and $D$ with hairline on CI scale and read answers under hairline on D scale. Example shown:

$$\frac{7.0}{2} = 3.5\ (R) \frac{7.0}{3.0} 2.33\ (S)$$

$$\frac{7.0}{4.0} = 1.75\ (T)$$

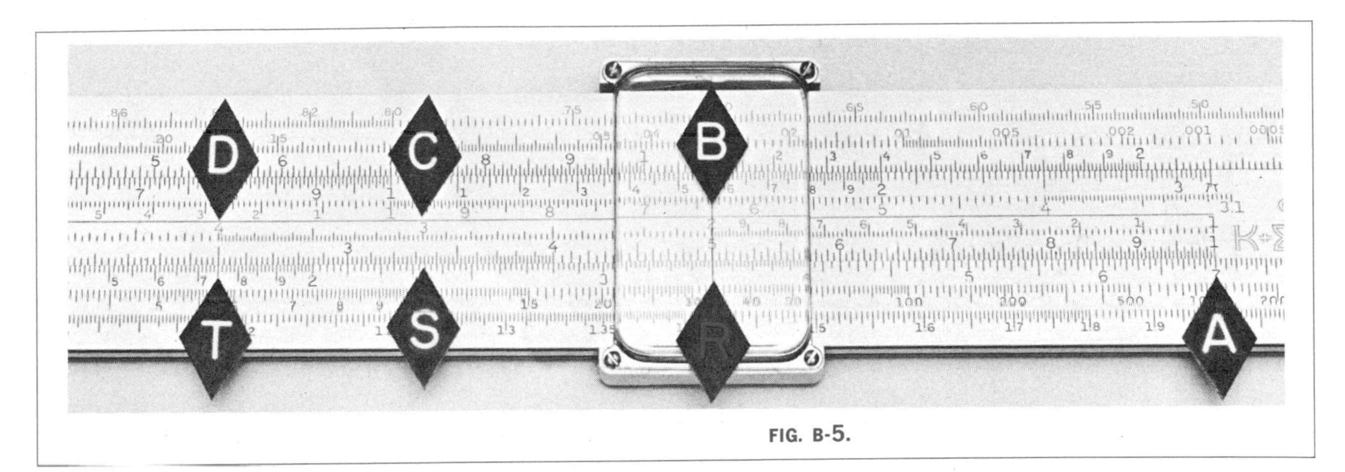


FIG. B-5.

## Multiplication by Using the CF and DF Scales

The "Folded" scales, CF and DF, are used whenever a value on the C scale falls beyond the D scale. In the example shown, value $B$ on the C scale cannot be set. Therefore, using the CF and DF scales:

$A \times C = R$

Set $A$ on D scale with index of C scale (Fig. B-6). CF and DF scales *have moved in exactly the same proportion as* C scale relative to D scale. Therefore, when value $B$ falls off end of C scale, set $C$ on CF scale and read answer on DF scale under hairline. Example shown:

$3.1 \times 4.5 = 13.95$

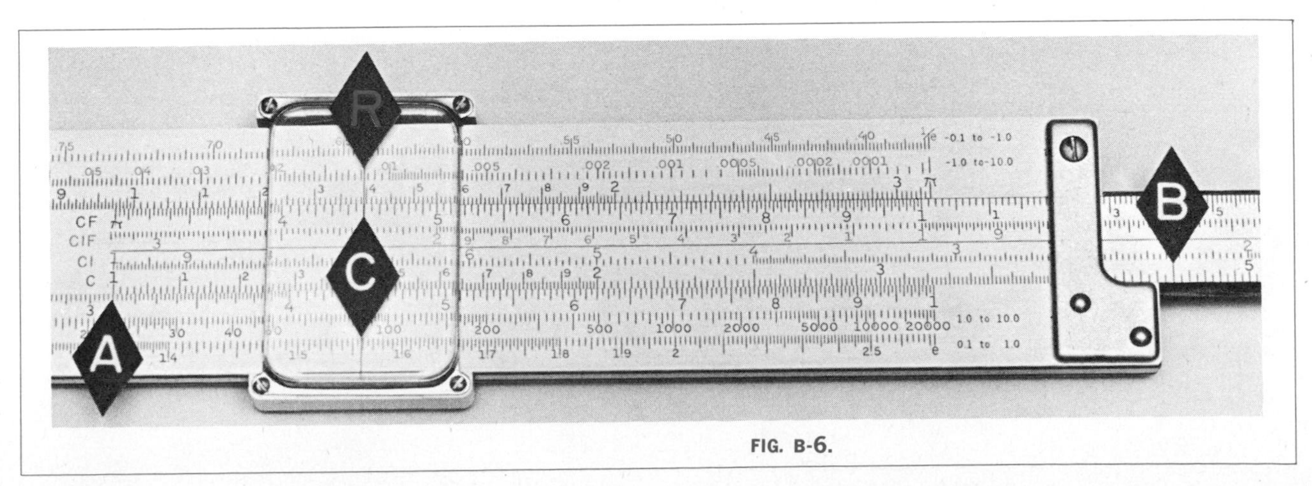


FIG. B-6.

## Division by Using the CIF Scale

The "folded and inverted" scale, CIF, is a folded scale to augment the CI scale. In successive division, using the CI scale, some values will be off the end of the scale. The CIF scale can then be used:

$\frac{A}{B} = R$ or $\frac{A}{C} = S$

Set $A$ on D scale with hairline (Fig. B-7). Set index of CIF scale under hairline. Then $\frac{A}{B} = R$, and $\frac{A}{C} = S$. Answer is read on D scale. Example shown:

$\frac{2.0}{8.0} = 0.25\ (R)$

$\frac{2.0}{5.0} = 0.40\ (S)$

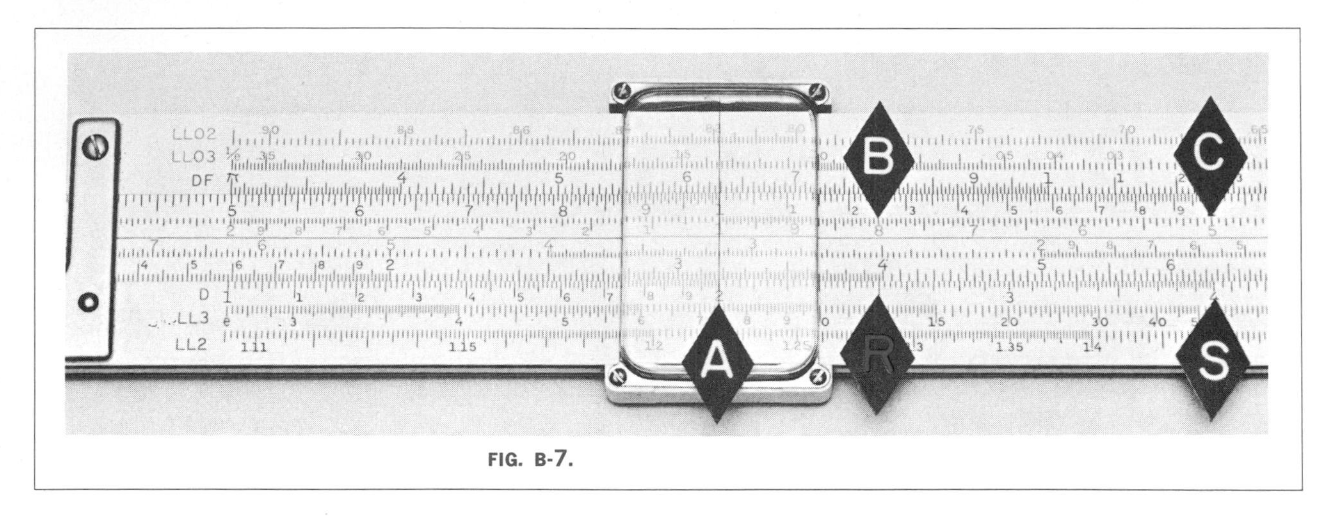


FIG. B-7.

## Multiplication of Three Values (D and CI Scales)

$A \times B \times C = S$

Set $A$ on D scale, using hairline (Fig. B-8). Set $B$ on CI scale under hairline. The answer of this operation $(A \times B)$ can then be read at index of C scale at $R$ on D scale. The third value $C$, now set on C scale, gives answer of $A \times B \times C$ at $S$ on D scale. Example shown:

$2.5 \times 2 \times 3.6 = 18.0$

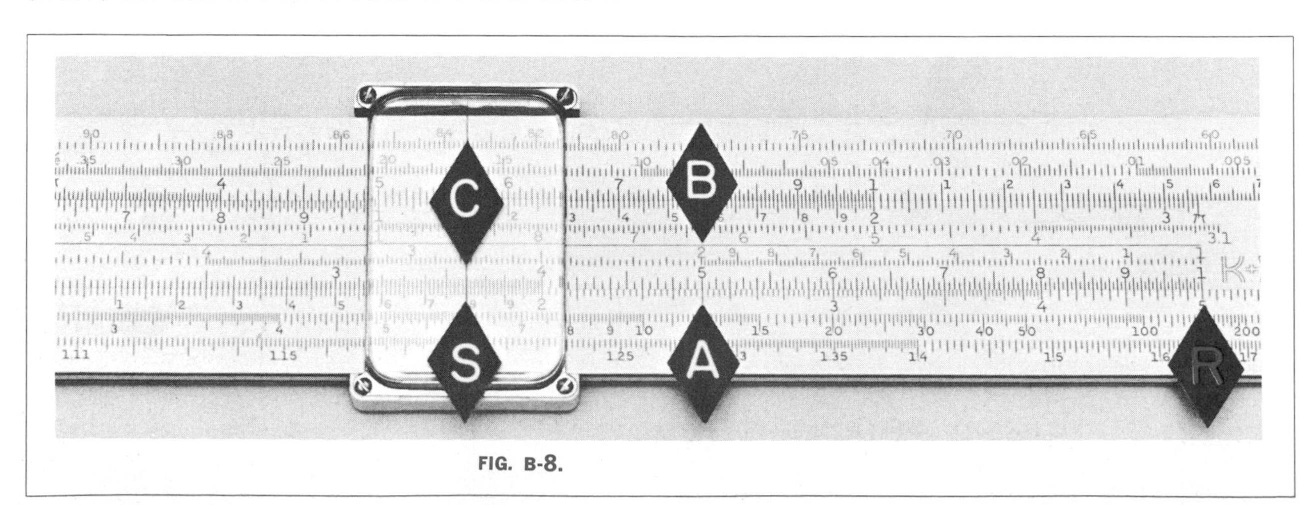


FIG. B-8.

## Combined Multiplication and Division (D and CI Scales)

$$\frac{A \times B}{C} = S$$

Set $A$ on D scale with hairline (Fig. B-9). Set $B$ on CI scale under hairline. The answer of $A \times B$ is at index of CI scale at $R$ on D scale. Then the third value $C$, on CI scale, gives answer of $\frac{A \times B}{C}$ at $S$ on D scale. Example shown:

$$\frac{1.5 \times 2.0}{15.0} = 0.20$$

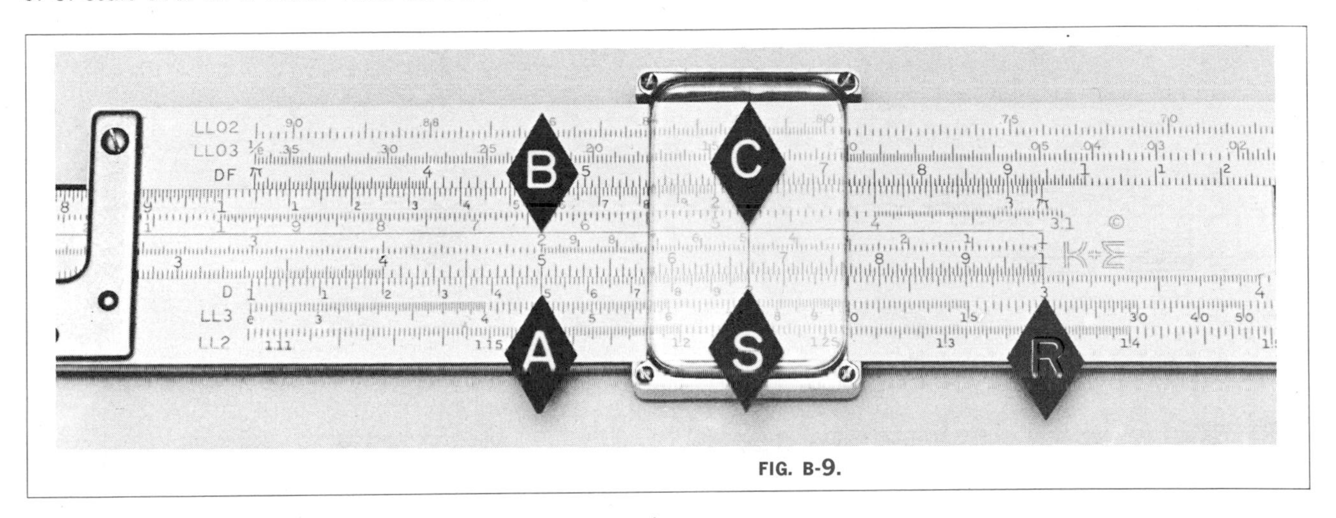


FIG. B-9.

## Combined Multiplication and Division (C, D, and CI Scales)

$$\frac{A}{B \times C} = S$$

Set $A$ on D scale with hairline (Fig. B-10). Set $B$ on C scale under hairline. The answer of $\frac{A}{B}$ is at index of CI scale at $R$ on D scale. Then the third value $C$, on the CI scale, gives the answer of $\frac{A}{B \times C}$ at $S$ on D scale. Example shown:

$$\frac{3.0}{2.0 \times 7.0} = 0.214$$

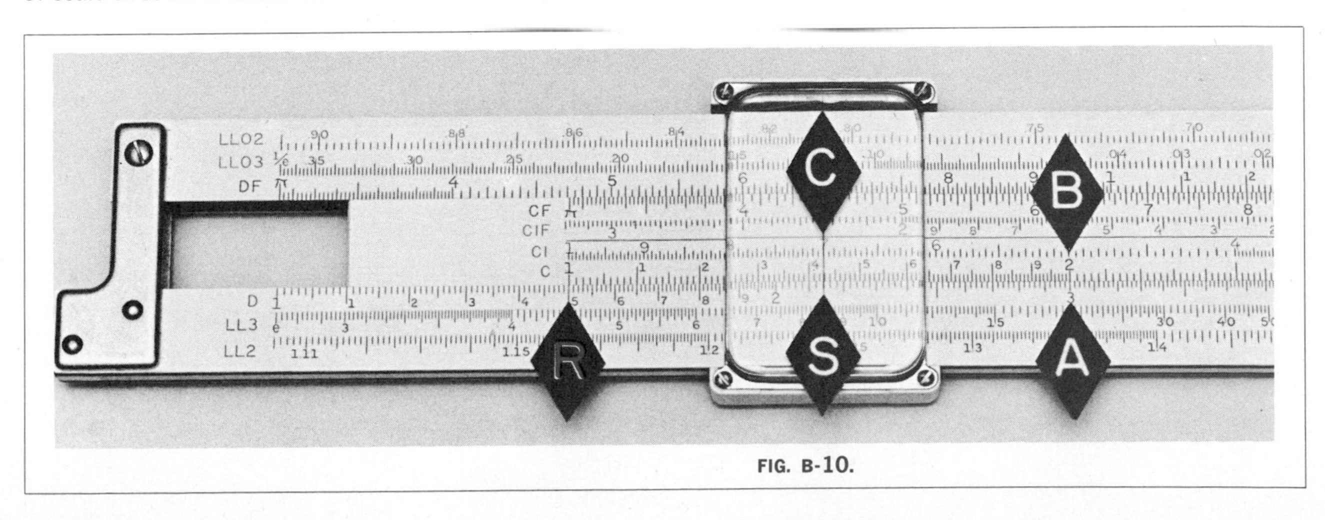


FIG. B-10.

## Combined Multiplication and Division (C, D, CF, DF, and CIF Scales)

$$\frac{A}{B \times C} = S$$

Sometimes in combined multiplication and division, values needed will be inaccessible on a scale. To illustrate, set $A$ on D scale with hairline (Fig. B-11). Set $B$ on C scale under hairline. Then the answer of $\frac{A}{B}$ is at index of C scale at $R$ on D scale. If $C$ is a value that cannot be set on CI scale, set $C$ on CIF scale and read answer at $S$ on DF scale. To prove this, note that CF scale has moved in exactly the same relationship to DF scale as C scale has moved in relation to D scale. Example shown:

$$\frac{15.0}{4.0 \times 6.0} = 0.625$$

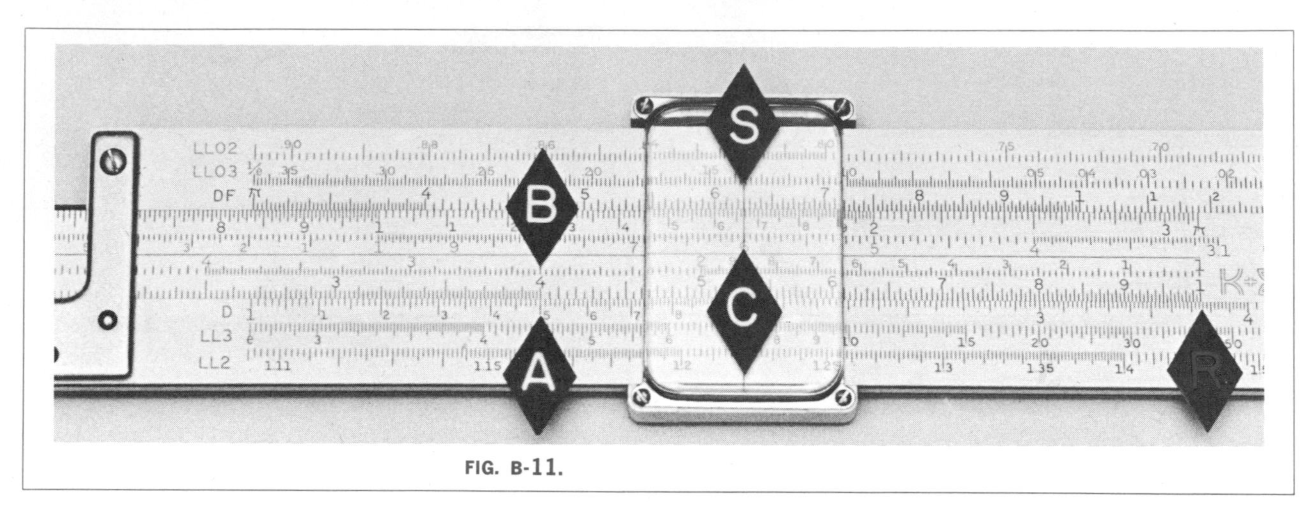


FIG. B-11.

## Squares and Square Roots (D and A Scales)

$R^2 = A$ and $\sqrt{A} = R$

The A scale is plotted as the square of the D scale.

To determine $R^2 = A$, set $R$ on D scale and read square on A scale (Fig. B-12). Example shown:

$$2.0^2 = 4.0$$

To determine $\sqrt{A} = R$, set $A$ on A scale and read square root on D scale. Example shown:

$$\sqrt{4.0} = 2.0$$

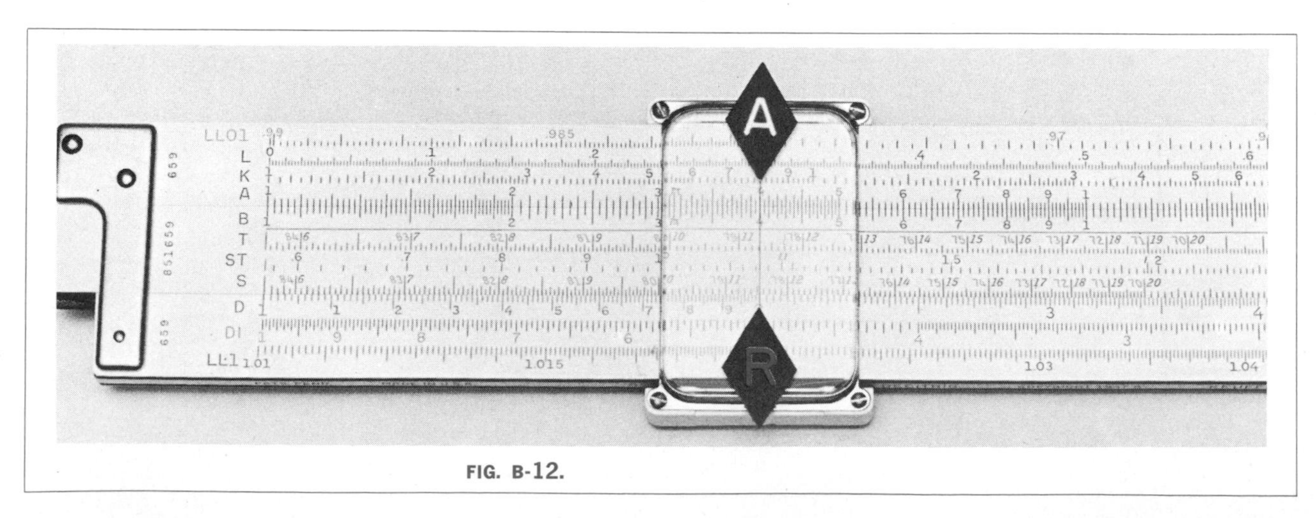


FIG. B-12.

## Cubes and Cube Roots (D and K Scales)

$A^3 = R$ and $\sqrt[3]{R} = A$

The K scale is plotted as the cube of the D scale.

To determine $A^3 = R$, set $A$ on D scale and read cube on K scale (Fig. B-13). Example shown:

$2.0^3 = 8.0$

To determine $\sqrt[3]{R} = A$, set $R$ on K scale and read cube root on D scale. Example shown:

$\sqrt[3]{8.0} = 2.0$

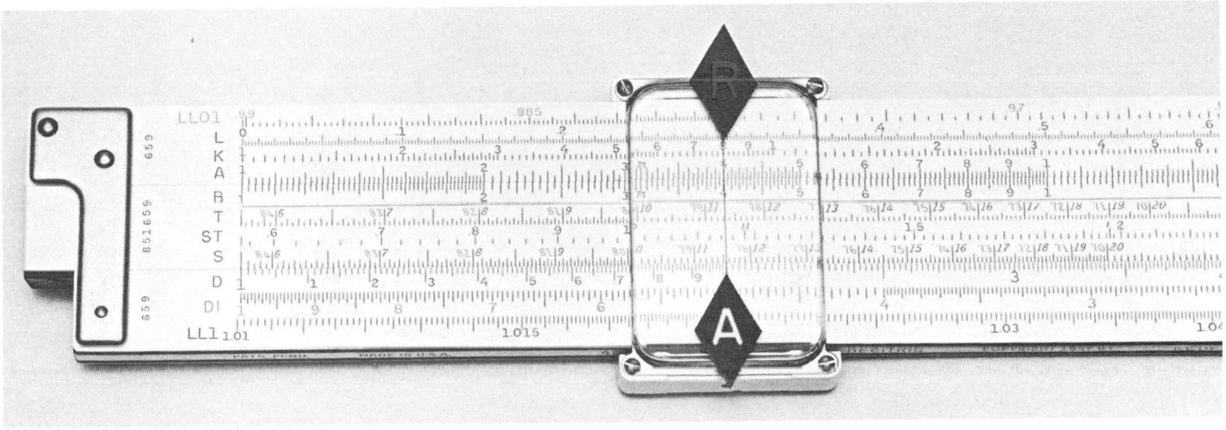

FIG. B-13.

## Use of the A and B Scales

The A and B scales can be used to multiply and divide, and are useful for performing these operations *after* a square has been obtained, *or* if the square root is needed *after* multiplication or division.

$A^2 \times B = S$

Set $A$ on D scale (Fig. B-14). $A^2$ can then be read on the A scale at $R$. Set $B$ on B scale with hairline. The answer $S$ is under hairline on A scale. Example shown:

$2.0^2 \times 4.0 = 16.0$

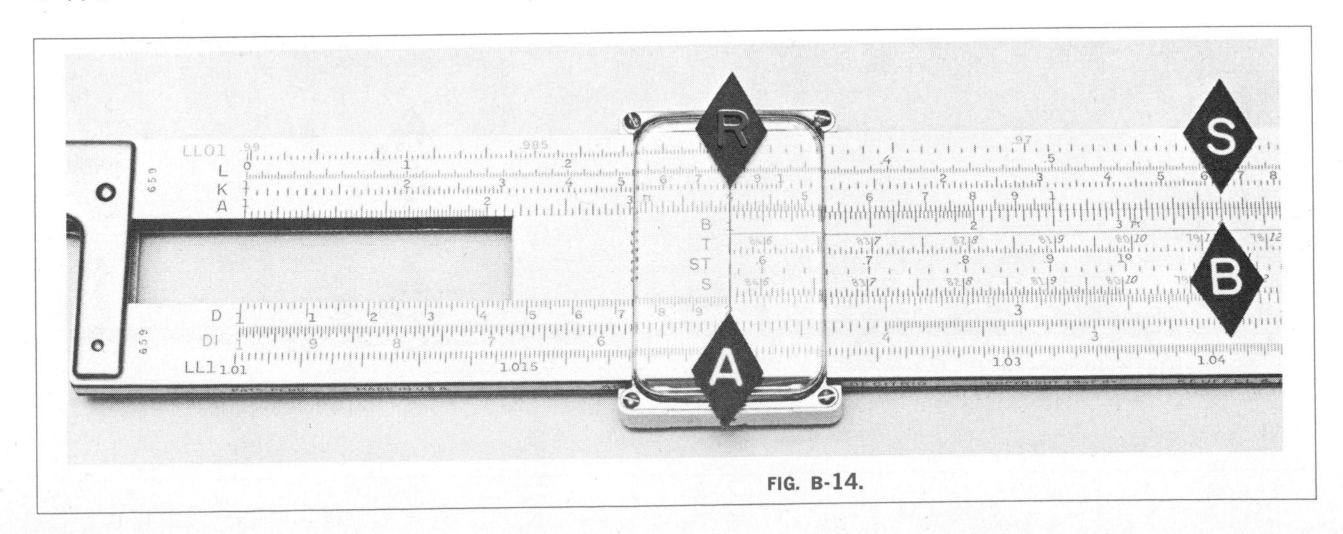


FIG. B-14.

## Mathematical Value of Sine or Tangent (ST Scale)

sin $A$ (or tan $A$) $= R$

The values of sines and tangents from zero degrees up to about $5\frac{1}{2}$ degrees are identical for the accuracy with which the scales can be read.

Index the ST scale with the D scale (Fig. B-15). Set angle ($A$) on ST scale with hairline and read mathematical value ($R$) on D scale under hairline. Example shown:

sin (or tan) 1.5° = 0.0262

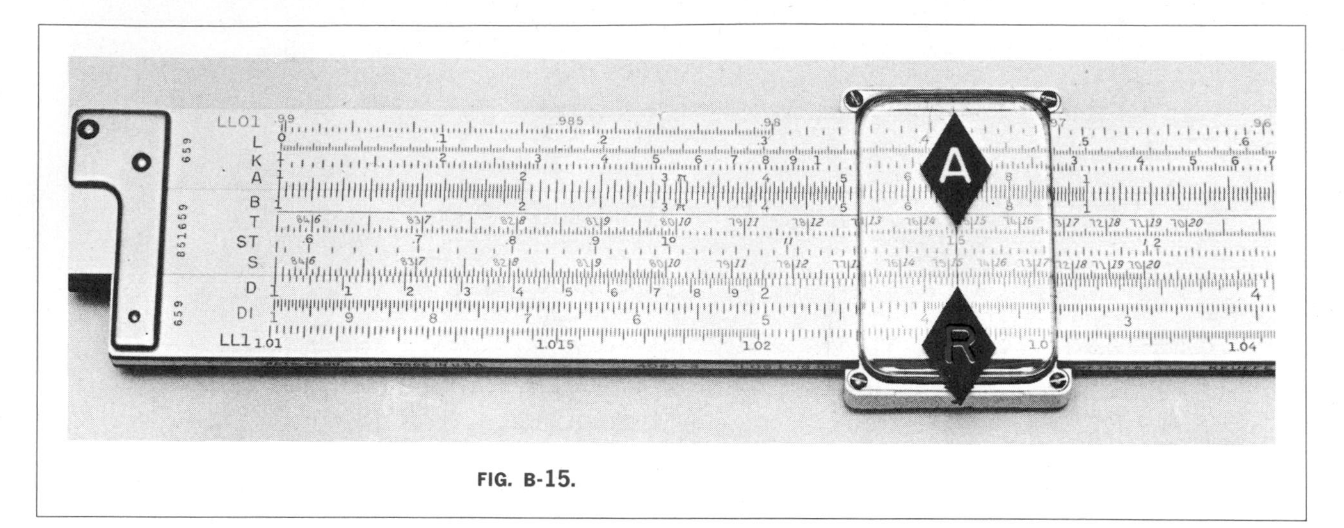


FIG. B-15.

## Mathematical Value of Sine (S Scale)

sin $A = R$

Index S scale with D scale (Fig. B-16). Set angle (degrees) ($A$) on S scale with hairline and read mathematical value ($R$) on D scale under hairline. The cosine (degrees) is printed on most rules on the S scale in red. Example shown:

sin 15° = 0.2588
cos 75° = 0.2588

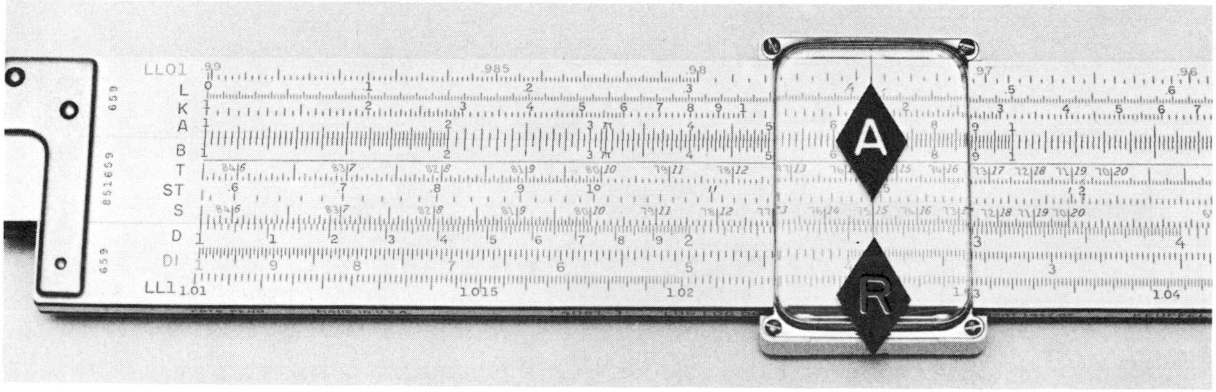

FIG. B-16.

## Angle of Sine, Corresponding to Mathematical Value (S Scale)

$A$ (sin) $= R$ (angle in degrees)

Index S scale with D scale (Fig. B-17). Set $A$ on D scale with hairline. Read angle in degrees ($R$) on S scale under hairline. Example shown:

sin $= 0.176 = 10.16°$

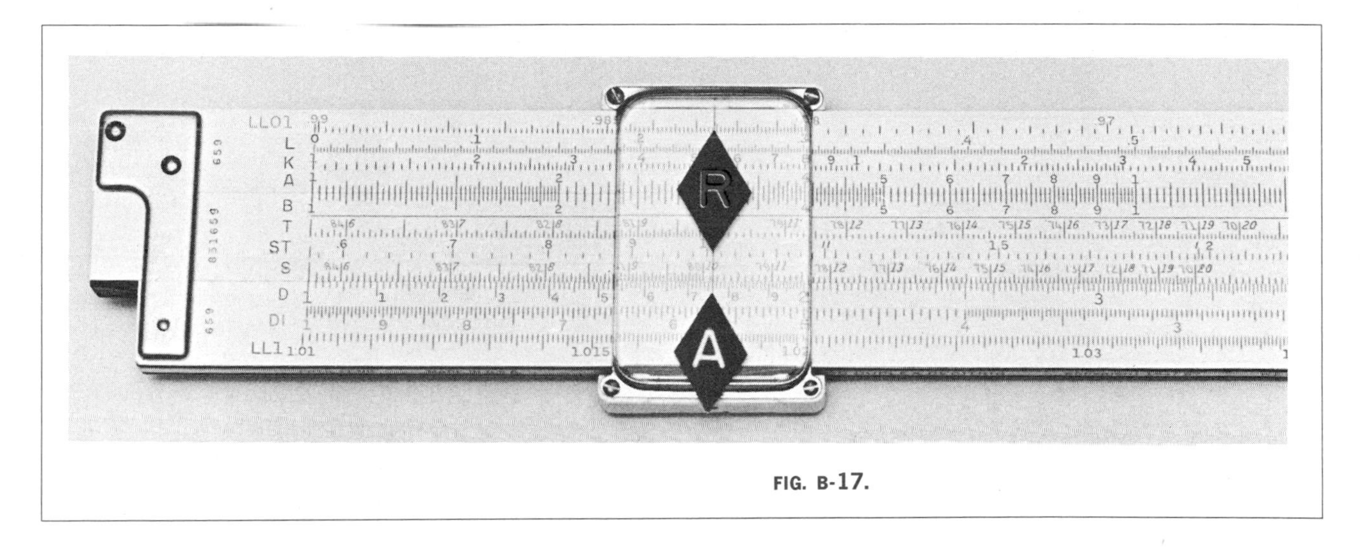


FIG. B-17.

## Mathematical Value of Tangent (T Scale) up to 45°

$\tan A = R$

Index T scale with D scale (Fig. B-18). Set $A$, angle in degrees, on T scale with hairline. Read mathematical value of tangent ($R$) on D scale under hairline. Example shown:

$\tan 10° = 0.176$

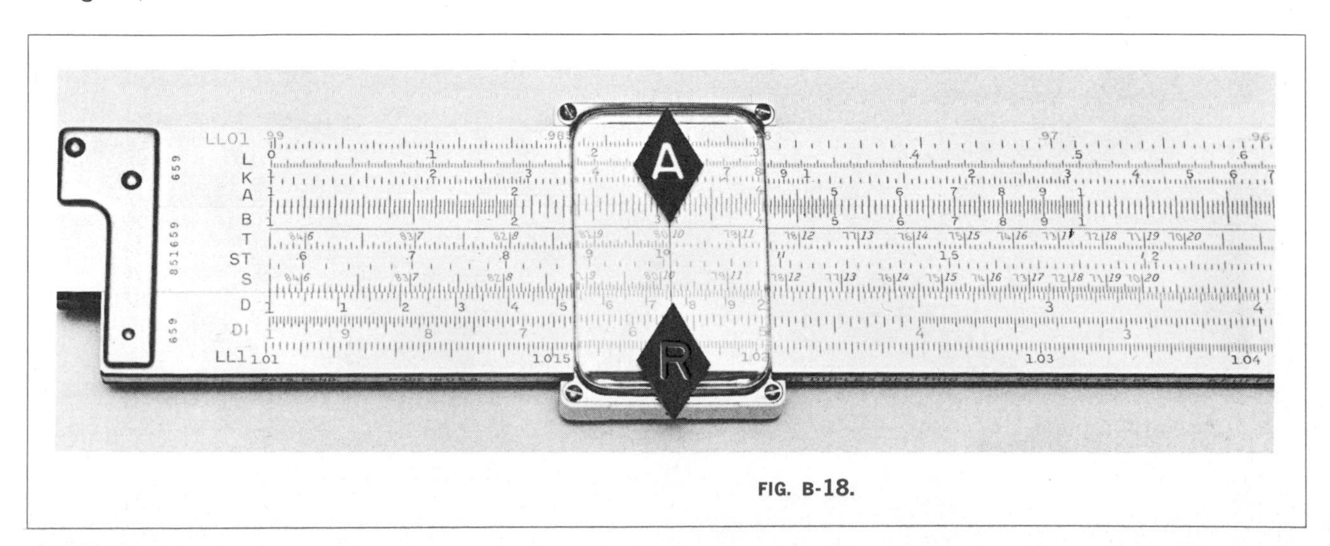


FIG. B-18.

## Mathematical Value of Tangent (T Scale) from 45° up

tan $A = R$

Index T scale with DI scale—inverted D scale (Fig. B-19). On most rules, the T scale above 45° and the DI scale are printed in red. Set angle in degrees ($A$) on T scale with hairline. Read mathematical value ($R$) on DI scale under hairline. Example shown:

tan 75° = 3.74

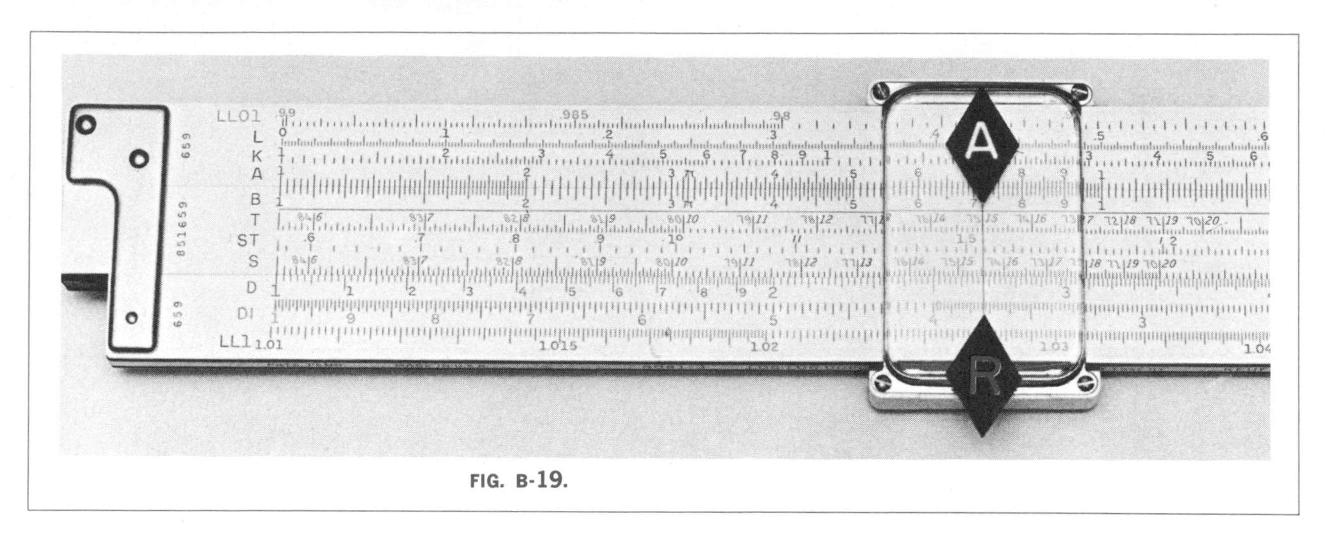


FIG. B-19.

## Multiplication with the Trigonometric Scales (ST Scale)

$A$ sin $B = R$

Set $A$ on D scale with index of ST scale (Fig. B-20). Set angle in degrees ($B$), corresponding to mathematical value of a sine or tangent, with hairline on ST scale. Read answer at $R$ on D scale. Example shown:

2 sin (tan) 1° = 0.0350

The S and T scales are employed for multiplication in the same manner.

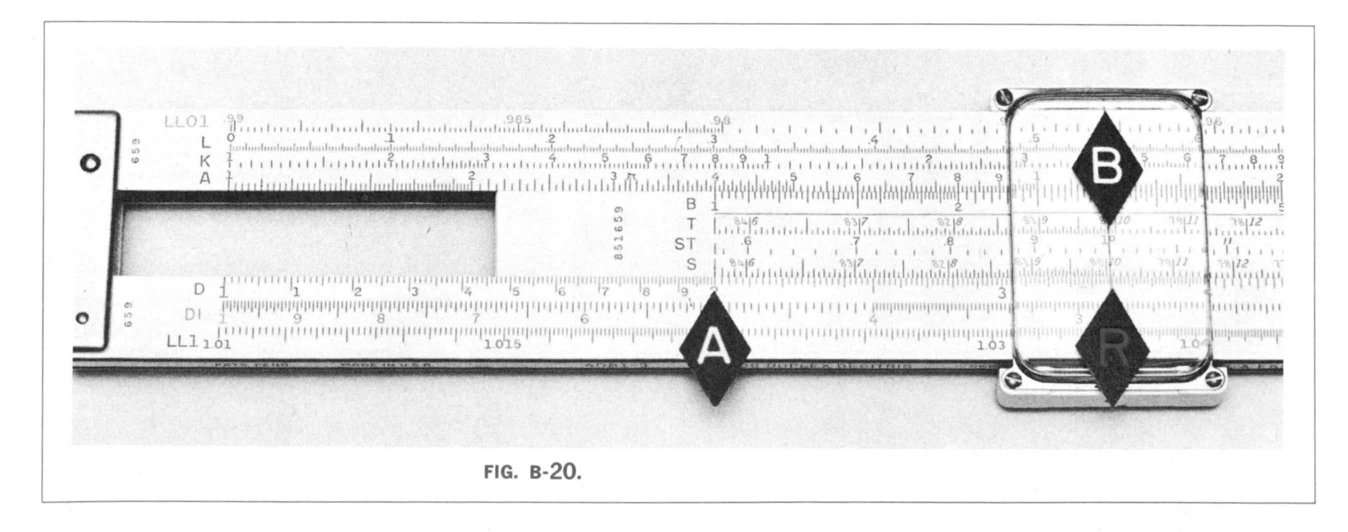


FIG. B-20.

## Division with the Trigonometric Scales (S Scale)

$$\frac{A}{\sin B} = R$$

Set $A$ on D scale with hairline (Fig. B-21). Set angle in degrees ($B$), corresponding to mathematical value of the sine, on S scale, under hairline. Read answer $R$, at index of S scale on D scale. Example shown:

$$\frac{3.24}{\sin 11^\circ} = 17.0$$

The ST and T scales are employed for division in the same manner.

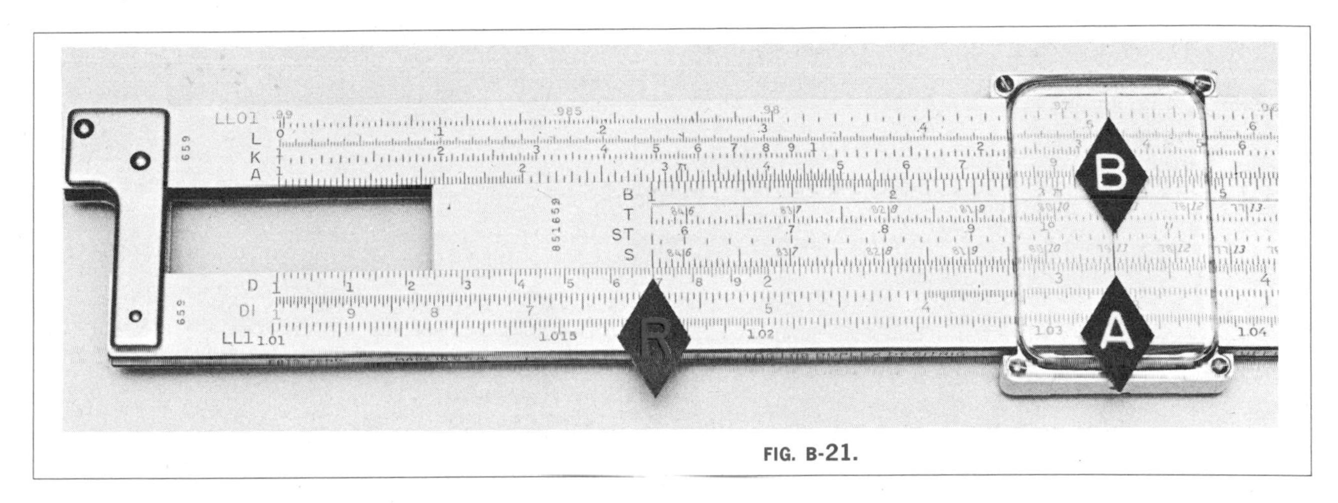


FIG. B-21.

## Logarithms (L Scale)

$$\log A = R$$

Set $A$ with hairline on D scale (Fig. B-22). Read logarithm, mantissa only, ($R$) under hairline on L scale. Example shown:

$$\log 2.10 = 0.322$$

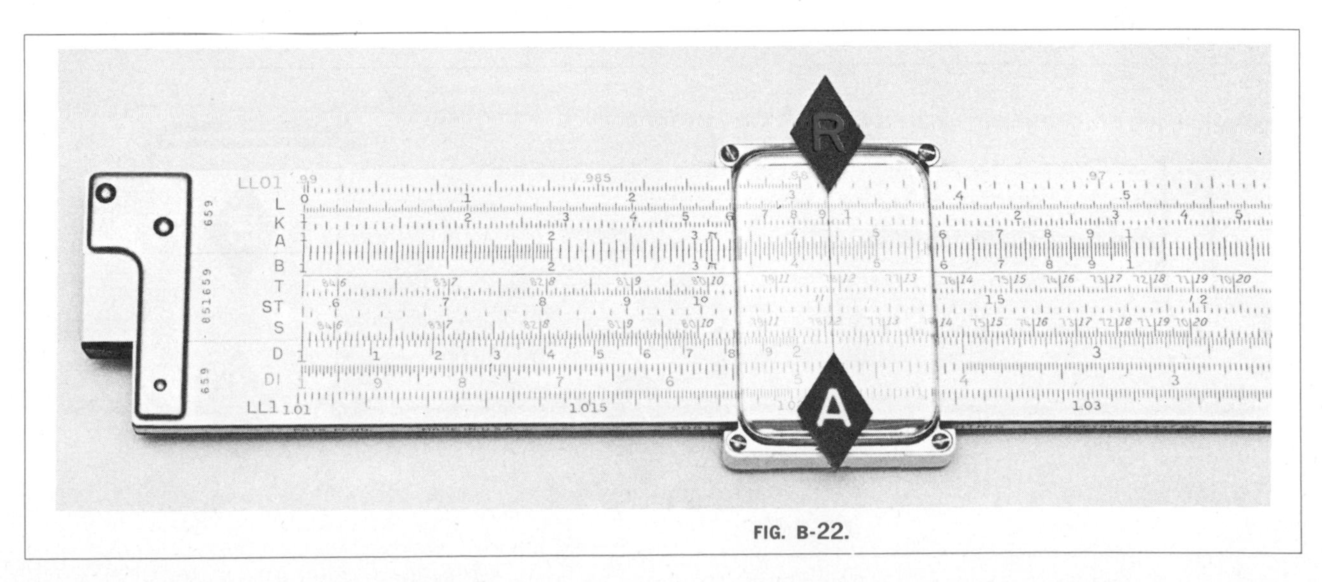


FIG. B-22.

## Antilogarithms (L Scale)

$A = \log R$

Set logarithms, mantissa only, ($A$) with hairline on L scale (Fig. B-23). Read antilog ($R$) under hairline on D scale. Example shown:

$0.330 = \log 2.14$

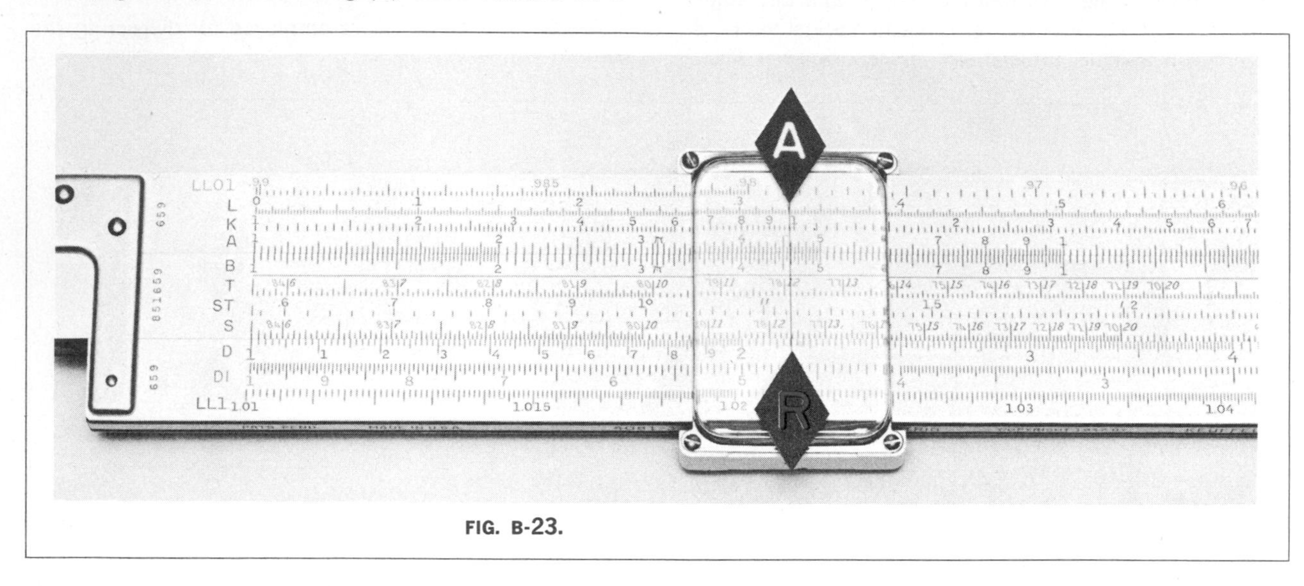


FIG. B-23.

## Powers (LL Scales)

$A^B = R$

Set $A$ with hairline on LL scale (depending upon the value) and match index of C scale under hairline (Fig. B-24). Set power $B$ with hairline on C scale. Read answer on LL scale at $R$. Example shown:

$6.0^{1.75} = 23.0$

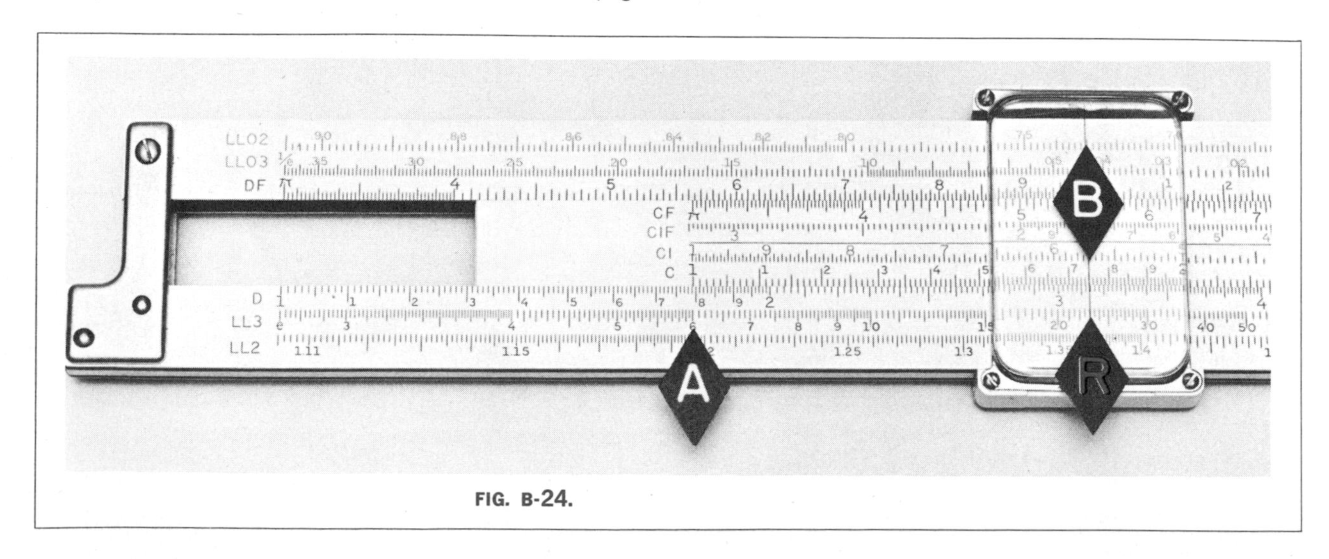


FIG. B-24.

## The LL Scales

The LL scales run continuously, for example, LL2 ends at $e$ at the right index, and LL3 begins with $e$ at the left index (Fig. B-25). Depending on the value set, and the power, the answer *may* be on the next scale in sequence.

$A^B = R$

Set as in previous example, but if left index of the C scale is used and value of the power is then beyond range of scale, change to right index and answer will appear on *next* LL scale. Example shown:

$2.0^4 = 16.0$

(The value 2.0 is set on LL2 and 16.0 is read on LL3.) The LL01, LL02, and LL03 scales are for numbers less than 1.0. Note that these scales are inverted.

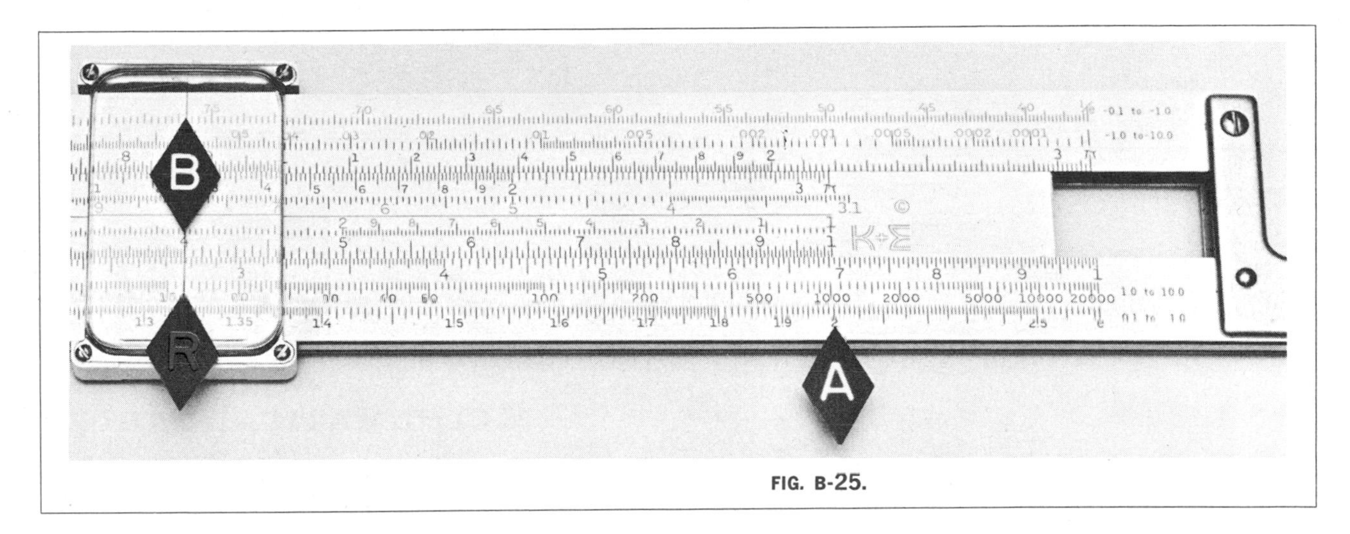


FIG. B-25.

## Roots (LL Scales)

$(A)^{1/B} = R$

Set $A$ with hairline on LL scale (depending upon the value) and set root ($B$), exponential denominator, on C scale under hairline (Fig. B-26). Read answer on LL scale at $R$. Example shown:

$49.0^{1/2} = 7.0$

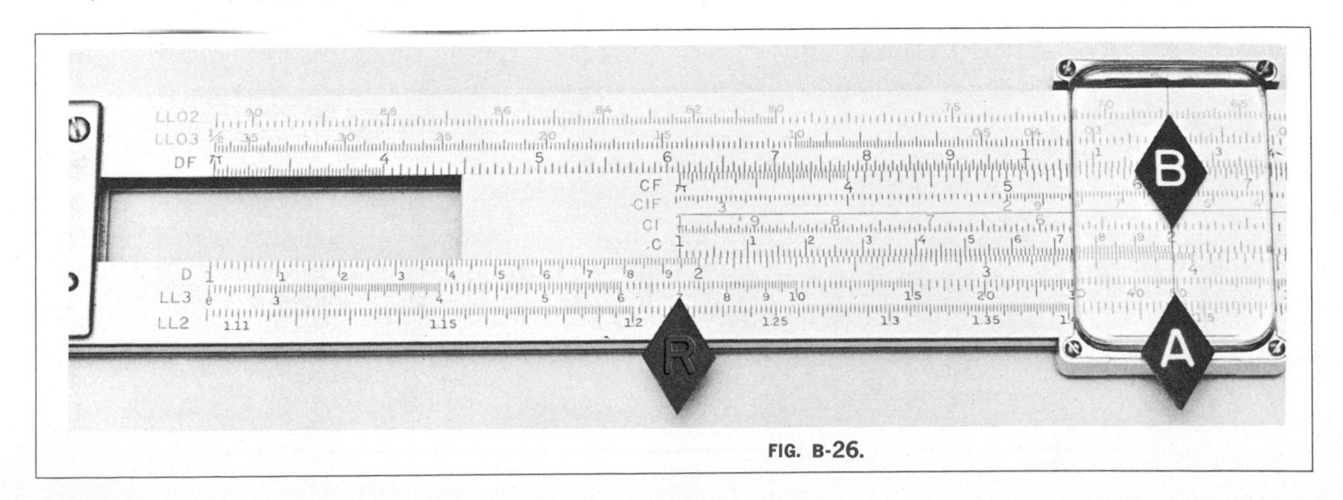


FIG. B-26.

# APPENDIX C

# MATHEMATICAL TABLES



## Selected Mathematical Values

| | | | | | |
|---|---|---|---|---|---|
| $\pi =$ | 3.141593 | $\sqrt{\pi} =$ | 1.772454 | $\sqrt{2} =$ | 1.414214 |
| $2\pi =$ | 6.283185 | $\sqrt[3]{\pi} =$ | 1.464592 | $\sqrt{3} =$ | 1.732055 |
| $3\pi =$ | 9.424778 | $\log_e \pi^* =$ | 1.144730 | $\frac{\pi}{\sqrt{g}} =$ | 0.55399 |
| $4\pi =$ | 12.566371 | $\frac{1}{\pi} =$ | 0.318310 | $\frac{1}{\sqrt{g}} =$ | 0.17634 |
| $\frac{\pi}{2} =$ | 1.570796 | $\frac{1}{\pi^2} =$ | 0.101321 | $\sqrt{2g} =$ | 8.01998 |
| $\frac{\pi}{3} =$ | 1.047198 | $\frac{1}{\sqrt{\pi}} =$ | 0.564190 | $\frac{1}{2g} =$ | 0.01555 |
| $\frac{\pi}{4} =$ | 0.785398 | $\log_{10} \pi =$ | 0.497150 | $g^2 =$ | 1034.266 |
| $\pi^2 =$ | 9.869604 | $\frac{1}{e} =$ | 0.367879 | $2g =$ | 64.32 |
| $\pi^3 =$ | 31.006277 | | | $g =$ | 32.16 |

*$e = 2.718282$

## Decimal Equivalents of Inch Fractions

| Fraction | Equiv. | Fraction | Equiv. | Fraction | Equiv. | Fraction | Equiv. |
|---|---|---|---|---|---|---|---|
| 1/64 | 0.015625 | 17/64 | 0.265625 | 33/64 | 0.515625 | 49/64 | 0.765625 |
| 1/32 | 0.03125 | 9/32 | 0.28125 | 17/32 | 0.53125 | 25/32 | 0.78125 |
| 3/64 | 0.046875 | 19/64 | 0.296875 | 35/64 | 0.546875 | 51/64 | 0.796875 |
| 1/16 | 0.0625 | 5/16 | 0.3125 | 9/16 | 0.5625 | 13/16 | 0.8125 |
| 5/64 | 0.078125 | 21/64 | 0.328125 | 37/64 | 0.578125 | 53/64 | 0.828125 |
| 3/32 | 0.09375 | 11/32 | 0.34375 | 19/32 | 0.59375 | 27/32 | 0.84375 |
| 7/64 | 0.109375 | 23/64 | 0.359375 | 39/64 | 0.609375 | 55/64 | 0.859375 |
| 1/8 | 0.1250 | 3/8 | 0.3750 | 5/8 | 0.6250 | 7/8 | 0.8750 |
| 9/64 | 0.140625 | 25/64 | 0.390625 | 41/64 | 0.640625 | 57/64 | 0.890625 |
| 5/32 | 0.15625 | 13/32 | 0.40625 | 21/32 | 0.65625 | 29/32 | 0.90625 |
| 11/64 | 0.171875 | 27/64 | 0.421875 | 43/64 | 0.671875 | 59/64 | 0.921875 |
| 3/16 | 0.1875 | 7/16 | 0.4375 | 11/16 | 0.6875 | 15/16 | 0.9375 |
| 13/64 | 0.203125 | 29/64 | 0.453125 | 45/64 | 0.703125 | 61/64 | 0.953125 |
| 7/32 | 0.21875 | 15/32 | 0.46875 | 23/32 | 0.71875 | 31/32 | 0.96875 |
| 15/64 | 0.234375 | 31/64 | 0.484375 | 47/64 | 0.734375 | 63/64 | 0.984375 |
| 1/4 | 0.2500 | 1/2 | 0.5000 | 3/4 | 0.7500 | 1 | 1.0000 |

## Metric Equivalents

| Mm | In.* | Mm | In. |
|---|---|---|---|
| 1 = | 0.0394 | 17 = | 0.6693 |
| 2 = | 0.0787 | 18 = | 0.7087 |
| 3 = | 0.1181 | 19 = | 0.7480 |
| 4 = | 0.1575 | 20 = | 0.7874 |
| 5 = | 0.1969 | 21 = | 0.8268 |
| 6 = | 0.2362 | 22 = | 0.8662 |
| 7 = | 0.2756 | 23 = | 0.9055 |
| 8 = | 0.3150 | 24 = | 0.9449 |
| 9 = | 0.3543 | 25 = | 0.9843 |
| 10 = | 0.3937 | 26 = | 1.0236 |
| 11 = | 0.4331 | 27 = | 1.0630 |
| 12 = | 0.4724 | 28 = | 1.1024 |
| 13 = | 0.5118 | 29 = | 1.1418 |
| 14 = | 0.5512 | 30 = | 1.1811 |
| 15 = | 0.5906 | 31 = | 1.2205 |
| 16 = | 0.6299 | 32 = | 1.2599 |

| In. | Mm † | In. | Mm |
|---|---|---|---|
| 1/32 = | 0.794 | 17/32 = | 13.493 |
| 1/16 = | 1.587 | 9/16 = | 14.287 |
| 3/32 = | 2.381 | 19/32 = | 15.081 |
| 1/8 = | 3.175 | 5/8 = | 15.875 |
| 5/32 = | 3.968 | 21/32 = | 16.668 |
| 3/16 = | 4.762 | 11/16 = | 17.462 |
| 7/32 = | 5.556 | 23/32 = | 18.256 |
| 1/4 = | 6.349 | 3/4 = | 19.050 |
| 9/32 = | 7.144 | 25/32 = | 19.843 |
| 5/16 = | 7.937 | 13/16 = | 20.637 |
| 11/32 = | 8.731 | 27/32 = | 21.431 |
| 3/8 = | 9.525 | 7/8 = | 22.225 |
| 13/32 = | 10.319 | 29/32 = | 23.018 |
| 7/16 = | 11.112 | 15/16 = | 23.812 |
| 15/32 = | 11.906 | 31/32 = | 24.606 |
| 1/2 = | 12.699 | 1 = | 25.400 |

* Calculated to *nearest* fourth decimal place.
† Calculated to *nearest* third decimal place.

## Logarithms of Numbers

| 10 | 0 | 1 | 2 | 3 | 4 | 5 | 6 | 7 | 8 | 9 |
|---|---|---|---|---|---|---|---|---|---|---|
| 10 | 0000 | 0043 | 0086 | 0128 | 0170 | 0212 | 0253 | 0294 | 0334 | 0374 |
| 11 | 0414 | 0453 | 0492 | 0531 | 0569 | 0607 | 0645 | 0682 | 0719 | 0755 |
| 12 | 0792 | 0828 | 0864 | 0899 | 0934 | 0969 | 1004 | 1038 | 1072 | 1106 |
| 13 | 1139 | 1173 | 1206 | 1239 | 1271 | 1303 | 1335 | 1367 | 1399 | 1430 |
| 14 | 1461 | 1492 | 1523 | 1553 | 1584 | 1614 | 1644 | 1673 | 1703 | 1732 |
| 15 | 1761 | 1790 | 1818 | 1847 | 1875 | 1903 | 1931 | 1959 | 1987 | 2014 |
| 16 | 2041 | 2068 | 2095 | 2122 | 2148 | 2175 | 2201 | 2227 | 2253 | 2279 |
| 17 | 2304 | 2330 | 2355 | 2380 | 2405 | 2430 | 2455 | 2480 | 2504 | 2529 |
| 18 | 2553 | 2577 | 2601 | 2625 | 2648 | 2672 | 2695 | 2718 | 2742 | 2765 |
| 19 | 2788 | 2810 | 2833 | 2856 | 2878 | 2900 | 2923 | 2945 | 2967 | 2989 |
| 20 | 3010 | 3032 | 3054 | 3075 | 3096 | 3118 | 3139 | 3160 | 3181 | 3201 |
| 21 | 3222 | 3243 | 3263 | 3284 | 3304 | 3324 | 3345 | 3365 | 3385 | 3404 |
| 22 | 3424 | 3444 | 3464 | 3483 | 3502 | 3522 | 3541 | 3560 | 3579 | 3598 |
| 23 | 3617 | 3636 | 3655 | 3674 | 3692 | 3711 | 3729 | 3747 | 3766 | 3784 |
| 24 | 3802 | 3820 | 3838 | 3856 | 3874 | 3892 | 3909 | 3927 | 3945 | 3962 |
| 25 | 3979 | 3997 | 4014 | 4031 | 4048 | 4065 | 4082 | 4099 | 4116 | 4133 |
| 26 | 4150 | 4166 | 4183 | 4200 | 4216 | 4232 | 4249 | 4265 | 4281 | 4298 |
| 27 | 4314 | 4330 | 4346 | 4362 | 4378 | 4393 | 4409 | 4425 | 4440 | 4456 |
| 28 | 4472 | 4487 | 4502 | 4518 | 4533 | 4548 | 4564 | 4579 | 4594 | 4609 |
| 29 | 4624 | 4639 | 4654 | 4669 | 4683 | 4698 | 4713 | 4728 | 4742 | 4757 |
| 30 | 4771 | 4786 | 4800 | 4814 | 4829 | 4843 | 4857 | 4871 | 4886 | 4900 |
| 31 | 4914 | 4928 | 4942 | 4955 | 4969 | 4983 | 4997 | 5011 | 5024 | 5038 |
| 32 | 5051 | 5065 | 5079 | 5092 | 5105 | 5119 | 5132 | 5145 | 5159 | 5172 |
| 33 | 5185 | 5198 | 5211 | 5224 | 5237 | 5250 | 5263 | 5276 | 5289 | 5302 |
| 34 | 5315 | 5328 | 5340 | 5353 | 5366 | 5378 | 5391 | 5403 | 5416 | 5428 |
| 35 | 5441 | 5453 | 5465 | 5478 | 5490 | 5502 | 5514 | 5527 | 5539 | 5551 |
| 36 | 5563 | 5575 | 5587 | 5599 | 5611 | 5623 | 5635 | 5647 | 5658 | 5670 |
| 37 | 5682 | 5694 | 5705 | 5717 | 5729 | 5740 | 5752 | 5763 | 5775 | 5786 |
| 38 | 5798 | 5809 | 5821 | 5832 | 5843 | 5855 | 5866 | 5877 | 5888 | 5899 |
| 39 | 5911 | 5922 | 5933 | 5944 | 5955 | 5966 | 5977 | 5988 | 5999 | 6010 |
| 40 | 6021 | 6031 | 6042 | 6053 | 6064 | 6075 | 6085 | 6096 | 6107 | 6117 |
| 41 | 6128 | 6138 | 6149 | 6160 | 6170 | 6180 | 6191 | 6201 | 6212 | 6222 |
| 42 | 6232 | 6243 | 6253 | 6263 | 6274 | 6284 | 6294 | 6304 | 6314 | 6325 |
| 43 | 6335 | 6345 | 6355 | 6365 | 6375 | 6385 | 6395 | 6405 | 6415 | 6425 |
| 44 | 6435 | 6444 | 6454 | 6464 | 6474 | 6484 | 6493 | 6503 | 6513 | 6522 |
| 45 | 6532 | 6542 | 6551 | 6561 | 6571 | 6580 | 6590 | 6599 | 6609 | 6618 |
| 46 | 6628 | 6637 | 6646 | 6656 | 6665 | 6675 | 6684 | 6693 | 6702 | 6712 |
| 47 | 6721 | 6730 | 6739 | 6749 | 6758 | 6767 | 6776 | 6785 | 6794 | 6803 |
| 48 | 6812 | 6821 | 6830 | 6839 | 6848 | 6857 | 6866 | 6875 | 6884 | 6893 |
| 49 | 6902 | 6911 | 6920 | 6928 | 6937 | 6946 | 6955 | 6964 | 6972 | 6981 |
| 50 | 6990 | 6998 | 7007 | 7016 | 7024 | 7033 | 7042 | 7050 | 7059 | 7067 |
| 51 | 7076 | 7084 | 7093 | 7101 | 7110 | 7118 | 7126 | 7135 | 7143 | 7152 |
| 52 | 7160 | 7168 | 7177 | 7185 | 7193 | 7202 | 7210 | 7218 | 7226 | 7235 |
| 53 | 7243 | 7251 | 7259 | 7267 | 7275 | 7284 | 7292 | 7300 | 7308 | 7316 |
| 54 | 7324 | 7332 | 7340 | 7348 | 7356 | 7364 | 7372 | 7380 | 7388 | 7396 |

**Logarithms of Numbers (Cont.)**

| 55 | 0 | 1 | 2 | 3 | 4 | 5 | 6 | 7 | 8 | 9 |
|---|---|---|---|---|---|---|---|---|---|---|
| 55 | 7404 | 7412 | 7419 | 7427 | 7435 | 7443 | 7451 | 7459 | 7466 | 7474 |
| 56 | 7482 | 7490 | 7497 | 7505 | 7513 | 7520 | 7528 | 7536 | 7543 | 7551 |
| 57 | 7559 | 7566 | 7574 | 7582 | 7589 | 7597 | 7604 | 7612 | 7619 | 7627 |
| 58 | 7634 | 7642 | 7649 | 7657 | 7664 | 7672 | 7679 | 7686 | 7694 | 7701 |
| 59 | 7709 | 7716 | 7723 | 7731 | 7738 | 7745 | 7752 | 7760 | 7767 | 7774 |
| 60 | 7782 | 7789 | 7796 | 7803 | 7810 | 7818 | 7825 | 7832 | 7839 | 7846 |
| 61 | 7853 | 7860 | 7868 | 7875 | 7882 | 7889 | 7896 | 7903 | 7910 | 7917 |
| 62 | 7924 | 7931 | 7938 | 7945 | 7952 | 7959 | 7966 | 7973 | 7980 | 7987 |
| 63 | 7993 | 8000 | 8007 | 8014 | 8021 | 8028 | 8035 | 8041 | 8048 | 8055 |
| 64 | 8062 | 8069 | 8075 | 8082 | 8089 | 8096 | 8102 | 8109 | 8116 | 8122 |
| 65 | 8129 | 8136 | 8142 | 8149 | 8156 | 8162 | 8169 | 8176 | 8182 | 8189 |
| 66 | 8195 | 8202 | 8209 | 8215 | 8222 | 8228 | 8235 | 8241 | 8248 | 8254 |
| 67 | 8261 | 8267 | 8274 | 8280 | 8287 | 8293 | 8299 | 8306 | 8312 | 8319 |
| 68 | 8325 | 8331 | 8338 | 8344 | 8351 | 8357 | 8363 | 8370 | 8376 | 8382 |
| 69 | 8388 | 8395 | 8401 | 8407 | 8414 | 8420 | 8426 | 8432 | 8439 | 8445 |
| 70 | 8451 | 8457 | 8463 | 8470 | 8476 | 8482 | 8488 | 8494 | 8500 | 8506 |
| 71 | 8513 | 8519 | 8525 | 8531 | 8537 | 8543 | 8549 | 8555 | 8561 | 8567 |
| 72 | 8573 | 8579 | 8585 | 8591 | 8597 | 8603 | 8609 | 8615 | 8621 | 8627 |
| 73 | 8633 | 8639 | 8645 | 8651 | 8657 | 8663 | 8669 | 8675 | 8681 | 8686 |
| 74 | 8692 | 8698 | 8704 | 8710 | 8716 | 8722 | 8727 | 8733 | 8739 | 8745 |
| 75 | 8751 | 8756 | 8762 | 8768 | 8774 | 8779 | 8785 | 8791 | 8797 | 8802 |
| 76 | 8808 | 8814 | 8820 | 8825 | 8831 | 8837 | 8842 | 8848 | 8854 | 8859 |
| 77 | 8865 | 8871 | 8876 | 8882 | 8887 | 8893 | 8899 | 8904 | 8910 | 8915 |
| 78 | 8921 | 8927 | 8932 | 8938 | 8943 | 8949 | 8954 | 8960 | 8965 | 8971 |
| 79 | 8976 | 8982 | 8987 | 8993 | 8998 | 9004 | 9009 | 9015 | 9020 | 9025 |
| 80 | 9031 | 9036 | 9042 | 9047 | 9053 | 9058 | 9063 | 9069 | 9074 | 9079 |
| 81 | 9085 | 9090 | 9096 | 9101 | 9106 | 9112 | 9117 | 9122 | 9128 | 9133 |
| 82 | 9138 | 9143 | 9149 | 9154 | 9159 | 9165 | 9170 | 9175 | 9180 | 9186 |
| 83 | 9191 | 9196 | 9201 | 9206 | 9212 | 9217 | 9222 | 9227 | 9232 | 9238 |
| 84 | 9243 | 9248 | 9253 | 9258 | 9263 | 9269 | 9274 | 9279 | 9284 | 9289 |
| 85 | 9294 | 9299 | 9304 | 9309 | 9315 | 9320 | 9325 | 9330 | 9335 | 9340 |
| 86 | 9345 | 9350 | 9355 | 9360 | 9365 | 9370 | 9375 | 9380 | 9385 | 9390 |
| 87 | 9395 | 9400 | 9405 | 9410 | 9415 | 9420 | 9425 | 9430 | 9435 | 9440 |
| 88 | 9445 | 9450 | 9455 | 9460 | 9465 | 9469 | 9474 | 9479 | 9484 | 9489 |
| 89 | 9494 | 9499 | 9504 | 9509 | 9513 | 9518 | 9523 | 9528 | 9533 | 9538 |
| 90 | 9542 | 9547 | 9552 | 9557 | 9562 | 9566 | 9571 | 9576 | 9581 | 9586 |
| 91 | 9590 | 9595 | 9600 | 9605 | 9609 | 9614 | 9619 | 9624 | 9628 | 9633 |
| 92 | 9638 | 9643 | 9647 | 9652 | 9657 | 9661 | 9666 | 9671 | 9675 | 9680 |
| 93 | 9685 | 9689 | 9694 | 9699 | 9703 | 9708 | 9713 | 9717 | 9722 | 9727 |
| 94 | 9731 | 9736 | 9741 | 9745 | 9750 | 9754 | 9759 | 9763 | 9768 | 9773 |
| 95 | 9777 | 9782 | 9786 | 9791 | 9795 | 9800 | 9805 | 9809 | 9814 | 9818 |
| 96 | 9823 | 9827 | 9832 | 9836 | 9841 | 9845 | 9850 | 9854 | 9859 | 9863 |
| 97 | 9868 | 9872 | 9877 | 9881 | 9886 | 9890 | 9894 | 9899 | 9903 | 9908 |
| 98 | 9912 | 9917 | 9921 | 9926 | 9930 | 9934 | 9939 | 9943 | 9948 | 9952 |
| 99 | 9956 | 9961 | 9965 | 9969 | 9974 | 9978 | 9983 | 9987 | 9991 | 9996 |

## Trigonometric Functions

| Angle | Sine | | Cosine | | Tangent | | Cotangent | | Angle |
|---|---|---|---|---|---|---|---|---|---|
| | Nat. | Log. | Nat. | Log. | Nat. | Log. | Nat. | Log. | |
| 0° 00′ | .0000 | ∞ | 1.0000 | 0.0000 | .0000 | ∞ | ∞ | ∞ | 90° 00′ |
| 10 | .0029 | 7.4637 | 1.0000 | 0000 | .0029 | 7.4637 | 343.77 | 2.5363 | 50 |
| 20 | .0058 | 7648 | 1.0000 | 0000 | .0058 | 7648 | 171.89 | 2352 | 40 |
| 30 | .0087 | 9408 | 1.0000 | 0000 | .0087 | 9409 | 114.59 | 0591 | 30 |
| 40 | .0116 | 8.0658 | .9999 | 0000 | .0116 | 8.0658 | 85.940 | 1.9342 | 20 |
| 50 | .0145 | 1627 | .9999 | 0000 | .0145 | 1627 | 68.750 | 8373 | 10 |
| 1° 00′ | .0175 | 8.2419 | .9998 | 9.9999 | .0175 | 8.2419 | 57.290 | 1.7581 | 89° 00′ |
| 10 | .0204 | 3088 | .9998 | 9999 | .0204 | 3089 | 49.104 | 6911 | 50 |
| 20 | .0233 | 3668 | .9997 | 9999 | .0233 | 3669 | 42.964 | 6331 | 40 |
| 30 | .0262 | 4179 | .9997 | 9999 | .0262 | 4181 | 38.188 | 5819 | 30 |
| 40 | .0291 | 4637 | .9996 | 9998 | .0291 | 4638 | 34.368 | 5362 | 20 |
| 50 | .0320 | 5050 | .9995 | 9998 | .0320 | 5053 | 31.242 | 4947 | 10 |
| 2° 00′ | .0349 | 8.5428 | .9994 | 9.9997 | .0349 | 8.5431 | 28.636 | 1.4569 | 88° 00′ |
| 10 | .0378 | 5776 | .9993 | 9997 | .0378 | 5779 | 26.432 | 4221 | 50 |
| 20 | .0407 | 6097 | .9992 | 9996 | .0407 | 6101 | 24.542 | 3899 | 40 |
| 30 | .0436 | 6397 | .9990 | 9996 | .0437 | 6401 | 22.904 | 3599 | 30 |
| 40 | .0465 | 6677 | .9989 | 9995 | .0466 | 6682 | 21.470 | 3318 | 20 |
| 50 | .0494 | 6940 | .9988 | 9995 | .0495 | 6945 | 20.206 | 3055 | 10 |
| 3° 00′ | .0523 | 8.7188 | .9986 | 9.9994 | .0524 | 8.7194 | 19.081 | 1.2806 | 87° 00′ |
| 10 | .0552 | 7423 | .9985 | 9993 | .0553 | 7429 | 18.075 | 2571 | 50 |
| 20 | .0581 | 7645 | .9983 | 9993 | .0582 | 7652 | 17.169 | 2348 | 40 |
| 30 | .0610 | 7857 | .9981 | 9992 | .0612 | 7865 | 16.350 | 2135 | 30 |
| 40 | .0640 | 8059 | .9980 | 9991 | .0641 | 8067 | 15.605 | 1933 | 20 |
| 50 | .0669 | 8251 | .9978 | 9990 | .0670 | 8261 | 14.924 | 1739 | 10 |
| 4° 00′ | .0698 | 8.8436 | .9976 | 9.9989 | .0699 | 8.8446 | 14.301 | 1.1554 | 86° 00′ |
| 10 | .0727 | 8613 | .9974 | 9989 | .0729 | 8624 | 13.727 | 1376 | 50 |
| 20 | .0756 | 8783 | .9971 | 9988 | .0758 | 8795 | 13.197 | 1205 | 40 |
| 30 | .0785 | 8946 | .9969 | 9987 | .0787 | 8960 | 12.706 | 1040 | 30 |
| 40 | .0814 | 9104 | .9967 | 9986 | .0816 | 9118 | 12.251 | 0882 | 20 |
| 50 | .0843 | 9256 | .9964 | 9985 | .0846 | 9272 | 11.826 | 0728 | 10 |
| 5° 00′ | .0872 | 8.9403 | .9962 | 9.9983 | .0875 | 8.9420 | 11.430 | 1.0580 | 85° 00′ |
| 10 | .0901 | 9545 | .9959 | 9982 | .0904 | 9563 | 11.059 | 0437 | 50 |
| 20 | .0929 | 9682 | .9957 | 9981 | .0934 | 9701 | 10.712 | 0299 | 40 |
| 30 | .0958 | 9816 | .9954 | 9980 | .0963 | 9836 | 10.385 | 0164 | 30 |
| 40 | .0987 | 9945 | .9951 | 9979 | .0992 | 9966 | 10.078 | 0034 | 20 |
| 50 | .1016 | 9.0070 | .9948 | 9977 | .1022 | 9.0093 | 9.7882 | 0.9907 | 10 |
| 6° 00′ | .1045 | 9.0192 | .9945 | 9.9976 | .1051 | 9.0216 | 9.5144 | 0.9784 | 84° 00′ |
| 10 | .1074 | 0311 | .9942 | 9975 | .1080 | 0336 | 9.2553 | 9664 | 50 |
| 20 | .1103 | 0426 | .9939 | 9973 | .1110 | 0453 | 9.0098 | 9547 | 40 |
| 30 | .1132 | 0539 | .9936 | 9972 | .1139 | 0567 | 8.7769 | 9433 | 30 |
| 40 | .1161 | 0648 | .9932 | 9971 | .1169 | 0678 | 8.5555 | 9322 | 20 |
| 50 | .1190 | 0755 | .9929 | 9969 | .1198 | 0786 | 8.3450 | 9214 | 10 |
| 7° 00′ | .1219 | 9.0859 | .9925 | 9.9968 | .1228 | 9.0891 | 8.1443 | 0.9109 | 83° 00′ |
| 10 | .1248 | 0961 | .9922 | 9966 | .1257 | 0995 | 7.9530 | 9005 | 50 |
| 20 | .1276 | 1060 | .9918 | 9964 | .1287 | 1096 | 7.7704 | 8904 | 40 |
| | Nat. | Log. | Nat. | Log. | Nat. | Log. | Nat. | Log. | |

## Trigonometric Functions (Cont.)

| Angle | Sine | | Cosine | | Tangent | | Cotangent | | Angle |
|---|---|---|---|---|---|---|---|---|---|
| | Nat. | Log. | Nat. | Log. | Nat. | Log. | Nat. | Log. | |
| 30 | .1305 | 1157 | .9914 | 9963 | .1317 | 1194 | 7.5958 | 8806 | 30 |
| 40 | .1334 | 1252 | .9911 | 9961 | .1346 | 1291 | 7.4287 | 8709 | 20 |
| 50 | .1363 | 1345 | .9907 | 9959 | .1376 | 1385 | 7.2687 | 8615 | 10 |
| 8° 00′ | .1392 | 9.1436 | .9903 | 9.9958 | .1405 | 9.1478 | 7.1154 | 0.8522 | 82° 00′ |
| 10 | .1421 | 1525 | .9899 | 9956 | .1435 | 1569 | 6.9682 | 8431 | 50 |
| 20 | .1449 | 1612 | .9894 | 9954 | .1465 | 1658 | 6.8269 | 8342 | 40 |
| 30 | .1478 | 1697 | .9890 | 9952 | .1495 | 1745 | 6.6912 | 8255 | 30 |
| 40 | .1507 | 1781 | .9886 | 9950 | .1524 | 1831 | 6.5606 | 8169 | 20 |
| 50 | .1536 | 1863 | .9881 | 9948 | .1554 | 1915 | 6.4348 | 8085 | 10 |
| 9° 00′ | .1564 | 9.1943 | .9877 | 9.9946 | .1584 | 9.1997 | 6.3138 | 0.8003 | 81° 00′ |
| 10 | .1593 | 2022 | .9872 | 9944 | .1614 | 2078 | 6.1970 | 7922 | 50 |
| 20 | .1622 | 2100 | .9868 | 9942 | .1644 | 2158 | 6.0844 | 7842 | 40 |
| 30 | .1650 | 2176 | .9863 | 9940 | .1673 | 2236 | 5.9758 | 7764 | 30 |
| 40 | .1679 | 2251 | .9858 | 9938 | .1703 | 2313 | 5.8708 | 7687 | 20 |
| 50 | .1708 | 2324 | .9853 | 9936 | .1733 | 2389 | 5.7694 | 7611 | 10 |
| 10° 00′ | .1736 | 9.2397 | .9848 | 9.9934 | .1763 | 9.2463 | 5.6713 | 0.7537 | 80° 00′ |
| 10 | .1765 | 2468 | .9843 | 9931 | .1793 | 2536 | 5.5764 | 7464 | 50 |
| 20 | .1794 | 2538 | .9838 | 9929 | .1823 | 2609 | 5.4845 | 7391 | 40 |
| 30 | .1822 | 2606 | .9833 | 9927 | .1853 | 2680 | 5.3955 | 7320 | 30 |
| 40 | .1851 | 2674 | .9827 | 9924 | .1883 | 2750 | 5.3093 | 7250 | 20 |
| 50 | .1880 | 2740 | .9822 | 9922 | .1914 | 2819 | 5.2257 | 7181 | 10 |
| 11° 00′ | .1908 | 9.2806 | .9816 | 9.9919 | .1944 | 9.2887 | 5.1446 | 0.7113 | 79° 00′ |
| 10 | .1937 | 2870 | .9811 | 9917 | .1974 | 2953 | 5.0658 | 7047 | 50 |
| 20 | .1965 | 2934 | .9805 | 9914 | .2004 | 3020 | 4.9894 | 6980 | 40 |
| 30 | .1994 | 2997 | .9799 | 9912 | .2035 | 3085 | 4.9152 | 6915 | 30 |
| 40 | .2022 | 3058 | .9793 | 9909 | .2065 | 3149 | 4.8430 | 6851 | 20 |
| 50 | .2051 | 3119 | .9787 | 9907 | .2095 | 3212 | 4.7729 | 6788 | 10 |
| 12° 00′ | .2079 | 9.3179 | .9781 | 9.9904 | .2126 | 9.3275 | 4.7046 | 0.6725 | 78° 00′ |
| 10 | .2108 | 3238 | .9775 | 9901 | .2156 | 3336 | 4.6382 | 6664 | 50 |
| 20 | .2136 | 3296 | .9769 | 9899 | .2186 | 3397 | 4.5736 | 6603 | 40 |
| 30 | .2164 | 3353 | .9763 | 9896 | .2217 | 3458 | 4.5107 | 6542 | 30 |
| 40 | .2193 | 3410 | .9757 | 9893 | .2247 | 3517 | 4.4494 | 6483 | 20 |
| 50 | .2221 | 3466 | .9750 | 9890 | .2278 | 3576 | 4.3897 | 6424 | 10 |
| 13° 00′ | .2250 | 9.3521 | .9744 | 9.9887 | .2309 | 9.3634 | 4.3315 | 0.6366 | 77° 00′ |
| 10 | .2278 | 3575 | .9737 | 9884 | .2339 | 3691 | 4.2747 | 6309 | 50 |
| 20 | .2306 | 3629 | .9730 | 9881 | .2370 | 3748 | 4.2193 | 6252 | 40 |
| 30 | .2334 | 3682 | .9724 | 9878 | .2401 | 3804 | 4.1653 | 6196 | 30 |
| 40 | .2363 | 3734 | .9717 | 9875 | .2432 | 3859 | 4.1126 | 6141 | 20 |
| 50 | .2391 | 3786 | .9710 | 9872 | .2462 | 3914 | 4.0611 | 6086 | 10 |
| 14° 00′ | .2419 | 9.3837 | .9703 | 9.9869 | .2493 | 9.3968 | 4.0108 | 0.6032 | 76° 00′ |
| 10 | .2447 | 3887 | .9696 | 9866 | .2524 | 4021 | 3.9617 | 5979 | 50 |
| 20 | .2476 | 3937 | .9689 | 9863 | .2555 | 4074 | 3.9136 | 5926 | 40 |
| 30 | .2504 | 3986 | .9681 | 9859 | .2586 | 4127 | 3.8667 | 5873 | 30 |
| 40 | .2532 | 4035 | .9674 | 9856 | .2617 | 4178 | 3.8208 | 5822 | 20 |
| 50 | .2560 | 4083 | .9667 | 9853 | .2648 | 4230 | 3.7760 | 5770 | 10 |
| | Nat. | Log. | Nat. | Log. | Nat. | Log. | Nat. | Log. | |

## Trigonometric Functions (Cont.)

| Angle | Sine | | Cosine | | Tangent | | Cotangent | | Angle |
|---|---|---|---|---|---|---|---|---|---|
| | Nat. | Log. | Nat. | Log. | Nat. | Log. | Nat. | Log. | |
| 15° 00′ | .2588 | 9.4130 | .9659 | 9.9849 | .2679 | 9.4281 | 3.7321 | 0.5719 | 75° 00′ |
| 10 | .2616 | 4177 | .9652 | 9846 | .2711 | 4331 | 3.6891 | 5669 | 50 |
| 20 | .2644 | 4223 | .9644 | 9843 | .2742 | 4381 | 3.6470 | 5619 | 40 |
| 30 | .2672 | 4269 | .9636 | 9839 | .2773 | 4430 | 3.6059 | 5570 | 30 |
| 40 | .2700 | 4314 | .9628 | 9836 | .2805 | 4479 | 3.5656 | 5521 | 20 |
| 50 | .2728 | 4359 | .9621 | 9832 | .2836 | 4527 | 3.5261 | 5473 | 10 |
| 16° 00′ | .2756 | 9.4403 | .9613 | 9.9828 | .2867 | 9.4575 | 3.4874 | 0.5425 | 74° 00′ |
| 10 | .2784 | 4447 | .9605 | 9825 | .2899 | 4622 | 3.4495 | 5378 | 50 |
| 20 | .2812 | 4491 | .9596 | 9821 | .2931 | 4669 | 3.4124 | 5331 | 40 |
| 30 | .2840 | 4533 | .9588 | 9817 | .2962 | 4716 | 3.3759 | 5284 | 30 |
| 40 | .2868 | 4576 | .9580 | 9814 | .2994 | 4762 | 3.3402 | 5238 | 20 |
| 50 | .2896 | 4618 | .9572 | 9810 | .3026 | 4808 | 3.3052 | 5192 | 10 |
| 17° 00′ | .2924 | 9.4659 | .9563 | 9.9806 | .3057 | 9.4853 | 3.2709 | 0.5147 | 73° 00′ |
| 10 | .2952 | 4700 | .9555 | 9802 | .3089 | 4898 | 3.2371 | 5102 | 50 |
| 20 | .2979 | 4741 | .9546 | 9798 | .3121 | 4943 | 3.2041 | 5057 | 40 |
| 30 | .3007 | 4781 | .9537 | 9794 | .3153 | 4987 | 3.1716 | 5013 | 30 |
| 40 | .3035 | 4821 | .9528 | 9790 | .3185 | 5031 | 3.1397 | 4969 | 20 |
| 50 | .3062 | 4861 | .9520 | 9786 | .3217 | 5075 | 3.1084 | 4925 | 10 |
| 18° 00′ | .3090 | 9.4900 | .9511 | 9.9782 | .3249 | 9.5118 | 3.0777 | 0.4882 | 72° 00′ |
| 10 | .3118 | 4939 | .9502 | 9778 | .3281 | 5161 | 3.0475 | 4839 | 50 |
| 20 | .3145 | 4977 | .9492 | 9774 | .3314 | 5203 | 3.0178 | 4797 | 40 |
| 30 | .3173 | 5015 | .9483 | 9770 | .3346 | 5245 | 2.9887 | 4755 | 30 |
| 40 | .3201 | 5052 | .9474 | 9765 | .3378 | 5287 | 2.9600 | 4713 | 20 |
| 50 | .3228 | 5090 | .9465 | 9761 | .3411 | 5329 | 2.9319 | 4671 | 10 |
| 19° 00′ | .3256 | 9.5126 | .9455 | 9.9757 | .3443 | 9.5370 | 2.9042 | 0.4630 | 71° 00′ |
| 10 | .3283 | 5163 | .9446 | 9752 | .3476 | 5411 | 2.8770 | 4589 | 50 |
| 20 | .3311 | 5199 | .9436 | 9748 | .3508 | 5451 | 2.8502 | 4549 | 40 |
| 30 | .3338 | 5235 | .9426 | 9743 | .3541 | 5491 | 2.8239 | 4509 | 30 |
| 40 | .3365 | 5270 | .9417 | 9739 | .3574 | 5531 | 2.7980 | 4469 | 20 |
| 50 | .3393 | 5306 | .9407 | 9734 | .3607 | 5571 | 2.7725 | 4429 | 10 |
| 20° 00′ | .3420 | 9.5341 | .9397 | 9.9730 | .3640 | 9.5611 | 2.7475 | 0.4389 | 70° 00′ |
| 10 | .3448 | 5375 | .9387 | 9725 | .3673 | 5650 | 2.7228 | 4350 | 50 |
| 20 | .3475 | 5409 | .9377 | 9721 | .3706 | 5689 | 2.6985 | 4311 | 40 |
| 30 | .3502 | 5443 | .9367 | 9716 | .3739 | 5727 | 2.6746 | 4273 | 30 |
| 40 | .3529 | 5477 | .9356 | 9711 | .3772 | 5766 | 2.6511 | 4234 | 20 |
| 50 | .3557 | 5510 | .9346 | 9706 | .3805 | 5804 | 2.6279 | 4196 | 10 |
| 21° 00′ | .3584 | 9.5543 | .9336 | 9.9702 | .3839 | 9.5842 | 2.6051 | 0.4158 | 69° 00′ |
| 10 | .3611 | 5576 | .9325 | 9697 | .3872 | 5879 | 2.5826 | 4121 | 50 |
| 20 | .3638 | 5609 | .9315 | 9692 | .3906 | 5917 | 2.5605 | 4083 | 40 |
| 30 | .3665 | 5641 | .9304 | 9687 | .3939 | 5954 | 2.5386 | 4046 | 30 |
| 40 | .3692 | 5673 | .9293 | 9682 | .3973 | 5991 | 2.5172 | 4009 | 20 |
| 50 | .3719 | 5704 | .9283 | 9677 | .4006 | 6028 | 2.4960 | 3972 | 10 |
| 22° 00′ | .3746 | 9.5736 | .9272 | 9.9672 | .4040 | 9.6064 | 2.4751 | 0.3936 | 68° 00′ |
| 10 | .3773 | 5767 | .9261 | 9667 | .4074 | 6100 | 2.4545 | 3900 | 50 |
| 20 | .3800 | 5798 | .9250 | 9661 | .4108 | 6136 | 2.4342 | 3864 | 40 |
| | Nat. | Log. | Nat. | Log. | Nat. | Log. | Nat. | Log. | |

## Trigonometric Functions (Cont.)

| Angle | Sine | | Cosine | | Tangent | | Cotangent | | Angle |
|---|---|---|---|---|---|---|---|---|---|
| | Nat. | Log. | Nat. | Log. | Nat. | Log. | Nat. | Log. | |
| 30 | .3827 | 5828 | .9239 | 9656 | .4142 | 6172 | 2.4142 | 3828 | 30 |
| 40 | .3854 | 5859 | .9228 | 9651 | .4176 | 6208 | 2.3945 | 3792 | 20 |
| 50 | .3881 | 5889 | .9216 | 9646 | .4210 | 6243 | 2.3750 | 3757 | 10 |
| 23° 00′ | .3907 | 9.5919 | .9205 | 9.9640 | 4245 | 9.6279 | 2.3559 | 0.3721 | 67° 00′ |
| 10 | .3934 | 5948 | .9194 | 9635 | .4279 | 6314 | 2.3369 | 3686 | 50 |
| 20 | .3961 | 5978 | .9182 | 9629 | .4314 | 6348 | 2.3183 | 3652 | 40 |
| 30 | .3987 | 6007 | .9171 | 9624 | .4348 | 6383 | 2.2998 | 3617 | 30 |
| 40 | .4014 | 6036 | .9159 | 9618 | .4383 | 6417 | 2.2817 | 3583 | 20 |
| 50 | .4041 | 6065 | .9147 | 9613 | .4417 | 6452 | 2.2637 | 3548 | 10 |
| 24° 00′ | .4067 | 9.6093 | .9135 | 9.9607 | .4452 | 9.6486 | 2.2460 | 0.3514 | 66° 00′ |
| 10 | .4094 | 6121 | .9124 | 9602 | .4487 | 6520 | 2.2286 | 3480 | 50 |
| 20 | .4120 | 6149 | .9112 | 9596 | .4522 | 6553 | 2.2113 | 3447 | 40 |
| 30 | .4147 | 6177 | .9100 | 9590 | .4557 | 6587 | 2.1943 | 3413 | 30 |
| 40 | .4173 | 6205 | .9088 | 9584 | .4592 | 6620 | 2.1775 | 3380 | 20 |
| 50 | .4200 | 6232 | .9075 | 9579 | .4628 | 6654 | 2.1609 | 3346 | 10 |
| 25° 00′ | .4226 | 9.6259 | .9063 | 9.9573 | .4663 | 9.6687 | 2.1445 | 0.3313 | 65° 00′ |
| 10 | .4253 | 6286 | .9051 | 9567 | .4699 | 6720 | 2.1283 | 3280 | 50 |
| 20 | .4279 | 6313 | .9038 | 9561 | .4734 | 6752 | 2.1123 | 3248 | 40 |
| 30 | .4305 | 6340 | .9026 | 9555 | .4770 | 6785 | 2.0965 | 3215 | 30 |
| 40 | .4331 | 6366 | .9013 | 9549 | .4806 | 6817 | 2.0809 | 3183 | 20 |
| 50 | .4358 | 6392 | .9001 | 9543 | .4841 | 6850 | 2.0655 | 3150 | 10 |
| 26° 00′ | .4384 | 9.6418 | .8988 | 9.9537 | .4877 | 9.6882 | 2.0503 | 0.3118 | 64° 00′ |
| 10 | .4410 | 6444 | .8975 | 9530 | .4913 | 6914 | 2.0353 | 3086 | 50 |
| 20 | .4436 | 6470 | .8962 | 9524 | .4950 | 6946 | 2.0204 | 3054 | 40 |
| 30 | .4462 | 6495 | .8949 | 9518 | .4986 | 6977 | 2.0057 | 3023 | 30 |
| 40 | .4488 | 6521 | .8936 | 9512 | .5022 | 7009 | 1.9912 | 2991 | 20 |
| 50 | .4514 | 6546 | .8923 | 9505 | .5059 | 7040 | 1.9768 | 2960 | 10 |
| 27° 00′ | .4540 | 9.6570 | .8910 | 9.9499 | .5095 | 9.7072 | 1.9626 | 0.2928 | 63° 00′ |
| 10 | .4566 | 6595 | .8897 | 9492 | .5132 | 7103 | 1.9486 | 2897 | 50 |
| 20 | .4592 | 6620 | .8884 | 9486 | .5169 | 7134 | 1.9347 | 2866 | 40 |
| 30 | .4617 | 6644 | .8870 | 9479 | .5206 | 7165 | 1.9210 | 2835 | 30 |
| 40 | .4643 | 6668 | .8857 | 9473 | .5243 | 7196 | 1.9074 | 2804 | 20 |
| 50 | .4669 | 6692 | .8843 | 9466 | .5280 | 7226 | 1.8940 | 2774 | 10 |
| 28° 00′ | .4695 | 9.6716 | .8829 | 9.9459 | .5317 | 9.7257 | 1.8807 | 0.2743 | 62° 00′ |
| 10 | .4720 | 6740 | .8816 | 9453 | .5354 | 7287 | 1.8676 | 2713 | 50 |
| 20 | .4746 | 6763 | .8802 | 9446 | .5392 | 7317 | 1.8546 | 2683 | 40 |
| 30 | .4772 | 6787 | .8788 | 9439 | .5430 | 7348 | 1.8418 | 2652 | 30 |
| 40 | .4797 | 6810 | .8774 | 9432 | .5467 | 7378 | 1.8291 | 2622 | 20 |
| 50 | .4823 | 6833 | .8760 | 9425 | .5505 | 7408 | 1.8165 | 2592 | 10 |
| 29° 00′ | .4848 | 9.6856 | .8746 | 9.9418 | .5543 | 9.7438 | 1.8040 | 0.2562 | 61° 00′ |
| 10 | .4874 | 6878 | .8732 | 9411 | .5581 | 7467 | 1.7917 | 2533 | 50 |
| 20 | .4899 | 6901 | .8718 | 9404 | .5619 | 7497 | 1.7796 | 2503 | 40 |
| 30 | .4924 | 6923 | .8704 | 9397 | .5658 | 7526 | 1.7675 | 2474 | 30 |
| 40 | .4950 | 6946 | .8689 | 9390 | .5696 | 7556 | 1.7556 | 2444 | 20 |
| 50 | .4975 | 6968 | .8675 | 9383 | .5735 | 7585 | 1.7437 | 2415 | 10 |
| | Nat. | Log. | Nat. | Log. | Nat. | Log. | Nat. | Log. | |

## Trigonometric Functions (Cont.)

| Angle | Sine | | Cosine | | Tangent | | Cotangent | | Angle |
|---|---|---|---|---|---|---|---|---|---|
| | Nat. | Log. | Nat. | Log. | Nat. | Log. | Nat. | Log. | |
| 30° 00′ | .5000 | 9.6990 | .8660 | 9.9375 | .5774 | 9.7614 | 1.7321 | 0.2386 | 60° 00′ |
| 10 | .5025 | 7012 | .8646 | 9368 | .5812 | 7644 | 1.7205 | 2356 | 50 |
| 20 | .5050 | 7033 | .8631 | 9361 | .5851 | 7673 | 1.7090 | 2327 | 40 |
| 30 | .5075 | 7055 | .8616 | 9353 | .5890 | 7701 | 1.6977 | 2299 | 30 |
| 40 | .5100 | 7076 | .8601 | 9346 | .5930 | 7730 | 1.6864 | 2270 | 20 |
| 50 | .5125 | 7097 | .8587 | 9338 | .5969 | 7759 | 1.6753 | 2241 | 10 |
| 31° 00′ | .5150 | 9.7118 | .8572 | 9.9331 | .6009 | 9.7788 | 1.6643 | 0.2212 | 59° 00′ |
| 10 | .5175 | 7139 | .8557 | 9323 | .6048 | 7816 | 1.6534 | 2184 | 50 |
| 20 | .5200 | 7160 | .8542 | 9315 | .6088 | 7845 | 1.6426 | 2155 | 40 |
| 30 | .5225 | 7181 | .8526 | 9308 | .6128 | 7873 | 1.6319 | 2127 | 30 |
| 40 | .5250 | 7201 | .8511 | 9300 | .6168 | 7902 | 1.6212 | 2098 | 20 |
| 50 | .5275 | 7222 | .8496 | 9292 | .6208 | 7930 | 1.6107 | 2070 | 10 |
| 32° 00′ | .5299 | 9.7242 | .8480 | 9.9284 | .6249 | 9.7958 | 1.6003 | 0.2042 | 58° 00′ |
| 10 | .5324 | 7262 | .8465 | 9276 | .6289 | 7986 | 1.5900 | 2014 | 50 |
| 20 | .5348 | 7282 | .8450 | 9268 | .6330 | 8014 | 1.5798 | 1986 | 40 |
| 30 | .5373 | 7302 | .8434 | 9260 | .6371 | 8042 | 1.5697 | 1958 | 30 |
| 40 | .5398 | 7322 | .8418 | 9252 | .6412 | 8070 | 1.5597 | 1930 | 20 |
| 50 | .5422 | 7342 | .8403 | 9244 | .6453 | 8097 | 1.5497 | 1903 | 10 |
| 33° 00′ | .5446 | 9.7361 | .8387 | 9.9236 | .6494 | 9.8125 | 1.5399 | 0.1875 | 57° 00′ |
| 10 | .5471 | 7380 | .8371 | 9228 | .6536 | 8153 | 1.5301 | 1847 | 50 |
| 20 | .5495 | 7400 | .8355 | 9219 | .6577 | 8180 | 1.5204 | 1820 | 40 |
| 30 | .5519 | 7419 | .8339 | 9211 | .6619 | 8208 | 1.5108 | 1792 | 30 |
| 40 | .5544 | 7438 | .8323 | 9203 | .6661 | 8235 | 1.5013 | 1765 | 20 |
| 50 | .5568 | 7457 | .8307 | 9194 | .6703 | 8263 | 1.4919 | 1737 | 10 |
| 34° 00′ | .5592 | 9.7476 | .8290 | 9.9186 | .6745 | 9.8290 | 1.4826 | 0.1710 | 56° 00′ |
| 10 | .5616 | 7494 | .8274 | 9177 | .6787 | 8317 | 1.4733 | 1683 | 50 |
| 20 | .5640 | 7513 | .8258 | 9169 | .6830 | 8344 | 1.4641 | 1656 | 40 |
| 30 | .5664 | 7531 | .8241 | 9160 | .6873 | 8371 | 1.4550 | 1629 | 30 |
| 40 | .5688 | 7550 | .8225 | 9151 | .6916 | 8398 | 1.4460 | 1602 | 20 |
| 50 | .5712 | 7568 | .8208 | 9142 | .6959 | 8425 | 1.4370 | 1575 | 10 |
| 35° 00′ | .5736 | 9.7586 | .8192 | 9.9134 | .7002 | 9.8452 | 1.4281 | 0.1548 | 55° 00′ |
| 10 | .5760 | 7604 | .8175 | 9125 | .7046 | 8479 | 1.4193 | 1521 | 50 |
| 20 | .5783 | 7622 | .8158 | 9116 | .7089 | 8506 | 1.4106 | 1494 | 40 |
| 30 | .5807 | 7640 | .8141 | 9107 | .7133 | 8533 | 1.4019 | 1467 | 30 |
| 40 | .5831 | 7657 | .8124 | 9098 | .7177 | 8559 | 1.3934 | 1441 | 20 |
| 50 | .5854 | 7675 | .8107 | 9089 | .7221 | 8586 | 1.3848 | 1414 | 10 |
| 36° 00′ | .5878 | 9.7692 | .8090 | 9.9080 | .7265 | 9.8613 | 1.3764 | 0.1387 | 54° 00′ |
| 10 | .5901 | 7710 | .8073 | 9070 | .7310 | 8639 | 1.3680 | 1361 | 50 |
| 20 | .5925 | 7727 | .8056 | 9061 | .7355 | 8666 | 1.3597 | 1334 | 40 |
| 30 | .5948 | 7744 | .8039 | 9052 | .7400 | 8692 | 1.3514 | 1308 | 30 |
| 40 | .5972 | 7761 | .8021 | 9042 | .7445 | 8718 | 1.3432 | 1282 | 20 |
| 50 | .5995 | 7778 | .8004 | 9033 | .7490 | 8745 | 1.3351 | 1255 | 10 |
| 37° 00′ | .6018 | 9.7795 | .7986 | 9.9023 | .7536 | 9.8771 | 1.3270 | 0.1229 | 53° 00′ |
| 10 | .6041 | 7811 | .7969 | 9014 | .7581 | 8797 | 1.3190 | 1203 | 50 |
| 20 | .6065 | 7828 | .7951 | 9004 | .7627 | 8824 | 1.3111 | 1176 | 40 |
| | Nat. | Log. | Nat. | Log. | Nat. | Log. | Nat. | Log. | |

## Trigonometric Functions (Cont.)

| Angle | Sine Nat. | Sine Log. | Cosine Nat. | Cosine Log. | Tangent Nat. | Tangent Log. | Cotangent Nat. | Cotangent Log. | Angle |
|---|---|---|---|---|---|---|---|---|---|
| 30 | .6088 | 7844 | .7934 | 8995 | .7673 | 8850 | 1.3032 | 1150 | 30 |
| 40 | .6111 | 7861 | .7916 | 8985 | .7720 | 8876 | 1.2954 | 1124 | 20 |
| 50 | .6134 | 7877 | .7898 | 8975 | .7766 | 8902 | 1.2876 | 1098 | 10 |
| 38° 00′ | .6157 | 9.7893 | .7880 | 9.8965 | .7813 | 9.8928 | 1.2799 | 0.1072 | 52° 00′ |
| 10 | .6180 | 7910 | .7862 | 8955 | .7860 | 8954 | 1.2723 | 1046 | 50 |
| 20 | .6202 | 7926 | .7844 | 8945 | .7907 | 8980 | 1.2647 | 1020 | 40 |
| 30 | .6225 | 7941 | .7826 | 8935 | .7954 | 9006 | 1.2572 | 0994 | 30 |
| 40 | .6248 | 7957 | .7808 | 8925 | .8002 | 9032 | 1.2497 | 0968 | 20 |
| 50 | .6271 | 7973 | .7790 | 8915 | .8050 | 9058 | 1.2423 | 0942 | 10 |
| 39° 00′ | .6293 | 9.7989 | .7771 | 9.8905 | .8098 | 9.9084 | 1.2349 | 0.0916 | 51° 00′ |
| 10 | .6316 | 8004 | .7753 | 8895 | .8146 | 9110 | 1.2276 | 0890 | 50 |
| 20 | .6338 | 8020 | .7735 | 8884 | .8195 | 9135 | 1.2203 | 0865 | 40 |
| 30 | .6361 | 8035 | .7716 | 8874 | .8243 | 9161 | 1.2131 | 0839 | 30 |
| 40 | .6383 | 8050 | .7698 | 8864 | .8292 | 9187 | 1.2059 | 0813 | 20 |
| 50 | .6406 | 8066 | .7679 | 8853 | .8342 | 9212 | 1.1988 | 0788 | 10 |
| 40° 00′ | .6428 | 9.8081 | .7660 | 9.8843 | .8391 | 9.9238 | 1.1918 | 0.0762 | 50° 00′ |
| 10 | .6450 | 8096 | .7642 | 8832 | .8441 | 9264 | 1.1847 | 0736 | 50 |
| 20 | .6472 | 8111 | .7623 | 8821 | .8491 | 9289 | 1.1778 | 0711 | 40 |
| 30 | .6494 | 8125 | .7604 | 8810 | .8541 | 9315 | 1.1708 | 0685 | 30 |
| 40 | .6517 | 8140 | .7585 | 8800 | .8591 | 9341 | 1.1640 | 0659 | 20 |
| 50 | .6539 | 8155 | .7566 | 8789 | .8642 | 9366 | 1.1571 | 0634 | 10 |
| 41° 00′ | .6561 | 9.8169 | .7547 | 9.8778 | .8693 | 9.9392 | 1.1504 | 0.0608 | 49° 00′ |
| 10 | .6583 | 8184 | .7528 | 8767 | .8744 | 9417 | 1.1436 | 0583 | 50 |
| 20 | .6604 | 8198 | .7509 | 8756 | .8796 | 9443 | 1.1369 | 0557 | 40 |
| 30 | .6626 | 8213 | .7490 | 8745 | .8847 | 9468 | 1.1303 | 0532 | 30 |
| 40 | .6648 | 8227 | .7470 | 8733 | .8899 | 9494 | 1.1237 | 0506 | 20 |
| 50 | .6670 | 8241 | .7451 | 8722 | .8952 | 9519 | 1.1171 | 0481 | 10 |
| 42° 00′ | .6691 | 9.8255 | .7431 | 9.8711 | .9004 | 9.9544 | 1.1106 | 0.0456 | 48° 00′ |
| 10 | .6713 | 8269 | .7412 | 8699 | .9057 | 9570 | 1.1041 | 0430 | 50 |
| 20 | .6734 | 8283 | .7392 | 8688 | .9110 | 9595 | 1.0977 | 0405 | 40 |
| 30 | .6756 | 8297 | .7373 | 8676 | .9163 | 9621 | 1.0913 | 0379 | 30 |
| 40 | .6777 | 8311 | .7353 | 8665 | .9217 | 9646 | 1.0850 | 0354 | 20 |
| 50 | .6799 | 8324 | .7333 | 8653 | .9271 | 9671 | 1.0786 | 0329 | 10 |
| 43° 00′ | .6820 | 9.8338 | .7314 | 9.8641 | .9325 | 9.9697 | 1.0724 | 0.0303 | 47° 00′ |
| 10 | .6841 | 8351 | .7294 | 8629 | .9380 | 9722 | 1.0661 | 0278 | 50 |
| 20 | .6862 | 8365 | .7274 | 8618 | .9435 | 9747 | 1.0599 | 0253 | 40 |
| 30 | .6884 | 8378 | .7254 | 8606 | .9490 | 9772 | 1.0538 | 0228 | 30 |
| 40 | .6905 | 8391 | .7234 | 8594 | .9545 | 9798 | 1.0477 | 0202 | 20 |
| 50 | .6926 | 8405 | .7214 | 8582 | .9601 | 9823 | 1.0416 | 0177 | 10 |
| 44° 00′ | .6947 | 9.8418 | .7193 | 9.8569 | .9657 | 9.9848 | 1.0355 | 0.0152 | 46° 00′ |
| 10 | .6967 | 8431 | .7173 | 8557 | .9713 | 9874 | 1.0295 | 0126 | 50 |
| 20 | .6988 | 8444 | .7153 | 8545 | .9770 | 9899 | 1.0235 | 0101 | 40 |
| 30 | .7009 | 8457 | .7133 | 8532 | .9827 | 9924 | 1.0176 | 0076 | 30 |
| 40 | .7030 | 8469 | .7112 | 8520 | .9884 | 9949 | 1.0117 | 0051 | 20 |
| 50 | .7050 | 8482 | .7092 | 8507 | .9942 | 9975 | 1.0058 | 0025 | 10 |
| 45° 00′ | .7071 | 9.8495 | .7071 | 9.8495 | 1.0000 | 0.0000 | 1.0000 | 0.0000 | 45° 00′ |
|  | Nat. | Log. | Nat. | Log. | Nat. | Log. | Nat. | Log |  |
|  | Cosine |  | Sine |  | Cotangent |  | Tangent |  | Angle |

## Length of Chord for Circle Arcs of 1-in. Radius

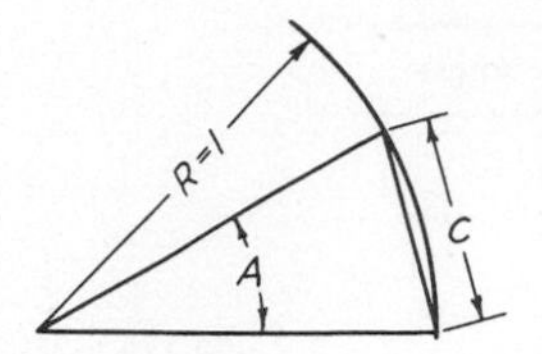


| ° | 0′ | 10′ | 20′ | 30′ | 40′ | 50′ |
|---|---|---|---|---|---|---|
| 0 | 0.0000 | 0.0029 | 0.0058 | 0.0087 | 0.0116 | 0.0145 |
| 1 | 0.0175 | 0.0204 | 0.0233 | 0.0262 | 0.0291 | 0.0320 |
| 2 | 0.0349 | 0.0378 | 0.0407 | 0.0436 | 0.0465 | 0.0494 |
| 3 | 0.0524 | 0.0553 | 0.0582 | 0.0611 | 0.0640 | 0.0669 |
| 4 | 0.0698 | 0.0727 | 0.0756 | 0.0785 | 0.0814 | 0.0843 |
| 5 | 0.0872 | 0.0901 | 0.0931 | 0.0960 | 0.0989 | 0.1018 |
| 6 | 0.1047 | 0.1076 | 0.1105 | 0.1134 | 0.1163 | 0.1192 |
| 7 | 0.1221 | 0.1250 | 0.1279 | 0.1308 | 0.1337 | 0.1366 |
| 8 | 0.1395 | 0.1424 | 0.1453 | 0.1482 | 0.1511 | 0.1540 |
| 9 | 0.1569 | 0.1598 | 0.1627 | 0.1656 | 0.1685 | 0.1714 |
| 10 | 0.1743 | 0.1772 | 0.1801 | 0.1830 | 0.1859 | 0.1888 |
| 11 | 0.1917 | 0.1946 | 0.1975 | 0.2004 | 0.2033 | 0.2062 |
| 12 | 0.2091 | 0.2119 | 0.2148 | 0.2177 | 0.2206 | 0.2235 |
| 13 | 0.2264 | 0.2293 | 0.2322 | 0.2351 | 0.2380 | 0.2409 |
| 14 | 0.2437 | 0.2466 | 0.2495 | 0.2524 | 0.2553 | 0.2582 |
| 15 | 0.2611 | 0.2639 | 0.2668 | 0.2697 | 0.2726 | 0.2755 |
| 16 | 0.2783 | 0.2812 | 0.2841 | 0.2870 | 0.2899 | 0.2927 |
| 17 | 0.2956 | 0.2985 | 0.3014 | 0.3042 | 0.3071 | 0.3100 |
| 18 | 0.3129 | 0.3157 | 0.3186 | 0.3215 | 0.3244 | 0.3272 |
| 19 | 0.3301 | 0.3330 | 0.3358 | 0.3387 | 0.3416 | 0.3444 |
| 20 | 0.3473 | 0.3502 | 0.3530 | 0.3559 | 0.3587 | 0.3616 |
| 21 | 0.3645 | 0.3673 | 0.3702 | 0.3730 | 0.3759 | 0.3788 |
| 22 | 0.3816 | 0.3845 | 0.3873 | 0.3902 | 0.3930 | 0.3959 |
| 23 | 0.3987 | 0.4016 | 0.4044 | 0.4073 | 0.4101 | 0.4130 |
| 24 | 0.4158 | 0.4187 | 0.4215 | 0.4244 | 0.4272 | 0.4300 |
| 25 | 0.4329 | 0.4357 | 0.4386 | 0.4414 | 0.4442 | 0.4471 |
| 26 | 0.4499 | 0.4527 | 0.4556 | 0.4584 | 0.4612 | 0.4641 |
| 27 | 0.4669 | 0.4697 | 0.4725 | 0.4754 | 0.4782 | 0.4810 |
| 28 | 0.4838 | 0.4867 | 0.4895 | 0.4923 | 0.4951 | 0.4979 |
| 29 | 0.5008 | 0.5036 | 0.5064 | 0.5092 | 0.5120 | 0.5148 |
| 30 | 0.5176 | 0.5204 | 0.5233 | 0.5261 | 0.5289 | 0.5317 |
| 31 | 0.5345 | 0.5373 | 0.5401 | 0.5429 | 0.5457 | 0.5485 |
| 32 | 0.5513 | 0.5541 | 0.5569 | 0.5597 | 0.5625 | 0.5652 |
| 33 | 0.5680 | 0.5708 | 0.5736 | 0.5764 | 0.5792 | 0.5820 |
| 34 | 0.5847 | 0.5875 | 0.5903 | 0.5931 | 0.5959 | 0.5986 |
| 35 | 0.6014 | 0.6042 | 0.6070 | 0.6097 | 0.6125 | 0.6153 |
| 36 | 0.6180 | 0.6208 | 0.6236 | 0.6263 | 0.6291 | 0.6319 |
| 37 | 0.6346 | 0.6374 | 0.6401 | 0.6429 | 0.6456 | 0.6484 |
| 38 | 0.6511 | 0.6539 | 0.6566 | 0.6594 | 0.6621 | 0.6649 |
| 39 | 0.6676 | 0.6704 | 0.6731 | 0.6758 | 0.6786 | 0.6813 |
| 40 | 0.6840 | 0.6868 | 0.6895 | 0.6922 | 0.6950 | 0.6977 |
| 41 | 0.7004 | 0.7031 | 0.7059 | 0.7086 | 0.7113 | 0.7140 |
| 42 | 0.7167 | 0.7195 | 0.7222 | 0.7249 | 0.7276 | 0.7303 |
| 43 | 0.7330 | 0.7357 | 0.7384 | 0.7411 | 0.7438 | 0.7465 |
| 44 | 0.7492 | 0.7519 | 0.7546 | 0.7573 | 0.7600 | 0.7627 |
| 45 * | 0.7654 | 0.7681 | 0.7707 | 0.7734 | 0.7761 | 0.7788 |

* For angles between 45° and 90°, draw 90° angle and lay off complement from 90° line.

## Conversions of Weights and Measures

| Given | Multiply by | To obtain |
|---|---|---|
| Absolute temperature (abs temp), centigrade | 1.0 | degrees Centigrade +273.160 ±.010 |
| Absolute temperature, Fahrenheit | 1.0 | degrees Fahrenheit +491.2 |
| Acceleration by gravity | 980.665 | centimeters per second |
| Acceleration by gravity | 32.16 | feet per second |
| Acres | 0.4047 | hectares |
| Acres | 10.0 | square chains |
| Acres | 43,560.0 | square feet |
| Acre | 0.00156 | square miles |
| Ares | 100.0 | centiares |
| Ares | 119.6 | square yards |
| Atmospheres (atm) | 29.921 | inches of mercury |
| Atmospheres | 33.934 | feet of water |
| Atmospheres | 1.033228 | kilograms per square centimeter |
| Atmospheres | 14.6959 | pounds per square inch |
| British thermal units (Btu) | 0.252 | calories |
| British thermal units | 778.0 | foot-pounds |
| British thermal units | 1,054.86 | watt-seconds |
| Bushels (bu), imperial | 1.032 | bushels, U.S. |
| Bushels, imperial | 1.2837 | cubic feet |
| Bushels, imperial | 2,218.19 | cubic inches |
| Bushels, imperial | 8.0 | gallons, imperial |
| Bushels, imperial | 36.368 | liters |
| Bushels, U.S. | 0.968 | bushels, imperial |
| Bushels, U.S. | 1.2445 | cubic feet |
| Bushels, U.S. | 2,150.42 | cubic inches |
| Bushels, U.S. | 35.2393 | liters |
| Bushels, U.S. | 4.0 | pecks |
| Bushels, U.S. | 64.0 | pints, dry |
| Bushels, U.S. | 32.0 | quarts, dry |
| Calories (cal) | 3.9682 | British thermal units |
| Calories | 3,088.4 | foot-pounds |
| Carats | 3.086 | grains |
| Carats | 200.0 | milligrams |
| Centiares | 1,549.997 | square inches |
| Centiares | 1.0 | square meters |
| Centigrade, degrees (°C) | $\frac{9}{5}$ °C + 32 | Fahrenheit, degrees |
| Centigrams (cg) | 0.1543 | grains |
| Centigrams | 0.01 | grams |
| Centiliters (cl) | 0.01 | liters |
| Centiliters | 0.0338 | ounces, fluid |

## Conversions of Weights and Measures (Cont.)

| Given | Multiply by | To obtain |
|---|---|---|
| Centimeters (cm) | 0.0328 | feet |
| Centimeters | 0.3937 | inches |
| Centimeters | 0.01 | meters |
| Chains | 0.10000 | furlongs |
| Chains | 0.01250 | miles, statute |
| Chains | 100.0 | links |
| Circle (angular) | 360.0 | degrees |
| Circular inch (cir in.) | 1.0 | area of a 1-in.-diameter circle |
| Circular inches | 1,000,000.0 | circular mils |
| Circular inches | 0.7854 | square inches |
| Circular mil | 1.0 | area of a 0.001-in.-diameter circle |
| Circular mils | 0.0000001 | circular inches |
| Circumference of the earth at the equator | 21,600.0 | miles, nautical |
| Circumference of the earth at the equator | 24,874.5 | miles, statute |
| Cord (cd), of wood, (4 × 4 × 8) | 128.0 | cubic feet |
| Cubic centimeters (cu cm) | 0.00003531 | cubic feet |
| Cubic centimeters | 0.06102 | cubic inches |
| Cubic centimeters | 0.0010 | liters |
| Cubic centimeters | 0.0000010 | cubic meter |
| Cubic decimeters | 1,000.0 | cubic centimeters |
| Cubic decimeters | 61.02 | cubic inches |
| Cubic feet (cu ft) | 0.7790 | bushels, imperial |
| Cubic feet | 0.80290 | bushels, U.S. |
| Cubic feet | 0.00781 | cords, of wood |
| Cubic feet | 28,317.08 | cubic centimeters |
| Cubic feet | 1,728.0 | cubic inches |
| Cubic feet | 0.0283 | cubic meters |
| Cubic feet | 0.0370 | cubic yards |
| Cubic feet | 7.4805 | gallons, U.S. |
| Cubic feet | 28.3163 | liters |
| Cubic feet | 0.04040 | perch, of masonry |
| Cubic feet of water at 39.1 degrees Fahrenheit (°F) | 28.3156 | kilograms |
| Cubic feet of water at 39.1 degrees Fahrenheit | 62.4245 | pounds |
| Cubic inches (cu in.) | 0.00045 | bushels, imperial |
| Cubic inches | 0.00046 | bushels, U.S. |
| Cubic inches | 16.3872 | cubic centimeters |
| Cubic inches | 0.00058 | cubic feet |
| Cubic inches | 0.000016 | cubic meters |
| Cubic inches | 0.0000214 | cubic yards |
| Cubic inches | 0.0036 | gallons, imperial |

## Conversions of Weights and Measures (Cont.)

| Given | Multiply by | To obtain |
|---|---|---|
| Cubic inches | 0.00432 | gallons, U.S. |
| Cubic inches | 0.0164 | liters |
| Cubic inches | 0.00186 | pecks |
| Cubic inches | 0.02976 | pints, dry |
| Cubic inches | 0.0346 | pints, liquid |
| Cubic inches | 0.01488 | quarts, dry |
| Cubic inches | 0.0173 | quarts, liquid |
| Cubic meters (cu m or $m^3$) | 1,000,000.0 | cubic centimeters |
| Cubic meters | 35.3133 | cubic feet |
| Cubic meters | 61,023.3753 | cubic inches |
| Cubic meters | 1.3079 | cubic yards |
| Cubic meters | 264.170 | gallons, U.S. |
| Cubic millimeters (cu mm or $mm^3$) | 0.001 | cubic centimeters |
| Cubic millimeters | 0.00006 | cubic inches |
| Cubic yards (cu yd) | 27.0 | cubic feet |
| Cubic yards | 46,656.0 | cubic inches |
| Cubic yards | 0.7646 | cubic meters |
| Decigrams | 1.5432 | grains |
| Decigrams | 0.1 | grams |
| Decimeters | 3.937 | inches |
| Decimeters | 0.01 | meters |
| Deciliter | 0.338 | ounces, fluid |
| Decagrams | 10.0 | grams |
| Decagrams | 0.3527 | ounces, avoirdupois |
| Decaliters | 0.284 | bushels, U.S. |
| Decaliters | 2.64 | gallons, U.S. |
| Decaliters | 10.0 | liters |
| Decameters | 393.7 | inches |
| Decameters | 10.0 | meters |
| Decileters | 0.1 | liters |
| Degrees (deg or °) | 60.0 | minutes |
| Degrees (arc) | 0.0175 | radians |
| Degrees (at the equator) | 60.0 | miles, nautical |
| Degrees (at the equator) | 69.168 | miles, statute |
| Dozens (doz) | 12.0 | units |
| Drams (dr), apothecaries | 60.0 | grains |
| Drams, apothecaries | 3.543 | grams |
| Drams, apothecaries | 3.0 | scruples |
| Drams, avoirdupois | 27.344 | grains |
| Drams, avoirdupois | 1.772 | grams |
| Drams, avoirdupois | 0.0625 | ounces, avoirdupois |
| Drams, fluid | 0.2256 | cubic inches |
| Drams, fluid | 3.6966 | milliliters |

## Conversions of Weights and Measures (Cont.)

| Given | Multiply by | To obtain |
|---|---|---|
| Drams, fluid | 60.0 | minims |
| Drams, fluid | 0.125 | ounces, U.S. fluid |
| Dynes | 0.00102 | grams |
| Ergs | 1.0 | dyne-centimeters |
| Fahrenheit | $\frac{5(°F-32)}{9}$ | centigrade, degrees |
| Fathoms | 6.0 | feet |
| Fathoms | 1.8288 | meters |
| Fathoms | 2.0 | yards |
| Feet (ft) | 30.4801 | centimeters |
| Feet | 0.16667 | fathom |
| Feet | 12.0 | inches |
| Feet | 0.66000 | links |
| Feet | 0.3048 | meters |
| Feet | 0.000189 | miles |
| Feet | 0.0001645 | miles, nautical |
| Feet | 0.06061 | rods |
| Feet | 0.3333 | yards |
| Feet of water at 62 degrees Fahrenheit | 304.442 | kilograms per square meter |
| Feet of water at 62 degrees Fahrenheit | 62.355 | pounds per square foot |
| Feet of water at 62 degrees Fahrenheit | 0.4334 | pounds per square inch |
| Feet per second (fps) | 0.5921 | knots |
| Feet per second | 0.6816 | miles per hour |
| Foot-pounds (ft-lb) | 0.00129 | British thermal units |
| Foot-pounds | 0.00032 | calories |
| Foot-pounds | 0.13835 | meter-kilograms |
| Foot-pounds per minute | 0.000003 | horsepower |
| Foot-pounds per second | 0.000018 | horsepower |
| Furlongs | 10.0 | chains |
| Furlongs | 660.0 | feet |
| Furlongs | 201.17 | meters |
| Furlongs | 0.12500 | miles, statute |
| Furlongs | 220.0 | yards |
| Gallons (gal), imperial | 0.125 | bushels, imperial |
| Gallons, imperial | 277.4176 | cubic inches |
| Gallons, imperial | 1.2009 | gallons, U.S. |
| Gallons, imperial | 4.54607 | liters |
| Gallons, U.S. | 0.1337 | cubic feet |
| Gallons, U.S. | 231.0 | cubic inches |
| Gallons, U.S. | 0.0038 | cubic meters |
| Gallons, U.S. | 0.8327 | gallons, imperial |
| Gallons, U.S. | 3.7878 | liters |

## Conversions of Weights and Measures (Cont.)

| Given | Multiply by | To obtain |
|---|---|---|
| Gallons, U.S. | 128.0 | ounces, U.S. fluid |
| Gallons, U.S., water | 8.5 | pounds |
| Gills | 0.25 | pints, liquid |
| Grains | 0.0366 | drams, avoirdupois |
| Grains | 0.0648 | grams |
| Grains | 64.7989 | milligrams |
| Grains | 0.00229 | ounces, avoirdupois |
| Grains | 0.00208 | ounces, troy and apothecaries' |
| Grains | 0.00014 | pounds, avoirdupois |
| Grains | 0.00017 | pounds, troy and apothecaries' |
| Grams (g) | 981 | dynes |
| Grams | 15.4475 | grains |
| Grams | 0.0010 | kilograms |
| Grams | 1,000.0 | milligrams |
| Grams | 0.0353 | ounces, avoirdupois |
| Grams | 0.0022 | pounds, avoirdupois |
| Grams per cubic centimeter | 1,000.0 | kilograms per cubic meter |
| Grams per cubic centimeter | 62.4 | pounds per cubic foot |
| Grams per cubic centimeter | 0.03613 | pounds per cubic inch |
| Gross | 12.0 | dozen |
| Gross, great | 12.0 | gross |
| Hands | 4.0 | inches |
| Hectares (ha) | 100.0 | ares |
| Hectograms | 100.0 | grams |
| Hectograms | 3.5274 | ounces, avoirdupois |
| Hectoliters | 26.417 | gallons |
| Hectoliters | 100.0 | liters |
| Hectometers | 328.083 | feet |
| Hectometers | 100.0 | meters |
| Horsepower (hp) | 76.042 | kilogram-meters per second |
| Horsepower | 550.0 | foot-pounds per second |
| Horsepower | 33,000.0 | foot-pounds per minute |
| Horsepower | 1.0139 | metric horsepower |
| Horsepower | 746.0 | watts per minute |
| Horsepower, metric | 0.9862 | horsepower |
| Horsepower, metric | 32,550.0 | foot-pounds per minute |
| Horsepower, metric | 542.5 | foot-pounds per second |
| Horsepower, metric | 75.0 | kilogram meters per second |
| Inches (in.) | 2.5400 | centimeters |
| Inches | 0.08333 | feet |
| Inches | 0.25000 | hands |
| Inches | 0.12626 | links |

## Conversions of Weights and Measures (Cont.)

| Given | Multiply by | To obtain |
|---|---|---|
| Inches | 0.0254 | meters |
| Inches | 1,000.0 | mils |
| Inches | 0.11111 | spans |
| Inches | 0.02778 | yards |
| Inches of mercury | 1.1341 | feet of water |
| Inches of mercury | 34.542 | grams per square centimeter |
| Inches of mercury | 13.6092 | inches of water |
| Inches of mercury | 0.49115 | pounds per square inch |
| Inches of water | 2.537 | grams per square centimeter |
| Inches of water | 0.07347 | inches of mercury |
| Inches of water | 5.1052 | pounds per square foot |
| Kilocycles | 1,000.0 | cycles per second |
| Kilogram-meters (kg-m) | 7.2330 | pound-feet |
| Kilogram-meters per second | 0.01305 | horsepower |
| Kilogram-meters per second | 0.01333 | horsepower, metric |
| Kilograms (kg) | 15,432.36 | grains |
| Kilograms | 1,000.0 | grams |
| Kilograms | 35.2740 | ounces, avoirdupois |
| Kilograms | 2.2046 | pounds, avoirdupois |
| Kilograms | 0.00110 | tons |
| Kilograms | 0.00098 | tons, long |
| Kilograms | 0.001 | tons, metric |
| Kilograms per cubic meter (kg per cu m or kg/m$^3$) | 0.06243 | pounds per cubic foot |
| Kilograms per meter | 0.6721 | pounds per foot |
| Kilograms per square centimeter | 14.22 | pounds per square inch |
| Kilograms per square meter | 0.2048 | pounds per square inch |
| Kilograms per square meter | 0.00142 | pounds per square inch |
| Kiloliters (kl) | 1,000.0 | liters |
| Kilometers (km) | 3,280.8330 | feet |
| Kilometers | 1,000.0 | meters |
| Kilometers | 0.5396 | miles, nautical |
| Kilometers | 0.6214 | miles, statute |
| Kilometers per hour | 0.5396 | knots |
| Kilometers per hour | 0.62138 | miles per hour |
| Kilowatts (kw) | 0.04426 | foot-pounds per minute |
| Kilowatt-hours (kwhr) | 3,412.75 | British thermal units per hour |
| Kilowatt-hours | 1.3414 | horsepower hours |
| Knots | 1.6889 | feet per second |
| Knots | 1.8532 | kilometers per hour |
| Knots | 0.5148 | meters per second |
| Knots | 1.1516 | miles per hour |
| Knots | 1.0 | nautical miles per hour |

## Conversions of Weights and Measures (Cont.)

| Given | Multiply by | To obtain |
|---|---|---|
| Leagues, land | 4.83 | kilometers |
| Leagues, land | 2.6050 | miles, nautical |
| Leagues, land | 3.0 | miles, statute |
| Leagues, marine | 5.56 | kilometers |
| Leagues, marine | 3.0 | miles, nautical |
| Leagues, marine | 3.45 | miles, statute |
| Links | 0.01 | chains |
| Links | 0.66 | feet |
| Links | 7.92 | inches |
| Links | 0.04 | rods |
| Links | 0.22 | yards |
| Liters (l) | 0.0284 | bushels, U.S. |
| Liters | 1,000.0 | cubic centimeters |
| Liters | 0.035313 | cubic feet |
| Liters | 61.02398 | cubic inches |
| Liters | 0.2199 | gallons, imperial |
| Liters | 0.2641 | gallons, U.S. |
| Liters | 0.1135 | pecks |
| Liters | 0.9081 | quarts, dry |
| Liters | 1.0567 | quarts, liquid |
| Long tons | 2,240.0 | pounds, avoirdupois |
| Megacycles | 1,000,000.0 | cycles per second |
| Megameters | 100,000.0 | meters |
| Meters (m) | 0.5460 | fathoms |
| Meters | 3.2808 | feet |
| Meters | 39.370 | inches |
| Meters | 0.000541 | miles, nautical |
| Meters | 0.000622 | miles, U.S. |
| Meters | 1.0936 | yards |
| Meter-kilograms (m-kg) | 7.2330 | foot-pounds |
| Meters per second | 1.9425 | knots |
| Meters per second | 2.2369 | miles per hour |
| Microns ($\mu$) | 0.000039 | inches |
| Microns | 0.000001 | meters |
| Microns | 0.03937 | mils |
| Miles, nautical | 6,080.20 | feet |
| Miles, nautical | 1.85325 | kilometers |
| Miles, nautical | 0.33333 | leagues, marine |
| Miles, nautical | 1,853.2486 | meters |
| Miles, nautical | 1.1516 | miles, statute |
| Miles, statute | 80.0 | chains |
| Miles, statute | 5,280.0 | feet |
| Miles, statute | 8.0 | furlongs |

## Conversions of Weights and Measures (Cont.)

| Given | Multiply by | To obtain |
|---|---|---|
| Miles, statute | 1.6093 | kilometers |
| Miles, statute | 0.33333 | leagues, land |
| Miles, statute | 1,609.35 | meters |
| Miles, statute | 0.86836 | miles, nautical |
| Miles, statute | 1,760.0 | yards |
| Miles per hour (mph) | 1.4667 | feet per second |
| Miles per hour | 1.6093 | kilometers per hour |
| Miles per hour | 0.8684 | knots |
| Miles per hour | 0.4470 | meters per second |
| Milligrams (mg) | 0.01543 | grains |
| Milligrams | 0.001 | grams |
| Milliliters (ml) | 0.2705 | drams, fluid |
| Milliliters | 0.001 | liters |
| Milliliters | 0.0338 | ounces, fluid |
| Millimeters (mm) | 0.03937 | inches |
| Millimeters | 0.001 | meters |
| Millimeters | 1,000.0 | microns |
| Millimeters | 39.37 | mils |
| Mils | 0.001 | inches |
| Mils | 25.4001 | microns |
| Mils | 0.0254 | millimeters |
| Minims | 0.01667 | drams, fluid |
| Minutes (min) | 60.0 | seconds |
| Myriagrams | 10,000.0 | grams |
| Myriameters | 10,000.0 | meters |
| Ounces (oz), apothecaries' | 8.0 | drams, apothecaries' |
| Ounces, avoirdupois | 16.0 | drams, avoirdupois |
| Ounces, avoirdupois | 437.5 | grains |
| Ounces, avoirdupois | 28.3495 | grams |
| Ounces, avoirdupois | 1.0971 | ounces, troy and apothecaries' |
| Ounces, avoirdupois | 0.0625 | pounds, avoirdupois |
| Ounces, British fluid | 28.382 | cubic centimeters |
| Ounces, British fluid | 1.732 | cubic inches |
| Ounces, fluid | 29.57 | milliliters |
| Ounces, troy | 20.0 | pennyweights |
| Ounces, troy and apothecaries' | 480.0 | grains |
| Ounces, troy and apothecaries' | 31.10348 | grams |
| Ounces, troy and apothecaries' | 0.91149 | ounces, avoirdupois |
| Ounces, U.S. fluid | 1.805 | cubic inches |
| Ounces, U.S. fluid | 8.0 | drams, fluid |
| Ounces, U.S. fluid | 0.00781 | gallons, U.S. |
| Ounces, U.S. fluid | 0.0296 | liters |
| Pecks (pk) | 0.25 | bushel, U.S. |

## Conversions of Weights and Measures (Cont.)

| Given | Multiply by | To obtain |
|---|---|---|
| Pecks | 537.61 | cubic inches |
| Pecks | 8.8096 | liters |
| Pecks | 8.0 | quarts, dry |
| Pennyweights (dwt) | 24.0 | grains |
| Perch (of masonry) | 24.75 | cubic feet |
| Pints (pt), dry | 0.015625 | bushels, U.S. |
| Pints, dry | 33.60 | cubic inches |
| Pints, dry | 0.5506 | liters |
| Pints, dry | 0.0625 | pecks |
| Pints, dry | 0.5 | quarts, dry |
| Pints, liquid | 28.875 | cubic inches |
| Pints, liquid | 4.0 | gills |
| Pints, liquid | 0.4732 | liters |
| Poundals | 0.03113 | pounds, avoirdupois |
| Pounds (lb), avoirdupois | 0.0160 | cubic feet of water |
| Pounds, avoirdupois | 7,000.0 | grains |
| Pounds, avoirdupois | 453.5924 | grams |
| Pounds, avoirdupois | 16.0 | ounces, avoirdupois |
| Pounds, avoirdupois | 32.1740 | poundals |
| Pounds, avoirdupois | 0.0311 | slugs |
| Pounds, avoirdupois | 0.00045 | tons, long |
| Pounds, avoirdupois | 0.0005 | tons, short |
| Pound-feet (lb-ft) | 0.1383 | kilogram-meters |
| Pounds per cubic foot (lb per cu ft) | 0.01602 | grams per cubic centimeter |
| Pounds per cubic foot | 16.0184 | kilograms per cubic meter |
| Pounds per cubic foot | 0.00058 | pounds per cubic inch |
| Pounds per foot | 1.4882 | kilograms per meter |
| Pounds per square foot (psf) | 0.1922 | inches of water |
| Pounds per square foot | 4.8824 | kilograms per square meter |
| Pounds per square foot | 0.00694 | pounds per square inch |
| Pounds per square inch (psi) | 0.0680 | atmospheres |
| Pounds per square inch | 2.3066 | feet of water |
| Pounds per square inch | 70.3067 | grams per square centimeter |
| Pounds per square inch | 27.7 | inches of water |
| Pounds per square inch | 2.0360 | inches of mercury |
| Pounds per square inch | 703.0669 | kilograms per square meter |
| Pounds per square inch | 144.0 | pounds per square foot |
| Pounds, troy and apothecaries' | 5,760.0 | grains |
| Pounds, troy and apothecaries' | 0.37324 | kilograms |
| Pounds, troy and apothecaries' | 12.0 | ounces, troy and apothecaries' |
| Pounds, troy and apothecaries' | 0.8229 | pounds, avoirdupois |
| Quadrants | 90.0 | degrees |
| Quarts (qt), dry, imperial | 1.032 | quarts, dry, U.S. |

## Conversions of Weights and Measures (Cont.)

| Given | Multiply by | To obtain |
|---|---|---|
| Quarts, dry, U.S. | 1.1012 | liters |
| Quarts, dry, U.S. | 0.125 | pecks |
| Quarts, dry, U.S. | 2.0 | pints, dry |
| Quarts, dry, U.S. | 0.968 | quarts, dry, imperial |
| Quarts, dry, U.S. | 0.03125 | bushel, U.S. |
| Quarts, dry, U.S. | 67.2 | cubic inches |
| Quarts, liquid | 57.75 | cubic inches |
| Quarts, liquid | 0.94636 | liters |
| Quintals | 100,000.0 | grams |
| Quintals | 220.46 | pounds, avoirdupois |
| Quire | 24.0 | sheets |
| Radians | 57.2958 | degrees, arc |
| Radians | 3,437.7468 | minutes, arc |
| Radians | 0.1591 | revolutions |
| Radians per second | 9.4460 | revolutions per minute |
| Ream | 480.0 | sheets |
| Ream, printing paper | 500.0 | sheets |
| Revolutions | 6.2832 | radians |
| Revolutions per minute (rpm) | 0.1059 | radians per second |
| Rods | 0.25 | chains |
| Rods | 16.5 | feet |
| Rods | 40.0 | furlongs |
| Rods | 25.0 | links |
| Rods | 5.029 | meters |
| Rods | 5.5 | yards |
| Score | 20.0 | units |
| Scruples | 20.0 | grains |
| Seconds | 0.01667 | minutes |
| Slugs | 32.1740 | pounds |
| Spans | 9.0 | inches |
| Square centimeters (sq cm or $cm^2$) | 0.001076 | square feet |
| Square centimeters | 0.1550 | square inches |
| Square centimeters | 100.0 | square millimeters |
| Square chains | 0.1 | acres |
| Square chains | 4,356.0 | square feet |
| Square chains | 404.7 | square meters |
| Square chains | 0.00016 | square miles |
| Square chains | 16.0 | square rods |
| Square chains | 484.0 | square yards |
| Square decameters | 100.0 | square meters |
| Square decimeters | 0.01 | square meters |
| Square feet (sq ft) | 0.000022988 | acres |

## Conversions of Weights and Measures (Cont.)

| Given | Multiply by | To obtain |
|---|---|---|
| Square feet | 929.0341 | square centimeters |
| Square feet | 0.00023 | square chains |
| Square feet | 144.0 | square inches |
| Square feet | 0.0929 | square meters |
| Square feet | 0.00368 | square rods |
| Square feet | 0.11111 | square yards |
| Square hectometers | 10,000.0 | square meters |
| Square inches (sq in.) | 1.27324 | circular inches |
| Square inches | 6.4516 | square centimeters |
| Square inches | 0.00694 | square feet |
| Square inches | 645.1625 | square millimeters |
| Square inches | 0.00077 | square yards |
| Square kilometers (sq km or km$^2$) | 100.0 | hectares |
| Square kilometers | 1,000,000.0 | square meters |
| Square kilometers | 0.3861 | square miles |
| Square links | 0.4356 | square feet |
| Square links | 0.0405 | square meters |
| Square links | 0.00160 | square rods |
| Square links | 0.04840 | square yards |
| Square meters (sq m or m$^2$) | 1.0 | centiares |
| Square meters | 10.7639 | square feet |
| Square meters | 1.1960 | square yards |
| Square miles | 2.590 | square kilometers |
| Square miles | 640.0 | acres |
| Square miles | 6,400.0 | square chains |
| Square millimeters (sq mm or mm$^2$) | 0.00155 | square inches |
| Square millimeters | 0.000001 | square meters |
| Square rods | 0.06250 | square chains |
| Square rods | 272.25 | square feet |
| Square rods | 625.0 | square links |
| Square rods | 25.29 | square meters |
| Square rods | 30.25 | square yards |
| Square yards | 0.00207 | square chains |
| Square yards | 9.0 | square feet |
| Square yards | 1,296.0 | square inches |
| Square yards | 20.66116 | square links |
| Square yards | 0.83613 | square meters |
| Square yards | 0.03306 | square rods |
| Tons, long | 1,016.0470 | kilograms |
| Tons, long | 2,240.0 | pounds, avoirdupois |
| Tons, metric | 1,000.0 | kilograms |
| Tons, metric | 2,204.62 | pounds, avoirdupois |
| Tons, metric | 10.0 | quintals |

## Conversions of Weights and Measures (Cont.)

| Given | Multiply by | To obtain |
| --- | --- | --- |
| Tons, register | 100.0 | cubic feet |
| Tons, shipping, British | 32.719 | bushels, imperial |
| Tons, shipping, British | 32.700 | bushels, U.S. |
| Tons, shipping, British | 42.0 | cubic feet |
| Tons, shipping, U.S. | 31.16 | bushels, imperial |
| Tons, shipping, U.S. | 32.143 | bushels, U.S. |
| Tons, shipping, U.S. | 40.0 | cubic feet |
| Tons, short | 907.18 | kilograms |
| Tons, short | 2,000.0 | pounds, avoirdupois |
| Watts (w) | 10,000,000.0 | ergs per second |
| Yards (yd) | 0.04545 | chains |
| Yards | 0.50000 | fathoms |
| Yards | 3.0 | feet |
| Yards | 0.004545 | furlongs |
| Yards | 36.0 | inches |
| Yards | 0.22000 | links |
| Yards | 0.9144 | meters |
| Yards | 0.000569 | miles, statute |
| Yards | 0.18182 | rods |

## Conversion Table: Fractions, Decimals, and Millimeters*

| inches | | mm | inches | | mm | inches | | mm | inches | | mm |
|---|---|---|---|---|---|---|---|---|---|---|---|
| fractions | decimals | | fractions | decimals | | fractions | decimals | | fractions | decimals | |
| – | .0004 | .01 | 25/32 | .781 | 19.844 | – | 2.165 | 55. | 3-11/16 | 3.6875 | 93.663 |
| – | .004 | .10 | – | .7874 | 20. | 2-3/16 | 2.1875 | 55.563 | – | 3.7008 | 94. |
| – | .01 | .25 | 51/64 | .797 | 20.241 | – | 2.2047 | 56. | 3-23/32 | 3.719 | 94.456 |
| 1/64 | .0156 | .397 | 13/16 | .8125 | 20.638 | 2-7/32 | 2.219 | 56.356 | – | 3.7401 | 95. |
| – | .0197 | .50 | – | .8268 | 21. | – | 2.244 | 57. | 3-3/4 | 3.750 | 95.250 |
| – | .0295 | .75 | 53/64 | .828 | 21.034 | 2-1/4 | 2.250 | 57.150 | – | 3.7795 | 96. |
| 1/32 | .03125 | .794 | 27/32 | .844 | 21.431 | 2-9/32 | 2.281 | 57.944 | 3-25/32 | 3.781 | 96.044 |
| – | .0394 | 1. | 55/64 | .859 | 21.828 | – | 2.2835 | 58. | 3-13/16 | 3.8125 | 96.838 |
| 3/64 | .0469 | 1.191 | – | .8661 | 22. | 2-5/16 | 2.312 | 58.738 | – | 3.8189 | 97. |
| – | .059 | 1.5 | 7/8 | .875 | 22.225 | – | 2.3228 | 59. | 3-27/32 | 3.844 | 97.631 |
| 1/16 | .062 | 1.588 | 57/64 | .8906 | 22.622 | 2-11/32 | 2.344 | 59.531 | – | 3.8583 | 98. |
| 5/64 | .0781 | 1.984 | – | .9055 | 23. | – | 2.3622 | 60. | 3-7/8 | 3.875 | 98.425 |
| – | .0787 | 2. | 29/32 | .9062 | 23.019 | 2-3/8 | 2.375 | 60.325 | – | 3.8976 | 99. |
| 3/32 | .094 | 2.381 | 59/64 | .922 | 23.416 | – | 2.4016 | 61. | 3-29/32 | 3.9062 | 99.219 |
| – | .0984 | 2.5 | 15/16 | .9375 | 23.813 | 2-13/32 | 2.406 | 61.119 | – | 3.9370 | 100. |
| 7/64 | .109 | 2.778 | – | .9449 | 24. | 2-7/16 | 2.438 | 61.913 | 3-15/16 | 3.9375 | 100.013 |
| – | .1181 | 3. | 61/64 | .953 | 24.209 | – | 2.4409 | 62. | 3-31/32 | 3.969 | 100.806 |
| 1/8 | .125 | 3.175 | 31/32 | .969 | 24.606 | 2-15/32 | 2.469 | 62.706 | – | 3.9764 | 101. |
| – | .1378 | 3.5 | – | .9843 | 25. | – | 2.4803 | 63. | 4 | 4.000 | 101.600 |
| 9/64 | .141 | 3.572 | 63/64 | .9844 | 25.003 | 2-1/2 | 2.500 | 63.500 | 4-1/16 | 4.062 | 103.188 |
| 5/32 | .156 | 3.969 | 1 | 1.000 | 25.400 | – | 2.5197 | 64. | 4-1/8 | 4.125 | 104.775 |
| – | .1575 | 4. | – | 1.0236 | 26. | 2-17/32 | 2.531 | 64.294 | – | 4.1338 | 105. |
| 11/64 | .172 | 4.366 | 1-1/32 | 1.0312 | 26.194 | – | 2.559 | 65. | 4-3/16 | 4.1875 | 106.363 |
| – | .177 | 4.5 | 1-1/16 | 1.062 | 26.988 | 2-9/16 | 2.562 | 65.088 | 4-1/4 | 4.250 | 107.950 |
| 3/16 | .1875 | 4.763 | – | 1.063 | 27. | 2-19/32 | 2.594 | 65.881 | 4-5/16 | 4.312 | 109.538 |
| – | .1969 | 5. | 1-3/32 | 1.094 | 27.781 | – | 2.5984 | 66. | – | 4.3307 | 110. |
| 13/64 | .203 | 5.159 | – | 1.1024 | 28. | 2-5/8 | 2.625 | 66.675 | 4-3/8 | 4.375 | 111.125 |
| – | .2165 | 5.5 | 1-1/8 | 1.125 | 28.575 | – | 2.638 | 67. | 4-7/16 | 4.438 | 112.713 |
| 7/32 | .219 | 5.556 | – | 1.1417 | 29. | 2-21/32 | 2.656 | 67.469 | 4-1/2 | 4.500 | 114.300 |
| 15/64 | .234 | 5.953 | 1-5/32 | 1.156 | 29.369 | – | 2.6772 | 68. | – | 4.5275 | 115. |
| – | .2362 | 6. | – | 1.1811 | 30. | 2-11/16 | 2.6875 | 68.263 | 4-9/16 | 4.562 | 115.888 |
| 1/4 | .250 | 6.350 | 1-3/16 | 1.1875 | 30.163 | – | 2.7165 | 69. | 4-5/8 | 4.625 | 117.475 |
| – | .2559 | 6.5 | 1-7/32 | 1.219 | 30.956 | 2-23/32 | 2.719 | 69.056 | 4-11/16 | 4.6875 | 119.063 |
| 17/64 | .2656 | 6.747 | – | 1.2205 | 31. | 2-3/4 | 2.750 | 69.850 | – | 4.7244 | 120. |
| – | .2756 | 7. | 1-1/4 | 1.250 | 31.750 | – | 2.7559 | 70. | 4-3/4 | 4.750 | 120.650 |
| 9/32 | .281 | 7.144 | – | 1.2598 | 32. | 2-25/32 | 2.781 | 70.6439 | 4-13/16 | 4.8125 | 122.238 |
| – | .2953 | 7.5 | 1-9/32 | 1.281 | 32.544 | – | 2.7953 | 71. | 4-7/8 | 4.875 | 123.825 |
| 19/64 | .297 | 7.541 | – | 1.2992 | 33. | 2-13/16 | 2.8125 | 71.4376 | – | 4.9212 | 125. |
| 5/16 | .312 | 7.938 | 1-5/16 | 1.312 | 33.338 | – | 2.8346 | 72. | 4-15/16 | 4.9375 | 125.413 |
| – | .315 | 8. | – | 1.3386 | 34. | 2-27/32 | 2.844 | 72.2314 | 5 | 5.000 | 127.000 |
| 21/64 | .328 | 8.334 | 1-11/32 | 1.344 | 34.131 | – | 2.8740 | 73. | – | 5.1181 | 130. |
| – | .335 | 8.5 | 1-3/8 | 1.375 | 34.925 | 2-7/8 | 2.875 | 73.025 | 5-1/4 | 5.250 | 133.350 |
| 11/32 | .344 | 8.731 | – | 1.3779 | 35. | 2-29/32 | 2.9062 | 73.819 | 5-1/2 | 5.500 | 139.700 |
| – | .3543 | 9. | 1-13/32 | 1.406 | 35.719 | – | 2.9134 | 74. | – | 5.5118 | 140. |
| 23/64 | .359 | 9.128 | – | 1.4173 | 36. | 2-15/16 | 2.9375 | 74.613 | 5-3/4 | 5.750 | 146.050 |
| – | .374 | 9.5 | 1-7/16 | 1.438 | 36.513 | – | 2.9527 | 75. | – | 5.9055 | 150. |
| 3/8 | .375 | 9.525 | – | 1.4567 | 37. | 2-31/32 | 2.969 | 75.406 | 6 | 6.000 | 152.400 |
| 25/64 | .391 | 9.922 | 1-15/32 | 1.469 | 37.306 | – | 2.9921 | 76. | 6-1/4 | 6.250 | 158.750 |
| – | .3937 | 10. | – | 1.4961 | 38. | 3 | 3.000 | 76.200 | – | 6.2992 | 160. |
| 13/32 | .406 | 10.319 | 1-1/2 | 1.500 | 38.100 | 3-1/32 | 3.0312 | 76.994 | 6-1/2 | 6.500 | 165.100 |
| – | .413 | 10.5 | 1-17/32 | 1.531 | 38.894 | – | 3.0315 | 77. | – | 6.6929 | 170. |
| 27/64 | .422 | 10.716 | – | 1.5354 | 39. | 3-1/16 | 3.062 | 77.788 | 6-3/4 | 6.750 | 171.450 |
| – | .4331 | 11. | 1-9/16 | 1.562 | 39.688 | – | 3.0709 | 78. | 7 | 7.000 | 177.800 |
| 7/16 | .438 | 11.113 | – | 1.5748 | 40. | 3-3/32 | 3.094 | 78.581 | – | 7.0866 | 180. |
| 29/64 | .453 | 11.509 | 1-19/32 | 1.594 | 40.481 | – | 3.1102 | 79. | – | 7.4803 | 190. |
| 15/32 | .469 | 11.906 | – | 1.6142 | 41. | 3-1/8 | 3.125 | 79.375 | 7-1/2 | 7.500 | 190.500 |
| – | .4724 | 12. | 1-5/8 | 1.625 | 41.275 | – | 3.1496 | 80. | – | 7.8740 | 200. |
| 31/64 | .484 | 12.303 | – | 1.6535 | 42. | 3-5/32 | 3.156 | 80.169 | 8 | 8.000 | 203.200 |
| – | .492 | 12.5 | 1-21/32 | 1.0502 | 42.060 | 3-3/16 | 3.1875 | 80.963 | – | 8.2677 | 210. |
| 1/2 | .500 | 12.700 | 1-11/16 | 1.6875 | 42.863 | – | 3.1890 | 81. | 8-1/2 | 8.500 | 215.900 |
| – | .5118 | 13. | – | 1.6929 | 43. | 3-7/32 | 3.219 | 81.756 | – | 8.6614 | 220. |
| 33/64 | .5156 | 13.097 | 1-23/32 | 1.719 | 43.656 | – | 3.2283 | 82. | 9 | 9.000 | 228.600 |
| 17/32 | .531 | 13.494 | – | 1.7323 | 44. | 3-1/4 | 3.250 | 82.550 | – | 9.0551 | 230. |
| 35/64 | .547 | 13.891 | 1-3/4 | 1.750 | 44.450 | – | 3.2677 | 83. | – | 9.4488 | 240. |
| – | .5512 | 14. | – | 1.7717 | 45. | 3-9/32 | 3.281 | 83.344 | 9-1/2 | 9.500 | 241.300 |
| 9/16 | .563 | 14.288 | 1-25/32 | 1.781 | 45.244 | – | 3.3071 | 84. | – | 9.8425 | 250. |
| – | .571 | 14.5 | – | 1.8110 | 46. | 3-5/16 | 3.312 | 84.1377 | 10 | 10.000 | 254.001 |
| 37/64 | .578 | 14.684 | 1-13/16 | 1.8125 | 46.038 | 3-11/32 | 3.344 | 84.9314 | – | 10.2362 | 260. |
| – | .5906 | 15. | 1-27/32 | 1.844 | 46.831 | – | 3.3464 | 85. | – | 10.6299 | 270. |
| 19/32 | .594 | 15.081 | – | 1.8504 | 47. | 3-3/8 | 3.375 | 85.725 | 11 | 11.000 | 279.401 |
| 39/64 | .609 | 15.478 | 1-7/8 | 1.875 | 47.625 | – | 3.3858 | 86. | – | 11.0236 | 280. |
| 5/8 | .625 | 15.875 | – | 1.8898 | 48. | 3-13/32 | 3.406 | 86.519 | – | 11.4173 | 290. |
| – | .6299 | 16. | 1-29/32 | 1.9062 | 48.419 | – | 3.4252 | 87. | – | 11.8110 | 300. |
| 41/64 | .6406 | 16.272 | – | 1.9291 | 49. | 3-7/16 | 3.438 | 87.313 | 12 | 12.000 | 304.801 |
| – | .6496 | 16.5 | 1-15/16 | 1.9375 | 49.213 | – | 3.4646 | 88. | 13 | 13.000 | 330.201 |
| 21/32 | .656 | 16.669 | – | 1.9685 | 50. | 3-15/32 | 3.469 | 88.106 | – | 13.7795 | 350. |
| – | .6693 | 17. | 1-31/32 | 1.969 | 50.006 | 3-1/2 | 3.500 | 88.900 | 14 | 14.000 | 355.601 |
| 43/64 | .672 | 17.066 | 2 | 2.000 | 50.800 | – | 3.5039 | 89. | 15 | 15.000 | 381.001 |
| 11/16 | .6875 | 17.463 | – | 2.0079 | 51. | 3-17/32 | 3.531 | 89.694 | – | 15.7480 | 400. |
| 45/64 | .703 | 17.859 | 2-1/32 | 2.03125 | 51.594 | – | 3.5433 | 90. | 16 | 16.000 | 406.401 |
| – | .7087 | 18. | – | 2.0472 | 52. | 3-9/16 | 3.562 | 90.4877 | 17 | 17.000 | 431.801 |
| 23/32 | .719 | 18.256 | 2-1/16 | 2.062 | 52.388 | – | 3.5827 | 91. | – | 17.7165 | 450. |
| – | .7283 | 18.5 | – | 2.0866 | 53. | 3-19/32 | 3.594 | 91.281 | 18 | 18.000 | 457.201 |
| 47/64 | .734 | 18.653 | 2-3/32 | 2.094 | 53.181 | – | 3.622 | 92. | 19 | 19.000 | 482.601 |
| – | .7480 | 19. | 2-1/8 | 2.125 | 53.975 | 3-5/8 | 3.625 | 92.075 | – | 19.6850 | 500. |
| 3/4 | .750 | 19.050 | – | 2.126 | 54. | 3-21/32 | 3.656 | 92.869 | 20 | 20.000 | 508.001 |
| 49/64 | .7656 | 19.447 | 2-5/32 | 2.156 | 54.769 | – | 3.6614 | 93. | | | |

* Adapted from "TRUARC Technical Manual of Retaining Rings and Assembly Tools."

## Dual (Inch-metric) Dimensioning*

| Conversion of tolerances | | | | |
|---|---|---|---|---|
| Drawing usage | | Reference data | | |
| | | Conversion error | | |
| Inch | Millimeter (roundoff) | Millimeter error, inches | Percent error | Millimeter equivalent |
| 0.00004 | 0.0010 | −0.00000062 | −1.57 | 0.001016 |
| 0.00005 | 0.0013 | +0.00000118 | +2.36 | 0.00127 |
| 0.0001 | 0.0025 | −0.00000157 | −1.57 | 0.00254 |
| 0.0002 | 0.0051 | +0.00000079 | +0.39 | 0.00508 |
| 0.0003 | 0.0076 | −0.00000079 | −0.26 | 0.00762 |
| 0.0004 | 0.010 | −0.00000629 | −1.57 | 0.01016 |
| 0.0005 | 0.013 | +0.00001181 | +2.36 | 0.01270 |
| 0.0006 | 0.015 | −0.00000945 | −1.57 | 0.01524 |
| 0.0007 | 0.018 | +0.00000866 | +1.24 | 0.01778 |
| 0.0008 | 0.020 | −0.00001260 | −1.57 | 0.02032 |
| 0.0009 | 0.023 | +0.00000551 | +0.62 | 0.02286 |
| 0.0010 | 0.025 | −0.00001575 | −1.57 | 0.0254 |
| 0.0015 | 0.038 | −0.00000394 | −0.26 | 0.0381 |
| 0.0020 | 0.051 | +0.00000787 | +0.39 | 0.0508 |
| 0.0025 | 0.064 | +0.00001968 | +0.79 | 0.0635 |
| 0.0030 | 0.076 | −0.00000787 | −0.26 | 0.0762 |
| 0.004 | 0.10 | −0.00006299 | −1.57 | 0.1016 |
| 0.005 | 0.13 | +0.00011811 | +2.36 | 0.1270 |
| 0.006 | 0.15 | −0.00009449 | −1.57 | 0.1524 |
| 0.007 | 0.18 | +0.00008661 | +1.24 | 0.1778 |
| 0.008 | 0.20 | −0.00012598 | −1.57 | 0.2032 |
| 0.009 | 0.23 | +0.00005511 | +0.62 | 0.2286 |
| 0.010 | 0.25 | −0.00015748 | −1.57 | 0.254 |
| 0.015 | 0.38 | −0.00003937 | −0.26 | 0.381 |
| 0.020 | 0.51 | +0.00007874 | +0.39 | 0.508 |
| 0.030 | 0.76 | −0.00007874 | −0.26 | 0.762 |
| 0.04 | 1.0 | −0.00062993 | −1.57 | 1.016 |
| 0.05 | 1.3 | +0.00118110 | +2.36 | 1.270 |
| 0.06 | 1.5 | −0.00094488 | −1.57 | 1.524 |
| 0.07 | 1.8 | +0.00086614 | +1.24 | 1.778 |
| 0.08 | 2.0 | −0.00125984 | −1.57 | 2.032 |
| 0.09 | 2.3 | +0.00055118 | +0.62 | 2.286 |
| 0.10 | 2.5 | −0.00157480 | −1.57 | 2.54 |
| 0.20 | 5.1 | +0.00078740 | +0.39 | 5.08 |
| 0.30 | 7.6 | −0.00078740 | −0.26 | 7.62 |
| 0.40 | 10.0 | −0.00629922 | −1.57 | 10.16 |

*Adapted from S.A.E. Aero-Space Standards.

# APPENDIX D

## STANDARD PARTS, SIZES, SYMBOLS, AND ABBREVIATIONS

## American Standard Unified and American Thread Series[a]

*Threads per inch for coarse, fine, extra-fine, 8-thread, 12-thread, and 16-thread series* [b] [*tap-drill sizes for approximately 75 per cent depth of thread* (*not American Standard*)]

| Nominal size (basic major diam.) | Coarse-thd. series UNC and NC[c] in classes 1A, 1B, 2A, 2B, 3A, 3B, 2, 3 | | Fine-thd. series UNF and NF[c] in classes 1A, 1B, 2A, 2B, 3A, 3B, 2, 3 | | Extra-fine thd. series UNEF and NEF[d] in classes 2A, 2B, 2, 3 | | 8-thd. series 8N[c] in classes 2A, 2B, 2, 3 | | 12-thd. series 12UN and 12N[d] in classes 2A, 2B, 2, 3 | | 16-thd. series 16UN and 16N[d] in classes 2A, 2B, 2, 3 | |
|---|---|---|---|---|---|---|---|---|---|---|---|---|
| | Thd./in. | Tap drill | Thd./in. | Tap drill | Thd./in. | Tap drill | Thd./in. | Tap drill | Thd./in. | Tap drill | Thd./in. | Tap drill |
| 0(0.060) | .... | ..... | 80 | 3/64 | | | | | | | | |
| 1(0.073) | 64 | No. 53 | 72 | No. 53 | | | | | | | | |
| 2(0.086) | 56 | No. 50 | 64 | No. 50 | | | | | | | | |
| 3(0.099) | 48 | No. 47 | 56 | No. 45 | | | | | | | | |
| 4(0.112) | **40** | No. 43 | 48 | No. 42 | | | | | | | | |
| 5(0.125) | 40 | No. 38 | 44 | No. 37 | | | | | | | | |
| 6(0.138) | **32** | No. 36 | 40 | No. 33 | | | | | | | | |
| 8(0.164) | **32** | No. 29 | 36 | No. 29 | | | | | | | | |
| 10(0.190) | **24** | No. 25 | **32** | No. 21 | | | | | | | | |
| 12(0.216) | 24 | No. 16 | 28 | No. 14 | 32 | No. 13 | | | | | | |
| 1/4 | **20** | No. 7 | **28** | No. 3 | 32 | 7/32 | | | | | | |
| 5/16 | **18** | Let. F | **24** | Let. I | 32 | 9/32 | | | | | | |
| 3/8 | **16** | 5/16 | **24** | Let. Q | 32 | 11/32 | | | | | | |
| 7/16 | **14** | Let. U | **20** | 25/64 | **28** | 13/32 | | | | | | |
| 1/2 | **13** | 27/64 | **20** | 29/64 | **28** | 15/32 | .. | ... | **12** | 27/64 | | |
| 9/16 | **12** | 31/64 | **18** | 33/64 | 24 | 33/64 | .. | ... | **12** | 31/64 | | |
| 5/8 | **11** | 17/32 | **18** | 37/64 | 24 | 37/64 | .. | ... | 12 | 35/64 | | |
| 11/16 | .... | ..... | .. | ..... | 24 | 41/64 | .. | ... | 12 | 39/64 | | |
| 3/4 | **10** | 21/32 | **16** | 11/16 | **20** | 45/64 | .. | ... | 12 | 43/64 | **16** | 11/16 |
| 13/16 | .... | ..... | .. | ..... | **20** | 49/64 | .. | ... | 12 | 47/64 | **16** | 3/4 |
| 7/8 | **9** | 49/64 | **14** | 13/16 | **20** | 53/64 | .. | ... | 12 | 51/64 | **16** | 13/16 |
| 15/16 | .... | ..... | .. | ..... | **20** | 57/64 | .. | ... | **12** | 55/64 | **16** | 7/8 |
| 1 | .... | ..... | 14 | 15/16 | .. | ..... | 8 | 7/8 | | | | |
| 1 | **8** | 7/8 | **12** | 59/64 | **20** | 61/64 | .. | ... | **12** | 59/64 | **16** | 15/16 |
| 1 1/16 | .... | ..... | .. | ..... | 18 | 1 | .. | ... | **12** | 63/64 | **16** | 1 |
| 1 1/8 | **7** | 63/64 | **12** | 1 3/64 | 18 | 1 5/64 | 8 | 1 | **12** | 1 3/64 | **16** | 1 1/16 |
| 1 3/16 | .... | ..... | .. | ..... | 18 | 1 9/64 | .. | ... | **12** | 1 7/64 | **16** | 1 1/8 |
| 1 1/4 | **7** | 1 7/64 | **12** | 1 11/64 | 18 | 1 3/16 | 8 | 1 1/8 | **12** | 1 11/64 | **16** | 1 3/16 |
| 1 5/16 | .... | ..... | .. | ..... | 18 | 1 17/64 | .. | ... | **12** | 1 15/64 | **16** | 1 1/4 |
| 1 3/8 | **6** | 1 7/32 | **12** | 1 19/64 | 18 | 1 5/16 | 8 | 1 1/4 | **12** | 1 19/64 | **16** | 1 5/16 |
| 1 7/16 | .... | ..... | .. | ..... | 18 | 1 3/8 | .. | ... | **12** | 1 23/64 | **16** | 1 3/8 |
| 1 1/2 | **6** | 1 11/32 | **12** | 1 27/64 | 18 | 1 7/16 | 8 | 1 3/8 | **12** | 1 27/64 | **16** | 1 7/16 |
| 1 9/16 | .... | ..... | .. | ..... | 18 | 1 1/2 | .. | ... | .. | ..... | 16 | 1 1/2 |
| 1 5/8 | .... | ..... | .. | ..... | 18 | 1 9/16 | 8 | 1 1/2 | 12 | 1 35/64 | 16 | 1 9/16 |
| 1 11/16 | .... | ..... | .. | ..... | 18 | 1 5/8 | .. | ... | .. | ..... | 16 | 1 5/8 |

# American Standard Unified and American Thread Series (Cont.)

| Nominal size (basic major diam.) | Coarse-thd. series UNC and NC[c] in classes 1A, 1B, 2A, 2B, 3A, 3B, 2, 3 | | Fine-thd. series UNF and NF[c] in classes 1A, 1B, 2A, 2B, 3A, 3B, 2, 3 | | Extra-fine thd. series UNEF and NEF[d] in classes 2A, 2B, 2, 3 | | 8-thd. series 8N[c] in classes 2A, 2B, 2, 3 | | 12-thd. series 12UN and 12N[d] in classes 2A, 2B, 2, 3 | | 16-thd. series 16UN and 16N[d] in classes 2A, 2B, 2, 3 | |
|---|---|---|---|---|---|---|---|---|---|---|---|---|
| | Thd. /in. | Tap drill | Thd. /in. | Tap drill | Thd. /in. | Tap drill | Thd. /in. | Tap drill | Thd. /in. | Tap drill | Thd. /in. | Tap drill |
| 1 3/4 | **5** | 1 9/16 | .. | ..... | **16** | 1 11/66 | 8[e] | 1 5/8 | **12** | 1 43/64 | **16** | 1 11/16 |
| 1 13/16 | .... | ..... | .. | ..... | .. | ..... | .. | ... | .. | ..... | 16 | 1 3/4 |
| 1 7/8 | .... | ..... | .. | ..... | .. | ..... | 8 | 1 3/4 | 12 | 1 51/64 | 16 | 1 13/16 |
| 1 15/16 | .... | ..... | .. | ..... | .. | ..... | .. | ... | .. | ..... | 16 | 1 7/8 |
| 2 | **4 1/2** | 1 25/32 | .. | ..... | **16** | 1 15/16 | 8[e] | 1 7/8 | **12** | 1 59/64 | **16** | 1 15/16 |
| 2 1/16 | .... | ..... | .. | ..... | .. | ..... | .. | ... | .. | ..... | 16 | 2 |
| 2 1/8 | .... | ..... | .. | ..... | .. | ..... | 8 | 2 | 12 | 2 3/64 | 16 | 2 1/16 |
| 2 3/16 | .... | ..... | .. | ..... | .. | ..... | .. | ... | .. | ..... | 16 | 2 1/8 |
| 2 1/4 | **4 1/2** | 2 1/32 | .. | ..... | .. | ..... | 8[e] | 2 1/8 | **12** | 2 11/64 | **16** | 2 3/16 |
| 2 5/16 | .... | ..... | .. | ..... | .. | ..... | .. | ... | .. | ..... | 16 | 2 1/4 |
| 2 3/8 | .... | ..... | .. | ..... | .. | ..... | .. | ... | 12 | 2 19/64 | 16 | 2 5/16 |
| 2 7/16 | .... | ..... | .. | ..... | .. | ..... | .. | ... | .. | ..... | 16 | 2 3/8 |
| 2 1/2 | **4** | 2 1/4 | .. | ..... | .. | ..... | 8[e] | 2 3/8 | **12** | 2 27/64 | **16** | 2 7/16 |
| 2 5/8 | .... | ..... | .. | ..... | .. | ..... | .. | ... | 12 | 2 35/64 | 16 | 2 9/16 |
| 2 3/4 | **4** | 2 1/2 | .. | ..... | .. | ..... | 8[e] | 2 5/8 | **12** | 2 43/64 | **16** | 2 11/16 |
| 2 7/8 | .... | ..... | .. | ..... | .. | ..... | .. | ... | 12 | 2 51/64 | 16 | 2 13/16 |
| 3 | **4** | 2 3/4 | .. | ..... | .. | ..... | 8[e] | 2 7/8 | **12** | 2 59/64 | **16** | 2 15/16 |
| 3 1/8 | .... | ..... | .. | ..... | .. | ..... | .. | ... | 12 | 3 3/64 | 16 | 3 1/16 |
| 3 1/4 | **4** | 3 | .. | ..... | .. | ..... | 8[e] | 3 1/8 | **12** | 3 11/64 | **16** | 3 3/16 |
| 3 3/8 | .... | ..... | .. | ..... | .. | ..... | .. | ... | 12 | 3 19/64 | 16 | 3 5/16 |
| 3 1/2 | **4** | 3 1/4 | .. | ..... | .. | ..... | 8[e] | 3 3/8 | **12** | 3 27/64 | **16** | 3 7/16 |
| 3 5/8 | .... | ..... | .. | ..... | .. | ..... | .. | ... | 12 | 3 35/64 | 16 | 3 9/16 |
| 3 3/4 | **4** | 3 1/2 | .. | ..... | .. | ..... | 8[e] | 3 5/8 | **12** | 3 43/64 | **16** | 3 11/16 |
| 3 7/8 | .... | ..... | .. | ..... | .. | ..... | .. | ... | 12 | 3 51/64 | 16 | 3 13/16 |
| 4 | **4** | 3 3/4 | .. | ..... | .. | ..... | 8[e] | 3 7/8 | **12** | 3 59/64 | **16** | 3 15/16 |
| 4 1/4 | .... | ..... | .. | ..... | .. | ..... | 8[e] | 4 1/8 | **12** | 4 11/64 | **16** | 4 3/16 |
| 4 1/2 | .... | ..... | .. | ..... | .. | ..... | 8[e] | 4 3/8 | **12** | 4 27/64 | **16** | 4 7/16 |
| 4 3/4 | .... | ..... | .. | ..... | .. | ..... | 8[e] | 4 5/8 | **12** | 4 43/64 | **16** | 4 11/16 |
| 5 | .... | ..... | .. | ..... | .. | ..... | 8[e] | 4 7/8 | **12** | 4 59/64 | **16** | 4 15/16 |
| 5 1/4 | .... | ..... | .. | ..... | .. | ..... | 8[e] | 5 1/8 | **12** | 5 11/64 | **16** | 5 3/16 |
| 5 1/2 | .... | ..... | .. | ..... | .. | ..... | 8[e] | 5 3/8 | **12** | 5 27/64 | **16** | 5 7/16 |
| 5 3/4 | .... | ..... | .. | ..... | .. | ..... | 8[e] | 5 5/8 | **12** | 5 43/64 | **16** | 5 11/16 |
| 6 | .... | ..... | .. | ..... | .. | ..... | 8[e] | 5 7/8 | **12** | 5 59/64 | **16** | 5 15/16 |

[a] ASA B1.1—1960. Dimensions are in inches. [b] Bold type indicates unified combinations.

[c] Limits of size for classes are based on a length of engagement equal to the nominal diameter.

[d] Limits of size for classes are based on a length of engagement equal to nine times the pitch.

[e] These sizes, with specified limits of size, based on a length of engagement of 9 threads in classes 2A and 2B, are designated UN.

*Note.* If a thread is in both the 8-, 12-, or 16-thread series and the coarse, fine, or extra-fine-thread series, the symbols and tolerances of the latter series apply.

## American Standard Square and Hexagon Bolts* and Hexagon Cap Screws†

| Nominal size (basic major diam.) | Regular bolts: Width across flats *W*, sq.‡ and hex. | Regular bolts: Height *H*, Unfaced sq. and hex. | Regular bolts: Height *H*, Semifin. hex., fin. hex., and hex. screw | Heavy bolts: Width across flats *W* | Heavy bolts: Height *H*, Unfaced hex. | Heavy bolts: Height *H*, Semifin. hex. and fin. hex. |
|---|---|---|---|---|---|---|
| 1/4 | 3/8(sq.), 7/16(hex.) | 11/64 | 5/32 | | | |
| 5/16 | 1/2 | 7/32 | 13/64 | | | |
| 3/8 | 9/16 | 1/4 | 15/64 | | | |
| 7/16 | 5/8 | 19/64 | 9/32 | | | |
| 1/2 | 3/4 | 11/32 | 5/16 | 7/8 | 7/16 | 13/32 |
| 9/16 | 13/16 | 25/64 | 23/64 | 15/16 | 15/32 | 7/16 |
| 5/8 | 15/16 | 27/64 | 25/64 | 1 1/16 | 17/32 | 1/2 |
| 3/4 | 1 1/8 | 1/2 | 15/32 | 1 1/4 | 5/8 | 19/32 |
| 7/8 | 1 5/16 | 37/64 | 35/64 | 1 7/16 | 23/32 | 11/16 |
| 1 | 1 1/2 | 43/64 | 39/64 | 1 5/8 | 13/16 | 3/4 |
| 1 1/8 | 1 11/16 | 3/4 | 11/16 | 1 13/16 | 29/32 | 27/32 |
| 1 1/4 | 1 7/8 | 27/32 | 25/32 | 2 | 1 | 15/16 |
| 1 3/8 | 2 1/16 | 29/32 | 27/32 | 2 3/16 | 1 3/32 | 1 1/32 |
| 1 1/2 | 2 1/4 | 1 | 15/16 | 2 3/8 | 1 3/16 | 1 1/8 |
| 1 5/8 | 2 7/16 | 1 1/16 | 1 | 2 9/16 | 1 9/32 | 1 7/32 |
| 1 3/4 | 2 5/8 | 1 5/32 | 1 3/32 | 2 3/4 | 1 3/8 | 1 5/16 |
| 1 7/8 | 2 13/16 | 1 7/32 | 1 5/32 | 2 15/16 | 1 15/32 | 1 13/32 |
| 2 | 3 | 1 11/32 | 1 7/32 | 3 1/8 | 1 9/16 | 1 7/16 |
| 2 1/4 | 3 3/8 | 1 1/2 | 1 3/8 | 3 1/2 | 1 3/4 | 1 5/8 |
| 2 1/2 | 3 3/4 | 1 21/32 | 1 17/32 | 3 7/8 | 1 15/16 | 1 13/16 |
| 2 3/4 | 4 1/8 | 1 13/16 | 1 11/16 | 4 1/4 | 2 1/8 | 2 |
| 3 | 4 1/2 | 2 | 1 7/8 | 4 5/8 | 2 5/16 | 2 3/16 |
| 3 1/4 | 4 7/8 | 2 3/16 | 2 | | | |
| 3 1/2 | 5 1/4 | 2 5/16 | 2 1/8 | | | |
| 3 3/4 | 5 5/8 | 2 1/2 | 2 5/16 | | | |
| 4 | 6 | 2 11/16 | 2 1/2 | | | |

* ASA B18.2–1960. All dimensions in inches.

† ASA B18.6.2–1956. Hexagon-head cap screw in sizes 1/4 to 1 1/2 only.

‡ Square bolts in (nominal) sizes 1/4 to 1 5/8 only.

For bolt-length increments, see table on page A68. Threads are coarse series, class 2A, except for finished hexagon bolt and hexagon cap screw, which are coarse, fine, or 8-pitch, class 2A. Minimum thread length: $2D + 1/4$ in. for bolts 6 in. or less in length; $2D + 1/2$ in. for bolts over 6 in. in length; bolts too short for formula, thread entire length.

## American Standard Square and Hexagon Nuts*

| Nominal size (basic major diam.) | Reg. nuts | | | | | | Heavy nuts | | | | | | | Reg. sq. nuts | |
|---|---|---|---|---|---|---|---|---|---|---|---|---|---|---|---|
| | Width across flats *W* | Thickness *T* | | | | | Width across flats *W* | Thickness *T* | | | | Slot | | Width across flats *W* | Thickness *T* |
| | | Reg. hex. | Reg. hex. jam | Semi-fin. and fin. hex. and hex. slotted | Semi-fin. and fin. hex. jam | Semi-fin. hex. thick, thick slotted,† and castle | | Reg. sq. and hex. | Reg. hex. jam | Semi-fin. hex. and hex. slotted | Semi-fin. hex. jam | Width | Depth | | |
| 1/4 | 7/16 | ..... | .... | 7/32 | 5/32 | 9/32 | 1/2 | 1/4 | 3/16 | 15/64 | 11/64 | 5/64 | 3/32 | 7/16 | 7/32 |
| 5/16 | 1/2 | ..... | .... | 17/64 | 3/16 | 21/64 | 9/16 | 5/16 | 7/32 | 19/64 | 13/64 | 3/32 | 3/32 | 9/16 | 17/64 |
| 3/8 | 9/16 | ..... | .... | 21/64 | 7/32 | 13/32 | 11/16 | 3/8 | 1/4 | 23/64 | 15/64 | 1/8 | 1/8 | 5/8 | 21/64 |
| 7/16 | 11/16 | ..... | .... | 3/8 | 1/4 | 29/64 | 3/4 | 7/16 | 9/32 | 27/64 | 17/64 | 1/8 | 5/32 | 3/4 | 3/8 |
| 1/2 | 3/4 | ..... | .... | 7/16 | 5/16 | 9/16 | 7/8 | 1/2 | 5/16 | 31/64 | 19/64 | 5/32 | 5/32 | 13/16 | 7/16 |
| 9/16 | 7/8 | ..... | .... | 31/64 | 5/16 | 39/64 | 15/16 | | | 35/64 | 21/64 | 5/32 | 3/16 | | |
| 5/8 | 15/16 | ..... | .... | 35/64 | 3/8 | 23/32 | 1 1/16 | 5/8 | 3/8 | 39/64 | 23/64 | 3/16 | 7/32 | 1 | 35/64 |
| 3/4 | 1 1/8 | 21/32 | 7/16 | 41/64 | 27/64 | 13/16 | 1 1/4 | 3/4 | 7/16 | 47/64 | 27/64 | 3/16 | 1/4 | 1 1/8 | 21/32 |
| 7/8 | 1 5/16 | 49/64 | 1/2 | 3/4 | 31/64 | 29/32 | 1 7/16 | 7/8 | 1/2 | 55/64 | 31/64 | 3/16 | 1/4 | 1 5/16 | 49/64 |
| 1 | 1 1/2 | 7/8 | 9/16 | 55/64 | 35/64 | 1 | 1 5/8 | 1 | 9/16 | 63/64 | 35/64 | 1/4 | 9/32 | 1 1/2 | 7/8 |
| 1 1/8 | 1 11/16 | 1 | 5/8 | 31/32 | 39/64 | 1 5/32 | 1 13/16 | 1 1/8 | 5/8 | 1 7/64 | 39/64 | 1/4 | 11/32 | 1 11/16 | 1 |
| 1 1/4 | 1 7/8 | 1 3/32 | 3/4 | 1 1/16 | 23/32 | 1 1/4 | 2 | 1 1/4 | 3/4 | 1 7/32 | 23/32 | 5/16 | 3/8 | 1 7/8 | 1 3/32 |
| 1 3/8 | 2 1/16 | 1 13/64 | 13/16 | 1 11/64 | 25/32 | 1 3/8 | 2 3/16 | 1 3/8 | 13/16 | 1 11/32 | 25/32 | 5/16 | 3/8 | 2 1/16 | 1 13/64 |
| 1 1/2 | 2 1/4 | 1 5/16 | 7/8 | 1 9/32 | 27/32 | 1 1/2 | 2 3/8 | 1 1/2 | 7/8 | 1 15/32 | 27/32 | 3/8 | 7/16 | 2 1/4 | 1 5/16 |
| 1 5/8 | 2 7/16 | ..... | .... | 1 25/64 | 29/32 | ..... | 2 9/16 | .... | ..... | 1 19/32 | 29/32 | 3/8 | 7/16 | | |
| 1 3/4 | 2 5/8 | ..... | .... | 1 1/2 | 31/32 | ..... | 2 3/4 | 1 3/4 | 1 | 1 23/32 | 31/32 | 7/16 | 1/2 | | |
| 1 7/8 | 2 13/16 | ..... | .... | 1 39/64 | 1 1/32 | ..... | 2 15/16 | .... | ..... | 1 27/32 | 1 1/32 | 7/16 | 9/16 | | |
| 2 | 3 | ..... | .... | 1 23/32 | 1 3/32 | ..... | 3 1/8 | 2 | 1 1/8 | 1 31/32 | 1 3/32 | 7/16 | 9/16 | | |
| 2 1/4 | 3 3/8 | ..... | .... | 1 59/64 | 1 13/64 | ..... | 3 1/2 | 2 1/4 | 1 1/4 | 2 13/64 | 1 13/64 | 7/16 | 9/16 | | |
| 2 1/2 | 3 3/4 | ..... | .... | 2 9/64 | 1 29/64 | ..... | 3 7/8 | 2 1/2 | 1 1/2 | 2 29/64 | 1 29/64 | 9/16 | 11/16 | | |
| 2 3/4 | 4 1/8 | ..... | .... | 2 23/64 | 1 37/64 | ..... | 4 1/4 | 2 3/4 | 1 5/8 | 2 45/64 | 1 37/64 | 9/16 | 11/16 | | |
| 3 | 4 1/2 | ..... | .... | 2 37/64 | 1 45/64 | ..... | 4 5/8 | 3 | 1 3/4 | 2 61/64 | 1 45/64 | 5/8 | 3/4 | | |
| 3 1/4 | ..... | ..... | .... | ..... | ..... | ..... | 5 | 3 1/4 | 1 7/8 | 3 3/16 | 1 13/16 | 5/8 | 3/4 | | |
| 3 1/2 | ..... | ..... | .... | ..... | ..... | ..... | 5 3/8 | 3 1/2 | 2 | 3 7/16 | 1 15/16 | 5/8 | 3/4 | | |
| 3 3/4 | ..... | ..... | .... | ..... | ..... | ..... | 5 3/4 | 3 3/4 | 2 1/8 | 3 11/16 | 2 1/16 | 5/8 | 3/4 | | |
| 4 | ..... | ..... | .... | ..... | ..... | ..... | 6 1/8 | 4 | 2 1/4 | 3 15/16 | 2 3/16 | 5/8 | 3/4 | | |

* ASA B18.2—1960.

† Slot dimensions for regular slotted nuts are same as for heavy slotted nuts.

*Thread.* Regular and heavy nuts: coarse series, class 2B; finished and semifinished nuts, regular and heavy (all): coarse, fine, or 8-pitch series, class 2B; thick nuts: coarse or fine series, class 2B.

## Sizes of Numbered and Lettered Drills

| No. | Size | No. | Size | No. | Size | Letter | Size |
|---|---|---|---|---|---|---|---|
| 80 | 0.0135 | 53 | 0.0595 | 26 | 0.1470 | *A* | 0.2340 |
| 79 | 0.0145 | 52 | 0.0635 | 25 | 0.1495 | *B* | 0.2380 |
| 78 | 0.0160 | 51 | 0.0670 | 24 | 0.1520 | *C* | 0.2420 |
| 77 | 0.0180 | 50 | 0.0700 | 23 | 0.1540 | *D* | 0.2460 |
| 76 | 0.0200 | 49 | 0.0730 | 22 | 0.1570 | *E* | 0.2500 |
| 75 | 0.0210 | 48 | 0.0760 | 21 | 0.1590 | *F* | 0.2570 |
| 74 | 0.0225 | 47 | 0.0785 | 20 | 0.1610 | *G* | 0.2610 |
| 73 | 0.0240 | 46 | 0.0810 | 19 | 0.1660 | *H* | 0.2660 |
| 72 | 0.0250 | 45 | 0.0820 | 18 | 0.1695 | *I* | 0.2720 |
| 71 | 0.0260 | 44 | 0.0860 | 17 | 0.1730 | *J* | 0.2770 |
| 70 | 0.0280 | 43 | 0.0890 | 16 | 0.1770 | *K* | 0.2810 |
| 69 | 0.0292 | 42 | 0.0935 | 15 | 0.1800 | *L* | 0.2900 |
| 68 | 0.0310 | 41 | 0.0960 | 14 | 0.1820 | *M* | 0.2950 |
| 67 | 0.0320 | 40 | 0.0980 | 13 | 0.1850 | *N* | 0.3020 |
| 66 | 0.0330 | 39 | 0.0995 | 12 | 0.1890 | *O* | 0.3160 |
| 65 | 0.0350 | 38 | 0.1015 | 11 | 0.1910 | *P* | 0.3230 |
| 64 | 0.0360 | 37 | 0.1040 | 10 | 0.1935 | *Q* | 0.3320 |
| 63 | 0.0370 | 36 | 0.1065 | 9 | 0.1960 | *R* | 0.3390 |
| 62 | 0.0380 | 35 | 0.1100 | 8 | 0.1990 | *S* | 0.3480 |
| 61 | 0.0390 | 34 | 0.1110 | 7 | 0.2010 | *T* | 0.3580 |
| 60 | 0.0400 | 33 | 0.1130 | 6 | 0.2040 | *U* | 0.3680 |
| 59 | 0.0410 | 32 | 0.1160 | 5 | 0.2055 | *V* | 0.3770 |
| 58 | 0.0420 | 31 | 0.1200 | 4 | 0.2090 | *W* | 0.3860 |
| 57 | 0.0430 | 30 | 0.1285 | 3 | 0.2130 | *X* | 0.3970 |
| 56 | 0.0465 | 29 | 0.1360 | 2 | 0.2210 | *Y* | 0.4040 |
| 55 | 0.0520 | 28 | 0.1405 | 1 | 0.2280 | *Z* | 0.4130 |
| 54 | 0.0550 | 27 | 0.1440 | | | | |

## Acme and Stub Acme Threads*

*ASA-preferred diameter-pitch combinations*

| Nominal (major) diam. | Threads/in. | Nominal (major) diam. | Threads/in. | Nominal (major) diam. | Threads/in. | Nominal (major) diam. | Threads/in. |
|---|---|---|---|---|---|---|---|
| 1/4 | 16 | 3/4 | 6 | 1 1/2 | 4 | 3 | 2 |
| 5/16 | 14 | 7/8 | 6 | 1 3/4 | 4 | 3 1/2 | 2 |
| 3/8 | 12 | 1 | 5 | 2 | 4 | 4 | 2 |
| 7/16 | 12 | 1 1/8 | 5 | 2 1/4 | 3 | 4 1/2 | 2 |
| 1/2 | 10 | 1 1/4 | 5 | 2 1/2 | 3 | 5 | 2 |
| 5/8 | 8 | 1 3/8 | 4 | 2 3/4 | 3 | | |

## Buttress Threads*

| Recommended diam., in. | Assoc. threads/in. |
|---|---|
| 1/2, 9/16, 5/8, 11/16 | 20, 16, 12 |
| 3/4, 7/8, 1 | 16, 12, 10 |
| 1 1/8, 1 1/4, 1 3/8, 1 1/2 | 16, 12, 10, 8, 6 |
| 1 3/4, 2, 2 1/4, 2 1/2, 2 3/4, 3, 3 1/2, 4 | 16, 12, 10, 8, 6, 5, 4 |
| 4 1/2, 5, 5 1/2, 6 | 12, 10, 8, 6, 5, 4, 3 |
| 7, 8, 9, 10 | 10, 8, 6, 5, 4, 3, 2 1/2, 2 |
| 11, 12, 14, 16 | 10, 8, 6, 5, 4, 3, 2 1/2, 2, 1 1/2, 1 1/4 |
| 18, 20, 22, 24 | 8, 6, 5, 4, 3, 2 1/2, 2, 1 1/2, 1 1/4, 1 |

* ASA B1.5—1952 and B1.9—1953. Diameters in inches.

## American Standard Round-head Bolts[a]

*Proportions for drawing purposes*

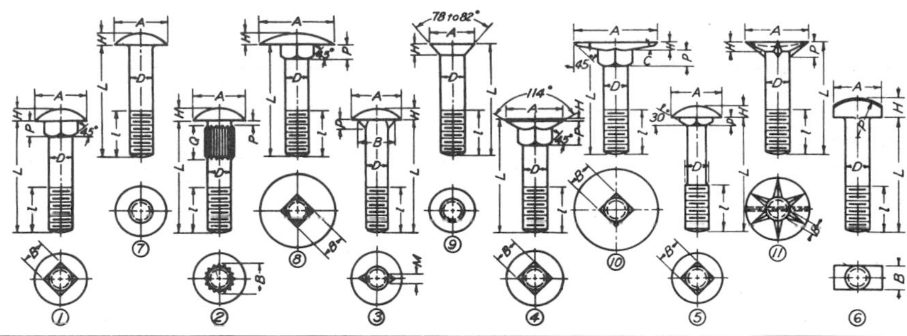


| Fastener name | Nominal diam.[b] (basic major diam.) | $A$ | $H$ | $P$ or $M$ | $B$ | |
|---|---|---|---|---|---|---|
| Carriage bolts: Round-head square-neck bolt[c] | No. 10, $\frac{1}{4}''$ to $\frac{1}{2}''$ by ($\frac{1}{16}''$), $\frac{5}{8}''$ to 1″ by ($\frac{1}{8}''$) | $2D + \frac{1}{16}$ | $\frac{D}{2}$ | $\frac{D}{2}$ | $D$ | |
| Carriage bolts: Round-head ribbed-neck bolt[d] | No. 10, $\frac{1}{4}''$ to $\frac{1}{2}''$ by ($\frac{1}{16}''$), $\frac{5}{8}''$ and $\frac{3}{4}''$ | $2D + \frac{1}{16}$ | $\frac{D}{2}$ | $\frac{1}{16}$ | $D + \frac{1}{16}$ | $Q$: $\frac{3}{16}''$ for $L = \frac{7}{8}''$ or less; $\frac{5}{16}''$ for $L = 1''$ and $1\frac{1}{8}''$; $\frac{1}{2}''$ for $L = 1\frac{1}{4}''$ or more |
| Carriage bolts: Round-head fin.-neck bolt[c] | No. 10, $\frac{1}{4}''$ to $\frac{1}{2}''$ by ($\frac{1}{16}''$) | $2D + \frac{1}{16}$ | $\frac{D}{2}$ | $\frac{3}{8}D$ | $1\frac{1}{2}D + \frac{1}{16}$ | |
| Carriage bolts: 114° counter-sunk square-neck bolt[c] | No. 10, $\frac{1}{4}''$ to $\frac{1}{2}''$ by ($\frac{1}{16}''$), $\frac{5}{8}''$ and $\frac{3}{4}''$ | $2D + \frac{1}{8}$ | $\frac{1}{32}$ | $D + \frac{1}{32}$ | $D$ | |
| Round-head short square-neck bolt[e] | $\frac{1}{4}''$ to $\frac{1}{2}''$ by ($\frac{1}{16}''$), $\frac{5}{8}''$ and $\frac{3}{4}''$ | $2D + \frac{1}{16}$ | $\frac{D}{2}$ | $\frac{D}{4} + \frac{1}{32}$ | $D$ | |
| T-head bolt[d] | $\frac{1}{4}''$ to $\frac{1}{2}''$ by ($\frac{1}{16}''$), $\frac{5}{8}''$ to 1″ by ($\frac{1}{8}''$) | $2D$ | $\frac{7}{8}D$ | $1\frac{5}{8}D$ | $D$ | |
| Round-head bolt[c] (buttonhead bolt) | No. 10, $\frac{1}{4}''$ to $\frac{1}{2}''$ by ($\frac{1}{16}''$), $\frac{5}{8}''$ to 1″ by ($\frac{1}{8}''$) | $2D + \frac{1}{16}$ | $\frac{D}{2}$ | | | |
| Step bolt[c] | No. 10, $\frac{1}{4}''$ to $\frac{1}{2}''$ by ($\frac{1}{16}''$) | $3D + \frac{1}{16}$ | $\frac{D}{2}$ | $\frac{D}{2}$ | $D$ | |
| Countersunk bolt[c] (may be slotted if so specified) | $\frac{1}{4}''$ to $\frac{1}{2}''$ by ($\frac{1}{16}''$), $\frac{5}{8}''$ to $1\frac{1}{2}''$ by ($\frac{1}{8}''$) | $1.8D$ | Obtain by projection | | | |
| Elevator bolt, flat head, countersunk[c] | No. 10, $\frac{1}{4}''$ to $\frac{1}{2}''$ by ($\frac{1}{16}''$) | $2\frac{1}{2}D + \frac{5}{16}$ | $\frac{D}{3}$ | $\frac{D}{2} + \frac{1}{16}$ | $D$ | Angle $C = 16D + 5°$ (approx.) |
| Elevator bolt, ribbed head[d] (slotted or unslotted as specified) | $\frac{1}{4}''$, $\frac{5}{16}''$, $\frac{3}{8}''$ | $2D + \frac{1}{16}$ | $\frac{D}{2} - \frac{1}{32}$ | $\frac{D}{2} + \frac{3}{64}$ | $\frac{D}{8} + \frac{1}{16}$ | |

[a] The proportions in this table are in some instances approximate and are intended for drawing purposes only. For exact dimensions see ASA B18.5—1959, from which this table was compiled. Dimensions are in inches.

[b] Fractions in parentheses show diameter increments, *e.g.*, $\frac{1}{4}$ in. to $\frac{1}{2}$ in. by ($\frac{1}{16}$ in.) includes the diameters $\frac{1}{4}$ in., $\frac{5}{16}$ in., $\frac{3}{8}$ in., $\frac{7}{16}$ in., and $\frac{1}{2}$ in

Threads are coarse series, class 2A.

Minimum thread length $l$: $2D + \frac{1}{4}$ in. for bolts 6 in. or less in length; $2D + \frac{1}{2}$ in. for bolts over 6 in. in length.

For bolt length increments see table page 720.

[c] Full-size body bolts furnished unless undersize body is specified.

[d] Only full-size body bolts furnished.

[e] Undersize body bolts furnished unless full-size body is specified.

## Bolt-length Increments*

| Bolt diameter | | 1/4 | 5/16 | 3/8 | 7/16 | 1/2 | 5/8 | 3/4 | 7/8 | 1 |
|---|---|---|---|---|---|---|---|---|---|---|
| Length increments | 1/4 | 3/4–3 | 3/4–4 | 3/4–6 | 1–3 | 1–6 | 1–6 | 1–6 | 1–4 1/2 | |
| | 1/2 | 3–4 | 4–5 | 6–9 | 3–6 | 6–13 | 6–10 | 6–15 | 4 1/2–6 | 3–6 |
| | 1 | 4–5 | .... | 9–12 | 6–8 | 13–24 | 10–22 | 15–24 | 6–20 | 6–12 |
| | 2 | .... | .... | ..... | ... | ..... | 22–30 | 24–30 | 20–30 | 12–30 |

* Compiled from manufacturers' catalogues.

*Example.* 1/4-in. bolt lengths increase by 1/4-in. increments from 3/4- to 3-in. length. 1/2-in. bolt lengths increase by 1/2-in. increments from 6- to 13-in. length. 1-in. bolt lengths increase by 2-in. increments from 12- to 30-in. length.

## American Standard Cap Screws[a]—Socket[b] and Slotted Heads[c]

*For hexagon-head screws, see page A64.*

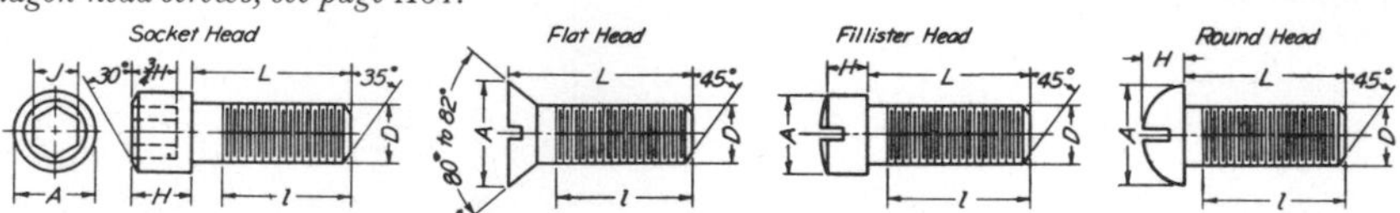


| Nominal diam. | Socket head[d] | | | Flat head[e] | Fillister head[e] | | Round head[e] | |
|---|---|---|---|---|---|---|---|---|
| | A | H | J | A | A | H | A | H |
| 0 | 0.096 | 0.060 | 0.050 | | | | | |
| 1 | 0.118 | 0.073 | 0.050 | | | | | |
| 2 | 0.140 | 0.086 | 1/16 | | | | | |
| 3 | 0.161 | 0.099 | 5/64 | | | | | |
| 4 | 0.183 | 0.112 | 5/64 | | | | | |
| 5 | 0.205 | 0.125 | 3/32 | | | | | |
| 6 | 0.226 | 0.138 | 3/32 | | | | | |
| 8 | 0.270 | 0.164 | 1/8 | | | | | |
| 10 | 5/16 | 0.190 | 5/32 | | | | | |
| 12 | 11/32 | 0.216 | 5/32 | | | | | |
| 1/4 | 3/8 | 1/4 | 3/16 | 1/2 | 3/8 | 11/64 | 7/16 | 3/16 |
| 5/16 | 7/16 | 5/16 | 7/32 | 5/8 | 7/16 | 13/64 | 9/16 | 15/64 |
| 3/8 | 9/16 | 3/8 | 5/16 | 3/4 | 9/16 | 1/4 | 5/8 | 17/64 |
| 7/16 | 5/8 | 7/16 | 5/16 | 13/16 | 5/8 | 19/64 | 3/4 | 5/16 |
| 1/2 | 3/4 | 1/2 | 3/8 | 7/8 | 3/4 | 21/64 | 13/16 | 11/32 |
| 9/16 | 13/16 | 9/16 | 3/8 | 1 | 13/16 | 3/8 | 15/16 | 13/32 |
| 5/8 | 7/8 | 5/8 | 1/2 | 1 1/8 | 7/8 | 27/64 | 1 | 7/16 |
| 3/4 | 1 | 3/4 | 9/16 | 1 3/8 | 1 | 1/2 | 1 1/4 | 17/32 |
| 7/8 | 1 1/8 | 7/8 | 9/16 | 1 5/8 | 1 1/8 | 19/32 | | |
| 1 | 1 5/16 | 1 | 5/8 | 1 7/8 | 1 5/16 | 21/32 | | |
| 1 1/8 | 1 1/2 | 1 1/8 | 3/4 | | | | | |
| 1 1/4 | 1 3/4 | 1 1/4 | 3/4 | | | | | |
| 1 3/8 | 1 7/8 | 1 3/8 | 3/4 | | | | | |
| 1 1/2 | 2 | 1 1/2 | 1 | | | | | |

[a] Dimensions in inches.

[b] ASA B18.3—1961.

[c] ASA B18.6.2—1956.

[d] Thread coarse or fine, class 3A. Thread length $l$: coarse thread, $2D + 1/2$ in.; fine thread, $1\frac{1}{2}D + 1/2$ in.

[e] Thread coarse, fine, or 8-pitch, class 2A. Thread length $l$: $2D + 1/4$ in.

Slot proportions vary with size of screw; draw to look well. All body-length increments for screw lengths 1/4 in. to 1 in. = 1/8 in., for screw lengths 1 in. to 4 in. = 1/4 in., for screw lengths 4 in. to 6 in. = 1/2 in.

## American Standard Machine Screws*

*Heads may be slotted or recessed*

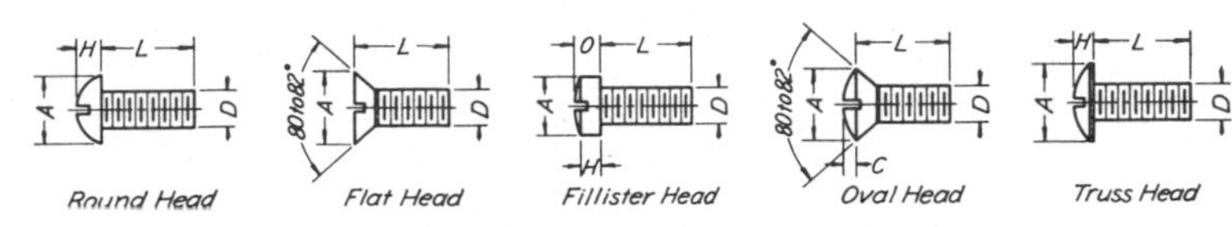


| Nominal diam. | Round head | | Flat head | Fillister head | | | Oval head | | Truss head | |
|---|---|---|---|---|---|---|---|---|---|---|
| | A | H | A | A | H | O | A | C | A | H |
| 0 | 0.113 | 0.053 | 0.119 | 0.096 | 0.045 | 0.059 | 0.119 | 0.021 | | |
| 1 | 0.138 | 0.061 | 0.146 | 0.118 | 0.053 | 0.071 | 0.146 | 0.025 | 0.194 | 0.053 |
| 2 | 0.162 | 0.069 | 0.172 | 0.140 | 0.062 | 0.083 | 0.172 | 0.029 | 0.226 | 0.061 |
| 3 | 0.187 | 0.078 | 0.199 | 0.161 | 0.070 | 0.095 | 0.199 | 0.033 | 0.257 | 0.069 |
| 4 | 0.211 | 0.086 | 0.225 | 0.183 | 0.079 | 0.107 | 0.225 | 0.037 | 0.289 | 0.078 |
| 5 | 0.236 | 0.095 | 0.252 | 0.205 | 0.088 | 0.120 | 0.252 | 0.041 | 0.321 | 0.086 |
| 6 | 0.260 | 0.103 | 0.279 | 0.226 | 0.096 | 0.132 | 0.279 | 0.045 | 0.352 | 0.094 |
| 8 | 0.309 | 0.120 | 0.332 | 0.270 | 0.113 | 0.156 | 0.332 | 0.052 | 0.384 | 0.102 |
| 10 | 0.359 | 0.137 | 0.385 | 0.313 | 0.130 | 0.180 | 0.385 | 0.060 | 0.448 | 0.118 |
| 12 | 0.408 | 0.153 | 0.438 | 0.357 | 0.148 | 0.205 | 0.438 | 0.068 | 0.511 | 0.134 |
| 1/4 | 0.472 | 0.175 | 0.507 | 0.414 | 0.170 | 0.237 | 0.507 | 0.079 | 0.573 | 0.150 |
| 5/16 | 0.590 | 0.216 | 0.635 | 0.518 | 0.211 | 0.295 | 0.635 | 0.099 | 0.698 | 0.183 |
| 3/8 | 0.708 | 0.256 | 0.762 | 0.622 | 0.253 | 0.355 | 0.762 | 0.117 | 0.823 | 0.215 |
| 7/16 | 0.750 | 0.328 | 0.812 | 0.625 | 0.265 | 0.368 | 0.812 | 0.122 | 0.948 | 0.248 |
| 1/2 | 0.813 | 0.355 | 0.875 | 0.750 | 0.297 | 0.412 | 0.875 | 0.131 | 1.073 | 0.280 |
| 9/16 | 0.938 | 0.410 | 1.000 | 0.812 | 0.336 | 0.466 | 1.000 | 0.150 | 1.198 | 0.312 |
| 5/8 | 1.000 | 0.438 | 1.125 | 0.875 | 0.375 | 0.521 | 1.125 | 0.169 | 1.323 | 0.345 |
| 3/4 | 1.250 | 0.547 | 1.375 | 1.000 | 0.441 | 0.612 | 1.375 | 0.206 | 1.573 | 0.410 |

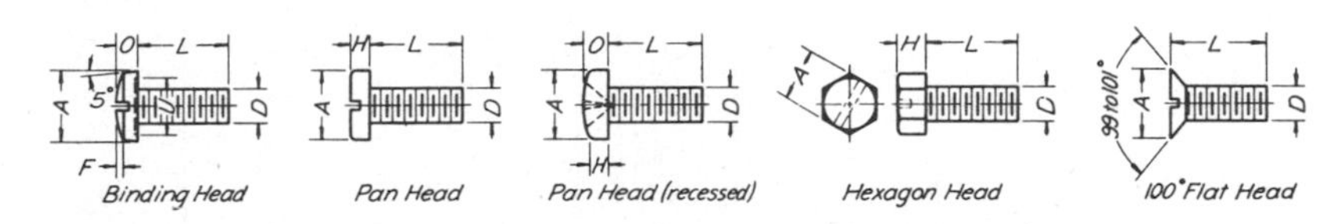


| Nominal diam. | Binding head | | | | Pan head | | | Hexagon head | | 100° flat head |
|---|---|---|---|---|---|---|---|---|---|---|
| | A | O | F | U | A | H | O | A | H | A |
| 2 | 0.181 | 0.046 | 0.018 | 0.141 | 0.167 | 0.053 | 0.062 | 0.125 | 0.050 | |
| 3 | 0.208 | 0.054 | 0.022 | 0.162 | 0.193 | 0.060 | 0.071 | 0.187 | 0.055 | |
| 4 | 0.235 | 0.063 | 0.025 | 0.184 | 0.219 | 0.068 | 0.080 | 0.187 | 0.060 | 0.225 |
| 5 | 0.263 | 0.071 | 0.029 | 0.205 | 0.245 | 0.075 | 0.089 | 0.187 | 0.070 | |
| 6 | 0.290 | 0.080 | 0.032 | 0.226 | 0.270 | 0.082 | 0.097 | 0.250 | 0.080 | 0.279 |
| 8 | 0.344 | 0.097 | 0.039 | 0.269 | 0.322 | 0.096 | 0.115 | 0.250 | 0.110 | 0.332 |
| 10 | 0.399 | 0.114 | 0.045 | 0.312 | 0.373 | 0.110 | 0.133 | 0.312 | 0.120 | 0.385 |
| 12 | 0.454 | 0.130 | 0.052 | 0.354 | 0.425 | 0.125 | 0.151 | 0.312 | 0.155 | |
| 1/4 | 0.513 | 0.153 | 0.061 | 0.410 | 0.492 | 0.144 | 0.175 | 0.375 | 0.190 | 0.507 |
| 5/16 | 0.641 | 0.193 | 0.077 | 0.513 | 0.615 | 0.178 | 0.218 | 0.500 | 0.230 | 0.635 |
| 3/8 | 0.769 | 0.234 | 0.094 | 0.615 | 0.740 | 0.212 | 0.261 | 0.562 | 0.295 | 0.762 |

* ASA B18.6.3—1962. Dimensions given are maximum values, all in inches.

Thread length: screws 2 in. long or less, thread entire length; screws over 2 in. long, thread length $l = 1\frac{3}{4}$ in. Threads are coarse or fine series, class 2. Heads may be slotted or recessed as specified, excepting hexagon form, which is plain or may be slotted if so specified. Slot and recess proportions vary with size of fastener; draw to look well.

## American Standard Machine-screw* and Stove-bolt† Nuts‡

| Nominal size | 0 | 1 | 2 | 3 | 4 | 5 | 6 | 8 | 10 | 12 | 1/4 | 5/16 | 3/8 |
|---|---|---|---|---|---|---|---|---|---|---|---|---|---|
| "W" | 5/32 | 5/32 | 3/16 | 3/16 | 1/4 | 5/16 | 5/16 | 11/32 | 3/8 | 7/16 | 7/16 | 9/16 | 5/8 |
| "T" | 3/64 | 3/64 | 1/16 | 1/16 | 3/32 | 7/64 | 7/64 | 1/8 | 1/8 | 5/32 | 3/16 | 7/32 | 1/4 |

* Machine-screw nuts are hexagonal and square.
† Stove-bolt nuts are square.
‡ ASA B18.6.3—1962. Dimensions are in inches.
Thread is coarse series for square nuts and coarse or fine series for hexagon nuts; class 2B.

## American Standard Hexagon Socket,* Slotted Headless,† and Square-head‡ Setscrews

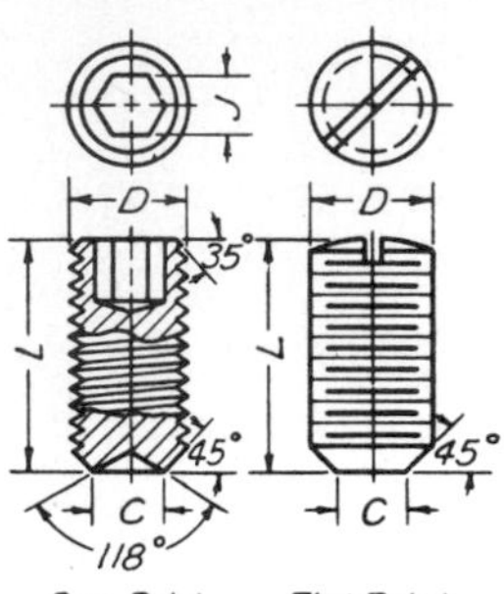

Cup Point Flat Point

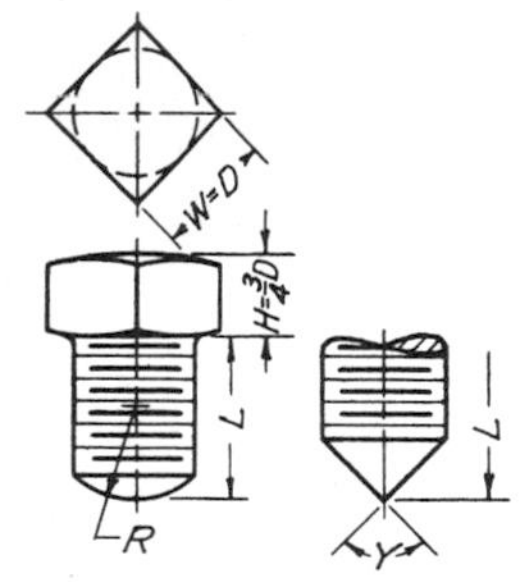

Oval Point Cone Point

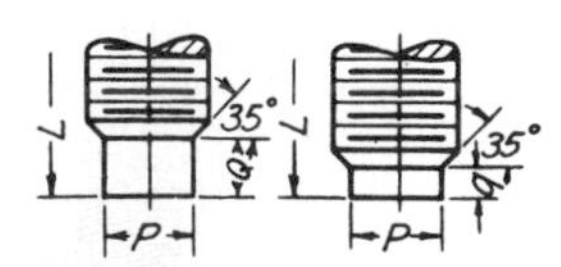

Full Dog Point Half Dog Point

(All six point types are available in all three head types)

| Diam. $D$ | Cup and flat-point diam. $C$ | Oval-point radius $R$ | Cone-point angle $Y$: 118° for these lengths and shorter | Cone-point angle $Y$: 90° for these lengths and longer | Full and half dog points: Diam. $P$ | Length: Full $Q$ | Length: Half $q$ | Socket width $J$ |
|---|---|---|---|---|---|---|---|---|
| 5 | 1/16 | 3/32 | 1/8 | 3/16 | 0.083 | 0.06 | 0.03 | 1/16 |
| 6 | 0.069 | 7/64 | 1/8 | 3/16 | 0.092 | 0.07 | 0.03 | 1/16 |
| 8 | 5/64 | 1/8 | 3/16 | 1/4 | 0.109 | 0.08 | 0.04 | 5/64 |
| 10 | 3/32 | 9/64 | 3/16 | 1/4 | 0.127 | 0.09 | 0.04 | 3/32 |
| 12 | 7/64 | 5/32 | 3/16 | 1/4 | 0.144 | 0.11 | 0.06 | 3/32 |
| 1/4 | 1/8 | 3/16 | 1/4 | 5/16 | 5/32 | 1/8 | 1/16 | 1/8 |
| 5/16 | 11/64 | 15/64 | 5/16 | 3/8 | 13/64 | 5/32 | 5/64 | 5/32 |
| 3/8 | 13/64 | 9/32 | 3/8 | 7/16 | 1/4 | 3/16 | 3/32 | 3/16 |
| 7/16 | 15/64 | 21/64 | 7/16 | 1/2 | 19/64 | 7/32 | 7/64 | 7/32 |
| 1/2 | 9/32 | 3/8 | 1/2 | 9/16 | 11/32 | 1/4 | 1/8 | 1/4 |
| 9/16 | 5/16 | 27/64 | 9/16 | 5/8 | 25/64 | 9/32 | 9/64 | 1/4 |
| 5/8 | 23/64 | 15/32 | 5/8 | 3/4 | 15/32 | 5/16 | 5/32 | 5/16 |
| 3/4 | 7/16 | 9/16 | 3/4 | 7/8 | 9/16 | 3/8 | 3/16 | 3/8 |
| 7/8 | 33/64 | 21/32 | 7/8 | 1 | 21/32 | 7/16 | 7/32 | 1/2 |
| 1 | 19/32 | 3/4 | 1 | 1 1/8 | 3/4 | 1/2 | 1/4 | 9/16 |
| 1 1/8 | 43/64 | 27/32 | 1 1/8 | 1 1/4 | 27/32 | 9/16 | 9/32 | 9/16 |
| 1 1/4 | 3/4 | 15/16 | 1 1/4 | 1 1/2 | 15/16 | 5/8 | 5/16 | 5/8 |
| 1 3/8 | 53/64 | 1 1/32 | 1 3/8 | 1 5/8 | 1 1/32 | 11/16 | 11/32 | 5/8 |
| 1 1/2 | 29/32 | 1 1/8 | 1 1/2 | 1 3/4 | 1 1/8 | 3/4 | 3/8 | 3/4 |
| 1 3/4 | 1 1/16 | 1 5/16 | 1 3/4 | 2 | 1 5/16 | 7/8 | 7/16 | 1 |
| 2 | 1 7/32 | 1 1/2 | 2 | 2 1/4 | 1 1/2 | 1 | 1/2 | 1 |

* ASA B18.3—1961. Dimensions are in inches. Threads coarse or fine, class 3A. Length increments: 1/4 in. to 5/8 in. by (1/16 in.); 5/8 in. to 1 in. by (1/8 in.); 1 in. to 4 in. by (1/4 in.); 4 in. to 6 in. by (1/2 in.). Fractions in parentheses show length increments; for example, 5/8 in. to 1 in. by (1/8 in.) includes the lengths 5/8 in., 3/4 in., 7/8 in., and 1 in.

† ASA B18.6.2—1956. Threads coarse or fine, class 2A. Slotted headless screws standardized in sizes No. 5 to 3/4 in. only. Slot proportions vary with diameter. Draw to look well.

‡ ASA B18.6.2—1956. Threads coarse, fine, or 8-pitch, class 2A. Square-head setscrews standardized in sizes No. 10 to 1 1/2 in. only.

## American Standard Socket-head Shoulder Screws[a]

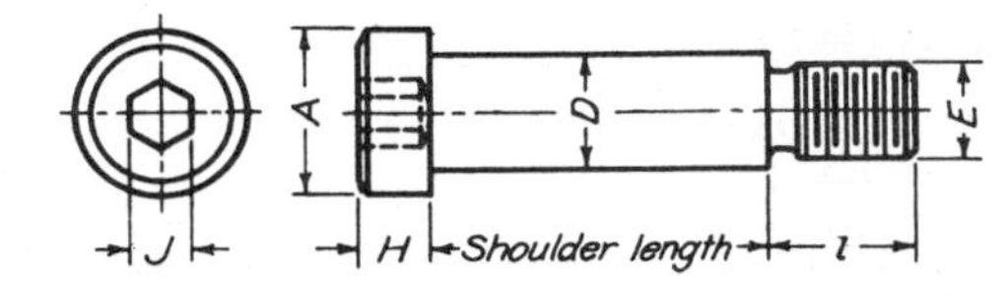


| Shoulder diameter $D$ | | | Head[b] | | | Thread | | Shoulder lengths[d] |
|---|---|---|---|---|---|---|---|---|
| Nominal | Max. | Min. | Diam. $A$ | Height $H$ | Hexagon[c] $J$ | Specification $E$ | Length $l$ | |
| 1/4 | 0.2480 | 0.2460 | 3/8 | 3/16 | 1/8 | 10-24NC-3 | 3/8 | 3/4–2 1/2 |
| 5/16 | 0.3105 | 0.3085 | 7/16 | 7/32 | 5/32 | 1/4-20NC-3 | 7/16 | 1 –3 |
| 3/8 | 0.3730 | 0.3710 | 9/16 | 1/4 | 3/16 | 5/16-18NC-3 | 1/2 | 1 –4 |
| 1/2 | 0.4980 | 0.4960 | 3/4 | 5/16 | 1/4 | 3/8-16NC-3 | 5/8 | 1 1/4–5 |
| 5/8 | 0.6230 | 0.6210 | 7/8 | 3/8 | 5/16 | 1/2-13NC-3 | 3/4 | 1 1/2–6 |
| 3/4 | 0.7480 | 0.7460 | 1 | 1/2 | 3/8 | 5/8-11NC-3 | 7/8 | 1 1/2–8 |
| 1 | 0.9980 | 0.9960 | 1 5/16 | 5/8 | 1/2 | 3/4-10NC-3 | 1 | 1 1/2–8 |
| 1 1/4 | 1.2480 | 1.2460 | 1 3/4 | 3/4 | 5/8 | 7/8-9NC-3 | 1 1/8 | 1 1/2–8 |

[a] ASA B18.3—1961. Dimensions are in inches.

[b] Head chamfer is 30°.

[c] Socket depth = 3/4 $H$.

[d] Shoulder-length increments: shoulder lengths from 3/4 in. to 1 in., 1/8-in. intervals; shoulder lengths from 1 in. to 5 in., 1/4-in. intervals; shoulder lengths from 5 in. to 7 in., 1/2-in. intervals; shoulder lengths from 7 in. to 8 in., 1-in. intervals. Shoulder-length tolerance ±0.005.

## American Standard Wood Screws*

| Nominal size | Basic diam. of screw $D$ | No. of threads per in.† | Slot width ‡ $J$ (all heads) | Round head | | Flat head | Oval head | |
|---|---|---|---|---|---|---|---|---|
| | | | | $A$ | $H$ | $A$ | $A$ | $C$ |
| 0 | 0.060 | 32 | 0.023 | 0.113 | 0.053 | 0.119 | 0.119 | 0.021 |
| 1 | 0.073 | 28 | 0.026 | 0.138 | 0.061 | 0.146 | 0.146 | 0.025 |
| 2 | 0.086 | 26 | 0.031 | 0.162 | 0.069 | 0.172 | 0.172 | 0.029 |
| 3 | 0.099 | 24 | 0.035 | 0.187 | 0.078 | 0.199 | 0.199 | 0.033 |
| 4 | 0.112 | 22 | 0.039 | 0.211 | 0.086 | 0.225 | 0.225 | 0.037 |
| 5 | 0.125 | 20 | 0.043 | 0.236 | 0.095 | 0.252 | 0.252 | 0.041 |
| 6 | 0.138 | 18 | 0.048 | 0.260 | 0.103 | 0.279 | 0.279 | 0.045 |
| 7 | 0.151 | 16 | 0.048 | 0.285 | 0.111 | 0.305 | 0.305 | 0.049 |
| 8 | 0.164 | 15 | 0.054 | 0.309 | 0.120 | 0.332 | 0.332 | 0.052 |
| 9 | 0.177 | 14 | 0.054 | 0.334 | 0.128 | 0.358 | 0.358 | 0.056 |
| 10 | 0.190 | 13 | 0.060 | 0.359 | 0.137 | 0.385 | 0.385 | 0.060 |
| 12 | 0.216 | 11 | 0.067 | 0.408 | 0.153 | 0.438 | 0.438 | 0.068 |
| 14 | 0.242 | 10 | 0.075 | 0.457 | 0.170 | 0.491 | 0.491 | 0.076 |
| 16 | 0.268 | 9 | 0.075 | 0.506 | 0.187 | 0.544 | 0.544 | 0.084 |
| 18 | 0.294 | 8 | 0.084 | 0.555 | 0.204 | 0.597 | 0.597 | 0.092 |
| 20 | 0.320 | 8 | 0.084 | 0.604 | 0.220 | 0.650 | 0.650 | 0.100 |
| 24 | 0.372 | 7 | 0.094 | 0.702 | 0.254 | 0.756 | 0.756 | 0.116 |

* ASA B18.6.1—1961. Dimensions given are maximum values, all in inches. Heads may be slotted or recessed as specified.

† Thread length $l$ = 2/3 $L$.

‡ Slot depths and recesses vary with type and size of screw; draw to look well.

## Small Rivets*

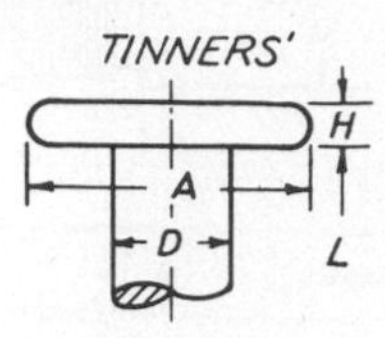


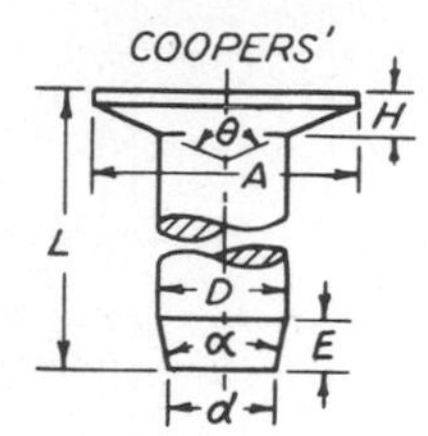


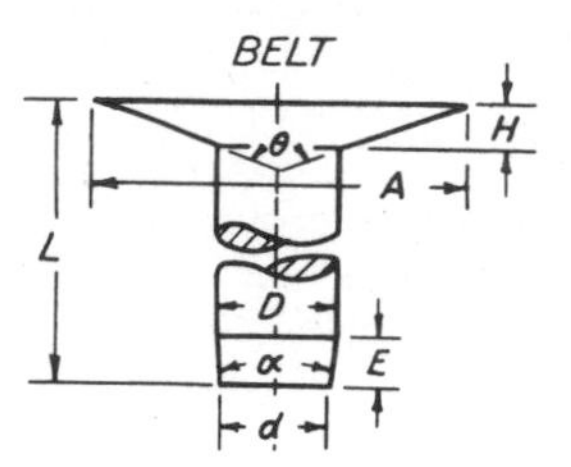


**Tinner's**

| Size No.† | D<br>Diam. body | L<br>Length |
|---|---|---|
| 8 oz | 0.089 | 0.16 |
| 12 | 0.105 | 0.19 |
| 1 lb | 0.111 | 0.20 |
| 1 1/2 | 0.130 | 0.23 |
| 2 | 0.144 | 0.27 |
| 2 1/2 | 0.148 | 0.28 |
| 3 | 0.160 | 0.31 |
| 4 | 0.176 | 0.34 |
| 6 | 0.203 | 0.39 |
| 8 | 0.224 | 0.44 |
| 10 | 0.238 | 0.47 |
| 12 | 0.259 | 0.50 |
| 14 | 0.284 | 0.52 |
| 16 | 0.300 | 0.53 |

Approx. proportions:
$A = 2.25 \times D$
$H = 0.30 \times D$

**Cooper's**

| Size No. | D<br>Diam. body | L<br>Length |
|---|---|---|
| 1 lb | 0.109 | 0.219 |
| 1 1/2 | 0.127 | 0.256 |
| 2 | 0.141 | 0.292 |
| 2 1/2 | 0.148 | 0.325 |
| 3 | 0.156 | 0.358 |
| 4 | 0.165 | 0.392 |
| 6 | 0.203 | 0.466 |
| 8 | 0.238 | 0.571 |
| 10 | 0.250 | 0.606 |
| 12 | 0.259 | 0.608 |
| 14 | 0.271 | 0.643 |
| 16 | 0.281 | 0.677 |

Approx. proportions:
$A = 2.25 \times D$, $d = 0.90 \times D$
$E = 0.40 \times D$, $H = 0.30 \times D$
Included $\angle\theta = 144°$
$\angle\alpha = 18°$

**Belt**

| Size No.‡ | D<br>Diam. body | L<br>Length |
|---|---|---|
| 7 | 0.180 | From 3/8 to 3/4 by 1/8" increments |
| 8 | 0.165 | |
| 9 | 0.148 | |
| 10 | 0.134 | |
| 11 | 0.120 | |
| 12 | 0.109 | |
| 13 | 0.095 | |

Approx. proportions:
$A = 2.8 \times D$, $d = 0.9 \times D$
$E = 0.4 \times D$, $H = 0.3 \times D$

Tolerances on the nominal diameter:
+0.002
−0.004

Finished rivets shall be free from injurious defects.

* ASA B18.1—1955. All dimensions given in inches.
† Size numbers refer to the Trade Name or weight of 1,000 rivets.
‡ Size number refers to the Stubs iron-wire gage number of the stock used in the body of the rivet.

## Dimensions of Standard Gib-head Keys, Square and Flat

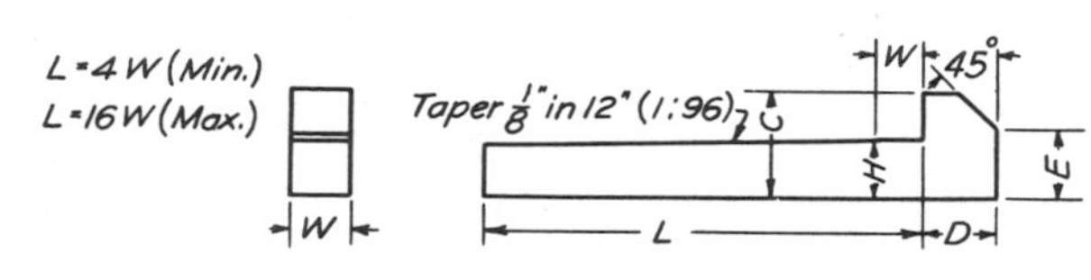


*Approved by ASA* *

| Diameters of shafts | Square type | | | | | Flat type | | | | |
|---|---|---|---|---|---|---|---|---|---|---|
| | Key | | Gib head | | | Key | | Gib head | | |
| | W | H | C | D | E | W | H | C | D | E |
| 1/2 – 9/16 | 1/8 | 1/8 | 1/4 | 7/32 | 5/32 | 1/8 | 3/32 | 3/16 | 1/8 | 1/8 |
| 5/8 – 7/8 | 3/16 | 3/16 | 5/16 | 9/32 | 7/32 | 3/16 | 1/8 | 1/4 | 3/16 | 5/32 |
| 15/16–1 1/4 | 1/4 | 1/4 | 7/16 | 11/32 | 11/32 | 1/4 | 3/16 | 5/16 | 1/4 | 3/16 |
| 1 5/16–1 3/8 | 5/16 | 5/16 | 9/16 | 13/32 | 13/32 | 5/16 | 1/4 | 3/8 | 5/16 | 1/4 |
| 1 7/16–1 3/4 | 3/8 | 3/8 | 11/16 | 15/32 | 15/32 | 3/8 | 1/4 | 7/16 | 3/8 | 5/16 |
| 1 13/16–2 1/4 | 1/2 | 1/2 | 7/8 | 19/32 | 5/8 | 1/2 | 3/8 | 5/8 | 1/2 | 7/16 |
| 2 5/16–2 3/4 | 5/8 | 5/8 | 1 1/16 | 23/32 | 3/4 | 5/8 | 7/16 | 3/4 | 5/8 | 1/2 |
| 2 7/8 –3 1/4 | 3/4 | 3/4 | 1 1/4 | 7/8 | 7/8 | 3/4 | 1/2 | 7/8 | 3/4 | 5/8 |
| 3 3/8 –3 3/4 | 7/8 | 7/8 | 1 1/2 | 1 | 1 | 7/8 | 5/8 | 1 1/16 | 7/8 | 3/4 |
| 3 7/8 –4 1/2 | 1 | 1 | 1 3/4 | 1 3/16 | 1 3/16 | 1 | 3/4 | 1 1/4 | 1 | 1 3/16 |
| 4 3/4 –5 1/2 | 1 1/4 | 1 1/4 | 2 | 1 7/16 | 1 7/16 | 1 1/4 | 7/8 | 1 1/2 | 1 1/4 | 1 |
| 5 3/4 –6 | 1 1/2 | 1 1/2 | 2 1/2 | 1 3/4 | 1 3/4 | 1 1/2 | 1 | 1 3/4 | 1 1/2 | 1 1/4 |

* ASA B17.1—1964. Dimensions in inches.

## Widths and Heights of Standard Square-and Flat-stock Keys with Corresponding Shaft Diameters*

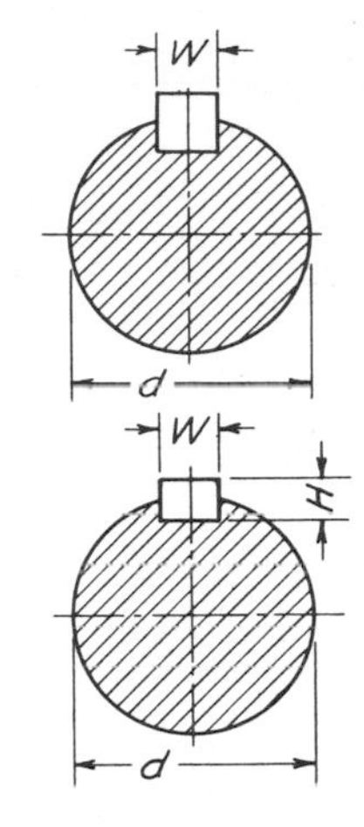


| Shaft diam. $d$ (inclusive) | Square-stock keys $W$ | Flat-stock keys, $W \times H$ | Shaft diam. $d$ (inclusive) | Square-stock keys $W$ | Flat-stock keys, $W \times H$ |
|---|---|---|---|---|---|
| 1/2 – 9/16 | 1/8 | 1/8 × 3/32 | 2 7/8–3 1/4 | 3/4 | 3/4 × 1/2 |
| 5/8 – 7/8 | 3/16 | 3/16 × 1/8 | 3 3/8–3 3/4 | 7/8 | 7/8 × 5/8 |
| 15/16–1 1/4 | 1/4 | 1/4 × 3/16 | 3 7/8–4 1/2 | 1 | 1 × 3/4 |
| 1 5/16–1 3/8 | 5/16 | 5/16 × 1/4 | | | |
| 1 7/16–1 3/4 | 3/8 | 3/8 × 1/4 | 4 3/4–5 1/2 | 1 1/4 | 1 1/4 × 7/8 |
| 1 13/16–2 1/4 | 1/2 | 1/2 × 3/8 | 5 3/4–6 | 1 1/2 | 1 1/2 × 1 |
| 2 5/16–2 3/4 | 5/8 | 5/8 × 7/16 | | | |

* Compiled from manufacturers' catalogues.

## Woodruff-key Dimensions

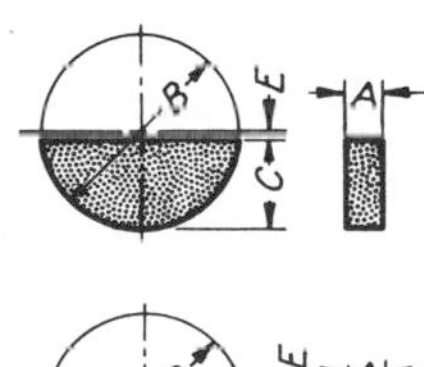


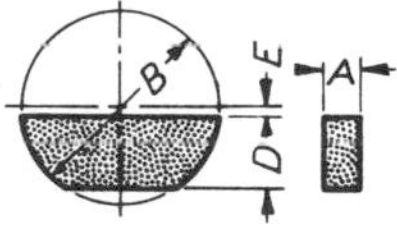


| Key* No. | Nominal size $A \times B$ | Max. width of key $A$ | Max. diam. of key $B$ | Max. height of key $C$ | Max. height of key $D$ | Distance below center $E$ |
|---|---|---|---|---|---|---|
| 204 | 1/16 × 1/2 | 0.0635 | 0.500 | 0.203 | 0.194 | 3/64 |
| 304 | 3/32 × 1/2 | 0.0948 | 0.500 | 0.203 | 0.194 | 3/64 |
| 305 | 3/32 × 5/8 | 0.0948 | 0.625 | 0.250 | 0.240 | 1/16 |
| 404 | 1/8 × 1/2 | 0.1260 | 0.500 | 0.203 | 0.194 | 3/64 |
| 405 | 1/8 × 5/8 | 0.1260 | 0.625 | 0.250 | 0.240 | 1/16 |
| 406 | 1/8 × 3/4 | 0.1260 | 0.750 | 0.313 | 0.303 | 1/16 |
| 505 | 5/32 × 5/8 | 0.1573 | 0.625 | 0.250 | 0.240 | 1/16 |
| 506 | 5/32 × 3/4 | 0.1573 | 0.750 | 0.313 | 0.303 | 1/16 |
| 507 | 5/32 × 7/8 | 0.1573 | 0.875 | 0.375 | 0.365 | 1/16 |
| 606 | 3/16 × 3/4 | 0.1885 | 0.750 | 0.313 | 0.303 | 1/16 |
| 607 | 3/16 × 7/8 | 0.1885 | 0.875 | 0.375 | 0.365 | 1/16 |
| 608 | 3/16 × 1 | 0.1885 | 1.000 | 0.438 | 0.428 | 1/16 |
| 609 | 3/16 × 1 1/8 | 0.1885 | 1.125 | 0.484 | 0.475 | 5/64 |
| 807 | 1/4 × 7/8 | 0.2510 | 0.875 | 0.375 | 0.365 | 1/16 |
| 808 | 1/4 × 1 | 0.2510 | 1.000 | 0.438 | 0.428 | 1/16 |
| 809 | 1/4 × 1 1/8 | 0.2510 | 1.125 | 0.484 | 0.475 | 5/64 |
| 810 | 1/4 × 1 1/4 | 0.2510 | 1.250 | 0.547 | 0.537 | 5/64 |
| 811 | 1/4 × 1 3/8 | 0.2510 | 1.375 | 0.594 | 0.584 | 3/32 |
| 812 | 1/4 × 1 1/2 | 0.2510 | 1.500 | 0.641 | 0.631 | 7/64 |
| 1008 | 5/16 × 1 | 0.3135 | 1.000 | 0.438 | 0.428 | 1/16 |
| 1009 | 5/16 × 1 1/8 | 0.3135 | 1.125 | 0.484 | 0.475 | 5/64 |
| 1010 | 5/16 × 1 1/4 | 0.3135 | 1.250 | 0.547 | 0.537 | 5/64 |
| 1011 | 5/16 × 1 3/8 | 0.3135 | 1.375 | 0.594 | 0.584 | 3/32 |
| 1012 | 5/16 × 1 1/2 | 0.3135 | 1.500 | 0.641 | 0.631 | 7/64 |
| 1210 | 3/8 × 1 1/4 | 0.3760 | 1.250 | 0.547 | 0.537 | 5/64 |
| 1211 | 3/8 × 1 3/8 | 0.3760 | 1.375 | 0.594 | 0.584 | 3/32 |
| 1212 | 3/8 × 1 1/2 | 0.3760 | 1.500 | 0.641 | 0.631 | 7/64 |

* Dimensions in inches. Key numbers indicate the nominal key dimensions. The last two digits give the nominal diameter $B$ in eighths of an inch, and the digits preceding the last two give the nominal width $A$ in thirty-seconds of an inch. Thus 204 indicates a key 2/32 by 4/8, or 1/16 by 1/2 in.

## Woodruff-key-seat Dimensions

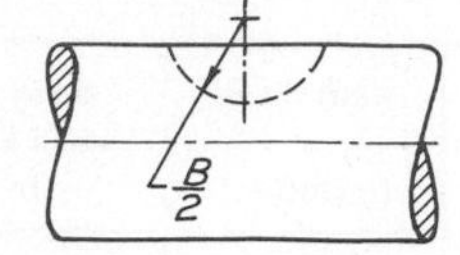


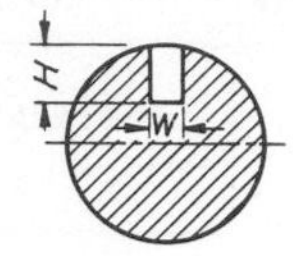


| Key* No. | Nominal size | Key slot | | | |
|---|---|---|---|---|---|
| | | Width $W$ | | Depth $H$ | |
| | | Max. | Min. | Max. | Min. |
| 204 | 1/16 × 1/2 | 0.0630 | 0.0615 | 0.1718 | 0.1668 |
| 304 | 3/32 × 1/2 | 0.0943 | 0.0928 | 0.1561 | 0.1511 |
| 305 | 3/32 × 5/8 | 0.0943 | 0 0928 | 0.2031 | 0.1981 |
| 404 | 1/8 × 1/2 | 0.1255 | 0.1240 | 0.1405 | 0.1355 |
| 405 | 1/8 × 5/8 | 0.1255 | 0.1240 | 0.1875 | 0.1825 |
| 406 | 1/8 × 3/4 | 0.1255 | 0.1240 | 0.2505 | 0.2455 |
| 505 | 5/32 × 5/8 | 0.1568 | 0.1553 | 0.1719 | 0.1669 |
| 506 | 5/32 × 3/4 | 0.1568 | 0.1553 | 0.2349 | 0.2299 |
| 507 | 5/32 × 7/8 | 0.1568 | 0.1553 | 0.2969 | 0.2919 |
| 606 | 3/16 × 3/4 | 0.1880 | 0.1863 | 0.2193 | 0.2143 |
| 607 | 3/16 × 7/8 | 0.1880 | 0.1863 | 0.2813 | 0.2763 |
| 608 | 3/16 × 1 | 0.1880 | 0.1863 | 0.3443 | 0.3393 |
| 609 | 3/16 × 1 1/8 | 0.1880 | 0.1863 | 0.3903 | 0.3853 |
| 807 | 1/4 × 7/8 | 0.2505 | 0.2487 | 0.2500 | 0.2450 |
| 808 | 1/4 × 1 | 0.2505 | 0.2487 | 0.3130 | 0.3080 |
| 809 | 1/4 × 1 1/8 | 0.2505 | 0.2487 | 0.3590 | 0.3540 |
| 810 | 1/4 × 1 1/4 | 0.2505 | 0.2487 | 0.4220 | 0.4170 |
| 811 | 1/4 × 1 3/8 | 0.2505 | 0.2487 | 0.4690 | 0.4640 |
| 812 | 1/4 × 1 1/2 | 0.2505 | 0.2487 | 0.5160 | 0.5110 |
| 1008 | 5/16 × 1 | 0.3130 | 0.3111 | 0.2818 | 0.2768 |
| 1009 | 5/16 × 1 1/8 | 0.3130 | 0.3111 | 0.3278 | 0.3228 |
| 1010 | 5/16 × 1 1/4 | 0.3130 | 0.3111 | 0.3908 | 0.3858 |
| 1011 | 5/16 × 1 3/8 | 0.3130 | 0.3111 | 0.4378 | 0.4328 |
| 1012 | 5/16 × 1 1/2 | 0.3130 | 0.3111 | 0.4848 | 0.4798 |
| 1210 | 3/8 × 1 1/4 | 0.3755 | 0.3735 | 0.3595 | 0.3545 |
| 1211 | 3/8 × 1 3/8 | 0.3755 | 0.3735 | 0.4060 | 0.4015 |
| 1212 | 3/8 × 1 1/2 | 0.3755 | 0.3735 | 0.4535 | 0.4485 |

* Dimensions in inches. Key numbers indicate the nominal key dimensions. The last two digits give the nominal diameter $B$ in eighths of an inch, and the digits preceding the last two give the nominal width $A$ in thirty-seconds of an inch. Thus 204 indicates a key 2/32 by 4/8, or 1/16 by 1/2 in.

## Dimensions of Pratt and Whitney Keys

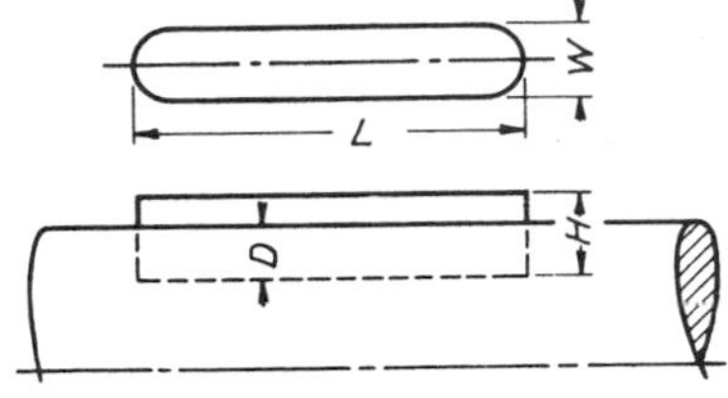


| Key No. | L | W | H | D | Key No. | L | W | H | D |
|---|---|---|---|---|---|---|---|---|---|
| 1 | 1/2 | 1/16 | 3/32 | 1/16 | 22 | 1 3/8 | 1/4 | 3/8 | 1/4 |
| 2 | 1/2 | 3/32 | 9/64 | 3/32 | 23 | 1 3/8 | 5/16 | 15/32 | 5/16 |
| 3 | 1/2 | 1/8 | 3/16 | 1/8 | F | 1 3/8 | 3/8 | 9/16 | 3/8 |
| 4 | 5/8 | 3/32 | 9/64 | 3/32 | 24 | 1 1/2 | 1/4 | 3/8 | 1/4 |
| 5 | 5/8 | 1/8 | 3/16 | 1/8 | 25 | 1 1/2 | 5/16 | 15/32 | 5/16 |
| 6 | 5/8 | 5/32 | 15/64 | 5/32 | G | 1 1/2 | 3/8 | 9/16 | 3/8 |
| 7 | 3/4 | 1/8 | 3/16 | 1/8 | 51 | 1 3/4 | 1/4 | 3/8 | 1/4 |
| 8 | 3/4 | 5/32 | 15/64 | 5/32 | 52 | 1 3/4 | 5/16 | 15/32 | 5/16 |
| 9 | 3/4 | 3/16 | 9/32 | 3/16 | 53 | 1 3/4 | 3/8 | 9/16 | 3/8 |
| 10 | 7/8 | 5/32 | 15/64 | 5/32 | 26 | 2 | 3/16 | 9/32 | 3/16 |
| 11 | 7/8 | 3/16 | 9/32 | 3/16 | 27 | 2 | 1/4 | 3/8 | 1/4 |
| 12 | 7/8 | 7/32 | 21/64 | 7/32 | 28 | 2 | 5/16 | 15/32 | 5/16 |
| A | 7/8 | 1/4 | 3/8 | 1/4 | 29 | 2 | 3/8 | 9/16 | 3/8 |
| 13 | 1 | 3/16 | 9/32 | 3/16 | 54 | 2 1/4 | 1/4 | 3/8 | 1/4 |
| 14 | 1 | 7/32 | 21/64 | 7/32 | 55 | 2 1/4 | 5/16 | 15/16 | 5/16 |
| 15 | 1 | 1/4 | 3/8 | 1/4 | 56 | 2 1/4 | 3/8 | 9/16 | 3/8 |
| B | 1 | 5/16 | 15/32 | 5/16 | 57 | 2 1/4 | 7/16 | 21/32 | 7/16 |
| 16 | 1 1/8 | 3/16 | 9/32 | 3/16 | 58 | 2 1/2 | 5/16 | 15/32 | 5/16 |
| 17 | 1 1/8 | 7/32 | 21/64 | 7/32 | 59 | 2 1/2 | 3/8 | 9/16 | 3/8 |
| 18 | 1 1/8 | 1/4 | 3/8 | 1/4 | 60 | 2 1/2 | 7/16 | 21/32 | 7/16 |
| C | 1 1/8 | 5/16 | 15/32 | 5/16 | 61 | 2 1/2 | 1/2 | 3/4 | 1/2 |
| 19 | 1 1/4 | 3/16 | 9/32 | 3/16 | 30 | 3 | 3/8 | 9/16 | 3/8 |
| 20 | 1 1/4 | 7/32 | 21/64 | 7/32 | 31 | 3 | 7/16 | 21/32 | 7/16 |
| 21 | 1 1/4 | 1/4 | 3/8 | 1/4 | 32 | 3 | 1/2 | 3/4 | 1/2 |
| D | 1 1/4 | 5/16 | 15/32 | 5/16 | 33 | 3 | 9/16 | 27/32 | 9/16 |
| E | 1 1/4 | 3/8 | 9/16 | 3/8 | 34 | 3 | 5/8 | 15/16 | 5/8 |

Dimensions in inches. Key is 2/3 in shaft; 1/3 in hub. Keys are 0.001 in. oversize in width to ensure proper fitting in keyway. Keyway size: width = $W$; depth = $H - D$. Length $L$ should never be less than $2W$.

## American Standard Plain Washers*

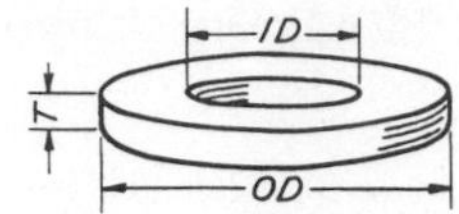


| Size | Light | | | Medium | | | Heavy | | | Extra heavy | | |
|---|---|---|---|---|---|---|---|---|---|---|---|---|
| | *ID* | *OD* | Thickness | *ID* | *OD* | Thickness | *ID* | *OD* | Thickness | *ID* | *OD* | Thickness |
| 0 | 5/64 | 3/16 | 0.020 | | | | | | | | | |
| 1 | 3/32 | 7/32 | 0.020 | | | | | | | | | |
| 2 | 3/32 | 1/4 | 0.020 | | | | | | | | | |
| 3 | 1/8 | 1/4 | 0.022 | | | | | | | | | |
| 4 | 1/8 | 1/4 | 0.022 | 1/8 | 5/16 | 0.032 | | | | | | |
| 5 | 5/32 | 5/16 | 0.035 | 5/32 | 3/8 | 0.049 | | | | | | |
| 6 | 5/32 | 5/16 | 0.035 | 5/32 | 3/8 | 0.049 | | | | | | |
| 7 | 11/64 | 13/32 | 0.049 | 3/16 | 3/8 | 0.049 | | | | | | |
| 8 | 3/16 | 3/8 | 0.049 | 3/16 | 7/16 | 0.049 | | | | | | |
| 9 | 13/64 | 15/32 | 0.049 | 7/32 | 1/2 | 0.049 | | | | | | |
| 3/16 | 7/32 | 7/16 | 0.049 | 7/32 | 1/2 | 0.049 | 1/4 | 9/16 | 0.049 | | | |
| 10 | 7/32 | 7/16 | 0.049 | 1/4 | 9/16 | 0.049 | 1/4 | 9/16 | 0.065 | | | |
| 11 | 15/64 | 17/32 | 0.049 | 1/4 | 9/16 | 0.049 | 1/4 | 9/16 | 0.065 | | | |
| 12 | 1/4 | 1/2 | 0.049 | 1/4 | 9/16 | 0.049 | 1/4 | 9/16 | 0.065 | | | |
| 14 | 17/64 | 5/8 | 0.049 | 5/16 | 3/4 | 0.065 | 5/16 | 7/8 | 0.065 | | | |
| 1/4 | 9/32 | 5/8 | 0.065 | 5/16 | 3/4 | 0.065 | 5/16 | 3/4 | 0.065 | 5/16 | 7/8 | 0.065 |
| 16 | 9/32 | 5/8 | 0.065 | 5/16 | 3/4 | 0.065 | 5/16 | 7/8 | 0.065 | 5/16 | 7/8 | 0.065 |
| 18 | 5/16 | 3/4 | 0.065 | 3/8 | 3/4 | 0.065 | 3/8 | 7/8 | 0.083 | 3/8 | 1 1/8 | 0.065 |
| 5/16 | 11/32 | 11/16 | 0.065 | 3/8 | 3/4 | 0.065 | 3/8 | 7/8 | 0.083 | 3/8 | 1 1/8 | 0.065 |
| 20 | 11/32 | 11/16 | 0.065 | 3/8 | 3/4 | 0.065 | 3/8 | 7/8 | 0.083 | 3/8 | 1 1/8 | 0.065 |
| 24 | 13/32 | 13/16 | 0.065 | 7/16 | 7/8 | 0.083 | 7/16 | 1 | 0.083 | 7/16 | 1 3/8 | 0.083 |
| 3/8 | 13/32 | 13/16 | 0.065 | 7/16 | 7/8 | 0.083 | 7/16 | 1 | 0.083 | 7/16 | 1 3/8 | 0.083 |
| 7/16 | 15/32 | 59/64 | 0.065 | 1/2 | 1 1/8 | 0.083 | 1/2 | 1 1/4 | 0.083 | 1/2 | 1 5/8 | 0.083 |
| 1/2 | 17/32 | 1 1/16 | 0.095 | 9/16 | 1 1/4 | 0.109 | 9/16 | 1 3/8 | 0.109 | 9/16 | 1 7/8 | 0.109 |
| 9/16 | 19/32 | 1 3/16 | 0.095 | 5/8 | 1 3/8 | 0.109 | 5/8 | 1 1/2 | 0.109 | 5/8 | 2 1/8 | 0.134 |
| 5/8 | 21/32 | 1 5/16 | 0.095 | 11/16 | 1 1/2 | 0.134 | 11/16 | 1 3/4 | 0.134 | 11/16 | 2 3/8 | 0.165 |
| 3/4 | 13/16 | 1 1/2 | 0.134 | 13/16 | 1 3/4 | 0.148 | 13/16 | 2 | 0.148 | 13/16 | 2 7/8 | 0.165 |
| 7/8 | 15/16 | 1 3/4 | 0.134 | 15/16 | 2 | 0.165 | 15/16 | 2 1/4 | 0.165 | 15/16 | 3 3/8 | 0.180 |
| 1 | 1 1/16 | 2 | 0.134 | 1 1/16 | 2 1/4 | 0.165 | 1 1/16 | 2 1/2 | 0.165 | 1 1/16 | 3 7/8 | 0.238 |
| 1 1/8 | ..... | ..... | ..... | 1 3/16 | 2 1/2 | 0.165 | 1 1/4 | 2 3/4 | 0.165 | | | |
| 1 1/4 | ..... | ..... | ..... | 1 5/16 | 2 3/4 | 0.165 | 1 3/8 | 3 | 0.165 | | | |
| 1 3/8 | ..... | ..... | ..... | 1 7/16 | 3 | 0.180 | 1 1/2 | 3 1/4 | 0.180 | | | |
| 1 1/2 | ..... | ..... | ..... | 1 9/16 | 3 1/4 | 0.180 | 1 5/8 | 3 1/2 | 0.180 | | | |
| 1 5/8 | ..... | ..... | ..... | 1 11/16 | 3 1/2 | 0.180 | 1 3/4 | 3 3/4 | 0.180 | | | |
| 1 3/4 | ..... | ..... | ..... | 1 13/16 | 3 3/4 | 0.180 | 1 7/8 | 4 | 0.180 | | | |
| 1 7/8 | ..... | ..... | ..... | 1 15/16 | 4 | 0.180 | 2 | 4 1/4 | 0.180 | | | |
| 2 | ..... | ..... | ..... | 2 1/16 | 4 1/4 | 0.180 | 2 1/8 | 4 1/2 | 0.180 | | | |
| 2 1/4 | ..... | ..... | ..... | ..... | .... | ..... | 2 3/8 | 4 3/4 | 0.220 | | | |
| 2 1/2 | ..... | ..... | ..... | ..... | .... | ..... | 2 5/8 | 5 | 0.238 | | | |
| 2 3/4 | ..... | ..... | ..... | ..... | .... | ..... | 2 7/8 | 5 1/4 | 0.259 | | | |
| 3 | ..... | ..... | ..... | ..... | .... | ..... | 3 1/8 | 5 1/2 | 0.284 | | | |

* ASA B27.2—1958. All dimensions in inches.

## American Standard Lock Washers*

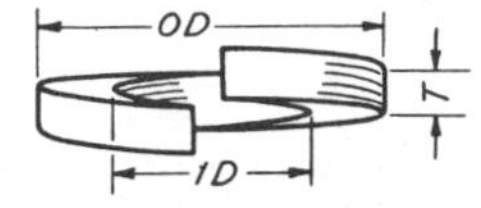


| Nominal size | Inside diam., min. | Light | | Medium | | Heavy | | Extra heavy | |
|---|---|---|---|---|---|---|---|---|---|
| | | Min. thickness | Outside diam., max. | Min. thickness | Outside diam., max. | Min. thickness | Outside diam., max. | Min. thickness | Outside diam., max. |
| 0.086 (No. 2) | 0.088 | 0.015 | 0.165 | 0.020 | 0.175 | 0.025 | 0.185 | 0.027 | 0.211 |
| 0.099 (No. 3) | 0.102 | 0.020 | 0.188 | 0.025 | 0.198 | 0.031 | 0.212 | 0.034 | 0.242 |
| 0.112 (No. 4) | 0.115 | 0.020 | 0.202 | 0.025 | 0.212 | 0.031 | 0.226 | 0.034 | 0.256 |
| 0.125 (No. 5) | 0.128 | 0.025 | 0.225 | 0.031 | 0.239 | 0.040 | 0.255 | 0.045 | 0.303 |
| 0.138 (No. 6) | 0.141 | 0.025 | 0.239 | 0.031 | 0.253 | 0.040 | 0.269 | 0.045 | 0.317 |
| 0.164 (No. 8) | 0.168 | 0.031 | 0.280 | 0.040 | 0.296 | 0.047 | 0.310 | 0.057 | 0.378 |
| 0.190 (No. 10) | 0.194 | 0.040 | 0.323 | 0.047 | 0.337 | 0.056 | 0.353 | 0.068 | 0.437 |
| 0.216 (No. 12) | 0.221 | 0.047 | 0.364 | 0.056 | 0.380 | 0.063 | 0.394 | 0.080 | 0.500 |
| 1/4 | 0.255 | 0.047 | 0.489 | 0.062 | 0.493 | 0.077 | 0.495 | 0.084 | 0.539 |
| 5/16 | 0.319 | 0.056 | 0.575 | 0.078 | 0.591 | 0.097 | 0.601 | 0.108 | 0.627 |
| 3/8 | 0.382 | 0.070 | 0.678 | 0.094 | 0.688 | 0.115 | 0.696 | 0.123 | 0.746 |
| 7/16 | 0.446 | 0.085 | 0.780 | 0.109 | 0.784 | 0.133 | 0.792 | 0.143 | 0.844 |
| 1/2 | 0.509 | 0.099 | 0.877 | 0.125 | 0.879 | 0.151 | 0.889 | 0.162 | 0.945 |
| 9/16 | 0.573 | 0.113 | 0.975 | 0.141 | 0.979 | 0.170 | 0.989 | 0.182 | 1.049 |
| 5/8 | 0.636 | 0.126 | 1.082 | 0.156 | 1.086 | 0.189 | 1.100 | 0.202 | 1.164 |
| 11/16 | 0.700 | 0.138 | 1.178 | 0.172 | 1.184 | 0.207 | 1.200 | 0.221 | 1.266 |
| 3/4 | 0.763 | 0.153 | 1.277 | 0.188 | 1.279 | 0.226 | 1.299 | 0.241 | 1.369 |
| 13/16 | 0.827 | 0.168 | 1.375 | 0.203 | 1.377 | 0.246 | 1.401 | 0.261 | 1.473 |
| 7/8 | 0.890 | 0.179 | 1.470 | 0.219 | 1.474 | 0.266 | 1.504 | 0.285 | 1.586 |
| 15/16 | 0.954 | 0.191 | 1.562 | 0.234 | 1.570 | 0.284 | 1.604 | 0.308 | 1.698 |
| 1 | 1.017 | 0.202 | 1.656 | 0.250 | 1.672 | 0.306 | 1.716 | 0.330 | 1.810 |
| 1 1/16 | 1.081 | 0.213 | 1.746 | 0.266 | 1.768 | 0.326 | 1.820 | 0.352 | 1.922 |
| 1 1/8 | 1.144 | 0.224 | 1.837 | 0.281 | 1.865 | 0.345 | 1.921 | 0.375 | 2.031 |
| 1 3/16 | 1.208 | 0.234 | 1.923 | 0.297 | 1.963 | 0.364 | 2.021 | 0.396 | 2.137 |
| 1 1/4 | 1.271 | 0.244 | 2.012 | 0.312 | 2.058 | 0.384 | 2.126 | 0.417 | 2.244 |
| 1 5/16 | 1.335 | 0.254 | 2.098 | 0.328 | 2.156 | 0.403 | 2.226 | 0.438 | 2.350 |
| 1 3/8 | 1.398 | 0.264 | 2.183 | 0.344 | 2.253 | 0.422 | 2.325 | 0.458 | 2.453 |
| 1 7/16 | 1.462 | 0.273 | 2.269 | 0.359 | 2.349 | 0.440 | 2.421 | 0.478 | 2.555 |
| 1 1/2 | 1.525 | 0.282 | 2.352 | 0.375 | 2.446 | 0.458 | 2.518 | 0.496 | 2.654 |

* ASA B27.1—1958. All dimensions in inches.

**Tapers.** Taper means the difference in diameter or width in 1 ft of length; see figure below. *Taper pins*, much used for fastening cylindrical parts and for doweling, have a standard taper of ¼ in. per ft.

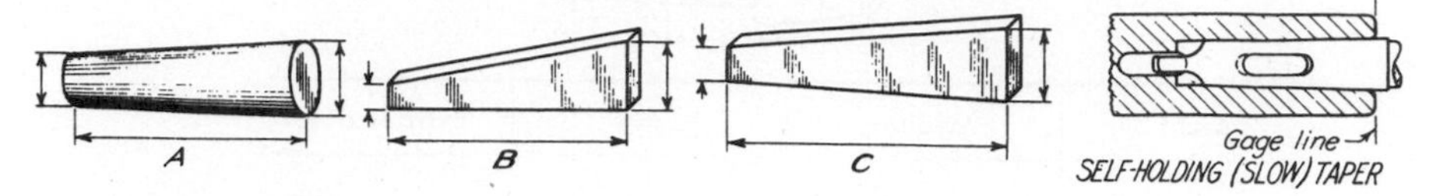


**Machine tapers.** The American Standard for self-holding (slow) machine tapers is designed to replace the various former standards. The table below shows its derivation. Detailed dimensions and tolerances for taper tool shanks and taper sockets will be found in ASA B5.10—1963.

## Dimensions of Taper Pins

*Taper ¼ in. per ft*

| Size No. | Diam., large end | Drill size for reamer | Max. length |
|---|---|---|---|
| 000000 | 0.072 | 53 | ⅝ |
| 00000 | 0.092 | 47 | ⅝ |
| 0000 | 0.108 | 42 | ¾ |
| 000 | 0.125 | 37 | ¾ |
| 00 | 0.147 | 31 | 1 |
| 0 | 0.156 | 28 | 1 |
| 1 | 0.172 | 25 | 1¼ |
| 2 | 0.193 | 19 | 1½ |
| 3 | 0.219 | 12 | 1¾ |
| 4 | 0.250 | 3 | 2 |
| 5 | 0.289 | ¼ | 2¼ |
| 6 | 0.341 | 9/32 | 3¼ |
| 7 | 0.409 | 11/32 | 3¾ |
| 8 | 0.492 | 13/32 | 4½ |
| 9 | 0 591 | 31/64 | 5¼ |
| 10 | 0.706 | 19/32 | 6 |
| 11 | 0.857 | 23/32 | 7¼ |
| 12 | 1.013 | 55/64 | 8¾ |
| 13 | 1.233 | 1 1/64 | 10¾ |

All dimensions in inches.

## American Standard Machine Tapers,* Self-holding (Slow) Taper Series

*Basic dimensions*

<table>
<tr><th>Origin of series</th><th>No. of taper</th><th>Taper /ft</th><th>Diam. at gage line</th><th colspan="4">Means of driving and holding</th></tr>
<tr><td rowspan="3">Brown and Sharpe taper series</td><td></td><td>0.500</td><td>0.239</td><td rowspan="13">Tongue drive with shank held in by friction</td><td rowspan="3" colspan="3"></td></tr>
<tr><td>0.299</td><td>0.500</td><td>0.299</td></tr>
<tr><td>0.375</td><td>0.500</td><td>0.375</td></tr>
<tr><td rowspan="6">Morse taper series</td><td>1</td><td>0.600</td><td>0.475</td><td rowspan="10">Tongue drive with shank held in by key</td><td rowspan="6" colspan="2"></td></tr>
<tr><td>2</td><td>0.600</td><td>0.700</td></tr>
<tr><td>3</td><td>0.602</td><td>0.938</td></tr>
<tr><td>4</td><td>0.623</td><td>1.231</td></tr>
<tr><td>4½</td><td>0.623</td><td>1.500</td></tr>
<tr><td>5</td><td>0.630</td><td>1.748</td></tr>
<tr><td rowspan="10">¾-in./ft taper series</td><td>200</td><td>0.750</td><td>2.000</td><td rowspan="10">Key drive with shank held in by key</td><td rowspan="10">Key drive with shank held in by drawbolt</td></tr>
<tr><td>250</td><td>0.750</td><td>2.500</td></tr>
<tr><td>300</td><td>0.750</td><td>3.000</td></tr>
<tr><td>350</td><td>0.750</td><td>3.500</td></tr>
<tr><td>400</td><td>0.750</td><td>4.000</td><td rowspan="6" colspan="2"></td></tr>
<tr><td>500</td><td>0.750</td><td>5.000</td></tr>
<tr><td>600</td><td>0.750</td><td>6.000</td></tr>
<tr><td>800</td><td>0.750</td><td>8.000</td></tr>
<tr><td>1,000</td><td>0.750</td><td>10.000</td></tr>
<tr><td>1,200</td><td>0.750</td><td>12.000</td></tr>
</table>

* ASA B5.10—1963. All dimensions in inches.

## Wire and Sheet-metal Gages

*Dimensions in decimal parts of an inch*

| No. of gage | American or Brown and Sharpe[a] | Washburn & Moen or American Steel & Wire Co.[b] | Birmingham or Stubs iron wire[c] | Music wire[d] | Imperial wire gage[e] | U.S. Std. for plate[f] |
|---|---|---|---|---|---|---|
| 0000000 | ..... | 0.4900 | .... | ..... | 0.5000 | 0.5000 |
| 000000 | 0.5800 | 0.4615 | ..... | 0.004 | 0.4640 | 0.4688 |
| 00000 | 0.5165 | 0.4305 | 0.500 | 0.005 | 0.4320 | 0.4375 |
| 0000 | 0.4600 | 0.3938 | 0.454 | 0.006 | 0.4000 | 0.4063 |
| 000 | 0.4096 | 0.3625 | 0.425 | 0.007 | 0.3720 | 0.3750 |
| 00 | 0.3648 | 0.3310 | 0.380 | 0.008 | 0.3480 | 0.3438 |
| 0 | 0.3249 | 0.3065 | 0.340 | 0.009 | 0.3240 | 0.3125 |
| 1 | 0.2893 | 0.2830 | 0.300 | 0.010 | 0.3000 | 0.2813 |
| 2 | 0.2576 | 0.2625 | 0.284 | 0.011 | 0.2760 | 0.2656 |
| 3 | 0.2294 | 0.2437 | 0.259 | 0.012 | 0.2520 | 0.2500 |
| 4 | 0.2043 | 0.2253 | 0.238 | 0.013 | 0.2320 | 0.2344 |
| 5 | 0.1819 | 0.2070 | 0.220 | 0.014 | 0.2120 | 0.2188 |
| 6 | 0.1620 | 0.1920 | 0.203 | 0.016 | 0.1920 | 0.2031 |
| 7 | 0.1443 | 0.1770 | 0.180 | 0.018 | 0.1760 | 0.1875 |
| 8 | 0.1285 | 0.1620 | 0.165 | 0.020 | 0.1600 | 0.1719 |
| 9 | 0.1144 | 0.1483 | 0.148 | 0.022 | 0.1440 | 0.1563 |
| 10 | 0.1019 | 0.1350 | 0.134 | 0.024 | 0.1280 | 0.1406 |
| 11 | 0.0907 | 0.1205 | 0.120 | 0.026 | 0.1160 | 0.1250 |
| 12 | 0.0808 | 0.1055 | 0.109 | 0.029 | 0.1040 | 0.1094 |
| 13 | 0.0720 | 0.0915 | 0.095 | 0.031 | 0.0920 | 0.0938 |
| 14 | 0.0641 | 0.0800 | 0.083 | 0.033 | 0.0800 | 0.0781 |
| 15 | 0.0571 | 0.0720 | 0.072 | 0.035 | 0.0720 | 0.0703 |
| 16 | 0.0508 | 0.0625 | 0.065 | 0.037 | 0.0640 | 0.0625 |
| 17 | 0.0453 | 0.0540 | 0.058 | 0.039 | 0.0560 | 0.0563 |
| 18 | 0.0403 | 0.0475 | 0.049 | 0.041 | 0.0480 | 0.0500 |
| 19 | 0.0359 | 0.0410 | 0.042 | 0.043 | 0.0400 | 0.0438 |
| 20 | 0.0320 | 0.0348 | 0.035 | 0.045 | 0.0360 | 0.0375 |
| 21 | 0.0285 | 0.0317 | 0.032 | 0.047 | 0.0320 | 0.0344 |
| 22 | 0.0253 | 0.0286 | 0.028 | 0.049 | 0.0280 | 0.0313 |
| 23 | 0.0226 | 0.0258 | 0.025 | 0.051 | 0.0240 | 0.0281 |
| 24 | 0.0201 | 0.0230 | 0.022 | 0.055 | 0.0220 | 0.0250 |
| 25 | 0.0179 | 0.0204 | 0.020 | 0.059 | 0.0200 | 0.0219 |
| 26 | 0.0159 | 0.0181 | 0.018 | 0.063 | 0.0180 | 0.0188 |
| 27 | 0.0142 | 0.0173 | 0.016 | 0.067 | 0.0164 | 0.0172 |
| 28 | 0.0126 | 0.0162 | 0.014 | 0.071 | 0.0148 | 0.0156 |
| 29 | 0.0113 | 0.0150 | 0.013 | 0.075 | 0.0136 | 0.0141 |
| 30 | 0.0100 | 0.0140 | 0.012 | 0.080 | 0.0124 | 0.0125 |
| 31 | 0.0089 | 0.0132 | 0.010 | 0.085 | 0.0116 | 0.0109 |
| 32 | 0.0080 | 0.0128 | 0.009 | 0.090 | 0.0108 | 0.0102 |
| 33 | 0.0071 | 0.0118 | 0.008 | 0.095 | 0.0100 | 0.0094 |
| 34 | 0.0063 | 0.0104 | 0.007 | 0.100 | 0.0092 | 0.0086 |
| 35 | 0.0056 | 0.0095 | 0.005 | 0.106 | 0.0084 | 0.0078 |
| 36 | 0.0050 | 0.0090 | 0.004 | 0.112 | 0.0076 | 0.0070 |
| 37 | 0.0045 | 0.0085 | ..... | 0.118 | 0.0068 | 0.0066 |
| 38 | 0.0040 | 0.0080 | ..... | 0.124 | 0.0060 | 0.0063 |
| 39 | 0.0035 | 0.0075 | ..... | 0.130 | 0.0052 | |
| 40 | 0.0031 | 0.0070 | ..... | 0.138 | 0.0048 | |

[a] Recognized standard in the United States for wire and sheet metal of copper and other metals except steel and iron.

[b] Recognized standard for steel and iron wire. Called the "U.S. steel wire gage."

[c] Formerly much used, now nearly obsolete.

[d] American Steel & Wire Co.'s music wire gage. Recommended by U.S. Bureau of Standards.

[e] Official British Standard.

[f] Legalized U.S. Standard for iron and steel plate, although plate is now always specified by its thickness in decimals of an inch.

Preferred thicknesses for uncoated thin flat metals (under 0.250 in.): ASA B32.1—1959 gives recommended sizes for sheets.

## Preferred Basic Sizes for Cylindrical Fits*

| | | | | | | | | |
|---|---|---|---|---|---|---|---|---|
| ... | 0.0100 | ... | ... | ... | 0.80 | 5½ | 5.5000 | 5.5 |
| ... | 0.0125 | ... | ⅞ | 0.8750 | ... | 5¾ | 5.7500 | 5.75 |
| 1/64 | 0.015625 | ... | ... | ... | 0.90 | 6 | 6.0000 | 6.0 |
| ... | 0.0200 | ... | 1 | 1.0000 | 1.0 | 6½ | 6.5000 | 6.5 |
| ... | 0.0250 | ... | ... | ... | 1.1 | 7 | 7.0000 | 7.0 |
| 1/32 | 0.03125 | ... | 1⅛ | 1.1250 | ... | 7½ | 7.5000 | 7.5 |
| ... | 0.0400 | 0.04 | ... | ... | ... | 8 | 8.0000 | 8.0 |
| ... | 0.0500 | ... | 1¼ | 1.2500 | 1.25 | 8½ | 8.5000 | 8.5 |
| ... | ... | 0.06 | 1⅜ | 1.3750 | ... | 9 | 9.0000 | 9.0 |
| 1/16 | 0.0625 | ... | ... | ... | 1.40 | 9½ | 9.5000 | 9.5 |
| ... | 0.0800 | ... | 1½ | 1.5000 | 1.50 | 10 | 10.0000 | 10.0 |
| 3/32 | 0.09375 | ... | 1⅝ | 1.6250 | ... | 10½ | 10.5000 | 10.5 |
| ... | 0.1000 | 0.10 | 1¾ | 1.7500 | 1.75 | 11 | 11.0000 | 11.0 |
| ⅛ | 0.1250 | ... | 1⅞ | 1.8750 | ... | 11½ | 11.5000 | 11.5 |
| ... | ... | 0.15 | 2 | 2.0000 | 2.0 | 12 | 12.0000 | 12.0 |
| 5/32 | 0.15625 | ... | 2⅛ | 2.1250 | ... | 12½ | 12.5000 | 12.5 |
| 3/16 | 0.1875 | ... | 2¼ | 2.2500 | 2.25 | 13 | 13.0000 | 13.0 |
| ... | ... | 0.20 | 2⅜ | 2.3750 | ... | 13½ | 13.5000 | 13.5 |
| ¼ | 0.2500 | 0.25 | 2½ | 2.5000 | 2.5 | 14 | 14.0000 | 14.0 |
| ... | ... | 0.30 | 2⅝ | 2.6250 | ... | 14½ | 14.5000 | 14.5 |
| 5/16 | 0.3125 | ... | 2¾ | 2.7500 | 2.75 | 15 | 15.0000 | 15.0 |
| ... | ... | 0.35 | 2⅞ | 2.8750 | ... | 15½ | 15.5000 | 15.5 |
| ⅜ | 0.3750 | ... | 3 | 3.0000 | 3.0 | 16 | 16.0000 | 16.0 |
| ... | ... | 0.40 | 3¼ | 3.2500 | 3.25 | 16½ | 16.5000 | 16.5 |
| 7/16 | 0.4375 | ... | 3½ | 3.5000 | 3.5 | 17 | 17.0000 | 17.0 |
| ½ | 0.5000 | 0.50 | 3¾ | 3.7500 | 3.75 | 17½ | 17.5000 | 17.5 |
| 9/16 | 0.5625 | ... | 4 | 4.0000 | 4.0 | 18 | 18.0000 | 18.0 |
| ... | ... | 0.60 | 4¼ | 4.2500 | 4.25 | 18½ | 18.5000 | 18.5 |
| ⅝ | 0.6250 | ... | 4½ | 4.5000 | 4.5 | 19 | 19.0000 | 19.0 |
| 11/16 | 0.6875 | ... | 4¾ | 4.7500 | 4.75 | 19½ | 19.5000 | 19.5 |
| ... | ... | 0.70 | 5 | 5 0000 | 5.0 | 20 | 20.0000 | 20.0 |
| ¾ | 0.7500 | 0.75 | 5¼ | 5.2500 | 5.25 | 20½ | 20.5000 | ... |
| | | | | | | 21 | 21.0000 | ... |

* Adapted from ASA B4.1—1955

# TABLES OF LIMITS FOR CYLINDRICAL FITS*

## Table 1. Running and Sliding Fits

| Nominal size, in. | Class RC 1 Clearance | Class RC 1 Std. lim. Hole | Class RC 1 Std. lim. Shaft | Class RC 2 Clearance | Class RC 2 Std. lim. Hole | Class RC 2 Std. lim. Shaft | Class RC 3 Clearance | Class RC 3 Std. lim. Hole | Class RC 3 Std. lim. Shaft | Class RC 4 Clearance | Class RC 4 Std. lim. Hole | Class RC 4 Std. lim. Shaft |
|---|---|---|---|---|---|---|---|---|---|---|---|---|
| 0.04–0.12 | 0.1 | +0.2 | −0.1 | 0.1 | +0.25 | −0.1 | 0.3 | +0.25 | −0.3 | 0.3 | +0.4 | −0.3 |
| | 0.45 | 0 | −0.25 | 0.55 | 0 | −0.3 | 0.8 | 0 | −0.55 | 1.1 | 0 | −0.7 |
| 0.12–0.24 | 0.15 | +0.2 | −0.15 | 0.15 | +0.3 | −0.15 | 0.4 | +0.3 | −0.4 | 0.4 | +0.5 | −0.4 |
| | 0.5 | 0 | −0.3 | 0.65 | 0 | −0.35 | 1.0 | 0 | −0.7 | 1.4 | 0 | −0.9 |
| 0.24–0.40 | 0.2 | +0.25 | −0.2 | 0.2 | +0.4 | −0.2 | 0.5 | +0.4 | −0.5 | 0.5 | +0.6 | −0.5 |
| | 0.6 | 0 | −0.35 | 0.85 | 0 | −0.45 | 1.3 | 0 | −0.9 | 1.7 | 0 | −1.1 |
| 0.40–0.71 | 0.25 | +0.3 | −0.25 | 0.25 | +0.4 | −0.25 | 0.6 | +0.4 | −0.6 | 0.6 | +0.7 | −0.6 |
| | 0.75 | 0 | −0.45 | 0.95 | 0 | −0.55 | 1.4 | 0 | −1.0 | 2.0 | 0 | −1.3 |
| 0.71–1.19 | 0.3 | +0.4 | −0.3 | 0.3 | +0.5 | −0.3 | 0.8 | +0.5 | −0.8 | 0.8 | +0.8 | −0.8 |
| | 0.95 | 0 | −0.55 | 1.2 | 0 | −0.7 | 1.8 | 0 | −1.3 | 2.4 | 0 | −1.6 |
| 1.19–1.97 | 0.4 | +0.4 | −0.4 | 0.4 | +0.6 | −0.4 | 1.0 | +0.6 | −1.0 | 1.0 | +1.0 | −1.0 |
| | 1.1 | 0 | −0.7 | 1.4 | 0 | −0.8 | 2.2 | 0 | −1.6 | 3.0 | 0 | −2.0 |
| 1.97–3.15 | 0.4 | +0.5 | −0.4 | 0.4 | +0.7 | −0.4 | 1.2 | +0.7 | −1.2 | 1.2 | +1.2 | −1.2 |
| | 1.2 | 0 | −0.7 | 1.6 | 0 | −0.9 | 2.6 | 0 | −1.9 | 3.6 | 0 | −2.4 |
| 3.15–4.73 | 0.5 | +0.6 | −0.5 | 0.5 | +0.9 | −0.5 | 1.4 | +0.9 | −1.4 | 1.4 | +1.4 | −1.4 |
| | 1.5 | 0 | −0.9 | 2.0 | 0 | −1.1 | 3.2 | 0 | −2.3 | 4.2 | 0 | −2.8 |
| 4.73–7.09 | 0.6 | +0.7 | −0.6 | 0.6 | +1.0 | −0.6 | 1.6 | +1.0 | −1.6 | 1.6 | +1.6 | −1.6 |
| | 1.8 | 0 | −1.1 | 2.3 | 0 | −1.3 | 3.6 | 0 | −2.6 | 4.8 | 0 | −3.2 |

| Nominal size, in. | Class RC 5 Clearance | Class RC 5 Hole | Class RC 5 Shaft | Class RC 6 Clearance | Class RC 6 Hole | Class RC 6 Shaft | Class RC 7 Clearance | Class RC 7 Hole | Class RC 7 Shaft | Class RC 8 Clearance | Class RC 8 Hole | Class RC 8 Shaft | Class RC 9 Clearance | Class RC 9 Hole | Class RC 9 Shaft |
|---|---|---|---|---|---|---|---|---|---|---|---|---|---|---|---|
| 0.04–0.12 | 0.6 | +0.4 | −0.6 | 0.6 | +0.6 | −0.6 | 1.0 | +1.0 | −1.0 | 2.5 | +1.6 | −2.5 | 4.0 | +2.5 | −4.0 |
| | 1.4 | 0 | −1.0 | 1.8 | 0 | −1.2 | 2.6 | 0 | −1.6 | 5.1 | 0 | −3.5 | 8.1 | 0 | −5.6 |
| 0.12–0.24 | 0.8 | +0.5 | −0.8 | 0.8 | +0.7 | −0.8 | 1.2 | +1.2 | −1.2 | 2.8 | +1.8 | −2.8 | 4.5 | +3.0 | −4.5 |
| | 1.8 | 0 | −1.3 | 2.2 | 0 | −1.5 | 3.1 | 0 | −1.9 | 5.8 | 0 | −4.0 | 9.0 | 0 | −6.0 |
| 0.24–0.40 | 1.0 | +0.6 | −1.0 | 1.0 | +0.9 | −1.0 | 1.6 | +1.4 | −1.6 | 3.0 | +2.2 | −3.0 | 5.0 | +3.5 | −5.0 |
| | 2.2 | 0 | −1.6 | 2.8 | 0 | −1.9 | 3.9 | 0 | −2.5 | 6.6 | 0 | −4.4 | 10.7 | 0 | −7.2 |
| 0.40–0.71 | 1.2 | +0.7 | −1.2 | 1.2 | +1.0 | −1.2 | 2.0 | +1.6 | −2.0 | 3.5 | +2.8 | −3.5 | 6.0 | +4.0 | −6.0 |
| | 2.6 | 0 | −1.9 | 3.2 | 0 | −2.2 | 4.6 | 0 | −3.0 | 7.9 | 0 | −5.1 | 12.8 | 0 | −8.8 |
| 0.71–1.19 | 1.6 | +0.8 | −1.6 | 1.6 | +1.2 | −1.6 | 2.5 | +2.0 | −2.5 | 4.5 | +3.5 | −4.5 | 7.0 | +5.0 | −7.0 |
| | 3.2 | 0 | −2.4 | 4.0 | 0 | −2.8 | 5.7 | 0 | −3.7 | 10.0 | 0 | −6.5 | 15.5 | 0 | −10.5 |
| 1.19–1.97 | 2.0 | +1.0 | −2.0 | 2.0 | +1.6 | −2.0 | 3.0 | +2.5 | −3.0 | 5.0 | +4.0 | −5.0 | 8.0 | +6.0 | −8.0 |
| | 4.0 | 0 | −3.0 | 5.2 | 0 | −3.6 | 7.1 | 0 | −4.6 | 11.5 | 0 | −7.5 | 18.0 | 0 | −12.0 |
| 1.97–3.15 | 2.5 | +1.2 | −2.5 | 2.5 | +1.8 | −2.5 | 4.0 | +3.0 | −4.0 | 6.0 | +4.5 | −6.0 | 9.0 | +7.0 | −9.0 |
| | 4.9 | 0 | −3.7 | 6.1 | 0 | −4.3 | 8.8 | 0 | −5.8 | 13.5 | 0 | −9.0 | 20.5 | 0 | −13.5 |
| 3.15–4.73 | 3.0 | +1.4 | −3.0 | 3.0 | +2.2 | −3.0 | 5.0 | +3.5 | −5.0 | 7.0 | +5.0 | −7.0 | 10.0 | +9.0 | −10.0 |
| | 5.8 | 0 | −4.4 | 7.4 | 0 | −5.2 | 10.7 | 0 | −7.2 | 15.5 | 0 | −10.5 | 24.0 | 0 | −15.0 |
| 4.73–7.09 | 3.5 | +1.6 | −3.5 | 3.5 | +2.5 | −3.5 | 6.0 | +4.0 | −6.0 | 8.0 | +6.0 | −8.0 | 12.0 | +10.0 | −12.0 |
| | 6.7 | 0 | −5.1 | 8.5 | 0 | −6.0 | 12.5 | 0 | −8.5 | 18.0 | 0 | −12.0 | 28.0 | 0 | −18.0 |

## Table 2. Clearance Locational Fits

| Nominal size, in. | Class LC 1 Clearance | Class LC 1 Hole | Class LC 1 Shaft | Class LC 2 Clearance | Class LC 2 Hole | Class LC 2 Shaft | Class LC 3 Clearance | Class LC 3 Hole | Class LC 3 Shaft | Class LC 4 Clearance | Class LC 4 Hole | Class LC 4 Shaft | Class LC 5 Clearance | Class LC 5 Hole | Class LC 5 Shaft |
|---|---|---|---|---|---|---|---|---|---|---|---|---|---|---|---|
| 0.04–0.12 | 0 | +0.25 | +0 | 0 | +0.4 | +0 | 0 | +0.6 | +0 | 0 | +1.0 | +0 | 0.1 | +0.4 | −0.1 |
| | 0.45 | −0 | −0.2 | 0.65 | −0 | −0.25 | 1 | −0 | −0.4 | 2.0 | −0 | −1.0 | 0.75 | −0 | −0.35 |
| 0.12–0.24 | 0 | +0.3 | +0 | 0 | +0.5 | +0 | 0 | +0.7 | +0 | 0 | +1.2 | +0 | 0.15 | +0.5 | −0.15 |
| | 0.5 | −0 | −0.2 | 0.8 | −0 | −0.3 | 1.2 | −0 | −0.5 | 2.4 | −0 | −1.2 | 0.95 | −0 | −0.45 |
| 0.24–0.40 | 0 | +0.4 | +0 | 0 | +0.6 | +0 | 0 | +0.9 | +0 | 0 | +1.4 | +0 | 0.2 | +0.6 | −0.2 |
| | 0.65 | −0 | −0.25 | 1.0 | −0 | −0.4 | 1.5 | −0 | −0.6 | 2.8 | −0 | −1.4 | 1.2 | −0 | −0.6 |
| 0.40–0.71 | 0 | +0.4 | +0 | 0 | +0.7 | +0 | 0 | +1.0 | +0 | 0 | +1.6 | +0 | 0.25 | +0.7 | −0.25 |
| | 0.7 | −0 | −0.3 | 1.1 | −0 | −0.4 | 1.7 | −0 | −0.7 | 3.2 | −0 | −1.6 | 1.35 | −0 | −0.65 |
| 0.71–1.19 | 0 | +0.5 | +0 | 0 | +0.8 | +0 | 0 | +1.2 | +0 | 0 | +2.0 | +0 | 0.3 | +0.8 | −0.3 |
| | 0.9 | −0 | −0.4 | 1.3 | −0 | −0.5 | 2 | −0 | −0.8 | 4 | −0 | −2.0 | 1.6 | −0 | −0.8 |
| 1.19–1.97 | 0 | +0.6 | +0 | 0 | +1.0 | +0 | 0 | +1.6 | +0 | 0 | +2.5 | +0 | 0.4 | +1.0 | −0.4 |
| | 1.0 | −0 | −0.4 | 1.6 | −0 | −0.6 | 2.6 | −0 | −1 | 5 | −0 | −2.5 | 2.0 | −0 | −1.0 |
| 1.97–3.15 | 0 | +0.7 | +0 | 0 | +1.2 | +0 | 0 | +1.8 | +0 | 0 | +3 | +0 | 0.4 | +1.2 | −0.4 |
| | 1.2 | −0 | −0.5 | 1.9 | −0 | −0.7 | 3 | −0 | −1.2 | 6 | −0 | −3 | 2.3 | −0 | −1.1 |
| 3.15–4.73 | 0 | +0.9 | +0 | 0 | +1.4 | +0 | 0 | +2.2 | +0 | 0 | +3.5 | +0 | 0.5 | +1.4 | −0.5 |
| | 1.5 | −0 | −0.6 | 2.3 | −0 | −0.9 | 3.6 | −0 | −1.4 | 7 | −0 | −3.5 | 2.8 | −0 | −1.4 |
| 4.73–7.09 | 0 | +1.0 | +0 | 0 | +1.6 | +0 | 0 | +2.5 | +0 | 0 | +4 | +0 | 0.6 | +1.6 | −0.6 |
| | 1.7 | −0 | −0.7 | 2.6 | −0 | −1.0 | 4.1 | −0 | −1.6 | 8 | −0 | −4 | 3.2 | −0 | −1.6 |

| Nominal size, in. | Class LC 6 Clearance | Class LC 6 Hole | Class LC 6 Shaft | Class LC 7 Clearance | Class LC 7 Hole | Class LC 7 Shaft | Class LC 8 Clearance | Class LC 8 Hole | Class LC 8 Shaft | Class LC 9 Clearance | Class LC 9 Hole | Class LC 9 Shaft | Class LC 10 Clearance | Class LC 10 Hole | Class LC 10 Shaft | Class LC 11 Clearance | Class LC 11 Hole | Class LC 11 Shaft |
|---|---|---|---|---|---|---|---|---|---|---|---|---|---|---|---|---|---|---|
| 0.04–0.12 | 0.3 | +0.6 | −0.3 | 0.6 | +1.0 | −0.6 | 1.0 | +1.6 | −1.0 | 2.5 | +2.5 | −2.5 | 4 | +4 | −4 | 5 | +6 | −5 |
| | 1.5 | −0 | −0.9 | 2.6 | −0 | −1.6 | 3.6 | −0 | −2.0 | 7.5 | −0 | −5.0 | 12 | −0 | −8 | 17 | −0 | −11 |
| 0.12–0.24 | 0.4 | +0.7 | −0.4 | 0.8 | +1.2 | −0.8 | 1.2 | +1.8 | −1.2 | 2.8 | +3.0 | −2.8 | 4.5 | +5 | −4.5 | 6 | +7 | −6 |
| | 1.8 | −0 | −1.1 | 3.2 | −0 | −2.0 | 4.2 | −0 | −2.4 | 8.8 | −0 | −5.8 | 14.5 | −0 | −9.5 | 20 | −0 | −13 |
| 0.24–0.40 | 0.5 | +0.9 | −0.5 | 1.0 | +1.4 | −1.0 | 1.6 | +2.2 | −1.6 | 3.0 | +3.5 | −3.0 | 5 | +6 | −5 | 7 | +9 | −7 |
| | 2.3 | −0 | −1.4 | 3.8 | −0 | −2.4 | 5.2 | −0 | −3.0 | 10.0 | −0 | −6.5 | 17 | −0 | −11 | 25 | −0 | −16 |
| 0.40–0.71 | 0.6 | +1.0 | −0.6 | 1.2 | +1.6 | −1.2 | 2.0 | +2.8 | −2.0 | 3.5 | +4.0 | −3.5 | 6 | +7 | −6 | 8 | +10 | −8 |
| | 2.6 | −0 | −1.6 | 4.4 | −0 | −2.8 | 6.4 | −0 | −3.6 | 11.5 | −0 | −7.5 | 20 | −0 | −13 | 28 | −0 | −18 |
| 0.71–1.19 | 0.8 | +1.2 | −0.8 | 1.6 | +2.0 | −1.6 | 2.5 | +3.5 | −2.5 | 4.5 | +5.0 | −4.5 | 7 | +8 | −7 | 10 | +12 | −10 |
| | 3.2 | −0 | −2.0 | 5.6 | −0 | −3.6 | 8.0 | −0 | −4.5 | 14.5 | −0 | −9.5 | 23 | −0 | −15 | 34 | −0 | −22 |
| 1.19–1.97 | 1.0 | +1.6 | −1.0 | 2.0 | +2.5 | −2.0 | 3.0 | +4.0 | −3.0 | 5 | +6 | −5 | 8 | +10 | −8 | 12 | +16 | −12 |
| | 4.2 | −0 | −2.6 | 7.0 | −0 | −4.5 | 9.5 | −0 | −5.5 | 17 | −0 | −11 | 28 | −0 | −18 | 44 | −0 | −28 |
| 1.97–3.15 | 1.2 | +1.8 | −1.2 | 2.5 | +3.0 | −2.5 | 4.0 | +4.5 | −4.0 | 6 | +7 | −6 | 10 | +12 | −10 | 14 | +18 | −14 |
| | 4.8 | −0 | −3.0 | 8.5 | −0 | −5.5 | 11.5 | −0 | −7.0 | 20 | −0 | −13 | 34 | −0 | −22 | 50 | −0 | −32 |
| 3.15–4.73 | 1.4 | +2.2 | −1.4 | 3.0 | +3.5 | −3.0 | 5.0 | +5.0 | −5.0 | 7 | +9 | −7 | 11 | +14 | −11 | 16 | +22 | −16 |
| | 5.8 | −0 | −3.6 | 10.0 | −0 | −6.5 | 13.5 | −0 | −8.5 | 25 | −0 | −16 | 39 | −0 | −25 | 60 | −0 | −38 |
| 4.73–7.09 | 1.6 | +2.5 | −1.6 | 3.5 | +4.0 | −3.5 | 6 | +6 | −6 | 8 | +10 | −8 | 12 | +16 | −12 | 18 | +25 | −18 |
| | 6.6 | −0 | −4.1 | 11.5 | −0 | −7.5 | 16 | −0 | −10 | 28 | −0 | −18 | 44 | −0 | −28 | 68 | −0 | −43 |

* Adapted from ASA B4.1—1955. Limits for hole and shaft are applied algebraically to basic size to obtain limits of size for parts. Data in boldface in accordance with ABC agreements.

## Table 3. Transition Locational Fits

| Nominal size, in. | Class LT 1 | | | Class LT 2 | | | Class LT 3 | | | Class LT 4 | | | Class LT 6 | | | Class LT 7 | | |
|---|---|---|---|---|---|---|---|---|---|---|---|---|---|---|---|---|---|---|
| | Fit† | Std. lim. | | Fit | Std. lim. | | Fit | Std. lim. | | Fit | Std. lim. | | Fit | Std. lim. | | Fit | Std. lim. | |
| | | Hole | Shaft | | Hole | Shaft | | Hole | Shaft | | Hole | Shaft | | Hole | Shaft | | Hole | Shaft |
| 0.04–0.12 | −0.15 | +0.4 | +0.15 | −0.3 | +0.6 | +0.3 | | | | | | | −0.55 | +0.6 | +0.55 | −0.5 | +0.4 | +0.5 |
| | +0.5 | −0 | −0.1 | +0.7 | −0 | −0.1 | | | | | | | +0.45 | −0 | +0.15 | +0.15 | −0 | +0.25 |
| 0.12–0.24 | −0.2 | +0.5 | +0.2 | −0.4 | +0.7 | +0.4 | | | | | | | −0.7 | +0.7 | +0.7 | −0.6 | +0.5 | +0.6 |
| | +0.6 | −0 | −0.1 | +0.8 | −0 | −0.1 | | | | | | | +0.5 | −0 | +0.2 | +0.2 | −0 | +0.3 |
| 0.24–0.40 | −0.3 | +0.6 | +0.3 | −0.4 | +0.9 | +0.4 | −0.5 | +0.6 | +0.5 | −0.7 | +0.9 | +0.7 | −0.8 | +0.9 | +0.8 | −0.8 | +0.6 | +0.8 |
| | +0.7 | −0 | −0.1 | +1.1 | −0 | −0.2 | +0.5 | −0 | +0.1 | +0.8 | −0 | +0.1 | +0.7 | −0 | +0.2 | −0.2 | −0 | +0.4 |
| 0 40–0.71 | −0.3 | +0.7 | +0.3 | −0.5 | +1.0 | +0.5 | −0.5 | +0.7 | +0.5 | −0.8 | +1.0 | +0.8 | −1.0 | +1.0 | +1.0 | −0.9 | +0.7 | +0.9 |
| | +0.8 | −0 | −0.1 | +1.2 | −0 | −0.2 | +0.6 | −0 | +0.1 | +0.9 | −0 | +0.1 | +0.7 | −0 | +0.3 | +0.2 | −0 | +0.5 |
| 0.71–1.19 | −0.3 | +0.8 | +0.3 | −0.5 | +1.2 | +0.5 | −0.6 | +0.8 | +0.6 | −0.9 | +1.2 | +0.9 | −1.1 | +1.2 | +1.1 | −1.1 | +0.8 | +1.1 |
| | +1.0 | −0 | −0.2 | +1.5 | −0 | −0.3 | +0.7 | −0 | +0.1 | +1.1 | −0 | +0.1 | +0.9 | −0 | +0.3 | +0.2 | −0 | +0.6 |
| 1.19–1.97 | −0.4 | +1.0 | +0.4 | −0.6 | +1.6 | +0.6 | −0.7 | +1.0 | +0.7 | −1.1 | +1.6 | +1.1 | −1.4 | +1.6 | +1.4 | −1.3 | +1.0 | +1.3 |
| | +1.2 | −0 | −0.2 | +2.0 | −0 | −0.4 | +0.9 | −0 | +0.1 | +1.5 | −0 | +0.1 | +1.2 | −0 | +0.4 | +0.3 | −0 | +0.7 |
| 1.97–3.15 | −0.4 | +1.2 | +0.4 | −0.7 | +1.8 | +0.7 | −0.8 | +1.2 | +0.8 | −1.3 | +1.8 | +1.3 | −1.7 | +1.8 | +1.7 | −1.5 | +1.2 | +1.5 |
| | +1.5 | −0 | −0.3 | +2.3 | −0 | −0.5 | +1.1 | −0 | +0.1 | +1.7 | −0 | +0.1 | +1.3 | −0 | +0.5 | +0.4 | −0 | +0.8 |
| 3.15–4.73 | −0.5 | +1.4 | +0.5 | −0.8 | +2.2 | +0.8 | −1.0 | +1.4 | +1.0 | −1.5 | +2.2 | +1.5 | −1.9 | +2.2 | +1.9 | −1.9 | +1.4 | +1.9 |
| | +1.8 | −0 | −0.4 | +2.8 | −0 | −0.6 | +1.3 | −0 | +0.1 | +2.1 | −0 | +0.1 | +1.7 | −0 | +0.5 | +0.4 | −0 | +1.0 |
| 4.73–7.09 | −0.6 | +1.6 | +0.6 | −0.9 | +2.5 | +0.9 | −1.1 | +1.6 | +1.1 | −1.7 | +2.5 | +1.7 | −2.2 | +2.5 | +2.2 | −2.2 | +1.6 | +2.2 |
| | +2.0 | −0 | −0.4 | +3.2 | −0 | −0.7 | +1.5 | −0 | +0.1 | +2.4 | −0 | +0.1 | +1.9 | −0 | +0.6 | +0.4 | −0 | +1.2 |

† "Fit" represents the maximum interference (minus values) and the maximum clearance (plus values).

## Table 4. Interference Locational Fits

| Nominal size, in. | Class LN 2 | | | Class LN 3 | | |
|---|---|---|---|---|---|---|
| | Interference | Std. lim. | | Interference | Std. lim. | |
| | | Hole | Shaft | | Hole | Shaft |
| 0.04–0.12 | 0 | +0.4 | +0.65 | 0.1 | +0.4 | +0.75 |
| | 0.65 | −0 | +0.4 | 0.75 | −0 | +0.5 |
| 0.12–0.24 | 0 | +0.5 | +0.8 | 0.1 | +0.5 | +0.9 |
| | 0.8 | −0 | +0.5 | 0.9 | −0 | +0.6 |
| 0.24–0.40 | 0 | +0.6 | +1.0 | 0.2 | +0.6 | +1.2 |
| | 1.0 | −0 | +0.6 | 1.2 | −0 | +0.8 |
| 0.40–0.71 | 0 | +0.7 | +1.1 | 0.3 | +0.7 | +1.4 |
| | 1.1 | −0 | +0.7 | 1.4 | −0 | +1.0 |
| 0.71–1.19 | 0 | +0.8 | +1.3 | 0.4 | +0.8 | +1.7 |
| | 1.3 | −0 | +0.8 | 1.7 | −0 | +1.2 |
| 1.19–1.97 | 0 | +1.0 | +1.6 | 0.4 | +1.0 | +2.0 |
| | 1.6 | −0 | +1.0 | 2.0 | −0 | +1.4 |
| 1.97–3.15 | 0.2 | +1.2 | +2.1 | 0.4 | +1.2 | +2.3 |
| | 2.1 | −0 | +1.4 | 2.3 | −0 | +1.6 |
| 3.15–4.73 | 0.2 | +1.4 | +2.5 | 0.6 | +1.4 | +2.9 |
| | 2.5 | −0 | +1.6 | 2.9 | −0 | +2.0 |
| 4.73–7.09 | 0.2 | +1.6 | +2.8 | 0.9 | +1.6 | +3.5 |
| | 2.8 | −0 | +1.8 | 3.5 | −0 | +2.5 |

## Table 5. Force and Shrink Fits

| | Class FN 1 | | | Class FN 2 | | | Class FN 3 | | | Class FN 4 | | | Class FN 5 | | |
|---|---|---|---|---|---|---|---|---|---|---|---|---|---|---|---|
| 0.04–0.12 | 0.05 | +0.25 | +0.5 | 0.2 | +0.4 | +0.85 | | | | 0.3 | +0.4 | +0.95 | 0.5 | +0.4 | +1.3 |
| | 0.5 | −0 | +0.3 | 0.85 | −0 | +0.6 | | | | 0.95 | −0 | +0.7 | 1.3 | −0 | +0.9 |
| 0.12–0.24 | 0.1 | +0.3 | +0.6 | 0.2 | +0.5 | +1.0 | | | | 0.4 | +0.5 | +1.2 | 0.7 | +0.5 | +1.7 |
| | 0.6 | −0 | +0.4 | 1.0 | −0 | +0.7 | | | | 1.2 | −0 | +0.9 | 1.7 | −0 | +1.2 |
| 0.24–0.40 | 0.1 | +0.4 | +0.75 | 0.4 | +0.6 | +1.4 | | | | 0.6 | +0.6 | +1.6 | 0.8 | +0.6 | +2.0 |
| | 0.75 | −0 | +0.5 | 1.4 | −0 | +1.0 | | | | 1.6 | −0 | +1.2 | 2.0 | −0 | +1.4 |
| 0.40–0.56 | 0.1 | +0.4 | +0.8 | 0.5 | +0.7 | +1.6 | | | | 0.7 | +0.7 | +1.8 | 0.9 | +0.7 | +2.3 |
| | 0.8 | −0 | +0.5 | 1.6 | −0 | +1.2 | | | | 1.8 | −0 | +1.4 | 2.3 | −0 | +1.6 |
| 0.56–0.71 | 0.2 | +0.4 | +0.9 | 0.5 | +0.7 | +1.6 | | | | 0.7 | +0.7 | +1.8 | 1.1 | +0.7 | +2.5 |
| | 0.9 | −0 | +0.6 | 1.6 | −0 | +1.2 | | | | 1.8 | −0 | +1.4 | 2.5 | −0 | +1.8 |
| 0.71–0.95 | 0.2 | +0.5 | +1.1 | 0.6 | +0.8 | +1.9 | | | | 0.8 | +0.8 | +2.1 | 1.4 | +0.8 | +3.0 |
| | 1.1 | −0 | +0.7 | 1.9 | −0 | +1.4 | | | | 2.1 | −0 | +1.6 | 3.0 | −0 | +2.2 |
| 0.95–1.19 | 0.3 | +0.5 | +1.2 | 0.6 | +0.8 | +1.9 | 0.8 | +0.8 | +2.1 | 1.0 | +0.8 | +2.3 | 1.7 | +0.8 | +3.3 |
| | 1.2 | −0 | +0.8 | 1.9 | −0 | +1.4 | 2.1 | −0 | +1.6 | 2.3 | −0 | +1.8 | 3.3 | −0 | +2.5 |
| 1.19–1.58 | 0.3 | +0.6 | +1.3 | 0.8 | +1.0 | +2.4 | 1.0 | +1.0 | +2.6 | 1.5 | +1.0 | +3.1 | 2.0 | +1.0 | +4.0 |
| | 1.3 | −0 | +0.9 | 2.4 | −0 | +1.8 | 2.6 | −0 | +2.0 | 3.1 | −0 | +2.5 | 4.0 | −0 | +3.0 |
| 1.58–1.97 | 0.4 | +0.6 | +1.4 | 0.8 | +1.0 | +2.4 | 1.2 | +1.0 | +2.8 | 1.8 | +1.0 | +3.4 | 3.0 | +1.0 | +5.0 |
| | 1.4 | −0 | +1.0 | 2.4 | −0 | +1.8 | 2.8 | −0 | +2.2 | 3.4 | −0 | +2.8 | 5.0 | −0 | +4.0 |
| 1.97–2.56 | 0.6 | +0.7 | +1.8 | 0.8 | +1.2 | +2.7 | 1.3 | +1.2 | +3.2 | 2.3 | +1.2 | +4.2 | 3.8 | +1.2 | +6.2 |
| | 1.8 | −0 | +1.3 | 2.7 | −0 | +2.0 | 3.2 | −0 | +2.5 | 4.2 | −0 | +3.5 | 6.2 | −0 | +5.0 |
| 2.56–3.15 | 0.7 | +0.7 | +1.9 | 1.0 | +1.2 | +2.9 | 1.8 | +1.2 | +3.7 | 2.8 | +1.2 | +4.7 | 4.8 | +1.2 | +7.2 |
| | 1.9 | −0 | +1.4 | 2.9 | −0 | +2.2 | 3.7 | −0 | +3.0 | 4.7 | −0 | +4.0 | 7.2 | −0 | +6.0 |
| 3.15–3.94 | 0.9 | +0.9 | +2.4 | 1.4 | +1.4 | +3.7 | 2.1 | +1.4 | +4.4 | 3.6 | +1.4 | +5.9 | 5.6 | +1.4 | +8.4 |
| | 2.4 | −0 | +1.8 | 3.7 | −0 | +2.8 | 4.4 | −0 | +3.5 | 5.9 | −0 | +5.0 | 8.4 | −0 | +7.0 |
| 3.94–4.73 | 1.1 | +0.9 | +2.6 | 1.6 | +1.4 | +3.9 | 2.6 | +1.4 | +4.9 | 4.6 | +1.4 | +6.9 | 6.6 | +1.4 | +9.4 |
| | 2.6 | −0 | +2.0 | 3.9 | −0 | +3.0 | 4.9 | −0 | +4.0 | 6.9 | −0 | +6.0 | 9.4 | −0 | +8.0 |
| 4.73–5.52 | 1.2 | +1.0 | +2.9 | 1.9 | +1.6 | +4.5 | 3.4 | +1.6 | +6.0 | 5.4 | +1.6 | +8.0 | 8.4 | +1.6 | +11.6 |
| | 2.9 | −0 | +2.2 | 4.5 | −0 | +3.5 | 6.0 | −0 | +5.0 | 8.0 | −0 | +7.0 | 11.6 | −0 | +10.0 |
| 5.52–6.30 | 1.5 | +1.0 | +3.2 | 2.4 | +1.6 | +5.0 | 3.4 | +1.6 | +6.0 | 5.4 | +1.6 | +8.0 | 10.4 | +1.6 | +13.6 |
| | 3.2 | −0 | +2.5 | 5.0 | −0 | +4.0 | 6.0 | −0 | +5.0 | 8.0 | −0 | +7.0 | 13.6 | −0 | +12.0 |
| 6.30–7.09 | 1.8 | +1.0 | +3.5 | 2.9 | +1.6 | +5.5 | 4.4 | +1.6 | +7.0 | 6.4 | +1.6 | +9.0 | 10.4 | +1.6 | +13.6 |
| | 3.5 | −0 | +2.8 | 5.5 | −0 | +4.5 | 7.0 | −0 | +6.0 | 9.0 | −0 | +8.0 | 13.6 | −0 | +12.0 |

## American Standard Pipe[a,b]

*Welded wrought iron*

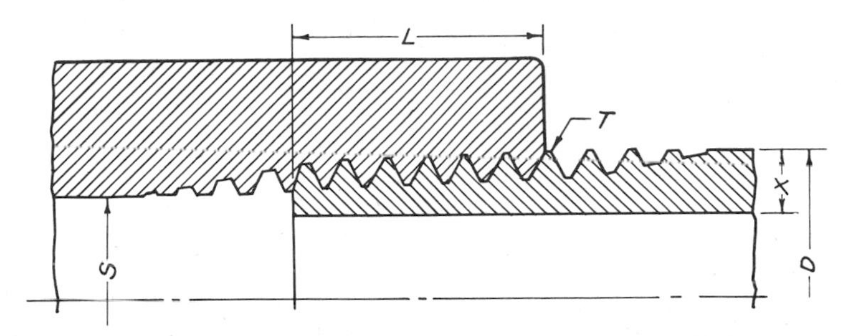


| Nominal pipe size | Actual outside diam. $D$ | Tap-drill size $S$ | Thd. /in. $T$ | Distance pipe enters fittings $L$ | Wall thickness $X$ | | | Weight, lb/ft[f] | | |
|---|---|---|---|---|---|---|---|---|---|---|
| | | | | | Standard 40[c] | Extra strong 80[d] | Double extra strong[e] | Standard 40[c] | Extra strong 80[d] | Double extra strong[e] |
| 1/8 | 0.405 | 11/32 | 27 | 3/16 | 0.070 | 0.098 | .... | 0.25 | 0.32 | |
| 1/4 | 0.540 | 7/16 | 18 | 9/32 | 0.090 | 0.122 | .... | 0.43 | 0.54 | |
| 3/8 | 0.675 | 37/64 | 18 | 19/64 | 0.093 | 0.129 | .... | 0.57 | 0.74 | |
| 1/2 | 0.840 | 23/32 | 14 | 3/8 | 0.111 | 0.151 | 0.307 | 0.86 | 1.09 | 1.714 |
| 3/4 | 1.050 | 59/64 | 14 | 13/32 | 0.115 | 0.157 | 0.318 | 1.14 | 1.48 | 2.440 |
| 1 | 1.315 | 1 5/32 | 11 1/2 | 1/2 | 0.136 | 0.183 | 0.369 | 1.68 | 2.18 | 3.659 |
| 1 1/4 | 1.660 | 1 1/2 | 11 1/2 | 35/64 | 0.143 | 0.195 | 0.393 | 2.28 | 3.00 | 5.214 |
| 1 1/2 | 1.900 | 1 47/64 | 11 1/2 | 9/16 | 0.148 | 0.204 | 0.411 | 2.72 | 3.64 | 6.408 |
| 2 | 2.375 | 2 7/32 | 11 1/2 | 37/64 | 0.158 | 0.223 | 0.447 | 3.66 | 5.03 | 9.029 |
| 2 1/2 | 2.875 | 2 5/8 | 8 | 7/8 | 0.208 | 0.282 | 0.565 | 5.80 | 7.67 | 13.695 |
| 3 | 3.5 | 3 1/4 | 8 | 1 5/16 | 0.221 | 0.306 | 0.615 | 7.58 | 10.3 | 18.583 |
| 3 1/2 | 4.0 | 3 3/4 | 8 | 1 | 0.231 | 0.325 | .... | 9.11 | 12.5 | |
| 4 | 4.5 | 4 1/4 | 8 | 1 1/16 | 0.242 | 0.344 | 0.690 | 10.8 | 15.0 | 27.451 |
| 5 | 5.563 | 5 5/16 | 8 | 1 5/32 | 0.263 | 0.383 | 0.768 | 14.7 | 20.8 | 38.552 |
| 6 | 6.625 | 6 5/16 | 8 | 1 1/4 | 0.286 | 0.441 | 0.884 | 19.0 | 28.6 | 53.160 |
| 8 | 8.625 | ..... | 8 | 1 15/32 | 0.329 | 0.510 | 0.895 | 28.6 | 43.4 | 72.424 |
| 10 | 10.75 | ..... | 8 | 1 43/64 | 0.372 | 0.606 | .... | 40.5 | 64.4 | |
| 12 | 12.75 | ..... | 8 | 1 7/8 | 0.414 | 0.702 | .... | 53.6 | 88.6 | |
| 14 OD | 14.0 | ..... | 8 | 2 | 0.437 | 0.750 | .... | 62.2 | 104. | |
| 16 OD | 16.0 | ..... | 8 | 2 13/64 | 0.500 | .... | .... | 82.2 | | |
| 18 OD | 18.0 | ..... | 8 | 2 13/32 | 0.562 | .... | .... | 103. | | |
| 20 OD | 20.0 | ..... | 8 | 2 19/32 | 0.562 | .... | .... | 115. | | |
| 24 OD | 24.0 | ..... | 8 | 3 | | | | | | |

[a] For welded and seamless steel pipe. See ASA B36.10—1959. Dimensions in inches.

[b] A pipe size may be designated by giving the nominal pipe size and wall thickness or by giving the nominal pipe size and weight per linear foot.

[c] Refers to American Standard schedule numbers, approximate values for the expression 1,000 $\times P/S$. Schedule 40—standard weight.

[d] Schedule 80—extra strong.

[e] Not American Standard, but commercially available in both wrought iron and steel.

[f] Plain ends.

## American Standard 150-lb Malleable-iron Screwed Fittings[a]

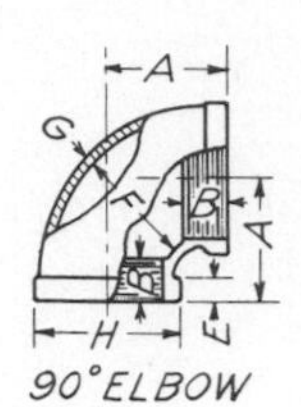
90° ELBOW

TEE

CROSS

| Nominal pipe size | A | B | C | E | F | G | H | J | K | L | M |
|---|---|---|---|---|---|---|---|---|---|---|---|
| 1/8 | 0.69 | 0.25 | .... | 0.200 | 0.405 | 0.090 | 0.693 | 1.00[b] | .... | 0.264 | |
| 1/4 | 0.81 | 0.32 | 0.73 | 0.215 | 0.540 | 0.095 | 0.844 | 1.19 | 0.94 | 0.402 | 1.00 |
| 3/8 | 0.95 | 0.36 | 0.80 | 0.230 | 0.675 | 0.100 | 1.015 | 1.44 | 1.03 | 0.408 | 1.13 |
| 1/2 | 1.12 | 0.43 | 0.88 | 0.249 | 0.840 | 0.105 | 1.197 | 1.63 | 1.15 | 0.534 | 1.25 |
| 3/4 | 1.31 | 0.50 | 0.98 | 0.273 | 1.050 | 0.120 | 1.458 | 1.89 | 1.29 | 0.546 | 1.44 |
| 1 | 1.50 | 0.58 | 1.12 | 0.302 | 1.315 | 0.134 | 1.771 | 2.14 | 1.47 | 0.683 | 1.69 |
| 1 1/4 | 1.75 | 0.67 | 1.29 | 0.341 | 1.660 | 0.145 | 2.153 | 2.45 | 1.71 | 0.707 | 2.06 |
| 1 1/2 | 1.94 | 0.70 | 1.43 | 0.368 | 1.900 | 0.155 | 2.427 | 2.69 | 1.88 | 0.724 | 2.31 |
| 2 | 2.25 | 0.75 | 1.68 | 0.422 | 2.375 | 0.173 | 2.963 | 3.26 | 2.22 | 0.757 | 2.81 |
| 2 1/2 | 2.70 | 0.92 | 1.95 | 0.478 | 2.875 | 0.210 | 3.589 | 3.86 | 2.57 | 1.138 | 3.25 |
| 3 | 3.08 | 0.98 | 2.17 | 0.548 | 3.500 | 0.231 | 4.285 | 4.51 | 3.00 | 1.200 | 3.69 |
| 3 1/2 | 3.42 | 1.03 | 2.39 | 0.604 | 4.000 | 0.248 | 4.843 | 5.09[b] | .... | 1.250 | 4.00 |
| 4 | 3.79 | 1.08 | 2.61 | 0.661 | 4.500 | 0.265 | 5.401 | 5.69 | 3.70 | 1.300 | 4.38 |
| 5 | 4.50 | 1.18 | 3.05 | 0.780 | 5.563 | 0.300 | 6.583 | 6.86[b] | .... | 1.406 | 5.12 |
| 6 | 5.13 | 1.28 | 3.46 | 0.900 | 6.625 | 0.336 | 7.767 | 8.03[b] | .... | 1.513 | 5.86 |

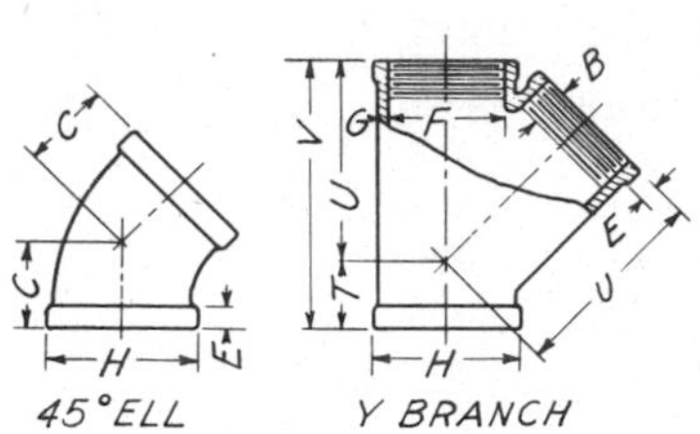
45° ELL Y BRANCH

| Nominal pipe size | N | P | T | U | V | W | X | Y | Z[c] | O[d] | Thickness of ribs on caps, couplings |
|---|---|---|---|---|---|---|---|---|---|---|---|
| 1/8 | 0.20 | .... | .... | .... | .... | 0.96 | 0.37 | 0.24 | 9/32 | .... | 0.090 |
| 1/4 | 0.26 | .... | .... | .... | .... | 1.06 | 0.44 | 0.28 | 3/8 | .... | 0.095 |
| 3/8 | 0.37 | .... | 0.50 | 1.43 | 1.93 | 1.16 | 0.48 | 0.31 | 7/16 | .... | 0.100 |
| 1/2 | 0.51 | 0.87 | 0.61 | 1.71 | 2.32 | 1.34 | 0.56 | 0.38 | 9/16 | 0.16 | 0.105 |
| 3/4 | 0.69 | 0.97 | 0.72 | 2.05 | 2.77 | 1.52 | 0.63 | 0.44 | 5/8 | 0.18 | 0.120 |
| 1 | 0.91 | 1.16 | 0.85 | 2.43 | 3.28 | 1.67 | 0.75 | 0.50 | 13/16 | 0.20 | 0.134 |
| 1 1/4 | 1.19 | 1.28 | 1.02 | 2.92 | 3.94 | 1.93 | 0.80 | 0.56 | 15/16 | 0.22 | 0.145 |
| 1 1/2 | 1.39 | 1.33 | 1.10 | 3.28 | 4.38 | 2.15 | 0.83 | 0.62 | 1 1/8 | 0.24 | 0.155 |
| 2 | 1.79 | 1.45 | 1.24 | 3.93 | 5.17 | 2.53 | 0.88 | 0.68 | 1 5/16 | 0.26 | 0.173 |
| 2 1/2 | 2.20 | 1.70 | 1.52 | 4.73 | 6.25 | 2.88 | 1.07 | 0.74 | 1 1/2 | 0.29 | 0.210 |
| 3 | 2.78 | 1.80 | 1.71 | 5.55 | 7.26 | 3.18 | 1.13 | 0.80 | 1 11/16 | 0.31 | 0.231 |
| 3 1/2 | 3.24 | 1.90 | .... | .... | .... | 3.43 | 1.18 | 0.86 | 1 7/8 | 0.34 | 0.248 |
| 4 | 3.70 | 2.08 | 2.01 | 6.97 | 8.98 | 3.69 | 1.22 | 1.00 | 2 1/8 | 0.37 | 0.265 |
| 5 | 4.69 | 2.32 | .... | .... | .... | .... | 1.31 | 1.00 | 2 5/16 | 0.46 | 0.300 |
| 6 | 5.67 | 2.55 | .... | .... | .... | .... | 1.40 | 1.25 | 2 1/2 | 0.52 | 0.336 |

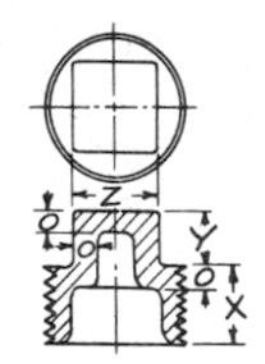
PLUG

[a] ASA B16.3—1963. Street tee not made in 1/8-in. size. Dimensions in inches. Left-hand couplings have four or more ribs. Right-hand couplings have two ribs. [b] Street ell only.

[c] These dimensions are the nominal size of wrench (ASA B18.2—1960). Square-head plugs are designed to fit these wrenches.

[d] Solid plugs are provided in sizes 1/8 to 3 1/2 in. inclusive; cored plugs 1/2 to 3 1/2 in. inclusive. Cored plugs have min. metal thickness at all points equal to dimension $O$ except at the end of thread.

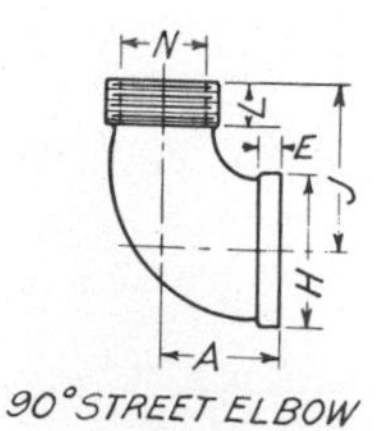
90° STREET ELBOW

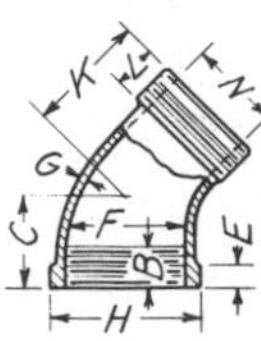

45° STREET ELBOW

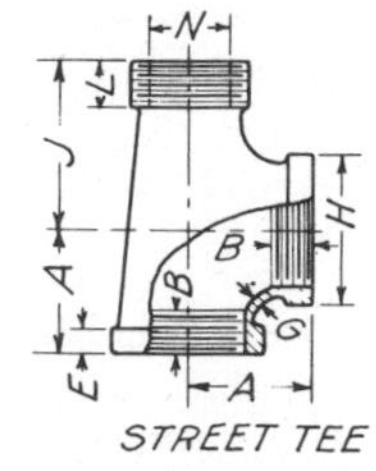
STREET TEE

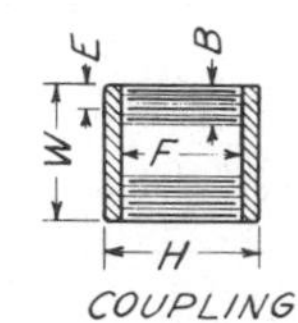
COUPLING

REDUCING COUPLING

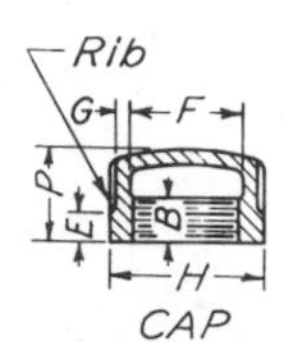
CAP

## American Standard 150-lb Malleable-iron Screwed Fittings: Dimensions of Close-, Medium-, and Open-pattern Return Bends*†

| Nominal pipe size | Length of thread, min. *B* | Width of band, min. *E* | Inside diam. of fittings *F* Max. | Inside diam. of fittings *F* Min. | Metal thickness ‡ *G* | Outside diam. of band, min. *H* | Center to center (close pattern) $R_1$ | Center to center (medium pattern) $R_2$ | Center to center (open pattern) $R_3$ |
|---|---|---|---|---|---|---|---|---|---|
| ½ | 0.43 | 0.249 | 0.840 | 0.897 | 0.116 | 1.197 | 1.000 | 1.250 | 1.50 |
| ¾ | 0.50 | 0.273 | 1.050 | 1.107 | 0.133 | 1.458 | 1.250 | 1.500 | 2.00 |
| 1 | 0.58 | 0.302 | 1.315 | 1.385 | 0.150 | 1.771 | 1.500 | 1.875 | 2.50 |
| 1¼ | 0.67 | 0.341 | 1.660 | 1.730 | 0.165 | 2.153 | 1.750 | 2.250 | 3.00 |
| 1½ | 0.70 | 0.368 | 1.900 | 1.970 | 0.178 | 2.427 | 2.188 | 2.500 | 3.50 |
| 2 | 0.75 | 0.422 | 2.375 | 2.445 | 0.201 | 2.963 | 2.625 | 3.000 | 4.00 |
| 2½ | 0.92 | 0.478 | 2.875 | 2.975 | 0.244 | 3.589 | .... | .... | 4.50 |
| 3 | 0.98 | 0.548 | 3.500 | 3.600 | 0.272 | 4.285 | .... | .... | 5.00 |
| 4 | 1.08 | 0.661 | 4.500 | 4.600 | 0.315 | 5.401 | .... | .... | 6.00 |

All dimensions given in inches.

* Adapted from ASA B16.3—1963.

† It is permissible to furnish close-pattern return bends not banded. Close-pattern return bends will not make up parallel coils, as the distance center to center of two adjacent bends is greater than the center to center of openings of a single bend.

‡ Patterns shall be designed to produce castings of metal thicknesses given in the table. Metal thickness at no point shall be less than 90 per cent of the thickness given in the table.

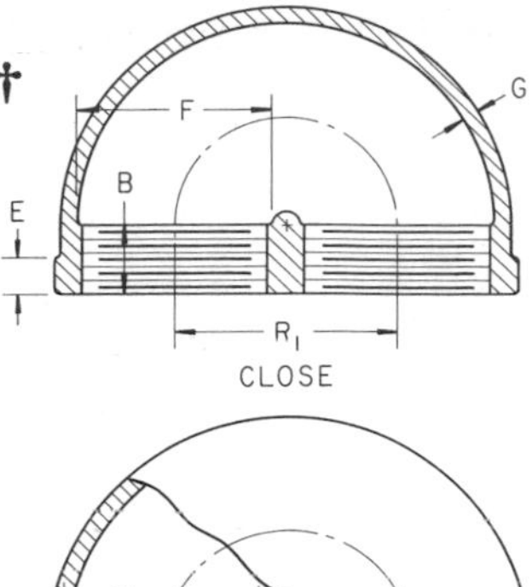


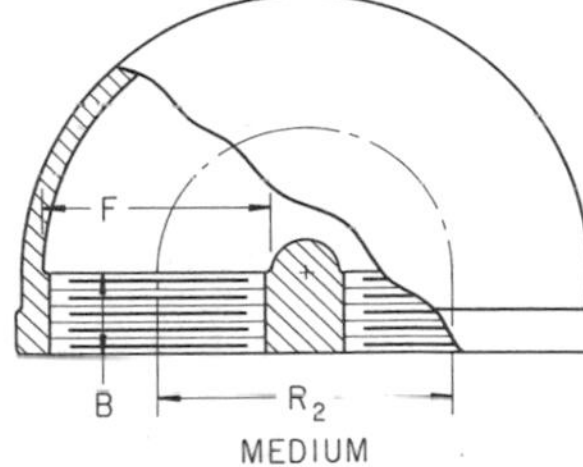


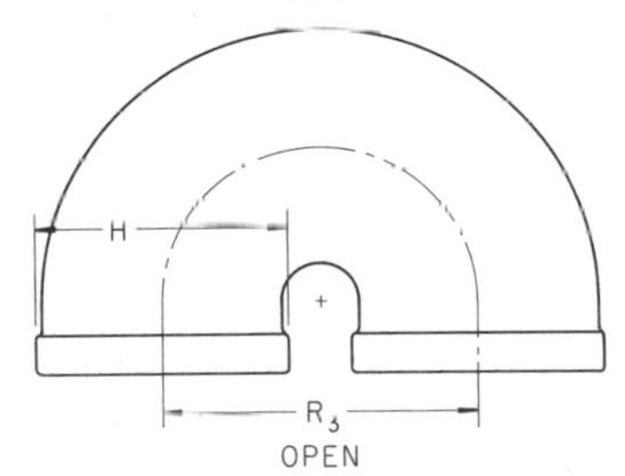


## American Standard Cast-iron Screwed Fittings*

*For maximum working saturated steam pressure of 125 and 250 lb per sq in.*

| Nominal pipe size | *A* | *B* min. | *C* | *E* min. | *F* Min. | *F* Max. | *G* min. | *H* min. |
|---|---|---|---|---|---|---|---|---|
| ¼ | 0.81 | 0.32 | 0.73 | 0.38 | 0.540 | 0.584 | 0.110 | 0.93 |
| ⅜ | 0.95 | 0.36 | 0.80 | 0.44 | 0.675 | 0.719 | 0.120 | 1.12 |
| ½ | 1.12 | 0.43 | 0.88 | 0.50 | 0.840 | 0.897 | 0.130 | 1.34 |
| ¾ | 1.31 | 0.50 | 0.98 | 0.56 | 1.050 | 1.107 | 0.155 | 1.63 |
| 1 | 1.50 | 0.58 | 1.12 | 0.62 | 1.315 | 1.385 | 0.170 | 1.95 |
| 1¼ | 1.75 | 0.67 | 1.29 | 0.69 | 1.660 | 1.730 | 0.185 | 2.39 |
| 1½ | 1.94 | 0.70 | 1.43 | 0.75 | 1.900 | 1.970 | 0.200 | 2.68 |
| 2 | 2.25 | 0.75 | 1.68 | 0.84 | 2.375 | 2.445 | 0.220 | 3.28 |
| 2½ | 2.70 | 0.92 | 1.95 | 0.94 | 2.875 | 2.975 | 0.240 | 3.86 |
| 3 | 3.08 | 0.98 | 2.17 | 1.00 | 3.500 | 3.600 | 0.260 | 4.62 |
| 3½ | 3.42 | 1.03 | 2.39 | 1.06 | 4.000 | 4.100 | 0.280 | 5.20 |
| 4 | 3.79 | 1.08 | 2.61 | 1.12 | 4.500 | 4.600 | 0.310 | 5.79 |
| 5 | 4.50 | 1.18 | 3.05 | 1.18 | 5.563 | 5.663 | 0.380 | 7.05 |
| 6 | 5.13 | 1.28 | 3.46 | 1.28 | 6.625 | 6.725 | 0.430 | 8.28 |
| 8 | 6.56 | 1.47 | 4.28 | 1.47 | 8.625 | 8.725 | 0.550 | 10.63 |
| 10 | 8.08 | 1.68 | 5.16 | 1.68 | 10.750 | 10.850 | 0.690 | 13.12 |
| 12 | 9.50 | 1.88 | 5.97 | 1.88 | 12.750 | 12.850 | 0.800 | 15.47 |

* ASA B16.4—1963. Dimensions in inches.

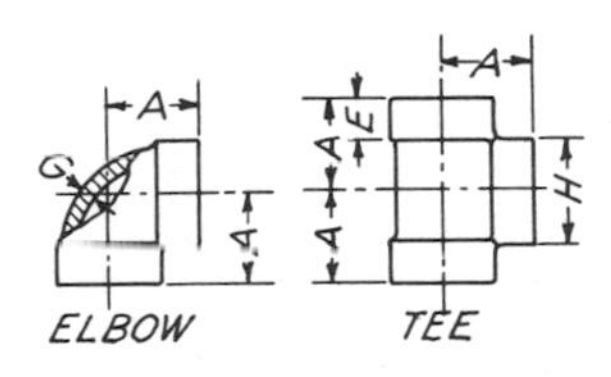


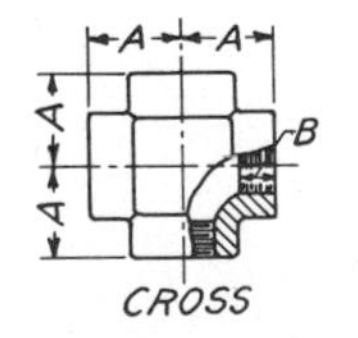

## Globe, Angle-globe, and Gate Valves*

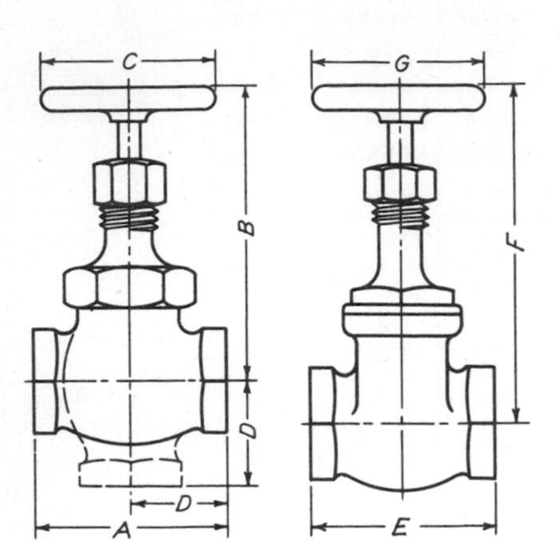


| Size | *A* (globe only) | *B* (open) | *C* | *D* (angle only) | *E* | *F* (open) | *G* |
|---|---|---|---|---|---|---|---|
| 1/8 | 2 | 4 | 1 3/4 | 1 | | | |
| 1/4 | 2 | 4 | 1 3/4 | 1 | 1 7/8 | 5 1/8 | 1 3/4 |
| 3/8 | 2 1/4 | 4 1/2 | 2 | 1 1/8 | 2 | 5 1/8 | 1 3/4 |
| 1/2 | 2 3/4 | 5 1/4 | 2 1/2 | 1 1/4 | 2 1/8 | 5 1/2 | 2 |
| 3/4 | 3 3/16 | 6 | 2 3/4 | 1 1/2 | 2 3/8 | 6 5/8 | 2 1/2 |
| 1 | 3 3/4 | 6 3/4 | 3 | 1 3/4 | 2 7/8 | 7 7/8 | 2 3/4 |
| 1 1/4 | 4 1/4 | 7 1/4 | 3 5/8 | 2 | 3 1/4 | 9 1/2 | 3 |
| 1 1/2 | 4 3/4 | 8 1/4 | 4 | 2 1/4 | 3 1/2 | 10 7/8 | 3 5/8 |
| 2 | 5 3/4 | 9 1/2 | 4 3/4 | 2 3/4 | 3 7/8 | 13 1/8 | 4 |
| 2 1/2 | 6 3/4 | 11 | 6 | 3 1/4 | 4 1/2 | 15 3/8 | 4 3/4 |
| 3 | 8 | 12 1/4 | 7 | 3 3/4 | 5 | 17 7/8 | 5 3/8 |

* Dimensions in inches and compiled from manufacturers' catalogues for drawing purposes.

## Lengths of Pipe Nipples*

| Size | Length | | Size | Length | | Size | Length | | Size | Length | |
|---|---|---|---|---|---|---|---|---|---|---|---|
| | Close | Short | | Close | Short | | Close | Short | | Close | Short |
| 1/8 | 3/4 | 1 1/2 | 1/2 | 1 1/8 | 1 1/2 | 1 1/4 | 1 5/8 | 2 1/2 | 2 1/2 | 2 1/2 | 3 |
| 1/4 | 7/8 | 1 1/2 | 3/4 | 1 3/8 | 2 | 1 1/2 | 1 3/4 | 2 1/2 | 3 | 2 5/8 | 3 |
| 3/8 | 1 | 1 1/2 | 1 | 1 1/2 | 2 | 2 | 2 | 2 1/2 | | | |

* Compiled from manufacturers' catalogues. Dimensions in inches.

Long-nipple lengths: from short-nipple lengths to 6 in. in 1/2 in. increments; from 6 in. nipple lengths to 12 in. in 1 in. increments; from 12 in. nipple lengths to 24 in. in 2 in. increments.

## Pipe Bushings*

*Dimensions of outside-head, inside-head, and face bushings in inches*

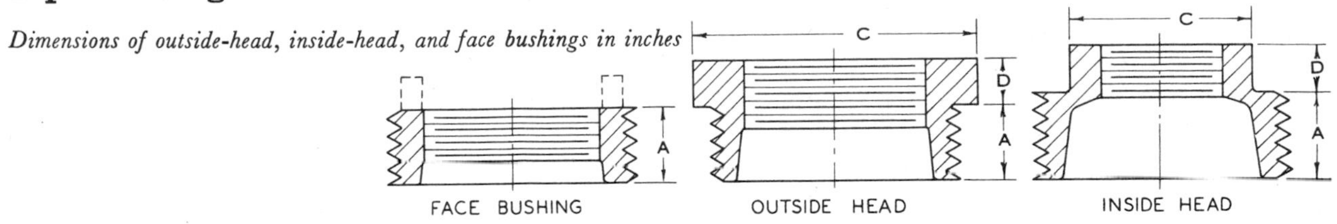


| Size | Length of external thread,† min., *A* | Height of head, min., *D* | Width of head,‡ min., *C* | |
|---|---|---|---|---|
| | | | Outside | Inside |
| ¼ × ⅛ | 0.44 | 0.14 | 0.64 | |
| ⅜ × ¼ | 0.48 | 0.16 | 0.68 | |
| ⅜ × ⅛ | 0.48 | 0.16 | 0.68 | |
| ½ × ⅜ | 0.56 | 0.19 | 0.87 | |
| ½ × ¼ | 0.56 | 0.19 | 0.87 | |
| ½ × ⅛ | 0.56 | 0.19 | 0.87 | |
| ¾ × ½ | 0.63 | 0.22 | 1.15 | |
| ¾ × ⅜ | 0.63 | 0.22 | 1.15 | |
| ¾ × ¼ | 0.63 | 0.22 | 1.15 | |
| ¾ × ⅛ | 0.63 | 0.22 | 1.15 | |
| 1 × ¾ | 0.75 | 0.25 | 1.42 | |
| 1 × ½ | 0.75 | 0.25 | 1.42 | |
| 1 × ⅜ | 0.75 | 0.30 | .... | 1.12 |
| 1 × ¼ | 0.75 | 0.30 | .... | 1.12 |
| 1 × ⅛ | 0.75 | 0.30 | .... | 1.12 |
| 1¼ × 1 | 0.80 | 0.28 | 1.76 | |
| 1¼ × ¾ | 0.80 | 0.28 | 1.76 | |
| 1¼ × ½ | 0.80 | 0.34 | .... | 1.34 |
| 1¼ × ⅜ | 0.80 | 0.34 | .... | 1.12 |
| 1¼ × ¼ | 0.80 | 0.34 | .... | 1.12 |
| 1½ × 1¼ | 0.83 | 0.31 | 2.00 | |
| 1½ × 1 | 0.83 | 0.31 | 2.00 | |
| 1½ × ¾ | 0.83 | 0.37 | .... | 1.63 |
| 1½ × ½ | 0.83 | 0.37 | .... | 1.34 |
| 1½ × ⅜ | 0.83 | 0.37 | .... | 1.12 |

| Size | Length of external thread,‡ min., *A* | Height of head, min., *D* | Width of head,† min., *C* | |
|---|---|---|---|---|
| | | | Outside | Inside |
| 1½ × ¼ | 0.83 | 0.37 | .... | 1.12 |
| 2 × 1½ | 0.88 | 0.34 | 2.48 | |
| 2 × 1¼ | 0.88 | 0.34 | 2.48 | |
| 2 × 1 | 0.88 | 0.41 | .... | 1.95 |
| 2 × ¾ | 0.88 | 0.41 | .... | 1.63 |
| 2 × ½ | 0.88 | 0.41 | .... | 1.34 |
| 2 × ⅜ | 0.88 | 0.41 | .... | 1.12 |
| 2 × ¼ | 0.88 | 0.41 | .... | 1.12 |
| 2½ × 2 | 1.07 | 0.37 | 2.98 | |
| 2½ × 1½ | 1.07 | 0.44 | 2.68 | |
| 2½ × 1¼ | 1.07 | 0.44 | .... | 2.39 |
| 2½ × 1 | 1.07 | 0.44 | .... | 1.95 |
| 2½ × ¾ | 1.07 | 0.44 | .... | 1.63 |
| 2½ × ½ | 1.07 | 0.44 | .... | 1.34 |
| 3 × 2½ | 1.13 | 0.40 | 3.86 | |
| 3 × 2 | 1.13 | 0.48 | 3.28 | |
| 3 × 1½ | 1.13 | 0.48 | .... | 2.68 |
| 3 × 1¼ | 1.13 | 0.48 | .... | 2.39 |
| 3 × 1 | 1.13 | 0.48 | .... | 1.95 |
| 3 × ¾ | 1.13 | 0.48 | .... | 1.63 |
| 3 × ½ | 1.13 | 0.48 | .... | 1.34 |

* ASA B16.14—1953.

† In the case of outside-head bushings, length *A* includes provisions for imperfect threads.

‡ Heads of bushings shall be hexagonal or octagonal, except that on the larger sizes of outside-head bushings the heads may be made round with lugs instead of hexagonal or octagonal.

## American Standard Cast-iron Pipe Flanges and Flanged Fittings*

*For maximum working saturated steam pressure of 125 lb per sq in. (gage)*

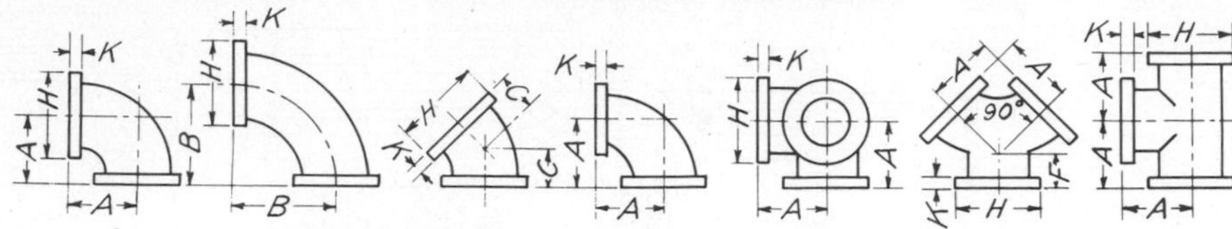


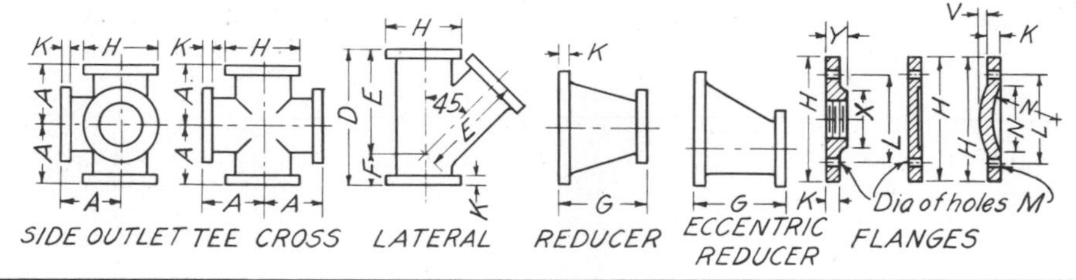


| Nominal pipe size $N$ | $A$ | $B$ | $C$ | $D$ | $E$ | $F$ | $G$ | $H$ | $K$ min. |
|---|---|---|---|---|---|---|---|---|---|
| 1 | 3½ | 5 | 1¾ | 7½ | 5¾ | 1¾ | .... | 4¼ | 7/16 |
| 1¼ | 3¾ | 5½ | 2 | 8 | 6¼ | 1¾ | .... | 4⅝ | ½ |
| 1½ | 4 | 6 | 2¼ | 9 | 7 | 2 | .... | 5 | 9/16 |
| 2 | 4½ | 6½ | 2½ | 10½ | 8 | 2½ | 5 | 6 | ⅝ |
| 2½ | 5 | 7 | 3 | 12 | 9½ | 2½ | 5½ | 7 | 11/16 |
| 3 | 5½ | 7¾ | 3 | 13 | 10 | 3 | 6 | 7½ | ¾ |
| 3½ | 6 | 8½ | 3½ | 14½ | 11½ | 3 | 6½ | 8½ | 13/16 |
| 4 | 6½ | 9 | 4 | 15 | 12 | 3 | 7 | 9 | 15/16 |
| 5 | 7½ | 10¼ | 4½ | 17 | 13½ | 3½ | 8 | 10 | 15/16 |
| 6 | 8 | 11½ | 5 | 18 | 14½ | 3½ | 9 | 11 | 1 |
| 8 | 9 | 14 | 5½ | 22 | 17½ | 4½ | 11 | 13½ | 1⅛ |
| 10 | 11 | 16½ | 6½ | 25½ | 20½ | 5 | 12 | 16 | 1 3/16 |
| 12 | 12 | 19 | 7½ | 30 | 24½ | 5½ | 14 | 19 | 1¼ |

| Nominal pipe size $N$ | $L$ | $M$ | No. of bolts | Diam. of bolts | Length of bolts | $X$ min. | $Y$ min. | Wall thickness | $V$ |
|---|---|---|---|---|---|---|---|---|---|
| 1 | 3⅛ | ⅝ | 4 | ½ | 1¾ | 1 15/16 | 1 1/16 | 5/16 | ⅜ |
| 1¼ | 3½ | ⅝ | 4 | ½ | 2 | 2 5/16 | 1 3/16 | 5/16 | 7/16 |
| 1½ | 3⅞ | ⅝ | 4 | ½ | 2 | 2 9/16 | ⅞ | 5/16 | ½ |
| 2 | 4¾ | ¾ | 4 | ⅝ | 2¼ | 3 1/16 | 1 | 5/16 | 9/16 |
| 2½ | 5½ | ¾ | 4 | ⅝ | 2½ | 3 9/16 | 1⅛ | 5/16 | ⅝ |
| 3 | 6 | ¾ | 4 | ⅝ | 2½ | 4¼ | 1 3/16 | ⅜ | 11/16 |
| 3½ | 7 | ¾ | 8 | ⅝ | 2¾ | 4 13/16 | 1¼ | 7/16 | ¾ |
| 4 | 7½ | ¾ | 8 | ⅝ | 3 | 5 5/16 | 1 5/16 | ½ | ⅞ |
| 5 | 8½ | ⅞ | 8 | ¾ | 3 | 6 7/16 | 1 7/16 | ½ | ⅞ |
| 6 | 9½ | ⅞ | 8 | ¾ | 3¼ | 7 9/16 | 1 9/16 | 9/16 | 15/16 |
| 8 | 11¾ | ⅞ | 8 | ¾ | 3½ | 9 11/16 | 1¾ | ⅝ | 1 1/16 |
| 10 | 14¼ | 1 | 12 | ⅞ | 3¾ | 11 15/16 | 1 15/16 | ¾ | 1⅛ |
| 12 | 17 | 1 | 12 | ⅞ | 3¾ | 14 1/16 | 2 3/16 | 13/16 | |

* ASA B16.1—1960. Dimensions in inches.

## Lengths of Malleable-iron Unions*

*Ground joint*

| Nominal size | 1/8 | 1/4 | 3/8 | 1/2 | 3/4 | 1 | 1 1/4 | 1 1/2 | 2 | 2 1/2 | 3 |
|---|---|---|---|---|---|---|---|---|---|---|---|
| End to end | 1 1/2 | 1 9/16 | 1 5/8 | 1 13/16 | 2 1/16 | 2 1/4 | 2 1/2 | 2 5/8 | 3 | 3 9/16 | 3 15/16 |

* Compiled from manufacturers' catalogues. Dimensions in inches.

## American Standard Steel Butt-welding Fittings*,†

*Elbows, tees, caps, and stub ends*

| Nominal pipe size | Outside diam. at bevel | Center to end | | | Welding caps $E$ | Lapped-joint stub ends | | |
|---|---|---|---|---|---|---|---|---|
| | | 90° welding elbow $A$ | 45° welding elbow $B$ | Of run, welding tee $C$ | | Lengths $F$ | Radius of fillet $R$ | Diameter of lap $G$ |
| 1 | 1.310 | 1 1/2 | 7/8 | 1 1/2 | 1 1/2 | 4 | 1/8 | 2 |
| 1 1/4 | 1.660 | 1 7/8 | 1 | 1 7/8 | 1 1/2 | 4 | 3/16 | 2 1/2 |
| 1 1/2 | 1.900 | 2 1/4 | 1 1/8 | 2 1/4 | 1 1/2 | 4 | 1/4 | 2 7/8 |
| 2 | 2.375 | 3 | 1 3/8 | 2 1/2 | 1 1/2 | 6 | 5/16 | 3 5/8 |
| 2 1/2 | 2.875 | 3 3/4 | 1 3/4 | 3 | 1 1/2 | 6 | 5/16 | 4 1/8 |
| 3 | 3.500 | 4 1/2 | 2 | 3 3/8 | 2 | 6 | 3/8 | 5 |
| 3 1/2 | 4.000 | 5 1/4 | 2 1/4 | 3 3/4 | 2 1/2 | 6 | 3/8 | 5 1/2 |
| 4 | 4.500 | 6 | 2 1/2 | 4 1/8 | 2 1/2 | 6 | 7/16 | 6 3/16 |

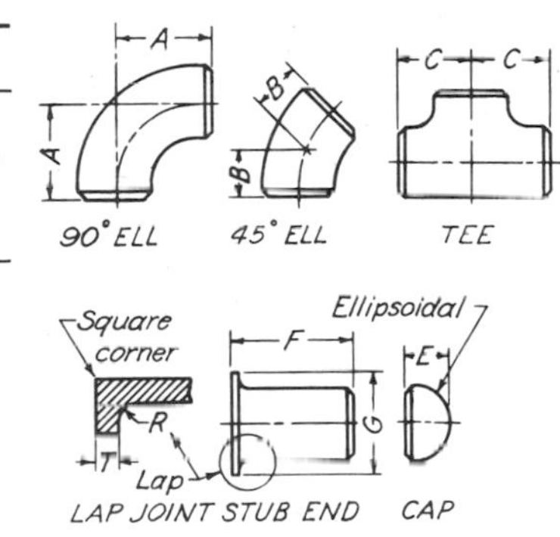


*Butt-welding reducers*

| Nominal pipe size | Outside diam. at bevel, Large end | Outside diam. at bevel, Small end | End to end $H$ |
|---|---|---|---|
| 1 × 3/4 | 1.315 | 1.050 | 2 |
| 1 × 1/2 | | 0.840 | |
| 1 × 3/8 | | 0.675 | |
| 1 1/4 × 1 | 1.660 | 1.315 | 2 |
| 1 1/4 × 3/4 | | 1.050 | |
| 1 1/4 × 1/2 | | 0.840 | |
| 1 1/2 × 1 1/4 | 1.900 | 1.660 | 2 1/2 |
| 1 1/2 × 1 | | 1.315 | |
| 1 1/2 × 3/4 | | 1.050 | |
| 1 1/2 × 1/2 | | 0.840 | |
| 2 × 1 1/2 | 2.375 | 1.900 | 3 |
| 2 × 1 1/4 | | 1.660 | |
| 2 × 1 | | 1.315 | |
| 2 × 3/4 | | 1.050 | |
| 2 1/2 × 2 | 2.875 | 2.375 | 3 1/2 |
| 2 1/2 × 1 1/2 | | 1.900 | |
| 2 1/2 × 1 1/4 | | 1.660 | |
| 2 1/2 × 1 | | 1.315 | |

| Nominal pipe size | Outside diam. at bevel, Large end | Outside diam. at bevel, Small end | End to end $H$ |
|---|---|---|---|
| 3 × 2 1/2 | 3.500 | 2.875 | 3 1/2 |
| 3 × 2 | | 2.375 | |
| 3 × 1 1/2 | | 1.900 | |
| 3 × 1 1/4 | | 1.660 | |
| 3 1/2 × 3 | 4.000 | 3.500 | 4 |
| 3 1/2 × 2 1/2 | | 2.875 | |
| 3 1/2 × 2 | | 2.375 | |
| 3 1/2 × 1 1/2 | | 1.900 | |
| 3 1/2 × 1 1/4 | | 1.660 | |
| 4 × 3 1/2 | 4.500 | 4.000 | 4 |
| 4 × 3 | | 3.500 | |
| 4 × 2 1/2 | | 2.875 | |
| 4 × 2 | | 2.375 | |
| 4 × 1 1/2 | | 1.900 | |
| 5 × 4 | 5.563 | 4.500 | 5 |
| 5 × 3 1/2 | | 4.000 | |
| 5 × 3 | | 3.500 | |
| 5 × 2 1/2 | | 2.875 | |

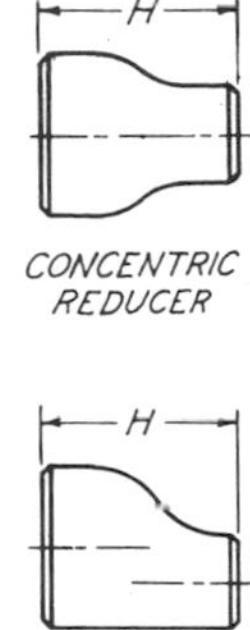


* For larger sizes, see ASA B16.3—1963.
† ASA B16.9—1964. Dimensions in inches.

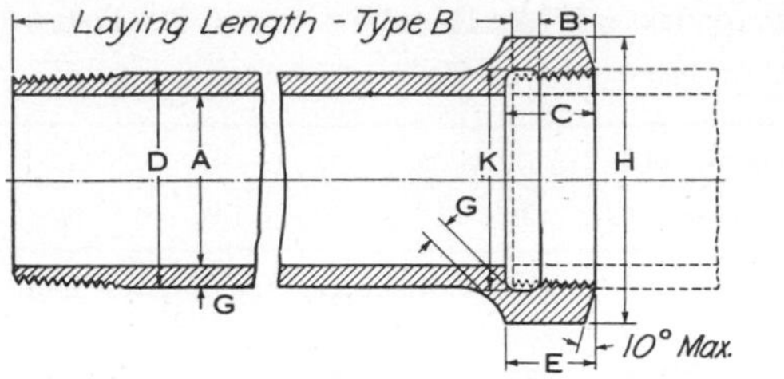

## Threaded Cast-iron Pipe*

*Dimension of pipe and drainage hubs*

| Pipe size | Pipe | | | Drainage hubs | | | | | Nominal weights | |
|---|---|---|---|---|---|---|---|---|---|---|
| | Nominal diam. | | Wall thick-ness, min., *G* | Thread length* *B* | Diam. of groove, max., *K* | End to shoulder † *C* | Min. band | | Type *A* and barrel of type *B* per foot | Addi-tional weight of hubs for type *B* |
| | Outside *D* | Inside *A* | | | | | Diam. *H* | Length *E* | | |
| 1¼ | 1.66 | 1.23 | 0.187 | 0.42 | 1.73 | 0.71 | 2.39 | 0.71 | 3.033 | 0.60 |
| 1½ | 1.90 | 1.45 | 0.195 | 0.42 | 1.97 | 0.72 | 2.68 | 0.72 | 3.666 | 0.90 |
| 2 | 2.38 | 1.89 | 0.211 | 0.43 | 2.44 | 0.76 | 3.28 | 0.76 | 5.041 | 1.00 |
| 2½ | 2.88 | 2.32 | 0.241 | 0.68 | 2.97 | 1.14 | 3.86 | 1.14 | 7.032 | 1.35 |
| 3 | 3.50 | 2.90 | 0.263 | 0.76 | 3.60 | 1.20 | 4.62 | 1.20 | 9.410 | 2.80 |
| 4 | 4.50 | 3.83 | 0.294 | 0.84 | 4.60 | 1.30 | 5.79 | 1.30 | 13.751 | 3.48 |
| 5 | 5.56 | 4.81 | 0.328 | 0.93 | 5.66 | 1.41 | 7.05 | 1.41 | 19.069 | 5.00 |
| 6 | 6.63 | 5.76 | 0.378 | 0.95 | 6.72 | 1.51 | 8.28 | 1.51 | 26.223 | 6.60 |
| 8 | 8.63 | 7.63 | 0.438 | 1.06 | 8.72 | 1.71 | 10.63 | 1.71 | 39.820 | 10.00 |
| 10 | 10.75 | 9.75 | 0.438 | 1.21 | 10.85 | 1.92 | 13.12 | 1.93 | 50.234 | |
| 12 | 12.75 | 11.75 | 0.438 | 1.36 | 12.85 | 2.12 | 15.47 | 2.13 | 60.036 | |

* ASA A40.5—1943. All dimensions are given in inches, except where otherwise stated. Type *A* has external threads both ends. Type *B* as shown.

† The length of thread *B* and the end to shoulder *C* shall not vary from the dimensions shown by more than plus or minus the equivalent of the pitch of one thread.

## American Standard Class-125 Cast-iron and 150-lb Steel Valves: Valves with Class-125 or 150-lb End Flanges, or with Welding Ends—Face-to-face and End-to-end Dimensions[a]

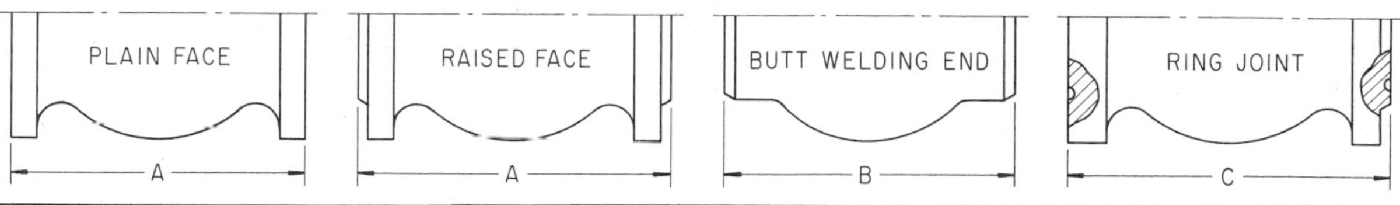


| Nominal valve size | Class-125 cast iron | | | | | | | | 150-lb steel | | | | | | | |
|---|---|---|---|---|---|---|---|---|---|---|---|---|---|---|---|---|
| | Flanged end—plain face | | | | | | | | Flanged end (1/16 in. raised face) and welding end | | | | | | | |
| | Gate | | Plug | | | Globe and lift check | Swing check[b] | Control | Gate | | | | Plug | | | |
| | Solid wedge | Double disc | Short pattern | Regular | Venturi | | | | Solid wedge | Double disc | Solid wedge | Double disc | Short pattern | Regular | Venturi | Round port full bore |
| | A | A | A | A | A | A | A | A | A | A | B | B | A | A | A | A |
| ¼ | ... | ... | ... | ... | ... | ... | ... | ... | 4 | 4 | 4 | 4 | ... | ... | ... | ... |
| ⅜ | ... | ... | ... | ... | ... | ... | ... | ... | 4 | 4 | 4 | 4 | ... | ... | ... | ... |
| ½ | ... | ... | ... | ... | ... | ... | ... | ... | 4¼ | 4¼ | 4¼ | 4¼ | ... | ... | ... | ... |
| ¾ | ... | ... | ... | ... | ... | ... | ... | ... | 4⅝ | 4⅝ | 4⅝ | 4⅝ | ... | ... | ... | ... |
| 1 | ... | ... | 5½ | 5½ | ... | ... | ... | 7¼ | 5 | 5 | 5 | 5 | 5½ | ... | ... | 7 |
| 1¼ | ... | ... | ... | 6½ | ... | ... | ... | ... | 5½ | 5½ | 5½ | 5½ | ... | ... | ... | ... |
| 1½ | ... | ... | 6½ | 6½ | ... | ... | ... | 8¾ | 6½ | 6½ | 6½ | 6½ | 6½ | ... | ... | 8¾ |
| 2 | 7 | 7 | 7 | 7½ | ... | 8 | 8 | 10 | 7 | 7 | 8½ | 8½ | 7 | ... | ... | 10½ |
| 2½ | 7½ | 7½ | 7½ | 8¼ | ... | 8½ | 8½ | 10⅞ | 7½ | 7½ | 9½ | 9½ | 7½ | ... | ... | 11¾ |
| 3 | 8 | 8 | 8 | 9 | ... | 9½ | 9½ | 11¾ | 8 | 8 | 11⅛ | 11⅛ | 8 | ... | ... | 13½ |
| 3½ | 8½ | 8½ | ... | ... | ... | ... | ... | ... | ... | ... | ... | ... | ... | ... | ... | ... |
| 4 | 9 | 9 | 9 | 9 | ... | 11½ | 11½ | 13⅞ | 9 | 9 | 12 | 12 | 9 | ... | ... | 17 |
| 5 | 10 | 10 | 10 | 14 | ... | 13 | 13 | ... | 10 | 10 | 15 | 15 | ... | ... | ... | ... |
| 6 | 10½ | 10½ | 10½ | 15½ | 15½ | 14 | 14 | 17¾ | 10½ | 10½ | 15⅞ | 15⅞ | 10½ | 15½ | ... | 21 |
| 8 | 11½ | 11½ | 11½ | 18 | 18 | 19½ | 19½ | 21⅜ | 11½ | 11½ | 16½ | 16½ | 11½ | 18 | ... | 25 |

| Nominal valve size | 150-lb steel | | | | | | | | | | | |
|---|---|---|---|---|---|---|---|---|---|---|---|---|
| | Flanged end (1/16 in. raised face) and welding end | | | | Flanged end (ring joint) | | | | | | | |
| | Globe and lift check | Swing check | "Y" pattern globe | Control | Gate | | Plug | | | Round port full bore | Globe and lift check | Swing check |
| | | | | | Solid wedge | Double disc | Short pattern | Regular | Venturi | | | |
| | A&B | A&B | A&B | A | C | C | C | C | C | C | C | C |
| ¼ | 4 | 4 | ... | ... | ... | ... | ... | ... | ... | ... | ... | ... |
| ⅜ | 4 | 4 | ... | ... | ... | ... | ... | ... | ... | ... | ... | ... |
| ½ | 4¼ | 4¼ | 5½ | ... | 4 11/16 | 4 11/16 | ... | ... | ... | ... | 4 11/16 | 4 11/16 |
| ¾ | 4⅝ | 4⅝ | 6 | ... | 5⅛ | 5⅛ | ... | ... | ... | ... | 5⅛ | 5⅛ |
| 1 | 5 | 5 | 6½ | 7¼ | 5½ | 5½ | 6 | ... | ... | 7½ | 5½ | 5½ |
| 1¼ | 5½ | 5½ | 7¼ | ... | 6 | 6 | ... | ... | ... | ... | 6 | 6 |
| 1½ | 6½ | 6½ | 8 | 8¾ | 7 | 7 | 7 | ... | ... | 9¼ | 7 | 7 |
| 2 | 8 | 8 | 9 | 10 | 7½ | 7½ | 7½ | ... | ... | 11 | 8½ | 8½ |
| 2½ | 8½ | 8½ | 11 | 10⅞ | 8 | 8 | 8 | ... | ... | 12¼ | 9 | 9 |
| 3 | 9½ | 9½ | 12½ | 11¾ | 8½ | 8½ | 8½ | ... | ... | 14 | | 10 |
| 3½ | ... | ... | ... | ... | ... | ... | ... | ... | ... | ... | ... | ... |
| 4 | 11½ | 11½ | 14½ | 13⅞ | 9½ | 9½ | 9½ | ... | ... | 17½ | 12 | 12 |
| 5 | 14 | 13 | ... | ... | 10½ | 10½ | ... | ... | ... | ... | 14½ | 13½ |
| 6 | 16 | 14 | 18½ | 17¾ | 11 | 11 | 11 | 16 | ... | 21½ | 16½ | 14½ |
| 8 | 19½ | 19½ | 23½ | 21⅜ | 12 | 12 | 12 | 18½ | ... | 25½ | 20 | 20 |

[a] Adapted from ASA B16.10—1957. API Standards 6D and 600 conform to the dimensions shown for corresponding sizes, valve type and flange class or welding end.

[b] These dimensions are not intended to cover the type of check valve having the seat angle at approximately 45 deg to the run of the valve or the "Underwriter Pattern" or other patterns where large clearances are required.

## Beam Connections

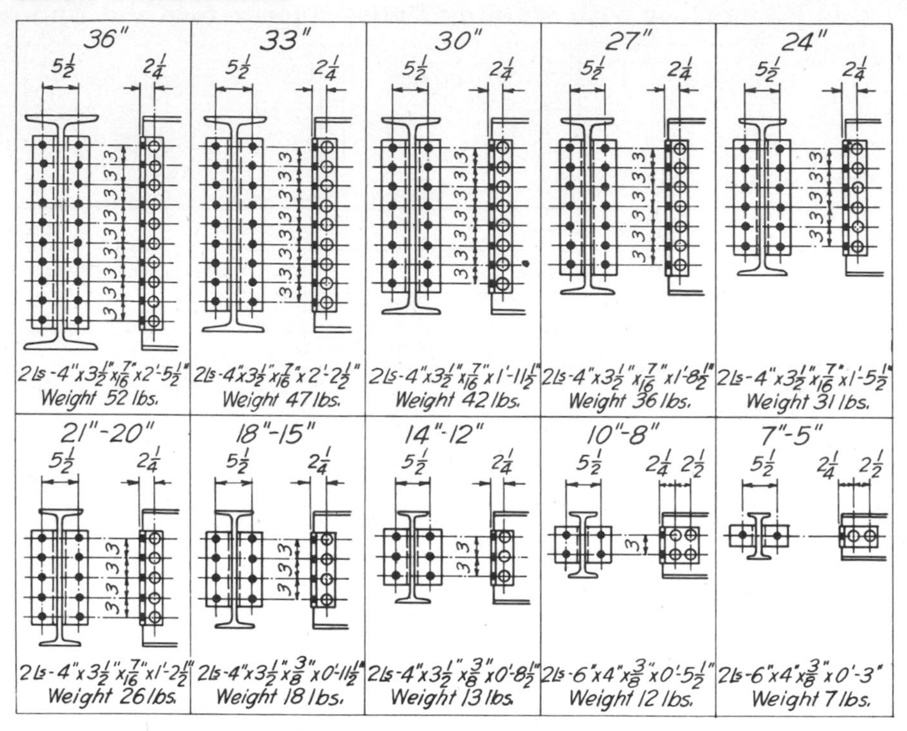


## Driving Clearances for Riveting

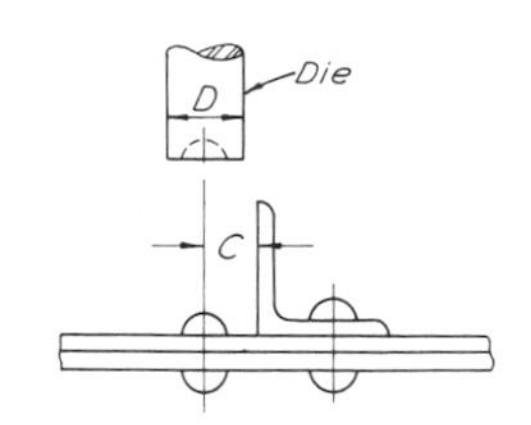


| | Diam. of rivet | | | | | | | | |
|---|---|---|---|---|---|---|---|---|---|
| | ½ | ⅝ | ¾ | ⅞ | 1 | 1⅛ | 1¼ | 1⅜ | 1½ |
| *D* | 1¾ | 2 | 2¼ | 2½ | 2¾ | 3 | 3¼ | 3½ | 3¾ |
| *C* | 1 | 1⅛ | 1¼ | 1⅜ | 1½ | 1⅝ | 1¾ | 1⅞ | 2 |

Dimensions in inches.

## Gage and Maximum Rivet Size for Angles

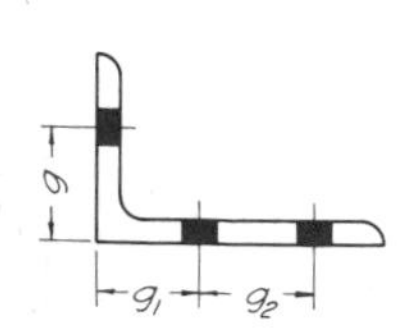


| Leg | 8 | 7 | 6 | 5 | 4 | 3½ | 3 | 2½ | 2 | 1¾ | 1½ | 1⅜ | 1¼ | 1 |
|---|---|---|---|---|---|---|---|---|---|---|---|---|---|---|
| $g$ | 4½ | 4 | 3½ | 3 | 2½ | 2 | 1¾ | 1⅜ | 1⅛ | 1 | ⅞ | ⅞ | ¾ | ⅝ |
| $g_1$ | 3 | 2½ | 2¼ | 2 | | | | | | | | | | |
| $g_2$ | 3 | 3 | 2½ | 1¾ | | | | | | | | | | |
| Max. rivet | 1⅛ | 1⅛ | 1 | 1 | ⅞ | ⅞ | ⅞ | ¾ | ⅝ | ½ | ⅜ | ⅜ | ⅜ | ¼ |

Dimensions in inches.

## Selected Structural Shapes*

*Dimensions for detailing*

| Name | Depth of section, in. | Weight per foot, lb | Flange | | Web | | Distances | | | | Grip, in. | Max. flange rivet, in. | Usual gage g, in. | Clearance m |
|---|---|---|---|---|---|---|---|---|---|---|---|---|---|---|
| | | | Width, in. | Mean thickness, in. | Thickness, in. | Half thickness, in. | T, in. | k, in. | $g_1$, in. | c, in. | | | | |
| Channels | 18 | 58.0 | 4 1/4 | 5/8 | 11/16 | 3/8 | 15 3/8 | 1 5/16 | 2 3/4 | 3/4 | 5/8 | 1 | 2 1/2 | |
| | 15 | 40.0 | 3 1/2 | 5/8 | 9/16 | 1/4 | 12 3/8 | 1 5/16 | 2 3/4 | 5/8 | 5/8 | 1 | 2 | |
| | 12 | 30.0 | 3 1/8 | 1/2 | 1/2 | 1/4 | 9 7/8 | 1 1/16 | 2 1/2 | 9/16 | 1/2 | 7/8 | 1 3/4 | |
| | 10 | 15.3 | 2 5/8 | 7/16 | 1/4 | 1/8 | 8 1/8 | 15/16 | 2 1/2 | 5/16 | 7/16 | 3/4 | 1 1/2 | |
| | 9 | 13.4 | 2 3/8 | 7/16 | 1/4 | 1/8 | 7 1/4 | 7/8 | 2 1/2 | 5/16 | 3/8 | 3/4 | 1 3/8 | |
| | 8 | 18.75 | 2 1/2 | 3/8 | 1/2 | 1/4 | 6 3/8 | 13/16 | 2 1/4 | 9/16 | 3/8 | 3/4 | 1 1/2 | |
| | 7 | 12.25 | 2 1/4 | 3/8 | 5/16 | 3/16 | 5 3/8 | 13/16 | 2 | 3/8 | 3/8 | 5/8 | 1 1/4 | |
| | 6 | 10.5 | 2 | 3/8 | 5/16 | 3/16 | 4 1/2 | 3/4 | 2 | 3/8 | 3/8 | 5/8 | 1 1/8 | |
| | 5 | 9.0 | 1 7/8 | 5/16 | 5/16 | 3/16 | 3 5/8 | 11/16 | 2 | 3/8 | 5/16 | 1/2 | 1 1/8 | |
| | 4 | 7.25 | 1 3/4 | 5/16 | 5/16 | 3/16 | 2 3/4 | 5/8 | 2 | 3/8 | 5/16 | 1/2 | 1 | |
| | 3 | 6.0 | 1 5/8 | 1/4 | 3/8 | 3/16 | 1 3/4 | 5/8 | . . . | 7/16 | 5/16 | 1/2 | 7/8 | |
| WF shapes | 21†(21 1/4) | 127 | 13 | 1 | 9/16 | 5/16 | 17 3/4 | 1 3/4 | 3 | 3/8 | ..... | ..... | 5 1/2 | 25 |
| | 16 (16 3/8) | 96 | 11 1/2 | 7/8 | 9/16 | 5/16 | 13 1/8 | 1 5/8 | 2 3/4 | 3/8 | ..... | ..... | 5 1/2 | 20 |
| | 14 (14 1/8) | 84 | 12 | 3/4 | 7/16 | 1/4 | 11 3/8 | 1 3/8 | 2 3/4 | 5/16 | ..... | ..... | 5 1/2 | 18 5/8 |
| | 14 (13 3/4) | 48 | 8 | 9/16 | 3/8 | 3/16 | 11 3/8 | 1 3/16 | 2 1/2 | 1/4 | ..... | ..... | 5 1/2 | 16 |
| | 12 (12 1/4) | 50 | 8 1/8 | 5/8 | 3/8 | 3/16 | 9 3/4 | 1 1/4 | 2 1/2 | 1/4 | ..... | ..... | 5 1/2 | 14 5/8 |
| | 10 (10) | 49 | 10 | 9/16 | 3/8 | 3/16 | 7 7/8 | 1 1/16 | 2 1/2 | 1/4 | ..... | ..... | 5 1/2 | 14 1/8 |
| | 10 (9 3/4) | 33 | 8 | 7/16 | 5/16 | 3/16 | 7 7/8 | 15/16 | 2 1/4 | 1/4 | ..... | ..... | 5 1/2 | 12 5/8 |
| | 8 (8) | 28 | 6 1/2 | 7/16 | 5/16 | 1/8 | 6 3/8 | 13/16 | 2 1/4 | 3/16 | ..... | ..... | 3 1/2 | 10 1/2 |
| Beams | 24 | 120.0 | 8 | 1 1/8 | 13/16 | 7/16 | 20 1/8 | 1 15/16 | 3 1/4 | 1/2 | 1 1/8 | 1 | 4 | |
| | 20 | 85.0 | 7 | 15/16 | 11/16 | 5/16 | 16 1/2 | 1 3/4 | 3 1/4 | 3/8 | 7/8 | 1 | 4 | |
| | 18 | 70.0 | 6 1/4 | 11/16 | 3/4 | 3/8 | 15 1/4 | 1 3/8 | 2 3/4 | 7/16 | 11/16 | 7/8 | 3 1/2 | |
| | 15 | 50.0 | 5 5/8 | 5/8 | 9/16 | 5/16 | 12 1/2 | 1 1/4 | 2 3/4 | 3/8 | 9/16 | 3/4 | 3 1/2 | |
| | 12 | 31.8 | 5 | 9/16 | 3/8 | 3/16 | 9 3/4 | 1 1/8 | 2 1/2 | 1/4 | 1/2 | 3/4 | 3 | |
| | 10 | 35.0 | 5 | 1/2 | 5/8 | 5/16 | 8 | 1 | 2 1/2 | 3/8 | 1/2 | 3/4 | 2 3/4 | |
| | 8 | 23.0 | 4 1/8 | 7/16 | 7/16 | 1/4 | 6 1/4 | 7/8 | 2 1/4 | 5/16 | 7/16 | 3/4 | 2 1/4 | |
| | 7 | 20.0 | 3 7/8 | 3/8 | 7/16 | 1/4 | 5 3/8 | 13/16 | 2 | 5/16 | 3/8 | 5/8 | 2 1/4 | |
| | 6 | 17.25 | 3 5/8 | 3/8 | 1/2 | 1/4 | 4 1/2 | 3/4 | 2 | 5/16 | 3/8 | 5/8 | 2 | |
| | 5 | 10.0 | 3 | 5/16 | 1/4 | 1/8 | 3 5/8 | 11/16 | 2 | 3/16 | 5/16 | 1/2 | 1 3/4 | |
| | 4 | 9.5 | 2 3/4 | 5/16 | 5/16 | 3/16 | 2 3/4 | 5/8 | 2 | 1/4 | 5/16 | 1/2 | 1 1/2 | |
| | 3 | 7.5 | 2 1/2 | 1/4 | 3/8 | 3/16 | 1 7/8 | 9/16 | . . . | 1/4 | 1/4 | 3/8 | 1 1/2 | |

* From Steel Construction Handbook.

† Nominal depth; (   ) indicates actual depth.

CHANNEL

WF SHAPE

BEAM

## Standard Jig Bushings*

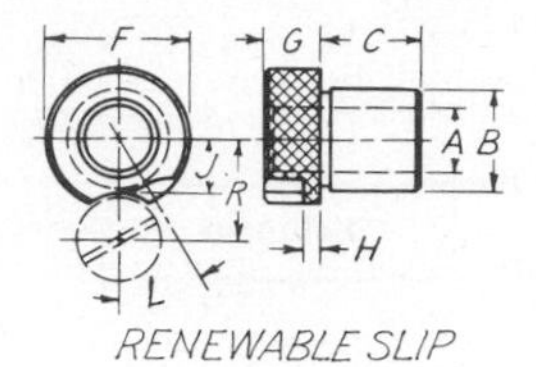

RENEWABLE SLIP

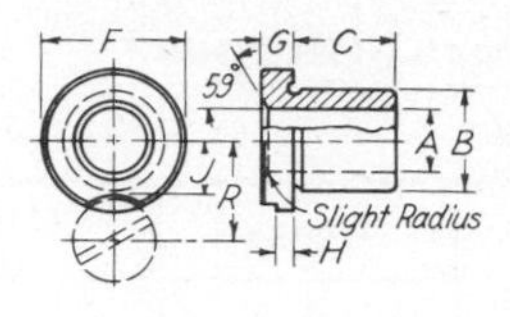

RENEWABLE FIXED

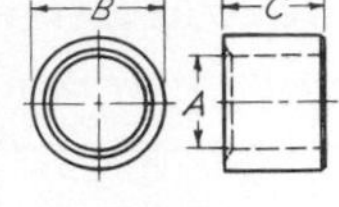

LINER (used with Renewable Type bushings)

| Renewable Slip Type and Renewable Fixed Type | | | | | | | | | | | | | Liners | | |
|---|---|---|---|---|---|---|---|---|---|---|---|---|---|---|---|
| Slip type<br>Hole size *A* | Fixed type<br>Hole size *A* | Tolerance on hole | Body diameter limits *B* | Lengths available *C* | Head dimensions *F* | *G* Slip type | *G* Fixed type | *H* | *J* | *L* | *R* | Lock screw no. | Hole limits *A* | OD limits *B* | Lengths |
| 0.052 to 0.089<br>0.0935 to 0.1562 | 0.055 to 0.089<br>0.0935 to 0.1562 | +0.0004<br>+0.0001 | 0.3125<br>0.3123 | 5/16, 1/2<br>3/4, 1<br>5/16, 1/2<br>3/4, 1 | 35/64 | 3/8 | 1/4 | 1/8 | 11/64 | 65° | 1/2 | 1 | 0.3126<br>0.3129 | 0.5017<br>0.5014 | Same as the bushing |
| 0.1406 to 0.3437 | 0.1570 to 0.3125 | Incl 1/4<br>+0.0004<br>+0.0001<br>Over 1/4<br>+0.0005<br>+0.0001 | 0.5000<br>0.4998 | 5/16, 1/2, 3/4<br>1, 1 3/8, 1 3/4 | 51/64 | 7/16 | 1/4 | 1/8 | 19/64 | 65° | 5/8 | 1 | 0.5002<br>0.5005 | 0.7518<br>0.7515 | Same as the bushing |
| 0.2812 to 0.5312 | 0.3160 to 0.5000 | +0.0005<br>+0.0001 | 0.7500<br>0.7498 | 1/2, 5/8, 1<br>1 3/8, 1 3/4, 2 1/8 | 1 3/64 | 7/16 | 1/4 | 1/8 | 27/64 | 50° | 3/4 | 1 | 0.7503<br>0.7506 | 1.0015<br>1.0018 | Same as the bushing |
| 0.4687 to 0.7812 | 0.5156 to 0.750 | +0.0005<br>+0.0001 | 1.0000<br>0.9998 | 3/4, 1, 1 3/8<br>1 3/4, 2 1/8, 2 1/2 | 1 27/64 | 7/16 | 3/8 | 3/16 | 19/32 | 35° | 59/64 | 2 | 1.0004<br>1.0007 | 1.3772<br>1.3768 | Same as the bushing |
| 0.7817 to 1.0312 | 0.7656 to 1.0000 | +0.0006<br>+0.0002 | 1.3750<br>1.3747 | 3/4, 1, 1 3/8<br>1 3/4, 2 1/8, 2 1/2 | 1 51/64 | 7/16 | 3/8 | 3/16 | 25/32 | 30° | 1 7/64 | 2 | 1.3756<br>1.3760 | 1.7523<br>1.7519 | Same as the bushing |
| 0.9687 to 1.4062 | 1.0156 to 1.3750 | +0.0006<br>+0.0002 | 1.7500<br>1.7497 | 1, 1 3/8, 1 3/4<br>2 1/8, 2 1/2, 3 | 2 19/64 | 5/8 | 3/8 | 3/16 | 1 | 30° | 1 25/64 | 3 | 1.7508<br>1.7512 | 2.2521<br>2.2525 | Same as the bushing |
| 1.3437 to 1.7812 | 1.3906 to 1.750 | Incl 1 1/2<br>+0.0006<br>+0.0002<br>Over 1 1/2<br>+0.0007<br>+0.0003 | 2.2500<br>2.2496 | 1, 1 3/8, 1 3/4<br>2 1/8, 2 1/2, 3 | 2 51/64 | 5/8 | 3/8 | 3/16 | 1 1/4 | 25° | 1 41/64 | 3 | 2.2510<br>2.2515 | 2.7526<br>2.7522 | Same as the bushing |

Dimensions in inches.
*ASA B5.6—1962.
Head design in accordance with manufacturer's practice; slip type usually knurled.

## Lock Screws

| Screw No. | *A* | *B* | *C* | *D* | *E* | *F* | ASA thd |
|---|---|---|---|---|---|---|---|
| 1 | 5/8 | 3/8 | 5/8 | 1/16 | 1/4 | .138<br>.132 | 5/16–18 |
| 2 | 7/8 | 3/8 | 5/8 | 3/32 | 3/8 | .200<br>.194 | 5/16–18 |
| 3 | 1 | 7/16 | 3/4 | 1/8 | 3/8 | .200<br>.194 | 3/8–16 |

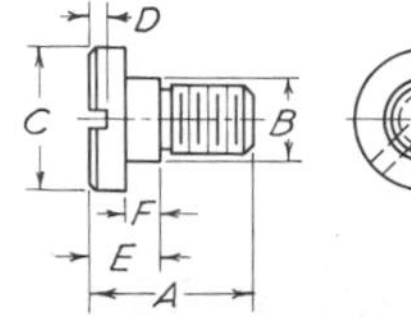

## Standard Jig Bushings (Cont.)

**Press-fit Headless and Press-fit Head Types**

| Hole size $A$ | Tolerance on hole | OD limits $B$ | Lengths available $C$ | Head type dimension $F$ | Head type dimension $G$ |
|---|---|---|---|---|---|
| 0.055 to 0.0995 | +0.0004 +0.0001 | 0.2046 / 0.2043 | 5⁄16, ½ | 19⁄64 | 3⁄32 |
| 0.1015 to 0.1360 | | 0.2516 / 0.2513 | 5⁄16, ½ | 23⁄64 | 3⁄32 |
| 0.1405 to 0.1875 | | 0.3141 / 0.3138 | 5⁄16, ½, ¾, 1 | 27⁄64 | ⅛ |
| 0.1890 to 0.2500 | | 0.4078 / 0.4075 | 5⁄16, ½, ¾, 1, 1⅜, 1¾ | ½ | 5⁄32 |
| 0.2570 to 0.3125 | +0.0005 +0.0001 | 0.5017 / 0.5014 | | 39⁄64 | 7⁄32 |
| 0.316 to 0.4219 | | 0.6267 / 0.6264 | ½, ¾, 1, 1⅜, 1¾, 2⅛ | 51⁄64 | 7⁄32 |
| 0.4375 to 0.500 | | 0.7518 / 0.7515 | | 59⁄64 | 7⁄32 |
| 0.5156 to 0.625 | | 0.8768 / 0.8765 | ¾, 1 | 1 7⁄64 | ¼ |
| 0.6406 to 0.7500 | | 1.0018 / 1.0015 | 1⅜, 1¾ | 1 15⁄64 | 5⁄16 |
| 0.7656 to 1.0000 | +0.0006 +0.0002 | 1.3772 / 1.3768 | 2⅛, 2½ | 1 39⁄64 | ⅜ |
| 1.0156 to 1.3750 | | 1.7523 / 1.7519 | 1, 1⅜, 1¾, 2⅛, 2½, 3 | 1 63⁄64 | ⅜ |
| 1.3906 to 1.7500 | Incl 1½ +0.0006 +0.0002 Over 1½ +0.0007 +0.0003 | 2.2525 / 2.2521 | | 2 31⁄64 | ⅜ |

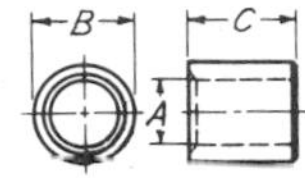


PRESS FIT HEADLESS TYPE

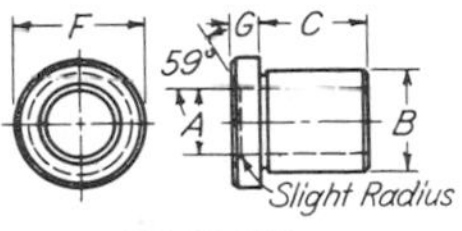


PRESS FIT HEAD TYPE

## American Standard Brass Solder-joint Fittings*

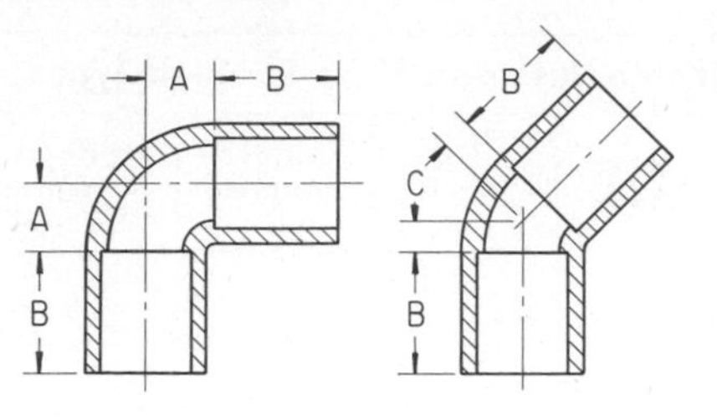

45° ELL

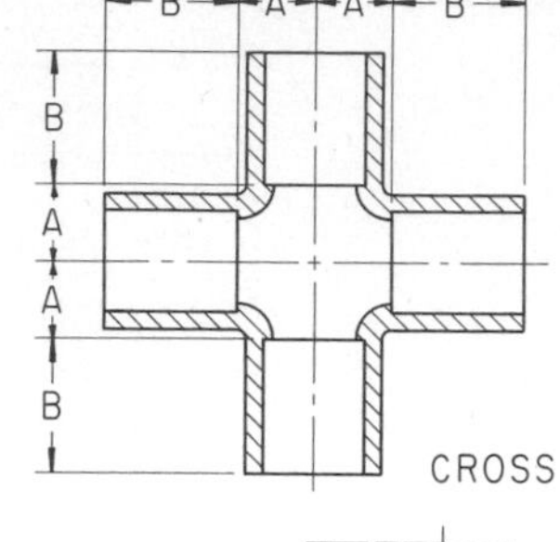


| Tubing size | A | B | C | D | Reducing size | F | E | G |
|---|---|---|---|---|---|---|---|---|
| 3/8 | 5/16 | 11/16 | 3/16 | 1/8 | 3/4 × 1/2 | 5/8 | 1 | 13/16 |
| 1/2 | 7/16 | 13/16 | 3/16 | 1/8 | 1 × 3/4 | 11/16 | 1 1/16 | 1 |
| 3/4 | 9/16 | 1 | 1/4 | 1/8 | 1 × 1/2 | 1/2 | 1 1/16 | 13/16 |
| 1 | 3/4 | 1 1/16 | 5/16 | 1/8 | 2 × 1 1/2 | 1 1/8 | 1 3/8 | 1 3/16 |
| 1 1/2 | 1 | 1 3/16 | 1/2 | 1/8 | 2 1/2 × 2 | 1 3/16 | 1 5/8 | 1 3/8 |
| 2 | 1 1/4 | 1 3/8 | 9/16 | 3/16 | 2 1/2 × 1 1/2 | 15/16 | 1 5/8 | 1 3/16 |
| 2 1/2 | 1 1/2 | 1 5/8 | 5/8 | 3/16 | 3 × 2 1/2 | 1 1/4 | 1 7/8 | 1 5/8 |
| 3 | 1 3/4 | 1 7/8 | 3/4 | 3/16 | 3 1/2 × 3 | 1 1/8 | 2 1/16 | 1 7/8 |
| 3 1/2 | 2 | 2 1/16 | 7/8 | 3/16 | 3 1/2 × 2 1/2 | 1 5/16 | 2 1/16 | 1 5/8 |
| 4 | 2 1/4 | 2 1/4 | 15/16 | 1/4 | 4 × 3 1/2 | 1 3/16 | 2 1/4 | 2 1/16 |

* Adapted from ASA B16.18—1963.

B C B — COUPLING

E F G — ECCENTRIC COUPLING

A B B A A B — TEE

SAE STANDARD COTTER PINS

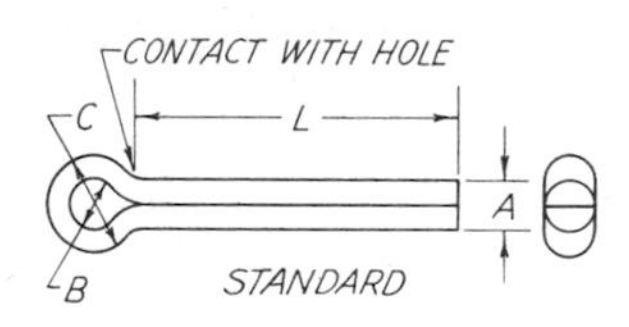


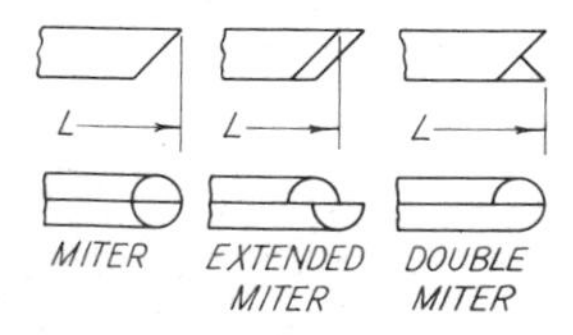


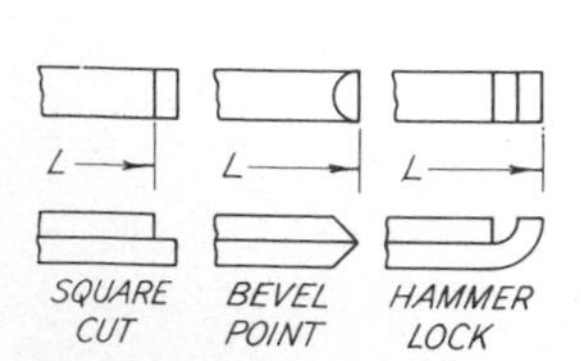


| Pin diameter A | | | Eye diameter, min. | | Recommended hole diam., drill size |
|---|---|---|---|---|---|
| Nominal | Max. | Min. | Inside B | Outside C | |
| 1/32 (0.031) | 0.032 | 0.028 | 1/32 | 1/16 | 3/64 (0.0469) |
| 3/64 (0.047) | 0.048 | 0.044 | 3/64 | 3/32 | 1/16 (0.0625) |
| 1/16 (0.062) | 0.060 | 0.056 | 1/16 | 1/8 | 5/64 (0.0781) |
| 5/64 (0.078) | 0.076 | 0.072 | 5/64 | 5/32 | 3/32 (0.0937) |
| 3/32 (0.094) | 0.090 | 0.086 | 3/32 | 3/16 | 7/64 (0.1094) |
| 1/8 (0.125) | 0.120 | 0.116 | 1/8 | 1/4 | 9/64 (0.1406) |
| 5/32 (0.156) | 0.150 | 0.146 | 5/32 | 5/16 | 11/64 (0.1719) |
| 3/16 (0.188) | 0.176 | 0.172 | 3/16 | 3/8 | 13/64 (0.2031) |
| 7/32 (0.219) | 0.207 | 0.202 | 7/32 | 7/16 | 15/64 (0.2344) |
| 1/4 (0.250) | 0.225 | 0.220 | 1/4 | 1/2 | 17/64 (0.2656) |
| 5/16 (0.312) | 0.280 | 0.275 | 5/16 | 5/8 | 5/16 (0.3125) |
| 3/8 (0.375) | 0.335 | 0.329 | 3/8 | 3/4 | 3/8 (0.3750) |
| 1/2 (0.500) | 0.473 | 0.467 | 1/2 | 1 | 1/2 (0.5000) |

# American Standard Graphic Symbols for Piping and Heating*

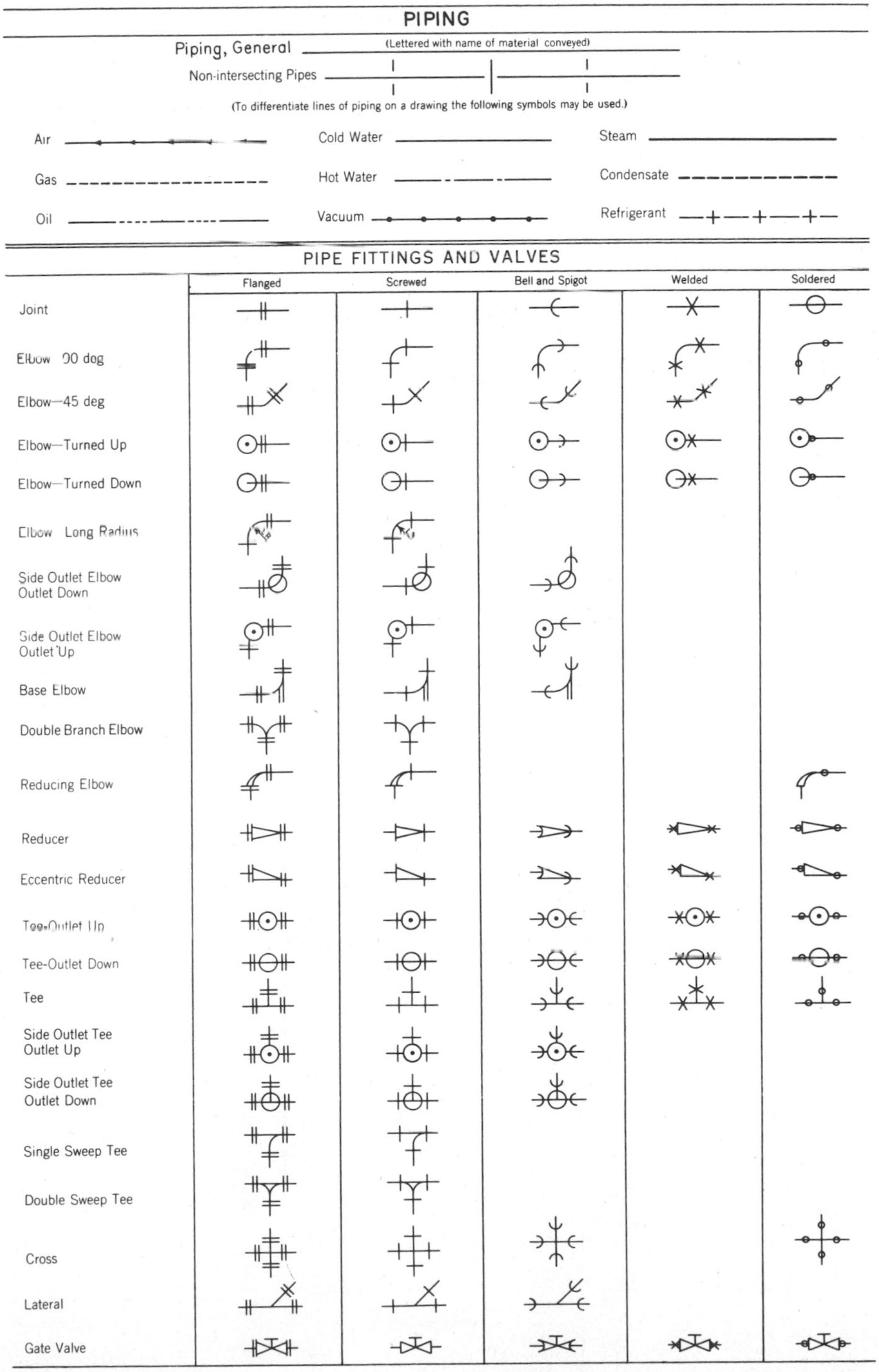

## PIPING

Piping, General (Lettered with name of material conveyed)

Non-intersecting Pipes

(To differentiate lines of piping on a drawing the following symbols may be used.)

| | | |
|---|---|---|
| Air | Cold Water | Steam |
| Gas | Hot Water | Condensate |
| Oil | Vacuum | Refrigerant |

## PIPE FITTINGS AND VALVES

| | Flanged | Screwed | Bell and Spigot | Welded | Soldered |
|---|---|---|---|---|---|
| Joint | | | | | |
| Elbow—90 deg | | | | | |
| Elbow—45 deg | | | | | |
| Elbow—Turned Up | | | | | |
| Elbow—Turned Down | | | | | |
| Elbow—Long Radius | | | | | |
| Side Outlet Elbow Outlet Down | | | | | |
| Side Outlet Elbow Outlet Up | | | | | |
| Base Elbow | | | | | |
| Double Branch Elbow | | | | | |
| Reducing Elbow | | | | | |
| Reducer | | | | | |
| Eccentric Reducer | | | | | |
| Tee-Outlet Up | | | | | |
| Tee-Outlet Down | | | | | |
| Tee | | | | | |
| Side Outlet Tee Outlet Up | | | | | |
| Side Outlet Tee Outlet Down | | | | | |
| Single Sweep Tee | | | | | |
| Double Sweep Tee | | | | | |
| Cross | | | | | |
| Lateral | | | | | |
| Gate Valve | | | | | |

* ASA Y32.2.3—1953.

## Symbols for Piping and Heating (Cont.)

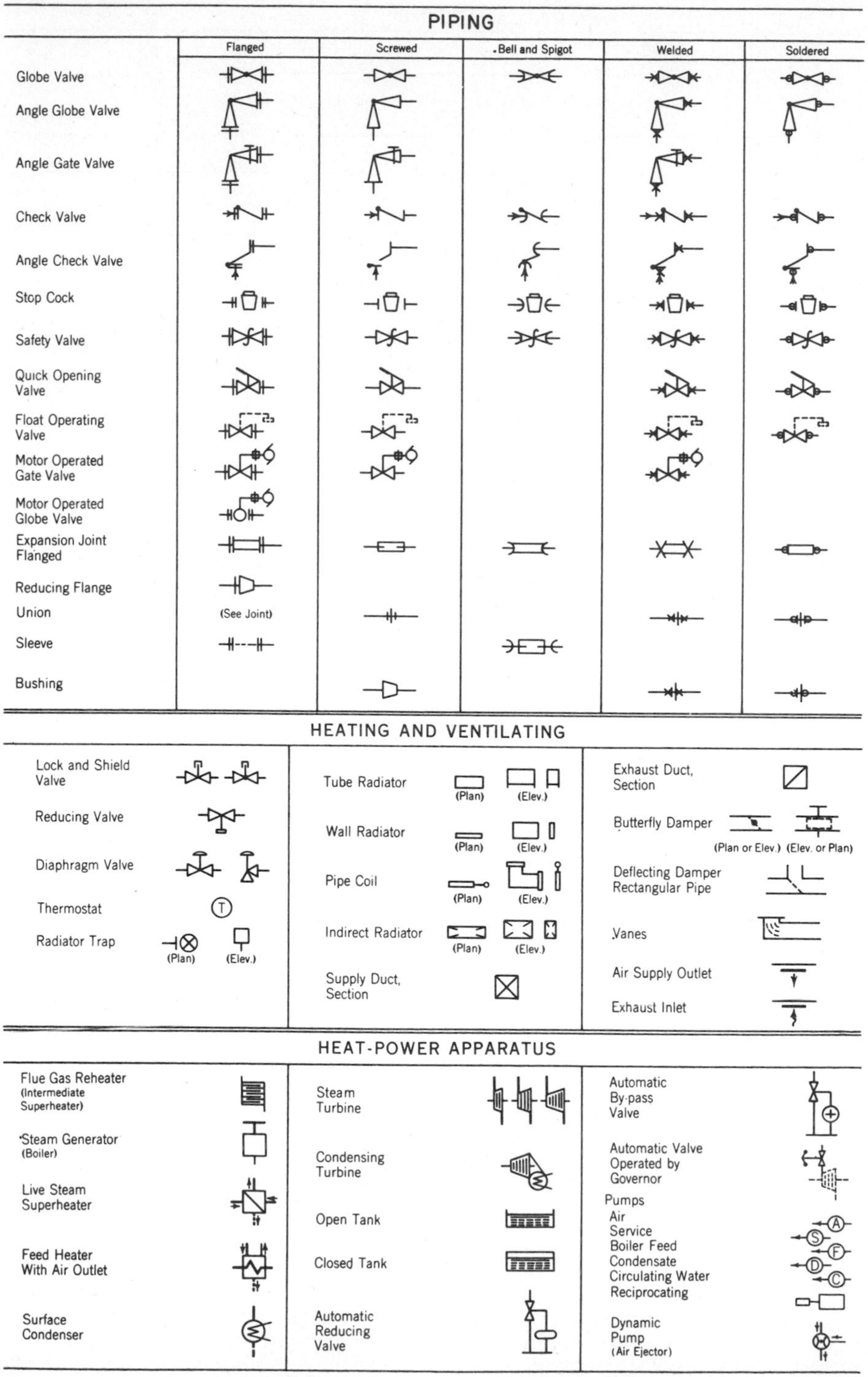

### PIPING

| | Flanged | Screwed | Bell and Spigot | Welded | Soldered |
|---|---|---|---|---|---|
| Globe Valve | | | | | |
| Angle Globe Valve | | | | | |
| Angle Gate Valve | | | | | |
| Check Valve | | | | | |
| Angle Check Valve | | | | | |
| Stop Cock | | | | | |
| Safety Valve | | | | | |
| Quick Opening Valve | | | | | |
| Float Operating Valve | | | | | |
| Motor Operated Gate Valve | | | | | |
| Motor Operated Globe Valve | | | | | |
| Expansion Joint Flanged | | | | | |
| Reducing Flange | | | | | |
| Union | (See Joint) | | | | |
| Sleeve | | | | | |
| Bushing | | | | | |

### HEATING AND VENTILATING

Lock and Shield Valve

Reducing Valve

Diaphragm Valve

Thermostat

Radiator Trap (Plan) (Elev.)

Tube Radiator (Plan) (Elev.)

Wall Radiator (Plan) (Elev.)

Pipe Coil (Plan) (Elev.)

Indirect Radiator (Plan) (Elev.)

Supply Duct, Section

Exhaust Duct, Section

Butterfly Damper (Plan or Elev.) (Elev. or Plan)

Deflecting Damper Rectangular Pipe

Vanes

Air Supply Outlet

Exhaust Inlet

### HEAT-POWER APPARATUS

Flue Gas Reheater (Intermediate Superheater)

Steam Generator (Boiler)

Live Steam Superheater

Feed Heater With Air Outlet

Surface Condenser

Steam Turbine

Condensing Turbine

Open Tank

Closed Tank

Automatic Reducing Valve

Automatic By-pass Valve

Automatic Valve Operated by Governor

Pumps
- Air
- Service
- Boiler Feed
- Condensate
- Circulating Water
- Reciprocating

Dynamic Pump (Air Ejector)

## American Standard Plumbing Symbols*

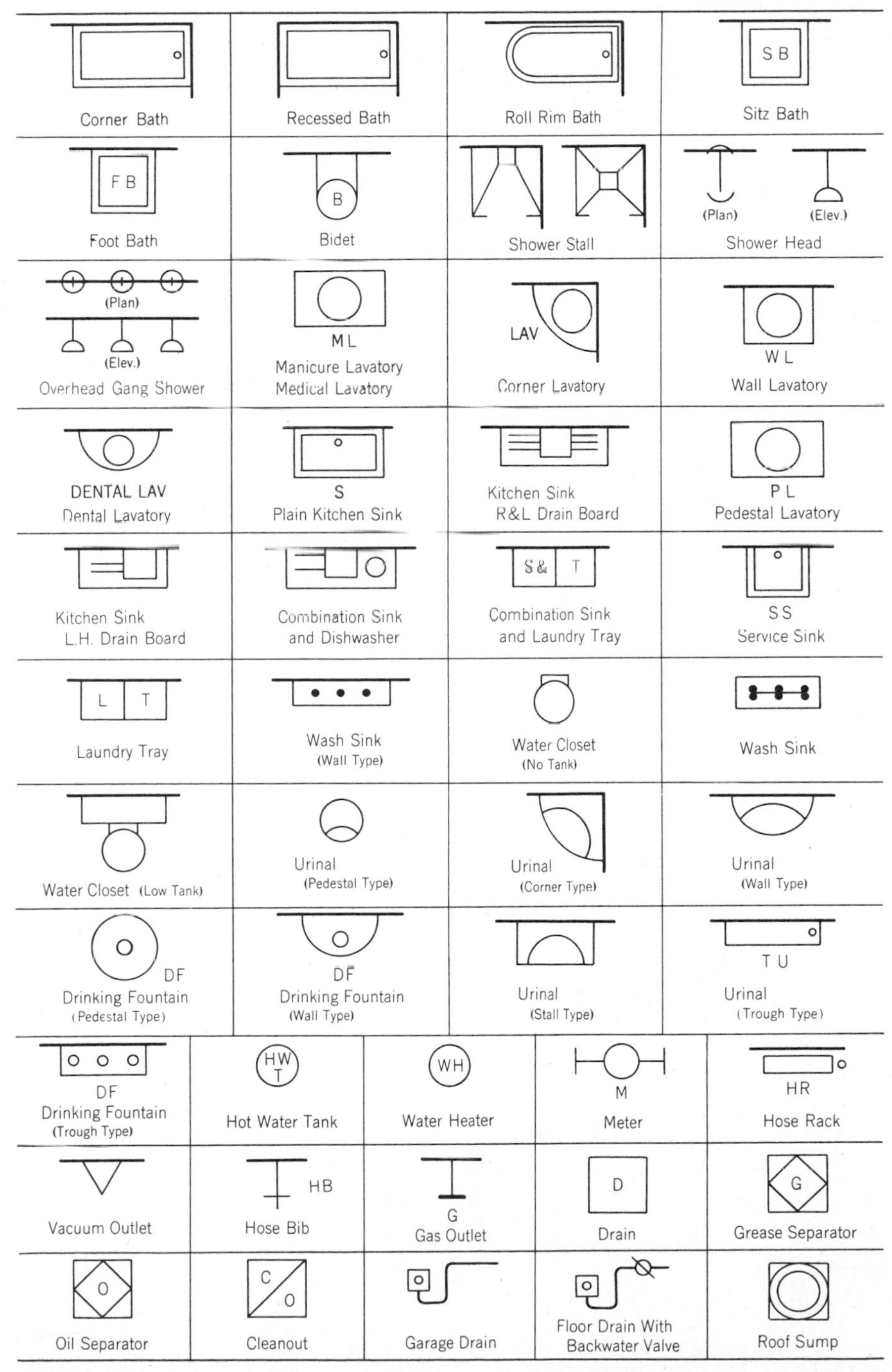
Corner Bath
Recessed Bath
Roll Rim Bath
S B
Sitz Bath
F B
Foot Bath
B
Bidet
Shower Stall
(Plan)
(Elev.)
Shower Head
(Plan)
(Elev.)
Overhead Gang Shower
M L
Manicure Lavatory
Medical Lavatory
LAV
Corner Lavatory
W L
Wall Lavatory
DENTAL LAV
Dental Lavatory
S
Plain Kitchen Sink
Kitchen Sink
R&L Drain Board
P L
Pedestal Lavatory
Kitchen Sink
L.H. Drain Board
Combination Sink
and Dishwasher
S&
T
Combination Sink
and Laundry Tray
S S
Service Sink
L
T
Laundry Tray
Wash Sink
(Wall Type)
Water Closet
(No Tank)
Wash Sink
Water Closet (Low Tank)
Urinal
(Pedestal Type)
Urinal
(Corner Type)
Urinal
(Wall Type)
DF
Drinking Fountain
(Pedestal Type)
DF
Drinking Fountain
(Wall Type)
Urinal
(Stall Type)
T U
Urinal
(Trough Type)
DF
Drinking Fountain
(Trough Type)
HW
T
Hot Water Tank
WH
Water Heater
M
Meter
HR
Hose Rack
Vacuum Outlet
HB
Hose Bib
G
Gas Outlet
D
Drain
G
Grease Separator
O
Oil Separator
C
O
Cleanout
Garage Drain
Floor Drain With
Backwater Valve
Roof Sump

* ASA Y32.4—1955.

## Wiring Symbols for Architecture

Ceiling Outlet
" " for Extensions
" Lamp Receptacle, Specifications to describe type, as Key, Keyless or Pull Chain
Ceiling Fan Outlet
Pull Switch — PS
Drop Cord — D
Wall Bracket
" Outlet for Extensions — E
" Lamp Receptacle, as specified — R
" Fan Outlet
Single Convenience Outlet
Double " " " — 2
Junction Box — J
Special Purpose Outlets Lighting, Heating and Power as described in specifications
Exit Light
Floor Outlet
Floor Elbow — E, Floor Tee — T
Local Switch, Single Pole — $S^1$
Double Pole — $S^2$, 3-Way — $S^3$, 4-Way — $S^4$
Automatic Door Switch — $S^D$
Key Push Button Switch — $S^K$
Electrolier Switch — $S^E$
Push Button Switch and Pilot — $S^P$
Remote Control Push Button Switch — $S^R$
Tank Switch — T.S.
Motor, Motor Controller — M.C.
Lighting Panel
Power Panel
Heating Panel
Pull Box
Cable Supporting Box
Meter
Transformer
Push Button
Pole Line
Buzzer, Bell
Annunciator

Branch Circuit, Run Exposed
Run Concealed Under Floor
" " " Floor Above
Feeder Run Exposed
Run Concealed Under Floor
" " " Floor Above
Telephone, Interior, Public
Clock, Secondary, Master
Time Stamp
Electric Door Opener
Local Fire Alarm Gong — F
City Fire Alarm Station
Local " " " — F
Fire Alarm Central Station — FA
Speaking Tube
Nurse's Signal Plug — N
Maid's Plug — M
Horn Outlet
District Messenger Call
Watchman Station — W
Watchman Central Station Detector — W
Public Telephone-P.B.X. Switchboard — P.B.X.
Interior Telephone Central Switchboard — IX
Interconnection Cabinet
Telephone Cabinet
Telegraph "
Special Outlet for Signal System as Specified
Battery
Signal Wires in Conduit Under Floor
" " " " " Floor Above

This Character Marked on Tap Circuits Indicates

| | | | | |
|---|---|---|---|---|
| 2 No. 14 Conductors in | $\frac{1}{2}$" Conduit | | | II |
| 3 " 14 " | " $\frac{1}{2}$" " | | | III |
| 4 " 14 " | " $\frac{3}{4}$" " | (Unless Marked $\frac{1}{2}$") | | IIII |
| 5 " 14 " | " $\frac{3}{4}$" " | | | IIIII |
| 6 " 14 " | " 1" " | (Unless Marked $\frac{3}{4}$") | | IIIIII |
| 7 " 14 " | " 1" " | | | IIIIIII |
| 8 " 14 " | " 1" " | | | IIIIIIII |

(Radio Outlet)
(Public Speaker Outlet)

# American Standard Graphic Symbols for Electrical Diagrams*

*Single-line symbols are shown at the left, complete symbols at the right, and symbols for both purposes are centered in each column. Listing is alphabetical.*

**ADJUSTABLE**
**CONTINUOUSLY ADJUSTABLE** (Variable)
*The shaft of the arrow is drawn at about 45 degrees across the body of the symbol.*

**AMPLIFIER**
See also MACHINE, ROTATING

General
The triangle is pointed in the direction of transmission.

Amplifier type may be indicated in the triangle by words, standard abbreviations, or a letter combination from the following list.

| | | | |
|---|---|---|---|
| BDG | Bridging | MON | Monitoring |
| BST | Booster | PGM | Program |
| CMP | Compression | PRE | Preliminary |
| DC | Direct Current | PWR | Power |
| EXP | Expansion | TRQ | Torque |
| LIM | Limiting | | |

Applications

Booster amplifier with two inputs

BST BST

Monitoring amplifier with two outputs

MON MON

Amplifier with associated power supply

PS PS

**ANTENNA**

General
Types or functions may be indicated by words or abbreviations adjacent to the symbol.

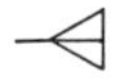

Dipole

**ARRESTER** (Electric Surge, Lightning, etc.)
**GAP**

General

Carbon block
*The sides of the rectangle are to be approximately in the ratio of 1 to 2 and the space between rectangles shall be approximately equal to the width of a rectangle.*

Electrolytic or aluminum cell
*This symbol is not composed of arrowheads.*

Protective gap
*These arrowheads shall not be filled.*

Sphere gap

Multigap, general

**ATTENUATOR**
See also PAD

General

Balanced, general

Unbalanced, general

**BATTERY**
The long line is always positive, but polarity may be indicated in addition.
Example:

Generalized direct-current source

**BREAKER, CIRCUIT**
If it is desired to show the condition causing the breaker to trip, the relay-protective-function symbols may be used alongside the breaker symbol.

General
Note 1—Use appropriate number of single-line diagram symbols.

SEE NOTE 1

Air or, if distinction is needed, for alternating-current circuit breaker rated at 1,500 volts or less and for direct-current circuit breaker.

SEE NOTE 1

**CAPACITOR**
See also TERMINATION

General
If it is necessary to identify the capacitor electrodes, the curved element shall represent the outside electrode in fixed paper-dielectric and ceramic-dielectric capacitors, the negative electrode in electrolytic capacitors, the moving element in adjustable and variable capacitors, and the low-potential element in feed-through capacitors.

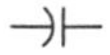

Application: shielded capacitor

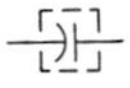

Application: adjustable or variable capacitor

If it is necessary to identify trimmer capacitors, the letter T should appear adjacent to the symbol.

Application: adjustable or variable capacitors with mechanical linkage of units

Shunt capacitor

Feed-through capacitor (with terminals shown on feed-through element)
Commonly used for bypassing high-frequency currents to chassis.

Application: feed-through capacitor between 2 inductors with third lead connected to chassis

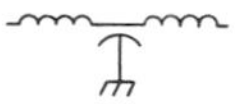

**CELL, PHOTOSENSITIVE** (Semiconductor)

λ indicates that the primary characteristic of the element within the circle is designed to vary under the influence of light.

Asymmetrical photoconductive transducer (resistive)
*This arrowhead shall be solid.*

Symmetrical photoconductive transducer; selenium cell

**CHASSIS**
**FRAME**
See also GROUND
The chassis or frame is not necessarily at ground potential.

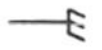

**COIL, BLOWOUT**

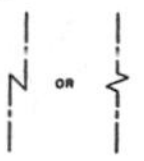

OR

**COIL, OPERATING**
See also INDUCTOR; WINDING

Note 3—The asterisk is not a part of the symbol. Always replace the asterisk by a device designation.

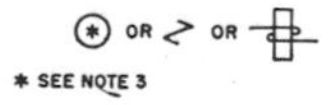

OR OR

* SEE NOTE 3

**CONNECTION, MECHANICAL — MECHANICAL INTERLOCK**
The preferred location of the mechanical connection is as shown in the various applications, but other locations may be equally acceptable.

Mechanical connection (*short dashes*)

* Adapted from ASA Y32.2—1962.

## Symbols for Electrical Diagrams (Cont.)

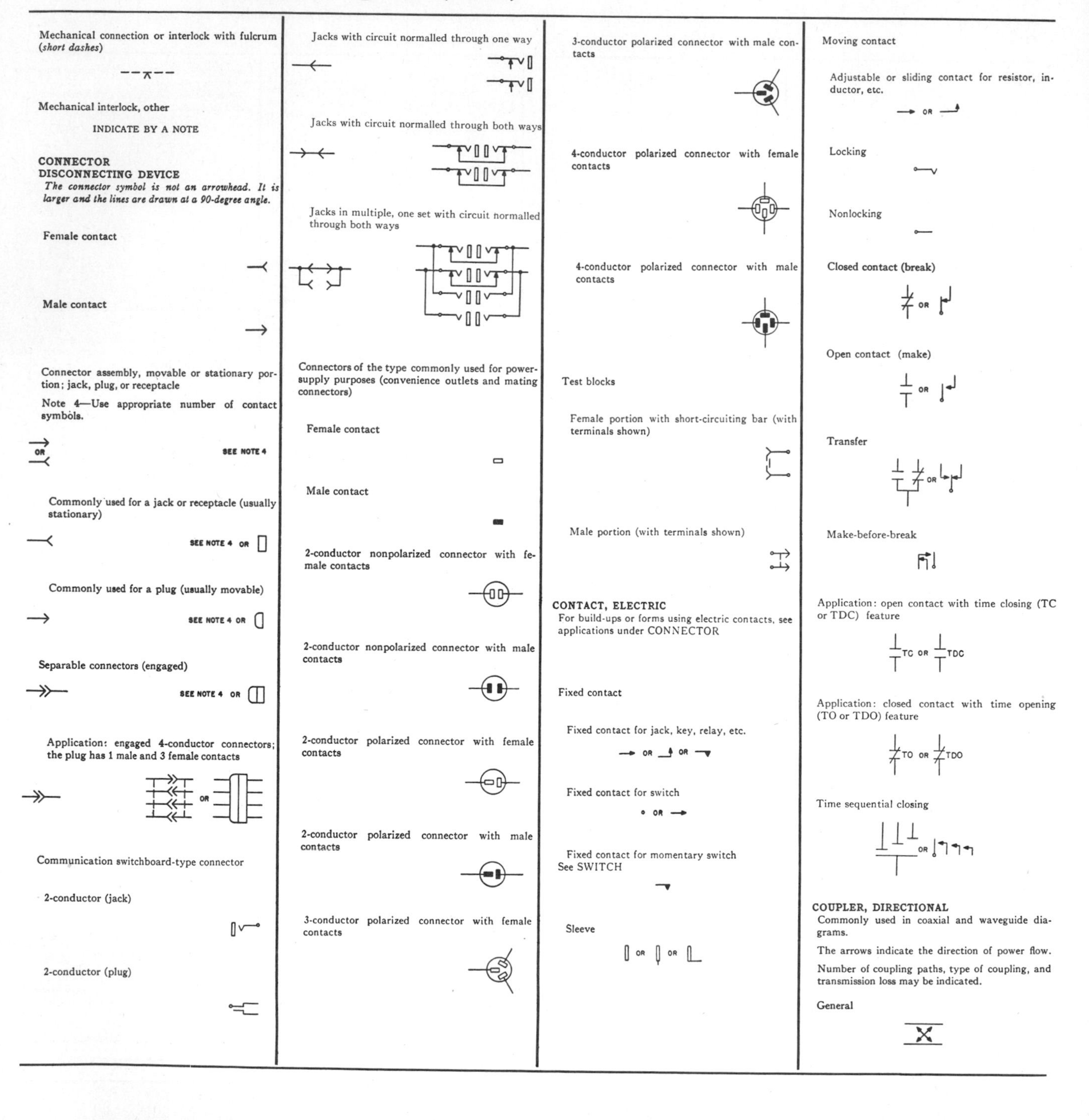

Mechanical connection or interlock with fulcrum (*short dashes*)

Mechanical interlock, other

INDICATE BY A NOTE

**CONNECTOR**
**DISCONNECTING DEVICE**
*The connector symbol is not an arrowhead. It is larger and the lines are drawn at a 90-degree angle.*

Female contact

Male contact

Connector assembly, movable or stationary portion; jack, plug, or receptacle

Note 4—Use appropriate number of contact symbols.

OR SEE NOTE 4

Commonly used for a jack or receptacle (usually stationary)

SEE NOTE 4 OR

Commonly used for a plug (usually movable)

SEE NOTE 4 OR

Separable connectors (engaged)

SEE NOTE 4 OR

Application: engaged 4-conductor connectors; the plug has 1 male and 3 female contacts

OR

Communication switchboard-type connector

2-conductor (jack)

2-conductor (plug)

Jacks with circuit normalled through one way

Jacks with circuit normalled through both ways

Jacks in multiple, one set with circuit normalled through both ways

Connectors of the type commonly used for power-supply purposes (convenience outlets and mating connectors)

Female contact

Male contact

2-conductor nonpolarized connector with female contacts

2-conductor nonpolarized connector with male contacts

2-conductor polarized connector with female contacts

2-conductor polarized connector with male contacts

3-conductor polarized connector with female contacts

3-conductor polarized connector with male contacts

4-conductor polarized connector with female contacts

4-conductor polarized connector with male contacts

Test blocks

Female portion with short-circuiting bar (with terminals shown)

Male portion (with terminals shown)

**CONTACT, ELECTRIC**
For build-ups or forms using electric contacts, see applications under CONNECTOR

Fixed contact

Fixed contact for jack, key, relay, etc.

OR OR

Fixed contact for switch

OR

Fixed contact for momentary switch
See SWITCH

Sleeve

OR OR

Moving contact

Adjustable or sliding contact for resistor, inductor, etc.

OR

Locking

Nonlocking

Closed contact (break)

OR

Open contact (make)

OR

Transfer

OR

Make-before-break

Application: open contact with time closing (TC or TDC) feature

TC OR TDC

Application: closed contact with time opening (TO or TDO) feature

TO OR TDO

Time sequential closing

OR

**COUPLER, DIRECTIONAL**
Commonly used in coaxial and waveguide diagrams.

The arrows indicate the direction of power flow.

Number of coupling paths, type of coupling, and transmission loss may be indicated.

General

## Symbols for Electrical Diagrams (Cont.)

Applications

*E*-plane aperture coupling, 30-decibel transmission loss

30DB

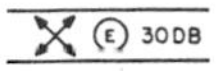

Loop coupling, 30-decibel transmission loss

30DB

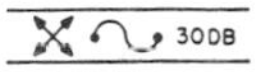

Probe coupling, 30-decibel transmission loss

30DB

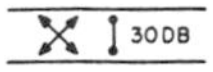

Resistance coupling, 30-decibel transmission loss

30DB

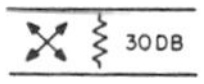

**DIRECTION OF FLOW OF POWER, SIGNAL, OR INFORMATION**

One-way

Note 10—The lower symbol is used if it is necessary to conserve space. *The arrowhead in the lower symbol shall be filled.*

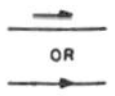

OR

SEE NOTE 10

Both ways

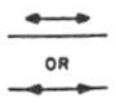

OR

SEE NOTE 10

**DISCONTINUITY**

A component that exhibits throughout the frequency range of interest the properties of the type of circuit element indicated by the symbol within the triangle.

Commonly used for coaxial and waveguide transmission.

Equivalent series element, general

Z

Capacitive reactance

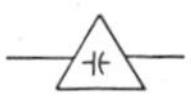

**ELEMENT, CIRCUIT** (General)

Note 12—The asterisk is not a part of the symbol. Always indicate the type of apparatus by appropriate words or letters in the rectangle.

*

* SEE NOTE 12

Accepted abbreviations in the latest edition of American Standard Z32.13 may be used in the rectangle.

The following letter combinations may be used in the rectangle.

| | | | |
|---|---|---|---|
| CB | Circuit breaker | NET | Network |
| DIAL | Telephone dial | PS | Power supply |
| EQ | Equalizer | RU | Reproducing unit |
| FAX | Facsimile set | | |
| FL | Filter | RG | Recording unit |
| FL-BE | Filter, band elimination | TEL | Telephone station |
| FL-BP | Filter, band pass | TPR | Teleprinter |
| FL-HP | Filter, high pass | TTY | Teletypewriter |
| FL-LP | Filter, low pass | | |

**ELEMENT, THERMAL**

Thermomechanical transducer

Note 13—Use appropriate number of single-line diagram symbols.

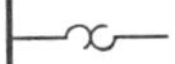

SEE NOTE 13

**FUSE**

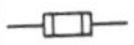

SEE NOTE 14

Note 14—Use appropriate number of single-line diagram symbols.

Fusible element

SEE NOTE 14

**GROUND**

See also CHASSIS; FRAME

**INDUCTOR**

**WINDING**

General

Either symbol may be used in the following subparagraphs.

OR

If it is desired especially to distinguish magnetic-core inductors

**LAMP**

Ballast lamp; ballast tube

The primary characteristic of the element within the circle is designed to vary nonlinearly with the temperature of the element.

B

Fluorescent lamp

2-terminal

4-terminal

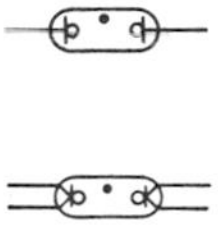

Glow lamp; cold-cathode lamp; neon lamp

Alternating-current type

Direct-current type

See also TUBE, ELECTRON

Incandescent-filament illuminating lamp

**MACHINE, ROTATING**

Basic

Generator, general

GEN

Motor, general

MOT

Motor, multispeed

USE BASIC MOTOR SYMBOL AND NOTE SPEEDS

Rotating armature with commutator and brushes

Wound rotor

Field, generator or motor

Compensating or commutating

Series

Shunt, or separately excited

Permanent magnet

PM

Winding symbols

Motor and generator winding symbols may be shown in the basic circle using the following representations.

1-phase

2-phase

3-phase wye (ungrounded)

3-phase wye (grounded)

3-phase delta

6-phase diametrical

6-phase double-delta

**MOTION, MECHANICAL**

Translation, one direction

Translation, both directions

Rotation, one direction

Rotation, both directions

## Symbols for Electrical Diagrams (Cont.)

**NETWORK**

General

NET

**OSCILLATOR GENERALIZED ALTERNATING-CURRENT SOURCE**

**PAD**

See also ATTENUATOR

General

Balanced, general

Unbalanced, general

**PATH, TRANSMISSION**

Air or space path

Dielectric path other than air

Commonly used for coaxial and waveguide transmission.

DIEL

Crossing of paths or conductors not connected
*The crossing is not necessarily at a 90-degree angle.*

Junction of paths or conductors

Junction (if desired)

Application: junction of different-size cables

Junction of connected paths, conductors, or wires

OR

OR ONLY IF REQUIRED BY SPACE LIMITATION

**RELAY**

See also CONTACT

Fundamental symbols for contacts, mechanical connections, coils, etc., are the basis of relay symbols and should be used to represent relays on complete diagrams.

Basic

Relay coil

Note 21—The asterisk is not a part of the symbol. Always replace the asterisk by a device designation.

* SEE NOTE 21

Application: 2-pole double-make

**RESISTOR**

Note 22—The asterisk is not a part of the symbol. Always add identification within or adjacent to the rectangle.

General

* SEE NOTE 22

Tapped resistor

* SEE NOTE 22

Application: with adjustable contact

* SEE NOTE 22

Application: adjustable or continuously adjustable (variable) resistor

* SEE NOTE 22

Heating resistor

* SEE NOTE 22

**SWITCH**

Fundamental symbols for contacts, mechanical connections, etc., may be used for switch symbols.

The standard method of showing switches is in a position with no operating force applied. For switches that may be in any one of two or more positions with no operating force applied and for switches actuated by some mechanical device (as in air-pressure, liquid-level, rate-of-flow, etc., switches), a clarifying note may be necessary to explain the point at which the switch functions.

Single throw, general

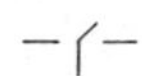

Double throw, general

Application: 2-pole double-throw switch with terminals shown

Knife switch, general

Switch, nonlocking; momentary or spring return

The symbols to the left are commonly used for spring buildups in key switches, relays, and jacks.

The symbols to the right are commonly used for toggle switches.

Circuit closing (make)

Circuit opening (break)

Switch, locking

The symbols to the left are commonly used for spring buildups in key switches, relays, and jacks.

The symbols to the right are commonly used for toggle switches.

Circuit closing (make)

Circuit opening (break)

Transfer, 2-position

**TERMINAL, CIRCUIT**

Terminal board or terminal strip with 4 terminals shown; group of 4 terminals

Number and arrangement as convenient.

**TRANSFORMER**

General

Either winding symbol may be used.

Additional windings may be shown or indicated by a note.

For power transformers, use polarity marking $H_1$-$X_1$, etc., from American Standard C6.1.

In coaxial and waveguide circuits, this symbol will represent a taper or step transformer without mode change.

If it is desired especially to distinguish a magnetic-core transformer

**TUBE, ELECTRON**

Tube-component symbols are shown first. These are followed by typical applications showing the use of these specific symbols in the various classes of devices such as thermionic, cold-cathode, and photoemissive tubes of varying structures and combinations of elements (triodes, pentodes, cathode-ray tubes, magnetrons, etc.).

Lines outside of the envelope are not part of the symbol but are electrical connections thereto.

***Connections between the external circuit and electron tube symbols within the envelope may be located as required to simplify the diagram.***

Emitting electrode

Directly heated (filamentary) cathode

Note—Leads may be connected in any convenient manner to ends of the ∧ provided the identity of the ∧ is retained.

Indirectly heated cathode

Lead may be connected to either extreme end of the ⌐¬ or, if required, to both ends, in any convenient manner.

Cold cathode (including ionically heated cathode)

Photocathode

Pool cathode

Ionically heated cathode with provision for supplementary heating

## Symbols for Electrical Diagrams (Cont.)

Controlling electrode

Grid (including beam-confining or beam-forming electrodes)

Deflecting electrodes (used in pairs); reflecting or repelling electrode (used in velocity-modulated tube)

Ignitor (in pool tubes) (should extend into pool)
Starter (in gas tubes)

Excitor (contactor type)

Collecting electrode

Anode or plate (including collecting electrode and fluorescent target)

Target or X-ray anode
*Drawn at about a 45-degree angle.*

Collecting and emitting electrode

Dynode

Alternately collecting and emitting

Composite anode-photocathode

Composite anode-cold cathode

Composite anode-ionically heated cathode with provision for supplementary heating

Heater

Envelope (shell)

The general envelope symbol identifies the envelope or enclosure regardless of evacuation or pressure. When used with electron-tube component symbols, the general envelope symbol indicates a vacuum enclosure unless otherwise specified. A gas-filled electron device may be indicated by a dot within the envelope symbol.

General

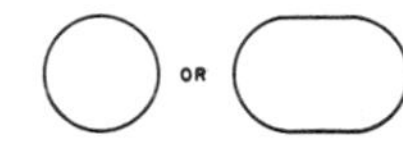


Split envelope
If necessary, envelope may be split.

Gas-filled
*The dot may be located as convenient.*

Shield

This is understood to shield against electric fields unless otherwise noted.

Any shield against electric fields that is within the envelope and that is connected to an independent terminal

Outside envelope of X-ray tube

Coupling by loop (electromagnetic type)
Coupling loop may be shown inside or outside envelope as desired, but if inside it should be shown grounded.

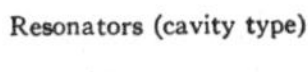

Resonators (cavity type)

Single-cavity envelope and grid-type associated electrodes

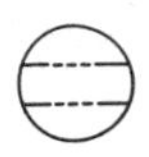

Double-cavity envelope and grid-type associated electrodes

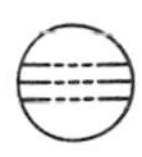

Multicavity magnetron anode and envelope

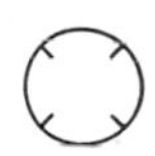

Associated parts of a circuit, such as focusing coils, deflecting coils, field coils, etc., are not a part of the tube symbol but may be added to the circuit in the form of standard symbols. For example, resonant-type magnetron with permanent magnet may be shown:

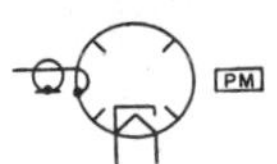


External and internal shields, whether integral parts of tubes or not, shall be omitted from the circuit diagram unless the circuit diagram requires their inclusion.

In line with standard drafting practice, straight-line crossovers are recommended.

Typical applications

Triode with directly heated filamentary cathode and envelope connection to base terminal

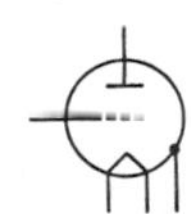

Equipotential-cathode pentode showing use of elongated envelope

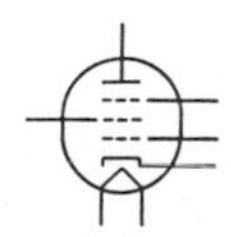

Typical wiring figure
This figure illustrates how tube symbols may be placed in any convenient position in a circuit.

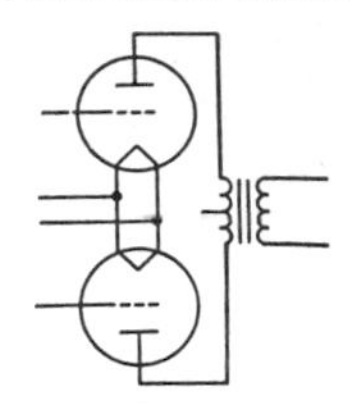

For tubes with keyed bases
Explanatory word and arrow are not a part of the symbol shown.

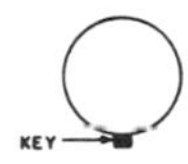


For tubes with bayonets, bosses, and other reference points

Base terminals
Explanatory words and arrows are not a part of the symbol.

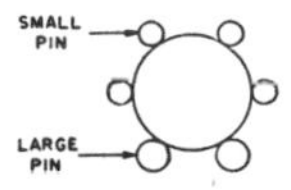


Envelope terminals
Explanatory words and arrows are not a part of the symbol.

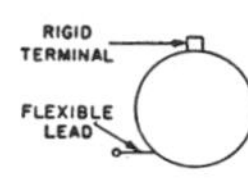


Applications

Triode with indirectly heated cathode and envelope connected to base terminal

Triode-heptode with rigid envelope connection

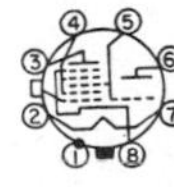


## Transistor Symbols (MIL-STD-15-1A, 22 May 1963)

PNP transistor

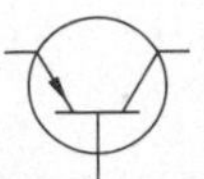

NPN transistor

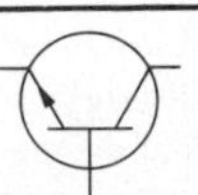

## Symbols for Materials (Exterior)

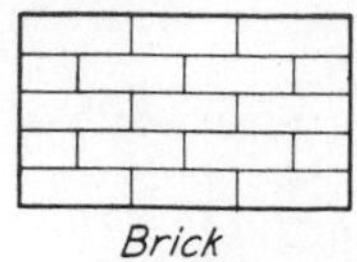
Brick

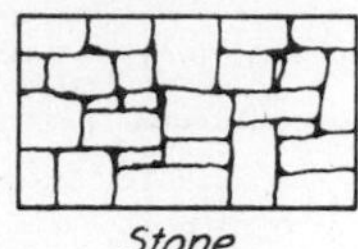
Stone

Transparent Material Glass. Celluloid. Etc.

Wood

## Symbols for Materials (Section)

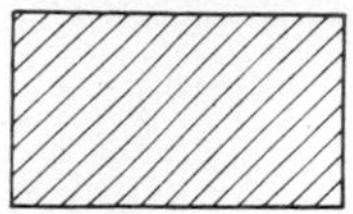
Cast Iron

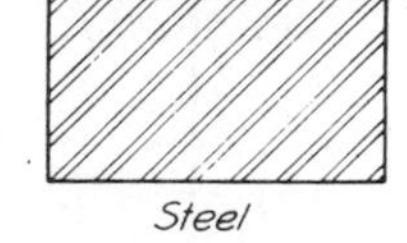
Steel

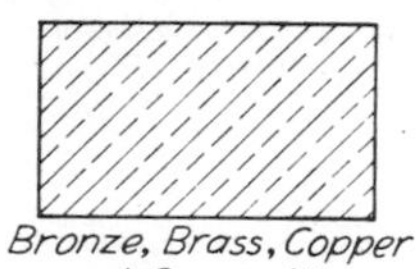
Bronze, Brass, Copper and Composition

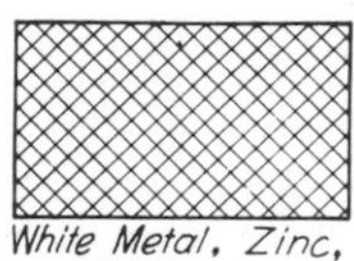
White Metal. Zinc, Lead, Babbitt & Alloys

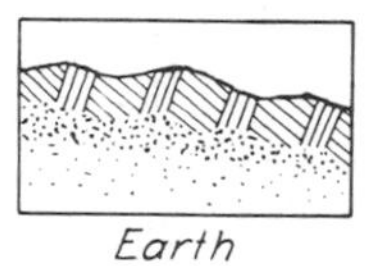
Earth

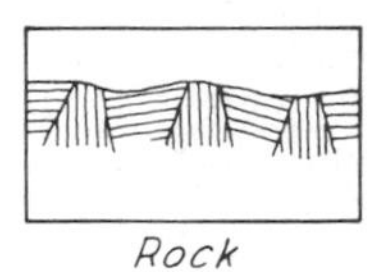
Rock

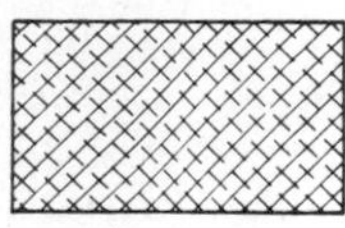
Aluminum

(Show solid for narrow sections)

Electric Insulation, Mica, Fibre, Vulcanite, Bakelite, Etc.

Sound or Heat Insulation Cork, Asbestos, Packing, Etc.

Flexible Material Fabric, Rubber, Etc.

Sand

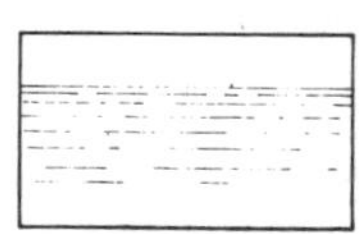
Water & Other Liquids

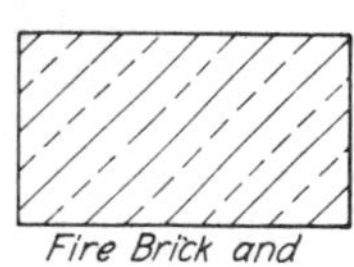
Fire Brick and Refractory Material

Concrete

Brick or Stone Masonry

Marble, Slate, Glass, Porcelain, Etc.

## Weights of Materials

| Metals | lb/cu in. |
|---|---|
| Aluminum alloy, cast | 0.099 |
| Aluminum, cast | 0.094 |
| Aluminum, wrought | 0.097 |
| Babbitt metal | 0.267 |
| Brass, cast or rolled | 0.303–0.313 |
| Brass, drawn | 0.323 |
| Bronze, aluminum cast | 0.277 |
| Bronze, phosphor | 0.315–0.321 |
| Chromium | 0.256 |
| Copper, cast | 0.311 |
| Copper, rolled, drawn or wire | 0.322 |
| Dowmetal *A* | 0.065 |
| Duralumin | 0.101 |
| Gold | 0.697 |
| Iron, cast | 0.260 |
| Iron, wrought | 0.283 |
| Lead | 0.411 |
| Magnesium | 0.063 |
| Mercury | 0.491 |
| Monel metal | 0.323 |
| Silver | 0.379 |
| Steel, cast or rolled | 0.274–0.281 |
| Steel, tool | 0.272 |
| Tin | 0.263 |
| Zinc | 0.258 |

| Wood | lb/cu in. |
|---|---|
| Ash | 0.024 |
| Balsa | 0.0058 |
| Cedar | 0.017 |
| Cork | 0.009 |
| Hickory | 0.0295 |
| Maple | 0.025 |
| Oak (white) | 0.028 |
| Pine (white) | 0.015 |
| Pine (yellow) | 0.025 |
| Poplar | 0.018 |
| Walnut (black) | 0.023 |

| Miscellaneous materials | lb/ cu ft |
|---|---|
| Asbestos | 175 |
| Bakelite | 79.5 |
| Brick, common | 112 |
| Brick, fire | 144 |
| Celluloid | 86.4 |
| Earth, packed | 100 |
| Fiber | 89.9 |
| Glass | 163 |
| Gravel | 109 |
| Limestone | 163 |
| Plexiglass | 74.3 |
| Sandstone | 144 |
| Water | 62.4 |

## Abbreviations for Use on Drawings

Abbreviations are shortened forms of words or expressions and their use on drawings is entirely for the purpose of conserving space and drafting time. Since they must be interpreted by shopmen, assemblers, construction men, and others, abbreviations should not be used where the meaning may not be clear. In case of doubt, spell out. Only words that are abbreviated on drawings are included in this list, which is selected from the more complete list given in the SAE Aerospace-Automotive Drawing Standards (Section Z.1)

The same abbreviation is used for all forms of a given word. Periods are used only to avoid misinterpretation of an abbreviation. Spaces between letters are for clarity only; the abbreviation of word combinations or phrases is sometimes spaced. The use of hyphens and slant bars has been avoided where practicable. Upper-case letters are used for abbreviations. Subscripts are not used in abbreviations. Whenever a special abbreviation is used (one not appearing in the SAE Aerospace-Automotive standard), it should be explained in a table on the drawing. For metals, chemicals, formulas, and equations, the abbreviations, symbols, and rules in general use and established by long acceptance are employed.

Abbreviations for colors and a partial list of chemical symbols are given after the general abbreviations.

| Word | Abbreviation | Word | Abbreviation |
|---|---|---|---|
| Abbreviate | ABBR | American Wire Gage | AWG |
| Absolute | ABS | Ammeter | AM |
| Accelerate | ACCEL | Amount | AMT |
| Acceleration due to gravity | G | Ampere | AMP |
| Access panel | AP | Ampere hour | AMP HR |
| Accessory | ACCESS. | Amplifier | AMPL |
| Actual | ACT. | Anneal | ANL |
| Adapter | ADPT | Antenna | ANT. |
| Addendum | ADD. | Apparatus | APP |
| Adjust | ADJ | Approved | APPD |
| Advance | ADV | Approximate | APPROX |
| After | AFT. | Arc weld | ARC/W |
| Aggregate | AGGR | Area | A |
| Aileron | AIL | Armature | ARM. |
| Air-break switch | ABS | Arrange | ARR. |
| Air-circuit breaker | ACB | Arrester | ARR. |
| Aircraft | ACFT | Asbestos | ASB |
| Airplane | APL | Assemble | ASSEM |
| Airtight | AT | Assembly | ASSY |
| Alarm | ALM | Atomic | AT |
| Allowance | ALLOW | Attach | ATT |
| Alloy | ALY | Audio-frequency | AF |
| Alteration | ALT | Automatic | AUTO |
| Alternate | ALT | Auto-transformer | AUTO TR |
| Alternating current | AC | Auxiliary | AUX |
| Alternator | ALT | Average | AVG |
| Altitude | ALT | | |
| Aluminum | AL | Babbitt | BAB |
| American Standard | AMER STD | Back to back | B to B |

| Word | Abbreviation | Word | Abbreviation |
|---|---|---|---|
| Baffle | BAF | Cast (used with other materials) | C |
| Balance | BAL | Cast iron | CI |
| Ball bearing | BB | Cast-iron pipe | CIP |
| Base line | BL | Cast steel | CS |
| Base plate | BP | Casting | CSTG |
| Battery | BAT. | Castle nut | CAS NUT |
| Bearing | BRG | Cement | CEM |
| Bent | BT | Center | CTR |
| Between | BET. | Center line | CL |
| Between centers | BC | Center to center | C to C |
| Between perpendiculars | BP | Centering | CTR |
| Bevel | BEV | Centigrade | C |
| Bill of material | B/M | Centigram | CG |
| Birmingham Wire Gage | BWG | Centiliter | CL |
| Blank | BLK | Centimeter | CM |
| Block | BLK | Centrifugal | CENT. |
| Blueprint | BP | Centrifugal force | CF |
| Bolt circle | BC | Ceramic | CER |
| Bottom | BOT | Chain | CH |
| Bottom chord | BC | Chamfer | CHAM |
| Brake | BK | Change | CHG |
| Brass | BRS | Change notice | CN |
| Brazing | BRZG | Change order | CO |
| Break | BRK | Channel | CHAN |
| Breaker | BKR | Check | CHK |
| Brinnell hardness | BH | Check valve | CV |
| British Standard | BR STD | Chemical | CHEM |
| British thermal units | BTU | Chord | CHD |
| Broach | BRO | Chrome molybdenum | CR MOLY |
| Bronze | BRZ | Chromium plate | CR PL |
| Brown & Sharp | B&S | Chrome vanadium | CR VAN |
| Brush | BR | Circle | CIR |
| Burnish | BNH | Circuit | CKT |
| Bushing | BUSH. | Circular | CIR |
| Bypass | BYP | Circular pitch | CP |
| | | Circulate | CIRC |
| Cadmium plate | CD PL | Circumference | CIRC |
| Calculate | CALC | Clamp | CLP |
| Calibrate | CAL | Class | CL |
| Calking | CLKG | Clear | CLR |
| Capacitor | CAP | Clearance | CL |
| Capacity | CAP | Clockwise | CW |
| Cap screw | CAP. SCR | Closing | CL |
| Carburize | CARB | Clutch | CL |
| Caseharden | CH | Coated | CTD |
| Casing | CSG | Coaxial | COAX |

| Word | Abbreviation | Word | Abbreviation |
|---|---|---|---|
| Coefficient | COEF | Counterbore | CBORE |
| Cold drawn | CD | Counterdrill | CDRILL |
| Cold-drawn steel | CDS | Counterpunch | CPUNCH |
| Cold rolled | CR | Countersink | CSK |
| Cold-rolled steel | CRS | Countersink other side | CSK-O |
| Column | COL | Coupling | CPLG |
| Combination | COMB. | Cover | COV |
| Combustion | COMB | Crank | CRK |
| Communication | COMM | Cross connection | XCONN |
| Commutator | COMM | Cross section | XSECT |
| Complete | COMPL | Cubic | CU |
| Composite | CX | Current | CUR |
| Composition | COMP | Cyanide | CYN |
| Compressor | COMPR | Cycle | CY |
| Concentric | CONC | Cycles per minute | CPM |
| Concrete | CONC | Cycles per second | CPS |
| Condition | COND | Cylinder | CYL |
| Conduct | COND | | |
| Conductor | COND | Decibel | DB |
| Conduit | CND | Decimal | DEC |
| Connect | CONN | Dedendum | DED |
| Constant | CONST | Deep drawn | DD |
| Contact | CONT | Deflect | DEFL |
| Container | CNTR | Degree | (°) DEG |
| Continue | CONT | Density | D |
| Continuous wave | CW | Describe | DESCR |
| Contract | CONT | Design | DSGN |
| Contractor | CONTR | Designation | DESIG |
| Control | CONT | Detail | DET |
| Control relay | CR | Detector | DET |
| Control switch | CS | Detonator | DET |
| Controller | CONT | Develop | DEV |
| Convert | CONV | Diagonal | DIAG |
| Conveyor | CNVR | Diagram | DIAG |
| Cooled | CLD | Diameter | DIA |
| Copper oxide | CUO | Diametral pitch | DP |
| Copper plate | COP. PL | Diaphragm | DIAPH |
| Cord | CD | Differential | DIFF |
| Correct | CORR | Dimension | DIM. |
| Corrosion resistant | CRE | Diode | DIO |
| Corrosion-resistant steel | CRES | Direct current | DC |
| Corrugate | CORR | Directional | DIR |
| Cotter | COT | Discharge | DISCH |
| Counter | CTR | Disconnect | DISC. |
| Counterclockwise | CCW | Distance | DIST |
| Counterbalance | CBAL | Distribute | DISTR |

| Word | Abbreviation | Word | Abbreviation |
|---|---|---|---|
| Ditto | DO. | Fahrenheit | F |
| Double | DBL | Fairing | FAIR. |
| Dovetail | DVTL | Farad | F |
| Dowel | DWL | Far side | FS |
| Down | DN | Feed | FD |
| Drafting | DFTG | Feeder | FDR |
| Draftsman | DFTSMN | Feet | (′) FT |
| Drain | DR | Feet per minute | FPM |
| Drawing | DWG | Feet per second | FPS |
| Drawing list | DL | Female | FEM |
| Drill | DR | Fiber | FBR |
| Drill rod | DR | Field | FLD |
| Drive | DR | Figure | FIG. |
| Drive fit | DF | Filament | FIL |
| Drop | D | Fillet | FIL |
| Drop forge | DF | Filling | FILL. |
| Duplex | DX | Fillister | FIL |
| Duplicate | DUP | Filter | FLT |
| Dynamic | DYN | Finish | FIN. |
| Dynamo | DYN | Finish all over | FAO |
| | | Fireproof | FPRF |
| Each | EA | Fitting | FTG |
| Eccentric | ECC | Fixture | FIX. |
| Effective | EFF | Flange | FLG |
| Electric | ELEC | Flashing | FL |
| Elevation | EL | Flat | F |
| Enclose | ENCL | Flat head | FH |
| End to end | E to E | Flexible | FLEX. |
| Envelope | ENV | Float | FLT |
| Equal | EQ | Floor | FL |
| Equation | EQ | Fluid | FL |
| Equipment | EQUIP. | Fluorescent | FLUOR |
| Equivalent | EQUIV | Flush | FL |
| Estimate | EST | Focus | FOC |
| Evaporate | EVAP | Foot | (′) FT |
| Excavate | EXC | Force | F |
| Exhaust | EXH | Forging | FORG |
| Expand | EXP | Forward | FWD |
| Exterior | EXT | Foundation | FDN |
| External | EXT | Foundry | FDRY |
| Extra heavy | X HVY | Fractional | FRAC |
| Extra strong | X STR | Frame | FR |
| Extrude | EXTR | Freezing point | FP |
| | | Frequency | FREQ |
| Fabricate | FAB | Frequency, high | HF |
| Face to face | F to F | Frequency, low | LF |

| Word | Abbreviation |
|---|---|
| Frequency, medium | MF |
| Frequency modulation | FM |
| Frequency, super high | SHF |
| Frequency, ultra high | UHF |
| Frequency, very high | VHF |
| Frequency, very low | VLF |
| Friction horsepower | FHP |
| From below | FR BEL |
| Front | FR |
| Fuel | F |
| Furnish | FURN |
| Fusible | FSBL |
| Fusion point | FNP |
| | |
| Gage or Gauge | GA |
| Gallon | GAL |
| Galvanize | GALV |
| Galvanized iron | GI |
| Galvanized steel | GS |
| Galvanized steel wire rope | GSWR |
| Gas | G |
| Gasket | GSKT |
| Gasoline | GASO |
| General | GEN |
| Glaze | GL |
| Government | GOVT |
| Government furnished equipment | GFE |
| Governor | GOV |
| Grade | GR |
| Graduation | GRAD |
| Gram | G |
| Graphic | GRAPH. |
| Graphite | GPH |
| Grating | GRTG |
| Gravity | G |
| Grid | G |
| Grind | GRD |
| Groove | GRV |
| Ground | GRD |
| | |
| Half hard | ½H |
| Half round | ½RD |
| Handle | HDL |
| Hanger | HGR |
| Hard | H |
| Hard-drawn | HD |
| Harden | HDN |
| Hardware | HDW |
| Head | HD |
| Headless | HDLS |
| Heat | HT |
| Heat treat | HT TR |
| Heater | HTR |
| Heavy | HVY |
| Height | HGT |
| Henry | H |
| Hexagon | HEX |
| High | H |
| High frequency | HF |
| High point | H PT |
| High pressure | HP |
| High speed | HS |
| High-speed steel | HSS |
| High tension | HT |
| High voltage | HV |
| Highway | HWY |
| Holder | HLR |
| Hollow | HOL |
| Horizontal | HOR |
| Horsepower | HP |
| Hot rolled | HR |
| Hot-rolled steel | HRS |
| Hour | HR |
| Hydraulic | HYD |
| | |
| Identify | IDENT |
| Ignition | IGN |
| Illuminate | ILLUM |
| Illustrate | ILLUS |
| Impact | IMP |
| Impedance | IMP. |
| Inch | (″) IN. |
| Inches per second | IPS |
| Include | INCL |
| Increase | INCR |
| Indicate | IND |
| Inductance or induction | IND |
| Industrial | IND |
| Information | INFO |
| Injection | INJ |
| Inlet | IN |
| Inspect | INSP |

| Word | Abbreviation |
|---|---|
| Install | INSTL |
| Instantaneous | INST |
| Instruct | INST |
| Instrument | INST |
| Insulate | INS |
| Interchangeable | INTCHG |
| Interior | INT |
| Interlock | INTLK |
| Intermediate | INTER |
| Intermittent | INTMT |
| Internal | INT |
| Interrupt | INTER |
| Interrupted continuous wave | ICW |
| Interruptions per minute | IPM |
| Interruptions per second | IPS |
| Intersect | INT |
| Inverse | INV |
| Invert | INV |
| Iron | I |
| Iron-pipe size | IPS |
| Irregular | IRREG |
| Issue | ISS |
| Jack | J |
| Job order | JO |
| Joint | JT |
| Junction | JCT |
| Kelvin | K |
| Key | K |
| Keyseat | KST |
| Keyway | KWY |
| Kilo | K |
| Kilocycle | KC |
| Kilocycles per second | KC |
| Kilogram | KG |
| Kiloliter | KL |
| Kilometer | KM |
| Kilovolt | KV |
| Kilovolt-ampere | KVA |
| Kilovolt-ampere hour | KVAH |
| Kilowatt | KW |
| Kilowatt-hour | KWH |
| Kip (1000 lb) | K |
| Knots | KN |
| Laboratory | LAB |
| Lacquer | LAQ |
| Laminate | LAM |
| Lateral | LAT |
| Lead-coated metal | LCM |
| Lead covered | LC |
| Leading edge | LE |
| Left | L |
| Left hand | LH |
| Length | LG |
| Length over all | LOA |
| Letter | LTR |
| Light | LT |
| Limit | LIM |
| Line | L |
| Linear | LIN |
| Link | LK |
| Liquid | LIQ |
| Liter | L |
| Locate | LOC |
| Long | LG |
| Longitude | LONG. |
| Low explosive | LE |
| Low frequency | LF |
| Low pressure | LP |
| Low tension | LT |
| Low voltage | LV |
| Low speed | LS |
| Low torque | LT |
| Lubricate | LUB |
| Lubricating oil | LO |
| Lumen | L |
| Lumens per watt | LPW |
| Machine | MACH |
| Magnct | MAG |
| Main | MN |
| Male and female | M&F |
| Malleable | MALL |
| Malleable iron | MI |
| Manual | MAN. |
| Manufacture | MFR |
| Manufactured | MFD |
| Manufacturing | MFG |
| Material | MATL |
| Material list | ML |
| Maximum | MAX |

| Word | Abbreviation | Word | Abbreviation |
|---|---|---|---|
| Maximum working pressure | MWP | Mounted | MTD |
| Mean effective pressure | MEP | Mounting | MTG |
| Mechanical | MECH | Multiple | MULT |
| Mechanism | MECH | Multiple contact | MC |
| Medium | MED | | |
| Mega | M | National | NATL |
| Megacycles | MC | Natural | NAT |
| Megawatt | MW | Near face | NF |
| Megohm | MEG | Near side | NS |
| Melting point | MP | Negative | NEG |
| Metal | MET. | Network | NET |
| Meter (instrument or measure of length) | M | Neutral | NEUT |
| Micro | $\mu$ or U | Nickel-silver | NI-SIL |
| Microampere | $\mu$A or UA | Nipple | NIP. |
| Microfarad | $\mu$F or UF | Nominal | NOM |
| Microhenry | $\mu$H or UH | Normal | NOR |
| Micro-inch-root-mean square | $\mu$ IN-RMS or U-IN-RMS | Normally closed | NC |
| Micrometer | MIC | Normally open | NO |
| Micron | $\mu$ or U | Not to scale | NTS |
| Microvolt | $\mu$V or UV | Number | NO. |
| Microwatt | $\mu$W or UW | | |
| Miles | MI | Obsolete | OBS |
| Miles per gallon | MPG | Octagon | OCT |
| Miles per hour | MPH | Ohm | Ω |
| Milli | M | Oil-circuit breaker | OCB |
| Milliampere | MA. | Oil insulated | OI |
| Milligram | MG | Oil switch | OS |
| Millihenry | MH | On center | OC |
| Millimeter | MM | One pole | 1 P |
| Milliseconds | MS | Opening | OPNG |
| Millivolt | MV | Operate | OPR |
| Milliwatt | MW | Opposite | OPP |
| Minimum | MIN | Optical | OPT |
| Minute | (′) MIN | Ordnance | ORD |
| Miscellaneous | MISC | Orifice | ORF |
| Mixture | MIX. | Original | ORIG |
| Model | MOD | Oscillate | OSC |
| Modify | MOD | Ounce | OZ |
| Modulated continuous wave | MCW | Out to Out | O to O |
| Modulator | MOD | Outlet | OUT. |
| Molecular weight | MOL WT | Output | OUT. |
| Monument | MON | Outside diameter | OD |
| Morse taper | MOR T | Outside face | OF |
| Motor | MOT | Outside radius | OR |
| | | Over-all | OA |
| | | Overhead | OVHD |

| Word | Abbreviation | Word | Abbreviation |
|---|---|---|---|
| Overload | OVLD | Pounds per cubic foot | PCF |
| Overvoltage | OVV | Pounds per square foot | PSF |
| Oxidized | OXD | Pounds per square inch | PSI |
| | | Pounds per square inch absolute | PSIA |
| Pack | PK | Power | PWR |
| Packing | PKG | Power amplifier | PA |
| Painted | PTD | Power directional relay | PDR |
| Pair | PR | Power factor | PF |
| Panel | PNL | Preamplifier | PREAMP |
| Parallel | PAR. | Precast | PRCST |
| Part | PT | Prefabricated | PREFAB |
| Pattern | PATT | Preferred | PFD |
| Perforate | PERF | Premolded | PRMLD |
| Permanent | PERM | Prepare | PREP |
| Permanent magnet | PM | Press | PRS |
| Perpendicular | PERP | Pressure | PRESS. |
| Phase | PH | Pressure angle | PA. |
| Phosphor bronze | PH BRZ | Primary | PRI |
| Photograph | PHOTO | Process | PROC |
| Physical | PHYS | Production | PROD |
| Piece | PC | Profile | PF |
| Piece mark | PC MK | Project | PROJ |
| Pierce | PRC | Punch | PCH |
| Pipe Tap | PT | Purchase | PUR |
| Pitch | P | Push-pull | P-P |
| Pitch circle | PC | | |
| Pitch diameter | PD | Quadrant | QUAD |
| Plastic | PLSTC | Quality | QUAL |
| Plate | PL | Quantity | QTY |
| Plotting | PLOT. | Quart | QT |
| Pneumatic | PNEU | Quarter | QTR |
| Point | PT | Quarter hard | ¼ H |
| Point of compound curve | PCC | Quarter round | ¼ RD |
| Point of curve | PC | Quartz | QTZ |
| Point of intersection | PI | | |
| Point of reverse curve | PRC | Radial | RAD |
| Point of switch | PS | Radio frequency | RF |
| Point of tangent | PT | Radius | R |
| Polar | POL | Reactive | REAC |
| Pole | P | Reactive kilovolt-ampere | KVAR |
| Polish | POL | Reactive volt-ampere | VAR |
| Port | P | Reactive voltmeter | RVM |
| Position | POS | Reactor | REAC |
| Positive | POS | Ream | RM |
| Potential | POT. | Reassemble | REASM |
| Pound | LB | Received | RECD |

| Word | Abbreviation | Word | Abbreviation |
|---|---|---|---|
| Receiver | REC | Saddle | SDL |
| Receptacle | RECP | Safe working pressure | SWP |
| Recriprocate | RECIP | Safety | SAF |
| Recirculate | RECIRC | Sand blast | SD BL |
| Reclosing | RECL | Saturate | SAT. |
| Record | REC | Schedule | SCH |
| Rectangle | RECT | Schematic | SCHEM |
| Rectifier | RECT | Scleroscope hardness | SH |
| Reduce | RED. | Screen | SCRN |
| Reference | REF | Screw | SCR |
| Reference line | REF L | Second | SEC |
| Regulator | REG | Section | SECT |
| Reinforce | REINF | Segment | SEG |
| Relay | REL | Select | SEL |
| Release | REL | Semifinished | SF |
| Relief | REL | Semifixed | SFXD |
| Remove | REM | Semisteel | SS |
| Repair | REP | Separate | SEP |
| Replace | REPL | Sequence | SEQ |
| Reproduce | REPRO | Serial | SER |
| Require | REQ | Series | SER |
| Required | REQD | Serrate | SERR |
| Resistance | RES | Service | SERV |
| Resistor | RES | Set screw | SS |
| Retainer | RET. | Shaft | SFT |
| Retard | RET. | Shield | SHLD |
| Return | RET. | Shipment | SHPT |
| Reverse | REV | Shop order | SO |
| Revise | REV | Short wave | SW |
| Revolution | REV | Shunt | SH |
| Revolutions per minute | RPM | Side | S |
| Revolutions per second | RPS | Signal | SIG |
| Rheostat | RHEO | Sink | SK |
| Right | R | Sketch | SK |
| Right hand | RH | Sleeve | SLV |
| Ring | R | Slide | SL |
| Rivet | RIV | Slotted | SLOT. |
| Rockwell hardness | RH | Small | SM |
| Roller bearing | RB | Smoke | SMK |
| Root diameter | RD | Smokeless | SMKLS |
| Root mean square | RMS | Socket | SOC |
| Rotary | ROT. | Soft | S |
| Rotate | ROT. | Solder | SLD |
| Rough | RGH | Solenoid | SOL |
| Round | RD | Sound | SND |
| Rubber | RUB. | South | S |

| Word | Abbreviation | Word | Abbreviation |
|---|---|---|---|
| Space | SP | Tangent | TAN. |
| Spare | SP | Taper | TPR |
| Speaker | SPKR | Technical | TECH |
| Special | SPL | Tee | T |
| Specific | SP | Teeth per inch | TPI |
| Specific gravity | SP GR | Television | TV |
| Specific heat | SP HT | Temperature | TEMP |
| Specification | SPEC | Template | TEMP |
| Speed | SP | Tensile strength | TS |
| Spherical | SPHER | Tension | TENS. |
| Spindle | SPDL | Terminal | TERM. |
| Split phase | SP PH | Terminal board | TB |
| Spot-faced | SF | That is | IE |
| Spring | SPG | Theoretical | THEO |
| Square | SQ | Thermal | THRM |
| Stabilize | STAB | Thermostat | THERMO |
| Stainless | STN | Thick | THK |
| Standard | STD | Thousand | M |
| Static pressure | SP | Thread | THD |
| Station | STA | Throttle | THROT |
| Stationary | STA | Through | THRU |
| Steel | STL | Time | T |
| Stiffener | STIFF. | Time delay | TD |
| Stock | STK | Time-delay closing | TDC |
| Storage | STG | Time-delay opening | TDO |
| Straight | STR | Tinned | TD |
| Strip | STR | Tobin bronze | TOB BRZ |
| Structural | STR | Toggle | TGL |
| Substitute | SUB | Tolerance | TOL |
| Suction | SUCT | Tongue and groove | T&G |
| Summary | SUM. | Tool steel | TS |
| Supervise | SUPV | Tooth | T |
| Supply | SUP | Total | TOT |
| Surface | SUR | Total indicator reading | TIR |
| Survey | SURV | Trace | TR |
| Switch | SW | Tracer | TCR |
| Symbol | SYM | Transfer | TRANS |
| Symmetrical | SYM | Transformer | TRANS |
| Synchronous | SYN | Transmission | XMSN |
| Synthetic | SYN | Transmitter | XMTR |
| System | SYS | Transmitting | XMTG |
| | | Transportation | TRANS |
| Tabulate | TAB. | Transverse | TRANSV |
| Tachometer | TACH | Trimmer | TRIM. |
| Tandem | TDM | Triode | TRI |

| Word | Abbreviation | Word | Abbreviation |
|---|---|---|---|
| True air speed | TAS | Voltmeter | VM |
| Truss | T | Volts per mil | VPM |
| Tubing | TUB | Volume | VOL |
| Tuned radio frequency | TRF | | |
| Turbine | TURB | Washer | WASH |
| Typical | TYP | Water | W |
| | | Water line | WL |
| Ultimate | ULT | Watertight | WT |
| Ultra-high frequency | UHF | Watt | W |
| Under voltage | UV | Watt-hour | WHR |
| Unit | U | Watt-hour meter | WHM |
| United States Gage | USG | Wattmeter | WM |
| United States Standard | USS | Weight | WT |
| Universal | UNIV | West | W |
| | | Wet bulb | WB |
| Vacuum | VAC | Width | W |
| Vacuum tube | VT | Wind | WD |
| Valve | V | Winding | WDG |
| Vapor proof | VAP PRF | Wire | W |
| Variable | VAR | With | W/ |
| Variable-frequency oscillator | VFO | With equipment and spare parts | W/E&SP |
| Velocity | V | Without | W/O |
| Ventilate | VENT. | Without equipment and spare parts | W/O E&SP |
| Versed sine | VERS | Wood | WD |
| Versus | VS | Woodruff | WDF |
| Vertical | VERT | Working point | WP |
| Very-high frequency | VHF | Working pressure | WP |
| Very-low frequency | VLF | Wrought | WRT |
| Video-frequency | VDF | Wrought iron | WI |
| Vibrate | VIB | | |
| Viscosity | VISC | Yard | YD |
| Vitreous | VIT | Year | YR |
| Voice frequency | VF | Yield point | YP |
| Volt | V | Yield strength | YS |
| Volt-ampere | VA | | |

## Abbreviations for Colors

| Word | Abbreviation | Word | Abbreviation |
|---|---|---|---|
| Amber | AMB | Green | GRN |
| Black | BLK | Orange | ORN |
| Blue | BLU | White | WHT |
| Brown | BRN | Yellow | YEL |

## Partial List of Chemical Symbols

| Word | Abbreviation | Word | Abbreviation |
|---|---|---|---|
| Aluminum | Al | Molybdenum | Mo |
| Antimony (stibium) | Sb | Neon | Ne |
| Barium | Ba | Nickel | Ni |
| Beryllium | Be | Nitrogen | N |
| Bismuth | Bi | Oxygen | O |
| Boron | B | Phosphorus | P |
| Bromine | Br | Platinum | Pt |
| Cadmium | Cd | Potassium (kalium) | K |
| Calcium | Ca | Radium | Ra |
| Carbon | C | Rhodium | Rh |
| Chlorine | Cl | Ruthenium | Ru |
| Chromium | Cr | Selenium | Se |
| Cobalt | Co | Silicon | Si |
| Copper | Cu | Silver (argentum) | Ag |
| Fluorine | F | Sodium (natrium) | Na |
| Gold (aurium) | Au | Strontium | Sr |
| Helium | He | Sulfur | S |
| Hydrogen | H | Tantalum | Ta |
| Indium | In | Tellurium | Te |
| Iodine | I | Thallium | Tl |
| Iridium | Ir | Tin (stannum) | Sn |
| Iron (ferrum) | Fe | Titanium | Ti |
| Lead (plumbum) | Pb | Tungsten (wolframium) | W |
| Lithium | Li | Uranium | U |
| Magnesium | Mg | Vanadium | V |
| Manganese | Mn | Zinc | Zn |
| Mercury (hydrargyrum) | Hg | Zirconium | Zr |

## Blind riveting is easier when you follow these simple **TIPS ON INSTALLATION**

### BEFORE YOU START THE JOB

**Measure your total material thickness. Select the correct grip length rivet from the tables on the next three pages by comparing your sheet thickness with the min. and max. grips shown in the column headed GRIP LENGTH.**

### IN METAL

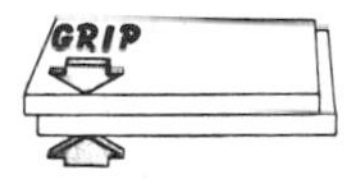


GRIP LENGTH = Total thickness of sheets to be fastened

### IN WOOD

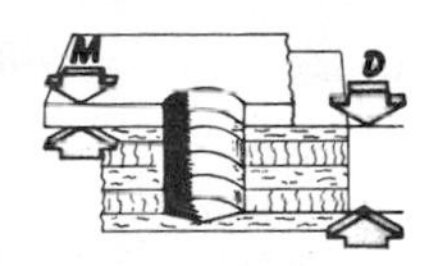


USE "L" DIMENSION (length under head) instead of grip length. L = M (thickness of metal) + D (hole depth in wood.)

### DRILL A HOLE

Select drill size according to your drilling method (hand or jig drilling)

| RIVET DIA. | JIG DRILLING | | | | | HAND DRILLING | | | | |
|---|---|---|---|---|---|---|---|---|---|---|
| | 1/8 | 5/32 | 3/16 | 1/4 | 5/16 | 1/8 | 5/32 | 3/16 | 1/4 | 5/16 |
| DRILL SIZE | 30 | 20 | 11 | F | P | 1/8 | 5/32 | 3/16 | 1/4 | 5/16 |
| HOLE DIA. | .128 | .161 | .191 | .257 | .323 | | | | | |

### INSERT A RIVET

seat head firmly against outer surface

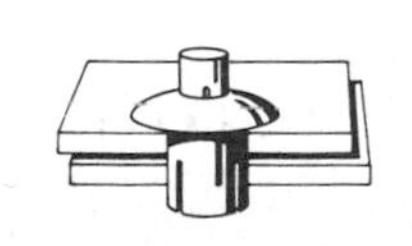

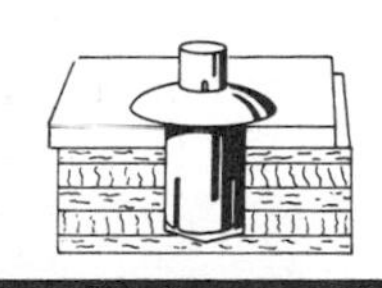

### HIT THE PIN

Drive pin flush with rivet head

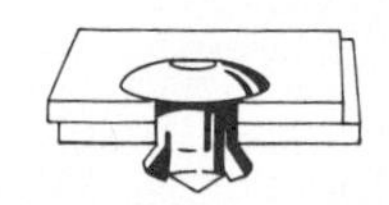

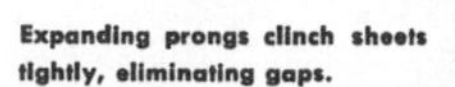
Expanding prongs clinch sheets tightly, eliminating gaps.

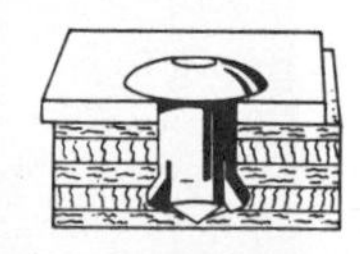

Metal and wood pulled tightly together. Nothing protrudes through wood.

# APPENDIX

## SUPPLY CATALOGUES AND TECHNICAL MANUALS

The following compilation shows representative pages from suppliers' and manufacturers' catalogues and technical manuals. These publications offer varied technical information, engineering specifications, and design data in many fields, material which is of great value both to students and to practicing engineers, designers, draftsmen, architects, and builders. In addition to their first-line role as sources of technical information and data, these publications exemplify interesting solutions to the graphic design problems involved in presenting large amounts of detail in readable form.

The student should take every opportunity to start and maintain a collection of such catalogues and manuals. He will find it a valuable addition to his technical library.

# STEEL DRIVE RIVETS

SEE PAGE 3 FOR INSTALLATION DATA

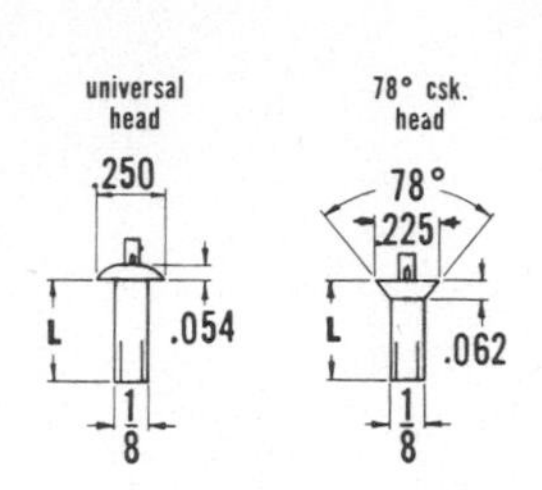


universal head, 78° csk. head, full brazier head: .375, 78°, .339, .469, .080, .094, .094, 3/16

## 1/8" DIA.

## 3/16" DIA.

USE THESE COLUMNS FIRST TO LOCATE YOUR CORRECT

## GRIP LENGTH

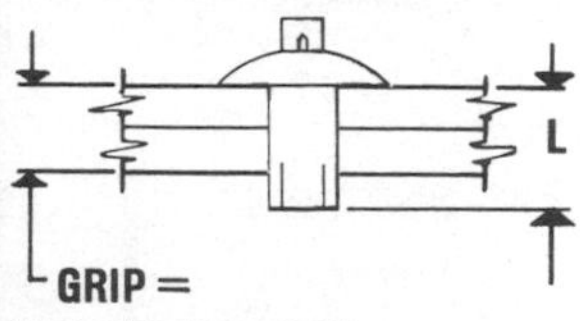


GRIP = TOTAL THICKNESS OF ALL SHEETS FASTENED TOGETHER

| length under head L | 1/8" DIA. PART NUMBERS: UNIVERSAL HEAD | 1/8" DIA. PART NUMBERS: 78° CSK. HEAD | length under head L | 3/16" DIA. PART NUMBERS: UNIVERSAL HEAD | 3/16" DIA. PART NUMBERS: 78° CSK. HEAD | 3/16" DIA. PART NUMBERS: FULL BRAZIER HEAD | MINIMUM GRIP | NOMINAL GRIP | MAXIMUM GRIP |
|---|---|---|---|---|---|---|---|---|---|
| 5/32 | 38-404-01-93* | | | | | | .015 | 1/32 | .046 |
| 5/32 | 38-104-02-93 | | 7/32 | | | 38-206-02-91 | .046 | 1/16 | .078 |
| 3/16 | 38-104-03-93 | 38-604-03-93‡ | 7/32 | 38-106-03-91 | | 38-206-03-91 | .078 | 3/32 | .109 |
| 7/32 | 38-104-04-93 | 38-604-04-93 | 1/4 | 38-106-04-91 | | 38-206-04-91 | .109 | 1/8 | .140 |
| 1/4 | 38-104-05-93 | 38-604-05-93 | 9/32 | 38-106-05-91 | | 38-206-05-91 | .140 | 5/32 | .171 |
| 9/32 | 38-104-06-93 | 38-604-06-93 | 5/16 | 38-106-06-91 | 38-606-06-91 | 38-206-06-91 | .171 | 3/16 | .203 |
| 5/16 | 38-104-07-93 | 38-604-07-93 | 11/32 | 38-106-07-91 | 38-606-07-91 | 38-206-07-91 | .203 | 7/32 | .234 |
| 11/32 | 38-104-08-93 | 38-604-08-93 | 3/8 | 38-106-08-91 | 38-606-08-91 | 38-206-08-91 | .234 | 1/4 | .265 |
| 3/8 | 38-104-09-93 | 38-604-09-93 | 13/32 | 38-106-09-91 | 38-606-09-91 | 38-206-09-91 | .265 | 9/32 | .296 |
| 13/32 | 38-104-10-93 | 38-604-10-93 | 7/16 | 38-106-10-91 | 38-606-10-91 | 38-206-10-91 | .296 | 5/16 | .328 |
| 7/16 | 38-104-11-93 | 38-604-11-93 | 15/32 | 38-106-11-91 | 38-606-11-91 | | .328 | 11/32 | .359 |
| 15/32 | 38-104-12-93 | 38-604-12-93 | 1/2 | 38-106-12-91 | 38-606-12-91 | | .359 | 3/8 | .390 |
| 1/2 | 38-104-13-93 | 38-604-13-93 | 17/32 | 38-106-13-91 | 38-606-13-91 | 38-206-13-91 | .390 | 13/32 | .421 |
| | | | 9/16 | 38-106-14-91 | 38-606-14-91 | | .421 | 7/16 | .453 |
| | * SPECIAL CONE HEAD | | 19/32 | 38-106-15-91 | 38-606-15-91 | | .453 | 15/32 | .484 |
| | | | 5/8 | 38-106-16-91 | 38-606-16-91 | 38-206-16-91 | .484 | 1/2 | .515 |
| | ‡ L = 7/32" | | 21/32 | 38-106-17-91 | 38-606-17-91 | | .515 | 17/32 | .546 |
| | | | 11/16 | 38-106-18-91 | 38-606-18-91 | | .546 | 9/16 | .578 |
| | | | 23/32 | 38-106-19-91 | 38-606-19-91 | | .578 | 19/32 | .609 |
| | | | 3/4 | 38-106-20-91 | 38-606-20-91 | 38-206-20-91 | .609 | 5/8 | .640 |

## New: handy kit for experimental & short run jobs.

### DRIVE RIVET SHOP ASSORTMENT

Contains approximately 1,000 Aluminum Drive Rivets, 3/16" & 1/4" diameter, with Full Brazier Heads. Grip lengths are carefully selected to handle the fastening of almost every common sheet thickness.

The package is a permanent metal-edge kit with refillable inner compartments. Complete instructions for drive riveting are a part of the box itself.

Part No. 38-88-101-10

**MATERIAL**

**Steel, cadmium plated. 1/8 dia. rivets have stainless steel pins. 3/16 & 1/4 dia. rivets have carbon steel pins*, cad. plated**

*Stainless steel pins may be substituted at Manufacturer's option.

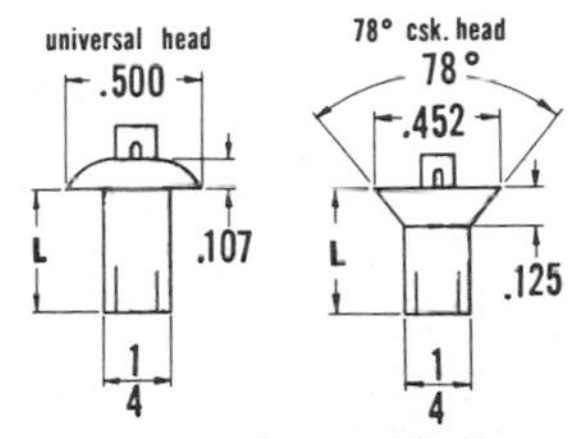


## 1/4" DIA.

| length under head L | PART NUMBERS UNIVERSAL HEAD | 78° CSK. HEAD |
|---|---|---|
| | | |
| | | |
| 7/32 | 38-108-03-91 | |
| 1/4 | 38-108-04-91 | |
| 9/32 | 38-108-05-91 | |
| 5/16 | 38-108-06-91 | 38-608-06-91 |
| 11/32 | 38-108-07-91 | 38-608-07-91 |
| 3/8 | 38-108-08-91 | 38-608-08-91 |
| 13/32 | 38-108-09-91 | 38-608-09-91 |
| 7/16 | 38-108-10-91 | 38-608-10-91 |
| 15/32 | 38-108-11-91 | 38-608-11-91 |
| 1/2 | 38-108-12-91 | 38-608-12-91 |
| 17/32 | 38-108-13-91 | 38-608-13-91 |
| 9/16 | 38-108-14-91 | 38-608-14-91 |
| 19/32 | 38-108-15-91 | 38-608-15-91 |
| 5/8 | 38-108-16-91 | 38-608-16-91 |
| 21/32 | 38-108-17-91 | 38-608-17-91 |
| 11/16 | 38-108-18-91 | 38-608-18-91 |
| 23/32 | 38-108-19-91 | 38-608-19-91 |
| 3/4 | 38-108-20-91 | 38-608-20-91 |

# *New . . . Larger diameter Steel Drive Rivets*

SEE PAGE 3 FOR INSTALLATION DATA

USE THESE COLUMNS FIRST TO LOCATE YOUR CORRECT

## GRIP LENGTH

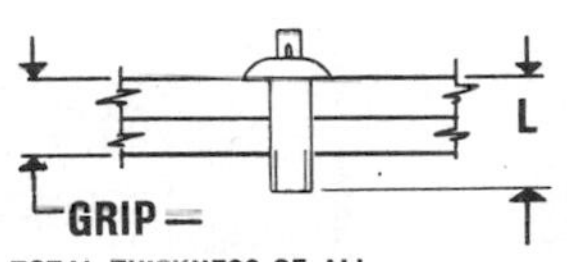


TOTAL THICKNESS OF ALL SHEETS FASTENED TOGETHER

universal head
.625
L
.138
5
16

## 5/16" DIA.

| MINIMUM GRIP | NOMINAL GRIP | MAXIMUM GRIP | length under head L | PART NUMBER UNIVERSAL HEAD |
|---|---|---|---|---|
| .094 | 1/8 | .156 | 5/16 | 38-110-04-91 |
| .157 | 3/16 | .218 | 3/8 | 38-110-06-91 |
| .219 | 1/4 | .281 | 7/16 | 38-110-08-91 |
| .282 | 5/16 | .343 | 1/2 | 38-110-10-91 |
| .344 | 3/8 | .406 | 9/16 | 38-110-12-91 |
| .407 | 7/16 | .468 | 5/8 | 38-110-14-91 |
| .469 | 1/2 | .531 | 11/16 | 38-110-16-91 |
| .532 | 9/16 | .593 | 3/4 | 38-110-18-91 |
| .594 | 5/8 | .656 | 13/16 | 38-110-20-91 |
| .657 | 11/16 | .718 | 7/8 | 38-110-22-91 |
| .719 | 3/4 | .781 | 15/16 | 38-110-24-91 |
| .782 | 13/16 | .843 | 1" | 38-110-26-91 |
| .844 | 7/8 | .906 | 1-1/16 | 38-110-28-91 |
| .907 | 15/16 | .968 | 1-1/8 | 38-110-30-91 |
| .969 | 1 | 1.031 | 1-3/16 | 38-110-32-91 |

**MATERIAL—RIVET and PIN**
**Steel, cadmium plated**

## Repair Truck Bodies in Minutes with RIVETPATCH™

Each Kit contains an aircraft type, Alclad Aluminum patch with formed edges, rounded corners, pilot holes. Sealant, Drive Rivets and instructions included. Only a drill and hammer needed to do a professional job in ten minutes!

4 sizes available: 3 x 5 (Part 38-99-337-11), 5 x 8 (Part 38-99-338-11), 8 x 11 5/8 (Part 38-99-339-11), and 11 5/8 x 24 (Part 38-99-353-11).

**OPTIONAL TOOLS AND ACCESSORIES**

| For Rivet Dia. | PART NUMBERS Rivet Set Tools | 100° Countersinks | Southco Rivet Selectors |
|---|---|---|---|
| 1/8 | 29-1-504-10 | 29-13-101-11 | 29-12-101-23 |
| 5/32 | 29-1-505-10 | | |
| 3/16 | 29-1-506-10 | 29-13-102-11 | |
| 1/4 | 29-1-508-10 | | |

# ALUMINUM

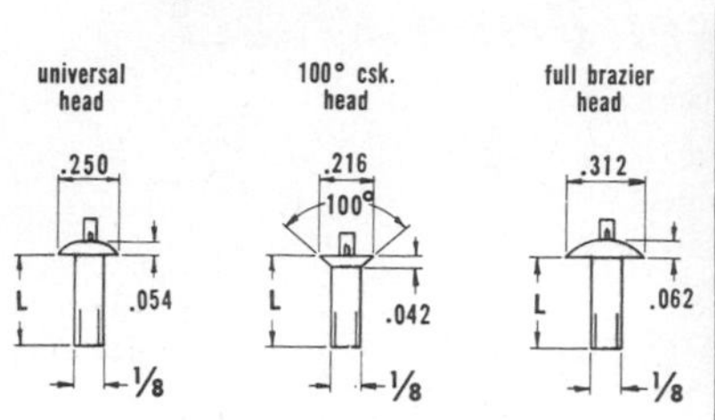


universal head .312 L .067 5/32 — 100° csk. head .278 100° L .055 5/32

USE THESE COLUMNS FIRST TO LOCATE YOUR CORRECT

## GRIP LENGTH

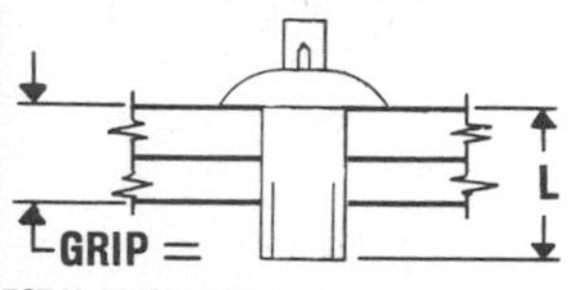


GRIP = TOTAL THICKNESS OF ALL SHEETS FASTENED TOGETHER

## 1/8" DIA.

| length under head L | PART NUMBERS UNIVERSAL HEAD | 100° CSK. HEAD | FULL BRAZIER HEAD |
|---|---|---|---|
| 5/32 | 38-404-01-13* | | |
| 5/32 | 38-104-02-13 | | |
| 3/16 | 38-104-03-13 | 38-504-03-13‡ | 38-204-03-13 |
| 7/32 | 38-104-04-13 | 38-504-04-13 | 38-204-04-13 |
| 1/4 | 38-104-05-13 | 38-504-05-13 | 38-204-05-13 |
| 9/32 | 38-104-06-13 | 38-504-06-13 | 38-204-06-13 |
| 5/16 | 38-104-07-13 | 38-504-07-13 | 38-204-07-13 |
| 11/32 | 38-104-08-13 | 38-504-08-13 | |
| 3/8 | 38-104-09-13 | 38-504-09-13 | |
| 13/32 | 38-104-10-13 | 38-504-10-13 | |
| 7/16 | 38-104-11-13 | 38-504-11-13 | |
| 15/32 | 38-104-12-13 | 38-504-12-13 | 38-204-12-13 |
| 1/2 | 38-104-13-13 | 38-504-13-13 | |
| 17/32 | 38-104-14-13 | | |
| 23/32 | 38-104-20-13 | | |

* SPECIAL CONE HEAD ‡ L = 7/32"

## 5/32" DIA.

| length under head L | PART NUMBERS UNIVERSAL HEAD | 100° CSK. HEAD |
|---|---|---|
| 3/16 | 38-105-01-13 | |
| 3/16 | 38-105-02-13 | |
| 7/32 | 38-105-03-13 | |
| 1/4 | 38-105-04-13 | 38-505-04-13 |
| 9/32 | 38-105-05-13 | 38-505-05-13 |
| 5/16 | 38-105-06-13 | 38-505-06-13 |
| 11/32 | 38-105-07-13 | 38-505-07-13 |
| 3/8 | 38-105-08-13 | 38-505-08-13 |
| 13/32 | 38-105-09-13 | 38-505-09-13 |
| 7/16 | 38-105-10-13 | 38-505-10-13 |
| 15/32 | 38-105-11-13 | 38-505-11-13 |
| 1/2 | 38-105-12-13 | 38-505-12-13 |
| 17/32 | 38-105-13-13 | 38-505-13-13 |
| 9/16 | 38-105-14-13 | 38-505-14-13 |
| 19/32 | 38-105-15-13 | 38-505-15-13 |
| 5/8 | 38-105-16-13 | 38-505-16-13 |
| 21/32 | 38-105-17-13 | 38-505-17-13 |
| 11/16 | 38-105-18-13 | 38-505-18-13 |
| 23/32 | 38-105-19-13 | 38-505-19-13 |
| 3/4 | 38-105-20-13 | 38-505-20-13 |

| MINIMUM GRIP | NOMINAL GRIP | MAXIMUM GRIP |
|---|---|---|
| .015 | 1/32 | .046 |
| .046 | 1/16 | .078 |
| .078 | 3/32 | .109 |
| .109 | 1/8 | .140 |
| .140 | 5/32 | .171 |
| .171 | 3/16 | .203 |
| .203 | 7/32 | .234 |
| .234 | 1/4 | .265 |
| .265 | 9/32 | .296 |
| .296 | 5/16 | .328 |
| .328 | 11/32 | .359 |
| .359 | 3/8 | .390 |
| .390 | 13/32 | .421 |
| .421 | 7/16 | .453 |
| .453 | 15/32 | .484 |
| .484 | 1/2 | .515 |
| .515 | 17/32 | .546 |
| .546 | 9/16 | .578 |
| .578 | 19/32 | .609 |
| .609 | 5/8 | .640 |
| .671 | 11/16 | .703 |
| .734 | 3/4 | .765 |
| .796 | 13/16 | .828 |
| .859 | 7/8 | .890 |
| .921 | 15/16 | .953 |
| .984 | 1 | 1.015 |

## ALUMINUM EXTRA LONG

RIVETS FOR METAL OR MASONRY
(Head dimensions same as on page 5)

Subtract 1/8" from "L" to determine grip

| length under head L | 3/16" DIA. FULL BRAZIER HEAD | 1/4" DIA. FULL BRAZIER HEAD | 1/4" DIA. 100° CSK. HEAD |
|---|---|---|---|
| 1 1/8 | 38-206-32-13 | 38-208-32-13 | 38-508-32-13 |
| 1 1/4 | | 38-208-36-13 | |
| 1 5/16 | | 38-208-38-13 | |
| 1 3/8 | | 38-208-40-13 | |
| 1 7/16 | | 38-208-42-13 | |
| 1 1/2 | | 38-208-44-13 | 38-508-44-13 |
| 1 9/16 | | 38-208-46-13 | |
| 1 5/8 | | 38-208-48-13 | |
| 1 3/4 | | 38-208-52-13 | 38-508-52-13 |

# DRIVE RIVETS

**2117-T4 ALUMINUM RIVET BODY WITH 300 SERIES STAINLESS STEEL PINS.**

(Aluminum rivets of other alloys, and pins of aluminum available on special request)

SEE PAGE 3 FOR INSTALLATION DATA

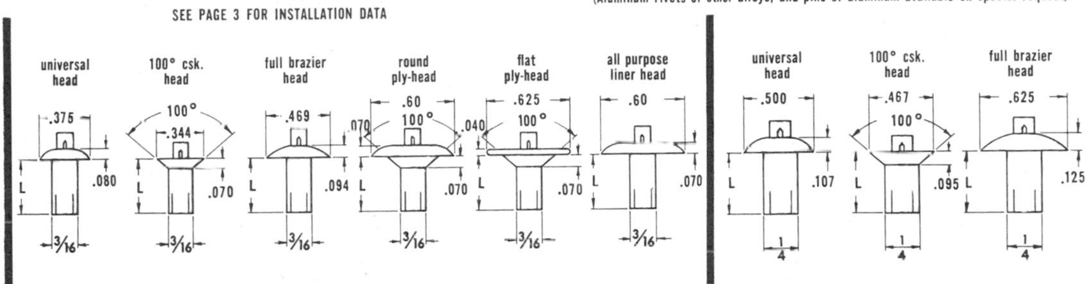


## 3/16" DIA.

### PART NUMBERS

| length under head L * | UNIVERSAL HEAD | 100° CSK. HEAD | FULL BRAZIER HEAD | ROUND PLY-HEAD® | FLAT PLY-HEAD® | All Purpose LINER HEAD |
|---|---|---|---|---|---|---|
| | | | | | | |
| 7/32 | 38-106-02-13 | | 38-206-02-13 | | | |
| 7/32 | 38-106-03-13 | | 38 206-03-13 | | | 38-99-269 03 |
| 1/4 | 38-106-04-13 | | 38-206-04-13 | | | 38-99-269 04 |
| 9/32 | 38-106-05-13 | | 38-206-05-13 | | | 38-99-269-05 |
| 5/16 | 38-106-06-13 | 38-506-06-13 | 38-206-06-13 | | | 38-99-269-06 |
| 11/32 | 38-106-07-13 | 38-506-07-13 | 38-206-07-13 | | | 38-99-269-07 |
| 3/8 | 38-106-08-13 | 38-506-08-13 | 38-206-08-13 | 38-906-08-13 | | 38-99-269-08 |
| 13/32 | 38-106-09-13 | 38-506-09-13 | 38-206-09-13 | 38-906-09-13 | | 38-99-269-09 |
| 7/16 | 38-106-10-13 | 38-506-10-13 | 38-206-10-13 | 38-906-10-13 | 38-806-10-13 | 38-99-269-10 |
| 15/32 | 38-106-11-13 | 38-506-11-13 | 38-206-11-13 | 38-906-11-13 | | 38-99-269-11 |
| 1/2 | 38-106-12-13 | 38-506-12-13 | 38-206-12-13 | 38-906-12-13 | 38-806-12-13 | 38-99-269-12 |
| 17/32 | 38-106-13-13 | 38-506-13-13 | 38-206-13-13 | | 38-806-13-13 | 38-99-269-13 |
| 9/16 | 38-106-14-13 | 38-506-14-13 | 38-206-14-13 | 38-906-14-13 | 38-806-14-13 | 38-99-269-14 |
| 19/32 | 38-106-15-13 | 38-506-15-13 | 38-206-15-13 | | | |
| 5/8 | 38-106-16-13 | 38-506-16-13 | 38-206-16-13 | 38-906-16-13 | 38-806-16-13 | 38-99-269-16 |
| 21/32 | 38-106-17-13 | 38-506-17-13 | 38-206-17-13 | | | |
| 11/16 | 38-106-18-13 | 38-506-18-13 | 38-206-18-13 | 38-906-18-13 | 38-806-18-13 | 38-99-269-18 |
| 23/32 | 38-106-19-13 | 38-506-19-13 | 38-206-19-13 | | | |
| 3/4 | 38-106-20-13 | 38-506-20-13 | 38-206-20-13 | 38-906-20-13 | 38-806-20-13 | 38-99-269-20 |
| | | | | | | |
| 13/16 | 38-106-22-13 | | 38-206-22-13 | | | |
| | | | | | | |
| 7/8 | 38-106-24-13 | | 38-206-24-13 | | | 38-99-269-24 |
| | | | | | | |
| 15/16 | 38-106-26-13 | | 38-206-26-13 | | | |
| | | | | | | |
| 1 | 38-106-28-13 | 38-506-28-13 | 38-206-28-13 | | | 38-99-269-28 |
| | | | | | | |
| 1-1/16 | 38-106-30-13 | | 38-206-30-13 | | | |
| | | | | | | |
| 1-1/8 | 38-106-32-13 | | 38-206-32-13 | | | |

## 1/4" DIA.

### PART NUMBERS

| length under head L * | UNIVERSAL HEAD | 100° CSK. HEAD | FULL BRAZIER HEAD |
|---|---|---|---|
| | | | |
| 7/32 | 38-108-02-13 | | |
| 7/32 | 38-108-03-13 | | 38-208-03-13 |
| 1/4 | 38-108-04-13 | | 38-208-04-13 |
| 9/32 | 38-108-05-13 | | 38-208-05-13 |
| 5/16 | 38-108-06-13 | 38-508-06-13 | 38-208-06-13 |
| 11/32 | 38-108-07-13 | 38-508-07-13 | 38-208-07-13 |
| 3/8 | 38-108-08-13 | 38-508-08-13 | 38-208-08-13 |
| 13/32 | 38-108-09-13 | 38-508-09-13 | 38-208-09-13 |
| 7/16 | 38-108-10-13 | 38-508-10-13 | 38-208-10-13 |
| 15/32 | 38-108-11-13 | 38-508-11-13 | 38-208-11-13 |
| 1/2 | 38-108-12-13 | 38-508-12-13 | 38-208-12-13 |
| 17/32 | 38-108-13-13 | 38-508-13-13 | 38-208-13-13 |
| 9/16 | 38-108-14-13 | 38-508-14-13 | 38-208-14-13 |
| 19/32 | 38-108-15-13 | 38-508-15-13 | 38-208-15-13 |
| 5/8 | 38-108-16-13 | 38-508-16-13 | 38-208-16-13 |
| 21/32 | 38-108-17-13 | 38-508-17-13 | 38-208-17-13 |
| 11/16 | 38-108-18-13 | 38-508-18-13 | 38-208-18-13 |
| 23/32 | 38-108-19-13 | 38-508-19-13 | 38-208-19-13 |
| 3/4 | 38-108-20-13 | 38-508-20-13 | 38-208-20-13 |
| | | | |
| 13/16 | 38-108-22-13 | 38-508-22-13 | 38-208-22-13 |
| 27/32 | | | 38-208-23-13 |
| 7/8 | 38-108-24-13 | 38-508-24-13 | 38-208-24-13 |
| | | | |
| 15/16 | 38-108-26-13 | 38-508-26-13 | 38-208-26-13 |
| | | | |
| 1 | 38-108-28-13 | 38-508-28-13 | 38-208-28-13 |
| | | | |
| 1-1/16 | 38-108-30-13 | 38-508-30-13 | 38-208-30-13 |
| | | | |
| 1-1/8 | 38-108-32-13 | 38-508-32-13 | 38-208-32-13 |

* NOTE: The L dimension (length under head) is of importance only as a means of dimensional description. This dimension is undergoing modification in some of our rivets. In cases where actual shank length differs from that given in the L column above, a note both inside the rivet package and on its label gives corrected information. The grip length of the rivet is not affected by shank length modification.

# PALNUT® Self-Threading Nuts

## SPECIFIC ADVANTAGES OF OTHER TYPES

### Regular Type

This is the lowest-cost PALNUT Self-threading Nut. Especially adapted to assemblies where available space is at a premium. Uses shorter studs. Competitive with push types, yet assembles fast and assures better, tighter assemblies. Shape of nut makes it possible to assemble with internal wrench. (See wrench illustration below).

### Acorn Type

All the cost and assembly advantages of the PALNUT Self-threading principle are gained in this decorative, crowned nut. It covers up the ends of studs for pleasing appearance and protection against scratching.

## EASY, FAST ASSEMBLY

PALNUT Self-threading Nuts are formed to standard hex shape. They may be fastened with any standard tools, manual or power. *However, by using the PALNUT Magnetic Socket Wrench, which fits all standard tools, greater convenience and higher-speed assembly are obtained.*

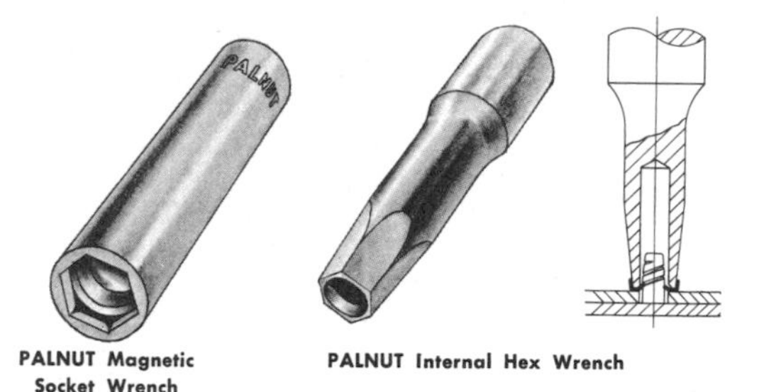

PALNUT Magnetic Socket Wrench

PALNUT Internal Hex Wrench

### Power Tool Assembly

Air or electric tools, equipped with the PALNUT Magnetic Socket Wrench, permit starting, running on and tightening in one simple, fast operation.

### Manual Assembly

PALNUT Self-threading Nuts may be applied with hand tools, but power tools are recommended for production runs. The PALNUT Magnetic Socket Wrench facilitates manual assembly.

Write for Bulletin WR-516 giving full details on PALNUT Wrenches and Tools

## STUD INFORMATION

Studs of any malleable material may be used, including zinc or aluminum die cast, steel, brass or aluminum wire and high tensile plastic materials.

Permissible tolerance on stud diameter is plus or minus 0.003″. Tapered studs should be within this tolerance at the point where the nut grips the stud at completion of assembly. (See illustration).

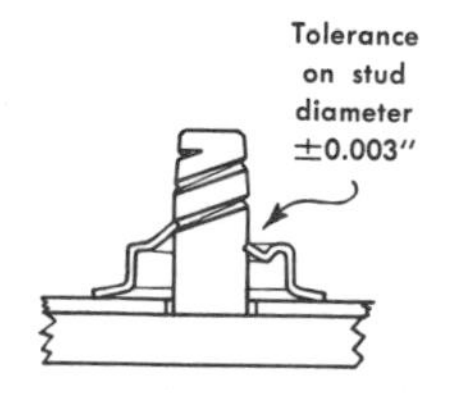


PALNUT Self-threading Nuts will perform satisfactorily on blunt end studs or wire, but a 1/32″ x 45° chamfered end is preferred for easier starting. Full data on stud specifications is available on request.

## TORQUE-TENSILE DATA

Recommended assembly torque and resulting stud tension for each Self-Threading Nut is shown on page 7. The nominal torque figures given can be used for preliminary adjustment of power tools.

Power tools with ratcheting (slip) type clutch or the "one shot" quick-breaking type clutch may be used. For Self-Threaders with Sealer we specifically recommend power tools with the ratcheting type clutch.

Where assembly conditions deviate from the normal, a downward or upward revision of the recommended assembly torque setting on the power tool clutch may be necessary to meet the current conditions. Examples given below.

1. Dry, flat, non-slippery surfaces and stud diameters on the high side of the tolerance will require higher assembly torque.
2. Irregular, slippery surfaces and stud diameters on the low side of the tolerance will require lower assembly torque.

# PALNUT® Self-Threading Nuts

## DIMENSIONS AND TORQUE-TENSILE DATA

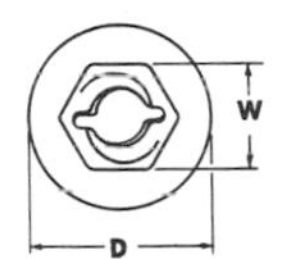


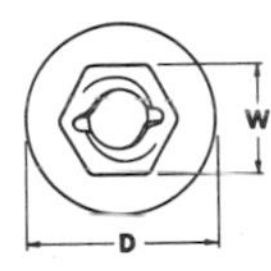


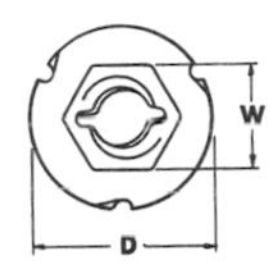

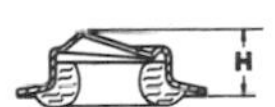


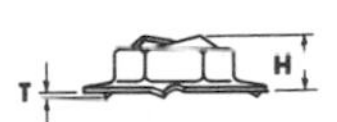

**WASHER TYPE**

**WASHER TYPE**
**with SEALER**
(See Sealer Information on Page 5)

**WASHER TYPE**
**STYLE SG**

**REGULAR TYPE**

**ACORN TYPE**

| Type | Stud Size | PALNUT Part No. | Hex Width "W" | PALNUT Height "H" | Washer Diameter "D" | Teeth Depth "T" | Net Wgt. Lbs./M Pieces | TORQUE-TENSILE DATA (See page 6 for details) ON ZINC DIE CAST STUDS Recommended Torque (Inch Lbs.) | ON ZINC DIE CAST STUDS Stud Tension Lbs. | ON STEEL STUDS Recommended Torque (Inch Lbs.) | ON STEEL STUDS Stud Tension (Lbs.) |
|---|---|---|---|---|---|---|---|---|---|---|---|
| WASHER TYPE | 1/8" | SD125007 | 5/16" | .186" | 7/16" | | 1.2 | 15 | 90 | 26 | 130 |
| | 1/8 | SD125085 | 5/16 | .194 | 17/32 | | 1.5 | 20 | 80 | 30 | 110 |
| | 5/32 | SD156009 | 11/32 | .212 | 9/16 | | 1.9 | 30 | 100 | 45 | 240 |
| | 3/16 | SD188008 | 3/8 | .217 | 1/2 | | 1.7 | 38 | 210 | 50 | 330 |
| | 3/16 | SD188009 | 3/8 | .225 | 9/16 | | 2.0 | 42 | 180 | 55 | 300 |
| | 3/16 | SD188010 | 3/8 | .235 | 5/8 | | 2.3 | 46 | 160 | 60 | 280 |
| | 1/4 | SD250095 | 7/16 | .232 | 19/32 | | 2.5 | 65 | 260 | 95 | 370 |
| | 1/4 | SD250011 | 7/16 | .247 | 11/16 | | 3.1 | 75 | 240 | 90 | 340 |
| WASHER TYPE STYLE SG | 1/8 | SG125085 | 5/16 | .194 | 17/32 | .013" | 1.5 | 20 | 80 | 32 | 110 |
| | 3/16 | SG188008 | 3/8 | .207 | 1/2 | .013 | 1.6 | 38 | 210 | 55 | 330 |
| | 3/16 | SG188010 | 3/8 | .235 | 5/8 | .013 | 2.3 | 46 | 160 | 60 | 280 |
| | 1/4 | SG250095 | 7/16 | .219 | 19/32 | .015 | 2.4 | 65 | 260 | 95 | 370 |
| REGULAR TYPE | 3/32 | SR094005 | 5/16 | .100 | | | 0.7 | 5 | 40 | 8 | 80 |
| | 1/8 | SR125004 | 1/4 | .008 | | | 0.4 | 8 | 50 | 10 | 110 |
| | 1/8 | SR125 | 5/16 | .100 | | | 0.7 | 11 | 65 | 16 | 130 |
| | 1/8 | SR125009 | 9/16 | .114 | | | 2.0 | 18 | 90 | 30 | 190 |
| | 5/32 | SR156 | 11/32 | .110 | | | 0.9 | 18 | 100 | 30 | 250 |
| | 3/16 | SR188006 | 3/8 | .116 | | | 1.1 | 26 | 140 | 32 | 280 |
| | 3/16 | SR188 | 1/2 | .129 | | | 2.1 | 30 | 140 | 38 | 250 |
| TYPE ACORN | 1/8 | SC125 | 5/16 | .265 | | | 1.6 | 11 | 65 | 18 | 130 |
| | 1/8 | SC125006 | 3/8 | .324 | | | 2.2 | 14 | 80 | 25 | 160 |
| | 1/8 | SC125007 | 7/16 | .372 | | | 3.2 | 16 | 90 | 28 | 210 |
| | 3/16 | SC188 | 7/16 | .380 | | | 3.7 | 26 | 140 | 45 | 320 |
| | 3/16 | SC188008 | 1/2 | .437 | | | | 30 | 140 | 55 | 350 |
| | 3/16 | SC188009 | 9/16 | .474 | | | 6.3 | 32 | 140 | 65 | 400 |
| | 1/4 | SC250009 | 9/16 | .484 | | | | 53 | 180 | 77 | 500 |

Standard Finish: Cadmium.* Material: Carbon Spring Steel. Optional Finishes: Phosphate, and Mechanical Zinc plus Chromate. (Except Acorn Type)

*Cadmium Plated PALNUT Self-Threading Nuts are usually colored yellow to differentiate them from PALNUT Lock Nuts. An exception is the Acorn Type which is plated with bright cadmium for appearance.

**WRITE FOR FREE SAMPLES, STATING TYPE, SIZE AND APPLICATION**

**Detailed information can be supplied on recommended assembly torques, stud tension, stud dimensions and types of sealers.**

# PUSHNUT® FASTENERS—FLAT ROUND TYPE

## USER ADVANTAGES

Flat Round PUSHNUT Fasteners provide the advantages common to push-on type fasteners generally, which are made by several manufacturers. These include • Exceptionally fast assembly • Elimination of more expensive screws and nuts, washers or retaining rings • Elimination of costly threading, hole drilling or annular grooving.

Compared with push-on nuts generally, Flat Round PUSHNUT Fasteners offer these *specific advantages*

**SUPERIOR GRIP** Means less slippage **GREATER HOLDING POWER** Means higher axial load strength
The number, shape, length and angularity of teeth in each part represent a carefully-designed balance between relative ease of assembly and ultimate strength. Steel thickness and relation of outside diameter to stud size are also part of this "balanced design".

**LOWER COST** Production in large quantities on ultra-modern equipment, plus advanced finishing techniques, assure a per-thousand cost that is invariably below competitive designs.

## STUD RECOMMENDATIONS

**DIAMETER TOLERANCE:** +.002″ —.003″, INCLUDING PLATING. Chamfered (45°) stud ends are preferred, but if cut or sheared from longer rod or wire, square ends should be free of burrs or mushrooming.

**STUD MATERIALS:** May be mild steel, aluminum, die cast zinc, or other malleable materials.

Subject to verification by your own tests, Style PV parts, and Styles PS and PD parts in the thinner gauges, are likely to perform satisfactorily on plastics having good tensile strength and toughness.

**STUD FINISHES:** Steel studs—any commercial finish is satisfactory, except nickel or chrome plating.
Die cast zinc—any finish is satisfactory, except nickel-chrome plating should not exceed .003″ maximum thickness.

**STUD HARDNESS:** See comments under PERFORMANCE DATA.

## PERFORMANCE DATA

Push-on Force and Removal Resistance values are "averages" obtained when parts are applied to low carbon, cold drawn steel rod of hardness not exceeding 78 on Rockwell 30T scale (this is corrected reading allowing for curvature error on round rod).

| PUSHNUT PART NO. | PUSH-ON FORCE LBS. | REMOV. RESIST. LBS. |
|---|---|---|
| PS062032 | 20 | 100 |
| PS094032 | 25 | 150 |
| PS125306 | 10 | 100 |
| PS125006 | 20 | 250 |
| PD156307 | 10 | 250 |
| PD156007 | 20 | 300 |
| PS188307 | 10 | 250 |
| PS188007 | 25 | 420 |
| PD219385 | 20 | 300 |

| PUSHNUT PART NO. | PUSH-ON FORCE LBS. | REMOV. RESIST. LBS. |
|---|---|---|
| PZ001514 | 130 | 600 |
| PS240085 | 30 | 600 |
| PS250385 | 25 | 400 |
| PS250085 | 45 | 650 |
| PV250015 | 10 | 140 |
| PR312075 | 40 | 300 |
| PS312310 | 30 | 650 |
| PS312010 | 50 | 900 |
| PV312015 | 15 | 300 |

| PUSHNUT PART NO. | PUSH-ON FORCE LBS. | REMOV. RESIST. LBS |
|---|---|---|
| PR375010 | 50 | 400 |
| PS375312 | 50 | 650 |
| PS375612 | 70 | 800 |
| PS375012 | 85 | 1000 |
| PV375015 | 15 | 450 |
| PS438014 | 70 | 1400 |
| PV438014 | 20 | 600 |
| PS500016 | 160 | 2000 |

**NOTE:** Application of PUSHNUT Fasteners to studs or rods exceeding specified size tolerances, or maximum hardness, can result in considerable variation in REMOVAL RESISTANCE values. Nevertheless, careful pre-testing may show the resulting strength is still more than adequate, depending on degree of performance needed.

## PUSHNUT® FASTENERS—FLAT ROUND TYPE

### AVAILABLE SIZES AND STYLES

DESIGN 1 STYLE PS

DESIGN 2 STYLE PS

STYLE PD AND PART PZ001514

DESIGN 3 STYLE PS

STYLE PV

PART PR312075

PART PR375010

| STUD DIAM. | PUSHNUT PART NO. | DES # | STEEL THICK. | TOTAL HEIGHT H | WASH. DIAM. INSIDE F | WASH. DIAM. OUTSIDE D | WGT. LBS/M. |
|---|---|---|---|---|---|---|---|
| .062" (1/16") | PS062032 | 3 | .009" | .038" | .142" | .195" | 0.09 |
| .094" (3/32") | PS094032 | 3 | .009" | .038" | .142" | .194" | 0.09 |
| .125" (1/8") | PS125306 | 1 | .009" | .045" | .228" | 3/8" | 0.3 |
| | PS125006 | 1 | .013" | .052" | .228" | 3/8" | 0.4 |
| .156" (5/32") | PD156307 | 2 | .010" | .047" | .320" | 7/16" | 0.4 |
| | PD156007 | 2 | .013" | .058" | .320" | 7/16" | 0.5 |
| .188" (3/16") | PS188307 | 2 | .010" | .056" | .320" | 7/16" | 0.4 |
| | PS188007 | 2 | .015" | .064" | .320" | 7/16" | 0.5 |
| .219" (7/32") | PD219385 | 2 | .012" | .067" | .388" | 17/32" | 0.7 |
| .237" | PZ001514 | 2 | .021" | .050" | .420" | 3/4" | 2.5 |
| .240" | PS240085 | 2 | .017" | .069" | .388" | 17/32" | 0.9 |
| .250" (1/4") | PS250385 | 2 | .012" | .057" | .388" | 17/32" | 0.6 |
| | PS250085 | 2 | .017" | .066" | .388" | 17/32" | 1.0 |
| | PV250015 | | .015" | .083" | .750" | 15/16" | 2.5 |
| .312" (5/16") | PR312075 | | .013" | .040" | .385" | 15/32" | 0.35 |
| | PS312310 | 2 | .015" | .059" | .456" | 5/8" | 1.1 |
| | PS312010 | 2 | .021" | .070" | .456" | 5/8" | 1.5 |
| | PV312015 | | .015" | .097" | .750" | 15/16" | 2.7 |
| .375" (3/8") | PR375010 | | .015" | .056" | .500" | 5/8" | 0.87 |
| | PS375312 | 2 | .017" | .061" | .546" | 3/4" | 1.7 |
| | PS375612 | 2 | .021" | .068" | .546" | 3/4" | 2.1 |
| | PS375012 | 2 | .027" | .081" | .546" | 3/4" | 2.8 |
| | PV375015 | | .015" | .093" | .750" | 15/16" | 2.5 |
| .438" (7/16") | PS438014 | 2 | .030" | .097" | .638" | 7/8" | 4.1 |
| | PV438015 | | .015" | .075" | .750" | 15/16" | 2.5 |
| .500" (1/2") | PS500016 | 2 | .035" | .112" | .730" | 1" | 6.4 |

**MATERIALS:** Spring Steel (some parts available in stainless steel on special order).

**FINISHES:** Mechanical Zinc, Cadmium or Electro-Zinc (see note), Phosphate and Oil, Plain. Other finishes on special order.

**NOTE:** Style PS and PD parts, and part PZ001514, are not available with Cadmium, Electro-Zinc or any other electroplated finish.

For aluminum, and some plastic studs, or wooden dowels, Style PV, Styles PS and PD in the thinner gauges, or Style PR parts are recommended.

# RIVNUT ENGINEERING DATA PREPARATION PROCEDURES

## FOR FLAT HEAD INSTALLATION

Flat head Rivnuts require only a punched or drilled hole of the proper size for their installation. Use a lead drill if desired and follow with a sharp finish drill held at right angles to the work. Always see that dirt or metal particles are removed from between metal sheets and burrs removed wherever possible. Sheets should be clamped or pressed into contact to reduce air gap to a minimum.

## FOR MACHINE COUNTERSINK INSTALLATION

Machine countersinking can be used only in metal thicker than the head thickness of the Rivnut. A precision hole and countersink can best be obtained by following these simple steps:

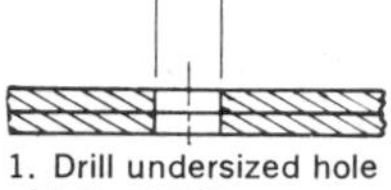

1. Drill undersized hole with lead drill

2. Countersink

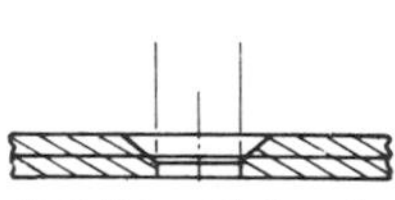

3. Drill correct diameter hole with sharp finish drill

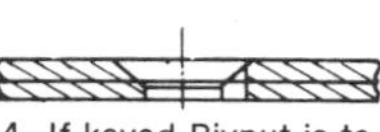

4. If keyed Rivnut is to be used cut keyway with key cutter tool or by the use of a guided drill

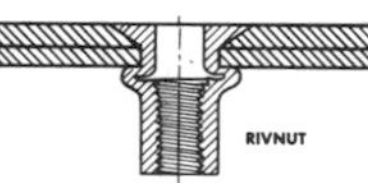


5. Install Rivnut

## FOR DIMPLE COUNTERSINK INSTALLATION

Metal thinner than Rivnut head thickness requires a dimple countersink.

The ideal bulge on any Rivnut application will always be formed against a flat under surface. The bell-mouth that results from ordinary dimpling will not permit proper formation of the bulge. Rivnuts upset against the sharp edge will form a weak bulge, a spread shank, or possibly shear.

To provide a flat surface in the dimpling operation, a ledge at the bottom of the dimpling die can be used. The "flat" on the dimple will save costly deburring before dimpling and enables the Rivnut bulge to form normally, providing maximum strength.

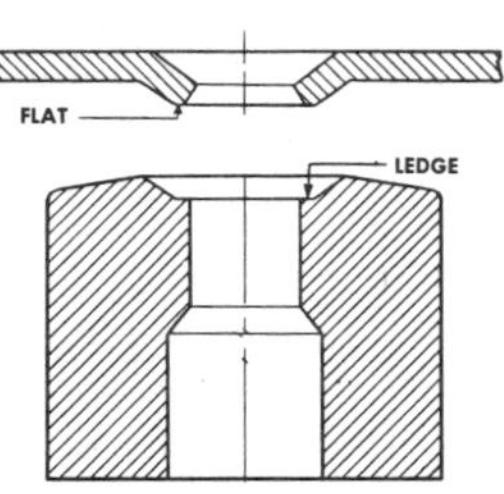


Dimple or press countersink hole. Note ledge at bottom of dimpling die

## FOR KEYED RIVNUTS

Keyways can be cut with a B.F.Goodrich Key Cutter tool. Standard tool will cut $\frac{3}{32}''$ aluminum, $\frac{1}{16}''$ mild steel, $\frac{1}{32}''$ stainless steel. To cut keyways in metal too thick for this tool, use a small, round file or guided drill.

## TORQUE WITHOUT KEYS

A hexagonal hole with a keyless Rivnut is an excellent substitute for keyed Rivnuts and eliminates the need to match the key to the keyway. This type installation offers torque resistance approximating that of keyed Rivnuts.

## HOW TO REMOVE RIVNUTS

Should it be necessary to remove an installed Rivnut, it can be drilled out by using the same size drill as used for the original hole. Drill through the head of the Rivnut and then punch out the shank. The counterbore will act as a guide for the drill. The same size Rivnut can then be installed in the same hole.

# ADJUSTING ANVIL TO SUIT RIVNUT LENGTH

Rivnut threads may be deformed or stripped if the pull-up stud does not engage all threads in the Rivnut. Speed Headers and all power tools have pull-up studs or anvils which can be adjusted easily to suit the Rivnut blank length. The following sketches demonstrate proper relation between face of anvil and end of pull-up stud.

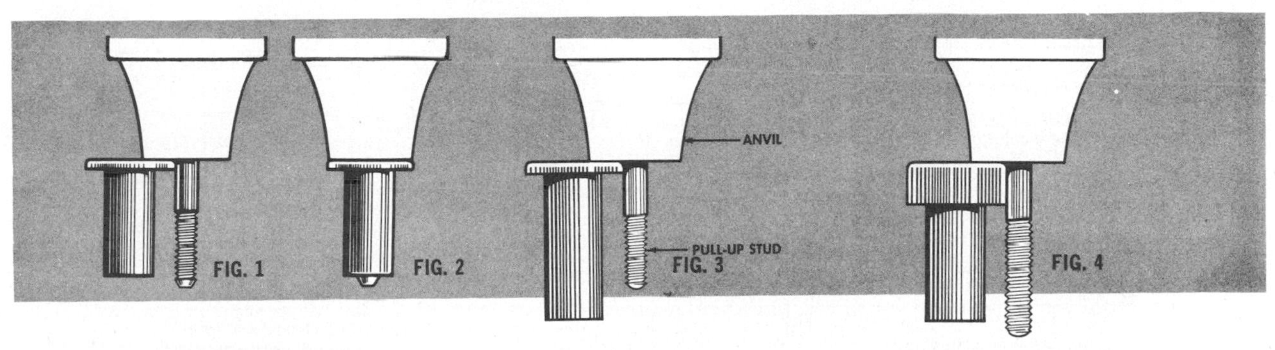


### OPEN END RIVNUT

Point of the pull-up stud should extend just beyond the end of the Rivnut, as in Fig. 1. Fig. 2 shows Rivnut head tight against anvil, ready for upset.

### CLOSED END RIVNUT

Thread Rivnut on pull-up stud all the way to bottom of threads. Back Rivnut off one complete turn, then adjust anvil so it contacts Rivnut head, as in Fig. 3.

### FILLISTER HEAD RIVNUT

Adjust stud to same reference points as conventional types.

# FINDING GRIP RANGE

### GRIP RANGE

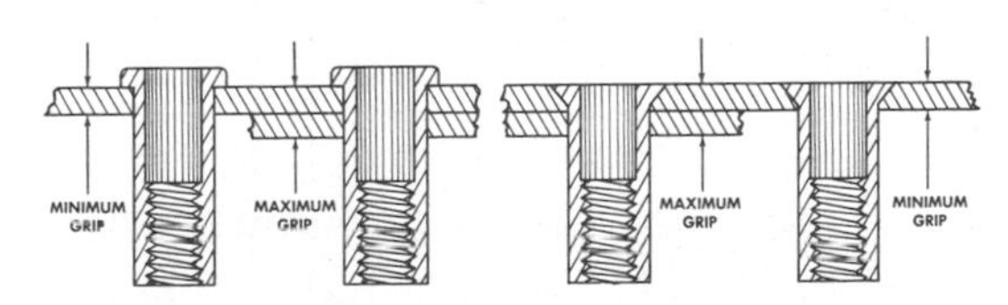


Maximum grip (above) represents the greatest material thickness in which a specific Rivnut should be properly installed. Minimum grip represents material of least thickness in which a specific Rivnut can be properly installed.

"Grip Range" is that area between maximum and minimum —the zone of thickness best suited to the installation of a specific Rivnut.

The grip ranges for various Rivnuts can be found on pages 8 and 9.

IMPORTANT: When material thickness (grip) approaches minimum or maximum for a given size Rivnut, a trial installation should be made.

### MEASURING "GRIP"

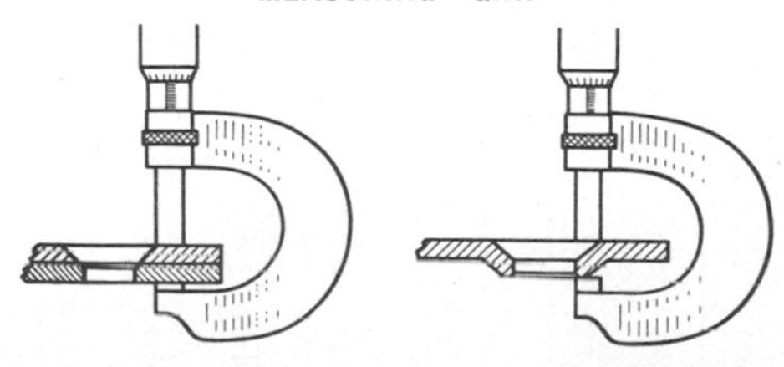

In order to select the correct Rivnut, physical measurements must be made. When installing flat head Rivnuts in a surface installation or countersunk types in machine countersunk holes, "grip" is the same as metal thickness (left above).

For dimpled or press countersunk holes, "grip" is the measurement from the metal surface to the underside of the dimpled hole (right above).

IMPORTANT: Measurements should include air gaps, paint and any burrs which cannot be removed.

# RIVNUT ENGINEERING DATA—FLAT HEAD

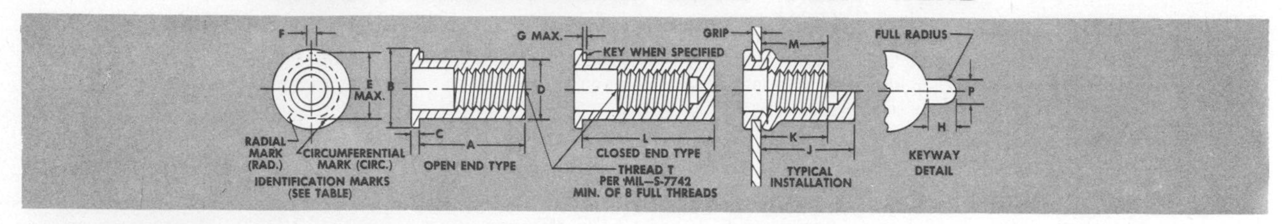


| First No. of Type No. | Thread Size* | B | C | D | E | F | G | Install Drill Size (Ref.) | Install Hole Size | | Keyway Dimensions | |
|---|---|---|---|---|---|---|---|---|---|---|---|---|
| | | ±.015 | Nom. | +.000 −.004 | Max. | +.005 −.000 | Max. | | Min. | Max. | P+.003 −.000 | H |
| 4 | # 4-40 UNC-3B | .270 | .025 | .155 | .198 | .054 | .023 | 5/32 | .155 | .157 | .062 | .046-.048 |
| 6 | # 6-32 UNC-3B | .325 | .032 | .189 | .240 | .054 | .023 | #12 | .189 | .193 | .062 | .056-.058 |
| 8 | # 8-32 UNC-3B | .357 | .032 | .221 | .271 | .054 | .023 | # 2 | .221 | .226 | .062 | .056-.058 |
| 10 | #10-32 UNF-3B | .406 | .038 | .250 | .302 | .054 | .023 | E | .250 | .256 | .062 | .056-.058 |
| 25 | ¼-20 UNC-3B | .475 | .058 | .332 | .382 | .054 | .035 | Q | .332 | .338 | .062 | .056-.058 |
| 31 | 5/16-18 UNC-3B | .665 | .062 | .413 | .505 | .120 | .040 | Z | .413 | .423 | .128 | .097-.102 |
| 37 | 3/8-16 UNC-3B | .781 | .088 | .490 | .597 | .120 | .040 | 12.5 MM | .490 | .500 | .128 | .110-.115 |
| 50 | ½-13 UNC-3B | 1.000 | .125 | .640 | .772 | .151 | .040 | 41/64 | .640 | .650 | .159 | .135-.140 |

*Both UNC and UNF threads are available in No. 10 and larger thread sizes.

CODE: Diameter and grip range as tabulated. First letter of type number indicates material and finish: "A" for aluminum alloy, "S" for C-1108 or C-1110 steel, "SS" for Type 430 corrosion resistant steel, "CH" for heat treated C-4037 steel and "BR" for brass. Letter between dash numbers indicate type "—" for keyless open end, "K" for keyed open end, "B" for keyless closed end, "KB" for keyed closed end.

EXAMPLES: A25K80 = Aluminum alloy keyed open end ¼-20 internal thread .020 to .080 grip range. SS6KB200 = Corrosion resistant steel keyed closed end No. 6-32 internal thread .160 to .200 grip range.

WEIGHTS — For Brass Rivnuts multiply weight of aluminum Rivnuts by 3.31. Weights for "CH" Rivnuts (C-4037 steel) and "SS" Rivnuts (Type 430 corrosion resistant steel ) same as for "S" Rivnuts.

| TYPE NUMBER | GRIP RANGE | IDENT. MARK | OPEN END KEYED AND KEYLESS | | | | CLOSED END KEYLESS | | | | | CLOSED END KEYED | | | | |
|---|---|---|---|---|---|---|---|---|---|---|---|---|---|---|---|---|
| | | | A ±.015 | M REF. | Wt. (LBS./1000) ALUM. | Wt. (LBS./1000) STEEL | L ±.015 | J REF. | K REF. | WT. (LBS./1000) ALUM. | WT. (LBS./1000) STEEL | L ±.015 | J REF. | K REF. | WT. (LBS./1000) ALUM. | WT. (LBS./1000) STEEL |
| 4-60 | .010-.060 | Blank | .345 | .230 | .4 | 1.3 | .500 | .385 | .230 | .6 | 1.9 | .500 | .385 | .230 | .6 | 1.9 |
| 4-85 | .060-.085 | 1-Rad. | .370 | .230 | .4 | 1.4 | .525 | .385 | .230 | .7 | 2.0 | .525 | .385 | .230 | .7 | 2.0 |
| 4-110 | .085-.110 | 2-Rad. | .400 | .230 | .5 | 1.4 | .555 | .390 | .230 | .7 | 2.0 | .555 | .390 | .230 | .7 | 2.0 |
| 4-135 | .110-.135 | 3-Rad. | .425 | .230 | .5 | 1.5 | .580 | .385 | .230 | .7 | 2.1 | .580 | .385 | .230 | .7 | 2.1 |
| 4-160 | .135-.160 | 4-Rad. | .450 | .230 | .5 | 1.5 | .605 | .385 | .230 | .7 | 2.1 | .605 | .385 | .230 | .7 | 2.1 |
| 4-185 | .160-.185 | 5-Rad. | .480 | .230 | .5 | 1.6 | .635 | .385 | .230 | .7 | 2.2 | .635 | .385 | .230 | .7 | 2.2 |
| 6-75 | .010-.075 | 1-Rad. | .438 | .300 | .8 | 2.4 | .625 | .490 | .305 | 1.2 | 3.5 | .750 | .615 | .405 | 1.4 | 4.1 |
| 6-120 | .075-.120 | 3-Rad. | .500 | .315 | .9 | 2.6 | .625 | .440 | .255 | 1.1 | 3.4 | .750 | .565 | .355 | 1.3 | 4.0 |
| 6-160 | .120-.160 | 5-Rad. | .500 | .270 | .9 | 2.6 | .750 | .520 | .260 | 1.3 | 4.0 | .750 | .520 | .310 | 1.3 | 4.0 |
| 6-200 | .160-.200 | 1-Circ. | .562 | .290 | .9 | 2.8 | .750 | .480 | .260 | 1.3 | 3.9 | .750 | .480 | .260 | 1.3 | 3.9 |
| 6-240 | .200-.240 | 2-Circ. | .625 | .310 | 1.0 | 3.0 | .750 | .435 | .260 | 1.3 | 3.8 | .750 | .435 | .260 | 1.3 | 3.8 |
| 6-280 | .240-.280 | 3-Circ. | .687 | .330 | 1.1 | 3.3 | .812 | .455 | .265 | 1.3 | 4.1 | .812 | .455 | .265 | 1.3 | 4.1 |
| 8-75 | .010-.075 | 1-Rad. | .438 | .300 | 1.0 | 3.0 | .625 | .490 | .305 | 1.5 | 4.5 | .750 | .615 | .405 | 1.7 | 5.3 |
| 8-120 | .075-.120 | 3-Rad. | .500 | .315 | 1.1 | 3.3 | .625 | .440 | .255 | 1.4 | 4.4 | .750 | .565 | .355 | 1.7 | 5.2 |
| 8-160 | .120-.160 | 5-Rad. | .500 | .270 | 1.1 | 3.2 | .750 | .520 | .260 | 1.7 | 5.1 | .750 | .520 | .310 | 1.7 | 5.1 |
| 8-200 | .160-.200 | 1-Circ. | .625 | .350 | 1.3 | 3.9 | .750 | .475 | .265 | 1.6 | 5.0 | .750 | .475 | .265 | 1.6 | 5.0 |
| 8-240 | .200-.240 | 2-Circ. | .625 | .305 | 1.2 | 3.8 | .875 | .555 | .310 | 1.9 | 5.6 | .875 | .555 | .310 | 1.9 | 5.6 |
| 8-280 | .240-.280 | 3-Circ. | .687 | .340 | 1.3 | 4.1 | .875 | .530 | .290 | 1.8 | 5.6 | .875 | .530 | .290 | 1.8 | 5.6 |
| 10-80 | .010-.080 | Blank | .531 | .380 | 1.5 | 4.5 | .781 | .630 | .380 | 2.3 | 6.8 | .781 | .630 | .380 | 2.3 | 6.8 |
| 10-130 | .080-.130 | 1-Rad. | .594 | .390 | 1.6 | 4.9 | .843 | .640 | .390 | 2.4 | 7.2 | .843 | .640 | .390 | 2.4 | 7.2 |
| 10-180 | .130-.180 | 2-Rad. | .641 | .390 | 1.7 | 5.1 | .891 | .640 | .390 | 2.4 | 7.4 | .891 | .640 | .390 | 2.4 | 7.4 |
| 10-230 | .180-.230 | 3-Rad. | .703 | .395 | 1.8 | 5.4 | .953 | .645 | .395 | 2.6 | 7.8 | .953 | .645 | .395 | 2.6 | 7.8 |
| 10-280 | .230-.280 | 4-Rad. | .750 | .395 | 1.9 | 5.7 | 1.000 | .645 | .395 | 2.6 | 8.0 | 1.000 | .645 | .395 | 2.6 | 8.0 |
| 10-330 | .280-.330 | 5-Rad. | .797 | .385 | 1.9 | 5.9 | 1.047 | .630 | .385 | 2.7 | 8.2 | 1.047 | .630 | .385 | 2.7 | 8.2 |
| 25-80 | .020-.080 | Blank | .625 | .450 | 3.2 | 9.7 | .937 | .760 | .440 | 4.9 | 15.1 | .937 | .760 | .440 | 5.0 | 15.1 |
| 25-140 | .080-.140 | 1-Rad. | .687 | .450 | 3.4 | 10.3 | 1.000 | .760 | .440 | 5.1 | 15.7 | 1.000 | .760 | .440 | 5.1 | 15.7 |
| 25-200 | .140-.200 | 2-Rad. | .750 | .450 | 3.6 | 10.9 | 1.062 | .760 | .440 | 5.3 | 16.2 | 1.062 | .760 | .440 | 5.3 | 16.3 |
| 25-260 | .200-.260 | 3-Rad. | .812 | .445 | 3.8 | 11.5 | 1.125 | .755 | .445 | 5.5 | 16.8 | 1.125 | .755 | .445 | 5.5 | 16.8 |
| 25-320 | .260-.320 | 4-Rad. | .875 | .445 | 4.0 | 12.0 | 1.187 | .755 | .445 | 5.7 | 17.4 | 1.187 | .755 | .445 | 5.7 | 17.4 |
| 25-380 | .320-.380 | 5-Rad. | .937 | .445 | 4.1 | 12.6 | 1.250 | .755 | .445 | 5.9 | 18.0 | 1.250 | .755 | .445 | 5.9 | 18.0 |
| 31-125 | .030-.125 | Blank | .750 | .505 | 6.0 | 18.2 | 1.187 | .940 | .550 | 9.6 | 29.1 | 1.187 | .940 | .550 | 9.6 | 29.2 |
| 31-200 | .125-.200 | 1-Rad. | .875 | .555 | 6.7 | 20.3 | 1.281 | .960 | .555 | 10.1 | 30.6 | 1.281 | .960 | .555 | 10.1 | 30.7 |
| 31-275 | .200-.275 | 2-Rad. | .937 | .540 | 6.9 | 21.1 | 1.343 | .950 | .560 | 10.3 | 31.4 | 1.343 | .950 | .560 | 10.3 | 31.5 |
| 31-350 | .275-.350 | 3-Rad. | 1.032 | .560 | 7.4 | 22.6 | 1.437 | .965 | .570 | 10.8 | 32.9 | 1.437 | .965 | .570 | 10.8 | 32.9 |
| 31-425 | .350-.425 | 4-Rad. | 1.125 | .580 | 7.9 | 24.0 | 1.531 | .985 | .575 | 11.3 | 34.3 | 1.531 | .985 | .575 | 11.3 | 34.4 |
| 31-500 | .425-.500 | 5-Rad. | 1.187 | .565 | 8.2 | 24.9 | 1.593 | .975 | .580 | 11.5 | 35.1 | 1.593 | .975 | .580 | 11.6 | 35.2 |
| 37-115 | .030-.115 | Blank | .844 | .585 | 9.7 | 29.7 | 1.281 | 1.020 | .660 | 14.8 | 45.0 | 1.281 | 1.020 | .660 | 14.8 | 45.1 |
| 37-200 | .115-.200 | 1-Rad. | .938 | .595 | 10.3 | 31.4 | 1.375 | 1.030 | .670 | 15.4 | 46.8 | 1.375 | 1.030 | .670 | 15.4 | 46.9 |
| 37-285 | .200-.285 | 2-Rad. | 1.031 | .605 | 10.9 | 33.2 | 1.468 | 1.040 | .680 | 15.9 | 48.5 | 1.468 | 1.040 | .680 | 16.0 | 48.6 |
| 37-370 | .285-.370 | 3-Rad. | 1.125 | .615 | 11.5 | 34.9 | 1.562 | 1.050 | .690 | 16.5 | 50.3 | 1.562 | 1.050 | .690 | 16.5 | 50.4 |
| 37-455 | .370-.455 | 4-Rad. | 1.218 | .630 | 12.0 | 36.7 | 1.656 | 1.065 | .710 | 17.1 | 52.1 | 1.656 | 1.065 | .710 | 17.1 | 52.2 |
| 37-540 | .455-.540 | 5-Rad. | 1.312 | .635 | 12.6 | 38.5 | 1.750 | 1.075 | .715 | 17.7 | 53.8 | 1.750 | 1.075 | .715 | 17.7 | 53.9 |
| 50-145 | .025-.145 | Blank | 1.062 | .730 | 20.8 | 63.5 | 1.656 | 1.325 | .855 | 31.9 | 97.3 | 1.656 | 1.325 | .855 | 32.0 | 97.5 |
| 50-265 | .145-.265 | 1-Rad. | 1.188 | .735 | 22.2 | 67.5 | 1.781 | 1.330 | .865 | 33.2 | 101.3 | 1.781 | 1.330 | .865 | 33.3 | 101.4 |
| 50-385 | .265-.385 | 2-Rad. | 1.312 | .740 | 23.4 | 71.4 | 1.906 | 1.335 | .875 | 34.5 | 105.1 | 1.906 | 1.335 | .875 | 34.5 | 105.2 |
| 50-505 | .385-.505 | 3-Rad. | 1.453 | .765 | 24.9 | 76.0 | 2.156 | 1.465 | .895 | 37.8 | 115.1 | 2.156 | 1.465 | .895 | 37.8 | 115.3 |
| 50-625 | .505-.625 | 4-Rad. | 1.578 | .770 | 26.2 | 80.0 | 2.297 | 1.485 | .905 | 39.3 | 119.9 | 2.297 | 1.485 | .905 | 39.4 | 120.0 |
| 50-745 | .625-.745 | 5-Rad. | 1.719 | .790 | 27.8 | 84.6 | 2.438 | 1.510 | .920 | 40.9 | 124.5 | 2.438 | 1.510 | .920 | 40.9 | 124.6 |

# RIVNUT ENGINEERING DATA—COUNTERSUNK HEAD

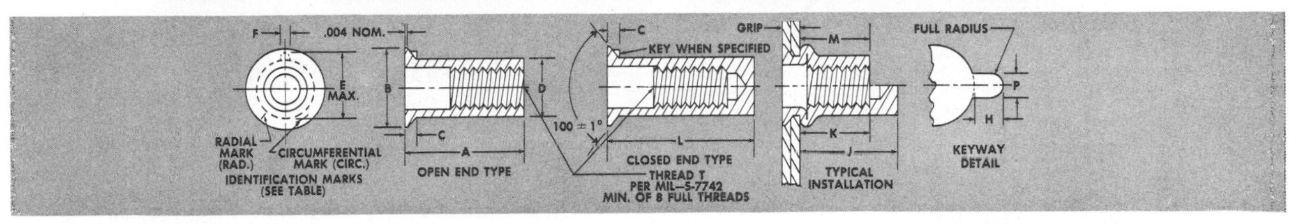


| First No. of Type No. | Thread Size* | B Ref. | C Max. | D +.000 −.004 | E Max. | F +.005 −.000 | Install Drill Size (Ref.) | Install Hole Size Min. | Install Hole Size Max. | Keyway Dimensions P +.003 −.000 | Keyway Dimensions H |
|---|---|---|---|---|---|---|---|---|---|---|---|
| 4 | # 4-40 UNC-3B | .263 | .051 | .155 | .198 | .054 | 5/32 | .155 | .157 | .062 | .046-.048 |
| 6 | # 6-32 UNC-3B | .323 | .063 | .189 | .240 | .054 | #12 | .189 | .193 | .062 | .056-.058 |
| 8 | # 8-32 UNC-3B | .355 | .063 | .221 | .271 | .054 | # 2 | .221 | .226 | .062 | .056-.058 |
| 10 | #10-32 UNF-3B | .391 | .065 | .250 | .302 | .054 | E | .250 | .256 | .062 | .056-.058 |
| 25 | 1/4-20 UNC-3B | .529 | .089 | .332 | .382 | .054 | Q | .332 | .338 | .062 | .056-.058 |
| 31 | 5/16-18 UNC-3B | .656 | .104 | .413 | .505 | .120 | Z | .413 | .423 | .128 | .097-.102 |
| 37 | 3/8-16 UNC-3B | .770 | .124 | .490 | .597 | .120 | 12.5 MM | .490 | .500 | .128 | .110-.115 |
| 50 | 1/2-13 UNC-3B | .990 | .154 | .640 | .772 | .151 | 41/64 | .640 | .650 | .159 | .135-.140 |

*Both UNC and UNF threads are available in No. 10 and larger thread sizes.

CODE: Diameter and grip range as tabulated. First letter of type number indicates material and finish: "A" for aluminum alloy, "S" for C-1108 or C-1110 steel, "SS" for Type 430 corrosion resistant steel, "CH" for heat treated C-4037 steel and "BR" for brass. Letter between dash numbers indicate type "—" for keyless open end, "K" for keyed open end, "B" for keyless closed end, "KB" for keyed closed end.

EXAMPLES: A25K151 = Aluminum alloy keyed open end 1/4-20 internal thread .089 to .151 grip range. SS6KB241 = Corrosion resistant steel keyed closed end No. 6-32 internal thread .201 to .241 grip range.

WEIGHTS — For Brass Rivnuts multiply weight of aluminum Rivnuts by 3.31. Weights for "CH" Rivnuts (C-4037 steel) and "SS" Rivnuts (Type 430 corrosion resistant steel) same as for "S" Rivnuts.

| TYPE NUMBER | GRIP RANGE | IDENT. MARK | OPEN END KEYED AND KEYLESS | | | | CLOSED END KEYLESS | | | | | CLOSED END KEYED | | | | |
|---|---|---|---|---|---|---|---|---|---|---|---|---|---|---|---|---|
| | | | A ±.015 | M REF. | Wt. (LBS./1000) ALUM. | Wt. (LBS./1000) STEEL | L ±.015 | J REF. | K REF. | WT. (LBS./1000) ALUM. | WT. (LBS./1000) STEEL | L ±.015 | J REF. | K REF. | WT. (LBS./1000) ALUM. | WT. (LBS./1000) STEEL |
| 4-81 | .050-.081 | Blank | .370 | .235 | .4 | 1.3 | .525 | .390 | .235 | .6 | 1.9 | .525 | .390 | .235 | .6 | 1.9 |
| 4-106 | .081-.106 | 1-Rad. | .395 | .235 | .4 | 1.3 | .550 | .390 | .235 | .6 | 1.9 | .550 | .390 | .235 | .6 | 1.9 |
| 4-131 | .106-.131 | 2-Rad. | .420 | .235 | .4 | 1.4 | .575 | .390 | .235 | .7 | 2.0 | .575 | .390 | .235 | .7 | 2.0 |
| 4-156 | .131-.156 | 3-Rad. | .450 | .235 | .5 | 1.4 | .600 | .390 | .235 | .7 | 2.0 | .600 | .390 | .235 | .7 | 2.0 |
| 4-181 | .156-.181 | 4-Rad. | .475 | .235 | .5 | 1.5 | .625 | .390 | .235 | .7 | 2.1 | .625 | .390 | .235 | .7 | 2.1 |
| 4-206 | .181-.206 | 5-Rad. | .500 | .235 | .5 | 1.5 | .650 | .390 | .235 | .7 | 2.1 | .650 | .390 | .235 | .7 | 2.1 |
| 6-106 | .065-.106 | Blank | .500 | .325 | .8 | 2.5 | .687 | .510 | .325 | 1.2 | 3.6 | .812 | .635 | .425 | 1.4 | 4.2 |
| 6-161 | .106-.161 | 2-Rad. | .500 | .280 | .8 | 2.4 | .687 | .465 | .280 | 1.2 | 3.5 | .812 | .590 | .380 | 1.3 | 4.1 |
| 6-201 | .161-.201 | 4-Rad. | .562 | .295 | .9 | 2.6 | .687 | .420 | .260 | 1.1 | 3.4 | .812 | .545 | .335 | 1.3 | 4.0 |
| 6-241 | .201-.241 | 1-Circ. | .625 | .315 | .9 | 2.9 | .812 | .505 | .295 | 1.3 | 4.0 | .812 | .505 | .295 | 1.3 | 4.0 |
| 6-281 | .241-.281 | 2-Circ. | .625 | .270 | .9 | 2.8 | .812 | .465 | .265 | 1.3 | 3.9 | .812 | .465 | .265 | 1.3 | 3.9 |
| 6-321 | .281-.321 | 3-Circ. | .687 | .290 | 1.0 | 3.0 | .844 | .455 | .265 | 1.3 | 4.0 | .844 | .455 | .265 | 1.3 | 4.0 |
| 8-106 | .065-.106 | Blank | .500 | .325 | 1.0 | 3.1 | .687 | .510 | .325 | 1.5 | 4.6 | .812 | .635 | .425 | 1.8 | 5.4 |
| 8-161 | .106-.161 | 2-Rad. | .500 | .280 | 1.0 | 3.0 | .687 | .465 | .280 | 1.5 | 4.5 | .812 | .590 | .380 | 1.7 | 5.3 |
| 8-201 | .161-.201 | 4-Rad. | .562 | .290 | 1.1 | 3.3 | .687 | .415 | .255 | 1.4 | 4.4 | .812 | .540 | .330 | 1.7 | 5.2 |
| 8-241 | .201-.241 | 1-Circ. | .625 | .310 | 1.2 | 3.6 | .875 | .560 | .290 | 1.8 | 5.5 | .875 | .560 | .290 | 1.8 | 5.5 |
| 8-281 | .241-.281 | 2-Circ. | .687 | .325 | 1.1 | 3.2 | .875 | .515 | .290 | 1.8 | 5.4 | .875 | .515 | .290 | 1.8 | 5.4 |
| 8-321 | .281-.321 | 3-Circ. | .687 | .295 | 1.2 | 3.8 | .875 | .485 | .300 | 1.7 | 5.2 | .875 | .485 | .300 | 1.7 | 5.2 |
| 10-116 | .065-.116 | Blank | .578 | .395 | 1.4 | 4.3 | .828 | .645 | .395 | 2.2 | 6.7 | .828 | .645 | .395 | 2.2 | 6.7 |
| 10-166 | .116-.166 | 1-Rad. | .625 | .385 | 1.5 | 4.6 | .875 | .635 | .385 | 2.3 | 6.9 | .875 | .635 | .385 | 2.3 | 6.9 |
| 10-216 | .166-.216 | 2-Rad. | .687 | .400 | 1.6 | 4.9 | .938 | .650 | .400 | 2.4 | 7.2 | .938 | .650 | .400 | 2.4 | 7.2 |
| 10-266 | .216-.266 | 3-Rad. | .734 | .390 | 1.7 | 5.1 | .984 | .640 | .390 | 2.5 | 7.5 | .984 | .640 | .390 | 2.5 | 7.5 |
| 10-316 | .266-.316 | 4-Rad. | .781 | .385 | 1.8 | 5.4 | 1.031 | .635 | .385 | 2.5 | 7.7 | 1.031 | .635 | .385 | 2.5 | 7.7 |
| 10-366 | .316-.366 | 5-Rad. | .844 | .400 | 1.9 | 5.7 | 1.094 | .650 | .400 | 2.6 | 8.0 | 1.094 | .650 | .400 | 2.6 | 8.0 |
| 25-151 | .089-.151 | Blank | .687 | .440 | 3.2 | 9.8 | 1.000 | .750 | .435 | 5.0 | 15.1 | 1.000 | .750 | .435 | 5.0 | 15.1 |
| 25-211 | .151-.211 | 1-Rad. | .750 | .440 | 3.4 | 10.3 | 1.062 | .750 | .435 | 5.2 | 15.7 | 1.062 | .750 | .435 | 5.2 | 15.7 |
| 25-271 | .211-.271 | 2-Rad. | .812 | .440 | 3.6 | 10.9 | 1.125 | .750 | .435 | 5.4 | 16.3 | 1.125 | .750 | .435 | 5.4 | 16.3 |
| 25-331 | .271-.331 | 3-Rad. | .875 | .435 | 3.8 | 11.5 | 1.187 | .750 | .435 | 5.5 | 16.9 | 1.187 | .750 | .435 | 5.5 | 16.9 |
| 25-391 | .331-.391 | 4-Rad. | .937 | .435 | 4.0 | 12.1 | 1.250 | .750 | .435 | 5.7 | 17.5 | 1.250 | .750 | .435 | 5.7 | 17.5 |
| 25-451 | .391-.451 | 5-Rad. | 1.000 | .445 | 4.2 | 12.7 | 1.312 | .760 | .445 | 5.9 | 18.1 | 1.312 | .760 | .445 | 5.9 | 18.1 |
| 31-181 | .106-.181 | Blank | .844 | .540 | 5.9 | 17.8 | 1.218 | .915 | .540 | 9.0 | 27.5 | 1.218 | .915 | .540 | 9.0 | 27.5 |
| 31-256 | .181-.256 | 1-Rad. | .937 | .560 | 6.3 | 19.3 | 1.312 | .935 | .560 | 9.5 | 28.9 | 1.312 | .935 | .560 | 9.5 | 29.0 |
| 31-331 | .256-.331 | 2-Rad. | 1.000 | .550 | 6.6 | 20.1 | 1.406 | .955 | .550 | 10.0 | 30.4 | 1.406 | .955 | .550 | 10.0 | 30.5 |
| 31-406 | .331-.406 | 3-Rad. | 1.093 | .565 | 7.1 | 21.5 | 1.468 | .940 | .565 | 10.2 | 31.1 | 1.468 | .940 | .565 | 10.2 | 31.2 |
| 31-481 | .406-.481 | 4-Rad. | 1.156 | .555 | 7.3 | 22.3 | 1.562 | .960 | .555 | 10.7 | 32.6 | 1.562 | .960 | .555 | 10.8 | 32.7 |
| 31-556 | .481-.556 | 5-Rad. | 1.250 | .575 | 7.8 | 23.7 | 1.625 | .950 | .575 | 10.9 | 33.3 | 1.625 | .950 | .575 | 11.0 | 33.4 |
| 37-211 | .125-.211 | Blank | .938 | .580 | 8.9 | 27.0 | 1.375 | 1.020 | .655 | 13.9 | 42.3 | 1.375 | 1.020 | .655 | 13.9 | 42.4 |
| 37-296 | .211-.296 | 1-Rad. | 1.031 | .590 | 9.4 | 28.7 | 1.468 | 1.030 | .655 | 14.5 | 44.1 | 1.468 | 1.030 | .655 | 14.5 | 44.1 |
| 37-381 | .296-.381 | 2-Rad. | 1.125 | .600 | 10.0 | 30.5 | 1.562 | 1.040 | .675 | 15.0 | 45.8 | 1.562 | 1.040 | .675 | 15.1 | 45.9 |
| 37-466 | .381-.466 | 3-Rad. | 1.219 | .615 | 10.6 | 32.3 | 1.656 | 1.050 | .690 | 15.6 | 47.6 | 1.656 | 1.050 | .690 | 15.7 | 47.7 |
| 37-551 | .466-.551 | 4-Rad. | 1.312 | .625 | 11.2 | 34.0 | 1.750 | 1.065 | .705 | 16.2 | 49.4 | 1.750 | 1.065 | .705 | 16.2 | 49.5 |
| 37-636 | .551-.636 | 5-Rad. | 1.422 | .650 | 11.9 | 36.2 | 1.859 | 1.090 | .715 | 16.9 | 51.6 | 1.859 | 1.090 | .715 | 17.0 | 51.7 |
| 50-276 | .156-.276 | Blank | 1.188 | .725 | 18.4 | 56.1 | 1.781 | 1.320 | .850 | 29.5 | 89.9 | 1.781 | 1.320 | .850 | 29.5 | 90.0 |
| 50-396 | .276-.396 | 1-Rad. | 1.312 | .730 | 19.7 | 60.0 | 1.906 | 1.325 | .865 | 30.8 | 93.7 | 1.906 | 1.325 | .865 | 30.8 | 93.9 |
| 50-516 | .396-.516 | 2-Rad. | 1.438 | .735 | 21.0 | 63.9 | 2.031 | 1.330 | .880 | 32.0 | 97.6 | 2.031 | 1.330 | .880 | 32.1 | 97.7 |
| 50-636 | .516-.636 | 3-Rad. | 1.578 | .750 | 22.5 | 68.5 | 2.172 | 1.350 | .890 | 33.6 | 102.3 | 2.172 | 1.350 | .890 | 33.6 | 102.4 |
| 50-756 | .636-.756 | 4-Rad. | 1.718 | .765 | 24.0 | 73.1 | 2.312 | 1.360 | .900 | 35.1 | 106.9 | 2.312 | 1.360 | .900 | 35.1 | 107.0 |
| 50-876 | .756-.876 | 5-Rad. | 1.843 | .780 | 25.3 | 77.1 | 2.437 | 1.375 | .930 | 36.3 | 110.8 | 2.437 | 1.375 | .930 | 36.4 | 110.9 |

# GENERAL DESIGN

## for Utilizing the unique,

**DOUBLE SHEAR STRENGTH** will be the determining factor in specifying the Rollpin size for a given stress in numerous applications. These values, for carbon steel and corrosion resistant steel Rollpins, are given in the table on Pages 14 and 16.

In soft metals ultimate failure depends more upon the strength of the hole material than upon shear strength of the Rollpin. Under increased load, hole edges are deformed by compressive forces. At ultimate loads the soft material may be expected to fail before the Rollpin.

In many cases, particularly with smaller mechanisms, the choice of Rollpin sizes will be influenced by space considerations and by general proportions of related members. Recommended practices are given in the accompanying tables.

As with other types of pins, good Rollpin design practice is to avoid conditions where the direction of vibration parallels the axis of the pin. For most applications the Rollpin's capacity to withstand the loosening effects of vibration can be consistently relied upon; however, unusual conditions can result in an axial vibration component that should be carefully evaluated. In such instances accelerated vibration tests should be made of prototypes of the proposed design. ESNA will be glad to provide sample Rollpins for such tests.

### EFFECT OF GAP ORIENTATION ON ROLLPIN SHEAR STRENGTH

LOAD

90°

Data obtained in tests made to determine the effect of gap orientation on Rollpin shear strength revealed that random slot positioning is satisfactory for all but the most extreme Rollpin applications. For special applications maximum shear strength is assured by inserting the Rollpin so the gap is in line with the direction of the load.

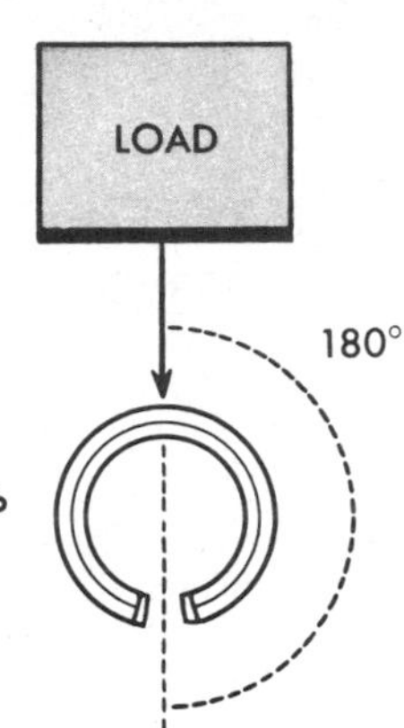


**SHEAR VALUE — 100%**

**SHEAR VALUE — 106%**

Double shear strength figures listed on Page 14 are based on tests made in accordance with Spec. MIL-P-10971 with pins in 90° position fastened in hardened steel fixtures.

### RECOMMENDED ROLLPIN SIZES FOR VARIOUS SHAFT DIAMETERS

**When used as a Transverse Pin**

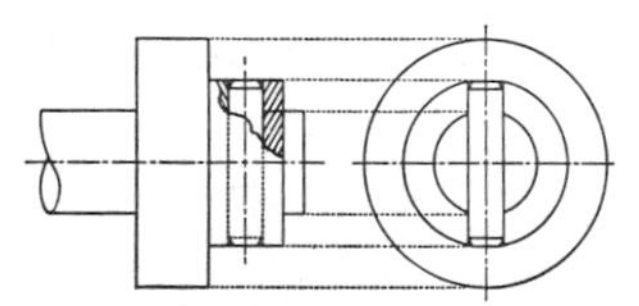

| D Shaft Diameter Inches | d Rollpin Nominal Diameter, Inches | Rollpin Size Number |
|---|---|---|
| 3/16 | 1/16 | 062 |
| 7/32 | 5/64 | 078 |
| 1/4 | 3/32 | 094 |
| 5/16-3/8 | 1/8 | 125 |
| 7/16-1/2 | 5/32 | 156 |
| 9/16 & 5/8 | 3/16 | 187 |
| 11/16 & 3/4 | 7/32 | 219 |
| 13/16 | 1/4 | 250 |
| 7/8 | 5/16 | 312 |
| 1 | 3/8 | 375 |
| 1-1/4 | 7/16 | 437 |
| 1-1/2 | 1/2 | 500 |

**When used as a Longitudinal Key**

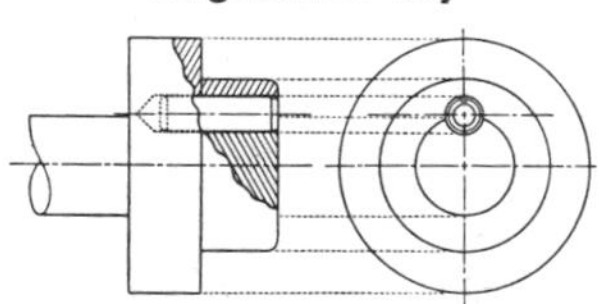

| D Shaft Diameter, Inches | d Rollpin Nominal Diameter, Inches | Rollpin Size Number |
|---|---|---|
| 1/4 | 1/16 | 062 |
| 5/16 | 5/64 | 078 |
| 3/8 | 3/32 | 094 |
| 7/16-1/2 | 1/8 | 125 |
| 9/16-5/8 | 5/32 | 156 |
| 11/16-3/4 | 3/16 | 187 |
| 7/8 | 7/32 | 219 |
| 15/16-1 | 1/4 | 250 |
| 1-1/4 | 5/16 | 312 |
| 1-3/8 | 3/8 | 375 |
| 1-1/2 - 1-5/8 | 7/16 | 437 |
| 2 | 1/2 | 500 |

**When used as a Replacement for Taper Pins**

### EQUIVALENT TAPER PINS

| Taper Pin Number | Equivalent Rollpin | |
|---|---|---|
| | Nominal Diameter, Inches | Rollpin Size Number |
| 7/0 | 1/16 | 062 |
| 6/0 | 5/64 | 078 |
| 5/0 | 3/32 | 094 |
| 4/0 | 1/8 | 125 |
| 3/0 | 1/8 | 125 |
| 2/0 | 5/32 | 156 |
| 0 | 5/32 | 156 |
| 1 | 3/16 | 187 |
| 2 | 3/16 | 187 |
| 3 | 7/32 | 219 |
| 4 | 1/4 | 250 |
| 5 | 5/16 | 312 |
| 6 | 3/8 | 375 |
| 7 | 7/16 | 437 |
| 8 | 1/2 | 500 |

# PRINCIPLES

## inherent ROLLPIN advantages

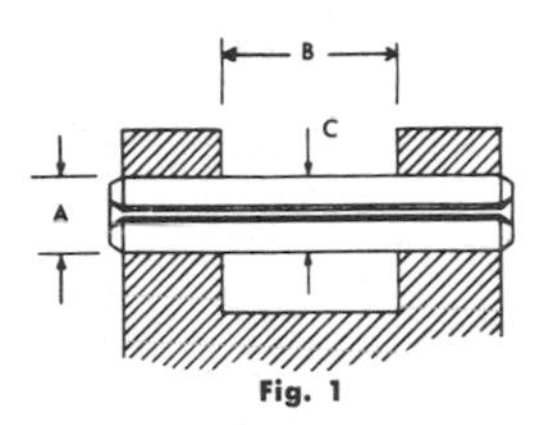


Fig. 1

Because of the inherent flexibility of the Rollpin it has found wide acceptance as a clevis joint pin. It is recommended, for greater bearing area, that the Rollpin be held by the outer members of the clevis (Fig. 1). However, the design may require that the inner member of the clevis (Fig. 2) be used to retain the pin. This alternate technique will also provide satisfactory performance results. The table below gives the average Rollpin spring back data relative to the span length.

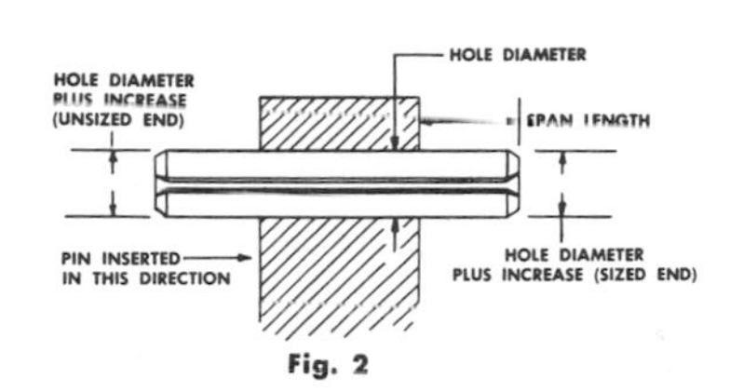


Fig. 2

BLACK — Minimum Hole | Fig. 1 | RED — Maximum Hole

| NOMINAL ROLLPIN DIAMETER | (A) HOLE DIAMETER | C — Rollpin Diameter at Center of Clevis Span —Thousandths<br>B — Span Length — Inches | | | | | | | | | | | | | |
|---|---|---|---|---|---|---|---|---|---|---|---|---|---|---|---|
| | | ⅛ | ¼ | ⅜ | ½ | ⅝ | ¾ | ⅞ | 1 | 1¼ | 1½ | 1¾ | 2 | 2½ | 3 |
| .062 | .062 | .063 | .065 | .065 | .065 | .065 | | | | | | | | | |
| | .065 | .066 | .067 | .067 | .067 | .067 | | | | | | | | | |
| .078 | .078 | .079 | .079 | .079 | .079 | .079 | .079 | | | | | | | | |
| | .081 | .083 | .084 | .084 | .084 | .084 | .084 | | | | | | | | |
| .094 | .094 | .096 | .097 | .098 | .098 | .098 | .098 | | | | | | | | |
| | .097 | .097 | .099 | .099 | .099 | .099 | .099 | | | | | | | | |
| .125 | .125 | .127 | .129 | .129 | .130 | .130 | .130 | .131 | .131 | | | | | | |
| | .129 | .129 | .130 | .130 | .130 | .131 | .131 | .131 | .131 | | | | | | |
| .156 | .156 | .158 | .159 | .160 | .161 | .162 | .162 | .163 | .163 | .163 | .163 | | | | |
| | .160 | .160 | .161 | .161 | .162 | .162 | .162 | .163 | .163 | .163 | .163 | | | | |
| .187 | .187 | .188 | .189 | .190 | .191 | .192 | .193 | .194 | .195 | .195 | .195 | .195 | .195 | | |
| | .192 | .192 | .193 | .193 | .194 | .194 | .194 | .195 | .195 | .195 | .195 | .195 | .195 | | |
| .219 | .219 | .220 | .221 | .222 | .223 | .223 | .224 | .225 | .225 | .226 | .226 | .226 | .226 | | |
| | .224 | .224 | .225 | .225 | .225 | .225 | .226 | .226 | .226 | .226 | .226 | .226 | .226 | | |
| .250 | .250 | .252 | .254 | .255 | .256 | .256 | .257 | .258 | .258 | .259 | .260 | .260 | .260 | | |
| | .256 | .257 | .258 | .258 | .259 | .259 | .260 | .260 | .260 | .260 | .260 | .260 | .260 | | |
| .312 | .312 | .313 | .313 | .313 | .314 | .315 | .315 | .316 | .317 | .317 | .318 | .319 | .320 | .321 | .321 |
| | .318 | .319 | .319 | .319 | .319 | .319 | .319 | .320 | .320 | .320 | .320 | .320 | .321 | .321 | .321 |
| .375 | .375 | .376 | .376 | .376 | .376 | .377 | .378 | .379 | .380 | .382 | .383 | .384 | .384 | .385 | .385 |
| | .382 | .383 | .383 | .383 | .383 | .383 | .383 | .384 | .384 | .384 | .384 | .385 | .385 | .385 | .385 |
| .500 | .500 | .501 | .502 | .502 | .503 | .504 | .504 | .505 | .506 | .507 | .508 | .510 | .512 | .514 | .514 |
| | .510 | .511 | .511 | .511 | .512 | .512 | .512 | .512 | .513 | .513 | .513 | .514 | .514 | .514 | .514 |

Black-unsized end | Fig. 2 | Red-Sized end

| | | ⅛ | | ¼ | | ⅜ | | ½ | | ⅝ | | ¾ | | ⅞ | | 1 | | 1¼ | | 1½ | | 2 | |
|---|---|---|---|---|---|---|---|---|---|---|---|---|---|---|---|---|---|---|---|---|---|---|---|
| .062 | .062 | .066 | .063 | .068 | .064 | .068 | .064 | .068 | .064 | .068 | .064 | .068 | .064 | | | | | | | | | | |
| | .065 | .067 | .066 | .067 | .067 | .067 | .067 | .067 | .067 | .067 | .067 | .067 | .067 | | | | | | | | | | |
| .078 | .078 | .081 | .080 | .083 | .080 | .083 | .080 | .083 | .080 | .083 | .080 | .083 | .080 | | | | | | | | | | |
| | .081 | .083 | .083 | .084 | .083 | .084 | .083 | .084 | .083 | .084 | .083 | .084 | .083 | | | | | | | | | | |
| .094 | .094 | .098 | .095 | .100 | .097 | .101 | .098 | .101 | .098 | .101 | .098 | .101 | .098 | | | | | | | | | | |
| | .097 | .099 | .098 | .101 | .099 | .101 | .099 | .101 | .099 | .101 | .099 | .101 | .099 | | | | | | | | | | |
| .125 | .125 | .128 | .126 | .130 | .127 | .131 | .127 | .131 | .128 | .132 | .128 | .132 | .128 | | | | | | | | | | |
| | .129 | .130 | .130 | .131 | .131 | .131 | .131 | .132 | .132 | .132 | .132 | .132 | .132 | | | | | | | | | | |
| .156 | .156 | .160 | .159 | .161 | .159 | .161 | .160 | .161 | .160 | .162 | .160 | .162 | .161 | .163 | .161 | .163 | .161 | | | | | | |
| | .160 | .162 | .162 | .163 | .163 | .163 | .163 | .163 | .163 | .163 | .163 | .163 | .163 | .163 | .163 | .163 | .163 | | | | | | |
| .187 | .187 | .190 | .188 | .192 | .189 | .194 | .190 | .195 | .191 | .195 | .191 | .195 | .191 | .195 | .191 | .195 | .191 | | | | | | |
| | .192 | .194 | .193 | .195 | .194 | .195 | .194 | .195 | .195 | .195 | .195 | .195 | .195 | .195 | .195 | .195 | .195 | | | | | | |
| .219 | .219 | .223 | .221 | .225 | .222 | .226 | .224 | .227 | .225 | .227 | .225 | .228 | .226 | .228 | .226 | .228 | .226 | | | | | | |
| | .224 | .225 | .225 | .226 | .226 | .227 | .226 | .227 | .227 | .227 | .227 | .227 | .227 | .227 | .227 | .227 | .227 | | | | | | |
| .250 | .250 | .252 | .251 | .253 | .252 | .255 | .253 | .257 | .254 | .258 | .255 | .258 | .255 | .258 | .256 | .259 | .256 | .259 | .256 | .259 | .256 | .259 | |
| | .256 | .257 | .257 | .258 | .258 | .259 | .259 | .259 | .259 | .260 | .259 | .260 | .260 | .260 | .260 | .260 | .260 | .260 | .260 | .260 | .260 | .260 | |
| .312 | .312 | .316 | .314 | .319 | .316 | .320 | .318 | .322 | .319 | .323 | .320 | .324 | .321 | .325 | .321 | .325 | .321 | .325 | .321 | .325 | .321 | .325 | |
| | .318 | .320 | .319 | .322 | .320 | .323 | .321 | .324 | .321 | .324 | .322 | .325 | .322 | .325 | .322 | .325 | .322 | .325 | .322 | .325 | .322 | .325 | |
| .375 | .375 | .378 | .376 | .380 | .378 | .384 | .380 | .385 | .381 | .386 | .383 | .387 | .384 | .388 | .385 | .388 | .385 | .389 | .385 | .389 | .385 | .389 | .385 |
| | .382 | .384 | .383 | .385 | .384 | .386 | .385 | .386 | .385 | .387 | .386 | .387 | .386 | .388 | .387 | .388 | .387 | .389 | .387 | .389 | .387 | .389 | .387 |
| .500 | .500 | .502 | .502 | .503 | .502 | .505 | .504 | .506 | .505 | .508 | .506 | .509 | .507 | .511 | .508 | .512 | 509 | .513 | .510 | .514 | .510 | .515 | .511 |
| | .510 | .511 | .511 | .512 | .512 | .513 | .513 | .514 | .514 | .514 | .514 | .514 | .514 | .515 | .515 | .515 | .515 | .515 | .515 | .515 | .515 | .515 | .515 |

ESNA FASTENER DIVISION

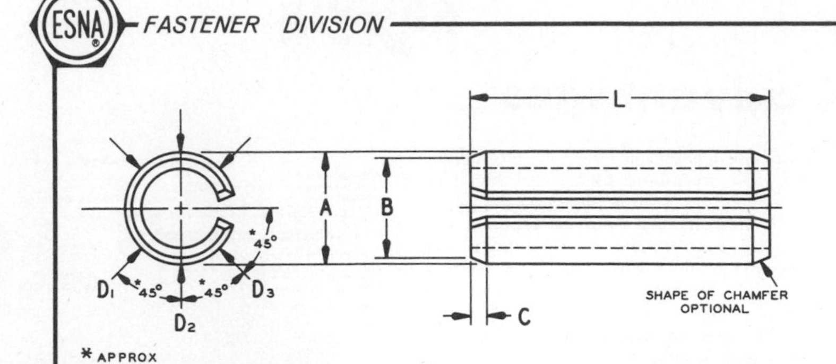


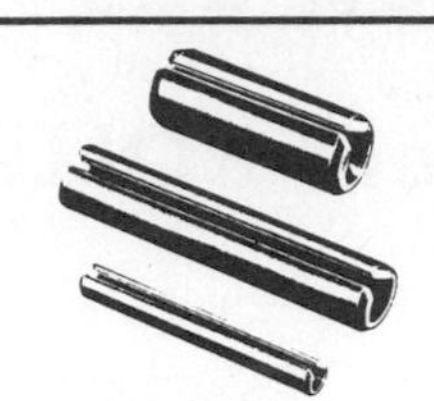

TOLERANCE ON SPECIFIED LENGTH "L"
0.187 TO 1.000 ±.015
1.001 TO 2.000 ±.020
2.001 TO 3.000 ±.025
3.001 TO 4.000 ±.030
4.001 & ABOVE ±.035

| A | | | B | C | | STOCK THICKNESS | RECOMMENDED HOLE SIZE | | MINIMUM DOUBLE SHEAR STRENGTH POUNDS |
|---|---|---|---|---|---|---|---|---|---|
| NOMINAL | MAXIMUM (GO RING GAGE) | MINIMUM $1/3(D_1+D_2+D_3)$ | MAX | MIN | MAX | | MIN | MAX | |
| .062 | .069 | .066 | .059 | .007 | .028 | .012 | .062 | .065 | 425 |
| .078 | .086 | .083 | .075 | .008 | .032 | .018 | .078 | .081 | 650 |
| .094 | .103 | .099 | .091 | .008 | .038 | .022 | .094 | .097 | 1,000 |
| .125 | .135 | .131 | .122 | .008 | .044 | .028 | .125 | .129 | 2,100 |
| .140 | .149 | .145 | .137 | .008 | .044 | .028 | .140 | .144 | 2,200 |
| .156 | .167 | .162 | .151 | .010 | .048 | .032 | .156 | .160 | 3,000 |
| .187 | .199 | .194 | .182 | .011 | .055 | .040 | .187 | .192 | 4,400 |
| .219 | .232 | .226 | .214 | .011 | .065 | .048 | .219 | .224 | 5,700 |
| .250 | .264 | .258 | .245 | .012 | .065 | .048 | .250 | .256 | 7,700 |
| .312 | .328 | .321 | .306 | .014 | .080 | .062 | .312 | .318 | 11,500 |
| .375 | .392 | .385 | .368 | .016 | .095 | .077 | .375 | .382 | 17,600 |
| .437 | .456 | .448 | .430 | .017 | .095 | .077 | .437 | .445 | 20,000 |
| .500 | .521 | .513 | .485 | .025 | .110 | .094 | .500 | .510 | 25,800 |
| | | | | | | | | | |
| | | | | | | | | | |

CODE: PARTS ARE IDENTIFIED BY A SERIES OF DASH NUMBERS IN ACCORDANCE WITH THE EXAMPLE SHOWN BELOW.

EXAMPLE: 5 9 - 028 - 125 - 0750 = ROLLPIN, .125 DIA., .023 WALL, 3/4 LONG, PLAIN CARBON STEEL.

THIRD DASH NUMBER INDICATES LENGTH IN THOUSANDTHS OF AN INCH. PREFIX "O" IF LESS THAN ONE INCH.

SECOND DASH NUMBER INDICATES NOMINAL DIAMETER IN THOUSANDTHS OF AN INCH AND IS DETERMINED FROM "A-NOMINAL" COLUMN IN TABULATION ABOVE.

FIRST DASH NUMBER INDICATES WALL THICKNESS IN THOUSANDTHS OF AN INCH AND IS DETERMINED FROM "STOCK THICKNESS" COLUMN IN TABULATION ABOVE.

SECOND DIGIT INDICATES FINISH:
9 = PLAIN.
(7) 3 = PHOSPHATE COATING, MIL SPEC MIL-C-16232(N Ord) TYPE Z.
2 = CADMIUM PLATE, FEDERAL SPEC QQ-P-416, TYPE 1, CLASS 3.
1 = ZINC PLATE, FEDERAL SPEC QQ-Z-325, TYPE 1, CLASS 3.

FIRST DIGIT INDICATES MATERIAL:
5 = HIGH CARBON STEEL.

FINISH: CARBON STEEL PINS ARE NORMALLY FURNISHED WITH A BLACK OILED FINISH.

HARDNESS: ROCKWELL "C" 46-53 OR EQUIVALENT.

ISSUED: 20 APR 50 REVISED: (7) 9 JAN 62

| PERFORMANCE SPECIFICATION | ROLLPIN | RP-A |
|---|---|---|
| ESNA SPEC 401 (SEE NOTE: 1) | HIGH CARBON STEEL | PAGE 1 OF 2 |

*FASTENER DIVISION*

NOTE 1. CARBON STEEL ROLLPINS MEET THE REQUIREMENTS OF THE FOLLOWING:
SPECIFICATIONS: MIL-P-10971, AMS 7205.
STANDARDS: ARMY ORDNANCE DRAWINGS BFSX2 AND BFSX2.1
NAVY DRAWINGS NAVORD OSTD 600, PAGES 5-6.11 TO 5-6.13
MS9047(ASG), MS9048(ASG), MS16562, NAS561

PLAIN CARBON STEEL ROLLPIN AVAILABILITY

NOMINAL DIAMETER

| LENGTH | .062 | .078 | .094 | .125 | .140 | .156 | .187 | .219 | .250 | .312 | .375 | .437 | .500 |
|---|---|---|---|---|---|---|---|---|---|---|---|---|---|
| 0.125" | * | | | | | | | | | | | | |
| 0.187 | * | * | * | | | | | | | | | | |
| 0.250 | * | * | * | * | Δ | * | | | | | | | |
| 0.312 | * | * | * | * | Δ | * | * | | | | | | |
| 0.375 | * | * | * | * | Δ | * | * | * | * | | | | |
| 0.437 | * | * | * | * | Δ | * | * | * | * | * | | | |
| 0.500 | * | * | * | * | Δ | * | * | * | * | * | * | | |
| 0.562 | * | * | * | * | Δ | * | * | * | * | * | * | | |
| 0.625 | * | * | * | * | Δ | * | * | * | * | Δ | * | | |
| 0.687 | * | * | * | * | Δ | * | * | * | * | Δ | Δ | | |
| 0.750 | * | * | * | * | * | * | * | * | * | * | * | Δ | * |
| 0.812 | * | * | * | * | * | * | * | * | * | * | Δ | Δ | Δ |
| 0.875 | * | * | * | * | * | * | * | * | * | * | * | Δ | * |
| 0.937 | * | * | * | * | * | * | * | * | * | * | * | Δ | Δ |
| 1.000 | * | * | * | * | * | * | * | * | * | * | * | * | * |
| 1.125 | * | * | * | * | Δ | * | * | * | * | * | * | Δ | * |
| 1.250 | Δ | * | * | * | * | * | * | * | * | * | * | * | * |
| 1.375 | | * | * | * | * | * | * | * | * | * | * | Δ | * |
| 1.500 | | * | * | * | Δ | * | * | * | * | * | * | * | * |
| 1.625 | | Δ | Δ | * | Δ | * | * | * | * | * | * | Δ | * |
| 1.750 | | Δ | Δ | * | Δ | * | * | * | * | * | * | * | * |
| 1.875 | | Δ | Δ | Δ | Δ | * | * | * | * | * | * | Δ | * |
| 2.000 | | Δ | Δ | * | Δ | * | * | * | * | * | * | * | * |
| 2.250 | | | | * | Δ | * | * | * | * | * | * | * | * |
| 2.500 | | | | Δ | Δ | * | * | * | * | * | * | * | * |
| 2.750 | | | | | | * | Δ | Δ | * | * | * | * | * |
| 3.000 | | | | | | Δ | Δ | Δ | * | * | * | * | * |
| 3.250 | | | | | | Δ | Δ | Δ | * | * | * | Δ | * |
| 3.500 | | | | | | Δ | Δ | Δ | * | * | * | * | * |
| 3.750 | | | | | | Δ | Δ | Δ | Δ | * | * | Δ | * |
| 4.000 | | | | | | Δ | Δ | Δ | * | * | * | * | * |
| 4.250 | | | | | | | Δ | Δ | Δ | Δ | Δ | Δ | Δ |
| 4.500 | | | | | | | Δ | Δ | Δ | Δ | Δ | Δ | * |
| 4.750 | | | | | | | Δ | Δ | Δ | Δ | Δ | Δ | Δ |
| 5.000 | | | | | | | Δ | Δ | Δ | Δ | * | Δ | * |
| 5.250 | | | | | | | | | | | Δ | Δ | Δ |
| 5.500 | | | | | | | | | | | Δ | Δ | Δ |

ESNA AVAILABILITY CODE:
* - STANDARD PARTS NORMALLY CARRIED IN STOCK.
Δ - STANDARD PARTS NOT ALWAYS CARRIED IN INVENTORY AND ON WHICH IT MAY BE NECESSARY TO REQUIRE MINIMUM ECONOMICAL RUNS.

IN ADDITION TO THE STOCK LENGTHS APPEARING IN THIS TABLE, ROLLPINS LESS THAN 1" LONG CAN BE MADE AVAILABLE IN LENGTH INCREMENTS OF 1/32". ROLLPINS LONGER THAN 1" CAN BE MADE AVAILABLE IN LENGTH INCREMENTS OF 1/16", BUT IN BOTH CASES SUCH PARTS ARE NON-STOCK ITEMS AND MUST BE MADE TO ORDER. ROLLPINS OF SPECIAL DECIMAL LENGTHS ARE SUBJECT TO SPECIAL ORDER REQUIREMENTS.

ISSUED: 20 APR 50 REVISED: (7) 9 JAN 62

| PERFORMANCE SPECIFICATION | ROLLPIN | RP-A |
|---|---|---|
| ESNA SPEC 401 (SEE NOTE: 1) | HIGH CARBON STEEL | PAGE 2 OF 2 |

BASIC
internal series
N5000
WALDES
TRUARC
RETAINING
RINGS


U. S. Pat. No. 2,861,824

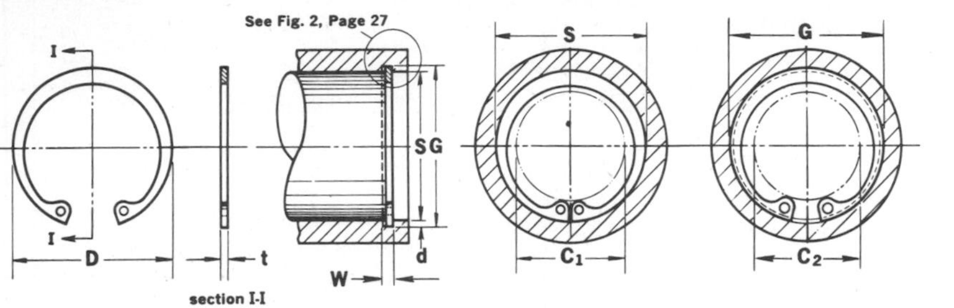


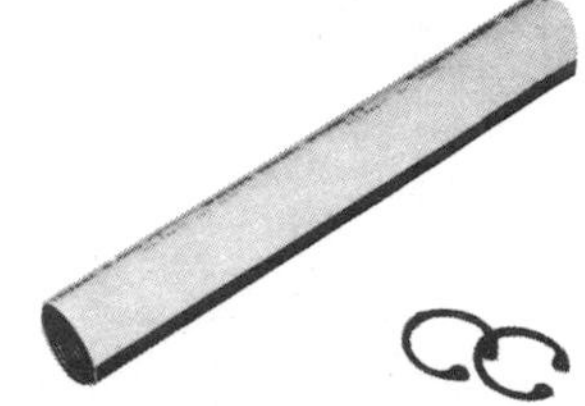

• Sizes identified by this symbol are available in tape-wrapped **Rol-Pak®** cartridges.

| HOUSING DIA. | | | MIL-R-21248 MS 16625 NAS 669 INTERNAL SERIES **N5000** | TRUARC RING DIMENSIONS (Thickness t applies only to unplated rings. For plated and stainless steel (Type H) rings, add .002" to the listed maximum thickness. Maximum ring thickness will be at least .0002" less than the listed minimum groove width (W).) | | | | | GROOVE DIMENSIONS (T.I.R. (total indicator reading) is the maximum allowable deviation of concentricity between groove and housing.) | | | | | APPLICATION DATA | | | |
|---|---|---|---|---|---|---|---|---|---|---|---|---|---|---|---|---|---|
| Dec. equiv. inch | Approx. fract. equiv. inch | Approx. mm. | | FREE DIA. | | THICKNESS | | Approx. weight per 1000 pieces | DIAMETER | | WIDTH | | Nominal groove depth | CLEARANCE DIAMETER: When sprung into housing S | CLEARANCE DIAMETER: When sprung into groove G | ALLOW. THRUST LOAD (lbs.) Sharp Corner Abutment: RINGS Safety factor=4 Important! See Page 19 | ALLOW. THRUST LOAD (lbs.) Sharp Corner Abutment: GROOVES Safety factor=2 Important! See Page 19 |
| S | S | S | size—no. | D | tol. | t | tol. | lbs. | G | tol. | W | tol. | d | $C_1$ | $C_2$ | $P_r$ | $P_g$ |
| .250 | 1/4 | 6.4 | • N5000-25 | .280 | +.010 −.005 | .015 | ±.002 | .08 | .268 | ±.001 .0015 ±T.I.R. | .018 | +.002 −.000 | .009 | .115 | .133 | 420 | 190 |
| .312 | 5/16 | 7.9 | • N5000-31 | .346 | | .015 | | .11 | .330 | | .018 | | .009 | .173 | .191 | 530 | 240 |
| .375 | 3/8 | 9.5 | • N5000-37 | .415 | | .025 | | .25 | .397 | ±.002 .002 T.I.R. | .029 | +.003 −.000 | .011 | .204 | .226 | 1050 | 350 |
| .438 | 7/16 | 11.1 | • N5000-43 | .482 | | .025 | | .37 | .461 | | .029 | | .012 | .23 | .254 | 1220 | 440 |
| .453 | 29/64 | 11.5 | • N5000-45 | .498 | | .025 | | .43 | .477 | | .029 | | .012 | .25 | .274 | 1280 | 460 |
| .500 | 1/2 | 12.7 | • N5000-50 | .548 | | .035 | | .70 | .530 | ±.002 .004 T.I.R. | .039 | | .015 | .26 | .29 | 1980 | 510 |
| .512 | — | 13.0 | • N5000-51 | .560 | | .035 | | .77 | .542 | | .039 | | .015 | .27 | .30 | 2030 | 520 |
| .562 | 9/16 | 14.3 | • N5000-56 | .620 | | .035 | | .86 | .596 | | .039 | | .017 | .275 | .305 | 2220 | 710 |
| .625 | 5/8 | 15.9 | • N5000-62 | .694 | | .035 | | 1.0 | .665 | | .039 | | .020 | .34 | .38 | 2470 | 1050 |
| .688 | 11/16 | 17.5 | • N5000-68 | .763 | | .035 | | 1.2 | .732 | | .039 | | .022 | .40 | .44 | 2700 | 1280 |
| .750 | 3/4 | 19.0 | • N5000-75 | .831 | | .035 | | 1.3 | .796 | | .039 | | .023 | .45 | .49 | 3000 | 1460 |
| .777 | — | 19.7 | • N5000-77 | .859 | +.015 −.010 | .042 | | 1.7 | .825 | ±.003 .004 T.I.R. | .046 | | .024 | .475 | .52 | 4550 | 1580 |
| .812 | 13/16 | 20.6 | • N5000-81 | .901 | | .042 | | 1.9 | .862 | | .046 | | .025 | .49 | .54 | 4800 | 1710 |
| .866 | — | 22.0 | • N5000-86 | .961 | | .042 | | 2.0 | .920 | | .046 | | .027 | .54 | .59 | 5100 | 1980 |
| .875 | 7/8 | 22.2 | • N5000-87 | .971 | | .042 | | 2.1 | .931 | | .046 | | .028 | .545 | .60 | 5150 | 2080 |
| .901 | — | 22.9 | • N5000-90 | 1.000 | | .042 | | 2.2 | .959 | | .046 | | .029 | .565 | .62 | 5350 | 2200 |
| .938 | 15/16 | 23.8 | • N5000-93 | 1.041 | | .042 | | 2.4 | 1.000 | | .046 | | .031 | .61 | .67 | 5600 | 2450 |
| 1.000 | 1 | 25.4 | • N5000-100 | 1.111 | | .042 | | 2.7 | 1.066 | | .046 | | .033 | .665 | .73 | 5950 | 2800 |
| 1.023 | — | 26.0 | • N5000-102 | 1.136 | | .042 | | 2.8 | 1.091 | | .046 | | .034 | .69 | .755 | 6050 | 3000 |
| 1.062 | 1 1/16 | 27.0 | • N5000-106 | 1.180 | +.025 −.020 | .050 | | 3.7 | 1.130 | ±.004 .005 T.I.R. | .056 | +.004 −.000 | .034 | .685 | .75 | 7450 | 3050 |
| 1.125 | 1 1/8 | 28.6 | • N5000-112 | 1.249 | | .050 | | 4.0 | 1.197 | | .056 | | .036 | .745 | .815 | 7900 | 3400 |
| 1.181 | — | 30.0 | • N5000-118 | 1.319 | | .050 | | 4.3 | 1.255 | | .056 | | .037 | .79 | .86 | 8400 | 3700 |
| 1.188 | 1 3/16 | 30.2 | • N5000-118 | 1.319 | | .050 | | 4.3 | 1.262 | | .056 | | .037 | .80 | .87 | 8400 | 3700 |
| 1.250 | 1 1/4 | 31.7 | • N5000-125 | 1.388 | | .050 | | 4.8 | 1.330 | | .056 | | .040 | .875 | .955 | 8800 | 4250 |
| 1.259 | — | 32.0 | • N5000-125 | 1.388 | | .050 | | 4.8 | 1.339 | | .056 | | .040 | .885 | .965 | 8800 | 4250 |
| 1.312 | 1 5/16 | 33.3 | • N5000-131 | 1.456 | | .050 | | 5.0 | 1.396 | | .056 | | .042 | .93 | 1.01 | 9300 | 4700 |
| 1.375 | 1 3/8 | 34.9 | • N5000-137 | 1.526 | | .050 | | 5.1 | 1.461 | | .056 | | .043 | .99 | 1.07 | 9700 | 5050 |
| 1.378 | — | 35.0 | • N5000-137 | 1.526 | | .050 | | 5.1 | 1.464 | | .056 | | .043 | .99 | 1.07 | 9700 | 5050 |
| 1.438 | 1 7/16 | 36.5 | • N5000-143 | 1.596 | | .050 | | 5.8 | 1.528 | | .056 | | .045 | 1.06 | 1.15 | 10200 | 5500 |
| 1.456 | — | 37.0 | • N5000-145 | 1.616 | | .050 | | 6.4 | 1.548 | | .056 | | .046 | 1.08 | 1.17 | 10300 | 5700 |
| 1.500 | 1 1/2 | 38.1 | • N5000-150 | 1.660 | | .050 | | 6.5 | 1.594 | | .056 | | .047 | 1.12 | 1.21 | 10550 | 6000 |
| 1.562 | 1 9/16 | 39.7 | • N5000-156 | 1.734 | +.035 −.025 | .062 | ±.003 | 8.1 | 1.658 | ±.005 .005 T.I.R. | .068 | | .048 | 1.14 | 1.23 | 13700 | 6350 |
| 1.575 | — | 40.0 | • N5000-156 | 1.734 | | .062 | | 8.1 | 1.671 | | .068 | | .048 | 1.15 | 1.24 | 13700 | 6350 |
| 1.625 | 1 5/8 | 41.3 | • N5000-162 | 1.804 | | .062 | | 10.0 | 1.725 | | .068 | | .050 | 1.15 | 1.25 | 14200 | 6900 |
| 1.653 | — | 42.0 | • N5000-165 | 1.835 | | .062 | | 10.4 | 1.755 | | .068 | | .051 | 1.17 | 1.27 | 14500 | 7200 |
| 1.688 | 1 11/16 | 42.9 | • N5000-168 | 1.874 | | .062 | | 10.8 | 1.792 | | .068 | | .052 | 1.21 | 1.31 | 14800 | 7450 |

Additional sizes appear on Pages 28 and 30

**FIG. 1:**
**MAXIMUM ALLOWABLE CORNER RADIUS ($R_{max.}$) AND CHAMFER ($Ch_{max.}$)**

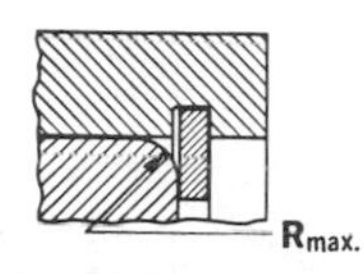


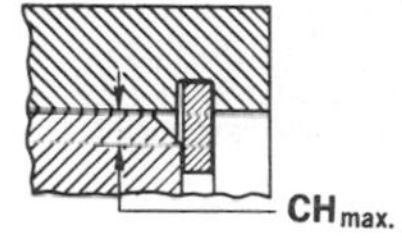


**FIG. 2:**
**ENLARGED DETAIL OF GROOVE PROFILE AND EDGE MARGIN (Z)**

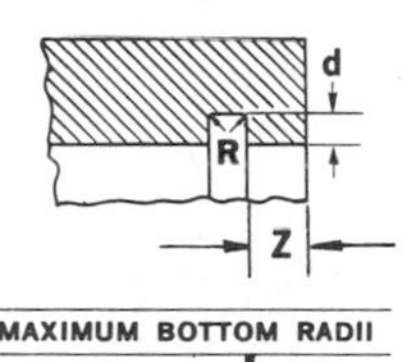


| MAXIMUM BOTTOM RADII | |
|---|---|
| Ring Size | R |
| -25 thru -100 | .005 |
| -102 thru -1000 | .010 |

**FIG. 3:**
**SUPPLEMENTARY RING DIMENSIONS**

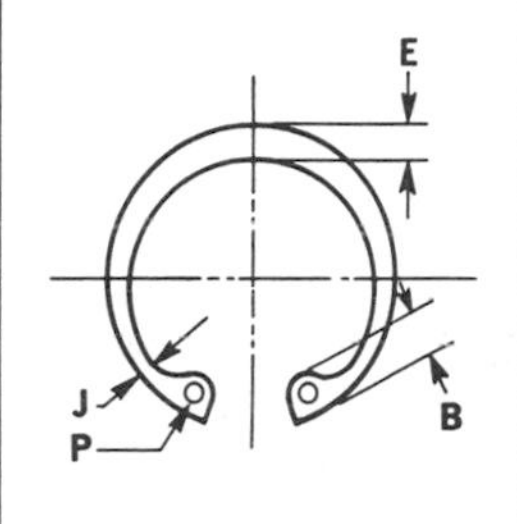


**FIG. 4:**
**MINIMUM GAP WIDTH**
(Ring installed in groove)

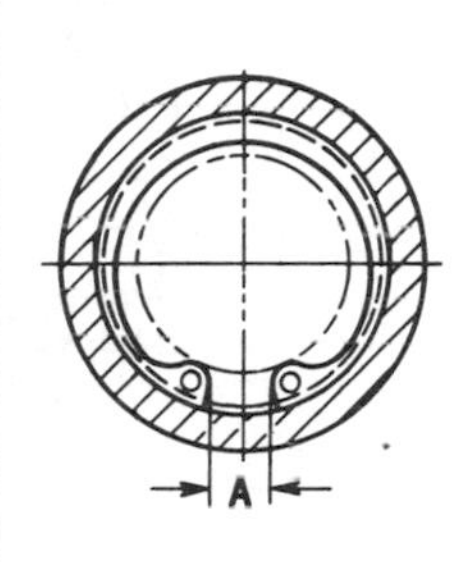


| SUPPLEMENTARY APPLICATION DATA | | | | | SUPPLEMENTARY RING DIMENSIONS | | | | | | | | |
|---|---|---|---|---|---|---|---|---|---|---|---|---|---|
| INTERNAL SERIES **N5000** | Maximum allowable corner radii and chamfers of retained parts (Fig. 1). See Page 21. | | Allowable assembly load with $R_{max.}$ or $Ch_{max.}$ | Edge margin (Fig. 2) See Page 19 | LUG | | LARGE SECTION | | SMALL SECTION | | HOLE DIAMETER | | MIN GAP WIDTH (Fig. 4) Ring installed in groove |
| size—no. | $R_{max.}$ | $Ch_{max.}$ | $P'_r$(lbs.) | Z | B | tol. | E | tol. | J | tol. | P | tol. | A |
| • N5000-25 | .011 | .0085 | 190 | .027 | .065 | | .025 | ±.002 | .015 | ±.002 | .031 | | .047 |
| • N5000-31 | .016 | .013 | 190 | .027 | .066 | | .033 | | .018 | | .031 | | .055 |
| • N5000-37 | .023 | .018 | 530 | .033 | .082 | | .040 | | .028 | | .041 | | .063 |
| • N5000-43 | .027 | .021 | 530 | .036 | .098 | ±.003 | .049 | ±.003 | .029 | ±.003 | .041 | | .063 |
| • N5000-45 | .027 | .021 | 530 | .036 | .098 | | .050 | | .030 | | .047 | | .071 |
| • N5000-50 | .027 | .021 | 1100 | .045 | .114 | | .053 | | .035 | | .047 | | .090 |
| • N5000-51 | .027 | .021 | 1100 | .045 | .114 | | .053 | | .035 | | .047 | | .092 |
| • N5000-56 | .027 | .021 | 1100 | .051 | .132 | | .053 | | .035 | | .047 | | .095 |
| • N5000-62 | .027 | .021 | 1100 | .060 | .132 | | .060 | ±.004 | .035 | ±.004 | .062 | | .104 |
| • N5000-68 | .027 | .021 | 1100 | .066 | .132 | | .063 | | .036 | | .062 | +.010 −.002 | .118 |
| • N5000-75 | .032 | .025 | 1100 | .069 | .142 | | .070 | | .040 | | .062 | | .143 |
| • N5000-77 | .035 | .028 | 1650 | .072 | .146 | | .074 | | .044 | | .062 | | .145 |
| • N5000-81 | .035 | .028 | 1650 | .075 | .155 | | .077 | | .044 | | .062 | | .153 |
| • N5000-86 | .035 | .028 | 1650 | .081 | .155 | | .081 | | .045 | | .062 | | .172 |
| • N5000-87 | .035 | .028 | 1650 | .084 | .155 | | .084 | ±.005 | .045 | ±.005 | .062 | | .179 |
| • N5000-90 | .038 | .030 | 1650 | .087 | .155 | | .087 | | .047 | | .062 | | .188 |
| • N5000-93 | .038 | .030 | 1650 | .093 | .155 | | .091 | | .050 | | .062 | | .200 |
| • N5000-100 | .042 | .034 | 2400 | .099 | .155 | | .104 | | .052 | | .062 | | .212 |
| • N5000-102 | .042 | .034 | 2400 | .102 | .155 | | .106 | | .054 | | .062 | | .220 |
| • N5000-106 | .044 | .035 | 2400 | .102 | .180 | | .110 | | .055 | | .078 | | .213 |
| • N5000-112 | .047 | .036 | 2400 | .108 | .180 | | .116 | | .057 | | .078 | | .232 |
| (S=1.181) • N5000-118 | .047 | .036 | 2400 | .111 | .180 | ±.005 | .120 | | .058 | | .078 | | .226 |
| (S=1.188) • N5000-118 | .047 | .036 | 2400 | .111 | .180 | | .120 | | .058 | | .078 | | .245 |
| (S=1.250) • N5000-125 | .048 | .038 | 2400 | .120 | .180 | | .124 | | .062 | | .078 | | .265 |
| (S=1.259) • N5000-125 | .048 | .038 | 2400 | .120 | .180 | | .124 | ±.006 | .062 | ±.006 | .078 | | .290 |
| • N5000-131 | .048 | .038 | 2400 | .126 | .180 | | .130 | | .062 | | .078 | | .284 |
| (S=1.375) • N5000-137 | .048 | .038 | 2400 | .129 | .180 | | .130 | | .063 | | .078 | +.015 −.002 | .297 |
| (S=1.378) • N5000-137 | .048 | .038 | 2400 | .129 | .180 | | .130 | | .063 | | .078 | | .305 |
| • N5000-143 | .048 | .038 | 2400 | .135 | .180 | | .133 | | .065 | | .078 | | .313 |
| • N5000-145 | .048 | .038 | 2400 | .138 | .180 | | .133 | | .065 | | .078 | | .320 |
| • N5000-150 | .048 | .038 | 2400 | .141 | .180 | | .133 | | .066 | | .078 | | .340 |
| (S=1.562) • N5000-156 | .064 | .050 | 3900 | .144 | .202 | | .157 | | .078 | | .078 | | .338 |
| (S=1.575) • N5000-156 | .064 | .050 | 3900 | .144 | .202 | | .157 | | .078 | | .078 | | .374 |
| • N5000-162 | .064 | .050 | 3900 | .150 | .227 | | .164 | ±.007 | .082 | ±.007 | .078 | | .339 |
| • N5000-165 | .064 | .050 | 3900 | .153 | .227 | | .167 | | .083 | | .078 | | .348 |
| • N5000-168 | .064 | .050 | 3900 | .156 | .227 | | .170 | | .085 | | .078 | | .357 |

U. S. Pat. No. 2,861,824

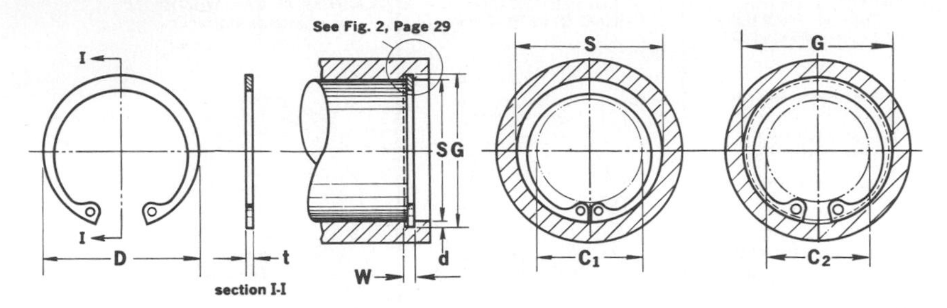


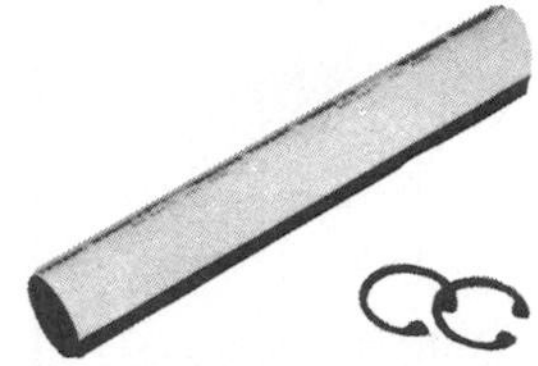

• Sizes identified by this symbol are available in tape-wrapped **Rol-Pak®** cartridges.

| HOUSING DIA. | | | MIL-R-21248 MS 16625 NAS 669 INTERNAL SERIES N5000 | TRUARC RING DIMENSIONS | | | | | GROOVE DIMENSIONS | | | | | APPLICATION DATA | | | |
|---|---|---|---|---|---|---|---|---|---|---|---|---|---|---|---|---|---|
| Dec. equiv. inch | Approx. fract. equiv. inch | Approx. mm. | | Thickness t applies only to unplated rings. For plated and stainless steel (Type H) rings, add .002" to the listed maximum thickness. Maximum ring thickness will be at least .0002" less than the listed minimum groove width (W). FREE DIA. | | THICKNESS | | Approx. weight per 1000 pieces | T.I.R. (total indicator reading) is the maximum allowable deviation of concentricity between groove and housing. DIAMETER | | WIDTH | | Nominal groove depth | CLEARANCE DIAMETER: When sprung into housing S | CLEARANCE DIAMETER: When sprung into groove G | ALLOW. THRUST LOAD (lbs.) Sharp Corner Abutment: RINGS Safety factor=4 Important! See Page 19 | ALLOW. THRUST LOAD (lbs.) Sharp Corner Abutment: GROOVES Safety factor=2 Important! See Page 19 |
| S | S | S | size—no. | D | tol. | t | tol. | lbs. | G | tol. | W | tol. | d | $C_1$ | $C_2$ | $P_r$ | $P_g$ |
| 1.750 | 1¾ | 44.4 | • N5000-175 | 1.942 | | .062 | | 10.3 | 1.858 | | .068 | | .054 | 1.26 | 1.36 | 15350 | 8050 |
| 1.812 | 1 13/16 | 46.0 | • N5000-181 | 2.012 | | .062 | | 11.5 | 1.922 | | .068 | | .055 | 1.32 | 1.43 | 15900 | 8450 |
| 1.850 | — | 47.0 | • N5000-185 | 2.054 | +.035 −.025 | .062 | | 12.8 | 1.962 | ±.005 .005 T.I.R. | .068 | +.004 −.000 | .056 | 1.36 | 1.47 | 16200 | 8750 |
| 1.875 | 1⅞ | 47.6 | • N5000-187 | 2.072 | | .062 | | 12.8 | 1.989 | | .068 | | .057 | 1.39 | 1.50 | 16450 | 9050 |
| 1.938 | 1 15/16 | 49.2 | • N5000-193 | 2.141 | | .062 | | 13.3 | 2.056 | | .068 | | .059 | 1.45 | 1.56 | 17000 | 9700 |
| 2.000 | 2 | 50.8 | • N5000-200 | 2.210 | | .062 | | 14.0 | 2.122 | | .068 | | .061 | 1.50 | 1.62 | 17500 | 10300 |
| 2.047 | — | 52.0 | N5000-206 | 2.280 | | .078 | | 18.0 | 2.171 | | .086 | | .062 | 1.52 | 1.64 | 22750 | 10850 |
| 2.062 | 2 1/16 | 52.4 | N5000-206 | 2.280 | | .078 | | 18.0 | 2.186 | | .086 | | .062 | 1.54 | 1.66 | 22750 | 10850 |
| 2.125 | 2⅛ | 54.0 | N5000-212 | 2.350 | | .078 | | 19.4 | 2.251 | | .086 | | .063 | 1.58 | 1.70 | 23400 | 11350 |
| 2.165 | — | 55.0 | N5000-218 | 2.415 | | .078 | | 19.6 | 2.295 | | .086 | | .065 | 1.61 | 1.74 | 24100 | 12050 |
| 2.188 | 2 3/16 | 55.6 | N5000-218 | 2.415 | | .078 | | 19.6 | 2.318 | | .086 | | .065 | 1.64 | 1.77 | 24100 | 12050 |
| 2.250 | 2¼ | 57.1 | N5000-225 | 2.490 | | .078 | | 21.8 | 2.382 | | .086 | | .066 | 1.69 | 1.82 | 24850 | 12600 |
| 2.312 | 2 5/16 | 58.7 | N5000-231 | 2.560 | | .078 | | 22.6 | 2.450 | | .086 | | .069 | 1.75 | 1.88 | 25450 | 13550 |
| 2.375 | 2⅜ | 60.3 | N5000-237 | 2.630 | | .078 | | 23.2 | 2.517 | | .086 | | .071 | 1.81 | 1.95 | 26150 | 14300 |
| 2.440 | 2 7/16 | 62.0 | N5000-244 | 2.702 | | .078 | | 25.4 | 2.584 | | .086 | | .072 | 1.86 | 2.00 | 26900 | 14900 |
| 2.500 | 2½ | 63.5 | N5000-250 | 2.775 | +.040 −.030 | .078 | | 25.5 | 2.648 | | .086 | | .074 | 1.91 | 2.05 | 27600 | 15650 |
| 2.531 | 2 17/32 | 64.3 | N5000-250 | 2.775 | | .078 | | 25.5 | 2.681 | | .086 | | .075 | 1.94 | 2.09 | 27600 | 15650 |
| 2.562 | 2 9/16 | 65.1 | N5000-256 | 2.844 | | .093 | | 34.0 | 2.714 | | .103 | | .076 | 1.95 | 2.10 | 33700 | 16500 |
| 2.625 | 2⅝ | 66.7 | N5000-262 | 2.910 | | .093 | ±.003 | 34.5 | 2.781 | | .103 | | .078 | 2.02 | 2.17 | 34550 | 17350 |
| 2.677 | — | 68.0 | N5000-268 | 2.980 | | .093 | | 35.0 | 2.837 | | .103 | | .080 | 2.05 | 2.21 | 35400 | 18250 |
| 2.688 | 2 11/16 | 68.3 | N5000-268 | 2.980 | | .093 | | 35.0 | 2.848 | ±.006 .006 T.I.R. | .103 | | .080 | 2.06 | 2.22 | 35400 | 18250 |
| 2.750 | 2¾ | 69.8 | N5000-275 | 3.050 | | .093 | | 35.5 | 2.914 | | .103 | +.005 −.000 | .082 | 2.12 | 2.28 | 36100 | 19200 |
| 2.812 | 2 13/16 | 71.4 | N5000-281 | 3.121 | | .093 | | 36.0 | 2.980 | | .103 | | .084 | 2.18 | 2.34 | 36950 | 20050 |
| 2.835 | — | 72.0 | N5000-281 | 3.121 | | .093 | | 36.0 | 3.006 | | .103 | | .085 | 2.21 | 2.38 | 36950 | 20050 |
| 2.875 | 2⅞ | 73.0 | N5000-287 | 3.191 | | .093 | | 41.0 | 3.051 | | .103 | | .088 | 2.22 | 2.39 | 37800 | 21500 |
| 2.953 | — | 75.0 | N5000-300 | 3.325 | | .093 | | 42.5 | 3.135 | | .103 | | .091 | 2.30 | 2.48 | 39500 | 23150 |
| 3.000 | 3 | 76.2 | N5000-300 | 3.325 | | .093 | | 42.5 | 3.182 | | .103 | | .091 | 2.35 | 2.53 | 39500 | 23150 |
| 3.062 | 3 1/16 | 77.8 | N5000-306 | 3.418 | | .109 | | 53.0 | 3.248 | | .120 | | .093 | 2.41 | 2.59 | 47100 | 24100 |
| 3.125 | 3⅛ | 79.4 | N5000-312 | 3.488 | | .109 | | 56.0 | 3.315 | | .120 | | .095 | 2.47 | 2.66 | 48100 | 25200 |
| 3.149 | — | 80.0 | N5000-315 | 3.523 | | .109 | | 57.0 | 3.341 | | .120 | | .096 | 2.49 | 2.68 | 48600 | 25700 |
| 3.156 | 3 5/32 | 80.2 | N5000-315 | 3.523 | | .109 | | 57.0 | 3.348 | | .120 | | .096 | 2.50 | 2.69 | 48600 | 25700 |
| 3.250 | 3¼ | 82.5 | N5000-325 | 3.623 | | .109 | | 60.0 | 3.446 | | .120 | | .098 | 2.54 | 2.73 | 50000 | 27000 |
| 3.346 | 3 11/32 | 85.0 | N5000-334 | 3.734 | ±.055 | .109 | | 65.0 | 3.546 | | .120 | | .100 | 2.63 | 2.83 | 51600 | 28300 |
| 3.469 | 3 15/32 | 88.1 | N5000-347 | 3.857 | | .109 | | 69.0 | 3.675 | | .120 | | .103 | 2.76 | 2.96 | 53400 | 30200 |
| 3.500 | 3½ | 88.9 | N5000-350 | 3.890 | | .109 | | 71.0 | 3.710 | | .120 | | .105 | 2.79 | 3.00 | 53900 | 31200 |
| 3.543 | — | 90.0 | N5000-354 | 3.936 | | .109 | | 72.0 | 3.755 | | .120 | | .106 | 2.83 | 3.04 | 54600 | 31800 |
| 3.562 | 3 9/16 | 90.5 | N5000-354 | 3.936 | | .109 | | 72.0 | 3.776 | | .120 | | .107 | 2.85 | 3.06 | 54600 | 31800 |

Additional sizes appear on Pages 26 and 30

**FIG. 1: MAXIMUM ALLOWABLE CORNER RADIUS ($R_{max.}$) AND CHAMFER ($Ch_{max.}$)**

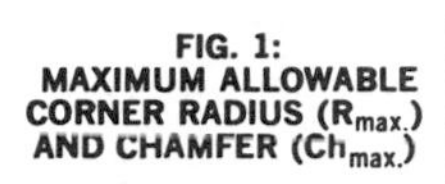

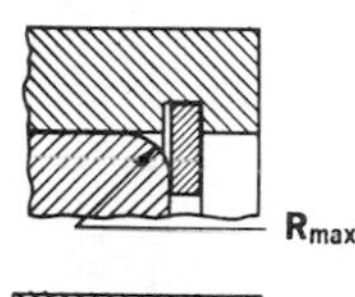


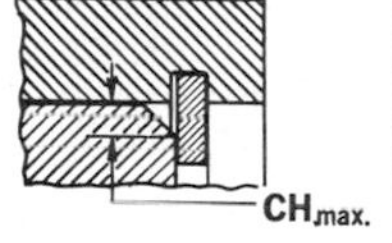


**FIG. 2: ENLARGED DETAIL OF GROOVE PROFILE AND EDGE MARGIN (Z)**

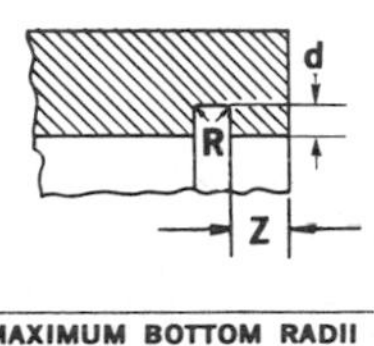


| MAXIMUM BOTTOM RADII | |
|---|---|
| Ring Size | R |
| -25 thru -100 | .005 |
| -102 thru -1000 | .010 |

**FIG. 3: SUPPLEMENTARY RING DIMENSIONS**

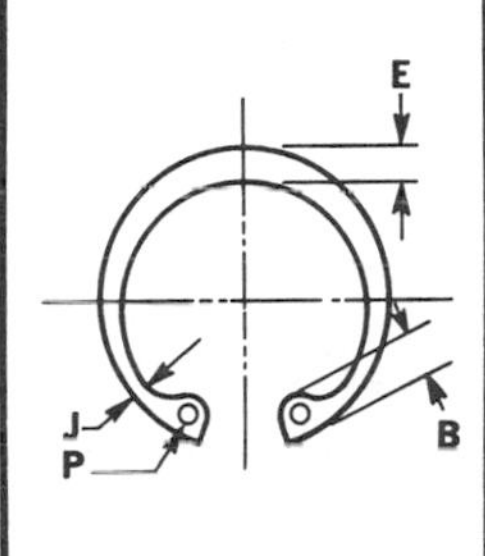


**FIG 4: MINIMUM GAP WIDTH**
(Ring installed in groove)

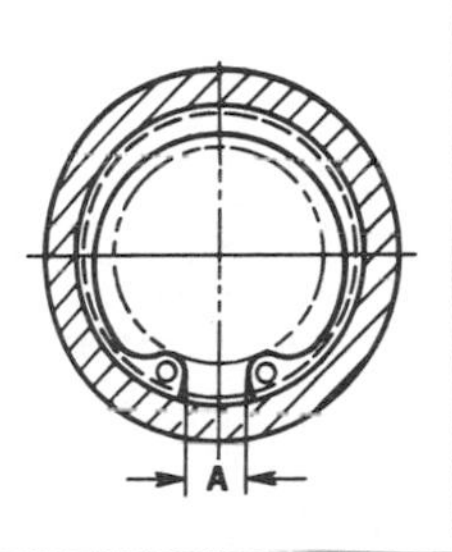


**FIG. 5: LUG DESIGN:**
' SIZES -206 THRU -275 AND -300 THRU -1000

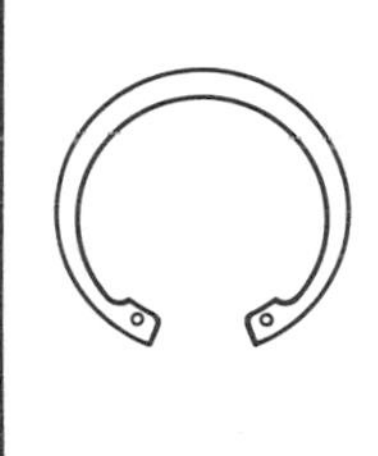

| INTERNAL SERIES N5000 | SUPPLEMENTARY APPLICATION DATA: Maximum allowable corner radii and chamfers of retained parts (Fig. 1). See Page 21. | | Allowable assembly load with $R_{max.}$ or $Ch_{max.}$ | Edge margin (Fig. 2) See Page 19 | SUPPLEMENTARY RING DIMENSIONS: LUG | | LARGE SECTION | | SMALL SECTION | | HOLE DIAMETER | | MIN. GAP WIDTH (Fig. 4) Ring installed in groove |
|---|---|---|---|---|---|---|---|---|---|---|---|---|---|
| size—no. | $R_{max.}$ | $Ch_{max.}$ | $P'_r$(lbs.) | Z | B | tol. | E | tol. | J | tol. | P | tol. | A |
| • N5000-175 | .064 | .050 | 3900 | .162 | .234 | | .171 | | .083 | | .078 | | .372 |
| • N5000-181 | .064 | .050 | 3900 | .165 | .234 | | .170 | | .084 | | .093 | | .382 |
| • N5000-185 | .064 | .050 | 3900 | .168 | .234 | | .170 | | .085 | | .093 | | .392 |
| • N5000-187 | .064 | .050 | 3900 | .171 | .234 | | .170 | | .085 | | .093 | | .420 |
| • N5000-193 | .064 | .050 | 3900 | .177 | .234 | | .170 | | .085 | | .093 | | .438 |
| • N5000-200 | .064 | .050 | 3900 | .183 | .240 | | .170 | | .085 | | .093 | | .453 |
| (S=2.047) N5000-206 | .076 | .061 | 6200 | .186 | .250 | | .186 | | .091 | | .093 | | .428 |
| (S=2.062) N5000-206 | .078 | .062 | 6200 | .186 | .250 | | .186 | | .091 | | .093 | | .468 |
| N5000-212 | .078 | .062 | 6200 | .189 | .260 | | .195 | | .096 | | .093 | | .460 |
| (S=2.165) N5000-218 | .078 | .062 | 6200 | .195 | .264 | | .199 | | .098 | | .093 | | .439 |
| (S=2.188) N5000-218 | .078 | .062 | 6200 | .195 | .264 | | .199 | | .098 | | .093 | | .489 |
| N5000-225 | .078 | .062 | 6200 | .198 | .270 | | .203 | | .099 | | .093 | | .478 |
| N5000-231 | .078 | .062 | 6200 | .207 | .270 | | .206 | | .100 | | .093 | | .486 |
| N5000-237 | .078 | .062 | 6200 | .213 | .270 | ±.005 | .207 | ±.007 | .102 | ±.007 | .093 | | .504 |
| N5000-244 | .078 | .062 | 6200 | .216 | .280 | | .209 | | .103 | | .110 | | .518 |
| (S=2.500) N5000-250 | .078 | .062 | 6200 | .222 | .280 | | .210 | | .103 | | .110 | | .532 |
| (S=2.531) N5000-250 | .078 | .062 | 6200 | .225 | .280 | | .210 | | .103 | | .110 | | .597 |
| N5000-256 | .088 | .070 | 9000 | .228 | .290 | | .222 | | .109 | | .110 | | .540 |
| N5000-262 | .088 | .070 | 9000 | .234 | .290 | | .226 | | .111 | | .110 | +.015 | .558 |
| (S=2.677) N5000-268 | .090 | .072 | 9000 | .240 | .300 | | .230 | | .113 | | .110 | −.002 | .539 |
| (S=2.688) N5000-268 | .090 | .072 | 9000 | .240 | .300 | | .230 | | .113 | | .110 | | .568 |
| N5000-275 | .092 | .074 | 9000 | .246 | .300 | | .234 | | .115 | | .110 | | .590 |
| (S=2.812) N5000-281 | .088 | .070 | 9000 | .252 | .300 | | .230 | | .115 | | .110 | | .615 |
| (S=2.835) N5000-281 | .088 | .070 | 9000 | .255 | .300 | | .230 | | .115 | | .110 | | .676 |
| N5000-287 | .092 | .074 | 9000 | .264 | .310 | | .240 | | .120 | | .110 | | .626 |
| (S=2.953) N5000-300 | .092 | .074 | 9000 | .273 | .310 | | .250 | | .122 | | .110 | | .619 |
| (S=3.000) N5000-300 | .092 | .074 | 9000 | .273 | .310 | | .250 | | .122 | | .110 | | .738 |
| N5000-306 | .097 | .078 | 12000 | .279 | .310 | | .254 | | .126 | | .125 | | .651 |
| N5000-312 | .099 | .079 | 12000 | .285 | .310 | | .259 | | .129 | | .125 | | .655 |
| (S=3.149) N5000-315 | .100 | .080 | 12000 | .288 | .310 | | .262 | | .129 | | .125 | | .650 |
| (S=3.156) N5000-315 | .100 | .080 | 12000 | .288 | .310 | | .262 | | .129 | | .125 | | .669 |
| N5000-325 | .104 | .083 | 12000 | .294 | .342 | | .269 | | .135 | | .125 | | .698 |
| N5000-334 | .108 | .086 | 12000 | .300 | .342 | ±.008 | .276 | ±.008 | .140 | ±.008 | .125 | | .705 |
| N5000-347 | .108 | .086 | 12000 | .309 | .342 | | .286 | | .144 | | .125 | | .763 |
| N5000-350 | .110 | .088 | 12000 | .315 | .342 | | .289 | | .142 | | .125 | | .774 |
| (S=3.543) N5000-354 | .110 | .088 | 12000 | .318 | .342 | | .292 | | .142 | | .125 | | .788 |
| (S=3.562) N5000-354 | .110 | .088 | 12000 | .321 | .342 | | .292 | | .142 | | .125 | | .842 |

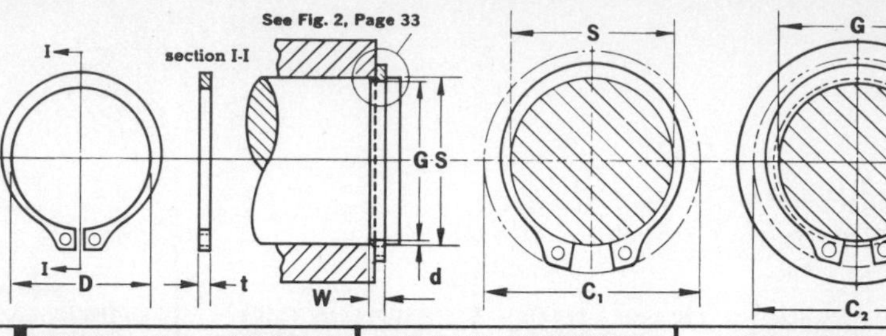


• Sizes identified by this symbol are available in tape-wrapped Rol-Pak® cartridges

| SHAFT DIAMETER | | | MIL-R-21248 MS 16624 NAS 670 EXTERNAL SERIES 5100 | TRUARC RING DIMENSIONS | | | | | GROOVE DIMENSIONS | | | | | APPLICATION DATA | | | |
|---|---|---|---|---|---|---|---|---|---|---|---|---|---|---|---|---|---|
| | | | | Thickness **t** applies only to **unplated** rings. For **plated** and **stainless steel** (Type H) rings, add .002" to the listed maximum thickness. Maximum ring thickness will be at least .0002" **less** than the listed minimum groove width (**W**). | | | | | T.I.R. (total indicator reading) is the maximum allowable deviation of concentricity between groove and shaft. | | | | | CLEARANCE DIAMETER | | ALLOW. THRUST LOAD (lbs.) Sharp Corner Abutment | |
| Dec. equiv. inch | Approx fract. equiv. inch | Approx. mm | | FREE DIA. | | THICKNESS | | Approx. weight per 1000 pieces | DIAMETER | | WIDTH | | Nominal groove depth | When sprung over shaft S | When sprung into groove G | RINGS Safety factor = 4 Important! See Page 19 | GROOVES Safety factor = 2 Important! See Page 19 |
| S | S | S | size — no. | D | tol. | t | tol. | lbs. | G | tol. | W | tol. | d | $C_1$ | $C_2$ | $P_r$ | $P_g$ |
| .125 | 1/8 | 3.2 | 5100-12 | .112 | | .010 | ±.001 | .018 | .117 | | .012 | | .004 | .222 | .214 | 110 | 35 |
| .156 | 5/32 | 4.0 | 5100-15 | .142 | | .010 | | .037 | .146 | | .012 | | .005 | .270 | .260 | 130 | 55 |
| .188 | 3/16 | 4.8 | 5100-18 | .168 | | .015 | | .059 | .175 | ±.0015 .0015 T.I.R. | .018 | +.002 −.000 | .006 | .298 | .286 | 240 | 80 |
| .197 | — — | 5.0 | 5100-19 | .179 | +.002 −.004 | .015 | | .063 | .185 | | .018 | | .006 | .319 | .307 | 250 | 85 |
| .219 | 7/32 | 5.6 | 5100-21 | .196 | | .015 | | .074 | .205 | | .018 | | .007 | .338 | .324 | 280 | 110 |
| .236 | 15/64 | 6.0 | 5100-23 | .215 | | .015 | | .086 | .222 | | .018 | | .007 | .355 | .341 | 310 | 120 |
| .250 | 1/4 | 6.4 | • 5100-25 | .225 | | .025 | | .21 | .230 | | .029 | | .010 | .45 | .43 | 590 | 175 |
| .276 | — — | 7.0 | 5100-27 | .250 | | .025 | | .23 | .255 | | .029 | | .010 | .48 | .46 | 650 | 195 |
| .281 | 9/32 | 7.1 | • 5100-28 | .256 | | .025 | | .24 | .261 | | .029 | | .010 | .49 | .47 | 660 | 200 |
| .312 | 5/16 | 7.9 | • 5100-31 | .281 | | .025 | | .27 | .290 | | .029 | | .011 | .54 | .52 | 740 | 240 |
| .344 | 11/32 | 8.7 | 5100-34 | .309 | | .025 | | .31 | .321 | ±.002 .002 T.I.R. | .029 | | .011 | .57 | .55 | 800 | 265 |
| .354 | — — | 9.0 | 5100-35 | .320 | | .025 | | .35 | .330 | | .029 | | .012 | .59 | .57 | 820 | 300 |
| .375 | 3/8 | 9.5 | • 5100-37 | .338 | +.002 −.005 | .025 | | .39 | .352 | | .029 | | .012 | .61 | .59 | 870 | 320 |
| .394 | — — | 10.0 | 5100-39 | .354 | | .025 | | .42 | .369 | | .029 | | .012 | .62 | .60 | 940 | 335 |
| .406 | 13/32 | 10.3 | 5100-40 | .366 | | .025 | | .43 | .382 | | .029 | | .012 | .63 | .61 | 950 | 350 |
| .438 | 7/16 | 11.1 | • 5100-43 | .395 | | .025 | | .50 | .412 | | .029 | | .013 | .66 | .64 | 1020 | 400 |
| .469 | 15/32 | 11.9 | 5100-46 | .428 | | .025 | | .54 | .443 | | .029 | | .013 | .68 | .66 | 1100 | 450 |
| .500 | 1/2 | 12.7 | • 5100-50 | .461 | | .035 | | .91 | .468 | ±.002 .004 T.I.R. | .039 | | .016 | .77 | .74 | 1650 | 550 |
| .551 | — — | 14.0 | 5100-55 | .509 | | .035 | | .90 | .519 | | .039 | | .016 | .81 | .78 | 1800 | 600 |
| .562 | 9/16 | 14.3 | • 5100-56 | .521 | | .035 | | 1.1 | .530 | | .039 | +.003 −.000 | .016 | .82 | .79 | 1850 | 650 |
| .594 | 19/32 | 15.1 | 5100-59 | .550 | | .035 | | 1.2 | .559 | | .039 | | .017 | .86 | .83 | 1950 | 750 |
| .625 | 5/8 | 15.9 | • 5100-62 | .579 | | .035 | ±.002 | 1.3 | .588 | | .039 | | .018 | .90 | .87 | 2060 | 800 |
| .669 | — — | 17.0 | 5100-66 | .621 | | .035 | | 1.4 | .629 | | .039 | | .020 | .93 | .89 | 2200 | 950 |
| .672 | 43/64 | 17.1 | 5100-66 | .621 | | .035 | | 1.4 | .631 | | .039 | | .020 | .93 | .89 | 2200 | 950 |
| .688 | 11/16 | 17.5 | • 5100-68 | .635 | +.005 −.010 | .042 | | 1.8 | .646 | ±.003 .004 T.I.R. | .046 | | .021 | 1.01 | .97 | 3400 | 1000 |
| .750 | 3/4 | 19.0 | • 5100-75 | .693 | | .042 | | 2.1 | .704 | | .046 | | .023 | 1.09 | 1.05 | 3700 | 1200 |
| .781 | 25/32 | 19.8 | 5100-78 | .722 | | .042 | | 2.2 | .733 | | .046 | | .024 | 1.12 | 1.08 | 3900 | 1300 |
| .812 | 13/16 | 20.6 | 5100-81 | .751 | | .042 | | 2.5 | .762 | | .046 | | .025 | 1.15 | 1.10 | 4000 | 1450 |
| .875 | 7/8 | 22.2 | 5100-87 | .810 | | .042 | | 2.8 | .821 | | .046 | | .027 | 1.21 | 1.16 | 4300 | 1650 |
| .938 | 15/16 | 23.8 | 5100-93 | .867 | | .042 | | 3.1 | .882 | | .046 | | .028 | 1.34 | 1.29 | 4650 | 1850 |
| .984 | 63/64 | 25.0 | 5100-98 | .910 | | .042 | | 3.5 | .926 | | .046 | | .029 | 1.39 | 1.34 | 4850 | 2000 |
| 1.000 | 1 | 25.4 | 5100-100 | .925 | | .042 | | 3.6 | .940 | | .046 | | .030 | 1.41 | 1.35 | 4950 | 2100 |
| 1.023 | — — | 26.0 | 5100-102 | .946 | | .042 | | 3.9 | .961 | | .046 | | .031 | 1.43 | 1.37 | 5050 | 2250 |
| 1.062 | 1 1/16 | 27.0 | 5100-106 | .982 | | .050 | | 4.8 | .998 | | .056 | | .032 | 1.50 | 1.44 | 6200 | 2400 |
| 1.125 | 1 1/8 | 28.6 | 5100-112 | 1.041 | | .050 | | 5.1 | 1.059 | | .056 | | .033 | 1.55 | 1.49 | 6600 | 2600 |
| 1.188 | 1 3/16 | 30.2 | 5100-118 | 1.098 | | .050 | | 5.6 | 1.118 | | .056 | | .035 | 1.61 | 1.54 | 7000 | 2950 |
| 1.250 | 1 1/4 | 31.7 | 5100-125 | 1.156 | +.010 −.015 | .050 | | 5.9 | 1.176 | ±.004 .005 T.I.R | .056 | | .037 | 1.69 | 1.62 | 7350 | 3250 |
| 1.312 | 1 5/16 | 33.3 | 5100-131 | 1.214 | | .050 | | 6.8 | 1.232 | | .056 | | .040 | 1.75 | 1.67 | 7750 | 3700 |
| 1.375 | 1 3/8 | 34.9 | 5100-137 | 1.272 | | .050 | | 7.2 | 1.291 | | .056 | | .042 | 1.80 | 1.72 | 8100 | 4100 |
| 1.438 | 1 7/16 | 36.5 | 5100-143 | 1.333 | | .050 | | 8.1 | 1.350 | | .056 | | .044 | 1.87 | 1.79 | 8500 | 4500 |
| 1.500 | 1 1/2 | 38.1 | 5100-150 | 1.387 | | .050 | | 9.0 | 1.406 | | .056 | +.004 −.000 | .047 | 1.99 | 1.90 | 8800 | 5000 |
| 1.562 | 1 9/16 | 39.7 | 5100-156 | 1.446 | | .062 | | 12.4 | 1.468 | | .068 | | .047 | 2.10 | 2.01 | 11400 | 5200 |
| 1.625 | 1 5/8 | 41.3 | 5100-162 | 1.503 | | .062 | | 13.2 | 1.529 | | .068 | | .048 | 2.17 | 2.08 | 11850 | 5500 |
| 1.688 | 1 11/16 | 42.9 | 5100-168 | 1.560 | | .062 | | 14.8 | 1.589 | | .068 | | .049 | 2.24 | 2.15 | 12350 | 5850 |
| 1.750 | 1 3/4 | 44.4 | 5100-175 | 1.618 | | .062 | ±.003 | 15.3 | 1.650 | ±.005 .005 T.I.R. | .068 | | .050 | 2.31 | 2.21 | 12800 | 6200 |
| 1.772 | — — | 45.0 | 5100-177 | 1.637 | +.013 −.020 | .062 | | 15.4 | 1.669 | | .068 | | .051 | 2.33 | 2.23 | 12950 | 6400 |
| 1.812 | 1 13/16 | 46.0 | 5100-181 | 1.675 | | .062 | | 16.2 | 1.708 | | .068 | | .052 | 2.38 | 2.28 | 13250 | 6650 |
| 1.875 | 1 7/8 | 47.6 | 5100-187 | 1.735 | | .062 | | 17.3 | 1.769 | | .068 | | .053 | 2.44 | 2.34 | 13700 | 7000 |
| 1.969 | 1 31/32 | 50.0 | 5100-196 | 1.819 | | .062 | | 18.0 | 1.857 | | .068 | | .056 | 2.54 | 2.43 | 14350 | 7800 |
| 2.000 | 2 | 50.8 | 5100-200 | 1.850 | | .062 | | 19.0 | 1.886 | | .068 | | .057 | 2.55 | 2.44 | 14600 | 8050 |

(For continuation of sizes, see Page 34)

WALDES TRUARC RETAINING RINGS

**BASIC** *external series* **5100**

**FIG. 1: MAXIMUM ALLOWABLE CORNER RADIUS ($R_{max.}$) AND CHAMFER ($Ch_{max.}$)**

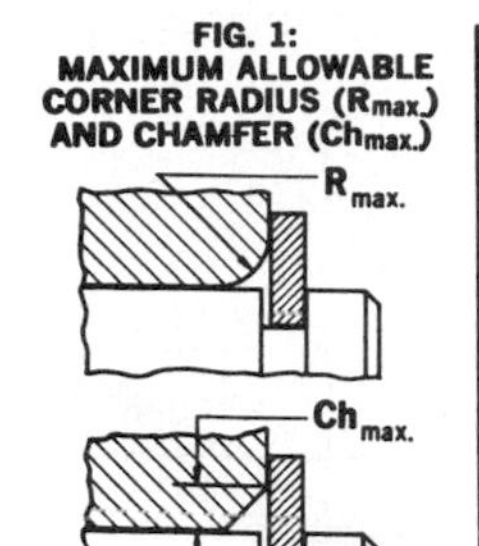


**FIG. 2: ENLARGED DETAIL OF GROOVE PROFILE AND EDGE MARGIN (Z)**

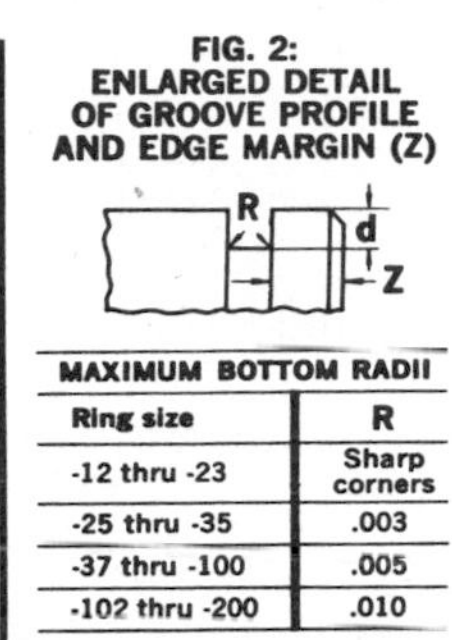


| MAXIMUM BOTTOM RADII | |
|---|---|
| Ring size | R |
| -12 thru -23 | Sharp corners |
| -25 thru -35 | .003 |
| -37 thru -100 | .005 |
| -102 thru -200 | .010 |

**FIG. 3: SUPPLEMENTARY RING DIMENSIONS**

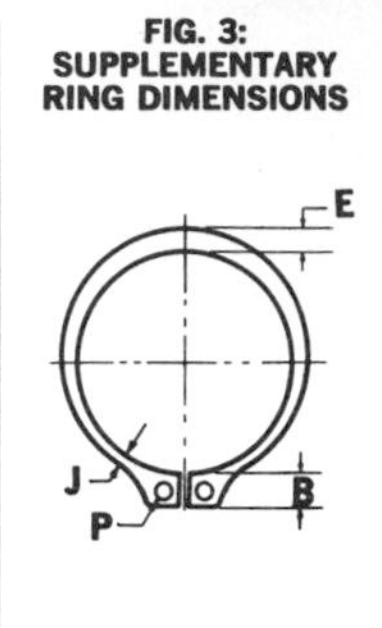


**FIG. 4: MAXIMUM GAGING DIAMETER** (Ring installed in groove)

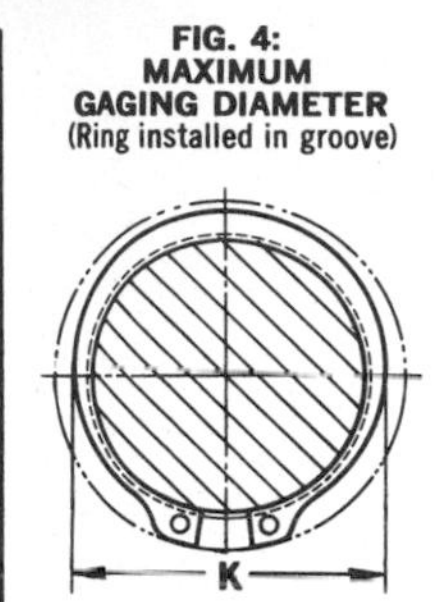


**FIG. 5: LUG DESIGN Sizes -12 thru -23**

| EXTERNAL SERIES 5100 | SUPPLEMENTARY APPLICATION DATA: Max. allow. corner radii and chamfers of retained parts. (Fig. 1). See Pg. 21 | | Allow. assembly load with $R_{max.}$ or $Ch_{max.}$ | Edge margin (Fig. 2) See Page 19 | Safe rotational speeds | SUPPLEMENTARY RING DIMENSIONS (Fig. 3): LUG | | LARGE SECTION | | SMALL SECTION | | HOLE DIAMETER | | MAX. GAGING DIA. Ring installed in groove (Fig. 4) |
|---|---|---|---|---|---|---|---|---|---|---|---|---|---|---|
| size — no. | $R_{max.}$ | $Ch_{max.}$ | P′ (lbs.) | Z | rpm | B | tol. | E | tol. | J | tol. | P | tol. | K |
| 5100-12 | .010 | .006 | 45 | .012 | 80000 | .046 | | .018 | ±.0015 | .011 | ±.0015 | .026 | | .148 |
| 5100-15 | .015 | .009 | 45 | .015 | 80000 | .054 | | .026 | | .016 | | .026 | | .189 |
| 5100-18 | .014 | .0085 | 105 | .018 | 80000 | .050 | ±.002 | .025 | | .016 | | .025 | | .218 |
| 5100-19 | .0145 | .009 | 105 | .018 | 80000 | .056 | | .026 | ±.002 | .016 | ±.002 | .026 | | .229 |
| 5100-21 | .015 | .009 | 105 | .021 | 80000 | .056 | | .028 | | .017 | | .026 | | .252 |
| 5100-23 | .0165 | .010 | 105 | .021 | 80000 | .056 | | .030 | | .019 | | .026 | | .272 |
| • 5100-25 | .018 | .011 | 470 | .030 | 80000 | .080 | | .035 | | .025 | | .041 | | .290 |
| 5100-27 | .0175 | .0105 | 470 | .031 | 76000 | .081 | | .035 | | .024 | | .041 | | .315 |
| • 5100-28 | .020 | .012 | 470 | .030 | 74000 | .080 | | .038 | | .0255 | | .041 | | .326 |
| • 5100-31 | .020 | .012 | 470 | .033 | 70000 | .087 | | .040 | | .026 | | .041 | | .357 |
| 5100-34 | .021 | .0125 | 470 | .033 | 64000 | .087 | | .042 | | .0265 | | .041 | | .390 |
| 5100-35 | .023 | .014 | 470 | .036 | 62000 | .087 | | .046 | ±.003 | .029 | ±.003 | .041 | | .405 |
| • 5100-37 | .026 | .0155 | 470 | .036 | 60000 | .088 | | .050 | | .0305 | | .041 | | .433 |
| 5100-39 | .027 | .016 | 470 | .037 | 56500 | .087 | | .052 | | .031 | | .041 | | .452 |
| 5100-40 | .0285 | .017 | 470 | .036 | 55000 | .087 | | .054 | | .033 | | .041 | +.010 −.002 | .468 |
| • 5100-43 | .029 | .0175 | 470 | .039 | 50000 | .088 | ±.003 | .055 | | .033 | | .041 | | .501 |
| 5100-46 | .031 | .018 | 470 | .039 | 42000 | .088 | | .060 | | .035 | | .041 | | .540 |
| • 5100-50 | .034 | .020 | 910 | .048 | 40000 | .108 | | .065 | | .040 | | .047 | | .574 |
| 5100-55 | .027 | .0165 | 910 | .048 | 36000 | .108 | | .053 | | .036 | | .047 | | .611 |
| • 5100-56 | .038 | .023 | 910 | .048 | 35000 | .108 | | .072 | | .041 | | .047 | | .644 |
| 5100-59 | .0395 | .0235 | 910 | .052 | 32000 | .109 | | .076 | ±.004 | .043 | ±.004 | .047 | | .680 |
| • 5100-62 | .0415 | .025 | 910 | .055 | 30000 | .110 | | .080 | | .045 | | .047 | | .715 |
| (S=.669) 5100-66 | .040 | .024 | 910 | .060 | 29000 | .110 | | .082 | | .043 | | .047 | | .756 |
| (S=.672) 5100-66 | .040 | .024 | 910 | .060 | 29000 | .110 | | .082 | | .043 | | .047 | | .758 |
| • 5100-68 | .042 | .025 | 1340 | .063 | 28000 | .136 | | .084 | | .048 | | .052 | | .779 |
| • 5100-75 | .046 | .0275 | 1340 | .069 | 26500 | .136 | | .092 | | .051 | | .052 | | .850 |
| 5100-78 | .047 | .028 | 1340 | .072 | 25500 | .136 | | .094 | | .052 | | .052 | | .883 |
| 5100-81 | .047 | .028 | 1340 | .075 | 24500 | .136 | | .096 | | .054 | | .052 | | .914 |
| 5100-87 | .051 | .0305 | 1340 | .081 | 23000 | .137 | | .104 | ±.005 | .057 | ±.005 | .052 | | .987 |
| 5100-93 | .055 | .033 | 1340 | .084 | 21500 | .166 | | .110 | | .063 | | .078 | | 1.054 |
| 5100-98 | .056 | .0335 | 1340 | .087 | 20500 | .167 | | .114 | | .0645 | | .078 | | 1.106 |
| 5100-100 | .057 | .034 | 1340 | .090 | 20000 | .167 | | .116 | | .065 | | .078 | | 1.122 |
| 5100-102 | .058 | .035 | 1340 | .093 | 19500 | .168 | | .118 | | .066 | | .078 | | 1.147 |
| 5100-106 | .060 | .036 | 1950 | .096 | 19000 | .181 | | .122 | | .069 | | .078 | | 1.192 |
| 5100-112 | .063 | .038 | 1950 | .099 | 18800 | .182 | | .128 | | .071 | | .078 | | 1.261 |
| 5100-118 | .064 | .0385 | 1950 | .105 | 18000 | .182 | ±.004 | .132 | | .072 | | .078 | | 1.325 |
| 5100-125 | .068 | .041 | 1950 | .111 | 17000 | .183 | | .140 | | .076 | | .078 | | 1.396 |
| 5100-131 | .068 | .041 | 1950 | .120 | 16500 | .183 | | .146 | | .0765 | | .078 | | 1.458 |
| 5100-137 | .072 | .043 | 1950 | .126 | 16000 | .184 | | .152 | | .082 | | .078 | | 1.529 |
| 5100-143 | .076 | .045 | 1950 | .132 | 15000 | .184 | | .160 | | .086 | | .078 | +.015 −.002 | 1.600 |
| 5100-150 | .079 | .047 | 1950 | .141 | 14800 | .214 | | .168 | | .091 | | .120 | | 1.668 |
| 5100-156 | .082 | .049 | 3000 | .141 | 14000 | .235 | | .172 | ±.006 | .093 | ±.006 | .125 | | 1.740 |
| 5100-162 | .087 | .052 | 3000 | .144 | 13200 | .235 | | .180 | | .097 | | .125 | | 1.812 |
| 5100-168 | .090 | .054 | 3000 | .148 | 13000 | .235 | | .184 | | .099 | | .125 | | 1.877 |
| 5100-175 | .091 | .054 | 3000 | .150 | 12200 | .237 | | .188 | | .101 | | .125 | | 1.945 |
| 5100-177 | .092 | .055 | 3000 | .154 | 11700 | .237 | | .190 | | .102 | | .125 | | 1.967 |
| 5100-181 | .092 | .055 | 3000 | .156 | 11500 | .238 | | .192 | | .102 | | .125 | | 2.010 |
| 5100-187 | .094 | .056 | 3000 | .159 | 11000 | .239 | | .196 | | .104 | | .125 | | 2.076 |
| 5100-196 | .094 | .056 | 3000 | .168 | 10500 | .245 | | .200 | | .106 | | .125 | | 2.170 |
| 5100-200 | .096 | .057 | 3000 | .171 | 10000 | .239 | | .204 | | .108 | | .125 | | 2.205 |

E-RING
external series
5133
X5133 • Y5133
WALDES
TRUARC
RETAINING
RINGS


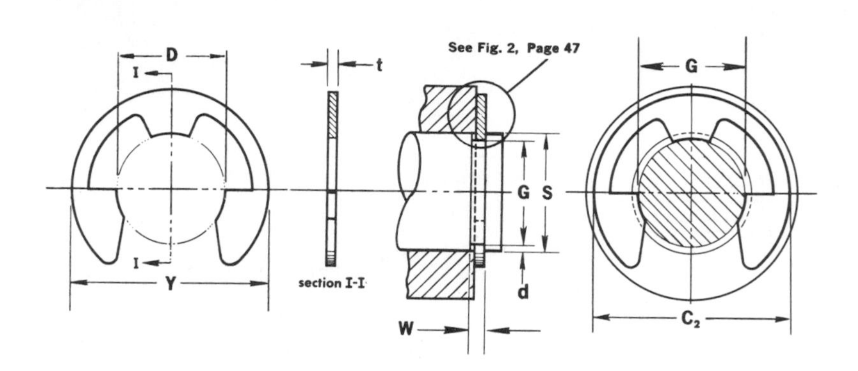

• Sizes identified by this symbol are available in tape-wrapped **Rol-Pak**® cartridges for use with Truarc Applicators, Dispensers and other assembly tools

| SHAFT DIA. | | | MIL-R-21248 MS 16633 | TRUARC RING DIMENSIONS | | | | | GROOVE DIMENSIONS | | | | | APPLICATION DATA | | | |
|---|---|---|---|---|---|---|---|---|---|---|---|---|---|---|---|---|---|
| | | | **EXTERNAL SERIES 5133** | Thickness t applies only to **unplated** rings. For **plated** and **stainless steel** (Type H) rings, add .002" to the listed maximum thickness. Maximum ring thickness will be at least .0002" **less** than the listed minimum groove width (**W**). | | | | Approx. weight per 1000 pieces | T.I.R. (total indicator reading) is the maximum allowable deviation of concentricity between groove and shaft. | | | | Nom-inal groove depth | CLEARANCE DIAMETER | | ALLOW. THRUST LOAD (lbs.) Sharp Corner Abutment | |
| Dec. equiv. inch | Approx. fract. equiv. inch | Approx. mm. | X5133 Y5133 | FREE DIA. | | THICKNESS | | | DIAMETER | | WIDTH | | | Free outside dia. (REF.) | When sprung into groove G | RINGS Safety factor = 3 **Important!** See Page 19 | GROOVES Safety factor = 2 **Important!** See Page 19 |
| S | S | S | size—no. | D | tol. | t | tol. | lbs. | G | tol. | W | tol. | d | Y | $C_2$ | $P_r$ | $P_g$ |
| .040 | — | 1.0 | X5133-4 | .025 | | .010 | ±.001 | .009 | .026 | | .012 | | .007 | .079 | .090 | 13 | 6 |
| .062 | 1/16 | 1.6 | • X5133-6 | .051 | +.001 | .010 | | .028 | .052 | | .012 | | .005 | .140 | .150 | 20 | 7 |
| .062 | 1/16 | 1.6 | • Y 5133-6 | .051 | −.003 | .020 | ±.002 | .094 | .052 | | .023 | | .005 | .187 | .200 | 40 | 7 |
| .062 | 1/16 | 1.6 | • 5133-6 | .051 | | .010 | ±.001 | .030 | .052 | +.002 | .012 | | .005 | .156 | .165 | 20 | 7 |
| .094 | 3/32 | 2.4 | • X5133-9 | .069 | +.002 −.003 | .015 | | .10 | .074 | −.000 | .018 | +.002 | .010 | .230 | .245 | 45 | 20 |
| .094 | 3/32 | 2.4 | • 5133-9 | .073 | | .015 | | .058 | .074 | .0015 T.I.R. | .018 | −.000 | .010 | .187 | .200 | 45 | 20 |
| .110 | 7/64 | 2.8 | • X5133-11 | .076 | | .015 | | .31 | .079 | | .018 | | .015 | .375 | .390 | 60 | 40 |
| .125 | 1/8 | 3.2 | • 5133-12 | .094 | | .015 | | .087 | .095 | | .018 | | .015 | .230 | .240 | 65 | 45 |
| .140 | 9/64 | 3.6 | • X5133-14 | .100 | | .015 | | .060 | .102 | | .018 | | .019 | .203 | .215 | 75 | 60 |
| .140 | 9/64 | 3.6 | • Y 5133-14 | .108 | | .015 | | .10 | .110 | | .018 | | .015 | .250 | .265 | 75 | 45 |
| .140 | 9/64 | 3.6 | • 5133-14 | .102 | +.001 | .025 | | .21 | .105 | +.002 | .029 | | .017 | .270 | .285 | 170 | 60 |
| .156 | 5/32 | 4.0 | • 5133-15 | .114 | −.003 | .025 | | .21 | .116 | −.000 | .029 | | .020 | .282 | .295 | 175 | 75 |
| .172 | 11/64 | 4.4 | • X5133-17 | .125 | | .025 | | .24 | .127 | .002 | .029 | | .022 | .312 | .325 | 180 | 90 |
| .188 | 3/16 | 4.8 | • X5133-18 | .122 | | .025 | | .45 | .125 | T.I.R. | .029 | | .031 | .375 | .39 | 200 | 135 |
| .188 | 3/16 | 4.8 | • 5133-18 | .145 | | .025 | | .29 | .147 | | .029 | | .020 | .335 | .35 | 190 | 90 |
| .219 | 7/32 | 5.6 | • X5133-21 | .185 | | .025 | | .47 | .188 | | .029 | | .015 | .437 | .45 | 225 | 75 |
| .250 | 1/4 | 6.3 | • 5133-25 | .207 | | .025 | ±.002 | .76 | .210 | | .029 | | .020 | .527 | .54 | 255 | 115 |
| .312 | 5/16 | 7.9 | • X5133-31 | .243 | | .025 | | .57 | .250 | | .029 | | .031 | .500 | .52 | 325 | 225 |
| .375 | 3/8 | 9.5 | • 5133-37 | .300 | | .035 | | 1.5 | .303 | | .039 | +.003 | .036 | .660 | .68 | 690 | 315 |
| .438 | 7/16 | 11.1 | • 5133-43 | .337 | +.002 | .035 | | 1.5 | .343 | | .039 | −.000 | .047 | .687 | .71 | 830 | 480 |
| .438 | 7/16 | 11.1 | • X5133-43 | .375 | −.004 | .035 | | 1.0 | .380 | | .039 | | .029 | .600 | .62 | 800 | 280 |
| .500 | 1/2 | 12.7 | • 5133-50 | .392 | | .042 | | 2.5 | .396 | +.003 | .046 | | .052 | .800 | .82 | 1110 | 600 |
| .625 | 5/8 | 15.9 | • 5133-62 | .480 | | .042 | | 3.2 | .485 | −.000 | .046 | | .070 | .940 | .96 | 1420 | 1050 |
| .744 | — | 18.9 | X5133-74 | .616 | | .050 | | 4.3 | .625 | .004 | .056 | | .059 | 1.000 | 1.02 | 1900 | 1050 |
| .750 | 3/4 | 19.0 | X5133-74 | .616 | +.003 | .050 | | 4.3 | .625 | T.I.R. | .056 | | .062 | 1.000 | 1.02 | 1950 | 1100 |
| .750 | 3/4 | 19.0 | • 5133-75 | .574 | −.005 | .050 | | 5.8 | .580 | | .056 | | .085 | 1.120 | 1.14 | 2000 | 1500 |
| .875 | 7/8 | 22.2 | 5133-87 | .668 | | .050 | | 7.6 | .675 | | .056 | | .100 | 1.300 | 1.32 | 2350 | 2050 |
| .984 | 63/64 | 25.0 | X5133-98 | .822 | | .050 | | 9.2 | .835 | | .056 | | .074 | 1.500 | 1.53 | 2600 | 1750 |
| 1.000 | 1 | 25.4 | X5133-98 | .822 | | .050 | | 9.2 | .835 | | .056 | | .082 | 1.500 | 1.53 | 2650 | 1900 |
| 1.188 | 1 3/16 | 30.2 | X5133-118 | 1.066 | +.006 | .062 | ±.003 | 11.3 | 1.079 | +.005 −.000 | .068 | +.004 | .054 | 1.626 | 1.67 | 3450 | 1500 |
| 1.375 | 1 3/8 | 34.9 | X5133-137 | 1.213 | −.010 | .062 | | 15.4 | 1.230 | .005 T.I.R. | .068 | −.000 | .072 | 1.875 | 1.92 | 4100 | 2350 |

**FIG. 1:**
**MAXIMUM ALLOWABLE CORNER RADIUS ($R_{max.}$) AND CHAMFER ($Ch_{max.}$)**

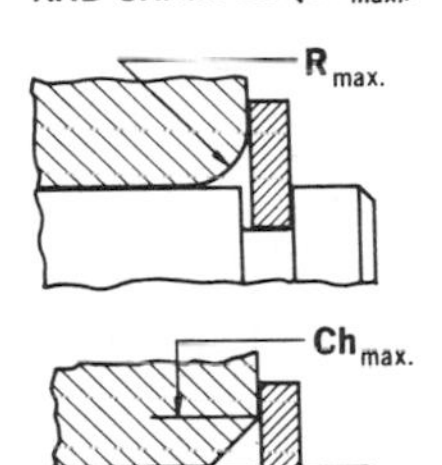


**FIG. 2:**
**ENLARGED DETAIL OF GROOVE PROFILE AND EDGE MARGIN (Z)**

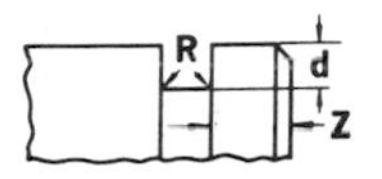


| MAXIMUM BOTTOM RADII | |
|---|---|
| **Ring size** | **R** |
| X-4 thru -6 | Sharp Corners |
| X-9 thru -25 | .005 |
| X-31 thru X-43 | .010 |
| -50 thru X-137 | .015 |

## SUPPLEMENTARY APPLICATION DATA

| EXTERNAL SERIES **5133** X5133 Y5133 | Maximum allowable corner radii and chamfers of retained parts (Fig. 1). See Page 21. | | Allow. assembly load with $R_{max.}$ or $Ch_{max.}$ | **Edge margin** (Fig. 2) See Page 19 | Safe rotational speeds |
|---|---|---|---|---|---|
| **size—no.** | $R_{max.}$ | $Ch_{max.}$ | $P'_{r(lbs.)}$ | **Z** | **rpm** |
| X5133-4 | .015 | .010 | 13 | .014 | 40000 |
| • X5133-6 | .030 | .020 | 20 | .010 | 40000 |
| • Y 5133-6 | .035 | .025 | 40 | .010 | 40000 |
| • 5133-6 | .030 | .020 | 20 | .010 | 40000 |
| • X5133-9 | .053 | .040 | 45 | .020 | 36000 |
| • 5133-9 | .040 | .030 | 45 | .020 | 36000 |
| • X5133-11 | .080 | .060 | 60 | .030 | 35000 |
| • 5133-12 | .040 | .030 | 65 | .030 | 35000 |
| • X5133-14 | .029 | .022 | 75 | .038 | 32000 |
| • Y 5133-14 | .040 | .030 | 75 | .030 | 32000 |
| • 5133-14 | .060 | .045 | 170 | .034 | 32000 |
| • 5133-15 | .060 | .045 | 175 | .040 | 31000 |
| • X5133-17 | .060 | .045 | 180 | .044 | 30000 |
| • X5133-18 | .060 | .045 | 200 | .062 | 30000 |
| • 5133-18 | .060 | .045 | 190 | .040 | 30000 |
| • X5133-21 | .060 | .045 | 225 | .030 | 26000 |
| • 5133-25 | .060 | .045 | 255 | .040 | 25000 |
| • X5133-31 | .060 | .045 | 325 | .062 | 22000 |
| • 5133-37 | .065 | .050 | 690 | .072 | 20000 |
| • 5133-43 | .065 | .050 | 830 | .094 | 16500 |
| • X5133-43 | .050 | .035 | 800 | .058 | 16500 |
| • 5133-50 | .080 | .060 | 1110 | .104 | 14000 |
| • 5133-62 | .080 | .060 | 1420 | .140 | 12000 |
| X5133-74 | .060 | .045 | 1900 | .118 | 11000 |
| X5133-74 | .057 | .042 | 1900 | .124 | 11000 |
| • 5133-75 | .085 | .065 | 2000 | .170 | 10500 |
| 5133-87 | .085 | .065 | 2350 | .200 | 9000 |
| X5133-98 | .085 | .065 | 2700 | .148 | 6500 |
| X5133-98 | .077 | .057 | 2700 | .164 | 6500 |
| X5133-118 | .090 | .070 | 3450 | .108 | 5500 |
| X5133-137 | .090 | .070 | 4100 | .144 | 4000 |

# INDEX

*Page references preceded by* A *indicate Appendix material.*